Food Microbiology

FUNDAMENTALS AND FRONTIERS

2nd Edition

Food Microbiology

FUNDAMENTALS AND FRONTIERS
2nd Edition

EDITED BY ————————————————————

Michael P. Doyle
Center for Food Safety, The University of Georgia, Griffin, Georgia 30223-1797

Larry R. Beuchat
Center for Food Safety, The University of Georgia, Griffin, Georgia 30223-1797

Thomas J. Montville
Department of Food Science, Rutgers–The State University of New Jersey, New Brunswick, NJ 08901-8520

ASM
PRESS   WASHINGTON, D.C.

Copyright © 2001   ASM Press
American Society for Microbiology
1752 N Street, N.W.
Washington, DC 20036-2904

Library of Congress Cataloging-in-Publication Data

Food microbiology : fundamentals and frontiers / edited by Michael P. Doyle, Larry R.
Beuchat, Thomas J. Montville.—2nd ed.
    p.   cm.
Includes bibliographical references and index.
ISBN 1-55581-208-2
    1. Food–Microbiology.   I. Doyle, Michael P.   II. Beuchat, Larry R.   III. Montville,
Thomas J.
QR115 .F654   2001
664′.001′579—dc21   2001022702

ISBN 1-55581-208-2

10   9   8   7   6   5   4   3   2   1

Address editorial correspondence to: ASM Press, 1752 N St., N.W., Washington, DC
20036-2904, U.S.A.

Send orders to: ASM Press, P.O. Box 605, Herndon, VA 20172, U.S.A.
Phone: 800-546-2416; 703-661-1593
Fax: 703-661-1501
Email: books@asmusa.org
Online: www.asmpress.org

Contents

Contributors

GARY R. ACUFF
Department of Animal Science, 2471 TAMU, Texas A&M University, College Station, TX 77843-2471

JOHN W. AUSTIN
Microbiology Research Division, Bureau of Microbial Hazards, Food Directorate, Health Canada, Banting Research Centre, Tunney's Pasture, PL 2204A2, Ottawa, Ontario K1A 0L2, Canada

J. STAN BAILEY
Russell Research Center, U.S. Department of Agriculture, Agricultural Research Service, P.O. Box 5677, Athens, GA 30606

DANE BERNARD
National Food Processors Association, 1350 I St., N.W., Suite 300, Washington, DC 20005-3305

LARRY R. BEUCHAT
Center for Food Safety and Quality Enhancement, Department of Food Science and Technology, University of Georgia, 1109 Experiment St., Griffin, GA 3022-1797

GREGORY A. BOHACH
Department of Microbiology, Molecular Biology and Biochemistry, University of Idaho, Moscow, ID 83843

ROBERT E. BRACKETT
Office of Plant and Dairy Foods and Beverages, U.S. Food and Drug Administration, 200 C St., S.W., Washington, DC 20204

ROBERT L. BUCHANAN
Center for Food Safety and Applied Nutrition, HFS-006, U.S. Food and Drug Administration, 200 C St., S.W., Washington, DC 20204

HERBERT J. BUCKENHÜSKES
Gewürzmüller GmbH, Klagenfurter Strasse 1-3, D-70469 Stuttgart, Germany

LLOYD B. BULLERMAN
Department of Food Science and Technology, 143 H. C. Filey Hall, East Campus, P.O. Box 830919, University of Nebraska, Lincoln, NE 68583-0919

IAIN CAMPBELL
Department of Biological Sciences, Heriot-Watt University, Riccarton, Edinburgh EH14 4AS, Scotland

MICHAEL L. CHIKINDAS
Department of Food Science, Rutgers-The State University of New Jersey, New Brunswick, NJ 08901-8520

DEAN O. CLIVER
Department of Population Health and Reproduction, School of Veterinary Medicine, University of California, Davis, Davis, CA 95616-8743

JEAN-YVES D'AOUST
Food Directorate, Health Products & Food Branch, Health Canada, Sir F. G. Banting Research Centre, Postal Locator 22.04.A2, Tunney's Pasture, Ottawa, Ontario K1A 0L2, Canada

P. MICHAEL DAVIDSON
Department of Food Science and Technology, University of Tennessee, 2509 River Rd., Knoxville, TN 37996

JAMES S. DICKSON
Department of Microbiology, Iowa State University, Ames, IA 50011-3211

MICHAEL P. DOYLE
Center for Food Safety, University of Georgia, Griffin, GA 30223

JÓZSEF FARKAS
Department of Refrigeration and Livestock Products' Technology, Szent István University, Ménesi út 45, Budapest H-1118, Hungary

PETER FENG
Division of Microbiological Studies, U.S. Food and Drug Administration, HFS-516, CFSAN, 200 C St., S.W., Washington, DC 20204

GRAHAM H. FLEET
Department of Food Science and Technology, The University of New South Wales, Sydney, New South Wales 2052, Australia

JOSEPH F. FRANK
Department of Food Science and Technology, Food Science Building, University of Georgia, Athens, GA 30602-7610

H. RAY GAMBLE
Parasite Biology and Epidemiology Laboratory, U.S. Department of Agriculture, Agricultural Research Service, Bldg 1040, Room 103, BARC-East, Beltsville, MD 20705

PER EINAR GRANUM
Department of Pharmacology, Microbiology and Food Hygiene, Norwegian School of Veterinary Science, PO Box 8146 Dep., Oslo N-0033, Norway

PAUL A. HARTMAN
Department of Microbiology, Immunology and Preventive Medicine, Iowa State University, Ames, IA 50014 [deceased]

EUGENE G. HAYUNGA
National Institutes of Health, Bldg. 31, Room 5B50, Bethesda, MD 20892

CRAIG W. HEDBERG
Division of Environmental and Occupational Health, School of Public Health, University of Minnesota, Minneapolis, MN 55455

AILSA D. HOCKING
Food Science Australia, P.O. Box 52, North Ryde, New South Wales 1670, Australia

LYNN M. JABLONSKI
Integrated Genomics, Inc., 2201 W. Campbell Park Dr., Chicago, IL 60612

TIMOTHY C. JACKSON
Nestle USA Quality Assurance Laboratory, Dublin, OH 43017

ERIC A. JOHNSON
Food Research Institute, University of Wisconsin-Madison, Madison, WI 53706

MARK E. JOHNSON
Center for Dairy Research, Department of Food Science, University of Wisconsin-Madison, 1605 Linden Dr., Madison, WI 53706-1565

JAMES B. KAPER
Center for Vaccine Development, Division of Geographic Medicine, Department of Medicine, University of Maryland School of Medicine, 685 West Baltimore St., Baltimore, MD 21201

JIMMY T. KEETON
Department of Animal Science, Room 338, Kleberg Animal and Food Science Center, Texas A&M University, College Station, TX 77843-2471

CHARLES W. KIM
Division of Infectious Diseases, Department of Medicine and Division of Microbiology, State University of New York, Stony Brook, NY 11794-8153

SYLVIA M. KIROV
Discipline of Pathology, Clinical School, University of Tasmania, GPO Box 252-29, Hobart, Tasmania 7001, Australia

TODD R. KLAENHAMMER
Department of Food Science, Schaub Hall, Box 7624, North Carolina State University, Raleigh, NC 27695-7624

KEITH A. LAMPEL
Center for Food Safety and Applied Nutrition, U.S. Food and Drug Administration, 200 C St., S.W., Washington, DC 20204

ALEX S. LOPEZ
c/o Malaysian Cocoa Board, Jen. Tunka Abd. Rahman, Beg Berkunci 211, Kota Kimbalu 88999, Sabah, Malaysia

DOUGLAS L. MARSHALL
Department of Food Science and Technology, Box 110 Herzer Building, Mississippi State University, Mississippi State, MS 39762

KARL R. MATTHEWS
Department of Food Science, Food Science Building, 65 Dudley Rd., Rutgers-The State University of New Jersey, New Brunswick, NJ 08901-8520

ANTHONY T. MAURELLI
Department of Microbiology and Immunology, F. Hébert School of Medicine, Uniformed Services University of the Health Sciences, 4301 Jones Bridge Rd., Bethesda, MD 20814-4799

JOHN MAURER
Department of Avian Medicine, University of Georgia, Athens, GA 30602

BRUCE A. MCCLANE
Department of Molecular Genetics and Biochemistry, University of Pittsburgh School of Medicine, E1240 Biomedical Science Tower, Pittsburgh, PA 15261-2072

JIANGHONG MENG
Department of Nutrition and Food Science, University of Maryland, College Park, MD 20742

KENNETH B. MILLER
Microbiology Research, Technical Center, Hershey Foods Corporation, 1025 Reese Ave., Hershey, PA 17033-0805

THOMAS J. MONTVILLE
Department of Food Science, Food Science Building, 65 Dudley Rd., Rutgers-The State University of New Jersey, New Brunswick, NJ 08901-8520

IRVING NACHAMKIN
Department of Pathology & Laboratory Medicine, University of Pennsylvania School of Medicine, Philadelphia, PA 19104-4283

JAMES D. OLIVER
Department of Biology, University of North Carolina at Charlotte, 9201 University City Blvd., Charlotte, NC 28223

YNES R. ORTEGA
Center for Food Safety and Quality Enhancement, Department of Food Science and Technology, University of Georgia, 1109 Experiment St., Griffin, GA 30223-1797

MERLE D. PIERSON
Department of Food Science and Technology, Virginia Polytechnic Institute and State University, Blacksburg, VA 24061

JOHN I. PITT
Food Science Australia, P.O. Box 52, North Ryde, New South Wales 1670, Australia

STEVEN C. RICKE
Department of Poultry Science, Room 338E, Kleberg Animal and Food Science Center, Texas A&M University, College Station, TX 77843-2472

ROY M. ROBINS-BROWNE
Department of Microbiology and Immunology, University of Melbourne, and Microbiological Research Unit, Murdoch Children's Research Institute, Parkville, Victoria 3052, Australia

PETER SETLOW
Department of Biochemistry, University of Connecticut Health Center, Farmington, CT 06030-3305

L. MICHELE SMOOT
Silliker Laboratories of Ohio, 1224 Kinnear Rd., Suite 114, Columbus, OH 43212

JAMES L. STEELE
Department of Food Science, University of Wisconsin-Madison, 1605 Linden Dr., Madison, WI 53706-1565

BALA SWAMINATHAN
Foodborne and Diarrheal Diseases Branch, Centers for Disease Control and Prevention, Mailstop C-03, 1600 Clifton Rd., Atlanta, GA 30333

STERLING S. THOMPSON
Microbiology Research, Technical Center, Hershey Foods Corporation, 1025 Reese Ave., Hershey, PA 17033-0805

RICHARD C. WHITING
Center for Food Safety and Applied Nutrition, HFS-300, U.S. Food and Drug Administration, 200 C St., S.W., Washington, DC 20204

KAREN WINKOWSKI
Department of Food Science, Rutgers-The State University of New Jersey, New Brunswick, NJ 08901-8520

IRENE ZABALA DIAZ
Department of Poultry Science, Room 327, Kleberg Animal and Food Science Center, Texas A&M University, College Station, TX 77843-2472

TONG ZHAO
Center for Food Safety and Quality Enhancement, University of Georgia, Griffin, GA 30223

SHAOHUA ZHAO
Division of Animal and Food Microbiology, Center for Veterinary Medicine/ Office of Research, U.S. Food & Drug Administration, Laurel, MD 20708

Reviewers

James N. Bacus
Saumya Bhaduri
J. Russell Bishop
Kathryn J. Boor
Scott L. Burnett
Robert G. Cassens
Fun S. Chu
Guy Cornelis
Tibor Deak
Andy DePaola
Siobain Duffy
Mel W. Eklund
Charles P. Gerba
Marcia Goldoft
Eugene G. Hayunga

Steven C. Ingham
Lee-Ann Jaykus
Michael G. Johnson
Sam W. Joseph
Ronald Labbe
Cathy Lewus
John E. Linz
Robert T. Marshall
Richard Meinersman
Aubrey F. Mendonca
Marguerite A. Neill
Ynes R. Ortega
Randall Phebus
Morrie Potter
N. Rukma Reddy

Donald Schaffner
Patrick M. Schlievert
Greg Siragusa
James L. Smith
John N. Sofos
Myron Solberg
Gerard N. Stelma, Jr.
Nancy Strockbine
Susan S. Sumner
Peter J. Taormina
Sita R. Tatini
Mikhail Tchikindas
Gloria L. Tetteh
Amy Wong

Preface to the Second Edition

The field of food microbiology is among the most diverse of the areas of study within the discipline of microbiology. Its scope encompasses a wide variety of microorganisms including spoilage, probiotic, fermentative, and pathogenic bacteria, molds, yeasts, viruses, and parasites; a diverse composition of foods; a broad spectrum of environmental factors that influence microbial survival and growth; and a multitude of research approaches that range from very applied studies of survival and growth of foodborne microorganisms to basic studies of the mechanisms of pathogenicity of harmful foodborne microorganisms. Several excellent books address many different aspects of food microbiology. The purpose of *Food Microbiology: Fundamentals and Frontiers* is to complement these books by providing new, state-of-the-art information that emphasizes the molecular and mechanistic aspects of food microbiology, and not to dwell on other aspects well covered in food microbiology texts. The second edition provides new information regarding recent advances in all aspects of food microbiology. In particular, major revisions have been made to chapters addressing foodborne pathogens, which is an area exploding with new findings.

This advanced text fulfills the need of research microbiologists, graduate students, and professors of food microbiology courses for an in-depth treatment of food microbiology. It provides current, definitive factual material written by experts on each subject. The book is written at a level which presupposes a general background in microbiology and biochemistry needed to understand the "how and why" of food microbiology at a basic scientific level.

The book is composed of nine major sections that address each of the major areas of the field. "Factors of Special Significance to Food Microbiology" provides a brief history of food microbiology; a perspective on and description of the basic principles that affect the growth, survival, and death of microbes;

coverage of spores; and the use of indicator microorganisms and microbiological criteria. "Microbial Spoilage of Foods" covers the principles of spoilage, dominant microorganisms, and spoilage patterns for each of three major food categories. The 13 chapters in the "Foodborne Pathogenic Bacteria" section provide a current molecular understanding of foodborne bacterial pathogens in the context of their pathogenic mechanisms, tolerance to preservation methods, and underlying epidemiology, as well as basic information about each microorganism's metabolic characteristics, symptoms of illness, and common food reservoirs. Similar perspectives are given by chapters in the sections "Mycotoxigenic Molds," "Viruses," and "Foodborne and Waterborne Parasites."

"Preservatives and Preservation Methods" presents information on mechanisms, models, and kinetics in three chapters which elucidate physical, chemical, and biological methods of food preservation. The "Food Fermentations" section emphasizes the genetics and physiology of microorganisms involved in fermentation of foods and beverages. The influence of fermentation on product characteristics is also examined.

Since rapid, genetic, and immunological methods for detecting foodborne microorganisms, the use of probiotics in promoting health, predictive modeling and quantitative risk assessment, and hazard analysis and critical control points are key issues to the future of food microbiology, it is appropriate that these topics are covered in the closing section, "Advanced Techniques in Food Microbiology."

We are grateful to all of our coauthors for their dedication to producing a book that is at the cutting edge of food microbiology, and to our reviewers whose critical evaluations enabled us to fine-tune each chapter.

Michael P. Doyle
Larry R. Beuchat
Thomas J. Montville

Factors of Special
Significance to
Food Microbiology

I

Food Microbiology: Fundamentals and Frontiers, 2nd Ed.
Edited by M. P. Doyle et al.
© 2001 ASM Press, Washington, D.C.

Paul A. Hartman

The Evolution of Food Microbiology

1

For centuries, humans have taken advantage of, as well as been plagued by, microorganisms and their activities in foods. This book describes the current state of food microbiology, whose ultimate goal is to provide an adequate, organoleptically satisfying, wholesome, and safe food supply. By necessity, this book represents the "state of the science" at a certain point in time. The current practice of food microbiology is, however, the culmination of decades of research and experience. Thus, food microbiology must be understood as an evolving science connected to both a history and a future. The evolution of food microbiology is an exciting story which is outlined in Fig. 1.1 and presented below.

EARLY DEVELOPMENTS IN FOOD PRODUCTION AND PRESERVATION

Food spoilage and food poisoning caused by microorganisms were problems that must have continually preoccupied early humans. Imagine the predicament of a Neanderthal (Europe, 50,000 B.C.) or Cro-Magnon (North America, 40,000 B.C.) who had just slaughtered a woolly mammoth and needed to preserve a ton of meat (20, 22). During the retreat of the last ice age, 10,000 to 20,000 years ago, what were formerly nomadic populations of hunters and gatherers domesticated both food crops and animals and turned to production agriculture in four major river civilizations—Egypt on the Nile, Sumer on the Tigris and Euphrates in Mesopotamia (now Iraq), India on the Indus, and China on the Yellow (21).

Studies continue even today to determine more precisely when food production originated. A recent study (24) of archaeological sites dated between 18,300 and 17,000 years ago revealed that barley flourished in the Nile Valley, near Aswan in Egypt. The practice of animal husbandry originated about 8,000 to 10,000 years ago. Evidence based on nucleic acid sequences indicates that there were two independent domestications of cattle (15) and a single domestication of chickens (10). The spread of agricultural practices from one society to another made it more possible to assure a stable food supply and encouraged community development. By 3,000 B.C., the people of Sumer (now Iraq) had developed a sophisticated agricultural economy. They constructed irrigation canals and grew a wide variety of crops and livestock (13, 20). The herders of Sumeria did not face food preservation problems as pressing as did the hunters that preceded them. Most livestock could be moved from

Reprinted from *Food Microbiology: Fundamentals and Frontiers* (first edition), 1997.

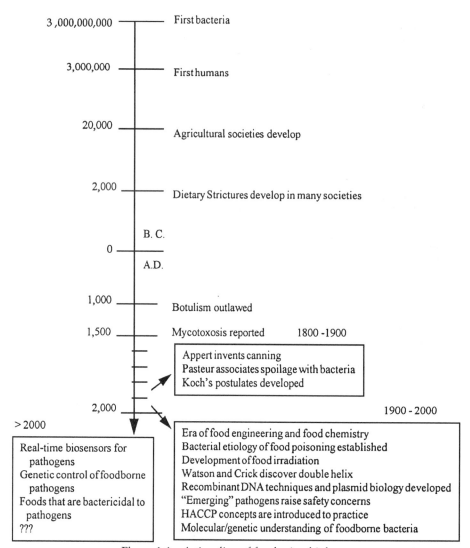

Figure 1.1 A time line of food microbiology.

place to place and slaughtered when needed. Livestock also had monetary value and could be considered as an early form of currency, as was salt. The ability to preserve food and store it from times of plenty to times of need was prerequisite for the shift from hunter-gatherers to an agricultural society.

The production of bread, alcoholic beverages (chapters 36 and 37), and a variety of acid-fermented foods (chapters 31 through 34), the preservation of meat and fish products by drying or adding salt (chapters 28 and 29), and the production of other indigenous foods (chapters 34 and 35) were critical to the development of stable societies. These microbial processes had multiple origins in different parts of the world (19). People also began to understand (although they did not yet know why) that foods should be kept away from contact with air, light,

and moisture. Some foods were preserved in early times by coating them with honey or clay and later with olive oil (22). Salt became an especially valuable commodity because it was essential to human health and useful for food preservation. However, salt was not available in quantity in many locales. According to some historians (22), the availability of salt, which was recognized as an important nutrient and food preservative, influenced the course of history. The salt of the Dead Sea was one reason for the Romans' interest in Palestine. If they had not colonized that country, the story of Christ and Christianity would have been quite different.

For thousands of years, people recognized that diseases could be spread by foods. Prohibitions on eating pork—in the Jewish and Muslim religions, for example—had their origins in medical doctrine.

Certainly, pork is a dangerous meat in a hot climate, and this must have been realized when dietary regulations were being formulated (20). Nevertheless, although the peoples of the Near East knew all about pork and its dangers, no taboos were placed on it until about 1,800 B.C. By the middle of the first millennium B.C., religious laws in India had also begun to list many "impure" items of diet. "Unclean food" included meat which had been cut with a sword, dog meat, human meat, and the meat of carnivorous animals, locusts, camels, and hairless or excessively hairy animals. Rice which had turned sour through being left to stand overnight, ready-made food from the market, and dishes which had been sullied by insects or mice or sniffed by a dog, cat, or human were all regarded as unfit for eating. Although most of these edicts carried the weight of religious sanction, they were fundamentally laws of simple hygiene and reflected similar practices derived in other countries (20).

AWARENESS OF BOTULISM AND OTHER FATAL FOODBORNE DISEASES

It was not until the 10th century A.D. that microbiological food poisoning was recognized in civil law. In 900 A.D., Emperor Leo VI of Byzantium issued an edict that forbade eating blood sausage prepared by stuffing blood into a pig stomach and preserving it by smoking. Those caught preparing blood sausage lost all their property and were exiled. Also, to keep civic authorities vigilant, the edict stated that the chief magistrate of the city would be fined 10 pounds of gold (about $35,000 in 1995 United States dollars). Emperor Leo's edict was proclaimed because of an association between botulism and blood sausage (20). The exact cause was unknown. At about that time, a common belief was that miasmas (infectious vapors) transmitted disease.

Poisoning by spoiled grains was recognized by the ancient Greeks and Romans, and many epidemics of major proportions occurred through the Middle Ages in Europe, Russia, and elsewhere. Ergotism, caused by growth of a mold, *Claviceps purpurea,* in grain, was recognized in 1582 and reported again around 1600. In the mid-16th century, epidemics were associated with scabrous rye and other grains infested with *C. purpurea,* and precautions were taken to avoid the contaminated grains. With few exceptions, these precautions were effective; the last major outbreak of ergot poisoning in the United States was in 1825.

Because its underlying causes were unknown, microbiological food poisoning was recurrent. Botulism reappeared many times. For example, in 1793 (9 centuries after Emperor Leo's edict), 13 were affected and 6 died in Germany after eating blood sausage. It was believed that the illness was caused by a fatty acid. The disease was made reportable and the product again was regulated (18).

THE ADVENT OF THERMAL PROCESSING

Because the difficulties with transportation and storage of foods were aggravated by war, in 1795 the French government offered a substantial award for a new preservation method. Nicholas Appert, a Paris confectioner, accepted the challenge and developed a method whereby wide-mouth glass bottles were filled with food, corked, and heated in boiling water. Appert won the prize in 1805 and, at the insistence of Napoleon I, published a book describing his process in 1810. That same year P. Durand of England patented the use of tin cans for thermally processed foods. Neither Appert nor Durand understood *why* thermally processed food did not spoil. Appert only recognized that the container must be sealed and heated to eliminate what he called "agents of putrefaction" and "fermentable principles." In an improvement on Appert's original canning process, I. Solomon, a Baltimore canner, developed in 1860 a simple process that enabled his cannery to increase its output from 2,500 to 20,000 cans per day. In this process, calcium chloride was added to the water in which cans were heated to increase the water temperature from boiling temperature (100°C) to 116°C. This modest increase in temperature reduced the cooking time from 6 h to only 25 to 40 min. R. Chevallier-Appert was issued a patent in 1853 for sterilization of food at even higher temperatures by using steam under pressure in an autoclave, and by 1874 commercial retorts had been introduced.

During this period, the causes of other types of food poisoning became known. For example, the cause of trichinosis in animals (chapter 25) was discovered in 1835 by J. Paget, and the first human case was described in 1859 by F. A. von Zenker. Little did anyone realize that Pasteur and other scientists would soon lay the foundation leading to the discovery of many microbes that caused food poisoning.

FOOD MICROBIOLOGY BECOMES A SCIENCE

Between 1854 and 1864, Louis Pasteur placed heat preservation methods on a scientific basis. He made many discoveries, including experimental proof that certain bacteria were associated with food spoilage and caused specific diseases. Thus, this food scientist became the father of food microbiology and the infant science was born. The first use of what we now know as "pasteurization," the heating of wine to destroy undesirable organisms, was introduced commercially in 1867–68. Meanwhile, in the 1800s, methods to cultivate microorganisms in pure culture and to associate specific bacteria as the causative agents of specific diseases were developed by R. Koch (who, in 1884, first isolated the cholera vibrio during a worldwide pandemic), J. Lister, and others (3, 13). The isolation and study of pure bacterial monocultures in the laboratory remained at the center of food microbiology for the next hundred years.

Although the cause of botulism, *Clostridium botulinum,* was discovered by E. Van Ermengem in 1896, botulism remained a major problem in thermally processed foods until the 1920s (18). The United States Public Health Service appointed a Botulism Commission, composed of a combination of industrial, academic, and federal experts, to recommend appropriate thermal processes for the prevention of botulism. Their report, published in 1922, emphasized the importance of heating canned foods sufficiently to destroy the most thermal-resistant spores. Consumer education for safely preparing home-canned foods also reduced the incidence of botulism. Nonetheless, to this day, the majority of botulism cases are caused by home canning. The importance of consumer education to the microbial safety of foods cannot be overemphasized. The Food and Drug Administration's general consumer education program through fiscal year 1998 targets the prevention of foodborne illness as the highest priority for its Center for Food Safety and Applied Nutrition.

In 1963, botulism again won notoriety when several outbreaks were attributed to smoked fish from the Great Lakes. Extensive studies by E. M. Foster and colleagues at the University of Wisconsin detailed the ecology of fish-borne *C. botulinum.* One result of the 1963 outbreaks was the realization that foods encased in plastic wraps that exclude oxygen to prevent the growth of aerobic spoilage bacteria and extend shelf life provide ideal conditions for the growth of anaerobic *C. botulinum* (chapter 15).

OTHER PIONEERING DEVELOPMENTS IN FOOD PRESERVATION

Many other pioneering developments on other food preservation processes occurred in the 1800s (13). In 1842, an English patent was issued to H. Benjamin for a process in which an ice-salt brine mixture was used to depress the freezing point to freeze foods more rapidly. An 1861 United States patent on freezing fish was issued to E. Piper of Maine. These patents were not used exten-

sively because refrigeration was in its infancy and there were problems in keeping the foods frozen. Clarence Birdseye, a Massachusetts inventor, when on an expedition to Labrador in 1915 noticed that fish and meat left out at −40 to −50°F (−40 to −46°C) froze solid almost immediately and tasted fresh when thawed months later. Birdseye sold his quick-freeze process to General Foods Corp. in 1929, but it was not until the 1950s that rapid-frozen foods gained popular acceptance. Powdered milk was produced in England in 1855. Pasteurized milk was sold in Germany by 1880 and in the United States by 1890. Commercially dried fruits and vegetables appeared in 1886.

Studies on the use of ionizing energy to preserve foods were initiated in 1925 by F. Ludwig and H. Hopf in Germany, but it was not until after World War II that intensive research commenced at the Massachusetts Institute of Technology and three industry locations. This consortium disbanded within a few years, however, when it became apparent that the project was becoming too expensive. The United States Army started, in 1948, an extensive study in collaboration with industry on the wholesomeness of irradiated foods. The Atomic Energy Commission joined the program in 1953. This National Food Irradiation Program made much progress, but it received a setback upon passage in 1958 of the Delaney Clause (described later), which defined ionizing energy as a food additive rather than a process. Some petitions for food-irradiation processes have been approved by the Food and Drug Administration, and several are presently in use (chapter 28). However, 70 years after its inception, food irradiation remains an underutilized technology.

If the period from the 1890s to the 1940s is described as the era of food preservation, then the era extending from the 1950s through the 1980s can be characterized as the era of "food science," based on chemistry and engineering. As the importance of water activity became clear, intermediate-moisture foods were introduced. Spray-dried and freeze-dried foods soon appeared in many pantries.

RECOGNITION OF OTHER BIOLOGICAL AGENTS OF FOOD POISONING (5)

A. A. Gärtner, in 1888, isolated from meat incriminated in a large food poisoning outbreak a bacterium subsequently named *Salmonella enteritidis.* The genus *Salmonella* was named in 1900 after a U.S. Department of Agriculture (USDA) bacteriologist, Dr. Salmon, who first described a member of the group, *Salmonella cholerae-suis,* which he thought caused hog cholera. (It was later discovered that a virus caused hog cholera and

Salmon's bacterium was an incidental isolation. Nevertheless, Salmon's name remains a part of the history of microbiology!) Salmonellosis remains a major problem (chapter 8). For example, one of the worst food poisoning incidents in the history of the United States occurred in 1985, when 16,284 cases and 7 deaths were documented when pasteurized milk somehow became recontaminated with a potent strain of *Salmonella typhimurium* (now termed *Salmonella enterica* serovar Typhimurium). In 1994, this was exceeded by a national outbreak of *S. enteritidis* affecting 225,000 people who consumed contaminated ice cream products (12). K. Shiga discovered in 1898 an enteric pathogen closely related to the salmonellae. Shiga's bacterium, which causes bacillary dysentery (chapter 12), was later named *Shigella dysenteriae*.

In 1906, an aerobic, spore-forming bacillus (chapter 17) was recognized as a cause of food poisoning. The true significance of this discovery did not become apparent until the taxonomy of the genus *Bacillus* was clarified by N. R. Smith and R. E. Gordon in 1946 and S. Hauge identified from among the numerous *Bacillus* species that the pathogen was always *Bacillus cereus*. Staphylococci (chapter 19) were first recognized by Pasteur, who found them in pus. They were not associated with food poisoning until studies by T. Denys in 1894 and M. A. Barber in 1914. Barber used himself as a test subject to show that milk that he obtained on a farm in the Philippines contained staphylococci that induced vomiting. The existence of an exotoxin was confirmed by G. M. Dack and coworkers, who studied cream-filled Christmas cakes in 1930.

In 1939, J. Schleifstein and M. B. Coleman described gastroenteritis caused by a bacterium that, in 1965, was named *Yersinia enterocolitica* by R. Sakazaki (chapter 11). However, it was not until 1969 that B. Nilehn documented a foodborne outbreak and 1971 when T. Dadisman and coworkers documented the first United States outbreak, which occurred in New York. Major attention was focused on yersiniosis in 1982, when pasteurized milk from a plant in Tennessee was implicated in a large interstate outbreak; 172 culture-positive infections were identified. Most patients required hospitalization, and 17 underwent appendectomies when their symptoms were mistakenly diagnosed as appendicitis.

Clostridium perfringens (formerly *Clostridium welchii*), the causative agent of human gas gangrene infections, was first implicated as a cause of foodborne illness by E. Klein in 1885. This type of food poisoning (chapter 16) went almost unrecognized until R. Knox and E. K. Macdonald in England in 1943 and L. S. McClung in the United States in 1945 alerted the scientific community about perfringens food poisoning. In 1953, B. Hobbs and coworkers in England reported that perfringens food poisoning was common but had been overlooked by most laboratories because anaerobic techniques were not used in food microbiology unless botulism was suspected. C. L. Duncan and D. W. Strong demonstrated, in 1969, that perfringens food poisoning was caused by an enterotoxin.

Milk had been implicated in the transmission of several diseases, so it was not surprising that G. Jubb, in 1915, reported milk to be the vehicle responsible for a small outbreak of poliomyelitis in England. Raw milk also was implicated in the transmission of infectious hepatitis (hepatitis A) virus by M. D. Campbell in 1943 in England, and an outbreak caused by contaminated shellfish in Sweden was described by B. Roos in 1956. In 1961, several large outbreaks of hepatitis A caused by clams and oysters occurred in Mississippi, Alabama, New Jersey, and Connecticut, focusing attention on foodborne viruses and especially the shellfish problem (see chapter 24).

In 1951, T. Fujino showed that *Vibrio parahaemolyticus* (chapter 13) caused food poisoning associated with seafood consumption in Japan. Within only a few years it became apparent that *V. parahaemolyticus* was responsible for over 70% of Japan's investigated cases of foodborne gastroenteritis. Twenty years after Fujino's work, in 1971, the first United States outbreak was documented in Maryland.

Food poisoning caused by molds again attracted attention when, in 1960–61, a massive outbreak of lethal hepatic "turkey X disease" occurred in England, involving over 100,000 turkey poults at 500 locations. *Aspergillus flavus* was isolated from Brazilian ground nut (peanut) meal and was shown by K. Sargeant and coworkers in 1961 to produce a toxin that caused acute hepatitis at high levels. Later, it was shown that the toxin was carcinogenic at lower levels. Within less than 10 years many different mycotoxigenic molds (chapters 21 to 23) were identified.

Various nematodes and cestodes (chapter 27) had been known for many decades to be transmitted by foods. Antonie van Leeuwenhoek, in 1681, examined his own stools during a bout with diarrhea and observed a protozoan, *Giardia lamblia*, in large numbers. He subsequently observed similar microbes in the guts of rodents and frogs, but did not associate animal reservoirs with disease transmission. This association was not fully appreciated until 1965, when a major waterborne outbreak of giardiasis occurred in Aspen, Colorado, and a series of other waterborne outbreaks occurred in the 1970s. Subsequently, other parasites have been confirmed as causes of major outbreaks of water- and foodborne infections and intoxications.

EMERGING FOODBORNE AND WATERBORNE PATHOGENS

The discovery (or sometimes rediscovery) of foodborne pathogens continues to this day. One of these pathogens is *Cryptosporidium*. This parasite was first described in 1907, and the first case was diagnosed in a human in 1976. Several small outbreaks occurred during the next 17 years, but the true danger of cryptosporidia was revealed when, in 1993, an estimated 403,000 persons in Milwaukee, Wisconsin, had prolonged diarrhea and approximately 4,400 required hospitalization during a waterborne outbreak.

Four relatively unknown bacteria became associated with foodborne illness in the 1980s (5, 6). One of these, *Vibrio fetus*, was identified as a cause of abortion in cattle and sheep in 1913 by J. McFadyean and S. Stockman in England. Although sporadic outbreaks of enteritis from milk had been described as early as 1938, *V. fetus* (now named *Campylobacter jejuni*) was considered mostly a veterinary pathogen. This status changed in 1977 following the publication of a report in the *British Medical Journal* by M. B. Skirrow, who described appropriate methods to isolate the bacterium. Once workers knew what to look for and how to conduct the search, campylobacteriosis was recognized as a leading cause of acute bacterial diarrhea worldwide (chapter 9). Another bacterium, enterohemorrhagic *Escherichia coli* O157:H7 (chapter 10), was first recognized as a pathogen in 1982. It causes bloody diarrhea and kidney failure. By 1987, *E. coli* O157:H7 was recognized as more common than *Shigella* spp. in the United States. In 1993 a large multistate outbreak from eating fast-food hamburgers occurred on the West Coast. The outbreak received wide media coverage, which greatly influenced public attitudes about meat sanitation and preparation. The reverberations of these outbreaks are still being felt. The public no longer accept the presence of pathogens in raw meat but are slow to accept their own role in assuring the microbial safety of food "from farm to fork."

Beginning in the 1970s, workers began to associate additional marine vibrios with severe tissue infections and death. *Vibrio vulnificus* (chapter 13), named in 1979, was identified as a cause of seafood-related illness with a high mortality rate in people who are immunologically compromised or have a preexisting hepatic disease. Foodborne *Listeria monocytogenes* (chapter 18) also affects the immunologically compromised. The first recognized outbreak of listeriosis occurred in Nova Scotia in 1981. The vehicle was coleslaw made from cabbage grown in a field in Maine that had been fertilized with sheep manure. After several other outbreaks, a large outbreak that occurred in California in 1985 focused attention on the unique epidemiology of listeriosis. Of 86 confirmed cases of listeriosis among people who had eaten Mexican-style soft cheese, over half were mother-infant pairs. The adoption of a zero tolerance in ready-to-eat foods and renewed attention to plant sanitation appear to have ended multicase listeriosis outbreaks, although sporadic cases still occur.

There have been no large confirmed outbreaks of food poisoning caused by *Aeromonas hydrophila* (named by M. Y. Popoff and coworkers in 1976) or *Plesiomonas shigelloides* (first described by W. W. Ferguson and N. D. Henderson in 1947 and named by H. Habs and R. H. W. Schubert in 1962). Nevertheless, much indirect evidence exists (chapter 14) that at least some strains cause gastroenteritis ranging from mild to severe, especially in the very young, aged, or otherwise immunologically compromised. An aging population and increasing prevalence of immunodeficiency diseases suggest that additional opportunistic pathogens will be discovered and have led to renewed interest in foodborne infections and intoxications. The struggle between humans and microbes is a continuing feature of the evolution of food microbiology. On the one hand, we become more knowledgeable about bacteria and develop better tools with which to identify and combat them. On the other hand, the current age of microbiology is barely a moment in the evolutionary history of bacteria, which extends 3 billion years into prehistory. Recent developments such as the spread of multiple drug resistance among clinically important pathogens warn against complacence.

MODERN LEGISLATION AND THE DEVELOPMENT OF PASTEURIZATION

Although the first British Food and Drugs Act was passed in 1860 and was revised and strengthened in 1872, it was not until 1890 that the United States passed its first food law, a National Meat Inspection Law. This law, which required the inspection of meats for export, was strengthened in 1895. Then, in 1906, a comprehensive Federal Food and Drug Act was approved by Congress; in 1939 a revised version became the Food, Drug and Cosmetics Act. Probably the most rigorous of the early United States food ordinances was the first U.S. Pasteurized Milk Ordinance, published in 1924 (25). It was a culmination of a century of experimentation and development; exactly 100 years earlier (in 1824), William Dewees recommended heating milk just to the boiling point to increase shelf life. In the United States, G. Borden, a Texas inventor, obtained a patent in 1853 for a process wherein milk was condensed under vacuum and heated to 50 to 60°C and sugar was added. The first commercial milk pasteurizer was made in Germany in 1882, and a variety of equipment and pasteurization times and temperatures ensued. In 1900, Russell and Hastings had defined the thermal death point of tubercle bacilli under commercial conditions, but their recommendations often were not followed or were unattainable because of faulty or poorly designed equipment. This oversight was corrected by the 1924 U.S. Pasteurized Milk Ordinance, which specified not only time and temperature, but most importantly, that the process must be conducted in approved equipment. Also during the 1920s, three associations representing sanitarians, manufacturers, and the milk industry worked with regulatory agents to formulate what are known as the "3-A" (for three associations) standards for the performance (such as cleanability) of dairy equipment (2). Included in a revised 1933 U.S. Public Health Service Milk Ordinance and Code was provision for high-temperature short-time (HTST) treatment of milk (chapter 28). Subsequently, in the 1950s, ultra-high-temperature (UHT) processing with aseptic packaging (chapter 28) became available.

During the past 70 years or so, milk pasteurization, together with a cattle-testing program, was so successful in reducing the incidence of tuberculosis in developed countries that it has been called "man's greatest victory over tuberculosis." This is attested to by the fact that tuberculosis is not included as a major food pathogen in this book (14). North America has also achieved freedom from brucellosis (undulant fever), another disease that once was transmitted by meat and milk. These are prime examples of what has been accomplished in food protection by cooperation and collaboration among farmers, veterinarians, food microbiologists, food scientists, food technologists, engineers, industry, and regulatory officials.

A U.S. Compulsory Poultry and Poultry Products Law was passed in 1957, a comprehensive Food Additives Amendment (often referred to as the Delaney Clause) was added to the Food, Drug and Cosmetics Act in 1958, a Wholesome Meat Act was enacted in 1967, and a Poultry Inspection Bill was passed in 1968.

THE "INDICATOR" CONCEPT AND THE DEVELOPMENT OF ANALYTICAL METHODS

When microbiological standards and regulations were being formulated, it was realized that tests could not be conducted for each and every enteric pathogen that might be present in a sample. Instead, a surrogate must be selected. A. von Fritsch of Germany suggested, in 1880, that certain klebsiellae were characteristic of human contamination of water supplies. Five years later, T. Escherich described a fecal bacterium, *Bacillus coli* (now named *Escherichia coli*), and in 1892, F. Schardinger suggested that *E. coli* would be useful as an indicator of fecal pollution (see chapter 4 on indicators and microbial criteria). At that time, methods to detect *E. coli* among a large group of related bacteria termed "coliforms" were not readily available. C. Eijkman, in 1904, determined that incubating test samples at 46°C would differentiate "fecal coliforms" that grow at high temperatures from coliforms that arose from other environments and do not grow at 46°C. Nevertheless, the U.S. Public Health Service, in 1914, changed the indicator standard from *E. coli* to the coliform group. Elaborate and time-consuming tests were conducted on isolates from positive coliform tests to determine if *E. coli*, the coliform bacterium whose presence correlated most closely with fecal pollution, was present. It was not until the 1980s, however, that relatively rapid, simple, and reliable methods for the simultaneous detection of both total coliforms and *E. coli* became available. Similarly, rapid methods were developed to detect and identify other indicator organisms such as enterococci, another bacterial group primarily of intestinal origin first described by M. E. Thiercelin in 1899. The need for indicator organisms was born at a time when the isolation and culture of specific pathogens was difficult, if not impossible. With the plethora of rapid and automated methods now available to test for specific pathogens and toxins it is difficult to justify continued reliance on indicators.

STANDARDIZATION OF METHODS

The promulgation of laws governing water and foods required uniform and efficient analytical methods, such as those described in chapter 38. At the turn of the century, workers became acutely aware that analytical results were almost useless for comparative purposes because of large laboratory-to-laboratory variations in methods. In 1895, the American Public Health Association recognized this need and appointed a committee to draft uniform procedures for the determination of several chemical attributes of water. In 1899, another committee was charged with extending standard procedures to all methods, including bacteriological, involved in water analysis. This committee's report, published in 1905, constituted the first edition of a manual, *Standard Methods of Water Analysis*. The book is now entitled *Standard Methods for the Examination of Water and Wastewater* (11). Also in 1905, S. C. Prescott of the Massachusetts Institute of Technology proposed that differences in composition of culture media, temperature and time of incubation, and other variables involved in the bacteriological examination of milk needed to be reconciled. A committee was formed and a report on *Standard Methods of Bacterial Milk Analysis* was published; this report evolved into another manual, *Standard Methods for the Examination of Dairy Products* (16). Another manual, *Recommended Methods for the Microbiological Examination of Foods,* was published in 1958. This manual eventually became the *Compendium of Methods for the Microbiological Examination of Foods* (23). Other organizations also publish compilations of microbiological procedures applicable to water and foods. One of these is the Association of Official Analytical Chemists International (AOAC), which since 1916 has validated methods used for regulatory purposes by subjecting the methods to collaborative studies to substantiate their accuracy and reproducibility. Approved methods are published in AOAC's *Official Methods of Analysis* (1, 4). In addition, the U.S. Food and Drug Administration publishes a *Bacteriological Analytical Manual*, which is now in its 8th edition (8). Also, throughout the years, as methodology has evolved, governmental agencies have published in the *Federal Register* methods acceptable for regulatory purposes (7, 9).

FROM DETECTION METHODS TO IN-PROCESS PREVENTION

In 1959, H. E. Baumann and others at the Pillsbury Co., in cooperation with the National Aeronautics and Space Agency, the U.S. Army Natick Laboratories, and the U.S. Air Force Space Laboratory Group, set out to produce almost completely safe foods for the space program. However, the sample-consumptive nature of food analysis was fundamentally different from quality assurance in electronics, where every item could be tested and then used. The amount of sample required to assure the desired safety level using statistically based microbial testing of finished products would require so much product that little would be left for use on the space flights. An alternative, preventive system of quality control, now known as the Hazard Analysis-Critical Control Point (HACCP) system (chapter 41), was developed. HACCP involves complete control over raw materials, process, environment, personnel, distribution, and storage (17). All of these had previously received attention, but not as an integrated system. Although the original emphasis was placed on raw materials from the time they arrived at the manufacturing plant on through to the consumer level, more recently the HACCP concept has included production agriculture, starting with the plant seed or individual animal production unit.

The HACCP system was first made public in 1971, but it was not seriously considered by the food industry until it was recommended in 1985 by the National Academy of Sciences Subcommittee on Microbiological Criteria for Foods and Food Ingredients. In 1989, the Subcommittee published a pamphlet (revised in 1992) titled "HACCP Principles for Food Production." HACCP has become widely accepted; for example, in 1995 the USDA Food Safety Inspection Service published in the *Federal Register* proposed regulations under which all slaughter and processing plants will be required to develop and implement a HACCP program within 3 years. Previously, in 1993, HACCP became international in scope when it was approved by the Committee of Food Hygiene of the Codex Alimentarius Commission. The Codex Alimentarius Commission is an international body responsible for the execution of the Joint FAO/WHO Food Standards Program, which was created in 1962 and is aimed at protecting the health of consumers and facilitating international trade in foods. The Codex consists of a large collection of food standards presented in a uniform manner. A well-documented HACCP plan also is required for certification of a food company under the recently formed European International Standards Organization (ISO) 9000 standards. The ISO 9000 is a voluntary process in which a third-party assessment certifies that a high-level quality assurance system is in place. Because of increased international trade, United States companies are discovering that ISO 9000 certification is necessary.

The International Commission on Microbiological Specifications for Foods (ICMSF), a standing committee of the International Association of Microbiological Societies, was formed in 1962. The ICMSF establishes internationally acceptable microbiological criteria and attempts to reach agreement on the essential supporting methods from among the plethora of methods in the literature.

THE ERA OF MOLECULAR BIOLOGY AND GENETICS

Until the 1960s, the practice of food microbiology was relatively unchanged since the time of Pasteur. It was a descriptive, qualitative science that focused on *what* happens with relatively little emphasis on, or understanding of, the *why* or underlying mechanisms. This observation is not meant to denigrate those pioneers of microbiology on whose foundations we build. One must know "what" before one can ask "why." The knowledge of molecular biology and modern experimental tools did not exist prior to the 1960s. Molecular biology was born in the 1940s and 1950s, when scientists from the physical sciences entered the field of biology with the express intent of applying the methods of physical science to biological phenomena. They brought with them a more quantitative and mechanistic approach to science which now permeates all areas of biology, including food microbiology. This second edition of *Food Microbiology: Fundamentals and Frontiers* has been written from this perspective. Wherever possible, the detailed mechanisms responsible for the topic at hand are stressed. It is clear, however, that in many cases, we have just begun to understand "what" happens, let alone "why."

Only within the last 30 years have fundamental concepts of biology, such as the genetic code, the structure-function relationship of proteins, the chemiosmotic coupling of energy-generating and -requiring reactions, and the transfer of genetic information been developed. Some of these are reviewed in chapter 2. In most cases, the general principles have been developed using relatively simple, well-studied bacteria such as *E. coli*. Frequently, foodborne microbes turn out to be quite different. Lactose catabolism is a case in point. The genes for lactose catabolism by *E. coli* were among the first to be studied in detail. The resultant *lac* operon is frequently used to teach the concepts of induction, derepression, carbon catabolite repression, and the role of protein kinases in the synthesis of the β-galactosidase which is ultimately excreted and cleaves lactose to glucose and galactose. However, from a practical standpoint, the most important application of lactose catabolism is in the dairy industry, where the lactose in milk is fermented by lactic acid bacteria such as *Lactococcus lactis*. Lactose catabolism by *L. lactis* is by a completely different mechanism from the *lac* operon model. *L. lactis* phosphorylates lactose as it is translocated across the cell membrane by the phosphoenolpyruvate:phosphotransferase system. The intracellular lactose phosphate is then hydrolyzed by an intracellular phospho-β-galactosidase.

The evolution of the dairy industry from a farm-based "art" to a highly technological industry provides other excellent examples of how basic science affects food microbiology. A fundamental understanding of the plasmid biology of fermentative organisms has reduced the incidence of "stuck" fermentations that have lost the ability to metabolize lactose. An understanding of the complex process by which bacteriophage attack and kill starter culture bacteria has generated many strategies for development of phage-resistant fermentations. Through the use of recombinant DNA technology, *E. coli* is rapidly replacing the fourth stomach of a milk-fed calf as the source of the rennet (chymotrypsin) used to make many cheeses.

Advances in molecular biology and genetics have revolutionized analytical food microbiology. Pasteur would be lost in a modern food microbiology laboratory. "Plate and count" microbiology is rapidly giving way to thermocyclers, gel boxes, microtiter plates, and ELISA readers which allow direct quantification of pathogens and their toxins (see chapter 38). "Rapid" salmonella tests have reduced analysis time from 5 days to less than 48 h. Methods being developed by companies from all over the world will consecutively break the 24-, 12-, 8-, and probably 4-h barriers. The ideal of a real-time biosensor for microbial contamination may one day be achieved.

CONCLUSION

The history of food microbiology is rich and exciting. It has taken us from the slow realization that certain diseases are caused by microorganisms that grow in foods to the empirical control of these microbes using physical, chemical, and biological manipulation. A mechanistic understanding of microbial physiology and metabolism has provided new approaches to food preservation and laid the foundation for genetic control of foodborne pathogens. We may one day be able to control foodborne pathogens by direct regulation of their genes. Control of insect pests in plants has progressed from the use of extrinsic chemical insecticides, to biocontrol using *Bacillus thuringiensis*, to the cloning of *B. thuringiensis* genes directly into plants to make them intrinsically insecticidal. Perhaps one day there will be salmonella-resistant chicken or listeria-resistant milk. Real-time biosensors may ultimately replace postprocess sampling.

Food microbiology stands on a scientific footing not more than 100 years old. Its current practice is being transformed by knowledge and tools generated by molecular biology and genetics. Its future, though hard to predict, will certainly be bright.

References

1. Andrews, W. H. 1994. Update on validation of microbiological methods by AOAC International. *J. AOAC Int.* 77:925–931.
2. Atherton, H. V. 1986. The 3-A story. *Dairy Food Sanit.* 6:96–98.

3. **Chung, K.-T., S. F. Stevens, and D. H. Ferris.** 1995. A chronology of events and pioneers of microbiology. *SIM News* 45:3–13.

4. **Cunniff, P. A. (ed.).** 1995. *Official Methods of Analysis of AOAC International,* 16th ed., vol. I and II. AOAC International, Arlington, Va.

5. **Doyle, M. P. (ed.).** 1989. *Foodborne Bacterial Pathogens.* Marcel Dekker, Inc., New York, N.Y.

6. **Doyle, M. P.** 1994. The emergence of new agents of foodborne disease in the 1980s. *Food Res. Int.* 27:219–226.

7. **Environmental Protection Agency.** 1986–1994. National primary drinking water regulations. *Fed. Regist.* 51:37608–37612; 52:42224–42245; 53:16348–16358; 54:27544–27568, 29998–30002; 55:22752–22756; 56:636–643, 49153–49154; 57:1850–1852, 24744–24747; 58:65622–65632; 59:6332–6444, 62456–62471.

8. **Food and Drug Administration.** 1992. *FDA Bacteriological Analytical Manual,* 7th ed. AOAC International, Arlington, Va.

9. **Food and Drug Administration.** 1993. Quality standards for foods with no identity standards: bottled water. *Fed. Regist.* 58:52042–52050.

10. **Fumihito, A., T. Miyake, S.-I. Sumi, M. Takada, S. Ohno, and N. Kondo.** 1994. One subspecies of the red junglefowl (*Gallus gallus gallus*) suffices as the matriarchic ancestor of all domestic breeds. *Proc. Natl. Acad. Sci. USA* 91:12505–12509.

11. **Greenberg, A. E., L. S. Clesceri, and A. D. Eaton (ed.).** 1992. *Standard Methods for the Examination of Water and Wastewater,* 18th ed. American Public Health Association, Washington, D.C.

12. **Hennessy, T. W., C. W. Hedberg, L. Slutsker, K. E. White, J. M. Besser-Wiek, M. E. Moen, J. Feldman, W. W. Coleman, L. M. Edmonson, K. L. MacDonald, M. T. Osterholm, and the Investigation Team.** 1996. A national outbreak of *Salmonella enteritidis* infections from ice cream. *N. Engl. J. Med.* 334:1281–1286.

13. **Jay, J. M.** 1992. *Modern Food Microbiology,* 4th ed. Van Nostrand Reinhold, New York, N.Y.

14. **Kapur, V., T. S. Whittam, and J. M. Musser.** 1994. Is *Mycobacterium tuberculosis* 15,000 years old? *J. Infect. Dis.* 170:1348–1349.

15. **Loftus, R. T., D. E. MacHugh, D. G. Bradley, P. M. Sharp, and P. Cunningham.** 1994. Evidence for two independent domestications of cattle. *Proc. Natl. Acad. Sci. USA* 91:2757–2761.

16. **Marshall, R. T. (ed.).** 1993. *Standard Methods for the Examination of Dairy Products,* 16th ed. American Public Health Association, Washington, D.C.

17. **Pierson, M. D., and D. A. Corlett, Jr.** 1992. *HACCP Principles and Applications.* Chapman & Hall, New York, N.Y.

18. **Smith, L. D. S.** 1977. *Botulism: the Organism, Its Toxins, the Disease.* Charles C Thomas Publishers, Springfield, Ill.

19. **Steinkraus, K. H. (ed.).** 1983. *Handbook of Indigenous Fermented Foods.* Marcel Dekker, Inc., New York, N.Y.

20. **Tannahill, R.** 1973. *Food in History.* Stein and Day Publishers, New York, N.Y.

21. **Thomas, H.** 1979. *A History of the World.* Harper & Row, Publishers, Inc., New York, N.Y.

22. **Toussaint-Samat, M. (translated by A. Bell).** 1992. *History of Food.* Blackwell Publishers, Cambridge, Mass.

23. **Vanderzant, C., and D. F. Splittstoesser (ed.).** 1992. *Compendium of Methods for the Microbiological Examination of Foods,* 3rd ed. American Public Health Association, Washington, D.C.

24. **Wendorf, F., R. Schild, N. El Hadidi, A. E. Close, M. Kobusiewicz, H. Wieckowska, B. Issawi, and H. Haas.** 1979. Use of barley in the Egyptian late paleolithic. *Science* 205:1341–1348.

25. **Westhoff, D. C.** 1978. Heating milk for microbial destruction: a historical outline and update. *J. Food Prot.* 41:122–130.

Food Microbiology: Fundamentals and Frontiers, 2nd Ed.
Edited by M. P. Doyle et al.
© 2001 ASM Press, Washington, D.C.

Thomas J. Montville
Karl R. Matthews

Principles Which Influence Microbial Growth, Survival, and Death in Foods

2

Food microbiologists must understand the basic biophysical principles of microbiology, must have a firm knowledge of food systems, and must be able to integrate both areas to solve the microbiological problems that occur in extremely complex food ecosystems. The first part of this chapter examines foods as ecosystems and discusses those intrinsic and extrinsic environmental factors that control the bacterial growth in foods. Ecology teaches that the *interaction* of environmental factors determines which organisms can or cannot grow in that environment. The manipulation of multiple environmental factors (e.g., pH, salt concentration, temperature) to inhibit microbial growth is the essence of multiple "hurdle" technology (64), which is increasingly being adopted as a technology for food preservation. Faced with changing environments, cells must maintain homeostasis in a variety of vital functions, such as internal pH and membrane fluidity. The ability of a cell to maintain homeostasis may ultimately determine its fate in foods (48).

Because many microbial processes are governed by first-order kinetics, fundamental kinetic concepts are explained in the second part of this chapter. The log phase of microbial growth and many types of lethality follow first-order or pseudo first-order kinetics. The doubling time, D values, and z values used by food microbiologists are kinetic constants based on first-order kinetics. Since food science is interdisciplinary and encompasses aspects of chemistry, engineering, and biology, food microbiologists should understand how microbiological rate constants relate to the kinetic constants used by chemists and engineers.

The third part of this chapter focuses on physiology and metabolism of foodborne microbes. The pathways for carbohydrate catabolism are especially important. The catabolic pathways used by the lactic acid bacteria dictate the characteristics of many fermented foods. In nonfermented foods, spoilage is usually associated with the appearance of a specific catabolic product(s). Lactic acid makes milk sour; carbon dioxide causes cans to swell. Foodborne microbes have a variety of catabolic pathways for processes as simple as glucose catabolism. The pathway used is determined by genetic and environmental parameters. The ability of bacteria to use different biochemical pathways which generate different amounts of ATP influences their ability to grow under adverse conditions in foods. For example, the pump that removes protons from the cytoplasm to maintain pH

Thomas J. Montville and Karl R. Matthews, Department of Food Science, Food Science Bldg., 65 Dudley Rd., Rutgers–The State University of New Jersey, New Brunswick, NJ 08901-8520.

homeostasis is fueled by ATP. Because facultative anaerobes generate more ATP by aerobic respiration than by anaerobic fermentation (see below), organisms like *Staphylococcus aureus* can grow at a lower pH and at lower water activities under aerobic conditions than under anaerobic conditions. The generation and utilization of energy, "bioenergenics," are critically important to the cell.

The limitations of classical microbiology are reviewed in the last section of this chapter. The power of techniques for genetic manipulation and development of improved instrumentation has changed our perception of the microbial world. The well-established concept that culturability reflects viability is under scrutiny. The inability of injured cells and cells that are "viable but not culturable" to form colonies on petri dishes does not mean that these cells do not exist. Indeed, there is a wealth of information suggesting that foodborne pathogens exist in the viable but not culturable state. Recent advances in microscopy have completely changed concepts of biofilm architecture. Bacteria living in biofilms on processing equipment or attached to surfaces are also physiologically different from the free-living cells studied in the laboratory.

FOOD ECOSYSTEMS, HOMEOSTASIS, AND HURDLE TECHNOLOGY

Foods as Ecosystems

Foods are complex ecosystems. This is so widely accepted that the International Commission on Microbiological Specifications for Foods chose *The Microbial Ecology of Foods* as the title of their books subtitled *Factors Affecting the Life and Death of Microorganisms* and *Food Commodities* (53, 54). Ecosystems are composed of the environment and the organisms that live in it. The food environment is composed of intrinsic factors inherent to the food (e.g., pH, water activity, and nutrients) and extrinsic factors external to it (e.g., temperature, gaseous environment, the presence of other bacteria). Intrinsic and extrinsic factors can be manipulated to preserve food, and food preservation can be viewed as "the ecology of zero growth" (14).

When applied to microbiology, ecology can be defined as "the study of the *interactions* between the chemical, physical, and structural aspects of a niche and the composition of its specific microbial population" (81). "Interactions" is emphasized to highlight the multivariable nature of ecosystems. The complex relationship between multiple environmental parameters in foods almost dictates a computer modeling approach. A very complete set of reviews about food ecosystems, from which this chapter draws heavily, has been published by the Society for Applied Bacteriology (13).

Foods can be heterogeneous on a micrometer scale. Heterogeneity and its associated gradients of pH, oxygen, nutrients, etc., are key ecological factors in foods (14). Foods may contain several distinct microenvironments. This is well illustrated by the food poisoning outbreaks in "aerobic" foods caused by the "obligate anaerobe" *Clostridium botulinum*. Growth of *C. botulinum* in potatoes, sautéed onions, and coleslaw exposed to air has caused botulism outbreaks (69). The oxygen in these foods is driven out during cooking and diffuses back in so slowly that the bulk of the product remains anaerobic. The growth of certain fungi and bacilli at the surface of tomato products can produce oxygen and pH gradients such that *C. botulinum* growth and toxin production can occur in a very limited area of the product (52, 80). This is also a good example of how ecosystems are nested, one inside the next, inside the next.

Intrinsic Factors That Influence Microbial Growth

Those factors inherent to the food itself are considered intrinsic factors. These include naturally occurring compounds that may stimulate or retard microbial growth, compounds added as preservatives, the oxidation-reduction potential, water activity, and pH. Most of these factors are covered separately in the chapters on physical and chemical methods of food preservation. The influence of pH on gene expression is a relatively new area. Genes encoding amino acid decarboxylases, lactate dehydrogenase, outer membrane proteins, and virulence factors are influenced by pH.

The expression of genes governing proton transport, amino acid degradation, adaptation to acidic or basic conditions, and even virulence can be regulated by the external pH (pH$_o$) (90). Cells sense changes in pH$_o$ through several different mechanisms. pH-induced protonation/deprotonation of amino acids can change the protein's secondary or tertiary protein structure, altering the protein's function which signals the change. The cell may respond to only one form of small signal molecules. For example, organic acids cross the cytoplasmic membrane only in the protonated form. An increased intracellular concentration would indicate increased environmental acidity. The transmembrane proton gradient (i.e., the ΔpH) itself can serve as a sensor and up- or down-regulate energy-dependent processes.

Intracellular pH (pH$_i$) must be maintained above some critical pH$_i$ at which intracellular proteins become irreversibly denatured. In *Salmonella enterica* serovar Typhimurium, where it has been studied most extensively, there are three progressively more stringent mechanisms to maintain a pH_i consistent with viability (44, 45). These three mechanisms are the homeostatic response, the acid tolerance response, and the synthesis of acid-shock proteins.

At $pH_o > 6.0$, salmonella cells adjust their pH_i through the homeostatic response. The homeostatic response maintains pH_i by allosterically modulating the activity of proton pumps, antiports, and symports to increase the rate at which protons are expelled from the cytoplasm. The homeostatic mechanism is constitutive and functions in the presence of protein synthesis inhibitors. In *Escherichia coli* it has been difficult to elucidate mechanisms of pH homeostasis; electron transport components pump hydrogen ions and contribute to the proton gradient, but the role they play in regulating pH_i has not yet been determined. During growth, *E. coli* does not require ATPase to maintain pH_i. However, in streptococci the proton-translocating ATPase regulates pH_i.

The acid tolerance response (ATR) is triggered by a pH_o of 5.5 to 6.0 (43, 44). This mechanism is sensitive to protein synthesis inhibitors; at least 18 ATR-induced proteins have been identified. ATR appears to involve the membrane-bound ATPase proton pump and maintains $pH_i > 5.0$ at pH_o values as low as 4.0. The loss of ATPase activity caused by gene disruption mutations or metabolic inhibitors abolishes the ATR but not the pH homeostatic mechanism described above. In enterobacteria at least four regulatory systems, an alternative sigma factor, two-component signal transduction system (PhoPQ), the major iron regulatory protein Fur, and Ada (involved in adaptive response to alkylating agents), are involved with acid survival (6). These systems may be activated depending on whether the stress is from an inorganic or organic acid (5). Acid adaptation occurs in *E. coli* O157:H7 and *Listeria monocytogenes* (66, 117). Acid adaptation (exposure to pH 5.5 for 1 h) of *L. monocytogenes* enhances its survival in yogurt, cottage cheese, and cheddar cheese (47). Loss of the gene encoding the general transcription factor σ^B in *L. monocytogenes* diminishes acid tolerance but has no effect on virulence, as characterized in a mouse model (117). The ATR may confer cross-protection to other environmental stressors. The exposure of *S. enterica* serovar Typhimurium cells to pH 5.8 for a few doublings induces 12 proteins, represses 6 proteins, and renders the cells less sensitive to salt and heat (65). Exposure of *S. enterica* serovar Typhimurium to short-chain fatty acids increases acid resistance (62). Following exposure to nisin, survival of acid-adapted *L. monocytogenes* is approximately 10-fold greater than that of nonadapted cells (113). Acid-adapted *L. monocytogenes* has increased resistance against heat shock, osmotic stress, and alcohol stress (94). Acid

adaptation of *E. coli* O157:H7 enhances thermotolerance (34).

The synthesis of acid-shock proteins is the third way that cells regulate pH_i. The synthesis of these proteins is triggered by pH_o from 3.0 to 5.0. They constitute a set of *trans*-acting regulatory proteins distinct from the ATR proteins. The majority of acid-induced proteins in *L. monocytogenes* are common for the responses to acid adaptation and acid stress (94). An array of genes are involved in acid resistance in gram-negative and gram-positive bacteria. Three stationary-phase-dependent acid resistance systems protect *E. coli* O157:H7 under extreme acid (pH 2.5 or less). These include the oxidative or glucose-repressed system, the glutamate decarboxylase system, and the arginine decarboxylase system. These systems have various levels of control and differing requirements for activity (23). DNA binding proteins interact with DNA to form stable complexes which protect the DNA from acid-mediated damage. Survival of an *E. coli* O157:H7 *dps* mutant is significantly less (4 log CFU/ml reduction) compared with the parent strain (1.0 log CFU/ml reduction) after acid (pH 1.8) challenge (26).

Other phenotypes can also be regulated by pH. The expression of the *Yersinia enterocolitica inv* gene in laboratory media at 23°C but not at 37°C seems paradoxical since its expression is required for infection of warm-blooded animals. However, at the pH of the small intestine (5.5), the *inv* gene is expressed at 37°C (93). The *yst* gene, which codes for a heat-stable enterotoxin in *Y. enterocolitica*, is regulated similarly (76). The *toxR* gene, which controls expression of cholera toxin in *Vibrio cholerae* is regulated in part by pH (83). In *Salmonella* spp., exposure to low pH enhances survival in macrophages, and in *Salmonella enterica* serovar Dublin genes on the resident virulence plasmid are induced by low pH (102).

Extrinsic Factors That Influence Microbial Growth

Temperature and gas composition are the primary extrinsic factors influencing microbial growth. Controlled and modified atmospheres are covered at length in chapter 29. The influence of temperature on microbial growth and physiology cannot be overemphasized. While the influence of temperature on growth kinetics is obvious and covered here in some detail, the influence of temperature on gene expression is equally important. Cells grown at refrigerated temperature are not just "slower" than those grown at ambient temperature, they express different genes and are physiologically different. Later chapters provide organism-specific detail about how temperature regulates phenotypes ranging from motility to virulence.

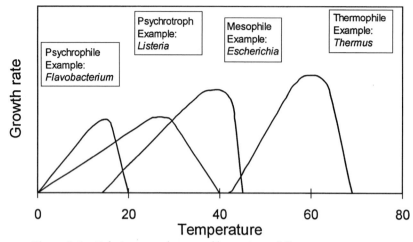

Figure 2.1   Relative growth rates of bacteria at different temperatures.

Influence of Temperature on Growth

A "rule of thumb" in chemistry suggests that reaction rates double with every 10°C increase in temperature. This simplifying assumption is valid for bacterial growth rates only over a limited range of organism-dependent temperatures (Fig. 2.1). Above the optimal growth temperature, the growth rates decrease precipitously. Below the optimum, growth rates also decrease, but more gradually. Bacteria can be classified as psychrophiles, psychrotrophs, mesophiles, and thermophiles according to how temperature influences their growth.

Both psychrophiles and psychrotrophs grow, albeit slowly, at 0°C. True psychrophiles have optimum growth rates at 15°C and cannot grow above 25°C. Psychrotrophs, such as *L. monocytogenes* and *C. botulinum* type E, have an optimum temperature of about 25°C and cannot grow above 40°C. Because these foodborne pathogens, and even some mesophilic *S. aureus,* strains, can grow at <10°C, conventional refrigeration is not sufficient to ensure the safety of food (92). Additional barriers to microbial growth should be incorporated into refrigerated foods containing no other inhibitors (84).

Several metabolic capabilities are important for growth in the cold. Homeoviscous adaptation enables cells to maintain membrane fluidity at low temperatures. As temperature decreases, the cell synthesizes increasing amounts of mono- and diunsaturated fatty acids (28, 100). The "kinks" caused by the double bonds prevent tight packing of the fatty acids into a more crystalline array. The accumulation of compatible solutes at low temperatures (59) is analogous to their accumulation under conditions of low water activity, as discussed in chapters 28 and 29. The membrane's physical state can influence and/or control expression of genes, particularly those that respond to temperature (114). The production of cold shock proteins (CSPs) contributes to an

organism's ability to grow at low temperatures. CSPs appear to function as RNA chaperones, minimizing the folding of mRNA, thereby facilitating the translation process. *Streptococcus thermophilus* CSPs are maximally expressed at 20°C. Northern blot analysis revealed a ninefold induction of *csp* mRNA and showed that its regulation takes place at the transcriptional level (119). Pretreatment at 20°C increases survival approximately 1,000-fold compared with nonadapted cells. *E. coli* CSPs have been categorized into two groups: class I proteins, which are expressed at low levels at 37°C and increase dramatically after shift to low temperature, and class II CSPs, which show a smaller increase after downshift in temperature (111). Cold-shocking *L. monocytogenes* from 37 to 5°C induces 12 CSPs with molecular weights ranging from 48,000 to 14,000 (4).

Temperature regulates the expression of virulence genes in several pathogens. The expression of 16 proteins on seven distinct operons on the *Y. enterocolitica* virulence plasmid is high at 37°C, weak at 22°C, and undetectable at 4°C (101). Similarly, the gene(s) required for virulence of *Shigella* spp. is expressed at 37°C but not at 30°C. The expression of genes required for *L. monocytogenes* virulence is also regulated by temperature (63). Cells grown at 4, 25, and 37°C all synthesize internalin, a protein required for penetration of the host cell. Cells grown at 37°C produce hemolytic activity, whereas this gene product is not produced at 4 or 25°C. However, the hemolytic activity is restored during the infection process (27). Temperature influences expression of *V. cholerae* *toxT* and *toxR* genes, which are essential for cholera toxin production. Maximal expression occurs at 30°C, whereas at 37°C expression is significantly decreased or abolished (83, 99). In enterohemorrhagic *E. coli,* temperature modulates transcription of the *esp* genes; synthesis of Esp proteins is enhanced when bacteria are grown at

37°C. Esp proteins are required for signal transduction events leading to the formation of attaching and effacing lesions linked to virulence of the pathogen (8).

The growth temperature can also influence a cell's thermal sensitivity. *L. monocytogenes* cells preheated at 48°C have increased thermal resistance (41). Holding *Listeria* cells at 48°C for 2 h in sausages increases their *D* values at 64°C by 2.4-fold. This thermotolerance is maintained for 24 h at 4°C (39). Subjecting *E. coli* O157:H7 cells to sublethal heating at 46°C increases their *D* value at 60°C by 1.5-fold. The increase of two proteins putatively, GroEL and DnaK, increased following heat shock (56). The role of heat shock proteins (HSPs) in increased thermal resistance is discussed further in chapter 28. In short, the heat shock response and regulated synthesis of HSPs in gram-negative bacteria can differ markedly from those is gram-positive bacteria. Many HSPs are molecular chaperones (e.g., DnaK and GroEL) or ATP-dependent proteases (e.g., Lon and ClpAP) and function in protein folding, assembly, transport, and repair under stress and nonstress conditions (120). Shock proteins synthesized in response to one stressor may provide cross-protection against other stressors (65). Exposing *Bacillus subtilis* to mild heat stress enables the organism to survive not only otherwise lethal temperatures but also exposure to toxic concentrations of NaCl (115). Three HSPs were identified based on their N-terminal amino acid sequence. Heat-adapted *Listeria* spp. (50°C for 45 min) have increased resistance to acid shock (94). Similarly, sublethal heat treatment of *E. coli* O157:H7 cells increases their tolerance to acidic conditions (116).

Influence of Quorum Sensing on Cell Growth

The discovery of cell-to-cell communication, or quorum sensing, has changed the belief that bacteria exist as individual cells in a population. Quorum sensing may provide bacteria protection from deleterious environments by precipitating adoption of new modes of growth, such as sporulation or biofilm formation. The phenomenon of quorum sensing is based on release by bacteria of autoinducers, which activate or repress target genes, when a threshold level is reached (reviewed in reference 33). In gram-negative bacteria (e.g., *Vibrio vulnificus*, *E. coli*, *Salmonella* spp.), *N*-acyl homoserine lactones generally act as signaling molecules (109). When in high enough concentration, these molecules can bind to and activate a transcriptional activator, which in turn induces expression of target genes. Signal molecules of gram-positive bacteria (e.g., *S. aureus* and *B. subtilis*) are usually small posttranslationally processed peptide signals. Quorum sensing has not been extensively studied in gram-positive bacteria.

In most bacteria, signaling molecules are at their greatest concentration during stationary phase; however, for *E. coli* O157:H7 and *S. enterica* serovar Typhimurium, quorum sensing is critical for regulating behavior in the prestationary phase of growth (108). In *E. coli*, the signal molecule exerts inhibitory activity during initiation of DNA replication rather than elongation (118). Moreover, unlike other gram-negative bacteria, the inhibitory activity does not require transcription or translation to be effective (108).

Quorum sensing likely has a role in food safety in association with behavior of a bacterium in a food matrix. Control of sporulation, germination of spores, formation of biofilm on food surfaces and food contact equipment, and production of virulence factors (toxins) may be regulated through quorum sensing.

Homeostasis and Hurdle Technology

Consumer trends toward blander foods have decreased the use of intrinsic factors such as acidity and salt as the sole means of providing food safety (48). Many food products use multiple hurdle technology to inhibit microbial growth. Instead of setting one environmental parameter to the extreme limit for growth, hurdle technology "deoptimizes" a variety of factors (64). For example, a limiting water activity of 0.85 or a limiting pH of 4.6 prevents the growth of foodborne pathogens. Hurdle technology might obtain similar inhibition at pH 5.2 and a water activity of 0.92. Mechanistically, hurdle technology assaults multiple distinct homeostatic processes (49). Intracellular pH must be maintained within relatively narrow limits. This is done by using energy to pump out protons, as described below. In low-water-activity environments, cells must use energy to accumulate compatible solutes, as described in chapter 29. Membrane fluidity must be maintained through homeoviscous adaptation (100), another energy-requiring process. The expenditure of energy to maintain homeostasis is fundamental to microbial life. When cells channel the energy needed for biosynthesis into maintenance of homeostasis, their growth is inhibited. When the energy demands of homeostasis exceed the cell's energy-producing capacity, the cell dies.

THE IMPORTANCE OF FIRST-ORDER KINETICS

Growth Kinetics

Food microbiology is concerned with all four phases of microbial growth. Growth curves showing the lag, exponential, stationary, and death phases of a culture are normally plotted as the number of cells on a log scale or $\log_{10}$ cell number versus time. These plots represent the state of microbial populations rather than individual microbes. Thus, both the lag phase and stationary phase of growth represent periods when the growth rate equals the death rate to produce no net change in cell numbers.

During the lag phase, cells adjust to their new environment by inducing or repressing enzyme synthesis and activity, initiating chromosome and plasmid replication, and, in the case of spores, differentiating into vegetative cells (see chapter 3). The length of the lag phase depends on the current temperature, the inoculum size (larger inocula usually have shorter lag phases), and the physiological history of the organism. If actively growing cells are inoculated into an identical fresh medium at the same temperature, the lag phase may vanish. Conversely, these factors can be manipulated to extend the lag phase beyond the time when some other food quality attribute (such as proteolysis or browning) becomes unacceptable. Foods are generally considered microbially safe if obvious spoilage precedes microbial growth. However, "spoiled" is a subjective and culturally biased concept. It is safer to manipulate factors to produce conditions in which the cell cannot grow regardless of time (such as reduction of pH to <4.6 to inhibit botulinal growth).

During the log, or exponential, phase of growth, bacteria reproduce by binary fission. One cell divides into two cells, which divide into four cells, which divide into eight cells, and so on. Thus, during exponential growth, first-order reaction kinetics can be used to describe the change in cell numbers. Food microbiologists often use doubling times as the kinetic constant to describe the rate of logarithmic growth. Doubling times (t_d), which are also referred to as generation times (t_{gen}), are related to classical kinetic constants, as shown in Table 2.1.

The influence of different parameters on a food's final microbial load can be illustrated by manipulating the equations in Table 2.1. Equation 1a states that the number of organisms (N) at any time is directly proportional to the initial number of organisms (N_0). Thus, decreasing the initial microbial load 10-fold will reduce the cell number at any time by 10-fold, although at extended times the population from the lower inoculum may reach the same final number. Because the instantaneous specific growth rate (μ) and time are in the power function of the equation, they have more marked effects on N. Consider a food in which $N_0 = 1 \times 10^4$ CFU/g and $\mu = 0.2$ h^{-1} at 37°C. After 24 h, the cell number would be 1.2×10^6 CFU/g. Reducing the initial number by 10-fold will reduce the number after 24 h by 10-fold to 1.2×10^5 CFU/g. However, reducing the temperature from 37 to 7°C has a much more profound effect. If one makes the simplifying assumption that the growth rate decreases twofold with every 10°C decrease in temperature, then μ will be decreased eightfold to 0.025 h^{-1} at 7°C. When equation 1a is solved using these values (e.g., $N = 10^4 e^{0.025 \times 24}$), then N at 24 h is 1.8×10^4 CFU/g. Both time and temperature have much greater influence over the final cell number than does the initial microbial load.

Equation 3a can be used to determine how long it will take a microbial population to reach a certain level. Consider the case of ground meat manufactured with an $N_0 = 1 \times 10^4$ CFU/g. How long can it be held at 7°C before reaching a level of 10^8 CFU/g? According to equation 3, $t = [2.3(\log 10^8/10^4)]/0.025$, or 368 h.

Food microbiologists frequently use t_d to describe growth rates of foodborne microbes. The relationship between t_d and μ is more obvious if equation 2a is written using natural logs (i.e., $\ln[N/N_0] = \mu \Delta t$) and solved for the condition where $t = t_d$ and $N = 2N_0$. Since the natural log of 2 is 0.693, the solution for equation 2a is $0.693/\mu = t_d$ (equation 4a). The average rate constant, k, defined as the number of generations per unit time (i.e., $1/t_{gen}$), is also used by applied microbiologists. The instantaneous growth rate constant, μ, is related to k by the equation $\mu = 0.693k$. Both rate constants characterize populations in the exponential phase of growth.

Table 2.1 First-order kinetics can be used to describe exponential growth and inactivation

Growth[a]	Thermal inactivation[b]	Irradiation[c]
1a. $N = N_0 e^{\mu t}$	1b. $N = N_0 e^{-kt}$	1c. $N = N_0 e^{-D/D_0}$
2a. $2.3\log(N/N_0) = \mu \Delta t$	2b. $2.3\log(N/N_0) = -(k\Delta t)$	
3a. $\Delta t = [2.3\log(N/N_0)]/\mu$	3b. $\Delta t = -[2.3\log(N/N_0)]/k$	
4a. $t_d = 0.693/\mu$	4b. $D = 2.3/k$	
	5b. $E_a = \frac{2.3 R T_1 T_2}{Z} \times \frac{9}{5}$	

[a] N, cell number (CFU/g); N_0, initial cell number (CFU/g); t, time (h); μ, specific growth rate (h^{-1}); t_d, doubling time (h).
[b] k, rate constant (h^{-1}); D, decimal reduction time (h); E_a, activation energy (kcal/mol); $T_1 T_2$, reference temperature and test temperature (Kelvin).
[c] D_0, rate constant (h^{-1}); D, dose (Gy).

Table 2.2 Representative specific growth rates and doubling times of microorganisms

Organism	μ (h^{-1})	t_d (h)
Bacteria		
Optimal conditions	2.3	0.3
Limited nutrients	0.20	3.46
Psychrotroph, 5°C	0.023	30
Molds, optimal	0.1–0.3	6.9–20

A more detailed explanation of the differences between these rate constants is given by Brock and Madigan (18). Some typical specific growth rates and doubling times are given in Table 2.2.

Death Kinetics

The killing of microbes by energy input (equations 1b and 1c), acid, bacteriocins, and other lethal agents is also governed by first-order kinetics. If one knows the initial microbial number, the first-order rate constant, and the time of exposure, one can predict the number of viable cells remaining. In food microbiology, the D value (amount of time required to reduce N_o by 90%) is the most frequently used kinetic constant. The use of D values in thermobacteriology is more fully covered in chapter 28. D values are inversely proportional to the rate constant, k, as shown in equation 4b. Both D and k values are defined for a given temperature (T). The relationship between k and T is related to the activation energy, E_a, as determined by the Arrhenius equation $k = s^{-E_a/RT}$, where s is the frequency constant, R is the ideal gas constant, and T is Kelvin. In thermobacteriology, the relationship between D and T is given by the z value. The z value is defined as the number of degrees Fahrenheit required to change the D value by a factor of 10. The z value is related to the E_a by the equation $Z = 2.3RT_1T_2/E_a \times (9/5)$, where T_1 and T_2 are actual and reference temperatures. A z value of 18°F equals an E_a of about 40 kcal/mol.

MICROBIAL PHYSIOLOGY AND METABOLISM

The Second Law of Thermodynamics dictates that all things progress to the state of maximum randomness in the absence of energy input. Since life is a fundamentally ordered process, all living things must generate energy to maintain their ordered state. Foodborne bacteria do this through the oxidation of reduced compounds. Oxidation only occurs in a chemical couple where the oxidation of one compound is linked to the reduction of another. In the case of aerobic bacteria, the initial carbon source, glucose, is oxidized to carbon dioxide, oxygen is reduced to water, and 38 ATPs are generated. Most of the ATP is generated through oxidative phosphorylation in the electron transport chain. In oxidative phosphorylation, the energy of the electrochemical gradient generated when oxygen is used as the terminal electron acceptor drives the formation of a high-energy bond between inorganic phosphate and an adenine nucleotide. Anaerobic bacteria, which lack functional electron transport chains, must reduce an internal compound through the process of fermentation and generate only 1 or 2 mol of ATP per mol of hexose catabolized. In this case, ATP is formed by substrate-level phosphorylation, and the phosphate group is transferred from an organic compound to the adenine nucleotide.

Glycolytic Pathways—Carbon Flow and Substrate-Level Phosphorylation

Embden-Meyerhof-Parnas Pathway

The most commonly used pathway for glucose catabolism (glycolysis) is the Embden-Meyerhof-Parnas (EMP) pathway (Fig. 2.2). In many organisms, the pathway is bidirectional (i.e., amphibolic) and can work in the direction of glucose, glycogen, and starch synthesis. The overall rate of glycolysis in this pathway is regulated by the activity of phosphofructokinase. This enzyme converts fructose-6-phosphate to fructose-1,6-bisphosphate. Phosphofructokinase activity is subject to allosteric regulation, whereby the binding of AMP or ATP at one site inhibits or stimulates (respectively) the phosphorylation of fructose-6-phosphate at the enzyme's active site. Fructose-1,6-bisphosphate activates lactate dehydrogenase (see below) so that the flow of carbon to pyruvate is tightly linked to the regeneration of NAD when pyruvate is reduced to lactic acid.

Another key enzyme of the EMP pathway is aldolase. The ultimate fermentation end products generated by the catabolism of pentoses and hexoses is partially determined by which enzyme converts the sugars to smaller units. Aldolase cleaves one molecule of fructose-1,6-bisphosphate to two three-carbon units, dihydroxyacetone phosphate and glyceraldehyde-3-phosphate. Other glycolytic pathways use keto-deoxyphosphogluconate aldolase to make two three-carbon units or phosphoketolase to produce one two-carbon compound and one three-carbon unit. Substrate-level phosphorylation generates a net gain of two ATPs when 1,3-diphosphoglycerate and phosphoenolpyruvate donate phosphoryl groups to ADP.

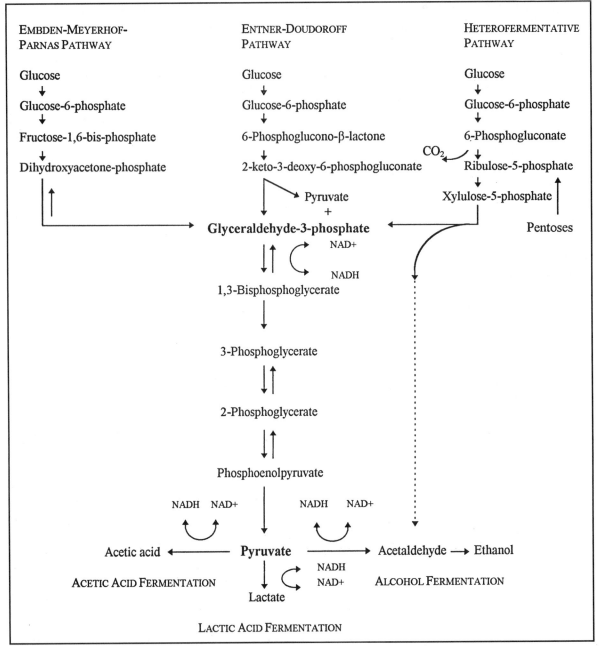

Figure 2.2 Major catabolic pathways used by foodborne bacteria.

Entner-Doudoroff Pathway

The Entner-Doudoroff pathway is an alternate glycolytic pathway that yields one ATP per molecule of glucose and diverts one three-carbon unit to biosynthetic pathways. In aerobes that use this pathway, such as *Pseudomonas* species, the difference between forming one ATP by this pathway versus the two ATPs formed by the EMP pathway is inconsequential compared with the 34 ATPs formed from oxidative phosphorylation. In the Entner-Doudoroff pathway, glucose is converted to 2-keto-3-deoxy-6-phosphogluconate. The enzyme keto-deoxy phosphogluconate aldolase cleaves this to one molecule of pyruvate (directly, without the generation of an ATP) and one molecule of 3-phosphoglyceraldehyde. The 3-phosphoglyceraldehyde is then catabolized by the same enzymes used in the EMP pathway, with the generation of one ATP by substrate-level phosphorylation using phosphoenolpyruvate as the phosphoryl-group donor.

Heterofermentative Catabolism

Heterofermentative bacteria, such as leuconostoc and some lactobacilli, have neither aldolases nor keto-deoxy-phosphogluconate aldolase. The heterofermentative pathway is based on pentose catabolism. The pentose can be obtained by transport into the cell or by intracellular decarboxylation of hexoses. In either case, the pentose is converted to xylulose-5-phosphate with ribulose-5-phosphate as an intermediate. The xylulose-5-phosphate is cleaved by phosphoketolase to a glyceraldehyde-3-phosphate and a two-carbon unit which can be converted to acetaldehyde, acetate, or ethanol. Although this pathway yields only one ATP, it offers cells a competitive advantage by allowing them to utilize pentoses which homolactic organisms cannot catabolize.

Homofermentative Catabolism

Homofermentative bacteria in the genera *Lactococcus,* *Pediococcus,* and some *Lactobacillus* species produce lactic acid as the sole fermentation product. The EMP pathway is used to produce pyruvate, which is then reduced by lactate dehydrogenase, forming lactic acid and regenerating NAD. Some lactobacillus species, such as *Lactobacillus plantarum* (112), are characterized as "facultatively heterofermentive." Hexoses are their preferred carbon source and are metabolized by the homofermentative pathway. If only pentoses are available, the cell shifts to a heterofermentative mode. When grown at low hexose concentrations, these bacteria do not make enough fructose-1,6-bisphosphate to activate their lactate dehydrogenase. This also causes them to shift to heterofermentative catabolism.

The Tricarboxylic Acid Cycle

The tricarboxylic acid (TCA) cycle links gylcolytic pathways to respiration. It generates $NADH_2$ and FADH as substrates for oxidative phosphorylation while providing additional ATP through substrate-level phosphorylation. With each turn of the TCA cycle, 2 pyruvate + 2 ADP + 2 FAD + 8 NAD → 6 CO_2 + 2 ATP + 2 $FADH_2$ + 8 NADH. Succinic acid, oxaloacetate, and α-ketoglutarate link the TCA cycle to amino acid biosynthesis. The TCA cycle is used by all aerobes, but some anaerobes lack all of the enzymes required to have a functional TCA cycle.

The TCA cycle is also the basis for two industrial fermentations important to the food industry. The industrial fermentations for the acidulant citric acid and

the flavor enhancer glutamic acid have a similar biochemical basis (31). Both fermentations take advantage of impaired TCA cycles. The production of citric acid uses mutants of *Aspergillus niger* and *Aspergillus wenti,* which do not express α-ketoglutarate dehydrogenase and have decreased activities of isocitrate dehydrogenase and aconitase during the stationary phase of growth. This causes citric acid to accumulate. Other reactions must then regenerate oxalacetate so that the flow of carbon to citrate continues. For example, pyruvate carboxylase mediates a condensation reaction between pyruvate and CO_2 to provide oxalacetate.

The production of glutamic acid is similar in principle. Glutamic acid producers such as *Corynebacterium glutamicum* are also deficient in α-ketoglutarate dehydrogenase. They have increased levels of glutamate dehydrogenase, so glutamic acid accumulates while other shunts are used to continue the flow of carbon into the TCA cycle. Because membranes are impermeable to charged compounds, glutamic acid accumulates and inhibits its own synthesis unless steps to permeabilize the membrane are taken. The membrane can be permeabilized by using biotin or oleic acid auxotrophs impaired in membrane synthesis. When fed biotin or oleic acid in limiting amounts, these cells have leaky membranes. Addition of saturated fatty acids or penicillin also permeabilizes membranes.

Aerobes, Anaerobes, the Regeneration of NAD, and Respiration

The flow of carbon to pyruvate always consumes NAD. Regardless of the pathway used to generate pyruvate, NAD must be regenerated for continued catabolism. When $NADH_2$ is oxidized to NAD, another compound must be reduced, i.e., serve as an electron acceptor. Aerobes having functional electron transport chains use molecular oxygen as the terminal electron acceptor during the oxidative phosphorylation that is the hallmark of respiration. As electrons travel down the electron transport chain, protons are pumped out, forming a proton gradient across the membrane. This proton gradient can be converted to ATP by the action of the BF_0F_1 ATPase. Oxidation of $NAD(P)H_2$ yields three ATPs. Oxidation of $FADH_2$ yields two ATPs. ATP and NADH are thus, in a sense, interconvertible. Sulfur and nitrite can also serve as terminal electron acceptors in "anaerobic respiration."

Anaerobes, in contrast, have a fermentative metabolism. Fermentations oxidize carbohydrates in the absence of an external electron acceptor. The final electron acceptor is an organic compound produced from the degradation of the carbohydrate. In the most obvious case, pyruvic acid is the terminal electron acceptor when it accepts an electron from NADH and is reduced to lactic acid. Paradoxically, some anaerobes are aerotolerant and can generate more energy in the presence of low levels of oxygen than in its absence. For example, some lactic acid bacteria have inducible NADH oxidases that regenerate NAD by reducing molecular oxygen to H_2O_2 (112). This spares the use of pyruvate as an electron acceptor and allows it to be converted to acetic acid with the concomitant generation of an additional ATP. Catalase-negative cells are inhibited by H_2O_2, but some lactic acid bacteria have NADH peroxidase, which detoxifies the H_2O_2. Obligate anaerobes, such as *C. botulinum*, have no means to detoxify H_2O_2, and they die rapidly when exposed to air.

Bioenergetics

All catabolic pathways generate energy with which the bacteria can perform useful work. Energy generation and utilization play a critical role in the life of bacteria. Several excellent reviews (57, 71) and books (51, 85) on bioenergetics provide additional depth and clarity on this topic. The preceding section on microbial biochemistry stressed the role of ATP in the cell's energy economy, but transmembrane gradients of other compounds play an equally important role. Transmembrane gradients release energy when one compound moves from high concentration to low concentration (i.e., with the gradient). This energy can be coupled to the transport of a second compound from a low concentration to a high concentration (i.e., against the gradient).

According to Mitchell's chemiosmotic theory (77), the proton motive force (PMF) has two components. An electrical component, the membrane potential ($\Delta\Psi$), represents the charge potential across the membrane. The pH gradient (ΔpH) across the membrane is the second component. Together, these compose the PMF, as stated by the equation PMF $= \Delta\Psi - z\Delta pH$. In this equation, $z = 2.3RT/F$, where R is the gas constant, T is the absolute temperature, and F is the Faraday constant. The factor z converts the pH gradient into millivolts and has a value of 59 mV at 25°C. The PMF is defined as being interior negative and alkaline, resulting in a negative value. (In the equation above, $z\Delta pH$ is not being subtracted from $\Delta\Psi$ but makes this negative term more negative.) There also is some interconversion of the $\Delta\Psi$ and the ΔpH components of PMF. If, for example, the ΔpH component decreases when an organism is transferred to a more neutral environment, the cell compensates by increasing $\Delta\Psi$ so that the total PMF remains relatively constant. PMF values can be as high as −200 mV

for aerobes, or in the range of −100 mV to −150 mV for anaerobes. Protein phosphorylation, flagellar synthesis and rotation, reversed electron transfer, and protein transport use PMF as an energy source (51).

PMF is generated by several mechanisms (Fig. 2.3). The translocation of protons down the electrochemical gradient during respiration generates a proton gradient when oxygen is used as the terminal electron acceptor. The oxidation of NADH is accompanied by the export of enough protons to make three ATPs. Proton gradients are also established during anaerobic respiration when nitrate or sulfate serves as the electron acceptor. The proton gradient is converted to ATP by the BF_0F_1 ATPase when it is driven in the direction of ATP synthesis (71). The bacterial BF_0F_1 ATPase is nearly identical to chloroplast and mitochondrial BF_0F_1 ATPases.

The BF_0F_1 ATPase is reversible. Aerobes use it to convert PMF to ATP. In anaerobes, it converts ATP to PMF. Maintaining internal pH homeostasis may be the principal role of the BF_0F_1 ATPase in anaerobes (57). Internal pH not only influences the activity of cytoplasmic enzymes but also regulates the expression of genes responsible for functions ranging from amino acid degradation to virulence (90). Anaerobes deacidify their cytoplasm using the BF_0F_1 ATPase to pump protons out. The proton pumping is driven by ATP hydrolysis. Some of the energy lost from ATP hydrolysis can be recovered if the resultant proton gradient is used to perform useful work, such as transport (see below). Most bacteria maintain their internal pH (pH_i) near neutrality, but lactic acid bacteria can tolerate lower pH_i values and expend less ATP on pH homeostasis. Acid-induced death is the direct result of an excessively low pH_i (44).

Given their limited capacity for ATP generation, it is not unexpected that some lactic acid bacteria can also generate ΔpH by ATP-independent mechanisms. The electropositive excretion of protons with acidic end products (75) has been demonstrated for lactate and acetate. For example, under some conditions, *L. plantarum* excretes three protons per molecule of acetate, thus sparing one ATP (112). The antiport (see below) exchange of precursor and product in anion-degrading systems, such as the malate^{2-}:lactate^{1-} exchange of the malolactic fermentation, might contribute to the generation of $\Delta\Psi$ (71).

Bacteria have evolved several mechanisms to achieve similar ends. The accumulation of compounds against a gradient (i.e., transport) is work and requires energy. In the case of primary transport systems and group translocation, this work is done by phosphoryl group transfer (Fig. 2.4). Secondary transport systems are fueled by the energy stored in the gradients that comprise the PMF.

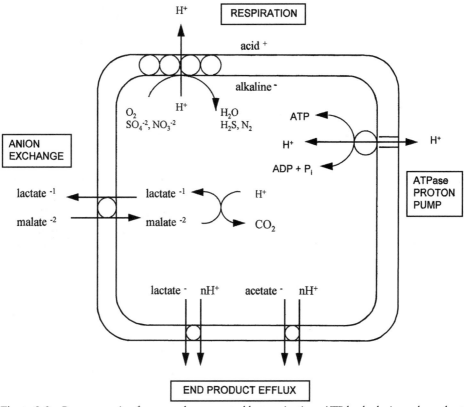

Figure 2.3 Proton motive force can be generated by respiration, ATP hydrolysis, end-product efflux, or anion exchange mechanisms. Modified from reference 73.

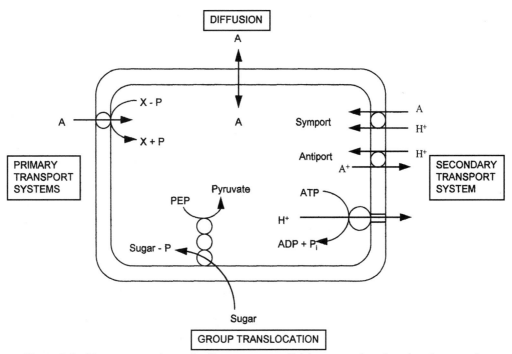

Figure 2.4 Transport can be at the direct expense of high-energy phosphate bonds or can be linked to the proton gradient of the proton motive force.

LIMITATIONS OF CLASSICAL MICROBIOLOGY

Limitations of Plate Counts

All methods based on the plate count and pure culture microbiology have the same limitations. The plate count is based on the assumptions that every cell forms one colony and that every colony originates from one cell. Accepting for the moment the underlying assumption that the free monocultured cell is an appropriate unit of study, the ability of a given cell to form a colony depends on a myriad of factors, including the physiological state of the cell, the medium used for enumeration, the incubation temperature, and so on. Table 2.3 (68) illustrates these points by providing D values at $55°C$ for *L. monocytogenes* with different thermal histories (heat shocked for 10 min at $80°C$ or not heat shocked), on selective (McBride's) or nonselective (TSAY) medium under aerobic or anaerobic atmospheres. The D values for *E. coli* O157:H7 are affected in a similar fashion (82). Injured cells and cells that are "viable but nonculturable" pose special problems to food microbiologists as discussed below.

Injury

Microorganisms may be injured rather than killed by sublethal levels of stressors such as heat, radiation, acid, or sanitizers. This injury is characterized by decreased resistance to selective agents or by increased nutritional requirements (55). Molecular level events associated with injury are likely complex and have yet to be defined. Food microbiologists became aware of the importance of microbial injury in the 1970s. There are excellent reviews from this pre-computer-database era dealing with injury in general (19) and injury in bacterial spores (1, 43), yeasts and molds (12, 107), and gram-negative bacteria (11).

Table 2.3 The influence of thermal history and enumeration protocols on experimentally determined D values at $55°C$ for *L. monocytogenes* (44)

Protocol	D_{55} values (min)	
	Aerobic atmosphere	Anaerobic atomosphere
TSAY medium		
With heat shock	18.7	26.4
Without heat shock	8.8	12.0
McBride's medium		
With heat shock	9.5	No growth
Without heat shock	6.6	No growth

Injury is a complex process influenced by time, temperature, concentration of injurious agent, strain of target pathogen, and experimental methodology. For example, while a standard sanitizer test indicates that several sanitizers kill listeria, viable cells can be recovered using listeria repair broth (98). The degree of injury decreases and the extent of lethality increases as the time and sanitizer concentration increase. For example, *L. monocytogenes* cells grown at $<28°C$ undergo a 3- to 4-log kill when exposed to $52°C$. However, if grown at 37 or $42°C$ there is little death, but 2 to 3 logs of injury occur when cells are heated to $52°C$ (103).

Data suggestive of injury are illustrated in Fig. 2.5. Cells subjected to a mild stress are plated on a rich nonselective medium or a medium containing 6% salt. The difference between the population of cells able to form colonies on each medium represents injured cells. (If 10^7 CFU/ml of a population are found on the selective medium and 10^4 CFU/ml can grow on the selective medium, then 10^3 CFU/ml are injured.) Media for recovery of injured cells are often specific for a given bacteria or stress condition. A shoulder on a lethality curve may represent injury that is easily repaired. As the time of heating increases, the extent of injury increases. But during the poststress period, the injured population undergoes repair and regains its ability to form colonies on the selective medium, and the number of injured cells decreases.

Injury is important to food safety for several reasons. (i) If injured cells are mistakenly classified as dead during the determination of thermal resistance, the organism's thermal sensitivity will be overestimated, and the resultant D values will be errantly low. (ii) Injured cells that escape detection at the time of postprocessing sampling may repair prior to consumption and present a safety or spoilage problem. A brief review on optimization of heat treatments that takes into account injuries of surviving bacteria is available (70). (iii) The "selective agent" may be common food ingredients such as salt, organic acids, or humectants or can even be suboptimal temperature. For example, *S. aureus* cells injured by acid during sausage fermentation can grow on tryptic soy agar but not on the same medium containing 7.5% salt. These injured cells can repair in sausage if held at $35°C$ (but not if held at $5°C$) and then grow and produce enterotoxin (104). Approximately 99% of *E. coli* O157:H7 cells are freeze-injured following freezing at $-20°C$ for 24 h (50). The type of food contaminated with *E. coli* O157:H7 also influences injury and subsequent recovery of the pathogen.

Injury in spores is complex. The many biochemical steps of sporulation, germination, and outgrowth

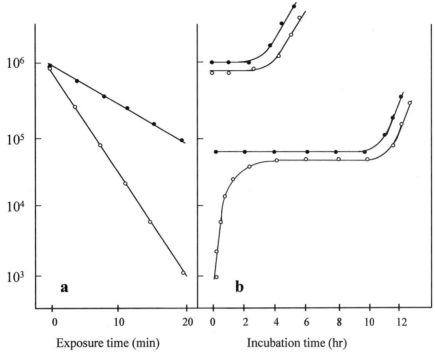

Figure 2.5 Data indicative of injury and repair. When bacteria are plated on selective (○) or nonselective (●) media during exposure to some stressor (a) (e.g., heat), the decrease in CFU on a nonselective medium represents the true lethality, while the difference between the values obtained on each medium is defined as "injury." During "repair" (b) resistance to selective agents is regained, and the value obtained on the selective medium approaches that of the nonselective medium. Unstressed controls are shown on the top of panel b. Modified and redrawn from reference 11.

explained in chapter 3 provide a plethora of targets that can be damaged. Thermal injury is the most well-studied form of injury and can occur during extrusion as well as during conventional thermal processing (67). Spores can also be injured by chemicals and irradiation. Extensive reviews of spore injury (1, 40, 43) indicate that DNA, RNA, enzymes, and membranes may be damaged during injury. Spore injury can be manifested as increased sensitivity to salt, surfactants, acidity, incubation temperature, and oxygen toxicity (42). Irradiation-induced injury of spores is primarily caused by single-strand breaks in DNA and is also manifested as increased sensitivity to pH, salt, and heat (40). *rec* systems can repair injury caused by single-strand DNA breaks. Radiation-induced heat sensitivity is caused by damage to the spore cortex peptidoglycan and can last for several weeks to several months (41, 42).

Injured spores can be revived using several different methods. The addition of soluble starch or charcoal to media binds surfactants that would otherwise enter the spore through the damaged cytoplasmic membrane. Increasing the osmolarity of the recovery medium stabilizes injured spores by minimizing leakage of spore constituents. Lysozyme enhances the recovery of spores whose germination systems have been damaged.

Vegetative cells injured by heat, freezing, and detergents usually leak intracellular constituents from damaged membranes (105). The reestablishment of membrane integrity is an important event during repair. Osmoprotectants can prevent or minimize freeze injury in *L. monocytogenes* (36, 37). Oxygen toxicity also causes injury. Recovery of injured cells is often enhanced by adding peroxide detoxifying agents such as catalase or pyruvate to the recovery medium or by excluding oxygen through the use of anaerobic incubation conditions or adding Oxyrase (which enzymatically reduces oxygen) to the recovery medium.

Repair is the process by which cells recover from injury. Repair requires de novo synthesis of RNA and protein (18, 95) and often is manifested as an extension of the lag phase of growth. The extent and rate of repair are influenced by a variety of environmental factors. *L. monocytogenes* cells injured at 55°C for 20 min start to repair immediately at 37°C and are completely recovered by 9 h (74). Heat-injured *L. monocytogenes* do not replicate in milk at 4°C. Repair at 4°C is delayed for 8 to 10

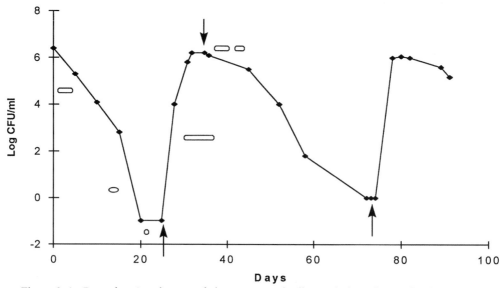

Figure 2.6 Data showing changes of plate count and cell morphology during development of the VNC state induced by temperature downshifts (at time 0 and ↓) and resuscitation of VNC cells by temperature upshifts (↑). Reprinted from reference 86 with permission.

days, and full recovery requires 16 to 19 days (30). Studies of cured luncheon meat (8) confirm that its microbial stability is due to the extended lag period required for the repair of spoilage organisms. The magnitude of the extension is determined by both the severity of the thermal process and the product's salt concentration.

Viable but Nonculturable

Salmonella, Campylobacter, Escherichia, Shigella, and *Vibrio* species, and species from other genera, can exist in a state where they are viable but cannot be cultured by normal microbiological methods. This differentiation of vegetative cells into a dormant "viable but nonculturable" (VNC) state is a survival strategy for many nonsporulating species. The VNC state is morphologically different from the "normal" vegetative cell. During the transition to the VNC state, rod-shaped cells shrink and become small spherical bodies that are totally different from bacillus and clostridial spores (60, 86). It takes from 2 days to several weeks for an entire population of vegetative cells to become VNC (86, 97).

Although VNC cells cannot be cultured, their viability can be determined through cytological methods (2, 58). Fluorescent reagents made popular by "molecular probes" are used to determine structural integrity of the bacterial cytoplasmic membrane. Cell integrity is based on differential permeability of intact cells for nucleic acid stains. Bacteria with intact cell membranes stain fluorescent green, whereas bacteria with damaged membranes stain fluorescent red (7). Iodonitrotetrazolium

violet can also identify VNC cells. Respiring cells reduce iodonitrotetrazolium violet to form an insoluble compound detectable by microscopic observation (86). Unculturable (<10 CFU/ml) *Salmonella* Enteritidis populations starved at 7°C have been quantified as 10^4 viable cells per ml using these methods (24). Experimental data (86) in Fig. 2.6 illustrate a *V. vulnificus* population that appears to have died off (i.e., gone through a 6-log reduction in CFU/ml) at a time when almost all of the cells (>10^5 per ml) are quantified as viable.

VNC cells can also be identified by their substrate-responsive metabolism. When VNC cells are incubated with yeast extract (as a nutrient) and nalidixic acid (an inhibitor of cell division), their elongation can be quantified microscopically. Because this widely used method does not work with gram-positive bacteria (which are insensitive to nalidixic acid), one might infer from the literature that the VNC state is limited to gram-negative species. In fact, researchers recently modified the method, replacing nalidixic acid with ciprofloxacin, an antibiotic efficient against gram-positive bacteria. The method was optimized to detect viable cells of *L. monocytogenes* (10). Other methods have been used to demonstrate the VNC state for *Streptococcus faecalis, Micrococcus flavus,* and *B. subtilis* (20).

Powerful new methods for detection of VNC cells are being developed as our understanding of bacteria at the molecular level and techniques for genetic manipulation advance. The detection of specific RNA by reverse transcriptase PCR is logical given the generally short half-life of RNA in bacteria (91). Alternatively, reporter genes can be used to determine activity in nonculturable cells. Green fluorescent protein-tagged and lux-tagged methods are used to mark cells and provide evidence of protein synthesis (25).

Because the VNC state is most often induced by nutrient limitation in aquatic environments, it might appear irrelevant to the nutrient-rich milieu of food. However, the VNC state can also be induced by changes in salt concentration, exposure to hypochlorite, and shifts in temperature (76, 89). *V. vulnificus* populations shifted to refrigeration temperatures are still lethal to mice when the entire population of 10^5 viable cells becomes nonculturable (<0.04 CFU/ml) (88). The bacteria resuscitate in the mice and can be isolated postmortem using culture methods. However, *S. enterica* serovar Typhimurium stressed with exposure to UV-C and seawater lost pathogenicity and virulence, as tested in a mouse model, but maintained viability (21). Increases or decreases in temperature can induce the VNC state in different organisms. When starved at 4 or 30°C for more than a month, *Vibrio harveyi* becomes VNC at 4°C but remains culturable at 30°C. In contrast, *E. coli* enters the VNC state at 30°C, but dies at 4°C (35). Foodborne pathogens in nutritionally rich media can become VNC when shifted to refrigerated temperature (73, 86, 88). This has chilling implications for the safety of refrigerated foods.

Resuscitation of VNC cells is demonstrated by an increase in culturability that is not accompanied by an increase in the total number of cells. The return to culturability can be induced by temperature shifts or gradual return of nutrients. It can take several days for the population to fully recover its culturability. The same population of bacteria can go through multiple cycles of the VNC and culturable states in the absence of growth (86). No one specific factor can be identified as preventing the resuscitation of VNC cells. Amending media with catalase or sodium pyruvate restores culturability of VNC cells of *E. coli* O157 and *Vibrio parahaemolyticus* (78, 79), suggesting that transfer of cells to nutrient-rich media initiates an imbalance in metabolism and rapid production of superoxide and free radicals. Moreover, autoclaving of rich media generates hydrogen peroxide in the media. Catalase enzymatically breaks down H_2O_2, and sodium pyruvate degrades H_2O_2 through decarboxylation of the α-keto acid to form acetic acid and CO_2 (38). Researchers resuscitated strains of *Campylobacter* sp. by injection of VNC cells into fertilized chicken eggs and incubation at 37°C for 48 h (110).

Increased awareness of the VNC state should lead to a reexamination of our concept of "viability," our dependence on "enrichment culture" to isolate pathogens, and our reliance on established cultural methods to monitor microbes in the environment (3, 73, 96). The mechanisms of VNC formation, the mechanisms that make VNC cells resistant to environmental stress, and the nature of the event that signals resuscitation are largely unknown. The relationship between viability and culturability may need to be redefined and, indeed, has spawned extensive discussion (2, 58, 72, 87). Some investigators question the ability of bacteria to enter or resuscitate from a VNC state and caution against blind acceptance of methods developed to determine viability (2, 15, 16). Clearly, food microbiologists need to do more research in the area of VNC foodborne pathogens.

Biofilms

To suggest that more research is needed about biofilms would be a gross understatement. Although planktonic (i.e., free, single) cells are easy to study, and pure culture is the foundation of microbiology as we know it, "in all natural habitats studied to date bacteria prefer to reproduce on any available surface rather than in the liquid phase" (22). Furthermore, most definitions of biofilms reveal that they exist as *communities* of microbial species embedded in a biopolymer matrix on some substratum. Biofilms are also heterogeneous in time and space, frequently appearing as collections of mushroom-shaped microcolonies with moving water channels between them (22, 121). Foodborne pathogens *E. coli* O157:H7, *L. monocytogenes*, *Y. enterocolitica*, and *Campylobacter jejuni* form biofilms on food surfaces and food contact equipment, leading to serious health problems and economic losses due to spoilage of food (61).

Biofilm formation is a multistep process in which the substratum first undergoes a conditioning process that allows cells to be adsorbed by weak reversible electrostatic forces. Adhesion or anchoring of the cells via some biopolymer follows rapidly. The synthesis of the matrix polymer may be upregulated by adsorption of the cell. In *Pseudomonas aeruginosa*, transcription of alginate biosynthetic genes is activated by response regulators that increase synthesis of a sigmalike factor, which regulates transcription of the *algD* promoter (17). The sensing part of this two-component signaling system has not yet been identified. The *algD* promoter regulates virtually all of the alginate biosynthetic genes, which are in a single operon. This system also contains an alginate lyase that allows for cell dispersion when the environment threatens communal life. The microcolonies have defined boundaries that allow fluid channels to run through the biomatrix. This requires higher-level differentiation, quorum sensing, or some kind of cell-to-cell communication to prevent undifferentiated growth from filling in these channels that bring nutrients and remove wastes. Costerton (29) paints a vivid picture of this system, concluding "that the highly structured biofilm mode of growth provide bacteria with a measure of homeostasis, a primitive circulatory system, a framework for the development of cooperative and specialized cell functions, and a large measure of protection from antimicrobial agents." Two special issues (vol. 15, no. 3 and 4) of the *Journal of Industrial Microbiology* provide an up-to-date understanding of biofilms in medical, dental, agricultural, and environmental settings. A recent report suggests a link between biofilm formation and quorum sensing in *P. aeruginosa* (32). *P. aeruginosa* mutants deficient in the production of the *las* signal molecule, 3-oxo-C_{12}-HSL, form a biofilm that is thinner and lacks the three-dimensional structure observed in the parent. The use of reporter gene technology can allow one to determine altered gene expression in target bacteria under the range of physicochemical conditions that occur in biofilms.

Cells in biofilms are more resistant to heat, chemicals, and sanitizers. This has been attributed to the diffusional barrier created by the biomatrix. The confocal scanning laser microscopy images of circulatory channels in hydrated biofilms shatter the barrier hypothesis. The increased resistance is now attributed to the very slow growth rates of cells in biofilms (29). Indeed, cells in the nutrient-depleted interior of the microcolony may be in the VNC state; however, the existence of VNC state in biofilm bacteria has not been conclusively established. From a pragmatic standpoint, reviews on biofilms in the food industry (22, 121) emphasize the importance of cleaning prior to sanitation of process equipment. Mean reduction values in viable counts of *L. monocytogenes* following treatment with a combination of sodium hypochlorite and heat are approximately 100 times lower in biofilms than for planktonic cells (46). Trisodium phosphate is effective toward *E. coli* O157:H7, *C. jejuni*, and *S. enterica* serovar Typhimurium planktonic and biofilm cells (106). While there are no materials inherently resistant to biofouling, true biofilms take days to weeks to reach equilibrium. Proper cleaning ensures that the cells in the nascent biofilm can be reached by sanitizers. Newer methods for control of biofilms include superhigh magnetic fields, ultrasound treatment, and high-pulsed electric fields (61). The design of equipment with smooth highly polished surfaces also impedes biofilm formation by making the initial adsorption step more difficult.

CONCLUSION

Microbial growth in foods is a complex process governed by genetic, biochemical, and environmental factors. Much of what we "know" about foodborne microbes must be held with the detached objectivity required of an unproven hypothesis. Developments in molecular biology and microbial ecology will change or deepen our perspective about the growth of microbes in foods. Some of these developments are detailed in this book. Other developments will unfold over the coming decades, perhaps by readers whose journey begins now.

Research in our laboratory and preparation of this manuscript were supported by state appropriations and U.S. Hatch Act Funds provided through the New Jersey Agricultural Experiment Station.

References

1. **Adams, D. M.** 1978. Heat injury of bacterial spores. *Adv. Appl. Microbiol.* **23**:245–261.

2. **Barer, M., and C. R. Harwood.** 1999. Bacterial viability and culturability. *Adv. Microb. Physiol.* **41**:93–137.

3. **Barer, M. R., L. T. Gribbon, C. R. Harwood, and C. E. Nwaguh.** 1993. The viable but non-culturable hypothesis and medical bacteriology. *Rev. Med. Microbiol.* **4**:183–191.

4. **Bayles, D. O., B. A. Annous, and B. J. Wilkinson.** 1996. Cold stress proteins induced in *Listeria monocytogenes* in response to temperature downshock and growth at low temperatures. *Appl. Environ. Microbiol.* **62**:1116–1119.

5. **Bearson, B. L., L. Wilson, and J. W. Foster.** 1998. A low pH-inducible, PhoPQ-dependent acid tolerance response protects *Salmonella typhimurium* against inorganic acid stress. *J. Bacteriol.* **180**:2409–2417.

6. **Bearson, S., B. Bearson, and J. W. Foster.** 1997. Acid stress responses in enterobacteria. *FEMS Microbiol. Lett.* **147**:173–180.

7. **Beeuwer, P., and T. Abee.** 2000. Assessment of viability of microorganisms employing fluorescence techniques. *Int. J. Food Microbiol.* **55**:193–200.

8. **Bell, R. G., and K. M. De Lacy.** 1984. Heat injury and recovery of *Streptococcus faecium* associated with the souring of chub-packed luncheon meat. *J. Appl. Bacteriol.* **57**:229–236.

9. **Beltrametti, F., A. U. Kresse, and C. A. Guzman.** 1999. Transcriptional regulation of *esp* genes of enterohemorrhagic *Escherichia coli*. *J. Bacteriol.* **181**:3409–3418.

10. **Besnard, V., M. Federighi, and J. M. Cappelier.** 2000. Development of a direct viable count procedure for the investigation of VBNC state in *Listeria monocytogenes*. *Lett. Appl. Microbiol.* **31**:77–81.

11. **Beuchat, L. R.** 1978. Injury and repair of Gram-negative bacteria with special consideration of the involvement of the cytoplasmic membrane. *Adv. Appl. Microbiol.* **23**:219–243.

12. **Beuchat, L. R.** 1984. Injury and repair of yeasts and moulds. *J. Appl. Bacteriol.* **12**:293–308.

13. **Board, R. G., D. Jones, R. G. Kroll, and G. L. Pettipher** (ed.). 1992. *Ecosystems: Microbes: Food. J. Appl. Bacteriol.* **73**:1S–178S.

14. **Boddy, L., and J. W. T. Wimpenny.** 1992. Ecological concepts in food microbiology. *J. Appl. Bacteriol.* **73**:23S–38S.

15. **Bogosian, G., N. D. Aardema, E. V. Bourneauf, P. J. Morris, J. P. O'Neil.** 2000. Recovery of hydrogen peroxide-sensitive culturable cells of *Vibrio vulnificus* gives the appearance of resuscitation from a viable but nonculturable state. *J. Bacteriol.* **182**:5070–5075.

16. **Bogosian, G., P. J. L. Morris, and J. P. O'Neil.** 1998. A mixed culture recovery method indicates that enteric bacteria do not enter the viable but nonculturable state. *Appl. Environ. Microbiol.* **64**:1736–1742.

17. **Boyd, A., and A. M. Chakrabarty.** 1995. *Pseudomonas aeruginosa* biofilms: role of the alginate exopolysaccharide. *J. Ind. Microbiol.* **15**:162–168.

18. **Brock, T. D., and M. T. Madigan.** 1988. *Biology of Microorganisms,* p. 793–795. Prentice-Hall, Inc., Englewood Cliffs, N.J.

19. **Busta, F. F.** 1978. Introduction to injury and repair of microbial cells. *Adv. Appl. Microbiol.* **23**:195–201.

20. **Byrd, J. J., H.-S. Xu, and R. R. Colwell.** 1991. Viable but nonculturable bacteria in drinking water. *Appl. Environ. Microbiol.* **57**:875–878.

21. **Caro, A., P. Got, J. Lense, S. Binard, and B. Baleux.** 1999. Viability and virulence of experimentally stressed nonculturable *Salmonella typhimurium. Appl. Environ. Microbiol.* **65**:3229–3232.

22. **Carpentier, B., and O. Cerf.** 1993. Biofilms and their consequences, with particular reference to hygiene in the food industry. *J. Appl. Bacteriol.* **75**:499–511.

23. **Castanie-Cornet, M., T. A. Penfound, D. Smith, J. F. Elliot, and J. W. Foster.** 1999. Control of acid resistance in *Escherichia coli. J. Bacteriol.* **181**:3525–3535.

24. **Chmielewski, R., and J. F. Frank.** 1995. Formation of viable but nonculturable *Salmonella* during starvation in chemically defined solutions. *Lett. Appl. Microbiol.* **20**:380–384.

25. **Cho, J. C., and S. J. Kim.** 1999. Green fluorescent protein-based direct viable count to verify a viable but nonculturable state of *Salmonella typhi* in environmental samples. *J. Microbiol. Methods* **36**:227–235.

26. **Choi, S. H., D. J. Baumler, and C. W. Kasper.** 2000. Contribution of *dps* to acid stress tolerance and oxidative stress tolerance in *Escherichia coli* O157:H7. *Appl. Environ. Microbiol.* **66**:3911–3916.

27. **Conte, M. P., C. Longhi, G. Petrone, M. Polidoro, P. Valenti, and L. Seganti.** 1994. *Listeria monocytogenes* infection of Caco-2 cells: role of growth temperature. *Res. Microbiol.* **145**:677–682.

28. **Cossins, A. R., and M. Sinensky.** 1984. Adaptation of membranes to temperature, pressure and exogenous lipids, p. 1–20. *In* M. Shinitzky (ed.), *Physiology of Membrane Fluidity.* CRC Press, Inc., Boca Raton, Fla.

29. **Costerton, J. W.** 1995. Overview of microbial biofilms. *J. Ind. Microbiol.* **15**:137–140.

30. **Crawford, R. W., C. M. Belizeau, T. J. Poeler, C. W. Donnelly, and U. K. Bunning.** 1989. Comparative recovery of uninjured and heat-injured *Listeria monocytogenes* cells from bovine milk. *Appl. Environ. Microbiol.* **55**:1490–1494.

31. **Crueger, W., and A. Crueger.** 1984. *Biotechnology: a Textbook of Industrial Microbiology.* Sinauer Associates, Inc., Sunderland, Mass.

32. **Davies, D. G., M. R. Parsek, J. P. Pearson, B. H. Iglewski, J. W. Costerton, and E. P. Greenberg.** 1998. The involvement of cell-to-cell signals in the development of a bacterial biofilm. *Science* **280**:295–298.

33. **de Kievit, T. R., and B. H. Iglewski.** 2000. Bacterial quorum sensing in pathogenic relationships. *Infect. Immun.* **68**:4839–4849.

34. **Duffy, G., D. C. Riordan, J. J. Sheridan, J. E. Call, R. C. Whiting, I. S. Blair, and D. A. McDowell.** 2000. Effect of pH on survival, thermotolerance, and verotoxin production of *Escherichia coli* O157:H7 during simulated fermentation and storage. *J. Food Prot.* **63**:12–18.

35. **Duncan, S., L. A. Glover, K. Killham, and J. I. Prosser.** 1994. Luminescence-based detection of activity of starved and viable but nonculturable bacteria. *Appl. Environ. Microbiol.* **60**:1308–1316.

36. **El-Kest, S. E., and E. H. Marth.** 1991. Injury and death of frozen *Listeria monocytogenes* as affected by glycerol and milk components. *J. Dairy Sci.* **74**:1201–1208.

37. **El-Kest, S. E., and E. H. Marth.** 1991. Strains and suspending menstrua as factors affecting death and injury of *Listeria monocytogenes* during freezing and frozen storage. *J. Dairy Sci.* **74**:1209–1213.

38. **Elstner, E. F., and A. Heupel.** 1973. On the decarboxylation of α-keto acid by isolated chloroplasts. *Biochim. Biophys. Acta* **352**:182–188.

39. **Farber, J. M., and B. E. Brown.** 1990. Effect of prior heat shock on heat resistance of *Listeria monocytogenes* in meat. *Appl. Environ. Microbiol.* **56**:1584–1587.

40. **Farkas, J.** 1994. Tolerance of spores to ionizing radiation: mechanisms of inactivation, injury, and repair. *J. Appl. Bacteriol.* **76**:81S–90S.

41. **Fedio, W. M., and H. Jackson.** 1989. Effect of tempering on the heat resistance of *Listeria monocytogenes. Lett. Appl. Microbiol.* **9**:157–160.

42. **Feeherry, F. E., D. T. Munsey, and D. B. Rowley.** 1987. Thermal inactivation and injury of *Bacillus stearothermophilus* spores. *Appl. Environ. Microbiol.* **53**:365–370.

43. **Foegoding, P. M., and F. F. Busta.** 1981. Bacterial spore injury—an update. *J. Food Prot.* **44**:776–786.

44. **Foster, J. W., and H. K. Hall.** 1991. Inducible pH homeostasis and the acid tolerance response of *Salmonella typhimurium. J. Bacteriol.* **173**:5129–5135.

45. **Foster, J. W., Y. K. Park, L. S. Bang, K. Karem, H. Betts, H. K. Hall, and E. Shaw.** 1994. Regulatory circuits involved with pH-regulated gene expression in *Salmonella typhimurium. Microbiology* **140**:341–352.

46. Frank, J. F., and R. A. Koffi. 1990. Surface-adherent growth of *Listeria monocytogenes* is associated with increased resistance to surfactant sanitizers and heat. *J. Food Prot.* **48:**740–742.

47. Gahan, C. G. M., B. O'Driscoll, and C. Hill. 1996. Acid adaptation of *Listeria monocytogenes* can enhance survival in acidic foods and during milk fermentation. *Appl. Environ. Microbiol.* **62:**3128–3132.

48. Gould, G. W. 1992. Ecosystems approaches to food microbiology. *J. Appl. Bacteriol.* **73:**58S–68S.

49. Gould, G. W. 1995. Homeostatic mechanisms during food preservation by combined methods, p. 397–410. *In* G. V. Barbosa-Canovas and J. Welti-Chanes (ed.), *Food Preservation by Moisture Control.* Technomic Publishing Co., Inc., Lancaster, Pa.

50. Hara-Kudo, Y., M. Ifedo, H. Kodaka, H. Nakagawa, K. Goto, T. Masuda, H. Konuma, T. Kojima, and S. Kumagai. 2000. Selective enrichment with resuscitation step for isolation of freeze-injured *Escherichia coli* O157:H7 from foods. *Appl. Environ. Microbiol.* **66:**2866–2872.

51. Harold, F. M. 1981. *The Vital Force: a Study of Bioenergetics.* W. H. Freeman & Co., New York, N.Y.

52. Huhtanen, C. N., J. Naghski, C. S. Custer, and R. W. Russel. 1976. Growth and toxin production by *Clostridium botulinum* in moldy tomato juice. *Appl. Environ. Microbiol.* **32:**711–715.

53. International Commission on Microbiological Specifications for Foods. 1980. *In The Microbial Ecology of Foods,* vol. 1, *Factors Affecting the Life and Death of Microorganisms.* Academic Press, Inc., New York, N.Y.

54. International Commission on Microbiological Specifications for Foods. 1980. *In The Microbial Ecology of Foods,* vol. 2, *Food Commodities.* Academic Press, Inc., New York, N.Y.

55. International Commission on Microbiological Specifications for Foods. 1980. Injury and its effect on recovery, p. 205–214. *In The Microbial Ecology of Foods,* vol. 1, *Factors Affecting the Life and Death of Microorganisms.* Academic Press, Inc., New York, N.Y.

56. Juneja, V. K., P. G. Klein, and B. S. Marmer. 1998. Heat shock and thermotolerance of *Escherichia coli* O157:H7 in a model beef gravy system and ground beef. *J. Appl. Microbiol.* **84:**677–684.

57. Kashket, E. R. 1987. Bioenergetics of lactic acid bacteria: cytoplasmic pH and osmotolerances. *FEMS Microbiol. Rev.* **46:**233–244.

58. Kell, D. B., A. S. Kaprelyants, D. H. Weichart, C. R. Harwood, and M. R. Barer. 1998. Viability and activity in readily culturable bacteria: a review and discussion of the practical issues. *Antonie Leeuwenhoek* **73:**169–187.

59. Ko, R., L. T. Smith, and G. M. Smith. 1994. Glycine betaine confers enhanced osmotolerance and cryotolerance in *Listeria monocytogenes. J. Bacteriol.* **176:**426–431.

60. Kondo, K., A. Takade, and K. Amako. 1994. Morphology of the viable but nonculturable *Vibrio cholerae* as determined by the freeze fixation technique. *FEMS Microbiol. Lett.* **123:**179–184.

61. Kumar, C. G., and S. K. Anand. 1998. Significance of microbial biofilms in food industry: a review. *Int. J. Food Microbiol.* **42:**9–27.

62. Kwon, Y. M., and S. C. Ricke. 1998. Induction of acid resistance of *Salmonella typhimurium* by exposure to short-chain fatty acids. *Appl. Environ. Microbiol.* **64:**3458–3463.

63. Leimeister-Wachter, M., E. Donnan, and T. Chakraborty. 1992. The expression of virulence genes in *Listeria monocytogenes* is thermal regulated. *J. Bacteriol.* **174:**947–952.

64. Leistner, L. 1994. Principles and applications of hurdle technology, p. 1–21. *In* G. W. Gould (ed.), *New Methods of Food Preservation.* Blackie Academic and Professional, Glasgow, Scotland.

65. Leyer, G. J., and E. A. Johnson. 1993. Acid adaptation induces cross-protection against environmental stresses in *Salmonella typhimurium. Appl. Environ. Microbiol.* **59:**1842–1847.

66. Leyer, G. J., L.-L. Wang, and E. A. Johnson. 1995. Acid adaptation of *Escherichia coli* O157:H7 increases survival in acid foods. *Appl. Environ. Microbiol.* **61:**3752–3755.

67. Likimani, T. A., and J. N. Sofos. 1990. Bacterial spore injury during extrusion cooking of corn/soybean mixtures. *Int. J. Food Microbiol.* **11:**243–249.

68. Linton, R. H., J. B. Webster, M. D. Pierson, J. R. Bishop, and C. R. Hackney. 1992. The effect of sublethal heat shock and growth atmosphere on the heat resistance of *Listeria monocytogenes* Scott A. *J. Food. Prot.* **55:**84–87.

69. Lund, B. M. 1992. Ecosystems in vegetable foods. *J. Appl. Bacteriol.* **73:**115S–126S.

70. Mafart, P. 2000. Taking injuries of surviving bacteria into account for optimizing heat treatments. *Int. J. Food Microbiol.* **55:**175–179.

71. Maloney, P. C. 1990. Microbes and membrane biology. *FEMS Microbiol. Rev.* **87:**91–102.

72. McDougald, D. S., A. Rice, D. Weichart, and S. Kjelleberg. 1998. Nonculturability: adaptation or debilitation? *FEMS Microbiol. Ecol.* **25:**1–9.

73. McKay, A. M. 1992. Viable but non-culturable forms of potentially pathogenic bacteria in water. *Lett. Appl. Microbiol.* **14:**129–135.

74. Meyer, D. H., and C. W. Donnelly. 1992. Effect of incubation temperature on repair of heat-injured *Listeria* in milk. *J. Food Prot.* **55:**579–582.

75. Michels, P. A. M., J. P. J. Michels, J. Boonstra, and W. L. Konings. 1979. Generation of an electrochemical proton gradient in bacteria by the excretion of metabolic end products. *FEMS Microbiol. Lett.* **5:**357–364.

76. Mikulskis, A. V., I. Delor, V. H. Thi, and G. R. Cornelis. 1994. Regulation of *Yersinia enterocolitica* enterotoxin *Yst* gene. Influence of growth phase, temperature, osmolarity, pH and bacterial host factors. *Mol. Microbiol.* **14:**905–915.

77. Mitchell, P. 1966. Chemiosmotic coupling in oxidative and photosynthetic phosphorylation. *Biol. Rev. Camb. Philos. Soc.* **41:**445–502.

78. **Mizunoe, Y., S. N. Wai, A. Takade, and S. Yoshida.** 1999. Restoration of culturability of starvation-stressed and low temperature-stressed *Escherichia coli* O157 cells by using H_2O_2-degrading compounds. *Arch. Microbiol.* **172:**63–67.

79. **Mizunoe, Y., S. N. Wai, T. Ishikawa, A. Takade, and S. Yoshida.** 2000. Resuscitation of viable but nonculturable cells of *Vibrio parahaemolyticus* induced at low temperature under starvation. *FEMS Microbiol. Lett.* **186:**115–120.

80. **Montville, T. J.** 1982. Metabiotic effect of *Bacillus licheniformis* on *Clostridium botulinum:* implications for home-canned tomatoes. *Appl. Environ. Microbiol.* **44:**334–338.

81. **Mossel, D. A. A., and C. B. Struijk.** 1992. The contribution of microbial ecology to management and monitoring of the safety, quality and acceptability (SQA) of foods. *J. Appl. Bacteriol.* **73:**1S–22S.

82. **Murano, E. A., and M. O. Pierson.** 1993. Effect of heat shock and incubation atmosphere on injury and recovery of *Escherichia coli* O157:H7. *J. Food Prot.* **56:**568–572.

83. **Murley, Y. M., P. A. Carrol, K. Skorupski, R. K. Taylor, and S. B. Calderwood.** 1999. Differential transcription of the *tcpPH* operon confers biotype-specific control of the *Vibrio cholerae* ToxR virulence regulon. *Infect. Immun.* **67:**5117–5123.

84. **National Food Processors Association.** 1988. Factors to be considered in establishing good manufacturing practices for the production of refrigerated food. *Dairy Food Sanit.* **8:**288–291.

85. **Nicholls, D. G., and S. J. Ferguson.** 1992. *Bioenergetics 2.* Academic Press, Inc., San Diego, Calif.

86. **Nilsson, L., J. D. Oliver, and S. Kjelleberg.** 1991. Resuscitation of *Vibrio vulnificus* from the viable but nonculturable state. *J. Bacteriol.* **173:**5054–5059.

87. **Nystrom, T.** 1998. To be or not to be: the ultimate decision of the growth-arrested bacterial cell. *FEMS Microbiol. Rev.* **21:**283–290.

88. **Oliver, J. D., and R. Bocklan.** 1995. In vivo resuscitation, and virulence towards mice, of viable but nonculturable cells of *Vibrio vulnificus. Appl. Environ. Microbiol.* **61:**2620–2623.

89. **Oliver, J. D., F. Hite, D. McDougald, N. L. Andon, and L. M. Simpson.** 1995. Entry into, and resuscitation from, the viable but nonculturable state by *Vibrio vulnificus* in an estuarine environment. *Appl. Environ. Microbiol.* **61:**2624–2630.

90. **Olson, E. R.** 1993. Influence of pH on bacterial gene expression. *Mol. Microbiol.* **8:**5–14.

91. **Pai, S., J. K. Actor, E. Sepulveda, R. L. Hunter, and C. Jegannath.** 2000. Identification of viable and non-viable *Mycobacterium tuberculosis* in mouse organs by directed RT-PCR for antigen 85B mRNA. *Microb. Pathog.* **28:**335–342.

92. **Palumbo, S. A.** 1986. Is refrigeration enough to restrain foodborne pathogens? *J. Food Prot.* **49:**1003–1009.

93. **Pepe, J. C., J. L. Badger, and V. L. Miller.** 1994. Growth phase and low pH affect the thermal regulation of the *Yersinia enterocolitica inv* gene. *Mol. Microbiol.* **11:**123–135.

94. **Phan-Thanh, L., F. Mahouin, and S. Alige.** 2000. Acid responses of *Listeria monocytogenes. Int. J. Food Microbiol.* **55:**121–126.

95. **Pierson, M. D., R. F. Gomez, and S. E. Martin.** 1978. The involvement of nucleic acids in bacterial injury. *Adv. Appl. Microbiol.* **23:**263–285.

96. **Rollins, D. M., and R. R. Colwell.** 1986. Viable but nonculturable stage of *Campylobacter jejuni* and its role in survival in the natural aquatic environment. *Appl. Environ. Microbiol.* **52:**531–538.

97. **Roszak, D. B., D. J. Grimes, and R. R. Colwell.** 1984. Viable but nonrecoverable stage of *Salmonella enteritidis* in aquatic systems. *Can. J. Microbiol.* **30:**334–338.

98. **Sallam, S. S., and C. W. Donnelly.** 1992. Destruction, injury and repair of *Listeria* species exposed to sanitizing compounds. *J. Food Prot.* **59:**771–776.

99. **Schuhmacker, D. A., and K. E. Klose.** 1999. Environmental signals modulate ToxT-dependent virulence factor expression in *Vibrio cholerae. J. Bacteriol.* **181:**1508–1514.

100. **Sinensky, M.** 1974. Homeoviscous adaptation — a homeostatic process that regulates the viscosity of membrane lipids in *Escherichia coli. Proc. Natl. Acad. Sci. USA* **71:**522–525.

101. **Skurnik, M.** 1985. Expression of antigens encoded by the virulence plasmid of *Yersinia enterocolitica* under different growth conditions. *Infect. Immun.* **47:**183–190.

102. **Slonczewski, J. L., and J. W. Foster.** 1999. pH-regulated genes and survival at extreme pH. *In* F. C. Neidhardt, R. Curtiss III, J. L. Ingraham, E. C. C. Lin, K. B. Low, Jr., B. Magasanik, W. S. Reznikoff, M. Riley, M. Schaechter, and H. E. Umbarger (ed.), *Escherichia coli and Salmonella: Cellular and Molecular Biology,* 2nd ed. (CD-ROM). ASM Press, Washington, D.C.

103. **Smith, J. L., B. S. Marmer, and R. C. Benedict.** 1991. Influence of growth temperature on injury and death of *Listeria monocytogenes* Scott A during a mild heat treatment. *J. Food Prot.* **54:**166–169.

104. **Smith, J. L., and S. A. Palumbo.** 1978. Injury to *Staphylococcus aureus* during sausage fermentation. *Appl. Environ. Microbiol.* **36:**857–860.

105. **Smith, J. L., and S. A. Palumbo.** 1982. Microbial injury reviewed for the sanitarian. *Dairy Food Sanit.* **2:**57–63.

106. **Somers, E. B., J. L. Schoeni, A. C. L. Wong.** 1994. Effect of trisodium phosphate on biofilm and planktonic cells of *Campylobacter jejuni, Escherichia coli* O157:H7, *Listeria monocytogenes,* and *Salmonella typhimurium. Int. J. Food Microbiol.* **22:**269–276.

107. **Stevenson, K. E., and T. R. Graumlich.** 1978. Injury and recovery of yeasts and molds. *Adv. Appl. Microbiol.* **23:**203–217.

108. **Surette, M. G., and B. L. Bassler.** 1998. Quorum sensing in *Escherichia coli* and *Salmonella typhimurium. Proc. Natl. Acad. Sci. USA* **95:**7046–7050.

109. **Surette, M. G., M. B. Miller, and B. L. Bassler.** 1999. Quorum sensing in *Escherichia coli, Salmonella typhimurium,*

and *Vibrio harveyi*: a new family of genes responsible for autoinducer production. *Proc. Natl. Acad. Sci. USA* **96**:1639–1644.

110. **Talibart, R., M. Denis, A. Castillo, J. M. Cappelier, and G. Ermel.** 2000. Survival and recovery of viable but noncultivable forms of *Campylobacter* in aqueous microcosm. *Int. J. Food Microbiol.* **55**:263–267.

111. **Thieringer, H. A., P. G. Jones, and M. Inouye.** 1998. Cold shock and adaptation. *Bioessays* **20**:49–57.

112. **Tseng, C.-P., and T. J. Montville.** 1993. Metabolic regulation of end product distribution in lactobacilli: causes and consequences. *Biotechnol. Prog.* **9**:113–121.

113. **Van Schaik, W., C. G. Gahan, and C. Hill.** 1999. Acid-adapted *Listeria monocytogenes* displays enhanced tolerance against the lantibiotics nisin and lactin 3147. *J. Food Prot.* **62**:536–539.

114. **Vigh, V., B. Maresca, and J. L. Harwood.** 1998. Does the membrane's physical state control the expression of heat shock and other genes? *Trends Biochem. Sci.* **23**:369–372.

115. **Volker, U., H. Mach, R. Schmid, and M. Hecker.** 1992. Stress proteins and cross-protection by heat shock and salt stress in *Bacillus subtilis*. *J. Gen. Microbiol.* **138**:2125–2135.

116. **Wang, G., and M. P. Doyle.** 1998. Heat shock response enhances acid tolerance of *Escherichia coli* O157:H7. *Lett. Appl. Microbiol.* **26**:31–34.

117. **Wiedmann, M., T. J. Arvik, R. J. Hurley, and K. J. Boor.** 1998. General stress transcription factor σ^B and its role in acid tolerance and virulence of *Listeria monocytogenes*. *J. Bacteriol.* **180**:3650–3656.

118. **Withers, H. L., and K. Nordstrom.** 1998. Quorum-sensing acts at initiation of chromosomal replication in *Escherichia coli*. *Proc. Natl. Acad. Sci. USA* **95**:15694–15699.

119. **Wouters, J. A., F. M. Rombouts, W. M. deVos, O. P. Kuipers, and T. Abee.** 1999. Cold shock proteins and low-temperature response of *Streptococcus thermophilus* CNRZ302. *Appl. Environ. Microbiol.* **65**:4436–4442.

120. **Yura, T., and K. Nakahigashi.** 1999. Regulation of the heat-shock response. *Curr. Opin. Microbiol.* **2**:153–158.

121. **Zottola, E. A., and K. C. Sasahara.** 1994. Microbial biofilms in the food processing industry—should they be a concern? *Int. J. Food Microbiol.* **23**:125–148.

Food Microbiology: Fundamentals and Frontiers, 2nd Ed.
Edited by M. P. Doyle et al.
© 2001 ASM Press, Washington, D.C.

Peter Setlow
Eric A. Johnson

Spores and Their Significance

3

Members of the gram-positive *Bacillus* and *Clostridium* spp. and some closely related genera respond to slowed growth or starvation by initiating the process of sporulation. The molecular biology of sporulation (93, 134, 212, 222) and spore resistance (10, 53, 61, 66, 142, 214, 215) has been extensively and elegantly studied in the genus *Bacillus* (80, 126, 172, 211, 213–215). Spores formed by certain species in the genera *Alicyclobacillus*, *Bacillus*, *Clostridium*, *Desulfotomaculum*, and *Sporolactobacillus* present practical problems in food microbiology. This chapter describes the fundamental basis of sporulation and problems that spores present to the food industry.

The first notable morphological event in sporulation is an unequal cell division. This creates the smaller nascent spore or forespore compartment and the larger mother cell compartment. As sporulation proceeds, the mother cell engulfs the forespore, resulting in a cell (the forespore) within a cell (the mother cell), each with a complete genome (64, 211). Since the spore is formed within the mother cell, it is termed an endospore.

Throughout sporulation there is a defined pattern of gene expression which is ordered not only temporally but also spatially, as some genes are expressed only in the mother cell or forespore (56, 126, 229). The gene expression pattern is controlled by the ordered synthesis and activation of new sigma (σ; specificity) factors for RNA polymerase, as at least four of these are synthesized specifically for modulation of gene expression during sporulation; a number of DNA binding proteins, both repressors and activators, also regulate gene expression during sporulation (56, 126, 229). As sporulation proceeds, there are striking morphological and biochemical changes in the developing spore. It becomes encased in two novel layers, a peptidoglycan layer (the spore cortex) and a number of layers of spore coats which contain the proteins unique to the spore (19, 52, 53, 212). The spore also accumulates a huge depot ($\geq 10\%$ of dry weight) of pyridine-2,6-dicarboxylic acid (dipicolinic acid [DPA]) (Fig. 3.1), found only in spores, as well as a large amount of divalent cations (157). The developing forespore also synthesizes a large amount of small, acid-soluble proteins (SASP), some of which coat the spore chromosome and protect the DNA from damage (212–215). In addition, the spore becomes both metabolically dormant and extremely resistant to a variety of harsh treatments, including heat, radiation, and chemicals (66, 71, 127, 215). As a consequence of these

Peter Setlow, Department of Biochemistry, University of Connecticut Health Center, Farmington, CT 06030-3305. **Eric A. Johnson,** Food Research Institute, University of Wisconsin-Madison, Madison, WI 53706.

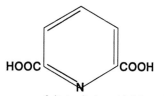

Figure 3.1 Structure of dipicolinic acid. Note that at physiological pH both carboxyl groups will be ionized.

physiological changes, spores can survive extremely long periods in the absence of exogenous nutrients (25, 26, 106).

Despite the spore's extreme dormancy, if given the appropriate stimulus (often a nutrient such as a sugar or amino acid) the spore can rapidly return to life via spore germination (70, 207). Within minutes of exposure to a germinant, spores lose most of their unique characteristics, including loss of DPA by excretion and loss of the cortex and SASP by degradation. The spore's resistance is also lost in the first minutes of germination, when active metabolism of both endogenous and exogenous compounds begins and macromolecular synthesis is initiated. Eventually, the germinated spore is converted back to a growing vegetative cell through the process of outgrowth.

Detailed study of sporulation, spores, and their "return to life" via spore germination and outgrowth has been motivated by a number of factors, one of which is the intrinsic attraction of this model developmental system. This attraction has led to an increasingly detailed understanding of mechanisms responsible for the elaborate networks that modulate gene expression during sporulation, as well as factors involved in spore resistance. An additional motivating factor behind work on this system is the major role played by sporeformers in food spoilage and foodborne diseases. While a tremendous amount of knowledge has been gained in studies motivated by these two disparate factors, the amount of cross talk between the workers in these two fields has never been optimal, to the detriment of both fields. Thus, one purpose of this chapter is to highlight the present state of knowledge concerning the molecular mechanisms behind sporulation, spore resistance and dormancy, and spore germination and outgrowth. Hopefully, it will provide a counterpoint to more applied aspects of this system. This review will focus on molecular mechanisms, most of which have been examined in *Bacillus subtilis*. This organism is neither an important pathogen nor an important agent of food spoilage. However, its natural transformability, as well as an abundance of molecular biological and genetic information, including the complete sequence of its genome (119), has

made *B. subtilis* the organism of choice for mechanistic studies on sporulation, spore germination, and spore resistance. Although much less mechanistic work has been carried out with other sporeformers, those studies indicate that the fundamental mechanisms regulating gene expression during sporulation, bringing about spore resistance and dormancy, and involved in spore germination are probably similar to those in *B. subtilis*. Determination of the genome sequences of at least five other gram-positive sporeformers (*Bacillus anthracis*, *Bacillus halodurans*, *Bacillus stearothermophilus*, *Clostridium acetobutylicum*, and *Clostridium difficile*) is at or near completion, and comparison of these data with those for *B. subtilis* has indicated a tremendous degree of conservation among genes initially involved in these processes.

SPORULATION

Distribution of Sporeformers

The sporulating organisms discussed in this chapter form heat-resistant endospores which contain DPA and are refractile or phase bright under phase-contrast microscopy. Most studies on sporulation, spores, and spore germination have been carried out with species of either the aerobic bacilli or the anaerobic clostridia. However, other genera form similar spores; among these are the genera *Sporosarcina*, *Sporolactobacillus*, and *Thermoactinomyces* (219). Studies using rRNA sequence analysis to determine evolutionary relatedness have shown that the genera *Bacillus*, *Sporosarcina*, *Sporolactobacillus*, and *Thermoactinomyces* (219, 225) are quite closely related (252) but are clearly derived from a common ancestor, most likely a sporeformer. A number of other genera, including *Staphylococcus*, are also derived from the same common ancestor (252) yet cannot sporulate. Sporulation-specific genes whose sequences are highly conserved among sporeformers appear to have disappeared from these latter organisms (18). Some of these nonsporing species, for example, *Planococcus citreus*, are more closely related to present-day sporeformers than are other sporeformers (225). Unfortunately, the genetic changes causing the loss of sporulation ability in these organisms have not been well characterized.

Induction of Sporulation

The commonest mechanism for inducing sporulation in the laboratory is limitation for one or more nutrients, including carbon or nitrogen. This is achieved either by exhausting one or more nutrients during cell growth, by shifting cells from a rich to a poor medium, or by adding an inhibitor (decoyinine) of guanine nucleotide biosynthesis. Although these laboratory methods cause

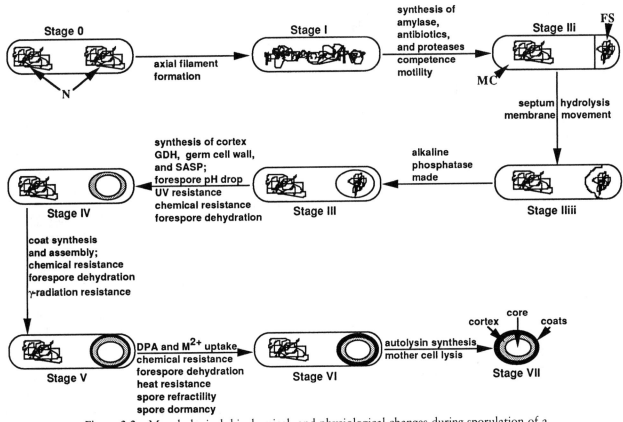

Figure 3.2 Morphological, biochemical, and physiological changes during sporulation of a rod-shaped *Bacillus* cell. In stage 0, a cell with two nucleoids (N) is shown; in stage IIi the mother cell and forespore are designated MC and FS, respectively. Note that the forespore nucleoid is more condensed than that in the mother cell. Stage IIii is not shown in this scheme, and for clarity the forespore nucleoid is not shown after stage III. The time of some biochemical and physiological events, such as forespore dehydration and acquisition of types of resistance to different chemicals (all lumped together as "chemical resistance"), stretches over a number of stages. The data for this figure are taken from references 56, 64, 158, 176, and 213.

the great majority of cells in a culture to sporulate, this is undoubtedly not the way sporulation is induced in nature. Indeed, even cells in a growing culture produce a finite number of spores, with the percentage of spores increasing as the culture's growth rate decreases (45). The time from initiation of sporulation to its completion may take as little as 8 h.

Sporulation is sensitive to repression by good carbon sources. This suggests that catabolite repression may regulate sporulation. Unfortunately, the precise mechanism for modulation of carbon catabolite repression of sporulation is not clear, despite significant insight in recent years (30, 47). The precise intracellular signal that triggers initiation of sporulation is unknown. Molecules such as cyclic AMP and cyclic GMP have been ruled out, and a possible role for guanine nucleotides has been suggested. Indeed, despite the impressive elucidation of the pathways regulating gene expression during sporulation

(see below), the physiological signals which initiate it remain generally obscure. However, induction of synthesis of enzymes of the tricarboxylic acid cycle is required (89). Sporulation is also modulated in a cell density-dependent fashion by small molecules secreted into the medium (22, 172, 244).

Sporulation-Associated Phenomena

In addition to sporulation, the cessation or slowing of growth of sporeformers can be associated with (i) synthesis of degradative enzymes such as amylases and proteases (118); (ii) synthesis of antibiotics such as gramicidin, bacitracin, or surfactin (258); (iii) in some species, synthesis of protein toxins active against insects (5), animals, or humans; (iv) development of motility (165); and (v) in a few species (e.g., *B. subtilis*), development of genetic competence (54) (Fig. 3.2). Although these phenomena are not necessary for sporulation, in most cases they are regulated at least partially by the mechanisms that modulate gene expression during sporulation (54, 118, 165, 258). In some cases, for example, protease synthesis and competence development in *B. subtilis*, the expression of the genes involved is under multiple controls by both positive and negative effectors (54, 118). Like sporulation, competence is also regulated in a cell density-dependent manner, in this case via the secretion of small peptides termed competence pheromones (141, 222).

There are drastic changes in the metabolism of the sporulating cell. Some of these changes include catabolism of previously formed polymers, such as poly-β-hydroxybutyrate, and initiation of oxidative metabolism due to synthesis of tricarboxylic acid cycle enzymes (22, 223). Many of these latter enzymes are not present in the developing forespore (207). Consequently, the mother cell and forespore have different metabolic capabilities. However, the precise source(s) of energy in the forespore is not clear, although the mother cell may provide much of the forespore's energy, either directly or indirectly.

Morphological, Biochemical, and Physiological Changes during Sporulation

Sporulation is generally divided into seven stages based primarily on the morphological characteristics of cells throughout the developmental process (Fig. 3.2). Growing cells are in stage 0, although sporulation is initiated only following completion of chromosome replication, i.e., only in binucleate cells (88). Older literature had suggested that the first morphological feature unique to sporulation was the presence of the two nucleoids in an axial filament which could be observed in electron micrographs; this was defined as stage I. Because of concerns that this axial filament was an artifact of preparation and because no single mutant has ever been found which is blocked in the transition between stage 0 and stage I, stage I has often been ignored. However, there is recent evidence that stage I is a discrete stage in sporulation (176), although its significance remains obscure.

The first morphological feature of sporulation is the formation of an asymmetric septum. It divides the sporulating cell (now at stage IIi [56, 64, 126]) into the larger mother cell and smaller forespore compartments. Despite the isolation of many genes involved in septum formation (126), the mechanisms for placement of this septum (midcell for normal cell division or asymmetric for sporulation) and its formation are not clear. Similarly, although the sporulation septum differs in some respects from the normal cell division septum, the nature of the

differences is not known. After the asymmetric sporulation division only ~30% of the chromosome is present in the forespore; the remainder is subsequently transferred into the forespore, and then both mother cell and forespore compartments contain complete and apparently identical chromosomes (253). However, the forespore chromosome is initially more condensed than the mother cell chromosome (200, 211), although the significance of this difference is presently unclear. A biochemical marker for late stage II is the synthesis of high levels of alkaline phosphatase. After stage IIi, genes in the two compartments of the sporulating cell may be expressed differentially.

Following septum formation, the mother cell membrane grows around and eventually engulfs the forespore. Thus, the forespore is surrounded by two complete membranes termed the inner and outer forespore membranes (see below). These two membranes have opposite polarities (55, 64, 249). Progression to this stage of sporulation, termed stage III, has been finely characterized by electron microscopy. There are a number of specific changes which occur in the transition from stage II to stage III, leading to the subdivision of stage II into three substages (56), only two of which are shown in Fig. 3.2.

In the transition from stage III to stage IV, a large peptidoglycan structure termed the cortex is laid down between the inner and outer forespore membranes. The cortex has a structure similar to that of cell wall peptidoglycan, but with a number of differences (discussed later). The spore's nascent germ cell wall is also made at about the same time as the cortex but appears to have the same structure as cell wall peptidoglycan and is formed between the cortex and the inner forespore membrane, probably by forespore enzymes (53, 157). During the stage III to stage IV transition, the forespore also synthesizes two biochemical markers, glucose dehydrogenase and SASP. Although the function of glucose dehydrogenase in spores is not known, a number of the SASP are involved in spore resistance. The developing forespore acquires full UV resistance and some chemical resistance at this time (213), and the forespore chromosome adopts a ringlike shape owing to the binding of specific SASP (176). Late in stage III, the forespore pH falls by 1 to 1.3 units, and forespore dehydration begins (139, 158).

In the stage IV to stage V transition, a series of proteinaceous coat layers are laid down outside the outer forespore membrane (52, 53, 212). Although coat synthesis is normally ascribed only to the stage IV to stage V transition, this is too narrow a distinction, as some coat proteins are clearly made in the stage III to stage IV transition and even after stage V (52, 56). The amount and variety of the spore coat proteins appear to vary considerably between species. *B. subtilis* has a complex coat structure, and other species have somewhat simpler coats (52, 53, 223). The reason for this is not clear, but many *B. subtilis* coat proteins can be lost with no apparent phenotypic effect. Forespore γ-radiation resistance begins to be acquired during this period, as is further chemical resistance. Forespore dehydration also continues during this time (158, 213). During the stage V to stage VI transition, the spore core's depot of DPA is accumulated following DPA synthesis in the mother cell (56). DPA uptake is paralleled by uptake of an enormous amount of divalent cations, predominantly Ca^{2+}, but with much Mg^{2+} and Mn^{2+} as well (157). The great majority of these cations are also in the spore core, presumably associated with DPA (157). The precise state of these compounds in the spore core is not known, although the amount of DPA accumulated exceeds its solubility. During this period the spore's central region or core also undergoes the final process of dehydration (158). As a consequence of the high ratio of solids to water in the spore core at this stage, the spore appears phase bright or white under the phase-contrast microscope. In addition, because of permeability changes in the spore membranes, the spore stains poorly if at all with common bacteriological stains. The spore also becomes metabolically dormant during this period and acquires further γ-radiation and chemical resistance (213, 215). Finally, in the transition to stage VII, one or more autolysins are produced in the mother cell (120), resulting in its lysis and release of the spore.

The seven stages are useful in characterizing various asporogenous or *spo* mutants which have no defect in growth but are blocked in a particular stage in sporulation. These *spo* mutants are given an added designation denoting the stage in which they are blocked. Thus *spo0* and *spoII* mutants are blocked in stages 0 and II, respectively. The various stages have also allowed facile correlation of biochemical changes with various morphological changes (Fig. 3.2). However, the sporulation scheme outlined above is an oversimplification for several reasons. First, since the various stages are intermediates in a continuous developmental process, it is probably misleading to think of the stages as discrete entities. Second, the scheme given is only for rod-shaped organisms which sporulate without terminal swelling of the forespore compartment. There are many sporeformers in which the forespore compartment swells considerably and the mother cell elongates (64). Similarly, some sporeformers grow as cocci (219). These differences undoubtedly modify the sporulation process. However, detailed knowledge of sporulation in these of organisms is limited.

Regulation of Gene Expression during Sporulation

Much of our knowledge about the regulation of gene expression during sporulation is derived from *spo* mutants. Asporogeny can be caused by mutations in any one of more than 75 distinct genetic loci (56, 229). The identification of biochemical or physiological markers for various stages of sporulation provides another major aid in understanding gene regulation during sporulation (Fig. 3.2). Analysis of these markers in *spo* mutants provides strong evidence that sporulation is primarily a linear process of sequential events. Thus, *spo0* mutants generally exhibit no sporulation-specific marker events, *spoII* mutants generally demonstrate only stage 0-specific marker events, and so on. Analysis of *spo* mutants has furnished a broad outline of regulation of gene expression during sporulation. However, the advent of molecular cloning technology was necessary to provide detailed understanding of this process, as the genes in which mutations block sporulation have been cloned and sequenced. In addition, many sporulation-specific genes in which mutation has little or at most a subtle effect on sporulation have been identified and analyzed. For most of these genes, their precise functions, the location of their expression (mother cell or forespore), and the factors involved in regulation of their expression are known (56, 126, 229).

It was hypothesized that sporulation requires transcription of many new genes and that this change in transcriptional pattern was modulated by changes in the specificity of the cell's RNA polymerase owing to synthesis of new σ (specificity) factors (133). This hypothesis has been expanded over the past 20 years because changes in gene expression during sporulation require modulation of transcription by many mechanisms, including (i) synthesis of new RNA polymerase σ factors with altered promoter specificity at various times and in the different compartments of the sporulating cell, (ii) activation or inactivation of new σ factors by both covalent and noncovalent modification, (iii) intricate regulatory communication between mother cell and forespore to ensure coordinate development in both compartments, (iv) synthesis of new DNA binding proteins to activate or repress transcription, (v) degradation of repressors, and (vi) modulation of transcription factor activity by phosphorylation. The information explosion from this work has now provided an extremely detailed picture of the control of gene expression during sporulation. This

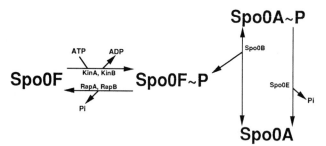

Figure 3.3 Gene products and reactions which affect levels of Spo0A~P. Spo0E is a phosphatase specific for Spo0A~P; RapA and RapB are phosphatases specific for Spo0F~P and are different names for Spo0L and Spo0P, respectively (80, 163, 172).

picture is striking not only in its complexity but also in the redundancy of its control mechanisms. The following discussion of these control mechanisms has been simplified and concentrates on major regulatory gene products.

Initiation of sporulation requires expression of *spo0* gene products at a significant level in vegetative cells (56, 77). The most important of these *spo0* gene products is Spo0A. This protein is the response-regulator half of a two-component (signal transduction) regulatory system.

These signal transduction systems transmit signals (often by binding to DNA and affecting transcription) when an aspartyl residue in the response regulator is phosphorylated by a sensor protein which has kinase activity (21). In growing cells, Spo0A is primarily in the dephosphorylated state. However, under conditions which initiate sporulation, the degree of Spo0A phosphorylation rises through action of multiple sensor kinases. In *B. subtilis* the majority of phosphate on Spo0A appears to be derived via a phosphorelay initiated by phosphorylation of Spo0F (Fig. 3.3 and 3.4). The phosphate is then transferred from Spo0F~P to Spo0A by the Spo0B protein. There are at least two major kinases (KinA and KinB) which can initiate the phosphorelay. These enzymes can also phosphorylate Spo0A directly but with low efficiency (80). Mutations in either *kinA* or *kinB* generally have little or no effect on sporulation, but a *kinA kinB* double mutant causes asporogeny. A third kinase (KinC) acts preferentially on Spo0A, but its precise function in wild-type cells is unclear (114, 124, 125). In addition to the many kinases feeding phosphate into Spo0A, there are at least three phosphatases that can dephosphorylate Spo0A~P. It can be dephosphorylated either directly or

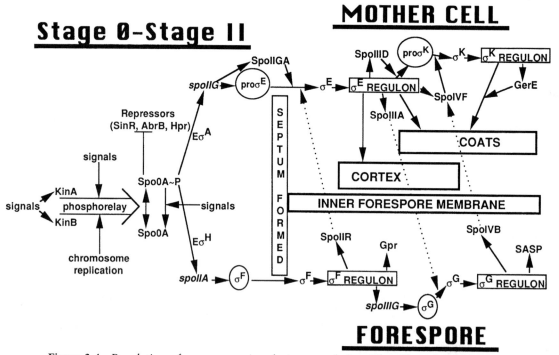

Figure 3.4 Regulation of gene expression during sporulation. The effect of Spo0A~P on repressors is negative; other effects of regulatory molecules on reactions are generally positive, although the effect of signals may be positive or negative. The enclosure of the pro-σ factors and σ factors denotes that at this time these factors are inactive. This figure is adapted from that of Piggot et al. (176) and includes data from references 56, 80, 104, 117, 118, 130, 172, 176, 228, and 229.

through the dephosphorylation of Spo0F∼P (163, 172) (Fig. 3.3). This multiplicity of kinases and phosphatases allows a variety of environmental signals to be integrated so that the precise intracellular level of Spo0A∼P can be set in response to the environment in toto. A specific threshold concentration of Spo0A∼P is probably needed to initiate sporulation (35). A number of regulators of both the kinases and phosphatases that modulate the level of Spo0A∼P have been identified (80, 172), although the complete picture of this regulation is not yet clear.

The phosphorylation of Spo0A greatly increases its affinity for binding to sites upstream of several key genes, although different sites have much different affinities for Spo0A∼P (80, 169). The *abrB* gene has a very high affinity for Spo0A∼P, and its binding of Spo0A∼P greatly decreases *abrB* transcription. Since AbrB is a labile protein, the decrease in *abrB* transcription rapidly decreases the intracellular AbrB concentration. AbrB is a repressor of several genes normally expressed in the stage 0 to stage II transition, including the *spo0H* gene, which codes for a minor σ factor for RNA polymerase termed σ^H. AbrB is also an activator for synthesis of a second repressor, termed Hpr, which represses additional stage 0-expressed genes, especially the genes for several proteases. In addition to AbrB and Hpr, there is a third repressor termed SinR. SinR binds to and represses other genes expressed in the stage 0 to stage II transition, including those of the *spoIIG* and *spoIIA* operons (22, 80). SinR action is blocked by synthesis of a protein termed SinI, which binds to SinR and blocks its action. Synthesis of SinI is stimulated by Spo0A∼P and probably repressed by AbrB and Hpr. Thus, early in the stage 0 to stage II transition, many genes normally repressed during vegetative growth are derepressed through action of Spo0A∼P (Fig. 3.4).

Important among the stage 0-II genes is *spo0H*, which encodes a minor sigma factor, σ^H. There is some transcription of *spo0H* during vegetative growth by RNA polymerase containing the cell's main sigma factor ($E\sigma^A$). However, levels of active σ^H during vegetative growth are maintained at a low level, probably by a posttranscriptional mechanism (22, 80). Early in sporulation, rising levels of functional σ^H promote transcription by $E\sigma^H$. As Spo0A∼P levels increase even further, concentrations become high enough to bind to the promoter regions of the *spoIIG* and *spoIIA* operons which have been derepressed by removal of SinR as a SinI-SinR complex (56, 80). The binding of Spo0A∼P stimulates transcription of these genes by RNA polymerase containing σ^A or σ^H, respectively (9, 22, 56, 80). $E\sigma^H$ has a different promoter specificity than does $E\sigma^A$, as do RNA polymerases

with all other sporulation-specific σ factors (76, 117). This allows these different RNA polymerases to recognize and transcribe different sets of genes. The *spoIIA* operon transcribed by $E\sigma^H$ encodes three proteins, with the third cistron (*spoIIAC*) encoding another sigma factor termed σ^F. The *spoIIAB* gene encodes an inhibitor of σ^F function, while *spoIIAA* encodes an antagonist of SpoIIAB (48, 117, 126). Prior to septation, σ^F is inactive in the sporulating cell due to its interaction with SpoIIAB. Interaction of SpoIIAB with σ^F rather than with SpoIIAA is promoted by the high ATP-to-ADP ratio in the preseptation sporulating cell (2), and phosphorylation of SpoIIAA by SpoIIAB also prevents SpoIIAA-SpoIIAB interaction (117, 126).

The *spoIIG* operon transcribed by $E\sigma^A$ encodes two proteins, with the second cistron, *spoIIGB*, also encoding an alternative σ factor for RNA polymerase, termed σ^E (56, 76, 117). However, unlike σ^F, σ^E is made as a precursor termed pro-σ^E, which is inactive in directing transcription. The first gene of the *spoIIG* operon (*spoIIGA*) is the protease responsible for processing pro-σ^E to σ^E (117, 126, 228). However, action of SpoIIGA on pro-σ^E requires gene expression under σ^F control in the forespore compartment (56, 117, 126, 216) (Fig. 3.4).

Although the *spoIIG* operon is transcribed before septum formation, σ^E is not generated until after septation, and then only in the mother cell (117, 126, 128). Similarly, σ^F is also produced before septum formation and is maintained in an inactive state by interaction with inhibitory SpoIIAB. However, following septation, σ^F becomes active in the forespore, as SpoIIAB leaves σ^F and interacts with SpoIIAA. This "partner switching" is promoted by dephosphorylation of SpoIIAA∼P by the SpoIIE phosphatase which resides in the sporulation septum at this time, and possibly by changes in the ATP-to-ADP ratio in the forespore (2). However, it is presently unclear how σ^F activation or activity is confined only to the forespore, although several models for this regulation have been presented (113, 117, 126, 128, 254). In any event, $E\sigma^F$ now initiates transcription of several genes in the forespore, including *spoIIR* (or *csfX*), *gpr*, and *spoIIIG* (104, 117, 126, 130, 210) (Fig. 3.4).

The *spoIIR* gene product acts by promoting pro-σ^E processing by SpoIIGA, although the precise mechanism of action of SpoIIR is not clear. SpoIIR can promote pro-σ^E processing not only in the mother cell, but also in the forespore (98, 104, 216). However, pro-σ^E is normally not present in the forespore shortly after septation. Although the reasons for this are not yet clear (117, 126), the absence of pro-σ^E from the forespore likely is dependent on the fact that pro-σ^E is associated specifically with the sporulation septum as is SpoIIGA (62, 81, 97).

During sporulation, pro-σ^E processing requires synthesis of SpoIIR in the forespore (104, 130). This requirement for an event in the forespore as a prerequisite to events in the mother cell ensures coordinate development of the mother cell and forespore (117, 134) and is the first of three examples of regulatory cross talk between the two compartments of the sporulating cell. Of the other two Eσ^F-transcribed genes, *gpr* encodes a protease that acts on SASP in the first minutes of spore germination, and *spoIIIG* encodes another σ factor termed σ^G, which is responsible for the bulk of forespore-specific transcription (210). However, σ^G is not active immediately following its synthesis in stage II. Instead, some mother cell-specific event(s), including transcription of the *spoIIIA* operon by Eσ^E, is required for generation of active Eσ^G, which now acts in stage III (56, 85, 183) (Fig. 3.3 and 3.4). This is the second example of regulatory cross talk between mother cell and forespore. The precise mechanism for regulation of σ^G activity is not yet clear but may involve regulation by SpoIIAB in the same manner as that for σ^F (183). The generation of active σ^E only in the mother cell, and the synthesis and activation of σ^G only in the forespore, firmly establishes the basis for compartment-specific transcription during sporulation.

As noted above, Eσ^G transcribes several genes expressed only in the forespore (56, 210). Some of these are genes essential for sporulation, but their function is unknown. As the third example of regulatory cross talk, at least one such gene (*spoIVB*) is also responsible for communicating with the mother cell and regulating gene expression in this compartment (117, 126) (Fig. 3.4) (see below). The *ssp* genes are a large set of genes transcribed by Eσ^G. These genes encode the SASP that are major protein components of the spore core. Many of the SASP are of the α/β-type and are DNA binding proteins which confer upon spore DNA resistance to a variety of treatments (213–215) (see below).

In contrast to Eσ^G, Eσ^E transcribes genes only in the mother cell, including genes needed specifically for spore cortex biosynthesis and some genes (termed *cot* genes) needed for spore coat formation and assembly (56). Eσ^E also transcribes the *spoIIID* gene, which encodes a DNA binding protein that modulates Eσ^E action, resulting in different classes of Eσ^E-transcribed genes. SpoIIID is needed for transcription of the *sigK* gene by Eσ^E (56, 117, 229). The *sigK* gene encodes a final new σ factor termed σ^K (76, 117, 229). In *B. subtilis*, the *sigK* gene has a large intervening sequence which is removed only from the mother cell genome by a recombinase. Expression of the recombinase is regulated such that generation

of an intact *sigK* gene just precedes *sigK* transcription. However, this intervening sequence is absent in the *sigK* genes of other *Bacillus* species (1). Consequently, removal of the intervening sequence does not appear to be a general mechanism for regulating σ^K function. As is the case with σ^E, σ^K is also synthesized as an inactive pro-σ^K and is processed proteolytically to σ^K about 1 h after its synthesis (117, 126). Conversion of pro-σ^K to σ^K in the mother cell requires expression of one σ^G-controlled gene (*spoIVB*) in the forespore (42, 134) (the third example of regulatory cross talk) and participation of the σ^E-controlled *spoIVF* operon in the mother cell (41) (Fig. 3.4). SpoIVFB is the protease which processes pro-σ^K, and SpoIVFA is an inhibitor of SpoIVFB function (117, 126, 185, 189). However, the precise mechanism whereby SpoIVB stimulates SpoIVFB action (or relieves SpoIVFA inhibition) is not clear. Eσ^K itself also transcribes *sigK* (in conjunction with SpoIIID) (76, 117), the gene for DPA synthase (43), other *cot* genes (52, 53, 229), and one gene termed *gerE* which encodes a DNA binding protein that modulates Eσ^K activity, causing a large change in the pattern of *cot* gene expression (56) (Fig. 3.4). Essentially all of our detailed knowledge of gene expression during sporulation ends at this point. There is little definitive information about genes which may be required for mother cell lysis. However, synthesis of one autolytic enzyme (CwlC) is known to be transcribed by Eσ^K (120).

The preceding picture of regulation of gene expression during sporulation is derived predominantly from studies with *B. subtilis*. Is the picture similar in organisms from other species or genera, including *Clostridium*? Although there is no definitive answer to this question, a number of key regulatory genes, including *spo0A*, *spoIIGA*, *spoIIGB*, *spoIIIG*, and *sigK*, have been analyzed in other sporeformers, including *Clostridium* spp. (1, 18, 246). The striking conservation of both the sequences of the encoded proteins and the organization of these genes strongly suggests that regulation of sporulation across species is by similar if not identical mechanisms.

THE SPORE

Spore Structure

The spore that is released from the mother cell at the end of sporulation is biochemically and physiologically different from the vegetative cell. The spore structure is also extremely different from that of a cell (Fig. 3.5). Many spore structures, including the exosporium and coats, have no counterparts in the vegetative cell. The outermost spore layer, the exosporium, varies significantly in

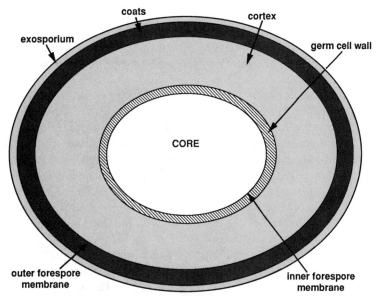

Figure 3.5 Structure of a dormant spore. The various structures are not drawn precisely to scale, especially the exosporium, whose size varies tremendously between spores of different species. The relative size of the germ cell wall is also generally smaller than that shown. The positions of the inner and outer forespore membranes, between the core and the germ cell wall and between the cortex and coats, respectively, are also noted.

size between species, and its various components are not well characterized (52, 53, 157). Underlying the exosporium are the spore coats. The complexity of this structure varies significantly between species. In *B. subtilis* spores, a number of distinct coat layers can be easily seen in electron micrographs (52, 53). Where individual coat proteins have been analyzed, they are unique to the spore stage of bacterial existence. Some coat proteins have an extremely unusual amino acid composition (52, 53). Several of these coat proteins appear active primarily in assembly of the final coat structure. Posttranslational modification of some coat proteins, in particular, formation of dityrosine cross-links, may be important in spore coat structure. The spore coats protect the spore cortex from attack by lytic enzymes. The coats also may provide an initial barrier to chemicals such as oxidizing agents (15, 52, 53, 66, 186, 190, 213, 215). However, the spore coats play no significant role in maintenance of spore resistance to heat or radiation (52, 66, 115, 213, 215).

Underlying the spore coats is the outer forespore membrane. This is a complete functional membrane in the developing forespore and may also be so in the spore (53). If it is a complete functional membrane, it may play a large role in the extreme impermeability of the spore to small molecules. The protein composition of this membrane is different from that of the inner forespore membrane (53).

Underlying the outer forespore membrane is the spore's cortex. This large peptidoglycan layer is structurally similar to cell wall peptidoglycan, but with several differences (19, 53). The cortical peptidoglycan always contains diaminopimelic acid, even if the vegetative cell wall peptidoglycan contains lysine. A significant percentage ($\sim$65%) of the muramic acid residues in cortical peptidoglycan also lack any peptide residues. Although a few muramic acid residues contain a single D-alanine, the majority are present as muramic acid lactam, which is not present in cell wall peptidoglycan. A number of genes involved in cortex synthesis have been identified and characterized (42, 53). Although the precise mechanism of cortex synthesis is not known, the required gene products are made primarily in the mother cell. Spore cortical peptidoglycan exhibits a much lower degree of peptide cross-linking compared with vegetative cell peptidoglycan (53). However, there is no knowledge of the timing of cross-link formation in the cortex, nor of the distribution of cross-links throughout the cortical volume, as the three-dimensional structure of the spore cortex is not known. This is a major deficiency, as the spore cortex is thought to be the structure responsible for the dehydration of the spore core and thus much of spore resistance (see below).

Between the cortex and the inner forespore membrane is the germ cell wall. Its structure may be identical to that in vegetative cells. The next structure, the inner forespore membrane, is a complete membrane and is an extremely strong permeability barrier to hydrophilic molecules and to most molecules of >150 molecular weight (67). This membrane's phospholipid content is similar to that of growing cells (55). However, the spore core volume surrounded by the inner forespore membrane appears smaller than is predicted based on this latter membrane's phospholipid content. Since the core's volume can expand significantly upon spore germination in the absence of membrane synthesis, the "excess" membrane in the inner forespore membrane may allow for this expansion.

Finally, the central region, or core, contains the spore's DNA, ribosomes, and most enzymes, as well as the depots of DPA and divalent cations. There are also many unique gene products in the dormant spore, including the large SASP pool (10 to 20% of spore protein), much of which is bound to spore DNA (209, 214, 215). A notable feature of the spore core is its low water content (66). While vegetative cells have ~4 g of water per g of dry weight, the spore core has only 0.4 to 1 g of water per g of dry weight. The core's low water content is thought to play a major role in spore dormancy and in spore resistance to a variety of agents (66, 213). In contrast to the low water content in the spore core, other regions of the spore have a more normal water content (66). Although the spore cortex is clearly essential for generating and maintaining spore core dehydration, the precise mechanism(s) involved is not clear.

Spore Macromolecules

Spores are biochemically different from vegetative cells. Some proteins found in the spore coats and core are not present in the vegetative cell. The SASP are particularly noteworthy, as some of these proteins play a major role in spore resistance (209, 210, 213–215). With the exception of several minor species, SASP are synthesized in the forespore during stage III of sporulation, when their coding genes are transcribed by Eσ^G, and are located in the spore core. There are three kinds of SASP in *Bacillus* spp., termed γ-type, α/β-type, and minor. There are at least nine minor SASP in *B. subtilis* spores, with sizes ranging from 34 to 71 amino acids (8, 23, 24). The function of these minor proteins is generally unclear. The γ-type SASP compose ~5% of spore protein and are 75- to 105-residue proteins which do not bind to any other spore macromolecule. Their only known function is to be degraded during germination, thus providing amino acids for the germinating spore. Their degradation is initiated by the sequence-specific protease GPR (germination protease, the product of the *gpr* gene). The γ-type SASP are encoded by a single gene in *Bacillus* spp. However, neither this gene nor γ-type SASP have been found

in clostridial spores. The sequences of γ-type SASP are not homologous to other proteins in currently available databases, and the proteins' sequence has diverged significantly during evolution. The α/β-type SASP compose 3 to 5% of total spore protein and are named for the two major proteins of this type in *B. subtilis* spores. In contrast to γ-type SASP, α/β-type SASP are coded for by at least seven genes. All of these are expressed, although in most species two proteins (major α/β-type SASP) are expressed at high levels and the remainder at much lower levels. However, all α/β-type SASP have similar properties in vitro and in vivo (see below).

The α/β-type SASP are also found in clostridial spores (82, 213–215). The amino acid sequences of these small (60- to 75-residue) proteins are highly conserved both within and across species, although they too have no homology to other proteins in current databases (209, 214, 215). The α/β-type SASP, like the γ-type, are also degraded to amino acids in the first minutes of spore germination. Their degradation is initiated by GPR (209). However, the α/β-type SASP are also DNA binding proteins, both in vivo and in vitro (178, 213–215). In vivo these proteins saturate the spore chromosome and provide much of the spore DNA's resistance to various treatments. Binding of α/β-type SASP to DNA is not sequence specific but exhibits a preference for G + C-rich regions, although A + T-rich regions are bound. These proteins bind to the outside of the DNA helix, interacting primarily with the sugar phosphate backbone. This binding alters the DNA structure from a B-like helix to an A-like helix, although this is not the classical A-like helix of a poorly hydrated DNA fiber. The precise structure of the α/β-type SASP-DNA complex is unknown. The binding of these proteins to DNA has striking effects on DNA properties, including providing great resistance to chemical and enzymatic cleavage of the DNA backbone, altering the DNA's UV photochemistry, and greatly slowing DNA depurination (see below). In addition to α/β-type SASP, the spore nucleoid also contains HBsu, the major protein found on the vegetative cell nucleoid; this protein covers ~5% of spore DNA and modulates the effect of α/β-type SASP on spore DNA properties (188).

Not only do spores contain unique proteins, but also a number of proteins present in vegetative cells are absent from spores. These include amino acid and nucleotide biosynthetic enzymes, which are degraded during sporulation and then synthesized at defined times during spore outgrowth (207). Other enzymes, in particular, the enzymes of amino acid and carbohydrate catabolism, are present at similar levels in spores and cells (207). Spores also contain all of the enzymes needed for RNA and protein synthesis and many DNA repair enzymes (207). However, at least one enzyme needed for initiation of

DNA replication may be absent from spores (207). Enzyme activities common to both cell and spore proteins are invariably the same gene product. In general, spore and cell rRNAs and tRNAs are similar if not identical, although much tRNA in spores lacks the 3'-terminal A residue, and little if any spore tRNA is aminoacylated (207). However, spores lack most if not all functional mRNA. Spore DNA appears identical to cell DNA but has different proteins associated with it.

Spore Small Molecules

Spores are quite different from cells in their content of small molecules (Table 3.1), which are located in the core. The low amount of spore core water and the huge depot of DPA and divalent cations have already been noted (Table 3.1). Physical methods have been used to show that the ions in the spore core are quite immobile (27), consistent with a dearth of free water in spores. The pH in the spore core is 1 to 1.5 units lower than that in a growing cell (139, 140, 202). The pH of the forespore changes relative to that of the mother cell during the stage

III to stage IV transition (Fig. 3.2). Spores of most species have a large depot of 3-phosphoglyceric acid (3PGA) (Table 3.1), which is accumulated shortly after and in response to the forespore pH decrease (140, 202). In contrast to growing cells, spores have little if any of the common "high-energy" compounds found in cells, including deoxynucleoside triphosphates, ribonucleoside triphosphates, reduced pyridine nucleotides, and acyl coenzyme A (acyl-CoA) (Table 3.1). However, spores do contain significant amounts of ribonucleoside mono- and diphosphates (but not deoxynucleotides), oxidized pyridine nucleotides, and CoA (Table 3.1). The high-energy forms of these latter compounds are lost from the forespore late in sporulation, as the spore becomes dormant at about the time of DPA uptake (Fig. 3.2). Interestingly, much of the CoA in spores is in disulfide linkage, some as a CoA disulfide and some linked to protein (Table 3.1). The function of these disulfides is not known, but they are reduced in the first minutes of spore germination (201). In addition to lacking many amino acid biosynthetic enzymes, spores have very low levels of most free amino acids (132) but often have high levels of glutamic acid (Table 3.1).

Spore Dormancy

Spores are metabolically dormant, catalyzing no detectable metabolism of endogenous or exogenous compounds and having no high-energy compounds (127, 207). The major cause of the spore's metabolic dormancy is undoubtedly the low water content of the core, which may preclude enzyme action (213). In a further reflection of this metabolic dormancy, the spore core contains at least two enzyme-substrate pairs which are stable in the dormant spore for months to years but which interact in the first 15 to 30 min of spore germination, resulting in complete substrate degradation. These two enzyme-substrate pairs are 3PGA-phosphoglycerate mutase (PGM) and SASP-GPR (213). Although special regulatory mechanisms other than dehydration stabilize 3PGA and SASP in the developing forespore despite the presence of PGM and GPR (212), these mechanisms do not explain the extreme stability of 3PGA and SASP in dormant spores. The reason for the lack of PGM and GPR action on their substrates in spores may again be enzyme inactivity caused by spore core dehydration. Although this explanation seems reasonable, in the absence of precise data on the amount of free water in the spore core, it is difficult to test this idea definitively.

Spore Resistance

The spore's metabolic dormancy is undoubtedly one factor in its ability to survive extremely long periods in the

Table 3.1 Small molecules in cells and spores of *Bacillus* species

Molecule	Content (mmol/g dry wt) in:	
	Cells[a]	Spores[b]
ATP	3.6	≤0.005
ADP	1	0.2
AMP	1	1.2–1.3
Deoxynucleotides	0.59[c]	<0.025[d]
NADH	0.35	<0.002[e]
NAD	1.95	0.11[e]
NADPH	0.52	<0.001[e]
NADP	0.44	0.018[e]
Acyl-CoA	0.6	<0.01[e]
CoASH[f]	0.7	0.26[e]
CoASSX[g]	<0.1	0.54[e]
3-PGA	<0.2	5–18
Glutamic acid	38	24–30
DPA	<0.1	410–470
Ca²⁺		380–916
Mg²⁺		86–120
Mn²⁺		27–56
H⁺	7.6–8.1[b]	6.3–6.9[b]

[a] Values for *B. megaterium* in mid-log phase are from references 46, 139, 201, 202, 206, 207, 213, and 218.
[b] Values are the range from spores of *B. cereus*, *B. subtilis*, and *B. megaterium* and are from references 139, 201, 202, 206, 207, and 213.
[c] Value is the total of all four deoxynucleoside triphosphates.
[d] Value is the sum of all four deoxynucleotides.
[e] Values are for *B. megaterium* only.
[f] CoASH, free CoA.
[g] CoASSX, CoA in disulfide linkage to CoA or a protein.
[b] Values are expressed as pH and are the range in *B. cereus*, *B. megaterium*, and *B. subtilis*.

absence of nutrients. A second factor in long-term spore survival is the spore's extreme resistance to potentially lethal treatments, including heat, radiation, chemicals, and desiccation (15, 61, 66, 214, 215). Spores are much more resistant than vegetative cells to a variety of killing treatments. Table 3.2 presents representative data for

Table 3.2 Killing and mutagenesis of spores and cells of *B. subtilis* by various treatments[a]

A. Freeze-drying

No. of freeze-drying cycles	Survival (%)		
	Cells[b]	Wild-type spores	$\alpha^-\beta^-$ spores[c]
1	2		
3		100 (<0.5)	7 (14)

B. 10% Hydrogen peroxide[a]

Time of treatment (min)	Survival (%)		
	Cells[b]	Wild-type spores	$\alpha^-\beta$ spores[c]
2.5	0.3	92	
5		88	50
10			10 (14)
20		50	0.1
60		6 (<0.5)	

C. UV[d]

	Dose to kill 90% of the population (J/m^2)		
	Cells[b]	Wild-type spores	$\alpha^-\beta^-$ spores[c]
	40	315	25

D. Moist heat

Treatment temp (°C)	D value		
	Cells[b]	Wild-type spores	$\alpha^-\beta^-$ spores[c]
95		14 min (≤0.5)	
85		360 min (≤0.5)	15 min (13)
65	<15 s	105 h	10 h
22		2.5 yr (≤0.5)	2.8 mo (18)

E. Dry heat

Treatment temp (°C)	D value		
	Cells[b]	Wild-type spores	$\alpha^-\beta^-$ spores[c]
120		33 min (12)	
90	5 min		5.5 min (12)

[a]Data taken from references 58, 59, 143, 203, 204, and 208. Values in parentheses are the percentage of survivors with asporogenous or auxotrophic mutations when spores undergo 30 to 99% killing.

[b]Cells in the log phase of growth. Similar results have been obtained with wild-type or $\alpha^-\beta^-$ cells.

[c]These spores lack the two major α/β-type SASP and thus ~70% of the α/β-type SASP pool.

[d]UV irradiation with light predominantly at 254 nm.

cells and spores of *B. subtilis*. Note, however, that some organisms form much more resistant spores than does *B. subtilis*. Spore resistance is due to a variety of factors, with factors such as spore core dehydration and α/β-type SASP involved in many types of resistance, while factors such as spore impermeability may be involved in only one. Since different and sometimes multiple factors contribute to different types of spore resistance, it is not surprising that spore resistance to different treatments is acquired at different times in sporulation (Fig. 3.2). Because resistance of spores to one treatment is often caused by multiple factors, elucidation of detailed mechanisms of spore resistance is difficult. However, in recent years the mechanisms of spore heat, UV, and H_2O_2 resistance have been elucidated. The following detailed discussion of spore resistance concentrates on *B. subtilis* because of the copious detailed mechanistic data available for this organism. However, studies with spores of other organisms, in particular, on the mechanism of heat resistance, have indicated that factors involved in resistance of *B. subtilis* spores are also involved in resistance of spores from other species and genera. However, the relative importance of particular factors in spore resistance may vary significantly between species.

Spore Freezing and Desiccation Resistance

Growing bacteria generally incur some lethality during freezing and greater lethality during desiccation, unless special precautions are taken to prevent killing. The precise mechanism(s) of killing is not clear, but one cause may be DNA damage; freeze-drying cells can cause significant mutagenesis (6). However, neither the precise type of DNA damage nor the reason for DNA damage caused by freeze-drying is clear. In contrast to the sensitivity of cells to freeze-drying, spores are resistant to multiple cycles of freeze-drying (71) (Table 3.2). A complete explanation for spore desiccation resistance is not yet available. However, the α/β-type SASP provide one component of spore desiccation resistance by preventing DNA damage caused by freeze-drying (59). Spores lacking these proteins (termed $\alpha^-\beta^-$ spores) are much more sensitive to killing by freeze-drying than are wild-type spores (Table 3.2), with the killing of $\alpha^-\beta$ spores due in large part to DNA damage. However, $\alpha^-\beta^-$ spores are still significantly more resistant to freeze-drying than are vegetative cells, indicating that spore resistance to freeze-drying has causes in addition to α/β-type SASP. Accumulation of various carbohydrates, such as the disaccharide trehalose, is often important in the resistance of yeasts or other fungal spores to freezing or desiccation (40). However, bacterial spores do not accumulate such sugars (207).

Spore Pressure Resistance

Spores are much more resistant to high pressures ($\geq$ 12,000 atm) than are cells (221). Although spores are more resistant than cells to lower pressures, spores are killed more rapidly at lower pressures than at higher pressures (36, 73). This apparent anomaly is caused by the promotion of spore germination at lower pressures; the germinated spores are then rapidly killed by the pressure treatment. In contrast, spore germination is not promoted by very high pressures. Although there is no complete understanding of the factors involved in spore resistance to high pressure or the precise mechanism whereby lower pressures promote spore germination, spore germination induced by pressure uses at least some of the components of normal germination pathways (255).

Spore γ-Radiation Resistance

Spores are generally more resistant to γ-radiation than are vegetative cells (61). In the few organisms where it has been studied, γ-radiation resistance is acquired 1 to 2 h before acquisition of heat resistance. The precise factors involved in spore γ-radiation resistance are not known, although SASP do not appear to be involved (214, 215). The low water content in the spore core would be expected to provide protection against γ-radiation. However, no analysis has correlated the degree of spore core dehydration with spore γ-radiation resistance. One significant impediment to the understanding of spore γ-radiation resistance is the lack of knowledge about the damage caused by γ-radiation that results in spore death. Presumably, this damage is to spore DNA. However, the precise nature of this damage and whether it is similar to that generated in vegetative cells is not known. Given the very different environments inside the vegetative cell and dormant spore, it is certainly possible that the γ-radiation damage to DNA differs significantly in these two states.

Spore UV-Radiation Resistance

Spores of many species are 7 to 50 times more resistant than are vegetative cells to UV radiation at 254 nm (208, 213–215) (Table 3.2), the wavelength giving maximal killing. Spores are also more resistant than cells at both longer and shorter UV wavelengths. UV resistance is acquired by the developing forespore 2 h before acquisition of heat resistance (Fig. 3.2), in parallel with synthesis of α/β-type SASP. These latter proteins are essential, and may also be sufficient, for spore UV resistance. Coats, cortex, and core dehydration are not necessary for spore UV resistance.

The major reason for spore UV resistance is the different UV photochemistry of DNA in spores and in cells.

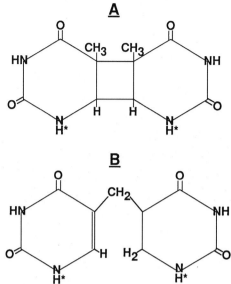

Figure 3.6 Structures of (A) cyclobutane-type TT dimer and (B) 5-thyminyl-5,6-dihydrothymine adduct (spore photoproduct). The positions of the hydrogens noted by the asterisks are the locations of the glycosylic bond in DNA.

The major photoproducts formed upon UV irradiation of cells or purified DNA are cyclobutane-type dimers between adjacent pyrimidines (63). The most abundant of these are between adjacent thymine residues (TT) (Fig. 3.6A), with smaller amounts between adjacent cytosine and thymine residues (CT) and adjacent cytosine residues (CC). In addition, UV irradiation of cells or purified DNA generates the various 6-4 photoproducts which are also formed between adjacent pyrimidines (247). All of these photoproducts can be lethal as well as mutagenic, although 6-4 photoproducts may be the most mutagenic (68). In contrast to the UV photochemistry of purified DNA or DNA in cells, UV irradiation of spores generates few cyclobutane-type pyrimidine dimers, and 6-4 photoproduct formation, if it takes place, is with a much lower yield as a function of UV fluence (50, 208). However, UV irradiation of spores does generate large amounts of a thyminyl-thymine adduct initially termed "spore photoproduct" (SP) (50) (Fig. 3.6B). The yield of SP as a function of UV fluence in spores is similar to the yield of TT as a function of UV fluence in cells. SP is a potentially lethal photoproduct (208, 214, 215). Thus, the difference in UV photochemistry between DNA in cells and spores is, by itself, not sufficient to explain spore UV resistance, as there must be a difference in the capacity of cells and spores to efficiently repair TT and SP, respectively. Indeed, spores have at least two mechanisms for SP repair, both of which operate in the first minutes of spore germination. One mechanism is via the same excision repair system that repairs TT and other lesions in growing cells (156). Spores lacking this repair system are two- to threefold more sensitive to UV than are wild-type spores (60, 208, 215). The second repair system, unique to both spores and SP, monomerizes SP to two thymines without excision of the lesion and is more error free than is excision repair (248). Spores lacking this SP-specific repair system are 5- to 10-fold more UV sensitive than are wild-type spores. Spores lacking both repair systems are 20- to 40-fold more UV sensitive than are wild-type spores (60, 208, 215). The gene in which mutation abolishes SP-specific repair has been cloned and sequenced (60). The product of this gene, termed spore photoproduct lyase (Spl), is an iron-sulfur protein (184) and shows some sequence homology to DNA-photoreactivation enzymes which cleave TT using light energy. However, SP monomerization does not require light, and Spl lacks at least one residue conserved in photoreactivating enzymes that is thought to be involved in transfer of light energy. In contrast to enzymes of excision repair which are present in both growing cells and spores, Spl is synthesized only in the developing forespore, with transcription of the spl gene directed by Eσ^G (169).

The major factor causing the altered UV photochemistry of spore DNA is the saturation of spore DNA with α/β-type SASP (208, 213–215). Spores lacking ~80% of these proteins ($\alpha^-\beta^-$ spores) are more UV sensitive than are vegetative cells (Table 3.2), and UV irradiation of $\alpha^-\beta^-$ spores generates significant amounts of TT and reduced amounts of SP. Generation of TT is the reason for the UV sensitivity of $\alpha^-\beta^-$ spores. The interchangeable role of α/β-type SASP in spore UV resistance has been shown by the restoration of UV resistance to $\alpha^-\beta^-$ spores by synthesis of sufficient levels of either major or minor α/β-type SASP from the same or a different species. Strikingly, UV irradiation of a complex between DNA and any of a number of purified α/β-type SASP generates SP and no TT. The yield of SP as a function of UV fluence in these complexes is ~10-fold lower than the yield of SP in vivo as a function of UV fluence. This difference is due to the large DPA depot in spores which acts as a photosensitizer. Binding of α/β-type SASP to DNA also blocks formation of CT and CC, as well as a variety of 6-4 photoproducts. The change in the UV photochemistry of DNA upon binding of α/β-type SASP strongly indicates that the DNA in this complex is in an altered structure. Indeed, DNA complexed with α/β-type SASP appears to be in an A-like helix, as has been suggested to be the case in spores (213–215).

During spore germination, α/β-type SASP are degraded to free amino acids which can then support protein synthesis. SASP degradation is initiated by the sequence-specific protease termed GPR (213–215). However, SASP degradation takes place more slowly than DPA release. Because of the photosensitizing action of DPA, spores early in germination are actually more UV resistant than are dormant spores owing to release of DPA prior to significant SASP degradation. However, as SASP degradation proceeds, this elevated UV resistance falls to that of the vegetative cell.

Spore Chemical Resistance

Spores are much more resistant than cells to a variety of chemical compounds, including cross-linking agents such as glutaraldehyde, oxidizing agents (Table 3.2), phenols, formaldehyde, chloroform, octanol, alkylating agents including ethylene oxide, iodine, and detergents, as well as to pH extremes and lytic enzymes such as lysozyme (15, 144, 190, 215). Resistance of spores to these various agents is acquired at different times in sporulation (Fig. 3.2). For some compounds, spore coats play a role in chemical resistance, possibly by providing an initial barrier against attack. This is clearly true for lytic enzymes, as spores with coats removed by chemical treatment or with coats altered owing to mutation can be sensitive to lysozyme degradation of their cortex. The very low permeation of most molecules into the spore core also plays an important role in spore resistance to many chemicals. For many chemicals, good data are not available for the temporal correlation between acquisition of spore resistance and other biochemical changes in the developing spore, nor for the correlation of spore chemical resistance with parameters such as core dehydration and cortex size, nor on the mechanism(s) whereby various chemicals cause spore death. However, some chemicals, such as formaldehyde and alkylating agents, kill spores at least in part by DNA damage, while other chemicals do not, and at least in some cases kill spores by inactivating the germination mechanism in some fashion (15, 131, 144, 190, 205, 237).

Some detailed information also is available about the points raised above for oxidizing agents, in particular, hydrogen peroxide. Data on hypochlorite are more limited. Hydrogen peroxide kills cells by several mechanisms, but a major one is the generation of hydroxyl radicals which can cause mutagenic or lethal DNA damage (86). However, spore killing by hydrogen peroxide or hypochlorite is not accompanied by significant DNA damage or mutagenesis (Table 3.2). Consequently, DNA in spores must be extremely well protected against damage by these agents. One reason for the high resistance of spores to these agents may be that spore coats form a protective barrier (15, 52, 53, 142, 186, 190, 217).

The spore core may also be relatively impermeable to these hydrophilic compounds (67), and the decreased spore core water content also plays a role in spore hydrogen peroxide resistance (181). However, these mechanisms do not specifically protect spore DNA against these agents, as this is accomplished through the saturation of the spore chromosome with α/β-type SASP (203). In contrast to wild-type spores, which are extremely resistant to hydrogen peroxide and hypochlorite, $\alpha^- \beta^-$ spores are much more sensitive to these agents (203, 215) (Table 3.2). Furthermore, while there is no increase in mutation frequency among survivors of wild-type spores killed 90 to 99% by hydrogen peroxide or hypochlorite, $\alpha^- \beta^-$ spores treated similarly have a 5 to 15% frequency of obvious mutations among the survivors (203, 215) (Table 3.2). DNA from $\alpha^- \beta^-$ spores killed by hydrogen peroxide has a high frequency of single-strand breaks, whereas DNA from wild-type spores killed similarly exhibits no such damage (203). These data suggest that, in wild-type spores, the saturation of DNA with α/β-type SASP provides such good protection against the DNA damage caused by oxidizing agents that spore killing by these agents is by other mechanisms. However, in $\alpha^- \beta^-$ spores in which much of the spore DNA is no longer covered with α/β-type SASP, the rates of DNA damage caused by such agents are greatly increased, and DNA damage is a significant cause of spore death. In support of this simple model, it has been shown that one component of spore hydrogen peroxide resistance is acquired during sporulation in parallel with accumulation of α/β-type SASP. This component of resistance is not acquired during sporulation of an $\alpha^- \beta^-$ strain (203). In addition, saturation of DNA with α/β-type SASP blocks hydrogen peroxide cleavage of the DNA backbone in vitro (203). However, α/β-type SASP binding to DNA is not the only factor involved in spore resistance to hydrogen peroxide. Indeed, one additional component of resistance is acquired during sporulation at about the time of final spore core dehydration (203). This may reflect a role for spore core dehydration in hydrogen peroxide resistance or, possibly, a mechanism in which the loss of reduced pyridine nucleotides from the developing spore (which also occurs at this time) blocks extensive production of hydroxyl radicals through the Fenton reaction (86, 203). The mechanism(s) of inducible resistance of growing cells of *B. subtilis* to hydrogen peroxide has been studied, and roles for catalases in this process have been demonstrated (16). However, such enzymes have no role in dormant spore resistance to hydrogen peroxide, nor does the MrgA protein, a DNA binding protein which plays a role in resistance of growing cells to hydrogen peroxide (28).

Spore Heat Resistance

Heat resistance, probably the spore resistance most familiar to food microbiologists, has the most implications for the food industry and is probably the best-studied form of resistance in spores. Spore heat resistance is truly remarkable, as spores of many species can withstand 100°C for several minutes. Heat resistance is often quantified as a D_t value, which is the time in minutes at temperature (t) needed to kill 90% of a cell or spore population. Generally, D values for spores at a temperature of $t+40$°C are approximately equal to those for their vegetative cell counterparts at temperature t. An often overlooked feature of spore heat resistance is that the extended survival of spores at elevated temperatures is paralleled by even longer survival times at lower temperatures. Spore D values increase 4- to 10-fold for each 10°C fall in temperature (66). Consequently, a spore with a D value at 90°C ($D_{90°C}$) of 30 min may have a $D_{20°C}$ value of many years. Indeed, there are several reports of spores surviving over periods of 70 to 100 years and others suggesting survival after millions of years (25, 26, 106). The mechanisms causing spore survival at elevated temperatures are presumed to be the same as those extending spore survival at lower temperatures. However, this assumption has not been rigorously tested.

One deficiency in our understanding of the spore heat resistance mechanism is the identity of the target(s) whose damage results in heat killing of spores. There are good data suggesting that this target is not spore DNA, as spore heat killing is associated with neither DNA damage nor mutagenesis (58) (Table 3.2). There are also data consistent with a protein or proteins being the target of spore heat killing (13). However, the identity of the protein(s) is not clear. It has not been proved that the changes in proteins associated with spore heat killing are the cause rather than the effect of heat killing. Sublethal heat treatment can also damage spores in some way, with this damage being repairable during spore germination and outgrowth (83). Again, the nature of this damage, including what macromolecule is damaged, remains obscure. In contrast to our lack of knowledge about the mechanism(s) of spore heat killing, there is much more information on factors which modulate spore heat resistance. Several of these factors are discussed below.

Sporulation Temperature

Elevated sporulation temperatures increase spore heat resistance (10, 66). Indeed, spores of thermophiles generally have much higher heat resistance than spores of mesophiles. Since spore macromolecules are generally

Table 3.3 Heat resistance of *B. subtilis* spores prepared at different temperatures with different ions and with or without α/β-type SASP[a]

Spore	Prepn temp (°C)	Mineralization	H_2O (g/g of spore core wet wt)	$D_{100°C}$
B. subtilis	50	Native	0.335	45
	37	Ca^{2+}	0.425	37
	37	Native	0.50	8.9
	37	H^+	0.571	2.7
	20	Native	0.55	4.9
B. subtilis 168 wild type	37	Native	0.37	360[b]
B. subtilis 168 $\alpha^-\beta^-$	37	Native	0.37	15[b]

[a]Data from references 10 and 58.
[b]$D_{85°C}$.

identical to cell macromolecules, spore macromolecules are not intrinsically heat resistant. Presumably, the total macromolecular content of spores from thermophiles is more heat stable than that from mesophiles, accounting for the higher heat resistance of spores from thermophiles. However, spores of the same strain prepared at various temperatures are most heat resistant when prepared at the highest temperature (Table 3.3). This is probably not due to temperature-related changes in total macromolecular composition and may be due to reduced core water content in spores prepared at higher temperatures (10) (Table 3.3). However, the mechanism whereby sporulation temperature affects spore core water content is not known.

In growing bacteria, adaptation to heat stress involves the proteins of the heat shock response. The levels of these proteins would be expected to increase with increasing sporulation temperature. Although heat shock proteins could play some role in spore heat resistance, this appears not to be the case (146).

α/β-Type SASP

One striking finding about the killing of spores by heat in water is that neither general mutagenesis (Table 3.2) (except possibly in the *gly* region of the *B. subtilis* chromosome [102]) nor DNA damage (213–215) occurs. This is despite the fact that the elevated temperatures would be expected to cause significant DNA depurination. Therefore, spore DNA must be remarkably well protected against heat damage, and the thermal inactivation of spores must be due to mechanisms other than DNA damage. The major cause of spore DNA protection against heat damage appears to be the saturation of spore DNA by α/β-type SASP. Consequently, $\alpha^-\beta^-$ spores of *B. subtilis* have D values 5 to 10% of those for wild-type spores (Tables 3.2 and 3.3). In addition, while heat killing of wild-type spores generates <1% obvious mutations in the survivors, heat killing of $\alpha^-\beta^-$ spores

produces ~20% obvious mutations among survivors, including auxotrophic, asporogenous, and colony morphology mutations (Table 3.2). Killing of $\alpha^-\beta^-$ spores by heat is also accompanied by a large amount of DNA damage. This DNA damage includes generation of abasic sites which may be the primary heat-induced lesion (214, 215), as well as single-strand breaks which may be generated by secondary cleavage reactions at abasic sites (58). As would be predicted from these latter findings, α/β-type SASP slow DNA depurination in vitro at least 20-fold (58).

Spores treated with dry heat are somewhat different from the spores heated in water discussed above. First, wild-type spores are much more resistant to dry heat than to aqueous heat, as D values are 2 to 3 orders of magnitude higher in dry versus hydrated spores (Table 3.2) (34, 214, 215). Second, wild-type spores exhibit a rather high level of mutagenesis (~12% of survivors) upon killing by dry heat (34, 214, 215) (Table 3.2), and this mutagenesis is associated with generation of damage in spore DNA (204). $\alpha^-\beta^-$ spores are much more sensitive to dry heat than are wild-type spores (with survivors of dry heat killing of $\alpha^-\beta^-$ spores also exhibiting a high percentage of mutations) and exhibit heat resistance similar to that of dry vegetative cells (204) (Table 3.2). These latter findings suggest that (i) α/β-type SASP are a major factor increasing the dry heat resistance of wild-type spores over that of vegetative cells; (ii) α/β-type SASP provide significant DNA protection against dry heat damage; but (iii) dry spores are so well protected against mechanisms of heat killing other than DNA damage that at elevated temperatures DNA damage does eventually kill dry spores. In support of these findings, α/β-type SASP slow DNA depurination caused by dry heat in vitro. However, in contrast to the $\geq$20-fold decrease in DNA depurination caused by α/β-type SASP in solution, these proteins only slow dry heat-induced depurination ~3-fold (204).

Spore Mineralization

Spores accumulate large amounts of divalent cations late in sporulation, approximately in parallel with DPA. DPA itself was thought initially to be involved in spore heat resistance. However, the isolation of DPA$^-$ spores which were heat resistant, and the removal of the majority of a spore's DPA with retention of heat resistance, has refuted this idea (11, 66, 214, 215). DPA may be required for the stability of the dormant state; DPA$^-$ spores are unstable and germinate readily. In contrast to the lack of a role for DPA in spore heat resistance, spore mineralization is implicated in heat resistance. Both the amount and type of mineral ions accumulated affect spore heat resistance (10, 66). The best data come from analyses in which mineral ions of spores of several species have been removed by titration with acid and the spores are then back-titrated with a mineral hydroxide. Analyses of such spores give the order of spore heat resistance with different cations as: $H^+ < Na^+ < K^+ < Mg^{2+} < Mn^{2+} < Ca^{2+} <$ untreated. Despite the clear role for spore mineral ions in spore heat resistance, the complete mechanism of this effect remains unclear. Alteration of spore mineralization can alter spore core water content (Table 3.3), which presumably causes a significant effect on heat resistance. However, mineralization may also affect spore heat resistance independently of effects on core water content (Table 3.3).

Spore Core Water Content

Low core water content is a major factor causing spore heat resistance. The dehydration of the spore begins in the stage III to stage IV transition and continues throughout stages IV and V, with final dehydration taking place approximately in parallel with acquisition of spore heat resistance (158). Synthesis of the spore cortex is essential both for affecting this dehydration and for maintaining the dehydrated state of the spore core. This is undoubtedly due to the ability of peptidoglycan to change its volume markedly upon changes in ionic strength and/or pH. If an expansion in cortex volume is restricted to one direction, i.e., toward the spore core, core water would be extruded via mechanical action. Although the precise mechanism of this process is unclear, the important role of the cortex in heat resistance is shown by the inverse correlation between spore heat resistance and the volume occupied by the spore cortex relative to that of the core (115) (Fig. 3.7). This correlates not only across species but also in a single species in which spore cortex biosynthesis has been altered by mutation (180). Presumably, the volume of the spore cortex influences the degree of core dehydration. It is also possible that the amount of cortex and thus its mechanical strength are also cru-

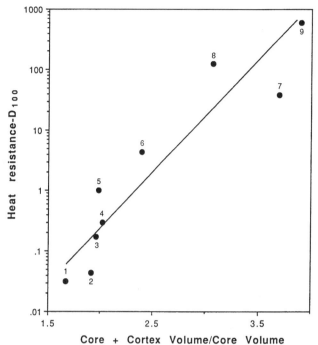

Figure 3.7 Correlation of spore heat resistance and the ratio of the core + cortex volume/core volume. Data were replotted from the work of Koshikawa et al. (115). The numbers correspond to the following spores: 1, *Bacillus megaterium* exosporiumless variant, coat stripped; 2, *B. megaterium* exosporiumless variant; 3, *B. megaterium*, coat stripped; 4, *B. megaterium*; 5, *Bacillus cereus* T, low calcium; 6, *B. cereus* T, high calcium; 7, *B. subtilis* niger; 8, *B. stearothermophilus* smooth; and 9, *B. stearothermophilus* rough.

cial to the spore's ability to maintain core dehydration during treatment.

Studies of spores from a large number of species have revealed a good correlation between spore core water content and heat resistance over a 20-fold range of D values (10, 66) (Fig. 3.8). However, at the extremes of core water contents, D values vary widely, presumably reflecting the importance of other factors, such as sporulation temperature, cortex structure, etc., in modulating spore heat resistance (66). One value that is unfortunately missing from analyses of spore water content is the amount of core water that is free water. Studies of ion movement in spores have indicated that core ions are immobile, consistent with the presence of little if any free water in the core. Presumably, low water content in the spore core causes heat resistance as well as long-term spore survival by slowing water-driven chemical reactions, such as DNA depurination, protein deamidation, and so on. Low water content also stabilizes macromolecules such as proteins against denaturation by restricting their molecular motion. It would be most

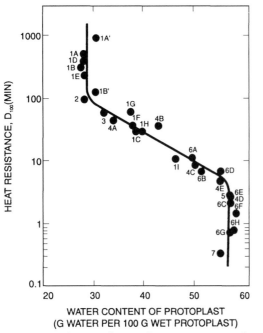

Figure 3.8 Correlation of spore heat resistance and protoplast (core) water content of lysozyme-sensitive spore types from seven *Bacillus* species which vary in thermal adaptation and mineralization. Reprinted from Beaman and Gerhardt (10) with permission. The numbers refer to spores of various species: 1, *B. stearothermophilus*; 2, "*B. caldolyticus*"; 3, *B. coagulans*; 4, *B. subtilis*; 5, *B. thuringiensis*; 6, *B. cereus*; and 7, *B. macquariensis*. The letters denote the sporulation temperature or the mineralization of the spores of various species as described in the original publication.

informative to know the precise amount of water associated with spore core macromolecules in order to calculate the degree of their stabilization by this process.

SPORE ACTIVATION, GERMINATION, AND OUTGROWTH

Activation

Although spores are metabolically dormant and can remain in this state for many years, if given the proper stimulus they can return to active metabolism within minutes through the process of spore germination (Fig. 3.9). A spore population will often initiate germination more rapidly and completely if activated prior to addition of a germinant (107). However, the requirement for activation varies widely among spores of different species. A number of agents cause spore activation, including low pH and many chemicals, although the most widely used agent is sublethal heat. The precise changes induced by spore activation are not clear, although, in some species, the activation process is reversible. In most species, heat

activation released a small amount of the spore's DPA. However, in *B. stearothermophilus* spores, heat activation releases essentially all of the DPA (11). There is no clear picture of the mechanism of this spore activation.

Germination

The precise period encompassed by spore germination, as distinguished from spore outgrowth, has been given a number of different definitions. For the purposes of this review, we consider spore germination to occur during the first 20 to 30 min following mixing of spores and germinant. During this period, a resistant dormant spore with a cortex and a large pool of DPA, minerals, and SASP is transformed to a sensitive, actively metabolizing germinated spore in which the cortex and SASP have been degraded and DPA and most minerals have been excreted (70, 207). These changes, along with initiation of RNA and protein synthesis, can take place in the complete absence of exogenous nutrients. However, further conversion of this germinated spore into a growing cell via the process of outgrowth requires exogenous nutrients.

The initiation of spore germination in different species can be triggered by a wide variety of compounds, including nucleosides, amino acids, sugars, salts, DPA, and long-chain alkylamines (70), although within a species the requirements are more specific. The precise mechanism whereby these compounds trigger spore germination is not clear. Metabolism of the germinant is not required. Indeed, spores of some species are germinated by inorganic salts (71). Detailed analyses of spores of several *B. megaterium* strains provide strong evidence that triggering of spore germination does not require metabolism of endogenous compounds (49, 194, 195). However, the stereospecificity exhibited by germinants (e.g., L-alanine is a germinant, whereas D-alanine often inhibits germination) and the alteration of germinant specificity by mutation strongly suggest that at least some germinants interact directly with a specific protein (70, 149, 150, 167). Analysis of *B. subtilis* mutants whose spores are defective or altered in their response to specific germinants has indicated that these proteins are likely the products of one of three tricistronic *ger* operons, *gerA*, *gerB*, or *gerK* (38, 90, 259). These operons are expressed in the developing forespore during sporulation, and mutations in any cistron of any of these *ger* operons results in spores defective or altered in initiating spore germination in response to one or more, but not all, germinants (149, 150, 167). However, spores from mutants lacking all three operons germinate extremely poorly in all nutrients (168). The products of these *ger* operons are probably membrane proteins, and it appears likely that the products of any

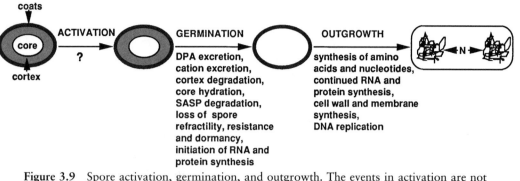

Figure 3.9 Spore activation, germination, and outgrowth. The events in activation are not known, hence the question mark. The loss of the spore cortex and the hydration and swelling of the core are shown in the germinated spore. Data are taken from references 70 and 207.

one operon form a complex which may function as a germinant receptor (149, 150, 167). The precise location of this receptor complex is not known, as there are data consistent with the location being either the inner or outer forespore membrane (149, 256). One interesting problem faced by a putative germinant receptor in the inner or outer forespore membrane is that at least part of the protein would have to be outside the spore core. How is this protein protected against heat, given its location outside the inner forespore membrane where the water content is similar to that of growing cells? In addition to the tricistronic *ger* operons noted above, there are several other *ger* genes in which mutations alter spore germination (149, 150). For some of these additional *ger* genes their function is known, while for others it is not (12, 150, 187, 257).

In the first minutes of spore germination, the earliest events are release of protons and some divalent cations, in particular Zn^{2+} (65, 96). Release of DPA, loss of spore refractility, and cortex degradation follow (65, 70, 96, 207). In this process, the spore excretes up to 30% of its dry weight, and the core increases its water content to that of the vegetative cell (70). These events clearly require large changes in the permeability of the inner forespore membrane. However, neither the nature nor the mechanism of these changes is understood. The time for the changes accompanying the initiation of spore germination may be extremely fast; an individual spore can lose refractility in as little as 30 s to 2 min. However, for a spore population the time can be much longer, as individual spores initiate germination after widely different lag times. Indeed, some spores in a population appear superdormant and may require special conditions or treatment to induce their germination (70). One concern about precise determination of the timing of various events during spore germination is that some of the methods used may overestimate the time for a particular event.

In particular, degradation of spore cortex has often been measured by the appearance of cortical fragments in the germination supernatant fluid, a measurement that may greatly underestimate the true rate of cortex degradation. Whatever the precise order of these events, initiation of cortex degradation is a crucial event, as it allows a rapid two- to threefold increase in spore core volume and eventual spore outgrowth. However, cortex hydrolysis is not needed for DPA excretion, and this excretion appears to be balanced by significant water uptake (179).

The importance of cortex lysis in spore germination has focused attention on lytic (lysozymelike) enzymes as key players in spore germination. A number of lytic enzymes have been isolated from spores of several species, including CwlJ, SleB, SleC, SleL, and SleM (32, 33, 91, 148, 152, 153). These enzymes generally have activity only on intact spore cortex or cortex fragments and appear to be specific for peptidoglycan containing muramic acid lactam. This is consistent with the lack of degradation of the cortex upon initiation of germination of spores of a *B. subtilis cwlD* mutant, as CwlD is needed in some fashion for formation of muramic acid lactam in spore cortex peptidoglycan (7, 179, 199). This specificity of cortex lytic enzymes will prevent these enzymes from hydrolyzing the germ cell wall during spore germination.

The function of the various cortex lytic enzymes in vivo has been studied best in *B. subtilis*, where CwlJ and SleB have been shown to play redundant but overlapping roles in cortex degradation, with SleB likely playing a larger role (91, 153). However, even a *cwlJ sleB* mutant undergoes early events in spore germination, indicating that these events do not require significant cortex degradation (91). Despite the establishment of the roles of CwlJ and SleB in spore germination, it is not yet clear how these enzymes, which are present in spores in a mature, potentially active form, are regulated. *Clostridium perfringens* spores contain a lytic enzyme, SleC, which is present in spores as a zymogen that is activated by proteolysis early in spore germination (148, 241). A similar enzyme has been suggested to be present in spores of at least one *Bacillus* sp., and there are reports of inhibition of spore germination by protease inhibitors (65, 96). However, the precise identity of the target for these protease inhibitors is not known.

Following the initial events in spore germination, a number of enzymatic reactions begin in the spore core (207). These include utilization of the spore's depot of 3PGA to generate ATP and NADH, degradation of SASP initiated by GPR action and completed by peptidases, catabolism of much of the amino acids produced by SASP degradation for production of high-energy phosphate and reduced pyridine nucleotides, and initiation of catabolism of exogenous compounds. The spore lacks a number of enzymes of the citric acid cycle, so most metabolism in germination and early outgrowth uses glycolysis and/or the hexose monophosphate shunt. In *B. megaterium*, endogenous energy reserves completely support ATP production during the first 10 to 15 min of spore germination, but exogenous catabolites are needed subsequently.

RNA synthesis is initiated in the first few minutes of germination, using nucleotides stored in the spore or generated by breakdown of preexisting spore RNA. The identity of the first RNAs made during germination is not clear, but they are probably mRNA. The precise structure of RNA polymerase acting during germination and early outgrowth is also not clear. Protein synthesis begins shortly after RNA synthesis, but the identity of the first proteins made is not clear. All the components of the protein-synthesizing machinery stored in the dormant spore appear functional, with the exception of some tRNA lacking the 3'-terminal adenosine residue. However, this tRNA is rapidly repaired by tRNA nucleotidyltransferase in the first minutes of spore germination, when aminoacyl-tRNA is also generated. As was the case with RNA synthesis early in germination, endogenous amino acids derived largely from α/β-type SASP breakdown support most protein synthesis in the first 20 to 30 min of germination.

Outgrowth

Although the transition from spore germination to spore outgrowth is not absolutely distinct, we will consider this as the time from ~25 min after the initiation of spore germination until the first cell division. Note that most processes during germination are supported from endogenous reserves, while exogenous sources are required to support outgrowth. Thus, outgrowth requires exogenous nutrients and can take as little as 90 min in a rich medium (207). Exogenous reserves of carbon, nitrogen, etc., are required to support spore outgrowth. During outgrowth, the spore regains the ability to synthesize amino acids, nucleotides, and other small molecules owing to synthesis of biosynthetic enzymes at defined times. However, the factors regulating gene expression during spore outgrowth are not well characterized.

DNA replication is not generally initiated until at least 60 min after the start of germination. However, the mechanisms controlling initiation of chromosomal replication during outgrowth have not been studied in detail. DNA repair can occur well before DNA replicative synthesis, and even in the first minutes of germination, spores contain deoxynucleoside triphosphates. However, the importance of DNA repair in the first minutes of germination, other than of UV damage, has not been well studied. During spore outgrowth, the volume of the outgrowing spore continues to increase, requiring the synthesis of membrane and cell wall components.

One question that remains intriguing is whether there are genes which are needed for outgrowth but no other stage of growth. A number of mutations affecting outgrowth (termed *out* mutants) have been isolated in *B. subtilis* (160, 197). Some of the genes targeted by these mutations have been cloned, sequenced, and analyzed. While the functions of some of these genes have been established, to date no outgrowth-specific gene has been definitely identified.

PRACTICAL PROBLEMS OF SPORES IN THE FOOD INDUSTRY

Spore-forming bacteria and heat-resistant fungi pose specific problems for the food industry. Three species of sporeformers, *Clostridium botulinum*, *C. perfringens*, and *B. cereus*, are infamous for producing toxins that can cause illness in humans and animals (39, 78). Many species of sporeformers cause spoilage of foods (87, 230). Strains of *Clostridium butyricum* and *Clostridium barati* have also been found to produce botulinal neurotoxin, but fortunately these isolates are rare and their spores appear to have a lower heat resistance than their nontoxigenic counterparts (78, 95). Sporeformers causing foodborne illness and spoilage are particularly important in low-acid foods (equilibrium pH >4.6) packaged in cans, bottles, pouches, or other hermetically sealed containers ("canned" foods), which are processed by heat (230). Certain sporeformers also cause various types of spoilage of high-acid foods (equilibrium pH ≤4.6). Psychrotrophic sporeformers are increasingly recognized as causing spoilage of refrigerated foods. Fungi that produce heat-resistant ascospores are also an important cause of spoilage of acidic foods and beverages, such as fruits and fruit products (177, 238).

Food spoilage by spore-forming bacteria was discovered by Pasteur during investigations of butyric acid fermentation in wines (20, 251). Pasteur was able to isolate an organism he termed *Vibrion butyrique*, which is probably the same organism now referred to as *C. butyricum*, the type species of the genus *Clostridium*. Endospores were discovered independently by Ferdinand Cohn and Robert Koch in 1876, soon after Pasteur had made microbiology famous (108). Investigations by Pasteur and Koch led to the association of microbial activity with the safety and quality of foods. During investigation of the anthrax bacillus, *B. anthracis*, the famous Koch's postulates were born; these proved that a disease is caused by a specific microorganism and also led to the development of pure culture techniques. Diseases and spoilage problems caused by sporeformers have traditionally been associated with foods that are thermally processed, since heat selects for survival and subsequent growth of spore-forming organisms.

The process of appertizing, or preserving food sterilized by heat in a hermetically sealed container, was invented in the late 1700s by Appert (3), who believed that the elimination of air was responsible for the long shelf lives of thermally processed foods. The empirical use of thermal processing gradually developed into modern-day thermal processing industries. In the late 1800s and early 1900s, several scientists in the United States were instrumental in developing scientific principles to ensure the safety and prevent spoilage of thermally processed foods (69). As a result of these studies, thermal processing of foods in hermetically sealed containers became an important industry in Europe and the United States (182, 192). Prescott and Underwood at the Massachusetts Institute of Technology and Russell at the University of Wisconsin found that endospore-forming bacilli caused the spoilage of thermally processed clams, lobsters, and corn (182, 192). The classic study of Esty and Meyer (57) in California provided definitive values of the heat resistance of *C. botulinum* type A and B spores and helped rescue the U.S. canning industry from its near demise from botulism in commercially canned olives and other foods. Quantitative thermal processes, understanding of spore heat resistance, aseptic processing, and implementation of the HACCP (Hazard Analysis and Critical Control Point) concept also resulted from developments in the canned food industry.

Low-Acid Canned Foods

The U.S. Food and Drug Administration (FDA) and the U.S. Department of Agriculture (USDA) Food Safety Inspection Service define a low-acid canned food as one with a finished equilibrium pH >4.6 and a water activity (a_w) >0.85. The regulations regarding thermal processing of canned foods are described in the U.S. *Code of Federal Regulations* (21 CFR, parts 108–114). USDA has regulatory oversight of products that contain at least 3% raw red meat or 2% cooked poultry, and the FDA supervises other products. A description of the process of the facility, equipment, and formulations must be filed with the proper governmental agency prior to commercial processing. The processor must report process deviations or instances of spoilage to the FDA or USDA.

Low-acid foods are packaged in hermetically sealed containers, often cans or glass jars but also plastic pouches and other types of containers. These "cans" (as collectively defined in this chapter) containing low-acid food products must be processed by heat to achieve commercial sterility, which is a condition achieved by application of heat that inactivates microorganisms of public health significance, as well as any microorganisms of non-health significance capable of reproducing in the food under normal nonrefrigerated conditions of storage and distribution. Commercial sterility is an empirical term to indicate a low level of microbial survival and provision of shelf stability (174). Preservation procedures such as acidification or lowering water activity by brining or other means can be used to attain commercial sterility. These preservation procedures are often combined with a reduced heat treatment.

In canned low-acid foods, the primary goal is to inactivate spores of *C. botulinum*, since this organism has the highest heat resistance of microbial pathogens. The degree of heat treatment applied varies considerably according to the class of food, its spore content, pH, storage conditions, and other factors. For example, canned low-acid vegetables and uncured meats usually receive a 12D process (see below) or "botulinum cook." Lesser heat treatments are applied to shelf-stable canned cured meats as well as to foods with reduced water activity or other antimicrobial factors which inhibit growth of spore-forming bacteria. In practice, the spore content of ingredients and the cleanliness of the cannery environment are of major importance for successful heat treatment of low-acid canned foods. Certain foods and food ingredients such as mushrooms, potatoes, spices, sugars, and starches may contain high levels of *C. botulinum* and spores of other sporeformers and can be monitored for their spore input to a process. Honey and certain other foods that can harbor *C. botulinum* spores should not be fed to infants less than one year of age, since only a few spores may be sufficient to cause infant botulism (234).

Researchers in the early and mid-1900s described thermal processes to prevent botulism and spoilage in canned foods (reviewed in references 69 and 230). Pflug (173–175) refined the semilogarithmic microbial destruction model and designed a strategy to achieve the required heat process F_T value. He also introduced the concept of the probability of a nonsterile unit (PNSU) on a one-container basis. This reasoning logically explains the traditional $12D$ term, which designates the time required in a thermal process for a 12-log reduction of $C. botulinum$ spores. In thermal processing, two values have traditionally been used to describe the thermal inactivation of an organism: the D value is the time required for a 1-log reduction of a microorganism, and the z value is the temperature change required to change the D value by a factor of 10 (230). While the D value represents the resistance of an organism to a specific temperature, the z value represents the relative resistance of an organism to inactivation at different temperatures (136, 230). The thermal process for the 10^{-11} to 10^{-13} level of probability of a botulism incident occurring will depend on the initial number of $C. botulinum$ spores present in a container. This value can be quite high, for example,

10^4 for a container of mushrooms, or very low, such as 10^{-1} or less for a container of meat product (74). Pflug (173–175) emphasized that the thermal processing industry should prioritize process design to protect against (i) public health hazard (botulism) from $C. botulinum$ spores, (ii) spoilage from mesophilic spore-forming organisms, and (iii) spoilage from thermophilic organisms in containers stored in warm climates or environments. Generally, low-acid foods in hermetically sealed containers are heated to achieve 3 to 6 min at a temperature of 121°C (250°F) or equivalent at the center or most heat-impermeable region of a food. This ensures the inactivation of the most heat-resistant $C. botulinum$ spores with a $D_{121°C(250°F)} = 0.21$ min and a z of 10°C (18°F). Economic spoilage is also avoided by achieving $\sim 5D$ killing of mesophilic spores that typically have a $D_{121°C}$ of ~ 1 min (44). Foods that are distributed in warm climates of tropical or desert areas require a particularly severe thermal treatment of ~ 20 min at 121°C to achieve a $5D$ killing of $Clostridium thermosaccharolyticum$, $B. stearothermophilus$, and $Desulfotomaculum nigrificans$ since these organisms have a $D_{121°C}$ of ~ 3 to 4 min (Table 3.4). Such severe treatment can have a detrimental

Table 3.4 Heat resistance of sporeformers of importance in foods[a]

Organism	Approx. D value[b] at temp:						
	80°C	85°C	90°C	95°C	100°C	110°C	120°C
Spores of public health significance							
Group I *Clostridium botulinum* types A and B			50		7–30	1–3	0.1–0.2
Group II *C. botulinum* type B		1–30	0.1–3	0.03–2			
C. botulinum type E	0.3–3				0.01		
Bacillus cereus					3–200	0.03–2.4	
Clostridium perfringens			3–145		0.3–18	2.3–5.2	
Mesophilic aerobes							
Bacillus subtilis					7–70	6.9	0.5
Bacillus licheniformis					13.5	0.5	
Bacillus megaterium					1		
Bacillus polymyxa			4–5		0.1–0.5		
Bacillus thermoacidurans			11–30		2–3		
Alicylcobacillus acidoterrestris			16	2.6			
Thermophilic aerobes							
Bacillus stearothermophilus					100–1,600		1–6
Bacillus coagulans					20–300		2–3
Mesophilic anaerobes							
Clostridium butyricum	4–5	0.4–0.8					
Clostridium sporogenes					80–100	21	0.1–1.5
Clostridium tyrobutyricum	13						
Thermophilic anaerobes							
Desulfotomaculum nigrificans					≤480		2–3
Clostridium thermosaccharolyticum					400		3–4

[a]Data are from references 69, 92, 112, 116, 135, 138, 155, 196, and 239.
[b]At pH ca. 7 and $a_w > 0.95$.

impact on nutrient content and organoleptic qualities, but it ensures a shelf-stable food. A predictive model has been proposed to simulate growth of proteolytic *C. botulinum* during cooling of cooked meat (100).

Recently, the concept of semilogarithmic or first-order kinetics of heat inactivation of spores and the use of *D* and *z* values has been carefully reevaluated (170, 171). It was concluded that most populations of spores differ in heat sensitivities and that non-log-linear death kinetics more accurately explain spore inactivation. Analysis using non-log kinetics can explain the tailing and curving that is commonly observed in heat inactivation of spore populations. The kinetics of heat inactivation of spore populations closely followed a cumulative Weibull distribution, i.e., $\log S = b(T)^{tn(T)}$, where S is the survival ratio and $b(T)$ and $n(T)$ are temperature-dependent coefficients (171).

Heat treatments are now commonly applied to aseptically processed low-acid canned foods, whereby commercially sterilized cooled product is poured into presterilized containers followed by aseptic hermetic sealing with a closure in a sterile environment. This technology was initially used for commercial sterilization of milk and creams in the 1950s and then encompassed other food products, such as soups, eggnog, cheese spreads, sour cream dips, puddings, and high-acid products such as fruit and vegetable drinks (44). Aseptic processing and packaging systems have the potential to reduce energy and packaging and distribution costs. Developments in aseptic packaging have also renewed interest in thermal processing and in the mechanisms of spore, cell, and enzyme inactivation (44). Other treatments are being evaluated for inactivation of spores, such as hydrostatic pressure (147), UV light (186), and combinations of physical and chemical treatments (227, 245).

Bacteriology of Sporeformers of Public Health Significance

Three species of sporeformers, *C. botulinum*, *C. perfringens*, and *B. cereus*, are well known to cause foodborne illness (Table 3.5). Certain other species of *Bacillus*, such

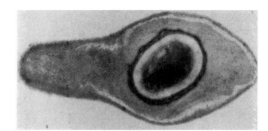

Figure 3.10 Transmission electron micrograph (×50,000) of a longitudinal section through a spore and sporangium of *C. botulinum* type A, showing the characteristic club-shaped morphology. Reprinted from reference 226.

as *B. licheniformis*, *B. subtilis*, and *B. pumilus*, have also been reported to sporadically cause foodborne disease (116), and rare strains of *C. butyricum* and *C. barati* produce type E and F botulinal toxins, respectively (77, 78, 95). Devastating incidents of intestinal anthrax caused by ingestion of contaminated raw or poorly cooked meat have also been reported (75).

The principal microbial hazard in heat-processed foods and in minimally processed refrigerated foods is *C. botulinum*. The genus *Clostridium* consists of grampositive, anaerobic, spore-forming bacilli that obtain energy by fermentation (29, 78, 79). The species *C. botulinum* is a heterogeneous collection of strains that differ widely in genetic relatedness and phenotypic properties but that all have the property of producing a characteristic neurotoxin of extraordinary potency (233). *C. botulinum* and other pathogenic bacteria produce spores that swell the mother sporangium, giving a "tennis racket" or club-shaped appearance (Fig. 3.10). Spores of *C. botulinum* types B and E frequently possess an exosporium, and type E characteristically produces appendages (Fig. 3.11). *C. botulinum* is commonly divided into four physiological groups (I through IV) on the basis of phenotypic properties (220). Group I (strongly proteolytic strains producing neurotoxin types A, B, and F) and group II (nonproteolytic strains producing neurotoxin serotypes B, E, and F) are the two groups of concern in food safety. Strains in groups I and II have certain properties that affect their ability to grow in foods

Table 3.5 Growth requirements of sporeformers of public health significance

Organism	Minimum pH	NaCl concn (%)	Minimum a_w	Temp range for growth (°C)
Group I *C. botulinum*	4.6	10	0.94	10–50
Group II *C. botulinum*	5.0	5	0.97	3.3–45
B. cereus	4.35–4.9	~10	0.91–0.95	5–50
C. perfringens	5.0	~7	0.95–0.97	15–50

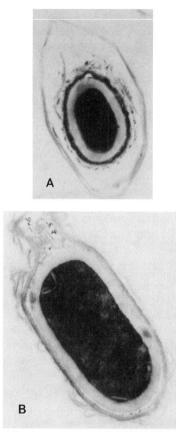

Figure 3.11 Electron micrographs of *C. botulinum* type B (A) and type E (B) showing characteristic exosporium in types B and E and appendages in type E. Micrographs courtesy of Philipp Gerhardt from spores produced in E.A.J.'s laboratory.

(138, 220). The spores of group II have considerably less heat resistance than do group I spores, but they can grow and produce toxin at refrigerator temperatures. The ability of *C. botulinum* to grow at low temperatures has generated considerable concern that refrigerated food products could lead to botulism outbreaks. It is recommended for refrigerated products with a shelf life of greater than 5 days, and that receive a heat-treatment killing of less than 6 logs of psychrotrophic spores of *C. botulinum*, that additional intrinsic preservation factors should be included to ensure the botulinal safety (72).

C. perfringens is widespread in soils and is a normal resident of the intestinal tracts of humans and certain animals (78). *C. perfringens* can grow extremely rapidly in high-protein foods, such as meats that have been cooked to eliminate competitors and are inadequately cooled (101), allowing it to produce an enterotoxin that causes diarrheal disease. It also produces a variety of other extracellular toxins and degradative enzymes, but these are mainly of significance in gas gangrene and diseases in an-

imals (77, 224). *C. perfringens* differs from many other clostridia in being nonmotile, reducing nitrate, and carrying out a stormy fermentation of lactose in milk. Contributing to the ability of *C. perfringens* to cause foodborne illness is its ubiquitous distribution in foods and food environments, formation of resistant endospores that survive cooking of foods, and an extremely rapid growth rate in warm foods (6 to 9 min at 43 to 45°C) (121).

C. perfringens is the cause of a relatively common type of food poisoning in the United States (39) and in several other countries where surveillance has been conducted. Many if not most foods contain spores of *C. perfringens*. In addition to vegetables and fruits that acquire spores from soil, foods of animal origin are contaminated during slaughtering with spores in the environment or spores residing in the intestinal tract. Dried foods such as spices are a common source of *C. perfringens* and other sporeformers (145). Sporulation of vegetative cells is often difficult to obtain in many laboratory media. When spores do occur, they are large, oval, and centrally or subterminally located and swell the cells. The optimum temperature for growth of vegetative cells is about 43 to 45°C (109 to 113°F), and in rich media or in certain foods at the optimum temperature for growth, doubling times as short as 6 to 9 min have been observed (121). Germination of spores and growth of cells will take place at up to 50°C (ca. 122°F). *C. perfringens* does not generally grow below 20°C (68°F), and true psychrotrophic strains of the organism have not been isolated. *C. perfringens* will grow over the pH range 5.5 to 8.0 or 8.5, and the optimum is ca. 6.5. At temperatures below 45°C (113°F), certain strains will grow at pH 5. Most strains are inhibited by 5 to 6% salt (Table 3.5).

The heat resistance of *C. perfringens* spores varies considerably among strains. In general, two classes of heat sensitivity are found. Heat-resistant spores have $D_{90°C}$ ($D_{194°F}$) values of 15 to 145 min and z values of 9 to 16°C (16 to 29°F), compared with heat-sensitive spores which have $D_{90°C}$ values of 3 to 5 min and z values of 6 to 8°C (11 to 14°F). The spores of the heat-resistant class generally require a heat shock of 75 to 100°C (167 to 212°F) for 5 to 20 min in order to germinate. The basis of the wide variation in heat resistance is not currently understood. The spores of both classes may survive cooking of foods and may be stimulated by heat shock for germination during the heating procedures. Both classes can cause diarrheal foodborne illness, although it would be expected that the heat-resistant forms would be a more frequent cause of illness.

Food poisoning by *C. perfringens* nearly always involves temperature abuse of a cooked food, and the great majority of food poisonings caused by *C. perfringens* could be avoided if cooked foods were eaten immediately after cooking or rapidly chilled and reheated before consumption to inactivate vegetative cells. The objective in prevention of food poisonings is to limit the multiplication of vegetative cells in the food. Since the spores are widespread and are resistant to heat, they will often survive the cooking procedure, germinate, and rapidly outgrow to large vegetative cell populations if the rate of cooling is inadequate. This property has led to new USDA performance standards for cooling in the production of certain ready-to-eat meat and poultry products (http://www.fsis.usda.gov/OA/fr/95033F-b.htm). If foods are not rapidly cooled, they should be held at 60°C (140°F) or higher. For cooled foods, the maximum internal temperature should not remain between 130 and 80°F for more than 1.5 h and not between 80 and 40°F for more than 5 h. For products that receive a pasteurization step, the entire process must not allow greater than 1 log growth in the product during processing and cooling.

The genus *Bacillus* contains only two species, *B. anthracis* and *B. cereus*, that are recognized as definitive human pathogens. *B. cereus* can produce a heat-labile enterotoxin causing diarrheal illness and a heat-stable toxin giving an emetic response in humans (116). Generally, the organism must grow to very high numbers (>10^6/g of food) to cause human illness. *B. cereus* is closely related to *B. megaterium*, *B. thuringiensis*, and *B. anthracis*, but *B. cereus* can be distinguished from these species by biochemical tests and the absence of toxin crystals. Other bacilli, including *B. licheniformis*, *B. subtilis*, and *B. pumilus*, have been reported to cause foodborne outbreaks, primarily in the United Kingdom to date (155).

B. cereus spores occur widely in foods and are commonly found in milk, cereals, starches, herbs, spices, and other dried foodstuffs. They are also frequently found on the surfaces of meats and poultry, probably because of soil or dust contamination. Investigators in Sweden reported isolation of *B. cereus* from 47.8% of 3,888 different food samples. In the United Kingdom, *B. cereus* was isolated from 98/108 (91%) of rice samples. The organism causes spoilage of raw and unpasteurized milk (31), and foods containing dried milks, such as infant formulas, may possess fairly high levels of spores or cells.

The organism grows over the temperature range of approximately 10 to 48°C (50 to 118.4°F), with an optimum of 28 to 35°C (82.4 to 95°F) (Table 3.5). Psychrotrophic strains that produce enterotoxin in milk have been isolated (31, 243). The doubling time at the optimum temperature in a rich medium is 18 to 27 min. Several strains can grow slowly in sodium chloride con-

centrations of 7.5%. The minimum water activity for growth is 0.95. The organism grows over a pH range of approximately 4.9 to 9.3, but these environmental limits for growth are dependent on water activity, temperature, and other interrelated parameters.

Spores of *B. cereus* are ellipsoidal and central to subterminal and do not distend the sporangium. Spore germination can occur over the temperature range of 8 to 30°C (46.4 to 86°F). Spores from strains associated with food poisoning have a heat resistance of $D_{95°C}$ ($D_{203°F}$) of ~24 min. Other strains have been shown to have a wider range of heat resistances. It has been suggested that the strains involved in food poisoning have higher heat resistances and therefore will be more apt to survive cooking. Spores are hydrophobic and attach to food contact surfaces (84).

Since *B. cereus* is widespread in nature and survives extended storage in dried food products, eliminating low numbers of its spores from foods is not practical. Control against food poisoning should be directed at preventing germination of spores and preventing multiplication of large populations of the bacterium. Cooked foods should be rapidly and efficiently cooled to less than 7°C (45°F) or maintained above 60°C (140°F) and should be thoroughly reheated before serving.

Heat Resistance of *C. botulinum* Spores

Group I *C. botulinum* type A and B strains can produce spores of remarkable heat resistance and are the most important sporeformers in the public health safety of thermally processed foods. The classic investigation on their heat resistance was carried out by Esty and Meyer (57) in California as a result of commercial outbreaks of botulism in canned olives and certain other canned vegetables. Esty and Meyer examined 109 type A and B strains at five heating temperatures over the range 100 to 120°C (212 to 248°F). They found that the inactivation rate is logarithmic between 100 and 120°C and that the inactivation rate depends on the spore concentration, the pH, and the heating menstruum. They demonstrated that 0.15 M phosphate buffer (Sorensen's buffer), pH 7.0, gave the most consistent heat-inactivation results, and their use of a standardized system enables comparisons of heat resistance by researchers today. The use of a reproducible system is valuable in periodically determining the heat resistance of new spore crops (92). The data of Esty and Meyer can be extrapolated to give a maximum value of $D_{121.1°C} = 0.21$ min for *C. botulinum* type A and B spores in phosphate buffer (92, 239). The thermal processing industries have used $D_{121°C}$ as a standard in calculating process requirements. Proteolytic type F *C. botulinum* spores have a heat resistance of $D_{98.9°C} = 12.2$ to 23.2 min and $D_{110°C} = 1.45$ to 1.82 min (138), which is much lower than that of type A spores. Spores of nonproteolytic type B and E *C. botulinum* have much lower heat resistance than proteolytic A and B strains. Ito et al. (92) reported that type E spores have a $D_{70°C(158°F)}$ varying from 29 to 33 min and a $D_{80°C(176°F)}$ from 0.3 to 2 min depending on the strain. The z value ranges from 13 to 15°F. These values are comparable to those of Ohye and Scott (164), who obtained $D_{80°C}$ values of 3.3 and 0.4 min for two type E strains. Spores of nonproteolytic *C. botulinum* type B can have heat resistance considerably higher than that of type E. Scott and Bernard (196) showed that the $D_{82.2°C}$ of nonproteolytic type B strains ranges from 1.5 to 32.3 min as compared with $D_{82.2°C} = 0.33$ min for a type E strain. Media containing lysozyme can significantly enhance recovery of group II *C. botulinum* spores, since lysozyme substitutes for spore lytic enzymes that are inactivated by heat (135). D values at 85 and 95°C are 100 and 4.4 min, respectively, for strain 17B and 45.6 and 2.8 min, respectively, for strain Beluga E on medium plus lysozyme. The thermal resistance of *C. botulinum* spores is strongly dependent on environmental and recovery conditions. Heat resistance is markedly affected by acidity (57, 100). Esty and Meyer found that spores have maximum resistance at pH 6.3 and 6.9, and resistance decreases markedly at pH values below 5 or above 9. Increased levels of sodium chloride (57) or sucrose (231) and decreased a_w increase the heat resistance of *C. botulinum* spores. Sugiyama (232) found that spores grown in media containing fatty acids increased their heat resistance. *C. botulinum* spores coated in oil are more resistant to heat (220). In common with *Bacillus* spp., sporulation of *C. botulinum* at higher temperatures results in spore crops with greater heat resistance, possibly by acquired thermotolerance through the formation of heat shock proteins (240).

Little is known of the compositional factors contributing to heat resistance of *C. botulinum* spores. The metal composition of purified group I *C. botulinum* spores is different from that of spores of *Bacillus* (110). The minerals required for sporulation and mechanisms of heat resistance of *C. botulinum* and *Bacillus* spp. are probably not the same (109, 110). Unlike *Bacillus* spp., which require manganese for sporulation, type B *C. botulinum* sporulation is enhanced by zinc and inhibited by copper (109). During sporulation, *C. botulinum* accumulated relatively high concentrations of transition metals, particularly zinc (~1% of cell dry weight) and iron and copper (0.05 to 1%). Spores containing increased contents of iron or copper are more rapidly inactivated by heat than are native spores or spores containing increased man-

ganese or zinc (110). In the anaerobic growth environment of clostridia, transition metals would be expected to have important roles in the sporulation and resistance properties of spores. Metals that undergo redox changes, including copper, iron, and manganese, tend to precipitate as the hydroxides or oxides in aerobic environments, but they are more soluble and biologically available at low redox potentials. The mechanisms by which iron and copper accelerate heat inactivation and those by which zinc and manganese protect *C. botulinum* spores against thermal energy have not been elucidated. Iron and copper are redox-active transition metals and may catalyze hydrolytic reactions (250) and may also spontaneously react with oxygen, generating toxic oxygen species that inflict mutations in DNA (129). Manganese and zinc can associate with nucleic acids, and Mn has been demonstrated to provide protection against heat denaturation (242) and also to act as an effective scavenger of free radicals in biological systems (4). Studies have also indicated that heat inactivation of *C. botulinum* spores is accelerated in modified gas atmospheres (111).

C. botulinum spores of groups I and II are highly resistant to irradiation compared with vegetative cells of most microorganisms, and it is probably not practical to inactivate them in foods by irradiation. *C. botulinum* spores have a $D = 0.1$ to 0.45 mrad (2.0 to 4.5 kGy). Irradiation resistance depends on *C. botulinum* type; proteolytic types of A, B, and F appear to be most resistant (112). *C. botulinum* spores are also highly resistant to ethylene oxide but are inactivated by halogen sanitizers and by hydrogen peroxide (112). Hydrogen peroxide, is commonly used for sanitizing surfaces in aseptic packaging, and halogen sanitizers are used in cannery cooling waters. Alternatives to hydrogen peroxide such as peracetic acid are receiving renewed consideration owing to the deleterious effects of hydrogen peroxide on packaging equipment and materials (14). The assessment of biocides and food preservatives in sporicidal efficacy has been reviewed (190, 191).

Incidence of Foodborne Illness Caused by *C. botulinum*

The epidemiology of botulism has been thoroughly reviewed (77, 95), and only aspects pertaining to spore survival and outgrowth are presented in this chapter. Fortunately, the incidence of botulism in commercial foods is very low. It has been estimated that about 30 billion cans, bottles, and pouches of low-acid foods are consumed in the United States each year. Since 1940, when heat-processing principles were firmly established, through 1975, less than 10 botulism outbreaks and fewer than four deaths were caused by inadequately commercially canned foods in the United States (137, 161). From 1971 through 1982, however, botulinum toxin was detected in several commercial canned foods, such as mushrooms, salmon, soups, peppers, tuna fish, beef stew, and tomatoes. Survival of spores and toxin production during this period were caused mainly by underprocessing or by container leakage following processing. The detection of botulinal toxin in canned mushrooms and in canned salmon in the 1970s and 1980s prompted U.S. regulatory agencies to recommend that chlorine or sanitizers be used in cooling water. Recently, botulism has been transmitted in several commercial low-acid foods, including chopped garlic in oil, cheese sauce, bean dip, and clam chowder (77, 78, 95). These incidents resulted from temperature abuse of products labeled "keep refrigerated" and the absence of inhibitory conditions other than temperature. Following an outbreak of botulism in the United States transmitted in garlic in oil, the FDA mandated that such products be acidified (154).

Botulism in commercial foods can have enormous medical and economic consequences. Outbreaks of botulism in commercial foods have been estimated to cost ca. $30 million per human case, which is a much higher cost estimate than for other foodborne illnesses, such as caused by *Salmonella* (ca. $10,000) and *Listeria monocytogenes* (ca. $12,000) (39). Outbreaks of botulism also generate considerable media publicity that negatively affects the food industry. Although commercial botulism has been quite rare owing to excellent control during thermal processing, botulism in home-canned and improperly fermented products occurs relatively frequently throughout the world (95). The heat resistance of *C. botulinum* spores is often not appreciated by home canners, and 20 to 30 cases occur each year in the United States, with a current case-fatality rate of about 10%. Botulism is more common in certain countries, such as Poland and China, where improper home canning of meats and poor fermentation of soybean curd occur relatively frequently (94). In recent years, the United States and certain other countries have seen a resurgence of botulism in restaurant-prepared foods, most often caused by poor temperature control of the prepared foods (95).

Inadvertent temperature abuse of foods has resulted in botulism outbreaks. Botulism has occurred from potatoes that were wrapped in foil, baked, and then held at room temperature until they were used for preparing salads (95, 198). During cooking, vegetative organisms are killed, but the spores of *C. botulinum* survive and grow in the anaerobic environment created by wrapping the potatoes in foil (235). In April 1994, an outbreak of botulism in Texas affected 23 individuals, 17 of whom were hospitalized (95). This was the largest botulism outbreak in the United States since 1983. The food vehicle was a potato-based dip (skordalia), which was prepared using foil-wrapped potatoes that were left at room temperature after baking. These examples illustrate ways in which changes in food processing, elimination of antimicrobials, and sole reliance on refrigeration can result in incidences of botulism. Changes in formulation, processing, or packaging can lead to botulism. To ensure the botulinogenic safety of a food with potential of supporting *C. botulinum* growth, it is recommended that laboratory challenge tests be performed. Guidelines have been recommended for such challenge studies (51, 105, 162).

HACCP and Prevention of Foodborne Disease by Sporeformers

The safety of thermally processed low-acid foods is enhanced by application of a HACCP program. HACCP entails a systematic and quantitative risk assessment program to ensure the safety of foods. The system was designed to have strict control over all aspects of the safety of food production, including raw materials, processing methods, the food plant environment, personnel, storage, and distribution. In practice, the identification of potential hazards for a given process and meticulous control of critical control points are required. Methods for HACCP and quality assurance programs for thermally processed foods have been outlined (44) and are the subjects of a chapter in this book (chapter 41).

Spoilage of Acid and Low-Acid Canned and Vacuum-Packaged Foods by Sporeformers

Thermally processed low-acid foods receive a heat treatment adequate to kill spores of *C. botulinum* but not sufficient to kill more heat-resistant spores of mesophiles and thermophiles (193). Acid and acidified foods with an equilibrium pH of ≤4.6 are not processed sufficiently to inactivate all spores, since most species of sporeformers do not grow under acid conditions, and inactivation of all spores would be detrimental to food quality and nutritional composition. Certain foods, such as cured meats and hams, do not receive a thermal process sufficient to inactivate sporeformers and thus must be kept under refrigerated conditions for microbial stability. These classes of foods present opportunities for the growth of sporeformers that do not present a public health hazard but which can cause economic spoilage (159). Most of the spoilers and their characteristic spoilage patterns have been recognized for many years (159, 230). In recent years, however, spoilage of vegetable and fruit products by *Alicyclobacillus acidoterrestris* has

Table 3.6 Spoilage of canned foods by sporeformers[a]

Type of spoilage	pH	Major sporeformers responsible	Spoilage defects
Flat-sour	≥5.3	*B. coagulans, B. stearothermophilus*	No gas, pH lowered. May have abnormal odor and cloudy liquor.
Thermophilic anaerobe	≥4.8	*C. thermosaccharolyticum*	Can swells, may burst. Anaerobic anaerobe and products give sour, fermented, or butyric odor. Typical foods are spinach, corn.
Sulfide spoilage	≥5.3	*D. nigrificans, Clostridium bifermentans*	Hydrogen sulfide produced, giving rotten egg odor. Iron sulfide precipitate gives blackened appearance. Typical foods are corn, peas.
Putrefactive anaerobe	≥4.8	*Clostridium sporogenes*	Plentiful gas. Disgusting putrid odor. pH often increased. Typical foods are corn, asparagus.
Psychrotrophic clostridia	≥4.6		Spoilage of vacuum packaged chilled meats. Production of gas, off flavors and odors, discoloration.
Aerobic sporeformers	≥4.8	*Bacillus* spp.	Gas usually absent except for cured meats; milk is coagulated. Typical foods are milk, meats, beets.
Butyric spoilage	≥4.0	*C. butyricum, Clostridium tertium*	Gas, acetic and butyric odor. Typical foods are tomatoes, peas, olives, cucumbers.
Acid spoilage	≥4.2	*Bacillus thermoacidurans*	Flat (*Bacillus*) or gas (butyric anaerobes). Off odors depend on organism. Common foods are tomatoes, tomato products, other fruits.
	<4	*A. acidoterrestris*	Flat spoilage with off flavors. Most common in fruit juices, acid vegetables, and also reported to spoil iced tea.

[a]Data are from references 17, 31, 103, 122, 123, and 159.

captured the interest of many microbiologists owing to its ability to grow under highly acidic conditions and to cause spoilage at low spore levels (166). Psychrotrophic clostridia have also been increasingly observed to spoil vacuum-packaged chilled meats, and certain strains can produce botulinal toxin (17, 95, 103, 151).

In practice, the inherent spore contamination of foods and food ingredients and of the cannery environment contributes to spoilage problems. Dry ingredients such as sugar, starches, flours, and spices often contain high levels of sporeformers. Spore populations can also accumulate in a food plant, such as thermophilic spores on heated equipment and saccharolytic clostridia in plants processing sugar-rich foods such as fruits.

The principal spoilage organisms and spoilage manifestations are presented in Table 3.6. The major classes of sporeformers causing spoilage are thermophilic flat-sour organisms, thermophilic anaerobes not producing hydrogen sulfide, thermophilic anaerobes forming hydrogen sulfide, putrefactive anaerobes, facultative *Bacillus* mesophiles, butyric clostridia, lactobacilli, and heat-resistant molds and yeasts (159). *A. acidoterrestris* and psychrotrophic clostridia have recently been implicated in spoilage of meat and fruit products, respectively (166).

Practical control of these organisms includes monitoring of raw foods entering the cannery, particularly sugars, starches, spices, onions, mushrooms, and dried foods, to limit the initial spore load in a food product; adequate thermal processing depending on subsequent storage and distribution conditions; rapid cooling of products; chlorination of cooling water; and implementing and maintaining good manufacturing practices within the food plant.

Certain species of spore-forming psychrophiles have the ability to spoil refrigerated foods and have caused spoilage of meats and dairy products in recent years. Psychrophilic strains of *Bacillus* spp. have been isolated from spoiled dairy products (31). Psychrophilic clostridia have also been associated with the spoilage of meats (37, 123).

Heat-resistant fungi are economic spoilers of acidic foods, particularly fruit products (177, 238). While most filamentous fungi and yeasts are killed by heating for a few minutes at 60 to 75°C, heat-resistant fungi produce thick-walled ascospores that survive heating at 85°C for 5 min. The most common genera of heat-resistant fungi causing spoilage are *Byssochlamys, Neosartorya, Talaromyces*, and *Eupenicillium* (177, 238). Certain heat-resistant fungi also produce toxic secondary metabolites collectively referred to as mycotoxins (238). To prevent spoilage of heat-treated foods, raw materials should be screened for heat-resistant fungi, and strict Good Manufacturing Practices and sanitation programs should be followed during processing. Additionally, manipulation of a_w and oxygen tension and the application of antimycotic agents can be used to prevent fungal growth.

Modeling Growth of Sporeformers in Foods

Abundant information exists on the behavior of microorganisms in foods, and it is often useful to generate statistical models to quantify safety risks in foods. Two general types of models have mainly been used in food microbiology: (i) those analyzing experimental growth and survival data using simple and higher-order polynominals and (ii) theoretic models derived from basic scientific principles and computer analysis, used to predict microbial survival. Statistical models can be particularly useful to define important variables and predict microbial behavior in advance of practical testing (162). Certain companies and institutions are beginning to use models involving neural nets. These are valuable since they are adaptive and improve in precision and predictive capability over time. An example of a model used extensively in the dairy industry is that of Tanaka et al. (236) for preventing growth of *C. botulinum* in processed cheese. Predictive models have also been developed for *C. botulinum* in cooked meat (99). Although models can provide guidelines for food safety, assurance of sterility generally needs to be ascertained by laboratory challenge studies.

CONCLUSION

The scientific investigation of sporeformers has greatly contributed to the development of microbiology for enhancement of food safety and quality. The fundamental understanding of sporulation in *B. subtilis* provides an elegant model of cellular differentiation. Advances in the understanding of the mechanisms of spore heat resistance have contributed to a greater knowledge of dormancy and the ecological success of sporeformers. The remarkable resistance properties of spores and their impact on human disease, particularly botulism, tetanus, and anthrax, have led to the development of microbiology and its importance in medicine and industry.

Work in the laboratory of P.S. has received generous support from both the Army Research Office and the National Institutes of Health (GM-19698). Research in the laboratory of E.J. has been supported by the USDA, NIH, and industrial sponsors of the Food Research Institute.

References

1. Adams, L. F., K. L. Brown, and H. R. Whiteley. 1991. Molecular cloning and characterization of two genes encoding sigma factors that direct transcription from a *Bacillus thuringiensis* crystal protein gene promoter. *J. Bacteriol.* **173:**3846–3854.

2. Alper, S., L. Duncan, and R. Losick. 1994. An adenosine nucleotide switch controlling the activity of a cell type-specific transcription factor in *B. subtilis*. *Cell* **77:**195–205.

3. Appert, N. 1810. L'Art de conserver pendant plusieurs années toutes les substances animales et végétables (translated by K. G. Bitting, 1920). *In* S. A. Goldblith, M. A. Joslyn, and J. T. R. Nickerson (ed.), *Introduction to the Thermal Processing of Foods.* AVI Publishing Co., Westport, Conn., 1961.

4. Archibald, F. S., and I. Fridovich. 1981. Manganese and defenses against oxygen toxicity in *Lactobacillus plantarum*. *J. Bacteriol.* **145:**442–451.

5. Aronson, A. I. 1993. Insecticidal toxins, p. 953–964. *In* A. L. Sonenshein, J. A. Hoch, and R. Losick (ed.), *Bacillus subtilis and Other Gram-Positive Bacteria: Biochemistry, Physiology, and Molecular Genetics.* American Society for Microbiology, Washington, D.C.

6. Ashwood-Smith, M. J., and E. Grant. 1976. Mutation induction in bacteria by freeze-drying. *Cryobiology* **13:**206–213.

7. Atrih, A., P. Zollner, G. Allmaier, and S. J. Foster. 1996. Structural analysis of *Bacillus subtilis* 168 endospore peptidoglycan and its role during differentiation. *J. Bacteriol.* **178:**6173–6183.

8. Bagyan, I., B. Setlow, and P. Setlow. 1998. New small, acid soluble proteins unique to spores of *Bacillus subtilis*: identification of the coding genes and studies of the regulation and function of two of these genes. *J. Bacteriol.* **180:**6704–6712.

9. Baldus, J. M., B. D. Green, P. Youngman, and C. P. Moran, Jr. 1994. Phosphorylation of *Bacillus subtilis* transcription factor SpoOA stimulates transcription from the *spoIIG* promoter by enhancing binding to weak 0A boxes. *J. Bacteriol.* **176:**296–306.

10. Beaman, T. C., and P. Gerhardt. 1986. Heat resistance of bacterial spores correlated with protoplast dehydration, mineralization, and thermal adaptation. *Appl. Environ. Microbiol.* **52:**1242–1246.

11. Beaman, T. C., H. S. Pankratz, and P. Gerhardt. 1988. Heat shock affects permeability and resistance of *Bacillus stearothermophilus* spores. *Appl. Environ. Microbiol.* **54:**2515–2520.

12. Behravan, J., H. Chirakkal, A. Masson, and A. Moir. 2000. Mutations in the *gerP* locus of *Bacillus subtilis* and *Bacillus cereus* affect access of germinants to their targets in spores. *J. Bacteriol.* **182:**1987–1994.

13. **Belliveau, B. H., T. C. Beaman, H. S. Pankratz, and P. Gerhardt.** 1992. Heat killing of bacterial spores analyzed by differential scanning calorimetry. *J. Bacteriol.* **174:**4463–4474.

14. **Blakistone, B., R. Chuyate, D. Kautter, Jr., J. Charbonneau, and K. Suit.** 1999. Efficacy of oxonia active against selective spore formers. *J. Food Prot.* **62:**262–267.

15. **Bloomfield, S. F., and M. Arthur.** 1994. Mechanisms of inactivation and resistance of spores to chemical biocides. *J. Appl. Bacteriol.* **76:**91S–104S.

16. **Bol, D. K., and R. Yasbin.** 1994. Analysis of the dual regulatory mechanisms controlling expression of the vegetative catalase gene of *Bacillus subtilis. J. Bacteriol.* **176:**6744–6748.

17. **Broda, D. M., D. J. Saul, P. A. Lawson, R. G. Bell, and D. R. Musgrave.** 2000. *Clostridium gasigenes* sp nov, a psychrophilic *Clostridium* causing spoilage of vacuum packaged meat. *Int. J. Syst. Bacteriol.* **50:**107–118.

18. **Brown, D. P., L. Ganova-Raeva, B. D. Green, S. R. Wilkinson, M. Young, and P. Youngman.** 1994. Characterization of *spo0A* homologues in diverse *Bacillus* and *Clostridium* species identifies a probable DNA-binding domain. *Mol. Microbiol.* **14:**411–426.

19. **Buchanan, C. E., A. O. Henriques, and P. J. Piggot.** 1994. Cell wall changes during bacterial endospore formation, p. 167–186. *In* J.-M. Ghuysen and R. Hakenbeck (ed.), *Bacterial Cell Wall.* Elsevier Science Publishers, New York, N.Y.

20. **Bulloch, W.** 1938. *The History of Bacteriology.* Oxford University Press, Oxford, United Kingdom.

21. **Burbulys, D., K. A. Trach, and J. A. Hoch.** 1991. Initiation of sporulation in *B. subtilis* is controlled by a multicomponent phosphorelay. *Cell* **64:**545–552.

22. **Burkholder, W. F., and A. D. Grossman.** 2000. Regulation of the initiation of endospore formation in *Bacillus subtilis*, p. 151–166. *In* L. J. Shimkets and Y. V. Brun (ed.), *Prokaryotic Development.* American Society for Microbiology, Washington, D.C.

23. **Cabrera-Hernandez, A., M. Paidhungat, and P. Setlow.** 1999. Analysis of the regulation of four genes encoding small, acid-soluble spore proteins in *Bacillus subtilis. Gene* **232:**1–10.

24. **Cabrera-Hernandez, A., and P. Setlow.** 2000. Analysis of the regulation and function of five genes encoding small, acid-soluble spore proteins of *Bacillus subtilis. Gene* **248:**169–181.

25. **Cano, R. J.** 1994. *Bacillus* DNA in amber: a window to ancient symbiotic relationships. *ASM News* **60:**129–134.

26. **Cano, R. J., and M. K. Borucki.** 1995. Revival and identification of bacterial spores in 25–40 million year old Dominican amber. *Science* **268:**1060–1064.

27. **Carstensen, E. L., and R. E. Marquis.** 1975. Dielectric and electrochemical properties of bacterial spores, p. 563–571. *In* P. Gerhardt, R. N. Costilow, and H. L. Sadoff (ed.), *Spores VI.* American Society for Microbiology, Washington, D.C.

28. **Casillas-Martinez, L., and P. Setlow.** 1997. Alkyl hydroperoxide reductase, catalase, MrgA, and superoxide dismutase are not involved in resistance of *Bacillus subtilis* spores to heat or oxidizing agents. *J. Bacteriol.* **179:**7420–7425.

29. **Cato, E. P., W. L. George, and S. M. Finegold.** 1986. The genus *Clostridium*, p. 1141–1200. *In* H. A. Sneath, N. S. Mair, and M. E. Sharpe (ed.), *Bergey's Manual of Systematic Bacteriology*, vol. 2. The Williams & Wilkins Co., Baltimore, Md.

30. **Chambliss, G. H.** 1993. Carbon source-mediated catabolite repression, p. 213–219. *In* A. L. Sonenshein, R. Losick, and J. A. Hoch (ed.), *Bacillus subtilis and Other Gram-Positive Bacteria: Biochemistry, Physiology, and Molecular Genetics.* American Society for Microbiology, Washington, D.C.

31. **Champagne, C. P., R. R. Laing, D. Roy, A. A. Mafu, and M. W. Griffiths.** 1994. Psychrotrophs in dairy products: their effects and their control. *Crit. Rev. Food Sci. Nutr.* **34:**1–30.

32. **Chen, Y., S. Fukuoka, and S. Makino.** 2000. A novel peptidoglycan hydrolase of *Bacillus cereus*: biochemical characterization and nucleotide sequence of the corresponding gene, *sleL. J. Bacteriol.* **182:**1499–1506.

33. **Chen, Y., S. Miyata, S. Makino, and R. Moriyama.** 1997. Molecular characterization of germination-specific muramidase from *Clostridium perfringens* S40 spores and nucleotide sequence of the corresponding gene. *J. Bacteriol.* **179:**3181–3187.

34. **Chiasson, L. P., and S. Zamenhof.** 1966. Studies on induction of mutations by heat in spores of *Bacillus subtilis. Can. J. Microbiol.* **12:**43–46.

35. **Chung, J. D., G. Stephanopoulos, K. Ireton, and A. D. Grossman.** 1994. Gene expression in single cells of *Bacillus subtilis*: evidence that a threshold mechanism controls the initiation of sporulation. *J. Bacteriol.* **176:**1977–1984.

36. **Clouston, J. G., and P. A Wills.** 1969. Initiation of germination and inactivation of *Bacillus pumilus* spores by hydrostatic pressure. *J. Bacteriol.* **97:**684–690.

37. **Collins, M. D., U. M. Rodrigues, R. A. Edwards, and T. A. Roberts.** 1992. Taxonomic studies on a psychrophilic *Clostridium* from vacuum-packed beef: description of *Clostridium estertheticum* sp. Nov. *FEMS Microbiol. Lett.* **96:**235–240.

38. **Corfe, B. M., R. L. Sammons, D. A. Smith, and C. Mauel.** 1994. The *gerB* region of the *Bacillus subtilis* 168 chromosome encodes a homologue of the *gerA* spore germination operon. *Microbiology* **140:**471–478.

39. **Council for Agricultural Science and Technology.** 1994. *Foodborne Pathogens: Risks and Consequences.* Task Force Report no. 122. Council for Agricultural Science and Technology.

40. **Crowe, J. H., E. A. Hoekstra, and L. M. Crowe.** 1992. Anhydrobiosis. *Annu. Rev. Physiol.* **54:**579–599.

41. **Cutting, S., S. Roels, and R. Losick.** 1991. Sporulation operon *spoIVF* and the characterization of mutations that uncouple mother-cell from forespore gene expression in *Bacillus subtilis. J. Mol. Biol.* **221:**1237–1256.

42. **Daniel, R. A., S. Drake, C. E. Buchanan, R. Scholle, and J. Errington.** 1994. The *Bacillus subtilis spoVD* gene

encodes a mother-cell-specific penicillin-binding protein required for spore morphogenesis. *J. Mol. Biol.* **235:**209–220.

43. **Daniel, R. A., and J. Errington.** 1993. Cloning, DNA sequence, functional analysis and transcriptional regulation of the genes encoding dipicolinic acid synthetase required for sporulation in *Bacillus subtilis. J. Mol. Biol.* **232:**468–483.

44. **David, J. R. D., R. H. Graves, and V. R. Carlson.** 1996. *Aseptic Processing and Packaging of Foods. A Food Industry Perspective.* CRC Press, Boca Raton, Fla.

45. **Dawes, I. W., and J. Mandelstam.** 1970. Sporulation of *Bacillus subtilis* in continuous culture. *J. Bacteriol.* **103:**529–535.

46. **Decker, S. J., and D. R. Lang.** 1978. Membrane bioenergetic parameters in uncoupler-resistant mutants of *Bacillus megaterium. J. Biol. Chem.* **253:**6738–6743.

47. **Deutscher, J., E. Kuster, U. Bergstedt, V. Charrier, and W. Hillen.** 1995. Protein kinase-dependent Hpr/CcpA interaction links glycolytic activity to carbon catabolite repression in Gram-positive bacteria. *Mol. Microbiol.* **15:**1049–1053.

48. **Diederich, B., J. F. Wilkinson, T. Magnin, S. M. A. Najafi, J. Errington, and M. D. Yudkin.** 1994. Role of interactions between SpoIIAA and SpoIIAB in regulating cell-specific transcription factor σ^F of *Bacillus subtilis. Genes Dev.* **8:**2653–2663.

49. **Dills, S. S., and J. C. Vary.** 1978. An evaluation of respiratory chain associated functions during initiation of germination of *Bacillus megaterium* spores. *Biochim. Biophys. Acta* **541:**301–311.

50. **Donnellan, J. E., Jr., and R. B. Setlow.** 1965. Thymine photoproducts but not thymine dimers are found in ultraviolet irradiated bacterial spores. *Science* **149:**308–310.

51. **Doyle, M. P.** 1991. Evaluating the potential risk from extended shelf-life refrigerated foods by *Clostridium botulinum* inoculation studies. *Food Technol.* **45:**154–156.

52. **Driks, A.** 1999. The *Bacillus subtilis* spore coat. *Microbiol. Mol. Biol. Rev.* **63:**1–20.

53. **Driks, A., and P. Setlow.** 2000. Morphogenesis and properties of the bacterial spore, p. 191–218. *In* L. J. Shimkets and Y. V. Brun (ed.), *Prokaryotic Development.* American Society for Microbiology, Washington, D.C.

54. **Dubnau, D.** 1991. The regulation of genetic competence in *Bacillus subtilis. Mol. Microbiol.* **5:**11–18.

55. **Ellar, D. J.** 1978. Spore specific structures and their functions, p. 295–325. *In* R. Y. Stanier, H. J. Rogers, and J. B. Ward (ed.), *Relations between Structure and Function in the Prokaryotic Cell.* Cambridge University Press, London, England.

56. **Errington, J.** 1993. *Bacillus subtilis* sporulation: regulation of gene expression and control of morphogenesis. *Microbiol. Rev.* **57:**1–33.

57. **Esty, J. R., and K. F. Meyer.** 1922. The heat resistance of the spores of *Bacillus botulinus* and allied anaerobes. XI. *J. Infect. Dis.* **31:**650–663.

58. **Fairhead, H., B. Setlow, and P. Setlow.** 1993. Prevention of DNA damage in spores and in vitro by small,

acid-soluble proteins from *Bacillus* species. *J. Bacteriol.* **175:**1367–1374.

59. **Fairhead, H., B. Setlow, W. M. Waites, and P. Setlow.** 1994. Small, acid-soluble proteins bound to DNA protect *Bacillus subtilis* spores from killing by freeze-drying. *Appl. Environ. Microbiol.* **60:**2647–2649.

60. **Fajardo-Cavazos, P., C. Salazar, and W. L. Nicholson.** 1993. Molecular cloning and characterization of the *Bacillus subtilis* spore photoproduct lyase (*spl*) gene, which is involved in the repair of UV radiation-induced DNA damage during spore germination. *J. Bacteriol.* **175:**1735–1744.

61. **Farkas, J.** 1994. Tolerance of spores to ionizing radiation: mechanisms of inactivation, injury and repair. *J. Appl. Bacteriol.* **76:**81S–90S.

62. **Fawcett, P., A. Melnikov, and P. Youngman.** 1997. The *Bacillus* SpoIIGA protein is targeted to sites of spore septum formation in a SpoIIE-independent manner. *Mol. Microbiol.* **28:**931–943.

63. **Fisher, G. J., and H. E. Johns.** 1976. Pyrimidine photodimers, p. 225–294. *In* S. Y. Wang (ed.), *Photochemistry and Photobiology of Nucleic Acids,* vol. I. Academic Press, Inc., New York, N.Y.

64. **Fitz-James, P., and E. Young.** 1969. Morphology of sporulation, p. 39–72. *In* G. W. Gould and A. Hurst (ed.), *The Bacterial Spore.* Academic Press, Ltd., London, England.

65. **Foster, S. J., and K. Johnstone.** 1989. The trigger mechanism of bacterial spore germination, p. 89–108. *In* I. Smith, R. Slepecky, and P. Setlow (ed.), *Regulation of Procaryotic Development.* American Society for Microbiology, Washington, D.C.

66. **Gerhardt, P., and R. E. Marquis.** 1989. Spore thermoresistance mechanisms, p. 17–63. *In* I. Smith, R. Slepecky, and P. Setlow (ed.), *Regulation of Procaryotic Development.* American Society for Microbiology, Washington, D.C.

67. **Gerhardt, P., R. Scherrer, and S. H. Black.** 1972. Molecular sieving by dormant spore structures, p. 68–74. *In* H. O. Halvorson, R. Hanson, and L. L. Campbell (ed.), *Spores V.* American Society for Microbiology, Washington, D.C.

68. **Glickman, B. W., R. M. Schaaper, W. A. Haseltine, R. L. Dunn, and D. E. Brash.** 1986. The C-C (6-4) UV photoproduct is mutagenic in *Escherichia coli. Proc. Natl. Acad. Sci. USA* **83:**6945–6949.

69. **Goldblith, S. A., M. A. Joslyn, and J. T. R. Nickerson.** 1961. *An Anthology of Food Science,* vol. 1. *Introduction to the Thermal Processing of Foods.* The AVI Publishing Co., Westport, Conn.

70. **Gould, G. W.** 1969. Germination, p. 397–444. *In* G. W. Gould and A. Hurst (ed.), *The Bacterial Spore.* Academic Press, Ltd., London, England.

71. **Gould, G. W.** 1983. Mechanisms of resistance and dormancy, p. 173–210. *In* A. Hurst and G. W. Gould (ed.), *The Bacterial Spore,* vol. 2. Academic Press, Ltd., London, England.

72. **Gould, G. W.** 1999. Sous vide foods: conclusions of an ECFF botulinum working party. *Food Control* **10:**47–51.

73. **Gould, G. W., and A. J. H. Sole.** 1970. Initiation of germination of bacterial spores by hydrostatic pressure. *J. Gen. Microbiol.* **60**:335–346.

74. **Greenberg, R. A., R. B. Tompkin, B. O. Blade, R. S. Kittaka, and A. Anelis.** 1966. Incidence of mesophilic spores in raw pork, beef, and chicken in processing plants in the United States and Canada. *Appl. Microbiol.* **14**:789–793.

75. **Guillemin, J.** 1999. *Anthrax. The Investigation of a Deadly Outbreak.* University of California Press, Berkeley.

76. **Haldenwang, W. G.** 1995. The sigma factors of *Bacillus subtilis*. *Microbiol. Rev.* **59**:1–30.

77. **Hatheway, C. L.** 1995. Botulism: the present status of the disease, p. 55–75. *In* C. Montecucco (ed.), *Clostridial Neurotoxins*. Springer Verlag, Berlin, Germany.

78. **Hatheway, C. L., and E. A. Johnson.** 1998. *Clostridium*: the spore-bearing anaerobes, p. 732–782. *In* W. J. Hausler and M. Sussman (ed.), *Topley and Wilson's Microbiology and Microbial Infections*, 9th ed., vol. 3. Edward Arnold, London, England.

79. **Hippe, H., J. R. Andreesen, and G. Gottschalk.** 1999. The genus *Clostridium*—nonmedical, p. 1800–1866. *In* M. Dworkin (ed. in chief), *The Prokaryotes: an Evolving Electronic Resource for the Microbiological Community* [1st electronic ed.]. Springer-Verlag, Berlin, Germany.

80. **Hoch, J. A.** 1995. Control of cellular development in sporulating bacteria by the phosphorelay two-component signal transduction system, p. 129–144. *In* J. A. Hoch and T. J. Silhavy (ed.), *Two-Component Signal Transduction*. American Society for Microbiology, Washington, D.C.

81. **Hofmeister, A.** 1998. Activation of the proprotein transcription factor pro-σ^E is associated with its progression through three patterns of subcellular localization during sporulation in *Bacillus subtilis*. *J. Bacteriol.* **180**:2426–2433.

82. **Holck, A., H. Blom, and P. E. Granum.** 1990. Cloning and sequencing of the genes encoding acid-soluble spore proteins from *Clostridium perfringens*. *Gene* **91**:107–111.

83. **Hurst, A.** 1983. Injury, p. 255–274. *In* A. Hurst and G. W. Gould (ed.), *The Bacterial Spore*, vol. 2. Academic Press Ltd., London, England.

84. **Husmark, U., and U. Ronner.** 1992. The influence of hydrophobic, electrostatic and morphologic properties on the adhesion of *Bacillus* spores. *Biofouling* **5**:335–344.

85. **Illing, N., and J. Errington.** 1991. The *spoIIIA* operon of *Bacillus subtilis* defines a new temporal class of mother-cell-specific sporulation genes under the control of the σ^E form of RNA polymerase. *Mol. Microbiol.* **5**:1927–1940.

86. **Imlay, J. A., and S. Linn.** 1988. DNA damage and oxygen radical toxicity. *Science* **240**:1302–1309.

87. **Ingram, M.** 1969. Sporeformers as food spoilage organisms, p. 549–610. *In* G. W. Gould and A. Hurst (ed.), *The Bacterial Spore*. Academic Press Ltd., London, England.

88. **Ireton, K., and A. D. Grossman.** 1994. A developmental checkpoint couples the initiation of sporulation to DNA replication in *Bacillus subtilis*. *EMBO J.* **13**:1566–1573.

89. **Ireton, K., S. Jin, A. D. Grossman, and A. L. Sonenshein.** 1995. Krebs cycle function is required for activation of the Spo0A transcription factor in *Bacillus subtilis*. *Proc. Natl. Acad. Sci. USA* **92**:2845–2849.

90. **Irie, R., Y. Fujita, and T. Okamoto.** 1993. Cloning and sequencing of the *gerK* spore germination gene of *Bacillus subtilis* 168. *J. Gen. Appl. Microbiol.* **39**:453–465.

91. **Ishikawa, S., K. Yamane, and J. Sekiguchi.** 1998. Regulation and characterization of a newly deduced cell wall hydrolase gene (*cwlJ*) which affects germination of *Bacillus subtilis* spores. *J. Bacteriol.* **180**:1375–1380.

92. **Ito, K. A., M. L. Seeger, C. W. Bohrer, C. B. Denny, and M. K. Bruch.** 1970. The thermal and germicidal resistance of *Clostridium botulinum* types A, B, and E spores, p. 410–415. *In* M. Herzberg (ed.), *Proceedings of the First U.S.-Japan Conference on Toxin Micro-Organisms*. UJNR Joint Panels on Toxic Micro-Organisms and the U.S. Department of the Interior, Washington, D.C.

93. **Jenal, U., and C. Stephens.** 1996. Bacterial differentiation: sizing up sporulation. *Curr. Biol.* **6**:111–114.

94. **Johnson, E. A.** 1991. Microbiological safety of fermented foods, p. 135–169. *In* J. G. Zeikus and E. A. Johnson (ed.), *Mixed Cultures in Biotechnology*. McGraw-Hill Book Co., New York, N.Y.

95. **Johnson, E. A., and M. C. Goodnough.** 1998. Botulism, p. 724–741. *In* W. J. Hausler and M. Sussman (ed.), *Topley and Wilson's Microbiology and Microbial Infections*, 9th ed., vol. 3. Edward Arnold, London, England.

96. **Johnstone, K.** 1994. The trigger mechanism of spore germination: current concepts. *J. Appl. Bacteriol.* **76**:175S–245S.

97. **Ju, J., T. Luo, and W. G. Haldenwang.** 1997. *Bacillus subtilis* pro-σ^E fusion protein localizes to the forespore septum and fails to be processed when synthesized in the forespore. *J. Bacteriol.* **179**:4888–4893.

98. **Ju, J., T. Luo, and W. G. Haldenwang.** 1998. Forespore expression and processing of the SigE transcription factor in wild-type and mutant *Bacillus subtilis*. *J. Bacteriol.* **180**:1673–1681.

99. **Juneja, V. K., and H. M. Marks.** 1999. Proteolytic *Clostridium botulinum* growth at 12–48°C simulating the cooling of cooked meat: development of a predictive model. *Food Microbiol.* **16**:583–592.

100. **Juneja, V. K., B. S. Marmer, J. G. Phillips, and A. J. Miller.** 1995. Influence of the intrinsic properties of food on thermal inactivation of spores of nonproteolytic *Clostridium botulinum*: development of a predictive model. *J. Food Safety* **15**:349–364.

101. **Juneja, V. K., R. C. Whiting, H. M. Marks, and O. P. Snyder.** 1999. Predictive model for growth of *Clostridium perfringens* at temperatures applicable to cooling of cooked meat. *Food Microbiol.* **16**:335–349.

102. **Kadota, H., A. Uchida, Y. Sako, and K. Harada.** 1978. Heat-induced DNA injury in spores and vegetative cells of *Bacillus subtilis*, p. 27–30. *In* G. Chambliss and J. C. Vary (ed.), *Spores VII*. American Society for Microbiology, Washington, D.C.

103. **Kalinowski, R. M., and R. B. Tompkin.** 1999. Psychrotrophic clostridia causing spoilage in cooked meat and poultry products. *J. Food Prot.* **62**:766–772.

104. Karow, M. L., P. Glaser, and P. J. Piggot. 1995. Identification of a gene, *spoIIR*, which links the activation of σ^E to the transcriptional activity of σ^F during sporulation in *Bacillus subtilis*. *Proc. Natl. Acad. Sci. USA* **92**:2012–2016.

105. Kautter, D. A., R. K. Lynt, Jr., and H. M. Solomon. 1981. Evaluation of the botulism hazard from nitrogen-packed sandwiches. *J. Food Prot.* **44**:59–61.

106. Kennedy, M. J., S. L. Reader, and L. M. Swierczynski. 1994. Preservation records of micro-organisms: evidence of the tenacity of life. *Microbiology* **140**:2513–2529.

107. Keynan, A., and Z. Evenchik. 1969. Activation, p. 359–396. *In* G. W. Gould and A. Hurst (ed.), *The Bacterial Spore*. Academic Press Ltd., London, England.

108. Keynan, A., and N. Sandler. 1984. Spore research in historical perspective, p. 1–48. *In* A. Hurst and G. W. Gould (ed.), *The Bacterial Spore*, vol. 2. Academic Press Ltd., London, England.

109. Kihm, D. J., M. T. Hutton, J. H. Hanlin, and E. A. Johnson. 1988. Zinc stimulates sporulation in *Clostridium botulinum* 113B. *Curr. Microbiol.* **17**:193–198.

110. Kihm, D. J., M. T. Hutton, J. H. Hanlin, and E. A. Johnson. 1990. Influence of transition metals added during sporulation on heat resistance of *Clostridium botulinum* 113B spores. *Appl. Environ. Microbiol.* **56**:681–685.

111. Kihm, D. J., and E. A. Johnson. 1990. Hydrogen gas accelerates thermal inactivation of *Clostridium botulinum* spores. *Appl. Microbiol. Biotechnol.* **33**:705–708.

112. Kim, J., and P. M. Foegeding. 1993. Principles of control, p. 121–176. *In* A. H. W. Hauschild and K. L. Dodds (ed.), *Clostridium botulinum. Ecology and Control in Foods*. Marcel Dekker, Inc., New York, N.Y.

113. King, N., O. Dreesen, P. Stragier, K. Pogliano, and R. Losick. 1999. Septation, dephosphorylation and the activation of sigma F during sporulation in *Bacillus subtilis*. *Genes Dev.* **13**:1156–1167.

114. Kobayashi, K., K. Shoji, T. Shimizu, K. Nakano, T. Sato, and Y. Kobayashi. 1995. Analysis of a suppressor mutation *ssb* (*kinC*) of *sur0B20* (*spo0A*) mutation in *Bacillus subtilis* reveals that *kinC* encodes a histidine protein kinase. *J. Bacteriol.* **177**:176–182.

115. Koshikawa, T., T. C. Beaman, H. S. Pankratz, S. Nakashio, T. R. Corner, and P. Gerhardt. 1984. Resistance, germination, and permeability correlates of *Bacillus megaterium* spores successively divested of integument layers. *J. Bacteriol.* **159**:624–632.

116. Kramer, J. H., and R. J. Gilbert. 1989. *Bacillus cereus* and other *Bacillus* species, p. 21–70. *In* M. P. Doyle (ed.), *Foodborne Bacterial Pathogens*. Marcel Dekker, Inc., New York, N.Y.

117. Kroos, L., B. Zhang, H. Ichikawa, and Y.-T. Yu. 1999. Control of σ factor activity during *Bacillus subtilis* sporulation. *Mol. Microbiol.* **31**:1285–1294.

118. Kunst, F., T. Msadek, and G. Rapoport. 1994. Signal transduction network controlling degradative enzyme synthesis and competence in *Bacillus subtilis*, p. 1–20. *In* P. J. Piggot, C. P. Moran, Jr., and P. Youngman (ed.), *Regulation of Bacterial Differentiation*. American Society for Microbiology, Washington, D.C.

119. Kunst, F., et al. 1997. The complete genome sequence of the gram-positive bacterium *Bacillus subtilis*. *Nature* **290**:249–256.

120. Kuroda, A., Y. Asami, and J. Sekiguchi. 1993. Molecular cloning of a sporulation-specific cell wall hydrolase gene of *Bacillus subtilis*. *J. Bacteriol.* **175**:6260–6268.

121. Labbe, R. G., and T. H. Huang. 1995. Generation times and modeling of enterotoxin-positive and enterotoxin-negative strains of *Clostridium perfringens* in laboratory media and ground beef. *J. Food Prot.* **58**:1303–1306.

122. Landry, W. L., A. H. Schwab, and G. A. Lancette. 1995. Examination of canned foods, p. 21.01–21.29. *In Bacteriological Analytical Manual*, 8th ed. Food and Drug Administration, AOAC International, Gaithersburg, Md.

123. Lawson, P., R. H. Dainty, N. Kristiansen, J. Berg, and M. D. Collins. 1994. Characterization of a psychrotrophic *Clostridium* causing spoilage in vacuum-packed cooked pork: description of *Clostridium algidicarnis* sp. nov. *Lett. Appl. Microbiol.* **19**:153–157.

124. LeDeaux, J. R., and A. D. Grossman. 1995. Isolation and characterization of *kinC*, a gene that encodes a sensor kinase homologous to the sporulation sensor kinases KinA and KinB in *Bacillus subtilis*. *J. Bacteriol.* **177**:166–175.

125. LeDeaux, J. R., N. Yu, and A. D. Grossman. 1995. Different roles for KinA, KinB, and KinC in the initiation of sporulation in *Bacillus subtilis*. *J. Bacteriol.* **177**:861–863.

126. Levin, P. A., and R. Losick. 2000. Asymmetric division and cell fate during sporulation in *Bacillus subtilis*, p. 167–190. In L. J. Shimkets and Y. V. Brun (ed.), *Prokaryotic Development*. American Society for Microbiology, Washington, D.C.

127. Lewis, J. C. 1969. Dormancy, p. 301–358. *In* G. W. Gould and A. Hurst (ed.), *The Bacterial Spore*. Academic Press Ltd., London, England.

128. Lewis, P. J., S. R. Partridge, and J. Errington. 1994. σ factors, asymmetry, and the determination of cell fate in *Bacillus subtilis*. *Proc. Natl. Acad. Sci. USA* **91**:3849–3853.

129. Loeb, L. A., E. A. James, A. M. Waltersdorph, and S. J. Klebanoff. 1988. Mutagenesis by the autoxidation of iron with isolated DNA. *Proc. Natl. Acad. Sci. USA* **85**:3918–3922.

130. Londono-Vallejo, J.-A., and P. Stragier. 1995. Cell-cell signalling pathway activating a developmental transcription factor in *Bacillus subtilis*. *Genes Dev.* **9**:503–508.

131. Loshon, C. A., P. C. Genest, B. Setlow, and P. Setlow. 1999. Formaldehyde kills spores of *Bacillus subtilis* by DNA damage, and small, acid-soluble spore proteins of the α/β-type protect spores against this DNA damage. *J. Appl. Microbiol.* **87**:8–14.

132. Loshon, C. A., and P. Setlow. 1993. Levels of small molecules in dormant spores of *Sporosarcina* species and comparison with levels in spores of *Bacillus* and *Clostridium* species. *Can. J. Microbiol.* **39**:259–262.

133. Losick, R., and J. Pero. 1981. Cascades of sigma factors. *Cell* **25**:582–584.

134. **Losick, R., and P. Stagier.** 1992. Crisscross regulation of cell-type specific gene expression during development in *Bacillus subtilis. Nature* (London) **355:**601–604.

135. **Lund, B. M., and M. W. Peck.** 1994. Heat resistance and recovery of spores of nonproteolytic *Clostridium botulinum* in relation to refrigerated, processed foods with extended shelf-life. *J. Appl. Bacteriol. Symp.* **76:**115S–128S.

136. **Lund, D.** 1975. Thermal processing, p. 31–92. *In* M. Karel, O. R. Fennema, and D. B. Lund (ed.), *Principles of Food Science*, part II. *Physical Principles of Food Preservation.* Marcel Dekker, Inc., New York, N.Y.

137. **Lynt, R. K., D. A. Kautter, and R. B. Read, Jr.** 1975. Botulism in commercially canned foods. *J. Milk Food Technol.* **38:**546–550.

138. **Lynt, R. K., D. A. Kautter, and H. M. Solomon.** 1982. Differences and similarities among proteolytic strains of *Clostridium botulinum* types A, B, E and F: a review. *J. Food Prot.* **45:**466–474.

139. **Magill, N. G., A. E. Cowan, D. E. Koppel, and P. Setlow.** 1996. The internal pH of the forespore compartment of *Bacillus megaterium* decreases by about 1 pH unit during sporulation. *J. Bacteriol.* **176:**2252–2258.

140. **Magill, N. G., A. E. Cowan, M. A. Leyva-Vazquez, M. Brown, D. E. Koppel, and P. Setlow.** 1996. Analysis of the relationship between the decrease in pH and accumulation of 3-phosphoglyceric acid in developing forespores of *Bacillus* species. *J. Bacteriol.* **178:**2204–2210.

141. **Magnuson, R., J. Solomoon, and A. D. Grossman.** 1994. Biochemical and genetic characterization of a competence pheromone from *B. subtilis. Cell* **77:**207–216.

142. **Marquis, R. E., J. Sim, and S. Y. Shin.** 1994. Molecular mechanisms of resistance to heat and oxidative damage. *J. Appl. Bacteriol.* **70:**40S–48S.

143. **Mason, J. M., and P. Setlow.** 1986. Essential role of small, acid-soluble spore proteins in resistance of *Bacillus subtilis* spores to UV light. *J. Bacteriol.* **167:**174–178.

144. **McDonnell, G., and A. D. Russell.** 1999. Antiseptics and disinfectants: activity, action and resistance. *Clin. Microbiol. Rev.* **12:**147–179.

145. **McKee, L. H.** 1995. Microbial contamination of spices and herbs: a review. *Lebensm.-Wiss. Technol.* **28:**1–11.

146. **Melly, E., and P. Setlow.** 2001. Heat shock proteins do not influence wet heat resistance of *Bacillus subtilis* spores. *J. Bacteriol.* **183:**779–784.

147. **Mills, G., R. Earnshwaw, and M. F. Patterson.** 1998. Effects of high hydrostatic pressure on *Clostridium botulinum* spores. *Lett. Appl. Microbiol.* **26:**227–230.

148. **Miyata, S., R. Moriyama, N. Miyahara, and S. Makino.** 1995. A gene (*sleC*) encoding a spore-cortex lytic enzyme from *Clostridium perfringens* S40 spores: cloning, sequence analysis and molecular characterization. *Microbiology* **141:**2643–2650.

149. **Moir, A., E. N. Kemp, G. Robinson, and B. M. Corfe.** 1994. The genetic analysis of bacterial spore germination. *J. Appl. Bacteriol.* **176:**9–16.

150. **Moir, A., and D. A. Smith.** 1990. The genetics of bacterial spore germination. *Annu. Rev. Microbiol.* **44:**531–533.

151. **Moorhead, S. M., and R. G. Bell.** 1999. Psychrotrophic clostridia mediated gas and botulinal toxin production in vacuum-packed chilled meat. *Lett. Appl. Microbiol.* **28:**108–112.

152. **Moriyama, R., H. Fukuoka, S. Miyata, S. Kudoh, A. Hattori, S. Kozuka, Y. Yasuda, K. Tochikubo, and S. Makino.** 1999. Expression of a germination-specific amidase, SleB, of *Bacilli* in the forespore compartment of sporulating cells and its localization on the exterior side of the cortex in dormant spores. *J. Bacteriol.* **181:**2373–2378.

153. **Moriyama, R., A. Hattori, S. Miyata, S. Kudoh, and S. Makino.** 1996. A gene (*sleB*) encoding a spore cortex-lytic enzyme from *Bacillus subtilis* and response of the enzyme to L-alanine mediated germination. *J. Bacteriol.* **178:**6059–6063.

154. **Morse, D. L., L. K. Leonard, J. J. Guzewich, B. D. Devine, and M. Shaygani.** 1990. Garlic-in-oil associated botulism: episode leads to product modification. *Am. J. Public Health* **80:**1372–1373.

155. **Mossel, D. A. A., J. E. L. Corry, C. B. Struik, and R. M. Baird** (ed.). 1995. *Essentials of the Microbiology of Foods: a Textbook for Advanced Studies.* John Wiley & Sons, Inc., New York, N.Y.

156. **Munakata, N., and C. S. Rupert.** 1974. Dark repair of DNA containing "spore photoproduct" in *Bacillus subtilis. Mol. Gen. Genet.* **130:**239–250.

157. **Murrell, W. G.** 1969. Chemical composition of spores and spore structures, p. 215–273. *In* G. W. Gould and A. Hurst (ed.), *The Bacterial Spore.* Academic Press Ltd., London, England.

158. **Murrell, W. G.** 1981. Biophysical studies on the molecular mechanisms of spore heat resistance and dormancy, p. 64–77. *In* H. S. Levinson, A. L. Sonenshein, and D. J. Tipper (ed.), *Sporulation and Germination.* American Society for Microbiology, Washington, D.C.

159. **Murrell, W. G.** 1987. Microbiology of canned foods. *Food Res. Q.* **45:**73–89.

160. **Nessi, C., A. M. Albertini, M. L. Speranza, and A. Galizzi.** 1995. The *outB* gene of *Bacillus subtilis* codes for NAD synthetase. *J. Biol. Chem.* **270:**6181–6185.

161. **NFPA/CMI Container Integrity Task Force, Microbiological Assessment Group Report.** 1984. Botulism risk from post-processing contamination of commercially canned foods in metal containers. *J. Food Prot.* **47:**801–816.

162. **Notermans, S., P. in't Veld, T. Wijtzes, and G. C. Mead.** 1993. A user's guide to microbial challenge testing for ensuring the safety and stability of food products. *Food Microbiol.* **10:**145–157.

163. **Ohlsen, R. L., J. K. Grimsley, and J. A. Hoch.** 1994. Deactivation of the sporulation transcription factor Spo0A by the Spo0E protein phosphatase. *Proc. Natl. Acad. Sci. USA* **91:**1756–1760.

164. **Ohye, D. F., and W. J. Scott.** 1957. Studies in the physiology of *Clostridium botulinum* type E. *Aust. J. Biol. Sci.* **10:**85–94.

165. **Ordal, G. W., L. Marquez-Magana, and M. J. Chamberlin.** 1993. Motility and chemotaxis, p. 765–784. *In* A. L. Sonenshein, J. A. Hoch, and R. Losick (ed.), *Bacillus*

subtilis and Other Gram-Positive Bacteria: Biochemistry, Physiology, and Molecular Genetics. American Society for Microbiology, Washington, D.C.

166. **Orr, R. V., and L. R. Beuchat.** 2000. Efficacy of disinfectants in killing of spores of *Alicyclobacillus acidoterrestris* and performance of media of supporting colony development by survivors. *J. Food Prot.* **63:**1117–1122.

167. **Paidhungat, M., and P. Setlow.** 1999. Isolation and characterization of mutations in *Bacillus subtilis* that allow spore germination in the novel germinant D-alanine. *J. Bacteriol.* **181:**3341–3350.

168. **Paidhungat, M., and P. Setlow.** 2000. Role of Ger-proteins in nutrient and non-nutrient triggering of spore germination in *Bacillus subtilis. J. Bacteriol.* **182:**2513–2519.

169. **Pedraza-Reyes, M., F. Gutierrez-Corona, and W. L. Nicholson.** 1994. Temporal regulation and forespore-specific expression of the spore photoproduct lyase gene by sigma-G RNA polymerase during *Bacillus subtilis* sporulation. *J. Bacteriol.* **176:**3983–3991.

170. **Peleg, M., and M. B. Cole.** 1998. Reinterpretation of microbial survival curves. *Crit. Rev. Food Sci.* **38:**353–380.

171. **Peleg, M., and M. B. Cole.** 2000. Estimating the survival of *Clostridium botulinum* spores during heat treatments. *J. Food Prot.* **63:**190–195.

172. **Perego, M.** 1999. Self-signalling by Phr peptides modulates *Bacillus subtilis* development, p. 243–258. *In* G. M. Dunny and S. C. Winans (ed.), *Cell-Cell Signaling in Bacteria.* American Society for Microbiology, Washington, D.C.

173. **Pflug, I. J.** 1987. Endpoint of a preservation process. *J. Food Prot.* **50:**347–351.

174. **Pflug, I. J.** 1987. Factors important in determining the heat process value, F_T, for low acid canned foods. *J. Food Prot.* **50:**528–533.

175. **Pflug, I. J.** 1987. Calculating F_T-values for heat preservation of shelf-stable, low acid canned foods using the straight-line semilogarithmic model. *J. Food Prot.* **50:**608–615.

176. **Piggot, P. J., J. E. Bylund, and M. L. Higgins.** 1994. Morphogenesis and gene expression during sporulation, p. 113–138. *In* P. J. Piggot, C. P. Moran, Jr., and P. Youngman (ed.), *Regulation of Bacterial Differentiation.* American Society for Microbiology, Washington, D.C.

177. **Pitt, J. I., and A. D. Hocking (ed.).** 1997. *Fungi and Food Spoilage,* 2nd ed. Blackie Academic & Professional, London, England.

178. **Pogliano, K., E. Harry, and R. Losick.** 1995. Visualization of the subcellular location of sporulation proteins in *Bacillus subtilis* using immunofluorescence microscopy. *Mol. Microbiol.* **18:**459–470.

179. **Popham, D. L., J. Helin, C. E. Costello, and P. Setlow.** 1996. Muramic lactam in peptidoglycan of *Bacillus subtilis* spores is required for spore outgrowth but not for spore dehydration or heat resistance. *Proc. Natl. Acad. Sci. USA* **93:**15405–15410.

180. **Popham, D. L., B. Illades-Aguiar, and P. Setlow.** 1995. The *Bacillus subtilis dacB* gene, encoding penicillin-binding protein 5*, is part of a three-gene operon required

for proper spore cortex synthesis and spore core dehydration. *J. Bacteriol.* **177:**4721–4729.

181. **Popham, D. L., S. Sengupta, and P. Setlow.** 1995. Heat, hydrogen peroxide, and UV resistance of *Bacillus subtilis* spores with increased core water content with or without major DNA binding proteins. *Appl. Environ. Microbiol.* **61:**3633–3638.

182. **Prescott, S. C., and W. L. Underwood.** 1897. Microorganisms and sterilizing processes in the canning industries. *Technol. Q.* **10:**183–199.

183. **Rather, P. N., R. Coppolecchia, H. DeGrazia, and C. P. Moran, Jr.** 1990. Negative regulator of σ^G-controlled gene expression in stationary-phase *Bacillus subtilis. J. Bacteriol.* **172:**709–715.

184. **Rebeil, R., Y. Sun, L. Chooback, M. Pedraza-Reyes, C. Kinsland, T. P. Begley, and W. L. Nicholson.** 1998. Spore photoproduct lyase from *Bacillus subtilis* spores is a novel-sulfur DNA repair enzyme which shares features with proteins such as class III anaerobic ribonucleotide reductases and pyruvate-formate lyases. *J. Bacteriol.* **180:**4879–4885.

185. **Resnekov, O., and R. Losick.** 1998. Negative regulation of the proteolytic activation of a developmental transcription factor in *Bacillus subtilis. Proc. Natl. Acad. Sci. USA* **95:**3162–3167.

186. **Riesenman, P. J., and W. L. Nicholson.** 2000. Role of the spore coat layers in *Bacillus subtilis* resistance to hydrogen peroxide, artificial UV-C, UV-B, and solar radiation. *Appl. Environ. Microbiol.* **66:**620–626.

187. **Robinson, C., C. Rivolta, D. Karamata, and A. Moir.** 1998. The product of the *yvoC* (*gerF*) gene of *Bacillus subtilis* is required for spore germination. *Microbiology* **144:**3105–3109.

188. **Ross, M. A., and P. Setlow.** 2000. The *Bacillus subtilis* HBsu protein modifies the effects of α/β-type small, acid-soluble spore proteins on DNA. *J. Bacteriol.* **182:**1942–1948.

189. **Rudner, D. Z., P. Fawcett, and R. Losick.** 1999. A family of membrane-embedded metalloproteases involved in regulated proteolysis of membrane-associated transcription factors. *Proc. Natl. Acad. Sci. USA* **96:**14765–14770.

190. **Russell, A. D.** 1982. *The Destruction of Bacterial Spores,* p. 169–231. Academic Press Ltd., London, England.

191. **Russell, A. D.** 1998. Assessment of sporicidal efficacy. *Int. Biodeterior. Biodegr.* **41:**281–287.

192. **Russell, H. L.** 1896. Gaseous fermentations in the canning industry, p. 227–231. *In Twelfth Annual Report of the Agricultural Experiment Station of the University of Wisconsin.* University of Wisconsin, Madison.

193. **Schmitt, H. P.** 1966. Commercial sterility in canned foods, its meaning and determination. *Assoc. Food Drug Off. U. S. Q. Bull.* **30:**141–151.

194. **Scott, I. R., and D. J. Ellar.** 1978. Metabolism and the triggering of germination of *Bacillus megaterium,* use of L-[³H] alanine and tritiated water to detect metabolism. *Biochem. J.* **174:**635–640.

195. **Scott, I. R., G. S. A. B. Stewart, M. A. Koncewicz, D. J. Ellar, and A. Crafts-Lighty.** 1978. Sequence of

biochemical events during germination of *Bacillus megaterium* spores, p. 95–103. *In* G. Chambliss and J. C. Vary (ed.), *Spores VII.* American Society for Microbiology, Washington, D.C.

196. **Scott, V. N., and D. T. Bernard.** 1982. Heat resistance of spores of non-proteolytic type B *Clostridium botulinum. J. Food Prot.* **45:**909–912.

197. **Scotti, C., M. Piatti, A. Cuzzoni, P. Perani, A. Tognoni, G. Grandi, A. Galizzi, and A. M. Albertini.** 1993. A *Bacillus subtilis* large ORF coding for a polypeptide highly similar to polyketide synthases. *Gene* **130:**65–71.

198. **Seals, J. E., J. E. Snijder, T. A. Edell, C. L. Hatheway, C. J. Johnson, R. C. Swanson, and J. M. Hughes.** 1981. Restaurant associated type A botulism: transmission by potato salad. *Am. J. Epidemiol.* **113:**436–444.

199. **Sekiguchi, J., K. Akeo, H. Yamamoto, F. K. Khasanou, J. C. Alonso, and A. Kuroda.** 1995. Nucleotide sequence and regulation of a new putative cell wall hydrolase gene, *cwlD*, which affects germination in *Bacillus subtilis. J. Bacteriol.* **177:**5582–5589.

200. **Setlow, B., N. Magill, P. Febbroriello, L. Nakhimovsky, D. E. Koppel, and P. Setlow.** 1991. Condensation of the forespore nucleoid early in sporulation of *Bacillus* species. *J. Bacteriol.* **173:**6270–6278.

201. **Setlow, B., and P. Setlow.** 1977. Levels of acetyl coenzyme A, reduced and oxidized coenzyme A, and coenzyme A in disulfide linkage to protein in dormant and germinated spores and growing and sporulating cells of *Bacillus megaterium. J. Bacteriol.* **132:**444–452.

202. **Setlow, B., and P. Setlow.** 1980. Measurements of the pH within dormant and germinated bacterial spores. *Proc. Natl. Acad. Sci. USA* **77:**2474–2476.

203. **Setlow, B., and P. Setlow.** 1993. Binding of small, acid-soluble spore proteins to DNA plays a significant role in the resistance of *Bacillus subtilis* spores to hydrogen peroxide. *Appl. Environ. Microbiol.* **59:**3418–3423.

204. **Setlow, B., and P. Setlow.** 1995. Small, acid-soluble proteins bound to DNA protect *Bacillus subtilis* spores from killing by dry heat. *Appl. Environ. Microbiol.* **61:**2787–2790.

205. **Setlow, B., K. J. Tautvydas, and P. Setlow.** 1998. Small, acid-soluble spore proteins of the α/β-type do not protect the DNA in *Bacillus subtilis* spores against base alkylation. *Appl. Environ. Microbiol.* **64:**1958–1962.

206. **Setlow, P.** 1973. Deoxyribonucleic acid synthesis and deoxynucleotide metabolism during bacterial spore germination. *J. Bacteriol.* **114:**1099.

207. **Setlow, P.** 1983. Germination and outgrowth, p. 211–254. *In* A. Hurst and G. W. Gould (ed.), *The Bacterial Spore,* vol. 2. Academic Press Ltd., London, England.

208. **Setlow, P.** 1988. Resistance of bacterial spores to ultraviolet light. *Commun. Mol. Cell. Biophys.* **5:**253–264.

209. **Setlow, P.** 1988. Small, acid-soluble spore proteins of *Bacillus* species: structure, synthesis, genetics, function and degradation. *Annu. Rev. Microbiol.* **42:**319–338.

210. **Setlow, P.** 1989. Forespore-specific genes of *Bacillus subtilis*: function and regulation of expression, p. 211–221.

In I. Smith, R. Slepecky, and P. Setlow (ed.), *Regulation of Procaryotic Development.* American Society for Microbiology, Washington, D.C.

211. **Setlow, P.** 1993. DNA structure, spore formation and spore properties, p. 181–194. *In* P. J. Piggot, P. Youngman, and C. P. Moran, Jr. (ed.), *Regulation of Bacterial Differentiation.* American Society for Microbiology, Washington, D.C.

212. **Setlow, P.** 1993. Spore structural proteins, p. 801–809. *In* J. A. Hoch, R. Losick, and A. L. Sonenshein (ed.), *Bacillus subtilis and Other Gram-Positive Bacteria: Biochemistry, Physiology, and Molecular Genetics.* American Society for Microbiology, Washington, D.C.

213. **Setlow, P.** 1994. Mechanisms which contribute to the long-term survival of spores of *Bacillus* species. *J. Appl. Bacteriol.* **176:**49S–60S.

214. **Setlow, P.** 1995. Mechanisms for the prevention of damage to the DNA in spores of *Bacillus* species. *Annu. Rev. Microbiol.* **49:**29–54.

215. **Setlow, P.** 2000. Resistance of bacterial spores, p. 217–230. *In* G. Storz and R. Hengge-Aronis (ed.), *Bacterial Stress Responses.* American Society for Microbiology, Washington, D.C.

216. **Shazand, K., N. Frandsen, and P. Stragier.** 1995. Cell-type specificity during development in *Bacillus subtilis*: the molecular and morphological requirements for σ^E activation. *EMBO J.* **14:**1439–1445.

217. **Shin, S.-Y., E. G. Calvisi, T. C. Beaman, H. S. Pankratz, P. Gerhardt, and R. E. Marquis.** 1994. Microscopic and thermal characterization of hydrogen peroxide killing and lysis of spores and protection by transition metal ions, chelators, and antioxidants. *Appl. Environ. Microbiol.* **60:**3192–3197.

218. **Shioi, J.-I., S. Matsuura, and Y. Imae.** 1980. Quantitative measurements of proton motive force and motility in *Bacillus subtilis. J. Bacteriol.* **144:**891–897.

219. **Slepecky, R. A., and E. R. Leadbetter.** 1994. Ecology and relationships of endospore forming bacteria: Changing perspectives, p. 195–206. *In* P. J. Piggot, C. P. Moran, Jr., and P. Youngman (ed.), *Regulation of Bacterial Differentiation.* American Society for Microbiology, Washington, D.C.

220. **Smith, L. D. S., and H. Sugiyama.** 1988. *Botulism. The Organism, Its Toxins, the Disease,* 2nd ed. Charles C Thomas Publisher, Springfield, Ill.

221. **Sole, A. J. H., G. W. Gould, and W. A. Hamilton.** 1970. Inactivation of bacterial spores by hydrostatic pressure. *J. Gen. Microbiol.* **60:**323–334.

222. **Solomon, J. M., R. Magnuson, A. Srivastava, and A. D. Grossman.** 1995. Convergent sensing pathways mediate response to two extracellular competence factors in *Bacillus subtilis. Genes Dev.* **9:**547–558.

223. **Sonenshein, A. L.** 2000. Endospore-forming bacteria—an overview, p. 133–150. *In* L. J. Shimkets and Y. V. Brun (ed.), *Prokaryotic Development.* American Society for Microbiology, Washington, D.C.

224. **Songer, J. G.** 1996. Clostridial enteric diseases of domestic animals. *Clin. Microbiol. Rev.* **9:**216–234.

225. Stackebrandt, E., W. Ludwig, M. Weizenegger, S. Dorn, T. J. McGill, G. E. Fox, C. R. Woese, W. Schubert, and K.-H. Schleifer. 1987. Comparative 16S rRNA oligonucleotide analyses and murein types of round-spore-forming bacilli and non-spore-forming relatives. *J. Gen. Microbiol.* **133:**2523–2529.

226. Stevenson, K. E., and R. H. Vaughn. 1972. Exosporium formation in sporulating cells of *Clostridium botulinum* 78A. *J. Bacteriol.* **112:**618–621.

227. Stewart, C. M., C. P. Dunne, A. Sikes, and D. G. Hoover. 2000. Sensitivity of spores of *Bacillus subtilis* and *Clostridium sporogenes* PA 3679 to combinations of high hydrostatic pressure and other processing parameters. *Innov. Food Sci. Emerg. Technol.* **1:**49–56.

228. Stragier, P., C. Bonamy, and C. Karmazyn-Campelli. 1988. Processing of a sporulation sigma factor in *Bacillus subtilis*: how morphological structure could control gene expression. *Cell* **52:**697–704.

229. Stragier, P., and R. Losick. 1996. Molecular genetics of sporulation in *Bacillus subtilis. Annu. Rev. Genet.* **30:**297–341.

230. Stumbo, C. R. 1973. *Thermobacteriology in Food Processing*, 2nd ed. Academic Press, Inc., New York, N.Y.

231. Sugiyama, H. 1951. Studies on factors affecting the heat resistance of spores of *Clostridium botulinum. J. Bacteriol.* **62:**81–96.

232. Sugiyama, H. 1952. Effect of fatty acids on the heat resistance of *Clostridium botulinum* spores. *Bacteriol. Rev.* **16:**125–126.

233. Sugiyama, H. 1980. *Clostridium botulinum* neurotoxin. *Microbiol. Rev.* **44:**419–448.

234. Sugiyama, H. 1986. Mouse models for infant botulism. *In* O. Zak and M. A. Sande (ed.), *Experimental Models in Antimicrobial Chemotherapy*. Academic Press, Inc., New York, N.Y.

235. Sugiyama, H., M. Woodburn, K. H. Yang, and C. Movroydis. 1983. Production of botulinum toxin in inoculated pack studies of foil-wrapped potatoes. *J. Food Prot.* **44:**896–898.

236. Tanaka, N., E. Traisman, P. Plantinga, L. Finn, W. Flom, L. Meske, and J. Guffisberg. 1986. Evaluation of factors involved in antibotulinal properties of pasteurized process cheese spreads. *J. Food Prot.* **49:**526–531.

237. Tennen, R., B. Setlow, K. L. Davis, C. A. Loshon, and P. Setlow. 2000. Mechanisms of killing of spores of *Bacillus subtilis* by iodine, glutaraldehyde and nitrous acid. *J. Appl. Microbiol.* **89:**330–338.

238. Tournas, V. 1994. Heat-resistant fungi of importance to the food and beverage industry. *Crit. Rev. Microbiol.* **20:**243–263.

239. Townsend, C. T., J. R. Esty, and F. C. Baselt. 1938. Heat-resistance studies on spores of putrefactive anaerobes in relation to the determination of safe processes for canned foods. *Food Res.* **3:**323–346.

240. Trent, J. D., M. Gabrielson, B. Jensen, J. Neuhard, and J. Olsen. 1994. Acquired thermotolerance and heat shock proteins in thermophiles from the three phylogenetic domains. *J. Bacteriol.* **176:**6148–6152.

241. Urakami, K., S. Miyata, R. Moriyama, K. Sugimoto, and S. Makino. 1999. Germination-specific cortex-lytic enzymes from *Clostridium perfringens* spores: time of synthesis, precursor structure and regulation of enzyme activity. *FEMS Microbiol. Lett.* **173:**467–473.

242. Vamvakopoulos, N. C., J. N. Vournakis, and S. J. Marcus. 1977. The effect of magnesium and manganous ions on the structure and template activity for reverse transcriptase of polyribocytidine and its 2′-O-methyl derivative. *Nucleic Acids Res.* **4:**3589–3597.

243. Van Netton, P., A. Van de Moosdijk, P. Van de Hoensel, D. A. A. Mossel, and I. Perales. 1990. Psychrotrophic strains of *Bacillus cereus* producing enterotoxin. *J. Appl. Bacteriol.* **69:**73–79.

244. Waldberger, C., D. Gonzalez, and G. H. Chambliss. 1993. Characterization of a new sporulation factor in *Bacillus subtilis. J. Bacteriol.* **175:**6321–6327.

245. Wandling, L. R., B. W. Sheldon, and P. M. Foegeding. 1999. Nisin in milk sensitizes *Bacillus* spores to heat and prevents recovery of survivors. *J. Food Prot.* **62:**492–498.

246. Wang, J., C. Sass, and G. N. Bennett. 1995. Sequence and arrangement of genes encoding sigma factors in *Clostridium acetobutylicum. Gene* **153:**89–92.

247. Wang, S. Y. 1976. Pyrimidine bimolecular photoproducts, p. 295–396. *In* S. Y. Wang (ed.), *Photochemistry and Photobiology of Nucleic Acids*, vol. 1. Academic Press, Inc., New York, N.Y.

248. Wang, T.-C., and C. S. Rupert. 1977. Evidence for the monomerization of spore photoproduct to two thymines by the light-independent "spore repair" process in *Bacillus subtilis. Photochem. Photobiol.* **25:**123–127.

249. Wilkinson, B. J., J. A. Deans, and D. J. Ellar. 1975. Biochemical evidence for the reversed polarity of the outer membrane of the bacterial forespore. *Biochem. J.* **152:**561–569.

250. Williams, R. J. P. 1981. The Bakerian lecture. Natural selection of the chemical elements. *Proc. R. Soc. Lond. Ser. B Biol. Sci.* **213:**361–397.

251. Willis, A. T. 1969. *Clostridia of Wound Infection*. Butterworths, London, England.

252. Woese, C. 1987. Bacterial evolution. *Microbiol. Rev.* **51:**221–271.

253. Wu, L. J., and J. Errington. 1994. *Bacillus subtilis* SpoIIIE protein required for DNA segregation during asymmetric cell division. *Science* **264:**572–575.

254. Wu, L. J., A. Heucht, and J. Errington. 1998. Prespore-specific gene expression in *Bacillus subtilis* is driven by sequestration of SpoIIE phosphatase to the prespore side of the asymmetric septum. *Genes Dev.* **12:**1371–1380.

255. Wuytack, E. Y., J. Soons, F. Poschet, and C. W. Michiels. 2000. Comparative study of pressure- and nutrient-induced germination of *Bacillus subtilis* spores. *Appl. Environ. Microbiol.* **66:**257–261.

256. Yasuda, Y., Y. Sakae, and K. Tochikubo. 1996. Immunological detection of the GerA spore germination proteins in the spore integuments of *Bacillus subtilis* using scanning electron microscopy. *FEMS Microbiol. Lett.* **139:**235–238.

257. **Zhang, Y. W., T. Koyama, and K. Ogura.** 1997. Two cistrons of the *gerC* operon of *Bacillus subtilis* encode the two subunits of heptaprenyl diphosphate synthase. *J. Bacteriol.* **179:**1417–1419.

258. **Zuber, P., M. M. Nakano, and M. A. Marahiel.** 1993. Peptide antibiotics, p. 897–916. *In* A. L. Sonenshein, J. A. Hoch, and R. Losick (ed.), *Bacillus subtilis and Other Gram-Positive Bacteria: Biochemistry, Physiology, and Molecular Genetics.* American Society for Microbiology, Washington, D.C.

259. **Zuberi, A. R., I. M. Feavers, and A. Moir.** 1987. The nucleotide sequence and gene organization of the *gerA* spore germination operon of *Bacillus subtilis* 168. *Gene* **162:**756–762.

Food Microbiology: Fundamentals and Frontiers, 2nd Ed.
Edited by M. P. Doyle et al.
© 2001 ASM Press, Washington, D.C.

Merle D. Pierson
L. Michele Smoot

Indicator Microorganisms and Microbiological Criteria

4

INTRODUCTION

Purpose of Microbiological Criteria

Microbiological criteria are used to distinguish between an acceptable and an unacceptable product, or between acceptable and unacceptable food processing and handling practices. The numbers and types of microorganisms present in or on a food product may be used to judge the microbiological safety and quality of that product. Safety is determined by the presence or absence of pathogenic microorganisms or their toxins, the number of pathogens, and the expected control or destruction of these agents. The level of spoilage microorganisms reflects the microbiological quality, or wholesomeness, of a food product as well as the effectiveness of measures used to control or destroy such microorganisms. Indicator organisms may be used to assess either the microbiological quality or safety. The use of indicator organisms to assess microbiological safety of a food requires that a relationship between the occurrence of the indicator organism and the likely presence of a pathogen or toxin has been established. Specifically, microbiological criteria are used to assess (i) the safety of food, (ii) adherence to Good Manufacturing Practices (GMPs), (iii) the keeping quality (shelf life) of certain perishable foods, and (iv) the utility (suitability) of a food or ingredient for a particular purpose (41). When appropriately applied, microbiological criteria can be a useful means of ensuring the safety and quality of foods, which in turn elevates consumer confidence. Microbiological criteria provide the food industry and regulatory agencies with guidelines for control of food processing systems and are an underlying component of any critical control point that addresses a microbiological hazard in Hazard Analysis and Critical Control Point (HACCP) systems. In addition, internationally accepted criteria can advance free trade though standardization of food safety/quality requirements.

Need To Establish Microbiological Criteria

A microbiological criterion should be established and implemented only when there is a need and when it can be shown to be both effective and practical. There are many considerations to be taken into account when establishing meaningful microbiological criteria. Listed below are factors considered important when assessing whether or not there is a need to establish a microbiological criterion (41).

Merle D. Pierson, Department of Food Science and Technology, Virginia Polytechnic Institute and State University, Blacksburg, VA 24061.
L. Michele Smoot, Silliker Laboratories Group, Inc., 1139 East Dominguez, Suite 1, Casson, CA 90746.

- evidence of a hazard to health based on epidemiological data or a hazard analysis
- the nature of the natural and commonly acquired microflora of the food and the ability of the food to support microbial growth
- the effect of processing on the microflora of the food
- the potential for microbial contamination and/or growth during processing, handling, storage, and distribution
- the category of consumers at risk
- the state in which the food is distributed
- the potential for abuse at the consumer level
- spoilage potential, utility, and GMPs
- the manner in which the food is prepared for ultimate consumption
- reliability of methods available to detect and/or quantify the microorganism(s) and toxins of concern
- the costs/benefits associated with the application of the criterion

Definitions

The National Research Council of the National Academy of Sciences addressed the issue of microbiological criteria in their 1985 report, *An Evaluation of the Role of Microbiological Criteria for Foods and Food Ingredients* (41). In that report, it was established that a microbiological criterion will stipulate that a type of microorganism, group of microorganisms, or toxin produced by a microorganism must either not be present at all, be present in only a limited number of samples, or be present as no less than a specified number or amount in a given quantity of a food or food ingredient. In addition, a microbiological criterion should include the following information (41):

- statement describing the identity of the food or food ingredient
- statement identifying the contaminant of concern
- analytical method to be used for the detection, enumeration, or quantification of the contaminant of concern
- sampling plan
- microbiological limits considered appropriate to the food and commensurate with the sampling plan

Criteria may be either mandatory or advisory. A mandatory criterion is a criterion that may not be exceeded, and food that does not meet the specified limit is required to be subjected to some action, including rejection, destruction, reprocessing, or diversion. An advisory

criterion permits acceptability judgments to be made, and it should serve as an alert to deficiencies in processing, distribution, storage, or marketing. For application purposes, the three categories of criteria that are employed include standards, guidelines, and specifications. The following definitions were recommended by the National Research Council's Subcommittee on Microbiological Criteria for Foods and Food Ingredients (41). Codex Alimentarius (12) does not provide specific definitions; however, they are implied in the "Application of Microbiological Criteria" section of their criteria principles document, as follows.

> **Standard:** A microbiological criterion that is part of a law, ordinance, or administrative regulation. A standard is a mandatory criterion. Failure to comply constitutes a violation of the law, ordinance, or regulation and will be subject to the enforcement policy of the regulatory agency having jurisdiction (41).... [A] criterion contained in a Codex Alimentarius standard. Wherever possible it should contain limits only for pathogenic microorganisms of public health significance in the food concerned. Limits for non-pathogenic microorganisms may be necessary when the methods of detection for the pathogens of concern are cumbersome or unreliable. Standards based on fixed numbers of nonpathogenic microorganisms may result in the recall or down grading of otherwise wholesome food. To minimize the effect of this approach penalty provisions could be applied when a lot is rejected. Such penalties would result in suspension in the privilege to process food only after repeated violations occur over a specified time period (12).
>
> **Guideline:** A microbiological criterion often used by the food industry or regulatory agency to monitor a manufacturing process. Guidelines function as alert mechanisms to signal whether microbiological conditions prevailing at critical control points or in the finished product are within the normal range. Hence, they are used to assess processing efficiency at critical control points and conformity with Good Manufacturing Practices. A microbiological guideline is advisory (41).... is intended to increase assurance that the provisions of hygienic significance in the Code have been met. It may include microorganisms which are not of direct public health significance (12).
>
> **Specifications:** A microbiological criterion that is used as a purchase requirement whereby conformance becomes a condition of purchase between buyer and vendor of a food ingredient. A microbiological specification may be advisory or mandatory (41).... is applied at the establishment at a specified point during or after processing to monitor hygiene. It is intended to guide the manufacturer and is not intended for official control purposes (12).

The Codex use of specification only refers to end products and does not include raw materials, ingredients, or foods in contractual agreement between two parties.

Who Establishes Microbiological Criteria?

Different scientific organizations have been involved in developing general principles for application of microbiological criteria by regulatory agencies and the food industry. The scientific organizations that have influenced the U.S. food industry the most include the Joint Food and Agriculture Organization (FAO) and World Health Organization (WHO) Codex Alimentarius International Food Standards Programme, the International Commission on Microbiological Specifications of Foods (ICMSF), the U.S. National Academy of Sciences, and the U.S. National Advisory Committee on Microbiological Criteria for Foods. The Codex Alimentarius Program first formulated "General Principles for the Establishment and Application of Microbiological Criteria" in 1981 (12). These microbiological principles have been well accepted internationally. In 1984, the NRC Subcommittee on Microbiological Criteria for Foods and Food Ingredients formulated general principles for the application of microbiological criteria to food and food ingredients as requested by four U.S. regulatory agencies (41). Recently, the Codex Committee published a revision of their earlier document (17). Codex Alimentarius contains standards for all principal foods in the form (i.e., processed, semiprocessed, or raw) in which they are delivered to the consumer. Fresh perishable commodities not traded internationally are excluded from these standards (35, 45). When establishing microbiological criteria for those foods intended for international trade, materials provided by the ICMSF (25–28) and the Codex Alimentarius (12) should be consulted.

The World Trade Organization provides a framework for ensuring fair trade and harmonizing standards and import requirements on food traded, through the Agreements on Sanitary and Phytosanitary Measures and Technical Barriers to Trade. Countries are required to base their standards on science, to base programs on risk analysis methodologies, and to develop ways of achieving equivalence between methods of inspection, analysis, and certification between trading countries (35, 45). The World Trade Organization recommends the use of standards, guidelines, and recommendations developed by Codex Alimentarius to facilitate harmonization of standards. Improvements in the development and execution of microbiological criteria will continue to be made with international acceptance and widespread implementation of Points HACCP, a science-based preventative system for food control. Several countries have mandated HACCP requirements and have established specific HACCP requirements for public sectors of their domestic food industries.

SAMPLING PLANS

Introduction

A sampling plan includes both the sampling procedure and the decision criteria. To examine a food for the presence of microorganisms, a representative sample is examined by defined methods. A "lot" is that quantity of product produced, handled, and stored within a limited time period under uniform conditions. Since it is impractical to examine the entire lot, statistical concepts of population probability and sampling must be used to determine the number and size of sample units from the lot and to provide conclusions drawn from the analytical results. The sampling plan is designed so that inferior lots within a set level of confidence are rejected. Detailed information regarding statistical concepts of population probabilities and sampling, choice of sampling procedures, decision criteria, and practical aspects of application as applied to microorganisms in food can be found in a publication by ICMSF (26).

A simplified example of a sampling plan described in the aforementioned publication is given below. Assume 10 samples were taken and analyzed for the presence of a particular microorganism. Based on the decision criterion, only a certain number of the sample units could be positive for the presence of that microorganism for the lot to be considered acceptable. If in the criterion the maximum allowable positive units had been set at 2 ($c = 2$), then a positive result for more than 2 of the 10 sample units ($n = 10$) would result in rejection of the lot. Ideally, the decision criterion is set to accept lots that are of the desired quality and to reject lots that are not. However, since the entire lot is not examined, there is always the risk that an acceptable lot may be rejected or an unacceptable lot may be accepted. The more samples examined, or the larger n is, the lower the risk of making an incorrect decision about the lot quality. However, as n increases, sampling becomes more time consuming and costly. Generally, a compromise is made between the size of n and the level of risk that is acceptable.

The level of risk that is associated with a particular sampling plan can be determined by an operating characteristic curve, as seen in Fig. 4.1. The two vertical scales in the graph show (i) the probability of acceptance, P_a, or the ratio of the number of times that the results will indicate a lot should be accepted to the number of times a lot of the given quality is sampled for a decision and (ii) the probability of rejection, P_r, or the ratio of the number of times that the result will indicate a lot should be rejected to the number of times a lot of the given quality is sampled for a decision. The horizontal axis shows the percentage of defective sample units (p) that are in

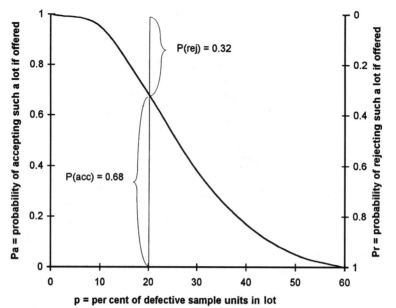

Figure 4.1 Operating characteristic curve for $n = 10$, $c = 2$. From reference 26 with permission.

the lot. This probability is usually expressed as percentages of "defectives." The operating characteristic curve in Fig. 4.1, in which $n = 10$ and $c = 2$, shows that as p increases, P_a will decrease. In other words, if the probability that a test unit will yield a positive result is great, then the probability that the lot will be accepted becomes small. The risks to be considered include those for the consumers or buyers and those for the vendors or producers. The vendors' risk is the probability that a lot of acceptable quality is rejected, while the probability that a lot of defective quality is accepted is referred to as the consumers' risk. The level of risk that a vendor is willing to accept relative to himself or herself and the consumer is then used to design the sampling plan.

Types of Sampling Plans

Sampling plans are divided into two main categories: variables and attributes. A variables plan depends on the frequency distribution of organisms in the food. For correct application of a variables plan, the organisms must be distributed log normal (i.e., counts transformed to logarithms are normally distributed). When the food is from a common source and is known to be produced or processed under uniform conditions, log-normal distribution of the organisms present is reasonably assumed (31, 32). Attributes sampling is the preferred plan when microorganisms are not homogeneously distributed throughout the food or when the target microorganism is present at low levels, which is often the case

with pathogenic microorganisms. Attributes plans are also widely used to determine the acceptance or rejection of products at ports or other points of entry, because there is little or no knowledge of how the food was processed, or past performance records are not available. Attributes sampling plans may also be used to monitor performance relative to accepted GMPs. Attributes sampling, however, is not appropriate when there is no defined lot or when random sampling is not possible, as might occur when monitoring cleaning practices.

Variables Sampling Plans

Because attributes sampling plans are widely used over variables sampling, this chapter will not cover variables sampling in detail. For more detailed information on variables sampling plans, the reader is referred to publications by the ICMSF (26), Kilsby (31), Kilsby et al. (32), and Malcolm (36).

Attributes Sampling Plans

Two-Class Plans

The two-class attributes sampling plan assigns the concentration of microorganisms of the sample units tested to a particular attribute class depending on whether the microbiological counts are above or below some preset concentration, represented by the letter m. The decision criterion is based on (i) the number of sample units tested, n, and (ii) the maximum allowable number of sample units yielding unsatisfactory tests results, c. For example, when $n = 5$ and $c = 2$ in a two-class sampling plan designed to make a presence/absence decision on the lot (i.e., $m = 0$), the lot is rejected if more than two of the five sample units tested are positive. As n increases for the set number c, the stringency of the sampling plan also increases. Conversely, for a set sample size n, as c increases, the stringency of the sampling plan decreases, allowing for a higher probability of acceptance, P_a, of the food lots of a given quality. Two-class plans are applied most often in qualitative (semiquantitative) pathogen testing in which the results are expressed as the presence or absence of the specific pathogen per sample weight analyzed.

Three-Class Plans

Three-class sampling plans use the concentration of microorganisms in the sample units to determine levels of quality and/or safety. Counts above a preset concentration M for any of the n sample units tested are considered unacceptable, and the lot is rejected. The level of the test organism acceptable in the food is denoted by m. This concentration in a three-class attribute plan separates acceptable lots (i.e., counts less than m) from marginally acceptable lots (i.e., counts greater than m but not exceeding M). Counts above m and up to and including M are not desirable, but the lot can be accepted provided the number of n samples that exceed m is no greater than the preset number, c. Thus, in a three-class sampling plan, the food lot will be rejected if any one of the sample units exceeds M or if the number of sample units with contamination levels above m exceeds c. Similar to the two-class sampling plan, the stringency of the three-class sampling plan is also dependent on the two numbers denoted by n and c. The larger the value of n for a given value of c, the better the food quality must be to have the same chance of passing, and vice versa. From n and c it is then possible to find the probability of acceptance, P_a, of a food lot of a given microbiological quality.

ESTABLISHING LIMITS

Microbiological limits, as defined in a criterion, represent the level above which action is required. Levels should be realistic and should be determined based on knowledge of the raw materials and the effects of processing, product handling, storage, and end use of the product. Limits should also take into account the likelihood of uneven distribution of microorganisms in the food, the inherent variability of the analytical procedure, the risk associated with the organisms, and the conditions under which the food is expected to be handled and consumed. Microbiological limits should include sample weight to

be analyzed, method reference, and confidence limits of the referenced method where applicable.

The shelf life of a perishable product is often determined by the number of microorganisms initially present. As a general rule, a food containing a large population of spoilage organisms will have a shorter shelf life than the same food containing fewer numbers of the same spoilage organisms. However, the relationship between total counts and shelf life is not absolute. Some types of microorganisms have a greater impact on the organoleptic characteristics of a food than others owing to the presence of different enzymes acting upon the food constituents. In addition to the effect of certain levels and/or types of spoilage microorganisms, changes in perceptible quality characteristics will also vary depending on the food and the conditions of storage, such as temperature and gaseous atmosphere. All of these parameters need to be considered when establishing limits for microbiological criteria used to determine product quality and/or shelf life.

Foods produced and stored under GMPs may be expected to have a different microbiological profile than those foods produced and stored under poor conditions. The use of poor quality materials, improper handling, or unsanitary conditions may result in higher bacterial counts in the finished product. However, low counts in the finished product do not necessarily mean that GMPs were adhered to. Processing steps such as heat treatments, fermentation, freezing, or frozen storage can reduce the counts of bacteria that have resulted from noncompliance with GMPs. Other products, such as ground beef, may normally contain high microbial counts even under the best conditions of manufacture owing to the growth of psychrotrophic bacteria during refrigeration. Limits set for microbiological criteria used to assess adherence to GMPs require a working knowledge of the types and levels of microorganisms present at the different processing steps to establish a relationship between the microbiology of the food and adherence to GMPs.

Microbiological criteria are also an underlying component of any critical control point in a HACCP system that addresses a microbiological hazard (9). In a HACCP system, a critical limit is set for each critical control point identified for an operation. A critical limit is the maximum and/or minimum value to which a biological, chemical, or physical parameter must be controlled at a critical control point to prevent, eliminate, or reduce to an acceptable level the occurrence of a food safety hazard. Limits assigned to microbiological criteria are not necessarily microbiological in nature but are often a physical or chemical attribute. For example, high-acid foods not thermally processed are required to have a pH of less than 4.6 to prevent growth of *Clostridium botulinum* spores.

Setting limits for a microbiological criterion allows for a decision to be made about the operation of a critical step in a process. The criterion may be either absolute or performance based. An absolute criterion is one in which a specific upper level or frequency of a microorganism, group of microorganisms, or product of microbial metabolism has been established. A negative result for *Escherichia coli* O157:H7 in 25 g of ground beef is an example of a critical limit established for an absolute microbiological criterion. Alternatively, a performance criterion refers to a specified change that a process is expected to exert on the level or frequency of a microorganism or microbial metabolite (9). For example, low-acid canned foods are required to be heat treated at a specific temperature and time to cause a reduction of *C. botulinum* spores by a factor of 10^{12}.

The critical limits associated with critical control points dictate the level of stringency of a food production process. For a limit to be effective, it must be well understood how that limit is related to the microbiological criterion, how the criterion is related to the microbiological hazard identified in the hazard analysis, and societal and regulatory expectations. The use of quantitative risk assessment techniques to scientifically determine the probability of occurrence and severity of known human exposure to foodborne hazards is strongly advocated. The process consists of (i) hazard identification, (ii) hazard characterization, (iii) exposure assessment, and (iv) risk characterization (33). Though quantitative risk assessment techniques are well established for chemical agents, application to food safety microbiology is relatively new. The application of risk assessment principles to biological agents has been complicated by the inherent differences between chemical and biological agents. Levels of microorganisms are often not uniformly distributed and are subject to change during processing and storage. Although systematic application of risk analysis methodology is still needed to establish microbiological criteria, quantitative data on microbial hazards may be replaced with current qualitative data within this same framework to facilitate establishment of microbiological limits.

INDICATORS OF MICROBIOLOGICAL QUALITY

Introduction

Estimation of a product for indicator organisms can provide simple, reliable, and rapid information about process failure, postprocessing contamination, contam-

ination from the environment, and the general level of hygiene under which the food was processed and stored. However, these methods cannot replace examination for specific pathogens where suitable methods exist and where such testing is appropriate, but usually provide information in a shorter time than that required for isolation and identification of specific organisms or pathogens.

It has been suggested that the ideal indicators of product quality or shelf life should meet the following criteria (29):

- They should be present and detectable in all foods whose quality is to be assessed.
- Their growth and numbers should have a direct negative correlation with product quality.
- They should be easily detected and enumerated and be clearly distinguishable from other organisms.
- They should be numerable in a short period of time, ideally within a working day.
- Their growth should not be affected adversely by other components of the food flora.

Indicator Microorganisms

Indicator microorganisms can be used in microbiological criteria. These criteria might be used to address existing product quality or to predict shelf life of the food. Some examples of indicators and the products in which they are used are shown in Table 4.1.

Those microorganisms listed in Table 4.1 are the primary spoilage organism corresponding to that specific product. Loss of quality in other products may be limited not to one organism but to a variety of organisms owing to the unrestricted environment of the food. In

Table 4.1 Organisms highly correlated with product quality[a]

Organisms	Products
Acetobacter spp.	Fresh cider
Bacillus spp.	Bread dough
Byssochlamys spp.	Canned fruits
Clostridium spp.	Hard cheeses
Flat-sour spores	Canned vegetables
Lactic acid bacteria	Beers, wines
Lactococcus lactis	Raw milk (refrigerated)
Leuconostoc mesenteroides	Sugar (during refining)
Pectinatus cerevisiiphilus	Beers
"*Pseudomonas putrefaciens*"	Butter
Yeasts	Fruit juice concentrates
Zygosaccharomyces bailii	Mayonnaise, salad dressing

[a]Reprinted with permission from reference 29.

those types of products, it is often more practical to determine the counts of groups of microorganisms most likely to cause spoilage in that particular food.

Aerobic plate count (APC) or standard plate count is commonly used to determine "total" numbers of microorganisms in a food product. By modifying the environment of incubation or the medium used, APC can be used to preferentially screen for groups of microorganisms, such as those that are thermodurics, mesophiles, psychrophiles, thermophiles, proteolytic, and lipolytic. APC may be a component of microbiological criteria for assessing product quality when those criteria are used to (i) monitor foods for compliance with standards or guidelines set by various regulatory agencies, (ii) monitor foods for compliance with purchase specifications, and (iii) monitor adherence to GMPs (41).

Microbiological criteria as a specification is used to determine the usefulness of a food or food ingredient for a particular purpose. For example, specifications are set for thermophilic spores in sugar and spices intended for use in the canning industry. Lots of sugar failing to meet specifications may not be suitable for use in the low-acid canning but could be diverted for other uses. The APC of refrigerated perishable foods such as milk, meat, poultry, and fish may be used to indicate the condition of equipment and utensils used, as well as the time/temperature profile of storage and distribution of the food.

When evaluating results of APC for a particular food, it is important to remember that (i) APCs only measure live cells and, therefore, it would not be of value, for example, to determine the quality of raw materials used for a heat-processed food, (ii) APCs are of little value in assessing organoleptic quality since high microbial counts are generally required prior to organoleptic quality loss, and (iii) since different bacteria vary in their biochemical activities, quality loss may also occur at low total counts depending on the predominant organisms present. With any food, specific causes of unexpectedly high counts can be identified by examination of samples at control points and by plant inspection. Reliable interpretation of the APC of a food requires knowledge of the expected microbial population at the point in the process or distribution at which the sample is collected. If counts are higher than expected, this will point to the need to determine why there has been a violation of the criterion. A review by Silliker (46) provides a detailed discussion of the use of total counts as an index of sanitary quality, organoleptic quality, and safety.

The direct microscopic count (DMC) is used to give an estimate of both viable and nonviable cells in samples containing a large number of microorganisms (i.e., $>10^5$ CFU/ml). Considering that the DMC does not differentiate between live and dead cells (unless a fluorescent dye such as acridine orange is employed) and requires that the total cell count exceed 10^5, use of the DMC is of limited value as part of microbiological criteria for quality issues. The use of DMC as part of microbiological criteria for foods or ingredients is restricted to a few products, such as raw, non-grade A milk; dried milks; liquid and frozen eggs; and dried eggs (41).

Other methods commonly used to indicate quality of different food products include the Howard mold count, yeast and mold count, heat-resistant mold count, and thermophilic spore count. The Howard mold count is used to detect the inclusion of moldy material in canned fruit and tomato products (21) as well as to evaluate the sanitary condition of processing machinery in vegetable canneries (19). Yeasts and molds frequently become predominant on foods when conditions for bacterial growth are less favorable. Therefore, they can potentially be a problem in fermented dairy products, fruits, fruit beverages, and soft drinks. Yeasts and mold counts are used as part of microbiological standards of various dairy products, such as cottage cheese and frozen cream (50), and sugar (42). Heat-resistant molds, such as *Byssochlamys fulva* and *Aspergillus fisheri*, that may survive the thermal processes applied to fruit and fruit products may need limits in purchase specifications for ingredients such as fruit concentrates. Concern for thermophilic spores in ingredients used in the canning industry is related to their ability to cause defects in foods held at elevated temperatures because of inadequate cooling and/or storage at too-high temperatures. Purchase specifications and verification criteria are often used for thermophilic spore counts in ingredients intended for use in low-acid, heat-processed canned foods.

Metabolic Products

In certain cases, bacterial populations in a food product can also be estimated by testing for metabolic products produced by the microorganisms present in the food. When a correlation is established between the presence of a metabolic product and product quality loss, tests for the metabolite may be a part of a microbiological criterion.

An example of the use of metabolic products as part of a microbiological criterion is the organoleptic evaluation of imported shrimp. Trained personnel are able to classify the degree of decomposition (i.e., quality loss) into one of three classes through organoleptic examination. The shrimp are placed into one of the following quality classes: class 1, passable; class 2, decomposed (slight but definite); class 3, decomposed (advanced). Limits of acceptability of a lot are based on the number of shrimp

Table 4.2 Some microbial metabolic products that correlate with food quality[a]

Metabolites	Applicable food product
Cadaverine and putrescine	Vacuum-packaged beef
Diacetyl	Frozen juice concentrate
Ethanol	Apple juice, fishery products
Histamine	Canned tuna
Lactic acid	Canned vegetables
Trimethylamine (TMA)	Fish
Total volatile bases (TVB), total volatile nitrogen (TVN)	Seafood
Volatile fatty acids	Butter, cream

[a] Reprinted with permission from references 26.

in a sample that are placed into each of the three classes (4). Other commodities in which organoleptic examination is used to determine quality deterioration include raw milk, meat, poultry, and fish and other seafood. The food industry also uses these examinations to classify certain foods into quality grades. For additional information about organoleptic examination of foods, the reader is referred to Amerine et al. (3) and Larmond (34).

Other examples of metabolic products used to assess product quality are listed in Table 4.2.

INDICATORS OF FOODBORNE PATHOGENS AND TOXINS

Introduction

Microbiological criteria, as they apply to product safety, should only be developed when the application of the criterion can reduce or eliminate a potential foodborne hazard. Each food type should be carefully evaluated through risk assessment to determine the potential hazards and their significance to consumers. When a food is repeatedly implicated as a vehicle in foodborne disease outbreaks, application of microbiological criteria may be useful. Public health officials and the dairy industry responded to widespread outbreaks of milk-borne disease occurring around the turn of the 20th century in the United States. By imposing controls on milk production, developing safe and effective pasteurization procedures, and setting microbiological criteria, the safety of commercial milk supplies was greatly improved. Epidemiological evidence alone, however, does not necessitate imposing microbiological criteria. The criteria should only be applied when its use results in a safer food (1).

Food products frequently subject to contamination by harmful microorganisms, such as shellfish, may benefit from the application of microbiological criteria. The National Shellfish Sanitation Program utilizes microbiological criteria in this manner to prevent use of shellfish from polluted waters that may contain various intestinal pathogens (2). Depending on the type and level of contamination anticipated, imposition of microbiological criteria may or may not be justified. Food contaminated with pathogens that do not have the opportunity to grow to levels that would potentially result in a health hazard do not warrant microbiological criteria. Though fresh vegetables are often contaminated with small numbers of C. botulinum, Clostridium perfringens, and Bacillus cereus, epidemiological evidence indicates that this contamination presents no health hazard. Thus, imposing microbiological criteria would not be beneficial for these microorganisms. However, criteria may be appropriate for enteric pathogens on produce, since there have been several outbreaks of foodborne illness resulting from fresh produce contaminated with enteric pathogens (7, 49).

Often, food processors alter the intrinsic or extrinsic parameters of a food (nutrients, pH, water activity, inhibitory chemicals, gaseous atmosphere, temperature of storage, and the presence of competing organisms) to prevent growth of undesirable microorganisms. If control over one or more of these parameters is lost, then there may be a risk of a health hazard. For example, in the manufacture of cheese or fermented sausage, a lactic acid starter culture is relied upon to produce acid quickly enough to inhibit the growth of Staphylococcus aureus to levels that are potentially harmful. Process critical control points, such as rate of acid formation, are implemented to prevent growth of contamination of the food by harmful microorganisms and to ensure that control of the process is maintained (43).

Depending on the pathogen, low levels of the microorganism in the food product may or may not be of concern. Some microorganisms have such a low infective dose that their mere presence in a food presents a significant public health risk. For such organisms, the concern is not whether the pathogen is able to grow in the food, but whether the microorganism could survive for any length of time in the food. Foods having intrinsic or extrinsic factors sufficient to prevent survival of pathogens or toxigenic microorganisms of concern may not be candidates for microbiological criteria related to safety. For example, the acidity of certain foods such as fermented meat products might be assumed sufficient for pathogen control. In fact, the growth and toxin production of S. aureus might be prevented; however, enteric pathogens such as E. coli O157:H7 could survive and result in a product unsafe for consumption.

Table 4.3 Plan stringency (case) in relation to degree of health hazard and conditions of use[a]

Type of hazard	Conditions in which food is expected to be handled and consumed after sampling, in the usual course of events[b]		
No direct health hazard	Increased shelf life	No change	Reduced shelf life
Utility (e.g., general contamination, reduced shelf life, and spoilage)	Case 1 Three-class $n = 5, c = 3$	Case 2 Three-class $n = 5, c = 2$	Case 3 Three-class $n = 5, c = 1$
Health hazard	Reduced hazard	No change	Increased hazard
Low, indirect (indicator)	Case 4 Three-class $n = 5, c = 3$	Case 5 Three-class $n = 5, c = 2$	Case 6 Three-class $n = 5, c = 1$
Moderate, direct, limited spread[c]	Case 7 Three-class $n = 5, c = 2$	Case 8 Three-class $n = 5, c = 1$	Case 9 Three-class $n = 5, c = 0$
Moderate, direct, potentially extensive spread	Case 10 Two-class $n = 5, c = 0$	Case 11 Two-class $n = 10, c = 0$	Case 12 Two-class $n = 20, c = 0$
Severe, direct	Case 13 Two-class $n = 15, c = 0$	Case 14 Two-class $n = 30, c = 0$	Case 15 Two-class $n = 60, c = 0$

[a] Reprinted with permission from reference 25.
[b] More stringent plans would generally be used for sensitive foods destined for susceptible populations.
[c] n, number of sample units drawn from lot; c, maximum allowable number of positive results.

An important and sometimes overlooked consideration when evaluating the potential microbiological risks associated with a food is the consumer. More rigid microbiological requirements may be needed if the food is intended for use by infants, elderly, or immunocompromised people, since they are more susceptible to infectious agents than are healthy adults. The sampling plan specified in a microbiological criterion should be appropriate to the hazard expected to be associated with the food, the consumer of the food, and the severity of the illness. The hazard associated with a food is determined by (i) the type of organism expected to be encountered and (ii) the expected conditions of handling and consumption after sampling. A more stringent sampling plan is desired for products expected to contain higher degrees of hazards. ICMSF (26) proposed a system for classification of foods according to risk into 15 hazard categories called cases, with suggested appropriate sampling plans (Table 4.3).

The stringency of sampling plans for foods is based either on the hazard to the consumer from pathogenic microorganisms and their toxins or toxic metabolites, or on the potential for quality deterioration to an unacceptable state, and it should take account the types of microorganisms present and their numbers (26). Foodborne pathogens are grouped into one of three categories based on the severity of the potential hazard (i.e.,

severe hazards, moderate hazards with potentially extensive spread, and moderate hazards with limited spread) (Table 4.4). Pathogens with potential for extensive spread are often initially associated with specific foods; however, secondary spread to other foods commonly occurs from environmental contamination and cross contamination within processing plants and food preparation areas, including homes. An example is fresh beef that is contaminated with E. coli O157:H7. One or a few contaminated pieces of meat can lead to widespread contamination of product during processing, such as grinding to produce ground beef. There can also be cross contamination if the fresh beef is improperly stored with ready-to-eat foods. Microbial pathogens in the lowest risk group (moderate hazards, limited spread) are found in many foods, usually in small numbers. Generally, illness is caused only when ingested foods contain large numbers of the pathogen, e.g., C. perfringens, or have at some time contained large enough numbers to produce sufficient toxin to cause illness, e.g., S. aureus. Outbreaks are usually restricted to consumers of a particular meal or a particular kind of food (41). A summary of those vehicles associated with outbreaks of foodborne diseases and the microorganisms involved that have occurred in the United States in recent years has been prepared by Bryan (8) (1977 to 1984) and by Bean et al. (6) (1983 to 1987).

Table 4.4 Hazardous microorganisms and parasites grouped on the basis of risk severity[a]

I.   Severe hazards
      Clostridium botulinum types A, B, E, and F
      Shigella dysenteriae
      Salmonella typhi: paratyphi A, B
      Enterohemorrhagic *Escherichia coli* (EHEC)
      Hepatitis A and E
      Brucella abortus: B suis
      Vibrio cholerae O1
      Vibrio vulnificus
      Taenia solium

II.   Moderate hazards: potentially extensive spread[b]
      Listeria monocytogenes
      Salmonella spp.
      Shigella spp.
      Other enterovirulent *Escherichia coli* (EEC)
      Streptococcus pyogenes
      Rotavirus
      Norwalk virus group
      Entamoeba histolytica
      Diphyllobothrium latum
      Ascaris lumbricoides
      Cryptosporidium parvum

III.   Moderate hazards: limited spread
      Bacillus cereus
      Campylobacter jejuni
      Clostridium perfringens
      Staphylococcus aureus
      Vibrio cholerae, non-O1
      Vibrio parahaemolyticus
      Yersinia enterocolitica
      Giardia lamblia
      Taenia saginata

[a] Reprinted with permission from reference 44.
[b] Although classified as moderate hazards, complications and sequelae may be severe in certain susceptible populations.

Indicator Organisms

Microbiological criteria for food safety may use tests for indicator organisms that suggest the possibility of a microbial hazard. *E. coli* in drinking water, for example, indicates possible fecal contamination and, therefore, the potential presence of enteric pathogens. Jay (29) suggested that indicators used to assess food safety should ideally meet the following criteria:

- be easily and rapidly detectable
- be easily distinguishable from other members of the food flora
- have a history of constant association with the pathogen whose presence it is to indicate
- always be present when the pathogen of concern is present

- be an organism whose number ideally should correlate with those of the pathogen of concern
- Possess growth requirements and a growth rate equaling those of the pathogen
- have a die-off rate that at least parallels that of the pathogen and ideally persists slightly longer than the pathogen of concern
- be absent from foods that are free of the pathogen except perhaps at certain minimum numbers

Buttiaux and Mossel (11) suggested additional criteria for fecal indicators used in food safety:

- Ideally, the bacteria selected should demonstrate specificity, occurring only in intestinal environments.
- They should occur in very high numbers in feces so as to be encountered in high dilutions.
- They should possess a high resistance to the external environment, the pollution of which is to be assessed.
- They should permit relatively easy and fully reliable detection even when present in low numbers.

With the exception of *Salmonella* spp. and *S. aureus*, most tests for ensuring safety use indicator organisms rather than direct tests for the specific hazard. An overview of some of the more common indicator microorganisms used for ensuring food safety is given below.

Fecal Coliforms and *E. coli*

Fecal coliforms, including *E. coli*, are easily destroyed by heat and may die during freezing and frozen storage of foods. Microbiological criteria involving *E. coli* are useful in those cases where it is desirable to determine if fecal contamination may have occurred. Contamination of a food with *E. coli* implies a risk that other enteric pathogens may be present in the food. Fecal coliform bacteria are used as a component of microbiological standards to monitor the wholesomeness of shellfish and the quality of shellfish-growing waters (23, 53). The purpose is to reduce the risk of harvesting shellfish from waters polluted with fecal material. The fecal coliforms have a higher probability of containing organisms of fecal origin than do coliforms that are composed of organisms of both fecal and nonfecal origins. Fecal coliforms can become established on equipment and utensils in the food processing environments and can contaminate processed foods. At present, *E. coli* is the most widely used indicator of fecal contamination. The failure to detect *E. coli* in a food, however, does not ensure the absence of enteric pathogens (23, 47). In many raw foods of animal origin, small numbers of *E. coli* can be expected because of the close association of these foods with the animal environment and the likelihood of contamination of carcasses from fecal material, hides, or feathers during slaughter-dressing procedures. Rapid, direct plating methods for *E. coli* now exist, and in some foods it may be advantageous to use *E. coli* rather than fecal coliforms as a component of microbiological criteria for foods.

The presence of *E. coli* in a heat-processed food means either process failure or, more commonly, postprocessing contamination from equipment or employees or from contact with contaminated raw foods. In the case of refrigerated ready-to-eat products, such as shrimp and crabmeat, coliforms are recommended as indicators of process integrity with regard to reintroduction of pathogens from the environmental sources and maintenance of adequate refrigeration (10). The source of coliforms after thermal processing, in these types of products, appears to be the processing environment, resulting from inadequate sanitation procedures and/or temperature control. Coliforms were recommended over *E. coli* and APC, because coliforms are often present in higher numbers than *E. coli*, and the levels of coliforms do not increase over time when the product is stored properly.

Enterococci

Sources of enterococci include fecal material from both warm-blooded and cold-blooded animals and plants (38). Enterococci differ from coliforms in that they are salt tolerant (grow in the presence of 6.5% NaCl) and relatively resistant to freezing (29). Certain enterococci (*Enterococcus faecalis* and *E. faecium*) are also relatively heat resistant and may survive usual milk pasteurization temperatures. Enterococci can establish themselves and persist in the food processing establishment for long periods. Because many foods contain small numbers of enterococci, a thorough understanding of the role and significance of enterococci in a food is required before any meaning can be attached to their presence and population numbers. Enterococci counts have few useful applications in microbiological criteria for food safety. This indicator may be applicable in specific cases to identify poor manufacturing practices.

Metabolic Products

Certain microbiological criteria related to safety rely on tests for metabolites to indicate a potential hazard rather than direct tests for pathogenic or indicator microorganisms. Examples of metabolites as a component of microbiological criteria include (i) tests for thermonuclease or thermostable deoxyribonuclease in foods containing or suspected of containing $\geq 10^6$ per ml or per g of *S. aureus*, (ii) illuminating grains under UV light to detect the presence of aflatoxin produced by *Aspergillus* spp., and (iii) assaying for the enzyme alkaline phosphatase, a natural constituent of milk that is inactivated during pasteurization, to detect postpasteurization contamination or contamination pasteurized milk with raw milk (37, 41).

APPLICATION AND SPECIFIC PROPOSALS FOR MICROBIOLOGICAL CRITERIA FOR FOOD AND FOOD INGREDIENTS

The application of useful microbiological criteria should address the following issues: (i) the sensitivity of the food product(s) relative to safety and quality, (ii) the needs for a microbiological standard(s) and/or guideline(s), (iii) assessment of information necessary for establishment of a criterion if one seems to be indicated, and (iv) where the criterion should be applied (39). In this section, examples of the application of microbiological criteria to various foods and food ingredients are presented. No general criteria suitable for all food product groups exist. The relevant background literature that relates to the quality and safety of a specific food product should be consulted prior to implementing microbiological criteria. Such information may be found in ICMSF publications, industry codes of practice, legislation, and peer-reviewed technical publications. Suggested microbiological limits for a wide range of products and product groups were recently published (48) as a guidance document for all involved in producing, using, and interpreting microbiological criteria in the food and catering industries.

The utilization of microbiological criteria may be either mandatory (standards) or advisory (guidelines, specifications). One obvious example of a mandatory criterion (standard) as it relates to food safety is the "zero tolerance" set for *Salmonella* spp. in all ready-to-eat foods. The Food Safety Inspection Service of the U.S. Department of Agriculture (USDA/FSIS) has mandated zero tolerance for *E. coli* O157:H7 in fresh ground beef. Since the emergence of *L. monocytogenes* as a foodborne pathogen the U.S. Food and Drug Administration (FDA) and USDA have also applied zero tolerance for this organism in all ready-to-eat foods. *Listeria monocytogenes* is one of the four microbiological criteria included for verification for cooked ready-to-eat shrimp and crabmeat (Table 4.5). However, there is considerable debate as to whether zero tolerance is warranted for *L. monocytogenes* (18). ICMSF (25) has recommended that the international community allow for tolerance of *L. monocytogenes* as outlined in Table 4.6. These recommendations are an example of advisory criteria. Other

Table 4.5 Microbiological criteria for verification of cooked, ready-to-eat shrimp and cooked, ready-to-eat crabmeat[a]

| Microorganism | Criteria | | Explanation |
	Shrimp	Crabmeat	
Salmonella sp.	$n = 30$	$n = 30$	Analytical unit = 25 g
	$c = 0$	$c = 0$	
	$m = M = 0$	$m = M = 0$	
Listeria monocytogenes	$n = 5$	$n = 5$	Sample unit = 50 g;
	$c = 0$	$c = 0$	analytical unit = 25 g
	$m = M = 0$	$m = M = 0$	through compositing of 5-g
			portions from 5 sample units
Staphylococcus aureus	$n = 5$	$n = 5$	
	$c = 2$	$c = 2$	
	$m = 50/g$	$m = 100/g$	
	$M = 500/g$	$M = 1,000/g$	
Thermal tolerant coliforms[b]	$n = 5$	$n = 5$	
	$c = 2$	$c = 2$	
	$m = 100/g$	$m = 500/g$	
	$M = 1,000/g$	$M = 5,000/g$	

[a] Reprinted with permission from reference 10.
[b] "Thermal tolerant coliforms" is used in lieu of the more traditional designation, "fecal coliforms."

advisory criteria, which have been established to address the safety of foods, are shown in Table 4.7.

Mandatory criteria in the form of standards have also been employed for quality issues. Presented in Table 4.8 are examples of foods and food ingredients for which federal, state, and city as well as international microbiological standards have been developed. Some of the advisory criteria that have been developed and applied to foods for quality monitoring are given in Table 4.7.

The USDA/FSIS coordinates a pathogen reduction program intended to reduce the level of pathogenic microorganisms in meat and poultry products (52). As part of this program, microbiological testing of carcasses, ground meat, and poultry for the presence of *Salmonella*

spp. would be required. The results of daily *Salmonella* testing, presence or absence, are evaluated over a specific time period using a statistical procedure known as the "moving window sum" to determine if the process is in control. The presence of *Salmonella* sp. is compared with a target frequency by examining a group of consecutive days (the window) of a specific duration in relation to a prespecified acceptable limit. The microbiological targets are summarized in Table 4.9.

If the number of positive samples is less than or equal to the acceptable limit, the process is in control. As each new sampling day is added, the window moves 1 day. Changing the size of the window, the acceptable limit, or the number of samples to be taken each day can alter the

Table 4.6 Summary of cases and sampling plans for *Listeria monocytogenes*[a]

| Intended consumer | Health hazard | Conditions in which food is expected to be handled and consumed after sampling in the usual course of events | | |
		Reduce degree of hazard	Cause no change in hazard	May increase hazard
Normal individuals	Moderate, direct, potentially extensive spread	Case 10 $n = 5$ $\leq 100/g$	Case 11 $n = 10$ $\leq 100/g$	Case 12 $n = 20$ $< 100/g$
Highly susceptible individuals	Severe, direct	Case 13 $n = 15$ $< 1/375$ g	Case 14 $n = 30$ $< 1/750$ g	Case 15 $n = 60$ $< 1/1,500$ g

[a] Reprinted with permission from reference 25.

Table 4.7 Examples of various food products for which advisory microbiological criteria have been established

Product category	Test parameters	Case	Plan class	n	c	Limit per g m	M	Reference
Roast beef	Salmonella sp.	12	2	20	0	0		26
Pâté	Salmonella sp.	12	2	20	0	0		
Raw chicken	APC	1	3	5	3	5×10^5	10^7	26
Cooked poultry, frozen; ready to eat	S. aureus	8	3	5	1	10^3	10^4	26
Cooked poultry, frozen; to be reheated	S. aureus	8	3	5	1	10^3	10^4	26
	Salmonella sp.	10	2	5	0	0		
Chocolate/confectionery	Salmonella sp.	11	2	10^a	0	0		26
Dried milk	APC	2	3	5	2	3×10^4	3×10^5	26
	Coliforms	5	3	5	1	10	10^2	
	Salmonella sp.^{b}	10	2	5	0	0		
	(normal routine)	11	2	10	0	0		
		12	2	20	0	0		
	Salmonella sp.^{b}	10	2	15	0	0		
	(high-risk	11	2	30	0	0		
	populations)	12	2	60	0	0		
Fresh cheese^c	S. aureus			5	2	10^2	10^3	13
	Coliforms			5	2	10^2	10^3	
	Yeasts and molds							
Soft cheese^c	S. aureus			5	2	10^2	10^3	13
Coliforms				5	2	10^2	10^3	
Pasteurized liquid, frozen and dried egg products	APC	2	3	5	2	5×10^4	10^6	26
	Coliforms	5	3	5	2	10^3	10^3	
	Salmonella sp.^{a}	10	2	5	0	0		
	(normal routine)	11	2	10	0	0		
		12	2	20	0	0		
	Salmonella sp.^{a}	10	2	15	0	0		
	(high-risk	11	2	30	0	0		
	populations)	12	2	60	0	0		
Fresh and frozen fish; to be cooked before eating	APC	1	3	5	3	5×10^3	10^7	26
	E. coli		4		3	11	500	
	Salmonella sp.^{d}	10	2	5	0	0		
	V. parahaemolyticus^d	7	3	5	2	10^2	10^3	
	S. aureus^a	7	3	5	2	10^3	10^4	
Coconut	Salmonella	1	3	5	3	5×10^5	10^7	26
	(growth not expected)	11	2	10	0	0		
	(growth expected)	12	2	20	0	0		

^{a} The 25-g analytical unit may be composited.
^{b} The case is to be chosen based on whether the hazard is expected to be reduced, unchanged, or increased.
^{c} Requirements are only for fresh and soft cheese made from pasteurized milk.
^{d} For fish known to derive from inshore or inland waters of doubtful bacteriological quality, or where fish are to be eaten raw, additional tests may be desirable.

stringency and sensitivity of the evaluation. The current acceptable limits are based on the criterion that there is an 80% probability that the plant is actually exceeding the target value if it exceeds the acceptable limit. As in any statistical approach, one needs to develop a program that has (i) a low probability of exceeding the limit when the producer is meeting the target and (ii) a high probability of exceeding the limit when the producer is not meeting the target. The 80% level was deemed to be a reasonable balance between the two, thus providing reasonable decision criteria. This verification procedure gives meat and poultry producers an opportunity to

Table 4.8 Examples of various food products for which mandatory microbiological criteria have been established

Product category	Test parameters	Comments	Reference
United States			
Dairy products			
Raw milk	Aerobic bacteria	Recommendations of U.S. Public Health Service (USPHS)	54
Grade A pasteurized	Aerobic bacteria Coliforms	USPHS	
Grade A pasteurized (cultured)	Aerobic bacteria Coliforms	USPHS	54
Dry milk (whole)	Standard plate count Coliforms	USPHS	54
Dry milk (nonfat)	Standard plate count Coliforms	Standards of Agriculture Marketing Service (USDA)	47
Frozen desserts	Standard plate count Coliforms	USPHS	54
Starch and sugars	Total thermophilic spore count Flat-sour spores Thermophilic anaerobic spores Sulfide spoilage spores	National Canners Assoc. (NFPA)	40
Breaded shrimp	Aerobic plate counts *E. coli* *S. aureus*	FDA *Compliance Policy Guides*	22
International			
Caseins and caseinates	Total bacterial count Thermophilic organisms Coliforms	Europe	41
Natural mineral waters	Aerobic mesophilic count Coliforms *E. coli* Fecal streptococci Sporulating sulfite-reducing anaerobes *Pseudomonas aeruginosa*, parasites, and pathogenic organisms	Codex	41
Hot meals served by airlines	*E. coli* *S. aureus* *B. cereus* *C. perfringens* *Salmonella* sp.	Europe	5
Tomato juice	Mold count	Canada	41
Fish protein	Total plate count *E. coli*	Canada	41
Gelatin	Total plate count Coliform *Salmonella* sp.	Canada	41

measure their process performance in relation to baseline data and national targets for pathogen reduction. Establishments whose production process exceeds the national targets are required to reevaluate their processing controls and, with USDA/FSIS oversight, initiate corrective actions (40).

CURRENT STATUS

ICMSF recommends that a series of steps be taken to manage microbiological hazards for foods intended for international trade (27). These steps include (i) conducting a risk assessment and an assessment of risk management options, (ii) establishing a food safety

Table 4.9 Moving sum rules for various commodities[a]

Commodity	Target (% positive Salmonella)	Window size (days)	Acceptable limit[b]
Steers/heifers	1	82	1
Cows/bulls	1	82	1
Raw ground beef	4	38	2
Fresh sausage	12	19	3
Turkeys	18	15	3
Hogs	18	17	4
Broilers	25	16	5

[a] Reprinted with permission from reference 51.
[b] There is approximately an 80% probability of meeting the acceptable limit when the process % positive equals the target.

objective (FSO), and (iii) confirming that the FSO is achievable by application of GMPs and HACCP. This stepwise approach is based on the following Codex Alimentarius documents: *Principles and Guidelines for the Conduct of Microbiological Risk Assessment* (15), *Risk Management and Food Safety* (20), *General Principles of Food Hygiene* (16), and *Hazard Analysis and Critical Control Point (HACCP) System and Guidelines for Its Application* (14).

An FSO is a statement of the frequency or maximum concentration of a microbiological hazard in a food considered acceptable for consumer protection. Representative member governments who serve as risk managers working within the Codex framework develop FSOs for foods in international commerce. Examples of an FSO might include the following: staphylococcal enterotoxin levels in cheese must not exceed 1 μg/100 g, or the aflatoxin concentration in peanuts should not exceed 15 μg/kg (55). FSOs are broader in scope than microbiological criteria and are intended to communicate the level of hazard acceptable. FSOs do not specify the control measures that should be taken to ensure that the food falls within the acceptable level. Control measures are actions and activities that can be used to prevent or eliminate a food safety hazard or reduce it to an acceptable level. The food industry is responsible for applying GMPs and establishing appropriate control measures and critical control points in their processes and HACCP plans (24, 55). Industry and regulatory authorities may use performance criteria, process criteria, and end product criteria associated with the food to meet the FSO. Performance criterion is the required outcome of one or more control measures at a step or combination of steps that will ensure the safety of the food. The control parameters (e.g., time, temperature, pH, water activity) at a step or a combination of steps applied to achieve a

performance criterion are the process criteria. For foods in international commerce, quantitative FSOs provide an objective basis to identify what an importing country is willing to accept in relation to the microbiological safety of its food supply and to assess, or demonstrate, the equivalence of the microbiological outcome of different control measures (30). The use of FSOs should facilitate the harmonization of international trade whereby practices between countries differ, but both practices provide safe food products.

SUMMARY

Microbiological criteria are most effectively applied as part of quality assurance programs in which HACCP and other prerequisite programs are in place. Here, criteria provide a means of determining the effectiveness of those control measures used to eliminate, reduce, or control the presence, survival, and growth of microorganisms. The establishment and application of microbiological criteria for foods will continue to evolve as new pathogens emerge and as new food vehicles are implicated as a result of changing industrial ecology of food production and consumption. Further development of quantitative risk assessment techniques will facilitate the establishment of microbiological criteria for foods in international trade and guidelines for national standards and policies.

References

1. Acuff, G. R. 1993. Microbiological criteria, p. A6.01–A6.07. *In Proceedings of the World Congress on Meat and Poultry Inspection*, October 10–14, 1993. Food Safety Inspection Service, U.S. Department of Agriculture, Washington, D.C.

2. Ahmed, F. E. (ed.). 1991. *Seafood Safety*. National Academy Press, Washington, D.C.

3. Amerine, M. A., R. M. Pangborn, and E. B. Roessler. 1965. *Principles of Sensory Evaluation of Food*. Academic Press, Inc., New York, N.Y.

4. Anonymous. 1979. Shrimp Decomposition Workshop. National Shrimp Breaders and Processors Association, National Fisheries Institute, and U.S. Food and Drug Administration, Tampa, Fla.

5. Association of European Airlines. 1996. *Hygiene Guidelines, Routine Microbiological Standards for Aircraft Ready Food*. 9-13. AEA, Brussels, Belgium.

6. Bean, N. H., P. M. Griffin, M. D., J. S. Goulding, and C. B. Ivey. 1990. Foodborne disease outbreaks, 5-year summary, 1983–1987. *J. Food Prot.* 53:711–728.

7. Beuchat, L. R. 1996. Pathogenic microorganisms associated with fresh produce. *J. Food Prot.* 59:204–216.

8. Bryan, F. L. 1988. Risks associated with vehicles of foodborne pathogens and toxins. *J. Food Prot.* 51:498–508.

9. Buchanan, R. L. 1995. The role of microbiological criteria and risk assessment in HACCP. *Food Microbiol.* **12**: 421–424.

10. Buchanan, R. L. 1991. Microbiological criteria for cooked, ready-to-eat shrimp and crabmeat. *Food Technol.* **1991**:157–160.

11. Buttiaux, R., and D. A. A. Mossel. 1961. The significance of various organism of faecal origin in foods and drinking water. *J. Appl. Bacteriol.* **24**:353–364.

12. Codex Alimentarius Commission, 14th Session. 1981. *Report of the 17th Session of the Codex Committee on Food Hygiene.* Alinorm. 81/13. Food and Agriculture Organization, Rome, Italy.

13. Codex Alimentarius Commission, 20th Session. 1993. *Report of the 20th Session of the Codex Commission on Food Hygiene.* Alinorm 93/13A. Food and Agriculture Organization, Rome, Italy.

14. Codex Alimentarius Commission. 1997. Joint FAO/WHO Food Standards Programme, Codex Committee on Food Hygiene, Supplement to Volume 1B-1997, *Hazard Analysis and Critical Control Point (HACCP) System and Guidelines for Its Application.* Annex to CAC/RCP 1-1969, Rev. 3.

15. Codex Alimentarius Commission. 1997. Joint FAO/WHO Food Standards Programme, Codex Committee on Food Hygiene. *Proposed Draft Principles and Guidelines for the Conduct of Microbiological Risk Assessment.* CX/FH 97/4.

16. Codex Alimentarius Commission. 1997. Joint FAO/WHO Food Standards Programme, Codex Committee on Food Hygiene, Supplement to Volume 1B-1997, *Recommended International Code of Practices, General Principles of Food Hygiene.* CAC/RCP 1-1969, Rev. 3.

17. Codex Alimentarius Commission. 1997. Joint FAO/WHO Food Standards Programme, Codex Committee on Food Hygiene, Supplement to Volume 1B-1997, *Principles for the Establishment and Application of Microbiological Criteria for Foods.* CAC/GL 21-997.

18. Doyle, M. P. 1991. Should regulatory agencies reconsider the policy of zero-tolerance of *Listeria monocytogenes* in all ready-to-eat foods? *Food Safety Notebook* **2**:98.

19. Eisenberg, W. V., and S. M. Cichowicz. 1977. Machinery mold-indicator organism in food. *Food Technol.* **31**(2):52–56.

20. Food and Agriculture Organization, United Nations/World Health Organization (FAO/WHO). 1997. *Risk Management and Food Safety, Report of the Joint FAO/WHO Consultation, Rome, Italy.* FAO Food and Nutrition Paper 65. FAO/WHO, Rome, Italy.

21. Food and Drug Administration. 1978. *The Food Defect Action Levels.* HFF-342. Food and Drug Administration, Washington, D.C.

22. Food and Drug Administration. 1989. Raw breaded shrimp—microbiological criteria for evaluating compliance with current good manufacturing practice regulations. *In Compliance Policy Guides.* 7108.25. Food and Drug Administration, Washington, D.C.

23. Hackney, C. R., and M. D. Pierson (ed.). 1994. *Environmental Indicators and Shellfish Safety.* Chapman & Hall, New York, N.Y.

24. Hathaway, S. 1999. Management of food safety in international trade. *Food Control* **10**:247–253.

25. International Commission on Microbiological Specifications of Foods (ICMSF). 1993. *Choice of Sampling Plan and Criteria for Listeria monocytogenes.* Papaendal, The Netherlands.

26. International Commission on Microbiological Specifications of Foods (ICMSF). 1986. *Microorganisms in Foods 2. Sampling for Microbiological Analysis: Principles and Applications,* 2nd ed. University of Toronto Press, Toronto, Ontario, Canada.

27. International Commission on Microbiological Specifications of Foods (ICMSF). 1997. Establishment of microbiological safety criteria for foods in international trade. *World Health Statist. Q.* **50**:119–123.

28. International Commission on Microbiological Specifications of Foods (ICMSF). 1998. *Microorganisms in Foods 6. Microbial Ecology of Food Commodities.* Blackie Academic and Professional, London, England.

29. Jay, J. M. 1992. *Modern Food Microbiology,* 4th ed. Chapman & Hall, New York, N.Y.

30. Jouve, J.-L. 1999. Establishment of food safety objectives. *Food Control* **10**:303–305.

31. Kilsby, D. 1982. Sampling schemes and limits, p. 387–421. *In* M. H. Brown (ed.), *Meat Microbiology.* Applied Science Publishers, London, England.

32. Kilsby, D., L. J. Aspinall, and A. C. Baird-Parker. 1979. A system for setting numerical microbiological specifications for foods. *J. Appl. Bacteriol.* **46**:591–599.

33. Lammerding, A. M. 1997. An overview of microbial food safety risk assessment. *J. Food Prot.* **60**:1420–1425.

34. Larmond, E. 1977. *Laboratory Methods for Sensory Evaluation of Food.* Pub. No. 1937. Research Branch, Canada Department of Agriculture.

35. Lupien, J. R., and M. F. Kenny. 1998. Tolerance limits and methodology: effect on international trade. *J. Food Prot.* **61**:1571–1578.

36. Malcolm, S. 1984. A note on the use of the non-central distribution in setting numerical specifications for foods. *J. Appl. Bacteriol.* **57**:175–177.

37. Marshall, R. T. (ed.). 1992. *Standard Methods for the Examination of Dairy Products,* 16th ed. American Public Health Association, Washington, D.C.

38. Mundt, J. O. 1970. Lactic acid bacteria associated with raw plant food materials. *J. Milk Food Technol.* **33**:550–553.

39. National Advisory Committee on Microbiological Criteria for Foods. 1993. Generic HACCP for raw beef. *Food Microbiol.* **10**:449–488.

40. National Canners Association. 1968. *Laboratory Manual for Food Canners and Processors,* vol. 1. AVI Publishing, Westport, Conn.

41. National Research Council. 1985. *An Evaluation of the Role of Microbiological Criteria for Foods and Food Ingredients.* National Academy Press, Washington, D.C.

42. **National Soft Drink Association.** 1975. *Quality Specifications and Test Procedures for "Bottler's Granulated and Liquid Sugar."* National Soft Drink Association, Washington, D.C.

43. **Pierson, M. D.** 1996. Critical limits, significance and determination of critical limits, p. 72–78. *In Proceedings of the 2nd Australian HACCP Conference.*

44. **Pierson, M. D., and D. A. Corlett, Jr. (ed.).** 1992. *HACCP: Principles and Applications.* Van Nostrand Reinhold, New York, N.Y.

45. **Randell, A. W., and A. J. Whitehead.** 1997. Codex Alimentarius: food quality and safety standards for international trade. *Rev. Sci. Tech. Off. Int. Epiz.* **16**:313–321.

46. **Silliker, J. H.** 1963. Total counts as indexes of food quality, p. 102–112. *In* L. W. Slanetz, C. O. Chichester, A. R. Gaufin, and Z. J. Ordal (ed.), *Microbiological Quality of Foods*, Academic Press, Inc., New York, N.Y.

47. **Silliker, J. H., and D. A. Gabis.** 1976. ICMSF method studies. VII. Indicator tests as substitutes for direct testing of dried foods and feeds for *Salmonella. Can. J. Microbiol.* **22**:971–974.

48. **Stannard, C.** 1997. Development and use of microbiological criteria for foods. *Food Sci. Technol. Today* **11**:137–177.

49. **Tauxe, R., H. Kruse, C. Hedberg, M. Potter, J. Madden,** and K. Wachsmuth. 1997. Microbial hazards and emerging issues associated with produce: a preliminary report to the National Advisory Committee on Microbiological Criteria for Foods. *J. Food Prot.* **60**:1400–1408.

50. **U.S. Department of Agriculture.** 1975. General specifications for approved dairy plants and standards for grades of dairy products. *Fed. Regist.* **40**:47910–47940.

51. **U.S. Department of Agriculture.** 1995. *Moving Sum Procedures for Microbial Testing in Meat and Poultry Establishment.* Food Safety Inspection Service, Science and Technology Program, Washington, D.C.

52. **U.S. Department of Agriculture.** 1995. Pathogen reduction; hazard analysis critical control point (HACCP) systems; proposed rule. *Fed. Regist.* **60**:6774–6889.

53. **U.S. Department of Health, Education and Welfare.** 1965. *National Shellfish Sanitation Program, Manual Operations*, part 1. *Sanitation of Shellfish Growing Areas.* U.S. Government Printing Office, Washington, D.C.

54. **U.S. Public Health Service/Food and Drug Administration.** 1978. *Grade A Pasteurized Milk Ordinance. 1978 Recommendations.* PHS/FDA publ. no. 229. U.S. Government Printing Office, Washington, D.C.

55. **van Schothorst, M.** 1998. Principles for the establishment of microbiological food safety objectives and related control measures. *Food Control* **9**:379–384.

Microbial Spoilage
of Foods

II

Food Microbiology: Fundamentals and Frontiers, 2nd Ed.
Edited by M. P. Doyle et al.
© 2001 ASM Press, Washington, D.C.

Timothy C. Jackson
Douglas L. Marshall
Gary R. Acuff
James S. Dickson

Meat, Poultry, and Seafood

5

Meat, poultry, and seafood (muscle foods) are described as spoiled if sensory changes make them unacceptable to the consumer. Factors associated with spoilage may include color defects or changes in texture, the development of off flavors, off odors, slime, or any other characteristic making the food undesirable for consumption. While enzymatic activity within muscle tissues contributes to changes during storage, organoleptically detectable spoilage is generally a result of decomposition and the formation of metabolites resulting from the growth of microorganisms.

The onset of spoilage may vary owing to differences in subjective judgments by the consumer, being influenced by cultural and economic factors, as well as the sensory acuity of the individual and the intensity of the defect. Despite variations in expectations, most consumers would agree that gross discoloration, strong off odors, and the development of slime would constitute spoilage (53). The types of spoilage defects produced in muscle foods will vary with the type of microflora, muscle type, product composition, and storage environment.

ECOLOGY OF THE SPOILAGE MICROFLORA OF MUSCLE FOODS

Origin of Microflora in Red Meats

Grau (61) stated that meat animals might be regarded as a source of edible tissue sandwiched between two regions that are heavily contaminated with microorganisms. While there has been some debate as to the sterility of muscle tissue, the relative ease with which sterile muscle tissue can be obtained (42, 52) would suggest that, if present, populations of bacteria in muscle tissues of healthy live animals are extremely low (66, 67). High numbers of bacteria are present on the hide, hair, and hooves of red meat animals, as well as in the gastrointestinal tract (104). Microorganisms on the hide include bacteria such as *Staphylococcus*, *Micrococcus*, and *Pseudomonas* species, and fungi such as yeasts and molds, which are normally associated with skin microflora, as well as species contributed by fecal material and soil (66). The population and composition of this microflora are influenced by environmental conditions.

Timothy C. Jackson, Microbiology, Nestle USA Quality Assurance Laboratory, Dublin, OH 43017. Douglas L. Marshall, Department of Food Science and Technology, Box 110 Herzer Building, Mississippi State University, Mississippi State, MS 39762. Gary R. Acuff, Department of Animal Science, 2471 TAMU, Texas A&M University, College Station, TX 77843-2471. James S. Dickson, Department of Microbiology, Iowa State University, Ames, IA 50011-3211.

Wet or muddy hides may contain high populations of bacteria indigenous to soil, and contamination of the hide with fecal material may increase the proportion of microorganisms of fecal origin (66).

It is generally agreed that the majority of bacteria on a dressed red meat carcass originate from the hide (61, 96). Initially, the tissue surface beneath the hide is essentially free of bacteria; however, once exposed, this tissue may be inoculated with bacteria from processing activities and the environment. During hide removal, bacteria are carried from the hide onto the underlying tissue with the initial incision. Further transfer of bacteria can occur from aerosols and dust generated from the hide during removal, from contact with workers' hands, or from contact of the hide or fleece with the exposed tissue surface (61).

Unlike cattle and sheep, the skin of hogs is usually not removed but is scalded and left on the carcass (61). While scalding reduces the numbers of microorganisms on the skin, recontamination can occur during dehairing owing to the presence of debris in dehairing machines (61, 72). The singeing process, used to burn hair remaining on the carcass after the dehairing process, kills a portion of microorganisms that are naturally present; however, some areas of the carcass surface may receive inadequate heat treatment (61). Microorganisms that may be present in deeper layers of the surface tissue are also protected from the heat applied in the singeing process.

Microorganisms may also be introduced to the carcass surface during the evisceration process. Contamination can occur if the intestinal tract is pierced or if fecal material is introduced from the rectum during the removal of abdominal contents, and handling can result in cross-contamination to other carcasses. Careful evisceration will reduce the potential for contamination (61). In addition to the hide and viscera, bacteria may originate from the processing environment such as floors, walls, contact surfaces, knives, and workers' hands. Rapid chilling of carcasses at low temperatures, low relative humidities, and high air velocities may result in a reduction of bacterial populations, whereas milder chilling conditions may allow the growth of psychrotrophs, increasing their population compared with mesophiles. Mesophiles may grow if carcasses are chilled at temperatures $\geq 15°C$ (66).

During the fabrication of meat into subprimal and retail cuts, bacteria present on the tissue surface are transferred by knives and workers' hands onto newly exposed meat surfaces. This is especially true in comminuted meats, where considerable new surface areas are created, and microorganisms are distributed throughout the product from the grinding equipment and the use of trimmings. Microorganisms in processed meats originate not only from the meat itself, but also from ingredients such as spices, salt, and dried milk powder. Of concern in the processing of injected meats, such as hams, is the microbiological quality of the brine solutions used.

Origin of Microflora in Poultry

As with other meat animals, the internal tissues of healthy poultry are essentially free from bacteria. The skin, feathers, and feet of the bird harbor microorganisms resident to the skin, as well as from litter and feces. Although present on the skin, psychrotrophic bacteria consisting primarily of *Acinetobacter* and *Moraxella* spp. are primarily associated with the feathers (61). Contamination and cross-contamination with fecal material may occur during transportation of birds from growing houses to slaughter facilities. Crowding and poor weather conditions during transport may stress the birds, leading to more frequent excretion of fecal material and cecal contents (91). Frequently, birds are withheld from food for several hours before slaughter to minimize intestinal contents and reduce the extent of fecal contamination during slaughter (91, 131).

Additional cross-contamination may occur as birds are hung and bled; the flapping of wings may generate dust and aerosols transferred onto nearby birds or carcasses (15). To facilitate the removal of feathers, carcasses are scalded, usually by immersion in a continuous flow water tank at 60 to 63°C. During this operation, bacteria from the carcasses are washed into the scald water. Many bacteria are removed from the carcasses, and the elevated temperature of the scald water kills some. Psychrotrophs are readily inactivated in the scald water (61). The defeathering process may spread microorganisms between carcasses or from the defeathering equipment and contribute to an increase in the numbers of psychrotrophs and aerobic mesophiles on the carcasses (61, 125). As with other meat animals, the evisceration process provides an opportunity for cross-contamination from knives, equipment, and workers' hands (15, 67). Spray washing of carcasses after picking and evisceration, but before chilling, will remove organic material and some of the bacteria introduced during evisceration (15, 67). Bacteria firmly attached to the carcass surface will remain (61, 91).

Freshly eviscerated carcasses are chilled rapidly to limit the multiplication of spoilage microorganisms and restrict the growth of pathogens. Slush ice chilling, continuous immersion chilling, spray chilling, air chilling, and carbon dioxide chilling systems have been utilized or proposed for this purpose (15, 67). The characteristics of each may influence the microbial population.

Origin of Microflora in Finfish

The population and composition of microflora on finfish are influenced by the environment from which they are taken, the season, conditions of harvesting, handling, and processing. Water temperature has a significant influence on the initial number and types of bacteria on the surface of fish. Higher numbers of bacteria are generally present on fish from warm subtropical or tropical waters compared with fish from colder waters (114). Fish taken from temperate waters harbor predominantly psychrotrophic bacteria, while mesophilic bacteria predominate on fish taken from tropical areas (67, 81). In addition, populations of bacteria may be influenced by water quality, with higher numbers occurring on the surfaces of fish taken from polluted waters. The activity of fish will influence populations in the intestine, with higher numbers present in feeding fish compared with nonfeeding fish (80). Bacteria from the genera *Acinetobacter*, *Aeromonas*, *Cytophaga*, *Flavobacterium*, *Moraxella*, *Pseudomonas*, *Shewanella* (formerly *Alteromonas*), and *Vibrio* dominate on fish and shellfish taken from temperate waters, while *Bacillus*, coryneforms, and *Micrococcus* species frequently predominate on fish taken from subtropical and tropical waters (81, 114). The microflora composition of fish from freshwater environments is also influenced by temperature and will vary from that in marine environments owing to the influence of bacteria present in the surrounding terrestrial environment (81).

The initial microflora on fish are influenced by the method of harvesting. Trawled fish generally have larger microbial populations than those that are line caught. In trawling, the dragging of fish and debris along the ocean bottom stirs up mud that contaminates the fish. In addition, the compaction of fish in trawling nets may cause expression of intestinal contents, with subsequent contamination of the fish surface (114). Handling and storage of fish aboard the fishing vessel will also affect bacterial populations. Fish are often stored in ice or chilled seawater or frozen while awaiting transportation to the processing plant (67, 81). The microbiological quality of chilled seawater and ice is a concern, as is the variation in temperature of the fish (114). A delay in chilling fish also enhances the possibility of rapid microbial growth, which for some *Scrombridae* and *Scomberesocidae* fish (tuna, mackerel, and skipjack) can lead to generation of toxic levels of histamine by *Morganella* (*Proteus*) *morganii* and related gram-negative bacteria (*Klebsiella pneumoniae* and *Hafnia alvei*) that produce histidine decarboxylase (119). *Photobacterium phosphorum* may produce histamine during low-temperature storage of these fish (119).

Nets and rough handling may result in penetration or bruising of the fish, which provides an avenue for penetration of bacteria into the muscle tissue. Large catches require more time to stow and, while awaiting storage, may be subject to physical abuse and high temperatures on deck. Liston (81) noted that, unlike the control over time of death possible in meat and poultry processing, methods of harvesting and shipboard working conditions allow little control over those variables in fish and shellfish. Additionally, contamination may occur from equipment and handling during processing.

Origin of Microflora in Shellfish

Unlike other crustacean shellfish (lobster, crab, or crayfish) that are kept alive until heat processed, shrimp die soon after harvesting. Decomposition begins soon after death and involves bacteria on the shrimp surface that originate from the marine environment or from contamination during handling and washing (45). Removal of heads before storage reduces the overall bacterial population (45); however, in the process, bacteria may be transferred to the tail meat (92). Crabs and crayfish are cooked before meat is removed by hand picking, but bacteria are reintroduced from the environment or from raw crabs during picking and processing. Molluscan shellfish (oysters, clams, scallops, and mussels) are sessile and filter feeders and, as such, their microflora depend greatly on the quality of water in which they reside (70), the quality of wash water, and other factors.

Bacterial Attachment

Because the edible muscle tissue of most healthy animals is sterile prior to processing, contaminating microorganisms are usually found on the surfaces of whole muscle. Therefore, spoilage of meat, poultry, and seafood generally occurs as a result of the growth of bacteria that have colonized muscle surfaces. The first stage in colonization and growth involves the attachment of microbial cells to the surface (13, 36, 47). An exception is some shellfish consumed with their intestinal tracts intact, resulting in the presence of contaminating microorganisms encased within the edible muscle tissue.

Bacterial attachment to muscle surfaces involves two stages (47). The first is a loose, reversible sorption that may be related to van der Waals forces or other physicochemical factors (85). One of the factors that influence attachment at this point is the population of bacteria in the water film (24, 47). The second stage consists of an irreversible attachment to surfaces involving the production of an extracellular polysaccharide layer known as a glycocalyx (26).

In addition to cell density, factors such as surface characteristics, growth phase, temperature, and motility may also influence bacterial attachment to muscle surfaces (16, 24, 34, 44, 79, 98). Bacteria already present on surfaces may influence the ability of other bacteria to attach to surfaces (88). However, Farber and Idziak (44) concluded that minimal competition occurs between meat spoilage bacteria during attachment to beef *longissimus dorsi* muscle. Chung et al. (24) observed an absence of significant competition in the attachment of spoilage and pathogenic bacteria to fat and lean beef surfaces. Differences in the rate of attachment of certain bacterial strains may be a contributing factor to the composition of the initial microbial flora. For example, *Pseudomonas* sp. has been reported to attach more rapidly to meat surfaces than several other types of spoilage bacteria (16, 48).

MICROBIAL PROGRESSION DURING STORAGE

The initial microflora of muscle foods are highly variable, arising from the resident microorganisms in and on the live animal; environmental sources such as vegetation, water, and soil; ingredients used in meat products; workers' hands; and contact surfaces in processing facilities. Despite this variability and differences between muscle tissues of different species, the characteristics of microbial spoilage are remarkably similar. A large portion of marketed perishable meat, poultry, and seafood products are stored at refrigeration temperatures to prolong their shelf life. Refrigeration will restrict the growth of mesophiles, generally a major component of the initial microflora, and allow psychrotrophic microorganisms to grow and eventually dominate the microflora. Fish and shellfish harvested from cold waters can have psychrophilic bacteria dominant at harvest (86). As microbial growth occurs during storage, the composition of the microflora is altered so that it is dominated by a few, or often a single, microbial species, usually of the genera *Pseudomonas*, *Lactobacillus*, *Moraxella*, or *Acinetobacter*, or *Brochothrix thermosphacta* (54). Although these bacteria often compose only a small proportion of the initial microflora, the types of bacteria that ultimately predominate during storage are reflective of these genera as well as characteristics of the muscle tissue and skin and characteristics of the storage environment.

Gill (54) noted that the final composition of the spoilage microflora may be influenced by the proportion of specific spoilage microorganisms in the initial population. High initial numbers of a slow-growing species may enable them to successfully compete with lower num-

bers of a species with a faster growth rate. If the number of spoilage microorganisms in the initial population is high, a slower growth rate may not be an important factor, since less growth may be necessary before spoilage occurs (54).

Pseudomonas species are able to compete successfully on aerobically stored refrigerated muscle foods for several reasons (75). *Pseudomonas* spp. are characterized by a competitive growth rate, even at refrigeration temperatures (53). In addition, pseudomonads are able to grow within the usual pH range of muscle foods (5.5 to 7.0), while many bacteria, e.g., *Moraxella* and *Acinetobacter* species, are less capable of competing under refrigeration temperatures at the lower pH in this range. Muscle foods of higher pH may allow for bacteria such as *Moraxella* and *Acinetobacter* spp. to compose a higher proportion of the microflora (53, 58). Because pseudomonads are highly oxidative, they are able to utilize low-molecular-weight nitrogen compounds as a source of energy. This is a clear competitive advantage for pseudomonads, since meats contain relatively low levels of simple sugars, and more complex energy sources, such as protein and fat, do not serve as significant substrates for growth until later in spoilage when high bacterial populations are attained.

Dominance of certain types of aerobic spoilage bacteria during spoilage is likely a result of their ability to utilize meat constituents under certain storage conditions rather than due to direct interactions between the competing species themselves. Gill and Newton (55) concluded that interaction only occurs when high population densities of *Pseudomonas* spp. are able to restrict the growth of competitors, perhaps by sequestering available oxygen and/or iron.

The growth of the aerobic spoilage microflora is suppressed during storage under vacuum and modified atmospheres (Table 5.1). Under these conditions, lactic acid bacteria and *B. thermosphacta* predominate. Lactic acid bacteria are favored by their growth rate, their fermentative metabolism, and their ability to grow at the pH range of meat (40). Some strains produce substances (bacteriocins) that inhibit the growth of competitors (93, 108). At higher pH, *B. thermosphacta* may grow and contribute to spoilage; however, it cannot grow anaerobically when the pH is below 5.8 (18, 60). The frequent occurrence of high populations of *B. thermosphacta* in vacuum-packaged pork and lamb has been attributed to a higher fat content in muscle tissue and more frequent occurrence of high-pH meat (40, 113). A higher pH may also allow *Shewanella putrefaciens*, which does not grow below pH 6.0, to contribute to the spoilage microflora (94).

Table 5.1 Percentage distribution of microflora of beef knuckles vacuum packaged and stored for 21 days at 0 to 2°C in packages with oxygen transmission rates ranging from 1 to 400 cm^3/m^2/24 h[a]

Microbial types	Percentage distribution[b] of microflora of beef knuckles packaged in films with oxygen transmission rates (cm^3/m^2/24 h)					
	1 cm^3	10 cm^3	12 cm^3	13 cm^3	30 cm^3	400 cm^{3c}
Micrococcus					2.6	
Lactobacillus coryneformis				6.1		
Lactobacillus plantarum	11.9		5.6	15.1	14.0	
Lactobacillus cellobiosus	45.7	57.0	65.7	5.5	6.6	3.9
All *Lactobacillus* spp.	**57.6**	**57.0**	**71.3**	**26.7**	**20.6**	**3.9**
Leuconostoc mesenteroides	1.1	38.3	15.0	39.8	51.3	12.4
Leuconostoc paramesenteroides	35.9		12.5	28.5	6.7	
All *Leuconostoc* spp.	**37.0**	**38.3**	**27.5**	**68.3**	**58.0**	**12.4**
Streptococcus spp.		2.3				
Brochothrix thermosphacta			1.0	0.1	4.6	3.8
Coryneform bacteria	1.1			1.5		
Staphylococcus spp.				1.8		
Moraxella-Acinetobacter spp.		0.3	0.2	0.1		
Flavobacterium spp.	2.2					
Pseudomonas spp.		1.6		1.5	14.2	79.9
Erwinia herbicola	1.1					
Aeromonas spp.	1.0	0.3				
All gram-negative rods	**4.3**	**2.2**	**0.2**	**1.6**	**14.2**	**79.9**

[a] From Savell et al. (111), with permission of the International Association for Food Protection, Des Moines, Iowa.
[b] Data points are averages of three knuckles.
[c] Data points are for microflora at 14 days; storage of knuckles packaged and stored in this film was not extended beyond 14 days.

An unusual spoilage of vacuum-packaged refrigerated fresh and cooked beef has been reported that is due to *Clostridium laramie* (71), which has the ability to grow at 0°C or below and can sporulate and germinate at 2°C. Occurring in normal pH product during storage at 2°C or below, spoilage is characterized by an initial color of pinkish-red, changing to green and production of large quantities of hydrogen sulfide and purge with considerable proteolysis of the meat. Vacuum or modified atmosphere packaging of refrigerated raw seafood is generally not acceptable owing to the potential outgrowth and toxin development by nonproteolytic *Clostridium botulinum* type E prior to spoilage, particularly during mild temperature abuse (>4°C) (106).

The water activity (a$_w$) of some types of processed meats is lowered by dehydration or the addition of solutes such as salt or sugar. As a$_w$ is lowered, the growth of some spoilage microorganisms is restricted, and characteristic spoilage microflora are altered. Below a$_w$ 0.98, growth of gram-negative spoilage bacteria is restricted (21). Lactic acid bacteria may grow at a$_w$ as low as 0.93 to 0.94, and micrococci may grow at even lower a$_w$ (127). When low a$_w$ restricts the growth of bacteria, growth of fungi may occur. Molds and yeasts may be involved in the spoilage of products in the a$_w$ range of 0.85 to 0.93, and below a$_w$ 0.85, xerophilic molds or osmophilic yeasts may be involved (21). Microbial growth does not occur on products with a$_w$ below 0.60 (21).

MUSCLE TISSUE AS A GROWTH MEDIUM

Fresh muscle tissue is a highly favorable environment for microbial growth and is subject to rapid spoilage unless modified or stored in an environment designed to retard microbial activity and reproduction (129). The usable water content of fresh muscle tissue is high, and readily available glycogen, peptides, and amino acids, as well as metal ions and soluble phosphorous, contribute to the suitability of muscle tissue as a substrate to support microbial growth. Detailed reviews of the composition and spoilage processes of postmortem muscle tissues have been published (52, 53, 66, 67, 70, 75, 77, 99).

Because the onset of spoilage of meats, poultry, and seafood is a subjective judgment (83), there has been some disagreement as to the number of bacteria present at the point spoilage is detected. It is generally agreed that spoilage defects in meat become evident when the number of spoilage bacteria at the surface reaches 10^7 CFU/cm^2 (66, 67). During aerobic spoilage, off odors are first detected when populations reach 10^7 CFU/cm^2.

When numbers reach 10^8 CFU/cm^2, the muscle tissue surface will begin to feel tacky, representing the first stage in slime formation (65). Slime formation is attributable to the growth of bacteria and synthesis of polysaccharide that gradually form a confluent, sticky layer on the surface of the tissue (75). Since spoilage characteristics do not become evident until amino acids are degraded, the concentration of glucose present in the tissue is a primary factor governing the time necessary for onset of aerobic spoilage (52, 75).

Composition and Spoilage of Red Meats

The a_w of red meat lean muscle tissue is 0.99, with a corresponding water content of 74 to 80%. The protein content may vary from 15 to 22% on a wet weight basis. The lipid content of intact red meats varies from 2.5 to 37%; carbohydrate composition ranges from 0 to 1.2% (75). While the bulk composition of muscle tissue does not change as a result of muscle rigor, significant changes in the concentration of some low-molecular-weight compounds do occur (58). The cessation of muscle cell respiration results in an end to ATP synthesis. Glycolysis leads to the accumulation of lactic acid, and, as a result, the pH of the muscle tissue decreases. The final pH and residual glycogen content of the tissue are influenced by the initial glycogen content in the muscle. In tissue with a high initial glycogen store, the pH may decrease to 5.5 before enzymatic activity associated with glycolysis ends owing to an inability to maintain ATP concentration. Where low amounts of glycogen are initially present, there is a direct correlation between glycogen content and final pH down to 5.8; however, tissue with a lower ultimate pH always contains some residual glycogen when glycolytic activity ceases, since ATP concentration is the limiting factor. In addition to glycogen, glycolytic intermediates such as glucose-6-phosphate and glucose are reduced to low levels following rigor (58).

The decrease in muscle pH and accumulation of various metabolites following rigor facilitates denaturation of some proteins. The release of proteolytic enzymes, such as cathepsins, from lysosomes results in a small amount of protein breakdown (58, 77). Soluble low-molecular-weight compounds constitute 1.2 to 3.5% of muscle tissue.

Spoilage characteristics of meat products will be influenced by the type of microflora involved. The composition of such microflora is related to intrinsic characteristics of the product, as well as extrinsic factors, such as temperature and the composition of the atmosphere in the storage environment. It is well established that the shelf life of perishable meat and other muscle tissues is extended by storage at refrigeration temperatures, and, consequently, much of the research

on meat spoilage has evaluated products stored at these temperatures. Storage temperature will influence the type and rate of growth of microorganisms that develop and, consequently, their patterns of substrate utilization and production of metabolites.

The spoilage of meats stored at ambient temperature results from the growth of mesophiles, predominantly *Clostridium perfringens* and members of the *Enterobacteriaceae* (65, 70). Spoilage deep within muscle tissues, known as "sours" or "bone taint," has been attributed to a slow cooling of carcasses, resulting in the growth of anaerobic mesophiles said to be already present in muscle tissues (17, 99). The natural presence of bacteria within meat tissues has been debated. Some have considered that low populations of bacteria are universally present in muscle tissue freshly removed from the carcass and that this is the reason bone taint develops. Ingram and Dainty (65) suggested that at ambient temperature spoilage occurs as a result of the growth of bacteria within muscle tissue that is so rapid that it precedes spoilage at the meat surface. Because these bacteria are seldom psychrotrophic, spoilage of meats stored at refrigeration temperatures is caused by bacteria on the meat surface. Others have argued that populations of bacteria isolated from tissues involved in bone taint are often too low to be directly responsible for such spoilage, and other processes may be involved (112). Gill (52) asserted that the spoilage of meats at ambient temperatures is a surface phenomenon, since the ease with which sterile muscle tissue can be obtained would indicate that the interior of intact meat is normally sterile. While possible, spoilage of meat by bacteria present in deep muscle tissues is rare under hygienic practices. A condition similar to bone taint in pork has been attributed to the introduction of spoilage bacteria into tissues in the curing brine, since the condition is most often observed in cured meats (99).

Storage at reduced temperatures will restrict the growth of mesophiles, allowing psychrotrophs to grow and dominate the spoilage microflora. If the surface of whole carcasses or fresh meat cuts becomes dry, bacterial growth may be restricted, and fungal spoilage may occur. The growth of molds in the genera *Thamnidium*, *Mucor*, and *Rhizopus* may result in the production of a whiskery, airy, or cottony gray-to-black growth on beef owing to the presence of mycelia. Other fungal genera may produce discolored areas on superficial layers of connective tissue or fat layers covering the muscle tissue (7). Black spot has been attributed to the growth of *Cladosporidium*, white spot to the growth of *Sporotrichum* and *Chrysosporium*, and green patches to *Penicillium* (82). Molds will not grow on beef held at temperatures below $-5°C$ (82).

Spoilage of high-moisture me... conditions of h... re... sult of ... tions, ... of *Pse*... and *Ae*...

Most ... preferen... glucose is u... from the bu... the supply f... diffusion fro... amino acids ar... by the spoilage ... ammonia, hydro... and other compo... flavors, and colors ... .es detectable to human ... *cter* and *Moraxella* spp. a... ...tle to spoilage defects, since, ...es, they apparently lack the abi... ...e off-odor compounds from the br... ...acids (54).

Composition and Sp... ...ltry Muscle

The nutritive compositio... ...muscle is similar to that of red meats, an... ..., the mechanism of microbial spoilage is als... .. The composition of raw poultry muscle tissue ... with age, sex, anatomy, and species (15). Fat is no... istributed throughout the muscle tissue, as in red meat, but is present in the abdominal cavity and beneath the skin. Spoilage of poultry meat has been associated with the growth of *Pseudomonas*, *S. putrefaciens*, *Acinetobacter*, and *Moraxella* spp. (10, 89, 90). Poultry skin and muscle provide excellent growth substrates for spoilage microorganisms. Spoilage is generally restricted to the outer surfaces of the skin and cuts and has been characterized by off odors and sliminess, as well as various types of discolorations. Skin may provide a barrier to the introduction of spoilage microorganisms to underlying muscle tissue.

The pH of poultry muscle varies with muscle type and age of the bird. Breast muscle typically has a pH of 5.7 to 5.9 (15, 67), and its spoilage patterns have been equated to those in fresh red meats. Under aerobic storage, *Pseudomonas* spp. make up the dominant spoilage microflora (10, 89, 94). The spoilage microflora of leg muscle, typically having a pH of 6.4 to 6.7 (15, 65), have been compared to those of dark, firm, dry (DFD) meat (94). Pseudomonads are also the predominant spoilage microorganisms in leg muscle; however, the high pH also facilitates the growth of *Aeromonas* spp. and *S. putrefaciens* to populations where they can contribute to spoilage (10, 90).

Unlike differences in spoilage patterns between DFD ... normal red meat, Clark (25) reported that there was ... ifference in the rate of spoilage of breast and leg mus... ...his observation led Newton and Gill (94) to con... ...at, despite a high pH, chicken leg muscle must ... ...ufficient available glucose for bacterial growth. ... ...v carcasses packaged in oxygen-impermeable ... ...age may be caused by *Shewanella* spp., *B. ther*... ...*ta*, and atypical lactobacilli. The potential for ... ...oduction of sulfide compounds, such as hydrogen ... ...de, dimethyl sulfide, and methyl mercaptan, makes ... *putrefaciens* an important component of the spoilage ...icroflora (49, 94).

New York-dressed poultry is stored uneviscerated before cooking. Under refrigeration, such carcasses have a longer shelf life than those that have been eviscerated. If the carcass skin is not broken and remains dry, the growth of psychrotrophic spoilage microorganisms on the surface is restricted. Instead, spoilage results from the production of hydrogen sulfide by bacteria in the intestines, which diffuses into the muscle tissue. A reaction occurs between the hydrogen sulfide and blood and muscle pigments when exposed to air to form sulphmyoglobin, a green pigment (91).

Composition and Spoilage of Finfish

The composition of fish muscle is highly variable between species and may fluctuate widely, depending upon size, season, fishing grounds, and diet, in the case of aquacultured species (114). The average composition of nonfatty fish, such as cod, has been characterized as 80% water, 18% protein, <1% lipid, and 1% carbohydrate (75). In contrast, the fat and water content of fatty fish, such as herring, varies widely. Lipid content may range from 1 to 30%, with water content varying so that fat and water compose approximately 80% of the muscle tissue (75). According to Shewan (114), non-protein-soluble components constitute about 1.5% of fish muscle; their composition and concentration vary with species and within species may vary with size, season, and fishing ground. Such components consist of sugars, minerals, vitamins, and nonprotein nitrogen compounds, such as free amino acids, ammonia, trimethylamine oxide, creatine, taurine, anserine, uric acid, betaine, carnosine, and histamine (14, 70, 114). Elasmobranchs (sharks, rays) contain about twice the percentage of soluble components as other fish.

As with other muscle foods, the spoilage microflora of fresh ice-stored fish consist largely of *Pseudomonas* species. *S. putrefaciens* may also contribute to the spoilage of seafood (19, 63, 67, 81, 94), and *Acinetobacter* and *Moraxella* spp. may compose a portion of the spoilage population (2, 63). Shewan (115) outlined the development of spoilage characteristics of fresh fish, such as cod, stored in ice. During the first phase (0 to 6 days), no marked spoilage occurs. The second phase (7 to 10 days) is marked by a lack of odor in the muscle tissue. During the third phase (11 to 14 days), some sourness or slightly sweet to fruity odors become evident, while during the fourth phase (>14 days) sulfur odors, such as hydrogen sulfide, are apparent. Fecal or potent ammonia odors may also develop. The low a_w of dried and salted fish will suppress the growth of bacteria; therefore, any spoilage that occurs with these products will likely be the result of mold growth (70).

The early stages of spoilage involve utilization of non-protein nitrogen, resulting in the formation and accumulation of fatty acids, ammonia, and volatile amines (81, 115). An important example is trimethylamine, which is produced when trimethylamine oxide in the muscle tissue is reduced by bacterial activity. Trimethylamine oxide has widespread occurrence in marine animals but is not commonly found in freshwater fish (14). The compound can be converted by microbial activity into trimethylamine, a major contributor to the fishy odor characteristic of some seafood spoilage (14). As proteolysis proceeds, spoilage occurs. Hydrogen sulfide and other sulfur compounds, such as mercaptans and dimethyl sulfide, produced by *S. putrefaciens* and some pseudomonads may contribute to spoilage (75, 114). While autolytic activities of enzymes present in seafood muscle tissue may contribute to spoilage, such contributions are difficult to estimate, since the activities of these enzymes and those of spoilage bacteria are not easily distinguishable (63, 114).

The internal muscle tissue of a healthy, live fish is generally considered to be sterile. Bacteria are present on the outer slime layer of the skin, gill surfaces, and, in feeding fish, the intestines (87, 114). During the spoilage of intact fish, bacterial activity is greater in the gill region than elsewhere on the carcass (114). Disagreement exists regarding the role of intestinal bacteria of uneviscerated fish in spoilage of muscle tissue. Jay (70) stated that bacteria from the intestinal tract of uneviscerated fish would penetrate the intestinal walls and enter the tissues surrounding the intestinal cavity. However, Liston (81) concluded that there is little evidence that adjacent tissues or blood vessels are invaded by intestinal bacteria.

Composition and Spoilage of Shellfish

Crustacean and molluscan shellfish generally contain larger amounts of free amino acids than do finfish (45, 117). Trimethylamine oxide is present in crustacean shellfish but, with the exception of cephalopods, scallops, and cockles, is absent in molluscan tissue (117).

Crustaceans possess potent cathepticlike enzymes, which rapidly break down proteins, leading to tissue softening and development of volatile off odors (27, 45, 70). These protease enzymes are primarily found in the hepatopancreas organ of crabs, shrimp, lobsters, freshwater prawns, and crayfish. Removal of the "head" after harvest can eliminate this organ and extend shelf life. Keeping these shellfish alive after harvesting is the primary way to prevent muscle liquefaction. Microbial spoilage of crustaceans occurs in a similar manner to fish flesh; however, the higher quantity of free amino acids and other soluble nitrogenous compounds facilitates rapid bacterial spoilage, accompanied by a production of large quantities of volatile base nitrogen (45, 70). Some crustacean meats (primarily shrimp) suffer from a visual defect known as black spot melanosis, which is due to polyphenol oxidase activity and not due to microbial action (132).

Molluscan shellfish contain a lower total nitrogen concentration in their flesh than do fish or crustacean shellfish. Additionally, while crustacean shellfish contain approximately 0.5% carbohydrate, molluscan shellfish contain considerably larger quantities of carbohydrate, e.g., 3.4% in clam meat and scallops and 5.6% in oysters, mostly in the form of glycogen. As a result, the spoilage pattern of molluscan shellfish differs from that of other seafood and is generally fermentative (70), the pH of tissues declining as spoilage progresses, yielding a predominance of lactobacilli and streptococci (75).

FACTORS INFLUENCING SPOILAGE

Proteolytic and Lipolytic Activity

Although *Pseudomonas* and other aerobic spoilage bacteria are able to produce proteolytic enzymes, the production of such enzymes is delayed until the late logarithmic phase of growth. There is general agreement that proteolysis occurs only at populations greater than 10^8 CFU/cm^2, when spoilage is well advanced and bacteria are approaching their maximum cell density (30, 52).

Oxidative rancidity of fat occurs when unsaturated fatty acids react with oxygen from the storage environment. Stable compounds such as aldehydes, ketones, and short-chain fatty acids are produced, resulting in the eventual development of rancid flavors and odors (52). Autoxidation, independent of microbial activity, occurs in muscle foods stored under aerobic environments, the rate being influenced by the proportion of unsaturated fatty acids in the fat (49). Autoxidation of fat is of particular importance in the deterioration of fatty fish and pork, which contain highly unsaturated lipids (75). The phospholipid component of muscle tissue membranes is also rich in unsaturated fatty acids that are susceptible to oxidation. Many spoilage microorganisms are able to produce lipases, which catalyze the hydrolysis of triacylglycerol into glycerol and free fatty acids that may, along with their oxidation products, contribute to rancidity. The production of lipases will be limited or inhibited by the presence of carbohydrates, lipids, and proteins in the medium (3, 4, 75). With a restriction of lipase production while carbohydrate substrates in muscle tissue are being utilized, it is unlikely that microbial lipolytic activity would occur until glucose at the muscle surface is depleted. At this point, amino acids would also be degraded, and the resulting spoilage characteristics would likely mask the effects of rancidity (49).

Spoilage of Adipose Tissue

Adipose tissue consists predominantly of insoluble fat, which cannot be effectively utilized as a substrate for microbial growth until it is broken down and emulsified. As previously noted, lipolytic enzymes are not produced until carbohydrates are exhausted. Spoilage of adipose tissue, therefore, is not dependent upon the ability of the microflora to produce lipases but is a process similar to the spoilage of muscle tissue (100) in which glucose and glycolytic intermediates are first utilized, followed by degradation of amino acids, resulting in spoilage. However, the amount of soluble components in adipose tissue is lower than that in muscle tissue. Adipose tissue lacks significant levels of glycolytic intermediates, and soluble nutrients from the underlying tissue are not readily replenished by diffusion (52, 100). While the spoilage process and rate of growth of spoilage bacteria are similar on adipose and muscle tissues, the low level of carbohydrates in adipose tissue means that spoilage odors will be detected when lower numbers of bacteria are present; most, if not all, available glucose is depleted when populations exceed 10^6 CFU/cm^2. In addition to containing limited amounts of carbohydrates, adipose tissue has a lower lactic acid content compared with muscle tissue and, therefore, the surface pH is higher, approaching 7.0. As a result, the spoilage of adipose tissue has been compared to that of DFD meat (100), and spoilage bacteria such as *S. putrefaciens* may potentially grow. The growth rate of some psychrotrophic microorganisms, e.g., *H. alvei, Serratia liquefaciens*, and *Lactobacillus plantarum*, has been reported to be higher on fat than on lean beef and pork tissues, but there is little difference in growth of others (128). In practice, spoilage of fat before lean tissue is unlikely, given the presence of muscle tissue fluids in vacuum-packaged

cuts and the restriction of bacterial growth due to the drying of carcass surfaces (100).

Spoilage under Anaerobic Conditions

The spoilage microflora of muscle foods are dominated by lactic acid bacteria when oxygen is excluded from the storage environment. If the pH of the muscle tissue is high or residual amounts of oxygen are present, other microorganisms, such as *B. thermosphacta* and *S. putrefaciens*, may make substantial contributions to product spoilage. The growth rate of bacteria under anaerobic conditions is considerably reduced when compared with growth under aerobic conditions. In addition, the maximum cell density achieved under anaerobic conditions (about 10^8 CFU/cm^2) is considerably less than that achieved under aerobic conditions ($>10^9$ CFU/cm^2) (52). Gill (51) reported that maximum populations of spoilage microorganisms under anaerobic conditions are determined by the rate of diffusion of fermentable substrates, such as glucose and arginine, from underlying tissues. It is not until the maximum density of microflora is reached that obvious spoilage slowly develops (53, 95).

The sour, acid, cheesy odors or cheesy and dairy flavors that develop in muscle tissue under anaerobic conditions can be attributed, at least in part, to the accumulation of short-chain fatty acids and amines (121). The presence of *B. thermosphacta* in significant numbers will cause more rapid spoilage than lactic acid bacteria. As with lactic acid bacteria, the spoilage characteristics primarily result from the production of organic acids (29, 53).

DFD Meats

Animals subjected to excessive stress or exercise before slaughter can have depleted levels of muscle glycogen. This results in the production of lower amounts of lactic acid and a higher ultimate pH. Muscles with a pH >6.0 will appear darker than normal red meat. This is due to a higher respiration rate that reduces the depth of oxygen penetration and, therefore, reduces the level of visible oxymyoglobin (62). The resulting condition is referred to as DFD muscle, and, although it occurs most often in beef, it can also occur in the muscle of pigs and other meat animals. In addition to a higher ultimate pH resulting from lower lactic acid content, such tissue is also deficient in glucose and glycolytic intermediates (94).

Spoilage of DFD meat occurs more quickly than spoilage of meat with normal pH. Rapid spoilage is a function of the absence of glucose in the tissues. With glucose unavailable, there is little delay in degradation of amino acids by pseudomonads. Spoilage will occur at lower bacterial cell densities than in normal meat

($>10^6$ CFU/cm^2) when a sufficient quantity of metabolites is produced to result in noticeable defects (94).

DFD meat stored vacuum packaged or in modified atmospheres will also spoil rapidly, typically resulting in the development of green discoloration (56). The high pH of the meat and the absence of glucose and glucose-6-phosphate have been reported to allow *Enterobacter liquefaciens* and *S. putrefaciens* to compete successfully with the normally dominant lactic acid bacteria to compose a significant portion of the spoilage microflora (56, 103). The green discoloration observed in DFD meat results from the production of hydrogen sulfide from cysteine or glutathione by *S. putrefaciens* (56). Hydrogen sulfide reacts with myoglobin in the muscle tissue to form sulphmyoglobin, a green pigment (97). *E. liquefaciens* does not contribute to greening even though it produces small quantities of hydrogen sulfide; however, its presence is significant, since low numbers produce spoilage odors on DFD meat (56, 94, 103).

PSE Meats

Pale, soft, exudative (PSE) muscle tissue is a condition that occurs in pork and turkey, and to a lesser extent in beef, in which accelerated postmortem glycolysis decreases the muscle pH to its ultimate pH while the muscle temperature is still high (52, 62). The condition, occurring in 5 to 20% of pig carcasses, is characterized by the development of a pale color, soft texture, and exudation of fluids from the muscle (52, 62). The occurrence of PSE is directly related to porcine stress syndrome, in which animals may die as a result of mild stress, or to malignant hyperthermia, in which death may be caused by exposure to certain anesthetics (62). There is debate over whether PSE meats spoil more slowly than meat with normal pH. Rey et al. (107) reported that the rate of bacterial growth was higher for DFD beef and slower for PSE pork, suggesting that pH affects the growth of spoilage microorganisms under various conditions. Gill (52) concluded that, even though an ultimate pH of 5.1 or below has been reported for PSE pork, there is little difference in the ultimate pH and chemical composition of PSE and normal meat. Since the concentration of low-molecular-weight soluble compounds is likely similar in PSE and normal meats, the development and characteristics of spoilage would likely be similar (52).

Comminuted Products

The limited shelf life of comminuted muscle foods has been attributed to both a higher initial number of bacteria due to use of a poorer-quality product for grinding and contamination during processing, as well as to the effects of the comminution of the muscle tissue.

Comminution ruptures tissue cells, releasing fluids and nutrients that provide a ready source of nutrients for bacteria. Additionally, bacteria that have been restricted to the surface of meats are distributed throughout the mass by the process (66, 67, 122). Although initial bacterial populations are higher, the type of microflora and spoilage resembles that of noncomminuted intact tissue. On the surface of aerobically stored comminuted products, *Pseudomonas*, *Acinetobacter*, and *Moraxella* species compose the dominant microflora, while in the interior, owing to the limited availability of oxygen, lactic acid bacteria are dominant (66). Occasional contaminants such as *Aeromonas* spp. or members of the *Enterobacteriaceae* occur more often on comminuted products than on intact tissue.

As with other comminuted meats, fresh sausage undergoes rapid spoilage. The addition of salt and spices is not sufficient to delay spoilage, which results from the activity of many of the same microorganisms involved in spoilage of ground beef. The spoilage microflora of refrigerated pork sausage often consist of *B. thermosphacta* (122).

Cooked Products

Cooking muscle foods results in destruction of vegetative cells of resident bacteria, although endospores may survive. For perishable, cooked, uncured meats, spoilage is dependent on the microflora surviving heat processing or the contribution of contaminants after cooking that are able to grow under storage conditions. Microorganisms responsible for spoilage of these products may include psychrotrophic micrococci, streptococci, lactobacilli, and *B. thermosphacta* (66). Recontamination is a concern if handling occurs following cooking. Where recontamination is minimal, spoilage is usually caused by nonproteolytic bacteria and involves the development of a sour odor. Recontamination with proteolytic microorganisms results in amino acid breakdown and a putrid odor (66). Retorted muscle food products are subject to spoilage resulting from defects produced before they are heat processed, inadequate thermal processing that enables the survival of heat-resistant mesophilic sporeformers, slow cooling that allows for proliferation of thermophilic sporeformers, or reintroduction of microorganisms from postprocessing leakage (66).

Processed Products

As with fresh muscle foods, microbial spoilage of processed products depends upon the nature of the product and the ingredients used. Growth of microorganisms in processed products is also affected by the conditions of processing and storage. Spoilage of these products is influenced by whether or not they are cured, heat processed, or fermented, and on the a_w and pH.

Jay (70) characterized the spoilage of processed meats, such as frankfurters, bologna, sausage, and luncheon meats, as belonging to three types: slimy spoilage, souring, and greening. Slimy spoilage, which is usually confined to the outside surfaces of product casings, may develop as discrete colonies that then expand to form a uniform gray slime layer. Slime formation usually requires the presence of moisture. Microorganisms implicated in slime formation include yeasts, two genera of lactic acid bacteria, *Lactobacillus* and *Enterococcus*, and *B. thermosphacta* (70). Souring occurs, typically beneath casings, when bacteria such as lactobacilli, enterococci, and *B. thermosphacta* utilize lactose and other sugars to produce acids. While greening of fresh meats may be a result of hydrogen sulfide production by certain bacteria, greening of cured meat products may also develop in the presence of hydrogen peroxide, which may form on the surface of vacuum- or modified atmosphere-packaged meats when exposed to air. Hydrogen peroxide may then react with nitrosohemochrome to produce choleglobin, which has a greenish color (59). Greening occurs in cured meats as a result of growth of bacteria. Catalase-negative bacteria are involved, and muscle catalase may be inactivated by the cooking process or the presence of nitrite in the cured meat. As a result, and because of the low oxidation/reduction potential in the meat interior, hydrogen peroxide may accumulate. As with surface greening, reaction of the hydrogen peroxide with nitrosohemochrome may result in the production of a green pigment. *Lactobacillus viridescens* is the most common cause of this type of greening, but species of *Streptococcus* and *Leuconostoc* may also produce this defect. The presence of these bacteria often is a result of poor sanitation before or following processing (59, 70).

Detailed reviews of the microbiology of cured meats are provided by Tompkin (126), Ingram and Simonsen (66), Kraft (75), and Gardner (50). Putrefaction of cured meats usually does not occur, mainly because of reduced a_w, which inhibits spoilage by gram-negative psychrotrophic bacteria at refrigeration temperatures (66). Cured meats are spoiled by microorganisms that tolerate low a_w, such as lactobacilli or micrococci (67, 127). According to Ingram and Dainty (65), aerobic micrococci will predominate on cured meats stored aerobically, while lactobacilli will predominate in the spoilage microflora of vacuum- or modified atmosphere-packaged products. If sucrose is added to cured meats, a slimy dextran layer may form owing to the activity of *Leuconostoc* species or other bacteria, such as *L. viridescens*. *B. thermosphacta* may also be involved in the

spoilage of these products (126). While some differences may exist in the rate of spoilage of cooked and cured products, vacuum-packaged bacon has a rate of spoilage similar to that of many cooked cured products, involving many of the same microorganisms. Ingram and Dainty (65) noted that different spoilage characteristics may develop on fat and lean tissue of aerobically stored bacon; a smoked fish flavor may develop on the lean tissue, and a rancid, cheesy flavor may develop on fat. Dry-cured products owe their stability to low a_w, the presence of nitrite, and smoke added before drying. As a result, bacterial growth is suppressed, and spoilage occurs mainly as a result of yeast and mold growth if products are stored in a humid environment (126). Spoilage of dry-cured meats may occur if spoilage microorganisms are allowed to grow in the product before salt penetration restricts their growth. Tompkin (126) reported that this also is particularly true if hams are not sufficiently cold at the time they are packed in salt and cured.

If properly prepared and stored, the low a_w of dried meats renders them microbiologically stable. To restrict the growth of some species of bacteria, yeasts, and molds that can grow at low a_w, the water content of dried meat products must be reduced to about 20%. To inhibit the growth of xerotolerant molds, the water content must be further reduced to about 15% (66). Microbial spoilage of these products should not occur unless exposure to high relative humidity or other high-moisture conditions results in an uptake of moisture. Addition of antifungal agents may be valuable to retard growth of spoilers in these products (70).

CONTROL OF SPOILAGE OF MUSCLE FOODS

The growth of spoilage microorganisms on muscle foods can be controlled by modifying intrinsic characteristics of products or extrinsic characteristics of the storage environment. The shelf life of processed meats can be extended by the use of processing procedures and ingredients that either prevent the growth of spoilage microorganisms or select for a specific spoilage microflora that produce less objectionable or slower development of spoilage. For fresh meats, the extension of shelf life has become especially important as the food industry moves toward centralization of processing activities, with products distributed to more distant domestic and international markets. Intrinsic and extrinsic factors influencing growth are discussed in chapter 2, and various methods of food preservation are presented in section VII. Some specific methods used to prevent spoilage and prolong the shelf life of muscle foods are noted here.

In his review of the control and evaluation of spoilage, Dainty (28) categorized methods used to control the spoilage of muscle foods into three strategies that could be used separately or in combination. Strategies are focused on prevention of initial microbial contamination, inactivation of microorganisms that are present on the product, or the use of storage conditions to prevent or slow the growth of microorganisms present in or on the product.

Initial Populations

Higher numbers of bacteria on products before storage result in shorter product shelf life. If the initial population of spoilage microorganisms on a product is high, less time will be required before high numbers are attained and spoilage defects become evident. High populations of microorganisms on carcass surfaces may also allow for attachment of larger numbers of bacteria (24). Attached bacteria are less likely to be removed by washing or other decontamination procedures and may be more resistant to processing conditions. If microbial populations are sufficiently high, spoilage microorganisms, and also pathogens, may survive processing treatments and, if storage conditions permit, proliferate and produce defects.

Many of the factors that influence populations of microorganisms during processing of muscle foods were noted earlier in the discussion of sources of microorganisms. Effective Good Manufacturing Practices will result in lower populations on products entering storage and distribution.

Water Rinsing

Several methods have been evaluated for their effectiveness in reducing initial microbial populations on muscle foods. Rinsing with water, immersion, and spray systems have been used to remove physical and microbial contaminants from carcasses (36). While the use of both hot and cold water has been shown to reduce microbial populations, greater reductions have been demonstrated when the temperature of water is elevated (5). At lower temperatures, reductions may be primarily a result of removal of cells, while at higher temperatures inactivation of cells by heat is also involved. High-pressure water sprays will also promote the removal of bacteria; however, there is some concern that higher pressures will drive bacteria into carcass tissues (33).

Spray washing is utilized in poultry processing after defeathering and evisceration to remove organic material and reduce bacterial populations before carcasses enter immersion chillers (15, 61, 67). High-pressure washes have been demonstrated to be effective in reducing bacte-

rial populations on the surface of whole fish by removing the slime layer (87).

Antimicrobial Treatments

In addition to washing, treatments with antimicrobial compounds such as chlorine and organic acids have been used to sanitize muscle foods. Various concentrations and degrees of effectiveness of chlorine compounds have been reported. In poultry processing, further reductions have been demonstrated during spray washing by the addition of 40 to 60 μg of chlorine per ml to the spray water (110). Chlorine (5 to 20 μg/ml) may also be added to poultry chill water in slush ice systems to inhibit multiplication of psychrotrophs (9). For finfish, the effectiveness of chlorine in dips and sprays has been evaluated; however, Mayer and Ward (87) concluded that the limited reduction of spoilage microflora on the surface of finfish treated with chlorine makes it only marginally effective in increasing shelf life. Similar observations have been made on the use of chlorination to extend the shelf life of meats (54, 73, 120). Kelly et al. (73) reported an additive antimicrobial effect of chlorine and elevated water temperature ($>80°$C).

Short-chain fatty acids have been utilized in sprays to reduce microbial populations on carcasses and extend the shelf life. The use of lactic, acetic, propionic, and citric acids has been investigated, although lactic and acetic acids are most often used to reduce microbial populations on carcass surfaces. The effectiveness of organic acids is influenced by such factors as type of acid, concentration, temperature, and point of application in processing (36). The point of application of organic acids in beef carcass processing is important. Organic acids cause the greatest reductions in populations when applied to hot carcass surfaces following dehiding and evisceration (105) and are relatively ineffective when applied to individual cuts following fabrication (1, 38). Organic acid decontamination of seafoods has also been reported (8).

Although some washing and decontamination treatments have been demonstrated to be effective in reducing bacterial populations on animal carcasses, the use of such treatments cannot replace effective sanitation. The use of strict sanitary procedures, including organic acid rinses and control of handling, vacuum packaging, and storage temperature, has been shown to increase the shelf life of carcasses and subprimal cuts over those produced from animals slaughtered and processed using conventional procedures (20, 39, 118). Samuels et al. (109) suggested that the shelf life and quality of finfish products could be improved by using an effective sanitation program in addition to an initial decontamination step before processing.

Refrigeration

Temperature is the most important environmental parameter influencing the growth of microorganisms in muscle foods. As temperature is decreased below the optimum for growth of microorganisms, generation times and lag times are extended, and growth is therefore slowed (Fig. 5.1). As temperature is reduced to the minimum for growth, extension of the lag time continues to increase until multiplication ceases (75). Changes in growth rate as the temperature is reduced, as well as the minimum temperature for growth, differ for each species of microorganism (54, 75).

The temperature at which muscle foods are stored ultimately influences the type of microorganisms that will grow and eventually cause spoilage, as well as the rate at which growth will occur. Under aerobic storage conditions, the comparatively rapid growth rate of *Pseudomonas* species allows successful competition with mesophiles and other psychrotrophs at temperatures below 20°C. Likewise, under anaerobic conditions at temperatures below 20°C, the rapid growth rate of psychrotrophic lactobacilli enables successful competi-

tion with other psychrotrophic spoilage microorganisms (9, 54, 57).

Storage of foods at temperatures at which mesophiles will grow allows these microorganisms to contribute to spoilage. Higher storage temperatures may also allow the growth of psychrotrophic spoilage microorganisms that would otherwise be restricted by the pH or lactic acid concentration of muscle tissue at reduced temperatures (54, 57). Likewise, the minimum temperature for growth of psychrotrophic spoilage microorganisms may be increased by other factors influencing growth, e.g., pH, a_w, or oxidation/reduction potential, or by inhibitory agents such as food additives or increased carbon dioxide (75).

Modified Atmosphere Storage and Vacuum Packaging

The shelf life of muscle foods can be extended by storage under vacuum or modified atmospheres. Modified atmosphere packaging involves the storage of products under high-oxygen barrier film with a headspace of different gaseous composition than air. Typically, this headspace contains elevated amounts of carbon dioxide and may also contain nitrogen and oxygen in various proportions. The use of carbon monoxide, nitrous oxide, and sulfur dioxide in the gaseous atmosphere of packaged muscle foods has also been suggested as a way to control microbial growth. Much of the extended shelf life of such products results from a modification of the spoilage microflora from an aerobic psychrotrophic population consisting of bacteria such as *Pseudomonas*, *Moraxella*, and *Acinetobacter* spp. to one consisting predominantly of lactic acid bacteria and *B. thermosphacta* (Fig. 5.2).

In vacuum-packaged products, exclusion of air from the package and slow diffusion of atmospheric oxygen through the high-oxygen barrier film leaves only residual oxygen remaining. In fresh muscle foods, respiration of the tissue utilizes the remaining oxygen in the package and generates carbon dioxide (12, 54). While the aerobic or facultative component of the bacterial microflora also utilizes residual oxygen in the package, its impact on oxygen content is insignificant compared to activity in the surrounding muscle, since such organisms are initially present in relatively low numbers (54). Residual oxygen remaining within vacuum-packaged products is a function of the rate of oxygen diffusion through the packaging film as well as the rate at which oxygen is depleted by the muscle tissue (54, 111).

Carbon dioxide is commonly used in modified atmosphere environments owing to its bacteriostatic activity. The presence of carbon dioxide will result in an increase in the lag phase and generation times of

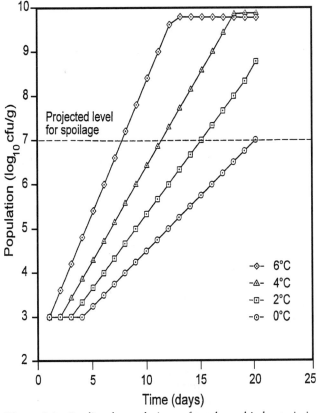

Figure 5.1 Predicted populations of total aerobic bacteria in ground beef as affected by storage temperature (35).

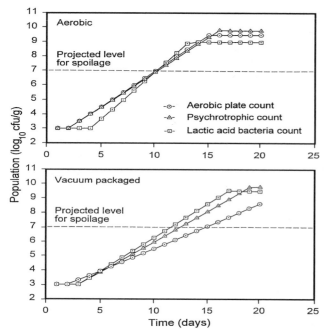

Figure 5.2 Populations of total aerobic, psychrotrophic, and lactic acid bacteria in aerobically packaged and vacuum-packaged ground pork stored at 0°C. The growth curves are based on the derived values for lag-phase duration and generation time (129).

many microorganisms (31). While the specific mechanism of bacteriostatic activity of carbon dioxide is unknown, several mechanisms have been suggested, including exclusion of oxygen from the atmosphere, thus restricting the growth of aerobes, altering cell membrane properties, altering enzyme systems, and acidifying intracellular pH (31, 37).

The inhibitory activity of carbon dioxide is influenced by many factors (43, 133). The temperature at which packaged products are stored will significantly influence the bacteriostatic effect of carbon dioxide (84), and solubility decreases as storage temperature is increased. Therefore, storage temperature should be maintained as low as possible, because elevated temperatures limit the effectiveness of carbon dioxide (31, 41, 46, 93, 101).

Oxygen is occasionally included in modified atmospheres used to package fresh red meats to maintain the oxymyoglobin responsible for the fresh red or "bloomed" color desired by the consumer (123, 134). While the inclusion of oxygen in modified atmospheres may improve the appearance of products, it may also reduce the shelf life. If significant amounts of oxygen are present, aerobic microorganisms and *B. thermosphacta* may grow, although the concentration of oxygen necessary for growth varies among microorganisms (22, 23, 68, 95, 116).

While nitrogen has no antimicrobial activity, its inclusion displaces oxygen in the package headspace, resulting in a delay in oxidative rancidity and an inhibition of the growth of aerobic microorganisms (43). As stated previously, fresh seafood can contain nonproteolytic *C. botulinum* that could outgrow and produce toxin in modified atmosphere-packaged refrigerated products (106). However, vacuum packaging of seafood is acceptable if it is to be frozen.

Cook-In Bag and Postpasteurization

Meat products can be cooked in flexible plastic packaging materials and distributed in the same packaging. Such a packaging approach avoids the reintroduction of spoilage microorganisms following cooking, and, as a result, the shelf life of the product is extended (32). A shortened shelf life may occur if cooked meats are removed from packaging and further processed or portioned for distribution (32). In some cases, surface pasteurization of muscle products following repackaging may be used to inactivate reintroduced spoilage bacteria and thereby increase shelf life. Such a process typically involves exposure of the product surface to temperatures of 82 to 96°C for periods of 30 s to 6 min (32).

Irradiation

Irradiation of meats enhances microbiological safety and quality by significantly reducing populations of pathogenic and spoilage bacteria. The effect of irradiation is dose dependent, with higher doses resulting in a greater reduction of bacterial populations. At present, most interest is in relatively low doses of radiation, i.e., less than 5 kGy.

The doses of radiation required to inactivate 90% of the vegetative cells of pathogenic bacteria vary somewhat with processing conditions (temperature, presence or absence of air) but typically range from a low of <0.2 kGy for *Aeromonas* (102) and *Campylobacter* (76) spp. to as high as 0.77 kGy for *Listeria* (64) and *Salmonella* (124) spp. Although there is some variation among species, doses required to inactivate 90% of the vegetative cells of spoilage bacteria typically fall in the range of 0.2 to 0.8 kGy. In contrast, the dose required for a 90% inactivation of *Clostridium* endospores is as high as 3.5 kGy (6).

Low-dose radiation (<5 kGy) has been demonstrated to be effective in extending the shelf life of fresh meats. Several studies have reported shelf life extensions of 14 to 21 days for irradiated meats, based on microbiological and organoleptic assays (11, 74, 78, 130). Because of significant reductions in both pathogenic and spoilage bacteria, low-dose radiation has been suggested as a pasteurization process for fresh meats.

References

1. Acuff, G. R., C. Vanderzant, J. W. Savell, D. K. Jones, D. B. Griffin, and J. G. Ehlers. 1987. Effect of acid decontamination of beef subprimal cuts on the microbiological and sensory characteristics of steaks. *Meat Sci.* **19**:217–226.

2. Adams, R., L. Farber, and P. Lerke. 1964. Bacteriology of spoilage of fish muscle. II. Incidence of spoilers during storage. *Appl. Microbiol.* **12**:277–279.

3. Alford, J. A., and L. E. Elliott. 1960. Lipolytic activity of microorganisms at low and intermediate temperatures. I. Action of *Pseudomonas fluorescens* on lard. *Food Res.* **25**:296–303.

4. Alford, J. A., J. L. Smith, and H. D. Lilly. 1971. Relationship of microbial activity to changes in lipids in foods. *J. Appl. Bacteriol.* **34**:133–146.

5. Anderson, M. E., R. T. Marshall, W. C. Stringer, and H. D. Naumann. 1979. Microbial growth on plate beef during extended storage after washing and sanitizing. *J. Food Prot.* **42**:389–392.

6. Anellis, A., E. Shattuck, M. Morin, B. Srisara, S. Qvale, D. B. Rowley, and E. W. Ross. 1977. Cryogenic gamma irradiation of prototype pork and chicken and antagonistic effect between *Clostridium botulinum* types A and B. *Appl. Microbiol.* **34**:823–831.

7. Ayres, J. C., J. O. Mundt, and W. E. Sandine. 1980. *Microbiology of Foods.* W. H. Freeman & Co., San Francisco, Calif.

8. Bal'a, M. F. A., and D. L. Marshall. 1998. Organic acid dipping of catfish fillets: effect on color, microbial load, and *Listeria monocytogenes. J. Food Prot.* **61**:1470–1474.

9. Barnes, E. M. 1976. Microbiological problems of poultry at refrigerator temperatures—a review. *J. Sci. Food Agric.* **27**:777–782.

10. Barnes, E. M., and C. S. Impey. 1968. Psychrotrophic spoilage bacteria of poultry. *J. Appl. Bacteriol.* **31**:97–107.

11. Basker, D., I. Klinger, M. Lapidot, and E. Eisenberg. 1986. Effect of chilled storage of radiation-pasteurized chicken carcasses on the eating quality of the resultant cooked meat. *J. Food Technol.* **21**:437–441.

12. Bendall, J. R., and A. A. Taylor. 1972. Consumption of oxygen by the muscles of beef animals and related species. *J. Sci. Food Agric.* **23**:707–719.

13. Benedict, R. C. 1988. Microbial attachment to meat surfaces. *Reciprocal Meat Conf. Proc.* **41**:1–6.

14. Brown, W. D. 1986. Fish muscle as food, p. 405–451. *In* P. J. Bechtel (ed.), *Muscle as Food.* Academic Press, Inc., Orlando, Fla.

15. Bryan, F. L. 1980. Poultry and poultry meat products, p. 410–469. *In* J. H. Silliker, R. P. Elliot, A. C. Baird-Parker, F. L. Bryan, J. H. B. Christian, D. S. Clark, J. C. Olsen, Jr., and T. A. Roberts (ed.), *Microbial Ecology of Foods,* vol. 2. *Food Commodities.* Academic Press, Inc., New York, N.Y.

16. Butler, J. L., J. C. Stewart, C. Vanderzant, Z. L. Carpenter, and G. C. Smith. 1979. Attachment of microorganisms to pork skin and surfaces of beef and lamb carcasses. *J. Food Prot.* **42**:401–406.

17. Callow, E. H., and M. Ingram. 1955. Bone-taint. *Food* **24**:52–55.

18. Campbell, R. J., A. F. Egan, F. H. Grau, and B. J. Shay. 1979. The growth of *Microbacterium thermosphactum* on beef. *J. Appl. Bacteriol.* **47**:505–509.

19. Chai, T., C. Chen, A. Rosen, and R. E. Levin. 1968. Detection and incidence of specific species of spoilage bacteria on fish. II. Relative incidence of *Pseudomonas putrefaciens* and fluorescent pseudomonads on haddock fillets. *Appl. Microbiol.* **16**:1738–1741.

20. Chandran, S. K., J. W. Savell, D. B. Griffin, and C. Vanderzant. 1986. Effect of slaughter-dressing, fabrication and storage conditions on the microbiological and sensory characteristics of vacuum-packaged beef steaks. *J. Food Sci.* **51**:37–53.

21. Christian, J. H. B. 1980. Reduced water activity, p. 70–111. *In* J. H. Silliker, R. P. Elliot, A. C. Baird-Parker, F. L. Bryan, J. H. B. Christian, D. S. Clark, J. C. Olsen, Jr., and T. A. Roberts (ed.), *Microbial Ecology of Foods,* vol. 1. *Factors Affecting the Life and Death of Microorganisms.* Academic Press, Inc., New York, N.Y.

22. Christopher, F. M., S. C. Seideman, Z. L. Carpenter, G. C. Smith, and C. Vanderzant. 1979. Microbiology of beef packaged in various gas atmospheres. *J. Food Prot.* **42**:240–244.

23. Christopher, F. M., C. Vanderzant, Z. L. Carpenter, and G. C. Smith. 1979. Microbiology of pork packaged in various gas atmospheres. *J. Food Prot.* **42**:323–327.

24. Chung, K.-T., J. S. Dickson, and J. D. Crouse. 1989. Attachment and proliferation of bacteria on meat. *J. Food Prot.* **52**:173–177.

25. Clark, D. S. 1970. Growth of psychrotolerant pseudomonads and achromobacteria on various chicken tissues. *Poult. Sci.* **49**:1315–1318.

26. Costerson, J. W., R. T. Irvin, and K.-J. Cheng. 1981. The bacterial glycocalyx in nature and disease. *Annu. Rev. Microbiol.* **35**:299–324.

27. Cotton, L. N., and D. L. Marshall. 1998. Rapid impediometric method to determine crustacean food freshness, p. 147–160. *In* M. H. Tunick, S. A. Palumbo, and P. M. Fratamico (ed.), *New Techniques in the Analysis of Foods.* Plenum Publishing Corp., New York, N.Y.

28. Dainty, R. H. 1971. The control and evaluation of spoilage. *J. Food Technol.* **6**:209–224.

29. Dainty, R. H., and C. M. Hibbard. 1980. Aerobic metabolism of *Brochothrix thermosphacta* growing on meat surfaces and in laboratory media. *J. Appl. Bacteriol.* **48**:387–396.

30. Dainty, R. H., B. G. Shaw, K. A. De Boer, and E. S. J. Scheps. 1975. Protein changes caused by bacterial growth on beef. *J. Appl. Bacteriol.* **39**:73–81.

31. Daniels, J. A., R. Krishnamurthi, and S. S. H. Rizvi. 1985. A review of effects of carbon dioxide on microbial growth and food quality. *J. Food Prot.* 48:532–537.

32. DeMasi, T. W., and K. R. Deily. 1990. Cooked meats packaging technology. *Reciprocal Meat Conf. Proc.* 43:117–121.

33. De Zuniga, A. G., M. E. Anderson, R. T. Marshall, and E. L. Iannotti. 1991. A model system for studying the penetration of microorganisms into meat. *J. Food Prot.* 54:256–258.

34. Dickson, J. S. 1991. Attachment of *Salmonella typhimurium* and *Listeria monocytogenes* to beef tissue: effects of inoculum level, growth temperature and bacterial culture age. *Food Microbiol.* 8:143–151.

35. Dickson, J. S. Unpublished data.

36. Dickson, J. S., and M. E. Anderson. 1992. Microbiological decontamination of food animal carcasses by washing and sanitizing systems: a review. *J. Food Prot.* 55:133–140.

37. Dixon, N. M., and D. B. Kell. 1989. The inhibition by CO_2 of the growth and metabolism of micro-organisms. *J. Appl. Bacteriol.* 67:109–136.

38. Dixon, Z. R., C. Vanderzant, G. R. Acuff, J. W. Savell, and D. K. Jones. 1987. Effect of acid treatment of beef strip loin steaks on microbiological and sensory characteristics. *Int. J. Food Microbiol.* 5:181–186.

39. Dixon, Z. R., G. R. Acuff, L. M. Lucia, C. Vanderzant, J. B. Morgan, S. G. May, and J. W. Savell. 1991. Effect of degree of sanitation from slaughter through fabrication on the microbiological and sensory characteristics of beef. *J. Food Prot.* 54:200–207.

40. Egan, A. F. 1983. Lactic acid bacteria of meat and meat products. *Antonie Leeuwenhoek* 49:327–336.

41. Enfors, S. O., and G. Molin. 1980. Effect of high concentrations of carbon dioxide on growth rate of *Pseudomonas fragi*, *Bacillus cereus*, and *Streptococcus cremoris*. *J. Appl. Bacteriol.* 48:409–416.

42. Fapohunda, A. O., K. W. McMillin, D. L. Marshall, and W. M. Waites. 1993. A simple and effective method for obtaining aseptic raw beef tissues. *J. Food Prot.* 56:543–544.

43. Farber, J. M. 1991. Microbiological aspects of modified-atmosphere packaging technology—a review. *J. Food Prot.* 54:58–70.

44. Farber, J. M., and E. S. Idziak. 1984. Attachment of psychrotrophic meat spoilage bacteria to muscle surfaces. *J. Food Prot.* 47:92–95.

45. Feiger, E. A., and A. F. Novak. 1961. Microbiology of shellfish deterioration, p. 561–611. *In* G. Borgstrom (ed.), *Fish as Food*, vol. 1. *Production, Biochemistry, and Microbiology*. Academic Press, Inc., New York, N.Y.

46. Finne, G. 1982. Modified- and controlled-atmosphere storage of muscle foods. *Food Technol.* 36:128–133.

47. Firstenberg-Eden, R. 1981. Attachment of bacteria to meat surfaces: a review. *J. Food Prot.* 44:602–607.

48. Firstenberg-Eden, R., S. Notermans, and M. Van Schothorst. 1978. Attachment of certain bacterial strains to chicken and beef meat. *J. Food Safety* 1:217–228.

49. Freeman, L. R., G. J. Silverman, P. Angelini, C. Merritt, Jr., and W. B. Esselen. 1976. Volatiles produced by microorganisms isolated from refrigerated chicken at spoilage. *Appl. Environ. Microbiol.* 32:222–231.

50. Gardner, G. A. 1982. Microbiology of processing: bacon and ham, p. 129–178. *In* M. H. Brown (ed.), *Meat Microbiology*. Applied Science Publishers, Inc., New York, N.Y.

51. Gill, C. O. 1976. Substrate limitation of bacterial growth at meat surfaces. *J. Appl. Bacteriol.* 41:401–410.

52. Gill, C. O. 1982. Microbial interaction with meats, p. 225–264. *In* M. H. Brown (ed.), *Meat Microbiology*. Applied Science Publishers, Inc., New York, N.Y.

53. Gill, C. O. 1983. Meat spoilage and evaluation of the potential storage life of fresh meat. *J. Food Prot.* 46:444–452.

54. Gill, C. O. 1986. The control of microbial spoilage in fresh meats, p. 49–88. *In* A. M. Pearson and T. R. Dutson (ed.), *Advances in Meat Research*, vol. 2. *Meat and Poultry Microbiology*. AVI Publishing Co., Inc., Westport, Conn.

55. Gill, C. O., and K. G. Newton. 1977. The development of aerobic spoilage flora on meat stored at chill temperatures. *J. Appl. Microbiol.* 43:189–195.

56. Gill, C. O., and K. G. Newton. 1979. Spoilage of vacuum-packaged dark, firm, dry meat at chill temperatures. *Appl. Environ. Microbiol.* 37:362–364.

57. Gill, C. O., and K. G. Newton. 1980. Growth of bacteria on meat at room temperatures. *J. Appl. Bacteriol.* 49:315–323.

58. Gill, C. O., and K. G. Newton. 1982. Effect of lactic acid concentration of growth on meat of gram-negative psychrotrophs from a meatworks. *Appl. Environ. Microbiol.* 43:284–288.

59. Grant, G. F., A. R. McCurdy, and A. D. Osborne. 1988. Bacterial greening in cured meats: a review. *Can. Inst. Food Sci. Technol. J.* 21:50–56.

60. Grau, F. H. 1980. Inhibition of the anaerobic growth of *Brochothrix thermosphacta* by lactic acid. *Appl. Environ. Microbiol.* 40:433–436.

61. Grau, F. H. 1986. Microbial ecology of meat and poultry, p. 1–47. *In* A. M. Pearson and T. R. Dutson (ed.), *Advances in Meat Research*, vol. 2. *Meat and Poultry Microbiology*. AVI Publishing Co., Inc., Westport, Conn.

62. Greaser, M. L. 1986. Conversion of muscle to meat, p. 37–102. *In* P. J. Bechtel (ed.), *Muscle as Food*. Academic Press, Inc., New York, N.Y.

63. Herbert, R. A., M. S. Hendrie, D. M. Gibson, and J. M. Shewan. 1971. Bacteria active in the spoilage of certain sea foods. *J. Appl. Bacteriol.* 34:41–50.

64. Huhtanen, C. N., R. K. Jenkins, and D. W. Thayer. 1989. Gamma radiation sensitivity of *Listeria monocytogenes*. *J. Food Prot.* 52:610–613.

65. Ingram, M., and R. H. Dainty. 1971. Changes caused by microbes in spoilage of meats. *J. Appl. Bacteriol.* 34:21–39.

66. **Ingram, M., and B. Simonsen.** 1980. Meats and meat products, p. 333–409. *In* J. H. Silliker, R. P. Elliot, A. C. Baird-Parker, F. L. Bryan, J. H. B. Christian, D. S. Clark, J. C. Olsen, Jr., and T. A. Roberts (ed.), *Microbial Ecology of Foods,* vol. 2. *Food Commodities.* Academic Press, Inc., New York, N.Y.

67. **International Commission on Microbiological Specification of Foods (ICMSF).** 1998. *Microorganisms in Foods 6. Microbial Ecology of Food Commodities.* Blackie Academic and Professional, London, England.

68. **Jackson, T. C., G. R. Acuff, C. Vanderzant, T. R. Sharp, and J. W. Savell.** 1992. Identification and evaluation of volatile compounds of vacuum and modified atmosphere packaged beef strip loins. *Meat Sci.* 31:175–190.

69. **Jay, J. M.** 1971. Mechanism and detection of microbial spoilage in meats at low temperatures: a status report. *J. Milk Food Technol.* 35:467–471.

70. **Jay, J. M.** 2000. *Modern Food Microbiology,* 6th ed. Aspen Publishers, Gaithersburg, Md.

71. **Kalchayanand, N., B. Ray, and R. A.** Field. 1993. Characteristics of psychrotrophic *Clostridium laramie* causing spoilage of vacuum-packaged refrigerated fresh and roasted beef. *J. Food Prot.* 56:13–17.

72. **Kampelmacher, E. H., P. A. M. Guinée, K. Hofstra, and A. van Keulen.** 1961. Studies on *Salmonella* in slaughterhouses. *Zentbl. Veterinarmed.* 8:1025–1042.

73. **Kelly, C. A., J. F. Dempster, and A. J. McLoughlin.** 1981. The effect of temperature, pressure and chlorine concentration of spray washing water on numbers of bacteria on lamb carcasses. *J. Appl. Bacteriol.* 51:415–424.

74. **Klinger, I., V. Fuchs, D. Basker, B. J. Juven, M. Lapidot, and E. Eisenberg.** 1986. Irradiation of broiler chicken meat. *Isr. J. Vet. Med.* 42:181–192.

75. **Kraft, A. A.** 1992. *Psychrotrophic Bacteria in Foods: Disease and Spoilage.* CRC Press, Inc., Boca Raton, Fla.

76. **Lambert, J. D., and R. B. Maxcy.** 1984. Effect of gamma radiation on *Campylobacter jejuni. J. Food. Sci.* 49:665–667.

77. **Lawrie, R. A.** 1985. *Meat Science,* 4th ed. Pergamon Press, Inc., New York, N.Y.

78. **Lescano, G., P. Narvaiz, E. Kairiyama, and N. Kaupert.** 1991. Effect of chicken breast irradiation on microbiological, chemical and organoleptic quality. *Lebensm.-Wiss.-Technol.* 24:130–134.

79. **Lillard, H. S.** 1986. Role of fimbriae and flagella in the attachment of *Salmonella typhimurium* to poultry skin. *J. Food Sci.* 51:54–56, 65.

80. **Liston, J.** 1956. Quantitative variations in the bacterial flora of flatfish. *J. Gen. Microbiol.* 15:305–314.

81. **Liston, J.** 1980. Fish and shellfish and their products, p. 567–605. *In* J. H. Silliker, R. P. Elliot, A. C. Baird-Parker, F. L. Bryan, J. H. B. Christian, D. S. Clark, J. C. Olsen, Jr., and T. A. Roberts (ed.), *Microbial Ecology of Foods,* vol. 2. *Food Commodities.* Academic Press, Inc., New York, N.Y.

82. **Lowry, P. D., and C. O. Gill.** 1984. Temperature and water activity minima for growth of spoilage moulds from meat. *J. Appl. Bacteriol.* 56:193–199.

83. **Marshall, D. L., and P. L. W. Lehigh.** 1993. Nobody's nose knows. *CHEMTECH* 23:38–42.

84. **Marshall, D. L., L. S. Andrews, J. H. Wells, and A. J. Farr.** 1992. Influence of modified atmosphere packaging on the competitive growth of *Listeria monocytogenes* and *Pseudomonas fluorescens* on precooked chicken. *Food Microbiol.* 9:303–309.

85. **Marshall, K. C., R. Stout, and R. Mitchell.** 1971. Mechanism of the initial events in the sorption of marine bacteria to surfaces. *J. Gen. Microbiol.* 68:337–348.

86. **Matches, J. R.** 1982. Effects of temperature on the decomposition of Pacific Coast shrimp. *J. Food Sci.* 47:1044–1047.

87. **Mayer, B. K., and D. R. Ward.** 1991. Microbiology of finfish and finfish processing, p. 3–17. *In* D. R. Ward and C. R. Hackney (ed.), *Microbiology of Marine Food Products.* Van Nostrand Reinhold, New York, N.Y.

88. **McEldowney, S., and M. Fletcher.** 1987. Adhesion of bacteria from mixed cell suspension to solid surfaces. *Arch. Microbiol.* 148:57–62.

89. **McMeekin, T. A.** 1975. Spoilage association of chicken breast muscle. *Appl. Microbiol.* 29:44–47.

90. **McMeekin, T. A.** 1977. Spoilage association of chicken leg muscle. *Appl. Microbiol.* 33:1244–1246.

91. **Mead, G. C.** 1982. Microbiology of poultry and game birds, p. 67–101. *In* M. H. Brown (ed.), *Meat Microbiology.* Applied Science Publishers, Inc., New York, N.Y.

92. **Miget, R. J.** 1991. Microbiology of crustacean processing: shrimp, crawfish, and prawns, p. 65–87. *In* D. R. Ward and C. R. Hackney (ed.), *Microbiology of Marine Products.* Van Nostrand Reinhold, New York, N.Y.

93. **Newton, K. G., and C. O. Gill.** 1978. The development of the anaerobic spoilage flora of meat stored at chill temperatures. *J. Appl. Bacteriol.* 44:91–95.

94. **Newton, K. G., and C. O. Gill.** 1980. The microbiology of DFD fresh meats: a review. *Meat Sci.* 5:223–232.

95. **Newton, K. G., J. C. L. Harrison, and K. M. Smith.** 1977. The effect of storage in various gaseous atmospheres on the microflora of lamb chops held at −1°C. *J. Appl. Bacteriol.* 43:53–59.

96. **Newton, K. G., J. C. L. Harrison, and A. M. Wauters.** 1978. Sources of psychrotrophic bacteria on meat at the abattoir. *J. Appl. Bacteriol.* 45:75–82.

97. **Nichol, D. J., M. K. Shaw, and D. A. Ledward.** 1970. Hydrogen sulfide production by bacteria and sulfmyoglobin formation in prepacked chilled beef. *Appl. Microbiol.* 19:937–939.

98. **Notermans, S., and E. H. Kampelmacher.** 1974. Attachment of some bacterial strains to the skin of broiler chickens. *Br. Poult. Sci.* 15:573–585.

99. **Nottingham, P. M.** 1982. Microbiology of carcass meats, p. 13–65. *In* M. H. Brown (ed.), *Meat Microbiology.* Applied Science Publishers, Inc., New York, N.Y.

100. **Nottingham, P. M., C. O. Gill, and K. G. Newton.** 1981. Spoilage at fat surfaces of meat, p. 183–190. *In* T. A. Roberts, G. Hobbs, J. H. B. Christian, and N. Skovgaard (ed.), *Psychrotrophic Microorganisms in Spoilage and Pathogenicity.* Academic Press, Inc., New York, N.Y.

101. Ogrydziak, D. M., and W. D. Brown. 1982. Temperature effects in modified-atmosphere storage of seafoods. *Food Technol.* **36**:86–96.

102. Palumbo, S. A., R. K. Jenkins, R. L. Buchanan, and D. W. Thayer. 1986. Determination of irradiation D-values for *Aeromonas hydrophila. J. Food Prot.* **49**:189–191.

103. Patterson, J. T., and P. A. Gibbs. 1977. Incidence and spoilage potential of isolates from vacuum-packaged meat of high pH value. *J. Appl. Bacteriol.* **43**:25–38.

104. Patterson, J. T., and P. A. Gibbs. 1978. Sources and properties of some organisms isolated in two abattoirs. *Meat Sci.* **2**:263–273.

105. Prasai, R. K., G. R. Acuff, L. M. Lucia, D. S. Hale, J. W. Savell, and J. B. Morgan. 1991. Microbiological effects of acid decontamination of beef carcasses at various locations in processing. *J. Food Prot.* **54**:868–872.

106. Reddy, N. R., H. M. Solomon, and E. J. Rhodehamel. 1999. Comparison of margin of safety between sensory spoilage and onset of *Clostridium botulinum* toxin development during storage of modified atmosphere (MA)-packaged fresh marine cod fillets with MA-packaged aquacultured fish fillets. *J. Food Safety* **19**:171–183.

107. Rey, C. R., A. A. Kraft, D. G. Topel, F. C. Parrish, Jr., and D. K. Hotchkiss. 1976. Microbiology of pale, dark and normal pork. *J. Food Sci.* **41**:111–116.

108. Roth, L. A., and D. S. Clark. 1975. Effect of lactobacilli and carbon dioxide on the growth of *Microbacterium thermosphactum* on fresh beef. *Can. J. Microbiol.* **21**:629–632.

109. Samuels, R. D., A. DeFeo, G. J. Flick, D. R. Ward, T. Rippen, J. K. Riggens, C. Coale, and C. Smith. 1984. *Demonstration of a Quality Maintenance Program for Fresh Fish Products. A Report Submitted to Mid-Atlantic Fisheries Development Foundation.* Virginia Polytechnic Institute and State University, Sea Grant, VPI-SG-84-04R.

110. Sanders, D. H., and C. D. Blackshear. 1971. Effect of chlorination in the final washer on bacterial counts of broiler chicken carcasses. *Poult. Sci.* **50**:215–219.

111. Savell, J. W., D. B. Griffin, C. W. Dill, G. R. Acuff, and C. Vanderzant. 1986. Effect of film oxygen transmission rate on lean color and microbiological characteristics of vacuum-packaged beef knuckles. *J. Food Prot.* **49**:917–919.

112. Shank, J. L., J. H. Silliker, and P. A. Goeser. 1962. The development of a nonmicrobial off-condition in fresh meat. *Appl. Microbiol.* **10**:240–246.

113. Shaw, B. G., C. D. Harding, and A. A. Taylor. 1980. The microbiology and storage stability of vacuum packed lamb. *J. Food Technol.* **15**:397–405.

114. Shewan, J. M. 1961. The microbiology of sea-water fish, p. 487–560 *In* G. Borgstrom (ed.), *Fish as Food*, vol. 1. *Production, Biochemistry, and Microbiology.* Academic Press, Inc., New York, N.Y.

115. Shewan, J. M. 1971. The microbiology of fish and fishery products—a progress report. *J. Appl Bacteriol.* **34**:299–315.

116. Silliker, J. H., R. E. Woodruff, J. R. Lugg, S. K. Wolfe, and W. D. Brown. 1977. Preservation of refrigerated meats with controlled atmospheres: treatment and post-treatment effects of carbon dioxide on pork and beef. *Meat Sci.* **1**:195–204.

117. Simidu, W. 1961. Nonprotein nitrogenous compounds, p. 353–384. *In* G. Borgstrom (ed.), *Fish as Food*, vol. 1. *Production, Biochemistry, and Microbiology.* Academic Press, Inc., New York, N.Y.

118. Smulders, F. J. M., and C. H. J. Woolthuis. 1983. Influence of two levels of hygiene on the microbiological condition of veal as a product of two slaughtering/processing sequences. *J. Food Prot.* **46**:1032–1035.

119. Stratton, J. E., and S. L. Taylor. 1991. Scombroid poisoning, p. 331–351. *In* D. R. Ward and C. R. Hackney (ed.), *Microbiology of Marine Food Products.* Van Nostrand Reinhold, New York, N.Y.

120. Stevenson, K. E., R. A. Merkel, and H. C. Lee. 1978. Effects of chilling rate, carcass fatness and chlorine spray on microbiological quality and case-life of beef. *J. Food Sci.* **43**:849–852.

121. Sutherland, J. P., P. A. Gibbs, J. T. Patterson, and J. G. Murray. 1976. Biochemical changes in vacuum packaged beef occurring during storage at 0–2°C. *J. Food Technol.* **11**:171–1180.

122. Sutherland, J. P., and A. Varnam. 1982. Fresh meat processing, p. 103–128. *In* M. H. Brown (ed.), *Meat Microbiology.* Applied Science Publishers, Inc., New York, N.Y.

123. Taylor, A. A., N. F. Down, and B. G. Shaw. 1990. A comparison of modified atmosphere and vacuum skin packing for the storage of red meats. *Int. J. Food Sci. Technol.* **25**:98–109.

124. Thayer, D. W., G. Boyd, W. S. Muller, C. A. Lipson, W. C. Hayne, and S. H. Bayer. 1990. Radiation resistance of *Salmonella. J. Ind. Microbiol.* **5**:383–390.

125. Thomas, C. J., and T. A. McMeekin. 1980. Contamination of broiler carcass skin during commercial processing procedures: an electron microscopic study. *Appl. Environ. Microbiol.* **40**:133–144.

126. Tompkin, R. B. 1986. Microbiology of ready-to-eat meat and poultry products, p. 89–121. *In* A. M. Pearson and T. R. Dutson (ed.), *Advances in Meat Research*, vol. 2. *Meat and Poultry Microbiology.* AVI Publishing Co., Inc., Westport, Conn.

127. Troller, J. A. 1979. Food spoilage by microorganisms tolerating low-a_w environments. *Food Technol.* **33**:72–75.

128. Vanderzant, C., J. W. Savell, M. O. Hanna, and V. Potluri. 1986. A comparison of growth of individual meat bacteria on the lean and fatty tissue of beef, pork and lamb. *J. Food Sci.* **51**:5–8, 11.

129. van Laak, R. L. J. M. 1994. Spoilage and preservation of muscle foods, p. 378–405. *In* D. M. Kinsman, A. W. Kotula, and B. C. Breidenstien (ed.), *Muscle Foods: Meat, Poultry and Seafood Technology.* Chapman & Hall, New York, N.Y.

130. **Venugopal, R. J., and J. S. Dickson.** 1999. Growth rates of mesophilic bacteria, aerobic psychrotrophic bacteria, and lactic acid bacteria in low-dose-irradiated pork. *J. Food Prot.* **62**:1297–1302.

131. **Wabek, C. J.** 1972. Feed and water withdrawal time relationship to processing yield and potential fecal contamination of broilers. *Poult. Sci.* **51**:1119–1121.

132. **Wiese-Lehigh, P. L., and D. L. Marshall.** 1993. Determination of seafood freshness using impedance technology, p. 248–261. *In* A. M. Spanier, H. Okai, and M. Tamura (ed.), *Food Flavor and Safety: Molecular Analysis and Design.* ACS Symposium Series No. 528. American Chemical Society, Washington, D.C.

133. **Wodzinski, R. J., and W. C. Frazier.** 1961. Moisture requirements of bacteria. IV. Influence of temperature and increased partial pressure of carbon dioxide on requirements of three species of bacteria. *J. Bacteriol.* **81**:401–418.

134. **Young, L. L., R. D. Reviere, and A. B. Cole.** 1988. Fresh red meats: a place to apply modified atmospheres. *Food Technol.* **42**:65–69.

Food Microbiology: Fundamentals and Frontiers, 2nd Ed.
Edited by M. P. Doyle et al.
© 2001 ASM Press, Washington, D.C.

Joseph F. Frank

Milk and Dairy Products

6

Being both highly perishable and nutritious, milk has since prehistoric times been subject to a variety of preservation treatments. Modern dairy processing utilizes pasteurization, heat sterilization, fermentation, dehydration, refrigeration, and freezing as preservation treatments. The result, when combined with component separation processes (i.e., churning, filtration, centrifugation, coagulation), is an assortment of dairy foods having vastly different tastes and textures and a complex variety of spoilage microflora. Spoilage of dairy foods is manifested as off flavors and odors and as changes in texture and appearance. Some defects of milk and cheese caused by microorganisms are listed in Tables 6.1 and 6.2. In this chapter, the discussion of spoilage is organized by types of microorganisms associated with various defects. These include gram-negative psychrotrophic microorganisms, coliform and lactic acid bacteria, spore-forming bacteria, and yeasts and molds. The major objective of this chapter is to describe the interactions of these microorganisms with dairy foods that lead to commonly encountered product defects.

MILK AND DAIRY PRODUCTS AS GROWTH MEDIA

Milk

Milk is a good growth medium for many microorganisms because of its high water content, near neutral pH, and variety of available nutrients. Milk, however, is not an ideal growth medium since, for example, the addition of yeast extract or protein hydrolysates often increases growth rates. Table 6.3 lists the major nutritional components of milk and their normal concentrations. These components consist of lactose, fat, protein, minerals, and various nonprotein nitrogenous compounds. Many microorganisms cannot utilize lactose and therefore must rely on proteolysis or lipolysis to obtain carbon and energy. In addition, freshly collected raw milk contains various growth inhibitors that decrease in effectiveness with storage.

Carbon and Nitrogen Availability

Carbon sources in milk include lactose, protein, and fat. The citrate in milk can be utilized by many microorganisms but is not present in sufficient amounts to support significant growth. A sufficient amount of glucose is present in milk to allow initiation of growth by some microorganisms, but for fermentative microorganisms to continue growth they must have the appropriate sugar transport system and hydrolytic enzymes for lactose utilization. These are described in chapter 32. Other spoilage microorganisms may oxidize lactose to lactobionic acid. The amount of lactose in milk is in great excess of that needed to support extensive microbial growth.

Although milk has a high fat content, few spoilage microorganisms utilize it as a carbon or energy source. This is because the fat is in the form of globules surrounded by a protective membrane composed of glycoproteins,

---

Joseph F. Frank, Department of Food Science and Technology, Food Science Building, University of Georgia, Athens, GA 30602-7610.

Table 6.1 Some defects of fluid milk that result from microbial growth

Defect	Associated microorganisms	Type of enzyme	Metabolic product	References
Bitter flavor	Psychrotrophic bacteria, *Bacillus* spp.	Protease, peptidase	Bitter peptides	77, 82
Rancid flavor	Psychrotrophic bacteria	Lipase	Free fatty acids	102
Fruity flavor	Psychrotrophic bacteria	Esterase	Ethyl esters	89
Coagulation	*Bacillus* spp.	Protease	Casein destabilization	77
Sour flavor	Lactic acid bacteria	Glycolytic	Lactic, acetic acids	38, 101
Malty flavor	Lactic acid bacteria	Oxidase	3-Methyl butanal	79
Ropy texture	Lactic acid bacteria	Polymerase	Exopolysaccharides	14

lipoproteins, and phospholipids. Milk fat is available for microbial metabolism only if the globule membrane is physically damaged or enzymatically degraded (2). Generally, milk will be spoiled by other mechanisms before this occurs.

There are primarily two types of proteins in milk, caseins and whey proteins. Caseins are present in the form of highly hydrated micelles and are readily susceptible to proteolysis. Whey proteins (β-lactoglobulin, α-lactalbumin, serum albumin, and immunoglobulins) remain soluble in the milk after precipitation of casein. They are less susceptible than caseins to microbial proteolysis. Milk contains nonprotein nitrogenous compounds such as urea, peptides, and amino acids that are readily available for microbial utilization (Table 6.3). These compounds are present in insufficient quantity to support the extensive growth required for spoilage.

Minerals and Micronutrients

Milk is a good source of B vitamins and mineral nutrients. The major salt cations and anions present in milk are listed in Table 6.3. Although milk contains many trace mineral nutrients, such as iron, cobalt, copper, and molybdenum, some of these, such as iron, may not be present in a readily usable form. Supplementation of milk with trace elements may be necessary to achieve maximum microbial growth rates. Milk also contains growth stimulants, of which orotic acid (a metabolic precursor for pyrimidines) is the most significant.

Natural Inhibitors

The major microbial inhibitors in raw milk are lactoferrin and the lactoperoxidase system. Natural inhibitors of lesser importance include lysozyme, specific immunoglobulins, and folate and vitamin B_{12} binding systems. Lactoferrin, a glycoprotein, acts as an antimicrobial agent by binding iron. Human milk contains over 2 mg of lactoferrin per ml, but it is of lesser importance in cow milk, which contains only 20 to 200 μg/ml (72). Psychrotrophic aerobes that commonly spoil refrigerated milk are inhibited by lactoferrin, but the presence of citrate in cow's milk limits its effectiveness, as the citrate competes with lactoferrin for binding the iron (8).

The most effective natural microbial inhibitor in cow's milk is the lactoperoxidase system. Lactoperoxidase catalyzes the oxidation of thiocyanate and simultaneous reduction of hydrogen peroxide, resulting in the accumulation of hypothiocyanite. Two mechanisms for this reaction are illustrated in Fig. 6.1. Hypothiocyanite oxidizes sulfhydryl groups of proteins, resulting in enzyme inactivation and structural damage to the microbial cytoplasmic membrane (116). Lactoperoxidase and thiocyanate are present in milk during synthesis, whereas hydrogen peroxide is formed in milk when oxygen is metabolized by lactic acid bacteria. Hydrogen peroxide is the limiting substrate of the reaction, so the effective use of this inhibitor system for preserving milk relies on adding a hydrogen peroxide-generating system or, less effectively, hydrogen peroxide to the milk (27). Lactic

Table 6.2 Some defects of cheese that result from microbial growth

Defect	Associated microorganisms	Metabolic product	Reference
Open texture, fissures	Heterofermentative lactobacilli	Carbon dioxide	62
Early gas	Coliforms, yeasts	Carbon dioxide, hydrogen	71
Late gas	*Clostridium* spp.	Carbon dioxide, hydrogen	25
Rancid	Psychrotrophic bacteria	Free fatty acids	65
Fruity	Lactic acid bacteria	Ethyl esters	9
White crystalline surface deposits	*Lactobacillus* spp.	Excessive D-lactate	91
Pink discoloration	*Lactobacillus delbrueckii* subsp. *bulgaricus*	High redox potential	99

Table 6.3 Approximate concentrations of some nutritional components of milk[a]

Component	Amount
	(g/100 g)
Water	87.3
Lactose	4.8
Fat	3.7
Casein	2.6
Whey protein	0.6
Salt cations	(mg/100 g)
Sodium	58
Potassium	140
Calcium	118
Magnesium	12
Salt anions	
Citrate	176
Chloride	104
Phosphorus	74
Nonprotein nitrogen (NPN)	(mg/liter)
Total NPN	296
Urea N	142
Peptide N	32
Amino acid N	44
Creatine N	25

[a] Adapted from Jenness (56).

acid bacteria, coliforms, and various pathogens are inhibited by this system (116).

Effect of Heat Treatments

The minimum required heat treatment for milk to be sold for fluid consumption in the United States is 72°C for 15 s, though most processors use slightly higher temperatures and longer holding times. Pasteurization affects the growth rate of the spoilage microflora by destroying inhibitor systems. More severe heat treatments affect microbial growth by increasing available nitrogen through protein hydrolysis and by the liberation of inhibitory sulfhydryl compounds. Lactoperoxidase is only partially inactivated by normal pasteurization treatments (116).

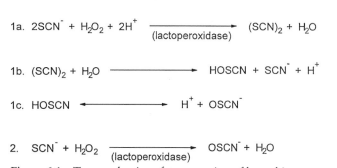

1a. $2SCN^- + H_2O_2 + 2H^+ \xrightarrow{\text{(lactoperoxidase)}} (SCN)_2 + H_2O$

1b. $(SCN)_2 + H_2O \longrightarrow HOSCN + SCN^- + H^+$

1c. $HOSCN \longleftrightarrow H^+ + OSCN^-$

2. $SCN^- + H_2O_2 \xrightarrow{\text{(lactoperoxidase)}} OSCN^- + H_2O$

Figure 6.1 Two mechanisms for generation of hypothiocyanite ($OSCN^-$) inhibitor in milk. Adapted from reference 116.

Dairy Products

Dairy products provide substantially different growth environments than fluid milk, because these products have nutrients removed or concentrated or have lower pH or water activity (a_w). The composition, pH, and a_w of selected dairy products are presented in Table 6.4. Yogurt is essentially acidified milk, and therefore provides a nutrient-rich low-pH environment. Cheeses are less acidified than yogurt, but they have added salt and less water, resulting in lower a_w. In addition, the solid nature of cheeses limits mobility of spoilage microorganisms. Liquid milk concentrates such as evaporated skim milk do not have sufficiently low a_w to inhibit spoilage and must be canned or refrigerated for preservation. Milk-derived powders have sufficiently low a_w to completely inhibit microbial growth. Butter is a water-in-oil emulsion, so microorganisms are trapped within serum droplets. If butter is salted, the mean salt content of the water droplets will be 6 to 8%, sufficient to inhibit gram-negative spoilage organisms that could grow during refrigeration. However, individual droplets will have significantly higher or lower salt content if the salt is not uniformly distributed during manufacture. This can result in the growth of psychrotrophic bacteria in droplets of low salt content. Unsalted butter is usually made from acidified cream and relies on low pH and refrigeration for preservation.

Table 6.4 Approximate composition, pH, and a_w of selected dairy products[a]

Product	Component (g/100 g)				a_w	pH
	Water	Fat	Protein	Carbohydrate		
Butter	16.0	81.0	3.6	0.06		6.3
Cheddar cheese	37.0	32.8	24.9	1.3	0.90–0.95	5.2
Swiss cheese	37.2	27.4	28.4	3.4		5.6
Nonfat dried milk	3.2	0.8	36.2	52.0	0.2	
Evaporated skim milk	79.4	0.2	7.5	11.3	0.93–0.98	
Yogurt	89	1.7	3.5	5.1		4.3

[a] Compiled from references 6, 7, 20, and 58.

PSYCHROTROPHIC SPOILAGE

Preservation of fluid milk relies on effective sanitation, timely marketing, pasteurization, and refrigeration. Raw milk is rapidly cooled after collection and is kept cold until pasteurized, after which it is kept cold until consumption. There is often sufficient time between milk collection and consumption for psychrotrophic bacteria to grow. Many of the flavor defects detected by milk consumers result from this growth. Pasteurized milk is expected to have a shelf life of 14 to 20 days, so contamination of the contents of a container with even one rapidly growing psychrotrophic microorganism can lead to spoilage. Growth of psychrotrophic bacteria in raw milk can lead to defects in products made from that milk because of the residual activity of degradative enzymes.

Psychrotrophic Bacteria in Milk

Psychrotrophic bacteria that spoil raw and pasteurized milk are primarily aerobic gram-negative rods in the family *Pseudomonadaceae*, with occasional representatives from the family *Neisseriaceae* and the genera *Flavobacterium* and *Alcaligenes*. It is typical that 65 to 70% of psychrotrophic isolates from raw milk are in the genus *Pseudomonas* (41). Although representatives of other genera, including *Bacillus*, *Micrococcus*, *Aerococcus*, and *Staphylococcus*, and the family *Enterobacteriaceae* may be present in raw milk and may be psychrotrophic, they are usually outgrown by the gram-negative obligate aerobes (especially *Pseudomonas* spp.) when milk is held at its typical 3 to 7°C storage temperature. The psychrotrophic spoilage microflora of milk are generally proteolytic, with many isolates able to produce extracellular lipases, phospholipase, and other hydrolytic enzymes but unable to utilize lactose. The bacterium most often associated with flavor defects in refrigerated milk is *Pseudomonas fluorescens* (30), with *Pseudomonas fragi*, *Pseudomonas putida*, and *Pseudomonas lundensis* (105) also commonly encountered. Jaspe et al. (55) observed that incubating raw milk for 3 days at 7°C selected for a population of *Pseudomonas* spp. that had a 10-fold higher growth rate at 7°C and a lower growth rate at 21°C than the initial *Pseudomonas* population. These strains also exhibited 1,000-fold greater proteolytic activity and 280-fold greater lipolytic activity than the initial *Pseudomonas* population. Psychrotrophic bacteria commonly found in raw milk are inactivated by pasteurization.

Sources of Psychrotrophic Bacteria in Milk

Soil, water, animals, and plant material constitute the natural habitat of psychrotrophic bacteria found in milk (21). Plant materials, such as grass and hay used for animal feed, may contain over 10^8 psychrotrophs per gram (106). Water used on the dairy farm has been found to contain psychrotrophic bacteria (107). Although farm water usually contains only low populations of psychrotrophic microorganisms, its use to clean and rinse milking equipment provides a direct means for their entry into milk. Psychrotrophic bacteria isolated from water are often very active producers of extracellular enzymes and grow rapidly at refrigeration temperatures (21). Consequently, water is an important source of milk spoilage bacteria. The teat and udder area of the cow can harbor high levels of psychrotrophic bacteria, even after washing and sanitizing (80). These psychrotrophs probably originate from soil.

Milking equipment, utensils, and storage tanks are the major source of psychrotrophic contamination of raw milk (21). Proper cleaning and sanitizing procedures can effectively reduce contamination from these sources. Milk residues on unclean equipment provide a growth niche for psychrotrophic bacteria, which enter milking machines, pipelines, and holding tanks with water rinses or milk. Milking equipment is generally fabricated from stainless steel, which is readily cleaned and sanitized; however, for some parts, rubber or other nonmetal materials must be used. These materials are difficult to sanitize, since only moderate use results in the formation of microscopic pores or cracks. Bacteria attached to these parts are difficult to inactivate by chemical sanitization (81).

Pasteurized milk products become contaminated with psychrotrophic bacteria by exposure to contaminated equipment or air. Schröder (98) determined that filling equipment was most often the source of psychrotrophs in packaged milk. Although the levels of psychrotrophic bacteria in air are generally quite low, only one viable cell per container is required to spoil the product. Product is often exposed to contaminated air during the packaging process. Pasteurized milk packaged using aseptic systems has a greatly extended shelf life compared with conventionally packaged milk.

Growth Characteristics and Defects

Generation times in milk of the most rapidly growing psychrotrophic *Pseudomonas* spp. isolated from raw milk are 8 to 12 h at 3°C and 5.5 to 10.5 h at 3 to 5°C (104). These growth rates are sufficient to cause spoilage within 5 days if the milk initially contains only one cell/ml. However, most psychrotrophic pseudomonads present in raw milk grow much more slowly, causing refrigerated milk to spoil in 10 to 20 days.

Defects of fluid milk associated with the growth of psychrotrophic bacteria are related to the production of extracellular enzymes. Sufficient enzyme to cause defects is usually present when the population of psychrotrophs reaches 10^6 to 10^7 CFU/ml (31), but this depends on the specific product; for example, the shelf life of ultra-high-temperature (UHT) milk is limited by the presence of less enzyme than is the shelf life of pasteurized milk. Bitter and putrid flavors and coagulation result from proteolysis. Rancid and fruity flavors result from lipolysis. The production of extracellular enzymes by psychrotrophic bacteria in raw milk also has implications for the quality of products produced from that milk.

Proteases

Factors Affecting Protease Production

P. fluorescens and other psychrotrophs that may be present in raw milk generally produce protease during the late exponential and stationary phases of growth (42, 95, 112), although there are reports of protease production throughout the growth phase (68). This discrepancy may be due to strain and species differences. Increases in protease activity associated with stationary and decline phases of growth may result from the release of preformed enzyme from the cells (42).

The effect of temperature on protease production does not parallel its effect on growth. The temperature for optimum production of protease by psychrotrophic *Pseudomonas* spp. is lower than the temperature for optimum rate of growth (74). Relatively high amounts of protease are produced at temperatures as low as 5°C (73). Rowe (95) and Griffiths (42) indicate that protease production by *Pseudomonas* spp. is inhibited in milk held at 2°C.

Since *Pseudomonas* spp. are obligate aerobes, it is expected that oxygen is required for protease synthesis. Myhara and Skura (84) reported optimum protease production for *P. fragi* in a medium containing 7.4 μg of dissolved oxygen per ml, slightly less than the 8.4 μg/ml contained in saturated water.

The effect of calcium and iron ions on protease production by *Pseudomonas* spp. is relevant to dairy spoilage. Ionic calcium is required for protease synthesis. McKellar and Cholette (75) reported that in the absence of ionic calcium, an inactive precursor to proteinase was produced by *P. fluorescens*. This precursor could not be activated, indicating that calcium is required to stabilize the enzyme. Iron, which may be at a growth-limiting concentration in milk, will repress protease production by *Pseudomonas* spp. when added to milk (32). Maximum

protease production will occur only if iron is growth limiting, though this may be an indirect effect of reduced cytochrome synthesis and decreased energy levels (74). The presence of pyoverdine, an iron-chelating pigment produced by *P. fluorescens*, stimulates protease production (76).

Most evidence indicates that *P. fluorescens* regulates protease production as a means to provide carbon to the cell rather than amino acids for protein synthesis (31). Protease production is induced by various low-molecular-weight protein degradation products and is subject to end product and catabolite repression. Asparagine is the most effective amino acid inducer, and citric acid is an effective inhibitor of synthesis (74). Protease production by *P. fluorescens* in raw milk is preceded by the depletion of glucose, galactose, lactate, glutamine, and glutamic acid. Supplementation of milk with these compounds delays or inhibits protease production (54).

Properties of Proteases from Psychrotrophic *Pseudomonas*

Fox et al. (35) summarized properties of proteinases produced by *P. fluorescens*. Properties most relevant to dairy product spoilage include temperature optima from 30 to 45°C with significant activity at 4°C and pH optima near neutrality or at alkaline conditions; all are metalloenzymes containing either Zn^{2+} or Ca^{2+}. Perhaps the most important characteristic of these enzymes in regard to dairy product spoilage is their extreme heat stability. Decimal reduction times at 140°C range from 50 to 200 s, sufficient to retain significant activity after UHT milk processing (60). This degree of heat stability is surprising, considering that the enzymes are produced and active at refrigeration temperatures. Another unexpected property of these enzymes is their susceptibility to autodegradation at 55 to 60°C, apparently due to an unfolding of the protein chain into a more sensitive conformation (103). This low-temperature inactivation phenomenon is highly dependent on pH (21) and varies with species biotype. Milk with high populations of psychrotrophs that has been subjected to a dual heat treatment (low temperature treatment for enzyme inactivation followed by a UHT treatment) has been reported to have a longer shelf life than conventional UHT milk (11, 46), but this effect is not consistent.

Protease-Induced Product Defects

Proteases of psychrotrophic bacteria cause product defects either at the time they are produced in the product or as a result of enzyme surviving a heat process. Most investigators have observed that these proteases preferentially hydrolyze κ-casein, although some show preference for β- or α_{s-1}-casein (22). No specific cleavage sites have been identified, since specificity depends on reaction conditions and the specific enzyme tested. Degradation of casein in milk by enzymes produced by psychrotrophs results in the liberation of bitter peptides. Bitterness is a common off flavor in pasteurized milk that has been subject to postpasteurization contamination with psychrotrophic bacteria. Continued proteolysis results in putrid off flavors associated with lower-molecular-weight degradation products such as ammonia, amines, and sulfides. Bitterness in UHT (commercially sterile) milk develops when sufficient psychrotrophic bacterial growth occurs in raw milk (estimated at 10^5 to 10^7 CFU/ml) to leave residual enzyme after heat treatment (82). Low-level protease activity in UHT milk can also result in coagulation or sediment formation. UHT milk appears to be more sensitive to protease-induced defects than raw milk, probably a result of either heat-induced changes in casein micelle structure or heat inactivation of protease inhibitors (90).

The effect of proteases of psychrotrophic origin on quality of cheese and other cultured dairy products is minimal, because the combination of low pH and low storage temperature inhibits their activity (63). In addition, proteases are removed with the whey fraction during cheese manufacture. However, growth of proteolytic bacteria in raw milk lowers cheese yield, because proteolytic products of casein degradation are lost to the whey rather than becoming part of the cheese (117).

Lipases

Psychrotrophic *P. fluorescens* isolated from milk often produces extracellular lipase in addition to protease. Other commonly found lipase-producing psychrotrophs include *P. fragi* and *Pseudomonas aeruginosa*.

Factors Affecting Lipase Production

Lipases of psychrotrophic pseudomonads, like proteases, are produced in the late log or stationary phase of growth (1, 74). As with protease, optimal synthesis of lipase generally occurs below the optimum temperature for growth. For example, Andersson (3) reported optimum production of lipase at 8°C by a *P. fluorescens* strain that exhibited optimum growth at 20°C. Milk is an excellent medium for lipase production by pseudomonads. Although *Pseudomonas* spp. can utilize inorganic nitrogen, lipase production requires an organic nitrogen source. However, some amino acids repress lipase production, especially those that serve as nitrogen but not carbon sources. This phenomenon has been related to accumulation of metabolic intermediates (33).

Ionic calcium is required for lipase activity, as activity inhibited by EDTA is reversed by addition of Ca^{2+} (1). *P. fluorescens* can produce lipase in calcium-free media, although the amount produced is less than that in the presence of calcium (75). Lipase production by *P. fluorescens*, to a greater degree than protease production, is stimulated by limiting iron availability (76). Supplementation of milk with iron delays the onset of lipase production by *P. fluorescens* in raw milk (32).

Although there is conflicting evidence, most reports indicate that lipase production is subject to catabolite repression (74, 102). Lipases are produced by *P. fluorescens* and *P. fragi* in the absence of triglycerides. The presence of triglycerides can either inhibit or stimulate production of lipase, depending on the specific strain, triglyceride concentration, and growth conditions. Surface-active agents such as polysaccharides and lecithin often stimulate the release of lipase from the cell surface (102).

Properties of Lipase from Psychrotrophic *Pseudomonas* spp.

Properties of lipases produced by *Pseudomonas* spp. have been summarized by Stead (102) and Fox et al. (35). Temperatures for optimal activity of these lipases range from 22 to 70°C with most between 30 and 40°C. Optimal pH of activity is from 7.0 to 9.0 with most having optima between 7.5 and 8.5. *P. fluorescens* lipase in milk is active at refrigeration temperatures, and significant activity remains at subfreezing temperatures and at low a_w (4).

Heat stability of lipases from psychrotrophic pseudomonads is similar to that of their proteases. Decimal reduction times at 140°C range from 48 to 437 s (60), sufficient to provide residual activity after UHT treatment. Most of these lipases are also subject to accelerated irreversible inactivation at temperatures less than 100°C. The optimal conditions for low-temperature inactivation vary considerably with strain, with optimal inactivation between 52 and 80°C being reported. Low-temperature inactivation is markedly affected by milk components, with milk salts increasing susceptibility and casein increasing stability (61). Histidine and $MgCl_2$ were required for the low-temperature inactivation of a heat-stable lipase isolated from *P. fluorescens* (19). The mechanism for low-temperature inactivation of lipase in milk is unknown, but the phenomenon may involve proteolysis of unfolding lipase molecules (60).

Lipase-Induced Product Defects

The triglycerides in raw milk are present in globules that are protected from enzymatic degradation by a membrane. Milk becomes susceptible to lipolysis if this membrane is disrupted by excessive shear force (from pumping, agitation, freezing, etc.). Raw milk contains a mammalian lipase (milk lipase) which will rapidly act on the fat if the globule membrane is disrupted. Most cases of rancidity in raw and pasteurized milk are a result of this process, rather than from the growth of lipase-producing microorganisms. Phospholipase C and protease produced by psychrotrophic bacteria can degrade the fat globule membrane, resulting in the enhancement of milk lipase activity (2, 18). Milk lipase is heat labile, so most milk products will not have residual activity. The exceptions are some hard cheeses made from unheated milk or milk given a subpasteurization heat treatment. These cheeses depend on milk lipase activity to develop a characteristic flavor.

Sufficient bacterial lipase can be produced in raw milk to cause defects in products manufactured from that milk. Since residual activities are usually low, and the reaction environment is less than optimum, usually only products with long storage times or high storage temperatures are affected. These include UHT milk, some cheeses, butter, and whole milk powder. Lipase-producing bacteria can also recontaminate pasteurized milk and cream and grow in these products during refrigerated storage. The rancid flavor and odor resulting from lipase action is usually from the liberation of C_4 to C_8 fatty acids. Fatty acids of higher molecular weight produce a flavor described as soapy. Low levels of unsaturated fatty acids liberated by enzymatic activity may be oxidized to ketones and aldehydes to produce oxidized or "cardboardy" off flavor (26). *P. fragi* produces a fruity off flavor in milk by esterifying free fatty acids with ethanol (89). Ethyl butyrate and ethyl hexanoate are the esters produced at highest amounts, with low levels of ethyl esters of acetate, propionate, and isovalerate also produced. Low levels of ethanol in milk, present as a result of microbial activity, stimulate ester production.

Residual activity from heat-stable microbial lipases can cause off flavors in UHT milk, but lipase-induced defects are not as common as those resulting from microbial protease (82). Lipolytic off flavors in UHT products generally take several weeks to months to develop owing to the low amounts of lipase present, or appear in products made from raw milk with excessively high populations of psychrotrophs ($>10^6$ CFU/ml) (102).

Rancid defect in butter may result from growth of lipolytic microorganisms during storage, residual heat-stable microbial lipase originating from the growth of psychrotrophic bacteria in the milk or cream, or milk lipase activity in the raw milk or cream. When butter is manufactured from rancid cream, low-molecular-weight free fatty acids are removed with the watery portion of the cream (buttermilk), so the resulting butter will not have a typical rancid flavor but a less pronounced soapy off flavor associated with C_{10} to C_{12} free fatty acids. However, the typical odor of rancid butter is associated with lower-molecular-weight fatty acids (C_4 to C_8). Microbial lipases present in butter will exhibit activity even if the product is stored at $-10°C$ (85). Growth of psychrotrophic bacteria in butter occurs only if the product is made from sweet rather than ripened (sour) cream. Sweet cream butter is preserved by salt and refrigeration. Butter is a water-in-fat emulsion, so moisture and salt will not equilibrate during storage. Consequently, if salt and moisture are not evenly distributed in the product during manufacture, then lipolytic psychrotrophs will have pockets of high a_w in which to grow (26).

Cheese is more susceptible to defects caused by bacterial lipases than those caused by proteases, because lipases, unlike most proteases, are concentrated along with the fat in the curd. The acidic environment of most cheeses limits, but may not eliminate, lipase activity. Some cheeses, such as Camembert and Brie, increase in pH to near neutrality during ripening. Camembert is, in fact, susceptible to defects associated with microbial lipase (29). More acidic cheeses, e.g., cheddar, are susceptible if cured for several months or if high amounts of lipase are present (64). Law et al. (65) reported that cheddar cheese made from milk containing over 10^6 CFU of *P. fluorescens* per ml before pasteurization developed a rancid flavor in 6 to 8 months. Psychrotrophic bacteria do not grow in cured cheeses because of the low pH and salt content but may grow in high-moisture fresh products, such as cottage cheese.

Whole milk powder containing bacterial lipase will develop rancidity, indicating a potential problem if milk supports growth of lipolytic psychrotrophs before processing. Low-fat powders, such as nonfat dried milk, whey, and whey protein concentrate, may contain residual lipase which becomes active when these products are used in fat-containing food formulations (102).

Control of Product Defects Associated with Psychrotrophic Bacteria

Raw Milk

Preventing product defects that result from growth of psychrotrophic bacteria in raw milk involves limiting contamination levels, rapid cooling immediately after milking, and maintenance of cold storage temperatures. Limiting populations of bacteria primarily involves cleaning, sanitizing, and drying cows' teats and udders before milking, and using cleaned and sanitized equipment. Removal of residual milk solids from milk contact surfaces is critical to psychrotroph control, since these residues protect cells from the action of chemical sanitizers and provide nutrients for growth. Subsequent growth over a period of days results in a biofilm which, in addition to containing high numbers of bacteria, is highly resistant to chemical sanitizers (36). Rapid cooling of milk after collection is important, because contamination of the product with psychrotrophic bacteria is unavoidable. Milk, fresh from the cow, enters the farm storage tank at 30 to 37°C. Sanitary standards in the United States require raw milk to be cooled to 7°C within 2 h after milking (5), but most farm systems achieve rapid cooling to less than 4°C. Since milk is often picked up from the farm every 48 h, three additional milkings will be added to the previously collected milk. The second milking should not warm the previously collected milk to over 10°C. As previously indicated, psychrotrophic activity in milk is inhibited at 2°C, but freezing of milk causes disruption of the fat globule membrane, making it highly susceptible to lipolysis. Therefore, the challenge of farm storage systems is to rapidly cool milk to as low a temperature as possible while avoiding ice formation.

Pasteurized Products

Preventing contamination of pasteurized dairy products with psychrotrophic bacteria is primarily a matter of equipment cleaning and sanitation, although airborne psychrotrophs may also limit product shelf life. Even when filling equipment is effectively cleaned and sanitized, it can still become a source of psychrotrophic microorganisms which accumulate during normal hours of continuous use (98). These microorganisms probably enter the filler through the vacuum system or from containers. Tanks used for holding pasteurized products before packaging can also be a source of psychrotrophic microorganisms. Tank walls may have microscopic fissures or pits that protect microorganisms from cleaning and sanitizing procedures. Complete elimination of psychrotrophic microorganisms from products can only be achieved by using aseptic packaging technologies whereby equipment and air that contact the pasteurized product are maintained free of microorganisms.

SPOILAGE BY FERMENTATIVE NONSPOREFORMERS

Spoilage of milk and dairy products resulting from growth of acid-producing fermentative bacteria occurs when storage temperatures are sufficiently high for these microorganisms to outgrow psychrotrophic bacteria, or when product composition is inhibitory to gram-negative aerobic organisms. For example, the presence

of lactic acid in fluid milk is a good indication that the product was exposed to an unacceptably high storage temperature that allowed growth of lactic acid bacteria. Fermented dairy foods, though manufactured using lactic acid bacteria, can be spoiled by the growth of "wild" lactics that produce unwanted gas, off flavors, or appearance defects. Fluid milk, cheese, and cultured milks are the major dairy products susceptible to spoilage by non-spore-forming fermentative bacteria.

Non-spore-forming bacteria responsible for fermentative spoilage of dairy products are mostly in either the lactic acid-producing or coliform groups. Lactic acid bacteria involved in dairy fermentations can spoil fluid milk, but the strains involved are often environmental types that produce defect-inducing metabolites in addition to lactic acid. Genera of lactic acid bacteria involved in spoilage of milk and fermented products include *Lactococcus*, *Lactobacillus*, *Leuconostoc*, *Enterococcus*, *Pediococcus*, and *Streptococcus* (16). Coliforms can spoil milk, but this is seldom a problem since they are usually outgrown by either the lactic acid or psychrotrophic bacteria. Coliform-induced spoilage is more common with fermented products, especially certain cheese varieties. Members of the *Enterobacter* and *Klebsiella* genera are most often associated with coliform spoilage, while *Escherichia* spp. only occasionally exhibit sufficient growth to produce a defect.

Sources of Fermentative Spoilage Bacteria

Lactic acid-producing bacteria are normal inhabitants on the skin of the cow's teat. Consequently, all raw milk contains at least low numbers of these microorganisms. Lactic acid bacteria are also associated with silage and other animal feeds and feces. Coliform bacteria are present on udder skin as a result of fecal contamination, so ineffective cleaning of this area before milking will contribute to high coliform populations in milk. Coliform bacteria in raw milk are also associated with milk residue buildup on inadequately cleaned milking equipment (10).

Defects of Fluid Milk Products

The most common fermentative defect in fluid milk products is souring caused by the growth of lactic acid bacteria. Lactic acid by itself has a clean pleasant acid flavor and no odor. The unpleasant "sour" odor and taste of spoiled milk is a result of small amounts of acetic and propionic acids (101). Sour odor can be detected before a noticeable acid flavor develops. For a discussion of lactic acid production in milk, see chapter 32. Other defects may occur in combination with acid production. A malty flavor results from growth of *Lactococcus lactis* subsp. *lactis* var. maltigenes. This strain is unique among lactococci in its ability to produce 2-methylpropanal, 3-methylbutanal, and the corresponding alcohols (79). The aldehydes are produced by decarboxylation of α-ketoisocaproic and α-ketoisovaleric acids. These keto acids are also concurrently used to synthesize leucine and valine by transamination with glutamic acid. Alcohols corresponding to the aldehydes are formed by the action of alcohol dehydrogenase in the presence of $NADH_2$. Malty flavor is primarily from 3-methylbutanal.

Another defect associated with growth of lactic acid bacteria in milk is ropy texture. Most dairy-associated species of lactic acid bacteria have strains that produce exocellular polymers which increase the viscosity of milk, causing the ropy defect (14). Some of these strains are used to produce high-viscosity fermented products such as yogurt and Scandinavian ropy milk (vilia, skyr). The defect in noncultured fluid milk products is usually caused by growth of specific strains of lactococci. The polymer produced by these organisms is a polysaccharide containing glucose and galactose with small amounts of mannose, rhamnose, and pentose (15). The polysaccharide is possibly associated with protein (67).

Defects in Cheese

Lactic Acid Bacteria

Some strains of lactic acid bacteria produce flavor and appearance defects in cheese. Lactobacilli are a normal part of the dominant microflora of aged cheddar cheese. If heterofermentative lactobacilli predominate, the cheese is prone to develop an "open" texture or fissures, a result of gas production during aging (62). Off flavors are also associated with the growth of these organisms (111). Gassy defects in aged cheddar cheese are more often associated with growth of lactobacilli than with growth of coliforms, yeasts, or sporeformers. The use of elevated ripening temperatures for cheddar cheese, e.g., 15 rather than 8°C, encourages growth of heterofermentative lactobacilli but not that of non-lactic acid bacteria (24). This phenomenon limits the use of high-temperature storage to accelerate ripening. *Lactobacillus brevis* and *Lactobacillus casei* subsp. *pseudoplantarum* have been associated with gas production in retail mozzarella cheese (53). *Lactobacillus casei* subsp. *casei* produces a soft body defect in mozzarella cheese (52). The softened cheese cannot be readily sliced or grated and does not melt properly.

Some cheese varieties occasionally exhibit a pink discoloration. Pink spots in Swiss-type varieties result from the growth of pigmented strains of propionibacteria. In Italian cheese varieties, a pink discoloration may occur either in a band near the surface or throughout the whole cheese. This defect is associated with certain strains of *Lactobacillus delbrueckii* subsp. *bulgaricus* that fail to lower the redox potential of the cheese (99). Another common defect of aged cheddar cheese is the appearance of white crystalline deposits on the surface. Although they do not affect flavor, these deposits reduce consumer acceptability. Rengpipat and Johnson (91) observed an atypical strain of a facultatively heterofermentative *Lactobacillus* sp. associated with the deposits. This strain produces an unusually high amount of D-lactic acid during cheese aging, resulting in the formation of insoluble calcium lactate crystals, the primary component of the white deposits. *Lactobacillus casei* subsp. *alactosus* and *L. casei* subsp. *rhamnosus* have been associated with the development of a phenolic flavor in cheddar cheese, described as being similar to that of horse urine (53). The flavor develops after 2 to 6 months of aging.

Fruity off flavor in cheddar cheese is usually not caused by growth of psychrotrophic bacteria, as it is in milk, but rather a result of growth of lactic acid bacteria (usually *Lactococcus* spp.) that produce esterase. Fruity-flavored cheeses contain high levels of ethanol, a substrate for esterification (9). The major esters contributing to fruity flavor in cheese are ethyl hexanoate and ethyl butyrate.

Coliform Bacteria

Coliform bacteria were recognized as causing gassy defect in cheddar and related cheese varieties as early as 1885 (96). Growth usually occurs during the cheese manufacture process or during the first few days of storage and therefore is referred to as early gas (or early blowing) defect. In hard cheeses, such as cheddar, this defect occurs when a slow lactic acid fermentation fails to rapidly lower pH, or when highly contaminated raw milk is used. Cheese varieties in which acid production is purposely delayed by washing the curds are highly susceptible to coliform growth (40). Soft, mold-ripened cheeses, such as Camembert, increase in pH during ripening, with a resulting susceptibility to coliform growth (39, 97). Frank and Marth (37) reported that 17% of commercial soft and semisoft cheeses tested contained over 10^4 coliforms per g. In a more recent survey, 13 and 17% of Brie and Camembert samples, respectively, contained greater than 10^7 CFU of *Enterobacteriaceae* per g on their mold rind (86). Gas formation in retail mozzarella cheese has been associated with growth of *Klebsiella pneumoniae* (71). Coliform growth in retail cheese is often manifested as a swelling of the plastic package. Approximately 10^7 CFU of coliform per g is needed to produce a gassy defect.

Defects in Fermented Milk Products

Fermented milk products such as cultured buttermilk, sour cream, and cottage cheese rely on diacetyl produced during fermentation for their typical "buttery" flavor and aroma. These products lose consumer appeal when this flavor is lost owing to reduction of diacetyl to acetoin and 2,3-butanediol (38). Lactococci, capable of growing at 7°C, may produce sufficient diacetyl reductase to destroy diacetyl in cultured milks (50). Other psychrotrophic contaminants in cultured milks, including yeasts and coliforms, may also be involved in diacetyl reduction (113).

Control of Defects Caused by Lactic Acid and Coliform Bacteria

Defects in fluid milk caused by coliforms and lactic acid bacteria are controlled by good sanitation practices during milking, maintaining raw milk at temperatures below 7°C, pasteurization, and refrigeration of pasteurized products. These microorganisms seldom grow to significant levels in refrigerated pasteurized milk because of their slow growth rates compared with psychrotrophic bacteria. Control of coliform growth in cheese is achieved by using pasteurized milk, encouraging rapid fermentation of lactose, and good sanitation during manufacture. Controlling defects produced by undesirable lactic acid bacteria in cheese and fermented milks is more difficult, since growth of lactic acid bacteria must be encouraged during manufacture, and the final products often provide suitable growth environments. Undesirable strains of lactic acid bacteria are readily isolated from the manufacturing environment, so their control requires attention to plant cleanliness and protecting the product during manufacture.

SPORE-FORMING BACTERIA

Spoilage by spore-forming bacteria can occur in low-acid fluid milk products that are preserved by substerilization heat treatments and packaged with little chance for recontamination with vegetative cells. Products in this category include aseptically packaged milk and cream and sweetened and unsweetened concentrated canned milks. Nonaseptic packaged refrigerated fluid milk may spoil owing to growth of psychrotrophic *Bacillus cereus* and *Bacillus polymyxa* in the absence of more rapidly growing gram-negative psychrotrophs (87, 105). Hard cheeses, especially those with low interior salt concentrations, are also susceptible to spoilage by spore-forming bacteria.

Spore-forming bacteria that spoil dairy products usually originate in the raw milk. Populations present in raw milk are generally quite low (<5,000 CFU/ml), and the occurrence of a sporeformer-induced defect does not always correlate with initial numbers of sporeformers in the raw product (78). This is because products prone to support sporeformer growth are stored for sufficiently long periods of time that outgrowth of small numbers of cells can eventually cause a defect. Sporeforming bacteria in raw milk are predominantly *Bacillus* spp., with *Bacillus licheniformis*, *B. cereus*, *Bacillus subtilis*, and *Bacillus megaterium* most commonly isolated (70, 100). *Clostridium* spp. are present in raw milk at such low levels that enrichment and most probable number techniques must be used for quantification (93). Populations of spore-forming bacteria in raw milk vary seasonally. In temperate climates, *Bacillus* and *Clostridium* spp. are at higher levels in raw milk collected in the winter than in the summer, because in the winter, cows lie on spore-contaminated bedding materials and are likely to consume spore-laden silage (10). The occurrence of psychrotrophic *Bacillus* spp. in raw milk does not follow this seasonal pattern (77), but there may be a seasonal occurrence of germination factors (88).

Defects in Fluid Milk Products

Pasteurized milk packaged under conditions that limit recontamination can spoil owing to the growth of psychrotrophic *B. cereus*. This topic has been reviewed by Meer et al. (77). Psychrotrophic *B. cereus* is present in over 80% of raw milk samples. There is also evidence that psychrotrophic *Bacillus* spp. are introduced into the milk at the processing plant as postpasteurization contaminants (43). Psychrotrophic *B. cereus* can reach populations exceeding 10^6 CFU/ml in milk held for 14 days at 7°C, although slower growth is more common (78). Germination of spores in raw milk occurs soon after pasteurization, indicating that they were heat activated. The defect produced by subsequent growth is described as sweet curdling, since it first appears as coagulation without significant acid or off flavor being formed. Coagulation is caused by a chymosinlike protease (17). Eventually, the enzyme degrades casein sufficiently to produce a bitter-flavored product. Growth may become visible as "buttons" at the bottom of the carton; these are actually bacterial colonies. Psychrotrophic *B. cereus* also produces phospholipase C (lecithinase), which degrades the fat globule membrane, resulting in the aggregation of the fat in cream (77). The result is described as "bitty" cream defect.

Psychrotrophic *Bacillus* spp. other than *B. cereus* are also capable of spoiling heat-treated milk. Cromie et al. (23) observed that psychrotrophic *Bacillus circulans* was the predominant spoilage organism in aseptically packaged heat-treated milk. This microorganism produces acid from lactose, giving the milk a sour flavor. *Bacillus mycoides* is another frequently isolated psychrotrophic sporeformer in milk (88).

Most bacterial spores present in raw milk are moderately heat labile and are destroyed by UHT treatments. The major heat-resistant species in milk is *Bacillus stearothermophilus* (83). Other, less heat-resistant *Bacillus* spp. have been isolated from UHT milk, especially *B. subtilis* and *B. megaterium* (47).

Defects in Canned Condensed Milk

Canned condensed milk may be either sweetened with sucrose and glucose to lower the a_w or left unsweetened. The unsweetened product must be sterilized by heat treatment. Defects associated with growth of surviving spore-forming organisms in this product have been described by Carić (13). "Sweet coagulation" is caused by growth of *Bacillus coagulans*, *B. stearothermophilus*, or *B. cereus*. This defect is similar to the sweet curdling defect caused by psychrotrophic *B. cereus* in pasteurized milk. Protein destruction, in addition to curdling, can also occur and is usually caused by growth of *B. subtilis* or *B. licheniformis*. Swelling or bursting of cans can be caused by growth of *Clostridium sporogenes*. "Flat sour" defect (acidification without gas production) can result from growth of *B. stearothermophilus*, *B. licheniformis*, *B. coagulans*, *Bacillus macerans*, and *B. subtilis* (57). Sweetened condensed milk should have sufficiently low a_w to inhibit bacterial spore germination. However, if a_w is not well controlled, *Bacillus* spp. may produce acid, or acid-proteolytic spoilage.

Control of Sporeformer-Associated Defects in Fluid Products

Methods for controlling growth of sporeformers in fluid products mainly involve the use of appropriate heat treatments. UHT treatments produce products microbiologically stable at room temperature. However, when sub-UHT heat treatments are more severe than that required for pasteurization, the shelf life of cream and milk can actually decrease, a phenomenon attributed to spore activation (83). Use of a double heat treatment, the first to activate spores and the second to destroy them, does not result in the expected increase in shelf life (45). Adding hippuric acid, a naturally occurring germinant to raw milk, does not germinate sufficient spores before heat treatment to provide a consistent increase in shelf life (44). Neither does adding lysozyme to milk sufficiently inhibit outgrowth of *Bacillus* spores to be of practical significance. A practical means to prevent sporeformers from spoiling nonfermented liquid dairy products given sub-UHT heat treatments has not been developed.

Defects in Cheese

The major defect in cheese caused by spore-forming bacteria is gas formation, usually resulting from growth of *Clostridium tyrobutyricum* and occasionally from growth of *C. sporogenes* and *Clostridium butyricum*. This defect is often called "late blowing" or "late gas" because it occurs after the cheese has aged for several weeks. Emmental, Swiss, Gouda, and Edam cheeses are most often affected because of their relatively high pH and moisture content plus their low interior salt levels. The defect can also occur in cheddar and Italian cheeses. Processed cheeses are susceptible to late blowing because spores are not inactivated during heat processing (58). Late gas defect results from the fermentation of lactate to butyric acid, acetic acid, carbon dioxide, and hydrogen gas. Flavor as well as appearance are affected. Populations of *C. tyrobutyricum* spores of less than 1 per ml of milk can produce the defect, because the spores are concentrated in the cheese curd during manufacture (94). The number of spores in raw milk needed to cause a defect varies with size and shape of the cheese in addition to pH and moisture, because cheese size and shape determine the extent of salt penetration after brining. The number of spores required to cause late blowing in 9-kg wheels of rinded Swiss cheese was estimated at >100 per liter of raw milk (25). The presence of *C. tyrobutyricum* spores in milk has been traced to the consumption of contaminated silage, which increases levels in the cow's feces (25). Contaminated silage generally has a high pH and is of low quality.

Control of Sporeformer-Associated Defects in Cheese

Ideally, control of late blowing defect would occur at the farm by instituting feeding and management practices that would reduce the number of spores entering the milk supply (48). In practice, this approach has not achieved the required results, so cheese manufacturers have tried to control the defect by removing spores from the milk at the plant or inhibiting their growth in the cheese. Numbers of bacterial spores can be reduced in milk by a centrifugation process known as bactofugation (58). Germination of spores in the cheese can be inhibited by addition of nitrate and/or lysozyme (25). Nitrate as a cheese additive is prohibited in many countries, and lysozyme by itself does not provide complete protection. Bacteriocins produced by lactic acid bacteria may provide a highly specific means of inhibiting anaerobic spore germination (108).

YEASTS AND MOLDS

Growth of yeasts and molds is a common cause of spoilage of fermented dairy products, because these microorganisms are able to grow well at low pH. Yeast spoilage is manifested as fruity or yeasty odor and/or gas formation. Hard (or cured) cheeses, when properly made, have very low amounts of lactose, thus limiting the potential for yeast growth. Cultured milks, such as yogurt and buttermilk, and fresh cheeses, such as cottage, normally contain fermentable levels of lactose and therefore are prone to yeast spoilage. A "fermented/yeasty" flavor observed in cheddar cheese spoiled by growth of a *Candida* sp. was associated with elevated ethanol, ethyl acetate, and ethyl butyrate (51). The affected cheese had a high moisture content (associated with low starter activity and therefore high residual lactose) and low salt content, which contributed to yeast growth. Yeast spoilage can also occur in dairy foods with low a_w, such as sweetened condensed milk and butter. The most common yeasts present in dairy products are *Kluyveromyces marxianus* and *Debaromyces hansenii* (the teleomorph) and their asporegenous counterparts (the anamorph), *Candida famata*, *Candida kefyr*, and other *Candida* species (34, 61). Also prevalent are *Rhodotorula mucilaginsoa*, *Yarrowia lipolytica*, and *Torulospora* and *Pichia* spp. (92, 115). Fermented dairy products provide a highly specialized ecological niche for yeasts, selecting for those that can utilize lactose or lactic acid and that tolerate high salt concentrations (34). Yeasts able to produce proteolytic or lipolytic enzymes may also have a selective advantage for growth in dairy products.

Mold Spoilage of Cheese

Growth of spoilage molds on cheese is a problem that still has significance, though it dates back to prehistory. The most common molds found on cheese are *Penicillium* spp. (12, 109), with others occasionally found, including *Aspergillus*, *Alternaria*, *Mucor*, *Fusarium*, *Cladosporium*, *Geotrichum*, and *Hormodendrum* spp. Mold species commonly isolated from processed cheese include *Penicillium roqueforti*, *Penicillium cyclopium*, *Penicillium viridicatum*, and *Penicillium crustosum* (110). Vacuum-packaged cheddar cheese supports the growth of *Cladosporium cladosporiodes*, *Penicillium commune*, *Cladosporium herbarum*, *Penicillium glabrum*, and *Phoma* species (49). *Candida* yeasts have also been isolated from vacuum-packaged cheese.

Controlling Mold Spoilage

Yeasts and molds that spoil dairy products can usually be isolated in the processing plant on packaging equipment, in the air, in salt brine, on manufacturing equipment, and in the general environment (floors, walls, ventilation ducts, etc.). Successful control efforts must start with limiting exposure of pasteurized products to these sources. Mold spores do not survive pasteurization (28). If the initial contamination level is limited, strategies to inhibit growth are more likely to succeed. These include packaging to reduce oxygen (and/or increase carbon dioxide), cold storage, and the use of antimycotic chemicals such as sorbate, propionate, and natamycin (pimaracin). Added liquid smoke is also a potent mold inhibitor (114). None of these control measures is completely effective. Vacuum-packaged cheese is susceptible to thread mold defect, whereby the fungi grow in the wrinkles of the plastic film (49). Some molds are resistant to antimycotic additives. Sorbate-resistant molds are commonly isolated from sorbate-treated cheese but not from untreated cheese (66). Some *Penicillium* spp. are not only resistant to sorbate but will degrade it by decarboxylation, producing 1,3-pentadiene (69). This imparts a kerosenelike odor to the cheese. Some *Mucor* spp. degrade sorbate to 4-hexenol, and some *Geotrichum* spp. degrade it to 4-hexenoic acid. Sorbate can also be used as a carbon source or can be oxidized to carbon dioxide and water (66). The ability of some molds to degrade sorbate explains why cheeses with high levels of mold contamination are not effectively preserved by this additive.

References

1. Abad, P., A. Villafafila, J. D. Frias, and C. Rodriguez-Fernandez. 1993. Extracellular lipolytic activity from *Pseudomonas fluorescens* biovar I *(Pseudomonas fluorescens* NC1). *Milchwissenschaft* **48**:680–683.

2. Alkanhal, H. A., J. F. Frank, and G. L. Christen. 1985. Microbial protease and phospholipase C stimulate lipolysis of washed cream. *J. Dairy Sci.* **68**:3162–3170.

3. Andersson, R. E. 1980. Lipase production, lipolysis and formation of volatile compounds by *Pseudomonas fluorescens* in fat containing media. *J. Food Sci.* **45**:1694–1701.

4. Andersson, R. E. 1980. Microbial lipolysis at low temperatures. *Appl. Environ. Microbiol.* **39**:36–40.

5. Anonymous. 1993. *Grade A Pasteurized Milk Ordinance.* U.S. Dept. of Health, Education and Welfare, Public Health Service/Food and Drug Administration, Washington, D.C.

6. Banwart, G. J. 1981. *Basic Food Microbiology.* AVI Publishing Co., Westport, Conn.

7. Bassette, R., and J. S. Acosta. 1988. Composition of milk products, p. 39–79. *In* N. P. Wong (ed.), *Fundamentals of Dairy Chemistry.* Van Nostrand Reinhold Co., New York, N.Y.

8. Batish, V. K., H. Chander, K. C. Zumdegni, K. L. Bhatia, and R. S. Singh. 1988. Antibacterial activity of lactoferrin against some common food-borne pathogenic organisms. *Aust. J. Dairy Technol.* **43**:16–18.

9. Bills, D. D., M. E. Morgan, L. M. Reddy, and E. A. Day. 1965. Identification of compounds responsible for fruit flavor defect of experimental Cheddar cheeses. *J. Dairy Sci.* **48**:1168–1170.

10. Bramley, A. J., and C. H. McKinnon. 1990. The microbiology of raw milk, p. 163–208. *In* R. K. Robinson (ed.), *Dairy Microbiology*, vol. 1. Elsevier Applied Science, New York, N.Y.

11. Bucky, A. R., P. R. Hayes, and D. S. Robinson. 1988. Enhanced inactivation of bacterial lipases and proteinases in whole milk by a modified ultra high temperature treatment. *J. Dairy Res.* **56**:373–380.

12. Bullerman, L. B., and F. J. Olivigni. 1974. Mycotoxin producing potential of molds isolated from Cheddar cheese. *J. Food Sci.* **39**:1166–1168.

13. Carić, M. 1994. *Concentrated and Dried Dairy Products.* VCH Publishers, Inc., New York, N.Y.

14. Cerning, J. 1990. Exocellular polysaccharides produced by lactic acid bacteria. *FEMS Microbiol. Rev.* **87**:113–130.

15. Cerning, J., C. Bouillanne, M. Landon, and M. Desmazeaud. 1992. Isolation and characterization of exopolysaccharides from slime-forming mesophilic lactic acid bacteria. *J. Dairy Sci.* **75**:692–699.

16. Chapman, H. R., and M. E. Sharpe. 1990. Microbiology of cheese, p. 203–289. *In* R. K. Robinson (ed.), *Dairy Microbiology*, vol. 2. Elsevier Applied Science, New York, N.Y.

17. Choudhery, A. K., and E. M. Mikolajcik. 1971. Activity of *Bacillus cereus* proteinases in milk. *J. Dairy Sci.* **53**:363–366.

18. Chrisope, G. L., and R. T. Marshall. 1976. Combined action of lipase and microbial phospholipase C on a model fat emulsion and raw milk. *J. Dairy Sci.* **59**:2024–2030.

19. Christen, G. L., and R. T. Marshall. 1985. Effect of histidine on thermostability of lipase and protease of *Pseudomonas fluorescens* 27. *J. Dairy Sci.* **68**:594–604.

20. Christian, J. H. B. 1980. Reduced water activity, p. 70–91. *In Microbial Ecology of Foods*, vol. 1. Academic Press, Inc., New York, N.Y.

21. Cousin, M. A. 1982. Presence and activity of psychrotrophic microorganisms in milk and dairy products: a review. *J. Food Prot.* **45**:172–207.

22. Cousin, M. A. 1989. Physical and biochemical effects of milk components, p. 205–225. *In* R. C. McKellar (ed.), *Enzymes of Psychrotrophs in Raw Food.* CRC Press, Inc., Boca Raton, Fla.

23. Cromie, S. J., T. W. Dommett, and D. Schmidt. 1989. Changes in the microflora of milk with different pasteurization and storage conditions and aseptic packaging. *Aust. J. Dairy Technol.* **44**:74–77.

24. Cromie, S. J., J. E. Giles, and J. R. Dulley. 1987. Effect of elevated ripening temperatures on the microflora of Cheddar cheese. *J. Dairy Res.* **54**:69–76.

25. Dasgupta, A. R., and R. R. Hull. 1989. Late blowing of Swiss cheese: incidence of *Clostridium tyrobutyricum* in manufacturing milk. *Aust. J. Dairy Technol.* **44**:82–87.

26. Deeth, H. C., and C. H. Fitz-Gerald. 1983. Lipolytic enzymes and hydrolytic rancidity in milk and milk products, p. 195–239. *In* P. F. Fox (ed.), *Developments in Dairy Chemistry*, part II. Applied Science, London, England.

27. Dionysius, D. A., P. A. Grieve, and A. C. Vos. 1992. Studies on the lactoperoxidase system: reaction kinetics and antibacterial activity using two methods for hydrogen peroxide generation. *J. Appl. Bacteriol.* **72**:146–153.

28. Doyle, M. P., and E. H. Marth. 1975. Thermal inactivation of conidia from *Aspergillus flavus* and *Aspergillus parasiticus*. I. Effects of moist heat, age of conidia, and sporulation medium. *J. Milk Food Technol.* **38**:678–682.

29. Dumont, J. P., G. Delespaul, B. Miquot, and J. Adda. 1977. Influence des bactéries psychrotrophs sur les qualités organoleptiques de fromages à pâte molle. *Lait* **57**:619–630.

30. Ewings, K. N., R. E. O'Conner, and G. E. Mitchell. 1984. Proteolytic microflora of refrigerated raw milk in South East Queensland. *Aust. J. Dairy Technol.* **39**:65–68.

31. Fairbairn, D. J., and B. A. Law. 1987. The effect of nitrogen and carbon sources on proteinase production by *Pseudomonas fluorescens*. *J. Appl. Bacteriol.* **62**:105–113.

32. Fernandez, L., J. A. Alvarez, P. Palacios, and C. San Jose. 1992. Proteolytic and lipolytic activities of *Pseudomonas fluorescens* grown in raw milk with variable iron content. *Milchwissenschaft* **47**:160–163.

33. Fernandez, L., C. San Jose, and R. C. McKellar. 1990. Repression of *Pseudomonas fluorescens* extracellular lipase secretion by arginine. *J. Dairy Res.* **57**:69–78.

34. Fleet, G. H. 1990. Yeasts in dairy products. *J. Appl. Bacteriol.* **68**:199–211.

35. Fox, P. F., P. Power, and T. M. Cogan. 1989. Isolation and molecular characteristics, p. 57–120. *In* R. C. McKellar (ed.), *Enzymes of Psychrotrophs in Raw Food.* CRC Press, Inc., Boca Raton, Fla.

36. Frank, J. F., and R. A. Koffi. 1990. Surface-adherent growth of *Listeria monocytogenes* is associated with increased resistance to surfactant sanitizers and heat. *J. Food Prot.* **53**:560–564.

37. Frank, J. F., and E. H. Marth. 1978. Survey of soft and semisoft cheese for presence of fecal coliforms and serotypes of enteropathogenic *Escherichia coli*. *J. Food Prot.* **41**:198–200.

38. Frank, J. F., and E. H. Marth. 1988. Fermentations, p. 656–738. *In* N. P. Wong (ed.), *Fundamentals of Dairy Chemistry*, 3rd ed. Van Nostrand Reinhold Co., New York, N.Y.

39. Frank, J. F., E. H. Marth, and N. F. Olson. 1977. Survival of enteropathogenic and non-pathogenic *Escherichia coli* during the manufacture of Camembert cheese. *J. Food Prot.* **40**:835–842.

40. Frank, J. F., E. H. Marth, and N. F. Olson. 1978. Behavior of enteropathogenic *Escherichia coli* during manufacture and ripening of brick cheese. *J. Food Prot.* **41**:111–115.

41. Garcia, M. L., B. Sanz, P. Garcia-Collia, and J. A. Ordonez. 1989. Activity and thermostability of the extracellular lipases and proteinases from pseudomonads isolated from raw milk. *Milchwissenschaft* **44**:547–560.

42. Griffiths, M. W. 1989. Effect of temperature and milk fat on extracellular enzyme synthesis by psychrotrophic bacteria during growth in milk. *Milchwissenschaft* **44**:539–543.

43. Griffiths, M. W., and J. D. Phillips. 1990. Incidence, source and some properties of psychrotrophic *Bacillus* spp. found in raw and pasteurized milk. *J. Soc. Dairy Technol.* **43**:62–70.

44. Griffiths, M. W., and J. D. Phillips. 1990. Strategies to control the outgrowth of spores of psychrotrophic *Bacillus* spp. in dairy products. I. Use of naturally occurring materials. *Milchwissenschaft* **45**:621–625.

45. Griffiths, M. W., and J. D. Phillips. 1990. Strategies to control the outgrowth of spores of psychrotrophic *Bacillus* spp. in dairy products. II. Use of heat treatments. *Milchwissenschaft* **45**:719–721.

46. Guamis, H., T. Huerta, and E. Garay. 1987. Heat-inactivation of bacterial proteases in milk before UHT-treatment. *Milchwissenschaft* **42**:651–653.

47. Hassan, A. N., A. S. Zahran, N. H. Metwalli, and S. I. Shalabi. 1993. Aerobic sporeforming bacteria isolated from UHT milk produced in Egypt. *Egypt. J. Dairy Sci.* **21**:109–121.

48. Herlin, A. H., and A. Christansson. 1993. Cheese-blowing anaerobic spores in bulk milk from loose-housed and tied dairy cows. *Milchwissenschaft* **48**:686–689.

49. Hocking, A. D., and M. Faedo. 1992. Fungi causing thread mould spoilage of vacuum packaged Cheddar cheese during maturation. *Int. J. Food Microbiol.* **16**:123–130.

50. Hogarty, S. L., and J. F. Frank. 1982. Low-temperature activity of lactic streptococci isolated from cultured buttermilk. *J. Food Prot.* **43**:1208–1211.

51. Horwood, J. F., W. Stark, and R. R. Hull. 1987. A "fermented, yeasty" flavour defect in Cheddar cheese. *Aust. J. Dairy Technol.* **42**:25:26.

52. Hull, R. R., A. V. Roberts, and J. J. Mayes. 1983. The association of *Lactobacillus casei* with a soft-body effect in commercial Mozzarella cheese. *Aust. J. Dairy Technol.* **22**:78–80.

53. Hull, R., S. Toyne, I. Haynes, and F. Lehman. 1992. Thermoduric bacteria: a reemerging problem in cheesemaking. *Aust. J. Dairy Technol.* **47**:91–94.

54. Jaspe, A., P. Palacios, P. Matias, L. Fernandez, and C. Sanjose. 1994. Proteinase activity of *Pseudomonas fluorescens* grown in cold milk supplemented with nitrogen and carbon sources. *J. Dairy Sci.* **77**:923–929.

55. Jaspe, A., P. Oviedo, L. Fernandez, P. Palacios, and C. Sanjose. 1995. Cooling raw milk: change in the spoilage potential of contaminating *Pseudomonas*. *J. Food Prot.* **58**:915–921.

56. Jenness, R. 1988. Composition of milk, p. 1–38. *In* N. P. Wong (ed.), *Fundamentals of Dairy Chemistry*. Van Nostrand Reinhold Co., New York, N.Y.

57. Kalogridou-Vassiliadou, D. 1992. Biochemical activities of *Bacillus* species isolated from flat sour evaporated milk. *J. Dairy Sci.* **75**:2681–2686.

58. Kosikowski, F. V. 1982. *Cheese and Fermented Milk Foods*, 2nd ed. F. V. Kosikowski and Associates, Brooktondale, N.Y.

59. Kosse, D., H. Seiler, R. Amann, W. Ludwig, and S. Scherer. 1997. Identification of yoghurt-spoiling yeasts with 18s rRNA oligonucleotide probes. *Syst. Appl. Microbiol.* **20**:468–480.

60. Kroll, S. 1989. Thermal stability, p. 121–152. *In* R. C. McKellar (ed.), *Enzymes of Psychrotrophs in Raw Food.* CRC Press, Inc., Boca Raton, Fla.

61. Kumura, H., K. Mikawa, and Z. Saito. 1993. Influence of milk proteins on the thermostability of the lipase from *Pseudomonas fluorescens* 33. *J. Dairy Sci.* **76**:2164–2167.

62. Lalaye, L. C., R. E. Simard, B.-H. Lee, R. A. Holley, and R. N. Giroux. 1987. Involvement of heterofermentative lactobacilli in development of open texture in cheeses. *J. Food Prot.* **50**:1009–1012.

63. Law, B. A. 1979. Reviews of the progress of dairy science: enzymes of psychrotrophic bacteria and their effects on milk and milk products. *J. Dairy Res.* **46**:573–588.

64. Law, B. A., C. M. Cousins, M. E. Sharpe, and F. L Davies. 1979. Psychrotrophs and their effects on milk and dairy products, p. 137–152. *In* A. D. Russell and R. Fuller (ed.), *Cold Tolerant Microbes in Spoilage and the Environment.* Academic Press, Inc., New York, N.Y.

65. Law, B. A., M. E. Sharpe, and H. R. Chapman. 1976. Effect of lipolytic Gram negative psychrotrophs in stored milk on the development of rancidity in Cheddar cheese. *J. Dairy Res.* **43**:459–468.

66. Liewen, M. B., and E. H. Marth. 1985. Growth and inhibition of microorganisms in the presence of sorbic acid: a review. *J. Food Prot.* **48**:364–375.

67. Macura, D., and P. M. Townsley. 1984. Scandinavian ropy milk—identification and characterization of endogenous ropy lactic streptococci and their extracellular excretion. *J. Dairy Sci.* **67**:734–744.

68. Malik, R. K., R. Prasad, and D. K. Mathur. 1985. Effect of some nutritional and environmental factors on extracellular protease production by *Pseudomonas* sp. B-25. *Lait* **65**:169–183.

69. Marth, E. H., C. M. Capp, L. Hasenzahl, H. W. Jackson, and R. V. Hussong. 1966. Degradation of potassium sorbate by *Penicillium* species. *J. Dairy Sci.* **49**:1197–1205.

70. Martin, J. H., D. P. Stahly, W. J. Harper, and I. A. Gould. 1962. Sporeforming microorganisms in selected milk supplies. *Proceedings of the XVI International Dairy Congress* **C**:295–304.

71. Massa, S., F. Gardini, M. Sinigaglia, and M. E. Guerzoni. 1992. *Klebsiella pneumoniae* as a spoilage organism in Mozzarella cheese. *J. Dairy Sci.* **75**:1411–1414.

72. Masson, P. L., and J. F. Heremans. 1971. Lactoferrin in milk from different species. *Comp. Biochem. Physiol.* **39B**:119–129.

73. **McKellar, R. C.** 1982. Factors influencing the production of extracellular proteinase by *Pseudomonas fluorescens*. *J. Appl. Bacteriol.* **53:**305–316.

74. **McKellar, R. C.** 1989. Regulation and control of synthesis, p. 153–172. *In* R. C. McKellar (ed.), *Enzymes of Psychrotrophs in Raw Foods.* CRC Press, Inc., Boca Raton, Fla.

75. **McKellar, R. C., and H. Cholette.** 1986. Possible role of calcium in the formation of active extracellular proteinase by *Pseudomonas fluorescens*. *J. Appl. Bacteriol.* **60:**37–44.

76. **McKellar, R. C., K. Shamsuzzaman, C. San Jose, and H. Cholette.** 1987. Influence of iron (iii) and pyoverdine, a siderophore produced by *Pseudomonas fluorescens* B52, on its extracellular proteinase and lipase production. *Arch. Microbiol.* **147:**225–230.

77. **Meer, R. R., J. Baker, F. W. Bodyfelt, and M. W. Griffiths.** 1991. Psychrotrophic *Bacillus* spp. in fluid milk products: a review. *J. Food Prot.* **54:**969–979.

78. **Mikolojcik, E. M., and N. T. Simon.** 1978. Heat resistant psychrotrophic bacteria in raw milk and their growth at 7°C. *J. Food Prot.* **41:**93–95.

79. **Morgan, M. E.** 1976. The chemistry of some microbially induced flavor defects in milk and dairy foods. *Biotechnol. Bioeng.* **18:**953–965.

80. **Morse, P. M., H. Jackson, C. H. McNaughton, A. G. Leggatt, G. B. Landerkin, and C. K. Johns.** 1968. Investigation of factors contributing to the bacteria count of bulk tank milk. II. Bacteria in milk from individual cows. *J. Dairy Sci.* **51:**1188–1191.

81. **Mosteller, T. M., and J. R. Bishop.** 1993. Sanitizer efficacy against attached bacteria in milk biofilm. *J. Food Prot.* **56:**34–41.

82. **Mottar, J. F.** 1989. Effect on the quality of dairy products, p. 227–243. *In* R. C. McKellar (ed.), *Enzymes of Psychrotrophs in Raw Food.* CRC Press, Inc., Boca Raton, Fla.

83. **Muir, D. D.** 1989. The microbiology of heat treated fluid milk products, p. 209–270. *In* R. K. Robinson (ed.), *Dairy Microbiology*, vol. 1. Elsevier Applied Science, New York, N.Y.

84. **Myhara, R. M., and B. Skura.** 1990. Centroid search optimization of cultural conditions affecting the production of extracellular proteinase by *Pseudomonas fragi* ATCC 4973. *J. Appl. Bacteriol.* **69:**530–538.

85. **Nashif, S. A., and F. E. Nelson.** 1953. The lipase of *Pseudomonas fragi*. III. Enzyme action in cream and butter. *J. Dairy Sci.* **36:**481–488.

86. **Nooitgedagt, A. J., and B. J. Hartog.** 1988. A survey of the microbiological quality of Brie and Camembert cheese. *Neth. Milk Dairy J.* **42:**57–72.

87. **Overcast, W. W., and K. Atmaran.** 1974. The role of *Bacillus cereus* in sweet curdling of fluid milk. *J. Milk Food Technol.* **37:**233–236.

88. **Phillips, J. D., and M. W. Griffiths.** 1986. Factors contributing to the seasonal variation of *Bacillus* species in pasteurized products. *J. Appl. Bacteriol.* **61:**275–285.

89. **Reddy, M. C., D. D. Bills, R. C. Lindsay, and L. M. Libbey.** 1968. Ester production by *Pseudomonas fragi*. I. Identification and quantification of some esters produced in milk cultures. *J. Dairy Sci.* **51:**656–659.

90. **Reimerdes, E. H.** 1982. Changes in the proteins of raw milk during storage, p. 271. *In* P. F. Fox (ed.), *Developments in Dairy Chemistry*, part I. Applied Science, London, England.

91. **Rengpipat, S., and E. A. Johnson.** 1989. Characterization of a *Lactobacillus* strain producing white crystals on Cheddar cheese. *Appl. Environ. Microbiol.* **56:**2579–2582.

92. **Rohm, H., F. Eliskases-Lechner, and M. Bräuer.** 1992. Diversity of yeasts in selected dairy products. *J. Appl. Bacteriol.* **72:**370–376.

93. **Rosen, B., U. Merin, and I. Rosenthal.** 1989. Evaluation of clostridia in raw milk. *Milchwissenschaft* **44:**356–357.

94. **Rosen, B., G. Popel, and I. Rosenthal.** 1990. The affinity of *Clostridium tyrobutyricum* to casein in raw milk. *Milchwissenschaft* **45:**152–154.

95. **Rowe, M. T.** 1990. Growth and extracellular enzyme production by psychrotrophic bacteria in raw milk stored at low temperature. *Milchwissenschaft* **45:**495–499.

96. **Russell, H. L.** 1885. Gas producing bacteria and the relation of the same to cheese, p. 139–150. *Wis. Agr. Expt. Sta. 12th Annual Report.*

97. **Rutzinski, J. L., E. H. Marth, and N. F. Olson.** 1979. Behavior of *Enterobacter aerogenes* and *Hafnia* species during the manufacture and ripening of Camembert cheese. *J. Food Prot.* **42:**790–793.

98. **Schröder, M. J. A.** 1984. Origins and levels of post pasteurization contamination of milk in the dairy and their effects on keeping quality. *J. Dairy Res.* **51:**59–67.

99. **Shannon, E. L., N. F. Olson, and J. H. von Elbe.** 1969. Effect of lactic starter culture on pink discoloration and oxidation-reduction potential in Italian cheese. *J. Dairy Sci.* **52:**1567–1561.

100. **Shehata, A. E., M. N. I. Magdoub, N. E. Sultan, and Y. A. El-Samragy.** 1983. Aerobic mesophilic and psychrotrophic sporeforming bacteria in buffalo milk. *J. Dairy Sci.* **66:**1228–1231.

101. **Shipe, W. F., R. Bassette, D. D. Deane, W. L. Dinkley, E. G. Hammond, W. J. Harper, D. H. Klein, M. E. Morgan, J. H. Nelson, and R. A. Scanlan.** 1978. Off flavors in milk: nomenclature, standards, and bibliography. *J. Dairy Sci.* **61:**856–869.

102. **Stead, D.** 1986. Microbial lipases: their characteristics, role in food spoilage and industrial uses. *J. Dairy Res.* **53:**481–505.

103. **Stepaniak, L., E. Zakrzewski, and T. Sorhaug.** 1991. Inactivation of heat-stable proteinase from *Pseudomonas fluorescens* P1 at pH 4.5 and 56°C. *Milchwissenshaft* **46:**139–142.

104. **Suhren, G.** 1989. Producer microorganisms, p. 3–34. *In* R. C. McKellar (ed.), *Enzymes of Psychrotrophs in Raw Foods.* CRC Press, Inc., Boca Raton, Fla.

105. **Ternstrom, A., M. A. Lindberg, and G. Molin.** 1993. Classification of the spoilage flora of raw and pasteurized bovine milk, with special reference to *Pseudomonas* and *Bacillus*. *J. Appl. Bacteriol.* **75:**25–34.

106. Thomas, S. B. 1966. Sources, incidence, and significance of psychrotrophic bacteria in milk. *Milchwissenschaft* **21**:270–275.

107. Thomas, S. B., and B. F. Thomas. 1973. Psychrotrophic bacteria in refrigerated bulk-collected raw milk. Part I. *Dairy Ind.* **38**:11–15.

108. Thualt, D., E. Beliard, J. Je Guern, and C.-M. Bourgeois. 1991. Inhibition of *Clostridium tyrobutyricum* by bacteriocin-like substances produced by lactic acid bacteria. *J. Dairy Sci.* **74**:1145–1150.

109. Torrey, G. S., and E. H. Marth. 1977. Isolation and toxicity of molds from foods stored in homes. *J. Food Prot.* **40**:187–190.

110. Tsai, W.-Y. J., M. B. Liewen, and L. Bullerman. 1988. Toxicity and sorbate sensitivity of molds isolated from surplus commodity cheese. *J. Food Prot.* **51**:457–462.

111. Turner, K. W., and T. D. Thomas. 1980. Lactose fermentation in Cheddar cheese and the effect of salt. *N. Z. J. Dairy Sci. Technol.* **15**:265–276.

112. Vilafafila, A., J. D. Frias, P. Abad, and C. Rodriguez-Fernandez. 1993. Extracellular proteinase activity from psychrotrophic *Pseudomonas fluorescens* biovar 1 (*Ps. fluorescens* NC1). *Milchwissenschaft* **48**:435–438.

113. Wang, J. J., and J. F. Frank. 1981. Characterization of psychrotrophic bacterial contamination of commercial buttermilk. *J. Dairy Sci.* **64**:2154–2160.

114. Wendorff, W. L, W. E. Riha, and E. Muehlenkamp. 1993. Growth of molds on cheese treated with heat or liquid smoke. *J. Food Prot.* **56**:963–966.

115. Westall, S., and O. Liltenborg. 1998. Spoilage yeasts of decorated soft cheese packed in modified atmosphere. *Food Microbiol.* **15**:243–249.

116. Wolfson, L. M., and S. S. Sumner. 1993. Antibacterial activity of the lactoperoxidase system: a review. *J. Food Prot.* **56**:887–892.

117. Yan, L., B. E. Langlois, J. O'Leary, and C. Hicks. 1983. Effect of storage conditions of grade A raw milk on proteolysis and cheese yield. *Milchwissenschaft* **38**:715–719.

Food Microbiology: Fundamentals and Frontiers, 2nd Ed.
Edited by M. P. Doyle et al.
© 2001 ASM Press, Washington, D.C.

Robert E. Brackett

Fruits, Vegetables, and Grains

7

INTRODUCTION

Foods of plant origin are diverse in composition. Consequently, patterns of microbiological spoilage differ substantially, sometimes dramatically. Despite their differences, plant products share some fundamental characteristics. They are all horticultural or agronomic products whose structural integrity depends on cellulose and pectin. This chapter aims to discuss the basic microflora and mechanisms involved in the spoilage of plant products while providing information on specific commodities.

Scientific literature describing the spoilage of fruits, vegetables, and grains is often confusing to readers not familiar with the topic. This is because spoilage of foods of plant origin often crosses the interface between traditional plant pathology and food microbiology. Both subdisciplines have distinct viewpoints and philosophies and use unique jargon. These differences often confuse readers as to the role of microflora in the spoilage of plant products and appropriate means of prevention.

For the purpose of this chapter, plant pathology as it relates to degradation of plant materials will be restricted to spoilage problems that arise before harvest, whereas food microbiology will deal with spoilage after harvest.

However, the reader should understand that scientific literature makes no such distinction, and differences between plant pathology and food microbiology are usually subtle. Consequently, some definition of terms often found in the literature is worth discussing. Although scientific definitions and differences exist between fruits and vegetables, some confusion does occasionally occur, especially in the minds of consumers. Fruits are defined as the seed-bearing organs of plants and include not only well-known commodities such as apples, citrus fruits, and berries, but also items sometimes thought of as vegetables, such as tomatoes, bell peppers, and cucumbers. In contrast, vegetables are defined as all other edible portions of plants including leaves, roots, and seeds.

Pathogens

Although most microbiologists recognize that a pathogen is a microorganism capable of causing disease, the host upon which the disease is inflicted is often assumed. Most clinical, veterinary, and food microbiologists understand pathogens to mean microorganisms that cause illness in humans or animals. In contrast, plant pathologists and other agricultural scientists who work with horticultural and agronomic commodities

---

Robert E. Brackett, Office of Plant and Dairy Foods and Beverages, U.S. Food and Drug Administration, 200 C St., S.W., Washington, DC 20204.

often think of pathogens as microorganisms that cause disease or decay of plants. Such assumptions are evident in the scientific literature and can confuse readers not familiar with this specialty area. However, confusion can easily be avoided by specifying the type of pathogen, i.e., plant pathogen or human pathogen, being discussed.

Plant pathogens can also be subdivided into those that are true plant pathogens and those that are opportunistic plant pathogens. Although exceptions exist, true plant pathogens are usually understood to be microorganisms which possess the ability to actively infect plant tissues. The ability to infect is derived from the ability of microorganisms to produce one of several degradative enzymes which enable them to penetrate protective external layers of cells. In contrast, opportunistic plant pathogens infect tissues only when normal defenses of the plant have been compromised in some way, e.g., by mechanical damage or insects.

Field diseases, storage diseases, and market diseases are descriptive terms for spoilage used in older or nonscientific literature. Field diseases include defects which are related to microbiological spoilage which occur either before or soon after harvest. In contrast, storage or market diseases refer to spoilage which is manifested some time after harvest, particularly during storage or marketing (39). Both terms are inadequate for scientific discussion because they describe the time at which spoilage becomes apparent but do not address specific etiologies or microorganisms involved. Moreover, some microorganisms that can cause field diseases often do not manifest themselves until the fruit or vegetable is harvested. One example involves the mold *Phytophthora infestans*, which causes late blight of potatoes. This microorganism can spread to potatoes either via airborne spores or by infecting neighboring potato plants. In most cases, infection causes death of the plant before potato tubers are harvested and, thus, can be defined as a storage disease. However, *P. infestans* can remain dormant in the tubers, only to rot the tubers during storage or reinfect plants of a new crop (14).

Spoilage

The term "spoilage" connotes different meanings to different people. In its broadest sense, spoilage refers to any change that occurs in a food whereby the food is made unacceptable for human consumption. This definition can include safety- as well as quality-related defects. Although microbiological safety is an important concern, this aspect of spoilage will be discussed only briefly here, since it will be dealt with in subsequent chapters. This chapter will focus on changes in color, flavor, texture, or aroma brought about by the growth of microorganisms on fruits, vegetables, and grains.

Spoilage is typically something that has a negative connotation, and a large part of research and development in the food industry is focused on extending shelf life, primarily by reducing spoilage. However, spoilage is not always undesirable. Undesirable colors, odors, and flavors often serve as a warning to consumers that foods have been improperly handled or stored too long. For example, Berrang et al. (3) observed that modified-atmosphere storage extended the shelf life of asparagus long enough to allow *Listeria monocytogenes* to reach higher populations than would have occurred with ambient-air storage. However, some types of spoilage can also increase food safety risks by making the microenvironment more suitable for human pathogens. Wells and Butterfield (56) reported that *Salmonella enterica* serovar Typhimurium grew better on tomatoes, potatoes, and onions in the presence of the spoilage molds *Botrytis* or *Rhizopus* than when the molds were absent. Growth of *Escherichia coli* O157:H7 is known to occur in bruised areas of apples (24). Hence, food microbiologists and technologists should always consider whether the benefits of reduced spoilage outweigh safety concerns.

Types of Spoilage

In general, three broad types of spoilage exist in plant products. The first type is active spoilage, caused by plant pathogenic microorganisms actually initiating infection of otherwise healthy and uncompromised products, resulting in reduction in sensory quality. A second type of spoilage is passive or wound-induced spoilage, in which opportunistic microorganisms gain access to internal tissues via damaged epidermal tissue, i.e., peels or skins. This type of spoilage often occurs soon after the product has been damaged by harvesting or processing equipment or by insects. For example, the common fruit fly (*Drosophila melanogaster*) can inoculate vegetables with *Rhizopus* species as it deposits eggs in wounds or other breaks in the epidermis, i.e., the outer protective tissue system of plant parts (13). Similarly, passive spoilage can occur when opportunistic spoilage microorganisms gain entry to internal tissues via lesions caused by plant pathogens or via natural openings such as lenticels, stomata, or hydathodes (37).

Spoilage of plant products can be manifested in a variety of ways, depending on the specific product, the environment, and the microorganisms involved. Traditionally, spoilage has been described by symptoms most often associated with a particular product. Listed in Table 7.1 are examples of some common types of fungal spoilage

Table 7.1 Types of fungal spoilage of fruits and vegetables[a]

Product	Type of spoilage	Mold responsible	Product involved
Fruits	*Alternaria*	*Alternaria* sp.	Citrus fruits
	Anthracnose (bitter rot)	*Colletotrichum musae*	Bananas
	Black rot	*Aspergillus niger, Ceraocystis fimriata*	Onions, sweet potatoes
	Brown rot	*Monilinia fructicola*	Peaches
	Crown rot	*Colletotrichum musae, Fusarium roseum, Verticillium theobromae, Ceratocystis paradoxa*	Bananas
	Gray mold rot	*Botrytis cinerea*	Grapes
	Pineapple black rot	*Ceratocystis paradoxa*	Pineapples
	Sour rot	*Geotrichum candidum*	Tomatoes, citrus fruits
	Lenticel rot	*Cryptosporiopsis malicorticus, Phylctaena vagabunda*	Apples, pears
	Green mold rot	*Penicillium digitatum*	Citrus fruits
	Blue rot	*Penicillium* sp.	Oranges
	Cladosporium rot	*Cladosporium herbarum*	Peaches, cherries
Vegetables	Black mold rot	*Aspergillus* sp.	Onions
	Black rot	*Alternaria* sp.	Carrots, cauliflower
	Downy mildew	*Bremia, Phytophthora* spp.	Lettuce, spinach
	Fusarium rot	*Fusarium* sp.	Asparagus
	Gray mold rot	*Botrytis* sp.	Cabbage
	Rhizopus soft rot	*Rhizopus* sp.	Green beans
	Smudge (anthracnose)	*Colletotrichum* sp.	Onions
	Tuber rot	*Fusarium* sp.	Potatoes
	Watery soft rot	*Sclerotinia* sp.	Celery
	Wilt	*Pythium* sp.	Green beans
	Blue rot	*Penicillium* sp.	Oranges
	Blight	*Phomopsis* sp.	Eggplant
	Finger rot	*Pestalozzia, Fusarium, Gleosporium* spp.	Bananas
	Pink rot	*Trichothecium* sp.	Peaches

[a]Adapted from Jay (34).

of fruits and vegetables and the molds normally associated with them. More comprehensive lists of microorganisms associated with spoilage of fruits and vegetables can be found in recent reviews (25, 41, 43). For some fruits or vegetables, this system is sufficient to adequately describe a spoilage problem. In other cases, however, the system is inadequate because more than one type of microorganism may produce identical or similar symptoms. For example, the best-known type of spoilage in vegetables is soft rot, which is usually evidenced by obvious softening and deliquescence of the plant tissue, especially around an initial point of infection. However, the term soft rot often does not fully describe the disease or spoilage problem because softening can be caused by various plant pathogenic bacteria, most notably *Erwinia carotovora* and several species of *Pseudomonas* (37). However, bacteria such as

Bacillus and *Clostridium* spp., and yeasts and molds, are also occasionally implicated in soft rot spoilage (40).

Types of spoilage are also known by the names of the microorganisms which cause them. For example, *Fusarium* rot and *Rhizopus* soft rot describe not only the symptoms of the spoilage but also the mold that causes them. This system of describing spoilage is preferable to naming the spoilage based on symptoms alone. However, to date no organized system of applying names to specific types of spoilage has been adopted.

Mechanism of Spoilage

Intact healthy plant cells possess a variety of defense mechanisms to resist microbial invasion. Thus, before microbial spoilage can occur, these defense mechanisms must be overcome. Agrios (1) provides a good

Table 7.2 Some compounds produced by microorganisms responsible for off flavors or odors in fruit and vegetable products[a]

Product	Microorganisms responsible	Chemical compound
Apple juice	*Alicyclobacillus* sp.	2,6-Dibromophenol, 2,6-dichlorophenol
Apples, pears, cherries	*Penicillium* sp.	Geosmin
Dried coconut	*Eurotium* sp.	Methyl ketones
	Bacillus subtilis	2,3,5,6-Tetramethylpyrazine, 2,3,5-trimethylpyrazine
Dried fruit, coffee	Various molds	2,4,6-Trichloroanisole
Orange juice	Lactic acid bacteria	Diacetyl, acetoin, 2,3,-hydroxybutane
	Penicillium sp.	4-Vinylguaiacol
Papaya puree	Bacteria	Methyl esters, short-chain fatty acids (e.g., butanoic acid)
Canned champignons	Actinomycetes	2-Methylisoborneol
Navy beans	Actinomycetes	Geosmin
Potatoes	*Erwinia carotovora*	Skatole
	Clostridium scatologenes	Indole, *p*-cresol

[a] Data from Whitfield (57).

general overview of the physical and chemical mechanisms by which microorganisms are able to invade plant tissues.

Fruits, vegetables, grains, and legumes have an epidermal layer of cells, i.e., skin, peel, or testa, that provides a protective barrier against infection of internal tissues. The composition of the epidermal tissue varies, but walls of cells in this tissue usually consist of cellulose and pectic materials, and the outermost cells are covered by a layer of waxes (cutin). This tissue forms a barrier which is resistant to penetration by most microorganisms. If this barrier is compromised in some way, however, the possibility exists for easy entry of microorganisms to internal tissues.

External barriers can be compromised in a number of ways. One obvious way is if the epidermal tissue is damaged by external sources. Damage due to relatively uncontrollable factors such as insect infestation, windblown sand, or rubbing against neighboring surfaces can occur before harvest. After harvest, the product is most often damaged by harvesting or processing equipment, particularly those that are poorly designed or maintained. In addition, some microorganisms, especially plant pathogenic molds, possess mechanisms to penetrate external tissues of plants. Once external barriers are penetrated, not only do these microorganisms quickly invade internal tissues, but other opportunistic microorganisms often take advantage of the availability of nutrients present in damaged tissue and cause additional spoilage.

Once microorganisms penetrate the outer tissues of fruits, vegetables, or grains, they must still gain access to internal areas and individual cells to extract nutrients. Plant tissues are held together by the middle lamella, which is composed primarily of pectic substances. These substances, composed of pectic acid, pectinic acid, and pectin, are polymers of α-1,4-linked D-galactopyranosyluronic acid units interspersed with 1,2-linked rhamnopyranose units (18). Any degradation of the middle lamella results in detachment of cells from one another. The integrity of individual plant cells is due to the cell wall, which consists of two main layers, viz., the primary cell wall layer, composed of cellulose and pectates, and the secondary cell wall layer, which consists almost entirely of cellulose.

Changes in the physiological state of plants can also result in openings in external tissues. For example, dehydration of vegetables can result in separation of cells, thereby allowing cracks to develop and facilitating the entry of microorganisms to internal tissues. In such cases, microorganisms that normally would not cause spoilage problems can utilize fermentable carbohydrates and produce metabolites which cause undesirable changes in flavor, aroma, or color. The sources of these objectionable flavors and odors are often chemically complex molecules produced by the spoilage organisms (Table 7.2).

Degradative Enzymes

Degradative enzymes play an important role in postharvest spoilage of plant products. Five classes of microbial enzymes are primarily responsible for degradation of plant materials (37). These are pectinases, cellulases, proteases, phosphatidases, and dehydrogenases. Because pectin and cellulose constitute the main structural components of plant cells, pectinases and cellulases are the most important degradative enzymes involved in spoilage.

Pectinases are enzymes that cause depolymerization of the pectin chain. Although pectinases are often discussed as a single enzyme, three main types are recognized for their role in plant spoilage (18, 20). Differences exist primarily in the type and site of reaction on the pectin polymer. Pectin methyl esterase (PME; EC 3.1.1.11) (31) hydrolyzes ester groups from pectin chains with subsequent production of methyl alcohol (1, 19). Although PME does not directly act to decrease chain length, it does affect pectin solubility and the rate at which other types of pectinases react. PME is produced by several plant pathogens, including *Botrytis cinerea*, *Monilinia fructicola*, *Penicillium citrinum*, and *E. carotovora* (18). However, PME is also produced by some microorganisms which are not considered plant pathogens (20).

In contrast to PME, polygalacturonase (PG; EC 3.2.1.17) and pectin lyase (PL; EC 4.2.2.2) (31) are chain-splitting pectinases which reduce the overall length of the pectin chain. Their mechanism of action differs primarily in the way in which pectin chains are cleaved. PG cleaves the pectin chain by hydrolyzing the linkage between two galacturonan molecules, whereas PL depolymerizes by β-elimination of the linkage. Both PG and PL can exist as endopectinases, which act on middle portions of the pectin chain, or as exopectinases, which act on terminal ends of the chains. The ultimate degradation of the pectin chain results in liquefaction of the pectin and complete maceration of plant tissues. The production of pectic enzymes by microorganisms does not necessarily mean that spoilage will result. For example, *Flavobacterium* species produce pectic enzymes without causing marked softening of plant tissues (39). In such cases the enzymes themselves may not be active in the target tissues. In addition, some plant tissues contain pectic enzyme inhibitors (20). Specifics describing regulation and genetics of pectinase production by bacteria have been published elsewhere (20, 36, 47).

Cellulases constitute the second major class of degradative enzymes that can lead to spoilage. Cellulases function by degrading cellulose, which is a glucose polymer, to glucose. Like pectinases, several types of cellulases exist, some of which attack native cellulose by cleaving cross-linkages between chains, while others act by breaking the cellulose into shorter chains (1). Cellulase activity is important in plant product spoilage not only by contributing to tissue softening and maceration, but also by producing glucose, which can be used by microorganisms devoid of degradative enzymes. Although cellulases are important in enabling plant pathogens to initiate infection, they are of lesser importance in

postharvest spoilage of plant products than are pectinases (37).

Influence of Physiological State

The physiological state of plant products, especially those consisting of fruits or vegetables, can have a dramatic effect on susceptibility to microbiological spoilage (32). Fruits, vegetables, and grains usually possess some sort of defense mechanism to resist infection by microorganisms. Usually these mechanisms are most effective when the plant is at peak physiologic health. Once plant tissues begin to age or are in a suboptimal physiologic state, resistance to infection diminishes (46).

Fruits and vegetables differ in the way in which they change physiologically after detachment from the plant. Nonclimacteric fruits and vegetables, such as strawberries, beans, and lettuce, cease to ripen once they have been harvested. In contrast, climacteric fruits and vegetables such as bananas and tomatoes continue to mature and ripen after harvest. Most consumers note development of color or softening of texture as the most obvious indications of ripening. However, ripening can continue to the point where fruits and vegetables are overripe, i.e., when normal cell integrity begins to diminish and tissues deteriorate (46). This process, known as senescence, arises from the accumulation of degradative enzymes produced by the fruit or vegetable and is unrelated to microbial decay. However, loss of cellular integrity brought about by senescence makes fruits and vegetables even more susceptible to microbial infection and spoilage. Consequently, climacteric fruits and vegetables are usually among the most perishable of plant products.

MICROBIOLOGICAL SPOILAGE OF VEGETABLES

Although vegetables, fruits, grains, and legumes are all plant-derived products, they possess inherent differences which influence both the type of microorganisms that form the natural microflora and the type of spoilage encountered. Important intrinsic factors that influence the microflora that develop on plant products include pH and water activity (a_w). The a_w of fresh fruits and vegetables is high enough to support growth of most bacteria and fungi and is therefore not considered a limiting factor. Similarly, with the exception of tomatoes, the pH of most vegetables is in the 5.0 to 6.0 range, which does not inhibit the growth of most microorganisms.

Natural Microflora

In general, the natural microflora of vegetables includes bacteria, yeasts, and molds representing many genera.

Table 7.3 Psychrotrophic and lactic acid bacteria and yeasts identified in stored vegetable salads[a]

Psychrotrophic bacteria	% of isolates	Lactic acid bacteria	% of isolates	Yeasts	% of isolates
Enterobacter intermedium	1	*Lactobacillus curvatus*	2	*Candida valida*	4
Pasteurella haemolytica biovar T	1	*Lactobacillus lactis*	2	*Geotrichum* sp.	4
Pseudomonas spp.	1	*Lactobacillus fermentum*	4	*Hansenula anomala*	4
Pseudomonas aeruginosa	1	*Lactobacillus plantarum*	7	*Rhodotorula splutinis*	4
Pseudomonas picketti biovar 1	1	*Lactobacillus paracaski*	11	*Torulopsis* spp.	4
Pseudomonas putrefaciens	1	*Lactobacillus mesenteroides*	16	*Candida lambica*	11
Pseudomonas stutz/mendo group	1	*Lactobacillus brevis*	22	*Trichosporon* sp.	18
Salmonella choleraesuis	1	ND	36	ND	50
Acinetobacter sp.	3				
Enterobacter agglomerans	3				
Pasteurella urae	3				
Plesiomonas shigelloides	3				
Pseudomonas maltophilia	3				
Staphylococcus cohnii	4				
Acinetobacter anitratus	5				
Chromobacterium violaceum	7				
Pseudomonas putida	8				
Pseudomonas fluorescens	19				
ND	33				

[a] Adapted from Garcia-Gimeno and Zurera-Cosano (29). ND, not determined.

An example of the wide variety of microorganisms associated with fruit and vegetable products is shown in Table 7.3 (29). However, microflora can vary considerably depending on the type of vegetable, environmental considerations, seasonality, and whether the vegetables were grown in close proximity to the soil. Bacteria normally present on vegetables at the time of harvest include both gram-positive and gram-negative forms. However, the manner in which vegetables are stored will often influence subsequent development of particular groups of microorganisms. For example, refrigeration tends to select for psychrotrophic bacteria, such as *Pseudomonas* species (14, 15). Many indigenous microorganisms are not considered to be plant pathogens but rather opportunistic microorganisms that are frequently associated with spoilage.

Spoilage Microflora

A variety of microorganisms can cause spoilage of vegetables. Table 7.1 lists some molds that cause spoilage and the types of spoilage associated with them. Yeasts, molds, and bacteria cause spoilage of vegetables; however, bacteria are more frequently isolated from initial spoilage defects (14,15, 37–39, 43). The reason for this is that bacteria are able to grow faster than yeasts or molds in most vegetables and therefore have a competitive advantage, particularly at refrigeration temperatures.

Extensive postharvest spoilage of raw vegetables occurs in the absence of other treatments. Spoilage can be influenced by the history of the land on which vegetables are grown (14). For example, repeated planting of one type of vegetable on the same land over several seasons can lead to the accumulation of plant pathogens in the soil and increased potential for spoilage (39). Likewise, soil contaminated by floodwater or poor-quality irrigation water may expose vegetables to spoilage microorganisms or human pathogens (11).

Virtually all vegetables receive at least some type of processing or handling before they are consumed. It is important to realize that many common processing steps increase the likelihood or chances for spoilage (42). Vegetables normally undergo washing and rinsing treatments to remove surface debris. Once washing is complete, vegetables are typically allowed to air dry or are partially dried in centrifugal spin dryers. In addition to removing surface soil and debris, the rinsing step also often reduces the number of microorganisms on the surface of vegetables (16).

In most cases, processors of vegetables destined to be sold as fresh produce add from 5 to 250 μl of chlorine per liter to wash water. Chlorine is an effective antimicrobial agent. However, washing with chlorinated water and other disinfectants has only a limited antimicrobial effect on the microflora of the produce (5). Senter et al. (49) observed that 90 to 280 μg of chlorine per liter had a minimal effect on the microflora of tomatoes. In contrast, Beuchat and Brackett (8) reported that dipping tomatoes in a 200- to 250-μg/liter chlorine solution significantly reduced populations of aerobic mesophilic microorganisms but not psychrotrophs or fungi. However, differences in microflora on treated and untreated tomatoes were not evident after 4 days of storage at 10°C. It has also been observed that carrots washed in water containing 200 to 260 μg of chlorine per liter contained about 10-fold fewer total aerobic microorganisms than carrots washed in water containing no chlorine (7). In this case, however, chlorine-treated carrots developed significantly higher populations of mesophiles, psychrotrophs, and fungi than did untreated carrots. The role of chlorine in reduction of populations of microorganisms and its effectiveness in extending the shelf life of fresh-cut vegetables has been demonstrated (23, 50).

Although many processors consider washing fresh produce with chlorinated water to be a disinfection step, this is not the real purpose for the inclusion of chlorine. Rather, chlorine is more effective in killing microorganisms in the water and minimizing contamination of the vegetables by the rinse water. Washing can actually contribute to spoilage problems if the rinse water contains a large amount of organic matter or if chlorine concentrations are not closely monitored and maintained (14).

Cutting, slicing, chopping, and mixing are other important processing steps to which fresh vegetables are commonly exposed. These operations are becoming even more important as the demand for ready-to-eat products has increased. These processes can sometimes have a dramatic influence on populations of microorganisms and their growth on fresh vegetables. For example, Splittstoesser (51) showed that cutting corn, slicing green beans, and chopping spinach increased populations of microorganisms by about 1 $\log_{10}$ CFU/g. Priepke et al. (45) similarly observed that populations of microorganisms increased faster on cut than on intact salad vegetables.

Processing operations that damage vegetable tissues can lead to increased microbial populations in several ways. First, poorly sanitized equipment can harbor contaminants that are transferred to the vegetables during the cutting operation. For example, *Geotrichum candidum*, sometimes called machinery mold or dairy mold, can accumulate on processing equipment and contaminate vegetables upon contact (14). Second, cutting, and slicing allow vegetable tissue fluids to be expressed onto outer surfaces of the vegetable as well as the processing equipment. These fluids can then serve as substrates for growth of microorganisms, allowing them to accumulate to higher populations (14, 28, 33, 54) and result in biofilm development (17, 27).

Storage, packaging—particularly modified-atmosphere packaging (MAP)—and transportation are other important processing steps that can influence the development of microbiological spoilage of vegetables. Most modified-atmosphere techniques involve reducing the concentration of oxygen while increasing the concentration of carbon dioxide. MAP functions by reducing respiration and the senescence process, thereby delaying undesirable changes in sensory quality. The specific composition of gas used for MAP varies depending on the product. Reduction of senescence in produce usually requires carbon dioxide concentrations of at least 5%; however, concentrations in excess of 20% can be detrimental to quality (14).

Microorganisms respond to MAP differently, depending upon their tolerance to oxygen and carbon dioxide. MAP can affect obligate aerobic microorganisms by displacing needed oxygen. In addition, carbon dioxide can directly affect microorganisms by causing a reduction in pH of the cytoplasm and interfering with normal cellular metabolism (22). In general, aerobic gram-negative bacteria are most sensitive to carbon dioxide, whereas obligately and facultatively anaerobic microorganisms are more resistant (2). Likewise, molds are more sensitive than are fermentative yeasts to carbon dioxide (21). However, the specific effects that modified atmosphere will have on the complex microbial ecology of fresh produce are not easy to predict (2).

Despite the ease with which one can often demonstrate the influence of carbon dioxide on microbial activity in the laboratory, the same is not always true in MAP produce. For example, storage of lettuce (6), asparagus, broccoli, cauliflower (4), and tomatoes (8) resulted in extension of shelf life but did not have an appreciable effect on populations of microorganisms. Similarly, Priepke et al. (45) observed that 10.5% carbon dioxide had only minimal effects on populations of aerobic microorganisms on salad vegetables stored in MAP. In contrast, Berrang et al. (4) reported that modified atmosphere (10% carbon dioxide, 11% oxygen) significantly inhibited the growth of total aerobic microorganisms in fresh broccoli. Similarly, Garcia-Gimeno and Zurera-Cosano (29) attributed enhanced growth of lactic acid bacteria and absence of molds in vegetable salads to the use of modified-atmosphere storage.

More extensive processing techniques such as thermal processing (canning) and freezing have been commonly used for many years to preserve vegetables. Microbiological spoilage of frozen vegetables is primarily caused by improper storage temperatures. The microflora in frozen vegetables is essentially the same as in raw products. Therefore, spoilage patterns can be expected to be similar to those of raw products if the products are not properly maintained in a frozen state. However, some molds have been reported to grow at temperatures as low as $-5°C$ (13). Thus, it is at least theoretically possible for frozen vegetable products to succumb to microbiological spoilage.

Preservation of vegetables by canning is accomplished by placing vegetables in hermetically sealed containers and then heating sufficiently to destroy microorganisms. For most vegetables, processing temperatures frequently exceed 120°C, which eliminates all but the most heat-resistant bacterial endospores. Consequently, spoilage of canned vegetables is usually caused by thermophilic spore-forming bacteria unless container integrity has been compromised in some way (30). Acidification without gas production is one of the most common types of spoilage observed in canned vegetables. This defect, called a "flat sour," is caused by *Bacillus stearothermophilus* or *Bacillus coagulans*. The defect ordinarily occurs when cans receive a marginally adequate thermal treatment or if cans are stored at temperatures above 40°C. Another type of spoilage observed with canned vegetables is evidenced by gas production and consequent swelling of cans. Swelling is caused when thermophilic spore-forming anaerobes, such as *Clostridium thermosaccharolyticum*, grow and produce large amounts of hydrogen and carbon dioxide. The production of hydrogen sulfide in some canned vegetables can lead to a spoilage problem known as "sulfide stinker." This type of spoilage usually occurs in the absence of can swelling (30).

MICROBIOLOGICAL SPOILAGE OF FRUITS

Fresh fruits are similar to vegetables in that they usually have a high enough a_w to support growth of all but the most xerophilic or osmophilic fungi. However, most fruits differ from vegetables in that they have a more acidic pH (<4.4), the exception being melons, as well as a higher sugar content. In addition, fruits usually possess more effective defense mechanisms such as thicker epidermal tissues and higher concentrations of antimicrobial organic acids.

Normal and Spoilage Microflora

As with vegetables, the normal microflora of fruits is varied and includes both bacteria and fungi (30). Sources of microorganisms include all those mentioned for vegetables, including air, soil, and insects. Unlike vegetables, however, spoilage of fruits is most often due to yeasts or molds, although yeasts are most important (22). However, bacteria can also be involved in the spoilage of fruits. Examples include *Erwinia* rots of pears (34) or

production in orange juice of buttermilk-like flavors by lactic acid bacteria (57).

Many of the same techniques involved in the processing and handling of fresh vegetables also apply to fresh fruits. Fruits are usually washed or rinsed immediately after harvest and then usually packed into shipping cartons. The washing step can play an important role in reducing microbial contamination. Splittstoesser (52) reported that rinsing and scrubbing can reduce populations of microorganisms on apples by more than 99.9%.

Fresh-cut or minimally processed fruit products have become increasingly popular in recent years and will likely become more so due to their convenience in preparation. Despite their popularity, however, cut fruits offer some additional challenges that are less often associated with whole fruits. For example, cutting and slicing will eliminate the protection normally offered by peels and skins. Moreover, high concentrations of sugars in fluids expressed from internal tissues of cut fruits enhance the growth of microorganisms that can tolerate acidic environments. Although both molds and yeasts have this capability, the latter are more often associated with spoilage of cut fruits due to their ability to grow faster than molds (34). However, Pao and Petracek (44) determined that gram-negative bacteria, specifically *Enterobacter agglomerans* and *Pseudomonas* spp., were predominant spoilage organisms of peeled oranges. As expected, yeasts (*Cryptococcus albidus*, *Rhodotorula glutinis*, and *Saccharomyces cerevisiae*) were also important in the spoilage of peeled oranges.

Several fruits are dried to yield intermediate-moisture products which normally rely on low a_w as the main mechanism for preservation. Raisins, prunes, dates, and figs are most commonly consumed as dried fruit products. Dried apples, apricots, and peaches are also popular with consumers. Depending on the means by which fruits are dehydrated, moisture levels can range from less than 5% to 35% (52). In general, most dried fruits have a sufficiently low a_w to inhibit the growth of most bacteria. Spoilage is usually limited to osmophilic yeasts or xerotolerant molds (52). Populations of yeasts and molds on dried fruit usually average under 10^3 CFU/g. Yeasts normally associated with spoilage of dried fruits include *Zygosaccharomyces rouxii* and *Hanseniaspora*, *Candida*, *Debaryomyces*, and *Pichia* species. Molds capable of growth below a_w 0.85 include several *Penicillium* and *Aspergillus* species (especially the *A. restrictus* series), *Eurotium* species (the *Aspergillus glaucus* series), and *Wallemia sebi*. However, spoilage may not always arise from direct contamination by microorganisms. Whitfield (57) described a situation in which a musty defect of dried fruits was traced to volatile chemicals produced as a result of fungal degradation of fiberboard in which the fruits were packaged.

Fruit concentrates, jellies, jams, preserves, and syrups owe their resistance to spoilage to low a_w. Unlike dried fruits, however, the reduced a_w of these products is achieved by adding sufficient sugar to achieve a_w values of 0.82 to 0.94. In addition to reduced a_w, these products are also usually heated to temperatures of 60 to 82°C, which kills most xerotolerant fungi. Consequently, spoilage of these products usually occurs when containers have been improperly sealed or after they are opened by consumers.

Various heat treatments are used to preserve fruit products. Fruits generally require less severe treatment than vegetables due to the enhanced lethality brought about by their acidic pH. Many canned fruits, whether halved, sliced, or diced, are processed by heating the products to a can center temperature of 85 to 90°C (53). Some processed fruit juices and nectars are rapidly heated to 93–110°C and then aseptically filled into containers. In either case, processes are sufficient to kill most vegetative bacteria, yeasts, and molds. However, several genera of molds produce heat-resistant ascospores or sclerotia. Molds typically associated with spoilage of thermally processed fruits products include *Byssochlamys fulva* (anamorph: *Paecilomyces fulvus*), *Byssochlamys nivea* (anamorph: *Paecilomyces niveus*), *Neosartorya fischeri* (anamorph: *Aspergillus fischeri*), and *Talaromyces flavus* (anamorphs: *Penicillium vermiculatum* and *Penicillium dangeardii*) (9, 53). Symptoms of spoilage typically observed in heat-processed fruit products include visible mold growth, off odors, breakdown of fruit texture, or solubilization of starch or pectin in the suspending medium (10).

In recent years, food microbiologists have become aware that the spore-forming bacterium *Alicyclobacillus acidoterrestris* can be a problem in pasteurized juices (55). This bacterium is sufficiently acid and heat tolerant to survive the pasteurization conditions that exclude survival of most other bacteria. *A. acidoterrestris* produces 2-methoxyphenol (guaiacol), which imparts a medicine-like or phenolic off flavor in several types of juices, most notably apple and orange juice (57).

MICROBIOLOGICAL SPOILAGE OF GRAINS AND GRAIN PRODUCTS

Although grains are similar to fruits and vegetables in that they are of plant origin, they differ in many ways. Unlike fruits and vegetables, grains are typically thought of as primarily agronomic or field crops rather than horticultural crops. Grains are planted and cultivated on a larger scale than fruits and vegetables and are harvested only after they have reached full maturity and often have dried to a desired moisture level. After harvest, grains are stored in bins or silos that facilitate additional drying and protection from the weather. Much of the grain produced in developed countries is used as animal feed. Grains destined for human food are ground into flour or meal for bakery or pasta products, or further processed into snacks or breakfast cereals. Final products, except doughs, often have a_w values (<0.65) below which most microorganisms will not grow.

Natural Microflora

Grains and grain products normally contain several genera of bacteria, molds, and yeasts, the specific species being dependent on conditions encountered during production, harvesting, storage, and processing. Although the low a_w of grains and grain products might lead one to believe that fungi, especially molds, are the predominant natural microflora, this is not always the case. Seiler (48), for example, surveyed microbial populations in wheat over a 2-year period and observed that bacterial populations were generally about 10-fold higher than mold populations. Of the many different types of microorganisms present on grains, only a few invade the kernel itself. Molds such as *Alternaria*, *Fusarium*, *Helminthosporium*, and *Cladosporium* are primarily responsible for invading wheat in the field. However, infection by these molds is minimal and does little damage unless the grains are allowed to become too moist. For additional discussions of molds naturally occurring on grains, see chapters 21, 22, and 23.

Effects of Processing

Grains intended for human consumption are rarely used in their native state but rather undergo various processing treatments. For example, grains destined for milling are washed, tempered, screened, and aspirated before being milled. The pretreated grains are then subjected to milling and sifting to separate the hull (bran), germ, and endosperm. The endosperm is then crushed to produce flour. Each step in the pretreatment and milling operations reduces populations of microorganisms such that the flours usually contain lower populations than do the grains from which they are made (26). Aside from fewer microorganisms being present in flour, the profile of microflora in milled grains is similar to that in whole grains. The final processing step to which most grain products, specifically flours, are subjected is the addition of liquids (e.g., water or milk) and other ingredients to produce doughs. The type and amount of these ingredients differ depending on the desired dough and final product.

Types of Spoilage

Properly dried and stored grains and grain products are inherently resistant to spoilage due to their low a_w. However, despite attempts to protect grains from uptake of water during storage, the a_w can increase to a level that enables xerotolerant molds to grow. Moreover, it should be kept in mind that reduced a_w often will only slow the growth of fungi and that spoilage will ultimately occur, given enough time. For example, a moisture content of 11 to 14% may be low enough to prevent spoilage of grains stored for less than 1 year, but a maximum of 10 to 12% may be required for long-term storage. However, temperature differentials of 0.5 to 1°C in large bulk quantities of grains can cause some areas of the bulk to reach moisture levels sufficient to allow mold growth. This happens when moisture evaporates from warm areas of bulk-stored grain and then condenses in cooler areas. In such cases, molds such as *Aspergillus*, *Eurotium* sp., and *W. sebi* may grow and cause spoilage.

Molds are the primary spoilage microorganisms in grains due to their ability to grow at reduced a_w. The primary concern with mold spoilage is the potential production of mycotoxins. However, molds can also affect the sensory quality of grain products. One of the first and most obvious indications of mold growth on grains is a change in appearance. Molded grains may have a powdery appearance. In addition, many molds produce colored spores that can discolor grain-based foods such as breads, crackers, or cakes.

Molds can also cause adverse changes in flavor and aroma of grain products due to the production of aromatic volatile compounds. Examples of some flavor compounds typically produced by molds in grains include 3-methyl-butanol, 3-octanone, 3-octanol, 1-octen-3-ol, 1-octanol, and 2-octen-one-ol (35). Although the specific compounds produced are somewhat affected by the substrate, production of volatile metabolites depends more on fungal species than on grain type (12).

Another important adverse affect of mold growth on grains is an increase in the free fatty acid value (FAV). Increases in FAV can result from fungal lipase activity. Because some of these lipases are heat stable, FAV is sometimes used as an indication of fungal spoilage of grains.

Spoilage of grain-based products often differs significantly from that of whole grains. For the most part, properly processed refrigerated doughs have few spoilage problems. However, improper storage conditions may result in growth of yeasts or lactic acid bacteria and cause splitting of bread loaves or the development of slimy texture or undesirable aroma or flavor. Elliott (26)

found that the microflora of spoiled doughs consisted of 53% *Leuconostoc dextranicum* and 35% *Leuconostoc mesenteroides*. *Lactobacillus*, *Streptococcus*, *Micrococcus*, and *Bacillus* spp. and gram-negative rods made up 1.1 to 4.5% of the microflora.

Most grain-based foods receive some type of heat treatment before they are considered edible. These treatments are usually sufficient to inactivate all but the most heat-resistant microorganisms. For example, during baking the internal temperature of most breads and cakes reaches or slightly exceeds 100°C. Consequently, spoilage microorganisms usually consist of airborne molds which contaminate the product after baking. A notable exception to this is the development of a spoilage condition known as "ropiness." This defect is caused by polysaccharide-producing strains of *Bacillus subtilis* or *Bacillus licheniformis*. Spores of these bacteria survive the baking process, germinate, and produce a stringy brown mass within the bread loaf. This defect usually only occurs when heavily contaminated loaves are improperly cooled. However, modern bread-making methods make ropy bread a rare occurrence. A defect referred to as "bloody bread" results from the growth of the red-pigmented bacterium *Serratia*. This defect likewise occurs only rarely in commercially produced breads.

References

1. **Agrios, G. N.** 1988. How pathogens attack plants, p. 63–86. *In Plant Pathology*, 3rd ed. Academic Press, Inc., San Diego, Calif.

2. **Bennik, M. H. J., W. Vorstman, E. J. Smid, and L. G. M. Gorris.** 1998. The influence of oxygen and carbon dioxide on the growth of prevalent *Enterobacteriaceae* and *Pseudomonas* species isolated from fresh and controlled-atmosphere-stored vegetables. *Food Microbiol.* **15:**459–469.

3. **Berrang, M. E., R. E. Brackett, and L. R. Beuchat.** 1989. Growth of *Listeria monocytogenes* on fresh vegetables stored under controlled atmosphere. *J. Food Prot.* **52:**702–705.

4. **Berrang, M. E., R. E. Brackett, and L. R. Beuchat.** 1990. Microbial, color and textural qualities of fresh asparagus, broccoli, and cauliflower stored under controlled atmosphere. *J. Food Prot.* **53:**391–395.

5. **Beuchat, L. R.** 1998. Surface decontamination of fruits and vegetables eaten raw: a review. WHO/FSE/98.2. Food Safety Issues, Food Safety Unit, WHO, Geneva, Switzerland.

6. **Beuchat, L. R., and R. E. Brackett.** 1990. Growth of *Listeria monocytogenes* on lettuce as influenced by shredding, chlorine treatment, modified atmosphere packaging, temperature and time. *J. Food Sci.* **55:**755–758, 870.

7. **Beuchat, L. R., and R. E. Brackett.** 1990. Inhibitory effects of raw carrots on *Listeria monocytogenes. Appl. Environ. Microbiol.* 56:1734–1742.

8. **Beuchat, L. R., and R. E. Brackett.** 1991. Behavior of *Listeria monocytogenes* inoculated into raw tomatoes and processed tomato products. *Appl. Environ. Microbiol.* 57:1367–1371.

9. **Beuchat, L. R., and J. I. Pitt.** 1992. Detection and enumeration of heat-resistant molds, p. 251–263. *In* C. Vanderzant and D. F. Splittoesser (ed.), *Compendium of Methods for the Microbiological Examination of Foods.* American Public Health Association, Washington, D.C.

10. **Beuchat, L. R., and S. L. Rice.** 1979. *Byssochlamys* spp. and their importance in processed fruits. *Adv. Food Res.* 25:237–288.

11. **Beuchat, L. R., and J.-H. Ryu.** 1997. Produce handling and processing practices. *Emerg. Infect. Dis.* 3:459–465.

12. **Börjesson, T., U. Stöllman, and J. Schnürer.** 1992. Volatile metabolites produced by six fungal species compared with other indicators of fungal growth on cereal grains. *Appl. Environ. Microbiol.* 58:2599–2605.

13. **Brackett, R. E.** 1987. Vegetables and related products, p. 129–154. *In* L. R. Beuchat (ed.), *Food and Beverage Mycology.* Van Nostrand Reinhold, New York, N.Y.

14. **Brackett, R. E.** 1993. Microbial quality, p. 125–148. *In* R. L. Shewfelt and S. E. Prussia (ed.), *Postharvest Handling: a Systems Approach.* Academic Press, Inc., New York, N.Y.

15. **Brackett, R. E.** 1994. Microbiological spoilage and pathogens in minimally processed fruits and vegetables, p. 269–312. *In* R. C. Wiley (ed.), *Minimally Processed Refrigerated Fruits and Vegetables.* Van Nostrand Reinhold, New York, N.Y.

16. **Brackett, R. E., and D. L. Splittstoesser.** 1992. Fruits and vegetables, p. 287–293. *In* C. Vanderzant and D. L. Splittstoesser (ed.), *Compendium of Methods for the Microbiological Examination of Food.* American Public Health Association, Washington, D.C.

17. **Carmichael, I., I. S. Harper, M. J. Coventry, P. W. J. Taylor, J. Wan, and M. W. Hickey.** 1999. Bacterial colonization and biofilm development on minimally processed vegetables. *J. Appl. Microbiol.* 28:45S–51S.

18. **Cheeson, A.** 1980. Maceration in relation to the postharvest handling and processing of plant material. *J. Appl. Bacteriol.* 48:1–45.

19. **Codner, R. C.** 1971. Pectinolytic and cellulytic enzymes in the microbial modification of plant tissues. *J. Appl. Bacteriol.* 34:147–160.

20. **Collmer, A., and N. T. Keen.** 1986. The role of pectic enzymes in plant pathogenesis. *Annu. Rev. Phytopathol.* 24:383–409.

21. **Daniels, J., A. R. Krishnamurthi, and S. S. H. Rizvi.** 1985. A review of effects of carbon dioxide on microbial growth and food quality. *J. Food Prot.* 48:532–537.

22. **Deak, T., and L. R. Beuchat.** 1996. Yeasts in specific types of foods, p. 61–96. *In Handbook of Food Spoilage Yeasts.* CRC Press, Boca Raton, Fla.

23. **Delaquis, P. J., S. Stewart, P. M. A. Toivonen, and A. L. Moyls.** 1999. Effect of warm, chlorinated water on the microbial flora of shredded iceberg lettuce. *Food Res. Int.* 32:7–14.

24. **Dingman, D. W.** 2000. Growth of *Escherichia coli* O157:H7 in bruised apple (*Malus domestica*) tissue as influenced by cultivar, date of harvest, and source. *Appl. Environ. Microbiol.* 66:1077–1083.

25. **Doan, C. H., and P. M. Davidson.** 2000. Microbiology of potatoes and potato products: a review. *J. Food Prot.* 5:668–683.

26. **Elliott, R. P.** 1980. Cereals and cereal products, p. 669–730. *In* J. H. Silliker, R. P. Elliot, A. C. Baird-Parker, F. L. Bryan, J. H. B. Christian, D. S. Clark, J. C. Olson, Jr., and T. A. Roberts (ed.), *Microbial Ecology of Foods,* vol. II. Academic Press, Inc., New York, N.Y.

27. **Fett, W. F.** 2000. Naturally occurring biofilms on alfalfa and other types of sprouts. *J. Food Prot.* 63:625–632.

28. **Francis, G. A., C. Thomas, and D. O'Beirne.** 1999. The microbiological safety of minimally processed vegetables. *Int. J. Food Sci. Technol.* 34:1–22.

29. **Garcia-Gimeno, R. M., and G. Zurera-Cosano.** 1997. Determination of ready-to-eat vegetable salad shelf-life. *Int. J. Food Microbiol.* 36:31–38.

30. **Goepfert, J. M.** 1980. Vegetables, fruits, nuts and their products, p. 606–642. *In* J. H. Silliker, R. P. Elliot, A. C. Baird-Parker, F. L. Bryan, J. H. B. Christian, D. S. Clark, J. C. Olson, Jr., and T. A. Roberts (ed.), *Microbial Ecology of Foods,* vol. II. Academic Press, Inc., New York, N.Y.

31. **Hanklin, L., and G. H. Lacy.** 1992. Pectinolytic microorganisms, p. 176–183. *In* C. Vanderzant and D. F. Splittstoesser (ed.), *Compendium of Methods for the Microbiological Examination of Foods.* American Public Health Association, Washington, D.C.

32. **Hao, Y. Y., and R. E. Brackett.** 1994. Pectinase activity of vegetable spoilage bacteria in modified atmosphere. *J. Food Sci.* 59:175–178.

33. **Heard, G.** 1999. Microbiological safety of ready-to-eat salads and minimally processed vegetables and fruits. *Food Austral.* 51:414–420.

34. **Jay, J. M.** 1992. Spoilage of fruis and vegetables, p. 187–198. *In Modern Food Microbiology,* 4th ed. Van Nostrand Reinhold, New York, N.Y.

35. **Kinderlerer, J. L.** 1989. Volatile metabolites of filamentous fungi and their role in food flavor. *J. Appl. Bacteriol Symp.* 67(Suppl.):133S–144S.

36. **Kotoujansky, A.** 1987. Molecular genetics of pathogenesis by soft-rot erwinias. *Annu. Rev. Phytopathol.* 25:405–430.

37. **Lund, B. M.** 1971. Bacterial spoilage of vegetables and certain fruits. *J. Appl. Bacteriol.* 34:9–20.

38. **Lund, B. M.** 1982. The effect of bacteria on post-harvest quality of vegetables and fruits, with particular reference to spoilage, p. 133–153. *In* M. E. Rhodes-Roberts and F. A. Skinner (ed.), *Bacteria and Plants.* Soc. Appl. Bacteriol. Symposium Series, no. 10. Academic Press. Inc., New York, N.Y.

39. **Lund, B. M.** 1983. Bacterial spoilage, p. 219–257. *In* C. Dennis (ed.), *Post-Harvest Pathology of Fruits and Vegetables*. Academic Press, London, United Kingdom.

40. **Lund, B. M.** 1992. Ecosystems in vegetable foods. *J. Appl. Bacteriol. Symp.* **1992**(Suppl.):115S–126S.

41. **Lund, B. M., and A. L. Snowdon.** 2000. Fresh and processed fruits, p. 738–758. *In* B. M. Lund, T. C. Baird-Parker, and G. W. Gould (ed.), *The Microbiological Safety and Quality of Food*, vol. I. Aspen Publ., Gaithersburg, Md.

42. **Nguyen-the, C., and F. Carlin.** 1994. The microbiology of minimally processed fresh fruits and vegetables. *Crit. Rev. Food Sci. Nutr.* **34**:371–401.

43. **Nguyen-the, C., and F. Carlin.** 2000. Fresh and processed vegetables, p. 620–684. *In* B. M. Lund, T. C. Baird-Parker, and G. W. Gould (ed.), *The Microbiological Safety and Quality of Food*, vol. I. Aspen Publ., Gaithersburg, Md.

44. **Pao, S., and P. D. Petracek.** 1997. Shelf life extension of peeled oranges by citric acid treatment. *Food Microbiol.* **14**:485–491.

45. **Priepke, P. E., L. Wei, and A. I. Nelson.** 1976. Refrigerated storage of prepackaged salad vegetables. *J. Food Sci.* **41**:379–382.

46. **Rolle, R. S., and G. W. Chism III.** 1987. Physiological consequences of minimally processed fruits and vegetables. *J. Food Qual.* **10**:157–177.

47. **Rombouts, F. M., and W. Pilnik.** 1980. Pectin enzymes, p. 227–282. *In* A. H. Rose (ed.), *Microbial Enzymes and Bioconversions*. Academic Press, London, United Kingdom.

48. **Seiler, D. A. L.** 1986. The microbial content of wheat and flour, p. 241–255. *In* B. Flannigan (ed.), *Spoilage and Mycotoxins of Cereals and Other Stored Products*. Crown, New York, N.Y.

49. **Senter, S. D., N. A. Cox, J. S. Bailey, and W. R. Forbus, Jr.** 1985. Microbiological changes in fresh market tomatoes during packing operations. *J. Food Sci.* **50**:254–255.

50. **Sinigaglia, M., M. Albenzio, and M. R. Corbo.** 1999. Influence of process operations on shelf-life and microbial populations of fresh-cut vegetables. *J. Ind. Microbiol. Biotechnol.* **23**:484–488.

51. **Splittstoesser, D. F.** 1973. The microbiology of frozen vegetables. *Food Technol.* **27**(1):54–60.

52. **Splittstoesser, D. F.** 1987. Fruits and vegetable products, p. 101–128. *In* L. R. Beuchat (ed.), *Fruit and Beverage Mycology*, 2nd ed. Van Nostrand Reinhold, New York, N.Y.

53. **Splittstoesser, D. F.** 1991. Fungi of importance in processed fruits, p. 201–219. *In* D. K. Arora, K. G. Mukerji, and E. H. Marth (ed.), *Handbook of Applied Mycology*. Marcel Dekker, Inc., New York, N.Y.

54. **Thomas, C., and D. O'Beirne.** 2000. Evaluation of the impact of short-term temperature abuse on the microbiology and shelf life of a model ready-to-eat vegetable combination product. *Int. J. Food Microbiol.* **59**:47–57.

55. **Walls, I., and R. Chuyate.** 2000. Spoilage of fruit juices by *Alicyclobacillus acidoterrestris*. *Food Austral.* **52**:286–288.

56. **Wells, J. M., and J. E. Butterfield.** 1999. Incidence of *Salmonella* on fresh fruits and vegetables affected by fungal rots or physical injury. *Plant Dis.* **83**:722–726.

57. **Whitfield, F. B.** 1998. Microbiology of food taints. *Int. J. Food Sci. Technol.* **33**:31–51.

Foodborne
Pathogenic
Bacteria

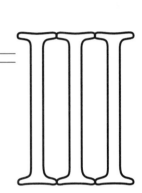

Food Microbiology: Fundamentals and Frontiers, 2nd Ed.
Edited by M. P. Doyle et al.
© 2001 ASM Press, Washington, D.C.

Jean-Yves D'Aoust
John Maurer
J. Stan Bailey

Salmonella Species

8

CHARACTERISTICS OF THE ORGANISM

Historical Considerations

In the early 19th century, clinical pathologists in France first documented the association of human intestinal ulceration with a contagious agent; the disease was later identified as typhoid fever. Further investigations by European workers led to the isolation and characterization of the typhoid bacillus responsible for typhoid fever and to the development of a serodiagnostic test for the detection of this serious human disease agent (65, 169). Differential clinical and serological traits were used subsequently to identify the closely related paratyphoid organisms. In the United States, contemporary work by Salmon and Smith (1885) led to the isolation of *Bacillus cholerae-suis*, now known as *Salmonella enterica* serovar Choleraesuis, from swine suffering from hog cholera (169). During the first quarter of the 20th century, great advances occurred in the serological detection of somatic and flagellar antigens within the *Salmonella* group, a generic term coined by Lignières in 1900 (169). An antigenic scheme for the classification of salmonellae

was first proposed by White (1926) and subsequently expanded by Kauffmann (1941) into the Kauffmann-White scheme, which currently includes more than 2,400 serovars (219).

Taxonomy

Salmonella spp. are facultatively anaerobic, gram-negative rod-shaped bacteria belonging to the family *Enterobacteriaceae*. Although members of this genus are motile by peritrichous flagella, nonflagellated variants, such as *Salmonella enterica* serovar Pullorum and *Salmonella enterica* serovar Gallinarum, and nonmotile strains resulting from dysfunctional flagella do occur. Salmonellae are chemoorganotrophic, with an ability to metabolize nutrients by both respiratory and fermentative pathways. The bacteria grow optimally at 37°C and catabolize D-glucose and other carbohydrates with the production of acid and gas. Salmonellae are oxidase negative and catalase negative, grow on citrate as a sole carbon source, generally produce hydrogen sulfide, decarboxylate lysine and ornithine, and do not hydrolyze

Jean-Yves D'Aoust, Food Directorate, Health Products & Food Branch, Health Canada, Sir F. G. Banting Research Centre, Postal Locator 22.04.A2, Tunney's Pasture, Ottawa, Ontario, Canada K1A 0L2.
John Maurer, Department of Avian Medicine, University of Georgia, Athens, GA 30602.
J. Stan Bailey, USDA, Agricultural Research Service, Russell Research Center, P.O. Box 5677, Athens, GA 30606.

urea. Many of these traits have formed the basis for the presumptive biochemical identification of *Salmonella* isolates. According to a contemporary definition, a typical *Salmonella* isolate would produce acid and gas from glucose in triple sugar iron agar medium and would not utilize lactose or sucrose in triple sugar iron or in differential plating media, such as brilliant green, xylose lysine deoxycholate, and Hektoen enteric agars. Additionally, typical salmonellae readily produce an alkaline reaction from the decarboxylation of lysine to cadaverine in lysine iron agar, generate hydrogen sulfide gas in triple sugar iron and lysine iron media, and fail to hydrolyze urea (9, 70). The dynamics of genetic variability arising from bacterial mutations and conjugative intra- and intergeneric exchange of plasmids encoding determinant biochemical traits continue to reduce the proportion of typical *Salmonella* biotypes. From the early studies of Le Minor and colleagues confirming that *Salmonella* utilization of lactose and sucrose was plasmid mediated (170, 171), many studies have since emphasized the occurrence of lac+ and/or suc+ biotypes in clinical specimens and food materials. This situation is of public health concern because biochemically atypical salmonellae could easily escape detection on disaccharide-dependent plating media, which are commonly used in hospital and food industry laboratories. Bismuth sulfite agar remains a medium of choice for isolating salmonellae because, in addition to its high level of selectivity, it responds solely and most effectively to the production of extremely low levels of hydrogen sulfide gas (65). The diagnostic hurdles encountered by the changing patterns of disaccharide utilization by *Salmonella* spp. are being further confounded by the increasing occurrence of biotypes that cannot decarboxylate lysine, that possess urease activity, that produce indole, and that readily grow in the presence of KCN. Clearly, the recognition of *Salmonella* as a biochemically homogeneous group of microorganisms is becoming obsolete. The one serovar-one species concept is untenable, because most serovars cannot be separated by biochemical tests. The situation has led to a reassessment of the diagnostic value of these and other biochemical traits and to their likely replacement with molecular technologies targeted at the identification of stable genetic loci and/or their products that are unique to the *Salmonella* genus.

Nomenclature of the *Salmonella* group has progressed through a succession of taxonomical schemes based on biochemical and serological characteristics and on principles of numerical taxonomy and DNA homology (Table 8.1). In the early development of taxonomic

Table 8.1 Taxonomic schemes for *Salmonella* spp.

Diagnostic basis	Salient features	Serovar designation	Reference
1. Biochemical	Five subgenera (I–V) Serovar = species status *S. arizonae*	*S. typhimurium*	156
2. Biochemical	Three species *S. typhi* *S. choleraesuis* *S. enteritidis* *Arizona* = separate genus	*S. enteritidis* serovar Typhimurium	83

3. Phenetic/DNA homology	Single species (*S. choleraesuis*)
	Seven subspecies
	  choleraesuis
	  salamae
	  arizonae
	  diarizonae
	  houtenae
	  bongori
	  indica
	Type strain = *S. choleraesuis*
		S. choleraesuis subsp. *choleraesuis* serovar Typhimurium	173

4. Phenetic/DNA homology	Single species (*S. enterica*) Seven subspecies (see above) Type strain = *S. typhimurium* LT2	*S. enterica* subsp. *enterica* serovar Typhimurium	172
5. Multilocus enzyme electrophoresis	Two species *S. enterica* (six subspecies) *S. bongori*		225

schemes, determinant biochemical reactions were used to separate salmonellae into subgroups. The Kauffmann-White scheme was the first attempt to systematically classify salmonellae using these scientific parameters. This major undertaking culminated in the identification of five biochemically defined subgenera (I to V) wherein individual serovars were afforded species status (156). Subgenus III included members of the *Arizona* species (*S. arizonae*). Subsequently, a three-species nomenclatural system was proposed using 16 discriminating tests to identify *S. typhi* (single serovar), *S. choleraesuis* (single serovar), and *S. enteritidis* (all other salmonellae serovars). The latter scheme recognized members of the *Arizona* group as a distinct genus (83). A defining development in *Salmonella* taxonomy occurred in 1973 when Crosa et al. (60) determined by DNA-DNA hybridization that all serotypes and subgenera of *Salmonella* and "*Arizona*" were related at the species level. The single exception to this observation was the subsequent description of *Salmonella bongori*, previously known as subspecies V, as a distinct species. Another system based on numerical taxonomy and DNA relatedness proposed a single species (*Salmonella choleraesuis*) consisting of seven subspecies (173). In numerical taxonomy, a statistical comparison of morphological and biochemical attributes of strains (phenetic analysis) measures the taxonomical proximity of test strains and allows for their separation into distinct taxa. In DNA homology, a high degree of hybridization of *Salmonella* reference DNA with extracts of test strains confirms the genetic relatedness of nucleic acid reactants and supports the inclusion of the test microorganisms into the *Salmonella* genus.

Subsequent modification of the former scheme has been proposed to change the type species from *S. choleraesuis* to *S. enterica* while retaining the names of the seven recognized subspecies. In 1987, Le Minor and Popoff (172) formally made a proposal as a "request for an opinion" to the Judicial Commission of the International Committee of Systematic Bacteriology to officially change the type strain of *Salmonella* from *S. choleraesuis* to *S. enterica* subsp. *enterica* serotype Typhimurium LT2 (172). The recommendation was denied by the Judicial Commission because its members believed that the status of *Salmonella* serotype Typhi was not adequately addressed in the request for opinion. The Judicial Commission therefore ruled that *S. choleraesuis* be retained as the legitimate type species pending an amended request for an opinion (276). Euźeby (82) made an amended request in 1999, which is pending, to adopt *S. enterica* as the type species of *Salmonella* while retaining the species "*S. typhi*" as an exception. Another proposal petitions for the elevation of *S. enterica* subsp. *bongori* to a

Table 8.2 Species within of the *Salmonella* genus[a]

Salmonella species and subspecies	No. of serovars
S. enterica subsp. *enterica* (I)	1,454
S. enterica subsp. *salamae* (II)	489
S. enterica subsp. *arizonae* (IIIa)	94
S. enterica subsp. *diarizonae* (IIIb)	324
S. enterica subsp. *houtenae* (IV)	70
S. enterica subsp. *indica* (VI)	12
S. bongori (V)	20
TOTAL	2,463

[a] From references 37 and 219.

new species based on multilocus enzyme electrophoretic patterns (225). The new species would be designated *S. bongori*. According to the World Health Organization (WHO) Collaborating Centre for Reference and Research on *Salmonella* (Institut Pasteur, Paris), *S. enterica* and *S. bongori* currently include 2,443 and 20 serovars, respectively (Table 8.2). Even though the Judicial Commission has not yet formally recognized *S. enterica* subsp. *enterica* serovar Typhimurium LT2 as the official type strain, most researchers and clinicians are now using this nomenclature.

The Centers for Disease Control and Prevention has adopted as its official nomenclature the following scheme. The genus *Salmonella* contains two species, each of which contains multiple serovars (Table 8.2). The two species are *S. enterica*, the type species, and *S. bongori* (formerly subspecies V). *S. enterica* is divided into six subspecies, which are referred to by a Roman numeral and a name (I, *S. enterica* subsp. *enterica*; II, *S. enterica* subsp. *salamae*; IIIa, *S. enterica* subsp. *arizonae*; IIIb, *S. enterica* subsp. *diarizonae*; IV, *S. enterica* subsp. *houtenae*; and VI, *S. enterica* subsp. *indica*) (37). *S. enterica* subspecies are differentiated biochemically and by genomic relatedness.

The biochemical identification of foodborne and clinical *Salmonella* isolates is generally coupled to serological confirmation, a complex and labor-intensive technique involving the agglutination of bacterial surface antigens with *Salmonella*-specific antibodies. These include somatic (O) lipopolysaccharides (LPS) on the external surface of the bacterial outer membrane, flagellar (H) antigens associated with the peritrichous flagella, and the capsular (Vi) antigen, which occurs only in *Salmonella* serovars Typhi, Paratyphi C, and Dublin (169). The heat-stable O antigens are classified as major or minor antigens. The former category consists of antigens such as the somatic factors O:4 and O:3, which are specific determinants for the somatic groups B and E, respectively. In contrast, minor somatic antigenic components, such as O:12, are nondiscriminatory, as evidenced by their presence in different somatic groups. Smooth (S) variants relate to strains with well-developed serotypic LPS that readily agglutinate with specific antibodies, whereas rough (R) variants exhibit incomplete LPS antigens, resulting in weak or no agglutination with *Salmonella* somatic antibodies. Flagellar (H) antigens are heat-labile proteins, and individual *Salmonella* strains may produce one (monophasic) or two (diphasic) sets of flagellar antigens. Although serovars such as Dublin produce a single set of flagellar (H) antigen, most serovars can alternatively elaborate two sets of antigens, i.e., phase 1 and phase 2 antigens. These homologous surface antigens are chromosomally encoded by the H_1 (phase 1) and H_2 (phase 2) genes and are transcribed under the control of the vh_2 locus (169). Capsular (K) antigens commonly encountered in members of *Enterobacteriaceae* are limited to the Vi antigen in the *Salmonella* genus. Thermal solubilization of the Vi antigen is necessary for the immunological identification of underlying serotypic LPS.

The aim of serological testing procedures is to determine the complete antigenic formula of individual *Salmonella* isolates. We shall use serovar Infantis (6,7:r:1,5) as an example. Commercially available polyvalent somatic antisera each consist of a mixture of antibodies specific for a limited number of major antigens; e.g., polyvalent B antiserum (Difco Laboratories, Detroit, Mich.) recognizes somatic (O) groups C_1, C_2, F, G, and H. Following a positive agglutination with polyvalent B antiserum, single grouping antisera representing the five somatic groupings included in the polyvalent B reagent would be used to define the serogroup of the isolate. The test isolate would react with the C_1 grouping antiserum, indicating that antigens 6,7 are present. Flagellar (H) antigens would then be determined by broth agglutination reactions using polyvalent H antisera or the Spicer-Edwards series of antisera. In the former assay, a positive agglutination reaction with one of the five polyvalent antisera (Poly A to E; Difco Laboratories) would lead to testing with single-factor antisera to specifically identify the phase 1 and/or phase 2 flagellar antigens present. Agglutination in Poly C flagellar antiserum and subsequent reaction of the isolate with single grouping H antisera would confirm the presence of the r antigen (phase 1). The empirical antigenic formula of the isolate would then be 6,7:r. Phase reversal in semisolid agar supplemented with r antiserum would immobilize phase 1 salmonellae at or near the point of inoculation, thereby facilitating the recovery of phase 2 cells from the edge of the zone of migration. Serological testing of phase 2 cells with Poly E and 1-complex antisera would confirm the presence of the flagellar 1 factor. Confirmation of the

flagellar 5 antigen with single-factor antiserum would yield the final antigenic formula 6,7:r:1,5, which corresponds to serovar Infantis. A similar analytical approach would be used with the Spicer-Edwards polyvalent H antisera; the identification of flagellar (H) antigens would arise from the pattern of agglutination reactions among the four Spicer-Edwards antisera and with three additional polyvalent antisera, including the L, 1, and e,n complexes.

Physiology

Growth

The *Salmonella* group consists of resilient microorganisms that readily adapt to extreme environmental conditions. Some *Salmonella* strains can grow at elevated temperatures ($\leq 54°C$), and others can exhibit psychrotrophic properties by their ability to grow in foods stored at 2 to 4°C (67) (Table 8.3). Moreover, preconditioning of cells to low temperatures can markedly increase the growth and survival of salmonellae in refrigerated food products (4). Such growth characteristics raise concerns regarding the efficacy of chill temperatures to ensure food safety through bacteriostasis. These concerns are further heightened by the widespread refrigerated storage of foods packaged under vacuum or modified atmosphere to prolong shelf life. Gaseous mixtures consisting of 60 to 80% (vol/vol) CO_2 with varying proportions of N_2 and/or O_2 can inhibit the growth of aerobic spoilage microorganisms such as *Pseudomonas* spp. without promoting the growth of *Salmonella* spp. (67). However, the proliferation of salmonellae in inoculated raw minced beef and cooked crabmeat stored at 8 to 11°C under modified atmospheres containing low levels of CO_2 (20 to 50% [vol/vol]) warrants caution in the general application of this novel processing technology (30, 147).

Studies on the maximum temperature for growth of *Salmonella* spp. in foods are generally lacking. Notwithstanding an early report of growth of salmonellae in inoculated custard and chicken à la king at 45.6°C (10), more recent evidence indicates that prolonged exposure of mesophilic strains to thermal stress conditions results in mutants of serovar Typhimurium capable of growth at 54°C (77). Although the mechanism of this phenomenon has yet to be elucidated, preliminary findings indicate that two separate mutations enable serovar Typhimurium to actively grow at 48°C (*ttl*) and 54°C (*mth*).

The physiological adaptability of *Salmonella* spp. is further demonstrated by their ability to proliferate at pH values ranging from 4.5 to 9.5, with an optimum pH for

Table 8.3 Physiological limits for the growth of *Salmonella* spp. in foods and bacteriological media

Parameter	Limits		Product	*Salmonella* serovar	Reference
	Minimum	Maximum			
Temperature	2°C (24 h)		Minced beef[a]	Typhimurium	44
	2°C (2 days)		Minced chicken[b]	Typhimurium	19
	4°C (≤10 days)		Shell eggs[b]	Enteritidis	158
		54°C[c]	Agar medium	Typhimurium	77
pH	3.99[d]		Tomatoes	Infantis	18
	4.05[e]		Liquid medium	Anatum	52
				Tennessee	
				Senftenberg	
		9.5	Egg wash water[b]	Typhimurium	139
Water activity	0.93[f]		Rehydrated dried soup[b]	Oranienburg	268

[a] Naturally contaminated.
[b] Artificially contaminated.
[c] Mutants selected to grow at elevated temperature.
[d] Growth within 24 h at 22°C.
[e] Acidified with HCl or citric acid; growth within 24 h at 30°C.
[f] Growth within 3 days at 30°C.

growth of 6.5 to 7.5 (Table 8.3). It is well established that the bacteriostatic or antibacterial effects of acidic conditions are acidulant dependent (65). Of the many organic acids produced by starter cultures in meat, dairy, and other fermented foods, and of the various organic and inorganic acids used in product acidification, propionic and acetic acids are more bactericidal than the common food-associated lactic and citric acids. Interestingly, the antibacterial action of organic acids decreases with increasing length of the fatty acid chain (65). Early research on the propensity for the growth of *Salmonella* spp. in acidic environments revealed that wild-type strains preconditioned on pH gradient plates could grow in liquid and solid media at considerably lower pH values than the parent strains (144). These findings raise concerns regarding the safety of fermented foods, such as cured sausages and fermented raw milk products, in which the progressive starter culture-dependent acidification of fermented foods could provide a favorable environment for the elevation of endogenous salmonellae to a state of increased acid tolerance. The growth and/or enhanced survival of salmonellae during the fermentative process would result in a contaminated ready-to-eat product. Leyer and Johnson (175) demonstrated the increased survival of acid-adapted *Salmonella* spp. in fermented milk and during storage (5°C) of cheddar, Swiss, and mozzarella cheese derived from the milk. The presence of acid-tolerant salmonellae in such foods further heightens the level of public health risk, because this acquired physiological trait could minimize the antimicrobial activity of gastric acidity (pH 2.5) and promote the survival

of salmonellae within the acidic cytoplasm of mononuclear and polynuclear phagocytes of the human host (66). Bearson et al. (25) identified four regulator genes, two of which, *ada* and *phoP*, are associated with virulence and two, *rpoS* and *fur*, with the development of acid tolerance. The complexity of the acid tolerance stress response suggests that acid tolerance and virulence are interrelated.

High salt concentrations have long been recognized for their ability to extend the shelf life of foods by inhibiting the growth of endogenous microflora (214). This bacteriostatic effect, which can also engender cell death, results from a dramatic decrease in water activity (a_w) and from bacterial plasmolysis commensurate with the hypertonicity of the suspending medium. Studies have revealed that foods with a_w values of ≤0.93 do not support the growth of salmonellae (65) (Table 8.3). Although *Salmonella* spp. are generally inhibited in the presence of 3 to 4% NaCl, bacterial salt tolerance increases with increasing temperature in the range of 10 to 30°C. However, the latter phenomenon is associated with a protracted log phase and a decreased rate of growth. Evidence further suggests that the magnitude of this adaptive response is food and serovar specific (65). A recent report on anaerobiosis and its potentiation of greater salt tolerance in *Salmonella* raises concerns regarding the safety of modified-atmosphere and vacuum-packaged foods that contain high levels of salt (12).

The pH, salt concentration, and temperature of the microenvironment can exert profound effects on the growth kinetics of *Salmonella* spp. Several studies revealed the increased ability of salmonellae to grow under acidic (pH ≤ 5.0) conditions or in environments of high salinity ($\geq 2\%$ NaCl) with increasing temperature (65, 86, 259). Similar research on the interrelationship between pH and NaCl at 20 to 30°C has underscored the dominance of medium pH on the growth of *Salmonella* spp. (259). Interestingly, the presence of salt in acidified foods can reduce the antibacterial action of organic acids where low concentrations of NaCl or KCl stimulate the growth of serovar Enteritidis in broth medium acidified to pH 5.19 with acetic acid (220). Such findings and other reports on the enhanced salt-dependent survival of salmonellae in rennet whey (pH 4.8 to 5.6) and mayonnaise (259) indicate that low levels of salt can undermine the preservative action of organic acids and potentially compromise the safety of fermented and acidified foods. Although the mechanisms for these salt phenomena remain elusive, the role of salinity in restoring cellular homeostasis may be linked to the Na^+/K^+-proton antiport systems (92, 93, 205).

Studies of the interactive forces generated by temperature, pH, and salt on the growth and survival of *Salmonella* spp. have led to the development and application of mathematical models to predict the fate of salmonellae in foods. In this approach, the growth, survival, or inactivation of a target microorganism under varying conditions of pH, NaCl, temperature, or other environmental factors of interest is laboriously characterized in laboratory media. The generated data are then used to derive mathematical models that depict the response of the microorganism under various combinations of environmental factors (106). Predictive models are not without limitations. For example, the use of a model to predict the growth of salmonellae in a food whose salt content lies beyond the range of values originally studied for the derivation of the mathematical model would likely lead to erroneous conclusions. Moreover, models are generally based on the behavior of a few *Salmonella* strains under selected environmental conditions. The physiological diversity among the large number of serovars (Table 8.2) may unduly challenge the reliability of models in predicting bacterial growth responses. Additionally, the lot-to-lot variations in the composition of a given food and in the types and numbers of background microflora could seriously undermine the predictive capability of mathematical models. A special issue on predictive modeling published in the *International Journal of Food Microbiology* (vol. 23, 1994) provides invaluable information on the merits and limitations of the modeling approach for *Salmonella* and other foodborne bacterial pathogens and should be consulted for an in-depth review of the subject.

Survival

Although the potential growth of foodborne *Salmonella* spp. is of primary importance in safety assessments, the propensity for these pathogens to persist in hostile environments further heightens public health concerns. The survival of *Salmonella* spp. for prolonged periods of time in foods stored at freezer and ambient temperatures is well documented (65). It is noteworthy that the composition of the freezing menstruum, the kinetics of the freezing process, the physiological state of foodborne salmonellae, and the serovar-specific responses to extreme temperatures determine the fate of salmonellae during freezer storage of foods (57). The viability of salmonellae in dry foods stored at $\geq 25°C$ decreases with increasing storage temperature and with increasing moisture content (58, 65).

Heat is widely used in food manufacturing processes to control the bacterial quality and safety of end products. Factors that potentiate the greater heat resistance of *Salmonella* spp. and other foodborne bacterial pathogens in food ingredients and finished products have been studied extensively (58). Although the heat resistance of *Salmonella* spp. increases as the a_w of the heating menstruum decreases, detailed studies have revealed that the nature of solutes used to alter the a_w of the heating menstruum play a determinant role in the level of acquired heat resistance (65). For example, the heating of serovar Typhimurium in menstrua adjusted to an a_w of 0.90 with sucrose and glycerol conferred different levels of heat resistance, as evidenced by $D_{57.2}$ values of 40 to 55 min and 1.8 to 8.3 min, respectively (112). The D value represents the amount of time required to effect a 90% kill (1.0 log_{10} reduction) in the number of viable cells upon heating at a specified constant temperature. Other important features associated with this adaptive response include the greater heat resistance of salmonellae grown in nutritionally rich medium than in minimal media, of cells derived from stationary- rather than logarithmic-phase cultures, and of salmonellae previously stored in a dry environment (112, 160, 199). The ability of *Salmonella* spp. to acquire greater heat resistance following exposure to sublethal temperatures is equally notable. The phenomenon stems from a rapid adaptation of the organism to rising temperatures in the microenvironment to a level of enhanced thermotolerance quite distinct from that described in conventional time-temperature curves of thermal lethality. This adaptive response has potentially serious implications for the safety of thermal processes that expose or maintain food products at marginally lethal temperatures. Exposure of salmonellae to sublethal temperatures ($\leq 50°C$) for 15 to 30 min enhances heat resistance through a rapid chloramphenicol-sensitive synthesis of heat shock proteins (145, 180, 181). Changes in the fatty acid composition of cell membranes in heat-stressed salmonellae to provide a greater proportion of saturated membrane phospholipids reduces the fluidity of the bacterial cell membrane with an attendant increase in membrane resistance to heat damage (145). The likelihood that other protective cellular functions are triggered by heat shock stimuli cannot be discounted.

The complexities of the foregoing considerations can best be illustrated by the following scenario involving *Salmonella* contamination of a chocolate confectionery product. Experience has shown that survival of salmonellae in dry roasted cocoa beans can lead to contamination of in-line and finished products. Thermal inactivation of salmonellae in molten chocolate is most difficult because the time-temperature conditions that would be required to effectively eliminate the pathogen in this sucrose-containing product of low a_w would likely result in an organoleptically unacceptable product. The problem is further compounded by the ability of salmonellae to survive for many years in the finished product when stored at ambient temperature (64). Clearly, effective decontamination of raw cocoa beans and stringent in-plant control measures to prevent cross-contamination of in-line products are of prime importance to this food industry.

Recent evidence indicates that brief exposure of serovar Typhimurium to mild acid environments of pH 5.5 to 6.0 (preshock) followed by exposure of the adapted cells to pH ≤4.5 (acid shock) triggers a complex acid tolerance response (ATR) that potentiates the survival of the microorganism under extreme acid environments (pH 3.0 to 4.0). The response translates into an induced synthesis of 43 acid shock and outer membrane proteins, reduced growth rate, and pH homeostasis, as demonstrated by the bacterial maintenance of internal pH values of 7.0 to 7.1 and 5.0 to 5.5 upon sequential exposure of cells to external pH of 5.0 and 3.3, respectively (93, 138). In the ATR response, bacterial Mg^{2+}-dependent proton-translocating ATPase encoded by the *atp* operon plays an important role in maintaining cellular pH at ≥5.0 through an energy-dependent transport of intracellular protons to the cell exterior (93). Other transmembrane mechanisms that putatively operate in pH homeostasis include the H^+-coupled ion-transport systems (antiport) for the intra- and extracellular transfer of K^+, Na^+, and H^+ ions, the electron transport chain-dependent efflux of H^+, and transport systems committed to the symport of H^+ and solutes (205). The Fe^{+2}-binding regulatory protein encoded by the *fur* (ferric uptake regulator) gene also impacts on

bacterial acid tolerance, as evidenced by the inability of *fur* mutant strains to survive under highly acidic conditions (103). Acid-induced activation of amino acid decarboxylases in *Salmonella* spp. provides an additional protective mechanism whereby cadaverine and putrescine from the enzymic breakdown of lysine and ornithine, respectively, potentiate acid neutralization and enhanced bacterial survival (205). Further characterization of the *Salmonella* response to acid stress has led to the identification of two additional protective mechanisms that operate in salmonellae in the stationary phase of growth and which are distinct from the previously discussed ATR response that prevails in log-phase cells (93, 166). One of these pH-dependent responses, designated stationary-phase ATR, provides greater acid resistance than the log-phase ATR; it is induced at pH <5.5 and functions maximally at pH 4.3. This stationary-phase ATR induces the synthesis of only 15 shock proteins, is not affected by mutations in the *atp* and *fur* genes, and is *rpoS* independent (166). The induction of the remaining acid-protective mechanism associated with *Salmonella* spp. in the stationary phase is independent of external pH and dependent on the alternative sigma factor (σ^S) encoded by the *rpoS* locus. The mechanism seemingly reinforces the ability of stationary-phase cells to survive in hostile environmental conditions. We have thus seen that three possibly overlapping cellular systems confer acid tolerance in *Salmonella* spp. These include (i) the pH-dependent, *rpoS*-independent log-phase ATR, (ii) the pH-dependent, *rpoS*-independent stationary-phase ATR, and (iii) the pH-independent, *rpoS*-dependent stationary-phase acid resistance. These systems likely operate in the acidic environments that prevail in fermented and in acidified foods and in phagocytic cells of the infected host.

Acid stress can also trigger enhanced bacterial resistance to other adverse environmental conditions. The growth of serovar Typhimurium at pH 5.8 engendered an increased thermal resistance at 50°C, an enhanced tolerance to high osmotic stress (2.5 M NaCl) ascribed to the induced synthesis of the OmpC outer membrane proteins, a greater surface hydrophobicity, and an increased resistance to the antibacterial lactoperoxidase system and to surface-active agents such as crystal violet and polymyxin B (176).

Reservoirs

The widespread occurrence of *Salmonella* spp. in the natural environment, coupled with the intensive husbandry practices used in the meat, fish, and shellfish industries and the recycling of offal and inedible raw materials into animal feeds, have favored the continued prominence of this human bacterial pathogen in the global food chain (65, 68). Of the many sectors within the meat industry, poultry and eggs remain a predominant reservoir of *Salmonella* spp. in many countries and tend to overshadow the importance of meats, such as pork, beef, and mutton, as potential vehicles of infection (65, 125, 126, 267, 281). In an effort to actively address the problem of *Salmonella* spp. in meat products, the USDA Food Safety Inspection Service (FSIS) published the "Final Rule on Pathogen Reduction and Hazard Analysis and Critical Control Point (HACCP) Systems" in July 1996. This Final Rule requires the meat and poultry industry to implement HACCP plans in all their plants and requires them to systematically sample final products for the indicator organism *Escherichia coli* biotype 1. At the same time, FSIS will conduct *Salmonella* testing to verify that the implemented HACCP is helping to control *Salmonella* spp. on finished products. Baseline studies prior to the implementation of the Final Rule indicated a *Salmonella* contamination rate of about 24% in U.S. broiler chickens. In 1999, the rate of *Salmonella* contamination on processed broiler chickens had been reduced to about 11%. Similar reductions were also observed with other meat animals. Future improvement in the *Salmonella* status of meat animals hinges on coordinated and enduring efforts from all sectors of the meat industry toward the implementation of stringent control measures at the farm level and within the processing, distribution, and retailing sectors (281).

The continuing pandemic of human serovar Enteritidis phage type 4 (Europe) and type 8 (North America) infections associated with the consumption of raw or lightly cooked shell eggs and egg-containing products further emphasizes the importance of poultry as vehicles of human salmonellosis and the need for sustained and stringent bacteriological control of poultry husbandry practices. This egg-related pandemic is of particular concern because the problem arises from transovarian transmission of the infective agent into the interior of the egg prior to shell deposition. The viability of internalized serovar Enteritidis remains unaffected by the egg surface-sanitizing practices currently applied in egg-grading stations. The significant public health and societal costs associated with egg-borne outbreaks of serovar Enteritidis infection together with the economic losses sustained by the agricultural sectors as a result of depopulation of infected layer flocks and mandatory pasteurization of shell eggs from infected commercial layer flocks will undoubtedly result in major financial losses. Increased efforts to intervene and reduce the prevalence of serovar Enteritidis have led to significant reductions in egg-borne outbreaks in the United Kingdom and the United States in 1998 and 1999.

Rapid depletion of feral stocks of fish and shellfish in recent years has greatly increased the importance of the international aquaculture industry as an alternate source for these popular food items. The high-density farming conditions required to maximize biological yields and to satisfy growing market demand open gateways to the widespread infection of species reared in earthen ponds and other unprotected facilities that are continuously exposed to environmental contamination. It is noteworthy that many of the currently available aquacultural products originate from the Asiatic, African, and South American continents. The feeding of raw meat scraps and offal, of night soil potentially contaminated with typhoid and paratyphoid salmonellae, and of animal feeds that may harbor a variety of *Salmonella* serovars to reared species is not uncommon in these geographic areas. Such practices create a situation that could lead to the perpetual bacterial contamination of rearing facilities and severe economic losses through suboptimal growth or death of aquacultural populations (68). Accordingly, aquaculture farmers are relying heavily on antibiotics applied at subtherapeutic levels to safeguard the vigor of farmed fish and shellfish. A serious public health concern arises from the use of antimicrobial agents such as ampicillin, chloramphenicol, sulfa drugs, and quinolones, which currently provide the mainstay for the clinical management of systemic salmonellosis in humans (13a, 68, 237). This antibiotic-dependent husbandry practice is short-sighted and irresponsible when one considers that the misuse of medically important therapeutic agents in the livestock industry continues to select for antibiotic-resistant *Salmonella* spp. (13a, 65, 68, 71). The health consequences of consuming contaminated aquacultural fish in a sushi dish or lightly cooked fish or shellfish that contains viable, antibiotic-resistant salmonellae are predictable.

Fruits and vegetables have gained notoriety in recent years as vehicles of human salmonellosis. The situation has developed from the increased global export of fresh and dehydrated fruits and vegetables from countries that enjoy tropical and subtropical climates. The prevailing hygienic conditions during the production, harvesting, and distribution of products in these countries do not always meet minimum standards, and they facilitate product contamination. More specifically, the fertilization of crops with untreated sludge or sewage effluents potentially contaminated with antibiotic-resistant *Salmonella* spp., the irrigation of garden plots and fields and the washing of fruits and vegetables with contaminated waters, the repeated handling of product by local workers, and the propensity for environmental contamination of spices and other condiments during drying

Table 8.4 Epidemiology of foodborne bacterial pathogens

Country	Year(s)	Mean no. of reported incidents[a]							
		Salmonella spp.	*Staphylococcus aureus*	*Clostridium perfringens*	*Campylobacter* spp.	*Bacillus cereus*	*Escherichia coli*	*Listeria monocytogenes*	Reference
Canada	1993	39.0	6.0	10.0	7.0	8.0	14.0[b]	2.0	126
Denmark	1985–89	5.2	2.4	5.8	0.4	2.0	0.2[c]	0.4	282
England and	1986–89	438.3	10.0	53.0	53.3	19.3	0.0	0.3	282
Wales	1989–91	922.0	8.3	50.7	5.0	24.7	1.0	0.0	248
Finland	1985–89	7.8	5.6	8.8	0.2	3.8	0.0	0.0	282
France	1997	201.0	32.0	13.0	0.0	1.0	0.0	0.0	39a
Germany	1992	177.0	1.0	6.0	0.0	2.0	1.0[c,d]	1.0	282a
Hungary	1992	113.0	13.0	4.0	0.0	2.0	0.0	0.0	282a
Japan	1987–90	110.8	120.3	19.0	31.8	10.3	20.0	0.0	258
	1991	159.0	95.0	21.0	24.0	9.0	30.0[c]	0.0	191
Netherlands	1985–89	8.0	1.6	2.4	2.0	4.8	0.0	0.0	282
Romania	1992	25.0	13.0	0.0	0.0	1.0	1.0[c]	0.0	282a
Scotland	1992	174.0	0.0	1.0	11.0	0.0	0.0	0.0	282a
Spain	1994	379.0	39.0	13.0	2.0	1.0	9.0[e]	0.0	211a
Sweden	1990–92	10.3	2.7	3.7	0.7	0.7	0.0	0.0	282a
United States	1992	81.0	6.0	12.0	6.0	3.0	3.0	0.0	24

[a] Mean annual number of reported incidents, where applicable.
[b] Enterohemorrhagic *E. coli* O157:H7.
[c] Generic reporting only.
[d] Human disease caused by *E. coli* and another bacterial pathogen.
[e] Principal cause of foodborne incidents remains *Vibrio parahaemolyticus* (mean = 303.5 incidents).

in unprotected facilities constitute as weak production links that undermine food safety. Operational changes in favor of field irrigation with treated effluents, washing of fruits and vegetables with disinfected waters, education of local workers on the hygienic handling of fresh produce, and greater protection of products from environmental contamination during all phases of production and marketing would markedly enhance the bacterial quality and safety of these frequently ready-to-eat products.

Foodborne Outbreaks

National epidemiologic registries continue to underscore the importance of *Salmonella* spp. as a leading cause of foodborne bacterial illnesses in humans. Reported incidents of foodborne salmonellosis tend to dwarf those associated with most other foodborne pathogens (Table 8.4). It is noteworthy that the problem of human salmonellosis from the consumption of contaminated foods generally remains on the increase worldwide (Table 8.5). Notwithstanding recent improvements in procedures for the epidemiologic investigation of foodborne outbreaks in many countries (267), the global increases in foodborne salmonellosis are considered real and not a result of enhanced surveillance programs and/or greater resourcefulness of medical and public health officials. Events such as the ongoing pandemic of

egg-borne *Salmonella* serovar Enteritidis clearly impact on current disease statistics. Major outbreaks of foodborne salmonellosis in the last few decades are of interest because they underline the multiplicity of foods and *Salmonella* serovars that have been implicated in human illness (Table 8.6). In 1974, temperature abuse of egg-containing potato salad served at an outdoor barbecue led to an estimated 3,400 human cases of serovar Newport infection; cross-contamination of the salad by an infected food handler was suspected (142). The large Swedish outbreak of serovar Enteritidis PT4 in 1977 was attributed to the consumption of a mayonnaise dressing in a school cafeteria (132). In 1984, Canada experienced its largest outbreak of foodborne salmonellosis, which was attributed to the consumption of cheddar cheese manufactured from heat-treated and pasteurized milk. The outbreak resulted in no fewer than 2,700 confirmed cases of serovar Typhimurium PT10 infection. Manual override of the flow diversion valve reportedly led to the entry of raw milk into vats of thermized and pasteurized cheese-milk (72). The following year witnessed the largest outbreak of foodborne salmonellosis in the United States, involving 16,284 confirmed cases of illness (164, 235). Although the cause of this outbreak was never ascertained, a cross-connection between raw and pasteurized milk lines seemingly was at fault. A large outbreak of salmonellosis affecting more than 10,000

Table 8.5 Trends in foodborne salmonellosis

Country	Annual no. of reported incidents					No. of cases[a]	Reference(s)
	1985	1987	1989	1991	1993		
Austria	124	151	440	963	—[b]	NA[c]	282, 282a
Bulgaria	15	10	12	13	—	2.1–13.9	282, 282a
Canada	59	53	51	28	43	0.5–3.8	68a, 125, 126
Czechoslovakia	94	120	258	—	—	94.2–258.0	282
Denmark	12	5	5	11	—	27.0–69.0	282, 282a
England and Wales	372	421	935	936	—	22.0–45.0	248, 282
Finland	11	3	2	4	—	89.1–153.2	282, 282a
France	7	178	462	477	626	NA	174, 282
Germany	—[d]	—[d]	23	62	—	103.5–242.7	282
Hungary	116	122	131	110	—	86.7	282, 282a
Italy	NA	NA	15	132	—	19.7—33.5	282a
Japan	—	90	146	159	—	2.9–8.2	258
Poland	380	690	709	625[e]	—	44.1–91.8	282, 282a
Scotland	133	180	151	115	—	35.0–46.0	282, 282a
United States	79	52	117	122	—	1.1–2.5	24, 45

[a] Range of cases per 100,000 population within the review period (1985–1993).
[b] —, data not published.
[c] NA, not available.
[d] German unification in 1989.
[e] Estimated from graphic representation of data.

Japanese consumers was attributed to a cooked egg dish (191). In 1993, paprika imported from South America was incriminated as the contaminated ingredient in the manufacture of potato chips distributed in Germany (168). A major outbreak of foodborne salmonellosis in the United States involved ice cream contaminated with serovar Enteritidis. The transportation of ice cream mix in an unsanitized truck that had previously carried raw eggs was identified as the source of contamination (13). Despite the general perception that meat and egg products are the primary sources of human *Salmonella* infections, many outbreaks in recent years have been associated with tomatoes (13e, 127), orange juice (18a, 34a, 42a, 209), and vegetable sprouts (1, 13d, 112a, 146, 193a, 217).

CHARACTERISTICS OF THE DISEASE

Symptoms and Treatment
Human *Salmonella* infections can lead to several clinical conditions, including enteric (typhoid) fever, uncomplicated enterocolitis, and systemic infections by nontyphoid microorganisms. Enteric fever is a serious human disease associated with the typhoid and paratyphoid strains, which are particularly well adapted for invasion and survival within host tissues. Clinical manifestations of enteric fever appear after a period of incubation ranging from 7 to 28 days and may include diarrhea, pro-

longed and spiking fever, abdominal pain, headache, and prostration (66). Diagnosis of the disease relies on the isolation of the infective agent from blood or urine samples in the early stages of the disease, or from stools after the onset of clinical symptoms (65). An asymptomatic chronic carrier state commonly follows the acute phase of enteric fever. The treatment of enteric fever is based on supportive therapy and/or the use of chloramphenicol, ampicillin, or trimethoprim-sulfamethoxazole to eliminate the systemic infection. Marked global increases in the resistance of typhoid and paratyphoid organisms to these antibacterial drugs in the last decade have greatly undermined their efficacy in human therapy. The problem is particularly serious in developing countries where multiply antibiotic-resistant salmonellae are frequently implicated in outbreaks of enteric fever and the cause of unusually high fatality rates (43, 222). The widespread use of fluoroquinolones in these countries is hampered by the high cost of these drugs (17, 38).

Human infections with nontyphoid *Salmonella* spp. commonly result in enterocolitis, which appears 8 to 72 h after contact with the invasive pathogen. The clinical condition is generally self-limiting, and remission of the characteristic nonbloody diarrheal stools and abdominal pain usually occurs within 5 days of onset of symptoms. The successful treatment of uncomplicated cases of enterocolitis may require only supportive therapy, such as fluid and electrolyte replacement. The use of antibiotics in such episodes is contraindicated, because

Table 8.6 Examples of major foodborne outbreaks of salmonellosis

Year	Country	Vehicle	*Salmonella* serovar	Cases[a]	Deaths	Reference(s)
				No.		
1973	Canada, United States	Chocolate	Eastbourne	217	0	59, 69
1973	Trinidad	Milk powder	Derby	3,000[b]	NS[c]	277
1974	United States	Potato salad	Newport	3,400[b]	0	142
1976	Spain	Egg salad	Typhimurium	702	6	11
1976	Australia	Raw milk	Typhimurium PT9	>500	NS	240
1977	Sweden	Mustard dressing	Enteritidis PT4	2,865	0	132
1981	The Netherlands	Salad base	Indiana	600[b]	0	26
1981	Scotland	Raw milk	Typhimurium PT204	654	2	54
1984	Canada	Cheddar cheese	Typhimurium PT10	2,700	0	72
1984	France, England	Liver pâté	Goldcoast	756	0	35, 208
1984	International	Aspic glaze	Enteritidis PT4	766	2	42
1984	United States	Salad bars	Typhimurium	751	0	267a
1985	United States	Pasteurized milk	Typhimurium	16,284	7	164
1987	China (People's Republic)	Egg drink	Typhimurium	1,113	NS	288
1987	Norway	Chocolate	Typhimurium	361	0	155
1988	Japan	Cuttlefish	Champaign	330	0	203
1988	Japan	Cooked eggs	*Salmonella* spp.	10,476	NS	191
1991	United States, Canada	Cantaloupes	Poona	>400	NS	94
1991	Germany	Fruit soup	Enteritidis	600	NS	105
1993	France	Mayonnaise	Enteritidis	751	0	228
1993	United States	Tomatoes	Montevideo	100	0	127
1993	Germany	Paprika chips	Saint-paul Javiana Rubislaw	>670	0	168
1994	Finland, Sweden	Alfalfa sprouts	Bovismorbificans	492	0	217
1994	United States	Ice cream	Enteritidis PT8	740	0	13, 134a
1995	United States	Orange juice	Hartford Gaminara	62	0	45, 209
1995	United States	Restaurant foods	Newport	850	0	13a
1996	United States	Alfalfa sprouts	Montevideo Meleagridis	481	1	193a
1997	United States	Stuffed ham	Heidelberg	746	1	26a
1998	United States	Toasted oat cereal	Agona	209	0	45
1998	Canada	Cheddar cheese	Enteritidis PT8	700	0	224a
1998	United States	Tomatoes	Baildon	>85	3	13e
1999	Japan	Dried squid	*Salmonella* spp.	>453	0	13c
1999	Australia	Orange juice	Typhimurium	427	NS	18a
1999	Japan	Cuttlefish chips	Oranienburg, Chester	>1,500	NS	269a
1999	Canada	Alfalfa sprouts	Paratyphi B (Java)	>53	NS	112a
1999	Japan	Peanut sauce	Enteritidis PT1	644	NS	152a
1999	United States, Canada	Orange juice	Muenchen	>220	0	34a
2000	United States, Canada	Cantaloupe	Poona	>39	0	13f
2000	United States	Orange juice	Enteritidis	>74	0	42a
2000	United States	Mung bean sprouts	Enteritidis	>45	0	13d

[a] Confirmed cases, unless stated otherwise.
[b] Estimated number of cases.
[c] NS, not specified.

it tends to prolong the carrier state and the intermittent excretion of salmonellae (65). This asymptomatic persistence of salmonellae in the gut likely results from a marked antibiotic-dependent repression of native gut microflora that normally compete with salmonellae for nutrients and intestinal binding sites. Human infections with nontyphoid strains can also degenerate into systemic infections and can precipitate various chronic conditions. In addition to serovar Dublin and serovar Choleraesuis, which exhibit a predilection toward septicemia, similarly high levels of virulence have been observed with other nontyphoid strains. Preexisting physiological, anatomical, and immunological disorders in human hosts can also favor severe and protracted illness through the inability of host defense mechanisms to respond effectively to the presence of invasive salmonellae (65). More frequent reports, in recent years, on the chronic and debilitating sequelae of nontyphoid systemic infections are of concern, because this emerging pattern may be linked to increased levels of virulence among nontyphoid salmonellae, increased susceptibility of human populations to chronic bacterial diseases, or synergy between both factors (51).

Salmonella spp.-induced chronic conditions, such as aseptic reactive arthritis, Reiter's syndrome, and ankylosing spondylitis, are noteworthy. Bacterial prerequisites for the onset of these chronic diseases include the ability of the bacterial strain to infect mucosal surfaces, the presence of outer membrane LPS, and a propensity to invade host cells (41, 247). Recent evidence suggests that these arthropathies may be linked to a genetic predisposition in individuals who carry the class I HLA-B27 histocompatibility antigen. Other antigens encoded by the HLA-B locus, such as the B7, B22, B40, and B60 antigens, which serologically cross-react with B27 antiserum, have also been associated with reactive arthritis (41, 184, 260). It is disconcerting that therapeutic eradication of a human *Salmonella* infection in the intestinal tract does not preclude the onset of chronic rheumatoid diseases in a distal, noninfected limb (260). Although the underlying mechanisms for these chronic conditions have yet to be fully elucidated, one of several theories suggests that phagocytosis of salmonellae results in the chemical alteration and slow phagocyte release of bacterial LPS. The altered LPS moieties are then reinserted into the cell membranes of synoviocytes and monocytes associated with articulations and synovial fluids, thereby rendering these surface-altered host cells susceptible to lysis by antibody and complement. This autoimmune response results in rheumatoid tissue damage caused by a localized inflammation of host tissues, and release of cytokines and enzymes from antigen-specific activated phagocytes (41, 247). Another suggested mechanism for the bacterial induction of arthropathies involves bacterial heat shock (stress) proteins. These molecular entities are synthesized in response to adverse environmental conditions and to the release of antibacterial substances, such as the reactive oxygen metabolites produced during the oxidative burst of activated phagocytes (66). Limited data indicate that heat shock proteins from invasive pathogens trig-

ger a proliferative response in synovial T lymphocytes of the host, which, in turn, engenders rheumatoid disease (247).

Prophylactics

The major global impact of typhoid and paratyphoid salmonellae on human health led to the early development of parenteral vaccines consisting of heat-, alcohol-, or acetone-killed cells (76). The phenol heat-killed typhoid vaccine, which continues to be widely used in protective immunity, can precipitate adverse reactions, including fever, headache, pain, and swelling at the site of injection. Although the need for immunogenic preparations against nontyphoid *Salmonella* spp. could be argued, the development of prophylactics against the multiplicity of serovars and bacterial surface antigens in the global ecosystem, the rapid succession of serovars in human populations and in reservoirs of infection, and the unpredictable pathogenicity of infective strains have dampened research interest and hampered significant advances in this field of clinical sciences. Nevertheless, the pandemic of poultry- and egg-borne serovar Enteritidis infections that continues to afflict consumers in many countries underscores the potential benefit of specific vaccines for human and veterinary applications (65, 183, 231).

Live attenuated vaccines continue to generate great research interest, because such preparations induce strong and durable humoral and cell-mediated responses in vaccinees (46, 63). The identification of bacterial genes wherein mutations lead to the attenuation of the carrier strain has greatly accelerated the development of live oral vaccines (46, 76). The avirulent mutant strain of serovar Typhi (Ty21a) is a product of chemical mutagenesis, in which the underlying molecular basis for attenuation is poorly understood. The absence of Vi antigens in the Ty21a strain and mutation in the *galE* locus, which significantly reduces the number of biosynthetic LPS enzymes, have been proposed as determinants for attenuation (46, 76). More recent evidence suggests that a mutation in the *rpoS* gene also contributes to the safety of this strain in humans (230). The *rpoS* gene, a component of the *Salmonella* virulence arsenal, engages the synthesis of numerous protective proteins in response to environmental stress conditions. The *rpoS* mutation in Ty21a is suggestively linked to a single nucleotide change in genomic DNA and results in a frameshift and gene transcription into an unstable RNA polymerase designated the σ^S factor (230). The Ty21a vaccine in the form of enteric-coated capsules is manufactured by the Swiss Serum and Vaccine Institute (Berne) under the trade name Vivotif Berna. This preparation, which has successfully met the challenge of intense field testing in Chile and Egypt (60 and 96% efficacy, respectively), currently enjoys wide usage in European and North American medical communities. Immunization results from the administration of three capsules on alternate days and provides up to 3 years of protection.

An injectable vaccine prepared from the Vi capsular polysaccharide antigen of serovar Typhi was recently released by Pasteur Mérieux Sérums et Vaccins (Lyon, France) under the trade name Typhim Vi. A single dose of phenol-preserved antigen provides for a rapid rise in serum levels of Vi antibodies and confers protection for up to 3 years. In the presence of an active typhoid infection, the Vi antibodies facilitate the bactericidal action of serum complement (classical pathway). Two large-scale clinical studies in Nepal and Eastern Transvaal (South Africa) confirmed vaccine efficacy that ranged from 65 to 75%, with low levels of adverse reactions (157). In contrast to the Ty21a vaccine, whose heterogeneous molecular composition predisposes recipients to stronger and varied secondary reactions, the use of a single, chemically defined antigen in Typhim Vi minimizes the risks of undesirable side effects. Developmental research may soon culminate in the marketing of new typhoid vaccines exhibiting greater levels of immunogenicity, biological safety, and storage stability (54). For example, a temperature-sensitive (*ts*) strain of serovar Typhi that grows optimally at 29°C but whose proliferation ceases at 37°C shows great potential as the immunogen in a novel live oral typhoid vaccine (29). Research at the National Institutes of Health (Bethesda, Md.) has also yielded promising results in the development of a conjugate vaccine consisting of O-acetylated pectin linked to a tetanus toxoid that raises the level of typhoid antibodies in mice (254a). The preparation is of particular interest as a human prophylactic because it is strongly immunogenic, is divalent, and carries no endotoxic moiety. Although beyond the scope of this chapter, another rapidly evolving and productive field of research involves the use of attenuated *Salmonella* strains as vectors for the delivery of heterologous antigens to mammalian immune systems (46, 76). For example, the administration of a live attenuated *Salmonella* strain harboring the gene encoding fragment C of the tetanus toxin would provoke a potent immunological response against the carrier *Salmonella* strain and the tetanus toxin (46). Clearly, the field of vaccine development has witnessed great advances in recent years, and such success will undoubtedly continue as new knowledge on the molecular complexities of bacterial attenuation facilitates the delivery of defined, effective, and stable prophylactic preparations. The subject of live attenuated

vaccines has been thoroughly reviewed (243). A live avirulent Food and Drug Administration (FDA)-approved serovar Typhimurium vaccine has been developed (123) for use in broiler and breeder chickens, whereas serovar Enteritidis vaccines are being used extensively in the egg layer industry to help control the problem of serovar Enteritidis.

Antibiotic Resistance

Current global trends depicting sustained increases in the number of antibiotic-resistant *Salmonella* spp. in humans and farm animals are most disquieting (167, 194, 207, 261). The major public health concern is that *Salmonella* spp. will become resistant to antibiotics used in human medicine, thereby greatly reducing therapeutic options and threatening the lives of infected individuals (256). The liberal administration of antimicrobial agents in hospitals and other treatment centers contributes significantly to the emergence and persistence of many resistant strains. The emergence of antimicrobial resistance in bacterial pathogens, many of which can be traced to food-producing animals, has generated much controversy and public media attention. The problem stems from the widespread use and abuse of antibiotics in the meat animal and aquaculture industries (148, 202, 281, 289). The imprudent use of subtherapeutic levels of antibiotics for growth promotion and prophylaxis in reared species has captured the attention of policymakers worldwide.

Because evolutionary processes offer genetic and phenotypic variability for resistance genes, there will always be some level of background resistance in bacterial populations (287). However, without selective pressures, resistance levels are low, and so are the chances for therapy failures. Antimicrobial agents are used in animal husbandry for three purposes: therapy, prophylaxis, and growth enhancement. In each of these instances, antibiotic residues are released into the environment, thereby increasing selective pressures for antibiotic-resistant bacteria. In particular, the widespread administration of prophylactic mixtures of antimicrobials, antimicrobials mixed into animal feeds, and therapy without diagnosis may lead to strong selective pressures on *Salmonella* spp. in intensive animal husbandry production units (287). Increased levels of resistant organisms have followed from newly licensed use of antimicrobials such as apramycin and quinolones (81, 133, 213, 265, 286). Interestingly, a low prevalence of antibiotic-resistant bacteria has been reported in Sweden, where the use of antibiotics in animal husbandry is restricted (31, 95, 279).

Antibiotic resistance in *Salmonella* spp. has been reported since the early 1960s (40, 272), when most of the reported resistance was to a single antibiotic (40, 49). In the mid-1970s a disturbing trend of multiply antibiotic-resistant-strains emerged. A serovar Typhimurium clone that carried resistance against chloramphenicol, streptomycin, sulfonamides, and tetracycline was documented in 1979 to move from calves through the food chain to humans (233). Many other multiply resistant clones were subsequently identified. Some of these clones appeared to be localized, but others, such as serovar Typhimurium DT204 and DT193, were found in several countries (264, 265).

The appearance of serovar Typhimurium DT104 in the late 1980s raised major concerns because of its common pentaresistance to ampicillin, chloramphenicol, streptomycin, sulfonamides, and tetracycline. Additional resistance to gentamicin, trimethoprim, and fluoroquinolones has been reported (262, 266). Although DT104 was originally detected in cattle (178, 274), this phagovar was later identified in livestock including horses, goats, and sheep as well as in cats, dogs, elk, mice, coyotes, squirrels, raccoons, chipmunks, and pigeons (287).

Fluoroquinolones belong to a class of newer antimicrobial agents with a broad spectrum of activity, high efficiency, and widespread application in human and veterinary medicine (284). Because fluoroquinolones are often recommended therapeutics for life-threatening Salmonella infections (17), there is a particular concern that all reasonable efforts should be made to keep resistance in Salmonella spp. and other foodborne bacterial pathogens as low as possible (278). An increase in the resistance of DT104 strains to fluoroquinolones was recently documented in England and Wales (266) and in Germany (262). As a result, several European countries have banned the nonhuman use of fluoroquinolones. The U.S. FDA recently recommended that the approval for therapeutic use of fluoroquinolones in poultry be withdrawn in January 2001.

Several antibiotics, such as bacitracin, virginiamycin, and apramycin, are considered bona fide veterinary drugs because they are reportedly devoid of any therapeutic value in human medicine. However, recent evidence indicates that the agricultural use of apramycin may promulgate the emergence of gentamicin-resistant Salmonella spp. in treated livestock (215, 261). The phenomenon which stems from the linkage of the apramycin and gentamicin resistance codons on a single R plasmid leads to a new realization that on-farm use of any antibiotics may soon engender major public health problems. The consideration that genetic determinants for Salmonella virulence and antibiotic resistance can occur on the same plasmid dramatically increases the disease

potential of this emerging bacterial elite, precursors to the "super bug" era (236, 263). In the next few decades (or perhaps sooner), we may well witness the international scientific community feverishly engaged in the development of new drugs to replace those that are currently in use and whose spectra of activity would then have slipped into obsolescence in the face of a rising tide of highly virulent and multiply antibiotic-resistant Salmonella spp. Most of the foregoing considerations also apply to the aquaculture industry, whose prominence in the global fish and shellfish market has soared as a result of the rapid depletion of feral stocks and increasing consumer demand for these products (68, 252). The aqueous environment and warm climates generally associated with aquacultural operations promote the proliferation of Salmonella spp. in rearing ponds and basins and provide favorable conditions for the exchange of R plasmids and cross-contamination of reared species. Moreover, the addition of subtherapeutic doses of antibiotics of human significance into holding facilities together with the sedimentation of detritus and medicated feeds provide a uniquely favorable habitat for the selection and growth of antibiotic-resistant salmonellae.

The issues that regulators are addressing presently are not new. The discovery of transferable drug resistance combined with an outbreak of salmonellosis in both cattle and people in the United Kingdom in the 1960s prompted the British Parliament to commission a study on the possibility that antimicrobial resistance was passing from animals to humans. This large-scale study and subsequent publication of the Swann Report had a two-part recommendation (254). First, it suggested that antibiotics used for treatment of human infections not be used in animal husbandry. Second, it recommended a ban on the subtherapeutic use (<200 g/ton) of antibiotics in animal feed. The issues raised in this report have yet to be fully resolved.

The potential link between the use of antimicrobial agents in animal feed and the emergence of drug-resistant human pathogens was examined in a 1980 report of the National Research Council (197). The conclusions were equivocal, and the report stated, "...the postulated hazards to human health from the subtherapeutic use of antimicrobials in animal feeds were neither proven nor disproven." The report cited problems with even doing the research because of the impossibility of determining the antimicrobial history of an animal from which a particular piece of meat was derived.

The Institute of Medicine of the National Academy of Sciences formed a committee to examine several cases of antibiotic resistance in humans and to determine possible relatedness of this phenomenon to the use of antimicrobials in animal feed (148). The committee could find no positive link between the subtherapeutic use of antibiotics in animals and human health hazards. A mathematical projection of disease probability was made, but the need for more data to make the projections useful was recognized.

An American Society for Microbiology task force report issued in 1995 (8) on the issue of antimicrobial resistance identified several areas of concern, including over-the-counter sale of antibiotics in developing countries, the empirical prescription of antibiotics in the absence of diagnostic test results, inappropriate pediatric prescriptions of drugs arising from parent pressure, and the use of antibiotics in the fruit, vegetable, and meat industries. The report further acknowledged the lack of information on the use of antimicrobial agents in agriculture and aquaculture.

The pressing need for data led to the joint development in 1996 of the National Antimicrobial Resistance Monitoring System (NARMS) by the FDA's Center for Veterinary Medicine, the Centers for Disease Control and Prevention, and the U.S. Department of Agriculture. The purpose of this program is to provide baseline data so that changes in the antimicrobial resistance of background microflora can be identified.

In 1997, a WHO workshop held in Berlin focused on the subtherapeutic use of antibiotics, especially fluoroquinolones (283). Recommendations from this meeting included phasing out subtherapeutic use of antimicrobials in animal feed, especially those agents that might be used in the treatment of human infections.

A workshop report from the Institute of Medicine on antimicrobial resistance published in 1998 recommended expanding collaborative research between human and animal medicine, with a suggestion that exploration of competitive exclusion as a means of controlling pathogenic bacteria be increased (149). The report highlighted the dire consequences of depleting the world's armamentarium of effective antimicrobials but acknowledged the difficulty of implementing a public policy that would undoubtedly have negative economic repercussions.

The most comprehensive report has been the National Research Council's publication in 1999 of *The Use of Drugs in Food Animals: Benefits and Risks*, by the Committee on Drug Use in Food Animals (198). This volume represents almost a decade of preparation and data gathering. The panel charged with this endeavor concluded that there were still insufficient data available to draw conclusions. They recommended a four-part effort—international harmonization, research into the development of new antimicrobials, development of a database,

and promotion of interdisciplinary work on antimicrobial use in animals and its link to human disease.

In 1999, the FDA issued a document entitled *A Proposed Framework for Evaluating and Assuring the Human Safety of the Microbial Effects of Antimicrobial New Animal Drugs Intended for Use in Food-Producing Animals* (91). This document is expected to be ratified in 2001 or 2002. The concepts described in the framework are designed to assist the agency in assessing not only new antibiotics but also previously approved antibiotics. The framework sets out three categories of antimicrobial drugs. The categories are based on the drug's unique or relative importance to human medicine. Category 1 drugs would include those that are (i) essential for treating a serious or life-threatening disease in humans, and for which there is no satisfactory alternative treatment, (ii) important for the treatment of foodborne diseases in humans when resistance to alternative antimicrobial drugs may limit therapeutic options, or (iii) in a class of drugs for which the mechanism of action and/or the nature of induced resistance is unique, resistance to antimicrobial drug is rare among human pathogens, and the drug holds potential for long-term therapy in human medicine. Any antimicrobial agent that can induce or select for cross-resistance to a Category 1 drug would be considered a Category 1 drug. Category 2 drugs are drugs of choice or are important in the treatment of a potentially serious disease, whether foodborne or otherwise, but for which there are satisfactory alternative therapies. Any antimicrobial agent that can induce or select for cross-resistance to a Category 2 drug would be considered a Category 2 drug. However, if an antimicrobial is not used in human medicine, then it would not be considered a Category 2 drug. Category 3 drugs are those that do not meet the criteria for Category 1 or Category 2, have little or no current use in human medicine, and are not the drug of choice or a significant alternative for treating human infections, including foodborne infections. As part of the framework, the likelihood of a drug to induce resistance in bacteria, the likelihood that use of a drug in food-producing animals will promote resistance, and the likelihood that such transfer will result in loss of availability of human antimicrobial therapies will be characterized. The likelihood of human exposure to resistant foodborne pathogens as a result of agricultural or aquacultural use of the specified antimicrobial agent will be classified as high, medium, or low. The amount of data required before and after approval of a new antimicrobial drug will depend upon the category of the drug and the potential for human exposure to bacteria resistant to the drug. For example, approval of a Category 1 drug with high potential for human exposure would require the greatest amount of data, i.e., preapproval studies on resistance and pathogen load, and the establishment of resistance and monitoring thresholds. On the other end of the spectrum, a Category 3 drug with low potential for human exposure would not require preapproval data on antimicrobial resistance. The ultimate success of this or any other program will be measured by NARMS, which will enable a prospective monitoring of changes in the antimicrobial susceptibility of zoonotic enteric pathogens.

Additionally, in June 2000, WHO released a set of global principles aimed at mitigating the risks related to the use of antimicrobial agents in food animals (285). The new recommendations are designed for use by governments, veterinary and other professional societies, industry, and academia. Some of the most important measures included in the new "Global Principles for the Containment of Antimicrobial Resistance due to Antimicrobial Use in Animals Intended for Food" are (i) obligatory prescriptions for all antimicrobials used for disease control in food animals, (ii) termination or rapid phasing out of the use of antimicrobials for growth promotion if they are also used for treatment of humans, even in the absence of a public health safety evaluation, (iii) creation of national systems to monitor antimicrobial usage in food animals, (iv) prelicensing safety evaluation of antimicrobials with consideration of potential resistance to human drugs, (v) monitoring of resistance to identify emerging health problems and timely corrective actions to protect human health, and (vi) guidelines for veterinarians to reduce overuse and misuse of antimicrobials in food animals.

Despite numerous investigations into this area during the last 30 years, the problem of there being insufficient data to accurately determine the risks persists. Hopefully, the recent flurry of attention and the development of appropriate databases will move the controversy closer to an equitable resolution.

Infectious Dose

It is well established that newborns, infants, the elderly, and immunocompromised individuals are more susceptible to *Salmonella* infections than healthy adults (65). The incompletely developed immune system in newborns and in infants, the frequently weak and/or delayed immunological responses in the elderly and debilitated persons, and the generally low gastric acid production in infants and seniors facilitate the intestinal colonization and systemic spread of salmonellae in these segments of the population (34, 66). Antibiotic treatment of subjects before their encounter with *Salmonella* spp. enhances

Table 8.7 Human infectious dose of *Salmonella*[a]

Food	*Salmonella* serovar	Infectious dose	Reference
Eggnog	Meleagridis	10^4–10^7	185
	Anatum	10^5–10^7	
Goat cheese	Zanzibar	10^5–10^{11}	226
Carmine dye	Cubana	10^4	163
Imitation ice cream	Typhimurium	10^4	15
Chocolate	Eastbourne	10^2	69
Hamburger	Newport	10^1–10^2	89
Cheddar cheese	Heidelberg	10^2	90
Chocolate	Napoli	10^1–10^2	113
Cheddar cheese	Typhimurium	10^0–10^1	72
Chocolate	Typhimurium	$\leq 10^1$	155
Paprika potato chips	Saint-paul	$\leq 4.5 \times 10^1$	168
	Javiana		
	Rubislaw		
Alfalfa sprouts	Newport	$\leq 4.6 \times 10^2$	1
Ice cream	Enteritidis	$\leq 2.8 \times 10^1$	273a

[a] Adapted from reference 68.

bacterial virulence through an antibiotic-mediated clearance of native gut microflora, which reduces the level of bacterial competition for nutrients and attachment sites in the intestinal tract of the host (140, 249).

Detailed investigations of foodborne outbreaks have indicated that the ingestion of only a few *Salmonella* cells can be infectious (Table 8.7). Although early reports that large numbers of salmonellae inoculated into eggnog and fed to human volunteers produced overt disease (185), more recent evidence suggests that 1 to 10 cells can constitute a human infectious dose (72, 155). Determinant factors in salmonellosis are not limited to the immunological heterogeneity within human populations and to the virulence of infecting strains; they may also include the chemical composition of incriminated food vehicles. A common denominator of the foods associated with low infectious doses (Table 8.7) is the high fat content in chocolate (cocoa butter), cheese (milk fat), and meat (animal fat). Suggestively, entrapment of salmonellae within hydrophobic lipid micelles would afford protection against the bactericidal action of gastric acidity. Following a bile-mediated dispersion of the lipid moieties in the duodenum, the viable salmonellae would resume their infectious course in search of suitable points of attachment in the lower portion of the small intestine (colonization). The rapid emptying of gastric contents could also provide an alternate mechanism for the successful infection of susceptible hosts. The swift passage of a liquid bolus through an empty stomach would minimize bacterial exposure to gastric acidity and sustain the migration of viable salmonellae in the intestinal tract (34).

Recent publications on the dynamics of human *Salmonella* infections are of singular interest (110, 111). An in-depth epidemiologic study of a large outbreak of serovar Typhimurium infection involving chicken served to delegates at a medical conference revealed that the clinical course in patients was directly related to the number of salmonellae ingested (111). The incubation period for the onset of symptoms was inversely related to the infectious dose. Patients with short (≤22 h) periods of incubation suffered more frequent diarrheal bowel movements, higher maximum body temperatures, greater persistence of clinical symptoms, and greater frequency of hospitalization. Interestingly, no association between the age of infected individuals and the length of the incubation period was noted. Similar findings were reported in retrospective dose-response studies of foodborne salmonellosis (110, 190).

The compelling evidence that ingestion of a few *Salmonella* cells can develop into a variety of clinical conditions and deteriorate into septicemia and even death underlines the unpredictable pathogenicity of this large and heterogeneous group of human bacterial pathogens. Food producers, processors, and distributors need to be reminded that low levels of salmonellae in a finished food product could lead to serious public health consequences and undermine the reputation and economic viability of the incriminated food manufacturer (65).

PATHOGENICITY AND VIRULENCE FACTORS

Specific and Nonspecific Human Responses

The presence of viable salmonellae in the human intestinal tract confirms the successful evasion of ingested organisms from nonspecific host defenses. Antibacterial lactoperoxidase in saliva, gastric acidity, mucoid secretions from intestinal goblet cells, intestinal peristalsis, and sloughing of luminal epithelial cells synergistically oppose bacterial colonization of the intestinal mucosa. In addition to these constitutive hurdles to bacterial infection, the antibacterial action of nonspecific phagocytic cells (neutrophils, macrophages, monocytes) coupled with the immune responses associated with specific T and B lymphocytes, the epithelio-lymphoid tissues (Peyer's patches), and the classical or alternative pathways for complement inactivation of invasive pathogens mount a formidable defense against the systemic spread of *Salmonella* spp. The interplay between these immunological defense factors was recently reviewed (66).

The human diarrheagenic response to foodborne salmonellosis results from the migration of the pathogen in the oral cavity to intestinal tissues and mesenteric lymph follicles (enterocolitis). The event coincides with extensive leukocyte influx into the infected tissues, increased mucus secretion by goblet cells, and mucosal inflammation triggered by the leukocytic release of prostaglandins. The latter occurrence also activates the adenyl cyclase in intestinal epithelial cells, resulting in increased fluid secretion into the intestinal lumen (88, 216). The failure of host defense systems to hold the invasive salmonellae in check can degenerate into septicemia and other chronic clinical conditions.

Attachment and Invasion

The establishment of a human *Salmonella* infection depends on the ability of the bacterium to attach (colonize) and enter (invade) intestinal columnar epithelial cells (enterocytes) and specialized M cells overlying Peyer's patches (Fig. 8.1). Salmonellae must successfully compete with indigenous gut microflora for suitable attachment sites on the luminal surface of the intestinal wall and evade capture by secretory immunoglobulin A that may also be present on the surface of epithelial cells. Colonization of enterocytes arises, in part, from the interaction of bacterial type 1 (mannose-sensitive) or type 3 (mannose-resistant) fimbriae, surface adhesins, nonfimbriate (mannose-resistant) hemagglutinins, or enterocyte-induced polypeptides with host glycoprotein receptors located on the microvilli or glycocalyx of the intestinal surface (66, 216). Although the role of *Salmonella* motility in the adherence and invasion processes is uncertain, motility may not be essential for bacterial internalization and possibly would simply increase the frequency of productive contacts between the pathogen and its targeted epithelial cell (99). An authoritative review on the morphology of interactions between salmonellae and the enterocytes, M cells, and the immune apparatus in the mucosal and submucosal layers of the host intestine is recommended as additional reading (216). Fine structural studies have also revealed that proteinaceous appendages develop on the surface of salmonellae upon contact with epithelial cells (99, 108). These invasion appendages are 0.3 to 1.0 nm in length and ca. 60 nm in diameter, which is considerably thicker than flagella (ca. 20 nm) and type 1 fimbriae (ca. 7 nm). The assembly of these *Salmonella* appendages is energy dependent but independent of de novo bacterial protein synthesis (56, 108). The appendages are short-lived and shed concomitantly with the appearance of membrane ruffles on colonized epithelial cells (108). Following bacterial attachment, signal transduction between the pathogen and host cell culminates in an energy-dependent *Salmonella* invasion of enterocytes and M

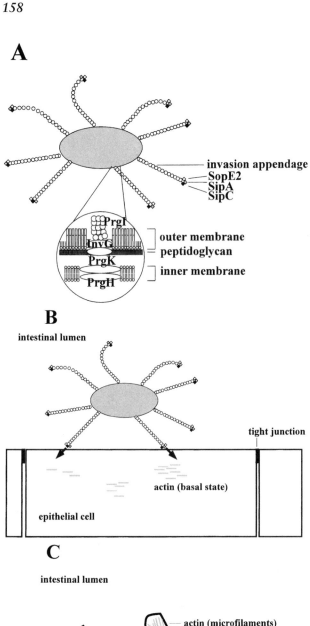

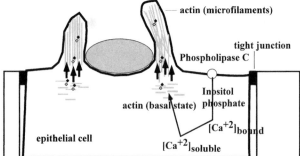

Figure 8.1 Mechanism for the *Salmonella* invasion of host epithelial cells. (A) Architecture of *Salmonella* invasion complex. (B) Translocation of effector proteins SipA, SipC, and SopE2 from invasion appendage into host cell cytoplasm. (C) Polymerization of actin into microfilaments, with SipA and SipC serving as the catalysts, and activation of host cell's signal transduction pathway (e.g., phospolipase C activation, Ca^{2+} influx) (101).

cells (102, 108). Although bacterial protein synthesis is not a prerequisite for the onset of invasion, its prolonged inhibition compromises the ability of salmonellae to enter cultured epithelial cells (179). The role of the *inv* (invasion) locus in *Salmonella* spp. is of singular importance in the invasion mechanism because it triggers two profound changes in enterocytes and M cells, namely, a Ca^{2+} influx and cytoskeleton rearrangement in the targeted host cells. The cytoskeleton of eukaryotic cells is a highly organized network of contractile and supportive filaments consisting of actin or other protein elements that define the apical-basal polarity of intestinal epithelial cells and control the movement of intracellular organelles (88, 216). Adherent *Salmonella* cells promote the influx of luminal Ca^{2+} into the mammalian cell, an important signal that influences actin polymerization (234). Salmonellae deliver invasion proteins SipA, SipB, SipC, SptP, SopE2, and SopB to eukaryotic cells (55). These proteins in turn cause polymerization of host cell actin into microfilaments in the vicinity of the invading pathogen (101, 109). *Salmonella* invasion proteins SipA and SipC act as catalysts in the nucleation and subsequent polymerization of F-actin into microfilaments (124, 291, 292). Another invasion protein, SopE2, activates signal transduction cascade involved in cytoskeletal rearrangement (250). Histologically, *Salmonella* invasion is seen as an evagination of the apical cytoplasm of epithelial cells (i.e., membrane ruffle) around the adherent salmonellae that subsequently mediate the pinocytotic uptake of the bacterial pathogen (153).

We have thus far seen that salmonellae attach to the microvilli of the intestinal epithelium, an intercellular contact that activates salmonellae into the production of transient, proteinaceous appendages deemed essential for *Salmonella* invasion. These events are closely followed by actin polymerization and formation of membrane ruffles. The completed internalization of salmonellae into the epithelial cell precipitates a reversion of actin microfilaments and membrane ruffles to their basal states. This final event may be mediated by the *Salmonella* tyrosine phosphatase SptP, a protein that causes disruption of the actin cytoskeleton and disappearance of stress fibers (97). *Salmonella* invasion of the epithelial cells also has profound effects on the cell's physiology. Epithelial cells produce cytokines, including chemoattractants, in response to the invading bacteria (78). The invading salmonellae also cause apoptosis (programmed cell death) of epithelial cells (159), neutrophils, and macrophages (137, 193, 229). Diarrhea associated with salmonellae now appears to be in response to bacterial invasion of enterocytes and M cells instead of the action of a putative enterotoxin, Stn (269, 275).

Table 8.8 Characteristics of *Salmonella* virulence genes

Position	Gene/locus	Regulation	Identity	Function	Reference(s)
59	SPI1[a]			Cell invasion	

	sipA	invF, hilA	ipaA[b]	Binds F-actin and T-plastin and induces actin bundling	80, 290, 291
	sipB	invF, hilA	ipaB[b]	Membrane fusion; induces apoptosis in macrophages	124, 137
	sipC	invF, hilA	ipaC[b]	Mediates actin condensation and cytoskeletal rearrangements	124
	sipD	invF, hilA	ipaD[b]		80
	prgH	phoP/Q		Type III secretion; outer membrane base	27
	prgI	phoP/Q	mxiH,[b] yscF[c]	Invasion appendage	210
	prgJ	phoP/Q	mxiI[b]	Type III secretion	210
	prgK	phoP/Q	mxiJ,[b] yscJ,[c] lipoprotein	Type III secretion; outer membrane base	210
	invA	hilA	flhA[c]	Type III secretion	80,100
	invB	hilA		Type III secretion; chaperone of SspA	80
	invC	hilA	fliI,[a] ATPase	Type III secretion	79, 80
	invD	hilA			80
	invE	hilA			80
	invF	hilA	AraC-type transcriptional activator	Regulation of sicA, sipBCDA operon	73, 80
	invG	hilA		Type III secretion; outer membrane secretion channel	80
	invH	hilA	Outer membrane lipoprotein	Type III secretion	7, 80
	spaM	hilA	ipaB[b]	Type III secretion	56, 80
	spaN	hilA	spa32,[b] eaeB[d]	Type III secretion	56, 80
	spaO	hilA	yscQ[c]	Type III secretion	55, 80
	spaP-S	hilA		Type III secretion	55, 80
	sptP	invF, hilA	yopE[c] (cytotoxin), yopH[c] (phosphatase)	Disruption of actin cytoskeleton	80, 154
	sitA,B,C,D	fur	yfe[c] ABC iron transport system	Iron transport	152, 292
	hilA	sirA	Transcriptional activator		3

(Continued)

Salmonella invasion genes are organized into contiguous and functionally related loci located within a 40- to 50-kb segment of chromosomal DNA (i.e., *Salmonella* pathogenicity island SPI1) that encode determinant factors for the facilitated entry of salmonellae into host cells (98, 115, 188; Table 8.8). The *inv* pathogenicity island is a multigenic locus consisting of 30 genes. Many of these genes code for enzymes and transcriptional activators responsible for the regulation, expression, and translocation of important effectors like SipA to the surface

Table 8.8 Characteristics of *Salmonella* virulence genes *(Continued)*

Position	Gene/locus	Regulation	Identity	Function	Reference(s)
30	SPI2[a]			Macrophage survival; cell invasion	

	ssaJ		*ycsJ*[c]	Type III secretion	135
	ssaK		*ycsL*[c]	Type III secretion	135
	ssaV		*lcrD*[c]	Type III secretion	135
	ssaN		*ycsN*[c]	Type III secretion	135
	ssaO		*ycsO*[c]	Type III secretion	135
	ssaQ		*ycsQ*[c]	Type III secretion	135
	ssaR		*ycsR*[c]	Type III secretion	135
	ssaS		*ycsS*[c]	Type III secretion	135
	ssaT		*ycsT*[c]	Type III secretion	135
	ssaU		*ycsU*[c]	Type III secretion	135
	ssaP				
	ssrA		*ompR*	Two-component regulatory element	165
	ssrB		*ompR*	Two-component regulatory element	53, 165
	sseA		*espA*[d]		53
	sseB				53
	sseC		*yopB*,[c] *espD*[d]	Translocation of proteins into eukaryotic membrane	53
	sseD		*espB*[d]		53
	sseE		*lcrR*[c]		53
	sseF				
	sseG				
	sscA		*lcrH*,[c] chaperone	Type III secretion	53
	sscB		*lcrH*,[c] *sicA*, *ipgC*,[c] chaperone	Type III secretion	
	sscC		*sicA*, *ipgC*,[c] chaperone	Type III secretion	53
	spiA		*sepC*,[d] *yscC*[c]	Type III secretion	
	spiB		*yscD*[c]	Type III secretion	
	spiC			Inhibit fusion of phagosomes with lysosome	269

of host cells (Table 8.8). Most of these genes encode type III secretion components involved in the assembly of the invasion appendages and the export and insertion of *Salmonella* invasion proteins into the epithelial cell (101, 143). This invasion-associated type III secretion apparatus structurally resembles the flagellar basal body of gram-negative bacteria (143). Elucidation of the role of many SPI1 invasion genes in the export and

assembly of the invasion complex is currently of major research interest (Table 8.8). Characterization of a second *Salmonella* pathogenicity island, SPI2, has identified additional genes necessary for the export of membrane fusion protein SipB (135). SPI1 invasion genes are evolutionarily conserved among *Salmonella* spp. (36), making these genes useful targets for PCR-based detection of *Salmonella* spp. in foods (48). Although this locus is

Table 8.8 *(Continued)*

Position	Gene/locus	Regulation	Identity	Function	Reference(s)
20	SPI5[a]			Enteropathogenicity (diarrhea)	

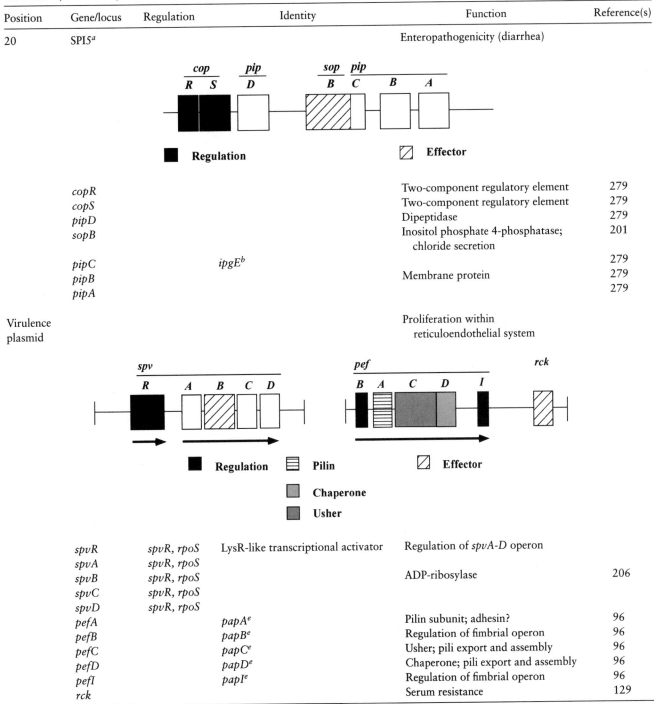

	copR			Two-component regulatory element	279
	copS			Two-component regulatory element	279
	pipD			Dipeptidase	279
	sopB			Inositol phosphate 4-phosphatase; chloride secretion	201
	pipC		*ipgE*[b]		279
	pipB			Membrane protein	279
	pipA				279

Virulence plasmid — Proliferation within reticuloendothelial system

	spvR	*spvR, rpoS*	LysR-like transcriptional activator	Regulation of *spvA-D* operon	
	spvA	*spvR, rpoS*			
	spvB	*spvR, rpoS*		ADP-ribosylase	206
	spvC	*spvR, rpoS*			
	spvD	*spvR, rpoS*			
	pefA		*papA*[e]	Pilin subunit; adhesin?	96
	pefB		*papB*[e]	Regulation of fimbrial operon	96
	pefC		*papC*[e]	Usher; pili export and assembly	96
	pefD		*papD*[e]	Chaperone; pili export and assembly	96
	pefI		*papI*[e]	Regulation of fimbrial operon	96
	rck			Serum resistance	129

[a] *Salmonella* pathogenicity island.
[b] *Shigella* invasion gene.
[c] *Yersinia* invasion gene.
[d] Enteropathogenic *E. coli* (EPEC) secreted protein.
[e] Uropathogenic *E. coli* P-pili.

not essential for the salmonellae to traverse the intestinal barrier and spread systemically (195), it does appear to be one of several factors that mediate enteritis associated with *Salmonella* infection (269).

Environmental factors such as high osmolarity and low pO_2 enhance bacterial invasiveness by altering the superhelicity of chromosomal DNA, which consequently impacts on the level of transcription of invasion-related genes such as *invA* (84, 102). There are several key transcriptional regulatory elements involved in the regulation of the *inv* locus (2, 27, 73, 80). In fact, the genes of *Salmonella* spp. are coordinately regulated as the local infection becomes systemic (131, 242).

Growth and Survival within Host Cells

In contrast to several bacterial pathogens, such as *Yersinia* and *Shigella* spp. and enteroinvasive *E. coli*, which replicate within the cytoplasm of host cells, salmonellae are confined to endocytotic vacuoles in which bacterial replication begins within hours following internalization (102). The infected vacuoles move from the apical to the basal pole of the host cell, where salmonellae are released into the lamina propria (150, 216). During their proliferation within epithelial vacuoles, salmonellae also induce the formation of stable filamentous structures within the epithelial cytosol (104). These organelles, which contain lysosomal membrane glycoproteins and acid phosphatase, are seemingly connected to the infected vacuoles. The formation of these filaments requires viable salmonellae within the membranous vacuoles and is blocked by inhibitors of vacuolar acidification. Although the role of these induced filamentous structures is poorly understood, preliminary findings point to their involvement in the intravacuolar replication of *Salmonella* spp. A bacterial surface mechanism has been tentatively identified that facilitates the migration of salmonellae into deeper layers of tissues upon their release into the lamina propria. The demonstrated ability of thin aggregate fimbriae on the outer surface of *Salmonella* cells to bind host plasminogen and the tissue-type plasminogen activator could markedly increase the invasiveness of infecting strains. Suggestively, the zymogen would be converted into its proteolytic (plasmin) form on the bacterial surface, thereby providing salmonellae with an effective tool to breach host tissue barriers and facilitate transcytosis into deeper tissues (244).

The systemic migration of salmonellae exposes the organism to phagocytosis by macrophages, monocytes, and polymorphonuclear leukocytes and to the antibacterial conditions that prevail in the cytoplasm of these host defense cells (46). Survival of the bacterial cell within the hostile confines of the macrophage or neutrophil determines the host's fate. Whether or not the host develops enteric fever from *Salmonella* infection is determined partly by the genetics of the host (245) and the salmonellae (134, 239). Intracellular pathogens have developed different strategies to survive within professional phagocytic cells, including (i) escaping from the phagosomes (21), (ii) inhibiting acidification of the phagosomes (141), (iii) preventing fusion of phagosomes and lysosomes (14), and (iv) withstanding the toxic environment of the phagolysosome (227). Although there is evidence to support the ability of *Salmonella* to prevent acidification of the phagosomes (6), fusion with the lysosome (39, 270), or maturation of the phagolysosome (223), *Salmonella* is a hardy bacterium that can survive and replicate within the bactericidal milieu of the acidified phagolysosome (204, 224). For example, the membrane oxidase-dependent formation of toxic oxygen products, such as singlet oxygen, superoxide anion, hydrogen peroxide, and hydroxyl radicals, during the oxidative metabolic burst of phagocytes is countered by the protective activity of several bacterial enzymes, including superoxide dismutase, peroxidase, and catalase. Recent findings have revealed that 30 bacterial proteins are synthesized in response to the toxic oxygen products of macrophages. Of these, nine are dependent upon the synthesis of a nucleic acid activator encoded by the chromosomal *oxyR* locus that induces the transcription of genetic loci responsible for the synthesis of protective proteins (218).

The *phoP/phoQ* regulon is a two-component transcriptional regulator system that enables *Salmonella* spp. to survive within the hostile environment of phagocytes, notably, the high acidity within phagolysosomes (host cell construct of an infected phagosome with incorporated lysosomal granules) and release of antibacterial defensins by phagocytic cells (186, 218). Defensins, which also occur in epithelial cells, are small, nonspecific cationic peptides that inactivate salmonellae by inserting into the outer bacterial membrane, thereby creating transmembrane channels that increase bacterial permeability to ions and precipitate cell death (87). The antidefensin gene product(s) encoded by the *Salmonella phoP/phoQ* system appears to involve a chemical alteration of the LPS core (271) and production of a cell surface protease that degrades host defensin (117), thereby conferring greater bacterial resistance to neutrophil antimicrobial peptides (241). The *phoP/phoQ* locus is part of a transcriptional cascade of activators and repressors (117, 121) necessary for expression of a subset of genes essential to the survival of the bacterial cell within the macrophage (25, 117, 121). PhoQ is a kinase that reportedly senses the hostile phagolysosomal environment by means of a short chain of 20 of its 487 amino acid residues that extends into the periplasmic space of the *Salmonella* cell envelope. Interaction between an external signal and the *phoQ* kinase anionic box triggers the autophosphorylation of the kinase sensor at a histidine residue, followed by a kinase-dependent transfer of phosphate to an aspartate residue in the amino terminus of the PhoP transcriptional activator protein (186, 187). The phosphorylated PhoP protein activates the transcription of several PhoP-activated genes (*pag*), including *pagA*, *pagB*, *pagC*, *psiD*, and the *phoN* locus, which encodes the periplasmic nonspecific acid phosphatase. The PhoP-activated genes *pagA* and the *psiD* loci are not involved in *Salmonella* virulence. The *Salmonella pagC* gene encodes an outer membrane protein that promotes survival within macrophages. Although the amino acid sequence of PagC is similar to the *ail* gene product in *Yersinia enterocolitica*, this structural homology between both gene products does not extend to functional homology, as evidenced by the loss of invasiveness in *ail* but not *pagC* mutants (218). The transcription of *pagC* within the macrophage is maximal at pH 4.9, and its protein product reportedly contributes to serum resistance but affords no protection against defensins or highly acidic environments (186, 273). Interestingly, transcription of the *pag* genes is induced within acidified *Salmonella*-infected phagocytes but not in infected epithelial vacuoles (6). The PhoP-activated gene *pagB* is part of an operon that includes polymyxin-resistance locus *prmA,B* (182), genes that encode a two-component regulatory system (232) similar to *phoP/phoQ* and *ompR/envZ* sensor kinase-activators involved in the regulation of multiple factors (20, 25, 107, 245), including virulence (75, 187). The *prmA,B* locus appears to regulate the *prmH-FIJKLM* operon, which contains genes for the synthesis of 4-aminoarabinose lipid A, a determinant for the resistance of *Salmonella* to polymyxin and defensins (122). In addition to the five previously mentioned PhoP-activated genes, 13 more positively regulated loci (*pagD* to *pagP*) have been identified (28). These additional *pag* genes do not contribute to the *phoP/phoQ*-dependent resistance to defensins, whereas *pagD*, *pagJ*, *pagK*, and *pagM* participate in mouse virulence and in bacterial survival within macrophages.

In contrast to the *pag* genes that are expressed under adverse environmental conditions, such as low pH, nutrient deficiency, and stationary phase, the *phoP/phoQ* system also regulates the expression of PhoP-repressed genes (*prg*), which are induced under nonstress conditions (27). Stated differently, conditions that activate *pag* genes generally repress *prg* expression. The *prgH*

operon (SPI1) plays a determinant role in the export of bacterial proteins necessary for invasion of epithelial cells (27, 210). Other proteins arising from the transcription of *prg* genes are required for the diffuse membrane ruffling of macrophages, increased pinocytosis, and formation within macrophages of spacious phagosomes in which slow acidification favors greater bacterial survival and attendant virulence (5). It is noteworthy that only a limited number of PhoP-activated genes contribute to *Salmonella* virulence (273) and that mutations in this regulon (*pagC*, -*D*, -*J*, -*K*, and -*M*) result in phenotypes exhibiting decreased survival within macrophages, increased susceptibility to acidic pH, serum complement (*pagC*) and defensins, and reduced invasiveness of epithelial cells (*prgH*).

As with invasion, several bacterial genes important to macrophage survival and proliferation have been identified as part of pathogenicity islands in the *Salmonella* chromosome (32, 135). Pathogenicity island SPI2 encodes a factor(s) necessary for intracellular survival and growth in epithelial cells and macrophages (Table 8.8; 53, 136). *Salmonella* spp. produce an effector, SpiC, that is exported through a type III secretion system to the host cell cytosol. This bacterial protein inhibits endosome-endosome fusion and prevents fusion between the phagosome and lysosome (270). A third pathogenicity island, SPI3, is also required for intracellular growth of *Salmonella* spp. within a macrophage (32). The genes *mgtCB* within SPI3 are required for growth under magnesium-limiting conditions and are essential for intracellular growth (32). The other genes identified in this locus do not contribute to *Salmonella* pathogenesis in a mouse typhoid model. Their functions are currently unknown (33).

Other Virulence Factors

The virulence of *Salmonella* spp., as reflected in their ability to cause acute and chronic diseases in humans and in a variety of animal hosts, stems from structural and physiological attributes that act synergistically or independently in promoting bacterial adhesion and invasiveness (65, 66). The previously discussed acid tolerance response (ATR) is an example of the physiological responsiveness of salmonellae to adverse environmental conditions, which concurrently potentiates greater bacterial virulence in hosts and greater acid tolerance in fermented foods. In addition to the *inv* locus and the *phoP/phoQ* system, which, respectively, encode determinants for *Salmonella* invasiveness and resistance to the antibacterial conditions within phagocytes, several other virulence factors impact on *Salmonella* pathogenicity.

Virulence Plasmids

Virulence plasmids are large cytoplasmic DNA structures that replicate in synchrony with the bacterial chromosome. These autonomous organelles contain many virulence loci ranging from 30 to 60 MDa in size and occur with a frequency of one to two copies per chromosome (118). The presence of virulence plasmids within the *Salmonella* genus is limited and has been confirmed in *Salmonella* serovars Typhimurium, Dublin, Gallinarum-Pullorum, Enteritidis, Choleraesuis, and Abortusovis (118). The absence of a virulence plasmid in the host-adapted and highly infectious serovar Typhi is noteworthy (66, 118). Although limited in its distribution in nature, the plasmid is self-transmissible (3). Gene products from the transcription of this plasmid potentiate systemic spread and infection of extraintestinal tissues but exert no effect on *Salmonella* adhesion and invasion of epithelial and M cells (65). More specifically, it is suggested that virulence plasmids enable carrier strains to rapidly multiply within host cells and overwhelm host defense mechanisms (162, 257). These plasmids also confer to salmonellae the ability to induce lysis of macrophages, elicit an inflammatory response (116), and induce enteritis in animal hosts (177). The *Salmonella* virulence plasmids contain highly conserved nucleotide sequences and exhibit functional homology when plasmid transfer from a wild-type serovar to another plasmid-cured serovar restores the virulence of the recipient strain (66, 118).

The *Salmonella* plasmid virulence (*spv*) region consists of a gene cluster that encodes products for the prolific growth of salmonellae in host reticuloendothelial tissues. This genetic entity was formerly identified as *mka* (mouse-killing agent), *mkf* (mouse-killing factor), or *vir* (virulence) (218). Signals that trigger *spv* transcription include the hostile environment within host phagocytes, iron limitation, elevated temperatures, low pH, and nutrient deprivation associated with the stationary phase of growth (130, 257). The *spv* regulon is approximately 8.0 kb in size and contains at least five genes (*spvR*, -*A*, -*B*, -*C*, and -*D*) and two principal promoters, one for the *spvR* locus and another for the *spvABCD* transcriptional unit (119). These loci are transcribed in a single direction under the combined regulatory action of *spvR* and a chromosomal *rpoS* that encodes a sigma (σ^S) factor (130). The sigma factor, a subunit within the RNA polymerase, recognizes the *spvR* promoter and activates transcription of the *spv* regulon (162). The SpvR protein positively regulates the expression of the downstream *spvABCD* transcriptional unit through the *spvA* promoter sequence (118, 253). Mutations in *spvR* severely affect the ability of salmonellae to proliferate in intestinal

tissues and in extraintestinal sites (120, 177). The biological function of the gene products encoded by the *spvABCD* in *Salmonella* virulence remains elusive (118), although recent evidence suggests that SpvB functions as an ADP-ribosyltransferase (206). The nucleotide sequences of the *spv* loci are highly conserved within the *Salmonella* genus (253, 257). Mutations in this region strongly attenuate or inactivate the ability of salmonellae to establish deep-seated infections. Minor differences in the nucleotide sequence of the *spvR* locus in serovars Dublin and Typhimurium markedly alter the capacity of the *spvR* gene to induce the *spvA* promoter and transcription of the *spvABCD* operon. These findings provide some insight into the molecular basis for the comparatively greater virulence of serovar Dublin in humans (257). The *Salmonella* virulence plasmid also contains a fimbrial operon (*pef*) that encodes adhesin, which is involved in colonization of the small intestine and enteropathogenicity of the organism (23, 96). Distribution of this and other adhesins may explain the narrow to broad host range within the *Salmonella* genus (22).

Iron Acquisition

Siderophores are yet another facet of the *Salmonella* virulence armamentarium. These elements retrieve essential iron from host tissues to drive key cellular functions, such as the electron transport chain and enzymes with iron cofactors (61). To this end, *Salmonella* spp. must compete with host transferrin, lactoferrin, and ferritin ligands for available iron (61, 66). For example, transferrin scavenges tissue fluids for Fe^{3+} ions to form Fe^{3+}-transferrin complexes that bind to surface host cell receptors. Upon internalization, the complexes dissociate, and the released Fe^{3+} is complexed with ferritin for intracellular storage (Fig. 8.2). In response to a limited availability of Fe^{3+} in host tissue, *Salmonella* spp. sequester Fe^{3+} ions by means of a high-affinity phenolate enterochelin (also designated enterobactin) consisting of a cyclic trimer of dihydroxybenzoic acid and L-serine and/or a low-affinity hydroxamate aerobactin chelator, an anabolic product derived from one citrate residue and two lysine (one hydroxylated, one acetylated) residues (196, 218). The *fur* (ferric uptake regulator) gene regulates the synthesis of these bacterial siderophores (61). The *Salmonella* binding of trivalent iron proceeds with the interaction of a ferri-siderophore complex with an outer membrane protein receptor that was induced in response to limiting concentrations of intracellular iron. The complex is then transposed into the bacterial cytoplasm, where the ferric moiety is reduced to the ferrous state. The low affinity of the siderophore for Fe^{2+} results in the release of the divalent ion into the bacterial

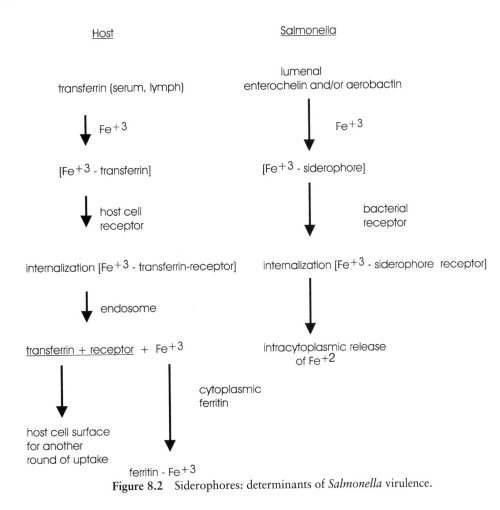

Figure 8.2 Siderophores: determinants of *Salmonella* virulence.

cytoplasm for subsequent use in key metabolic functions. It is notable that the degree of *Salmonella* virulence is directly related to the enterochelin content of the infecting strain. Experimental insight into this relationship follows from the reduced virulence of auxotrophic strains deficient in the aromatic pathway which is responsible for the formation of dihydroxybenzoic acid, the precursor of enterochelin (65). Siderophores do not appear to be the only mechanism by which salmonellae can acquire iron from the host. A new, *fur*-regulated operon was identified within SPI1 that can compete with chelators like 2,2'-dipyridyl for iron and influence *Salmonella* growth in vivo (152, 292). These iron transport proteins have considerable homology to the *yfe* ABC iron transport system in *Yersinia pestis*. The *sitABCD* operon is evolutionarily conserved among the salmonellae (61, 152).

Toxins

Diarrheagenic enterotoxin is a putative *Salmonella* virulence factor. The release of toxin into the cytoplasm of infected host cells precipitates an activation of adenyl cyclase localized in the epithelial cell membrane and a marked increase in the cytoplasmic concentration of cyclic AMP in host cells. The concurrent fluid exsorption into the intestinal lumen results from a net secretion of Cl^- ions in the crypt regions of the intestinal mucosa and depressed Na^+ absorption at the level of the intestinal villi (66, 251). Enterotoxigenicity is a virulence phenotype that prevails in *Salmonella* serovars, including serovar Typhi, and is expressed within hours following bacterial contact with the targeted host cells (65, 85).

One putative *Salmonella* enterotoxin is a thermolabile protein with a molecular mass of 90 to 110 kDa (66). The narrow pH range (6.0 to 8.0) for active enterotoxin suggests that the delivery of a functional toxin may be impaired or inhibited in acidic phagolysosomes (pH 4.5 to 5.0) and in epithelial endosomes (pH 5.0 to 6.5). The ability of enterotoxin to bind to GM_1 ganglioside receptors on host cell surfaces remains a controversial issue (65, 66, 85, 221). The *Salmonella* enterotoxin is encoded by a 6.3-kb chromosomal gene (*stx*) which regulates the synthesis of three proteins of 45, 26, and 12 kDa (50).

In addition to enterotoxin, *Salmonella* strains generally elaborate a thermolabile cytotoxic protein, which is localized in the bacterial outer membrane (16, 66). The cytotoxin is not inactivated with antisera raised against Shiga toxin or *E. coli* Shiga toxins 1 and 2, as determined in green monkey kidney (Vero) and HeLa cell bioassays (16, 74). Maximum production of cytotoxin in laboratory media occurs at pH 7.0 and 37°C during the early stationary phase of growth. Hostile environments, such as acidic pH and elevated (42°C) temperature, precipitate an extracellular release of toxin, possibly as a result of induced bacterial lysis (74). The virulence attribute of cytotoxin stems from its inhibition of protein synthesis and lysis of host cells, thereby promoting the dissemination of viable salmonellae into host tissues (66, 161). The ability of added Ca^{2+} ions to block the cytotoxin-dependent disruption of cell monolayers suggests that cytotoxin may function as a chelator of divalent cations that normally contribute to the structural integrity of monolayers (211). This cytotoxin may actually be one of the *Salmonella* invasion proteins reported to activate apoptosis in epithelial and phagocytic cell types (137, 159, 193, 229).

A new *Salmonella* pathogenicity island, SPI5, has been identified that appears to be responsible for the ability of salmonellae to induce diarrhea (Table 8.8). This conserved locus contains *pipABCD* and *sopB*, which are implicated in intestinal secretion and inflammatory response but which are not essential for systemic infection (280). SopB appears to function as an inositol phosphate phosphatase and hydrolyzes phosphatidylinositol 3,4,5-triphosphate, an inhibitor of Ca^{2+}-dependent chloride secretion. Mutation affecting the phosphatase activity of SopB diminished the ability of *Salmonella* spp. to cause diarrhea (201). The SPI1 locus has also emerged as an important factor in the enteropathogenicity of *Salmonella* spp. (269, 275). New evidence now questions the role, if any, of enterotoxin in diarrhea (275).

Vi Antigen, LPS, and Porins

To conclude, three virulence determinants located within or on the external surface of the *Salmonella* outer membrane will be discussed briefly. The capsular polysaccharide Vi (virulence) antigen occurs in most strains of serovar Typhi, in a few strains of serovar Paratyphi C, and rarely in serovar Dublin (218). The Vi antigen significantly increases the virulence of carrier strains by inhibiting the opsonization of the C3b host complement factor to surface LPS, a critical event in the induction of macrophage phagocytosis of invasive salmonellae (66).

Two genes (*viaA* and *viaB*) are associated with the formation of the Vi antigen. However, the *viaB* locus, which encodes no fewer than six proteins, appears to be the key genetic element in the control of Vi synthesis, as evidenced by the presence of *viaA* in members of *Enterobacteriaceae* and in *Salmonella* spp. that do not express the Vi antigen (218). Recent data further indicate that the *ompR* locus also impacts on Vi capsule formation in serovar Typhi (212).

The length of serotypic LPS that protrudes from the bacterial outer membrane not only defines the rough (short LPS) and smooth (complete LPS) phenotypes but also plays an important role in repelling the potentially lytic attack of the host complement system (66). In effect, the LPS of smooth variants sterically hinders the stable insertion of the C5b-9 complement factor into the inner cytoplasmic membrane, which otherwise would precipitate bacteriolysis. The general inability of the short LPS in rough variants to protect against the C5b-9 lytic insertion renders such variants more susceptible to lysis and consequently less virulent (66). The primary carbohydrate composition of LPS can also affect the level of serum complement activation. For example, an isogenic pair of *Salmonella* spp. containing either the somatic B (antigens 4,12) or C (antigens 6,7) LPS antigens activated serum complement at a slow or rapid rate, respectively (238). Correspondingly, serogroup B exhibits greater virulence than serogroup C.

Porins are outer membrane proteins that function as transmembrane (outer membrane) channels in regulating the influx of nutrients, antibiotics, and other small molecular species (62, 200). To date, four porins encoded by the *ompF*, *ompC*, *ompD*, and *phoE* genes have been described in *Salmonella* serovar Typhimurium (200). Gene transcription of the *ompF* and *ompC* loci is induced by changes in the microenvironment. Low osmolarity, low nutrient availability, and low temperature can trigger the transcription of the *ompF* gene with a concomitant repression of the *ompC* locus. Conversely, favorable environmental conditions induce transcription of the *ompC* gene and repression of the *ompF* locus (62, 192, 200). This is analogous to the previously described *phoP/phoQ*-dependent activation of *pag* genes under adverse conditions and the concomitant repression of *prg* genes (27). The transmembrane channels formed by OmpF and OmpC are 1.1 to 1.3 nm in diameter and consist of trimers of monomeric subunits. EnvZ activates OmpR through a phosphorylation-dependent mechanism, whereas the activated *ompR* locus regulates transcription of *ompF* and *ompC* (62, 192, 246). This mechanism recalls the previously discussed *phoP/phoQ* regulon where phosphorylation of *phoP* by the *phoQ*-encoded kinase activates the transcription of *pag* genes. Mutations in the *ompR/envZ* (designated *ompB*) regulon dramatically attenuate the virulence of carrier strains; in contrast, mutations in the *ompC*, *-F*, or *-D* locus exert little effect on *Salmonella* virulence (75). However, mutations in both *ompC* and *ompF* reduced the virulence of *Salmonella* spp. in mice and protected vaccinated mice upon challenge with wild-type serovar Typhimurium (47). Like *phoP/phoQ*, the two-component regulatory system *ompB* controls a series of genes important to the physiology (20, 107, 245) and pathogenesis (165, 189) of *Salmonella* spp. Interestingly, reports on the ability of *Salmonella* porins to elicit immunological host defense responses suggest that porins are of limited importance in *Salmonella* pathogenicity (66, 255). Recent evidence indicates that the *rck* (resistance to complement killing) locus, which is located on the *Salmonella* virulence plasmid, encodes gene products that protect both smooth and rough variants against the lytic C5b-9 complement factor (128, 129). Nucleotide sequencing of *rck* and amino acid analysis of the Rck protein showed homology with the *phoP/phoQ*-dependent *pagC* locus and gene product (129).

CONCLUSION

Salmonella spp. continue to be a leading cause of foodborne bacterial illnesses. This situation has endured because of the widespread occurrence of salmonellae in the natural environment and their prevalence in many sectors of the global food chain (65). Because of intense animal husbandry practices and major difficulties in controlling the spread of *Salmonella* spp. in vertically integrated meat and poultry production and processing industries, raw poultry and meats rank as principal vehicles of human foodborne salmonellosis. The widespread administration of prophylactic doses of medically important antibiotics to reared animal species may promote on-farm selection of antibiotic-resistant strains and markedly increase the human health risks associated with the handling and consumption of contaminated meat products. A similar scenario can be drawn for the expansive aquacultural industry where intensive rearing of fish and shellfish in generally unprotected facilities, together with a liberal use of subtherapeutic regimens of antibiotic treatment, is common practice. The growing worldwide popularity of fluoroquinolones as prophylactic drugs in the agricultural and aquacultural industries is of serious concern because such a husbandry approach may undermine the clinical efficacy of these novel and invaluable drugs (65, 66). Reports on the acquired resistance of foodborne *Salmonella* and

Campylobacter spp. to fluoroquinolones are disquieting (65). Moreover, the propensity for bacterial cross-resistance to fluoroquinolones adds yet another critical dimension to this emerging problem (114, 151). The importance of *Salmonella* vehicles other than raw poultry and meats and derived products cannot be minimized. Human salmonellosis associated with the consumption of fresh fruits and vegetables, spices, chocolate, and milk products (Table 8.6) reiterates the importance of sanitary practices during the harvesting, processing, and distribution of raw foods and food ingredients (64, 68).

The arsenal of virulence factors that enable *Salmonella* spp. to evade the various antibacterial host defense mechanisms is remarkable yet disconcerting. The physiological adaptability of salmonellae to hostile conditions in the natural environment safeguards their survival and infectious potential (Table 8.3). In addition, the pathogen benefits from chromosome- and plasmid-encoded virulence determinants that provide for facilitated attachment and invasion of host cells (*inv*, *prgH*) (Fig. 8.1); resistance to intraphagocyte acidity (*phoP/phoQ*), complement lysis (Vi, LPS, *phoP/phoQ*, *rck*), and antibacterial substances (*ompF*, *ompC*, *ompD*, *phoE*); widespread invasion of deep host tissues (*spv*); competition for available Fe^{3+} (siderophores); and induction of diarrhea.

It is clear that the prevalence of *Salmonella* in the global food chain and its current and projected repercussions on human health is cause for concern. Unless changes in agricultural and aquacultural practices are implemented, the notoriety of human foodborne salmonellosis will prevail.

References

1. **Aabo, S., and D. L. Baggesen.** 1997. Growth of *Salmonella* Newport in naturally contaminated alfalfa sprouts and estimation of infectious dose in a Danish *Salmonella* Newport outbreak due to alfalfa sprouts, p. 425–426. *Proceedings of Salmonella and Salmonellosis*, Ploufragan, France.

2. **Ahmer, B. M. M., J. van Reeuwijk, P. R. Watson, T. S. Wallis, and F. Heffron.** 1999. *Salmonella* SirA is a global regulator of genes mediating enteropathogenesis. *Mol. Microbiol.* 31:971–982.

3. **Ahmer, B. M., M. Tran, and F. Heffron.** 1999. The virulence plasmid of *Salmonella typhimurium* is self-transmissible. *J. Bacteriol.* 181:1364–1368.

4. **Airoldi, A. A., and E. A. Zottola.** 1988. Growth and survival of *Salmonella typhimurium* at low temperature in nutrient deficient media. *J. Food Sci.* 53:1511–1513.

5. **Alpuche-Aranda, C. M., E. L. Racoussin, J. A. Swanson, and S. I. Miller.** 1994. *Salmonella* stimulate macrophage macropinocytosis and persist within spacious phagosomes. *J. Exp. Med.* 179:601–608.

6. **Alpuche-Aranda, C. M., J. A. Swanson, W. P. Loomis, and S. I. Miller.** 1992. *Salmonella typhimurium* activates virulence gene transcription within acidified macrophage/phagosomes. *Proc. Natl. Acad. Sci. USA* **89:**10079–10083.

7. **Altmeyer, R. M., J. K. McNern, J. C. Bossio, I. Rosenshine, B. B. Finlay, and J. E. Galan.** 1993. Cloning and molecular characterization of a gene involved in *Salmonella* adherence and invasion of cultured epithelial cells. *Mol. Microbiol.* **7:**89–98.

8. **American Society for Microbiology.** 1995. *Report of the ASM Task Force on Antibiotic Resistance.* American Society for Microbiology, Washington, D.C.

9. **Andrews, W. H., V. R. Bruce, G. A. June, P. Sherrod, T. S. Hammack, and R. M. Amaguana.** 1995. *Salmonella. Bacteriological Analytical Manual*, 8th ed. AOAC International, Arlington, Va.

10. **Angelotti, R., M. J. Foter, and K. H. Lewis.** 1961. Time-temperature effects on salmonellae and staphylococci in foods. *Am. J. Public Health* **51:**76–88.

11. **Anonymous.** 1977. *Aviation Catering.* Report of Working Group. World Health Organization (WHO) Regional Office for Europe, Copenhagen, Denmark.

12. **Anonymous.** 1986. Microbiological safety of vacuum-packed, high salt foods questioned. *Food Chem. News* **28:**30–31.

13. **Anonymous.** 1995. Ice-cream firm reaches tentative *Salmonella* case agreement. *Food Chem. News* **36:**53.

13a. **Anonymous.** 1995. Aquaculture cited by ASM as problem for antibiotic resistance. *Food Chem. News* **37:**20–21.

13b. **Anonymous.** 1995. Florida *Salmonella* outbreak caused changes in State inspections. *Food Chem. News* **37:**27.

13c. **Anonymous.** 1999. Salmonellosis, dried squid—Japan (Tokyo). *Mainichi Daily News*, April 25.

13d. **Anonymous.** 2000. Salmonellosis outbreak associated with raw mung bean sprouts. News release (22-00), April. California Department of Health Services, Sacramento, Calif.

13e. **Anonymous.** 1999. *Food Chem. News* **41:**39.

13f. **Anonymous.** 2000. *Food Chem. News* **42:**12.

14. **Arenas, G. N., A. S. Staskevich, A. Aballay, and L. S. Mayorga.** 2000. Intracellular trafficking of *Brucella abortus* in J774 macrophages. *Infect. Immun.* **68:**4255–4263.

15. **Armstrong, R. W., T. Fodor, G. T. Curlin, A. B. Cohen, G. K. Morris, W. T. Martin, and J. Feldman.** 1970. Epidemic *Salmonella* gastroenteritis due to contaminated imitation ice cream. *Am. J. Epidemiol.* **91:**300–307.

16. **Ashkenazi, S., T. G. Cleary, B. E. Murray, A. Wanger, and L. K. Pickering.** 1988. Quantitative analysis and partial characterization of cytotoxin production by *Salmonella* strains. *Infect. Immun.* **56:**3089–3094.

17. **Asperilla, M. A., R. A. Smego, Jr., and L. K. Scott.** 1990. Quinolone antibiotics in the treatment of *Salmonella* infections. *Rev. Infect. Dis.* **12:**873–889.

18. **Asplund, K., and E. Nurmi.** 1991. The growth of salmonellae in tomatoes. *Int. J. Food Microbiol.* **13:**177–182.

18a. **Australian Health Commission.** 1999. End to *Salmonella* outbreak. Press release, March 25. Australian Health Commission, Adelaide, Australia.

19. **Baker, R. C., R. A. Qureshi, and J. H. Hotchkins.** 1986. Effect of an elevated level of carbon dioxide containing 5 Australian Health Commission, atmosphere on the growth of spoilage and pathogenic bacteria. *Poult. Sci.* **65:**729–737.

20. **Bang, I. S., B. H. Kim, J. W. Foster, and Y. K. Park.** 2000. OmpR regulates the stationary-phase acid tolerance response of *Salmonella enterica* serovar Typhimurium. *J. Bacteriol.* **182:**2245–2252.

21. **Barry, R. A., H. G. Bouwer, D. A. Portnoy, and D. J. Hinrichs.** 1992. Pathogenicity and immunogenicity of *Listeria monocytogenes* small-plaque mutants defective for intracellular growth and cell-to-cell spread. *Infect. Immun.* **60:**1625–1632.

22. **Baumler, A. J., A. J. Gilde, R. M. Tsolis, A. W. M. Van Der Velden, B. M. M. Ahmer, and F. Heffron.** 1997. Contribution of horizontal gene transfer and deletion events to development of distinctive patterns of fimbrial operons during evolution of *Salmonella* serotypes. *J. Bacteriol.* **179:**317–322.

23. **Baumler, A. J., R. M. Tsolis, F. A. Bowe, J. G. Kusters, S. Hoffmann, and F. Heffron.** 1996. The *pef* fimbrial operon of *Salmonella typhimurium* mediates adhesion to murine small intestine and is necessary for fluid accumulation. *Infect. Immun.* **64:**61–68.

24. **Bean, N. H., J. S. Goulding, C. Lao, and F. J. Angulo.** 1996. Surveillance for foodborne disease outbreaks— United States, 1988–1992. CDC Surveillance Summaries. *Morb. Mortal. Wkly. Rep.* **45(SS-5):**1–66.

25. **Bearson, B. L., L. Wilson, and J. W. Foster.** 1998. A low pH-inducible, PhoPQ-dependent acid tolerance response protects *Salmonella typhimurium* against inorganic acid stress. *J. Bacteriol.* **180:**2409–2417.

26. **Beckers, H. J., M. S. M. Daniels-Bosman, A. Ament, J. Daenen, A. W. J. Hanekamp, P. Knipschild, A. H. H. Schuurmann, and H. Bijkerk.** 1985. Two outbreaks of salmonellosis caused by *Salmonella indiana*. A survey of the European Summit outbreak and its consequences. *Int. J. Food Microbiol.* **2:**185–195.

26a. **Beers, A.** 1997. Maryland church dinner *Salmonella* outbreak blamed on faulty food preparation. *Food Chem. News* **39:**19–20.

27. **Behlau, I., and S. I. Miller.** 1993. A PhoP-repressed gene promotes *Salmonella typhimurium* invasion of epithelial cells. *J. Bacteriol.* **175:**4475–4484.

28. **Belden, W. J., and S. I. Miller.** 1994. Further characterization of the PhoP regulon: identification of new PhoP-activated virulence loci. *Infect. Immun.* **62:**5095–5101.

29. **Bellanti, J. A., B. J. Zeligs, S. Vetro, Y. H. Pung, S. Luccioli, M. J. Malavasic, A. M. Hooke, T. R. Ubertini, R. Vanni, and L. Nencioni.** 1993. Studies of safety, infectivity and immunogenicity of a new temperature-sensitive (ts) 51-1 strain of *Salmonella typhi* as a new live

oral typhoid fever vaccine candidate. *Vaccine* **11**:587–590.

30. **Bergis, H., G. Poumeyrol, and A. Beaufort.** 1994. Etude de développement de la flore saprophyte et de *Salmonella* dans les viandes hachées conditionnées sous atmosphère modifiée. *Sci. Aliments* **14**:217–228.

31. **Björnerot, L., A. Franklin, and E. Tysen.** 1996. Usage of antibacterial and antiparasitic drugs in animals in Sweden between 1988 and 1993. *Vet. Rec.* **21**:282–286.

32. **Blanc-Potard, A. B., and E. A. Groisman.** 1997. The *Salmonella selC* locus contains a pathogenicity island mediating intramacrophage survival. *EMBO J.* **16**:5376–5385.

33. **Blanc-Potard, A. B., F. Solomon, J. Kayser, and E. A. Groisman.** 1999. The SPI-3 pathogenicity island of *Salmonella enterica*. *J. Bacteriol.* **181**:998–1004.

34. **Blaser, M. J., and L. S. Newman.** 1982. A review of human salmonellosis. 1. Infective dose. *Rev. Infect. Dis.* **4**:1096–1106.

34a. **Boase, J., S. Lipsky, P. Simani, et al.** 1999. Outbreak of *Salmonella* serotype Muenchen infections associated with unpasteurized orange juice—United States and Canada. *JAMA* **282**:726–728.

35. **Bouvet, E., C. Jestin, and R. Ancelle.** 1986. Importance of exported cases of salmonellosis in the revelation of an epidemic, p. 303. *In Proceedings of the Second World Congress on Foodborne Infections and Intoxications*, Berlin, Germany.

36. **Boyd, E. F., J. Li, H. Ochman, and R. K. Selander.** 1997. Comparative genetics of the *inv-spa* invasion gene complex of *Salmonella enterica*. *J. Bacteriol.* **179**:1985–1991.

37. **Brenner, F. W., R. G. Villar, F. J. Angulo, R. Tauxe, and B. Swaminathan.** 2000. *Salmonella* nomenclature. *J. Clin. Microbiol.* **38**:2465–2467.

38. **Bryan, J. P., H. Rocha, and W. M. Scheld.** 1986. Problems in salmonellosis: rationale for clinical trials with newer β-lactam agents and quinolones. *Rev. Infect. Dis.* **8**:189–203.

39. **Buchmeier, N. A., and F. Heffron.** 1991. Inhibition of macrophage phagosomes-lysosome fusion by *Salmonella typhimurium*. *Infect. Immun.* **59**:2232–2238.

39a. **Bulletin Épidémiologique Annuel.** 1999. Epidémiologie des maladies infectieuses en France. Situation en 1997 et tendances évolutives recentes. Réseau National de Santé Publique, Saint-Maurice, France.

40. **Bulling, E., R. Stephan, and V. Sebek.** 1973. The development of antibiotic resistance among *Salmonella* bacteria of animal origin in the Federal Republic of Germany and West Berlin: 1st communication: a comparison between the years of 1961 and 1970–1971. *Zentbl. Bakteriol. Mikrobiol. Hyg. 1 Abt. Orig. A* **225**:245–256.

41. **Bunning, V. K., R. B. Raybourne, and D. L. Archer.** 1988. Foodborne enterobacterial pathogens and rheumatoid disease. *J. Appl. Bacteriol. Symp. Suppl.* **17**:87S–107S.

42. **Burslem, C. D., M. J. Kelly, and F. S. Preston.** 1990. Food poisoning—a major threat to airline operations. *J. Soc. Occup. Med.* **40**:97–100.

42a. **Butler, M. A.** 2000. *Salmonella* outbreak leads to juice recall in western states. *Food Chem. News* **42**:19–20.

43. **Carmen Palomino, W., R. Lucia Aguad, L. Manuel Rodriguez, G. Graciela Cofre, and J. Villanueva.** 1986. Clinical course of infections by *Salmonella typhi*, paratyphus A and paratyphus B in relation to the sensitivity of the etiological agent to chloramphenicol. *Rev. Méd. Chile* **114**:919–927.

44. **Catsaras, M., and D. Grebot.** 1984. Multiplication des *Salmonella* dans la viande hachée. *Bull. Acad. Vét. France* **57**:501–512.

45. **Centers for Disease Control and Prevention.** 2000. 1999 surveillance results. *http://www.cdc.gov/ncidod/dbmd/foodnet*.

46. **Chatfield, S., M. Roberts, P. Londono, I. Cropley, G. Douce, and G. Dougan.** 1993. The development of oral vaccines based on live attenuated *Salmonella* strains. *FEMS Immunol. Med. Microbiol.* **7**:1–8.

47. **Chatfield, S. N., C. J. Dorman, C. Hayward, and G. Dougan.** 1991. Role of *ompR*-dependent genes in *Salmonella typhimurium* virulence: mutants deficient in both *ompC* and *ompF* are attenuated in vivo. *Infect. Immun.* **59**:449–452.

48. **Chen, S., A. Yee, M. Griffiths, C. Larkin, C. T. Yamashiro, R. Behari, C. Paszko-Kolva, K. Rahn, and S. A. De Grandis.** 1997. The evaluation of a fluorogenic polymerase chain reaction assay for the detection of *Salmonella* species in food commodities. *Int. J. Food Microbiol.* **35**:239–250.

49. **Cherubin, C. E.** 1981. Antibiotic resistance of *Salmonella* in Europe and the United States. *Rev. Infect. Dis.* **3**:1105–1125.

50. **Chopra, A. K., C. W. Houston, J. W. Peterson, R. Prasad, and J. J. Mekalanos.** 1987. Cloning and expression of the *Salmonella* enterotoxin gene. *J. Bacteriol.* **169**:5095–5100.

51. **Christmann, D., T. Staub, and Y. Hansmann.** 1992. Manifestations extra-digestives des salmonelloses. *Méd. Mal. Infect.* **22**:289–298.

52. **Chung, K. C., and J. M. Goepfert.** 1970. Growth of *Salmonella* at low pH. *J. Food Sci.* **35**:326–328.

53. **Cirillo, D. M., R. H. Valdivia, D. M. Monack, and S. Falkow.** 1998. Macrophage-dependent induction of the *Salmonella* pathogenicity island 2 type III secretion system and its role in intracellular survival. *Mol. Microbiol.* **30**:175–188.

54. **Cohen, D. R., I. A. Porter, T. M. S. Reid, J. C. M. Sharp, G. I. Forbes, and G. M. Paterson.** 1983. A cost benefit study of milk-borne salmonellosis. *J. Hyg.* **91**:17–23.

55. **Collazo, C. M., and J. E. Galan.** 1997. The invasion-associated type III system of *Salmonella typhimurium* directs the translocation of Sip proteins into the host cell. *Mol. Microbiol.* **24**:747–756.

56. **Collazo, C. M., M. K. Zierler, and J. E. Galan.** 1995. Functional analysis of the *Salmonella typhimurium* invasion genes *inv I* and *inv J* and identification of a target of the protein secretion apparatus encoded in the *inv* locus. *Mol. Microbiol.* **15**:25–38.

57. **Corry, J. E. L.** 1971. The water relations and heat resistance of microorganisms, p. 1–42. *Scientific and Technical Surveys, no. 73.* British Food Manufacturing Industries Research Association, Leatherhead, Surrey, United Kingdom.

58. **Corry, J. E. L.** 1976. The safety of intermediate moisture foods with respect to *Salmonella*, p. 215–238. *In* R. Davies, G. G. Birch, and K. J. Parker (ed.), *Intermediate Moisture Foods.* Applied Science Publishers Ltd., London, England.

59. **Craven, P. C., D. C. Mackel, W. B. Baine, W. H. Barker, E. J. Gangarosa, M. Goldfield, H. Rosenfeld, R. Altman, G. Lachapelle, J. W. Davies, and R. C. Swanson.** 1975. International outbreak of *Salmonella eastbourne* infection traced to contaminated chocolate. *Lancet* i:788–793.

60. **Crosa, J. H., D. J. Brenner, W. H. Ewing, and S. Falkow.** 1973. Molecular relationships among the salmonellae. *J. Bacteriol.* **115:**307–315.

61. **Crosa, J. H.** 1989. Genetics and molecular biology of siderophore-mediated iron transport in bacteria. *Microbiol. Rev.* **53:**517–530.

62. **Csonka, L. N.** 1989. Physiological and genetic responses of bacteria to osmotic stress. *Microbiol. Rev.* **53:**121–147.

63. **Curtiss, R., III, S. M. Kelly, and J. O. Hassan.** 1993. Live oral avirulent *Salmonella* vaccines. *Vet. Microbiol.* **37:**397–405.

64. **D'Aoust, J.-Y.** 1977. *Salmonella* and the chocolate industry. A review. *J. Food Prot.* **40:**718–727.

65. **D'Aoust, J.-Y.** 1989. *Salmonella*, p. 327-445. *In* M.P. Doyle (ed.), *Foodborne Bacterial Pathogens.* Marcel Dekker, Inc., New York, N.Y.

66. **D'Aoust, J.-Y.** 1991. Pathogenicity of foodborne *Salmonella*. *Int. J. Food Microbiol.* **12:**17–40.

67. **D'Aoust, J.-Y.** 1991. Psychrotrophy and foodborne *Salmonella*. *Int. J. Food Microbiol.* **13:**207–216.

68. **D'Aoust, J.-Y.** 1994. *Salmonella* and the international food trade. *Int. J. Food Microbiol.* **24:**11–31.

68a. **D'Aoust, J.-Y.** 2000. *Salmonella*, p. 1233–1299. *In* B. M. Lund, A. C. Baird-Parker, and G. W. Gould (ed.), *The Microbiological Safety and Quality of Food.* Aspen Publishers, Inc., Gaithersburg, Md.

69. **D'Aoust, J.-Y., B. J. Aris, P. Thisdele, A. Durante, N. Brisson, D. Dragon, G. Lachapelle, M. Johnston, and R. Laidley.** 1975. *Salmonella eastbourne* outbreak associated with chocolate. *Can. Inst. Food Sci. Technol. J.* **8:**181–184.

70. **D'Aoust, J.-Y., and U. Purvis.** 1998. *Isolation and Identification of Salmonella from Foods.* MFHPB-20. Health Protection Branch, Health Canada, Ottawa, Canada.

71. **D'Aoust, J.-Y., A. M. Sewell, E. Daley, and P. Greco.** 1992. Antibiotic resistance of agricultural and foodborne *Salmonella* isolates in Canada: 1986-1989. *J. Food Prot.* **55:**428–434.

72. **D'Aoust, J.-Y., D. W. Warburton, and A. M. Sewell.** 1985. *Salmonella typhimurium* phage-type 10 from cheddar cheese implicated in a major Canadian foodborne outbreak. *J. Food Prot.* **48:**1062–1066.

73. **Darwin, K. H., and V. L. Miller.** 1999. InvF is required for expression of genes encoding proteins secreted by the SPI1 type III secretion apparatus in *Salmonella typhimurium*. *J. Bacteriol.* **181:**4949–4954.

74. **Dewanti, R., and M. P. Doyle.** 1992. Influence of cultural conditions on cytotoxin production by *Salmonella enteritidis*. *J. Food Prot.* **55:**28–33.

75. **Dorman, C. J., S. Chatfield, C. F. Higgins, C. Hayward, and G. Dougan.** 1989. Characterization of porin and *ompR* mutants of a virulent strain of *Salmonella typhimurium*: *ompR* mutants are attenuated in vivo. *Infect. Immun.* **57:**2136–2140.

76. **Dougan, G.** 1994. The molecular basis for the virulence of bacterial pathogens: implications for oral vaccine development. *Microbiology* **140:**215–224.

77. **Droffner, M. L., and N. Yamamoto.** 1992. Procedure for isolation of *Escherichia, Salmonella*, and *Pseudomonas* mutants capable of growth at the refractory temperature of 54°C. *J. Microbiol. Methods* **14:**201–206.

78. **Eckmann, L., J. R. Smith, M. P. Housley, M. B. Dwinell, and M. F. Kagnoff.** 2000. Analysis by high-density cDNA arrays of altered gene expression in human intestinal epithelial cells in response to infection with the invasive enteric bacteria *Salmonella*. *J. Biol. Chem.* **275:**14084–14094.

79. **Eichelberg, K., C. C. Ginocchio, and J. E. Galan.** 1994. Molecular and functional characterization of the *Salmonella typhimurium* invasion genes *invB* and *invC*: homology of *invC* to the F_0F_1 ATPase family of proteins. *J. Bacteriol.* **176:**4501–4510.

80. **Eichelberg, K., and J. E. Galan.** 1999. Differential regulation of *Salmonella typhimurium* type III secreted proteins by pathogenicity island 1 (SPI-1)-encoded transcriptional activators InvF and HilA. *Infect. Immun.* **67:**4099–4105.

81. **Endtz, H. P., G. J. Ruijs, B. van Klingeren, W. H. Jansen, T. van der Reyden, and R. P. Mouton.** 1991. Quinolone resistance in *Campylobacter* isolated from man and poultry following the introduction of fluoroquinolones in veterinary medicine. *J. Antimicrob. Chemother.* **27:**199–208.

82. **Euzéby, J. P.** 1999. Revised *Salmonella* nomenclature: designation of *Salmonella enterica* (exk1 Dauffmann and Edwards 1952) Le Minor and Popoff 1987 sp. nom., nom. rev. as the neotype species of the genus *Salmonella* Lignières 1900 (approved lists 1980), rejection of the name *Salmonella choleraesuis* (Smith 1894) Weldin 1927 (approved lists 1980), and conservation of the name *Salmonella typhi* (Schroeter 1886) Warren and Scott 1930 (approved lists 1980). Request for an opinion. *Int. J. Syst. Bacteriol.* **49:**927–930.

83. **Ewing, W. H.** 1972. The nomenclature of *Salmonella*, its usage, and definitions for the three species. *Can. J. Microbiol.* **18:**1629–1637.

84. **Falkow, S., R. R. Isberg, and D. A. Portnoy.** 1992. The interaction of bacteria with mammalian cells. *Annu. Rev. Cell Biol.* **8:**333–363.

85. **Fernandez, M., J. Sierra-Madero, H. de la Vega, M. Vazquez, Y. Lopez-Vidal, G. M. Ruiz-Palacios, and E. Calva.** 1988. Molecular cloning of a *Salmonella*

typhi LT-like enterotoxin gene. *Mol. Microbiol.* 2:821–825.

86. **Ferreira, M. A. S. S., and B. M. Lund.** 1987. The influence of pH and temperature on initiation of growth of *Salmonella* spp. *Lett. Appl. Microbiol.* 5:67–70.

87. **Fields, P. I., E. A. Groisman, and F. Herron.** 1989. A *Salmonella* locus that controls resistance to microbicidal proteins from phagocytic cells. *Science* 243:1059–1060.

88. **Finlay, B. B.** 1994. Molecular and cellular mechanisms of *Salmonella* pathogenesis. *Curr. Top. Microbiol. Immunol.* 192:163–185.

89. **Fontaine, R. E., S. Arnon, W. T. Martin, T. M. Vernon, E. J. Gangarosa, J. J. Farmer, A. B. Moran, J. H. Silliker, and D. L. Decker.** 1978. Raw hamburger: an interstate common source of human salmonellosis. *Am. J. Epidemiol.* 107:36–45.

90. **Fontaine, R. E., M. L. Cohen, W. T. Martin, and T. M. Vernon.** 1980. Epidemic salmonellosis from Cheddar cheese: surveillance and prevention. *Am. J. Epidemiol.* 111:247–253.

91. **Food and Drug Administration.** 1999. *A Proposed Framework for Evaluating and Assuring the Human Safety of the Microbial Effects of Antimicrobial New Animal Drugs Intended for Use in Food-Producing Animals.* U.S. Food and Drug Administration, Washington, D.C.

92. **Foster, J. W., and B. Bearson.** 1994. Acid-sensitive mutants of *Salmonella typhimurium* identified through a dinitrophenol lethal screening strategy. *J. Bacteriol.* 176:2596–2602.

93. **Foster, J. W., and H. K. Hall.** 1991. Inducible pH homeostasis and the acid tolerance response of *Salmonella typhimurium. J. Bacteriol.* 173:5129–5135.

94. **Francis, B. J., J. V. Altamirano, M. G. Stobierski, W. Hall, B. Robinson, S. Dietrich, R. Martin, F. Downes, K. R. Wilcox, C. Hedberg, R. Wood, M. Osterholm, G. Genese, M. J. Hung, S. Paul, K. C. Spitalny, C. Whalen, and J. Spika.** 1991. Multistate outbreak of *Salmonella poona* infections—United States and Canada, 1991. *Morb. Mortal. Wkly. Rep.* 40:549–552.

95. **Franklin, A.** 1997. Current status of antibiotic resistance in animal production in Sweden, p. 229–235. *In Report of a WHO Meeting.* WHO/EMC/ZOO/97.4. World Health Organization, Geneva, Switzerland.

96. **Friedrich, M. J., N. E. Kinsey, J. Vila, and R. J. Kadner.** 1993. Nucleotide sequence of a 13.9 kb segment of the 90 kb virulence plasmid of *Salmonella typhimurium*: the presence of fimbrial biosynthetic genes. *Mol. Microbiol.* 8:543–558.

97. **Fu Y., and J. E. Galan.** 1998. The *Salmonella typhimurium* tyrosine phosphatase SptP is translocated into host cells and disrupts the actin cytoskeleton. *Mol. Microbiol.* 27:359–368.

98. **Galan, J. E., and R. Curtiss III.** 1991. Distribution of the *invA, -B, -C,* and *-D* genes of *Salmonella typhimurium* among other *Salmonella* serovars: *invA* mutants of *Salmonella typhi* are deficient for entry into mammalian cells. *Infect. Immun.* 59:2901–2908.

99. **Galan, J. E., and C. Ginocchio.** 1994. The molecular genetic bases of *Salmonella* entry into mammalian cells. *Biochem. Soc. Trans.* 22:301–306.

100. **Galan, J. E., C. Ginocchio, and P. Costeas.** 1992. Molecular and functional characterization of the *Salmonella* invasion gene *invA*: homology of InvA to members of a new protein family. *J. Bacteriol.* 174:4338–4349.

101. **Galan, J. E., and D. Zhou.** 2000. Striking a balance: modulation of the actin cytoskeleton by *Salmonella. Proc. Natl. Acad. Sci. USA* 95:8754–8761.

102. **Garcia-del-Portillo, F., and B. B. Finlay.** 1994. Invasion and intracellular proliferation of *Salmonella* within non-phagocytic cells. *Microbiol. SEM* 10:229–238.

103. **Garcia-del Portillo, F., J. W. Foster, and B. B. Finlay.** 1993. Role of acid tolerance response genes in *Salmonella typhimurium* virulence. *Infect. Immun.* 61:4489–4492.

104. **Garcia-del Portillo, F., M. B. Zwick, K. Y. Leung, and B. B. Finlay.** 1994. Intracellular replication of *Salmonella* within epithelial cells is associated with filamentous structures containing lysosomal membrane glycoproteins. *Infect. Agents Dis.* 2:227–231.

105. **Geiss, H. K., I. Ehrhard, A. Rösen-Wolff, H. G. Sonntag, J. Pratsch, A. Wirth, D. Krüger, I. Knollmann-Schanbacher, H. Kühn, and C. Treiber-Klötzer.** 1993. Foodborne outbreak of a *Salmonella enteritidis* epidemic in a major pharmaceutical company. *Gesundheitswesen* 55:127–132.

106. **Gibson, A. M., N. Bratchell, and T. A. Roberts.** 1988. Predicting microbial growth: growth responses of salmonellae in laboratory medium as affected by pH, sodium chloride and storage temperature. *Int. J. Food Microbiol.* 6:155–178.

107. **Gibson, M. M., E. M. Ellis, K. A. Graeme-Cook, and C. F. Higgins.** 1987. OmpR and EnvZ are pleiotropic regulatory proteins: positive regulation of the tripeptide permease (*tppB*) of *Salmonella typhimurium. Mol. Gen. Genet.* 207:120–129.

108. **Ginocchio, C. C., S. B. Olmsted, C. L. Wells, and J. E. Galan.** 1994. Contact with epithelial cells induces the formation of surface appendages on *Salmonella typhimurium. Cell* 76:717–724.

109. **Ginocchio, C., J. Pace, and J. E. Galan.** 1992. Identification and molecular characterization of a *Salmonella typhimurium* gene involved in triggering the internalization of salmonellae into cultured epithelial cells. *Proc. Natl. Acad. Sci. USA* 89:5976–5980.

110. **Glynn, J. R., and D. J. Bradley.** 1992. The relationship of infecting dose and severity of disease in reported outbreaks of *Salmonella* infections. *Epidemiol. Infect.* 109:371–388.

111. **Glynn, J. R., and S. R. Palmer.** 1992. Incubation period, severity of disease, and infecting dose: evidence from a *Salmonella* outbreak. *Am. J. Epidemiol.* 136:1369–1377.

112. **Goepfert, J. M., I. K. Iskander, and C. H. Amundson.** 1970. Relation of the heat resistance of salmonellae to the water activity of the environment. *Appl. Microbiol.* 19:429–433.

112a. Gordenker, A. 1999. Common seed source identified in Canadian outbreak. *Food Chem. News* **41**:23.

113. Greenwood, M. H., and W. L. Hooper. 1983. Chocolate bars contaminated with *Salmonella napoli*: an infectivity study. *Br. Med. J.* **286**:1394.

114. Griggs, D. J., M. C. Hall, Y. F. Jin, and L. J. V. Piddock. 1994. Quinolone resistance in veterinary isolates of *Salmonella*. *J. Antimicrob. Chemother.* **33**:1173–1189.

115. Groisman, E. A., and H. Ochman. 1993. Cognate gene clusters govern invasion of host epithelial cells by *Salmonella typhimurium* and *Shigella flexneri*. *EMBO J.* **12**:3779–3787.

116. Guilloteau, L. A., T. S. Wallis, A. V. Gautier, S. McIntyre, D. J. Platt, and A. J. Lax. 1996. The *Salmonella* virulence plasmid enhances *Salmonella*-induced lysis of macrophages and influences inflammatory responses. *Infect. Immun.* **64**:3385–3393.

117. Guina, T., E. C. Yi, H. Wang, M. Hackett, and S. I. Miller. 2000. A *phoP*-regulated outer membrane protease of *Salmonella enterica* serovar Typhimurium promotes resistance to alpha-helical antimicrobial peptides. *J. Bacteriol.* **182**:4077–4086.

118. Guiney, D. G., F. C. Fang, M. Krause, and S. Libby. 1994. Plasmid-mediated virulence genes in non-typhoid *Salmonella* serovars. *FEMS Microbiol. Lett.* **124**:1–10.

119. Gulig, P. A., H. Danbar, D. G. Guiney, A. J. Lax, F. Norel, and M. Rhen. 1993. Molecular analysis of *spv* virulence genes of the *Salmonella* virulence plasmids. *Mol. Microbiol.* **7**:825–830.

120. Gulig, P. A., T. J. Doyle, J. A. Hughes, and H. Matsui. 1998. Analysis of host cells associated with the *spv*-mediated increased intracellular growth rate of *Salmonella typhimurium* in mice. *Infect. Immun.* **66**:2471–2485.

121. Gunn, J. S., and S. I. Miller. 1996. PhoP-PhoQ activates transcription of *pmrAB*, encoding a two component regulatory system involved in *Salmonella typhimurium* antimicrobial peptide resistance. *J. Bacteriol.* **178**:6857–6864.

122. Gunn, J. S., S. S. Ryan, J. C. van Velkinburgh, R. K. Ernst, and S. I. Miller. 2000. Genetic and functional analysis of a PmrA-PmrB-regulated locus necessary for lipopolysaccharide modification, antimicrobial peptide resistance, and oral virulence of *Salmonella enterica* serovar Typhimurium. *Infect. Immun.* **68**:6139–6146.

123. Hassan, J. O., and R. Curtiss. 1994. Development and evaluation of an experimental vaccination program using a live avirulent *Salmonella typhimurium* strain to protect immunized chickens against challenge with homologous and heterologous *Salmonella* serotypes. *Infect. Immun.* **62**:5519–5527.

124. Hayward, R. D., and V. Koronakis. 1999. Direct nucleation and bundling of actin by the SipC protein of invasive *Salmonella*. *EMBO J.* **18**:4926–4934.

125. Health Canada. 1991. *Foodborne and Waterborne Disease in Canada. Annual Summary 1985–86*. Polyscience Publications, Inc., Morin Heights, Québec, Canada.

126. Health Canada. 1998. *Foodborne and Waterborne Disease in Canada. Annual Summary 1992–93*. Poly-

science Publications, Inc., Morin Heights, Québec, Canada.

127. Hedberg, C. W., F. J. Angulo, K. E. White, C. W. Langkop, W. L. Schell, M. G. Stobierski, A. Schuchat, J. M. Besser, S. Dietrich, L. Helsel, P. M. Griffin, J. W. McFarland, and M. T. Osterholm. 1999. Outbreaks of salmonellosis associated with eating uncooked tomatoes: implications for public health. The Investigation Team. *Epidemiol. Infect.* **122**:385–393.

128. Hefferman, E. J., J. Harwood, J. Fierer, and D. Guiney. 1992. The *Salmonella typhimurium* virulence plasmid complement resistance gene *rck* is homologous to a family of virulence-related outer membrane protein genes, including *pagC* and *ail*. *J. Bacteriol.* **174**:84–91.

129. Hefferman, E. J., L. Wu, J. Louie, S. Okamoto, J. Fierer, and D. G. Guiney. 1994. Specificity of the complement resistance and cell association phenotypes encoded by the outer membrane protein genes *rck* from *Salmonella typhimurium* and *ail* from *Yersinia enterocolitica*. *Infect. Immun.* **62**:5183–5186.

130. Heiskanen, P., S. Taira, and M. Rhen. 1994. Role of *rpoS* in the regulation of *Salmonella* plasmid virulence (*spv*) genes. *FEMS Microbiol. Lett.* **123**:125–130.

131. Heithoff, D. M., C. P. Conner, P. C. Hanna, S. M. Julio, U. Hentschel, and M. J. Mahan. 1997. Bacterial infection as assessed by in vivo gene expression. *Proc. Natl. Acad. Sci. USA* **94**:934–939.

132. Hellström, L. 1980. Food-transmitted *S. enteritidis* epidemic in 28 schools, p. 397–400. *In Proceedings of the World Congress on Foodborne Infections and Intoxications*, Berlin, Germany.

133. Helmuth, R., and D. Protz. 1997. How to modify conditions limiting resistance in bacteria in animals and other reservoirs. *Clin. Infect. Dis.* **24**:8136–8138.

134. Henderson, S. C., D. I. Bounous, and M. D. Lee. 1999. Early events in the pathogenesis of avian salmonellosis. *Infect. Immun.* **67**:3580–3586.

134a. Henkel, J. 1995. Ice cream firm linked to *Salmonella* outbreak. *FDA Consumer* **29**:30–31.

135. Hensel, M., J. E. Shea, B. Raupach, D. Monack, S. Falkow, C. Gleeson, T. Kubo, and D. W. Holden. 1997. Functional analysis of *ssaJ* and the *ssaK/U* operon, 13 genes encoding components of the type III secretion apparatus of *Salmonella* pathogenicity island 2. *Mol. Microbiol.* **24**:155–167.

136. Hensel, M., J. E. Shea, S. R. Waterman, R. Mundy, T. Nikolaus, G. Banks, A. Vazquez-Torres, C. Gleeson, F. C. Fang, and D. W. Holden. 1998. Genes encoding putative effector proteins of the type III secretion system of *Salmonella* pathogenicity island 2 are required for bacterial virulence and proliferation in macrophages. *Mol. Microbiol.* **30**:163–174.

137. Hersh, D., D. M. Monack, M. R. Smith, N. Ghori, S. Falkow, and A. Zychlinsky. 1999. The *Salmonella* invasin SipB induces macrophage apoptosis by binding to caspase-1. *Proc. Natl. Acad. Sci. USA* **96**:2396–2401.

138. Hickey, E. W., and I. N. Hirshfield. 1990. Low pH-induced effects of patterns of protein synthesis and on internal pH in *Escherichia coli* and *Salmonella*

typhimurium. Appl. Environ. Microbiol. **56:**1038–1045.

139. **Holley, R. A., and M. Proulx.** 1986. Use of egg washwater pH to prevent survival of *Salmonella* at moderate temperatures. *Poult. Sci.* **65:**922–928.

140. **Holmberg, S. D., M. T. Osterholm, K. A. Senger, and M. L. Cohen.** 1984. Drug-resistant *Salmonella* from animal fed antimicrobials. *N. Engl. J. Med.* **311:**617–622.

141. **Horwitz, M. A., and F. R. Maxfield.** 1984. *Legionella pneumophila* inhibits acidification of its phagosome in human monocytes. *J. Cell Biol.* **99:**1936–1943.

142. **Horwitz, M. A., R. A. Pollard, M. H. Merson, and S. M. Martin.** 1977. A large outbreak of foodborne salmonellosis on the Navajo nation Indian reservation, epidemiology and secondary transmission. *Am. J. Public Health* **67:**1071–1076.

143. **Hueck, C. J.** 1998. Type III secretion systems in bacterial pathogens of animals and plants. *Microbiol. Mol. Biol. Rev.* **62:**379–433.

144. **Huhtanen, C. N.** 1975. Use of pH gradient plates for increasing the acid tolerance of salmonellae. *Appl. Microbiol.* **29:**309–312.

145. **Humphrey, T. J., N. P. Richardson, K. M. Statton, and R. J. Rowbury.** 1993. Effects of temperature shift on acid and heat tolerance in *Salmonella enteritidis* phage type 4. *Appl. Environ. Microbiol.* **59:**3120–3122.

146. **Inami, G. B., and S. E. Moler.** 1999. Detection and isolation of *Salmonella* from naturally contaminated alfalfa seeds following and outbreak investigation. *J. Food Prot.* **62:**662–664.

147. **Ingham, S. C., R. A. Alford, and A. P. McCown.** 1990. Comparative growth rates of *Salmonella typhimurium* and *Pseudomonas fragi* on cooked crab meat stored under air and modified atmosphere. *J. Food Prot.* **53:**566–567.

148. **Institute of Medicine.** 1988. *Report of a Study. Human Health Risks with the Subtherapeutic Use of Penicillin and Tetracyclines in Animal Feed.* National Academy Press, Washington, D.C.

149. **Institute of Medicine.** 1998. *Antimicrobial Resistance: Issues and Options. Forum on Emerging Infections.* National Academy Press, Washington, D.C.

150. **Isberg, R. R., and G. T. V. Nhieu.** 1994. Two mammalian cell internalization strategies used by pathogenic bacteria. *Annu. Rev. Genet.* **27:**395–422.

151. **Jacobs-Reitsma, W. F., P. M. F. J. Koenraad, N. M. Bolder, and R. W. A. W. Mulder.** 1994. In vitro susceptibility of *Campylobacter* and *Salmonella* isolates from broilers to quinolones, ampicillin, tetracycline, and erythromycin. *Vet. Q.* **16:**206–208.

152. **Janakiraman, A., and J. M. Slauch.** 2000. The putative iron transport system SitABCD encoded on SPI1 is required for full virulence of *Salmonella typhimurium. Mol. Microbiol.* **35:**1146–1155.

152a. **Japan Ministry of Health and Welfare.** 1999. *Departmental Health Report.* Japan Ministry of Health and Welfare, Tokyo, Japan.

153. **Jones, B. D., H. F. Paterson, A. Hall, and S. Falkow.** 1993. *Salmonella typhimurium* induces membrane ruffling by a growth factor-receptor-independent mechanism. *Proc. Natl. Acad. Sci. USA* **90:**10390–10394.

154. **Kaniga, K., J. C. Bossio, and J. E. Galan.** 1994. The *Salmonella typhimurium* invasion genes *inv* F and *inv* G encode homologues of the AraC and PulD family of proteins. *Mol. Microbiol.* **13:**555–568.

155. **Kapperud, G., S. Gustavsen, I. Hellesnes, A. H. Hansen, J. Lassen, J. Hirn, M. Jahkola, M. A. Montenegro, and R. Helmuth.** 1990. Outbreak of *Salmonella typhimurium* infection traced to contaminated chocolate and caused by a strain lacking the 60-megadalton virulence plasmid. *J. Clin. Microbiol.* **28:**2597–2601.

156. **Kauffmann, F.** 1966. *The Bacteriology of Enterobacteriaceae.* Munksgaard, Copenhagen, Denmark.

157. **Keitel, W. A., N. L. Bond, J. M. Zahradnik, T. A. Cramton, and J. B. Robbins.** 1994. Clinical and serological responses following primary and booster immunization with *Salmonella typhi* Vi capsular polysaccharide vaccines. *Vaccine* **12:**195–199.

158. **Kim, C. J., D. A. Emery, H. Rinke, K. V. Nagaraja, and D. A. Halvorson.** 1989. Effect of time and temperature on growth of *Salmonella enteritidis* in experimentally inoculated eggs. *Avian Dis.* **33:**735–742.

159. **Kim, J. M., L. Eckmann, T. C. Savidge, D. C. Lowe, T. Witthoft, and M. F. Kagnoff.** 1998. Apoptosis of human intestinal epithelial cells after bacterial invasion. *J. Clin. Invest.* **102:**1815–1823.

160. **Kirby, R. M., and R. Davies.** 1990. Survival of dehydrated cells of *Salmonella typhimurium* LT 2 at high temperatures. *J. Appl. Microbiol.* **68:**241–246.

161. **Koo, F. C. W., J. W. Peterson, C. W. Houston, and N. C. Molina.** 1984. Pathogenesis of experimental salmonellosis: inhibition of protein synthesis by cytotoxin. *Infect. Immun.* **43:**93–100.

162. **Kowarz, L., C. Coynault, V. Robbe-Saule, and F. Norel.** 1994. The *Salmonella typhimurium katF (rpoS)* gene: cloning, nucleotide sequence, and regulation of *spvR* and *spvABCD* virulence plasmid genes. *J. Bacteriol.* **176:**6852–6860.

163. **Lang, D. J., L. J. Kunz, A. R. Martin, S. A. Schroeder, and L. A. Thomson.** 1967. Carmine as a source of nosocomial salmonellosis. *N. Engl. J. Med.* **276:**829–832.

164. **Lecos, C.** 1986. Of microbes and milk: probing America's worst *Salmonella* outbreak. *Dairy Food Sanit.* **6:**136–140.

165. **Lee, A. K., C. S. Detweiler, and S. Falkow.** 2000. OmpR regulates the two-component system SsrA-SsrB in *Salmonella* pathogenicity island 2. *J. Bacteriol.* **182:**771–781.

166. **Lee, I. S., J. L. Slonczewski, and J. W. Foster.** 1994. A low-pH-inducible, stationary phase acid tolerance response in *Salmonella typhimurium. J. Bacteriol.* **176:**1422–1426.

167. **Lee, L. A., N. D. Puhr, E. K. Maloney, N. H. Bean, and R. V. Tauxe.** 1994. Increase in antimicrobial-resistant *Salmonella* infections in the United States, 1989–1990. *J. Infect. Dis.* **170:**128–134.

168. Lehmacher, A., J. Bockemühl, and S. Aleksic. 1995. Nationwide outbreak of human salmonellosis in Germany due to contaminated paprika and paprika-powdered potato chips. *Epidemiol. Infect.* **115**:501–511.

169. Le Minor, L. 1981. The genus *Salmonella*, p. 1148–1159. *In* M. P. Starr, H. Stolp, H. G. Truper, A. Balows, and H. G. Schlegel (ed.), *The Prokaryotes*. Springer-Verlag, New York, N.Y.

170. Le Minor, L., C. Coynault, and G. Pessoa. 1974. Déterminisme plasmidique du caractère atypique "lactose positif" de souches de *S. typhimurium* et de *S. oranienburg* isolées au Brésil lors d'épidémies de 1971 à 1973. *Ann. Microbiol.* (Paris) **125A**:261–285.

171. Le Minor, L., C. Coynault, R. Rhode, B. Rowe, and S. Aleksic. 1973. Localisation plasmidique de déterminant génétique du caractère atypique "saccharose+" des *Salmonella*. *Ann. Microbiol. (Paris).* **124B**:295–306.

172. Le Minor, L., and M. Y. Popoff. 1987. Request for an opinion. Designation of *Salmonella enterica* sp. nov., nom. rev., as the type and only species of the genus *Salmonella*. *Int. J. Syst. Bacteriol.* **37**:465–468.

173. Le Minor, L., M. Y. Popoff, B. Laurent, and D. Hermant. 1986. Individualisation d'une septième sous-espèce de *Salmonella*: *S. choleraesuis* subsp. *indica* subsp. nov. *Ann. Inst. Pasteur/Microbiol.* **137B**:211–217.

174. Lepoutre, A., J. Salomon, C. Charley, and F. LeQuerrec. 1994. Les toxi-infections alimentaires collectives en 1993. *Bull. Epidémiol. Hebd.* **52**:245–247.

175. Leyer, G. J., and E. A. Johnson. 1992. Acid adaptation promotes survival of *Salmonella* spp. in cheese. *Appl. Environ. Microbiol.* **58**:2075–2080.

176. Leyer, G. J., and E. A. Johnson. 1993. Acid adaptation induces cross-protection against environmental stresses in *Salmonella typhimurium*. *Appl. Environ. Microbiol.* **59**:1842–1847.

177. Libby, S. J., L. G. Adams, T. A. Ficht, C. Allen, H. A. Whitford, N. A. Buchmeier, S. Bossie, and D. G. Guiney. 1997. The *spv* genes on the *Salmonella dublin* virulence plasmid are required for severe enteritis and systemic infection in the natural host. *Infect. Immun.* **65**:1786–1792.

178. Low, J. C., G. Hopkins, T. King, and D. Munro. 1996. Antibiotic resistant *Salmonella typhimurium* DT104 in cattle. *Vet. Rec.* **138**:650–651.

179. MacBeth, K. J., and C. A. Lee. 1993. Prolonged inhibition of bacterial protein synthesis abolishes *Salmonella* invasion. *Infect. Immun.* **61**:1544–1546.

180. Mackey, B. M., and C. M. Derrick. 1986. Elevation of the heat resistance of *Salmonella typhimurium* by sublethal heat shock. *J. Appl. Bacteriol.* **61**:389–393.

181. Mackey, B. M., and C. Derrick. 1990. Heat shock protein synthesis and thermotolerance in *Salmonella typhimurium*. *J. Appl. Bacteriol.* **69**:373–383.

182. Makela, P. H., M. Sarvas, S. Calcagno, and K. Lounatmaa. 1978. Isolation and characterization of polymyxin-resistant mutants of *Salmonella typhimurium*. *FEMS Microbiol. Lett.* **3**:323–326.

183. Mason, J. 1994. *Salmonella enteritidis* control programs in the United States. *Int. J. Food Microbiol.* **21**:155–169.

184. Mattila, L. M. Leirisalo-Repo, S. Koskimies, K. Granfors, and A. Sütonen. 1994. Reactive arthritis following an outbreak of *Salmonella* infection in Finland. *Br. J. Rheumatol.* **33**:1136–1141.

185. McCullough, N. B., and C. W. Eisele. 1951. Experimental human salmonellosis. 1. Pathogenicity of strains of *Salmonella meleagridis* and *Salmonella anatum* obtained from spray-dried whole egg. *J. Infect. Dis.* **88**:278–290.

186. Miller, S. I. 1991. PhoP/PhoQ: macrophage-specific modulators of *Salmonella* virulence. *Mol. Microbiol.* **5**:2073–2078.

187. Miller, S. I., A. M. Kukral, and J. J. Mekalanos. 1989. A two-component regulatory system (phoP/phoQ) controls *Salmonella typhimurium* virulence. *Proc. Natl. Acad. Sci. USA* **86**:5054–5058.

188. Mills, D. M., V. Bajaj, and C. A. Lee. 1995. A 40 kb chromosomal fragment encoding *Salmonella typhimurium* invasion genes is absent from the corresponding region of the *Escherichia coli* K-12 chromosome. *Mol. Microbiol.* **15**:749–759.

189. Mills, S. D., S. R. Ruschkowski, M. A. Stein, and B. B. Finlay. 1998. Trafficking of porin-deficient *Salmonella typhimurium* mutants inside HeLa cells: *ompR* and *envZ* mutants are defective for the formation of *Salmonella*-induced filaments. *Infect. Immun.* **66**:1806–1811.

190. Mintz, E. D., M. L. Cartter, J. L. Hadler, J. T. Wassell, J. A. Zingeser, and R. V. Tauxe. 1994. Dose-response effects in an outbreak of *Salmonella enteritidis*. *Epidemiol. Infect.* **112**:13–23.

191. Miyagawa, S., and A. Miki. 1992. The epidemiological data of food poisoning in 1991. *Food Sanit. Res.* **42**:78–104.

192. Mizuno, T., and S. Mizushima. 1990. Signal transduction and gene regulation through the phosphorylation of two regulatory components:the molecular basis for the osmotic regulation of the porin genes. *Mol. Microbiol.* **4**:1077–1082.

193. Monack, D. M., B. Raupach, A. E. Hromockyj, and S. Falkow. 1996. *Salmonella typhimurium* invasion induces apoptosis in infected macrophages. *Proc. Natl. Acad. Sci. USA* **93**:9833–9838.

193a. Mouzin, E., S. B. Werner, R. G. Bryant, et al. 1997. When a health food becomes a hazard: a large outbreak of salmonellosis associated with alfalfa sprouts—California. Presented at Epidemic Intelligence Service Conference, Centers for Disease Control and Prevention, April.

194. Murray, B. E. 1986. Resistance of *Shigella*, *Salmonella*, and other selected enteric pathogens to antimicrobial agents. *Rev. Infect. Dis.* **8**(Suppl. 2):S172–S181.

195. Murray, R. A., and C. A. Lee. 2000. Invasion genes are not required for *Salmonella enterica* serovar Typhimurium to breach the intestinal epithelium: evidence that *Salmonella* pathogenicity island 1 has alternative functions during infection. *Infect. Immun.* **68**:5050–5055.

196. **Nassif, X., and P. Sansonetti.** 1987. Les systèmes bactériens de captation du fer: leur rôle dans la virulence. *Bull. Inst. Pasteur* **85:**307–327.

197. **National Research Council.** 1980. *Effects on Human Health of Subtherapeutic Use of Antimicrobials in Animal Feeds.* National Academy Press, Washington, D.C.

198. **National Research Council.** 1999. *The Use of Drugs in Food Animals: Benefits and Risks.* National Academy Press, Washington, D.C.

199. **Ng, H., H. G. Bayne, and J. A. Garibaldi.** 1969. Heat resistance of *Salmonella*: the uniqueness of *Salmonella senftenberg* 775W. *Appl. Microbiol.* **17:**78–82.

200. **Nguyen Van, J. C., and L. Gutman.** 1994. Résistance aux antibiotiques par diminution de la perméabilité chez les bactéries à Gram négatif. *Presse Méd.* **23:**522–531.

201. **Norris, F. A., M. P. Wilson, T. S. Wallis, E. E. Galyov, and P. W. Majerus.** 1998. SopB, a protein required for virulence of *Salmonella dublin*, is an inositol phosphate phosphatase. *Proc. Natl. Acad. Sci. USA* **95:**14057–14059.

202. **Novick, R. P.** 1981. The development and spread of antibiotic-resistant bacteria as a consequence of feeding antibiotics to livestock. *Ann. N.Y. Acad. Sci.* **368:**23–59.

203. **Ogawa, H., H. Tokunou, M. Sasaki, T. Kishimoto, and K. Tamura.** 1991. An outbreak of bacterial food poisoning caused by roast cuttlefish "yaki-ika" contaminated with *Salmonella* spp. Champaign. *Jpn. J. Food Microbiol.* **7:**151–157.

204. **Oh, Y., C. Alpuche-Aranda, E. Berthiaume, T. Jinks, S. I. Miller, and J. A. Swanson.** 1996. Rapid and complete fusion of macrophage lysosomes with phagosomes containing *Salmonella typhimurium*. *Infect. Immun.* **64:**3877–3883.

205. **Olson, E. R.** 1993. MicroReview. Influence of pH on bacterial gene expression. *Mol. Microbiol.* **8:**5–14.

206. **Otto, H., D. Tezcan-Merdol, R. Girisch, F. Haag, M. Rhen, and F. Koch–Nolte.** 2000. The *spvB* gene-product of the *Salmonella enterica* virulence plasmid is a mono(ADP-ribosyl)transferase. *Mol. Microbiol.* **37:**1106–1115.

207. **Pacer, R. E., J. S. Spika, M. C. Thurmond, N. Hargrett-Bean, and M. E. Potter.** 1989. Prevalence of *Salmonella* and multiple antimicrobial-resistant *Salmonella* in California dairies. *J. Am. Vet. Med. Assoc.* **195:**59–63.

208. **Palmer, S. R., and B. Rowe.** 1986. Trends in *Salmonella* infections. *PHLS Microbiol. Digest* **3:**18–21.

209. **Parish, M. E.** 1998. Coliforms, *Escherichia coli* and *Salmonella* serovars associated with a citrus-processing facility implicated in a salmonellosis outbreak. *J. Food Prot.* **61:**280–284.

210. **Pengues, D. A., M. J. Hantman, I. Behlau, and S. I. Miller.** 1995. PhoP/PhoQ transcriptional repression of *Salmonella typhimurium* invasion genes: evidence for a role in protein secretion. *Mol. Microbiol.* **17:**169–181.

211. **Peterson, J. W., and D. W. Niesel.** 1988. Enhancement by calcium of the invasiveness of *Salmonella* for HeLa cell monolayers. *Rev. Infect. Dis.* **10:**S319-S322.

211a. **Pezzi, G. H., I. M. Gallardo, S. M. Ontanon, et al.** 1995. Vigilancia de brotes de infecciones e intoxica ciones de origen alimentario, Espana, Ans 1994 (excluye brotes hidricos). *Bol. Epidemiol. Semanal.* **3:**293–299.

212. **Pickard, D., J. Li, M. Roberts, D. Maskell, D. Hone, M. Levine, G. Dougan, and S. Chatfield.** 1994. Characterization of defined *ompR* mutants of *Salmonella typhi*: *ompR* is involved in the regulation of Vi polysaccharide expression. *Infect. Immun.* **62:**3984–3993.

213. **Piddock, L. J. V., C. Wray, I. McLaren, and R. Wise.** 1990. Quinolone resistance in *Salmonella* species: veterinary pointers. *Lancet* **336:**125.

214. **Pivnick, H.** 1980. Curing salts and related materials, p. 136–159. *In* J. H. Silliker, R. P. Elliott, A. C. Baird-Parker, F. L. Bryan, J. H. B. Christian, D. S. Clark, J. C. Olson, Jr., and T. A. Roberts (ed.), *Microbial Ecology of Foods*, vol. 1. *Factors Affecting Life and Death of Microorganisms*. Academic Press, Inc., New York:, N.Y.

215. **Pohl, P., Y. Glupczynski, M. Marin, G. Van Robaeys, P. Lintermans, and M. Couturier.** 1993. Replicon typing characterization of plasmids encoding resistance to gentamicin and apramycin in *Escherichia coli* and *Salmonella typhimurium* isolated from human and animal sources in Belgium. *Epidemiol. Infect.* **111:**229–238.

216. **Polotsky, Y., E. Dragunsky, and T. Khavkin.** 1994. Morphologic evaluation of the pathogenesis of bacterial enteric infections. *Crit. Rev. Microbiol.* **20:**161–208.

217. **Pönkä, A., Y. Andersson, A. Sütonen, B. de Jong, M. Jahkola, O. Haikala, A. Kuhmonen, and P. Pakkala.** 1995. *Salmonella* in alfalfa sprouts. *Lancet* **345:**462–463.

218. **Popoff, M. Y., and F. Norel.** 1992. Bases moléculaires de la pathogénicité des *Salmonella*. *Méd. Mal. Infect.* **22:**310–324.

219. **Popoff, M. Y., J. Bockemuhl, and F. W. Brenner.** 2000. Supplement 1998 (no. 42) to the Kauffmann-White scheme. *Res. Microbiol.* **151:**63–65.

220. **Radford, S. A., and R. G. Board.** 1995. The influence of sodium chloride and pH on the growth of *Salmonella enteritidis* PT 4. *Lett. Appl. Microbiol.* **20:**11–13.

221. **Rahman, H., V. B. Singh, and V. D. Sharma.** 1994. Purification and characterization of enterotoxic moiety present in cell-free culture supernatant of *Salmonella typhimurium*. *Vet. Microbiol.* **39:**245–254.

222. **Rajajee, S., T. B. Anandi, S. Subha, and B. R. Vatsala.** 1995. Patterns of resistant *Salmonella typhi* infection in infants. *J. Trop. Pediatr.* **41:**52–54.

223. **Rathman, M., L. P. Barker, and S. Falkow.** 1997. The unique trafficking pattern of *Salmonella typhimurium*-containing phagosomes in murine macrophages is independent of the mechanism of bacterial entry. *Infect. Immun.* **65:**1475–1485.

224. **Rathman, M., M. D. Sjaastad, and S. Falkow.** 1996. Acidification of phagosomes containing *Salmonella typhimurium* in murine macrophages. *Infect. Immun.* **64:**2765–2773.

224a. **Ratman, S., F. Stratton, C. O'Keefe, et al.** 1999. *Salmonella* Enteritidis outbreak due to contaminated cheese—Newfoundland. *Can. Commun. Dis. Rep.* **25:**17–20.

225. Reeves, M. W., G. M. Evins, A. A. Heiba, B. D. Plikaytis, and J. J. Farmer III. 1989. Clonal nature of *Salmonella typhi* and its genetic relatedness to other salmonellae as shown by multilocus enzyme electrophoresis, and proposal of *Salmonella bongori* comb. nov. *J. Clin. Microbiol.* 27:313–320.

226. Reitler, R., D. Yarom, and R. Seligmann. 1960. The enhancing effect of staphylococcal enterotoxin on *Salmonella* infection. *Med. Officer* 104:181.

227. Rhoades, E. R., and H. J. Ullrich. 2000. How to establish a lasting relationship with your host: lessons learned from *Mycobacterium* spp. *Immunol. Cell Biol.* 78:301–310.

228. Richard, F., E. Pons, B. Lelore, V. Bleuze, B. Grandbastien, C. Collinet, R. Mathis, J.-F. Diependale, and P. Legrand. 1994. Toxi-infection alimentaire collective du 8 juin 1993 à Douai. *Bull. Epidemiol. Hebd.* 3:9–11.

229. Richter-Dahlfors, A., A. M. J. Buchan, and B. B. Finlay. 1997. Murine salmonellosis studied by confocal microscopy: *Salmonella typhimurium* resides intracellularly inside macrophages and exerts a cytotoxic effect on phagocytes in vivo. *J. Exp. Med.* 186:569–580.

230. Robbe-Saule, V., C. Coynault, and F. Norel. 1995. The live oral typhoid vaccine Ty21a is a rpoS mutant and is susceptible to various environmental stresses. *FEMS Microbiol. Lett.* 126:171–176.

231. Roberts, J. A., and P. N. Sockett. 1994. The socioeconomic impact of human *Salmonella enteritidis* infection. *Int. J. Food Microbiol.* 21:117–129.

232. Roland, K. L., L. E. Martin, C. R. Esther, and J. K. Spitznagel. 1993. Spontaneous *pmrA* mutants of *Salmonella typhimurium* LT2 define a new two-component regulatory system with a possible role in virulence. *J. Bacteriol.* 175:4154–4164.

233. Rowe, B., E. J. Threlfall, L. R. Ward, and A. S. Ashley. 1979. International spread of multiresistant strains of *Salmonella typhimurium* phage types 204 and 193 from Britain to Europe. *Vet. Rec.* 105:468–469.

234. Ruschkowski, S., I. Rosenshine, and B. B. Finlay. 1992. *Salmonella typhimurium* induces an inositol phosphate flux in infected epithelial cells. *FEMS Microbiol. Lett.* 95:121–126.

235. Ryan, C. A., M. K. Nickels, N. T. Hargrett-Bean, M. E. Potter, T. Endo, L. Mayer, C. W. Langkop, C. Gibson, R. C. McDonald, R. T. Kenney, N. D. Puhr, P. J. McDonnell, R. J. Martin, M. L. Cohen, and P. A. Blake. 1987. Massive outbreak of antimicrobial-resistant salmonellosis traced to pasteurized milk. *JAMA* 258:3269–3274.

236. Sameshima, T., H. Ito, I. Uchida, H. Danbara, and N. Terakado. 1993. A conjugative plasmid pTE195 coding for drug resistance and virulence phenotypes from *Salmonella naestved* strain of calf origin. *Vet. Microbiol.* 36:197–203.

237. Sandaa, R. A., V. L. Torsvik, and J. Goksoyr. 1992. Transferable drug resistance in bacteria from fish-farm sediments. *Can. J. Microbiol.* 38:1061–1065.

238. Saxen, H., I. Reima, and P. H. Mäkelä. 1987. Alternative complement pathway activation by *Salmonella* O polysaccharide as a virulence determinant in the mouse. *Microb. Pathog.* 2:15–28.

239. Schwan, W. R., X. Huang, L. Hu, and D. J. Kopecko. 2000. Differential bacterial survival, replication, and apoptosis-inducing ability of *Salmonella* serovars within human and murine macrophages. *Infect. Immun.* 68:1005–1013.

240. Seglenieks, Z., and S. Dixon. 1977. Outbreak of milk-borne *Salmonella* gastroenteritis—South Australia. *Morb. Mortal. Wkly. Rep.* 26:127.

241. Shafer, W. M., L. E. Martin, and J. K. Spitznagel. 1984. Cationic antimicrobial proteins isolated from human neutrophil granulocytes in the presence of diisopropyl fluorophosphates. *Infect. Immun.* 45:29–35.

242. Shea, J. E., M. Hensel, C. Gleeson, and D. W. Holden. 1996. Identification of a virulence locus encoding a second type III secretion system in *Salmonella typhimurium*. *Proc. Natl. Acad. Sci. USA* 93:2593–2597.

243. Sirard, J.-C., F. Niedergang, and J.-P. Kraehenbuhl. 1999. Live attenuated *Salmonella*: a paradigm of mucosal vaccines. *Immunol. Rev.* 171:5–26.

244. Sjöbring, U., G. Pohl, and A. Olsen. 1994. Plasminogen, adsorbed by *Escherichia coli* expressing curli or by *Salmonella enteritidis* expressing thin aggregative fimbriae, can be activated by simultaneously captured tissue-type plasminogen activator (t-PA). *Mol. Microbiol.* 14:443–452.

245. Skamene, E., E. Schurr, and P. Gros. 1998. Infection genomics: *nramp1* as a major determinant of natural resistance to intracellular infections. *Annu. Rev. Med.* 49:275–287.

246. Slauch, J. M., S. Garrett, D. E. Jackson, and T. J. Silhavy. 1988. EnvZ functions through OmpR to control porin gene expression in *Escherichia coli* K-12. *J. Bacteriol.* 170:439–441.

247. Smith, J. L. 1994. Arthritis and foodborne bacteria. *J. Food Prot.* 57:935–941.

248. Sockett, P. N., J. M. Cowden, S. LeBaigne, D. Ross, G. K. Adak, and H. Evans. 1993. Foodborne disease surveillance in England and Wales: 1989–1991. *Commun. Dis. Rep.* 3:R159–R173.

249. Spika, J. S., S. H. Waterman, G. W. Soo Hoo, M. E. St.Louis, R. E. Pacer, S. M. James, M. L. Bissett, L. W. Mayer, J. Y. Chiu, B. Hall, K. Greene, M. E. Potter, M. L. Cohen, and P. A. Blake. 1987. Chloramphenicol-resistant *Salmonella newport* traced through hamburger to dairy farms. *N. Engl. J. Med.* 316:565–570.

250. Stender, S., A. Friebel, S. Linder, M. Rohde, S. Mirold, and W. D. Hardt. 2000. Identification of SopE2 from *Salmonella typhimurium*, a conserved guanine nucleotide exchange factor for Cdc42 of the host cell. *Mol. Microbiol.* 36:1206–1221.

251. Stephen, J., T. S. Wallis, W. G. Starkey, D. C. A. Candy, M. P. Osborne, and S. Haddon. 1985. Salmonellosis: in retrospect and prospect, p. 175–192. *Microbial Toxins and Diarrhoeal Disease* (Ciba Foundation Symposium 112). Pitman, London, England.

252. Stickney, R. R. 1990. A global overview of aquaculture production. *Food Rev. Int.* 6:299–315.

253. Suzuki, S., K. Komase, H. Matsui, A. Abe, K. Kawahara, Y. Tamura, M. Kijima, H. Danbara, M. Nakamura, and S. Sato. 1994. Virulence region of plasmid pN2001 of *Salmonella enteritidis*. *Microbiology* **140:**1307–1318.

254. Swann, M. M. 1969. *Report of the Joint Committee on the Use of Antibiotics in Animal Husbandry and Veterinary Medicine*. Cmnd. 4190. Her Majesty's Stationery Office, London. England.

254a. Szu, S. C. 1994. Presented at Asia-Pacific Symposium on Typhoid Fever, Bangkok, Thailand.

255. Tabaraie, B., B. K. Sharma, P. R. Sharma, R. Sehgal, and N. K. Ganguly. 1994. Stimulation of macrophage oxygen free radical production and lymphocyte blastogenic response by immunization with porins. *Microbiol. Immunol.* **38:**561–565.

256. Tacket, C. O., L. B. Dominguez, H. J. Fisher, and M. L. Cohen. 1985. An outbreak of multiple drug-resistant *Salmonella* enteritis from raw milk. *JAMA* **253:**2058–2060.

257. Taira, S., P. Heiskanen, R. Hurme, H. Heikkilä, P. Riikonen, and M. Rhen. 1995. Evidence for functional polymorphism of the *spvR* gene regulating virulence gene expression in *Salmonella*. *Mol. Gen. Genet.* **246:**437–444.

258. Tanaka, N. 1993. Food hygiene in Japan—Japanese food hygiene regulations and food poisoning incidents. *Dairy Food Environ. Sanit.* **13:**152–156.

259. Thomas, L. V., J. W. T. Wimpenny, and A. C. Peters. 1992. Testing multiple variables on the growth of a mixed inoculum of *Salmonella* strains using gradient plates. *Int. J. Food Microbiol.* **15:**165–175.

260. Thomson, G. T. D., D. A. DeRubeis, M. A. Hodge, C. Rajanayagam, and R. D. Inman. 1995. Post-*Salmonella* reactive arthritis: late clinical sequelae in point source cohort. *Am. J. Med.* **98:**13–21.

261. Threlfall, E. J. 1992. Antibiotics and the selection of food-borne pathogens. *J. Appl. Bacteriol. Symp. Suppl.* **73:**96S–102S.

262. Threlfall, E. J., F. J. Angulo, and P. G. Wall. 1998. Ciprofloxacin-resistant *Salmonella typhimurium* DT104. *Vet. Rec.* **142:**255.

263. Threlfall, E. J., M. D. Hampton, H. Chart, and B. Rowe. 1994. Identification of a conjugative plasmid carrying antibiotic resistance and *Salmonella* plasmid virulence (*spv*) genes in epidemic strains of *Salmonella typhimurium* phagetype 193. *Lett. Appl. Microbiol.* **18:**82–85.

264. Threlfall, E. J., B. Rowe, J. L. Ferguson, and L. R. Ward. 1985. Increasing incidence of resistance to gentamicin and related aminoglycosides in *Salmonella typhimurium* phage type 204c in England, Wales and Scotland. *Vet. Rec.* **117:**355–357.

265. Threlfall, E. J., B. Rowe, J. L. Ferbuson, and L. R. Ward. 1986. Characterization of plasmids conferring resistance to gentamicin and apramycin in strains of *Salmonella typhimurium* phage type 204c isolated in Britain. *J. Hyg. Camb.* **97:**419–426.

266. Threlfall, E. J., L. R. Ward, J. A. Skinner, and B. Rowe. 1997. Increase in multiple antibiotic resistance in non-typhoidal Salmonellas from humans in England and Wales: a comparison of data for 1994 and 1996. *Microb. Drug Resist.* **3:**263–266.

267. Todd, E. C. D. 1994. Surveillance of foodborne disease, p. 461–536. *In* Y. H. Hui, J. R. Gorham, K. D. Murrell, and D. O. Cliver (ed.), *Foodborne Disease Handbook*, vol. 1. *Diseases Caused by Bacteria*. Marcel Dekker, Inc., New York, N.Y.

267a. Torok, T. J., R. V. Tauxe, R. P. Wise, et al. 1997. A large community outbreak of salmonellosis caused by intentional contamination of restaurant salad bars. *JAMA* **278:**389–395.

268. Troller, J. A. 1986. Water relations of foodborne bacterial pathogens—an updated review. *J. Food Prot.* **49:**656–670.

269. Tsolis, R. M., L. G. Adams, T. A. Ficht, and A. J. Baumler. 1999. Contribution of *Salmonella typhimurium* virulence factors to diarrheal disease in calves. *Infect. Immun.* **67:**4879–4885.

269a. Tsujii, H., and K. Hamada. 1999. Outbreak of salmonellosis caused by ingestion of cuttlefish chips contaminated by both *Salmonella* Chester and *Salmonella* Oranienburg. *Jpn. J. Infect. Dis.* **52:**138–139.

270. Uchiya, K., M. A. Barbieri, K. Funato, A. H. Shah, P. D. Stahl, and E. A. Groisman. 1999. A *Salmonella* virulence protein that inhibits cellular trafficking. *EMBO J.* **18:**3924–3933.

271. Vaara, M., T. Vaara, M. Jenson, I. Helander, M. Nurminen, E. T. Rietschel, and P. H. Makela. 1981. Characterization of the lipopolysaccharide from the polymixin-resistant *pmrA* mutants of *Salmonella typhimurium*. *FEBS Lett.* **129:**145–149.

272. van Leeuwen, W. J., J. D. A. van Embden, P. A. M. Guinee, E. H. Kampelmacher, A. Manten, M. van Schothorst, and C. E. Voogd. 1979. Decrease of drug resistance in *Salmonella* in The Netherlands. *Antimicrob. Agents Chemother.* **16:**237–239.

273. Vescovi, E. G., F. C. Soncini, and E. A. Groisman. 1994. The role of the PhoP/PhoQ regulon in *Salmonella* virulence. *Res. Microbiol.* **145:**473–480.

273a. Vought, K. J., and S. R. Tatini. 1998. *Salmonella enteritidis* contamination of ice cream associated with a 1994 multistate outbreak. *J. Food Prot.* **61:**5–10.

274. Wall, P. G., D. Morgan, K. Lamden, M. Ryan, M. Griffin, E. J. Threlfall, L. R. Ward, and B. Rowe. 1994. A case control study of infection with an epidemic strain of multiple resistant *Salmonella typhimurium* DT104 in England and Wales. *Commun. Dis. Rep.* **4:**R130–R135.

275. Watson, P. R., E. E. Galyov, S. M. Paulin, P. W. Jones, and T. S. Wallis. 1998. Mutation of *invH*, but not *stn*, reduces *Salmonella*-induced enteritis in cattle. *Infect. Immun.* **66:**1432–1438.

276. Wayne, L. G. 1991. Judicial Commission of the International Committee on Systematic Bacteriology. *Int. J. Syst. Bacteriol.* **41:**185–187.

277. Weissman, J. B., R. M. A. D. Deen, M. Williams, N. Swanton, and S. Ali. 1977. An island-wide epidemic of salmonellosis in Trinidad traced to contaminated powdered milk. *West Indies Med. J.* **26:**135–143.

278. **Wiedemann, B., and P. Heisig.** 1994. Mechanisms of quinolone resistance. *Infection* **22**(Suppl. 2):S73–S79.

279. **Wierup, M.** 1997. Ten years without antibiotic growth promoters—results from Sweden with special reference to production results, alternative disease preventive methods and the usage of antibacterial drugs, p. 229–235 *In Report of a WHO Meeting.* WHO/EMC/ ZOO/97.4. World Health Organization, Geneva, Switzerland.

280. **Wood, M. W., M. A. Jones, P. R. Watson, S. Hedges, T. S. Wallis, and E. E. Galyov.** 1998. Identification of a pathogenicity island required for *Salmonella* enteropathogenicity. *Mol. Microbiol.* **29**:883–891.

281. **World Health Organization.** 1988. *Salmonellosis Control: the Role of Animal and Product Hygiene.* Technical Report Series 774. World Health Organization, Geneva, Switzerland.

282. **World Health Organization.** 1992. *WHO Surveillance Programme for Control of Foodborne Infections and Intoxications in Europe. Fifth Report 1985–1989.* Institute of Veterinary Medicine, Robert von Ostertag Institute, Berlin, Germany.

282a. **World Health Organization.** 1995. *WHO Surveillance Programme for Control of Foodborne Infections and Intoxications in Europe,* p. 1–340. Federal Institute for Health Protection of Consumers and Veterinary Medicine, Berlin, Germany.

283. **World Health Organization.** 1997. *The Medical Impact of the Use of Antimicrobials in Food Animals: Report of a WHO Meeting.* WHO/E.C./ZOO/97.4. World Health Organization, Geneva, Switzerland.

284. **World Health Organization.** 1998. Use of quinolones in food animals and potential impact on human health, p. 1–24. *In Report of a WHO Meeting.* WHO/EMC/ZDI/98.10. World Health Organization, Geneva, Switzerland.

285. **World Health Organization.** 2000. WHO issues new recommendations to protect human health from antimicrobial use in food animals. http//www.who.int.

286. **Wray, C., R. W. Hedges, K. P. Shannon, and D. E. Bradley.** 1986. Apramycin and gentamicin resistance in *Escherichia coli* and Salmonellas isolated from farm animals. *J. Hyg. Camb.* **97**:445–456.

287. **Wray, C., and A. Wray. (ed.).** 2000. Antibiotic resistance in Salmonella. *In Salmonella in Domestic Animals.* CABI, Oxon, United Kingdom.

288. **Ye, X. L., C. C. Yan, H. H. Xie, X. P. Tan, Y. Z. Wang, and L. M. Ye.** 1990. An outbreak of food poisoning due to *Salmonella typhimurium* in the People's Republic of China. *J. Diarrheal Dis. Res.* **8**:97–98.

289. **Young, H. K.** 1994. Do nonclinical uses of antibiotics make a difference? *Infect. Control Hosp. Epidemiol.* **15**:484–487.

290. **Zhou, D., M. S. Mooseker, and J. E. Galan.** 1999. An invasion-associated *Salmonella* protein modulates the actin-bundling activity of plastin. *Proc. Natl. Acad. Sci. USA* **96**:10176–10181.

291. **Zhou, D., M. S. Mooseker, and J. E. Galan.** 1999. Role of the S. typhimurium actin-binding protein SipA in bacterial internalization. *Science* **283**:2092–2095.

292. **Zhou, D., W. D. Hardt, and J. E. Galan.** 1999. *Salmonella typhimurium* encodes a putative iron transport system within the centisome 63 pathogenicity island. *Infect. Immun.* **67**:1974–1981.

Food Microbiology: Fundamentals and Frontiers, 2nd Ed.
Edited by M. P. Doyle et al.
© 2001 ASM Press, Washington, D.C.

Irving Nachamkin

Campylobacter jejuni

9

CHARACTERISTICS OF THE ORGANISM

Campylobacter jejuni subsp. *jejuni* (hereafter referred to as *C. jejuni*) is one of many species and subspecies within the genus *Campylobacter*, family *Campylobacteraceae*, and has recently been reviewed (54, 94). During the 1980s there was an explosion of information published on these bacteria, as evident from the 1984 edition of *Bergey's Manual of Systematic Bacteriology* (86), published at a time when there were only eight species and subspecies within the genus *Campylobacter*. Since then, many investigators have become interested in studying the taxonomy and clinical importance of campylobacters and have greatly expanded the number of genera and species associated with this group of bacteria.

Campylobacter and *Arcobacter* are included in the family *Campylobacteraceae* (96–98). The family *Campylobacteraceae* includes, at present, 18 species and subspecies within the genus *Campylobacter* and four species in the genus *Arcobacter* (Table 9.1). *C. hyoilei*, associated with porcine proliferative enteritis and previously thought to be a distinct species (1), is now considered to be identical to *C. coli* (99). *Bacteroides gracilis* was reclassified as *C. gracilis* (95), and *Bacteroides ureolyticus* is thought to be closely related to the genus

Campylobacter; however, this classification is still uncertain. Some phenotypically unusual isolates of *C. gracilis* were recently reclassified into a new genus and species, *Sutterella wadsworthensis* (104). The classification of *C. sputorum* has been amended recently to include three biovars (65). Lawson et al. recently detected 16S rRNA sequences representing a new putative noncultivatable *Campylobacter* sp. from the stool samples of healthy individuals (41). Strains originally described as free-living *Campylobacter* are now known as *Sulfurospirillum* sp. An extensive description of the various species within the family *Campylobacteraceae* was recently given by Vandamme (94).

Campylobacter is the type genus within the family *Campylobacteraceae*. Organisms are curved, S-shaped, or spiral rods that are 0.2 to 0.9 μm wide and 0.5 to 5 μm long. They are gram-negative, non-spore-forming rods that may form spherical or coccoid bodies in old cultures or cultures exposed to air for prolonged periods. Organisms are motile by means of a single polar unsheathed flagellum at one or both ends. The various species are microaerobic with a respiratory type of metabolism. Some strains grow aerobically or anaerobically. An atmosphere containing increased hydrogen may be required

Irving Nachamkin, Department of Pathology & Laboratory Medicine, University of Pennsylvania School of Medicine, Philadelphia, PA 19104-4283.

Table 9.1 Reservoirs and disease-associated species in the family *Campylobacteraceae*[a]

Organism	Reservoirs	Disease, sequelae, or comments	
		Human	Animals
Campylobacter jejuni subsp. *jejuni*	Human, other mammals, birds	Diarrhea, systemic illness, Guillain-Barré syndrome	Diarrhea in primates
C. jejuni subsp. *doylei*	Unknown	Diarrhea	
C. fetus subsp. *fetus*	Cattle, sheep	Systemic illness, diarrhea	Abortion
C. fetus subsp. *venerealis*	Cattle		Infertility
C. coli	Pigs, birds	Diarrhea	
C. lari	Birds, dogs	Diarrhea	
C. upsaliensis	Domestic pets	Diarrhea	Diarrhea
C. hyointestinalis subsp. *hyointestinalis*	Cattle, pigs, hamsters, deer	Rare, proctitis, diarrhea	Proliferative enteritis
C. hyointestinalis subsp. *lawsonii*	Pigs		
C. mucosalis	Pigs	Rare, diarrhea	Proliferative enteritis
C. hyoilei	Pigs		
C. sputorum biovar sputorum	Humans	Oral cavity, abscesses	Genital tract of bulls, abortion in sheep
C. sputorum biovar paraureolyticus	Cattle	Diarrhea	
C. sputorum biovar faecalis	Cattle, sheep		Enteritis
C. concisus	Humans	Periodontal disease	
C. curvus	Humans	Periodontal disease	
C. rectus	Humans	Periodontal disease, pulmonary infections	
C. showae	Humans	Periodontal disease	
C. helveticus	Domestic pets		Diarrhea
C. gracilis	Humans	Infections of head, neck, other sites	
Arcobocter butzleri	Cattle, pigs	Diarrhea, other	Diarrhea, abortion
A. cryaerophilus	Cattle, sheep, pigs	Diarrhea, bacteremia	Abortion
A. skirrowii	Cattle, sheep, pigs		Abortion, diarrhea; isolated from genital tract of bulls
A. nitrofigilis	Plants		

[a] Data from references 84, 95, and 99.

by some species for microaerobic growth. *C. jejuni* and *C. coli* have genomes approximately 1.7 Mb in size, as determined by pulsed-field gel electrophoresis, which is about one-third the size of the *Escherichia coli* genome (91). The genome sequence of *C. jejuni* NCTC 11168 was recently completed and confirms the size of genome of 1,641,481 bp (30.6% G+C) (67).

Arcobacters are gram-negative, curved, S-shaped, or helical non-spore-forming rods that are 0.2 to 0.9 μm wide and 1 to 3 μm long. Organisms are motile with a single polar unsheathed flagellum. Arcobacters grow at 15, 25, and 30°C but have variable growth at 37 and 42°C. Organisms are microaerobic and do not require hydrogen for growth. Arcobacters may grow aerobically at 30°C and anaerobically at 35 to 37°C. Most strains are nonhemolytic. *Arcobacter skirrowii* may be alpha-hemolytic. Most strains are susceptible to nalidixic acid but variable in susceptibility to cephalothin.

ENVIRONMENTAL SUSCEPTIBILITY

C. jejuni is susceptible to a variety of environmental conditions that make it unlikely to survive for long periods of time outside the host. The organism does not grow at temperatures below 30°C, is microaerobic, and is sensitive to drying, high-oxygen conditions, and low pH (19). The decimal reduction time for campylobacters varies and depends upon the food source and temperature (Table 9.2). Thus, the organism should not survive in food products brought to adequate cooking temperatures. Organisms are susceptible to gamma radiation (1 kGy), but the rate of killing is dependent on the type of product being processed. Irradiation is less effective for frozen materials than for refrigerated or room temperature meats. Early-log-phase cells are more susceptible than cells grown to log or stationary phase (74). Recent studies suggest that *Campylobacter* spp. are more radiation sensitive than other foodborne pathogens, such as salmonellae and *Listeria monocytogenes*, and that

Table 9.2 *D* values for various food sources[a]

Source	Temp (°C)	*D* value range (min)
Skim milk	48	7.2–12.8
	55	0.74–1.0
Red meat	50	5.9–6.3
	60	<1
Ground chicken	49	20
	57	<1

[a] Data from reference 34.

Table 9.3 Isolation of *Campylobacter* sp. from various food sources[a]

Product	% Positive samples
Chicken	14–98
Turkey	3–25
Duck	48
Goose	38
Cow's milk	0–12.3
Goat's milk	0
Ewe's milk	0
Beef	0–23.6
Pork	1–23.5
Lamb	0–15.5
Sheep	3
Offal	47
Mussels	47–69
Oysters	6–27
Vegetables	<5

[a] Data from reference 34.

irradiation treatment that is effective against the latter organisms should be sufficient to kill *Campylobacter* spp. as well (68).

Disinfectants such as sodium hypochlorite (Clorox), *o*-phenylphenol (Amphyl), iodine-polyvinylpyrrolidine (Betadine), alkylbenzyl dimethylammonium chloride (Zephiran), glutaraldehyde (Cidex), formaldehyde, and ethanol have antibacterial activity at commonly used concentrations (100). Ascorbic acid at 0.05% inhibits the organism and at 0.09% is bactericidal (34).

Campylobacter spp. are susceptible to low pH and are killed readily at pH 2.3 (11). Campylobacters remain viable and multiply in bile at 37°C and survive better in feces, milk, water, and urine held at 4°C than in material held at 25°C. The maximum periods of viability of *Campylobacter* spp. at 4°C were 3 weeks in feces, 4 weeks in water, and 5 weeks in urine (11). Freezing reduces the cell populations of *Campylobacter* sp. in contaminated poultry, but even after freezing to −20°C, low levels of *Campylobacter* can be recovered (34).

RESERVOIRS AND FOODBORNE OUTBREAKS

C. jejuni is zoonotic, with many animals serving as reservoirs for human disease. Reservoirs for infection include rabbits, rodents, wild birds, sheep, horses, cows, pigs, poultry, and domestic pets (3, 19, 34, 89) (Table 9.3). Contaminated vegetables and shellfish may also be vehicles of infection (3, 19, 34). Campylobacters are frequently isolated from water, and water supplies have been sources of infection in some reported outbreaks. *C. jejuni* can remain dormant in water in a state that has been termed viable but nonculturable (80); that is, under unfavorable conditions, the organisms essentially remain dormant and cannot be easily recovered on artificial media. Under favorable conditions, campylobacters are able to multiply (47). The role of these forms as a source of infection for humans is not clear; Medema et al. (47) determined that viable but non-

culturable *C. jejuni* isolates were not able to colonize chicks.

From 1978 to 1996, there were 111 outbreaks of *Campylobacter* enteritis reported to the Centers for Disease Control and Prevention, affecting 9,913 individuals (24) (Table 9.4). The vehicles of *Campylobacter* outbreaks have changed over the past 2 decades. Water and unpasteurized milk were responsible for over half of

Table 9.4 Foodborne and waterborne outbreaks of *Campylobacter* infections reported in the United States, by vehicle, 1978 to 1996[a]

Origin	No. of outbreaks	No. of outbreak-associated cases
Foodborne		
Milk	30	1,212
Chicken	2	16
Turkey	1	11
Beef	1	24
Other meat	2	30
Eggs	1	26
Fruits	4	227
Other foods	4	251
Multiple foods	10	411
Unknown food	42	2,775
Waterborne		
Community water supply	8	5,068
Other water supply	4	104
Total	111	10,115

[a] Table from reference 24.

the outbreaks between 1978 and 1987, whereas other foods accounted for over 80% of outbreaks between 1988 and 1996 (24). In a 10-year review of outbreaks from 1981 to 1990, 20 outbreaks occurred that affected 1,013 individuals who drank raw milk. The attack rate was 45%. At least one outbreak occurred each year, and most of the outbreaks occurred in children who had gone on field trips to dairy farms (107).

The seasonal distribution of outbreaks is somewhat different from that of sporadic cases. Milk-borne and waterborne outbreaks tend to occur in the spring and fall but do not occur frequently in the summer months (24) (Fig. 9.1). Sporadic cases are usually most frequent during the summertime (24).

In contrast to the relatively low occurrence of outbreaks caused by campylobacters, campylobacters have been identified in many studies as among the most common causes of sporadic bacterial enteritis in the United States (13, 22, 24, 90). In 1995, the Centers for Disease Control and Prevention, the U.S. Department of Agriculture, and the U.S. Food and Drug Administration developed an active surveillance system for foodborne diseases, including *Campylobacter* sp., called the Foodborne Diseases Active Surveillance Network, also known as FoodNet. Using a variety of data, including FoodNet incidence data, U.S. Census Bureau data, and data from other sources, it is estimated that there are 2.4 million *Campylobacter* infections in the United States each year (24). This is similar to the incidence of sporadic infection in the United Kingdom and other developed nations. Sporadic cases occur more often during the summer months and usually follow ingestion of improperly handled or cooked food, primarily poultry products

Table 9.5 Food sources of sporadic *Campylobacter* infections[a]

Source	Estimated proportion of infections (%)
Poultry	10–70
Raw milk	5
Contact with pets	6–30
Drinking water	8
International travel	9

[a] Adapted from reference 24.

(24, 89). Other exposures include drinking raw milk, contaminated surface water, overseas travel, and contact with domestic pets (Table 9.5).

CHARACTERISTICS OF DISEASE
C. jejuni and *C. coli*
Most *Campylobacter* species are associated with lower gastrointestinal tract infection; however, extraintestinal infections are common with some species, such as *C. fetus* subsp. *fetus* (hereafter referred to as *C. fetus*), and sequelae of *Campylobacter* infection are being more frequently recognized. *C. jejuni* and *C. coli* have been recognized since the 1970s as agents of gastrointestinal tract infection. Campylobacters have been identified as common causes of sporadic bacterial enteritis in the United States (24).

C. jejuni and *C. coli* are the most common *Campylobacter* species associated with diarrheal illness and are clinically indistinguishable. The relative ratio of *C. jejuni* to *C. coli* is not known, because most laboratories do not routinely distinguish between these organisms. In the United States, an estimate of approximately 5 to 10% of cases attributed to *C. jejuni* are actually due to *C. coli*; this figure may be higher in other parts of the world (64).

A spectrum of illness can occur during *C. jejuni* or *C. coli* infection, and patients may be asymptomatic to severely ill. Symptoms and signs usually include fever, abdominal cramping, and diarrhea (with or without blood or fecal leukocytes) that last several days to more than 1 week. Symptomatic infections are usually self-limited, but relapses may occur in 5 to 10% of untreated patients. *Campylobacter* infection may mimic acute appendicitis and result in unnecessary surgery. Extraintestinal infections and sequelae do occur, including bacteremia, bursitis, urinary tract infection, meningitis, endocarditis, peritonitis, erythema nodosum, pancreatitis, abortion and neonatal sepsis, reactive arthritis, and Guillain-Barré syndrome (GBS). Deaths directly attributable

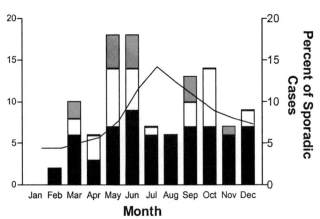

Figure 9.1 Outbreaks (gray bars, waterborne; open bars, milk borne; solid bars, other food) of *Campylobacter* infections, 1978 to 1996, and distribution of sporadic cases (line) by month, 1982 to 1995, in the United States. From reference 24.

to *C. jejuni* infection have been reported but rarely occur (10).

C. jejuni and *C. coli* are susceptible to a variety of antimicrobial agents, including macrolides, fluoroquinolones, aminoglycosides, chloramphenicol, and tetracycline. Erythromycin has been the drug of choice for treating *C. jejuni* gastrointestinal tract infections, but ciprofloxacin is a good alternative drug. Early therapy of *Campylobacter* infection with erythromycin or ciprofloxacin is effective in eliminating the campylobacters from stool and may also reduce the duration of symptoms associated with infection (10).

C. jejuni is generally susceptible to erythromycin, with resistance rates of less than 5%. Rates of erythromycin resistance in *C. coli* vary considerably, with up to 80% of strains having resistance in some studies. Although ciprofloxacin has been effective in treating *Campylobacter* infections, the emergence of fluoroquinolone resistance during therapy has been reported. Several in vitro studies indicate there is increasing resistance to fluoroquinolones (20, 76, 78, 88). While controversial, there is evidence that fluoroquinolone resistance in *Campylobacter* sp. is related to the use of these compounds in poultry. In a recent review on the subject, Smith et al. presented the following arguments (87):

1. Poultry is a major food vehicle of *Campylobacter* and is infrequently transmitted from person to person.
2. Experimental treatment of chickens with enrofloxacin does not eradicate the bacteria and selects for fluoroquinolone-resistant bacteria.
3. Fluoroquinolone use in animals is widespread throughout the world.
4. There is a direct relationship between the licensure of fluoroquinolone use in food animals and the subsequent increase in fluoroquinolone-resistant *Campylobacter* sp. in human infections.
5. Fluoroquinolone-resistant *Campylobacter* sp. can be isolated from retail poultry, and the prevalence of resistance roughly parallels the resistance rates observed in human infections.
6. In studies by Smith et al. (87), molecular typing using flagellin gene typing (58) identified fluoroquinolone-resistant *Campylobacter* sp. from human infections and poultry with the same molecular types.
7. Fluoroquinolone use in humans occurred well before the emergence of resistant strains in humans and cannot account for the rapid increase in resistance seen in human infections.

Other *Campylobacter* and *Arcobacter* Species

In contrast to *C. jejuni*, *C. fetus* is primarily associated with bacteremia and extraintestinal infections in patients with underlying diseases and may be associated with a poor outcome in some patients. *C. fetus* is also associated with septic abortions, septic arthritis, abscesses, meningitis, endocarditis, mycotic aneurysm, thrombophlebitis, peritonitis, and salpingitis. Although gastroenteritis does occur with this species, the incidence is probably underestimated, because the organism does not grow well at 42°C and is usually susceptible to cephalothin, an antimicrobial agent used in some common selective media for stool culture. *C. fetus* subsp. *venerealis* is not associated with human infection (54).

C. upsaliensis is a recently described thermotolerant species that is an important cause of diarrhea and bacteremia. *C. lari* (formerly *C. laridis*) is a thermophilic species isolated first from gulls (of the genus *Larus*) and subsequently from other avian species, dogs, cats, and chickens. *C. lari* has been isolated infrequently from humans with bacteremia and gastrointestinal and urinary tract infections (54). A waterborne outbreak of infection affecting over 100 individuals occurred in 1985 (14). Other *Campylobacter* species have been isolated from clinical specimens of patients with a variety of diseases, but their pathogenic role has not been determined (54). *C. jejuni* subsp. *doylei* is a nitrate-negative subspecies of *C. jejuni* rarely isolated from patients with upper gastrointestinal tract infections and gastroenteritis. *C. hyointestinalis* has been associated occasionally with proctitis and diarrhea in human infection. *C. concisus* is associated primarily with periodontal disease but has also been isolated from patients with bacteremia, foot ulcer, and upper and lower gastrointestinal tract infections. *C. sputorum* biovar sputorum and biovar bubulus have associated with lung, axillary, scrotal, and groin abscesses. *C. mucosalis* was isolated from two children with enteritis. *C. helveticus* is a newly described species recovered from domestic cats and dogs but not from humans. The species is phenotypically most closely related to *C. upsaliensis*. *C. rectus* is primarily isolated from patients with active periodontal infections but was also isolated from a patient with pulmonary infection. *C. showae* is a recently described species isolated from human gingival crevice that is somewhat distinctive in its morphologic characteristics from other *Campylobacter* species, appearing as straight rods and multiple flagella. *C. hyoilei* is associated with proliferative enteritis in pigs (1). *C. gracilis*, formerly considered an anaerobic bacterium, has been isolated from gingival crevices of humans and is involved in visceral, head, and neck infections (95).

Arcobacters are aerotolerant, *Campylobacter*-like organisms frequently isolated from bovine and porcine abortion and enteritis. Two of the four *Arcobacter* species have been associated with human infection. *A. butzleri* has been isolated from patients with bacteremia, endocarditis, peritonitis, and diarrhea. In addition, *A. butzleri* has been associated with diarrheal disease in monkeys (*Macaca mulatta*). *A. cryaerophilus* group 1B has been isolated from patients with bacteremia and diarrhea (54).

EPIDEMIOLOGIC SUBTYPING SYSTEMS USEFUL FOR INVESTIGATING FOODBORNE ILLNESSES

Many typing systems have been devised to study epidemiology of *Campylobacter* infections; such systems vary in complexity and ability to discriminate between strains. These methods include biotyping, serotyping, bacteriocin sensitivity, detection of preformed enzymes, auxotyping, lectin binding, phage typing, multilocus enzyme electrophoresis, and molecular-based methods such as pulsed-field gel electrophoresis, ribotyping, and restriction fragment length polymorphism combined with PCR methods (60, 69).

The most frequently used phenotypic systems are biotyping and serotyping (69). Several biotyping schemes that have been published can, on the basis of only a few biochemical tests, group *C. jejuni*, *C. coli*, and *C. lari* into major categories. While the discrimination of strains is low with such schemes, biotyping may be useful as a first step for epidemiologic investigation.

The two major serotyping schemes used worldwide detect heat-labile (45) and heat-stable (HS) (70) antigens. The heat-labile serotyping scheme originally described by Lior et al. (45) can detect over 100 serotypes of *C. jejuni*, *C. coli*, and *C. lari*. Uncharacterized bacterial surface antigens and, in some serotypes, flagella are the serodeterminants for this serotyping system (2). The more widely used Penner HS serotyping scheme (70) detects over 60 types of *C. jejuni* and *C. coli* (69). Previously thought to be based on detection of a high-molecular-weight O antigen, the HS serodeterminant was recently determined to be a capsular polysaccharide (36).

Both serotyping systems, while simple to perform, have good ability to discriminate between strains. These systems, however, are available in only a few reference laboratories because of the time and expense needed to maintain quality serotyping antisera. Several attempts have been made by commercial enterprises to market limited serotyping antisera for *Campylobacter* species; however, either too few antisera were included or the quality was poor (62).

During the past 5 years, molecular-based methods have come to the forefront for studying the epidemiology of *Campylobacter* infections and were recently reviewed by Newell and colleagues (60, 101). Pulsed-field gel electrophoresis is one of the most powerful methods for molecular analysis of *Campylobacter* species and has excellent discriminatory power. A variety of PCR-based methods, including flagellin gene typing, ribotyping, random amplified polymorphic DNA, amplified fragment length polymorphism, and multiplex PCR-restriction fragment length polymorphism have all been used in various applications to study *Campylobacter* sp.

INFECTIVE DOSE AND SUSCEPTIBLE POPULATIONS

C. jejuni is susceptible to low pH, and hence the gastric environment is sufficient to kill most campylobacters (11). The infective dose of *C. jejuni*, however, does not appear to be high, with <1,000 organisms being capable of causing illness (9). The only controlled study to determine the infective dose of *C. jejuni* was conducted by Black and colleagues (9). In a study using two different strains of *C. jejuni*, only 18% of human volunteers became ill when infected with 10^8 CFU of strain A3249; however, 46% of the volunteers became ill when infected with another strain, 81-176. In an interesting experiment, Robinson ingested 500 CFU of *C. jejuni* in milk, which resulted in abdominal cramps and nonbloody diarrhea occurring 4 days after ingestion and lasting 3 days (79). Although not studied directly, there appeared to be a dose-related effect on both rate of infection and severity of illness in individuals involved in an outbreak of *Campylobacter* infection after ingesting raw milk (12).

Young children and young adults, 20 to 40 years of age, have the highest incidence of sporadic infections in the United States (90). The incidence of *Campylobacter* infection in developing countries such as Mexico and Thailand may be orders of magnitude higher than in the United States (90, 92). In contrast to developed countries, campylobacters are frequently isolated from individuals who may or may not have diarrheal disease. Most symptomatic infections occur in infancy and early childhood, and the incidence decreases with age (16, 92, 93). Age-related increases in humoral immune responses to *Campylobacter* antigens are associated with a decrease in symptomatic illness (46, 93). Travelers to developing countries may acquire *Campylobacter* infection, with isolation rates from 0 to 39% reported in different studies (92).

Bacteremia reportedly occurs at a rate of 1.5 per 1,000 intestinal infections, with the highest rate in the elderly (85). Persistent diarrheal illness and bacteremia may occur in immunocompromised hosts, such as in patients with human immunodeficiency virus infection or hypogammaglobulinemia, and are difficult to treat (10).

VIRULENCE FACTORS AND MECHANISMS OF PATHOGENCITY

Little is known about the mechanism by which *C. jejuni* causes human disease. *C. jejuni* can cause an enterotoxigeniclike illness with loose or watery diarrhea or an inflammatory colitis with fever and the presence of fecal blood and leukocytes and occasionally bacteremia that suggests an invasive mechanism of disease. A major problem in elucidating the pathogenesis of *Campylobacter* infection has been the lack of suitable animal models (108).

Cell Association and Invasion

The interaction of *Campylobacter* sp. with eukaryotic cells has recently been reviewed (31, 38). A proposed model of the initial interaction of *Campylobacter* sp. with cells was recently described by Konkel et al. (38) as follows.

Uptake of *C. jejuni* by host cells is mediated by intimate binding with the cell surface, using multiple adhesins, including CadF and PEB1. Binding of *Campylobacter* sp. to fibronectin leads to aggregation of host cell integrin receptors to initiate a cascade of host cell signaling events. Following contact, *C. jejuni* synthesizes a minimum of 14 new proteins, of which synthesis corresponded to a rapid increase in internalization and rearrangement of the host cell cytoskeletal components. A subset of newly synthesized proteins is secreted and translocated into the cytoplasm of target cells, augmenting the host cell signaling events required for *C. jejuni* uptake. Following binding and internalization, *Campylobacter*-infected cells release inflammatory cytokines, including interleukin-8, that promote the recruitment of lymphocytes and professional phagocytic cells to the site of the infection. Induction of apoptosis by *C. jejuni* in host cells may promote the survival and transmission of the pathogen.

Flagella and Motility

Campylobacter species are motile and have a single polar, unsheathed flagellum at one or both ends. Motility and flagella are to be important determi-

nants for the invasion-translocation process (27, 103). *Campylobacter* colonization and/or infection in a variety of animal models are dependent on intact motility and full-length flagella (59); however, other colonization factors may be involved (49). Two genes, *flaA* and *flaB*, are involved in the expression of the flagellar filament and are arranged in tandem in both *C. jejuni* and *C. coli* (23, 28, 63). Motility and *flaA* are essential for colonization (27, 59, 103). Components of intestinal mucin, particularly L-fucose, are chemotactic for *C. jejuni*, and motility toward these components may be important in the pathogenesis of infection (32). The structure, function, and role of *Campylobacter* flagella in immunity to infection were recently reviewed by Guerry et al. (29).

Toxins

C. jejuni has been reported to produce a choleralike enterotoxin (18, 42, 82); however, other studies have refuted the significance of these findings (39, 71). Daikoku et al. (18) reported the isolation of a toxin with cholera toxin-like properties, having three subunits with molecular masses of 68, 54, and 43 kDa. The toxin enhanced adenylate cyclase activity in HeLa cells, and the 68-kDa protein had immunologic cross-reactivity with cholera toxin. Further, it was reported that both the 68- and 54-kDa proteins had putative ganglioside binding activity (18). Small amounts of this toxin were detected in strains studied using a rabbit ileal loop model. These strains produced mucosal hemorrhage, inflammation with a polymorphonuclear leukocyte infiltrate, tissue edema, cell damage, and submucosal bleeding. Cholera toxin activity was not detected in the tissues of infected animals (21).

The lack of genetic evidence for the presence of this putative toxin, as determined by hybridization with probes directed against the A and B subunits of cholera toxin and *E. coli* heat-labile toxin genes (7, 66) or by low-stringency hybridization (15), raises serious doubts about the existence of a classical enterotoxin. *C. jejuni* infection of rabbit ileal loops caused an elevation of cyclic AMP, prostaglandin E_2, and leukotriene B_4 levels in tissue and fluids. This fluid caused elevated cellular cyclic AMP in CaCo-2 cells and was inhibited by antiserum against prostaglandin E_2 (15). This finding suggests that inflammatory mediators elicited by *C. jejuni* could be involved in the pathogenesis of infection.

Investigations on the cytolethal distending toxin (CDT) have received increased attention over the past few years, and the biology of this toxin was recently reviewed by Pickett (72). First reported by Johnson and Lior in 1988 (35), the role of CDT in pathogenesis is unknown, but the toxin was reported to be active in CHO, Vero, HeLa, and HEp-2 cells but not in Y-1 cells. Forty-one percent of over 700 strains produced the toxin, which produced a hemorrhagic response in rat ligated ileal loops. *cdt* genes were characterized by Pickett et al. (73), and three adjacent genes, *cdtA*, *cdtB*, and *cdtC*, were determined to encode this toxin and were similar to *E. coli* CDT proteins. Most strains of *C. jejuni* have *cdt* genes, but the amount of toxin produced by strains assayed in vitro varies (72). CDT appears to cause a block in the G_2 phase of the cell cycle in eukaryotic cells (105).

Cover et al. (17) studied the presence of cytotoxic activity in fecal filtrates of patients with *Campylobacter* enteritis. Cytotoxic activity was detected in both patients with disease and healthy asymptomatic subjects, raising doubts about the clinical relevance of the toxin and its role in pathogenicity (19). The production of a hepatotoxin described by Kita et al. (37) is interesting in that a strain of *C. jejuni* (GIFU 8734) produced hepatitis in mice, and specific toxin activity could be isolated from the bacteria. Other toxins such as Shiga toxins (52) and hepatoxins (37) have been described. Several recent reviews on *Campylobacter* toxins have been published (72, 102).

The completion of the genome sequence of *C. jejuni* NCTC 11168 by Parkhill and colleagues has enabled a reexamination of the presence of putative toxin genes (67). Genes *cdtA-C* were identified in the genome sequence; however, choleralike toxin genes were notably absent (67). Genes with similarity to contact-dependent hemolysins present in pathogenic *Serpulina* and *Mycobacterium* species, integral membrane protein with a hemolysin domain and phospholipase (*pldA*), also were identified in NCTC 11168 (67). Genes for other putative toxins described in *Campylobacter* sp. were not detected.

Other Factors

Environmentally regulated gene expression is becoming an important area of research in *Campylobacter* sp. Campylobacters are microaerobic, and *C. jejuni* has a higher temperature for optimal growth (42°C) than other bacterial pathogens. Understanding the impact of environmental signals on the growth, metabolism, and pathogenicity of *Campylobacter* sp. will have a major impact on the ability to control campylobacters in the environment and food chain. Such pathways include response to iron, oxidative stress, temperature regulation, including cold and heat shock responses, and starvation. The environmental regulatory pathways of *Campylobacter* sp. were recently reviewed by Park (66).

Autoimmune Sequelae

It is now clearly established that *Campylobacter* infection is a major trigger of GBS, an acute, immune-mediated paralytic disorder affecting the peripheral nervous system (55).

Mishu and Blaser (50) estimated that the annual incidence of GBS preceded by *C. jejuni* infection ranges from 0.17 to 0.51 cases per 100,000 population and accounts for 425 to 1,272 cases per year in the United States. The pathogenesis of GBS induced by *C. jejuni* is not clear, but molecular mimicry between bacterial lipopolysaccharide and relevant target epitopes in peripheral nerve tissue is likely a major mechanism of *Campylobacter*-induced GBS (55).

Campylobacter infection has been recognized as the most common identifiable event preceding GBS and has been estimated to occur in up to 40% of GBS patients (33). Although serologic and culture studies revealed that some patients with GBS had evidence of infection, an important study by Kuroki and colleagues solidified the association of *Campylobacter* and GBS (40). In a study of Japanese patients, Kuroki et al. isolated *C. jejuni* from 14 of 46 GBS patients (30.4%), compared with only 6 (1.2%) of 503 in a healthy control population. Using HS serotyping to characterize the isolates, 10 of 12 available isolates had the same HS serotype, HS:19. This serotype, however, only occurred in 1.7% of 1,150 *C. jejuni* isolates from patients with uncomplicated gastrointestinal infection. A recent analysis of 31 strains of *C. jejuni* from Japanese GBS patients by Yuki et al. (110) also revealed that 52% of strains were HS:19, but this serotype occurred in only 5% of 215 strains from patients with uncomplicated gastroenteritis. In isolates from patients with Miller-Fisher syndrome, a less common form of GBS, HS:2 strains were overrepresented compared with control isolates (71 versus 38%), although only seven patients were studied.

Aspinall et al. (4, 5) and Yuki et al. (109, 111) have revealed that strains of serotype HS:19 as well as certain other serotypes (HS:4, HS:1) of *Campylobacter* sp. have core oligosaccharide lipopolysacchamide structures that mimic ganglioside structures such as GM1 and GD1a, a component of motor neurons. Ganglioside epitopes are produced by strains involved in GBS as well as by enteritis-associated isolates (57). A variety of anti-ganglioside antibodies are produced by patients with GBS, and these antibodies likely play an important role in the pathogenesis of the disease (106).

The genetics of ganglioside epitope expression in *C. jejuni* is only now being understood. *C. jejuni* produces a number of sugar transferases and sialyltransferases involved in sialyation of the outer core lipo-oligosaccharide (25, 26, 43, 44, 53). Host genetic factors such HLA type clearly play an important role in the development of disease (51, 77, 109).

IMMUNITY

Protective immunity appears after infection with *Campylobacter* species and is likely antibody mediated. Black et al. (9) determined that rechallenge of homologous strains 28 days after the initial volunteer challenge resulted in protection against illness but not necessarily against colonization by *C. jejuni*. Blaser et al. (12) revealed that 76% of acutely exposed individuals who had not been exposed previously to raw milk became acutely ill, compared with none of 10 individuals who were regular milk drinkers and who drank the implicated milk. Humoral immunity is likely to be an important component of protective immunity, as suggested by studies on persistent infection in immunocompromised hosts with human immunodeficiency virus infection or hypogammaglobulinemia (10).

In developing countries where *Campylobacter* infections are endemic, immunity to *Campylobacter* infection appears to be age dependent. In a cohort of Mexican children studied by Calva et al. (16), the ratio of symptomatic to asymptomatic infection decreased from infancy to 6 years of age. An inverse relationship between serum antiflagellin antibodies and diarrheal illness in another cohort of children also supports the role of antibodies in protective immunity (46). Production of cytokines during intestinal infection may play some role in immunity and immune responses. Using a mouse colonization model, Baqar et al. (8) determined that oral administration of interleukin-5 and interleukin-6 reduced the level of gut colonization with *C. jejuni*.

Flagellin is an important immunogen during *Campylobacter* infection (29), and antibodies against this protein correlate to some degree with protective immunity. Breast feeding provides protection against *Campylobacter* infection in developing countries (46, 48, 56, 81). Antibodies against flagellin present in breast milk appear to be associated with protection of infants against infection (56).

Several researchers are working toward developing a vaccine for *Campylobacter* infection. Baqar et al. (7) tested an oral whole-cell killed vaccine coadministered with *E. coli* heat-labile enterotoxin as an immunoadjuvant to rhesus monkeys. The vaccine elicited both humoral and cellular responses; however, the ability to protect against infection was not studied. Killed whole-cell vaccine provided colonization protection in a mouse model (6). Guerry et al. (30) described the production of a *recA* mutant of *C. jejuni* 81-176 that colonized rabbits and induced colonization protection. Whether this strain was attenuated is not known, but *recA* mutants of other bacteria, such as *Salmonella enterica* serovar Typhimurium, are avirulent (30). Strategies for human vaccine development were recently reviewed by Scott and Tribble (83).

Other strategies to prevent the transmission of infection to humans include improved hygiene practices during broiler production, such as decontamination of water supplies, use of competitive exclusion flora that may prevent *C. jejuni* colonization of young chicks, and immunological approaches through the use of animal vaccines (61).

CONCLUSIONS

Campylobacter species are among the most important human bacterial enteric pathogens, yet little is known about how these intriguing bacteria cause disease. Difficult challenges remain in identifying suitable in vitro and in vivo models of infection that will enable investigators to study *Campylobacter* species at both the biological and genetic levels. Elucidation of the genetic basis for pathogenesis of *Campylobacter* infection is still in the embryonic stage; however, with the recent completion of the *C. jejuni* genome project, a rich source of information is now available to study this intriguing pathogen. Finally, *Campylobacter* sp. has received renewed attention by governmental agencies that has already led to infusion of resources for research in the future (75).

References

1. **Alderton, M. R., V. Korolik, P. J. Coloe, F. E. Dewhirst, and B. J. Paster.** 1995. *Campylobacter hyoilei* sp. nov., associated with porcine proliferative enteritis. *Int. J. Syst. Bacteriol.* 45:61–66.

2. **Alm, R. A., P. Guerry, M. E. Power, H. Lior, and T. J. Trust.** 1991. Analysis of the role of flagella in the heat-labile Lior serotyping scheme of thermophilic campylobacters by mutant allele exchange. *J. Clin. Microbiol.* 29:2438–2445.

3. **Altekruse, S. F., J. M. Hunt, L. K. Tollefson, and J. M. Madden.** 1994. Food and animal sources of human *Campylobacter jejuni* infection. *J. Am. Vet. Med. Assoc.* 204:57–61.

4. **Aspinall, G. O., A. G. McDonald, H. Pang, L. A. Kurjanczyk, and J. L. Penner.** 1994. Lipopolysaccharides of *Campylobacter jejuni* serotype O:19: structures of core oligosaccharide regions from the serostrain and two

bacterial isolates from patients with the Guillain-Barré syndrome. *Biochemistry* **33**:241–249.

5. **Aspinall, G. O., A. G. McDonald, T. S. Raju, H. Pang, L. A. Kurjanczyk, J. L. Penner, and A. P. Moran.** 1993. Chemical structure of the core region of *Campylobacter jejuni* serotype O:2 lipopolysaccharide. *Eur. J. Biochem.* **213**:1029–1037.

6. **Baqar, S., L. A. Applebee, and A. L. Bourgeois.** 1995. Immunogenicity and protective efficacy of a prototype *Campylobacter* killed whole-cell vaccine in mice. *Infect. Immun.* **63**:3731–3735.

7. **Baqar, S., A. L. Bourgeois, P. J. Schultheiss, R. I. Walker, D. M. Rollins, R. L. Haberberger, and O. R. Pavlovskis.** 1995. Safety and immunogenicity of a prototype oral whole-cell killed *Campylobacter* vaccine administered with a mucosal adjuvant in non-human primates. *Vaccine* **13**:22–28.

8. **Baqar, S., N. D. Pacheco, and F. M. Rollwagen.** 1993. Modulation of mucosal immunity against *Campylobacter jejuni* by orally administered cytokines. *Antimicrob. Agents Chemother.* **37**:2688–2692.

9. **Black, R. E., M. M. Levine, M. L. Clements, T. P. Hughs, and M. J. Blaser.** 1988. Experimental *Campylobacter jejuni* infections in humans. *J. Infect. Dis.* **157**:472–480.

10. **Blaser, M. J.** 2000. *Campylobacter jejuni* and related species, p. 2276–2285. *In* G. L. Mandell, J. E. Bennett, and R. Dolin (ed.), *Principles and Practice of Infectious Diseases.* Churchill Livingstone, Philadelphia, Pa.

11. **Blaser, M. J., H. L. Hardesty, B. Powers, and W. L. Wang.** 1980. Survival of *Campylobacter fetus* subsp. *jejuni* in biological milieus. *J. Clin. Microbiol.* **11**:309–313.

12. **Blaser, M. J., E. Sazie, and P. Williams.** 1987. The influence of immunity on raw milk-associated *Campylobacter* infection. *JAMA* **257**:43–46.

13. **Borczyk, A., S. C. Rosa, and H. Lior.** 1991. Enhanced recognition of *Campylobacter cryaerophila* in clinical and environmental specimens, abstr. C-267, p. 386. *In Abstr. 91st Gen. Meet. Am. Soc. Microbiol. 1991.* American Society for Microbiology, Washington, D.C.

14. **Borczyk, A., S. Thompson, D. Smith, and H. Lior.** 1987. Water-borne outbreak of *Campylobacter laridis*-associated gastroenteritis. *Lancet* **i**:164–165.

15. **Calva, E., J. Torres, M. Vazquez, V. Angeles, H. de la Vega, and G. M. Ruiz-Palacios.** 1989. *Campylobacter jejuni* chromosomal sequences that hybridize to *Vibrio cholerae* and *Escherichia coli* LT enterotoxin genes. *Gene* **75**:243–251.

16. **Calva, J. J., G. M. Ruiz-Palacios, A. B. Lopez-Vidal, A. Ramos, and R. Bojalil.** 1988. Cohort study of intestinal infection with *Campylobacter* in Mexican children. *Lancet* **i**:503–505.

17. **Cover, T. L., G. I. Perez-Perez, and M. J. Blaser.** 1990. Evaluation of cytotoxic activity in fecal filtrates from patients with *Campylobacter jejuni* or *Campylobacter coli* enteritis. *FEMS Microbiol. Lett.* **58**:301–304.

18. **Daikoku, T., M. Kawaguchi, K. Takama, and S. Susuki.** 1990. Partial purification and characterization of the enterotoxin produced by *Campylobacter jejuni*. *Infect. Immun.* **58**:2414–2419.

19. **Doyle, M. P., and D. M. Jones.** 1992. Food-borne transmission and antibiotic resistance of *Campylobacter jejuni*, p. 45–48. *In* I. Nachamkin, M. J. Blaser, and L. S. Tompkins (ed.), *Campylobacter jejuni: Current Status and Future Trends.* American Society for Microbiology, Washington, D.C.

20. **Everest, P. H., A. T. Cole, C. J. Hawkey, S. Knutton, H. Goossens, J. P. Butzler, J. M. Ketley, and P. H. Williams.** 1993. Role of leukotriene B4, prostaglandin E2, and cyclic AMP in *Campylobacter jejuni* induced intestinal fluid secretion. *Infect. Immun.* **61**:4885–4887.

21. **Everest, P. H., H. Goossens, P. Sibbons, D. R. Lloyd, S. Knutton, R. Leece, J. M. Ketley, and P. H. Williams.** 1993. Pathological changes in the rabbit ileal loop model caused by *Campylobacter jejuni* from human colitis. *J. Med. Microbiol.* **38**:316–321.

22. **Finch, M. J., and L. W. Riley.** 1984. *Campylobacter* infections in the United States: results of an 11-state surveillance. *Arch. Intern. Med.* **144**:1610–1612.

23. **Fischer, S. H., and I. Nachamkin.** 1991. Common and variable domains of the flagellin gene, flaA, in *Campylobacter jejuni*. *Mol. Microbiol.* **5**:1151–1158.

24. **Friedman, C. R., J. Neimann, H. C. Wegener, and R. V. Tauxe.** 2000. Epidemiology of *Campylobacter jejuni* infections in the United States and other industrialized nations, p. 121–138. *In* I. Nachamkin and M. J. Blaser (ed.), *Campylobacter*, 2nd ed. ASM Press, Washington, D.C.

25. **Fry, B. N., N. J. Oldfield, V. Korolik, P. J. Coloe, and J. M. Ketley.** 2000. Genetics of *Campylobacter* lipopolysaccharide biosynthesis, p. 381–403. *In* I. Nachamkin and M. J. Blaser (ed.), *Campylobacter*, 2nd ed. ASM Press, Washington, D.C.

26. **Gilbert, M., J. R. Brisson, M. F. Karwaski, J. Michniewicz, A. M. Cunningham, Y. Wu, N. M. Young, and W. W. Wakarchuk.** 2000. Biosynthesis of ganglioside mimics in *Campylobacter jejuni* OH4384. *J. Biol. Chem.* **275**:3896–3906.

27. **Grant, C. C. R., M. E. Konkel, W. Cieplak, and L. S. Tompkins.** 1993. Role of flagella in adherence, internalization, and translocation of *Campylobacter jejuni* in nonpolarized and polarized epithelial cells. *Infect. Immun.* **61**:1764–1771.

28. **Guerry, P., R. A. Alm, M. E. Power, and T. J. Trust.** 1992. Molecular and structural analysis of *Campylobacter* flagellin, p. 267–281. *In* I. Nachamkin, M. J. Blaser, and L. S. Tompkins (ed.), *Campylobacter jejuni: Current Status and Future Trends.* American Society for Microbiology, Washington, D.C.

29. **Guerry, P., R. A. Alm, C. Szymanski, and T. J. Trust.** 2000. Structure, function and antigenicity of *Campylobacter* flagella, p. 405–421. *In* I. Nachamkin and M. J. Blaser (ed.), *Campylobacter*, 2nd ed. ASM Press, Washington, D.C.

30. **Guerry, P., P. M. Pope, D. H. Burr, J. Leifer, S. W. Joseph, and A. L. Bourgeois.** 1994. Development and characterization of recA mutants of *Campylobacter jejuni* for inclusion in vaccines. *Infect. Immun.* **62**:426–432.

31. **Hu, L., and D. J. Kopecko.** 2000. Interactions of *Campylobacter* with eukaryotic cells: gut luminal colonization and mucosal invasion mechanisms, p. 191–215. *In* I. Nachamkin and M. J. Blaser (ed.), *Campylobacter*, 2nd ed. ASM Press, Washington, D.C.

32. **Hugdahl, M. B., J. T. Beery, and M. P. Doyle.** 1988. Chemotactic behavior of *Campylobacter jejuni*. *Infect. Immun.* **56:**1560–1566.

33. **Hughes, R. A. C., and J. H. Rees.** 1997. Clinical and epidemiologic features of Guillain-Barré syndrome. *J. Infect. Dis.* **176**(Suppl. 2)**:**92–98.

34. **Jacobs-Reitsma, W.** 2000. *Campylobacter* in the food supply, p. 467–481. *In* I. Nachamkin and M. J. Blaser (ed.), *Campylobacter*, 2nd ed. ASM Press, Washington, D.C.

35. **Johnson, W. M., and H. Lior.** 1988. A new heat-labile cytolethal distending toxin (CLDT) produced by *Campylobacter* spp. *Microb. Pathog.* **4:**115–126.

36. **Karlyshev, A. V., D. Linton, N. A. Gregson, A. J. Lastovica, and B. W. Wren.** 2000. Genetic and biochemical evidence of a *Campylobacter jejuni* capsular polysaccharide that accounts for Penner serotype specificity. *Mol. Microbiol.* **35:**529–541.

37. **Kita, E., D. Oku, A. Hamuro, F. Nishikawa, M. Emoto, Y. Yagyu, N. Katsui, and S. Kashiba.** 1990. Hepatotoxic activity of *Campylobacter jejuni*. *J. Med. Microbiol.* **33:**171–182.

38. **Konkel, M. E., L. A. Joens, and P. F. Mixter.** 2000. Molecular characterization of *Campylobacter jejuni* virulence determinants, p. 217–240. *In* I. Nachamkin and M. J. Blaser (ed.), *Campylobacter*, 2nd ed. ASM Press, Washington, D.C.

39. **Konkel, M. E., Y. Lobet, and W. Cieplak.** 1992. Examination of multiple isolates of *Campylobacter jejuni* for evidence of cholera toxin-like activity, p. 193–198. *In* I. Nachamkin, M. J. Blaser, and L. S. Tompkins (ed.), *Campylobacter jejuni: Current Status and Future Trends.* American Society for Microbiology, Washington, D.C.

40. **Kuroki, S., T. Saida, M. Nukina, T. Haruta, M. Yoshioka, Y. Kobayashi, and H. Nakanishi.** 1993. *Campylobacter jejuni* strains from patients with Guillain-Barre syndrome belong mostly to Penner serogroup 19 and contain β-N-acetylglucosamine residues. *Ann. Neurol.* **33:**243–247.

41. **Lawson, A. J., D. Linton, and J. Stanley.** 1998. 16S rRNA gene sequences of "*Candidatus* Campylobacter hominis," a novel uncultivated species, are found in the gastrointestinal tract of healthy humans. *Microbiology* **144:**2063–2071.

42. **Lindblom, G. B., M. Johny, K. Khalil, K. Mazhar, G. M. Ruiz-Palacios, and B. Kaijser.** 1990. Enterotoxigenicity and frequency of *Campylobacter jejuni*, *C. coli* and *C. laridis* in human and animal stool isolates from different countries. *FEMS Microbiol. Lett.* **54:**163–167.

43. **Linton, D., M. Gilbert, P. G. Hitchen, A. Dell, H. R. Morris, W. W. Wakarchuk, N. A. Gregson, and B. W. Wren.** 2000. Phase variation of a β-1,3 galactosyltransferase involved in generation of the ganglioside GM1-like lipo-

oligosaccharide of *Campylobacter jejuni*. *Mol. Microbiol.* **37:**501–514.

44. **Linton, D., A. V. Karlyshev, P. G. Hitchen, H. R. Morris, A. Dell, N. A. Gregson, and B. W. Wren.** 2000. Multiple N-acetyl neuramic acid synthetase (*neuB*) genes in *Campylobacter jejuni*: identification and characterization of the gene involved in sialyation of lipo-oligosaccharide. *Mol. Microbiol.* **35:**1120–1134.

45. **Lior, H., D. L. Woodward, J. A. Edgar, L. J. Laroche, and P. Gill.** 1982. Serotyping of *Campylobacter jejuni* by slide agglutination based on heat-labile antigenic factors. *J. Clin. Microbiol.* **15:**761–768.

46. **Martin, P. M. V., J. Mathiot, J. Ipero, M. Kirimat, A. J. Georges, and M. C. Georges-Courbot.** 1989. Immune response to *Campylobacter jejuni* and *Campylobacter coli* in a cohort of children from birth to 2 years of age. *Infect. Immun.* **57:**2542–2546.

47. **Medema, G. J., F. M. Schets, A. W. van de Giessen, and A. H. Gavelaar.** 1992. Lack of colonization of 1 day old chicks by viable, non-culturable *Campylobacter jejuni*. *J. Appl. Bacteriol.* **72:**512–516.

48. **Megraud, F., G. Boudraa, K. Bessaoud, S. Bensid, F. Dabis, R. Soltana, and M. Touhami.** 1990. Incidence of *Campylobacter* infection in infants in western Algeria and the possible protective role of breast feeding. *Epidemiol. Infect.* **105:**73–78.

49. **Meinersmann, R. J., W. E. Rigsby, N. J. Stern, L. C. Kelley, J. E. Hill, and M. P. Doyle.** 1991. Comparative study of colonizing and noncolonizing *Campylobacter jejuni*. *Am. J. Vet. Res.* **52:**1518–1522.

50. **Mishu, B., and M. J. Blaser.** 1993. Role of infection due to *Campylobacter jejuni* in the initiation of Guillain-Barré syndrome. *Clin. Infect. Dis.* **17:**104–108.

51. **Monos, D. S., M. Papaioakim, T. W. Ho, C. Y. Li, and G. M. McKhann.** 1997. Differential distribution of HLA alleles in two forms of Guillain-Barré syndrome. *J. Infect. Dis.* **176**(Suppl.)**:**180–182.

52. **Moore, M. A., M. J. Blaser, G. I. Perez-Perez, and A. D. O'Brien.** 1988. Production of Shiga-like cytotoxin by *Campylobacter*. *Microb. Pathog.* **4:**455–462.

53. **Moran, A. P., J. L. Penner, and G. O. Aspinall.** 2000. *Campylobacter* lipopolysaccharides, p. 241–257. *In* I. Nachamkin and M. J. Blaser (ed.), *Campylobacter*, 2nd ed. ASM Press, Washington, D.C.

54. **Nachamkin, I.** 1999. *Campylobacter* and *Arcobacter*, p. 716–726. *In* P. R. Murray, E. J. Baron, M. A. Pfaller, F. C. Tenover, and R. H. Yolken (ed.), *Manual of Clinical Microbiology*, 7th ed. ASM Press, Washington, D.C.

55. **Nachamkin, I., B. M. Allos, and T. W. Ho.** 1998. *Campylobacter* and Guillain-Barré syndrome. *Clin. Microbiol. Rev.* **11:**555–567.

56. **Nachamkin, I., S. H. Fischer, X. H. Yang, O. Benitez, and A. Cravioto.** 1994. Immunoglobulin A antibodies directed against *Campylobacter jejuni* flagellin present in breast-milk. *Epidemiol. Infect.* **112:**359–365.

57. **Nachamkin, I., H. Ung, A. P. Moran, D. Yoo, M. M. Prendergast, M. A. Nicholson, K. Sheikh, T. W. Ho, A. K. Asbury, G. M. McKhann, and J. W. Griffin.** 1999.

Ganglioside GM1 mimicry in *Campylobacter* strains from sporadic infections in the United States. *J. Infect. Dis.* **179**:1183–1189.

58. Nachamkin, I., H. Ung, and C. M. Patton. 1996. Analysis of HL and O serotypes of *Campylobacter* strains by the flagellin gene typing system. *J. Clin. Microbiol.* **34**:277–281.

59. Nachamkin, I., X. H. Yang, and N. J. Stern. 1993. Role of *Campylobacter jejuni* flagella as colonization factors for three-day-old chicks: analysis with flagellar mutants. *Appl. Environ. Microbiol.* **59**:1269–1273.

60. Newell, D. G., J. A. Frost, B. Duim, J. A. Wagenaar, R. H. Madden, J. van der Plas, and S. L. W. On. 2000. New developments in the subtyping of *Campylobacter* species, p. 27–44. *In* I. Nachamkin and M. J. Blaser (ed.), *Campylobacter*, 2nd ed. ASM Press, Washington, D.C.

61. Newell, D. G., and J. A. Wagenaar. 2000. Poultry infections and their control at the farm level, p. 497–509. *In* I. Nachamkin and M. J. Blaser (ed.), *Campylobacter*, 2nd ed. ASM Press, Washington, D.C.

62. Nicholson, M. A., and C. M. Patton. 1993. Evaluation of commercial antisera for serotyping heat-labile antigens of *Campylobacter jejuni* and *Campylobacter coli*. *J. Clin. Microbiol.* **31**:900–903.

63. Nuijten, P. J., C. Bartels, N. M. Bleumink-Pluym, W. Gaastra, and B. A. van der Zeijst. 1990. Size and physical map of the *Campylobacter jejuni* chromosome. *Nucleic Acids Res.* **18**:6211–6214.

64. Oberhelman, R. A., and D. N. Taylor. 2000. *Campylobacter* infections in developing countries, p. 139–153. *In* I. Nachamkin and M. J. Blaser (ed.), *Campylobacter*, 2nd ed. ASM Press, Washington, D.C.

65. On, S. L. W., H. I. Atabay, J. E. L. Corry, C. S. Harrington, and P. Vandamme. 1998. Emended description of *Campylobacter sputorum* and revision of its infrasubspecific (biovar) divisions, including *C. sputorum* biovar paraureolyticus, a urease-producing variant from cattle and humans. *Int. J. Syst. Bacteriol.* **48**:195–206.

66. Park, S. F. 2000. Environmental regulatory genes, p. 423–440. *In* I. Nachamkin and M. J. Blaser (ed.), *Campylobacter*, 2nd ed. ASM Press, Washington, D.C.

67. Parkhill, J., B. W. Wren, K. Mungall, J. M. Ketley, C. Churcher, D. Basham, T. Chillingworth, R. M. Davies, T. Feltwell, S. Holroyd, K. Jagels, A. V. Karlyshev, S. Moule, M. J. Pallen, C. W. Penn, M. A. Quail, M. A. Rajandream, K. M. Rutherford, A. H. M. van Vliet, S. Whitehead, and B. G. Barrell. 2000. The genome sequence of the food-borne pathogen *Campylobacter jejuni* reveals hypervariable sequences. *Nature* **403**:665–668.

68. Patterson, M. F. 1995. Sensitivity of *Campylobacter* spp. to irradiation in poultry meat. *Lett. Appl. Microbiol.* **20**:338–340.

69. Patton, C. M., and I. K. Wachsmuth. 1992. Typing schemes: are current methods useful?, p. 110–128. *In* I. Nachamkin, M. J. Blaser, and L. S. Tompkins (ed.), *Campylobacter jejuni: Current Status and Future Trends.* American Society for Microbiology, Washington, D.C.

70. Penner, J. L., and J. N. Hennessy. 1980. Passive hemagglutination technique for serotyping *Campylobacter fetus* subsp. *jejuni* on the basis of soluble heat-stable antigens. *J. Clin. Microbiol.* **12**:732–737.

71. Perez-Perez, G. I., D. N. Taylor, P. D. Echeverria, and M. J. Blaser. 1992. Lack of evidence of enterotoxin involvement in pathogenesis of *Campylobacter* diarrhea, p. 184–192. *In* I. Nachamkin, M. J. Blaser, and L. S. Tompkins (ed.), *Campylobacter jejuni: Current Status and Future Trends.* American Society for Microbiology, Washington, D.C.

72. Pickett, C. L. 2000. *Campylobacter* toxins and their role in pathogenesis, p. 179–190. *In* I. Nachamkin and M. J. Blaser (ed.), *Campylobacter*, 2nd ed. ASM Press, Washington, D.C.

73. Pickett, C. L., E. C. Pesci, D. L. Cottle, G. Russell, N. Erdem, and H. Zeytin. 1996. Prevalence of cytolethal distending toxin production in *Campylobacter jejuni* and relatedness of *Campylobacter cdtB* genes. *Infect. Immun.* **64**:2070–2078.

74. Radomyski, T., E. A. Murano, D. G. Olson, and P. S. Murano. 1994. Elimination of pathogens of significance in food by low-dose irradiation: a review. *J. Food Prot.* **57**:73–86.

75. Ransom, G. M., B. Kaplan, A. M. McNamara, and I. Wachsmuth. 2000. *Campylobacter* prevention and control: the USDA-Food Safety and Inspection Service role and new food safety approaches, p. 511–528. *In* I. Nachamkin and M. J. Blaser (ed.), *Campylobacter*, 2nd ed. ASM Press, Washington, D.C.

76. Rautelin, H., O. V. Renkonen, and T. U. Kosunen. 1991. Emergence of fluoroquinolone resistance in *Campylobacter jejuni* and *Campylobacter coli* in subjects from Finland. *Antimicrob. Agents Chemother.* **35**:2065–2069.

77. Rees, J. H., R. W. Vaughan, E. Kondeatis, and R. A. C. Hughes. 1995. HLA class II alleles in Guillain-Barre syndrome and Miller Fisher syndrome and their associations with preceding *Campylobacter jejuni* infection. *J. Neuroimmunol.* **38**:53–57.

78. Reina, J., N. Borrell, and A. Serra. 1992. Emergence of resistance to erythromycin and fluoroquinolones in thermotolerant *Campylobacter* strains isolated from feces 1987–1991. *Eur. J. Clin. Microbiol. Infect. Dis.* **11**:1163–1166.

79. Robinson, D. A. 1981. Infective dose of *Campylobacter jejuni* in milk. *Br. Med. J.* **282**:1584.

80. Rollins, D. M., and R. R. Colwell. 1986. Viable but nonculturable stage of *Campylobacter jejuni* and its role in the survival in the natural aquatic environment. *Appl. Environ. Microbiol.* **52**:531–538.

81. Ruiz-Palacios, G. M., J. J. Calva, L. K. Pickering, Y. Lopez-Vidal, P. Volkow, H. Pezzarossi, and M. S. West. 1990. Protection of breast-fed infants against *Campylobacter* diarrhea by antibodies in human milk. *J. Pediatr.* **116**:707–713.

82. Ruiz-Palacios, G. M., N. I. Torres, B. R. Ruiz-Palacios, J. Torres, E. Escamilla, and J. Tamayo. 1983. Cholera-like enterotoxin produced by *Campylobacter jejuni*. *Lancet* **ii**:250–253.

83. Scott, D. A., and D. R. Tribble. 2000. Protection against *Campylobacter* infection and vaccine development,

p. 303–319. *In* I. Nachamkin and M. J. Blaser (ed.), *Campylobacter*, 2nd ed. ASM Press, Washington, D.C.

84. Skirrow, M. B. 1994. Diseases due to *Campylobacter*, *Helicobacter* and related bacteria. *J. Comp. Pathol.* 111:113–149.

85. Skirrow, M. B., D. M. Jones, E. Sutcliffe, and J. Benjamin. 1993. *Campylobacter* bacteremia in England and Wales, 1981–1991. *Epidemiol. Infect.* 110:567–573.

86. Smibert, R. M. 1984. Genus *Campylobacter* Sebald and Veron 1963, p. 111–118. *In* N. R. Krieg and J. G. Holt (ed.), *Bergey's Manual of Systematic Bacteriology*. The Williams & Wilkins co., Baltimore, Md.

87. Smith, K. E., J. B. Bender, and M. T. Osterholm. 2000. Antimicrobial resistance in animals and relevance to human infections, p. 483–495. *In* I. Nachamkin and M. J. Blaser (ed.), *Campylobacter*, 2nd ed. ASM Press, Washington, D.C.

88. Smith, K. E., J. M. Besser, C. W. Hedberg, F. T. Ieano, J. B. Bender, J. H. Wicklund, B. P. Johnson, K. A. Moore, and M. T. Osterholm. 1999. Quinolone-resistant *Campylobacter jejuni* infections in Minnesota, 1992–1998. *N. Engl. J. Med.* 340:1525–1532.

89. Stern, N. J. 1992. Reservoirs for *Campylobacter jejuni* and approaches for intervention in poultry, p. 49–60. *In* I. Nachamkin, M. J. Blaser, and L. S. Tompkins (ed.), *Campylobacter jejuni: Current Status and Future Trends*. American Society for Microbiology, Washington, D.C.

90. Tauxe, R. V. 1992. Epidemiology of *Campylobacter jejuni* infections in the United States and other industrialized nations, p. 9–19. *In* I. Nachamkin, M. J. Blaser, and L. S. Tompkins (ed.), *Campylobacter jejuni: Current Status and Future Trends*. American Society for Microbiology, Washington, D.C.

91. Taylor, D. E. 1992. Genetic analysis of *Campylobacter* spp., p. 255–266. *In* I. Nachamkin, M. J. Blaser, and L. S. Tompkins (ed.), *Campylobacter jejuni: Current Status and Future Trends*. American Society for Microbiology, Washington, D.C.

92. Taylor, D. N. 1992. *Campylobacter* infections in developing countries, p. 20–30. *In* I. Nachamkin, M. J. Blaser, and L. S. Tompkins (ed.), *Campylobacter jejuni: Current Status and Future Trends*. American Society for Microbiology, Washington, D.C.

93. Taylor, D. N., P. Echeverria, C. Pitarangsi, J. Seriwatana, L. Bodhidatta, and M. J. Blaser. 1988. Influence of strain characteristics and immunity on the epidemiology of *Campylobacter* infections in Thailand. *J. Clin. Microbiol.* 26:863–868.

94. Vandamme, P. 2000. Taxonomy of the family *Campylobacteraceae*, p. 3–26. *In* I. Nachamkin and M. J. Blaser (ed.), *Campylobacter*, 2nd ed. ASM Press, Washington, D.C.

95. Vandamme, P., M. I. Daneshvar, F. E. Dewhirst, B. J. Paster, K. Kersters, H. Goossens, and C. W. Moss. 1995. Chemotaxonomic analyses of *Bacteroides gracilis* and *Bacteroides ureolyticus* and reclassification of *B. gracilis* as *Campylobacter gracilis* comb. nov. *Int. J. Syst. Bacteriol.* 45:145–152.

96. Vandamme, P., and J. De Ley. 1991. Proposal for a new family, *Campylobacteraceae*. *Int. J. Syst. Bacteriol.* 41:451–455.

97. Vandamme, P., E. Falsen, R. Rossau, B. Hoste, P. Segers, R. Tytgat, and J. De Ley. 1991. Revision of *Campylobacter*, *Helicobacter*, and *Wolinella* taxonomy: emendation of generic descriptions and proposal of *Arcobacter* gen. nov. *Int. J. Syst. Bacteriol.* 41:81–103.

98. Vandamme, P., M. Vancanneyt, B. Pot, L. Mels, B. Hoste, D. Dewettinck, L. Vlaes, C. Van den Borre, R. Higgins, and J. Hommez. 1992. Polyphasic taxonomic study of the emended genus *Arcobacter* with *Arcobacter butzleri* comb. nov. and *Arcobacter skirrowii* sp. nov., an aerotolerant bacterium isolated from veterinary specimens. *Int. J. Syst. Bacteriol.* 42:344–356.

99. Vandamme, P., L. J. VanDoorn, S. T. Alrashid, W. G. V. Quint, J. VanderPlas, V. L. Chan, and S. L. W. On. 1997. *Campylobacter hyoilei* Alderton et al. 1995 and *Campylobacter coli* Veron and Chatelain 1973 are subjective synonyms. *Int. J. Syst. Bacteriol.* 47:1055–1060.

100. Wang, W. L., B. W. Powers, N. W. Luechtefeld, and M. J. Blaser. 1980. Effects of disinfectants on *Campylobacter jejuni*. *Appl. Environ. Microbiol.* 45:1202–1205.

101. Wassenaar, T., and D. G. Newell. 2000. Genotyping of *Campylobacter* spp. *Appl. Environ. Microbiol.* 66:1–9.

102. Wassenaar, T. M. 1997. Toxin production by *Campylobacter* spp. *Clin. Microbiol. Rev.* 10:466–476.

103. Wassenaar, T. M., N. M. Bleumink-Pluym, and B. A. van der Zeijst. 1991. Inactivation of *Campylobacter jejuni* flagellin genes by homologous recombination demonstrates that flaA but not flaB is required for invasion. *EMBO J.* 10:2055–2061.

104. Wexler, H. M., D. Reeves, P. H. Summanen, E. Molitoris, M. McTeague, J. Duncan, K. H. Wilson, and S. M. Finegold. 1996. *Sutterella wadsworthensis* gen. nov., sp. nov., bile-resistant microaerophilic *Campylobacter gracilis*-like clinical isolates. *Int. J. Syst. Bacteriol.* 46:252–258.

105. Whitehouse, C. A., P. B. Balbo, E. C. Pesci, D. L. Cottle, P. M. Mirabito, and C. L. Pickett. 1998. *Campylobacter jejuni* cytolethal distending toxin causes a G2-phase cell cycle block. *Infect. Immun.* 66:1934–1940.

106. Willison, H. J., and G. M. O'Hanlon. 2000. Antiglycosphingolipid antibodies and Guillain-Barre syndrome, p. 259–285. *In* I. Nachamkin and M. J. Blaser (ed.), *Campylobacter*, 2nd ed. ASM Press, Washington, D.C.

107. Wood, R. C., K. L. MacDonald, and M. T. Osterholm. 1992. *Campylobacter* enteritis outbreaks associated with drinking raw milk during youth activities. A 10-year review of outbreaks in the United States. *JAMA* 268:3228–3230.

108. Young, V. B., D. B. Schauer, and J. G. Fox. 2000. Animal models of *Campylobacter* infection, p. 287–301. *In* I. Nachamkin and M. J. Blaser (ed.), *Campylobacter*, 2nd ed., ASM Press, Washington, D.C.

109. **Yuki, N., S. Sato, T. Itoh, and T. Miyatake.** 1991. HLA-B35 and acute axonal polyneuropathy following *Campylobacter* infection. *Neurology* **41**:1561–1563.

110. **Yuki, N., M. Takahashi, Y. Tagawa, K. Kashiwase, K. Tadokoro, and K. Saito.** 1997. Association of *Campylobacter jejuni* serotype with antiganglioside antibody in Guillain-Barre syndrome and Fisher's syndrome. *Ann. Neurol.* **42**:28–33.

111. **Yuki, N., T. Taki, F. Inagaki, T. Kasama, M. Takahashi, K. Saito, S. Handa, and T. Miyatake.** 1993. A bacterium lipopolysaccharide that elicits Guillain-Barre syndrome has a GM1 ganglioside structure. *J. Exp. Med.* **178**:1771–1775.

Food Microbiology: Fundamentals and Frontiers, 2nd Ed.
Edited by M. P. Doyle et al.
© 2001 ASM Press, Washington, D.C.

Jianghong Meng
Michael P. Doyle
Tong Zhao
Shaohua Zhao

10

Enterohemorrhagic *Escherichia coli*

Escherichia coli is a common part of the normal facultative anaerobic microflora in the intestinal tract of humans and warm-blooded animals. Most *E. coli* strains are harmless; however, some are pathogenic and cause diarrheal disease. *E. coli* isolates are serologically differentiated based on three major surface antigens, which enable serotyping: the O (somatic), H (flagella), and K (capsule) antigens. At present, a total of 167 O antigens, 53 H antigens, and 74 K antigens have been identified (80). Currently, it is considered necessary only to determine the O and the H antigens to serotype strains of *E. coli* associated with diarrheal disease. The O antigen identifies the serogroup of a strain, and the H antigen identifies its serotype. The application of serotyping to isolates associated with diarrheal disease has shown that particular serogroups often fall into one category of diarrheagenic *E. coli*. However, some serogroups, such as O55, O111, O126, and O128, appear in more than one category.

Diarrheagenic *E. coli* isolates are categorized into specific groups based on virulence properties, mechanisms of pathogenicity, clinical syndromes, and distinct O:H serotypes. These categories include enteropathogenic *E. coli* (EPEC), enterotoxigenic *E. coli* (ETEC), enteroinvasive *E. coli* (EIEC), diffuse-adhering *E. coli* (DAEC), enteroaggregative *E. coli* (EAEC), and enterohemorrhagic *E. coli* (EHEC). This chapter will focus on EHEC, which, among the *E. coli* isolates that cause foodborne illness, is the most significant group based on severity of illness. More information on other diarrheagenic *E. coli* strains is available in a recent review article by Nataro and Kaper (92).

EPEC

Enteropathogenic *E. coli* isolates can cause severe diarrhea in infants, especially in developing countries. EPEC previously were also associated with outbreaks of diarrhea in nurseries in developed countries. The major O serogroups associated with illness include O55, O86, O111ab, O119, O125ac, O126, O127, O128ab, and O142. Humans are an important reservoir. The original definition of EPEC is "diarrheagenic *E. coli* belonging to serogroups epidemiologically incriminated as pathogens

Jianghong Meng, Department of Nutrition and Food Science, University of Maryland, College Park, MD 20742.
Michael P. Doyle and Tong Zhao, Center for Food Safety and Quality Enhancement, University of Georgia, Griffin, GA 30223.
Shaohua Zhao, Division of Animal and Food Microbiology, Center for Veterinary Medicine/Office of Research, Food & Drug Administration, Laurel, MD 20708.

but whose pathogenic mechanism has not been proven to be related to either enterotoxins, or *Shigella*-like invasiveness." However, EPEC have been determined to induce the attaching and effacing lesions in cells to which they adhere and can invade epithelial cells.

ETEC

Enterotoxigenic *E. coli* isolates are a major cause of infantile diarrhea in developing countries. They are also the agents most frequently responsible for traveler's diarrhea. ETEC colonize the proximal small intestine by fimbrial colonization factors (e.g., CFA/I and CFA/II) and produce heat-labile or heat-stable enterotoxin that elicits fluid accumulation and a diarrheal response. The most frequent ETEC serogroups include O6, O8, O15, O20, O25, O27, O63, O78, O85, O115, O128ac, O148, O159, and O167. Humans are the principal reservoir of ETEC strains that cause human illness.

EIEC

Enteroinvasive *E. coli* isolates cause nonbloody diarrhea and dysentery similar to that caused by *Shigella* spp. by invading and multiplying within colonic epithelial cells. As for *Shigella*, the invasive capacity of EIEC is associated with the presence of a large plasmid (ca. 140 MDa) which encodes several outer membrane proteins involved in invasiveness. The antigenicity of these outer membrane proteins and the O antigens of EIEC are closely related. The principal site of bacterial localization is the colon, where EIEC invade and proliferate in epithelial cells, causing cell death. Humans are a major reservoir, and the serogroups most frequently associated with illness include O28ac, O29, O112, O124, O136, O143, O144, O152, O164, and O167. O124 is the serogroup most commonly encountered.

DAEC

Diffuse-adhering *E. coli* isolates have been associated with diarrhea, primarily in young children who are older than infants. The relative risk of DAEC-associated diarrhea increases with age from 1 to 5 years. The basis for this age-related infection is unknown. DAEC are most commonly of serogroups O1, O2, O21, and O75. Typical symptoms of DAEC infection are mild diarrhea without blood or fecal leukocytes, and DAEC strains produce a characteristic diffuse-adherent pattern of attachment to HEp-2 or HeLa cell lines. DAEC generally do not elaborate heat-labile or heat-stable enterotoxin or elevated

levels of Shiga toxin, nor do they possess EPEC adherence factor plasmids or invade epithelial cells.

EAEC

Enteroaggregative *E. coli* isolates recently have been associated with persistent diarrhea in infants and children in several countries worldwide (92). EAEC are different from the other types of pathogenic *E. coli* because of their ability to produce a characteristic pattern of aggregative adherence on HEp-2 cells. EAEC adhere in an appearance of stacked bricks to the surface of HEp-2 cells. Serogroups associated with EAEC include O3, O15, O44, O77, O86, O92, O111, and O127. A gene probe derived from a plasmid associated with EAEC strains has been developed to identify *E. coli* isolates of this type; however, more epidemiologic information is needed to elucidate the significance of EAEC as an agent of diarrheal disease.

EHEC

Enterohemorrhagic *E. coli* isolates were first recognized as human pathogens in 1982 when *E. coli* O157:H7 was identified as the cause of two outbreaks of hemorrhagic colitis. Since then, other serotypes of *E. coli*, such as O26, O111, and sorbitol-fermenting O157:NM, also have been associated with cases of hemorrhagic colitis and have been classified as EHEC. However, serotype O157:H7 is the predominant cause of EHEC-associated disease in the United States and many other countries. All EHEC strains produce factors cytotoxic to African green monkey kidney (Vero) cells, which hence have been named verotoxins or Shiga toxins (Stx) because of similarity to the Shiga toxin produced by *Shigella dysenteriae* type 1 (15). Production of Shiga toxins by *E. coli* O157:H7 was first reported by Johnson et al. (64). Karmali et al. (66) subsequently associated Shiga toxin-producing *E. coli* infections with a severe and sometimes fatal condition, hemolytic-uremic syndrome (HUS). Many different serotypes of *E. coli* subsequently were determined to produce Shiga toxins; hence they have been named Shiga toxin-producing *E. coli* (STEC). More than 200 serotypes of STEC have been isolated from humans (Table 10.1) (118); however, only those strains that cause hemorrhagic colitis are considered to be EHEC. Major non-O157:H7 EHEC serotypes include O26:H11, O111:H8, and O157:NM. Since *E. coli* O157:H7 is the most common serotype of the EHEC, and because more is known about this serotype than other serotypes of EHEC, this chapter will focus on *E. coli* O157:H7.

Table 10.1 Serotypes of non-O157 STEC isolated from humans[a,b]

O antigen	H antigen	O antigen	H antigen	O antigen	H antigen	O antigen	H antigen
O1	H-, H1, H2, H7, H20, HNT	O38	H21	O91	H-, H10, H14, **H21**, HNT	O126	H-, **H2**, H8, **H21**, H27
O2	**K1:H2**, **H5**, **H6**, H7, **H27**, **H29**, **H44**	O39	H4, H8	O98	H-, H8	O128	H-, **H8**, H12, **H25**
O4	H-, **H5**, H10, H40, H40	O45	H-, **H2**, H7	O100	H25, H32	O128ab	**H2**
O5	H-, **H16**	O48	**H21**	O101	H-, H9	O132	H-
O6	H-, H1, **H2**, **H4**, **H28**, **H29**, H31	O49	H-, **H10**	O103	H-, **H2**, H4, **H25**, HNT	O137	**H41**
O7	H4, H8	O50	H-, **H7**	O104	H-, **H2**, H7, H21	O141	H-
O8	H-, **H14**, **H21**	O52	H23	O105	H20	O144	H-
O9ab	H-	O55	H-, **H6**, **H7**, **H10**, H?	O105ac	**H18**	O145	H-, **H16**, **H25**, **H28**, HNT
O11	H49	O60	H-	O109	H2	O146	H21, H28
O14	H-	O65	H16	O110	H-, H19	O150	H10
O15	H-, **H2**, H27	O69	H-	O111	H-, **H2**, **H8**, H30, H34	O153	H2, H11, H12, **H25**
O16	H-, **H6**	O70	H11	O112	H21	O154	H-, **H4**, **H19/20**
O17	H18	O73	H34	O113	H2, H4, H7, **H21**	O161	H-
O18	H-, H7, **H12**, H?	O75	H-, **H5**	O114	H4	O163	H19
O20	H7	O76	H19	O115	H10, **H18**	O165	H-, **H10**, **H19**, **H25**
O21	**H5**, H?	O77	H4, H18	O116	H19	O166	H12, H15, **H28**
O22	H-, **H1**, **H8**, H16, H40	O79	H7	O117	H-, **H4**, H7, K1:H7, H19	O168	H-
O23	H7, H16	O80	H-	O118	**H16**, H30	O169	H-
O25	H-, **K2:H2**	O82	H-, **H5**, H8	O119	H-, **H5**, H6	O171	H-, **H2**
O26	H-, **H8**, **H11**, H21, H32	O83	H11	O121	H-, **H8**, **H19**	O172	H-
O27	H-	O84	H2	O123	H49	ONT	H-, **H2**, H18, H21, H25, H47
O30	H2, H21, H23	O85	H-, **H10**, H23	O124	H-	O-rough	H-, **H5**, **H11**, **H16**, H20, H21
O37	H41	O90	H-	O125	H-, H8	OX3	H2, H21

[a] Data from reference 118.
[b] Serotypes in boldface type have been isolated from patients with hemolytic-uremic syndrome.

Non-O157 EHEC will be included where information is available.

CHARACTERISTICS OF *E. COLI* O157:H7 AND NON-O157 EHEC

E. coli O157:H7 was first identified as a foodborne pathogen in 1982 (116). There had been prior isolation of the organism, identified retrospectively among isolates at the Centers for Disease Control and Prevention; the isolate was from a California woman with bloody diarrhea in 1975 (56). In addition to production of Shiga toxin(s), most strains of *E. coli* O157:H7 also possess several characteristics uncommon to most other *E. coli*: inability to grow well, if at all, at temperatures $\geq$44.5°C in *E. coli* broth, inability to ferment sorbitol within 24 h, inability to produce β-glucuronidase (i.e., inability to hydrolyze 4-methylumbelliferyl-D-glucuronide), possession of an attaching and effacing (*eae*) gene, and carriage of a 60-MDa plasmid. Non-O157 EHEC do not share the previously described growth and metabolic characteristics, although they all produce Shiga toxin(s), and most carry the *eae* gene and large plasmid (54).

Acid Tolerance

Unlike most foodborne pathogens, many strains of *E. coli* O157:H7 are unusually tolerant to acidic environments. The minimum pH for *E. coli* O157:H7 growth is 4.0 to 4.5 but is dependent upon the interaction of pH with other growth factors. Studies with organic acid sprays on beef using acetic, citric, or lactic acid at concentrations up to 1.5% revealed that *E. coli* O157:H7 populations were not appreciably affected by any of the treatments (10). *E. coli* O157:H7, when inoculated at high populations, survived fermentation, drying, and storage in fermented sausage (pH 4.5) for up to 2 months at 4°C (51), in mayonnaise (pH 3.6 to 3.9) for 5 to 7 weeks at 5°C and for 1 to 3 weeks at 20°C (120), and in apple cider (pH 3.6 to 4.0) for 10 to 31 days or 2 to 3 days at 8 or 25°C (121), respectively.

Three systems in *E. coli* O157:H7 are reported that are involved in acid tolerance, including an acid-induced oxidative system, an acid-induced arginine-dependent system, and a glutamate-dependent system (77). The oxidative system is less effective in protecting the organism from acid stress than are the arginine-dependent and glutamate-dependent systems. The alternate sigma factor RpoS is required for oxidative acid tolerance but is only partially involved with the other two systems. Once induced, the acid-tolerant state can persist for a prolonged period of time (≥ 28 days) at refrigeration temperature. More important, induction of acid tolerance in *E. coli* O157:H7 also can increase tolerance to other environmental stresses. Acid tolerant-induced cells also can have increased tolerance to heating, radiation, and antimicrobials (13, 100).

Antibiotic Resistance

Initially, when *E. coli* O157:H7 was first associated with human illness, the pathogen was susceptible to most antibiotics affecting gram-negative bacteria (9, 99). However, more recent studies indicate a trend toward increasing resistance to antibiotics among *E. coli* O157:H7 isolates (45, 73, 89). *E. coli* O157:H7 strains isolated from humans, animals, and food have developed resistance to multiple antibiotics, with streptomycin-sulfisoxazole-tetracycline being the most common resistance profile. Non-O157 EHEC strains isolated from humans and animals also have acquired antibiotic resistance, and some are resistant to multiple antimicrobials commonly used in human and veterinary medicine (45, 106). Antibiotic-resistant EHEC strains possess a selective advantage over other bacteria colonizing the gastrointestinal tract of animals that are treated with antibiotics (therapeutically or subtherapeutically). Hence, antibiotic-resistant EHEC strains may become the predominant *E. coli* present under antibiotic selective pressure and subsequently be more prevalent in fecal excretions.

Inactivation by Heat and Irradiation

Studies on the thermal sensitivity of *E. coli* O157:H7 in ground beef revealed that the pathogen has no unusual resistance to heat, with *D* values of 270, 45, 24, and 9.6 s at 57.2, 60, 62.8, and 64.3°C, respectively (41). Heating ground beef sufficiently to kill typical strains of *Salmonella* sp. will also kill *E. coli* O157:H7 (Table 10.2)

Table 10.2 Comparison of *D* values for *E. coli* O157:H7 and *Salmonella* spp. in ground beef

	D value (min)		
	E. coli O157:H7		
Temp (°C)	30.5% fat[a]	17–20% fat[b]	*Salmonella* spp.[c]
51.7	115.5	ND[d]	54.3
57.2	5.3	4.5	5.43
62.8	0.47	0.40	0.54

[a] Data from reference 78.
[b] Data from reference 41.
[c] Data from reference 52.
[d] ND, not done.

(52). The presence of fat protects *E. coli* O157:H7 in ground beef, with *D* values for lean (2.0% fat) and fatty (30.5% fat) ground beef of 4.1 and 5.3 min, respectively, at 57.2°C, and 0.3 and 0.5 min, respectively, at 62.8°C, (78). Pasteurization of milk (72°C, 16.2 s) is an effective treatment that will kill more than 10^4 *E. coli* O157:H7 cells per ml (34). Proper heating of foods of animal origin, e.g., heating foods to an internal temperature of at least 68.3°C for several seconds, is an important critical control point to ensure inactivation of *E. coli* O157:H7.

The use of irradiation to eliminate foodborne pathogens in food has been approved by many countries. Unlike other processing technologies, irradiation eliminates foodborne pathogens yet maintains the raw character of foods. In the United States, 4.5 kGy is approved for refrigerated raw ground beef and 7.5 kGy for frozen. Bacterial foodborne pathogens are relatively sensitive to irradiation. D_{10} values for *E. coli* O157:H7 in raw ground beef patties range from 0.241 to 0.307 kGy, depending on temperature, with D_{10} values significantly higher for patties irradiated at -16°C than at 4°C (30). Hence, a radiation dose of 1.5 kGy should be sufficient to eliminate *E. coli* O157:H7 at the cell concentrations likely to occur in ground beef.

RESERVOIRS OF *E. COLI* O157:H7 AND NON-O157 EHEC

Cattle

Undercooked ground beef and, less frequently, unpasteurized milk are well-recognized vehicles associated with outbreaks of *E. coli* O157:H7 infection; hence, cattle have been the focus of many studies on their role as a reservoir of *E. coli* O157:H7.

Detection of *E. coli* O157:H7 and STEC on Farms

The first reported isolation of *E. coli* O157:H7 from cattle was from a <3-week-old calf with colibacillosis in Argentina in 1977 (95). Prevalence rates of STEC as high as 60% have been found in bovine herds in many countries, but in most cases, the rates range from 10 to 25%. The isolation rates of *E. coli* O157:H7 are much lower than those of non-O157 STEC. A study on STEC carriage by dairy cattle on farms in Canada revealed that 36% of cows and 57% of calves were STEC positive in all of 80 herds tested (117). On four farms, the same STEC serotypes were identified in cattle and humans. Seven animals (0.45%) on four farms (5%) were positive for *E. coli* O157:H7. Results of two major surveys in the United States revealed that 31 (3.2%) of 965 dairy calves and 191 (1.6%) of 11,881 feedlot cattle were positive for *E. coli* O157:H7 (122). An additional 0.4% of feedlot cattle were positive for *E. coli* O157:NM. *E. coli* O157:H7 levels in calf feces range from detectable by enrichment culture (but <10^2 CFU/g) to 10^5 CFU/g. Young, weaned animals more frequently carry *E. coli* O157:H7 than do adult cattle. Studies in some locations indicate increased prevalence of *E. coli* O157:H7 in cattle during warmer months of the year, which correlates with the seasonal variation in human disease. For example, a 1997 study in England revealed that *E. coli* O157:H7 was isolated from 38% of cattle presented for slaughter in the spring, but from only 4.8% of cattle during the winter months. A U.S. study also revealed a high prevalence of *E. coli* O157:H7 in beef cattle presented for slaughter from July to August 1999 (42). *E. coli* O157:H7 was present in 28% (91 of 327) of fecal samples and on 11% (38 of 355) of hide samples. All but two of the 30 lots of cattle or carcasses tested were positive for *E. coli* O157:H7.

Factors Associated with Bovine Carriage of *E. coli* O157:H7

Several tentative associations between fecal shedding of *E. coli* O157:H7 and feed or environmental factors have been made from epidemiologic studies of dairy herds (101). For example, some calf starter feed regimens or environmental factors and some feed components, such as whole cottonseed, were associated with reduced prevalence of *E. coli* O157:H7 (50). In contrast, grouping calves before weaning, sharing among calves feeding utensils without sanitation, and early feeding of grain were associated with increased carriage of *E. coli* O157:H7. Controversial results have been obtained from studies on the effect of grain and hay feeding on *E. coli* O157:H7 carriage by cattle. An early study suggested that grain feeding increased both the number and degree of acid resistance of *E. coli* compared with hay feeding (38). However, a later study determined that hay-fed cattle shed *E. coli* O157:H7 longer than grain-fed animals, and the pathogen was equally acid resistant in the bovine host irrespective of diet (61).

Environmental factors, such as water and feed sources, or farm management practices, such as manure handling, may play important roles in influencing the prevalence of *E. coli* O157:H7 on dairy farms (57). The pathogen was frequently found in water troughs on farms. Several studies have revealed that *E. coli* O157:H7 can survive for weeks and months in bovine feces and water. Commercial feeds often contained detectable *E. coli*, indicating widespread fecal contamination, although limited sampling has failed to detect naturally occurring *E. coli* O157:H7 (83).

Susceptibility of cattle to subsequent gastrointestinal colonization with *E. coli* O157:H7 is largely a function of age. Young animals are more likely to be positive compared with older ones of the same herd. For many cattle, *E. coli* O157:H7 is transiently carried in the gastrointestinal tract and is intermittently excreted for a few weeks to months in young calves and heifers. Carriage of more than one strain of *E. coli* O157:H7 has been described. Other influences that disrupt the normal flora of the gastrointestinal tract, such as temporary withholding of feed, can alter the pattern of fecal shedding. Fasting calves for 48 h can increase the populations of *E. coli* O157:H7 in the gastrointestinal tract of some animals.

Cattle Model for Infection of *E. coli* O157:H7

E. coli O157:H7 is not a pathogen of weaned calves and adult cattle; hence, animals that carry the pathogen are not ill. However, there is evidence that *E. coli* O157:H7 can cause diarrhea and attaching and effacing lesions in neonatal calves (33, 35). Studies by Brown et al. (11) revealed that the initial sites of localization of *E. coli* O157:H7 in cattle are the forestomachs (rumen, omasum, and reticulum).

Domestic Animals and Wildlife

Although cattle are thought to be the main source of STEC in the food chain, STEC have also been isolated from other domestic animals and wildlife, such as sheep, goats, deer, dogs, horses, swine, and cats. *E. coli* O157:H7 also was isolated from seagulls and rats (29, 114). Prevalence of *E. coli* O157:H7 and STEC in sheep is generally higher than in other animals. A 6-month study of healthy, naturally infected ewes revealed that fecal shedding of the pathogen was transient and seasonal, with 31% of sheep positive in June, 5.7% positive in August, and none positive in November (75). In a survey of seven animal species in Germany, STEC were isolated most frequently from sheep (66.6%), goats (56.1%), and cattle (21.1%) (8), with lower prevalence rates in chickens (0.1%), pigs (7.5%), cats (13.8%), and dogs (4.8%).

Humans

Fecal shedding of *E. coli* O157:H7 by patients with hemorrhagic colitis or HUS usually lasts for no more than 13 to 21 days following onset of symptoms. However, in some instances, the pathogen can excrete in feces for weeks (65). A child infected during a day care center-associated outbreak continued to excrete the pathogen for 62 days after the onset of diarrhea (94). Studies of persons living on dairy farms to determine carriage of *E. coli* O157:H7 by farm families revealed elevated antibody titers against the surface antigens of *E. coli* O157; however, the pathogen was not isolated from feces. An asymptomatic long-term carrier state has not been identified. The significance of fecal carriage of *E. coli* O157:H7 by humans is the potential for person-to-person dissemination of the pathogen, a situation which has been observed repeatedly in outbreak settings. A contributing factor to person-to-person transmission of the pathogen is its extraordinarily low infectious dose: <100 cells and possibly as few as 10 can produce illness in highly susceptible populations. Inadequate attention to personal hygiene, especially after using the bathroom, can transfer the pathogen to other persons through contaminated hands, resulting in secondary transmission.

DISEASE OUTBREAKS

Geographic Distribution

E. coli O157:H7 has been the cause of many major outbreaks of severe illness worldwide. At least 30 countries on six continents have reported *E. coli* O157:H7 infection in humans. In the United States, 196 outbreaks or clusters of *E. coli* O157:H7 infection were documented through 1998 (55). The number of outbreaks has increased from an average of 2 per year between 1982 and 1992 to 29 annually between 1993 and 1998. This dramatic increase may be due in part to improved recognition of *E. coli* O157:H7 infection following the publicity of a large multistate outbreak in the western United States in 1993. Since January 1, 1994, individual cases of *E. coli* O157:H7 infection are reportable to the National Notifiable Diseases Surveillance System. The number of reported outbreaks and clusters and the number of cases of infection in the United States between 1982 and 1998 are presented in Table 10.3. However, the precise incidence of *E. coli* O157:H7 foodborne illness in the United States is not known, because infected persons presenting mild or no symptoms and persons with nonbloody diarrhea are less likely to seek medical attention than patients with bloody diarrhea; hence, such cases would not be reported. The Foodborne Diseases Active Surveillance Network (FoodNet; http://www.cdc.gov/ncidod/dbmd/foodnet/) reports that the annual rate of *E. coli* O157:H7 infection at several surveillance sites in the United States ranges from 2.1 to 2.8 per 100,000 population for 1996 to 1999 (27). The Centers for Disease Control and Prevention estimates that *E. coli* O157:H7 causes 73,480 illnesses and

Table 10.3 Number of reported outbreaks and clusters and number of cases of *E. coli* O157:H7 infection in the United States, 1982–1998[a]

Year	No. of outbreaks	No. of cases
1982	2	47
1984	2	70
1986	2	52
1987	1	52
1988	3	153
1989	2	246
1990	2	75
1991	4	54
1992	4	75
1993	17	1,000
1994	32	543
1995	32	455
1996	29	488
1997	22	298
1998	42	477
Total	196	4,085

[a] Data from reference 55.

61 deaths annually in the United States, and non-O157 STEC account for an additional 37,740 cases, with 30 deaths (86). Eighty-five percent of these cases are attributed to foodborne transmission.

Large outbreaks of *E. coli* O157:H7 infections involving hundreds of cases also have been reported in Canada, Japan, and the United Kingdom. The largest multiple outbreaks reported worldwide occurred in May to December 1996 in Japan, involving more than 11,000 reported cases (90). In the same year, 21 elderly people died in a large outbreak involving 501 cases in central Scotland (2). Although *E. coli* O157:H7 is still the predominant serotype of EHEC in the United States, Canada, the United Kingdom, and Japan, an increasing number of outbreaks and sporadic cases related to EHEC of serotypes other than O157:H7 have been reported. A large epidemic involving several thousand cases of *E. coli* O157:NM infection occurred in Swaziland and South Africa, following consumption of contaminated surface water (62). In continental Europe, Australia, and Latin America, non-O157 EHEC infections are more common than *E. coli* O157:H7 infections. Details of many reported foodborne and waterborne outbreaks of EHEC infections are provided in Table 10.4. There are no distinguishing biochemical phenotypes for non-O157 EHEC, making screening for these bacteria problematic and labor intensive, and for this reason many clinical laboratories do only limited testing for them. Therefore, the prevalence of non-O157 EHEC infections may be underestimated.

Table 10.4 Representative foodborne and waterborne outbreaks of *E. coli* O157:H7 and other EHEC infections[a]

Year	Month	Location[b]	No. of cases (deaths)	Setting	Vehicle	References
1982	2	Oregon	26	Community	Ground beef	116
1982	5	Michigan	21	Community	Ground beef	116
1985		Canada	73 (17)	Nursing home	Sandwiches	17
1987	6	Utah	51	Custodial institution	Ground beef/person to person	96
1988	10	Minnesota	54	School	Precooked ground beef	5
1989	12	Missouri	243	Community	Water	109
1990	7	North Dakota	65	Community	Roast beef	21
1991	11	Massachusetts	23	Community	Apple cider	7
1991	7	Oregon	21	Community	Swimming water	69
1992[c]		France	>4	Community	Goat cheese	36
1992	12	Oregon	9	Community	Raw milk	68
1993	1	California, Idaho, Nevada, Washington	732 (4)	Restaurant	Ground beef	4, 28, 55
1993	3	Oregon	47	Restaurant	Mayonnaise?	55
1993	7	Washington	16	Church picnic	Pea salad	55
1993	8	Oregon	27	Restaurant	Cantaloupe	55
1994[d]	2	Montana	18	Community	Milk	23
1994	11	Washington, California	19	Home	Salami	20, 111
1995[e]	2	Adelaide, Australia	>200	Community	Semidry sausage	18
1995	10	Kansas	21	Wedding	Punch/fruit salad	55
1995	11	Oregon	11	Home	Venison jerky	70
1995	7	Montana	74	Community	Leaf lettuce	1
1995	9	Maine	37	Camp	Lettuce	55
1996[f]		Komatsu, Japan	126	School	Luncheon	58
1996	5, 6	Connecticut, Illinois	47	Community	Mesclun lettuce	59
1996	7	Osaka, Japan	7,966 (3)	Community	White radish sprouts	90
1996	10	California, Washington, Colorado	71 (1)	Community	Apple juice	26
1996	11	Central Scotland, U.K.	<501 (21)	Community	Cooked meat	2
1997	5	Illinois	3	School	Ice cream bar	55
1997	7	Michigan	60	Community	Alfalfa sprouts	55
1997	11	Wisconsin	13	Church banquet	Meatballs/coleslaw	55
1998	6	Wisconsin	63	Community	Cheese curds	25
1998	6	Wyoming	114	Community	Water	3
1998	7	North Carolina	142	Restaurant	Coleslaw	55
1998	7	California	28	Prison	Milk	55
1998	8	New York	11	Deli	Macaroni salad	55
1998	9	California	20	Church	Cake	55
1999[g]	7	Texas	58	Camp	Ice?	19
1999	8	New York	900 (2)	Fair	Well water	24

[a] *E. coli* O157:H7, unless otherwise noted.
[b] State of the United States unless otherwise noted.
[c] *E. coli* O119.
[d] *E. coli* O104:H21; bloody diarrhea was not observed.
[e] *E. coli* O111:NM.
[f] *E. coli* O118:H2.
[g] *E. coli* O111:H8.

Seasonality of *E. coli* O157:H7

Outbreaks and clusters of *E. coli* O157:H7 infections peak during the warmest months of the year. Approximately 86% (169/196) of outbreaks and clusters reported in the United States occurred from May to October. FoodNet data indicate the same trend. The reasons for this seasonal pattern are unknown, but may include (i) an increased prevalence of the pathogen in cattle or other livestock or vehicles of transmission during the summer, (ii) greater human exposure to ground beef or other *E. coli* O157:H7-contaminated foods during the "cookout" months, and/or (iii) greater improper handling (temperature abuse and cross-contamination) or incomplete cooking of products such as ground beef during warm months.

Age of Patients

All age groups can be infected by *E. coli* O157:H7, but the very young and the elderly most frequently experience severe illness with complications. HUS usually occurs in children, whereas thrombotic thrombocytopenic purpura (TTP), which is uncommon, occurs in adults (54). Population-based studies have suggested that the highest age-specific incidence of *E. coli* O157:H7 infection occurs in children 2 to 10 years of age. The high rate of infection in this age group likely is the result of increased exposure to contaminated foods, contaminated environments, and infected animals; more opportunities for person-to-person spread between infected children with relatively undeveloped hygiene skills; and the lack of adequate protective antibodies to Shiga toxins. The fairly high rate of infection among young women suggests that sociological factors influence acquisition of illness; e.g., women are more likely to prepare food and are the primary care providers for infected young children.

Locations of Outbreaks

Outbreaks have occurred in communities at large, schools, homes, day care centers, camps, and custodial and chronic care institutions; community outbreaks are often associated with restaurant exposure, but waterborne outbreaks have also occurred.

Transmission of *E. coli* O157:H7

A variety of foods have been identified as vehicles of *E. coli* O157:H7 infections. Examples include ground beef, roast beef, cooked meats, venison jerky, salami, raw milk, pasteurized milk, yogurt, cheese, lettuce, unpasteurized apple cider/juice, cantaloupe, handling potatoes, radish sprouts, alfalfa sprouts, fruit/vegetable salad, and coleslaw (88). Ice cream bars and cake also have been associated with *E. coli* O157:H7 outbreaks (55). Among the 196 outbreaks reported in the United States for which a vehicle has been identified, 48 (33.1%) were associated with ground beef, 4 (2.8%) with raw milk, and 3 (2.1%) with roast beef (Table 10.5). Contact of foods with meat or feces (human or bovine) contaminated with *E. coli* O157:H7 is a likely primary source of cross-contamination. Outbreaks attributed to person-to-person transmission, including secondary transmission (25.5%), and waterborne (particularly recreational water) transmission (12.4%) have been increasingly reported. In contrast to *E. coli* O157:H7 outbreaks in which a food is most often identified as a vehicle, the

Table 10.5 Leading vehicles or mode of transmission associated with *E. coli* O157:H7 outbreaks in the United States, 1982–1998[a]

Rank	Vehicle	No. of outbreaks/% of known modes of transmission
1	Ground beef	48/33.1
2	Person to person	37/25.5
3	Vegetables/salad bars	18/12.4
3	Water/swimming water	18/12.4
5	Raw milk/milk	4/2.8
5	Apple cider/juice	4/2.8
7	Roast beef	3/2.1
	Others	13/9.0
	Unknown	51

[a] Data from reference 55.

modes of transmission of most outbreaks caused by non-O157 EHEC are not known (63). Only a few outbreaks of non-O157 EHEC have been clearly associated with foods (Table 10.4).

Examples of Foodborne and Waterborne Outbreaks

The Original Outbreaks

The first documented outbreak of *E. coli* O157:H7 infection occurred in Oregon in 1982, with 26 cases and 19 persons hospitalized (116). All patients had bloody diarrhea and severe abdominal pain. The median age was 28 years, with a range of 8 to 76 years. The duration of illness ranged from 2 to 9 days, with a median of 4 days. This outbreak was associated with eating undercooked hamburgers from a specific fast-food restaurant chain. *E. coli* O157:H7 was recovered from stools of patients. A second outbreak followed 3 months later and was associated with the same fast-food restaurant chain in Michigan, with 21 cases and 14 persons hospitalized. The median age was 17 years, with a range of 4 to 58 years. Contaminated hamburgers again were implicated as the vehicle, and *E. coli* O157:H7 was isolated both from patients and from a frozen ground beef patty. That *E. coli* O157:H7, a heretofore unknown human pathogen, was the causative agent was established by its association with the food and recovery of the bacterium with identical microbiologic characteristics from both the patients and the meat from the implicated supplier.

Waterborne Outbreaks

Reported waterborne outbreaks of *E. coli* O157:H7 infection have increased substantially in recent years, being associated with swimming water, drinking water, well water, and ice. Investigations of lake-associated outbreaks revealed that in some instances the water was likely contaminated with *E. coli* O157:H7 by toddlers defecating while swimming and that swallowing lake water was subsequently identified as the risk factor. A 1995 outbreak in Illinois involved 12 children ranging in age from 2 to 12 years (22). Although *E. coli* O157:H7 was not recovered from water samples, high levels of *E. coli* were detected, indicating likely fecal contamination. A large waterborne outbreak of *E. coli* O157:H7 among attendees of a county fair in New York occurred in August 1999 (24). More than 900 persons were infected, of whom 65 were hospitalized. *Campylobacter jejuni* also was identified in some patients. Two persons died, including a 3-year-old girl from HUS and a 79-year-old man from HUS/TTP. Unchlorinated well water used to make beverages and ice was the vehicle, and *E. coli* O157:H7 was isolated from samples of well water.

Waterborne outbreaks of *E. coli* O157 infections also have been reported in other locations of the world. Drinking water, which was probably contaminated with bovine feces, was implicated in outbreaks in Scotland (37) and southern Africa (62). *E. coli* O157:NM was isolated from water associated with the latter outbreak. Outbreaks associated with drinking contaminated well water also have been reported in Japan.

Apple Cider/Juice Outbreaks

The first confirmed outbreak of *E. coli* O157:H7 infection associated with apple cider occurred in Massachusetts in 1991, involving 23 cases. In 1996, three outbreaks of *E. coli* O157:H7 infection associated with unpasteurized apple juice/cider were reported in the United States. The largest of the three occurred in three western states (California, Colorado, and Washington) and British Columbia, Canada, with 71 confirmed cases and one death (26). *E. coli* O157:H7 was isolated from the implicated apple juice. An outbreak also occurred in Connecticut, with 14 confirmed cases. Manure contamination of apples was the suspected source of *E. coli* O157:H7 in several of the outbreaks. Using apple drops (i.e., apples picked up from the ground) for making apple cider is a common practice, and apples can become contaminated by resting on soil contaminated with manure. Apples also can become contaminated if transported or stored in areas that contain manure or if treated with contaminated water. Investigation of the 1991 outbreak in Massachusetts revealed that the implicated cider press processor also raised cattle that grazed in a field adjacent to the cider mill (7). Fecal droppings from deer also were found in the orchard where apples used to make the cider were harvested.

Large Multistate Outbreak

A large multistate outbreak of *E. coli* O157:H7 infection in the United States occurred in Washington, Idaho, California, and Nevada in early 1993 (4). Approximately 90% of primary cases were associated with eating at a single fast-food restaurant chain (chain A), from which *E. coli* O157:H7 was isolated from hamburger patties. Transmission was amplified by secondary spread (48 patients in Washington alone) via person-to-person transmission. In total, 732 cases were identified, with 629 in Washington, 13 in Idaho, 58 in Las Vegas, Nevada, and 32 in Southern California. The median age of patients was 11 years, with a range of 4 months to 88 years. One hundred seventy-eight were hospitalized, 56 developed HUS, and 4 children died. Because neither specific laboratory testing nor surveillance for *E. coli* O157:H7 was carried out for earlier cases in Nevada, Idaho, and California, the outbreak went unrecognized until a sharp increase in cases of HUS was identified and investigated in the state of Washington.

The outbreak resulted from insufficient cooking of hamburgers by chain A restaurants. Epidemiologic investigation revealed that 10 of 16 hamburgers cooked according to chain A's cooking procedures in Washington had internal temperatures below 60°C, which is substantially less than the minimum internal temperature of 68.3°C required by the state of Washington. Cooking patties to an internal temperature of 68.3°C would have been sufficient to kill the low populations of *E. coli* O157:H7 detected in the contaminated ground beef.

Outbreaks of EHEC O111

Several outbreaks of EHEC *E. coli* O111 infection have been reported worldwide; however, an outbreak in early 1995 in South Australia was one in which the vehicle of transmission was identified (18). Twenty-three cases of HUS among children <12 years of age (median age was 4 years) were reported after consumption of an uncooked, semidry fermented sausage product. Of 10 sausage samples obtained from the homes of nine patients (eight homes), 8 were positive for Shiga toxin genes by PCR, and *E. coli* O111:NM was isolated from 4 of these samples. Eighteen (39%) of 47 additional sausage samples from the same manufacturer and retail stores were PCR positive; 3 yielded *E. coli* O111:NM. In June 1999, an outbreak of *E. coli* O111:H8 involving 58 cases occurred at a teenage cheerleading camp in Texas (19). Contaminated ice was the implicated vehicle.

Outbreaks Associated with Vegetables

Raw vegetables, particularly lettuce and vegetable and alfalfa sprouts, have been implicated in several outbreaks of *E. coli* O157:H7 infection in North America, Europe, and Japan. Several outbreaks associated with lettuce have been reported in the United States. In May 1996, mesclun lettuce was associated with a multistate outbreak in which 47 cases were identified in Illinois and Connecticut (59). Isolates of *E. coli* O157:H7 from patients in the two states were indistinguishable by DNA subtyping. Traceback studies implicated one grower as the likely source of the contaminated lettuce and revealed that cattle were present in the vicinity of the lettuce-growing and -processing areas.

Between May and December 1996, multiple outbreaks of *E. coli* O157:H7 infection occurred in Japan, involving 11,826 cases and 12 deaths (90). The largest outbreak affected 7,892 schoolchildren and 74 teachers and staff in Osaka in July 1996; 606 individuals were hospitalized, 106 had HUS, and 3 died. Epidemiologic investigations revealed that white radish sprouts were the vehicle of transmission.

CHARACTERISTICS OF DISEASE

The spectrum of human illness of *E. coli* O157:H7 infection includes nonbloody diarrhea, hemorrhagic colitis, HUS, and TTP. Some persons may be infected but asymptomatic. Ingestion of the organism is followed typically by a 3- to 4-day incubation period (range 2 to 12 days) during which colonization of the large bowel occurs. Illness begins with nonbloody diarrhea and severe abdominal cramps for 1 to 2 days, progressing in the second or third day of illness to bloody diarrhea that lasts for 4 to 10 days (6). Many outbreak investigations revealed that more than 90% of microbiologically documented cases of diarrhea caused by *E. coli* O157:H7 were bloody, but in some outbreaks there have been 30% of cases with nonbloody diarrhea. Symptoms usually resolve after a week, but about 6% of patients progress to HUS; half of these require dialysis, and 75% require transfusions of erythrocytes and/or platelets (32). The case-fatality rate from *E. coli* O157:H:7 infection is about 1% (6). Similar symptoms have been observed in infections with non-O157 EHEC. However, the incubation period and the proportion of persons that develop HUS have not yet been determined.

HUS largely affects children, for whom it is the leading cause of acute renal failure (53). The syndrome is characterized by a triad of features: acute renal insufficiency, microangiopathic hemolytic anemia, and thrombocytopenia. Significant pathological changes include swelling of endothelial cells, widened subendothelial regions, and hypertrophied mesangial cells between glomerular capillaries. These changes combine to narrow the lumina of the glomerular capillaries and afferent arterioles and result in thrombosis of the arteriolar and glomerular microcirculation. Complete obstruction of renal microvessels can produce glomerular and tubular necrosis, with an increased probability of subsequent hypertension or renal failure (91). TTP mainly affects adults and resembles HUS histologically. It is accompanied by distinct neurological abnormalities resulting from blood clots in the brain (74).

INFECTIOUS DOSE

Retrospective analyses of foods associated with outbreaks of EHEC infection revealed that the infectious dose is very low. For example, between 0.3 to 15 CFU of *E. coli* O157:H7 per gram was enumerated in lots of frozen ground beef patties associated with a 1993 multistate outbreak in the western United States. Similarly, 0.3 to 0.4 *E. coli* O157:H7 per gram was detected in several intact packages of salami that were associated with a foodborne outbreak. These data suggest that the infectious dose of *E. coli* O157:H7 may be fewer than 100 cells. In an outbreak of *E. coli* O111:NM infection in Australia, the implicated salami was estimated to contain fewer than 1 cell per 10 g. Additional evidence for a low infectious dose is the capability for person-to-person and waterborne transmission of EHEC infection.

MECHANISMS OF PATHOGENICITY

The precise mechanism of pathogenicity of EHEC has not been fully elucidated. Significant virulence factors, however, have been identified based on histopathology of tissues of HUS and hemorrhagic colitis patients, studies with tissue culture and animal models, and studies using genetic approaches. A general body of knowledge of the pathogenicity of EHEC has been developed that indicates the bacteria cause disease by their ability to adhere to the host cell membrane and colonize the large intestine, and then producing one or more Stx. However, the mechanisms of colon colonization are not well understood. The role of various other potential virulence factors and host factors remains to be determined. Virulence factors involved in the pathogenesis of EHEC are summarized in Table 10.6.

Attaching and Effacing

Substantial progress has been made in characterizing adherence factors and their genes, but the mechanisms of adherence and colonization by *E. coli* O157:H7 and other EHEC remain to be completely elucidated. Studies in animal models revealed that the pathogen can

Table 10.6 Proteins and their genes involved in the pathogenesis of EHEC[a]

Genetic locus	Protein description	Gene	Function
Chromosome (locus of enterocyte effacement)	Intimin	*eae*	Adherence
	Tir	*tir*	Intimin receptor
	Secretion proteins	*espA, espB, espD*	Induces signal transduction
	Type III secretion system	*escC, escD, escF, escJ, escN, escR, escS, escT, escU, escV, sepQ, sepZ*	Apparatus for extracellular protein secretion
Phage	Shiga toxins	*stx1, stx2, stx2c, stx2d*	Inhibits protein synthesis
Plasmid	EHEC-hemolysin	EHEC-*hlyA*	Disrupts cell membrane permeability
	Catalase-peroxidase	*katP*	

[a] Data from references 14, 93, and 98.

colonize the cecum and colon of orally infected gnotobiotic piglets, chickens, infant rabbits, and mice by an attaching-and-effacing (AE) mechanism (Fig. 10.1). The AE lesion is characterized by intimate attachment of the bacteria to intestinal cells, with effacement of the underlying microvilli and accumulation of filamentous actin (F-actin) in the subjacent cytoplasm. The molecular genetics of the AE lesion were first studied in EPEC and subsequently in *E. coli* O157:H7. A three-stage model of adhesion and AE lesion formation by EPEC

has been proposed by Donnenberg and Kaper (39); it includes (i) localized adherence, (ii) signal transduction, and (iii) intimate adherence. Although *E. coli* O157:H7 produces an AE lesion in the large intestine similar to that induced by EPEC, which in contrast occurs predominantly in the small intestine, the identity of the bacterial factors and the genes responsible for each stage of AE lesion formation have not been fully elucidated. It is believed that similar mechanisms, but factors and genes different from those of EPEC, regulate the processes

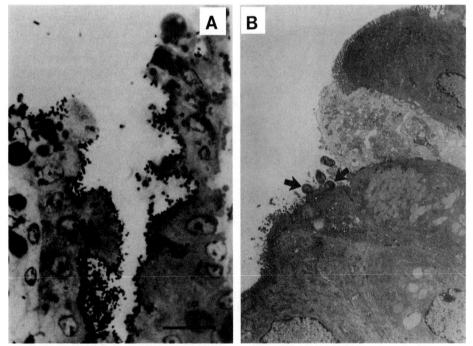

Figure 10.1 AE lesion caused by *E. coli* O157:H7. (A) Irregularly shaped, cuboidal to low columnar crypt neck cells with bacteria adherent to luminal plasma membrane of a gnotobiotic pig (toluidine blue stain). Reproduced with permission from reference 46. (B) Transmission electron photomicrograph of rabbit ileum with adherent bacteria (arrow). Reproduced with permission from reference 107.

of AE formation by *E. coli* O157:H7. A pathogenicity island, involved in AE formation, known as the locus of enterocyte effacement (LEE), has been identified in EPEC and EHEC.

Locus of Enterocyte Effacement

Studies have revealed that genes on the LEE of the EPEC chromosome are involved in producing the AE lesion (84). This 35-kb region is not present in *E. coli* constituting the normal intestinal flora, *E. coli* K-12, or ETEC, but is found in those EPEC and EHEC strains that produce the AE lesion. Such a large block of DNA that encodes multiple virulence factors and is absent from nonpathogenic members of a species is termed a pathogenicity island. The complete DNA sequences of the LEE from EPEC strain E2348/69 and EHEC O157:H7 strain EDL 933 have been determined (43, 98). The EHEC LEE is larger, containing ca. 43 kb compared with ca. 35 kb for EPEC. Most of the size difference is due to a 7.5-kb putative prophage present on the EHEC LEE. The G+C contents of the EPEC LEE and EHEC LEE are 38.3% and 39.6%, respectively, which are significantly lower than the 50.8% G+C content of the total *E. coli* genome. This suggests that EPEC and EHEC may have acquired this pathogenicity island by horizontal transfer from another species. The LEE pathogenicity island is necessary for expression of the AE phenotype in EPEC. A recombinant plasmid clone containing the entire LEE of EPEC is sufficient to confer the AE phenotype when cloned into *E. coli* K-12 or *E. coli* from the normal intestinal flora (85). However, the EHEC LEE is unable to confer the AE phenotype upon *E. coli* K-12, suggesting functional and/or regulatory differences between the LEE pathogenicity islands of EPEC and EHEC (44).

The LEE region consists of three segments with genes of known function. The middle segment includes the *eae* gene, which encodes intimin, and the *tir* gene, which encodes a translocated receptor for intimin. Downstream of *eae* are located the *esp* genes, which encode secreted proteins responsible for inducing epithelial cell signal transduction events leading to the AE lesion. Upstream of *eae* and *tir* are several genes (*esc* and *sep*) that encode a type III secretion system involved in extracellular secretion of proteins encoded by *esp* genes (see below). The nucleotide sequences of EPEC and EHEC LEE display significant divergence, although the average nucleotide identity for the 41 shared genes is 93.9%. The most remarkable feature of the divergent genes is that they include all of those encoding proteins that are known to interact directly with the host, such as *eae*, *espA*, *espB*, *espD*, and *tir*. The *eae* genes are 87.2%

identical, with most of the heterogeneity at the 3′ end that encodes the putative receptor-binding portion. The *tir* genes have only 66.5% identity. The *esp* genes are also divergent: *espB* (26.0% difference), *espA* (15.4%), and *espD* (19.6%). These variations between the EPEC and EHEC LEEs may have resulted from natural selection for adaptation to either host specificity or evasion for the host immune system. In contrast to the hypervariability that occurs in genes encoding proteins known to interact directly with the host, other LEE-encoded proteins that do not interact with the host are highly conserved. The *esc* genes encoding the type III secretion system have 98 to 100% nucleotide identity.

Intimin

Intimin is a 94- to 97-kDa outer membrane protein encoded by *eae* (which stands for *E. coli* attaching and effacing) (40). The protein is produced by those enteric pathogens, including EPEC, EHEC, *Hafnia alvei*, and *Citrobacter rodentium*, that can produce AE lesions. To date, intimin is the sole described adherence factor of *E. coli* O157:H7. Intimin has been identified as important for intestinal colonization in animal models. Mutants with insertions in *eae* are defective in intimate adherence to epithelial cells but still induce the signal transduction activities in cells infected with AE pathogens. The importance of intimin in intestinal colonization also has been directly determined in animal studies with *eae* mutants of *E. coli* O157:H7 and in volunteer studies with *eae* mutants of EPEC. *E. coli* O157:H7 strains mutated in *eae* do not produce AE lesions in piglets, nor do they colonize any intestinal site. Also, there is a difference in the location of intestinal colonization between EHEC and EPEC. EPEC strains produce AE lesions in both the small and large intestines, whereas EHEC strains produce AE lesions only in the large intestine. EHEC and EPEC intimins share only 49% identity (119). Most of the amino acid sequence differences are in the C-terminal region, i.e., the putative receptor binding portion. Volunteer studies have not been done with EHEC, but antibodies to intimin have been identified in patients with HUS.

Tir

The Tir protein was initially identified as a 90-kDa eukaryotic cytoskeletal protein (Hp90) that became phosphorylated on a tyrosine residue in response to EPEC infection. Recently, Hp90 was identified as a bacterial protein that is translocated from EPEC to the host cell, where it is phosphorylated and serves as a receptor for intimin (108). Hp90 has been renamed Tir, for translocated intimin receptor. Tir is produced in the bacterial cell as a 78-kDa unphosphorylated protein and is subsequently translocated to the host cell, where it is phosphorylated, increasing its molecular mass to 90 kDa. Translocation of Tir is dependent upon a functional type III secretion system. Tyrosine phosphorylation of Tir may vary by EHEC serotypes; e.g., it has not been observed in EHEC O157:H7, but it has been observed in EHEC O26:NM. Studies revealed that tyrosine protein kinase inhibitors did not inhibit AE lesion formation. Hence, the physiological significance of Tir phosphorylation is not clear.

Tir consists of at least three functional regions, i.e., an extracellular domain that interacts with intimin, a transmembrane domain, and a cytoplasmic domain that can induce focusing of the polymerized actin and other cytoskeletal proteins beneath intimately attached bacteria to produce the characteristic pedestallike structure of the AE lesion (71). As most bacterial pathogens exploit existing host cell components as receptors, the ability of EPEC and EHEC to supply their own mucosal receptors may provide additional advantages for initiating the process of adhesion, motility, and signal transduction.

Type III Secretion System and Secreted Proteins

The type III secretion system, present in many gram-negative bacterial pathogens, is induced upon contact with the host cells and specifically exports virulence factors delivered directly to eukaryotic cells. In pathogenic *E. coli*, at least 10 *esc* (for *E. coli* secretion) genes encoding proteins homologous to type III system proteins in other gram-negative pathogens have been identified (98). Two LEE *sep* (for secretion of EPEC proteins) genes, *sepQ* and *speZ*, are also involved in type III secretion but are heterologous to systems of other pathogens. Using the type III secretion system, EPEC and EHEC secrete several proteins (Esps, for *E. coli*-secreted proteins) responsible for signal transduction events during AE lesion formation (76). These proteins include EspA (25 kDa), EspB (38 kDa), and EspD (40 kDa). EspA is a structural protein and a major component of a large filamentous organelle that is transiently present on the bacterial surface and interacts with the host cell during the early stage of AE lesion formation. The protein also is a component of a translocation apparatus and is essential for the translocation of EspB and Tir to a host cell. EspA filaments have a hollow cylindrical structure that forms a channel through which proteins are delivered into the host cell. EspD, whose function is unknown, may also be a component of the EspA filaments. EspB is involved in forming the translocation channel in the host cell membrane.

Mutation of the genes encoding any of these proteins abolishes signal transduction in epithelial cells by wild-type EPEC and the AE lesion histopathology.

60-MDa Plasmid (pO157)

E. coli O157:H7 isolates possess a plasmid (pO157) of approximately 60 MDa (unrelated to the 60-MDa plasmid present in EPEC) that contains DNA sequences common to plasmids present in other serotypes of EHEC isolated from patients with hemorrhagic colitis (60). Because of the consistent association of the 60-MDa plasmid with several EHEC serotypes, the plasmid has been implicated in the pathogenesis of EHEC infections. However, its exact role in virulence has not been fully elucidated.

Based on DNA sequence analysis, pO157 is a 92-kb F-like plasmid composed of segments of putative virulence genes in a framework of replication and maintenance regions, with seven insertion sequence elements located largely at the boundaries of the virulence segments (14). There are 100 open reading frames (ORFs), of which 19 have been previously sequenced and implicated as potential virulence genes, including those encoding EHEC-hemolysin and catalase-peroxidase. The term EHEC-hemolysin is used to distinguish it from alpha-hemolysin, to which it is related but not identical. EHEC-hemolysin belongs to the repeats-in-toxin family of exoproteins. Four gene products of the *hlyCABD* operon encode a pore-forming cytolysin and its secretion apparatus (104). Toxicity results from the insertion of HlyA into the cytoplasmic membrane of target mammalian cells, thereby disrupting permeability. The EHEC catalase-peroxidase is encoded by *katP*, whose product is a bifunctional periplasmic enzyme that protects the bacterium against oxidative stress, a possible defense strategy of mammalian cells during bacterial infection (12). The occurrence of *katP* in EHEC strains is associated with the EHEC-hemolysin operon.

One newly identified large ORF encodes a putative cytotoxin active site shared with a large clostridial toxin family (14). The unusually large ORF of 3,169 amino acids on pO157 has strong sequence similarity within the first 700 residues to the activity-containing N-terminal domain of the large clostridial toxins which includes ToxA (2,710 amino acids) and ToxB (2,366 amino acids) of *Clostridium difficile*. These AB cytotoxins share a C-terminal domain that functions in toxin entry into the cell and an N-terminal glucosyltransferase that modifies G proteins, resulting in the altered regulation of the cytoskeleton. This is a significant finding, because in cases of *E. coli* O157:H7 colitis, intestinal lesions are produced that are similar to the toxin-mediated damage to intestinal cells seen in *C. difficile* infections.

Shiga Toxins

EHEC produce one or two Stx. The nomenclature of the Stx family and their important characteristics are listed in Table 10.7. Molecular studies on Stx1 from different *E. coli* strains revealed that Stx1 is either completely identical to the Stx of *S. dysenteriae* type 1 or differs by only one amino acid. Unlike Stx1, toxins of the Stx2 group are not neutralized by serum raised against Stx and do not cross-hybridize with Stx1-specific DNA probes. There is sequence and antigenic variation within toxins of the Stx2 family produced by *E. coli* O157:H7 and other STEC. The subgroups of Stx2 include Stx2, Stx2c, Stx2d, and Stx2e (87). The Stx2c subgroup is approximately 97% related to the amino acid sequence of the B subunits of Stx2, whereas the A subunit of Stx2c shares 98 to 100% amino acid sequence homology with Stx2. Another subtype of the Stx2 family is Stx2e, which is associated with edema disease that principally occurs in piglets (115). Stx2e shares 93% and 84% amino acid sequence homology with the A and B subunits, respectively, of Stx2. Hence, the Stx2-related toxins have only partial serological reactivity with anti-Stx2 serum. Additional Stx2 variants, such as Stx2f of STEC strains isolated from feral pigeons, have been described recently, suggesting there is great heterogeneity among the Stx2-related toxins (105).

Structure of the Stx Family

Shiga toxins are holotoxins composed of a single enzymatic A subunit of approximately 32 kDa in association with a pentamer of receptor-binding B subunits of 7.7 kDa (93). The Stx A subunit can be split by trypsin into an enzymatic A1 fragment (approximately 27 kDa) and a carboxyl-terminal A2 fragment (approximately 4 kDa) that links A1 to the B subunits. The A1 and A2 subunits remain linked by a single disulfide bond until the enzymatic fragment is released and enters the cytosol of a susceptible mammalian cell. Each B subunit is composed of six antiparallel strands forming a closed barrel, capped by a single helix between strands 3 and 4. The A subunit lies on the side of the B subunit pentamer, nearest to the C-terminal end of the B subunit helices. The A subunit interacts with the B subunit pentamer through a hydrophobic helix which extends half of the 2.0-nm length of the pore in the B pentamer. This pore is lined with the hydrophobic side chains of the B subunit helices. The A subunit also interacts with the B subunit via a four-stranded mixed sheet composed of residues of both the A2 and A1 fragments.

Genetics of Stx

While all *stx1* operons examined thus far are essentially identical and located on the genomes of lysogenic lambdoid bacteriophages, there is considerable heterogeneity in the *stx2* family (93). Unlike the genes of other Stx2 that are located on bacteriophage that integrate into the chromosome, Stx of *S. dysenteriae* type 1 and Stx2e are encoded on chromosomal genes (115). Subsequently, the genes of additional variants of Stx2 have been isolated from STEC. A sequence comparison of the growing *stx2* family indicates that genetic recombination among the B subunit genes, rather than base substitutions, has given rise to the variants of Stx2 present in human and animal strains of *E. coli* (48, 49). However, the operons for every member of the Stx subgroups are organized identically; the A and B subunit genes are arranged in tandem and separated by a 12- to 15-nucleotide gap. The operons

Table 10.7 Nomenclature and biological characteristics of Shiga toxins (Stx)[a]

		Biological characteristics				
		Amino acid homology to Stx2 (%)			Activated by	
Nomenclature	Genetic loci	A subunit	B subunit	Receptor	intestinal mucus	Disease
Stx	Chromosome	55	57	Gb$_3$	No	Human diarrhea, HC,[b] HUS[c]
Stx1	Phage	55	57	Gb$_3$	No	Human diarrhea, HC, HUS
Stx2	Phage	100	100	Gb$_3$	No	Human diarrhea, HC, HUS
Stx2c	Phage	100	97	Gb$_3$	No	Human diarrhea, HC, HUS
Stx2d	Phage	99	97	Gb$_3$	Yes	Human diarrhea, HC, HUS
Stx2e	Chromosome	93	84	Gb$_4$	No	Pig edema disease

[a] Data from reference 87.
[b] HC, hemorrhagic colitis.
[c] HUS, hemolytic-uremic syndrome.

are transcribed from a promoter that is located 5' to the A subunit gene, and each gene is preceded by a putative ribosome-binding site. The existence of an independent promoter for the B subunit genes has been suggested. The holotoxin stoichiometry suggests that expression of the A and B subunit genes is differentially regulated, permitting overproduction of the B polypeptides. Finally, Stx and Stx1 production are negatively regulated at the transcriptional level by an iron-Fur protein corepressor complex that binds at the *stx1* promoter but is unaffected by temperature, whereas Stx2 production is neither iron nor temperature regulated (16, 93).

Receptors

All members of the Stx family bind to globoseries glycolipids on the eukaryotic cell surface; Stx, Stx1, Stx2, Stx2c, and Stx2d bind to glycolipid globotriaosylceramide (Gb$_3$), whereas Stx2e primarily binds to glycolipid globotetraosylceramide (Gb$_4$) (31, 87). The alteration of binding specificity between Stx2e and the rest of the Stx family is related to carbohydrate specificity of receptors (79). The amino acid composition of B subunits of Stx2 and Stx2e differ at only 11 positions, yet Stx2e binds primarily to Gb$_4$, whereas Stx2 binds only to Gb$_3$ (112). High-affinity binding also depends on multivalent presentation of the carbohydrate, as would be provided by glycolipids in a membrane. Studies by Kiarash et al. (72) revealed that the affinity of Stx1 for Gb$_3$ isoforms is influenced by fatty acyl chain length and by its level of saturation. Stx1 binds preferentially to Gb$_3$ containing C20:1 fatty acid, whereas Stx2c prefers Gb$_3$ containing C18:1 fatty acid (97). The basis for these findings may be related to the ability of different Gb$_3$ isoforms to present multivalent sugar-binding sites in the optimal orientation and position at the membrane surface. It is also possible that different fatty acyl groups affect the conformation of individual receptor epitopes on the sugar.

Mode of Action of the Stx

STX act by inhibiting protein synthesis. Each of the B subunits is capable of binding with high affinity to an unusual disaccharide linkage (galactose 1-4 galactose) in the terminal trisaccharide sequence of Gb$_3$ (or Gb$_4$) (110). Following binding to the glycolipid receptor, the toxin is endocytosed from clathrin-coated pits and transferred first to the *trans* Golgi network and subsequently to the endoplasmic reticulum and nuclear envelope (102). While it appears that transfer of the toxin to the Golgi apparatus is essential for intoxication, the mechanism of entry of the A subunit from the endosome to the cytosol, and particularly the role of the B subunit in the process, remains unclear. In the cytosol,

the A subunit undergoes partial proteolysis and splits into a 27-kDa active intracellular enzyme (A1) and a 4-kDa fragment (A2) bridged by a disulfide bond. Although the entire toxin is necessary for its toxic effect on whole cells, the A1 subunit is capable of cleaving the N-glycoside bond in one adenosine position of the 28S rRNA that comprises 60S ribosomal subunits (103). This elimination of a single adenine nucleotide inhibits the elongation factor-dependent binding to ribosomes of aminoacyl-bound transfer RNA molecules. Peptide chain elongation is truncated, and overall protein synthesis is suppressed, resulting in cell death.

The Role of Stx in Disease

The precise role of Stx in mediating colonic disease, HUS, and neurological disorders has not been fully elucidated. There is no satisfactory animal model for hemorrhagic colitis or HUS, and the severity of disease precludes study of experimental infections in humans. Therefore, our present understanding of the role of Stx in causing disease is obtained from a combination of studies, including histopathology of diseased human tissues, animal models, and endothelial tissue culture cells. Results of recent studies support the concept that Stx contribute to pathogenesis by directly damaging vascular endothelial cells in certain organs, thereby disrupting the homeostatic properties of these cells (67).

The involvement of Stx in enterocolitis was demonstrated when fluid accumulation and histological damage occurred after purified Stx was injected into ligated rabbit intestinal loops. The fluid secretion may be due to the selective killing of absorptive villus tip intestinal epithelial cells by Stx. However, intravenous administration of Stx to rabbits can produce nonbloody diarrhea, suggesting other mechanisms of diarrhea are possible. Studies with genetically mutated STEC strains also indicate that Stx has a role in intestinal disease, but the significance of Stx in provoking a diarrheal response differs depending upon the animal model used.

Many epidemiologic studies have identified a correlation between enteric infection with *E. coli* O157:H7 and development of HUS in humans. Histopathologic examination of kidney tissue from HUS patients revealed profound structural alterations in the glomeruli, the basic filtration unit of the kidney (82). Glomerular endothelial cells were swollen and were often detached from the glomerular basement membrane. Hence, subendothelial matrix components may be exposed and serve as sites for platelet adherence and activation. The damage caused by Stx is often not limited to the glomeruli. Arteriolar damage involving internal cell proliferation, fibrin thrombi deposition, and perivascular inflammation occurs (113).

Cortical necrosis also occurs in a small number of HUS cases. In addition, human glomerular endothelial cells are sensitive to the direct cytotoxic action of bacterial endotoxin. Endotoxin in the presence of Stx also can activate macrophage and polymorphonuclear neutrophils to synthesize and release cytokines, superoxide radicals, or proteinases and amplify endothelial cell damage (81).

Neurological symptoms in patients and laboratory animals infected with *E. coli* O157:H7 also have been described and may be caused by secondary neuron disturbances that result from endothelial cell damage by Stx. Studies in mice perorally administered an *E. coli* O157:H− strain revealed that Stx2v impaired the blood-brain barrier and damaged neuron fibers, resulting in death (47). Presence of the toxin in neurons was verified by immunoelectron microscopy.

Epidemiologic studies also have revealed that *E. coli* O157:H7 strains isolated from patients with hemorrhagic colitis usually produce both Stx1 and Stx2 or Stx2 only; isolates producing only Stx 1 are uncommon. Patients infected with *E. coli* O157:H7 producing only Stx2 or Stx2 in combination with Stx1 were more likely to develop serious renal or circulatory complications in comparison with patients infected with EHEC strains producing Stx1 only.

CONCLUDING REMARKS

The serious nature of the symptoms of hemorrhagic colitis and HUS caused by *E. coli* O157:H7 places this pathogen in a category apart from other foodborne pathogens that typically cause only mild symptoms. The severity of the illness it causes combined with its apparent low infectious dose (<100 cells) qualifies *E. coli* O157:H7 to be among the most serious of known foodborne pathogens. Although the pathogen has been isolated from a variety of domestic animals and wildlife, cattle are a major reservoir of *E. coli* O157:H7, with undercooked ground beef being the single most frequently implicated vehicle of transmission. An important feature of this pathogen is its acid tolerance. Outbreaks have been associated with consumption of contaminated high-acid foods, including apple juice and fermented dry salami. Many other foods and recreational and drinking water also have been identified as vehicles of transmission of *E. coli* O157:H7 infections. The mechanisms of pathogenicity of *E. coli* O157:H7 have not been fully elucidated; however, production of one or more Shiga toxins and AE adherence are important virulence factors.

EHEC other than O157:H7 have been increasingly associated with cases of HUS. Over 200 non-O157 STEC serotypes have been isolated from humans, but not all of these serotypes have been shown to cause illness. Some STEC may have a low potential to cause HUS; other non-O157 STEC isolates, found in healthy individuals, may not be pathogens. Furthermore, multiple STEC serotypes have been isolated from a single patient. The contribution of each serotype to the pathogenesis of disease is difficult to determine. *E. coli* O157:H7 is still by far the most important serotype of STEC in North America. Isolation of non-O157:H7 STEC requires techniques not generally used in clinical laboratories; hence, these bacteria are rarely sought or detected in routine practice. Recognition of non-O157 EHEC strains in foodborne illness necessitates identification of serotypes of EHEC other than O157:H7 in persons with bloody diarrhea and/or HUS and preferably in implicated food. The increased availability in clinical laboratories of techniques such as testing for Stx or their genes and identification of other virulence markers unique for EHEC may enhance the detection of disease attributable to non-O157 EHEC.

References

1. Ackers, M. L., B. E. Mahon, E. Leahy, B. Goode, T. Damrow, P. S. Hayes, W. F. Bibb, D. H. Rice, T. J. Barrett, L. Hutwagner, P. M. Griffin, and L. Slutsker. 1998. An outbreak of *Escherichia coli* O157:H7 infections associated with leaf lettuce consumption. *J. Infect. Dis.* 177:1588–1593.

2. Ahmed, S., and M. Donaghy. 1998. An outbreak of *Escherichia coli* O157:H7 in central Scotland, p. 59–65. *In* J. Kaper and A. O'Brien (ed.), *Escherichia coli O157:H7 and Other Shiga Toxin-Producing E. coli Strains.* ASM Press, Washington, D.C.

3. Barwick, R. S., D. A. Levy, G. F. Craun, M. J. Beach, and R. L. Calderon. 2000. Surveillance for waterborne-disease outbreaks—United States, 1997–1998. *Morb. Mortal. Wkly. Rep. CDC Surveill. Summ.* 49:1–21.

4. Bell, B. P., M. Goldoft, P. M. Griffin, M. A. Davis, D. C. Gordon, P. I. Tarr, C. A. Bartleson, J. H. Lewis, T. J. Barrett, J. G. Wells, et al. 1994. A multistate outbreak of *Escherichia coli* O157:H7-associated bloody diarrhea and hemolytic uremic syndrome from hamburgers. The Washington experience. *JAMA* 272:1349–1353.

5. Belongia, E. A., K. L. MacDonald, G. L. Parham, K. E. White, J. A. Korlath, M. N. Lobato, S. M. Strand, K. A. Casale, and M. T. Osterholm. 1991. An outbreak of *Escherichia coli* O157:H7 colitis associated with consumption of precooked meat patties. *J. Infect. Dis.* 164:338–343.

6. Besser, R. E., P. M. Griffin, and L. Slutsker. 1999. *Escherichia coli* O157:H7 gastroenteritis and the hemolytic uremic syndrome: an emerging infectious disease. *Annu. Rev. Med.* 50:355–367.

7. Besser, R. E., S. M. Lett, J. T. Weber, M. P. Doyle, T. J. Barrett, J. G. Wells, and P. M. Griffin. 1993. An

outbreak of diarrhea and hemolytic uremic syndrome from *Escherichia coli* O157:H7 in fresh-pressed apple cider. *JAMA* **269**:2217–2220.

8. **Beutin, L., D. Geier, H. Steinruck, S. Zimmermann, and F. Scheutz.** 1993. Prevalence and some properties of verotoxin (Shiga-like toxin)-producing *Escherichia coli* in seven different species of healthy domestic animals. *J. Clin. Microbiol.* **31**:2483–2488.

9. **Bopp, C. A., K. D. Greene, F. P. Downes, E. G. Sowers, J. G. Wells, and I. K. Wachsmuth.** 1987. Unusual verotoxin-producing *Escherichia coli* associated with hemorrhagic colitis. *J. Clin. Microbiol.* **25**:1486–1489.

10. **Brackett, R., Y. Hao, and M. Doyle.** 1994. Ineffectiveness of hot acid sprays to decontaminate *Escherichia coli* O157:H7 on beef. *J. Food Prot.* **57**:198–203.

11. **Brown, C. A., B. G. Harmon, T. Zhao, and M. P. Doyle.** 1997. Experimental *Escherichia coli* O157:H7 carriage in calves. *Appl. Environ. Microbiol.* **63**:27–32.

12. **Brunder, W., H. Schmidt, and H. Karch.** 1996. KatP, a novel catalase-peroxidase encoded by the large plasmid of enterohaemorrhagic *Escherichia coli* O157:H7. *Microbiology* **142**:3305–3315.

13. **Buchanan, R. L., S. G. Edelson, and G. Boyd.** 1999. Effects of pH and acid resistance on the radiation resistance of enterohemorrhagic *Escherichia coli*. *J. Food Prot.* **62**:219–228.

14. **Burland, V., Y. Shao, N. T. Perna, G. Plunkett, H. J. Sofia, and F. R. Blattner.** 1998. The complete DNA sequence and analysis of the large virulence plasmid of *Escherichia coli* O157:H7. *Nucleic Acids Res.* **26**:4196–4204.

15. **Calderwood, S., D. Acheson, G. Keusch, T. Barrett, P. Griffin, N. Strockbine, B. Swaminathan, J. Kaper, M. Levine, B. Kaplan, H. Karch, A. O'Brien, T. Obrig, Y. Takeda, P. Tarr, and I. Wachsmuth.** 1996. Proposed new nomenclature for SLT(VT) family. *ASM News* **62**:118–119.

16. **Calderwood, S. B., and J. J. Mekalanos.** 1987. Iron regulation of Shiga-like toxin expression in *Escherichia coli* is mediated by the fur locus. *J. Bacteriol.* **169**:4759–4764.

17. **Carter, A. O., A. A. Borczyk, J. A. Carlson, B. Harvey, J. C. Hockin, M. A. Karmali, C. Krishnan, D. A. Korn, and H. Lior.** 1987. A severe outbreak of *Escherichia coli* O157:H7-associated hemorrhagic colitis in a nursing home. *N. Engl. J. Med.* **317**:1496–1500.

18. **Centers for Disease Control and Prevention.** 1995. Community outbreak of hemolytic uremic syndrome attributable to *Escherichia coli* O111:NM—South Australia, 1995. *Morb. Mortal. Wkly. Rep.* **44**:550–558.

19. **Centers for Disease Control and Prevention.** 2000. *Escherichia coli* O111:H8 outbreak among teenage campers—Texas, 1999. *Morb. Mortal. Wkly. Rep.* **49**:321–324.

20. **Centers for Disease Control and Prevention.** 1995. *Escherichia coli* O157:H7 outbreak linked to commercially distributed dry-cured salami—Washington and California, 1994. *Morb. Mortal. Wkly. Rep.* **44**:157–160.

21. **Centers for Disease Control and Prevention.** 1991. Foodborne outbreak of gastroenteritis caused by *Escherichia*

coli O157:H7—North Dakota, 1990. *Morb. Mortal. Wkly. Rep.* **40**:265–267.

22. **Centers for Disease Control and Prevention.** 1996. Lake-associated outbreak of *Escherichia coli* O157:H7—Illinois, 1995. *Morb. Mortal. Wkly. Rep.* **45**:437–439.

23. **Centers for Disease Control and Prevention.** 1995. Outbreak of acute gastroenteritis attributable to *Escherichia coli* serotype O104:H21—Helena, Montana, 1994. *Morb. Mortal. Wkly. Rep.* **44**:501–503.

24. **Centers for Disease Control and Prevention.** 1999. Outbreak of *Escherichia coli* O157:H7 and *Campylobacter* among attendees of the Washington County Fair—New York, 1999. *Morb. Mortal. Wkly. Rep.* **48**:803–805.

25. **Centers for Disease Control and Prevention.** 2000. Outbreak of *Escherichia coli* O157:H7 infection associated with eating fresh cheese curds—Wisconsin, June 1998. *Morb. Mortal. Wkly. Rep.* **49**:911–913.

26. **Centers for Disease Control and Prevention.** 1996. Outbreak of *Escherichia coli* O157:H7 infections associated with drinking unpasteurized commercial apple juice—British Columbia, California, Colorado, and Washington, October 1996. *Morb. Mortal. Wkly. Rep.* **45**:975.

27. **Centers for Disease Control and Prevention.** 2000. Preliminary FoodNet data on the incidence of foodborne illnesses—selected sites, United States, 1999. *Morb. Mortal. Wkly. Rep.* **49**:201–205.

28. **Centers for Disease Control and Prevention.** 1993. Preliminary report: foodborne outbreak of *Escherichia coli* O157:H7 infections from hamburgers—western United States, 1993. *Morb. Mortal. Wkly. Rep.* **42**:85–86.

29. **Cizek, A., P. Alexa, I. Literak, J. Hamrik, P. Novak, and J. Smola.** 1999. Shiga toxin-producing *Escherichia coli* O157 in feedlot cattle and Norwegian rats from a large-scale farm. *Lett. Appl. Microbiol.* **28**:435–439.

30. **Clavero, M., J. Monk, L. Beuchat, M. Doyle, and R. Brackett.** 1994. Inactivation of *Escherichia coli* O157:H7, salmonellae, and *Campylobacter jejuni* in raw ground beef by gamma irradiation. *Appl. Environ. Microbiol.* **60**:2069–2075.

31. **Cohen, A., G. E. Hannigan, B. R. Williams, and C. A. Lingwood.** 1987. Roles of globotriosyl- and galabiosyl-ceramide in verotoxin binding and high affinity interferon receptor. *J. Biol. Chem.* **262**:17088–17091.

32. **Cohen, M. B.** 1996. *Escherichia coli* O157:H7 infections: a frequent cause of bloody diarrhea and the hemolytic-uremic syndrome. *Adv. Pediatr.* **43**:171–207.

33. **Cray, W. C., Jr., and H. W. Moon.** 1995. Experimental infection of calves and adult cattle with *Escherichia coli* O157:H7. *Appl. Environ. Microbiol.* **61**:1586–1590.

34. **D'Aoust, J., C. Park, R. Szabo, E. Todd, B. Emmons, and R. McKellar.** 1988. Thermal inactivation of *Campylobacter* species, *Yersinia enterocolitca*, and hemorrhagic *Escherichia coli* O157:H7 in fluid milk. *J. Dairy Sci.* **71**:3230–3236.

35. **Dean-Nystrom, E. A., B. T. Bosworth, W. C. Cray, Jr., and H. W. Moon.** 1997. Pathogenicity of *Escherichia coli* O157:H7 in the intestines of neonatal calves. *Infect. Immun.* **65**:1842–1848.

36. Deschenes, G., C. Casenave, F. Grimont, J. C. Desenclos, S. Benoit, M. Collin, S. Baron, P. Mariani, P. A. Grimont, and H. Nivet. 1996. Cluster of cases of haemolytic uraemic syndrome due to unpasteurised cheese. *Pediatr. Nephrol.* **10**:203–205.

37. Dev, V. J., M. Main, and I. Gould. 1991. Waterborne outbreak of *Escherichia coli* O157. *Lancet* **337**:1412.

38. Diez-Gonzalez, F., T. R. Callaway, M. G. Kizoulis, and J. B. Russell. 1998. Grain feeding and the dissemination of acid-resistant *Escherichia coli* from cattle. *Science* **281**:1666–1668.

39. Donnenberg, M. S., and J. B. Kaper. 1992. Enteropathogenic *Escherichia coli*. *Infect. Immun.* **60**:3953–3961.

40. Donnenberg, M. S., S. Tzipori, M. L. McKee, A. D. O'Brien, J. Alroy, and J. B. Kaper. 1993. The role of the eae gene of enterohemorrhagic *Escherichia coli* in intimate attachment in vitro and in a porcine model. *J. Clin. Invest.* **92**:1418–1424.

41. Doyle, M. P., and J. L. Schoeni. 1984. Survival and growth characteristics of *Escherichia coli* associated with hemorrhagic colitis. *Appl. Environ. Microbiol.* **48**:855–856.

42. Elder, R. O., J. E. Keen, G. R. Siragusa, G. A. Barkocy-Gallagher, M. Koohmaraie, and W. W. Laegreid. 2000. From the cover: correlation of enterohemorrhagic *Escherichia coli* O157 prevalence in feces, hides, and carcasses of beef cattle during processing. *Proc. Natl. Acad. Sci. USA* **97**:2999–3003.

43. Elliott, S. J., L. A. Wainwright, T. K. McDaniel, K. G. Jarvis, Y. K. Deng, L. C. Lai, B. P. McNamara, M. S. Donnenberg, and J. B. Kaper. 1998. The complete sequence of the locus of enterocyte effacement (LEE) from enteropathogenic *Escherichia coli* E2348/69. *Mol. Microbiol.* **28**:1–4.

44. Elliott, S. J., J. Yu, and J. B. Kaper. 1999. The cloned locus of enterocyte effacement from enterohemorrhagic *Escherichia coli* O157:H7 is unable to confer the attaching and effacing phenotype upon *E. coli* K-12. *Infect. Immun.* **67**:4260–4263.

45. Farina, C., A. Goglio, G. Conedera, F. Minelli, and A. Caprioli. 1996. Antimicrobial susceptibility of *Escherichia coli* O157 and other enterohaemorrhagic *Escherichia coli* isolated in Italy. *Eur. J. Clin. Microbiol. Infect. Dis.* **15**:351–353.

46. Francis, D. H., J. E. Collins, and J. R. Duimstra. 1986. Infection of gnotobiotic pigs with an *Escherichia coli* O157:H7 strain associated with an outbreak of hemorrhagic colitis. *Infect. Immun.* **51**:953–956.

47. Fujii, J., T. Kita, S. Yoshida, T. Takeda, H. Kobayashi, N. Tanaka, K. Ohsato, and Y. Mizuguchi. 1994. Direct evidence of neuron impairment by oral infection with verotoxin-producing *Escherichia coli* O157:H⁻ in mitomycin-treated mice. *Infect. Immun.* **62**:3447–3453.

48. Gannon, V. P., and C. L. Gyles. 1990. Characteristics of the Shiga-like toxin produced by *Escherichia coli* associated with porcine edema disease. *Vet. Microbiol.* **24**:89–100.

49. Gannon, V. P., C. Teerling, S. A. Masri, and C. L. Gyles. 1990. Molecular cloning and nucleotide sequence of an-

50. Garber, L. P., S. J. Wells, D. D. Hancock, M. P. Doyle, J. Tuttle, J. A. Shere, and T. Zhao. 1995. Risk factors for fecal shedding of *Escherichia coli* O157:H7 in dairy calves. *J. Am. Vet. Med. Assoc.* **207**:46–49.

51. Glass, K., J. Loeffelholz, J. Ford, and M. Doyle. 1992. Fate of *Escherichia coli* O157:H7 as affected by pH or sodium chloride in fermented, dry sausage. *Appl. Environ. Microbiol.* **58**:2513–2516.

52. Goodfellow, S., and W. Brown. 1987. Fate of *Salmonella* inoculated into beef for cooking. *J. Food. Prot.* **41**:598–605.

53. Gransden, W. R., M. A. Damm, J. D. Anderson, J. E. Carter, and H. Lior. 1986. Further evidence associating hemolytic uremic syndrome with infection by verotoxin-producing *Escherichia coli* O157:H7. *J. Infect. Dis.* **154**:522–524.

54. Griffin, P. M. 1995. *Escherichia coli* O157:H7 and other enterohemorrhagic *Escherichia coli*, p. 739–761. *In* M. J. Blaser, P. D. Smith, J. I. Ravdin, H. B. Greenberg, and R. L. Guerrant (ed.), *Infections of Gastrointestinal Tract*. Raven Press, New York, N.Y.

55. Griffin, P. M. 2000. Personal communication.

56. Griffin, P. M., and R. V. Tauxe. 1991. The epidemiology of infections caused by *Escherichia coli* O157:H7, other enterohemorrhagic *E. coli*, and the associated hemolytic uremic syndrome. *Epidemiol. Rev.* **13**:60–98.

57. Hancock, D. D., T. E. Besser, and D. H. Rice. 1998. Ecology of *Escherichia coli* O157:H7 in cattle and impact of management practices, p. 85–91. *In* J. Kaper and A. O'Brien (ed.), *Escherichia coli* O157:H7 and Other Shiga Toxin-Producing E. coli Strains. ASM Press, Washington, D.C.

58. Hashimoto, H., K. Mizukoshi, M. Nishi, T. Kawakita, S. Hasui, Y. Kato, Y. Ueno, R. Takeya, N. Okuda, and T. Takeda. 1999. Epidemic of gastrointestinal tract infection including hemorrhagic colitis attributable to Shiga toxin 1-producing *Escherichia coli* O118:H2 at a junior high school in Japan. *Pediatrics* **103**:E2.

59. Hilborn, E. D., J. H. Mermin, P. A. Mshar, J. L. Hadler, A. Voetsch, C. Wojtkunski, M. Swartz, R. Mshar, M. A. Lambert-Fair, J. A. Farrar, M. K. Glynn, and L. Slutsker. 1999. A multistate outbreak of *Escherichia coli* O157:H7 infections associated with consumption of mesclun lettuce. *Arch. Intern. Med.* **159**:1758–1764.

60. Hofinger, C., H. Karch, and H. Schmidt. 1998. Structure and function of plasmid pColD157 of enterohemorrhagic *Escherichia coli* O157 and its distribution among strains from patients with diarrhea and hemolytic-uremic syndrome. *J. Clin. Microbiol.* **36**:24–29.

61. Hovde, C. J., P. R. Austin, K. A. Cloud, C. J. Williams, and C. W. Hunt. 1999. Effect of cattle diet on *Escherichia coli* O157:H7 acid resistance. *Appl. Environ. Microbiol.* **65**:3233–3235.

62. Isaacson, M., P. Canter, P. Effler, L. Arntzen, P. Bomans, and R. Heenon. 1993. Haemorrhagic colitis epidemic in Africa. *Lancet* **341**:961.

other variant of the *Escherichia coli* Shiga-like toxin II family. *J. Gen. Microbiol.* **136**:1125–1135.

63. Johnson, R., R. Clarke, J. Wilson, S. Read, K. Rahn, S. Renwick, K. Sandhu, D. Alves, M. Karmali, H. Lior, S. Mcewen, J. Spika, and C. Gyles. 1996. Growing concerns and recent outbreaks involving non-O157:H7 serotypes of verotoxigenic *Escherichia coli. J. Food Prot.* 59:1112–1122.

64. Johnson, W. M., H. Lior, and G. S. Bezanson. 1983. Cytotoxic *Escherichia coli* O157:H7 associated with haemorrhagic colitis in Canada. *Lancet* i:76.

65. Karch, H., H. Russmann, H. Schmidt, A. Schwarzkopf, and J. Heesemann. 1995. Long-term shedding and clonal turnover of enterohemorrhagic *Escherichia coli* O157 in diarrheal diseases. *J. Clin. Microbiol.* 33:1602–1605.

66. Karmali, M. A., M. Petric, C. Lim, P. C. Fleming, G. S. Arbus, and H. Lior. 1985. The association between idiopathic hemolytic uremic syndrome and infection by verotoxin-producing *Escherichia coli. J. Infect. Dis.* 151:775–782.

67. Karpman, D., A. Andreasson, H. Thysell, B. S. Kaplan, and C. Svanborg. 1995. Cytokines in childhood hemolytic uremic syndrome and thrombotic thrombocytopenic purpura. *Pediatr. Nephrol.* 9:694–699.

68. Keene, W. E., K. Hedberg, D. E. Herriott, D. D. Hancock, R. W. McKay, T. J. Barrett, and D. W. Fleming. 1997. A prolonged outbreak of *Escherichia coli* O157:H7 infections caused by commercially distributed raw milk. *J. Infect. Dis.* 176:815–8.

69. Keene, W. E., J. M. McAnulty, F. C. Hoesly, L. P. Williams, Jr., K. Hedberg, G. L. Oxman, T. J. Barrett, M. A. Pfaller, and D. W. Fleming. 1994. A swimming-associated outbreak of hemorrhagic colitis caused by *Escherichia coli* O157:H7 and *Shigella sonnei. N. Engl. J. Med.* 331:579–584.

70. Keene, W. E., E. Sazie, J. Kok, D. H. Rice, D. D. Hancock, V. K. Balan, T. Zhao, and M. P. Doyle. 1997. An outbreak of *Escherichia coli* O157:H7 infections traced to jerky made from deer meat. *JAMA* 277:1229–1231.

71. Kenny, B., R. DeVinney, M. Stein, D. J. Reinscheid, E. A. Frey, and B. B. Finlay. 1997. Enteropathogenic *E. coli* (EPEC) transfers its receptor for intimate adherence into mammalian cells. *Cell* 91:511–520.

72. Kiarash, A., B. Boyd, and C. A. Lingwood. 1994. Glycosphingolipid receptor function is modified by fatty acid content. Verotoxin 1 and verotoxin 2c preferentially recognize different globotriaosyl ceramide fatty acid homologues. *J. Biol. Chem.* 269:11138–11146.

73. Kim, H. H., M. Samadpour, L. Grimm, C. R. Clausen, T. E. Besser, M. Baylor, J. M. Kobayashi, M. A. Neill, F. D. Schoenknecht, and P. I. Tarr. 1994. Characteristics of antibiotic-resistant *Escherichia coli* O157:H7 in Washington State, 1984–1991. *J. Infect. Dis.* 170:1606–1609.

74. Kovacs, M. J., J. Roddy, S. Gregoire, W. Cameron, L. Eidus, and J. Drouin. 1990. Thrombotic thrombocytopenic purpura following hemorrhagic colitis due to *Escherichia coli* O157:H7. *Am. J. Med.* 88:177–179.

75. Kudva, I. T., P. G. Hatfield, and C. J. Hovde. 1996. *Escherichia coli* O157:H7 in microbial flora of sheep. *J. Clin. Microbiol.* 34:431–433.

76. Lai, L. C., L. A. Wainwright, K. D. Stone, and M. S. Donnenberg. 1997. A third secreted protein that is encoded by the enteropathogenic *Escherichia coli* pathogenicity island is required for transduction of signals and for attaching and effacing activities in host cells. *Infect. Immun.* 65:2211–2217.

77. Lin, J., M. Smith, K. Chapin, H. Baik, G. Bennett, and J. Foster. 1996. Mechanisms of acid resistance in enterohemorrhagic *Escherichia coli. Appl. Environ. Microbiol.* 62:3094–3100.

78. Line, J., A. Fain, A. Moran, L. Martin, R. Lechowich, J. Carosella, and W. Brown. 1991. Lethality of heat to *Escherichia coli* O157:H7: D-value and z-value determinations in ground beef. *J. Food Prot.* 54:762–766.

79. Lingwood, C. A. 1996. Role of verotoxin receptors in pathogenesis. *Trends Microbiol.* 4:147–153.

80. Lior, H. 1994. Classification of *Escherichia coli*, p. 31–72. *In* C. Gyles (ed.), *Escherichia coli in Domestic Animals and Humans.* CAB International, Wallingford, United Kingdom.

81. Louise, C. B., and T. G. Obrig. 1992. Shiga toxin-associated hemolytic uremic syndrome: combined cytotoxic effects of Shiga toxin and lipopolysaccharide (endotoxin) on human vascular endothelial cells in vitro. *Infect. Immun.* 60:1536–1543.

82. Louise, C. B., and T. G. Obrig. 1995. Specific interaction of *Escherichia coli* O157:H7-derived Shiga-like toxin II with human renal endothelial cells. *J. Infect. Dis.* 172:1397–1401.

83. Lynn, T. V., D. D. Hancock, T. E. Besser, J. H. Harrison, D. H. Rice, N. T. Stewart, and L. L. Rowan. 1998. The occurrence and replication of *Escherichia coli* in cattle feeds. *J. Dairy Sci.* 81:1102–1108.

84. McDaniel, T. K., K. G. Jarvis, M. S. Donnenberg, and J. B. Kaper. 1995. A genetic locus of enterocyte effacement conserved among diverse enterobacterial pathogens. *Proc. Natl. Acad. Sci. USA* 92:1664–1668.

85. McDaniel, T. K., and J. B. Kaper. 1997. A cloned pathogenicity island from enteropathogenic *Escherichia coli* confers the attaching and effacing phenotype on *E. coli* K-12. *Mol. Microbiol.* 23:399–407.

86. Mead, P. S., L. Slutsker, V. Dietz, L. F. McCaig, J. S. Bresee, C. Shapiro, P. M. Griffin, and R. V. Tauxe. 1999. Food-related illness and death in the United States. *Emerg. Infect. Dis.* 5:607–625.

87. Melton-Celsa, A., and A. O'Brien. 1998. Structure, biology, and relative toxicity of Shiga toxin family members for cells and animals, p. 121–128. *In* J. Kaper and A. O'Brien (ed.), *Escherichia coli O157:H7 and Other Shiga Toxin-Producing E. coli Strains.* ASM Press, Washington, D.C.

88. Meng, J., and M. P. Doyle. 1998. Microbiology of Shiga toxin-producing *Escherichia coli* in foods, p. 92–111. *In* J. Kaper and A. O'Brien (ed.), *Escherichia coli O157:H7 and Other Shiga Toxin-Producing E. coli Strains.* ASM Press, Washington, D.C.

89. Meng, J., S. Zhao, M. Doyle, and S. Joseph. 1998. Antibiotic resistance of *Escherichia coli* O157:H7 and

O157:NM isolated from animals, food and humans. *J. Food Prot.* **61**:1511–1514.

90. **Michino, H., K. Araki, S. Minami, T. Nakayama, Y. Ejima, K. Hiroe, H. Tanaka, N. Fujita, S. Usami, M. Yonekawa, K. Sadamoto, S. Takaya, and N. Sakai.** 1998. Recent outbreaks of infections caused by *Escherichia coli* O157:H7 in Japan, p. 73–81. *In* J. Kaper and A. O'Brien (ed.), *Escherichia coli O157:H7 and Other Shiga Toxin-Producing E. coli Strains.* ASM Press, Washington, D.C.

91. **Moake, J. L.** 1994. Haemolytic-uraemic syndrome: basic science. *Lancet* **343**:393–397.

92. **Nataro, J. P., and J. B. Kaper.** 1998. Diarrheagenic *Escherichia coli. Clin. Microbiol. Rev.* **11**:142–201.

93. **O'Brien, A. D., V. L. Tesh, A. Donohue-Rolfe, M. P. Jackson, S. Olsnes, K. Sandvig, A. A. Lindberg, and G. T. Keusch.** 1992. Shiga toxin: biochemistry, genetics, mode of action, and role in pathogenesis. *Curr. Top. Microbiol. Immunol.* **180**:65–94.

94. **Orr, P., D. Milley, D. Coiby, and M. Fast.** 1994. Prolonged fecal excretion of verotoxin-producing *Escherichia coli* following diarrheal illness. *Clin. Infect. Dis.* **19**:796–797.

95. **Orskov, F., I. Orskov, and J. A. Villar.** 1987. Cattle as reservoir of verotoxin-producing *Escherichia coli* O157:H7. *Lancet* **ii**:276.

96. **Pavia, A. T., C. R. Nichols, D. P. Green, R. V. Tauxe, S. Mottice, K. D. Greene, J. G. Wells, R. L. Siegler, E. D. Brewer, D. Hannon, et al.** 1990. Hemolytic-uremic syndrome during an outbreak of *Escherichia coli* O157:H7 infections in institutions for mentally retarded persons: clinical and epidemiologic observations. *J. Pediatr.* **116**:544–51.

97. **Pellizzari, A., H. Pang, and C. A. Lingwood.** 1992. Binding of verocytotoxin 1 to its receptor is influenced by differences in receptor fatty acid content. *Biochemistry* **31**:1363–1370.

98. **Perna, N. T., G. F. Mayhew, G. Posfai, S. Elliott, M. S. Donnenberg, J. B. Kaper, and F. R. Blattner.** 1998. Molecular evolution of a pathogenicity island from enterohemorrhagic *Escherichia coli* O157:H7. *Infect. Immun.* **66**:3810–3817.

99. **Ratnam, S., S. B. March, R. Ahmed, G. S. Bezanson, and S. Kasatiya.** 1988. Characterization of *Escherichia coli* serotype O157:H7. *J. Clin. Microbiol.* **26**:2006–2012.

100. **Rowbury, R. J.** 1995. An assessment of environmental factors influencing acid tolerance and sensitivity in *Escherichia coli, Salmonella* spp. and other enterobacteria. *Lett. Appl. Microbiol.* **20**:333–337.

101. **Russell, J. B., F. Diez-Gonzalez, and G. N. Jarvis.** 2000. Invited review: effects of diet shifts on *Escherichia coli* in cattle. *J. Dairy Sci.* **83**:863–873.

102. **Sandvig, K., S. Olsnes, J. E. Brown, O. W. Petersen, and B. van Deurs.** 1989. Endocytosis from coated pits of Shiga toxin: a glycolipid-binding protein from *Shigella dysenteriae* 1. *J. Cell Biol.* **108**:1331–1343.

103. **Sandvig, K., and B. van Deurs.** 1996. Endocytosis, intracellular transport, and cytotoxic action of Shiga toxin and ricin. *Physiol. Rev.* **76**:949–966.

104. **Schmidt, H., E. Maier, H. Karch, and R. Benz.** 1996. Pore-forming properties of the plasmid-encoded hemolysin of enterohemorrhagic *Escherichia coli* O157:H7. *Eur. J. Biochem.* **241**:594–601.

105. **Schmidt, H., J. Scheef, S. Morabito, A. Caprioli, L. H. Wieler, and H. Karch.** 2000. A new Shiga toxin 2 variant (Stx2f) from *Escherichia coli* isolated from pigeons. *Appl. Environ. Microbiol.* **66**:1205–1208.

106. **Schmidt, H., J. von Maldeghem, M. Frosch, and H. Karch.** 1998. Antibiotic susceptibilities of verocytotoxin-producing *Escherichia coli* O157 and non-O157 strains isolated from patients and healthy subjects in Germany during 1996. *J. Antimicrob. Chemother.* **42**:548–550.

107. **Sherman, P., R. Soni, and M. Karmali.** 1988. Attaching and effacing adherence of Vero cytotoxin-producing *Escherichia coli* to rabbit intestinal epithelium in vivo. *Infect. Immun.* **56**:756–761.

108. **Stein, M., B. Kenny, M. A. Stein, and B. B. Finlay.** 1996. Characterization of EspC, a 110-kilodalton protein secreted by enteropathogenic *Escherichia coli* which is homologous to members of the immunoglobulin A protease-like family of secreted proteins. *J. Bacteriol.* **178**:6546–6554.

109. **Swerdlow, D. L., B. A. Woodruff, R. C. Brady, P. M. Griffin, S. Tippen, H. D. Donnell, Jr., E. Geldreich, B. J. Payne, A. Meyer, Jr., J. G. Wells, et al.** 1992. A waterborne outbreak in Missouri of *Escherichia coli* O157:H7 associated with bloody diarrhea and death. *Ann. Intern. Med.* **117**:812–819.

110. **Tesh, V. L., and A. D. O'Brien.** 1991. The pathogenic mechanisms of Shiga toxin and the Shiga-like toxins. *Mol. Microbiol.* **5**:1817–1822.

111. **Tilden, J., Jr., W. Young, A. M. McNamara, C. Custer, B. Boesel, M. A. Lambert-Fair, J. Majkowski, D. Vugia, S. B. Werner, J. Hollingsworth, and J. G. Morris, Jr.** 1996. A new route of transmission for *Escherichia coli*: infection from dry fermented salami. *Am. J. Public Health* **86**:1142–1145.

112. **Tyrrell, G. J., K. Ramotar, B. Toye, B. Boyd, C. A. Lingwood, and J. L. Brunton.** 1992. Alteration of the carbohydrate binding specificity of verotoxins from Gal alpha 1-4Gal to GalNAc beta 1-3Gal alpha 1-4Gal and vice versa by site-directed mutagenesis of the binding subunit. *Proc. Natl. Acad. Sci. USA* **89**:524–528.

113. **van Setten, P. A., L. A. Monnens, R. G. Verstraten, L. P. van den Heuvel, and V. W. van Hinsbergh.** 1996. Effects of verocytotoxin-1 on nonadherent human monocytes: binding characteristics, protein synthesis, and induction of cytokine release. *Blood* **88**:174–183.

114. **Wallace, J., S. Cheasty, and K. Jones.** 1997. Isolation of Vero cytotoxin-producing *Escherichia coli* O157 from wild birds. *J. Appl. Microbiol.* **82**:399–404.

115. **Weinstein, D. L., M. P. Jackson, J. E. Samuel, R. K. Holmes, and A. D. O'Brien.** 1988. Cloning and sequencing of a Shiga-like toxin type II variant from *Escherichia coli* strain responsible for edema disease of swine. *J. Bacteriol.* **170**:4223–4230.

116. **Wells, J. G., B. R. Davis, I. K. Wachsmuth, L. W. Riley, R. S. Remis, R. Sokolow, and G. K. Morris.** 1983. Laboratory investigation of hemorrhagic colitis outbreaks associated with a rare *Escherichia coli* serotype. *J. Clin. Microbiol.* **18:**512–520.

117. **Wilson, J. B., R. C. Clarke, S. A. Renwick, K. Rahn, R. P. Johnson, M. A. Karmali, H. Lior, D. Alves, C. L. Gyles, K. S. Sandhu, S. A. McEwen, and J. S. Spika.** 1996. Vero cytotoxigenic *Escherichia coli* infection in dairy farm families. *J. Infect. Dis.* **174:**1021–1027.

118. **World Health Organization.** 1998. Zoonotic non-O157 Shiga toxin-producing *Escherichia coli* (STEC). Report of a WHO Scientific Working Group Meeting, 23-26 June 1998, Berlin, Germany.

119. **Yu, J., and J. B. Kaper.** 1992. Cloning and characterization of the *eae* gene of enterohaemorrhagic *Escherichia coli* O157:H7. *Mol. Microbiol.* **6:**411–417.

120. **Zhao, T., and M. Doyle.** 1994. Fate of enterohemorrhagic *Escherichia coli* O157:H7 in commercial mayonnaise. *J. Food Prot.* **57:**780–783.

121. **Zhao, T., M. Doyle, and R. Besser.** 1993. Fate of enterohemorrhagic *Escherichia coli* O157:H7 in apple cider with and without preservatives. *Appl. Environ. Microbiol.* **59:**2526–2530.

122. **Zhao, T., M. P. Doyle, J. Shere, and L. Garber.** 1995. Prevalence of enterohemorrhagic *Escherichia coli* O157:H7 in a survey of dairy herds. *Appl. Environ. Microbiol.* **61:**1290–1293.

Food Microbiology: Fundamentals and Frontiers, 2nd Ed.
Edited by M. P. Doyle et al.
© 2001 ASM Press, Washington, D.C.

Roy M. Robins-Browne

Yersinia enterocolitica

11

CHARACTERISTICS OF THE ORGANISM

Introduction

The genus *Yersinia* comprises 11 species within the family *Enterobacteriaceae* (Table 11.1) (18, 254). Like other members of the family, yersiniae are gram-negative, oxidase-negative, rod-shaped facultative anaerobes which ferment glucose. The genus includes three well-characterized pathogens of mammals, one of fish, and several other species whose etiologic roles in disease are uncertain (for a review see reference 230). The four known pathogenic species are *Yersinia pestis*, the causative agent of bubonic and pneumonic plague (the black death); *Yersinia pseudotuberculosis*, a rodent pathogen which occasionally causes mesenteric lymphadenitis, septicemia, and immune-mediated diseases in humans; *Yersinia ruckeri*, a cause of enteric redmouth disease in salmonids and other freshwater fish; and *Yersinia enterocolitica*, a versatile intestinal pathogen which is the most prevalent *Yersinia* species among humans.

Y. pestis is transmitted to its host via the bites of fleas or respiratory aerosols, whereas *Y. pseudotuberculosis* and *Y. enterocolitica* are foodborne pathogens. Never-

theless, these three species share a number of essential virulence determinants which enable them to overcome the innate defenses of their hosts. Analogs of these virulence determinants occur in several other enterobacteria, such as enteropathogenic and enterohemorrhagic *Escherichia coli* and *Salmonella* and *Shigella* species, as well as in various pathogens of animals (e.g., *Pseudomonas aeruginosa* and *Bordetella* species) and plants (e.g., *Erwinia amylovora*, *Xanthomonas campestris*, and *Pseudomonas syringae*), thus providing evidence for horizontal transfer of virulence genes among diverse bacterial pathogens. In addition, yersiniae may have acquired a number of genetic elements of eukaryotic origin which enable them to undermine key aspects of the physiological response to infection.

Classification

Y. enterocolitica first emerged as a human pathogen during the 1930s (28, 230). After several unsuccessful attempts to allocate it to a suitable taxonomic position, *Y. enterocolitica* was assigned to the family *Enterobacteriaceae*. *Y. enterocolitica* shares between 10 and 30% DNA homology with members of other genera in the

Roy M. Robins-Browne, Department of Microbiology and Immunology, University of Melbourne, and Microbiological Research Unit, Murdoch Children's Research Institute, Parkville, Victoria 3052, Australia.

Table 11.1 Biochemical tests used to differentiate *Yersinia* species[a]

			Result[b]									
			Y. enterocolitica									
			Biovars 1									
Test	*Y. aldovae*	*Y. bercovieri*	through 4	Biovar 5	*Y. frederiksenii*	*Y. intermedia*	*Y. kristensenii*	*Y. pestis*	*Y. mollaretii*	*Y. pseudotuberculosis*	*Y. robdei*	*Y. ruckeri*
Indole	−	−	D	−	+	+	D	−	−	−	−	−
Voges-Proskauer	+	−	+	+[c]	D	+	−	−	−	−	−	−
Citrate (Simmons)	D	−	−	−	D	+	−	−	−	−	+	−
L-Ornithine	+	+	+	−	+	+	+	+	−	−	+	+
Mucate, acid	D	+	−	−	D	D	−	+	−	−	−	−
Pyrazinamidase	+	+	D	−	+	+	+[c]	+	−	−[c]	+	ND
Sucrose	−	+	+	D	+	+	−	+	−	−	+	−
Cellobiose	−	+	+	+	+	+	+	+	−	−	+	−
L-Rhamnose	+	−	−	−	−	+	−	−	−	+	−	−
Melibiose	−	−	D	−	−	+	−	+	D	+	D	−
L-Sorbose	−	−	D	D	+	+	+	−	−	−	ND	ND
L-Fucose	D	+	D	−	+	D	D	−	ND	−	ND	ND

[a] Adapted from reference 254.
[b] +, positive; −, negative; D, different reactions; ND, not determined.
[c] Some reactions may be delayed or weakly positive.

Table 11.2 Biotyping scheme of *Y. enterocolitica*[a]

	Reaction of biovar[b]					
Test	1A	1B	2	3	4	5
Lipase (Tween hydrolysis)	+	+	−	−	−	−
Esculin hydrolysis	D	−	−	−	−	−
Indole production	+	+	(+), −	−	−	−
D-Xylose fermentation	+	+	+	+	−	D
Voges-Proskauer reaction	+	+	+	+	+	(+)
Trehalose fermentation	+	+	+	+	+	−
Nitrate reduction	+	+	+	+	+	−
Pyrazinamidase	+	−	−	−	−	−
β-D-Glucosidase	+	−	−	−	−	−
Proline peptidase	D	−	−	−	−	−

[a] Adapted from reference 255.
[b] +, positive; (+), delayed positive; −, negative; D, different reactions.

family *Enterobacteriaceae*, and is approximately 50% related to *Y. pseudotuberculosis* and *Y. pestis*. The last two species share greater than 90% DNA homology; genetic analysis has revealed that *Y. pestis* is a clone of *Y. pseudotuberculosis* which evolved some 1,500 to 20,000 years ago, shortly before the first known pandemics of human plague (2).

Y. enterocolitica is a heterogenous species that is divisible into a large number of subgroups, largely according to biochemical activity and lipopolysaccharide (LPS) O antigens (Tables 11.2 and 11.3). Biotyping is based on the ability of *Y. enterocolitica* to metabolize selected organic substrates and provides a convenient means to

Table 11.3 Relationship between O serogroup and pathogenicity of *Y. enterocolitica* and related species

Species and biovar	Serogroup(s)[a]
Y. enterocolitica	
Biovar 1A	O:4; O:5; O:6,30; O:6,31; O:7,8; O:7,13; O:10; O:14; O:16; O:21; O:22; O:25; O:37; O:41,42; O:46; O:47; O:57; NT
Biovar 1B	**O:4,32; O:8; O:13a,13b; O:16; O:18; O:20; O:21;** O:25; O:41,42; NT
Biovar 2	**O:5,27; O:9;** O:27
Biovar 3	**O:1,2,3; O:3; O:5,27**
Biovar 4	**O:3**
Biovar 5	**O:2,3**
Y. bercovieri	O:8; O:10; O:58,16; NT
Y. frederiksenii	O:3; O:16; O:35; O:38; O:44; NT
Y. intermedia	O:17; O:21,46; O:35; O:37; O:40; O:48; O:52; O:55; NT
Y. kristensenii	O:11; O:12,25; O:12,26; O:16; O:16,29; O:28,50; O:46; O:52; O:59; O:61; NT
Y. mollaretii	O:3; O:6,30; O:7,13; O:59; O:62,22; NT

[a] NT, not typeable. Serogroups which include strains considered to be primary pathogens are in boldface.

subdivide the species into subtypes of various degrees of clinical and epidemiologic significance (Tables 11.2 and 11.3) (255). Most primary pathogenic strains of humans and domestic animals occur within biovars 1B, 2, 3, 4, and 5. By contrast, *Y. enterocolitica* strains of biovar 1A are commonly obtained from terrestrial and freshwater ecosystems. For this reason, they are often referred to as environmental strains, although some of them may be responsible for intestinal infections (79). Not all isolates of *Y. enterocolitica* obtained from soil, water, or unprocessed foods can be assigned to a biovar. These unassigned strains lack the characteristic virulence determinants of biovars 1B though 5 (see below) and may represent novel nonpathogenic subtypes or even new *Yersinia* species (67, 105).

The most frequent *Y. enterocolitica* biovar obtained from human clinical material worldwide is biovar 4. Biovar 1B bacteria are usually isolated from patients in the United States and are referred to as "American" strains, although they have also been detected in a number of countries in Europe, Africa, Asia, and Australasia. Although not common anywhere, biovar 1B yersiniae are inherently more virulent for mice (and possibly for humans) than strains in the other pathogenic categories and have been identified as the cause of several foodborne outbreaks of yersiniosis in the United States.

Serogroups of *Y. enterocolitica*, based on LPS surface O antigens, coincide to some extent with the biovars (Table 11.3), and serotyping provides a useful additional tool to subdivide the species in a way that relates to pathologic significance (253). Serogroup O:3 is the variety most frequently isolated from humans. Almost all of these isolates belong to biovar 4. Other serogroups commonly obtained from humans, particularly in northern Europe, include O:9 (biovar 2) and O:5,27 (biovar 2 or 3). The usefulness of serotyping is limited to some extent by the fact that the overwhelming majority of human infections are due to strains of serogroup O:3 and by the presence of cross-reacting O antigens in *Y. enterocolitica* strains of varying pathological and epidemiologic significance. In addition, some bacteria that were originally allocated to O serogroups of *Y. enterocolitica* were later reclassified as separate species (253).

At least 18 flagellar (H) antigens of *Y. enterocolitica*, designated by lowercase letters (a,b; b,c; b,c,e,f,k; m, etc.), have been also identified. Although there is little overlap between the H antigens of *Y. enterocolitica* sensu stricto and those of the related species, antigenic characterization of isolates by complete O and H serotyping is seldom attempted (253).

Other schemes for subtyping *Yersinia* species include bacteriophage typing, multienzyme electrophoresis, and the demonstration of restriction fragment length poly-morphisms of chromosomal and plasmid DNAs (116). These techniques can be used to facilitate epidemiologic investigations of outbreaks or to trace the sources of sporadic infections (13, 74).

Susceptibility and Tolerance

Y. enterocolitica is unusual among pathogenic enterobacteria in being psychrotrophic, as evidenced by its ability to replicate at temperatures below 4°C. The doubling time at the optimum growth temperature (ca. 28 to 30°C) is around 34 min, which increases to 1 h at 22°C, 5 h at 7°C, and approximately 40 h at 1°C (Figure 11. 1) (211). *Y. enterocolitica* readily withstands freezing and can survive in frozen foods for extended periods, even after repeated freezing and thawing (244). Studies of the ability of *Y. enterocolitica* to survive and grow in artificially contaminated foods under various conditions of storage have shown that it generally survives better at room temperature and refrigeration temperature than at intermediate temperatures. *Y. enterocolitica* persists longer in cooked foods than in raw foods, probably due to an increased availability of nutrients in cooked foods and the fact that the presence of other psychrotrophic bacteria, including nonpathogenic strains of *Y. enterocolitica*, in unprocessed food may restrict growth (211). The number of viable *Y. enterocolitica* organisms may increase more than a millionfold on cooked beef or pork within

24 h at 25°C or within 10 days at 7°C (89). Growth rates are lower on raw beef and pork. *Y. enterocolitica* can grow at refrigeration temperature in vacuum-packed meat, boiled eggs, boiled fish, pasteurized liquid eggs, pasteurized whole milk, cottage cheese, and tofu (soybean curd) (66, 211). Proliferation also occurs in refrigerated seafoods, such as oysters, raw shrimp, and cooked crab meat, but at a lower rate than in pork or beef (173). Yersiniae can also persist for extended periods in refrigerated vegetables and cottage cheese, particularly in the presence of chicken meat (218). The psychrotrophic nature of *Y. enterocolitica* also poses problems for the blood transfusion industry, mainly because of its ability to proliferate and release endotoxin in blood products stored at 4°C without manifestly altering their appearance (12).

Y. enterocolitica and *Y. pseudotuberculosis* can grow over a pH range of approximately 4 to 10, with an optimum pH of ca. 7.6 (207). They tolerate alkaline conditions extremely well, but their acid tolerance is less pronounced and depends on the acidulent used, the environmental temperature, the composition of the medium, and the growth phase of the bacteria (4). The acid tolerance of *Y. enterocolitica* is enhanced by the production of urease, which hydrolyzes urea to release ammonia that elevates the cytoplasmic pH (55, 261).

Y. enterocolitica and *Y. pseudotuberculosis* are susceptible to heat and are readily destroyed by pasteurization at 71.8°C for 18 s or 62.8°C for 30 min (50, 244). Exposure of surface-contaminated meat to hot water (80°C) for 10 to 20 s reduces bacterial viability by at least 99.9% (225). *Y. enterocolitica* is also readily inactivated by ionizing and UV irradiation (35, 61) and by sodium nitrate and nitrite when they are added to food (52). The pathogen is relatively resistant to these salts in solution, however, and can also tolerate NaCl at concentrations of up to 5% (52, 227). *Y. enterocolitica* is generally susceptible to organic acids, such as lactic and acetic acids, and to chlorine (63). However, some resistance to chlorine occurs among yersiniae grown under conditions that approximate natural aquatic environments or when they are cocultivated with predatory aquatic protozoa (91, 130).

Sutherland et al. formulated models to predict the influence of temperature, pH, and the concentrations of sodium chloride and lactic acid on the survival and growth of *Y. enterocolitica* in foods and determined that the models correlated well with published data for meat and milk products (121, 232). Bhaduri et al. developed a model to assess the likely safety of foods contaminated with *Y. enterocolitica* as a result of manufacturing problems or storage abuse (20). Nevertheless, mathematical models should not be relied upon to predict food

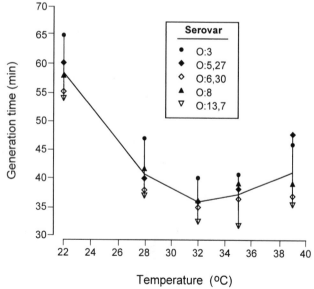

Figure 11.1 Generation times at different temperatures of *Y. enterocolitica* strains of five different serogroups grown in 4% tryptone–1% mannitol salt broth, pH 7.6. (Adapted from reference 209.)

safety, because they may overestimate bacterial numbers in some circumstances and underestimate them in others (81).

CHARACTERISTICS OF INFECTION

Infections with *Y. enterocolitica* typically manifest as nonspecific, self-limiting diarrhea but may lead to a variety of suppurative and autoimmune complications (Table 11.4), the risk of which is determined partly by host factors, in particular age and underlying immune status.

Acute Infection

Y. enterocolitica enters the gastrointestinal tract after ingestion in contaminated food or water. The median infective dose for humans is not known, but it is likely to exceed 10^4 CFU. Gastric acid is a significant barrier to infection with *Y. enterocolitica*, and in individuals with gastric hypoacidity, the infectious dose may be lower (55, 70).

Most symptomatic infections with *Y. enterocolitica* occur in children, especially in those less than 5 years of age. In these patients, yersiniosis presents as diarrhea, often accompanied by low-grade fever and abdominal pain (102, 147). The character of the diarrhea varies from

Table 11.4 Clinical manifestations of infection with *Y. enterocolitica*

Common manifestations
    Diarrhea ("gastroenteritis"), especially in young children
    Enterocolitis
    Pseudoappendicitis syndrome due to terminal ileitis;
        acute mesenteric lymphadenitis
    Pharyngitis
    Postinfection autoimmune sequelae
        Arthritis, especially associated with HLA-B27
        Erythema nodosum
        Uveitis, associated with HLA-B27
        Glomerulonephritis (uncommon)
        Myocarditis (uncommon)
        Thyroiditis (uncertain)
Less common manifestations
    Septicemia
    Visceral abscesses (for example, in liver, spleen, lung)
    Skin infection: pustules, wound infection, pyomyositis, etc.
    Pneumonia
    Endocarditis
    Osteomyelitis
    Peritonitis
    Meningitis
    Intussusception
    Eye infections: conjunctivitis, panophthalmitis

watery to mucoid. A small proportion of children (generally less than 10%) have frankly bloody stools. Children with *Y. enterocolitica*-induced diarrhea often complain of abdominal pain and headache. Sore throat is a frequent accompaniment and may dominate the clinical picture in older patients (234). The illness typically lasts from a few days to 3 weeks, although some patients develop chronic enterolitis, which may persist for several months (203). Occasionally, acute enteritis progresses to intestinal ulceration and perforation or to ileocolic intussusception, toxic megacolon, or mesenteric vein thrombosis (48). On rare occasions, patients may present with peritonitis in the absence of intestinal perforation (187).

In children older than 5 years of age and adolescents, acute yersiniosis often presents as a pseudoappendicular syndrome due to acute inflammation of the terminal ileum or the mesenteric lymph nodes. The usual features of this syndrome are abdominal pain and tenderness localized to the right lower quadrant. These symptoms are usually accompanied by fever, with little or no diarrhea. The importance of this form of the disease lies in its close resemblance to appendicitis (48). Of those patients with this syndrome who undergo laparotomy, approximately 60 to 80% have terminal ileitis, with or without mesenteric adenitis, and a normal or slightly inflamed appendix (184, 249). *Y. enterocolitica* can be cultured from the distal ileum and the mesenteric lymph nodes. The pseudoappendicular syndrome appears to be more frequent in patients infected with the relatively more virulent strains of *Y. enterocolitica*, notably strains of biovar 1B. *Y. enterocolitica* is rarely found in patients with true appendicitis (249).

Although *Y. enterocolitica* is seldom isolated from extraintestinal sites, there appears to be no tissue in which it will not grow. In adults, pharyngitis, sometimes with cervical lymphadenitis, is the dominant clinical presentation (234). Focal disease, in the absence of obvious bacteremia, may present as cellulitis, subcutaneous abscess, pyomyositis, suppurative lymphadenitis, septic arthritis, osteomyelitis, urinary tract infection, renal abscess, sinusitis, pneumonia, lung abscess, or empyema (48).

Bacteremia is a rare complication of infection, except in patients who are immunocompromised or in an iron overload state (70). Factors which predispose to the development of *Yersinia* bacteremia include immunosuppression, blood dyscrasias, malnutrition, chronic renal failure, cirrhosis, alcoholism, diabetes mellitus, and acute and chronic iron overload states, particularly when managed by chelation therapy with desferrioxamine B (48). Bacteremic dissemination of *Y. enterocolitica* can lead to various manifestations, including splenic, hepatic, and lung abscesses, catheter-associated infections, osteomyelitis, panophthalmitis, endocarditis, mycotic aneurysm, and meningitis (48). *Yersinia* bacteremia has a case fatality rate of between 30 and 60%.

Bacteremia may also result from direct inoculation of *Y. enterocolitica* into the circulation during blood transfusion (77, 206). *Y. enterocolitica* is the single most important cause of fatal bacteremia following transfusion with packed red blood cells or platelets (77). Patients infused with contaminated blood may develop symptoms of a severe transfusion reaction minutes to hours after exposure, depending on the number of bacteria and the amount of endotoxin administered with the blood (12). The varieties of *Y. enterocolitica* responsible for transfusion-acquired yersiniosis are the same serobiovars as those associated with enteric infections. The probable source of these infections are blood donors with low-grade, subclinical bacteremia. A small number of bacteria in donated blood will increase during storage at refrigeration temperature without manifestly altering the appearance of the blood (241).

Autoimmune Complications

Although most episodes of yersiniosis remit spontaneously without long-term sequelae, infections with *Y. enterocolitica* are noteworthy for the large variety of immunological complications, such as reactive arthritis, erythema nodosum, uveitis, glomerulonephritis, carditis, and thyroiditis, which may follow acute infection (48). Of these, reactive arthritis is the most widely recognized (6, 27, 137). This manifestation of infection is infrequent before the age of 10 years and occurs most often in Scandinavian countries, where serotype O:3 strains and the human leukocyte antigen HLA-B27 are especially prevalent. Men and women are affected equally. Arthritis typically follows the onset of diarrhea or the pseudoappendicular syndrome by 1 to 2 weeks, with a range of from 1 to 38 days. The joints most commonly involved are the knees, ankles, toes, tarsal joints, fingers, wrists, and elbows. Synovial fluid from affected joints contains large numbers of inflammatory cells, principally polymorphonuclear leukocytes, and is invariably sterile, although it generally contains bacterial antigens (78). The duration of arthritis is typically less than 3 months, and the long-term prognosis in terms of joint destruction is excellent, although some patients may have symptoms that persist for several years (100, 137). Many patients with arthritis also have extra-articular symptoms, including urethritis, uveitis, and erythema nodosum (27).

Y. enterocolitica-induced erythema nodosum occurs predominantly in women and is not associated with HLA-B27. Other autoimmune complications of yersiniosis, including Reiter's syndrome, uveitis, acute proliferative glomerulonephritis, collagenous colitis, and rheumatic-like carditis, have been reported, mostly from Scandinavian countries (134). Yersiniosis has also been linked to various thyroid disorders, including Graves' disease hyperthyroidism, nontoxic goiter, and Hashimoto's thyroiditis, although the causative role of yersiniae in these conditions is uncertain (242). In Japan, *Y. pseudotuberculosis* has been implicated in the etiology of Kawasaki's disease (245).

RESERVOIRS

Infections with *Yersinia* species are zoonoses. The subgroups of *Y. enterocolitica* which commonly occur in humans also occur in domestic animals, whereas those which are infrequent in humans generally reside in wild rodents. *Y. enterocolitica* can occupy a broad range of environments and has been isolated from the intestinal tracts of many different mammalian species, as well as from birds, frogs, fish, flies, fleas, crabs, and oysters (28, 48, 245).

Foods that may harbor *Y. enterocolitica* include pork, beef, lamb, poultry, and dairy products, notably milk, cream, and ice cream (66, 124, 211). *Y. enterocolitica* is also commonly present in a variety of terrestrial and freshwater ecosystems, including soil, vegetation, lakes, rivers, wells, and streams, and can persist for extended periods in soil, vegetation, streams, lakes, wells, and spring water, particularly at low environmental temperatures (37, 124). Many environmental isolates of *Y. enterocolitica* lack markers of bacterial virulence and are of uncertain significance for human or animal health (157).

Although *Y. enterocolitica* has been recovered from a variety of wild and domesticated animals, pigs are the only animal species from which *Y. enterocolitica* of biovar 4 serogroup O:3 (the variety most commonly associated with human disease) has been isolated with any frequency (124). Pigs may also carry *Y. enterocolitica* of serogroups O:9 and O:5,27, particularly in regions where human infections with these varieties are comparatively common. In countries with a high incidence of human yersiniosis, *Y. enterocolitica* is commonly isolated from pigs at slaughterhouses (9, 51, 74). The tissue most frequently culture positive at slaughter is the tonsils, which appear to be the preferred site of *Y. enterocolitica* infection in pigs. Other tissues which frequently yield yersiniae include the tongue, cecum, rectum, feces, and gut-associated lymphoid tissue. *Y. enterocolitica* is seldom isolated from meat offered for retail sale, however, apart from pork tongue (124, 211), although standard methods of bacterial isolation and detection may underestimate the true incidence of contamination (164).

Although some domesticated farm animals, notably sheep, cattle, and deer, can suffer symptoms as a result of infection with *Y. enterocolitica* or *Y. pseudotuberculosis* (119, 224), in most cases the biovars and serotypes of these bacteria differ from those responsible for human infection, indicating a lack of transmission of these particular bacteria between animals and humans. By contrast, individual isolates of *Y. enterocolitica* from pigs and humans are indistinguishable from each other in terms of serotype, biovar, restriction fragment length polymorphism of chromosomal and plasmid DNA, and carriage of virulence determinants (124). Further evidence that pigs are a significant reservoir of human infections is provided by epidemiologic studies identifying the ingestion of raw or undercooked pork as a major risk factor for the acquisition of yersiniosis (169, 239). Infection also occurs after handling contaminated pig intestines while preparing chitterlings (129, 135, 228).

Food animals are seldom infected with biovar 1B strains of *Y. enterocolitica*, the reservoir of which remains unknown (211). The relatively low incidence of human yersiniosis caused by these strains, despite their comparatively high virulence, indicates a lack of significant contact between their reservoir and humans. Since yersiniae of this biovar are primary pathogens of rodents, it is conceivable that rats or mice are the natural reservoir of these strains (95).

FOODBORNE OUTBREAKS

Considering the widespread occurrence of *Y. enterocolitica* in nature, and its ability to colonize food animals, to persist within animals and the environment, and to pro-

liferate at refrigeration temperature, outbreaks of yersiniosis are surprisingly uncommon. Most foodborne outbreaks in which a source was identified have been traced to milk (Table 11.5). *Y. enterocolitica* is rapidly destroyed by pasteurization, and hence, infection results from the consumption of raw milk or milk that is contaminated after pasteurization (62, 209). During the mid-1970s, two outbreaks of yersiniosis caused by *Y. enterocolitica* O:5,27 occurred among 138 Canadian schoolchildren who had consumed raw milk, but the organism was not recovered from the suspected source (127). In 1976, serogroup O:8 *Y. enterocolitica* was responsible for an outbreak in New York State which affected 217 people, 38 of whom were culture positive (22). The source of infection was chocolate-flavored milk, which evidently became contaminated after pasteurization.

In 1981, an outbreak of infection with *Y. enterocolitica* O:8 affected 35% of 455 individuals at a diet camp in New York State (161). Seven patients were hospitalized as a result of the infection, five of whom underwent appendectomy. The source of the infection was reconstituted powdered milk and/or chow mein, which probably became contaminated during preparation by an infected food handler. During 1982, 172 cases of infection with *Y. enterocolitica* O:13a,13b occurred in an area which included parts of Tennessee, Arkansas, and Mississippi (235). The suspected source was pasteurized milk, which may have become contaminated with pig manure during transport (62).

More recently, an outbreak of infection with *Y. enterocolitica* O:3 affected 15 infants and children in metropolitan Atlanta (135). In this instance, bacteria were transmitted from raw chitterlings to the affected children on the hands of food handlers. Other foods which have been responsible for outbreaks of yersiniosis include "pork cheese" (a type of sausage prepared from

Table 11.5 Selected foodborne outbreaks of infection with *Y. enterocolitica*

Location	Yr	Mo	No. of cases	Serogroup	Source[a]	Reference
Canada	1976	April	138	O:5,27	Raw milk (?)	127
New York	1976	September	38	O:8	Chocolate-flavored milk	22
Japan	1980	April	1,051	O:3	Milk	149
New York	1981	July	159	O:8	Powdered milk, chow mein	215
Washington	1981	December	50	O:8	Tofu, spring water	233
Pennsylvania	1982	February	16	O:8	Bean sprouts, well water	48
Southern United States	1982	June	172	O:13a,13b	Milk (?)	235
Hungary	1983	December	8	O:3	Pork cheese (sausage)	146
Georgia, United States	1989	November	15	O:3	Pork chitterlings	136
Northeastern United States	1995	October	10	O:8	Pasteurized milk (?)	3

[a] (?), bacteria were not isolated from the incriminated source.

chitterlings), bean sprouts, and tofu (48, 146, 233). In the outbreaks associated with bean sprouts and tofu, contaminated well or spring water was the probable source of yersiniae. Water was also the putative source of infection in a case of sporadic *Y. enterocolitica* bacteremia in a 75-year-old man in New York State (128) and a small family outbreak in Ontario, Canada (240). Several outbreaks of presumed foodborne infection with *Y. enterocolitica* O:3 have also been reported from the United Kingdom and Japan, but in most cases the source of the outbreak was not identified.

MECHANISMS OF PATHOGENICITY

Y. enterocolitica is an invasive enteric pathogen whose virulence determinants have been the subject of intensive investigation (93), but not all strains of *Y. enterocolitica* are equally virulent (Table 11.6). *Y. enterocolitica* strains of biovars 1B, 2, 3, 4, and 5 possess a multitude of interactive virulence determinants, including a chromosomally encoded invasin and a ca. 70-kb virulence plasmid, termed pYV (plasmid for *Yersinia* virulence) (93). In addition, biovar 1B strains of *Y. enterocolitica* carry a pathogenicity island, which is associated with enhanced virulence for mice and probably humans. All pYV-bearing clones of *Y. enterocolitica* have the capacity to invade epithelial cells in large numbers in vitro—a feature which distinguishes them from clones that do not carry pYV (79, 189). Surprisingly, however, this highly invasive phenotype is not specified by genes within pYV and is maximally expressed by bacteria in which pYV has been cured.

Until recently, weakly invasive, pYV-negative strains of *Y. enterocolitica*, most of which belong to biovar 1A, were regarded as avirulent because they never carry pYV or any of the other well-characterized virulence-associated genes of the species. There is now persuasive epidemiologic evidence, however, indicating that at least some of these strains may cause gastrointestinal symptoms clinically indistinguishable from those due to pYV-bearing strains (34, 160). This is supported by the laboratory demonstration that biovar 1A strains fall into two categories, including one which comprises strains recovered from symptomatic patients that have the ability

to penetrate epithelial cells in moderate numbers and to resist killing by macrophages to a significantly greater extent than biovar 1A strains obtained from nonclinical sources (79, 80). Because the mechanisms by which biovar 1A strains cause disease are unknown, the remainder of this section will focus chiefly on the virulence determinants of the classical pathogenic, i.e., pYV-bearing, highly invasive strains of *Y. enterocolitica*.

Pathological Changes

Examination of surgical specimens from patients with yersiniosis reveals that *Y. enterocolitica* is an invasive pathogen which displays a tropism for lymphoid tissue (33, 43). The distal ileum, in particular the gut-associated lymphoid tissue, bears the brunt of the infection, although adjacent regions of the intestine and the mesenteric lymph nodes are also frequently involved.

As investigations with volunteers are precluded by the risk of autoimmune sequelae, most information regarding the pathogenesis of yersiniosis in vivo has been obtained from animal models, in particular mice and rabbits (96, 141, 247). Although these animals are not the natural hosts of the serotypes of *Y. enterocolitica* that commonly infect humans, they have provided important insights into the probable pathogenesis of human disease. Nevertheless, some data derived from animal studies should be interpreted with caution, particularly where death is used as the end point of infection, since this is not the usual outcome of human infection (175).

After oral inoculation of mice with a virulent strain of serogroup O:8, biovar 1B, most bacteria remain within the intestinal lumen, while a small number adhere to the mucosal epithelium, having no particular preference for any cell type (84). However, invasion of the epithelium takes place almost exclusively through M cells (Fig. 11.2) (84). These are specialized epithelial cells that overlie intestinal lymphoid follicles (Peyer's patches), where they play a major role in antigen sampling (118, 205). Studies of experimentally infected rabbits and pigs have revealed that after penetrating the epithelium, *Y. enterocolitica* traverses the basement membrane to reach the gut-associated lymphoid tissue and the lamina propria, where it causes localized tissue destruction

Table 11.6 Characteristics of the pathogenic subgroups of *Y. enterocolitica*

Subgroup	Capacity to invade epithelial cells in vitro	Biovar(s)	Virulence-associated determinants
Classical	High (lower if pYV present)	1B, 2, 3, 4, 5	Invasin, Ail, Myf, *Yersinia* heat-stable enterotoxin (Yst), virulence plasmid (pYV), high-pathogenicity island (biovar 1B only)
Atypical	Low to moderate	1A	Unknown

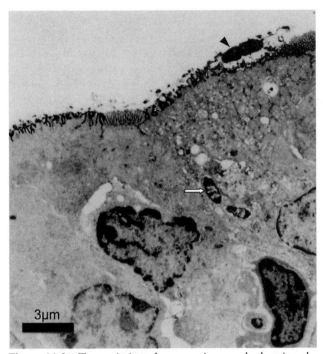

Figure 11.2 Transmission electron micrograph showing the initial interaction (arrowhead) and transport (arrow) of *Y. enterocolitica* through an intestinal M cell 60 min after inoculation into mouse ileum. (Reprinted with permission from reference 84.)

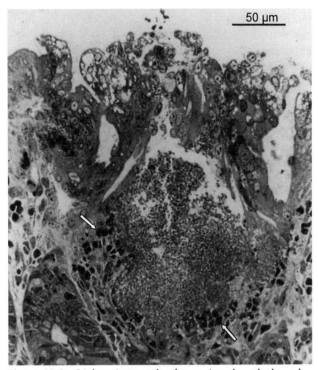

Figure 11.3 Light micrograph of a section though the colon of a gnotobiotic piglet 3 days after inoculation with a virulent strain of *Y. enterocolitica* O:3. Note the microabscess, comprising mostly bacteria; the surrounding inflammatory cells (arrows); and the disrupted epithelium with vacuolated and necrotic cells. An epoxy section with methylene blue stain is shown. (Reprinted with permission from reference 246.)

leading to the formation of microabscesses (Fig. 11.3) (141, 195, 246). These lesions occur chiefly within intestinal crypts but may extend as far as the crypt-villus junction. *Y. enterocolitica* often spreads via the lymph to the draining mesenteric lymph nodes, where it may also lead to microabscess formation. If the bacteria circumvent the lymph nodes to enter the bloodstream, they can disseminate to any organ, but they continue to show a tropism for lymphoid tissue by preferentially localizing in the reticuloendothelial tissues of the liver and spleen. Although *Y. enterocolitica* is often regarded as a facultative intracellular pathogen because of its innate resistance to being killed by macrophages (54), most of the bacteria observed in histological sections are located extracellularly (90). Nevertheless, macrophages containing viable bacteria may play an important role in the dissemination of yersiniae throughout the body (167, 195).

Virulence Determinants
Chromosomal Determinants of Virulence

Invasin

All strains of *Y. enterocolitica* which carry pYV also produce a 91-kDa surface-expressed protein termed invasin. This outer membrane protein was first identified in *Y. pseudotuberculosis* as a 102-kDa protein product of the chromosomal *inv* gene (114). When introduced into an innocuous laboratory strain of *E. coli*, such as K-12, *inv* imbues the recipient with the ability to penetrate mammalian cells, including epithelial cells and macrophages (54, 114). Despite the difference in size of invasins from *Y. enterocolitica* and *Y. pseudotuberculosis*, the two proteins are functionally highly conserved. Invasins are also related to intimin, an essential virulence determinant of enteropathogenic and enterohemorrhagic strains of *E. coli*, which require this protein to produce the distinctive attaching-effacing lesions that characterize infection with these bacteria (120).

The amino terminus of invasin is inserted into the bacterial outer membrane, while the carboxyl terminus is exposed on the surface, where it mediates binding to host cell integrins (204). Integrins are heterodimeric transmembrane proteins which communicate extracellular signals to the cytoskeleton. They comprise α and β subunits, which form the basis of their classification into families. The β_1 class integrins, which are the principal receptors for invasin, occur on many cell types, including epithelial cells, macrophages, T lymphocytes, and Peyer's patch M cells (42). Their physiological role is to act as receptors for fibronectin, laminin, and related host proteins, which may bear conformational similarities to invasin (87). However, the affinity of invasin for integrins $\alpha_3\beta_1$, $\alpha_4\beta_1$, $\alpha_5\beta_1$, and $\alpha_6\beta_1$ is much greater than that of fibronectin. Accordingly, when invasin binds to these integrins, it causes them to cluster and initiate a sequence of events, including the activation of focal adhesion kinase, which results in reorganization of the host cell cytoskeleton and internalization of the bacteria (8, 60). Inhibitors of actin polymerization and tyrosine kinases block invasin-mediated invasion (69, 198), indicating that uptake of *Y. enterocolitica* by eukaryotic cells requires both an intact cytoskeleton and signal transduction pathways involving tyrosine phosphorylation. The internalization process is governed entirely by the host cell, because nonviable bacteria and even latex particles coated with invasin are internalized in the same way as living bacteria (115).

Although DNA sequences homologous to *inv* occur in all *Yersinia* species (except *Yersinia ruckeri*), this gene is functional only in *Y. pseudotuberculosis* and the classical pathogenic biovars (1B through 5) of *Y. enterocolitica*, suggesting that invasin plays a key role in virulence (153). Nevertheless, although *inv* mutants of *Y. enterocolitica* show a pronounced reduction in their ability to invade epithelial cells in vitro, their virulence for orally inoculated mice is barely affected (175). There is evidence, however, that invasin contributes to gastrointestinal tract colonization of mice, as *inv* mutants of *Y. enterocolitica* show diminished translocation into Peyer's patches compared with that of wild-type strains (148, 175). The fact that *inv* mutants show only marginally reduced virulence for mice suggests that M cells may be able to internalize the bacteria to some extent in the absence of a specific stimulus or that *Y. enterocolitica* expresses supplementary factors, such as YadA, which can compensate for the absence of invasin.

Ail

Classical pathogenic strains of *Y. enterocolitica* produce an outer membrane protein, unrelated to invasin, which also confers invasive ability on *E. coli* (152). This 17-kDa peptide is specified by a chromosomal *ail* (attachment-invasion) locus, so called because it mediates bacterial attachment to some cultured epithelial cell lines and invasion of others. In concert with YadA, a pYV-encoded protein, Ail may also enable yersiniae to persist extracellularly by protecting them from complement-mediated lysis via its ability to interfere with the binding of com-

plement proteins to the bacterial surface (24). The *ail* gene occurs only in the classical pathogenic varieties of *Y. enterocolitica* (189) and in vitro is optimally expressed at 37°C (unlike *inv*, which is optimally expressed at 25°C, unless the pH is lowered to 5.5 [174]). Further circumstantial evidence supporting a role for Ail in virulence stems from the finding that *ail* mutants fail to adhere to or invade cultured cells when the *inv* gene is not expressed (113). Surprisingly, however, an *ail* mutant of *Y. enterocolitica* showed no reduction in virulence for perorally inoculated mice, indicating that Ail is not required to establish infection or even to cause systemic infection in these animals (251). However, as *Y. enterocolitica* is inherently resistant to complement-mediated killing by murine serum (in contrast to its pronounced susceptibility to killing by human serum [170]), the mouse model may not be well suited to investigate the contribution of anticomplement factors to virulence.

Outer membrane proteins which share significant amino acid homology with Ail have been identified in other members of the family *Enterobacteriaceae*. One of these is PagC, a virulence-associated, surface-expressed protein of *Salmonella enterica* serovar Typhimurium which is required for bacterial survival within macrophages (183). Despite these structural similarities, however, Ail and PagC are functionally distinct, because PagC does not contribute to the invasive ability of salmonella and Ail is not required by yersiniae to resist being killed by macrophages (152).

Heat-Stable Enterotoxins

When first isolated from clinical material, most strains of *Y. enterocolitica* secrete a heat-stable enterotoxin, known as Yst (or Yst-a), which is reactive in infant mice (171). Yst comprises 30 amino acids (Fig. 11.4). Its carboxyl terminus is homologous to those of heat-stable enterotoxins from enterotoxigenic *E. coli*, *Citrobacter freundii*, and non-O1 serotypes of *Vibrio cholerae* and also to that of guanylin, an intestinal paracrine hormone (Fig. 11.4) (7, 49, 85, 103, 186, 236–238). These polypeptides also share a common mechanism of action which involves binding to and activation of cell-associated guanylate cyclase, with subsequent elevation of intracellular concentrations of cyclic GMP (193). This in turn causes perturbation of fluid and electrolyte transport in intestinal absorptive cells, which may result in diarrhea.

Yst is encoded by the chromosomal *yst* gene and is synthesized as a 71-amino-acid polypeptide, the carboxyl terminus of which becomes the mature toxin (59). The 19 amino acids at the amino terminus contain a signal sequence, which is cleaved and removed during secretion, as is the 23-amino-acid central region. This

Y. enterocolitica Yst	Q A C D P P S P P A E V S S D W D	C C	D V	C C N P A C	A G C	
Y. enterocolitica Yst-b	K A C D T Q T P S P S E E N D D W	C C	E V	C C N P A C	A G C	
Y. enterocolitica Yst-c	ᵃ- E C G T Q S A T T Q G E N D W D W	C C	E L	C C N P A C	A G C	
E. coli STh	N S S N Y	C C	E L	C C N P A C	T G C	Y
E. coli STp	N T F Y	C C	E L	C C N P A C	A G C	Y
C. freundii ST	N T T Y	C C	E L	C C N P A C	A G C	
V. cholerae non-O1 ST	I D	C C	E I	C C N P A C	F G C	L N
Guanylin	P N T	C E	I	C A Y A	A C T G C	

Figure 11.4 Amino acid sequences of the mature heat-stable enterotoxins produced by *Y. enterocolitica* (103, 186, 238), enterotoxigenic *E. coli* of human (STh) and porcine (STp) subtypes (7, 236), *C. freundii* (85), *V. cholerae* non-O1 (237), and the intestinal hormone guanylin (49). Amino acid residues which are shaded are common to all seven peptides. The first 23 amino acids at the N terminus of the Yst-c mature toxin (denoted by superscript "a") are not included in the sequence alignments.

maturation process resembles that of *E. coli* STa, although these toxins have no significant homology at the amino terminus.

Despite its similarity to known virulence factors and the fact that production of Yst is largely restricted to the classical pathogenic biovars of *Y. enterocolitica* (59), the contribution of Yst to the pathogenesis of diarrhea associated with yersiniosis is uncertain. Doubts regarding the role of Yst in virulence stem from the following observations: (i) the toxin is generally not detectable in bacterial cultures incubated at temperatures above 30°C, (ii) production of Yst has not been observed in vitro, and (iii) strains of *Y. enterocolitica* that have spontaneously lost the ability to produce Yst retain full virulence for experimental animals (195, 208). However, Delor and Cornelis have determined that a *yst* mutant of a serogroup O:9 strain of *Y. enterocolitica* caused milder diarrhea in infant rabbits than the wild-type strain (58). This observation suggests that the reason why *yst* (and *inv*) are not normally expressed at 37°C is that the conditions used to study expression of these genes in vitro do not reflect those to which the bacteria are exposed in vivo. In this regard, the finding that *Y. enterocolitica* can produce Yst at 37°C if the bacteria are grown in medium with a high osmolarity and pH resembling that present in the intestinal lumen is significant, although the precise stimulus to Yst synthesis in vivo is not yet known (151).

After repeated passage or prolonged storage, Yst-secreting strains of *Y. enterocolitica* frequently become toxin negative. This phenomenon is not caused by mutation of the *yst* gene but is due to the silencing of this gene by YmoA (*Yersinia* modulator). YmoA is an 8-kDa protein which resembles histones and evidently downregulates gene expression in yersiniae by altering DNA topology (47). Expression of *yst* is also regulated by RpoS, an alternative sigma factor of RNA polymerase, which is involved in regulating the expression of a number of stationary-phase genes in other enterobacteria (110).

Toxins that resemble Yst in terms of heat stability and reactivity in infant mice, but with a different structure, molecular weight, and/or mechanism of action, have been detected in various *Yersinia* species, including biovar 1A strains of *Yersinia enterolitica* and "avirulent" *Yersinia* species, such as *Yersinia bercovieri* and *Yersinia mollaretii* (186, 194, 231, 260). These bacteria may be responsible for diarrhea in some patients; hence, the contribution of these toxins to bacterial virulence cannot be discounted (230). Some of these toxins have been characterized, and the genes encoding their production have been cloned and sequenced (Fig. 11.4). The availability of diagnostic probes for the genes encoding Yst-b and Yst-c has indicated that biovar 1A *Y. enterocolitica* organisms commonly carry the genes for Yst-b, whereas Yst-c, which was originally identified in a serogroup O:3 strain of biotype 4, is rare (79, 186, 260).

Some strains of *Y. enterocolitica* can elaborate Yst or the other enterotoxins over a wide range of temperatures from 4 to 37°C (123, 194). Since these toxins are relatively acid stable, they could resist inactivation by stomach acid and thus conceivably cause foodborne illness if they were ingested preformed in food. In artificially inoculated foods, however, these toxins are optimally synthesized at 25°C during the stationary phase of bacterial growth (210). Accordingly, the storage conditions required for their production in food would generally result in severe spoilage, making the possibility of ingestion of preformed Yst unlikely.

Myf Fibrillae

Many intestinal pathogens possess distinctive colonization factors on their surface which mediate their adherence to specific targets on the intestinal epithelium. In noninvasive, enterotoxin-secreting bacteria, such as enterotoxigenic *E. coli*, these factors frequently are surface fimbriae, which enable the bacteria to deliver their toxins close to epithelial cells while resisting removal by peristalsis (75). In enteroinvasive bacteria, surface adhesins may augment virulence by enabling the bacteria to home in on cells, such as M cells, which they preferentially invade. The key intestinal-colonization factors of *Y. enterocolitica* are invasin and YadA, which mediate binding to M cells. In addition, some strains of *Y. enterocolitica* produce a fimbrial adhesin, named Myf (for mucoid *Yersinia* fibrillae) because it bestows a mucoid appearance on bacterial colonies which express it (112). Myf consists of narrow flexible fimbriae which resemble CS3, an essential colonization factor of some human strains of enterotoxigenic *E. coli*. MyfA, the major structural subunit of Myf, has some homology to the PapG protein of pyelonephritis-associated strains of *E. coli* and is 44% identical at the DNA level to the pH6 antigen of *Y. pseudotuberculosis* and *Y. pestis*, which also has a fibrillar structure and mediates thermoinducible binding of *Y. pseudotuberculosis* to tissue culture cells (143, 259). Like *ail* and *yst*, *myf* occurs predominantly in *Y. enterocolitica* strains of the classical pathogenic varieties commonly associated with disease (112). The pH6 antigen of *Y. pseudotuberculosis* is synthesized within the acidic phagolysosomes of macrophages and may play a role in the interaction between bacteria and phagocytic cells, although it apparently does not contribute to bacterial survival in these cells. Its main role in virulence may be related to its ability to mediate binding of bacteria to intestinal mucus before the bacteria contact epithelial cells (148). Although Myf may contribute to the colonizing ability of *Y. enterocolitica*, direct proof of this is lacking.

The *myf* operon in *Y. enterocolitica* comprises three genes, *myfA*, *myfB*, and *myfC*, corresponding to *psaA*, *psaB*, and *psaC* in *Y. pseudotuberculosis* (112, 259). *myfA* encodes the major 21-kDa pilus subunit, whereas *myfB* and *myfC* appear to be involved in fimbrial assembly (112). Two regulatory genes, *myfE* and *myfF* (*psaE* and *psaF*), are located upstream of *myfA* (106). In vitro, *myf* and *psa* are optimally expressed at 37°C in stationary phase and low pH but are not subject to regulation by RpoS (106, 110).

LPS

As with other enterobacteria, *Y. enterocolitica* can be classified as smooth or rough depending on the amount of O side chain polysaccharide attached to the inner core region of the cell wall LPS. Synthesis of the O side chain by *Y. enterocolitica* is specified by the chromosomal *rfb* locus and is regulated by temperature, such that colonies are smooth when grown at temperatures below 30°C but rough at 37°C (222). *Y. enterocolitica* strains of serogroups O:3 and O:8 which carry a mutation in the *rfb* locus display reduced virulence for mice, indicating that the bacteria require smooth LPS for the full expression of virulence (222, 264). The outer core region of LPS plays a role in maintaining outer membrane integrity and may contribute to the resistance of *Y. enterocolitica* to bactericidal peptides in host tissues (223). Smooth LPS may also enhance virulence by increasing bacterial hydrophilicity and thus facilitating the passage of yersiniae through the mucous secretions which line the intestinal epithelium. Additionally, elongated O side chains could conceal surface virulence-associated proteins, such as Ail and YadA, which may enhance the survival of bacteria which have entered the extracellular milieu. Hence, repression of O side chain synthesis at 37°C may increase the likelihood of bacteria surviving in tissues by enabling them to express essential virulence determinants on their surfaces at the appropriate stage of infection (180).

Flagella

In vitro, *Y. enterocolitica* is motile when grown at 25°C but not at 37°C. The expression of the genes encoding flagellin and invasin appear to be coordinately regulated at the level of transcription in such a way that mutants of *Y. enterocolitica* which are defective in the expression of invasin are hypermotile (15). Although flagellar proteins are evidently synthesized in vivo, as evidenced by the antibody responses of patients with systemic yersiniosis to these proteins, motility does not appear to contribute to the virulence of *Y. enterocolitica* in mice (111).

Iron Acquisition

Iron is an essential micronutrient of almost all bacteria. Despite the nutrient-rich environment provided to bacteria by mammalian tissues, the availability of iron in many extracellular locations is limited (256). This is because most iron in tissues is bound to high-affinity transport glycoproteins, such as transferrin and lactoferrin, or is incorporated into organic molecules, such as hemoglobin. Several species of pathogenic bacteria produce low-molecular-weight, high-affinity iron chelators known as siderophores (163). These compounds are secreted by the bacteria into the surrounding medium, where they form a complex with ferric iron. The resultant ferri-siderophore complex then binds to specific receptors on the bacterial surface and is taken up by cells. The observation that patients suffering from iron overload have increased susceptibility to severe infections with *Y. enterocolitica* suggests that the availability of iron in tissues may determine the outcome of yersiniosis (192).

Investigation of the relationship of yersiniae to iron has revealed that these bacteria employ a wide array of processes to acquire iron from inorganic and organic sources (177, 178). The fact that most clinical isolates of *Y. enterocolitica* do not produce siderophores accounts for their reliance on abnormally high concentrations of iron for growth in tissues. Interestingly, however, biovar 1B strains (which are intrinsically more virulent than strains of biovars 2 to 5), as well as wild-type strains of *Y. pestis* and *Y. pseudotuberculosis*, carry genes for the biosynthesis, transport, and regulation of a 482-Da, catechol-containing siderophore known as yersiniabactin. The ca. 40-kbp *ybt* locus which contains these genes has a higher G+C content (57.5 mol%) than that of the *Y. enterocolitica* chromosome (47 mol%), is flanked on one side by an *asn* tRNA gene, and carries the gene for a putative integrase (185). These features, which are typical of a pathogenicity island, have led to the *ybt* locus also being described as the *Yersinia* high-pathogenicity island. The designation "high" alludes to the observation that bacteria which carry this locus are more virulent for mice infected perorally (median lethal dose, $<10^3$ CFU) than strains which lack it (median lethal dose, typically $>10^6$ CFU). Interestingly, the *Yersinia* high-pathogenicity island is also present in some enterovirulent strains of *E. coli*, indicating the possible lateral transfer of these virulence genes between different species of enteric pathogens (126, 213).

The complete nucleotide sequence of the high-pathogenicity island of a serogroup O:8 strain of *Y. enterocolitica* has been determined (185). It contains 22 open reading frames within 43.4 kb, approximately 30.5 kb of which are conserved among *Y. enterocolitica*, *Y. pestis*, and *Y. pseudotuberculosis* (Fig. 11.5). The functions of several genes in the pathogenicity island are still to be elucidated, but it is known that the synthesis of yersiniabactin requires the *irp* (iron-regulated protein) genes *irp1* to *irp5*, as well as *ybtA*, which encodes an AraC-like regulator. The ortholog of *irp9* in *Y. pestis* may also be involved in the synthesis of yersiniabactin, whereas those of *irp6*, *irp7*, and *irp8* may contribute to the uptake of the ferri-yersiniabactin complex. The major receptor for this complex, however, is a 65-kDa outer membrane protein, named FyuA, which also serves as a receptor for pesticin, a bacteriocin produced by *Y. pestis*. For this reason, strains of *Yersinia* species which carry the high-pathogenicity island are susceptible to pesticin (97). Transport of ferri-yersiniabactin complexes across the cell wall of *Y. enterocolitica* resembles the analogous pathway in *E. coli* in that it is an energy-dependent process which requires TonB. The latter protein couples energy provided by inner membrane metabolism to outer membrane protein receptors, such as FyuA. TonB-, FyuA-, and yersiniabactin-deficient mutants of *Y. enterocolitica* all have reduced virulence for mice, presumably because of their limited capacities to acquire sufficient iron to grow in tissues (16). The fact that *aroA* is also required for yersiniabactin synthesis may partly explain the attenuation of *aroA* mutants of *Y. enterocolitica* for mice (30).

Although biovars of *Y. enterocolitica* other than 1B do not produce yersiniabactin, they are able to acquire iron from a number of sources, including ferri-siderophore complexes in which the siderophore, such as desferrioxamine B, is synthesized by another microorganism

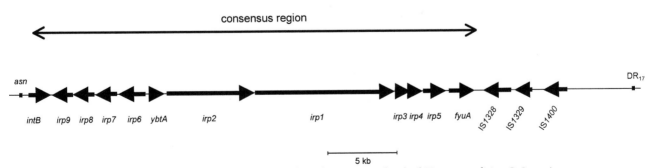

Figure 11.5 Representation of the high-pathogenicity island of *Y. enterocolitica* O:8 strain WA-C. The arrows indicate the positions of the open reading frames and the direction of transcription. The region that is conserved in *Y. pestis* and *Y. pseudotuberculosis* is indicated by a double-headed arrow. (Adapted from reference 185.)

(16, 191). The resultant iron-desferrioxamine complex (known as ferrioxamine) binds to FoxA, a 76-kDa outer membrane protein which shares 33% amino acid homology with FhuA, the high-affinity ferrichrome receptor of *E. coli* (17). The ability of *Y. enterocolitica* to acquire iron from ferrioxamine may have important clinical implications, because desferrioxamine B is used therapeutically to reduce iron overload in patients with hemosiderosis and other forms of iron intoxication. When administered to patients, desferrioxamine B forms a ferri-siderophore complex, which *Y. enterocolitica* can utilize as a growth factor (190). Accordingly, If patients undergoing iron chelation therapy with desferrioxamine B become infected with *Y. enterocolitica*, the bacteria may proliferate in tissues in which, under normal circumstances, the poor availability of iron would limit their growth. Apart from its effects on microbial iron metabolism, desferrioxamine B may also increase susceptibility to systemic yersiniosis by interfering with host immune responses (14).

Phospholipase

Some isolates of *Y. enterocolitica* are hemolytic due to the production of phospholipase A. A strain of *Y. enterocolitica* in which the *yplA* gene encoding this enzyme was deleted had diminished virulence for perorally inoculated mice (212). Interestingly, this mutant induced less inflammation and necrosis in intestinal and lymphoid tissues than the wild type, suggesting that phospholipase contributes to microabscess formation by *Y. enterocolitica*. YplA is secreted by *Y. enterocolitica* via the same type III export apparatus that is used for flagellar proteins (262).

Urease

All enteric pathogens must negotiate the acid barrier of the stomach to cause disease. In *Y. enterocolitica*, acid tolerance relies on the production of urease, which catalyzes the release of ammonia from urea and enables the bacteria to resist pHs as low as 2.5 (55, 261). Urease also contributes to the survival of *Y. enterocolitica* in host tissues, but the mechanism by which this occurs is not known (83).

Although the urease gene complex of *Y. enterocolitica*, comprising seven open reading frames, resembles that of other enterobacteria, analysis of the complex suggests that *Y. enterocolitica* acquired it from an unrelated organism or that *Yersinia* diverged from the other enterobacteria some considerable time ago (57). The urease of *Y. enterocolitica* is also unusual in that it displays optimal activity at pH 3.5 to 4.5, suggesting a physiological role in protecting the bacteria from acid (56). The production of urease by *Y. enterocolitica* is maximal at 28°C during the stationary phase of growth, but expression of the urease gene complex does not depend on RpoS.

The Virulence Plasmid (pYV)

All fully virulent, highly invasive strains of *Y. enterocolitica* carry a ca. 70-kb plasmid, termed pYV (plasmid for *Yersinia* virulence), which is highly conserved among *Y. enterocolitica*, *Y. pestis*, and *Y. pseudotuberculosis* (for a review, see reference 46). pYV functions mainly as an antihost genome that permits bacteria which carry it to resist phagocytosis and complement-mediated lysis, thus enabling them to proliferate extracellularly in tissues (Table 11.7). Yersiniae which carry pYV exhibit a

Table 11.7 Major pYV-encoded determinants of *Y. enterocolitica* and their roles in virulence

Determinant(s) Yops	Role in virulence
Yops	
Effectors	
YopE, YopT	Interference with phagocytosis by disruption of actin filaments
YopH	Interference with phagocytosis by dephosphorylation of focal adhesion kinase, etc.
YopM	Not known
YopO (YpkA)[a]	Disruption of cytoskeletal actin
YopP (YopJ)[a]	Reduction of inflammation by suppressing TNF-α; induction of macrophage apoptosis by suppressing host kinases
YopB, YopD, LcrV	Yop translocation; possible antihost effects and regulation of *yop* gene expression
Syc proteins	
SycE, -H, -D, -N, -T	Yop chaperones
Ysc complex (product of *virC*, *virG*, *virA*, *virB*)	Type III secretory apparatus for Yops
YadA	Bacterial adhesion; reduction of opsonization by interfering with binding of complement proteins
VirF	Regulation of expression of *yop*, *virC*, and *yadA* genes

[a] Designation in *Y. pseudotuberculosis*.

distinctive phenotype known as "calcium dependency," or "the low-calcium response," because it manifests only when pYV-bearing bacteria are grown in media containing low concentrations of Ca^{2+}. The principal features of this response are the cessation of bacterial growth after one or two generations and the appearance of at least 12 new proteins (Yops) on the bacterial surface or in the culture medium. The expression of pYV-encoded proteins imbues *Y. enterocolitica* with novel properties, such as autoagglutination, resistance to killing by normal human serum, and an ability to bind Congo red and crystal violet (29). Some of these characteristics have been exploited in the design of culture media, such as calcium-depleted agar containing Congo red (21, 65), to facilitate the isolation of pYV-bearing colonies of yersiniae. The inherent instability of pYV, however, particularly when yersiniae are grown in low-calcium media at 37°C (139), indicates that caution is needed in using these media if the presence of pYV is required for further testing.

Yops are so named because they were once thought to be outer membrane proteins, but they are now known to be secreted by the bacteria via the type III secretory path-

way. Although Yops are highly conserved among *Yersinia* species, there is little homology between the individual Yops of a single species. The 12 Yops encoded by pYV are characterized by their common mode of secretion and their regulation by a pYV-encoded regulator known as VirF.

The genes carried on pYV include those for (i) an outer membrane protein adhesin, YadA; (ii) a secretion apparatus (Ysc) whereby Yops are transported across the *Yersinia* inner and outer membranes via the type III secretory pathway; (iii) at least six distinct antihost effector Yops; (iv) a translocation apparatus, comprising certain Yops, which the effector Yops require to gain access to the host cell cytosol; and (v) factors for the regulation of Yop biosynthesis, secretion, and translocation.

Genes for the effector Yops are scattered around pYV, whereas those required for Yop secretion and translocation are clustered together (Fig. 11.6). Homologs of several of these genes occur in *P. aeruginosa* (73, 94). pYV also encodes YlpA, a 29-kDa lipoprotein related to the TraT protein of plasmids in various enterobacteria, and, in *Y. enterocolitica* strains of biovars 2 to 5 (but not

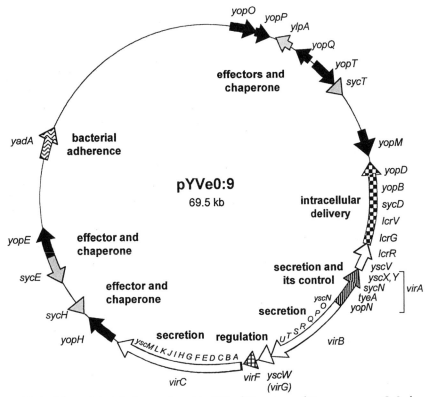

Figure 11.6 Map of the virulence plasmid pYV of *Y. enterocolitica* serogroup O:9 showing the locations and directions of transcription (arrows) of the genes encoding (i) YadA; (ii) YlpA; (iii) YopB, -D, -E, -H, -M, -N, -O, -P, -Q, and -T and LcrV; (iv) specific Yop chaperones SycD, -E, -H, and -T; (v) secretion elements VirA, -B, -C, and -G; and (vi) the regulatory element VirF. (Adapted from reference 108.)

biovar 1B or *Y. pseudotuberculosis*), an operon which specifies resistance to arsenic (166).

YadA: a pYV-Encoded Adhesin

YadA, formerly known as Yop1 or P1, is a 44- to 47-kDa outer membrane protein which contains a typical 25-amino-acid signal sequence, indicating that it is transported across the bacterial membrane via the general (type II) secretory pathway. Individual YadA monomers aggregate in solution to form oligomers with an apparent molecular mass of ca. 200 kDa. Early studies of the morphology of YadA suggested that it polymerized to form fibrils on the bacterial surface (125), but more recent studies have revealed that YadA comprises lollipop-shaped oligomers which envelop the entire outer membrane as a densely packed array (5).

YadA mediates bacterial adhesion to intestinal mucus and to certain extracellular matrix proteins, including collagen, laminin, and cellular fibronectin (221). These proteins in turn may bind to β_1 integrins, including $\alpha_1\beta_1$, $\alpha_2\beta_1$, $\alpha_3\beta_1$, and $\alpha_5\beta_1$, on epithelial cells and stimulate bacterial internalization by endocytosis in a manner similar to that mediated by invasin. Thus, YadA may contribute to bacterial invasion, even though binding to integrins via matrix proteins generally does not stimulate bacterial uptake to the same extent as that mediated by invasin (25). YadA may also promote bacterial invasion by binding to integrins directly (23, 258). Studies involving the deletion of portions of YadA and the substitution of individual amino acids have revealed that YadA contains a hydrophobic C-terminal outer membrane anchor, a central rodlike region (formed by a coiled coil) which is responsible for binding to matrix proteins, and an N-terminal domain which is required for binding to polymorphonuclear leukocytes (5, 46).

Apart from its role as an adhesin and invasin, YadA also contributes to virulence by conveying resistance to complement-mediated opsonization. It achieves this by binding factor H, thereby reducing deposition of C3b on the bacterial surface (41). As a consequence, YadA is associated with resistance of *Y. enterocolitica* complement-mediated lysis and phagocytosis and an ability to inhibit the respiratory burst of polymorphonuclear leukocytes, all of which require the bacteria to be preopsonized (40, 41). Given the pluripotential capacity of YadA to increase the likelihood of bacterial survival in host tissues, it is not surprising that YadA mutants of *Y. enterocolitica* show significantly reduced virulence for mice (176, 196). This is in contrast to YadA mutants of *Y. pseudotuberculosis*, which are not attenuated, and *Y. pestis*, which is extremely virulent for mice but is naturally defective in YadA production due to a single-base-

pair deletion resulting in a shift of the reading frame of the gene (88, 201).

Synthesis of YadA is regulated at the transcriptional level by temperature, but not by the concentration of Ca^{2+}. This is in contrast to *ylpA*, the *yop* genes, and the *virC* operon, which are regulated by both temperature and Ca^{2+}. Expression of *yadA* is controlled by VirF, a pYV-encoded, DNA-binding protein of the AraC family (44). Other members of this family include VirF from *Shigella flexneri*, Rns from enterotoxigenic *E. coli*, and ExsA from *P. aeruginosa*, all of which are involved in the regulation of expression of virulence in their respective bacteria. VirF plays a central role in the virulence of *Y. enterocolitica* by governing the transcriptional activation of *yadA*, *ylpA*, all of the *yop* genes, and the *virC* operon. Because these genes are coregulated, they have been named the Yop regulon (45). Transcription of *virF* itself is regulated by temperature (but not by Ca^{2+}), probably as a result of combined effects of YmoA and the influence of elevated temperature on DNA supercoiling (197).

Yop Secretion and the Ysc Secretion Apparatus

The inner and outer membranes of gram-negative bacteria are major barriers to protein export. The term "secretion" is used to denote the active transport of proteins from the cytoplasm across these membranes. Secretion of Yops occurs via the type III secretory pathway in which protein transport is governed by an amino-terminal sequence. In contrast to proteins exported via the general or type II secretory pathway, however, the amino termini of type III-secreted proteins have no resemblance to each other and are not cleaved during export (for reviews, see references 104 and 182). Another feature of type III secretion is the presence of structurally conserved chaperone proteins, which bind specifically to individual secreted proteins and guide them to the secretion apparatus while preventing their premature interaction with other proteins (252). Chaperone proteins for Yops are denoted by the prefix Syc (for specific Yop chaperone) and include SycE (for YopE), SycH (for YopH), SycD (for YopB and YopD), SycN (for YopN), and SycT (for YopT). The genes encoding these chaperones are located on pYV close to the corresponding *yop* gene (Fig. 11.6). Although they share no significant homology with each other, all Yop chaperones identified to date are low-molecular-mass (14- to 19-kDa) proteins with a C-terminal amphipathic α-helix and a pI of around 4.5 (46).

The N-terminal signal required to secrete a Yop generally resides within the first 20 amino acid residues of

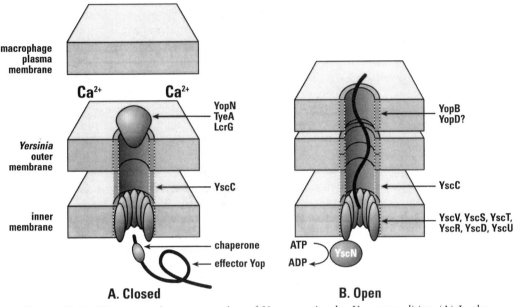

Figure 11.7 Diagrammatic representation of Yop secretion by *Y. enterocolitica*. (A) In the absence of contact with host cells or in the presence of elevated Ca^{2+}, the secretory apparatus (Ysc) is closed by a plug comprising YopN, TyeA, and LcrG. (B) When Ca^{2+} is removed (in vitro) or contact with host cells is established, a continuous channel which links the bacterial cytoplasm to that of the host cell is established. This allows effector Yops to traverse the inner and outer membranes of the bacteria (via the Ysc type III secretory machine) and to enter the host cell (via the translocation apparatus comprising YopB and -D).

the protein. Nevertheless, Yop secretion domains are not similar to each other with respect to amino acid sequence, hydropathy profile, distribution of charged structures, or predicted secondary structure (46). Studies by Anderson and Schneewind (10, 11) involving mutagenesis of the signal sequences of YopE and YopN and frameshift mutations that completely altered the peptide sequence suggested that the signal for Yop secretion lies in the messenger RNA rather than the peptide sequence. Subsequent work from the same laboratory, however, confirmed a second secretion signal in YopE that corresponds to the SycE binding site (38, 39). Thus, it appears that there are two distinct signals for the secretion of YopE, one dependent on the nucleotide sequence at the 5′ end of the messenger RNA and the other determined by the SycE-YopE complex. Similar processes seem likely to govern the secretion of YopH and YopT. For Yops without specific chaperones, however, it is not known if secretion is determined by an amino or 5′ signal.

The Ysc secretion apparatus is a paradigm of all type III secretory machines (46). The 29 genes encoding this apparatus are contained within four contiguous loci named *virC* (comprising *yscABCDEFGHIJKLM*), *virG* (which encodes YscW), *virA* (encoding YopN,

TyeA, SycN, YscX, YscY, YscV, and YscR [also known as LcrD]), and *virB* (comprising *yscNOPQRSTU*) (Fig. 11.6) (108). Many Ysc proteins have not been characterized, and their contributions to Yop secretion are obscure. Nevertheless, some components of Ysc have been localized to the bacterial cell envelope, where they are envisaged to form a channel that spans the inner and outer membranes (Fig. 11.7) (46).

YscC is an outer membrane protein of the secretin family which is envisaged to form a ringlike structure with an external diameter of 200 Å and an apparent central pore around 50 Å in diameter (132). It is stabilized in the outer membrane by YscW, a lipoprotein product of the *virG* gene (132). At least four Ysc proteins span the inner membrane, namely, YscD, YscR, YscU, and YscV. YscN is a 48-kDa protein with ATP-binding motifs (Walker boxes A and B) resembling those of many ATPases; it is hypothesized to provide the energy required for Yop secretion.

Several Ysc proteins are secreted, including YscM, YopR (the product of *yscH*), YopN, YscO, and YscP. YopN is thought to plug the channel together with TyeA, a surface protein, which binds to YopN and is required for the translocation of YopE and YopH (109). The roles of the other secreted proteins are unknown.

Yop Translocation

Although the type III secretory pathway appears to have evolved from the process used to assemble flagella (104), many type III systems, including that of the enteropathogenic yersiniae, are used by bacteria to inject (translocate) proteins into the cytosol of eukaryotic cells and thus to facilitate pathogenesis (104, 200, 226). Yops are classified into those which are translocated into host cells to exert an antihost action (effector Yops) and those which are primarily involved in the translocation process. There is evidence, however, that some Yops required for translocation may also act as antihost effectors.

The transport of Yops from the bacterial cytoplasm (via Ysc) into the host cell cytosol (via the translocation apparatus) is envisaged to occur in one step from bacteria which are closely bound to the host cell (via invasin and/or YadA) (Fig. 11.7) (136). Two of the proteins required for translocation are YopB (42 kDa) and YopD (33 kDa,) both of which have hydrophobic domains, suggesting that they could interact directly with host cell membranes (86, 165). YopB resembles members of the RTX toxin family and can evidently form a pore in the plasma membranes of eukaryotic cells (165). YopD associates with YopB and may contribute to the formation of the putative pore (165). The observation that YopD is also translocated into cells, however, suggests that it may fulfill more than one function (72). YopB and YopD are encoded by the *lcrGV-sycD-yopBD* operon, which also encodes LcrV, LcrG, and SycD, the chaperone for YopB and YopD (Fig. 11.6). LcrV, also known as the V antigen, is a versatile Yop which acts as an essential component of the translocation apparatus, as a regulator of Yop expression, and possibly as an effector Yop by inhibiting neutrophil chemotaxis and cytokine production (68, 257). Its key role in the virulence of *Y. pestis* is evidenced by the fact that antibodies to LcrV protect mice from infection with this species (248).

Other proteins which appear to contribute to the translocation process include YopQ (known as YopK in *Y. pseudotuberculosis*), the gene for which lies outside the *lcrGV-sycD-yopBD* operon (Fig. 11.6). Although LcrG is not a component of the translocation apparatus per se, it acts in concert with YopN and TyeA to control Yop release in vitro and may also mediate binding between the Syc translocation channel and a heparinlike receptor on host cells (Fig. 11.7) (31, 109).

Mechanism of Action of Effector Yops

YopE is a 25-kDa protein which is translocated into host cells and contributes to the ability of *Y. enterocolitica* to resist phagocytosis. Direct microinjection of YopE into mammalian cells disrupts actin microfilaments and leads to cytotoxic changes (199). YopE does not act on actin directly, and although its target in host cells is not known, there is evidence that it may act on Rho GTPases which contribute to actin integrity (172). YopT is a 35-kDa protein which exerts an effect on actin microfilaments similar to that of YopE (107). Its intracellular target appears to be a Rho GTPase known as RhoA (266).

YopH, a 51-kDa protein, is a potent protein tyrosine phosphatase. Protein tyrosine phosphorylation is a key element of signal transduction in host cells which affects many fundamental processes, including phagocytosis and cell division. YopH dephosphorylates several phosphotyrosine residues, such as those on focal adhesion kinase and the focal adhesion proteins, paxillin and p130^{Cas} (179). Focal adhesions are sites where an integrin receptor acts as a transmembrane bridge between extracellular matrix proteins and intracellular signaling proteins. The fact that autophosphorylation of focal adhesion kinase is directly involved in invasin-mediated uptake of yersiniae by epithelial cells (8) provides an explanation for the antiphagocytic effects of YopH.

YopM is a 41-kDa protein which contains a succession of 12 repeated structures that resemble leucine-rich repeat motifs (138). For this reason, YopM shows weak homology to a large number of proteins, including the α chain of membrane glycoprotein 1b (GP1bα), a platelet-specific receptor for thrombin, and the von Willebrand factor (138). Although YopM can to bind to thrombin in vitro, the finding that it is translocated into cells and traffics to the nucleus by means of a vesicle-associated, microtubule-dependent pathway (219) indicates that its principal mechanism of action is probably unrelated to its antithrombin activity.

YopO is an 82-kDa protein which is known as YpkA (*Yersinia* protein kinase A) in *Y. pseudotuberculosis*. It is secreted by the bacteria in an inactive form, which is activated by actin in the target cell (122). The functional kinase then acts on actin to disrupt the cytoskeleton and presumably retard phagocytosis.

YopP (known as YopJ in *Y. pseudotuberculosis*) is a 32-kDa protein that is encoded by the same operon as YopO. YopP interferes with the mitogen-activated protein kinase activities of c-Jun N-terminal kinase, p38, and extracellular signal-regulated kinase (202). Its inhibitory action on these enzymes is evidently linked to the ability of YopP to induce apoptosis in macrophages. YopP also acts on various cell types to inhibit release of tumor necrosis factor alpha (TNF-α), a proinflammatory cytokine which plays a central role in inflammatory and immune responses (26). This effect of YopP may be due to an inhibitory action on NF-κB. Accordingly, YopP may contribute to the establishment of bacterial infection by eliminating macrophages while retarding inflammatory responses (158).

Regulation of Yop Production and Secretion

All Yops and YlpA are produced in vitro at 37°C (but not at temperatures below 30°C) when the concentration of Ca^{2+} is sufficiently low to induce bacteriostasis. The mechanism of regulation by temperature involves VirF (which itself is regulated by temperature) and probably DNA supercoiling (197). The mechanism of regulation by Ca^{2+} is not known, but there is evidence that Ca^{2+} chelation may affect the interaction among YopN, TyeA, and LcrG, which in the presence of millimolar concentrations of Ca^{2+} act together to plug the Ysc channel (Fig. 11.7).

There is little doubt that Yops are produced in vivo, because animals and humans infected with virulent *Yersinia* species develop antibodies to these proteins during the course of infection (Fig. 11.8). Although the host

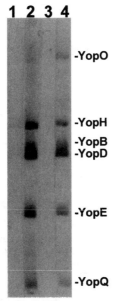

Figure 11.8 Antibody response of sheep infected with *Y. enterocolitica* or *Y. pseudotuberculosis* to Yops. Yops were prepared from *Y. enterocolitica* serogroup O:3, separated by polyacrylamide gel electrophoresis, transferred to a nitrocellulose membrane, and reacted with preimmune (lanes 1 and 3) or immune (lanes 2 and 4) sera from lambs with naturally acquired infection with pYV-bearing *Y. enterocolitica* (lanes 1 and 2) or *Y. pseudotuberculosis* (lanes 3 and 4). (Reprinted with permission from reference 188.)

temperature (37°C) is likely to be a key stimulus for the production of Yops in vivo, the second signal, equating to low Ca^{2+}, is obscure. Originally, when yersiniae were thought to localize within phagocytes, the intraphagocytic environment, which is relatively low in Ca^{2+}, was thought to provide this stimulus (181). It is now clear, however, that *Y. enterocolitica* is located mainly extracellularly in tissues and that extracellular bacteria are the major source of intracellular Yops (117, 136). Forsberg et al. (71) determined that extracellular yersiniae are cytotoxic for HeLa cells even in the presence of millimolar concentrations of Ca^{2+}, provided the bacteria can attach to the target cells. This observation indicates that in tissues, contact between *Y. enterocolitica* and host cells permits Yop release in the same way that Ca^{2+} depletion achieves it in vitro. The nature of the stimulus which host cells provide to the bacteria to synthesize and release Yops is not known.

Antihost Action of pYV

When strains of *Y. enterocolitica* cured of pYV are incubated with host epithelial cells or phagocytes, they typically penetrate them in large numbers without causing significant cytopathology (Fig. 11.9). By contrast, pYV-bearing bacteria generally remain extracellular but exert a range of effects which are attributable to the action of effector Yops (Fig. 11.9). Some of these effects can be ascribed to one particular Yop, whereas others are due to a number of Yops acting collaboratively.

As macrophages and polymorphonuclear leukocytes are major participants in the first line of defense against invading microbes, the effect of *Y. enterocolitica* on these cells is of particular interest in explaining the contribution of pYV to virulence. When pYV-positive strains of *Y. enterocolitica* are incubated with macrophages in vitro, they resist phagocytosis, inhibit the respiratory burst, interfere with the release of TNF-α and gamma interferon, and induce apoptosis (19, 92, 142, 155, 159). Bacteria cured of plasmids exhibit none of this antimacrophage activity and are readily phagocytosed, but they are able to resist killing, possibly due to the action of HtrA (also known as DegP), a chromosomally encoded, heat shock-induced serine protease (140). If pYV-bearing bacteria are preopsonized by antibodies, resistance to phagocytosis is mediated largely by YadA, whereas unopsonized bacteria resist ingestion by phagocytes due to the action of YopH, possible acting in concert with YopE and YopT. These same Yops are involved in inhibition of the respiratory burst, whereas the inhibition of cytokine release and the induction of apoptosis are primarily due to YopP and YopJ. The ability of pYV-bearing *Y. enterocolitica* to inhibit the release and/or production

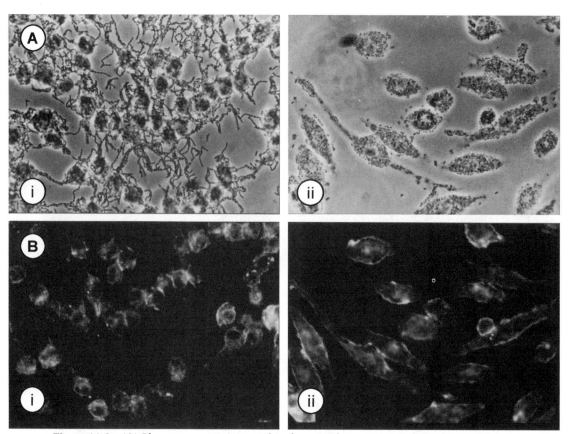

Figure 11.9   (A) Phase-contrast micrographs of murine bone marrow-derived macrophages incubated for 3 h with *Y. enterocolitica* bearing (i) pYV or (ii) no plasmid. (B) The same preparations shown in panel A stained with fluorescein-conjugated phalloidin (to demonstrate filamentous actin) and visualized by fluorescence microscopy. Note the predominantly extracellular location of the bacteria, rounded macrophages, and disrupted actin filaments in panels A i and B i compared with panels A ii and B ii. Magnification, ×400. (Reprinted with permission from reference 91.)

of proinflammatory cytokines, such as TNF-α, gamma interferon, and interleukin 8, results in a reduction of inflammatory responses (26, 162, 214).

pYV-bearing strains of *Y. enterocolitica* resist ingestion and killing by polymorphonuclear leukocytes, whereas strains cured of the plasmid are killed (40). The principal virulence determinants involved in this process are YadA, which interferes with opsonization, and YopH and YopE, which inhibit the respiratory burst and retard phagocytosis (41). Yops of *Y. enterocolitica* also inhibit the production of receptor-dependent superoxide anion by human granulocytes, but the mechanism by which this occurs is not known (250). Should pYV-bearing bacteria be ingested by polymorphonuclear leukocytes, they can resist being killed due to their reduced susceptibility to the antimicrobial peptides produced by these cells. This effect is attributable to YadA and smooth LPS (223).

Pathogenesis of *Yersinia*-Induced Autoimmunity

Arthritis

Following an acute infection with pYV-bearing *Y. enterocolitica*, a small proportion of patients develop autoimmune (reactive) arthritis. A similar syndrome can also occur after infection with *Campylobacter, Salmonella, Shigella,* or *Chlamydia* species (131, 217). The overwhelming majority of individuals with postinfective reactive arthritis caused by *Y. enterocolitica* carry the class I human leukocyte antigen HLA-B27 (217). In addition, HLA-B27-positive individuals have more severe arthritic symptoms and a more prolonged course than individuals with different HLA antigens.

The pathogenesis of reactive arthritis is poorly understood. The synovial fluid from affected joints of patients with *Yersinia*-induced arthritis is culture negative but generally contains bacterial antigens, such as LPS and heat shock proteins, within inflammatory cells (76, 78). Explanations for the link between yersiniosis and autoimmunity include antigen persistence, molecular mimicry, and infection-induced presentation of normally cryptic cell antigens (98, 242). Although patients with reactive arthritis display higher levels of serum immunoglobulin A antibodies to *Yersinia* antigens than individuals without arthritis, these antibodies are unlikely to contribute to the development of arthritis. Instead, they probably reflect enhanced stimulation of the mucosal immune system due to the persistence of bacterial antigens in the intestine or other tissues (53).

The observation that rats transgenic for HLA-B27 have a weaker cytotoxic T-cell response against *Y. pseudotuberculosis* than nontransgenic syngeneic rats provides a link between prolonged infection, antigen persistence, and HLA-B27 (64). Moreover, Hermann et al. (99) have determined that T-cell clones derived from the synovial fluid of patients with *Yersinia*-triggered reactive arthritis are selectively cytotoxic for HLA-B27-bearing cells infected with *Y. enterocolitica*. This suggests that $CD8^+$ class I-restricted cytotoxic T cells can recognize specific *Yersinia*-derived peptides when they are presented together with HLA-B27. The nature of these peptides is not known.

Support for the molecular-mimicry hypothesis comes from the demonstration of autoreactive T cells in the synovial tissue of patients with reactive arthritis (99). Among the bacterial antigens which may provoke autoreactivity are heat shock proteins, which are somewhat conserved among bacteria and mammals and share a number of antigenic determinants. Accordingly, an immune response to selected epitopes on bacterial heat shock proteins may lead to an autoimmune response at sites where bacterial antigens accumulate. In keeping with this suggestion is the observation that synovial fluid from patients with reactive arthritis may contain $CD4^+$ major histocompatibility complex class II-restricted T lymphocytes which recognize epitopes that are shared by a 60-kDa *Yersinia* heat shock protein and its human counterpart (99). Further evidence for the molecular-mimicry hypothesis comes from the finding that an immunodominant epitope from GroEL, a heat shock protein of *S. enterica* serovar Typhimurium (which is almost identical to the homologous protein in *Y. enterocolitica*), can be presented by a major histocompatibility complex class I molecule of mice (144). This complex can subsequently be recognized by $CD8^+$ cytotoxic T cells that cross-react with a peptide derived from a heat shock protein of mice (144). Other bacterial antigens which have been claimed to contribute to autoimmunity via

specific interactions with the immune system are YadA and the β subunit of urease (82, 220). However, early suggestions that YadA shared epitopes with the peptide-binding groove of the HLA-B27 antigen have been discounted (133). The β subunit of urease is a cationic protein which (i) is recognized by $CD4^+$ T cells from patients with *Yersinia*-induced reactive arthritis and (ii) produces arthritis when injected into the joints of rats (220). The finding that a urease-deficient mutant of *Y. enterocolitica* retains its capacity to induce arthritis in a rat model, however, casts doubt on the role of urease in this condition (83).

Y. enterocolitica and *Y. pseudotuberculosis* may also induce polyclonal T-cell stimulation by virtue of their ability to secrete toxins which resemble superantigens (1, 229). YpmA and its variant, YpmB, are 14- to 15-kDa proteins produced by approximately 20% of *Y. pseudotuberculosis* strains that activate human T cells of $V\beta$ phenotypes 3, 9, 31.1, and 13.2 (36, 156). The *Y. enterocolitica* superantigen is not as well characterized as that of *Y. pseudotuberculosis*, but evidently it can stimulate T cells with $V\beta$ phenotypes 3, 7, 8.1, 9, and 11 (229). Yersiniae may also provoke nonspecific immune stimulation when invasin binds to β_1 integrins on T lymphocytes, thus providing a costimulatory signal to these cells (32).

Thyroid Diseases

Y. enterocolitica has been implicated in the etiology of various thyroid disorders, including autoimmune thyroiditis and Graves' disease hyperthyroidism (242, 243). The latter is an immunologic disorder mediated by autoantibodies to the thyrotropin (TSH) receptor. The chief link between *Y. enterocolitica* and thyroid diseases is that patients with these disorders frequently have elevated titers of serum agglutinins to *Y. enterocolitica* O:3 (216). As there is no clear relationship between the incidence or geographic distribution of yersiniosis and that of autoimmune thyroid diseases; however, the presence of circulating antibodies to *Y. enterocolitica* in such patients is likely to reflect a fortuitous cross-reaction between *Yersinia* and thyroid antigens rather than a causal relationship (242). In addition, follow-up of patients many years after infection with *Y. enterocolitica* or *Y. pseudotuberculosis* has revealed no increased frequency of thyroid disease, nor is there any evidence that hyperthyroidism is exacerbated by infection with *Y. enterocolitica* (242). However, observations that the outer membrane of *Y. enterocolitica* carries binding sites for TSH which are recognized by immunoglobulins from patients with Graves' disease (101) and that mice immunized with a component of the human TSH receptor develop antibodies that recognize bacterial surface proteins (145, 263) suggest that infection with *Y. enterocolitica* may trigger an autoimmune response which leads to the development of Graves' disease in susceptible individuals.

Although few studies have been conducted of the pathogenesis of *Yersinia*-induced erythema nodosum or glomerulonephritis, experience with other infective agents suggests that these manifestations are caused by the deposition of immune complexes in affected organs.

SUMMARY AND CONCLUSIONS

Y. enterocolitica is a versatile foodborne pathogen with a remarkable ability to adapt to a wide range of environments within and outside its mammalian hosts. *Y. enterocolitica* typically accesses its hosts via food or water in which it has grown to stationary phase at ambient temperature. Under these circumstances, *Y. enterocolitica* expresses factors, such as urease, flagella, and smooth LPS, which facilitate its passage through the stomach and the mucus layer of the small intestine. Bacteria in this state may also carry Myf fibrillae and invasin that may promote adherence to and penetration of the dome epithelium overlying the Peyer's patches. The higher infectivity of *Y. enterocolitica* when grown at ambient temperature compared with that at 37°C may account for the relatively small number of reports of human-to-human transmission of yersiniosis (168).

Once *Y. enterocolitica* begins to replicate in the intestine at 37°C, LPS becomes rough, exposing Ail and YadA on the bacterial surface. These factors may promote further invasion while protecting the bacteria from complement-mediated opsonization. When *Y. enterocolitica* makes contact with host cells in lymphoid tissue, it is stimulated to synthesize, secrete, and translocate Yops, notably the effector Yops, YopE, -T, -H, and -P, which further frustrate the efforts of phagocytes to ingest and remove the bacteria. Subsequent bacterial replication may lead to tissue damage and the formation of microabscesses. If strains of *Y. enterocolitica* which possess the *Yersinia* high-pathogenicity island gain access to sites where iron supplies are growth limiting, they may produce yersiniabactin so that replication can proceed. Eventually, the cycle is completed when the bacteria rupture through microabscesses in intestinal crypts to reenter the intestine and regain access to the environment. This well-defined life cycle of *Y. enterocolitica*, with its distinctive temperature-induced phases is reminiscent of the flea-rat-flea cycle of *Y. pestis*.

Although much remains to be learned about *Y. enterocolitica*, investigations into the pathogenesis of yersiniosis to date have provided fascinating new insights into bacterial pathogenesis as a whole and its genetic control. *Y. enterocolitica* and *Y. pseudotuberculosis* were the first invasive human pathogens in which plasmid-mediated virulence was documented (265), from which internalins (invasin, Ail, and YadA) were cloned and characterized (113, 154), in which the relationship between iron limitation and ferri-siderophore uptake assumed clinical significance (191), and in which type III protein secretion was identified (150). Future research in this area will no doubt lead to new and unexpected discoveries of bacterial strategies to evade host immunity that will further advance our understanding of the interface between microbes and the animal world.

References

1. Abe, J., M. Onimaru, S. Matsumoto, S. Noma, K. Baba, Y. Ito, T. Kohsaka, and T. Takeda. 1997. Clinical role for a superantigen in *Yersinia pseudotuberculosis* infection. *J. Clin. Investig.* **99:**1823–1830.

2. Achtman, M., K. Zurth, G. Morelli, G. Torrea, A. Guiyoule, and E. Carniel. 1999. *Yersinia pestis*, the cause of plague, is a recently emerged clone of *Yersinia pseudotuberculosis. Proc. Natl. Acad. Sci. USA* **96:**14043–14048.

3. Ackers, M. L., S. Schoenfeld, J. Markman, M. G. Smith, M. A. Nicholson, W. DeWitt, D. N. Cameron, P. M. Griffin, and L. Slutsker. 2000. An outbreak of *Yersinia enterocolitica* O:8 infections associated with pasteurized milk. *J. Infect. Dis.* **181:**1834–1837.

4. Adams, M. R., C. L. Little, and M. C. Easter. 1991. Modelling the effect of pH, acidulant and temperature on the growth rate of *Yersinia enterocolitica. J. Appl. Bacteriol.* **71:**65–71.

5. Aepfelbacher, M., R. Zumbihl, K. Ruckdeschel, C. A. Jacobi, C. Barz, and J. Heesemann. 1999. The tranquilizing injection of *Yersinia* proteins: a pathogen's strategy to resist host defense. *Biol. Chem.* **380:**795–802.

6. Ahvonen, P., K. Sievers, and K. Aho. 1969. Arthritis associated with *Yersinia enterocolitica. Acta Rheumatol. Scand.* **15:**232–255.

7. Aimoto, S., T. Takao, Y. Shimonishi, S. Hara, T. Takeda, Y. Takeda, and T. Miwatani. 1982. Amino-acid sequence of a heat-stable enterotoxin produced by human enterotoxigenic *Escherichia coli. Eur. J. Biochem.* **129:**257–263.

8. Alrutz, M. A., and R. R. Isberg. 1998. Involvement of focal adhesion kinase in invasin-mediated uptake. *Proc. Natl. Acad. Sci. USA* **95:**13658–13663.

9. Andersen, J. K., R. Sorensen, and M. Glensbjerg. 1991. Aspects of the epidemiology of *Yersinia enterocolitica*: a review. *Int. J. Food Microbiol.* **13:**231–237.

10. Anderson, D. M., and O. Schneewind. 1997. A mRNA signal for the type III secretion of Yop proteins by *Yersinia enterocolitica. Science* **278:**1140–1143.

11. Anderson, D. M., and O. Schneewind. 1999. *Yersinia enterocolitica* type III secretion: an mRNA signal that

couples translation and secretion of YopQ. *Mol. Microbiol.* **31**:1139–1148.

12. Arduino, M. J., L. A. Bland, M. A. Tipple, S. M. Aguero, M. S. Favero, and W. R. Jarvis. 1989. Growth and endotoxin production of *Yersinia enterocolitica* and *Enterobacter agglomerans* in packed erythrocytes. *J. Clin. Microbiol.* **27**:1483–1485.

13. Asplund, K., T. Johansson, and A. Siitonen. 1998. Evaluation of pulsed-field gel electrophoresis of genomic restriction fragments in the discrimination of Yersinia enterocolitica O:3. *Epidemiol. Infect.* **121**:579–586.

14. Autenrieth, I. B., R. Reissbrodt, E. Saken, R. Berner, U. Vogel, W. Rabsch, and J. Heesemann. 1994. Desferrioxamine-promoted virulence of *Yersinia enterocolitica* in mice depends on both desferrioxamine type and mouse strain. *J. Infect. Dis.* **169**:562–567.

15. Badger, J. L., and V. L. Miller. 1998. Expression of invasin and motility are coordinately regulated in *Yersinia enterocolitica*. *J. Bacteriol.* **180**:793–800.

16. Baumler, A., R. Koebnik, I. Stojiljkovic, J. Heesemann, V. Braun, and K. Hantke. 1993. Survey on newly characterized iron uptake systems of *Yersinia enterocolitica*. *Int. J. Med. Microbiol. Virol. Parasitol. Infect. Dis.* **278**:416–424.

17. Baumler, A. J., and K. Hantke. 1992. Ferrioxamine uptake in *Yersinia enterocolitica*: characterization of the receptor protein FoxA. *Mol. Microbiol.* **6**:1309–1321.

18. Bercovier, H., and H. H. Mollaret. 1984. Genus XIV. *Yersinia* Van Loghem 1944, 15[AL], p. 498–506. *In* N. R. Krieg and J. G. Holt (ed.), *Bergey's Manual of Systematic Bacteriology*, vol. 1. Williams & Wilkins, Baltimore, Md.

19. Beuscher, H. U., F. Rödel, Å. Forsberg, and M. Röllinghoff. 1995. Bacterial evasion of host immune defense: *Yersinia enterocolitica* encodes a suppressor for tumor necrosis factor alpha expression. *Infect. Immun.* **63**:1270–1277.

20. Bhaduri, S., R. L. Buchanan, and J. G. Phillips. 1995. Expanded response surface model for predicting the effects of temperature, pH, sodium chloride contents and sodium nitrite concentrations on the growth rate of *Yersinia enterocolitica*. *J. Appl. Bacteriol.* **9**:163–170.

21. Bhaduri, S., B. Cottrell, and A. R. Pickard. 1997. Use of a single procedure for selective enrichment, isolation, and identification of plasmid-bearing virulent *Yersinia enterocolitica* of various serotypes from pork samples. *Appl. Environ. Microbiol.* **63**:1657–1660.

22. Black, R. E., R. J. Jackson, T. Tsai, M. Medvesky, M. Shayegani, J. C. Feeley, K. I. E. MacLeod, and A. M. Wakelee. 1978. Epidemic *Yersinia enterocolitica* infection due to contaminated chocolate milk. *N. Engl. J. Med.* **298**:76–79.

23. Bliska, J. B., M. C. Copass, and S. Falkow. 1993. The *Yersinia pseudotuberculosis* adhesin YadA mediates intimate bacterial attachment to and entry into HEp-2 cells. *Infect. Immun.* **61**:3914–3921.

24. Bliska, J. B., and S. Falkow. 1992. Bacterial resistance to complement killing mediated by the Ail protein of *Yersinia enterocolitica*. *Proc. Natl. Acad. Sci. USA* **89**:3561–3565.

25. Bliska, J. B., and S. Falkow. 1994. Interplay between determinants of cellular entry and cellular disruption in the enteropathogenic *Yersinia*. *Curr. Opin. Infect. Dis.* **7**:323–328.

26. Boland, A., and G. R. Cornelis. 1998. Role of YopP in suppression of tumor necrosis factor alpha release by macrophages during *Yersinia* infection. *Infect. Immun.* **66**:1878–1884.

27. Borg, A. A., J. Gray, and P. T. Dawes. 1992. *Yersinia*-related arthritis in the United Kingdom. A report of 12 cases and review of the literature. *Q. J. Med.* **84**:575–582.

28. Bottone, E. J. 1977. *Yersinia enterocolitica*: a panoramic view of a charismatic microorganism. *Crit. Rev. Microbiol.* **5**:211–241.

29. Bottone, E. J. 1997. *Yersinia enterocolitica*: the charisma continues. *Clin. Microbiol. Rev.* **10**:257–276.

30. Bowe, F., P. O'Gaora, D. Maskell, M. Cafferkey, and G. Dougan. 1989. Virulence, persistence, and immunogenicity of *Yersinia enterocolitica* O:8 *aroA* mutants. *Infect. Immun.* **57**:3234–3236.

31. Boyd, A. P., M. P. Sory, M. Iriarte, and G. R. Cornelis. 1998. Heparin interferes with translocation of Yop proteins into HeLa cells and binds to LcrG, a regulatory component of the Yersinia Yop apparatus. *Mol. Microbiol.* **27**:425–436.

32. Brett, S. J., A. V. Mazurov, I. G. Charles, and J. P. Tite. 1993. The invasin protein of *Yersinia* spp. provides costimulatory activity to human T cells through interaction with beta 1 integrins. *Eur. J. Immunol.* **23**:1608–1614.

33. Brubaker, R. R. 1991. Factors promoting acute and chronic diseases caused by yersiniae. *Clin. Microbiol. Rev.* **4**:309–324.

34. Burnens, A. P., A. Frey, and J. Nicolet. 1996. Association between clinical presentation, biogroups and virulence attributes of *Yersinia enterocolitica* strains in human diarrhoeal disease. *Epidemiol. Infect.* **116**:27–34.

35. Butler, R. C., V. Lund, and D. A. Carlson. 1987. Susceptibility of *Campylobacter jejuni* and *Yersinia enterocolitica* to UV radiation. *Appl. Environ. Microbiol.* **53**:375–378.

36. Carnoy, C., H. Müeller-Alouf, S. Haentjens, and M. Simonet. 1998. Polymorphism of *ypm*, *Yersinia pseudotuberculosis* superantigen encoding gene. *Zentbl. Bakteriol.* **29**(Suppl.):397–398.

37. Chao, W. L., R. J. Ding, and R. S. Chen. 1988. Survival of *Yersinia enterocolitica* in the environment. *Can. J. Microbiol.* **34**:753–756.

38. Cheng, L. W., D. M. Anderson, and O. Schneewind. 1997. Two independent type III secretion mechanisms for YopE in *Yersinia enterocolitica*. *Mol. Microbiol.* **24**:757–765.

39. Cheng, L. W., and O. Schneewind. 1999. *Yersinia enterocolitica* type III secretion. On the role of SycE in targeting YopE into HeLa cells. *J. Biol. Chem.* **274**:22102–22108.

40. China, B., B. T. N'Guyen, M. de Bruyere, and G. R. Cornelis. 1994. Role of YadA in resistance of *Yersinia enterocolitica* to phagocytosis by human polymorphonuclear leukocytes. *Infect. Immun.* **62**:1275–1281.

41. China, B., M. P. Sory, B. T. N'Guyen, M. de Bruyere, and G. R. Cornelis. 1993. Role of the YadA protein in

prevention of opsonization of *Yersinia enterocolitica* by C3b molecules. *Infect. Immun.* **61:**3129–3136.

42. Clark, M. A., B. H. Hirst, and M. A. Jepson. 1998. M-cell surface beta1 integrin expression and invasin-mediated targeting of *Yersinia pseudotuberculosis* to mouse Peyer's patch M cells. *Infect. Immun.* **66:**1237–1243.

43. Cornelis, G., Y. Laroche, G. Balligand, M. P. Sory, and G. Wauters. 1987. *Yersinia enterocolitica*, a primary model for bacterial invasiveness. *Rev. Infect. Dis.* **9:**64–87.

44. Cornelis, G., C. Sluiters, C. L. de Rouvroit, and T. Michiels. 1989. Homology between *virF*, the transcriptional activator of the *Yersinia* virulence regulon, and AraC, the *Escherichia coli* arabinose operon regulator. *J. Bacteriol.* **171:**254–262.

45. Cornelis, G. R., T. Biot, C. Lambert de Rouvroit, T. Michiels, B. Mulder, C. Sluiters, M. P. Sory, M. Van Bouchaute, and J. C. Vanooteghem. 1989. The *Yersinia* yop regulon. *Mol. Microbiol.* **3:**1455–1459.

46. Cornelis, G. R., A. Boland, A. P. Boyd, C. Geuijen, M. Iriarte, C. Neyt, M.-P. Sory, and I. Stainier. 1998. The virulence plasmid of *Yersinia*, an antihost genome. *Microbiol. Mol. Biol. Rev.* **62:**1315–1352.

47. Cornelis, G. R., C. Sluiters, I. Delor, D. Geib, K. Kaniga, C. Lambert de Rouvroit, M. P. Sory, J. C. Vanooteghem, and T. Michiels. 1991. ymoA, a *Yersinia enterocolitica* chromosomal gene modulating the expression of virulence functions. *Mol. Microbiol.* **5:**1023–1034.

48. Cover, T. L., and R. C. Aber. 1989. *Yersinia enterocolitica.* *N. Engl. J. Med.* **321:**16–24.

49. Currie, M. G., K. F. Fok, J. Kato, R. J. Moore, F. K. Hamra, K. L. Duffin, and C. E. Smith. 1992. Guanylin: an endogenous activator of intestinal guanylate cyclase. *Proc. Natl. Acad. Sci. USA* **89:**947–951.

50. D'Aoust, J. Y., C. E. Park, R. A. Szabo, E. C. Todd, D. B. Emmons, and R. C. McKellar. 1988. Thermal inactivation of *Campylobacter* species, *Yersinia enterocolitica*, and hemorrhagic *Escherichia coli* O157:H7 in fluid milk. *J. Dairy Sci.* **71:**3230–3236.

51. de Boer, E., and J. F. Nouws. 1991. Slaughter pigs and pork as a source of human pathogenic *Yersinia enterocolitica.* *Int. J. Food Microbiol.* **12:**375–378.

52. de Giusti, M., and E. de Vito. 1992. Inactivation of *Yersinia enterocolitica* by nitrite and nitrate in food. *Food Addit. Contam.* **9:**405–408.

53. de Koning, J., J. Heesemann, J. A. Hoogkamp-Korstanje, J. J. Festen, P. M. Houtman, and P. L. van Oijen. 1989. *Yersinia* in intestinal biopsy specimens from patients with seronegative spondyloarthropathy: correlation with specific serum IgA antibodies. *J. Infect. Dis.* **159:**109–112.

54. de Koning-Ward, T. F., T. Grant, F. Oppedisano, and R. M. Robins-Browne. 1998. Effect of bacterial invasion of macrophages on the outcome of assays to assess bacterium-macrophage interactions. *J. Immunol. Methods* **215:**39–44.

55. de Koning-Ward, T. F., and R. M. Robins-Browne. 1995. Contribution of urease to acid tolerance in *Yersinia enterocolitica.* *Infect. Immun.* **63:**3790–3795.

56. de Koning-Ward, T. F., and R. M. Robins-Browne. 1997.

A novel mechanism of urease regulation in *Yersinia enterocolitica.* FEMS Microbiol. Lett. **147:**221–226.

57. de Koning-Ward, T. F., A. C. Ward, and R. M. Robins-Browne. 1994. Characterisation of the urease-encoding gene complex of *Yersinia enterocolitica.* *Gene* **145:**25–32.

58. Delor, I., and G. R. Cornelis. 1992. Role of *Yersinia enterocolitica* Yst toxin in experimental infection of young rabbits. *Infect. Immun.* **60:**4269–4277.

59. Delor, I., A. Kaeckenbeeck, G. Wauters, and G. R. Cornelis. 1990. Nucleotide sequence of *yst*, the *Yersinia enterocolitica* gene encoding the heat-stable enterotoxin, and prevalence of the gene among pathogenic and nonpathogenic yersiniae. *Infect. Immun.* **58:**2983–2988.

60. Dersch, P., and R. R. Isberg. 1999. A region of the *Yersinia pseudotuberculosis* invasin protein enhances integrin-mediated uptake into mammalian cells and promotes self-association. *EMBO J.* **18:**1199–1213.

61. Dion, P., R. Charbonneau, and C. Thibault. 1994. Effect of ionizing dose rate on the radioresistance of some food pathogenic bacteria. *Can. J. Microbiol.* **40:**369–374.

62. Doyle, M. P. 1990. Pathogenic *Escherichia coli*, *Yersinia enterocolitica*, and *Vibrio parahaemolyticus*. *Lancet* **336:**1111–1115.

63. Escudero, M. E., L. Velazquez, M. S. Di Genaro, and A. M. de Guzman. 1999. Effectiveness of various disinfectants in the elimination of *Yersinia enterocolitica* on fresh lettuce. *J. Food Prot.* **62:**665–669.

64. Falgarone, G., H. S. Blanchard, B. Riot, M. Simonet, and M. Breban. 1999. Cytotoxic T-cell-mediated response against *Yersinia pseudotuberculosis* in HLA-B27 transgenic rat. *Infect. Immun.* **67:**3773–3779.

65. Farmer, J. J., III, G. P. Carter, V. L. Miller, S. Falkow, and I. K. Wachsmuth. 1992. Pyrazinamidase, CR-MOX agar, salicin fermentation-esculin hydrolysis, and D-xylose fermentation for identifying pathogenic serotypes of *Yersinia enterocolitica.* *J. Clin. Microbiol.* **30:**2589–2594.

66. Feng, P., and S. D. Weagant. 1993. *Yersinia*, p. 427–460. *In* Y. H. Hui, J. R. Gorham, K. D. Murrell, and D. O. Cliver (ed.), *Foodborne Disease Handbook*, vol. 1. Marcel Dekker, New York, N.Y.

67. Fenwick, S. G., G. Wauters, J. Ursing, and M. Gwozdz. 1996. Unusual *Yersinia enterocolitica* strains recovered from domestic animals and people in New Zealand. *FEMS Immunol. Med. Microbiol.* **16:**241–245.

68. Fields, K. A., M. L. Nilles, C. Cowan, and S. C. Straley. 1999. Virulence role of V antigen of *Yersinia pestis* at the bacterial surface. *Infect. Immun.* **67:**5395–5408.

69. Finlay, B. B., and S. Falkow. 1988. Comparison of the invasion strategies used by *Salmonella cholerae-suis*, *Shigella flexneri* and *Yersinia enterocolitica* to enter cultured animal cells: endosome acidification is not required for bacterial invasion or intracellular replication. *Biochimie* **70:**1089–1099.

70. Foberg, U., A. Fryden, E. Kihlstrom, K. Persson, and O. Weiland. 1986. *Yersinia enterocolitica* septicemia: clinical and microbiological aspects. *Scand. J. Infect. Dis.* **18:**269–279.

71. **Forsberg, A., R. Rosqvist, and H. Wolf-Watz.** 1994. Regulation and polarized transfer of the *Yersinia* outer proteins (Yops) involved in antiphagocytosis. *Trends Microbiol.* **2:**14–19.

72. **Francis, M. S., and H. Wolf-Watz.** 1998. YopD of *Yersinia pseudotuberculosis* is translocated into the cytosol of HeLa epithelial cells: evidence of a structural domain necessary for translocation. *Mol. Microbiol.* **29:**799–813.

73. **Frank, D. W.** 1997. The exoenzyme S regulon of *Pseudomonas aeruginosa. Mol. Microbiol.* **26:**621–629.

74. **Fredriksson-Ahomaa, M., T. Korte, and H. Korkeala.** 2000. Contamination of carcasses, offals, and the environment with *yadA*-positive *Yersinia enterocolitica* in a pig slaughterhouse. *J. Food Prot.* **63:**31–35.

75. **Gaastra, W., and A. M. Svennerholm.** 1996. Colonization factors of human enterotoxigenic *Escherichia coli* (ETEC). *Trends Microbiol.* **4:**444–452.

76. **Gaston, J. S., C. Cox, and K. Granfors.** 1999. Clinical and experimental evidence for persistent *Yersinia* infection in reactive arthritis. *Arthritis Rheum.* **42:**2239–2242.

77. **Gottlieb, T.** 1993. Hazards of bacterial contamination of blood products. *Anaesth. Intensive Care* **21:**20–23.

78. **Granfors, K., S. Jalkanen, R. von Essen, R. Lahesmaa-Rantala, O. Isomaki, K. Pekkola-Heino, R. Merilahti-Palo, R. Saario, H. Isomaki, and A. Toivanen.** 1989. *Yersinia* antigens in synovial-fluid cells from patients with reactive arthritis. *N. Engl. J. Med.* **320:**216–221.

79. **Grant, T., V. Bennett-Wood, and R. M. Robins-Browne.** 1998. Identification of virulence-associated characteristics in clinical isolates of *Yersinia enterocolitica* lacking classical virulence markers. *Infect. Immun.* **66:**1113–1120.

80. **Grant, T., V. Bennett-Wood, and R. M. Robins-Browne.** 1999. Characterization of the interaction between *Yersinia enterocolitica* biotype 1A and phagocytes and epithelial cells in vitro. *Infect. Immun.* **67:**4367–4375.

81. **Greer, G. G., C. O. Gill, and B. D. Dilts.** 1995. Predicting the aerobic growth of *Yersinia enterocolitica* on pork fat and muscle tissues. *Food Microbiol.* **12:**463–469.

82. **Gripenberg-Lerche, C., M. Skurnik, L. Zhang, K.-O. Söderström, and P. Toivanen.** 1994. Role of YadA in arthritogenicity of *Yersinia enterocolitica* serotype O:8: experimental studies with rats. *Infect. Immun.* **62:**5568–5575.

83. **Gripenberg-Lerche, C., L. Zhang, P. Ahtonen, P. Toivanen, and M. Skurnik.** 2000. Construction of urease-negative mutants of *Yersinia enterocolitica* serotypes O:3 and O:8: role of urease in virulence and arthritogenicity. *Infect. Immun.* **68:**942–947.

84. **Grützkau, A., C. Hanski, H. Hahn, and E. O. Riecken.** 1990. Involvement of M cells in the bacterial invasion of Peyer's patches: a common mechanism shared by *Yersinia enterocolitica* and other enteroinvasive bacteria. *Gut* **31:**1011–1015.

85. **Guarino, A., R. Giannella, and M. R. Thompson.** 1989. *Citrobacter freundii* produces an 18-amino-acid heat-stable enterotoxin identical to the 18-amino-acid *Escherichia coli* heat-stable enterotoxin (STIa). *Infect. Immun.* **57:**649–652.

86. **Håkansson, S., T. Bergman, J. C. Vanooteghem, G. Cornelis, and H. Wolf-Watz.** 1993. YopB and YopD constitute a novel class of *Yersinia* Yop proteins. *Infect. Immun.* **61:**71–80.

87. **Hamburger, Z. A., M. S. Brown, R. R. Isberg, and P. J. Bjorkman.** 1999. Crystal structure of invasin: a bacterial integrin-binding protein. *Science* **286:**291–295.

88. **Han, Y. W., and V. L. Miller.** 1997. Reevaluation of the virulence phenotype of the *inv yadA* double mutants of *Yersinia pseudotuberculosis. Infect. Immun.* **65:**327–330.

89. **Hanna, M. O., J. C. Stewart, D. L. Zink, Z. L. Carpenter, and C. Vanderzant.** 1977. Development of *Yersinia enterocolitica* on raw and cooked beef and pork at different temperatures. *J. Food Sci.* **42:**1180–1184.

90. **Hanski, C., U. Kutschka, H. P. Schmoranzer, M. Naumann, A. Stallmach, H. Hahn, H. Menge, and E. O. Riecken.** 1989. Immunohistochemical and electron microscopic study of interaction of *Yersinia enterocolitica* serotype O:8 with intestinal mucosa during experimental enteritis. *Infect. Immun.* **57:**673–678.

91. **Harakeh, M. S., J. D. Berg, J. C. Hoff, and A. Matin.** 1985. Susceptibility of chemostat-grown *Yersinia enterocolitica* and *Klebsiella pneumoniae* to chlorine dioxide. *Appl. Environ. Microbiol.* **49:**69–72.

92. **Hartland, E. L., S. P. Green, W. A. Phillips, and R. M. Robins-Browne.** 1994. Essential role of YopD in inhibition of the respiratory burst of macrophages by *Yersinia enterocolitica. Infect. Immun.* **62:**4445–4453.

93. **Hartland, E. L., and R. M. Robins–Browne.** 1998. Infections with enteropathogenic *Yersinia* species: paradigms of bacterial pathogenesis. *Rev. Med. Microbiol.* **9:**191–205.

94. **Hauser, A. R., S. Fleiszig, P. J. Kang, K. Mostov, and J. N. Engel.** 1998. Defects in type III secretion correlate with internalization of *Pseudomonas aeruginosa* by epithelial cells. *Infect. Immun.* **66:**1413–1420.

95. **Hayashidani, H., Y. Ohtomo, Y. Toyokawa, M. Saito, K. Kaneko, J. Kosuge, M. Kato, M. Ogawa, and G. Kapperud.** 1995. Potential sources of sporadic human infection with *Yersinia enterocolitica* serovar O:8 in Aomori Prefecture, Japan. *J. Clin. Microbiol.* **33:**1253–1257.

96. **Heesemann, J., K. Gaede, and I. B. Autenrieth.** 1993. Experimental *Yersinia enterocolitica* infection in rodents: a model for human yersiniosis. *APMIS* **101:**417–429.

97. **Heesemann, J., K. Hantke, T. Vocke, E. Saken, A. Rakin, I. Stojiljkovic, and R. Berner.** 1993. Virulence of *Yersinia enterocolitica* is closely associated with siderophore production, expression of an iron-repressible outer membrane polypeptide of 65,000 Da and pesticin sensitivity. *Mol. Microbiol.* **8:**397–408.

98. **Hermann, E.** 1993. T cells in reactive arthritis. *APMIS* **101:**177–186.

99. **Hermann, E., D. T. Yu, K. H. Meyer zum Buschenfelde, and B. Fleischer.** 1993. HLA-B27-restricted CD8 T cells derived from synovial fluids of patients with reactive

arthritis and ankylosing spondylitis. *Lancet* **342**:646–650.

100. **Herrlinger, J. D., and J. U. Asmussen.** 1992. Long term prognosis in *Yersinia* arthritis: clinical and serological findings. *Ann. Rheum. Dis.* **51**:1332–1334.

101. **Heyma, P., L. C. Harrison, and R. Robins-Browne.** 1986. Thyrotrophin (TSH) binding sites on *Yersinia enterocolitica* recognized by immunoglobulins from humans with Graves' disease. *Clin. Exp. Immunol.* **64**:249–254.

102. **Hoogkamp-Korstanje, J. A. A., and V. M. M. Stolk-Engelaar.** 1995. *Yersinia enterocolitica* infection in children. *Pediatr. Infect. Dis. J.* **14**:771–775.

103. **Huang, X., K. Yoshino, H. Nakao, and T. Takeda.** 1997. Nucleotide sequence of a gene encoding the novel *Yersinia enterocolitica* heat-stable enterotoxin that includes a proregion-like sequence in its mature toxin molecule. *Microb. Pathog.* **22**:89–97.

104. **Hueck, C. J.** 1998. Type III protein secretion systems in bacterial pathogens of animals and plants. *Microbiol. Mol. Biol. Rev.* **62**:379–433.

105. **Ibrahim, A., W. Liesack, A. G. Steigerwalt, D. J. Brenner, E. Stackebrandt, and R. M. Robins-Browne.** 1997. A cluster of atypical *Yersinia* strains with a distinctive 16S rRNA signature. *FEMS Microbiol. Lett.* **146**:73–78.

106. **Iriarte, M., and G. R. Cornelis.** 1995. MyfF, an element of the network regulating the synthesis of fibrillae in *Yersinia enterocolitica*. *J. Bacteriol.* **177**:738–744.

107. **Iriarte, M., and G. R. Cornelis.** 1998. YopT, a new *Yersinia* Yop effector protein, affects the cytoskeleton of host cells. *Mol. Microbiol.* **29**:915–929.

108. **Iriarte, M., and G. R. Cornelis.** 1999. Identification of SycN, YscX, and YscY, three new elements of the *Yersinia yop* virulon. *J. Bacteriol.* **181**:675–680.

109. **Iriarte, M., M. P. Sory, A. Boland, A. P. Boyd, S. D. Mills, I. Lambermont, and G. R. Cornelis.** 1998. TyeA, a protein involved in control of Yop release and in translocation of Yersinia Yop effectors. *EMBO J.* **17**:1907–1918.

110. **Iriarte, M., I. Stainier, and G. R. Cornelis.** 1995. The *rpoS* gene from *Yersinia enterocolitica* and its influence on expression of virulence factors. *Infect. Immun.* **63**:1840–1847.

111. **Iriarte, M., I. Stainier, A. V. Mikulskis, and G. R. Cornelis.** 1995. The *fliA* gene encoding σ^{28} in *Yersinia enterocolitica*. *J. Bacteriol.* **177**:2299–2304.

112. **Iriarte, M., J. C. Vanooteghem, I. Delor, R. Diaz, S. Knutton, and G. R. Cornelis.** 1993. The Myf fibrillae of *Yersinia enterocolitica*. *Mol. Microbiol.* **9**:507–520.

113. **Isberg, R. R.** 1990. Pathways for the penetration of enteroinvasive *Yersinia* into mammalian cells. *Mol. Biol. Med.* **7**:73–82.

114. **Isberg, R. R., and S. Falkow.** 1985. A single genetic locus encoded by *Yersinia pseudotuberculosis* permits invasion of cultured animal cells by *Escherichia coli* K-12. *Nature* **317**:262–264.

115. **Isberg, R. R., and G. T. Van Nhieu.** 1994. Two mammalian cell internalization strategies used by pathogenic bacteria. *Annu. Rev. Genet.* **28**:395–422.

116. **Iteman, I., A. Guiyoule, and E. Carniel.** 1996. Compari-

son of three molecular methods for typing and subtyping pathogenic *Yersinia enterocolitica* strains. *J. Med. Microbiol.* **45**:48–56.

117. **Jacobi, C. A., A. Roggenkamp, A. Rakin, R. Zumbihl, L. Leitritz, and J. Heesemann.** 1998. In vitro and in vivo expression studies of yopE from *Yersinia enterocolitica* using the gfp reporter gene. *Mol. Microbiol.* **30**:865–882.

118. **Jepson, M. A., and M. A. Clark.** 1998. Studying M cells and their role in infection. *Trends Microbiol.* **9**:359–365.

119. **Jerrett, I. V., K. J. Slee, and B. I. Robertson.** 1990. Yersiniosis in farmed deer. *Aust. Vet. J.* **67**:212–214.

120. **Jerse, A. E., J. Yu, B. D. Tall, and J. B. Kaper.** 1990. A genetic locus of enteropathogenic *Escherichia coli* necessary for the production of attaching and effacing lesions on tissue culture cells. *Proc. Natl. Acad. Sci. USA* **87**:7839–7843.

121. **Jones, J. E., S. J. Walker, J. P. Sutherland, M. W. Peck, and C. L. Little.** 1994. Mathematical modelling of the growth, survival and death of *Yersinia enterocolitica*. *Int. J. Food Microbiol.* **23**:433–447.

122. **Juris, S. J., A. E. Rudolph, D. Huddler, K. Orth, and J. E. Dixon.** 2000. A distinctive role for the *Yersinia* protein kinase: actin binding, kinase activation, and cytoskeleton disruption. *Proc. Natl. Acad. Sci. USA* **97**:9431–9436.

123. **Kapperud, G.** 1982. Enterotoxin production at 4°, 22°, and 37° among *Yersinia enterocolitica* and *Y. enterocolitica*-like bacteria. *APMIS* **90B**:185–189.

124. **Kapperud, G.** 1991. *Yersinia enterocolitica* in food hygiene. *Int. J. Food Microbiol.* **12**:53–65.

125. **Kapperud, G., E. Namork, and H. J. Skarpeid.** 1985. Temperature-inducible surface fibrillae associated with the virulence plasmid of *Yersinia enterocolitica* and *Yersinia pseudotuberculosis*. *Infect. Immun.* **47**:561–566.

126. **Karch, H., S. Schubert, D. Zhang, W. Zhang, H. Schmidt, T. Olschlager, and J. Hacker.** 1999. A genomic island, termed high-pathogenicity island, is present in certain non-O157 Shiga toxin-producing *Escherichia coli* clonal lineages. *Infect. Immun.* **67**:5994–6001.

127. **Kasatiya, S. S.** 1976. *Yersinia enterocolitica* gastroenteritis outbreak—Montreal. *Can. Dis. Wkly. Rep.* **2**:73–74.

128. **Keet, E. E.** 1974. *Yersinia enterocolitica* septicemia: source of infection and incubation period identified. *N. Y. State J. Med.* **74**:2226–2229.

129. **Kellogg, C. M., E. A. Tarakji, M. Smith, and P. D. Brown.** 1995. Bacteremia and suppurative lymphadenitis due to *Yersinia enterocolitica* in a neutropenic patient who prepared chitterlings. *Clin. Infect. Dis.* **21**:236–237.

130. **King, C. H., E. B. Shotts, Jr., R. E. Wooley, and K. G. Porter.** 1988. Survival of coliforms and bacterial pathogens within protozoa during chlorination. *Appl. Environ. Microbiol.* **54**:3023–3033.

131. **Kingsley, G., and G. Panayi.** 1992. Antigenic responses in reactive arthritis. *Rheum. Dis. Clin. N. Am.* **18**:49–66.

132. **Koster, M., W. Bitter, H. de Cock, A. Allaoui, G. R. Cornelis, and J. Tommassen.** 1997. The outer membrane component, YscC, of the Yop secretion machinery of *Yersinia enterocolitica* forms a ring-shaped multimeric complex. *Mol. Microbiol.* **26**:789–797.

133. Lahesmaa, R., M. Skurnik, K. Granfors, T. Mottonen, R. Saario, A. Toivanen, and P. Toivanen. 1992. Molecular mimicry in the pathogenesis of spondyloarthropathies. A critical appraisal of cross-reactivity between microbial antigens and HLA-B27. *Br. J. Rheumatol.* **31:**221–229.

134. Larsen, J. H. 1980. *Yersinia enterocolitica* infection and rheumatic diseases. *Scand. J. Rheumatol.* **9:**129–137.

135. Lee, L. A., A. R. Gerber, D. R. Lonsway, J. D. Smith, G. P. Carter, N. D. Puhr, C. M. Parrish, R. K. Sikes, R. J. Finton, and R. V. Tauxe. 1990. *Yersinia enterocolitica* O:3 infections in infants and children, associated with the household preparation of chitterlings. *N. Engl. J. Med.* **322:**984–987.

136. Lee, V. T., D. M. Anderson, and O. Schneewind. 1998. Targeting of *Yersinia* Yop proteins into the cytosol of HeLa cells: one-step translocation of YopE across bacterial and eukaryotic membranes is dependent on SycE chaperone. *Mol. Microbiol.* **28:**593–601.

137. Leirisalo-Repo, M. 1987. *Yersinia* arthritis. Acute clinical picture and long-term prognosis. *Contrib. Microbiol. Immunol.* **9:**145–154.

138. Leung, K. Y., and S. C. Straley. 1989. The *yopM* gene of *Yersinia pestis* encodes a released protein having homology with the human platelet surface protein GPIbα. *J. Bacteriol.* **171:**4623–4632.

139. Li, H., S. Bhaduri, and W. E. Magee. 1998. Maximizing plasmid stability and production of released proteins in *Yersinia enterocolitica*. *Appl. Environ. Microbiol.* **64:**1812–1815.

140. Li, S.-R., N. Dorrell, P. H. Everest, G. Dougan, and B. W. Wren. 1996. Construction and characterization of a *Yersinia enterocolitica* O:8 high-temperature requirement (*htrA*) isogenic mutant. *Infect. Immun.* **64:**2088–2094.

141. Lian, C. J., W. S. Hwang, J. K. Kelly, and C. H. Pai. 1987. Invasiveness of *Yersinia enterocolitica* lacking the virulence plasmid: an in-vivo study. *J. Med. Microbiol.* **24:**219–226.

142. Lian, C. J., W. S. Hwang, and C. H. Pai. 1987. Plasmid-mediated resistance to phagocytosis in *Yersinia enterocolitica*. *Infect. Immun.* **55:**1176–1183.

143. Lindler, L. E., and B. D. Tall. 1993. *Yersinia pestis* pH 6 antigen forms fimbriae and is induced by intracellular association with macrophages. *Mol. Microbiol.* **8:**311–324.

144. Lo, W. F., A. S. Woods, A. DeCloux, R. J. Cotter, E. S. Metcalf, and M. J. Soloski. 2000. Molecular mimicry mediated by MHC class Ib molecules after infection with gram-negative pathogens. *Nat. Med.* **6:**215–218.

145. Luo, G., J. L. Fan, G. S. Seetharamaiah, R. K. Desai, J. S. Dallas, N. Wagle, R. Doan, D. W. Niesel, G. R. Klimpel, and B. S. Prabhakar. 1993. Immunization of mice with *Yersinia enterocolitica* leads to the induction of antithyrotropin receptor antibodies. *J. Immunol.* **151:**922–928.

146. Marjai, E., M. Kalman, I. Kajary, A. Belteky, and M. Rodler. 1987. Isolation from food and characterization by virulence tests of *Yersinia enterocolitica* associated with an outbreak. *Acta Microbiol. Hung.* **34:**97–109.

147. Marks, M. I., C. H. Pai, L. Lafleur, L. Lackman, and O. Hammerberg. 1980. *Yersinia enterocolitica* gastroenteritis: a prospective study of clinical, bacteriologic, and epidemiologic features. *J. Pediatr.* **96:**26–31.

148. Marra, A., and R. R. Isberg. 1997. Invasin-dependent and invasin-independent pathways for translocation of *Yersinia pseudotuberculosis* across the Peyer's patch intestinal epithelium. *Infect. Immun.* **65:**3412–3421.

149. Maruyama, T. 1987. *Yersinia enterocolitica* infection in humans and isolation of the microorganism from pigs in Japan. *Contrib. Microbiol. Immunol.* **9:**48–55.

150. Michiels, T., P. Wattiau, R. Brasseur, J. M. Ruysschaert, and G. Cornelis. 1990. Secretion of Yop proteins by yersiniae. *Infect. Immun.* **58:**2840–2849.

151. Mikulskis, A. V., I. Delor, V. H. Thi, and G. R. Cornelis. 1994. Regulation of the *Yersinia enterocolitica* enterotoxin Yst gene. Influence of growth phase, temperature, osmolarity, pH and bacterial host factors. *Mol. Microbiol.* **14:**905–915.

152. Miller, V. L. 1992. *Yersinia* invasion genes and their products. *ASM News* **58:**26–33.

153. Miller, V. L., and S. Falkow. 1988. Evidence for two genetic loci in *Yersinia enterocolitica* that can promote invasion of epithelial cells. *Infect. Immun.* **56:**1242–1248.

154. Miller, V. L., B. B. Finlay, and S. Falkow. 1988. Factors essential for the penetration of mammalian cells by *Yersinia*. *Curr. Top. Microbiol. Immunol.* **138:**15–39.

155. Mills, S. D., A. Boland, M. P. Sory, P. van der Smissen, C. Kerbourch, B. B. Finlay, and G. R. Cornelis. 1997. *Yersinia enterocolitica* induces apoptosis in macrophages by a process requiring functional type III secretion and translocation mechanisms and involving YopP, presumably acting as an effector protein. *Proc. Natl. Acad. Sci. USA* **94:**12638–12643.

156. Miyoshi-Akiyama, T., W. Fujimaki, X. J. Yan, J. Yagi, K. Imanishi, H. Kato, K. Tomonari, and T. Uchiyama. 1997. Identification of murine T cells reactive with the bacterial superantigen *Yersinia pseudotuberculosis*-derived mitogen (YPM) and factors involved in YPM-induced toxicity in mice. *Microbiol. Immunol.* **41:**345–352.

157. Mollaret, H. H., H. Bercovier, and J. M. Alonso. 1979. Summary of the data received at the WHO Reference Centre for *Yersinia enterocolitica*. *Contrib. Microbiol. Immunol.* **5:**174–184.

158. Monack, D. M., J. Mecsas, D. Bouley, and S. Falkow. 1998. *Yersinia*-induced apoptosis *in vivo* aids in the establishment of a systemic infection of mice. *J. Exp. Med.* **188:**2127–2137.

159. Monack, D. M., J. Mecsas, N. Ghori, and S. Falkow. 1997. *Yersinia* signals macrophages to undergo apoptosis and YopJ is necessary for this cell death. *Proc. Natl. Acad. Sci. USA* **94:**10385–10390.

160. Morris, J. G., Jr., V. Prado, C. Ferreccio, R. M. Robins-Browne, A. M. Bordun, M. Cayazzo, B. A. Kay, and M. M. Levine. 1991. *Yersinia enterocolitica* isolated from two cohorts of young children in Santiago, Chile: incidence of and lack of correlation between illness and proposed virulence factors. *J. Clin. Microbiol.* **29:**2784–2788.

161. Morse, D. L., M. Shayegani, and R. J. Gallo. 1984. Epidemiologic investigation of a *Yersinia* camp outbreak

linked to a food handler. *Am. J. Public Health* **74**:589–592.

162. **Nakajima, R., and R. R. Brubaker.** 1993. Association between virulence of *Yersinia pestis* and suppression of gamma interferon and tumor necrosis factor alpha. *Infect. Immun.* **61**:23–31.

163. **Neilands, J. B.** 1981. Microbial iron compounds. *Annu. Rev. Biochem.* **50**:715–731.

164. **Nesbakken, T., G. Kapperud, K. Dommarsnes, M. Skurnik, and E. Hornes.** 1991. Comparative study of a DNA hybridization method and two isolation procedures for detection of *Yersinia enterocolitica* O:3 in naturally contaminated pork products. *Appl. Environ. Microbiol.* **57**:389–394.

165. **Neyt, C., and G. R. Cornelis.** 1999. Insertion of a Yop translocation pore into the macrophage plasma membrane by *Yersinia enterocolitica*: requirement for translocators YopB and YopD, but not LcrG. *Mol. Microbiol.* **33**:971–981.

166. **Neyt, C., M. Iriarte, V. H. Thi, and G. R. Cornelis.** 1997. Virulence and arsenic resistance in yersiniae. *J. Bacteriol.* **179**:612–619.

167. **Nikolova, S., H. Najdenski, D. Wesselinova, A. Vesselinova, D. Kazatchca, and P. Neikov.** 1997. Immunological and electronmicroscopic studies in pigs infected with *Yersinia enterocolitica* O:3. *Zentbl. Bakteriol.* **286**:503–510.

168. **Nilehn, B.** 1969. Studies on *Yersinia enterocolitica* with special reference to bacterial diagnosis and occurrence in human acute enteric disease. *Acta Pathol. Microbiol. Scand. Suppl.* **206**:1–48.

169. **Ostroff, S. M., G. Kapperud, L. C. Hutwagner, T. Nesbakken, N. H. Bean, J. Lassen, and R. V. Tauxe.** 1994. Sources of sporadic *Yersinia enterocolitica* infections in Norway: a prospective case-control study. *Epidemiol. Infect.* **112**:133–141.

170. **Pai, C. H., and L. De Stephano.** 1982. Serum resistance associated with virulence in *Yersinia enterocolitica*. *Infect. Immun.* **35**:605–611.

171. **Pai, C. H., V. Mors, and S. Toma.** 1978. Prevalence of enterotoxigenicity in human and nonhuman isolates of *Yersinia enterocolitica*. *Infect. Immun.* **22**:334–338.

172. **Pederson, K. J., A. J. Vallis, K. Aktories, D. W. Frank, and J. T. Barbieri.** 1999. The amino-terminal domain of *Pseudomonas aeruginosa* ExoS disrupts actin filaments via small molecular weight GTP-binding proteins. *Mol. Microbiol.* **32**:393–401.

173. **Peixotto, S. S., G. Finne, M. O. Hanna, and C. Vanderzant.** 1979. Presence, growth and survival of *Yersinia enterocolitica* in oyster, shrimp and crab. *J. Food Prot.* **42**:974–981.

174. **Pepe, J. C., J. L. Badger, and V. L. Miller.** 1994. Growth phase and low pH affect the thermal regulation of the *Yersinia enterocolitica inv* gene. *Mol. Microbiol.* **11**:123–135.

175. **Pepe, J. C., and V. L. Miller.** 1993. *Yersinia enterocolitica* invasin: a primary role in the initiation of infection. *Proc. Natl. Acad. Sci. USA* **90**:6473–6477.

176. **Pepe, J. C., M. R. Wachtel, E. Wagar, and V. L. Miller.** 1995. Pathogenesis of defined invasion mutants of *Yersinia enterocolitica* in a BALB/c mouse model of infection. *Infect. Immun.* **63**:4837–4848.

177. **Perry, R. D.** 1993. Acquisition and storage of inorganic iron and hemin by the yersiniae. *Trends Microbiol.* **1**:142–147.

178. **Perry, R. D., and J. D. Fetherston.** 1997. *Yersinia pestis*—etiologic agent of plague. *Clin. Microbiol. Rev.* **10**:35–66.

179. **Persson, C., N. Carballeira, H. Wolf-Watz, and M. Fallman.** 1997. The PTPase YopH inhibits uptake of *Yersinia*, tyrosine phosphorylation of p130Cas and FAK, and the associated accumulation of these proteins in peripheral focal adhesions. *EMBO J.* **16**:2307–2318.

180. **Pierson, D. E.** 1994. Mutations affecting lipopolysaccharide enhance ail-mediated entry of *Yersinia enterocolitica* into mammalian cells. *J. Bacteriol.* **176**:4043–4051.

181. **Pollack, C., S. C. Straley, and M. S. Klempner.** 1986. Probing the phagolysosomal environment of human macrophages with a Ca2+-responsive operon fusion in *Yersinia pestis*. *Nature* **322**:834–836.

182. **Pugsley, A. P.** 1993. The complete general secretory pathway in gram-negative bacteria. *Microbiol. Rev.* **57**:50–108.

183. **Pulkkinen, W. S., and S. I. Miller.** 1991. A *Salmonella typhimurium* virulence protein is similar to a *Yersinia enterocolitica* invasion protein and a bacteriophage lambda outer membrane protein. *J. Bacteriol.* **173**:86–93.

184. **Puylaert, J. B., R. J. Vermeijden, S. D. van der Werf, L. Doornbos, and R. K. Koumans.** 1989. Incidence and sonographic diagnosis of bacterial ileocaecitis masquerading as appendicitis. *Lancet* **2**:84–86.

185. **Rakin, A., C. Noelting, S. Schubert, and J. Heesemann.** 1999. Common and specific characteristics of the high-pathogenicity island of *Yersinia enterocolitica*. *Infect. Immun.* **67**:5265–5274.

186. **Ramamurthy, T., K. Yoshino, X. Huang, G. B. Nair, E. Carniel, T. Maruyama, H. Fukushima, and T. Takeda.** 1997. The novel heat-stable enterotoxin subtype gene (ystB) of *Yersinia enterocolitica*: nucleotide sequence and distribution of the yst genes. *Microb. Pathog.* **23**:189–200.

187. **Reed, R. P., R. M. Robins-Browne, and M. L. Williams.** 1997. *Yersinia enterocolitica* peritonitis. *Clin. Infect. Dis.* **25**:1468–1469.

188. **Robins-Browne, R. M., A. M. Bordun, and K. J. Slee.** 1993. Serological response of sheep to plasmid-encoded proteins of *Yersinia* species following natural infection with *Y. enterocolitica* and *Y. pseudotuberculosis*. *J. Med. Microbiol.* **39**:268–272.

189. **Robins-Browne, R. M., M. D. Miliotis, S. Cianciosi, V. L. Miller, S. Falkow, and J. G. Morris, Jr.** 1989. Evaluation of DNA colony hybridization and other techniques for detection of virulence in *Yersinia* species. *J. Clin. Microbiol.* **27**:644–650.

190. **Robins-Browne, R. M., and J. K. Prpic.** 1985. Effects of iron and desferrioxamine on infections with *Yersinia enterocolitica*. *Infect. Immun.* **47**:774–779.

191. Robins-Browne, R. M., J. K. Prpic, and S. J. Stuart. 1987. Yersiniae and iron. A study in host-parasite relationships. *Contrib. Microbiol. Immunol.* **9:**254–258.

192. Robins-Browne, R. M., A. R. Rabson, and H. J. Koornhof. 1979. Generalised infection with *Yersinia enterocolitica* and the role of iron. *Contrib. Microbiol. Immunol.* **5:**277–282.

193. Robins-Browne, R. M., C. S. Still, M. D. Miliotis, and H. J. Koornhof. 1979. Mechanism of action of *Yersinia enterocolitica* enterotoxin. *Infect. Immun.* **25:**680–684.

194. Robins-Browne, R. M., T. Takeda, A. Fasano, A.-M. Bordun, S. Dohi, H. Kasuga, G. Fong, V. Prado, R. L. Guerrant, and J. G. Morris, Jr. 1993. Assessment of enterotoxin production by *Yersinia enterocolitica* and identification of a novel heat-stable enterotoxin produced by a noninvasive *Y. enterocolitica* strain isolated from clinical material. *Infect. Immun.* **61:**764–767.

195. Robins-Browne, R. M., S. Tzipori, G. Gonis, J. Hayes, M. Withers, and J. K. Prpic. 1985. The pathogenesis of *Yersinia enterocolitica* infection in gnotobiotic piglets. *J. Med. Microbiol.* **19:**297–308.

196. Roggenkamp, A., H.-R. Neuberger, A. Flugel, T. Schmoll, and J. Heesemann. 1995. Substitution of two histidine residues in YadA protein of *Yersinia enterocolitica* abrogates collagen binding, cell adherence and mouse virulence. *Mol. Microbiol.* **16:**1207–1219.

197. Rohde, J. R., J. M. Fox, and S. A. Minnich. 1994. Thermoregulation in *Yersinia enterocolitica* is coincident with changes in DNA supercoiling. *Mol. Microbiol.* **12:**187–199.

198. Rosenshine, I., V. Duronio, and B. B. Finlay. 1992. Tyrosine protein kinase inhibitors block invasin-promoted bacterial uptake by epithelial cells. *Infect. Immun.* **60:**2211–2217.

199. Rosqvist, R., A. Forsberg, and H. Wolf-Watz. 1991. Intracellular targeting of the *Yersinia* YopE cytotoxin in mammalian cells induces actin microfilament disruption. *Infect. Immun.* **59:**4562–4569.

200. Rosqvist, R., K. E. Magnusson, and H. Wolf-Watz. 1994. Target cell contact triggers expression and polarized transfer of *Yersinia* YopE cytotoxin into mammalian cells. *EMBO J.* **13:**964–972.

201. Rosqvist, R., M. Skurnik, and H. Wolf-Watz. 1988. Increased virulence of *Yersinia pseudotuberculosis* by two independent mutations. *Nature* **334:**522–524.

202. Ruckdeschel, K., J. Machold, A. Roggenkamp, S. Schubert, J. Pierre, R. Zumbihl, J. P. Liautard, J. Heesemann, and B. Rouot. 1997. *Yersinia enterocolitica* promotes deactivation of macrophage mitogen-activated protein kinases extracellular signal-regulated kinase-1/2, p38, and c-Jun NH2-terminal kinase. Correlation with its inhibitory effect on tumor necrosis factor-a production. *J. Biol. Chem.* **272:**15920–15927.

203. Saebo, A., and J. Lassen. 1992. Acute and chronic gastrointestinal manifestations associated with *Yersinia enterocolitica* infection. A Norwegian 10-year follow-up study on 458 hospitalized patients. *Ann. Surg.* **215:**250–255.

204. Saltman, L. H., Y. Lu, E. M. Zaharias, and R. R. Isberg.

1996. A region of the *Yersinia pseudotuberculosis* invasin protein that contributes to high affinity binding to integrin receptors. *J. Biol. Chem.* **271:**23438–23444.

205. Sansonetti, P. J., and A. Phalipon. 1999. M cells as ports of entry for enteroinvasive pathogens: mechanisms of interaction, consequences for the disease process. *Semin. Immunol.* **11:**193–203.

206. Sazama, K. 1994. Bacteria in blood for transfusion. A review. *Arch. Pathol. Lab. Med.* **118:**350–365.

207. Schiemann, D. A. 1980. *Yersinia enterocolitica*: observations on some growth characteristics and response to selective agents. *Can. J. Microbiol.* **26:**1232–1240.

208. Schiemann, D. A. 1981. An enterotoxin-negative strain of *Yersinia enterocolitica* serotype O:3 is capable of producing diarrhea in mice. *Infect. Immun.* **32:**571–574.

209. Schiemann, D. A. 1987. *Yersinia enterocolitica* in milk and dairy products. *J. Dairy. Sci.* **70:**383–391.

210. Schiemann, D. A. 1988. Examination of enterotoxin production at low temperatures by *Yersinia* spp. in culture media and foods. *J. Food Prot.* **51:**571–573.

211. Schiemann, D. A. 1989. *Yersinia enterocolitica* and *Yersinia pseudotuberculosis*, p. 601–672. *In* M. P. Doyle (ed.), *Foodborne Bacterial Pathogens*. Marcel Dekker, New York, N.Y.

212. Schmiel, D. H., E. Wagar, L. Karamanou, D. Weeks, and V. L. Miller. 1998. Phospholipase A of *Yersinia enterocolitica* contributes to pathogenesis in a mouse model. *Infect. Immun.* **66:**3941–3951.

213. Schubert, S., A. Rakin, H. Karch, E. Carniel, and J. Heesemann. 1998. Prevalence of the "high-pathogenicity island" of *Yersinia* species among *Escherichia coli* strains that are pathogenic to humans. *Infect. Immun.* **66:**480–485.

214. Schulte, R., P. Wattiau, E. L. Hartland, R. M. Robins-Browne, and G. R. Cornelis. 1996. Differential secretion of interleukin-8 by human epithelial cell lines upon entry of virulent or non-virulent strains of *Yersinia enterocolitica*. *Infect. Immun.* **64:**2106–2113.

215. Shayegani, M., D. Morse, I. DeForge, T. Root, L. M. Parsons, and P. S. Maupin. 1983. Microbiology of a major foodborne outbreak of gastroenteritis caused by *Yersinia enterocolitica* serogroup O:8. *J. Clin. Microbiol.* **17:**35–40.

216. Shenkman, L., and E. J. Bottone. 1976. Antibodies to *Yersinia enterocolitica* in thyroid disease. *Ann. Intern. Med.* **85:**735–739.

217. Simonet, M. L. 1999. Enterobacteria in reactive arthritis: *Yersinia*, *Shigella*, and *Salmonella*. *Rev. Rhum. Engl. Ed.* **66:**14S–18S.

218. Sims, G. R., D. A. Glenister, T. F. Brocklehurst, and B. M. Lund. 1989. Survival and growth of food poisoning bacteria following inoculation into cottage cheese varieties. *Int. J. Food Microbiol.* **9:**173–195.

219. Skrzypek, E., C. Cowan, and S. C. Straley. 1998. Targeting of the *Yersinia pestis* YopM protein into HeLa cells and intracellular trafficking to the nucleus. *Mol. Microbiol.* **30:**1051–1065.

220. Skurnik, M., S. Batsford, A. Mertz, E. Schiltz, and P. Toivanen. 1993. The putative arthritogenic cationic

19-kilodalton antigen of *Yersinia enterocolitica* is a urease beta-subunit. *Infect. Immun.* **61**:2498–2504.

221. **Skurnik, M., Y. el Tahir, M. Saarinen, S. Jalkanen, and P. Toivanen.** 1994. YadA mediates specific binding of enteropathogenic *Yersinia enterocolitica* to human intestinal submucosa. *Infect. Immun.* **62**:1252–1261.

222. **Skurnik, M., and P. Toivanen.** 1993. *Yersinia enterocolitica* lipopolysaccharide: genetics and virulence. *Trends Microbiol.* **1**:148–152.

223. **Skurnik, M., R. Venho, J. A. Bengoechea, and I. Moriyon.** 1999. The lipopolysaccharide outer core of *Yersinia enterocolitica* serotype O:3 is required for virulence and plays a role in outer membrane integrity. *Mol. Microbiol.* **31**:1443–1462.

224. **Slee, K. J., and N. W. Skilbeck.** 1992. Epidemiology of *Yersinia pseudotuberculosis* and *Y. enterocolitica* infections in sheep in Australia. *J. Clin. Microbiol.* **30**:712–715.

225. **Smith, M. G.** 1992. Destruction of bacteria on fresh meat by hot water. *Epidemiol. Infect.* **109**:491–496.

226. **Sory, M.-P., and G. R. Cornelis.** 1994. Translocation of a hybrid YopE-adenylate cyclase from *Yersinia enterocolitica* into HeLa cells. *Mol. Microbiol.* **14**:583–594.

227. **Stern, N. J., M. D. Pierson, and A. W. Kotula.** 1980. Effects of pH and sodium chloride on *Yersinia enterocolitica* growth at room and refrigeration temperatures. *J. Food Sci.* **45**:64–67.

228. **Stoddard, J. J., D. S. Wechsler, J. P. Nataro, and J. F. Casella.** 1994. *Yersinia enterocolitica* infection in a patient with sickle cell disease after exposure to chitterlings. *Am. J. Pediatr. Hematol. Oncol.* **16**:153–155.

229. **Stuart, P. M., and J. G. Woodward.** 1992. *Yersinia enterocolitica* produces superantigenic activity. *J. Immunol.* **148**:225–233.

230. **Sulakvelidze, A.** 2000. Yersiniae other than *Y. enterocolitica*, *Y. pseudotuberculosis*, and *Y. pestis*: the ignored species. *Microbes Infect.* **2**:497–513.

231. **Sulakvelidze, A., A. Kreger, A. Joseph, R. M. Robins-Browne, A. Fasano, G. Wauters, N. Harnett, L. DeTolla, and J. G. Morris, Jr.** 1999. Production of enterotoxin by *Yersinia bercovieri*, a recently identified *Yersinia enterocolitica*-like species. *Infect. Immun.* **67**:968–971.

232. **Sutherland, J. P., and A. J. Bayliss.** 1994. Predictive modelling of growth of *Yersinia enterocolitica*: the effects of temperature, pH and sodium chloride. *Int. J. Food Microbiol.* **21**:197–215.

233. **Tacket, C. O., J. Ballard, N. Harris, J. Allard, C. Nolan, T. Quan, and M. L. Cohen.** 1985. An outbreak of *Yersinia enterocolitica* infections caused by contaminated tofu (soybean curd). *Am. J. Epidemiol.* **121**:705–711.

234. **Tacket, C. O., B. R. Davis, G. P. Carter, J. F. Randolph, and M. L. Cohen.** 1983. *Yersinia enterocolitica* pharyngitis. *Ann. Intern. Med.* **99**:40–42.

235. **Tacket, C. O., J. P. Narain, R. Sattin, J. P. Lofgren, C. Konigsberg, Jr., R. C. Rendtorff, A. Rausa, B. R. Davis, and M. L. Cohen.** 1984. A multistate outbreak of infections caused by *Yersinia enterocolitica* transmitted by pasteurized milk. *JAMA* **251**:483–486.

236. **Takao, T., T. Hitouji, S. Aimoto, Y. Shimonishi, S. Hara,**

T. Takeda, Y. Takeda, and T. Miwatani. 1983. Amino acid sequence of a heat–stable enterotoxin isolated from enterotoxigenic *Escherichia coli* strain 18D. *FEBS Lett.* **152**:1–5.

237. **Takao, T., Y. Shimonishi, M. Kobayashi, O. Nishimura, M. Arita, T. Takeda, T. Honda, and T. Miwatani.** 1985. Amino acid sequence of heat-stable enterotoxin produced by *Vibrio cholerae* non-O1. *FEBS Lett.* **193**:250–254.

238. **Takao, T., N. Tominaga, S. Yoshimura, Y. Shimonishi, S. Hara, T. Inoue, and A. Miyama.** 1985. Isolation, primary structure and synthesis of heat-stable enterotoxin produced by *Yersinia enterocolitica*. *Eur. J. Biochem.* **152**:199–206.

239. **Tauxe, R. V., J. Vandepitte, G. Wauters, S. M. Martin, V. Goossens, P. de Mol, R. Van Noyen, and G. Thiers.** 1987. *Yersinia enterocolitica* infections and pork: the missing link. *Lancet* **1**:1129–1132.

240. **Thompson, J. S., and M. J. Gravel.** 1986. Family outbreak of gastroenteritis due to *Yersinia enterocolitica* serotype O:3 from well water. *Can. J. Microbiol.* **32**:700–701.

241. **Tipple, M. A., L. A. Bland, J. J. Murphy, M. J. Arduino, A. L. Panlilio, J. J. Farmer III, M. A. Tourault, C. R. Macpherson, J. E. Menitove, and A. J. Grindon.** 1990. Sepsis associated with transfusion of red cells contaminated with *Yersinia enterocolitica*. *Transfusion* **30**:207–213.

242. **Toivanen, P., and A. Toivanen.** 1994. Does *Yersinia* induce autoimmunity? *Int. Arch. Allergy Immunol.* **104**:107–111.

243. **Tomer, Y., and T. F. Davies.** 1993. Infection, thyroid disease, and autoimmunity. *Endocrine Rev.* **14**:107–120.

244. **Toora, S., E. Budu-Amoako, R. F. Ablett, and J. Smith.** 1992. Effect of high-temperature short-time pasteurization, freezing and thawing and constant freezing, on the survival of *Yersinia enterocolitica* in milk. *J. Food Prot.* **55**:803–805.

245. **Tsubokura, M., K. Otsuki, K. Sato, M. Tanaka, T. Hongo, H. Fukushima, T. Maruyama, and M. Inoue.** 1989. Special features of distribution of *Yersinia pseudotuberculosis* in Japan. *J. Clin. Microbiol.* **27**:790–791.

246. **Tzipori, S., R. Robins-Browne, and J. K. Prpic.** 1987. Studies on the role of virulence determinants of *Yersinia enterocolitica* in gnotobiotic piglets. *Contrib. Microbiol. Immunol.* **9**:233–238.

247. **Une, T.** 1977. Studies on the pathogenicity of *Y. enterocolitica*. I. Experimental infection in rabbits. *Microbiol. Immunol.* **21**:349–363.

248. **Une, T., and R. R. Brubaker.** 1984. Roles of V antigen in promoting virulence and immunity in yersiniae. *J. Immunol.* **133**:2226–2230.

249. **Van Noyen, R., R. Selderslaghs, J. Bekaert, G. Wauters, and J. Vandepitte.** 1991. Causative role of *Yersinia* and other enteric pathogens in the appendicular syndrome. *Eur. J. Clin. Microbiol. Infect. Dis.* **10**:735–741.

250. **Visser, L. G., E. Seijmonsbergen, P. H. Nibbering, P. J. van den Broek, and R. van Furth.** 1999. Yops of *Yersinia enterocolitica* inhibit receptor-dependent superoxide anion production by human granulocytes. *Infect. Immun.* **67**:1245–1250.

251. Wachtel, M. R., and V. L. Miller. 1995. In vitro and in vivo characterization of an *ail* mutant of *Yersinia enterocolitica*. *Infect. Immun.* **63:**2541–2548.

252. Wattiau, P., B. Bernier, P. Deslée, T. Michiels, and G. R. Cornelis. 1994. Individual chaperones required for Yop secretion by *Yersinia*. *Proc. Natl. Acad. Sci. USA* **91:**10493–10497.

253. Wauters, G., S. Aleksic, J. Charlier, and G. Schulze. 1991. Somatic and flagellar antigens of *Yersinia enterocolitica* and related species. *Contrib. Microbiol. Immunol.* **12:**239–243.

254. Wauters, G., M. Janssens, A. G. Steigerwalt, and D. J. Brenner. 1988. *Yersinia mollaretii* sp. nov. and *Yersinia bercovieri* sp. nov., formerly called *Yersinia enterocolitica* biogroups 3A and 3B. *Int. J. Syst. Bacteriol.* **38:**424–429.

255. Wauters, G., K. Kandolo, and M. Janssens. 1987. Revised biogrouping scheme of *Yersinia enterocolitica*. *Contrib. Microbiol. Immunol.* **9:**14–21.

256. Weinberg, E. D. 1984. Iron withholding: a defense against infection and neoplasia. *Physiol. Rev.* **64:**65–102.

257. Welkos, S., A. Friedlander, D. McDowell, J. Weeks, and S. Tobery. 1998. V antigen of *Yersinia pestis* inhibits neutrophil chemotaxis. *Microb. Pathog.* **24:**185–196.

258. Yang, Y., and R. R. Isberg. 1993. Cellular internalization in the absence of invasin expression is promoted by the *Yersinia pseudotuberculosis yadA* product. *Infect. Immun.* **61:**3907–3913.

259. Yang, Y., and R. R. Isberg. 1997. Transcriptional regulation of the *Yersinia pseudotuberculosis* pH6 antigen adhesin by two envelope-associated components. *Mol. Microbiol.* **24:**499–510.

260. Yoshino, K., T. Takao, X. Huang, H. Murata, H. Nakao, T. Takeda, and Y. Shimonishi. 1995. Characterization of a highly toxic, large molecular size heat-stable enterotoxin produced by a clinical isolate of *Yersinia enterocolitica*. *FEBS Lett.* **362:**319–322.

261. Young, G. M., D. Amid, and V. L. Miller. 1996. A bifunctional urease enhances survival of pathogenic *Yersinia enterocolitica* and *Morganella morganii* at low pH. *J. Bacteriol.* **178:**6487–6495.

262. Young, G. M., D. H. Schmiel, and V. L. Miller. 1999. A new pathway for the secretion of virulence factors by bacteria: the flagellar export apparatus functions as a protein-secretion system. *Proc. Natl. Acad. Sci. USA* **96:**6456–6461.

263. Zhang, H., I. Kaur, D. W. Niesel, G. S. Seetharamaiah, J. W. Peterson, B. S. Prabhakar, and G. R. Klimpel. 1997. Lipoprotein from *Yersinia enterocolitica* contains epitopes that cross-react with the human thyrotropin receptor. *J. Immunol.* **158:**1976–1983.

264. Zhang, L., J. Radziejewska-Lebrecht, D. Krajewska-Pietrasik, P. Toivanen, and M. Skurnik. 1997. Molecular and chemical characterization of the lipopolysaccharide O-antigen and its role in the virulence of *Yersinia enterocolitica* serotype O:8. *Mol. Microbiol.* **23:**63–76.

265. Zink, D. L., J. C. Feeley, J. G. Wells, C. Vanderzant, J. C. Vickery, W. D. Roof, and G. A. O'onovan. 1980. Plasmid-mediated tissue invasiveness in *Yersinia enterocolitica*. *Nature* **283:**224–226.

266. Zumbihl, R., M. Aepfelbacher, A. Andor, C. A. Jacobi, K. Ruckdeschel, B. Rouot, and J. Heesemann. 1999. The cytotoxin YopT of *Yersinia enterocolitica* induces modification and cellular redistribution of the small GTP-binding protein RhoA. *J. Biol. Chem.* **274:**29289–29293.

Food Microbiology: Fundamentals and Frontiers, 2nd Ed.
Edited by M. P. Doyle et al.
© 2001 ASM Press, Washington, D.C.

Keith A. Lampel
Anthony T. Maurelli

Shigella Species

12

Bacillary dysentery or shigellosis is caused by members of the *Shigella* species. Dysentery was the term used by Hippocrates to describe an illness characterized by frequent passage of stools containing blood and mucus accompanied by painful abdominal cramps. Perhaps one of the greatest historical impacts of this disease has been its powerful influence in military operations. Long, protracted military campaigns and sieges almost always spawned epidemics of dysentery causing large numbers of military and civilian casualties. With a low infectious dose required to cause disease coupled with oral transmission via fecally contaminated food and water, it is not surprising that dysentery caused by *Shigella* spp. follows in the wake of many natural (earthquakes, floods, famine) and man-made (war) disasters. Apart from these special circumstances, shigellosis remains an important disease in developed countries as well as in developing countries.

Foodborne shigellosis is a neglected area of study. With this chapter we hope to bring the members of the *Shigella* spp. and the disease they cause to the attention of food microbiologists. Modes of transmission and examples of recent foodborne outbreaks will also be high-

lighted. Finally, our most current understanding of the genetics of *Shigella* pathogenesis, the genes involved in causing disease and how they are regulated, will be discussed. Since no single review can be completely comprehensive, the reader is encouraged to refer to several excellent recent reviews for additional information (25, 35, 67, 83, 97).

CHARACTERISTICS OF THE ORGANISM

Classification and Biochemical Characteristics

There are four species of the genus *Shigella* serologically grouped (41 serotypes) based on their somatic O antigens: *Shigella dysenteriae* (group A), *S. flexneri* (group B), *S. boydii* (group C), and *S. sonnei* (group D). As members of the family *Enterobacteriaceae*, they are nearly genetically identical to the *Escherichieae* family and closely related to the salmonellae (80). *Shigella* spp. are nonmotile, oxidase-negative, gram-negative rods. Some important biochemical characteristics that distinguish these bacteria from other enterics are their inability to ferment lactose, although some strains of *S. sonnei* may ferment lactose slowly, or utilize citric acid as

Keith A. Lampel, Center for Food Safety and Applied Nutrition, Food and Drug Administration, 200 C St. S.W., Washington, DC 20204.
Anthony T. Maurelli, Department of Microbiology and Immunology, Uniformed Services University of the Health Sciences, F. Hébert School of Medicine, 4301 Jones Bridge Rd., Bethesda, MD 20814–4799.

a sole carbon source. They do not produce H_2S, except for *S. flexneri* 6 and *S. boydii* serotypes 13 and 14, and do not produce gas from glucose. *Shigella* spp. are inhibited by potassium cyanide and do not synthesize lysine decarboxylase (28). Enteroinvasive *Escherichia coli* strains (EIEC) have pathogenic and biochemical properties that are similar to those of *Shigella* spp. This similarity poses a problem in distinguishing these pathogens. For example, EIEC are nonmotile and are unable to ferment lactose. Some serotypes of EIEC also share identical O antigens with *Shigella* spp. (93).

Shigella spp. are not particularly fastidious in their growth requirements, and in most cases the organisms are routinely cultivated in the laboratory on artificial medium. Cultures of *Shigella* spp. are easily isolated and grown from analytical samples, including water and clinical specimens. In the latter case, *Shigella* spp. are present in fecal specimens in large numbers (10^3 to 10^9 per gram of stool) during the acute phase of infection; therefore, identification is readily accomplished using culture media, biochemical analysis, and serological typing. Shigellae are shed and continue to be detected from convalescent patients (10^2 to 10^3 per gram of stool) for weeks or longer after the initial infection. Isolation of *Shigella* spp. at this stage of infection is more difficult, because a selective enrichment broth for shigellae is not available; therefore, shigellae can be outgrown by resident bacterial fecal flora.

Isolation of *Shigella* spp. from foods is not as facile as from other sources. Foods have many different physical attributes that may affect the successful recovery of shigellae. These factors include composition, such as fat content of the food; physical parameters, such as pH and salt; and natural microbial flora of the food. In the latter case, other microbes in a sample may overgrow shigellae during culture in broth media. The amount of time from the clinical report of a suspected outbreak to the analysis of the food samples can be considerable, and therefore it can lessen the chances of identifying the causative agent. The physiological state of shigellae present in the food is a contributing factor in the successful recovery of this pathogen. *Shigella* spp. may be present in low numbers or in poor physiological state in the suspected food samples. Under these conditions, special enrichment procedures are required for successful detection of shigellae (6).

Shigella in Foods

Shigella spp. are not associated with any specific foods. Common foods that have been implicated in outbreaks caused by shigellae include potato salad, chicken, tossed salad, and shellfish. Establishments where contaminated foods have been served include homes, restaurants, camps, picnics, schools, airlines, sorority houses, and military mess halls (102). In many cases, the source (food) was not identified. From 1983 to 1987, 2,397 foodborne outbreaks representing 54,453 cases were reported to the Centers for Disease Control and Prevention. In only 38% of the cases was the source of the etiological agent identified (8). Whereas epidemiologic methods may strongly imply a common food source, *Shigella* spp. are not often recovered and identified from foods using standard bacteriological methods. Also, since shigellae are not commonly associated with any particular food, routine testing of foods to identify these pathogens is not usually performed.

The traditional approach to address the problem of microbially contaminated foods in the processing plant is to inspect the final product. There are several drawbacks to this approach (42). Current bacteriological methods are often time-consuming and laborious. An alternative to end product testing is the Hazard Analysis and Critical Control Point (HACCP; http://vm.cfsan.fda.gov/~lrd/haccp.html) system, which identifies certain points of the processing system that may be most vulnerable to microbial contamination and chemical and physical hazards. This approach would be instituted as a preventive program with less reliance on end product testing. A monitoring system such as HACCP may be well suited for pathogenic bacteria such as *E. coli* O157:H7 and *Salmonella* spp., which are known to be associated with specific foods, e.g., meats and egg products, respectively.

In contrast, establishing specific critical control points for preventing *Shigella* contamination of foods is not always suitable for the HACCP concept. This pathogen is usually introduced into the food supply by an infected person, such as a food handler with poor personal hygiene. In some cases, this may occur at the manufacturing site but more likely happens at a point between the processing plant and the consumer. Another factor is that foods, such as vegetables (lettuce is a good example), can be contaminated at the site of collection and shipped directly to market. Although HACCP is a method for controlling food safety and preventing foodborne outbreaks, pathogens such as *Shigella* that are not indigenous to, but rather introduced into, foods are most likely to be undetected.

Survival and Growth in Foods

Depending upon growth conditions, *Shigella* spp. can survive in media with a pH range of 2 to 3 for several hours. Acid resistance is modulated by a sigma factor encoded by the *rpoS* (*katF*) gene (101). However, under acidic conditions, shigellae do not usually survive well, either in foods or in stool samples. In foods, studies using citric juices (orange, grape, and lemon), carbonated beverages, and wine revealed that shigellae were recovered after 1 to 6 days (49). In neutral pH foods, such as butter or margarine, shigellae were recovered after 100 days when stored frozen or at 6°C.

Shigella spp. can survive a temperature range of −20°C to room temperature. Survival of shigellae was longer in foods stored frozen or at refrigeration temperature than at room temperature. In foods such as salads containing mayonnaise and some cheese products, *Shigella* spp. survived for 13 to 92 days. Shigellae can survive for an extended period of time on dried surfaces and in foods, e.g., shrimp, ice cream, and minced pork meat, stored frozen. Growth of *Shigella* spp. is impeded in the presence of 3.8 to 5.2% NaCl at pH 4.8 to 5.0, in 300 to 700 mg of $NaNO_2$ per liter, and in 0.5 to 1.5 mg of sodium hypochlorite (NaClO) per liter of water at 4°C. *Shigella* spp. are sensitive to ionizing radiation, with a reduction of 10^7 CFU/g at 3 kGy.

FOODBORNE OUTBREAKS

It is estimated there are 76 million cases of foodborne illness in the United States annually (9, 69) and that the actual number of foodborne outbreaks may be greatly underreported (103). The adverse impact on the public health caused by foodborne pathogens is reflected by their notable morbidity and mortality (87). In the United States, foodborne diseases are responsible for an estimated 5,000 deaths and 325,000 hospitalizations annually. Although reported foodborne outbreaks of shigellosis in the United States have declined recently, shigellosis continues to be a major public health concern. In compiling data from 1982 to 1997, which includes reported cases (from 1983 to 1992) and the number of cases from passive (1992 to 1997) and active surveillance via FoodNet (http://www.cdc.gov/ncidod/dbmd/foodnet) (20), the Centers for Disease Control and Prevention estimated that there were approximately 448,000 cases of shigellosis in the United States, the third leading cause of foodborne outbreaks by bacterial pathogens (69). Worldwide, the World Health Organization estimates that *Shigella* spp. are responsible for 164.7 million cases of shigellosis annually in developing countries and 1.5 million cases in developed countries (55).

Transmission and Susceptible Populations

Human-to-human transmission of *Shigella* spp. is by the fecal-oral route. Most cases of shigellosis result from the ingestion of fecally contaminated food or water, and with foods, the major contribution to contamination is poor personal hygiene of food handlers. From infected carriers, shigellae are spread by several routes, including food, fingers, feces, and flies. Flies transmit shigellae from fecal matter to foods. The highest incidence of shigellosis occurs during the warmer months of the year. Improper storage of contaminated foods is the second most common factor contributing to foodborne outbreaks of shigellosis (102). Other contributing factors are inadequate cooking, contaminated equipment, and food obtained from unsafe sources (8). To reduce the spread of shigellosis, infected patients should be monitored until stool samples are negative for *Shigella* spp.

Shigella is a serious pathogen capable of causing disease in otherwise healthy individuals. Certain populations, however, may be more predisposed to infection and disease owing to the nature of transmission of the organism. The greatest frequency of illness occurs among children less than 6 years of age. In the United States, outbreaks of shigellosis and other diarrheal diseases in day care centers is increasing as more single-parent and two-working-parent families use these facilities to care for their children (59, 84). Typical toddler behavior, such as oral exploration of the environment and inadequate personal hygiene habits, creates conditions ideally suited to the transmission of bacterial, protozoal, and viral pathogens that are spread by fecal contamination. Transmission of *Shigella* spp. in this population is very efficient, and the low infectious dose for causing disease increases the risk for shigellosis. Increased risk also extends to family contacts of day care attendees (114).

Shigellosis can be endemic in other institutional settings, such as prisons, mental hospitals, and nursing homes, where crowding and/or insufficient hygienic conditions create an environment for direct fecal-oral contamination. Crowded conditions and poor sanitation contribute to shigellosis being endemic in developing countries as well.

When natural or man-made disasters destroy the sanitary waste treatment and water purification infrastructure, developed countries assume the conditions of developing countries. These conditions place a population at risk for diarrheal diseases such as cholera and shigellosis. Recent examples include famine and political upheaval in Somalia and the war in Bosnia-Herzegovina (59). Massive population displacement (e.g., refugees fleeing from Rwanda into Zaire in 1994) can also lead to explosive epidemics of diarrheal disease caused by *Vibrio cholerae* and *S. dysenteriae* 1 (39).

The low infectious dose and oral route of transmission also make *Shigella* spp. a potent biological weapon. The literature includes at least one report of deliberate contamination of food with *Shigella* that resulted in a dozen cases of shigellosis (54).

Reservoirs and Vehicles of Infection

Humans are the natural reservoir of *Shigella* spp., although several cases of shigellosis transmitted by monkeys have been reported (52). In one instance, three animal caretakers at a monkey house complained of having diarrhea, and *S. flexneri* 1b was isolated from their stool specimens. Further investigation revealed that four monkeys were also shedding the identical serotype. The disease was spread by direct contact of the caretakers with excrement from the infected monkeys.

Asymptomatic carriers of *Shigella* spp. may exacerbate the maintenance and spread of this pathogen in developing countries. Two studies, one in Bangladesh (47) and another in Mexico (41), revealed that *Shigella* was isolated from stool samples collected from asymptomatic children under the age of 5 years. *Shigella* spp. were rarely isolated from infants under the age of 6 months.

Examples of Foodborne Outbreaks

Through the cases described below, it becomes obvious that a wide range of foods can be contaminated with shigellae. Disease is caused by the ingestion of these contaminated foods and in some instances subsequently leads to rapid dissemination through contaminated feces.

1987—Rainbow Family Gathering

As many as half of the 12,700 people in attendance at an annual gathering of the Rainbow Family may have had shigellosis (116). *S. sonnei* was isolated from stool cultures of tested attendees. Spread of the organism most likely occurred through the fecal-oral route in a crowded environment by contamination of food or water or both. Subsequent outbreaks reported in three other states were due to attendees returning home and infecting others.

1989 and 1994—Shigellosis aboard Cruise Ships

In October 1989, 14% of passengers and 3% of crew members aboard a cruise ship reported having gastrointestinal symptoms (60). A multiple-antibiotic-resistant strain of *S. flexneri* 2a was isolated from several ill passengers and crew. The vehicle of the outbreak was German potato salad. Contamination was introduced by infected food handlers, initially in the country where the food was originally prepared and subsequently by a member of the galley crew on the cruise ship. Another outbreak of shigellosis occurred in August 1994 on the cruise ship SS *Viking Serenade* (19). Thirty-seven percent (*n* = 586) of the passengers and 4% (*n* = 24) of

the crew reported having diarrhea, and one death occurred. *S. flexneri* 2a was isolated from patients, and the suspected vehicle was spring onions.

1990—Operation Desert Shield

Diarrheal diseases during a military operation can be a major factor in reducing troop readiness. Enteric pathogens were isolated from 214 U.S. soldiers in Operation Desert Shield, and of those, 113 cases were diagnosed as shigellosis; *S. sonnei* was the most prevalent species isolated (48). Shigellosis accounted for more time lost from military duties and was responsible for more severe morbidity than enterotoxigenic *E. coli*, the most common enteric pathogen isolated from U.S. troops in Saudi Arabia (48). The suspected vehicle was contaminated fresh vegetables, specifically lettuce. Twelve heads of lettuce were tested, and enteric pathogens were isolated from all.

1991—Alaska Moose Soup

In September 1991 in Galena, Alaska, 25 people who had participated in a gathering of local residents contracted shigellosis associated with eating homemade moose soup. One of five women who made the soup reported having gastroenteritis while preparing the soup. *S. sonnei* was isolated from a hospitalized patient.

1994—Contaminated Lettuce in Norway and the United Kingdom

An outbreak in Norway of 110 culture-confirmed cases of shigellosis caused by *S. sonnei* was reported in 1994 (50). Iceberg lettuce from Spain, served in a salad bar, was suspected as the source of the outbreak in Norway and likely responsible for increases in shigellosis in other European countries, including the United Kingdom (34) and Sweden. *S. sonnei* was isolated from patients from several northwest European countries but was not isolated from any foods. Epidemiologic evidence indicated that imported lettuce was the vehicle of these outbreaks.

1994—Green Onions

An outbreak of *S. flexneri* serotype 6 (mannitol-negative) infection occurred in the midwest of the United States in 1994 (18). Although not confirmed, the suspected vehicle was Mexican green onions (scallions, spring onions). Seventeen cases of shigellosis occurred at a church potluck meal in Indiana, 29 cases occurred at an anniversary reception in Indiana, and 26 culture-confirmed mannitol-negative *S. flexneri* or *Shigella* cases were reported to the Illinois State Department of Health. *S. flexneri* serotype 6 was also isolated from patients in Missouri, Minnesota, Wisconsin, Michigan, and Kentucky. Ingestion of green onions was implicated as the vehicle of infection. An infected worker most likely contaminated the onions at the time of harvest or packing.

1998—Fresh Parsley

Several health departments in the United States and Canada were notified during August 1998 that a number of people had shigellosis. Most of the patients ate at restaurants that served chopped, uncooked parsley. The causative agent was identified as *S. sonnei*, and based on pulsed-field gel electrophoresis, epidemiologic traceback, and data from other investigations, one farm in Mexico was implicated as the source of the contaminated parsley. Further investigations revealed that the water supply used for chilling the parsley in a hydrocooler and making ice was unchlorinated and prone to bacterial contamination (21).

2000—5-Layer Bean Dip

An outbreak of shigellosis associated with contaminated 5-layer (bean, salsa, guacamole, nacho cheese, and sour cream) party dip occurred in three West Coast states (22). The causative agent, *S. sonnei*, was isolated from at least 30 patients. The pathogen was isolated from only one layer (cheese) of the dip and was initially detected by a PCR assay targeting shigellae. *Shigella* was isolated subsequently by enrichment followed by plating on selective agar (115).

One of the striking features regarding foodborne outbreaks of shigellosis is that contamination of foods usually is not at the processing plant, but rather the source can be traced to a food handler. As evident from the examples above and in Table 12.1, these incidents can occur from contamination of foods by infected food handlers at small town gatherings and picnics to larger-scale outbreaks such as those on cruise ships and at institutions.

CHARACTERISTICS OF DISEASE

Clinical Presentation

Shigellosis is distinguished from disease caused by most other foodborne pathogens described in this book in at least two important aspects: (i) the production of bloody diarrhea or dysentery and (ii) the low infectious dose that can cause clinical symptoms. Bloody diarrhea refers to diarrhea in which stools contain visible red blood. Dysentery also involves bloody diarrhea, but the passage of bloody mucoid stools is accompanied by severe abdominal and rectal pain, cramps, and fever. While abdominal pain and diarrhea are experienced by nearly all

Table 12.1 Examples of foodborne outbreaks caused by *Shigella* spp.

Year	Location; source of contamination[a]	Isolate	Reference
1986	Texas; shredded lettuce	*S. sonnei*	24
1987	Rainbow Family gathering; food handlers	*S. sonnei*	116
1988–1989	Monroe, N.Y.; multiple sources	*S. sonnei*	117
1988	Outdoor music festival, Michigan; food handlers	*S. sonnei*	58
1988	Commercial airline; cold sandwiches	*S. sonnei*	44
1989	Cruise ship; potato salad	*S. flexneri*	60
1990	Operation Desert Shield (U.S. troops); fresh produce	*Shigella* spp.	48
1991	Alaska; moose soup	*S. sonnei*	37
1992–1993	Operation Restore Hope, Somalia, U.S. troops	*Shigella* spp.	98
1994	Europe; shredded lettuce from Spain	*S. sonnei*	50
1994	Midwest U.S.; green onions	*S. flexneri*	18
1994	Cruise ship	*S. flexneri*	19
1998	Fresh parsley	*S. sonnei*	21
2000	Bean dip	*S. sonnei*	22

[a] The source of contamination is listed when known.

patients with shigellosis, fever occurs in about one-third and gross blood in the stools occurs in about 40% of cases (26).

The clinical features of shigellosis range from a mild watery diarrhea to severe dysentery. The dysentery stage caused by *Shigella* spp. may or may not be preceded by watery diarrhea. This stage reflects the transient multiplication of bacteria as they pass through the small bowel. Jejunal secretions probably are not effectively reabsorbed in the colon owing to transport abnormalities caused by bacterial invasion and destruction of the colonic mucosa (53). The dysentery stage of disease correlates with extensive bacterial colonization of the colonic mucosa. The bacteria invade epithelial cells of the colon and spread from cell to cell but penetrate only as far as the lamina propria. Foci of individually infected cells produce microabscesses that coalesce, forming large abscesses and mucosal ulcerations. As the infection progresses, dead cells of the mucosal surface slough off, subsequently leading to the presence of blood, pus, and mucus in the stools.

The incubation period for shigellosis is 1 to 7 days, but the symptoms usually begin within 3 days. Strains of *S. dysenteriae* 1 produce the most severe symptoms of shigellosis, whereas *S. sonnei* produces the mildest consequences. *S. flexneri* and *S. boydii* infections can be either mild or severe. Despite the severity of the illness, shigellosis is self-limiting. If left untreated, clinical illness usually persists for 1 to 2 weeks (although it may be as long as a month), and the patient recovers.

Infectious Dose

An important aspect of *Shigella* pathogenesis is the extremely low 50% infectious dose (ID_{50}), i.e., the experimentally determined oral dose that causes disease in 50% of volunteers challenged with a virulent strain of the cells. The ID_{50} for *S. flexneri*, *S. sonnei*, and *S. dysenteriae* is approximately 5,000 cells. Volunteers have become ill when administered doses as low as 200 bacteria (27). The low infectious dose of *Shigella* underlies the high rate of transmission of bacillary dysentery and the great explosive potential for person-to-person spread as well as foodborne and waterborne outbreaks of diarrhea.

Complications

Shigellosis can be a very painful and incapacitating disease and is more likely to require hospitalization than other bacterial diarrheas. It is not usually life-threatening, and mortality is rare except in malnourished children, immunocompromised individuals, and the elderly (11). However, complications arising from the disease include severe dehydration, intestinal perforation, toxic megacolon, septicemia, seizures, hemolytic-uremic syndrome (HUS), and Reiter's syndrome (10). These last two syndromes are receiving increasing research attention. HUS is a rare but potentially fatal complication associated with infection by *S. dysenteriae* 1 (85). The syndrome is characterized by hemolytic anemia, thrombocytopenia, and acute renal failure. Epidemiologic studies indicate that Shiga toxin produced by *S. dysenteriae* 1 may be the cause of HUS (61). This hypothesis is supported by the fact that HUS is also caused by strains of *E. coli* O157:H7, which produce high levels of Shiga toxin (51). It has been suggested that Shiga toxins may cause HUS by entering the bloodstream and damaging vascular endothelial cells such as those in the kidney (51, 61, 104). Reiter's syndrome, a form of reactive arthritis, is a postinfection sequela to shigellosis that is strongly associated with individuals of the HLA-B27 histocompatibility group (100). The syndrome is composed of three symptoms—urethritis, conjunctivitis, and arthritis—with the latter being the most dominant symptom. Infections caused by several other gram-negative enteric pathogens also can lead to this type of sterile inflammatory polyarthropathy (16).

Treatment and Prevention

Although stool fluid losses are not as massive as with other bacterial diarrheas, the diarrhea associated with shigellosis combined with water loss due to fever and decreased water intake due to anorexia may result in severe dehydration (11). Oral intake can generally replace fluid losses, although intravenous rehydration may be required in very young and elderly patients.

The antibiotic of choice for treatment of shigellosis is trimethoprim-sulfamethoxazole (26). However, there is some controversy regarding the use of antibiotics in treating shigellosis. Since the infection is self-limited in normally healthy patients, and full recovery occurs without the use of antibiotics, drug therapy is usually not indicated. In addition, multiple drug resistance among isolates of *Shigella* is becoming more common. Clinical isolates resistant to sulfonamides, ampicillin, trimethoprim-sulfamethoxazole, tetracycline, chloramphenicol, and streptomycin have been obtained (12, 46). Extensive use of antibiotics selects for drug-resistant organisms, and, therefore, many believe that antimicrobial therapy for shigellosis should be reserved only for the most severely ill patients. On the other hand, there are persuasive public health arguments for the antibiotic management of shigellosis. Antibiotic treatment limits the duration of disease and shortens the period of fecal excretion of bacteria (43). An infected person or asymptomatic carrier can be an index case for person-to-person and food- and waterborne spread; hence, antibiotic treatment of these individuals can be a significant public health tool to contain the spread of shigellosis. However, antibiotics are not a substitute for improved hygienic conditions to contain secondary spread of shigellosis. The single most effective means of preventing secondary transmission is hand washing. Food handling and preparation are important processes that also deserve attention, and persons with diarrhea should be excluded from handling food.

Despite many years of intensive effort, an effective vaccine against shigellosis has not been developed. Attenuated, oral vaccine strains of *S. flexneri* are currently being tested. One such strain is SC602. This strain has mutations that block iron uptake and abolish the intracellular and intercellular motility phenotypes (7). A limited challenge study showed protection among those who were vaccinated. However, one major drawback still to be resolved is the transient fever and mild diarrhea associated with administration of the vaccine (23). This study highlights one of the persistent problems impeding the development of a safe *Shigella* vaccine: designing a strain that can induce a protective immune response without producing unacceptable side effects.

VIRULENCE FACTORS

Hallmarks of Virulence

Shigella spp. and EIEC are the principal agents of bacillary dysentery and as such belong to the group of enteric pathogens that cause disease by overt invasion of epithelial cells in the large intestine. The clinical symptoms of shigellosis can be directly attributed to the hallmarks of *Shigella* virulence: the ability to induce diarrhea, invade epithelial cells of the intestine, multiply intracellularly, and spread from cell to cell.

Shigella spp. colonize the small intestine only transiently and cause little tissue damage (88). Production of enterotoxins by the bacteria while they are in the small bowel probably results in the diarrhea that generally precedes onset of dysentery (31, 77). The jejunal secretions elicited by these toxins may facilitate passage of the bacteria through the small intestine and into the colon, where they colonize and invade the epithelium.

Formal and colleagues (56) established the essential role of epithelial cell invasion in *Shigella* pathogenesis in a landmark study that employed in vitro tissue culture assays for both invasion and animal models. Spontaneous colonial variants of *S. flexneri* 2a that are unable to invade epithelial cells in tissue culture do not cause disease in monkeys. Gene transfer studies using *E. coli* K-12 donors and *S. flexneri* 2a recipients established the third hallmark of *Shigella* virulence. An *S. flexneri* 2a recipient that inherits the *xyl-rha* region of the *E. coli* K-12 chromosome retains the ability to invade epithelial cells but has a reduced ability to multiply within these cells (30). This hybrid strain fails to cause a fatal infection in the opium-treated guinea pig model and is unable to cause disease when fed to rhesus monkeys (33).

It is necessary but not sufficient for *Shigella* to multiply within the host epithelial cell after invasion. The bacterium must also spread through the epithelial layer of the colon by cell-to-cell spread that does not require the bacterium to leave the intracellular environment and be reexposed to the intestinal lumen. Mutants of *Shigella* spp. that are competent for invasion and multiplication but unable to spread between cells in this fashion have been isolated. These mutants established intracellular spread as the fourth hallmark of *Shigella* virulence and will be discussed further below.

Along with the ability to colonize and cause disease, an intrinsic part of a bacterium's pathogenicity is its mechanism for regulating expression of the genes involved in virulence. Virulence in *Shigella* spp. is regulated by growth temperature. After growth at 37°C, virulent strains of *Shigella* are able to invade mammalian cells, but when cultivated at 30°C, they are noninvasive. This noninvasive phenotype is reversible by shifting the growth temperature to 37°C. The temperature change enables the bacteria to reexpress its virulence properties (65). Temperature regulation of virulence gene expression is a characteristic that *Shigella* spp. share with other human pathogens, such as *E. coli*, *Salmonella enterica* serovar Typhimurium, *Bordetella pertussis*, *Yersinia* spp., and *Listeria monocytogenes* (see reference 63 for review). Regulation of gene expression in response to environmental temperature is a useful bacterial strategy. By sensing the ambient temperature of the mammalian host (e.g., 37°C for humans) to trigger gene expression, this strategy permits *Shigella* spp. to economize energy that would be expended on the synthesis of virulence products when the bacteria are outside the host. The system also permits the bacteria to coordinately regulate expression of multiple unlinked genes that are required for the full virulence phenotype. Temperature regulation in *S. flexneri* 2a operates at the level of gene transcription and is mediated by both positive and negative transcription factors. A chromosomal gene, *virR(hns)*, encodes a repressor of virulence gene expression (68), whereas two other genes, *virF* and *virB*, encode positive activators (1, 89). These genes will be discussed in a later section. A more thorough treatment of virulence gene regulation in *Shigella* spp. can be found in several recent review articles (25, 81).

Genetics

Virulence-Associated Plasmid Genes

Given the complexity of the interactions between host and pathogen, it is not surprising that *Shigella* virulence is multigenic, involving both chromosomal and plasmid-encoded genes (Table 12.2). Another landmark report on the pathogenicity of *Shigella* spp. was the demonstration of the indispensable role for a large plasmid in invasion. A 180-kb plasmid in *S. sonnei* and a 220-kb plasmid in *S. flexneri* are essential for invasion (94, 95). Other *Shigella* spp. as well as some EIEC contain homologous plasmids which are functionally interchangeable and share significant degrees of DNA homology (93). Hence, it is probable that the plasmids of *Shigella* spp. and EIEC are derived from a common ancestor (92).

A 37-kb region of the invasion plasmid of *S. flexneri* 2a contains all of the genes necessary to permit the bacteria to penetrate into tissue culture cells. This DNA segment was identified as the minimal region of the virulence

Table 12.2 Virulence-associated loci of *Shigella*

Locus	Product	Role in virulence
Chromosomal		
cpxR	Response regulator of CpxA-CpxR two-component system	Activator of *virF*
iuc	Synthesis of aerobactin and receptor	Acquisition of iron in the host
rfa; rfb	Enzymes for core and O-antigen biosynthesis	Correct polar localization of IcsA
set[a]	ShET1	Enterotoxin
she	Putative hemagglutinin and mucinase	Unknown
sodB	Superoxide dismutase	Inactivation of superoxide radicals; defense against oxygen-dependent killing in host
stx[b]	Shiga toxin	Destruction of vascular tissue
vacB(rnr)	Exoribonuclease RNase R	Posttranscriptional regulation of virulence gene expression
virR(hns)	Histonelike protein	Repressor of virulence gene expression
Plasmid		
icsA(virG)	120-kDa cell-bound and secreted protein	Actin polymerization for intracellular motility and intercellular spread
ipaA	70-kDa protein	Efficient invasion; binds to vinculin and promotes F-actin
ipaB	62-kDa protein	Invasion; lysis of vacuole; induction of apoptosis
ipaC	43-kDa protein	Invasion; induces cytoskeletal reorganization
ipaD	38-kDa protein	Invasion; antisecretion plug (with IpaB) depolymerization
ipgC	17-kDa protein	Chaperone for IpaB and IpaC
mxi/spa	20 proteins	Type III system for secretion of Ipa and other virulence proteins
sen	ShET2	Enterotoxin
virB	Transcriptional activator	Temperature regulation of virulence genes
virF	Transcriptional activator	Temperature regulation of virulence genes

[a] The *set* locus and production of ShET1 are found almost exclusively in *S. flexneri*.
[b] The *stx* locus and production of Shiga toxin are observed only in *S. dysenteriae* 1.

plasmid needed to allow a plasmid-cured derivative of *S. flexneri* (and *E. coli* K-12) to invade tissue culture cells (64). The nucleotide sequence of this part of the virulence plasmid from *S. flexneri* and *S. sonnei* is known (see reference 35 for summary; GenBank accession no. D50601 for *S. sonnei*). The region encodes about 33 genes contained in two groups of genes transcribed in opposite orientation (Fig. 12.1). Although a precise transcription map of these genes has not been defined, available evidence and the DNA sequence of the region suggest a multiple operon organization.

The genes composing the *ipaBCDA* (invasion plasmid antigens) cluster the immunodominant antigens detected with sera from convalescent patients and experimentally challenged monkeys (78). *ipaBCD* have been experimentally demonstrated to be absolutely required for invasion of mammalian cells (71). An *ipaA* mutant is still invasive but has a 10-fold reduced ability compared with wild type (107). The Ipa products are found associated with the outer membrane of *Shigella* spp. IpaB and IpaC (and probably IpaA) form a complex on the bacterial cell surface and are responsible for transducing the signal that leads to entry of *Shigella* into the host cells via bacterium-

directed phagocytosis (73). When coated onto latex beads, IpaB and IpaC can form a complex that promotes uptake of the beads by HeLa cells (70). This complex of Ipa proteins also binds to cell surface receptors such as $\alpha5$-$\beta1$ integrins (113). Purified IpaC induces cytoskeletal reorganization, including formation of filopodia and lamellipodial extensions on permeabilized cells (108). IpaA binds to vinculin and promotes F-actin depolymerization (14, 107). This step is thought to facilitate reorganization of the host cell surface structures induced by contact with *Shigella* spp. and modulate bacterial entry.

Although the Ipa proteins have no typical signal sequence, these proteins are secreted into the extracellular medium. Contact of the bacterium with epithelial cells causes increased secretion of the cytoplasmic pool of Ipa products (4, 72). IpaD forms an antisecretion complex, or plug, with IpaB. Consequently, *ipaD* mutants are hypersecreters of the Ipa products (72). *ipgC* (invasion plasmid gene) is required for invasion and acts as a cytoplasmic chaperon which prevents IpaB and IpaC from forming complexes while in the bacterial cytoplasm (73). In the absence of IpgC, IpaB and IpaC are rapidly degraded.

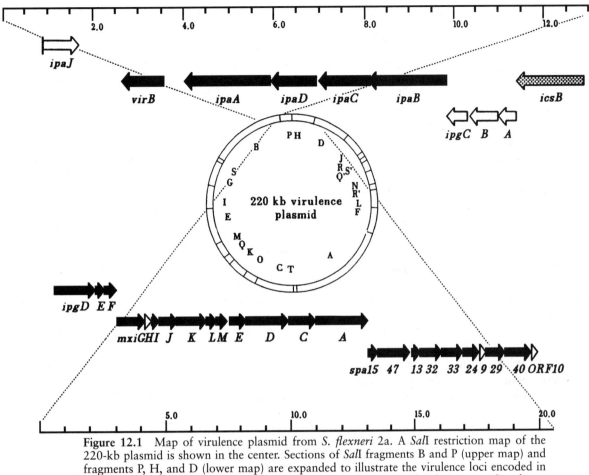

Figure 12.1 Map of virulence plasmid from *S. flexneri* 2a. A *Sal*I restriction map of the 220-kb plasmid is shown in the center. Sections of *Sal*I fragments B and P (upper map) and fragments P, H, and D (lower map) are expanded to illustrate the virulence loci encoded in these regions. The expanded regions are contiguous and cover 32 kb. The open reading frame for *icsB* is separated from that of *ipgD* by 314 bp. The entire sequence of the virulence plasmid of *S. flexneri* 5 has been determined (15a) (GenBank accession no. AL391753).

The product of *ipaB* has also been postulated to be the "contact hemolysin" which is responsible for lysis of the phagocytic vacuole minutes after entry of the bacterium into the host cell (45). The ability of *S. flexneri* to induce apoptosis in infected macrophages is an additional property assigned to IpaB (118).

As mentioned above, the products of the *ipa* genes are actively secreted into the extracellular medium even though they contain no signal sequence for recognition by the usual gram-negative bacterial transport system. Ipa secretion is mediated by a type III secretion pathway (97) and requires a dedicated apparatus composed of gene products from the *mxi/spa* loci (Fig. 12.1). The *mxi* (membrane expression of invasion plasmid antigens) genes compose an operon that encodes several lipoproteins (MxiJ and MxiM), a transmembrane protein (MxiA), and proteins containing signal sequences (MxiD, MxiJ, and MxiM) (2, 3, 5). MxiH, MxiJ, MxiD, and MxiA share homology with proteins involved in

secretion of virulence proteins (Yops) in *Yersinia* spp. (2, 3, 5). The *spa* (surface presentation of Ipa antigens) genes encode proteins that share significant homologies with proteins involved in flagellar synthesis in *E. coli*, *S. enterica* serovar Typhimurium, *Bacillus subtilis*, and *Caulobacter crescentus* (96, 110). Included among these genes is *spa47*, which encodes a protein that likely functions as the energy-generating component of the secretion apparatus because it has sequence similarities with ATPases of the flagellar assembly machinery of other bacteria (111). Nonpolar null mutations in all of the *mxi/spa* genes tested so far result in loss of the ability to secrete the Ipas and loss of invasive capacity for cultured cells. Hence, the type III secretion system is an essential component of *Shigella* virulence.

Apart from the resemblance to genes involved in flagellar synthesis and secretion of Yops, there is an even more striking similarity both in gene organization and predicted protein sequence between the *mxi/spa* region of the *Shigella* virulence plasmid and a virulence-associated chromosomal region from *S. enterica* serovar Typhimurium (35, 40). The *Salmonella spa* region encodes homologues of the *Shigella spa* genes in the same gene order. Sequence identities between the protein homologues range as high as 86% (Spa9 versus SpaQ). The relatedness of the *spa* regions strongly suggests that these two human pathogens evolved similar mechanisms for secretion of the virulence proteins required for signal transduction with the mammalian host. Interestingly, plant pathogens such as *Erwinia carotovora*, *Xanthomonas campestris*, and *Pseudomonas solanacearum* also contain genes that encode homologues of the *mxi/spa*-encoded proteins (109). It is now recognized that the type III secretion system for transport of virulence proteins is a critical element of both plant and animal bacterial pathogenesis.

Secretion of the Ipa proteins via the Mxi/Spa apparatus is induced when *Shigella* cells contact the host cell (72, 112). Other compounds that mimic this signal and induce Ipa secretion are fibronectin, laminin, collagen type IV, Congo red, bile salts, and fetal bovine serum (82, 113). A similar phenomenon of contact-induced secretion is observed in other bacterial pathogens that utilize a type III secretion pathway for extracellular transport of virulence effectors (97).

A plasmid-encoded virulence gene which is unlinked to the 37-kb region shown in Fig. 12.1 is not required for invasion but is crucial for intra- and intercellular motility. This gene, known as *virG* or *icsA* (intracellular spread), encodes a protein that catalyzes the polymerization of actin in the cytoplasm of the infected host cell (13, 62). The IcsA protein is unusual in that it is expressed asymmetrically on the bacterial surface, being found only at one pole (38). The polymerization of actin monomers by IcsA forms a tail leading from the pole and provides the force that propels the bacterium through the cytoplasm. Hence, unipolar expression of IcsA imparts directionality of movement to the bacterium. Although the mechanism of unipolar localization of IcsA is unknown, it is dependent on synthesis of a complete lipopolysaccharide (LPS) (90). LPS mutants of *S. flexneri* 2a express surface IcsA in a circumferential fashion, and while the protein is still capable of polymerizing actin, movement is restricted as the bacterium becomes encased in a shell of actin. A plasmid-encoded protease, SopA/IcsP, cleaves IcsA and is proposed to play a role in unipolar localization of IcsA (29, 99). However, *E. coli* and plasmid-cured derivatives of *S. flexneri* transformed with a cloned *icsA* gene localize IcsA normally (91). These results suggest that IcsA localization does not require any other virulence plasmid-encoded gene and that motifs within IcsA itself contain the information that directs the protein to the pole.

Chromosomal Virulence Loci

In contrast to the genes of the virulence plasmid that are responsible for invasion of mammalian tissues, most of the chromosomal loci associated with *Shigella* virulence are involved in regulation or survival within the host. Mutations that alter O antigen and core synthesis or assembly lead to a "rough" phenotype and render *Shigella* spp. avirulent. Synthesis of a complete LPS, which is crucial for correct unipolar localization of IcsA (see above), requires chromosomal loci such as *rfa* and *rfb*. In the case of *S. sonnei* and *S. dysenteriae* 1, plasmid-encoded genes are also necessary for synthesis of the LPS O side chain (15).

Although *Shigella* spp. and *E. coli* are very closely related at the genetic level, there are significant differences beyond the presence of the virulence plasmid in *Shigella* spp. Two pathogenicity islands have been identified in the chromosome of *S. flexneri* (Table 12.2). SHI-1 (*Shigella* pathogenicity island 1) contains the *set* gene, which encodes an enterotoxin (31). It is contained within the open reading frame of another gene, *she*, which encodes a protein with putative hemagglutinin and mucinase activity (86). The *iuc* locus, which contains the genes for aerobactin synthesis and transport, is present in SHI-2 (74, 111). Aerobactin is a hydroxamate siderophore that *S. flexneri* uses to scavenge iron. When the *iuc* locus is inactivated, the aerobactin-deficient mutants retain their capacity to invade host cells but are altered in virulence as measured in animal models. These results suggest that aerobactin synthesis is important for bacterial growth within the mammalian host (57, 76).

The *stx* locus in *S. dysenteriae* 1 encodes Shiga toxin (for a review see reference 79). A mutation in this locus does not alter the ability of the organism to invade epithelial cells or cause keratoconjunctivitis in the Sereny test. However, when tested in macaque monkeys, the mutant strain caused less vascular damage in the colonic tissue than did the toxin-producing parent (32). Hence, production of Shiga toxin may account for the fact that infections caused by *S. dysenteriae* 1 are generally more severe than those caused by other species of *Shigella*.

In addition to extra genes in the *Shigella* chromosome, there are genes that are present in the closely related *E. coli* but that are missing from the *Shigella* chromosome. *ompT* is part of a cryptic prophage in the *E. coli* chromosome and encodes an outer membrane protease. This prophage is missing in *Shigella* spp. When the *ompT* gene is introduced into *Shigella* spp. by conjugation, the strain loses the ability to spread from cell to cell and is attenuated. This phenotype is due to degradation of

IcsA by the *ompT* protease (75). Another example of a missing genetic locus or "black hole" in the *Shigella* genome is *cadA*, the gene for lysine decarboxylase. Although lysine decarboxylase activity is present in >85% of *E. coli* strains, it is missing in all strains of *Shigella* spp. and EIEC. When *cadA* is reintroduced into *S. flexneri*, the production of cadaverine (by the decarboxylation of lysine) acts to inhibit the action of the *Shigella* enterotoxins (66). Hence, these examples of "antivirulence genes" illustrate another way that pathogens evolve from their nonpathogenic commensal relatives by both acquiring genes (e.g., the virulence plasmid) that contribute to virulence and deleting genes that are incompatible with expression of these new virulence traits.

Virulence Gene Regulation

An important feature of *Shigella* pathogenesis is the ability of the organism to modulate expression of its virulence genes in response to growth temperature. Both activators and a repressor control the virulence regulon of *Shigella* spp. The product of the chromosomal *virR/hns* locus is a histonelike protein, H-NS, which acts as a repressor of *Shigella* virulence gene expression (68). Mutations in *virR/hns* cause deregulation of temperature control such that genes in the virulence regulon are expressed even at the nonpermissive temperature of 30°C. The *virR/hns* locus is allelic with regulatory loci in other enteric bacteria and, like *virR/hns*, these alleles act as repressors of their respective regulons (references cited in reference 25). Several different models to explain how VirR/H-NS acts as a transcriptional repressor have been proposed and are analyzed in a recent review (25). However, because VirR/H-NS is involved in gene regulation in response to diverse environmental stimuli, such as osmolarity, pH, and temperature, a comprehensive model to explain its activity has been elusive.

One of the targets of VirR/H-NS regulation is the plasmid-encoded transcriptional activator *virB* (106). Expression of genes in the *ipa* and *mxi/spa* clusters is dependent on VirB, and mutations in *virB* abolish the bacterium's ability to invade tissue culture cells (1, 17). Transcription of *virB* is dependent on growth temperature and VirF (105). VirB is likely a DNA-binding protein, as suggested by the homology it shares with the plasmid-partitioning proteins ParB of bacteriophage P1 and SopB of plasmid F. However, targets for VirB binding in the promoter regions of the *ipa* and *mxi/spa* genes have not yet been identified, and the mechanism by which VirB activates virulence gene expression is unknown.

The product of the *virF* locus is a key element in temperature regulation of the *Shigella* virulence regulon. A helix-turn-helix motif in the carboxyl-terminal portion of VirF is characteristic of members of the AraC family of transcriptional activators (36). Consistent with its predicted role as a DNA-binding protein, VirF binds to sequences upstream of *virB* (106). The binding of VirF may act as an antagonist to binding by VirR/H-NS and thereby provides a mechanism for responding to temperature. However, expression of *virF* itself is apparently not subject to temperature regulation, leaving open the possibility that an unknown regulator may be involved in activating VirF. Alternatively, temperature may induce conformational changes in VirF that influence its binding affinity and hence its ability to activate promoters.

CONCLUSIONS

While foodborne infections due to members of the *Shigella* spp. may not be as frequent as those caused by other foodborne pathogens, they have the potential for explosive spread owing to the extremely low infectious dose that can cause overt clinical disease. In addition, cases of bacillary dysentery frequently require medical attention (even hospitalization), resulting in lost time from work, as the severity and duration of symptoms can be incapacitating. There is no effective vaccine against dysentery caused by *Shigella* spp. These features, coupled with the wide geographical distribution of the strains and sensitivity of the human population to *Shigella* infection, make *Shigella* spp. a formidable public health threat.

Research on the genetics of Shigella virulence in the laboratory of A.T.M. is supported by USUHS protocol RO-7385 and Public Health Service grant AI24656 from the National Institute of Allergy and Infectious Diseases.

References

1. **Adler, B., C. Sasakawa, T. Tobe, S. Makino, K. Komatsu, and M. Yoshikawa.** 1989. A dual transcriptional activation system for the 230 kb plasmid genes coding for virulence-associated antigens of *Shigella flexneri*. Mol. Microbiol. 3:627–635.
2. **Allaoui, A., P. J. Sansonetti, and C. Parsot.** 1992. MxiJ, a lipoprotein involved in secretion of *Shigella* Ipa invasins, is homologous to YscJ, a secretion factor of the *Yersinia* Yop proteins. J. Bacteriol. 174:7661–7669.
3. **Allaoui, A., P. J. Sansonetti, and C. Parsot.** 1993. MxiD, an outer membrane protein necessary for the secretion of the *Shigella flexneri* Ipa invasins. Mol. Microbiol. 7:59–68.
4. **Andrews, G. P., A. E. Hromockyj, C. Coker, and A. T. Maurelli.** 1991. Two novel virulence loci, *mxiA* and *mxiB*, in *Shigella flexneri* 2a facilitate excretion of invasion plasmid antigen. Infect. Immun. 59:1997–2005.
5. **Andrews, G. P., and A. T. Maurelli.** 1992. *mxiA* of *Shigella flexneri* 2a, which facilitates export of invasion plasmid antigens, encodes a homologue of the low-calcium response protein, LcrD, of *Yersinia pestis*. Infect. Immun. 60:3287–3295.

6. **Andrews, W. H.** 1989. Methods for recovering injured classical enteric pathogenic bacteria (*Salmonella*, *Shigella* and enteropathogenic *Escherichia coli*) from foods, p. 55–113. *In* B. Ray (ed.), *Injured Index and Pathogenic Bacteria: Occurrence and Detection in Foods, Water and Feeds.* CRC Press, Inc., Boca Raton, Fla.

7. **Barzu, S., A. Fontaine, P. Sansonetti, and A. Phalipon.** 1996. Induction of a local anti-IpaC antibody response in mice by use of a *Shigella flexneri* 2a vaccine candidate: implications for use of IpaC as a protein carrier. *Infect. Immun.* 64:1190–1196.

8. **Bean, N. H., and P. M. Griffin.** 1990. Foodborne disease outbreaks in the United States, 1973–1987: pathogens, vehicles, and trends. *J. Food Prot.* 53:804–817.

9. **Bennett, J. V., S. D. Holmberg, M. F. Rogers, and S. L. Solomon.** 1987. Infectious and parasitic diseases, p. 102–114. *In* R.W. Amlet and H.B. Dull (ed.), *Closing the Gap: the Burden of Unnecessary Illness.* Oxford Press, New York, N.Y.

10. **Bennish, M. L.** 1991. Potentially lethal complications of shigellosis. *Rev. Infect. Dis.* 13(Suppl. 4):S319–S324.

11. **Bennish, M. L., J. R. Harris, B. J. Wojtynaik, and M. Struelens.** 1990. Death in shigellosis: incidence and risk factors in hospitalized patients. *J. Infect. Dis.* 161:500–506.

12. **Bennish, M. L., and M. A. Salam.** 1992. Rethinking options for the treatment of shigellosis. *J. Antimicrob. Chemother.* 30:243–247.

13. **Bernardini, M. L., J. Mounier, H. d'Hauteville, M. Coquis-Rondon, and P. J. Sansonetti.** 1989. Identification of *icsA*, a plasmid locus in *Shigella flexneri* that governs bacterial intra- and intercellular spread through interaction with F-actin. *Proc. Natl. Acad. Sci. USA* 86:3867–3871.

14. **Bourdet-Sicard, R., M. Rudiger, B. M. Jockusch, P. Gounon, P. J. Sansonetti, and G. T. Nhieu.** 1999. Binding of the *Shigella* protein IpaA to vinculin induces F-actin depolymerization. *EMBO J.* 18:5853–5862.

15. **Brahmbhatt, H. N., A. A. Lindberg, and K. N. Timmis.** 1992. *Shigella* lipopolysaccharide: structure, genetics, and vaccine development. *Curr. Top. Microbiol. Immunol.* 180:45–64.

15a. **Buchrieser, C., P. Glaser, C. Rusniok, H. Nedjari, H. d'Hauteville, F. Kunst, P. Sansonetti, and C. Parsot.** 2000. The virulence plasmid pWR100 and the repertoire of proteins secreted by the type III secretion apparatus of *Shigella flexneri*. *Mol. Microbiol.* 38:760–771.

16. **Bunning, V. K., R. B. Raybourne, and D. L. Archer.** 1988. Foodborne enterobacterial pathogens and rheumatoid disease. *J. Appl. Bacteriol. Symp. Suppl.* 65:87S–107S.

17. **Buysse, J. M., M. M. Venkatesan, J. Mills, and E. V. Oaks.** 1990. Molecular characterization of a transacting, positive effector (*ipaR*) of invasion plasmid antigen synthesis in *Shigella flexneri* serotype 5. *Microb. Pathog.* 8:197–211.

18. **Centers for Disease Control and Prevention.** 1994. Personal communication.

19. **Centers for Disease Control and Prevention.** 1994. Outbreak of *Shigella flexneri* 2a infections on a cruise ship. *Morb. Mortal. Wkly. Rep.* 43:657.

20. **Centers for Disease Control and Prevention.** 1996. The Foodborne Diseases Active Surveillance Network. *Morb. Mortal. Wkly. Rep.* 46:258–261.

21. **Centers for Disease Control and Prevention.** 1999. Outbreaks of *Shigella sonnei* infection associated with eating fresh parsley—United States and Canada, July-August, 1998. *Morb. Mortal. Wkly. Rep.* 48:285–289.

22. **Centers for Disease Control and Prevention.** 2000. Outbreaks of *Shigella sonnei* infections associated with eating a nationally distributed dip—California, Oregon, and Washington, January 2000. *Morb. Mortal. Wkly. Rep.* 49:60–61.

23. **Coster, T. S., C. W. Hoge, L. L. VanDeVerg, A. B. Hartman, E. V. Oaks, M. M. Venkatesan, D. Cohen, G. Robin, A. Fontaine-Thompson, P. J. Sansonetti, and T. L. Hale.** 1999. Vaccination against shigellosis with attenuated *Shigella flexneri* 2a strain SC602. *Infect. Immun.* 67:3437–3443.

24. **Davis, H., J. P. Taylor, J. N. Perdue, G. N. Stelma, Jr., J. M. Humphreys, Jr., R. Rowntree III, and K. D. Greene.** 1988. A shigellosis outbreak traced to commercially distributed lettuce. *Am. J. Epidemiol.* 128:1312–1321.

25. **Dorman, C. J., and M. E. Porter.** 1998. The *Shigella* virulence gene regulatory cascade: a paradigm of bacterial gene control mechanisms. *Mol. Microbiol.* 29:677–684.

26. **DuPont, H. L.** 1995. *Shigella* species (bacillary dysentery), p. 2033–2039. *In* G. L. Mandell, J. E. Bennett, and R. Dolin (ed.), *Principles and Practice of Infectious Diseases.* Churchill Livingstone, Inc., New York, N.Y.

27. **DuPont, H. L., M. M. Levine, R. B. Hornick, and S. B. Formal.** 1989. Inoculum size in shigellosis and implications for expected mode of transmission. *J. Infect. Dis.* 159:1126–1128.

28. **Edwards, P. R., and W. H. Ewing.** 1972. *Identification of Enterobacteriaceae.* Burgess Publishing, Co., Minneapolis, Minn.

29. **Egile, C. H. d'Hauteville, C. Parsot, and P. J. Sansonetti.** 1997. SopA, the outer membrane protease responsible for polar localization of IcsA in *Shigella flexneri*. *Mol. Microbiol.* 23:1063–1073.

30. **Falkow, S., H. Schneider, L. Baron, and S. B. Formal.** 1963. Virulence of *Escherichia-Shigella* genetic hybrids for the guinea pig. *J. Bacteriol.* 86:1251–1258.

31. **Fasano, A., F. R. Noriega, D. R. Maneval, Jr., S. Chanasongcram, R. Russell, S. Guandalini, and M. M. Levine.** 1995. *Shigella* enterotoxin 1: an enterotoxin of *Shigella flexneri* 2a active in rabbit small intestine in vivo and in vitro. *J. Clin. Invest.* 95:2853–2861.

32. **Fontaine, A., J. Arondel, and P. J. Sansonetti.** 1988. Role of the Shiga toxin in the pathogenesis of bacillary dysentery studied by using a Tox⁻ mutant of *Shigella dysenteriae* 1. *Infect. Immun.* 56:3099–3109.

33. **Formal, S. B., E. H. LaBrec, T. H. Kent, and S. Falkow.** 1965. Abortive intestinal infection with an *Escherichia coli-Shigella flexneri* hybrid strain. *J. Bacteriol.* 89:1374–1382.

34. **Frost, J. A., M. B. McEvoy, C. A. Bentley, Y. Andersson, and B. Rowe.** 1995. An outbreak of *Shigella sonnei* infection associated with consumption of iceberg lettuce. *Emerg. Infect. Dis.* 1:26–29.

35. Galán, J. E., and P. J. Sansonetti. 1995. Molecular and cellular bases of *Salmonella* and *Shigella* interactions with host cells, p. 2757–2773. *In* F. C. Neidhardt, R. Curtiss III, C. A. Gross, J. I. Ingraham, E. C. C. Lin, K. B. Low, Jr., B. Magasanik, W. Reznikoff, M. Riley, M. Schaechter, and H. E. Umbarger (ed.), *Escherichia coli and Salmonella: Cellular and Molecular Biology*, 2nd ed. American Society for Microbiology, Washington, D.C.

36. Gallegos, M. T., C. Michan, and J. L. Ramos. 1993. The XylS/AraC family of regulators. *Nucleic Acids Res.* 21:807–810.

37. Gessner, B. D., and M. Beller. 1994. Moose soup shigellosis in Alaska. *West. J. Med.* 160:430–433.

38. Goldberg, M. B., O. Barzu, C. Parsot, and P. J. Sansonetti. 1993. Unipolar localization and ATPase activity of IcsA, a *Shigella flexneri* protein involved in intracellular movement. *J. Bacteriol.* 175:2189–2196.

39. Goma Epidemiology Group. 1995. Public health impact of Rwandan refugee crisis: what happened in Goma, Zaire, in July, 1994? *Lancet* 345:339–344.

40. Groisman, E. A., and H. Ochman. 1993. Cognate gene clusters govern invasion of host epithelial cells by *Salmonella typhimurium* and *Shigella flexneri*. *EMBO J.* 12:3779–3787.

41. Guerrero, L., J. J. Calva, A. L. Morrow, F. R. Velazquez, F. Tuz-Dzib, Y. Lopez-Vidal, H. Ortega, H. Arroyo, T. G. Cleary, L. K. Pickering, and G. M. Ruiz-Palacios. 1994. Asymptomatic *Shigella* infections in a cohort of Mexican children younger than two years of age. *Pediatr. Infect. Dis. J.* 13:597–602.

42. Hall, P. A. 1994. Scope for rapid microbiological methods in modern food production, p. 255–267. *In* P. Patel (ed.), *Rapid Analysis Techniques in Food Microbiology*. Blackie Academic & Professional, New York, N.Y.

43. Haltalin, K., J. Nelson, and R. Ring. 1967. Double-blind treatment study of shigellosis comparing ampicillin, sulfadiazone and placebo. *J. Pediatr.* 70:970–981.

44. Hedberg, C. W., W. C. Levine, K. E. White, R. H. Carlson, D. K. Winsor, D. N. Cameron, K. L. MacDonald, and M. T. Osterholm. 1992. An international foodborne outbreak of shigellosis associated with a commercial airline. *JAMA* 268:3208–3212.

45. High, N., J. Mounier, M. C. Prevost, and P. J. Sansonetti. 1992. IpaB of *Shigella flexneri* causes entry into epithelial cells and escape from the phagocytic vacuole. *EMBO J.* 11:1991–1999.

46. Hoge, C. W., J. M. Gambel, A. Srijan, C. Pitarangsi, and P. Echeverria. 1998. Trends in antibiotic resistance among diarrheal pathogens isolated in Thailand over 15 years. *Clin. Infect. Dis.* 26:341–345.

47. Hossain, M. A., K. Z. Hasan, and M. J. Albert. 1994. *Shigella* carriers among nondiarrhoeal children in an endemic area of shigellosis in Bangladesh. *Trop. Geogr. Med.* 46:40–42.

48. Hyams, K. C., A. L Bourgeois, B. R. Merrell, P. Rozmajzl, J. Escamilla, S. A. Thornton, G. M. Wasserman, A. Burke, P. Echeverria, K. Y. Green, A. Z. Kapikian, and J. N. Woody. 1991. Diarrheal disease during Operation Desert Shield. *N. Engl. J. Med.* 325:1423–1428.

49. International Commission on Microbiological Spec-

ifications for Foods. 1996. *Shigella*, p. 280–298. *In Microorganisms in Foods*, vol. 5. *Microbiological Specification of Food Pathogens*. Blackie Academic & Professional, New York, N.Y.

50. Kapperud, G., L. M. Rorvik, V. Hasseltvedt, E. A. Hoiby, B. G. Iversen, K. Staveland, G. Johnsen, J. Leitao, H. Herikstad, Y. Andersson, G. Langeland, B. Gondrosen, and J. Lassen. 1995. Outbreak of *Shigella sonnei* infection traced to imported iceberg lettuce. *J. Clin. Microbiol.* 33:609–614.

51. Karmali, M. A., M. Petric, C. Lim, P. C. Fleming, G. S. Arbus, and H. Lior. 1985. The association between idiopathic hemolytic syndrome and infection by Verotoxin-producing *Escherichia coli*. *J. Infect. Dis.* 151:775–782.

52. Kennedy, F. M., J. Astbury, J. R. Needham, and T. Cheasty. 1993. Shigellosis due to occupational contact with non-human primates. *Epidemiol. Infect.* 110:247–257.

53. Kinsey, M. D., S. B. Formal, G. J. Dammin, and R. A. Giannella. 1976. Fluid and electrolyte transport in rhesus monkeys challenged intracecally with *Shigella flexneri* 2a. *Infect. Immun.* 14:368–371.

54. Kolavic, S. A., A. Kimura, S. L. Simons, L. Slutsker, S. Barth, and C. E. Haley. 1997. An outbreak of *Shigella dysenteriae* type 2 among laboratory workers due to intentional food contamination. *JAMA* 278:396–398.

55. Kotloff, K. L., J. P. Winickoff, B. Ivanoff, J. D. Clemens, D. L. Swerdlow, P. J. Sansonetti, G. K. Adak, and M. M. Levine. 1999. Global burden of *Shigella* infections: implications for vaccine development and implementation of control strategies. *Bull. W. H. O.* 77:651–666.

56. LaBrec, E. H., H. Schneider, T. J. Magnani, and S. B. Formal. 1964. Epithelial cell penetration as an essential step in the pathogenesis of bacillary dysentery. *J. Bacteriol.* 88:1503–1518.

57. Lawlor, K. M., P. A. Daskeleros, R. E. Robinson, and S. M. Payne. 1987. Virulence of iron-transport mutants of *Shigella flexneri* and utilization of host iron compounds. *Infect. Immun.* 55:594–599.

58. Lee, L. A., S. M. Ostroff, H. B. McGee, D. R. Johnson, F. P. Downes, D. N. Cameron, N. H. Bean, and P. M. Griffin. 1991. An outbreak of shigellosis at an outdoor music festival. *Am. J. Epidemiol.* 133:608–615.

59. Levine, M. M., and O. S. Levine. 1994. Changes in human ecology and behavior in relation to the emergence of diarrheal diseases, including cholera. *Proc. Natl. Acad. Sci. USA* 91:2390–2394.

60. Lew, J. F., D. L. Swerdlow, M. E. Dance, P. M. Griffin, C. A. Bopp, M. J. Gillenwater, T. Mercatante, and R. I. Glass. 1991. An outbreak of shigellosis aboard a cruise ship caused by a multiple-antibiotic-resistant strain of *Shigella flexneri*. *Am. J. Epidemiol.* 134:413–420.

61. Lopez, E. L., M. Diaz, S. Grinstein, S. Dovoto, F. Mendila-Harzu, B. E. Murray, S. Ashkenazi, E. Rubeglio, M. Woloj, M. Vasquez, M. Turco, L. K. Pickering, and T. G. Cleary. 1989. Hemolytic uremic syndrome and diarrhea in Argentine children: the role of Shiga-like toxins. *J. Infect. Dis.* 160:469–475.

62. Makino, S., C. Sasakawa, K. Kamata, T. Kurata, and M. Yoshikawa. 1986. A genetic determinant required for

continuous reinfection of adjacent cells on large plasmid in *Shigella flexneri* 2a. *Cell* **46**:551–555.

63. **Maurelli, A. T.** 1989. Temperature regulation of virulence genes in pathogenic bacteria: a general strategy for human pathogens? *Microb. Pathog.* **7**:1–10.

64. **Maurelli, A. T., B. Baudry, H. d'Hauteville, T. L. Hale, and P. J. Sansonetti.** 1985. Cloning of virulence plasmid DNA sequences involved in invasion of HeLa cells by *Shigella flexneri. Infect. Immun.* **49**:164–171.

65. **Maurelli, A. T., B. Blackmon, and R. Curtiss III.** 1984. Temperature-dependent expression of virulence genes in *Shigella* species. *Infect. Immun.***43**:195–201.

66. **Maurelli, A. T., R. E. Fernández, C. A. Bloch, C. K. Rode, and A. Fasano.** 1998. "Black holes" and bacterial pathogenicity: a large genomic deletion that enhances the virulence of *Shigella* spp. and enteroinvasive *Escherichia coli. Proc. Natl. Acad. Sci. USA* **95**:3943–3948.

67. **Maurelli, A. T., and K. A. Lampel.** 2001. *Shigella,* p. 323–343. *In* Y. H. Hui, M. D. Pierson, and J. R. Gorman (ed.), *Foodborne Disease Handbook,* vol. I, 2nd ed. Marcel Dekker, Inc., New York, N.Y.

68. **Maurelli, A. T., and P. J. Sansonetti.** 1988. Identification of a chromosomal gene controlling temperature regulated expression of *Shigella* virulence. *Proc. Natl. Acad. Sci. USA* **85**:2820–2824.

69. **Mead, P. S., L. Slutsker, V. Dietz, L. F. McCaig, J. S. Bresee, C. Shapiro, P. M. Griffin, and R. V. Tauxe.** 1999. Food-related illness and death in the United States. *Emerg. Infect. Dis.* **5**:607–625.

70. **Ménard, R., M.-C. Prévost, P. Gounon, P. Sansonetti, and C. Dehio.** 1996. The secreted Ipa complex of *Shigella flexneri* promotes entry into mammalian cells. *Proc. Natl. Acad. Sci. USA* **93**:1254–158.

71. **Ménard, R., P. J. Sansonetti, and C. Parsot.** 1993. Nonpolar mutagenesis of the *ipa* genes defines IpaB, IpaC, and IpaD as effectors of *Shigella flexneri* entry into epithelial cells. *J. Bacteriol.* **175**:5899–5906.

72. **Ménard, R., P. J. Sansonetti, and C. Parsot.** 1994. The secretion of the *Shigella flexneri* Ipa invasins is induced by the epithelial cell and controlled by IpaB and IpaD. *EMBO J.* **13**:5293–5302.

73. **Ménard, R., P. J. Sansonetti, C. Parsot, and T. Vasselon.** 1994. The IpaB and IpaC invasins of *Shigella flexneri* associate in the extracellular medium and are partitioned in the cytoplasm by a specific chaperon. *Cell* **76**:829–839.

74. **Moss, J. E., T. J. Cardozo, A. Zychlinsky, and E. A. Groisman.** 1999. The *selC*-associated SHI-2 pathogenicity island of *Shigella flexneri. Mol. Microbiol.* **33**:74–83.

75. **Nakata, N., T. Tobe, I. Fukuda, T. Suzuki, K. Komatsu, M. Yoshikawa, and C. Sasakawa.** 1993. The absence of a surface protease, OmpT, determines the intercellular spreading ability of *Shigella*: the relationship between the *ompT* and *kcpA* loci. *Mol. Microbiol.* **9**:459–468.

76. **Nassif, X., M. C. Mazert, J. Mounier, and P. J. Sansonetti.** 1987. Evaluation with an *iuc*::Tn*10* mutant of the role of aerobactin production in the virulence of *Shigella flexneri. Infect. Immun.* **55**:1963–1969.

77. **Nataro, J. P., J. Seriwatana, A. Fasano, D. R. Maneval, L. D. Guers, F. Noriega, F. Dubovsky, M. M. Levine,**

and J. G. Morris, Jr. 1995. Identification and cloning of a novel plasmid-encoded enterotoxin of enteroinvasive *Escherichia coli* and *Shigella* strains. *Infect. Immun.* **63**:4721–4728.

78. **Oaks, E. V., T. L. Hale, and S. B. Formal.** 1986. Serum immune response to *Shigella* protein antigens in rhesus monkeys and humans infected with *Shigella* spp. *Infect. Immun.* **53**:57–63.

79. **O'Brien, A. D., and R. K. Holmes.** 1995. Protein toxins of *Escherichia coli* and *Salmonella,* p. 2788–2802. *In* F. C. Neidhardt, R. Curtiss III, C. A. Gross, J. I. Ingraham, E. C. C. Lin, K. B. Low, Jr., B. Magasanik, W. Reznikoff, M. Riley, M. Schaechter, and H. E. Umbarger (ed.), *Escherichia coli and Salmonella: Cellular and Molecular Biology,* 2nd ed. American Society for Microbiology, Washington, D.C.

80. **Ochman, H., T. S. Whittam, D. A. Caugant, and R. K. Selander.** 1983. Enzyme polymorphism and genetic population structure in *Escherichia coli* and *Shigella. J. Gen. Microbiol.* **129**:2715–2726.

81. **O'Connell, C. M. C., R. C. Sandlin, and A. T. Maurelli.** 1995. Signal transduction and virulence gene regulation in *Shigella* spp.: temperature and (maybe) a whole lot more, p. 111–127. *In* R. Rappuoli (ed.), *Signal Transduction and Bacterial Virulence.* R. G. Landes Company, Austin, Tex.

82. **Parsot, C., R. Ménard, P. Giunon, and P. J. Sansonetti.** 1995. Enhanced secretion through the *Shigella flexneri* Mxi-Spa translocon leads to assembly of extracellular proteins into macromolecular structures. *Mol. Microbiol.* **16**:291–300.

83. **Parsot, C., and P. J. Sansonetti.** 1996. Invasion and the pathogenesis of *Shigella* infections. *Curr. Top. Microbiol. Immunol.* **209**:25–42.

84. **Pickering, L. K., A. V. Bartlett, and W. E. Woodward.** 1986. Acute infectious diarrhea among children in day-care: epidemiology and control. *Rev. Infect. Dis.* **8**:539–547.

85. **Raghupathy, P., A. Date, J. C. M. Shastry, A. Sudarsanam, and M. Jadhav.** 1978. Haemolytic-uraemic syndrome complicating shigella dysentery in south Indian children. *Br. Med. J.* **1**:1518–1521.

86. **Rajakumar, K., C. Sasakawa, and B. Adler.** 1997. Use of a novel approach, termed island probing, identifies the *Shigella flexneri she* pathogenicity island which encodes a homolog of the immunoglobulin A protease-like family of proteins. *Infect. Immun.* **65**:4606–4614.

87. **Roberts, T.** 1989. Human illness costs of foodborne bacteria. *Am. J. Agric. Econ.* **71**:468–474.

88. **Rout, W. R., S. B. Formal, R. A. Giannella, and G. J. Dammin.** 1975. Pathophysiology of *Shigella* diarrhea in the rhesus monkey: intestinal transport, morphological, and bacteriological studies. *Gastroenterology* **68**:270–278.

89. **Sakai, T., C. Sasakawa, and M. Yoshikawa.** 1988. Expression of four virulence antigens of *Shigella flexneri* is positively regulated at the transcriptional level by the 30 kilodalton *virF* protein. *Mol. Microbiol.* **2**:589–597.

90. Sandlin, R. C., K. A. Lampel, S. P. Keasler, M. B. Goldberg, A. L. Stolzer, and A. T. Maurelli. 1995. Avirulence of rough mutants of *Shigella flexneri*: requirement of O-antigen for correct unipolar localization of IcsA in bacterial outer membrane. *Infect. Immun.* **63**:229–237.

91. Sandlin, R. C., and A. T. Maurelli. 1999. Establishment of unipolar localization of IcsA in *Shigella flexneri* 2a is not dependent on virulence plasmid determinants. *Infect. Immun.* **67**:350–356.

92. Sansonetti, P. J., H. d'Hauteville, C. Ecobichon, and C. Pourcel. 1983. Molecular comparison of virulence plasmids in *Shigella* and enteroinvasive *Escherichia coli*. *Ann. Inst. Pasteur Microbiol.* **134A**:295–318.

93. Sansonetti, P. J., T. L. Hale, and E. V. Oaks. 1985. Genetics of virulence in enteroinvasive *Escherichia coli*, p. 74–77. *In* D. Schlessinger (ed.), *Microbiology—1985*. American Society for Microbiology, Washington, D.C.

94. Sansonetti, P. J., D. J. Kopecko, and S. B. Formal. 1981. *Shigella sonnei* plasmids: evidence that a large plasmid is necessary for virulence. *Infect. Immun.* **34**:75–83.

95. Sansonetti, P. J., D. J. Kopecko, and S. B. Formal. 1982. Involvement of a plasmid in the invasive ability of *Shigella flexneri*. *Infect. Immun.* **35**:852–860.

96. Sasakawa, C., K. Komatsu, T. Tobe, T. Suzuki, and M. Yoshikawa. 1993. Eight genes in region 5 that form an operon are essential for invasion of epithelial cells by *Shigella flexneri* 2a. *J. Bacteriol.* **175**:2334–2346.

97. Schuch, R., and A. T. Maurelli. 2000. The type III secretion pathway: dictating the outcome of bacterial-host interactions, p. 203–223. *In* K. A. Brogden, J. A. Roth, T. B. Stanton, C. A. Bolin, F. C. Minion, and M. J. Wannemuehler (ed.), *Virulence Mechanisms of Bacterial Pathogens*, 3rd ed. American Society for Microbiology, Washington, D.C.

98. Sharp, T. W., S. A. Thornton, M. R. Wallace, R. F. Defraites, J. L. Sanchez, R. A. Batchelor, P. J. Rozmajzl, R. K. Hanson, P. Echeverria, A. Z. Kapikian, X. J. Xiang, M. K. Estes, J. P. Burans. 1995. Diarrheal disease among military personnel during Operation Restore Hope, Somalia, 1992–1993. *Am J. Trop. Med. Hyg.* **52**:188–193.

99. Shere, K. D., S. Sallustio, A. Manessis, T. G. D'Aversa, and M. B. Goldberg. 1997. Disruption of IcsP, the major *Shigella* protease that cleaves IcsA, accelerates actin-based motility. *Mol. Microbiol.* **25**:451–462.

100. Simon, D. G., R. A. Kaslow, J. Rosenbaum, R. L. Kaye, and A. Calin. 1981. Reiter's syndrome following epidemic shigellosis. *J. Rheumatol.* **8**:969–973.

101. Small, P., D. Blankenhorn, D. Welty, E. Zinser, J. L. Slonczewski. 1994. Acid and base resistance in *Escherichia coli* and *Shigella flexneri*: role of *rpoS* and growth pH. *J. Bacteriol.* **176**:1729–1737.

102. Smith, J. L. 1987. *Shigella* as a foodborne pathogen. *J. Food Prot.* **50**:788–801.

103. Snyder, O. P. 1992. HACCP—an industry food safety self-control program. Part IV. *Dairy Food Environ. Sanit.* **12**:230–232.

104. Tesh, V. L., and A. D. O'Brien. 1991. The pathogenic mechanisms of Shiga toxin and the Shiga-like toxins. *Mol. Microbiol.* **5**:1817–1822.

105. Tobe, T., S. Nagai, N. Okada, B. Adler, M. Yoshikawa, and C. Sasakawa. 1991. Temperature-regulated expression of invasion genes in *Shigella flexneri* is controlled through the transcriptional activation of the *virB* gene on the large plasmid. *Mol. Microbiol.* **5**:887–893.

106. Tobe, T., M. Yoshikawa, T. Mizuno, and C. Sasakawa. 1993. Transcriptional control of the invasion regulatory gene *virB* of *Shigella flexneri*: activation by VirF and repression by H-NS. *J. Bacteriol.* **175**:6142–6149.

107. Tran Van Nhieu, G., A. Ben-Ze'ev, and P. J. Sansonetti. 1997. Modulation of bacterial entry into epithelial cells by association between vinculin and the *Shigella* IpaA invasin. *EMBO J.* **16**:2717–2729.

108. Tran Van Nhieu, G., E. Caron, A. Hall, and P. J. Sansonetti. 1999. IpaC induces actin polymerization and filopodia formation during *Shigella* entry into epithelial cells. *EMBO J.* **18**:3249–3262.

109. Van Gijsegem, F., S. Genin, and C. Boucher. 1993. Conservation of secretion pathways for pathogenicity determinants of plant and animal bacteria. *Trends Microbiol.* **1**:175–180.

110. Venkatesan, M. M., J. M. Buysse, and E. V. Oaks. 1992. Surface presentation of *Shigella flexneri* invasion plasmid antigens requires the products of the *spa* locus. *J. Bacteriol.* **174**:1990–2001.

111. Vokes, S. A., S. A. Reeves, A. G. Torres, and S. M. Payne. 1999. The aerobactin iron transport system genes in *Shigella flexneri* are present within a pathogenicity island. *Mol. Microbiol.* **33**:63–73.

112. Watarai, M., T. Tobe, M. Yoshikawa, and C. Sasakawa. 1995. Contact of *Shigella* with host cells triggers release of Ipa invasins and is an essential function of invasiveness. *EMBO J.* **14**:2461–2470.

113. Watarai, M., S. Funato, and C. Sasakawa. 1996. Interaction of Ipa proteins of *Shigella flexneri* with $\alpha 5$-$\beta 1$ integrin promotes entry of the bacteria into mammalian cells. *J. Exp. Med.* **183**:991–999.

114. Weissman, J. B., A. Schmerler, P. Weiler, G. Filice, N. Godby, and I. Hansen. 1974. The role of preschool children and day-care centers in the spread of shigellosis in urban communities. *J. Pediatr.* **84**:797–802.

115. Wetherington, J., J. L. Bryant, K. A. Lampel, and J. M. Johnson. 2000. PCR screening and isolation of *Shigella sonnei* from layered party dip. *Lab. Inform. Bull.* **16**(4215):1–8.

116. Wharton, M., R. A. Spiegel, J. M. Horan, R. V. Tauxe, J. G. Wells, N. Barg, J. Herndon, R. A. Meriwether, J. N. MacCormack, and R. H. Levine. 1990. A large outbreak of antibiotic-resistant shigellosis at a mass gathering. *J. Infect. Dis.* **162**:1324–1328.

117. Yagupsky, P., M. Loeffelholz, K. Bell, and M. A. Menegus. 1991. Use of multiple markers for investigation of an epidemic of *Shigella sonnei* infections in Monroe County, New York. *J. Clin. Microbiol.* **29**:2850–2855.

118. Zychlinsky, A., B. Kenny, R. Ménard, M. C. Prevost, I. B. Holland, and P. J. Sansonetti. 1994. IpaB mediates macrophage apoptosis induced by *Shigella flexneri*. *Mol. Microbiol.* **11**:619–627.

Food Microbiology: Fundamentals and Frontiers, 2nd Ed.
Edited by M. P. Doyle et al.
© 2001 ASM Press, Washington, D.C.

James D. Oliver
James B. Kaper

13

Vibrio Species

Whereas the 8th edition of *Bergey's Manual* listed five *Vibrio* species, with two recognized as human pathogens, over 20 species have now been described, including at least 12 capable of causing infection in humans. A number of excellent reviews have been published about pathogenic vibrios (12, 62, 67, 70, 94, 104, 115, 158, 162, 169), although with the exception of *V. cholerae* and *V. parahaemolyticus*, little is known about the virulence mechanisms they employ. Of the 12 pathogens, 8 are known to be directly food associated, and they are the subject of this review. (*V. carchariae* and *V. damsela* infections appear to result solely from wound infections, and the route of infection for *V. metschnikovii* and *V. cincinnatiensis* is unclear.)

INCIDENCE OF VIBRIOS IN SEAFOOD

One of the most consistent aspects of vibrio infections is a recent history of seafood consumption. Vibrios, which are generally the predominant bacterial genus in estuarine waters, are associated with a great variety of seafoods. To cite a few recent studies, Prasad and Rao (127) determined that 30% of 129 samples of fresh, frozen, and iced prawns were contaminated with vibrios, including *V. parahaemolyticus*, *V. vulnificus*, *V. metschnikovii*, *V. cholerae* non-O1, and *V. fluvialis*. In another study, potentially pathogenic vibrios were present on 13 to 36% of shrimp and fish samples assayed and also were present in high numbers in freshwater clams (170, 171). Studies by Buck (16) revealed 36 to 60% of finfish and shellfish obtained from supermarkets contained *Vibrio* spp., with *V. parahaemolyticus* and *V. alginolyticus* most commonly isolated. Vibrios are most frequently isolated from molluscan shellfish during the summer months. Lowry et al. (92) determined that 100% of raw oysters they examined were contaminated with *V. parahaemolyticus*, and 67% contained *V. vulnificus*. Interestingly, 50% of cooked oysters tested also contained *V. parahaemolyticus*, and 25% contained *V. vulnificus*. A recent survey of frozen raw shrimp imported from Mexico, China, and Ecuador revealed that more than 63% were contaminated with *Vibrio* species, including *V. vulnificus* and *V. parahaemolyticus* (8). A study of the distribution of vibrios in oysters (*Crassostrea virginica*) originating from the coast of Brazil revealed *Vibrio* sp. contamination in the following order of incidence:

James D. Oliver, Department of Biology, University of North Carolina at Charlotte, 9201 University City Blvd., Charlotte, NC 28223.
James B. Kaper, Center for Vaccine Development, Department of Microbiology and Immunology, University of Maryland School of Medicine, 685 West Baltimore St., Baltimore, MD 21201.

V. alginolyticus (81%), *V. parahaemolyticus* (77%), *V. cholerae* non-O1 (31%), *V. fluvialis* (27%), *V. furnissii* (19%), *V. mimicus* (12%), and *V. vulnificus* (12%) (96).

ISOLATION

Various enrichment broths have been described for the isolation of vibrios, although alkaline peptone water remains the most commonly used (118). These broths are frequently coupled with thiosulfate-citrate-bile salts-sucrose (TCBS) agar or other plating media. With sucrose as a differentiating trait, the 12 human pathogenic vibrios can be separated on TCBS into 6 that are generally sucrose positive (*V. cholerae*, *V. metschnikovii*, *V. fluvialis*, *V. furnissii*, *V. alginolyticus*, and *V. carchariae*) and 5 that are generally sucrose negative (*V. mimicus*, *V. hollisae*, *V. damsela*, *V. parahaemolyticus*, and *V. vulnificus*). Most vibrios grow well on TCBS agar, although *V. hollisae* grows very poorly or not at all on this medium, which is also the case for *V. damsela* when incubated at 37°C (40). Further, the oxidase test, a crucial test for distinguishing vibrios from the *Enterobacteriaceae*, may give erroneous results when colonies are obtained directly from TCBS agar. Therefore, sucrose-positive colonies on TCBS should be subcultured by heavy inoculation onto a nonselective medium, such as blood agar, and allowed to grow for 5 to 8 h before being tested for oxidase activity. For a recent review of enrichment and plating media for vibrios, see reference 118.

IDENTIFICATION

Salient differentiating features of the 12 vibrios currently recognized as human pathogens and those tests that are of special significance in differentiating these pathogens are provided in Tables 13.1 and 13.2, respectively. However, although the determination of phenotypic traits for the speciation of vibrios remains routine, identifications based on these classic methods have always been problematic. Some researchers have developed immunological assays to overcome this problem, and molecular techniques to identify the presence of vibrios in foods are becoming more common and are proving to be a very powerful adjunct to more traditional taxonomic methods.

EPIDEMIOLOGY

The cell populations of most vibrios in both surface waters and shellfish correlate with seasonality, generally being greater during the warm weather months between April and October. Similarly, vibrios are more commonly isolated from the warmer waters of the Gulf and East Coasts than those of the West and Pacific Northwest. Seasonality is most notable for *V. vulnificus* and *V. parahaemolyticus* infections, whereas those of some vibrios, such as *V. fluvialis*, occur throughout the year.

There is considerable variation in the severity of the various vibrio diseases, and even with infections caused by the same *Vibrio* species; the most severely ill patients generally suffer preexisting underlying illnesses, with chronic liver disease being one of the most common. An exception to this generalization is *V. cholerae* O1/O139, which can readily cause illness in noncompromised individuals. Most cases are associated with a recent history of seafood consumption, especially eating raw oysters (34). *V. cholerae* O1/O139 is also exceptional in having a broad range of vehicles of infection, although seafood is important in the transmission of many cases of cholera. A 10-year compilation of vibrio isolates recovered in Alabama indicates that 87% of patients had a recent history of seawater-associated activity (13). A study by Lowry et al. (92) revealed that all of 51 persons with diarrhea who had a *Vibrio* species present in a stool specimen had eaten raw or cooked seafood. Unfortunately, because vibrios are part of the normal estuarine microflora and not a result of fecal contamination, vibrio infections will not likely be controlled through shellfish sanitation programs (12, 169). It is thus essential that raw seafood be adequately refrigerated or iced to prevent significant bacterial growth, and be properly cooked before consumption.

INCIDENCE OF VIBRIO INFECTIONS

The routine use of TCBS agar for screening stool samples for vibrios generally has revealed a low incidence of vibrio infections. For example, a 15-year survey of stool specimens obtained in a hospital adjacent to the Chesapeake Bay of Maryland and assayed with TCBS agar resulted in only 40 vibrio isolations from 32 patients, for an overall incidence of 1.6 isolations per 100,000 patients per year (59). Similarly, vibrios were isolated from only 55 (17%) of 3,250 diarrheal stools received between 1989 and 1991 at a clinical laboratory in Brazil using enrichment in APW supplemented with 2% NaCl before culture on TCBS (95). Bonner et al. (13) concluded that vibrios were not a major cause of infectious diarrhea after identifying only one case of vibrio gastroenteritis during the 4 years TCBS was used to isolate vibrios from stool specimens in a Gulf Coast survey. However, the routine use of TCBS for screening stool and wound samples for vibrios may be warranted in some

Table 13.1 Biochemical and other characteristics of 12 *Vibrio* species found in human clinical specimens[a]

	% Positive[b]											
Test	*V. cholerae*	*V. mimicus*	*V. metschnikovii*	*V. cincinnatiensis*	*V. hollisae*	*V. damsela*	*V. fluvialis*	*V. furnissii*	*V. alginolyticus*	*V. parahaemolyticus*	*V. vulnificus*	*V. carchariae*
Indole production (HIB, 1% NaCl)	99	98	20	8	97	0	13	11	85	98	97	100
Methyl red (1% NaCl)	99	99	96	93	0	100	96	100	75	80	80	100
Voges-Proskauer[c] (1% NaCl, Barritt)	75	9	96	0	0	95	0	0	95	0	0	50
Citrate, Simmons	97	99	75	21	0	0	93	100	1	3	75	0
H$_2$S on TSI	0	0	0	0	0	0	0	0	0	0	0	0
Urea hydrolysis	0	1	0	0	0	0	0	0	0	15	1	0
Phenylalanine deaminase	0	0	0	0	0	0	0	0	1	1	35	NG
Arginine, Moeller[c] (1% NaCl)	0	0	60	0	0	95	93	100	0	0	0	0
Lysine, Moeller[c] (1% NaCl)	99	100	35	57	0	50	0	0	50	95	55	0
Ornithine, Moeller[c] (1% NaCl)	99	99	0	0	0	0	0	0	50	95	55	0
Motility (36°C)	99	98	74	86	0	25	70	89	99	99	99	0
Gelatin hydrolysis (1% NaCl, 22°C)	90	65	65	0	0	6	85	86	90	95	75	0
KCN test (% that grow)	10	2	0	0	0	5	65	89	15	20	1	0
Malonate utilization	1	0	0	0	0	0	0	11	0	0	0	0
D-Glucose acid production[c]	100	100	100	100	100	100	100	100	100	100	100	50
D-Glucose gas production[c]	0	0	0	0	0	10	0	100	0	0	0	0
Acid production from:												
D-Adonitol	0	0	0	0	0	0	0	0	1	0	0	0
L-Arabinose[c]	0	1	0	100	97	0	93	100	1	1	80	0
D-Arabitol[c]	0	0	0	0	0	0	65	89	0	0	0	0
Cellobiose[c]	8	0	9	100	0	0	30	11	3	5	99	50
Dulcitol	0	0	0	0	0	0	0	0	0	3	0	0
Erythritol	0	0	0	0	0	0	0	0	0	0	0	0
D-Galactose	90	82	45	100	100	90	96	100	20	92	96	0
Glycerol	30	13	100	100	0	0	7	55	80	50	1	0
myo-Inositol	0	0	40	100	0	0	0	0	0	0	0	0
Lactose[c]	7	21	50	0	0	0	3	0	0	1	85	0
Maltose[c]	99	99	100	100	0	100	100	100	100	99	100	100
D-Mannitol[c]	99	99	96	100	0	0	97	100	100	100	45	50
D-Mannose	78	99	100	100	100	100	100	100	99	100	98	50
Melibiose	1	0	0	7	0	0	3	11	1	1	40	0
α-Methyl-D-glucoside	0	0	25	57	0	5	0	0	1	0	0	0
Raffinose	0	0	0	0	0	0	0	11	0	0	0	0
D-Rhamnose	0	0	0	0	0	0	0	45	0	1	0	0
Salicin[c]	1	0	9	100	0	0	0	0	4	1	95	0
D-Sorbitol	1	0	45	0	0	0	3	0	1	1	0	0
Sucrose[c]	100	0	100	100	0	5	100	100	99	1	15	50
Trehalose	99	94	100	100	0	86	100	100	100	99	100	50
D-Xylose	0	0	0	43	0	0	0	0	0	0	0	0
Mucate, acid production	1	0	0	0	0	0	0	0	0	0	0	0
Tartrate, Jordan	75	12	35	0	65	0	35	22	95	93	84	50
Esculin hydrolysis	0	0	60	0	0	0	8	0	3	1	40	0
Acetate utilization	92	78	25	14	0	0	70	65	0	1	7	0
Nitrate→nitrate[c]	99	100	0	100	100	100	100	100	100	100	100	100

(Continued)

Table 13.1 Biochemical and other characteristics of 12 *Vibrio* species found in human clinical specimens[a] *(Continued)*

Test	% Positive[b]											
	V. cholerae	*V. mimicus*	*V. metschnikovii*	*V. cincinnatiensis*	*V. hollisae*	*V. damsela*	*V. fluvialis*	*V. furnissii*	*V. alginolyticus*	*V. parahaemolyticus*	*V. vulnificus*	*V. carchariae*
Oxidase[c]	100	100	0	100	100	95	100	100	100	100	100	100
DNase, 25°C	93	55	50	79	0	75	100	100	95	92	50	100
Lipase (corn oil)[c]	92	17	100	36	0	0	90	89	85	90	92	0
ONPG test[c]	94	90	50	86	0	0	40	35	0	5	75	0
Yellow pigment at 25°C	0	0	0	0	0	0	0	0	0	0	0	0
Tyrosine clearing	13	30	5	0	3	0	65	45	70	77	75	0
Growth in nutrient broth with:												
0% NaCl[d]	100	100	0	0	0	0	0	0	0	0	0	0
1% NaCl[d]	100	100	100	100	99	100	99	99	99	100	99	100
6% NaCl[d]	53	49	78	100	83	95	96	100	100	99	65	100
8% NaCl[d]	1	0	44	62	0	0	71	78	94	80	0	0
10% NaCl[d]	0	0	4	0	0	0	4	0	69	2	0	0
12% NaCl[d]	0	0	0	0	0	0	0	0	17	1	0	0
Swarming (marine agar, 25°C)	−	−	−	+	−	−	−	−	+	+	−	100
String test	100	100	100	80	100	80	100	100	91	64	100	100
O/129, zone of inhibition[c]	99	95	90	25	40	90	31	0	19	20	98	100
Polymyxin B, zone of inhibition	22	88	100	92	100	85	100	89	63	54	3	100

[a] HIB, heart infusion broth; 1% NaCl, 1% NaCl added to the standard medium to enhance growth: TSI, triple sugar iron agar.
[b] After 48 h of incubation at 36°C (unless conditions are indicated). Most positive reactions occur during the first 24 h. NG, no growth (probably because NaCl concentration is too low); +, most strains (generally about 90 to 100% positive); −, most strains negative (generally about 0 to 10% positive).
[c] Rest is recommended as part of the routine set for *Vibrio* identification.
[d] Disk potency, 150 μg.

Table 13.2 Eight key differential tests to divide the 12 clinically significant *Vibrio* species into six groups[a]

Test	Reactions of the species in:											
	Group 1		Group 2: *V. metschnikovii*	Group 3: *V. cincinnatiensis*	Group 4: *V. hollisae*	Group 5			Group 6			
	V. cholerae	*V. mimicus*				*V. damsela*	*V. fluvialis*	*V. furnissii*	*V. alginolyticus*	*V. parahaemolyticus*	*V. vulnificus*	*V. carchariae*
Growth in nutrient broth												
With no NaCl added	+	+	−	−	−	−	−	−	−	−	−	−
With 1% NaCl added	+	+	+	+	+	+	+	+	+	+	+	+
Oxidase			−	+	+	+	+	+	+	+	+	+
Nitrate → nitrite		−	−	+	+	+	+	+	+	+	+	+
myo-Inositol fermentation	−	−	−	+	−	−	−	−	−	−	−	−
Arginine dihydrolase					−	+	+	+	−	−	−	−
Lysine decarboxylase					−				+	+	+	+
Ornithine decarboxylase					−							

[a] All data are for reactions within 2 days at 35 to 37°C unless otherwise specified. Symbols; +, most strains (generally about 90 to 100%) positive; −, most strains negative (generally about 0 to 100% positive). Key test result are boxed.

instances. Desenclos et al. (34) reported an overall annual incidence for any vibrio illness of 11.3 per million for raw oyster-eating populations in Florida, but only 2.2 for persons who did not eat raw oysters. These rates were greatly affected by several host factors. For example, the annual incidence rate of vibrio illness for raw oyster eaters with liver disease was 95.4, compared to a rate of 9.2 for those without liver disease who ate raw oysters. Similarly, Levine et al. (87) suggested that, while the reported incidence rate of infections yielding *Vibrio* species is low (0.7/100,000 population) compared to other enteric pathogens (e.g., 7.7/100,000 population for *Shigella* species), the high proportion (45%) of patients with vibrio gastroenteritis who are hospitalized suggests that many milder infections may occur that are not reported. They noted that 86% of patients with *V. fluvialis* and 35% of those with *V. parahaemolyticus* had bloody stools as an additional indication of the severity of these diseases.

SUSCEPTIBILITY TO PHYSICAL/CHEMICAL TREATMENTS

With the exception of *V. cholerae* and *V. parahaemolyticus*, relatively little is known regarding the effect of various food preservation methods on the inactivation of vibrios. Summarized here are some of the studies relevant to this point.

Cold

Although reports generally indicate that vibrios are sensitive to cold, seafoods can be protective for vibrios at refrigeration temperatures. Several psychrotrophic strains of *V. mimicus*, *V. fluvialis*, and *V. parahaemolyticus* have been isolated from seafoods, and they survive well at 4, 10, and 30°C (170). *V. parahaemolyticus* was rapidly inactivated (ca. 2 log CFU/g) when incubated on whole shrimp at 3, 7, 10, or −18°C, although survivors were detected at the end of the 8-day study. *V. parahaemolyticus* survived storage in shellstock oysters for at least 3 weeks at 4°C, and subsequently multiplied after incubation at 35°C for 2 to 3 days (64). Similarly, *V. parahaemolyticus* populations were reduced in cooked fish mince and surimi held at 5°C for 48 h, but growth occurred when the product was held at 25°C. *V. vulnificus* did not grow in oysters held at 13°C and below, but grew well in oysters stored at 18°C and higher (28). Hence, naturally occurring vibrios will likely grow in unchilled shellstock oysters. Similarly, studies involving temperature abuse of octopus, cooked shrimp, and crabmeat have documented growth of *V. parahaemolyticus* to very large numbers when held for even short periods of time under improper refrigeration.

The persistence of *V. vulnificus* in oysters following freezing and storage at −20°C, with or without vacuum packaging, was dependent on the length of frozen storage time for cells packaged without vacuum. Vacuum-packaged oysters had significantly less *V. vulnificus* over a 70-day study period as compared to film-wrapped samples (124). "Individually quick-frozen" oyster technology can extend the shelf life of oysters to more than 18 months, during which time *V. vulnificus* populations decline to nondetectable levels (55).

Heat

All of the vibrios are sensitive to heat, although a wide range of thermal inactivation rates have been reported. Heating of *V. parahaemolyticus* cells at 60, 80, or 100°C for 1 min is lethal to small (5×10^2 CFU/g) populations, although some cells survived heating at 60 and even 80°C for 15 min when populations of 2×10^5 were present (165). Cells did not survive boiling for 1 min, however. D values at 47°C for 52 strains of *V. vulnificus* averaged 78 s (29). Low-temperature pasteurization (e.g., 50°C for 10 min) of oysters reduced *V. vulnificus* by 6 log CFU/g (3). Heating oysters for 10 min in water at 50°C was sufficient to reduce the *V. vulnificus* populations to nondetectable levels (29). Heating a strain of *V. cholerae* O1 inoculated into white shrimp (*Penaeus schimitti*) to boiling resulted in complete inactivation within 1 to 2 min (109). The U.S. Food and Drug Administration (FDA) recommends steaming shellstock oysters, clams, and mussels for 4 to 9 min, frying shucked oysters for 10 min at 375°C, or baking oysters for 10 min at 450°C. Thorough heating of shellfish to an internal temperature of at least 60°C for several minutes should be sufficient to eliminate pathogenic vibrios (169).

Irradiation

Doses of 3 kGy of gamma irradiation eliminated vibrios from frozen shrimp (130). Treatment of fresh and frozen frog legs with >50 krad of ^{60}Co eliminated *V. cholerae* (137). A dose of 1 kGy decreased *V. vulnificus* populations in shellstock oysters by greater than 5 log CFU/g, with no mortality of the oysters. Higher doses, e.g., 1.5 kGy, completely inactivated *V. vulnificus* but increased oyster mortalities by up to 16% (36).

High Pressure

High hydrostatic pressure (30,000 to 50,000 lb/in^2) applied to oysters reduced *V. vulnificus* populations by 6 log CFU/g after 10 min and up to 9 log CFU of *V. parahaemolyticus* cells per g for 30 s (19, 76). Such treatments also killed the oysters.

Miscellaneous Inhibitors

A large variety of dried spices, the oils of several herbs, tomato sauce, and several organic acids are bactericidal to *V. parahaemolyticus*, and many of these substances are highly toxic at low levels (133). *V. parahaemolyticus* is highly sensitive to 50 ppm of butylated hydroxyanisole and is inhibited by 0.1% sorbic acid. Diacetyl is inhibitory to *V. vulnificus* in shellstock oysters (151). Although polyphosphates are highly inhibitory to a variety of foodborne pathogens, we observed (unpublished) that 1% tripolyphosphate has no lethal effect on *V. vulnificus*.

Studies on the effect of pH revealed that vibrios are highly acid sensitive, although growth in media with a pH as low as 4.8 has been reported for *V. parahaemolyticus*.

Depuration

Depuration, wherein filter-feeding bivalves are purified of certain bacteria by pumping bacteria-free water through the bivalves' tissues, is of considerable value in removing *Salmonella* spp. and *Escherichia coli* (17, 131). However, depuration of oysters and clams, by itself, offers little hope of significantly reducing the naturally occurring *Vibrio* microflora present in bivalves (66, 154). In contrast, depuration is effective in eliminating vibrios from artificially infected oysters under laboratory conditions (38, 47, 134). In related studies, Motes et al. (108) determined that relaying oysters to offshore waters (30 to 34‰) resulted in a reduction in *V. vulnificus* levels from 10^3 to 10^4 to <10 most probable number per g within 7 to 17 days.

V. CHOLERAE

V. cholerae O1 is the causative agent of cholera, one of the few foodborne illnesses with epidemic and pandemic potential. *V. cholerae* is a well-defined species on the basis of biochemical tests and DNA homology studies, but as reviewed elsewhere (70), this species is not homogeneous with regard to pathogenic potential. Specifically, important distinctions within the species are made on the basis of production of cholera enterotoxin (cholera toxin, or CT), serogroup, and potential for epidemic spread. Until recently, the public health distinction was simple, that is, *V. cholerae* strains of the O1 serogroup that produced CT were associated with epidemic cholera and all other members of the species were either nonpathogenic or only occasional pathogens. However, with the emergence of cholera due to strains of the O139 serogroup (see below), such previous distinctions

are no longer valid. Two serogroups, O1 and O139, have been associated with epidemic disease, but there are also strains of these serogroups that do not produce CT, do not cause cholera, and are not involved in human illness. Conversely, there are occasional strains of serogroups other than O1 or O139 that are clearly pathogenic, by the production of either CT or other virulence factors (see below); however, none of these other serogroups has caused large epidemics or pandemics. Therefore, in assessing the public health significance of an isolate of *V. cholerae*, there are two critical properties to be determined beyond the biochemical identification of the species *V. cholerae*. The first of these properties is production of CT, which is the toxin responsible for severe, cholera-like disease in epidemic and sporadic forms. The second property is possession of the O1 or O139 antigen, which, since the actual determinant of epidemic or pandemic potential is not known, is at least a marker of such potential. The subject of cholera has been reviewed elsewhere by Kaper et al., and much of this section is drawn from that review (70). Due to space constraints, most primary references for *V. cholerae* O1/O139 will not be repeated here but can be found in this review. Furthermore, unless otherwise stated, all information about this species will refer to *V. cholerae* O1 or O139 strains capable of causing cholera.

Classification

Serogroups

V. cholerae *O1*

V. cholerae strains of the O1 serogroup that produce CT have long been associated with epidemic and pandemic cholera. Strains isolated from environmental samples in nonepidemic areas are usually CT negative and are considered to be nonpathogenic, based on volunteer studies (see "Reservoir"). However, CT-negative *V. cholerae* O1 strains have been isolated from occasional cases of diarrhea or extraintestinal infections. This serogroup can be further subdivided into serotypes of the O1 serogroup called Ogawa and Inaba. *V. cholerae* O1 can also be divided into two biotypes, classical and El Tor, which differ in several characteristics. The El Tor biotype is currently the most important biotype; in recent decades, occasional strains of the classical biotype have only been isolated in Bangladesh.

V. cholerae *Non-O1/Non-O139*

Nearly 200 serogroups of *V. cholerae* have been described. In recent years, until the emergence of the O139 serogroup, all isolates that were identified as *V. cholerae* on the basis of biochemical tests but that were negative for the O1 serogroup were referred to as non-O1 *V. cholerae*. In earlier years, the non-O1 *V. cholerae* was referred to as NCV (non-cholera vibrios) or NAG (nonagglutinable) vibrios. The basis for serotyping in *V. cholerae* is the lipopolysaccharide (LPS) somatic antigen; H antigens are not useful in serotyping. The great majority of these strains do not produce CT and are not associated with epidemic diarrhea (reviewed in reference 103). These strains are occasionally isolated from cases of diarrhea (usually associated with consumption of shellfish) and have been isolated from a variety of extraintestinal infections. They are regularly present in estuarine environments, and infections due to these strains are commonly of environmental origin. While the great majority of these strains do not produce CT, some strains may produce other toxins (see below); however, for many strains of *V. cholerae* non-O1/non-O139 isolated from cases of gastroenteritis, the pathogenic mechanisms are unknown.

V. cholerae *O139 Bengal*

The simple distinction between *V. cholerae* O1 and *V. cholerae* non-O1 was rendered obsolete in early 1993 when the first reports appeared of a new epidemic of severe, cholera-like disease emerging from eastern India and Bangladesh (reviewed in reference 2). Further investigations revealed that this bacterium did not belong to the O serogroups previously described for *V. cholerae* but to a new serogroup, which was given the designation O139 and a synonym "Bengal," in recognition of the origin of this strain. This organism appears to be a hybrid of the O1 strains and the non-O1 strains. In important virulence characteristics, specifically, cholera enterotoxin and toxin coregulated pilus (TCP), *V. cholerae* O139 is indistinguishable from typical El Tor *V. cholerae* O1 strains (2). However, this organism does not produce the O1 LPS and lacks several kilobases of the genetic material necessary for production of the O1 antigen (27). Furthermore, like many strains of non-O1 *V. cholerae* and unlike *V. cholerae* O1, this organism produces a polysaccharide capsule (see below).

Biochemical and Serological Identification

Suspected *V. cholerae* isolates can be transferred from primary isolation plates to a standard series of biochemical media used to identify *Enterobacteriaceae* and *Vibrionaceae*. Both conventional tube tests and commercially available enteric identification systems are suitable for identifying this species. Several key characteristics for distinguishing *V. cholerae* from other species are given in Tables 13.1 and 13.2.

The key confirmation for identification of *V. cholerae* O1 is agglutination in polyvalent antisera raised against the O1 antigen. Polyvalent antiserum for *V. cholerae* O1 and O139 is commercially available and can be used in slide agglutination or coagglutination tests. Monoclonal antibody-based, coagglutination tests suitable for testing isolated colonies or diarrheal stool samples for O1 or O139 are also available commercially (Cholera SMART and Bengal SMART, New Horizons Diagnostics Corp., Columbia, Md.). Oxidase-positive organisms that agglutinate in O1 or O139 antisera can be reported presumptively as *V. cholerae* O1 or O139 and then forwarded to a public health reference laboratory for confirmation. Antisera for serogroups other than O1 or O139 are not commercially available.

DNA Probes and PCR Techniques

Nucleic acid probes are not routinely employed for the identification of *V. cholerae* due to the ease of identifying this species by conventional methods. Where DNA probes have been extremely useful is in distinguishing those strains of *V. cholerae* that contain genes encoding CT (*ctx*) from those that do not contain these genes. This distinction is particularly important in examining environmental isolates of *V. cholerae* because the great majority of these strains lack *ctx* sequences.

A number of DNA fragment probes and synthetic oligonucleotide probes have been developed to detect *ctx* sequences (reviewed in reference 126). An oligonucleotide probe labeled with alkaline phosphatase was reported to identify *ctx*-positive strains in just 3 h from picking the colonies from TCBS agar to final hybridization results. The PCR technique has also been used to detect *ctx* sequences, and PCR has been used to detect toxigenic *V. cholerae* O1 in food samples. A study of fruit, vegetables, and shellfish that had been seeded with *V. cholerae* O1 revealed that amplification of the desired fragment was readily observed with fruit and vegetable samples, but seeded shellfish homogenates often inhibited the PCR assay. By diluting the shellfish homogenate to 1%, the desired fragment was readily amplified (79). PCR was used to investigate a small outbreak of cholera in New York from crabs imported from Ecuador. Although the crabs implicated in the outbreak did not yield *V. cholerae* O1 after culture, the *ctx* PCR technique detected *ctx* sequences in one of four crab samples examined. Restriction fragment length polymorphism (RFLP) analysis of variations in *ctx* genes and in flanking DNA sequences has yielded significant insights into the molecular epidemiology of *V. cholerae*. For example, RFLP analysis demonstrated that toxigenic O1 strains isolated from various states bordering the U.S.

Gulf Coast are clonal. RFLP analysis also revealed that a strain isolated from a patient with cholera in Maryland was identical to isolates from Louisiana and Texas, which concurred with epidemiologic investigations indicating that the crabs eaten by the Maryland patient were harvested along the Texas coast. The most commonly used variation of RFLP currently employed to differentiate strains of *V. cholerae* is pulsed-field gel electrophoresis (PFGE), which has been used in many studies throughout the world. A study by the U.S. Centers for Disease Control and Prevention (CDC) revealed that PFGE could distinguish strains that were identical when examined by multilocus enzyme electrophoresis or ribotyping (20).

Reservoirs

Environment

V. cholerae is part of the normal, free-living (autochthonous) bacterial flora in estuarine areas. Non-O1/non-O139 strains are much more commonly isolated from the environment than are O1 strains, even in epidemic settings in which fecal contamination of the environment might be expected. Outside of epidemic areas (and away from areas that may have been contaminated by patients with cholera), O1 environmental isolates are almost always CT negative. However, it is clear that CT-producing *V. cholerae* O1 can persist in the environment in the absence of known human disease. An environmental reservoir is the most likely explanation for the persistence of a single strain along the U.S. Gulf Coast for over 20 years. Periodic introduction of such environmental isolates into the human population through ingestion of uncooked or undercooked shellfish appears to be responsible for isolated foci of cholera endemicity along the U.S. Gulf Coast and in Australia (11).

V. cholerae O1 strains are capable of colonizing the surfaces of zooplankton such as copepods with between 10^4 and 10^5 *V. cholerae* cells attached to a single copepod. Other aquatic biota, such as water hyacinths from Bangladeshi waters, also can be colonized by *V. cholerae* and promote its growth. *V. cholerae* produces a chitinase and is able to bind to chitin, which is the principal component of crustacean shells; the bacterium can grow in media with chitin as the sole carbon source.

Persistence of *V. cholerae* within the environment may be facilitated by its ability to assume survival forms, including a viable but nonculturable state and a rugose survival form. The viable but nonculturable state is a dormant state in which *V. cholerae* is still viable but not culturable in conventional laboratory media (reviewed in references 25, 116, and 117). In this dormant state, the

cells are reduced in size and become ovoid. The continued viability of the nonculturable *V. cholerae* can be assessed by a direct viable count procedure in which cells are incubated in the presence of yeast extract and nalidixic acid and examined microscopically for cell elongation. The viable but nonculturable state can be induced in the laboratory by incubating a culture of *V. cholerae* in phosphate-buffered saline at 4°C for several days. Although these cells are not culturable with nonselective enrichment broth or plates, nonculturable *V. cholerae* O1 injected into ligated rabbit ileal loops or ingested by volunteers has yielded culturable *V. cholerae* O1 in intestinal contents or stool specimens (26).

V. cholerae can also assume a "rugose" or "wrinkled" colonial morphology when passaged on a nonselective medium such as Luria agar (reviewed in reference 70). Upon laboratory passage, one in ca. 1,000 smooth colonies will shift to a rugose form; when a rugose strain is passaged, approximately 1 in 1,000 of the resultant colonies will shift back to a smooth morphology. The rate of shift to rugose forms can be increased by stress such as passage in APW. When examined by light and electron microscopy, cells in a rugose culture are small and spherical and are embedded in an amorphous matrix material composed primarily of carbohydrate. The matrix material, or exopolymer, appears to aid aggregation of bacteria in clusters of up to 100 bacteria. In this state, the cells are protected against adverse environmental conditions. Notably, rugose variants survive in the presence of chlorine and other disinfectants and are still capable of causing diarrhea in volunteers.

Humans and Animals

Long-term carriage of *V. cholerae* in humans is extremely rare and is not considered to be significant in transmission of disease. However, short-term carriage of *V. cholerae* by humans is quite important in transmission of disease. Persons with acute cholera excrete 10^7 to 10^8 *V. cholerae* cells per gram of stool; for patients who have 5 to 10 liters of diarrheal stool, total output of *V. cholerae* can be in the range of 10^{11} to 10^{13} CFU. Even after cessation of symptoms, patients who have not been treated with antibiotics may continue to excrete vibrios for 1 to 2 weeks. Furthermore, a high percentage of persons infected with *V. cholerae* in areas of endemicity have inapparent illness and can still excrete vibrios, although excretion generally lasts for less than a week. Asymptomatic carriers are most commonly identified among household members of persons with acute illness: in various studies, the rate of asymptomatic carriage in this group has ranged from 4% to almost 22%. Studies also have revealed that *V. cholerae* O1 can be sporadically carried by household animals, including cows, dogs, and chickens, but no animal species consistently carries the bacterium.

In the 1970s it was widely accepted that asymptomatic and convalescent human (and, possibly, animal) carriers were the primary reservoir for *V. cholerae*. With the recognition that *V. cholerae* can live and multiply in the environment, much greater attention has been given to the identification and characterization of environmental reservoirs (reviewed in reference 25). Nonetheless, CT-producing *V. cholerae* O1 strains (i.e., disease-causing strains) continue to be isolated almost exclusively from areas that have been contaminated by human feces or sewage from persons or groups of persons known to have had cholera. Similarly, in areas of endemicity such as Lima, Peru, rates of isolation from the environment correlate primarily with degree of sewage contamination. However, even areas where cholera is not endemic may be contaminated by ballast water from ships originating in areas of cholera endemicity, such as was seen with oysters in Mobile Bay, Ala., that contained toxigenic *V. cholerae* O1 strains with PFGE patterns that were identical to those of toxigenic *V. cholerae* strains isolated from ballast water from ships docked at Gulf of Mexico ports (107). A dynamic relationship between human and environmental sources of the organism is apparent, with carriage and amplification by human populations playing a critical role in epidemic spread of CT-producing *V. cholerae* (70).

Foodborne Outbreaks

The critical role of water in transmission of cholera has been recognized for more than a century, ever since the London physician and epidemiologist John Snow determined in 1854 that illness was associated with consumption of water from a water system that drew its water from the Thames at a point below major sewage inflows. In developing countries, ingestion of contaminated water and food is probably the major vehicle for transmission of cholera whereas in developed countries, foodborne transmission is more important (45). Such distinctions are often difficult to make because contaminated water is frequently used in food preparation. For example, rice prepared with water contaminated with *V. cholerae* O1 has been implicated in outbreaks from Bangladesh as well as from the U.S. Gulf Coast. Fruit juices diluted with contaminated water and vegetables irrigated with untreated sewage have been associated with disease in South America. Seafood may acquire the organism from environmental sources and may serve as a vehicle in both endemic and epidemic disease, particularly if it is uncooked or only partially cooked.

The role of food in transmitting *V. cholerae* O1 has been extensively reviewed elsewhere (72, 101), and the reader is referred to these reviews for more detail and primary references. The spectrum of food items implicated in transmission of cholera includes crabs, shrimp, raw fish, mussels, cockles, squid, oysters, clams, rice, raw pork, millet gruel, cooked rice, street vendor food, frozen coconut milk, and raw vegetables and fruit (101). One shared characteristic of the implicated foods is their neutral or nearly neutral pH. Thus, in investigating a suspected foodborne outbreak with many possible vehicles, one can eliminate the foods with an acid pH and concentrate on neutral or alkaline foods (72). This predilection for neutral foods was demonstrated in an epidemiologic study in West Africa, where boiled rice is commonly prepared in the morning, held unrefrigerated, and eaten with sauce at the midday and evening meals. In a case-control study of illness, tomato sauces with a pH of 4.5 to 5.0 were protective against illness, whereas less acidic sauces (pH 6.0 to 7.0) prepared from ground peanuts were associated with illness due to *V. cholerae* O1 (101). Alkaline conditions are employed in enrichment cultures of food and water samples wherein the use of APW (pH 8.5) followed by plating on TCBS agar is a common isolation method (72). Survival and growth of *V. cholerae* O1 in food are also enhanced by low temperatures, high organic content, high moisture, and absence of competing flora (101). Survival is increased when foods are cooked before contamination; cooking eliminates competing organisms and may destroy some heat-labile growth inhibitors and produce denatured proteins that the bacterium uses for growth (101).

Food buffers *V. cholerae* O1 against killing by gastric acid. Although many different food items can provide this buffering capacity, the protection provided by chitin is noteworthy because crustaceans are a frequent vehicle of disease. In dilute hydrochloric acid solutions of approximately the same pH as human gastric acid, survival of *V. cholerae* absorbed to chitin was enhanced compared to *V. cholerae* in the absence of chitin (101).

In the United States, both domestically acquired and imported cases of cholera occur. Crabs, shrimp, and oysters have been the most frequently implicated vehicles of domestic cases, although the largest single outbreak (16 cases) was due to ingestion of contaminated rice. In this outbreak, which occurred on a Gulf Coast oil rig in 1981, cooked rice was moistened with water contaminated by human feces and then held for 8 h after cooking (101). Although most domestic cases occur in states bordering the Gulf Coast, seafood shipped from this area has caused disease in both Maryland and Colorado (101). The risk of imported cholera has greatly increased since the establishment of endemic cholera in South America in 1991. The largest such outbreak (75 cases) involved crab salad served on an airplane flying from Peru to California. A smaller outbreak of eight cases occurred in New Jersey attributed to consumption of contaminated crabs purchased in Ecuador and carried to the United States in an individual's luggage. Importation from Asia can also occur, even in commercially imported food. A small outbreak of four cases in Maryland was attributed to frozen coconut milk imported from Thailand that was subsequently used in a topping for a rice pudding.

Nearly all cases of gastroenteritis caused by *V. cholerae* of serogroups other than O1 or O139 have been associated with the consumption of raw oysters (reviewed in reference 103). Both disease incidence and isolation rate of non-O1/O139 serogroups from oysters are highest in the summer. Such strains were isolated from up to 14% of freshly harvested oysters in one study conducted by the FDA. Outside the United States, outbreaks have also been linked to consumption of contaminated potatoes, chopped eggs, pre-prepared gelatin, vegetables, and meat samples (56, 103). As with *V. cholerae* O1, survival of non-O1/non-O139 *V. cholerae* is enhanced in foods of alkaline pH.

Characteristics of Disease

The explosive, potentially fatal dehydrating diarrhea that is characteristic of cholera actually occurs in only a minority of persons infected with CT-producing *V. cholerae* O1/O139. Most infections with *V. cholerae* O1 are mild or even asymptomatic. It has been estimated that 11% of patients with classical infections develop severe disease, compared with 2% of those with El Tor infections. An additional 5% of El Tor infections and 15% of classical infections result in moderate illness (70). Symptoms of persons infected with *V. cholerae* O139 Bengal are virtually identical to those of persons infected with O1 strains.

The incubation period of cholera can range from several hours to 5 days and is dependent in part on inoculum size. Onset of illness may be sudden, with profuse, watery diarrhea, or there can be premonitory symptoms such as anorexia, abdominal discomfort, and simple diarrhea. Initially the stool is brown with fecal matter, but soon thereafter diarrhea becomes a pale gray color with an inoffensive, slightly fishy odor. Mucus in the stool imparts the characteristic "rice water" appearance. Vomiting is often present, occurring a few hours after the onset of diarrhea.

In its most severe form, termed cholera gravis, the rate of diarrhea can quickly reach 500 to 1,000 ml/h, leading rapidly to tachycardia, hypotension, and vascular collapse due to dehydration. Peripheral pulses may be absent, and blood pressure may be unobtainable. Skin turgor is poor, giving the skin a doughy consistency; the eyes are sunken; and hands and feet become wrinkled, as after long immersion ("washerwoman's hands"). Such severe dehydration can lead to death within hours of the onset of symptoms unless fluids and electrolytes are rapidly replaced. Although cholera gravis is a striking clinical entity, milder illnesses are not readily differentiated from other causes of gastroenteritis in areas of cholera endemicity.

Gastroenteritis associated with *V. cholerae* non-O1/non-O139 is generally of mild to moderate severity, although severe, cholera-like illness has also been seen occasionally. Besides nonbloody and occasionally bloody diarrhea, symptoms can also include abdominal cramps, and fever with nausea and vomiting occurring in a minority of patients (103). Non-O1/O139 *V. cholerae* strains are also frequently isolated from extraintestinal infections such as septicemia, wound infections, and ear infections; these infections usually involve exposure to fresh or brackish water (103).

Infectious Dose and Susceptible Population

In healthy North American volunteers, doses of 10^{11} CFU of *V. cholerae* were required to consistently cause diarrhea when the inoculum was administered in buffered saline (pH 7.2). When stomach acidity was neutralized with 2 g of sodium bicarbonate immediately before administration of the inoculum, attack rates of 90% occurred with an inoculum of 10^6 CFU. Food has a buffering capacity comparable to that of sodium bicarbonate. Ingestion of 10^6 vibrios with food such as fish and rice resulted in the same high attack rate (100%) as when this inoculum is administered with buffer (85). Further studies revealed that most volunteers who received as few as 10^3 to 10^4 vibrios with buffer develop diarrhea, although lower inocula correlated with a longer incubation period and diminished severity. The incubation time in volunteers between ingestion of vibrios and onset of diarrhea ranged from 8 to 96 h.

The inoculum size in naturally occurring infections is not known with certainty. A study in Bangladesh revealed that it was almost always necessary to use enrichment techniques to isolate *V. cholerae* from household water samples, suggesting that counts were <500 CFU/ml (149). Attack rates in the households surveyed were only 11%, and only half of infected persons had clinical illness, consistent with the hypothesis that inoculum sizes were relatively low. Although conditions will obviously vary widely from one community to another, studies such as these have led to the suggestion that the inoculum in nature is in the range of 10^2 to 10^3 CFU (45).

The volunteer data on the effect of buffer on infectious dose are consistent with epidemiologic data showing that people who are achlorhydric because of surgery, medication (e.g., antacids), or other reasons are at increased risk for cholera. Individuals of blood group O are at increased risk of more severe cholera, which has been shown for natural infection as well as with experimental infection; the mechanism of this increased susceptibility is unknown. In addition to these factors, additional host factors, as yet poorly defined, play a role in disease susceptibility to cholera. The effect of other host factors is illustrated by a study in which the identical inoculum that caused 44 liters of diarrhea in one volunteer caused little or no illness in other individuals. The various factors affecting host susceptibility have been reviewed elsewhere (132).

Strains of non-O1/non-O139 *V. cholerae* have also been studied in volunteers. Of three strains fed to volunteers that did not produce CT, only one strain caused diarrhea. This strain produced a heat-stable enterotoxin (ST)-like toxin and caused diarrhea in six of eight volunteers at doses of 10^6 to 10^9 CFU (after neutralization of stomach acid with sodium bicarbonate) (103). The severity of disease was generally mild, but in one volunteer, diarrheal stool volume exceeded 5 liters.

Virulence Mechanisms

Infection due to *V. cholerae* O1/O139 begins with the ingestion of food or water contaminated with the pathogen. After passage through the acid barrier of the stomach, vibrios colonize the epithelium of the small intestine by means of one or more adherence factors. Invasion into epithelial cells or the lamina propria does not occur. Production of cholera enterotoxin (and possibly other toxins) disrupts ion transport by intestinal epithelial cells. The subsequent loss of water and electrolytes leads to the severe diarrhea characteristic of cholera.

Cholera Toxin

Volunteer studies revealed that the massive, dehydrating diarrhea characteristic of cholera is induced by cholera enterotoxin, also referred to as cholera toxin or choleragen. CT is among the most extensively studied of bacterial toxins, and several reviews have been published on this topic (69, 70, 147).

Structure

The structure of CT is typical of the A-B subunit group of toxins in which each of the subunits has a specific function. The B subunit serves to bind the holotoxin to the eukaryotic cell receptor, and the A subunit possesses a specific enzymatic function that acts intracellularly. CT consists of five identical B subunits and a single A subunit, and neither of the subunits individually has significant secretogenic activity in animal or intact cell systems. The mature B subunit contains 103 amino acids with a subunit weight of 11.6 kDa. The mature A subunit has a mass of 27.2 kDa and is proteolytically cleaved to yield two polypeptide chains, a 195-residue A1 peptide of 21.8 kDa and a 45-residue A2 peptide of 5.4 kDa. After proteolytic cleavage, the A1 and A2 peptides are still linked by a disulfide bond before internalization. The *ctxA* and *ctxB* genes encoding the A and B subunits reside on a filamentous bacteriophage (CTXΦ), which is capable of transducing *ctx* genes to nontoxigenic strains (166a).

Receptor Binding

The receptor for CT is the ganglioside GM1. Binding of CT to epithelial cells is enhanced by a neuraminidase (NANase) produced by *V. cholerae*. This 83-kDa enzyme enhances the effect of CT by catalyzing the conversion of higher order gangliosides to GM1, thereby enhancing the binding of CT and leading to greater fluid secretion. When culture supernatant fluids from a CT-positive, NANase-positive *V. cholerae* strain were compared to supernatant fluids from an isogenic CT-positive, NANase-negative strain, binding of CT to mouse fibroblasts was increased five- to eightfold in the presence of NANase (44). Furthermore, the short circuit current measured in Ussing chambers (a measure of secretory activity) increased 65% with NANase-positive filtrates compared with NANase-negative filtrates. These results indicate that NANase plays a subtle but significant role in the binding and uptake of CT, although this enzyme is not a primary virulence factor of *V. cholerae*.

Enzymatic Activity

The intracellular target of CT is adenylate cyclase, which mediates the transformation of ATP to cyclic AMP (cAMP), a crucial intracellular messenger for a variety of cellular pathways. Regulation of adenylate cyclase occurs via G proteins, which serve to link many cell surface receptors to effector proteins at the plasma membrane. The specific G protein involved is the $G_{s\alpha}$ protein, activation of which leads to increased adenylate cyclase activity. CT (specifically, the A1 peptide) catalyzes the transfer of the ADP-ribose moiety of NAD to a specific arginine residue in the $G_{s\alpha}$ protein, resulting in the activation of adenylate cyclase and subsequent increases in intracellular levels of cAMP. cAMP

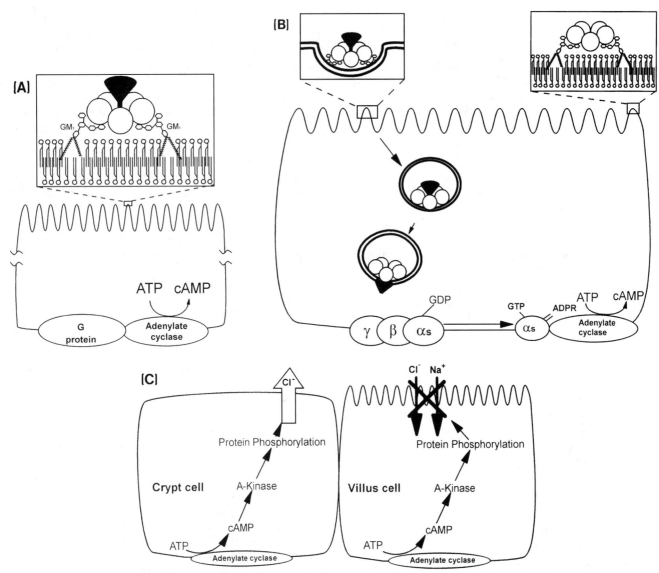

Figure 13.1 Classic model of CT mode of action involving cAMP. More recent evidence indicates that prostaglandins and the enteric nervous system are also involved in the response to CT (see text for details). (A) Adenylate cyclase, located in the basolateral membrane of intestinal epithelial cells, is regulated by G proteins. CT binds via the B subunit pentamer (shown as open circles with the A subunit as the inverted solid triangle) to the GM1 ganglioside receptor inserted into the lipid bilayer. (B) The toxin enters the cell via endosomes, and the A1 peptide ADP-ribosylates $G_{s\alpha}$ located in the basolateral membrane. (C) Increased cAMP activates protein kinase A, leading to protein phosphorylation. In crypt cells, the protein phosphorylation leads to increased Cl$^-$ secretion; in villus cells, it leads to decreased NaCl absorption. Adapted from reference 69.

activates a cAMP-dependent protein kinase, leading to protein phosphorylation, alteration of ion transport, and ultimately to diarrhea (Fig. 13.1). The alpha subunit of G_s contains a GTP-binding site and an intrinsic GTPase activity. Binding of GTP to the alpha subunit leads to dissociation of the alpha and the beta-gamma subunits and subsequent increased affinity of alpha for adenylate cyclase. The resulting activation of adenylate cyclase continues until the intrinsic GTPase activity hydrolyzes GTP to GDP, thereby inactivating the G protein and adenylate cyclase. ADP-ribosylation of the alpha subunit by the A1 peptide of CT inhibits the hydrolysis of GTP to GDP, thus leaving adenylate cyclase constitutively activated, probably for the life of the cell. The ADP-ribosylation activity of A1 is stimulated in vitro by a family of proteins termed ADP-ribosylation factors (ARFs) (reviewed in reference 106). The ARF proteins are ca. 20-kDa GTP-binding proteins and constitute a distinct family within the larger group of ca. 20-kDa guanine nucleotide-binding proteins that include the *ras* oncogene protein. At least in vitro, ARFs serve as allosteric activators of A1 and increase the ADP-ribosyltransferase activity of the proteins.

Cellular Response

The increased intracellular cAMP concentrations resulting from the activation of adenylate cyclase by CT leads to increased Cl^- secretion by intestinal crypt cells and decreased NaCl-coupled absorption by villus cells (Fig. 13.1C). The net movement of electrolytes into the lumen results in a transepithelial osmotic gradient that causes water flow into the lumen. The massive volume of water overwhelms the absorptive capacity of the intestine, resulting in diarrhea. The steps between increased levels of cAMP and secretory diarrhea are not known in their entirety, but one crucial step resulting from increased cAMP levels is activation of protein kinase A, which subsequently phosphorylates numerous substrates in the cell.

Regulation of chloride channels by cAMP-dependent protein kinases is well known, and the crucial chloride channel involved in diarrhea due to cholera is the cystic fibrosis (CF) gene product, CFTR, a Cl^- channel with multiple potential substrate sequences for kinase A. In the CF mouse model, mice that expressed no CFTR protein did not secrete fluid in response to CT whereas heterozygotes expressed 50% of the normal amount of CFTR and secreted 50% of the normal fluid and chloride ion in response to CT (43).

Alternate Mechanisms of Action

The activation of adenylate cyclase leading to increased cAMP and subsequent altered ion transport is the "classic" mode of action of CT. However, it has recently been appreciated that the increased levels of cAMP and subsequent protein kinase A activation may not explain all of the secretory effects of CT. There is compelling evidence that prostaglandins and the enteric nervous system are involved in the response to CT in addition to the mechanism described above. Most studies presenting these alternative mechanisms do not conclude that the above scenario is wrong; rather, they suggest that this scenario simply does not explain all of the secretion due to CT.

Prostaglandins. Several reports have implicated prostaglandins in the pathogenesis of cholera (69, 70). Although the exact mechanisms are unclear, the role of prostaglandins, leukotrienes, and other metabolites of arachidonic acid in causing intestinal secretion has been well documented. Cholera patients in the active secretory disease stage have elevated jejunal concentrations of prostaglandin E_2 (PGE_2) compared to patients in the convalescent stage. In one animal study, addition of cAMP induced only a small, transient fluid accumulation in rabbit intestinal loops whereas addition of PGE_2 caused a much stronger fluid accumulation in rabbit loops. Addition of CT led to increases in both cAMP and PGE_2 in rabbit loops and in Chinese hamster ovary cells, resulting in the release of arachidonic acid from membrane phospholipids. A model has been suggested in which cAMP levels increased by CT serve not only to activate protein kinase A but to also regulate transcription of a phospholipase or a phospholipase-activating protein. The activated phospholipase could act on membrane phospholipids to produce arachidonic acid, a precursor of prostaglandins and leukotrienes. Consistent with this model, it has been reported that 40 to 60% of the short current response to CT is inhibited by relatively low concentrations of the phospholipase A2 inhibitor mepacrine. The implication of platelet-activating factor, a stimulator of phospholipase A2, in the response to CT (48) further supports a key role for prostaglandin synthesis in the mode of action of CT.

Enteric nervous system. The enteric nervous system plays an important role in intestinal secretion and absorption. The intestine also contains a variety of cells that can produce hormones and neuropeptides such as vasoactive intestinal peptide (VIP) and serotonin (5-hydroxytryptamine or 5-HT) that can affect secretion. It has been estimated that ca. 60% of the effect of CT on intestinal fluid transport could be attributed to nervous mechanisms (69, 70). One proposed mechanism is that cholera toxin binds to "receptor cells," namely enterochromaffin cells, which release a substance such as serotonin, which activates dendrite-like structures located beneath the intestinal epithelium. This leads to the release of VIP, resulting in electrolyte and fluid secretion. This model is supported by a variety of studies using receptor antagonists, ganglionic or neurotransmitter blockers, as well as direct measurements of increased levels of serotonin (5-HT) and VIP. Perhaps the most convincing evidence comes from a study in which fluid communication between proximal and distal regions of rabbit intestine is prevented by ligation and fluid secretion is measured in each isolated area (114). Addition of CT to the proximal loop results in changes in fluid secretion in both the proximal and distal loop. However, when the enteric nervous system connecting the two regions is disrupted, addition of CT to the proximal loop has no effect on fluid secretion in the distal region.

Other Toxins Produced by *V. cholerae*

When the first recombinant *V. cholerae* vaccine strains specifically deleted of genes encoding CT were tested in volunteers, it was somewhat surprising that mild to moderate diarrhea was still observed in ca. 50% of volunteers (86). The volume of diarrhea was not the severe, dehydrating diarrhea seen with wild-type strains, which can exceed 40 liters in volume, but a much milder diarrhea that ranged from 0.3 to 2.1 liters in volume. In addition, some volunteers also experienced abdominal cramps, anorexia, and low-grade fever when fed Δ*ctx* *V. cholerae* strains. These results prompted a search for additional toxins produced by *V. cholerae*, and it is now known that *V. cholerae* produces a variety of extracellular products that have deleterious effects on eukaryotic cells. In addition to the well-characterized toxins described below, *V. cholerae* can produce a number of other toxic factors, but the responsible proteins and genes have not yet been purified or cloned.

The role of toxins other than CT in the pathogenesis of disease due to *V. cholerae* is largely unknown. These toxins clearly cannot cause cholera gravis because the diarrhea seen with Δ*ctx* strains presumably still producing these toxins is not the severe purging seen with wild-type *V. cholerae* strains. Δ*ctx* *V. cholerae* vaccine candidate strains lacking genes encoding zonula occludenotoxin (Zot), accessory cholera enterotoxin (Ace), hemolysin/cytolysin, or RtxA toxins still caused mild to moderate diarrhea as well as fever and abdominal cramps in volunteers. Only mutation of the *hap* gene encoding soluble hemagglutinin/protease reduced the reactogenicity observed with Δ*ctx* *V. cholerae* vaccine strains (7). Toxins other than CT may contribute in part to the diarrhea and other symptoms observed with *V. cholerae* strains, perhaps serving as a secondary secretogenic mechanism when conditions for producing cholera toxin are not optimal.

Zot

Zot increases the permeability of the small intestinal mucosa by affecting the structure of the intercellular tight junction, or zonula occludens. In rabbit ileal tissue mounted in Ussing chambers, Zot causes a decreased tissue resistance that reflects modification of tissue permeability through the intercellular space, i.e., the paracellular pathway (42). By increasing intestinal permeability, Zot might cause diarrhea by leakage of water and electrolytes into the lumen under the force of hydrostatic pressure.

Ace

Ace causes fluid accumulation in rabbit ligated ileal loops, and like CT, and in contrast to Zot, this toxin increases potential difference rather than tissue conductivity in Ussing chambers (160). The *ace* and *zot* genes are located immediately upstream of the *ctx* genes and encode predicted proteins of 11.3 kDa and 45 kDa, respectively. In addition to their toxic activities, Ace and Zot are also believed to be components of the CTXΦ filamentous phage encoding CT (166a).

Soluble Hemagglutinin/Protease

V. cholerae produces a soluble factor that possesses zinc-dependent metalloprotease activity. This factor was first characterized by its ability to agglutinate certain erythrocytes and is therefore named hemagglutinin/protease (HA/P). HA/P is capable of nicking and activating the A subunit of CT as well as cleaving mucin, fibronectin, and lactoferrin. More recently, it has been shown to perturb the barrier function of epithelial cells by affecting tight junctions. Specifically, HA/P affects two major tight junction components, occludin and ZO-1, by digesting occludin and rearranging the distribution of ZO-1 (173). The in vivo relevance of HA/P was observed in volunteer studies where Δ*ctx* HA/P-positive *V. cholerae* strains caused mild diarrhea, whereas as an isogenic HA/P-negative strain did not (7).

Hemolysin/Cytolysin

Hemolysis of sheep red blood cells was traditionally used to distinguish between the El Tor and classical biotypes of *V. cholerae*, although more recent El Tor isolates are only poorly hemolytic on sheep erythrocytes. The hemolysin/cytolysin is made initially as an 82-kDa protein and processed in two steps to a 65-kDa active cytolysin. The purified hemolysin is cytolytic for a variety of erythrocytes and cultured mammalian cells, is rapidly lethal for mice, and causes bloody fluid accumulation in ligated rabbit ileal loops.

RtxA

V. cholerae O1 El Tor and O139 strains produce a toxin (RtxA) that has substantial sequence similarity to members of the RTX (repeats in toxin) toxin family, a group that includes the hemolysin of uropathogenic *Escherichia coli* and adenylate cyclase of *Bordetella pertussis*. The huge RtxA toxin, with a predicted unprocessed size of ca. 500 kDa, is responsible for the cytotoxicity of El Tor strains to HEp-2 cells (88). The gene encoding RtxA is located downstream of the *ctx* genes in El Tor strains but is not part of CTXΦ. Classical strains contain a deletion of ca. 8 kb in this area and thus do not produce RtxA.

Toxins of *V. cholerae* Non-O1 or Non-O139 Strains

No toxins unique to *V. cholerae* non-O1/non-O139 serogroups have been identified, but some strains of these serogroups may produce one or more toxins that have been characterized previously in *V. cholerae* O1 or other species (60). Most *V. cholerae* non-O1/non-O139 strains produce a cytotoxic protein apparently identical to the El Tor hemolysin (hemolysin-cytolysin) of *V. cholerae* O1. Most non-O1/O139 strains do not contain genes encoding CT, Zot, or Ace, but a few strains possess genes for these toxins.

Some strains of *V. cholerae* non-O1/O139 produce a 17-amino-acid heat-stable enterotoxin (designated NAG-ST for nonagglutinable *Vibrio* ST) that has 50% sequence homology with the STa of enterotoxigenic *E. coli*. (A single strain of *V. cholerae* O1 produces an ST-like toxin but this toxin is more common among non-O1 strains.) In a volunteer study, one subject who ingested a CT-negative *V. cholerae* non-O1 strain producing NAG-ST purged over 5 liters of diarrheal stool (103). This toxin is present only in a minority of *V. cholerae* strains. In one study, 6.8% of *V. cholerae* non-O1 strains from Thailand and none of the strains from the United States or Mexico produced this toxin. In another study, 2.3% of all non-O1 *V. cholerae* isolated from Calcutta, India, contained genes for this toxin. The *tdh* gene encoding the thermostable direct hemolysin (TDH) of *V. parahaemolyticus* is also present in *V. cholerae* non-O1/non-O139. In one strain, the *tdh* gene was present on a plasmid, suggesting that this toxin gene could be readily transferred to other species.

Colonization Factors

TCP

TCP is the best-characterized intestinal colonization factor of *V. cholerae*. TCP consists of long filaments 7 nm in diameter that are laterally associated in bundles. The name of the pilus results from the fact that expression of the pilus is correlated with expression of CT (156). The predicted amino acid sequence of the 20.5-kDa *tcpA* subunit has significant homology with type IV pili of *Pseudomonas aeruginosa*, *Neisseria gonorrhoeae*, *Moraxella*

bovis, and *Bacteroides nodosus*. Epitope differences occur in pili produced by classical and El Tor strains, and the predicted protein sequences share 82 to 83% homology between the biotypes. El Tor strains produce less TCP than classical strains, and culture conditions for optimal TCP expression differ between biotypes. Synthesis of TCP is complex and incompletely understood, and up to 15 open reading frames are present in the *tcp* gene cluster. The sequence and regulation of TCP in the O139 serogroup are identical to those observed in O1 El Tor strains.

TCP is the only colonization factor of *V. cholerae* whose importance in human disease has been proven. Volunteers ingesting *V. cholerae* strains specifically mutated in the *tcpA* gene did not experience diarrhea and no vibrios were recovered from stools of the volunteers (70). The *tcp* gene cluster along with *toxT* and *acf* are encoded on a 40-kb pathogenicity island termed *Vibrio* pathogenicity island (VPI) that is present in all O1 and O139 clinical isolates and absent from nearly all environmental isolates (71).

Accessory Colonization Factor

Mutations in a ToxR-regulated locus called *acf* for accessory colonization factor also diminish colonization in mice but not to the extent as mutations in *tcp* (i.e., ca. 10- to 50-fold decrease for *acf* compared to 1,000-fold for *tcp*). The *acf* gene cluster apparently does not encode a fimbrial colonization factor but may be involved in intestinal colonization via motility and/or chemotaxis functions.

Mannose-Fucose-Resistant Hemagglutinin

Mutations in a gene encoding a cell-associated mannose-fucose-resistant hemagglutinin (MFRHA) also lead to decreased colonization in mice (500- to 1,300-fold relative to the parent strain). The exact nature of the MFRHA, expressed by both biotypes, is not known, but it was suggested that the MFRHA is a 27-kDa cationic outer membrane protein that is held on the cell surface primarily by charge interactions with the LPS.

Mannose-Sensitive Hemagglutinin

The mannose-sensitive hemagglutinin (MSHA) of *V. cholerae* is expressed by strains of the El Tor biotype but is only rarely expressed by strains of the classical biotype. The MSHA is a thin, flexible pilus composed of subunits with a molecular mass of ca. 17 kDa. MSHA does not play a role in human disease but does appear to be involved in biofilm formation (168).

OmpU

The 38-kDa outer membrane protein OmpU was originally described as being regulated by ToxR. There is evidence indicating that this protein can serve as an adherence factor because antibodies raised against OmpU completely inhibit adherence of *V. cholerae* O1 to cultured epithelial cells and protect mice against challenge with either El Tor or classical strains (148).

Motility/Flagella

V. cholerae is motile by means of a single, polar, sheathed flagellum. Motility is an important virulence property, with nonmotile, fully enterotoxinogenic mutants being diminished in virulence. In several animal and in vitro models, motile *V. cholerae* rapidly enters the mucous gel overlying the intestinal epithelium and enters intervillous spaces within minutes to a few hours. In addition to its role in motility, it has also been suggested that the flagellum serves as an adhesin, and recent evidence suggests that flagella and/or motility may be involved in regulation of virulence gene expression (50).

LPS

The LPS of *V. cholerae* O1 is the major protective antigen of this pathogen, and its importance in protective immunity greatly outweighs that of the CT. The importance of this antigen was seen in India and Bangladesh when the O139 serogroup caused widespread disease in individuals who were presumably immune to the O1 serogroup. There is also evidence to suggest that LPS is involved in adherence of *V. cholerae* O1 to the intestinal mucosa. In one study, purified Inaba LPS significantly inhibited attachment of *V. cholerae* Inaba to rabbit mucosa; in other in vitro and in vivo studies, antibodies against Ogawa or Inaba LPS prevented adhesion of *V. cholerae* to intestinal mucosa.

Polysaccharide Capsule

Although *V. cholerae* O1 is unencapsulated, strains of *V. cholerae* O139 produce a polysaccharide capsule (63), termed an O-antigen capsule. The genetic change involved in the conversion of the O1 serogroup to O139 involves the deletion of ca. 22 kb of the *rfb* genes encoding the O1 LPS and the substitution of a ca. 35-kb insert encoding the O139 LPS and polysaccharide capsule (27). The polysaccharide capsule of an O139 strain could mediate adherence to tissue culture epithelial cells (148).

Adherence Factors of Non-O1/Non-O139 V. cholerae

Intestinal colonization factors of non-O1/non-O139 *V. cholerae* strains are poorly characterized. Sequences encoding TCP are not usually found in *V. cholerae* non-O1 strains or in environmental isolates of *V. cholerae* O1, which do not produce CT. A variety of fimbria hemagglutinins have been described for these strains, but their role in intestinal adherence is unclear. There is evidence that the capsular polysaccharide may also mediate adherence to intestinal epithelial cells for non-O1 strains. Approximately 70% of such strains produce a polysaccharide capsule. In addition to a potential role in adherence, such a capsule could facilitate septicemia often seen with infections due to non-O1/non-O139 strains.

Regulation

Multiple systems are involved in regulation of virulence in *V. cholerae*. The ToxR regulon controls expression of several critical virulence factors and has been the most extensively characterized. Regulation in response to iron concentration is a distinct regulatory system that controls additional putative virulence factors. Other putative virulence factors such as neuraminidase and various hemagglutinins are apparently not controlled by either regulatory system. There is also a set of poorly characterized genes that are expressed only in vivo and do not belong to the ToxR or iron regulatory systems. These different regulatory systems no doubt allow *V. cholerae* to vary expression of its genes to optimize survival in a variety of environments, from the human intestine to the estuarine environment.

The ToxR Regulon

Expression of several virulence genes in *V. cholerae* O1 and O139 is coordinately regulated so that multiple genes respond in a similar fashion to environmental conditions (35). The "master switch" for control of these factors is ToxR, a 32-kDa transmembrane protein that shares homology with other sensory transduction proteins seen in various bacterial pathogens. It has been proposed that the ToxR protein senses environmental conditions and transmits this information to other genes in the ToxR regulon by signal transduction. At least 17 distinct genes are regulated by ToxR, including those encoding CT, TCP, ACF, OmpU, OmpT, and three lipoproteins. The effect of ToxR on expression of most of these factors is to increase expression, but expression of OmpT is decreased in the presence of ToxR. These genes comprise the ToxR regulon. The importance of ToxR in human disease was demonstrated in volunteer studies wherein a ToxR mutant *V. cholerae* O1 strain was fed to volunteers who subsequently suffered no diarrheal symptoms and did not shed the strain in their stools.

Several other proteins interact with ToxR to control gene expression in *V. cholerae*. ToxS is a 19-kDa transmembrane protein that helps assemble or stabilize ToxR monomers into a dimeric form. Expression of many genes of the ToxR regulon, including *ctx* and *tcpA*, is controlled by another regulatory factor, ToxT, a 32-kDa protein that has significant sequence homology with the AraC family of transcriptional activators. Expression of *toxT* is regulated by the TcpP and TcpH proteins that are encoded by genes located on the VPI pathogenicity island upstream of *toxT* (51). ToxR acts in concert with TcpPH to activate *toxT* expression. Expression of the *tcpPH* genes is in turn regulated by the AphAB proteins, which are encoded outside the VPI (145). Thus, the major virulence factors of *V. cholerae*, CT and TCP, are regulated in a cascade fashion in which AphAB controls expression of TcpPH, which acts together with ToxRS to activate expression of ToxT, which then activates expression of CT and TCP.

Iron Regulation

Growth of *V. cholerae* under low-iron conditions induces the expression of several new outer membrane proteins that are not seen in cells grown in iron-rich media. Many of these proteins are similar to proteins induced by in vivo growth of *V. cholerae*, indicating that the intestinal site of *V. cholerae* is a low-iron environment. In addition, expression of some OMPs decreases under iron-limiting conditions. Several proteins whose expression is increased under low-iron conditions have been identified, including the hemolysin/cytolysin, vibriobactin, and IrgA.

V. cholerae has at least two high-affinity systems for acquiring iron. The first system involves a phenolate-like siderophore, vibriobactin, which is produced under low-iron conditions. Vibriobactin binds iron extracellularly and transports it into the cell through a specific receptor. A second system for acquiring iron uses heme and hemoglobin. The *hutA* gene (53) encodes a 77-kDa OMP that serves as a receptor for heme.

Iron-regulated gene expression in *V. cholerae* involves a protein named Fur, which binds to a 21-bp operator sequence present in the promoter of iron-regulated genes, thereby repressing transcription (18). In *V. cholerae*, Fur acts as a repressor for >20 genes, including the vibriobactin gene (*viuA*) and *irgA*, which encodes an OMP whose mutation results in a ca. 10-fold decrease in mouse colonization.

V. MIMICUS

Prior to 1981, *V. mimicus* was known as sucrose-negative *V. cholerae* non-O1. This species was determined to be a distinctly different species on the basis of biochemical reactions and DNA hybridization studies, and the name *mimicus* was given because of its similarity to *V. cholerae* (32). This organism is chiefly isolated from cases of gastroenteritis, but occasional strains have been isolated from ear infections as well.

Reservoir

As with other *Vibrio* species, the reservoir of *V. mimicus* is the aquatic environment. One very interesting study compared the ecology of *V. mimicus* in the tropic, polluted environment of Bangladesh to the cleaner, more temperate environment of Okayama, Japan (22). This species was isolated from Bangladeshi waters throughout the year whereas it was not isolated in Okayama when the water temperature decreased below 10°C. In Japan, *V. mimicus* was present both in freshwater and in brackish waters with a salinity optimum of 4 ppt; the bacterium was not recovered from waters with salinity >10 ppt. Besides being present free in the water column, *V. mimicus* was also isolated from the roots of aquatic plants, sediments, and plankton, at levels up to 6×10^4 CFU per 100 g of plankton. *V. mimicus* has also been isolated from fish (80) and freshwater prawns (24).

Foodborne Outbreaks

Gastroenteritis due to *V. mimicus* has only been associated with consumption of seafood. In initial patient studies in the United States (139), consumption of raw oysters was significantly associated with disease due to this species; one patient reported eating only shrimp and crab, not oysters, in the week before onset of disease. Cases are usually sporadic, rather than associated with common-source outbreaks, but three patients in Louisiana became ill with *V. mimicus* infection after attending a company banquet where the foods served included crawfish (139). Disease in Japan is associated with consumption of raw fish and at least two outbreaks of seafood-borne disease have involved *V. mimicus* of serogroup O41 (80).

Characteristics of Disease

Disease due to *V. mimicus* is characterized by diarrhea, nausea, vomiting, and abdominal cramps in most patients. In a minority of patients, fever, headache, and bloody diarrhea also occur. In one group of patients, diarrhea lasted a median of 6 days, with a range of 1.5 to 10 days (139). In this report, disease was associated with consumption of seafood, particularly raw oysters, and the median interval from the time of consumption to onset of illness was 24 h (range, 3 to 72 h). In a 1989 survey of *Vibrio* infections on the Gulf Coast, *V. mimicus* was isolated from 4 cases of gastroenteritis, compared to

26 cases of *V. parahaemolyticus* and 18 cases of non-O1 *V. cholerae* (87).

Infectious Dose and Susceptible Population

There are no volunteer or epidemiologic data to enable estimation of an infectious dose for *V. mimicus*. There is not a particularly susceptible population, other than people who eat raw oysters. Most patients who developed gastroenteritis with *V. mimicus* were in good health before onset of illness (139).

Virulence Factors

V. mimicus appears not to produce unique enterotoxins, but many strains produce toxins that were first described in other *Vibrio* species, including CT, TDH, Zot, and a heat-stable enterotoxin apparently identical to the NAG-ST produced by *V. cholerae* non-O1/O139 strains. One study in Bangladesh (22) revealed that 10% of clinical isolates produced a CT-like toxin compared to less than 1% of environmental strains producing such a toxin. Another study in Japan revealed that 94% of clinical isolates produced TDH, whereas none of the environmental strains studied produced this toxin. The *tdh* gene from *V. mimicus* has 97% homology with the prototypic *tdh* gene from *V. parahaemolyticus* (112).

There is little information about potential intestinal colonization factors of *V. mimicus*. Like *V. cholerae* non-O1/non-O139, *V. mimicus* does not appear to express TCP or possess *tcp* genes. One study revealed that the levels of expression of a cell-associated hemagglutinin (HA) correlated with adherence to human intestinal epithelial cells (163). Pili were also observed on the surface of *V. mimicus* cells, but expression of these pili did not correlate with adherence. The levels of adherence to epithelial cells were much lower than those observed with *V. cholerae* O1, thus suggesting that reduced adherence is responsible for the reduced virulence of *V. mimicus* relative to *V. cholerae* O1. Two other HAs, one with protease activity, the other without, have been identified in the culture supernatant fluids, but no relationship to intestinal adherence has been suggested.

V. PARAHAEMOLYTICUS

Along with *V. cholerae*, *V. parahaemolyticus* is the best described of the pathogenic vibrios, with numerous studies reported since the first description of its involvement in a major outbreak of food poisoning in 1950. Between 1973 and 1998, a total of 40 outbreaks of *V. parahaemolyticus* in the United States were reported to the CDC, with over 1,000 persons involved (31).

Classification

The taxonomy of *V. parahaemolyticus* has been described in reviews by Joseph et al. (67) and Twedt (162), which include the minimal characteristics for identification of *V. parahaemolyticus*, as well as tests that serve to distinguish this species from other pathogenic vibrios. As with many of the vibrios, however, strain variation is common, and phenotypic testing is often insufficient for speciation. A significant percentage of isolates of *V. parahaemolyticus* that are sucrose positive, for example, have been described. In addition, some traits, such as H_2S production, are dependent on the medium or assay method employed, and caution must be exercised in their determination.

V. parahaemolyticus is serotyped according to both its somatic O and capsular polysaccharide K antigens, based on a scheme developed by Sakazaki et al. (135) from a study of 2,720 strains. There are presently 12 O (LPS) antigens and 59 K (acidic polysaccharide) antigens recognized. Although many environmental and some clinical isolates are untypeable by the K antigen, most clinical strains can be classified to their O type. Until the recent description of serotype O3:K6, there was no correlation between serotype and virulence. This serotype, however, has recently been involved in epidemics of gastroenteritis in Southeast Asia, Japan, and North America (110).

A special consideration in the taxonomy of *V. parahaemolyticus* is the ability of certain strains to produce a hemolysin, termed the TDH or Kanagawa hemolysin, which is correlated with virulence in this species (see "Virulence Mechanisms"). The production of this hemolysin is determined on Wagatsuma agar, which contains yeast extract (5 g/liter), peptone (10 g/liter), mannitol (5 g/liter), K_2HPO_4 (0.5 g/liter), NaCl (70 g/liter), agar (15 g/liter), and crystal violet (1 ml of a 0.1% solution). Freshly drawn and washed human blood cells are added to the cooled medium. The use of Wagatsuma agar for determining the production of beta hemolysis by Kanagawa phenomenon-positive (KP+) strains, however, has been replaced to a great extent by in vitro DNA amplification and gene probe hybridization methods to differentiate KP+ and KP− cells (6).

Although a correlation between urea hydrolysis and the ability to produce the Kanagawa hemolysin has been reported (74), Osawa et al. (123) subsequently examined 132 strains of *V. parahaemolyticus* and determined that urea hydrolysis was not a reliable marker for the production of the TDH. However, they did find that urea hydrolysis may be a marker for the TDH-related hemolysin, an additional virulence factor described under "Virulence Mechanisms."

Reservoirs

V. parahaemolyticus occurs in estuarine waters throughout the world and is easily isolated from coastal waters of the entire United States, as well as from sediment, suspended particles, plankton, and a variety of fish and shellfish. The latter include at least 30 different species, among them clams, oysters, lobster, scallops, shrimp, and crab. In a study conducted by the FDA, 86% of the 635 seafood samples examined positive for this species. *V. parahaemolyticus* counts as high as 1,300/g of oyster tissue and 1,000/g of crabmeat have been reported, although cell populations of 10/g are more typical for seafood products. Hackney et al. (49), in a 3-year survey of 716 seafood samples obtained in North Carolina, revealed 46% were positive for *V. parahaemolyticus*. Notable were unshucked oysters (79% positive), unshucked clams (83% positive), unpeeled shrimp (60% positive), and live crabs (100% positive). *V. parahaemolyticus* was isolated in another study from ca. 69 to 100% of the commercially obtained or cultured oysters, clams, and shrimps tested, but only 42% of the crabs.

Hackney et al. (49), as well as others, have observed no correlation between fecal coliforms or other indicator microorganisms and the occurrence of *V. parahaemolyticus*, but *V. parahaemolyticus* counts varied with the season, with samples analyzed in January and February often free of *V. parahaemolyticus*. The ecology of *V. parahaemolyticus* is greatly influenced by water temperature, salinity, and association with certain plankton, with highest numbers occurring in warmer months and in waters of intermediate salinity. The relationship between each of these parameters has been demonstrated by Kaneko and Colwell (68) in their study of the occurrence of this species in Chesapeake Bay waters. *V. parahaemolyticus* has been occasionally isolated from freshwater sites, but only at extremely low levels (<5 CFU/liter) and only during the warmest periods of the year.

Although KP+ strains of *V. parahaemolyticus* are of principal importance in human disease, occasional KP− strains also have been isolated from diarrheal stools. KP+ strains constitute a very small percentage (typically <1%) of the *V. parahaemolyticus* strains present in aquatic environments and seafoods. Thus, the simple isolation of this species from water or foodstuffs does not, in itself, indicate a health hazard.

Isolation of *V. parahaemolyticus* generally involves an enrichment procedure, and many enrichment broths have been developed (67, 162). APW provides superior recovery of *V. parahaemolyticus* from a variety of fish and shellfish, even when samples have been chilled or frozen (118). Many plating media also have been developed for isolating this species. Eighteen such media are described and reviewed by Joseph et al. (67) and Twedt (162), of which TCBS agar remains the most commonly employed (118). The 7th edition of the *Bacteriological Analytical Manual* of the FDA (164) prescribed a 16-h enrichment at 35 to 37°C in APW, from which a loopful is streaked onto TCBS agar to obtain isolated colonies of *V. parahaemolyticus*. The newest edition of the manual indicates direct plating of samples onto TCBS agar and employing DNA probes for the *tlh* and *tdh* genes. The former is to obtain isolates and identify as *V. parahaemolyticus*, and the latter as pathogenic *V. parahaemolyticus* strains.

Foodborne Outbreaks

Gastroenteritis with *V. parahaemolyticus* is almost exclusively associated with seafood that is consumed raw, inadequately cooked, or cooked but recontaminated. In Japan, *V. parahaemolyticus* is the major cause of foodborne illness, with as great as 70% of all bacterial foodborne illnesses in that country attributable to this species in the 1960s (162). Outbreaks of *V. parahaemolyticus* gastroenteritis in the United States occurring between 1973 and 1987 have been summarized by Daniels et al. (31). Whereas most Japanese outbreaks involve fish, crab, shrimp, lobster, and oysters are primarily associated with U.S. outbreaks. Shellfish were associated with 33 of 42 outbreaks described by Beuchat (9). The first major outbreak (320 persons ill) in the United States occurred in Maryland in 1971, a result of eating improperly steamed crabs (30). Subsequent outbreaks occurred at all U.S. coasts and Hawaii. The largest U.S. outbreak of *V. parahaemolyticus* occurred during the summer of 1978 and affected 1,133 of 1,700 persons attending a dinner in Port Allen, La. (4). All stool isolates of *V. parahaemolyticus* were KP+. Boiled shrimp, which yielded positive cultures of *V. parahaemolyticus*, was the implicated vehicle. The raw shrimp had been purchased and shipped in standard wooden seafood boxes. They were boiled the morning of the dinner but returned back to the same boxes in which they had been shipped. The warm shrimp were then transported 40 miles in an unrefrigerated truck to the site of the dinner and held an additional 7 to 8 h until served that evening. Another major outbreak, involving 1,416 cases, was associated with consumption of raw oysters from Galveston Bay, Tex. (31).

In a survey of four Gulf Coast states (87), *V. parahaemolyticus* was the most common cause of gastroenteritis (37% of 71 cases) in that area. Similarly, Desenclos et al. (34) determined that *V. parahaemolyticus* was the second leading cause (over 26%) of gastroenteritis in persons who had consumed raw oysters in Florida. In a 15-year survey of *Vibrio* infections reported by a hospital adjacent to the Chesapeake Bay, 9 (69%) of 13 *Vibrio*-positive stool specimens were determined to contain *V. parahaemolyticus* as the sole pathogen (59). In the largest survey reported to date on the epidemiology of 690 *Vibrio* infections in Florida during a 13-year period, 68% of gastroenteritis cases were associated with raw oyster consumption (57).

Characteristics of Disease

V. parahaemolyticus has a remarkable ability for rapid growth, and generation times as short as 8 to 9 min at 37°C have been reported. Even in seafoods, generation times of 12 to 18 min have been observed. Hence, *V. parahaemolyticus* has the ability to grow rapidly, both in vitro and in vivo, and this is evidenced in the characteristics of illness it produces. Symptoms reported in the Maryland outbreak cited above began 4 to more than 30 h after consumption of the implicated food, with a mean time of onset of 23.6 h. The primary symptoms were diarrhea (100%) and abdominal cramps (86%), along with nausea and vomiting (26%) and fever (23%). Symptoms subsided in 3 to 5 days in most individuals, although they lasted 5 to 7 days in 30% and for more than 7 days in another 20% of cases. In the more severe cases, diarrhea was watery with mucus, blood, and tenesmus.

In the Louisiana outbreak of 1978, a mean incubation period of 16.7 h was reported, with a range of 3 to 76 h (4). The duration of illness was from less than 1 day to more than 8 days, with a mean of 4.6 days. Hospitalization was required for more than 7% of the victims. Symptoms included diarrhea (95.1%), cramps (91.5%), weakness (90.2%), nausea (71.9%), chills (54.9%), headache (47.7%), fever (47.5%), and vomiting (12.2%).

Infectious Dose and Susceptible Population

V. parahaemolyticus and *V. cholerae* are the only *Vibrio* spp. for which experimental evidence exists regarding dosages required to initiate gastroenteritis. Studies using human volunteers have revealed that ingestion of 2×10^5 to 3×10^7 CFU of KP$^+$ cells will lead to the rapid development of gastrointestinal disease. Conversely, volunteers receiving as many as 1.6×10^{10} CFU of KP$^-$ cells did not experience diarrheal disease (138). These observations are consistent for all serotypes of *V. parahaemolyticus*, and except for the newly described O3:K6 strains (110), no specific serotype predominates in human illness or environments. On the basis of typical numbers of *V. parahaemolyticus* present in fish and shellfish, and the low

incidence of KP$^+$ cells in these natural samples, a meal of raw shellfish would likely contain no greater that 10^4 KP$^+$ cells. Hence, for disease to result from consumption of contaminated food, mishandling at temperatures allowing growth of the cells likely would be necessary (162). Indeed, Gooch et al. (45a) determined that incubation of shellstock oysters at 26°C for 24 h postharvest resulted in an average 3 log CFU/g increase of *V. parahaemolyticus*.

Virulence Mechanisms

Although the epidemiologic linkage between human virulence and the ability of *V. parahaemolyticus* isolates to produce the Kanagawa hemolysin has long been established, the molecular mechanisms by which this factor can cause diarrhea have only been recently elucidated. Sakazaki et al. (136) observed that 96.5% (2,655 of 2,720) of *V. parahaemolyticus* strains isolated from human patients were KP$^+$. Surprisingly, however, such hemolysin-producing strains are rarely found in the environment. Sakazaki et al. (136) determined that only 1% (7 of 650) of environmental isolates were KP$^+$. Others have reported even lower frequencies (e.g., 4 of 2,218, or 0.18%, were KP$^+$ strains in Galveston Bay, Tex.). It is now believed that such observations can be explained through a natural selection of KP$^+$ strains in the intestinal tract and better survival of KP$^-$ strains in the estuarine environment, and some experimental evidence exists for this explanation (67).

V. parahaemolyticus possesses at least three hemolytic components: a thermolabile hemolysin gene (*tlh*), a thermostable direct hemolysin gene (*tdh*), and a thermostable direct hemolysin-related gene (*trh*). At present, only the direct hemolysins (i.e., those whose lysis of erythrocytes occurs without additional substituents) have been extensively studied. The Kanagawa hemolysin (TDH) is a protein that is only partially inactivated at 100°C for 30 min at pH 6.0. TDH produces edema, erythema, and induration in skin and has capillary permeability activity. TDH is lethal for mice, with a minimum lethal dose of ca. 0.6 μg of protein. Intravenous injections of 5 μg of TDH kill mice rapidly, whereas intraperitoneal injections may produce only signs of cramping. TDH lyses erythrocytes from a large variety of animals, but not those of horses. Hemolytic activity is temperature dependent and is not enhanced by the addition of lecithin, but is inhibited by various gangliosides.

Although the hemolysis is a convenient assay of TDH activity, it does not explain how this toxin causes diarrheal disease. When administered orally to suckling mice at low levels, TDH causes diarrhea, and at higher levels (e.g., 50 μg) TDH produces diarrhea and death. When

100 μg of TDH was administered to ligated rabbit ileal loops, fluid accumulation was not observed, in contrast to CT for which this response occurs following inoculation of only 0.2 μg. Doses of 200 μg of TDH induce a positive rabbit ileal loop response, but with erosive lesions and desquamation of necrotic intestinal mucosa. Such histological changes were not observed when whole bacteria expressing more physiologically relevant TDH levels are tested in rabbit ileal loops. Using a KP$^+$ strain and an isogenic TDH$^-$ mutant, it was determined that only the KP$^+$ parent strain was capable of inducing fluid accumulation in the rabbit ileal loop assay (111). When the complete *tdh* gene was returned to the isogenic mutant, restoration of activity was observed. Similar results were obtained when culture supernatant fluids of these strains were tested on rabbit ileal tissue mounted in Ussing chambers, a more sensitive measure of secretory activity. In this assay, the ability of TDH to alter ion transport in the intestinal tract was observed at nanogram levels, with no histological changes. Raimondi et al. (129) further investigated the effect of purified TDH in Ussing chambers and determined that TDH induces intestinal chloride ion secretion and that the trisialoganglioside GT1b is the cellular receptor. Furthermore, rather than cAMP or cGMP, TDH uses Ca^{2+} as an intracellular second messenger, thereby being the first bacterial enterotoxin for which the linkage between changes in intracellular calcium and secretory activity has been established. More recently (128), human and rat cell cultures were used to determine that the ion influx pathway triggered by TDH is not selective for calcium, with sodium and manganese ions also being admitted. The hemolysin was also determined to trigger a calcium-dependent chloride secretion.

TDH is encoded by one or more nonidentical copies of the *tdh* gene in *V. parahaemolyticus*. Initial studies revealed that KP$^+$ isolates usually contain two copies of the *tdh* gene but that many KP$^-$ or weakly positive isolates have only one gene copy. The two gene copies (*tdh1* and *tdh2*) of KP$^+$ strains are not identical and the predicted protein products differ in seven amino acid residues, although the proteins are immunologically indistinguishable. While both gene products contribute to the KP phenotype, >90% of the protein of TDH can be attributed to high-level expression of the *tdh2* gene (112). Recent studies have revealed that the level of TDH production may be under the control of a regulator similar to the ToxR of *V. cholerae*. A homolog of the *toxRS* genes in *V. parahaemolyticus*, which promotes the expression of the *tdh2* gene but not of the *tdh1* gene, has been reported. As occurs in *V. cholerae*, the extent of the *V. parahaemolyticus* ToxR-stimulated increase in

tdh expression is dependent on the culture medium. The reader is referred to the review by Nishibuchi and Kaper (112) regarding the regulation of *tdh* gene expression and its involvement in the pathogenesis of *V. parahaemolyticus*.

During an outbreak of gastroenteritis, KP$^-$ isolates of *V. parahaemolyticus* were obtained that produced a TDH-related hemolysin, termed TRH, but not TDH. The *trh* gene has 69% identity to the *tdh2* gene, and the biological, immunological, and physicochemical characteristics of TRH are similar, but not identical, to those of TDH (113). A survey of 285 strains of *V. parahaemolyticus* revealed that both *tdh*-positive and *trh*-positive strains were strongly associated with gastroenteritis. As determined by a gene probe, *trh* occurs in more than 35% of 214 clinical strains, including 24% of those lacking the *tdh* gene (142). This suggests that TRH may be an important virulence factor, and possibly the cause of diarrhea in those patients from whom only KP$^-$ strains of *V. parahaemolyticus* are isolated from stools.

A novel siderophore, termed vibrioferrin, has been described for *V. parahaemolyticus*, which sequesters iron from 30% iron-saturated human transferrin. Its importance in pathogenesis is unknown, although the ability to utilize such a source of host iron could enhance the bacterium's survival and proliferation in vivo.

While much is understood regarding the toxins of *V. parahaemolyticus*, little is known of the adherence process, which is an essential step in the pathogenesis of most enteropathogens. Several adhesive factors have been proposed, including the outer membrane, lateral flagella, pili, and a mannose-resistant, cell-associated hemagglutinin, but the importance of any of these factors in human disease is unknown.

Recently, Nasu et al. (110) determined that serotype O3:K6 strains of *V. parahaemolyticus* contain a filamentous phage, f237, which may be exclusively associated with strains isolated since 1996. The involvement of this phage in the virulence of these strains is yet to be determined.

V. VULNIFICUS

V. vulnificus is the most serious of all of the pathogenic vibrios in the United States, solely responsible for 95% of all seafood-borne deaths in this country. This bacterium is the leading cause of reported deaths from foodborne illness in Florida (58). Among that portion of the population that is at risk to infection by this bacterium, primary septicemia cases resulting from raw oyster consumption typically have fatality rates of 60%. This is the highest death rate of any foodborne disease agent in the United

States (98, 159). CDC estimates there are approximately 50 foodborne cases of *V. vulnificus* annually (98, 159). The bacterium can produce wound infections in addition to gastroenteritis and primary septicemias. Wound infections carry a 20 to 25% fatality rate, are also seawater and/or shellfish associated, and generally require surgical debridement of the affected tissue or amputation. The biology of *V. vulnificus*, as well as the clinical manifestations of both the primary septicemic and wound forms, has been described in recent reviews (89, 115, 150). Discussion here is limited to the primary septicemic (foodborne) form of infection caused by this bacterium.

Classification

The first detailed taxonomic study of this species was by researchers at the CDC who in 1976 described 38 strains submitted by clinicians from around the country. Originally termed the "lactose-positive" vibrio, its current name was suggested after a series of phenotypic and genetic studies by several laboratories (115). In 1982, a second biotype of *V. vulnificus* was described that can easily be differentiated from biotype 1 by its negative indole reaction. Biotype 2 strains are a major source of fatalities in eels, and although not generally considered to be pathogenic for humans, can lead to human infection in isolated instances. In 1999, a third biotype was described, which differs from biotype 1 and 2 isolates in being negative for citrate and *o*-nitrophenyl-β-D-galactopyranoside tests and in lack of fermentation of salicin, cellobiose, and lactose. The biotype 3 strains have been isolated only in Israel, and all cases were wound infections. The fish *Tilapia* spp. were implicated as the source of these infections in 98% of the 62 cases. The discussions here will focus on the originally described biotype 1, which is the major human pathogen.

The phenotypic traits of this species have been fully described in several studies (40, 115). The isolation of *V. vulnificus* from blood samples is straightforward, as the bacterium grows readily on TCBS, MacConkey, and blood agars. Isolation from the environment, however, has proven much more problematic. Vibrios comprise 50% or more of estuarine bacterial populations, and most have not been characterized. Further, considerable variation exists in the phenotypic traits ascribed to *V. vulnificus*, including lactose and sucrose fermentation, considered among the most important characteristics in identifying this species. This problem is best exemplified by the isolation from a clinical sample of a bioluminescent strain of this species (120).

Attempts to characterize the distribution of this species have routinely employed TCBS agar for its isolation, although the value of this medium for isolating environmental vibrios has been frequently questioned. Various other media have been proposed for isolating *V. vulnificus*, but the best appears to be colistin-polymyxin B-cellobiose (CPC) agar or one of several more recently described derivatives (118). This medium has been tested in the field for isolating *V. vulnificus* from oysters and clams (119), and Sun and Oliver (153) determined that >80% of 1,000 colonies of appropriate morphology that grew on this medium following plating of oyster homogenates could be identified as *V. vulnificus*. A similar conclusion was reported by Sloan et al. (146) following their study of five selective enrichment broths and two selective agar media for isolating *V. vulnificus* from oysters. The 7th edition of the FDA *Bacteriological Analytical Manual* (164) prescribes a 16-h enrichment at 35 to 37°C in APW, from which a loopful is streaked onto CPC agar to obtain isolated colonies of *V. vulnificus*. Identification of this species is best determined by using a probe against its hemolysin gene, such as that developed by Morris et al. (105), and the latest FDA *Bacteriological Analytical Manual* recommends the use of this probe.

Susceptibility to Control Methods

Cook and Ruple (29) determined that *V. vulnificus* cells naturally occurring in oysters undergo a time-dependent inactivation when either shellstock oysters or shucked oyster meats are held at 4, 0, or −1.9°C. Cook (28) determined that, after harvest, *V. vulnificus* failed within 30 h to grow in shellstock oysters stored at 13°C or below. In oysters held at 18°C or higher for 12 or 30 h, however, *V. vulnificus* populations increased. Susceptibility of *V. vulnificus* to freezing, low-temperature pasteurization, high hydrostatic pressure, and ionizing radiation has been discussed in the introduction of this chapter.

In response to a report that *V. vulnificus* could be killed by applying cocktail or Tabasco sauces to raw oysters, Sun and Oliver (152) determined the sensitivity of this species to a commercial horseradish-based sauce and Tabasco sauce. Results revealed that, while Tabasco sauce (but not cocktail sauce) was highly effective in reducing the *V. vulnificus* populations on the oyster meat surface, little reduction occurred in cell populations within the oysters for either sauce. The in vitro bactericidal activity of a variety of compounds generally recognized as safe against *V. vulnificus* has also been determined. Whereas several of these have been found to kill *V. vulnificus* at low levels, only diacetyl had any significant in vivo effects against this bacterium when naturally present within oysters (151).

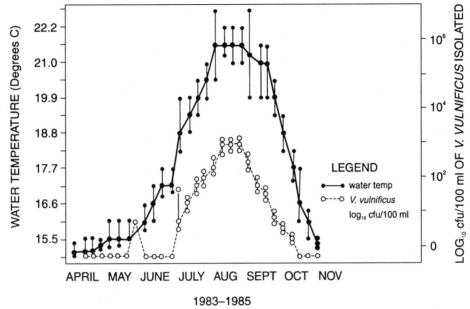

Figure 13.2 Numbers of *V. vulnificus* isolated from northeastern United States coastal waters as a function of time of year and water temperature. From reference 157 with permission of the publisher.

Reservoirs

V. vulnificus is a widespread inhabitant of estuarine environments, having been isolated from the Gulf, East, and Pacific coasts of the United States and from around the world (75, 108, 115, 122, 155). One of the most comprehensive studies on the distribution and ecology of *V. vulnificus* involved a three-summer study of the entire East Coast of the United States (121). Over 6,000 sucrose-negative isolates were examined, and on the basis of phenotypic and DNA-DNA hybridization analyses, it was concluded that approximately 1% of the culturable vibrios were identified as *V. vulnificus*. Isolates were obtained from Cape Cod, Mass., to Miami, Fla., from seawater, plankton, fish, and shellfish. Only ca. 7 CFU of *V. vulnificus* per ml were detected in seawater; however, an average of 6×10^4 CFU/g were detected in oysters. Similar results were obtained by Tamplin et al. (155), who reported from 0.3 to 7,000 CFU of *V. vulnificus* per 100 ml of seawater obtained from two estuarine bays in Florida. *V. vulnificus* has also been isolated from crabs, clams, ark shells, tarbos, and seawater surface samples (115). DePaola et al. (33) have isolated *V. vulnificus* from the intestinal tracts of a variety of bottom-feeding coastal fish and have suggested that such fish may represent a major reservoir of this species.

No correlation could be made between the presence of *V. vulnificus* and the presence of fecal coliforms (121), and this observation has been repeatedly confirmed by other investigators. However, a strong correlation between water temperature and the presence of *V. vulnificus* has been observed (Fig. 13.2) (75, 115, 157), which agrees with epidemiologic studies on infections caused by *V. vulnificus*. Because of this seasonality and the inability to isolate *V. vulnificus* from water or oysters when water temperatures are low, many studies have been done on the apparent die-off of this pathogen during cold weather months. This is now attributed to a cold-induced "viable but nonculturable" state, wherein the cells remain viable but are no longer culturable on the routine media normally employed for their isolation. This phenomenon, which has now been demonstrated in at least 16 genera, has been the subject of several reviews (116, 117). It is also possible, however, that *V. vulnificus* overwinters in certain environments. It is interesting in this context that DePaola et al. (33) isolated the pathogen during the winter from sheepshead fish at higher densities than in sediment or seawater.

Foodborne Outbreaks

There is no report of more than one person developing *V. vulnificus* infection following consumption of the same lot of oysters. There is an occasional report on consumption of raw oysters with subsequent illness in one family member, but others in the family consuming oysters exhibited no symptoms. Further, because raw oysters are usually eaten whole, there rarely exists any remains of the implicated oyster to sample. Raw oysters from the same lot, or even the same serving, may be available;

however, studies have indicated that two oysters, taken from the same estuarine location, may have vastly different cell populations of *V. vulnificus*. Hence, it is difficult to trace back to the source the *V. vulnificus* strain isolated from the patient. It is possible that newer methods, such as randomly amplified polymorphic DNA PCR, may enable molecular epidemiology to be used to aid in tracking infections. However, Buchrieser et al. (15), employing clamped homogeneous electric field gel electrophoresis to examine 118 strains of *V. vulnificus* isolated from three oysters, determined that no two isolates had the same profile. Studies using arbitrarily primed PCR, ribotyping, PFGE, and AFLP all revealed that no two *V. vulnificus* isolates of those evaluated had the same chromosomal arrangement (167). Recently, Vickery et al. (166) employed arbitrarily primed PCR to analyze the DNA profiles obtained from 10 *V. vulnificus* strains isolated from patients who became infected and died following raw oyster consumption. They concluded ". . . that multiple pathogenic strains with greatly diverse genomic arrangements, rather than a single type of infective strain or serogroup, caused these infections". Whether all strains are capable of causing infection or whether only certain strains are pathogenic remains to be determined (15). It is interesting that this species, like other *Vibrio* species, contains two distinct chromosomes (161).

Infections due to *V. vulnificus* correlate distinctly with water temperature, with most cases occurring during the months of April through October. Nearly all cases of *V. vulnificus* infection result from consumption of raw oysters, and most of these infections result in primary septicemias. In a study of 422 infections occurring in 23 states, Shapiro et al. (141) reported that 96% of those patients developing primary septicemia had consumed raw oysters.

A prospective study to determine the incidence of vibrios in symptomatic or asymptomatic infections among persons eating raw shellfish revealed that only 3 of 479 persons had *V. vulnificus* in their stools, and none was ill (92). Coupled with their finding was that two-thirds of the raw oysters tested were culture positive for *V. vulnificus*, indicating that exposure to this species might be relatively high and underscores the need for avoidance of raw seafood by at-risk persons. Inland laboratories rarely encounter *Vibrio* species, with most isolates coming from coastal areas. Even in coastal areas, however, the frequency of *V. vulnificus* infection is not high. In a collaborative 10-year study conducted by five hospitals in Mobile, Ala., only 12 of the vibrios isolated were identified as *V. vulnificus* and none was obtained from stool specimens.

Characteristics of Disease

Several major studies of *V. vulnificus* infections have been reported. In a review of 57 cases of primary septicemia, 85% of which resulted from consumption of raw oysters, the incubation period was determined to be short, ranging from 7 h to several days, with a median of 26 h (115). The most significant symptoms included fever (94%), chills (86%), nausea (60%), and hypotension (systolic pressure <85 mm Hg, 43%). Although frequently present, symptoms typical of gastroenteritis were not as common: abdominal pain (44%), vomiting (35%), and diarrhea (30%). An unusual symptom that generally (69%) occurs in these cases is the development of secondary lesions. These usually occur on the extremities, frequently develop into necrotizing fasciitis or vasculitis, and often necessitate surgical debridement or limb amputation.

Surprisingly, gastrointestinal disease with associated diarrhea is relatively rare. Johnston et al. (65) were the first to provide evidence of such a syndrome, reporting on three males who presented with abdominal cramps; *V. vulnificus* was isolated from their diarrheal stools. All three had a history of alcohol abuse, routinely consumed antacids or cimetidine, and had eaten raw oysters during the week before their illness. Although symptoms in all three subsided without antibiotic treatment, diarrhea continued for a month in one case. A survey of *Vibrio* infections in raw oyster eaters in Florida revealed eight cases of *V. vulnificus*-induced gastroenteritis, six of which involved consumption of raw oysters (34). A survey of *Vibrio* infections occurring during one year in four Gulf Coast states identified three cases of gastroenteritis from *V. vulnificus* (87).

The time to death in the fatal cases of *V. vulnificus* varies considerably, ranging from 2 h post-hospital admission to as long as 6 weeks (115). Most deaths occur within a few days. Hospital stays of several weeks are the norm for those surviving primary septicemic disease. Survival depends to a great extent directly on prompt antibiotic administration.

Infectious Dose and Susceptible Population

In almost all *V. vulnificus* infections that follow ingestion of raw oysters, the patient has an underlying chronic disease (15). The most common (80%) of these is a liver- or blood-related disorder, with liver cirrhosis secondary to alcoholism or alcohol abuse being the most typical (34, 115). These diseases typically result in elevated serum iron levels, and laboratory studies of experimental *V. vulnificus* infections have revealed that elevated serum iron plays a major role in this disease (172). Other risk factors include hematopoietic disorders, chronic renal disease, gastric disease, use of immunosuppressive agents, and diabetes.

Although cases of *V. vulnificus* infection have been reported in children, most infections occur in males (82% of those reviewed by Oliver [115]) whose average age exceeds 50 years. The reason for this gender specificity is that estrogen protects against *V. vulnificus* endotoxin, a virulence factor critical to infection by this pathogen (Huet-Hudson and Oliver, unpublished).

The infectious dose of *V. vulnificus* is not known. A mouse model, however, may offer some insight to this question. When mice were injected with 16 μg of iron to produce serum iron overload, the 50% lethal dose (LD$_{50}$) of *V. vulnificus* decreased from ca. 10^6 to a single cell (172). In a variation of these studies, the administration of small amounts of CCl$_4$ to produce short-term liver necrosis increased serum iron levels, with an inverse correlation between serum iron levels and LD$_{50}$ values observed. Such data agree with epidemiologic studies indicating that liver damage and sometimes immunocompromising diseases are major underlying factors in the development of *V. vulnificus* infections and suggest that extremely small populations of this pathogen may be sufficient to initiate potentially fatal infections.

V. vulnificus is susceptible to most antibiotics and, while a large variety of antibiotics have been used clinically, laboratory studies indicate tetracycline may be especially effective against this species.

Virulence Mechanisms

Capsule

The polysaccharide capsule, which is produced by nearly all strains of *V. vulnificus*, is essential to the bacterium's ability to initiate infection. A study of 38 strains of *V. vulnificus*, including both clinical and environmental isolates and virulent and avirulent strains, revealed that all virulent strains were of the "opaque" (encapsulated) colony type, whereas isogenic cells obtained from "translucent" (acapsular) colonies were avirulent (144) (Fig. 13.3). It was further observed that encapsulated cells produce nonencapsulated cells, and this loss of capsule correlates with loss of virulence in otherwise isogenic strains. Only the encapsulated cells were able to utilize transferrin-bound iron, and only these cells were iron responsive, i.e., were virulent at an inoculum of 10^3 in iron-overloaded mice. The primary reason for the avirulence of translucent cells likely resides in the observation

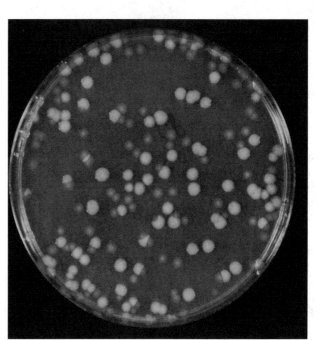

Figure 13.3 Opaque and translucent colonies of *V. vulnificus* C7184. From reference 144.

that the capsule (Fig. 13.4) enables these cells to resist phagocytosis.

At least 10 distinct serotypes were identified using polyclonal antibodies to study the capsule of *V. vulnificus* (143). Of those that were associated with human infection, all 10 serotypes were represented, with 34.6%

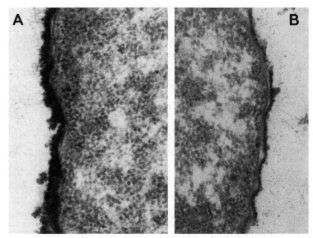

Figure 13.4 Electron micrographs of cells from opaque and translucent colonies of *V. vulnificus* C7184. Cells were stained for acidic polysaccharide with ruthenium red. (A) Opaque cell type, showing thick polysaccharide layer outside the cell envelope. (B) Translucent cell type, demonstrating thin layer of ruthenium red-staining polysaccharide. From reference 144.

being of types 2 and 4. In contrast, 84.6% of the type-able environmental strains were of serotype 3 or 5, but only 7.7% were of type 2 or 4. Similarly, Hayat et al. (52) reported that 19% of 21 clinical strains, but none of the 67 environmental isolates examined, agglutinated with antiserum prepared against a clinical isolate. Whether such differences play a role in the epidemiology of *V. vulnificus* infections remains to be been determined.

Iron

Elevated levels of serum iron appear to be essential for *V. vulnificus* to multiply in the human host (115). The effect of elevated serum iron on reduction of LD_{50} values in mice has been described above. Normal human serum does not permit growth of *V. vulnificus*, suggesting that this bacterium may only produce septicemia in those with elevated serum iron levels (172). Although *V. vulnificus* simultaneously produces both hydroxymate and phenolate siderophores, it is unable to compete with serum transferrin for iron, and this is likely a major factor in its inability to initiate infections in healthy individuals (172). A similar result has been reported for other iron-binding proteins, such as lactoferrin and ferritin. *V. vulnificus* is able to overcome the binding of haptoglobin to hemoglobin, however, and this may represent another aspect of the importance of iron in the pathogenesis of these infections.

LPS

Symptoms that occur during *V. vulnificus* septicemia, including fever, tissue edema, hemorrhage, and especially the significant hypotension, are those classically associated with gram-negative endotoxic shock. Another product of *V. vulnificus*, which may be critical to its virulence, is the endotoxic LPS. Intravenous injections in rodents of extracted and partially purified LPS from *V. vulnificus* caused decreased arterial blood pressure within 10 min, which further declined, leading to a decrease in heart rate and death within 30 to 60 min (97). Subsequent studies revealed that an inhibitor of nitric oxide synthase (the LPS-induced enzyme responsible for release of nitric oxide and subsequent host tissue damage) administered 10 min after LPS injection reversed this lethal effect. These results indicate that the classic symptoms of endotoxic shock observed following *V. vulnificus* infection are likely due to the stimulation by LPS of nitric oxide synthase and that inhibition of this enzyme is a possible treatment for endotoxic shock produced by this pathogen. The gender specificity of *V. vulnificus* infections is likely due to an estrogen-induced protection of females against this endotoxin.

Five LPS serological varieties in *V. vulnificus* have been identified using monoclonal antibodies (143). Of these, LPS antigens 1 and/or 5 were expressed by 25% of clinical isolates, whereas only 0.3% of environmental isolates were of these serotypes. While the chemical composition of LPS extracted from one strain of *V. vulnificus* has been determined, little is known regarding the relative virulence of the different serotypes.

Other Toxins

In addition to the role of capsule, iron, and endotoxin in the pathogenesis of *V. vulnificus* infections, this bacterium produces a large number of extracellular compounds, including hemolysin, protease, elastase, collagenase, DNase, lipase, phospholipase, mucinase, chondroitin sulfatase, hyaluronidase, and fibrinolysin (115). To date, none of these putative virulence factors has definitively been shown to be involved in pathogenesis (89). The most studied of these is a potent heat-stable hemolysin/cytotoxin that possesses cytolytic activity against a variety of mammalian erythrocytes, cytotoxic activity against CHO cells, vascular permeability factor activity against guinea pig skin, and lethal activity for mice. Production of the hemolysin, a metalloprotease, is regulated by a transmembrane virulence regulator, homologous to ToxRS of *V. cholerae*. The toxin has been purified, having a molecular weight of ca. 56,000 and an LD_{50} value for mice of ca. 3 μg/kg following intravenous injection. It enhances vascular permeability through the release of bradykinin and can also specifically degrade type IV collagen, thereby destroying the basal membrane layer of capillary vessels. Mutants for this protease, however, have no reduction in LD_{50} for mice following intraperitoneal injection (140), and the role the protease as a virulence factor remains unclear.

An elastolytic protease has been described that lacks hemolytic activity but degrades albumin, immunoglobulin G, complement factors C3 and 4, and elastin. Minimum lethal doses in mice, regardless of route of injection, are ca. 25 μg, with extensive hemorrhagic necrosis, edema, and muscle tissue destruction occurring.

Although none of these many putative virulence factors is known to be essential for virulence of *V. vulnificus*, they may play a role in pathogenesis or might be essential for wound infections also produced by this species.

V. FLUVIALIS

Classification

V. fluvialis was originally described by Lee et al. (84) and referred to as group F. Subsequently, it was determined that this group was identical to that referred to as Group EF-6 by the CDC, and as a result of further taxonomic study, it was concluded that these isolates were a new species, *V. fluvialis*. Two "biogroups" of group F vibrios were originally described, of which biogroup I was anaerogenic and isolated from aquatic environments and diarrheal cases, whereas biogroup II was aerogenic and not associated with disease. Subsequent studies indicated that the aerogenic strains were a unique species, and these were subsequently reclassified as *V. furnissii* (14).

The possibility of confusion with *Aeromonas* strains (especially *A. hydrophila*) has been recognized as both are arginine dihydrolase positive. The simplest differentiation is the inability of *V. fluvialis*, being halophilic, to grow in media lacking NaCl. The lack of production of indole by *V. fluvialis* also differentiates this species from *Aeromonas* spp. (67).

Reservoirs

V. fluvialis has been isolated frequently from brackish and marine waters and sediments in the United States (67) as well as other countries. It also has been isolated from fish and shellfish from the Pacific Northwest and Gulf Coast. Approximately 65 to 79% of oysters, hard clams, and freshwater clams tested harbored *V. fluvialis*, whereas only 25% of crabs and 6% of shrimp were *V. fluvialis* positive (171). It has only rarely been detected in freshwaters.

Foodborne Outbreaks

V. fluvialis was isolated from more than 500 patients with diarrheal stools at the Cholera Research Laboratory of Bangladesh during a 9-month period between 1976 and 1977 (61). Approximately half of the patients were less than 5 years of age. Since that epidemic, however, *V. fluvialis* has only occasionally been reported to be an enteric pathogen.

In the United States, *V. fluvialis* accounted for 10% of clinical cases in a survey of *Vibrio* infections along the Gulf Coast (87). All seven of these cases resulted in gastroenteritis, with three requiring hospitalization. Consumption of raw oysters was implicated in at least three of the seven, and shrimp in one other.

In the largest clinical series of *V. fluvialis* cases described in the United States, Klontz and Desenclos (78) described 12 persons in Florida from whom this species was recovered between 1982 and 1988. This species was cultured from the stools of 10 of the 12 who presented with gastroenteritis. Eight of the 10 who reported eating seafood during the week before became ill, with raw oysters implicated in five cases, shrimp in two, and cooked fish in one case. In a subsequent survey of vibrioses resulting from raw oyster consumption in Florida during an 8-year period, Desenclos et al. (34) reported that 5.6% of the 125 gastroenteritis cases were caused by *V. fluvialis*.

Characteristics of Disease

Gastroenteritis with *V. fluvialis* is similar to that of cholera, with diarrhea and vomiting (97%), moderate to severe dehydration (67%), abdominal pain (75%), and fever (35%) being common symptoms (61). Passage of 10 to 12 stools per day has been reported, with individual stool outputs of 0.5 to 7 liters. Diarrhea typically lasts from 16 h to more than 3 days. Stools collected during the original Bangladesh outbreak revealed an average of 10^6 *V. fluvialis* cells per ml. A notable difference from cholera is the frequent occurrence of bloody stools in infections due to *V. fluvialis*. While not always observed, Huq et al. (61) determined that 75% of the stool specimens from patients contained erythrocytes or leukocytes, and frank bloody stools were reported for several of the cases. Bloody stools were also reported in 86% of the seven cases reported by Levine et al. (87).

All 10 cases of gastroenteritis described by Klontz and Desenclos (78) had diarrhea (generally watery), with 2 to 20 stools per day (median of 7). Five patients reported at least one episode of bloody stools. Other symptoms included nausea, vomiting, and abdominal cramps. None had hypotension. The median age of the patients was 37 (range of 1 month to 67 years), with eight being male. The median incubation period for six of the patients who had eaten seafood was 39 h (range of 16 to 60 h). The median duration of the illness was 6 days (range of 1 to 60 days); five patients were hospitalized for a median of 4 days (range of 3 to 11 days).

Infectious Dose and Susceptible Population

The infectious dose for *V. fluvialis* is not known. Persons developing gastroenteritis range from 1 month to >80 years of age. Underlying disease does not appear to play a major role in *V. fluvialis* infection. Klontz and Desenclos (78) reported that only 4 of the 10 patients with *V. fluvialis* gastroenteritis they studied had underlying medical conditions, including diabetes, alcohol abuse, and ulcerative colitis. One patient had a history of cardiopulmonary disease and peptic ulcers and was taking antacids at the time of hospitalization. The patient described by Klontz et al. (77) did not have a history of alcoholism or liver disease but had extensive coronary artery disease.

Although successful resolution of infection without antibiotic therapy has been reported, such treatment, often along with intravenous fluids, is generally required. Antibiotic therapy was not successful in at least one case.

Virulence Mechanisms

Most early studies on the virulence of *V. fluvialis* isolates failed to detect evidence of production of an enterotoxin. This might have been due to the growth conditions employed, as it was subsequently determined that production of a heat-labile enterotoxin, which induces fluid accumulation in the intestines of suckling mice, is culture medium dependent. It also was determined that, unlike clinical isolates, most environmental isolates do not induce fluid accumulation.

Three potential enterotoxins have been described for *V. fluvialis*, a CHO cell cytotoxin, a CHO cell rounding toxin, and a CHO cell elongation factor (90). Crude preparations of all three heat-labile toxins cause fluid accumulation in infant mice, but only the CHO cell rounding toxin stimulates secretion in ligated ileal loops.

In possibly the most comprehensive study on the putative virulence factors produced by *V. fluvialis*, Chikahira and Hamada (21) studied cell filtrates and extracts of 39 environmental and clinical isolates. An effect on Chinese hamster ovary cells was observed in 84% and 75% of environmental and clinical isolates, respectively, and 45% and 12.5%, respectively, caused cell elongation. The cell elongation factor was appreciably neutralized by anti-CT serum, but the cell killing factor was not. Live cultures of all of the human and half of the environmental isolates also caused fluid accumulation in rabbit ileal loops. This activity was not neutralized by anti-CT serum. In addition, enzyme-linked immunosorbent assays revealed that 64% of environmental and 67% of human isolates produced CT. None of the culture filtrates produced fluid accumulation in suckling mice, however. Culture filtrates of 15 of the 16 strains tested were lethal for mice (one human isolate was not lethal), with death occurring within 20 min. Less than a third of the isolates were hemolytic for rabbit erythrocytes.

In addition to the above factors, Oliver et al. (120) reported that the strain of *V. fluvialis* they studied produced elastase, mucinase, protease, lipase, lecithinase, chondroitin sulfatase, hyaluronidase, DNase, fibrinolysin, and a hemolysin. In this range of exoproducts, *V. fluvialis* was comparable to *V. vulnificus* and *V. mimicus*.

V. FURNISSII

Classification

The taxonomy of *V. furnissii* has been described by Brenner et al. (14) and Farmer et al. (41). Their studies included biochemical reactions and DNA-DNA hybridization relatedness to *V. fluvialis*, as well as to *Aeromonas* and *Alteromonas* species and to other vibrios. *V. furnissii* is similar to *A. hydrophila*, from which it can easily be distinguished by its ability to grow in 6% NaCl. *V. furnissii* can be differentiated from *V. fluvialis* by, among other traits, its production of gas from glucose.

Reservoirs

V. furnissii has been isolated from river and estuarine water, marine molluscs, and crustacea from throughout the world. Wong et al. (171) isolated *V. furnissii* from a relatively small percentage (ca. 7 to 12%) of oysters, clams, shrimps, and crabs they examined.

Foodborne Outbreaks

The largest documented outbreaks of *V. furnissii* were in 1969, when this bacterium was isolated during an investigation of two outbreaks of acute gastroenteritis in American tourists returning from the Orient. In the first outbreak, 23 of 42 elderly passengers returning from Tokyo developed gastroenteritis; one woman died and two other persons required hospitalization. Food histories implicated shrimp and crab salad and/or the cocktail sauce served with the salads. *V. furnissii* was isolated from seven stool specimens, two of which also contained *V. parahaemolyticus*. The second outbreak affected 24 of 59 persons returning from Hong Kong. Nine persons were hospitalized. A food vehicle was not identified, but *V. furnissii* was isolated from at least five fecal specimens. However, because several other potentially enteropathogenic bacteria also were isolated from these stool specimens, an absolute causal role of *V. furnissii* could not be determined.

Characteristics of Disease

Symptoms of the gastroenteritis outbreaks described above included diarrhea (91 to 100%), abdominal cramps (79 to 100%), nausea (65 to 89%), and vomiting (39 to 78%) (14). There were no reports of fever. Onset of symptoms occurred between 5 and 20 h, with the patients recovering within 24 h.

Infectious Dose and Susceptible Population

These are not known.

Virulence Mechanisms

Chikahira and Hamada (21) provided an extensive description of the toxic products produced by nine environmental strains of *V. furnissii*. Unlike strains of *V. fluvialis*, they found culture filtrates or cell extracts of only one (11%) caused cell elongation of CHO cells; however, 100% caused cell death, an activity which is distinct from the elongation activity (90). They found 86% of the *V. furnissii* strains produced a factor cross-reacting with CT; 50% caused fluid accumulation in rabbit ileal loops and produced a hemolysin active against rabbit erythrocytes. The culture supernatant fluids of all *V. furnissii* strains caused mouse lethality.

V. HOLLISAE

Classification

V. hollisae was described as a new species in 1982 (54). Of the original 16 strains obtained and characterized by CDC, 15 were from stool specimens or intestinal contents, and many of these patients had diarrhea. The phenotypic traits of this species are described by Hickman et al. (54).

V. hollisae is unusual among the vibrios in its inability to grow on TCBS agar or MacConkey agar. As these two media are routinely employed for the examination of stool samples for vibrios, it is possible that many infections caused by this species are missed in clinical laboratories. *V. hollisae* does grow well on blood agar; however, xylose-lysine-deoxycholate agar recovers this species. The API20E system generally can properly identify *V. hollisae*, although the system failed to identify one of the isolates described by Abbott and Janda (1).

Reservoirs

The distribution of *V. hollisae* is not well documented, although it is likely a marine species. It appears that the bacterium prefers warm waters (1).

Foodborne Outbreaks

Thirty cases of *V. hollisae* infection have been reported in the literature since the original description of this species (1). Most (87%) have been cases of gastroenteritis in adults, with only 13% causing extraintestinal disease. Three cases of septicemia associated with *V. hollisae* have been reported.

There is a strong correlation between *V. hollisae* infections and consumption of raw seafood. However, cases associated with consumption of fried catfish and of dried and salted fish have been reported. In addition, Abbott and Janda (1) cited a case of *V. hollisae*-associated diarrhea that involved a 61-year-old male who denied recent travel or seafood consumption. This suggests that additional vehicles may remain to be identified.

A study in 1989 of 121 *Vibrio* infections in four Gulf Coast states identified 9 cases due to *V. hollisae* (87). Eight were cases of gastroenteritis and one was a wound infection. Two of the patients required hospitalization. The most commonly consumed foods by the patients with gastroenteritis were raw oysters, raw clams, crabs, and shrimp. A study of 333 adult cases of *Vibrio* illnesses in Florida during an 8-year period revealed 32 cases were due to *V. hollisae* (34). Of these, 20 (62.5%) occurred following ingestion of raw oysters. Seventeen of these 20 cases resulted in gastroenteritis and 3 developed septicemia. Morris et al. (102) described nine cases of diarrhea that were culture positive for *V. hollisae*, with no other enteric pathogen identified. All had diarrhea and abdominal pain, and all but one were hospitalized. Raw oysters or clams were implicated as vehicles in six cases, and raw shrimp in another. A seventh patient had eaten seafood but denied eating it raw. Abbott and Janda (1) described two incidences of severe gastrointestinal disease caused by *V. hollisae*. In both cases, one a 42-year-old male and the other a 25-year-old female, ingestion of raw oysters preceded the infections. In one of the cases, steamed oysters and cooked crab were also consumed.

Although *V. hollisae* previously had been isolated from a case of septicemia (102), Lowry et al. (93) were the first to describe a case of septicemia from which *V. hollisae* solely was isolated from blood cultures. A 65-year-old male was hospitalized with a 12-h history of night fever, vomiting, and abdominal pain. He had passed two loose stools. On the day before hospitalization, the patient ate fried Mississippi River catfish for lunch and again at dinner. His illness began at midnight, and on admission *V. hollisae* was isolated from a blood culture. The patient was given antibiotics and was discharged 8 days after admission. This case is unusual not only in being a septicemia, but that the infection occurred following consumption of a freshwater fish that had been fried. There was no history of exposure to saltwater or other consumption of seafood by the patient. Apparently, catfish are capable of adapting to low salinities, and the Mississippi River in the area where the fish was obtained often has salt concentrations up to 0.5%. It is likely that incomplete cooking or recontamination of the fish may have occurred. An additional case of foodborne septicemia has been reported that was unusual in that the vehicle appeared to be dried and salted fish obtained from a Southeast Asian food store and which was eaten uncooked.

In the first case of *V. hollisae* infection reported in Europe, Gras-Rouzet et al. (46) described a case of gastroenteritis and bacteremia in a previously healthy 76-year-old man who ate cockles from Brittany, France.

Characteristics of Disease

Symptoms of gastroenteritis caused by *V. hollisae* are similar to those caused by non-O1 strains of *V. cholerae* (102) and typically include severe abdominal cramping, vomiting, fever, and watery diarrhea (1). In cases reported by Morris et al. (102), the median duration of diarrhea was 1 day (range, 4 h to 13 days), with occasional reports of bloody diarrhea. Eight of these patients were hospitalized (median duration, 5 days; range, 2 to 9 days), and all recovered.

Infectious Dose and Susceptible Population

Of the nine gastroenteritis cases described by Morris et al. (102), six were male with an average age of 35 years (range, 31 to 59 years). Other reports do not suggest any significant preference for males or females, or for age differences (1). No seasonality for infections is apparent, with cases occurring in winter and summer months. Most cases of gastroenteritis are in otherwise healthy individuals (1, 102). Abnormal liver function, secondary to alcohol abuse, was present in one case reported by Morris et al. (102). In the septicemia case described by Lowry et al. (93), the patient had consumed a fifth of wine daily for 20 years and had previously undergone pancreatectomy and splenectomy as well as other surgical procedures. Unlike cases of gastroenteritis, underlying disease appears to be the norm for septicemic cases (1, 102).

The organism is highly susceptible to most antimicrobial agents (1, 102), and tobramycin and cefamandole have been used successfully in a septicemic case (93).

Virulence Mechanisms

Gene sequences in *V. hollisae* that are homologous to the TDH gene of *V. parahaemolyticus* have been reported, although strain-to-strain variation exists (112). The hemolysin has been purified and partially characterized, and is related to *V. parahaemolyticus* TDH.

Intragastric administration of *V. hollisae* cells into infant mice elicits intestinal fluid accumulation. An enterotoxin that elongates CHO cells and causes fluid accumulation in mice has been purified from *V. hollisae* (81). It also has been detected in extracts of infected mice and in culture fluids from various growth media.

A hydroxamate siderophore, identified as aerobactin, is produced by *V. hollisae* in response to iron limitation. This iron-binding protein may play a role in the ability of this bacterium to cause infection, and a bacteremic patient had improperly dosed himself with ferrous sulfate as a supplement for chronic anemia, an act that may have exacerbated the disease process.

V. hollisae can adhere to and invade cultured epithelial cells, with internalization involving both eukaryotic and prokaryotic factors (99). In addition to toxin production, the ability to invade epithelial cells is consistent with the invasive disease produced by *V. hollisae* in some patients (93).

V. ALGINOLYTICUS

Classification

V. alginolyticus was originally classified as a biotype of *V. parahaemolyticus*; however, the two are genetically quite similar. They can easily be differentiated phenotypically, most readily by the fermentation of sucrose by *V. alginolyticus*. Some differences in taxonomic traits between clinical and environmental isolates of *V. alginolyticus* have been suggested, but whether these are significant has not been determined.

Reservoirs

V. alginolyticus inhabits, often in large cell populations, seawater and seafood obtained from throughout the world (67). It is easily isolated from fish, clams, crabs, oysters, mussels, and shrimp, as well as water. Many surveys have identified this species of *Vibrio* to be one of the most commonly isolated vibrios. Several studies have reported a temperature correlation with its isolation, with success greatest during the warm water months.

Foodborne Outbreaks

Most *V. alginolyticus* infections are associated with the marine environment and generally remain superficial. Farmer et al. (40), for example, determined only 5.4% of 74 isolates studied at the *Vibrio* Reference Laboratory of the CDC were of fecal or intestinal origin. Only rarely has *V. alginolyticus* been implicated as a foodborne pathogen. This bacterium has been isolated from 0.5% of healthy people in Japan, with no clinically associated intestinal disease evident. *V. alginolyticus* has been isolated from rice water diarrheal stool of a female patient with acute enterocolitis and also from the trout roe she had consumed. *V. alginolyticus* also has been isolated from blood of a leukemic, 49-year-old female who had consumed raw oysters one week earlier. Desenclos et al. (34) reported a case of *V. alginolyticus*-induced gastroenteritis in Florida during their 8-year survey, which represented only 1 of 333 cases of bacteriologically confirmed *Vibrio* illnesses reported in that study.

Characteristics of Disease

There are few reports describing symptoms of gastroenteric disease caused by *V. alginolyticus*. A leukemic patient with *V. alginolyticus* infection was hospitalized with recent confusion, shock, and anemia. She had a temperature of 40°C and systolic blood pressure of 80 mm Hg. Despite antibiotic administration and treatment for shock, the patient died 12 days after admission. The role of *V. alginolyticus* in this case is unknown and was confounded by the presence of *Pseudomonas aeruginosa* in the patient's blood, which was likely a major contributing factor to the fatal outcome.

Infectious Dose and Susceptible Population

Extraintestinal *V. alginolyticus* infections are usually self-limiting and relatively mild, whereas systemic infections are generally severe. Most cases involve patients who are immunocompromised due to severe burns or cancer. A history of alcohol abuse may also be important in the development of *V. alginolyticus* infections.

Virulence Mechanisms

Virtually nothing is known regarding the pathogenic mechanisms of *V. alginolyticus*. Some strains produce lipase, lecithinase, chondroitin sulfatase, DNase, and hemolysin but not elastase, mucinase, protease, or hyaluronidase (120). The role, if any, these putative virulence factors may play in infection has not been elucidated.

CONCLUSIONS

The presence of vibrios, especially *V. cholerae*, *V. vulnificus*, and *V. parahaemolyticus*, in foods represents a serious and growing public health hazard. These species, in addition to seven other recognized human pathogenic vibrios present in foods, are present in estuarine waters and occur frequently in a variety of fish and shellfish. Studies routinely report 30 to 100% of fresh, frozen, or iced fish and shellfish to harbor vibrios, with *V. parahaemolyticus*, *V. cholerae* non-O1, *V. vulnificus*, *V. alginolyticus*, and *V. fluvialis* predominating. The isolation of vibrios from the environment and foodstuffs and their association with infections are seasonal, being generally greater during warm months. Although the overall incidence of *Vibrio* infection has been estimated to be relatively low, many mild infections likely occur that are not reported. Symptoms of *Vibrio* infections range from only mild gastrointestinal upset to death; fatalities are primarily associated with two species, *V. cholerae* and *V. vulnificus*. With some exceptions, e.g., *V. cholerae* O1/O139, most patients who become severely ill have preexisting underlying illnesses, with chronic liver disease being the most common.

Cholera produced by *V. cholerae* is one of the few foodborne diseases with epidemic and endemic potential. The seventh pandemic of cholera, which began in 1961, has involved more than 100 countries, affecting more than 3 million persons and killing many thousands. As more has been learned about the pathogenesis of this bacterium in recent years, it has become clear that its pathogenic potential is quite heterogeneous. Although a number of toxins are elaborated by *V. cholerae*, it is evident that production of CT is essential to its disease production and that possession of either the O1 or O139 antigen usually correlates with this potential. Disease due to *V. cholerae* often involves seafood, but a wide variety of nonacidic foods have been involved in disease transmission.

V. cholerae and *V. parahaemolyticus* are the best described of the pathogenic vibrios. *V. parahaemolyticus* is present in estuarine waters throughout the world; hence, it is generally found in a wide variety of seafoods. As great as 70% of reported bacterial foodborne illnesses in Japan are attributable to this *Vibrio* sp. Outbreaks of *V. parahaemolyticus* infection affecting more than 1,100 people have occurred in the United States, and in all cases, seafood (especially shrimp and oysters) has been identified as the vehicles of illness. Disease is generally associated with strains producing the so-called Kanagawa hemolysin (TDH), which is not produced by most environmental isolates of *V. parahaemolyticus*.

V. vulnificus is the most serious of all the pathogenic vibrios in the United States, being responsible for 95% of all seafood-borne deaths in this country. Like *V. cholerae* and *V. parahaemolyticus*, *V. vulnificus* is present in estuarine waters and in shellfish that inhabit those environments. Development of human infection usually follows consumption of raw oysters harvested from the Gulf Coast but is generally restricted to persons having certain underlying diseases of the liver such as cirrhosis. Fatality rates are approximately 60%.

Human infections caused by other vibrios such as *V. mimicus*, *V. fluvialis*, *V. furnissii*, *V. hollisae*, and *V. alginolyticus* are less common and usually less severe, although deaths have been reported. Seafood is usually the source of infection.

Minimal information is available regarding the susceptibility of vibrios to food preservation methods. Cold is an effective defense against proliferation of vibrios, although seafoods can be protective of some *Vibrio* spp. at refrigeration temperature. Thorough heating of shellfish is adequate to kill all *Vibrio* spp. and is among the best protective measures currently available.

References

1. **Abbott, S. L., and J. M. Janda.** 1994. Severe gastroenteritis associated with *Vibrio hollisae* infection: report of two cases and review. *Clin. Infect. Dis.* **18:**310–312.
2. **Albert, M. J.** 1994. *Vibrio cholerae* O139 Bengal. *J. Clin. Microbiol.* **32:**2345–2349.

3. **Andrews, L., D. L. Park, and Y.-P. Chen.** 2000. Low temperature pasteurization to reduce the risk of *Vibrio* infections in raw-shellstock oysters. *Abstr. 25th Annu. Meet. Seafood Sci. Technol. Soc.*, Longboat Key, Fla.

4. **Anonymous.** 1978. *V. parahaemolyticus* foodborne outbreak—Louisiana. *Morb. Mortal. Wkly. Rep.* **27:**345–346.

5. **Beane, N. H., and P. M. Griffin.** 1990. Foodborne disease outbreaks in the United States, 1973–1987: pathogens, vehicles, and trends. *J. Food Prot.* **53:**804–817.

6. **Bej, A. K., D. P. Patterson, C. W. Brasher, M. C. Vickery, D. D. Jones, and C. A. Kaysner.** 1999. Detection of total and hemolysin-producing *Vibrio parahaemolyticus* in shellfish using multiplex PCR amplification of *tl*, *tdh* and *trh*. *J. Microbiol. Methods* **36:**215–225.

7. **Benitez, J. A., L. Garcia, A. Silva, H. Garcia, R. Fando, B. Cedre, A. Perez, J. Campos, B. L. Rodriguez, J. L. Perez, T. Valmaseda, O. Perez, A. Perez, M. Ramirez, T. Ledon, M. D. Jidy, M. Lastre, L. Bravo, and G. Sierra.** 1999. Preliminary assessment of the safety and immunogenicity of a new CTXϕ-negative, hemagglutinin/protease-defective El Tor strain as a cholera vaccine candidate. *Infect. Immun.* **67:**539–545.

8. **Berry, T. M., D. L. Park, and D. V. Lightner.** 1994. Comparison of the microbial quality of raw shrimp from China, Ecuador, or Mexico at both wholesale and retail levels. *J. Food Prot.* **57:**150–153.

9. **Beuchat, L. R.** 1982. *Vibrio parahaemolyticus*: public health significance. *Food Technol.* **36:**80–83.

10. **Bisharat, N., V. Agmon, R. Finkelstein, R. Raz, G. Ben-Dror, L. Lerner, S. Soboh, R. Coldner, D. N. Cameron, D. L. Wykstra, D. L. Swerdlow, and J. J. Farmer III.** 1999. Clinical, epidemiological, and microbiological features of *Vibrio vulnificus* biogroup 3 causing outbreaks of wound infection and bacteraemia in Israel. *Lancet* **354:**1421–1424.

11. **Blake, P. A.** 1994. Endemic cholera in Australia and the United States, p. 309–319. *In* I. K. Wachsmuth, P. A. Blake, and Ø. Olsvik (ed.), *Vibrio cholerae and Cholera: Molecular to Global Perspectives.* ASM Press, Washington, D.C.

12. **Blake, P. A., R. E. Weaver, and D. G. Hollis.** 1980. Diseases of humans (other than cholera) caused by vibrios. *Annu. Rev. Microbiol.* **34:**341–367.

13. **Bonner, J. R., A. S. Coker, C. R. Berryman, and H. M. Pollock.** 1983. Spectrum of *Vibrio* infections in a Gulf Coast community. *Ann. Intern. Med.* **99:**464–469.

14. **Brenner, D. J., F. W. Hickman-Brenner, J. V. Lee, A. G. Steigerwalt, G. R. Fanning, D. G. Hollis, J. J. Farmer III, R. E. Weaver, S. W. Joseph, and R. J. Seidler.** 1983. *Vibrio furnissii* (formerly aerogenic biogroup of *Vibrio fluvialis*), a new species isolated from human feces and the environment. *J. Clin. Microbiol.* **18:**816–824.

15. **Buchrieser, C., V. V. Gangar, R. L. Murphree, M. L. Tamplin, and C. W. Kaspar.** 1995. Multiple *Vibrio vulnificus* strains in oysters as demonstrated by clamped homogeneous electric field gel electrophoresis. *Appl. Environ. Microbiol.* **61:**1163–1168.

16. **Buck, J. D.** 1998. Potentially pathogenic *Vibrio* spp. in market seafood and natural habitats from Southern New England and Florida. *J. Aquat. Food Prod. Technol.* **7:**53–62.

17. **Burkhardt, W., III, S. R. Rippey, and W. D. Watkins.** 1992. Depuration rates of Northern quahogs, *Mercenaria mercenaria* (Linnaeus, 1758) and Eastern oysters *Crassostrea virginica* (Gmelin, 1791) in ozone- and ultraviolet light-disinfected seawater systems. *J. Shellfish Res.* **11:**105–109.

18. **Butterton, J. R., J. A. Stoebner, S. M. Payne, and S. B. Calderwood.** 1992. Cloning, sequencing, and transcriptional regulation of *viuA*, the gene encoding the ferric vibriobactin receptor of *Vibrio cholerae*. *J. Bacteriol.* **174:**3729–3738.

19. **Calik, H., M. T. Morrissey, P. Reno, R. Adams, and H. An.** 2000. Low temperature pasteurization to reduce the risk of *Vibrio* infections in raw-shellstock oysters. *Abstr. 25th Annu. Meet. Seafood Sci. Technol. Soc.*, Longboat Key, Fla.

20. **Cameron, D. N., F. M. Khambaty, I. K. Wachsmuth, R. V. Tauxe, and T. J. Barrett.** 1994. Molecular characterization of *Vibrio cholerae* O1 by pulsed-field gel electrophoresis. *J. Clin. Microbiol.* **32:**1685–1690.

21. **Chikahira, M., and K. Hamada.** 1988. Enterotoxigenic substances and other toxins produced by *Vibrio fluvialis* and *Vibrio furnissii*. *Jpn. J. Vet. Sci.* **50:**865–873.

22. **Chowdhury, M. A. R., K. M. S. Aziz, B. A. Kay, and Z. Rahim.** 1987. Toxin production by *Vibrio mimicus* strains isolated from human and environmental sources in Bangladesh. *J. Clin. Microbiol.* **25:**2200–2203.

23. **Chowdhury, M. A. R., R. T. Hill, and R. R. Colwell.** 1994. A gene for the enterotoxin zonula occludens toxin is present in *Vibrio mimicus* and *Vibrio cholerae* O139. *FEMS Microbiol. Lett.* **119:**377–380.

24. **Chowdhury, M. A. R., H. Yamanaka, S. Miyoshi, K. M. S. Aziz, and S. Shinoda.** 1989. Ecology of *Vibrio mimicus* in aquatic environments. *Appl. Environ. Microbiol.* **55:**2073–2078.

25. **Colwell, R. R., and A. Huq.** 1994. Vibrios in the environment: viable but nonculturable *Vibrio cholerae*, p. 117–133. *In* K. Wachsmuth, P. A. Blake, and Ø. Olsvik (ed.), *Vibrio cholerae and Cholera: Molecular to Global Perspectives.* ASM Press, Washington, D.C.

26. **Colwell, R. R., M. L. Tamplin, P. R. Brayton, A. L. Gauzens, B. D. Tall, D. Herrington, M. M. Levine, S. Hall, A. Huq, and D. A. Sack.** 1990. Environmental aspects of *Vibrio cholerae* in transmission of cholera, p. 327–343. *In* R. B. Sack and Y. Zinnaka (ed.), *Advances in Research on Cholera and Related Diarrheas*, vol. 7. KTK Scientific Publishers, Tokyo, Japan.

27. **Comstock, L. E., J. A. Johnson, J. M. Michalski, J. G. Morris, Jr., and J. B. Kaper.** 1996. Cloning and sequence of a region encoding surface polysaccharide of *Vibrio cholerae* O139 and characterization of the insertion site in the chromosome of *Vibrio cholerae* O1. *Mol. Microbiol.* **19:**815–826.

28. **Cook, D. W.** 1994. Effect of time and temperature on multiplication of *Vibrio vulnificus* in postharvest

Gulf Coast shellstock oysters. *Appl. Environ. Microbiol.* **60:**3483–3484.

29. Cook, D. W., and A. D. Ruple. 1992. Cold storage and mild heat treatment as processing aids to reduce the numbers of *Vibrio vulnificus* in raw oysters. *J. Food Prot.* **55:**985–989.

30. Dadisman, T. A., R. Nelson, J. R. Molenda, and H. J. Garber. 1972. *Vibrio parahaemolyticus* gastroenteritis in Maryland. I. Clinical and epidemiological aspects. *Am. J. Epidemiol.* **96:**414–426.

31. Daniels, N. A., L. MacKinnon, R. Bishop, S. Altekruse, B. Ray, R. M. Hammond, S. Thompson, S. Wilson, N. H. Bean, P. M. Griffin, and L. Slutsker. 2000. *Vibrio parahaemolyticus* infections in the United States, 1973–1998. *J. Infect. Dis.* **181:**1661–1666.

32. Davis, B. R., G. R. Fanning, J. M. Madden, A. G. Steigerwalt, H. B. Bradford, Jr., H. L. Smith, Jr., and D. J. Brenner. 1981. Characterization of biochemically atypical *Vibrio cholerae* strains and designation of a new pathogenic species, *Vibrio mimicus*. *J. Clin. Microbiol.* **14:**631–639.

33. DePaola, A., G. M. Capers, and D. Alexander. 1994. Densities of *Vibrio vulnificus* in the intestines of fish from the U.S. Gulf Coast. *Appl. Environ. Microbiol.* **60:**984–988.

34. Desenclos, J.-C. A., K. C. Klontz, L. E. Wolfe, and S. Hoecherl. 1991. The risk of *Vibrio* illness in the Florida raw oyster eating population, 1981–1988. *Am. J. Epidemiol.* **134:**290–297.

35. DiRita, V. J. 1992. Co-ordinate expression of virulence genes by ToxR in *Vibrio cholerae*. *Mol. Microbiol.* **6:**451–458.

36. Dixon, W. D. 1992. The effects of gamma radiation (^{60}Co) upon shellstock oysters in terms of shelf life and bacterial reduction, including *Vibrio vulnificus* levels. M.S. thesis. University of Florida, Gainesville.

37. Everiss, K. D., K. J. Hughes, M. E. Kovach, and K. M. Peterson. 1994. The *Vibrio cholerae* acfB colonization determinant encodes an inner membrane protein that is related to a family of signal-transducing proteins. *Infect. Immun.* **62:**3289–3298.

38. Eyles, M. J., and G. R. Davey. 1984. Microbiology of commercial depuration of the Sydney rock oyster, *Crassostrea commercialis*. *J. Food Prot.* **47:**703–706.

39. Farmer, J. J., III, F. W. Hickman-Brenner, G. R. Fanning, C. M. Gordon, and D. J. Brenner. 1988. Characterization of *Vibrio metschnikovii* and *Vibrio gazogenes* by DNA-DNA hybrization and phenotype. *J. Clin. Microbiol.* **26:**1993–2000.

40. Farmer, J. J., III, F. W. Hickman-Brenner, and M. T. Kelly. 1995. *Vibrio*, p. 282–301. *In* P. Murray et al. (ed.), *Manual of Clinical Microbiology*, 6th ed. American Society for Microbiology, Washington, D.C.

41. Farmer, R. E., III, Weaver, S. W. Joseph, and R. J. Seidler. 1983. *Vibrio furnissii* (formerly aerogenic biogroup of *Vibrio fluvialis*), a new species isolated from human feces and the environment. *J. Clin. Microbiol.* **18:**816–824.

42. Fasano, A., B. Baudry, D. W. Pumplin, S. S. Wasserman, B. D. Tall, J. M. Ketley, and J. B. Kaper. 1991. *Vibrio cholerae* produces a second enterotoxin, which affects intestinal tight junctions. *Proc. Natl. Acad. Sci. USA* **88:**5242–5246.

43. Gabriel, S. E., K. N. Brigman, B. H. Koller, R. C. Boucher, and M. J. Stutts. 1994. Cystic fibrosis heterozygote resistance to cholera toxin in the cystic fibrosis mouse model. *Science* **266:**107–109.

44. Galen, J. E., J. M. Ketley, A. Fasano, S. H. Richardson, S. S. Wasserman, and J. B. Kaper. 1992. Role of *Vibrio cholerae* neuraminidase in the function of cholera toxin. *Infect. Immun.* **60:**406–415.

45. Glass, R. I., and R. E. Black. 1992. The epidemiology of cholera, p. 129–154. *In* D. Barua and W. B. Greenough III (ed.), *Cholera*. Plenum Medical Book Co., New York, N.Y.

45a. Gooch, J. A., A. DePaola, C. A. Kaysner, and D. L. Marshall. 1999. Postharvest growth and survival of *Vibrio parahaemolyticus* in oysters stored at 26° and 3°C. *Abstr. Gen. Meet. Am. Soc. Microbiol.* P-52, p. 521. American Society for Microbiology, Washington, D.C.

46. Gras-Rouzet, S., P. Y. Donnio, F. Juguet, P. Plessis, J. Minet, and J. L. Avril. 1996. First European case of gastroenteritis and bacteremia due to *Vibrio hollisae*. *Eur. J. Clin. Microbiol. Infect. Dis.* **15:**864–866.

47. Groubert, T. N., and J. D. Oliver. 1994. Interaction of *Vibrio vulnificus* and the Eastern oyster, *Crassostrea virginica*. *J. Food Prot.* **57:**224–228.

48. Guerrant, R. L., G. D. Fang, N. M. Thielman, and M. C. Fonteles. 1994. Role of platelet activating factor (PAF) in the intestinal epithelial secretory and Chinese hamster ovary (CHO) cell cytoskeletal responses to cholera toxin. *Proc. Natl. Acad. Sci. USA* **91:**9655–9658.

49. Hackney, C. R., B. Ray, and M. L. Speck. 1980. Incidence of *Vibrio parahaemolyticus* in and the microbiological quality of seafood in North Carolina. *J. Food Prot.* **43:**769–773.

50. Häse, C. C., and J. J. Mekalanos. 1999. Effects of changes in membrane sodium flux on virulence gene expression in *Vibrio cholerae*. *Proc. Natl. Acad. Sci. USA* **96:**3183–3187.

51. Häse, C. C., and J. J. Mekalanos. 1998. TcpP protein is a positive regulator of virulence gene expression in *Vibrio cholerae*. *Proc. Natl. Acad. Sci. USA* **95:**730–734.

52. Hayat, U., G. P. Reddy, C. A. Bush, J. A. Johnson, A. C. Wright, and J. G. Morris, Jr. 1993. Capsular types of *Vibrio vulnificus*: an analysis of strains from clinical and environmental sources. *J. Infect. Dis.* **168:**758–762.

53. Henderson, D. P., and S. M. Payne. 1994. Characterization of the *Vibrio cholerae* outer membrane heme transport protein HutA: sequence of the gene, regulation of expression, and homology to the family of TonB-dependent proteins. *J. Bacteriol.* **176:**3269–3277.

54. Hickman, F. W., J. J. Farmer III, D. G. Hollis, G. R. Fanning, A. G. Steigerwalt, R. E. Weaver, and D. J. Brenner. 1982. Identification of *Vibrio hollisae* sp. nov. from patients with diarrhea. *J. Clin. Microbiol.* **15:**395–401.

55. **Hillman, C.** 2000. Low temperature pasteurization to reduce the risk of *Vibrio* infections in raw-shellstock oysters. *Abstr. 25th Annu. Meet. Seafood Sci. Technol. Soc.*, Longboat Key, Fla.

56. **Hlady, W. G.** 1997. *Vibrio* infections associated with raw oyster consumption in Florida, 1981–94. *J. Food Prot.* **60**:353–357.

57. **Hlady, W. G., and K. C. Klontz.** 1996. The epidemiology of *Vibrio* infections in Florida, 1981–1993. *J. Infect. Dis.* **173**:1176–1183.

58. **Hlady, W. G., R. C. Mullen, and R. S. Hopkin.** 1993. *Vibrio vulnificus* from raw oysters. Leading cause of reported deaths from foodborne illness in Florida. *J. Fla. Med. Assoc.* **80**:536–538.

59. **Hoge, C. W., D. Watsky, R. N. Peeler, J. P. Libonati, E. Israel, and J. G. Morris, Jr.** 1989. Epidemiology and spectrum in *Vibrio* infections in a Chesapeake Bay USA community. *J. Infect. Dis.* **160**:985–993.

60. **Honda, T., and T. Miwatani.** 1988. Multi-toxigenicity of *Vibrio cholerae* non-O1, p. 23–32. *In* N. Ohtomo and R. B. Sack (ed.), *Advances in Research on Cholera and Related Diarrheas*, vol. 6. KTK Scientific Publishers, Tokyo, Japan.

60a. **Howard, R. J., and N. T. Bennett.** 1993. Infections caused by halophilic marine *Vibrio* bacteria. *Ann. Surg.* **217**:525–530.

61. **Huq, M. I., A. K. M. J. Alam, D. J. Brenner, and G. K. Morris.** 1980. Isolation of *Vibrio*-like group, EF-6, from patients with diarrhea. *J. Clin. Microbiol.* **11**:621–624.

62. **Janda, J. M., C. Powers, R. G. Bryant, and S. L. Abbott.** 1988. Current perspectives on the epidemiology and pathogenesis of clinically significant *Vibrio* spp. *Clin. Microbiol. Rev.* **1**:245–267.

63. **Johnson, J. A., C. A. Salles, P. Panigrahi, M. J. Albert, A. C. Wright, R. J. Johnson, and J. G. Morris, Jr.** 1994. *Vibrio cholerae* O139 synonym Bengal is closely related to *Vibrio cholerae* El Tor but has important differences. *Infect. Immun.* **62**:2108–2110.

64. **Johnson, W. G., Jr., A. C. Salinger, and W. C. King.** 1973. Survival of *V. parahaemolyticus* in oyster shellstock at two different storage temperatures. *Appl. Microbiol.* **26**:122–123.

65. **Johnston, J. M., S. F. Becker, and L. M. McFarland.** 1986. Gastroenteritis in patients with stool isolations of *Vibrio vulnificus*. *Am. J. Med.* **80**:336–338.

66. **Jones, S. H., T. L. Howell, and K. R. O'Neill.** 1991. Differential elimination of indicator bacteria and pathogenic *Vibrio* sp. from Eastern oysters (*Crassostrea virginica gmelin*, 1791) in a commercial purification facility in Maine. *J. Shellfish Res.* **10**:105–112.

67. **Joseph, S. W., R. R. Colwell, and J. B. Kaper.** 1982. *Vibrio parahaemolyticus* and related halophilic vibrios. *Crit. Rev. Microbiol.* **10**:77–124.

68. **Kaneko, T., and R. R. Colwell.** 1978. The annual cycle of *Vibrio parahaemolyticus* in Chesapeake Bay. *Microb. Ecol.* **4**:135–155.

69. **Kaper, J. B., A. Fasano, and M. Trucksis.** 1994. Toxins of *Vibrio cholerae*, p. 145–176. *In* I. K. Wachsmuth, P.

Blake, and Ø. Olsvik (ed.), *Vibrio cholerae and Cholera*. American Society for Microbiology, Washington, D.C.

70. **Kaper, J. B., J. G. Morris, Jr., and M. M. Levine.** 1995. Cholera. *Clin. Microbiol. Rev.* **8**:48–86.

71. **Karaolis, D. K. R., J. A. Johnson, C. C. Bailey, E. C. Boedeker, J. B. Kaper, and P. R. Reeves.** 1998. A *Vibrio cholerae* pathogenicity island associated with epidemic and pandemic strains. *Proc. Natl. Acad. Sci. USA* **95**:3134–3139.

72. **Kaysner, C. A., and W. E. Hill.** 1994. Toxigenic *Vibrio cholerae* O1 in food and water, p. 27–39. *In* I. K. Wachsmuth, P. A. Blake, and Ø. Olsvik (ed.), *Vibrio cholerae and Cholera: Molecular to Global Perspectives*. ASM Press, Washington, D.C.

73. **Kaysner, C. A., C. Abeyta, M. M. Wekell, A. DePaola, R. F. Stott, and J. M. Leitch.** 1987. Virulent strains of *Vibrio vulnificus* from estuaries of the U.S. West Coast. *Appl. Environ. Microbiol.* **53**:1349–1351.

74. **Kaysner, C. A., C. Abeyta, Jr., P. A. Trost, J. H. Wetherington, K. C. Jinneman, W. E. Hill, and M. M. Wekell.** 1994. Urea hydrolysis can predict the potential pathogenicity of *Vibrio parahaemolyticus* strains isolated in the Pacific Northwest. *Appl. Environ. Microbiol.* **60**:3020–3022.

75. **Kelly, M. T.** 1982. Effect of temperature and salinity on *Vibrio (Beneckea) vulnificus* occurrence in a Gulf Coast environment. *Appl. Environ. Microbiol.* **44**:820–824.

76. **Kilgen, M. B.** 2000. Low temperature pasteurization to reduce the risk of *Vibrio* infections in raw-shellstock oysters. *Abstr. 25th Annu. Meet. Seafood Sci. Technol. Soc.*, Longboat Key, Fla.

77. **Klontz, K. C., D. E. Cover, F. N. Hyman, and R. C. Mullen.** 1994. Fatal gastroenteritis due to *Vibrio fluvialis* and nonfatal bacteremia due to *Vibrio mimicus*: unusual vibrio infections in two patients. *Clin. Infect. Dis.* **19**:541–542.

78. **Klontz, K. C., and J.-C. A. Desenclos.** 1990. Clinical and epidemiological features of sporadic infections with *Vibrio fluvialis* in Florida USA. *J. Diarrh. Dis. Res.* **8**:1–2.

79. **Koch, W. H., W. L. Payne, B. A. Wentz, and T. A. Cebula.** 1993. Rapid polymerase chain reaction method for detection of *Vibrio cholerae* in foods. *Appl. Environ. Microbiol.* **59**:556–560.

80. **Kodama, H., Y. Gyobu, N. Tokuman, H. Uetake, T. Shimada, and R. Sakazaki.** 1988. Ecology of non-O1 *Vibrio cholerae* and *Vibrio mimicus* in Toyama prefecture, p. 79–88. *In* N. Ohtomo and R. B. Sack (ed.), *Advances in Research on Cholera and Related Diarrheas*, vol. 6. KTK Scientific Publishers, Tokyo, Japan.

81. **Kothary, M. H., E. F. Claverie, M. D. Miliotis, J. M. Madden, and S. H. Richardson.** 1995. Purification and characterization of a Chinese hamster ovary cell elongation factor of *Vibrio hollisae*. *Infect. Immun.* **63**:2418–2423.

82. **Kreger, A., and D. Lockwood.** 1981. Detection of extracellular toxin(s) produced by *Vibrio vulnificus*. *Infect. Immun.* **33**:583–590.

83. Lee, J. V., T. J. Donovan, and A. Furniss. 1978. Characterisation, taxonomy and emended description of *Vibrio metschnikovii. Int. J. Syst. Bacteriol.* 28:99–111.

84. Lee, J. V., P. Shread, and A. L. Furniss. 1978. The taxonomy of group F organisms: relationships to *Vibrio* and *Aeromonas. J. Appl. Bacteriol.* 45:ix.

85. Levine, M. M., R. E. Black, M. L. Clements, D. R. Nalin, L. Cisneros, and R. A. Finkelstein. 1981. Volunteer studies in development of vaccines against cholera and enterotoxigenic *Escherichia coli*: a review, p. 443–459. *In* T. Holme, J. Holmgren, M. H. Merson, and R. Mollby (ed.), *Acute Enteric Infections in Children. New Prospects for Treatment and Prevention.* Elsevier/North-Holland Biomedical Press, Amsterdam, The Netherlands.

86. Levine, M. M., J. B. Kaper, D. Herrington, G. Losonsky, J. G. Morris, M. L. Clements, R. E. Black, B. Tall, and R. Hall. 1988. Volunteer studies of deletion mutants of *Vibrio cholerae* O1 prepared by recombinant techniques. *Infect. Immun.* 56:161–167.

87. Levine, W. C., P. M. Griffin, and the Gulf Coast Vibrio Working Group. 1993. *Vibrio* infections on the Gulf Coast: results of first year of regional surveillance. *J. Infect. Dis.* 167:479–483.

88. Lin, W., K. J. Fullner, R. Clayton, J. A. Sexton, M. B. Rogers, K. E. Calia, S. B. Calderwood, C. Fraser, and J. J. Mekalanos. 1999. Identification of a *Vibrio cholerae* RTX toxin gene cluster that is tightly linked to the cholera toxin prophage. *Proc. Natl. Acad. Sci. USA* 96:1071–1076.

89. Linkous, D. A., and J. D. Oliver. 1999. Pathogenesis of *Vibrio vulnificus. FEMS Microbiol. Lett.* 174:207–214.

90. Lockwood, D. E., A. S. Kreger, and S. H. Richardson. 1982. Detection of toxins produced by *Vibrio fluvialis. Infect. Immun.* 35:702–708.

91. Lockwood, D. E., S. H. Richardson, A. S. Kreger, M. Aiken, and B. McCreedy. 1983. In vitro and in vivo biologic activities of *Vibrio fluvialis* and its toxic products, p. 87–99. *In* S. Kuwahara and N. F. Pierce (ed.), *Advances in Research on Cholera and Related Diarrheas*, vol. 1. KTK Scientific Publishers, Tokyo, Japan.

92. Lowry, P. W., L. M. McFarland, B. H. Peltier, N. C. Roberts, H. B. Bradford, J. L. Herndon, D. F. Stroup, J. B. Mathison, P. A. Blake, and R. A. Gunn. 1989. *Vibrio* gastroenteritis in Louisiana: a prospective study among attendees of a scientific congress in New Orleans. *J. Infect. Dis.* 160:978–984.

93. Lowry, P. W., L. M. McFarland, and H. K. Threefoot. 1986. *Vibrio hollisae* septicemia after consumption of catfish. *J. Infect. Dis.* 154:730–731.

94. Madden, J. M., B. A. McCardell, and J. G. Morris, Jr. 1989. *Vibrio cholerae*. p. 525–542. *In* M. P. Doyle (ed.), *Foodborne Bacterial Pathogens.* Marcel Dekker, Inc., New York, N.Y.

95. Magalhaes, V., A. Castello Filho, M. Magalhaes, and T. T. Gomes. 1993. Laboratory evaluation on pathogenic potentialities of *Vibrio furnissii. Mem. Inst. Oswaldo Cruz Rio de Janeiro* 88:593–597.

96. Matté G. R., M. H. Matté, I. G. Rivera, and M. T. Martins. 1994. Distribution of pathogenic vibrios in oysters from a tropical region. *J. Food Prot.* 57:870–873.

97. McPherson, V. L., J. A. Watts, L. M. Simpson, and J. D. Oliver. 1991. Physiological effects of the lipopolysaccharide of *Vibrio vulnificus* on mice and rats. *Microbios* 67:141–149.

98. Mead, P. S., L. Slutsker, V. Dietz, L. F. McCaig, J. S. Bresee, C. Shapiro, P. M. Griffin, and R. B. V. Tauxe. 1999. Food-related illness and death in the United States. *Emerg. Infect. Dis.* 5:607–625.

99. Miliotis, M. D., B. D. Tall, and R. T. Gray. 1995. Adherence to and invasion of tissue culture cells by *Vibrio hollisae. Infect. Immun.* 63:4959–4963.

100. Minami, A., S. Hashimoto, H. Abe, M. Arita, T. Taniguchi, T. Honda, T. Miwatani, and M. Nishibuchi. 1991. Cholera enterotoxin production in *Vibrio cholerae* O1 strains isolated from the environment and from humans in Japan. *Appl. Environ. Microbiol.* 57:2152–2157.

101. Mintz, E. D., T. Popovic, and P. A. Blake. 1994. Transmission of *Vibrio cholerae* O1, p. 345–356. *In* I. K. Wachsmuth, P. A. Blake, and Ø. Olsvik (ed.), *Vibrio cholerae and Cholera: Molecular to Global Perspectives.* ASM Press, Washington, D.C.

102. Morris, G. J., R. Wilson, D. G. Hollis, R. E. Weaver, H. G. Miller, C. O. Tacket, F. W. Hickman, and P. A. Blake. 1982. Illness caused by *Vibrio damsela* and *Vibrio hollisae. Lancet* 1:1294–1296

103. Morris, J. G., Jr. 1990. Non-O group 1 *Vibrio cholerae*: a look at the epidemiology of an occasional pathogen. *Epidemiol. Rev.* 12:179–191.

104. Morris, J. G., Jr. 1995. "Noncholera" *Vibrio* species, p. 671–685. *In* M. J. Blaser, P. D. Smith, J. I. Ravdin, H. B. Greenberg, and R. L. Guerrant (ed.), *Infections of the Gastrointestinal Tract.* Raven Press, Ltd., New York, N.Y.

105. Morris, J. G., Jr., A. C. Wright, D. M. Roberts, P. K. Wood, L. M. Simspon, and J. D. Oliver. 1987. Identification of environmental *Vibrio vulnificus* isolates with a DNA probe for the cytotoxin-hemolysin gene. *Appl. Environ. Microbiol.* 53:193–195.

106. Moss, J., and M. Vaughan. 1991. Activation of cholera toxin and *Escherichia coli* heat-labile enterotoxins by ADP-ribosylation factors, a family of 20-kDa guanine nucleotide-binding proteins. *Mol. Microbiol.* 5:2621–2627.

107. Motes, M., A. DePaola, S. Zywno-Van Ginkel, and M. McPhearson. 1994. Occurrence of toxigenic *Vibrio cholerae* O1 in oysters in Mobile Bay, Alabama: an ecological investigation. *J. Food Prot.* 57:975–980.

108. Motes, M. L., A. DePaola, D. W. Cook, J. E. Veazey, J. C. Hunsucker, W. E. Garthright, R. J. Blodgett, and S. Chirtel. 1998. Influence of water temperature and salinity on *Vibrio vulnificus* in Northern Gulf and Atlantic Coast oysters (*Crassostrea virginica*). *Appl. Environ. Microbiol.* 64:1459–1465.

109. Nascumento, D. R., R. H. Vieira, H. B. Almeida, T. R. Patel, and S. T. Iaria. 1998. Survival of *Vibrio cholerae* O1 strains in shrimp subjected to freezing and boiling. *J. Food Prot.* **61:**1317–1320.

110. Nasu, H., I. Tetsuya, T. Sugahara, Y. Yamaichi, K.-S. Park, K. Yokoyama, K. Makino, H. Shinagawa, and T. Honda. 2000. A filamentous phage associated with recent pandemic *Vibrio parahaemolyticus* O3:K6 strains. *J. Clin. Microbiol.* **38:**2156–2161.

111. Nishibuchi, M., A. Fasano, R. G. Russell, and J. B. Kaper. 1992. Enterotoxigenicity of *Vibrio parahaemolyticus* with and without genes encoding thermostable direct hemolysin. *Infect. Immun.* **60:**3539–3545.

112. Nishibuchi, M., and J. B. Kaper. 1995. Thermostable direct hemolysin gene of *V. parahaemolyticus*: a virulence gene acquired by a marine bacterium. *Infect. Immun.* **63:**2093–2099.

113. Nishibuchi, M., T. Taniguchi, T. Misawa, V. Khaeomanee-iam, T. Honda, and T. Miwatani. 1989. Cloning and nucleotide sequence of the gene (*trh*) encoding the hemolysin related to the thermostable direct hemolysin of *Vibrio parahaemolyticus*. *Infect. Immun.* **57:**2691–2697.

114. Nocerino, A., M. Iafusco, and S. Guandalini. 1995. Cholera toxin-induced small intestinal secretion has a secretory effect on the colon of the rat. *Gastroenterology* **108:**34–39.

115. Oliver, J. D. 1989. *Vibrio vulnificus*, p. 569–600. *In* M. P. Doyle (ed.), *Foodborne Bacterial Pathogens*. Marcel Dekker, New York, N.Y.

116. Oliver, J. D. 1993. Formation of viable but nonculturable cells, p. 239–272. *In* S. Kjelleberg (ed.), *Starvation in Bacteria*. Plenum Press, New York, N.Y.

117. Oliver, J. D. 2000. Public health significance of viable but nonculturable bacteria, p. 277–300. *In* R. R. Colwell and D. J. Grimes (ed.), *Nonculturable Microorganisms in the Environment*. ASM Press, Washington, D.C.

118. Oliver, J. D. Culture media for the isolation and enumeration of pathogenic *Vibrio* species in foods and environmental samples. *In* J. E. L. Corry, G. D. W. Curtis, and R. M. Baird (ed.), *Culture Media for Food Microbiology*, 2nd ed., in press. Elsevier Science, Amsterdam, The Netherlands.

119. Oliver, J. D., K. Guthrie, J. Preyer, A. Wright, L. M. Simpson, R. Siebeling, and J. G. Morris, Jr. 1992. Use of colistin-polymyxin B-cellobiose agar in the isolation of *Vibrio vulnificus* from the environment. *Appl. Environ. Microbiol.* **58:**737–739.

120. Oliver, J. D., M. B. Thomas, and J. Wear. 1986. Production of extracellular enzymes and cytotoxicity by *Vibrio vulnificus*. *Diagn. Microbiol. Infect. Dis.* **5:**99–111.

121. Oliver, J. D., R. A. Warner, and D. R. Cleland. 1983. Distribution of *Vibrio vulnificus* and other lactose-fermenting vibrios in the marine environment. *Appl. Environ. Microbiol.* **45:**985–998.

122. O'Neill, K. R., S. H. Jones, and D. J. Grimes. 1992. Seasonal incidence of *Vibrio vulnificus* in the Great Bay estuary of New Hampshire and Maine. *Appl. Environ. Microbiol.* **58:**3257–3262.

123. Osawa, R., T. Okitsu, H. Morozumi, and S. Yamai. 1996. Occurrence of urease-positive *Vibrio parahaemolyticus* in Kanagawa, Japan, with specific reference to presence of thermostable direct hemolysin (TDH) and the TDH-related-hemolysin genes. *Appl. Environ. Microbiol.* **62:**725–727.

124. Parker, R. W., E. M. Maurer, A. B. Childers, and D. H. Lewis. 1994. Effect of frozen storage and vacuum-packaging on survival of *Vibrio vulnificus* in Gulf Coast oysters (*Crassostrea virginica*). *J. Food Prot.* **57:**604–606.

125. Pearson, G. D. N., A. Woods, S. L. Chiang, and J. J. Mekalanos. 1993. CTX genetic element encodes a site-specific recombination system and an intestinal colonization factor. *Proc. Natl. Acad. Sci. USA* **90:**3750–3754.

126. Popovic, T., P. I. Fields, and O. Ølsvik. 1994. Detection of cholera toxin genes, p. 41–52. *In* I. K. Wachsmuth, P. A. Blake, and Ø. Olsvik (ed.), *Vibrio cholerae and Cholera: Molecular to Global Perspectives*. ASM Press, Washington, D.C.

127. Prasad, M. M., and C. C. P. Rao. 1994. Pathogenic vibrios associated with seafoods in and around Kakinada, India. *Fish. Technol.* **31:**185–187.

128. Raimondi, F., J. P. Y. Kao, C. Fiorentini, A. Fabbri, G. Donelli, N. Gasparini, A. Rubino, and A. Fasano. 2000. Enterotoxicity and cytotoxicity of *Vibrio parahaemolyticus* thermostable direct hemolysin in in vitro systems. *Infect. Immun.* **68:**3180–3185.

129. Raimondi, D., J. P. Y. Kao, J. B. Kaper, S. Guandalini, and A. Fasano. 1995. Calcium-dependent intestinal chloride secretion by *Vibrio parahaemolyticus* thermostable direct hemolysin in a rabbit model. *Gastroenterology* **109:**381–386.

130. Rashid, H. O., H. Ito, and I. Ishigaki. 1992. Distribution of pathogenic vibrios and other bacteria in imported frozen shrimps and their decontamination by gamma-irradiation. *World J. Microbiol. Biotechnol.* **8:**494–499.

131. Richards, G. P. 1988. Microbial purification of shellfish: a review of depuration and relaying. *J. Food Prot.* **51:**218–251.

132. Richardson, S. H. 1994. Host susceptibility, p. 273–289. *In* I. K. Wachsmuth, P. A. Blake, and Ø. Olsvik (ed.), *Vibrio cholerae and Cholera: Molecular to Global Perspectives*. ASM Press, Washington, D.C.

133. Robach, M. C., and C. S. Hickey. 1978. Inhibition of *Vibrio parahaemolyticus* by sorbic acid in crab meat and flounder homogenates. *J. Food Prot.* **41:**699–702.

134. Rodrick, G. E., K. R. Schneider, F. A. Steslow, N. J. Blake, and W. S. Otwell. 1988. Uptake, fate and ultraviolet depuration of vibrios in *Mercenaria campechiensis*. *Mar. Technol. Soc. J.* **23:**21–26.

135. Sakazaki, R., S. Iwanami, and K. Tamura. 1968. Studies on the enteropathogenic, facultatively halophilic bacteria, *Vibrio parahaemolyticus*. II. Serological characteristics. *Jpn. J. Med. Sci. Biol.* 21:313–324.

136. Sakazaki, R., K. Tamura, T. Kato, Y. Obara, S. Yamai, and K. Hobo. 1968. Studies on the enteropathogenic, facultatively halophilic bacteria, *Vibrio parahaemolyticus*. III. Enteropathogenicity. *Jpn. J. Med. Sci. Biol.* 21:325–331.

137. Sang, F. C., M. E. Hugh-Jones, and H. V. Hagstad. 1987. Viability of *Vibrio cholerae* O1 on frog legs under frozen and refrigerated conditions and low dose radiation treatment. *J. Food Prot.* 50:662–664.

138. Sanyal, S. C., and P. C. Sen. 1974. Human volunteer study on the pathogenicity of *Vibrio parahaemolyticus*, p. 227–230. *In* T. Fujino, G. Sakaguchi, R. Sakazaki, and Y. Takeda (ed.), *International Symposium on Vibrio parahaemolyticus*. Saikon Publishing Co., Ltd., Tokyo, Japan.

139. Shandera, W. X., J. M. Johnston, B. R. Davis, and P. A. Blake. 1983. Disease from infection with *Vibrio mimicus*, a newly recognized *Vibrio* species. *Ann. Intern. Med.* 99:169–171.

140. Shao, C.-P., and L.-I. Hor. 2000. Metalloprotease is not essential for *Vibrio vulnificus* virulence in mice. *Infect. Immun.* 68:3569–3573.

141. Shapiro, R. L., S. Altekruse, L. Hutwagner, R. Bishop, R. Hammond, S. Wilson, B. Ray, S. Thompson, R. V. Tauxe, and P. M. Griffin. 1998. The role of Gulf Coast oysters harvested in warmer months in *Vibrio vulnificus* infections in the United States, 1988–1996. *J. Infect. Dis.* 178:752–759.

142. Shirai, H., H. Ito, T. Kirayama, Y. Nakamoto, N. Nakabayashi, K. Kumagni, Y. Takeda, and M. Nishibuchi. 1990. Molecular epidemiologic evidence for association of thermostable direct hemolysin (TDH) and TDH-related hemolysin of *Vibrio parahaemolyticus* with gastroenteritis. *Infect. Immun.* 58:3568–3573.

143. Simonson, J. G., P. Danieu, A. B. Zuppardo, R. J. Siebeling, R. L. Murphree, and M. L. Tamplin. 1995. Distribution of capsular and lipopolysaccharide antigens among clinical and environmental *Vibrio vulnificus* isolates. *Abstr. Annu. Meet. Am. Soc. Microbiol.*, abstr. B-286, p. 215.

144. Simpson, L. M., V. K. White, S. F. Zane, and J. D. Oliver. 1987. Correlation between virulence and colony morphology in *Vibrio vulnificus*. *Infect. Immun.* 55:269–272.

145. Skorupski, K., and R. K. Taylor. 1999. A new level in the *Vibrio cholerae* ToxR virulence cascade: AphA is required for transcriptional activation of the *tcpPH* operon. *Mol. Microbiol.* 31:763–771.

146. Sloan, E. M., C. J. Hagen, G. A. Lancette, J. T. Peeler, and J. N. Sofos. 1992. Comparison of five selective enrichment broths and two selective agars for recovery of *Vibrio vulnificus* from oysters. *J. Food Prot.* 55:356–359.

147. Spangler, B. D. 1992. Structure and function of cholera toxin and the related *Escherichia coli* heat-labile enterotoxin. *Microbiol. Rev.* 56:622–647.

148. Sperandio, V., J. A. Girón, W. D. Silveira, and J. B. Kaper. 1995. The OmpU outer membrane protein, a potential adherence factor of *Vibrio cholerae*. *Infect. Immun.* 63:4433–4438.

149. Spira, W. M., M. U. Khan, Y. A. Saeed, and M. A. Sattar. 1980. Microbiological surveillance of intra-neighbourhood El Tor cholera transmission in rural Bangladesh. *Bull. WHO* 58:731–740.

150. Strom, M. S., and R. N. Paranjpye. 2000. Epidemiology and pathogenesis of *Vibrio vulnificus*. *Microbes Infect.* 2:177–188.

151. Sun, Y., and J. D. Oliver. 1994. Effects of GRAS compounds on natural *Vibrio vulnificus* populations in oysters. *J. Food Prot.* 57:921–923.

152. Sun, Y., and J. D. Oliver. 1995. Hot sauce: no elimination of *Vibrio vulnificus* in oysters. *J. Food Prot.* 58:441–442.

153. Sun, Y., and J. D. Oliver. 1995. The value of CPC agar for the isolation of *Vibrio vulnificus* from oysters. *J. Food Prot.* 58:439–440.

154. Tamplin, M. L., and G. M. Capers. 1992. Persistence of *Vibrio vulnificus* in tissues of Gulf Coast oysters, *Crassostrea virginica*, exposed to seawater disinfected with UV light. *Appl. Environ. Microbiol.* 58:1506–1510.

155. Tamplin, M., G. E. Rodrick, N. J. Blake, and T. Cuba. 1982. Isolation and characterization of *Vibrio vulnificus* from two Florida estuaries. *Appl. Environ. Microbiol.* 44:1466–1470.

156. Taylor, R. K., V. L. Miller, D. B. Furlong, and J. J. Mekalanos. 1987. Use of *phoA* gene fusions to identify a pilus colonization factor coordinately regulated with cholera toxin. *Proc. Natl. Acad. Sci. USA* 84:2833–2837.

157. Tilton, R. C., and R. W. Ryan. 1987. Clinical and ecological characteristics of *Vibrio vulnificus* in the Northeastern United States. *Diagn. Microbiol. Infect. Dis.* 6:109–117.

158. Tison, D. L., and M. T. Kelly. 1984. *Vibrio* species of medical importance. *Diagn. Microbiol. Infect. Dis.* 2:263–276.

159. Todd, E. C. D. 1989. Preliminary estimates of costs of foodborne disease in the United States. *J. Food Prot.* 52:595–601.

160. Trucksis, M., J. E. Galen, J. Michalski, A. Fasano, and J. B. Kaper. 1993. Accessory cholera enterotoxin (Ace), the third toxin of a *Vibrio cholerae* virulence cassette. *Proc. Natl. Acad. Sci. USA* 90:5267–5271.

161. Trucksis, M., J. Michalski, Y. K. Denk, and J. B. Kaper. 1998. The *Vibrio cholerae* genome contains two unique circular chromosomes. *Proc. Natl. Acad. Sci. USA* 95:14464–14469.

162. Twedt, R. M. 1989. *Vibrio parahaemolyticus*, p. 543–568. *In* M. P. Doyle (ed.). *Foodborne Bacterial Pathogens*. Marcel Dekker, Inc., New York, N.Y.

163. **Uchimura, M., and T. Yamamoto.** 1992. Production of hemagglutinins and pili by *Vibrio mimicus* and its adherence to human and rabbit small intestines in vitro. *FEMS Microbiol. Lett.* **91:**73–78.

164. **U.S. Food and Drug Administration.** 1992. *Bacteriological Analytical Manual*, 7th ed. Office of Analytical Chemistry, Arlington, Va.

165. **Vanderzant, C., and R. Nickelson.** 1972. Survival of *Vibrio parahaemolyticus* in shrimp tissue under various environmental conditions. *Appl. Microbiol.* **23:**34–37.

166. **Vickery, M. C., N. Harold, and A. K. Bej.** 2000. Cluster analysis of AP-PCR generated DNA fingerprints of *V. vulnificus* isolates from patients fatally infected after consumption of raw oysters. *Lett. Appl. Microbiol.* **30:**258–262.

166a. **Waldor, M. K., and J. J. Mekalanos.** 1996. Lysogenic conversion by a filamentous phage encoding cholera toxin. *Science* **272:**1910–1914.

167. **Warner, J. M., and J. D. Oliver.** 1999. Randomly amplified polymorphic DNA analysis of clinical and environmental isolates of *Vibrio vulnificus* and other *Vibrio* species. *Appl. Environ. Microbiol.* **65:**1141–1144.

168. **Watnick, P. I., K. J. Fullner, and R. Kolter.** 1999. A role for the mannose-sensitive hemagglutinin in biofilm formation by *Vibrio cholerae* El Tor. *J. Bacteriol.* **181:**3606–3609.

169. **West, P. A.** 1989. The human pathogenic vibrios — a public health update with environmental perspectives. *Epidemiol. Infect.* **103:**1–34.

170. **Wong, H.-C., L.-L. Chen, and C.-M. Yu.** 1994. Survival of psychrotrophic *Vibrio mimicus*, *Vibrio fluvialis* and *Vibrio parahaemolyticus* in culture broth at low temperatures. *J. Food Prot.* **57:**607–610.

171. **Wong, H.-C., S.-H. Ting, and W.-R. Shieh.** 1992. Incidence of toxigenic vibrios in foods available in Taiwan. *J. Appl. Bacteriol.* **73:**197–202.

172. **Wright, A. C., L. M. Simpson, and J. D. Oliver.** 1981. Role of iron in the pathogenesis of *Vibrio vulnificus* infections. *Infect. Immun.* **34:**503–507.

173. **Wu, Z., P. Nbom, and K.-E. Magnusson.** 2000. Distinct effects of *Vibrio cholerae* haemagglutinin/protease on the structure and localization of the tight junction-associated proteins occludin and ZO-1. *Cell. Microbiol.* **2:**11–17.

Food Microbiology: Fundamentals and Frontiers, 2nd Ed.
Edited by M. P. Doyle et al.
© 2001 ASM Press, Washington, D.C.

Sylvia M. Kirov

Aeromonas and *Plesiomonas* Species

14

The genera *Aeromonas* and *Plesiomonas* comprise gram-negative, facultatively anaerobic, oxidase-positive, glucose-fermenting, rod-shaped bacteria that are generally motile by polar flagella. (The species *Aeromonas salmonicida* and *Aeromonas media* are reported to be nonmotile.) *Aeromonas* spp. (aeromonads) usually have a single polar flagellum, but up to 50% of isolates can produce numerous lateral flagella that are distinct from the polar flagellum; *Plesiomonas shigelloides* has two to seven polar flagella and may also produce lateral flagella with a shorter wave length. Until recently, both genera were classified in the family *Vibrionaceae*. They are commonly found in aquatic environments and have been incriminated epidemiologically as enteropathogens. Thus, they have traditionally been considered together. However, molecular genetic evidence (including 16S rRNA catalog, 5S rRNA sequence, and rRNA-DNA hybridization) suggests they are not closely related to each other or to other *Vibrio* species. In the latest edition of *Bergey's Manual of Systematic Bacteriology*, therefore, the proposal that *Aeromonas* spp. form a separate family, the *Aeromonadaceae*, has been adopted. *P. shigelloides*, which is more closely related to bacteria within the family *Enterobacteriaceae*, is now also classified in a separate family, the *Plesiomonadaceae* (45, 126, 163).

The histories of the two genera have been reviewed by Farmer and colleagues (66). Aeromonads were discovered more than 100 years ago, but their role in human illness dates to 1954, when the organism was associated with the death of a 40-year-old Jamaican woman with acute fulminating metastatic myositis (47). In 1964, the first well-described case associating *Aeromonas* spp. with diarrhea was described (209). The organism now known as *P. shigelloides* was first described in 1947 by Ferguson and Henderson, who isolated it from a fecal specimen (66). Although it is now well accepted that members of both genera can cause serious extraintestinal infections (some of which may be acquired via the oral route), particularly in immunocompromised hosts, *Aeromonas* and *Plesiomonas* species are still regarded by some as controversial gastrointestinal pathogens. This is because definite proof (conclusive human volunteer trials and animal models) of enteropathogenicity is lacking. Nevertheless, substantial clinical and microbiological evidence now supports the epidemiologic evidence that at least some strains of certain *Aeromonas* species can cause gastroenteritis in some individuals. The situation with *Aeromonas* spp. may be analogous to that which exists for the various pathotypes of *Escherichia coli* and *Yersinia enterocolitica* for which combinations of

Sylvia M. Kirov, Discipline of Pathology, University of Tasmania, GPO Box 252-29, Hobart, Tasmania 7001, Australia.

virulence determinants render particular strains pathogenic. At present, however, as *Aeromonas* virulence mechanisms are not well understood, it is not possible to distinguish the strains which pose a threat to human health from the diversity of strains present in foods and water, or even to determine the significance of a given isolate in the clinical laboratory. This chapter reviews existing knowledge of putative *Aeromonas* virulence factors and outlines possible future research directions that may eventually be able to resolve this dilemma. Even less is known about *P. shigelloides* and its possible virulence determinants than is known for *Aeromonas* spp. The role of *P. shigelloides* in gastroenteric disease is, therefore, still uncertain.

CLASSIFICATION AND IDENTIFICATION

Aeromonas Species

The taxonomy of the *Aeromonas* genus is complex and has undergone constant change over the last decade. The number of recognized species has increased markedly in that period (66, 118, 121, 126). *Aeromonas* species can be differentiated from the halophilic vibrios because they are unable to grow in 6% NaCl and from the *Vibrio cholerae* group by their resistance to 150 μg of vibriostatic agent (O/129). Other biochemical tests and antibiotic susceptibility testing may also be required for correct genus assignment because of emerging worldwide resistance of *V. cholerae* to O/129 and susceptibility of some aeromonads to this concentration of O/129 (40, 66, 120, 121).

The genus *Aeromonas* was originally divided into a single mesophilic species, *A. hydrophila*, and a single psychrophilic species, *A. salmonicida*. Thus, in the older literature (and unfortunately in some more recent reports), the name *A. hydrophila* is used to include all aeromonads except *A. salmonicida*, a fish pathogen. DNA-DNA hybridization experiments performed by Popoff in the early 1980s revealed that there were at least seven distinct hybridization groups (HGs) among the *A. hydrophila* strains. These fell into three main phenotypic groups which could be identified based on the reactions of isolates in 8 to 18 biochemical tests. These "phenospecies" were named *A. hydrophila*, *A. sobria*, and *A. caviae* (201). Hence, the name *A. hydrophila* began to be limited to HG1, HG2, and HG3. Most clinical laboratories adopted this simplified phenotypic classification. The genetic complexity of the group, however, meant that not only the *A. hydrophila* phenospecies but also the other two phenospecies each encompassed several HGs (*A. sobria* HG7, HG8, HG9, and HG10; *A. caviae* HG4,

HG5, and HG6). The number of DNA hybridization groups ("genospecies"/"genomospecies") has since expanded to >14, of which 8 have been associated with human illness. As DNA hybridization technology is not routinely carried out in diagnostic laboratories, more comprehensive biotyping schemes (19 to 24 biochemical tests) have been proposed in recent years which will identify strains to the genospecies level (1, 13). Key tests have been identified so that biochemical typing schemes, such as Aerokey II, now permit preliminary grouping of most clinical isolates (13, 42, 120, 121, 126). Currently recognized genospecies and phenospecies of the genus *Aeromonas* are shown in Table 14.1. More than 10 phenotypic species have been named, with HG8 and HG10 (now combined as HG8/10) having two biovars. The name *A. hydrophila* is thus now being further limited in some of the recent literature to mean only HG1,

Table 14.1 Currently recognized genospecies and phenospecies of the genus *Aeromonas*[a]

DNA group	Genospecies	Phenospecies
1	*A. hydrophila*	*A. hydrophila*
2	*A. bestiarum*	*A. hydrophila*-like
3	*A. salmonicida*[b]	*A. salmonicida* subsp. *salmonicida* subsp. *achromogenes* subsp. *masoucida*
	A. salmonicida	*A. smithia*
	Unnamed	*A. hydrophila*
4	*A. caviae*	*A. caviae*
5A	*A. media*	*A. caviae*-like
5B	*A. media*	*A. media*
6	*A. eucrenophila*	*A. eucrenophila*[c]
7	*A. sobria*	*A. sobria*[c]
8	*A. veronii*	*A. veronii* biovar sobria[d]
9	*A. jandaei*	*A. jandaei*
10	*A. veronii*	*A. veronii* biovar veronii[d]
11	Unnamed	*Aeromonas* sp. (ornithine positive)
12	*A. schubertii*	*A. schubertii*
13	*Aeromonas* group 501	*A. schubertii*-like
14	*A. trota*	*A. trota*
15?	*A. allosaccharophila*	*A. allosaccharophila*
16?	*A. encheleia*	*A. encheleia*
17?	*A. popoffii*	*A. popoffii*

[a] The new family, *Aeromonadaceae*, includes 14 validated species. For comprehensive coverages, see references 45 and 126.

[b] Includes psychrophilic strains originating from fish. These strains are biochemically distinct (nonmotile and indole negative) from HG3 isolates recovered from clinical material (1, 120).

[c] HG6 and HG7 have not yet been recovered from clinical material (1, 118).

[d] HG8 and HG10 are genetically identical, but the type strains are biochemically distinguishable; hence, they are now called biovars of the first described species, *A. veronii* (45, 118, 126).

Table 14.2 Relevant biochemical properties of clinical *Aeromonas* phenospecies and of *Plesiomonas shigelloides*[a,b]

Test	A. hydrophila[c]	A. caviae[d]	A. veronii biovar sobria (HG8/10)	A. veronii biovar veronii (HG8/10)	A. jandaei (HG9)	A. schubertii (HG12)	A. trota (HG14)	P. shigelloides
O/129, 10 μg/150 μg[e]	R/R	R/R	R/R	R/R	R/R	R/R	R/R	S/S
Indole	+	+	+	+	+	−	+	+
Glucose (gas)	+	−	+	+	+	+	+	−
Voges-Proskauer	+	−	+	+	+	−	−	−
Ornithine decarboxylase	−	−	−	+	−	−	−	+
Acid from								
L-Arabinose	+	+	−	−	−	−	−	−
Sucrose	+	+	+	+	−	−	−	−
m-Inositol	−	−	−	−	−	−	−	+
D-Mannitol	+	+	+	+	+	−	+	−
Salicin	+	+	−	+	−	−	−	V
Esculin hydrolysis[f]	+	+	−	+	−	−	−	−
H2S[g]	+	−	+	+	+	−	+	
Cephalothin 30 μg[h]	R	R	S	S	R	S	R	
Beta-hemolysis (sheep blood)	+	V	+	+	+	+	V	−

[a] Adapted from references 122 and 126.
[b] R, resistant; S, susceptible; +, positive for more than 80% of strains; −, negative for more than 80% of strains; V, variable (20 to 80% positive). Biochemical tests are read daily for 3 to 4 days (126).
[c] Refers to phenospecies HG1, HG2, and HG3; additional biochemical tests are required to separate these HGs (13, 141).
[d] Refers to both *A. caviae* (HG4) and *A. media* (HG5A and HG5B), which have proven difficult to separate biochemically (1, 118, 126).
[e] 99% of strains are resistant to 10 μg; >80% are resistant to 150 μg.
[f] Agar formulation only (42).
[g] Gelatin-cysteine thiosulfate medium.
[h] Bauer-Kirby disk diffusion. Susceptibility for this method requires a zone size of ≥18 mm.

which includes the type strain for this species. More than 85% of gastroenteritis-associated isolates belong to three main species, *A. hydrophila* HG1, *A. caviae* HG4, and *A. veronii* biovar sobria HG8/10 (29, 89, 120, 145). Salient features in the identification of *Aeromonas* phenospecies and genospecies in the clinical laboratory are summarized in Table 14.2.

In the discussion to follow, where the name *A. hydrophila* has been used in the literature in its broad historical sense to refer to the entire group of motile mesophilic aeromonads, such organisms will generally be referred to as *Aeromonas* spp. or aeromonads or as an unqualified *A. hydrophila* (Tables 14.1, 14.2, and 14.4). The name *A. hydrophila* (HG1, HG2, and HG3) will be used to refer to the more limited group of strains defined by phenotype (criteria of Popoff [201]) or DNA-DNA hybridization. Similarly, *A. caviae* refers to strains belonging to DNA HG4, HG5, and HG6, and *A. sobria* refers to strains belonging to HG7, HG8, HG9, and HG10. *A. sobria* is now the designation for HG7 and should only be used for this species in future studies. Hence, where the correct phenospecies is now known or can be inferred from published data, this will be given in brackets. For more recent studies in which species

have been identified to hybridization group, either by genetic or biochemical means, names such as *A. hydrophila* (HG1), *A. caviae* (HG5), *A. veronii* biovar sobria (HG8/10), and *A. jandaei* (HG9) will be used.

The taxonomy of the *Aeromonas* genus is still evolving. It is important that strains are accurately identified, as antibiotic susceptibilities and pathogenic mechanisms (see below) may vary between strains. Misidentification of unusual *Aeromonas* species as members of the genus *Vibrio* is a continuing problem (2). Comprehensive biochemical typing schemes that will accurately identify all *Aeromonas* isolates, including all environmental strains, are, however, not yet available. Those that have been proposed for environmental isolates involve large numbers of tests (>40) and, hence, are beyond the scope of most routine laboratories (134). The fact that some biochemical characteristics are temperature dependent is a further complication for such biochemical typing schemes (126). There is a need for further work in this area to develop schemes that will accurately and easily identify all *Aeromonas* isolates. Until then, food isolates of *Aeromonas* species should at least be identified to a complex (*A. hydrophila*, *A. caviae*, or *A. veronii*) using key phenotypic features (121).

Possible new approaches for the identification of aeromonads based on genetic methods include PCR assays with 16S rDNA-targeted species-specific oligonucleotide primers, determination of 16S rRNA gene restriction patterns, or a recently developed genomic fingerprinting technique (AFLP; European Patent Office no. 534858A1) which detects DNA polymorphisms by selective amplifications of restriction fragments (30, 63, 113, 163). Such genetic techniques and other typing techniques, such as multilocus enzyme electrophoresis, electrophoretic typing of esterases, phage typing, gas-liquid chromatography of cell wall fatty acid methyl esters, and polyacrylamide gel electrophoresis of total soluble proteins (11, 197), have already proved useful in the research setting for identifying *Aeromonas* strains and tracing possible sources of infection (12, 199). A serogrouping scheme (97 serogroups) for aeromonads has been developed in the United Kingdom (225). Thus far, however, it has been minimally used in epidemiologic investigations.

Media used for isolation of aeromonads usually exploit their resistance to ampicillin. Blood ampicillin agar (10 to 30 μg of ampicillin per ml) is often used for clinical specimens, and starch ampicillin agar is used for environmental specimens. (However, *A. trota* is ampicillin sensitive.) Further discussion of these and other selective media for the isolation of aeromonads from foods is provided elsewhere (66, 121, 123, 126, 134).

P. shigelloides

In contrast to the genus *Aeromonas*, *Plesiomonas* consists of a homogeneous species with a single DNA hybridization group (*P. shigelloides*). Like many *Vibrio* spp., *P. shigelloides* is susceptible to O/129. Like *Aeromonas* spp., it is unable to grow in 6% NaCl. Separation of *P. shigelloides* from aeromonads is based on simple phenotypic tests, exoenzyme production, and a range of carbohydrate fermentations (32, 66, 172). Useful biochemical reactions for differentiating *P. shigelloides* from *Aeromonas* spp. (Table 14.2) include myoinositol fermentation and decarboxylation of ornithine. The L-histidine decarboxylase test (conventional Moeller's tube test) may also be a potentially useful diagnostic feature for differentiating *P. shigelloides* from members of the family *Vibrionaceae* (186). More than 100 serovars of *P. shigelloides* have been described. Several react with *Shigella* antisera. At present, 76 O and 41 H antigens are included in a proposed international typing scheme (7–9, 32). Not all *P. shigelloides* strains, however, are groupable using this international typing system (121). Ampicillin-containing selective *Aeromonas* media are not suitable for the isolation of *P. shigelloides*. Suitable isolation media include xylose-sodium deoxycholate-citrate and inositol-

brilliant green bile salts and *Plesiomonas* agars (32, 66, 109, 123, 172).

TOLERANCE OR SUSCEPTIBILITY TO PRESERVATION METHODS

Temperature

Most aeromonads have an optimal growth temperature of 28°C; however, they constitute a very heterogeneous group, including strains with a very wide temperature growth range (<5 to 45°C). Hence, the optimal temperature is not 28°C for all strains. The source of strains in foods can influence the rate at which they grow at low temperatures (151). While most strains in foods are mesophiles, many can grow at refrigeration temperatures. Populations of *Aeromonas* species naturally occurring in foods of all types can increase 10- to 1,000-fold during 7 to 10 days of storage at 5°C (25, 41, 190). Psychrotrophic strains that grow very rapidly have also been described. One of these, isolated from goat milk, has a theoretical minimum temperature (T_{min}) for growth of -5.3°C. The T_{min} is determined from growth patterns in broth cultures placed in a temperature gradient incubator. Growth rates are determined as the time required for an increase in optical density of 0.3. The T_{min} is then estimated by plotting the square root of the growth rate versus temperature (135, 205). Most strains isolated from clinical specimens can grow at refrigeration temperature (64, 136, 190). Temperature alone, therefore, cannot be relied on to control the growth of *Aeromonas* strains in foods (26, 132).

Most strains of *P. shigelloides* do not grow below 8°C, but at least one strain has been reported to grow at 0°C, and *P. shigelloides* may occur in the aquatic environment of cold climates (162, 172, 173). Their optimal temperature for growth, however, is considered to be 38 to 39°C, with a maximum around 45°C. About 25% of strains grow at 45°C (66, 172). Temperatures of 42 to 44°C have been recommended for the isolation of *P. shigelloides* from environmental specimens in which the presence of aeromonads and other organisms poses a problem (109).

Strains of *Aeromonas* spp. and *P. shigelloides* can be recovered from foods stored at -20°C for considerable periods (years) and are successfully stored in culture collections at -70°C.

pH and Salt Concentration

Aeromonas spp. are fairly sensitive to low pH (<5.5). Studies have revealed that *Aeromonas* spp. are unlikely to grow in foods with more than 3 to 3.5% (wt/wt) NaCl and pH values below 6.0 when foods are stored at low temperature (132). Acetic acid, lactic acid, tartaric acid, citric acid, sulfuric acid, and hydrochloric acid, in that order, are effective at restricting growth (3). Polyphosphates also can control the growth of aeromonads in certain foods (189).

P. shigelloides grows at pH 5 to 8 but may be especially susceptible to low pH. Cells are killed rapidly at pH ≤4 (32, 172). Tolerance to salt is similar to that of *Aeromonas* spp. (173).

Atmosphere

Overall, aeromonads grow as well anaerobically as they do aerobically. However, they are more sensitive to $NaNO_2$ under anaerobic than aerobic conditions. *Aeromonas* growth under modified atmospheres depends on the nature and number of competing microflora. The use of modified atmospheres to extend the shelf life of packaged meats and fresh vegetables may enable aeromonads to grow to high populations (25, 26). The type of vegetable has more influence on *Aeromonas* growth rate than the type of atmosphere present, with more rapid growth occurring on shredded endives and iceberg lettuce than on brussels sprouts or grated carrots (116). *Aeromonas* spp. contribute to the spoilage of many foods, but in others, such as milk, they can reach large populations (up to 10^8) without detectable organoleptic changes (142).

P. shigelloides is extremely susceptible to modified atmospheric storage (80% CO_2, no O_2), with no growth under such conditions as in cooked crayfish tails held at 11 and 14°C (114).

Destructive Methods: Heat, Irradiation, Disinfectants, and Chlorine

Aeromonas spp. are readily killed by either heat or irradiation (3, 183, 191, 204). Palumbo et al. concluded that the thermal resistance of aeromonads was similar to that of other gram-negative organisms (191), D values at 48°C for different clinical and food isolates heated in raw milk ranging from 3.2 to 6.2 min. However, one report indicates that aeromonads are more sensitive to heat than other foodborne pathogens, such as *E. coli* O157:H7, *Staphylococcus aureus*, and *Salmonella enterica* serovar Typhimurium. Aeromonads in peptone water were killed within 2 min at 55°C, compared with 15 min for the other pathogens, and were also less resistant to heat in hamburger steaks than the other pathogens (183). Some *Aeromonas* toxins, however, are reportedly heat stable (56°C, 20 min, to 100°C, 30 min) (50, 56).

Aeromonas strains are as susceptible to disinfectants, including chlorine, as other gram-negative bacteria associated with foods, yet recovery of aeromonads from chlorinated water supplies is commonly reported. This could be the result of posttreatment recontamination, inactivation of chlorine by organic matter, the presence of very high initial numbers of aeromonads, or possibly the persistence of the organism in a "viable but nonculturable" state following treatment (3, 132).

More research is needed to determine the effects of the factors discussed above on plesiomonads, including their ability to survive and grow under storage conditions commonly used for many foods, especially those of aquatic origin. Studies to date indicate that adequately cooked foods will not contain viable *P. shigelloides*, and proper refrigeration should control its growth in stored foods (173).

RESERVOIRS

Members of both genera are primarily aquatic organisms. The motile, mesophilic aeromonads are found in fresh, stagnant, estuarine, or brackish water worldwide. As they are also commonly present in drinking water, they are found in sinks, drainpipes, and household effluents. A significant correlation between organic matter content and total numbers of aeromonads in waters has been reported (16). Temperature-dependent seasonal variations in cell populations have been observed, with numbers highest during the summer months (37, 94, 127, 157). *A. caviae* uniformly self-destructs ("suicide phenomenon") in the presence of low pH induced under the action of acetic acid accumulation (178). This could account for its absence in acidic waters, although it is found in alkaline environments. Thus, different phenospecies (Popoff criteria) of aeromonads may predominate in different water sources. *A. hydrophila* (HG1, HG2, and HG3) is prevalent in springwater; *A. caviae* is prevalent in marine samples; and *A. sobria* (*A. veronii* biovar sobria) predominates in recreational lakes and river water. All three phenospecies are found in sewage-contaminated waters (18, 66). The distribution of species and the proportion of strains producing toxins can vary in different geographic regions. Clinical strains reflect differences in the distribution of environmental strains expressing virulence-associated properties (139, 140).

Motile aeromonads may associate with or colonize many water-dwelling plants and animals (e.g., healthy fish, leeches, and frogs). They are recognized as the causal agent of red-leg disease in amphibians and are responsible for diseases in reptiles, fish, shellfish, and snails. In many fish species, they cause a hemorrhagic septicemia (126). They can be found as minor components of fecal flora of a wide variety of animals, including domestic animals used for food (pigs, cows, sheep, and poultry).

They have been implicated in outbreaks of bovine abortion and diarrhea in piglets, as well as other infections in a variety of animals (83, 201).

Aeromonads are not generally considered to be normal inhabitants of the gastrointestinal tract of humans. However, fecal carriage rate can approach 3% in asymptomatic persons in temperate climates (129, 174). In the tropics or developing regions, carriage rates may reach 30% or more (129, 200). Aeromonads are widespread in foods and are readily isolated from meat, raw milk, poultry, fish, shellfish, and vegetables. They also are detected in prepared foods, such as pasteurized milk, but less frequently (69, 132, 142). As for water, there are geographic differences in the species isolated and the proportion of isolates possessing putative virulence properties. *A. sobria* (*A. veronii* biovar sobria) is most frequently isolated from poultry, raw meat and offal, and meat products (76, 87, 135). *A. caviae* is the most common (~60%) phenospecies isolated in Japan from sea fish, vegetables, and their products (182), whereas *A. hydrophila* (HG1, HG2, and HG3) and *A. caviae* are predominant in vegetables in the United States, comprising 48 and 26%, respectively, of *Aeromonas* isolates (41). Kühn et al. proposed that certain *Aeromonas* strains may be especially suited to survive and multiply in particular foods (156).

P. shigelloides is found in fresh and estuarine waters, as well as seawater in warm weather (7–9, 66, 165). Its limited temperature range for growth influences the more frequent occurrence of *P. shigelloides* in tropical and subtropical climates and also accounts for the seasonal variation (summer incidence) that occurs in its isolation from river water in temperate climates. However, *P. shigelloides* has been isolated from aquatic environments in cold climates, such as northern Europe (155). *P. shigelloides* also has been isolated from warm- and cold-blooded animals, including dogs, cats, cattle, pigs, snakes, shellfish, and tropical fish. It has been found in healthy humans at very low rates (0.0078%; 3 carriers among 38,454 food handlers and schoolchildren in Japan) (15). A much higher isolation rate (24%; 12 of 51 adults in Thailand) has been reported in developing countries (32). Few systematic studies of the prevalence of *P. shigelloides* in foods have been conducted; to date, *P. shigelloides* has been isolated predominantly from fish and seafood (66, 173).

FOODBORNE OUTBREAKS

The primary source of *Aeromonas* infection is likely water. Studies revealed that increased populations of aeromonads in drinking water coincided with an increased incidence of *Aeromonas*-associated gastroenteri-tis (37). Drinking untreated water has also been identified as a significant risk factor for *Aeromonas*-associated gastroenteric disease (176). Similarly, not only gastroenteric but also septicemic and surgical infections in immunocompromised patients hospitalized for more than 2 days were associated with the hospital water supply (198, 199). For the latter, infection rates increased in the summer and were correlated with the increased populations of aeromonads detected in the water storage tanks (198). Moreover, the implementation control measures, such as the use of mineral water for drinking and sterile water for medical preparations used for such things as gastric lavage and radiopaque solutions, reduced the number of cases of septicemia (199). Increased isolation of aeromonads from human stools also has been correlated with large populations of aeromonads in foods (particularly minced beef, pork, and chicken) in Japan (182). However, typing methods have revealed there is little similarity between most water-related *Aeromonas* strains and diarrhea-associated isolates (93, 199). Strains able to colonize or infect the human gastrointestinal tract are probably only a small fraction of environmental strains (93, 188).

There are few published cases in which *Aeromonas* species have been associated with foodborne gastroenteritis, and the few suspected outbreaks of foodborne *Aeromonas* infections have not been documented satisfactorily. Cases that have been described are summarized elsewhere (134). In most instances, suspect foods (oysters and other seafood, edible land snails, egg salad) had been prefrozen and presumably inadequately cooked before consumption or were foods that were consumed after receiving minimal cooking or being improperly stored, which allowed rapid growth of aeromonads (31, 132). In only one case (38-year-old male) was the isolate from food (shrimp cocktail) and feces typed (rRNA gene restriction patterns) in a manner to establish definitively the source of the strain (12).

Two outbreaks in Japan attributed to *Aeromonas* species have been reported (152). More recently, *Aeromonas* spp. also have been implicated in food-poisoning outbreaks in Norway, Sweden, and France (82, 91, 154). In Norway, three of four people who had eaten raw fermented fish became ill. The fish contained up to 10^7 CFU of *A. hydrophila* (HG1, HG2, and HG3) (82). In the Swedish incident, 22 of 27 persons became ill 20 to 30 h after eating from a smorgasbord containing shrimp, smoked sausage, liver paté, and boiled ham, all of which contained large populations of aeromonads (10^6 to >10^7/g of food sample) in virtually pure culture (154). *A. caviae* was implicated in an outbreak of foodborne illness in France. This outbreak involved 10 hospital staff members who became ill after eating

Vietnamese food. *A. caviae* was isolated at high populations (10^6/ml of sauce) from the implicated food (fish sauce). Fecal specimens from only two patients were examined, and *A. caviae* was not isolated from them. Although no other potential enteropathogens were isolated from food or fecal specimens, the role of *A. caviae* in this outbreak is still unproven (91).

The ability of many strains to grow and produce toxins (some of which are heat stable and can induce fluid accumulation in animal intestines) at low temperatures has led to speculation that illness from eating *Aeromonas*-contaminated foods might be possible by intoxication (caused by preformed toxins in foods) (64, 161). Considerable growth and toxin production occurs in bacteriological media under refrigeration conditions; however, considerably less toxin is produced in foods. Those psychrotrophic strains that can produce high levels of toxins in foods are not common. Moreover, toxins are inactivated by some foods, such as milk (136, 138, 142). Thus, illness resulting from intoxication is an unlikely risk compared with the possible risk posed by strains able to colonize the human intestine and express virulence properties in vivo.

P. shigelloides has been implicated in at least two large waterborne epidemics (involving ~1,000 people) (230). In one, a single predominant serotype (O17:H2) was isolated from patients, and this serotype also was isolated from tap water from different locations in the area where the outbreak occurred (230). Other, more recent water-associated outbreaks have been reported (159). Outbreaks of *Plesiomonas*-associated gastroenteritis have been attributed to *Plesiomonas*-contaminated oysters, chicken, fish (salt mackerel, cuttlefish salad), and shrimp (32, 172, 210). Oysters have been the major food incriminated in the United States (66). Thirty-one patients with gastroenteric symptoms from whom *P. shigelloides* had been isolated were studied by Holmberg et al. (104). Consumption of raw seafood and foreign travel (particularly to Mexico) were identified as major risk factors for these patients with *Plesiomonas*-positive stools (104). *P. shigelloides* was isolated from 5.6% of 23,215 travelers with diarrhea returning to Japan from developing countries (238), and in Nigeria, a strong association of the bacterium with diarrhea has been reported for patients from rural areas (185).

CHARACTERISTICS OF DISEASE

Extraintestinal Infections

Although extraintestinal infections with *Aeromonas* spp. and *P. shigelloides* are uncommon, they tend to be severe and often fatal, particularly in patients with sep-

sis and meningitis. *Aeromonas* extraintestinal disease is more common and varied (peritonitis, endocarditis, pneumonia, conjunctivitis, and urinary tract infections) than that caused by *P. shigelloides* (cellulitis, arthritis, endophthalmitis, and cholecystitis) (32, 68, 129, 194). The gastrointestinal tract may be the source of some disseminating infections, although some may originate from infected wounds and trauma. *Aeromonas* spp. are well documented as causative agents of wound infections, usually linked to water-associated injuries or aquatic recreational activities (78).

Gastroenteritis

Aeromonas- and *Plesiomonas*-associated cases of gastroenteritis are distinctly seasonal (sharp summer peak). The most common (approximately three-quarters of reported cases) clinical presentation with which aeromonads have been associated is watery diarrhea accompanied by a mild fever. Vomiting may also occur in children under 2 years of age. Generally, *Aeromonas* gastrointestinal infection is a self-limiting disease lasting less than 1 week. In more than one-third of patients, however, diarrhea may last for over 2 weeks, and cases of longer than 17 months have been documented (206). The remaining one-quarter of incidents are dysenterylike cases with blood and mucus present in the stools. In some cases, the clinical features are suggestive of ulcerative colitis (79). Isolated cases of a choleralike illness (rice water stools) also are reported in the literature (52).

Patients with stools positive for *P. shigelloides* also have either watery diarrhea or diarrhea with blood and mucus. The secretory form is usually reported to last from 1 to 7 days but can be prolonged (~3 weeks), with many (up to 30) bowel movements per day at the peak of the disease (32, 172). Holmberg et al. (104) determined that most patients (adults) from whom *P. shigelloides* was isolated had symptoms of invasive disease (bloody, mucus-containing stools with polymorphonuclear leukocytes). Patients often had severe abdominal cramps, vomiting, and some degree of dehydration. Three culture-positive patients had another known concurrent infection (with *Entamoeba histolytica* or *Campylobacter jejuni*) and were excluded from the above analysis.

The lack of suitable animal models that reproduce the features of *Aeromonas*- and *Plesiomonas*-associated diarrhea has made it difficult to definitively establish the enteropathogenicity of these bacteria. The intestinal secretory immunoglobulin A (sIgA) response among adults in Mexico with diarrhea was studied as an indicator of enteropathogenicity. Of 12 subjects shedding the phenospecies *A. sobria* (Popoff classification) or *A. hydrophila* (HG1, HG2, and HG3), 11 had a fourfold or greater sIgA titer rise against the infecting strain. However, no sIgA titer rises were detected in 14 patients shedding *P. shigelloides*. Thus, although this study provided further evidence for the significance of *Aeromonas* species as pathogens in acute diarrhea, it raised questions about the role of *P. shigelloides*, at least in adults with traveler's diarrhea (124).

There is a species-associated disease spectrum for *Aeromonas*. *A. veronii* biovar sobria (HG8/10) is more frequently associated with bacteremia than the other species. *A. schubertii* (HG12) is associated with aquatic wound infections but as yet has not been documented in association with gastroenteritis (42, 44). *A. veronii* biovar sobria also occurs in wound infections and following the use of medicinal leeches (81, 222). *A. veronii* biovar sobria (HG8/10), *A. hydrophila* (HG1 and HG3), *A. caviae* (HG4), and, to a lesser extent, *A. veronii* biovar veronii (HG8/10), *A. trota* (HG 14) and *A. jandaei* (HG9) are the species which have been associated with gastroenteritis (42, 43, 118, 126). *A. veronii* biovar sobria has been associated with the dysenteric presentation more frequently than the other species, although studies of invasive *Aeromonas* gastroenteritis are few (158, 233). *A. caviae* (HG4) is most common in pediatric diarrhea (178). *A. media* (HG5) is also occasionally isolated from human feces, but its significance is not known (42, 118).

INFECTIOUS DOSE AND SUSCEPTIBLE POPULATIONS

Extraintestinal Infections

Aeromonas opportunistic extraintestinal infections occur particularly in children and adolescents, who account for approximately one-quarter of all patients presenting with bacteremia. Most patients have chronic underlying disorders, such as leukemia, solid tumors, aplastic anemia, hemoglobinopathies, cirrhosis, or renal failure (68, 129, 194). *Plesiomonas* meningitis has been reported in neonates, particularly those whose deliveries have been complicated by various medical conditions, including prolonged rupture of the mother's membranes. *Plesiomonas* septicemia has been reported in splenectomized adults (32, 66, 68).

Gastroenteritis

Aeromonas-associated gastroenteritis occurs most commonly in children, the elderly, and the immunocompromised (19, 79, 211). Predisposing risk factors in adults include hospitalization, antimicrobial therapy, neutralization of gastric acid or inhibition of acid secretion,

hepatic diseases, and underlying enteric conditions, such as gastric and colonic surgery, gastrointestinal tract bleeding, and idiopathic inflammatory bowel disease (75). Cases of traveler's diarrhea have also been attributed to *Aeromonas* spp. (80, 88). Formula-fed infants, or those with altered gastrointestinal tract flora as a consequence of disease or antibiotic administration, have been reported to favor the survival and colonization of *A. caviae* (178).

The infectious dose of aeromonads is not known. The one human trial, conducted in 1985, was largely unsuccessful. Diarrhea was observed in only 2 of 57 healthy adult volunteers, with doses ranging from 10^4 to 10^{10} CFU. One person experienced mild diarrhea with 10^9 CFU (enterotoxin-positive strain, isolated from a healthy person); a second person developed moderate diarrhea with 10^7 CFU of another strain (enterotoxin-positive strain, isolated from diarrheal stool) (175).

The essentially negative results of the human trial should not be considered evidence negating the enteropathogenicity of *Aeromonas* spp. First, the trial was conducted in healthy, human adults who may have acquired considerable immunity to *Aeromonas* spp. More important, a number of other considerations regarding the temperature-dependent expression of possible virulence factors, such as intestinal colonization factors, have since emerged. Virtually nothing was known in 1985 about these virulence determinants (140), or the suicide phenomenon (the acetic acid self-killing phenomenon among mesophilic aeromonads) (177), which could account for the complete lack of recovery of aeromonads from the stools of most challenged volunteers (175). A laboratory accident in the early 1990s in which a stock culture broth of *A. trota* (HG14) ($\sim 10^9$ CFU) was accidentally ingested by a 28-year-old laboratory worker with no preexisting health problems did correlate with clinical illness. He developed severe gastroenteric symptoms (rice water stools for 2 days) within 24 h, and an *Aeromonas* strain identical to the culture stock was isolated from his stools on two occasions (43).

P. shigelloides has been isolated from the diarrheal feces of both adults and children. It may pose an occupational hazard for veterinarians, zookeepers, fish handlers, and athletes in water sports. Symptomatic individuals from whom *P. shigelloides* has been isolated often have a history of freshwater contact, exposure to fish, amphibia, or reptiles, or travel to developing countries. Thus, ~70% of persons who present with plesiomonad-associated diarrhea have either an underlying illness (e.g., cancer or cirrhosis) or an identifiable risk factor (foreign travel, consumption of raw seafood) (32, 104). If *P. shigelloides* is responsible for the gastroenteritis in these individuals, its infectious dose is unknown. In a human volunteer study, none of 33 volunteers (given ampicillin to reduce normal intestinal flora) developed diarrhea, even though some received over 10^9 cells of a strain originally isolated from a 4-year-old patient with diarrhea whose stool had blood, mucus, and leukocytes. A similar dose in piglets, however, caused moderate to mild diarrhea in three of three infected animals (66, 104, 172).

Most Aeromonas- and Plesiomonas-associated gastrointestinal infections are self-limiting. However, for persistent or severe cases (e.g., bloody diarrhea in children, enterocolitis in adults), trimethoprim-sulfamethoxazole is suggested as a first choice, with tetracycline or doxycycline as an alternative for adults with allergies to sulfa drugs. For extraintestinal infections, quinolones or aminoglycosides are effective (121).

PATHOGENICITY AND VIRULENCE FACTORS OF AEROMONAS SPECIES

Most research into the possible virulence factors of aeromonads has been done on strains isolated from patients with gastroenteric disease. Less is known of the factors that may be significant in the pathogenesis of extraintestinal infections. Some virulence determinants, such as the cytotoxic enterotoxin (Act/aerolysin), may be common to both forms of infections, whereas others, such as S-layers, may be more important in extraintestinal infections. However, as the roles of the different putative virulence factors so far described have not been established for any human Aeromonas infections, the proposed virulence determinants will be considered together.

The varied clinical picture of Aeromonas infections, and gastroenteric illness in particular, suggests that complex pathogenic mechanisms occur in aeromonads. For the different gastroenteric disease manifestations, it may be that strains possess multiple virulence factors in different combinations which may interact with different levels of the intestine. Putative Aeromonas virulence factors, inferred from comparisons with better-studied enteropathogenic bacteria, are listed in Table 14.3. These inferred virulence factors tend to be found more frequently, or are better expressed, in clinical strains than in environmental isolates. However, despite significant advances in this area, there is still no group of virulence factors that can be used to identify a given clinical or environmental isolate that has the capacity to cause human illness. Enterotoxins have been the most studied, and evidence is accumulating that the cytotoxic enterotoxin (Act/aerolysin) plays a key role in the pathogenesis

Table 14.3 Potential virulence factors of Aeromonas spp.[a]

Extracellular enzymes (e.g., proteases, lipases, elastase)
Siderophores (enterobactin, amonabactin)
Exotoxins
Cytotoxic (cytolytic enterotoxin, aerolysin, Asao toxin, β-hemolysins)[b]
Cytotonic
Heat stable (56°C, 10–20 min), cholera toxin cross-reacting, and non-cross-reacting
Heat labile, non-cholera toxin cross-reacting
Shiga-like toxins
Endotoxins (lipopolysaccharide)
Invasins (invasion of HEp-2 and Caco-2 cells)
Adhesins
Type IV pili
Outer membrane proteins (lectinlike, possibly porins)
S-layers
Lateral flagella

[a] See references in text for possible roles of these factors in Aeromonas extraintestinal and gastroenteritis-associated infections.
[b] Synonyms.

of Aeromonas infections. Knowledge of invasins and intestinal adhesins is still very scant.

The lack of a suitable animal model of Aeromonas diarrhea has slowed progress in the identification of critical Aeromonas virulence determinants involved in gastroenteric disease. The removable intestinal tie adult rabbit diarrhea (RITARD) model is currently the most promising model available, but it involves moderately complex surgery (195). In this model, histological changes consistent with colonization and infection of the ileum are seen. Rats pretreated with clindamycin and fed aeromonads reportedly can develop diarrhea and histopathologic evidence of enteritis (85). However, this model has not been validated by other investigators (128). An adult streptomycin mouse model (fecal shedding of aeromonads monitored) and 50% lethal dose experiments using infant mice are not diarrheal models but may nevertheless measure the ability of strains to bind to the intestinal mucosa (146). The medicinal leech Hirudo medicinalis maintains a pure culture of A. veronii biovar sobria in its digestive tract and has been proposed as an experimental model for studying Aeromonas-host interaction (81). Additional research to establish appropriate animal models for studying virulence determinants is of crucial importance if we are to understand Aeromonas pathogenicity (118, 146).

S-Layers

S-layers are macromolecular arrays of protein subunits (49 to 58 kDa) found on the bacterial cell surface. Strains of the psychrophilic aeromonad *A. salmonicida* produce an S-layer, previously known as A-layer. S-layer has many biological activities. It contributes to protection against the bactericidal activity of both immune and nonimmune serum, influences the outcome of interaction of the bacterium with macrophages, protects against the action of proteases, and binds immunoglobulins. S-layer is also capable of binding a variety of extracellular matrix proteins (e.g., basement membrane components collagen IV and laminin) and may play a role in colonization. Transposon mutants with altered ability to produce S-layer lose their ability to kill fish (228).

The biological functions of the S-layers of the mesophilic aeromonads are not yet known. However, one study suggests that for *A. hydrophila*, reduction in virulence caused by the loss of the S-layer is less significant than is the case for *A. salmonicida* (184). S-layer-positive strains of the mesophilic aeromonads *A. hydrophila* (HG1) and *A. veronii* biovar sobria (HG8/10) have been identified. These strains were from extraintestinal infections, such as wounds and peritonitis. S-layer-producing strains have been particularly associated with a single lipopolysaccharide serogroup, O:11, which contains some highly virulent strains, although there are possibly other groups with S-layers. Like *A. salmonicida*, these S-layer-positive, mesophilic strains are more refractory to the bactericidal activity of pooled human sera than strains that do not possess S-layers, although the effect is probably not directly due to the presence of S-layer. Serogroup O:11 strains, as well as serogroup O:34 strains, which reportedly do not possess S-layers but which also occur in septicemic disease, produce capsular polysaccharide. This may be involved in the virulence of these strains (164, 169). *A. hydrophila* (HG1, HG2, and HG3) serogroup O:34 strains adhered better and were more invasive in two fish cell lines when strains were grown under conditions promoting capsule formation (167). The S-layer may possess antiphagocytic activity which could facilitate systemic dissemination after invasion through the gastrointestinal mucosa. The virulence of S-layer-negative strains for mice following intraperitoneal injection increased when the bacteria were mixed with purified S-layer (153). An enzyme-linked immunosorbent assay using specific polyclonal antibodies against S-layer has been developed for the detection of O:11 aeromonads in foods (170).

Extracellular Enzymes

Most aeromonads elaborate a variety of extracellular enzymes: proteases (at least four or five, including a thermolabile serine protease and a thermostable metalloprotease), DNase, RNase, elastase, lecithinase, amylase, lipases, gelatinase, and chitinase. Several of these enzymes have now been cloned, as have two *exe* operons required for extracellular secretion of proteins by *Aeromonas* species (14, 53, 77, 117, 166, 207). The roles of the enzymes in virulence have yet to be determined. Proteases may contribute to pathogenicity by causing direct tissue damage or enhanced invasiveness. In addition, at least three toxins produced by *Aeromonas* spp. (Act/aerolysin, hemolysin, and glycerophospholipid:cholesterol acyltransferase [GCAT]) as well as Shiga-like toxins (which some *Aeromonas* strains may produce) are activated by proteolytic cleavage. Lipases may be important for bacterial nutrition. They may also constitute virulence factors by interacting with human leukocytes, or by affecting several immune system functions by free fatty acids generated by lipolytic activity.

GCAT is a novel lipase (37.4-kDa protein) that produces cholesterol esters (33). This enzyme, which is found in *Aeromonas* and *Vibrio* species, can catalyze the hydrolysis of a number of lipids and destroys host cells by digesting their cell membranes. For this reason, GCAT is also a hemolytic toxin. The structural gene encoding GCAT has been cloned and sequenced (227). GCAT is believed to be a critical virulence determinant in the pathogenesis of furunculosis in fish. However, its role in human disease is considered less significant, because GCAT is much less efficient at lysing mammalian cells.

Siderophores

Efficient mechanisms for iron acquisition from the host during an infection are considered essential for virulence. Siderophores are low-molecular-weight compounds with high affinities for various forms of iron. Mesophilic aeromonads produce either of two siderophores, enterobactin or amonabactin. Enterobactin is produced by various gram-negative bacteria; amonabactin has so far been detected only in *Aeromonas* spp. Two biologically active forms of this latter siderophore exist. The type of siderophore produced by aeromonads varies with hybridization group. Amonabactin is the predominant siderophore in *A. hydrophila* (HG1, HG2, and HG3), *A. caviae* (HG4), *A. media* (HG5), *A. schubertii* (HG12), and *A. trota* (HG14), whereas *A. veronii* biovar sobria (HG8/10) and *A. jandaei* (HG9) produce enterobactin. Amonabactin (*amoA*) and enterobactin (*aebC*) biosynthetic genes have been cloned. This will facilitate determination of the role(s) of these siderophores in *Aeromonas* virulence (21, 38, 39).

Exotoxins

Although the toxins are the best characterized of the putative virulence determinants of *Aeromonas* species, their role in human illness is unknown because few have been purified to homogeneity to enable investigation of their involvement in *Aeromonas*-induced diseases. Reports on "toxins" are usually based on the measurement of biological activities in vitro and sometimes in vivo. In recent years, molecular genetic analysis has confirmed that pathogenesis of some *Aeromonas* infections is due to molecules analogous to the toxins of other, better-studied enteropathogenic bacteria. However, for *Aeromonas* species, the direct role of these molecules as virulence determinants is still essentially unknown because of the lack of case-control human studies. There is little evidence that production of a specific toxin correlates with the ability of a given strain to cause gastrointestinal disease in humans. Nevertheless, an adhesive-enterotoxic mechanism has been widely postulated as the one most likely to result in the commonest gastroenteric presentation (watery diarrhea) attributed to *Aeromonas* spp. A cytotoxic enterotoxin from *A. hydrophila* (HG1) and aerolysins from *A. bestiarum* (HG2) and *A. trota* (HG14) are possibly the best characterized of the bacterial enterotoxins/hemolysins in terms of their molecular properties (34, 51, 61, 232). Moreover, a recent case-control human study at the International Centre for Diarrhoeal Disease Research, Bangladesh, has revealed for the first time an association of diarrhea with the production of different enterotoxins (6).

Factors that have added to the confusion include the difficulties with the taxonomy of the species in this genus and the variety of toxins that may be produced by different strains of aeromonads. Not all strains produce all the toxins described to date. Even if strains do possess specific toxin genes, these may be expressed only under certain growth conditions. Also, at least one toxin (i.e., Act/aerolysin) possesses multiple biological activities. Traditionally, the major measurable effects of exotoxins have been their lethality for mice and their enterotoxigenic, hemolytic, and cytotoxic activities. In the laboratory, relatively simple tests have been devised to measure such activities. Those most commonly used are illustrated in Fig. 14.1. Bacterium-free broth culture supernatant fluids are tested for enterotoxic activity (fluid accumulation) in a suckling mouse or rat/rabbit ligated ileal loop assay; cytotoxic activity is measured by observing, or quantifying with dye, disruption of a Vero or CHO cell monolayer following addition of the broth culture supernatant fluid; hemolysin production is quantified by titration of broth supernatant fluids against a suspension of rabbit erythrocytes (136, 142). A significant correlation in the production of these three exotoxin activities has often been observed. However, this has not always been found, particularly for environmen-

tal strains, and the different activities peak at different time periods during bacterial growth (36). Attempts to purify toxins led to reports of both heat-stable (56°C, 10 to 20 min) and heat-labile cytotonic and cytotoxic enterotoxins, some associated and some not associated with hemolytic activity and some related and some unrelated to cholera toxin (CT).

The different types of enterotoxins produced by *Aeromonas* spp. can cause a problem with the interpretation of toxic activity, especially when studies are performed with tissue culture cell lines. If a toxin has "cytotoxic" activity, the "cytotonic" enterotoxins from the same strain will be difficult to detect because cytotoxicity is the dominant effect. Proteolytic activities expressed by individual strains may also complicate assays, as these activities can lead to rounding up of cells. In addition, some of the above-mentioned toxins (e.g., Act/aerolysin) may require enzymatic activation (106). Hence, the so-called heat lability of a particular toxin may be the result of inactivation of a protease required for its activation. These factors may help to explain some of the conflicting reports in the literature. A PCR-based method for the detection of Act/aerolysin genes in *Aeromonas* species has recently been described as an alternative to the biological assays described above (131). PCR also has been used to detect heat-labile cytotonic enterotoxin in *Aeromonas* isolates from food and water (82). The increased use of such genetic detection methods is helping to clarify the occurrence of enterotoxins among *Aeromonas* isolates and their distribution in *Aeromonas* species.

Cytotoxic/Cytolytic Enterotoxin

Cytotoxic enterotoxin (Act) has associated hemolytic, cytotoxic, and enterotoxic activities and is often produced by diarrheal isolates. The toxin is lethal to mice when injected intravenously. This molecule also has been referred to as aerolysin and β-hemolysin. Several groups have cloned and sequenced Act/aerolysin-encoding genes (Table 14.4) (48, 58, 70, 98, 107, 111, 130), and it has now been determined conclusively that the products of these genes have biological activities that include lethality for mice, hemolysis, cytotoxicity, and enterotoxicity (58, 70, 237). Act increases cyclic AMP (cAMP) levels in macrophages; hence, it is correctly recognized as an enterotoxin (61). Alignment of the predicted amino acid sequences of the Act/aerolysin gene products suggests that there are at least two different variants of this toxin. The toxins designated AHH3 and AHH5 (96), Act (58), and aerolysin (108) exhibit a high degree of identity. In contrast, those designated ASA1 (96), AerA (111), and ASH3 (97) exhibit considerable sequence divergence in their N- and C-terminal regions from the above

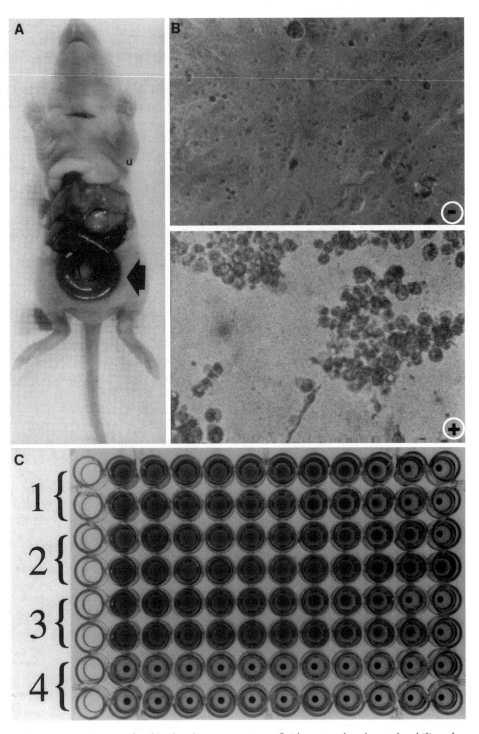

Figure 14.1 Bacterium-free broth culture supernatant fluids are used to detect the ability of an *Aeromonas* isolate to elaborate toxins. (A) Suckling mouse assay to detect enterotoxic activity. Note fluid accumulation in the intestine (arrow). (B) Cytotoxin assay (Vero cells). Cytotoxin causes healthy cells (top) to round up and lift from the monolayer (bottom). (C) Hemolysin titration against rabbit erythrocytes. Strains 1 to 3 have hemolytic titers >128; strain 4 has not produced detectable levels of hemolysin. Adapted from reference 147.

Table 14.4 Molecular masses of the enterotoxins of *Aeromonas* spp.

Factor	Molecular mass (kDa)	*Aeromonas* spp. from which gene(s) was cloned[a]	Source of strain	Reference
Cytotoxic enterotoxin (aerolysin, cytolysin, Asao toxin, β-hemolysin)	49–54	*A. hydrophila* = *A. bestiarum*	Rainbow trout	107
		A. sobria = *A. trota*[b]	Human diarrheal feces	111
		A. hydrophila (HG1, HG2, and HG3)	Eel	98
		A. hydrophila (HG1, HG2, and HG3)	Human	98
		A. sobria	Human	98
		A. hydrophila	Human diarrheal feces	58
		A. trota	ATCC 49659	130
		A. sobria	Human diarrheal feces	70
Hemolysin	~60–64	*A. hydrophila* (HG1)	ATCC 7966	96
Cytotonic enterotoxin		*A. hydrophila* = *A. trota*[b]	Human diarrheal feces	50
Heat labile	35–40	*A. hydrophila*	Human diarrheal feces	60
Heat stable	33 and 67[c]	Possibly similar to gene from A. *trota* above		50

[a] In most instances, the data in the papers reviewed were unclear on the method of classification to the species level. It is assumed that when details were not given, *A. hydrophila* was used in its broad sense to refer to a strain of the motile mesophilc aeromonads. That is its meaning unless otherwise stated in the table. *A. sobria* refers to the phenospecies defined by Popoff (201).

[b] In some studies, strains designated *A. hydrophila* or *A. sobria* were subsequently determined to be *A. trota* (HG14) as indicated.

[c] Two polypeptides have been identified, but the relationship between the two is unknown (54).

family of toxins but overall have a high degree of identity with each other. The "hemolysin" with enterotoxigenic activity described by Fujii et al. also appears to belong to this latter group (70, 71). The flanking sequences of aerolysins and Act structural genes are highly divergent (58). Khan et al. (130) concluded that aerolysin and aerolysinlike genes fell into three phylogenetic groups based on protein sequences available in GenBank. All the above Act/aerolysinlike gene products are heat-labile (56°C, 10 min) proteins with molecular masses of 49 kDa to 54 kDa.

The primary amino acid sequence of Act/aerolysin is fairly unique, but a short (31-amino-acid) region does exhibit homology with *Pseudomonas aeruginosa* cytotoxin, *S. aureus* alpha-toxin, *Clostridium perfringens* epsilon-toxin, and perfringolysin O (51, 55, 58, 193). The primary structure of *Clostridium septicum* alpha-toxin exhibits 27% identity and 72% similarity over a 387-residue region with the primary structure of *Aeromonas* aerolysin (20).

Act/aerolysin proteins are channel-forming proteins which ultimately produce a transmembrane channel that destroys cells by breaking their permeability barriers (51, 62, 67, 193, 232, 234, 235). There are two inactive precursor forms of the toxin. The first, preprotoxin, contains a typical 23-amino-acid signal sequence that directs transport across the inner bacterial membrane. This sequence is removed during transit. The resulting protoxin is then exported and activated by proteolysis of about 25 to 48 amino acids from the carboxy-terminal end. The active aerolysin binds to the receptor glycophorin

on eukaryotic cells (74). However, Act does not bind to glycophorin but uses cholesterol as one of the receptors on the host cell (67). Aerolysin has a novel protein fold as a result of its relatively hydrophobic primary structure, which lacks extended regions of hydrophobic residues, such as those normally present in transmembrane proteins. After binding, both Act and aerolysin oligomerize, which is an essential step in channel formation (34, 51, 62, 192, 234, 235). Oligomerization precedes membrane insertion. Purified aerolysin at very low concentrations forms pores in artificial lipid bilayers. These pores are ~1.5 nm and exhibit weak anion selectivity. This could explain the enterotoxic properties of the protein at low concentrations (51, 232). However, it also has been suggested that aerolysin does not cause cell death simply by allowing leakage of intracellular components. One study has determined that channel formation leads to selective permeabilization of the target cell to small ions, such as potassium, which causes depolarization of the plasma membrane. This in turn results in vacuolation of the endoplasmic reticulum and selective alteration of the early biosynthetic membrane pathway by inhibiting transport of newly synthesized membrane proteins to the plasma membrane (4). Recently, a hemolysin with enterotoxic activity from *A. sobria* (Popoff phenospecies) and Act from *A. hydrophila* (HG1) have been shown to increase cAMP in cultured cells (61, 71). Act also increases the level of prostaglandin E_2 in macrophages and in the fluids of rat ligated loops treated with the toxin (61). This provides a conclusive mechanism by which Act and possibly aerolysins could produce fluid secretion in the intestine.

In addition to its roles as a cytotoxic enterotoxin and in increasing the lethality of *A. hydrophila* (HG1) for mice, Chopra and colleagues have determined that Act inhibits the phagocytic ability of mouse phagocytes both in vitro and in vivo and stimulates the chemotactic activity of human leukocytes. Act also increases the production of proinflammatory cytokines, such as tumor necrosis factor-α, interleukin-1β, and interleukin-6, in a macrophage cell line. These activities could contribute to the pathology observed in the small intestine following *Aeromonas* infection (56, 61, 125).

A. veronii biovar sobria (HG8/10) is in standard exotoxin assays the species that most often contains the highest proportion of toxigenic strains and the strains producing the highest titers of toxins. Strains of *A. hydrophila* (HG1, HG2, and HG3) also are frequently positive for these activities. Interestingly, strains of *A. caviae* (HG4) tend not to produce exotoxins under the same conditions; however, genetic analysis has shown that ~50% of *A. caviae* strains carry the aerolysin gene (110). Husslein et al. suggested that the gene in these strains is silenced in some way (mutation, insertion, or deletion) (112). In glucose-free, double-strength Trypticase broth, and particularly in the late stationary phase of growth (that is, under culture conditions different from those used in a standard assay), some clinical strains of *A. caviae* produce detectable cytotoxic and cytotonic enterotoxin activities (178). The production of Act by *A. hydrophila* (HG1) and aerolysin by *A. bestiarum* is reduced when the bacteria are grown in medium containing glucose and iron (56). Singh and Sanyal (221) have determined that passage through rabbit ileal loops of strains of all of the three phenospecies described above increased the proportion of strains able to produce detectable enterotoxin. The conclusion from studies to date is that most strains of *A. veronii* biovar sobria (HG8/10), *A. hydrophila* (HG1, HG2, and HG3), and *A. caviae* (HG4), whether from clinical or environmental sources, are potentially enterotoxigenic.

Hemolysins

Aeromonads produce at least two major classes of hemolysins, α and β. Act/aerolysinlike molecules (β-hemolysins) are enterotoxigenic cytolysins. Genes encoding other β-hemolysins distinct from Act/aerolysin also have been cloned from several *A. hydrophila* (HG1, HG2, and HG3) strains, including ATCC 7966 *A. hydrophila* (HG1) (96, 236). These toxins, termed hemolysin (HlyA), are larger (~64 kDa) than Act/aerolysin (~54 kDa) and have only 17.8 to 19.6% amino acid sequence identity with Act/aerolysin (58, 108). They do, however, exhibit homology with HlyA

from *V. cholerae* El Tor (10). In one study, the gene for this hemolysin was detected in 43 of 62 (69%) hemolysin-producing *Aeromonas* strains by colony hybridization analysis (96). A more recent survey of the distribution of *hlyA* and *aerA* genes in clinical and environmental *Aeromonas* isolates examined 61 *A. hydrophila* (HG1, HG2, and HG3), 83 *A. veronii* biovar sobria, and 34 *A. caviae* strains. The *hlyA* gene was detected in 96, 12, and 35% of these species, respectively, whereas 78, 97, and 41%, respectively, were *aerA* positive (95). Studies with mutant strains in a suckling mouse model revealed that all virulent *A. hydrophila* (HG1, HG2, and HG3) isolates were *hlyA⁺aerA⁺*, and a two-hemolytic toxin model of virulence for *A. hydrophila* (HG1, HG2, and HG3) was therefore proposed (95, 236). This model adds further support to other evidence, such as that obtained in adhesion and animal model studies, that there are likely to be species-associated differences in *Aeromonas* virulence mechanisms.

The second class of hemolysins, α-hemolysin (molecular mass, 65 kDa), is elaborated during the late stationary phase of growth and causes incomplete lysis of erythrocytes (double-zone lysis on blood agar). It is not expressed at temperatures above 30°C and has not been associated with enterotoxic properties.

Cytotonic Enterotoxins

A variety of cytotonic enterotoxins have been described, and several have now been cloned (Table 14.4). They cause elongation of and increased levels of cAMP in CHO cells. They are all thought to act similarly to CT, despite their differing molecular characteristics and variable reactivity with cholera antitoxin. Several non-CT-reactive cytotonic enterotoxins have been described for *Aeromonas* spp., including a 15- to 20-kDa heat-stable protein by Ljungh et al. (160) and a 44-kDa heat-labile protein, referred to as Alt by Chopra and Houston (56, 57). Other cytotonic enterotoxins, such as that purified from a strain of *A. sobria* (Popoff phenospecies) by Potomski et al. (202), have exhibited cross-reactivity to CT. This latter purified toxin exhibited bands of 43.5, 29.5, and 26 kDa on sodium dodecyl sulfate-polyacrylamide gel electrophoresis. Schultz and McCardell similarly demonstrated that in cell lysates of *Aeromonas* strains, some strains had three protein bands (sizes of 89, 37, and 11 kDa) that reacted with CT antitoxin on an immunoblot (215). Chakraborty et al. cloned a cytotonic enterotoxin gene from an isolate [now known to be *A. trota* (HG14)] and determined conclusively that this strain produced a cytotonic enterotoxin that was distinct from cytotoxic/hemolytic enterotoxin (50). Chopra et al. cloned the gene encoding Alt and subsequently purified the recombinant protein from *E. coli*. The recombinant Alt was slightly smaller (35 to 38 kDa) compared with the native Alt (44 kDa, mentioned above) purified from *A. hydrophila* (HG1) (56, 60). At the DNA level there was no homology with the *A. trota* cytotonic toxin gene studied by Chakraborty and colleagues (50), but at the amino acid level, Alt exhibited significant homology with lipase and phospholipase of *A. hydrophila* (HG1, HG2, and HG3), although Alt did not exhibit phospholipase activity (59, 166). This toxin increases Ca^{2+} levels in CHO cells through host cell phospholipase C activation (54). Further, active immunization of mice with Alt reduced the fluid secretory capacity of *A. hydrophila* (HG1) in a mouse ligated ileal loop model (59). A second cytotonic enterotoxin gene (designated *ast*) also was identified in the Alt⁺ *A. hydrophila* strain by Chopra and colleagues, and a 4.8-kb *Sal*I-*Bam*HI DNA fragment encoding this heat-stable toxin hybridized to a 3.5-kb *Bam*HI DNA fragment of a plasmid that contained the *A. trota* heat-stable cytotonic enterotoxin gene (60). This toxin elevated cAMP in cells through a unique mechanism (54).

Aeromonas spp., therefore, may produce different types of cytotonic enterotoxins that are functionally similar. However, the mechanism(s) involved in elevating second messengers (e.g., cAMP) might be quite different. In a recent study, the *alt* gene was detected by PCR in 75% (23 of 31) of food and water *Aeromonas* isolates, including the *A. hydrophila* (HG1, HG2, and HG3) strain implicated in the outbreak of foodborne illness caused by the ingestion of raw, fermented fish (82). However, cytotonic enterotoxins may be produced by only a minority of strains. Seidler et al. reported their production in <6% (20 of 330) of strains (216). Shimada et al. determined that a CT-like enterotoxin was produced by *Aeromonas* spp. in ~4.5% (8 of 179) of strains tested (219).

In summary, despite significant advances, much still remains to be done to clarify the picture regarding the relative roles of the above *Aeromonas* toxins in gastrointestinal infections. For extraintestinal infections (septicemia), it has been determined in a mouse infection model that Act/aerolysin is required for both the initiation and maintenance of infection (49, 237). Jin et al. (125) have determined that intraperitoneal pretreatment of mice with Act predisposed them to nonlethal quantities of *Aeromonas* that was translocated into liver, spleen, and heart blood, whereas in mice not given toxin, *Aeromonas* did not spread to these organs. The toxin impairs the phagocytic ability of mouse phagocytes and is also a potent chemoattractant, inducing leukocyte infiltration in vivo, resulting in inflammation. The discovery

of these pathogenic roles for Act helps to explain why *Aeromonas* species are capable of causing wound infections, septicemia, and/or diarrhea.

Shiga-Like toxins

Aeromonas infection has been associated with hemolytic-uremic syndrome (27, 65, 120, 208). Hemolytic-uremic syndrome is caused by Shiga toxin (Stx)-producing bacteria. Hence, it is possible that some strains of *Aeromonas* have acquired the ability to produce Shiga-like toxins. One published study reported Stx1 production in 3 of 28 (11%) clinical *Aeromonas* isolates and 1 of 11 (9%) environmental isolates. Stx-producing bacteria were identified on the basis of neutralization of cytotoxicity and PCR amplification of isolated DNA using oligonucleotide primers from the Stx1 gene of *E. coli* O157:H7. The gene encoding Stx1 was found on a small 2.14-kb plasmid (90). Additional studies are required to confirm these findings.

Invasins

The Sereny test for tissue invasiveness has been universally negative for aeromonads (66). Only a few studies have examined the ability of strains to penetrate epithelial cell lines, such as HEp-2 of epipharyngeal origin and the intestinal Caco-2 (Fig. 14.2) (158, 181, 233). Invasive ability has been reported most commonly for strains of *A. sobria* (*A. veronii* biovar sobria), followed by *A. hydrophila* (HG1, HG2, and HG3), particularly strains isolated from patients with symptoms of dysentery (158, 233). Invasive strains of *A. caviae* also have been described, however (217). The possible link between tissue

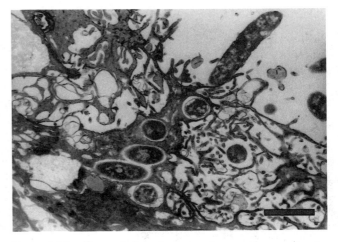

Figure 14.2 Transmission electron micrograph revealing invasion of Caco-2 cells by aeromonads. Bacteria are enclosed within membrane-bound vacuoles inside Caco-2 cell. Bar = 1 μm. Reprinted from reference 181 with permission.

culture cell invasiveness and dysentery needs to be confirmed with larger numbers of patients. The mechanisms of *Aeromonas* cell line adhesion and invasion remain to be elucidated. Invasin determinants are probably chromosomally located (158). Nishikawa et al. determined that the DNA of four invasive *Aeromonas* strains that they studied did not hybridize with probes to the genes associated with the attaching and effacing (*eae*) and invasion (*ipaB*) abilities of *E. coli* (181).

Plasmids

Plasmids carry important virulence determinants in many bacterial enteropathogens, but little is known of their role (if any) in *Aeromonas* virulence. Most strains of *A. salmonicida* carry at least one large plasmid (60 to 150 kb) and two low-molecular-weight plasmids (5.2 and 5.4 kb); additional plasmids are also frequently encountered (180). Plasmids are detected less commonly (17 to 27% of strains) in human diarrhea-associated *Aeromonas* isolates but have been detected in up to 95% of isolates from patients with bacteremia (5, 28, 231). Plasmids of clinical isolates are predominantly cryptic (function not known), but some are involved in antibiotic resistance and possibly siderophore production (28). Mini pilin, a putative adhesin, has been localized to a 7.6-kb plasmid in a strain of *A. hydrophila* (HG1, HG2, and HG3) (100). Shiga toxin genes were localized on a small 2.14-kb plasmid (90). There has also been a report that adherence and hemolytic properties of a strain of *Aeromonas* spp. may be regulated by a plasmid (40 MDa, ∼62 kb) that also codes for antibiotic resistance (86). Further study is needed to confirm this finding and to determine whether such regulation occurs in other *Aeromonas* strains.

Adhesins

Adhesion is likely to be an essential virulence factor for aeromonads that infect through mucosal surfaces or cause gastroenteric disease (133). Most studies of *Aeromonas* adhesion to cultured mammalian cells, such as mouse Y_1 adrenal, HEp-2, and human intestinal cell lines, such as INT407 (embryonic human intestinal cell line) and Caco-2 (colon carcinoma cells which differentiate in culture to express characteristics of small intestinal enterocytes), have revealed that strains of *A. veronii* biovar sobria are the most adhesive (the highest proportion of adherent strains and the strains with the greatest number of adherent bacteria per cell) (46, 84, 133). Clinical strains of *A. caviae* (>30%) also have been reported to be adherent (84, 178, 179). *A. hydrophila* (HG1, HG2, and HG3) is less adherent to cell lines than are the above two species. Limited research with animal models also indicates possible species-related differences in the

mechanisms by which strains adhere to the mucosal surface (146). A recent study has investigated the interaction of *Aeromonas* species with mucins. Species-associated differences in mucin interactions were observed, with *A. hydrophila* (HG1, HG2, and HG3) revealing greater binding than *A. sobria* (*A. veronii* biovar sobria) and *A. caviae* species (17). Hence, there are differences in adhesive mechanisms between species.

Receptors in the host are usually carbohydrate moieties which may also be expressed on erythrocytes. Hemagglutination has therefore been used as a screening model to detect adhesins. The hemagglutinins of *Aeromonas* species are numerous. Some have been associated with filamentous appendages (fimbriae or pili); others have been associated with outer membrane proteins. Some hemagglutination patterns (e.g., not inhibited by fucose, galactose, or mannose) correlate with diarrhea, but hemagglutination screening has not proved useful for the identification of potentially significant strains (35).

Filamentous Adhesins

Aeromonas filamentous surface structures are diverse. Their number and type vary depending on the source of the isolate and the bacterial growth conditions (99, 133, 143). Environmental strains of *A. veronii* biovar sobria are heavily piliated when isolated (143). Short, rigid pili are the predominant type expressed on heavily piliated aeromonads (Fig. 14.3A). These pili cause autoaggregation of bacteria but are not hemagglutinating and do not bind to intestinal cells (105). The amino acid sequence of pili purified from a clinical *A. hydrophila* (HG1, HG2, and HG3) isolate revealed homology with *E. coli* type I and Pap pilin (99). The role of the short, rigid pili in colonization is unknown.

In contrast, many *Aeromonas* isolates (in particular, strains of *A. veronii* biovar sobria and *A. caviae*) from diarrheal stools are poorly piliated (<10 per cell). Short, rigid pili can be induced on some clinical strains grown in liquid medium at <22°C (143), but long, thin (4- to 7-nm), flexible, or wavy pili are the predominant type present on strains isolated from feces (Fig. 14.3B) (46, 143, 148, 149). On some strains, bundles of flexible pili and thin (<2-nm) filamentous networks are present on small numbers of bacteria (<10%), especially when strains are grown in liquid medium at temperatures below 22°C (Fig. 14.4) (143, 146). Lateral flagella have been observed on 20 to 50% of strains grown on solid medium for 6 h at both 22 and 37°C (218, 220). These flagella have N-terminal homology (first 20 amino acids) with the lateral flagella of *Vibrio parahaemolyticus* (226). Approximately 50% of strains overall possess *laf* genes (150). At least some of the surface structures described above are colonization factors.

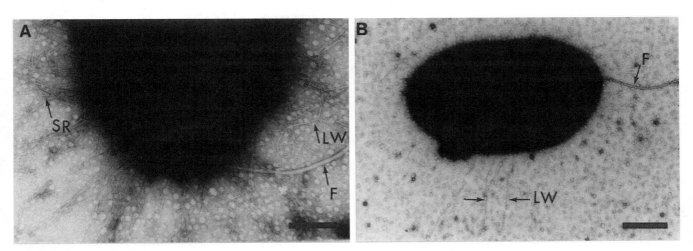

Figure 14.3 Transmission electron micrographs of negatively stained *A. veronii* biovar sobria strains revealing pili. (A) A heavily piliated bacterium with pili of short, rigid (SR) and long, flexible, or wavy (LW) morphology and a flagellum (F). Environmental strains are heavily piliated. Bar = 0.2 μm. (B) Fecal isolates of this species are poorly piliated. Generally a few long, flexible (LW) pili (arrows) are seen. Bar = 0.4 μm.

Long, thin, flexible pili have been purified from all *Aeromonas* species associated with diarrheal infection (101–103, 115, 148). They form a family of type IV pili whose structural subunits share N-terminal amino acid homology and have molecular masses of 19 to 23 kDa. They have been designated bundle-forming pili (Bfp) because the bundles of pili that link cells are of this pilus type (Fig. 14.4) (148). Despite this tendency to bundle formation, the pilins of this family have closer N-terminal homology to the classical type IVA pilins, such as the mannose-sensitive hemagglutinin pilus of *V. cholerae*, than they do to the type IVB pilins, such as the bundle-forming pilus of enteropathogenic *E. coli* and the toxin-coregulated pilus of *V. cholerae*. There is no published genetic analysis of this *Aeromonas* pilus type. However, a second family of type IV pili has been identified in the *Aeromonas* species following the cloning of a type IV pilus biogenesis gene cluster, *tapABCD*. This gene cluster was originally cloned from a strain of "*A. hydrophila*" (strain Ah65 = *A. bestiarum* HG2), and the product was designated Tap, for type IV *Aeromonas* pilus (196). The cluster was subsequently cloned from a strain of *A. veronii* biovar sobria from which a Bfp had been purified. It was identified subsequently in other

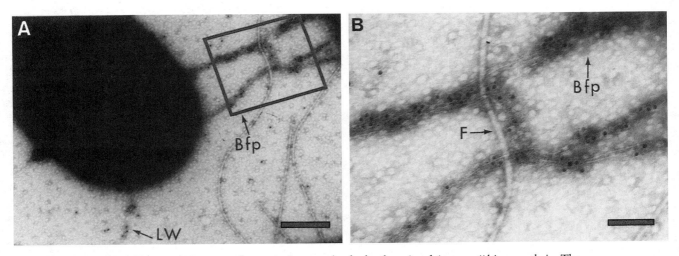

Figure 14.4 (A) Immunoelectron micrograph of a fecal strain of *A. veronii* biovar sobria. The gold particles bind specifically to pili and stain bundles of long, flexible pili (Bfp) as well as isolated long, flexible (LW) pili (arrow). Bar = 0.3 μm. (B) Enlargement of Bfp and flagellum (F) shown in panel A. Bar = 0.1 μm.

Bfp-positive strains and is widely conserved in all *Aeromonas* species (22, 23). Tap pilins have a predicted molecular mass of ~17 kDa and are more closely related to the classical type IVA pilin proteins of *Dichelobacter nodosus*, *P. aeruginosa*, and *Neissria gonorrhoeae* than they are to *Aeromonas* Bfp pilin proteins (22).

There is substantial evidence that the Bfp are intestinal colonization factors. Purified Bfp bind to epithelial and intestinal cell lines, freshly isolated human enterocytes, and fresh and fixed human and rabbit intestinal tissues, as determined by light and electron microscopy and immunohistochemical detection. Removal of Bfp by mechanical means decreased adhesion to cell lines by up to 80% (140). Purified Bfp blocked (to varying degrees) adhesion of *Aeromonas* bacteria to intestinal cells in a dose-dependent manner. Bacterial adhesion also could be blocked by the Fab fraction of anti-Bfp immunoglobulin (101–103, 115, 144). Moreover, ultrastructural studies (ruthenium red staining and transmission and scanning electron microscopy) revealed pilus-mediated adhesion to enterocytes (144). Bfp also may promote colonization by forming bacterium-to-bacterium linkages (144, 148).

By contrast, inactivation of the *tapA* gene which encodes the Tap structural subunit protein had no effect on bacterial adherence to epithelial and intestinal cell lines or to human intestinal tissue. When a *tapA* mutant strain of *A. veronii* biovar sobria was compared with the wild-type strain in the RITARD model, no significant effect on either the duration or incidence of diarrhea was observed. The ability of the mutant strain to colonize infant mice also was unaffected (137). A homologous type IV pilus cluster, *pilABCD*, present in *V. cholerae* recently has been determined not to be important for adhesion to HEp-2 cells or for the colonization of infant mice (72). Hence, Tap pili do not appear to be as significant as Bfp for *Aeromonas* intestinal colonization, but their widespread conservation suggests they do play an important role in the biology of the bacterium.

Not all *Aeromonas* pili of flexible morphology are type IV pili. The plasmid-coded flexible pilus mentioned above was characterized by Ho et al. and is composed of a novel 46-amino-acid polypeptide (mini pilin) (99). The mini pilin structural gene (*fxp*) has not, however, been detected in any other strains (66 strains tested in original report; >100 strains tested in our laboratory). Its significance as a colonization factor is unclear. Mini pilin is also environmentally regulated and is maximally expressed in liquid medium at 22°C in conditions of iron depletion (99, 100).

Lateral flagella may facilitate intestinal colonization and biofilm formation at the intestinal mucosal surface by mediating swarming motility (92). They may also form bacterium-to-bacterium linkages to facilitate these processes (218), as has been described for *V. para-haemolyticus* (24).

Quorum sensing, a mechanism for controlling gene expression in response to an expanding bacterial population, has been reported in *Aeromonas* species, as in many other gram-negative bacteria (223, 224). It is likely to be an important factor in intestinal colonization and the establishment of persistent *Aeromonas* infections.

Nonfilamentous Adhesins

Outer membrane proteins and a lipopolysaccharide-protein complex also have been implicated as adhesins of *Aeromonas* species (133, 171, 203). A nonfimbrial hemagglutinin associated with a 43-kDa outer membrane protein of a strain of *A. hydrophila* has been purified. The receptor for this latter protein was human blood group H antigen, an oligosaccharide that lines the enteric tract. H-antigen terminal trisaccharides were used to prepare by affinity chromatography carbohydrate-reactive outer membrane proteins from solubilized outer membranes. Carbohydrate-reactive outer membrane proteins of this strain were indistinguishable from a porin subset isolated from the same strain. It has been proposed that porins may act as lectinlike adhesins for attachment of this strain to carbohydrate-rich surfaces, such as erythrocytes, and possibly the human gut (203).

Virulence and Temperature

Bacterial growth temperature affects the expression of several virulence-associated factors. Numbers of flexible pili (Bfp intestinal colonization factors) were greater when bacteria were grown at 22°C and 7°C than when they were grown at 37°C, and adhesive ability to cell lines was also optimal at 22°C (or 7°C for some environmental strains) (140, 146). The titers of extracellular products (hemolysin, cytotoxin, and proteases) reportedly increased for some strains, serum sensitivity decreased, lipopolysaccharide levels increased, and virulence for animals (mice and fish) was greater in strains grown at 20°C than in those grown at 37°C (168, 188, 229). Increased expression of virulence determinants at low temperature may have implications for the pathogenicity of aeromonads in stored foods and requires further investigation.

PATHOGENICITY AND VIRULENCE FACTORS OF *PLESIOMONAS* SPECIES

There are no virulence factors that are widely accepted as being important for *Plesiomonas*-associated infections. None is tested for routinely (66, 119, 187). Potential virulence factors that have been described are summarized

Table 14.5 Potential virulence factors for *P. shigelloides*[a]

Extracellular enzymes (e.g., elastase)
Enterotoxins/cytotoxins
    SMA, RIL, lysis of Y_1 cells (after passage in RIL)
    CHO elongation factor (when grown in iron-poor medium)
β-Hemolysin (cell associated)
Endotoxin
Invasins (invasion of HeLa cells)
Adhesins (glycocalyx?)

[a] See references in text. SMA, suckling mouse assay; RIL, rabbit ileal loop.

in Table 14.5. Very few have been investigated in any detail.

A glycocalyx has been detected on the outer surface of *P. shigelloides* by ruthenium red staining and transmission electron microscopy. This material may be involved in the attachment of plesiomonads to epithelial cells. Endotoxin may play a role in *Plesiomonas* virulence, as it does for other pathogens (32). Plesiomonads do produce extracellular enzymes, such as elastase, which may be involved in connective tissue degradation. However, whether strains produce toxins or not has been a controversial issue. It may be that toxin assays used for *Aeromonas* species are not suitable for detecting *Plesiomonas* toxins. Several investigators concluded that plesiomonads do not produce toxins because rabbit ileal loop and suckling mouse assays, cell culture assays (Y_1, Vero, CHO) and rabbit skin permeability toxin assays all yielded negative results (32, 66, 172). However, the production of both heat-stable and heat-labile cytotoxic enterotoxins from all strains, irrespective of origin or serotype, has been reported (213). Manorama et al. attempted to purify and characterize these toxins (162). The key to detecting *Plesiomonas* enterotoxins may be serial passage of isolates in vivo (32). Bioassays and enzyme-linked immunosorbent assays used to detect *E. coli* enterotoxins have been negative for *P. shigelloides*, as have nucleic acid probes for such toxins (187). Iron regulates production of a factor that causes elongation of CHO cells similar to that produced by *V. cholerae* enterotoxin (73). More than 90% of 36 *Plesiomonas* strains tested produced a β-hemolysin, as judged by the results of agar overlay and contact-dependent hemolysis assays. The hemolysin was cell associated and was produced at both 25° and 35°C (119). It may play a role in iron acquisition in vivo or have another, as yet unknown, role in gastrointestinal disease. Hemolytic and elastolytic activities of *P. shigelloides* are enhanced when bacteria are grown in an iron-depleted medium (212).

Little is known about intestinal colonization and mucus interactions by plesiomonads. Some isolates from

clinical material reportedly adhere to HeLa, HEp-2, and INT407 cells. However, adhesion was low compared with that of clinical *Aeromonas* isolates (214). Invasiveness in the Sereny test does not occur, but invasion of the cell lines described above occurs. With freshly isolated strains from children with acute diarrhea, invasion of HeLa cells was comparable with that of *Shigella sonnei* (32, 66, 187, 213). As mentioned, the patients from whom plesiomonads were isolated in a study by Holmberg et al. had symptoms typical of an invasive pathogen (104). Twenty-three of 27 isolates from these patients contained plasmids. A single plasmid of very high molecular mass (>150 MDa, ~230 kb) was identified in 12 isolates, but it was not the same as the large virulence plasmid described for *Shigella* species or enteroinvasive *E. coli*. Large plasmids also have been detected in strains from patients with *Plesiomonas*-associated colitis. It is possible that unstable virulence plasmids may be involved in *Plesiomonas* pathogenicity.

SUMMARY AND CONCLUSION

Aeromonas species are commonly isolated from foods and water. Molecular genetic studies have led to considerable advances in the taxonomy of these organisms in recent years. Good typing schemes for accurately identifying clinical isolates now exist, but there is a need for additional studies to develop schemes that will identify all environmental isolates. *Aeromonas* spp. are the etiologic agent in a variety of extraintestinal infections involving immunocompetent as well as immunocompromised individuals. There is also strong evidence that some strains are enteropathogenic despite the essentially negative result obtained in one human trial. Disease-causing strains appear to be only a selection from the diversity of strains present in the environment. There is still much to be learned about *Aeromonas* virulence determinants and how they may combine to result in the virulent subsets within each *Aeromonas* species that cause disease. At present, it is not possible to identify the disease-causing strains because of our lack of knowledge about *Aeromonas* virulence mechanisms. Some of the confusion surrounding *Aeromonas* enterotoxins has been resolved. It is likely that ongoing molecular genetic investigations will further elucidate the mechanisms of adhesion and invasiveness. We may then be able to develop tests which will identify those strains that pose a risk to human health.

The case for *Plesiomonas* being an enteric pathogen is less convincing than that for *Aeromonas* spp. Although *Plesiomonas* has been isolated from diarrheal patients and has been incriminated in several large water- and foodborne outbreaks, no definite virulence mechanism has been identified in most strains associated with gastrointestinal infections. The negative results obtained in human volunteer studies and the lack of an intestinal sIgA response in infected patients cast further doubt on the enteropathogenicity of the species. Establishment of the role of *P. shigelloides* in human disease therefore awaits further studies.

References

1. **Abbott, S. L., W. K. W. Cheung, S. Kroske-Bystrom, T. Malekzadeh, and J. M. Janda.** 1992. Identification of *Aeromonas* strains to the genospecies level in the clinical laboratory. *J. Clin. Microbiol.* 30:1262–1266.

2. **Abbott, S. L., L. S. Seli, M. Catino, Jr., M. A. Hartley, and J. M. Janda.** 1998. Misidentification of unusual *Aeromonas* species as members of the genus *Vibrio*: a continuing problem. *J. Clin. Microbiol.* 36:1103–1104.

3. **Abeyta, C, Jr., S. A. Palumbo, and G. N. Stelma.** 1994. *Aeromonas hydrophila* group, p. 1–27. *In* Y. H. Hui, J. R. Gorham, K. D. Murrel, and D. O. Cliver (ed.), *Foodborne Disease Handbook*, vol. 1. Marcel Dekker, Inc., New York, N.Y.

4. **Abrami, L., M. Fivaz, P.-E. Glauser, R. G. Parton, and F. van der Goot.** 1998. A pore-forming toxin interacts with a GPI-anchored protein and causes vacuolation of the endoplasmic reticulum. *J. Cell Biol.* 140:525–540.

5. **Alabi, S. A., and T. Odugbemi.** 1990. Plasmid screening amongst *Aeromonas* species and *Plesiomonas shigelloides* isolated from subjects with diarrhoea in Lagos, Nigeria. *Afr. J. Med. Med. Sci.* 19:303–306.

6. **Albert, M. J., and A. K. Chopra.** Personal communication.

7. **Aldová, E.** 1997. New serovars of *Plesiomonas shigelloides* 1996. *Cent. Eur. J. Public Health* 5:21–23.

8. **Aldová, E., D. Danesová, J. Postupa, and T. Shimada.** 1994. New serovars of *Plesiomonas shigelloides* 1992. *Cent. Eur. J. Public Health* 1:32–36.

9. **Aldová, E., and R. H. W. Schubert.** 1996. Serotyping of *Plesiomonas shigelloides*—a tool for understanding ecological relationships. *Med. Microbiol. Lett.* 5:33–39.

10. **Alm, R. A., U. H. Stroeher, and P. A. Manning.** 1988. Extracellular proteins of *Vibrio cholerae*: nucleotide sequence of the structural gene (*hlyA*) for the haemolysin of the haemolytic El Tor 017 and characterization of the *hlyA* mutation in the non-haemolytic classical strain 569B. *Mol. Microbiol.* 2:481–488.

11. **Altwegg, M.** 1993. A polyphasic approach to the classification and identification of *Aeromonas* strains. *Med. Microbiol. Lett.* 2:200–205.

12. **Altwegg, M., G. Martinetti-Lucchini, J. Lüthy-Hottenstein, and M. Rohrbach.** 1991. *Aeromonas*-associated gastroenteritis after consumption of contaminated shrimp. *Eur. J. Clin. Microbiol. Infect. Dis.* 10:44–45.

13. **Altwegg, M., A. G. Steigerwalt, R. Altwegg-Bissig, J. Lüthy-Hottenstein, and D. J. Brenner.** 1990. Biochemi-

14. **Anguita, J., L. B. Rodriguez Aparicio, and G. Naharro.** 1993. Purification, gene cloning, amino acid sequence analysis, and expression of an extracellular lipase from an *Aeromonas hydrophila* isolate. *Appl. Environ. Microbiol.* 59:2411–2417.

15. **Arai, T., N. Ikejima, T. Itoh, S. Sakai, T. Shimada, and R. Sakazaki.** 1980. A survey of *Plesiomonas shigelloides* from aquatic environments, domestic animals, pets, and humans. *J. Hyg. Camb.* 84:203–211.

16. **Araujo, R. M., R. M. Arribas, F. Lucena, and R. Pares.** 1989. Relation between *Aeromonas* and fecal coliforms in fresh waters. *J. Appl. Bacteriol.* 67:213–217.

17. **Ascencio, F., W. Martinez-Arias, M. J. Romero, and T. Wadstrom.** 1998. Analysis of the interaction of *Aeromonas caviae*, *A. hydrophila*, and *A. sobria* with mucins. *FEMS Immunol. Med. Microbiol.* 20:219–229.

18. **Ashbolt, N. J., A. Ball, M. Dorsch, C. Turner, P. Cox, A. Chapman, and S. M. Kirov.** 1995. The identification and human health significance of environmental aeromonads. *Water Sci. Technol.* 31:263–269.

19. **Ashdown, L. R., and J. M. Koehler.** 1993. The spectrum of *Aeromonas*-associated diarrhea in tropical Queensland, Australia. *Southeast Asian J. Trop. Med. Public Health* 24:347–353.

20. **Ballard, J., J. Crabtree, B. A. Roe, and R. K. Tweten.** 1995. The primary structure of *Clostridium septicum* alpha-toxin exhibits similarity with that of *Aeromonas hydrophila* aerolysin. *Infect. Immun.* 63:340–344.

21. **Bargouthi, S., S. M. Payne, J. E. L. Arceneaux, and B. R. Byers.** 1991. Cloning, mutagenesis, and nucleotide sequence of a siderophore biosynthetic gene (*amoA*) from *Aeromonas hydrophila*. *J. Bacteriol.* 173:5121–5128.

22. **Barnett, T. C., S. M. Kirov, M. S. Strom, and K. Sanderson.** 1997. *Aeromonas* spp. possess at least two distinct types of type IV pilus families. *Microb. Pathog.* 23:241–247.

23. **Barnett, T. C., and S. M. Kirov.** 1999. The type IV *Aeromonas* pilus (Tap) gene cluster is widely conserved in *Aeromonas* species. *Microb. Pathog.* 26:77–84.

24. **Belas, M. R., and R. R. Colwell.** 1982. Scanning electron microscope observation of the swarming phenomenon of *Vibrio parahaemolyticus*. *J. Bacteriol.* 150:956–959.

25. **Berrang, M. E., R. E. Brackett, and L. R. Beuchat.** 1989. Growth of *Aeromonas hydrophila* on fresh vegetables grown under a controlled atmosphere. *Appl. Environ. Microbiol.* 55:2167–2171.

26. **Beuchat, L. R.** 1991. Behaviour of *Aeromonas* species at refrigeration temperatures. *Int. J. Food Microbiol.* 13:217–224.

27. **Bogdanovic, R., M. Cobeljic, M. Markovic, V. Nikolic, M. Ognjanovic, L. Sarjanovic, and D. Makic.** 1991. Haemolytic-uraemic syndrome associated with *Aeromonas hydrophila* enterocolitis. *Pediatr. Nephrol.* 5:293–295.

28. **Borrego, J. J., M. A. Morinigo, E. Martinez-Manzanares, M. Bosca, D. Castro, J. L. Barja, and A. E. Toranzo.** 1991.

Plasmid associated virulence properties of environmental isolates of *Aeromonas hydrophila*. *J. Med. Microbiol.* **35:**264–269.

29. **Borrell, N., M.-J. Figueras, and J. Guarro.** 1998. Phenotypic identification of *Aeromonas* genomospecies from clinical and environmental sources. *Can. J. Microbiol.* **44:**103–108.

30. **Borrell, N., S. G. Acinas, M.-J. Figueras, and A. J. Martínez-Murcia.** 1997. Identification of *Aeromonas* clinical isolates by restriction fragment length polymorphism of PCR-amplified 16S rRNA genes. *J. Clin. Microbiol.* **35:**1671–1674.

31. **Bottone, E. J.** 1993. Correlation between known exposure to contaminated food or surface water and development of *Aeromonas hydrophila* and *Plesiomonas shigelloides* diarrheas. *Med. Microbiol. Lett.* **2:**217–225.

32. **Brenden, R. A., M. A. Miller, and J. M. Janda.** 1988. Clinical disease spectrum and pathogenic factors associated with *Plesiomonas shigelloides* infections in humans. *Rev. Infect. Dis.* **10:**303–316.

33. **Buckley, J. T.** 1982. Substrate specificity of bacterial glycerophospholipid:cholesterol acyltransferase. *Biochemistry* **21:**6699–6703.

34. **Buckley, J. T.** 1991. Secretion and mechanism of action of the hole-forming toxin aerolysin. *Experientia* **47:**418–419.

35. **Burke, V., M. Cooper, J. Robinson, M. Gracey, M. Lesmana, P. Echeverria, and J. M. Janda.** 1984. Hemagglutination patterns of *Aeromonas* spp. in relation to biotype and source. *J. Clin. Microbiol.* **19:**39–43.

36. **Burke, V., J. Robinson, H. M. Atkinson, M. Dibley, R. J. Berry, and M. Gracey.** 1981. Exotoxins of *Aeromonas hydrophila*. *Aust. J. Exp. Biol. Med. Sci.* **59:**753–761.

37. **Burke, V., J. Robinson, M. Gracey, D. Peterson, and K. Partridge.** 1984. Isolation of *Aeromonas hydrophila* from a metropolitan water supply: seasonal correlation with clinical isolates. *Appl. Environ. Microbiol.* **48:**361–366.

38. **Byers, B. R., and J. E. L. Arceneaux.** 1993. Siderophore diversity in the genus *Aeromonas*. *Med. Microbiol. Lett.* **2:**281–285.

39. **Byers, B. R., and J. E. L. Arceneaux.** 1994. Iron acquisition and virulence of the bacterial genus *Aeromonas*, p. 29–40. *In* J. A. Mantley, D. E. Crowley, and D. G. Luster (ed.), *Biochemistry of Metal Micronutrients in the Rhizosphere*. CRC Press, Inc., Boca Raton, Fla.

40. **Cahill, M. M., and I. C. MacRae.** 1992. Characteristics of O/129-sensitive motile *Aeromonas* strains isolated from freshwater on starch-ampicillin agar. *Microb. Ecol.* **24:**215–226.

41. **Callister, S. M., and W. A. Agger.** 1987. Enumeration and characterization of *Aeromonas hydrophila* and *Aeromonas caviae* isolated from grocery store produce. *Appl. Environ. Microbiol.* **53:**249–253.

42. **Carnahan, A. M., S. Behram, and S. W. Joseph.** 1991. Aerokey II: a flexible key for identifying clinical *Aeromonas* species. *J. Clin. Microbiol.* **29:**2843–2849.

43. **Carnahan, A. M., T. Chakraborty, G. R. Fanning, D. Verma, A. Ali, J. M. Janda, and S. W. Joseph.** 1991. *Aeromonas trota* sp. nov., an ampicillin-susceptible species isolated from clinical specimens. *J. Clin. Microbiol.* **29:**1206–1210.

44. **Carnahan, A. M., M. A. Marii, G. R. Fanning, M. A. Pass, and S. W. Joseph.** 1989. Characterization of *Aeromonas schubertii* strains recently isolated from traumatic wound infections. *J. Clin. Microbiol.* **27:**1826–1830.

45. **Carnahan, A. M., and S. W. Joseph.** *In* G. Garrity (ed.), *Bergey's Manual of Systematic Bacteriology*, vol. 2, 2nd ed., in press. Springer-Verlag, New York, N.Y.

46. **Carrello, A., K. A. Silburn, J. R. Budden, and B. J. Chang.** 1988. Adhesion of clinical and environmental *Aeromonas* isolates to HEp-2 cells. *J. Med. Microbiol.* **26:**19–27.

47. **Caselitz, F.-H.** 1996. How the *Aeromonas* story started in medical microbiology. *Med. Microbiol. Lett.* **5:**46–54.

48. **Chakraborty, T., B. Huhle, H. Bergbauer, and W. Goebel.** 1986. Cloning, expression, and mapping of the *Aeromonas hydrophila* aerolysin gene determinant in *Escherichia coli* K-12. *J. Bacteriol.* **167:**368–374.

49. **Chakraborty, T., B. Huhle, H. Hof, H. Bergbauer, and W. Goebel.** 1987. Marker exchange mutagenesis of the aerolysin determinant in *Aeromonas hydrophila* demonstrates the role of aerolysin in *A. hydrophila*-associated systemic infections. *Infect. Immun.* **55:**2274–2280.

50. **Chakraborty, T., M. A. Montenegro, S. C. Sanyal, R. Helmuth, E. Bulling, and K. N. Timmis.** 1984. Cloning of enterotoxin gene from *Aeromonas hydrophila* provides conclusive evidence of production of a cytotonic enterotoxin. *Infect. Immun.* **46:**435–441.

51. **Chakraborty, T., A. Schmid, S. Notermans, and R. Benz.** 1990. Aerolysin of *Aeromonas sobria*: evidence for formation of iron-permeable channels and comparison with alpha-toxin of *Staphylococcus aureus*. *Infect. Immun.* **58:**2127–2132.

52. **Champsaur, H., A. Andremont, D. Mathieu, E. Rottman, and P. Auzepy.** 1982. Cholera-like illness due to *Aeromonas sobria*. *J. Infect. Dis.* **145:**248–254.

53. **Chang, M. C., S. Y. Chang, S. L. Chen, and S. M. Chuang.** 1992. Cloning and expression in *Escherichia coli* of the gene encoding an extracellular deoxyribonuclease (DNase) from *Aeromonas hydrophila*. *Gene* **122:**175–180.

54. **Chopra, A. K.** Personal communication.

55. **Chopra, A. K., M. R. Ferguson, and C. W. Houston.** 1993. Molecular characterization of enterotoxins from *Aeromonas hydrophila*. *Med. Microbiol. Lett.* **2:**261–268.

56. **Chopra, A. K., and C. W. Houston.** 1999. Enterotoxins in *Aeromonas*-associated gastroenteritis. *Microb. Infect.* **1:**1129–1137.

57. **Chopra, A. K., and C. W. Houston.** 1989. Purification and partial characterization of a cytotonic enterotoxin produced by *Aeromonas hydrophila*. *Can. J. Microbiol.* **35:**719–727.

58. **Chopra, A. K., C. W. Houston, J. W. Peterson, and G.-F. Jin.** 1993. Cloning, expression and sequence analysis of a cytolytic enterotoxin gene from *Aeromonas hydrophila*. *Can. J. Microbiol.* **39:**513–523.

59. **Chopra, A. K., J. W. Peterson, X. J. Xu, D. H. Coppenhaver, and C. W. Houston.** 1996. Molecular and

biochemical characterization of a heat-labile enterotoxin from *Aeromonas hydrophila. Microb. Pathog.* 21:357–377.

60. Chopra, A. K., R. Pham, and C. W. Houston. 1994. Cloning and expression of putative cytotonic enterotoxin-encoding genes from *Aeromonas hydrophila. Gene* 139:87–91.

61. Chopra, A. K., X.-J. Xu, D. Ribardo, M. Gonzales, K. Kuhl, J. W. Peterson, and C. W. Houston. 2000. The cytotoxic enterotoxin of *Aeromonas hydrophila* induces proinflammatory cytokine production and activates arachidonic acid metabolism in macrophages. *Infect. Immun.* 68:2808–2818.

62. Diep, D. B., T. S. Lawrence, J. Ausio, S. P. Howard, and J. T. Buckley. 1998. Secretion and properties of the large and small lobes of the channel-forming toxin aerolysin. *Mol. Microbiol.* 30:341–352.

63. Dorsch, M., N. J. Ashbolt, P. T. Cox, and A. E. Goodman. 1994. Rapid identification of *Aeromonas* using 16S rDNA targeted oligonucleotide primers: a molecular approach based on screening of environmental isolates. *J. Appl. Bacteriol.* 77:722–726.

64. Eley, A., I. Geary, and M. H. Wilcox. 1993. Growth of *Aeromonas* spp. at 4°C and related toxin production. *Lett. Appl. Microbiol.* 16:36–39.

65. Fang, J.-S., J.-B. Chen, W.-J. Chen, and K.-T. Hsu. 1999. Hemolytic-uremic syndrome in an adult male with *Aeromonas hydrophila* enterocolitis. *Nephrol. Dial. Transplant.* 14:439–440.

66. Farmer, J. J., III, M. J. Arduino, and F. W. Hickman-Brenner. 1992. The genera *Aeromonas* and *Plesiomonas*, p. 3012–3028. *In* A. Balows, H. G. Trüper, M. Dworkin, W. Harder, and K.-H. Schleifer (ed.), *The Prokaryotes*, 2nd ed. Springer-Verlag, New York, N.Y.

67. Ferguson, M. R., X.-J. Xu, C. W. Houston, J. W. Peterson, D. H. Coppenhaver, V. L. Popov, and A. K. Chopra. 1997. Hyperproduction, purification, and mechanism of action of the cytotoxic enterotoxin produced by *Aeromonas hydrophila. Infect. Immun.* 65:4299–4308.

68. Freij, B. J. 1987. Extraintestinal *Aeromonas* and *Plesiomonas* infections of humans. *Experientia* 43:359–360.

69. Fricker, C. R., and S. Tompsett. 1989. *Aeromonas* spp. in foods: a significant cause of food poisoning? *Int. J. Food Microbiol.* 9:17–23.

70. Fujii, Y., T. Nomura, H. Kanzawa, M. Kameyama, H. Yamanaka, M. Aikita, K. Setsu, and K. Okamoto. 1998. Purification and characterization of enterotoxin produced by *Aeromonas sobria. Microbiol. Immunol.* 42:703–714.

71. Fujii, Y., T. Nomura, and K. Okamoto. 1999. Hemolysin of *Aeromonas sobria* stimulates production of cyclic AMP by cultured cells. *FEMS Microbiol. Lett.* 176:67–72.

72. Fullner, K. J., and J. J. Mekalanos. 1999. Genetic characterization of a new type IV-A pilus gene cluster found in both classical and El Tor biotypes of *Vibrio cholerae. Infect. Immun.* 67:1393–1404.

73. Gardner, S. E., S. E. Fowlston, and W. L. George. 1990. Effect of iron on production of a possible virulence factor

by *Plesiomonas shigelloides. J. Clin. Microbiol.* 28:811–813.

74. Garland, W. J., and J. T. Buckley. 1988. The cytolytic toxin aerolysin must aggregate to disrupt erythrocytes, and aggregation is stimulated by human glycophorin. *Infect. Immun.* 56:1249–1253.

75. George, W. L., M. M. Nakata, J. Thompson, and M. L. White. 1985. *Aeromonas*-related diarrhea in adults. *Arch. Intern. Med.* 145:2207–2211.

76. Gobat, P.-F., and T. Jemmi. 1993. Distribution of mesophilic *Aeromonas* species in raw and ready-to-eat fish and meat products in Switzerland. *Int. J. Food Microbiol.* 20:117–120.

77. Gobius, K. S., and J. M. Pemberton. 1988. Molecular cloning, characterization, and nucleotide sequence of an extracellular amylase gene from *Aeromonas hydrophila. J. Bacteriol.* 170:1325–1332.

78. Gold, W. L., and I. E. Salit. 1993. *Aeromonas hydrophila* infections of skin and soft tissue: report of 11 cases and review. *Clin. Infect. Dis.* 16:69–74.

79. Gracey, M., V. Burke, and J. Robinson. 1982. *Aeromonas*-associated gastroenteritis. *Lancet* ii:1304–1306.

80. Gracey, M., V. Burke, J. Robinson, P. L. Masters, J. Stewart, and J. Pearman. 1984. *Aeromonas* spp. in travellers' diarrhoea. *Br. Med. J.* 289:658.

81. Graf, J. 1999. Symbiosis of *Aeromonas veronii* biovar sobria and *Hirudo medicinalis*, the medicinal leech: a novel model for digestive tract associations. *Infect. Immun.* 67:1–7.

82. Granum, P. E., K. O'Sullivan, J. M. Tomás, and O. Ormen. 1998. Possible virulence factors of *Aeromonas* spp. from food and water. *FEMS Immunol. Med. Microbiol.* 21:131–137.

83. Gray, S. J., and D. J. Stickler. 1989. Some observations on the faecal carriage of mesophilic *Aeromonas* in cows and pigs. *Epidemiol. Infect.* 103:523–537.

84. Grey, P. A., and S. M. Kirov. 1993. Adherence to HEp-2 cells and enteropathogenic potential of *Aeromonas* spp. *Epidemiol. Infect.* 110:279–287.

85. Haberberger, R. L., Jr., W. P. Yonushonis, R. L. Daise, I. A. Mikhail, and E. A. Ishak. 1991. Re-examination of *Rattus norvegicus* as an animal model for *Aeromonas*-associated gastroenteritis in man. *Experientia* 47:426–429.

86. Hanes, D. E., and D. K. F. Chandler. 1993. The role of a 40-megadalton plasmid in the adherence and hemolytic properties of *Aeromonas hydrophila. Microb. Pathog.* 15:313–317.

87. Hänninen, M.-L. 1993. Occurrence of *Aeromonas* spp. in samples of ground meat and chicken. *Int. J. Food Microbiol.* 18:339–342.

88. Hänninen, M.-L., S. Salmi, L. Mattila, R. Taipalinen, and A. Siitonen. 1995. Association of *Aeromonas* spp. with travellers' diarrhoea in Finland. *J. Med. Microbiol.* 42:26–31.

89. Hänninen, M.-L., and A. Siitonen. 1995. Distribution of *Aeromonas* phenospecies and genospecies among strains isolated from water, foods, or from human clinical specimens. *Epidemiol. Infect.* 115:39–50.

90. **Haque, Q. M., A. Sugiyama, Y. Iwade, Y. Midorikawa, and T. Yamauchi.** 1996. Diarrheal and environmental isolates of *Aeromonas* spp. produce a toxin similar to Shiga-like toxin 1. *Curr. Microbiol.* **32**:239–245.

91. **Harf-Monteil, C., B. Jaulhac, J. M. Scheftel, Y. Hansmann, and H. Monteil.** 1999. Clinical significance of *Aeromonas* isolates: predisposing factors. 6th International *Aeromonas-Plesiomonas* Symposium, Chicago, Ill. http://www.aeromonas.com.

92. **Harshey, R. M.** 1994. Bees aren't the only ones: swarming in gram-negative bacteria. *Mol. Microbiol.* **13**:389–394.

93. **Havelaar, A. H., F. M. Schets, A. van Silhouf, W. H. Jansen, G. Wieten, and D. van der Kooij.** 1992. Typing of *Aeromonas* strains from patients with diarrhoea and from drinking water. *J. Appl. Bacteriol.* **72**:435–444.

94. **Havelaar, A. H., J. F. M. Versteegh, and M. During.** 1990. The presence of *Aeromonas* in drinking water supplies in the Netherlands. *Zentbl. Hyg.* **190**:236–256.

95. **Heuzenroeder, M. W., C. Y. F. Wong, and R. L. P. Flower.** 1999. Distribution of two hemolytic toxin genes in clinical and environmental isolates of *Aeromonas* spp.: correlation with virulence in a suckling mouse model. *FEMS Microbiol. Lett.* **174**:131–136.

96. **Hirono, I., and T. Aoki.** 1991. Nucleotide sequence and expression of an extracellular hemolysin gene of *Aeromonas hydrophila. Microb. Pathog.* **11**:189–197.

97. **Hirono, I., and T. Aoki.** 1993. Cloning and characterization of three hemolysin genes from *Aeromonas salmonicida. Microb. Pathog.* **15**:269–282.

98. **Hirono, I., T. Aoki, T. Asao, and S. Kozaki.** 1992. Nucleotide sequences and characterization of haemolysin genes from *Aeromonas hydrophila* and *Aeromonas sobria. Microb. Pathog.* **13**:433–446.

99. **Ho, A. S. Y., T. A. Mietzner, A. J. Smith, and G. K. Schoolnik.** 1990. The pili of *Aeromonas hydrophila*: identification of an environmentally regulated "mini pilin." *J. Exp. Med.* **172**:795–806.

100. **Ho, A. S. Y., I. Sohel, and G. K. Schoolnik.** 1992. Cloning and characterization of *fxp*, the flexible pilin gene of *Aeromonas hydrophila. Mol. Microbiol.* **6**:2725–2732.

101. **Hokama, A., Y. Honma, and N. Nakasone.** 1990. Pili of an *Aeromonas hydrophila* strain as a possible colonization factor. *Microbiol. Immunol.* **34**:901–915.

102. **Hokama, A., and M. Iwanaga.** 1991. Purification and characterization of *Aeromonas sobria* pili, a possible colonization factor. *Infect. Immun.* **59**:3478–3483.

103. **Hokama, A., and M. Iwanaga.** 1992. Purification and characterization of *Aeromonas sobria* Ae24 pili: a possible new colonization factor. *Microb. Pathog.* **13**:325–334.

104. **Holmberg, S. D., I. K. Wachsmuth, F. W. Hickman-Brenner, P. A. Blake, and J. J. Farmer III.** 1986. *Plesiomonas* enteric infections in the United States. *Ann. Intern. Med.* **105**:690–694.

105. **Honma, Y., and N. Nakasone.** 1990. Pili of *Aeromonas hydrophila*: purification, characterization and biological role. *Microbiol. Immunol.* **34**:83–98.

106. **Howard, S. P., and J. T. Buckley.** 1985. Activation of the hole-forming toxin aerolysin and extracellular processing. *J. Bacteriol.* **163**:336–340.

107. **Howard, S. P., and J. T. Buckley.** 1986. Molecular cloning and expression in *Escherichia coli* of the structural gene for the hemolytic toxin aerolysin from *Aeromonas hydrophila. Mol. Gen. Genet.* **204**:289–295.

108. **Howard, S. P., W. J. Garland, M. J. Green, and J. T. Buckley.** 1987. Nucleotide sequence of the gene for the hole-forming toxin aerolysin of *Aeromonas hydrophila. J. Bacteriol.* **169**:2869–2871.

109. **Huq, A., A. Akhtar, M. A. R. Chowdhury, and D. A. Sack.** 1991. Optimal growth temperature for the isolation of *Plesiomonas shigelloides*, using various selective and differential agars. *Can. J. Microbiol.* **37**:800–802.

110. **Husslein, V., T. Chakraborty, A. Carnahan, and S. W. Joseph.** 1992. Molecular studies on the aerolysin gene of *Aeromonas* species and discovery of a species-specific probe for *Aeromonas trota* species nova. *Clin. Infect. Dis.* **14**:1061–1068.

111. **Husslein, V., B. Huhle, T. Jarchau, R. Lurz, W. Goebel, and T. Chakraborty.** 1988. Nucleotide sequence and transcriptional analysis of the *aerCaerA* region of *Aeromonas sobria* encoding aerolysin and its regulatory region. *Mol. Microbiol.* **2**:507–517.

112. **Husslein, V., S. H. E. Notermans, and T. Chakraborty.** 1988. Gene probes for the detection of aerolysin in *Aeromonas* spp. *J. Diarrhoeal Dis. Res.* **6**:124–130.

113. **Huys, G., R. Coopman, P. Janssen, and K. Kersters.** 1996. High-resolution genotypic analysis of the genus *Aeromonas* by AFLP fingerprinting. *Int. J. Syst. Bacteriol.* **46**:572–580.

114. **Ingham, S. C.** 1990. Growth of *Aeromonas hydrophila* and *Plesiomonas shigelloides* on cooked crayfish tails during cold storage under air, vacuum, and a modified atmosphere. *J. Food Prot.* **53**:665–667.

115. **Iwanaga, M., and A. Hokama.** 1992. Characterization of *Aeromonas sobria* TAP 13 pili: a possible new colonization factor. *J. Gen. Microbiol.* **138**:1913–1919.

116. **Jacxsens, L., F. Devlieghere, P. Falcato, and J. Debevere.** 1999. Behavior of *Listeria monocytogenes* and *Aeromonas* spp. on fresh-cut produce packaged under equilibrium-modified atmosphere. *J. Food Prot.* **62**:1128–1135.

117. **Jahagirdar, R., and P. Howard.** 1994. Isolation and characterization of a second *exe* operon required for extracellular protein secretion in *Aeromonas hydrophila. J. Bacteriol.* **176**:6819–6826.

118. **Janda, J. M.** 1991. Recent advances in the study of the taxonomy, pathogenicity, and infectious syndromes associated with the genus *Aeromonas. Clin. Microbiol. Rev.* **4**:397–410.

119. **Janda, J. M., and S. L. Abbott.** 1993. Expression of hemolytic activity by *Plesiomonas shigelloides. J. Clin. Microbiol.* **31**:1206–1208.

120. **Janda, J. M., and S. L. Abbott.** 1998. Evolving concepts regarding the genus *Aeromonas*: an expanding panorama of species, disease presentations, and unanswered questions. *Clin. Infect. Dis.* **27**:332–344.

121. **Janda, J. M., and S. L. Abbott.** 1999. Unusual foodborne pathogens: *Listeria monocytogenes, Aeromonas,*

Plesiomonas and *Edwardsiella* species. *Clin. Lab. Med.* **19:**553–582.

122. **Janda, J. M., S. L. Abbott, and A. M. Carnahan.** 1995. *Aeromonas* and *Plesiomonas*, p. 477–482. *In* P. R. Murray, E. J. Baron, M. A. Pfaller, F. C. Tenover, and R. H. Yolken (ed.), *Manual of Clinical Microbiology*, 6th ed. American Society for Microbiology, Washington, D.C.

123. **Jeppesen, C.** 1995. Media for *Aeromonas* spp., *Plesiomonas shigelloides* and *Pseudomonas* spp. from food and environment. *Int. J. Food Microbiol.* **26:**25–41.

124. **Jiang, Z. D., A. C. Nelson, J. J. Mathewson, C. D. Ericsson, and H. L. DuPont.** 1991. Intestinal secretory immune response to infection with *Aeromonas* species and *Plesiomonas shigelloides* among students from the United States in Mexico. *J. Infect. Dis.* **164:**979–982.

125. **Jin, G.-F., A. K. Chopra, and C. W. Houston.** 1992. Stimulation of neutrophil leukocyte chemotaxis by a cloned cytolytic enterotoxin of *Aeromonas hydrophila*. *FEMS Microbiol. Lett.* **98:**285–290.

126. **Joseph, S. W., and A. Carnahan.** 1994. The isolation, identification and systematics of the motile *Aeromonas* species. *Annu. Rev. Fish Dis.* **4:**315–343.

127. **Kaper, J. B., H. Lockman, R. R. Colwell, and S. W. Joseph.** 1981. *Aeromonas hydrophila*: ecology, and toxigenicity of isolates from an estuary. *J. Appl. Bacteriol.* **50:**359–377.

128. **Kelleher, A., and S. M. Kirov.** 2000. *Rattus norvegicus*: not an animal model for *Aeromonas*-associated gastroenteritis in man. *FEMS Immunol. Med. Microbiol.* **28:**313–318.

129. **Kelly, K. A., M. Koehler, and L. R. Ashdown.** 1993. Spectrum of extraintestinal disease due to *Aeromonas* species in tropical Queensland, Australia. *Clin. Infect. Dis.* **16:**574–579.

130. **Khan, A. A., E. Kim, and C. E. Cerniglia.** 1998. Molecular cloning, nucleotide sequence, and expression in *Escherichia coli* of a hemolytic toxin (aerolysin) gene from *Aeromonas trota*. *Appl. Environ. Microbiol.* **64:**2473–2478.

131. **Kingombe, C. I., G. Huys, M. Tonolla, M. J. Albert, J. Swings, R. Peduzzi, and T. Jemmi.** 1999. PCR detection, characterization, and distribution of virulence genes in *Aeromonas* spp. *Appl. Environ. Microbiol.* **65:**5293–5302.

132. **Kirov, S. M.** 1993. The public health significance of *Aeromonas* spp. in foods. *Int. J. Food Microbiol.* **20:**179–198.

133. **Kirov, S. M.** 1993. Adhesion and piliation of *Aeromonas* spp. *Med. Microbiol. Lett.* **2:**274–280.

134. **Kirov, S. M.** 1997. *Aeromonas*, p. 473–492. *In* A. D. Hocking, G. Arnold, I. Jenson, K. Newton, and P. Sutherland (ed.), *Foodborne Microorganisms of Public Health Significance*, 5th ed. AIFST (NSW Branch) Food Microbiology Group, North Sydney, NSW, Australia.

135. **Kirov, S. M., M. J. Anderson, and T. A. McMeekin.** 1990. A note on *Aeromonas* spp. from chickens as possible food-borne pathogens. *J. Appl. Bacteriol.* **68:**327–334.

136. **Kirov, S. M., E. K. Ardestani, and L. J. Hayward.** 1993. The growth and expression of virulence factors at refrig-

eration temperature by *Aeromonas* strains isolated from foods. *Int. J. Food Microbiol.* **20:**159–168.

137. **Kirov, S. M., T. C. Barnett, C. P. Pepe, M. S. Strom, and M. J. Albert.** 2000. Investigation of the role of type IV *Aeromonas* pilus (Tap) in the pathogenesis of *Aeromonas* gastrointestinal infection. *Infect. Immun.* **68:**4040–4048.

138. **Kirov, S. M., and F. Brodribb.** 1993. Exotoxin production by *Aeromonas* spp. in foods. *Lett. Appl. Microbiol.* **17:**208–211.

139. **Kirov, S. M., and L. J. Hayward.** 1993. Virulence traits of *Aeromonas* in relation to species and geographic region. *Aust. J. Med. Sci.* **14:**54–58.

140. **Kirov, S. M., L. J. Hayward, and M. A. Nerrie.** 1995. Adhesion of *Aeromonas* sp. to cell lines used as models for intestinal colonization. *Epidemiol. Infect.* **115:**465–473.

141. **Kirov, S. M., J. A. Hudson, L. J. Hayward, and S. J. Mott.** 1994. Distribution of *Aeromonas hydrophila* hybridization groups and their virulence properties in Australasian clinical and environmental strains. *Lett. Appl. Microbiol.* **18:**71–73.

142. **Kirov, S. M., D. S. Hui, and L. J. Hayward.** 1993. Milk as a potential source of *Aeromonas* gastrointestinal infection. *J. Food Prot.* **56:**306–312.

143. **Kirov, S. M., I. Jacobs, L. J. Hayward, and R. Hapin.** 1995. Electron microscopic examination of factors influencing the expression of filamentous surface structures on clinical and environmental isolates of *Aeromonas veronii* biotype sobria. *Microbiol. Immunol.* **39:**329–338.

144. **Kirov, S. M., L. A. O'Donovan, and K. Sanderson.** 1999. Functional characterization of type IV pili expressed on diarrhea-associated isolates of *Aeromonas* species. *Infect. Immun.* **67:**5447–5454.

145. **Kirov, S. M., B. Rees, R. C. Wellock, J. M. Goldsmid, and A. D. van Galen.** 1986. Virulence characteristics of *Aeromonas* spp. in relation to source and biotype. *J. Clin. Microbiol.* **24:**827–834.

146. **Kirov, S. M., and K. Sanderson.** 1995. *Aeromonas* cell line adhesion, surface structures and in vivo models of intestinal colonization. *Med. Microbiol. Lett.* **4:**305–315.

147. **Kirov, S. M., and K. Sanderson.** 1995. *Aeromonas*: recognising the enemy. *Today's Life Sci.* **7:**30–35.

148. **Kirov, S. M., and K. Sanderson.** 1996. Characterization of a type IV bundle-forming pilus (SFP) from a gastroenteritis-associated strain of *Aeromonas veronii* biovar sobria. *Microb. Pathog.* **21:**23–34.

149. **Kirov, S. M., K. Sanderson, and T. C. Dickson.** 1998. Characterisation of a type IV pilus produced by *Aeromonas caviae*. *J. Med. Microbiol.* **47:**527–531.

150. **Kirov, S. M., and A. B. T. Semmler.** Unpublished observations.

151. **Knøchel, S.** 1990. Growth characteristics of motile *Aeromonas* spp. isolated from different environments. *Int. J. Food Microbiol.* **10:**235–244.

152. **Kohbayashi, K., and T. Ohnaka.** 1989. Food poisoning due to newly recognized pathogens. *Asian Med. J.* **32:**1–12.

153. **Kokka, R. P., N. A. Vedros, and J. M. Janda.** 1992. Immunochemical analysis and possible biological role of an *Aeromonas hydrophila* surface array protein in septicaemia. *J. Gen. Microbiol.* **138:**1229–1236.

154. **Krovacek, K., S. Dumontet, E. Eriksson, and S. B. Baloda.** 1995. Isolation, and virulence profiles, of *Aeromonas hydrophila* implicated in an outbreak of food poisoning in Sweden. *Microbiol. Immunol.* **39:**655–661.

155. **Krovacek, K., L. M. Eriksson, C. Gonzalez-Rey, J. Rozinsky, and I. Ciznar.** 2000. Isolation, biochemical and serological characterisation of *Plesiomonas shigelloides* from freshwater in Northern Europe. *Comp. Immunol. Microbiol. Infect. Dis.* **23:**45–51.

156. **Kühn, I. T. Lindberg, K. Olsson, and T. Stentström.** 1992. Biochemical fingerprinting for typing of *Aeromonas* strains from food and water. *Lett. Appl. Microbiol.* **15:**261–265.

157. **Kuijper, E. J., P. Bol, M. F. Peeters, A. G. Steigerwalt, H. C. Zanen, and J. G. Brenner.** 1989. Clinical and epidemiologic aspects of members of *Aeromonas* DNA hybridization groups isolated from human feces. *J. Clin. Microbiol.* **27:**1531–1537.

158. **Lawson, M. A., V. Burke, and B. J. Chang.** 1985. Invasion of HEp-2 cells by fecal isolates of *Aeromonas hydrophila*. *Infect. Immun.* **47:**680–683.

159. **Levy, D. A., M. S. Bens, G. F. Craun, R. L. Calderon, and B. L. Herwaldt.** 1998. Surveillance for waterborne-disease outbreaks—United States. *Morb. Mortal. Wkly. Rep.* **47:**1–34.

160. **Ljungh, Å., P. Enroth, and T. Wadström.** 1982. Cytotonic enterotoxin from *Aeromonas hydrophila*. *Toxicon* **20:**787–794.

161. **Majeed, K. N., and I. C. MacRae.** 1991. Experimental evidence for toxin production by *Aeromonas hydrophila* and *Aeromonas sobria* in a meat extract at low temperatures. *Int. J. Food Microbiol.* **12:**181–188.

162. **Manorama, V. Taneja, R. K. Agarwal, and S. C. Sanyal.** 1983. Enterotoxins of *Plesiomonas shigelloides*: partial purification and characterization. *Toxicon* **3(Suppl.):**269–272.

163. **Martinez-Murcia, A. J., S. Benlloch, and M. D. Collins.** 1992. Phylogenetic interrelationships of members of the genera *Aeromonas* and *Plesiomonas* as determined by 16S ribosomal DNA sequencing: lack of congruence with results of DNA-DNA hybridizations. *Int. J. Syst. Bacteriol.* **42:**412–421.

164. **Martínez, M. J., D. Simon-Pujol, F. Congregado, S. Merino, X. Rubires, and J. M. Tomás.** 1995. The presence of capsular polysaccharide in mesophilic *Aeromonas hydrophila* serotypes O:11 and O:34. *FEMS Microbiol. Lett.* **128:**69–74.

165. **Medema, G., and C. Schets.** 1993. Occurrence of *Plesiomonas shigelloides* in surface water: relationship with faecal pollution and trophic state. *Zentbl. Hyg.* **194:**398–404.

166. **Merino, S., A. Anguilar, M. M. Nogueras, M. Swift, and J. M. Tomás.** 1999. Cloning, sequencing, and role in virulence of two phospholipases (A1 and C) from mesophilic *Aeromonas* sp. serogroup O:34. *Infect. Immun.* **67:**4008–4013.

167. **Merino, S., A. Anguilar, X. Rubires, N. Abitiu, M. Regue, and J. M. Tomás.** 1997. The role of the capsular polysaccharide of *Aeromonas hydrophila* serogroup O:34 in the adherence to and invasion of fish cell lines. *Res. Microbiol.* **148:**625–631.

168. **Merino, S., A. Anguilar, X. Rubires, and J. M. Tomás.** 1998. Mesophilic *Aeromonas* strains from different serogroups: the influence of growth temperature and osmolarity on lipopolysaccharide and virulence. *Res. Microbiol.* **149:**407–416.

169. **Merino, S., S. Camprubí, and J. M. Tomás.** 1993. Incidence of *Aeromonas* sp. serotypes O:34 and O:11 among clinical isolates. *Med. Microbiol. Lett.* **2:**48–55.

170. **Merino, S., S. Camprubí, and J. M. Tomás.** 1993. Detection of *Aeromonas hydrophila* in food with an enzyme-linked immunosorbent assay. *J. Appl. Bacteriol.* **74:**149–154.

171. **Merino, S., X. Rubires, A. Aguillar, J. F. Guillot, and J. M. Tomás.** 1996. The role of O-antigen lipopolysaccharide on the colonization in vivo of germfree chicken gut by *Aeromonas hydrophila* serogroup O:34. *Microb. Pathog.* **20:**325–333.

172. **Miller, M. L., and J. A. Kohburger.** 1985. *Plesiomonas shigelloides*: an opportunistic food and waterborne pathogen. *J. Food Prot.* **48:**449–457.

173. **Miller, M. L., and J. A. Kohburger.** 1986. Tolerance of *Plesiomonas shigelloides* to pH, sodium chloride and temperature. *J. Food Prot.* **49:**877–879.

174. **Millership, S. E., S. R. Curnow, and B. Chattopadhyay.** 1983. Faecal carriage rate of *Aeromonas hydrophila*. *J. Clin. Pathol.* **36:**920–923.

175. **Morgan, D. R., P. C. Johnson, H. L. DuPont, T. K. Satterwhite, and L. V. Wood.** 1985. Lack of correlation between known virulence properties of *Aeromonas hydrophila* and enteropathogenicity for humans. *Infect. Immun.* **50:**62–65.

176. **Moyer, N. P.** 1987. Clinical significance of *Aeromonas* species isolated from patients with diarrhea. *J. Clin. Microbiol.* **25:**2044–2048.

177. **Namdari, H., and E. J. Bottone.** 1988. Correlation of the suicide phenomenon in *Aeromonas* species with virulence and enteropathogenicity. *J. Clin. Microbiol.* **26:**2615–2619.

178. **Namdari, H., and E. J. Bottone.** 1991. *Aeromonas caviae*: ecologic adaptation in the intestinal tract of infants coupled to adherence and enterotoxin production as factors in enteropathogenicity. *Experientia* **47:**434–436.

179. **Neves, M. S., M. P. Nunes, and A. M. Milhomen.** 1994. *Aeromonas* species exhibit aggregative adherence to HEp-2 cells. *J. Clin. Microbiol.* **32:**1130–1131.

180. **Nielsen, B., J. E. Olsen, and J. L. Larsen.** 1993. Plasmid profiling as an epidemiological marker within *Aeromonas salmonicida*. *Dis. Aquat. Org.* **15:**129–135.

181. **Nishikawa, Y., A. Hase, J. Ogawasara, S. M. Scotland, H. R. Smith, and T. Kimura.** 1994. Adhesion to and invasion of human colon carcinoma Caco-2 cells by *Aeromonas* strains. *J. Med. Microbiol.* **40:**55–61.

182. **Nishikawa, Y., and T. Kishi.** 1988. Isolation and characterization of motile *Aeromonas* from human, food and

environmental specimens. *Epidemiol. Infect.* **101**:213–223.

183. **Nishikawa, Y., J. Ogasawara, and T. Kimura.** 1993. Heat and acid sensitivity of motile *Aeromonas*: a comparison with other food-poisoning bacteria. *Int. J. Food Microbiol.* **18**:271–278.

184. **Noonan, B., and T. J. Trust.** 1997. The synthesis, secretion and role in virulence of the paracrystalline surface protein layers of *Aeromonas salmonicida* and *A. hydrophila. FEMS Microbiol. Lett.* **154**:1–7.

185. **Obi, C. L., A. O. Coker, J. Epoke, and R. N. Ndip.** 1997. Enteric bacterial pathogens in stools of residents of urban and rural regions in Nigeria: a comparison of patients with and without diarrhoea and controls without diarrhoea. *J. Diarrhoeal Dis. Res.* **15**:241–247.

186. **O'Brien, M., and M. Tandy.** 1994. L-Histidine decarboxylase: a new diagnostic test for *Plesiomonas shigelloides. Aust. Microbiol.* **15**:A–69.

187. **Olsvik, Ø., K. Wachsmuth, B. Kay, K. A. Birkness, A. Yi, and B. Sack.** 1990. Laboratory observations on *Plesiomonas shigelloides* strains isolated from children with diarrhea in Peru. *J. Clin. Microbiol.* **28**:886–889.

188. **Palumbo, S. A.** 1993. The occurrence and significance of organisms of the *Aeromonas hydrophila* group in food and water. *Med. Microbiol. Lett.* **2**:339–346.

189. **Palumbo, S. A., J. E. Call, P. H. Cooke, and A. C. Williams.** 1995. Effect of polyphosphates and NaCl on *Aeromonas hydrophila* K144. *J. Food Safety* **15**:77–87.

190. **Palumbo, S. A., F. Maxino, A. C. Williams, R. L. Buchanan, and D. W. Thayer.** 1985. Starch-ampicillin agar for the quantitative detection of *Aeromonas hydrophila. Appl. Environ. Microbiol.* **50**:1027–1030.

191. **Palumbo, S. A., A. C. Williams, R. L. Buchanan, and J. G. Philips.** 1987. Thermal resistance of *Aeromonas hydrophila. J. Food Prot.* **50**:761–764.

192. **Parker, M. W., J. T. Buckley, J. P. M. Postma, A. D. Tucker, K. Leonard, F. Pattus, and D. Tsernoglou.** 1994. Structure of *Aeromonas* toxin proaerolysin in its water-soluble and membrane-channel states. *Nature* (London) **367**:292–295.

193. **Parker, M. W., F. G. van der Goot, and T. J. Buckley.** 1996. Aerolysin—the ins and outs of a model channel-forming toxin. *Mol. Microbiol.* **19**:205–212.

194. **Parras, F., M. D. Díaz, J. Reina, S. Moreno, C. Guerrero, and E. Bouza.** 1993. Meningitis due to *Aeromonas* species: case report and review. *Clin. Infect. Dis.* **17**:1058–1060.

195. **Pazzaglia, G., R. B. Sack, A. L. Bourgeois, J. Froehlich, and J. Eckstein.** 1990. Diarrhea and intestinal invasiveness of *Aeromonas* strains in the removable intestinal tie rabbit model. *Infect. Immun.* **58**:1924–1931.

196. **Pepe, C. M., M. W. Eklund, and M. S. Strom.** 1996. Cloning of an *Aeromonas hydrophila* type IV pilus biogenesis gene cluster: complementation of pilus assembly functions and characterization of a type IV leader peptidase/N-methyltransferase required for extracellular protein secretion. *Mol. Microbiol.* **19**:857–869.

197. **Picard, B., and P. Goullet.** 1985. Comparative electrophoretic profiles of esterases, and of glutamate, lac-

198. **Picard, B., and P. Goullet.** 1987. Seasonal prevalence of nosocomial *Aeromonas hydrophila* infection related to aeromonas in hospital water. *J. Hosp. Infect.* **10**:152–155.

199. **Picard, B., and P. Goullet.** 1987. Epidemiological complexity of hospital *Aeromonas* infections revealed by electrophoretic typing of esterases. *Epidemiol. Infect.* **98**:5–14.

200. **Pitarangsi, C., P. Echeverria, R. Whitmire, C. Tiripat, S. Formal, G. J. Dammin, and M. Tingtalapong.** 1982. Enteropathogenicity of *Aeromonas hydrophila* and *Plesiomonas shigelloides*: prevalence among individuals with and without diarrhea in Thailand. *Infect. Immun.* **35**:666–673.

201. **Popoff, M.** 1984. Genus III. *Aeromonas* Kluyver and Van Neil 1936, 398, p. 545–548. *In* N. R Krieg and J. G. Holt (ed.), *Bergey's Manual of Systematic Bacteriology*, vol. 1. The Williams & Wilkins Co., Baltimore, Md.

202. **Potomski, J., V. Burke, J. Robinson, D. Fumarola, and G. Miragliotta.** 1987. *Aeromonas* cytotonic enterotoxin cross–reactive with cholera toxin. *J. Med. Microbiol.* **23**:179–186.

203. **Quinn, D. M., H. M. Atkinson, A. H. Bretag, M. Tester, T. J. Trust, C. Y. F. Wong, and R. L. P. Flower.** 1994. Carbohydrate-reactive, pore-forming outer membrane proteins of *Aeromonas hydrophila. Infect. Immun.* **62**:4054–4058.

204. **Radomyski, T., E. A. Murano, D. G. Olson, and P. S. Murano.** 1994. Elimination of pathogens of significance in food by low dose irradiation: a review. *J. Food Prot.* **57**:73–86.

205. **Ratkowsky, D. A., J. Olley, T. A. McMeekin, and A. Ball.** 1982. Relationship between temperature and growth rate of bacterial cultures. *J. Bacteriol.* **149**:1–5.

206. **Rautelin, H., M. L. Hänninen, A. Sivonen, U. Turunen, and V. Valtonen.** 1995. Chronic diarrhea due to a single strain of *Aeromonas caviae. Eur. J. Clin. Microbiol. Infect. Dis.* **14**:51–53.

207. **Rivero, O., J. Anguta, C. Paniagua, and G. Naharro.** 1990. Molecular cloning and characterization of an extracellular protease gene from *Aeromonas hydrophila. J. Bacteriol.* **172**:3905–3908.

208. **Robson, W. L., A. K. Leung, and C. L. Trevenen.** 1992. Haemolytic-uraemic syndrome associated with *Aeromonas hydrophila* enterocolitis. *Pediatr. Nephrol.* **6**:221. (Letter.)

209. **Rosner, R.** 1964. *Aeromonas hydrophila* as the etiologic agent in a case of severe gastroenteritis. *Am. J. Clin. Pathol.* **42**:402–404.

210. **Rutala, W. A., F. A. Sarubbi, Jr., C. S. Finch, J. N. MacCormack, and G. E. Steinkraus.** 1982. Oyster-associated outbreak of diarrhoeal disease possibly caused by *Plesiomonas shigelloides. Lancet* **i**:739.

211. **San Joaquin, V. H., and D. A. Pickett.** 1988. *Aeromonas*-associated gastroenteritis in children. *Pediatr. Infect. Dis. J.* **7**:53–57.

212. Santos, J. A., C. J. Gonzalez, T. M. Lopez, A. Otero, and M. L. Garcia-Lopez. 1999. Hemolytic and elastolytic activities influenced by iron in *Plesiomonas shigelloides*. *J. Food Prot.* 62:1475–1477.

213. Saraswathi, B., R. K. Agarwal, and S. C. Sanyal. 1983. Further studies on enteropathogenicity of *Plesiomonas shigelloides*. *Indian J. Med. Res.* 78:12–18.

214. Schubert, R. H., and A. Holz-Bremer. 1999. Cell adhesion of *Plesiomonas shigelloides*. *Zentbl. Hyg. Umweltmed.* 202:383–388.

215. Schultz, A. J., and B. A. McCardell. 1988. DNA homology and immunological cross-reactivity between *Aeromonas hydrophila* cytotonic enterotoxin and cholera toxin. *J. Clin. Microbiol.* 26:57–61.

216. Seidler, R. J., D. A. Allen, H. Lockman, R. R. Colwell, S. W. Joseph, and O. P. Daily. 1980. Isolation, enumeration and characterization of *Aeromonas* from polluted waters encountered in diving operations. *Appl. Environ. Microbiol.* 39:1010–1018.

217. Shaw, J. G., J. P. Thornley, L. Palmer, and I. Geary. 1995. Invasion of tissue culture cells by *Aeromonas caviae*. *Med. Microbiol. Lett.* 4:316–323.

218. Shaw, K. N., and S. M. Kirov. 1999. Diarrhoea-associated *Aeromonas* species produce lateral flagella: accessory colonization factors? Presented at 6th International *Aeromonas-Plesiomonas* Symposium, Chicago, Ill. http://www.aeromonas.com.

219. Shimada, T., R. Sakazaki, K. Horigome, Y. Ueska, and K. Niwano. 1984. Production of cholera-like enterotoxin by *Aeromonas hydrophila*. *Jpn. J. Med. Sci. Biol.* 37:141–144.

220. Shimada, T., R. Sakazaki, and K. Suzuki. 1985. Peritrichous flagella in mesophilic strains of *Aeromonas*. *Jpn. J. Med. Sci. Biol.* 38:141–145.

221. Singh, D. V., and S. C. Sanyal. 1992. Enterotoxicity of clinical and environmental isolates of *Aeromonas* spp. *J. Med. Microbiol.* 36:269–272.

222. Snower, D. P., C. Ruef, A. P. Kuritza, and S. C. Edberg. 1989. *Aeromonas hydrophila* infection associated with the use of medicinal leeches. *J. Clin. Microbiol.* 27:1421–1422.

223. Swift, S., A. V. Karlyshev, L. Fish, E. L. Durant, M. K. Winson, S. R. Chhabra, P. Williams, S. Macintyre, and G. S. A. B. Stewart. 1997. Quorum sensing in *Aeromonas hydrophila* and *Aeromonas salmonicida*: identification of the LuxRI homologs AhyRI and AsaRI and their cognate N-acylhomoserine lactone signal molecules. *J. Bacteriol.* 179:5721–5281.

224. Swift, S., M. J. Lynch, L. Fish, D. F. Kirke, J. M. Tomás, G. S. Stewart, and P. Williams. 1999. Quorum sensing-dependent regulation and blockade of exoprotease production in *Aeromonas hydrophila*. *Infect. Immun.* 67:5192–5199.

225. Thomas, L. V., R. J. Gross, T. Cheasty, and B. Rowe. 1990. Extended serogrouping scheme for motile, mesophilic *Aeromonas* species. *J. Clin. Microbiol.* 28:980–984.

226. Thornley, J. P., J. G. Shaw, I. A. Gryllos, and A. Eley. 1997. Virulence properties of clinically significant *Aeromonas* species: evidence for pathogenicity. *Rev. Med. Microbiol.* 8:61–72.

227. Thornton, J., S. P. Howard, and J. T. Buckley. 1988. Molecular cloning of a phospholipid-cholesterol acyltransferase from *Aeromonas hydrophila*. Sequence homologies with lecithin-cholesterol acyltransferase and other lipases. *Biochim. Biophys. Acta* 959:153–159.

228. Trust, T. J. 1993. Molecular, structural and functional properties of *Aeromonas* S-layers, p. 159–171. *In* T. J. Beveridge and S. F. Koval (ed.), *Advances in Bacterial Paracrystalline Surface Layers*. Plenum Press, Inc., New York, N.Y.

229. Tso, M. D., and J. S. G. Dooley. 1995. Temperature-dependent protein and lipopolysaccharide expression in clinical *Aeromonas* strains. *J. Med. Microbiol.* 42:32–38.

230. Tsukamoto, T., Y. Kinoshita, T. Shimada, and R. Sakazaki. 1978. Two epidemics of diarrheal disease possibly caused by *Plesiomonas shigelloides*. *J. Hyg. Camb.* 80:275–280.

231. Vadivelu, J., S. D. Puthucheary, M. Phipps, and Y. M. Chee. 1995. Possible virulence factors involved in bacteraemia caused by *Aeromonas hydrophila*. *J. Med. Microbiol.* 42:171–174.

232. van der Goot, F. G., F. Pattus, M. Parker, and J. T. Buckley. 1994. The cytolytic toxin aerolysin: from the soluble form to the transmembrane channel. *Toxicology* 87:19–28.

233. Watson, I. M., J. O. Robinson, V. Burke, and M. Gracey. 1985. Invasiveness of *Aeromonas* spp. in relation to biotype, virulence factors and clinical features. *J. Clin. Microbiol.* 22:48–51.

234. Wilmsen, H. U., J. T. Buckley, and F. Pattus. 1991. Site-directed mutagenesis at histidines of aerolysin from *Aeromonas hydrophila*: a lipid planar bilayer study. *Mol. Microbiol.* 5:2745–2751.

235. Wilmsen, H. U., K. R. Leonard, W. Tichelaar, J. T. Buckley, and F. Pattus. 1992. The aerolysin membrane channel is formed by heptamerization of the monomer. *EMBO J.* 11:2457–2463.

236. Wong, C. Y. F., M. W. Heuzenroeder, and R. F. L. P. Flower. 1998. Inactivation of two haemolytic toxin genes in *Aeromonas hydrophila* attenuates virulence in a suckling mouse model. *Microbiology* 144:291–298.

237. Xu, X. J., M. R. Ferguson, V. L. Popov, C. W. Houston, J. W. Peterson, and A. K. Chopra. 1998. Role of a cytotoxic enterotoxin in *Aeromonas*-mediated infections: development of transposon and isogenic mutants. *Infect. Immun.* 66:3501–3509.

238. Yamada, S., S. Matsushita, S. Dejsirilert, and Y. Kudoh. 1997. Incidence and clinical symptoms of *Aeromonas*-associated travellers' diarrhoea in Tokyo. *Epidemiol. Infect.* 119:121–126.

Food Microbiology: Fundamentals and Frontiers, 2nd Ed.
Edited by M. P. Doyle et al.
© 2001 ASM Press, Washington, D.C.

John W. Austin

Clostridium botulinum

15

Since the first recognition of botulism as a foodborne disease in the late 1800s (193), it has been a major concern of food processors and consumers alike. Currently, four categories of human botulism are recognized. Foodborne botulism is caused by eating food contaminated with preformed botulinum neurotoxin (BoNT). Infant botulism is caused by ingestion of viable spores of *Clostridium botulinum* which germinate, colonize, and produce neurotoxin in the intestinal tract of infants under 1 year of age. Infant botulism was first recognized in 1976 (142) and is now the most common form of botulism in the United States (29). Wound botulism is due to infection of a wound with spores of *C. botulinum* which grow and produce neurotoxin in the wound. While wound botulism is rare, it is being increasingly associated with intravenous drug use (133). The fourth category, adult infectious botulism, includes adult cases with intestinal colonization and toxemia (31, 70, 106, 112). Animal botulism affects cattle and birds worldwide and occurs to a lesser extent in sheep, horses, zoo animals, and various other animals (201, 203).

As a result of the severity of the disease, botulism is extensively reported. In the United States, approximately 9.4 outbreaks of botulism occur annually, with an average of 2.5 cases per outbreak. From 1950 to 1996 there were 444 foodborne botulism outbreaks reported in the United States (29). Since 1973, a median of 24 cases of foodborne botulism, 3 cases of wound botulism, and 71 cases of infant botulism have been reported annually to the Centers for Disease Control and Prevention (170). In Canada, since 1985, approximately 4.4 outbreaks of foodborne botulism have occurred annually, with an average of 2.75 cases per outbreak (9–11, 24a, 147). Seven outbreaks, involving 18 cases with one fatality, were reported in Canada in 1997 (10).

CHARACTERISTICS OF *C. BOTULINUM*

Classification

C. botulinum is a gram-positive, anaerobic, rod-shaped bacterium (73). Subterminal oval endospores are formed in stationary-phase cultures (Fig. 15.1). Originally, all clostridia known to produce BoNT were included in this species (146). There are seven types of *C. botulinum*, A through G, based on the serological specificity of the neurotoxin produced. Human botulism, including foodborne, wound, and infant botulism, is associated with types A, B, E, and, very rarely, F. Types C and D cause

John W. Austin, Microbiology Research Division, Bureau of Microbial Hazards, Food Directorate, Health Canada, Banting Research Centre, PL 2204A2, Ottawa, Ontario K1A 0L2, Canada.

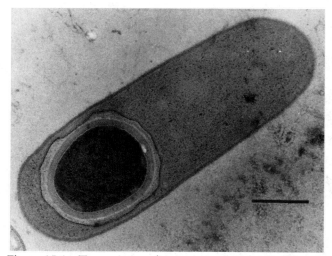

Figure 15.1 Transmission electron micrograph of a thin section of a sporulating cell of *C. botulinum*. Bar, 500 nm.

botulism in animals. To date, there is no direct evidence linking type G to disease.

The species is also divided into four groups based on physiological differences (Table 15.1) (73), as follows: group I, all type A strains and proteolytic strains of types B and F; group II, all type E strains and nonproteolytic strains of types B and F; group III, type C and D strains; and group IV, *C. botulinum* type G, for which the new name *C. argentinense* has been proposed (181). This grouping agrees with results of DNA homology studies and of 16S and 23S rRNA sequence studies (82, 83, 103, 149) that show a high degree of relatedness among strains within each group, but little relatedness between groups.

Group I strains are proteolytic and are typified by strains that produce neurotoxin type A (73). The optimal temperature for growth is 37°C, with growth occurring between 10 and 48°C. High levels of neurotoxin (10^6 mouse LD_{50}/ml)(1 LD_{50} is the amount of neurotoxin required to kill 50% of injected mice within 4 days) are typically produced in cultures. Spores have a high heat resistance, with $D_{100°C}$ values of approximately 25 min (the D value is the time required to inactivate 90% of the population at a given temperature). To inhibit growth, the pH must be below 4.6, the salt concentration above 10%, or the water activity (a_w) below 0.94 (89).

Group II strains are nonproteolytic, have a lower optimum growth temperature (30°C), and will grow at temperatures as low as 3°C (69). The spores have a much lower heat resistance, with $D_{100°C}$ values less than 0.1 min. Group II strains are inhibited by a pH below 5.0, salt concentrations above 5%, or a_w below 0.97 (89). Since these bacteria lack proteolytic enzymes, the toxicity of cultures is usually increased by treating with trypsin, which activates the neurotoxin.

Group III includes types C and D strains, which are not involved in human botulism but cause animal botulism. Consequently, they have been studied in less detail. These strains are nonproteolytic and grow optimally at 40°C and at temperatures only as low as 15°C. Group IV strains, which produce type G neurotoxin, grow optimally at 37°C and have a minimal growth temperature of 10°C. Spores are rarely seen and have a fairly low resistance to heat, with $D_{104°C}$ values of 0.8 to 1.12 min.

These four groups, plus neurotoxin-producing strains of *Clostridium butyricum* and *Clostridium baratii*, compose a total of six distinct genomic groups that produce BoNT. The focus in this chapter will be on groups I and II, because they include the strains most commonly involved in human illness.

Tolerance to Preservation Methods

Temperature, pH, a_w, redox potential (E_h), added preservatives, and the presence of other microorganisms are

Table 15.1 Grouping and characteristics of strains of *C. botulinum*

Characteristic	Group			
	I	II	III	IV
Neurotoxin types	A, B, F	B, E, F	C, D	G
Minimum temperature for growth	10°C	3°C	15°C	ND[a]
Optimum temperature for growth	35–40°C	18–25°C	40°C	37°C
Minimum pH for growth	4.6	ca. 5	ND	ND
Inhibitory [NaCl]	10%	5%	ND	ND
Minimum a_w for growth	0.94	0.97	ND	ND
$D_{100°C}$ of spores	25 min	<0.1 min	0.1–0.9 min	0.8–1.12 min
$D_{121°C}$ of spores	0.1–0.2 min	<0.001 min	ND	ND

[a] ND, not determined.

the main factors controlling growth of *C. botulinum* in foods. Historically, studies have established maximum and/or minimum limits for these parameters that control growth of *C. botulinum* (Table 15.1). These factors seldom function independently; usually they act in concert, often having synergistic effects.

Low Temperature

Refrigerated storage is used to prevent growth of *C. botulinum*. The minimum temperatures allowing growth have been determined, largely to assess the impact of refrigeration as a control. The established lower limits are 10°C for group I and 3.0°C for group II (69, 73). However, these limits apply to few strains and depend on otherwise optimal growth conditions. Irrespective of the actual minimum growth temperature, production of neurotoxin generally requires weeks at the lower temperature limits for group I and group II clostridia. The optimum growth temperatures are between 35 and 40°C for group I clostridia and between 25 and 30°C for group II clostridia.

Thermal Inactivation

Thermal processing is used to inactivate spores of *C. botulinum* and is the most common method of producing shelf-stable foods. *C. botulinum* spores of group I, which are very heat resistant, are the target organisms for most thermal processes. D values vary considerably among *C. botulinum* strains (89). D values depend on how the spores are produced and treated, the heating environment, and the recovery system (89, 109, 135, 136). Spores of group I strains are the most heat resistant, having $D_{121°C}$ values between 0.1 and 0.2 min. These spores are of particular concern in the commercial sterilization of canned low-acid foods. The canning industry has adopted a D value of 0.2 min at 121°C as a standard for calculating thermal processes. The z value (the temperature change necessary to cause a 10-fold change in the D value) for the most resistant strains is approximately 10°C, which has also been adopted as a standard. Despite variations in D and z values, the adoption of a 12-D process as the minimum thermoprocess applied to commercial canned low-acid foods by the canning industry has ensured the production of safe products (74).

Strains of group II are considerably less heat resistant ($D_{100°C}$ < 0.1 min) than those of group I. Spores of *C. botulinum* group II can be inactivated at moderate temperatures (40 to 50°C) when heating is combined with high pressures of up to 827 mPa (148). However, survival of spores of *C. botulinum* group II in pasteurized, refrigerated products is of concern because of their ability to grow at refrigeration temperatures (86,

108, 134). $D_{82°C}$ values of *C. botulinum* type E in neutral phosphate buffer are generally in the range of 0.2 to 1.0 min. Values ranging from 0.15 to greater than 4.90 min have been reported for type E strains, depending on the heating menstruum, strain, recovery medium, and the presence of lysozyme (137). Various regulations and guidelines for the safe production, distribution, and sale of refrigerated foods of extended durability have been published (108, 134). The European Chilled Food Federation "Botulinum Working Group" has produced guidelines for sous vide (vacuum-packed) foods depending upon the specified shelf life of the food. For products with a short shelf life (i.e., less than 5 days) safety can be ensured by a minimum heat process and strict limitation of chill shelf life. For products intended for longer shelf life (i.e., greater than 5 days), a thermal process ensuring a 6-D reduction in numbers of spores of psychrotrophic strains of *C. botulinum* is required, provided that the temperature is kept below 10°C to prevent the growth of the more heat-resistant proteolytic strains of *C. botulinum*. If a thermal process resulting in less than a 6-D reduction in nonproteolytic spores is applied to a food product with an intended shelf life greater than 5 days, evidence must be provided that other preservation factors (e.g., lowered pH or a_w, strict control of temperature below 3°C) are effective in controlling growth of nonproteolytic *C. botulinum* using modeling or inoculated pack/challenge tests (68). American recommendations from the National Advisory Committee on Microbiological Criteria for Foods recommend inoculated pack studies with *C. botulinum* to decide shelf life.

pH

The minimum pH allowing growth of *C. botulinum* group I is 4.6; for group II, it is approximately pH 5. Hence, many fruits and vegetables are sufficiently acidic to inhibit *C. botulinum* by their pH alone, whereas acidulents are used to preserve other products. Substrate, temperature, nature of the acidulent, the presence of preservatives, a_w, and E_h are all factors that influence the acid tolerance of *C. botulinum*. Acid-tolerant microorganisms, such as yeasts and molds, may grow in acidic products and raise the pH in their immediate vicinity to a level that allows growth of *C. botulinum* (80). *C. botulinum* can also grow in some acidified foods if excessively slow pH equilibration occurs.

Salt and a_w

Sodium chloride is one of the most important factors used to control *C. botulinum* in foods. It acts primarily by decreasing the a_w. Consequently, its concentration in the aqueous phase, called the brine concentration

(% brine = % NaCl × 100/% H_2O + % NaCl), is critical. The growth-limiting brine concentrations are about 10% for group I and 5% for group II under otherwise optimal conditions. These concentrations correspond well to the limiting a_w of 0.94 for group I and 0.97 for group II in foods where NaCl is the main a_w depressant. The solute used to control a_w may influence these limits. Generally, NaCl, KCl, glucose, and sucrose show similar effects, whereas growth occurs in glycerol at lower a_w (74). The limiting a_w may be increased substantially by other factors, such as increased acidity or the use of preservatives.

Atmosphere, Redox Potential

Modified atmosphere packaging (MAP) is being increasingly used to extend the shelf life of foods. MAP has been a concern because conditions might promote growth and toxin formation by *C. botulinum* while suppressing the growth of spoilage microorganisms. The high incidence of type E spores in fish (85) has raised concerns regarding the safety of MAP fish. *C. botulinum* can grow and produce neurotoxin in MAP fish, and, depending on conditions, neurotoxin can be present before the fish is considered spoiled (46, 66, 110, 111, 143). Growth of *C. botulinum* has been observed in several other types of MAP foods, including meats (98, 99), vegetables (12, 72, 101, 102, 176, 177), and bakery products (139–141).

While it is commonly assumed that *C. botulinum* cannot grow in foods exposed to O_2, the E_h of most foods is usually low enough to allow its growth (89). *C. botulinum* grows optimally at an E_h of -350 mV, but growth initiation may occur in the E_h range of $+30$ to $+250$ mV (89). The presence of other inhibitory factors lowers this upper limit. Once growth is initiated, the E_h declines rapidly.

Initial atmospheres containing 20% O_2 did not delay neurotoxin production by *C. botulinum* in inoculated pork compared with samples packaged with 100% N_2, and some toxic samples contained 15% residual O_2 (100). CO_2 is used in MAP to inhibit spoilage and pathogenic microorganisms, but CO_2 can stimulate growth of *C. botulinum* (58). Initial levels of 15 to 30% CO_2 did not inhibit *C. botulinum* in inoculated pork; only 75% CO_2 provided significant inhibition (99). Toxin production in English-style crumpets inoculated with *C. botulinum* spores types A and proteolytic B was delayed 1.5 to 3 days when packaged under 100% CO_2 (140). Pressurized CO_2 is lethal to *C. botulinum*, with lethality increasing with the pressure of CO_2 (45). Inclusion of an ethanol vapor-generating sachet in the package delays toxigenesis (141). The safety of different atmospheres with respect to *C. botulinum* in packaging foods should be carefully investigated before use.

Preservatives

Nitrite has several functions in cured meat products; an important role is the inhibition of *C. botulinum*, whereas its effects on color and flavor are important organoleptic considerations. The exact mechanism of botulinal inhibition by nitrite is not known. Its effectiveness is dependent upon complex interactions among pH, sodium chloride, heat treatment, time and temperature of storage, and the composition of the food (89). Nitrite is depleted from cured foods, and the depletion rate is dependent on product formulation, pH, and time and temperature during processing and storage. However, a significant contribution of nitrite to the inhibition of *C. botulinum* continues even when nitrite is no longer detectable (74). Nitrite reacts with many cellular constituents and appears to inhibit *C. botulinum* by more than one mechanism, including reaction with essential iron-sulfur proteins to inhibit energy-yielding systems in the cell (89). The reaction of nitrite, or nitric oxide, with secondary amines in meats to produce nitrosamines, some of which are carcinogenic, has led to regulations limiting the amount of nitrite used.

Sorbates, parabens, nisin, phenolic antioxidants, polyphosphates, ascorbates, EDTA, metabisulfite, *n*-monoalkyl maleates and fumarates, and lactate salts are also active against *C. botulinum* (89). The use of natural or liquid smoke has a significant inhibitory effect against *C. botulinum* in fish but has an insignificant effect in meats.

Competitive and Growth-Enhancing Microorganisms

The growth of competitive and growth-promoting microorganisms in foods has a very significant effect on the fate of *C. botulinum* (33, 110, 122, 127, 128). Acid-tolerant molds such as *Cladosporium* sp. or *Penicillium* sp. can provide an environment that enhances the growth of *C. botulinum* (80). Other microorganisms may inhibit *C. botulinum*, either by changing the environment or by producing specific inhibitory substances, or both. Growth of *C. botulinum* type E was inhibited in cooked surimi nuggets by naturally occurring *Bacillus* spp. (110). Lactic acid bacteria, including *Lactobacillus*, *Pediococcus*, and *Streptococcus* spp., can inhibit growth of *C. botulinum* in foods, largely by reducing the pH but also by the production of bacteriocins (33, 84, 175). The use of lactic acid bacteria and a fermentable carbohydrate, the "Wisconsin process," is permitted for producing bacon with a decreased level of nitrite in the United States (183). The growth of other microorganisms can protect consumers by causing spoilage of food in which botulinum toxin is produced.

Table 15.2 Incidence of *C. botulinum* in soils and sediments

Location	Sample size (g)	% Positive samples	MPN[a] per kg	Type (% of types identified)				
				A	B	C/D	E	F
Eastern U.S., soil	10	19	21	12	64	12	12	0
Western U.S., soil	10	29	33	62	16	14	8	0
Green Bay, U.S., sediment	1	77	1,280	0	0	0	100	0
Alaska, U.S., soil	1	41	660	0	0	0	100	0
Britain, soil	50	6	2	0	100	0	0	0
British coast, sediment	2	4	18	0	100	0	0	0
Scandinavian coast, sediment	6	100	>780	0	0	0	100	0
Finland, Baltic offshore	500	88	1,020	0	0	0	100	0
Finland, trout farm sediments	200	68	2,020	0	0	0	100	0
Netherlands, soil	0.5	94	2,500	0	22	46	32	0
Switzerland, soil	12	44	48	28	83	6	0	27
Rome, Italy, soil	7.5	1	2	86	14	0	0	0
Iran, Caspian sea, sediment	2	17	93	0	8	0	92	0
China, Sinkiang, soil	10	70	25,000	47	32	19	2	0
Japan, Hokkaido, soil	5–10	4	4	0	0	0	100	0
Japan, Ishikawa, soil	40–50	56	16	0	0	100	0	0
Brazil, soil	5	35	86	57	7	29	0	7
Paraguay, soil	5	24	10	14	0	14	0	71
South Africa, soil	30	3	1	0	100	0	0	0
Thailand, sediment	10	3	3	0	0	83	17	0
New Zealand, sediment	20	55	40	0	0	100	0	0

[a] MPN, most probable number.

Inactivation by Irradiation

C. botulinum spores are probably the most radiation-resistant spores of public health concern. *D* values (irradiation dose required to inactivate 90% of the population) of group I strains at −50 to −10°C are between 2.0 and 4.5 kGy in neutral buffers and in foods (89). Spores of type E are more sensitive, having *D* values between 1 and 2 kGy. Radappertization is designed to reduce the number of viable spores of the most radiation-resistant *C. botulinum* strains by 12 log cycles. Different environmental conditions, such as the presence of O_2, change in irradiation temperature, and irradiation and recovery environments, can affect the *D* values of spores. Generally, spores in the presence of O_2 or preservatives and at temperatures above 20°C have greater sensitivity to irradiation.

RESERVOIRS

Occurrence of *C. botulinum* in the Environment

Because contamination of food largely depends on the incidence of *C. botulinum* in the environment, many surveys of different environments for spores of *C. botulinum*

have been undertaken worldwide. Results revealed that spores of *C. botulinum* are commonly present in soils and sediments, but their numbers and types vary depending on the location (Table 15.2) (42).

In North America, *C. botulinum* spores are widely distributed in nature, but the spore load varies considerably, as does the predominating type. Soils in the United States east of the rise of the Rocky Mountains usually contain spores of type B proteolytic *C. botulinum*. Type A spores predominate in the western United States. Overall, type E is found infrequently, and only in damp or wet locations. However, type E predominates and high numbers are found in the region around the Great Lakes, particularly around Green Bay of Lake Michigan (24), and in the coastal areas of Washington and Alaska. The distribution of types on the Pacific coast changes with latitude; south of 36°N, the prevalent types shift from E to A and B.

In Europe, surveys show that type B predominates in the terrestrial environments of Britain, Ireland, Iceland, Denmark, and Switzerland, and in the aquatic environments of the United Kingdom. Disease outbreaks also confirm its occurrence in Spain, Portugal, Italy, France, Belgium, Germany, Poland, the former Czechoslovakia, Hungary, and the former Yugoslavia. Most European

Table 15.3 Prevalence of *C. botulinum* spores in food

Product	Origin	Sample size (g)	% Positive samples	MPN[a] per kg	Type(s) identified
Eviscerated whitefish chubs	Great Lakes	10	12	14	E, C
Vacuum-packed frozen flounder	Atlantic Ocean	1.5	10	70	E
Dressed rockfish	California	10	100	2,400	A, E
Salmon	Alaska	24–36	100	190	A
Vacuum-packed fish	Viking Bank		42	63	E
Smoked salmon	Denmark	20	2	<1	B
Salted carp	Caspian Sea	2	63	490	E
Fish and seafood	Osaka, Japan	30	8	3	C, D
Raw meat	North America	3	<1	0.1	C
Cured meat	Canada	75	2	0.2	A
Raw pork	U.K.	30	0–14	<0.1–5	A, B, C
Cooked, vacuum-packed potatoes	The Netherlands		0	0.63	
Mushrooms	Canada			2,100	B
Random honey samples	U.S.	30	1	0.4	A, B
Honey samples associated with infant botulism	U.S.	30	100	8×10^4	A, B

[a]MPN, most probable number.

type B strains are nonproteolytic. Type E is the predominant serotype from other aquatic environments, with the highest numbers found in Scandinavian waters, particularly in the Sound and Kattegat between Denmark and Sweden. In the former USSR, type E spores also predominate, except in the central region, where type B spores predominate.

Surveys within Asia reveal lower numbers of spores, with exceptions of a high incidence of type E spores around the Caspian Sea (151) and a high incidence of all types in the Xinjiang region of China (65). In Japan, the incidence of type E spores is high on the northern island of Hokkaido, whereas type C predominates in other areas. In the tropical regions of Asia, types C and D replace type E as the predominant type in aquatic environments.

Fewer surveys have been done in the southern hemisphere. Type A spores predominate in Brazilian and Argentine soils, but types B, C, F, and G are also present. In Paraguay, the prevalent spore type is F, followed by A and C (204). Few surveys have been done in Africa, but one following an outbreak of type A botulism detected type A and C spores in Kenyan soil (205). Type B spores were found in South Africa in soil, and type D spores were found in sediment (93). In Indonesia, types C and D predominate, but types A, B, and E have also been detected. In Australia and New Zealand, the environmental incidence is very low, with both types A and B spores found.

In summary, type A spores predominate in soils in the western United States, China, Brazil, and Argentina. Type B spores predominate in soils in the eastern United States, the United Kingdom, and much of continental Europe. Most American type B strains are proteolytic, whereas most European strains are nonproteolytic. Type E predominates in northern regions and in most temperate aquatic regions and their surroundings, whereas types C and D are present more frequently in warmer environments.

Occurrence of *C. botulinum* in Foods

Many surveys have determined the incidence of *C. botulinum* spores in foods (Table 15.3) (41). However, considering that the risk of foodborne botulism is directly related to the contamination of foods, surprisingly there have been fewer surveys of foods than of the environment. Food surveys have focused largely on fish, meats, and infant foods, especially honey.

C. botulinum type E spores are commonly present in fish and aquatic animals. In North America, the incidence and level of contamination are highest in samples from the Pacific coast, followed by samples from the Great Lakes and then from the Atlantic seaboard. Fish in Europe have a lower level of contamination, except for fish from Scandinavia and from the Caspian Sea. In Indonesia, most positive fish samples contain either type C or D spores, but types A, B, and F are also present.

Meat and meat products have been studied less intensively than fish and fishery products, and the level of contamination is generally low. These products are less likely than fish to be contaminated with spores, because there is considerably less contamination of the farm environment than of the aquatic environment. In North America, very low levels of *C. botulinum* spore contamination occur in raw pork, beef, and chicken in processing plants. Studies have either failed to detect or detected a very low incidence of *C. botulinum* spores in cured meats and meat trimmings and in vacuum-packed sliced processed meat, such as bologna, smoked beef, turkey, chicken, liver sausage, luncheon loaf, salami, and pastrami. In the United Kingdom, the incidence in both raw and finished product varies considerably between sampling occasions. In North America, the average most probable number (MPN) is ca. 0.1 spore per kg of meat products, whereas in Europe, the average MPN is ca. 2.5 spores per kg. The spore types most often associated with meats are A and B.

C. botulinum spores, usually type A or B, can contaminate fruits and vegetables, particularly those in close contact with the soil. Different agricultural practices, such as the use of manure for fertilizer, may affect the level of contamination. Products in which contamination has often been detected include asparagus, beans, cabbage, carrots, celery, corn, onions, potatoes, turnips, olives, apricots, cherries, peaches, and tomatoes. The overall incidence of *C. botulinum* spores in commercially available precut MAP vegetables is low, approximately 0.36%; however, spores of both *C. botulinum* types A and B have been detected in these products (107). Cultivated mushrooms are of special concern, with up to 2.1×10^3 type B spores per kg being detected (76).

The potential presence of spores in honey and other infant foods is problematic because, in some infants, the spores can colonize the intestines, produce neurotoxin, and cause infant botulism. Honey is the only food that has ever been implicated in infant botulism. Molecular typing, using pulsed-field gel electrophoresis, ribotyping, and randomly amplified polymorphic DNA typing, has associated clinical isolates of *C. botulinum* from infant feces with isolates from honey implicated in cases of infant botulism (11a). Surveys of random honey samples revealed that approximately 2 to 7% of honey samples contain *C. botulinum* (87, 166), with spore levels between 1 and 10 spores per kg (41). The level is higher in honey samples associated with infant botulism, approximately 10^4 spores per kg (9, 41). While spores have been detected in other infant foods, namely, corn syrup and rice cereal, these foods do not appear to present the same risk as honey because the level of contamination is low, and spores are unlikely to increase during production and storage of such products. A very low occurrence of *C. botulinum* spores has been found in other foods, including dairy products, vacuum-packed products, and convenience foods (41).

FOODBORNE OUTBREAKS

Although botulism from commercial foods is rare, many countries report relatively frequent outbreaks (Table 15.4). Unrecognized and misdiagnosed cases of botulism do occur, as revealed by a 1985 outbreak in Vancouver, Canada, in which the initial diagnoses for 28 patients included psychiatric illness, viral syndrome, and a variety of other maladies (180). In large areas of the

Table 15.4 Reported foodborne botulism cases

Country(ies)	Period	No. of cases	Usual type(s)	Usual food(s)	Reference(s)
Argentina	1979–1997	277	A	Preserved vegetables	194
Belgium	1988–1998	10	B	Meats	185
Canada	1985–1999	183	E	Traditional Inuit fermented marine mammals	24a
China	1958–1989	2,861	A, B	Fermented bean products	65
Denmark (incl. Greenland and the Faroe Islands)	1984–1989	16	B, E	Mutton, fish	75, 185, 197
England and Wales	1922–1998	58	A, B	Various	26
France	1988–1998	72	B	Home-cured ham	185
Germany	1988–1998	177	B	Meats	185
Iran	1972–1974	314	E	Fish	75
Italy	1988–1998	412	B	Vegetables preserved in oil or water	6
Japan	1951–1987	479	E	Fish or fish products	75
Norway	1975–1997	26	E	"Rakfisk," traditional fermented fish	96
Poland	1988–1998	1,995	B	Home-preserved meats	64
Russia	1988–1992	2,300	B	Home-preserved mushrooms, fish	88
Spain	1988–1998	92	B	Vegetables	185
United States	1950–1996	1,087	A	Vegetables	29

world, particularly where botulism occurs infrequently, epidemiologic data are scarce.

In many northern areas, including the Canadian north, Alaska, Scandinavia, and northern Japan, most botulism outbreaks involve fish, especially traditional native dishes, such as raw and parboiled meats from sea mammals; fermented meats such as muktuk (meat, blubber, and skin of the beluga whale); and fermented salmon eggs (75). *C. botulinum* type E has been implicated in the vast majority of outbreaks involving northern native foods (9–11, 147, 168, 195, 196). Many of these foods are fermented products, but the level of fermentable carbohydrates is too low to ensure a pH reduction sufficiently rapid to prevent growth of *C. botulinum*.

In Poland and several other European countries, including France, Germany, Hungary, Portugal, the former Czechoslovakia, and Belgium, the foods most often implicated in botulism outbreaks are home-preserved meats, such as ham, fermented sausages, and canned products, with the predominant type being B (75). Poland reports the highest number of botulism cases annually, reflecting both a high local incidence of botulism and a very thorough surveillance system. In Poland, the use of "weck jars" (weckglas) to hermetically seal cooked meats (mostly pork) under a vacuum led to a significant number of botulism cases (64). The incidence of botulism was highest in Poland during the period of social change in the 1980s, reaching a peak of 738 cases in 1982. Commercial foods, usually canned meat and fish, were responsible for 26% of botulism cases in Poland (64).

Uneviscerated, salt-cured fish (ribyetz or kapchunka) was the vehicle of several outbreaks of botulism. In 1981, a California man became ill and, in 1985, two Russian immigrants died after eating uneviscerated, salt-cured fish (13). In 1987, an international outbreak involving eight people in Israel and the United States, with one fatality, was caused by consumption of kapchunka distributed in New York City (184). The largest outbreak of type E botulism ever reported, and the first recorded outbreak of botulism in Egypt, was associated with eating uneviscerated salted mullet fish (198). Ninety-one patients were hospitalized in Cairo after consuming faseikh purchased from the same shop. Eighteen of the patients died. The viscera of the ungutted, salted fish may have provided a low-salt environment for *C. botulinum* to germinate and produce neurotoxin.

Commercial products generally have a good safety record since the early days of canning, the 1920s. Bottled garlic in oil, which was the vehicle of an outbreak in Canada and another in the United States, is a recently implicated commercial product (23, 123, 180). As a result of these two outbreaks, garlic in oil can be sold in

North America only if a second barrier, such as acidification, is present in addition to refrigeration (123). Two outbreaks of type B botulism in Italy were associated with commercially prepared sliced roasted eggplant in oil (34). Additional commercial products implicated in outbreaks in Italy include canned tuna fish in oil, mascarpone cheese, and low-acid canned vegetables (6).

In the United Kingdom, hazelnut yogurt was the vehicle of a recent outbreak (129). Type B neurotoxin was produced in an underprocessed hazelnut purée that was then used to flavor a yogurt produced by a local dairy. Commercial mascarpone cheese was the vehicle of an eight-case outbreak of type A botulism in Italy in 1996 (6, 8). Two outbreaks in Japan were associated with commercial food. In one, vacuum-packaged, stuffed lotus rhizome was the vehicle of a type A outbreak, which involved 36 cases with 11 deaths (130). In the other, imported bottled caviar was the vehicle, with 21 cases and 3 deaths due to type B neurotoxin. Taiwan reported an unusual botulism outbreak involving commercially canned peanuts (32).

Several outbreaks have also occurred at food-service establishments (40). Temperature abuse of either food ingredients or the final product is often the contributing factor. Leftover baked potatoes were the vehicles of two reported outbreaks, one in which the potatoes were used for potato salad (167) and another in which the potatoes were used in a Greek food known as skordalia (5). During baking, the temperature of the potatoes would not exceed 100°C, below the temperature required to kill spores of proteolytic *C. botulinum*. The baked potatoes were then held at room temperature after baking for a time sufficient to allow germination and growth of *C. botulinum* spores. Potatoes stuffed with meat and cheese sauce and served at a delicatessen were the vehicle for an outbreak of type A botulism involving eight cases and one death. In this outbreak, a commercial cheese sauce used to fill the potatoes became toxic after contamination with *C. botulinum* and subsequent storage at room temperature (189).

Not all foodborne botulism outbreaks have been caused by *C. botulinum*. In Guanyun, Jiangsu Province, China, in January 1994, six clinical cases of botulism were reported after consumption of salted and fermented paste made of soybeans and wax gourds (114, 115). Type E toxin was detected in the implicated food, and *C. butyricum*, but not *C. botulinum*, was isolated. Subsequently, type E neurotoxin-producing *C. butyricum* was isolated from soil at four sites in an area near the locations where the type E outbreak occurred (115). *C. butyricum* has also been isolated from sevu (crisp made of graham flour), which was associated with a

34-case outbreak of botulism in India in September 1996 (30).

CHARACTERISTICS OF DISEASE

Foodborne botulism can range from a mild illness, which may be disregarded or misdiagnosed, to a serious disease that can be fatal within 24 h. The clinical symptoms of botulism were the subject of a recent review (170). Symptoms typically appear 12 to 36 h after ingestion of neurotoxin but may appear within a few hours or not until up to 14 days. The earlier that symptoms appear, the more serious the illness. The first symptoms are generally nausea and vomiting, followed by neurological signs and symptoms, including visual impairments (blurred or double vision, ptosis, fixed and dilated pupils), loss of normal mouth and throat functions (difficulty in speaking and swallowing; dry mouth, throat, and tongue; sore throat), general fatigue and lack of muscle coordination, and respiratory impairment. Other gastrointestinal symptoms may include abdominal pain, diarrhea, or constipation. Nausea and vomiting appear more often in cases of botulism types B and E than in those of type A. Dysphagia and muscle weakness are more common in outbreaks of types A and B than in outbreaks of type E. Dry mouth, tongue, and throat are observed most frequently in type B cases. Respiratory failure and airway obstruction are the main causes of death. Fatality rates in the first half of the century were about 50% or higher, but with the availability today of antisera and modern respiratory support systems, they have decreased to about 10%.

Botulism may be confused with other illnesses, including other forms of food poisoning, stroke, myasthenia gravis, and carbon monoxide poisoning, but most commonly with Guillain-Barré syndrome (170). The neurological signs in botulism appear first in the cranial nerve area (eyes, mouth, and throat) and then descend; in Guillain-Barré, the syndrome progresses in an ascending fashion, beginning in the extremities. Electrodiagnostic testing, including nerve conduction studies and electromyography, is useful for differentiating botulism from several other paralytic disorders, including Guillain-Barré syndrome, tick paralysis, poliomyelitis, porphyria, myasthenia gravis, Lambert-Eaton myasthenic syndrome, hypermagnesemia, diphtheria, and organophosphate intoxication (169).

The initial symptoms of infant botulism are less clearcut (43, 116). The most common, and usually the earliest, symptom is constipation. Medical attention is generally requested several days to a week later. The infants usually present a generalized weakness and a weak cry. Other symptoms may include feeding difficulty and poor sucking, lethargy, lack of facial expression, irritability, and progressive "floppiness." Respiratory arrests occur frequently but are seldom fatal.

Initial treatment of foodborne botulism involves removing or inactivating the neurotoxin by (i) neutralization of circulating neurotoxin with antiserum, (ii) use of enema to remove residual neurotoxin from the bowel, and (iii) gastric lavage or treatment with emetics if the potential food exposure was recent. Antiserum is most effective in the early stages of illness (182). The impact of antiserum is obvious from the Chinese experience: before the availability of antisera in 1960, the death rate in China was approximately 50%, but it was reduced to only 8% for the nearly 4,000 patients who have received antitoxin since then (171). Subsequent treatment is mainly to counteract paralysis of the respiratory muscles by mechanical ventilation. Optimal treatment for infant botulism consists primarily of high-quality supportive care (43). Approximately 25% of all affected infants require mechanical ventilation, and many require gavage feeding. The use of equine antitoxin and antibiotics is not recommended and does not appear to change the course or outcome of infant botulism. Botulism immune globulin has been used to treat infant botulism and has demonstrated significant reductions in the length of hospital stay and cost (5a).

TOXIC/INFECTIOUS DOSE AND SUSCEPTIBLE POPULATIONS

Little is known regarding the minimum toxic (infective) dose of *C. botulinum* and its neurotoxins. From a food safety perspective, there is no tolerance for the presence of neurotoxin or for conditions permitting growth of *C. botulinum*. The parenteral LD_{50} for BoNT in monkeys and mice is approximately 0.4 ng/kg (67). A minimum infective dose has not been established for infant or wound botulism. Two separate studies revealed that the spore levels in honey samples implicated in infant botulism were quite high, 8×10^3 and 8×10^4 spores per kg, which were approximately 1,000-fold higher than those in random samples of honey (43). However, this does not indicate a high infective dose, because the babies affected consumed very little honey.

While infants are most susceptible to germination of spores in their intestines, with subsequent growth, toxigenesis, and illness, some adults are also susceptible (31, 70, 106, 112). Susceptibility of infants is related to their diet and indigenous intestinal microflora. Susceptibility of adults appears to be associated with major perturbations in their intestinal microflora caused by such treatments as chemotherapy and antibiotic therapy.

The only effective means of preventing foodborne botulism is by preventing neurotoxin production in foods. Immunization of high-risk populations with botulinal toxoids has been considered, but it is not considered effective; at present, only laboratory workers at risk are normally immunized. Usually, the preservation of high-moisture foods is designed to prevent growth of *C. botulinum*. In shelf-stable canned foods, the thermal process generally ensures destruction of *C. botulinum* spores. Control of *C. botulinum* in minimally processed foods is achieved by inhibition, typically by the hurdle effect using a combination of factors. Controlling the growth of *C. botulinum* generally also ensures control of other foodborne pathogens and of many spoilage microorganisms.

VIRULENCE FACTORS/MECHANISMS OF PATHOGENICITY

Eight antigenically distinct toxins, designated types A, B, C_1, C_2, D, E, F, and G, are produced by *C. botulinum* (73). All of the toxins, except C_2, are neurotoxins. C_2 and exoenzyme C_3 have been identified as ADP-ribosylating enzymes (4).

C. botulinum is categorized into four broad groups (73). Group I strains produce types A, B, and F neurotoxins or combinations of A_B, A_F, B_A, and B_F neurotoxins, where the minor component is designated by the subscript letter. Group II strains produce type B, E, and F neurotoxins. Group III strains produce C_1 and D neurotoxins and C_2 toxin. These toxins can be produced separately or in combinations of C_1 and C_2 or D and C_2 toxins. Group IV strains produce type G neurotoxin. Other species of *Clostridium* also can produce neurotoxins. *C. butyricum* can produce type E neurotoxin (7, 113), *C. baratii* can produce type F neurotoxin (71), and *C. novyi* can produce type C_1 or D neurotoxin (53).

C. botulinum C_2 toxin and exoenzyme C_3 are produced by type C and D serotypes and belong to a class of bacterial ADP-ribosylating toxins which modify actin or the small GTP-binding proteins of the Rho family (4). Other ADP-ribosylating toxins include diphtheria toxin, *Pseudomonas aeruginosa* exotoxin A, exoenzyme S of *P. aeruginosa*, cholera toxin, pertussis toxin, and the heat-labile enterotoxin of *Escherichia coli* (132). These toxins are all capable of transferring the ADP-ribosyl moiety of NAD to specific protein substrates.

C. botulinum C_2 toxin is a binary toxin, composed of a binding component and an enzyme component with actin ADP-ribosylating activity (4). Monomeric G-actin is ADP-ribosylated at arginine 177 by the enzymatic component of C_2 toxin (C2-I) (192). The ribosylated actin is no longer capable of polymerization, resulting in destabilization of the cytoskeleton.

C. botulinum exoenzyme C3 has been purified as a protein with a molecular mass of 23 kDa. Exoenzyme C3 ADP-ribosylates a specific asparagine residue of the low-molecular-mass GTP-binding proteins of the Rho family. The Rho proteins are involved in regulation of the actin cytoskeleton (4). No role in pathogenesis has been attributed to either C_2 toxin or exoenzyme C3.

Neurotoxins

All seven of the neurotoxins are similar in structure and mode of action. *C. botulinum* neurotoxins are high-molecular-mass (150 kDa) two-chain proteins which are among the most toxic substances known. The neurotoxins block neurotransmission at peripheral motor nerve terminals by selectively hydrolyzing proteins involved in the fusion of synaptic vesicles with the presynaptic plasma membrane, thereby preventing acetylcholine release (120, 138, 150, 188).

Structure of the Neurotoxins

BoNTs are water-soluble proteins produced as a single polypeptide with an approximate M_r of 150,000. They are cleaved by a protease approximately one-third of the distance from the N terminus to produce the active neurotoxin, which is composed of one heavy chain (H; $M_r = 100,000$) and one light chain (L; $M_r = 50,000$) linked by a single disulfide bond (36, 37).

Endogenous bacterial proteases or proteases such as trypsin can cause the proteolytic cleavage. The two chains individually are nontoxic. The L chain remains bound to the N-terminal half of the H chain by a disulfide bond between Cys-429 and Cys-453 (94) and noncovalent bonds (35).

Native gel electrophoresis and chemical cross-linking experiments have been used to determine the quaternary structure of the BoNTs (104). These studies indicate that BoNT type A exists in several forms, including a dimer, a trimer, and a larger species; BoNT type E exists as a monomer and primarily as a dimer; and BoNT type B exists as a dimer in aqueous solution. The oligomerization of BoNT molecules may be involved in possible channel formation and translocation of the chain into the cytoplasm. BoNT type A was first crystallized in 1991 (179), and the complete three-dimensional structure of BoNT type A has been recently determined at 3.3-Å resolution (97).

BoNTs form complexes with nontoxic proteins in naturally contaminated foods and culture supernatant fluids to form what is referred to as progenitor toxin (152). The nontoxic proteins can be dissociated from

BoNT by pH values greater than 7.2, and they spontaneously reassociate when the pH is lowered. Three forms of progenitor neurotoxin have been distinguished and are referred to as M (medium-sized), L (large), and LL (extra large) (152). M toxin is produced by all strains producing neurotoxin except those producing type G neurotoxin, has a molecular mass of approximately 300 kDa, and is composed of the neurotoxin and a nontoxic nonhemagglutinin (NTNH) protein with a size similar to that of the neurotoxin. L toxin has a molecular mass of approximately 500 kDa and is produced by strains that produce types A, B, C, D, and G neurotoxins. L toxin contains hemagglutinins with molecular masses of 33 kDa (HA-33, Hn-33) and 17 kDa (HA-II) in addition to BoNT and NTNH. The nontoxic proteins consist of several polypeptides with different molecular masses (27, 35, 55, 115, and 120 kDa or 33, 53, and 130 kDa) (190, 191). Hn-33 displays complete resistance to proteolysis when treated with trypsin, chymotrypsin, pepsin, and subtilisin (59), suggesting that Hn-33 and perhaps the other nontoxic proteins are important for protection of the neurotoxin from low pH and proteases during passage through the stomach (59, 174). The HA proteins function as adhesins enabling BoNTs to bind to the microvilli of the small intestine, leading to adsorption of BoNT (61, 63).

Genetic Regulation of the Neurotoxins

The neurotoxins are arranged as part of a transcriptional unit which includes the genes encoding BoNT as well as genes encoding NTNH components and hemagglutinins. This transcriptional unit may be referred to as the BoNT gene complex (49, 51).

Complete gene sequences have been determined for the neurotoxins produced by *C. botulinum* types A, B (proteolytic and nonproteolytic), C, D, E, F, and G; *C. baratii* type F; and *C. butyricum* type E (17–19, 27, 50, 78, 81, 90, 91, 145, 186, 187, 199, 200). The degree of relatedness of the various neurotoxins has been determined on the basis of sequence homologies (27). Different serotype neurotoxins display less sequence homology than the same serotypes, even if the same serotypes are from different species. Type E BoNTs from *C. botulinum* and *C. butyricum* share 97% identical amino acids, suggesting gene transfer between *C. botulinum* and *C. butyricum* (79, 145). Type F BoNTs from nonproteolytic *C. botulinum* and *C. baratii* have approximately 70% identity (187). Type E and type F BoNTs have approximately 63% identity (187), whereas botulinum toxins type G and type B are approximately 58% identical (27).

The locations of the genes coding for BoNTs and the associated nontoxic proteins vary depending upon the serotype. The genes coding for BoNTs A, B, E, and F and the associated nontoxic proteins are located on the bacterial chromosome (19, 186, 199). The genes coding for botulinum toxins C_1 and D and the associated nontoxic proteins are encoded by bacteriophages (52, 55, 56, 77, 78, 190), whereas the genes coding for BoNT G and the associated nontoxic proteins are located on a plasmid (54, 207).

The organization of the gene clusters encoding components of the BoNT complexes have now been determined for all serotypes of *C. botulinum* (15, 47, 48, 51, 77, 95, 126, 153, 190, 191). The gene encoding NTNH is located immediately upstream of that of BoNT in all toxin types (49). Genes for other components of the BoNT complexes are clustered upstream of the NTNH gene in strains encoding types A and B (51), type C (62, 77, 191), type D (126), and type G (15). The six genes encoding BoNT C_1 and its associated nontoxic proteins (Antp) are organized into three transcriptional units (77). One cluster encodes the structural gene for BoNT C_1 and an associated nontoxic protein (Antp 139/C1 or NTNH). A second cluster contains three genes, two of which encode proteins similar to HA-33 and HA-II. The third cluster consists of a single gene which displays homology with a regulatory protein of *Clostridium perfringens* (*uviA* gene product) and could regulate the expression of the genes in the other two clusters (77).

Mode of Action of the Neurotoxins

BoNTs block the exocytic release of the excitatory neurotransmitter acetylcholine from synaptic vesicles at peripheral motor nerve terminals. This results in the flaccid paralysis observed in botulism poisoning and has been utilized in the therapy of several neurologic disorders (28). BoNTs are also being used as tools to study the molecular mechanisms of vesicle docking and membrane fusion mechanisms involved in exocytosis (25, 161).

The mechanism of action of BoNTs can be divided into three steps: binding, internalization, and intracellular action (172). A review on the mode of action of BoNT by Montecucco et al. (121) includes membrane translocation in addition to these three steps. The H chains are responsible for selective binding of the neurotoxin to neurons, internalization of the entire neurotoxin, intraneuronal sorting, and translocation of the L chains into the cytosol. The L chains block exocytosis as soon as they are released into the cytoplasm (3, 118, 124).

Binding

The potency of the BoNTs results, in part, from the specificity of the toxins for neurons. Neurotoxin receptors are located at the motor neuron plasma membrane at the neuromuscular junction (44). The nature of the cell surface receptors for BoNTs has been recently reviewed (154). Receptors may be different for the various neurotoxins (20, 206). Binding experiments with ^{125}I-labeled BoNTs and synaptosomes indicate type A, B, C_1, and D neurotoxins all have a small number of high-affinity binding sites and a large number of sites with much lower affinity (2, 57, 202).

While the neurotoxins appear not to share receptors, the receptors may all possess at least one sialic acid residue. Gangliosides, which are sialic acid-containing glycosphingolipids, bind to the BoNTs (92, 155, 173). Sialic acid-specific lectins from the plants *Triticum vulgaris* and *Limax flavus* are able to block binding of BoNTs to brain membranes (14). Putative receptors for BoNT A and tetanus neurotoxin have been identified from blots of synaptosomal proteins (156). Both neurotoxins adhered to 80-kDa and 116-kDa glycoproteins. The H chain of BoNT A specifically adhered to the same proteins, further evidence that attachment to target cell membranes is mediated through the C-terminal half of the H chain (131, 144, 156).

A double-receptor model for binding of botulinum and tetanus neurotoxins has been used to explain the high-affinity binding (117). In this model, the neurotoxins first bind to the negatively charged surface of the presynaptic membrane, which contains large amounts of acidic lipids. After binding to the negatively charged lipids, the neurotoxin may diffuse laterally in the membrane to bind to a protein receptor. The high binding affinity of the neurotoxins for the presynaptic plasma membrane is explained in this model because the association constant is the product of the association constants of the neurotoxin with the negatively charged lipids and to the protein receptor (118).

Internalization

After the neurotoxin has bound to its receptors at the neuromuscular junctions, the entire neurotoxin is internalized by receptor-mediated endocytosis. Once the neurotoxin has been internalized, it can no longer be neutralized by antibodies. Studies using chemical cross-linking and high-resolution electron microscopy indicate that the H chain aggregates and forms channels in the endosomal membrane to allow the L chain to pass into the cytoplasm. Cross-linking studies have revealed that, in aqueous solution, type A neurotoxin exists as a dimer, a trimer, and a larger species; type E neurotoxin exists as a monomer and dimer; and type B neurotoxin exists as a dimer (104). Three-dimensional reconstructions of ordered arrays of type B neurotoxin on surfaces of ganglioside/phosphocholine lipid vesicles reveal that four neu-

rotoxin molecules combine to form channels which cross the vesicle membrane (165). In addition, synthetic peptides of the predicted transmembrane sequence of BoNT A form cation-selective channels in planar lipid bilayers (125). Thus, it appears that the L chain exits the endosome by passing through channels created by the H chain.

Intracellular Action

The L chain is capable of inhibiting neurotransmitter release independently of the H chain (124). This has been demonstrated by intracellular injection of isolated L chain (144), by intracellular delivery of the L chain by liposome fusion with motor nerve endings (39), and by exposure of permeabilized cells to the L chain (178).

Comparison of the nucleotide sequences of the genes encoding the BoNTs with that of the gene encoding tetanus neurotoxin revealed a conserved segment in the central region of the L chain, which contained the His-Glu-x-x-His zinc-binding motif of metalloendoproteinases (38, 60, 118, 119, 160, 162). L chains act as zinc-dependent endopeptidases, whose substrates are components of the synaptic vesicle docking and fusion complex (Fig. 15.2). BoNT types B, F, D, and G and tetanus neurotoxin cause selective degradation of VAMP (also called synaptobrevin) (157–159). Type A and E neurotoxins degrade the synaptosome-associated protein SNAP-25 (16, 21, 163). Type C_1 neurotoxin degrades syntaxin/HPC-1 (22, 164).

The zinc-dependent protease activity of the BoNTs is very specific. BoNT type B and tetanus neurotoxin hydrolyze a Gln-Phe peptide bond in VAMP-2, BoNT type F cleaves VAMP at a Gln-Lys bond present in both VAMP-1 and VAMP-2, and BoNT type G hydrolyzes VAMP at a single Ala-Ala peptide bond (158). VAMP, syntaxin, and SNAP-25 form the core of a multicomponent complex which mediates fusion of carrier vesicles to target membranes in eukaryotic cells (Fig. 15.2). Proteolysis of these proteins involved in docking and fusion of synaptic vesicles blocks neuroexocytosis and subsequent neurotransmitter release.

Concluding Remarks

C. botulinum is a spore-forming bacterium that is widely distributed, occurring in soils, freshwater, and marine sediments and the intestinal tracts of many animals. Control of foodborne botulism is based almost entirely on thermal destruction of the spores or inhibition of spore germination and bacterial cell growth in foods. Outbreaks continue to occur in home-preserved foods (including vegetables and meats), commercial foods, and restaurant foods. Some traditional northern

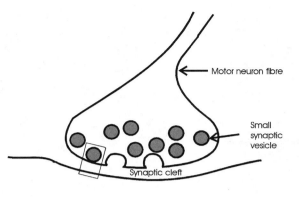

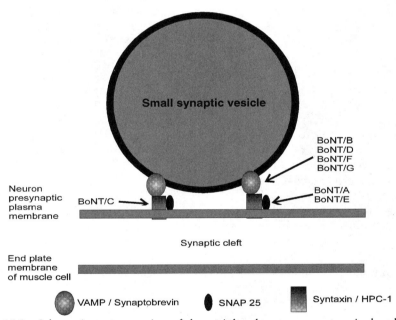

Figure 15.2 Schematic representation of the peripheral motor nerve terminal and synapse, depicting sites of action of BoNTs. The upper diagram depicts the anatomy of the presynaptic terminal and synapse. The boxed area is enlarged beneath, revealing details of the small synaptic vesicle. Synaptobrevin (also known as VAMP) (open ovals), SNAP-25 (solid ovals), and syntaxin (also known as HPC-1) (striped boxes) are proteins involved in the fusion of synaptic vesicles with the presynaptic plasma membrane. Synaptobrevin is embedded in the synaptic vesicle membrane, whereas SNAP-25 and syntaxin are located on the cytoplasmic surface of the neuron plasma membrane. The arrows indicate the protein which is the specific target of the different BoNTs.

native foods, including whale, seal, and walrus meat and salmon eggs, are responsible for a high proportion of cases of botulism, especially in Canada.

The increased consumption of ready-to-eat, minimally processed foods, especially those packaged in modified atmospheres, is of concern. The combina-

tion of lack of a heat treatment sufficient to destroy *C. botulinum* spores, lack of heating prior to consumption, and packaging in an atmosphere with reduced or no oxygen increases the risk of botulism from these foods. The reliance on refrigeration for storage of these foods does not preclude the growth of, and toxin production by, nonproteolytic *C. botulinum*. Use of novel preservatives or bacteriocin-producing bacteria to inhibit the germination and growth of *C. botulinum* in these foods is a possible alternative for controlling this pathogen.

Complete nucleotide sequences are now known for all the BoNTs. In addition, several accessory genes, including those encoding the hemagglutinins, have been sequenced and localized into transcriptional units. The isolation of neurotoxin-producing strains of *C. butyricum* and *C. baratii* and the high sequence similarity between BoNTs produced in these strains and *C. botulinum* strains suggest lateral transfer of the genes encoding BoNTs and their associated polypeptides. Future research in this area may include elucidation of possible transcriptional or translational control of synthesis of the neurotoxins, organization of gene clusters in unusual strains (153), the role of the nontoxic proteins, and transfer of the genes encoding neurotoxin and associated proteins, between species of *Clostridium*.

The modes of action of the BoNTs have recently been elucidated. BoNT types B, D, F, and G cleave VAMP/synaptobrevin, an integral membrane protein of synaptic vesicles, whereas BoNT types A and E cleave SNAP-25, and BoNT type C cleaves syntaxin. Both SNAP-25 and syntaxin are synaptic proteins located on the plasma membrane of the neuron. Identification of zinc-dependent endopeptidase activity of BoNTs and discovery of the substrates of neurotoxins have facilitated the study of synaptic vesicle docking, exocytosis, and neurotransmitter release. Future research will address selective alteration of neurotoxins by site-directed mutagenesis to determine which amino acid residues are involved in binding, uptake, proteolysis (105), and antigenicity. Elucidation of the mode of action of the BoNTs, combined with the recent determination of the three-dimensional structure of the neurotoxin (97), should facilitate design and development of pharmacological treatments for botulinum intoxication (1).

References

1. **Adler, M., R. E. Dinterman, and R. W. Wannemacher.** 1997. Protection by the heavy metal chelator N,N,N′,N′-tetrakis (2-pyridylmethyl)ethylenediamine (TPEN) against the lethal action of botulinum neurotoxin A and B. *Toxicon* **35**:1089–100.

2. **Agui, T., B. Syuto, K. Oguma, H. Iida, and S. Kubo.** 1983. Binding of *Clostridium botulinum* type C neurotoxin to rat brain synaptosomes. *J. Biochem.* (Tokyo) **94**:521–725.

3. **Ahnert-Hilger, G., and H. Bigalke.** 1995. Molecular aspects of tetanus and botulinum neurotoxin poisoning. *Prog. Neurobiol.* **46**:83–96.

4. **Aktories, K.** 1994. Clostridial ADP-ribosylating toxins: effects on ATP and GTP-binding proteins. *Mol. Cell. Biochem.* **138**:167–176.

5. **Angulo, F. J., J. Getz, J. P. Taylor, K. A. Hendricks, C. L. Hatheway, S. S. Barth, H. M. Solomon, A. E. Larson, E. A. Johnson, L. N. Nickey, and A. A. Ries.** 1998. A large outbreak of botulism: the hazardous baked potato. *J. Infect. Dis.* **178**:172–77.

5a. **Arnon, S.** Personal communication.

6. **Aureli, P., L. Fenica, and G. Franciosa.** 1999. Classic and emergent forms of botulism: the current status in Italy. *Eurosurveillance* **4**:7–9.

7. **Aureli, P., L. Fenicia, B. Pasolini, M. Gianfranceschi, L. M. McCroskey, and C. L. Hatheway.** 1986. Two cases of type E infant botulism caused by neurotoxigenic *Clostridium butyricum* in Italy. *J. Infect. Dis.* **154**:207–211.

8. **Aureli, P., G. Franciosa, and M. Pourshaban.** 1996. Food-borne botulism in Italy. *Lancet* **348**:1594.

9. **Austin, J.** 1996. Botulism in Canada—summary for 1995. *Can. Commun. Dis. Rep.* **22**:182–183.

10. **Austin, J., B. Blanchfield, E. Ashton, M. Lorange, J. F. Proulx, A. Trinidad, and W. Winther.** 1999. Botulism in Canada—summary for 1997. *Can. Commun. Dis. Rep.* **25**:121–122.

11. **Austin, J., B. Blanchfield, J.-F. Proulx, and E. Ashton.** 1997. Botulism in Canada—summary for 1996. *Can. Commun. Dis. Rep.* **23**:132.

11a. **Austin, J. W., and S. Hielm.** Unpublished results.

12. **Austin, J. W., K. L. Dodds, B. Blanchfield, and J. M. Farber.** 1998. Growth and toxin production by *Clostridium botulinum* on inoculated fresh-cut packaged vegetables. *J. Food Prot.* **61**:324–328.

13. **Badhey, H., D. J. Cleri, R. F. D'Amato, J. R. Vernaleo, V. Veinni, J. Tessler, A. A. Wallman, A. J. Mastellone, M. Giuliani, and L. Hochstein.** 1986. Two fatal cases of type E adult food-borne botulism with early symptoms and terminal neurologic signs. *J. Clin. Microbiol.* **23**:616–618.

14. **Bakry, N., Y. Kamata, and L. L. Simpson.** 1991. Lectins from *Triticum vulgaris* and *Limax flavus* are universal antagonists of botulinum neurotoxin and tetanus toxin. *J. Pharmacol. Exp. Ther.* **258**:830–836.

15. **Bhandari, M., K. D. Campbell, M. D. Collins, and A. K. East.** 1997. Molecular characterization of the clusters of genes encoding the botulinum neurotoxin complex in *Clostridium botulinum* (*Clostridium argentinense*) type G and nonproteolytic *Clostridium botulinum* type B. *Curr. Microbiol.* **35**:207–214.

16. **Binz, T., J. Blasi, S. Yamasaki, A. Baumeister, E. Link, T. C. Sudhof, R. Jahn, and H. Niemann.** 1994. Proteolysis of SNAP-25 by types E and A botulinal neurotoxins. *J. Biol. Chem.* **269**:1617–1620.

17. **Binz, T., U. Eisel, H. Kurazono, I. T. Binschek, H. Bigalke, and H. Niemann.** 1990. Tetanus and botulinum A toxins: comparison of sequences and microinjection of tetanus toxin A subunit-specific mRNA into bovine chromaffin cells. *Zentbl. Bakteriol.* **S19**:105–108.

18. **Binz, T., H. Kurazono, M. R. Popoff, M. W. Eklund, G. Sakaguchi, S. Kozaki, K. Krieglstein, A. Henschen,**

D. M. Gill, and H. Niemann. 1990. Nucleotide sequence of the gene encoding *Clostridium botulinum* neurotoxin type D. *Nucleic Acids Res.* **18:**5556.

19. Binz, T., H. Kurazono, M. Wille, J. Frevert, K. Wernars, and H. Niemann. 1990. The complete sequence of botulinum neurotoxin type A and comparison with other clostridial neurotoxins. *J. Biol. Chem.* **265:**9153–9158.

20. Black, J. D., and J. O. Dolly. 1986. Interaction of ^{125}I-labeled botulinum neurotoxins with nerve terminals. I. Ultrastructural autoradiographic localization and quantitation of distinct membrane acceptors for types A and B on motor nerves. *J. Cell Biol.* **103:**521–534.

21. Blasi, J., E. R. Chapman, E. Link, T. Binz, S. Yamasaki, P. De Camilli, T. C. Sudhof, H. Niemann, and R. Jahn. 1993. Botulinum neurotoxin A selectively cleaves the synaptic protein SNAP-25. *Nature* **365:**160–163.

22. Blasi, J., E. R. Chapman, S. Yamasaki, T. Binz, H. Niemann, and R. Jahn. 1993. Botulinum neurotoxin C1 blocks neurotransmitter release by means of cleaving HPC-1/syntaxin. *EMBO J.* **12:**4821–4828.

23. Blatherwick, F. J., S. H. Peck, G. B. Morgan, M. E. Milling, G. D. Kettyls, T. J. Johnstone, D. Bowering, and M. St. Louis. 1985. An international outbreak of botulism associated with a restaurant in Vancouver, British Columbia. *Can. Dis. Wkly. Rep.* **11:**177–178.

24. Bott, T. L., J. Johnson, Jr., E. M. Foster, and H. Sugiyama. 1968. Possible origin of the high incidence of *Clostridium botulinum* type E in an inland bay (Green Bay of Lake Michigan). *J. Bacteriol.* **95:**1542–1547.

24a. Botulism Reference Serrice for Canda. Unpublished data.

25. Boyd, R. S., M. J. Duggan, C. C. Shone, and K. A. Foster. 1995. The effect of botulinum neurotoxins on the release of insulin from the insulinoma cell lines HIT-15 and RINm5F. *J. Biol. Chem.* **270:**18216–18218.

26. Brett, M. 1999. Botulism in the United Kingdom. *Eurosurveillance* **4:**9–10.

27. Campbell, K., M. D. Collins, and A. K. East. 1993. Nucleotide sequence of the gene coding for *Clostridium botulinum* (*Clostridium argentinense*) type G neurotoxin: genealogical comparison with other clostridial neurotoxins. *Biochim. Biophys. Acta* **1216:**487–491.

28. Cardoso, F., and J. Jankovic. 1995. Clinical use of botulinum neurotoxins. *Curr. Top. Microbiol. Immunol.* **195:**123–141.

29. Centers for Disease Control and Prevention. 1998. Botulism in the United States, 1899–1996. *In Handbook for Epidemiologists, Clinicians, and Laboratory Workers.* Centers for Disease Control and Prevention, Atlanta, Ga.

30. Chaudhry, R., B. Dhawan, D. Kumar, R. Bhatia, J. C. Gandhi, R. K. Patel, and B. C. Purohit. 1998. Outbreak of suspected *Clostridium butyricum* botulism in India. *Emerg. Infect. Dis.* **4:**506–507.

31. Chia, J. K., J. B. Clark, C. A. Ryan, and M. Pollack. 1986. Botulism in an adult associated with food-borne intestinal infection with *Clostridium botulinum*. *N. Engl. J. Med.* **315:**239–241.

32. Chou, J. H., P. H. Hwang, and M. D. Malison. 1988. An outbreak of type A foodborne botulism in Taiwan due to commercially preserved peanuts. *Int. J. Epidemiol.* **17:**899–902.

33. Crandall, A. D., and T. J. Montville. 1993. Inhibition of *Clostridium botulinum* growth and toxigenesis in a model gravy system by coinoculation with bacteriocin-producing lactic acid bacteria. *J. Food Prot.* **56:**485.

34. D'Argenio, P., F. Palumbo, R. Ortolani, R. Pizzuti, M. Russo, R. Carducci, M. Soscia, P. Aureli, L. Fenicia, G. Franciosa, A. Parella, and V. Scala. 1995. Type B botulism associated with roasted eggplant in oil Italy, 1993. *Morb. Mortal. Wkly. Rep.* **44:**33–36.

35. DasGupta, B. R. 1990. Structure and biological activity of botulinum neurotoxin. *J. Physiol.* (Paris) **84:**220–228.

36. DasGupta, B. R., and H. Sugiyama. 1972. Isolation and characterization of a protease from *Clostridium botulinum* type B. *Biochim. Biophys. Acta* **268:**719–729.

37. DasGupta, B. R., and H. Sugiyama. 1972. A common subunit structure in *Clostridium botulinum* type A, B and E toxins. *Biochem. Biophys. Res. Commun.* **48:**108–112.

38. de Paiva, A., A. C. Ashton, P. Foran, G. Schiavo, C. Montecucco, and J. O. Dolly. 1993. Botulinum A like type B and tetanus toxins fulfils criteria for being a zinc-dependent protease. *J. Neurochem.* **61:**2338–2341.

39. de Paiva, A., and J. O. Dolly. 1990. Light chain of botulinum neurotoxin is active in mammalian motor nerve terminals when delivered via liposomes. *FEBS Lett.* **277:**171–174.

40. Dodds, K. L. 1990. Restaurant-associated botulism outbreaks in North America. *Food Control* **1:**139–142.

41. Dodds, K. L. 1993. *Clostridium botulinum* in foods, p. 53–68. *In* A. H. W. Hauschild and K. L. Dodds (ed.), *Clostridium botulinum: Ecology and Control in Foods.* Marcel Dekker, Inc., New York, N.Y.

42. Dodds, K. L. 1993. *Clostridium botulinum* in the environment, p. 21–52. *In* A. H. W. Hauschild and K. L. Dodds (ed.), *Clostridium botulinum: Ecology and Control in Foods.* Marcel Dekker, Inc., New York, N.Y.

43. Dodds, K. L. 1993. Worldwide incidence and ecology of infant botulism, p. 105–120. *In* A. H. W. Hauschild and K. L. Dodds (ed.), *Clostridium botulinum: Ecology and Control in Foods.* Marcel Dekker, Inc., New York, N.Y.

44. Dolly, J. O., J. Black, R. S. Williams, and J. Melling. 1984. Acceptors for botulinum neurotoxin reside on motor nerve terminals and mediate its internalization. *Nature* **307:**457–460.

45. Doyle, M. P. 1983. Effect of carbon dioxide on toxin production by *Clostridium botulinum*. *Eur. J. Appl. Microbiol. Biotechnol.* **17:**53–56.

46. Dufresne, I., J. P. Smith, J.-N. Liu, I. Tarte, B. Blanchfield, and J. W. Austin. Effect of headspace oxygen and films of different oxygen transmission rates on toxin production by *Clostridium botulinum* type E in rainbow trout fillets stored under modified atmospheres. *J. Food Safety*, in press.

47. East, A. K., M. Bhandari, S. Hielm, and M. D. Collins. 1998. Analysis of the botulinum neurotoxin type F gene clusters in proteolytic and nonproteolytic *Clostridium botulinum* and *Clostridium barati*. *Curr. Microbiol.* **37:**262–268.

48. East, A. K., M. Bhandari, J. M. Stacey, K. D. Campbell, and M. D. Collins. 1996. Organization and phylogenetic interrelationships of genes encoding components of the botulinum toxin complex in proteolytic *Clostridium botulinum* types A, B, and F: evidence of chimeric sequences in the gene encoding the nontoxic nonhemagglutinin component. *Int. J. Syst. Bacteriol.* **46:**1105–1112.

49. East, A. K., and M. D. Collins. 1994. Conserved structure of genes encoding components of botulinum neurotoxin complex M and the sequence of the gene coding for the nontoxic component in nonproteolytic *Clostridium botulinum* type F. *Curr. Microbiol.* **29:**69–77.

50. East, A. K., P. T. Richardson, D. Allaway, M. D. Collins, T. A. Roberts, and D. E. Thompson. 1992. Sequence of the gene encoding type F neurotoxin of *Clostridium botulinum*. *FEMS Microbiol. Lett.* **75:**225–230.

51. East, A. K., J. M. Stacey, and M. D. Collins. 1994. Cloning and sequencing of a hemagglutinin component of the botulinum neurotoxin complex encoded by *Clostridium botulinum* types A and B. *Syst. Appl. Microbiol.* **17:**306–312.

52. Eklund, M. W., F. T. Poysky, and W. H. Habig. 1989. Bacteriophages and plasmids in *Clostridium botulinum* and *Clostridium tetani* and their relationship to production of toxins, p. 25–51. *In* L. L. Simpson (ed.), *Botulinum Neurotoxin and Tetanus Toxin*. Academic Press, Inc., New York, N.Y.

53. Eklund, M. W., F. T. Poysky, J. A. Meyers, and G. A. Pelroy. 1974. Interspecies conversion of *Clostridium botulinum* type C to *Clostridium novyi* type A by bacteriophage. *Science* **186:**456–458.

54. Eklund, M. W., F. T. Poysky, L. M. Mseitif, and M. S. Strom. 1988. Evidence for plasmid-mediated toxin and bacteriocin production in *Clostridium botulinum* type G. *Appl. Environ. Microbiol.* **54:**1405–1408.

55. Eklund, M. W., F. T. Poysky, and S. M. Reed. 1972. Bacteriophage and the toxigenicity of *Clostridium botulinum* type D. *Nat. New Biol.* **235:**16–17.

56. Eklund, M. W., F. T. Poysky, S. M. Reed, and C. A. Smith. 1971. Bacteriophage and the toxigenicity of *Clostridium botulinum* type C. *Science* **172:**480–482.

57. Evans, D. M., R. S. Williams, C. C. Shone, P. Hambleton, J. Melling, and J. O. Dolly. 1986. Botulinum neurotoxin type B. Its purification, radioiodination and interaction with rat-brain synaptosomal membranes. *Eur. J. Biochem.* **154:**409–416.

58. Foegeding, P. M., and F. A. Busta. 1983. Effect of carbon dioxide, nitrogen and hydrogen gases on germination of *Clostridium botulinum* spores. *J. Food Prot.* **46:**987–989.

59. Fu, F. N., S. K. Sharma, and B. R. Singh. 1998. A protease-resistant novel hemagglutinin purified from type A *Clostridium botulinum*. *J. Protein Chem.* **17:**53–60.

60. Fujii, N., K. Kimura, N. Yokosawa, K. Tsuzuki, and K. Oguma. 1992. A zinc-protease specific domain in botulinum and tetanus neurotoxins. *Toxicon* **30:**1486–1488.

61. Fujinaga, Y., K. Inoue, T. Nomura, J. Sasaki, J. C. Marvaud, M. R. Popoff, S. Kozaki, and K. Oguma. 2000. Identification and characterization of functional subunits of *Clostridium botulinum* type A progenitor toxin involved in binding to intestinal microvilli and erythrocytes. *FEBS Lett.* **467:**179–183.

62. Fujinaga, Y., K. Inoue, S. Shimazaki, K. Tomochika, K. Tsuzuki, N. Fujii, T. Watanabe, T. Ohyama, K. Takeshi, and K. Inoue. 1994. Molecular construction of *Clostridium botulinum* type C progenitor toxin and its gene organization. *Biochem. Biophys. Res. Commun.* **205:**1291–1298.

63. Fujinaga, Y., K. Inoue, S. Watanabe, K. Yokota, Y. Hirai, E. Nagamachi, and K. Oguma. 1997. The haemagglutinin of *Clostridium botulinum* type C progenitor toxin plays an essential role in binding of toxin to the epithelial cells of guinea pig small intestine, leading to the efficient absorption of the toxin. *Microbiology* **143:**3841–3847.

64. Galazka, A., and A. Przybylska. 1999. Surveillance of foodborne botulism in Poland: 1960–1998. *Eurosurveillance* **4:**69–72.

65. Gao, Q. Y., Y. F. Huang, J. G. Wu, H. D. Liu, and H. Q. Xia. 1990. A review of botulism in China. *Biomed. Environ. Sci.* **3:**326–336.

66. Garcia, G. W., C. Genigeorgis, and S. Lindroth. 1987. Risk of growth and toxin production by *Clostridium botulinum* nonproteolytic types B, E and F in salmon fillets stored under modified atmospheres at low and abused temperatures. *J. Food Prot.* **50:**330–336.

67. Gill, D. M. 1982. Bacterial toxins: a table of lethal amounts. *Microbiol. Rev.* **46:**86–94.

68. Gould, G. W. 1999. Sous vide foods: conclusions of an ECFF Botulinum Working Party. *Food Control* **10:**47–51.

69. Graham, A. F., D. R. Mason, F. J. Maxwell, and M. W. Peck. 1997. Effect of pH and NaCl on growth from spores of non-proteolytic *Clostridium botulinum* at chill temperature. *Lett. Appl. Microbiol.* **24:**95–100.

70. Griffin, P. M., C. L. Hatheway, R. B. Rosenbaum, and R. Sokolow. 1997. Endogenous antibody production to botulinum toxin in an adult with intestinal colonization botulism and underlying Crohn's disease. *J. Infect. Dis.* **175:** 633–637.

71. Hall, J. D., L. M. McCroskey, B. J. Pincomb, and C. L. Hatheway. 1985. Isolation of an organism resembling *Clostridium baratii* which produces type F botulinal toxin from an infant with botulism. *J. Clin. Microbiol.* **21:**654–655.

72. Hao, Y. Y., R. E. Brackett, L. R. Beuchat, and M. P. Doyle. 1999. Microbiological quality and production of botulinal toxin in film-packaged broccoli, carrots, and green beans. *J. Food Prot.* **62:**499–508.

73. Hatheway, C. L. 1993. *Clostridium botulinum* and other clostridia that produce botulinum neurotoxin, p. 3–20. *In* A. H. W. Hauschild and K. L. Dodds (ed.), *Clostridium botulinum: Ecology and Control in Foods*. Marcel Dekker, Inc., New York, N.Y.

74. Hauschild, A. H. W. 1989. *Clostridium botulinum*, p. 111–189. *In* M. P. Doyle (ed.), *Bacterial Foodborne Pathogens*. Marcel Dekker, Inc., New York, N.Y.

75. Hauschild, A. H. W. 1993. Epidemiology of human foodborne botulism, p. 69–104. *In* A. H. W. Hauschild and K. L. Dodds (ed.), *Clostridium botulinum: Ecology*

and Control in Foods. Marcel Dekker, Inc., New York, N.Y.

76. **Hauschild, A. H. W., B. J. Aris, and R. Hilsheimer.** 1975. *Clostridium botulinum* in marinated products. *Can. Inst. Food Sci. Technol. J.* 8:84–87.

77. **Hauser, D., M. W. Eklund, P. Boquet, and M. R. Popoff.** 1994. Organization of the botulinum neurotoxin C_1 gene and its associated nontoxic protein genes in *Clostridium botulinum* C 468. *Mol. Gen. Genet.* 243:631–640.

78. **Hauser, D., M. W. Eklund, H. Kurazono, T. Binz, H. Niemann, D. M. Gill, P. Boquet, and M. R. Popoff.** 1990. Nucleotide sequence of *Clostridium botulinum* C1 neurotoxin. *Nucleic Acids Res.* 18:4924.

79. **Hauser, D., M. Gibert, P. Boquet, and M. R. Popoff.** 1992. Plasmid localization of a type E botulinal neurotoxin gene homologue in toxigenic *Clostridium butyricum* strains, and absence of this gene in nontoxigenic *C. butyricum* strains. *FEMS Microbiol. Lett.* 78:251–255.

80. **Huhtanen, C. N., J. Naghski, C. S. Custer, and R. W. Russell.** 1976. Growth and toxin production by *Clostridium botulinum* in moldy tomato juice. *Appl. Environ. Microbiol.* 32:711–715.

81. **Hutson, R. A., M. D. Collins, A. K. East, and D. E. Thompson.** 1994. Nucleotide sequence of the gene coding for non-proteolytic *Clostridium botulinum* type B neurotoxin: comparison with other clostridial neurotoxins. *Curr. Microbiol.* 28:101–110.

82. **Hutson, R. A., D. E. Thompson, and M. D. Collins.** 1993. Genetic interrelationships of saccharolytic *Clostridium botulinum* types B, E and F and related clostridia as revealed by small-subunit rRNA gene sequences. *FEMS Microbiol. Lett.* 108:103–110.

83. **Hutson, R. A., D. E. Thompson, P. A. Lawson, R. P. Schocken-Itturino, E. C. Bottger, and M. D. Collins.** 1993. Genetic interrelationships of proteolytic *Clostridium botulinum* types A, B, and F and other members of the *Clostridium botulinum* complex as revealed by small-subunit rRNA gene sequences. *Antonie Leeuwenhoek* 64:273–283.

84. **Hutton, M. T., P. A. Chehak, and J. H. Hanlin.** 1991. Inhibition of botulinum toxin production by *Pediococcus acidilactici* in temperature abused refrigerated foods. *J. Food Safety* 11:255–267.

85. **Hyytia-Trees, E., M. Lindstrom, B. Schalch, A. Stolle, and H. Korkeala.** 1999. *Clostridium botulinum* type E in Bavarian fish. *Archiv. Lebensmittelhyg.* 50:79–82.

86. **Hyytia-Trees, E., E. Skytta, M. Mokkila, A. Kinnunen, M. Lindstrom, L. Lahteenmaki, R. Ahvenainen, and H. Korkeala.** 2000. Safety evaluation of sous vide-processed products with respect to nonproteolytic *Clostridium botulinum* by use of challenge studies and predictive microbiological models. *Appl. Environ. Microbiol.* 66:223–229.

87. **Kautter, D. A., T. Lilly, H. M. Solomon, and R. K. Lynt.** 1982. *Clostridium botulinum* spores in infant foods: a survey. *J. Food Prot.* 45:1028–1029.

88. **Kazdobina, I. S., E. B. Terekhova, L. G. Podunova, and Y. V. Vertiev.** 1995. Botulism in the Russian Federation (1988-1992). *Zh. Mikrobiol.* (Moscow) 6:96–99.

89. **Kim, J., and P. M. Foegeding.** 1993. Principles of control, p. 121–176. *In* A. H. W. Hauschild and K. L. Dodds (ed.), *Clostridium botulinum: Ecology and Control in Foods.* Marcel Dekker, Inc., New York, N.Y.

90. **Kimura, K., N. Fujii, K. Tsuzuki, T. Murakami, T. Indoh, N. Yokosawa, and K. Oguma.** 1991. Cloning of the structural gene for *Clostridium botulinum* type C_1 toxin and whole nucleotide sequence of its light chain component. *Appl. Environ. Microbiol.* 57:1168–1172.

91. **Kimura, K., N. Fujii, K. Tsuzuki, T. Murakami, T. Indoh, N. Yokosawa, K. Takeshi, B. Syuto, and K. Oguma.** 1990. The complete nucleotide sequence of the gene coding for botulinum type C1 toxin in the C-ST phage genome. *Biochem. Biophys. Res. Commun.* 171:1304–1311.

92. **Kitamura, M., M. Iwamori, and Y. Nagai.** 1980. Interaction between *Clostridium botulinum* neurotoxin and gangliosides. *Biochim. Biophys. Acta* 628:328–335.

93. **Knock, G. G.** 1952. Survey of soils for spores of *Clostridium botulinum* (Union of South Africa and South West Africa). *J. Sci. Food Agric.* 3:86–91.

94. **Krieglstein, K. G., B. R. Dasgupta, and A. H. Henschen.** 1994. Covalent structure of botulinum neurotoxin type-A. Location of sulfhydryl groups, and disulfide bridges and identification of C-termini of light and heavy chains. *J. Protein Chem.* 13:49–57.

95. **Kubota, T., N. Yonekura, Y. Hariya, E. Isogai, H. Isogai, K. Amano, and N. Fujii.** 1998. Gene arrangement in the upstream region of *Clostridium botulinum* type E and *Clostridium butyricum* BL6340 progenitor toxin genes is different from that of other types. *FEMS Microbiol. Lett.* 158:215–221.

96. **Kuusi, M.** 1998. Botulisme. *Tidsskr. Nor Laegeforen* 118:4338.

97. **Lacy, D. B., W. Tepp, A. C. Cohen, B. R. Dasgupta, and R. C. Stevens.** 1998. Crystal structure of botulinum neurotoxin type A and implications for toxicity. *Nat. Struct. Biol.* 5:898–902.

98. **Lambert, A. D., J. P. Smith, and K. L. Dodds.** 1991. Combined effect of modified atmosphere packaging and low-dose irradiation on toxin production by *Clostridium botulinum* in fresh pork. *J. Food Prot.* 54:94–101.

99. **Lambert, A. D., J. P. Smith, and K. L. Dodds.** 1991. Effect of headspace carbon dioxide concentration on toxin production by *Clostridium botulinum* in MAP irradiated fresh pork. *J. Food Prot.* 54:588–592.

100. **Lambert, A. D., J. P. Smith, and K. L. Dodds.** 1991. Effect of initial O_2 and CO_2 and low-dose irradiation on toxin production by *Clostridium botulinum* in MAP fresh pork. *J. Food Prot.* 54:939–944.

101. **Larson, A. E., and E. A. Johnson.** 1999. Evaluation of botulinal toxin production in packaged fresh-cut cantaloupe and honeydew melons. *J. Food Prot.* 62:948–952.

102. **Larson, A. E., E. A. Johnson, C. R. Barmore, and M. D. Hughes.** 1997. Evaluation of the botulism hazard from vegetables in modified atmosphere packaging. *J. Food Prot.* 60:1208–1214.

103. **Lawson, P. A., P. Llop-Perez, R. A. Hutson, H. Hippe, and M. D. Collins.** 1993. Towards a phylogeny of the

clostridia based on 16S rRNA sequences. *FEMS Microbiol. Lett.* **113:**87–92.

104. **Ledoux, D. N., X. H. Be, and B. R. Singh.** 1994. Quaternary structure of botulinum and tetanus neurotoxins as probed by chemical cross-linking and native gel electrophoresis. *Toxicon* **32:**1095–1104.

105. **Li, L., T. Binz, H. Niemann, and B. R. Singh.** 2000. Probing the mechanistic role of glutamate residue in the zinc-binding motif of type A botulinum neurotoxin light chain. *Biochemistry* **39:**2399–2405.

106. **Li, L. Y., P. Kelkar, R. E. Exconde, J. Day, and G. J. Parry.** 1999. Adult-onset "infant" botulism: an unusual cause of weakness in the intensive care unit. *Neurology* **53:**891.

107. **Lilly, T., H. M. Solomon, and E. J. Rhodehamel.** 1996. Incidence of *Clostridium botulinum* in vegetables packaged under vacuum or modified atmosphere. *J. Food Prot.* **59:**59–61.

108. **Lund, B. M., and S. H. W. Notermans.** 1993. Potential hazards associated with REPFEDS, p. 279–304. *In* A. H. W. Hauschild and K. L. Dodds (ed.), *Clostridium botulinum: Ecology and Control in Foods.* Marcel Dekker, Inc., New York, N.Y.

109. **Lynt, R. K., D. A. Kautter, and H. M. Solomon.** 1979. Heat resistance of nonproteolytic *Clostridium botulinum* in phosphate buffer and crabmeat. *J. Food Sci.* **44:**108–111.

110. **Lyver, A., J. P. Smith, J. Austin, and B. Blanchfield.** 1998. Competitive inhibition of *Clostridium botulinum* type E by *Bacillus* species in a value-added seafood product packaged under a modified atmosphere. *Food Res. Int.* **31:**311–319.

111. **Lyver, A., J. P. Smith, F. M. Nattress, J. W. Austin, and B. Blanchfield.** 1998. Challenge studies with *Clostridium botulinum* type E in a value-added surimi product stored under a modified atmosphere. *J. Food Safety* **18:**1–23.

112. **McCroskey, L. M., and C. L. Hatheway.** 1988. Laboratory findings in four cases of adult botulism suggest colonization of the intestinal tract. *J. Clin. Microbiol.* **26:**1052–1054.

113. **McCroskey, L. M., C. L. Hatheway, L. Fenicia, B. Pasolini, and P. Aureli.** 1986. Characterization of an organism that produces type E botulinal toxin but which resembles *Clostridium butyricum* from the feces of an infant with type E botulism. *J. Clin. Microbiol.* **23:**201–202.

114. **Meng, X., T. Karasawa, K. Zou, X. Kuang, X. Wang, C. Lu, C. Wang, K. Yamakawa, and S. Nakamura.** 1997. Characterization of a neurotoxigenic *Clostridium butyricum* strain isolated from the food implicated in an outbreak of food-borne type E botulism. *J. Clin. Microbiol.* **35:**2160–2162.

115. **Meng, X., K. Yamakawa, K. Zou, X. Wang, X. Kuang, C. Lu, C. Wang, T. Karasawa, and S. Nakamura.** 1999. Isolation and characterisation of neurotoxigenic *Clostridium butyricum* from soil in China. *J. Med. Microbiol.* **48:**133–137.

116. **Midura, T. F.** 1996. Update: infant botulism. *Clin. Microbiol. Rev.* **9:**119.

117. **Montecucco, C.** 1986. How do tetanus and botulinum toxins bind to neuronal membranes. *Trends Biochem.* **11:**314–317.

118. **Montecucco, C., and G. Schiavo.** 1994. Mechanism of action of tetanus and botulinum neurotoxins. *Mol. Microbiol.* **13:**1–8.

119. **Montecucco, C., G. Schiavo, and F. Facchiano.** 1993. Tetanus and botulism neurotoxins: a new group of zinc proteases. *Trends Biol. Sci.* **18:**324–327.

120. **Montecucco, C., G. Schiavo, and O. Rossetto.** 1996. The mechanism of action of tetanus and botulinum neurotoxins. *Arch. Toxicol. Suppl.* **18:**342–354.

121. **Montecucco, C., G. Schiavo, and O. Rossetto.** 1996. The mechanism of action of tetanus and botulinum neurotoxins. *Arch. Toxicol. Suppl.* **18:**342–354.

122. **Montville, T. J.** 1982. Metabiotic effect of *Bacillus licheniformis* on *Clostridium botulinum:* implications for home-canned tomatoes. *Appl. Environ. Microbiol.* **44:**334–338.

123. **Morse, D. L., L. K. Pickard, J. J. Guzewich, B. D. Devine, and M. Shayegani.** 1990. Garlic-in-oil associated botulism: episode leads to product modification. *Am. J. Publ. Health* **80:**1372–1373.

124. **Niemann, H., T. Binz, O. Grebenstein, H. Kurazono, J. Thierer, S. Mochida, B. Poulain, and L. Tauc.** 1991. Clostridial neurotoxins: from toxins to therapeutic tools? *Behring Inst. Mitt.* 153–162.

125. **Oblatt-Montal, M., M. Yamazaki, R. Nelson, and M. Montal.** 1995. Formation of ion channels in lipid bilayers by a peptide with the predicted transmembrane sequence of botulinum neurotoxin A. *Protein Sci.* **4:**1490–1497.

126. **Ohyama, T., T. Watanabe, Y. Fujinaga, K. Inoue, H. Sunagawa, H. Fujii, K. Inoue, and K. Oguma.** 1995. Characterization of nontoxic-nonhemagglutinin component of the two types of progenitor toxin (M and L) produced by *Clostridium botulinum* type D CB-16. *Microbiol. Immunol.* **39:**457–465.

127. **Okereke, A., and T. J. Montville.** 1991. Bacteriocin inhibition of *Clostridium botulinum* spores by lactic acid bacteria. *J. Food Prot.* **54:**349–353, 356.

128. **Okereke, A., and T. J. Montville.** 1991. Bacteriocin-mediated inhibition of *Clostridium botulinum* spores by lactic acid bacteria at refrigeration and abuse temperatures. *Appl. Environ. Microbiol.* **57:**3423–3428.

129. **O'Mahony, M., E. Mitchell, R. J. Gilbert, D. N. Hutchinson, N. T. Begg, J. C. Rodhouse, and J. E. Morris.** 1990. An outbreak of foodborne botulism associated with contaminated hazelnut yoghurt. *Epidemiol. Infect.* **104:**389–395.

130. **Otofuji, T., H. Tokiwa, and K. Takahashi.** 1987. A food-poisoning incident caused by *Clostridium botulinum* toxin A in Japan. *Epidemiol. Infect.* **99:**167–172.

131. **Park, M. K., H. H. Jung, and K. H. Yang.** 1990. Binding of *Clostridium botulinum* type B toxin to rat brain synaptosome. *FEMS Microbiol. Lett.* **60:**243–247.

132. **Passador, L., and W. Iglewski.** 1994. ADP-ribosylating toxins. *Methods Enzymol.* **235:**617–631.

133. **Passaro, D. J., S. B. Werner, J. McGee, W. R. Mac Kenzie, and D. J. Vugia.** 1998. Wound botulism associated with black tar heroin among injecting drug users. *JAMA* **279:**859–863.

134. Peck, M. W. 1997. *Clostridium botulinum* and the safety of refrigerated processed foods of extended durability. *Trends Food Sci. Technol.* 8:186–192.

135. Peck, M. W., D. A. Fairbairn, and B. M. Lund. 1992. The effect of recovery medium on the estimated heat-inactivation of spores of non-proteolytic *Clostridium botulinum*. *Lett. Appl. Microbiol.* 15:146–151.

136. Peck, M. W., D. A. Fairbairn, and B. M. Lund. 1992. Factors affecting growth from heat-treated spores of non-proteolytic *Clostridium botulinum*. *Lett. Appl. Microbiol.* 15:152–155.

137. Peck, M. W., D. A. Fairbairn, and B. M. Lund. 1993. Heat-resistance of spores of non-proteolytic *Clostridium botulinum* estimated on medium containing lysozyme. *Lett. Appl. Microbiol.* 16:126–131.

138. Pellizzari, R., O. Rossetto, B. Schiavo, and C. Montecucco. 1999. Tetanus and botulinum neurotoxins: mechanism of action and therapeutic uses. *Philos. Trans. R. Soc. Lond. B Biol. Sci.* 354:259–268.

139. Phillips-Daifas, D., J. P. Smith, B. Blanchfield, and J. W. Austin. 1999. Growth and toxin production by *Clostridium botulinum* in English-style crumpets packaged under modified atmospheres. *J. Food Prot.* 62:349–355.

140. Phillips-Daifas, D., J. P. Smith, B. Blanchfield, and J. W. Austin. 1999. Effect of pH and CO_2 on growth and toxin production by *Clostridium botulinum* in English-style crumpets packaged under modified atmospheres. *J. Food Prot.* 62:1157–1161.

141. Phillips-Daifas, D., J. P. Smith, I. Tarte, B. Blanchfield, and J. W. Austin. 2000. Effect of ethanol vapor on growth and toxin production by *Clostridium botulinum* in a high moisture bakery product. *J. Food Safety* 20:111–125.

142. Pickett, J., B. Berg, E. Chaplin, and M. A. Brunstetter-Shafer. 1976. Syndrome of botulism in infancy: clinical and electrophysiologic study. *N. Engl. J. Med.* 295:770–772.

143. Post, L. S., D. A. Lee, M. Solberg, D. Furgang, J. Specchio, and C. Graham. 1985. Development of botulinal toxin and sensory deterioration during storage of vacuum and modified atmosphere packaged fish fillets. *J. Food. Sci.* 50:990–996.

144. Poulain, B., S. Mochida, U. Weller, B. Hogy, E. Habermann, J. D. Wadsworth, C. C. Shone, J. O. Dolly, and L. Tauc. 1991. Heterologous combinations of heavy and light chains from botulinum neurotoxin A and tetanus toxin inhibit neurotransmitter release in *Aplysia*. *J. Biol. Chem.* 266:9580–9585.

145. Poulet, S., D. Hauser, M. Quanz, H. Niemann, and M. R. Popoff. 1992. Sequences of the botulinal neurotoxin E derived from *Clostridium botulinum* type E (strain Beluga) and *Clostridium butyricum* (strains ATCC 43181 and ATCC 43755). *Biochem. Biophys. Res. Commun.* 183:107–113.

146. Prevot, A. R. 1953. Rapport d'introduction du President du Sous-Comite *Clostridium* pour l'unification de la nomenclature des types toxigeniques de *C. botulinum*. *Int. Bull. Bacteriol. Nomencl.* 3:120–123.

147. Proulx, J. F., V. Milot-Roy, and J. Austin. 1997. Four outbreaks of botulism in Ungava Bay, Nunavik, Quebec. *Can. Commun. Dis. Rep.* 23:30–32.

148. Reddy, N. R., H. M. Solomon, G. A. Fingerhut, E. J. Rhodehamel, V. M. Balasubramaniam, and S. Palaniappan. 1999. Inactivation of *Clostridium botulinum* type E spores by high pressure processing. *J. Food Safety* 19:277–288.

149. Ronner, S. G. E., and E. Stackebrandt. 1994. Further evidence for the genetic heterogeneity of *Clostridium botulinum* as determined by 23S rDNA oligonucleotide probing. *Syst. Appl. Microbiol.* 17:180–188.

150. Rossetto, O., F. Deloye, B. Poulain, R. Pellizzari, G. Schiavo, and C. Montecucco. 1995. The metalloproteinase activity of tetanus and botulism neurotoxins. *J. Physiol.* (Paris) 89:43–50.

151. Rouhbakhsh-Khaleghdoust, A. 1975. The incidence of *Clostridium botulinum* type E in fish and bottom deposits in the Caspian Sea coastal waters. *Pahlavi Med. J.* 6:550–556.

152. Sakaguchi, G. 1990. Molecular structure of *Clostridium botulinum* progenitor toxins, p. 173–180. *In* A. E. Pohland, V. R. Dowell, and J. L. Richard (ed.), *Microbial Toxins in Foods and Feeds. Cellular and Molecular Modes of Action*. Plenum Press, Inc., New York, N.Y.

153. Santos-Buelga, J. A., M. D. Collins, and A. K. East. 1998. Characterization of the genes encoding the botulinum neurotoxin complex in a strain of *Clostridium botulinum* producing type B and F neurotoxins. *Curr. Microbiol.* 37:312–318.

154. Schengrund, C. L. 1999. What is the cell surface receptor(s) for the different serotypes of botulinum neurotoxin? *J. Toxicol. Toxin Rev.* 18:35–44.

155. Schengrund, C. L., B. R. DasGupta, and N. J. Ringler. 1991. Binding of botulinum and tetanus neurotoxins to ganglioside GT1b and derivatives thereof. *J. Neurochem.* 57:1024–1032.

156. Schengrund, C. L., N. J. Ringler, and B. R. Dasgupta. 1992. Adherence of botulinum and tetanus neurotoxins to synaptosomal proteins. *Brain Res. Bull.* 29:917–924.

157. Schiavo, G., F. Benfenati, B. Poulain, O. Rossetto, P. Polverino de Laureto, B. R. DasGupta, and C. Montecucco. 1992. Tetanus and botulinum-B neurotoxins block neurotransmitter release by proteolytic cleavage of synaptobrevin. *Nature* 359:832–835.

158. Schiavo, G., C. Malizio, W. S. Trimble, P. Polverino de Laureto, G. Milan, H. Sugiyama, E. A. Johnson, and C. Montecucco. 1994. Botulinum G neurotoxin cleaves VAMP/synaptobrevin at a single Ala-Ala peptide bond. *J. Biol. Chem.* 269:20213–20216.

159. Schiavo, G., O. Rossetto, F. Benfenati, B. Poulain, and C. Montecucco. 1994. Tetanus and botulinum neurotoxins are zinc proteases specific for components of the neuroexocytosis apparatus. *Ann. N. Y. Acad. Sci.* 710:65–75.

160. Schiavo, G., O. Rossetto, F. Benfenati, B. Poulain, and C. Montecucco. 1994. Tetanus and botulinum neurotoxins are zinc proteases specific for components of the neuroexocytosis apparatus. *Ann. N. Y. Acad. Sci.* 710:65–75.

161. Schiavo, G., O. Rossetto, and C. Montecucco. 1994. Clostridial neurotoxins as tools to investigate the molecular events of neurotransmitter release. *Semin. Cell Biol.* 5:221–229.

162. **Schiavo, G., O. Rossetto, A. Santucci, B. R. DasGupta and C. Montecucco.** 1992. Botulinum neurotoxins are zinc proteins. *J. Biol. Chem.* **267:**23479–23483.

163. **Schiavo, G., A. Santucci, B. R. Dasgupta, P. O. Mehta, J. Jontes, F. Benfenati, M. C. Wilson, and C. Montecucco.** 1993. Botulinum neurotoxins serotypes A and E cleave SNAP-25 at distinct COOH-terminal peptide bonds. *FEBS Lett.* **335:**99–103.

164. **Schiavo, G., C. C. Shone, M. K. Bennett, R. H. Scheller, and C. Montecucco.** 1995. Botulinum neurotoxin type C cleaves a single Lys-Ala bond within the carboxyl-terminal region of syntaxins. *J. Biol. Chem.* **270:**10566–10570.

165. **Schmid, M. F., J. P. Robinson, and B. R. DasGupta.** 1993. Direct visualization of botulinum neurotoxin-induced channels in phospholipid vesicles. *Nature* **364:**827–830.

166. **Schocken-Iturrino, R. P., M. C. Carneiro, E. Kato, J. O. B. Sorbara, O. D. Rossi, and L. E. R. Gerbasi.** 1999. Study of the presence of the spores of *Clostridium botulinum* in honey in Brazil. *FEMS Immunol. Med. Microbiol.* **24:**379–382.

167. **Seals, J. E., J. D. Snyder, T. A. Edell, C. L. Hatheway, C. J. Johnson, R. C. Swanson, and J. M. Hughes.** 1981. Restaurant-associated type A botulism: transmission by potato salad. *Am. J. Epidemiol.* **113:**436–444.

168. **Shaffer, N., R. B. Wainwright, J. P. Middaugh, and R. V. Tauxe.** 1990. Botulism among Alaska natives. The role of changing food preparation and consumption practices. *West. J. Med.* **153:**390–393.

169. **Shapiro, B. E., O. Soto, S. Shafqat, and H. Blumenfeld.** 1997. Adult botulism. *Muscle Nerve* **20:**100–102.

170. **Shapiro, R. L., C. Hatheway, and D. L. Swerdlow.** 1998. Botulism in the United States: a clinical and epidemiologic review. *Ann. Intern. Med.* **129:**221–228.

171. **Shih, Y., and S. Y. Chao.** 1986. Botulism in China. *Rev. Infect. Dis.* **8:**984–990.

172. **Simpson, L. L.** 1986. Molecular pharmacology of botulinum toxin and tetanus toxin. *Annu. Rev. Pharmacol. Toxicol.* **26:**427–453.

173. **Simpson, L. L., and M. M. Rapport.** 1971. Ganglioside inactivation of botulinum toxin. *J. Neurochem.* **18:**1341–1343.

174. **Singh, B. R., B. Li, and D. Read.** 1995. Botulinum versus tetanus neurotoxins: why is botulinum neurotoxin but not tetanus neurotoxin a food poison? *Toxicon* **33:**1541–1547.

175. **Skinner, G. E., H. M. Solomon, and G. A. Fingerhut.** 1999. Prevention of *Clostridium botulinum* type A, proteolytic B and E toxin formation in refrigerated pea soup by *Lactobacillus plantarum*, ATCC 8014. *J. Food Sci.* **64:**724–727.

176. **Solomon, H. M., D. A. Kautter, T. Lilly, and E. J. Rhodehamel.** 1990. Outgrowth of *Clostridium botulinum* in shredded cabbage at room temperature under a modified atmosphere. *J. Food Prot.* **53:**831–833.

177. **Solomon, H. M., E. J. Rhodehamel, and D. A. Kautter.** 1998. Growth and toxin production by *Clostridium botulinum* on sliced raw potatoes in a modified atmosphere with and without sulfite. *J. Food Prot.* **61:**126–128.

178. **Stecher, B., U. Weller, E. Habermann, M. Gratzl, and G. Ahnert-Hilger.** 1989. The light chain but not the heavy chain of botulinum A toxin inhibits exocytosis from permeabilized adrenal chromaffin cells. *FEBS Lett.* **255:**391–394.

179. **Stevens, R. C., M. L. Evenson, W. Tepp, and B. R. DasGupta.** 1991. Crystallization and preliminary X-ray analysis of botulinum neurotoxin type A. *J. Mol. Biol.* **222:**877–880.

180. **St. Louis, M. E., S. H. Peck, D. Bowering, G. B. Morgan, J. Blatherwick, S. Banerjee, G. D. Kettyls, W. A. Black, M. E. Milling, A. H. Hauschild, R. Tauxe, and P. Black.** 1988. Botulism from chopped garlic: delayed recognition of a major outbreak. *Ann. Intern. Med.* **108:**363–368.

181. **Suen, J. C., C. L. Hatheway, A. G. Streigerwalt, and D. J. Brenner.** 1988. *Clostridium argentinense* sp. nov.: a genetically homogeneous group composed of all strains of *Clostridium botulinum* toxin type G and some nontoxigenic strains previously identified as *Clostridium subterminale* or *Clostridium hastiforme*. *Int. J. Syst. Bacteriol.* **38:**375–385.

182. **Tacket, C. O., W. X. Shandera, J. M. Mann, N. T. Hargrett, and P. A. Blake.** 1984. Equine antitoxin use and other factors that predict outcome in type A foodborne botulism. *Am. J. Med.* **76:**794–798.

183. **Tanaka, N., L. Meske, M. P. Doyle, E. Traisman, D. W. Thayer, and R. W. Johnston.** 1985. Plant trials of bacon made with lactic acid bacteria, sucrose and lowered sodium nitrite. *J. Food Prot.* **48:**679–686.

184. **Telzak, E. E., E. P. Bell, D. A. Kautter, L. Crowell, L. D. Budnick, D. L. Morse, and S. Schultz.** 1990. An international outbreak of type E botulism due to uneviscerated fish. *J. Infect. Dis.* **161:**340–342.

185. **Therre, H.** 1999. Botulism in the European Union. *Eurosurveillance* **4:**2–7.

186. **Thompson, D. E., J. K. Brehm, J. D. Oultram, T. J. Swinfield, C. C. Shone, T. Atkinson, J. Melling, and N. P. Minton.** 1990. The complete amino acid sequence of the *Clostridium botulinum* type A neurotoxin, deduced by nucleotide sequence analysis of the encoding gene. *Eur. J. Biochem.* **189:**73–81.

187. **Thompson, D. E., R. A. Hutson, A. K. East, D. Allaway, M. D. Collins, and P. T. Richardson.** 1993. Nucleotide sequence of the gene coding for *Clostridium barati* type F neurotoxin: comparison with other clostridial neurotoxins. *FEMS Microbiol. Lett.* **108:**175–182.

188. **Tonello, F., S. Morante, O. Rossetto, G. Schiavo, and C. Montecucco.** 1996. Tetanus and botulism neurotoxins: a novel group of zinc-endopeptidases. *Adv. Exp. Med. Biol.* **389:**251–260.

189. **Townes, J. M., P. R. Cieslak, C. L. Hatheway, H. M. Solomon, J. T. Holloway, M. P. Baker, C. F. Keller, L. M. McCroskey, and P. M. Griffin.** 1996. An outbreak of type A botulism associated with a commercial cheese sauce. *Ann. Intern. Med.* **125:**558–563.

190. **Tsuzuki, K., K. Kimura, N. Fujii, N. Yokosawa, T. Indoh, T. Murakami, and K. Oguma.** 1990. Cloning and complete nucleotide sequence of the gene for the main component of hemagglutinin produced by *Clostridium botulinum* type C. *Infect. Immun.* **58:**3173–3177.

191. Tsuzuki, K., K. Kimura, N. Fujii, N. Yokosawa, and K. Oguma. 1992. The complete nucleotide sequence of the gene coding for the nontoxic-nonhemagglutinin component of *Clostridium botulinum* type C progenitor toxin. *Biochem. Biophys. Res. Commun.* **183:**1273–1279.

192. Vandekerckhove, J., B. Schering, M. Barmann, and K. Aktories. 1988. Botulinum C_2 toxin ADP-ribosylates cytoplasmic beta/gamma-actin in arginine 177. *J. Biol. Chem.* **263:**696–700.

193. van Ermengem, E. 1979. Classics in infectious disease. A new anaerobic bacillus and its relation to botulism. *Rev. Infect. Dis.* **1:**701–719. (Originally published in 1897, as Ueber einen neuen anaeroben Bacillus und seine Beziehungen zum Botulismus. *Z. Hyg. Infektionskr.* **26:**1–56.)

194. Villar, R. G., R. L. Shapiro, S. Busto, C. Riva-Posse, G. Verdejo, M. I. Farace, F. Rosetti, J. A. San Juan, C. M. Julia, J. Becher, S. E. Maslanka, and D. L. Swerdlow. 1999. Outbreak of type A botulism and development of a botulism surveillance and antitoxin release system in Argentina. *JAMA* **281:**1334–1338, 1340.

195. Wainwright, R. B. 1993. Hazards from northern native foods, p. 305–322. *In* A. H. W. Hauschild and K. L. Dodds (ed.), *Clostridium botulinum: Ecology and Control in Foods.* Marcel Dekker, Inc., New York, N.Y.

196. Wainwright, R. B., W. L. Heyward, J. P. Middaugh, C. L. Hatheway, A. P. Harpster, and T. R. Bender. 1988. Foodborne botulism in Alaska, 1947–1985: epidemiology and clinical findings. *J. Infect. Dis.* **157:**1158–1162.

197. Wandall, E. P., and T. Videro. 1991. [Botulism on the Faeroe Islands]. *Ugeskr. Laeger* **153:**833–835. (In Danish.)

198. Weber, J. T., R. G. Hibbs, Jr., A. Darwish, B. Mishu, A. L. Corwin, M. Rakha, C. L. Hatheway, S. el Sharkawy, S. A. el-Rahim, M. F. al-Hamd, J. E. Sarn, P. A. Blake, and R. V. Tauxe. 1993. A massive outbreak of type E botulism associated with traditional salted fish in Cairo. *J. Infect. Dis.* **167:**451–454.

199. Whelan, S. M., M. J. Elmore, N. J. Bodsworth, T. Atkinson, and N. P. Minton. 1992. The complete amino acid sequence of the *Clostridium botulinum* type-E neurotoxin, derived by nucleotide-sequence analysis of the encoding gene. *Eur. J. Biochem.* **204:**657–667.

200. Whelan, S. M., M. J. Elmore, N. J. Bodsworth, J. K. Brehm, T. Atkinson, and N. P. Minton. 1992. Molecular cloning of the *Clostridium botulinum* structural gene encoding the type B neurotoxin and determination of its entire nucleotide sequence. *Appl. Environ. Microbiol.* **58:**2345–2354.

201. Whitlock, R. H., and C. Buckley. 1997. Botulism. *Vet. Clin. N. Am. Equine Pract.* **13:**107–128.

202. Williams, R. S., C. K. Tse, J. O. Dolly, P. Hambleton, and J. Melling. 1983. Radioiodination of botulinum neurotoxin type A with retention of biological activity and its binding to brain synaptosomes. *Eur. J. Biochem.* **131:**437–445.

203. Wobeser, G., K. Baptiste, E. G. Clark, and A. W. Deyo. 1997. Type C botulism in cattle in association with a botulism die-off in waterfowl in Saskatchewan. *Can. Vet. J.* **38:**782.

204. Yamakawa, K., T. Hayashi, S. Kamiya, K. Yoshimura, and S. Nakamura. 1990. *Clostridium botulinum* in the soil of Paraguay. *Jpn. J. Med. Sci. Biol.* **43:**19–21.

205. Yamakawa, K., S. Kamiya, K. Yoshimura, S. Nakamura, and T. Ezaki. 1990. *Clostridium botulinum* in the soil of Kenya. *Ann. Trop. Med. Parasitol.* **84:**201–203.

206. Yokosawa, N., K. Tsuzuki, B. Syuto, N. Fujii, K. Kimura, and K. Oguma. 1991. Binding of botulinum type C_1, D and E neurotoxins to neuronal cell lines and synaptosomes. *Toxicon* **29:**261–264.

207. Zhou, Y., H. Sugiyama, H. Nakano, and E. A. Johnson. 1995. The genes for the *Clostridium botulinum* type G toxin complex are on a plasmid. *Infect. Immun.* **63:**2087–2091.

Food Microbiology: Fundamentals and Frontiers, 2nd Ed.
Edited by M. P. Doyle et al.
© 2001 ASM Press, Washington, D.C.

Bruce A. McClane

Clostridium perfringens

16

Clostridium perfringens first became recognized in the 1940s and 1950s as an important cause of food-borne disease, following the pioneering work of Knox, McDonald, McClung, and Hobbs (see references 31, 43, and 56 for reviews). It gradually became apparent that *C. perfringens* causes two quite different human enteric diseases that can be transmitted by food, i.e., *C. perfringens* type A food poisoning and necrotic enteritis (also known as Darmbrand or Pig-Bel). Since foodborne necrotic enteritis is rare in industrialized societies, this chapter will focus on *C. perfringens* type A food poisoning; readers interested in necrotic enteritis are directed to the review in reference 46.

CHARACTERISTICS OF THE ORGANISM

General

C. perfringens is a gram-positive, rod-shaped, encapsulated, nonmotile bacterium that causes a broad spectrum of human and veterinary diseases (31, 43, 56). The virulence of *C. perfringens* largely results from its prolific toxin-producing ability, including at least two toxins, *C. perfringens* enterotoxin (CPE) and beta-toxin, that are active on the human gastrointestinal (GI) tract.

In addition to producing GI-active toxins, *C. perfringens* has several other characteristics that contribute to its ability to cause foodborne disease. First, its rapid doubling time (reportedly <10 min for vegetative cells [43]) allows *C. perfringens* to quickly multiply in food. Second, *C. perfringens* has the ability to form spores (Fig. 16.1) which are resistant to environmental stresses such as radiation, desiccation, and heat (43); the ability to form such resistant spores facilitates survival of *C. perfringens* in incompletely cooked or inadequately warmed foods.

C. perfringens is considered an anaerobe since it does not produce colonies on agar plates continuously exposed to air (43). However, this bacterium tolerates moderate exposure to air; compared with most other anaerobes, *C. perfringens* requires only relatively modest reductions in oxidation-reduction potential (E_h) for growth (43).

Classification: Toxin Typing of *C. perfringens*

At least 14 different *C. perfringens* toxins have now been identified (21, 56). However, each individual *C. perfringens* cell carries a gene(s) encoding only a defined subset of this vast toxin repertoire (17, 75). That limitation

Bruce A. McClane, Department of Molecular Genetics and Biochemistry, University of Pittsburgh School of Medicine, E1240 Biomedical Science Tower, Pittsburgh, PA 15261-2072.

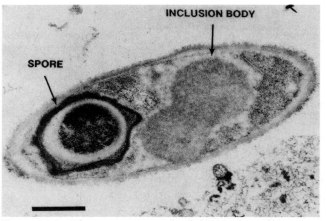

Figure 16.1 Electron micrograph of thin sections of a sporulating cell of *C. perfringens*. Magnification, ×40,000. Bar, 0.5 μm. Arrows indicate the endospore and a CPE-containing inclusion body in the cytoplasm of the mother cell. Reproduced with permission from reference 44.

provides the basis for a toxin typing system that classifies *C. perfringens* isolates into five types (A through E), depending upon an isolate's ability to express four (alpha, beta, epsilon, and iota) of the 14 *C. perfringens* toxins (Table 16.1). Toxin typing of *C. perfringens* traditionally involved laborious toxin-antiserum neutralization tests in mice (56), but the development of PCR-based schemes permitting toxin genotyping of *C. perfringens* isolates (17, 75) has greatly simplified toxin typing.

The two foodborne diseases caused by *C. perfringens* are each associated with a distinct *C. perfringens* type. Necrotic enteritis is caused by type C isolates, with the beta-toxin produced by those isolates considered the primary virulence factor for that disease (46). As implied by its name, *C. perfringens* type A food poisoning is almost always associated with type A isolates of *C. perfringens* (43, 56), even though other *C. perfringens* types sometimes express an enterotoxin with similar (or identical) physical, serologic, and biologic properties as CPE (43, 56), the toxin responsible for the characteristic symptoms of *C. perfringens* type A food poisoning. A

Table 16.1 Toxin typing of *C. perfringens*

C. perfringens type	Toxins produced			
	Alpha	Beta	Epsilon	Iota
A	+	−	−	−
B	+	+	+	−
C	+	+	−	−
D	+	−	+	−
E	+	−	−	+

possible explanation for the strong association between CPE-positive type A isolates and *C. perfringens* type A food poisoning is that CPE-producing type A isolates are much more widely distributed than other types of CPE-producing *C. perfringens* isolates (17, 75).

Tolerance/Susceptibility of *C. perfringens* to Preservation Methods

It is well established that pathogen growth in food is affected by factors such as temperature, E_h, pH, and water activity (a_w) levels. Therefore, the effect of those factors on the growth of *C. perfringens* will be briefly discussed below.

Temperature

The heat resistance of *C. perfringens* spores contributes to this bacterium's ability to cause food poisoning by enabling its survival in incompletely cooked foods. The heat resistance of *C. perfringens* spores is influenced by both environmental and genetic factors. With respect to environmental factors, the medium in which a *C. perfringens* spore is heated clearly affects heat resistance properties (43). Spores of some *C. perfringens* strains can survive boiling for an hour or longer in a relatively protective medium (e.g., cooked-meat medium) (43, 68). A genetic contribution to spore heat resistance is evident from studies demonstrating that spores of different *C. perfringens* strains exhibit considerable variation in their heat resistance properties (see reference 43 for a listing of decimal reduction values for spores made by representative *C. perfringens* strains). Perhaps owing to selective pressure, spores of food poisoning isolates generally have considerably greater heat resistance than spores of *C. perfringens* isolates from other sources (43, 68); this may be important for foodborne virulence. Finally, incomplete cooking of foods may not only fail to kill *C. perfringens* spores in foods but can actually favor development of *C. perfringens* type A food poisoning by inducing germination of *C. perfringens* spores (43).

Vegetative cells of *C. perfringens* are also somewhat heat tolerant. Although not truly thermophilic, *C. perfringens* vegetative cells possess a relatively high optimal growth temperature (43 to 45°C), and can often grow at temperatures up to 50°C or more (43). Recent studies suggest that at 55°C vegetative cells of food poisoning isolates are about twofold more heat resistant than vegetative cells of other *C. perfringens* isolates (68).

However, vegetative *C. perfringens* cells are not notably tolerant of cold temperatures, including either refrigeration or freezing conditions (43). Growth rates of

C. perfringens rapidly decrease at temperatures below ~15°C, with no growth occurring at 6°C (43). In contrast, spores of *C. perfringens* are cold resistant (43). Food poisoning can result if viable spores present in refrigerated or frozen foods later germinate when that food is warmed for serving.

Other Factors

Growth of *C. perfringens* in food is also affected by a_w, E_h, pH, and (probably) the presence of curing agents (31, 43). *C. perfringens* is less tolerant of low-a_w environments than is *Staphylococcus aureus*, another common gram-positive foodborne pathogen (43). The lowest a_w supporting vegetative growth of *C. perfringens* is 0.93 to 0.97, depending upon the solute used to control the a_w of the medium (31, 43).

C. perfringens does not require (for an anaerobe) an extremely reduced environment for growth. Provided that the environmental E_h is suitably low for initiating growth (the exact E_h value needed to initiate *C. perfringens* growth depends on environmental factors such as pH), *C. perfringens* can then modify the E_h of its surrounding environment (by producing reducing molecules such as ferredoxin) to produce more optimal growth conditions (43). As a practical guide for food microbiologists, the E_h of many common foods (e.g., raw meats and gravies) is often sufficient to support the growth of *C. perfringens* (43).

Growth of *C. perfringens* is also pH sensitive. Optimal growth of this bacterium occurs at pH 6 to 7, whereas *C. perfringens* grows poorly, if at all, at pH values ≤5 and ≥8.3 (31, 43).

The effectiveness of curing agents for limiting *C. perfringens* growth in foods is an unsettled subject (31, 43). Some studies (31) indicate that the concentration of curing salts needed to significantly affect *C. perfringens* growth exceeds commercially acceptable levels; e.g., growth inhibition of *C. perfringens* requires at least 6 to 8% NaCl and 10,000 ppm of $NaNO_3$ or 400 ppm of $NaNO_2$. However, more recent studies (43) revealed that curing salts can be at least partially effective at preventing *C. perfringens* growth in food, even when used at commercially acceptable levels, since (i) simultaneous coapplication of other preservation factors, such as heating and acidic pH values increases the sensitivity of *C. perfringens* to curing salts (43), (ii) simultaneous use of several curing agents often provides a synergistic effect to inhibit *C. perfringens* growth (43), and (iii) foods may contain lower initial populations of *C. perfringens* cells and spores than are used in laboratory studies evaluating the effectiveness of curing agents for inhibiting *C. perfringens* growth (31, 43). A more recent study

indicates that the presence of 3% NaCl delays the growth of *C. perfringens* in vacuum-packed beef (34). However, one argument for the ability of curing agents to influence *C. perfringens* growth is that commercially cured meat products are rarely associated with *C. perfringens* type A foodborne outbreaks (43).

Preservation factors such as pH, a_w, and, perhaps, curing agents also control *C. perfringens* populations in foods by inhibiting the outgrowth of germinating *C. perfringens* spores (43). However, ungerminated spores may remain viable in foods in the presence of preservation factors that prevent cell growth, and those spores may later undergo germination/outgrowth if the growth-limiting factor(s) is removed during food preparation.

RESERVOIRS FOR *C. PERFRINGENS* TYPE A

C. perfringens is present throughout the natural environment (31, 43), including soil (at levels of 10^3 to 10^4 CFU/g), foods (e.g., approximately 50% of raw or frozen meat contains some *C. perfringens*), dust, and the intestinal tract of humans and domestic animals (e.g., human feces usually contain 10^4 to 10^6 CFU/g). The widespread distribution of *C. perfringens* has been considered an important factor in the frequent occurrence of *C. perfringens* as an agent of foodborne illness. However, recent studies revealed that the ubiquitous distribution of *C. perfringens* in nature is not especially relevant for understanding the reservoir(s) for this foodborne pathogen. Notably, several independent surveys (17, 41, 75, 81) have revealed that <5% of global *C. perfringens* isolates harbor the *cpe* gene. Only CPE-positive *C. perfringens* isolates cause foodborne illness; hence, the fact that most *C. perfringens* isolates are *cpe* negative indicates that determining the specific locations where the small minority of enterotoxigenic *C. perfringens* isolates exist in the natural environment is essential for identifying the reservoirs of *C. perfringens* type A foodborne illness.

Moreover, it has become apparent that even identifying where *cpe*-positive isolates reside in nature is probably insufficient for determining *C. perfringens* type A food poisoning strains. Recent studies have revealed that the *cpe* gene can be located on either the chromosome or a plasmid (13, 14, 37); however, the *cpe* gene of most, if not all, human food poisoning isolates is located on the chromosome (Fig. 16.2). This strong association between foodborne illness and isolates that possess a chromosomal *cpe* provides a basis for more specifically identifying where food poisoning isolates reside in nature. The specificity of such testing of isolates will enable microbiologists to answer many critical questions regarding the ecology of *C. perfringens* food poisoning isolates.

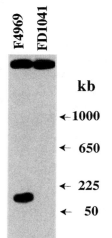

Figure 16.2 Pulsed-field gel electrophoresis distinguishes *C. perfringens* type A isolates carrying a chromosomal versus a plasmid *cpe* gene. Because of the large size of chromosomal DNA, *cpe*-containing DNA is unable to enter pulsed-field gels when an isolate carries its *cpe* gene on the chromosome. However, owing to the smaller size of plasmid DNA, some *cpe*-containing DNA does enter pulsed-field gels when an isolate carries its *cpe* gene on a plasmid. Shown are pulsed-field gel results with strain FD1041 (a food poisoning isolate carrying a chromosomal *cpe* gene) and F4969 (a nonfoodborne human gastrointestinal disease isolate carrying a plasmid *cpe* gene).

Questions to be answered include the following: Are those isolates present in low numbers in some healthy human carriers? Are they present in some food animals? Do they enter foods during processing? Do they enter foods during final handling, cooking, or holding? Unfortunately, such definitive surveys have not yet been conducted.

It is apparent that current knowledge regarding the principal reservoir(s) of *C. perfringens* type A food poisoning strains is severely deficient. Such information would be invaluable for controlling *C. perfringens* type A food poisoning, since it should provide insights into how and when food becomes contaminated with *C. perfringens* food poisoning isolates, thereby enabling a more focused strategy targeted toward intervention at locations in the food chain where strains responsible for foodborne illness exist.

C. PERFRINGENS TYPE A FOOD POISONING OUTBREAKS

Incidence
The most recent statistics from the Centers for Disease Control and Prevention (CDC) (63) indicate that *C. perfringens* type A food poisoning ranks as the third most commonly reported bacterial cause of foodborne disease

in the United States. From 1993 to 1997, 40 outbreaks (representing 4.1% of total bacterial foodborne disease outbreaks) of *C. perfringens* type A food poisoning occurred in the United States, involving 2,772 cases (6.3% of total cases of bacterial foodborne diseases). However, because most cases of *C. perfringens* type A food poisoning are not recognized or reported, the official CDC statistics significantly understate the true prevalence and impact of this foodborne disease. Recent estimates, which may be conservative, indicate that 250,000 cases of *C. perfringens* type A foodborne illness actually occur in the United States each year, causing an average of seven deaths per year (60). Economic costs associated with *C. perfringens* type A food poisoning are estimated to exceed $120 million (in 1989 dollars) (79, 80).

Identified outbreaks of *C. perfringens* type A foodborne illness are usually large (the average outbreak size is ~50 to 100 cases) and often occur in institutionalized settings (2). The large size of most recognized *C. perfringens* type A foodborne disease outbreaks is largely attributable to two factors. First, large institutions often prepare food in advance and then hold that food for later serving, which enables the growth of *C. perfringens* if the stored food is temperature abused. Second, given the relatively mild and nondistinguishing symptoms of most cases of *C. perfringens* type A foodborne illness, public health officials usually only become involved in investigating and reporting such foodborne illnesses when large numbers of people become ill. *C. perfringens* type A foodborne illness occurs throughout the year but is slightly more common during summer months (2), perhaps because higher ambient temperatures facilitate temperature abuse of foods during cooling and holding.

Food Vehicles for *C. perfringens* Foodborne Illness
Recent CDC statistics indicate that the most common food vehicles for *C. perfringens* type A foodborne illness in the United States are meats (notably beef and poultry) (63). Meat-containing products (e.g., gravies and stews) and Mexican foods also represent important vehicles for *C. perfringens* type A foodborne illness.

Contributing Factors Leading to *C. perfringens* Type A Foodborne Illness
C. perfringens type A foodborne illness usually results from temperature abuse during the cooking, cooling, or holding of foods. CDC reports that improper storage or holding temperatures were a contributor to 100% of all recent outbreaks of *C. perfringens* type A foodborne illness for which factors were identified (63). Improper cooking was identified as a factor in ~30% of

these outbreaks, whereas use of contaminated equipment contributed to ~15% of recent outbreaks.

The importance of temperature abuse in *C. perfringens* type A foodborne illness is not surprising, considering the relative heat tolerance of some *C. perfringens* vegetative cells and, more important, the heat resistance of spores produced by some *C. perfringens* isolates (68). Incomplete cooking can increase the likelihood of illness by inducing the germination of *C. perfringens* spores present in foods (43); if the vegetative cells resulting from those germinated spores are present in improperly cooled or stored foods, they can rapidly multiply to the populations necessary to cause foodborne illness.

Prevention and Control of *C. perfringens* Type A Foodborne Illness

Thoroughly cooking food is the best approach to prevent and control *C. perfringens* type A foodborne illness. This is particularly important for large roasts and turkeys, for which, because of their size, it is difficult to obtain the high internal temperatures needed to kill *C. perfringens* spores. The difficulty of cooking large roasts and whole poultry carcasses to such high internal temperatures in part explains why those foods are such common food vehicles for outbreaks of *C. perfringens* type A foodborne illness. A second, and perhaps even more important, step for preventing *C. perfringens* type A foodborne illness involves rapid cooling of cooked foods and then storing and serving these foods at nonpermissive conditions for vegetative growth of *C. perfringens* (e.g., either at refrigeration temperatures or at temperatures of >70°C).

Examples of Recent Outbreaks of *C. perfringens* Type A Foodborne Illness

In 1994, CDC reported the results of its investigation of two outbreaks of *C. perfringens* type A foodborne illness associated with St. Patrick's Day meals (11). The first outbreak occurred in Cleveland, Ohio, and involved 156 persons, whose illnesses were associated with ingesting corned beef prepared at a local delicatessen. During preparation, that corned beef had been boiled for 3 h; it was then allowed to cool slowly (at room temperature) before refrigeration. Four days later, portions of the corned beef were warmed to 48.8°C and served. Some sandwiches prepared with this corned beef were held at room temperature from the time when they were prepared (at 11:00 a.m.) until consumption (which occurred throughout the afternoon).

The second St. Patrick's Day outbreak occurred in Virginia, involving 86 persons who had attended a traditional St. Patrick's Day dinner, which also included corned beef. The corned beef, which was determined to contain large populations of *C. perfringens*, was a frozen, commercially prepared, brined product that had been thawed, cooked in large (~4.5-kg) pieces, stored in a refrigerator, and held for 90 min under a heat lamp before serving.

These two outbreaks illustrate the typical involvement of meat in *C. perfringens* type A foodborne illness, as well as the importance of temperature abuse as a contributing factor to this foodborne illness. In the outbreak associated with the Ohio delicatessen, the corned beef was cooled too slowly after cooking and was then reheated before serving to a temperature insufficient to kill *C. perfringens* cells. In the outbreak that occurred in Virginia, the beef was cooked in overly large portions and then was not adequately reheated prior to serving. The Virginia outbreak is atypical because it involved a commercially prepared, brined meat product.

An outbreak of *C. perfringens* type A foodborne illness in a Northern California juvenile detention center involved ~100 residents and staff who suffered the characteristic symptoms of diarrhea and cramping between 5 and 19 h after eating a Thanksgiving holiday meal (64). The meal consisted of turkey, gravy, mashed potatoes, melon, and milk, and the turkey was identified as the vehicle of illness. Further investigation revealed that the turkey was likely inadequately cooked, was stored under inadequate conditions after cooking, and was not adequately reheated before serving. This outbreak illustrates two important points: (i) *C. perfringens* type A foodborne illness outbreaks often involve multiple contributing factors, in particular, inadequate cooking and improper handling/storage conditions, and (ii) large poultry products are frequent vehicles for this type of foodborne illness.

Interestingly, in none of the three outbreaks described was CPE-positive *C. perfringens* isolated from either the implicated food or the feces of patients. In fact, investigators failed to attempt detection of CPE in the feces of any of the ill individuals. Identifying the presence of CPE in feces is particularly useful, because it can provide conclusive evidence that *C. perfringens* type A was responsible for illness.

Identification of *C. perfringens* Type A Foodborne Illness Outbreaks

Public health agencies traditionally have relied upon descriptive criteria, such as incubation time and symptoms of illness or type and history of food vehicles (e.g., is temperature-abused meat or poultry involved?), for identifying *C. perfringens* type A food-associated outbreaks. However, complete dependence on clinical and epidemiologic features for identifying such outbreaks is not well founded, considering the similarities that exist between the onset times and symptomology of *C. perfringens* type A foodborne illness and certain other foodborne illnesses, such as the diarrheal form of *Bacillus cereus* food poisoning.

Many public health agencies now include laboratory analyses to more reliably identify outbreaks of *C. perfringens* type A foodborne illness. Bacteriologic criteria used by CDC to identify an outbreak include demonstrating the presence of either (i) 10^5 *C. perfringens* organisms per gram of stool from two or more ill persons or (ii) 10^5 *C. perfringens* organisms per gram of epidemiologically implicated food (63). Methods for enumerating *C. perfringens* in food are described in the U.S. Food and Drug Administration's *Bacteriological Methods* (66).

Merely demonstrating the presence of *C. perfringens* in suspect food or feces from ill individuals often is not sufficient to unequivocally identify an outbreak of *C. perfringens* type A foodborne illness. *C. perfringens* is widely distributed in the environment, including food or feces, often at high populations (50). Most strains of *C. perfringens* in food or feces are *cpe* negative and hence are unable to cause foodborne illness.

CDC and the Food and Drug Administration now use the detection of CPE in feces of ill individuals as a diagnostic criterion to identify *C. perfringens* type A food poisoning outbreaks (18a, 63). Several commercially available serologic assays are available for fecal CPE detection (43, 50), including reverse-passive latex agglutination assay (Oxoid, Basingstoke, England) and a rapid enzyme-linked immunosorbent assay (Tech Lab, Blacksburg, Va.).

Since CPE also can be present in feces of people suffering from nonfoodborne GI diseases (such as antibiotic-associated diarrhea), determining the presence of CPE in feces of a single individual is not sufficient to identify *C. perfringens* type A food poisoning. However, detection of CPE in feces from several individuals does provide strong evidence for a *C. perfringens* type A food poisoning outbreak, particularly when those individuals consumed a common food, developed illness within typical incubation times, and presented with the characteristic symptoms of *C. perfringens* type A food poisoning. The use of fecal CPE detection approaches for identifying *C. perfringens* food poisoning outbreaks is limited by the fact that fecal samples must be collected soon after the onset of food poisoning symptoms to ensure meaningful results (1).

In theory, demonstrating the presence of *C. perfringens* food poisoning isolates in foods or feces represents an alternative or supplemental approach for diagnosing

C. perfringens type A food poisoning outbreaks. However, there are several factors that limit the use of this criterion for epidemiologic purposes. First, while CPE serologic detection assays can be used to evaluate the enterotoxigenicity of a food or fecal isolate, such assays are limited by the requirement that isolates must sporulate in vitro (CPE expression is sporulation associated), which is often difficult to achieve under laboratory conditions (12, 13, 41, 43). However, even if in vitro sporulation can be obtained, demonstrating CPE expression by a *C. perfringens* isolate does not establish that the isolate carries the chromosomal *cpe* gene that is specifically associated with food poisoning. Most simple *cpe* gene detection assays (e.g., simple PCR assays) are unable to distinguish between isolates with plasmid or chromosomal *cpe*. Furthermore, the recent discovery of silent, plasmid-borne *cpe* sequences in most or all *C. perfringens* type E isolates (3) raises additional concerns about relying upon simple *cpe* gene detection approaches for diagnosing food poisoning. Fortunately, two genotyping approaches, *cpe* Southern blot analyses of DNA subjected to either pulsed-field gel electrophoresis (PFGE) or restriction fragment length polymorphism (RFLP) analysis, can specifically identify isolates carrying the chromosomal *cpe* gene linked to food poisoning (12, 13). These approaches hold promise as powerful epidemiologic tools.

Public health agencies also have begun to use a related genetic approach to study the epidemiology of *C. perfringens* type A food poisoning events. DNAs obtained from different *C. perfringens* isolates yield different RFLP patterns when digested with rare base-cutting restriction endonucleases and stained with ethidium bromide; hence, protocols have recently been developed (47) which apply similar PFGE/RFLP analyses in the investigation of *C. perfringens* type A food poisoning outbreaks. For example, if several people with symptoms of *C. perfringens* type A food poisoning became ill from a similar isolate, this would imply they consumed the same contaminated food.

In summary, the laboratory plays an increasingly important role in identifying *C. perfringens* type A food poisoning outbreaks. Within the proper epidemiologic or clinical context, demonstrating that CPE is present in feces of several ill individuals provides compelling evidence for the occurrence of a *C. perfringens* type A food poisoning outbreak.

CHARACTERISTICS OF *C. PERFRINGENS* TYPE A FOODBORNE ILLNESS

Symptoms of *C. perfringens* type A food poisoning generally develop about 8 to 16 h after ingestion of contaminated food (43, 56) and then resolve spontaneously within the next 12 to 24 hours. Victims of *C. perfringens* type A food poisoning usually suffer only diarrhea and severe abdominal cramps; vomiting and fever are not typical features. While death rates from *C. perfringens* type A food poisoning are low, fatalities do occur in debilitated or elderly people.

The typical pathogenesis of *C. perfringens* type A food poisoning is illustrated in Fig. 16.3. Initially, as a result of temperature abuse, vegetative cells of enterotoxigenic *C. perfringens* rapidly multiply in food; these bacteria are then consumed with the contaminated food. Many of the ingested *C. perfringens* vegetative cells are likely killed when exposed to stomach acidity (31), but if the food vehicle is sufficiently contaminated, some vegetative cells survive passage through the stomach and remain viable when they enter the small intestine, where they multiply and sporulate. A low-M_r sporulation factor, produced by both CPE-negative and CPE-positive strains, may promote in vivo sporulation (70). CPE is expressed during the sporulation of *C. perfringens* cells in the small intestines. After being released into the intestinal lumen, CPE quickly binds to intestinal epithelial cells and exerts its action, which induces morphologic damage to intestinal epithelial cells (Fig. 16.3). This CPE-induced intestinal tissue damage initiates intestinal fluid loss, which is clinically manifested as diarrhea.

Two factors help explain why most cases of *C. perfringens* type A food poisoning are usually relatively mild and self-limited (69): (i) the diarrhea associated with *C. perfringens* type A food poisoning likely mitigates the severity of illness by flushing unbound CPE and many *C. perfringens* cells (containing additional unreleased CPE) from the small intestine, and (ii) CPE preferentially affects villus tip cells, which are the oldest intestinal cells and are rapidly replaced in young, healthy individuals by normal turnover of intestinal cells.

INFECTIOUS DOSE AND SUSCEPTIBLE POPULATIONS FOR *C. PERFRINGENS* TYPE A FOODBORNE ILLNESS

C. perfringens cells are susceptible to killing when exposed to stomach acidity (31); hence, cases of *C. perfringens* type A food poisoning usually develop only when a heavily contaminated food (i.e., a food containing >10^6 to 10^7 *C. perfringens* vegetative cells per gram of food) is consumed (50). The enterotoxin responsible for disease

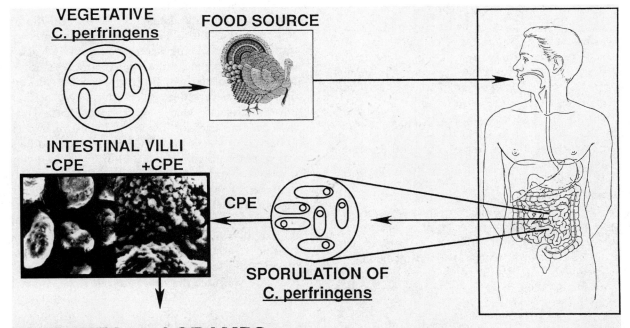

DIARRHEA and CRAMPS

Figure 16.3 The pathogenesis of *C. perfringens* type A food poisoning. Vegetative *C. perfringens* cells multiply rapidly in contaminated food (usually a meat or poultry product) and, after ingestion, sporulate in the small intestine. Sporulating *C. perfringens* cells produce an enterotoxin, CPE, that causes morphologic damage to the small intestine, resulting in diarrhea and abdominal cramps. Reproduced with permission from reference 50.

symptoms of *C. perfringens* type A food poisoning is usually produced in vivo when *C. perfringens* isolates sporulate in the intestines. Therefore, this illness is considered an infection, not an intoxication. A few reports that describe the onset of early symptoms are consistent with the possibility that preformed CPE in foods might occasionally contribute to this food poisoning (43). Nevertheless, the typically long incubation period of this type of food poisoning (despite the quick action of CPE) indicates that the involvement of preformed CPE in *C. perfringens* type A food poisoning symptoms must be rare.

Everyone is susceptible to *C. perfringens* type A food poisoning; however, this illness tends to be more serious in elderly or debilitated individuals (56). Many individuals develop at least a transient serum antibody response to CPE following illness (4), but there is no evidence to indicate that previous exposure to this type of food poisoning provides significant future protection (31).

VIRULENCE FACTORS CONTRIBUTING TO *C. PERFRINGENS* TYPE A FOODBORNE ILLNESS

Heat Resistance

Most (if not all) cases of food poisoning are caused by *C. perfringens* isolates carrying a chromosomal *cpe* gene. This association between chromosomal *cpe* isolates and food poisoning was initially puzzling, considering that *C. perfringens* isolates carrying chromosomal or plasmid *cpe* genes both expressed similar levels of an identical CPE protein, which is the toxin responsible for the GI symptoms of *C. perfringens* type A food poisoning (12). A recent study (68) suggests that the strong association between chromosomal *cpe* isolates and food poisoning may be largely attributable to the spores and vegetative cells of isolates with chromosomal *cpe* possessing greater heat resistance than the cells and spores of isolates with plasmid *cpe* . Since cooked meat products are the food vehicle causing most *C. perfringens* type A food poisoning outbreaks, the greater heat resistance of isolates with chromosomal *cpe* favors their survival during incomplete cooking or inadequate holding of foods, which are the two major factors contributing to *C. perfringens* type A foodborne illness.

CPE

Evidence that CPE Is Involved in *C. perfringens* Type A Foodborne Illness

Epidemiologic studies have provided the following strong evidence that CPE plays a major role in *C.*

perfringens type A foodborne illness. (i) A strong positive correlation exists between illness and the presence of CPE in a victim's feces. Depending on the sensitivity of the assay used and how quickly the fecal sample was collected after the onset of symptoms, 80 to 100% of feces from individuals ill with *C. perfringens* type A food poisoning test CPE positive, whereas virtually no feces from well individuals test CPE positive (1, 4). (ii) CPE is often present in feces of food poisoning victims at levels (1, 4) known to cause serious intestinal effects in laboratory animals (69). (iii) Human volunteers fed highly purified CPE develop the characteristic symptoms of *C. perfringens* type A foodborne illness (73). (iv) CPE-positive *C. perfringens* food poisoning isolates are dramatically more effective than CPE-negative *C. perfringens* isolates at producing either fluid accumulation in rabbit ileal loops or diarrhea in human volunteers (77). (v) Rabbit ileal loop effects caused by CPE-positive isolates can be neutralized with CPE-specific antisera (29).

More recently, the importance of CPE for the GI pathogenesis of *C. perfringens* food poisoning isolates received compelling support from experiments (67) that fulfilled molecular Koch's postulates. Those experiments (Fig. 16.4) revealed that sporulating (but not vegetative) culture lysates of a wild-type CPE-positive food poisoning strain could induce both fluid accumulation and histopathologic damage in rabbit ileal loops, a result consistent with CPE (whose expression is sporulation associated) being necessary for the GI activity of that food poisoning isolate. Furthermore, neither vegetative nor sporulating culture lysates of a *cpe* knockout mutant (constructed by allelic exchange) of that food poisoning isolate were able to induce either intestinal fluid accumulation or histopathologic damage. The *cpe* knockout mutant's loss of GI virulence could be specifically attributed to inactivation of its *cpe* gene, because it was shown that full GI virulence was restored when that mutant was complemented with a shuttle plasmid carrying the wild-type *cpe* gene.

C. perfringens type A food poisoning is not the only disease involving CPE. CPE-positive *C. perfringens* isolates are also responsible for several nonfoodborne human GI illnesses, including antibiotic-associated diarrhea and sporadic diarrhea, as well as some veterinary diarrheas (10, 52). Molecular Koch's postulate studies (67) using a *cpe* knockout mutant of a nonfoodborne human GI disease isolate have also confirmed the importance of CPE expression for the pathogenesis of nonfoodborne human GI diseases. However, the *cpe*-positive *C. perfringens* isolates causing nonfoodborne GI diseases are genotypically distinct from those causing food poisoning; i.e., *cpe* is located on the chromosome of food

SM101, FTG

SM101, DS

MRS101, DS

MRS101(pJRC200), DS

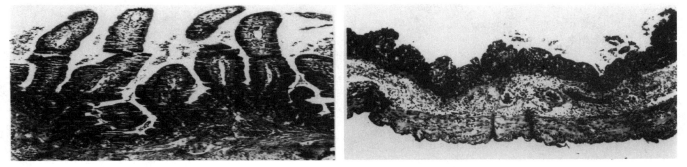

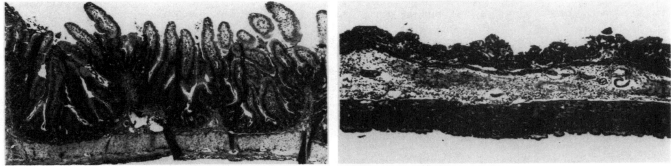

Figure 16.4 Fulfilling molecular Koch's postulates demonstrates that CPE is important for the gastrointestinal virulence of *C. perfringens* type A food poisoning isolates. Tissue specimens shown were collected from rabbit ileal loops treated with either concentrated vegetative (FTG) or sporulating (DS) culture lysates of *C. perfringens* strain SM101, an electroporatable derivative of food poisoning strain NCTC 8798; MRS101, which is a *cpe* knockout mutant of SM101; or MRS101(pJRC200), which is the MRS101 mutant complemented with a shuttle plasmid carrying the cloned wild-type *cpe* gene. Note that (i) tissue specimens treated with concentrated FTG lysates of SM101 (or its derivatives) were indistinguishable from control ileal loop specimens, and (ii) fluid accumulation was observed only in loops treated with DS culture lysates of SM101 or MRS(pJRC200). Reprinted with permission from reference 67.

poisoning isolates but is present on a plasmid in nonfoodborne disease isolates (13, 14, 37).

The Genetics of CPE

Cloning and sequencing of the intact *cpe* gene provided tools (e.g., *cpe*-specific gene probes) that have substantially increased knowledge of *cpe* genetics (16). An example of these advances was the discovery that the *cpe* gene is present in only ~5% of *C. perfringens* isolates, most of which are type A (17, 41, 75, 81). The presence of the *cpe* gene in only a small subset of global *C. perfringens* isolates suggests that *cpe* might be transferable by mobile genetic elements. Southern blot analyses of PFGE and RFLP gels (13, 14) have confirmed an association between mobile genetic elements and the *cpe* gene by demonstrating that the *cpe* gene of nonfoodborne human GI disease isolates, and of most animal

isolates, is located on a plasmid (Fig. 16.2). Recent studies (7) have revealed that the *cpe*-carrying plasmid can be transferred, apparently by conjugation, to naturally *cpe*-negative *C. perfringens* isolates.

Some evidence suggests that the chromosomal *cpe* of food poisoning isolates may also be associated with a mobile genetic element. For example, the *cpe* gene of food poisoning isolate NCTC 8798 was mapped by PFGE studies to a highly variable region of the chromosome, which is consistent with *cpe* of that strain being present on a phage or transposon that has integrated into a chromosomal "hot spot" (9). More recently, sequencing analysis of food poisoning strain NCTC 8239 (8) revealed that IS*1470* sequences are present upstream and downstream of the *cpe* open reading frame (ORF), suggesting that NCTC 8239 *cpe* is located on a 6.3-kb transposon with terminal IS*1470* elements. Interestingly, different IS elements are associated with the plasmid *cpe* gene. A follow-up study revealed that the putative 6.3-kb *cpe*-containing transposon, named Tn*5555*, may have a circular intermediate form (5).

While the *cpe* ORF sequences encoded by both the plasmid and chromosomal *cpe* gene of *C. perfringens* type A isolates appear to be identical (12), most, if not all, *C. perfringens* type E isolates carry *cpe* sequences that are not expressed owing to the presence of numerous mutations in the ORF, promoter(s), and ribosome binding site of their *cpe* sequences (3). The implication of these silent type E *cpe* sequences for simple *cpe* detection assays has been described previously.

Expression and Release of CPE

There are at least three interesting features of the expression and release of CPE from *C. perfringens*: (i) CPE expression is tightly regulated, i.e., this toxin is strongly expressed by sporulating, but not vegetative, *C. perfringens* cells; (ii) during sporulation, many CPE-positive isolates express extremely large amounts of this toxin; and (iii) CPE is not actually secreted by sporulating *C. perfringens* cells but is instead released into the intestines when the mother cell lyses at the completion of sporulation.

Regulation of CPE Synthesis

Classic studies by Duncan et al. in the 1960s and 1970s first established the relationship between CPE expression and sporulation (56). They determined that *C. perfringens* mutants blocked at stage 0 of sporulation completely lose their ability to produce CPE (18). This putative relationship between sporulation and CPE expression was later confirmed by using highly sensitive and specific Western immunoblots (Fig. 16.5) (16, 41), which revealed that vegetative *C. perfringens* cells produce only trace amounts of CPE (probably as a result of "leaky" gene regulation), but sporulating cells of that same *C. perfringens* strain express at least 1,500-fold more CPE.

The sporulation-associated pattern of CPE expression raises an interesting question: Is CPE expression repressed (negatively regulated) in vegetative *C. perfringens* cells or is it positively regulated (activated) during sporulation? One approach used to address this question has been to test CPE expression by an *Escherichia coli* (a nonsporulating organism) clone transformed with a shuttle plasmid carrying *cpe* (15). Although many nonsporulation-associated clostridial toxins are expressed by *E. coli*, an *E. coli* transformant carrying more than 50 copies of the *cpe*-containing shuttle plasmid did not express detectable levels of CPE. Presuming it unlikely that *E. coli* produces a *cpe*-specific repressor, these results provide evidence that CPE expression is not negatively

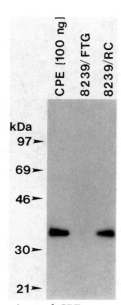

Figure 16.5 Comparison of CPE expression between vegetative and sporulating *C. perfringens* cultures. Western immunoblot results for CPE detection are shown. CPE, 100 ng of purified CPE; 8239/FTG, 80 μl of a lysate from an 8-h vegetative culture of enterotoxigenic *C. perfringens* strain NCTC 8239; 8239/RC, 2 μl of a lysate from an 8-h sporulating culture of NCTC 8239. Western immunoblotting was performed as described in reference 41. Note: In order to visualize trace CPE expression by vegetative cultures, special conditions (e.g., long autoradiography, lysate concentration) must be used (16).

regulated during *C. perfringens* vegetative growth and that CPE expression likely involves positive regulation during sporulation of *C. perfringens*.

The ability of naturally *cpe*-negative *C. perfringens* types A, B, and C isolates transformed with a *cpe*-containing shuttle plasmid to express CPE has also been evaluated (15). Although no CPE expression was detected during vegetative growth of the *C. perfringens* transformants, all three transformants produced CPE during sporulation. These results indicate that most, if not all, *C. perfringens* isolates produce some (or all) of the regulatory factor(s) involved in modulating normal, sporulation-associated CPE expression. The apparent widespread distribution of this regulatory factor(s) among *C. perfringens* isolates suggests that these regulators may function globally, i.e., they may help regulate transcription of other genes in *C. perfringens*.

One potential global regulator, Hpr, has already been associated with sporulation regulation of CPE expression (6). Hpr-like binding sequences have been identified upstream and downstream of the *cpe* ORF. Furthermore, DNA isolated from several *C. perfringens* strains hybridizes with an *hpr*-specific gene probe. Since Hpr is known to repress expression of some proteins during exponential growth of *Bacillus subtilis*, the putative involvement of Hpr in CPE expression may help explain the lack of CPE expression during vegetative growth.

Synthesis of CPE

Synthesis of CPE begins shortly after the induction of sporulation and progressively increases for at least the next 6 to 8 h (61, 74). After 6 to 8 h of sporulation, CPE can represent up to 15% (16), or even 30% (44), of the total cell protein present in sporulating cells of some *C. perfringens* isolates.

Why do some *C. perfringens* strains produce so much CPE during sporulation? The amount of CPE expressed by an isolate is not influenced by whether that isolate carries a chromosomal or plasmid *cpe* gene (12). Nor can the strong enterotoxin expression exhibited by some *C. perfringens* isolates be attributed to a gene dosage effect, because all *cpe*-positive isolates apparently carry a single copy of the *cpe* gene (13, 14). There is a general relationship between an isolate's sporulation ability and its CPE production; i.e., the better a *C. perfringens* isolate sporulates, the more CPE is produced. However, that correlation is not absolute, suggesting that, while sporulation is necessary, other unrecognized factors also modulate CPE expression levels (12).

Considerable attention has recently been focused on understanding how transcriptional regulation contributes to CPE expression levels. RNA slot blot studies and Northern blot analyses (15, 61, 90) have confirmed that CPE expression is regulated at the transcriptional level, with significant amounts of *cpe* mRNA produced during sporulation but little or no *cpe* mRNA produced during vegetative growth of *C. perfringens*. Northern blot studies (15) revealed that *cpe* mRNA is transcribed as a monocistronic message of ~1.2 kb, which would be consistent with primer extension analysis studies indicating that *cpe* mRNA transcription starts ~200 bp upstream of the CPE translation start site (61). Subsequent primer extension studies, RNase T$_2$ protection studies, and deletion mutagenesis analyses (90) identified at least three start sites (P1, P2, and P3) for the initiation of *cpe* mRNA transcription. Interestingly, P1 shares some homology with SigK-dependent promoters, whereas P2 and P3 share some homology with SigE-dependent promoters (SigE and SigK are sporulation-associated sigma factors active in mother cells during sporulation of *B. subtilis*).

Posttranscriptional effects may also help regulate CPE expression levels. For example, the functional half-life of *cpe* mRNA in sporulating *C. perfringens* cells is 58 min, which is unusually long-lived for a bacterial message (45). Such exceptional message stability could contribute to the abundant CPE expression noted for sporulating cells of some *C. perfringens* strains. Given that stem-loop structures are believed to contribute to message stability, the putative stability of *cpe* mRNA could result from a stem-loop structure that appears to lie 36 bp downstream of the 3' end of the *cpe* ORF (16). This stem-loop structure is followed by an oligo(dT) tract, suggesting it also functions as a rho-independent transcriptional terminator (16), which would be consistent with the transcriptional start sites identified (90) 200 bp upstream of the *cpe* initiation codon and the 1.2-kb size of *cpe* mRNA observed in Northern blot studies.

Release of CPE from C. perfringens

Unlike most *C. perfringens* toxins, CPE is not secreted; i.e., it is not an exotoxin in the classic sense (16). Consistent with that, *cpe* does not encode the 5' signal peptide often associated with secreted toxins (16). After synthesis, CPE accumulates in the cytoplasm of the mother cell, where it can reach sufficiently high concentrations to induce formation of cytoplasmic CPE-containing paracrystalline inclusion bodies (Fig. 16.1) (43). Intracellular CPE is eventually released into the intestines at the completion of sporulation, when the mother cell lyses to free its mature spore. Dependence upon mother cell lysis for CPE release explains, at least in part, why (despite CPE's quick intestinal action) *C. perfringens* type A food poisoning symptoms develop 8 to 24 h after ingestion of contaminated foods; i.e., sporulating *C. perfringens* cells must complete sporulation, which takes at least 8 to 12 h (43), before CPE can be released into the intestine to produce its effects.

CPE Biochemistry

CPE was initially purified and characterized in the early 1970s (56). Results from early studies revealed that the enterotoxin is a single polypeptide of ~35,000 Da, with an isoelectric point of 4.3. Later, *cpe* ORF sequencing studies demonstrated that the *cpe* protein is composed of 319 amino acids, with an M_r of 35,317 (16). Additional *cpe* ORF sequencing revealed that the CPE sequence is highly conserved among different CPE-positive *C. perfringens* type A isolates; i.e., the *cpe* ORF nucleotide sequences of eight different type A isolates, including five isolates carrying a chromosomal *cpe* gene and three isolates carrying a plasmid *cpe* gene, were determined to be identical. This highly conserved CPE consensus sequence lacks significant homology with other proteins, except for some limited homology with a nonneurotoxic protein produced by *Clostridium botulinum* (30). The significance of that limited homology remains unclear.

CPE is not a heat-stable enterotoxin; its biologic activity can be inactivated by heating for 5 min at 60°C (56).The toxin is also quite sensitive to pH extremes but is resistant to some proteolytic treatments (56). In fact, limited trypsinization or chymotrypsinization actually increases CPE activity by about two- to threefold (22, 23, 26), suggesting that intestinal proteases may activate CPE during food poisoning. However, no direct in vivo evidence supporting this hypothesis has been presented.

CPE Action

Considerable progress has been achieved toward understanding the action of CPE during *C. perfringens* type A food poisoning. These findings indicate that the in vivo and in vitro effects of CPE represent a novel mechanism of action for a bacterial enterotoxin.

CPE Effects on the GI Tract

CPE is classified as an enterotoxin because it induces fluid and electrolyte losses from the GI tract of many mammalian species (56). The principal target organ for CPE is believed to be the small intestine, with animal model studies indicating that the ileum is particularly sensitive to this toxin (56). Interestingly, the rabbit colon is insensitive to CPE, despite the fact that the enterotoxin binds well to rabbit colonic cells (56). Whether the insensitivity of rabbit colon to CPE treatment indicates that the human colon is also unaffected by this enterotoxin remains to be determined.

Several features distinguish the biologic activity of CPE from that of cholera and *E. coli* heat-labile enterotoxins (56): (i) CPE does not increase intestinal cyclic AMP levels, (ii) CPE inhibits glucose absorption, and (iii) CPE induces direct histopathologic damage to the small intestine (Fig. 16.3), with villus tips being particularly sensitive. While some other bacterial enterotoxins (e.g., Shiga toxin and *Clostridium difficile* toxins) are also cytotoxic and cause intestinal tissue damage, CPE is unique with respect to how quickly it elicits intestinal tissue damage; i.e., CPE-induced intestinal damage can develop within as little as 15 to 30 min (69).

Two observations strongly indicate that CPE-induced tissue damage (Fig. 16.4) plays a major role in initiating CPE-induced fluid/electrolyte intestinal transport alterations (58, 69). First, the onset of fluid transport changes closely coincides with the development of tissue damage in CPE-treated rabbit ileum. Second, only those CPE doses that produce tissue damage can induce intestinal fluid and electrolyte transport alterations in the rabbit ileum. Given those two observations, it appears that CPE's intestinal effects during *C. perfringens* type A foodborne illness occur when the enterotoxin induces tissue damage, which disrupts villus integrity and induces a breakdown of the normal intestinal secretion/absorption equilibrium. This effect is clinically manifested as diarrhea.

Recent studies suggest that CPE may also affect the paracellular permeability properties of the intestinal epithelium (76), which could contribute to intestinal fluid/electrolyte secretion in CPE-treated intestinal tissue. Finally, the ability of CPE to induce significant changes in levels of some proinflammatory cytokines (42, 82) raises the possibility that inflammation might be another contributor to CPE's intestinal effects, particularly later in the illness.

The Cellular and Molecular Basis of CPE Action: an Overview of Early Steps

Recent CPE studies have focused primarily on understanding the early events in CPE action, with the rationale that these should be the primary events leading to CPE's cytotoxic effects on mammalian cells, which appear to be responsible for the tissue damage that seemingly initiates CPE's intestinal effects (49). A model of the initial steps in CPE's biologic activity is presented in Fig. 16.6. This model predicts that CPE acts on the intestine

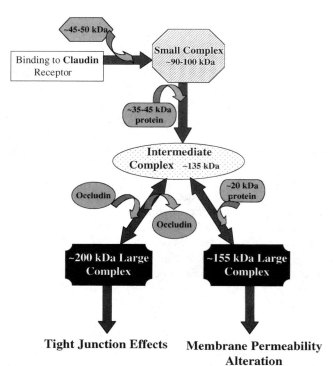

Figure 16.6 A current model for early events in CPE action. This model reflects the four known early events in CPE's action: (i) binding of CPE to its receptor, (ii) formation of a small (90-kDa) CPE-containing complex, which entraps CPE on the membrane surface, (iii) formation of several intermediate to large CPE-containing complexes, and (iv) consequences of that intermediate to large complex formation. Note that steps iii and iv do not occur at low temperature. See text for full discussion of each event.

in a multistep process that may involve at least five early events. CPE begins its enterotoxic activity when the toxin binds to its intestinal receptor(s); CPE binding induces formation of a 90-kDa small complex, which entraps CPE on the membrane surface. The CPE-containing small complex may then interact with other proteins to form an intermediate-size complex, which can interact with either occludin (an ~65-kDa tight junction protein) or another unidentified eucaryotic protein to form ~200-kDa and ~155-kDa complexes, respectively. Formation of the ~155-kDa complex causes the plasma membrane to lose its normal permeability characteristics, possibly because that complex has porelike properties. Those membrane permeability alterations collapse the colloid-osmotic equilibrium of the CPE-treated cell, leading to morphologic damage (Fig. 16.7) and cell death. Meanwhile, formation of the ~200-kDa complex might be inducing tight junction rearrangements, which could result in paracellular permeability changes in the CPE-treated intestinal epithelium.

This model emphasizes the uniqueness of CPE cellular/molecular activity by predicting that CPE closely interacts with eucaryotic proteins at every early step of its action. No other membrane-active toxin is known to be so interactive with different eucaryotic proteins in its mechanism of action.

Step 1 in the Molecular Action of CPE: Receptor Binding

The initial binding of CPE to mammalian plasma membranes exhibits characteristics of a receptor-mediated process. For example, CPE binding is specific; i.e., native CPE competes against binding of ^{125}I-CPE to membranes (55). Furthermore, CPE binding is saturable, with estimates indicating the existence of ~10^6 CPE receptors per cell (33, 49, 59, 86). The binding of CPE to its receptor(s) is rapid and temperature sensitive (less CPE binding occurs at 4°C than at 37°C [49]) and appears to be essential for the efficient initiation of CPE cytotoxicity; e.g., cell types (such as CHO cells) that do not have a receptor to specifically bind CPE also do not respond to treatment with physiologic doses of CPE (33, 49, 84). The small intestines of most mammalian species (49, 55, 78) can specifically bind CPE, which helps to explain why CPE

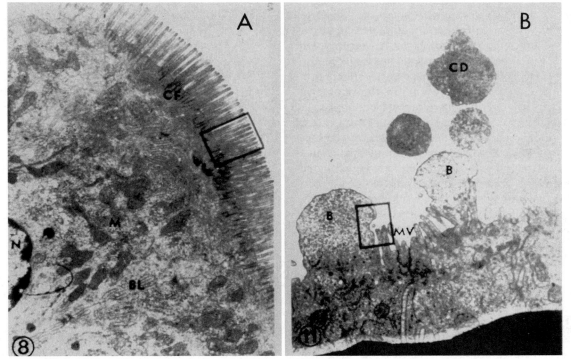

Figure 16.7 Morphologic damage to rabbit small intestinal cells in response to treatment with CPE. (A) Control specimen. An intestinal epithelial cell with normal brush border membranes (box). (B) CPE-treated intestinal cells show visible damage, including bleb (B) formation, to their brush border membranes. Brush border membrane damage precedes the development of damage to internal organelles in CPE-treated cells (57). Reproduced with permission from reference 57.

induces fluid and electrolyte losses from so many mammalian species. Many nonintestinal cells also express CPE receptors (33, 49, 59). However, the CPE receptor is not essential for survival of mammalian cells, because certain cell types do not bind CPE (33, 84). Finally, kinetic analyses of CPE binding have yielded contradictory conclusions as to the existence of single versus multiple CPE receptors (33, 54, 55, 59, 76).

Early studies revealed that protease pretreatment of cells or isolated intestinal brush border membranes (BBMs) destroys the subsequent ability of those cells or membranes to bind CPE (51, 55, 86), indicating that the CPE receptor(s) is proteinaceous. Identification of a protein(s) acting as a CPE receptor was initially studied using biochemical approaches, which implicated a ~45- to 50-kDa mammalian membrane protein as the CPE receptor in BBMs and Vero cells (a very CPE-sensitive cell line). For example, early affinity chromatography studies revealed that a ~45- to 50-kDa protein present in BBMs or Vero cell detergent extracts binds to a column containing CPE immobilized to Sepharose 4B (78, 87, 88). Subsequently, Wieckowski et al. (84) determined that a ~90-kDa complex (now known as small complex) forms upon binding of CPE to cell types capable of binding and responding to CPE, suggesting that small complex represents the binding of CPE to its functional receptor(s). Immunoprecipitation analysis of two CPE-sensitive cell lines revealed that this small complex contains, at a minimum, one CPE molecule and one ~45- to 50-kDa eucaryotic membrane protein (presumably the same protein isolates by affinity chromatography techniques).

More recently, Katahira et al. (35, 36) performed expression cloning studies that identified certain claudins as functional CPE receptors. They determined that transfection of human cDNAs encoding either of two related 22-kDa membrane proteins into L cells, which are normally CPE insensitive because they cannot bind CPE, confers both CPE binding capability and CPE sensitivity to the resultant transfectants. These two 22-kDa CPE receptors subsequently were identified as claudins-3 and -4 (20). The claudin family includes at least 18 related four-transmembrane-domain proteins which play important structural roles in epithelial tight junctions. Claudins-3, -4, -6, -7, -8, and -14, but not claudins-5 and -10, bind CPE (19). CPE binding to claudin-3 is mediated by the second extracellular loop of that tight junction protein.

How might the apparent conflict between biochemical versus expression cloning data regarding the identity of CPE receptors be resolved? One possible explanation could be the existence of coreceptors; i.e., CPE might bind to both a claudin and a ~45- to 50-kDa protein. The coreceptor hypothesis is supported by studies

revealing that antibodies directed against a FLAG epitope present on recombinant claudin-3 can immunoprecipitate a ~45- to 50-kDa protein from lysates of mouse L cells expressing the tagged claudin, a result consistent with the ~45- to 50-kDa protein interacting with certain claudins prior to CPE treatment (35).

Finally, it has been proposed (36, 76) that claudin-3 and claudin-4 correspond to the multiple CPE receptors putatively identified in some kinetic studies. However, that assertion remains debatable, because (i) multiple claudins are now known to bind CPE (19) and (ii) the kinetic values measured for CPE binding to claudin-3 and claudin-4 transfectants do not precisely correspond to those measured using intact Vero cells (59); i.e., the issue of which CPE receptor(s) or coreceptor(s) is physiologically important remains unresolved.

Step 2 in the Molecular Action of CPE: Small Complex Formation

The event(s) responsible for formation of the ~90-kDa CPE-containing small complex that forms upon CPE binding to sensitive cells remains unclear. As depicted in Fig. 16.6, formation of the 90-kDa small complex could involve CPE binding to a claudin receptor, followed by that CPE-receptor complex interacting with a ~45- to 50-kDa protein. Alternatively, CPE might bind independently to multiple receptors, e.g., claudins or a ~45- to 50-kDa protein, which then interact with other CPE receptor(s)/binding proteins. A final possibility, already alluded to, would involve small complex resulting from CPE binding to a coreceptor consisting of both claudins and a ~45- to 50-kDa protein.

When present in small complex formation, CPE is exposed on the membrane surface, as evident from experiments indicating that CPE localized in small complex remains fully accessible to externally applied antibodies and proteases (40, 83). Despite being exposed on the surface of membranes or cells, CPE localized in small complex does not readily dissociate from membranes or cells, even when small complex-containing cells or membranes are incubated for long periods in the presence of chaotropic salts or EDTA, which are agents known to facilitate release of peripherally bound proteins (49). Additionally, kinetic studies (51, 54) have revealed that the association of CPE with membranes at 4°C, at which all bound CPE is present in small complex, involves a rapid two-step process, i.e., binding followed rapidly by a second process. Collectively, these observations indicate that the process of small complex formation involves a postbinding conformational change to one or more small complex proteins, which entrap CPE on the membrane surface.

Step 3 in the Molecular Action of CPE: Formation of Intermediate and Large Complexes

It is now well established (35, 38–40, 53, 72, 84, 88) that, under physiologic conditions, e.g., at 37°C, CPE-containing small complex rapidly becomes associated with additional complexes larger than the ~90-kDa small complex. In Caco-2 cells, those other identified CPE-containing species can include an intermediate (~135 kDa) and two large (~155 kDa and ~200 kDa) complexes (Fig. 16.8) (72). Under physiologic conditions, both the small complex and some or all intermediate and large complexes can exist simultaneously in the same plasma membrane (84). However, small complex is easily distinguished from the intermediate and large complexes on the basis of size, stability in sodium dodecyl sulfate (SDS) (intermediate and large complexes are stable in SDS, whereas small complex dissociates in the presence of SDS), and temperature sensitivity (formation of intermediate and large complexes is blocked at 4°C, whereas small complex forms at both low and physiologic temperatures). The two large complex species, and likely the ~135-kDa intermediate species, are urea sensitive, suggesting these complexes are not held together by covalent bonds (72).

Small complex appears to be a precursor for formation of the intermediate and large complexes. For example, intermediate and large complex formation occurs quickly when Vero cells treated with CPE at 4°C, at which all toxin is in small complex, are quickly shifted to 37°C (53). At physiologic temperatures, Caco-2 cells rapidly form all four CPE-containing complexes, i.e., the ~90-kDa small complex, the ~135-kDa intermediate complex, and the ~155-kDa and ~200-kDa large complexes (72); after 30 min of CPE treatment at 37°C, most CPE becomes localized in intermediate and large complexes, with the ratio of the ~135-kDa, ~155-kDa, and ~200-kDa complexes being 10:65:25 (72).

Kinetic analyses (72) indicate that levels of the ~135-kDa intermediate complex slightly increase when Caco-2 cells are treated for an extended period of time with CPE at 37°C. The increased amount of ~135-kDa complex coincides with a slight decrease in the amount of the ~155-kDa and ~200-kDa large complex species present in those cells, consistent with the ~135-kDa complex species representing a possible intermediate for formation of both large complex species. The possibility that the ~135-kDa complex may serve as an intermediate complex for formation of the larger complexes is further supported by observations that CPE mutants containing a single point mutation are blocked for formation of all three intermediate and large complex species, even though those mutants can bind and form small complex (72).

The identity of the eucaryotic protein constituents of intermediate and large CPE-containing complexes is incompletely understood. However, recent studies (72) have revealed that the ~200-kDa complex contains occludin, which is a ~65-kDa tight junction protein. However, occludin is not present in the small complex,

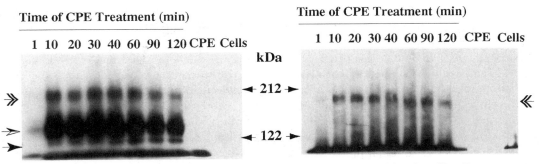

Figure 16.8 The kinetics of formation of CPE-containing intermediate to large complexes in Caco-2 cells. CPE was added to Caco-2 cells for the indicated times at 37°C. After removal of unbound enterotoxin, the cells were lysed with SDS, and cell lysates were analyzed by SDS-polyacrylamide gel electrophoresis (no sample boiling) using 4% acrylamide gels, followed by Western blotting with either CPE antibodies or occludin antibodies, as indicated. The time of CPE treatment is shown on top of the gel, and the migration of myosin (212-kDa) and β-galactosidase (122-kDa) markers is indicated in the center space between these two blots. The double, open, and closed arrows indicate the location of the ~200-kDa large complex, the ~155-kDa large complex, and the ~135-kDa intermediate complex, respectively. Reproduced with permission from reference 72.

intermediate complex, or ~155- kDa large complex (72). Furthermore, Katahira et al. (35) have determined that claudins may also be present in one or more of the intermediate and large complexes (35). Finally, a ~45- to 50-kDa membrane protein, presumably the same protein present in small complex, also appears to be associated with one or more of the intermediate and large complexes (88).

Considerable evidence indicates that at least some of the intermediate or large complexes in CPE induce cytotoxicity. For example, when Vero cells are treated with CPE at 4°C, they do not form any intermediate or large complex material and remain viable, despite the fact that CPE binds and forms small complex at low temperatures (84, 85). If unbound CPE is removed from the cultures and the cells are warmed to 37°C, both cytotoxicity and formation of intermediate and large complex material occur concurrently, as would be expected if some or all intermediate and large complexes were mediating cytotoxicity (53). Perhaps the strongest evidence for the involvement of some, if not all, intermediate and large complexes in CPE-induced cytotoxicity has been provided by recent studies using CPE deletion and point mutants (38, 39). These studies have clearly established a direct correlation between a mutant's cytotoxic activity and its ability to form intermediate and large complexes.

Step 4 in the Molecular Action of CPE: Consequences of Large Complex Formation

Many studies conducted in the late 1970s to early 1980s revealed that CPE rapidly affects the plasma membrane permeability properties of sensitive mammalian cells (48, 49, 51, 53, 56, 59). CPE-induced membrane permeability alterations are initially restricted to small molecules of <200 Da, which suggests that the initial CPE-induced membrane "lesion" is ~0.5 to 1 nm^2 (Fig. 16.9). Furthermore, this CPE-induced lesion appears to be neither directional nor discriminating; i.e., CPE induces a rapid increase in both influx and efflux of most or all small molecules, including ions and amino acids (49). These CPE-induced small-molecule permeability alterations severely perturb the normal cytoplasmic levels of ions and other small molecules, leading to cell death from either (i) a collapse in the cellular colloid-osmotic equilibrium, with subsequent cell lysis, or (ii) metabolic shutdown, e.g., inhibition of macromolecular synthesis (49). Similar time frames between the development of the cytotoxic effect of CPE and the onset of intestinal histopathologic damage, which appears to initiate CPE-induced intestinal fluid and electrolyte losses, suggest that CPE's cytotoxic activity is responsible, at least in large part, for initiating the enterotoxin's intestinal effects.

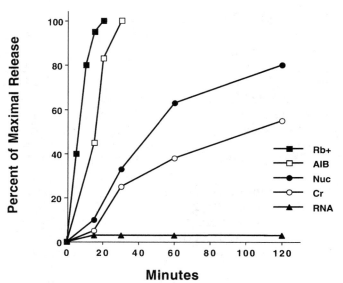

Figure 16.9 The effects of CPE on plasma membrane permeability in Vero cells. CPE effects on Vero cell membrane permeability were measured by specific CPE-induced release of radioactive cytoplasmic markers of defined sizes, including ^{86}Rb (Rb+, M_r ~100), ^{14}C-aminoisobutyric acid (AIB, M_r ~100), ^{3}H nucleotides (Nuc, M_r <1,000), ^{51}Cr (Cr, M_r ~3,000), and ^{3}H-RNA (RNA, M_r >25,000). Note the rapid release of small molecules, such as ions (Rb$^+$) and amino acids (AIB), following CPE treatment, whereas very large molecules (e.g., RNA) are never released from intact CPE-treated cells. Data compiled from studies discussed in references 49 and 56.

What mechanism does CPE use to induce changes in small-molecule membrane permeability? A close temporal correlation has been established between formation of CPE-containing intermediate and large complexes and the onset of CPE-induced small-molecule permeability alterations (53). If it is assumed that the ~135-kDa intermediate complex represents a precursor for formation of the ~155-kDa and ~200-kDa large complexes in CPE-treated Caco-2 cells, then the close correlation between formation of these large complexes and onset of membrane permeability changes suggests that one or both of these large complex species are responsible for CPE-induced membrane permeability effects. The putative involvement of one or both large complexes in CPE-induced membrane permeability changes is further supported by studies indicating that noncytotoxic CPE mutants fail to form either the ~155-kDa or ~200-kDa large complexes in Caco-2 cells (72). More recently, it has been determined that CPE-treated Transwell cultures of Caco-2 cells can exhibit membrane permeability alterations (Fig. 16.9) while forming only the ~155-kDa large complex, indicating that formation of the ~155-kDa large complex is probably sufficient for CPE to induce membrane permeability effects (71).

How might formation of the ~155-kDa large complex induce membrane permeability changes? Recent electrophysiology studies have detected "pore" formation in the apical membranes of Caco-2 cells treated with CPE under physiologic conditions in which the ~155-kDa large complex forms (28). Other recent observations suggest that some, if not all, CPE-containing large-complex material (presumably including the ~155-kDa complex) corresponds to a membrane "pore." For example, protease challenge studies have revealed that CPE localized in large complexes becomes very closely associated with membranes, which is consistent with part or all of those CPE-containing complex(s) being inserted into membranes (83). However, existing data do not preclude other possible explanations for CPE effects on membrane permeability; e.g., perhaps formation of the ~155-kDa complex leads to activation of an existing ion channel(s) in mammalian cells.

What role might the ~200-kDa large complex play in CPE's intestinal activity? A C-terminal CPE fragment, which is noncytotoxic and fails to form any large complex species, has been determined to bind to certain claudins and induce rearrangements in tight junctions of MDCK cells (76). These tight junction rearrangements can apparently cause paracellular permeability changes in MDCK cells (76). It is conceivable that native CPE might induce similar paracellular permeability effects in the intestinal epithelium, with those paracellular permeability effects contributing to CPE-induced intestinal secretion (i.e., diarrhea) during *C. perfringens* type A foodborne illness.

It has further been determined that after only 1 h of treatment, native CPE (but not a C-terminal CPE fragment) can induce tight junction rearrangements in rat liver (65). Since N-terminal CPE sequences appear to be important for formation of the ~200-kDa large complex containing the major tight junction structural protein occludin (72), the observation that tight junction structural rearrangements are induced more quickly by native CPE versus C-terminal CPE fragments may indicate that, in vivo, the ~200-kDa large complex species plays a key role in inducing tight junction changes and, by extension, intestinal paracellular permeability changes.

Although the ability of CPE to interact with tight junctions is interesting and perhaps contributes to CPE's intestinal enterotoxigenic activity, available data indicate that CPE's primary intestinal effect is its ability to induce histopathologic damage to the intestines through its cytotoxic action. For example, the close correlation between histopathologic damage and CPE-induced fluid and electrolyte losses has already been described. Furthermore, CPE-induced tight junction rearrangements were observed in both polarized monolayers of MDCK cells and rat liver only when the enterotoxin was applied to the basal side of those epithelia (65, 76). Since (i) CPE is present in the intestinal lumen and, therefore, initially interacts with the apical side of epithelial cells and (ii) CPE is not internalized inside mammalian cells (49), the ability of CPE to cause CPE-induced tight junction rearrangements only when applied from the basolateral surface suggests that tight junction rearrangements would only develop in the intestinal epithelium after CPE's cytotoxic effect has produced sufficient histopathologic damage to provide the enterotoxin with access to the basolateral surface of epithelial cells. In this regard, it deserves mention that the apical side of epithelial cells, such as Caco-2 cells, are sensitive to the cytotoxic effects of CPE (28).

CPE Structure-Function Relationships

There has been a substantial increase in our understanding of how the CPE protein mediates its pathophysiologic effects. Most notably, mapping studies utilizing CPE fragments and point mutants have localized several important functional regions on the CPE protein (22–27, 32, 38, 39). An overview of CPE functional regions, as currently understood, is shown in Fig. 16.10.

Recent structure-function mapping studies have determined that receptor binding activity is present in the extreme C terminus of the CPE protein (24–27, 32, 35, 36, 38, 39). For example, deleting the five C-terminal amino acids from native CPE completely abrogated the enterotoxin's ability to bind to mammalian cells or membranes (39). Furthermore, a synthetic peptide

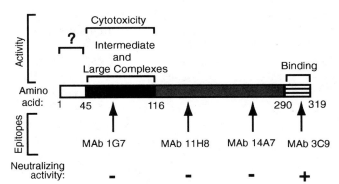

Figure 16.10 Map of CPE functional regions. CPE regions required for large complex formation, cytotoxicity, and binding to cells are depicted in this diagram (compiled from references 22, 23, 25–27, 32, 38, and 39), as are sequences required for presentation of MAb1G7, MAb11H8, MAb14A7, and MAb3C9 (24). The C-terminal region of CPE also appears to contain at least one other epitope, named RPC-1 (26). Antibodies that neutralize CPE's cytotoxicity are designated with a plus sign.

corresponding to amino acids 290 to 319 of native CPE (i.e., the 30 C-terminal CPE amino acids) was equally efficient as native CPE at competing against ^{125}I-CPE binding to cells or isolated membranes (25). Binding-capable C-terminal CPE fragments are not cytotoxic; i.e., C-terminal CPE fragments do not induce small molecule permeability alterations in mammalian cells (25, 27), indicating that mere occupancy of the CPE receptor is insufficient to trigger CPE-like cytotoxicity. That finding supports the importance of postbinding steps for CPE activity and also indicates that sequences present in the N-terminal half of the CPE molecule are necessary for cytotoxicity; i.e., like most bacterial toxins, CPE segregates its receptor binding and activity regions.

Deletion mutagenesis studies have revealed that the first 45 N-terminal amino acids of CPE are not required for cytotoxicity (39). Removal of the first 45 N-terminal CPE amino acids activates the cytotoxic activity of CPE (39). Similar activation may also occur during *C. perfringens* type A foodborne illness because the intestinal proteases trypsin and chymotrypsin remove the first 25 or 36 N-terminal amino acids, respectively, from CPE, thereby activating cytotoxic activity (22, 23, 26).

Why does the removal of N-terminal sequences activate CPE cytotoxicity? Results from recent deletion mutagenesis and random mutagenesis studies have revealed that activated CPE fragments lacking extreme N-terminal sequences can bind and form small complex similar to native CPE (38, 39). However, relative to native enterotoxin, those activated CPE fragments have enhanced ability to form intermediate and large complex species, which also provides additional support for the important role of intermediate and large complexes in CPE-induced cytotoxicity.

Studies with other CPE deletion fragments have revealed that removing additional N-terminal amino acids beyond residue 45 eliminates all cytotoxic activity. This loss of cytotoxic activity has been attributed to those deletion fragments being specifically blocked in their ability to form the intermediate and large CPE complexes. Recent studies used CPE point mutants to map the region responsible for formation of these intermediate and large complexes to amino acids 45 to 116 of the native enterotoxin (38). For example, a CPE Gly49Asp random mutant, which binds and forms small complex, was defective in its ability to form the ~135-kDa, ~155-kDa, or ~200- kDa complexes (72), a finding which also supports the concept that the ~135-kDa intermediate complex may be a precursor for formation of the ~155-kDa and ~200-kDa large complexes.

What regions of CPE are involved in inducing tight junction rearrangements and paracellular permeability changes? The noncytotoxic C-terminal half of CPE is sufficient to induce tight junction rearrangements, likely because C-terminal CPE fragments can interact with claudin receptors, which are also major structural proteins of tight junctions (35, 36, 76). However, native CPE seems to induce tight junction rearrangements more rapidly than C-terminal CPE fragments do (65), suggesting that N-terminal CPE sequences also contribute to CPE-induced tight junction rearrangements and, possibly, paracellular permeability changes. The apparently more rapid ability of native CPE versus C-terminal CPE fragments to induce tight junction rearrangements may be attributable to the presence of the amino acid 45 to 116 region in native CPE. That 45 to 116 region seems to be important for formation of the ~200-kDa complex, which contains the other major structural tight junction protein, occludin (72). Therefore, native CPE may disrupt tight junctions by simultaneously interacting with both claudins (via its C-terminal region) and occludin (directly or indirectly via its N-terminal 45 to 116 region). The fact that occludin becomes associated with native CPE, but not C-terminal CPE fragments that appear fully capable of binding claudins, indicates that the CPE protein plays a role in occludin's association with the ~200-kDa large complex; i.e., formation of that complex does not simply result from previously reported interactions between claudin and occludin (20).

Studies are currently ongoing to more precisely map each CPE functional region. Once that information is available, it will be combined with physical data regarding CPE's three-dimensional structure to determine precisely how this enterotoxin exerts its effects.

CPE Epitopes: Is a CPE Vaccine Possible?

The CPE fragments prepared for the structure-function mapping studies described above were reacted (26) with a series of CPE-specific monoclonal antibodies (MAbs) (89). These epitope mapping studies identified at least four or five disparate regions, scattered throughout the enterotoxin primary sequence, which appear to be involved in presentation of CPE epitopes.

Of greatest significance, the CPE epitope recognized by MAb 3C9 (89) has been mapped to the extreme C terminus of the enterotoxin protein (26). Since MAb 3C9 is a neutralizing monoclonal antibody that blocks CPE binding to cells (89), the presence of the MAb 3C9 epitope in the C terminus of CPE provides additional evidence that that region of the CPE molecule has receptor binding activity. Moreover, identifying the presence of a neutralizing linear epitope in C-terminal CPE fragments, which are not themselves cytotoxic (25, 27, 32, 39), suggests that those fragments might be potential CPE vaccine candidates. The possible use of C-terminal CPE fragments for developing a CPE vaccine was explored by preparing a 30-mer synthetic peptide corresponding to the extreme C-terminal CPE sequence (62) and chemically conjugating that peptide to a thyroglobulin carrier. When the resultant conjugate was administered intravenously to mice, the conjugate-immunized mice developed very high titers of serum antibodies capable of neutralizing the cytotoxicity of native CPE.

Of course, effective immunity to *C. perfringens* type A foodborne illness likely requires a secretory IgA response in the intestinal lumen. Therefore, should further development of a CPE food poisoning vaccine be deemed desirable, it would be necessary to pursue approaches that can specifically stimulate the development of intestinal IgA immunity against CPE. Furthermore, because recent studies have shown that, under certain circumstances, noncytotoxic C-terminal CPE fragments can still induce, although slowly, tight junction rearrangements (76), use of C-terminal CPE sequences for vaccine purposes might require fine-mapping the MAb 3C9 epitope to identify peptide sequences containing that neutralizing epitope but lacking claudin-binding activity.

CONCLUDING REMARKS

Since publication of the first edition of this book, considerable progress has been achieved toward understanding the unique pathogenesis of *C. perfringens* type A foodborne illness. Despite those advances, many future challenges remain for CPE researchers, e.g., identifying the reservoirs of the chromosomal *cpe C. perfringens* isolates responsible for foodborne illness; determining how *C. perfringens* regulates the timing and amount of CPE produced during sporulation; defining the role the 45- to 50-kDa eucaryotic protein, which appears to be a constituent of the small, intermediate, and large complexes, plays in CPE activity; characterizing that 45- to 50-kDa protein; determining what other eucaryotic proteins are present in the ~135-kDa, ~155-kDa, and ~200-kDa intermediate and large complexes; elucidating precisely how each of those complexes contributes to CPE's intestinal activity; revealing the importance of paracellular permeability effects versus cytotoxic effects of CPE-induced intestinal fluid and electrolyte loss; solving the three-dimensional structure of CPE; and locating the important functional regions within the native CPE structure.

Answering these and other questions through continued basic research should provide numerous practical applications, such as better approaches for controlling CPE-associated diseases, including *C. perfringens* type A foodborne illness. For example, identifying the specific reservoirs of *C. perfringens* foodborne illness isolates should enable public health agencies to design specific hygienic measures to reduce food contamination with these organisms. Additional studies may also lead to the development of agents capable of blocking CPE expression or activity. Finally, continued research on the mechanism of CPE activity should enable the use of this enterotoxin as a powerful probe of normal intestinal physiology.

Preparation of this chapter was supported by Public Health Service grant AI 19844-17 from the National Institute of Allergy and Infectious Diseases and grant 9802822 from the Ensuring Food Safety Program of the U.S. Department of Agriculture. The assistance of J. F. Kokai-Kun, M. R. Sarker, U. Singh, and E. Wieckowski with graphics is greatly appreciated.

References

1. **Bartholomew, B. A., M. F. Stringer, G. N. Watson, and R. J. Gilbert.** 1985. Development and application of an enzyme linked immunosorbent assay for *Clostridium perfringens* type A enterotoxin. *J. Clin. Pathol.* 38:222–228.

2. **Bean, N. H., J. S. Goulding, C. Lao, and F. J. Angulo.** 1996. Surveillance for foodborne-disease outbreaks—United States, 1988-1992. *Morbid. Mortal. Wkly. Rep.* 45:1–54.

3. **Billington, S. J., E. U. Wieckowski, M. R. Sarker, D. Bueschel, J. G. Songer, and B. A. McClane.** 1998. *Clostridium perfringens* type E animal enteritis isolates with highly conserved, silent enterotoxin sequences. *Infect. Immun.* 66:4531–4536.

4. **Birkhead, G., R. L. Vogt, E. M. Heun, J. T. Snyder, and B. A. McClane.** 1988. Characterization of an outbreak of *Clostridium perfringens* food poisoning by quantitative fecal culture and fecal enterotoxin measurement. *J. Clin. Microbiol.* 26:471–474.

5. **Brynestad, S., and P. E. Granum.** 1999. Evidence that Tn5565, which includes the enterotoxin gene in *Clostridium perfringens*, can have a circular form which may be a transposition intermediate. *FEMS Microbiol. Lett.* 170:281–286.

6. **Brynestad, S., L. A. Iwanejko, G. S. A. B. Stewart, and P. E. Granum.** 1994. A complex array of Hpr consensus DNA recognition sequences proximal to the enterotoxin gene in *Clostridium perfringens* type A. *Microbiology* 140:97–104.

7. **Brynestad, S., M. R. Sarker, B. A. McClane, P. E. Granum, and J. I. Rood.** The enterotoxin (CPE) plasmid from *Clostridium perfringens* is conjugative. *Infect. Immun.*, in press.

8. **Brynestad, S., B. Synstad, and P. E. Granum.** 1997. The *Clostridium perfringens* eterotoxin gene is on a transposable element in type A human food poisoning strains. *Microbiology* 143:2109–2115.

9. **Canard, B., B. Saint-Joanis, and S. T. Cole.** 1992. Genomic diversity and organization of virulence genes in the pathogenic anaerobe *Clostridium perfringens*. *Mol. Microbiol.* 6:1421–1429.

10. **Carman, R. J.** 1997. *Clostridium perfringens* in spontaneous and antibiotic-associated diarrhoea of man and other animals. *Rev. Med. Microbiol.* 8(Suppl. 1):S43–S45.

11. **Centers for Disease Control and Prevention.** 1994. *Clostridium perfringens* gastroenteritis associated with corned beef served at St. Patrick's day meals—Ohio and Virginia. 1993. *Morb. Mortal. Wkly. Rep.* 43:137–144.

12. **Collie, R. E., J. F. Kokai-Kun, and B. A. McClane.** 1998. Phenotypic characterization of enterotoxigenic *Clostridium perfringens* isolates from non-foodborne human gastrointestinal diseases. *Anaerobe* 4:69–79.

13. **Collie, R. E., and B. A. McClane.** 1998. Evidence that the enterotoxin gene can be episomal in *Clostridium perfringens* isolates associated with nonfoodborne human gastrointestinal diseases. *J. Clin. Microbiol.* 36:30–36.

14. **Cornillot, E., B. Saint-Joanis, G. Daube, S. Katayama, P. E. Granum, B. Carnard, and S. T. Cole.** 1995. The enterotoxin gene (*cpe*) of *Clostridium perfringens* can be chromosomal or plasmid-borne. *Mol. Microbiol.* 15:639–647.

15. **Czeczulin, J. R., R. E. Collie, and B. A. McClane.** 1996. Regulated expression of *Clostridium perfringens* enterotoxin in naturally *cpe*-negative type A, B, and C isolates of C. *perfringens. Infect. Immun.* 64:3301–3309.

16. **Czeczulin, J. R., P. C. Hanna, and B. A. McClane.** 1993. Cloning, nucleotide sequencing, and expression of the *Clostridium perfringens* enterotoxin gene in *Escherichia coli. Infect. Immun.* 61:3429–3439.

17. **Daube, G., P. Simon, B. Limbourg, C. Manteca, J. Mainil, and A. Kaeckenbeeck.** 1996. Hybridization of 2,659 *Clostridium perfringens* isolates with gene probes for seven toxins ($\alpha, \beta, \varepsilon, \iota, \theta, \mu$ and enterotoxin) and for sialidase. *Am. J. Vet. Res.* 57:496–501.

18. **Duncan, C. L., D. H. Strong, and M. Sebald.** 1972. Sporulation and enterotoxin production by mutants of *Clostridium perfringens.* J. Bacteriol. 110:378–391.

18a. **Food and Drug Administration.** 2000. Foodborne pathogenic microorganisms and natural toxins. http://vm.cfsan.fda.gov/~mow/intro.html.

19. **Fujita, K., J. Katahira, Y. Horiguchi, N. Sonoda, M. Furuse, and S. Tskuita.** 2000. *Clostridium perfringens* enterotoxin binds to the second extracellular loop of claudin-3, a tight junction membrane protein. *FEBS Lett.* 476:258–261.

20. **Furuse, M., K. Fujita, T. Hiiragi, K. Fujumoto, and S. Tsukita.** 1998. Claudin-1 and -2: novel integral membrane proteins localizing at tight junctions with no sequence similarity to occludin. *J. Cell Biol.* 141:1539–1550.

21. **Gibert, M., C. Jolivet-Reynaud, and M. R. Popoff.** 1997. Beta2 toxin, a novel toxin produced by *Clostridium perfringens.* Gene 203:65–73.

22. **Granum, P. E., and M. Richardson.** 1991. Chymotrypsin treatment increases the activity of *Clostridium perfringens* enterotoxin. *Toxicon* 29:445–453.

23. **Granum, P. E., J. R. Whitaker, and R. Skjelkvale.** 1981. Trypsin activation of enterotoxin from *Clostridium perfringens* type A. *Biochim. Biophys. Acta* 668:325–332.

24. **Hanna, P. C., and B. A. McClane.** 1991. A recombinant C-terminal toxin fragment provides evidence that membrane insertion is important for *Clostridium perfringens* enterotoxin cytotoxicity. *Mol. Microbiol.* 5:225–230.

25. **Hanna, P. C., T. A. Mietzner, G. K. Schoolnik, and B. A. McClane.** 1991. Localization of the receptor-binding region of *Clostridium perfringens* enterotoxin utilizing cloned toxin fragments and synthetic peptides. The 30 C-terminal amino acids define a functional binding region. *J. Biol. Chem.* 266:11037–11043.

26. **Hanna, P. C., E. U. Wieckowski, T. A. Mietzner, and B. A. McClane.** 1992. Mapping functional regions of *Clostridium perfringens* type A enterotoxin. *Infect. Immun.* 60:2110–2114.

27. **Hanna, P. C., A. P. Wnek, and B. A. McClane.** 1989. Molecular cloning of the 3′ half of the *Clostridium perfringens* enterotoxin gene and demonstration that this region encodes receptor-binding activity. *J. Bacteriol.* 171:6815–6820.

28. **Hardy, S. P., M. Denmead, N. Parekh, and P. E. Granum.** 1999. Cationic currents induced by *Clostridium perfringens* type A enterotoxin in human intestinal CaCo-2 cells. *J. Med. Microbiol.* 48:235–243.

29. **Hauschild, A. H., L. Niilo, and W. J. Dorward.** 1971. The role of enterotoxin in *Clostridium perfringens* type A enteritis. *Can. J. Microbiol.* 17:987–991.

30. **Hauser, D., M. W. Eklund, P. Boquet, and M. R. Popoff.** 1994. Organization of the botulinum neurotoxin C1 gene and its non-toxic protein genes in *Clostridium botulinum* C 468. *Mol. Gen. Genet.* 243:631–640.

31. **Hobbs, B. C.** 1979. *Clostridium perfringens* gastroenteritis, p. 131–167. *In* H. Riemann and F. L. Bryan (ed.), *Food-Borne Infections and Intoxications,* 2nd ed. Academic Press, Inc., New York, N.Y.

32. **Horiguchi, Y., T. Akai, and G. Sakaguchi.** 1987. Isolation and function of a *Clostridium perfringens* enterotoxin fragment. *Infect. Immun.* 55:2912–2915.

33. **Horiguchi, Y., T. Uemura, S. Kozaki, and G. Sakaguchi.** 1985. The relationship between cytotoxic effects and binding to mammalian cultures cells of *Clostridium perfringens* enterotoxin. *FEMS Microbiol. Lett.* 28:131–135.

34. **Juneja, V. K., and W. M. Majka.** 1995. Outgrowth of *Clostridium perfringens* spores in spores in cook-in-bag beef products. *J. Food Safety* 15:21–34.

35. **Katahira, J., N. Inoue, Y. Horiguchi, M. Matsuda, and N. Sugimoto.** 1997. Molecular cloning and functional characterization of the receptor for *Clostridium perfringens* enterotoxin. *J. Cell Biol.* 136:1239–1247.

36. **Katahira, J., H. Sugiyama, N. Inoue, Y. Horiguchi, M. Matsuda, and N. Sugimoto.** 1997. *Clostridium perfringens* enterotoxin utilizes two structurally related membrane proteins as functional receptors *in vivo. J. Biol. Chem.* 272:26652–26658.

37. **Katayama, S. I., B. Dupuy, G. Daube, B. China, and S. T. Cole.** 1996. Genome mapping of *Clostridium perfringens* strains with *I-Ceu I* shows many virulence genes to be plasmid-borne. *Mol. Gen. Genet.* 251:720–726.

38. **Kokai-Kun, J. F., K. Benton, E. U. Wieckowski, and B. A. McClane.** 1999. Identification of a *Clostridium perfringens* enterotoxin region required for large complex formation

and cytotoxicity by random mutagenesis. *Infect. Immun.* 67:6534–6541.

39. **Kokai-Kun, J. F., and B. A. McClane.** 1997. Deletion analysis of the *Clostridium perfringens* enterotoxin. *Infect. Immun.* 65:1014–1022 .

40. **Kokai-Kun, J. F., and B. A. McClane.** 1996. Evidence that region(s) of the *Clostridium perfringens* enterotoxin molecule remain exposed on the external surface of the mammalian plasma membrane when the toxin is sequestered in small or large complex. *Infect. Immun.* 64:1020–1025.

41. **Kokai-Kun, J. F., J. G. Songer, J. R. Czeczulin, F. Chen, and B. A. McClane.** 1994. Comparison of Western immunoblots and gene detection assays for identification of potentially enterotoxigenic isolates of *Clostridium perfringens*. *J. Clin. Microbiol.* 32:2533–2539.

42. **Krakauer, T., B. Fleischer, D. L. Stevens, B. A. McClane, and B. G. Stiles.** 1997. *Clostridium perfringens* enterotoxin lacks superantigenic activity but induces an interleukin-6 response from human peripheral blood mononuclear cells. *Infect. Immun.* 65:3485–3488.

43. **Labbe, R. G.** 1989. *Clostridium perfringens*, p. 192–234. *In* M. P. Doyle (ed.), *Foodborne Bacterial Pathogens*. Marcel Dekker, Inc., New York, N.Y.

44. **Labbe, R. G.** 1981. Enterotoxin formation by *Clostridium perfringens* type A in a defined medium. *Appl. Environ. Microbiol.* 41:315–317.

45. **Labbe, R. G., and C. L. Duncan.** 1977. Evidence for stable messenger ribonucleic acid during sporulation and enterotoxin synthesis by *Clostridium perfringens* type A. *J. Bacteriol.* 129:843–849.

46. **Lawrence, G. W.** 1997. The pathogenesis of enteritis necroticans, p. 198–207. *In* J. I. Rood, B. A. McClane, J. G. Songer, and R. W. Titball (ed.), *The Clostridia: Molecular Genetics and Pathogenesis*. Academic Press Inc., London, England.

47. **Maslanka, S. E., J. G. Kerr, G. Williams, J. M. Barbaree, L. A. Carson, J. M. Miller, and B. Swaminathan.** 1999. Molecular subtyping of *Clostridium perfringens* by pulsed-field gel electrophoresis to facilitate food-borne-disease outbreak investigations. *J. Clin. Microbiol.* 37:2209–2214.

48. **Matsuda, M., K. Ozutsumi, H. Iwashi, and N. Sugimoto.** 1986. Primary action of *Clostridium perfringens* type A enterotoxin on HeLa and Vero cells in the absence of extracellular calcium: rapid and characteristic changes in membrane permeability. *Biochem. Biophys. Res. Commun.* 141:704–710.

49. **McClane, B. A.** 1994. *Clostridium perfringens* enterotoxin acts by producing small molecule permeability alterations in plasma membranes. *Toxicology* 87:43–67.

50. **McClane, B. A.** 1992. *Clostridium perfringens* enterotoxin: structure, action and detection. *J. Food Safety* 12:237–252.

51. **McClane, B. A., P. C. Hanna, and A. P. Wnek.** 1988. *Clostridium perfringens* type A enterotoxin. *Microb. Pathogen.* 4:317–323.

52. **McClane, B. A., D. M. Lyerly, J. S. Moncrief, and T. D. Wilkins.** 2000. Enterotoxic clostridia: *Clostridium perfringens* type A and *Clostridium difficile*, p. 551-562. *In* V. A. Fischetti, R. P. Novick, J. J. Ferretti, D. A. Portnoy, and J. I. Rood (ed.), *Gram-Positive Pathogens*. ASM Press, Washington, D.C.

53. **McClane, B. A., and A. P. Wnek.** 1990. Studies of *Clostridium perfringens* enterotoxin action at different temperatures demonstrate a correlation between complex formation and cytotoxicity. *Infect. Immun.* 58:3109–3115.

54. **McClane, B. A., A. P. Wnek, K. I. Hulkower, and P. C. Hanna.** 1988. Divalent cation involvement in the action of *Clostridium perfringens* type A enterotoxin. *J. Biol. Chem.* 263:2423–2435.

55. **McDonel, J. L.** 1980. Binding of *Clostridium perfringens* ^{125}I-enterotoxin to rabbit intestinal cells. *Biochemistry* 21:4801–4807.

56. **McDonel, J. L.** 1986. Toxins of *Clostridium perfringens* types A, B, C, D, and E, p. 477–517. *In* F. Dorner and H. Drews (ed.), *Pharmacology of Bacterial Toxins*. Pergamon Press, Oxford, England.

57. **McDonel, J. L., L. W. Chang , J. L. Pounds, and C. L. Duncan.** 1978. The effects of *Clostridium perfringens* enterotoxin on rat and rabbit ileum: an electron microscopy study. *Lab. Invest.* 39:210–218.

58. **McDonel, J. L., and C. L. Duncan.** 1975. Histopathological effect of *Clostridium perfringens* enterotoxin in the rabbit ileum. *Infect. Immun.* 12:1214–1218.

59. **McDonel, J. L., and B. A. McClane.** 1979. Binding vs. biological activity of *Clostridium perfringens* enterotoxin in Vero cells. *Biochem. Biophys. Res. Commun.* 87:497–504.

60. **Mead, P. S., L. Slutsker, V. Dietz, L. F. McCaig, J. S. Bresee, C. Shapiro, P. M. Griffen, and R. V. Tauxe.** 1999. Food-related illness and death in the United States. *Emerg. Infect. Dis.* 5:607–625.

61. **Melville, S. B., R. Labbe, and A. L. Sonenshein.** 1994. Expression from the *Clostridium perfringens* cpe promoter in *C. perfringens* and *Bacillus subtilis*. *Infect. Immun.* 62:5550–5558.

62. **Mietzner, T. A., J. F. Kokai-Kun, P. C. Hanna, and B. A. McClane.** 1992. A conjugated synthetic peptide corresponding to the C-terminal region of *Clostridium perfringens* type A enterotoxin elicits an enterotoxin-neutralizing antibody response in mice. *Infect. Immun.* 60:3947–3951.

63. **Olsen, S. J., L. C. MacKinon, J. S. Goulding, N. H. Bean, and L. Slutsker.** 2000. Surveillance for foodborne-disease outbreaks—United States, 1993-97. *Morb. Mortal. Wkly. Rep.* 49:1–51.

64. **Parikh, A. I., M. T. Jay, D. Kassam, T. Kociemba, B. Dworkis, P. D. Bradley, and K. Takata.** 1997. *Clostridium perfringens* outbreak at a juvenile detention facility linked to a Thanksgiving holiday meal. *West. J. Med.* 166:417–419.

65. **Rahner, C., L. L. Mitic, B. A. McClane, and J. M. Anderson.** 1999. *Clostridium perfringens* enterotoxin impairs bile flow in the isolated perfused rat liver and induces fragmentation of tight junction fibrils. *Hepatology* 30:326A.

66. **Rhodehamel, J., and S. M. Harmon.** 1998. *Clostridium perfringens*, p. 16.01–16.06. *In FDA Bacteriologic Methods*, 8th ed. AOAC International, Gaithersburg, Md.

67. Sarker, M. R., R. J. Carman, and B. A. McClane. 1999. Inactivation of the gene (*cpe*) encoding *Clostridium perfringens* enterotoxin eliminates the ability of two *cpe*-positive *C. perfringens* type A human gastrointestinal disease isolates to affect rabbit ileal loops. *Mol. Microbiol.* **33:**946–958.

68. Sarker, M. R., R. P. Shivers, S. G. Sparks, V. K. Juneja, and B. A. McClane. 2000. Comparative experiments to examine the effects of heating on vegetative cells and spores of *Clostridium perfringens* isolates carrying plasmid versus chromosomal enterotoxin genes. *Appl. Environ. Microbiol.* **66:**3234–3240.

69. Sherman, S., E. Klein, and B. A. McClane. 1994. *Clostridium perfringens* type A enterotoxin induces concurrent development of tissue damage and fluid accumulation in the rabbit ileum. *J. Diarrheal Dis. Res.* **12:**200–207.

70. Shih, N. J., and R. G. Labbe. 1996. Sporulation-promoting ability of *Clostridium perfringens* culture fluids. *Appl. Environ. Microbiol.* **62:**1441–1443.

71. Singh, U., and B. A. McClane. Unpublished data.

72. Singh, U., C. M. Van Itallie, L. L. Mitic, J. M. Anderson, and B. A. McClane. 2000. CaCo-2 cells treated with *Clostridium perfringens* enterotoxin form multiple large complex species, one of which contains the tight junction protein occludin. *J. Biol. Chem.* **275:**18407–18417.

73. Skjelkvale, R., and T. Uemura. 1977. Experimental diarrhea in human volunteers following oral administration of *Clostridium perfringens* enterotoxin. *J. Appl. Bacteriol.* **46:**281–286.

74. Smith, W. P., and J. L. McDonel. 1980. *Clostridium perfringens* type A: *in vitro* systems for sporulation and enterotoxin synthesis. *J. Bacteriol.* **144:**306–311.

75. Songer, J. G., and R. M. Meer. 1996. Genotyping of *Clostridium perfringens* by polymerase chain reaction is a useful adjunct to diagnosis of clostridial enteric disease in animals. *Anaerobe* **2:**197–203.

76. Sonoda, N., M. Furuse, H. Sasaki, S. Yonemura, J. Katahira, Y. Horiguchi, and S. Tsukita. 1999. *Clostridium perfringens* enterotoxin fragments removes specific claudins from tight junction strands: evidence for direct involvement of claudins in tight junction barrier. *J. Cell Biol.* **147:**195–204.

77. Strong, D. H., C. L. Duncan, and G. Perna. 1971. *Clostridium perfringens* type A food poisoning. II. Response of the rabbit ileum as an indication of enteropathogenicity of strains of *Clostridium perfringens. Infect. Immun.* **3:**171–178.

78. Sugii, S., and Y. Horiguchi. 1988. Identification and isolation of the binding substance for *Clostridium perfringens* enterotoxin on Vero cells. *FEMS Microbiol. Lett.* **52:**85–90.

79. Todd, E. C. D. 1989. Cost of acute bacterial foodborne disease in Canada and the United States. *Int. J. Food Microbiol.* **9:**313–326.

80. Todd, E. C. D. 1989. Preliminary estimates of costs of foodborne disease in the United States. *J. Food Prot.* **52:**595–601.

81. Van Damme-Jongsten, M., M. J. Rodhouse, R. J. Gilbert, and S. Notermans. 1990. Synthetic DNA probes for detection of enterotoxigenic *Clostridium perfringens* strains isolated from outbreaks of food poisoning. *J. Clin. Microbiol.* **28:**131–133.

82. Wallace, F. M., A. S. Mach, A. M. Keller, and J. A. Lindsay. 1999. Evidence for *Clostridium perfringens* enterotoxin inducing a mitogenic and cytokine response in vitro and a cytokine response in vivo. *Curr. Microbiol.* **38:**96–100.

83. Wieckowski, E., J. F. Kokai-Kun, and B. A. McClane. 1998. Characterization of membrane-associated *Clostridium perfringens* enterotoxin following Pronase treatment. *Infect. Immun.* **66:**5897–5905.

84. Wieckowski, E. U., A. P. Wnek, and B. A. McClane. 1994. Evidence that an ∼50 kDa mammalian plasma membrane protein with receptor-like properties mediates the amphiphilicity of specifically-bound *Clostridium perfringens* enterotoxin. *J. Biol. Chem.* **269:**10838–10848.

85. Wnek, A. P., and B. A. McClane. 1990. *Clostridium perfringens* enterotoxin in complex with eucaryotic membrane proteins demonstrates amphiphilic properties, p. 78. *Abstr. 90th Annu. Meet. Am. Soc. Microbiol. 1990.* American Society for Microbiology, Washington, D.C.

86. Wnek, A. P., and B. A. McClane. 1986. Comparison of receptors for *Clostridium perfringens* type A and cholera enterotoxins in isolated rabbit intestinal brush border membranes. *Microb. Pathog.* **1:**89–100.

87. Wnek, A. P., and B. A. McClane. 1983. Identification of a 50,000 Mr protein from rabbit brush boarder membranes that binds *Clostridium perfringens* enterotoxin. *Biochem. Biophys. Res. Commun.* **112:**1099–1105.

88. Wnek, A. P., and B. A. McClane. 1989. Preliminary evidence that *Clostridium perfringens* type A enterotoxin is present in a 160,000-M$_r$ complex in mammalian membranes. *Infect. Immun.* **57:**574–581.

89. Wnek, A. P., R. J. Strouse, and B. A. McClane. 1985. Production and characterization of monoclonal antibodies against *Clostridium perfringens* type A enterotoxin. *Infect. Immun.* **50:**442–448.

90. Zhao, Y., and S. B. Melville. 1998. Identification and characterization of sporulation-dependent promoters upstream of the enterotoxin gene (*cpe*) of *Clostridium perfringens. J. Bacteriol.* **180:**136–142.

Food Microbiology: Fundamentals and Frontiers, 2nd Ed.
Edited by M. P. Doyle et al.
© 2001 ASM Press, Washington, D.C.

Per Einar Granum

Bacillus cereus

17

INTRODUCTION

Bacillus cereus causes two different types of foodborne illness: the diarrheal type, which was first recognized after a hospital outbreak associated with contaminated vanilla sauce in Oslo, Norway, in 1948 (34, 35), and the emetic type, described about 20 years later after several outbreaks associated with fried rice in London (48). The diarrheal type of foodborne illness is caused by enterotoxin(s) produced during vegetative growth of *B. cereus* in the small intestine (26), whereas the emetic toxin is produced by cells growing in the food (41). For both types of foodborne illness, the food involved has usually been heat treated, and surviving spores are the source of the food poisoning. Although *B. cereus* is not a competitive microorganism, it grows well after cooking and cooling (<48°C). Heat treatment causes spore germination, and in the absence of competing flora, *B. cereus* grows well. *B. cereus* is a common soil saprophyte and is easily spread to many types of foods, especially those of plant origin (rice and pasta), but it is also frequently isolated from meat, eggs, and dairy products (41). Increased numbers of psychrotolerant strains, specifically in the dairy industry, have led to greater surveillance of *B. cereus* in recent years (20, 28, 33, 42, 63, 64).

B. cereus foodborne illness is underreported, as both types of illness are relatively mild and usually last less than 24 h (41). However, more severe forms of the diarrheal type of *B. cereus* foodborne illness have occasionally been reported (25, 26). These reports include three deaths caused by a newly discovered necrotic enterotoxin (44) and one death resulting from the intake of large amounts of the emetic toxin (47).

CHARACTERISTICS OF THE ORGANISM

B. cereus is a gram-positive, spore-forming, motile, aerobic rod but grows well anaerobically. The genus *Bacillus*, as described in *Bergey's Manual of Systematic Bacteriology* (21), is too diverse for a single genus, and dramatic changes have been suggested for its classification (50). A pioneering analysis of 16S rRNA sequences from many *Bacillus* revealed that there are at least five genera or rRNA groups within the genus *Bacillus* (11). With the subsequent identification of many new species, the number of "genera" has increased to approximately 16.

Bacillus anthracis, B. cereus, B. mycoides, B. thuringiensis, and, more recently, *B. pseudomycoides* (49) and *B. weihenstephanensis* (42) make up the *B. cereus*

Per Einar Granum, Department of Pharmacology, Microbiology and Food Hygiene, Norwegian School of Veterinary Science, N-0033 Oslo, Norway.

Table 17.1   Criteria for differentiating members of the *B. cereus* group

Species	Colony morphology	Hemolysis	Motility	Susceptible to penicillin	Parasporal crystal inclusion
B. cereus	White	+	+	−	−
B. anthracis	White	−	−	+	−
B. thuringiensis	White/gray	+	+	−	+
B. mycoides	Rhizoid	(+)	−	−	−
B. weihenstephanensis	Differentiated from *B. cereus* based on growth at <7°C and not at 43°C; can be identified rapidly using rDNA or cspA (cold shock protein A) targeted PCR (42)				
B. pseudomycoides	Not distinguishable from *B. mycoides* by physiological and morphological characteristics; clearly differentiated based on fatty acid composition and 16S RNA sequences (49)				

group. These bacteria have highly similar 16S and 23S rRNA sequences, indicating that they have diverged from a common evolutionary line relatively recently (9, 10). Strains of *B. anthracis* have been related to the other species within the *B. cereus* group based on rRNA sequences; however, taxonomically and because of its highly virulent pathogenicity, *B. anthracis* is the most distinctive member of this group. In addition, extensive genomic studies of DNA from strains of *B. cereus* and *B. thuringiensis* have revealed that there is no taxonomic basis for separate species status (17, 18). Nevertheless, the name *B. thuringiensis* is retained for those strains that synthesize a crystalline inclusion (Cry protein) or δ-endotoxin that is highly toxic to specific insects. The *cry* genes are usually located on plasmids, and loss of the relevant plasmid(s) makes the bacterium indistinguishable from *B. cereus*.

Cells of the six species discussed above are all large (cell width, >0.9 μm) and produce central to terminal ellipsoid or cylindrical spores that do not distend the sporangia (21, 42, 49). Bacteria of these *Bacillus* species sporulate easily on most media after 2 to 3 days, and *B. cereus* and *B. thuringiensis* lose their motility during the early stages of sporulation. The six species can be differentiated by using the criteria cited in Table 17.1.

RESERVOIRS

B. cereus is widespread in nature and is frequently isolated from soil and growing plants (41). From this natural environment it is easily spread to foods, especially those of plant origin. Through cross-contamination, it may then be spread to other foods, such as meat products (41). The problems in milk and milk products are caused by *B. cereus* that is spread from soil and grass to the udders of cows and into raw milk. *B. cereus* spores survive milk pasteurization, and after germination, the cells are free from competition from other vegetative cells (7). According to the classic literature (21), *B. cereus* is unable to grow at temperatures below 10°C and cannot

grow in milk and milk products stored at temperatures between 4 and 8°C. However, the psychrotolerant strains that have evolved can grow at temperatures as low as 4 to 6°C (20, 28, 64). In addition to rice, pasta, and spices, dairy products are among the most common food vehicles for *B. cereus*.

The closely related *B. thuringiensis* is reported to produce enterotoxin(s) (22, 23, 51, 52) and has caused foodborne illness when administered to human volunteers (51). This characteristic may develop into a serious problem because several countries commonly spray this organism to protect crops against insect infestations. *B. thuringiensis* reportedly caused one outbreak of foodborne illness (40); however, because the procedures normally used to identify *B. cereus* do not differentiate between the two species of *Bacillus* (Table 17.1), other outbreaks caused by *B. thuringiensis* may have gone unrecognized. To ensure safe spraying of *B. thuringiensis*, the organism should not produce enterotoxins.

FOODBORNE OUTBREAKS

The number of outbreaks of *B. cereus* foodborne illness has been greatly underestimated. Reasons for this include the relatively short duration of both types of illness (usually <24 h) and the likelihood that foodborne illness involving *B. cereus*-contaminated milk is limited to one or two cases within a family and is thus not identified as an outbreak. Perhaps some people who drink milk are protected against *B. cereus* foodborne illness even though the milk they drink every day contains small numbers of *B. cereus* cells. Toward the end of its shelf life, milk frequently contains enough *B. cereus* cells to cause illness if the strains present are enterotoxin producers; however, *B. cereus* can produce a protease that results in off flavors, which might limit the consumption of milk containing high numbers of the organism.

The dominant type of illness caused by *B. cereus* differs from country to country. In Japan, the emetic type is reported about 10 times more frequently than

Table 17.2 Characteristics of the two types of illness caused by *B. cereus*[a]

Characteristic	Diarrheal syndrome	Emetic syndrome
Dose causing illness (no. of cells)	10^5–10^7 (total)	10^5–10^8/g
Toxin produced	In the small intestine of the host	Preformed in foods
Type of toxin	Protein; enterotoxin(s)	Cyclic peptide; emetic toxin
Incubation period (h)	8–16 (occasionally >24)	0.5–5
Duration of illness (h)	12–24 (occasionally several days)	6–24
Symptoms	Abdominal pain, watery diarrhea, occasionally nausea	Nausea, vomiting, malaise (sometimes followed by diarrhea due to production of enterotoxin)
Foods most frequently implicated	Meat products, soups, vegetables, puddings and sauces, milk and milk products	Fried and cooked rice, pasta, pastry, noodles

[a] Based on references 25, 26, 41, 56, and 62.

the diarrheal type (56), whereas in Europe and North America the diarrheal type is reported more frequently (1, 41, 54). Eating habits probably account for this difference, although contaminated milk was reported to have caused at least one large outbreak of the emetic type in Japan (56). Some patients simultaneously experience both types of *B. cereus* foodborne illness (41). In addition, many *B. cereus* strains can produce both types of toxins (28, 41).

Because surveillance of foodborne illnesses differs greatly among countries, it is not possible to directly compare the incidence of outbreaks reported. The percentages of outbreaks and cases attributed to *B. cereus* in Japan, North America, and Europe vary from approximately 1 to 47% of outbreaks and from about 0.7 to 33% of cases (reports from different periods between 1960 and 1992) (1, 41, 54). Iceland, The Netherlands, and Norway reported the greatest number of *B. cereus* outbreaks and cases. Relatively few outbreaks of salmonellosis and campylobacter enteritis have occurred in Norway and Iceland (1, 54), although they are the two most frequently reported causes of foodborne illness in the United Kingdom (60). In 1991, *B. cereus* was responsible for 27% of the outbreaks in The Netherlands for which the causative agent was identified. However, the actual incidence of *B. cereus* was only 2.8% of the total number of cases because most cases of foodborne illness were of unknown etiology (54).

CHARACTERISTICS OF THE DISEASE

As mentioned above, there are two types of *B. cereus* foodborne illness. The first type, which is caused by an emetic toxin, induces vomiting, whereas the second type, caused by enterotoxin(s), results in diarrhea (41).

In a small number of cases, both types of symptoms occur (41), likely due to production of both types of toxin. There has been some debate about whether enterotoxin(s) can be preformed in foods and cause an intoxication. A review of the literature shows that the incubation time (>6 h; average, 12 h) is likely too long for diarrheal illness to be caused by preformed enterotoxin(s) (41), and model experiments revealed that enterotoxin(s) is degraded as it proceeds to the ileum (26). There is no doubt that enterotoxin(s) can be preformed in food, but the number of *B. cereus* cells in the food is also at least 2 orders of magnitude higher than that necessary to cause food poisoning (19, 20, 26). Usually products with such large populations of *B. cereus* are no longer acceptable to the consumer, although food containing >10^7 *B. cereus*/ml may not always appear spoiled. The characteristics of the two types of *B. cereus* foodborne illness are described in Table 17.2.

Some recent reports of *B. cereus* foodborne illness have caused considerable concern, particularly two outbreaks with fatal outcomes. An outbreak in Norway associated with eating stew containing approximately 10^4 to 10^5 *B. cereus* per serving affected 17 people, 3 of whom were hospitalized, 1 for 3 weeks (25). For the hospitalized patients, the onset of the illness was late, greater than 24 h (see The Spore, below). In another case, the emetic toxin cereulide was responsible for the death of a 17-year-old Swiss boy, due to fulminant liver failure (47). A large amount of *B. cereus* emetic toxin was detected in residue in the pan used to reheat the implicated food (pasta) as well as in the boy's liver and bile. The newly discovered cytotoxin K (CytK) is similar to the β-toxin of *Clostridium perfringens* (and other related toxins) and was the cause of a severe outbreak of *B. cereus* foodborne illness in France in 1998 (44). In this outbreak, several people developed bloody diarrhea, and three died. This is the first recorded outbreak of *B. cereus* necrotic enteritis, although it was not nearly as severe as *C. perfringens* type C foodborne illness (24).

DOSE AND SUSCEPTIBLE POPULATIONS

After the first recognized outbreak of *B. cereus* foodborne illness in Oslo (caused by contaminated vanilla sauce), Hauge isolated the causative agent, grew it to 4×10^6/ml, and drank 200 ml of the culture (34). Approximately 13 h later he developed abdominal pain and watery diarrhea that lasted for about 8 h. The cell population of *B. cereus* ingested was approximately 8×10^8. Counts of *B. cereus* ranging from 200 to 10^9/g (or ml) (25, 26, 35, 41) have been reported in foods incriminated in outbreaks, indicating the total dose ranged from approximately 5×10^4 to 10^{11} cells/dose. The total number of *B. cereus* required to be ingested to produce illness is likely in the range of 10^5 to 10^8 viable cells or spores. This wide range is due in part to differences in the amount of enterotoxin produced by different strains (26). Hence, food containing more than 10^3 *B. cereus* per g cannot be considered completely safe for consumption.

Although little is known about susceptible populations, the more severe types of the illness have occasionally involved either young athletes (<19 years) or the elderly (>60 years) (25, 30, 44). More studies are needed to address this important question.

VIRULENCE FACTORS AND MECHANISMS OF PATHOGENICITY

The two types of *B. cereus* foodborne illness are caused by very different types of toxins. The emetic toxin, causing vomiting, recently has been isolated and characterized (3), whereas the diarrheal disease is caused by one or more enterotoxins (13, 14, 26, 27, 31, 41, 44, 59, 61, 62).

The Emetic Toxin

The structure of the emetic toxin (Table 17.3) has long been a mystery because the only detection system involved living primates (37, 41). However, the recent discovery that the toxin could be detected by HEp-2 cells (vacuolation activity) (37) has led to its isolation and determination of its structure (3). Although some uncertainty remains as to whether the emetic toxin and the vacuolating factor are the same component (56, 58), there is no doubt that they are identical (5, 57). The emetic toxin, named cereulide, consists of a ring structure of three repeats of four amino and/or oxy acids: [D-O-Leu-

Table 17.3 Properties of the emetic toxin cereulide[a]

Trait	Property or activity
Molecular mass	1.2 kDa
Structure	Ring-shaped peptide
Isoelectric point	Uncharged
Antigenic	No (?)
Biological activity in living primates	Vomiting
Receptor	5-HT$_3$ (stimulation of the vagus afferent)
Ileal loop tests (rabbit, mouse)	None
Cytotoxic	No
HEp-2 cells	Vacuolation activity
Stability to heat	90 min at 121°C
Stability to pH	Stable at pH 2–11
Effect of proteolysis (trypsin, pepsin)	None
Conditions under which toxin is produced	In food: rice and milk at 25–32°C
Mechanisms of production	Not known (probably enzymatically)

[a] Based on references 3, 5, 41, 56, and 57.

D-Ala-L-O-Val-L-Val]$_3$. This ring structure (dodecadepsipeptide) has a molecular mass of 1.2 kDa and is closely related chemically to the potassium ionophore, valinomycin (3). The biosynthetic pathway and mechanism of action of the emetic toxin have yet to be elucidated, although the emetic toxin has been shown to stimulate the vagus afferent through binding to the 5-HT$_3$ receptor (5). It is not clear if the toxin is a modified gene product or is enzymatically produced through modification of components in the growth medium. However, given its structure, cereulide is more likely to be an enzymatically synthesized peptide and not a gene product.

For many years it was speculated that the emetic toxin had a lipid structure (26, 41). The reason for this belief is probably because cereulide has a hydrophobic nature and is thus associated with lipid environments. Emetic toxin is resistant to heat, pH, and proteolysis and is not antigenic (41) (Table 17.3). The recent death of a 17-year-old Swiss boy from liver failure caused by emetic toxin has stimulated research on other aspects of cereulide besides food poisoning. Recently, mice were injected intraperitoneally with synthetic cereulide and the histopathological changes that developed were examined (65). High cereulide doses produced massive degeneration of hepatocytes. The serum values of hepatic enzymes were highest on days 2 and 3 after the inoculation of cereulide and decreased rapidly thereafter. General recovery from pathological changes and regeneration of hepatocytes were observed after 4 weeks.

Table 17.4 Toxins produced by *B. cereus*

Toxin	Type or size	Food poisoning	Reference(s)
Hemolysin BL (Hbl)	Protein, three components	Probably	13, 16, 36, 53
Nonhemolytic enterotoxin (Nhe)	Protein, three components	Yes	32, 45, 46
Enterotoxin T (BceT)	Protein, one component, 41 kDa	No (?)	4
Enterotoxin FM (EntFM)	Protein, one component, 45 kDa	No (?)	8
Cytotoxin K (CytK)	Protein, one component, 34 kDa	Yes, three deaths	44
Emetic toxin (cereulide)	Cyclic peptide, 1.2 kDa	Yes, one death	3, 5, 47, 65

Enterotoxins

Cloning and sequencing studies have shown that *B. cereus* produces at least five different proteins (or protein complexes) referred to as enterotoxins (4, 8, 29, 36, 44). There is no evidence, however, that the recently reported enterotoxin T (4) and enterotoxin FM (8) cause foodborne illness (Table 17.4). Perhaps through specific mutations or gene knockout studies the extent to which the different proteins are involved in foodborne illness can be determined.

Three of these enterotoxins likely are involved in *B. cereus* foodborne illness (Table 17.4). Two of them are multicomponent and related, whereas the third (enterotoxin K) is a single protein of 34 kDa. The three-component hemolysin (Hbl; consisting of three proteins: B, L_1, and L_2) with enterotoxin activity was the first to be fully characterized (14, 16). This toxin also has dermonecrotic and vascular permeability activities and causes fluid accumulation in ligated rabbit ileal loops. Hbl has been suggested to be a primary virulence factor in *B. cereus* diarrhea (13). Convincing evidence has shown that all three components are necessary for maximal enterotoxin activity (13). More recently, Lund and Granum characterized a nonhemolytic, three-component enterotoxin (Nhe) (45, 46). The three components of this toxin were different from the components of Hbl, although there are similarities. The three components of Nhe enterotoxin were first purified from a *B. cereus* strain isolated after a large food-associated outbreak in Norway in 1995. The strain used to characterize Nhe did not produce the L_2 component (27). We initially thought that the L_2 component was unnecessary for biological activity of the Hbl enterotoxin, but then we isolated and characterized another three-component enterotoxin from this strain. Binary combinations of the components of this enterotoxin possess some biological activity, but not nearly as much as when all the components are present (46).

Almost all strains of *B. cereus* and *B. thuringiensis* tested produce Nhe, and about 60% produce Hbl (unpublished results). It is not known how important each is in relation to foodborne illness. The enterotoxin T gene was found to be absent in 57 of 95 strains of *B. cereus* and in 5 of 7 strains involved in foodborne illness (27). Enterotoxin T does not have a signal sequence and is released only after *B. cereus* cell lysis. Little is known about the role of enterotoxin FM in foodborne illness, and no biological studies have been performed on this protein.

The newly discovered cytotoxin K (CytK) is similar to the β-toxin of *C. perfringens* (and other related toxins) and was the cause of symptoms in a severe outbreak of *B. cereus* foodborne illness in a nursing home in France in 1998 (44). In this outbreak several people developed bloody diarrhea and three died. This was an outbreak of *B. cereus* necrotic enteritis, although it was not nearly as severe as the necrotic enteritis caused by *C. perfringens* type C (24).

Studies of interactions of Hbl with erythrocytes have suggested that the B protein binds Hbl to the target cells and that L_1 and L_2 have lytic functions (12). More recently, a better model for the action of Hbl has been proposed, suggesting that the components of Hbl bind to target cells independently and then constitute a membrane-attacking complex resulting in a colloid osmotic lysis mechanism (16). Substantial heterogeneity has been observed in the components of Hbl (55).

There is significant sequence identity among the three proteins of Nhe and between the Nhe and Hbl proteins. The identity is greatest in the N-terminal third of the proteins. The most pronounced similarities occur between *nheA* and *hblC*, *nheB* and *hblD*, and *nheC* and *hblA*. This is not only in direct comparison of the sequences but also in predicted transmembrane helices for the six proteins. NheA and HblC have no predicted transmembrane helices, whereas NheB and HblD have two each. Finally, NheC and HblA each have one predicted transmembrane helix, in the same position of the two proteins (32).

Gene Organization of the Enterotoxins

All three proteins of Hbl are transcribed from one operon (*hbl*) (53), and Northern blot analysis has revealed an RNA transcript of 5.5 kb. *hblC* (transcribing

L₂) and *hblD* (transcribing L₁) are separated by only 37 bp and encode proteins of 447 amino acids (aa) and 384 aa. L₂ has a signal peptide of 32 aa, and L₁ has a signal peptide of 30 aa. The B protein, transcribed from *hblA*, consists of 375 aa, with a signal peptide of 31 aa (36). The exact spacing between *hblD* and *hblA* is at least 100 bp (overlapping sequence not published), but is claimed to be approximately 115 bp (53). The spacing between *hblA* and *hblB* is 381 bp, and the length of *hblB* is not known (36). However, based on the length of the Northern blot, a size similar to *hblA* could be suggested for *hblB*. The B and the putative B′ protein are very similar in the first 158 aa (the known sequence of B′ protein based on the DNA sequence). The function of this putative protein is not yet known, but it is possible that it may substitute for the B protein. The *hbl* operon is mapped to the unstable part of the *B. cereus* chromosome (18).

The *nhe* operon contains three open reading frames (Fig. 17.1): *nheA*, *nheB*, and *nheC* (32). The first two gene products have been addressed previously as the 45- and 39-kDa proteins, respectively. The last possible transcript from the *nhe* operon (*nheC*) has not been purified and its function is not known, although we do know that a third component is involved in Nhe activity. The three proteins transcribed from the *nhe* operon have properties described in Table 17.5. The correct sizes of the 45- and 39-kDa proteins are 41 kDa and 39.8 kDa, respectively. The size of NheC is 36.5 kDa. There is a gap of 40 bp between *nheA* and *nheB* and a gap of 109 bp between *nheB*

Table 17.5 Properties of the Nhe proteins[a]

Protein	Signal peptide (aa)	Active protein (aa)	Mol wt (active protein)	pI
NheA	26	360	41.019	5.13
NheB	30	372	39.820	5.61
NheC	30	329	36.481	5.28

[a] Based on reference 32.

and *nheC*. In the 109 bp between *nheB* and *nheC* there is an inverted repeat of 13 bp (32). This structure may result in little production of NheC, compared to that of NheA and NheB. This may be the reason for difficulties encountered in isolating the NheC protein.

The *B. cereus* strain producing CytK did not contain genes for other known *B. cereus* enterotoxins. CytK is a protein of 34 kDa and is also hemolytic. The *cytK* is genetically organized, as shown in Fig. 17.1.

For both Hbl and Nhe, maximal enterotoxin activity is found during the late exponential or early stationary phase, and both of these enterotoxin operons are regulated by *plcR* (2), a gene first described to regulate the *plcA* (phospholipase C) (43). According to Agaisse et al. (2), the binding sequence for this regulatory protein is (TATGNAN₄TNCATG).

The Spore

The spore of *B. cereus* is an important factor in foodborne illness, as it is more hydrophobic than any other

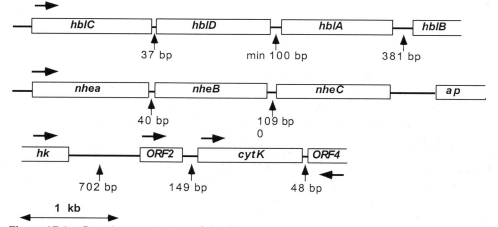

Figure 17.1 Genetic organization of the three enterotoxin genes: *hbl*, *nhe*, and *cytK* (2, 32, 36, 44, 53). The two three-component enterotoxins are regulated by PlcR, which binds upstream of *hblC* and *nheA*. There is an inverted repeat of 13 bp between *nheB* and *nheC*. An aminopeptidase gene (*ap*) is found about 600 bp downstream of *nheC*. *cytK* has an open reading frame (ORF2) upstream, encoding a protein of unknown function, and a histidine kinase (*hk*) further upstream. ORF4 encodes a long-chain fatty acid-coenzyme A ligase (orientated in the opposite direction to that of *cytK*). The arrowheads indicate the orientation of the genes. The gap between the different genes is indicated.

Bacillus spp. spores, which enables it to adhere to several types of surfaces (38, 39). Hence, it is not easily removed during cleaning and is a difficult target for disinfectants. *B. cereus* spores also contain appendages and/or pili (7, 39) that are at least partly involved in adhesion (39). These properties of the *B. cereus* spore not only enable it to withstand sanitation, and hence remain present on surfaces for contamination of different foods, but also aid in adherence to epithelial cells. Studies have revealed that spores of at least one strain associated with an outbreak (25) can adhere to Caco-2 cells in culture, and that these properties are associated with hydrophobicity and possibly the spore's appendages (6).

Commercial Methods for Detection of *B. cereus* Toxins

Commercial kits for detection of the emetic toxin (cereulide) are not yet available. Now that the structure of cereulide is known, however (3), it is likely that a kit will be available in the near future.

Neither of the two commercial immunoassays can quantify the toxicity of the *B. cereus* enterotoxins. The assay from Oxoid measures the presence of the HblC (L$_2$) component, whereas the Tecra kit mainly detects the NheA (45-kDa) component (15, 28, 29, 46). However, if one or both of the commercial kits react positively with proteins from *B. cereus* supernatant fluids, the strain is probably enterotoxin positive. If culture supernatant fluids are also cytotoxic, the strains can be regarded as enterotoxin positive. At present no commercial method is available for detecting CytK.

CONCLUDING REMARKS

B. cereus is a normal inhabitant of soil and is frequently isolated from a variety of foods, including vegetables, dairy products, and meat. It causes an emetic or a diarrheal type of food-associated illness that is becoming increasingly important in developed countries. The diarrheal type of illness is more prevalent in the Western Hemisphere, whereas the emetic type is more prevalent in Japan. Desserts, meat dishes, and dairy products are the foods most frequently associated with diarrheal illness, whereas rice and pasta are the most common vehicles of emetic illness.

A *B. cereus* emetic toxin has been isolated and characterized. Three types of *B. cereus* enterotoxins involved in outbreaks of foodborne illness have been identified. Two of these enterotoxins possess three components and are related, whereas the third is a one-component protein (CytK). Deaths have been caused by the emetic toxin and by a strain producing only CytK.

Some strains of the *B. cereus* group can grow at refrigeration temperature. These variants raise concern about the safety of cooked, refrigerated foods with extended shelf lives. *B. cereus* spores adhere to many surfaces and will survive normal cleaning and disinfection (except for hypochlorite and UVC) procedures. *B. cereus* foodborne illness is likely to be highly underreported because its relatively mild symptoms are of short duration. However, consumer interest in precooked, chilled food products with long shelf lives may lead to products well suited for *B. cereus* survival and growth. Such foods could increase the prominence of *B. cereus* as a foodborne pathogen.

References

1. Aas, N., B. Gondrosen, and G. Langeland. 1992. *Norwegian Food Control Authority's Report on Food Associated Diseases in 1990.* SNT-report 3, Norwegian Food Control Authority, Oslo, Norway.
2. Agaisse, H., M. Gominet, O. A. Okstad, A. B. Kolsto, and D. Lereclus. 1999. PlcR is a pleiotropic regulator of extracellular virulence factor gene expression in *Bacillus thuringiensis. Mol. Microbiol.* 32:1043–1053.
3. Agata, N., M. Mori, M. Ohta, S. Suwan, I. Ohtani, and M. Isobe. 1994. A novel dodecadepsipeptide, cereulide, isolated from *Bacillus cereus* causes vacuole formation in HEp-2 cells. *FEMS Microbiol. Lett.* 121:31–34.
4. Agata, N., M. Ohta, Y. Arakawa, and M. Mori. 1995. The *bceT* gene of *Bacillus cereus* encodes an enterotoxic protein. *Microbiology* 141:983–988.
5. Agata, N., M. Ohta, M. Mori, and M. Isobe. 1995. A novel dodecadepsipeptide, cereulide, is an emetic toxin of *Bacillus cereus. FEMS Microbiol. Lett.* 129:17–20.
6. Andersson, A., P. E. Granum, and U. Rönner. 1998. The adhesion of *Bacillus cereus* spores to epithelial cells might be an additional virulence mechanism. *Int. J. Food Microbiol.* 39:93–99.
7. Andersson, A., U. Rönner, and P. E. Granum. 1995. What problems does the food industry have with the sporeforming pathogens *Bacillus cereus* and *Clostridium perfringens? Int. J. Food Microbiol.* 28:145–156.
8. Asano, S. I., Y. Nukumizu, H. Bando, T. Iizuka, and T. Yamamoto. 1997. Cloning of novel enterotoxin genes from *Bacillus cereus* and *Bacillus thuringiensis. Appl. Environ. Microbiol.* 63:1054–1057.
9. Ash, C., and M. D. Collins. 1992. Comparative analysis of 23S ribosomal RNA gene sequence of *Bacillus anthracis* and emetic *Bacillus cereus* determined by PCR-direct sequencing. *FEMS Microbiol. Lett.* 73:75–80.
10. Ash, C., J. A. Farrow, M. Dorsch, E. Steckebrandt, and M. D. Collins. 1991. Comparative analysis of *Bacillus anthracis, Bacillus cereus,* and related species on the basis of reverse transcriptase sequencing of 16S rRNA. *Int. J. Syst. Bacteriol.* 41:343–346.
11. Ash, C., J. A. Farrow, S. Wallbanks, and M. D. Collins. 1991. Phylogenetic heterogeneity of the genus *Bacillus* revealed by comparative analysis of small subunit ribosomal RNA sequences. *Lett. Appl. Microbiol.* 13:202–206.

12. Beecher, D. J., and J. D. Macmillan. 1991. Characterization of the components of hemolysin BL from *Bacillus cereus*. *Infect. Immun.* **59**:1778–1784.

13. Beecher, D. J., J. L. Schoeni, and A. C. L. Wong. 1995. Enterotoxin activity of hemolysion BL from *Bacillus cereus*. *Infect. Immun.* **63**:4423–4428.

14. Beecher, D. J., and A. C. L. Wong. 1994. Improved purification and characterization of hemolysin BL, a hemolytic dermonecrotic vascular permeability factor from *Bacillus cereus*. *Infect. Immun.* **62**:980–986.

15. Beecher, D. J., and A. C. L. Wong. 1994. Identification and analysis of the antigens detected by two commercial *Bacillus cereus* diarrheal enterotoxin immunoassay kits. *Appl. Environ. Microbiol.* **60**:4614–4616.

16. Beecher, D. J., and A. C. L. Wong. 1997. Tripartite hemolysin BL from *Bacillus cereus*. Hemolytic analysis of component interaction and model for its characteristic paradoxical zone phenomenon. *J. Biol. Chem.* **272**:233–239.

17. Carlson, C. R., D. A. Caugant, and A.-B. Kolstø. 1994. Genotypic diversity among *Bacillus cereus* and *Bacillus thuringiensis* strains. *Appl. Environ. Microbiol.* **60**:1719–1725.

18. Carlson, C. R., T. Johansen, and A.-B. Kolstø. 1996. The chromosome map of *Bacillus thuringiensis* subsp. *canadensis* HD224 is highly similar to that of *Bacillus cereus* type strain ATCC 14579. *FEMS Microbiol. Lett.* **141**:163–167.

19. Christiansson, A. 1993. Enterotoxin production in milk by *Bacillus cereus*: a comparison of methods for toxin detection. *Bull. Int. Dairy Fed.* **287**:54–59.

20. Christiansson, A., A. S. Naidu, I. Nilsson, T. Wadström, and H.-E. Pettersson. 1989. Toxin production by *Bacillus cereus* dairy isolates in milk at low temperatures. *Appl. Environ. Microbiol.* **55**:2595–2600.

21. Claus, D., and R. C. W. Berkeley. 1986. Genus *Bacillus*, p. 1105–1139. *In* P. H. A. Sneath (ed.), *Bergey's Manual of Systematic Bacteriology*, vol. 2. The Williams and Wilkins Co., Baltimore, Md.

22. Damgaaerd, P. H., H. D. Larsen, B. M. Hansen, J. Bresciani, and K. Jørgensen. 1996. Enterotoxin-producing strains of *Bacillus thuringiensis* isolated from food. *Lett. Appl. Microbiol.* **23**:146–150.

23. Drobniewski, F. A. 1993. *Bacillus cereus* and related species. *Clin. Microbiol. Rev.* **6**:324–338.

24. Granum, P. E. 1990. *Clostridium perfringens* toxins involved in food poisoning. *Int. J. Food Microbiol.* **10**:101–112.

25. Granum, P. E. 1994. *Bacillus cereus* in food hygiene. *Norsk Vet. Tidskr.* **106**:911–915. (In Norwegian.)

26. Granum, P. E. 1994. *Bacillus cereus* and its toxins. *J. Appl. Bacteriol. Symp. Suppl.* **76**:61S–66S.

27. Granum, P. E., A. Andersson, C. Gayther, M. C. te Giffel, H. D. Larsen, T. Lund, and K. O'Sullivan. 1996. Evidence for a further enterotoxin complex produced by *Bacillus cereus*. *FEMS Microbiol. Lett.* **141**:145–149.

28. Granum, P. E., S. Brynestad, and J. M. Kramer. 1993. Analysis of enterotoxin production by *Bacillus cereus* from dairy products, food poisoning incidents and non-gastrointestinal infections. *Int. J. Food Microbiol.* **17**:269–279.

29. Granum, P. E., and T. Lund. 1997. *Bacillus cereus* enterotoxins. *FEMS Microbiol. Lett.* **157**:223–228.

30. Granum, P. E., A. Næstvold and K. N. Gundersby. 1995. *Bacillus cereus* food poisoning during the Norwegian Ski Championship for juniors. *Norsk Vet. Tidskr.* **107**:945–948. (In Norwegian.)

31. Granum, P. E., and H. Nissen. 1993. Sphingomyelinase is part of the "enterotoxin complex" produced by *Bacillus cereus*. *FEMS Microbiol. Lett.* **110**:97–100.

32. Granum, P. E., K. O'Sullivan, and T. Lund. 1999. The sequence of the non-haemolytic enterotoxin operon from *Bacillus cereus*. *FEMS Microbiol. Lett.* **177**:225–229.

33. Griffiths, M. W. 1990. Toxin production by psychrotrophic *Bacillus* spp. present in milk. *J. Food Prot.* **53**:790–792.

34. Hauge, S. 1950. Matforgiftninger framkalt av *Bacillus cereus*. *Nord. Hyg. Tidsk.* **31**:189–206. (In Norwegian.)

35. Hauge, S. 1955. Food poisoning caused by aerobic spore forming bacilli. *J. Appl. Bacteriol.* **18**:591–595.

36. Heinrichs, J. H., D. J. Beecher, J. M. MacMillan, and B. A. Zilinskas. 1993. Molecular cloning and characterization of the hblA gene encoding the B component of hemolysin BL from *Bacillus cereus*. *J. Bacteriol.* **175**:6760–6766.

37. Hughes, S., B. Bartholomew, J. C. Hardy, and J. M. Kramer. 1988. Potential application of a HEp-2 cell assay in the investigation of *Bacillus cereus* emetic-syndrome food poisoning. *FEMS Microbiol. Lett.* **52**:7–12.

38. Husmark, U. 1993. Adhesion mechanisms of bacterial spores to solid surfaces. Ph.D. thesis. Chalmers University of Technology and SIK, The Swedish Institute for Food Research, Göteborg, Sweden.

39. Husmark, U., and U. Rönner. 1992. The influence of hydrophobic, electrostatic and morphologic properties on the adhesion of *Bacillus* spores. *Biofouling* **5**:335–344.

40. Jackson, S. G., R. B. Goodbrand, R. Ahmed, and S. Kasatiya. 1995. *Bacillus cereus* and *Bacillus thuringiensis* isolated in a gastroenteritis outbreak investigation. *Lett. Appl. Microbiol.* **21**:103–105.

41. Kramer, J. M., and R. J. Gilbert. 1989. *Bacillus cereus* and other *Bacillus* species, p. 21–70. *In* M. P. Doyle (ed.), *Foodborne Bacterial Pathogens*. Marcel Dekker, New York, N.Y.

42. Lechner, S., R. Mayr, K. P. Francic, B. M. Prub, T. Kaplan, E. Wieber-Gunkel, G. A. S. B. Stewart, and S. Scherer. 1998. *Bacillus weihenstephanensis* sp. nov. is a new psychrotolerant species of the *Bacillus cereus* group. *Int. J. Syst. Bacteriol.* **48**:1373–1382.

43. Lereclus, D., H. Agaisse, M. Gominet, S. Salamitou, and V. Sanchis. 1996. Identification of a *Bacillus thuringiensis* gene that positively regulates transcription of the phosphatidylinositol-specific phospholipase C gene at the onset of the stationary phase. *J. Bacteriol.* **178**:2749–2756.

44. Lund, T., M. L. De Buyser, and P. E. Granum. 2000. A new cytotoxin from *Bacillus cereus* that may cause necrotic enteritis. *Mol. Microbiol.* **38**:254–261.

45. Lund, T., and P. E. Granum. 1996. Characterisation of a non-haemolytic enterotoxin complex from *Bacillus cereus*

isolated after a foodborne outbreak. *FEMS Microbiol. Lett.* **141:**151–156.

46. **Lund, T., and P. E. Granum.** 1997. Comparison of biological effect of the two different enterotoxin complexes isolated from three different strains of *Bacillus cereus. Microbiology* **143:**3329–3336.

47. **Mahler, H., A. Pasi, J. M. Kramer, P. Schulte, A. C. Scoging, W. Bar, and S. Krahenbuhl.** 1997. Fulminant liver failure in association with the emetic toxin of Bacillus cereus. *N. Engl. J. Med.* **336:**1142–1148.

48. **Mortimer, P. R., and G. McCann.** 1974. Food poisoning episodes associated with *Bacillus cereus* in fried rice. *Lancet* **i:**1043–1045.

49. **Nakamura, L. K.** 1998. *Bacillus pseudomycoides* sp. nov. *Int. J. Syst. Bacteriol.* **48:**1031–1035.

50. **Preist, F. G.** 1993. Systematics and ecology of *Bacillus*, p. 3–16. *In* A. L. Sonenshein, J. A. Hoch, and R. Losick (ed.), *Bacillus subtilis and Other Gram-Positive Bacteria.* American Society for Microbiology, Washington, D.C.

51. **Ray, D. E.** 1991. Pesticides derived from plants and other organisms, p. 585–636. *In* W. J. Hayes and E. R. Laws Jr. (ed.), *Handbook of Pesticide Toxology.* Academic Press, Inc., New York, N.Y.

52. **Rivera, A. M. G., P. E. Granum, and F. G. Priest.** 2000. Common occurrence of enterotoxin genes and enterotoxicity in *Bacillus thuringiensis. FEMS Microbiol. Lett.* **190:**151–155.

53. **Ryan, P. A, J. M. Macmillan, and B. A. Zilinskas.** 1997. Molecular cloning and characterization of the genes encoding the L_1 and L_2 components of hemolysin BL from *Bacillus cereus. J. Bacteriol.* **179:**2551–2556.

54. **Schmidt, K. (ed.).** 1995. *WHO Surveillance Programme for Control of Foodborne Infections and Intoxications in Europe.* Sixth Report—FAO/WHO Collaborating Centre for Research and Training in Food Hygiene and Zoonoses, Berlin, Germany.

55. **Schoeni, J. L., and A. C. L. Wong.** 1999. Heterogeneity observed in the components of hemolysin BL, an enterotoxin produced by Bacillus cereus. *Int. J. Food Microbiol.* **53:**159–167.

56. **Shinagawa, K.** 1993. Serology and characterization of *Bacillus cereus* in relation to toxin production. *Bull. Int. Dairy Fed.* **287:**42–49.

57. **Shinagawa, K., H. Konuma, H. Sekita, and S. Sugii.** 1995. Emesis of rhesus monkeys induced by intragastric administration with the HEp-2 vacuolation factor (cereulide) produced by *Bacillus cereus. FEMS Microbiol. Lett.* **130:**87–90.

58. **Shinagawa, K., S. Otake, N. Matsusaka, and S. Sugii.** 1992. Production of the vacuolation factor of *Bacillus cereus* isolated from vomiting-type food poisoning. *J. Vet. Med. Sci.* **54:**443–446.

59. **Shinagawa, K., S. Ueno, H. Konuma, N. Matsusaka, and S. Sugii.** 1991. Purification and characterization of the vascular permeability factor produced by *Bacillus cereus. J. Vet. Med. Sci.* **53:**281–286.

60. **Skirrow, M. B.** 1990. Foodborne illness: *Campylobacter. Lancet* **336:**921–923.

61. **Thompson, N. E., M. J. Ketterhagen, M. S. Bergdoll, and E. J. Schantz.** 1984. Isolation and some properties of an enterotoxin produced by *Bacillus cereus. Infect. Immun.* **43:**887–894.

62. **Turnbull, P. C. B.** 1986. *Bacillus cereus* toxins, p. 397–448. *In* F. Dorner and J. Drews (ed.), *Pharmacology of Bacterial Toxins. International Encyclopedia of Pharmacology and Therapeutics,* sec. 119. Pergamon Press, Oxford, United Kingdom.

63. **Väisänen, O. M., N. J. Mwaisumo, and M. S. Salkinoja-Salonen.** 1991. Differentiation of dairy strains of the *Bacillus cereus* group by phage typing, minimum growth temperature, and fatty acid analysis. *J. Appl. Bacteriol.* **70:**315–324.

64. **van Netten, P., A. van de Moosdijk, P., van Hoensel, D. A. A. Mossel, and I. Perales.** 1990. Psychrotrophic strains of *Bacillus cereus* producing enterotoxin. *J. Appl. Bacteriol.* **69:**73–79.

65. **Yokoyama, K., M. Ito, N. Agata, M. Isobe, K. Shibayama, T. Horii, and M. Ohta.** 1999. Pathological effect of synthetic cereulide, an emetic toxin of *Bacillus cereus,* is reversible in mice. *FEMS Immunol. Med. Microbiol.* **24:**115–120.

Food Microbiology: Fundamentals and Frontiers, 2nd Ed.
Edited by M. P. Doyle et al.
© 2001 ASM Press, Washington, D.C.

Bala Swaminathan

Listeria monocytogenes

18

Listeriosis has emerged as a major foodborne disease during the past 2 decades, after a 1981 outbreak of listeriosis in Nova Scotia, Canada, was traced to contaminated coleslaw (129). However, the causative agent of listeriosis, *Listeria monocytogenes*, was discovered more than 70 years ago by E. G. D. Murray and James Pirie, who were working independently of each other and provided exquisitely detailed descriptions of listeriosis in small animals (122). A fascinating account of the discovery of *L. monocytogenes* has been compiled by Jim McLauchlin (100). The first documented human listeriosis case was a soldier who suffered from meningitis at the end of World War I. However, there is a suggestion in the literature that listeriosis may have been the cause of Queen Anne's 17 unsuccessful pregnancies (128).

Between 1930 and 1950, a few human listeriosis cases were reported. However, there are now hundreds reported every year (124). The emergence of listeriosis is the result of complex interactions between various factors reflecting changes in social patterns. These factors include:

- improvements during the past 50 years in medicine, public health, sanitation, and nutrition that have resulted in increased life expectancy, particularly in developed countries
- the ongoing AIDS epidemic, as well as the widespread use of immunosuppressive medications for the treatment of malignancies and management of organ transplantations, which have greatly expanded the immunocompromised population at increased risk for listeriosis
- changes in food production practices, particularly the high degree of centralization and consolidation of food production and processing, the ever-expanding national and international distribution of foods, and increased use of refrigeration as a primary means of preservation of foods
- changes in food habits (increased consumer demand for convenience food that has a fresh-cooked taste, can be purchased ready-to-eat, refrigerated, or frozen, can be prepared rapidly, and requires essentially little cooking before consumption) and changes in handling and preparation practices

Listeriosis is an atypical foodborne illness of major public health concern because of the severity of the disease (meningitis, septicemia, and abortion), a high

Bala Swaminathan, Foodborne and Diarrheal Diseases Branch, Centers for Disease Control and Prevention, 1600 Clifton Road, Mailstop C-03, Atlanta, GA 30333.

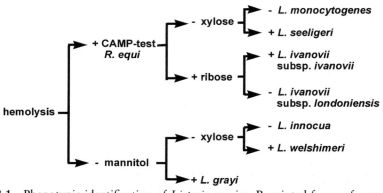

Figure 18.1 Phenotypic identification of *Listeria* species. Reprinted from reference 49 with permission.

case-fatality rate (approximately 20 to 30% of cases), a long incubation time, and a predilection for individuals who have an underlying condition which leads to impairment of T-cell-mediated immunity. *L. monocytogenes* differs in many respects from most other foodborne pathogens: it is widely distributed, resistant to diverse environmental conditions, including low pH and high NaCl concentrations, and is microaerobic and psychrotrophic. The various ways the bacterium can enter into food processing plants, its ability to survive for long periods of time in the environment (soil, plants, and water), on foods, and in food processing plants, and its ability to grow at very low temperatures (2 to 4°C) and to survive in or on food for prolonged periods under adverse conditions have made this bacterium a major concern for the agrifood industry during the last decade. The significance of *L. monocytogenes* as a foodborne pathogen is complex. The severity and case-fatality rate of the disease require appropriate preventive measures, but the characteristics of the microorganism are such that it is unrealistic to expect all food to be *Listeria*-free. This dilemma has generated animated debate and prompted research in various areas, such as conventional and rapid detection methods for *Listeria* in food, its behavior in food, the genetics of virulence and molecular typing of the organism, as well as epidemiologic investigations of the disease.

CHARACTERISTICS OF THE ORGANISM

Classification

The genus *Listeria* belongs to the *Clostridium* subbranch, together with *Staphylococcus*, *Streptococcus*, *Lactobacillus*, and *Brochothrix*. This phylogenetic position of *Listeria* is consistent with its low G+C DNA content (36 to 42%).

L. monocytogenes is one of six species in the genus *Listeria*. The other species are *L. ivanovii*, *L. innocua*,

L. seeligeri, *L. welshimeri*, and *L. grayi* (122). Two subspecies of *L. ivanovii* have been described: *L. ivanovii* subsp. *ivanovii* and *L. ivanovii* subsp. *londoniensis* (11). *L. murrayi*, which was a separate species in the genus *Listeria*, is now included in the species *L. grayi* (123). Based on results of DNA-DNA hybridization, multilocus enzyme analysis, and 16S rRNA sequencing, the six species in the genus *Listeria* are divided on two lines of descent: (i) *L. monocytogenes* and its closely related species, *L. innocua*, *L. ivanovii* (*L. ivanovii* subsp. *ivanovii* and *L. ivanovii* subsp. *londoniensis*), *L. welshimeri*, and *L. seeligeri*, and (ii) *L. grayi*. Within the genus *Listeria*, only *L. monocytogenes* and *L. ivanovii* are considered to be pathogenic, as evidenced by their 50% lethal dose (LD$_{50}$) in mice and their ability to grow in mouse spleen and liver. *L. monocytogenes* is a human pathogen of high public health concern; *L. ivanovii* is primarily an animal pathogen.

The identification of *Listeria* species is based on a limited number of biochemical markers, among which hemolysis is used to differentiate between *L. monocytogenes* and the most frequently encountered nonpathogenic *Listeria* species, *L. innocua* (136). The biochemical tests useful for discriminating between the species are acid production from D-xylose, L-rhamnose, alpha-methyl-D-mannoside, and D-mannitol (Fig. 18.1).

Further Characterization and Subtyping of *L. monocytogenes*

L. monocytogenes isolates are often characterized below the species level for the purposes of public health surveillance and to assist in outbreak investigations. Serotyping has proved its value over many years. There are 13 serotypes of *L. monocytogenes* which can cause disease, but 95% of human isolates belong to three serotypes: 1/2a, 1/2b, and 4b (62, 137). Because of the low discriminatory power of this method, phage-typing systems were developed and were the only means to distinguish between strains of the same serotype before the introduction of molecular typing methods. Since 1989, various molecular typing methods have been applied to *L. monocytogenes,* including multilocus enzyme electrophoresis, ribotyping, DNA microrestriction (high-frequency cutting enzymes and conventional electrophoresis) and macrorestriction (infrequently cutting enzymes and pulsed-field gel electrophoresis [PFGE]), and random amplification of polymorphic DNA (62). Because of their ability to type all strains and the high discriminatory power of some of them, these methods have become invaluable tools for epidemiologic investigations. In addition, unlike serotyping and phage typing, which require specialized reagents (typing sera and bacteriophages) and thus are available only in a few reference laboratories, these methods can be performed in any reasonably equipped laboratory. In the United States, the Centers for Disease Control and Prevention has established a network (PulseNet) of public health and food regulatory laboratories that are routinely subtyping foodborne pathogenic bacteria to rapidly detect foodborne disease clusters that may have a common source. PulseNet laboratories use highly standardized protocols for subtyping of bacteria by PFGE and are able to quickly compare PFGE patterns of foodborne pathogens from different locations within the country via the Internet. *L. monocytogenes* was added to PulseNet in 1999. Despite these advances in molecular subtyping methods, serotyping and phage typing continue to be used in public health laboratories. Serotyping is useful for first-level discrimination between isolates before more sensitive subtyping methods are applied. Bacteriophage typing is still the method of choice for rapidly typing large numbers of isolates in acute outbreak situations.

L. monocytogenes serotype 4b strains are responsible for 33 to 50% of sporadic human cases worldwide and caused all major foodborne outbreaks in Europe and North America in the 1980s. In contrast, isolates recovered from foods in many countries mostly belong to serotypes 1/2a and 1/2c. Although the reasons for this have not been fully elucidated, some recent observations provide intriguing clues. DNA from several epidemic-associated strains of *L. monocytogenes* 4b were resistant to restriction by *Sau*3A and by other restriction enzymes known to be sensitive to cytosine methylation at 5'-GATC-3' sites. This modification of *Sau*3A restriction appears to be host mediated (160). Also, a novel gene (*gtcA*) involved in the incorporation of galactose and glucose to the cell wall teichoic acid appears to be present only in serotype 4b and other serotype 4 isolates (119).

SUSCEPTIBILITY TO PHYSICAL AND CHEMICAL AGENTS

L. monocytogenes is able to initiate growth in the temperature range of 0 to 45°C. As would be expected, the bacterium grows more slowly at lower temperatures. The average generation times for 39 *L. monocytogenes* strains were 43, 6.6, and 1.1 h at 4, 10, and 37°C, and the respective lag times were 151, 48, and 7.3 h (7). Temperatures below 0°C preserve or moderately inactivate the bacterium. Survival and injury during frozen storage depend on the substrate and the rate of freezing. *L. monocytogenes* is inactivated by exposure to temperatures above 50°C.

Zheng and Kathariou (159) cloned three genes (*ltrA*, *ltrB*, and *ltrC*) of *L. monocytogenes* that are essential for low-temperature growth. When a 1.2-kb internal fragment of *ltrB* was used as a probe in Southern hybridizations of *Hin*dIII-digested *L. monocytogenes* DNA, a 9.5-kb DNA fragment that hybridized with the probe was found to be unique to epidemic-associated serotype 4b strains of *L. monocytogenes*.

The pH range for the growth of *L. monocytogenes* was thought to be 5.6 to 9.6, although recent investigations indicate that the organism can initiate growth in laboratory media at pH values as low as 4.4 (91). Growth at low pH values is influenced by the incubation temperature and the type of acid. At pH values below 4.3, the bacterial cells may survive but do not multiply. Experimentally, the presence of up to 0.1% acetic, citric, and lactic acids in tryptose broth inhibits the growth of *L. monocytogenes,* with inhibition increasing as the incubation temperature decreases (1). The antilisterial activity of these acids is related to their degree of dissociation, with citric and lactic acids being less detrimental for the pathogen than acetic acid at an equivalent pH.

L. monocytogenes grows optimally at water activity (a_w) ≥ 0.97. For most strains, the minimum a_w for growth is 0.93 (91), but some strains may grow at a_w values as low as 0.90. Furthermore, the bacterium may survive for long periods at a_w values as low as 0.83 (138). Also, an inverse relationship exists between thermal resistance of *L. monocytogenes* and the a_w of the medium in which it is suspended (91, 147). This should be a cause for concern for food manufacturers who rely on low a_w and thermal treatment for preservation of their food products.

L. monocytogenes is able to grow in the presence of 10 to 12% sodium chloride; it grows to high populations in moderate salt concentrations (6.5%). The bacterium survives for long periods in high salt concentrations; the survival in high-salt environments is significantly increased by lowering the temperature.

LISTERIOSIS AND READY-TO-EAT FOODS

Certain ready-to-eat processed foods pose a high risk of contracting listeriosis for susceptible populations, as determined by active surveillance for sporadic listeriosis and epidemiologic investigation of listeriosis outbreaks. These foods are usually preserved by refrigeration and offer an appropriate environment for the multiplication of *L. monocytogenes* during manufacture, aging, transportation, and storage. The foods in this category include unpasteurized milk and products prepared from unpasteurized milk, soft cheeses, frankfurters, delicatessen meats and poultry products, and some seafood. In Canada, the regulatory policy directs inspection and compliance action on ready-to-eat foods that can support the growth of *L. monocytogenes*. The highest priority is given to those foods that have been known to cause listeriosis and to those that have greater than 10 days of shelf life (45).

Fluid Milk Products

Raw milk is a well-documented source of *L. monocytogenes*. The 1985 Mexican-style cheese-associated outbreak in California was likely caused by the use of unpasteurized milk for cheesemaking (90). Thermal inactivation studies of *L. monocytogenes* in milk have produced conflicting results. Several investigators observed that cells of *L. monocytogenes* suspended in milk were effectively inactivated under high-temperature–short-time (HTST) pasteurization conditions (71°C for 15 s or equivalent), whereas others contended that the bacteria were not completely inactivated (8, 12, 37). The methods used for the determination of survival (open tubes versus sealed capillary tubes) were responsible for the discrepant results. The most likely explanation for the survival of *L. monocytogenes* in the open tubes was that the bacterial cells adhering to the wall of the test tubes during initial mixing of the suspensions were not subjected to the thermal inactivation temperatures because the entire tube was not submerged in the hot water.

When the thermal inactivation of freely suspended *L. monocytogenes* cells was compared with that of cells that were internalized with phagocytic leukocytes, once again discrepant results were reported. Differences in the physiological state of the bacterial cells (actively growing cells versus cells in stationary phase) and growth of bacterial cells at elevated temperatures (infected cows that may have developed fever) before exposure to pasteurization conditions are some of the potential factors that may have caused the cells to become more heat resistant. Lou and Yousef (91) indicate that the following factors should be considered in evaluating the effect of pasteur-

ization on *L. monocytogenes*: (i) the safety margin of pasteurization for inactivation of the bacterium may be lower than previously thought, (ii) the level of contamination in pooled milk is lower than that used in most inoculation studies, (iii) homogenization of milk destroys the integrity of phagocytic cells in milk, thus removing any protection offered to bacterial cells, and (iv) thermoduric spoilage microorganisms surviving HTST are likely to out-compete *L. monocytogenes*. Thus there appears to be general agreement with the observations of the World Health Organization informal study group, which concluded that "pasteurization is a safe process which reduces the number of *L. monocytogenes* in raw milk to levels that do not pose an appreciable risk to human health" (4). Recognizing the small margin of safety offered by the HTST process, most raw milk processors have adopted processes that employ temperatures well above the minimum legal requirements for pasteurized milk. *L. monocytogenes* grows in pasteurized milk, with the numbers increasing 10-fold in 7 days at 4°C; also, listeriae grow more rapidly in pasteurized milk than in raw milk when incubated at 7°C (111). Therefore, fluid milk that is contaminated after pasteurization and stored under refrigeration may attain very high populations of *L. monocytogenes* after 1 week; temperature abuse may further enhance the multiplication of bacterial cells, as was evidenced in a chocolate milk-associated outbreak in Illinois (31). Additional details may be obtained from a recent review of the prevalence, growth, and survival of *L. monocytogenes* in fluid milk and unfermented dairy products (126).

Cheeses

Because of its relative hardiness to temperature fluctuations, ability to multiply at refrigeration temperature, and salt tolerance, *L. monocytogenes* can survive the cheese manufacturing and ripening process. Its growth in cheese milk is retarded but not completely inhibited by lactic starter cultures. During the manufacturing process, *L. monocytogenes* is primarily concentrated in the cheese curd, with only a very small proportion of cells appearing in whey. The behavior of listeriae in the curd is influenced by the type of cheese, ranging from growth in feta cheese to significant inactivation during cottage cheese manufacture. During ripening of the cheese, the numbers of *L. monocytogenes* cells may increase (Camembert), decrease gradually (cheddar or Colby), or decrease rapidly during early ripening and then stabilize (blue cheese) (113). Consumption of soft cheeses by susceptible persons is a risk factor for sporadic and epidemic listeriosis in North America and Europe (10, 74, 90, 132, 134). Recently, French investigators concluded that 49% of sporadic listeriosis in that country could be attributed to the consumption of soft cheeses (34). In the United States, the Centers for Disease Control and Prevention recommends that pregnant women avoid the consumption of soft cheeses (21).

Meat and Poultry Products

The multiplication potential for *L. monocytogenes* in meat and poultry products depends on the type of meat, the pH, and the type and cell populations of competitive flora. Poultry supports the growth of *L. monocytogenes* better than other meats; roast beef and summer sausage support the least growth. Contamination of animal muscle tissue may occur either from symptomatic or asymptomatic carriage of *L. monocytogenes* by the food animal before slaughter or contamination of the carcass after slaughter. Because *L. monocytogenes* tends to concentrate and multiply in the kidney, mesenteric and mammary lymph nodes, and liver and spleen of infected animals, eating organ meat may be more hazardous than eating muscle tissue (48). Regardless of the route of contamination of meat, *L. monocytogenes* attaches strongly to the surface of raw meats and is difficult to remove or inactivate. *L. monocytogenes* multiplies readily in meat products, including vacuum-packaged beef, at pH values near 6.0, whereas there is very little or no multiplication at approximately pH 5.0 (48, 58). Ready-to-eat meat products that have received a heat treatment followed by cooling in brine before packaging may provide a particularly conducive environment for multiplication of *L. monocytogenes* because of the reduction in competitive flora and the high salt tolerance of the organism.

A monitoring program for *L. monocytogenes* in cooked ready-to-eat meat products instituted by the U.S. Department of Agriculture revealed the following incidence rates for 1993 to 1996: beef jerky, 0–2.2%; cooked sausages, 1.0–5.3%; salads and spreads, 2.2–4.7%; sliced ham and sliced luncheon meats, 5.1–81%. In Canada, the incidence of *L. monocytogenes* in domestic ready-to-eat, cooked meat products was 24% in 1989–90 but declined to 3% or less during 1991–92. However, the specimens obtained from the environments of establishments producing *Listeria*-positive foods remained constant at 12% between 1989 and 1992 (48). This observation is critically important because it suggests that once it is established in a food processing plant environment, *L. monocytogenes* may persist in that location for several years even if the products produced at that location are *Listeria*-free for several months.

Cooked, ready-to-eat meat and poultry products have been implicated as the source of sporadic and epidemic listeriosis on several occasions in North America and

Europe. Consumption of unreheated frankfurters and undercooked chicken was identified as a risk factor for sporadic listeriosis in a study conducted in the United States (135). The Centers for Disease Control and Prevention identified a contaminated turkey frankfurter product as the source of sporadic *L. monocytogenes* infection in a cancer patient in 1989 (18, 156). Two recent (1998 and 2000) multistate outbreaks of listeriosis in the United States were linked to contaminated frankfurters and turkey deli meat, respectively (19, 103).

Seafood

The role of seafood in human listeriosis was recently reviewed (125). Shrimp, smoked mussels, and imitation crabmeat have been implicated as the sources of infections for human listeriosis. Gravad rainbow trout and cold-smoked rainbow trout were implicated as the sources of infection in two outbreaks in Sweden and Finland, respectively (43, 107). In the United States, *L. monocytogenes* has been isolated from both domestic and imported, fresh, frozen, and processed seafood products, including crustaceans, molluscan shellfish, and finfish (77). A U.S. Food and Drug Administration survey of domestic and imported refrigerated or frozen cooked crabmeat in 1987–88 determined contamination levels of 4.1% for domestic products and 8.3% for imported products (42). Crab and smoked fish samples analyzed by the U.S. Food and Drug Administration between 1991 and 1996 showed that 7.5% and 13.6% of the samples, respectively, were contaminated with *L. monocytogenes*. A quantitative risk assessment for *L. monocytogenes* in ready-to-eat foods has been done in the United States (42).

Huss et al. (69) have classified the following seafoods as potential high-risk foods for listeriosis: (i) molluscs, including fresh and frozen mussels, clams, and oysters in shell or shucked, (ii) raw fish, (iii) lightly preserved fish products, including salted, marinated, fermented, cold-smoked, and gravad fish, and (iv) mildly heat-processed fish products and crustaceans.

Seafood consumption is still much less when compared with consumption of meats and cheeses. Also, the production of seafood products is done on a much smaller scale than meat and cheese manufacture. This may be the reason that large outbreaks of listeriosis due to seafoods have not been reported and that case-control studies have not identified this group of foods as a major risk for listeriosis (125).

Effect of Newer Methods of Food Preservation

New trends in food preservation have recently been developed, including the use of biopreservatives, vacuum-packaging, and modified atmospheres (92). The antagonistic effect of nisin on *L. monocytogenes* has been demonstrated, and its activity in foods is strongly dependent on the chemical composition of the food to which it is added. Pediocins (from *Pediococcus pentosaceus* and *Pediococcus acidilactici*) inhibit *L. monocytogenes* growth. Bacteriocin from *Lactobacillus bavaricus* transiently affects *L. monocytogenes* in various beef systems, especially at low temperature. A similar effect was observed with a bacteriocin from *Carnobacterium piscicola* in broth and skimmed milk (99). However, a drawback to the use of bacteriocins is the emergence of resistant mutants, as has been observed with bavaricin A and nisin (32, 86). Modi et al. (109) reported that a nisin-resistant strain of *L. monocytogenes* grew in the presence of nisin; it was more sensitive to heat than the wild-type strain.

A survey of vacuum-packaged processed meat in retail stores revealed that 53% of the samples tested were contaminated with *L. monocytogenes* and that 4% contained more than 1,000 CFU/g (60). This observation corroborates experimental evidence that the growth of *L. monocytogenes* (a facultative anaerobe) is not significantly affected by vacuum packaging. There has been considerable interest in modified-atmosphere packaging of meat products (low oxygen and high carbon dioxide concentrations) over recent years because of the increasing demand for refrigerated convenience foods with extended shelf life. Studies with meat juice, raw chicken, and precooked chicken nuggets revealed that such atmospheres do not significantly affect the growth of *L. monocytogenes*. However, *L. monocytogenes* is sensitive to low doses of irradiation.

A substantial portion of surviving cells are injured by heating to 54°C or above, freezing, or various other treatments. Heat-stressed *L. monocytogenes* may be considerably less pathogenic than nonstressed cells. On nonselective agar, injured cells can repair the damage induced by stress and grow, but they are subject to additional stress in selective agar, and variable recovery is observed. Thus, the presence of sublethally damaged cells in food samples may lead to differences in cell recovery rates.

RESERVOIRS

The role of silage in the transmission of animal disease was bacteriologically documented in 1960, and *L. monocytogenes* strains were isolated from natural decaying vegetation in 1968 and thereafter (70, 155). The bacterium can survive and grow in soil and water. *Listeria* species have been detected in various aqueous environments: surface water of canals and lakes, ditches of polders in The Netherlands, freshwater tributaries draining into a California bay, and sewage (24, 35, 64). Alfalfa plants and other crops grown on soil treated with sewage sludge are contaminated with *Listeria* spp. (3). Half the radish samples grown in soil inoculated with *L. monocytogenes* were confirmed positive 3 months later (151). Similarly, the presence of *L. monocytogenes* in pasture grasses and grass silages has often been documented (70). The widespread presence of *L. monocytogenes* in soil is likely due to contamination by decaying plant and fecal material, with the soil providing a cool, moist environment and the decaying material providing the nutrients (49).

L. monocytogenes has been isolated from the feces of many healthy animals and birds; listeriosis in many animal species has been recorded. Humans exhibiting symptoms of listeriosis and asymptomatic carriers shed the organism in their feces. Figure 18.2 illustrates the many ways in which *L. monocytogenes* can be spread from the environment to animals and humans and back to the environment.

Food Processing Plants

Entry of *L. monocytogenes* into food processing plants occurs through soil on workers' shoes and clothing and on transport equipment, animals which excrete the bacterium or have contaminated hides or surfaces, raw plant tissue, raw food of animal origin, and possibly healthy human carriers. Growth of listeriae is favored by high humidity and the presence of nutrients. *L. monocytogenes* is most often detected in moist areas such as floor drains, condensed and stagnant water, floors, residues, and processing equipment (30). *L. monocytogenes* can attach to various kinds of surfaces (including stainless steel, glass, and rubber), and biofilms have been described in meat and dairy processing environments (76). *Listeria* sp. can survive on fingers after hand washing and in aerosols. The presence of *L. monocytogenes* in the food processing chain is evidenced by the widespread distribution of the listeriae in processed products. Contaminated effluents from food processing plants increase the spread of *L. monocytogenes* in the environment. Sources of *L. monocytogenes* in dairy processing plants include the environment (floors and floor drains, especially in areas in and around coolers or places subject to outside contamination) and raw milk. Efforts to ensure that milk is safe from *L. monocytogenes* contamination should focus on promoting appropriate methods of pasteurization and on identifying and eliminating sources of postpasteurization contamination (154).

The presence of *L. monocytogenes* on carcasses is usually attributed to contamination by fecal matter during slaughter. A high percentage (11 to 52%) of animals

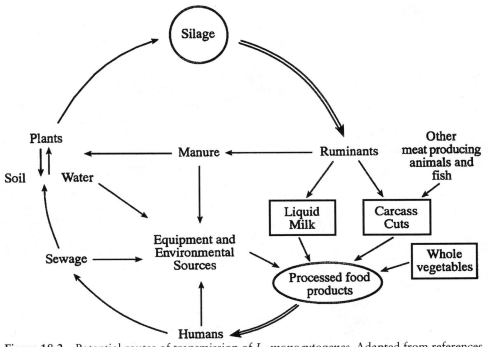

Figure 18.2 Potential routes of transmission of *L. monocytogenes*. Adapted from references 9 and 49. Circles or ovals indicate areas of greatest risk of *L. monocytogenes* multiplication. Boxes indicate where direct consumption of minimally processed products (e.g., whole fresh vegetables, cooked carcass cuts of meat and fish, and effectively pasteurized milk) presents a low risk. Double arrows indicate consumer is at risk.

are healthy fecal carriers. Up to 45% of pigs harbor *L. monocytogenes* in tonsils, and 24% of cattle have contaminated internal retropharyngeal nodes (17, 143). *L. monocytogenes* has been recovered from both unclean and clean zones (especially on workers' hands) in slaughterhouses; the most heavily contaminated working areas are cow dehiding and pig stunning and hoisting. Studies in turkey and poultry slaughterhouses failed to detect *L. monocytogenes* in feather samples, scalding tank water overflow, neck skin, livers, hearts, ceca, or large intestines. In contrast, *L. monocytogenes* was recovered from feather plucker drip water, chill water overflow, recycling water for cleaning gutters, and mechanically deboned meat. These findings demonstrate the importance of the defeathering machine, chillers, and recycled water in product cross-contamination (56).

Postprocessing contamination is the most likely route of contamination of processed foods by *L. monocytogenes*. To date, there is no evidence to indicate that *L. monocytogenes* can survive heat processing protocols used to render foods safe. *L. monocytogenes* is a particularly difficult bacterium to eliminate from the food processing plant because it has the propensity to adhere to food contact surfaces and to form biofilms, which makes implementing effective sanitation procedures

difficult (61). The refrigerated, moist environments in food processing plants provide a good growth environment for *L. monocytogenes*. Because *L. monocytogenes* is a frequent contaminant of raw materials used in food processing plants, there are ample opportunities for reintroduction of listeriae into food processing facilities (38). Extensive information on the problem of *L. monocytogenes* in various food processing environments and approaches to control is presented in a recent review by Gravani (61).

SURVEYS OF FOODS FOR PREVALENCE OF *L. MONOCYTOGENES* AND THE REGULATORY STATUS IN DIFFERENT COUNTRIES

Several recently completed surveys confirm that *L. monocytogenes* contamination of food is widespread in many parts of the world. Hutchins (71) summarized the data on the prevalence of *L. monocytogenes* in foods in the United States and the United Kingdom. The data indicate that contamination levels range from 0 (2% milk, desserts) to 16% (salami) in ready-to-eat foods. The contamination of foods to be cooked or heated is much higher, ranging from 2% in cook-chill foods in catering establishments to 60% in raw chicken. In a survey conducted in Denmark in 1994–1995, *L. monocytogenes* was isolated from 14.2% and 30.9% of raw fish and raw meats, respectively. Preserved, but not heat-treated, fish and meat products were more frequently contaminated (10.8% and 23.5%, respectively) than heat-treated meat products (5%). *L. monocytogenes* was present at levels exceeding 100 CFU/g in 1.3% of preserved (both heat-treated and not heat-treated) fish and meat products packed under vacuum or modified atmospheres for extended shelf life. In contrast, a survey of the same type of products that were not packed under vacuum or modified atmospheres revealed significantly lower contamination levels (0.3 to 0.6%) (110). In a survey of retail foods in Japan, *L. monocytogenes* was isolated from 12, 20, 37, and 25% of minced beef, minced pork, minced chicken, and minced pork-beef mixture, respectively; only five chicken samples had populations higher than 100 CFU/g. *L. monocytogenes* was isolated from 5.4% of smoked salmon samples and 3.3% of ready-to-eat uncooked seafood products (72). A survey conducted in Barcelona, Spain, revealed *L. monocytogenes* contamination in 9.3% of ready-to-eat foods and 2.9% of foods intended to be cooked before consumption (33). *L. monocytogenes* has been isolated from many vegetables, including bean sprouts, cabbage, cucumbers, leafy vegetables, potatoes, prepackaged salads, radishes, salad vegetables, and tomatoes in North America, Europe, and Asia (9).

Because of the frequent occurrence of *L. monocytogenes* in foods and in the food processing environment, food regulatory agencies in many countries have accepted the argument that it is impossible to produce *L. monocytogenes*-free foods and have established tolerance levels for *L. monocytogenes*. For example, the Canadian government policy directs inspection and compliance action to ready-to-eat foods that are capable of supporting the growth of *L. monocytogenes*. Foods are divided into three risk categories, with products causally linked to human listeriosis being placed in category 1 and regulated more stringently than foods in categories 2 and 3 (47). The French position directs that foods (25-g samples) should be *L. monocytogenes*-free, but when this is not possible, one must try to obtain the lowest level possible. In France, foods are divided into three groups as follows: (i) food for populations "at risk," (ii) foods heated in their wrapping or aseptically conditioned after treatment, and (iii) raw foods or foods susceptible to recontamination after treatment (85). In contrast, the United Kingdom and the United States, although they acknowledge the ubiquitous distribution of *L. monocytogenes* in the food supply and the difficulties

in producing *L. monocytogenes*-free ready-to-eat foods, have decided not to adopt tolerance levels for *L. monocytogenes* in these foods. Both countries argue that any "acceptable" levels for *L. monocytogenes* would require knowledge of the number of organisms unlikely to cause human infection. Because there are no scientific data to establish such tolerance levels, these two countries continue a "zero-tolerance" policy for *L. monocytogenes* in ready-to-eat foods (139, 141).

HUMAN CARRIAGE

Asymptomatic fecal carriage of *L. monocytogenes* has been studied in a variety of human populations, including healthy people, pregnant women, outpatients with gastroenteritis, slaughterhouse workers, laboratory workers handling *Listeria*, food handlers, and patients undergoing renal transplantation or hemodialysis (144). *L. monocytogenes* was isolated from 2 to 6% of fecal samples from healthy people. Patients with listeriosis often excrete high populations of *L. monocytogenes*; e.g., specimens from 21% of patients had $\geq 10^4$ *L. monocytogenes*/g of feces, and 18% of household contacts of patients with listeriosis fecally shed the same serotype and multilocus enzyme type of *L. monocytogenes* as the corresponding index case (75, 131). Among household contacts of 18 pregnant women, 8.3% asymptomatically shed *L. monocytogenes*, whereas no listeriae were isolated from 30 household contacts of age-, sex-, and hospital-matched control subjects (98). In addition, results of an investigation of an outbreak in California in 1985 revealed that community-acquired outbreaks might be amplified through secondary transmission by fecal carriers (98).

L. monocytogenes has not been isolated from oropharyngeal samples of healthy people, and the presence of listeriae in cervico-vaginal specimens is always associated with pregnancy-related listeriosis. The role of healthy carriers in the epidemiology of listeriosis is unclear and warrants further study.

FOODBORNE OUTBREAKS

Foodborne transmission of listeriosis was suggested early in the medical literature but was first definitively documented in 1981 during the investigation of an outbreak in Canada with the simultaneous use of a case-control study and strain typing (129). Since 1981, epidemiologic investigations have repeatedly revealed that the consumption of contaminated food is a primary mode of transmission of listeriosis. Food has been identified as the vehicle of several major (>30 cases) outbreaks of listeriosis investigated since 1981.

An outbreak of listeriosis in Massachusetts in 1979 may have been caused by raw produce, but the food source was not positively identified. Twenty patients with *L. monocytogenes* serotype 4b infection were hospitalized during a 2-month period; only nine cases had been detected in the previous 26 months. Ten of the patients were immunosuppressed adults, and five died. Fifteen patients are thought to have acquired the infection in the hospital. Consumption of tuna fish, chicken salad, or cheese was associated with illness, but no specific brand was implicated. It was postulated that the raw celery and lettuce, served as a garnish with the three foods, may have been contaminated with *L. monocytogenes* (68). Although the source of infection was not definitively identified in this outbreak, consumption of cimetidine or antacids was implicated as a risk factor for listeriosis. Decreased gastric acidity might have increased the survival of *L. monocytogenes* cells as they passed through the stomach.

The first confirmed foodborne outbreak of listeriosis occurred in 1981 in Nova Scotia, Canada. Thirty-four pregnancy-associated cases and seven cases in nonpregnant adults occurred during a 6-month period. A case-control study implicated a locally prepared coleslaw as the vehicle, and the epidemic strain was subsequently isolated from an unopened package of this product. Cabbage fertilized with manure from sheep suspected to have had *Listeria* meningitis was the probable source. Harvested cabbage was stored over the winter and spring in an unheated shed, providing a definite growth advantage for the psychrotrophic *L. monocytogenes* (129). Pasteurized milk was identified as the most likely source of infection in another large outbreak of listeriosis in Boston, Massachusetts, in 1983. Forty-nine cases occurred during a 2-month period: 42 in immunosuppressed adults and 7 in pregnant women; the overall case-fatality rate was 29% (50). A case-control study implicated pasteurized milk with 2% fat as the vehicle. Multiple serotypes of *L. monocytogenes* were isolated from raw milk at the implicated dairy, but none was the epidemic strain;

no deviations from approved pasteurization process was noted at the dairy, suggesting that the contamination occurred after pasteurization of the milk.

In 1985 in California, an outbreak of listeriosis with 142 cases during an 8-month period occurred in Los Angeles County and was traced to contaminated Mexican-style cheese (90). Pregnant women accounted for 93 cases, and the remaining 49 were nonpregnant adults; 48 of 49 nonpregnant adults had a predisposing condition for listeriosis. Among pregnancy-associated cases, 87% occurred in Hispanic women. The case-fatality rates were 32% for perinatal cases and 32% for nonpregnant adults. Inadequate pasteurization of milk that was used to prepare the cheese and mixing of raw milk with pasteurized milk likely resulted in the contaminated cheese.

L. monocytogenes-contaminated soft cheese was responsible for a 4-year (1983 to 1987) outbreak of 122 cases in Switzerland (10), and a contaminated paté was the vehicle of a 300-case outbreak in the United Kingdom in 1989–1990 (101). More recently in France, contaminated pork tongue in aspic was the principal vehicle of 279 cases of listeriosis in 10 months in 1992 (74), potted pork ("rillettes") was associated with 39 cases in 1993, and soft cheese was the vehicle of 33 cases in 1995 (59). Recalling the implicated food, advising the general population through the mass media to avoid eating contaminated products, and taking appropriate action to prevent *L. monocytogenes* contamination at product processing and handling facilities terminated these large outbreaks.

A large multistate outbreak of listeriosis occurred in the United States between August 1998 and March 1999. A total of 101 outbreak-associated cases (15 were perinatal cases) were identified in 22 states (Table 18.1). Fifteen adult deaths and six miscarriages or stillbirths were associated with this outbreak. A case-control study implicated frankfurters manufactured in one facility of a large food manufacturing company. *L. monocytogenes* serotype 4b of the epidemic PFGE subtype was isolated

Table 18.1 Invasive listeriosis outbreaks, 1990–2000

Year	Geographic location	No. of persons affected	Deaths	Vehicle	Serotype	Reference
1992	France	279	85	Pork tongue in jelly	4b	74
1994	Illinois	3	0	Chocolate milk	1/2b	118
1996	Ontario, Canada	2	0	Imitation crab (?)	1/2b	46
1998	U.S., multiple states	110	4	Frankfurters	4b	103
1998–1999	Finland	11	4	Butter	3a	93
1999–2000	France	26	7	Pork tongue in jelly	4b	25
2000	U.S., multiple states	29	7	Turkey deli meat	1/2a	19

from open and unopened packages of frankfurters from the implicated factory (20, 103).

Between December 1998 and February 1999, an increase in cases of listeriosis due to *L. monocytogenes* serotype 3a was recognized in Finland. A total of 25 persons, most of whom were in hematological or organ transplant patients, were identified as part of the outbreak; 6 patients died from the *Listeria* infection. Butter served at the tertiary care hospital was implicated as the source of infection. The epidemic strain was isolated from all 13 butter samples obtained from the hospital kitchen and from several lots from the dairy and wholesale store. One sample contained 11,000 *L. monocytogenes* CFU/g, but the other samples had lower counts (5 to 60 CFU/g) (93).

Between May and November 2000, a listeriosis outbreak was identified in the United States in 10 states. When subtyped, the *L. monocytogenes* isolates from these cases were all serotype 1/2a and were indistinguishable from each other by PFGE. Eight perinatal and 21 nonperinatal cases were reported. Among the 21 nonperinatal case-patients, the median age was 65 years (range, 29 to 92 years); 13 (62%) were female. This outbreak resulted in four deaths and three miscarriages or stillbirths. A case-control study implicated deli turkey meat as the probable source of infection (19).

Strains responsible for the major outbreaks between 1981 and 1992 (Canada, 1981; California, 1985; Switzerland, 1983 to 1987; France, 1992) were all of serotype 4b and belonged to a small number of closely related clones, as evidenced by ribotyping, multilocus enzyme electrophoresis, and DNA macrorestriction pattern analysis (15). In Southern hybridizations with DNA fragments of one of the genes responsible for facilitating low-temperature growth (*ltrB*), the epidemic-associated serotype 4b strains reveal a 9.5-kb fragment that is not present in nonepidemic strains (159). Also, DNA from the epidemic-associated serotype 4b strains is generally resistant to restriction by *Sau*3AI and other restriction endonucleases sensitive to cytosine methylation at 5′-GATC-3′ sites. This modification of *Sau*3AI restriction appears to be host mediated and is uncommon among other strains of serotype 4b that are not associated with epidemics (160). When multiple isolates from the Nova Scotia, Boston, and California outbreaks were characterized, 11 to 29% of the outbreak strains were antigenically unique, i.e., negative reaction with a serotype 4b-specific monoclonal antibody, absence of galactose in the cell wall teichoic acid, and resistance to serotype 4b-specific bacteriophage 2671 (23). Whether these characteristics are associated with the propensity of these strains to cause epidemics has not been elucidated.

Also, in the past decade, several outbreaks of febrile gastroenteritis caused by *L. monocytogenes* have been documented (Table 18.2). These gastroenteritis outbreaks differ from the invasive outbreaks described in several respects. They affect persons with no known predisposing risk factors for listeriosis. The infectious dose appears to be higher (1.9×10^5 to 1×10^9 CFU per g or per ml) than that for typical invasive listeriosis in the susceptible population. Finally, the symptoms appear within several hours (18 to 27 h) of exposure (similar to other bacterial enteric infections) in gastrointestinal listeriosis, in contrast to the several weeks of incubation observed for invasive listeriosis.

Data obtained during the past 10 years regarding the sources of outbreaks suggest that some foods are more hazardous vehicles of listeriosis than others. Highest-risk foods are ready-to-eat foods stored under refrigeration for a long period of time and those that are contaminated with *L. monocytogenes* at levels of >100 CFU per g or per ml. This observation was also made during investigations of sporadic listeriosis cases (72, 78). Among ready-to-eat foods that have been previously implicated as a source of infections in listeriosis cases are frankfurters, deli meats, soft cheeses, and smoked fish.

CHARACTERISTICS OF THE DISEASE

Human disease caused by *L. monocytogenes* usually occurs in certain well-defined high-risk groups, including pregnant women, neonates, and immunocompromised adults, but may occasionally occur in persons with no predisposing underlying conditions. In nonpregnant adults, *L. monocytogenes* primarily causes septicemia, meningitis, and meningoencephalitis, with a mortality rate of 20 to 25%. Other infrequent manifestations of listeriosis in this population include endocarditis in persons with underlying cardiac lesions (including prosthetic or porcine valves) and various types of focal infections, including endophthalmitis, septic arthritis, osteomyelitis, pleural infection, and peritonitis (144). Clinical conditions known to predispose persons to the serious manifestations of listeriosis include malignancy, organ transplants, immunosuppressive therapy, infection with the human immunodeficiency virus, and advanced age. Although pregnant women, particularly in the third trimester of pregnancy, may experience only mild flulike symptoms (fever, myalgias with or without diarrhea) as a result of *L. monocytogenes* infection, the infection has serious consequences for the fetus, leading to stillbirth or abortion. In neonates who are less than 7 days old, sepsis and pneumonia are predominant symptoms, whereas in neonates older than

Table 18.2 Outbreaks of gastrointestinal manifestation of listeriosis in humans with no known predisposing condition

Year	Geographic location	Incubation time	Persons affected	Symptoms	Vehicle	Serotype	Contamination level (CFU/g or ml)	Reference
1986—87	Pennsylvania	Unknown	Unknown	Fever, vomiting, and diarrhea in the week before positive culture	Unknown	Multiple serotypes	NA[a]	135
1994	Italy	18 h	18	Fever, diarrhea	Foods served at a dinner	1/2b	NA	127
1996	Illinois	20 h	80	Diarrhea, fever	Chocolate milk, temperature abused	1/2b	10^9	31
1998?	Finland	<27 h	5	Nausea, abdominal cramps, diarrhea, fever	Cold-smoked rainbow trout	1/2a	1.9×10^5	107
1997	Northern Italy	24 h	1,566	Headache, abdominal pain, diarrhea, fever	Cold salad of corn and tuna	4b	10^6	6
2000	Winnipeg, Canada	NA	>20	Gastrointestinal illness	Defective cans of aerosol-dispensed whipping cream	NA	NA	22

[a] NA, not available.

393

7 days, the infections manifest as meningitis and sepsis (144).

As previously stated, several recent investigations of listeriosis outbreaks suggest that *L. monocytogenes* also causes febrile gastroenteritis in normal hosts (Table 18.2). Interestingly, Murray and Pirie clearly stated in their original descriptions of cases from the 1920s that diarrhea was a common feature of listeriosis in small animals (100). The most compelling evidence for the gastrointestinal manifestation of *L. monocytogenes* infections comes from outbreak investigations by Dalton et al. (31) and Aureli et al. (6). Fever and diarrhea are the most consistent symptoms in gastrointestinal listeriosis. Also, the incubation time for the enteric form is rather short, in the range of 20 to 27 h. In the two investigations in which attempts were made to quantify the numbers of *L. monocytogenes* cells in the implicated foods, the contamination level was very high, indicating the ingestion of several million bacteria. From the data available thus far for gastrointestinal listeriosis, it appears that infection requires a high dose of *L. monocytogenes*. It is not known whether the strains involved in the gastrointestinal forms of listeriosis possess additional virulence factors similar to those of common enteric pathogens.

While listeriosis outbreaks attract the most attention, most cases of human listeriosis occur sporadically. However, some of these sporadic cases may be unrecognized common-source clusters. The source and route of infection of most of these cases remain unknown, although foodborne transmission is demonstrated in some cases. Many cases have not been associated with food because of the difficulties of prospectively investigating sporadic cases of the disease: long incubation times (up to 5 weeks) make accurate food histories difficult to obtain and make examination of incriminated foodstuffs difficult to accomplish because they are usually consumed or discarded. Understanding the epidemiology of sporadic cases is critical to the development of effective control strategies. In association with active surveillance of listeriosis, a case-control study of dietary risk factors undertaken in the United States revealed that patients with listeriosis were more likely than control subjects to have eaten soft cheeses or food purchased from store delicatessen counters. Data suggested that foodborne transmission could be responsible for about one-third of cases (115, 132).

Sporadic listeriosis is a rare disease with an incidence of between 2 and 7 cases per million population, evaluated over recent years by passive surveillance in some European countries and Canada. A more precise estimate of 7.4 cases per million population was obtained through an active surveillance study in the United States in 1989 and 1990 in a population base of 19 million people. A decrease in the number of cases (44%) and deaths (48%) attributable to sporadic listeriosis has been observed in the United States since 1990, suggesting that the measures taken in the food industry and the recommendations for people at greatest risk for the disease have been effective (149). The most recent estimates of the burden of human listeriosis in the United States indicate that 2,518 illnesses, 2,322 hospitalizations, and 504 deaths are attributable to listeriosis each year (104). Listeriosis is mainly reported from industrialized countries, and the prevalence in Africa, Asia, and South America is unknown or low. Whether this reflects different consumption patterns and dietary habits, different host susceptibility, different food processing and storage technologies, or lack of awareness or laboratory facilities is not known (124).

Although exposure to *L. monocytogenes* is common, invasive listeriosis is rare. It is unclear whether this is due to early acquired protection or to most strains' being only weakly virulent. The pathogenesis of human listeriosis is poorly understood. Many healthy individuals (2 to 6%) are asymptomatic fecal carriers of *L. monocytogenes*. The risk of clinical disease in those intestinal carriers of *L. monocytogenes* is unknown. Endogenous infection by *L. monocytogenes* in the gut is plausible, especially in patients receiving immunosuppressive therapy, which not only impairs resistance to infection but can also result in alterations of intestinal defense mechanisms favoring listerial invasion. Nevertheless, asymptomatic fecal carriage has been observed in pregnant women who proceed to normal birth at term; women who have given birth to infected infants do not necessarily suffer the same problem in later pregnancies. Similarly, recent transplant recipients may harbor *L. monocytogenes* in the gut without developing the disease.

Epidemiologic studies since 1981 have focused on the role of contaminated food in the transmission of listeriosis. However, two unusual transmission routes have been described. Hospital-acquired listeriosis is sporadically described, mainly in nursery mates, with equipment serving as the vehicle. Amniotic fluid during intrauterine infections could contain as many as 10^8 *L. monocytogenes* CFU/ml (102). Mineral oil has been implicated in an outbreak of neonatal listeriosis (133). Primary cutaneous infections without systemic involvement have been observed as an occupational disease in veterinarians and farmers; most cases are caused by manipulation of presumably infected bovine fetuses or cows.

INFECTIOUS DOSE AND SUSCEPTIBLE POPULATIONS

Infectious Dose

The infectious dose of *L. monocytogenes* depends on many factors, including the immunological status of the host. In addition to host factors and exposure to particular foods, it is likely that microbial characteristics are important risk factors for disease. The occurrence and the course of infection may depend on virulence factors and infective dose. Severity of the disease is such that tests with human volunteers are impossible. Studies in monkeys and in mice suggest that reducing levels of exposure will reduce clinical disease (44). However, these experiments do not help to determine the minimal infective dose for humans. Published data indicate that the populations of *L. monocytogenes* in contaminated food responsible for epidemic and sporadic foodborne cases were usually more than 100 CFU/g. However, the frankfurters implicated in the 1998 listeriosis outbreak in the United States contained less than 0.3 CFU/g (103). Because enumeration procedures are not fully reliable and because the time between consumption and analysis of the contaminated food can enable growth or death of *L. monocytogenes*, results may not always be indicative of the numbers consumed. More epidemiologic information is needed for an accurate assessment of the infectious dose.

Susceptible Populations

Most human cases of listeriosis occur in individuals who have a predisposing disease that leads to impairment of their T-cell-mediated immunity. The percentage of patients suffering from a known underlying condition varies greatly between studies, accounting for 70 to 85% of the cases in some surveys and nearly all cases in others (16, 142). The most commonly affected populations include extreme ages (neonates, the elderly), pregnant women, and those persons who are immunosuppressed by medication (corticosteroids, cytotoxic drugs), especially after organ transplantation or illness (hematologic malignancies, such as leukemia, lymphoma, and myeloma, as well as solid malignancies). Listeriosis is 300 times more frequent in people with AIDS than in the general population (78, 134). In addition to T-cell-immunity impairment, a small percentage of listeriosis patients suffer from chronic diseases not usually associated with immunosuppression, such as congestive heart failure, diabetes, cirrhosis, alcoholism, and systemic lupus erythematosus, alone or in association with known predisposing diseases.

A concurrent infection could influence susceptibility to listeriosis. This was exemplified by a cluster of cases in 1987 in Philadelphia that were characterized by a number of different strains isolated from patients. A single food vehicle could not be identified because of the diversity of strains. Clinical and epidemiologic investigations suggested that individuals who were previously asymptomatic for listerial infection but whose gastrointestinal tract harbored *L. monocytogenes* became symptomatic, possibly because of a coinfecting agent (135).

VIRULENCE FACTORS AND MECHANISMS OF PATHOGENICITY

L. monocytogenes can infect laboratory animals and mammalian cells in tissue culture, can grow well in culture media, and can be genetically manipulated and used to experimentally infect animals. Hence, during the last decade there have been major advances in studying and identifying the virulence factors of this highly invasive intracellular pathogen (27, 28, 39, 117, 140, 150).

Pathogenicity of *L. monocytogenes*

Many pathogenicity tests for studying *L. monocytogenes* have been developed, including a fertilized hen egg test, tissue culture assays, and tests using laboratory animals, particularly mice, which are either immunocompetent or immunocompromised (9, 71). In these studies, mice are infected intraperitoneally, intravenously or intragastrically, and virulence is evaluated either by comparing the LD_{50} or by enumerating bacteria in the spleen or liver. A broad range in levels of virulence of various *L. monocytogenes* strains has been observed with immunocompetent mice: LD_{50} values range from 10^3 to 10^7 CFU. No clear correlation between origin (human, animal, category of food, environment) or strain characteristics (serotype, phage type, multilocus enzyme type, or DNA micro- or macrorestriction patterns) and virulence has been observed (14, 114). Epidemic strains also have a wide range of virulence in the mouse model. Although rare nonpathogenic or weakly pathogenic *L. monocytogenes* isolates have been reported, all strains of *L. monocytogenes* are considered to be potentially capable of causing human disease.

Experimental Infection and Cell Biology of the Infectious Process

Experimental infection of rodents with *L. monocytogenes* has been widely used for the study of cell-mediated immunity. When mice are exposed to a sublethal inoculum of *L. monocytogenes*, the ensuing infection follows a well-defined course, lasting for approximately 1 week. Routinely, mice are injected intravenously, and bacterial growth kinetics are monitored in the spleen and liver. Within 10 min after intravenous injection, 90% of the inoculum is taken up by the liver, and 5 to 10% is taken up by the spleen. During the first 6 h, the number of viable listeriae in the liver decreases 10-fold, indicating rapid destruction of most of the bacteria. Surviving listeriae then multiply within susceptible macrophages and grow exponentially in the spleen and liver for the next 48 h, peaking at day 2 or 3 after infection (5). Rapid inactivation ensues during the next 3 to 4 days, indicating recovery of the host. Convalescent mice are resistant to challenge and have a delayed-type hypersensitivity with swelling of the footpads injected with crude cell preparations of *L. monocytogenes*.

L. monocytogenes crosses the intestinal barrier in animals infected by the oral route. The portal of entry has been a subject of controversy for quite some time. There is no preferential site for translocation between enterocytes and M cells (120), but the intestinal dendritic cells in the Peyer's patches appear to be the preferred site for invasion and multiplication (83). Bacteria are then internalized by resident macrophages, in which they can survive and replicate. They are subsequently transported via the blood to regional lymph nodes. When they reach the liver and the spleen, most listeriae are rapidly killed. In the initial phase of infection, infected hepatocytes are the target for neutrophils and later for mononuclear phagocytes, which are responsible for control and resolution of infection. Depending on the level of T-cell response induced in the first days following initial infection, further dissemination via the blood to the brain or, in the pregnant animal, the placenta may subsequently occur. Hence, infection is not localized at the site of entry but involves entry and multiplication in a wide variety of cell types and tissues. The principal site of infection is the liver. *L. monocytogenes* invades many cell lines of different types (macrophages, fibroblasts, hepatocytes, and epithelial cells) in vitro and is one of the most invasive bacteria known. It has presumably evolved specific strategies allowing entry into different cell types, but retaining some tropism for particular organs.

Invasion of brain cell endothelial cells is a prerequisite for meningeal pathogens that penetrate the central nervous system and cause meningitis, encephalitis, and brain abscesses. *L. monocytogenes* utilizes penetration of human brain microvascular endothelial cells (HBMEC) as a means of crossing the blood-brain barrier. As a first step, *L. monocytogenes* adheres to HBMEC through the microvilli. This step does not require the invasion-associated protein InlB. This is followed by InlB-dependent invasion of the cells, intracellular multiplication, and intracellular movement leading to the formation of *L. monocytogenes*-containing protrusions on the surface of the infected HBMEC cells (63).

L. monocytogenes enters both phagocytic cells and non-professional phagocytes by phagocytosis. This first step in infection is prevented by addition of cytochalasin D, a drug which inhibits actin polymerization and hence active participation of the mammalian cell. In the case of non-professional phagocytic cells, this process is triggered by the bacterium and is therefore called induced phagocytosis. Soon after entry, bacteria are internalized in membrane-bound vacuoles, which are lysed within 30 min. Intracellular bacteria are released to the cytosol and begin to multiply with a doubling time of about 1 h. These intracytoplasmic bacteria become progressively covered by a "cloud" of cell actin filaments that later rearranges into a polarized "comet tail" up to 40 μm in length (26, 27, 95). The actin comet tail is made of actin microfilaments that are continuously assembled in the vicinity of the bacterium, then released and cross-linked. The actin comet tail is stationary in the cytosol and left behind by moving bacteria. The length of the tail is thus proportional to the speed of movement, with faster-moving bacteria having longer tails. Their speed ranges from 0.1 to 1 μm/s. When bacteria reach the plasma membrane, they put out long protrusions, each with a bacterium at the tip. These protrusions are then internalized by a neighboring cell, giving rise to a two-membrane-bound vacuole. After lysis of this vacuole, a new cycle of replication, movement, and spreading of the bacteria begins. The entire cycle is completed in about 5 h. If cytochalasin D is added after entry, bacteria do not spread within the cytosol. They replicate and form microcolonies in the vicinity of the nucleus. Hence, actin polymerization is essential to intracellular movement and cell-to-cell spreading (Fig. 18.3).

The Role of Immune Response of the Host in Listeriosis

The strategy of direct cell-to-cell spread allows bacteria to disseminate within host tissues and induce the formation of infectious foci while being sheltered from host defenses, such as circulating antibodies. This may explain why antibodies play no role in recovery from infection or protection against secondary infection (116); during normal infection, the anti-*Listeria* antibody titers, including antibodies to listeriolysin O (LLO), are low. However, recently Edelson and Unanue (41) demonstrated that passive immunization of mice with a neutralizing

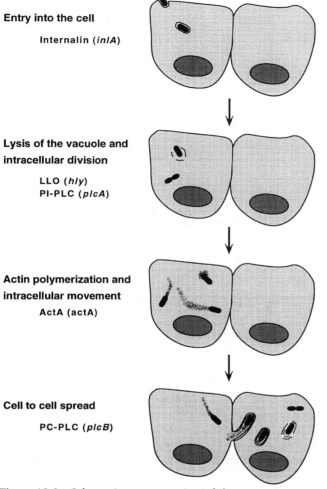

Entry into the cell

Internalin (*inlA*)

Lysis of the vacuole and intracellular division

LLO (*hly*)
PI-PLC (*plcA*)

Actin polymerization and intracellular movement

ActA (*actA*)

Cell to cell spread

PC-PLC (*plcB*)

Figure 18.3 Schematic representation of the successive steps in the infectious process.

monoclonal antibody to LLO increased the resistance to infection by *L. monocytogenes*. The investigators provided three possible explanations for this phenomenon: (i) opsonization of bacteria and/or complement activation through the binding of LLO, (ii) neutralization of LLO inside the phagosome of infected cells, thereby preventing escape of bacterial cells into the cytosol, and (iii) neutralization of exotoxic functions of LLO when the organism is present in the extracellular environment.

Genetic Approaches to the Study of Virulence Factors

Genetic manipulation of *L. monocytogenes* in vitro is now straightforward; however, some useful techniques applicable to other bacterial genera, e.g., transduction, are not available. The first transposon mutagenesis experiment with *L. monocytogenes* was conducted in

1985 (53). The conjugative transposons Tn*1545* and Tn*916*, or the nonconjugative transposon Tn*917* and its derivative Tn*917-lac*, can all be used for transposon mutagenesis. Transformation, originally involving protoplast formation, is now possible by electroporation. Efficiency varies, and complementation of spontaneous mutants with plasmid libraries is not possible. However, complementary mutants can be derived with cloned genes, and allelic exchange is available for several thermosensitive plasmids. Plasmids can also be introduced into *Listeria* spp. by conjugation. Vectors such as pAT18 have been very useful.

Adhesion to Mammalian Cells

Most of the research on the interaction of *L. monocytogenes* with mammalian cells has focused on invasion of and internalization in host cells. However, adhesion of *L. monocytogenes* to mammalian cells may be an important and critical aspect of pathogen-host interaction. Adhesion may promote colonization of the gastrointestinal tract and direct the bacteria to appropriate target cells or tissues, such as the central nervous system or the placenta, and may even facilitate cell invasion by activating cell signal transduction pathways or triggering the synthesis of a target cell receptor required for invasion (108). *L. monocytogenes* adheres to Hep-G2 hepatocytes via a lectin-substrate interaction, and a 104-kDa cell surface protein (p104) may be involved in adhesion to Caco-2 cells. Using insertional mutagenesis with Tn*1545* in the genetic background of a strain defective in the invasion genes, Milohanic et al. (108) identified four insertional mutants that were defective in adhesion to the human melanoma cell line SK-MEL 28. The most severely adhesion-deficient mutant had a Tn*1545* insertion upstream from *ami*, a gene that encodes a surface-exposed autolysin with a C terminus similar to that of InlB.

Entry into Mammalian Cells

Following invasion of the intestinal barrier, two types of cells are critical to infection, i.e., macrophages and hepatocytes. In vivo observations have led to conflicting conclusions concerning the primary site of entry (epithelial cells or M cells) of *L. monocytogenes* into the host. Entry into epithelial cells has been studied by genetic approaches (40), leading to the identification of internalin, a protein required for entry into epithelial cells and encoded by the gene *inlA* (52). *inlA* is part of a multigene family. A second gene of this family, *inlB*, contributes to entry into hepatocytes in vitro (39) (Fig. 18.4). Analysis of spontaneous rough mutants has led to the identification of a second protein, p60, associated with invasion.

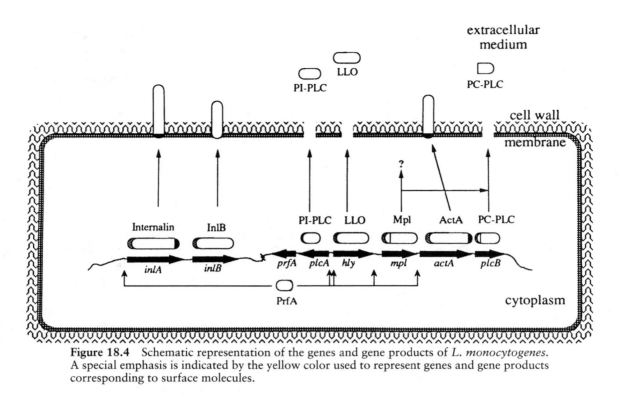

Figure 18.4 Schematic representation of the genes and gene products of *L. monocytogenes*. A special emphasis is indicated by the yellow color used to represent genes and gene products corresponding to surface molecules.

Its gene was named *iap* for invasion-associated protein (82).

Internalin and the *inl* Gene Family

L. monocytogenes cells are internalized by both phagocytic and non-professional host cells. The bacteria induce their own uptake into cells that are not normally phagocytic. The sequence of events that lead to the uptake of *L. monocytogenes* cells resembles the process of uptake of *Yersinia* cells, characterized by highly local apposition of the plasma membrane with the incoming bacterial cell. The bacterial cell appears to progressively sink into the mammalian cell surface (13, 73, 106).

Two surface proteins of *L. monocytogenes* directly participate in the entry into mammalian cells in culture. These two proteins, internalin A (InlA) and internalin B (InlB), mediate invasion in different types of cells. InlA is a 800-amino-acid polypeptide that facilitates entry of *L. monocytogenes* into Caco-2 or other cells that express its receptor, E-cadherin, a calcium-dependent cell adhesion molecule composed of five extracellular domains and a cytoplasmic tail. InlA has two regions of repeats: (i) 15 successive leucine-rich repeats (LRR) of 22 amino acids and (ii) two 70-amino-acid-residue repeats and a partial repeat of 49 amino acids. The C-terminal region of InlA contains an LPXTG motif that covalently links of the protein to the peptidoglycan. The amino-terminal

region that includes the LRR region and the second repeat region are necessary and sufficient to promote bacterial entry into cells expressing its receptor.

The InlA-mediated entry of *L. monocytogenes* into human cells has an absolute requirement for a proline residue at position 16 of E-cadherin (87). Interestingly, the E-cadherin of guinea pigs and rabbits, which were the first-described hosts of *L. monocytogenes* infections, also have proline at position 16, whereas the mouse E-cadherin has glutamic acid at position 16 and thus does not serve as a receptor for InlA.

InlB is a 630-amino-acid surface polypeptide that facilitates entry into a wider variety of cell types, including cultured hepatocytes and some epithelial, fibroblast, and endothelial cell lines, including Vero, HeLa, HEp-2, CHO, and L2 cells. InlB has eight tandem repeats similar to those of InlA and a carboxy-terminal cell surface anchoring region (Csa) made up of three tandem repeats of GW modules (repeats beginning with the dipeptide GW). The 213 N-terminal amino acids of InlB, which consist of eight LRRs, are necessary and sufficient for entry into Vero cells. The same LRR region of InlB also activates phosphoinositide (PI) kinase, a lipid kinase required for InlB-mediated invasion of mammalian cells. Also, it stimulates tyrosine phosphorylation of the three adapter proteins Gab1, Cb1, and Shc, which are also used by epidermal growth factor and insulin to stimulate

PI kinase. In addition, the LRR region of InlB stimulates changes in the actin cytoskeleton, giving rise to membrane ruffles (13). Recent X-ray crystallographic studies of the LRR domain of InlB reveal that InlB binds calcium and that calcium may play a critical role in receptor interactions (94). The Csa region of InlB is associated with the bacterial membrane component lipoteichoic acid. This InlB is partially exposed on the cell surface (N-terminal domain with LRR) and partially buried in the peptidoglycan layer (Csa domain).

Five other *inl* genes have been cloned and sequenced, and their properties are currently under investigation. Preliminary data suggest that the *inl* family of genes encodes surface proteins, each with a specific tropism for a different cell type. Internalin expression is greater at 37°C than at 20°C and is under the control of the pleiotropic activator gene *prfA* (see below).

p60 and *iap*

Spontaneously rough mutants (R-mutants) produce reduced levels of a 60-kDa extracellular protein termed p60. These R-mutants form long chains separated by double septa and have reduced virulence in mice. They also poorly invade 3T6 fibroblasts. Adding partially purified p60 to R-mutants disaggregates the chains, and the resulting bacteria become invasive. In contrast, if the chains are disrupted by ultrasonication, the single cells are not invasive. The gene coding for p60 (*iap*) has been cloned and sequenced. The p60 protein is 484 amino acids long, with a 27-amino-acid signal sequence but no membrane anchor (82). A peculiar sequence of 19 Thr-Asn repeats is near the C terminus of the protein. In cultures of wild-type *L. monocytogenes*, p60 is present both on the cell surface and in the culture supernatant fluid. However, the protein is present only on the surface of the R-mutant. p60 is an essential protein with bacteriolytic activity and may be a murein hydrolase involved in septum formation (158). The exact role of p60 in invasion is unclear, but it primarily affects entry into fibroblasts. R-mutants are thought to be noninvasive for fibroblasts, not because of their large size but rather because of their inability to adhere to these cells. In contrast, when R-mutants are in contact with Caco-2 cells, the bacteria are adherent but are not invasive. After addition of p60 or after disruption of the chains, these mutants invade Caco-2 cells. Hence, neither adherence nor uptake of *L. monocytogenes* by epithelial cells is dependent on p60.

A 45-kDa protein (P45), partially associated with the cell surface in *L. monocytogenes* and several other *Listeria* spp., has been described by Schubert et al. (130). The gene encoding P45 (*spl*, secreted protein with lytic activity) has been cloned and sequenced. The deduced amino acid sequence of the P45 protein is highly similar to that of the p60 protein in a 120-amino-acid stretch near the C terminus. However, P45 lacks the Thr-Asn repeats that are present in the C-terminal region of p60. Like p60, P45 possesses peptidoglycan hydrolase activity.

L. monocytogenes invasion of mammalian cells is inhibited by tyrosine kinase inhibitors (148, 153). In agreement with these data, entry into mammalian cells by *L. monocytogenes* induces phosphorylation of a 37-kDa host cell protein (148).

Escape from the Phagosome and Intracellular Growth

Once internalized in a vacuole, *L. monocytogenes* rapidly escapes into the cytosol, where it multiplies. The main factor involved in lysis of the vacuole is a protein that has a pore-forming activity, LLO. However, a second factor, a phosphatidylinositol phospholipase C (PI-PLC), may also be involved (117, 140).

LLO is a toxin that belongs to the class of thiol-activated toxins whose prototype is streptolysin O, which is produced by *Streptococcus pyogenes*. These thiol-activated toxins are produced by several gram-positive bacteria, including *Streptococcus*, *Clostridium*, and *Bacillus* species. They have close similarities in sequence and probably have a common mode of action. They are only active on cholesterol-containing membranes, with cholesterol likely acting as the receptor in the membrane. It has been suggested that after LLO binds to the surface of the cell membrane, the protein forms oligomers that lead to pore formation. The activity of pore-forming toxins is usually determined by their lytic activity on red blood cells; hence, they are termed hemolysins. Colonies of bacteria producing this type of toxin are hemolytic on blood agar plates, a phenotype that has been used to identify the gene-encoding LLO.

Nonhemolytic *L. monocytogenes* strains, which are very rare, are avirulent in experimentally infected mice. This stimulated study of the role of hemolysin in virulence. Using transposon Tn*1545*, a nonhemolytic mutant was obtained that was avirulent in mice. Both the wild type and the nonhemolytic strains were able to enter mammalian (Caco-2) cells; however, the nonhemolytic strain was impaired in its ability to grow intracellularly. Electron microscopy revealed that the nonhemolytic mutant, although invasive, was unable to escape from the phagocytic vacuole.

The locus containing the transposon was cloned and determined to be a structural gene, named *hlyA*, later shortened to *hly*. The insertion had no polar effect, because *hly* is a monocistronic unit. *hly* encodes a 58-kDa protein that, like all thiol-activated toxins, contains a single cysteine residue in the amino-acid sequence ECTGLAWEWWR located toward the C terminus of the protein. The cysteine residue is not essential for hemolysin activity, as previously thought. LLO is needed for the escape of *L. monocytogenes* from the primary vacuole and intracellular growth. Using six-His-tagged LLO and an LLO-negative mutant, Gedde et al. (55) have determined that LLO is also essential for the escape of *L. monocytogenes* from the double-membrane vacuole that encases the bacteria during cell-to-cell spread.

LLO is also a potent inflammatory stimulus, promoting the expression of adhesion molecules in human umbilical vein endothelial cells and chemokine secretion (79). The crossing of endothelial cells by *L. monocytogenes* may be a complex phenomenon involving the LLO-secreting bacteria and inflammatory cells producing tumor necrosis factor alpha and interleukin-1 resulting in increased leukocyte-endothelial cell interaction and leading to the influx of leukocytes through the endothelial barrier.

LLO has been purified from culture supernatant fluids. Unlike all other thiol-activated toxins, it has maximal activity at pH 5 and is inactive at pH 7 (57). It has been proposed that this property is particularly important in the acidic environment that results from phagosome-lysosome fusion. The most interesting feature of LLO is not its high activity at low pH but its low activity at high or neutral pH which prevents the deleterious effect of LLO on cellular membranes when the bacterium is free in the cytosol.

The gene *plcA* is upstream from *hly*. *plcA* encodes the PI-PLC that is present in culture supernatant fluids of *L. monocytogenes*. This 37-kDa protein hydrolyzes glycosylphosphatidylinositol anchors. Analysis of in-frame deletions of *plcA* revealed that this gene is not required for virulence in mice; however, *plcA* is involved in escape from the vacuole in primary macrophages. These results indicate that *L. monocytogenes* has evolved the use of specific genes not only for particular steps of the infection but also for specific cell types.

A priori it was believed that the cytoplasm of eucaryotic cells may not be permissive for the growth of bacteria and that intracellular pathogens would have to induce expression of several genes for survival. However, infecting macrophages with a *Bacillus subtilis* strain expressing *hly* has revealed that the cytosol is permissive for growth of at least this species. Additional studies revealed that most auxotrophic mutants of *L. monocytogenes* in cell culture have growth rates unaffected by the

intracellular environment and are only slightly affected in virulence (2, 80, 95).

Intracytoplasmic Movement and Cell-to-Cell Spreading

Induced phagocytosis, escape from the phagosome, and intracellular multiplication are essential steps for infection of individual cells but are not sufficient to achieve infection in mice. *L. monocytogenes* must efficiently invade its preferred tissues by direct cell-to-cell spreading to infect mice. Direct cell-to-cell spreading is the result of intracellular movement, protrusion formation, phagocytosis of the protrusion by a neighboring cell, formation of a double-membrane-bound vacuole, and lysis of the vacuole. Some of the bacterial genes involved in these steps are clustered in an operon located immediately downstream from *hly*.

Intracytoplasmic movement is strictly coupled to continuous actin assembly, which provides the force for bacterial propulsion (16) (Fig. 18.5). Actin polymerization requires expression of the *actA* gene (36, 81). *actA* mutants are invasive, escape from the phagosome, replicate, and form intracellular microcolonies but are not covered with actin, do not move intracellularly, and do not spread from cell to cell. *actA* encodes a 610-amino-acid surface protein (ActA) anchored in the bacterial cytoplasmic membrane by its hydrophobic C-terminal region. Half of the molecule protrudes from the cell wall and can interact with cellular components. Immunolocalization of the ActA protein in infected cells has revealed that ActA is

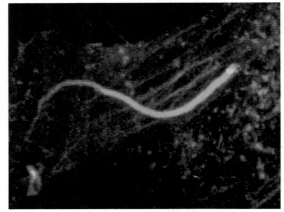

Figure 18.5 Directional actin assembly by *L. monocytogenes*, with filamentous actin assembled at one bacterial extremity. This is a double immunofluorescence confocal laser scanning micrograph of a Vero cell infected for 5 h, treated with anti-*L. monocytogenes* antiserum (bacterium, red), and fluorescein isothiocyanate-phalloidin (F-actin, green). Reprinted with permission from *Molecular Microbiology* (vol. 13, no. 3, 1994, front cover).

asymmetrically distributed on the bacterial surface such that it is weakly detectable at one pole, with an increasing concentration toward the other pole, which is the site of comet tail formation. Even before comet tail formation, ActA colocalizes with F-actin. This strongly suggests that ActA is involved in actin filament nucleation from the bacterial surface. In addition, the asymmetrical distribution of ActA appears to be essential for intracellular *L. monocytogenes* movement in the direction of the non-ActA-expressing pole. ActA seems to be sufficient to trigger actin-based motility, because expression of *actA* in *L. innocua* is sufficient to produce movement in cell extracts.

When the actin comet tails propel the bacterium into its host's plasma membrane, the membrane becomes distended, forming long protrusions into the extracellular space. When these bacterium-containing membrane protrusions from one host cell enter a neighboring host cell, intercellular spread occurs. However, at this stage, the bacterial cell is enclosed in a double-membrane secondary vacuole, one membrane coming from the donor cell and the other from the recipient cell. Lysis of the two membrane vacuoles requires LLO, a metalloprotease encoded by *mpl*, PI-PLC, and a broad-spectrum phospholipase (PC-PLC) encoded by *plcB*. Mutants lacking in either or both of the PLCs or Mpl form small plaques in confluent layers of mouse fibroblasts, presumably because they are less efficient in escaping from the secondary vacuole. *plcB* is a gene located immediately downstream from *actA* (152). PC-PLC is a zinc-requiring phospholipase of 29 kDa. Its optimum activity is between pH 5.5 and 7, and it can cleave phosphatidylcholine, phosphatidylserine, and phosphatidylethanolamine but not phosphatidylinositol. During intracellular growth of *L. monocytogenes*, PC-PLC is synthesized as a precursor that must be cleaved by Mpl and a cysteine protease of host lysosomal origin (96). Acidification of the host compartment in which the bacteria are confined triggers rapid Mpl-mediated activation and release of active PC-PLC (97). Because PC-PLC in the cytosol of the mammalian cell is cytotoxic, regulated release of PC-PLC just when it is needed is essential to the pathogenicity of *L. monocytogenes*.

Coregulation of Virulence Factors

The pathogenic mechanism of *L. monocytogenes* involves the following steps: adhesion to host cells, bacterially induced uptake into nonphagocytic host cells, lysis of the phagosome, replication in the host cytoplasm, actin-based intracellular movement and propulsion, and intercellular spread. These events are mediated by eight genes

of *L. monocytogenes*. Six of these eight genes are clustered in one operon on the chromosome: *prfA, plcA, hly, mpl, actA,* and *plcB. inlA* and *inlB* are part of a multigene family and form another distinct operon. The expression of these genes is modulated by the same environmental conditions, being repressed at low temperatures and maximally expressed at 37°C (88). In addition, they are coordinately regulated by PrfA, a 29-kDa transcriptional activator encoded by *prfA. prfA* is downstream from *plcA* and is cotranscribed with it either as a monocistronic or a bicistronic transcript (88, 89, 105). PrfA-mediated activation begins with its binding to a conserved 14-bp palindromic sequence (PrfA box) in PrfA target promoter regions. A hierarchy exists in PrfA activation, with more efficient transcription of *hly* and *plcA* (perfectly symmetrical PrfA box) than of *mpl*, which has base substitutions in its PrfA box. Substitution of the *mpl* promoter palindrome increases *mpl* expression. Interestingly, the expression of *actA*, which also has several base substitutions in its promoter palindrome, is unaffected by substitution with *hly* promoter palindrome (157). PrfA probably is converted from an inactive form to an active form by binding of a cofactor.

Expression of *prfA* is a complex phenomenon. *prfA* is the second gene of the *plcA-prfA* operon whose primary promoter is regulated by *prfA* (105). Hence, *prfA* autoregulates (activates) its own synthesis. *prfA* can also be transcribed from its own promoter region. Since *prfA* transcripts are fewer in stationary growth, it has been proposed that *prfA* may also downregulate its own synthesis. It has been determined in a *prfA* mutant that transcription at the *prfA* specific promoter is upregulated (51).

prfA is similar to CAP, the cyclic AMP receptor in *Escherichia coli* (140). It was suggested that this protein contains a helix-turn-helix motif that can interact with DNA. Mutagenesis in the region coding for this motif and gel shift assays strongly suggest that the product of the *prfA* gene is a DNA-binding protein belonging to the CAP/FNR family that mediates transcription activation by binding to a conserved palindromic region located in the −35 region of promoters under its control.

The mechanism of thermoregulation of *prfA* expression is unclear. At low temperatures, the small *prfA*-specific transcript is present (89). However, no apparent transcription of virulence genes is observed. For activation, the *prfA* gene product may act in concert with another regulatory protein or may be posttranscriptionally modified. It is also possible that oligomerization of *prfA* plays a role in activation, and the concentration of *prfA* may be critical for its activity. In addition, the various *prfA*-regulated promoters have different affinities for *prfA*. Clearly, the expression of virulence genes is highly regulated. There is evidence that heat shock (145, 146) or various nutrients in broth medium (112) can also affect virulence gene expression.

A group of Tn*917-lac* mutants was screened to identify genes specifically induced in the intracellular environment to test for strains in which *lacZ* expression was greater inside cells than in broth culture. Five genes were identified as preferentially expressed within cells (80). Three were genes involved in the biosynthesis of purines and pyrimidines, one encoded an arginine ABC transporter, and another was the *plcA* gene coding for the PI-PLC that is cotranscribed with *prfA*, the pleiotropic regulator of all virulence genes. Recently, Renzoni et al. (121) determined that PrfA synthesis is enhanced during adherence of *L. monocytogenes* to host cells. This enhanced synthesis is independent of InlB, suggesting that only a loose association with the host cell is required. Proteinase K-sensitive proteins of the host may play a role in upregulation of PrfA synthesis (121).

Stress Proteins of *L. monocytogenes*

Like many other bacteria, *L. monocytogenes* induces the synthesis of stress proteins in response to heat shock or oxidative stresses (67). The DnaK protein of *L. monocytogenes* belongs to the HSP70 family of heat shock proteins. The *dnaK* locus on the *L. monocytogenes* chromosome consists of the following six genes in the order given: *hrcA-grpE-dnaK-dnaJ-orf35-orf29* (66). A *dnaK* insertional mutant loses its ability to grow at temperatures above 37°C and forms filaments at these temperatures. The mutant also is much less efficiently phagocytosed by macrophages, probably owing to deficiency in binding of bacterial cells to macrophages. However, the mutant's ability to multiply in macrophages is unaffected (65).

A two-component signal transduction system involved in stress tolerance and virulence has been identified in *L. monocytogenes* (29). The genetic locus, *lisRK*, includes two genes: (i) *lisR*, which encodes a listerial response regulator protein that may also function as a transcriptional activator, and (ii) *lisK*, which encodes a histidine kinase. A nonpolar 498-bp deletion in *lisK* resulted in a mutant that was acid sensitive during log-phase growth and had reduced mouse virulence compared with the wild-type parent strain. ClpP is another stress-induced protein of *L. monocytogenes*. It is a 21.6-kDa protein that belongs to a family of proteases that are highly conserved in procaryotes and eucaryotes. A *clpP* deletion mutant showed significant reduction in survival inside macrophages and a loss of virulence in

mice. The activity of LLO in the mutant was greatly reduced under stress conditions. ClpP may be involved in rapid adaptive response of *L. monocytogenes* during infection (54).

Host Involvement in Virulence of *L. monocytogenes*

During its interaction with host cells, *L. monocytogenes* interferes with and modulates several host cell functions. These signal transduction events are just being recognized. In addition to its important role in escape of listeriae from primary and secondary vacuoles after invasion of host cells and cell-to-cell spread, respectively, LLO triggers several cellular events, including apoptosis and second-message generation in several mammalian cells. LLO induces the generation of second messengers inositol phosphate and prostaglandin in human umbilical vein endothelial cells (HUVEC). A synergism is observed between LLO and PI-PLC in stimulating these events. In HUVEC, *L. monocytogenes* infection leads to the activation of inducible transcription factor NF-κB, which is involved in the regulation of expression of many immunologically important genes. LLO is required and sufficient for NF-κB activation in HUVEC; also, a fraction of lipoteichoic acid induces strong transient activation of NF-κB (84). Also, LLO mediates the phosphorylation of mitogen-activated protein kinases which are part of signal transduction pathways that link extracellular signals to gene expression. LLO induces apoptosis in hepatocytes and dendritic cells. While the programmed cell death of dendritic cells may benefit the invading bacteria by limiting antigen presentation, the advantage to the bacteria of the programmed cell death of hepatocytes is not clear.

InlA and InlB mediate the invasion of nonphagocytic cells by *L. monocytogenes*. The signaling events caused by the interaction of InlA and its receptor E-cadherin have not been identified. InlB-mediated uptake of *L. monocytogenes* by Vero cells is associated with the activation of PI 3-kinase p85/p110. This activation also requires tyrosine phosphorylation in the host cell. After invasion of the host cell, *L. monocytogenes* modulates the fusion of phagosome with lysosomes by interfering with phagosome maturation. The mechanism for this is not known but does not involve LLO.

Signal transduction events that occur during the intracytoplasmic growth of *L. monocytogenes* in host cells have not been adequately investigated. PI-PLC and PC-PLC are the only known factors which may potentially interfere with host cell signaling to favor listerial growth. Both PLCs are required for permanent induction of NF-κB, but it has not been shown that NF-κB activation is a prerequisite for listerial intracytoplasmic growth. Also, signaling events during intercellular spread of *L. monocytogenes* have not been studied. The presence of PC-PLC in the membrane protrusions containing the bacterium may cause a specific alteration of the surrounding cell membrane which may be specifically recognized by the neighboring cell (84).

CONCLUDING REMARKS

Public health surveillance, outbreak investigations, and applied and basic research conducted during the past 2 decades have helped to characterize listeriosis, to define the magnitude of its public health problem and its impact on the food industry, to identify the risk factors associated with the disease, and to develop appropriate control strategies. The food processing industry has made progress in reducing the prevalence of *L. monocytogenes* in processing plant environments and in high-risk foods, and preventive measures have been developed and implemented for persons at increased risk of infection. In the United States, there was a large reduction in the incidence of illness (44%) and death (48%) due to sporadic listeriosis between 1989 and 1993; the decline was attributed to enhanced regulatory control and the implementation of preventive strategies by the food industry. Nevertheless, there are still many unanswered questions: Are all *L. monocytogenes* strains equally pathogenic for humans, and are all of them capable of causing outbreaks? What is the infectious dose of *L. monocytogenes* for nonpregnant adults, persons with immune system deficiencies, and pregnant women? Do the infectious doses vary significantly between strains? Are only a few strains capable of causing febrile gastroenteritis in humans, or are all strains equally capable of causing this gastrointestinal syndrome if they are present in high enough numbers? Because *L. monocytogenes* is primarily a soil microorganism and is ubiquitously distributed in the environment, is it practical to expect all ready-to-eat foods to be *Listeria*-free? How do we ensure that ready-to-eat prepackaged salads are not contaminated with *L. monocytogenes*? How do we prevent *Listeria* contamination of other ready-to-eat foods, such as frankfurters, soft cheeses, and smoked fish? Are some strains of *L. monocytogenes* better able to form biofilms, and, if yes, are these strains more likely to persist in food processing environments for a long time?

A multidisciplinary approach has enabled the identification and characterization of several virulence factors. Research during the past 15 years has led to (i) identification of the internalin receptor of mammalian cells, (ii) elucidation of the role of the internalin multigene functions, (iii) understanding of the functions of *actA* and of *prfA* and the global regulation of virulence, and (iv) elucidation of the modulation of host cell signaling by the pathogen. *L. monocytogenes* has become one of the best-understood intracellular pathogens at the cellular and molecular level. The immune response to this pathogen continues to be a topic of intensive investigations. The complete sequence of the genome of *L. monocytogenes* is being determined by a consortium of European researchers and research institutions and is expected to be available in 2001. It is widely expected that the availability of the complete genome sequence will stimulate further research and will help researchers answer additional questions about this fascinating bacterial pathogen.

References

1. **Ahamad, N., and E. Marth.** 1989. Behavior of *Listeria monocytogenes* at 7, 13, 21, and 35°C in tryptose broth acidified with acetic, citric, or lactic acid. *J. Food Prot.* **52:**688–695.

2. **Alexander, J., P. Andrew, D. Jones, and I. Roberts.** 1993. Characterization of an aromatic amino acid-dependent *Listeria monocytogenes* mutant: attenuation, persistence, and ability to induce protective immunity in mice. *Infect. Immun.* **61:**2245–2248.

3. **Al-Ghazali, M., and S. Al-Azawi.** 1990. *Listeria monocytogenes* contamination of crops grown on soil treated with sewage sludge cake. *J. Appl. Bacteriol.* **69:**642–674.

4. **Anonymous.** 1988. *Food Listeriosis—Report of the WHO Informal Working Group.* WHO/EHE/FOS/88.5. World Health Organization, Geneva, Switzerland.

5. **Audurier, A., P. Pardon, J. Marly, and F. Lantier.** 1980. Experimental infection of mice with *Listeria monocytogenes* and *L. innocua. Ann. Microbiol. (Paris)* **131B:**47–57.

6. **Aureli, P., G. C. Fiorucci, D. Caroli, G. Marchiaro, O. Novara, L. Leone, and S. Salmaso.** 2000. An outbreak of febrile gastroenteritis associated with corn contaminated with *Listeria monocytogenes. N. Engl. J. Med.* **342:**1236–1241.

7. **Barbosa, W. B., L. Cabedo, H. J. Wederquist, J. N. Sofos, and G. R. Schmidt.** 1994. Growth variation among species and strains of *Listeria monocytogenes. J. Food Prot.* **57:**765–769.

8. **Bearns, R. E., and K. F. Gerard.** 1958. The effect of pasteurization on *Listeria monocytogenes. Can. J. Microbiol.* **4:**55–61.

9. **Beuchat, L. R.** 1996. *Listeria monocytogenes*: incidence on vegetables. *Food Control* **7:**223–228.

10. **Bille, J.** 1989. Presented at "Foodborne Listeriosis," Wiesbaden, Germany, Sept. 7, 1998.

11. **Boerlin, P., J. Rocourt, F. Grimont, P. A. D. Grimont, C. Jacquet, and J. C. Piffaretti.** 1992. *Listeria ivanovii* subspecies *londoniensis. Int. J. Syst. Bacteriol.* **15:**42–46.

12. Bradshaw, J. G., J. T. Peeler, J. J. Corwin, J. M. Hunt, J. T. Tierney, E. P. Larkin, and R. M. Twedt. 1985. Thermal resistance of *Listeria monocytogenes* in milk. *J. Food Prot.* 48:743–745.

13. Braun, L., F. Nato, B. Payrastre, J.-C. Mazie, and P. Cossart. 1999. The 213-amino-acid leucine-rich repeat region of the *Listeria monocytogenes* InlB protein is sufficient for entry into mammalian cells, stimulation of PI 3-kinase and membrane ruffling. *Mol. Microbiol.* 34:10–23.

14. Brosch, R., B. Catimel, G. Milon, C. Buchrieser, E. Vindel, and J. Rocourt. 1993. Virulence heterogeneity of *Listeria monocytogenes* strains from various sources (food, human, animal) in immunocompetent mice and its association with typing characteristics. *J. Food Prot.* 56:301–312.

15. Buchrieser, C., R. Brosch, B. Catimel, and J. Rocourt. 1993. Pulsed-field electrophoresis applied for comparing *Listeria monocytogenes* strains involved in outbreaks. *Can. J. Microbiol.* 39:395–401.

16. Bula, C., J. Bille, and M. Glauser. 1995. An epidemic of food-borne listeriosis in Western Switzerland: description of 57 cases involving adults. *Clin. Infect. Dis.* 20:66–72.

17. Buncie, S. 1991. The incidence of *Listeria monocytogenes* in slaughtered animals, in meat, and in meat products in Yugoslavia. *Int. J. Food Microbiol.* 12:173–180.

18. Centers for Disease Control and Prevention. 1989. Listeriosis associated with consumption of turkey franks. *Morb. Mortal. Wkly. Rep.* 38:267–268.

19. Centers for Disease Control and Prevention. 2000. Multistate outbreak of listeriosis—United States. *Morb. Mortal. Wkly. Rep.* 49:1129–1130.

20. Centers for Disease Control and Prevention. 1998. Multistate outbreak of listeriosis—United States, 1998. *Morb. Mortal. Wkly. Rep.* 47:1085–86.

21. Centers for Disease Control and Prevention. 1996. *Preventing Foodborne Illness: Listeriosis*. U.S. Department of Health and Human Services, Public Health Service, Centers for Disease Control and Prevention, National Center for Infectious Diseases, Atlanta, Georgia.

22. Chief Medical Officer of Manitoba. 2000. *Faulty Food Product Identified as a Source of Listeria Foodborne Illness*. Manitoba Health-Office of CMOH-Media Bulletin (07.06.2000).

23. Clark, E., I. Wesley, F. Fiedler, N. Promadej, and S. Kathariou. 2000. Absence of serotype-specific surface antigen and altered teichoic acid glycosylation among epidemic-associated strains of *Listeria monocytogenes*. *J. Clin. Microbiol.* 38:3856–3859.

24. Colburn, K., C. Kaysner, C. Abeyta, Jr., and M. Wekell. 1990. *Listeria* species in a California estuarine environment. *Appl. Environ. Microbiol.* 56:2007–2011.

25. Communicable Disease Report. 2000. Outbreak of *Listeria monocytogenes* serotype 4b infection in France. *CDR Wkly.* 10:81, 84.

26. Cossart, P. 1995. Actin-based bacterial motility. *Curr. Top. Cell Biol.* 7:94–101.

27. Cossart, P., and C. Kocks. 1994. The actin-based motility of the intracellular pathogen *Listeria monocytogenes*. *Mol. Microbiol.* 13:395–402.

28. Cossart, P., and J. Mengaud. 1989. *Listeria monocytogenes*: a model system for the molecular study of intracellular parasitism. *Mol. Biol. Med.* 6:463–474.

29. Cotter, P., N. Emerson, C. Gahan, and C. Hill. 1999. Identification and disruption of *lisRK*, a genetic locus encoding a two-component signal transduction system involved in stress tolerance and virulence in *Listeria monocytogenes*. *J. Bacteriol.* 181:6840–6843.

30. Cox, L., T. Kleiss, J. Cordier, C. Cordellana, P. Konkel, C. Pedrazzini, R. Beumer, and A. Siebenga. 1989. *Listeria* spp. in food processing, non-food and domestic environments. *Food Microbiol.* 6:49–61.

31. Dalton, C. B., C. C. Austin, J. Sobel, P. S. Hayes, W. F. Bibb, L. M. Graves, B. Swaminathan, M. E. Procter, and P. M. Griffin. 1997. An outbreak of gastroenteritis and fever due to *Listeria monocytogenes* in milk. *N. Engl. J. Med.* 336:100–105.

32. Davies, E., and M. Adams. 1994. Resistance of *Listeria monocytogenes* to the bacteriocin nisin. *Int. J. Food Microbiol.* 21:341–347.

33. de Simon, M., and M. Dolores Ferrer. 1998. Initial numbers, serovars and phagovars of *Listeria monocytogenes* in prepared foods in the city of Barcelona (Spain). *Int. J. Food Microbiol.* 44:141–144.

34. de Valk, H., V. Vaillant, V. Pierre, J. Rocourt, C. Jacquet, F. Lequerrec, J.-C. Thomas, and V. Goulet. 1998. Risk factors for sporadic listeriosis in France. http://www.invs.sante.fr/epiet/seminar/1998/valk.html.

35. Dijkstra, R. 1982. The occurrence of *Listeria monocytogenes* in surface water of canals and lakes, in ditches of one big polder and in the effluents and canals of a sewage treatment plant. *Zentbl. Bakteriol. Hyg. Abt. 1 Orig. Reihe B* 176:202–205.

36. Domann, E., J. Wehland, M. Rohde, S. Pistor, M. Hartl, W. Goebel, M. Leimester-Wachter, M. Wuenscher, and T. Chakraborty. 1992. A novel bacterial gene in *Listeria monocytogenes* required for host cell microfilament interaction with homology to the proline-rich region of vinculin. *EMBO J.* 11:1981–1990.

37. Donnelly, C. W., E. H. Briggs, and L. S. Donnelly. 1987. Comparison of heat resistance of *Listeria monocytogenes*. *Abstr. 87th Annu. Meet. Am. Soc. Microbiol. 1987*. American Society for Microbiology, Washington, D.C.

38. Doyle, M. 1988. Effect of environmental and processing conditions on *Listeria monocytogenes*. *Food Technol.* 42:169–171.

39. Dramsi, S., I. Biswas, L. Braun, E. Maguin, P. Mastroenni, and P. Cossart. 1995. Entry into hepatocytes requires expression of the *inlB* gene product. *Mol. Microbiol.* 16:251–261.

40. Dramsi, S., M. Lebrun, and P. Cossart. 1996. Molecular and genetic determinants involved in invasion of mammalian cells by *Listeria monocytogenes*. *Curr. Top. Microbiol. Immunol.* 203:61–77.

41. Edelson, B., and E. Unanue. 2000. Immunity to *Listeria* infection. *Curr. Opin. Immunol.* 12:425–431.

42. Elliot, E. L., and J. E. Kvenberg. 2000. Risk assessment used to evaluate the US position on *Listeria monocytogenes* in seafood. *Int. J. Food Microbiol.* 62:253–260.

43. Ericsson, H., W. Eklow, M. L. Danielson-Tham, S. Loncarevic, L. O. Mentzing, I. Person, H. Unnerstad, and W. Tham. 1997. An outbreak of listeriosis suspected to have been caused by rainbow trout. *J. Clin. Microbiol.* 35:2904–2907.

44. Farber, J., E. Coates, N. Beausoleil, and J. Fournier. 1991. Feeding trials of *Listeria monocytogenes* with a nonhuman primate model. *J. Clin. Microbiol.* 29:2606–2608.

45. Farber, J. M. 2000. Present situation in Canada regarding *Listeria monocytogenes* and ready-to-eat seafood products. *Int. J. Food Microbiol.* 62:247–251.

46. Farber, J. M., E. M. Daley, M. T. Mackie, and B. Limerick. 2000. A small outbreak of listeriosis potentially linked to the consumption of crab meat. *Lett. Appl. Microbiol.* 31:100–104.

47. Farber, J. M., and J. Harwig. 1996. The Canadian position on *Listeria monocytogenes* in ready-to-eat foods. *Food Control* 7:253–258.

48. Farber, J. M., and P. I. Peterkin. 1999. Incidence and behavior of *Listeria monocytogenes* in meat products, p. 505–564. *In* E. T. Ryser and E. H. Marth (ed.), *Listeria, Listeriosis, and Food Safety,* 2nd ed. Marcel Dekker, Inc., New York, N.Y.

49. Fenlon, D. R. 1999. *Listeria monocytogenes* in the natural environment, p. 21–37. *In* E. T. Ryser and E. H. Marth (ed.), *Listeria, Listeriosis, and Food Safety.* Marcel Dekker, Inc., New York, N.Y.

50. Fleming, D., S. Cochi, K. MacDonald, J. Brondum, P. Hayes, B. Plikaytis, M. Holmes, A. Audurier, C. Broome, and A. Reingold. 1985. Pasteurized milk as a vehicle of infection in an outbreak of listeriosis. *N. Engl. J. Med.* 312:404–407.

51. Freitag, N., L. Rong, and D. Portnoy. 1993. Regulation of the *prfA* transcriptional activator of *Listeria monocytogenes:* multiple promoter elements contribute to intracellular growth and cell-to-cell spread. *Infect. Immun.* 61:2537–2544.

52. Gaillard, J., P. Berche, C. Frehel, E. Gouin, and P. Cossart. 1991. Entry of *L. monocytogenes* into cells is mediated by internalin, a repeat protein reminiscent of surface antigens from Gram-positive cocci. *Cell* 65:1127–1141.

53. Galliard, J., P. Berche, and P. Sansonetti. 1986. Transposon mutagenesis as a tool to study the role of hemolysin in the virulence of *Listeria monocytogenes. Infect. Immun.* 52:50–55.

54. Gaillot, O., E. Pelligrini, S. Bregenholt, S. Nair, and P. Berche. 2000. The ClpP serine protease is essential for the intracellular parasitism and virulence of *Listeria monocytogenes. Mol. Microbiol.* 35:1286–1294.

55. Gedde, M., D. Higgins, L. Tilney, and D. Portnoy. 2000. Role of listeriolysin O in cell-to-cell spread of *Listeria monocytogenes. Infect. Immun.* 68:999–1003.

56. Genigeorgis, C., D. Dutulescu, and J. Fernandez Garayzabal. 1989. Prevalence of *Listeria* spp. in poultry meat at the supermarket and slaughterhouse level. *J. Food Prot.* 52:618–624.

57. Geoffroy, C., J. Galliard, J. Alouf, and P. Berche. 1987. Purification, characterization, and toxicity of the sulfhydryl-activated hemolysin listeriolysin O from *Listeria monocytogenes. Infect. Immun.* 55:1641–1646.

58. Glass, K. A., and M. P. Doyle. 1989. Fate of *Listeria monocytogenes* in processed meat products during refrigerated storage. *Appl. Environ. Microbiol.* 55:1565–1569.

59. Goulet, V., C. Jacquet, V. Vaillant, I. Rebiere, E. Mouret, E. Lorente, F. Steiner, and J. Rocourt. 1995. Listeriosis from consumption of raw milk cheese. *Lancet* 345:1581–1582.

60. Grau, F., and P. Vanderlinde. 1992. Occurrence, numbers and growth of *Listeria monocytogenes* on some vacuum-packaged processed meats. *J. Food Prot.* 55:4–7.

61. Gravani, R. 1999. *Listeria* in food processing facilities, p. 657–709. *In* E. T. Ryser and E. H. Marth (ed.), *Listeria, Listeriosis, and Food Safety.* Marcel Dekker, Inc., New York, N.Y.

62. Graves, L. M., B. Swaminathan, and S. Hunter. 1999. Subtyping *Listeria monocytogenes,* p. 279–297. *In* E. T. Ryser and E. H. Marth (ed.), *Listeria, Listeriosis, and Food Safety,* 2nd ed. Marcel Dekker, Inc., New York, N.Y.

63. Greeiffenberg, L., W. Goebel, K. S. Kim, J. Daniels, and M. Kuhn. 2000. Interactions of *Listeria monocytogenes* with human brain microvascular endothelial cells: an electron microscopic study. *Infect. Immun.* 68:3275–3279.

64. Guenich, H., H. Muller, A. Schrettenbrunner, and H. Seeliger. 1985. The occurrence of different *Listeria* species in municipal waste water. *Zentbl. Bakteriol. Hyg. Abt. 1 Orig. Reihe B* 181:563–565.

65. Hanawa, T., M. Fukuda, H. Kawakami, H. Hirano, S. Kamiya, and T. Yamamoto. 1999. The *Listeria monocytogenes* DnaK chaperone is required for stress tolerance and efficient phagocytosis with macrophages. *Cell Stress Chaperones* 4:118–128.

66. Hanawa, T., M. Kai, S. Kamiya, and T. Yamamoto. 2000. Cloning, sequencing, and transcriptional analysis of the *dnaK* heat shock operon of *Listeria monocytogenes. Cell Stress Chaperones* 5:21–29.

67. Hanawa, T., Y. Yamamoto, and S. Kamiya. 1995. *Listeria monocytogenes* can grow in macrophages without the aid of proteins induced by environmental stresses. *Infect. Immun.* 63:4595–4599.

68. Ho, J. L., K. N. Shands, G. Friedland, P. Eckind, and D. W. Fraser. 1986. An outbreak of type 4b *Listeria monocytogenes* infection involving patients in eight Boston hospitals. *Arch. Intern. Med.* 146:520–524.

69. Huss, H. H., A. Reilly, and P. K. Ben Embarek. 2000. Prevention and control of safety hazards in cold smoked salmon production. *Food Control* 11:149–156.

70. Husu, J., S. Sivela, and A. Rauramaa. 1990. Prevalence of *Listeria* species as related to chemical quality of farm-ensiled grass. *Grass Forage Sci.* 45:309–314.

71. Hutchins, A. D. 1996. Assessment of alimentary exposure to *Listeria monocytogenes*. *Int. J. Food Microbiol.* **30:** 71–85.

72. Inoue, S., A. Nakama, Y. Arai, Y. Kokubu, T. Maruyama, A. Saito, T. Yoshida, M. Terao, S. Yamamoto, and S. Kumagai. 2000. Prevalence and contamination levels of *Listeria monocytogenes* in retail foods in Japan. *Int. J. Food Microbiol.* **59:**73–77.

73. Isberg, R. R. 1991. Discrimination between intracellular uptake and surface adhesion of bacterial pathogens. *Science* **252:**934–938.

74. Jacquet, C., B. Catimel, R. Brosch, C. Buchrieser, P. Dehaumont, V. Goulet, V. Lepoutre, P. Veit, and J. Rocourt. 1995. Investigations related to the epidemic strain involved in the French listeriosis outbreak in 1992. *Appl. Environ. Microbiol.* **61:**2242–2246.

75. Jensen, A. 1993. Excretion of *Listeria monocytogenes* in faeces after listeriosis: rate, quantity and duration. *Med. Microbiol. Lett.* **2:**176–182.

76. Jeong, D., and J. Frank. 1994. Growth of *Listeria monocytogenes* at 10°C in biofilms with microorganisms isolated from meat and dairy processing environments. *J. Food Prot.* **57:**576–586.

77. Jinneman, K. C., M. M. Wekell, and M. W. Eklund. 1999. Incidence and behavior of *Listeria monocytogenes* in fish and seafood, p. 601–630. *In* E. T. Ryser and E. H. Marth (ed.), *Listeria, Listeriosis, and Food Safety*, 2nd ed. Marcel Dekker, Inc., New York, N.Y.

78. Jurado, R., M. Farley, E. Pereira, R. Harvey, A. Schuchat, J. Wenger, and D. Stephens. 1993. Increased risk of meningitis and bacteremia due to *Listeria monocytogenes* in patients with human immunodeficiency virus infection. *Clin. Infect. Dis.* **17:**224–227.

79. Kayal, S., A. Lilienbaum, C. Poyart, S. Memet, A. Israel, and P. Berche. 1999. Listeriolysin O-dependent activation of endothelial cells during infection with *Listeria monocytogenes*: activation of NF-κB and upregulation of adhesion molecules and chemokines. *Mol. Microbiol.* **31:**1709–1722.

80. Klarsfeld, A., P. Goossens, and P. Cossart. 1994. Five *Listeria monocytogenes* preferentially expressed in mammalian cells. *Mol. Microbiol.* **13:**585–597.

81. Kocks, C., E. Gouin, M. Tabouret, P. Berche, H. Ohayon, and P. Cossart. 1992. *Listeria monocytogenes* induced actin assembly requires the ActA gene product, a surface protein. *Cell* **68:**521–531.

82. Kohler, S., M. Leimeister-Wachter, T. Chakraborty, F. Lottspeich, and W. Goebel. 1990. The gene coding for protein p60 of *Listeria monocytogenes* and its use as a specific probe for *Listeria monocytogenes*. *Infect. Immun.* **58:**1943–1950.

83. Kolb-Maurer, A., I. Gentschev, H.-W. Fries, F. Fiedler, E.-B. Brocker, E. Kampgen, and W. Goebel. 2000. *Listeria monocytogenes*-infected human dendritic cells: uptake and host cell response. *Infect. Immun.* **68:**3680–3688.

84. Kuhn, M., T. Pfeuffer, L. Greiffenberg, and W. Goebel. 1999. Host cell signal transduction during *Listeria monocytogenes* infection. *Arch. Biochem. Biophys.* **372:**166–172.

85. Lahellec, C. 1996. *Listeria monocytogenes* in foods: the French position. International Food Safety Conference. *Listeria*: the State of Science—29–30 June 1995—Rome, Italy. *Food Control* **7:**241–243.

86. Larsen, A., and B. Norrung. 1993. Inhibition of *Listeria monocytogenes* by bavaricin A, a bacteriocin produced by *Lactobacillus bavaricus* Ml401. *Lett. Appl. Microbiol.* **17:**132–134.

87. Lecuit, M., S. Dramsi, C. Gottardi, M. Fedor-Chaiken, B. Gumbiner, and P. Cossart. 1999. A single amino acid in E-cadherin responsible for host specificity towards the human pathogen *Listeria monocytogenes*. *EMBO J.* **18:**3956–3963.

88. Leimeister-Wachter, M., E. Domann, and T. Chakraborty. 1992. The expression of virulence genes in *L. monocytogenes* is thermoregulated. *J. Bacteriol.* **174:**947–952.

89. Leimeister-Wachter, M., C. Haffner, E. Domann, W. Goebel, and T. Chakraborty. 1990. Identification of a gene that positively regulates expression of listeriolysin, the major virulence factor of *Listeria monocytogenes*. *Proc. Natl. Acad. Sci. USA* **87:**8336–8340.

90. Linnan, M., L. Mascola, X. Lou, V. Goulet, S. May, C. Salminen, D. Hird, M. Yonekura, P. Hayes, R. Weaver, A. Audurier, B. Plikaytis, S. Fannin, A. Kleks, and C. Broome. 1988. Epidemic listeriosis associated with Mexican-style cheese. *N. Engl. J. Med.* **319:**823–828.

91. Lou, Y., and A. E. Yousef. 1999. Characteristics of *Listeria monocytogenes* important to food processors, p. 131–224. *In* E. T. Ryser and E. H. Marth (ed.), *Listeria, Listeriosis, and Food Safety*, 2nd ed. Marcel Dekker, Inc., New York, N.Y.

92. Luchansky, J. B. 1994. Use of biopreservatives to control pathogenic and spoilage microbes in food, p. 253–262. *In* A. Amgar (ed.), *Food Safety 94*. ASEPT, Laval, France.

93. Lyytikainen, O., T. Autio, R. Maijala, P. Ruutu, T. Honkanen-Buzalski, M. Miettinen, M. Hatakka, J. Mikkola, V.-J. Anttila, T. Johansson, L. Rantala, T. Aalto, H. Korkeala, and A. Siitonen. 2000. An outbreak of *Listeria monocytogenes* serotype 3a infections from butter in Finland. *J. Infect. Dis.* **181:**1838–41.

94. Marino, M., L. Braun, P. Cossart, and P. Ghosh. 1999. Structure of the InlB leucine-rich repeats, a domain that triggers host cell invasion by the bacterial pathogen *L. monocytogenes*. *Mol. Cell* **4:**1063–1072.

95. Marquis, H., H. Bouwer, D. Hinrichs, and D. Portnoy. 1993. Intracytoplasmic growth and virulence of *Listeria monocytogenes* auxotrophic mutants. *Infect. Immun.* **61:**3756–3760.

96. Marquis, H., H. Goldfine, and D. A. Portnoy. 1997. Proteolytic pathways of activation and degradation of a bacterial phospholipase C during intracellular infection by *Listeria monocytogenes*. *J. Cell Biol.* **137:**1381–1392.

97. Marquis, H., and E. Hager. 2000. pH-regulated activation and release of a bacteria-associated phospholipase C during intracellular infection by *Listeria monocytogenes*. *Mol. Microbiol.* **35:**289–298.

98. Mascola, L., F. Sorvillo, V. Goulet, B. Hall, R. Weaver, and M. Linnan. 1992. Fecal carriage of *Listeria monocytogenes*—observations during a community

wide, common-source outbreak. *Clin. Infect. Dis.* **15:** 557–558.

99. **Matthieu, F., M. Michel, A. Lebrihi, and G. Lefebvre.** 1994. Effect of the bacteriocin carnocin CP5 and of the producing strain *Carnobacterium piscicola* CP5 on the viability of *Listeria monocytogenes* ATCC 15313 in salt solution, broth and skimmed milk, at various incubation temperatures. *Int. J. Food Microbiol.* **22:**155–172.

100. **McLauchlin, J.** 1997. The discovery of *Listeria. PHLS Microbiol. Digest* **14:**76–78.

101. **McLauchlin, J., S. Hall, S. Velani, and R. Gilbert.** 1991. Human listeriosis and pate—a possible association. *Br. Med. J.* **303:**773–775.

102. **McLauchlin, J., and P. Hoffman.** 1989. Neonatal cross-infection from *Listeria monocytogenes. Commun. Dis. Rep.* **6:**3–4.

103. **Mead, P. S.** 1999. *Multistate Outbreak of Listeriosis Traced to Processed Meats, August 1998–March 1999.* Epidemiologic Investigation Report. Centers for Disease Control and Prevention, Atlanta, Ga.

104. **Mead, P. S., L. Slutsker, V. Dietz, L. F. McCaig, J. Bresee, C. Shapiro, P. M. Griffin, and R. V. Tauxe.** 1999. Food-related illness and death in the United States. *Emerg. Infect. Dis.* **5:**607–625.

105. **Mengaud, J., S. Dramsi, E. Gouin, J. Vasquez-Boland, G. Milon, and P. Cossart.** 1991. Pleiotropic control of *Listeria monocytogenes* virulence factors by a gene which is autoregulated. *Mol. Microbiol.* **5:**2273–2283.

106. **Mengaud, J., H. Ohayon, P. Gounon, R.-M. Mege, and P. Cossart.** 1996. E-cadherin is the receptor for internalin, a surface protein required for entry of *L. monocytogenes* into epithelial cells. *Cell* **84:**923–932.

107. **Miettinen, M. K., A. Siitonen, P. Heiskanen, H. Haajanen, K. J. Bjorkroth, and H. J. Korkeala.** 1999. Molecular epidemiology of an outbreak of febrile gastroenteritis caused by *Listeria monocytogenes* in cold-smoked rainbow trout. *J. Clin. Microbiol.* **37:**2358–2360.

108. **Milohanic, E., B. Pron, The European Listeria Genome Consortium, P. Berche, and J. Gaillard.** 2000. Identification of new loci involved in adhesion of *Listeria monocytogenes* to eukaryotic cells. *Microbiology* **146:**731–739.

109. **Modi, K., M. Chikindas, and T. Montville.** 2000. Sensitivity of nisin-resistant *Listeria monocytogenes* to heat and the synergistic action of heat and nisin. *Lett. Appl. Microbiol.* **30:**249–253.

110. **Norrung, B., J. K. Andersen, and J. Schlundt.** 1999. Incidence and control of *Listeria monocytogenes* in foods in Denmark. *Int. J. Food Microbiol.* **53:**195–203.

111. **Northolt, M. D., H. J. Beckers, U. Vecht, L. Toepoel, P. S. S. Soentoro, and H. J. Wisselink.** 1988. *Listeria monocytogenes:* heat resistance and behavior during storage of milk and whey and making of Dutch type of cheeses. *Neth. Milk Dairy J.* **42:**207–219.

112. **Park, S., and R. Kroll.** 1993. Expression of listeriolysin and phosphatidylinositol-specific phospholipase C is repressed by the plant-derived molecular cellobiose in *Listeria monocytogenes. Mol. Microbiol.* **8:**653–661.

113. **Pearson, L. J., and E. H. Marth.** 1990. *Listeria monocytogenes*—threat to food supply: a review. *J. Dairy Sci.* **73:**912–928.

114. **Pine, L., S. Kathariou, F. Quinn, G. V, J. Wenger, and R. Weaver.** 1991. Cytopathogenic effects in enterocytelike Caco-2 cells differentiate virulent from avirulent *Listeria* strains. *J. Clin. Microbiol.* **29:**990–996.

115. **Pinner, R., A. Schuchat, B. Swaminathan, P. Hayes, K. Deaver, R. Weaver, B. Plikaytis, M. Reeves, C. Broome, and J. Wenger.** 1992. Role of foods in sporadic listeriosis. 2. Microbiologic and epidemiologic investigation. *JAMA* **267:**2046–2050.

116. **Portnoy, D.** 1992. Innate immunity to a facultative intracellular bacterial pathogen. *Curr. Top. Immunol.* **4:**20–24.

117. **Portnoy, D., T. Chakraborty, W. Goebel, and P. Cossart.** 1992. Molecular determinants of *Listeria monocytogenes* pathogenesis. *Infect. Immun.* **60:**1263–1267.

118. **Proctor, M. E., R. Brosch, J. W. Mellen, L. A. Garrett, C. W. Kaspar, and J. B. Luchansky.** 1995. Use of pulsed-field gel electrophoresis to link sporadic cases of invasive listeriosis with recalled chocolate milk. *Appl. Environ. Microbiol.* **61:**3177–3179.

119. **Promadej, N., F. Fiedler, P. Cossart, S. Dramsi, and S. Kathariou.** 1999. Cell wall teichoic acid glycosylation in *Listeria monocytogenes* serotype 4b requires *gtcA*, a novel, serogroup-specific gene. *J. Bacteriol.* **181:**418–425.

120. **Pron, B., C. Boumaila, F. Jaubert, S. Sarnacki, J. P. Monnet, P. Berche, and J. L. Gaillard.** 1998. Comprehensive study of the intestinal stage of listeriosis in a rat ligated loop system. *Infect. Immun.* **66:**747–755.

121. **Renzoni, A., P. Cossart, and S. Dramsi.** 1999. PrfA, the transcriptional activator of virulence genes, is upregulated during interaction of *Listeria monocytogenes* with mammalian cells and in eukaryotic cell extracts. *Mol. Microbiol.* **34:**552–561.

122. **Rocourt, J.** 1999. The genus *Listeria* and *Listeria monocytogenes:* phylogenetic position, taxonomy, and identification, p. 1–20. *In* E. T. Ryser and E. H. Marth (ed.), *Listeria, Listeriosis, and Food Safety,* 2nd ed. Marcel Dekker, Inc., New York, N.Y.

123. **Rocourt, J., P. Boerlin, F. Grimont, C. Jacquet, and J. C. Piffaretti.** 1992. Assignment of *Listeria grayi* and *Listeria murrayi* to a single species, *Listeria grayi,* with a revised description of *Listeria grayi. Int. J. Syst. Bacteriol.* **42:**69–73.

124. **Rocourt, J., and R. Brosch.** 1992. *Human Listeriosis—1990.* WHO/HPP/FOS/92.3. World Health Organization, Geneva, Switzerland.

125. **Rocourt, J., C. Jacquet, and A. Reilly.** 2000. Epidemiology of human listeriosis and seafoods. *Int. J. Food Microbiol.* **62:**197–209.

126. **Ryser, E. T.** 1999. Incidence and behavior of *Listeria monocytogenes* in unfermented dairy products, p. 359–409. *In* E. T. Ryser and E. H. Marth (ed.), *Listeria, Listeriosis, and Food Safety,* 2nd ed. Marcel Dekker, Inc., New York, N.Y.

127. **Salamina, G., E. Dalle Donne, A. Niccolini, G. Poda, D. Cesaroni, M. Bucci, R. Fini, M. Maldini, A. Schuchat, B. Swaminathan, W. Bibb, J. Rocourt, N. Binkin, and S. Salmaso.** 1996. A foodborne outbreak of gastroenteritis involving *Listeria monocytogenes. Epidemiol. Infect.* **117:**429–436.

128. **Saxbe, W. B., Jr.** 1972. *Listeria monocytogenes* and Queen Anne. *Pediatrics* **49:**97–101.

129. **Schlech, W. I., P. M. Lavigne, R. A. Bortolussi, A. C. Allen, E. V. Haldane, A. J. Won, A. W. Hightower, S. E. Johnson, S. H. King, E. S. Nicholls, and C. V. Broome.** 1983. Epidemic listeriosis—evidence for transmission by food. *N. Engl. J. Med.* **308:**203–206.

130. **Schubert, K., A. Bichlmaier, E. Mager, K. Wolff, G. Ruhland, and F. Fiedler.** 2000. P45, an extracellular 45 kDa protein of *Listeria monocytogenes* with similarity to protein p60 and exhibiting peptidoglycan lytic activity. *Arch. Microbiol.* **173:**21–28.

131. **Schuchat, A., K. Deaver, P. Hayes, L. Graves, L. Mascola, and J. Wenger.** 1993. Gastrointestinal carriage of *Listeria monocytogenes* in household contacts of patients with listeriosis. *J. Infect. Dis.* **167:**1261–1262.

132. **Schuchat, A., K. Deaver, J. Wenger, B. Plikaytis, L. Mascola, R. Pinner, A. Reingold, and C. Broome.** 1992. Role of foods in sporadic listeriosis. 1. Case-control study of dietary risk factors. *JAMA* **267:**2041–2045.

133. **Schuchat, A., C. Lizano, C. Broome, B. Swaminathan, C. Kim, and K. Win.** 1991. Outbreak of neonatal listeriosis associated with mineral oil. *Pediatr. Infect. Dis. J.* **10:**183–189.

134. **Schuchatt, A., R. W. Pinner, K. Deaver, B. Swaminathan, R. Weaver, P. S. Hayes, M. Reeves, P. Pierce, J. D. Wenger, C. V. Broome, and The Listeria Study Group.** 1991. Epidemiology of listeriosis in the USA, p. 69–73. *In* A. Amgar (ed.), *Listeria and Food Safety.* ASEPT, Laval, France.

135. **Schwartz, B., D. Hexter, C. Broome, A. Hightower, R. Hischorn, J. Porter, P. Hayes, W. Bibb, B. Lorber, and D. Faris.** 1989. Investigation of an outbreak of listeriosis: new hypotheses for the etiology of epidemic *Listeria monocytogenes* infections. *J. Infect. Dis.* **159:**680–685.

136. **Seeliger, H., and D. Jones.** 1986. *Listeria,* p. 1235–1245. *In* P. H. A. Sneath, N. S. Mair, M. E. Sharpe, and J. G. Holt (ed.), *Bergey's Manual of Systematic Bacteriology,* vol. 2. Williams & Wilkins, Baltimore, Md.

137. **Seeliger, H. P. R., and K. Hohne.** 1979. Serotyping of *Listeria monocytogenes* and related species. *Methods Microbiol.* **13:**31–49.

138. **Shahamat, M., A. Seaman, and M. Woodbine.** 1980. Survival of *Listeria monocytogenes* in high salt concentrations. *Zentbl. Bakteriol. Hyg. Abt. 1 Orig. A* **246:**506–511.

139. **Shank, F., E. L. Elliot, I. K. Wachsmuth, and M. E. Losikoff.** 1996. US position on *Listeria monocytogenes* in foods. *Food Control* **7:**229–234.

140. **Sheehan, B., C. Kocks, S. Dramsi, E. Gouin, A. Klarsfeld, J. Mengaud, and P. Cossart.** 1994. Molecular and genetic determinants of the *Listeria monocytogenes* infectious process. *Curr. Top. Microbiol. Immunol.* **192:**187–216.

141. **Skinner, R.** 1996. *Listeria:* the State of Science. Rome 29-30 June 1995. Session IV: country and organizational postures on *Listeria monocytogenes* in food. *Listeria:* UK government's approach. *Food Control* **7:**245–247.

142. **Skogberg, K., J. Syrjanen, M. Jahkola, O. Renkonen, J. Paavonen, J. Ahonen, S. Kontiainen, P. Ruutu, and V. Valtonen.** 1992. Clinical presentation and outcome of listeriosis in patients with and without immunosuppressive therapy. *Clin. Infect. Dis.* **14:**815–821.

143. **Skovgaard, N., and B. Norrung.** 1989. The incidence of *Listeria* spp. in faeces of Danish pigs and in minced pork meat. *Int. J. Food Microbiol.* **8:**59–63.

144. **Slutsker, L., and A. Schuchat.** 1999. Listeriosis in humans, p. 75–95. *In* E. T. Rysaer and E. H. Marth (ed.), *Listeria, Listeriosis, and Food Safety,* 2nd ed. Marcel Dekker, Inc., New York, N.Y.

145. **Sokolovic, Z., A. Fuchs, and W. Goebel.** 1990. Synthesis of species-specific stress proteins by virulent strains of *Listeria monocytogenes. Infect. Immun.* **58:**3582–3587.

146. **Sokolovic, Z., and W. Goebel.** 1989. Synthesis of listeriolysin in *Listeria monocytogenes* under heat shock conditions. *Infect. Immun.* **57:**295–298.

147. **Sumner, S. S., T. M. Sandros, M. Harmon, V. N. Scott, and D. T. Bernard.** 1991. Heat resistance of *Salmonella typhimurium* and *Listeria monocytogenes* in sucrose solutions of various water activities. *J. Food Sci.* **56:**1741–1743.

148. **Tang, P., I. Rosenshine, and B. Finley.** 1994. *Listeria monocytogenes,* an invasive bacterium stimulates MAP kinase upon attachment to epithelial cells. *Mol. Biol. Cell* **5:**455–464.

149. **Tappero, J., A. Schuchat, K. Deaver, L. Mascola, and J. Wenger.** 1995. Reduction in the incidence of human listeriosis in the United States. Effectiveness of prevention efforts. *JAMA* **273:**1118–1122.

150. **Tilney, L., and M. Tilney.** 1993. The wily ways of a parasite: induction of actin assembly by *Listeria. Trends Microbiol.* **1:**25–31.

151. **Van Renterghem, B., F. Huysman, R. Rygole, and W. Verstraete.** 1991. Detection and prevalence of *Listeria monocytogenes* in the agricultural ecosystem. *J. Appl. Bacteriol.* **71:**211–217.

152. **Vasquez-Boland, J., C. Kocks, S. Dramsi, H. Ohayon, C. Geoffroy, J. Mengaud, and P. Cossart.** 1992. Nucleotide sequence of the lecithinase operon of *Listeria monocytogenes* and possible role of lecithinase in cell-to-cell spread. *Infect. Immun.* **60:**219–230.

153. **Velge, P., E. Bottreau, B. Kaeffer, N. Yurdusev, P. Pardon, and N. Van Langendonck.** 1994. Protein tyrosine kinase inhibitors block the entries of *Listeria monocytogenes* and *Listeria ivanovii* into epithelial cells. *Microb. Pathog.* **17:**37–50.

154. **Walker, R., L. Jensen, H. Kinde, A. Alexander, and L. Owen.** 1991. Environment survey for *Listeria* species in frozen milk plants in California. *J. Food Prot.* **54:**178–182.

155. **Weis, J., and H. Seeliger.** 1975. Incidence of *Listeria monocytogenes* in nature. *Appl. Microbiol.* **30:**29–32.

156. **Wenger, J. D., B. Swaminathan, P. S. Hayes, S. S. Green, M. Pratt, R. W. Pinner, A. Schuchat, and C. V. Broome.** 1990. *Listeria monocytogenes* contamination of turkey franks: evaluation of a production facility. *J. Food Prot.* **53:**1015–1019.

157. **Williams, J., C. Thayyullathil, and N. Freitag.** 2000. Sequence variations within Prfa DNA binding sites and effects on *Listeria monocytogenes* virulence gene expression. *J. Bacteriol.* **182:**837–841.

158. **Wuenscher, M., S. Kohler, A. Bubert, U. Gerike, and W. Goebel.** 1993. The *iap* gene of *Listeria monocytogenes* is essential for cell viability and its gene product, p60, has bacteriolytic activity. *J. Bacteriol.* **175:**3491–3501.

159. **Zheng, W., and S. Kathariou.** 1995. Differentiation of epidemic-associated strains of *Listeria monocytogenes* by restriction fragment length polymorphism in a gene region essential for growth at low temperatures (4°C). *Appl. Environ. Microbiol.* **61:**4310–4314.

160. **Zheng, W., and S. Kathariou.** 1997. Host-mediated modification of *Sau*3AI restriction in *Listeria monocytogenes*: prevalence in epidemic-associated strains. *Appl. Environ. Microbiol.* **63:**3085–3089.

Food Microbiology: Fundamentals and Frontiers, 2nd Ed.
Edited by M. P. Doyle et al.
© 2001 ASM Press, Washington, D.C.

Lynn M. Jablonski
Gregory A. Bohach

19

Staphylococcus aureus

CHARACTERISTICS OF THE ORGANISM

Historical Aspects and General Considerations

Staphylococcal food poisoning (SFP) is a common cause of gastroenteritis worldwide. In the United States, estimates of costs associated with SFP are higher than for many other foodborne pathogens (22). The etiologic agents of SFP are members of the genus *Staphylococcus*, predominantly *Staphylococcus aureus*. Unlike many other forms of gastroenteritis, SFP is not due to the ingestion of live microorganisms but rather results from one or more preformed staphylococcal enterotoxins (SEs) in staphylococci-contaminated food. This form of food poisoning is considered an intoxication because it does not require growth of the bacterium in the host.

The association of staphylococci with foodborne illness was made almost a century ago. In 1914, Barber (8) determined that repeated ingestion of contaminated milk produced symptoms of vomiting and diarrhea, thereby providing the first evidence that a soluble toxin was responsible for SFP. The next major advance in understanding SFP etiology was reported in 1930 by Dack et al. (28), who voluntarily consumed supernatant fluids from cultures of "a yellow hemolytic *Staphylococ-*

cus" grown from contaminated sponge cake. Upon ingestion of the filtrates, they became ill with vomiting, abdominal cramps, and diarrhea. SE was the first true enterotoxin described because, unlike botulinum toxin, it exerted an effect on the gastrointestinal tract. SE was particularly distinctive because its activity was "not entirely destroyed by heating even for 30 minutes at 100°C."

S. aureus has been extensively characterized. This organism produces a variety of extracellular products. (For a review of staphylococcal exotoxins, see references 20 and 30.) Many of these, including the SEs, are virulence factors that have been implicated in diseases of humans and animals. As a group, the SEs have biological properties that enable staphylococci to cause at least two common human diseases, toxic shock syndrome (TSS) and SFP. This chapter primarily addresses SFP; however, in regard to the SEs, there is significant overlap in the natural histories of both diseases. Hence, TSS is also discussed in sections where this overlap is most relevant.

Nomenclature, Characteristics, and Distribution of SE-Producing Staphylococci

The term "staphylococci" informally describes a group of small, spherical, gram-positive bacteria. Depending on

Lynn M. Jablonski, Integrated Genomics, Inc., 2201 W. Campbell Park Dr., Chicago, IL 60612.
Gregory A. Bohach, Department of Microbiology, Molecular Biology and Biochemistry, University of Idaho, Moscow, ID 83843.

Table 19.1 General characteristics of selected species of *Staphylococcus*

Characteristic	*S. aureus*	*S. chromogenes*	*S. hyicus*	*S. intermedius*	*S. epidermidis*	*S. saprophyticus*
Coagulase	+	−	+	+	−	−
Thermostable nuclease	+	−	+	+	+/−	−
Clumping factor	+	−	−	+	−	−
Yellow pigment	+	+	−	−	−	+/−
Hemolytic activity	+	−	−	+	+/−	−
Phosphatase	+	+	+	+	+/−	−
Lysostaphin	Sensitive	Sensitive	Sensitive	Sensitive	Slightly sensitive	ND[a]
Hyaluronidase	+	−	+	−	+/−	ND
Mannitol fermentation	+	+/−	−	+/−	−	+/−
Novobiocin resistance	−	−	−	−	−	+

[a] ND, not determined.

the species and culture conditions, their cells have a diameter ranging from approximately 0.5 to 1.5 µm. They are catalase-positive chemoorganotrophs with a DNA composition of 30 to 40 mol% guanine-plus-cytosine content. Staphylococci have a typical gram-positive cell wall containing peptidoglycan and teichoic acids. Except for clinical isolates and strains exposed to antimicrobial therapy, most staphylococci are sensitive to β-lactams, tetracyclines, macrolides, lincosamides, novobiocin, and chloramphenicol but are resistant to polymyxin and polyene. Some differential characteristics of *S. aureus* and several other selected species of staphylococci are summarized in Table 19.1.

There have been many useful schemes for classification of the staphylococci. According to *Bergey's Manual of Determinative Bacteriology*, staphylococci are grouped in the family *Micrococcaceae*. This family includes the genera *Micrococcus*, *Staphylococcus*, and *Planococcus*. The genus *Staphylococcus* is further subdivided into more than 23 species and subspecies. Many of these contaminate food from human, animal, or environmental sources. Several species of *Staphylococcus*, including both coagulase-negative and coagulase-positive isolates, can produce SEs. Although several species have the potential to cause gastroenteritis, with few exceptions, nearly all cases of SFP are attributed to *S. aureus*. This is a reflection of the relatively high incidence of SE production by *S. aureus* in comparison to other staphylococcal species. Although the reason for this has not been confirmed, the SEs act as superantigens (described below) and therefore are potential immunomodulating agents. Thus, SE production may provide a selective advantage to *S. aureus*, a species that is common to both humans and animals, the two most frequent sources of food contamination.

Enterotoxigenic strains of staphylococci have been well characterized on the basis of a number of genotypic and phenotypic characteristics. An extensive phage typing system is available for *S. aureus*. Most SE-producing isolates belong to phage group I or III or are nontypeable. Although SE production by other phage groups is less common, it has been documented. Hajek and Marsalek (38) were able to differentiate *S. aureus* into at least six biotypes by using a classification scheme based largely on the animal host of origin. By far, SE production was most prevalent among human isolates within biotype A. SE production by other biotypes is rare except for biotype C bovine and ovine mastitis isolates. SEs may also be produced by *Staphylococcus intermedius* and *Staphylococcus hyicus* (formerly *S. aureus* biotypes E and F, respectively), albeit less frequently.

Introduction and Nomenclature of the SEs
Structure-function and mechanisms of pathogenicity of the SEs are discussed later in this chapter. This section will introduce the nomenclature and evolution of SE.

Current Classification Scheme Based on Antigenicity
Major advances in characterization of SEs were made approximately 2 decades after Dack and colleagues (28) associated SFP with an exotoxin. Bergdoll and colleagues were the first investigators to produce purified SE preparations and develop specific antisera (reviewed in reference 10). They and others, using purified or partially purified toxins, determined that protective antibodies could be induced in several animal species. Unfortunately, immunity was strain specific and did not provide protection against strains other than the one that was used to induce the initial immune response (12). It soon became apparent that *S. aureus* can produce multiple toxins with similar molecular weights as well as biological and physicochemical properties.

Initially, differentiation between antigenic forms of SE was based on the observation that many food isolates produce one common antigenic type of toxin, tentatively designated the "F" toxin. Most other enterotoxigenic strains, such as those from enteritis patients, also produced a second antigenic form that was classified as the "E" toxin. The discovery of additional isolates that did not conform to this pattern prompted adoption of an improved nomenclature system. A committee was assembled in 1990 to establish an alphabetical nomenclature for the classification of SEs that is still used today (15). Based on this nomenclature, SEs are assigned a letter of the alphabet in the order of their discovery. Subsequently, the "F" and "E" toxins were designated SEA and SEB, respectively. Between 1962 and 1972, three additional SE serotypes (SEC, SED, and SEE) were reported (5, 11). Protein sequencing and recombinant DNA methods have resulted in our current detailed knowledge of the primary sequences of all of the classical SEs (5, 27, 45, 56, 65, 92) (Fig. 19.1). The most recent additions to the SE family include SEG, SEH, SEI, and SEJ (75, 88, 109). Initially, SEF was used in reference to an exotoxin commonly produced by isolates of *S. aureus* associated with TSS. This designation was later changed when it was confirmed that SEF was not emetic. To avoid confusion, SEF has been retired from use in the SE nomenclature system and is now referred to as TSS toxin 1 (TSST-1) (15).

The incidence of SE involvement in SFP appears to change with time. Prior to 1971, SEA was the predominant toxin identified in cases of SFP, followed in frequency by SED and SEC. SEB was only rarely associated with SFP (43). In other outbreaks of SFP, SEA either alone or together with other enterotoxins still remained the most common toxin implicated (51, 107). Interestingly, in one study SEA and SEB were the only enterotoxins associated with SFP. Holmberg and Blake suggested that the observed decrease in cases attributed to other SEs (such as SEC and SED) was due to improved conditions for the processing and storage of milk, which had been commonly contaminated by SEC- and SED-producing strains of *S. aureus* in the past. However, this model seems to be localized to the United States because SEA and SEC were implicated in an outbreak of food poisoning of 485 individuals in Thailand (99). A survey of food handlers in community or hospital-located kitchens revealed that SEC was the most common enterotoxin produced by *S. aureus* isolates (94). In France, 66% of the enterotoxigenic strains of *S. aureus* produced SEC (either singly or in combination with other enterotoxins) (88). A similar scenario was observed in Brazil, where SEC was the most frequent enterotoxin produced from isolates of *S. aureus* (94).

SE Antigenic Subtypes and Molecular Variants
Detailed immunological studies and sequence analysis have revealed that there are molecular variants and sub-

types of the SEs. The best example of this variation is observed with SEC. It had been known for some time that the SEC serological variant can be divided into at least three subtypes (SEC1, SEC2, and SEC3) based on minor differences in immunological reactivity (Fig. 19.2). Within each subtype, however, significant sequence variability may occur. For example, SEs produced by strains FRI-909 and FRI-913 were designated as SEC3 according to their immunological reactivity. However, it was later determined that their amino acid sequences actually differ by nine residues (65). SEC variants produced by bovine and ovine isolates of *S. aureus* have very similar sequences and are indistinguishable from SEC1 in immunological assays. Surprisingly, they behave very differently from SEC1 in biological assays. For example, although SEC-bovine differs from SEC1 by only three residues, the potency of the two toxins differs by several orders of magnitude in lymphocyte proliferation assays (65). Similarly, SEC-canine isolates of *S. intermedius* (previously classified as *S. aureus*) have 93 to 96% sequence identity to other SEC variants, with sequence similarity most closely related to SEC2 and SEC3. However, SEC-canine stimulated T lymphocytes in a profile resembling SEC1 rather than SEC3 (65).

Staphylococcal Genetics and Evolutionary Aspects of SE Production

SEs Are Superantigens and Belong to a Large Pyrogenic Toxin Family
SEs are part of a larger family of pyrogenic toxins (PTs) produced by *S. aureus* and *Streptococcus pyogenes* (17). Members of this family are grouped together based on shared biological and biochemical properties, including their unique ability to act as superantigens (SAgs) (66).

SAgs are bifunctional molecules that form a bridge between MHC class II molecules on antigen-presenting cells and T-cell receptors (TCRs) located on the surface of T cells (Fig. 19.3). Unlike conventional antigens, SAgs are not processed before binding to the MHC class II molecules. In addition, they do not bind in the peptide-binding groove but instead bind on the outer surface of the MHC class II molecules. Once formed, the MHC-SAg complex interacts with the TCR. This interaction with the TCR is also nonconventional and relatively nonspecific; it occurs at a variable (V) location on the TCR β-chain (the Vβ region). Since SAgs bind outside the area on the TCR used for antigen recognition, they activate a much higher percentage of T cells than can be activated by conventional antigens. Compared to mitogens that stimulate T cells in a indiscriminate manner, SAgs stimulate specific subsets of T cells bearing SAg-recognized Vβ sequences.

```
SEA         1 --MKKTAFTLLLFIALTLTTSPLVNGSEKSEEINEKDLRKKSELQGTALGN-LKQIYYYNEKAKTENKESHDQFLQHTIL
SEE         1 --MKKTAFILLLFIALTLTTSPLVNGSEKSEEINEKDLRKKSELQRNALSN-LRQIYYYNEKAITENKESDDQFLENTLL
SEJ         1 --MKKTIFILIFSLTLTLLITPLVYSDSKNETIKEKNLHKKSELSSITLNN-LRHIYFFNEKGISEKIMTEDQFLDYTLL
SED         1 -MKKFNILIALLFFTSLVISPLNVKANENIDSVKEKELHKKSELSSTALNN-MKHSYADKNPIIGENKSTGDQFLENTLL
SEC1        1 MNKSRFISCVILIFALILVLFTPNVLAESQPDPTPDELHKASKFTGLMEN---MKVLYDDHYVSATKVKSVDKFLAHDLI
SEC2        1 MNKSRFISCVILIFALILVLFTPNVLAESQPDPTPDELHKSSEFTGTMGN---MKYLYDDHYVSATKVKSVDKFLAHDLI
SEC3        1 MYKRLFISRVILIFALILVISTPNVLAESQPDPMPDDLHKSSEFTGTMGN---MKYLYDDHYVSATKVKSVDKFLAHDLI
SEB         1 MYKRLFISHVILIFALILVISTPNVLAESQPDPKPDELHKSSKFTGLMEN---MKVLYDDNHVSAINVKSIDQFLYFDLI
SEG         1 ---MKKLSTVIIILILEIVFHNMNYVN-AQPDPKLDELNKVSDYKNNKGTMGNVMNLYTSPPVEGRGVINSRQFLSHDLI
SEH         1 -----------MINKIKILFSFLALLLSFTSYAKAEDLHDKSELTDLALAN---AYGQYNHPFIKENIKSDEISGEKDLI
SEI         1 --MKKFKYSFILVFILLFNIKDLTYAQGDIGVGNLRNFYTKHDYIDLKGVT-------------------DKNLPIANQ
consensus   1      kk i  lilifallllis l  l  e n e    ddlhkksefs     n    h  y     ig   is dqfl hdli


SEA        78 FKGFFTDHSWYNDLLVDFDSKDIVDKYKGKKVDLYGAYYGYQCAGGTPN---------KTACMYGGVTLHDNNRLTEEK
SEE        78 FKGFFTGHPWYNDLLVDLGSKDATNKYKGKKVDLYGAYYGYQCAGGTPN---------KTACMYGGVTLHDNNRLTEEK
SEJ        78 FKSFFISHSQYNDLLVQFDSKETVNKFKGKQVDLYGSYYGFQCSGGKPN---------KTACMYGGVTLHENNQLYDTK
SED        79 YKKFFTDLINFEDLLINFNSKEMAQHFKSKNVDVYPIRYSINCYGGEID---------RTACTYGGVTPHEGNKLKERK
SEC1       78 YNISDKKLKNYDKVKTELLNEGLAKKYKDEVVDVYGSNYYVNCYFSSKD---NVGKVTGGKTCMYGGITKHEGNHFDNGN
SEC2       78 YNISDKKLKNYDKVKTELLNEDLAKKYKDEVVDVYGSNYYVNCYFSSKD---NVGKVTGGKTCMYGGITKHEGNHFDNGN
SEC3       78 YNISDKKLKNYDKVKTELLNEDLAKKYKDEVVDVYGSNYYVNCYFSSKD---NVGKVTGGKTCMYGGITKHEGNHFDNGN
SEB        78 YSIKDTKLGNYDNVRVEFKNKDLADKYKDKYVDVFGANYYYQCYFSKKTNDINSHQTDKRKTCMYGGVTEHNGNQLD--K
SEG        77 FPIEY-K--SYNEVKTELENTELANNYKDKKVDIFGVPYFYTCIIPKSEPDIN---QNFGGCCMYGGLTFNSSENERD--
SEH        67 FRNQG---DSGNDLRVKFATADLAQKFKNKNVDIYGASFYYKCEKISEN---------ISECLYGGTTLNSEKLAQER-
SEI        59 LEFSTG----TNDLISESNNWDEISKFKGKKLDIFGIDYNGPCKSKYMY-------------GGATLSGQYLNSAR---
consensus  81 f         qyndlkve n dla kyK k vDlyg  yyyqC   t           cmyggvt hd nr  e


SEA       148 KVPINLWLDGKQNTVPLETVKTNKKNVTVQELDLQARRYLQEKYNLYNSDVFD-----------GKVQRGLIVFHTSTE
SEE       148 KVPINLWIDGKQTTVPIDKVKTSKKEVTVQELDLQARHYLHGKFGLYNSDSFG-----------GKVQRGLIVFHSSEG
SEJ       148 KIPINLWIDSIRTVVPLDIVKTNKKKVTIQELDLQARYYLHKQYNLYNPSTFD----------GKIQKGLIVFHTSKE
SED       149 KIPINLWINGVQKEVSLDKVQTDKKNVTVQELDAQARRYLQKDLKLYNNDTLG----------GKIQRGKIEFDSSDG
SEC1      155 LQNVLIRVYENKRNTISFEVQTDKKSVTAQELDIKARNFLINKKNLYEFN--S-----------SPYETGYIKFIENNG
SEC2      155 LQNVLIRVYENKRNTISFEVQTDKKSVTAQELDIKARNFLINKKNLYEFN--S-----------SPYETGYIKFIENNG
SEC3      155 LQNVLVRVYENKRNTISFEVQTDKKSVTAQELDIKARNFLINKKNLYEFN--S-----------SPYETGYIKFIENNG
SEB       156 YRSITVRVFEDGKNLLSFDVQTNKKKVTAQELDYLTRHYLVKNKKLYEFN--N-----------SPYETGYIKFIENE-
SEG       149 -KLITVQVTIDNRQSLGFTITTNKNMVTIQELDYKARHWLTKEKKLYEFDG------------SAFESGYIKFTEKNN
SEH       133 VIGANVWVDGIQK--ETELIRTNKKNVTLQELDIKIRKILSDKYKIYYKD-------------SEISKGLIEFDMKTP
SEI       118 KIPINLWVNGKHKTISTDKIATNKKLVTAQEIDVKLRRYLQEEYNIYGHNNTGKGKEYGYKSKFYSGFNNGKVLFHLNNE
consensus 161 v inlwv     k v se v TnKk VTvQElDlka RryL  k nlY                 s  y  G i F


SEA       216 PSVNYDLFGAQGQYSN---TLLRIYRDNKTINSE-NMHIDIYLYTS-----------
SEE       216 STVSYDLFDAQGQYPD---TLLRIYRDNKTINSE-NLHIDLYLYTT-----------
SEJ       216 PLVSYDLFNVIGQYPD---KLLKIYQDNKIIESE-NMHIDIYLYTSLIVLISLPLVL
SED       217 SKVSYDLFDVKGDFPE---KQLRIYSDNKTLSTE-HLHIDIYLYEK-----------
SEC1      221 NTFWYDMMPAPGDKFDQ-SKYLMMYNDNKTVDSK-SVKIEVHLTTKNG--------
SEC2      221 NTFWYDMMPAPGDKFDQ-SKYLMMYNDNKTVDSK-SVKIEVHLTTKNG--------
SEC3      221 NTFWYDMMPAPGDKFDQ-SKYLMMYNDNKTVDSK-SVKIEVHLTTKNG--------
SEB       221 NSFWYDMMPAPGDKFDQ-SKYLMMYNDNKMVDSK-DVKIEVYLTTKKK--------
SEG       214 TSFWFDLFPKKELVPFVPYKFLNIYGDNKVVDSK-SIKMEVFLNTH----------
SEH       196 RDYSFDIYDLKGENDY---EIDKIYEDNKTLKSDDISHIDVNLYTKKKV-------
SEI       198 KSFSYDLFYTGDGLPVS---FLKIYEDNKIIESE-KFHLDVEISYVDSN-------
consensus 241 sf yDlf a gd  d   kylriY DNKtiese  mhidvyl tk
```

Figure 19.1 Alignment of primary sequences of predicted translation products for SE proteins (including the signal peptides) published in the current literature (5, 27, 45, 56, 65, 74, 87, 92, 108). Also shown are residue sequence numbers (on the left) and dashes (-) to indicate gaps in the sequences made by alignment. Sequence alignment and output were generated by using CLUSTAL W and BoxShade 3.21.

Figure 19.2 Immunodiffusion assays for demonstrating SE uniqueness. A precipitin line of partial identity forms when SEC1 and SEC2 diffuse from separate wells toward the center well containing hyperimmune SEC2 rabbit antiserum. While this demonstrates the highly related nature of the SEC subtypes, less related antigenic types such as SEB are more immunologically distinct and do not react.

The staphylococcal and streptococcal PTs are prototype microbial SAgs that exert a variety of immunomodulatory effects leading to shock, immunosuppression, and other systemic abnormalities associated with TSS. While the SEs are included with the PTs, they have the

unique distinction of possessing an additional ability to induce an emetic response upon oral ingestion and are thus solely responsible for SFP. It is generally agreed that many of the toxins in this family, including the SEs, arose from a common ancestral gene that was introduced into both *Staphylococcus* and *Streptococcus* genera. Evidence for this idea is strongly supported by the observation that the structural genes for some of the SEs, and related streptococcal PTs, are carried on mobile genetic elements (14, 55, 63).

Role of Mobile Genetic Elements in Generation and Dissemination of the SEs

Some staphylococcal and streptococcal toxins are encoded by structural genes located on bacteriophage genomes. Although the streptococcal toxins have been best studied, the genetic mobility of SEA and SEE appears to be similar to the situation in group A streptococci. Betley and Mekalanos confirmed that the SEA structural gene (*sea*) is carried by a lysogenic phage and can be cloned directly from the induced bacteriophage genome (14). Both the *sea* and *see* genes map near the *att* site on their respective phage genomes. In a high percentage of cases, the toxin-encoding phages cannot be induced to

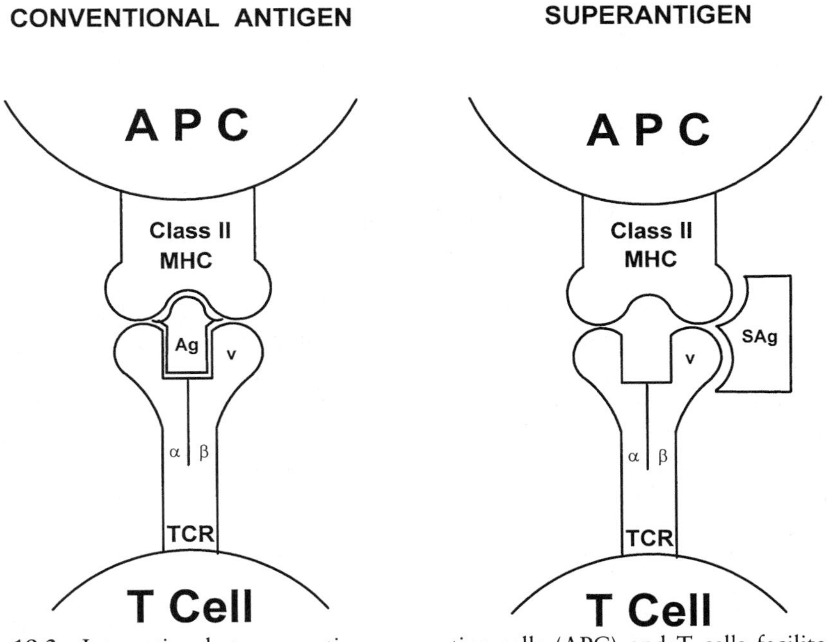

CONVENTIONAL ANTIGEN

SUPERANTIGEN

APC

Class II MHC

Ag

v

α | β

TCR

T Cell

APC

Class II MHC

v

SAg

α | β

TCR

T Cell

Figure 19.3 Interaction between antigen-presenting cells (APC) and T cells facilitated by conventional antigens (Ag) and superantigens (SAg). Following processing by the APC, conventional Ags are presented to highly specific T-cell receptors (TCR) in association with the antigen-binding groove of the MHC class II molecule. Superantigens interact with MHC class II molecules (without processing) outside the antigen-binding groove. The SAg/MHC bimolecular complex interacts with the TCR through specificity determined only by the variable (V) region of the receptor β chain.

a lytic cycle and appear to be defective. The most likely explanation for these observations is that the toxin genes were originally located on the bacterial genome, but were subsequently obtained by the phage upon abnormal excision from the chromosome. This phenomenon is documented for some bacterial toxins that are transferred by lysogenic conversion in other genera of bacteria such as *Streptococcus* and *Corynebacterium* (55).

Researchers have determined that *tst* (encoding for TSST-1) is part of a 15.2-kb pathogenicity island (63). Not to be confused with phages or transposons, the prototype of these elements (SaPI1) contains a putative integrase, VapE homolog, as well as a second putative enterotoxin. Flanking the element is a 17-bp direct repeat that may be used to help localize the unique chromosomal copy of the same repeat sequence for insertion. The element itself is mobilized by staphylococcal phages ϕ13 and 80α. Staphylococcal pathogenicity islands reported to date are mobilized by either ϕ13 or 80α but not both. It has been speculated that elements such as these may be responsible for the transfer of *tst* between strains of *S. aureus*.

The role of plasmids has received considerable attention in relation to transmission of PT genes. However, of the PTs, only SED and SEJ are plasmid encoded (5, 108). Iandolo and colleagues reported that in more than 20 characterized Sed⁺ isolates, the *sed* and *sej* structural genes are localized to a stable 27.6-kb plasmid (pIB485), which also encodes penicillin and cadmium resistance (5, 108). The literature also includes reports associating SEC and SEB genes with transmissible penicillin resistance plasmids. However, it is now generally agreed that both of these toxins are chromosomally encoded. This does not preclude the possibility that these genes may be harbored on a variable genetic element as predicted by several investigators. SEB is the better characterized of the two toxins in this regard. Evidence suggests that *seb* is located on a DNA element that is at least 26.8kb in length (54). It remains to be determined whether this element represents a phage or an integrated plasmid. While SEB production has not been associated with lysogenic conversion, based on what is known about SEA, it is possible that the *seb* gene may reside on a defective phage. Circumstantial evidence for this possibility includes variable upstream regions in different SEB-producing strains of *S. aureus*.

Other SEs may also be mobile. The structural genes of SEG and SEI (*seg* and *sei*, respectively) are located in very close proximity to one another on the chromosome (50, 71). In fact, they are separated by approximately 2 kb, with an intervening region that has a high degree of similarity to other known enterotoxins. Although *seg*⁺

sei⁺ strains of *S. aureus* produce lytic phages, they have not been shown to harbor *seg* or *sei*. However, this region may be mobile via an alternative mechanism similar to that identified with *sed* and *sej*. The genes coding for SED and SEJ are localized to the same plasmid and are present together on all SED-encoding plasmids (108). Like *seg* and *sei*, *sed* and *sej* are separated by a short intergenic region; however, the DNA sequence of this region has not been reported.

Some staphylococcal toxin genes are insertion sites for genetic elements carrying other virulence determinants (13). The expression of several SEs is often affected by this feature. For example, SEB and TSST-1 syntheses are mutually exclusive in *S. aureus*. However, the SEB and TSST-1 genes (*seb* and *tst*, respectively) often coexist in *S. aureus*. The lack of Seb⁺ Tst⁺ strains is due to the insertion of *tst* into the *seb* locus. Likewise, the phage that harbors *sea* utilizes the β-toxin locus, *hlb*, as its insertion site so that Sea⁺ isolates do not produce β-toxin.

Mechanisms and Rationale for Generation of SE Diversity

Currently known SEs can be divided into three groups based on their amino acid sequences (Fig. 19.4). Group 1 contains SEB, the SEC subtypes, and their molecular variants. Toxins in this group are highly related to each other (66 to 99% identical) and to several streptococcal PTs. Group 2 contains the highly related SEA and SEE, with 84% identity, as well as SED and SEJ. Of the remaining toxins, SEG has 38 to 42% identity to members of group 1 while SEH and SEI exhibit 26 to 38% identity with group 2 toxins.

Most staphylococcal and streptococcal PTs, even of those with no significant overall homology, contain four highly conserved stretches of primary sequence (42). This suggests that there is a selective advantage in host-parasite interactions for these organisms to maintain certain toxin characteristics. At the same time, modification of selected regions of the proteins could enable the microorganisms to broaden their host range. This molecular diversity may explain how a group of toxins with the same function but different host specificities could have arisen for the purpose of exploiting a broader repertoire of receptors.

Sequence comparison of members within the PT family has provided several examples in which diversity among the toxins appears to have arisen through gene duplication and/or homologous recombination. For example, SEC1 is most similar to SEC2 and SEC3. However, residues 14 through 26 of mature SEC1 are identical to the analogous region of SEB but significantly different

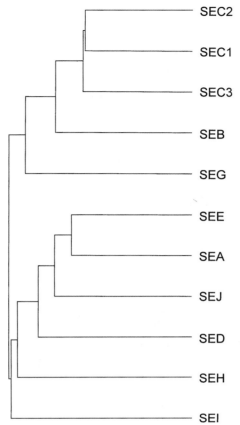

Figure 19.4 Tree representation demonstrating molecular relatedness of the currently known SE family. This tree was created with the clustering feature of the PileUp Program (GCG version 8, Genetics Computer Group, Madison, Wis.).

from the other SEC subtypes (65). Genetic recombination between *seb* and *sec* in a strain producing both SEB and SEC2 (or SEC3) could explain the generation of SEC1. Interestingly, the intergenic region between SEG and SEI has three open reading frames (ORFs 1 through 3) with high levels of identity to other SEs (71). ORF1 (if translated) would produce a polypeptide with 45% identity to the N terminus of SEB. In contrast, a polypeptide produced from ORF2 would have 63% identity with the C terminus of the SECs. ORF3 could produce a full-length polypeptide with 37 to 39% identity to SEA, SED, SEE, SEH, and/or SEJ.

Additional minor variabilities may have resulted by point mutations; even closely related SEs display some sequence differences. This may reflect fine-tuning of the toxin sequences for interacting with cells from a variety of hosts, as demonstrated with the SEC subtype variants. The sequences of SEC toxins produced by strains of *S. aureus* isolated from humans differ only slightly (>95%

identity) from sequences of SEC variants produced by bovine and ovine isolates (65).

Staphylococcal Regulation of SE Expression

General Considerations

SEs are produced in extremely low quantities throughout most of the exponential growth phase (8, 76). There is generally a large increase in expression during the late exponential or early stationary phase of growth, with SEA and SED accumulating somewhat earlier than other SEs (13). Production of SE is dependent on de novo synthesis within the cell. The quantity of toxin produced is strain dependent, with SEB and SEC being produced in the highest quantities, up to 350 μg/ml. SEA, SED, and SEE are easily detectable by gel diffusion assays, which detect as little as 100 ng of SE per ml of culture. Some strains produce very low levels of toxins and require more sensitive analysis methods for detection. Of these, SED and SEJ are most likely to be undetected in cultures of enterotoxigenic *S. aureus*. Transcriptional fusion studies of *sej* and *sed* revealed that *sej* was expressed at relatively low levels in an *agr*-independent manner (108).

Molecular Regulation of SE Production

Three loci implicated in regulation of virulence factor expression in *S. aureus* include *agr* (accessory gene regulator) (77), *sar* (staphylococcal accessory regulator) (23), and *sae* (*S. aureus* exoprotein expression) (37). The best characterized of these is *agr*. Mutations in the *agr* locus result in decreased expression of several SEs and other exoproteins. Regulation of gene expression by *agr* can be transcriptional or translational. It can regulate α-hemolysin at both the transcriptional and translational levels, whereas SEB and SEC are regulated only at the transcriptional level. Not all SEs are regulated by *agr*. SEA expression does not appear to be affected by *agr* mutations (101). Since at least 15 genes are under *agr* control, it fits into the general class of global regulators.

The *agr* locus maps to approximately 4 o'clock on the staphylococcal genome, near *purA*, *bla*, and *sea*, on the standard map of *S. aureus*. It contains two divergent operons separated by approximately 120 bp (Fig. 19.5). Transcription can be initiated from three promoters (P1, P2, and P3). P1 is weakly constitutive and transcribes *agrA*. P2 and P3 are induced strongly during the late exponential and early stationary phases but are only weakly expressed earlier. The P2 transcript, RNAII, encodes four proteins designated AgrA, AgrB, AgrC, and AgrD. Together these four proteins form a quorum-sensing apparatus that allows the bacterium to

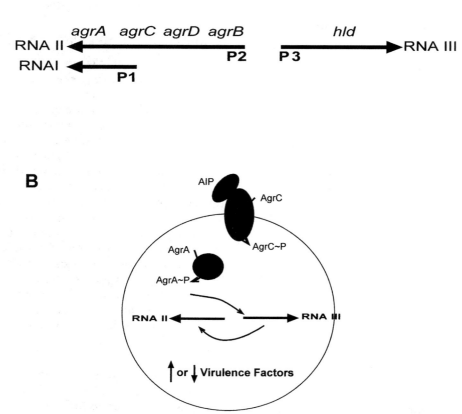

Figure 19.5 General characteristics of the accessory gene regulator in *S. aureus*. (A) Physical map of the *agr* locus indicating the relative locations of genes within the locus (not drawn to scale). (B) Potential role of AIPs and AgrC in environmental sensing and signal transduction as a classical two-component system.

respond to its environment. AgrB is thought to secrete the peptide pheromone derived from AgrD from the cell. Increased levels of pheromone are recognized by the membrane-bound receptor AgrC, which in turn initiates a signal transduction pathway. This pathway is thought to activate AgrA, thereby activating expression of both RNAII and RNAIII. Strains with mutations in AgrA, AgrB, AgrC, or AgrD have an Agr⁻ phenotype and do not initiate transcription from either P2 or P3.

The 514-nucleotide transcript from P3, designated RNAIII, encodes the 26-residue staphylococcal δ-hemolysin and contains a significant amount of untranslated sequence. Interestingly, Agr⁻ phenotypes can be complemented with plasmids encoding RNAIII under control of an inducible promoter, even when the δ-hemolysin gene, *hld*, has been inactivated (62). Therefore, it appears that RNAIII is a diffusible element that plays a key role in the *agr* regulation of exoprotein structural genes, including those for SEs (49, 62, 77). This RNA molecule

can replace the regulatory function of the entire *agr* locus and regulates expression predominantly at the transcriptional level. The exact mechanism by which transcriptional regulation occurs has not been determined, although it has been proposed to require an association with peptide factors and formation of an RNAIII-peptide complex.

Expression of both of the *agr* operons is dependent on a DNA-binding protein (*sarA*) encoded by the *sar* locus (23–26, 72, 73). Northern blot analysis of the *sar* locus revealed three overlapping transcripts (*sarA*, *sarB*, and *sarC*) coding for SarA, SarB, and SarC (4). DNase I footprinting experiments revealed that SarA binds to three dimeric regions on the intergenic region between P2 and P3 designated A1/A2, B1/B2, and C1/C2 (83). Rechtin et al. propose that binding of SarA actually induces changes in the local conformation or superhelicity of the DNA. This could explain why SarA might induce expression of some genes while repressing others

(16). *sar* mutants display phenotypes distinct from those of mutants produced by affecting the *agr* loci. In contrast to *agr* mutants, *sar* mutants express an increased amount of α-hemolysin and decreased amounts of δ- and β-hemolysins and collagen adhesin (16).

Additional studies have revealed that the *sae* locus contains two genes, *saeR* and *saeS*. *saeR* has high similarity to other bacterial response regulators, whereas *saeS* has high similarity to a histidine kinase, suggesting that these proteins function as part of a two-component regulatory system (36). The *sae* locus is a positive effector of cell-free β- and α-hemolysins, coagulase, DNase, and protein A, but apparently does not affect SEA, protease, lipase, staphylokinase, or cell-bound protein A. These results suggest that the *sae* locus acts differently from either the *agr* or *sar* locus in regulation of expression of SEs.

Sigma factors may also affect temporal expression of SEs. RNA polymerase, purified from exponential-phase cultures of *S. aureus*, contains a σ^{70}-related factor similar to that of *Escherichia coli* (81a). The holoenzyme containing this σ factor transcribed the *sea* more efficiently than *sec* or *agrP2*, suggesting there is additional complexity in the differential expression of certain SEs. While *agr* and most exoprotein genes (including *sec*) are expressed as the cells enter the stationary phase, *sea* is expressed earlier during the exponential phase, simultaneously with the σ^{70}-like factor. Hence, differential SE expression, at specific points of the bacterial growth phase, may coincide with availability of compatible σ factors.

Environmental Signaling in Regulation of SE Gene Expression

The dissection of the properties of *agr* has helped explain several characteristics of SE production. For example, *agr* expression coincides temporally with expression of SEB and SEC during the bacterial growth cycle. All are maximally expressed during late exponential and postexponential growth. SEA, which is not regulated by *agr*, is produced earlier (13). Furthermore, the production of several SEs is negatively regulated by growth in media containing glucose, which is now known to affect *agr*. Most research in this area has addressed SEC. SEC expression is affected by glucose through at least two different mechanisms. First, metabolism of glucose indirectly influences SEC production through *agr* by reducing pH (86). Since *agr* is maximally expressed at neutral pH, growth in a nonbuffered environment containing glucose lowers pH levels, which directly reduce *agr* expression. Consequently, expression of *sec* and other *agr* target genes is affected correspondingly (86). Glucose also reduces *sec* expression in *agr⁻* strains. This observation

indicates the existence of a second glucose-dependent mechanism for reduction of SE expression, independent of *agr* and apparently not influenced by pH. The tight regulation of *agr* expression also occurs under alkaline conditions. Regardless of glucose levels, RNAIII is not expressed efficiently above pH 8.0.

In many organisms, the expression of various proteins in response to their environment is coordinated by a process known as autoinduction. At the center of the autoinduction response of *S. aureus* is the *agr* locus. Components of the *agr* operon make up a quorum-sensing apparatus analogous to the signal transduction pathway of bacterial two-component systems (76). In this system, *agrD* encodes a propeptide that is processed and possibly secreted by an integral membrane protein, AgrB. The resulting autoinducer peptide (AIP) then acts as an activation ligand binding to the transmembrane protein receptor AgrC. AgrC is subsequently autophosphorylated and transmits the signal across the membrane to AgrA. AgrA then activates transcription of the P2 and P3 promoters of the *agr* operon through a mechanism that is not well understood.

Strains of *S. aureus* and other staphylococci can be subdivided into four groups or classes based on activation of RNAIII transcription by AIPs (53). Mature AIPs from *S. aureus*, *Staphylococcus epidermidis,* and *Staphylococcus lugdunensis* are small peptides, usually between 7 and 9 amino acids in length, with a conserved cysteine five residues from the carboxy terminus. A thiolactone bond between the conserved cysteine residue and the C-terminal carboxy residue is necessary for *agr* activation because mutants lacking this ringed structure are devoid of activity (67). Though not highly conserved themselves, AIPs are located between two conserved regions on AgrDs from different species. Interestingly, AIPs can induce expression of *agr* in other strains belonging to the same group. However, intergroup AIPs inhibit the expression of the *agr* locus. Mayville et al. (67) suggest that intragroup AIPs interact with AgrC from the same class, enabling a *trans*-acylation and signal-transducing conformational change in the receptor. AIPs exposed to receptors belonging to a different group cannot trigger the *trans*-acylation and subsequent signal transduction of AgrC. It has been proposed that AIPs may also regulate colonization factors (76).

Other Relevant Molecular Aspects of SE Expression

Although many chemical and physical factors selectively inhibit SE expression (34, 48, 59), their effects on regulation and signal transduction are only beginning to be defined. *agr* is not the only signal transduction mechanism for *S. aureus*, as suggested by investigations into the inhibitory effect of glucose. The negative regulatory effect of glucose described above cannot be attributed entirely to lower pH levels in cultures grown on glucose because cultures containing the carbohydrate produce less SE, even when the pH is stably maintained. Although this effect has some attributes of catabolite repression, there are major differences between the glucose inhibitory effect in *S. aureus* and catabolite repression in *E. coli*. For example, inhibition by glucose cannot be reversed by adding cyclic AMP (cAMP) to staphylococcal cultures. The significance of these differences is still unclear. Even the catabolite-repressible staphylococcal *lac* operon has features that differ significantly from the analogous operon in *E. coli*, and it is apparently unresponsive to cAMP.

S. aureus is an osmotolerant bacterium. While it is able to survive and grow in environments with low water activity (a_w), production of some SEs, especially SEB and SEC, is reduced when the bacterium is grown under osmotic stress. In experiments performed with SEC-producing strains, levels of *sec* mRNA and SEC protein are both reduced in response to high NaCl concentrations. However, addition of osmoprotectants reverses the effect. This reduced expression is also seen in Agr⁻ strains, indicating that the signal transduction pathway used in this mechanism occurs by an alternative pathway.

Low concentrations of the commonly used emulsifier glycerol monolaurate (GML) inhibit transcription of many exoprotein genes, including *sea*, without inhibiting *S. aureus* growth. Inhibition of SE production is not associated with a simultaneous effect on *agr* transcription and occurs in *agr* mutants as well as wild-type strains. These results, plus the finding that constitutive expression of some genes is not affected, suggest that GML interferes with non-*agr*-mediated signal transduction. It has been proposed that GML and a variety of related food additives exert this effect by inserting into the staphylococcal membrane, altering the membrane protein conformation, and thereby interfering with signal transduction.

RESERVOIRS

Sources of Staphylococcal Food Contamination

Humans are the main reservoir of staphylococci involved in human disease, including *S. aureus*. Although most species are considered to be normal inhabitants of the external regions of the body, *S. aureus* is a leading human pathogen. Colonized individuals are carriers and provide the main source for dissemination of staphylo-

cocci to others and to food. In humans, the anterior nares is the predominant site of colonization, although *S. aureus* also occurs on other sites such as the skin or perineum. Dissemination of *S. aureus* among humans and from humans to food can occur by direct contact, indirectly by skin fragments, or through respiratory tract droplet nuclei.

One of the difficulties associated with controlling SFP is the large number of human and animal reservoirs. *S. aureus* has a remarkable ability to persist in sites such as the mucosal surface because of its intracellular survival. *S. aureus* can bind to and be internalized by many different cell types, including bovine mammary epithelial cells (6, 106), chick osteoblast cells (46), and bovine aortic endothelial cells. Once bound, *S. aureus* triggers a series of host-specific changes that are very similar to other facultative intracellular pathogens, including membrane pseudopod formation, cytoskeletal rearrangement, phagosome formation, protein tyrosine kinase activation, and changes in cell morphology (6, 106). Inside the cell, *S. aureus* is localized temporarily in the membrane-bound endosome before finally residing in the cell's cytoplasm (31, 46) and triggering the host cells to go through apoptosis (or programmed cell death) (105).

Today, most sources of SFP are traced to humans who contaminate food during preparation. In addition to contamination by food preparers who are carriers, *S. aureus* may also be introduced into food by fomites used in food processing such as meat grinders, knives, storage utensils, cutting blocks, and saw blades. A survey of more than 700 outbreaks of foodborne illness outbreaks revealed the following conditions most often associated with foodborne illness: (i) inadequate refrigeration; (ii) preparation of foods far in advance; (iii) poor personal hygiene, e.g., not washing either hands or instruments properly; (iv) inadequate cooking or heating of food; or (v) prolonged use of warming plates when serving foods, a practice that promotes staphylococcal growth and SE production (21).

Animals, also an important source of *S. aureus*, are often heavily colonized with staphylococci. Predisposing factors that facilitate the survival of staphylococci are major concerns in the maintenance and processing of domestic animals and their products. For example, one very serious problem for the dairy industry is mastitis, an infectious disease often caused by *S. aureus*. The combined losses and expenses associated with bovine mastitis make it the single most costly disease of agriculture in the United States. Colonization of animals by this bacterium is also a public health concern because it may result in contamination of food and milk with *S. aureus* before or during processing.

Table 19.2 Prevalence of *S. aureus* in several common food products

Product	No. of samples tested	% Positive for *S. aureus*	*S. aureus* content/g[a]	Reference
Ground beef	74	57	≥100	96a
	1,830	8	≥1,000	22a
	1,090	9	>100	81
Big game	112	46	≥10	93a
Pork sausage	67	25	100	96a
Ground turkey	50	6	>10	37b
	75	80	>3.4	37a
Salmon steaks	86	2	>3.6	32
Oysters	59	10	>3.6	32
Blue crabmeat	896	52	≥3	104a
Peeled shrimp	1,468	27	≥3	97a
Lobster tail	1,315	24	≥3	97a
Assorted cream pies	465	1	≥25	99a
Tuna pot pies	1,290	2	≥10	104b
Delicatessen salads	517	12	≥3	77a

[a] Determined by either direct plate count or most-probable-number technique.

It has not always been possible to trace the source of staphylococcal food contamination to human or animal origin. Regardless of its source, many studies have revealed the common presence of *S. aureus* in many types of food products (Table 19.2). The cell populations of staphylococci are usually low initially. However, their widespread presence provides a potential source of staphylococci capable of inducing SFP if conditions appropriate for SE expression are provided.

Resistance to Adverse Environmental Conditions

Some unique resistance properties of *S. aureus* facilitate its contamination and growth in food. Outside the body, *S. aureus* is one of the most resistant non-spore-forming human pathogens and can survive for extended periods in a dry state. Its survival is facilitated by organic material, which is likely to be associated with the bacterium from an inflammatory lesion. Isolation of *S. aureus* from air, dust, sewage, and water is relatively easy, and environmental sources of food contamination have been documented in several outbreaks of SFP.

S. aureus is known for acquiring genetic resistance to heavy metals and antimicrobial agents used in clinical medicine. Generally, however, the resistance of *S. aureus* to common food preservative methods is unremarkable. One noteworthy exception is its osmotolerance, which permits growth in medium containing the equivalent of 3.5 M NaCl and survival at a_w less than 0.86. This is especially problematic because other microorganisms, with which *S. aureus* does not compete successfully, are likely to be inhibited under these conditions.

The molecular basis for staphylococcal osmotolerance has received much interest in recent years, although systems for responding to osmotic stress have been more intensively studied in less tolerant microorganisms. Considering the unique resistance of staphylococci to osmotic stress, it would not be surprising to find they have developed a highly efficient osmoprotectant system. As in other microorganisms, several compounds accumulate in the cell or enhance staphylococcal growth under osmotic stress. Glycine betaine may be the most important osmoprotectant for *S. aureus*. To varying degrees, other compounds, including L-proline, proline betaine, choline, and taurine, can also act as compatible solutes for this bacterium. Intracellular levels of proline and glycine betaine accumulate to very high levels in *S. aureus* in response to increased concentrations of NaCl in the environment. Although the signal transduction pathway is not known for staphylococci, in other microorganisms it involves a loss in turgor pressure in the cell and activation of required transport systems. High-affinity and low-affinity transport systems operate in *S. aureus* for both proline and glycine betaine (100, 104). The low-affinity systems are primarily stimulated by osmotic stress, have broad substrate specificity, and may be the same transporter shared by both osmoprotectants.

By itself, the demonstration of a stress response system in *S. aureus* does not explain the unusual osmotolerance of staphylococci. Other less tolerant microorganisms possess mechanisms for counteracting osmotic stress. The efficiency of the staphylococcal system may reflect an unusually high endogenous level of intracellular K^+ and lack of need for de novo transporter synthesis. For example, in other well-studied systems such as that of *E. coli*, changes in osmotic stress activate K$^+$ transport systems. Elevated intracellular K$^+$ levels that result are required for induction of *proU* and eventually lead to synthesis of transporters for glycine betaine and other osmoprotectants. In *S. aureus*, K$^+$ levels are high in unstressed cells. Therefore, the transport system is constitutively present in this bacterium and is preformed when high salt conditions are encountered. The net result is a very rapid and efficient response. *S. aureus* cells accumulate a 21-fold increase in proline after less than 3 min of exposure to high salt concentrations (100).

FOODBORNE OUTBREAKS

Incidence of SFP

SFP occurs as either isolated cases or outbreaks affecting a large number of individuals. Unlike other forms of foodborne illness, there has been less incentive to report cases of SFP because it is self-limiting and typically resolves itself within 24 to 48 h after onset. Although there is a national surveillance system for SFP, it is not an officially reportable disease. It has been estimated that only 1 to 5% of all SFP cases that occur in the United States are reported, usually at the state health department level. Most of these are highly publicized outbreaks. Isolated cases occurring in the home are not usually reported. SFP accounts for an estimated 14% of the total outbreaks of foodborne illness within the United States, ranking as the third most common confirmed bacterial foodborne illness caused by known bacterial pathogens (43, 68). There has been an average of approximately 25 major outbreaks of SFP annually within the United States. Occurrence of SFP is cyclical in nature. The highest incidence is typically in the late summer when temperatures are warm and food is more likely to be stored improperly. A second peak occurs in November and December. Approximately one-third of these outbreaks are associated with leftover holiday food.

SFP is likely a leading cause of foodborne illness worldwide, although reporting in other countries is even less complete than in the United States. In one study, 40% of outbreaks of foodborne gastroenteritis in Hungary were due to SFP (8). The percentage is slightly lower in Japan (approximately 20 to 25%), where contamination of rice balls during preparation is a potential problem (8). Outbreaks resulting from improper manufacturing of canned corned beef have been reported in England, Brazil, Argentina, Malta, northern Europe, and Australia. In Great Britain, 75% of the 359 cases of SFP reported between 1969 and 1990 were associated with meat, poultry, or their products (107). Other cases in Great Britain have been attributed to milk and cheese contamination resulting from sheep mastitis (51). In some countries, ice cream has been a major cause of SFP. Outbreaks of SFP have also occurred in Thailand, in which SEA and SEC were the most common enterotoxins produced (99).

Characteristics of a Large Typical SFP Outbreak

The following is a summary of an outbreak of SFP reported by the Food and Drug Administration (FDA) (3). Many aspects of this outbreak such as type of food involved, mechanism of contamination, inadequate safety measures in food handling, and clinical pictures are typical of the usual SFP outbreak. This particular outbreak originated from one meal that was fed to 5,824 elementary school children at 16 sites in Texas. A total of 1,364 of the children exposed developed typical signs of SFP. Investigation into the vehicle of the illness revealed that 95% of the children who became ill had eaten chicken salad. Large populations of *S. aureus* were cultured from the chicken salad implicated in the outbreak.

The series of events leading up to the outbreak was as follows. The meals were prepared in a centralized kitchen beginning on the preceding day. Frozen chickens used for the salad were boiled for 3 h. After cooking, the chicken meat was deboned, cooled to room temperature with a fan, ground into small pieces, placed in 30.5-cm-deep aluminum pans, and stored overnight in a walk-in refrigerator at 5.5 to 7°C. The following morning, the remaining ingredients were added to the salad and the mixture was blended with an electric mixer. The food was placed in thermal containers and transported by truck to the various schools between 9:30 and 10:30 a.m. It was kept at room temperature until served between 11:30 a.m. and noon. It is believed that the chicken became contaminated when it was deboned after cooking. Most likely, the storage of the warm chicken in the deep aluminum pans did not permit rapid cooling and provided an environment favorable for staphylococcal growth and SE production. Further growth of the bacteria probably occurred during the period when the food was kept in the warm classrooms. Prevention of the incident would have entailed screening food preparers to identify *S. aureus* carriers, more rapid cooling of the chicken, and refrigeration of the salad after preparation.

CHARACTERISTICS OF DISEASE

SFP is usually described as a self-limiting illness presenting with emesis following a short incubation period. However, vomiting is not the only symptom that is commonly observed. A significant number of patients with SFP do not vomit. Other common symptoms include nausea, abdominal cramps, diarrhea, headaches, muscular cramping, and/or prostration. In a summary of clinical symptoms involving 2,992 patients diagnosed with SFP, 82% complained of vomiting, 74% felt nauseated, 68% had diarrhea, and 64% exhibited abdominal pain (43). In all cases of diarrhea, vomiting was present. Diarrhea is usually watery but may contain blood as well. The absence of high fever is consistent with the absence of infection in this type of foodborne illness, although some patients present with low-grade fever. Other potential symptoms include headaches, general weakness, dizziness, chills, and perspiration.

Symptoms usually develop within 6 h after ingestion of contaminated food. In one report, 75% of the exposed individuals exhibited symptoms of SFP within 6 to 10 h post ingestion (64). The mean incubation period is 4.4 h, although incubation periods as short as 1 h have been reported. In another outbreak, symptoms lasted for 1 to 88 h, with a mean of 26.3 h. Death due to SFP is not common, but the fatality rate ranges from 0.03% for the general public to 4.4% for more susceptible populations such as children and the elderly (43). Approximately 10% patients with confirmed SFP seek medical attention. Treatment is minimal in most cases, although administration of fluids is indicated when diarrhea and vomiting are severe.

INFECTIVE DOSE AND SUSCEPTIBLE POPULATIONS

Numbers of Staphylococci Required

Since many variables affect the amount of SE produced, one cannot predict with certainty the number of *S. aureus* in food required to cause SFP. Factors contributing to the levels of toxin concentration have been extensively studied and include environmental conditions such as food composition, temperature, other physical and chemical parameters, and the presence of inhibitors. Also, bacterial factors to be considered include potential differences in the types, amounts, and numbers of different SEs that the strain in question has the physiological ability to produce. It is likely that these combined conditions are unique for each isolated case or outbreak of SFP. Despite this variability, there are several general guidelines that are useful for assessing general risk. According to the FDA, effective doses of SE may be achieved when populations of *S. aureus* are greater than 10^5 cells per g of contaminated food (3). In other studies, 10^5 to 10^8 cells were determined to be the typical range, despite the fact that lower cell populations were sometimes implicated (43).

Toxin Dose Required

Many studies have been conducted to assess SE potency and the amount of toxin in food required to initiate SFP symptoms. Perhaps the most valid information in this regard has come from analysis of food recovered from outbreaks of the illness. Although the SEs are quite potent, the amount required to induce symptoms is relatively large compared to many other exotoxins that are acquired through contaminated food. A basal level of approximately 1 ng of SE per g of contaminated food is sufficient to cause symptoms associated with SFP. Although levels of 1 to 5 μg of ingested toxin are usually associated with many outbreaks, the actual levels of detectable SE were even less (<0.01 μg) in 16 SFP outbreaks (35). One of the most useful studies for predicting the minimal oral dose of SE required to induce SFP in humans was a well-documented investigation of an outbreak caused by ingestion of contaminated chocolate milk (32). In that study, the minimal dose of SEA required to cause SFP in schoolchildren was 144 ± 50 ng.

Many factors contribute to the likelihood of developing symptoms of SFP and their severity. The most important include susceptibility of the individual to the toxin, the total amount of food ingested, and the overall health of the affected person. The toxin type may also influence the likelihood and severity of disease. Though SFP outbreaks attributed to ingestion of SEA are much more common, individuals exposed to SEB exhibit more severe symptoms. Forty-six percent of 2,291 individuals exposed to SEB exhibited SFP symptoms severe enough to be admitted to the hospital (43). Only 5% of 1,813 individuals exposed to SEA required such treatment. It is possible that these observations reflect differences in levels of toxin expression, because SEB is generally produced at higher levels than SEA.

Human volunteers and several species of macaque monkeys have been used to determine the minimal amount of purified SEs required to induce emesis when administered orally. Generally, monkeys are less susceptible to SE-induced enterotoxicity than humans. Studies with purified SEs have provided useful comparative information and have important research applications, but their direct relevance to SFP is uncertain because potential stabilization of unpurified SEs by food is an important consideration. Based on a study in which human volunteers ingested partially purified toxin, Raj and Bergdoll estimated that 20 to 25 μg of SEB (0.4 μg/kg) is sufficient to cause vomiting in humans (81). In the rhesus monkey model, the 50% emetic dose (ED_{50}) is between 5 and 20 μg per animal (or approximately 1 μg/kg) when administered intragastrically. Munson et al. (74) determined that four of six rhesus monkeys vomited when given a dose of 80 μg of SEG-containing culture supernatant fluid per kg of body weight, suggesting that the toxin may be less active than other SEs. Likewise, only one of four animals receiving 150 μg of SEI-containing culture supernatant fluid per kg of body weight experienced emesis. In our investigations, the minimal emetic dose of SEC1 for pigtail monkeys (*Macaca nemestrina*) is consistently between 0.1 and 1.0 μg of purified toxin per kg of body weight (91).

VIRULENCE FACTORS AND MECHANISMS OF PATHOGENICITY

SE Structure-Function Associations

Basic Structural and Biophysical Features

SEs are single polypeptides of approximately 25 to 28 kDa. Most are neutral or basic proteins with pIs ranging from 7 to 8.6, but they have a high degree of microheterogeneity when assessed by isoelectric focusing. For example, SEC2 focuses into at least eight bands ranging in pI from 5.50 to 7.35 (29). Although this has been attributed to enzymatic deamidation, the putative enzyme has not been identified. Based on what is known about their sequences, biochemistry, and functional aspects, all of the SE antigenic variants exist as monomeric proteins. Current evidence does not suggest that SE molecules are arranged functionally into a function-dependent A-B subunit organization. Typical of most other exotoxins, the SEs are translated as larger precursors with a classical signal peptide that is cleaved during export.

SE sequence analysis, relatedness, and diversity have been discussed above (Fig. 19.1 and 19.3). Structural studies have revealed that the molecular topology of all SEs is similar, indicative of their partial sequence conservation (97). Spectral techniques and computer-assisted structural predictions revealed that SEA, SEB, SECs, and SEE contain a low content of α-helix (<10%) compared to β-pleated sheets/β-turn structures (approximately 60 to 85%) (19, 93). These predictions have recently been confirmed with the reported three-dimensional crystal structures of SEA, SEB, SEC2, and SEC3 (19, 42, 78, 80, 89, 97).

Although detailed studies have not been performed with every SE, as a group they are stable molecules in many respects. Their recognition as heat-stable toxins arose from early studies in which enterotoxicity and antigenicity were not completely destroyed upon boiling crude preparations. Furthermore, less extreme elevations of temperature such as those used for pasteurization of milk had little or no effect on SE toxicity. The heat resistance of the SEs has been extensively studied. The general conclusion from this research is that SEs are difficult to inactivate by heating and are even more stable when present in high concentrations or in crude states such as in the environment of food. Since temperatures required to inactivate the SEs are much higher than those needed to kill *S. aureus* under the same environmental conditions, toxic food involved in many cases of SFP is devoid of viable organisms at the time of serving.

One additional property of SEs that has potential significance toward development of SFP is their resistance to inactivation by proteases found in the gastrointestinal tract. Resistance to pepsin, especially in a relatively low pH environment, is a key requirement for SE activity in vivo. All of the SEs have some resistance to pepsin, a property not shared by at least one nonemetic staphylococcal PT, TSST-1. SEB is susceptible to degradation by pepsin at very low pH levels, but partial neutralization of the gastric acidity by food intake is presumed to temporarily provide a protective environment for the toxin (8).

The SEs may be cleaved by other common proteases. However, unless the fragments generated are separated in the presence of denaturing agents, proteolysis alone may not be sufficient to cause a loss of biological activity. This is apparently representative of inherent SE molecular stability, which can be demonstrated by renaturing studies. For example, denaturation occurs only under strong denaturing conditions employing high concentrations of urea or guanidine hydrochloride. If the denaturing conditions are removed, the SEs may spontaneously renature and regain biological activity (95). Differences in stability do exist among the toxins. For example, SEB and SEC1 are approximately 50-fold more stable under denaturing conditions than SEA (102). There has been some speculation that the SE disulfide bond is responsible for an inherent molecular stability. Experiments by several investigators suggest that the closed disulfide bond does contribute at least some degree of conformational stabilization, but its disruption has only minimal effects on the overall stability and activity of the molecule (104).

Three-Dimensional Structure

The crystal structure of SEC3, determined at a resolution of 1.9 Å and depicted in Fig. 19.6, is very similar to that of SEA, SEB, SEC2, SED, and TSST-1 (17, 19, 78, 80, 89, 93, 97). The molecule has an overall ellipsoidal shape with maximal dimensions of 43 by 38 by 32 Å and is folded into two domains containing a mixture of α and β structures. Domain 1, the smaller of the two, contains residues near the N terminus but not the N-terminal residues themselves. The residues in SEC3 that are located in domain 1 include those in positions 35 through 120 (Fig. 19.1). The folding conformation

DOMAIN 2

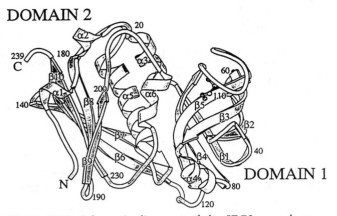

Figure 19.6 Schematic diagrams of the SEC3 crystal structure illustrating major structural features. Numerical designation of the location of select residues and each α- and β-strand is shown within the two major domains. Also indicated are the N and C termini. The intramolecular disulfide linkage between Cys residues 93 and 110 (ball-and-stick) connects the disulfide loop to the β5-strand containing the conserved residues (see Fig. 19.7) potentially important for emesis. The zinc atom bound by SEC3 faces toward the back of the SEC3 molecule between domains 1 and 2 and is coordinated by D83, H118, and H122. In contrast, the zinc in SEA is positioned on the opposite edge of domain 2. The conformational topology of domain 1 is the same as the OB fold domains of other proteins described in the text.

of this domain may have potential significance for the function of the toxin. Its topology, in which a Greek-key β-barrel structure is capped at one end by an α-helix, is known as the oligonucleotide/oligosaccharide binding (OB) fold. The internal portion of the β-barrel is rich in hydrophobic residues, and the potential oligomer binding surface is covered with mainly hydrophilic residues. This same conformational folding pattern is found in several bacterial enzymes and exotoxins. Although members of this diverse group of proteins are not related according to their amino acid sequences, they share the common feature of exerting their activity by interacting with either oligosaccharides or oligonucleotides. Staphylococcal nuclease and *E. coli* Shiga toxin are examples of proteins that interact through their OB fold domains with oligonucleotides or oligosaccharides, respectively. The other prominent feature of domain 1 is that it contains two cysteine residues responsible for forming the disulfide linkage characteristic of all SEs. This bond and the cystine loop are located at the end of the domain opposite its α-helix cap. Crystallographic data for SEA, SEB, and SEC3 indicate that the loop regions of all three toxins are quite flexible (20, 89, 97).

The larger domain 2 contains the N and C termini and encompasses residues 1 through 33 and 123 through 239 of SEC3. It can be described as a five-strand antiparallel

		Cystine loop or analogous region	Conserved Downstream Sequences

Enterotoxigenic PTs:

SEC1	93	CYFSSKDNVGKVTGG---KTC	111	M Y G G I T K H E G N H
SEC2	93	CYFSSKDNVGKVTGG---KTC	111	M Y G G I T K H E G N H
SEC3	93	CYFSSKDNVGKVTGG---KTC	111	M Y G G I T K H E G N H
SEB	93	CYFSKKTNDINSHQTDKRKTC	114	M Y G G V T E H N G N Q
SED	92	CYGGEIDRTAC	103	T Y G G V T P H E G N K
SEA	96	CAGGTPNKTAC	107	M Y G G V T L H D N N R
SEE	93	CAGGTPNKTAC	104	M Y G G V T L H D N N R
SEH	82	CEKISENISEC	93	L Y G G T T L - N S E K

Nonenterotoxigenic PTs:

SPEA	87	CYLCENAERSAC	99	I Y G G V T N H E G N H
SPEC	74	GLFYILNSHTGE	86	Y I Y G G I T P A Q N N
TSST-1	69	KRTKKSQHTSEGTYYH	85	Q I S G V T N T E K L P

Figure 19.7 Comparison of cysteine loop and adjacent sequences for selected SEs and the analogous regions of nonenterotoxin PTs. Numbers designate the position within the primary sequences of the mature proteins. Evidence suggests that proper positioning of the critical downstream residues by a stable disulfide bond is required for emesis. The streptococcal superantigen sequence is not included here because its enterotoxic ability has not been investigated. SPEA and SPEC refer to streptococcal pyrogenic exotoxins types A and C, respectively. SPEA is nonemetic (91) despite having the potential to form a disulfide bond in this region. Although SPEA also contains multiple cysteines, the nature of the putative SPEA disulfide bond is unclear because it has an additional cysteine at position 90.

β-sheet wall, overlaid with a group of α-helices. The N-terminal 20 residues of SEC3 form a loosely attached structure that folds over the edge of this domain. Residues immediately downstream from the N terminus form α-helices that mark the interfaces between both domains. Specifically, these α-helices form a long groove on one side of the molecule (α5 groove) and a shallow α3 cavity near the top of the molecule (42).

Binding of Zinc by SEs

The first evidence that a cation could affect SE structure or function was the demonstration that binding of SEA, SEE, and possibly SED to MHC class II requires zinc. Fraser et al. (33) determined that SEA and SEE bind zinc via a single site with a dissociation constant of 1 to 2 μM. The binding site was subsequently predicted by mutagenesis of SEA to be composed of a nonlinear stretch of residues (H187, H225, and D227). SED is also likely to bind zinc through the same site, but it contains aspartic acid at position 187 instead of histidine. The crystal structure for SEA reported by Schad et al. (89) has confirmed that these three residues bind zinc and also implicate the N-terminal serine in the metal coordination.

SEC2 and SEC3 also bind zinc, albeit through a different mechanism compared to SEA and SEE (17, 19). The α5 groove of SEC3 contains a zinc atom bound by the classical motif (H-E-X-X-H), typically present in the catalytic site of metalloenzymes such as thermolysin (41). The significance of a zinc-binding site with potential enzymatic activity in SEC3 is being actively investigated. Its presence raises several interesting questions. None of the SEs is known to possess protease activity. However, zinc-mediated metalloprotease activity dependent upon a conserved H-E-X-X-H motif has been demonstrated for other exotoxins produced by gram-positive bacteria (60). In the tetanus and botulinum neurotoxin group, protease activity involving this zinc-binding motif is central to the toxins' mechanism of action. Future work will determine whether similarities between botulism and SFP are coincidental or whether zinc binding represents a related mechanism of action. Interestingly, both diseases are usually acquired as intoxications. The responsible toxins are ingested orally but subsequently exert their effects through an action on nerves. In any regard, if this highly speculative possibility were true for SEC3, the heterogeneity in zinc binding among the staphylococcal toxins makes it unlikely that zinc-dependent protease activity is a uniform mechanism for all SEs.

Molecular Regions of SEs Responsible for Enterotoxicity

The structural aspects of SEs that enable them to survive degradation by pepsin and other enzymes in the gastrointestinal tract are required for the toxins to induce SFP. However, stability alone is not sufficient. SEs must also interact with the appropriate target, leading to emesis, diarrhea, and other gastrointestinal tract symptoms. Initial attempts to define molecular regions responsible for enterotoxicity involved testing biological activity of protease-generated fragments derived from SEA, SEB, or SEC1 (95). Three main conclusions were drawn from this work. First, only large toxin fragments containing central and C-terminal portions of SEs retained enough of the native structure to cause emesis. Second, N-terminal residues of SEs were not required for emesis. SEC1, modified by removal of the 59 N-terminal residues, retains the ability to cause emesis in monkeys. Smaller toxin fragments from SEC1 and other SEs were inactive. The third conclusion was that preservation of structure in the area of the conserved SE disulfide loop was necessary for emesis to occur.

The disulfide bond is a structural feature that is characteristic of most SEs but is not uniformly present in nonemetic staphylococcal and streptococcal exotoxins. With two known exceptions, SEs contain exactly two cysteine residues that could potentially form an intramolecular linkage and a spacer disulfide loop. Only SEI and the SEC molecular variant produced by staphylococcal isolates from cattle (SEC-bovine) deviate from this pattern. SEI has only one cysteine and is therefore unable to form a disulfide bond; however, it is emetic (at higher doses) and retains a highly conserved sequence (Fig. 19.1 and 19.7) that is directly downstream from the analogous cysteine loop region in the other staphylococcal enterotoxins (74). SEC from bovine sources possesses an additional third cysteine (Fig. 19.1) which may influence disulfide bond formation and overall structure. There is speculation about the role of the disulfide bond, which is located approximately in the middle of every SE regardless of its antigenic type, in emesis. Potential structural contributions from the Cys-Cys bond that could contribute to SE conformation necessary for emetic activity could include one or more of the following features: (i) proper positioning of cysteine residues upon formation of the disulfide bond, (ii) exposure and/or orientation of crucial residues in the loop formed between the two linked cysteines, (iii) exposure and/or orientation of residues immediately adjacent to the disulfide linkage but not contained within the cystine loop, and (iv) contributions to the overall SE conformation by the linkage of the two cysteine residues.

Each of these possibilities has been considered. The cysteine residues and the loop probably do not play a direct role in the emetic response. It has been possible to substitute the cysteine residues in several SEs by site-directed mutagenesis and show that neither of these two residues is absolutely critical. Although most of the SEs have a cystine loop, the lengths and composition of residues within the loops of different SE antigenic types vary greatly. This lack of consistency among SE loop properties suggests that they are unlikely to have a shared enterotoxigenic function (Fig. 19.7). Furthermore, proteolytic nicking of toxins in their loops has no effect on their ability to cause emesis (95). Warren et al. (103) determined that the disulfide bond contributes only minimally to overall protein conformation. Presumably, then, if the disulfide linkage is important in emesis, the effect is probably to provide a particular orientation of residues near the disulfide linkage, but not in the loop. The most convincing evidence in support of this possibility was provided by mutagenesis of SEC1 in which its cysteine residues were substituted by either serine or alanine (44). It was shown that mutants with serine substitutions were emetic, whereas the analogous mutants with alanine substitutions were nonemetic. Although serine and alanine are both conservative substitutions for cysteine, these two amino acids differ in that serine has the ability to hydrogen bond. Thus, hydrogen bonding by serine may be able to replace the disulfide linkage stabilization of local structure. The fact that SEI is also emetic and does not have a disulfide bond further supports this premise (74).

Which critical local residues require proper orientation by the disulfide bond (or hydrogen bonding at the same positions) in order for the SEs to induce emesis? Possible candidates are those within a highly conserved stretch of residues directly adjacent to, and downstream from, the disulfide loop (Fig. 19.7). These residues in SEC3 are located on the $\beta5$-strand. An attractive hypothesis is that in addition to stability in the gut, two other structural requirements are required for enterotoxicity. First, the appropriate conserved residues must be present in the toxin. Many PTs, emetic and nonemetic, have similar sets of highly conserved residues in a location analogous to the $\beta5$-strand of SEC3. Second, they must be positioned properly for interacting with their target in the gut. The unique SE disulfide bond may serve this function. Of the entire PT family, only two non-SE toxins produced by *S. pyogenes* (streptococcal pyrogenic exotoxin and streptococcal SAg) could potentially form a disulfide linkage (85). However, the presence of more than two cysteines in these toxins suggests that the structure in this area and the degree of local stabilization by their putative disulfide linkage may not be identical to those of the SEs.

Additional work with SEA revealed that single-site substitutions of residues closer to the N terminus also influence emesis (40). This was especially the case for mutants constructed by substitution with glycine. For

example, mutagenesis of residues 25, 47, and 48 causes a significant reduction in the emetic potency of SEA. Although these residues are far from the disulfide bond in the SEA primary sequence, they are located near or within domain 1 of SEA and could potentially influence the area around the disulfide bond.

SE Antigenic Epitopes

The need for reagents that could detect SEs in food and clinical samples, plus a desire to differentiate antigenically different toxins in the family, has been the impetus for considerable research on epitope characterization and mapping. Individually, each SE type and subtype has a sufficient degree of antigenic distinctness to enable its differentiation from other PTs by using highly specific polyclonal antisera and monoclonal antibodies. However, some degree of cross-reactivity can often be observed between several SEs. The level of cross-reactivity generally correlates with shared primary sequences. The type C SE subtypes and their molecular variants exhibit a substantial amount of cross-reactivity, as do SEA and SEE, the two major serological types with the greatest sequence relatedness. For these toxins, cross-reactivity can even be demonstrated in relatively insensitive assays such as immunodiffusion assays, in which these two toxins produce lines of partial identity with the heterologous antiserum. SEB and the SEC subtypes and molecular variants are also highly related at the amino acid level. Cross-reactivity between SEB and SEC may be demonstrated occasionally by immunodiffusion but more consistently by using sensitive methods such as radioimmunoassays or immunoblotting. Generally, it has not been possible to produce useful antibodies that cross-react among less-related SEs. Although one investigator has produced a monoclonal antibody that cross-reacts with all five major SE antigenic types A through E, this antibody has low affinity and cross-reacts with other staphylococcal proteins (9). The two most distantly related SEs that can be recognized by a common epitope are SEA and SED (11).

The mapping of conserved and specific antigenic epitopes on SEs, and their differentiation from potentially toxic regions, has potential applications for rational development of nontoxic vaccines. Considering the array of SE antigenic types, the most efficient toxoid would presumably contain one or more epitopes that are shared by multiple toxins. Several approaches have been used to partially localize antigenic epitopes on SEs and differentiate them from toxic regions. One of the earlier methods used for this purpose was to identify protease-generated toxin fragments from several SEs that bind to cross-reactive antibodies (18). These studies identified both N- and C-terminal toxin fragments that contain cross-reactive epitopes but were unable to define shorter stretches of residues.

There is some evidence that immunization with short, highly conserved peptides could provide protective immunity. For example, the use of synthetic peptides from highly conserved stretches of primary sequence has resulted in the production of neutralizing antibody for several of the SEs. Immunization with synthetic peptides corresponding to residues 130 to 160 of SEB or the same region of SEC1 (residues 148 to 162) induced antibodies that neutralized both native toxins (42). The highly conserved SE sequence K-K-X-V-T-X-Q-E-L-D (see Fig. 19.1), encompassed by both peptides, may represent part of an epitope that could be useful for protective immunity. It is unknown whether major epitopes identified on other toxins such as SEA can provide protective immunity (79).

SE Mode of Action in Induction of Emesis and Other Symptoms Related in SFP

SE-Induced Emesis Requires Nerve Stimulation

Except for rare SFP cases in which massive doses of SEs are consumed, systemic dissemination of the toxins does not contribute significantly to the illness. When fed to rodents, SEs do enter the circulation but are rapidly removed by the kidneys (7). Most studies using the simian model indicate that the SE site of action following ingestion is the abdominal viscera. Early studies into the mechanism of action of SEs tested the emetic responsiveness of animals to the toxins after disruption of well-defined neural systems or after visceral deafferation. The characteristic emetic response was observed to result from a stimulation of local neural receptors in the abdomen (96) that transmit impulses through the vagus and sympathetic nerves, ultimately stimulating the medullary emetic center.

Cellular Histopathology in the Gastrointestinal Tract

Information on the histological effects of oral doses of SEs in humans is extremely limited. Most of what is known has been derived from information obtained from experiments in rhesus monkeys. Upon ingestion of the toxin, pathological changes compatible with a definition of gastroenteritis are observed in several parts of the gastrointestinal tract (59).

The primate stomach becomes hyperemic and is marked by lesions that begin with the influx of neutrophils into the lamina propria and epithelium. A mucopurulent exudate in the gastric lumen is also typi-cally observed. Also characteristic are mucus-filled surface cells which eventually release their contents. Later in the illness, neutrophils are replaced by macrophages. Eventually, upon resolution of the symptoms, the cellular infiltrate clears.

A similar cellular infiltrate and lumen exudate occurs in the small intestine, although they decrease in severity in sections taken from lower, compared to upper, portions of the intestine. Clearly evident in the jejunum are extension of crypts, disruption or loss of the brush border, and an extensive infiltrate of neutrophils and macrophages into the lamina propria. Changes in the colon are minimal in the monkey model. Only a mild cellular exudate and mucus depletion are evident. The only other significant effect in monkeys is acute lymphadenitis in the mesenteric lymph nodes.

Search for the Gastrointestinal Tract Target

The specific cells with which SEs interact in the abdomen have not been clearly identified, nor have their receptors. Evidence suggests that the interaction of SEs with their target directly or indirectly causes production of inflammatory mediators that induce SFP symptoms. Jett et al. (52) determined that oral administration of SEB produced elevated levels of arachidonic acid cascade products. Specifically, they observed significant increases in prostaglandin E_2, leukotreine B4, and 5-hydroxyeicosatetraenoic acid. These three compounds are potent vasoactive inflammatory mediators that can also act as chemoattractants for neutrophils. Both of these activities are consistent with the histopathology described above in the SFP monkey model.

Scheuber et al. (90) could not demonstrate an effect of prostanoid inhibitors on SEB-induced emesis, but they were able to correlate gastrointestinal symptoms with cysteinyl leukotreine generation. Intoxication of animals with SEB resulted in a 10-fold increase in levels of leukotreine E4 in bile, plus an unidentified leukotreine in the urine. These investigators suggested a role for mast cells in pathogenesis of SFP. Although induction of histamine production by SEB was responsible for some secondary nonenteric immediate hypersensitivity skin reactions, it did not correlate with emesis.

Currently available data indicate that ingestion of SEs causes a stimulation of mast cells and possibly other inflammatory cells in the abdomen. Thus far, the abdominal receptor has not been identified. Experiments using anti-idiotype antibodies in binding assays and protection assays have provided circumstantial evidence for its existence on monkey mast cells (84). Komisar et al. determined that SEB can stimulate rodent peritoneal mast cells as well and provided evidence for a protein receptor

(61). SEs may act directly on their receptor and circumvent the typical two-stage mast cell IgE antibody-antigen interaction (90).

Despite these observations, Alber et al. (1) were unable to directly stimulate monkey or human skin and intestinal mast cells with SEB to release inflammatory mediators. It was suggested that stimulation of mast cells in vivo occurs through a nonimmunological mechanism requiring the generation of neuropeptides released from peripheral terminals of primary sensory nerves. At least one putative mast cell-stimulating peptide, substance P, was implicated in SEB-induced toxicity by use of antibodies and a variety of inhibitors. The attractiveness of this explanation is that it is consistent with earlier predictions of a neural involvement in the pathogenesis of SFP.

Researchers have determined that certain PTs can cross epithelial membranes (39, 91). SEB and TSST-1 were transported across the membrane more rapidly than a protein of similar size, horseradish peroxidase, suggesting that movement was due to facilitated transcytosis. Interestingly, SEA was transported at a rate similar to horseradish peroxidase, suggesting that this toxin was transported in a nonspecific manner. Site-directed mutants of SEB (N23>K and F44>S) had reduced rates of transcytosis. These residues are located in domain 1 of the protein. Previous studies by Kappler et al. have revealed that N23 interacts with TCR, whereas F44 is involved in binding to MHC (57). These residues may also influence the toxin's ability to move across the membrane. In vivo studies revealed that both SEA and SEB could cross the epithelial layer and enter the bloodstream, with SEB reaching an average level of 2,750 pg/ml and SEA approximately 100 pg/ml. The investigators suggest that the difference in levels of the enterotoxins in the bloodstream may be due to their differing modes of entry. In contrast, Schlievert et al. (91) determined that when comparing TSST-1, SEC1, and streptococcal pyrogenic exotoxin A (SPEA), only TSST-1 could consistently transverse the mucosal membranes of Dutch Belted rabbits. Oral or vaginal administration of TSST-1 induced endotoxic shock and death in rabbits, unlike either SEC1 or SPEA. This is intriguing considering that neither TSST-1 nor SPEA is emetic when administered to primates (91).

Is There a Relationship Between Superantigenicity, TSS, and SFP?

The discovery of the mechanism of SAg action and the unique properties of this class of proteins provided an explanation for the multiple systemic effects observed in TSS patients. The massive cellular stimulation induced by superantigenic PTs explained the long-recognized fact that TSS patients had elevated serum cytokine levels, which presumably mediate many symptoms of the disease. The realization that at least some of the pathogenesis of TSS could be attributed to immune cell stimulation led some investigators to predict that SFP could also be a reflection of SAg function. Consistent with this prediction was the fact that TSS patients often have a gastrointestinal tract component characterized by vomiting and diarrhea. Also, patients with endotoxin shock have elevated cytokine levels and similarly display vomiting and diarrhea. If superantigenicity is responsible for SFP, the SEs presumably act directly on T cells and antigen-presenting cells in the gut. Although some SEs enter the circulation, they appear to be rapidly cleared by the kidneys so that significant systemic concentrations are unlikely to occur (7). This, plus the fact that TSS-associated symptoms (i.e., shock and fever) are not observed in SFP patients, suggests that SEs do not mediate SFP through systemic cytokines.

Despite their similarities and the evidence cited above, several lines of evidence suggest that the partial overlap between SFP and TSS symptoms is probably coincidental and that superantigenicity is not directly responsible for SFP. First, nonimmunological mast cell stimulation has been linked to the release of inflammatory mediators resulting from nerve interactions. The second line of evidence has come through mutagenesis of several SEs. Studies have revealed that T-cell stimulation and induction of emesis are separable functions and are determined by distinct portions of the SE molecules (1, 40, 44, 45). It has been possible to construct SEA, SEB, or SEC1 mutants that are deficient in T-cell stimulatory activity but retain the ability to induce emesis and vice versa. Finally, although all PTs have superantigen function, only the SEs are emetic when ingested. The lack of emesis-inducing ability of some nonenterotoxic PTs has been attributed to instability in the gastrointestinal tract. However, at least one nonemetic PT, SPEA, is very stable in gastric fluid but does not cause emesis when ingested by monkeys (91).

If superantigenicity does not explain SFP, how do PTs cause vomiting and diarrhea in TSS? Several possibilities could explain how SEs and other PTs act on the gastrointestinal tract if they are not consumed through the oral route. First, the toxins' ability to induce TSS symptoms may be limited to the systemic circulation. If so, cytokines would need to enter the abdomen from the circulation and mediate gastroenteritis pathogenesis. Alternatively, the PTs could enter the gut from the circulation and act directly at the local level as superantigens or through other mechanisms as proposed for SFP. The latter possibility is less likely because even nonenterotoxic PTs, including those that are susceptible to degradation in the gut, cause gastrointestinal symptoms when they are associated with TSS.

These complex and unresolved issues are relevant to interpreting models for studying the enterotoxic activity of SEs. Oral administration of staphylococcal culture filtrates or suspected food to monkeys is regarded as the preferred method for detecting the enterotoxic properties of SEs. However, intravenous administration of SEs to primates and other animals, especially cats, has been used as an alternative to feeding. Considering that systemic exposure to SEs mimics the situation leading to TSS symptoms, intravenous administration may not accurately reflect the pathogenesis and toxin properties required to induce emesis in SFP.

CONCLUDING REMARKS AND FUTURE DIRECTIONS

Remarkable progress has been made toward understanding the molecular aspects relevant to SFP. *S. aureus* has some unique properties that promote its ability to induce foodborne illness. Further understanding of the molecular aspects of these unique properties should facilitate the implementation of more effective ways to selectively inhibit staphylococcal survival, growth, and SE production in food. For example, improved methods for detection of enterotoxin-producing strains of *S. aureus* such as multiplex PCR (71), trichloroacetic acid precipitation (69), and biosensors (2, 48, 82, 98) may aid both epidemiologic and therapeutic studies.

One issue that continues to puzzle the scientific community is the questionable rationale for staphylococci to produce an emetic toxin. Unlike enteric pathogens that inhabit the gut, the ability to induce emesis and diarrhea as a mechanism to promote exit from the host and dissemination does not appear to be important for staphylococci. One may ask a similar question regarding the ability of SEs to induce lethal shock in TSS, since it is generally agreed that lethality is not advantageous to microorganisms. Instead, long-term host-pathogen coexistence usually relies on adaptations that allow the organism to survive for extended periods of time without harming the host or being cleared by the immune response.

Based on what is now known regarding the superantigenic properties and proposed host-specific molecular adaptation of SEs, one may propose that SEs are produced for the purpose of immunomodulating the host. Exposure of animals and peripheral blood mononuclear cell cultures to SEs and other superantigens induces at least a transient immunosuppression. Similarly, TSS patients often fail to produce a significant immune response

to causative superantigenic PTs and remain susceptible to subsequent toxigenic illnesses. Thus, one may speculate that the harmful effects on the host (SFP and TSS) are merely secondary effects of the staphylococci's attempt to affect immune cell function, allowing the bacteria to survive and persistently colonize their many potential animal hosts.

Our preparation of this chapter was supported by grants from the Public Health Service (AI28401, RR15587, and RR00166), the U.S. Department of Agriculture (NRI-CGP), and the Idaho Agriculture Experiment Station. We also thank Patrick Schlievert, Richard Novick, Amy Wong, Cynthia Stauffacher, Sibyl Munson, Brian Wilkinson, and Scott Minnich for comments, suggestions, and critical opinions on certain topics covered in this chapter. Jana Joyce, Scott Callantine, and Claudia Deobald are acknowledged for computer analysis, artwork, and organizational aspects of this manuscript.

References

1. **Alber, G., P. H. Scheuber, B. Reck, B. Sailer-Kramer, A. Hartmann, and D. K. Hammer.** 1989. Role of substance P in immediate-type skin reactions induced by staphylococcal enterotoxin B in unsensitized monkeys. *J. Allergy Clin. Immunol.* **84:**880–885.

2. **Anderson, G. P., K. A. Breslin, and F. S. Ligler.** 1996. Assay development for a portable fiberoptic biosensor. *Asaio J.* **42:**942–946.

3. **Anonymous.** 1992. *Foodborne Pathogenic Microorganisms and Natural Toxins.* Center for Food Safety and Applied Nutrition, U.S. Food and Drug Administration, Washington, D.C.

4. **Bayer, M. G., J. H. Heinrichs, and A. L. Cheung.** 1996. The molecular architecture of the *sar* locus in *Staphylococcus aureus*. *J. Bacteriol.* **178:**4563–4570.

5. **Bayles, K. W., and J. J. Iandolo.** 1989. Genetic and molecular analyses of the gene encoding staphylococcal enterotoxin D. *J. Bacteriol.* **171:**4799–4806.

6. **Bayles, K. W., C. A. Wesson, L. E. Liou, L. K. Fox, G. A. Bohach, and W. R. Trumble.** 1998. Intracellular *Staphylococcus aureus* escapes the endosome and induces apoptosis in epithelial cells. *Infect. Immun.* **66:**336–342.

7. **Beery, J. T., S. L. Taylor, L. R. Schlunz, R. C. Freed, and M. S. Bergdoll.** 1984. Effects of staphylococcal enterotoxin A on the rat gastrointestinal tract. *Infect. Immun.* **44:**234–240.

8. **Bergdoll, M. S.** 1979. Staphylococcal intoxications, p. 443–494. *In* H. Riemann and F. L. Bryan (ed.), *Food-Borne Infections and Intoxications*, 2nd ed. Academic Press, New York, N.Y.

9. **Bergdoll, M. S.** 1985. The staphylococcal enterotoxins—an update, p. 247–254. *In* J. Jeljaszewicz (ed.), *The staphylococci. Zentralbl. Bakteriol.* Suppl. 14. Gustav Fischer Verlag, Stuttgart, Germany.

10. **Bergdoll, M. S.** 1989. *Staphylococcus aureus*, p. 247–254. *In* M. P. Doyle (ed.), *Foodborne Bacterial Pathogens*. Marcel Dekker, Inc, New York, N.Y.

11. **Bergdoll, M. S., C. R. Borja, R. N. Robbins, and K. F. Weiss.** 1971. Identification of enterotoxin E. *Infect. Immun.* **4:**593–595.

12. **Bergdoll, M. S., M. J. Surgalla, and G. M. Dack.** 1959. Staphylococcal enterotoxin. Identification of a specific neutralizing antibody with enterotoxin neutralizing properties. *J. Immunol.* **83:**334–338.

13. **Betley, M. J., D. W. Borst, and L. B. Regassa.** 1992. Staphylococcal enterotoxins, toxic shock syndrome toxin and streptococcal pyrogenic exotoxins: a comparative study of their molecular biology. *Chem. Immunol.* **55:** 1–35.

14. **Betley, M. J., and J. J. Mekalanos.** 1985. Staphylococcal enterotoxin A is encoded by phage. *Science* **229:**185–187.

15. **Betley, M. J., P. M. Schlievert, M. S. Bergdoll, G. A. Bohach, J. J. Iandolo, S. A. Khan, P. A. Pattee, and R. R. Reiser.** 1990. Staphylococcal gene nomenclature. *ASM News* **56:**182.

16. **Blevins, J. S., A. F. Gillaspy, T. M. Rechtin, B. K. Hurlburt, and M. S. Smeltzer.** 1999. The staphylococcal accessory regulator (*sar*) represses transcription of the *Staphylococcus aureus* collagen adhesin gene (*cna*) in an *agr*-independent manner. *Mol. Microbiol.* **33:**317–326.

17. **Bohach, G. A., D. J. Fast, R. D. Nelson, and P. M. Schlievert.** 1990. Staphylococcal and streptococcal pyrogenic toxins involved in toxic shock syndrome and related illnesses. *Crit. Rev. Microbiol.* **17:**251–272.

18. **Bohach, G. A., C. J. Hovde, J. P. Handley, and P. M. Schlievert.** 1988. Cross-neutralization of staphylococcal and streptococcal pyrogenic toxins by monoclonal and polyclonal antibodies. *Infect. Immun.* **56:**400–404.

19. **Bohach, G. A., L. M. Jablonski, C. F. Deobald, Y. I. Chi, and C. V. Stauffacher.** 1995. Functional domains of staphylococcal enterotoxins. *In* M. Ecklund, J. L. Richard, and K. Mise (ed.), *Molecular Approaches to Food Safety; Issues Involving Toxic Microorganisms.* Alaken Inc., Fort Collins, Colo.

20. **Bohach, G. A., C. V. Stauffacher, D. H. Ohlendorf, Y. I. Chi, G. M. Vath, and P. M. Schlievert.** 1996. The staphylococcal and streptococcal pyrogenic toxin family. *Adv. Exp. Med. Biol.* **391:**131–154.

21. **Bryan, F. L.** 1976. *Staphylococcus aureus*, p. 12–128. *In* M. P. deFigueiredo and D. F. Splittstoesser (ed.), *Food Microbiology: Public Health and Spoilage Aspects.* AVI Publishers, Westport, Conn.

22. **Buzby, J. C., T. Roberts, C. T. J. Lin, and J. M. MacDonald.** 1996. *Bacterial Foodborne Disease: Medical Costs and Productivity Losses.* Economic Research Service Report no. 741. USDA Economic Research Service, Washington, D.C.

22a. **Carl, K. E.** 1975. Oregon's experience with microbiological standards for meat. *J. Milk Food Technol.* **38:**483–486.

23. **Cheung, A. L., and S. J. Projan.** 1994. Cloning and sequencing of *sarA* of *Staphylococcus aureus*, a gene required for the expression of *agr*. *J. Bacteriol.* **176:**4168–4172.

24. **Chien, Y., and A. L. Cheung.** 1998. Molecular interactions between two global regulators, *sar* and *agr*, in *Staphylococcus aureus*. *J. Biol. Chem.* **273:**2645–2652.

25. **Chien, Y., A. C. Manna, and A. L. Cheung.** 1998. SarA level is a determinant of *agr* activation in *Staphylococcus aureus*. *Mol. Microbiol.* **30:**991–1001.

26. **Chien, Y., A. C. Manna, S. J. Projan, and A. L. Cheung.** 1999. SarA, a global regulator of virulence determinants in *Staphylococcus aureus*, binds to a conserved motif essential for *sar*-dependent gene regulation. *J. Biol Chem.* **274:**37169–37176.

27. **Couch, J. L., M. T. Soltis, and M. J. Betley.** 1988. Cloning and nucleotide sequence of the type E staphylococcal enterotoxin gene. *J. Bacteriol.* **170:**2954–2960.

28. **Dack, G. M., W. E. Cary, O. Woolper, and H. Wiggers.** 1930. An outbreak of food poisoning proved to be due to a yellow hemolytic *Staphylococcus*. *J. Prev. Med.* **4:**167–175.

29. **Dickie, N., Y. Yano, H. Robern, and S. Stavric.** 1972. On the heterogeneity of staphylococcal enterotoxin C 2. *Can. J. Microbiol.* **18:**801–804.

30. **Dinges, M. M., P. M. Orwin, and P. M. Schlievert.** 2000. Exotoxins of *Staphylococcus aureus*. *Clin. Microbiol. Rev.* **13:**16–34.

31. **Dziewanowska, K., J. M. Patti, C. F. Deobald, K. W. Bayles, W. R. Trumble, and G. A. Bohach.** 1999. Fibronectin binding protein and host cell tyrosine kinase are required for internalization of *Staphylococcus aureus* by epithelial cells. *Infect. Immun.* **67:**4673–4678.

32. **Everson, M. L., M. W. Hinds, R. S. Bernstein, and M. S. Bergdoll.** 1988. Estimation of human dose of staphylococcal enterotoxin A from a large outbreak of staphylococcal food poisoning involving chocolate milk. *Int. J. Food Microbiol.* **7:**311–316.

33. **Fraser, J. D., S. Lowe, M. J. Irwin, N. R. J. Gascoigne, and K. R. Hudson** 1993. Structural model of staphylococcal enterotoxin A interaction with MHC class II antigens, p. 7–30. *In* B. T. Huber and E. Palmer (ed.), *Superantigens.* Cold Spring Harbor Laboratory Press, Plainview, N.Y.

34. **Friedman, M. E.** 1966. Inhibition of staphylococcal enterotoxin B formation in broth cultures. *J. Bacteriol.* **92:**277–278.

35. **Gilbert, R. J., and A. A. Wieneke.** 1973. Staphylococcal food poisoning with special reference to the detection of enterotoxin in food, p. 273–285. *In* B. C. Hobbs and J. H. B. Christian (ed.), *The Microbiological Safety of Food.* Academic Press, New York, N.Y.

36. **Giraudo, A. T., A. Calzolari, A. A. Cataldi, C. Bogni, and R. Nagel.** 1999. The *sae* locus of *Staphylococcus aureus* encodes a two-component regulatory system. *FEMS Microbiol. Lett.* **177:**15–22. (Corrigendum, **180:**117, 1999.)

37. **Giraudo, A. T., C. G. Raspanti, A. Calzolari, and R. Nagel.** 1994. Characterization of a Tn551-mutant of *Staphylococcus aureus* defective in the production of several exoproteins. *Can. J. Microbiol.* **40:**677–681.

37a. **Gutherz, L. S., J. T. Fruin, R. L. Okoluk, and J. L. Fowler.** 1977. Microbial quality of frozen comminuted turkey meat. *J. Food Sci.* **42:**1444–1447.

37b. **Gutherz, L. S., J. T. Fruin, D. Spicer, and J. L. Fowler.** 1976. Microbial quality of fresh comminuted turkey meat. *J. Milk Food Technol.* **39:**823–829.

38. **Hajek, V., and E. Marsalek.** 1973. The occurrence of enterotoxigenic *Staphylococcus aureus* strains in hosts of different animal species. *Zentralbl. Bakteriol. Orig. Reihe A* **223:**63–68.

39. **Hamad, A. R., P. Marrack, and J. W. Kappler.** 1997. Transcytosis of staphylococcal superantigen toxins. *J. Exp. Med.* **185:**1447–1454.

40. **Harris, T. O., and M. J. Betley.** 1995. Biological activities of staphylococcal enterotoxin type A mutants with N-terminal substitutions. *Infect. Immun.* **63:**2133–2140.

41. **Hase, C. C., and R. A. Finkelstein.** 1993. Bacterial extracellular zinc-containing metalloproteases. *Microbiol. Rev.* **57:**823–837.

42. **Hoffmann, M. L., L. M. Jablonski, K. K. Crum, S. P. Hackett, Y. I. Chi, C. V. Stauffacher, D. L. Stevens, and G. A. Bohach.** 1994. Predictions of T-cell receptor- and major histocompatibility complex-binding sites on staphylococcal enterotoxin C1. *Infect. Immun.* **62:**3396–3407.

43. **Holmberg, S. D., and P. A. Blake.** 1984. Staphylococcal food poisoning in the United States. New facts and old misconceptions. *JAMA* **251:**487–489.

44. **Hovde, C. J., J. C. Marr, M. L. Hoffmann, S. P. Hackett, Y. I. Chi, K. K. Crum, D. L. Stevens, C. V. Stauffacher, and G. A. Bohach.** 1994. Investigation of the role of the disulphide bond in the activity and structure of staphylococcal enterotoxin C1. *Mol. Microbiol.* **13:**897–909.

45. **Huang, I. Y., and M. S. Bergdoll.** 1970. The primary structure of staphylococcal enterotoxin B. 3. The cyanogen bromide peptides of reduced and aminoethylated enterotoxin B, and the complete amino acid sequence. *J. Biol. Chem.* **245:**3518–3525.

46. **Hudson, M. C., W. K. Ramp, N. C. Nicholson, A. S. Williams, and M. T. Nousiainen.** 1995. Internalization of *Staphylococcus aureus* by cultured osteoblasts. *Microb. Pathogen.* **19:**409–419.

47. **Iandolo, J. J., and W. M. Shafer.** 1977. Regulation of staphylococcal enterotoxin B. *Infect. Immun.* **16:**610–616.

48. **James, E. A., K. Schmeltzer, and F. S. Ligler.** 1996. Detection of endotoxin using an evanescent wave fiber-optic biosensor. *Appl. Biochem. Biotechnol.* **60:**189–202.

49. **Janzon, L., and S. Arvidson.** 1990. The role of the delta-lysin gene (*hld*) in the regulation of virulence genes by the accessory gene regulator (*agr*) in *Staphylococcus aureus*. *EMBO J.* **9:**1391–1399.

50. **Jarraud, S., G. Cozon, F. Vandenesch, M. Bes, J. Etienne, and G. Lina.** 1999. Involvement of enterotoxins G and I in staphylococcal toxic shock syndrome and staphylococcal scarlet fever. *J. Clin. Microbiol.* **37:**2446–2449.

51. **Jay, J. M.** 1986. Staphylococcal gastroenteritis, p. 437–458. In *Modern Food Microbiology*, 3rd ed. Van Nostrand Reinhold Company, New York, N.Y.

52. **Jett, M., R. Neill, C. Welch, T. Boyle, E. Bernton, D. Hoover, G. Lowell, R. E. Hunt, S. Chatterjee, and P. Gemski.** 1994. Identification of staphylococcal enterotoxin B sequences important for induction of lymphocyte proliferation by using synthetic peptide fragments of the toxin. *Infect. Immun.* **62:**3408–3415.

53. **Ji, G., R. Beavis, and R. P. Novick.** 1997. Bacterial interference caused by autoinducing peptide variants. *Science* **276:**2027–2030.

54. **Johns, M. B., Jr., and S. A. Khan.** 1988. Staphylococcal enterotoxin B gene is associated with a discrete genetic element. *J. Bacteriol.* **170:**4033–4039.

55. **Johnson, L. P., and P. M. Schlievert.** 1983. A physical map of the group A streptococcal pyrogenic exotoxin bacteriophage T12 genome. *Mol. Gen. Genet.* **189:**251–255.

56. **Jones, C. L., and S. A. Khan.** 1986. Nucleotide sequence of the enterotoxin B gene from *Staphylococcus aureus*. *J Bacteriol.* **166:**29–33.

57. **Kappler, J. W., A. Herman, J. Clements, and P. Marrack.** 1992. Mutations defining functional regions of the superantigen staphylococcal enterotoxin B. *J. Exp. Med.* **175:**387–396.

58. **Katsuno, S., and M. Kondo.** 1973. Regulation of staphylococcal enterotoxin B synthesis and its relation to other extracellular proteins. *Jpn. J. Med. Sci. Biol.* **26:**26–29.

59. **Kent, T. H.** 1966. Staphylococcal enterotoxin gastroenteritis in rhesus monkeys. *Am. J. Pathol.* **48:**387–407.

60. **Klimpel, K. R., N. Arora, and S. H. Leppla.** 1994. Anthrax toxin lethal factor contains a zinc metalloprotease consensus sequence which is required for lethal toxin activity. *Mol. Microbiol.* **13:**1093–1100.

61. **Komisar, J., J. Rivera, A. Vega, and J. Tseng.** 1992. Effects of staphylococcal enterotoxin B on rodent mast cells. *Infect. Immun.* **60:**2969–2975.

62. **Kornblum, J., B. Kreiswirth, S. J. Projan, H. Ross, and R. P. Novick.** 1990. *agr*: a polycistronic locus regulating exoprotein synthesis in *Staphylococcus aureus*, p. 373–402. *In* R. P. Novick (ed.), *Molecular Biology of the Staphylococci.* VCH Publishers, New York, N.Y.

63. **Lindsay, J. A., A. Ruzin, H. F. Ross, N. Kurepina, and R. P. Novick.** 1998. The gene for toxic shock toxin is carried by a family of mobile pathogenicity islands in *Staphylococcus aureus*. *Mol. Microbiol.* **29:**527–543.

64. **Lovejoy, H. F. J.** 1991. Unpublished data.

65. **Marr, J. C., J. D. Lyon, J. R. Roberson, M. Lupher, W. C. Davis, and G. A. Bohach.** 1993. Characterization of novel type C staphylococcal enterotoxins: biological and evolutionary implications. *Infect. Immun.* **61:**4254–4262.

66. **Marrack, P., and J. Kappler.** 1990. The staphylococcal enterotoxins and their relatives. *Science* **248:**1066.

67. **Mayville, P., G. Ji, R. Beavis, H. Yang, M. Goger, R. P. Novick, and T. W. Muir.** 1999. Structure-activity analysis of synthetic autoinducing thiolactone peptides from *Staphylococcus aureus* responsible for virulence. *Proc. Natl. Acad. Sci. USA* **96:**1218–1223.

68. Mead, P. S., L. Slutsker, V. Dietz, L. F. McCaig, J. S. Bresee, C. Shapiro, P. M. Griffin, and R. V. Tauxe. 1999. Food-related illness and death in the United States. *Emerg. Infect. Dis.* 5:607–625.

69. Meyrand, A., V. Atrache, C. Bavai, M. P. Montet, and C. Vernozy-Rozand. 1999. Evaluation of an alternative extraction procedure for enterotoxin determination in dairy products. *Lett. Appl. Microbiol.* 28:411–415.

70. Monday, S. R., and G. A. Bohach. 1999. Use of multiplex PCR to detect classical and newly described pyrogenic toxin genes in staphylococcal isolates. *J. Clin. Microbiol.* 37:3411–3414.

71. Monday, S. R., and G. A. Bohach. Genes encoding staphylococcal entertoxin G and I are linked and separated by DNA related to other staphylococcal enterotoxins. *J. Nat. Toxins,* in press.

72. Morfeldt, E., I. Panova-Sapundjieva, B. Gustafsson, and S. Arvidson. 1996. Detection of the response regulator AgrA in the cytosolic fraction of *Staphylococcus aureus* by monoclonal antibodies. *FEMS Microbiol. Lett.* 143:195–201.

73. Morfeldt, E., K. Tegmark, and S. Arvidson. 1996. Transcriptional control of the *agr*-dependent virulence gene regulator, RNAIII, in *Staphylococcus aureus. Mol. Microbiol.* 21:1227–1237.

74. Munson, S. H., M. T. Tremaine, M. J. Betley, and R. A. Welch. 1998. Identification and characterization of staphylococcal enterotoxin types G and I from *Staphylococcus aureus. Infect. Immun.* 66:3337–3348.

75. Noleto, A. L., and M. S. Bergdoll. 1982. Production of enterotoxin by a *Staphylococcus aureus* strain that produces three identifiable enterotoxins. *J. Food Prot.* 45:1096–1097.

76. Novick, R. P., and T. W. Muir. 1999. Virulence gene regulation by peptides in staphylococci and other gram-positive bacteria. *Curr. Opin. Microbiol.* 2:40–45.

77. Novick, R. P., H. F. Ross, S. J. Projan, J. Kornblum, B. Kreiswirth, and S. Moghazeh. 1993. Synthesis of staphylococcal virulence factors is controlled by a regulatory RNA molecule. *EMBO J.* 12:3967–3975.

77a. Pace, P. J. 1975. Bacteriological quality of delicatessen foods. *J. Milk Food Technol.* 38:347–353.

78. Papageorgiou, A. C., K. R. Acharya, R. Shapiro, E. F. Passalacqua, R. D. Brehm, and H. S. Tranter. 1995. Crystal structure of the superantigen enterotoxin C2 from *Staphylococcus aureus* reveals a zinc-binding site. *Structure* 3:769–779.

79. Pontzer, C. H., J. K. Russell, and H. M. Johnson. 1989. Localization of an immune functional site on staphylococcal enterotoxin A using the synthetic peptide approach. *J. Immunol.* 143:280–284.

80. Prasad, G. S., C. A. Earhart, D. L. Murray, R. P. Novick, P. M. Schlievert, and D. H. Ohlendorf. 1993. Structure of toxic shock syndrome toxin 1. *Biochemistry* 32:13761–13766.

81. Raj, H. D., and M. S. Bergdoll. 1969. Effect of enterotoxin B on human volunteers. *J. Bacteriol.* 98:833–834.

81a. Rao, L., R. K. Karls, and M. J. Betley. 1995. In vitro transcription of pathogenesis-related genes by pruified

82. Rasooly, L., and A. Rasooly. 1999. Real time biosensor analysis of staphylococcal enterotoxin A in food. *Int. J. Food Microbiol.* 49:119–127.

83. Rechtin, T. M., A. F. Gillaspy, M. A. Schumacher, R. G. Brennan, M. S. Smeltzer, and B. K. Hurlburt. 1999. Characterization of the SarA virulence gene regulator of *Staphylococcus aureus. Mol. Microbiol.* 33:307–316.

84. Reck, B., P. H. Scheuber, W. Londong, B. Sailer-Kramer, K. Bartsch, and D. K. Hammer. 1988. Protection against the staphylococcal enterotoxin-induced intestinal disorder in the monkey by anti-idiotypic antibodies. *Proc. Natl. Acad. Sci. USA* 85:3170–3174.

85. Reda, K. B., V. Kapur, J. A. Mollick, J. G. Lamphear, J. M. Musser, and R. R. Rich. 1994. Molecular characterization and phylogenetic distribution of the streptococcal superantigen gene (*ssa*) from *Streptococcus pyogenes. Infect. Immun.* 62:1867–1874.

86. Regassa, L. B., J. L. Couch, and M. J. Betley. 1991. Steady-state staphylococcal enterotoxin type C mRNA is affected by a product of the accessory gene regulator (*agr*) and by glucose. *Infect. Immun.* 59:955–962.

87. Ren, K., J. D. Bannan, V. Pancholi, A. L. Cheung, J. C. Robbins, V. A. Fischetti, and J. B. Zabriskie. 1994. Characterization and biological properties of a new staphylococcal exotoxin. *J. Exp. Med.* 180:1675–1683.

88. Rosec, J. P., J. P. Guiraud, C. Dalet, and N. Richard. 1997. Enterotoxin production by staphylococci isolated from foods in France. *Int. J. Food Microbiol.* 35:213–221.

89. Schad, E. M., I. Zaitseva, V. N. Zaitsev, M. Dohlsten, T. Kalland, P. M. Schlievert, D. H. Ohlendorf, and L. A. Svensson. 1995. Crystal structure of the superantigen staphylococcal enterotoxin type A. *EMBO J.* 14:3292–3301.

90. Scheuber, P. H., C. Denzlinger, D. Wilker, G. Beck, D. Keppler, and D. K. Hammer. 1987. Staphylococcal enterotoxin B as a nonimmunological mast cell stimulus in primates: the role of endogenous cysteinyl leukotrienes. *Int. Arch. Allergy Appl. Immunol.* 82:289–291.

91. Schlievert, P. M., L. M. Jablonski, M. Roggiani, I. Sadler, S. Callantine, D. T. Mitchell, D. H. Ohlendorf, and G. A. Bohach. 2000. Pyrogenic toxin superantigen site specificity in toxic shock syndrome and food poisoning in animals. *Infect. Immun.* 68:3630–3634.

92. Schmidt, J. J., and L. Spero. 1983. The complete amino acid sequence of staphylococcal enterotoxin C1. *J. Biol. Chem.* 285:6300–6306.

93. Singh, B. R., and M. J. Betley. 1989. Comparative structural analysis of staphylococcal enterotoxins A and E. *J. Biol. Chem.* 264:4404–4411.

93a. Smith, F. C., R. A. Field, and J. C. Adams. 1974. Microbiology of Wyoming big game meat. *J. Milk Food Technol.* 37:129–131.

94. Soares, M. J., N. H. Tokumaru-Miyazaki, A. L. Noleto, and A. M. Figueiredo. 1997. Enterotoxin production by *Staphylococcus aureus* clones and detection of Brazilian

RNA polymerase from *Staphylococcus aureus. J. Bacteriol.* 177:2609–2614.

epidemic MRSA clone (III::B:A) among isolates from food handlers. *J. Med. Microbiol.* **46**:214–221.

95. **Spero, L., B. Y. Griffin, J. L. Middlebrook, and J. F. Metzger.** 1976. Effect of single and double peptide bond scission by trypsin on the structure and activity of staphylococcal enterotoxin C. *J. Biol. Chem.* **251**:5580–5588.

96. **Sugiyama, H., and T. Hayama.** 1965. Abdominal viscera as site of emetic action for staphylococcal enterotoxin in the monkey. *J. Infect. Dis.* **115**:330–336.

96a. **Surkiewicz, B. F., M. E. Harris, R. P. Elliott, J. F. Macaluso, and M. M. Strand.** 1975. Bacteriological survey of raw beef patties produced at establishments under federal inspection. *Appl. Microbiol.* **29**:331–334.

97. **Swaminathan, S., W. Furey, J. Pletcher, and M. Sax.** 1992. Crystal structure of staphylococcal enterotoxin B, a superantigen. *Nature* **359**:801–806.

97a. **Swartzentruber, A., A. H. Schwab, A. P. Duran, B. A. Wentz, and R. B. Read, Jr.** 1980. Microbiological quality of frozen shrimp and lobster tail in the retail market. *Appl. Environ. Microbiol.* **40**:765–769.

98. **Tempelman, L. A., K. D. King, G. P. Anderson, and F. S. Ligler.** 1996. Quantitating staphylococcal enterotoxin B in diverse media using a portable fiber-optic biosensor. *Anal. Biochem.* **233**:50–57.

99. **Thaikruea, L., J. Pataraarechachai, P. Savanpunyalert, and U. Naluponjiragul.** 1995. An unusual outbreak of food poisoning. *SE Asian J. Trop. Med. Public Health* **26**:78–85.

99a. **Todd, E. C. D., G. A. Jarvis, K. F. Weiss, G. W. Riedall, and S. Charbonneau.** 1983. Microbiological quality of frozen cream-type pies sold in Canada. *J. Food Protect.* **46**:34–40.

100. **Townsend, D. E., and B. J. Wilkinson.** 1992. Proline transport in *Staphylococcus aureus*: a high-affinity system and a low-affinity system involved in osmoregulation. *J. Bacteriol.* **174**:2702–2710.

101. **Tremaine, M. T., D. K. Brockman, and M. J. Betley.** 1993. Staphylococcal enterotoxin A gene (*sea*) expression is not affected by the accessory gene regulator (*agr*). *Infect. Immun.* **61**:356–359.

102. **Warren, J. R.** 1977. Comparative kinetic stabilities of staphylococcal enterotoxin types A, B, and C1. *J. Biol. Chem.* **252**:6831–6834.

103. **Warren, J. R., L. Spero, and J. F. Metzger.** 1974. Stabilization of native structure by the closed disulfide loop of staphylococcal enterotoxin B. *Biochim. Biophys. Acta* **359**:351–363.

104. **Wengender, P. A., and K. J. Miller.** 1995. Identification of a PutP proline permease gene homolog from *Staphylococcus aureus* by expression cloning of the high-affinity proline transport system in *Escherichia coli*. *Appl. Environ. Microbiol.* **61**:252–259.

104a. **Wentz, B. A., A. P. Duran, A. Swartzentruber, A. H. Schwab, and R. B. Read, Jr.** 1983. Microbiological quality of fresh blue crabmeat, clams and oysters. *J. Food Protect.* **46**:978–981.

104b. **Wentz, B. A., A. P. Duran, A. Swartzentruber, A. H. Schwab, and R. B. Read, Jr.** 1984. Microbiological quality of frozen onion rings and tuna pot pies. *J. Food Protect.* **47**:58–60.

105. **Wesson, C. A., J. Deringer, L. E. Liou, K. W. Bayles, G. A. Bohach, and W. R. Trumble.** 2000. Apoptosis induced by *Staphylococcus aureus* in epithelial cells utilizes a mechanism involving caspases 8 and 3. *Infect. Immun.* **68**:2998–3001.

106. **Wesson, C. A., L. E. Liou, K. M. Todd, G. A. Bohach, W. R. Trumble, and K. W. Bayles.** 1998. *Staphylococcus aureus agr* and *sar* global regulators influence internalization and induction of apoptosis. *Infect. Immun.* **66**:5238–5243.

107. **Wieneke, A. A., D. Roberts, and R. J. Gilbert.** 1993. Staphylococcal food poisoning in the United Kingdom, 1969–90. *Epidemiol. Infect.* **110**:519–531.

108. **Zhang, S., J. J. Iandolo, and G. C. Stewart.** 1998. The enterotoxin D plasmid of *Staphylococcus aureus* encodes a second enterotoxin determinant (*sej*). *FEMS Microbiol. Lett.* **168**:227–233.

Food Microbiology: Fundamentals and Frontiers, 2nd Ed.
Edited by M. P. Doyle et al.
© 2001 ASM Press, Washington, D.C.

Craig W. Hedberg

Epidemiology of Foodborne Illnesses

20

The epidemiology of foodborne illnesses is constantly changing (24). As new foods and sources of foods become available, new opportunities for transmission of foodborne illness often follow. For example, the exportation of raspberries from Guatemala was closely followed by widespread outbreaks of diarrheal illnesses caused by the coccidian parasite *Cyclospora cayetanensis* in the United States and Canada (29, 30). In addition, diets change in response to tastes and health concerns. Americans are now being encouraged to eat five servings of fresh fruits and vegetables daily to help prevent cancer, heart disease, obesity, and diabetes (52, 63). As a nation, we have increased consumption of these foods but continue to fall short of this goal (38, 60). However, the number and proportion of outbreaks of foodborne illness associated with fresh fruits and vegetables have increased (Fig. 20.1). Moreover, the U.S. population itself is changing. As the population ages, the number and proportion of people at high risk for serious complications of foodborne illnesses will increase. Thus, even if the incidence of foodborne illnesses remains constant, the public health burden of these illnesses will grow.

Not all change is bad, however. During the early 1990s, eating hamburgers at fast food restaurants was identified as an important risk factor for *Escherichia coli* O157:H7 infections (55). Following a large foodborne outbreak of *E. coli* O157:H7 in the Pacific Northwest in 1993, public health recommendations and industry actions addressed this problem (2). In the initial Foodborne Diseases Active Surveillance Network (FoodNet) case-control study of sporadic *E. coli* O157:H7 infections, conducted during 1996–1997, eating hamburgers at fast food restaurants was not associated with illness (35).

Does the constantly changing epidemiology of foodborne illnesses mean that food is getting safer or less safe? When we look beyond specific pathogens or foods, it is difficult to say. Food safety is the product of complex interactions among environmental, cultural, and socioeconomic factors (49). Before FoodNet was established as part of the Emerging Infection Program of the Centers for Disease Control and Prevention (CDC) in 1995, there was no systematic basis for even estimating the magnitude of foodborne illnesses in the United States (43). Although the estimate by Mead and colleagues (43) that 76 million foodborne illnesses occur in the United States each year was larger than previous estimates, it did serve as a baseline against which future trends would be measured.

Craig W. Hedberg, Division of Environmental and Occupational Health, School of Public Health, University of Minnesota, Minneapolis, MN 55455.

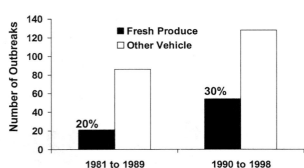

Figure 20.1 Confirmed outbreaks of foodborne illness associated with fresh produce, Minnesota, 1981–1998.

This chapter will provide an introduction to epidemiology and epidemiologic methods as they are applied to foodborne illnesses. An understanding of epidemiology is important because, despite our best efforts to prevent foodborne illnesses, humans remain the ultimate bioassay for low-level or sporadic contamination of our food supply (24). Epidemiologic methods of foodborne illness surveillance are needed to detect outbreaks, identify their causes, and assess the effectiveness of control measures. Epidemiologic data are also important in establishing food safety priorities, allocating food safety resources, stimulating public interest in food safety issues, establishing risk reduction strategies and public education campaigns, and evaluating the effectiveness of food safety programs (61). The examples in the chapter were drawn from the relevant literature and the author's own experiences as a foodborne disease epidemiologist. They were chosen to be illustrative, not necessarily representative. Epidemiologic data relevant to individual foodborne illnesses are presented in their respective chapters.

EPIDEMIOLOGY

Epidemiology is the study of events in populations, with events usually thought of in terms of illnesses. For example, FoodNet was established in part to determine actively how many laboratory-confirmed infections caused by *Campylobacter* spp., *E. coli* O157:H7, and *Salmonella* spp. occur in the populations under surveillance. A second major FoodNet activity has been a survey to determine how many diarrheal illnesses occur in the same populations that are under active surveillance for specific foodborne pathogens (7). By comparing the incidence and proportion of specific illnesses attributable to foodborne transmission with the incidence of occurrence and medical evaluation for clinical syndromes, such as diarrhea, it becomes possible to determine what propor-

tion of the diarrheal illnesses may be foodborne. This was the basis of the approach taken by Mead and colleagues in determining that 76 million foodborne illnesses occur each year in the United States (43).

Epidemiology also includes the study of factors associated with the occurrence of illnesses. The same population survey that FoodNet conducts to assess the frequency of diarrheal illness has also surveyed the population to determine how often people have potential exposures, such as eating undercooked hamburger, eating runny eggs, failing to wash their hands after handling raw chicken, or making a salad on the same cutting board they used to cut up raw chicken. Many of these same factors are likely to be assessed in the context of outbreak investigations to enable identification of the factors actually contributing to the occurrence of an outbreak.

The careful description of events in populations is generally called descriptive epidemiology. This process of determining characteristics of person, place, and time associated with illness occurrence forms the basis of all surveillance systems. The defining example of this process was the work of John Snow during the London cholera epidemic of 1860 (50). He located cases on a map of London, saw a cluster around the Broad Street pump, and pulled the pump handle. The contaminated water source was shut down and the outbreak ended. Such dramatic examples of consequential epidemiology are not common. However, surveillance of foodborne disease, based on the principles of descriptive epidemiology, remains a cornerstone on which other food safety activities are built.

Surveillance is one function of public health agencies. In mid-September 1994, a microbiologist at the Minnesota Department of Health's (MDH) Public Health Laboratory noted an increase in the number of *Salmonella enterica* serovar Enteritidis isolates being identified. MDH epidemiologists began to track the age, gender, and county of residence of the cases. After tracking new cases for several weeks, it was apparent that they were clustered in southeastern Minnesota. Epidemiologists concluded that a regional outbreak was occurring (28).

In contrast to the mere description of events, the comparison of different rates at which they occur between groups is known as analytical epidemiology. This involves determining a measure of association and a measure of the variability, or uncertainty, in the measurement of the point estimate of that association. To learn the cause of an outbreak, etiologic studies must determine the difference in risk of illness among those exposed to the "contaminated" food compared to those who were not exposed. This is customarily presented as an odds ratio or risk ratio, depending on the form of the epidemiologic study design.

The classic tool of analytical epidemiology for foodborne diseases is the use of a case-control study to identify the source of an outbreak. Case-control studies are frequently used in foodborne outbreak investigations because it is generally easier to identify ill persons (cases) than to identify the entire group at risk. Outbreak investigations initiated from pathogen-specific surveillance will typically include case-control studies. After concluding that a regional outbreak was occurring in southeastern Minnesota, epidemiologists constructed a questionnaire to ascertain what the individuals had eaten in the 5 days before they became ill. The same questionnaire was then used to interview healthy individuals of about the same age from the same communities. By comparing the responses of those who were ill to the responses of the healthy individuals who served as controls, the investigations determined that the outbreak was due to consumption of commercially manufactured ice cream (28). Ten (67%) of 15 cases but only 2 (13%) of 15 controls had eaten Schwan's ice cream (matched odds ratio [OR] = 10.0; 95% confidence interval [CI] = 1.4 to 434). The magnitude of the odds ratio is a measure of the strength of the association between illness and exposure. The strength of this association left little doubt that the ice cream was the source of the outbreak. A public health intervention was initiated days before *Salmonella* was actually isolated from the suspected product. In addition, the results of the case-control study suggested that the problem likely involved multiple ice cream flavors and days of production. In response a nationwide product recall was initiated. Further investigation led to the determination that the ice cream premix had been contaminated by raw egg residues in tanker trailers used to haul the premix from two other suppliers. What had initially appeared to be a regional outbreak in southeastern Minnesota ultimately involved an estimated 224,000 illnesses, with cases reported from 41 states (28).

Another closely related tool of analytical epidemiology is the cohort study. In outbreak investigations, retrospective cohort studies are conducted when an entire group (or population) with a common exposure can be identified. Individuals can be interviewed without prior knowledge of whether they were ill or not. Identifying groups based on their exposure status rather than their illness status enables researches to directly calculate a risk ratio for specific exposures. The risk ratio is the percentage of exposed persons who were ill divided by the percentage of unexposed persons who were ill. For example, if 100 people were exposed and 50 became ill,

the attack rate is 50%. However, if 5 of 100 people who were not exposed became ill, the risk ratio would be 50% divided by 5%. Thus, the risk ratio would be 10, meaning that exposed individuals were 10 times more likely to become ill than were individuals who were not exposed.

Typical settings for cohort studies involve banquets, wedding receptions, and events where a list of all persons attending can be obtained. For specific exposures, such as a particular food item, the percentage of persons who became ill after eating the food can be determined. Thus, in addition to implicating the food item, it is possible to determine the attack rate among persons who ate it. Information on attack rates may also help in evaluating the factors contributing to the occurrence of the outbreak. For example, for many bacterial foodborne agents, the attack rate may be related to the exposure dose. A high attack rate may imply high levels of contamination, such as may occur after prolonged temperature abuse.

In both case-control and cohort studies, much attention is paid to the precision of the calculated estimate, as measured by the P value or 95% confidence interval. Most conventions treat a P value of <0.05 as being significant. Similarly, 95% confidence intervals are expected to exclude 1.0 when a significant exposure occurs. While these criteria have served as useful guides to data analysis over the years, it must always be noted that the main contributor to the P value and confidence interval is the size of the sample. The larger the sample size, the more likely that a statistically significant result will be obtained. Investigators should not discard potentially important associations just because they fail to achieve a P value of <0.05. Conversely, "highly significant" results may be totally spurious. A greater concern is that bias in sampling may have introduced reasons other than contamination of food for the observed difference in risk of illness. When potential sources of bias are properly controlled and accounted for, epidemiologic methods have demonstrated an impressive ability to accurately identify the source and contributing causes of foodborne outbreaks.

As powerful as epidemiologic methods are, they do have limitations. In an outbreak setting where everyone ate a contaminated food item, there is no opportunity to demonstrate a difference in risk of illness between those who were exposed and those who were not. Alternative approaches, such as determining the amount eaten or when it was eaten, may allow some epidemiologic discrimination, but results are not likely to be "significant." For example, in outbreaks of shigellosis associated with parsley, parsley was associated with illness at one restaurant in Minnesota but not at a second restaurant. Chopped parsley was used on so many dishes at the second restaurant that most of the cases and the controls ate it (11).

In epidemiologic studies, the absence of an association does not necessarily mean that the food item or exposure in question is safe. For example, in the FoodNet case-control study of sporadic *E. coli* O157:H7 infections, there was no association between sporadic infections and consumption of unpasteurized apple cider or attendance at day care (35). Both have been associated with outbreaks of illness and remain important potential sources of infection. Although they were not identified as the primary sources of sporadic *E. coli* O157:H7, it would be misleading to interpret the results to imply that there is no risk.

The concepts and methods of epidemiology can be used to examine the relationships between disease and all levels of food safety, from production and distribution to preparation and consumption.

AGENTS AND TRANSMISSION

Information about specific agents is contained in their respective chapters in this volume. However, in discussing the epidemiology of foodborne illnesses it is important to keep in mind the chain of infection. This includes the agent, the reservoir that contains the agent, a means of escape from the reservoir, a mode of transmission to a susceptible host, and a means of entry to the host. These elements define the agent-host environment that results in the occurrence of illness.

Specific agents vary with respect to the types of illness they cause, from self-limited gastroenteritis associated with bacterial enterotoxins to invasive and life-threatening diseases associated with *Listeria monocytogenes*. Some agents, such as Norwalk-like virus, are rapidly cleared from the gastrointestinal tract, whereas others, such as *Giardia*, may cause prolonged carriage. An infectious dose may range from ingestion of a few *E. coli* O157:H7 to hundreds of thousands of *Clostridium perfringens*. Incubation periods range from a few hours for staphylococcal intoxication to weeks for viral hepatitis type A. Some agents, such as *Shigella* spp., infect only humans, while *Salmonella* spp., *Campylobacter*, spp., and *E. coli* O157:H7 primarily come from animal sources. Many agents are associated with typical food vehicles, such as *E. coli* O157:H7 with ground beef, and *Salmonella enterica* serovar Enteritidis with eggs.

The characteristics of specific agents affect the potential sources of contamination during food production and preparation and form the basis for hazard analysis for specific foods. They also help shape the conduct of foodborne illness investigations and enable the investigation of source and transmission to be based on particular features of the disease.

FOODBORNE ILLNESS SURVEILLANCE

Surveillance involves the systematic collection of data with analysis and dissemination of results. Surveillance systems may be passive or active, national or regional in scope, or based on a sentinel system of individual sites designed to provide estimates that can be generalized. Types of systems, their use, and their limitations are described below.

Traditionally, there have been four purposes for conducting foodborne illness surveillance:

    (i)   identify, control and prevent outbreaks of foodborne illness;
    (ii) determine the causes of foodborne illness;
    (iii) monitor trends in the occurrence of foodborne illness; and
    (iv) quantify the magnitude of foodborne illness.

"Pulling the pump handle" is the goal of all public health epidemiologists conducting foodborne illness surveillance. Thus it is necessary to identify, control, and prevent outbreaks of foodborne illness. For products with a long, shelf life, such as frankfurters, ice cream, or toasted oats cereal, even a relatively long investigation can identify a contaminated product quickly enough to remove it from the marketplace in time to prevent many illnesses (9, 10, 28). On the other hand, for highly perishable products such as fresh fruits and vegetables, the outbreak will often have run its course before the food item can be implicated (6, 11, 23).

Even if it is impossible to directly intervene to control the outbreak, however, there is still value in determining the causes of foodborne illness. Hazard analysis and critical control point (HACCP) systems for protecting the safety of foods rely on accurate knowledge of potential hazards (45). Many of these hazards, in terms of specific agents, specific food ingredients, or various agent-food interactions, were originally identified as a result of foodborne illness surveillance. Because food sources and agents of foodborne illness are constantly changing, hazard analysis is an ongoing process that requires continuous support from public health surveillance of foodborne illnesses. Additionally, Mead and colleagues determined that the cause of most foodborne illnesses is unknown and much research remains to be done in this area (43). As the experience with the *Cyclospora* outbreaks from

Guatemalan raspberries reveals, surveillance of food-borne illness may be the only way to identify both new causes of foodborne illness and potential foodborne hazards.

A third traditional purpose for conducting surveillance of foodborne illness is to monitor trends in occurrence of foodborne illness. This is important to identify priorities for new food safety activities and to monitor the effectiveness of existing programs. For example, during the 1980s, the increased occurrence of sporadic *Salmonella* serovar Enteritidis infections and outbreaks in New England led to the identification of a new problem with *Salmonella* serovar Enteritidis contamination of grade A shell eggs (57). The problem has taken on international dimensions (51). In the United States, the U.S. Department of Agriculture (USDA) and the Food and Drug Administration (FDA) have worked with the egg industry to develop and implement a number of control strategies (31). The incidence of serovar Enteritidis infections in FoodNet sites declined 48% from 1996 to 1999, suggesting that these control strategies are beginning to have an impact (13).

As part of the Food Safety Initiative, FoodNet was also set up to quantify the magnitude of foodborne illnesses. Surveys of the population, physicians, and clinical laboratories linked to active surveillance of laboratory-confirmed infections provide the most comprehensive data available from which to estimate the magnitude of foodborne illnesses. For the first time, a generally recognized estimate will be available to serve as a basis for setting public health priorities and for evaluating the effectiveness of prevention measures. However, significant gaps remain in our knowledge base. Eighty-two percent of estimated foodborne illnesses cannot be attributed to known agents, and no systematic surveillance exists for Norwalk-like viruses, which are responsible for an estimated two-thirds of foodborne illnesses caused by a known pathogen (Table 20.1) (43).

METHODS OF SURVEILLANCE OF FOODBORNE ILLNESS

Surveillance of foodborne illness requires close collaboration among acute-disease epidemiologists, public health laboratories, and environmental health specialists to determine the likely sources of exposure to pathogens or toxins and route of food contamination. There are four primary components of surveillance of foodborne illness (Table 20.2).

The first is identification and reporting of outbreaks associated with events and establishments, usually by ill persons who were part of the outbreak. There is a bias toward detecting outbreaks at events, such as a wedding, where a large group gathers with a common experience and has some reason to discuss it later. People eating at a restaurant may never know that their illness was also experienced by other patrons, who are unknown to them.

Outbreaks associated with events and establishments require prompt and thorough investigation to identify both the agent and the source. These outbreaks are usually recognized because common symptoms, such as diarrhea and vomiting, occur and are reported to public health officials before a specific diagnosis has been established. Unfortunately, in many outbreak investigations, laboratory testing is not sought, not available, or not adequate to identify the causative agent. Of 2,751 outbreaks of foodborne illness reported to the CDC from 1993 to 1997, 1,873 (68%) were classified as of unknown etiology (47). However, many of these were likely outbreaks of Norwalk-like viral gastroenteritis, which is not detected by routine testing. Fankhauser and colleagues at CDC retrospectively confirmed the presence of Norwalk-like viruses in 90% of outbreaks of acute nonbacterial gastroenteritis (17). In Minnesota,

Table 20.1 Leading known causes of foodborne illness in the United States, and ranking based on available surveillance systems[a]

Agent	Estimated cases	Rank by surveillance system		
		Active	Passive	Outbreak
Norwalk-like virus	9,200,000	NA[b]	NA	NA
Campylobacter spp.	1,963,141	1	2	7
Salmonella, nontyphoidal	1,341,873	2	1	1
Clostridium perfringens	248,520	NA	NA	3
Giardia lamblia	200,000	3	7	NA

[a] Adapted from reference 43.
[b] NA, not available.

Table 20.2 Primary components of surveillance of foodborne illness

1. Investigation of outbreaks associated with events and establishments.
2. Pathogen-specific surveillance to identify clusters of cases caused by the same organism.
3. Determination of risk factors for sporadic cases of infection with common foodborne pathogens.
4. Population surveillance to determine the frequency of gastrointestinal illnesses, health care-seeking behavior, food consumption, and personal prevention measures.

outbreaks whose etiology could not be confirmed by laboratory testing have been characterized by the clinical and epidemiologic features of the reported illnesses. From 1981 to 1998, 120 (41%) of 295 confirmed foodborne outbreaks had clinical and epidemiologic features consistent with Norwalk-like viruses and distinct from outbreaks caused by known bacterial infections or toxins (15). These include a median incubation period between 24 and 48 h, a 12- to 60-h duration of symptoms, and a relatively high proportion of cases experiencing vomiting (25, 33).

Such epidemiologic profiling of outbreaks is useful to guide laboratory testing as well. Clinical laboratories and many public health laboratories do not test for the presence of enterotoxigenic *E. coli* (ETEC). However, if a high proportion of individuals became ill several days after a common exposure and experienced diarrhea and cramps but little vomiting or fever, the investigator should seek specialized laboratory testing for ETEC (14). Just such a clinical and epidemiologic profile led to the identification of an atypical enteropathogenic strain of *E. coli* O39 as the cause of a food-associated outbreak at a restaurant (26). Similarly, investigation of a series of event-associated outbreaks with unusually long median incubation periods of a week led to the recognition of *Cyclospora* as a foodborne pathogen (29).

The investigation of outbreaks associated with events is the oldest form of foodborne illness surveillance and is frequently overlooked in discussions of how to improve such surveillance. However, it remains the primary way to identify "new" foodborne pathogens, such as *Cyclospora* or diarrheagenic *E. coli*, which are not routinely identified by clinical laboratories.

A major focus of the National Food Safety Initiative has been to improve pathogen-specific surveillance (59). Serotype-specific surveillance of *Salmonella* spp. has identified many large, multistate outbreaks of salmonellosis; many were associated with previously unrecognized vehicles, most notably fresh fruits and vegetables (6, 23, 40). Outbreaks of salmonellosis caused by cantaloupes, tomatoes, and alfalfa sprouts were all recognized because of unusual temporal clusters of cases with an uncommon serotype. Epidemiologic investigation of these cases identified the source.

Cases of *Salmonella* infections are electronically reported to CDC from state public health laboratories through the Public Health Laboratory Information System (PHLIS). An automated *Salmonella* outbreak detection algorithm (SODA) has been developed to look for unusual case clusters based on the 5-year mean number of cases from the same geographic area and week of the year (32). Similar schemes have been developed

in Europe and Australia (56). Detection of clusters by SODA has helped confirm the existence of multistate outbreaks of *S. enterica* serovar Stanley infections associated with alfalfa sprouts in 1995 and serovar Agona infections associated with toasted oats cereal in 1998 (9, 40). However, both outbreaks were initially recognized by individual state health departments. Because it compares current cases to 5-year means, SODA appears to be most effective at detecting case clusters of uncommon serotypes. As with other applications of serotype-specific surveillance, SODA is not likely to be especially sensitive to detect outbreaks caused by common serotypes, such as *S. enterica* serovar Typhimurium.

Molecular subtyping schemes, such as pulsed-field gel electrophoresis (PFGE), can improve the investigation of outbreaks by distinguishing unrelated sporadic cases from the main outbreak-associated strain. This improves the specificity of the case definition and makes it more likely that the source of the outbreak can be identified. PFGE has been used in outbreaks of *E. coli* O157:H7, *Salmonella*, and *Shigella* (8, 9, 11). Molecular subtyping can also facilitate outbreak detection by changing serotype-specific surveillance into molecular subtype-specific surveillance. The public health utility of incorporating molecular subtyping into routine surveillance of *E. coli* O157:H7 at the state level has been demonstrated (3).

The ability to distinguish specific subtypes among relatively common pathogens, such as *E. coli* O157:H7 and *S. enterica* serovar Typhimurium, is the basis of the National Molecular Subtyping Network (PulseNet). PulseNet takes advantage of the combined revolutions in molecular biology and information technology. Highly reproducible PFGE patterns are generated under standardized conditions, and the PFGE patterns can be transmitted electronically between participating laboratories. For example, an outbreak of *E. coli* O157:H7 infections in Colorado was associated with consumption of a nationally distributed ground beef product. Within days, it was possible to compare the outbreak strain to PFGE patterns of *E. coli* O157:H7 isolates throughout the United States (8). This gives PulseNet the potential to be the "backbone" of a public health surveillance system that can provide truly national surveillance for a variety of foodborne pathogens in a manner timely enough to be an early warning system for outbreaks of foodborne illness.

Molecular subtyping can also be used to enhance the investigation of outbreaks associated with events or establishments, such as when multiple, apparently independent outbreaks of *Shigella sonnei* infections across the United States and Canada were shown to have been caused by a common strain (11). Although the independent investigations did not identify the cause, the linkage of the outbreaks allowed investigators to identify parsley that had been imported from Mexico as the common source.

All outbreak investigations require close collaboration with environmental health specialists and field investigators to identify and trace the source of food items that may have been a vehicle for the causative agent. Although product tracebacks are frequently started only after a food item has been implicated, detailed product information is actually critical to epidemiologic analysis. In the outbreak of S. enterica serovar Enteritidis associated with Schwan's ice cream, for example, if cases and controls had initially been asked only if they had eaten ice cream, investigators would never have identified the source of the outbreak (28). Thirteen (87%) cases and 10 (67%) controls ate some type of ice cream (OR = 2.5; 95% CI = 0.4 to 26.3). These data would not have raised suspicions that ice cream could be the source of the outbreak. Thus, collecting product source information should begin as early in the investigation as possible.

Determining risk factors for sporadic cases of infection with common foodborne pathogens is a third major component of foodborne illness surveillance. This practice can aid in identifying targets for intervention and provide a basis to evaluate its effectiveness. Case-control studies of sporadic S. enterica serovar Enteritidis infections in the United States and Europe helped establish the role of grade A shell eggs in the epidemiology of these infections. Case-control studies of Campylobacter infections helped establish chickens as a primary source of Campylobacter spp. (34, 58). Similarly, case-control studies of sporadic E. coli O157:H7 infections confirmed the importance of ground beef, particularly undercooked ground beef, as a source of the pathogen (37, 42, 55).

FoodNet has conducted or is planning population-based case-control studies for E. coli O157:H7, Salmonella serogroups B and D, Campylobacter, Cryptosporidium, and Listeria. These promise to provide updated or original estimates of the proportion of these infections that are attributable to specific food items. As previously noted, results of the E. coli O157:H7 case-control study demonstrated that eating hamburgers at a fast food restaurant was not associated with illness, as was the case in previous case-control studies (35). This appears to be due to the fact that fast food restaurants now thoroughly cook the hamburgers they serve. At home or at table service restaurants, where consumers may choose to order an undercooked hamburger, eating undercooked hamburgers was associ-

ated with illness (35). Preliminary results of case-control studies for S. enterica serovars Enteritidis and Heidelberg also demonstrated associations with eating out (27, 36). Other interesting findings included associations between serovar Enteritidis infection and chicken consumption and serovar Heidelberg and egg consumption.

The final important component of foodborne illness surveillance is population surveillance to determine the frequency of gastrointestinal illnesses, health care-seeking behavior, food consumption, and personal prevention measures. This type of syndrome-specific surveillance is not usually conducted by state or local health departments. However, these measures formed the basis of CDC's recent estimates that 76 million foodborne illnesses occur annually in the United States. These same data are also useful for understanding the results of active surveillance. By describing when and why people seek medical care when they are ill, these data can help to estimate how many illnesses occur for each one that receives a specific diagnosis. Finally, population-based food consumption data are being used to estimate the proportion of specific diseases attributable to specific exposures in the FoodNet case-control studies.

SURVEILLANCE FOR FOOD HAZARDS

Risk assessments for specific pathogens such as E. coli O157:H7 in ground beef or S. enterica serovar Enteritidis in shell eggs require making measurements or assumptions about many parameters. At every step from farm to table, potential sources of contamination can be identified, measured, and modeled to determine whether they contribute to the overall risk of foodborne illness.

For many foodborne pathogens of public health importance, contamination of the finished product is a sporadic and rare occurrence. Hence, end-product testing is difficult to justify as a prevention measure. Given the scale of modern agriculture and food production systems, contamination levels below the statistical and microbiological threshold can still be large public health problems. For example, internal contamination of eggs with S. enterica serovar Enteritidis is rare—approximately 1 in 20,000 eggs on a national basis in the United States, with a range from 2.5 to 62.5 per 10,000 eggs from environmentally positive flocks (54). Despite this sporadic occurrence, shell eggs or foods made with shell eggs accounted for 137 (36%) of 380 outbreaks of serovar Enteritidis infections reported to CDC from 1985 to 1991 (44). Similarly, ground beef remains an important source of E. coli O157:H7 infections, although E. coli O157:H7 was isolated from fewer than 1 in 1,000 portions of ground beef sampled at the retail level by USDA from 1995 to 1998 (21). In 1999, the level of positive samples jumped to 2.5 per 1,000. This increase was likely due to more sensitive testing methods rather than more contamination of the finished product.

For this reason, surrogate measures of contamination, such as identifying infected hens or environmental contamination in egg laying houses, are more efficient and effective than testing individual eggs. Alternatively, measuring a relatively common microorganism such as commensal strains of *E. coli* may provide surveillance for control of operations that can be independently associated with degree of risk for disease transmission. For example, in a series of baseline surveys of beef slaughter plants and ground beef conducted by USDA in preparation for their introduction of HACCP regulation in the meat industry, 8.2% of steer and heifer carcasses and 15.8% of cow and bull carcasses were contaminated by commensal *E. coli* (18, 19). However, only 4 (0.2%) of 2,081 steer and heifer carcasses and none of 2,112 cow and bull carcasses were contaminated with *E. coli* O157:H7. Of 563 finished ground beef samples, 78.6% were contaminated by commensal *E. coli* but none by *E. coli* O157:H7 (20). Thus, microbiological testing for commensal *E. coli* can provide useful information for monitoring HACCP control points, even though preventing contamination by *E. coli* O157:H7 is the desired outcome.

In addition to surveys of food products conducted by USDA and FDA, USDA implemented the National Animal Health Monitoring System (NAHMS) in 1983 to conduct national studies and compile data from industry sources to address emerging issues such as the association between calf management practices and the presence of *E. coli* O157:H7 and *Salmonella* spp. in cattle herds (22, 39). The Veterinary Diagnostic Laboratory Reporting System compiles and analyzes reports from state veterinary diagnostic laboratories to assess trends in infectious diseases among food animals (53).

Behavioral risk factors for foodborne illness range from choosing to eat alfalfa sprouts or undercooked ground beef to failing to wash hands between using a toilet and making a salad. These factors are evaluated and can be identified during the outbreak investigations. The frequency of these behaviors in the general population can be assessed through surveys, such as that being conducted as part of FoodNet. For example, 7.7% of respondents to the FoodNet population survey conducted from 1998 to 1999 reported eating alfalfa sprouts during the 7 days before the interview (12). However, 10.5% of California and Oregon residents reported eating sprouts. Thus, it is not entirely sur-

prising that sprout-associated outbreaks of salmonellosis are more common on the West Coast than in the rest of the United States (62). Among other potentially important food exposures, 25% of persons who ate eggs, consumed runny eggs; 11% of persons who ate hamburgers, ate hamburgers that were pink; 4.4% of respondents drank unpasteurized apple juice or apple cider; 3.4% drank unpasteurized milk; and 2.5% ate fresh oysters (12). Among persons who handled raw chicken, 88.6% reported washing their hands with soap and water after touching the raw chicken, but only 48.3% of respondents had a meat thermometer in the home (12). Information of this type can provide important perspectives for understanding the occurrence of outbreaks of foodborne illness and sporadic infections. It also serves as the basis for developing and modifying public education campaigns to improve food safety, such as the Partnership for Food Safety Education's national public education campaign, Fight BAC! (5).

Factors contributing to the occurrence of outbreaks of foodborne illness are compiled by CDC from national surveillance data and by individual states. National data are typically published in summaries covering 5-year periods. These include detailed tables, by year, but very little synthesis or trend analysis. For example, in outbreaks for which contributing factors were reported between 1988 and 1997, the two most commonly reported contributing factors were improper holding temperatures (60%) and poor personal hygiene (34%) (1, 47). During the first 5 years of this period, poor personal hygiene declined as a contributing factor, from 44% of outbreaks in 1988 to 29% in 1992. During the second 5-year period, the role of poor personal hygiene increased from 27% of outbreaks in 1993 to 38% in 1997. It would be useful to know whether these represented actual trends in the occurrence of foodborne illness outbreaks, trends in the way outbreaks are investigated and reported, or mere variations in passively collected data that do not correlate with the actual occurrence of foodborne illnesses. Unfortunately, contributing factors were reported for only 58% of all outbreaks reported to CDC, and not all outbreaks were reported (1, 47). For example, there was a 24% decline in the number of outbreaks reported between 1995 and 1996. Minnesota reported 20 outbreaks in 1995 and 1 in 1996. Similarly, New York reported 89 outbreaks in 1995 and 9 in 1996. However, these apparent declines represent underreporting to the passive outbreak surveillance system rather than fewer actual foodborne illness outbreaks in these states. Bias in the type of information reported is the greatest systematic problem for interpreting national foodborne disease outbreak surveillance data.

The New York State Health Department has been a leader in attempting to quantify environmental and behavioral factors contributing to outbreaks of foodborne illness. In contrast to the five categories of contributing factors historically tracked by CDC, the New York State Department of Health's Bureau of Community Sanitation and Food Protection expanded the list to 19 contributing factors and began to distinguish foods based on both methods of preparation and significant ingredients (64). These have been further expanded and divided into 14 factors that could contribute to contamination of foods, 5 factors associated with survival or lack of inactivation, and 12 factors associated with proliferation and amplification (4). The CDC is now adapting this scheme for incorporation into the national foodborne illness outbreak reporting system; however, much information will be submitted on an optional basis.

These refined summaries should provide better information on conditions found to contribute to the occurrence of individual outbreaks. However, they will not provide an estimate of the proportion of outbreaks attributable to the contributing factor identified. For example, in investigating an outbreak of shigellosis in a restaurant, it may be observed that food preparers routinely failed to wash their hands after returning from the toilet. This would likely be considered a contributing factor to the occurrence of the outbreak. To put this in perspective, however, it would be useful to know in general how often food workers in restaurants fail to wash their hands properly. The proportion of *Shigella* outbreaks attributable to improper handwashing is not simply the sum of all outbreaks in which it was identified as a contributing factor. Although outbreaks of shigellosis have usually occurred because of errors in final food preparation, recent outbreaks associated with imported parsley highlight the importance of not drawing conclusions about the source before the investigation is complete.

In addressing likely food hazards, it is first necessary to determine the proportion of illnesses and outbreaks attributable to the hazard and then to estimate the predictive value of the hazard for causing disease. This is part of the risk assessment process and an important application of foodborne illness epidemiology (48). The frequency of egg contamination is x, the proportion of *Salmonella* serovar Enteritidis cases attributable to eggs is y. The likelihood of developing serovar Enteritidis infection after eating an undercooked egg is z. We need to provide estimates for x, y, and z. In a restaurant setting, we may know how many times failure to wash hands is identified as a contributing factor in an outbreak of foodborne illness, but what is the frequency of this failure in restaurants? The FDA began to address this data need in 1997 with the creation of the FDA Retail Food Program Database of Foodborne Illness Risk Factors. Baseline data were obtained from 895 inspections of institutional food service establishments, restaurants, and retail food stores conducted by FDA's Regional Retail Food Specialists (http://vm.cfsan.fda.gov/~dms/retrsk.html). What proportion of outbreaks are attributable to this problem? FoodNet is pilot testing a case-control study of outbreaks in restaurants to address this need for data. Finally, how likely is an outbreak to occur if an employee fails to wash his or her hands? These questions can all be addressed by using epidemiologic methods applied to surveillance for the hazards.

EVALUATING FOOD SAFETY SYSTEMS

Public health surveillance of foodborne illness is critical to the performance of food safety systems that are based on HACCP plans. Surveillance is required to identify new hazards. It also provides the ultimate feedback on the efficacy of HACCP plans. Recalling the outbreak of salmonellosis due to commercial ice cream, cross contamination during transportation was identified as a hazard that had not been addressed in the manufacturer's HACCP plan had. Epidemiologic data can also measure the success of food safety interventions. After implementation of the nationwide recall of Schwan's ice cream, a survey of Schwan's customers in Georgia revealed that 50% had not learned about the recall until 5 days after it was issued, and 9% did not learn about it at all (41). Furthermore, 26% of those who learned about the recall did not initially believe the products were unsafe, and many subsequently ate the ice cream. Of an estimated 11,000 cases of illness in Georgia associated with this outbreak, 16% were exposed after the recall (41).

As noted earlier, case-control studies of sporadic *E. coli* O157:H7 conducted before and after the Jack-in-Box outbreak of 1993 demonstrated a change in the risk of illness associated with eating hamburgers at fast food restaurants. Recommendations made following the outbreak appeared to reduce the risk of eating hamburgers in fast food restaurants. This appears to be due to the fact that these restaurants now cook hamburger more thoroughly, because cattle continue to be a primary reservoir for *E. coli* O157:H7. Moreover, the improved sensitivity of testing for *E. coli* O157:H7 has provided us a more complete picture of its abundance. In 1999, sampling of animals and carcasses at four midwestern beef processing 0plants identified *E. coli* O157:H7 in 27.8% of fecal samples and on 10.7% of hides (16). Carcass contamination decreased from 43.4% of pre-evisceration samples

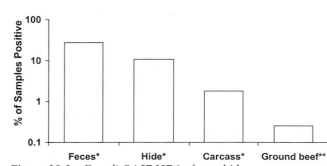

Figure 20.2 *E. coli* O157:H7 in feces, hides, carcasses, and retail ground beef samples. *, adapted from reference 16; **, adapted from reference 21.

to 1.8% of postprocessing samples. These results confirm the widespread distribution of *E. coli* O157:H7 among cattle and the effectiveness of the sanitary procedures used to reduce contamination in processing plants (16). However, these results, together with those of ground beef sampling at retail stores, and the continuing identification of undercooked ground beef as an important cause of sporadic human *E. coli* O157:H7 infections also demonstrate the limits of these procedures to prevent contamination of the finished product (Fig. 20.2). The consistency of available microbiological and epidemiologic data continues to support the need for terminal control treatments such as cooking or irradiation to further reduce the risks of *E. coli* O157:H7 infections.

The FoodNet surveillance network was established, in part, to provide a tool to evaluate the public health impact of the USDA's implementation of meat processing plant HACCP. Surveillance data from the five original FoodNet sites provide consistent, population-based measures of the incidence of *Campylobacter*, *E. coli* O157:H7, *Listeria*, *Salmonella*, *Shigella*, *Vibrio*, and

Yersinia infections. In aggregate, the incidence of these infections declined from 51.2 cases per 100,000 population in 1996 to 50.3 in 1997, 47.2 in 1998, and 40.7 cases per 100,000 population in 1999 (13). It has been noted that this declining incidence occurred at the same time as implementation of changes in meat and poultry processing plants, revisions to the FDA Food Code for restaurants, and increased attention to good agricultural practices for fresh fruits and vegetables and eggs on farms (13). In fact, these are just the types of effects that FoodNet was designed to measure.

However, determining cause and effect is not as simple as measuring the trends. To begin with, the FoodNet sites exhibit considerable regional variability in the incidence of specific infections. For example, in 1996, the reported incidence of *Campylobacter* infections ranged from 58 per 100,000 in California to 14 per 100,000 in Georgia (7). Similarly, the incidence of *E. coli* O157:H7 infections ranged from 5 per 100,000 in Minnesota to 0.6 per 100,000 in Georgia (7). Although the rates of *Salmonella* infections did not vary much between sites, the distribution of serotypes did; *S. enterica* serovar Enteritidis was the most common serotype in Connecticut but was rare in Georgia.

Overall, the incidence of *Campylobacter* infections in FoodNet declined from 23.5 per 100,000 in 1996 to 17.3 per 100,000 in 1999 (Fig. 20.3). However, a large portion of that overall trend was due to marked declines in the incidence of *Campylobacter* infections in California, from 58 per 100,000 in 1996 to 33 per 100,000 in 1999. Among the other sites, *Campylobacter* rates remained stable or declined slightly. Results of a FoodNet case-control study for sporadic *Campylobacter* infections will help provide

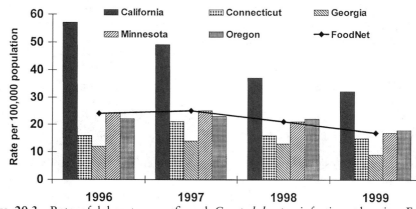

Figure 20.3 Rate of laboratory-confirmed *Campylobacter* infections, by site, FoodNet, 1996–1999.

answers to the question of whether the observed decreases in *Campylobacter* infections may actually be due to regulatory changes that were implemented by USDA. In the meantime, it is important to remember that the (unknown) underlying factors that contribute to the regional variability of these infections may also lead to regional differences in how they respond to prevention and control measures.

SUMMARY

Humans remain the ultimate bioassay for low-level or sporadic contamination of our food supply (24). Epidemiologic methods of surveillance of foodborne illness are powerful tools because they take advantage of events that are occurring throughout the population. This population-based lens, focused by advances in molecular subtyping and information technology available to public health laboratories, is particularly well suited to associating foodborne illnesses with mass-produced and widely distributed food products.

Epidemiologic methods of foodborne illness surveillance are needed to detect outbreaks. In outbreak settings, rapid determination of detailed information on food source and production is needed to confirm the cause of the outbreak and identify contributing factors. Epidemiologic data are also very important to establish food safety priorities, allocate food safety resources, stimulate public interest in food safety issues, establish risk reduction strategies and public education campaigns, and evaluate the effectiveness of food safety programs (61). The principles and methods of epidemiologic surveillance can be applied to surveillance for both food hazards and foodborne illnesses.

Although this chapter concentrated on experiences in the United States, the same methods of observation and analysis should form the basis of foodborne illness surveillance in developed and developing countries throughout the world. Since national food supplies are rapidly becoming global in origin, the need exists for an international system for foodborne illness surveillance as well. Models such as PulseNet provide opportunities to conduct multinational surveillance for at least the major bacterial agents of foodborne illness. Because foodborne illness problems imported into countries such as the United States may represent endemic disease problems in the producing country, a growing awareness of these problems could stimulate investment in interventions such as safe water distribution and sanitary waste disposal systems that would benefit both the local population and consumers thousands of miles away.

References

1. Bean, N. H., J. S. Goulding, C. Lao, and F. J. Angulo. 1996. Surveillance for foodborne-disease outbreaks—United States, 1988–1992. *Morbid. Mortal. Weekly Rep.* 45(SS-5):1–55.
2. Bell, B. P., M. Goldoft, P. M. Griffin, M. A. Davis, D. C. Gordon, P. I. Tarr, C. A. Bertelson, J. H. Lewis, T. J. Barrett, J. G. Wells, R. Baron, and J. Kobayashi. 1994. A multistate outbreak of *Escherichia coli* O157:H7-associated bloody diarrhea and hemolytic uremic syndrome from hamburgers: the Washington experience. *JAMA* 272:1349–1353.
3. Bender, J. B., C. W. Hedberg, J. M. Besser, D. J. Boxrud, K. L. MacDonald, and M. T. Osterholm. 1997. Surveillance for *Escherichia coli* O157:H7 infections in Minnesota by molecular subtyping. *N. Engl. J. Med.* 337:388–394.
4. Bryan, F. L., J. J. Guzewich, and E. C. D. Todd. 1997. Surveillance of foodborne disease. III. Summary and presentation of data on vehicles and contributory factors; their value and limitations. *J. Food Prot.* 60:701–714.
5. Carr, C. J., and F. C. Lu. 1998. Partnership for food safety education—"Fight BAC!" *Regul. Toxicol. Pharmacol.* 27:281–282.
6. Centers for Disease Control. 1991. Multistate outbreak of *Salmonella poona* infections—United States and Canada, 1991. *Morbid. Mortal. Weekly Rep.* 40:549–552.
7. Centers for Disease Control and Prevention. 1997. Foodborne diseases active surveillance network, 1996. *Morbid. Mortal. Weekly Rep.* 46:258–261.
8. Centers for Disease Control and Prevention. 1997. *Escherichia coli* O157:H7 infections associated with eating a nationally distributed commercial brand of frozen ground beef patties and burgers—Colorado, 1997. *Morbid. Mortal. Weekly Rep.* 46:777–778.
9. Centers for Disease Control and Prevention. 1998. Multistate outbreak of *Salmonella* serotype Agona infections linked to toasted oats cereal—United States, April–May, 1998. *Morbid. Mortal. Weekly Rep.* 47:462–464.
10. Centers for Disease Control and Prevention. 1998. Multistate outbreak of listeriosis—United States, 1998. *Morbid. Mortal. Weekly Rep.* 47:1085–1086.
11. Centers for Disease Control and Prevention. 1999. Outbreaks of *Shigella sonnei* infection associated with eating fresh parsley—United States and Canada, July–August, 1998. *Morbid. Mortal. Weekly Rep.* 48:285–289.
12. Centers for Disease Control and Prevention. 1999. Foodborne diseases active surveillance network (FoodNet): population survey atlas of exposures: 1998–1999. Centers for Disease Control and Prevention, Atlanta, Ga.
13. Centers for Disease Control and Prevention. 2000. Preliminary FoodNet data on the incidence of foodborne illnesses—selected sites, 1999. *Morbid. Mortal. Weekly Rep.* 49:201–205.
14. Dalton, C. B., E. D. Mintz, J. G. Wells, C. A. Bopp, and R. V. Tauxe. 1999. Outbreaks of enterotoxigenic *Escherichia coli* infection in American adults: a clinical and epidemiologic profile. *Epidemiol. Infect.* 123:9–16.

15. Deneen, V. C., J. M. Hunt, C. R. Paule, R. I. James, R. G. Johnson, M. J. Raymond, and C. W. Hedberg. 2000. The impact of foodborne calicivirus disease: the Minnesota experience. *J. Infect. Dis.* **181**(S2):S281–S283.

16. Elder, R. O., J. E. Keen, G. R. Siragusa, G. A. Barkocy-Gallagher, M. Koohmaraie, and W. W. Laegreid. 2000. Correlation of enterohemorrhagic *Escherichia coli* O157 prevalence in feces, hides, and carcasses of beef cattle during processing. *Proc. Natl. Acad. Sci. USA* **97**:2999–3003.

17. Fankhauser, R. L., J. S. Noel, S. S. Monroe, T. Ando, and R. I. Glass. 1998. Molecular epidemiology of "Norwalk-like viruses" in outbreaks of gastroenteritis in the United States. *J. Infect. Dis.* **178**:1571–1578.

18. Food Safety and Inspection Service. 1994. Nationwide beef microbiological data collection program: steers and heifers. U.S. Department of Agriculture, Washington, D.C.

19. Food Safety and Inspection Service. 1996. Nationwide beef microbiological data collection program: cows and bulls. U.S. Department of Agriculture, Washington, D.C.

20. Food Safety and Inspection Service. 1996. Nationwide federal plant ground beef microbiological survey. U.S. Department of Agriculture, Washington, D.C.

21. Food Safety and Inspection Service. 2000. Microbiological testing program for *Escherichia coli* O157:H7 in raw ground beef, 1995 through 1999 data. U.S. Department of Agriculture, Washington, D.C. http://www.fsis.usda.gov/OPHS/ecoltest.

22. Garber, L. P., S. J. Wells, D. D. Hancock, M. P. Doyle, J. Tuttle, J. A. Shere, and T. Zhao. 1995. Risk factors for fecal shedding of *Escherichia coli* O157:H7 in dairy calves. *J. Am. Vet. Med. Assoc.* **207**:46–49.

23. Hedberg, C. W., F. J. Angulo, K. E. White, C. W. Langkop, W. L. Schell, M. G. Stobierski, A. Schuchat, J. M. Besser, S. Dietrich, L. Helsel, P. M. Griffin, J. W. McFarland, M. T. Osterholm, and the Investigation Team. 1999. Outbreaks of salmonellosis associated with eating uncooked tomatoes: implications for public health. *Epidemiol. Infect.* **122**:385–393.

24. Hedberg, C. W., K. L. MacDonald, and M. T. Osterholm. 1994. Changing epidemiology of foodborne disease: a Minnesota perspective. *Clin. Infect. Dis.* **18**:671–682.

25. Hedberg, C. W., and M. T. Osterholm. 1993. Outbreaks of foodborne and waterborne viral gastroenteritis. *Clin. Microbiol. Rev.* **6**:199–210.

26. Hedberg, C. W., S. J. Savarino, J. M. Besser, C. J. Paulus, V. M. Thelen, L. J. Meyers, D. N. Cameron, T. J. Barrett, J. B. Kaper, M. T. Osterholm, and the Investigation Team. 1997. An outbreak of foodborne illness caused by *Escherichia coli* O39:NM, an agent not fitting into the existing scheme for classifying diarrheogenic *E. coli*. *J. Infect. Dis.* **176**:1625–1628.

27. Hennessy, T., L. Cheng, H. Kassenborg, S. Desai, J. Mohle-Boetani, R. Marcus, W. Shiferaw, F. Angulo, and the Food-Net Working Group. 1998. Eggs identified as a risk factor for sporadic *Salmonella* serotype Heidelberg infection: a case-control study in FoodNet sites. *36th Annual Meeting of the Infectious Disease Society of America,* Denver, Colo., November 12–15, 1998. Infectious Diseases Society of America, Alexandria, Va.

28. Hennessy, T. W., C. W. Hedberg, L. Slutsker, K. E. White, J. M. Besser-Wiek, M. E. Moen, J. Feldman, W. W. Coleman, L. M. Edmonson, K. L. MacDonald, M. T. Osterholm, and the Investigation Team. 1996. A national outbreak of *Salmonella enteritidis* infections from ice cream. *N. Engl. J. Med.* **334**:1281–1286.

29. Herwaldt, B. L., M. L. Ackers, and the *Cyclospora* Working Group. 1999. An outbreak in 1996 of cyclosporiasis associated with imported raspberries. *N. Engl. J. Med.* **336**:1548–1556.

30. Herwaldt, B. L., M. J. Beach, and the *Cyclospora* Working Group. 1999. The return of *Cyclospora* in 1997: another outbreak of cyclosporiasis in North America associated with imported raspberries. *Ann. Intern. Med.* **130**:210–220.

31. Hogue, A., P. White, J. Guard-Petter, W. Schlosser, R. Gast, E. Ebel, J. Farrar, T. Gomez, J. Madden, M. Madison, A. M. McNamara, R. Morales, D. Parham, P. Sparling, W. Sutherlin, and D. Swerdlow. 1997. Epidemiology and control of egg-associated *Salmonella enteritidis* in the United States of America. *Rev. Sci. Technol.* **16**:542–553.

32. Hutwagner, L. C., E. K. Maloney, N. H. Bean, L. Slutsker, and S. M. Martin. 1997. Using laboratory-based surveillance data for prevention: an algorithm for detecting *Salmonella* outbreaks. *Emerg. Infect. Dis.* **3**:395–400.

33. Kaplan, J. E., G. W. Gary, R. C. Baron, N. Singh, L. B. Schonberger, R. Feldman, and H. B. Greenberg. 1982. The epidemiology of Norwalk gastroenteritis and the role of Norwalk virus in outbreaks of acute nonbacterial gastroenteritis. *Ann. Intern. Med.* **96**:756–761.

34. Kapperud, G., L. M. Rorvik, V. Hasseltvedt, E. A. Hoiby, B. G. Iversen, K. Staveland, G. Johnson, J. Leitao, H. Herikstad, Y. Andersson, G. Langeland, B. Gondrosen, and J. Lassen. 1992. Risk factors for sporadic *Campylobacter* infections: results of a case-control study in southeastern Norway. *J. Clin. Microbiol.* **30**:3117–3121.

35. Kassenborg, H., C. Hedberg, M. Evans, G. Chin, T. Fiorentino, D. Vugia, M. Bardsley, L. Slutsker, and P. Griffin. 1998. Case-control study of sporadic *Escherichia coli* O157:H7 infections in 5 FoodNet sites (CA, CT, GA, MN, OR). *First International Conference on Emerging Infectious Diseases,* Atlanta, Ga., March 8–11, 1998. American Society for Microbiology, Washington, D.C. http://www.cdc.gov/foodnet/pub/iceid/1998/kassenborg_h.htm

36. Kimura, A., S. Reddy, R. Marcus, P. Cieslak, J. Mohle-Boetani, H. Kassenborg, S. Segler, D. Swerdlow, and the FoodNet Working Group. 1998. Chicken, a newly identified risk factor for sporadic *Salmonella* serotype Enteritidis infections in the United States: a case-control study in FoodNet sites. *36th Annual Meeting of the Infectious Disease Society of America,* Denver, Colo., November 12–15, 1998. Infectious Diseases Society of America, Alexandria, Va. http://www.cdc.gov/foodnet/pub/idsa/1998/kimura_a.htm

37. Le Saux, N. J. S. Spika, B. Friesen, I. Johnson, D. Mlnychuck, C. Anderson, R. Dion, M. Rahman, and W. Tostowarky. 1993. Ground beef consumption in

non-commercial settings is a risk factor for sporadic *Escherichia coli* O157:H7 infection in Canada. *J. Infect. Dis.* **167**: 500–502.

38. Li, R., M. Serdula, S. Bland, A. Mokdad, B. Bowman, and D. Nelson. 2000. Trends in fruit and vegetables consumption among adults in 16 U.S. states: behavioral risk factor surveillance system. *Am. J. Public Health* **90**:777–781.

39. Losinger, W. C., S. J. Wells, L. P. Garber, H. S. Hurd, and L. A. Thomas. 1995. Management factors related to *Salmonella* shedding by dairy heifers. *J. Dairy Sci.* **78**:2474–2472.

40. Mahon, B. E., A. Ponka, W. N. Hall, K. Komatsu, S. E. Dietrich, A. Siitonen, G. Cage, P. S. Hayes, M. A. Lambert-Fair, N. H. Bean, P. M. Griffin, and L. Slutsker. 1997. An international outbreak of *Salmonella* infections caused by alfalfa sprouts grown from contaminated seeds. *J. Infect. Dis.* **175**:876–882.

41. Mahon, B. E., L. Slutsker, L. Hutwagner, C. Drenzek, K. Maloney, K. Toomey, and P. M. Griffin. 1999. Consequences in Georgia of a nationwide outbreak of *Salmonella* infections: what you don't know might hurt you. *Am. J. Public Health* **89**:31–35.

42. Mead, P. S., L. Finelli, M. A. Lambert-Fair, D. Champ, J. Townes, L. Hutwagner, T. Barrett, K. Spitalny, and E. Mintz. 1997. Risk factors for sporadic infection with *Escherichia coli* O157:H7. *Arch. Intern. Med.* **157**:204–208.

43. Mead, P. S., L. Slutsker, V. Dietz, L. F. McCaig, J.S. Bresee, C. Shapiro, P. M. Griffin, and R. V. Tauxe. 1999. Food-related illness and death in the United States. *Emerg. Infect. Dis.* **5**:607–625.

44. Mishu, B., J. Koehler, L. A. Lee, D. Rodrigue, F. H. Brenner, P. Blake, and R. V. Tauxe. 1994. Outbreaks of *Salmonella enteritidis* infections in the United States, 1985–1991. *J. Infect. Dis.* **169**:547–552.

45. National Advisory Committee on Microbiological Criteria for Foods. 1998. Hazard analysis and critical control point principles and application guidelines. *J. Food Prot.* **61**:762–765.

46. National Advisory Committee on Microbiological Criteria for Foods. 1998. Principles of risk assessment for illness caused by foodborne biological agents. *J. Food Prot.* **61**:1071–1074.

47. Olsen, S. J., L. C. MacKinon, J. S. Goulding, N. H. Bean, and L. Slutsker. 2000. Surveillance for foodborne-disease outbreaks—United States, 1993–1997. *Morbid. Mortal. Weekly Rep.* **49**(SS-1):1–51.

48. Potter, M. E. 1996. Risk assessment terms and definitions. *J. Food Prot.* **57**(Suppl.):6–9.

49. Potter, M. E., S. Gonzalez Ayala, and N. Silarug. 1996. Epidemiology of foodborne diseases, p. 376–390. *In* M. P. Doyle, L. R. Beuchat, and T. J. Montville (ed.), *Food Microbiology: Fundamentals and Frontiers.* American Society for Microbiology, Washington, D.C.

50. Richardson, B. W., and W. H. Frost. 1936. *Snow on Cholera.* The Commonwealth Fund, New York, N.Y.

51. Rodrigue, D. C., R. V. Tauxe, and B. Rowe. 1990. International increase in *Salmonella enteritidis*: a new pandemic? *Epidemiol. Infect.* **105**:21–27.

52. Rolls, B. J., and E. A. Bell. 2000. Dietary approaches to the treatment of obesity. *Med. Clin. North Am.* **84**:401–418.

53. Salman, M. D., G. R. Frank, D. W. MacVean, J. S. Reif, J. K. Collins, and R. Jones. 1988. Validation of disease diagnoses reported to the National Animal Health Monitoring System from a large Colorado beef feedlot. *J. Am. Vet. Med. Assoc.* **192**:1069–1073.

54. Schlosser, W., D. Henzler, J. Mason, S. Hurd, S. Trock, W. Sischo, D. Kradel, and A. Hogue. 1995. *Salmonella enteritidis* pilot project progress report. U.S. Government Printing Office, Washington, D.C.

55. Slutsker, L., A. A. Ries, K. Maloney, J. G. Wells, K. D. Greene, and P. M. Griffin. 1998. A nationwide case-control study of *Escherichia coli* O157:H7 infection in the United States. *J. Infect. Dis.* **177**:962–966.

56. Stern, L., and D. Lightfoot. 1999. Automated outbreak detection: a quantitative retrospective analysis. *Epidemiol. Infect.* **122**:103–110.

57. St. Louis, M. E., D. L. Morse, M. E. Potter, T. M. Demelfig, J. J. Guzewich, R. V. Tauxe, P. A. Blake, and the *Salmonella enteritidis* Working Group. 1988. The emergence of grade A eggs as a major source of *Salmonella enteritidis* infections: new implications for the control of salmonellosis. *JAMA* **259**:2103–2107.

58. Tauxe, R. V. 1992. Epidemiology of *Campylobacter jejuni* infections in the United States and other industrialized nations, p. 9–19. *In* I. Nachamkin, M. J. Blaser, and L. S. Tompkins (ed.), *Campylobacter jejuni: Current Status and Future Trends.* American Society for Microbiology, Washington, D.C.

59. Tauxe, R. V. 1998. New approaches to surveillance and control of emerging foodborne diseases. *Emerg. Infect. Dis.* **4**:455–456.

60. Thompson B., W. Demark-Wahnefried, G. Taylor, J. W. McClelland, G. Stables, S. Havas, Z. Feng, M. Topor, J. Heimendinger, K. D. Reynolds, and N. Cohen. 1999. Baseline fruit and vegetable intake among adults in seven 5 a day study centers located in diverse geographic areas. *J. Am. Diet. Assoc.* **99**:1241–1248.

61. Todd, E. C. D. 1996. Worldwide surveillance of foodborne disease: the need to improve. *J. Food Prot.* **59**:82–92.

62. Van Beneden, C. A., W. E. Keene, R. A. Strang, D. H. Werker, A. S. King, B. Mahon, K. Hedberg, A. Bell, M. T. Kelly, V. K. Balan, W. R. MacKenzie, and D. Fleming. 1999. Multinational outbreak of *Salmonella enterica* Newport infections due to contaminated alfalfa sprouts. *JAMA* **281**:158–162.

63. van't Veer, P., M. C. Jansen, M. Klerk, and F. J. Kok. 2000. Fruits and vegetables in the prevention of cancer and cardiovascular disease. *Public Health Nutr.* **3**:103–107.

64. Weingold, S. E., J. J. Guzewich, and J. K. Fudala. 1994. Use of foodborne disease data for HACCP risk assessment. *J. Food Prot.* **57**:820–830.

Mycotoxigenic Molds

IV

Food Microbiology: Fundamentals and Frontiers, 2nd Ed.
Edited by M. P. Doyle et al.
© 2001 ASM Press, Washington, D.C.

Ailsa D. Hocking

Toxigenic *Aspergillus* Species

21

For centuries it has been known that eating particular species of mushrooms can cause illness and death, but the recognition that some common food spoilage molds are also responsible for animal and human diseases and deaths has come relatively recently. Although it has been known since 1940 that a *Penicillium* species was the source of toxicity in "yellow" rice in Japan (90; see chapter 22, this volume), mycotoxins were brought to the attention of scientists in the Western world in the early 1960s with the outbreak of turkey "X" disease in England, which killed about 100,000 turkeys and other farm animals. The cause of this disease was traced to peanut meal in the feed, which was heavily contaminated with *Aspergillus flavus*. Analysis of the feed revealed that a group of fluorescent compounds, later named aflatoxins, were responsible for this outbreak, the deaths of large numbers of ducklings in Kenya, and widespread hepatoma in hatchery-reared trout in California that occurred more or less simultaneously (4, 7, 124).

Aspergillus was first described almost 300 years ago and is an important genus in foods. Although a few species have been harnessed in production of food (e.g., *Aspergillus oryzae* in soy sauce manufacture), most *Aspergillus* species occur in foods as spoilage or biodeterioration fungi. They are extremely common in stored commodities such as grains, nuts, and spices and occur more frequently in tropical and subtropical than in temperate climates (110).

The genus *Aspergillus* contains several species capable of producing mycotoxins, although the mycotoxin literature of the past 35 years has been dominated by papers on aflatoxins: chemistry, detection, toxicology, production, occurrence in foods, and regulatory aspects. The existence of other toxigenic *Aspergillus* species means that correct identification of isolates from foods and knowledge of the ecology of these molds are of paramount importance.

TAXONOMY

Aspergillus is a large genus containing more than 100 recognized species, most of which grow well in laboratory culture. There are a number of teleomorphic (ascosporic) genera that have *Aspergillus* conidial states (anamorphs), but the only two of real importance in foods are the xerophilic genus *Eurotium* (previously known as the *Aspergillus glaucus* group) and *Neosartorya* species, which produce heat-resistant ascospores and cause spoilage in heat-processed foods, mainly fruit products (110).

Ailsa D. Hocking, Food Science Australia, P.O. Box 52, North Ryde, New South Wales 1670, Australia.

The most widely used taxonomy for *Aspergillus* is that of Raper and Fennell (117), although some of their concepts are now out of date (121, 122), and many new species have since been described (115). Gams et al. (50) proposed a nomenclaturally correct classification for species within the genus *Aspergillus*, grouping species into six subgenera that are subdivided into sections. A recent taxonomy to the most common *Aspergillus* species, including those important in foods, is provided by Klich and Pitt (68).

In addition to traditional morphological taxonomic techniques, chemical techniques such as isoenzyme patterns (30), secondary metabolites (45, 46), ubiquinone systems (76), and molecular techniques (39, 91, 92, 95, 137) have been used to clarify relationships within the genus *Aspergillus*.

ISOLATION, ENUMERATION, AND IDENTIFICATION

Techniques for the isolation and enumeration of *Aspergillus* species from foods have been described in detail (110, 120). Antibacterial media containing compounds to inhibit or reduce mold colony spreading, such as dichloran-rose bengal-chloramphenicol (DRBC) agar (65) or dichloran–18% glycerol (DG18) agar (63), are recommended for enumerating fungi in foods (67, 119).

As mentioned, the most recent complete taxonomy is that of Raper and Fennell (117), but keys and descriptions of the most common foodborne *Aspergillus* species can be found elsewhere (68, 72, 110, 120). Identification of *Aspergillus* species requires growth on media developed for this purpose, including Czapek agar, a defined medium based on mineral salts, or a derivative such as Czapek-yeast extract agar, and malt extract agar. Growth on Czapek–yeast extract–20% sucrose agar (CY20S) can be a useful aid in identifying species of *Aspergillus* (68).

Unlike *Penicillium* species, *Aspergillus* species are conveniently "color coded," and the color of the conidia can be a very useful starting point in identification, at least to the series level. Microscopic morphology is also important in identification. Phialides (cells producing conidia) may be produced directly from the swollen apex (vesicle) of long stalks (stipes), or there may be an intermediate row of supporting cells (metulae). Correct identification of *Aspergillus* species is an essential prerequisite to assessing the potential for mycotoxin contamination in a commodity, food, or feedstuff.

SIGNIFICANT *ASPERGILLUS* MYCOTOXINS

Observations on "significant *Penicillium* mycotoxins" (102) can also be applied to *Aspergillus*, although iden-

tification of *Aspergillus* species is less frequently incorrect. The exception to this is the confusion in the literature over the identity of *Aspergillus* isolates reported to produce aflatoxins, with *A. flavus* and *Aspergillus parasiticus* frequently misidentified or misreported (69). The close taxonomic relationship between *Penicillium* and *Aspergillus* is reflected in the common production of many mycotoxins.

Almost 50 species of *Aspergillus* have been listed as capable of producing toxic metabolites (26), but the *Aspergillus* mycotoxins of greatest significance in foods and feeds are aflatoxins (produced by *A. flavus*, *A. parasiticus*, and *Aspergillus nomius*); ochratoxin A from *Aspergillus ochraceus* and related species and from *Aspergillus carbonarius* and occasionally *Aspergillus niger*; sterigmatocystin, produced primarily by *Aspergillus versicolor* but also by *Emericella* species; and cyclopiazonic acid (*A. flavus* is the primary source, but it is also reported to be produced by *Aspergillus tamarii*). Citrinin, patulin, and penicillic acid may also be produced by certain *Aspergillus* species, and tremorgenic toxins are produced by *Aspergillus terreus* (territrems), *Aspergillus fumigatus* (fumitremorgens), and *Aspergillus clavatus* (tryptoquivaline) (44, 73, 126).

Like *Penicillium*, *Aspergillus* species produce toxins that exhibit a wide range of toxicities, with the most significant effects being long term. Aflatoxin B_1 is perhaps the most potent liver carcinogen known for a wide range of animal species, including humans. Ochratoxin A and citrinin both affect kidney function. Cyclopiazonic acid has a wide range of effects (28), and tremorgenic toxins such as territrems affect the central nervous system. Table 21.1 lists the most significant toxins produced by *Aspergillus* species and their toxic effects.

A. FLAVUS AND *A. PARASITICUS*

Undoubtedly, the most important group of toxigenic aspergilli are the aflatoxigenic molds, *A. flavus*, *A. parasiticus*, and the recently described but much less common species *A. nomius*, all of which are classified in *Aspergillus* section *Flavi* (50). Although these three species are closely related and share many similarities, a number of characteristics may be used in their differentiation (Table 21.2). The suite of toxins produced by these three species is species specific (69, 74). *A. flavus* can produce aflatoxins B_1 and B_2 and cyclopiazonic acid, but only a proportion of isolates are toxigenic. *A. parasiticus* produces aflatoxins B_1, B_2, G_1, and G_2, but not cyclopiazonic acid, and almost all isolates are toxigenic. *A. nomius* is morphologically similar to *A. flavus* but, like *A. parasiticus*, produces B and G aflatoxins without cyclopiazonic acid. Because this species

Table 21.1 Significant mycotoxins produced by *Aspergillus* species and their toxic effects

Mycotoxins	Toxicity	Species
Aflatoxins B_1 and B_2	Acute liver damage, cirrhosis, carcinogenic (liver), teratogenic, immunosuppressive	*A. flavus, A. parasiticus, A. nomius*
Aflatoxins G_1 and G_2	Similar effects to B aflatoxins: G_1 toxicity less than B_1 but greater than B_2	*A. parasiticus, A. nomius*
Cyclopiazonic acid	Degeneration and necrosis of various organs, tremorgenic, low oral toxicity	*A. flavus*
Ochratoxin A	Kidney necrosis (especially pigs), teratogenic, immunosuppressive, probably carcinogenic	*A. ochraceus* and related species, *A. carbonarius*
Sterigmatocystin	Acute liver and kidney damage, carcinogenic (liver)	*A. versicolor, Emericella* spp.
Fumitremorgens	Tremorgenic (rats and mice)	*A. fumigatus*
Territrems	Tremorgenic (rats and mice)	*A. terreus*
Tryptoquivalines	Tremorgenic	*A. clavatus*
Cytochalasins	Cytotoxic	*A. clavatus*
Echinulins	Feed refusal (pigs)	*Eurotium chevalieri, E. amstelodami*

appears to be uncommon, it has been little studied, so the potential toxigenicity of isolates is not known, and the practical importance of this species is hard to assess. *A. flavus* and *A. parasiticus* are closely related to *A. oryzae* and *Aspergillus sojae*, species that are used in the manufacture of fermented foods and do not produce toxins (39). Production of aflatoxin B and cyclopiazonic acid by *A. tamarii* has also been reported (55). Obviously, accurate differentiation of related species within section *Flavi* is important to determine the potential for toxin production and the types of toxins likely to be present.

Detection and Identification

The most effective medium for rapid detection of aflatoxigenic molds is Aspergillus flavus and parasiticus agar (AFPA) (113), a medium formulated specifically for this purpose. Under the incubation conditions specified for this medium (30°C for 42 to 48 h), *A. flavus, A. parasiticus*, and *A. nomius* produce a bright orange-yellow colony reverse that is diagnostic and readily recognized. *A. flavus* and related species also grow well on general-purpose yeast and mold enumeration media such as DRBC agar or DG18 agar, and they are relatively easily recognized by experienced personnel. Moderately deep, yellow-green colonies with moplike fruiting structures (best observed under a stereomicroscope) can be counted as *A. flavus* or *A. parasiticus*.

A. flavus and *A. parasiticus* (Fig. 21.1) are easily distinguished from other *Aspergillus* species by using appropriate media and methods (68, 110, 120). However, differentiating between these two species and *A. nomius* is more difficult, but it is important because of their differing mycotoxin profiles and toxin-producing potential. The texture of the conidial walls is a reliable differentiating feature: conidia of *A. flavus* (Fig. 21.1A and B) are usually smooth to finely roughened, while those of *A. parasiticus* (Fig. 21.1C and D) are clearly rough when observed under an oil immersion lens. Screening isolates for aflatoxin production can also be used to differentiate the species. Cultures can be grown on coconut-cream agar and observed under UV light (38), or a simple agar plug technique coupled with thin-layer chromatography (TLC) (43) can be used to screen cultures for aflatoxin production as an aid to identification. The combinations of characteristics most useful in differentiation among the three aflatoxigenic species are summarized in Table 21.2.

Aflatoxins

Aflatoxins are difuranocoumarin derivatives (16). Aflatoxins B_1, B_2, G_1, and G_2 are produced in nature by the molds discussed above. The letters B and G refer to the fluorescent colors (blue and green, respectively) observed under long-wave UV light, and the subscripts 1 and 2 refer to their separation patterns on TLC plates.

Table 21.2 Distinguishing features of *Aspergillus flavus, A. parasiticus*, and *A. nomius*

Species	Conidia	Sclerotia	Toxins
A. flavus	Smooth to moderately roughened, variable in size	Large, globose	Aflatoxins B_1 and B_2, cyclopiazonic acid
A. parasiticus	Conspicuously roughened, little variation in size	Large, globose	Aflatoxins B and G
A. nomius	Similar to *A. flavus*	Small, elongated (bullet shaped)	Aflatoxins B and G

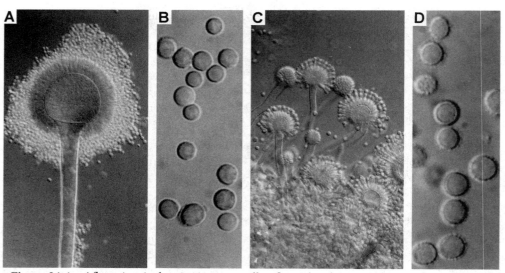

Figure 21.1 Aflatoxigenic fungi. (A) *Aspergillus flavus* head (×215); (B) *A. flavus* conidia (×1,350); (C) young *A. parasiticus* heads (×215); (D) *A. parasiticus* conidia (×1,350).

Aflatoxins M_1 and M_2 are produced from their respective B aflatoxins by hydroxylation in lactating animals and are excreted in milk at a rate of approximately 1.5% of ingested B aflatoxins (48).

Aflatoxins are synthesized through the polyketide pathway, beginning with condensation of an acetyl unit with two malonyl units and the loss of carbon dioxide (11). The resultant hexanoate is enzyme bound and reacts successively with seven malonate units to yield a polyketide intermediate that undergoes cyclization and aromatization to give norsolorinic acid. Norsolorinic acid undergoes several metabolic conversions to form aflatoxin B_1 (40, 147). G-group aflatoxins are formed from the same substrate as B-group aflatoxins, namely, O-methylsterigmatocystin, but by an independent pathway (152). Most of the genes involved in the biosynthesis of aflatoxins are contained within a single gene cluster in the genomes of *A. flavus* and *A. parasiticus*, and their regulation and expression are now relatively well understood (13, 84, 104, 151, 155).

Toxicity

Aflatoxins are both acutely and chronically toxic in animals and humans, producing acute liver damage, liver cirrhosis, tumor induction, and teratogenesis (132). Perhaps of greater significance to human health are the immunosuppressive effects of aflatoxins, either alone or in combination with other mycotoxins (107). Immunosuppression can increase susceptibility to infectious diseases, particularly in populations where aflatoxin ingestion is chronic (32), and can interfere with production of antibodies in response to immunization in animals (107) and perhaps also in children.

Acute aflatoxicosis in humans is rare (127); however, several outbreaks have been reported. In 1967, 26 people in two farming communities in Taiwan became ill with apparent food poisoning. Nineteen were children, three of whom died. Although postmortems were not performed, rice from affected households contained about 200 μg of aflatoxin B_1 per kg, which was probably responsible for the outbreak. An outbreak of hepatitis in India in 1974 that affected 400 people, 100 of whom died, almost certainly was caused by aflatoxins (75). The outbreak was traced to corn heavily contaminated with *A. flavus* and containing up to 15 mg of aflatoxins per kg. It was calculated that affected adults may have consumed 2 to 6 mg on a single day, implying that the acute lethal dose for adult humans is of the order of 10 mg. More recently, deaths of 13 Chinese children in the northwestern Malaysian state of Perak were reported (88) which were apparently due to ingestion of contaminated noodles. Aflatoxins were confirmed in postmortem tissue samples. A case of systemic aspergillosis caused by an aflatoxin-producing strain of *A. flavus* has been reported in which aflatoxins B_1, B_2 and M_1 were detected in lung lesions and were considered to have played a role in damaging the immune system of the patient (93).

Undoubtedly, the greatest direct impact of aflatoxins on human health is their potential to induce liver cancer (87). Human liver cancer has a high incidence in central Africa and parts of Southeast Asia, and studies in several African countries and Thailand have shown a correlation between aflatoxin intake and the occurrence of primary liver cancer (144). No such correlation could be demonstrated for populations in rural areas of the United States, despite the occurrence of considerable amounts of aflatoxins in corn (133, 134). This apparent anomaly may be explained by the relationship between the roles of hepatitis B virus and aflatoxins in induction of human liver cancer. Aflatoxins and hepatitis B virus are apparently cocarcinogens, and the occurrence of both predisposing factors greatly increases the probability of human liver cancer. However, evidence supports the hypothesis that high aflatoxin intakes are causally related to high incidences of cancer, even in the absence of hepatitis B (58, 60, 105).

In animals, aflatoxins have been shown to cause various syndromes, including cancer of the liver, colon, and kidneys in rats, mice, monkeys, ducklings, and trout (40). Regular low-level intake of aflatoxins can lead to poor feed conversion, low weight gain, and poor milk yields in cattle (14).

Mechanism of Action

Aflatoxin B_1 is metabolized by the microsomal mixed function oxidase system in the liver, leading to the formation of highly reactive intermediates, one of which is 2,3-epoxy-aflatoxin B_1 (94). Binding of these reactive intermediates to DNA results in disruption of transcription and abnormal cell proliferation, leading to mutagenesis or carcinogenesis (40). Aflatoxins also inhibit oxygen uptake in the tissues by acting on the electron transport chain and inhibiting various enzymes, resulting in decreased production of ATP (40).

Toxin Detection

Aflatoxins can be detected by chemical or biological methods. Chemical determination of aflatoxins has become fairly standardized (6, 96). Samples are extracted with organic solvents such as chloroform or methanol in combination with small amounts of water. The presence of fats, lipids, or pigments in extracts reduces the efficiency of the separation, and solvents such as hexane may be used to partition these components from the extract (40). Extracts are further cleaned up by passage through a silica gel column or proprietary separating cartridge. The extract is then concentrated, usually by evaporation under nitrogen, and then separated by TLC or high-performance liquid chromatography (HPLC). TLC is commonly used, and aflatoxins are visualized under UV light and quantified by visual comparison with known concentrations of standards or by fluorimetry. HPLC with detection by UV light or absorption spectrometry provides a more readily quantifiable (although not necessarily more accurate or sensitive) technique.

Immunoassay techniques, including enzyme-linked immunosorbent assays (ELISA) and dipstick tests for aflatoxin detection, have been developed (21, 79, 106, 125, 139), and a number of kits are now commercially available. Immunoaffinity with fluorimetric detection has been combined in developing biosensors able to detect aflatoxin down to 50 μg/kg in as little as 2 min (17, 18, 19). PCR has been applied to detect the presence of aflatoxigenic fungi in foods, targeting aflatoxin biosynthetic genes (51, 128).

Occurrence of Aflatoxigenic Molds and Aflatoxins

A. flavus is widely distributed in nature, but *A. parasiticus* is probably less widespread, the actual extent of its occurrence being complicated by the tendency for both species to be reported indiscriminately as *A. flavus*. In a wide-ranging survey of the mycoflora of commodities in Thailand (111, 112), *A. flavus* was one of the most commonly occurring molds in nuts and oilseeds, but *A. parasiticus* was rarely encountered. *A. flavus* was the most common species in peanuts and the second most common (after *Fusarium moniliforme*) in corn. *A. nomius* was reported from both commodities, the first published report of the occurrence of this species in food (111). Soybeans, mung beans, sorghum, and other commodities also contained considerable populations of *A. flavus*, but *A. parasiticus* was rare (112).

A. flavus and *A. parasiticus* have a strong affinity with nuts and oilseeds. Corn, peanuts, and cottonseed are the most important crops invaded by these molds, and in many instances, invasion takes place before harvest, not during storage as was once believed. Peanuts are invaded while still in the ground if the crop suffers drought stress or related factors (27, 109, 123). In corn, insect damage to developing kernels allows entry of aflatoxigenic molds, but invasion can also occur through the silks of developing ears (81). Cottonseeds are invaded through the nectaries (70).

Cereals and spices are common substrates for *A. flavus* (109), but aflatoxin production in these commodities is almost always a result of poor drying, handling, or storage, and aflatoxin levels are rarely significant. Significant amounts of aflatoxins can occur in peanuts, corn, and other nuts and oilseeds, particularly in some tropical countries where crops may be grown under marginal conditions and where drying and storage facilities are limited (5, 33, 87).

Factors Affecting Growth and Toxin Production

A. flavus and *A. parasiticus* have similar growth patterns. Both grow at temperatures ranging from 10 to 12°C to 42 to 43°C, with an optimum from 30 to 33°C (8, 57), with aflatoxins being produced from 12 to 40°C (34, 71,

100). The optimum water activity (a_w) for growth is near 0.996 (57), with minima reported as a_w 0.80 (8) to 0.82 to 0.83 (100, 114). Aflatoxins are generally produced in greater quantity at higher values (0.98 to 0.99), with toxin production apparently ceasing at or near a_w 0.85 (34, 71, 100). Although growth can take place over the pH range of just above 2.0 up to 10.5 (*A. parasiticus*) or 11.2 (*A. flavus*) (149), aflatoxin production has been reported for *A. parasiticus* only between pH 3.0 and 8.0, with an optimum near pH 6.0 (15). Reduction of available oxygen by modified atmosphere packaging of foods in barrier film or with oxygen scavengers can inhibit aflatoxin formation by *A. flavus* and *A. parasiticus* (41, 42).

Control and Inactivation

Control of aflatoxins in commodities generally relies on screening techniques that separate affected nuts, grains, or seeds. In corn, cottonseed, and figs, screening for aflatoxins can be done by examination under UV light: those particles that fluoresce may be contaminated. All peanuts fluoresce when exposed to UV light, so this method is not useful for detecting kernels that are contaminated with aflatoxins. Peanuts containing aflatoxins are segregated by electronic color-sorting machines that detect discolored kernels.

Aflatoxins can be partially destroyed by various chemical treatments. Oxidizing agents such as ozone and hydrogen peroxide have been demonstrated to remove aflatoxins from contaminated peanut meals (148). Although ozone was effective in removal of aflatoxins B_1 and G_1 under the conditions applied (100°C for 2 h), there was no effect on aflatoxin B_2, and the treatment decreased the lysine content of the meal (35). Treatment with hydrogen peroxide resulted in 97% destruction of aflatoxin in defatted peanut meal (131). The most practical chemical method of aflatoxin destruction appears to be the use of anhydrous ammonia gas at elevated temperatures and pressures, with a 95 to 98% reduction in total aflatoxin in peanut meal reported (143). This technique is used commercially for detoxification of animal feeds in Senegal, France, and the United States (94, 143).

The ultimate method of control of aflatoxins in commodities, particularly peanuts, is to prevent the plants from becoming infected with aflatoxigenic strains of molds. Progress toward this goal is being made by a biological control strategy: early infection of plants with nontoxigenic strains of *A. flavus* to prevent the subsequent entry of toxigenic strains (20, 28, 29, 31, 36).

Aflatoxins are one of the few mycotoxins covered by legislation. Some countries impose statutory limits on the amount of aflatoxin that can be present in particular foods. The limit imposed by most Western countries is 5 to 20 μg of aflatoxin B_1 per kg in several human foods, including peanuts and peanut products, with the amount allowed in animal feeds varying, but up to 300 μg/kg allowed in feedstuffs for beef cattle and sheep in the United States (52, 87). There are currently no statutory limits for aflatoxins in foods or feeds in Southeast Asia (87, 143).

Cyclopiazonic Acid

Cyclopiazonic acid is produced by *A. flavus* and has also been reported from *A. tamarii* (55, 129) and *A. versicolor* (26). Maximum production of cyclopiazonic acid by *A. flavus* occurs at a_w 0.98 at 20°C, with minimum production reported at a_w 0.90 at 30°C (56). The toxin is an indole tetramic acid that can occur in naturally contaminated agricultural commodities and compounded animal feeds. It is acutely toxic to rats and other test animals, causing severe gastrointestinal and neurological disorders (98). Degenerative changes and necrosis may occur in the digestive tract, liver, kidney, and heart (99). Cyclopiazonic acid may have been responsible for many of the symptoms observed in turkey "X" disease in the 1960s, originally attributed to aflatoxins (12). Cyclopiazonic acid can be detected by TLC with visualization by spraying with dimethylaminobenzaldehyde and 50% ethanolic H_2SO_4 followed by heating for 5 min at 100°C (26).

ASPERGILLUS OCHRACEUS

A. ochraceus (Fig. 21.2A) is a widely distributed mold, particularly common on dried foods (110). It is the most commonly occurring species in what was known as the "*Aspergillus ochraceus* group" of Raper and Fennell (117), now correctly known as *Aspergillus* section *Circumdati* (50). Ochratoxin was first isolated from *A. ochraceus* in 1965 (141), not as a result of a toxicosis but in a laboratory study on toxigenic molds. Three toxins were found; ochratoxin A was the major toxin, with minor components of lower toxicity designated as ochratoxins B and C. *A. ochraceus* also produces penicillic acid, a mycotoxin of lower toxicity and of uncertain importance in human health. Other reported toxic metabolites are xanthomegnin and viomellein (47).

Other *Aspergillus* species closely related to *A. ochraceus* can also produce ochratoxin A. *Aspergillus sclerotiorum*, *Aspergillus alliaceus*, *Aspergillus melleus*, and *Aspergillus sulphureus* have all been reported to produce ochratoxins (23, 145), but in *A. ochraceus* and these related species, only a proportion of isolates are toxigenic.

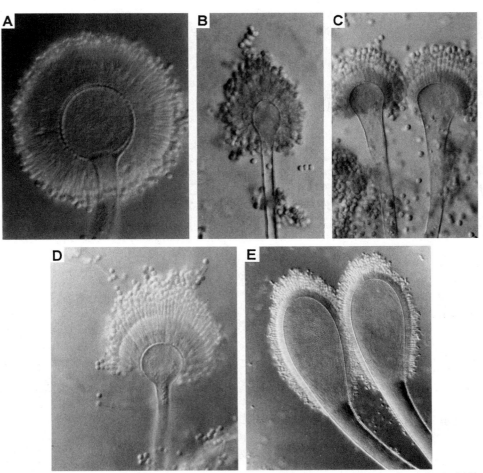

Figure 21.2 Some common mycotoxigenic *Aspergillus* species. (A) *A. ochraceus* (×540); (B) *A. versicolor* (×540); (C) *A. fumigatus* (×540); (D) *A. terreus* (×540); (E) *A. clavatus* (×215).

Isolation and Identification

A. ochraceus and other species in *Aspergillus* section *Circumdati* are xerophilic and are best isolated on reduced-a_w media such as DG18 agar (110, 119), although DRBC agar should also give satisfactory results. Colonies of *A. ochraceus* and related species are relatively deep ochre-brown to yellow-brown in color, with long stipes bearing radiate *Aspergillus* heads. The vesicles are spherical, bearing densely packed metulae and phialides with small, smooth, pale brown conidia. Differentiating these closely related species can be difficult (68, 117) but is rarely necessary, as *A. ochraceus* is by far the most commonly occurring.

Growth and Toxin Production

A. ochraceus is widely distributed in dried foods such as nuts (peanuts, pecans), beans, dried fruit, biltong, and dried fish (110). It is a xerophile, capable of growth down to a_w 0.79 and with an optimum near a_w 0.99 (110). *A. ochraceus* grows from 8 to 37°C (101, 102) and within a wide pH range (2.2 to 10.3) (149). Ochratoxin A is produced optimally at a_w 0.98 to 0.96 (116) and at quite high temperatures (between 25 to 30°C), compared with the optimum for penicillic acid production, which is 10 to 20°C (9, 23, 101, 102, 116).

Ochratoxin A contamination of foods is of greatest concern in Scandinavia and possibly in the Baltic states, where the source is *Penicillium verrucosum*. Ochratoxin A from *A. ochraceus* or related species would occur only in warmer climates, but ochratoxin is rarely reported from tropical commodities (154). More recently, ochratoxin A has been reported from figs (37). The toxic effects of ochratoxin A are discussed in chapter 22.

Ochratoxins can be assayed by routine chromatographic methods (TLC, HPLC, LC) with detection of green fluorescence under UV light at 333 nm (26, 138). Immunoassay techniques have been developed for detection of ochratoxin, including a monoclonal antibody capable of detecting ochratoxin A at the picogram level (59) and an ELISA for detection of ochratoxin A in swine kidneys (24).

A. CARBONARIUS AND ASPERGILLUS NIGER

Aspergillus species in section Nigri have only recently been recognized as a source of ochratoxin A. Production of ochratoxin A by Aspergillus niger var. niger was reported in 1994 (1), and since then ochratoxin A has also been reported from A. carbonarius (61, 64, 136, 150) and Aspergillus awamori (103) but not Aspergillus japonicus.

Little is known about A. carbonarius; however, its close relationship with A. niger would suggest similar physiological characteristics of optimal growth near 33 to 35°C and a minimum a_w for growth near 0.80 (110). A. niger and A. carbonarius are easily recognized in culture by the production of rapidly growing dark brown to black colonies (110). The two species also share habitats, and because of their high resistance to sunlight, both have been reported from sun-dried fruits such as sultanas, raisins, and figs (37, 66, 80). Ochratoxin A has been reported from these commodities as well as from coffee and wine (37, 89, 142, 156). The conditions under which A. carbonarius produces ochratoxin have not been reported.

A. VERSICOLOR

A. versicolor (Fig. 21.2B) is the most important food spoilage and toxigenic species in Aspergillus section Versicolores (50), previously known as the Aspergillus versicolor group (117). A. versicolor is the major producer of sterigmatocystin, a carcinogenic dihydrofurano-xanthone that is a precursor of the aflatoxins (26), but aflatoxins are not produced by this species. Sterigmatocystin is also produced by members of the Aspergillus section Nidulantes, including A. nidulans and a number of Emericella species (129). A. versicolor has been reported to produce ochratoxin A (2).

A. versicolor is a xerophile, with a minimum a_w for growth of 0.74 to 0.78 (110). It is very widely distributed in foods, particularly stored cereals, cereal products, nuts, spices, and dried meat products (110). A. versicolor grows slowly on all media, but because it is most commonly found in reduced a_w foods and feeds, a medium of reduced a_w, such as DG18 agar, is recommended for isolation. Small, gray-green colonies showing pinkish or reddish colors in the mycelium and/or reverse and with moplike heads are indicative of A. versi-

color. The reported minimum temperature for growth is 9°C at a_w 0.97, and the maximum temperature is 39°C at a_w 0.87, with optimum growth at 27°C at a_w 0.97 (130). Little is known about factors affecting production of sterigmatocystin.

Occurrence, Toxicity, and Detection of Sterigmatocystin

Natural occurrence of sterigmatocystin has been reported in rice in Japan, wheat and barley in Canada, cereal-based products in the United Kingdom (154), and Ras cheese (3). Sterigmatocystin has also been reported from mold-affected buildings and building materials (97, 140). Although the toxin has not been reported in significant quantities, its acute and chronic toxicity are such that it should be regarded as a major potential hazard in foods and the environment (135). Sterigmatocystin has low acute oral toxicity because it is relatively insoluble in water and gastric juices. However, even low doses can cause tumors in mice (49) and pathological changes in rat livers (132). As a liver carcinogen, sterigmatocystin appears to be only about 1/150 as potent as aflatoxin B_1, but it is still much more potent than most other known liver carcinogens. Doses as low as 15 μg/day fed continuously or a single dose of 10 mg caused liver cancer in ca. 30% of male Wistar rats (132).

Sterigmatocystin can be detected by TLC, with chloroform-methanol (98:2) as a solvent system. The toxin is visualized as an orange-red spot under UV light. A light yellow fluorescence develops after spraying with acetic acid (26). An ELISA method for detection of sterigmatocystin has been described (22).

A. FUMIGATUS

A. fumigatus, section Fumigati (Fig. 21.2C), is best recognized as a human pathogen, causing aspergillosis of the lung (73). It is thermophilic, with a temperature range for growth of between 10 and 55°C and an optimum of between 40 and 42°C. It is one of the least xerophilic of the common aspergilli, with a minimum a_w for growth of 0.85 (77). Its prime habitat is decaying vegetation, in which it causes spontaneous heating. A. fumigatus is frequently isolated from foods, particularly stored commodities, but is not regarded as a serious spoilage mold (110).

A. fumigatus is capable of producing several toxins that affect the central nervous system, causing tremors. Fumitremorgens A, B, and C are toxic cyclic dipeptides that are produced by A. fumigatus and Aspergillus caespitosus (26). When fed to laboratory rats and mice, fumitremorgens caused tremors and death in 70% of animals tested (54). Verruculogen, which is also produced by these species (26, 78), has a structure similar to the fumitremorgens but with a different side chain. Verruculogen is tremorgenic when fed to mice and day-old chicks (54), and it appears to cause inhibition of the alpha motor cells of the anterior horn (26). *A. fumigatus* also produces gliotoxin, a toxin with immunosuppressive activity that causes DNA damage and may play a role in the pathogenesis of this species (10, 53, 118).

A. TERREUS

A. terreus, section *Terrei* (Fig. 21.2D), produces rapidly growing pale brown colonies, with *Aspergillus* heads bearing densely packed metulae and phialides with minute conidia borne in long columns. *A. terreus* commonly occurs in soil and in foods, particularly stored cereals and cereal products, beans, pulses, and nuts, but is not regarded as an important spoilage mold (110). It is probably thermophilic, growing much more strongly at 37°C than at 25°C (110), but little information has been published on its physiology. The reported minimum a_w for growth is 0.78 at 37°C (5).

A. terreus can produce a group of tremorgenic toxins known as territrems (83). The most distinctive structural feature of these toxins is that they do not contain nitrogen. Territrems are acutely toxic, with an intraperitoneal dose of 1 mg of territrem B injected into Swiss mice (20 g body weight) causing whole body tremors within 5 min and other neurological symptoms within 20 to 30 min, all of which subside within 1 h (82). A 2-mg dose can cause death after convulsions and apnea. Territrem B appears to act by blocking acetylcholinesterase activity (82).

Territrems can be detected in chloroform extracts by TLC, exhibiting blue fluorescence under UV light. Natural contamination of food or feeds with territrems has not been reported (82).

A. CLAVATUS

A. clavatus (Fig. 21.2E) is found in soil and decomposing plant materials. The most common member of the section *Clavati*, subgenus *Clavati*, it is easily recognizable by its large blue-green clavate (club-shaped) heads (68, 110, 117). Although *A. clavatus* has been reported to be present in various stored grains (110), it is especially common in malting barley, an environment particularly suited to its growth and sporulation (44).

A. clavatus produces patulin, cytochalasins, and the tremorgenic mycotoxins tryptoquivaline, tryptoquivalone, and related compounds. Patulin and cytochalasins

may be formed in barley during the malting process (85, 86), and outbreaks of *A. clavatus*-associated mycotoxicoses have been reported in stock fed on culms from distillery maltings in Europe, the United Kingdom, and South Africa (44).

A. clavatus is a recognized health hazard to workers in the malting industry. Inhalation of large numbers of highly allergenic spores can cause respiratory diseases such as bronchitis, emphysema, or malt worker's lung, a serious occupational extrinsic allergic alveolitis (44).

EUROTIUM SPECIES

The genus *Eurotium* is an ascomycete genus, characterized by the formation of bright yellow cleistothecia, often enmeshed in yellow, orange, or red hyphae, overlayed by the gray-green (glaucus) *Aspergillus* heads of the anamorphic state. Members of this genus were (and often still are) referred to as the *Aspergillus glaucus* group (117). All *Eurotium* species are xerophilic. They are important spoilage molds in all types of stored commodities, often being the primary invaders of stored grains, spices, nuts, and animal feeds. The four most common species are *Eurotium chevalieri*, *Eurotium repens*, *Eurotium rubrum*, and *Eurotium amstelodami*.

Eurotium species grow poorly on all high-a_w media; thus, DG18 agar is recommended for their isolation and enumeration from foods (62, 110). Identification to species is based on colony color and ascospore morphology, as observed on CY20S agar (68, 110).

E. chevalieri and *E. amstelodami* have been reported to produce toxic alkaloid metabolites called echinulin and neoechinulins (26). Echinulin was identified as the compound responsible for refusal by swine of moldy feed containing high populations of *E. chevalieri* and *E. amstelodami*. The toxin was extracted in acetone and identified by TLC using ethyl acetate-hexane (8:2) as the solvent system. Echinulin turned blue in the presence of *p*-anisaldehyde reagent at 110°C (146). There is little published information on the toxicity of echinulins. Neoechinulin from *E. repens* has been reported as having strong antioxidant properties (153).

CONCLUSION

Aspergillus is one of the most important genera in the spoilage of foods and animal feeds, particularly in warm-temperate climates and the tropics. The genus contains a number of highly mycotoxigenic molds, the aflatoxigenic species clearly being the most important from the point of view of human health, although there are reports of the production of ochratoxin by an increasing number of *Aspergillus* species. There is still much to learn about the real significance of many of the other mycotoxins produced by *Aspergillus*, particularly their roles in cancer induction and immunosuppression. The interactive effects of naturally occurring mixtures of mycotoxins, e.g., aflatoxins and *Fusarium* toxins (particularly fumonisins) in corn, are poorly understood and may be of much greater significance to human and animal health than we yet realize.

References

1. **Abarca, M. L., M. R. Bragulat, G. Castellá, and F. J. Cabañes.** 1994. Ochratoxin production by strains of *Aspergillus niger* var. *niger*. *Appl. Environ. Microbiol.* **60:**2650–2652.

2. **Abarca, M. L., M. R. Bragulat, G. Castellá, and F. J. Cabañes.** 1997. New ochratoxigenic species in the genus *Aspergillus*. *J. Food Prot.* **60:**1580–1582.

3. **AbdaAlla, E. A. M., M. M. Metwally, A. M. Mehriz, and Y. H. AbuSree.** 1996. Sterigmatocystin: incidence, fate and production by *Aspergillus versicolor* in Ras cheese. *Nahrung* **40:**310–313.

4. **Allcroft, R., and R. B. A. Carnaghan.** 1963. Toxic products in groundnuts—biological effects. *Chem. Ind.* **1963:**50–53.

5. **Arim, R. H.** 1995. Present status of the aflatoxin situation in the Philippines. *Food Addit. Contam.* **12:**291–296.

6. **Association of Official Analytical Chemists International (AOAC).** 2000. Natural toxins, ch. 25. *In Official Methods of Analysis of the AOAC International*, 17th ed., vol. II. Association of Official Analytical Chemists International, Washington, D.C.

7. **Austwick, P. K. C., and G. Ayerst.** 1963. Groundnut mycoflora and toxicity. *Chem. Ind.* **1963:**55–61.

8. **Ayerst, G.** 1969. The effects of moisture and temperature on growth and spore germination in some fungi. *J. Stored Prod. Res.* **5:**127–141.

9. **Bacon, C. W., J. G. Sweeney, J. D. Robbins, and D. Burdick.** 1973. Production of penicillic acid and ochratoxin A on poultry feed by *Aspergillus ochraceus*: temperature and moisture requirements. *J. Food Prot.* **44:**450–454.

10. **Belkacemi, L., R. C. Barton, V. Hopwood, and E. G. V. Evans.** 1999. Determination of optimum growth conditions for gliotoxin production by *Aspergillus fumigatus* and development of a novel method for gliotoxin detection. *Med. Mycol.* **37:**227–233.

11. **Bennett, J. W., and L. S. Lee.** 1979. Mycotoxins: their biosynthesis in fungi—aflatoxins and other bisfuranoids. *J. Food Prot.* **42:**805–809.

12. **Bradburn, N., R. D. Coker, and G. Blunden.** 1994. The aetiology of turkey "X" disease. *Phytochemistry* **35:**817.

13. **Brown, M. P., C. S. Brown-Jenco, and G. A. Payne.** 1999. Genetic and molecular analysis of aflatoxin biosynthesis. *Fungal Genetics Biol.* **26:**81–98.

14. **Bryden, W. L.** 1982. Aflatoxins and animal production: an Australian perspective. *Food Technol. Aust.* **34:**216–223.

15. **Buchanan, R. L., and J. C. Ayres.** 1976. Effect of sodium acetate on growth and aflatoxin production in *Aspergillus parasiticus* NRRL 2999. *J. Food Sci.* **41:**128–132.

16. **Büchi, G., and I. D. Rae.** 1969. The structure and chemistry of aflatoxins, p. 55–75. *In* L. A. Goldblatt (ed.), *Aflatoxins*. Academic Press, New York, N.Y.

17. **Carlson, M. A., C. B. Bargeron, R. C. Benson, A. B. Fraser, T. E. Phillips. J. T. Velky, J. D. Groopman, P. T. Strickland, and H. W. Ko.** 2000. An automated hand-held biosensor for aflatoxin. *Biosensors Bioelectronics* **14:**841–848.

18. **Carman, A. S., S. S. Kuan, G. M. Ware, P. P. Umrigar, K. V. Miller, and H. G. Guerrero.** 1996. Robotic automated analysis of foods for aflatoxin. *J. AOAC Int.* **79:**456–464.

19. **Carter, R. M., M. B. Jacobs, G. J. Lubrano, and G. G. Guilbault.** 1997. Rapid detection of aflatoxin B$_1$ with immunochemical electrodes. *Anal. Lett.* **30:**1465–1482.

20. **Chourasia, H. K., and R. K. Sinha.** 1994. Potential of the biological control of aflatoxin contamination in developing peanut (*Arachis hypogaea* L.) by atoxigenic strains of *Aspergillus flavus*. *J. Food Sci. Technol. Mysore* **31:**362–366.

21. **Chu, F. S.** 1984. Immunoassays for analysis of mycotoxins. *J. Food Prot.* **47:**562–569.

22. **Chung, D. H., M. I. Abouzied, and J. J. Pestka.** 1989. Immunochemical assay applied to mycotoxin biosynthesis: ELISA comparison of sterigmatocystin production by *Aspergillus versicolor* and *Aspergillus nidulans*. *Mycopathologia* **107:**93–100.

23. **Ciegler, A.** 1972. Bioproduction of ochratoxin A and penicillic acid by members of the *Aspergillus ochraceus* group. *Can. J. Microbiol.* **18:**631–636.

24. **Clarke, J. R., R. R. Marquardt, A. A. Frohlich, and R. J. Pitura.** 1994. Quantification of ochratoxin A in swine kidneys by enzyme-linked immunosorbent assay using a simplified sample preparation procedure. *J. Food Prot.* **57:**991–995.

25. **Cole, R. J.** 1986. Etiology of turkey "X" disease in retrospect: a case for the involvement of cyclopiazonic acid. *Mycotoxin Res.* **2:**3–7

26. **Cole, R. J., and R. H. Cox.** 1981. *Handbook of Toxic Fungal Metabolites*. Academic Press, New York, N.Y.

27. **Cole, R. J., R. A. Hill, P. D. Blankenship, T. H. Sanders, and H. Garren.** 1982. Influence of irrigation and drought on invasion of *Aspergillus flavus* in corn kernels and peanut pods. *Dev. Ind. Microbiol.* **23:**299–326.

28. **Cotty, P. J.** 1994. Influence of field application of an atoxigenic strain of *Aspergillus flavus* on the populations of *A. flavus* infecting cotton bolls and on aflatoxin content of cottonseed. *Phytopathology* **84:**1270–1277.

29. **Cotty, P. J., P. Bayman, D. S. Engel, and K. S. Elias.** 1994. Agriculture, aflatoxins and *Aspergillus*, p. 1–27. *In* K. A. Powell, A. Renwick, and J. F. Peberdy (ed.), *The Genus Aspergillus: From Taxonomy and Genetics to Industrial Application*. Plenum Press, New York, N.Y.

30. **Cruickshank, R., and J. I. Pitt.** 1990. Isoenzyme patterns in *Aspergillus flavus* and closely related taxa, p. 259–265. *In* R. A. Samson and J. I. Pitt (ed.), *Modern Concepts in Penicillium and Aspergillus Classification*. Plenum Press, New York, N.Y.

31. **Daigle, D. J., and P. J. Cotty.** 1995. Formulating atoxigenic *Aspergillus flavus* for field release. *Biocontrol Sci. Tech.* 5:175–184.

32. **Denning, D. W., S. C. Quiepo, D. G. Altman, K. Makarananda, G. E. Neal, E. L. Camellere, M. R. A. Morgan, and T. E. Tupasi.** 1995. Aflatoxin and outcome from acute lower respiratory infection in children in the Philippines. *Ann. Trop. Paediatr.* 15:209–216.

33. **Dhavan, A. S., and M. R. Choudary.** 1995. Incidence of aflatoxins in animal feedstuff: a decade's scenario in India. *J. AOAC Int.* 78:693–698.

34. **Diener, U. L., and N. D. Davis.** 1967. Limiting temperature and relative humidity for growth and production of aflatoxins and free fatty acids by *Aspergillus flavus* in sterile peanuts. *J. Am. Oil Chem. Soc.* 44:259–263.

35. **Dollear, F. G., G. E. Mann, L. P. Codifer, H. K. Gardner, S. P. Koltun, and H. L. E. Vix.** 1986. Elimination of aflatoxin from peanut meal. *J. Am. Oil Chem. Soc.* 45:862–865.

36. **Dorner, J. W., R. J. Cole, and D. W. Wicklow.** 1999. Aflatoxin reduction in corn through field application of competitive fungi. *J. Food Prot.* 62:650–656.

37. **Doster, M. A., T. J. Michailides, and D. P. Morgan.** 1996. *Aspergillus* species and mycotoxins in figs from California. *Plant Dis.* 80:484–489.

38. **Dyer, S. K., and S. McCammon.** 1994. Detection of toxigenic isolates of *Aspergillus flavus* and related species on coconut cream agar. *J. Appl. Bacteriol.* 76:75–78.

39. **Egel, D. S., P. J. Cotty, and K. S. Elias.** 1994. Relationships among isolates of *Aspergillus* sect. *Flavi* that vary in aflatoxin production. *Phytopathology* 84:906–912.

40. **Ellis, W. O., J. P. Smith, B. K. Simpson, and J. H. Oldham.** 1991. Aflatoxins in food: occurrence, biosynthesis, effects on organisms, detection, and methods of control. *CRC Crit. Rev. Food Sci. Nutr.* 30:403–439.

41. **Ellis, W. O., J. P. Smith, B. K. Simpson, H. Ramaswamy, and G. Doyon.** 1994. Effect of gas barrier characteristics of films on aflatoxin production by *Aspergillus flavus* in peanuts packaged under modifed atmosphere packaging (MAP) conditions. *Food Res. Int.* 27:505–512.

42. **Ellis, W. O., J. P. Smith, B. K. Simpson, H. Ramaswamy, and G. Doyon.** 1994. Novel techniques for controlling growth of and aflatoxin production by *Aspergillus parasiticus* in packaged peanuts. *Food Microbiol.* 11:357–368.

43. **Filtenborg, O., J. C. Frisvad, and J. A. Svendsen.** 1983. Simple screening method for molds producing intracellular mycotoxins in pure culture. *Appl. Environ. Microbiol.* 45:581–585.

44. **Flannigan, B., and A. R. Pearce.** 1994. *Aspergillus* spoilage: spoilage of cereals and cereal products by the hazardous species *A. clavatus*, p. 115–127. *In* K. A. Powell, A. Renwick, and J. F. Peberdy (ed.), *The Genus Aspergillus: From Taxonomy and Genetics to Industrial Application.* Plenum Press, New York, N.Y.

45. **Frisvad, J. C.** 1985. Secondary metabolites as an aid to *Emericella* classification, p. 437–444. *In* R. A. Samson and J. I. Pitt (ed.), *Advances in Penicillium and Aspergillus Systematics.* Plenum Press, New York, N.Y.

46. **Frisvad, J. C.** 1989. The connection between the penicillia and aspergilli and mycotoxins with species emphasis on misidentified isolates. *Arch. Environ. Contam. Toxicol.* 18:452–467.

47. **Frisvad, J. C., and U. Thrane.** 1995. Mycotoxins production by food-borne fungi, p. 251–260. *In* R. A. Samson, E. S. Hoekstra, J. C. Frisvad, and O. Filtenborg (ed.), *Introduction to Food-Borne Fungi*, 4th ed. Centraabureau voor Schimmelcultures, Baarn, The Netherlands.

48. **Frobish, R. A., B. D. Bradley, D. D. Wagner, P. E. Long-Bradley, and H. Hairston.** 1986. Aflatoxin residues in milk of dairy cows after ingestion of naturally contaminated grain. *J. Food Prot.* 49:781–785.

49. **Fujii, K., H. Kurata, S. Odashirna, and Y. Hatsuda.** 1976. Tumour induction by a single subcutaneous injection of sterigmatocystin in newborn mice. *Cancer Res.* 36:1615–1618.

50. **Gams, W., M. Christensen, A. H. S. Onions, J. I. Pitt, and R. A. Samson.** 1985. Intrageneric taxa of *Aspergillus*, p. 55–62. *In* R. A. Samson and J. I. Pitt (ed.), *Advances in Penicillium and Aspergillus Systematics.* Plenum Press, New York, N.Y.

51. **Geisen, R.** 1996. Multiplex polymerase chain reaction for the detection of potential aflatoxin and sterigmatocystin producing fungi. *Syst. Appl. Microbiol.* 19:388–392.

52. **Gilbert, J.** 1991. Regulatory aspects of mycotoxins in the European Community and USA, p. 194–197. *In* B. R. Champ, E. Highley, A. D. Hocking, and J. I. Pitt (ed.), *Fungi and Mycotoxins in Stored Products: Proceedings of an International Conference, Bangkok, Thailand, 23–26 April 1991.* ACIAR Proceedings no. 36. Australian Centre for International Agriculture Research, Canberra, Australia.

53. **Golden, M. C., S. J. Hahm, R. E. Elessar, S. Saksonov, and J. J. Steinberg.** 1998. DNA damage by gliotoxin from *Aspergillus fumigatus*. An occupational and environmental propagule: adduct detection as measured by ^{32}P DNA radiolabelling and two-dimensional thin-layer chromatography. *Mycoses* 41:97–104.

54. **Golinski, P.** 1991. Secondary metabolites (mycotoxins) produced by fungi colonizing cereal grain in store—structure and properties, p. 355–403. *In* J. Chelkowski (ed.), *Cereal Grain. Mycotoxins, Fungi and Quality in Drying and Storage.* Elsevier, Amsterdam, The Netherlands.

55. **Goto, T., D. T. Wicklow, and Y. Ito.** 1996. Aflatoxin and cyclopiazonic acid production by a sclerotium-producing *Aspergillus tamarii* strain. *Appl. Environ. Microbiol.* 62:4036–4038.

56. **Gqaleni, N., J. E. Smith, J. Lacey, and G. Gettinby.** 1996. The production of cyclopiazonic acid by *Penicillium commune* and cyclopiazonic acid and aflatoxins by *Aspergillus flavus* as affected by water activity and temperature on maize grains. *Mycopathologia* 136:103–108.

57. **Gqaleni, N., J. E. Smith, J. Lacey, and G. Gettinby.** 1997. Effects of temperature, water activity, and incubation time on production of aflatoxins and cyclopiazonic acid by an isolate of *Aspergillus flavus* in surface agar culture. *Appl. Environ. Microbiol.* 63:1048–1053.

58. **Groopman, J. D., L. G. Cain, and T. W. Kensler.** 1988. Aflatoxin exposure in human populations: measurements

and relation to cancer. *CRC Crit. Rev. Toxicol.* **19:**113–145.

59. **Gyongyosihorvath, A., I. Barnavetro, and L. Solti.** 1996. A new monoclonal antibody detecting ochratoxin A at the picogram level. *Lett. Appl. Microbiol.* **22:**103–105.

60. **Hatch, M. C., C.-J. Chen, B. Levin, B.-T. Ji, G.-Y. Yang, S.-W. Hsu, L.-W. Wang, L.-L. Hsieh, and R. M. Santella.** 1993. Urinary aflatoxin levels, hepatitis B virus infection and hepatocellular carcinoma in Taiwan. *Int. J. Cancer* **54:**931–934.

61. **Heenan, C. N., K. J. Shaw, and J. I. Pitt.** 1998. Ochratoxin A production by *Aspergillus carbonarius* and *A. niger* and detection using coconut cream agar. *J. Food Mycol.* **1:**67–72.

62. **Hocking, A. D.** 1992. Collaborative study on media for enumeration of xerophilic fungi, p. 121–125. *In* R. A. Samson, A. D. Hocking, J. I. Pitt, and A. D. King (ed.), *Modern Methods in Food Mycology.* Elsevier, Amsterdam, The Netherlands.

63. **Hocking, A. D., and J. I. Pitt.** 1980. Dichloran-glycerol medium for enumeration of xerophilic fungi from low moisture foods. *Appl. Environ. Microbiol.* **39:**488–492.

64. **Horie, Y.** 1995. Productivity of ochratoxin A of *Aspergillus carbonarius* in *Aspergillus* section *Nigri. Nippon Kingakukai Kaiho* **36:**73–76.

65. **King, A. D., A. D. Hocking, and J. I. Pitt.** 1979. Dichloran-rose bengal medium for enumeration and isolation of molds from foods. *Appl. Environ. Microbiol.* **37:**959–964.

66. **King, A. D., A. D. Hocking, and J. I. Pitt.** 1981. Mycoflora of some Australian foods. *Food Technol. Aust.* **33:**55–60.

67. **King, A. D., J. I. Pitt, L. R. Beuchat, and J. E. L. Corry** (ed.). 1986. *Methods for the Mycological Examination of Food.* Plenum Press, New York, N.Y.

68. **Klich, M. A., and J. I. Pitt.** 1988. *A Laboratory Guide to Common Aspergillus Species and Their Teleomorphs.* CSIRO Division of Food Processing, North Ryde, NSW, Australia.

69. **Klich, M. A., and J. I. Pitt.** 1988. Differentiation of *Aspergillus flavus* from *A. parasiticus* and closely related species. *Trans. Br. Mycol. Soc.* **91:**99–108.

70. **Klich, M. A., S. H. Thomas, and J. E. Mellon.** 1984. Field studies on the mode of entry of *Aspergillus flavus* into cotton seeds. *Mycologia* **76:**665–669.

71. **Koehler, P. E., L. R. Beuchat, and M. S. Chinnan.** 1985. Influence of temperature and water activity on aflatoxin production by *Aspergillus flavus* in cowpea (*Vigna unguiculata*) seeds and meal. *J. Food Prot.* **48:**1040–1043.

72. **Kozakiewicz, Z.** 1989. *Aspergillus* species on stored products. *Mycol. Pap.* **161:**1–188.

73. **Kozakiewicz, Z.** 1994. *Aspergillus,* p. 575–616. *In* Y. H. Hui, J. R. Gorham, K. D. Murrell, and D. O. Cliver (ed.), *Foodborne Disease Handbook,* vol. 2. Marcel Dekker, New York, N.Y.

74. **Kozakiewicz, Z.** 1994. *Aspergillus* toxins and taxonomy, p. 303–311. *In* K. A. Powell, A. Renwick, and J. F. Peberdy (ed.), *The Genus Aspergillus: From Genetics to Industrial Application.* Plenum Press, New York, N.Y.

75. **Krishnamachari, K. A. V., R. V. Bhat, V. Nagarajan, and T. B. G. Tilak.** 1975. Investigations into an outbreak of hepatitis in parts of Western India. *Indian J. Med. Res.* **63:**1036–1048.

76. **Kuraishi, H., M. Ito, M. Tsuzaki, Y. Katayama, T. Yokohama, and J. Sugiyama.** 1990. Ubiquinone systems as a taxonomic tool in *Aspergillus* and its teleomorphs, p. 407–421. *In* R. A. Samson and J. I. Pitt (ed.), *Modern Concepts in Penicillium and Aspergillus Classification.* Plenum Press, New York, N.Y.

77. **Lacey, J.** 1994. Aspergilli in feeds and seeds, p. 73–92. *In* K. A. Powell, A. Renwick, and J. F. Peberdy (ed.), *The Genus Aspergillus: From Taxonomy and Genetics to Industrial Application.* Plenum Press, New York, N.Y.

78. **Land, C. J., H. Lundstrom, and S. Werner,** 1993. Production of tremorgenic mycotoxins by isolates of *Aspergillus fumigatus* from sawmills in Sweden. *Mycopathologia* **124:**87–93.

79. **Lawellin, D. W., D. W. Grant, and B. K. Joyce.** 1977. Enzyme-linked immunosorbent analysis of aflatoxin B$_1$. *Appl. Environ. Microbiol.* **34:**94–96.

80. **Leong, S.-L. L., J. I. Pitt, and A. D. Hocking.** 1999. Black Aspergilli on dried vine fruits, abstr. no. MP15.12. *IXth International Congress of Mycology, International Union of Microbiological Societies,* Sydney, NSW, Australia, 16–20 August 1999.

81. **Lillehoj, E. B., W. F. Kwolek, E. S. Horner, N. W. Widstrom, L. M. Josephson, A. O. Franz, and E. A. Catalano.** 1980. Aflatoxin contamination of preharvest corn: role of *Aspergillus flavus* inoculum and insect damage. *Cereal Chem.* **57:**255–257.

82. **Ling, K. H.** 1994. Territrems, tremorgenic mycotoxin isolated from *Aspergillus terreus. J. Toxicol. Toxin Rev.* **13:**243–252.

83. **Ling, K. H., C. K. Yang, and F. T. Peng.** 1979. Territrems, tremorgenic mycotoxins of *Aspergillus terreus. Appl. Environ. Microbiol.* **37:**355–357.

84. **Liu, B. H., and F. S. Chu.** 1998. Regulation of the *aflR* and its product AFLR, associated with aflatoxin biosynthesis. *Appl. Environ. Microbiol.* **64:**3718–3723.

85. **LopezDiaz, T. M., and B. Flannigan.** 1997. Mycotoxins of *Aspergillus clavatus:* toxicity of cytochalasin E, patulin, and extracts of contaminated barley malt. *J. Food Prot.* **60:**1381–1385.

86. **LopezDiaz, T. M., and B. Flannigan.** 1997. Production of patulin and cytochalasin E by *Aspergillus clavatus* during malting of barley and wheat. *Int. J. Food Microbiol.* **35:**129–136.

87. **Lubulwa, A. S. G., and J. S. Davis.** 1994. Estimating the social costs of the impacts of fungi and aflatoxins, p. 1017–1042. *In* E. Highley, E. J. Wright, H. J. Banks, and B. R. Champ (ed.), *Stored Product Protection. Proceedings of the 6th International Working Conference on Stored-Product Protection.* CAB International, Wallingford, Oxford, United Kingdom.

88. **Lye, M. S., A. A. Ghazali, J. Mohan, N. Alwin, and R. C. Nair.** 1995. An outbreak of acute hepatic encephalopathy due to severe aflatoxicosis in Malaysia. *Am. J. Trop. Med. Hyg.* **53:**67–82.

89. **Ministry of Agriculture, Fisheries and Food (MAFF).** 1999. 1998 survey of retail products for ochratoxin A. MAFF, Food Surveillance Information Sheet 185, p. 1–36. HMSO Publications Centre, London, United Kingdom.

90. **Miyake, I., H. Naito, and H. Sumeda.** 1940. *Rep. Res. Inst. Rice Improvement* **1:**1. (In Japanese.)

91. **Moody, S. F., and B. M. Tyler.** 1990. Restriction enzyme and mitochondrial DNA of the *Aspergillus flavus* group, *Aspergillus flavus, Aspergillus parasiticus* and *Aspergillus nomius. Appl. Environ. Microbiol.* **56:**2441–2452.

92. **Moody, S. F., and B. M. Tyler.** 1990. Use of DNA restriction length polymorphisms to analyse the diversity of the *Aspergillus flavus* group, *Aspergillus flavus, Aspergillus parasiticus* and *Aspergillus nomius. Appl. Environ. Microbiol.* **56:**2453–2461.

93. **Mori, T., M. Matsumura, K. Yamada, S. Irie, K. Oshimi, K. Suda, T. Oguri, and M. Ichinoe.** 1998. Systemic aspergillosis caused by an aflatoxin-producing strain of *Aspergillus flavus. Med. Mycol.* **36:**107–112.

94. **Moss, M. O., and J. E. Smith.** 1985. *Mycotoxins: Formation, Analysis and Significance.* John Wiley & Sons, Chichester, United Kingdom.

95. **Mullaney, E. J., and M. A. Klich.** 1990. A review of molecular biological techniques for systematic studies of *Aspergillus* and *Penicillium*, p. 301–307. *In* R. A. Samson and J. I. Pitt (ed.), *Modern Concepts in Penicillium and Aspergillus Classification.* Plenum Press, New York, N.Y.

96. **Nesheim, S., and M. Trucksess.** 1986. Thin-layer chromatography/high-performance thin-layer chromatography as a tool for mycotoxin determination. *In* R. J. Cole (ed.), *Modern Methods for Analysis and Structural Elucidation of Mycotoxins.* Academic Press, Orlando, Fla.

97. **Nielsen, K. F., S. Gravesen, P. A. Nielsen, B. Andersen, U. Thrane, and J. C. Frisvad.** 1999. Production of mycotoxins on artificially infested building materials. *Mycopathologia* **145:**43–56.

98. **Nishie, K., R. J. Cole, and F. W. Dorner.** 1985. Toxicity and neuropharmacology of cyclopiazonic acid. *Food Chem. Toxicol.* **23:**831–839.

99. **Norred, W. P., R. E. Morrisey, R. T. Riely, R. J. Cole, and L. W. Dorner.** 1985. Distribution, excretion and skeletal muscle effects of the mycotoxin (^{14}C) cyclopiazonic acid in rats. *Food Chem. Toxicol.* **23:**1069–1076.

100. **Northolt, M. D., H. P. van Egmond, and W. E. Paulsch.** 1977. Differences in *Aspergillus flavus* strains in growth and aflatoxin B_1 production in relation to water activity and temperature. *J. Food Prot.* **40:**778–781.

101. **Northolt, M. D., H. P. van Egmond, and W. E. Paulsch.** 1979. Ochratoxin A production by some fungal species in relation to water activity and temperature. *J. Food Prot.* **42:**485–490.

102. **Northolt, M. D., H. P. van Egmond, and W. E. Paulsch.** 1979. Penicillic acid production by some fungal species in relation to water activity and temperature. *J. Food Prot.* **42:**476–484.

103. **Ono, H., A. Kataoka, M. Koakutsu, K. Tanaka, S. Kawasugi, M. Wakazawa, Y. Ueno, and M. Manabe.** 1995. Ochratoxin A producibility by strains of *Aspergillus niger* group stored in IFO culture collection. *Mycotoxins* **41:**47–51.

104. **Payne, G. A., and M. P. Brown.** 1998. Genetics and physiology of aflatoxin biosynthesis. *Annu. Rev. Phytopathol.* **36:**329–362.

105. **Peers, F., X. Bosch, J. Kaldor, A. Linsell, and M. Pluumen.** 1987. Aflatoxin exposure, hepatitis B virus infection and liver cancer in Swaziland. *Int. J. Cancer* **39:**545–553.

106. **Pestka, J. J., M. N. Abouzied, and Sutikno.** 1995. Immunological assays for mycotoxin detection. *Food Technol.* **49**(2):120–128.

107. **Pier, A. C.** 1991. The influence of mycotoxins on the immune system, p. 489–497. *In* J. E. Smith and R. S. Henderson (ed.), *Mycotoxins and Animal Foods.* CRC Press, Boca Raton, Fla.

109. **Pitt, J. I., S. K. Dyer, and S. McCammon.** 1991. Systemic invasion of developing peanut plants by *Aspergillus flavus. Lett. Appl. Microbiol.* **13:**16–20.

110. **Pitt, J. I., and A. D. Hocking.** 1997. *Fungi and Food Spoilage.* 2nd ed. Aspen Publishers, Gaithersburg, Md.

111. **Pitt, J. I., A. D. Hocking, K. Bhudhasamai, B. F. Miscamble, K. A. Wheeler, and P. Tanboon-Ek.** 1993. The normal mycoflora of commodities from Thailand. 1. Nuts and oilseeds. *Int. J. Food Microbiol.* **20:**211–226.

112. **Pitt, J. I., A. D. Hocking, K. Bhudhasamai, B. F. Miscamble, K. A. Wheeler, and P. Tanboon-Ek.** 1994. The normal mycoflora of commodities from Thailand. 2. Beans, rice, small grains and other commodities. *Int. J. Food Microbiol.* **23:**35–53.

113. **Pitt, J. I., A. D. Hocking, and D. R. Glenn.** 1983. An improved medium for the detection of *Aspergillus flavus* and *A. parasiticus. J. Appl. Bacteriol.* **54:**109–114.

114. **Pitt, J. I., and B. F. Miscamble.** 1995. Water relations of *Aspergillus flavus* and closely related species. *J. Food Prot.* **58:**86–90.

115. **Pitt, J. I., and R. A. Samson.** 1993. Species names in current use in the *Trichocomaceae* (Fungi, Eurotiales). *Regnum Veg.* **128:**13–57.

116. **Ramos, A. J., N. la Bernia, S. Marin, V. Sanchis, and N. Magan.** 1998. Effect of water activity and temperature on growth and ochratoxin production by three strains of *Aspergillus ochraceus* on a barley extract medium and on barley grains. *Int. J. Food Microbiol.* **44:**133–140.

117. **Raper, K. B., and D. I. Fennell.** 1965. *The Genus Aspergillus.* The Williams and Wilkins Co., Baltimore, Md.

118. **Richards, J. L.** 1997. Gliotoxin, a mycotoxin associated with cases of avian aspergillosis. *J. Nat. Toxins* **6:**11–18.

119. **Samson, R. A., A. D. Hocking, J. I. Pitt, and A. D. King (ed.).** 1992. *Modern Methods in Food Mycology.* Elsevier, Amsterdam, The Netherlands.

120. **Samson, R. A., E. S. Hoekstra, J. C. Frisvad, and O. Filtenborg.** 1995. *Introduction to Food-Borne Fungi,* 4th ed. Centraalbureau voor Schimmelcultures, Baarn, The Netherlands.

121. **Samson, R. A., and J. I. Pitt.** 1985. *Advances in Penicillium and Aspergillus Systematics.* Plenum Press, New York, N.Y.

122. **Samson, R. A., and J. I. Pitt.** 1990. *Modern Concepts in Penicillium and Aspergillus Classification.* Plenum Press, New York, N.Y.

123. **Sanders, T. H., R. A. Hill, R. J. Cole, and P. D. Blankenship.** 1981. Effect of drought on the occurrence of *Aspergillus flavus* in maturing peanuts. *J. Am. Oil Chem. Soc.* 58:966A–970A.

124. **Sargeant, K., R. B. A. Carnaghan, and R. Allcroft.** 1963. Toxic products in groundnuts—chemistry and origin. *Chem. Ind.* 1963:53–55.

125. **Schneider, E., E. Usleber, E. Martl-Bauer, R. Dietrich, and G. Terplan.** 1995. Multimycotoxin dipstick enzyme immunoassay applied to wheat. *Food Addit. Contam.* 12:387–393.

126. **Scott, P. M.** 1994. *Penicillium* and *Aspergillus* toxins, p. 261–285. *In* J. D. Miller and H. L. Trenholm (ed.). *Mycotoxins in Grain. Compounds Other Than Aflatoxin.* Eagan Press, St. Paul, Minn.

127. **Shank, R. C.** 1978. Mycotoxicoses of man: dietary and epidemiological considerations, p. 1–12. *In* T. D. Wyllie and L. G. Morhouse (ed.), *Mycotoxigenic Fungi, Mycotoxins, Mycotoxicoses: an Encyclopedic Handbook.* Marcel Dekker, New York, N.Y.

128. **Shapira, R., N. Paster, O. Eyal, M. Menasherov, A. Mett, and R. Salomon.** 1996. Detection of aflatoxigenic molds in grains by PCR. *Appl. Environ. Microbiol.* 62:3270–3273.

129. **Smith, J. E., and K. Ross.** 1991. The toxigenic aspergilli, p. 101–118. *In* J. E. Smith and R. S. Henderson (ed.), *Mycotoxins and Animal Foods.* CRC Press, Boca Raton, Fla.

130. **Smith, S. L., and S. T. Hill.** 1982. Influence of temperature and water activity on germination and growth of *Aspergillus restrictus* and *A. versicolor. Trans. Br. Mycol. Soc.* 79:558–560.

131. **Sreenivasamurthy, V., H. A. B. Parpia, S. Srikanta, and A. Shankarmurti.** 1967. Detoxification of aflatoxin in peanut meal by hydrogen peroxide. *J. Assoc. Off. Anal. Chem.* 50:350–354.

132. **Stoloff, L.** 1977. Aflatoxins—an overview, p. 7–28. *In* J. V. Rodricks, C. W. Hesseltine, and M. A. Mehlman (ed.), *Mycotoxins in Human and Animal Health.* Pathatox Publishers, Park Forest South, Ill.

133. **Stoloff, L.** 1983. Aflatoxin as a cause of primary liver-cell cancer in the United States: a probability study. *Nutr. Cancer* 5:165–168.

134. **Stoloff, L., and L. Friedman.** 1976. Information bearing on the evaluation of the hazard to man from aflatoxin ingestion. *PAG Bull.* 6:21–32.

135. **Terao, K.** 1983. Sterigmatocystin—a masked potent carcinogenic mycotoxin. *J. Toxicol. Toxin Rev.* 2:77–100.

136. **Téren, J., J. Varga, Z. Hamari, E. Rinyu, and É. Kevei.** 1996. Immunochemical detection of ochratoxin A in black *Aspergillus* strains. *Mycopathologia* 134:171–176.

137. **Tran-Dinh, N., J. I. Pitt, and D. A. Carter.** 1999. Molecular genotype analysis of natural toxigenic and non-toxigenic isolates of *Aspergillus flavus* and *A. parasiticus. Mycol. Res.* 103:1485–1490.

138. **Trucksess, M. W., J. Giler, K. Young, K. D. White, and S. W. Page.** 1999. Determination and survey of ochratoxin A in wheat, barley and coffee. *J. AOAC Int.* 82:85–98.

139. **Trucksess, M. W., and M. E. Stack.** 1994. Enzyme-linked immunosorbent assay of total aflatoxins B_1, B_2 and G_1 in corn: follow-up collaborative study. *J. AOAC Int.* 77:655–658.

140. **Tuomi, T., K. Reijula, T. Johnsson, K. Hemminki, E. L. Hintikka, O. Lindroos, S. Kalso, P. Koukila-Kahkola, H. Mussalo-Rauhamaa, and T. Haahtela.** 2000. Mycotoxins in crude building materials from water-damaged buildings. *Appl. Environ. Microbiol.* 66:1899–1904.

141. **van der Merwe, K. J., P. S. Steyn, L. Fourie, D. B. Scott, and J. J. Theron.** 1965. Ochratoxin A, a toxic metabolite produced by *Aspergillus ochraceus* Wilh. *Nature* 205:1112–1113.

142. **van der Stegen G., U. J. Jörissen, A. Pittet, M. Saccon, W. Steiner, M. Vincenzi, M. Winkler, J. Zapp, and C. Schlatter.** 1997. Screening of European coffee final products for occurrence of ochratoxin A. *Food Addit. Contam.* 14:211–216.

143. **Van Egmond, H. P.** 1991. Regulatory aspects of mycotoxins in Asia and Africa, p. 198–204. *In* B. R. Champ, E. Highley, A. D. Hocking, and J. I. Pitt (ed.), *Fungi and Mycotoxins in Stored Products: Proceedings of an International Conference, Bangkok, Thailand, 23–26 April 1991.* ACIAR Proceedings no. 36. Australian Centre for International Agricultural Research (ACIAR), Canberra, Australia.

144. **van Rensburg, S. J.** 1977. Role of epidemiology in the elucidation of mycotoxin health risks, p. 699–711. *In* J. V. Rodricks, C. W. Hesseltine, and M. A. Mehlman (ed.), *Mycotoxins in Human and Animal Health.* Pathatox Publishers, Park Forest South, Ill.

145. **Varga, J., E., Kevei, É. Rinyu, J. Téren, and Z. Kozakiewicz.** 1996. Ochratoxin production by *Aspergillus* species. *Appl. Environ. Microbiol.* 62:4461–4464.

146. **Vesonder, R. F., R. Lambert, D. T. Wicklow, and M. L. Biehl.** 1988. *Eurotium* spp. and echinulin in feed refusal by swine. *Appl. Environ. Microbiol.* 54:830–831.

147. **Watanabe, C. M. H., and C. A. Townsend.** 1998. The *in vitro* conversion of norsolinic acid to aflatoxin B_1. An improved method of cell-free enzyme preparation and stabilization. *J. Am. Chem. Soc.* 120:6231–6239.

148. **Weng, C. Y., A. J. Martinez, and D. L. Park.** 1994. Efficacy and permanency of ammonia treatment in reducing aflatoxin levels in corn. *Food Addit. Contam.* 11:649–658.

149. **Wheeler, K. A., B. F. Hurdman, and J. I. Pitt.** 1991. Influence of pH on the growth of some toxigenic species of *Aspergillus, Penicillium* and *Fusarium. Int. J. Food Microbiol.* 12:141–150.

150. **Wicklow, D. T., P. F. Dowd, A. A. Alfatafta, and J. B. Gloer.** 1996. Ochratoxin A: an antiinsectan metabolite from the sclerotia of *Aspergillus carbonarius* NRRL 369. *Can. J. Microbiol.* 42:1100–1103.

151. **Woloshuk, C. P., and R. Prieto.** 1998. Genetic organization and function of the aflatoxin B$_1$ biosynthetic genes. *FEMS Microbiol. Lett.* **160:**169–176.

152. **Yabe, K., M. Nakamura, and T. Hamasaki.** 1999. Enzymatic formation of G-group aflatoxins and biosynthetic relationship between G- and B-group aflatoxins. *Appl. Environ. Microbiol.* **65:**3867–3872.

153. **Yagi, R., and M. Doi.** 1999. Isolation of an antioxidative substance produced by *Aspergillus repens. Biosci. Biotechnol. Biochem.* **63:**932–933.

154. **Yoshizawa, T.** 1991. Natural occurrence of mycotoxins in small grain cereals (wheat, barley, rye, oats, sorghum, millet, rice), p. 301–324. *In* J. E. Smith and R. S. Henderson (ed.), *Mycotoxins and Animal Foods.* CRC Press, Boca Raton, Fla.

155. **Yu, J. J., P. K. Chang, J. W. Cary, M. Wright, D. Bhatnagar, T. E. Cleveland, G. A. Payne, and J. E. Linz.** 1995. Comparative mapping and aflatoxin pathway gene clusters in *Aspergillus parasiticus* and *Aspergillus flavus. Appl. Environ. Microbiol.* **61:**2365–2371.

156. **Zimmerli, B., and R. Dick.** 1996. Ochratoxin A in table wine and grape-juice: occurrence and risk assessment. *Food Addit. Contam.* **13:**655–668.

Food Microbiology: Fundamentals and Frontiers, 2nd Ed.
Edited by M. P. Doyle et al.
© 2001 ASM Press, Washington, D.C.

John I. Pitt

Toxigenic *Penicillium* Species

22

Experimental evidence that common microfungi can produce toxins is popularly believed to date only from about 1960. However, the study of mycotoxicology began 100 years ago. In 1891, in Japan, Sakaki demonstrated that an ethanol extract from moldy, unpolished "yellow" rice was fatal to dogs, rabbits, and guinea pigs, with symptoms indicating paralysis of the central nervous system (99). In consequence, the sale of yellow rice was banned in Japan in 1910. In 1913, an extract from a *Penicillium puberulum* culture isolated from moldy corn in Nebraska was found to be toxic to animals when injected at 200 to 300 mg/kg of body weight (1). This was the first reliable account of toxin production by a mold in pure culture. This careful study sought to resolve the question of whether common molds or mold products could have an injurious effect on animals, produced positive results, and was far ahead of its time. Nevertheless, such direct evidence that common molds could be toxic was largely ignored.

In 1940, *Penicillium toxicarium* was isolated from yellow rice and described (67). It produced a highly toxic metabolite, subsequently named citreoviridin. Perhaps because this study was published in wartime, it also failed to alert the world to the potential or actual danger of the toxicity of common molds.

The discovery of penicillin in 1929 gave impetus to a search for other *Penicillium* metabolites with antibiotic properties and ultimately led to the recognition of citrinin, patulin, and griseofulvin as "toxic antibiotics" or, later, mycotoxins.

The literature on toxigenic penicillia is now quite vast. In a comprehensive review of literature on fungal metabolites, about 120 common mold species were demonstrably toxic to higher animals (18). Forty-two were reported to be produced by one or more *Penicillium* species. No fewer than 85 *Penicillium* species were listed as toxigenic (18). The literature on this subject has accumulated in a rather random fashion, with emphasis in the majority of papers on chemistry or toxicology rather than mycology. The impression gained from the literature is that toxin production by penicillia lacks species specificity, i.e., most toxins are produced by a variety of species. This viewpoint is commonly accepted. For example, citrinin has been reported to be produced by at least 22 species (18, 55, 88), but perhaps as few as three species are actual producers. Explanations include the fact that the taxonomy of toxigenic *Penicillium* species has been uncertain and that many reports on toxigenicity of particular species have been based on misidentifications.

John I. Pitt, Food Science Australia, P.O. Box 52, North Ryde, New South Wales 1670, Australia.

Recent developments in *Penicillium* taxonomy have profoundly changed the accuracy of identifications. Improvements in morphological classification (72), the development of simple screening methods for secondary metabolites and the recognition that such compounds are of taxonomic value (32, 34, 35), emphasis on accurate identification and mycotoxin production (81), and the use of electrophoretic fingerprinting of certain isoenzymes (4, 21) have all assisted in developing a more definitive taxonomy (85). Molecular studies have to date mainly confirmed the more traditional findings (60, 90) and permitted refinement of phylogeny (5, 59). A classification of the more common species has been published (75), and an accurate picture of the more important species-mycotoxin relationships has been developed (81). A list of "Names in Current Use" provides a complete list of accepted names in *Penicillium* and related genera (82).

TAXONOMY

Penicillium is a large genus. Although 150 species were recognized in the most recent complete taxonomy (72), subsequent studies indicate that this number is conservative. At least 50 species are of common occurrence (75). All common species grow and sporulate well on synthetic or semisynthetic media and are usually readily recognizable to genus level.

Classification within *Penicillium* is based primarily on microscopic morphology (Fig. 22.1). The genus is divided into subgenera based on the number and arrangement of phialides (elements producing conidia) and metulae and rami (elements supporting phialides) on the main stalk cells (stipes). The classification of Pitt (72) includes four subgenera: *Aspergilloides*, where phialides are borne directly on the stipes without intervening supporting elements; *Furcatum* and *Biverticillium*, where phialides are supported by metulae; and *Penicillium*, where both metulae and rami are usually present (Fig. 22.1). The majority of important toxigenic and food spoilage species are found in the subgenus *Penicillium*.

ENUMERATION

General enumeration procedures suitable for foodborne molds are effective for enumerating all common *Penicillium* species. Many antibacterial enumeration media can be expected to give satisfactory results. However, some *Penicillium* species grow rather weakly on very dilute media such as potato-dextrose agar. Dichloran-rose bengal-chloramphenicol agar (DRBC) (77) and

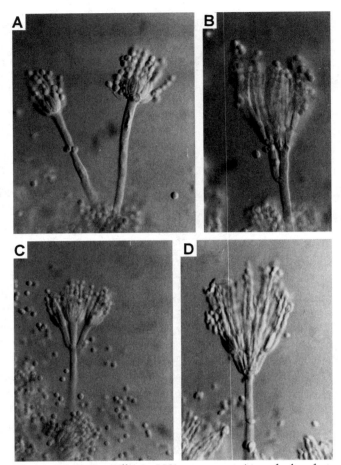

Figure 22.1 Penicilli (×550) representative of the four *Penicillium* subgenera. (A) *Penicillium* subgenus *Aspergilloides* (*P. glabrum*); (B) *Penicillium* subgenus *Penicillium* (*P. expansum*); (C) *Penicillium* subgenus *Furcatum* (*P. citrinum*); (D) *Penicillium* subgenus *Biverticillium* (*P. variabile*).

dichloran–18% glycerol (DG18) agar (77) are recommended (42, 48, 84).

IDENTIFICATION

For a comprehensive taxonomy of *Penicillium*, see Pitt (72); for keys and descriptions to common species, see Pitt (75); and for foodborne species, see Pitt and Hocking (77) and Samson et al. (84).

Identification of *Penicillium* isolates to species level is not easy and is preferably carried out under carefully standardized conditions of media, incubation time, and temperature. In addition to microscopic morphology, gross physiological features, including colony diameters and colors of conidia and colony pigments, are used to distinguish species (72, 75).

SIGNIFICANT *PENICILLIUM* MYCOTOXINS

As already noted, more than 80 *Penicillium* species have been reported as toxin producers. In assessing the relevance of these reports, several points need to be kept in mind: (i) many of the identifications reported in the literature are incorrect, and this has resulted in confusion; (ii) a number of genuinely toxic compounds are produced by species not usually occurring in commodities, foods, or feeds—such toxins are of little practical importance and are not discussed here; (iii) some compounds often cited as mycotoxins in the literature are of low toxicity and again are of little practical importance; and (iv) growth of mold does not always mean production of toxin. The conditions under which toxins are produced are often narrower than the conditions for growth. Appropriate information is provided where this is known. Taking these factors into account, a recent review listed 27 mycotoxins, produced by 32 species, that possessed demonstrated toxicity to humans or domestic animals (81).

The range of mycotoxin classes produced by *Penicillium* is broader than for any other fungal genus. Moreover, molecular composition is diverse in the extreme. Patulin is an unsaturated lactone with a molecular weight of 150, whereas penitrem A is a polycyclic ether with nine contiguous rings and a molecular weight of more than 650 (18, 22).

Toxicity due to *Penicillium* species is also very diverse. However, most toxins can be placed in two broad groups: those that affect liver and kidney function and those that are neurotoxins. In general terms, the *Penicillium* toxins that affect liver or kidney function are asymptomatic or cause generalized debility in humans or animals. In contrast, toxicity of the neurotoxins in animals is often characterized by sustained trembling. However, individual toxins show wide variations from these generalizations.

Table 22.1 lists nine mycotoxins, produced by 17 *Penicillium* species, that the author considers to be significant, or to have potential significance, in human health.

Table 22.1   Significant mycotoxins produced by *Penicillium* species

Mycotoxin	Toxicity, LD$_{50}$[a]	Species producing[b]
Citreoviridin	Mice, 7.5 mg/kg i.p.	*Penicillium citreonigrum* Dierckx
	Mice, 20 mg/kg oral	*Eupenicillium ochrosalmoneum* Scott & Stolk
Citrinin	Mice, 35 mg/kg i.p.	*P. citrinum* Thom
	Mice, 110 mg/kg oral	*P. expansum* Link
		P. verrucosum Dierckx
Cyclopiazonic acid	Rats, 2.3 mg/kg i.p.	*P. camemberti* Thom
	Male rats, 36 mg/kg oral	*P. commune* Thom
	Female rats, 63 mg/kg oral	*P. chrysogenum* Thom
		P. griseofulvum Dierckx
		P. hirsutum Dierckx
		P. viridicatum Westling
Ochratoxin A	Young rats, 22 mg/kg oral	*P. verrucosum*
Patulin	Mice, 5 mg/kg i.p.	*P. expansum*
	Mice, 35 mg/kg oral	*P. vulpinum* (Cooke & Massee) Seifert & Samson
		P. griseofulvum
		P. roqueforti Thom
Penitrem A	Mice, 1 mg/kg i.p.	*P. crustosum* Thom
		P. glandicola (Oudem.) Seifert & Samson
PR toxin	Mice, 6 mg/kg i.p.	*P. roqueforti*
	Rats, 115 mg/kg oral	
Roquefortine C	Mice, 340 mg/kg i.p.	*P. roqueforti*
		P. chrysogenum
		P. crustosum
Secalonic acid D	Mice, 42 mg/kg i.p.	*P. oxalicum* Currie & Thom

[a] References 13 and 75. LD$_{50}$, 50%, lethal dose. Dosing: i.p., intraperitoneal injection.
[b] Reference 23 and personal observations.

Relative toxicities of the various compounds are shown for comparative purposes. For this reason, toxicities in a standard test animal (mouse) and by a standard route of administration (oral) have been provided where possible. The list of species shown as producing each toxin results from the author's own experience and is probably conservative. Some general notes follow on these toxins and the species that produce them.

Ochratoxin A

Undoubtedly the most important toxin produced by a *Penicillium* species, ochratoxin A is derived from isocomarin linked to phenylalanine. The International Agency for Research on Cancer (IARC) has classified this compound as a possible human carcinogen (group 2B), based on sufficient evidence of carcinogenicity in experimental animal studies and inadequate evidence in humans (45). The target organ of toxicity in all mammalian species tested is the kidneys, in which lesions can be produced by both acute and chronic exposure (38).

The source of ochratoxin A in nature is relatively complex. Ochratoxin A was originally described as a metabolite of *Aspergillus ochraceus* (101), and it has recently been shown to be commonly produced by *Aspergillus carbonarius* (98). Later, ochratoxin A (and sometimes citrinin) was reported to be produced by *Penicillium viridicatum* (104), and this view prevailed for more than a decade. Aftere some confusion, it became clear that isolates regarded as *P. viridicatum* but producing ochratoxin and citrinin were more correctly classified in a separate species, *Penicillium verrucosum* (74). The majority of *P. verrucosum* isolates are ochratoxin A producers. In a comprehensive study, 48 of 84 isolates produced ochratoxin A in the laboratory, and 6 of these same isolates produced citrinin (74).

P. verrucosum grows in barley and wheat crops in cold climates, especially Scandinavia, central Europe, and western Canada. Ochratoxin A plays a major role in the etiology of nephritis (kidney disease) in pigs in Scandinavia (49, 50, 51) and indeed in much of northern Europe. In an extensive survey on 70 Danish barley samples from farms where pigs were suffering from nephritis, 67 contained high populations of *P. verrucosum* and 66 contained ochratoxin A (32, 36). It seems certain that *P. verrucosum* is the major source of ochratoxin A in Scandinavia and other cool temperate zone areas.

The major source of ochratoxin A in foods is bread made from barley or wheat in which *P. verrucosum* has grown. Because ochratoxin A is fat soluble and not readily excreted, it also accumulates in the depot fat of animals eating feeds containing ochratoxin A and from

there is ingested by humans eating pig meats. In consequence, ochratoxin A has been found to be widespread in Europeans' blood (7, 37) and breast milk (8, 65). Although clear evidence of human disease is still elusive, such levels indicate a widespread problem with ochratoxin A in Europe.

It has been suggested that ochratoxin A is a causal agent of Balkan endemic nephropathy, a kidney disease with a high mortality rate in certain areas of Bulgaria, Yugoslavia, and Romania. The symptoms in people are sufficiently similar to those of ochratoxin A toxicity in animals that a role for ochratoxin A in this disease has been postulated on a number of occasions (3, 52, 71). However, no conclusive epidemiological evidence has emerged to support this hypothesis, and its relevance to studies on ochratoxin A toxicity remains doubtful (63). Other toxins or contributing factors are probably involved (97).

Taxonomy

Described originally in 1901, *P. verrucosum* was ignored until revived by Samson et al. in 1976 (86). Their concept was very broad, encompassing several species previously considered distinct. *P. verrucosum* was restricted to a much narrower concept by Pitt (72). With the recognition that this concept is linked to ochratoxin production (74), it has now been widely accepted. *P. verrucosum* is classified in subgenus *Penicillium* section *Penicillium*, which includes many mycotoxigenic species that commonly occur in foods. *P. verrucosum* (Fig. 22.2A) is distinguished by slow growth on Czapek yeast extract agar (CYA) and malt extract agar (MEA) at 25°C (17 to 24 mm and 10 to 20 mm after 7 days, respectively), bright green conidia, clear to pale yellow exudate, and rough stipes (74). It is similar in general appearance to *P. viridicatum*, differing most obviously by slower growth, and to *Penicillium solitum*, from which it differs by having green rather than blue conidia (75).

Enumeration and Identification

The media specified above for general enumeration of *Penicillium* species are effective for *P. verrucosum*. On dichloran-rose bengal-yeast extract-sucrose (DRYS) agar, a selective medium for the enumeration of *P. verrucosum* and *P. viridicatum*, *P. verrucosum* produces a violet brown reverse coloration (31). Isolation and identification of *P. verrucosum* in pure culture are essential for confirmation.

Ecology

P. verrucosum has been reported almost exclusively in grain from temperate zones. It is associated with barley

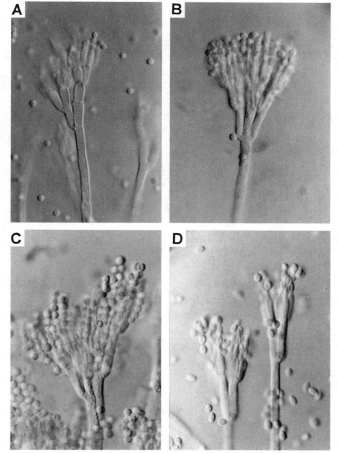

Figure 22.2 Some toxigenic *Penicillium* species (×550). (A) *P. verrucosum*; (B) *P. crustosum*; (C) *P. roqueforti*; (D) *P. oxalicum*.

watery diarrhea, increased water consumption, and reduced weight gain due to kidney degeneration (92). The importance of citrinin in human health is currently difficult to assess. When it is ingested in the absence of other toxins, significant effects appear unlikely. However, citrinin is produced in nature along with ochratoxin A by some isolates of *P. verrucosum* (51, 74) and with patulin by some isolates of *Penicillium expansum* (81). The possibility of synergy between these toxins should not be discounted.

Primarily recognized as a metabolite of *Penicillium citrinum*, citrinin has been reported from more than 20 other fungal species (18, 31, 88). However, apart from *P. citrinum*, only *P. expansum* and *P. verrucosum* are certain producers (81). These three species are among the more commonly occurring penicillia, so citrinin is probably the most widely occurring *Penicillium* toxin.

Taxonomy

Classified in subgenus *Furcatum* section *Furcatum* (72), *P. citrinum* is a very well-circumscribed species, accepted without controversy for many years. The most distinctive feature of *P. citrinum* is its penicillus (Fig. 21.1C), which consists of a cluster of three to five divergent metulae, usually apically swollen. Under the stereomicroscope, the phialides from each metula bear conidia as long columns, producing a distinctive pattern that can be of diagnostic value. Colonies of this species on CYA and MEA are of moderate size (25 to 30 mm and 14 to 18 mm, respectively), with the smaller size on MEA also a distinctive feature. Growth normally occurs at 37°C, but colonies seldom exceed 10 mm after 7 days (75).

Enumeration and Identification

P. citrinum can be enumerated effectively on any of the enumeration media mentioned above. In all cases, confirmation requires isolation and identification by the standard methods. No selective media have been developed for this species.

Ecology

P. citrinum is a ubiquitous mold and has been isolated from nearly every kind of food surveyed for fungi. The most common sources are cereals, especially rice, wheat, and corn, milled grains, and flour (77). Toxin production is also likely to be a common occurrence.

P. citrinum grows from 5 to 7 to 40°C, with an optimum near 30°C (77). It is xerophilic, with a minimum a_w for growth near 0.82 (41). Citrinin is produced over most of the temperature growth range, but the effect of a_w is unknown (46).

and wheat from Scandinavia (36) and Canada (56). It has also been isolated quite frequently from meat products in Germany and other European countries. It is uncommon elsewhere (77).

P. verrucosum grows most strongly at relatively low temperatures, down to 0°C, with a maximum near 31°C. It is xerophilic, growing down to water activity (a_w) 0.80. Maximum ochratoxin A production occurs at about 20°C and is possible down to about a_w 0.85 (46). Cereals in storage only marginally above levels considered safe from mold growth appear especially vulnerable.

Citrinin

Citrinin is a significant renal toxin affecting monogastric domestic animals such as pigs and dogs (13, 30). It is also an important toxin in domestic birds, where it produces

Patulin

Patulin is a lactone and produces teratogenic, neurological, and gastrointestinal effects in rodents (92). Early reports of carcinogenicity (23) have not been substantiated, but mutagenicity in a cell line has recently been reported (2). Like most other mycotoxins, patulin has no obvious immunological effects (58).

The most important *Penicillium* species producing patulin is *P. expansum*, best known as a fruit pathogen but also of widespread occurrence in other fresh and processed foods (77, 93). *P. expansum* produces patulin as it rots apples and pears. The use of rotting fruit in juice or cider manufacture can result in high concentrations of patulin, 350 μg/liter (6) or even 630 μg/liter (105) having been found in the resultant juice. Levels in commercial practice are usually much lower than this, and, given that patulin appears to lack chronic toxic effects in humans, low levels in juices are perhaps of little concern (68). However, because apple juice is widely consumed by children as well as adults, some countries have set an upper limit of 50 μg/liter for patulin in apple juice and other apple products (92, 102). It is also important as an indicator of the use of poor quality raw materials in juice manufacture.

Patulin is produced by several other *Penicillium* species besides *P. expansum* (Table 21.1), but the potential for production of unacceptable levels in foods appears to be much lower.

Taxonomy

One of the oldest described *Penicillium* species, *P. expansum* has been established as the principal cause of spoilage of pome fruits throughout this century. It belongs to subgenus *Penicillium* section *Penicillium* (72). *P. expansum* colonies usually grow quite rapidly on CYA and MEA (30 to 40 mm in diameter in 7 days) and show orange brown or cinnamon colors in exudate, soluble pigment, and reverse (75). Penicilli are terverticillate, with smooth stipes (Fig. 21.1B). Destructive rots of pomaceous fruits are almost always due to this species.

Enumeration and Identification

This species can be enumerated effectively on DRBC or DG18. No selective media have been developed; however, isolation from decaying apples and pears provides presumptive identification.

Ecology

The major habitat for *P. expansum* is as a destructive pathogen of pome fruits, but it is in reality a broad-spectrum pathogen on fruits, including avocados,

grapes, tomatoes, and mangoes (93). Growth occurs from around −3 to 35°C, with an optimum near 25°C, and down to 0.82 a_w. Patulin can be produced over the range 0 to 25°C, but not at 31°C, with the optimum being at 25°C. The minimum a_w for patulin production by *P. expansum* is 0.95 at 25°C (46). Patulin is quite stable in apple juice during storage; pasteurization at 90°C for 10 s caused less than 20% reduction (77).

Cyclopiazonic Acid

Cyclopiazonic acid, a highly toxic compound, causes fatty degeneration and hepatic cell necrosis in the liver and kidneys of domestic animals. Chickens are particularly susceptible (25). When the compound is administered by injection, central nervous system dysfunction occurs, and high doses may result in death of experimental animals (92). Mammals may be less affected (103).

Cyclopiazonic acid is produced by *Aspergillus flavus* and at least six *Penicillium* species (Table 21.1). Along with *A. flavus*, *Penicillium commune*, a common cause of cheese spoilage (76), is probably the most common source of cyclopiazonic acid in foods. It has been shown (76) that *P. commune* is the wild ancestor of *Penicillium camemberti*, used in the production of Camembert-type cheeses. Nearly all isolates of *P. camemberti* produce cyclopiazonic acid (95), but according to some reports, not under commercial cheese manufacturing conditions (70, 87). As others did not agree (53, 89), the search for nontoxigenic *P. camemberti* strains suitable for use as starter cultures continues (54).

Taxonomy

The *Penicillium* species producing cyclopiazonic acid are all classified in *Penicillium* subgenus *Penicillium* (Table 21.1).

Enumeration and Identification

All of the species producing cyclopiazonic acid can be enumerated effectively on the media discussed above. No selective media are available for any of these species. Identification of any of them requires specialist methods and information (75, 84).

Ecology

The principal *Penicillium* species producing cyclopiazonic acid, *P. commune*, is a common cause of cheese spoilage around the world (40, 62). Other species producing this toxin have a wide range of habitats.

All of the *Penicillium* species producing cyclopiazonic acid have similar physiology, including the ability to grow at or near 0°C, optima around 25°C, and maxima at or below 37°C. Most are able to grow at or below a_w 0.85 (41).

Citreoviridin

The role of citreoviridin in the human disease acute cardiac beriberi has been well documented (100). Acute cardiac beriberi was a common disease in Japan in the second half of the 19th century. Symptoms were heart distress, labored breathing, nausea, and vomiting, followed by anguish, pain, restlessness, and sometimes maniacal behavior. In extreme cases, progressive paralysis leading to respiratory failure occurred. It is notable that victims of this disease were often young, healthy adults. In 1910 the incidence of acute cardiac beriberi suddenly decreased. This coincided with implementation of a government inspection scheme that dramatically reduced the sale of moldy rice in Japan (99, 100).

The major source of citreoviridin is *Penicillium citreonigrum* (synonyms *P. citreoviride, P. toxicarium*), a species that usually occurs in rice, less commonly in other cereals, and rarely in other foods (72, 77). Recent studies (79, 80) failed to find more than occasional infections of *P. citreonigrum* in Southeast Asian rice. The threat of citreoviridin toxicosis at least in advanced Southeast Asian countries now appears to be very low. However, the possibility cannot be discounted that *P. citreonigrum* still occurs in less-developed African or Asian countries where adequate rice-drying systems and controls are not yet in place.

Citreoviridin is also produced by *Eupenicillium ochrosalmoneum* (anamorph *Penicillium ochrosalmoneum*), which is a relatively uncommon though widespread species, usually associated with cereals (72). The reported formation of citreoviridin by this species in corn in the United States is potentially of great concern (107, 108), but it has not been reported from Southeast Asian corn (78, 80). The ecology of this mold-commodity-mycotoxin relationship remains to be fully elucidated.

Taxonomy

Classified in subgenus *Aspergilloides, P. citreonigrum* was described in 1901 but subsequently ignored because the description was meager. In 1979, however, the name was revived (72) on the basis of the 1923 neotypification by Biourge, as it has priority over *P. citreoviride*, the more widely used name. *E. ochrosalmoneum* is a relatively recently described species, with no taxonomic complications.

When grown on standard identification media, *P. citreonigrum* is a distinctive species. Colonies grow quite slowly (after 7 days at 25°C, 20 to 28 mm in

diameter on CYA and 22 to 26 mm in diameter on MEA; 0 to 10 mm at 37°C); produce low numbers of pale gray-green conidia; and exhibit yellow mycelial, soluble pigment and reverse colors. Penicilli consist of small clusters of phialides only; stipes are slender and not apically enlarged, and conidia are spherical, smooth walled, and tiny (72, 75).

E. ochrosalmoneum forms slowly growing colonies (usually less than 30 mm in diameter) that are bright yellow on both CYA and MEA at 25°C. Growth on CYA at 37°C is similar to that at 25°C. Penicilli are biverticillate, with few and divergent metulae (75).

Enumeration and Identification

Little specific information exists on techniques for enumerating *P. citreonigrum* or *E. ochrosalmoneum*. Both would be expected to grow satisfactorily on media such as DRBC or DG18 agar. No selective media have been developed, however.

Ecology

The physiology of *P. citreonigrum* and of citreoviridin production has been little studied. Cardinal temperatures are below 5°C, 20 to 24°C, and 37 to 38°C (46). The minimum a_w for growth is not known, but this species is undoubtedly xerophilic (77). Citreoviridin is produced at temperatures from 10 to 37°C, with a maximum near 20°C (46).

Penitrem A

Chemicals capable of inducing a tremorgenic (trembling) response in vertebrate animals are regarded as rare—except for mold metabolites, where at least 20 such compounds have been reported (17). Tremorgens are neurotoxins; in low doses they appear to cause no adverse effects on animals, which are able to feed and function more or less normally while sustained trembling occurs, even over periods as long as 18 days (47). However, relatively small increases in dosage (5- to 20-fold) can be rapidly lethal (20, 43, 44). The presence of tremorgens is exceptionally difficult to diagnose postmortem because visible pathological effects are not produced.

Several tremorgenic mycotoxins are produced by *Penicillium* species, the most important being the highly toxic penitrem A. Verruculogen, equally toxic, is not produced by species of common occurrence in foods. Less toxic compounds include fumitremorgen B, paxilline, verrucosidin, and janthitrems (81). At doses around 1 mg/kg, penitrem A causes brain disturbance, brain damage, and death in rats, but the mechanism remains unclear (9, 14). The response in humans is also unknown: circumstantial evidence suggests that penitrem A may produce an emetic effect, which would render serious toxic effects much less likely.

The classification of *Penicillium* species producing penitrem A was particularly confusing, with penitrem production by several species described in early literature. It was later concluded that this confusion was due to misidentification and that the only common foodborne species producing penitrem A is *Penicillium crustosum* (73). Nearly all isolates of *Penicillium crustosum* produce penitrem A at high levels, so the presence of this species in food or feed is a warning signal (27, 28).

Taxonomy

The validity of *P. crustosum* as a species was in doubt until 1980 (72). However, it has been accepted without any confusion since. A member of *Penicillium* subgenus *Penicillium*, *P. crustosum* produces large penicilli, with rami, metulae, and phialides characteristic of this subgenus (Fig. 21.2B). It is one of the faster growing species in section *Penicillium*, producing dull green colonies with a granular texture on both CYA and MEA. Microscopically, *P. crustosum* is characterized by large, rough-walled stipes and smooth-walled, usually spherical conidia. However, the most distinctive feature of typical isolates is the production of enormous numbers of conidia on MEA, which become detached from the colony when the petri dish is jarred (72, 75).

Enumeration and Identification

P. crustosum grows relatively rapidly, so enumeration is best carried out by using a modern medium such as DRBC, which inhibits spreading of colonies (77). Confirmation of this species requires isolation and growth on standard identification media. No selective media have been developed for this species.

Ecology

P. crustosum is a ubiquitous spoilage mold, occurring almost universally in cereal and animal feed samples (77). *P. crustosum* causes spoilage of corn, processed meats, nuts, cheese, and fruit juices, as well as being a weak pathogen on pomaceous fruits and cucurbits (77, 93). The occurrence of penitrem A in animal feeds is well documented (44, 81). Its occurrence in human foods appears equally certain.

As has been noted earlier, *P. crustosum* has been poorly recognized until recently, so a lack of information about growth and toxin production is to be expected. Like nearly all species in *Penicillium* subgenus *Penicillium*, *P. crustosum* does not grow at 37°C. Penitrem A appears to be produced only at high a_w (46).

PR Toxin and Roquefortine

Treated together here are toxins produced by *Penicillium roqueforti*, a species used in cheese manufacture, but which can also be a spoilage mold. Roquefortine has a relatively high 50% lethal dose (Table 21.1), but it has been reported to be the cause of the deaths of dogs in Canada, with symptoms similar to strychnine poisoning (61). PR toxin is apparently much more toxic than roquefortine (Table 21.1), but has not been implicated in animal or human disease.

Apart from *P. roqueforti*, roquefortine has also been shown to be produced by *P. crustosum* (19) and *Penicillium chrysogenum* (28). PR toxin is produced only by *P. roqueforti* (33).

Taxonomy

P. roqueforti is classified in subgenus *Penicillium* section *Penicillium* (Fig. 21.2C). This species has been shown to comprise two distinct varieties: *P. roqueforti* var. *roqueforti* is widely used as a starter culture for mold-ripened cheeses, while *P. roqueforti* var. *carneum* occurs in meats and silage (35). *P. roqueforti* var. *roqueforti* produces PR toxin, *P. roqueforti* var. *carneum* produces patulin, and both varieties produce roquefortine (35).

P. roqueforti grows very rapidly on CYA and MEA and produces green reverse colors on one or both media. Stipes are often very rough; conidia are large, spherical, and smooth walled (75).

Enumeration and Identification

Like other similar species, *P. roqueforti* can be enumerated on DRBC or DG18. A valuable property in some circumstances is the fact that *P. roqueforti* is more tolerant of acetic acid than are other *Penicillium* species. In consequence, *P. roqueforti* can be selectively enumerated on a medium such as malt acetic agar (MEA–0.5% glacial acetic acid) (29, 77).

Ecology

P. roqueforti and the other two *Penicillium* species producing roquefortine are common in foods. As well as growing rapidly at refrigeration temperatures, *P. roqueforti* is able to grow in oxygen concentrations below 0.5%, even in the presence of up to 20% carbon dioxide (96). Hence, it is a common cause of spoilage in chilled meats, cheese, and other products (11, 12, 69). The presence of roquefortine in cheese and other foods is to be expected. Disease from PR toxin or roquefortine in association with cheese has not been reported

(77). However, roquefortine has been implicated in a case of human intoxication from moldy beer infected with *P. crustosum* (19).

Because of their potential occurrence in staple foods, PR toxin and roquefortine are of considerable significance from the public health viewpoint. PR toxin is unstable in cheese in storage, but roquefortine has been isolated from finished products (88). Extensive searches for nontoxic strains for use as cheese starter cultures have so far been largely unsuccessful (54, 64).

Like other species in subgenus *Penicillium*, *P. roqueforti* is psychrotrophic, growing vigorously at temperatures as low as 2°C. It is notably tolerant of weak acid preservatives such as sorbic acid and is able to grow in 0.5% acetic acid or more (29). This species also has the lowest oxygen requirement for growth of any *Penicillium* species, being capable of growth in less than 0.5% oxygen in the presence of 20% carbon dioxide (96).

Secalonic Acid D

Secalonic acids are dimeric xanthones produced by a range of taxonomically distant molds (18). Secalonic acid D, the only secalonic acid produced by *Penicillium* species, has significant animal toxicity (15, 94). Secalonic acid D is produced as a major metabolite of *Penicillium oxalicum*. It has been found in grain dusts at levels of up to 4.5 mg/kg (26). The possibility that such concentrations can be toxic to grain handlers by inhalation should not be ignored. However, the role of secalonic acid D in human or animal disease remains unclear.

Taxonomy

P. oxalicum is a distinctive species in subgenus *Furcatum* section *Furcatum* (Fig. 21.2D). It grows rapidly on CYA at 25 and 37°C, forming a continuous layer of conidia that, under the low power microscope, can be seen to lie in closely packed, readily fractured sheets, with a uniquely shiny appearance. Penicilli are terminal and biverticillate, with long phialides producing large ellipsoidal conidia (75).

Enumeration and Identification

This species can be enumerated without difficulty on standard media such as DRBC and DG18. Selective media have not been developed, however.

Ecology

A major habitat for *P. oxalicum* is corn at harvest, where it is the most common *Penicillium* species isolated (57, 66, 77). The occurrence of secalonic acid D in corn and hence in grain dusts is a potential hazard (26). *P. oxalicum* grows from about 8 to about 40°C. The minimum a_w for conidial germination is 0.86 (77).

QUANTIFYING MYCOTOXINS

Quantifying most *Penicillium* mycotoxins is relatively straightforward, but procedures vary with commodity and toxin. As mycotoxins are usually present in minute amounts, best practice is essential to obtain good results. Scrupulous attention to safety is necessary, in analyst training, in laboratory equipment and procedures, and in toxin containment and disposal.

Basic procedures for mycotoxin assays are sampling and subsampling, extraction and cleanup, and detection, quantification, and confirmation. Each of these areas is described in general terms below.

Sampling and Subsampling

The incidence of mycotoxins in a food commodity is usually heterogeneous, so sampling plays an important part in assay accuracy. Adequate (representative) sample sizes depend on particle size. While a 500-g sample is adequate for oils or milk powder, 3 kg of flour or peanut butter is necessary. To be representative, corn samples should be 10 kg and peanuts require 20 kg per batch (92).

Samples of raw materials should be taken with on-line samplers during processing or be composites from several sites in raw material lots. Much attention has been paid to sample plans for aflatoxins in peanuts (10, 16, 83, 106), corn (24), and figs (91). Such detailed studies have not been carried out on sampling for other mycotoxins, but the aflatoxin systems should have broad applicability.

To obtain a representative subsample for analysis, samples of particulate foods should be ground or blended. After thorough mixing, subsamples (20 to 50 g) are taken. To check on the adequacy of sampling procedures, assays are often performed in duplicate.

Extraction and Cleanup

To release toxins from the ground food material, extraction in a solvent system is necessary. This is usually carried out by blending the subsample with a suitable solvent in an explosion proof blender for 1 to 3 min or shaking on a wrist action shaker (30 min). Various solvent systems are used, depending on circumstances and likely toxins, the most common being methanol plus acidified water, dichloromethane, or chloroform (39).

The extract usually requires some form of cleanup procedure, in particular to remove lipids and pigments that would interfere with analysis. Dialysis, precipitation, chromatographic columns, solvent partitioning, and specific antibody layers are all in use (92). Concentration of the purified extract is then carried out by using a steam bath or vacuum rotary evaporator.

Detection and Quantification

One of the earliest and still the most common techniques for mycotoxin assays is thin-layer chromatography (TLC). Concentrated extracts are spotted (1 to 10 μl) on glass or aluminum plates coated with a thin layer of activated silica gel and then developed in tanks containing suitable solvents designed to separate toxins from interfering chemicals of all types. For some substrates, two-dimensional TLC (developing the plate twice at right angles) is of value. Appropriate standards are also run at the same time.

After development, toxins are visualized by comparison of spots on the chromatogram with standard spots. Many mycotoxins, including aflatoxins, fluoresce under UV light and are readily visualized. Others require reaction with spray reagents, such as sulfuric acid or aluminium chloride. Comparison of the R_f and color of unknown spots against standards enables separation of a wide variety of toxins. Intensity of spots when compared with suitable standards provides quantification. Confirmation is important and usually involves use of a spray reagent to produce a different color.

In recent years, high-performance liquid chromatography (HPLC) has increasingly become the analysis method of choice. Passage of extracts through a long column packed with an inert layer and a suitable adsorbent causes separation of molecules in a manner similar to TLC. Visualization is by spectrophotometric detectors. Sometimes compounds must be derivatized before analysis to improve sensitivity. HPLC is more sensitive than TLC and more suited to automation.

Rapid Methods of Analysis

Assay techniques such as TLC and HPLC are effective but slow. The search for more rapid methods continues. Observation of cracked corn kernels under UV light has long been used as a screening technique for aflatoxins. This method is also of value for aflatoxin in figs but is ineffective for peanuts, which autofluoresce under UV light. Peanuts are screened for aflatoxins by using minicolumns, in which aflatoxin is bound to an absorbent material in a small tube and then assayed by UV light.

The newest approach to rapid assays involves the use of antibodies developed to specific toxins. Immunoassays using either spot or minicolumn tests have been developed for several major toxins (92).

Specific assay methods exist for most *Penicillium* toxins, using the general principles outlined above (39, 92). However, the literature for the less common and well-known toxins is widely scattered.

CONCLUSIONS

Penicillium species produce a wide range of toxic compounds. The role of one toxin, citreoviridin, in a historical human health problem is well established. Of the other toxins discussed here, only ochratoxin is currently regarded as a serious threat to human health. However, *Penicillium* species are so widespread and abundant in foods and feeds that they must be considered to be a potential hazard to both human and animal health. Many species make several compounds known to be toxic, and the possibility of synergy also cannot be ignored. Much more research is needed to improve detection methods and understand the ecology of toxigenic *Penicillium* species, as well as to evaluate the significance of *Penicillium* toxins in human health.

References

1. **Alsberg, C. L., and O. F. Black.** 1913. Contributions to the study of maize deterioration: biochemical and toxicological investigations of *Penicillium puberulum* and *Penicillium stoloniferum. Bull. Bur. Anim. Ind. U.S. Dept. Agric.* **270:**1–47.

2. **Alves, I., N. G. Oliviera, A. Laires, A. S. Rodrigues, and J. Rueff.** 2000. Induction of micronuclei and chromosomal aberrations by the mycotoxin patulin in mammalian cells: role of ascorbic acid as a modulator of patulin clastogenicity. *Mutagenesis* **15:**229–234.

3. **Austwick, P. K. C.** 1981. Balkan nephropathy. *Practitioner* **225:**1031–1038.

4. **Banke, S., J. C. Frisvad, and S. Rosendahl.** 1997. Taxonomy of *Penicillium chrysogenum* and related xerophilic species, based on isozyme analysis. *Mycol. Res.* **101:**617–624.

5. **Boysen, M., P. Skouboe, J. Frisvad, and L. Rossen.** 1996. Reclassification of the *Penicillium roqueforti* group into three species on the basis of molecular genetic and biochemical profiles. *Microbiology* **142:**541–549.

6. **Brackett, R. E., and E. H. Marth.** 1979. Patulin in apple juice from roadside stands in Wisconsin. *J. Food Prot.* **42:**862–863.

7. **Breitholtz, A., M. Olsen, A. Dahlbäck, and K. Hult.** 1991. Plasma ochratoxin A levels in three Swedish populations surveyed using an ion-pair HPLC technique. *Food Addit. Contam.* **8:**183–92.

8. **Breitholtz-Emanuelsson, A., M. Olsen, A. Oskarsson, I. Palminger, and K. Hult.** 1993. Ochratoxin A in cow's

milk and human milk with corresponding human blood samples. *J. Assoc. Off. Anal. Chem. Int.* 76:842–846.

9. **Breton, P., J. C. Bizot, J. Buee, and I. Delamanche.** 1998. Brain neurotoxicity of penitrem A: electrophysiological, behavioral and histopathological changes. *Toxicon* 36:645–655.

10. **Brown, G. H.** 1982. Sampling for "needles in haystacks." *Food Technol. Aust.* 34:224–227.

11. **Bullerman, L. B.** 1980. Incidence of mycotoxic molds in domestic and imported cheeses. *J. Food Safety* 2:47–58.

12. **Bullerman, L. B.** 1981. Public health significance of molds and mycotoxins in fermented dairy products. *J. Dairy Sci.* 64:2439–2452.

13. **Carlton, W. W., G. Sansing, G. M. Szczech, and J. Tuite.** 1974. Citrinin mycotoxicosis in beagle dogs. *Food Cosmet. Toxicol.* 12:479–490.

14. **Cavanagh, J. B., J. L. Holton, C. C. Nolan, D. E. Ray, J. T. Naik, and P. G. Mantle.** 1998. The effects of the tremorgenic mycotoxin penitrem A on the rat cerebellum. *Vet. Pathol.* 35:53–63.

15. **Ciegler, A., A. W. Hayes, and R. F. Vesonder.** 1980. Production and biological activity of secalonic acid D. *Appl. Environ. Microbiol.* 39:285–287.

16. **Coker, R. D.** 1989. Control of aflatoxin in groundnut products with emphasis on sampling, analysis, and detoxification, p. 123–132. *In Aflatoxin Contamination of Groundnut: Proceedings of the International Workshop, 6–9 October, 1987, ICRISAT Centre, India.* ICRISAT, Patancheru, India.

17. **Cole, R. J.** 1981. Tremorgenic mycotoxins: an update, p. 17-33. *In* R. L. Ory (ed.), *Antinutrients and Natural Toxicants in Foods.* Food and Nutrition Press, Westport, Conn.

18. **Cole, R. J., and R. H. Cox.** 1981. *Handbook of Toxic Fungal Metabolites.* Academic Press, New York, N.Y.

19. **Cole, R. J., J. W. Dorner, R. H. Cox, and L. W. Raymond.** 1983. Two classes of alkaloid mycotoxins produced by *Penicillium crustosum* Thom isolated from contaminated beer. *J. Agric. Food Chem.* 31:655–657.

20. **Cole, R. J., J. W. Kirksey, J. H. Moore, B. R. Blankenship, U. L. Diener, and N. D. Davis.** 1972. Tremorgenic toxin from *Penicillium verruculosum.* *Appl. Microbiol.* 24:248–250.

21. **Cruickshank, R. H., and J. I. Pitt.** 1987. Identification of species in *Penicillium* subgenus *Penicillium* by enzyme electrophoresis. *Mycologia* 79:614–620.

22. **De Jesus, A. E., P. S. Steyn, F. R. van Heerden, R. Vleggaar, and P. L. Wessels.** 1981. Structure and biosynthesis of the penitrems A-F, six novel tremorgenic mycotoxins from *Penicillium crustosum.* *J. Chem. Soc. Chem. Commun.* 1981:289–291.

23. **Dickens, F., and H. E. H. Jones.** 1961. Carcinogenic activity of a series of reactive lactones and related substances. *Br. J. Cancer* 15:85–100.

24. **Dickens, J. W., and T. B. Whitaker.** 1983. Sampling, BGYF, and aflatoxin analysis in corn, p. 35–37. *In* U. L. Diener, R. L. Asquith, and J. W. Dickens (ed.) *Aflatoxin and Aspergillus flavus in Corn.* Alabama Agricultural Experiment Station, Auburn University, Auburn, Ala.

25. **Dorner, J. W., R. J. Cole, L. G. Lomax, H. S. Gosser, and U. L. Diener.** 1983. Cyclopiazonic acid production by *Aspergillus flavus* and its effect on broiler chickens. *Appl. Environ. Microbiol.* 46:698–703.

26. **Ehrlich, K. C., L. S. Lee, A. Ciegler, and M. S. Palmgren.** 1982. Secalonic acid D: natural contaminant of corn dust. *Appl. Environ. Microbiol.* 44:1007–1008.

27. **El-Banna, A. A., and L. Leistner.** 1988. Production of penitrem A by *Penicillium crustosum* from foodstuffs. *Int. J. Food Microbiol.* 7:9–17.

28. **El-Banna, A. A., J. I. Pitt, and L. Leistner.** 1987. Production of mycotoxins by *Penicillium* species. *Syst. Appl. Microbiol.* 10:42–46.

29. **Engel, G., and M. Teuber.** 1978. Simple aid for the identification of *Penicillium roqueforti* Thom. *Eur. J. Appl. Microbiol. Biotechnol.* 6:107–111.

30. **Friis, P., E. Hasselager, and P. Krogh.** 1969. Isolation of citrinin and oxalic acid from *Penicillium viridicatum* Westling and their nephrotoxicity in rats and pigs. *Acta Pathol. Microbiol. Scand.* 77:559–560.

31. **Frisvad, J. C.** 1983. A selective and indicative medium for groups of *Penicillium viridicatum* producing different mycotoxins in cereals. *J. Appl. Bacteriol.* 54:409–416.

32. **Frisvad, J. C.** 1986. Taxonomic approaches to mycotoxin identification, p. 415–457. *In* R. J. Cole (ed.), *Modern Methods in the Analysis and Structural Elucidation of Mycotoxins.* Academic Press, Orlando, Fla.

33. **Frisvad, J. C.** 1987. High performance liquid chromatographic determination of profiles of mycotoxins and other secondary metabolites. *J. Chromatogr.* 392:333–347.

34. **Frisvad, J. C., and O. Filtenborg.** 1983. Classification of terverticillate penicillia based on profiles of mycotoxins and other secondary metabolites. *Appl. Environ. Microbiol.* 46:1301–1310.

35. **Frisvad, J. C., and O. Filtenborg.** 1989. Terverticillate Penicillia: chemotaxonomy and mycotoxin production. *Mycologia* 81:837–861.

36. **Frisvad, J. C., and B. T. Viuf.** 1986. Comparison of direct and dilution plating for detection of *Penicillium viridicatum* in barley containing ochratoxin, p. 45–47. *In* A. D. King, J. I. Pitt, L. R. Beuchat, and J. E. L. Corry (ed.), *Methods for the Mycological Examination of Food.* Plenum Press, New York, N.Y.

37. **Hald, B.** 1991. Ochratoxin A in human blood in European countries, p. 159–164. *In* F. Castegnaro, R. Pleština, G. Dirheimer, I. N. Chernozemsky, and H. Bartsch (ed.), *Mycotoxins, Endemic Nephropathy and Urinary Tract Tumours.* International Agency for Research on Cancer, Lyon, France.

38. **Harwig, J., T. Kuiper-Goodman, and P. M. Scott.** 1983. Microbial food toxicants: ochratoxins, p. 193–238. *In* M. Reichcigl (ed.), *Handbook of Foodborne Diseases of Biological Origin.* CRC Press, Boca Raton, Fla.

39. **Helrich, K. (ed.).** 1990. *Official Methods of Analysis of the Association of Official Analytical Chemists,* 15th ed. Association of Official Analytical Chemists, Arlington, Va.

40. **Hocking, A. D., and M. Faedo.** 1992. Fungi causing thread mould spoilage of vacuum packaged cheddar

cheese during maturation. *Int. J. Food Microbiol.* **16:** 123–130.

41. Hocking, A. D., and J. I. Pitt. 1979. Water relations of some *Penicillium* species at 25°C. *Trans. Br. Mycol. Soc.* **73:**141–145.

42. Hocking, A. D., J. I. Pitt, R. A. Samson, and A. D. King. 1992. Recommendations from the closing session of SMMEF II, p. 359–364. *In* R. A. Samson, A. D. Hocking, J. I. Pitt, and A. D. King (ed.), *Modern Methods in Food Mycology.* Elsevier, Amsterdam, The Netherlands.

43. Hou, C. T., A. Ciegler, and C. W. Hesseltine. 1971. Tremorgenic toxins from Penicillia. II. A new tremorgenic toxin, tremortin B, from *Penicillium palitans. Can. J. Microbiol.* **17:**599–603.

44. Hou, C. T., A. Ciegler, and C. W. Hesseltine. 1971. Tremorgenic toxins from Penicillia. III. Tremortin production by *Penicillium* species on various agricultural commodities. *Appl. Microbiol.* **21:**1101–1103.

45. International Agency for Research on Cancer (IARC). 1994. Ochratoxin A, p. 489–521. *In IARC Monographs on the Evaluation of Carcinogenic Risks to Humans,* vol. 56. *Some Naturally Occurring Substances: Food Items and Constitutents, Heterocyclic Aromatic Amines and Mycotoxins.* International Agency for Research on Cancer, Lyon, France.

46. International Commission on Microbiological Specifications for Foods (ICMSF). 1996. Toxigenic fungi: *Penicillium,* p. 397–413. *In Microorganisms in Foods.* 5. *Characteristics of Food Pathogens.* Blackie Academic and Professional, London, United Kingdom.

47. Jortner, B. S., M. Ehrich, A. E. Katherman, W. R. Huckle, and M. E. Carter. 1986. Effects of prolonged tremor due to penitrem A in mice. *Drug Chem. Toxicol.* **9:**101–116.

48. King, A. D., J. I. Pitt, L. R. Beuchat, and J. E. L. Corry (ed.). 1986. *Methods for the Mycological Examination of Food.* Plenum Press, New York, N.Y.

49. Krogh, P., N. H. Axelsen, F. Elling, N. Gyrd-Hansen, B. Hald, J. Hyldgaard-Jensen, A. E. Larsen, A. Madsen, H. P. Mortensen, T. Møller, O. K. Petersen, U. Ravnskov, M. Rostgaard, and O. Aalund. 1974. Experimental porcine nephropathy. *Acta Pathol. Microbiol. Scand. Sect. A Suppl.* **246:**1–21.

50. Krogh, P., B. Hald, P. Englund, L. Rutqvist, and O. Swahn. 1974. Contamination of Swedish cereals with ochratoxin A. *Acta Pathol. Microbiol. Scand. Sect. B* **82:**301–302.

51. Krogh, P., B. Hald, and E. J. Pedersen. 1973. Occurrence of ochratoxin A and citrinin in cereals associated with mycotoxic porcine nephropathy. *Acta Pathol. Microbiol. Scand. Sect. B* **81:**689–695.

52. Krogh, P., B. Hald, R. Plestina, and S. Ceovic. 1977. Balkan (endemic) nephropathy and foodborn ochratoxin A: preliminary results of a survey of foodstuffs. *Acta Pathol. Microbiol. Scand. Sect B.* **85:**238–240.

53. Le Bars, J. 1979. Cyclopiazonic acid production by *Penicillium camemberti* Thom and natural occurrence of this mycotoxin in cheese. *Appl. Environ. Microbiol.* **38:**1052–1055.

54. Leistner, L., R. Geisen, and J. Fink-Gremmels. 1989. Mould-fermented foods of Europe: hazards and developments, p. 145–154. *In* S. Natori, K. Hashimoto, and Y. Ueno (ed.), *Mycotoxins and Phytotoxins '88.* Elsevier Science, Amsterdam, The Netherlands.

55. Leistner, L., and J. I. Pitt. 1977. Miscellaneous *Penicillium* toxins, p. 639–653. *In* J. V. Rodricks, C. W. Hesseltine, and M. E. Mehlman (ed.), *Mycotoxins in Human and Animal Health.* Pathotox Publishers, Park Forest South, Ill.

56. Li, S., R. R. Marquandt, A. A. Frolich, T. G. Vitti, and G. Crow. 1997. Pharmokinetics of ochratoxin A and its metabolites in rats. *Toxicol. Appl. Pharmacol.* **145:**90–92.

57. Lichtwardt, R. W., and L. H. Tiffany. 1958. Mold flora associated with shelled corn in Iowa. *Iowa State Coll. J. Sci.* **33:**1–11.

58. Llewellyn, G. C., J. A. McCay, R. D. Brown, D. L. Musgrove, L. F. Butterworth, A. E. Munson, and K. L. White. 1998. Immunological evaluation of the mycotoxin patulin in female B6C3F(1) mice. *Food Cosmet. Toxicol.* **36:**1107–1115.

59. LoBuglio, K. F., J. I. Pitt, and J. W. Taylor. 1994. Independent origin of the synnematous *Penicillium* species, *P. duclauxiii, P. clavigerum* and *P. vulpinum,* as assessed by two ribosomal DNA regions. *Mycol. Res.* **98:**250–256.

60. LoBuglio, K. F., and J. W. Taylor. 1994. Phylogeny and PCR identification of the human pathogenic fungus *Penicillium marneffei. J. Clin. Microbiol.* **33:**85–89.

61. Lowes, N. R., R. A. Smith, and B. E. Beck. 1992. Roquefortine in the stomach contents of dogs suspected of strychnine poisoning. *Can. Vet. J.* **33:**535–538.

62. Lund, F., O. Filtenborg, and J. C. Frisvad. 1995. Associated mycoflora of cheese. *Food Microbiol.* **12:**173–180.

63. Mantle, P. G., and K. M. Macgeorge. 1991. Nephrotoxic fungi in a Yugoslav community in which Balkan nephropathy is hyperendemic. *Mycol. Res.* **95:**660–664.

64. Medina, M., P. Gaya, and M. Nunez. 1985. Production of PR toxin and roquefortine by *Penicillium roqueforti* isolates from Cabrales blue cheese. *J. Food Prot.* **48:**118–121.

65. Miraglia, M., C. Brera, and M. Colatosti. 1996. Application of biomarkers to assessment of risk to human health from exposure to mycotoxins. *Microchem. J.* **54:**472–477.

66. Mislivec, P. B., and J. Tuite. 1970. Species of *Penicillium* occurring in freshly-harvested and in stored dent corn kernels. *Mycologia* **62:**67–74.

67. Miyake, I., H. Naito, and H. Sumeda. 1940. (Japanese title.) *Rep. Res. Inst. Rice Improvement* **1:**1.

68. Mortimer, D. N., I. Parker, M. J. Shepherd, and J. Gilbert. 1985. A limited survey of retail apple and grape juices for the mycotoxin patulin. *Food Addit. Contam.* **2:**165–170.

69. Northolt, M. D., H. P. van Egmond, P. Soentoro, and E. Deijll. 1980. Fungal growth and the presence of sterigmatocystin in hard cheese. *J. Assoc. Off. Anal. Chem.* **63:**115–119.

70. Orth, R. 1981. Mykotoxine von Pilzen der Käseherstellung, p. 273–296. *In* J. Reiss (ed.), *Mykotoxine*

in Lebensmitteln. Gustaf Fisher Verlag, Stuttgart, Germany.

71. **Pavlović, M., R. Plestina, and P. Krogh.** 1979. Ochratoxin A contamination of foodstuffs in an area with Balkan (endemic) nephropathy. *Acta Pathol. Microbiol. Scand. Sect. B* 87:243–246.

72. **Pitt, J. I.** 1979. *The Genus Penicillium and Its Teleomorphic States Eupenicillium and Talaromyces.* Academic Press, London, United Kingdom.

73. **Pitt, J. I.** 1979. *Penicillium crustosum* and *P. simplicissimum,* the correct names for two common species producing tremorgenic mycotoxins. *Mycologia* 71:1166–1177.

74. **Pitt, J. I.** 1987. *Penicillium viridicatum, Penicillium verrucosum,* and production of ochratoxin A. *Appl. Environ. Microbiol.* 53:266–269.

75. **Pitt, J. I.** 2000. *A Laboratory Guide to Common Penicillium Species,* 3rd ed. CSIRO Food Science Australia, North Ryde, N.S.W., Australia.

76. **Pitt, J. I., R. H. Cruickshank, and L. Leistner.** 1986. *Penicillium commune, P. camembertii,* the origin of white cheese moulds, and the production of cyclopiazonic acid. *Food Microbiol.* 3:363–371.

77. **Pitt, J. I., and A. D. Hocking.** 1997. *Fungi and Food Spoilage,* 2nd ed. Aspen Publishers, Gaithersburg, Md.

78. **Pitt, J. I., A. D. Hocking, K. Bhudhasamai, B. F. Miscamble, K. A. Wheeler, and P. Tanboon-Ek.** 1993. The normal mycoflora of commodities from Thailand. 1. Nuts and oilseeds. *Int. J. Food Microbiol.* 20:211–226.

79. **Pitt, J. I., A. D. Hocking, K. Bhudhasamai, B. F. Miscamble, K. A. Wheeler, and P. Tanboon-Ek.** 1994. The normal mycoflora of commodities from Thailand. 2. Beans, rice and other commodities. *Int. J. Food Microbiol.* 23:35–53.

80. **Pitt, J. I., A. D. Hocking, B. F. Miscamble, O. S. Dharmaputra, K. R. Kuswanto, E. S. Rahayu, and Sardjono.** 1998. The mycoflora of food commodities from Indonesia. *J. Food Mycol.* 1:41–60.

81. **Pitt, J. I., and L. Leistner.** 1991. Toxigenic *Penicillium* species, p. 91–99. *In* J. E. Smith and R. S. Henderson (ed.), *Mycotoxins and Animal Foods.* CRC Press, Boca Raton, Fla.

82. **Pitt, J. I., and R. A. Samson.** 1993. Species names in current use in the *Trichocomaceae* (Fungi, Eurotiales). *Regnum Veg.* 128:13–57.

83. **Read, M.** 1989. Removal of aflatoxin from the Australian groundnut crop, p. 133–140. *In Aflatoxin Contamination of Groundnut: Proceedings of the International Workshop, 6–9 October, 1987, ICRISAT Centre, India.* ICRISAT, Patancheru, India.

84. **Samson, R. A., E. S. Hoekstra, J. C. Frisvad, and O. Filtenborg.** 1995. *Introduction to Food-borne Fungi,* 4th ed. Centraalbureau voor Schimmelcultures, Baarn, Netherlands.

85. **Samson, R. A., and J. I. Pitt (ed.).** 1990. *Modern Concepts in Penicillium and Aspergillus Classification.* Plenum Press, New York, N.Y.

86. **Samson, R. A., A. C. Stolk, and R. Hadlok.** 1976. Revision of the subsection Fasciculata of *Penicillium* and some allied species. *Stud. Mycol. (Baarn)* 11:1–47.

87. **Schoch, U., J. Lüthy, and C. Schlatter.** 1984. Mutagenitätsprüfung industriell verwendeter *Penicillium camemberti-* und *P. roqueforti-*Stämme. *Z. Lebensm. Unters. Forsch.* 178:351–355.

88. **Scott, P. M.** 1977. *Penicillium* mycotoxins, p. 283–356. *In* T. D. Wyllie and L. G. Morehouse (ed.), *Mycotoxic Fungi, Mycotoxins, Mycotoxicoses, an Encyclopedic Handbook,* vol. 1. *Mycotoxigenic Fungi.* Marcel Dekker, New York, N.Y.

89. **Scott, P. M.** 1981. Toxins of *Penicillium* species used in cheese manufacture. *J. Food Prot.* 44: 702–710.

90. **Scouboe, P., J. C. Frisvad, J. W. Taylor, D. Lauritsen, M. Boysen, and L. Rossen.** 1999. Phylogenetic analysis of nucleotide sequences from the ITS region of terverticillate *Penicillium* species. *Mycol. Res.* 103:873–881.

91. **Sharman, M., S. Macdonald, A. J. Sharkey, and J. Gilbert.** 1994. Sampling bulk consignments of dried figs for aflatoxin analysis. *Food Addit. Contam.* 11:17–23.

92. **Smith, J. E., C. W. Lewis, J. G. Anderson, and G. L. Solomons.** 1994. *Mycotoxins in Human Nutrition and Health.* Report EUR 16048 EN. European Commission Directorate-General XII, Brussels, Belgium.

93. **Snowdon, A. L.** 1990. *A Colour Atlas of Post-Harvest Diseases and Disorders of Fruits and Vegetables. 1. General Introduction and Fruits.* Wolfe Scientific, London, United Kingdom.

94. **Sorenson, W. G., F. H. Y. Green, V. Vallyathan, and A. Ciegler.** 1982. Secalonic acid D toxicity in rat lung. *J. Toxicol. Environ. Health* 9:515–525.

95. **Still, P., C. Eckardt, and L. Leistner.** 1978. Bildung von Cyclopazonsaure durch *Penicillium camembertii-*isolate von Kase. *Fleischwirtschaft* 58:876–878.

96. **Taniwaki, M.** 1995. Growth and mycotoxin production by fungi under modified atmospheres. Ph.D. thesis. University of New South Wales, Kensington, N.S.W., Australia.

97. **Tatu, C. A., W. H. Orem, R. B. Finkelman, and G. L. Feder.** 1998. The etiology of Baltic endemic nephropathy: still more questions than answers. *Environ. Health Perspect.* 106:689–700.

98. **Téren, J., J. Varga, Z. Hamari, E. Rinyu, and É. Kevei.** 1996. Immunochemical detection of ochratoxin A in black *Aspergillus* strains. *Mycopathologia* 134:171–176.

99. **Ueno, Y., and I. Ueno.** 1972. Isolation and acute toxicity of citreoviridin, a neurotoxic mycotoxin of *Penicillium citreo-viride* Biourge. *Jpn. J. Exp. Med.* 42:91–105.

100. **Uraguchi, K.** 1969. Mycotoxic origin of cardiac beriberi. *J. Stored Prod. Res.* 5:227–236.

101. **Van der Merwe, K. J., P. S. Steyn, L. Fourie, D. B. Scott., and J. J. Theron.** 1965. Ochratoxin A, a toxic metabolite produced by *Aspergillus ochraceus* Wilh. *Nature* 205:1112–1113.

102. **Van Egmond, H. P.** 1989. Current situation on regulations for mycotoxins. Overview of tolerances and status of standard methods of sampling and analysis. *Food Addit. Contam.* 6:139–188.

103. Van Rensburg, S. J. 1984. Subacute toxicity of the mycotoxin cyclopiazonic acid. *Food Chem. Toxicol.* 22:993–
998.

104. Van Walbeek, W., P. M. Scott, J. Harwig, and J. W.
Lawrence. 1969. *Penicillium viridicatum* Westling: a new
source of ochratoxin A. *Can. J. Microbiol.* 15:1281–
1285.

105. Watkins, K. L., G. Fazekas, and M. V. Palmer. 1990.
Patulin in Australian apple juice. *Food Australia* 42:
438–439.

106. Whitaker, T. B., F. E. Dowell, W. M. Hagler, F. G.
Giesbrecht, and J. Wu. 1994. Variability associated with
sampling, sample preparation, and chemical testing for
aflatoxin in farmers' stock peanuts. *J. Assoc. Off. Anal.
Chem. Int.* 77:107–116.

107. Wicklow, D. T., and R. J. Cole. 1984. Citreoviridin
in standing corn infested by *Eupenicillium ochrosalmoneum. Mycologia* 76:959–961.

108. Wicklow, D. T., R. D. Stubblefield, B. W. Horn, and
O. L. Shotwell. 1988. Citreoviridin levels in *Eupenicil-
lium ochrosalmoneum*-infested maize kernels at harvest.
Appl. Environ. Microbiol. 54:1096–1098.

Food Microbiology: Fundamentals and Frontiers, 2nd Ed.
Edited by M. P. Doyle et al.
© 2001 ASM Press, Washington, D.C.

Lloyd B. Bullerman

Fusaria and Toxigenic Molds Other than Aspergilli and Penicillia

23

Toxigenic molds in genera other than *Aspergillus* and *Penicillium* are most often found as contaminants of plant-derived foods, especially cereal grains. As such, these molds and their toxic metabolites (mycotoxins) find their way into animal feeds and human foods. Animals, both food-producing animals and pets, are more often affected by the toxins of these molds than are humans because of the nature of animal feed and the way it is stored and handled. Human food is generally higher in quality and more carefully protected, particularly in the so-called developed countries in temperate climates. In other areas with more tropical and subtropical climates, this may not be the case. Nevertheless, there is evidence that grains and processed human foods are occasionally contaminated with mycotoxins produced by molds other than aspergilli and penicillia (31). Also, exposure of food-producing animals to mycotoxigenic molds may have an impact on the human food supply by causing the death of animals, reducing their rate of growth, or depositing toxins in meats, milk, and eggs. Diseases in animals caused by mycotoxins may also suggest that similar conditions can occur in humans.

The most important group of mycotoxigenic molds other than *Aspergillus* and *Penicillium* species are species of the genus *Fusarium* (44). Many *Fusarium* species are plant pathogens, while others are saprophytic; most can be found in the soil (15, 16). In terms of human foods, *Fusarium* species are most often encountered as contaminants of cereal grains, oil seeds, and beans. Corn, wheat, and products made from these grains are most commonly contaminated. However, barley, rye, triticale, millet, and oats can also be contaminated.

Another mold genus other than *Aspergillus* and *Penicillium* that can produce mycotoxins is *Alternaria* (80). *Alternaria* species are widely distributed in the environment and can be found in soil, decaying plant materials, and dust. Both plant pathogenic and saprophytic species occur. Toxigenic molds also belong to several other genera. These include species of *Acremonium, Chaetomium, Claviceps, Diplodia maydis, Myrothecium, Phoma herbarum, Phomopsis, Pithomyces chartarum, Rhizoctonia, Rhizopus, Stachybotrys,* and *Trichothecium roseum* (37).

DISEASES OF HUMANS ASSOCIATED WITH *FUSARIUM*

Alimentary Toxic Aleukia

During World War II a very severe human disease occurred in the former Soviet Union, particularly in the

Lloyd B. Bullerman, Department of Food Science and Technology, University of Nebraska, 349 Food Industry Building, East Campus, Lincoln, NE 68583-0919.

Orenburg area of Russia. The disease, known as alimentary toxic aleukia (ATA), is believed to be caused by T-2 and HT-2 toxins produced by *Fusarium sporotrichioides* and *Fusarium poae* of the *Sporotrichiella* section of *Fusarium* (33, 39). Because of the war, there was such a shortage of farm workers that much grain was not harvested in the fall and overwintered in the field. By spring, near-famine conditions existed and people were forced to eat the overwintered cereal grains that were milled into flour and made into bread. This resulted in a severe toxicosis manifested as alimentary toxic aleukia. The disease developed over several weeks, with increasingly severe symptoms with continued consumption of the toxic grain.

The disease actually has four stages (33). The first stage is characterized by symptoms that appear a short time after ingestion of the toxic grain. These symptoms include a burning sensation in the mouth, tongue, esophagus, and stomach (Table 23.1). The stomach and intestinal mucosa become inflamed, resulting in vomiting, diarrhea, and abdominal pain. This first stage may appear and disappear rather quickly, within 3 to 9 days. During the second stage, the individual experiences no outward signs of the disease and feels well. However, during this stage the hematopoietic system is being damaged or destroyed by progressive leukopenia, granulopenia, and lymphocytosis. The blood-making capacity of the bone marrow is being destroyed, the platelet count decreases, and anemia develops. The leukocyte count decreases and secondary bacterial infections occur. There are also disturbances in the central and autonomic nervous systems.

The third stage develops suddenly and is marked by petechial hemorrhage on the skin and mucous membranes. The hemorrhaging becomes more severe, with bleeding from the nose and gums and hemorrhaging in the stomach and intestines. Necrotic lesions also develop in the mouth, gums, mucosa, larynx, and vocal cords. At this stage the disease is highly fatal, with up to 60% mortality. If death does not occur, the fourth stage of the disease is recovery or convalescence. It takes 3 to 4 weeks of treatment for the necrotic lesions, hemorrhaging, and bacterial infections to clear up and 2 months or more for the blood-making capacity of the bone marrow to return to normal. Alimentary toxic aleukia has been reproduced in cats and monkeys administered T-2 toxin isolated from the strain of *F. sporotrichioides* involved in the fatal human outbreaks of the disease in Russia. The related compound, HT-2 toxin, may also contribute to the disease.

Urov or Kashin-Beck Disease

Urov or Kashin-Beck disease has been observed among the Cossack people of eastern Russia for well over 100 years (30). The disease has been endemic in areas along the Urov River in Siberia and was studied extensively by two Russian scientists, Kashin and Beck, thus the name Urov or Kashin-Beck disease (33). The disease is a deforming bone-joint osteoarthrosis that is manifested by a shortening of long bones and a thickening deformation of joints, plus muscular weakness and atrophy (Table 23.1). The disease occurs most commonly in preschool and school-age children.

Table 23.1 Human diseases that have been associated with *Fusarium* species and toxins

Disease	Food(s)	Organism(s)	Toxin(s)	Symptoms or effects
Alimentary toxic aleukia (ATA)	Cereal grains, wheat, rye, bread	*F. sporotrichioides, F. poae*	Possibly T-2 toxin	Burning sensation in mouth and throat, vomiting, diarrhea, abdominal pain, bone marrow destruction, hemorrhaging, death
Urov or Kasin-Beck disease	Cereal grains	*F. poae*	Unknown	Osteoarthritis, shortened long bones, deformed joints, muscular weakness
Drunken bread	Cereal grains, wheat, bread	*F. graminearum*	Unknown	Headache, dizziness, tinnitus, trembling, unsteady gait, abdominal pain, nausea, diarrhea
Akakabi-byo (scabby grain intoxication)	Cereal grains, wheat, barley, noodles	*F. graminearum*	Unknown, possibly deoxynivalenol	Anorexia, nausea, vomiting, headache, abdominal pain, diarrhea, chills, giddiness, convulsions
Foodborne illness outbreaks	Cereal grains, wheat, barley, corn, bread	*F. graminearum*	Deoxynivalenol, acetyl-deoxynivalenol, nivalenol, T-2 toxin	Irritation of throat, nausea, headaches, vomiting, abdominal pain, diarrhea
Esophageal cancer	Corn	*F. moniliforme*	Unknown, possibly fumonisins and other toxins	Precancerous and cancerous lesions in the esophagus

Several hypotheses have been advanced to describe possible causes of the disease, including nutritional deficiencies caused by low levels of calcium and other minerals in water and food crops of the region, and hereditary considerations. Russian scientists have discounted these ideas in favor of a mycotoxic origin of the disease (39). This conclusion is based on the fact that the disease has been reproduced in puppies and rats by feeding *F. poae* isolates obtained only from regions where the disease is endemic. No specific toxin has yet been isolated and proven to be the definite cause, so the actual etiology has not been completely resolved and evidence for involvement of a mycotoxin may be weak. The same disease has also been reported in North Korea and northern China.

Drunken Bread

Another human mycotoxicosis reported in the former Soviet Union is known as "drunken bread" (30). This syndrome is apparently caused by consumption of bread made from rye grain infected with *Fusarium graminearum*. The illness is milder than alimentary toxic aleukia and is a nonfatal self-limiting disorder. Symptoms associated with this disease are headache, dizziness, tinnitus, trembling, and shaking of the extremities, with an unsteady or stumbling gait, hence the name (Table 23.1). There is also flushing of the face and gastrointestinal symptoms, including abdominal pain, nausea, and diarrhea. Victims may appear to be euphoric and confused. The duration of the illness is 1 to 2 days after consumption of the toxic food has ceased. The infected grain may seem normal or appear shriveled and light in weight, with a white to pinkish coloration suggestive of fusarium head blight, or scab.

Akakabi-Byo

Akakabi-byo is also called scabby grain intoxication or red mold disease. It has been observed in Russia, Japan, and China (39). In China the disease is known as mi-chum. In all cases the disease is associated with eating bread made from scab-infested wheat, barley, or other grains infected with *F. graminearum*. Symptoms of this illness are anorexia, nausea, vomiting, headache, abdominal pain, diarrhea, chills, giddiness, and convulsions (Table 23.1). Clinical signs are similar to those of the so-called drunken bread syndrome discussed above. No specific mycotoxins have been shown to cause this illness, but deoxynivalenol and nivalenol naturally occur in scabby grain from the regions where it is endemic.

Foodborne Illness Outbreaks

Outbreaks of foodborne illness associated with *Fusarium* species have involved foods made from wheat or barley infected mainly with *F. graminearum*. These outbreaks resemble scabby grain intoxication or red mold disease and may be essentially the same thing. Foodborne illnesses reported in Japan, Korea, and China have involved foods, particularly noodles, made from scabby wheat (77). The onset of illness is usually rapid, from 5 to 30 min, suggesting the presence of a preformed toxin, most likely deoxynivalenol. The most common symptoms include nausea, vomiting, abdominal pain, diarrhea with headache, fever, chills, and throat irritation in some victims (Table 23.1). An outbreak in India was characterized by an onset time of 15 to 60 min, abdominal pain, irritation of the throat, vomiting, and diarrhea, some with blood in the stools (7). In addition, some victims had a facial rash, nausea, and flatulence. The food involved was bread made from molded wheat. Apparently, flour millers had mixed infected wheat with sound wheat and milled it into flour. *Fusarium* species were isolated from the wheat and flour, and trichothecene mycotoxins including deoxynivalenol, acetyl-deoxynivalenol, nivalenol, and T-2 toxin were detected in many samples.

Outbreaks of foodborne illness that have occurred in China were reviewed by Kuiper-Goodman (35). These involved corn and wheat contaminated with *Fusarium* species, deoxynivalenol, and zearalenone. An outbreak of precocious pubertal changes in thousands of young children in Puerto Rico in which zearalenone or its derivatives or other exogenous estrogenic substances were the suspected cause has also been reported (59, 60). Affected persons experienced premature pubarche, prepubertal gynecomastia, and precocious pseudopuberty. Zearalenone or a derivative was found in the blood of some of the patients, and food was believed to be the source of the estrogenic substances.

Esophageal Cancer

Fusarium verticillioides (moniliforme) has been associated with high rates of esophageal cancer in certain parts of the world, particularly the Transkei region of South Africa, northeastern Italy, and northern China (25, 35, 47, 53). In these regions, corn is a dietary staple and is the main or only food consumed. Corn and corn-based foods from these regions may contain significant amounts of fumonisins and possibly other metabolites of this fungus.

Immunotoxic Effects of *Fusarium* Toxins

The preceding discussions of human diseases that have been associated with various *Fusarium* species in grains and grain-based foods illustrate the range of illnesses caused by *Fusarium* toxins. Members of the genus

Fusarium produce a diverse array of biologically active compounds. These differ in chemical structure and activity. However, it is believed that many of these compounds, especially the trichothecenes, can affect the immune system by suppressing immune functions. T-2 toxin, for example, is known to be highly immunosuppressive (27). There are reports of so-called "sick houses," i.e., houses where individuals have contracted diseases such as leukemia, where *Fusarium* species have been detected in dust (40). In one report, a husband and wife both contracted leukemia while living in a house that contained dust-associated spores of *Fusarium equiseti* (85). In another report, *Fusarium* spores were detected in a house in which four cases of leukemia occurred (84). *F. equiseti* is a known producer of diacetoxyscirpenol, a trichothecene that has known immunosuppressive properties (40). Other *Fusarium* isolates from the house were toxic to ducklings, hamsters, and mice. The conclusion was that mycotoxigenic *Fusarium* species, including *F. equiseti*, were possibly involved in the development of leukemia by their immunosuppressive effects (85).

Immunotoxicity can cause two general types of adverse effects on the immune system (50). In the first type, a toxin or chemical suppresses one or more functions of the immune system. This can result in increased susceptibility to infection or neoplastic disease. In the second type of immunotoxicity, the toxin or chemical may stimulate an immune function, resulting in autoimmune types of disorders. Trichothecenes, for example, are known to inhibit protein and DNA synthesis and to interact with cell membranes, causing weakening and damage. Exposure to trichothecenes can cause damage to bone marrow, spleen, thymus, lymph nodes, and intestinal mucosa (50, 66). T-2 toxin and deoxynivalenol have been shown to affect B-cell and T-cell mitogen responses in lymphocytes. Dietary exposures to deoxynivalenol at concentrations as low as 2 μg/g for 5 weeks or 5 μg/g for 1 week cause decreased mitogen responses (50). T-2 toxin and diacetoxyscirpenol cause increased susceptibility to *Candida* infections, as well as to *Listeria, Salmonella, Mycobacterium,* and *Cryptococcus* infection, in experimental animals (50).

Dietary deoxynivalenol has been shown to stimulate immunoglobin production, causing elevated immunoglobin A (IgA) levels in mice. Among the harmful effects of this stimulation are kidney damage that is very similar to a common human kidney condition known as glomerulonephritis or IgA nephropathy (50). While the cause of this condition is unknown, there is an association with grain-based diets. Other *Fusarium* mycotoxins, e.g., zearalenone and fumonisins, may also have immunotoxic effects, but less information is available about these toxins. Of all the harmful effects of mycotoxins, immunotoxicity or immunomodulation may have the most significant impact on human health. It appears that relatively low levels of the toxins can cause these responses.

TOXIGENIC *FUSARIUM* SPECIES AND THEIR TOXINS

According to the key of Nelson et al. (49), there are twelve sections, or groupings, within the genus *Fusarium*. Only four sections, containing the most common toxic species, will be discussed here: *Sporotrichiella* (*F. sporotrichioides* and *F. poae*), *Gibbosum* (*F. equiseti*), *Discolor* (*F. graminearum* and *Fusarium culmorum*), and *Liseola* (*F. verticillioides* [*moniliforme*], *Fusarium proliferatum,* and *Fusarium subglutinans*). The major mycotoxins produced by these species are summarized in Table 23.2. The key of Nelson et al. (49) should be consulted for identifying *Fusarium* species, although this book will be revised in the future because of changes in

Table 23.2 Major mycotoxins that may be produced by *Fusarium* species of importance in cereal grains and grain-based foods

Section	Species	Potential mycotoxins
Sporotrichiella	*F. poae*	Type A trichothecenes, T-2 toxin, diacetoxyscirpenol
	F. sporotrichioides	T-2 toxin
Gibbosum	*F. equiseti*	Unknown
Discolor	*F. graminearum*	Deoxynivalenol (DON, vomitoxin), 3-acetyldeoxynivalenol, 15-acetyldeoxynivalenol, zearalenone, possibly others
	F. culmorum	Nivalenol, zearalenone
Liseola	*F. verticillioides*	Fumonisins and others
	F. proliferatum	Fumonisins, moniliformin, and others
	F. subglutinans	Moniliformin and others

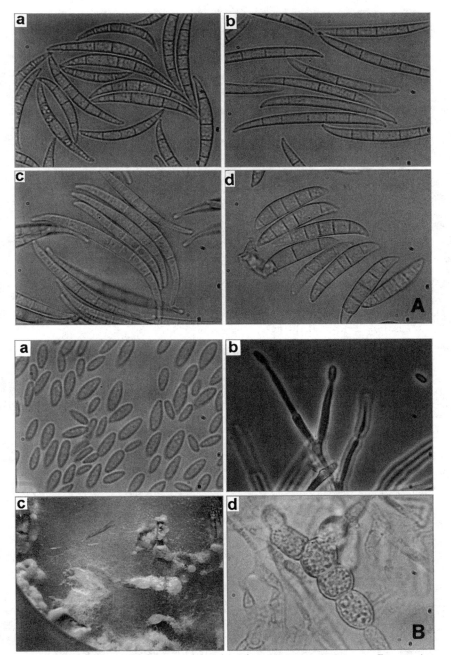

Figure 23.1 Examples of (A) macroconida of *Fusarium* species: a, *F. graminearum*; b, *F. verticillioides (moniliforme)*; c, *F. equiseti*; d, *F. culmorum* (all ×1,000) and (B) micro- and macroscopic structures of *Fusarium*: a, microconidia; b, monophialides; c, sporodochia; d, chlamydospores (a, b, and d, ×1,000; c, ×10).

Fusarium taxonomy. In addition, the book by Samson et al. (61) can be used to identify most isolates encountered in food. Another good general reference is the book by Pitt and Hocking (52).

The genus *Fusarium* is characterized by production of septate hyphae that generally range in color from white to pink, red, purple, or brown as a result of pigment production. The most common characteristic of the genus is the production of large septate, crescent-shaped, fusiform, or sickle-shaped spores known as macroconidia. The macroconidia exhibit a foot-shaped basal cell and beak-shaped or snout-like apical cell (Fig. 23.1A).

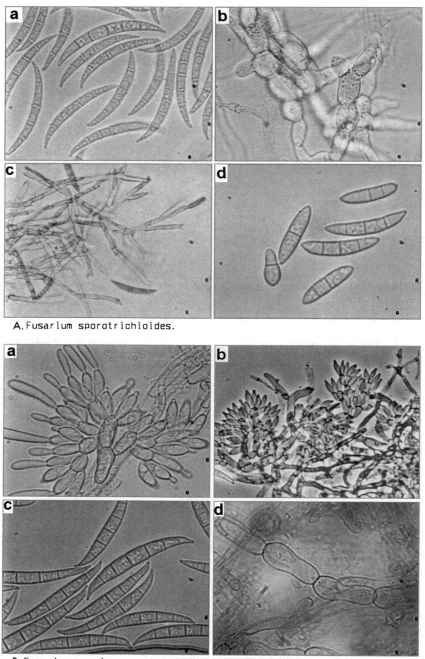

Figure 23.2 Microscopic structures. (A) *Fusarium sporotrichioides*: a, macroconidia; b, chlamydospores; c, phialides; d, microconidia (a, b, and d, ×1,000; c, ×550); (B) *Fusarium graminearum*: a, conidiophores (monophialides); b, monophialides in sporodochia; c, macroconidia; d, chlamydospores (a, c, and d, ×1,000; b, ×550). (C) *Fusarium verticillioides* (*moniliforme*): a, microconidia in chains; b, microconidia; c, monophialides producing microconidia; d, macroconidia (a, ×165; b, ×550; c and d, ×1,000). (D) *Fusarium proliferatum*: a, microconidia in chains; b, microconidia; c, polyphialides; d, macroconidia (a, ×165; b, c, and d, ×1,000).

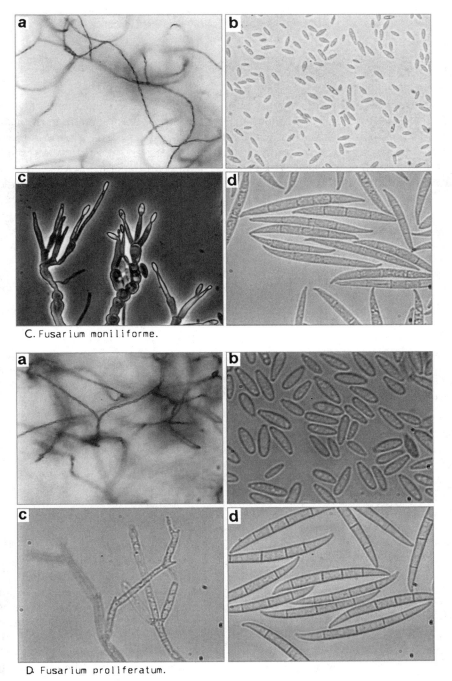

C. Fusarium moniliforme.

D. Fusarium proliferatum.

Figure 23.2   *(Continued)*

The macroconidia are produced from phialides in a stroma known as a sporodochium or in mucoid or slimy masses known as pionnotes. Macroconidia can also be produced in the hyphae, but these are less typical and more variable. Some species also produce smaller one- or two-celled conidia known as microconidia (Fig. 23.1B). Some species also produce swollen, thick-walled chlamydospores in the hyphae or in the macroconidia (Fig. 23.1B). *Fusarium* species are highly variable because of their genetic makeup and can undergo mutations and morphological changes in culture after isolation.

Section *Sporotrichiella*

F. poae

F. poae is widespread in soils of temperate climate regions and is found on grains such as wheat, corn, and barley (39, 52). It exists primarily as a saprophyte, but may be weakly parasitic, and is most commonly found in temperate regions of Russia, Europe, Canada, and northern United States; *F. poae* has also been found in warmer regions, such as Australia, India, Iraq, and South Africa. Its optimum growth temperature is 22 to 27°C, but it can grow in temperature as low as 2 to 3°C. *F. poae* can be isolated from overwintered grain. Diseases that have been associated with *F. poae* include alimentary toxic aleukia, a hemorrhagic syndrome, and Urov or Kashin-Beck disease. *F. poae* produces rapid, profuse white to pink mycelial growth on potato dextrose agar (PDA), with a red to very deep carmine red reverse coloration (49). The most distinctive feature of *F. poae* is its production of abundant globose to oval, almost pyriform (pear-shaped) microconidia, with few macroconidia. Conidia are produced on branched or unbranched monophialides. Chlamydospores are produced infrequently (49). The teleomorphic state has not been observed in *F. poae*. *F. poae* primarily produces type A trichothecenes such as T-2 toxin and diacetoxyscirpenol.

F. sporotrichioides

F. sporotrichioides is found in soil and a wide variety of plant materials. This mold is found in the temperate to colder regions of the world, including Russia, northern Europe, Canada, northern United States, and Japan (39). It can grow at low to very low temperatures, e.g., at −2°C, on grain overwintering in the field. Its optimum growth temperature is 22 to 27°C. Diseases that have been associated with *F. sporotrichioides* include alimentary toxic aleukia, a hemorrhagic syndrome, and Akakabi-byo. *F. sporotrichioides* produces T-2 toxin, diacetoxyscirpenol, zearalenone, and fusarin C.

F. sporotrichioides produces dense white to pink or brown mycelia on PDA, with a deep red reverse. The most distinctive feature of *F. sporotrichioides* is the production of branched and unbranched polyphialides that produce two kinds of microconidia, oval to pear shaped and multiseptate spindle shaped, that resemble macroconidia. Macroconidia are also produced but on monophialides (Fig. 23.2A). *F. sporotrichioides* produces abundant chlamydospores singly and in chains and bunches (49, 52). The teleomorphic state has not been observed in *F. sporotrichioides*. *Fusarium tricinctum* and *Fusarium chlamydosporum* are also members of the section *Sporotrichiella*, and both have

been reported to be toxigenic, producing moniliformin (49).

Section *Gibbosum*

F. equiseti

F. equiseti is a very cosmopolitan mold found in the soil. It is particularly common in tropical and subtropical areas but also occurs in temperate regions (39, 52). For the most part, *F. equiseti* is saprophytic but may be pathogenic to plants such as bananas, avocados, and curcubits. *F. equiseti* has been found in soils from Alaska to tropical regions, and it has been isolated from cereal grains and overwintered cereals in Europe, Russia, and North America. It has been suggested that *F. equiseti* may contribute to leukemia in humans by affecting the immune system (39, 84, 85).

Growth of *F. equiseti* on PDA is rapid, resulting in dense, cottony aerial mycelia that fill the petri dish. The colony is white with a pale salmon to almost brown reverse (49, 52). As cultures age, orange sporodochia may be produced. The most distinctive characteristic of *F. equiseti* is the shape of the macroconidia, which are long, slender, and curved, with five to seven septa. The apical cell is elongated, and the basal cell has a distinctive foot shape. Microconidia and chlamydospores are also produced (49, 52). The teleomorph of *F. equiseti* is *Gibberella intricans*, but its occurrence in nature is rare (49, 52). Other species in the section *Gibbosum* include *Fusarium scirpi*, *Fusarium acuminatum*, and *Fusarium longipes*, with *F. acuminatum* also being reported to be toxigenic (49).

Section *Discolor*

F. graminearum

F. graminearum, a plant pathogen found worldwide in the soil, is the most widely distributed toxigenic *Fusarium* species (31). It causes various diseases of cereal grains, including gibberella ear rots in corn and *Fusarium* head blight or scab in wheat and other small grains (39, 44). These two diseases are important to food microbiology and food safety because the mold and its major toxic metabolites, deoxynivalenol and zearalenone, may contaminate grain and subsequent food products made from the grain. Other *F. graminearum* mycotoxins are 3-acetyldeoxynivalenol, 15-acetyl-deoxynivalenol, diacetyldeoxynivalenol, butenolide, diacetoxyscirpenol, fusarenon-X (4-acetylnivalenol), monoacetoxy scirpenol, neosolaniol, nivalenol, or T-2 toxin that may be produced by some strains (39). Some confusion may surround the taxonomy of *F. graminearum*. In some of the older literature, the name *Fusarium roseum* was used for *F. graminearum* and other species, e.g., *F. roseum* Graminearum and *F. roseum* var. *graminearum*.

Growth of *F. graminearum* on PDA is rapid, with the formation of dense aerial mycelia that fill the petri dish (49, 52). The mycelia are grayish with yellow and brown tinges and white margins. The reverse is usually a deep carmine red. The formation of sporodochia on PDA is sparse and may take 30 days to form. Sporodochia appear red-brown to orange. The macroconidia of *F. graminearum* are distinctive of the species, in that they are almost cylindrical, with the central dorsal and ventral surfaces parallel (Fig. 23.2B). They most often have five septa. The ends of the macroconidia are slightly and unequally curved, and the apical cell is cone shaped and slightly bent. The basal cell has a distinct foot shape. Microconidia are not formed, and chlamydospores, when formed, most often occur in the macroconidia with some in the mycelia (49, 52). The teleomorphic state of *F. graminearum* is *Gibberella zeae*.

F. culmorum

F. culmorum is also widely distributed in the soil and causes diseases of cereal grains, one of which, ear rot in corn, is important in food microbiology and food safety, since the mold and its toxins may contaminate corn-based foods. *F. culmorum* is in many ways similar to *F. graminearum* and has previously been lumped with *F. graminearum* in the nonexistent species *F. roseum*, e.g., *F. roseum* Culmorum and *F. roseum* var. *culmorum* (49, 52). Mycotoxins reported to be produced by *F. culmorum* include deoxynivalenol, zearalenone, and acetyldeoxnivalenol (39, 44).

Growth of *F. culmorum* on PDA is rapid, with the formation of dense aerial mycelium that is white with some tinges of yellow and brown. The reverse of the colony is a deep carmine red (49). Sporodochia that are orange to red-brown may be produced in older cultures. While *F. culmorum* may resemble *F. graminearum* in macroscopic appearance when growing on PDA, the macroconidia are quite different. Macroconidia of *F. culmorum* are very short and stout, compared to other members of the section, and have three to five septa (49, 52). The macroconida have curved dorsal and ventral surfaces, and the basal cell varies from being foot shaped to having a notched appearance. Chlamydospores are readily formed in the macroconidia and mycelia. Microconidia are not formed. No teleomorphic state of *F. culmorum* has been observed. Other species in the section *Discolor* include *Fusarium heterosporum*, *Fusarium reticulatum*, *Fusarium sambucinum*, and *Fusarium crookswellense*. Of these, *F. heterosporum* and *F. sambucinum* have been reported to form toxins (49).

Section *Liseola*

F. moniliforme (verticillioides)

The organism formerly called *F. moniliforme* in much of the scientific literature is now divided into two species, *F. verticillioides* and *Fusarium thapsinum* (41). Within the section *Liseola* exist six distinct mating types (A through F), of which mating types A and F were within the old *F. moniliforme* anamorph. These mating types were very distinct and differed in their ability to produce the mycotoxin, fumonisin (36, 47). Mating type A, which is found in corn, is capable of producing high levels of fumonisins. The anamorph of this mating type, now designated *F. verticillioides*, is synonymous with *F. moniliforme* from corn. Mating type F is found in sorghum and produces little or no fumonisins. The anamorph of this mating type, now designated *F. thapsinum*, is synonymous with *F. moniliforme* from sorghum. The mating populations of the section *Liseola* with their anamorphs and teleomorphs are summarized in Table 23.3. *Fusarium fujikuroi* is also an older synonym for *F. verticillioides (moniliforme)*.

F. verticillioides is a soil-borne plant pathogen that is found in corn growing in all regions of the world. It is the most prevalent mold associated with corn. It often produces symptomless infections of corn plants, but may infect the grain as well, and has been found worldwide on food- and feed-grade corn. The presence of *F. verticillioides* in corn grain is often not discernible, e.g., the grain does not appear infected, yet it is not uncommon to find lots of shelled corn with 100% internal kernel infection (39). The presence of *F. verticillioides* in corn is a major concern in food microbiology and food safety because of the possible widespread contamination of corn and corn-based foods with its toxic metabolites, especially the fumonisins.

Table 23.3 Mating populations of the *Liseola* section of the genus *Fusarium* with corresponding anamorphic and teleomorphic names[a]

Mating population	Anamorph (*Fusarium*)	Teleomorph (*Gibberella*)
A	*F. verticillioides*	*G. moniliformis*
B	*F. sacchari*	*Gibberella* species
C	*F. fujikuroi*	*G. fujikuroi*
D	*F. proliferatum*	*G. intermedia*
E	*F. subglutinans*	*G. subglutinans*
F	*F. thapsinum*	*G. thapsina*

[a] Data from reference 41. The teleomorph of mating population B (anamorph *F. sacchari*) is an unnamed species of *Gibberella*. The teleomorphs of *F. globosum* and *F. anthophilum* have not been reported and are unknown.

F. verticillioides has long been suspected of being involved in animal and human diseases. In the 1880s, a mold found on corn in Italy, called *Oospora verticillioides*, was associated with pellagra (40). In the United States, *F. moniliforme (verticillioides)* growing on corn was linked to diseases of farm animals in Nebraska and other parts of the Midwest (51, 67). In more recent years, animal diseases associated with *F. verticillioides* have included equine leukoencephalomalacia (ELEM), a liquefactive necrosis of the brain of horses and other equidae (34); pulmonary edema and hydrothorax in swine (29); and experimental liver cancer in rats (25). In addition, *F. verticillioides* has been associated with abnormal bone development in chicks and pigs, manifested as leg deformities and ricketslike diseases (39). Experimental toxicity has been induced in animals by feeding culture material of *F. verticillioides*. These animals include baboons, in which acute congestive heart failure and cirrhosis of the liver were observed (39), monkeys, in which atherogenic and hypercholesterolemic responses occurred (22), chickens, donkeys, ducklings, geese, horses, mice, pigeons, pigs, rabbits, rats, and sheep (39).

The main human disease associated with *F. verticillioides* is esophageal cancer. Several studies have linked the presence of *F. verticillioides* and fumonisins in corn to high incidences of esophageal cancer in humans in certain regions of the world, including the Transkei of South Africa, northeastern Italy, northern China, and an area around Charleston, South Carolina, in the United States (24, 26, 42, 53). Mycotoxins produced by *F. verticillioides* include fumonisins, fusaric acid, fusarins, and fusariocins.

F. verticillioides grows rapidly and produces dense white mycelia that might be tinged with purple on PDA (49). The reverse of the colony can range from colorless to purple. Macroconidia are long, slender, and almost straight to slightly curved, especially near the ends (Fig. 23.2C). Macroconidia have three to five septa, a snout-shaped apical cell, and a foot-shaped basal cell. The most distinctive microscopic feature of *F. verticillioides* is the formation of long chains of oval, single-celled microconidia on monophialides (Fig. 23.2C). Microconidia can also be formed in false heads. Chlamydospores are not formed. *F. verticillioides* can grow over a wide temperature range, from 2.5 to 37°C, with an optimum range of 22 to 27°C, and at water activity (a_w) above 0.87 (52). Besides corn, *F. moniliforme (verticillioides* and/or *thapsinum)* has also been isolated from rice, sorghum, yams, hazelnuts, pecans, and cheeses (52).

F. proliferatum

F. proliferatum is closely related to *F. verticillioides*, yet less is known about this species, possibly because it is often misidentified as *F. moniliforme* (39). It is also frequently isolated from corn, where it probably occurs in much the same way as *F. verticillioides*. *F. proliferatum* is capable of producing fumonisins, but has not as yet been associated with animal or human diseases (39). In addition to producing fumonisins, *F. proliferatum* has been reported to produce moniliformin, fusaric acid, fusarin C, beauvericin, and fusaproliferin (41).

On PDA, *F. proliferatum* grows rapidly to produce heavy white aerial mycelia that may become tinged with purple. Macroconidia, which are produced abundantly, are long and thin, have three to five septa, and are only slightly curved to almost straight (Fig. 23.2D). The basal cell is foot shaped. Microconidia are single celled with a flattened base and are produced in short to varying length chains or in false heads from polyphialides (more than one opening), which distinguishes *F. proliferatum* from *F. verticillioides* (49). Chlamydospores are not produced. *F. proliferatum* is widely distributed in the soil and may contaminate several types of food grains. *F. proliferatum* equates with mating population D, and the teleomorphic state is *Gibberella intermedia*. *F. proliferatum* appears to be a significant producer of fumonisins (36, 47).

F. subglutinans

F. subglutinans is very similar to both *F. verticillioides* and *F. proliferatum*. It is widely distributed on corn and other grains. Little information is available about this mold, again probably because of its misidentification as *F. moniliforme* (45, 52). *F. subglutinans* has not been specifically associated with any reported animal or human diseases, but it has been found in corn from regions with high incidences of human esophageal cancer. Cultures of *F. subglutinans* have been shown to be acutely toxic to ducklings and rats and to be dermatoxic to rabbit skin (39). However, *F. subglutinans* does not produce fumonisins (48, 73). In these studies, isolates of *F. subglutinans* from the United States, Mexico, Nigeria, and South Africa were examined. Thus, toxicity attributed to *F. subglutinans* must be due to other toxic metabolites, such as moniliformin, beauvericin, and fusaproliferin (41).

F. subglutinans grows rapidly on PDA, forming white aerial mycelia that may be tinged with purple. The reverse of the colony may range from colorless to a dark purple (49, 52). Macroconidia are long, slender, and almost straight to slightly curved, with a foot-shaped basal cell and three to five septa. Microconidia are oval, usually single celled (but may be one to three septate), and are produced only in false heads. The teleomorphic state is *Gibberella subglutinans*, with an older synonym of *Gibberella fujikuroi* var. *subglutinans*; synonyms of the anamorphic state are *F. moniliforme* var. *subglutinans*, *F. moniliforme* Subglutinans, and *Fusarium sacchari* (36, 49). *F. sacchari* is now the anamorph of mating population B, and *F. subglutinans* is the anamorph of mating population E (41). Another species in the *Liseola* section is *Fusarium anthophilum*.

DETECTION, ISOLATION, AND IDENTIFICATION OF *FUSARIUM* SPECIES

Fusarium species are most often associated with cereal grains, seeds, milled cereal products such as flour and corn meal, barley malt, animal feeds, and necrotic plant tissue. These substrates may also contain or be colonized by many other microorganisms, and *Fusarium* species may be present in low numbers. To isolate *Fusarium* species from these products, it is necessary to use selective media. The basic techniques for detection and isolation of *Fusarium* species employ plating techniques, either as plate counts of serial dilutions of products or by the placement of seeds or kernels of grain directly on the surface of agar media in petri dishes, i.e., direct plating.

Several culture media have been used to detect and isolate *Fusarium* species. These include Nash Snyder (NS) medium (46), modified Czapek Dox (MCZ) agar (49), Czapek iprodione dichloran (CZID) agar (2), potato dextrose iprodione dichloran (PDID) agar (74), and dichloran chloramphenicol peptone agar (DCPA) (3, 52). The NS medium and MCZ agar contain pentachloronitrobenzene, a known carcinogen, and are not favored for routine use in food microbiology laboratories. However, these media can be useful for evaluating samples that are heavily contaminated with bacteria and other fungi. The CZID agar is being used more regularly for isolating *Fusarium* from foods, but rapid identification of *Fusarium* isolates to species level is difficult, if not impossible, on this medium. Isolates must be subcultured on other media such as carnation leaf agar (CLA) for identification. However, CZID agar is a good selective medium for *Fusarium*. While some other molds may not be completely inhibited on CZID, most are, and *Fusarium* species can be readily distinguished. Thrane et al. (74) reported that PDID agar is as selective as CZID agar for *Fusarium* species, with the advantage

that it supports *Fusarium* growth with morphological and cultural characteristics that are the same as on PDA. This facilitates more rapid identification, since various monographs and manuals for *Fusarium* identification describe characteristics of colonies grown on PDA. Thrane et al. (74) compared several media for their suitability to support colony development by *Fusarium* and found that PDID and CZID agars were better than DCPA. Growth rates were much higher on DCPA, making colony counts more difficult. Conner (18), however, modified DCPA by adding 0.5 µg/mL of crystal violet and reported increased selectivity by *Aspergillus* and *Penicillium* species, but not *Fusarium* species.

Identification of *Fusarium* species is based largely on the production and morphology of macroconidia and microconidia. Identification keys described by Nelson et al. (49) rely heavily on the morphology of conidia and conidiophores, the structures on which conidia are produced. *Fusarium* species do not readily form conidia on all culture media, and conidia formed on high-carbohydrate media such as PDA are often more variable and less typical. A medium that supports abundant and consistent spore production is CLA. Carnation leaves from actively growing, disbudded, young carnation plants free of pesticide residues are cut into small pieces (5 mm^2), dried in an oven at 45 to 55°C for 2 h, and sterilized by irradiation (49). CLA is prepared by placing a few pieces of carnation leaf on the surface of 2.0% water agar (23, 49). *Fusarium* isolates are then inoculated on the agar and leaf interface, where they form abundant and typical conidia and conidiophores in sporodochia, rather than mycelia. CLA is low in carbohydrates and rich in other complex naturally occurring substances that apparently stimulate spore production.

Since many *Fusarium* species are plant pathogens and all are found in fields where crops are grown, these molds respond to light. Growth, pigmentation, and spore production are most typical when cultures are grown in alternating light and dark cycles of 12 h each. Fluorescent light or diffuse sunlight from a north window is best. Fluctuating temperatures such as 25°C (day) and 20°C (night) also enhance growth and sporulation. For identification keys, refer to Nelson et al. (49), Samson et al. (61), Marasas et al. (40), and Marasas (38).

DETECTION AND QUANTITATION OF *FUSARIUM* TOXINS

Fusarium species produce several toxic or biologically active metabolites. The trichothecenes are a group of closely related compounds that are esters of sesquiterpene alcohols that possess a basic trichothecene skeleton and an epoxide group (71). The trichothecenes are divided into three groups: the type A trichothecenes, which include diacetoxyscirpenol, T-2 toxin, HT-2 toxin, and neosolaniol; the type B trichothecenes, which include deoxynivalenol, 3-acetydeoxynivalenol, 15-acetyldeoxynivalenol, nivalenol, and fusarenon-X; and the type C or so-called macrocyclic trichothecenes known as satratoxins. Of these, the toxin most commonly found in cereal grains or most often associated with human illness is deoxynivalenol (44). Other *Fusarium* toxins associated with diseases are zearalenone and the fumonisins. T-2 toxin rarely occurs in grain in the United States but has been associated with alimentary toxic aleukia in Russia in the 1940s and earlier. Moniliformin, fusarin C, and fusaric acid are also of interest and concern, but they have not been shown to commonly occur or be specifically associated with diseases.

If present, *Fusarium* toxins are usually found at low levels in cereal grains and processed grain-based foods. Their concentrations may range from less than nanogram to microgram quantities per gram (parts per billion to parts per million, respectively). *Fusarium* toxins vary in their chemical structures and properties, making it difficult to develop a single method for quantitating all toxins. The basic steps involved in detection of *Fusarium* mycotoxins are similar to those for other mycotoxins. These include sampling, size reduction and mixing, subsampling, extraction, filtration, cleanup, concentration, separation of components, detection, quantification, and confirmation (9, 71).

The first problem encountered in the analysis of grains for *Fusarium* toxins is the same as for other mycotoxins, i.e., sampling. Obtaining a representative sample from a large lot of cereal grain can be very difficult if the toxin is present in a relatively small percentage of the kernels, which may be the case with toxins such as deoxynivalenol and zearalenone. On the other hand, fumonisins appear to be more evenly distributed in corn. Processed grain-based foods may contain a more even distribution of toxins as a result of grinding and mixing. Samples are usually ground and mixed further, and a subsample of 50 to 100 g is taken for extraction. *Fusarium* toxins, like all mycotoxins, must be extracted from the matrix in which they are found. Most mycotoxins are more soluble in slightly polar organic solvents than water. The most commonly used extraction solvents consist of combinations of water with organics such as methanol, acetone, and acetonitrile. Following extraction, the extract is filtered to remove solids and subjected to a cleanup step to remove interfering substances. Cleanup can be done in several ways, but the most common method used for

Fusarium toxins is to pass the extract through a column packed with sorbent packing materials. In recent years, the use of small, prepacked, commercially available disposable columns or cartridges such as Sep-Pak, Bond Elut, and MycoSep has become common. After the extract has been cleaned, the sample may need to be concentrated before analysis in order to detect the toxin. This may be accomplished by mild heating such as in a water bath, heating block, or rotary evaporator under reduced pressure or a stream of nitrogen. Detection and quantification of the toxins are done after they are separated from other components by chromatographic means. The most common chromatographic separation techniques used are thin-layer chromatography (TLC) and high-performance liquid chromatography (HPLC). Gas chromatography (GC) also has some applications, particularly when coupled with mass spectrometry.

A commonly used method for quantitating deoxynivalenol is TLC (21, 54, 76). Gas chromatography is more sensitive than TLC but is also more laborious. While HPLC methods employing UV absorbance at 219 nm for detection are fairly sensitive, they require purification of deoxynivalenol by using high-capacity activated charcoal columns (13). The method of choice for quantitation of zearalenone is HPLC with fluorescence detection (6, 81). A TLC method for zearalenone has been tested collaboratively and is useful as a screening method (69). GC methods are most commonly used for T-2 toxin (14, 17, 55). Because type A trichothecenes lack a UV chromophore and are not fluorescent, both TLC and HPLC methods are unsuitable, resulting in the reliance on GC methods. The most widely used analytical methods for fumonisins are HPLC methods involving the formation of fluorescent derivatives (54). Methods have been developed using derivatizing agents, such as o-phthaldehyde (OPA) (56, 57, 68), fluorescamine (82), and naphthalene dicarboxaldehyde (5). A TLC method for fumonisins has also been developed but is used mainly for screening (54, 58). For discussions of methods most commonly used for analysis of *Fusarium* and other mycotoxins, see Richard et al. (54) and Steyn et al. (71).

Immunoassays have been developed for *Fusarium* toxins. Enzyme-linked immunosorbent assay (ELISA) kits for *Fusarium* toxins are commercially available (Neogen Corporation, Lansing, Mich.). Qualitative kits for screening as well as kits for quantitative analyses are available for deoxynivalenol, zearalenone, T-2 toxin, and fumonisins. A rapid screening TLC kit for deoxynivalenol is available from Romer Labs (Union, Mo.). This method uses a special cleanup column that requires only 10 s per sample. An antibody-based affinity column for fumonisins is also available (Vicam, Watertown, Mass.) for quantitation as well as for use as a cleanup tool.

OCCURRENCE OF *FUSARIUM* TOXINS IN FOODS

Fusarium toxins, particularly deoxynivalenol and fumonisins, have been found in finished human food products. Various *Fusarium* toxins have been found naturally occurring in numerous cereal grains, but most of these grains have been destined for animal feed. Deoxynivalenol is the most common trichothecene found in commodity grains; therefore, it is most likely to occur in finished foods. Food products such as bread, pasta, and beer may contain at least trace amounts of the toxin. Scott et al. (65) reported finding trace amounts (ca. 5 ng/ml) of deoxynivalenol in 29 of 50 samples of commercially available domestic and imported beer in Canada. In the United States, deoxynivalenol has been found in breakfast cereals in average amounts of 100 ng/g (75), while corn syrup and beer samples were negative. In another study (1), the toxin was found in corn breakfast cereals, wheat flour muffin mixes, wheat- and oat-based cookies, crackers, corn chips, popcorn, and mixed grain cereals. Deoxynivalenol concentrations ranged from 4 to 19 μg/g. Brumley et al. (8) found deoxynivalenol in wheat, flour, corn, corn meal, and snack foods in levels ranging from 0.08 to 0.3 μg/g. Thus, there is evidence that deoxynivalenol is a contaminant of processed human food products and that levels sometimes exceed the U.S. government guideline of 1.0 μg/g in finished food products. Deoxynivalenol is quite heat stable, probably tolerating most thermal processes to some degree (63, 64).

Fumonisins have also been found in processed or finished food products. Food products that have been examined include corn meal, corn grits, corn breakfast cereals, tortillas, tortilla chips, corn chips, popcorn, and hominy corn (10, 11, 78). The most consistently contaminated products with the greatest amounts of fumonisins are those foods that receive only physical processing, such as milled products, e.g., corn meal and corn muffin mixes. Sydenham et al. (72) determined fumonisin levels in corn meal obtained from Canada, Egypt, Peru, South Africa, and the United States. Corn meal from Egypt and the United States had the highest levels. Corn meal from the United States contained fumonisins in concentrations of less than 1.0 up to 2.8 μg/g of corn meal. In another survey, the highest levels of fumonisins in corn-based foods were found in corn meal and corn grits (70). Corn flakes and corn pops cereals, corn chips, and corn tortilla chips were negative for fumon-

isins, and very low levels were found in tortillas, popcorn, and hominy. Fumonisins have been detected in processed corn products in Germany, Italy, Japan, Spain, and Switzerland (20, 62, 78, 79, 86). Fumonisins and moniliformin have been found to co-occur in food-grade corn and corn-based foods in the United States (28).

OTHER TOXIC MOLDS

Other potentially toxic molds that may contaminate foods include species of the genera *Acremonium*, *Alternaria*, *Chaetomium*, *Cladosporium*, *Claviceps*, *Myrothecium*, *Phomopis*, *Rhizoctonia*, and *Rhizopus*. Molds such as *Diplodia maydis*, *Phoma herbarum*, *Pithomyces chartarum*, *Stachybotrys chartarum*, and *Trichothecium roseum* are also potentially toxic (37). However, most of these molds are more likely to be present in animal feeds, and their significance to food safety may be minimal. Some have been shown to produce toxic secondary metabolites in vitro that have yet to be found to occur naturally.

The ergot mold, *Claviceps purpurea*, is the cause of the earliest recognized human mycotoxicosis, ergotism (4). Ergotism has been reported in sporadic outbreaks in Europe since 857, with near-epidemic outbreaks occurring in the Middle Ages. The disease was known as St. Anthony's fire during the Middle Ages because it was believed that pilgrimages to a shrine of St. Anthony could lead to a cure for the disease. It is probable that as pilgrims traveled to the shrine they left areas where ergot was endemic and traveled through areas where it was not a problem. Consequently, when they stopped eating toxic bread from the endemic area and ate nontoxic bread obtained along the way, by the time they reached the shrine, their symptoms subsided. Ergot is a disease of rye in which the rye grains are replaced by ergot sclerotia that contain toxic alkaloids. If sclerotia are not removed when the rye is milled into flour, the flour and bread made from the flour become contaminated. Ergotism can be manifested as a convulsive condition or a necrotic gangrenous condition of the extremities. During the Middle Ages, necrotic gangrenous ergotism was characterized by swollen limbs and alternating cold and burning sensations in fingers, hands, and feet, hence the term "fire" in St. Anthony's fire. The main ergot alkaloid, ergotamine, has vasoconstrictive properties that cause these symptoms. In severe cases, the extremities, such as feet in humans and hooves in animals, were sloughed off. Convulsive ergotism may have been the reason for the Salem witchcraft trials of 1692 in Salem, Massachusetts (12, 43). In more recent times, outbreaks of ergotism have occurred in Russia in 1926, Ireland in

1929, France in 1953, India in 1958, and Ethiopia in 1973 (4).

Alternaria species may be especially significant as potential toxic contaminants of food. *Alternaria* species infect plants in the field and may contaminate wheat, sorghum, and barley (19). *Alternaria* species also infect various fruits and vegetables, including apples, pears, citrus fruits, peppers, tomatoes, and potatoes. *Alternaria* species can cause spoilage of these foods in refrigerated storage. Several *Alternaria* toxins have been described, including alternariol (AOH), alternariol monomethyl ether (AME), altenuene (ALT), tenuazonic acid (TA), and the altertoxins (ATX). Relatively little is known about the toxicity of these toxins; however, cultures of *Alternaria* that have been grown on corn or rice and fed to rats, chicks, turkey poults, and ducklings have been shown to be quite toxic. *Alternaria* was also implicated in the alimentary toxic aleukia toxicoses in Russia in the 1940s (32). Toxicity of various *Alternaria* toxins has not been studied extensively, but there is evidence that the toxins may have synergistic activity, i.e., mixtures of *Alternaria* toxins or culture extracts are more toxic than the individual toxins. Species of *Alternaria* that produce large amounts of tenuazonic acid and one or more of the other toxins include *A. tenuis*, *A. alternata*, *A. citri*, and *A. solani*. *A. alternata* f. sp. *lycopersici*, which is a pathogen of tomatoes, produces a host-specific phytotoxin known as (AAL) toxin that is nearly identical in structure to fumonisins. It has been shown that fumonisins can cause lesions in tomatoes identical to those caused by AAL toxin and that AAL toxin is toxic to animal cells in tissue culture (83).

References

1. **Abbouzied, M. M., J. I. Azcona, W. E. Braselton, and J. J. Pestka.** 1991. Immunochemical assessment of mycotoxins in 1989 grain foods: evidence for deoxynivalenol (vomitoxin) contamination. *Appl. Environ. Microbiol.* 57:672–677.

2. **Abildgren, M. P., F. Lund, U. Thrane, and S. Elmholt.** 1987. Czapek-Dox agar containing iprodione and dichloran as a selective medium for the isolation of *Fusarium* species. *Lett. Appl. Microbiol.* 5:83–86.

3. **Andrews, S., and J. I. Pitt.** 1986. Selective medium for isolation of *Fusarium* species and dematiaceous hyphomycetes from cereals. *Appl. Environ. Microbiol.* 51:1235–1238.

4. **Beardall, J. M., and J. D. Miller.** 1994. Diseases in humans with mycotoxins as possible causes, p. 487–539. *In* J. D. Miller and H. L. Trenholm (ed.), *Mycotoxins in Grain. Compounds Other than Aflatoxin.* Eagan Press, St. Paul, Minn.

5. **Bennett, G. A., and J. L. Richard.** 1992. High performance liquid chromatographic method for naphthalene dicarboxaldehyde derivative of fumonisins, p. 143. *In Proceeding of the AOAC International Meeting, Cincinnati, Ohio.*

6. **Bennett, G. A., O. L. Shotwell, and W. F. Kwolek.** 1985. Liquid chromatographic determination of α zearalenol and zearalenone in corn: collaborative study. *J. Assoc. Off. Anal. Chem.* 68:958–962.

7. **Bhat, R. V., S. R. Beedu, Y. Ramakrisna, and K. L. Munshi.** 1989. Outbreak of trichothecene mycotoxicosis associated with consumption of mould-damaged wheat products in Kashmir Valley, India. *Lancet* i:35–37.

8. **Brumley, W. C., M. W. Trucksess, S. H. Adler, C. K. Cohen, K. D. White, and J. A. Sphon.** 1985. Negative ion chemical ionization mass spectrometry of deoxynivalenol (DON): application to identification of DON in grains and snack foods after quantitation/isolation by thin-layer chromatography. *J. Agric. Food Chem.* 33:326–330.

9. **Bullerman, L. B.** 1987. Methods for detecting mycotoxins in foods and beverages, p. 571–598. *In* L. R. Beuchat (ed.), *Food and Beverage Mycology*, 2nd ed. Van Nostrand Reinhold Company Inc., New York, N.Y.

10. **Bullerman, L. B.** 1996. Occurrence of *Fusarium* and fumonisins on food grains and in foods, p. 27–38. *In* L. Jackson, J. DeVries, and L. Bullerman (ed.), *Fumonisins in Foods.* Plenum Publishing Corp., New York, N.Y.

11. **Bullerman, L. B., and W. Y. J. Tsai.** 1994. Incidence and levels of *Fusarium moniliforme*, *Fusarium proliferatum* and fumonisins in corn and corn-based foods and feeds. *J. Food Prot.* 57:541–546.

12. **Caporeal, L. R.** 1976. Ergotism: the Satan loosed in Salem? *Science* 192:21–26.

13. **Chang, H. L., J. W. Devries, P. A. Larson, and H. H. Patel.** 1984. Rapid determination of deoxynivalenol (vomitoxin) by liquid chromatography using modified Romer column clean-ups. *J. Assoc. Off. Anal. Chem.* 67:52–54.

14. **Chaytor, J. P., and M. J. Saxby.** 1982. Development of a method for the analysis of T-2 toxin in maize by gas-chromatography-mass spectrometry. *J. Chromatogr.* 237:107–111.

15. **Chelkowski, J.** 1989. Mycotoxins associated with corn cob fusariosis, p. 53–62. *In* J. Chelkowski (ed.), *Fusarium. Mycotoxins, Taxonomy and Pathogenicity.* Elsevier Science Publishing Company, New York, N.Y.

16. **Chelkowski, J.** 1989. Formation of mycotoxins produced by fusaria in heads of wheat triticale and rye, *In* J. Chelkowski (ed.), p. 63–84. *Fusarium. Mycotoxins, Taxonomy and Pathogenicity.* Elsevier Science Publishing Company, New York, N.Y.

17. **Cohen, H., and M. Lapointe.** 1984. Capillary gas chromatographic determination of T-2 toxin, HT-2 toxin and diacetoxyscirpenol in cereal grains. *J. Assoc. Off. Anal. Chem.* 67:1105–1109.

18. **Conner, D. E.** 1992. Evaluation of methods for selective enumeration of *Fusarium* species in feedstuffs, p. 299–302. *In* R. A. Samson, A. D. Hocking, J. I. Pitt, and A. D. King (ed.), *Modern Methods in Food Mycology.* Elsevier Scientific Publishers, Amsterdam, The Netherlands.

19. Coulombe, R. A. 1991. *Alternaria* toxins, p. 425–433. *In* R. P. Sharma and D. K. Salunke (ed.), *Mycotoxins and Phytoalexins*. CRC Press, Inc., Boca Raton, Fla.

20. Doko, M. B., and A. Visconti. 1994. Occurrence of fumonisins B_1 and B_2 in corn and corn-based human foodstuffs in Italy. *Food Addit. Contam.* 11:433–439.

21. Eppley, R. M., M. W. Trucksess, S. Nesheim, C. W. Thorpe, and A. E. Pohland. 1986. Thin layer chromatographic method for detection of deoxynivalenol in wheat: collaborative study. *J. Assoc. Off. Anal.Chem.* 69:37–40.

22. Fincham, J. E., W. F. O. Marasas, J. J. F. Taljaard, N. P. J. Kriek, C. J. Badenhorst, W. C. A. Gelderblom, J. V. Seier, C. M. Smuts, M. Faber, M. J. Weight, W. Slazus, C. W. Woodroof, M. J. van Wyk, M. Kruger, and P. G. Thiel. 1992. Atherogenic effects in a non-human primate of *Fusarium moniliforme* cultures added to a carbohydrate diet. *Atherosclerosis* 94:13–25.

23. Fisher, N. L., L. W. Burgess, T. A. Toussoun, and P. E. Nelson. 1982. Carnation leaves as a substrate and for preserving cultures of *Fusarium* species. *Phytopathology* 72:151–153.

24. Franceschi, S., E. Bidoli, A. E. Baron, and C. LaVecchia. 1990. Maize and the risk of cancers of the oral cavity, pharynx and esophagus in northeastern Italy. *J. Natl. Cancer Inst.* 82:1407–1411.

25. Gelderblom, W. C. A., N. P. J. Kriek, W. F. O. Marasas, and P. G. Thiel. 1991. Toxicity and carcinogenicity of the *Fusarium moniliforme* metabolite fumonisin B_1 in rats. *Carcinogenesis* 12:1247–1251.

26. Gelderblom, W. C. A., W. F. O. Marasas, R. Vleggaar, P. G. Thiel, and M. E. Cawood. 1992. Fumonisins: isolation, chemical characterization and biological effects. *Mycopathologia* 117:11–16.

27. Graveson, S., J. C. Frisvad, and R. A. Samson. 1994. *Microfungi*. Munksgaard International Publishers Ltd., Copenhagen, Denmark.

28. Gutema, T., C. Munimbazi, and L. B. Bullerman. 2000. Occurrence of fumonisins and moniliformin in corn and corn-based food products of U.S. origin. *J. Food Prot.* 63:1732–1737.

29. Harrison, L. R., B. M. Colvin, J. T. Greens, L. E. Newman, and J. R. Cole. 1990. Pulmonary edema and hydrothorax in swine produced by fumonisin B_1 a toxic metabolite of *Fusarium moniliforme*. *J. Vet. Diagn. Invest.* 2:217–221.

30. Hayes, A. W. 1981. Involvement of mycotoxins in animal and human health, p. 11–40. *In Mycotoxin Teratogenicity and Mutagenicity*. CRC Press, Inc., Boca Raton, Fla.

31. International Agency for Research on Cancer (IARC). 1993. *Some Naturally Occurring Substances: Food Items and Constituents, Heterocyclic Aromatic Amines and Mycotoxins*. Monograph 56. International Agency for Research on Cancer, Lyon, France.

32. Joffe, A. Z. 1960. The mycoflora of overwintered cereals and its toxicity. *Bull. Res. Counc. Isr.* 90:101–126.

33. Joffe, A. Z. 1986. Effects of fusariotoxins in humans, p. 225–298. *In Fusarium Species: Their Biology and Toxicology*. John Wiley & Sons, New York, N.Y.

34. Kellerman, T. S., W. F. O. Marasas, P. G. Thiel, W. C. A. Gelderblom, M. Cawood, and J. A. W. Coetzer. 1990. Leukoencephalomalacia in two horses induced by oral dosing of fumonisin B_1. *Onderstepoort J. Vet. Res.* 57:269–275.

35. Kuiper-Goodman, T. 1994. Prevention of human mycotoxicoses through risk assessment and risk management, p. 439–469. *In* J. D. Miller and H. L. Trenholm (ed.), *Mycotoxins in Grain. Compounds Other than Aflatoxin*. Eagan Press, St. Paul, Minn.

36. Leslie, J. F., R. D. Plattner, A. E. Desjardins, and C. J. R. Klittich. 1992. Fumonisin B_1 production by strains from different mating populations of *Gibberella fujikuroi* (*Fusarium* section Liseola). *Phytopathology* 82:341–345.

37. Mantle, P. G. 1991. Miscellaneous toxigenic fungi, p. 141–152. *In* J. E. Smith and R. S. Henderson (ed.), *Mycotoxins and Animal Foods*. CRC Press, Inc., Boca Raton, Fla.

38. Marasas, W. F. O. 1991. Toxigenic fusaria, p. 119–139. *In* J. E. Smith and R. S. Henderson (ed.), *Mycotoxins and Animal Foods*. CRC Press, Inc., Boca Raton, Fla.

39. Marasas, W. F. O., P. E. Nelson, and T. A. Tousson. 1984. *Toxigenic Fusarium Species: Identity and Mycotoxicology*. The Pennsylvania State University Press, University Park, Pa.

40. Marasas, W. F. O., P. E. Nelson, and T. A. Tousson. 1985. Taxonomy of toxigenic fusaria, p. 3–14. *In* J. Lacey (ed.), *Trichothecenes and Other Mycotoxins. Proceedings of the International Mycotoxin Symposium*, Sydney, Australia, 1984. John Wiley & Sons, New York, N.Y.

41. Marasas, W. F. O., and J. P. Reeder. 2000. *Sections Liseola, Sporotrichiella and Arthrosporiella. Fusarium Laboratory Workshop*, June 11–16. Kansas State University, Manhattan, Kans.

42. Marasas, W. F. O., F. C. Wehner, S. J. van Rensberg, and D. J. van Schalkwyk. 1981. Mycoflora of corn produced in human esophageal cancer areas in Transkei, Southern Africa. *Phytopathology* 71:792–796.

43. Matossian, M. K. 1982. Ergot and the Salem witchcraft affair. *Am. Sci.* 70:355–357.

44. Miller, J. D. 1995. Fungi and mycotoxins in grain: implications for stored product research. *J. Stored Prod. Res.* 31:1–16.

45. Mills, J. T. 1989. Ecology of mycotoxigenic *Fusarium* species on cereal seeds. *J. Food Prot.* 52:737–742.

46. Nash, S. M., and W. C. Snyder. 1962. Quantitative estimations by plate counts of propagules of the bean root rot *Fusarium* in field soils. *Phytopathology* 52:567–572.

47. Nelson, P. E., A. E. Desjardins, and R. D. Plattner. 1993. Fumonisins, mycotoxins produced by *Fusarium* species: biology, chemistry and significance. *Annu. Rev. Phytopathol.* 31:233–252.

48. Nelson, P. E., R. D. Plattner, D. D. Shackelford, and A. E. Desjardins. 1992. Fumonisin B1 production by *Fusarium* species other than *F. moniliforme* in section Liseola and by some related species. *Appl. Environ. Microbiol.* 58:984–989.

49. Nelson, P. E., T. A. Tousoun, and W. F. O. Marasas. 1983. *Fusarium Species. An Illustrated Manual for Identification*. The Pennsylvania State University Press, University Park, Pa.

50. Pestka, J. J., and G. S. Bondy. 1994. Immunotoxic effects

of mycotoxins, p. 339–359. *In* J. D. Miller and H. L. Trenholm (ed.), *Mycotoxins in Grain. Compounds Other than Aflatoxin.* Eagan Press, St. Paul, Minn.

51. **Peters, A. T.** 1904. A fungus disease in corn. *Agric. Exp. Stn. Nebr. Annu. Rep.* **17**:13–22.

52. **Pitt, J. I., and A. D. Hocking.** 1999. *Fungi and Food Spoilage,* 2nd ed. Academic Press, Inc., Sydney, Australia.

53. **Rheeder, J. P., W. F. O. Marasas, P. G. Thiel, E. W. Sydenham, G. S. Shepard, and D. J. van Schalkwyk.** 1992. *Fusarium moniliforme* and fumonisins in corn in relation to human esophageal cancer in Transkei. *Phytopathology* **82**:353–357.

54. **Richard, J. L., G. A. Bennett, P. F. Ross, and P. E. Nelson.** 1993. Analysis of naturally occurring mycotoxins in feedstuffs and foods. *J. Anim. Sci.* **71**:2563–2574.

55. **Romer, T. R., T. M. Boling, and J. L. McDonald.** 1978. Gas-liquid chromatographic determination of T-2 toxin and diacetoxyscirpenol in corn and mixed feeds. *J. Assoc. Off. Anal. Chem.* **61**:801–805.

56. **Ross, P. F., P. E. Nelson, J. L. Richard, G. D. Osweiler, L. G. Rice, R. D. Plattner, and T. M. Wilson.** 1990. Production of fumonisins by *Fusarium moniliforme* and *Fusarium proliferatum* isolates associated with equine leukoencephalomalacia and a pulmonary edema syndrome in swine. *Appl. Environ. Microbiol.* **56**:3225–3226.

57. **Ross, P. F., L. G. Rice, R. D. Plattner, G. O. Osweiler, T. M. Wilson, D. L. Owens, P. A. Nelson, and J. L. Richard.** 1991. Concentrations of fumonisin B_1 in feeds associated with animal health problems. *Mycopathologia* **114**:129–135.

58. **Rottinghaus, G. E., C. F. Coatney, and Harry C. Minoir.** 1992. A rapid, sensitive thin layer chromatography procedure for the detection of fumonisin B_1 and B_2. *J. Vet. Diagn. Invest.* **4**:326–329.

59. **Saenz de Rodriguez, C. A.** 1984. Environmental hormone contamination in Puerto Rico. *N. Engl. J. Med.* **310**:1741–1742.

60. **Saenz de Rodriguez, C. A., A. M. Bongiovanni, and L. Conde de Borrego.** 1985. An epidemic of precocious development in Puerto Rican children. *J. Pediatr.* **107**:393–396.

61. **Samson, R. A., E. S. Hoekstra, J. C. Frisvad, and O. Filtenborg (ed.).** 1995. *Introduction to Food-Borne Fungi.* Centraalbureau voor Schimmelcultures, Baarn, The Netherlands.

62. **Sanchis, V., M. Abadias, L. Oncins, N. Sala, I. Vinas, and R. Canela.** 1994. Occurrence of fumonisins B_1 and B_2 in corn-based products from the Spanish market. *Appl. Environ. Microbiol.* **60**:2147–2148.

63. **Scott, P. M.** 1984. Effects of food processing on mycotoxins. *J. Food Prot.* **47**:489–499.

64. **Scott, P. M., S. R. Kanhere, P.-Y. Lau, J. E. Dexter, and R. Greenhalgh.** 1983. Effects of experimental flour milling and bread baking on retention of deoxynivalenol (vomitoxin) in hard red spring wheat. *Cereal Chem.* **60**:421–424.

65. **Scott, P. M., S. R. Kanhere, and D. Weber.** 1993. Analysis of Canadian and imported beers for *Fusarium* mycotoxins by gas chromatography-mass spectrometry. *Food Addit. Contam.* **10**:381–389.

66. **Sharma, R. P., and Y. W. Kim.** 1991. Trichothecenes, p. 339–359. *In* R. P. Sharma and D. K. Salunkhe (ed.), *Mycotoxins and Phytoalexins.* CRC Press, Inc. Boca Raton, Fla.

67. **Sheldon, J. L.** 1904. A corn mold (*Fusarium moniliforme* n. sp.) *Agric. Exp. Stn. Nebr. Annu. Rep.* **17**:23–43.

68. **Shepard, G. W., E. W. Sydenham, P. G. Thiel, and W. C. A. Gelderblom.** 1990. Quantitative determination of fumonisins B_1 and B_2 by high performance liquid chromatography with fluorescence detection. *J. Liquid Chromatogr.* **13**:2077–2087.

69. **Shotwell, O. L., M. L. Goulden, and G. A. Bennett.** 1976. Determination of zearalenone in corn: collaborative study. *J. Assoc. Off. Anal. Chem.* **59**:666–669.

70. **Stack, M. E., and R. M. Eppley.** 1992. Liquid chromatographic determination of fumonisins B_1 and B_2 in corn and corn products. *J. Am. Offic. Anal. Chem. Int.* **75**:834–837.

71. **Steyn, P. S., P. G. Thiel, and D. W. Trinder.** 1991. Detection and quantification of mycotoxins by chemical analysis, p. 165–221. *In* J. E. Smith and R. S. Henderson (ed.), *Mycotoxins and Animal Foods.* CRC Press, Inc., Boca Raton, Fla.

72. **Sydenham, E. W., G. S. Shephard, P. G. Thiel, W. F. O. Marasas, and S. Stockenstrom.** 1991. Fumonisin contamination of commercial corn-based human foodstuffs. *J. Agric. Food Chem.* **25**:767–771.

73. **Thiel, P. G., W. F. O. Marasas, E. W. Sydenham, G. S. Shepard, W. C. A. Gelderblom, and J. J. Nieuwenhuis.** 1991. Survey of fumonisin production by *Fusarium* species. *Appl. Environ. Microbiol.* **57**:1089–1093.

74. **Thrane, U., O. Filtenborg, F. C. Frisvad, and F. Lund.** 1992. Improved methods for the detection and identification of toxigenic *Fusarium* species, p. 285–291. *In* R. A. Samson, A. D. Hocking, J. I. Pitt, and A. D. King (ed.), *Modern Methods in Food Mycology.* Elsevier Science Publishers, New York, N.Y.

75. **Trucksess, M. W., M. T. Flood, and S. W. Page.** 1986. Thin layer-chromatography determination of deoxynivalenol in processed grain products. *J. Assoc. Off. Anal. Chem.* **69**:35–36.

76. **Trucksess, M. W., S. Nesheim, and R. M. Eppley.** 1984. Thin layer chromatographic determination of deoxynivalenol in wheat and corn. *J. Assoc. Off. Anal. Chem.* **67**:40–44.

77. **Ueno, Y.** 1983. Toxicoses, natural occurrence and control, p. 195–307. *In Trichothecenes. Chemical, Biological and Toxicological Aspects. Developments in Food Science.* Elsevier Science Publishing Company, Inc., New York, N.Y.

78. **Ueno, Y., S. Aoyama, Y. Sugiura, D. S. Wang, U. S. Lee, E. Y. Hirooka, S. Hara, T. Karki, G. Chen, and S. Z. Yu.** 1993. A limited survey of fumonisins in corn and corn-based products in Asian countries. *Mycotoxin Res.* **9**:27–34.

79. **Usleber, E., M. Straka, and G. Terplan.** 1994. Enzyme immunoassay for fumonisin B_1 applied to corn-based food. *J. Agric. Food Chem.* **42**:1392–1396.

80. **Visconti, A., and A. Sibilia.** 1994. *Alternaria* toxins, p. 315–336. *In* J. D. Miller and H. L. Trenholm (ed.),

Mycotoxins in Grain. Compounds Other than Aflatoxin. Eagan Press, St. Paul, Minn.

81. **Ware, G. M., and C. W. Thorp.** 1978. Determination of zearalenone in corn by high-pressure liquid chromatography and fluorescence detection. *J. Assoc. Off. Anal. Chem.* **61:**1058–1061.

82. **Wilson, T. M., P. F. Ross, L. G. Rice, G. D. Osweiler, H. A. Nelson, D. L. Owens, R. D. Plattner, C. Reggiardo, T. H. Noon, and J. W. Pickrell.** 1990. Fumonisin B_1 levels associated with an epizootic of equine leukoencephalomalacia. *J. Vet. Diagn. Invest.* **2:**213–216.

83. **Winter, C. K., D. G. Gilchrist, M. B. Dickman, and C. J. Jones.** 1996.Chemistry and biological activity of AAL toxins, p. 307–316. *In* L. Jackson, J. DeVries, and L. Bullerman (ed.), *Fumonisins in Foods.* Plenum Publishing Corp., New York, N.Y.

84. **Wray, B. B., and K. G. O'Steen.** 1975. Mycotoxin-producing fungi from a house associated with leukemia. *Arch. Environ. Health* **30:**571–573.

85. **Wray, B. B., E. J. Rushings, R. C. Boyd, and A. M. Schindel.** 1979. Suppression of phytohemagglutinin response by fungi from a "leukemia" house. *Arch. Environ. Health* **34:**350–353.

86. **Zoller, O., F. Sager, and B. Zimmerli.** 1994. Occurrence of fumonisins in foods. *Mitt. Geb. Lebensm. Hyg.* **85:**81–99.

Viruses

V

Food Microbiology: Fundamentals and Frontiers, 2nd Ed.
Edited by M. P. Doyle et al.
© 2001 ASM Press, Washington, D.C.

Dean O. Cliver

Foodborne Viruses

24

CHARACTERISTICS OF VIRUSES—GENERAL

According to the passive system of recording foodborne disease outbreaks used by the U.S. Centers for Disease Control and Prevention (CDC), viruses had a significant role in causing foodborne illness during the most recent compilation period, 1993 to 1997 (10; Table 24.1), and it seems likely that a substantial portion of outbreaks of undetermined etiology were also caused by viruses. Indeed, viruses occupied 3 of the top 10 places (based on number of persons ill) among causes of foodborne disease outbreaks reported in the United States during this period (10; Table 24.2). The active system of surveillance for foodborne diseases, FoodNet, does not include viruses (9), even though hepatitis A, at least, is a reportable viral disease. CDC has estimated that the Norwalk virus (or related agents) causes two-thirds of foodborne illnesses in the United States (44; Table 24.3). A small, round, structured virus (Norwalk-like) was responsible for the only viral outbreak of waterborne illness in the United States in 1995–1996, but there were also eight outbreaks of acute gastrointestinal illness of unknown etiology, some of which might have been viral (8).

Viruses are transmitted as particles of submicroscopic size. Particles of viruses known to be foodborne (as-

troviruses, caliciviruses, picornaviruses, and rotaviruses) range generally from 25 to 75 nm in diameter (Table 24.4). Although they have icosahedral symmetry, they are approximately spherical, as seen in electron micrographs. Particles of foodborne viruses generally contain RNA (rather than DNA) genomes. With the exception of the rotaviruses, the genome is made up of a single strand with "plus sense"—which is to say, capable of serving as mRNA and being directly translated into protein. At least part of the virus-specific protein that results from this initial translation is an enzyme that catalyzes RNA synthesis on an RNA template; reverse transcription is unnecessary, in that DNA has no role in the replicative cycles of these viruses. Viral particles are totally inert and thus cannot multiply in food or anywhere outside of susceptible living host cells.

Foodborne viruses are generally *enteric*: they infect perorally (by ingestion) and are shed with feces. This means that they may also be transmitted by person-to-person "contact" and via water, as are other enteric infectious agents. By no means do all foodborne viral infections occur as part of outbreaks (defined by CDC as a cluster of two or more cases); but until now there has been no means of attributing single illnesses to food,

Dean O. Cliver, Department of Population Health and Reproduction, School of Veterinary Medicine, University of California, Davis, Davis, CA 95616-8743.

Table 24.1 Foodborne disease outbreaks, United States, 1993–1997[a]

Etiology	Outbreaks		Cases		No. of deaths
	No.	%	No.	%	
Bacterial	655	23.8	43,821	50.9	28
Chemical	148	5.4	576	0.7	0
Parasitic	19	0.7	2,325	2.7	0
Viral	56	2.0	4,066	4.7	0
Confirmed etiology	878	31.9	50,788	59.0	28
Unknown etiology	1,873	68.1	35,270	41.0	1
Totals	2,751	100.0	86,058	100.0	29

[a] Data from reference 10.

Table 24.3 Estimated foodborne illnesses, hospitalizations, and deaths in the United States from viral pathogens[a]

Viral pathogen	Illnesses	Hospitalizations	Deaths
Norwalk-like viruses	9,200,000	20,000	124
Rotavirus	39,000	500	0
Astrovirus	39,000	125	0
Hepatitis A	4,170	90	4
Subtotal, viral	9,282,170	20,715	128
Grand total foodborne[b]	13,814,924	60,854	1,809

[a] Data from reference 44.
[b] Foodborne illnesses, all causes.

except in a few studies in which shellfish consumption was recorded. A distinctive property of enteric foodborne viruses is that they are generally adapted exclusively to humans as hosts—alternative animal hosts are unknown or of no practical significance. With the exception of tick-borne encephalitis virus, to be discussed briefly, neither nonhuman reservoirs nor biological vectors are significant in the transmission of these agents. Mechanical vectors, such as flies, that transmit other feces-associated agents have been mentioned occasionally but have not been directly implicated in recent outbreaks.

After enteric viruses are ingested, some have their principal site of action in the lining of the small intestine. Others infect the liver, and still others may affect other parts of the host's body. Disease results from killing of infected cells by the viral replicative process or from destruction of infected cells by the host's immune response. Although the genomic organizations of some of the food-borne viruses are known, virulence factors as such have not been described. Common elements of the genome include a region that codes for the structural (coat) protein of the virus and another that codes for the RNA-dependent RNA polymerase that is required for replication of the viral nucleic acid. Other regions may not be translated, may code for additional virus-specific enzymes, may interfere with normal activities of the host cell, or may be of unknown function.

The peroral infectious dose of virus is subject to debate. A certain amount of research on this subject has involved human subjects, but doses were generally measured in cell culture infectious units rather than individual viral particles. There is little doubt that one mature viral particle contains all of the information needed to produce a human infection. Because there are some intrinsic inefficiencies in the process, however, ingestion of a single viral particle is unlikely to result in infection. Haas (30) has applied elegant analytical methods to the results of feeding trials with viruses and concludes that the possibility of infection from ingesting a single particle cannot be ruled out. The effect of vehicles (e.g., specific foods, drinking water) on the efficiency with which viruses produce infection has not been studied either. Given that peak rates of shedding of some viruses exceed 10^8 particles per g of feces, 10 μg of fecal contamination, which could occur by handling food with feces-soiled fingers, would contain 1,000 particles and would carry a very serious threat of infection. If the vehicle were

Table 24.2 Top 10 identified causes of foodborne disease outbreaks, United States, 1993 to 1997, ranked by numbers of illnesses[a]

Rank	Causative agent	Illnesses		Outbreaks	
		No.	%	No.	%
1	Salmonella spp.	32,610	37.9	357	13.0
2	Escherichia coli	3,260	3.8	84	3.1
3	Clostridium perfringens	2,772	3.2	57	2.1
4	Other parasitic	2,261	2.6	13	0.5
5	Other viral	2,104	2.4	24	0.9
6	Shigella spp.	1,555	1.8	43	1.6
7	Staphylococcus aureus	1,413	1.6	42	1.5
8	Norwalk virus	1,233	1.4	9	0.3
9	Hepatitis A virus	729	0.8	23	0.8
10	Bacillus cereus	691	0.8	14	0.5

[a] Data from reference 10.

Table 24.4 Some properties that define the major groups of human enteric viruses

Size	NA strands	RNA	DNA
25–35 nm	Single	Astro- Calici- Picorna-	Parvo-
70–85 nm	Double	Reo- Rota-	Adeno-

Table 24.5 Viruses of human hepatitis

Type	Former name	Mode of transmission	Remarks
A	Infectious hepatitis	Fecal-oral	"Killed" vaccine available
B	Serum hepatitis	Parenteral	Recombinant vaccine available
C	Non-A, non-B hepatitis	Parenteral	Now screened for in U.S. blood banks
D	Delta agent	Parenteral	Satellite of hepatitis B virus
E	Non-A, non-B hepatitis	Fecal-oral	Not yet in North America

100 g of food or water, the dilution factor would be 10^{-7}, which is why viruses are almost never detected in foods implicated in outbreaks.

CHARACTERISTICS OF SPECIFIC FOODBORNE VIRUSES

Hepatitis Viruses

The hepatitis viruses transmitted enterically are types A and E (18). These differ not only serologically but in many other ways as well. The viruses of human hepatitis generally have in common only that they all infect the liver (Table 24.5); other taxonomic properties and even modes of transmission vary widely. One or more variants of the hepatitis A virus that replicate and produce cytopathic effects in cell cultures have been identified; this has enabled research on many aspects of the virus and led to the development of effective vaccines. Replication of hepatitis E virus in cell cultures has been reported recently; this should enable important strides in research.

Size and Shape of Particle

The hepatitis A virus looks like other picornaviruses: diameter of ~28 nm, apparently spherical, with an approximately smooth surface, as seen in transmission electron micrographs. The hepatitis E virus looks like the caliciviruses; diameter of ~32 nm, apparently spherical, with visible cup-shaped concavities on its surface ("calicivirus" was coined from the Greek *kalyx*, or cup), but it has recently been excluded from the group on other taxonomic grounds (26). Thus, although a skilled electron microscopist could distinguish between hepatitis viruses A and E, neither could be distinguished morphologically from many other viruses that infect humans.

Characteristics of Disease; Shedding

Once either hepatitis A or E causes illness, the two diseases are not clinically distinguishable. Either of these viruses can produce illness that lasts several weeks and includes jaundice, anorexia, vomiting, and profound malaise. One clue that is sometimes observed is that hepatitis E is much more serious in pregnant women (17% to 33% mortality) than is hepatitis A. Diagnosis of hepatitis A is usually based on demonstration of immunoglobulin M (IgM)-class antibody against the virus in the blood of the affected person, using one of several commercial kits. Hepatitis A is more likely to cause asymptomatic or mild infections in young children than in adolescents or adults, whereas serologic surveys indicate that hepatitis E less often infects young children. The incubation period of hepatitis A ranges from 15 to 50 days, with a median of 28 or 30 days. The virus is often shed in feces for 10 to 14 days before the onset of illness, during which time an infected food handler has ample opportunity to contaminate food unless personal hygiene (specifically, hand washing) has been scrupulous. Shedding may continue for 1 to 2 weeks after onset of illness (18). The incubation period of hepatitis E is 22 to 60 days, with an average of 40 days; shedding during the incubation period has not been known to lead to transmission via food.

Mechanisms of Pathogenesis; Genome

Virulence factors as such are poorly characterized. The hepatitis A virus ordinarily does not kill the liver cells that it infects (18). Rather, the body's immune response to the infection eventually leads to destruction of infected hepatocytes by cytotoxic T cells, hence the long incubation period. The death rate is estimated at 0.3% (44); survivors may experience permanent sequelae or occasional relapses. Immunity is durable, probably lifelong. Pathogenesis in hepatitis E is probably similar. Genomic organizations are not remarkable among single (plus-sense)-strand RNA viruses: consistent features are an extensive 5' nontranslated region and specific sequences that code for structural (coat, or capsid) protein and for the required RNA-dependent RNA polymerase that was mentioned previously. The hepatitis A capsid comprises 60 copies of each of four different structural peptides; the genome also codes for one or more proteases that process the translated large polypeptide into four small units. Because many more molecules of coat protein than of the enzymes are needed during replication, translation of the capsid region of the genome is selectively enhanced.

Stability in Food

Food-associated outbreaks of hepatitis A have been recorded in great numbers. In most instances, these outbreaks have not appeared to depend on extreme stability of the virus in the food, in that the food was eaten uncooked or was thought to have been contaminated just before it was eaten. However, limited experiments to date have shown that hepatitis A virus is much more resistant to heat and drying than are other picornaviruses (18), whereas resistance to acid (e.g., pH 2 for short periods) and to gamma rays is common among picornaviruses. More research relating these properties to situations in food is needed. Hepatitis A virus is not remarkably resistant to chlorine in water. Comparable information regarding hepatitis E virus is not yet available: the virus is often transmitted via water, but under circumstances that chlorination could not have been expected to prevent. Transmission of hepatitis E via foods has been proposed on various occasions, but conclusive epidemiologic evidence has not resulted. Swine infection with this agent has been reported, but the epidemiologic significance of this is not known, and swine have not yet been used to study the stability of the virus as it might pertain to transmission via food or water.

Susceptible Populations

Susceptibility to hepatitis A is general. In countries where fecal-oral transmission of disease agents is common, most children have experienced infection and acquired active immunity by 5 years of age. Hepatitis A is more likely to cause symptoms in adults in developed countries, with more severe consequences. Other than age, host factors affecting the severity of hepatitis A are not well characterized; immune impairment may be less significant than with some other foodborne diseases, but liver abnormalities from other causes (e.g., cirrhosis) may be significant. Pooled human immune serum globulin has been used to afford passive immunity to susceptible persons who are going into "at risk" situations and to persons recently exposed to the virus (e.g., via food) if it can be administered within 2 weeks of probable exposure. One or more vaccines already in use in Europe have been licensed in the United States (11); food handlers have been proposed as an important target group for vaccination in countries where they are unlikely to have acquired immunity by earlier infection. Means of immunization against hepatitis E virus are not yet available. No vaccine is in prospect, and especially in North America, infection is evidently so rare that significant levels of antibody against hepatitis E virus are not found in pooled human immune serum globulin. Whereas hepatitis A is virtually ubiquitous, outbreaks of hepatitis E are seen principally in Asia, Africa, and Latin America.

Outbreaks

The reported annual incidence of hepatitis A in the United States declined from ~31,000 cases in 1990 to ~23,000 cases in 1992 (6); this rate of decline has not continued since. For the years 1990, 1991, and 1992, respectively, suspected food- or waterborne outbreaks were cited as possible sources of these infections at crude rates of 9.4, 6.0, and 8.0%, but at rates of 4.4, 3.0, and 4.7% when sources were assessed on a mutually exclusive (only one possible source per infection) basis. One report includes estimates of 4,800 to 35,000 cases of food-associated hepatitis A in the United States annually (17), whereas another estimates 27,797 foodborne cases (among 83,391 total cases) annually (44). The United States recorded 30,021 total cases of hepatitis A in 1997 (7).

Hundreds of outbreaks of food-associated hepatitis A have been reported since 1943 (12). Shellfish (bivalve mollusks) taken from waters contaminated with human feces have been the vehicle in a plurality of outbreaks, but any food handled by an infected person may become contaminated and, if it is not cooked before being eaten, may transmit the infection (13).

Four outbreaks will be summarized here to provide some idea of the scope of the problem:

- During the Christmas season of 1955, oysters that had been held in a harbor in Sweden awaiting sale were contaminated by feces from a privy that overhung the water. These oysters were eaten raw as part of a traditional holiday celebration and gave rise, over the next few weeks, to 629 illnesses among people in many cities and villages and some who had by then left the country. This was the first recorded instance of hepatitis A transmission via shellfish (22).
- In 1962, 28 cases of hepatitis A occurred among the staff of a hospital in St. Louis, Missouri (21). The vehicle was found to be orange juice that had been thawed and reconstituted by a worker who was inapparently infected by the virus. Although she never became ill, her husband, who had not been at the hospital, did become ill at the time of the staff outbreak. Much more conclusive diagnostic methods for such infections are now available.
- In 1978, outbreaks involving 82 cases in Arkansas and 58 cases in Texas were traced to a single worker (55). The man prepared sandwiches and salads in several vegetarian health food restaurants in the two states. He continued to work at the second location even while he was visibly ill.
- In 1988, clams taken from waters off Shanghai, China, and eaten raw were the source of nearly 300,000 cases of hepatitis A (31). The waters from which the shellfish

were taken had apparently become contaminated with human sewage.

Another five recent outbreaks of foodborne hepatitis A illustrate the variety of vehicles that may transmit the disease (the list is far from exhaustive):

- In late 1988 and early 1989, a cluster of 38 cases of hepatitis A in Italy was traced to consumption of raw mussels and to drinking a particular brand of mineral water (54).
- In 1989, 50 people in several villages in the United Kingdom acquired hepatitis A by eating foods from a single bake shop (53).
- In 1990, 110 people in Missouri acquired hepatitis A by eating lettuce apparently handled by an infected worker (5).
- A 1994 report tells of four people in Denmark who acquired hepatitis A by eating caviar that had been illegally imported from Latvia (23).
- In February and March 1997, a hepatitis A outbreak began in Michigan and extended to other states. Totals of 213 cases from 23 schools in Michigan and 29 cases from 13 schools in Maine were associated with eating frozen strawberries, largely distributed through the U.S. Department of Agriculture School Lunch Program. Genetic sequencing of virus genomes from cases that occurred in other states identified another five patients in Wisconsin, seven patients in Arizona, and two patients in Louisiana with related infections (34).

Norwalk-like Viruses

Despite difficulties of diagnosis, the Norwalk-like viruses are being recognized as a major cause of foodborne disease in the United States and elsewhere. During the period 1993 to 1997, the Norwalk virus ranked eighth as a cause of foodborne illnesses among outbreaks reported in the United States (10; Table 24.2). The CDC, moreover, estimates that this virus, or group of viruses, causes two-thirds of the foodborne disease in the United States (44; Table 24.3). The virus does not multiply in any known laboratory host. Historically, diagnoses were based on detection of the viral particles in patients' stool samples by electron microscopy, immune electron microscopy, or enzyme immunoassay (EIA) (57), or on demonstration of antibody against the virus in patients' serum samples by EIA. Viral antigen has been produced biosynthetically (37). Several serologic varieties, comprising at least two major genogroups, occur among the "Norwalk-like" viruses (1, 27). Another group of human caliciviruses, the "Sapporo-like viruses," differs in genomic organization from the Norwalk-like viruses (27, 47). In general, diagnostic tests have been directed to the Norwalk virus proper, with detection of novel types of Norwalk-like viruses most easily done by immune electron microscopy, using virus in acute-phase stool extract and antibody in convalescent-phase serum from the same patient (3). More recently, molecular methods of detection have become available for several types of Norwalk-like viruses (26, 44, 52).

Size and Shape of Particle

The Norwalk-like viruses have been assigned to the calicivirus group (2, 27). Although the Norwalk-like viruses are not uniformly reported as 32 nm in diameter, they do regularly reveal the pattern of surface depressions from which the name calicivirus derives. When detected by electron microscopy, they have been called "small, round, structured viruses" (2). Like most other foodborne viruses, they contain a single, plus-sense strand of RNA.

Characteristics of Disease; Shedding

The disease includes nausea, vomiting, diarrhea, and other symptoms common to gastroenteritis. If enough people are involved to make comparisons valid, vomiting is usually as common as diarrhea; before transmission via food and water had been described, "winter vomiting disease" was a commonly reported syndrome. Both the incubation period and the duration of symptoms usually range from 24 to 48 h. The ill person sheds the virus in both vomitus and feces; food contamination by emesis is relatively rare but not unknown. Now that more sensitive methods are available to detect the virus, it has been determined that fecal shedding may last a week or more after the gastroenteritis ends (25).

Mechanisms of Pathogenesis; Genome

Virulence factors as such have not been identified for these agents. The viruses infect and kill cells of the small intestinal mucosa (2). Progeny virus passes down the digestive tract and is shed in the feces. This is an exceptionally fast-acting virus. Although the usual incubation period is longer, volunteers have become ill as early as 10 h after ingesting the virus. Because the virus particle contains plus-sense RNA, it must code for an RNA-dependent RNA polymerase that can be translated early to permit replication. In the expectation that the genomic sequence coding for this polymerase would be highly conserved, sequences of many outbreak strains have been determined and compared. It was surprising to learn that each strain differed to some extent from the others, though at least two groups with substantial homology were identified (1, 27, 43, 58).

Stability in Food

Because the infectivity of the virus can be determined only by feeding it to people, its stability in food (e.g., in processing) has not been determined. However, qualitative studies in human volunteers have shown that infectivity persists during 3 h at pH 2.7 at room temperature and for 60 min at neutral pH at 60°C (20). Vehicles in reported outbreaks have been either shellfish eaten raw or slightly cooked or other foods that were not heated after having undergone hands-on contamination. Thus, there is no reason to suppose that the virus is exceptionally heat stable in food. Human volunteers have been used to study the depuration of Norwalk virus-contaminated oysters, which led to the determination that the virus persists longer than fecal coliform bacterial indicators in contaminated shellfish (29).

Susceptible Populations

Attack rates in reported outbreaks have often exceeded 50% of those at risk, which suggests that susceptibility to the viruses is widespread (45). Antibody produced in response to the infection is of diagnostic significance, but it evidently does not afford protection to the person who has been infected. Indeed, in volunteer studies, those who had preexisting antibody frequently became ill, and although their antibody levels often rose as a result of the infection, they were again susceptible to the same agent when challenged over a year later. Given the lack of protection afforded by antibody produced as a result of infection, the hope of producing a useful vaccine against these agents seems dim.

Outbreaks

The many recorded outbreaks of food-associated gastroenteritis that have been attributed to Norwalk-like viruses show that the pattern of transmission resembles that of hepatitis A. Shellfish are frequently vehicles, but foods handled by infected persons often transmit these viruses as well. According to Mead et al. (44), "No data are available to directly determine the proportion of cases of NLV-associated disease attributable to foodborne transmission." However, several recent reports have implicated human caliciviruses or small, round, structured viruses in a large proportion of food-associated outbreaks in Minnesota (19), Sweden (32), Japan (35), and the United Kingdom (57).

- A series of outbreaks of gastroenteritis associated with consuming cockles in the United Kingdom in late 1976 and early 1977 led to the detection of small, round viruses in the feces of several patients (3).
- The first fully documented foodborne outbreak of Norwalk virus gastroenteritis occurred in 1978 in New South Wales, Australia (45). More than 2,000 people

who ate oysters, principally from the Georges River, were affected. Some groups of people who ate oysters showed attack rates of 85%.
- Norwalk virus was the leading cause of reported foodborne disease in the United States in 1982, due largely to an outbreak that involved some 3,000 people in the area of Minneapolis and St. Paul, Minnesota (41). A baker's assistant who was ill apparently contaminated butter cream frosting that was later applied to a great number and variety of pastries.
- In 1987, ice made with contaminated water was used to cool drinks served at a college football game and a fundraising gathering; approximately 5,000 people became ill with gastroenteritis associated with a small, round virus in feces (4).
- The Snow Mountain agent, a small, round, structured virus that is antigenically distinct from the Norwalk virus, caused 155 illnesses (two deaths) among residents and 28 illnesses among employees of a retirement facility in the area of San Francisco Bay, California (24). The vehicles were evidently shrimp dishes.
- Small, round, structured viruses resembling the Norwalk agent were detected in the stools of persons in a large gastroenteritis outbreak in Toyota City, Japan (40). School lunches from a preparation center that served nine schools were implicated as vehicles. Those affected included 3,236 students and 117 teachers.
- Various outbreaks of viral gastroenteritis have been reported aboard cruise ships. A Hawaiian cruise ship was the site of one such outbreak in 1990 that affected 238 people (33); the vehicle was fresh cut fruit. Another Hawaiian cruise ship served contaminated ice in 1992 that led to infections in approximately 200 people (39).

Astroviruses

Like the rotaviruses to be discussed below, the astroviruses most often affect the very young and the elderly, but there have been significant recorded instances of transmission via foods. Mead et al. (44) have estimated that 39,000 illnesses from foodborne astroviruses occur annually in the United States (Table 24.3).

Size and Shape of Particle

The astroviruses are among the small, round, structured viruses seen with the electron microscope. Their size is similar to the human caliciviruses, but they are distinguished by an apparently raised, five- or six-pointed star pattern on the protein coat surface, whereas the caliciviruses have depressions (2). The nucleic acid is a single, plus-sense strand of RNA.

Characteristics of Disease; Shedding

The illness differs from that of the Norwalk-like viruses in that the incubation period is 3 to 4 days, diarrhea is more typical than vomiting (which is rare), and the duration of illness is generally 2 to 3 days but may be up to 7 to 14 days (2). Shedding of the virus in feces lasts at least during the period of diarrhea.

Mechanisms of Pathogenesis; Genome

The virus infects and kills mature enterocytes on the villi of the small intestine. These are replaced after a few days by new enterocytes generated in the crypts. An antibody response, which may be protective, results from infection. At least six serotypes have been identified, and several of them are able to infect laboratory cell cultures and cause cytopathic effects (2).

Stability in Food

Like other enteric viruses, the astroviruses are resistant to pH 3 for periods of time. They are resistant to lipid solvents and show residual infectivity after 30 min at 50°C (2). Decimal reduction times at 60°C are in the range of 2 to 3 min (2).

Susceptible Populations

Because of the high rate of infection in children under 1 year old, 70% of children in the United Kingdom are said to have antibody against astrovirus by 3 to 4 years of age. However, resistance is likely to be serotype specific, so susceptibility to some types of astroviruses continues. A foodborne outbreak of astrovirus gastroenteritis involving approximately 4,700 people was reported in Katano City, Osaka, Japan (48).

Rotaviruses

Rotaviral illness, especially that caused by serogroup A, is common in infants throughout the world and is a significant cause of infant death in developing countries (51). Mead et al. (44) have estimated that 39,000 illnesses from foodborne rotaviruses occur annually in the United States (Table 24.3).

Size and Shape of Particle

The rotavirus particle is 70 to 75 nm in diameter. Although its shape is demonstrably icosahedral, it appears roughly spherical in most electron micrographs. The coat protein is double layered, and the nucleic acid is double-stranded RNA in 11 segments.

Characteristics of Disease; Shedding

After an incubation period of 1 to 3 days, illness is characterized by fever, vomiting, and diarrhea that may last for 4 to 6 days. The virus is shed at least during illness but usually not longer than 8 days. Infection begins in the enterocytes of the proximal small intestine but eventually involves the jejunum and ileum as well.

Stability in Food

Rotavirus loses 99% of its infectivity titer during 30 min at 50°C (36). It is unstable outside the pH range of 3 to 10 (51). It survives for many days on vegetables at 4 or 20°C and has been shown experimentally to withstand the process of making soft cheese (51).

Susceptible Populations

Even though most rotaviral illness involves infants, outbreaks of foodborne and waterborne disease affecting a broad spectrum of ages have been recorded. Most foodborne outbreaks have been recorded in Japan and in New York State, whereas waterborne outbreaks have occurred in Germany, Israel, Russia, Sweden, and China. Serogroups other than A are sometimes involved, and animals may also be infected with some non-A serogroups but are not known to transmit their infections to humans. Shellfish growing in sewage-polluted waters undoubtedly become contaminated but have not yet been implicated as a vehicle in a recorded outbreak of human illness.

Other Viruses

A few other viruses have been reported to be transmitted via foods on occasion (14). Each will be discussed briefly here.

Tick-Borne Encephalitis Virus

In Slovakia and perhaps adjacent regions, dairy animals (particularly goats but possibly also cattle and sheep) bitten by infected ticks become infected with an encephalitis virus that is then shed in milk (28). This tick-borne encephalitis virus is perhaps the only known viral zoonosis that is transmissible via food. Although the virus is readily inactivated by pasteurizing the milk, it may be found in several milk products made with unpasteurized milk. Fortunately, infection is rarely contracted by this route because of the limited distribution of the virus; however, small outbreaks continue to occur even though the problem has been recognized for many years. For example, a group of seven people who drank goats' milk raw contracted the disease in Slovakia in 1993.

Enteroviruses

Poliomyelitis, caused by any of three serotypes of enteroviruses (a subset of the picornavirus group, which includes the hepatitis A virus as a separate subset), was the first viral disease reported to be foodborne. However, no outbreak of foodborne poliomyelitis has been reported in developed countries since 1949, *before* the introduction of preventive vaccines (12). Members of the enterovirus group other than the polioviruses include the coxsackieviruses and the echoviruses. One outbreak of food-associated coxsackieviruses B1, B3, and B5 infection has been reported from the then USSR, in association with widespread contamination in a kitchen that prepared food for children in a day-care facility (49). Echovirus outbreaks that have been reported include one of type 4 that occurred in Pennsylvania in 1976 (56) and one of type 5 that occurred in New York State in 1988 (46). Eighty people who ate contaminated coleslaw were affected in the former outbreak; the vehicle in the latter outbreak was not determined, but foods were the common element among the 161 persons affected.

Parvoviruses

The parvoviruses are perhaps the smallest of the enteric viruses, with diameters of 20 to 26 nm (2). Their protein coat appears smooth, and they contain single-stranded DNA. Several (e.g., cockle, Wollan, Ditchling, and Parramatta agents) have been detected in the stools of persons involved in food-associated outbreaks of gastroenteritis. However, Appleton (2) says that the role of these agents in causation of human gastroenteritis has yet to be confirmed and that parvoviruses are often detected in diarrheal stools in association with other viruses that are known to cause gastroenteritis. Replication of the viral genome from a single-stranded DNA template presents some special challenges. Parvoviruses certainly cause illnesses other than gastroenteritis in humans and cause gastroenteritis in other animal species, but these viruses may be just satellites in human gastroenteritis, as they are known to be in some other contexts.

Other Gastroenteritis Viruses

Other human enteric viruses, particularly adenoviruses 40 and 41 and coronaviruses, are known to cause gastroenteritis in humans, occasionally in significant outbreaks. The adenoviruses are ~75 nm in diameter and contain double-stranded DNA. They have not been known to be transmitted via food or water. The coronaviruses are among the very rare enteric viruses that have lipid-containing envelopes; in this case, the envelope surrounds a nucleocapsid of single-stranded RNA wrapped in protein. One gastroenteritis outbreak, among persons who ate at a sports club in Scotland, has been attributed to coronavirus (16). The specific food by which the virus was transmitted could not be identified.

Viruses Probably Not Transmitted via Foods

Because they represent a threat to human health that is quite serious in some instances, some other viruses have been suspected of being transmissible via foods. Despite the inability of these viruses to infect humans perorally, the perceived risk of their introduction into food has at times led to detention or destruction of food that presented no danger to consumer health. Included are the human immunodeficiency virus, herpesvirus, hantavirus, rabies virus, rhinoviruses, and the agent of bovine spongiform encephalopathy.

The human immunodeficiency virus (HIV) is not transmissible via food because the virus evidently does not infect perorally. Not only are HIV-positive persons not a risk as food handlers, but people with frank symptoms of AIDS are not a threat when handling food, as long as they do not have diarrhea (e.g., such as that caused by *Cryptosporidium*). Neither are the herpesviruses, the hantavirus, the rabies virus, and the rhinoviruses (which cause the "common cold") infectious perorally, with or without food. The agent (a prion) of bovine spongiform encephalopathy (BSE), which may infect cattle perorally, apparently can also infect humans who ingest certain bovine tissues, probably not including voluntary muscle (red meat). What is said here does not suggest that it is reasonable to eat meat from sick animals or to allow people to handle food (professionally or as amateurs) while ill. It simply means that food that is suspected of contamination with one of the viruses mentioned (e.g., milk from a cow that subsequently showed symptoms of rabies) is unlikely to injure human consumers.

OTHER CONSIDERATIONS

Once diagnostic methods for a foodborne viral illness are well established, it is reasonable to aspire to detect the virus in food. Whatever other considerations there are, one would rather be able to detect virus in food before the food was eaten than to detect virus in the feces of those who became ill by eating the food. When diagnostic methods for enteric viral infections were based on detection of the virus in a patient's stools, one might have aspired to increase the sensitivity of the same test by at least two orders of magnitude (assuming that feces never make up more than 1% of a contaminated food) and apply it to food testing. However, diagnostic tests have now evolved in the direction of detecting antiviral antibody (usually of the IgM class, which denotes recent or current infection) in the patient's serum. This approach offers no help in detecting viruses in food.

Viruses in food might be detected on the basis of their infectivity in living hosts (most likely laboratory cell cultures), their size and shape as seen by electron microscopy, the unique reactions of their coat antigens in some serologic test, or the specific reactions of their nucleic acids with selected probes or primers (15). The first two options have not worked well because the most important foodborne viruses do not cause demonstrable infections in cell cultures and are likely to be present at such low levels as to be undetectable by electron microscopy. Many methods for detecting viruses in food and water by specific reactions with the viral protein or nucleic acid have been described (26, 36, 38, 43, 52); however, none has yet gained general acceptance and all are exacting, time-consuming, and expensive.

Some tests that have been described might be applied to test for viruses in foods implicated in an outbreak, but these methods are unlikely to be used in routine monitoring of foods destined for human consumption. Indicators of fecal contamination have been used to monitor the sanitary quality of some foods (e.g., shellfish) and might be pertinent to virus contamination because viruses are shed in the feces of infected humans. However, the established indicator systems (e.g., fecal coliforms) are bacterial, and their presence or absence is not a good predictor of viruses for a number of reasons. Alternative tests, such as detection of bacteriophages that infect enteric bacteria (*Escherichia coli*, *Bacteroides fragilis*, etc.), have been under study for a few years, but none has yet emerged as a solution to the problem of routine monitoring of foods for possible virus contamination.

Avoiding fecal contamination of food is the surest means of preventing food-associated transmission of viruses. However, the epidemiologic record indicates that both fecal and viral contamination occur at times, either from an infected food handler or with wastewater. Now that a vaccine against hepatitis A is available, it might be well to immunize professional food handlers who are not already immune. Of course, this would not protect against contamination of food by amateurs infected with hepatitis A virus, nor by anyone infected with a gastroenteritis virus. Although it is extremely difficult to document, it seems likely that a large proportion of viral disease is transmitted via foods contaminated in the home. Proper hand washing and exclusion of ill persons from handling food are important preventive measures, as is avoidance of contaminating food with polluted water or wastewater. Proper use of chlorine or of UV light will inactivate virus in water and on surfaces, but virus within a food is best inactivated by cooking to temperatures that will kill vegetative bacterial pathogens. Enteric viruses are likely to persist for weeks in refrigerated foods and indefinitely in frozen foods.

SUMMARY

Viruses are a frequent cause of foodborne disease in the United States and probably in many other countries. The most important are the human caliciviruses (Norwalk-like viruses) and the hepatitis A virus. Almost all of the viruses transmissible to humans via foods are produced in the human body and shed with feces, and almost all of these enteric viruses make up RNA (usually single-stranded) coated only with protein. They are more acid stable, but not more heat stable, than most enteric bacteria. These viruses enter food as a result of fecal contamination. Because the particles are inert outside of living human cells, they cannot multiply in food. Viruses in food are inactivated by high temperatures and preserved by low temperatures.

References

1. **Ando, T., M. N. Mulders, D. C. Lewis, M. K. Estes, S. S. Monroe, and R. I. Glass.** 1994. Comparison of the polymerase region of small round structured virus strains previously classified in three antigenic types by solid-phase immune electron microscopy. *Arch. Virol.* **135**:217–226.

2. **Appleton, H.** 1994. Norwalk virus and the small round viruses causing foodborne gastroenteritis, p. 57–79. *In* Y. H. Hui, J. R. Gorham, K. D. Murrell, and D. O. Cliver (ed.), *Foodborne Disease Handbook*, vol. 2. *Diseases Caused by Viruses, Parasites, and Fungi*. Marcel Dekker, New York, N.Y.

3. **Appleton, H., and M. S. Pereira.** 1977. A possible virus aetiology in outbreaks of food-poisoning from cockles. *Lancet* **i**:780–781.

4. **Centers for Disease Control.** 1987. Outbreak of viral gastroenteritis—Pennsylvania and Delaware. *Morbid. Mortal. Weekly Rep.* **36**:709–711.

5. **Centers for Disease Control and Prevention.** 1993. Foodborne hepatitis A—Missouri, Wisconsin, and Alaska. *Morbid. Mortal. Weekly Rep.* **42**:526–529.

6. **Centers for Disease Control and Prevention.** 1994. Viral hepatitis surveillance program, 1990–1992, p. 19-34. *Hepatitis Surveillance Report No. 55.* Centers for Disease Control and Prevention, Atlanta, Ga.

7. **Centers for Disease Control and Prevention.** 1998. Summary of notifiable diseases, United States, 1997. *Morbid. Mortal. Weekly Rep.* **46**(54):1–87.

8. **Centers for Disease Control and Prevention.** 1998. Surveillance for waterborne-disease outbreaks—United States, 1995–1996. *Morbid. Mortal. Weekly Rep.* **47**(SS-5):1–34.

9. **Centers for Disease Control and Prevention.** 2000. Preliminary FoodNet data on the incidence of foodborne

illnesses—selected sites, United States, 1999. *Morbid. Mortal. Weekly Rep.* **49:**201–205.

10. **Centers for Disease Control and Prevention.** 2000. Surveillance for foodborne-disease outbreaks—United States, 1993–1997. *Morbid. Mortal. Weekly Rep.* **49**(SS-1): 1–62.

11. **Cimons, M.** 1995. Hepatitis A vaccine licensed, progress for rotavirus, parvovirus. *ASM News* **61:**324–326.

12. **Cliver, D. O.** 1983. *Manual on Food Virology.* VPH/83.46. World Health Organization, Geneva, Switzerland.

13. **Cliver, D. O.** 1985. Vehicular transmission of hepatitis A. *Public Health Rev.* **13:**235–292.

14. **Cliver, D. O.** 1994. Other foodborne viral diseases, p. 137–143. *In* Y. H. Hui, J. R. Gorham, K. D. Murrell, and D. O. Cliver (ed.), *Foodborne Disease Handbook*, vol. 2. *Diseases Caused by Viruses, Parasites, and Fungi.* Marcel Dekker, New York, N.Y.

15. **Cliver, D. O., R. D. Ellender, G. S. Fout, P. A. Shields, and M. D. Sobsey.** 1992. Foodborne viruses, p. 763–787. *In* C. Vanderzant and D. F. Splittstoesser (ed.), *Compendium of Methods for the Microbiological Examination of Foods*, 3rd ed. American Public Health Association, Washington, D.C.

16. **Communicable Diseases and Environmental Health Scotland Weekly Report.** 1991. Outbreak of diarrhoeal illness associated with coronovirus [sic] infection. *Commun. Dis. Environ. Health Scotland Weekly Rep.* **25**(46):1.

17. **Council for Agricultural Science and Technology.** 1994. *Foodborne Pathogens: Risks and Consequences.* Task Force Report No. 122, September 1994. Council for Agricultural Science and Technology (CAST), Ames, Iowa.

18. **Cromeans, T., O. V. Nainan, H. A. Fields, M. O. Favorov, and H. S. Margolis.** 1994. Hepatitis A and E viruses, p. 1–56. *In* Y. H. Hui, J. R. Gorham, K. D. Murrell, and D. O. Cliver (ed.), *Foodborne Disease Handbook*, vol. 2. *Diseases Caused by Viruses, Parasites, and Fungi.* Marcel Dekker, New York, N.Y.

19. **Deneen, V. C., J. M. Hunt, C. R. Paule, R. I. James, R. G. Johnson, M. J. Hedberg, and C. W. Hedberg.** 2000. The impact of foodborne calicivirus disease: the Minnesota experience. *J. Infect. Dis.* **181**(Suppl. 2):S281–S283.

20. **Dolin, R., N. R. Blacklow, H. DuPont, R. F. Buscho, R. G. Wyatt, J. A. Kasel, R. Hornick, and R. M. Chanock.** 1972. Biological properties of Norwalk agent of acute infectious nonbacterial gastroenteritis. *Proc. Soc. Exp. Biol. Med.* **140:**578–583.

21. **Eisenstein, A. B., R. D. Aach, W. Jacobson, and A. Goldman.** 1963. An epidemic of infectious hepatitis in a general hospital. *JAMA* **185:**171–174.

22. **Gard, S.** 1957. Discussion, p. 241–243. *In* F. W. Hartman (ed.), *Hepatitis Frontiers.* Little, Brown & Co., Boston, Mass.

23. **Glerup, H., H. T. Sorensen, A. Flyvbjerg, P. Stokvad, and H. Vilstrup.** 1994. A "mini epidemic" of hepatitis A after eating Russian caviar. *J. Hepatol.* **21:**479.

24. **Gordon, S. M., L. S. Oshiro, W. R. Jarvis, D. Donenfeld, M. S. Ho, F. Taylor, H. B. Greenberg, R. Glass, H. P. Madore, R. Dolin, and O. Tablan.** 1990. Foodborne Snow Mountain agent gastroenteritis with secondary person-to-

25. **Graham, D. Y., X. Jiang, T. Tanaka, A. R. Opekun, H. P. Madore, and M. K. Estes.** 1994. Norwalk virus infection of volunteers: new insights based on improved assays. *J. Infect. Dis.* **170:**34–43.

26. **Green, J., K. Henshilwood, C. I. Gallimore, D. G. Brown, and D. N. Lees.** 1998. A nested reverse transcriptase PCR assay for detection of small round-structured viruses in environmentally contaminated molluscan shellfish. *Appl. Environ. Microbiol.* **64:**858–863.

27. **Green, K. Y., T. Ando, M. S. Balayan, T. Berke, I. N. Clarke, M. K. Estes, D. O. Matson, S. Nakata, J. D. O'Neill, M. J. Studdert, and H.-J. Thiel.** 2000. Taxonomy of the caliciviruses. *J. Infect. Dis.* **181**(Suppl. 2):S322–S330.

28. **Grešíková, M.** 1994. Tickborne encephalitis, p. 113–135. *In* Y. H. Hui, J. R. Gorham, K. D. Murrell, and D. O. Cliver (ed.), *Foodborne Disease Handbook*, vol. 2. *Diseases Caused by Viruses, Parasites, and Fungi.* Marcel Dekker, New York, N.Y.

29. **Grohmann, G. S., A. M. Murphy, P. J. Christopher, E. Auty, and H. B. Greenberg.** 1981. Norwalk virus gastroenteritis in volunteers consuming depurated oysters. *Austral. J. Exp. Biol. Med.* **59**(Pt. 2):219–228.

30. **Haas, C. N.** 1983. Estimation of risk due to low doses of microorganisms: a comparison of alternate methodologies. *Am. J. Epidemiol.* **118:**573–582.

31. **Halliday, M. L., L.-Y. Kang, T.-K. Zhou, M.-D. Hu, Q.-C. Pan, T.-Y. Fu, Y.-S. Huang, and S.-L. Hu.** 1991. An epidemic of hepatitis A attributable to the ingestion of raw clams in Shanghai, China. *J. Infect. Dis.* **164:**852–859.

32. **Hedlund, K. O., E. Rubilar-Abreu, and I. Svensson.** 2000. Epidemiology of calicivirus infections in Sweden, 1994–1998. *J. Infect. Dis.* **181**(Suppl. 2):S275–S280.

33. **Herwaldt, B. L., J. F. Lew, C. L. Moe, D. C. Lewis, C. D. Humphrey, S. S. Monroe, E. W. Pon, and R. I. Glass.** 1994. Characterization of a variant strain of Norwalk virus from a food-borne outbreak of gastroenteritis on a cruise ship in Hawaii. *J. Clin. Microbiol.* **32:**861–866.

34. **Hutin, Y. J., V. Pool, E. H. Cramer, O. V. Nainan, J. Weth, I. T. Williams, S. T. Goldstein, K. F. Gensheimer, B. P. Bell, C. N. Shapiro, M. J. Alter, and H. S. Margolis.** 1999. A multistate, foodborne outbreak of hepatitis A. *N. Engl. J. Med.* **340:**595–602.

35. **Inouye, S., K. Yamashita, S. Yamadera, M. Yoshikawa, N. Kato, and N. Okabe.** 2000. Surveillance for viral gastroenteritis in Japan: pediatric cases and outbreak incidents. *J. Infect. Dis.* **181**(Suppl. 2):S270–S274.

36. **Jaykus, L., R. De Leon, and M. D. Sobsey.** 1996. A virion concentration method for detection of human enteric viruses in oysters by PCR and oligoprobe hybridization. *Appl. Environ. Microbiol.* **62:**2074–2080.

37. **Jiang, X., J. Wang, D. Y. Graham, and M. K. Estes.** 1992. Expression, self-assembly, and antigenicity of the Norwalk virus capsid protein. *J. Virol.* **66:**6517–6531.

38. **Jothikumar, N., D. O. Cliver, and T. W. Mariam.** 1998. Immunomagnetic capture PCR for rapid concentration and

person spread in a retirement community. *Am. J. Epidemiol.* **131:**702–710.

detection of hepatitis A virus from environmental samples. *Appl. Environ. Microbiol.* **64:**504–508.

39. **Khan, A. S., C. K. Moe, R. I. Glass, S. S. Monroe, M. K. Estes, L. E. Chapman, X. Jiang, C. Humphrey, E. Pon, J. K. Iskander, and L. B. Schonberger.** 1994. Norwalk virus-associated gastroenteritis traced to ice consumption aboard a cruise ship in Hawaii: comparison and application of molecular method-based assays. *J. Clin. Microbiol.* **32:**318–322.

40. **Kobayashi, S., T. Morishita, T. Yamashita, K. Sakae, O. Nishio, T. Miyake, Y. Ishihara, and S. Isomura.** 1991. A large outbreak of gastroenteritis associated with a small round structured virus among schoolchildren and teachers in Japan. *Epidemiol. Infect.* **107:**81–86.

41. **Kuritsky, J. N., M. T. Osterholm, H. B. Greenberg, J. A. Korlath, J. R. Godes, C. W. Hodberg, J. C. Forfang, A. Z. Kapikian, J. C. McCullough, and K. E. White.** 1984. Norwalk gastroenteritis: a community outbreak associated with bakery product consumption. *Ann. Intern. Med.* **100:**519–521.

42. **Lees, D. N., K. Henshilwood, J. Green, C. I. Gallimore, and D. W. G. Brown.** 1995. Detection of small round structured viruses in shellfish by reverse transcription-PCR. *Appl. Environ. Microbiol.* **61:**4418–4424.

43. **Lew, J. F., A. Z. Kapikian, J. Valdesuso, and K. Y. Green.** 1994. Molecular characterization of Hawaii virus and other Norwalk-like viruses: evidence for genetic polymorphism among human caliciviruses. *J. Infect. Dis.* **170:**535–542.

44. **Mead, P. S., L. Slutsker, V. Dietz, L. F. McCaig, J. S. Bresee, C. Shapiro, P. M. Griffin, and R. V. Tauxe.** 1999. Food-related illness and death in the United States. *Emerg. Infect. Dis.* **5:**607–625.

45. **Murphy, A. M., G. S. Grohmann, P. J. Christopher, W. A. Lopez, G. R. Davey, and R. H. Millsom.** 1979. An Australia-wide outbreak of gastroenteritis from oysters caused by Norwalk virus. *Med. J. Austral.* **2:**329–333.

46. **New York Department of Health.** 1989. *A Review of Foodborne Disease Outbreaks in New York State 1988.* Bureau of Community Sanitation and Food Protection, Albany, N.Y.

47. **Noel, J. S., B. L. Liu, C. D. Humphrey, E. M. Rodriguez, P. R. Lambden, I. N. Clarke, D. M. Dwyer, T. Ando, R. I. Glass, and S. S. Monroe.** 1997. Parkville virus: a novel genetic variant of human calicivirus in the Sapporo virus clade, associated with an outbreak of gastroenteritis in adults. *J. Med. Virol.* **52:**173–178.

48. **Oishi, I., K. Yamazaki, T. Kimoto, Y. Minekawa, E. Utagawa, S. Yamazaki, S. Inouye, G. S. Grohmann,** S. S. Monroe, S. E. Stine, C. Carcamo, T. Ando, and R. I. Glass. 1994. A large outbreak of acute gastroenteritis associated with astrovirus among students and teachers in Osaka, Japan. *J. Infect. Dis.* **170:**439–443.

49. **Osherovich, A. M., and G. S. Chasovnikova.** 1967. Study on the isolation of enteroviruses from environmental objects (in Russian), p. 89–90. *In* M. P. Chumakov (ed.), *Materialy Problemnoi Komissii Akad. Med. Nauk SSSR "Poliomyelit i Virusnye Entsefality,"* Vypusk 1, *Enterovirusy.* Akad. Med. Nauk SSSR, Moscow, USSR.

50. **Romalde, J. L., M. K. Estes, G. Szűcs, R. L. Atmar, C. M. Woodley, and T. G. Metcalf.** 1994. In situ detection of hepatitis A virus in cell cultures and shellfish tissues. *Appl. Environ. Microbiol.* **60:**1921–1926.

51. **Sattar, S. A., V. S. Springthorpe, and S. A. Ansari.** 1994. Rotavirus, p. 81–111. *In* Y. H. Hui, J. R. Gorham, K. D. Murrell, and D. O. Cliver (ed.), *Foodborne Disease Handbook,* vol. 2. *Diseases Caused by Viruses, Parasites, and Fungi.* Marcel Dekker, New York, N.Y.

52. **Shieh, Y. C., K. R. Calci, and R. S. Baric.** 1999. A method to detect low levels of enteric viruses in contaminated oysters. *Appl. Environ. Microbiol.* **65:**4709–4714.

53. **Sockett, P. N., J. M. Cowden, S. Le Baigue, D. Ross, G. K. Adak, and H. Evans.** 1993. Foodborne disease surveillance in England and Wales: 1989–1991. *Commun. Dis. Rep.* **3:**159–172.

54. **Stroffolini, T., W. Biagini, L. Lorenzoni, G. P. Palazzesi, M. Divizia, and R. Frongillo.** 1990. An outbreak of hepatitis A in young adults in central Italy. *Eur. J. Epidemiol.* **6:**156–159.

55. **U.S. Department of Health, Education, and Welfare.** 1978. Unpublished information.

56. **U.S. Department of Health, Education, and Welfare.** 1979. Aseptic meningitis outbreak at a military installation in Pennsylvania, p. 11. *Aseptic Meningitis Surveillance, Annual Summary 1976.* HEW-CDC publication no. 79–8231. U.S. Department of Health, Education, and Welfare, Atlanta, Ga.

57. **Vipond, I. B., E. Pelosi, J. Williams, C. R. Ashley, P. R. Lambden, I. N. Clarke, and E. O. Caul.** 2000. A diagnostic EIA for detection of the prevalent SRSV strain in the United Kingdom outbreaks of gastroenteritis. *J. Med. Virol.* **61:**132–137.

58. **Wang, J. X., X. Jiang, H. P. Madore, J. Gray, U. Desselberger, T. Ando, Y. Seto, I. Oishi, J. F. Lew, K. Y. Green, and M. Estes.** 1994. Sequence diversity of small, round-structured viruses in the Norwalk virus group. *J. Virol.* **68:**5982–5990.

Foodborne and Waterborne Parasites

Food Microbiology: Fundamentals and Frontiers, 2nd Ed.
Edited by M. P. Doyle et al.
© 2001 ASM Press, Washington, D.C.

Charles W. Kim
H. Ray Gamble

Helminths in Meat

25

Foodborne parasites have undoubtedly had an impact on human health throughout history. They are important from the standpoint of their direct effect on the well-being of humans, who almost universally consume animal meat as a source of protein and other nutrients. They also serve as a trade obstacle for countries where a high prevalence of zoonotic parasites in livestock may prevent trade with countries where these parasites are rare. There are three meat-borne helminths of medical significance: *Trichinella* spp. and *Taenia solium*, which occur primarily in pork, and *Taenia saginata*, which is found in beef.

Despite the availability of sensitive, specific diagnostic tests, veterinary public health programs (meat inspection), and effective chemotherapeutic agents for human tapeworm carriers, these parasites continue to be a threat to public health throughout the world. Reasons for this include animal management systems that perpetuate infection, inadequate or poorly enforced inspection requirements for slaughtered animals, new sources of infection, and demographic changes in human populations that introduce new culinary procedures for preparing meats. Thus, current control and preventive procedures are inadequate to varying degrees, and more effective control measures are needed to ensure safe meat for human consumption.

TRICHINELLOSIS

The history of trichinellosis is fascinating, going back to when infection by *Trichinella* species was presumed more than proven. Whether the commandment in the Bible (Leviticus 11 and Deuteronomy 14) not to eat the flesh of cloven-footed animals (swine) was due in part to the potential danger of contracting trichinellosis is only speculative. In the 7th century A.D., Mohammed prohibited the eating of pork. Hence, to this day, trichinellosis is rare in Jews and Moslems. The earliest known case of trichinellosis may be evidenced by the mummy of a person who probably lived near the River Nile circa 1200 B.C., an observation made only in 1980 when supposedly *Trichinella* larvae were identified in the mummy's intercostal muscle (63).

Trichinella spiralis was first observed by Paget in 1835 in the muscles of a man during postmortem dissection (12). A landmark in the history of clinical trichinellosis

Charles W. Kim, Division of Infectious Diseases, Department of Medicine and Division of Microbiology, State University of New York, Stony Brook, NY 11794-8153.
H. Ray Gamble, Parasite Biology and Epidemiology Laboratory, U.S. Department of Agriculture, Agricultural Research Service, Building 1040, Room 103, BARC-East, Beltsville, MD 20705.

was Zenker's demonstration in 1860 that encapsulated larvae in the arm muscle caused the illness and death of a young woman. Even after 1860, many cases of trichinellosis were undoubtedly not diagnosed because of the difficulty in recognizing the infection clinically.

Species of *Trichinella*

Trichinellosis is worldwide in its distribution because the etiologic agent, nematodes of the genus *Trichinella*, are ubiquitous in animals, both domestic and wild. Historically, it was assumed that there was only one species of *Trichinella*, *T. spiralis*. However, our understanding of the composition of this genus has changed remarkably in recent years. A large volume of information is now available on species within the genus and the molecular and biological characters that distinguish them (70, 78).

Initial knowledge of variation within the genus *Trichinella* was based primarily on observed biological differences. New species were proposed for an isolate of *Trichinella* from arctic carnivores, designated *Trichinella nativa*, and a species first obtained from a hyena in Kenya, designated *Trichinella nelsoni* (11). The distinguishing characteristic of *T. nativa* is its resistance to freezing. It is distributed throughout arctic and subarctic zones, where it is found in a variety of wild mammals, including walruses and seals. The role of walruses in transmission of *T. nativa* to humans is well documented (32). *T. nelsoni*, which occurs in equatorial Africa, is also found in a wide range of wild animals, with the primary hosts being the bush pig and warthog (71). This species has occasionally been linked to human disease (51). Both *T. nativa* and *T. nelsoni* have low infectivity for the domestic pig. Another species, *Trichinella pseudospiralis*, believed to be a parasite of birds and hence not pathogenic for humans, was reported to be smaller than other isolates and, more important, lacking in the formation of a capsule (38). Since its description, this species has been found in various birds and mammals throughout the world (50), and it has been implicated in human disease (2, 49).

With the application of molecular techniques to parasite systematics, new interest has developed in the genus *Trichinella*. Molecular differentiation has supported observed biological differences, and, as a result, several new species and types of *Trichinella* have emerged. In addition to those described above, others include *Trichinella britovi*, common in wild mammals and feral swine, and occasionally found in horses and humans in temperate regions of Europe and Asia. *T. britovi* has some intermediate characteristics of other species, including some resistance to freezing, moderate infectivity for swine, and slow capsule formation (larvae

have been confused for nonencapsulating species in some cases) (78). *Trichinella murrelli* (79) is a North American species found in wildlife and occasionally in horses and humans. It has low infectivity for domestic pigs but poses a risk to humans who eat game meats. *Trichinella papuae* (80) is a second species that does not form a capsule. To date, it has only been reported from Papua New Guinea. In addition to the species described, several other isolates are recognized. *Trichinella* T6 is found in North America. It is resistant to freezing, has low infectivity for pigs, is found in a variety of wild mammals, and has been implicated in human disease (21). *Trichinella* T8 is an isolate from Africa that is similar to *T. britovi* but differs by using molecular markers. Further analysis of this type is needed for taxonomic placement. Finally, *Trichinella* T9 from Japan is similar to *T. britovi* but differs based on molecular analysis. As molecular techniques are refined and additional comparative studies are performed on *Trichinella* isolates, further taxonomic resolution should be applied to this genus. Several excellent reviews of the current status of *Trichinella* species are available (70, 78).

Life Cycle

All species of *Trichinella* complete their life cycle within one host; no intermediate hosts or extrinsic development is required. When larvae, encysted in raw or inadequately cooked meat, are ingested, the muscle fibers and capsules that enclose the parasite are digested. The liberated larvae burrow into the lamina propria of the villi in the jejunum and ileum. Four molts occur within 48 h, and by the third day, the worms are sexually mature. The small, tapered head of the adult worm has a round, unarmed mouth that opens into a tubular esophagus. Approximately one-third of the anterior portion of the body is composed of the stichosome, consisting of stacks of discoid stichocyte cells. The secreted contents of these cells are important for serological detection of infection and also contain those antigens that confer protective resistance to the host. Immediately below the esophagus and the stichosome lies a thin-walled intestine, the hind portion of which terminates in the rectum, a muscular tube lined with chitin. The female worm measures about 3.5 mm in length and possesses a vulval opening about one-fifth the body length from the anterior end (Fig. 25.1). The male measures 1.3 to 1.6 mm in length and possesses a single testis that originates in the posterior portion of the body and extends anteriorly to near the posterior end of the esophagus, where it turns posteriorly to form the vas deferens, which becomes the enlarged vesicula seminalis. The vesicula seminalis becomes the ejaculatory duct to join

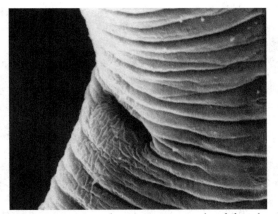

Figure 25.1 Scanning electron micrograph of female adult worm of *T. spiralis* with its prominent vulval opening (×2,450).

the copulatory tube in the cloaca. The copulatory tube forms the copulatory bell that is extruded during copulation (Fig. 25.2). There are two ventrally located copulatory appendages on each side of the cloacal opening that possibly serve to clasp the female in copula. Between these appendages lie four tubercles or papillae. Variations in size and differences in the cuticle have been noted for different isolates. For example, *T. pseudospiralis* is as much as one-fourth to one-third smaller than other isolates.

In the host, sexually mature adult worms reenter the lumen of the small intestine, where copulation takes place. The adult males die shortly after copulation. The female worms reburrow into the mucosa and begin to larviposit, usually into the central lacteals of the villi, about 7 days after infection and may continue to do so for a period up to a few weeks. Each female worm bears

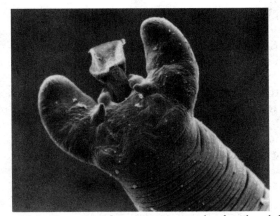

Figure 25.2 Scanning electron micrograph of male adult of *T. spiralis* with its copulatory bell (×1,400).

approximately 1,500 newborn larvae, but this varies based on *Trichinella* species and host.

The tiny newborn larvae (100 by 6 μm) are carried from the intestinal lymphatic vessels to the regional lymph nodes and into the thoracic duct and the venous blood, passing through the right side of the heart, through the pulmonary capillaries back to the left side of the heart, and into peripheral circulation. During migration, the larvae are known to enter many tissues, including those of the myocardium, the brain, and other sites, but here they either are destroyed or reenter the bloodstream. Generally, only larvae that reach striated muscles are able to continue development. They penetrate the sarcolemma of the fibers, where they mature, reaching approximately 1,250 pm in length. They become coiled within the fibers and, in the case of most species, are encapsulated as a result of the host's cellular response. This host-parasite complex, called the Nurse cell, is capable of supporting the infective larvae for months or even years. An increased vascular supply to the Nurse cell provides nutrients and oxygen vital to the parasite's survival. The encapsulated cyst eventually becomes calcified, and the larva dies as a result.

Infection in humans typically represents a dead end in the parasite's life cycle. However, in animal hosts, the carcass serves as a source of infection. Typically, infected animals are refractory to subsequent infection. However, molecular evidence has documented dual infection in animals (78), and it is possible that these dual infections result from multiple exposures.

Epidemiology

Trichinellosis, the human disease resulting from infection with species of the genus *Trichinella*, is considered a zoonosis because infection occurs as a result of ingestion of raw or poorly cooked meat from infected animals. Sources of human infection include domestic livestock, primarily pigs, and wild mammals. The transmission patterns of *Trichinella* spp. can be divided, based on hosts, into domestic (synanthropic) and sylvatic (wildlife) cycles. Humans can become involved in both of these cycles.

The domestic cycle is defined by habitat and, by definition, must involve domesticated (farm) animals. There are various patterns of transmission within the domestic cycle. Pigs may become infected through the deliberate feeding of uncooked or undercooked animal flesh containing infective larvae. Infected pigs may then serve as a source of infection to other pigs through cannibalism. Within the domestic cycle, rats may serve as a reservoir host, acquiring infection from wildlife, carrion, or pigs. Rats may transmit infection directly to pigs, which ingest them. Pigs may acquire infection directly from wildlife when allowed to wander under unrestricted conditions.

The species most frequently associated with the domestic cycle is *T. spiralis*. This species has a high infectivity for pigs and rats relative to other species of *Trichinella*. The lower infectivity of other species diminishes their importance in the domestic cycle. Although pigs have been reported to be infected with most known species, few cases of human infection with these species have been linked to pork. Because cooking, freezing, and many other processing methods kill trichinae in meat, most human infections have resulted from situations where meat preparation was not adequate. Pork products, such as fresh sausage, summer sausage, and dried or smoked sausage, have been implicated as vehicles of human trichinellosis in the United States. Over the past 5 years, only about 10 cases of human trichinellosis have been reported annually in the United States (65). Of these, less than one-half have been attributed to pork or pork products. However, pork and pork products continue to serve as a major source of infection in many parts of the world (20, 72, 96).

Another domesticated animal that has emerged as a major source of human trichinellosis is the horse. In 1975 and 1985, outbreaks in the southern suburbs of Paris were attributed to the consumption of raw horse meat in the form of steak tartare (9). It was initially difficult to believe that human infections resulted from ingestion of meat of an herbivorous animal, but a series of large outbreaks over the past 2 decades have documented the role of horsemeat as a cause of human trichinellosis. A review of the occurrence of horsemeat trichinellosis is found in reference 7. To date, more than 3,200 cases of human trichinellosis have been reported in France and Italy as a result of ingesting raw or undercooked horsemeat. Other countries where horsemeat is consumed have not had human cases, as meat is cooked thoroughly before consumption. Following the first outbreaks it was speculated that the horsemeat was contaminated with meat from other species (5). However, the recovery of larvae from the suspected meat (43) and identification of naturally infected horses (3) have solidified this source of human exposure. The route by which horses become infected largely remains an enigma. However, infected horses typically originate from countries or regions where a high incidence of *Trichinella* infection is known to occur in pigs, rats, and other animal species. It is likely that horses raised under conditions of minimal management and nutrition will ingest dead or dying rodents in the environment. It is also possible that nesting rodents are killed when hay bales are chopped

for feeding to horses. We know that three species have contributed to human trichinellosis from horsemeat: *T. spiralis, T. britovi,* and *T. murrelli* (7). Further studies are needed on the epidemiology of *Trichinella* infection in the horse.

Other "domestic" sources of trichinellosis have been reported sporadically. An outbreak in China was reported to be due to the consumption of mutton (16). An unusual outbreak in northern Germany was attributed to the consumption of air-dried meat known as pastyrma that had been purchased as camel meat in Egypt (8). It was not certain whether this delicacy was indeed camel meat, since pastyrma is usually made from beef. If beef was the source, it may have been adulterated with pork. In Thailand, dog meat has been implicated as a source of infection (19). Dog meat is prepared as a special dish and is favored by the Vietnamese, Chinese, and hill tribe Thais. Dog meat has also been implicated in an outbreak of trichinellosis in China (47). Most recently, farmed crocodile flesh was reported to harbor *Trichinella* larvae (66), but no human cases have been reported from this source.

The sylvatic cycle of *Trichinella* involves more than 100 species of mammals, and many of these species pose a risk to humans who eat raw or undercooked game meats. In the sylvatic cycle, wild carnivorous or omnivorous animals scavenge the carrion of dead animals or eat meat from prey animals. Higher infection rates typically occur in animals near the top of the food chain. For example, in the arctic, *Trichinella* is quite prevalent in polar bears, grizzly bears, foxes, and wolves. Transmission in the arctic environment is further facilitated by survival of larvae in frozen meat. Trichinous polar bear meat may conceivably have been responsible for the deaths of three Swedish explorers on an expedition to the North Pole in 1897. More than 50 years later, laboratory examination verified the presence of larvae in minute particles of bear meat still remaining on the explorers' equipment (84). In Alaska, all human cases of trichinellosis have been traced to the consumption of bear or walrus meat (90). Polar bear meat is considered a delicacy among the Eskimos. Among the Inuit Eskimos in northeastern Canada, there have been outbreaks resulting from the consumption of undercooked walrus meat (97) and probably other mammals (60). In the former Soviet Union, up to 96% of human trichinellosis cases have been traced to the consumption of wild animals, particularly boars (6). In the United States, bear meat is the most commonly incriminated game meat, and wild pig is second. An isolated case in Idaho was attributed to the consumption of cougar meat (21).

Prevalence of Human Disease

Human trichinellosis is essentially nonexistent in some countries of the world but remains a significant problem in many others. In countries of the European Union, where slaughter inspection of pigs for *Trichinella* infection has been required for more than 100 years, the parasite has been eradicated from domestic swine, and in some of these countries human trichinellosis has not been reported for several decades.

In the United States, a National Institutes of Health report, published in 1943, found 16.2% of the population to be infected with *Trichinella* (101). This information led to considerable publicity on the dangers of eating pork and was responsible for strict federal control of methods used to prepare ready-to-eat pork products. Human cases of clinical trichinellosis reported to the Centers for Disease Control and Prevention declined from about 500 per year in the 1940s to fewer than 50 per year in the 1980s and less than 10 per year from 1996 through 2000. Furthermore, many of these cases resulted from sources other than pork, such as bear and other game meats. Although there has been a general decline in incidence in the United States, trends in incidence and transmission patterns changed over the years, especially with a large influx of people from Southeast Asia (54). Historically, trichinellosis was more common in individuals of German, Italian, and Polish descent because of culinary preferences for raw or undercooked pork. Between 1975 and 1984, trichinellosis among Southeast Asians, who prepare some dishes containing essentially raw pork, was 25 times more frequent than in the general population (94). Of the 1,260 cases of trichinellosis reported to the Centers for Disease Control and Prevention during this period, 60 (4.8%) were among refugees of Southeast Asia. One of the largest outbreaks ever reported in the United States occurred in 1990 when Southeast Asian refugees from six states and Canada developed trichinellosis after eating pork sausage at a wedding held in Iowa (61).

Trichinellosis, acquired from eating raw or undercooked pork, remains a serious public health problem in many parts of the world and may be considered an emerging or a reemerging disease in some countries. Regions where trichinellosis continues to be a problem include parts of Eastern Europe, Asia, and some countries of Central and South America. In most cases, poor hygienic conditions for raising pigs and inadequate veterinary control at slaughter are the reasons for high rates of human infection. Pigs raised outdoors, pigs fed raw garbage, or pigs raised in contact with large rodent populations are at high risk of acquiring infection.

Inadequate veterinary inspection may result from insensitive or improperly performed testing methods or from pigs that are butchered outside of normal inspection channels (backyard pigs).

In addition to gaps in inspection programs, culinary habits (improper cooking) impact the risk of human exposure. This is exemplified in some Asian countries in which epidemics have been reported, including Thailand, Laos, Japan, China, and Hong Kong. A dish known as nahm is a common source among the inhabitants of northern Thailand (52). Other Thai dishes, such as lu (lahb [raw spiced meat] mixed with fresh blood) and satay (small pieces of spiced pork grilled rare to medium on a bamboo skewer), are also sources of infection. Sometimes satay is eaten raw. In both dishes, the larvae may remain alive. Outbreaks of trichinellosis in Laos have been due to the consumption of pork dishes known as som-mou, lap mou, and lap leuat. Laotian immigrants in the United States still prepare these traditional dishes, resulting in outbreaks (15).

It had been suspected for many years that trichinellosis can be contracted via contaminated utensils, such as butcher knives, meat grinders, butcher blocks, or even human hands after handling infected pork. Circumstantial evidence for this mode of transmission was reported for an outbreak that occurred in Liverpool, England (92). Trichinellosis associated with beef that had been in contact with the meat grinder or butcher block used for pork, or from mixing of beef and pork, has also been reported in the United States (4, 14) and Canada (30).

Pathogenesis and Pathology

Damage caused by *Trichinella* infection varies with the intensity of the infection and the tissues invaded. The in-and-out movements of the adult worms, especially the females in intestinal tissue, cause an acute inflammatory response and petechial hemorrhages. The cellular response primarily consists of neutrophils with eosinophils. This is followed by an infiltration of lymphocytes, plasma cells, and macrophages that peaks at about 12 days after infection, gradually declining thereafter. Thus, lesions in the intestine are due to the host's response to the adult worms or their protein products.

The newborn larvae cause an acute inflammatory response as they pass through or become lodged in various tissues and organs. The infiltration consists of lymphocytes, neutrophils, and especially eosinophils. Although there is myocarditis, viable larvae are more numerous in the pericardial fluid than in the myocardium. Pulmonary hemorrhage and bronchopneumonia may be observed during this stage of larval migration through the capillaries. Rarely, encephalitis may result if the larvae migrate through cerebral capillaries.

When larvae encyst in striated muscle fibers, there is an immediate tissue response consisting of inflammation of the sarcolemma of the involved muscle fibers. The disturbance of ultrastructure and metabolic processes in muscle fibers results in basophilic transformation. This is followed by destruction of the muscle fibers and the eventual formation of a capsule. The larvae gradually die, provoking an intense granulomatous reaction or foreign body cellular response that culminates in calcification. The most heavily parasitized striated muscles include the diaphragm, intercostals, ocular, and masseter muscles. Those larvae that enter tissues other than striated muscles disintegrate and are eventually absorbed.

Clinical Manifestations

The vast majority of individuals infected with *Trichinella* spp. are asymptomatic, probably because low numbers of larvae are ingested. Classical trichinellosis is usually described as a febrile disease with gastrointestinal symptoms, periorbital edema, myalgia, petechial hemorrhage, and eosinophilia (56). During the intestinal stage of infection, gastrointestinal symptoms such as nausea, vomiting, "toxic" diarrhea or dysentery, fever (over 38°C), and sweating may be observed. Malaise can be severe and lasts longer than fever. The onset of intestinal symptoms usually occurs within 72 h after infection and may last for 2 weeks or longer.

Generally, from the second week, when the newborn larvae are migrating and can reach almost any tissue within the body, a characteristic edema is noted around the eyes and in some cases around the sides of the nose, at the temples, and even on the hands. Periorbital edema is present in about 85% of patients and is often complicated by subconjunctival and subungual splinter hemorrhages that disappear within 2 weeks. Eosinophilia greater than 50% is the most consistent manifestation. In severe cases, however, the number of eosinophils may be low. As the larvae enter striated muscle fibers, a striking myositis and muscular pain are evident. In addition to inflammation and pain, eosinophilia is also observed during this phase. Serum creatine phosphokinase and transaminase can be elevated as a result of leakage of these enzymes from muscle fibers into the serum. Higher levels of serum enzymes may not correlate well with the severity of the clinical condition. Hypoalbuminemia, which is believed to be due to a demand for protein by the parasite, is usually accompanied by hypopotassemia. Muscle atrophy and contractures may also occur.

Respiratory symptoms, including dyspnea, cough, and hoarseness, either may be due to myositis of respiratory muscles or may be secondary to pulmonary congestion. Neurologic manifestations are rare but may result from invasion of the brain by migrating larvae. In severe cases, death may result following cardiac decompensation or respiratory failure, cyanosis, and coma.

Intimate contact between the parasite and host tissues stimulates the production of antibodies that can be demonstrated in the serum. The exact role of antibodies in acquired immunity is not clear. On the basis of experimental animal data, it would be safe to assume that in most cases in humans, the severity of infection is reduced considerably by the development of immunity from previous subclinical infections. Local gut immunity in mice has been shown to be T-cell dependent (42).

Diagnosis

It is difficult to clinically diagnose trichinellosis because it mimics so many other infections as a result of its wide distribution in the body. Diagnosis is based on a history of eating infected meat, symptoms, laboratory findings (including serology), and recovery of larvae from muscles (34). Classical symptoms of trichinellosis include fever, myalgia, and periorbital edema. The most characteristic feature of laboratory tests is a marked eosinophilia. Although eosinophilia is not restricted to trichinellosis, its presence in 30 to 85% of infected individuals is a constant and important diagnostic aid. Eosinophilia is initiated in the second week of infection, reaching its peak by about day 20. It may be absent not only in patients with very severe and fatal cases but also in individuals with secondary infection.

The most definitive diagnostic method is muscle biopsy to detect unencysted or encysted larvae. Diagnostic success depends on chance distribution of larvae in the particular striated muscle that is sampled. Gastrocnemius, pectoralis major, deltoid, or biceps muscles are commonly used because of easy accessibility. The muscle strip is compressed tightly between two microscope slides and examined under a microscope (Fig. 25.3). Part of the biopsied sample can be either digested or fixed and then sectioned, stained, and examined. The presence of active larvae following digestion with artificial gastric juice indicates a recent infection.

Numerous immunological tests are available for diagnosis of trichinellosis (53, 55, 68). With the introduction of marker assays such as immunofluorescence and enzyme immunoassays, the sensitivity of tests has greatly improved. Common methods in which fluorescence-labeled antibodies are used include direct, inhibition, and indirect staining techniques. The enzyme-linked immunosorbent assay (ELISA) has been shown to be both

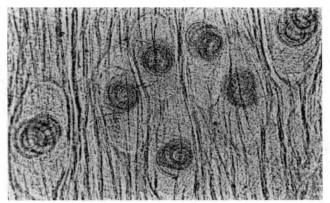

Figure 25.3 Pressed muscle containing *T. spiralis* larvae.

sensitive and specific. With the aid of class-specific conjugates, specific antibodies of various classes that are important for diagnosis of acute trichinellosis can be detected; an increase in immunoglobulin M and A titers is indicative of a recent infection. Various types of ELISA can be used for measuring antigen. The sandwich ELISA has been used to detect circulating antigens in sera (48). Other serologic tests, including complement fixation, precipitin, latex agglutination, and various flocculation tests, have been replaced by newer techniques that are more sensitive and specific. It is advisable to take multiple serum samples at intervals of several weeks to demonstrate seroconversion in patients whose sera were initially negative or to detect rising titers.

Treatment

The efficacy of treatment of trichinellosis depends on the intensity of infection, the species of *Trichinella*, the stage of infection, and the character and intensity of the host response. The purpose of treatment during the intestinal phase is to destroy adult worms and to interfere with the production of newborn larvae. The drug of choice is mebendazole (Vermox) at dosages of 200 to 400 mg three times a day for 3 days and then 400 to 500 mg three times a day for 10 days (34, 56). Mebendazole is also believed to be active against developing encysted larvae but at dosages higher than those used for adult worms. Mebendazole is administered during the first week of infection before newborn larvae migrate, in moderate or severe infections in combination with corticosteroids, and in infections with *T. nativa*, which responds poorly to treatment with nonbenzimidazole compounds. Mebendazole is not recommended for women in the first trimester of pregnancy. Thiabendazole (Mintezol) is no longer used because of adverse reactions. Albendazole (Valbazen, Zentel) is better absorbed and is probably as active as

or even more active than mebendazole (56). Pyrantel (Antiminth) at 10 mg/kg of body weight per day for 4 days and levamisole at 2.5 mg/kg/day (maximum of 1 to 50 mg) are active only against adult worms in the intestine.

To minimize the provocation of hypersensitivity, it is recommended that corticosteroids be given in combination with anthelmintic drugs. Corticosteroids are recommended for acute trichinellosis not only for antiallergic action but also for anti-inflammatory and antishock actions. Supportive therapy, such as bed rest, is very important.

Prevention and Control

Prevention of human trichinellosis resulting from pork and pork products is accomplished by one or more of several methods, including preventing infection on the farm, detecting infected carcasses at slaughter, processing meat to inactivate worms, and proper cooking.

Several measures to prevent human infection can be directed against the principal sources of human trichinellosis (pigs) before they are slaughtered for market. In the United States, Public Law 96-468 of 1980 requires treatment of meat-containing waste if it is to be fed to pigs (59). The Swine Health Protection Act of 1983 stipulates that all such garbage must be boiled for 30 min in a licensed facility before being fed to pigs (67). The other, more desirable alternative is to strictly prohibit feeding of garbage to pigs, a measure taken by many states. Unfortunately, these preventive measures do not apply to feeding of raw scraps of wild animal meat or scraps of raw household garbage to pigs. Some countries, including the United States, are now proposing Good Production Practices (GPPs) for pigs that preclude exposure to *Trichinella* from all identified sources (62). These GPPs include elements of biosecurity (i.e., eliminating exposure to wildlife and rats) and general good hygiene.

In the United States, as in many countries, all animals slaughtered for human consumption are examined by a federal or state inspector at the time of slaughter to determine whether the animal is healthy and suitable for use as human food. In many European countries, notably those members of the European Union, each pig carcass is inspected for *Trichinella* infection by using one of several approved direct methods of inspection. These methods include trichinoscope examination, in which a piece of muscle is compressed between two glass slides and scanned under a microscope, and several artificial digestion methods, where a muscle sample is digested in acidified pepsin and the product is examined for larvae. By applying this rigorous approach to slaughter testing, *Trichinella* infection has been virtually eliminated from domestic pigs in these countries. However, this effort is not without cost, with an estimated $570 million spent annually just for *Trichinella* inspection (77). In the United States, pigs are not inspected individually for larvae because of the large numbers of animals that are produced (69). In other countries, testing programs are in place, but not all pigs are processed through official slaughter facilities. These gaps in veterinary control are one reason that trichinellosis remains a significant problem.

In the United States, pork products that are considered ready-to-eat, such as cold-smoked sausage, must be processed to kill all larvae by heating pork muscle to at least 58°C, freezing to −35°C in its center, or freeze-drying (59, 67). Curing procedures are specified in the U.S. Department of Agriculture (USDA), Code of Federal Regulations (9CFR, §318.10) for most classes of pork products (59). The effectiveness of curing depends on a combination of salt concentration, temperature, and time (57).

The USDA and the Food and Drug Administration have approved irradiation of pork to destroy *Trichinella* larvae (68). However, commercially produced irradiated pork is not yet available. The success of application of irradiation will depend on cost-effectiveness for the meat industry and willingness on the part of consumers to purchase and eat irradiated meat.

The most effective control measure for fresh pork sold from areas of endemicity is education of the public. The responsibility lies with consumers to ensure that larvae are killed before the pork is eaten. It is very important that fresh pork be cooked or processed properly before consumption in the home. The USDA recommends that consumers assume that all pork is infected with *Trichinella* larvae. Thus, it is recommended that all pork be cooked to an internal temperature of 160°F (71°C) (68). This recommendation is based on observations that rapid cooking methods, such as the use of microwave ovens, may not heat the pork uniformly or for sufficient time to destroy the larvae (58). The USDA, Code of Federal Regulations (9CFR, §318.10) requires that pork be frozen for 20 days at 5°F (−15°C), 10 days at −10°F (−23°C), or 6 days at −20°F (−29°C) to kill the larvae, provided that the meat is less than about 6 in. (15 cm) thick (68). These freezing temperature and time requirements do not appear to be effective for killing the arctic isolate, *T. nativa*, or other freeze-resistant types of *Trichinella*. In fact, Alaskan bear meat has been shown to be infective for laboratory animals even after 35 days of storage at −15°C (100). Thus, freezing meat may not be adequate for eliminating the potential source of infection by some species.

TAENIA SAGINATA TAENIASIS

The history of taeniasis caused by *T. saginata* has been described in the ancient literature with intriguing theories as to the parasite's origin and nature. *T. saginata* was recognized as a distinct species by Goeze in 1782. It was not until 1861 that Leuckart established the relationship between the adult worm in humans and the larval bladder worm in cattle. Although some controversy has arisen regarding the generic term for the unarmed tapeworm of humans acquired from the consumption of beef, *T. saginata* is the currently accepted name.

Epidemiology

Taeniasis can result when raw or inadequately cooked beef is consumed. Raw or rare beef appears to be popular in many countries, particularly in the form of beef tartare (basterma) in the Middle East, shish kebab in India, lahb in Thailand, yuk hoe in Korea (64), and raw or slightly roasted meat in parts of Africa, especially Ethiopia (13), where the beef is diced into cubes, skewered, and then roasted or is served raw as kitfo, the equivalent of steak tartare. Although the meat of reindeer in northern Siberia, as well as meat of other herbivorous animals, has been reported to contain cysticerci of *T. saginata* and has probably been responsible for taeniasis in humans, the only evidence of wild intermediate hosts in the Americas has been a single report at the turn of the century concerning llama and pronghorn antelope (74). Thus, it can be assumed that at present beef is the primary source of *T. saginata* taeniasis in humans.

With respect to the prevalence of taeniasis in humans, geographic areas can be classified into three groups: countries or regions where infection is highly endemic, with a prevalence in the human population exceeding 10%; countries with moderate infection rates; and countries with a very low prevalence, below 0.1%, or even no infections caused by the parasite. The areas of high endemicity are Central and East African countries, such as Ethiopia, Kenya, and Zaire; Caucasian South Central Asian republics of the former Soviet Union; Middle Eastern countries, such as Syria and Lebanon; and parts of the former Yugoslavia. In the Asian republics of the former Soviet Union, the prevalence rate may reach 45% (76). The incidence is low in the United States, where only 443 patients were reportedly treated for taeniasis due to *T. saginata* in 1981 (69). Critical factors responsible for the persistence of and periodic increases in human *T. saginata* infections depend on multiple means that govern infection in cattle. Cattle have an increased risk of infection if they have access to pastures contaminated directly with human feces or sewage effluent, feed contaminated by laborers, and irrigation ditches and cattle pens directly contaminated by farm workers. Eggs of *T. saginata* are capable of surviving long periods in the environment and are known to be resistant to moderate desiccation, disinfectants, and low temperatures (4 to 5°C). A longevity of 71 days in liquid manure, 33 days in river water, and 154 days in pastures has been reported (93). The eggs can also survive many sewage treatment processes and remain viable in fluid effluent or dried sludge.

An interesting observation is that *T. saginata* is a problem not only in developing countries, where relatively lower standards of hygiene and sanitation are practiced, but also in economically developed countries, where people can afford to consume more beef and sewage treatment facilities are overtaxed. Thus, both the practice of defecating in the bush in poor countries and placing stress on sewage treatment facilities in rich countries ensure a continuity of the infection. Among affluent Americans, a high infection rate has been attributed to greater freedom of international travel, in addition to a preference for imaginative diets such as those including rare or raw beef. The penchant for raw beef and steak tartare is not limited to affluent Americans and Europeans but is also found among more affluent Ethiopians who can afford beef. In many countries of the Middle East and East Africa, a combination of poor sanitation and culinary preference for raw and rare beef has contributed to a higher risk of infection. It should be remembered that transmission can also result from tasting raw or undercooked meat during preparation and cooking.

Life Cycle
Following ingestion of raw or rare beef containing a viable cyst of the larval stage of *T. saginata* (often referred to as *Cysticercus bovis*), the cysticercus is released from its surrounding muscle tissue by digestion in the small intestine. The scolex of the cysticercus evaginates from the vesicle or bladder and attaches to the mucosa of the jejunum, where the larva develops into a mature adult worm in 8 to 10 weeks. A mature adult worm is characterized by having a scolex with four hemispherical suckers of 0.7 to 0.8 mm in diameter situated at the four angles of the head. The entire strobila measures 4 to 12 m and possesses 1,000 to 2,000 proglottids or segments. Immediately posterior to the neck is a series of immature, mature, and gravid proglottids. The mature and gravid proglottids are hermaphroditic, containing both male and female reproductive organs. Self-fertilization occurs when distal proglottids come together by folding of the worm's body.

The scolex attaches to the mucosal surface of the upper jejunum by means of the four suckers but, because of its length, the entire worm might extend down to the terminal ileum. The developing proglottids extend down the small intestine. The most distal gravid proglottids, which measure about 20 mm long and 5 to 7 mm wide, detach singly from the rest of the strobila and independently migrate through the rectum to the outside. Each gravid proglottid contains about 80,000 eggs, which are expressed from the proglottids and deposited on the perianal skin. When gravid proglottids come to rest on the ground, eggs are extruded. The eggs may also be present as a result of indiscriminate promiscuous defecation.

Cattle ingest *T. saginata* eggs during grazing. Eggs hatch in the duodenum, liberating a six-hooked embryo that penetrates mesenteric venules or the lymphatics and reaches skeletal muscles or the heart, where it develops into the cysticercus larva. The cysticercus is essentially a miniature scolex and neck invaginated into a fluid-filled bladder that measures about 10 by 6 mm. The bladder larva becomes infective within 8 to 10 weeks following ingestion of eggs by cattle and remains infective for more than a year. A person eating raw beef from infected cattle within this period is subject to infection.

Pathogenesis and Pathology
The scolex of the adult worm is generally lodged in the upper part of the jejunum. Usually, only a single worm is present, although multiple infections have been reported. The adult tapeworm does not cause pathologic changes. However, mucosal biopsies have shown minimal inflammatory reactions, suggesting that the worm can have irritative action that results in bowel distention or spasm.

Migration by the adult worm to unusual sites is rare, but complications such as appendicitis, invasion of pancreatic and bile ducts, intestinal obstruction, or perforation and vomiting of proglottids with aspiration have been reported.

Clinical Manifestations
Although most cases of taeniasis are asymptomatic, up to one-third of patients complain of nausea or abdominal "hunger" pain that is often relieved by eating. Epigastric pain may be accompanied by weakness, weight loss, increased appetite, headache, constipation, dizziness, and diarrhea. The patient usually becomes aware of the infection when a proglottid is passed in the stool or is found on the perianal area or even on underclothing.

The adult worm is weakly immunogenic, as manifested by a moderate eosinophilia and increased levels of serum immunoglobulin E. Allergic reactions, such as urticaria and pruritus, may be due to the worm and its

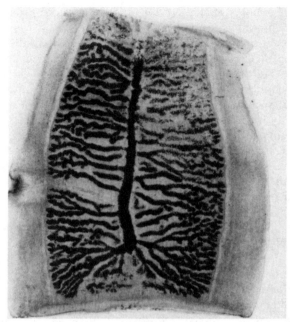

Figure 25.4 Gravid proglottid of *T. saginata*. Note at least 16 lateral uterine branches.

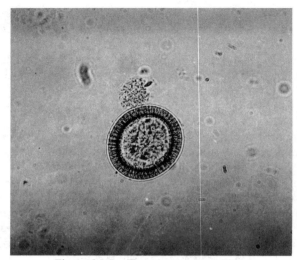

Figure 25.5 *T. saginata* egg (×590).

metabolites. The adult worms induce the production of antibodies. The persistence for years of a large, actively growing worm reflects lack of protective immunity to resident worms but is consistent with concomitant immunity observed in other helminth infections.

Diagnosis

Definitive diagnosis is based on identifying the proglottid, since the eggs of *T. saginata* cannot be distinguished from those of other species of *Taenia* or those of *Multiceps* and *Echinococcus* species. The gravid proglottid of *T. saginata* has 15 to 20 lateral branches of the uterus on each side of the main uterine stem, a characteristic feature (Fig. 25.4). If the gravid proglottid is treated with 10% formaldehyde and injected with India ink, the uterine branches are very prominent. Uterine branches can also be seen by gently pressing the proglottid between two microscope slides and holding them in front of a bright light. If the scolex is present, the four characteristic hookless suckers can be used as a distinguishing feature for identification.

The egg is nearly spherical in shape, measuring 30 to 40 μm in diameter, has characteristic radial striations on its thick shell, and contains a hexacanth embryo with delicate lancet-shaped hooklets (Fig. 25.5). Rather than looking for the eggs in the stool, it is better to use the commercial Scotch tape method to obtain the eggs or proglottids from the perianal region. Since detection of

eggs is nonspecific in terms of species of origin, DNA probes that allow differentiation of genomic DNA of tapeworm species by hybridization techniques have been developed (31, 44). PCR has been used to discriminate between various *Taenia* and *Echinococcus* species at the genomic level (41). The amplification of a 0.55-kb DNA fragment is a unique *T. saginata* genomic DNA template (40).

Although an antibody response is induced by the adult worm, specific serologic tests such as are used for diagnosis of trichinosis have not been routinely used for diagnosis of taeniasis. However, a highly specific immunoassay, using a <12-kDa antigen prepared from the cyst fluid of *Taenia hydatigena*, that is able to differentiate between the types of human taeniasis has been developed (45).

Treatment

The drug of choice in treating taeniasis is niclosamide (Niclocide, Yomesan), a taeniacide that is effective in damaging the worm to such an extent that a purge following therapy will often produce the scolex. Cure rates are very high. A single oral dose of 10 mg of praziquantel (Biltricide) per kg has also been reported to be highly effective (87).

Prevention and Control

In addition to effective drug therapy against the adult *Taenia* worms, certain measures can be taken to prevent cysticercosis in cattle. The most important measure is improved sanitation for adequate sewage disposal so that tapeworm eggs are unavailable to cattle. Since humans are the only definitive hosts of *T. saginata* and thus the only disseminators of eggs, which can number 480,000 to 720,000 daily, the success of this measure depends on educating livestock producers and their employees about modes of transmission, examining workers' stools for *Taenia* species, and providing adequate toilet facilities in cattle-feeding establishments (91). Where toilet facilities are available, overloading of systems should be avoided. Prevention includes sanitary protection of cattle feed. For example, a recent epidemic of cattle cysticercosis that led to extensive carcass condemnation resulted from feeding cottonseed harvested from an area where a municipal water treatment facility and hard rains had combined to contaminate the feed with *T. saginata* eggs.

Meat inspection to detect cysticercosis in slaughtered cattle can reduce the transmission of taeniasis; however, its effectiveness is limited even if practiced extensively. Even routine examination of the heart and masseter muscles may miss a significant percentage of infected cattle. It is estimated that as few as 25% of infected cattle are detected by methods currently employed for meat inspection (83). In the United States, the percentage of cattle inspected by state and municipal authorities varies from state to state. Cattle from small farms where sanitary conditions may often be poor are frequently slaughtered in local abattoirs without inspection requirements. All infected carcasses should be condemned, but approximately 70% of infected carcasses are trimmed of visible cysts and then passed unrestricted for consumption (91). However, the U.S. Code of Federal Regulations (9CFR, §311.23) requires infected carcasses to be frozen for a minimum of 10 days at 15°F (−9.4°C) or cooked to a temperature of 140°F (75).

An alternative approach for assessing cysticercosis in cattle might be to use a sensitive serologic method. The ELISA technique has been suggested (83, 86), but the efficacy of this method has not been proven to be superior to physical inspection; furthermore, correlation between direct and serological methods is poor. Serological testing might be useful for epidemiological purposes, especially if cysticercosis is suspected in a cattle herd. A positive ELISA would warrant gross inspection of the carcass for the larval bladder worm (beef measles) at slaughter; a negative result from direct examination would indicate that the carcass is very likely free of infection. The development of an effective vaccine for use in highly endemic areas would not only prevent initial infection but also protect cattle against challenge infections.

On the part of the consumer, mass public health education concerning risks of eating raw or inadequately cooked beef is important. The cysticerci in meat are inactivated by freezing the meat at −10°C for 10 days or −18°C for 5 days, heating to an internal temperature of 56°C, or salt curing under appropriate conditions.

Major obstacles to this approach have been the consumers' reluctance to modify preferred culinary habits and to break with long-established cultural traditions. Thus, attempts to eradicate *T. saginata* taeniasis have been unsuccessful.

TAENIA SOLIUM TAENIASIS

Taeniasis caused by *T. solium* was known at the time of Hippocrates. The Greeks described the larval stage in the tongue of swine as resembling a hailstone. In contrast to *T. saginata*, *T. solium*, the pork tapeworm, possesses an armed rostellum, has a smaller number of lateral uterine branches, and lacks a vaginal sphincter (81).

Epidemiology

Humans are the only known natural definitive hosts for *T. solium* taeniasis, which has a worldwide distribution. It is very difficult to evaluate the prevalence of infection caused by this species because the coproscopical methods used for surveys cannot differentiate between infection caused by *T. saginata* and that caused by *T. solium*. *T. solium* is important in all countries where pork is consumed, especially those with suboptimal practices of sanitation and pig husbandry. Human infection is practically nonexistent in Moslem countries (76). It is most common in Slavic countries, Mexico, Central and South America, Central and South Africa, Asia, non-Islamic Southeast Asia, and southern and eastern Europe. Prevalence rates have been reported at 0.6% (state of Michoacan) and 1.5% (state of Yucatan) in Mexico (85, 89), 1.0 to 2.8% in Guatemala (1, 37), 3.0 to 8.6% in Peru (35), 2.0% in Honduras (88), 11.5% in northern Nigeria (17), 0.7% in Bali, Indonesia (95), and 5.0% in aborigines of southeast Taiwan (28). Specific habits of eating raw or undercooked pork are linked to these foci of taeniasis. In developed societies, raw pork is eaten less frequently than raw beef, which may be responsible for the lower incidence of *T. solium* than of *T. saginata* taeniasis.

The consequences of *T. solium* taeniasis are much more serious than those of *T. saginata* taeniasis because the larval or cysticercus stage of *T. solium*, which infects pigs, can also infect humans (82). Cysticercosis in humans is an important public health problem in developing countries where hygiene and sanitary conditions are deficient or nonexistent and where swine-rearing practices are primitive (73). The relatively high prevalence of *T. solium* taeniasis in the Bantu population of South Africa has been attributed to a widespread practice of using tapeworm proglottids as a component of muti, a medication used by native herbalists for the treatment of intestinal worms (46). Infection of humans with larvae results from inadvertent ingestion of eggs in contaminated food, e.g., vegetables fertilized with night soil (human excrement) or contaminated water, from hands of individuals already infected with the adult worm (external autoinfection), and from direct contact with a tapeworm carrier. Another important mode of transmission involves patients infected with adult tapeworms, in whom the eggs are carried by reversed peristalsis back to the duodenum or stomach, where they are stimulated to hatch and subsequently invade extraintestinal tissues to become cysticerci (internal autoinfection). The prevalence of human neurocysticercosis is increasing in some parts of the world, and a considerable body of literature is devoted to the diagnosis, treatment, and prevention of this disease (36, 98).

Life Cycle

When meat, usually pork, containing a viable cyst of the larval stage of *T. solium* (often referred to as *Cysticercus cellulosae*) is ingested, the head of the larva evaginates from the fluid-filled milky white bladder. The scolex bears four suckers and an apical crown of hooklets. The cysticerci are referred to as pork measles and are larger (5 to 20 mm in diameter) than *Cysticercus bovis*. They attach to the wall of the small intestine and mature into adult worms in 5 to 12 weeks. The adult worm measures up to 8 m (usually 1.5 to 5 m) and has a scolex that is roughly quadrate, possessing a conspicuous rounded rostellum armed with a double row of large and small hooklets, numbering 22 to 32. A short cervical region is anterior to a series of proglottids or segments. Immature proglottids are broader than long, mature ones, being nearly square, while gravid ones are longer than broad. The total number of distinct proglottids is less than 1,000. The terminal proglottids become separated from the rest of the strobila and migrate out of the anus or are passively expelled in the stool. A single gravid proglottid contains fewer than 50,000 eggs. Upon escape from the ruptured uterus of the gravid proglottid and after deposition on the soil, the eggs may remain viable for many weeks. The eggs of *T. solium* are more apt than those of *T. saginata* to appear in the stool.

In the normal cycle, the eggs are ingested by pigs, the usual intermediate host. The hexacanth embryo hatches in the duodenum, migrates through the intestinal wall to reach the blood and lymphatic channels, and is carried to the skeletal muscles and the myocardium. The embryo develops in 8 to 10 weeks into a cysticercus or *Cysticercus cellulosae*. In the abnormal cycle, humans can serve as intermediate hosts and harbor cysticerci acquired by accidentally ingesting eggs or by the autochronus cycle (an infection that results from the movement of eggs or gravid proglottids from the intestine back into the stomach by reverse peristalsis). The cysticercus develops most commonly in striated muscles and subcutaneous tissues but also in the brain, eye, heart, lung, and peritoneum.

Pathogenesis and Pathology

As with *T. saginata* infection, there is usually only a single adult worm, and pathological changes are similar. There is a mild local inflammation of the intestinal mucosa as a result of attachment by the suckers and especially the hooklets. Because of its smaller size, however, *T. solium* is less likely to cause intestinal obstruction. Rare instances of intestinal perforation with secondary peritonitis and gallbladder infection have been reported.

Pathological changes due to cysticerci can be serious, depending on the tissue invaded and the number of cysticerci that become established. Damage results from pressure caused by encapsulated larvae on surrounding tissue, since the cysticerci produce occupying lesions. There may be no prodromal symptoms or only slight muscular pain and a mild fever during invasion of the muscles and subcutaneous tissues. In ocular cysticercosis, which accounts for about 20% of neurocysticercosis cases, there may be loss of vision. Invasion of the meninges, cortex, cerebral substance, and ventricles evokes tissue reactions leading to focal epileptic attacks or other motor or sensory involvement. The reasons for the predilection of cysticerci for the central nervous system are still obscure.

Clinical Manifestations

Since only a single adult worm is usually present in *T. solium* taeniasis, there are no symptoms of epigastric fullness. Patients are asymptomatic and become aware of the infection only when they find proglottids in their stools or on perianal skin. However, there may be vague abdominal discomfort, hunger pains, anorexia, and nervous disorders. Rare instances of intestinal perforation with secondary peritonitis and gallbladder infection have been reported. Eosinophilia, as high as 28%, and leukopenia can occur. The persistence for years of large, actively growing tapeworms does not appear to be consistent with the development of protective immunity.

Diagnosis

Diagnosis is based on stool examination and perianal scrapings. The sensitivity of both methods is increased by three examinations daily. Since *T. solium* eggs cannot be distinguished from those of *T. saginata*, specific diagnosis is based on the identification of the gravid proglottid, which has fewer lateral branches (7 to 12) on each side of the main uterine stem (Fig. 25.6) than that of *T. saginata*.

Figure 25.6 Gravid proglottid of *T. solium*. Note fewer lateral uterine branches than in *T. saginata* in Fig. 25.4.

If the scolex is obtained, it possesses hooklets in addition to four suckers. The development of DNA probes has made it possible to distinguish *T. solium* from *T. saginata* (44). Sensitivity of serologic tests varies, depending on the particular method and the clinical form of infection. Excretory-secretory antigens have been used with varying degrees of success for the serological detection of *T. solium* taeniasis (99). A highly specific immunoassay for diagnosis of cysticercosis, using the <12-kDa *T. hydatigena* antigen, is reported to be so specific as to distinguish between human clinical cases of cysticercosis and taeniasis. The assay is positive only with serum from cases of cysticercosis (45).

Treatment

Niclosamide is the drug of choice because of its effectiveness against the scolex and proliferative zone of strobila. A single dose of praziquantel is also highly effective. Mebendazole has also been reported to be effective. It is imperative to treat patients harboring the adult worm, since cysticercosis can occur from internal autoinfection. The major concerns in treating patients with the adult worm are to prevent vomiting and to ensure rapid expulsion of disintegrated proglottids from the intestine.

Prevention and Control

Prevention and eradication of *T. solium* adults in the intestine and of cysticerci in various tissues in humans and in pigs are a concern in areas of endemicity where economic, social, and sanitary conditions are substandard. Needed changes are monumental in scope. Without fundamental changes, the most scrupulous personal hygiene and eating habits will not prevent or eradicate infection in underdeveloped or developing countries. In developed countries, infection can be avoided by adherence to modern animal husbandry practices (39). The

best preventive measure in interrupting the transmission from humans to animals is to introduce and maintain proper sanitary facilities to dispose of contaminated feces. Even with proper toilet facilities, measures must be taken to make sure that sewage treatment is adequate to kill the eggs.

Since one infected individual can infect literally thousands of market hogs via contaminated feedlots, management personnel and employees must be educated regarding the parasite and means of avoiding transmission. Prospective employees for animal care should be examined for tapeworm infection before employment and semiannually during employment, and chemical toilets should be installed and properly maintained at convenient locations on slaughtering premises. The most important practice is to keep pigs in enclosures or indoors to prevent access to human fecal matter.

Meat inspection programs are partially effective for identifying and condemning infected carcasses. Meat inspection for measly pork is done in the United States only for interstate commerce, but inspection can miss lightly infected carcasses. Thus, the consumer should make sure that the meat is thoroughly cooked to an internal temperature of 56°C or higher or is frozen at −10°C for at least 14 days to render it incapable of transmitting cysticerci (69). Other means of inactivating cysticerci in pork, such as irradiation, have not been commercially applied. The best preventive measure is to avoid eating raw or uninspected pork.

ASIAN TAENIASIS

Asian taeniasis was first described in the aboriginal population in the mountainous regions of Taiwan (22). Initially, the etiologic agent was considered to be *T. saginata* because of morphologic similarities of the adult worms, although notable differences, including its shorter length, fewer number of proglottids, wider diameter of the scolex, and fewer number of testes in the mature proglottid, were found (22). Cloned ribosomal DNA fragments and sequence amplification by PCR showed that the Asian *Taenia* species is similar but genetically distinct from *T. saginata* (102). Also, sequence variation in the 28S rRNA and mitochondrial cytochrome c oxidase I genes and the cytochrome *c* oxidase I and ribosomal DNA internal transcribed spacer I PCR restriction fragment length polymorphism pattern in the Asian *Taenia* species have been used to identify it as an entity genetically distinct from other taeniid cestodes (10). Some authors now refer to this variant as a subspecies, *T. saginata asiatica* (25). Although further studies are needed, the fact that the Asian *Taenia* species is closely related to *T. saginata* suggests that it is unlikely to be an important cause of human cysticercosis (33).

Epidemiology

Asian *Taenia* infection has been found in several Asian countries, notably the mountainous regions of Taiwan, Cheju Island of Korea, and Samosir Island of Indonesia. It is estimated that public health costs in these areas alone exceed $35 million U.S. annually (25). In Taiwan, of 1,661 aboriginal cases of Asian taeniasis, the overall clinical infection rate was 76% among nine aboriginal tribes in 10 counties in mountainous areas (27). Pigs, cattle, goats, wild boars, and monkeys can serve as intermediate hosts. Of these, the wild boar appears to be the probable natural intermediate host in Taiwan. This may also be true in Indonesia, where people have become infected presumably from consuming pork (18). The infectivity for pigs, calves, goats, and monkeys has been confirmed experimentally, with the pig as the most favorable laboratory intermediate host (23).

The cysticercus is armed with tiny rostellar hooklets like that of *T. solium* and develops in a period shorter than that for either *T. saginata* or *T. solium*. Interestingly, the cysticercus of the Asian *Taenia* species is found mainly in the parenchyma of the liver (22), whereas cysticerci of *T. saginata* and *T. solium* are found primarily in the muscles of cattle and pigs, respectively. Thus, the custom of eating the viscera, especially the liver, of freshly killed animals appears to be a contributing factor in the infection.

Clinical Manifestations

Infected individuals can pass proglottids in their feces for 30 years or more, suggesting that the life span of this form of *Taenia* is very long (27). Common clinical manifestations include pruritis ani, nausea, abdominal pain, and dizziness. Abdominal pain is usually localized on the midline of the epigastrium or in the umbilical region and varies in intensity from a dull, aching, gnawing, or burning to intense colic-like, sharp pain. There may be headache, either an increased appetite or a lack of appetite, and diarrhea or constipation.

Treatment

Clinical trials in Taiwan have shown that a single dose (150 mg) of praziquantel (Biltricide) is highly effective against the Asian *Taenia* infection (24, 26).

Prevention and Control

The best preventive measure is to avoid eating raw, uninspected pork or other meat that contains the cysticerci. Since many cases of Asian taeniasis have been reported to result from the consumption of infected pork (29), the meat should be cooked thoroughly or frozen for at least 14 days at −10°C, to kill the cysticerci, the same procedures as are used to prevent *T. solium* taeniasis.

References

1. **Allan, J. C., M. Velasquez-Tohom, J. Garcia-Noval, R. Torres-Alvarez, P. Yurrita, C. Fletes, F. de Mata, H. Soto de Alfaro, and P. S. Craig.** 1996. Epidemiology of intestinal taeniasis in four rural Guatemalan communities. *Ann. Trop. Med. Parasitol.* **90:**157–165.
2. **Andrews, J. R. H., R. Ainsworth, and D. Abernethy.** 1994. *Trichinella pseudospiralis* in humans: description of a case and its treatment. *Trans. R. Soc. Trop. Med. Hyg.* **88:**200–203.
3. **Arriaga, C., L. Yépez-Mulia, N. Viveros, L. A. Adame, D. S. Zarlenga, J. R. Lichtenfels, E. Benitez, and M. G. Ortega-Pierres.** 1995. Detection of *Trichinella spiralis* muscle larvae in naturally infected horses. *J. Parasitol.* **81:**781–783.
4. **Bailey, T. M., and P. M. Schantz.** 1990. Trends in the incidence and transmission patterns of trichinosis in humans in the United States: comparison of the periods 1975–1981 and 1981–1986. *Rev. Infect. Dis.* **12:**5–11.
5. **Bellani, L., A. Mantovani, S. Pampiglione, and I. Fillippini.** 1978. Observations on an outbreak of human trichinellosis in northern Italy, p. 535–539. *In* C. W. Kim and Z. S. Pawlowski (ed.), *Trichinellosis.* University Press of New England, Hanover, N.H.
6. **Bessonov, A. S.** 1981. Changes in the epizootic and epidemic situation of trichinellosis in the USSR, p. 365–368. *In* C. W. Kim, E. J. Ruitenberg, and J. S. Tepperna (ed.), *Trichinellosis.* Reedbooks, Chertsey, England.
7. **Boireau, P., I. Vallee, T. Roman, C. Perret, L. Mingyuan, H. R. Gamble, and A. A. Gajadhar.** 2000. *Trichinella* in horses: a low frequency infection with high human risk. *Vet. Parasitol.* **93:**309–320.
8. **Bommer, W., H. Kaiser, W. Mannweiler, H. Mergerian, and G. Pottkamper.** 1985. An outbreak of trichinellosis in northern Germany caused by imported air-dried meat from Egypt, p. 314-317. *In* C. W. Kim (ed.), *Trichinellosis.* State University of New York Press, Albany, N.Y.
9. **Bouree, P., J. L. Leymarie, and C. Aube.** 1989. Epidemiological study of two outbreaks of trichinosis in France, due to horse meat, p. 382–386. *In* C. E. Tanner, A. R. Martinez-Fernandez, and F. Bolas-Fernandez (ed.), *Trichinellosis.* CSIC Press, Madrid, Spain.
10. **Bowles, I., and D. P. McManus.** 1994. Genetic characterization of the Asian *Taenia*, a newly described taeniid cestode of humans. *Am. J. Trop. Med. Hyg.* **50:**33–34.
11. **Britov, V. A., and S. N. Boev.** 1972. Taxonomic rank of various strains of *Trichinella* and their circulation in nature. *Vestn. Akad. Nauk SSSR* **28:**27–32.
12. **Campbell, W. C.** 1983. Historical introduction, p. 1–30. *In* W. C. Campbell (ed.), *Trichinella and Trichinosis.* Plenum, Publishing Corp., New York, N.Y.

13. **Carmichael, J.** 1952. Animal-man relationships in tropical disease in Africa. *Trans. R. Soc. Trop. Med. Hyg.* **46:**385–394.

14. **Centers for Disease Control.** 1976. Trichinosis surveillance annual summary—1975.

15. **Centers for Disease Control.** 1982. Common-source outbreaks of trichinosis—New York City; Rhode Island. *Morbid. Mortal. Weekly Rep.* **31:**161–164.

16. **Coordinating Group for Prevention and Treatment of Trichinosis, Harbin City.** 1981. *A Survey of Trichinosis Due to Eating Scalded Mutton.* WHO/HELM/82.5. (English abstr.) World Health Organization, Geneva, Switzerland.

17. **Dada, E. O., C. M. Adeiyongo, J. C. Anosike, V. O. Zaccheaus, S. N. Okoye, and E. E. Oto.** 1993. Observations on the epidemiology of human taeniasis amongst the Goemai tribe of northern Nigeria. *Appl. Parasitol.* **34:**251–257.

18. **Depary, A. A., and M. L. Kosman.** 1990. Taeniasis in Indonesia with special reference to Samosir Island, North Sumatra. *Southeast Asian J. Trop. Med. Public Health* **22**(Suppl.):239–241.

19. **Dissamarn, R., and P. Indrakamhang.** 1985. Trichinosis in Thailand during 1962–1983. *Int. J. Zoon.* **12:**257–266.

20. **Dupouy-Camet, J.** 2000. Trichinellosis: a worldwide zoonosis. *Vet. Parasitol.* **93:**191–200.

21. **Dworkin, M. S., H. R. Gamble, D. S. Zarlenga, and P. O. Tennican.** 1996. Outbreak of trichinellosis associated with eating cougar jerky. *J. Infect. Dis.* **174:**663–666.

22. **Fan, P. C.** 1988. Taiwan *Taenia* and taeniasis. *Parasitol. Today* **4:**86–88.

23. **Fan, P. C.** 1990. Asian *Taenia saginata*: species or strain? *Southeast Asian J. Trop. Med. Public Health* **22** (Suppl.):245–250.

24. **Fan, P. C.** 1995. Review of taeniasis in Asia. *Chung Hua Min Kuo Wei Sheng Wu Chi Mien I Hsueh Tsa Chih* **28:**79–94.

25. **Fan, P. C., and W. C. Chung.** 1997. Sociocultural factors and local customs related to taeniasis in east Asia. *Kao Hsiung I Hsueh Ko Hsueh Tsa Chih* **13:**647–652.

26. **Fan, P. C., W. C. Chung, C. H. Chan, Y. A. Chen, F. Y. Cheng, and M. C. Hsu.** 1986. Studies on taeniasis in Taiwan. V. Field trial on evaluation of therapeutic efficacy of mebendazole and praziquantel against taeniasis. *Southeast Asian J. Trop. Med. Public Health* **17:**82–90.

27. **Fan, P. C., W. C. Chung, C. Y. Lin, and C. H. Chan.** 1992. Clinical manifestations of taeniasis in Taiwan aborigines. *J. Helminthol.* **66:**118–123.

28. **Fan, P. C., W. C. Chung, C. Y. Lin, and C. C. Wu.** 1992. Studies of taeniasis in Taiwan. XIV. Current status of taeniasis among Yami aborigines on Lanyu Island, Taitung County, southeast Taiwan. *Kao Hsiung I Hsueh Ko Hsueh Tsa Chih* **8:**266.

29. **Fan, P. C., W. C. Chung, C. T. Soh, and M. L. Kosman.** 1992. Eating habits of east Asian people and transmission of taeniasis. *Acta Trop.* **50:**305–315.

30. **Faubert, G. M., J. C. Pechere, R. Delisle, H.-C. Smith, and Y. Brindle.** 1981. An outbreak of trichinellosis in Canada:

the enzyme linked immunosorbent assay (ELISA) and clinical findings, p. 269–273. *In* C. W. Kim, I. J. Ruitenberg, and J. S. Teppema (ed.), *Trichinellosis.* Reedbooks, Chertsey, England.

31. **Flisser, A., A. Reid, E. Garcia-Zepeda, and D. P. McManus.** 1988. Specific detection of *Taenia saginata* eggs by DNA hybridisation. *Lancet* **2:**1429–1430.

32. **Forbes, L. B.** 2000. The occurrence and ecology of *Trichinella* in marine mammals. *Vet. Parasitol.* **93:**321–334.

33. **Galan-Puchades, M. T., and M. V. Fuentes.** 2000. The Asian *Taenia* and the possibility of cysticercosis. *Korean J. Parasitol.* **38:**1–7.

34. **Gamble, H. R., and K. D. Murrell.** 1988. Trichinellosis, p. 1018–1024. *In* W. Balows (ed.), *Laboratory Diagnosis of Infectious Disease: Principles and Practice.* Springer-Verlag, New York, N.Y.

35. **Garcia, H. H., R. Araoz, R. H. Gilman, J. Valdez, A. E. Gonzalez, C. Gavidia, M. L. Bravo, and V. C. Tsang.** 1998. Increased prevalence of cysticercosis and taeniasis among professional fried pork vendors and the general population of a village in the Peruvian highlands. Cysticercosis Working Group in Peru. *Am. J. Trop. Med. Hyg.* **59:**902–905.

36. **Garcia, H. H., and O. H. Del Brutto.** 2000. *Taenia solium* cysticercosis. *Infect. Dis. Clin. North Am.* **14:**97–119.

37. **Garcia-Noval, J., J. C. Allan, C. Fletes, E. Moreno, F. DeMata, R. Torres-Alvarez, H. Soto de Alfaro, P. Yurrita, H. Higueros-Morales, F. Mencos, and P. S. Craig.** 1996. Epidemiology of *Taenia solium* taeniasis and cysticercosis in two rural Guatemalan communities. *Am. J. Trop. Med. Hyg.* **55:**282–289.

38. **Garkavi, B. L.** 1972. Species of *Trichinella* from wild carnivores. *Veterinariya* (Moscow) **49:**90–101.

39. **Gemmell, M. A.** 1987. A critical approach to the concepts of control and eradication of echinococcosis/hydatidosis and taeniasis/cysticercosis. *Int. J. Parasitol.* **17:**465–472.

40. **Gottstein, B., P. Deplazes, I. Tanner, and J. S. Skaggs.** 1991. Diagnostic identification of *Taenia saginata* with the polymerase chain reaction. *Trans. R. Soc. Trop. Med. Hyg.* **85:**248–249.

41. **Gottstein, B., and M. R. Mowatt.** 1991. Sequencing and characterization of an *Echinococcus multilocularis* DNA probe and its use in the polymerase chain reaction. *Mol. Biochem. Parasitol.* **44:**183–194.

42. **Grencis, R. K., and D. Wakelin.** 1985. Analysis of lymphocyte subsets involved in mediation of intestinal immunity to *Trichinella spiralis* in the mouse, p. 26–30. *In* C. W. Kim (ed.), *Trichinellosis.* State University of New York Press, Albany, N.Y.

43. **Haeghebaert, S., and E. Maillot.** 1999. Community-wide outbreak of trichinellosis, Tarn et Garonne, haute Garonne, Tarn districts, January–March 1998. *Réseau National de Santé Publique.*

44. **Harrison, L. J. S., J. Delgado, and R. M. E. Parkhouse.** 1990. Differential diagnosis of *Taenia saginata* and *Taenia solium* with DNA probes. *Parasitology* **100:**459–461.

45. **Hayunga, E. G., M. P. Sumner, M. L. Rhoads, K. D. Murrell, and R. S. Isenstein.** 1991. Development of a

serologic assay for cysticercosis, using an antigen isolated from *Taenia spp.* cyst fluid. *Am. J. Vet. Res.* 52:462–470.

46. **Heinz, H., and G. Macnab.** 1965. Cysticercosis in the Bantu of Southern Africa. *S. Afr. J. Med. Sci.* 30:19–31.

47. **Hou, H. W., et al.** 1983. *A Survey of an Outbreak of Trichinosis Caused by Eating Roasted Dog Meat.* WHO/HELM/84.15. (English abstr.) World Health Organization, Geneva, Switzerland.

48. **Ivanoska, D., K. Cuperlovic, H. R. Gamble, and K. D. Murrell.** 1989. Comparative efficacy of antigen and antibody detection tests for human trichinellosis. *J. Parasitol.* 75:38–41.

49. **Jongwutiwes S., N. Chantachum, P. Kraivichian, P. Siriyasatien, C. Putaporntip, A. Tamburrini G. La Rosa, C. Sreesunpasirikul, P. Yingyourd, and E. Pozio.** 1998. First outbreak of human trichinellosis caused by *Trichinella pseudospiralis. Clin. Infect. Dis.* 26:111–115.

50. **Kapel, C. M. O.** 2000. Host diversity and biological characteristics of the *Trichinella* genotypes and their effect on their transmission. *Vet. Parasitol.* 93:263–278.

51. **Kefenie, H., and G. Bero.** 1992. Trichinosis from wild boar meat in Gojjam, north-west Ethiopia. *Trop. Geogr. Med.* 44:278–280.

52. **Khamboonruang, C.** 1990. The present status of trichinellosis in Thailand. *Southeast Asian J. Trop. Med. Public Health* 22(Suppl.):312–315.

53. **Kim, C. W.** 1975. The diagnosis of parasitic diseases. *Prog. Clin. Pathol.* 6:267–288.

54. **Kim, C. W.** 1991. The significance of changing trends in trichinellosis. *Southeast Asian J. Trop. Med. Public Health* 22(Suppl.):316–320.

55. **Kim, C. W.** 1994. A decade of progress in trichinellosis, p. 35–47. *In* W. C. Campbell (ed.), *Trichinellosis.* Istituto Superiore di Sanita Press, Rome.

56. **Kocięcka, W.** 2000. Human disease, pathology, diagnosis and treatment. *Vet. Parasitol.* 93:265–283.

57. **Kotula, A. W.** 1983. Postslaughter control of *Trichinella spiralis. Food Technol.* 37:91–94.

58. **Kotula, A. W., K. D. Murrell, L. Acosta-Stein, L. Lamb, and L. Douglass.** 1983. Destruction of *Trichinella spiralis* during cooking. *J. Food Sci.* 48:765–768.

59. **Leighty, J. C.** 1983. Regulatory action to control *Trichinella spiralis. Food Technol.* 37:95–97.

60. **MacLean, J. D., J. Viallet, C. Law, and M. Staudt.** 1989. Trichinosis in the Canadian arctic: report of five outbreaks and a new clinical syndrome. *J. Infect. Dis.* 160:513–520.

61. **McAuley, J. B., M. K. Michelson, A. W. Hightower, S. Engeran, L. A. Wintermeyer, and P. M. Schantz.** 1992. A trichinosis outbreak among Southeast Asian refugees. *Am. J. Epidemiol.* 135:1404–1410.

62. **Miller, L. E., H. R. Gamble, and B. Lautner.** 1997. Use of risk factor evaluation and ELISA testing to certify herds for trichinae status in the United States, p. 687–690. *In* M. Ortega Pierres, H. R. Gamble, D. Wakelin, and F. van Knapen (ed.), *Proceedings of the 9th International Conference on Trichinellosis.* Centro de Investigacion y Estudios Avanzados del Instituto Politecnico National Mexico.

63. **Millet, N. B., G. D. Hart, T. A. Reyman, M. R. Zimmerman, and P. K. Lewin.** 1980. ROM I: mummification for the common people, p. 71–84. *In* A. Cockburn and E. Cockburn (ed.), *Mummies Disease and Ancient Cultures.* Cambridge University Press, Cambridge, United Kingdom.

64. **Moon, J. R.** 1976. Public health significance of zoonotic tapeworms in Korea. *Int. J. Zoon.* 3:1–18.

65. **Moorhead, A., P. E. Grunenwald, V. J. Diet, and P. M. Schantz.** 1999. Trichinellosis in the United States, 1991–1996: declining but not gone. *Am. J. Trop. Med. Hyg.* 60:66–69.

66. **Mukaratirwa, S., and C. M. Foggin.** 1999. Infectivity of *Trichinella* sp. isolated from *Crocodylus niloticus* to the indigenous Zimbabwean pig. *Int. J. Parasitol.* 29:1129–1131.

67. **Murrell, K. D.** 1985. Strategies for the control of trichinosis transmitted by pork. *Food Technol.* 39:65–68, 110–111.

68. **Murrell, K. D., and F. Bruschi.** 1994. Clinical trichinellosis. *Prog. Clin. Parasitol.* 4:117–150.

69. **Murrell, K. D., R. Fayer, and J. P. Dubey.** 1986. Parasitic organisms. *Adv. Meat Res.* 2:311–377.

70. **Murrell, K. D., J. R. Lichtenfels, D. S. Zarlenga, and E. Pozio.** 2000. The systematics of the genus *Trichinella* with a key to species. *Vet. Parasitol.* 93:293–307.

71. **Nelson, G. S., and J. Mikundi.** 1963. A strain of *Trichinella spiralis* from Kenya of low infectivity to rats and domestic pigs. *J. Helminthol.* 37:329–338.

72. **Ortega Pierres, M. G., C. Arriaga, and L. Yepez-Mulia.** 2000. Epidemiology of trichinellosis in Central and South America. *Vet. Parasitol.* 93:201–225.

73. **Pal, D. K., A. Carpio, and J. W. Sander.** 2000. Neurocysticercosis and epilepsy in developing countries. *J. Neurol. Neurosurg. Psychiatry* 68:137–143.

74. **Pawlowski, Z., and M. G. Schultz.** 1972. Taeniasis and cysticercosis (*Taenia saginata*). *Adv. Parasitol.* 10:269–343.

75. **Pawlowski, Z. S.** 1982. Taeniasis and cysticercosis, p. 313–348. *In* J. H. Steele (ed.), *Parasitic Zoonosis,* vol. 1. CRC I Handbook Series in Zoonoses. CRC Press, Boca Raton, Fla.

76. **Pawlowski, Z. S.** 1990. Cestodiasis, p. 490–504. *In* K. S. Warren and A. A. F. Mahmoud (ed.), *Tropical and Geographic Medicine,* 2nd ed. McGraw-Hill Book Co., New York, N.Y.

77. **Pozio, E.** 1998. Trichinellosis in the European Union: epidemiology, ecology and economic impact. *Parasitol. Today* 14:35–38.

78. **Pozio, E.** 2000. Factors affecting the flow among domestic, synanthropic and sylvatic cycles of *Trichinella. Vet. Parasitol.* 93:241–262.

79. **Pozio, E., and G. La Rosa.** 2000. *Trichinella murrelli* n. sp: etiological agent of sylvatic trichinellosis in temperate areas of North America. *J. Parasitol.* 86:134–139.

80. **Pozio, E., I. L. Owen, G. La Rosa, L. Sacchi, P. Rossi, and S. Corona.** 1999. *Trichinella papuae* n. sp. (Nematoda), a new non-encapsulated species from domestic and sylvatic

swine of Papua New Guinea. *Int. J. Parasitol.* 29:1825–1839.

81. Proctor, E. M. 1972. Identification of tapeworms. *S. Afr. Med. J.* 46:234–238.

82. Rhoads, M. L., and D. D. Murrell. 1988. Taeniasis and cysticercosis, p. 987–992. *In* W. Balows (ed.), *Laboratory Diagnosis of Infectious Disease: Principles and Practice.* Springer-Verlag, New York, N.Y.

83. Rhoads, M. L., K. D. Murrell, G. W. Dilling, M. M. Wong, and N. F. Baker. 1985. A potential diagnostic reagent for bovine cysticercosis. *J. Parasitol.* 71:779–787.

84. Roberts, D. 1986. The last trace. *Am. Photogr.* 16:64–68.

85. Rodriguez-Canul, R., A. Fraser, J. C. Allan, J. L. Dominguez-Alpizar, F. Argaez-Rodriguez, and P. S. Craig. 1999. Epidemiological study of *Taenia solium* taeniasis/cysticercosis in a rural village in Yucatan state, Mexico. *Ann. Trop. Med. Parasitol.* 93:57–67.

86. Ruitenberg, E. J., F. Van Knapen, and J. W. Weiss. 1979. Foodborne parasitic infections—a review. *Vet. Parasitol.* 5:1–10.

87. Ruiz-Perez, A., M. Santana-Ane, B. Villaverde-Ane, F. Bandera-Tirado, and N. Santana-Santos. 1995. The minimum dosage of praziquantel in the treatment of *Taenia saginata*, 1986–1993. *Rev. Cubana Med. Trop.* 47:219–220.

88. Sanchez, A. L., O. Gomez, P. Allebeck, H. Cosenza, and I. Ljungstrom. 1997. Epidemiological study of *Taenia solium* infections in a rural village in Honduras. *Ann. Trop. Med. Parasitol.* 91:163–171.

89. Sarti, E., P. M. Schantz, A. Plancarte, M. Wilson, O. I. Gutierrez, J. Aguilera, J. Roberts, and A. Flisser. 1994. Epidemiological investigation of *Taenia solium* taeniasis and cysticercosis in a rural village of Michoacan state, Mexico. *Trans. R. Soc. Trop. Med. Hyg.* 88:49–52.

90. Schantz, P. M. 1983. Trichinosis in the United States, 1947–1981. *Food Technol.* 37:83–86.

91. Schultz, M. G., J. A. Hermos, and J. H. Steele. 1970. Epidemiology of beef tapeworm infection in the United States. *Public Health Rep.* 85:169–176.

92. Semple, A. B., J. B. M. Davies, W. E. Kershaw, and C. A. St. Hill. 1954. An outbreak of trichinosis in Liverpool in 1953. *Br. Med. J.* 1:1002–1006.

93. Snyder, G. R., and K. D. Murrell. 1983. Bovine cysticercosis, p. 161–170. *In* G. Woods (ed.), *Practices in Veterinary Health and Preventive Medicine.* Iowa State University Press, Ames, Iowa.

94. Stehr-Green, J. K., and P. M. Schantz. 1986. Trichinosis in Southeast Asian refugees in the United States. *Am. J. Public Health* 76:1238–1239.

95. Sutisna, I. P., A. Fraser, I. N. Kapti, R. Rodriguez-Canul, D. Puta Widjana, P. S. Craig, and J. C. Allan. 1999. Community prevalence study of taeniasis and cysticercosis in Bali, Indonesia. *Trop. Med. Int. Health* 4:288–294.

96. Takahashi, Y., L. Mingyuan, and J. Waikagul. 2000. Epidemiology of trichinellosis in Asia and the Pacific rim. *Vet. Parasitol.* 93:227–239.

97. Viens, P., and P. Auger. 1981. Clinical and epidemiological aspects of trichinosis in Montreal, p. 275–277. *In* C. W. Kim, E. J. Ruitenberg, and J. S. Teppema (ed.), *Trichinellosis.* Reedbooks, Chertsey, England.

98. White, A. C., Jr. 2000. Neurocysticercosis: updates on epidemiology, pathogenesis, diagnosis, and management. *Annu. Rev. Med.* 51:187–206.

99. Wilkins, P. P., J. C. Allan, M. Verastegui, M. Acosta, A. G. Eason, H. H. Garcia, A. E. Gonzalez, R. H. Gilman, and V. C. Tsang. 1999. Development of a serologic assay to detect *Taenia solium* taeniasis. *Am. J. Trop. Med. Hyg.* 60:199–204.

100. Woods, G. T. 1986. Trichinosis surveillance in the United States and the swine industry's action plan, p. 190–198. *In* G. T. Woods (ed.), *Practices in Veterinary Public Health and Preventative Medicine in the United States.* Iowa State University Press, Ames, Iowa.

101. Wright, W. H., K. B. Kerr, and L. Jacobs. 1943. Studies on trichinosis. XV. Summary of the findings of *Trichinella spiralis* in a random sampling and other samplings of the population of the United States. *Public Health Rep.* 58:1293–1313.

102. Zarlenga, D. S., D. P. McManus, P. C. Fan, and J. H. Cross. 1991. Characterization and detection of a newly described Asian Taeniid using cloned ribosomal DNA fragments and sequence amplification by the polymerase chain reaction. *Exp. Parasitol.* 72:174–183.

103. Zarlenga, D. S., and K. D. Murrell. 1989. Molecular cloning of *Trichinella spiralis* ribosomal RNA genes: application as genetic markers for isolate classification, p. 35–40. In C. E. Tanner, A. R. Martinez-Fernandez, and F. Bolas-Fernandez (ed.), *Trichinellosis.* CSIC Press, Madrid, Spain.

Food Microbiology: Fundamentals and Frontiers, 2nd Ed.
Edited by M. P. Doyle et al.
© 2001 ASM Press, Washington, D.C.

Eugene G. Hayunga

26

Helminths Acquired from Finfish, Shellfish, and Other Food Sources

A variety of human helminthic infections can be acquired by eating food products from infected animals and plants, by the accidental ingestion of infected invertebrates in foodstuffs or drinking water, or as a result of inadvertent fecal contamination by humans or animals (Table 26.1). Effective prevention involves exploiting parasite vulnerabilities (34) and requires a sound understanding of parasite life cycles and modes of transmission. Unlike bacteria, the infective stages of helminths generally do not propagate. As a result, the critical control point for foodborne helminthiases is initial food preparation, not subsequent storage, reheating, or processing. Although these helminthic infections can readily be prevented, the reality is that safe water supplies, adequate sanitation, and reliable food handling simply do not exist for much of the world's population, a fact generally not appreciated by tourists "who explore tropical countries with a zeal undamped by any knowledge of preventive medicine" (44).

Foodborne helminths, although taxonomically diverse, share the common characteristic of requiring more than one host to complete their life cycles. The transmission of these helminths, termed biohelminths by Kisielewska (42), requires close behavioral contact between hosts. Typically, the definitive host of a biohelminth occupies the highest trophic level of the food chain. Prevention of biohelminth infections can be accomplished by avoiding the intermediate hosts or by adequate cooking of foods. In contrast, helminths with eggs or free-living stages that can survive a certain length of time in the external environment, termed geohelminths (42), are typically transmitted via contaminated water or foods and are best controlled by improved sanitation. In addition, any parasite capable of penetrating the skin can also penetrate the buccal epithelium.

HELMINTHS ACQUIRED FROM FINFISH AND SHELLFISH

Anisakis and Related Roundworms

Several related nematodes of the genera *Anisakis*, *Pseudoterranova*, and *Contracaecum* may be acquired by eating raw fish or squid in seafood dishes such as sushi, sashimi, ceviche, and lomi-lomi (61). The noninvasive form of anisakiasis is generally asymptomatic, resulting in "tingling throat syndrome" when worms are released from seafood following digestion and migrate up the esophagus into the pharynx, where they may be expectorated (62). The invasive form typically penetrates the mucosa or submucosa of the stomach or small intestine

Eugene G. Hayunga, National Institutes of Health, Bethesda, MD 20892.

Table 26.1 Sources of infection for some foodborne helminths of humans

Helminth	Beef or pork[a]	Finfish	Shellfish	Other invertebrates	Fruits or vegetables	Other food source	Fecal contamination
Alaria americana						X[b]	
Angiostrongylus sp.			X	X	X		
Anisakis sp.		X	X[c]				
Ascaris lumbricoides							X
Baylisascaris procyonis							X
Capillaria philippinensis		X					
Clonorchis sinensis		X	X				
Dicrocoelium dendriticum				X	X[d]		
Diphyllobothrium sp.		X					
Dipylidium caninum				X			
Dracunculus medinensis				X			
Echinococcus granulosus							X
Echinococcus multiocularis							X
Echinostomum sp.		X		X		X[e]	
Enterobius vermicularis							X
Eustrongylides sp.		X					
Fasciola hepatica					X	X[f]	
Fasciolopsis buski					X		
Gnathostoma spinigerum	X	X		X		X[g]	
Heterophyes heterophyes		X					
Hymenolepis diminuta				X			
Hymenolepis nana							X
Ligula intestinalis		X					
Macracanthorhynchus hirudinaceus				X			
Metagonimus yokogawai		X					
Moniliformis moniliformis				X			
Multiceps multiceps							X
Nanophyetus salmincola		X					
Nybelinia surmenicola			X[c]				
Opisthorchis sp.		X	X				
Paragonimus westermani			X			X[b]	
Phaeneropsolos bonnei				X			
Philometra sp.		X					
Prosthodendrium molenkampi				X			
Spirometra sp.	X	X		X		X[i]	
Strongyloides sp.						X[j]	
Taenia saginata	X						
Taenia solium	X						X
Toxocara canis							X
Trichinella spiralis	X					X[k]	
Trichostrongylus sp.							X
Trichuris trichiura							X

[a] Described in chapter 25.
[b] Frog, raccoon, and opossum.
[c] Squid.
[d] Acquired by ingestion of ants on unwashed herbs and vegetables.
[e] Frog and tadpole.
[f] The condition halzoun occurs when adult worms in uncooked sheep liver attach to the pharynx.
[g] Pork, chicken, duck, frog, eel, snake, and rat.
[h] Wild boar.
[i] Acquired by ingestion of procercoids in copepods and plerocercoids in frogs, tadpoles, lizards, snakes, birds, and mammals; infection has also been reported from eating undercooked pork.
[j] Transmammary infection has been reported.
[k] Reported from a variety of animals including bear, walrus, and horse.

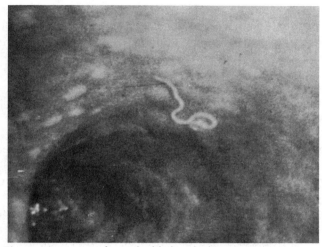

Figure 26.1 *Anisakis* embedded in the human gastric mucosa as visualized by gastroscopy. (Photograph contributed by Tomoo Oshima; illustration courtesy of the Armed Forces Institute of Pathology, AFIP 76-2118.)

(Fig. 26.1), resulting in epigastric pain, nausea, vomiting, and diarrhea, usually 12 h after eating the infected seafood. Chronic anisakiasis may mimic peptic ulcer, appendicitis, enteritis, Crohn's disease, or gastric carcinoma. The only effective treatment is surgical removal of the worms.

The life cycles of these helminths are not completely known. The adult worms are intestinal parasites of dolphins, whales, seals, and sea lions. Thus, the important reservoir hosts for this disease include several protected species, and there is evidence that the prevalence of such fishborne helminthiases is closely related to the local geographic distribution of the reservoir hosts (17, 20). Eggs passed in the feces of marine mammals embryonate in seawater and develop into larvae that are eaten by krill. The infected crustaceans are next eaten by fish or squid, and the larvae develop further. The life cycle is completed in marine mammals, but when fish or squid are eaten by the unsuitable human host, the parasites do not develop further or reproduce. Anisakidae larvae have been reported from rockfish, herring, cod, halibut, mackerel, and salmon. It has been estimated that 90% of the cod fillets sold in the Washington, D.C., area may be infected (23).

The prominent reddish-brown larvae of *Pseudoterranova* are readily visible in contrast to the whitish fish tissue, but the smaller, lighter-colored *Anisakis* larvae are more difficult to detect (61). Comparison of salmon steaks with salmon fillets indicates predilection sites for the larvae and suggests that certain cuts may pose a lesser risk for infection (19). In some fishes, most of the juvenile larvae are found in the viscera (16), suggesting that immediate evisceration would prevent their subsequent migration into the musculature. However, only thorough cooking or prolonged freezing will kill the parasite and completely eliminate the risk of infection. The U.S. Food and Drug Administration (FDA) recommends that all finfish and shellfish intended for raw, partly cooked, or marinated consumption be blast frozen to $-35°C$ or below for 15 h or regularly frozen to $-20°C$ or below for 7 days (67). Cold smoking and most methods of brining fish are not reliable preventive measures (63).

Most human infections of the so-called sushi parasite have been reported from Japan and The Netherlands. In Japan, hypochlorhydria or achlorhydria has been found in more than half of anisakiasis patients and may predispose them to infection (23). In the United States, the number of documented cases was reported to be between 25 and 50 in 1989 (63), and fewer than 10 cases are diagnosed annually (67). Considering the increasing popularity of raw seafood dishes and the proliferation of thousands of sushi restaurants in the last decade, the low rate of reported infection is remarkable. The rare occurrence of restaurant-acquired cases has been attributed to the training and expertise of professional sushi chefs (63), while the majority of human infections acquired in the United States have been associated with dishes prepared at home. The source of commercial fish harvesting with regard to the distribution of reservoir hosts may also contribute to the epidemiology of this disease, as geographic variation has been reported in infection rates of fish from different localities (20).

Capillaria

Capillaria philippinensis, a nematode found in freshwater fishes, causes a severe and potentially fatal infection in humans (9). In 1967 and 1968, the disease reached epidemic proportions in the Philippines, with more than 1,000 confirmed cases and over 100 deaths (11); cases have also been reported from Thailand. The infection appears to be acquired by eating raw fish that contain infective larvae. The freshwater fishes *Ambassis miops*, *Eleotris melanosoma*, and *Hypseleotris bipartita* have been implicated in experimental infections, but the life cycle of *C. philippinensis* has not been fully elucidated, nor have natural reservoir hosts been identified. Massive infections are believed to result from internal autoinfection, but it is not known whether the disease can be acquired by the ingestion of *C. philippinensis* eggs.

Both larvae and adult worms may be found embedded in the intestine (Fig. 26.2). The long, slender adult worms are dioecious. Males are approximately 2 to 3 mm long and 20 to 30 μm in diameter, while the larger females

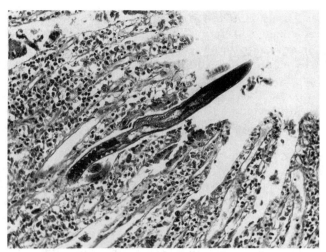

Figure 26.2 Transverse section of larval *Capillaria philippinensis* embedded in human intestinal glands (×95). (Illustration courtesy of the Armed Forces Institute of Pathology, AFIP 69-1066.)

are approximately 3 to 5 mm long and 30 to 50 µm in diameter. *C. philippinensis* eggs are distinctive, thick-shelled, capsule-shaped structures with bipolar plugs, and they measure approximately 45 by 21 µm, differing in appearance from *Trichuris* eggs of the same size that also have bipolar plugs. Diagnosis is made by identifying eggs in the feces. Symptoms of capillariasis include borborygmus, abdominal pain, nausea, vomiting, diarrhea, and anorexia during the acute phases. If untreated, intestinal malabsorption and intractable diarrhea lead to cachexia and possibly death. Treatment consists of a regimen of mebendazole with electrolyte and protein supplementation.

Gnathostoma

Roundworms of the genus *Gnathostoma* reside in the stomach wall of a variety of carnivorous mammals. The life cycle of this parasite typically involves two intermediate hosts, a copepod and a freshwater fish; however, a variety of animals may serve as paratenic or transport hosts. Unable to mature in human hosts, the parasite wanders aimlessly through the tissues, causing a severe larva migrans that may persist for years. Cases have been reported throughout the world, with the greatest prevalence in Thailand.

Gnathostoma spinigerum has been reported in tigers, leopards, lions, domestic cats, minks, and dogs. The stout, reddish female worms range in length from 25 to 54 mm and are characterized by spines covering the anterior half of the body and by a prominent cephalic bulb covered by rows of sharp hooklets. Male worms

are about half as long as females. Unembryonated eggs are passed in the feces and hatch in about a week, releasing motile first-stage larvae. When eaten by a copepod, the larva penetrates into the hemocoel and develops further. Infected copepods are next ingested by fish, frogs, or snakes, and the parasite continues its development into a third-stage larva, which is the infective stage for the final mammalian host.

Human infection typically results from eating raw, marinated, or poorly cooked freshwater fish, as once happened at a diplomatic dinner for visiting dignitaries (63). Recent cases in Mexico have been attributed to substituting marinated freshwater tilapia for the more expensive marine fish normally used in ceviche (60). In addition, infection can be acquired from pork, chicken, duck, frog, eel, snake, or rat, or by the accidental ingestion of infected copepods in drinking water. Larvae may also penetrate the skin during food handling (13). Larvae can be killed by cooking or by immersion in strong vinegar for 5 h or longer; immersion in lime juice or chilling at 4°C for 1 month is not effective (4). Raw foods, particularly fish and chicken, should be avoided in areas of endemicity, and drinking water should be filtered before consumption.

Nausea, abdominal pain, and vomiting usually develop between 24 and 48 h following the ingestion of infected meat or fish. Symptoms of larva migrans or creeping eruption include pruritus, urticaria, tenderness, and painful subcutaneous swelling. Invasion of the central nervous system may result in meningitis and neuropathy. Diagnosis is difficult and chemotherapy is of questionable value. Recently, albendazole has been reported as an alternative to surgical removal of the larvae (1).

Diphyllobothrium latum

The broad tapeworm, *D. latum*, occurs in northern temperate regions of the world where raw or undercooked freshwater fishes are eaten; infection is closely related to dietary and cultural practices in food preparation (21). Infective larvae may be found in whitefish, trout, pike, and salmon. Cases have been reported throughout Europe, particularly in the Baltic countries, and from the Great Lakes region of the United States, Canada, Japan, South America, and Australia. *D. latum* was introduced into the United States by European immigrants in the middle of the 19th century (17). Although it is the largest of the human tapeworms, measuring up to 9 m in length, infections may be mild or asymptomatic. Nausea, abdominal pain, diarrhea, and weakness are common manifestations. *D. latum* may also cause pernicious anemia and vitamin B_{12} deficiency because the worm is highly efficient in competing with its host for available vitamins.

D. latum eggs shed in feces require approximately 12 days to embryonate, at which time the ciliated coracidium exits the egg through the operculum. When eaten by *Cyclops* or *Diaptomus*, the coracidium penetrates the body cavity of the freshwater crustacean and develops into the procercoid stage of the parasite. When small fish such as minnows eat infected crustaceans, the procercoid penetrates into the viscera or muscles of the fish and develops into another unsegmented stage called the plerocercoid. The carnivorous fish that eat these fish are termed paratenic hosts because, although the plerocercoid penetrates into their viscera or muscles, it does not develop beyond this stage. Further development from plerocercoid into adult worm requires ingestion by the human definitive host.

Diagnosis is made on the basis of eggs in the feces. The distinctive *D. latum* egg is ovoid, approximately 45 by 70 μm, with a prominent operculum at one end and a characteristic knob at the abopercular end; proglottids may occasionally be found in the feces. Infection is best prevented by adequate cooking or freezing of fish before consumption. Improved sanitation measures can also help reduce prevalence by interrupting the parasite's life cycle. Praziquantel is the recommended anthelminthic (1). *Diphyllobothrium dendriticum* and *Ligula intestinalis*, tapeworms of piscivorous birds, and *Diphyllobothrium pacificum*, a tapeworm of seals, have also been reported in humans.

Clonorchis sinensis

Although the correct scientific name for the Chinese liver fluke is *Opisthorchis sinensis*, its original name, *Clonorchis sinensis*, remains more commonly accepted. Widespread throughout Asia, *C. sinensis* has been reported from Hong Kong, Japan, Korea, Taiwan, Vietnam, and large areas of the People's Republic of China. The parasite is typically acquired by eating infected freshwater fishes; the infective stage of *C. sinensis* has also been reported in crayfish (24). Reservoir hosts include dogs, cats, foxes, pigs, rats, mink, badgers, and tigers.

Adult worms, measuring approximately 1.2 to 2.4 cm in length and 0.3 to 0.5 cm in width, reside in the bile duct. When eggs passed in the feces are eaten by *Parafossarulus manchouricus* or other hydrobiid snails such as *Bulimus*, *Semisulcospira*, *Alocinma*, or *Melanoides*, the miracidium is released in the digestive tract and penetrates into the hemocoel, where it develops into first the sporocyst and then the redia stage. The redia gives rise to free-swimming cercariae that leave the snail and penetrate the second intermediate host, a cyprinid fish, where they encyst as metacercariae in the gills, fins, or muscles, or under the skin. When the definitive host eats an infected fish, metacercariae excyst in the duodenum, migrate into the bile duct, and develop into adult worms.

C. sinensis may live in the human host for as long as 25 to 30 years, and massive infections have been reported with as many as 500 to 1,000 parasites. The severity of symptoms is related to the intensity and duration of infection. Diarrhea, epigastric pain, and anorexia are typical manifestations of acute clonorchiasis. The adult worm produces localized tissue damage that may result in hyperplasia or metaplasia of the bile duct epithelium, duct thickening, fibrosis, bilary stasis, and secondary bacterial infection. Pancreatitis may occur when worms enter the pancreatic duct. An association between cholangiocarcinoma and *C. sinensis* infection has also been reported (37). Diagnosis is made by identifying eggs in the feces. The operculate *C. sinensis* eggs measure approximately 27 by 16 μm and are characterized by distinctive opercular shoulders and a small spinelike process at the abopercular end. Infection is best prevented by avoiding raw, undercooked, or pickled finfish and shellfish. Snail control and improved sanitation can also help reduce transmission. Praziquantel and albendazole are effective anthelminthics.

Two closely related bile duct flukes of dogs and cats may also be acquired by eating uncooked cyprinid fishes. *Opisthorchis felineus* occurs throughout Eastern Europe and portions of the former Soviet Union, while *Opisthorchis viverrini* has been reported from Southeast Asia. The clinical picture for human infection by these species is very similar to that seen for clonorchiasis.

The Human Lung Fluke, *Paragonimus westermani*

P. westermani is the best-known and most widely distributed lung fluke, although other species of this genus may parasitize humans (73). *P. westermani* infection is common throughout the Far East and occurs to a lesser extent in parts of Africa and the Indian subcontinent. It has been reported from Japan, Korea, the People's Republic of China, Manchuria, Taiwan, the Philippines, Indonesia, the Solomon Islands, both American and British Samoa, India, Sri Lanka, Cameroon, and Nigeria and in American troops in the South Pacific during World War II. It is a common parasite of mink in Canada and the eastern United States. Other reservoir hosts include a variety of animals that may eat crustaceans. Infection has been reported from dogs, cats, tigers, and cattle.

The reddish-brown, thick-bodied flatworms are found encapsulated in cystic structures adjacent to the bronchi or bronchioles. They measure approximately 0.8 to 1.6 cm in length by 0.4 to 0.8 cm in width and are 3.0 to 0.5 cm thick. Eggs are released by the parasite into the bronchioles, where they may be expectorated or swallowed and passed in the feces. Eggs require several weeks in an aqueous environment for development. Upon hatching, the free-swimming miracidium penetrates a snail of the genus *Brotia, Semisulcospira, Tarebia,* or *Thiara,* where it undergoes further development into a sporocyst, then two generations of rediae. Approximately 11 weeks after the snail is infected, cercariae are shed and then penetrate any of a variety of freshwater crabs and crayfish, where they become encysted in the muscles, gills, and other organs as metacercariae. Important shellfish hosts include the freshwater and brackish water crabs *Eriocheir, Potamon,* and *Sundathelphusa,* and the crayfish *Procambarus.*

Humans become infected by eating raw or improperly cooked freshwater crabs or crayfish. Dishes such as raw crayfish salad, jumping salad (live shrimp), drunken crab (live crabs in wine), and crayfish curd are popular throughout the Orient. Pickling does not kill the parasite. The infective metacercariae can be killed by boiling the crabs for several minutes until the meat has congealed and turned opaque (40). Infection may also be acquired from shellfish juices used in food dishes or folk remedies, from food prepared using contaminated utensils or chopping blocks, or by drinking water contaminated with metacercariae released from dead or injured crustaceans.

Migration of parasites through host tissues produces localized hemorrhage and infiltration of lymphocytes. Pulmonary symptoms include dyspnea, chronic cough, chest pain, night sweats, hemoptysis, and persistent rales. The severity of symptoms appears to be related to the number of parasites present. Pleural effusion and fibrosis may occur in long-standing infections, although there is also evidence that lesions may resolve without treatment. Neurological complications result from migration of the parasite into the spinal cord or brain. *P. westermani* has also been found in the intestinal wall, peritoneum, pleural cavity, and testes. *Paragonimus szechuanensis* and *Paragonimus hueitungensis,* two species that apparently are incapable of developing in the lungs, cause cutaneous larva migrans, characterized by eosinophilia, anemia, and low-grade fever. Diagnosis is made on the basis of eggs in expectorant or feces. The characteristic golden-brown eggs are 80 to 120 μm long by 48 to 60 μm wide. Chest X rays may show nodular shadows or calcified spots. Praziquantel and bithionol are effective anthelminthics.

Other Helminths Associated with Seafood

Heterophyes heterophyes, an intestinal fluke acquired from the mullet, has been found in Egypt, in Israel, and

throughout Asia. The parasite may cause nausea, diarrhea, and abdominal pain, but light infections are often asymptomatic. There have been reports of fatal myocarditis and neurological complications when helminth eggs penetrate the intestine and enter the circulatory system. The closely related trematode, *Metagonimus yokagawai,* acquired from salmonid fishes, has been reported from Asia, the Balkans, Israel, Spain, and portions of the former Soviet Union. Fish-eating mammals and birds, such as dogs, cats, and pelicans, serve as reservoir hosts. Infection by species of another trematode, *Echinostomum,* has been reported in Japan and has been attributed to eating raw freshwater fish, in particular sashimi (32). Diagnosis of these parasites is made by identifying eggs passed in the feces. Intestinal symptoms are mild and depend upon the number of worms present. Praziquantel is an effective anthelminthic.

Salmon poisoning, a severe and frequently fatal disease of dogs in the Pacific Northwest, is associated with the intestinal trematode *Nanophyetus* (=*Troglotrema*) *salmincola.* The disease occurs in dogs because *N. salmincola* serves as a vector for the rickettsial pathogen *Neorickettsia helmintheca.* Human *N. salmincola* infection, which is characterized by nausea, diarrhea, and intestinal discomfort, may be acquired by eating uncooked salmonid fishes and has also been attributed to handling freshly killed coho salmon (33).

The nematode *Eustrongylides* is a common parasite of piscivorous birds. In addition to fish, reptiles and amphibians may also serve as intermediate hosts and thus contain infective larvae. An infection in New York City was attributed to eating raw fish prepared at home (71). Cases in Maryland and New Jersey were reported in fishermen who had eaten live bait (8, 26). As with *Anisakis* infections, surgical removal of worms is the only effective treatment.

Philometra, a nematode closely related to *Dracunculus,* was identified in a fisherman in Hawaii who filleted a carangid fish (18). Infection by the tapeworm *Nybelinia surmenicola* has been attributed to eating raw squid (41). Sparganosis may be transmitted by a variety of animals, including fish. Although typically associated with snails, *Angiostrongylus* may also be acquired from freshwater prawns and land crabs.

HELMINTHS ACQUIRED FROM VEGETATION

The Sheep Liver Fluke, *Fasciola hepatica*
Sheep liver rot, caused by the digenetic trematode *F. hepatica,* was recognized as early as the 14th century (15). It was the first trematode for which a complete life cycle

was elucidated (45). Sheep, cattle, and other herbivores acquire the infection by eating metacercariae encysted on aquatic plants. *F. hepatica* has been found in goats, horses, deer, rabbits, camels, vicuna, swine, dogs, and squirrels. Human infection is prevalent in sheep-raising areas throughout the world. In the United States, human fascioliasis has been reported from California (51) and Puerto Rico (35).

F. hepatica is a large fluke, measuring approximately 3 cm in length by 1.5 cm in width, and is readily identified by its characteristic "cephalic zone," a distinct conical projection at the anterior end. Adult worms reside in the biliary passages and gallbladder. Eggs passed with feces into the water require 9 to 15 days to mature, but they may remain viable for several months in soil if they stay moist. Upon hatching, the free-swimming miracidium penetrates a lymnaeid snail of the genus *Lymnaea*, *Succinea*, *Fossaria*, or *Practicolella*, where it develops into a sporocyst and then two generations of rediae. Cercariae are shed from the snail, swim freely, and then attach to aquatic vegetation, where they encyst as metacercariae. Encysted metacercariae are susceptible to drying but can survive over winter (54).

Humans become infected by eating infested freshwater plants or free metacercariae in drinking water. Human cases are frequently traced to watercress *Nasturtium officinalis* (39). A pharyngeal form of disease, called halzoun, may occur following the ingestion of raw liver from infected animals, when worms present in liver attach to the pharynx. Transmission can be controlled by the use of molluscicides, by draining ponds, and by protecting crops and water supplies from contact with livestock.

Some inflammation is associated with migration of the parasite, but mild infections are often asymptomatic. Tissue destruction occurs when worms penetrate the liver. In sheep, liver rot causes massive damage. There may be mechanical obstruction of bile ducts, hyperplasia of biliary epithelium, and proliferation of the connective tissue. Worms may erode the walls of the bile ducts and invade the liver parenchyma. Secondary bacterial infection and portal cirrhosis have been reported, but liver calcification appears rare. Pain, bleeding, and edema of the face and neck are associated with halzoun.

Diagnosis is made on the basis of eggs in the feces. The relatively large, operculate *F. hepatica* eggs are approximately 130 to 150 μm long by 63 to 90 μm wide. Ingestion of liver from infected sheep or cattle may result in spurious infection when eggs present in the food are passed in feces. Such false-positive findings can be ruled out by subsequent stool examinations. A lengthy regimen of bithionol has been the treatment of choice,

although triclabendazole offers the advantage of a single dosage.

The Giant Intestinal Fluke, *Fasciolopsis buski*

F. buski is the largest trematode parasite of humans. Endemic throughout Asia, it has been reported from China, Taiwan, Korea, Vietnam, Cambodia, Laos, Myanmar, Thailand, Borneo, Sumatra, Malaysia, Indonesia, India, and Bangladesh. The pig is an important reservoir host. Infection has also been reported in dogs and rabbits.

Although only 0.8 to 3.0 mm thick, *F. buski* can grow to 7.5 cm in length by 2 cm in width. Adult worms attach to the bowel walls, primarily along the duodenum and jejunum. Eggs passed in the feces are unembryonated and require 3 to 7 weeks in freshwater to develop. Upon hatching, the free-swimming miracidium penetrates a snail of the genus *Segmentina* or *Hippeutis*, where it develops into a sporocyst and then two generations of rediae. Approximately 4 to 7 weeks after the snail is infected, cercariae are shed into the water and then attach to vegetation, where they encyst as metacercariae.

Humans become infected by eating infested water chestnuts, bamboo, caltrop, or lotus. Individuals may also acquire the parasite by peeling the hulls of plants with their teeth. The metacercariae excyst in the small intestine, attach to the mucosa, and develop into adult worms in about 3 months. Drying or cooking the plants before eating kills the metacercariae (5). Immersion of vegetables in boiling water for a few seconds, or even peeling and washing them in clear water, is sufficient to preclude infection (4). Poor sanitation and the use of human and swine feces as fertilizer are major factors in disease transmission (59).

Adult worms feed not only on intestinal contents but also on intestinal epithelium, leading to local ulceration and hemorrhage. Nausea, abdominal pain, diarrhea, and hunger pangs are common. In heavy infections, stools are profuse and light yellow in color, suggestive of malabsorption; intestinal obstruction and ascites have been reported. Diagnosis is made on the basis of eggs in the feces. The yellow-brown eggs have a small operculum and are approximately 130 to 140 μm long by 80 to 85 μm wide. Adult worms are seen in feces only following chemotherapy or purgation. Praziquantel is the anthelminthic of choice.

Other Helminths Associated with Vegetation

Fresh vegetables grown in areas where nightsoil is used as fertilizer are frequently contaminated and thus may facilitate transmission of any of a number of geohelminths. Human infection by *Dicrocoelium* is explained by the accidental ingestion of ants on vegetation, and

Angiostrongylus costaricensis is thought to be acquired by eating raw fruits and vegetables on which snails have left larvae in mucous deposits, or by accidentally ingesting infected snails on unwashed vegetation. *Trichostrongylus* and *Echinococcus granulosus* infections have also been attributed to contaminated vegetation (4).

HELMINTHS ACQUIRED FROM INVERTEBRATES IN DRINKING WATER

The Guinea Worm, *Dracunculus medinensis*

The long, threadlike roundworm *Dracunculus* has plagued humans since antiquity. It is believed to be the "fiery serpent" described in the Bible (52) and is symbolically depicted in the caduceus, the insignia of the medical profession. Within the last decade, prevalence has declined impressively as a result of an aggressive and highly effective worldwide eradication campaign. Dracunculiasis now occurs primarily in the Sudan, and to lesser degree among rural populations in parts of India, in Yemen, and in several African countries (36). Although rarely fatal, this parasite causes considerable discomfort and disability, particularly in areas where poverty is severe. The female nematode is almost a meter in length but less than 2 mm thick; males are inconspicuous and only 2 cm long. Worms develop to maturity in the body cavity or deep connective tissues. Females then migrate to the subcutaneous tissues, become gravid, and stimulate the formation of a blister that eventually ruptures to expose part of the worm. The uterus accounts for most of the parasite's volume and is distended with as many as 1 to 3 million larvae. Upon contact with fresh water, the worm bursts to release larvae. Larvae are ingested by copepods of the genera *Cyclops*, *Mesocyclops*, or *Thermocyclops* and mature into the infective form in approximately 3 weeks.

Humans become infected by swallowing infected copepods present in drinking water (Fig. 26.3). The larvae penetrate the digestive tract and take about a year to reach maturity. Transmission is clearly related to poverty, the quality of drinking water, and water contact by infected individuals. A measure as simple as sieving drinking water through a piece of cloth will remove copepods and prevent infection. The provision of safe drinking water supplies in a village is usually followed by disappearance of the disease (49, 56). *Dracunculus* infections have been reported in dogs in China and the former Soviet Union, but it is not clear whether canines play any significant role as reservoir hosts.

Subcutaneous migration of the female nematode results in localized erythema, tenderness, and pruritus. Other symptoms include nausea, vomiting, diarrhea,

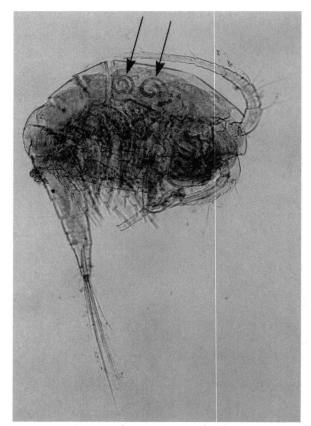

Figure 26.3 Arrows show *Dracunculus* larvae within the body cavity of the intermediate host, *Cyclops* (×60). (Specimen contributed by E. L. Schiller, Johns Hopkins University School of Public Health, Baltimore, Md.; illustration courtesy of the Armed Forces Institute of Pathology, AFIP 68-4629.)

generalized urticaria, and asthma. A papule forms under the skin, where the anterior end of the parasite lies, and becomes vesicular, then ulcerates, exposing the worm. There may be a painful local reaction when larvae are discharged. Secondary bacterial infection may occur when the parasite is resorbed and is a frequent complication when the worm breaks following unsuccessful attempts at removal. On rare occasions, nervous system involvement has been reported, resulting in paraplegia (47), quadriplegia (22), and death (57).

The traditional treatment of winding the worm around a small stick and slowly extruding can still be effective provided that asepsis is maintained. Surgical removal of *Dracunculus* has been practiced in India and Pakistan. Although not curative, metronidazole decreases inflammation and facilitates removal of the worm, but this compound may be mutagenic in the dosage recommended; mebendazole has been reported to kill the parasite directly (1).

Other Helminths Acquired from Copepods

There is evidence that the tissue-dwelling nematode *G. spinigerum* may be acquired directly from infected copepods (12). Sparganosis, although typically acquired from a variety of vertebrate hosts, may also result from ingesting copepods infected with procercoids.

HELMINTHS ACQUIRED FROM OTHER INVERTEBRATES

Snails

The rodent lungworm, *Angiostrongylus cantonensis*, is a slender roundworm about 25 mm in length that typically resides in the pulmonary arteries of rodents. Human cases have been reported from Taiwan, Thailand, India, the Philippines, Hawaii, several Pacific islands, Cuba, and the Ivory Coast. An autochthonous *A. cantonensis* infection was recently found in New Orleans in a child who had consumed a raw snail (50). Migration of the parasite through the brain causes meningitis, and fatal cases have been reported.

The adult nematodes lay eggs that hatch in the lungs. First-stage larvae migrate to the trachea, are swallowed, and pass in the feces, where they are ingested by any of a variety of slugs, land snails, or planarians. They then develop into infective third-stage larvae in about 2 weeks. When rodents eat infected molluscs, the larvae migrate to the brain, where they develop into adults in about 4 weeks, then the pulmonary arteries, where they begin to lay eggs after an additional 2 weeks.

Humans become infected by eating raw or undercooked molluscs such as the freshwater snail *Pila* or the giant African land snail *Achatina fulica*. Freshwater prawns, which are frequently eaten raw in Thailand, Vietnam, and Tahiti, have been implicated in human infection (58). Natural infections have also been found in the land crab *Cardisoma hirtipes* and the coconut crab *Birgus latro* (2). Shrimp, crabs, and frogs may serve as paratenic hosts. Infection can also be acquired by the accidental ingestion of slugs on lettuce or of parasite larvae left by snails in mucous deposits on fruits and vegetables; contaminated drinking water may be another source of infection. Transmission can be prevented by thorough cooking and appropriate attention to food-handling practices, particularly the careful washing of fruits and vegetables, and washing of hands after exposure to molluscs while gardening.

Angiostrongyliasis usually presents as a self-limiting meningitis during the migratory phase. Symptoms of meningitis or meningoencephalitis are characterized by abrupt onset and include headache, stiff neck, and sensorial changes. Cerebrospinal fluid may contain 100 to 2,000 white blood cells/mm^3, and there is marked eosinophilia. Nausea, vomiting, fever, abdominal pain, malaise, and constipation have also been reported (72). While larvae may be found in cerebrospinal fluid, diagnosis is often presumptive, based upon eosinophilia and symptoms of meningitis in an area of endemicity. Analgesics or corticosteroids may provide symptomatic relief. Mebendazole is now the recommended anthelminthic (1). Some have cautioned that anthelminthics may not be necessary because the disease is self-limiting, and killing of parasites while in the central nervous system could result in even greater complications from an increased inflammatory reaction (4).

Angiostrongylus costaricensis is a closely related roundworm found in the mesenteric arteries of the cotton rat *Sigmodon hispidus*. Human abdominal angiostrongyliasis has been reported from Costa Rica, Honduras, Panama, Mexico, Brazil, and Venezuela. The veronicellid slug, *Vaginulus plebius*, has been implicated in the transmission of *A. costaricensis*. Human infection is acquired from accidental ingestion of slugs or by contact with contaminated fruits, vegetables, or grass.

Several species of the intestinal trematode *Echinostomum* may be acquired by eating metacercariae encysted in snails. Human echinostomiasis has been reported from China, Taiwan, India, Korea, Malaysia, Indonesia, and the Philippines (32). Infections are generally mild and often asymptomatic. Adult worms are found in a variety of domestic animals and birds, and the Norway rat is believed to be an important reservoir host. Diagnosis is based on identification of unembryonated, operculate eggs, measuring approximately 83 to 116 μm in length by 58 to 69 μm in width, in stool specimens. Praziquantel appears to be an effective anthelminthic.

Ants

Dicrocoelium dendriticum is a trematode found in the biliary passages of sheep, deer, and other herbivores. The intermediate hosts are land snails and ants. Cercariae are shed from snails in slime balls that are deposited on the grass; when eaten by ants, they encyst as metacercariae. The mammalian host becomes infected by ingestion of infected ants. Reports of human infection are frequently spurious. Ingestion of liver from infected sheep can result in a false-positive diagnosis when eggs present in the food are passed in feces. However, genuine human cases have been reported in Europe, Asia, and Africa. Transmission of *Dicrocoelium* can best be prevented by careful washing of herbs and vegetables to remove ants.

Fleas

Dipylidium caninum is a common tapeworm of dogs and cats throughout the world. Tapeworm eggs are ingested by the flea *Ctenocephalides*, where they develop into the cysticercoid stage. Human infection appears limited to young children and results from accidental ingestion of infected fleas. Infection can best be prevented by controlling fleas on pets and by periodic worming of animals when necessary.

Beetles, Cockroaches, and Other Insects

Hymenolepis diminuta is a tapeworm of rats, mice, and other rodents. Eggs passed in rodent feces are ingested by flour beetles (*Tribolium, Tenebrio*), cockroaches (*Blattella, Periplaneta*), or fleas, where they develop into the cysticercoid stage. Human infection, acquired by the accidental ingestion of infected insects, is typically asymptomatic, although nausea, abdominal pain, and diarrhea may occur. Preventive measures include protecting grains and foodstuffs from insects and initiating rodent control.

The acanthocephalan *Moniliformis moniliformis* is an intestinal parasite of rats that also uses cockroaches and beetles as intermediate hosts. As with *H. diminuta*, human infection is usually found in young children. The giant leech-like acanthocephalan *Macracanthorhynchus hirudenaceus* is a cosmopolitan parasite of swine. The spiny proboscis embeds itself in the intestinal mucosa; female worms measure up to 65 cm in length. Human infections are rare but have been directly attributed to eating raw beetles. The trematodes *Prosthodendrium molenkampi* and *Phaneropsolos bonnei* may be acquired from ingesting dragon fly and damsel fly aquatic larvae.

HELMINTHS ACQUIRED FROM OTHER FOOD SOURCES

Tapeworms Causing Sparganosis

Sparganosis refers to infection by the plerocercoid stage of the diphyllobothrid tapeworm *Sparganum* or *Spirometra*. Human infection may be acquired by the accidental ingestion of copepods infected with the procercoid stage of the tapeworm. Alternatively, humans may serve as paratenic hosts by eating any of a variety of animals infected with the plerocercoid stage, such as frogs, tadpoles, lizards, snakes, birds, and mammals. Foodborne transmission has been attributed to eating raw pork (10). Infection may also result from the folk medicine practice of applying poultices of frog or snake to the eyes, skin, or vagina, as sparagana are capable of migrating out of the infected animal flesh and penetrating the lesion.

The sparganum is a wrinkled, ivory-white, ribbon-like flatworm that can grow up to 30 cm in length but is only about 3 mm wide (Fig. 26.4). Histological sections reveal

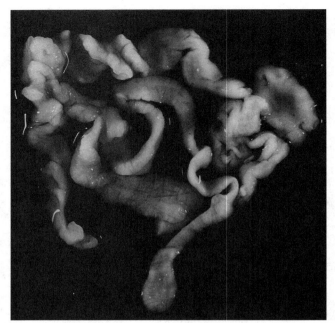

Figure 26.4 A sparganum removed from a subcutaneous nodule in the inguinal region (×1.7). (Illustration courtesy of the Armed Forces Institute of Pathology, AFIP 70-7392.)

a parenchymal tissue typical of undifferentiated cestode plerocercoids. Although usually found in subcutaneous tissue, spargana are highly motile and may migrate into muscle, viscera, brain, and eye. *Sparganum proliferum* is an especially hyperplastic species, exhibiting extensive branching and the capability of asexual reproduction by budding into separate organisms.

Migration of spargana is characterized by a painful, localized inflammatory reaction. Excessive lacrimation, periorbital edema, and swelling of the eyelids are associated with ocular sparganosis. Subcutaneous lesions may develop into abscesses; if the lesion ulcerates, the sparganum may be mistaken for a guinea worm. Diagnosis is presumptive and usually not confirmed until the worm is removed and identified in histological section. Surgical removal of spargana is the treatment of choice. Praziquantel has been proven to be an effective anthelminthic for laboratory animals.

Other Biohelminths

Alaria americana is a strigeid trematode typically found in the intestine of dogs and foxes. The normal life cycle involves sporocysts in snails and mesocercariae (a nonencysted stage) in tadpoles and frogs. Human infection has been attributed to eating inadequately cooked frog legs, raccoon, and opossum. Parasites have been recovered from the eye and from intradermal lesions, and at least one massive, fatal infection has been reported with

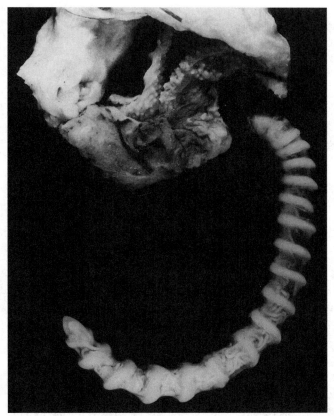

Figure 26.5 *Armillifer armillatus* adult female attached to the respiratory surface of the lung of a rock python from Zaire (×1.0). (Illustration courtesy of the Armed Forces Institute of Pathology, AFIP 72-881.)

mesocercariae disseminated into the lungs and other internal organs (28). Human infection by the trematode *Echinostomum* has also been attributed to eating crustaceans, tadpoles, and frogs (32).

Pentastomids or "tongue worms" (Fig. 26.5) are larval arthropods that have been recovered from liver, spleen, lungs, and eye. Human infection has been attributed to eating raw snake, lizard, goat, and sheep. Ocular involvement probably results from direct contact with pentastomid eggs in water. Most infections are asymptomatic, although respiratory discomfort and intestinal obstruction have been reported (45). Hoarseness and coughing due to young worms attached to the pharynx, a condition known as halzoun in the Near East, have been attributed to pentastomid infection (5). Halzoun may also be caused by *F. hepatica* in sheep liver. Ingestion of liver from sheep infected with adult *F. hepatica* or *Dicrocoelium dentriticum* may result in false-positive diagnoses of these parasites when their eggs are released into the alimentary canal. Although typically associated

with fish and copepods, *Gnathostoma* may also be acquired from pork, chicken, duck, frog, eel, snake, or rat.

Trichinella spiralis, described in detail in chapter 25, is typically associated with ingestion of raw or undercooked pork. However, human infection has also been acquired from sources as diverse as bear and walrus. An outbreak of trichinellosis in France was traced to horsemeat imported from the United States (3). Wild boar, a paratenic host for *P. westermani*, has been implicated as another potential source of human infection by that species (48). Prenatal and transmammary transmission of infective larval stages of helminths is more common than generally realized, and in some instances may be the major route of infection. It has been observed for cestodes, trematodes, and, most frequently, nematodes (46). Larvae of *Strongyloides fuelleborni* have been recovered from human milk (6), and *Strongyloides stercoralis* may be similarly transmitted (4). Strongyloidiasis is a common cause of infant death in Papua New Guinea (70).

HELMINTHS ACQUIRED FROM FECAL CONTAMINATION

Inadequate washing of produce or poor hygiene among food handlers can result in a variety of helminthic infections. Personal hygiene is critical because helminth eggs are often adherent, and contamination may be found not only on hands but also under fingernails, on clothing, and in washwater (53). Parasites can be acquired from both animal and human waste, and the use of nightsoil to fertilize crops represents a major source of infection. In addition to the direct transmission of zoonoses from animals to humans, experimental studies with seagulls and *Taenia* suggest the possibility that birds may play an important ancillary role in disseminating helminth eggs (64).

A variety of geohelminth species use the fecal-oral route for person-to-person and animal-to-person transmission. *Ascaris lumbricoides*, the largest intestinal roundworm of humans (Fig. 26.6), has been known since antiquity. *Ascaris eggs* have been found in mummified remains in Egypt and in archeological artifacts from the sites of Roman legion encampments. In parts of Central and South America, infection rates have been reported to approach 45% (4). It has been estimated that, at one time, in China alone, 18,000 tons of *Ascaris* eggs were produced each year (66). Unembryonated eggs shed in the feces are resistant to desiccation, moderate freezing temperatures, and chemical treatment of sewage, and they may remain dormant in the soil for years (7). The swine ascarid, *Ascaris suum*, can also infect humans but does not develop to maturity (14).

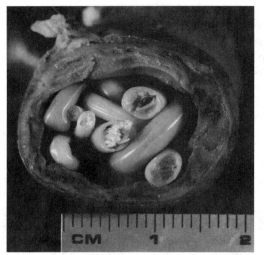

Figure 26.6 Numerous adult *Ascaris lumbricoides* obstructing the jejunum of a 13-year-old Zairian (×2.5). (Illustration courtesy of the Armed Forces Institute of Pathology, AFIP 72-13204.)

The whipworm, *Trichuris trichiura* (Fig. 26.7), occurs throughout the world but is most common in the tropics and in regions where sanitation is poor. Infections have been reported from the southeastern United States. *Trichuris* eggs can survive in the soil for several years.

Figure 26.7 *Trichuris trichiura* adult worms, showing the slender anterior ends threaded beneath the colonic epithelium (×3.7). (Illustration courtesy of the Armed Forces Institute of Pathology, AFIP 69-3583.)

There is no reservoir host for *T. trichiura*, but *Trichuris suis* and *Trichuris vulpis*, parasites of swine and dogs, respectively, have been reported in humans.

Trichostrongylus is a common intestinal nematode of herbivores that resembles the hookworm. Human trichostrongyliasis, caused by the accidental ingestion of larvae on contaminated vegetation or in drinking water, occurs primarily in Asia but has been reported throughout the world.

Oesophagostomum is a common intestinal nematode of primates, swine, and domestic animals throughout Asia, Africa, and South America. In human infections the accidentally ingested larvae produce nodular lesions and abscesses, approximately 1 to 2 cm in diameter, in the intestinal wall and sometimes other organs (31). As with anisakiasis, there is no effective anthelminthic, and surgical removal of the parasite is the only effective treatment. Definitive diagnosis is based upon identifying worms in surgical or biopsy specimens. In livestock, severe diarrhea may kill the animal, and parasitic nodules render the intestine unsuitable for making sausage casings, thus representing a serious economic loss to farmers (25).

The dwarf tapeworm, *Hymenolepis nana*, is the only human tapeworm that does not require an intermediate host; its life cycle is maintained by person-to-person contact and by autoinfection, although rodents may serve as reservoir hosts. Distributed throughout the world, *H. nana* is common in southern Europe, the former Soviet Union, India, and Latin America. Young children are most frequently infected. Eggs shed in feces are fully embryonated and immediately infective but have poor resistance outside the host. Internal autoinfection may occur when eggs hatch in the small intestine and penetrate the villi to repeat the developmental cycle without leaving the host. Eggs ingested by fleas and flour beetles develop into cysticercoids, but insects do not appear to play a major role in transmission. Mild infections are typically asymptomatic, and even large numbers of *H. nana* are well tolerated.

Toxocara canis, an intestinal roundworm of dogs and foxes, appears to be the principal cause of visceral larva migrans in the United States (55). *Toxocara cati* is a related species in domestic felines. Following the accidental ingestion of soil contaminated with embryonated *Toxocara* eggs, the larvae migrate throughout the body and may cause serious complications when they invade the central nervous system or the eye. Eggs may survive for years in the environment, and the typically high rates of contamination in municipal parks, playgrounds, and schoolyards pose a serious potential risk for young children (68). Appropriate preventive measures include animal control, particularly leash laws and waste disposal regulations, the periodic worming of pets, and scrupulous encouragement of hand washing before meals. Ocular involvement is a serious complication, and, tragically, a number of eyes with benign *Toxocara* inflammatory lesions have been unnecessarily enucleated because of suspected retinoblastoma (29). Ocular lesions may be more likely to occur in mild infections (30).

Baylisascaris procyonis, an ascarid of raccoons, can cause severe neurological complications in humans because this species is aggressive and grows considerably during migration. Fatal human infections have been reported in children believed to have been exposed to *Baylisascaris* eggs present on pieces of firewood or in raccoon feces deposited by animals nesting in unused hearths and chimneys (27, 38).

Echinococcus granulosus is a tapeworm of dogs and other canids. Human infection following the accidental ingestion of eggs in canine feces is common in sheep-raising areas throughout the world and among indigenous populations such as Eskimos, where there is a close ecological relationship between humans and dogs. *Echinococcus multilocularis* is a closely related species producing an alveolar cyst that grossly resembles an invading neoplasm. The course of the disease resembles that of a growing carcinoma, and it is among the most lethal of all helminthic infections (4). Risk factors for *Echinococcus* infection include exposure to infected dogs; consumption of contaminated water, ice, or snow; contact with infected foxes or their skins; and ingestion or handling of contaminated strawberries, huckleberries, cranberries, dandelions, or other vegetation (65, 69). Increases in grassland rodent and red fox populations in some parts of Europe, attributed to changes in land use as a result of agricultural policy, may further contribute to the risk of human infection (69).

Multiceps multiceps develops into a similar proliferative stage called a coenurus. In sheep, coenuriasis of the brain or spinal cord is relatively common and is known as gid or staggers. Human infections may occur in any organ but frequently involve subcutaneous tissue, brain, and eye. Cysticercosis, described in chapter 25 of this volume, results from the accidental ingestion of *Taenia solium* eggs. Racemose cysticercosis has been described as an aberrant cysticercus form of *T. solium* but may represent coenurus infection.

I thank Judy A. Sakanari, University of California, San Francisco, for sharing recent research findings about Anisakis and other fishborne parasites, and John H. Cross, Uniformed Services University of the Health Sciences, Bethesda, Md., for providing background information about capillariasis and other helminthic diseases in the Far East. I am especially indebted to John S. Mackiewicz, State University of New York at Albany, for encouraging attention to the importance of ecological factors when considering parasitic diseases. Illustrations for this chapter were obtained from the extensive slide collection of the Armed Forces Institute of Pathology, Washington, D.C.

References

1. **Abramowicz, M. (ed.).** 1998. Drugs for parasitic infections. *Med. Lett. Drugs Therapeut.* **40:**1–12.
2. **Alicata, J. E.** 1965. Notes and observations on murine angiostrongylosis and eosinophilic meningoencephalitis in Micronesia. *Can. J. Zool.* **43:**667–672
3. **Ancelle, T., J. Dupouy-Camet, M. E. Bougnoux, V. Fourestie, H. Petit, G. Mougeot, J. P. Nozais, and J. LaPierre.** 1988. Two outbreaks of trichinosis caused by horsemeat in France in 1985. *Am. J. Epidemiol.* **127:**1302–1311.
4. **Beaver, P. C., R. C. Jung, and E. W. Cupp.** 1984. *Clinical Parasitology*, 9th ed. Lea & Febiger, Philadelphia, Pa.
5. **Bogitsh, B. J., and T. C. Cheng.** 1990. *Human Parasitology*. Saunders College Publications, Holt, Rinehart & Winston, New York, N.Y.
6. **Brown, R. C., and M. H. F. Girardeau.** 1977. Transmammary passage of *Strongyloides* sp. larvae in the human host. *Am. J. Trop. Med. Hyg.* **26:**215–219.
7. **Bryan, F. D.** 1977. Diseases transmitted by foods contaminated by waste water. *J. Food Prot.* **40:**45–56.
8. **Centers for Disease Control.** 1982. Intestinal perforation caused by larval *Eustrongylides*—Maryland. *Morb. Mortal. Wkly. Rep.* **31:**383–389.
9. **Chitwood, M. B., C. Valesquez, and N. G. Salazar.** 1968. *Capillaria philippinensis* sp.n. (Nematoda:Trichinellida) from the intestine of man in the Philippines. *J. Parasitol.* **54:**368–371.
10. **Corkum, K. C.** 1966. Sparganosis in some vertebrates of Louisiana and observations on a human infection. *J. Parasitol.* **52:**444–448.
11. **Cross, J. H.** 1992. Intestinal capillariasis. *Clin. Microbiol. Rev.* **5:**120–129.
12. **Daengsvang, S.** 1971. Infectivity of *Gnathostoma spinigerum* larvae in primates. *J. Parasitol.* **57:**476–578.
13. **Daengsvang, S., B. Sermswatsri, P. Youngyi, and D. Guname.** 1970. Development of adult *Gnathostoma spinigerum* in the definitive host (cat and dog) by skin penetration of the advanced third-stage larvae. *Southeast Asia J. Trop. Med. Public Health* **1:**187–192.
14. **Davies, N. J., and J. M. Goldsmid.** 1978. Intestinal obstruction due to *Ascaris suum* infection. *Trans. R. Soc. Trop. Med. Hyg.* **72:**107.
15. **de Brie, J.** 1879. Le bon berger ou le vray reegime et gouvenement de bergers et bergeres: compose par le rustique Jehan de Brie le bon berger. Isidor Liseux, Paris, France.
16. **Deardorff, T. L., and R. M. Overstreet.** 1981. Larval *Hysterothylacium* (=*Thynnascaris*) (Nematoda:Anisakidae) from fishes and invertebrates in the Gulf of Mexico. *Proc. Helminthol. Soc. Wash.* **48:**113–126.
17. **Deardorff, T. L., and R. M. Overstreet.** 1990. Seafood-transmitted zoonoses in the United States: the fishes, the

dishes, and the worms, p. 211–265. *In* D. R. Ward and D. R. Hackney (ed.), *Microbiology of Marine Food Products*. Van Nostrand Reinhold, New York, N.Y.

18. **Deardorff, T. L., R. M. Overstreet, M. Okihiro, and R. Tam.** 1986. Piscine adult nematode invading an open lesion in a human hand. *Am. J. Trop. Med. Hyg.* **35:**827–830.

19. **Deardorff, T. L., and R. Throm.** 1988. Commercial blast freezing of third-stage *Anisakis simplex* larvae encapsulated in salmon and rockfish. *J. Parasitol.* **74:**600–603.

20. **Delaware Sea Grant.** 2000. Eating raw finfish: what are the risks, the benefits? University of Delaware. http://www.ocean.udel.edu/mas/seafood/raw.html.

21. **Desowitz, R. S.** 1981. *New Guinea Tapeworms and Jewish Grandmothers.* Norton and Company, New York, N.Y.

22. **Donaldson, J. R., and T. A. Angelo.** 1961. Quadriplegia due to guinea-worm abscess. *J. Bone Joint Surg.* **43A:**197–198.

23. **Dooley, J. R., and R. C. Neafie.** 1976. Anisakiasis, p. 475–481. *In* C. H. Binford and D. H. Connor (ed.), *Pathology of Tropical and Extraordinary Diseases*, vol. 2. Armed Forces Institute of Pathology, Washington, D.C.

24. **Dooley, J. R., and R. C. Neafie.** 1976. Clonorchiasis and opisthorchiasis, p. 509–516. *In* C. H. Binford and D. H. Connor (ed.), *Pathology of Tropical and Extraordinary Diseases*, vol. 2. Armed Forces Institute of Pathology, Washington, D.C.

25. **Dooley, J. R., and R. C. Neafie.** 1976. Oesophagostomiasis, p. 440–445. *In* C. H. Binford and D. H. Connor (ed.), *Pathology of Tropical and Extraordinary Diseases*, vol. 2. Armed Forces Institute of Pathology, Washington, D.C.

26. **Eberhard, M. L., H. Hurwitz, A. M. Sun, and D. Coletta.** 1989. Intestinal perforation caused by larval *Eustrongylides* (Nematode:Dioctophymatoidae) in New Jersey. *Am. J. Trop. Med. Hyg.* **40:**648–650.

27. **Fox, A. S., K. R. Kazacos, N. S. Gould, P. T. Heydemann, C. Thomas, and K. M. Boyer.** 1985. Fatal eosinophilic meningoencephalitis and visceral larva migrans caused by the raccoon ascarid *Baylisascaris procyonis*. *N. Engl. J. Med.* **312:**1619–1623.

28. **Freeman, R. S., P. F. Stuart, J. B. Cullen, A. C. Ritchie, A. Mildon, B. J. Fernandes, and R. Bonin.** 1976. Fatal human infection with mesocercariae of the trematode *Alaria americana*. *Am. J. Trop. Med. Hyg.* **25:**803–807.

29. **Glickman, L. T.** 1984. Toxocariasis. *In* K. S. Warren and A. A. F. Mahmoud (ed.), *Tropical and Geographical Medicine*. McGraw-Hill, Book Company, New York, N.Y.

30. **Glickman, L. T., P. M. Schantz, and R. H. Cypress.** 1979. Canine and human toxocariasis: review of transmission, pathogenesis and clinical disease. *J. Am. Vet. Med. Assoc.* **175:**1265–1269.

31. **Gordon, J. A., C. M. D. Ross, and H. Affleck.** 1969. Abdominal emergency due to an oesophagostome. *Ann. Trop. Med. Parasitol.* **63:**161–164.

32. **Graczyk, T. K., and B. Fried.** 1998. Echinostomiasis: a common but forgotten food-borne disease. *Am. J. Trop. Med. Hyg.* **58:**501–504.

33. **Harrell, L. W., and T. L. Deardorff.** 1990. Human nanophyetiasis: transmission by handling naturally infected coho salmon (*Oncorhynchus kisutch*). *J. Infect. Dis.* **161:**146–148.

34. **Hayunga, E. G.** 1989. Parasites and immunity: tactical considerations in the war against disease—or, how did the worms learn about Clausewitz? *Perspect. Biol. Med.* **32:**349–370.

35. **Hillyer, G. V.** 1981. Fascioliasis in Puerto Rico: a review. *Bol. Assoc. Med. P.R.* **73:**94–101.

36. **Hopkins, D. R., E. Ruiz-Tiben, and T. K. Ruebush III.** 1997. Dracunculiasis eradication: almost a reality. *Am. J. Trop. Med. Hyg.* **57:**252–259.

37. **Hou, P. C.** 1965. Hepatic clonorchiasis and carcinoma of the bile duct in a dog. *J. Pathol. Bacteriol.* **89:**365.

38. **Huff, D. S., R. C. Neafie, M. J. Binder, G. A. DeLeon, L. W. Brown, and K. R. Kazacos.** 1984. Case 4: the first fatal *Baylisascaris* infection in humans: an infant with eosinophilic meningoencephalitis. *Pediatr. Pathol.* **2:**345–352.

39. **Jones, E. A., J. M. Key, H. P. Milligan, and D. Owens.** 1977. Massive infection with *Fasciola hepatica* in man. *JAMA* **63:**836–842.

40. **Katz, M., D. D. Despommier, and R. W. Gwadz.** 1982. *Parasitic Diseases.* Springer-Verlag, New York, N.Y.

41. **Kikuchi, Y., T. Takenouchi, M. Kamiya, and H. Ozake.** 1981. Trypanorhynchiid cestode larva found on the human palatine tonsil. *Jpn. J. Parasitol.* **30:**3497–3499.

42. **Kisielewska, K.** 1970. Ecological organization of intestinal helminth groupings in *Clethrionomys glareolus* (Schreb.) (Rodentia). I. Structure and seasonal dyamics of helminth groupings in a host population in the Bialowieza National Park. *Acta Parasitol. Pol.* **18:**121–147.

43. **Leuckart, K. G.** 1882. Zur Entwickelungsgeschichte des Leberegels. *Zweite Mittheilung. Zool. Anz.* **5:**524–528.

44. **MacArthur, W. P.** 1933. Cysticercosis as seen in the British army, with special reference to the production of epilepsy. *Trans. R. Soc. Trop. Med. Hyg.* **27:**343–363.

45. **Meyers, W. M., R. C. Neafie, and D. H. Connor.** 1976. Pentastomiasis, p. 546–550. *In* C. H. Binford and D. H. Connor (ed.), *Pathology of Tropical and Extraordinary Diseases*, vol. 2. Armed Forces Institute of Pathology, Washington, D.C.

46. **Miller, G. C.** 1981. Helminths and the transmammary route of infection. *Parasitology* **82:**335–342.

47. **Mitra, A. K., and D. R. W. Haddock.** 1970. Paraplegia due to guinea worm infection. *Trans. Roy. Soc. Trop. Med. Hyg.* **64:**102–106.

48. **Miyazaki, I., and S. Habe.** 1976. A newly recognized mode of human infection with the lung fluke, *Paragonimus westermani* (Kerbert 1978). *J. Parasitol.* **62:**646–648.

49. **Muller, R.** 1971. *Dracunculus* and dracunculiasis. *Adv. Parasitol.* **9:**73–151.

50. **New, D., M. D. Little, and J. Cross.** 1995. *Angiostrongylus cantonensis* infection from eating raw snails. *N. Engl. J. Med.* **332:**1105–1106.

51. **Norton, R. A., and L. Monroe.** 1961. Infection by *Fasciola hepatica* acquired in California. *Gastroenterology* **41:**46–48.

52. **Numbers 21:**4–9.

53. Ockert, G., and J. Obst. 1973. Ausstreuung umhüllter Onkosphären durch Bandwurmträger. *Monat. Veterinaermed.* 28:97–98.

54. Ollerenshaw, C. B. 1980. Forecasting liver fluke disease, p. 33–52. *In* A. E. R. Taylor and R. Muller (ed.), *12th Annual Symposium of the British Society for Parasitology.* Blackwell, London, United Kingdom.

55. Paul, A. J., K. S. Todd, Jr., and J. A. Dipietro. 1988. Environmental contamination by eggs of *Toxocara* species. *Vet. Parasitol.* 26:339–342.

56. Rao, C. K., R. C. Paul, and M. I. D. Sharma. 1981. Guinea worm disease in India—control status and strategy of its eradication. *J. Commun. Dis.* 13:1–7.

57. Reddy, C. R. R. M., and V. V. Valli. 1967. Extradural guinea-worm abscess. *Am. J. Trop. Med. Hyg.* 16:23–25.

58. Rosen, L., G. Loison, J. Laigret, and G. D. Wallace. 1967. Studies on eosinophilic meningitis. 3. Epidemiologic and clinical observations on Pacific islands and the possible etiologic role of *Angiostrongylus cantonensis*. *Am. J. Epidemiol.* 85:17–44.

59. Sadun, E. H., and C. Maiphoom. 1953. Studies in the epidemiology of the human intestinal fluke, *Fasciolopsis buski* (Lankester) in Central Thailand. *Am. J. Trop. Med. Hyg.* 2:1070–1084.

60. Sakanari, J. A., M. Moser, and T. L. Deardorff. 1995. *Fish Parasites and Human Health, Epidemiology of Human Helminthic Infections.* Report No. T-CSGCP-034. California Sea Grant College, University of California, La Jolla.

61. Sakanari, J. A., H. M. Loinaz, T. L. Deardorff, R. B. Raybourne, H. H. McKerrow, and J. G. Frierson. Intestinal anisakiasis: a case diagnosed by morphologic and immunologic methods. *Am. J. Clin. Pathol.* 90:107–113.

62. Sakanari, J. A., and J. H. McKerrow. 1989. Anisakiasis. *Clin. Microbiol. Rev.* 2:278–284.

63. Schantz, P. M. 1989. The dangers of eating raw fish. *N. Engl. J. Med.* 320:1143–1145.

64. Silverman, P. H., and R. B. Griffiths. 1955. A review of methods of sewage disposal in Great Britain with special reference to the epizootiology of *Cysticercus bovis*. *Ann. Trop. Med. Parasitol.* 49:436–450.

65. Stehr-Green, J. K., P. A. Stehr-Green, P. M. Schantz, J. F. Wilson, and A. Lanier. 1988. Risk factors for infection with *Echinococcus multilocularis* in Alaska. *Am. J. Trop. Med. Hyg.* 38:380–385.

66. Stoll, N. R. 1947. This wormy world. *J. Parasitol.* 33:1–18.

67. U.S. Food and Drug Administration, Center for Food Safety and Applied Nutrition. 1999. *Anisakis simplex* and related worms. *In Foodborne Pathogenic Microorganisms and Natural Toxins Handbook (Bad Bug Book).* Available electronically at http://vm.cfsan.fda.gov/~mow/chap25.html.

68. Uga, S., and N. Kataoka. 1995. Measures to control *Toxocara* egg contamination in sandpits of public parks. *Am. J. Trop. Med. Hyg.* 52:21–24.

69. Viel, J. F., P. Giraudoux, V. Abrial, and S. Bresson-Hadni. 1999. Water vole (*Arvicola terrestris scherman*) density as risk factor for human alveolar echinococcosis. *Am. J. Trop. Med. Hyg.* 61:559–565.

70. Vince, J. D., R. W. Ashford, M. J. Gratten, and J. Bana-Koiri. 1979. *Strongyloides* species infestation in young infants of Papua New Guinea: association with generalized oedema. *Papua New Guinea Med. J.* 22:120–127.

71. Wittner, M., J. W. Turner, G. Jacquotte, L. R. Ash, M. P. Salgo, and H. B. Tanowitz. 1989. Eustrongylidiasis—a parasitic infection acquired by eating sushi. *N. Engl. J. Med.* 320:1124–1126.

72. Yii, C.-Y. 1976. Clinical observations of eosinophilic meningitis and meningoencephalitis caused by *Angiostrongylus cantonensis* in Taiwan. *Am. J. Trop. Med. Hyg.* 25:233–249.

73. Yokogawa, M. 1969. *Paragonimus* and paragonimiasis. *Adv. Parasitol.* 7:375–387.

Food Microbiology: Fundamentals and Frontiers, 2nd Ed.
Edited by M. P. Doyle et al.
© 2001 ASM Press, Washington, D.C.

Ynes R. Ortega

Protozoan Parasites

27

Protozoan parasites have long been associated with food-borne and waterborne outbreaks of disease in humans. Difficulties arise with the inactivation of these organisms because of size and resistance to environmental conditions. The following groups of protozoan parasites are relevant to transmission via food: Apicomplexa, Microspora, flagellates, ciliates, and amoebae (Table 27.1).

A major characterization of apicomplexan parasites is that a vertebrate host is required to complete their complex life cycle and produce infectious cysts. Of this group, *Cryptosporidium parvum*, *Cyclospora cayetanensis*, and *Isospora belli* inhabit the intestinal mucosa and produce diarrheal illness in humans. These coccidian parasites affect immunocompetent as well as immunocompromised individuals, causing more severe and prolonged symptoms in the latter. *Cryptosporidium parvum* can infect both animals and humans, and two genotypes have been described that seem to have host preference or specificity. *C. cayetanensis* has been isolated exclusively in humans. Another apicomplexan, *Toxoplasma gondii*, infects human tissues other than the intestinal mucosa and causes birth defects, blindness, and chorioretinitis. The life cycle stages of apicomplexan parasites are produced intracellularly in the host. In *Cyclospora*, *Toxoplasma*, and *Isospora*, sporogony typically occurs outside the host, requiring the passage of some time before oocysts are infective to a new host, whereas *Cryptosporidium* oocysts are excreted already sporulated and are infectious when shed.

Microsporidia, the second group of parasites implicated in human food and waterborne diseases, have to propagate and complete their life cycle intracellularly. Of the microsporidia, five genera have been implicated in human diseases: *Encephalitozoon*, *Enterocytozoon*, *Septata*, *Pleistophora*, and *Vittaforma*. None of these species is host, tissue, or organ specific, with the exception of *Enterocytozoon bieneusi*, which appears to infect only the human intestinal tract.

Flagellates, ciliates, and amoebae can propagate in more than one host. All three types of parasites can cause disease in humans and animals. The infectious stage in this group is the cyst. Once ingested by the susceptible host, the trophozoites or motile forms will be released from the cyst and proceed to colonize the host's intestinal cells. In contrast with the previously described protozoan parasites, this group of parasites do not present intracellular stages, but attach with specific structures to the luminal surface of the intestinal cells and feed off

Ynes R. Ortega, Center for Food Safety and Quality Enhancement, Department of Food Science and Technology, University of Georgia, 1109 Experiment St., Griffin, GA 30223-1797.

Table 27.1 Protozoa of medical importance acquired from food and water

Phylum	Protozoan	Infective stage
Apicomplexa	*Cryptosporidium parvum*	Oocyst
	Cyclospora cayentanensis	Oocyst
	Isospora belli	Oocyst
	Toxoplasma gondii	Oocyst/tissue cyst
	Sarcocystis hominis	Oocyst
	Sarcocystis suihominis	Oocyst
Ciliphora	*Balantidium coli*	Cyst
Microspora	*Enterocytozoon bieneusi*	Spore
	Septata intestinalis	Spore
Sarcomastigophora	*Acanthamoeba* spp.[a]	Trophozoite
	Dientamoeba fragilis	Trophozoite
	Entamoeba dispar[b]	Cyst
	Entamoeba histolytica	Cyst
	Giardia intestinilis	Cyst
	Naegleria fowleri[a]	Trophozoite

[a] Normally free-living amoeba. Invasion through mucosa and can in some instances infect the central nervous system of humans. Amoebic meningitis is usually fatal.

[b] Commensal, morphologically similar to *E. histolytica*, not normally capable of inducing disease.

the nutrients and cellular debris of the intestinal epithelium. *Giardia lamblia*, a flagellate originally considered a commensal, is well recognized as an etiological agent for acute or chronic diarrhea in humans. *Balantidium coli*, a ciliate, also causes diarrhea in humans and, if not treated, can cause ulcerative colitis. Various species of amoebae can infect humans. Most of them are commensal and do not produce disease; however, others can cause diarrhea, dysentery, or amebomas if not treated.

CRYPTOSPORIDIUM SPP.

Cryptosporidium was first isolated from the intestines of mice in 1910. Since then various species of *Cryptosporidium* that infect animals have been described. It was not until 1976 that *C. parvum* was first described in humans. After 1982, *C. parvum* was one of the most frequently diagnosed opportunistic pathogens associated with diarrhea and wasting syndrome in patients with AIDS. Cryptosporidiosis produces a life-threatening, prolonged cholera-like illness in immunocompromised patients (82), whereas in immunocompetent patients, *C. parvum* causes self-resolving acute diarrheal disease with variability in the severity of symptoms. Isolate variation and its association with severity of the disease has been a subject of intensive

investigation for some time. Recently, *C. parvum* isolates have been separated into two genotypes: type I and type II. Genotype I has been isolated exclusively from humans. Genotype II has been isolated from bovines and from humans who have been exposed to infected cattle and is readily infective to laboratory animals. Genotype differentiation was based on the sequencing of the thrombospondin-related adhesive protein (TRAP-C2) (68).

Cryptosporidiosis is acquired after ingesting food or water contaminated with infective *Cryptosporidium* oocysts. Once ingested, the oocysts excyst (Fig. 27.1). The released sporozoites proceed to invade the enterocytes. A parasitophorous vacuole is formed as the parasite enters the cell. This structure contains the intracellular parasites and communicates with the host cell via a "feeder organelle." The parasite goes through two asexual multiplication stages called merogony. Type I meronts contain eight merozoites that are released and proceed to invade other enterocytes to form type II meronts, each containing four merozoites. Once merozoites from the type II meronts are released, they infect enterocytes but differentiate into sexual stages identified as macro- (female) and microgametocytes (male). The union of a microgametocyte and a macrogametocyte produces a zygote (immature oocyst), which matures within its host into a fully sporulated oocyst (Fig. 27.2). Two types of oocysts are produced and excreted: thin-walled oocysts, which excyst endogenously, resulting in autoinfection, and thick-walled oocysts, which are environmentally resistant and are shed in the feces and become immediately infectious to other hosts (28, 94).

Generally, *C. parvum* infects the brush border of the intestinal epithelium and causes villous atrophy, although the exact mechanism for this pathological alteration is not yet fully understood. Immunocompetent patients develop a profuse, watery diarrhea accompanied by epigastric cramping, nausea, and anorexia, which is usually self-limiting and lasts for about 15 days.

However, extraintestinal dissemination can be observed in immunocompromised patients. *Cryptosporidium* may be found in epithelial cells from other organs such as those of the respiratory tract and the biliary tree. Immunocompromised patients (AIDS patients or those receiving immunosuppressive drugs) develop severe diarrhea (3 to 6 liters/day) which persists for several weeks to months or years. These cases are the most severe and life-threatening, with continuous shedding of oocysts. In patients with human immunodeficiency virus (HIV) infection, the CD4$^{\pm}$ cell count is the best marker for the ability of the immune system to self-resolve the infection. In the industrialized world, patients with CD4 counts of

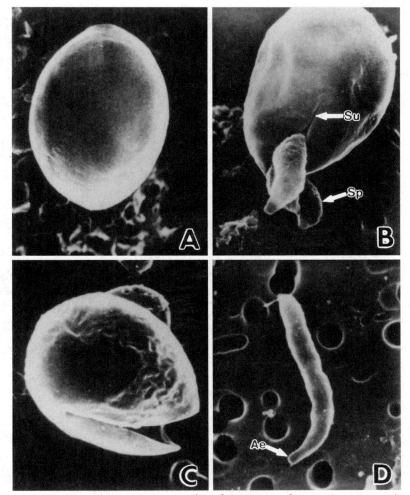

Figure 27.1 Scanning electron micrographs of oocysts and excysting sporozoites of *C. parvum*. (A) Intact oocyst prior to excystation (×11,200). (B). Three sporozoites (Sp) excysting from oocyst simultaneously via the cleaved suture (Su) (×11,200). (C) Empty oocyst (×11,200). (D) Excysted sporozoite; Ae, apical end (×9,800). From Reduker et al. (75).

180 cells/ml or more usually develop a self-limiting cryptosporidiosis, while those with counts below 180 cells/ml usually develop chronic and profuse diarrhea, which is exacerbated by the lack of an effective therapy (42). The pathogenesis of the diarrhea is not clear, and the presence of a toxin has been suggested but not demonstrated to date.

The mean prevalence of *C. parvum* in Europe and the United States is between 1 and 3%, and it is considerably higher in developing countries. Outbreaks of disease associated with *Cryptosporidium* have been reported in the United States as well as overseas (94).

Cryptosporidium infection is highly associated with travel abroad, exposure to farm animals, and person-to-person transmission in settings such as day care centers and medical institutions. A large number of waterborne outbreaks have been reported in the literature. Some of the most notable occurred in Milwaukee and Georgia, where more than 400,000 cases were reported. In 1995, *Cryptosporidium* was associated with cases of acute gastroenteritis among persons who attended a social event in Minnesota. This outbreak was epidemiologically associated with contaminated chicken salad (8). In 1993 in central Maine and in 1996 in New York, apple cider was associated with outbreaks of cryptosporidiosis (5, 59). In 1998, another foodborne outbreak affected about 50 people in Spokane, Washington, but it could not be traced to a specific type of food.

Although there is no effective and definite therapeutic agent for the treatment of cryptosporidiosis, spiramycin

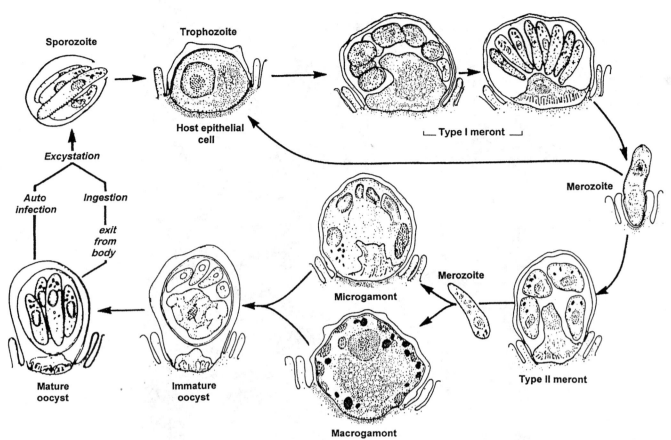

Figure 27.2 Life cycle of *Cryptosporidium*. From Dubey et al. (33).

has been reported to decrease diarrhea in early infections; however, it is not effective in advanced infections. Azithromycin, nitazoxanide, and paromomycin have been reported to be effective in AIDS patients with cryptosporidiosis. Alternative experimental therapies already evaluated in humans involve the use of passive immunotherapies such as hyperimmune bovine colostrum. Monoclonal antibodies and polyclonal hyperimmune hen yolk against *Cryptosporidium* have also been tested in animal models with some success (24, 30, 40, 43, 69).

Cryptosporidium oocysts are 4 to 6 μm in diameter. Because of their size, they can be overlooked in fecal examinations or confused with yeast cells. *Cryptosporidium* can be identified by light microscopy using acid-fast staining. Immunoassays (fluorescent antibody or enzyme immunoassay) are very sensitive and specific diagnostic procedures (10, 11, 25, 85, 87). *Cryptosporidium* can be mistaken for *Cyclospora*, which is also acid-fast positive but larger (Fig. 27.3). It is important that laboratories carefully measure the diameter of the organisms, particularly if they appear to be larger

than *Cryptosporidium*. New molecular diagnostic tests such as PCR are being developed and could improve the sensitivity in detecting *Cryptosporidium* oocysts in produce (13, 23).

CYCLOSPORA

Cyclospora was probably first reported from humans in 1979 by Ashford, who described it as an *Isospora*-like coccidian affecting humans in Papua New Guinea (12). Thereafter, other investigators found similar structures in fecal samples of patients with diarrhea, but because of the morphology of the unsporulated oocyst and its autofluorescence, it was considered to be a CLB (coccidian-like body or cyanobacterium-like body). In 1993, the conclusive identification was performed, and the CLBs were fully characterized as coccidian parasites and placed in the genus *Cyclospora* with the proposed species name *cayetanensis* (63, 66).

Cyclospora belongs to the family *Eimeriidae*, subphylum Apicomplexa. *Cyclospora* species infect moles, rodents, insectivores, snakes, and humans. In 1998,

Newly shed oocysts (Fig. 27.4A) of *C. cayetanensis* require 2 weeks to sporulate and become infectious (Fig. 27.4B) under optimal laboratory conditions (63). This period required for the oocyst to become infectious suggests that contamination of produce occurs with oocysts that are fully sporulated or almost fully sporulated. Otherwise, unsporulated oocysts would require optimal time and environmental conditions to induce oocyst sporulation, while produce would still have to remain edible.

In studies performed in areas of endemic disease, *Cyclospora* oocysts were isolated from produce by washing thoroughly with distilled water. The washes were then concentrated by centrifugation, and pellets were fixed and preserved in 10% formalin. Samples were examined directly by using epifluorescence microscopy and phase-contrast microscopy (66). Nested PCR targeted to amplify the 18S rDNA can also be used with these preparations; however, it should be noted that further testing using restriction fragment length polymorphism (RFLP) was required, since the described PCR cross-reacted with *Eimeria* species. *Eimeria* parasites are infectious to animals but not to humans and can be readily found in the environment (52).

Cyclosporiasis is characterized by mild to severe nausea, anorexia, abdominal cramping, mild fever, and watery diarrhea. Diarrhea alternating with constipation has been commonly reported. Some patients present with flatulent dyspepsia and, less frequently, joint pain and night sweats. Onset of illness in patients is usually sudden, and symptoms persist an average of 7 weeks (83).

C. cayetanensis infects epithelial cells of the duodenum and jejunum of humans. Merogony and gametogony occur intracytoplasmically within the parasitophorous vacuoles in intestinal cells (Fig. 27.4C). Duodenal and jejunal biopsies of patients with cyclosporiasis show varying degrees of jejunal villous blunting, atrophy, and crypt hyperplasia (26, 64). Extensive lymphocytic infiltration into the surface epithelium is present, especially at the tips of the shortened villi. The reactive inflammatory response of the host does not correlate with the number of intracellular parasites present in the tissues (64).

The routes of transmission for *Cyclospora* are still undocumented, although the fecal-oral route, either directly or via water and food, is probably the major one. In the United States, epidemiological evidence suggests that water has been responsible for sporadic cases of cyclosporiasis. In Utah, a man became infected after cleaning his basement, which was flooded with runoff from a nearby farm following heavy rains. In 1990, an outbreak involved residents of a physicians' dormitory in

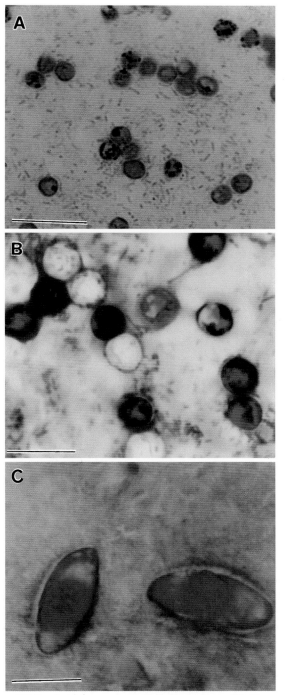

Figure 27.3   Acid-fast staining. (A) *C. parvum*; (B) *C. cayetanensis*; (C) *I. belli*. (Bar, 20 μm.)

not recovered from or detected in the produce associated with any of the reported outbreaks. In the United States, most of the cases have been reported from April to August, suggesting some possible seasonality.

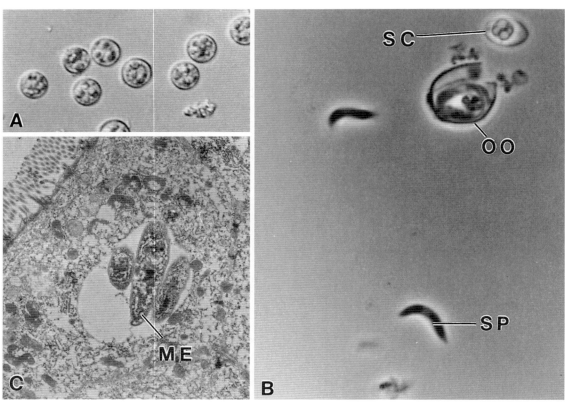

Figure 27.4 *C. cayetanensis* oocysts. (A) Phase-contrast microscopy of unsporulated oocysts. (B) Oocysts (OO) in process of excystation. Note the two sporozoites (SP) free of the sporocyst (SC). (C) Transmission electron microscopy of human small intestine showing *Cyclospora* intracellular stages.

sudden, and symptoms persist an average of 7 weeks (83).

C. cayetanensis infects epithelial cells of the duodenum and jejunum of humans. Merogony and gametogony occur intracytoplasmically within the parasitophorous vacuoles in intestinal cells (Fig. 27.4C). Duodenal and jejunal biopsies of patients with cyclosporiasis show varying degrees of jejunal villous blunting, atrophy, and crypt hyperplasia (26, 64). Extensive lymphocytic infiltration into the surface epithelium is present, especially at the tips of the shortened villi. The reactive inflammatory response of the host does not correlate with the number of intracellular parasites present in the tissues (64).

The routes of transmission for *Cyclospora* are still undocumented, although the fecal-oral route, either directly or via water and food, is probably the major one. In the United States, epidemiological evidence suggests that water has been responsible for sporadic cases of cyclosporiasis. In Utah, a man became infected after cleaning his basement, which was flooded with runoff from

a nearby farm following heavy rains. In 1990, an outbreak involved residents of a physicians' dormitory in a Chicago hospital. It was epidemiologically associated with tap water from unprotected reservoir tanks that served the building and which had a broken water pump (9). In Pokhara, Nepal, British Gurkha soldiers were confirmed to have cyclosporiasis, and oocysts were isolated from a water sample. The water was a mixture of river and municipal water that was routinely chlorinated and served the houses in the camp where the soldiers were stationed (72).

A few reports have described the isolation of *C. cayetanensis* oocysts in animals (chicken, ducks, and dogs). After several experimental studies attempting to infect those and additional animal species, it seems that *Cyclospora* species are host specific.

Foodborne outbreaks of cyclosporiasis have also been reported. In 1996 and 1997, *Cyclospora* infections in the United States were associated with imported raspberries. It was speculated that the raspberries could have become contaminated when they were sprayed with insecticide possibly diluted with contaminated surface water. Analysis of the irrigation water demonstrated the presence of *Cyclospora* oocysts (17). In Peru, *C. cayetanensis* oocysts have been isolated from vegetables from markets in areas of endemicity. In studies where vegetables were experimentally inoculated with *C. cayetanensis* oocysts, it was demonstrated that washing with water does not remove all the oocysts (65). The minimum infectious dose of oocysts, the oocyst sporulation rate, and their survival under different environmental conditions are unknown.

Analysis of the intervening transcribed spacer 1 region sequence of *Cyclospora* isolates from the 1996 outbreak demonstrated that all were identical, suggesting that a single source of contamination caused the infection (2).

There is strong evidence to suggest seasonality of *Cyclospora* infections. In Peru, in more than 6 years of prospective epidemiological studies investigating endemic *Cyclospora* infections, nearly all infections occurred between December and July (57). Rarely were infections documented at any other time of the year. In the United States, the major epidemics occurred from May to July. In Nepal, infection and illness occurred most frequently from May to August (48). The specific reasons for this marked seasonality have not been defined (49).

The only successful antimicrobial treatment for *Cyclospora* is trimethoprim-sulfamethoxazole (TMP-SMX). AIDS patients appear to have a higher parasite infestation than immunocompetent individuals infected with *Cyclospora*. However, the prevalence of *Cyclospora* in HIV patients is not higher than in immunocompetent populations. This is probably due to the frequent use of TMP-SMX for *Pneumocystis carinii* prophylaxis among HIV patients (56, 67, 73).

ISOSPORA

Isospora is another coccidian parasite that infects humans. It is more frequently identified in AIDS patients (29, 82). *Isospora* can be acquired by ingestion of contaminated food or water. Oocysts excyst in the intestine and sporocysts are released. These in turn release sporozoites that infect intestinal epithelial cells. Asexual and sexual life cycle stages occur in the cytoplasm of enterocytes. Unsporulated *Isospora* oocysts require 12 to 48 h to mature and become infectious outside the host.

Isospora can be diagnosed in stools from infected patients by observing unsporulated oocysts shed in feces. *Isospora* belli oocysts (10 to 19 by 20 to 30 μm) can be detected in direct wet mounts in heavy infections during ova and parasite examination (Fig. 27.5). However, most infections are not heavy, and shedding

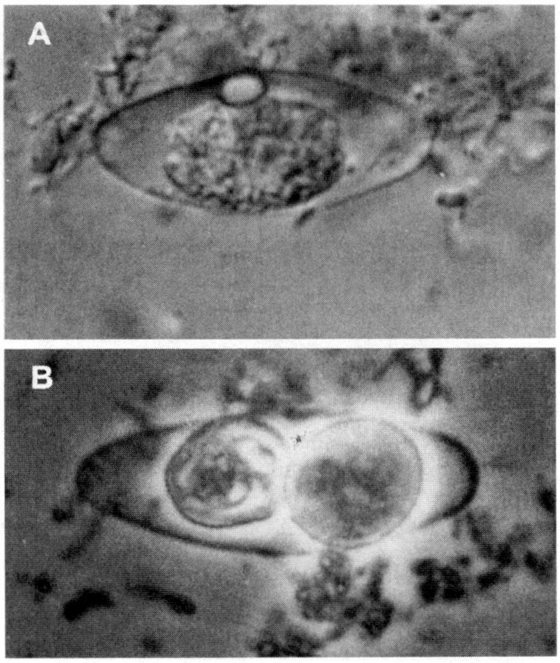

Figure 27.5 Bright-field photomicrograph of *I. belli* oocysts (A) unsporulated and (B) sporulated.

of oocysts may be variable. Therefore, it is necessary to examine a series of samples and perform modified acid-fast staining (Fig. 27.3C). *I. belli* infects the entire intestine and produces severe intestinal disease (84). Deaths from overwhelming infections have been reported, especially in immunocompromised patients. Symptoms include diarrhea, nausea, steatorrhea, headache, and weight loss. The disease may persist for months and even years.

Isospora is rare in immunocompetent people but occurs in 0.2 to 0.3% of the immunocompromised AIDS patients in the United States and 8 to 20% of AIDS patients in Africa and Haiti. *Isospora* is endemic in many parts of Africa, Asia, and South America. The treatment of choice is TMP-SMX. In HIV patients, recurrence is common after discontinuation of therapy (70).

TOXOPLASMA

Toxoplasma gondii is a coccidian parasite that infects a variety of warm-blooded hosts. Cats are the definitive hosts, and other warm-blooded animals can serve as intermediate hosts. Cats excrete oocysts in their feces. Oocysts are environmentally resistant and can survive for several years in moist, shaded conditions. Infections are acquired principally by ingestion of food or water containing oocysts, by ingestion of animal tissues

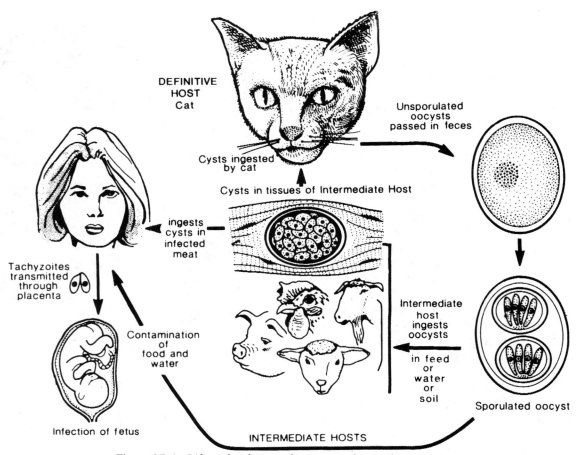

Figure 27.6 Life cycle of *T. gondii*. From Dubey and Beattie (31).

containing cystic forms (bradyzoites) (31), or by transplacental transmission.

The unsporulated oocysts require 24 h outside the host to differentiate and become infectious. When oocysts are ingested by the intermediate host, the oocyst walls are ruptured and the sporozoites are released (Fig. 27.6). They invade epithelial cells and rapidly multiply asexually (tachyzoites). Tachyzoites multiply by endodyogeny, a process in which the mother tachyzoite is consumed by the formation of two daughter zoites (Fig. 27.7). Eventually, the tachyzoites encyst in the brain, liver, skeletal muscle, and cardiac muscle. These cysts contain bradyzoites, which are slow, multiplying forms. Cysts persist for the duration of the life of the host. By encysting, the parasite evades the host's immune response and ensures its viability (77).

When the cat ingests animal tissues containing cysts, proteolytic enzymes digest the cyst wall and the bradyzoites are released. They infect and multiply in the intestinal epithelial cells, transforming into tachyzoites. These in turn disperse via blood and lymph. When tissues of other animal species are ingested by felines, tachyzoites or bradyzoites begin the enteroepithelial cycle and sexual multiplication also occurs (Fig. 27.6). Macro- and microgametocytes are produced. After fertilization, the zygote differentiates into oocysts, which are passed in the feces (31, 32).

Although most infections occur by ingestion of contaminated meat, other foods, and water, they can also be acquired by organ transplantation or by blood transfusion. Disseminated toxoplasmosis may occur in those who received organ transplants and are receiving immunosuppressive therapy. Chorioretinitis is frequently observed in adults who acquire the infection. In immunosupppressed patients, toxoplasmosis can reactivate from latent infections (89).

Toxoplasmosis can also be acquired vertically by transplacental transmission when a pregnant woman gets infected. After multiplying in the placenta, tachyzoites spread into the fetal tissues (92). Infection can occur at any stage of the pregnancy, but the fetus is most affected when infection occurs during the first months

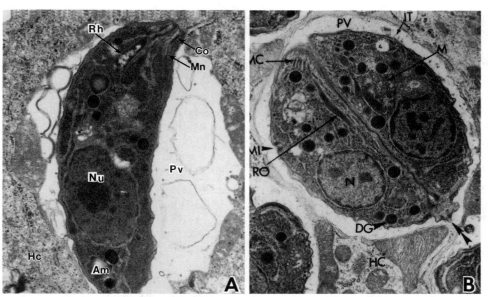

Figure 27.7 Transmission electron micrographs of *T. gondii*. (A) Sporozoite in parasitophorous vacuole (Pv) of host cell (Hc) at 24 h after inoculation; Am, amylopectin granule; Co, conoid; Mn, microneme; Nu, nucleus of sporozoite; Rh, rhoptry. (B) Final stage of endodyogeny to form two daughter tachyzoites that are still attached to the posterior ends (arrowheads); dense granules (DG); HC, host cell; IT, intravacular tubules; M, mitochondrion; MC, microneme; MI, micropore; N, nucleus; RO, rhoptry. From Dubey and Beattie (31).

of pregnancy. The majority of infected children do not show any signs of the disease until later in life, when they may present with chorioretinitis and mental retardation (31, 92).

The overall prevalence in humans and animals varies according to eating habits and lifestyle. The prevalence of *T. gondii* in swine is highest if they have outdoor access. Confined housing reduces the exposure to cat feces or infected rodents (36).

Toxoplasmosis can be acquired by ingestion of lamb, poultry, horse, and wild game. Cooking, freezing, or gamma irradiation will kill the *Toxoplasma* cysts and oocysts. Temperatures of 61°C or higher for 3.6 min will inactivate the parasites, and freezing at −13°C will result in nonviable cysts (54). Pyrimethamine in combination with folinic acid or trisulfapyrimidine is the treatment of choice for acute infections. TMP-SMX is effective and frequently used to prevent the recurrence of acute infections in AIDS patients (73).

MICROSPORIDIA

Most microsporidium infections have been reported in AIDS patients (37). Five genera of microsporidia have been associated with human infections: *Enterocytozoon*, *Septata*, *Pleistophora*, *Encephalitozoon*, and

Vittaforma. *Enterocytozoon bieneusi* is the only species that seems to be tissue specific, associated only with human enteric infections. The microsporidia spores, which are highly resistant environmental forms, vary in size among the different species. They are ovoid or piriform and 1 to 2 μm in diameter. *Enterocytozoon intestinalis* spores are bigger than those of *E. bieneusi*. These spores contain a polar filament that is used to eject the sporoplasm and penetrate the host cell cytoplasm. After entering the cell, the parasite multiplies asexually and eventually forms spores lysing the host cell and invading neighboring cells. Although the mechanisms of transmission are not clear, it is believed that it can be acquired by ingesting spores in contaminated water and produce (22).

Microsporidia infection is associated with watery and large-volume stools. These infections are occasionally associated with biliary tract disease and may potentially cause cholangiopathy observed in AIDS patients (15, 27).

Microsporidia spores can be histologically identified in tissue, but their identification in fecal samples requires special staining processes. Microsporidia can be identified in tissues by using various conventional staining procedures such as hematoxylin-eosin, Gram, and Giemsa. Identification in fecal samples can be achieved by using nonspecific staining techniques such as Calcofluor white and a modified trichrome with Chromotrope 2R; however, small structures and yeast spores also stain by using these procedures. Identification of microsporidia in the environment presents an additional challenge: most species of the animal kingdom can be parasitized by some other species of microsporidia which could be confused with those infectious to humans. Definite diagnosis is limited to transmission electron microscopy (16, 50, 71, 80). New tests using specific monoclonal antibodies and molecular tools such as PCR are being developed to aid in the diagnosis of microsporidia in tissue, fecal, and environmental samples (41, 62).

Infections caused by *Septata* can be treated with albendazole or with metronidazole and atovaquone. To date, there is no effective treatment for infections caused by *E. bieneusi* (18, 20, 58).

GIARDIA

Giardia is a protozoan flagellate that belongs to the phylum Zoomastigophora. *Giardia*, which was initially thought to be a commensal organism in humans, is now clearly recognized as a common cause of diarrhea and malabsorption. *Giardia* infects millions of people throughout the world in both epidemic and sporadic forms. Most human infections result from the ingestion of contaminated water or food or by direct fecal-oral transmission such as would occur in person-to-person contact in child care centers and in male homosexual activity (1).

Three species of *Giardia* have been described based on differences discernible in cysts and trophozoites by light microscopy: *G. agilis* from amphibians; *G. muris* from rodents, birds, and reptiles; and *G. lamblia* (also called *G. intestinalis* or *G. duodenalis*) from various mammals, including humans. Two additional species that are indistinguishable from *G. lamblia* by light microscopy, *Giardia ardeae* (heron) and *Giardia psittaci* (psittacine birds), have been identified based on ultrastructural morphologic differences (38, 88). *G. lamblia* does not appear to be host restricted, and wild animals, such as beavers and muskrats, have been implicated in waterborne outbreaks. More recently, molecular classification using small subunit rRNA has placed *Giardia* as one of the most primitive eukaryotic organisms (86).

Giardia can be observed in two forms: the trophozoite and the cyst. The cyst is the infectious form and is relatively inert and environmentally resistant. After cysts are ingested, excystation occurs in the duodenum after exposure to the acidic gastric pH and pancreatic enzymes, chemotrypsin and trypsin. Each cyst releases two vegetative trophozoites. The trophozoites replicate in the crypts of the duodenum and upper jejunum and reproduce asexually by binary fission. Some of the trophozoites then encyst in the ileum, possibly as a result of exposure to bile salts or from cholesterol starvation (1). The trophozoites and cysts are excreted in the feces.

Cysts are round or oval shaped. Each cyst measures 11 to 14 by 7 to 10 μm, has four nuclei, and contains axonemes and median bodies. Trophozoites have the shape of a teardrop (viewed dorsally or ventrally) and measure 10 to 20 μm in length by 5 to 15 μm in width. The trophozoite has a concave sucking disk with four pairs of flagella, two axonemes, two median bodies, and two nuclei. The ventral disks act as suction cups, allowing mechanical attachment to the surface of the intestine (Fig. 27.8) (38). Infections may result from the ingestion of 10 or fewer *Giardia* cysts (76). Boiling is very effective in inactivating *Giardia* cysts, but cysts can survive after freezing for a few days (76).

G. lamblia, which is prevalent worldwide, is especially common in areas where poor sanitary conditions and insufficient water treatment facilities prevail. Seasonality has been reported during late summer in the United Kingdom, the United States, and Mexico. The majority of cases of *Giardia* are asymptomatic, but they can present as chronic diarrhea. Travelers to areas of endemicity are at high risk for developing symptomatic giardiasis. In Leningrad (now St. Petersburg), Russia, 95% of travelers developed symptomatic giardiasis. Hikers and campers

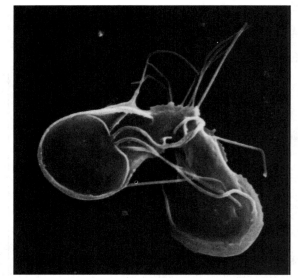

Figure 27.8 Scanning electron micrograph of *G. lamblia* trophozoites. One trophozoite shows the dorsal surface and the other shows the ventral surface with sucking disk and flagella.

are also at increased risk, since *Giardia* cysts, often of animal origin, can be found in freshwater lakes and streams. The prevalence of *Giardia* can be as high as 35% in children attending child care centers. Although these children are frequently asymptomatic, they may infect other family members who may develop symptomatic giardiasis (14, 19, 21, 51, 91).

In waterborne outbreaks of diarrhea in which the etiologic agent is identified, *Giardia* has been the most commonly identified agent. Waterborne transmission is commonly a result of inadequate water treatment or sewage contamination of drinking, well, or surface water. Giardiasis has also been associated with exposure to contaminated recreational water such as swimming pools. *Giardia* cysts are susceptible to inactivation by ozone and halogens; however, the concentration of chlorine used for drinking water may not inactivate *Giardia* cysts. Inactivation by chlorine requires prolonged contact time, and filtration is the recommended means for purifying water (51, 55, 60, 61, 90).

Symptomatic patients present with loose, foul-smelling stools and increased fat and mucus in fecal samples. Flatulence, abdominal cramps, bloating, and nausea are common, as are anorexia, malaise, and weight loss. Blood is not present in stools. Fever is occasionally present at the beginning of the infection. In contrast with most other forms of acute infectious diarrhea, *G. lamblia* infection results in prolonged symptoms. Although giardiasis may resolve spontaneously, the illness frequently lasts for several weeks if left untreated, and sometimes for months. Those with chronic giardiasis have profound malaise and diffuse epigastric and abdominal discomfort. Although diarrhea may persist, it may be replaced by constipation or even by normal bowel habits (1).

Malabsorption associated with giardiasis may be responsible for substantial weight loss. Even in asymptomatic infections, malabsorption of fats, carbohydrates, and vitamins may occur. Reduced intestinal disaccharidase activity may persist even after *Giardia* is eradicated. Lactase deficiency is the most common residual deficiency and occurs in 20 to 40% of cases (47).

Villous blunting, lymphocytic infiltration, and malabsorption are observed in biopsy samples of symptomatic cases. No tissue invasion is observed, and high numbers of trophozoites are sometimes present in the crypts without obvious pathology. To date, the presence of a toxin has not been demonstrated, and no other potential mechanisms by which *Giardia* causes diarrhea have been identified.

Giardia can be diagnosed by finding cysts or, less commonly, trophozoites in fecal specimens (Fig. 27.9A). *Giardia* can be detected in feces by enzyme immunoas-says, by indirect and direct immunofluorescent assays using monoclonal antibodies, or by PCR. All of these procedures are highly sensitive and specific for environmental and stool samples. In some patients with chronic diarrhea and malabsorption, stool examinations are repeatedly negative despite ongoing suspicion of giardiasis (3, 44, 93).

Effective treatment for patients with symptomatic giardiasis is mainly a single treatment course with metronidazole. In refractory cases, multiple or combination courses have occasionally been required. Tinidazole is widely used throughout the world, and a single dose is effective for treatment of giardiasis (45, 47).

BALANTIDIUM

Balantidium coli is a ciliate parasite that, although found worldwide, is not highly prevalent. It is a commensal parasite of pigs. The trophozoites reside in the large intestine and multiply by binary fission. In humans, it can cause ulcerative colitis and diarrhea. The ulcers differ from those caused by *Entamoeba histolytica* in that the epithelial surface is damaged but with more superficial lesions compared to those caused by amoebae. The parasite encysts and is excreted in the feces. The cyst, which is the environmentally resistant form, is large and oblong, 45 to 65 μm in diameter (Fig. 27.9B). Both cyst and trophozoite contain two nuclei. Trophozoites move via their cilia and rotate on their longitudinal axis (53). Treatment is preferentially with tetracycline and alternatively iodoquinol and metronidazole.

AMOEBIASIS

Various amoebae can infect humans as commensals; however, *E. histolytica* is pathogenic to humans. *Entamoeba dispar*, which is morphologically similar to *E. histolytica*, is not pathogenic. Cysts and trophozoites of *E. histolytica* are excreted in the feces of infected individuals (Fig. 27.9C), with cysts being environmentally resistant. Once cysts are ingested, the trophozoite excysts and colonizes the large intestine and multiplies by binary fission followed by encystation.

Most patients are asymptomatic even when shedding cysts in their feces. In other instances the parasites can invade the mucosa and cause an ulceration that goes from the luminal surface of the intestine, through the lamina propria and to the muscularis mucosa. The parasite then spreads laterally, forming a flask-shaped ulcer. The trophozoites feed on cell debris and red blood cells (78). Infection can progress, producing ulcerative amoebic colitis and causing perforation of the intestinal wall.

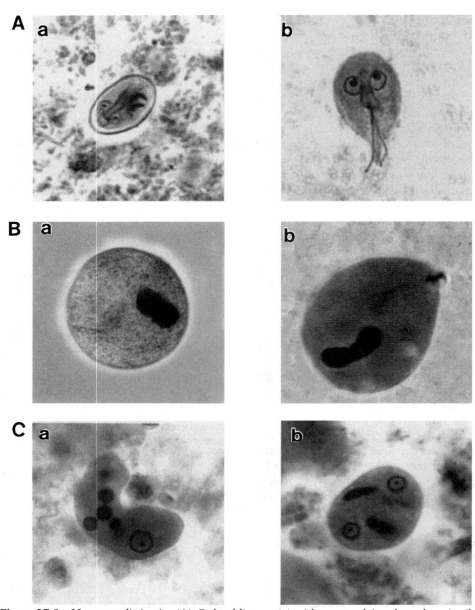

Figure 27.9 Hematoxylin/eosin. (A) *G. lamblia* cyst (a) with two nuclei and trophozoite (b) showing two nuclei and median body visible at one pole. (B) *B. coli* cyst (a) showing large macronucleus and cilia beneath cyst wall and trophozoite (b) showing oval macronucleus. (C) *E. histolytica* trophozoite (a) showing a nucleus and few red blood cells in the cytoplasm; cyst (b) showing two of four nuclei and rod-shaped inclusion bodies with rounded ends. (Courtesy of Lynne S. Garcia.)

Patients complain of diarrhea, with stools containing blood and mucus, back pain, tenesmus, dehydration, and abdominal tenderness. Fulminant colitis is characterized by severe bloody diarrhea, fever, and abdominal tenderness due to transmural necrosis of the bowel. Another presentation of amoebiasis is ameboma, which resembles a carcinoma and is not necessarily associated with pain. Amoebae can also disseminate to the liver. Complications with amoebiasis are observed when the liver parenchyma is gradually replaced with necrotic debris, inflammatory cells, and trophozoites. Patients present with hepatomegaly, weight loss, and anemia (39, 74).

Asymptomatic amoebiasis can be treated with iodoquinol, paromomycin, or diloxanide. If mild to severe and hepatic abcesses are present, amoebiasis can be treated with metronidazole or tinidazole followed by iodoquinol to treat asymptomatic amoebiasis.

CONCLUSION

A growing number of foodborne disease outbreaks caused by parasites are being observed. The consumption of fresh fruits and vegetables in developed countries has increased, and local production cannot meet this demand. Thus, importation of products from countries where sanitary standards are not controlled makes it even more difficult to control food-related diseases (4). In addition, inactivation of these parasites has been a challenging job.

We face new challenges to ensure the safety of produce. Enrichment media are not available for cultivating parasites; therefore, isolation and identification methodologies need to be extremely sensitive and specific to detect small numbers of contaminants that may be present in foods. Molecular tools such as PCR, restriction fragment length polymorphism, and variations of these techniques are being developed to improve the sensitivity and specificity of detection and identification processes.

Treatment of drinking water by chlorination at permissible concentrations is relatively ineffective at inactivating spores, cysts, and oocysts and cannot be recommended as a sole method for water treatment. Boiling can inactivate cysts, oocysts, and spores. Irradiation treatment of produce to inactivate pathogens that could be present has also been examined. This methodology effectively inactivates *Toxoplasma* cyst forms and oocysts. In addition, there is an increase in consumer acceptance of irradiated produce (34).

References

1. Adam, R. D. 1991. The biology of *Giardia* spp. *Microbiol. Rev.* 55:706–732.
2. Adam, R. D., Y. R. Ortega, R. H. Gilman, and C. R. Sterling. 2000. Intervening transcribed spacer region 1 variability in *Cyclospora cayetanensis*. *J. Clin. Microbiol.* 38:2339–2343.
3. Aldeen, W. E., D. Hale, A. J. Robison, and K. Carroll. 1995. Evaluation of a commercially available ELISA assay for detection of *Giardia lamblia* in fecal specimens. *Diagn. Microbiol. Infect. Dis.* 21:77–79.
4. Altekruse, S. F., M. L. Cohen, and D. L. Swerdlow. 1997. Emerging foodborne diseases. *Emerg. Infect. Dis.* 3:285–293.
5. Anonymous. 1997. From the Centers for Disease Control and Prevention. Outbreaks of *Escherichia coli* O157:H7 infection and cryptosporidiosis associated with drinking unpasteurized apple cider—Connecticut and New York, October 1996. *JAMA* 277:781–782.
6. Anonymous. 1997. Outbreak of cyclosporiasis—Northern Virginia–Washington, D.C.–Baltimore, Maryland, Metropolitan Area, 1997. *Morb. Mortal. Wkly. Rep.* 46:689–691.
7. Anonymous. 1997. Outbreaks of cyclosporiasis—United States, 1997. *Morb. Mortal. Wkly. Rep.* 46:451–452.
8. Anonymous. 1998. From the Centers for Disease Control and Prevention. Foodborne outbreak of cryptosporidiosis—Spokane, Washington, 1997. *JAMA* 280:595–596.
9. Anonymous. 2000. Outbreaks of diarrheal illness associated with Cyanobacteria (blue-green-algae)-like-bodies—Chicago and Nepal, 1989 and 1990. *Morb. Mortal. Wkly. Rep.* 40:325–327.
10. Anusz, K. Z., P. H. Mason, M. W. Riggs, and L. E. Perryman. 1990. Detection of *Cryptosporidium parvum* oocysts in bovine feces by monoclonal antibody capture enzyme-linked immunosorbent assay. *J. Clin. Microbiol.* 28:2770–2774.
11. Arrowood, M. J., and C. R. Sterling. 1989. Comparison of conventional staining methods and monoclonal antibody-based methods for *Cryptosporidium* oocyst detection. *J. Clin. Microbiol.* 27:1490–1495.
12. Ashford, R. W. 1979. Occurrence of an undescribed coccidian in man in Papua New Guinea. *Ann. Trop. Med. Parasitol.* 73:497–500.
13. Balatbat, A. B., G. W. Jordan, Y. J. Tang, and J. Silva, Jr. 1996. Detection of *Cryptosporidium parvum* DNA in human feces by nested PCR. *J. Clin. Microbiol.* 34:1769–1772.
14. Barbour, A. G., C. R. Nichols, and T. Fukushima. 1976. An outbreak of giardiasis in a group of campers. *Am. J. Trop. Med. Hyg.* 25:384–389.
15. Beaugerie, L., M. F. Teilhac, A. M. Deluol, J. Fritsch, P. M. Girard, W. Rozenbaum, Y. Le Quintrec, and F. P. Chatelet. 1992. Cholangiopathy associated with Microsporidia infection of the common bile duct mucosa in a patient with HIV infection. *Ann. Intern. Med.* 117:401–402.
16. Berlin, O. G., L. R. Ash, C. N. Conteas, and J. B. Peter. 1999. Rapid, hot chromotrope stain for detecting Microsporidia. *Clin. Infect. Dis.* 29:209.
17. Bern, C., B. Hernandez, M. B. Lopez, M. J. Arrowood, M. A. de Mejia, A. M. de Merida, A. W. Hightower, L. Venczel, B. L. Herwaldt, and R. E. Klein. 1999. Epidemiologic studies of *Cyclospora cayetanensis* in Guatemala. *Emerg. Infect. Dis.* 5:766–774.
18. Bicart-See, A., P. Massip, M. D. Linas, and A. Datry. 2000. Successful treatment with nitazoxanide of *Enterocytozoon bieneusi* microsporidiosis in a patient with AIDS. *Antimicrob. Agents Chemother.* 44:167–168.
19. Black, R. E. 1990. Epidemiology of travelers' diarrhea and relative importance of various pathogens. *Rev. Infect. Dis.* 12(Suppl. 1):S73–S79.
20. Blanshard, C., D. S. Ellis, D. G. Tovey, S. Dowell, and B. G. Gazzard. 1992. Treatment of intestinal microsporidiosis with albendazole in patients with AIDS. *AIDS* 6:311–313.
21. Brodsky, R. E., H. C. Spencer, Jr., and M. G. Schultz.

1974. Giardiasis in American travelers to the Soviet Union. *J. Infect. Dis.* **130:**319–323.

22. **Bryan, R. T.** 1995. Microsporidiosis as an AIDS-related opportunistic infection. *Clin. Infect. Dis.* **21**(Suppl. 1): S62–S65.

23. **Buchanan, R.** 1998. Principles of risk assessment for illness caused by foodborne biological agents. National Advisory Committee on Microbiological Criteria for Foods. *J. Food Prot.* **61:**1071–1074.

24. **Cama, V. A., and C. R. Sterling.** 1991. Hyperimmune hens as a novel source of anti-*Cryptosporidium* antibodies suitable for passive immune transfer. *J. Protozool.* **38:**42S–43S.

25. **Casemore, D. P., M. Armstrong, and R. L. Sands.** 1985. Laboratory diagnosis of cryptosporidiosis. *J. Clin. Pathol.* **38:**1337–1341.

26. **Connor, B. A., J. Reidy, and R. Soave.** 1999. Cyclosporiasis: clinical and histopathologic correlates. *Clin. Infect. Dis.* **28:**1216–1222.

27. **Conteas, C. N., E. S. Didier, and O. G. Berlin.** 1997. Workup of gastrointestinal microsporidiosis. *Dig. Dis.* **15:**330–345.

28. **Current, W. L., and P. H. Bick.** 1989. Immunobiology of *Cryptosporidium spp. Pathol. Immunopathol. Res.* **8:**141–160.

29. **Curry, A., and H. V. Smith.** 1998. Emerging pathogens: *Isospora, Cyclospora* and *Microsporidia. Parasitology* **117** (Suppl.):S143–S159.

30. **Danziger, L. H., T. P. Kanyok, and R. M. Novak.** 1993. Treatment of cryptosporidial diarrhea in an AIDS patient with paromomycin. *Ann. Pharmacother.* **27:**1460–1462.

31. **Dubey, J. P., and C. P. Beattie.** 1988. *Toxoplasmosis of Animals and Man.* CRC Press, Inc., Boca Raton, Fla.

32. **Dubey, J. P., D. S. Lindsay, and C. A. Speer.** 1998. Structures of *Toxoplasma gondii* tachyzoites, bradyzoites, and sporozoites and biology and development of tissue cysts. *Clin. Microbiol. Rev.* **11:**267–299.

33. **Dubey, J. P., C. A. Speer, and R. Fayer.** 1990. *Cryptosporidiosis of Man and Animals.* CRC Press, Inc, Boca Raton, Fla.

34. **Dubey, J. P., D. W. Thayer, C. A. Speer, and S. K. Shen.** 1998. Effect of gamma irradiation on unsporulated and sporulated *Toxoplasma gondii* oocysts. *Int. J. Parasitol.* **28:**369–375.

35. **Eberhard, M. L., A. J. da Silva, B. G. Lilley, and N. J. Pieniazek.** 1999. Morphologic and molecular characterization of new *Cyclospora* species from Ethiopian monkeys: *C. cercopitheci sp.n., C. colobi sp.n.,* and *C. papionis sp.n. Emerg. Infect. Dis.* **5:**651–658.

36. **Edelhofer, R.** 1994. Prevalence of antibodies against *Toxoplasma gondii* in pigs in Austria—an evaluation of data from 1982 and 1992. *Parasitol. Res.* **80:**642–644.

37. **Eeftinck Schattenkerk, J. K., and T. van Gool.** 1992. Clinical and microbiological aspects of microsporidiosis. *Trop. Geogr. Med.* **44:**287.

38. **Erlandsen, S. L., and W. J. Bemrick.** 1987. SEM evidence for a new species, *Giardia psittaci. J. Parasitol.* **73:**623–629.

39. **Espinosa-Cantellano, M., and A. Martinez-Palomo.** 2000. Pathogenesis of intestinal amebiasis: from molecules to disease. *Clin. Microbiol. Rev.* **13:**318–331.

40. **Fayer, R., M. Tilley, S. J. Upton, A. J. Guidry, D. W. Thayer, M. Hildreth, and J. Thomson.** 1991. Production and preparation of hyperimmune bovine colostrum for passive immunotherapy of cryptosporidiosis. *J. Protozool.* **38:**38S–39S.

41. **Fedorko, D. P., N. A. Nelson, and C. P. Cartwright.** 1995. Identification of Microsporidia in stool specimens by using PCR and restriction endonucleases. *J. Clin. Microbiol.* **33:**1739–1741.

42. **Furio, M. M., and C. J. Wordell.** 1985. Treatment of infectious complications of acquired immunodeficiency syndrome. *Clin. Pharm.* **4:**539–554.

43. **Galvagno, G., G. Cattaneo, and E. Reverso-Giovantin.** 1993. [Chronic diarrhea due to *Cryptosporidium*: the efficacy of spiramycin treatment]. *Pediatr. Med. Chir.* **15:**297–298.

44. **Green, E., D. Warhurst, J. Williams, T. Dickens, and M. Miles.** 1990. Application of a capture enzyme immunoassay in an outbreak of waterborne giardiasis in the United Kingdom. *Eur. J. Clin. Microbiol. Infect. Dis.* **9:**424–428.

45. **Guerreiro, N. M., P. M. Herrera, L. de Escalona, C. E. de Kolster, V. G. de Yanes, O. de Febres, O. Naveda, and M. de Naveda.** 1991. [*Giardia lamblia*: comparison of two diagnostic methods and evaluation of response to treatment with metronidazole]. *G.E.N.* **45:**105–110.

46. **Herwaldt, B. L., and M. L. Ackers.** 1997. An outbreak in 1996 of cyclosporiasis associated with imported raspberries. The Cyclospora Working Group. *N. Engl. J. Med.* **336:**1548–1556.

47. **Hill, D. R.** 1993. Giardiasis. Issues in diagnosis and management. *Infect. Dis. Clin. North Am.* **7:**503–525.

48. **Hoge, C. W., D. R. Shlim, M. Ghimire, J. G. Rabold, P. Pandey, A. Walch, R. Rajah, P. Gaudio, and P. Echeverria.** 1995. Placebo-controlled trial of cotrimoxazole for *Cyclospora* infections among travellers and foreign residents in Nepal. *Lancet* **345:**691–693. (Erratum, 345(8956):1060, 1995.)

49. **Hoge, C. W., D. R. Shlim, R. Rajah, J. Triplett, M. Shear, J. G. Rabold, and P. Echeverria.** 1993. Epidemiology of diarrhoeal illness associated with coccidian-like organisms among travellers and foreign residents in Nepal. *Lancet* **341:**1175–1179.

50. **Hollister, W. S., and E. U. Canning.** 1987. An enzyme-linked immunosorbent assay (ELISA) for detection of antibodies to *Encephalitozoon cuniculi* and its use in determination of infections in man. *Parasitology* **94**(Pt. 2):209–219.

51. **Hopkins, R. S., P. Shillam, B. Gaspard, L. Eisnach, and R. J. Karlin.** 1985. Waterborne disease in Colorado: three years' surveillance and 18 outbreaks. *Am. J. Public Health* **75:**254–257.

52. **Jinneman, K. C., J. H. Wetherington, W. E. Hill, A. M. Adams, J. M. Johnson, B. J. Tenge, N. L. Dang, R. L. Manger, and M. M. Wekell.** 1998. Template preparation for PCR and RFLP of amplification products for the detection and identification of *Cyclospora* sp. and *Eimeria* sp.

oocysts directly from raspberries. *J. Food Prot.* **61**:1497–1503.

53. Karanis, P., D. Schoenen, W. A. Maier, and H. M. Seitz. 1993. [Drinking water and parasites]. *Immun. Infekt.* **21**:132–136.

54. Kolychev, V. V. 1967. [Viability of *Toxoplasma* under various environmental conditions]. *Veterinariia* **44**:58–59.

55. Lopez, C. E., A. C. Dykes, D. D. Juranek, S. P. Sinclair, J. M. Conn, R. W. Christie, E. C. Lippy, M. G. Schultz, and M. H. Mires. 1980. Waterborne giardiasis: a community-wide outbreak of disease and a high rate of asymptomatic infection. *Am. J. Epidemiol.* **112**:495–507.

56. Madico, G., R. H. Gilman, E. Miranda, L. Cabrera, and C. R. Sterling. 1993. Treatment of *Cyclospora* infections with co-trimoxazole. *Lancet* **342**:122–123.

57. Madico, G., J. McDonald, R. H. Gilman, L. Cabrera, and C. R. Sterling. 1997. Epidemiology and treatment of *Cyclospora cayetanensis* infection in Peruvian children. *Clin. Infect. Dis.* **24**:977–981.

58. Marshall, M. M., D. Naumovitz, Y. Ortega, and C. R. Sterling. 1997. Waterborne protozoan pathogens. *Clin. Microbiol. Rev.* **10**:67–85.

59. Millard, P. S., K. F. Gensheimer, D. G. Addiss, D. M. Sosin, G. A. Beckett, A. Houck-Jankoski, and A. Hudson. 1994. An outbreak of cryptosporidiosis from fresh-pressed apple cider. *JAMA* **272**:1592–1596.

60. Moorehead, W. P., R. Guasparini, C. A. Donovan, R. G. Mathias, R. Cottle, and G. Baytalan. 1990. Giardiasis outbreak from a chlorinated community water supply. *Can. J. Public Health* **81**:358–362.

61. Navin, T. R., D. D. Juranek, M. Ford, D. J. Minedew, E. C. Lippy, and R. A. Pollard. 1985. Case-control study of waterborne giardiasis in Reno, Nevada. *Am. J. Epidemiol.* **122**:269–275.

62. Ombrouck, C., L. Ciceron, and I. Desportes-Livage. 1996. Specific and rapid detection of Microsporidia in stool specimens from AIDS patients by PCR. *Parasite* **3**:85–86.

63. Ortega, Y. R., R. H. Gilman, and C. R. Sterling. 1994. A new coccidian parasite (Apicomplexa: Eimeriidae) from humans. *J. Parasitol.* **80**:625–629.

64. Ortega, Y. R., R. Nagle, R. H. Gilman, J. Watanabe, J. Miyagui, H. Quispe, P. Kanagusuku, C. Roxas, and C. R. Sterling. 1997. Pathologic and clinical findings in patients with cyclosporiasis and a description of intracellular parasite life-cycle stages. *J. Infect. Dis.* **176**:1584–1589.

65. Ortega, Y. R., C. R. Roxas, R. H. Gilman, N. J. Miller, L. Cabrera, C. Taquiri, and C. R. Sterling. 1997. Isolation of *Cryptosporidium parvum* and *Cyclospora cayetanensis* from vegetables collected in markets of an endemic region in Peru. *Am. J. Trop. Med. Hyg.* **57**:683–686.

66. Ortega, Y. R., C. R. Sterling, R. H. Gilman, V. A. Cama, and F. Diaz. 1993. *Cyclospora* species—a new protozoan pathogen of humans. *N. Engl. J. Med.* **328**:1308–1312.

67. Pape, J. W., R. I. Verdier, M. Boncy, J. Boncy, and W. D. Johnson, Jr. 1994. *Cyclospora* infection in adults infected with HIV. Clinical manifestations, treatment, and prophylaxis. *Ann. Intern. Med.* **121**:654–657.

68. Peng, M. M., L. Xiao, A. R. Freeman, M. J. Arrowood, A. A. Escalante, A. C. Weltman, C. S. Ong, W. R.

MacKenzie, A. A. Lal, and C. B. Beard. 1997. Genetic polymorphism among *Cryptosporidium parvum* isolates: evidence of two distinct human transmission cycles. *Emerg. Infect. Dis.* **3**:567–573.

69. Perryman, L. E., and J. M. Bjorneby. 1991. Immunotherapy of cryptosporidiosis in immunodeficient animal models. *J. Protozool.* **38**:98S–100S.

70. Pollok, R. C., and M. J. Farthing. 1999. Managing gastrointestinal parasite infections in AIDS. *Trop. Doct.* **29**:238–241.

71. Punpoowong, B., P. Pitisuttithum, D. Chindanond, D. Phiboonnakit, and S. Leelasuphasri. 1995. Microsporidium: modified technique for light microscopic diagnosis. *J. Med. Assoc. Thai.* **78**:251–254.

72. Rabold, J. G., C. W. Hoge, D. R. Shlim, C. Kefford, R. Rajah, and P. Echeverria. 1994. *Cyclospora* outbreak associated with chlorinated drinking water. *Lancet* **344**:1360–1361. (Letter.)

73. Ramakrishna, B. S. 1999. Prevalence of intestinal pathogens in HIV patients with diarrhea: implications for treatment. *Indian J. Pediatr.* **66**:85–91.

74. Ravdin, J. I., and R. L. Guerrant. 1982. A review of the parasite cellular mechanisms involved in the pathogenesis of amebiasis. *Rev. Infect. Dis.* **4**:1185–1207.

75. Reduker, D. W., C. A. Speer, and J. A. Blixt. 1985. Ultrastructure of *Cryptosporidium parvum* oocysts and excysting sporozoites as revealed by high resolution scanning electron microscopy. *J. Protozool.* **32**:708–711.

76. Rose, J. B., C. N. Haas, and S. Regli. 1991. Risk assessment and control of waterborne giardiasis. *Am. J. Public Health* **81**:709–713.

77. Schwartzman, J. D., J. C. Boothroyd, and L. H. Kasper. 1994. *Toxoplasma* workshop overview. *J. Eukaryot. Microbiol.* **41**:19S–21S.

78. Sepulveda, B. 1982. Amebiasis: host-pathogen biology. *Rev. Infect. Dis.* **4**:1247–1253.

79. Shlim, D. R., C. W. Hoge, R. Rajah, R. M. Scott, P. Pandy, and P. Echeverria. 1999. Persistent high risk of diarrhea among foreigners in Nepal during the first 2 years of residence. *Clin. Infect. Dis.* **29**:613–616.

80. Sironi, M., C. Bandi, S. Novati, and M. Scaglia. 1997. A PCR-RFLP method for the detection and species identification of human Microsporidia. *Parasitologia* **39**:437–439.

81. Smith, H. V., C. A. Paton, R. W. Girdwood, and M. M. Mtambo. 1996. *Cyclospora* in non-human primates in Gombe, Tanzania. *Vet. Rec.* **138**:528.

82. Soave, R. 1988. Cryptosporidiosis and isosporiasis in patients with AIDS. *Infect. Dis. Clin. North Am.* **2**:485–493.

83. Soave, R. 1996. Cyclospora: an overview. *Clin. Infect. Dis.* **23**:429–435.

84. Soave, R., and W. D. Johnson, Jr. 1988. *Cryptosporidium* and *Isospora belli* infections. *J. Infect. Dis.* **157**:225–229.

85. Tee, G. H., A. H. Moody, A. H. Cooke, and P. L. Chiodini. 1993. Comparison of techniques for detecting antigens of *Giardia lamblia* and *Cryptosporidium parvum* in faeces. *J. Clin. Pathol.* **46**:555–558.

86. Thompson, R. C., and A. J. Lymbery. 1996. Genetic

variability in parasites and host-parasite interactions. *Parasitology* **112**(Suppl.):S7–22.

87. **Ungar, B. L.** 1990. Enzyme-linked immunoassay for detection of *Cryptosporidium* antigens in fecal specimens. *J. Clin. Microbiol.* **28:**2491–2495.

88. **van Keulen, H., S. R. Campbell, S. L. Erlandsen, and E. L. Jarroll.** 1991. Cloning and restriction enzyme mapping of ribosomal DNA of *Giardia duodenalis*, *Giardia ardeae* and *Giardia muris*. *Mol. Biochem. Parasitol.* **46:**275–284.

89. **Wong, B.** 1984. Parasitic diseases in immunocompromised hosts. *Am. J. Med.* **76:**479–486.

90. **Wright, M. S., and P. A. Collins.** 1997. Waterborne transmission of *Cryptosporidium*, *Cyclospora* and *Giardia*. *Clin. Lab Sci.* **10:**287–290.

91. **Wright, R. A., H. C. Spencer, R. E. Brodsky, and T. M. Vernon.** 1977. Giardiasis in Colorado: an epidemiologic study. *Am. J. Epidemiol.* **105:**330–336.

92. **Zardi, O., and B. Soubotian.** 1979. Biology of *Toxoplasma gondii*, its survival in body tissues and liquids, risks for the pregnant woman. *Biochem. Exp. Biol.* **15:**355–360.

93. **Zimmerman, S. K., and C. A. Needham.** 1995. Comparison of conventional stool concentration and preserved-smear methods with Merifluor Cryptosporidium/Giardia Direct Immunofluorescence Assay and ProSpecT Giardia EZ Microplate Assay for detection of *Giardia lamblia*. *J. Clin. Microbiol.* **33:**1942–1943.

94. **Zu, S. X., and R. L. Guerrant.** 1993. Cryptosporidiosis. *J. Trop. Pediatr.* **39:**132–136.

Preservatives and Preservation Methods

VII

Food Microbiology: Fundamentals and Frontiers, 2nd Ed.
Edited by M. P. Doyle et al.
© 2001 ASM Press, Washington, D.C.

József Farkas

Physical Methods of Food Preservation

28

Whenever the values of environmental stress factors discussed in chapter 2 are outside the vital range for growth and survival of a specific microorganism, cellular damage results. Depending on the severity of this effect, growth may be inhibited (microbistatic effect or reversible inhibition) or cells may be killed (microbicidal effect). Food spoilage can be prevented by applying one or more of the following strategies: inhibition of microbial growth, destruction (irreversible inactivation) of microbial cells, or mechanical removal of microorganisms from the food.

Physical methods of food preservation are those that utilize physical treatments to inhibit, destroy, or remove undesirable microorganisms without involving antimicrobial additives or products of microbial metabolism as preservative factors. Microbial growth can be inhibited by physical dehydration processes (drying, freeze-drying, and freeze concentration), cold storage, or freezing and frozen storage. Microorganisms can be destroyed (irreversibly inactivated) by established physical microbicide treatments, such as heating (including microwave heat treatment); UV or ionizing radiation; emerging methods of new nonthermal treatments, such as high hydrostatic pressure, pulsed electric fields, oscillating magnetic fields, or photodynamic effects; or a combination of physical processes, such as heat-irradiation, dehydro-irradiation,

or mano-thermo-sonication. Mechanical removal of microorganisms from food may be accomplished by membrane filtration of food liquids. This chapter discusses the microbiological fundamentals of the physical preservation methods outlined above, with the exception of mechanical removal. Wherever possible, the mechanisms and underlying principles are explained.

PHYSICAL DEHYDRATION PROCESSES

Water is one of the most important factors controlling the rate of deterioration of food, either by microbial or nonmicrobial effects. Knowledge of the moisture content alone is not sufficient to predict the stability of foods, because it is not the total moisture content but rather the availability of water that determines the shelf life of a food. A proportion of the total water in food is strongly bound to specific sites.

The availability of water is measured by the water activity (a_w) of a food, defined as the ratio of the vapor pressure of water in a food, P, to the vapor pressure of pure water, P_0, at the same temperature:

$$a_w = \frac{P}{P_0}$$

József Farkas, Department of Refrigeration and Livestock Products' Technology, Szent István University, Ménesi út 45, H-1118 Budapest, Hungary.

The relation of water activity to solute concentration is expressed by Raoult's law for "ideal" solutions:

$$a_w = \frac{P}{P_0} = \frac{n_2}{n_1 + n_2}$$

where n_1 and n_2 are the number of moles of solute and solvent, respectively. An ideal solution may be defined as one in which the molecules of a solute are affected by their environment in exactly the same manner as they are in a pure state at the same temperature and pressure as the solution. In reality, there are no such solutions, because interactions between molecules may reduce the effective number of particles in solution, or dissociation may greatly increase the number. Departures from the ideal are large for nonelectrolytes at concentrations above 1 M (molality is the number of gram moles per kilogram of water) and for electrolytes at all concentrations.

The movement of water vapor from a food to the surrounding air depends on the moisture content, the composition of the food, the temperature, and the humidity of the air. At constant temperature, the moisture content of food changes until it comes into equilibrium with water vapor in the surrounding air. This is called the equilibrium moisture content of the food. At the equilibrium moisture content, the food neither gains nor loses weight on storage under those conditions. The relative humidity of the surrounding air is then the equilibrium relative humidity (ERH). When different values of relative humidity versus equilibrium moisture content are plotted, a curve known as a water sorption isotherm is obtained. ERH is related to a_w by the expression ERH (%) = $a_w \times 100$.

Fresh, raw, high-moisture food materials such as fruits, vegetables, meats, and fish have a_w levels of 0.98 or higher. Food preservation by dehydration is based on the principle that microbial growth is inhibited if the available water required for the growth is removed, i.e., if the water activity is reduced (88). Technologies for drying solid foods remove water by hot air; freeze-drying does so by sublimation after freezing. Liquid foods can be dried by freeze concentration, whereby freezing is followed by mechanical removal of frozen water.

These processes reduce the a_w of both the food and the microbial cells to levels insufficient for microbial growth. Water activities of various raw materials and food products are compared in Table 28.1. The microbiological stability of many intermediate-moisture foods with a_w levels of 0.7 to 0.85 (jams, sausages, etc.) depends also on other preservative factors (reduced pH, preservatives, pasteurization, etc.); however, their reduced a_w is of major importance. Because of their chemical composition

Table 28.1 Water activities of various foods[a]

Food	a_w
Fresh, raw fruits, vegetables, meat, and fish	>0.98
Cooked meat, bread	0.98–0.95
Cured meat products, cheeses	0.95–0.91
Sausages, syrups	0.91–0.87
Flours, rice, beans, peas	0.87–0.80
Jams, marmalades	0.80–0.75
Candies	0.75–0.65
Dried fruits	0.65–0.60
Dehydrated vermicelli, spices, milk powder	0.60–0.20

[a] Adapted from references 26 and 162.

and the water binding capacity of their components, various foods may have very different moisture contents at the same a_w level, as is illustrated by Table 28.2.

The a_w requirements of various microorganisms vary significantly (70). In the vital range of growth, decreasing the a_w increases the lag phase of growth and decreases the growth rate. Foodborne microorganisms are grouped according to their minimal a_w requirements in Table 28.3.

In general, among bacteria, gram-negative species have the highest a_w requirement. Important gram-negative bacteria such as *Pseudomonas* spp. and most members of the family *Enterobacteriaceae* can usually grow only above a_w of 0.96 and 0.93, respectively. Less sensitive to reduced a_w are gram-positive non-spore-forming bacteria. While many members of the family *Lactobacillaceae* have minimum a_w near 0.94, some representatives of the family *Micrococcaceae* are capable of growth below a_w of 0.90. Staphylococci are unique among nonhalophilic bacteria in being capable of growing at high levels of NaCl. The generally recognized minimum a_w for growth of *Staphylococcus aureus* is about

Table 28.2 Moisture content of various dry or dehydrated food products when their a_w is 0.7 at 20°C[a]

Food	% Moisture content
Various seed grains	4–9
Milk powder	7–10
Cocoa powder	7–10
Whole egg powder	10–11
Skim milk powder	10–15
Dried, fat-free meat	10–15
Rice and legume seeds	12–15
Dehydrated vegetables	12–22
Dried soups	13–21
Dried fruits	18–25

[a] Adapted from references 22, 26, 109, 110, and 162.

Table 28.3 Minimal a_w levels required for growth of foodborne microorganisms at 25°C[a]

Group of microorganisms	Minimal a_w required
Most bacteria	0.91–0.88
Most yeasts	0.88
"Regular" molds	0.80
Halophilic bacteria	0.75
Xerotolerant molds	0.71
"Xerophilic" molds and "osmophilic" yeasts	0.62–0.60

[a] Adapted from references 26, 70, 109, and 162.

0.86 (70); however, under otherwise ideal conditions, its growth has been demonstrated at as low an a_w as 0.83. Production of staphylococcal enterotoxin in food slurries has not been observed below a_w of 0.93 (163). Most spore-forming bacteria do not grow below a_w of 0.93. Germination and outgrowth of spores of food-poisoning strains of *Bacillus cereus* are prevented at a_w of 0.97 to 0.93, depending on the nature of the bulk solute (72). Various serological types of *Clostridium botulinum* differ in their a_w requirements. The lower a_w limits in salt-adjusted media for growth from spore inocula are 0.95 for type A, 0.94 for type B, and 0.97 for type E (118). Toxin production occurs at a_w levels closely approaching these growth minima. The minimum a_w for the growth and spore germination of *Clostridium perfringens* is between 0.97 and 0.95 in complex media when sucrose or NaCl is used to adjust a_w (78). For both *C. botulinum* and *C. perfringens*, growth proceeds at lower a_w levels when it is controlled by glycerol instead of by salt (6).

Several species of yeasts have a_w requirements much lower than those of bacteria. Salt-tolerant species such as *Debaryomyces hansenii*, *Hansenula anomala*, and *Candida pseudotropicalis* may grow well on cured meats and pickles at NaCl concentrations of up to 11% ($a_w = 0.93$). Some "osmophilic" species (such as *Zygosaccharomyces rouxii*, *Zygosaccharomyces baillii*, and *Zygosaccharomyces bisporus*) are able to grow in food of high sugar content (jams, honey, syrups). Terms such as "xerotolerant" or "saccharotolerant" would be more appropriate for these "osmophilic" yeasts.

In general, molds have a_w requirements much lower than those of other groups of foodborne microorganisms (5). However, molds with sporangiospores have greater moisture requirements, and germination of their spores may be inhibited at $a_w = 0.93$. The most common xerotolerant molds belong to the genus *Eurotium* (their asexual forms are the members of the *Aspergillus glaucus* group). The minimal a_w for their growth is in the range of 0.71 to 0.77, while the optimal a_w is 0.96. True xerophilic molds (128), such as *Monascus (Xeromyces) bisporus* do not grow at a_w levels greater than 0.97 to 0.99. The relationship of a_w to mold growth and toxin formation is complex (142). It is of great public health importance that the minimal a_w requirement for mycotoxin production is generally higher than that for growth of toxigenic molds (Table 28.4). That is, under many conditions where molds grow, they do not produce toxins.

Differences in a_w limits for growth reflect differences in osmoregulatory capacities (51). Mechanisms of tolerance to low a_w are different in bacteria and fungi; the key point, however, is that the cell osmoregulation mechanism operates to maintain homeostasis with respect to water content (51). In general, the strategy employed by microorganisms as protection against osmotic stress is the intracellular accumulation of compatible solutes, i.e., compounds that have the general property of interfering minimally with the metabolic activities of the cell. While some bacteria accumulate K^+ ions and amino acids, such as proline, as a response to low a_w (12), halotolerant and xerotolerant fungi concentrate polyols such as glycerol, erythritol, and arabitol (32). These polyols not only act as osmoregulators but also prevent inhibition or

Table 28.4 Minimal a_w requirements for growth and mycotoxin production of some toxigenic molds[a]

Mycotoxins	Mold	Minimal a_w requirements for:	
		Growth	Toxin production
Aflatoxins	*Aspergillus flavus*	0.82	0.83–0.87
	Aspergillus parasiticus	0.82	0.87
Ochratoxin	*Aspergillus ochraceus*	0.77	0.85
	Penicillium cyclopium	0.82–0.85	0.87–0.90
Patulin	*Penicillium expansum*	0.81	0.99
	Penicillium patulum	0.81	0.95
Stachybotryn	*Stachybotrys atra*	0.94	0.94

[a] Adapted from references 22, 26, and 130.

inactivation of enzymes and probably serve as nutrient reserves (13). In the case of halophilic bacteria, KCl is a requirement, while osmophilic yeasts have a high tolerance for high solute concentrations (11).

Drying

During hot air drying of vegetables, vermicelli, and similar solid foods, the airstream has two functions; it transfers heat to evaporate the water from the raw materials, and it carries away the vapor produced. In the course of drying, microorganisms on the raw materials are affected by both the drying temperature and changes of a_w. Microbiological consequences of the drying technology are influenced by a number of other factors, e.g., size and composition of food pieces and time-temperature combinations.

During the warming-up period of drying, the temperature is still low and the relative humidity is high. The length of this phase depends mainly on the size of food particles. If this phase is long, microorganisms may even grow during the slow increase of temperature in the range of 20 to 40°C under the existing high a_w. During later phases of drying, the temperature exceeds 50 to 70°C. Although there is no opportunity for growth, heat destruction is not significant either, because the higher temperature develops parallel to decreased moisture content. Wet heat is more lethal than dry heat. If the drying lasts long enough (at least 30 min), time-temperature combinations which irreversibly damage microbial cells occur. The temperature and moisture distribution within the food are important with respect to the thermal inactivation of microorganisms. With certain drying technologies, the surface temperature may even reach 100°C, while the internal temperature remains lower. When microorganisms are mainly on the product surface, in the drying process their inactivation is enhanced. On the other hand, the moisture content of the food surface decreases rapidly with heating, thereby decreasing the heat sensitivity of microbial cells. Therefore, the lethal combination of high temperature and high humidity is rare in drying technologies. In fact, a major microbicidal effect is due to high temperature-low humidity conditions which last for long periods of time. Loss of viability of microorganisms continues during storage because reversibly damaged cells, being unable to regenerate at low a_w, gradually die. Further viability losses may occur during dehydration of dried products, especially when the rehydration is rapid, causing large differences in osmotic pressure.

The a_w of dried foods is usually considerably lower than the critical value for microbial growth. Therefore, dried products are stable microbiologically, provided that the relative humidity of storage atmosphere is less than 70%. If the ERH increases or the temperature changes, conditions on the product surface may change and permit the growth of xerotolerant molds (162). Of course, the microbiological quality of dried foods depends also on the microbiological quality of raw materials and their handling prior to drying (103).

Freeze-Drying

Freeze-drying (lyophilization) is a combined method of food preservation. Its principle is dehydration of the food in the frozen state through vacuum sublimation of its ice content (48). This method of water removal is very gentle. In other methods for dehydrating solid foods, the distribution of solutes in the food changes owing to a continuous transport of solutions toward the surface and increased solute loss when they reach the surface. During freeze-drying, the sublimation front moves and forwards the core, and the ice sublimates in the location where it is formed. Thus, solutes remain inside the food at their original location, their loss is insignificant, and the food retains its original form and structure very well. As a result, rehydration of freeze-dried foods is very fast and almost complete, and such foods regain 90 to 95% of their original moisture contents. On the other hand, the very large specific surface of freeze-dried products makes them vulnerable to other damage, such as lipid oxidation. To prevent spontaneous rehydration, freeze-dried foods should be kept in vapor-impermeable packaging. They should be packaged in an inert atmosphere to delay oxidation.

The effect of freeze-drying of foods on their microbial contaminants is a combination of decreased a_w by freezing and further reduction of a_w by sublimation of the ice. The extent of cell damage in these two phases may be different, depending on the conditions (temperatures and rates) of freezing and sublimation. Microbial survival also depends on the composition of freeze-dried food. Carbohydrates, proteins, and collodial substances, in general, have a protective effect on the microorganisms. During storage of freeze-dried products with 2 to 5% residual moisture content, a gradual, slow decrease of cell viability occurs, similar to that of conventionally dried foods, especially in the presence of atmospheric oxygen. Again, there may be additional microbial destruction during rehydration of freeze-dried products because large differences in solution concentrations develop suddenly.

Because of the mild manner of moisture removal, lyophilization conditions can be intentionally optimized for maximal survival of microbial cells (i.e., preservation of stock cultures). Low dehydration temperature, protective additives (such as glycerol) in the microbe suspension, and storage of lyophilized cultures in a nitrogen atmosphere or in vacuo increase culture viability. Gram-positive bacteria survive freeze-drying better than gram-negative bacteria do.

COOL STORAGE

Cool (or chill) storage generally refers to storage at temperatures above freezing, from about 16°C down to −2°C (135). While pure water freezes at 0°C, most foods remain unfrozen until a temperature of about −2°C or lower is reached. One should bear in mind, however, that many fresh fruits and vegetables may suffer from a chilling injury when they are kept below critical temperatures of 4 to 12°C. Depending on the inherent storability of raw foods, the duration of cool storage may vary from a few days to several weeks (Table 28.5).

The preservation effect of cool storage (refrigeration) is based on the fact that by reducing the temperature, the rates of chemical reactions and growth of microorganisms are decreased. The temperature proper for cool storage is generally less than the minimal growth temperature of most foodborne microorganisms (see chapter 2). The generation time of psychrotrophic microorganisms is considerably increased at low temperature. The refrigeration temperatures allow the growth of psychrophilic and psychrotrophic microorganisms (141). If their initial population is high, refrigerated foods can spoil in a short time. Since the temperature requirements for various microorganisms differ, refrigeration may considerably change the qualitative composition of the microbiota.

Because growth rates of microorganisms increase significantly with an increase in temperature of only a few degrees, variation of storage temperatures should be avoided. Psychrotrophic pathogens such as *Yersinia enterocolitica*, *Vibrio parahaemolyticus*, *Listeria monocytogenes*, and *Aeromonas hydrophila* (see section III of

Table 28.5 Shelf-life extension of raw foods by cool storage[a]

Food	Avg useful storage life (days) at:	
	0°C (32°F)	22°C (72°F)
Meat	6–10	1
Fish	2–7	1
Poultry	5–18	1
Fruits	2–180	1–20
Leafy vegetables	3–20	1–7
Root crops	90–300	7–50

[a] Adapted from reference 135.

this volume) are important, especially in "minimally processed, extended shelf-life" products (100).

There is some evidence that the minimum growth temperature of microorganisms is determined by the inhibition of solute transport. In this regard, it has been proposed that the growth temperature range of an organism depends on how well the organism can regulate its lipid fluidity within a given range (41). Psychrotrophs contain increased amounts of unsaturated fatty acid residues in their lipids when grown at low temperatures. This increase in the degree of unsaturation of fatty acids leads to a decrease in lipid melting point, suggesting that increased synthesis of unsaturated fatty acids at low temperatures acts to maintain the lipid in a liquid and mobile state, thereby allowing membrane proteins to continue to function (73). In addition, the transport permeases of psychrotrophs are apparently more active at low temperatures than those of mesophiles (9), and a cold-resistant transport system is characteric of psychrotrophic bacteria (167).

Controlled-Atmosphere Storage

The minimal temperature for microbial growth is influenced by many factors and is lowest when other growth conditions are optimal. Under unfavorable values of other environmental conditions, the minimal growth temperature needed for microbial growth increases considerably. Very important environmental factors include the pH, a_w, and oxygen concentration. Therefore, when suboptimal values of these factors are combined with low temperature, the refrigeration stability of many commodities increases greatly. Controlled atmosphere storage of certain fruits and vegetables is widely used. In this operation, a reduced (2 to 5%) oxygen content and generally an increased (8 to 10%) carbon dioxide content are established and maintained in airtight chilled storage rooms. This modified storage atmosphere is advantageous because it depresses the respiration and adverse changes of sensorial and textural quality of stored fruits and vegetables while inhibiting the growth of certain spoilage organisms.

Growth inhibition is evident both in the extension of the lag phase and the reduction of maximal biomass formed. The inhibitory effect of carbon dioxide is due to several factors. It reduces after-ripening, thereby maintaining greater phytoimmunity of plant tissues. When dissolved in aqueous phase, carbon dioxide reduces the pH. But the primary mechanism is direct inhibition of microbial respiration. The simultaneous reduction of oxygen concentration contributes significantly to the inhibitory effect of carbon dioxide. Psychrotrophic spoilage bacteria of flesh foods, such as the aerobic genera *Pseudomonas* and *Acinetobacter*, are particularly sensitive to carbon dioxide, while lactic acid bacteria and yeasts are not (23).

MAP

Modified atmosphere packaging (MAP) operates by a principle similar to that for controlled atmosphere storage. In the case of vacuum-packaged (VP) meat products, it is not the composition but the pressure of air that is changed (the residual air pressure is only 0.3 to 0.4 bar [1 bar = 10^5 Pa] instead of 1 bar for atmospheric air), thereby reducing the amount of available oxygen. If packaging films with very low gas permeability are used for fresh meats, fruits, and vegetables, the composition of the carbon dioxide will change owing to the enzymatic activity of the tissue and its associated microorganisms; this causes the oxygen level to decrease as the carbon dioxide level increases. This modified atmosphere is very inhibitory to certain microorganisms and greatly increases the keeping quality of foods. However, with regard to vacuum packaging of chilled foods, one must remember that *C. botulinum* (and some other pathogens too) grow well in the absence of O_2. Thus, a maximum temperature of 3.3°C is advisable for some VP foods, especially pasteurized foods.

Almost any combination of carbon dioxide, nitrogen, and oxygen may be used in MAP to sustain visual appearance and/or to extend shelf life of meat and meat products (47, 117). In meats, aerobic spoilage bacteria such as the *Pseudomonas/Acinetobacter/Moraxella* group are inhibited (34). The lactic acid bacteria then become the dominant components of the microbiota, but they grow more slowly than the former group and produce less offensive sensory changes (47).

MAP for respiring ("live") product, i.e., fresh fruit and vegetables, is rather complicated since a proper gas permeability of the packaging film is required to achieve an equilibrium atmosphere in the package while both the respiration rate and the gas permeability change with temperature (81). Despite an increasing interest in the use of MAP to extend the shelf life of many perishable products, the concern about the potential growth of pathogenic bacteria at refrigeration temperatures (122) remains the limiting factor to further expansion of the method. Therefore, recent studies have examined the effect of VP/MAP on the growth and survival of psychrotrophic pathogens and the effect on physicochemical changes occurring in VP/MAP foods (52, 116).

The effects of epiphytic flora on the growth of *L. monocytogenes* Scott A were assessed on different types of ready-to-eat leafy vegetables under modified atmosphere at 10°C (49). The highest growth rate was obtained on butterhead lettuce, and no growth occurred on lamb lettuce. Strain Scott A grows faster on young, yellow leaves than on older, green leaves of broad leaf chicory.

There is a positive correlation between the extent of spoilage of the leaves and the number of listeriae as well as the number of epiphytes after storage. *L. monocytogenes* inoculated on leaves disinfected with H_2O_2 (10%), which reduced the number of epiphytic bacteria by 1 to 2 log units, grew faster and to higher number than on leaves rinsed with water only.

Survival and growth of *B. cereus* and *L. monocytogenes* were determined on mung bean sprouts and chicory endive (49). *B. cereus* proliferated during storage at 10°C under atmospheric pressure conditions but diminished in a moderate vacuum system (reduced oxygen tension, 40 kPa of pressure) system. In the latter system, *L. monocytogenes* expired on mung bean sprouts but grew to significant levels on chicory endive.

Historically, fish and fish products are the greatest concern for *C. botulinum* growth in MAP products. The U.S. National Academy of Sciences recommends that fish not be packed under modified atmospheres until the safety of the system has been established. In European Community projects, the microbiological safety of MAP fishes (cod and rainbow trout) was studied at 0, 5, and up to 12°C. In no instance was the growth or survival of *Aeromonas* spp., *Y. enterocolitica*, and *Salmonella enterica* serovar Typhimurium greater under MAP than in the aerobically stored control, and frequently the growth was reduced under MAP (18). However, these investigations did not include studies of *C. botulinum*.

FREEZING AND FROZEN STORAGE

Freezing normally lowers the temperature of a food to −18°C, and the food is stored at that temperature or below. Many convenience food products are made possible by this technology, which preserves foods without causing major changes in their sensory quality. Freezing and frozen transportation and storage are, however, highly energy intensive. There are three basic freezing methods in commercial use: (i) in cold air, (ii) by placing the food or the food package in contact with a surface that is cooled by a refrigerant (indirect contact freezing), and (iii) by submerging the food or spraying the cold refrigerant liquid onto the food or package surface (135).

The freezing of foods does not occur at one defined temperature, but over a broad temperature range. Depending on the composition of the foods, the water starts freezing at −1 to −3°C. The freezing increases the solute concentration in the water not yet frozen, which decreases further the freezing point of the solution. At the so-called eutectic temperature, the concentration of solutes reaches the solubility limit (the point at which they precipitate), and the residual water freezes. A totally frozen state, which is a complex system of ice crystals and crystallized soluble substances, results at −15 to −20°C for fruits and vegetables and at less than −40°C for meats.

As an effect of freezing, both the temperature and the water activity decrease. Thus, in frozen foods, only those microorganisms which are cold tolerant and xerotolerant can grow. While the spoilage of refrigerated non-frozen meat is due mainly to bacteria, on frozen meats those molds which are able to grow at lower a_w levels may be problematic.

During freezing, microorganisms suffer multiple damage which may cause their inactivation directly or at a later time (92). This is the outcome of several effects, the magnitude of which is influenced by, for example, the rate and temperature of freezing and the composition of the food. The rapid cooling of mesophilic bacteria from the normal growth temperature brings about immediate death to a proportion of the culture. Freezing also involves a temperature shock and may cause metabolic injury, presumably by virtue of damage to the plasma membrane. Gram-negative bacteria, particularly mesophiles, appear to be more susceptible to this cold shock than are gram-positive bacteria (67, 149). Rapid chilling results in membrane phase transitions from liquid crystal to gel status without allowing lateral phase separation of phospholipid and protein domains. This results in loss of permeability control of cytoplasmic and outer membranes (98). Cold shock sensitizes cells to various forms of oxidative stress (46, 94).

During freezing of water, an osmotic shock occurs, and intercellular ice crystal formation causes mechanical injury. The concentration of cellular liquids changes the pH and ionic strength, thereby inactivating enzymes, denaturing proteins, and hampering the function of DNA, RNA, and cellular organelles. The main targets of injury are cell membranes. Not only do injured cells die gradually during frozen storage, but surviving cells are reexposed to osmotic effects during thawing. The injury of microbial cells may be reversible or irreversible. In the former case, cells are able to repair if nutrients, energy sources, and specific ions (mainly Mg^{2+} and phosphate) are available and metabolism can commence. The extent of injury and repair and the death and survival rates vary according to the conditions of freezing, frozen storage, and thawing.

Freezing rate is an important determinant of microbial viability. During slow freezing, crystallization occurs extracellularly, as the cell cytoplasmic fluid becomes supercooled at −5 to −10°C. Because their vapor pressure is greater than that of ice crystals, cells lose water, which freezes extracellularly. As a result of crystallization, the concentration of the external solution increases, which also removes water from the cells. On the whole, microorganisms are exposed to osmotic effects for a relatively long time. This causes increased injury. With increased freezing rates, the duration of osmotic effects decreases, thereby increasing survival to a maximum. When freezing rates are too high, crystal formation also occurs intracellularly, injuring the cells drastically and causing a decreasing survival rate.

The most important factor influencing the effect of freezing on microbial cells is the composition of the suspending media. Certain compounds enhance, while others diminish, the lethal effects of freezing. Sodium chloride is very important in this regard because it reduces the freezing point of solutions, thereby extending the time period during which cells are exposed to high solute concentrations before freezing occurs. Other compounds, e.g., glycerol, saccharose, gelatin, and proteins, generally have a cryoprotective effect.

At temperatures below about −8°C, there is, in practice, no microbial growth (68). The fate of microorganisms surviving freezing may vary during frozen storage. Often the death of survivors is fast initially and then slows gradually, and finally the survival level stabilizes. During frozen storage, the death rate of microbial populations is usually lower than that during the freezing process. Death during frozen storage is probably due to the unfrozen, very concentrated residual solution formed by freezing. The concentration and composition of this residual solution, may change during the course of storage, and the size of the ice crystals may increase, especially at fluctuating storage temperatures. These processes also influence the survival of microorganisms. In general, there is less loss of viability during frozen storage when the storage temperature is stable rather than fluctuating. The lower the temperature of frozen storage, the slower the death rate of survivors (68). Gram-positive microorganisms survive frozen storage better than gram-negative bacteria.

Although freezing and frozen storage reduce considerably the viable cell counts in food, freezing preservation cannot be considered a sterilization process; the extent of microbial destruction may be significantly reduced by the components of food. Under some conditions, survivors can grow during the thawing of frozen food. Their populations may then equal or exceed the level prior to destruction during freezing and frozen storage. The cells and the tissue structure of foods are also damaged by freezing, frozen storage, and thawing. During thawing, a solution rich in nutrients is released from the food cells, the microorganisms can penetrate the damaged tissue more easily, and a liquid film condenses on the surface of the product. These conditions favor microbial growth. Therefore, thawed products may be vulnerable to quick microbial spoilage. Refreezing of thawed products may be a dangerous practice and should be avoided.

PRESERVATION BY HEAT TREATMENTS

The most effective and most widely used method for destroying microorganisms and inactivating enzymes is heat treatment. The heat resistance of various microorganisms is, therefore, one of their most important characteristics from an industrial point of view.

The heat processing method called pasteurization (named after Louis Pasteur) is a relatively mild heat treatment aimed at inactivating enzymes and destroying a large population (99 to 99.9%) of vegetative bacterial cells. The main objective of this treatment is to eliminate non-spore-forming pathogenic bacteria. Because a significant proportion of the spoilage microorganisms are also destroyed by pasteurization, pasteurized products have an extended shelf life. For safety and keeping quality, the pasteurization must be complemented with packaging which prevents recontamination. Pasteurized food must be stored at low temperature to prevent the growth of sporeformers whose spores survive the heat treatment. The shelf life of pasteurized products depends on the type of food and conditions of pasteurization and storage.

Sterilization refers to the complete destruction of microorganisms. Industrially, the emphasis is not on absolute sterility but rather on freedom from pathogenic microorganisms and shelf stability of hermetically packaged products. Shelf-stable and microbiologically safe products may contain low numbers of viable, but dormant, bacterial spores. These products are "commercially sterile."

Technological Fundamentals

Heat preservation practices can be divided into two broad categories: those that heat foods in their final containers, and those that use heat prior to packaging. The latter technology requires separate sterilization of the food and the packaging material before the food is placed aseptically (in a sterile environment) into the package (aseptic technology) (140, 143). Commercial sterilization of food in hermetically closed containers (cans or bottles) is called canning, or appertization (named after

Nicholas Appert, who invented the basic process in the early 1800s).

Major technological developments in heat processing are the use of ultrahigh-temperature, short-time treatments (rapid heating to temperatures about 140°C, holding for several seconds, and then rapid cooling) for the production of foods that are shelf stable at ambient temperature. High-temperature short-time processes provide possibilities for increasing the flexibility of packaging and product types as well as improving product quality and processing efficiency (90). The use of higher temperatures for shorter times results in less damage to important nutrients and functional ingredients. The thermobacteriology of ultrahigh-temperature-processed foods has been reviewed (14). Aseptic technology is widely used for fruit juices, dairy products, creams, and sauces; continuous heat treatment with heat exchangers enables an efficient process with improved product quality. Significant progress has also been made for the adoption of this technology to low-acid soups, including particulate products (143).

Ohmic heating (148) and microwave energy treat foods in unique fashions because these processes generate heat throughout the food mass simultaneously, which causes very rapid heating.

Recently, a combined process of microbial inactivation, including heat (112°C) and ultrasonication (20 kHz) under pressure (300 kPa) (mano-thermosonication), was proposed (137) for some liquid foods. The lethality of this combined process is much greater than that of heat treatments at the same temperature.

There has been a recent expansion in the use of the combination of mild heating of packaged foods with well-controlled chill storage. These products are also known as REPFEDs (i.e., "refrigerated processed foods of extended durability") (115) or cook-chill products and include "sous vide" meals (foods mildly heated within vacuum packs) (111). For these chilled pasteurized foods, specific attention is given to the control of nonproteolytic (psychrotrophic) C. botulinum because of its ability to grow at chilled temperatures as low as 3.0°C (53). To ensure safety, some form of demonstrable intrinsic preservation of products (i.e., an additional hurdle) is generally regarded as necessary (99).

Fundamentals of Thermobacteriology

Heat in the presence of water (so-called wet-heat treatment) kills microorganisms by the denaturation of nucleic acids, structural proteins, and enzymes. Identification of primary sites for lethal heat-induced damage is difficult because of many changes which heat brings about in cells. Nevertheless, there is substantial evidence that DNA damage, directly or indirectly, may be the key lethal event in heated cells (52). However, heat also damages many other, less critical sites. In general, the thermal stability of ribosomes corresponds to the maximal growth temperature of microorganisms (73), and cytoplasmic membranes seem to be major sites of injury of mild heating (42) with consequent increase of sensitivity to environmental stress factors. However, degradation of rRNA in membrane-damaged, heated cells often precedes loss of viability. Therefore, it is not thought to be a prime cause of cell death (65). In bacterial spores, some part of the germination system (i.e., a specific enzyme[s]) may be the least heat-resistant component and thus a key site of heat inactivation (52). Dry heat is less lethal and kills the microorganisms by dehydration and oxidation. Dry heat necessitates higher temperatures and longer heating times to generate the same lethality as wet heat.

The calculation of the heat treatment required to kill a given number of bacteria is based on the assumption that the destruction of a pure bacteria culture by wet heat at constant temperature exhibits a negative exponential kinetics (the order of death is logarithmic), $N = N_0 \cdot e^{-k \cdot t}$, where N is the actual number of viable cells, N_0 is the initial number of viable cells, k is the rate constant (time^{-1}) of heat destruction at constant temperature T, and t is the heat processing time. That is, equal periods at lethal temperatures result in equal percentage of destruction regardless of the microbial population size. A theoretical treatment of this relationship is given in chapter 2.

The destruction rate is inversely related to the decimal reduction time (D value), $D_T = \ln 10 / k_T$. The D value is defined as the time in minutes required to destroy 90% of the population, as shown in Fig. 28.1 (survival curve). Survival curves provide data on the rate of destruction of a specific organism in a specific medium or food at a specific temperature. Under given conditions, the rate of death is constant at any given temperature and independent of the initial viable cell number. The logarithms of D values plotted against exposure temperatures give the thermal death time curve (Fig. 28.2). The thermal death time is the time necessary to kill a given number of organisms at a specified temperature.

While the D value represents the measure of heat resistance of the given microorganism at a given temperature, the thermal death time curve indicates the relative resistance at different temperatures. This enables one to calculate the lethal effect of different temperatures. The slope of the thermal death time curve is expressed by the z value, which is the temperature increase required to reduce the thermal death time by a factor of 10.

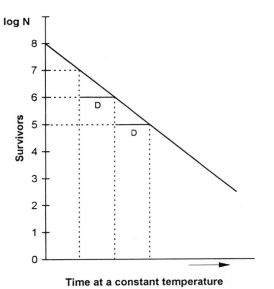

Figure 28.1 Bacterial survival curve showing logarithmic order of death and the concept of the D value.

Heat resistances (D values) of various microorganisms and the effect of the temperature changes on the death rate are different and are influenced by many factors (68). There are inherent differences among species, among strains within the same species, and between spores and vegetative cells. Bacterial spores are more heat resistant than vegetative cells of the same strain, some being capable of surviving for minutes at 120°C or for several hours

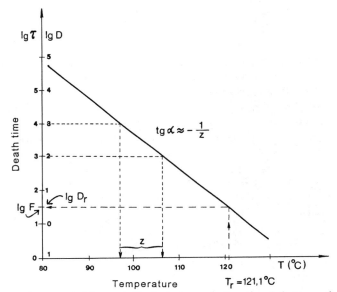

Figure 28.2 Thermal death time curve showing definition of z value and F value. T_r = reference temperature; τ = death time.

Table 28.6 Relative heat resistance of some vegetative bacteria[a]

Organism	Heating medium	D value (min) at:				z value (°C)
		70°C	65°C	60°C	55°C	
Escherichia coli	Ringer solution, pH 7.0				4	
Lactobacillus planatarum	TSB[b] + sucrose (a_w = 0.95)		4.7–8.1			
	Tomato juice, pH 4.5	11.0				12.5
Pseudomonas aeruginosa	Nutrient agar				1.9	
Pseudomonas fluorescens	Nutrient agar				1–2	
Salmonella senftenberg (775 W)	Skim milk			10.8		6.0
	Phosphate buffer (0.1 M), pH 6.5		0.29			
	Phosphate buffer plus sucrose (%w/v) 30:70		1.4:4.3			
	Phosphate buffer plus glucose (%w/v) 30:70		2.0:17.0			
	Milk chocolate	440				18.0
Salmonella senftenberg	Heart infusion broth (a_w = 0.99), pH 7.4			6.1		6.8
	Heart infusion broth + NaCl (a_w = 0.90), pH 7.4			2.7		13
	Heart infusion broth + sucrose (a_w = 0.90), pH 7.4			75.2		8.9
Salmonella enterica serovar Typhimurium	Phosphate buffer (0.1 M), pH 6.5		0.056			
	Milk chocolate	816				19.0
Staphylococcus aureus	Custard or pea soup			7.8		4.5

[a] Adapted from reference 161.
[b] TSB, tryptic soy broth.

at 100°C. Less heat-resistant spores may have $D_{100°C}$ values of less than 1 min (106). Vegetative cells of bacteria, yeasts, and molds are killed after a few minutes at 70 to 80°C.

The age of the cells and their stage of growth, the temperature at which the microorganisms were grown, and the composition of growth medium also affect heat resistance. Physicochemical factors, such as pH, a_w, salt content, and organic composition of the suspension medium in which microorganisms are heated, also profoundly influence heat resistance (21, 106, 153).

Tables 28.6 and 28.7 illustrate selected heat resistance data for vegetative bacteria and bacterial endospores, respectively. Vegetative cells of spore-forming bacteria are as heat sensitive as other vegetative cells. The high heat resistance of spores relates to their specific structure and composition, and it is basically the consequence of the dehydrated state of the spore core (52). The molecular basis of heat resistance in spores is covered more fully in chapter 3.

Bacterial spores suspended in oils are more heat resistant than those suspended in aqueous systems. This increased heat resistance was ascribed to the low a_w of oils,

and the difference in z values suggests that the mechanism of inactivation differs for spores suspended in lipids and in aqueous systems (1).

Vegetative forms of yeasts can usually be killed by 10- to 20-min heat treatment at 55 to 60°C (136). Given the same period of heat treatment, yeast ascospores require only 5 to 10°C higher temperature than that required by the mother cells to effect similar lethality (25, 74). Vegetative propagules of most molds and conidiospores can be inactivated by 5 to 10 min of wet-heat treatment at 60°C (30). Several mold species forming sclerotia or ascospores are much more heat resistant (3, 31, 154). Mold spores survive dry-heat treatment for 30 min at 120°C.

Calculating Heat Processes for Foods

Containers of food do not heat instantaneously, and since all temperatures (above a minimum value) contribute to the microbial destruction, a method to determine the relative effect of changing temperature while a food is being heated and cooled during thermal processing is necessary.

For the determination of heat processes for canned foods, the F value was introduced. The F value is defined

Table 28.7 Approximate heat resistance (*D* values) of some bacterial spores

Types of food and typical organisms	*D* value (min) at:		*z* value (°C)
	121°C	100°C	
Low-acid foods (pH > 4.6)			
Thermophilic aerobes			
Bacillus stearothermophilus	4.0–4.5	3,000	7
Thermophilic anaerobes			
Clostridium thermosaccharolyticum	3.0–4.0		12–18
Desulfotomaculum nigrificans	2.0–3.0		
Mesophilic anaerobes			
Clostridium sporogenes	0.1–1.5		9–13
Clostridium botulinum types A and B	0.1–0.2	50	10
Clostridium perfringens		0.3–20	10–30
Mesophilic aerobes			
Bacillus licheniformis		13	6
Bacillus subtilis		11	7
Bacillus cereus		5	10
Bacillus megaterium		1	9
Acid foods (pH ≤ 4.6)			
Thermotolerant aerobes			
Bacillus coagulans	0.01–0.1		
Mesophilic aerobes			
Bacillus polymyxa		0.1–0.5	
Bacillus macerans		0.1–0.5	
Clostridium butyricum (or *Clostridium pasteurianum*)		0.1–0.5	

[a] Adapted from references 68 and 158.

as the number of minutes at a specific temperature required to destroy a specific number of viable cells having a specific *z* value (Fig. 28.2).

Since *F* values represent the number of minutes to diminish a homogeneous population with a specific *z* value at a specific temperature, and because *z* values as well as temperatures vary, it is necessary to designate a reference *F* value. This reference is the F_0 value, which is the number of minutes at 121°C (250°F) required to destroy a specific number of cells whose *z* value is 10°C (18°F). The F_0 value of a heat treatment is called its "sterilization value" and is a measure of the lethality of a given heat treatment. It is related to τ, the time required for identical lethality at temperature *T*, shown by the equation

$$\lg F - \lg \tau = \frac{T - 121.1}{z}$$

F_0 requirements of various foods differ. Since heat sensitivities of microorganisms and the characteristics of the thermal death curve are affected by many factors, thermal death curves should be established in the specific food for which a heat process is designed. Because the spores of *C. botulinum* type A and B are the most heat-resistant spores of a foodborne pathogen (Table 28.7), commercial sterilization of low-acid (pH > 4.6) foods must be sufficient to destroy these spores. (The targets for the heat processing of acid foods [pH ≤ 4.6] are heat-resistant fungi [126].)

A "botulinum cook" is a heat process which reduces the population of the most resistant *C. botulinum* spores by an arbitrarily established factor of 12 decimal values, the 12*D* concept. This provides a substantial margin of safety in low-acid canned foods. In practice, 2.45 min is the 12*D* value of botulinal spores in phosphate buffer (68). The botulinum cook will produce a product that is safe but not necessarily commercially sterile. Other heat-resistant spores, which may cause spoilage but are not pathogenic, may be present and viable (Table 28.7). Thus, their higher heat resistance often is the determining factor in establishing commercial heat processes. On the other hand, *C. botulinum* can be controlled by pasteurization in some cured meat products, low a_w, a reduction in pH, and the addition of nitrite. This multiple-hurdle preservation system allows long, safe stability (87).

Detailed guidance for calculating heat processes on the basis of semilog-linear kinetics is provided by several monographs (126, 157, 158). Recently, kinetic models taking into account the concurrent activation of dormant spores and distinct inactivation of dormant and activated spores during heat treatments have been developed for advanced process design (89, 147). In the case of vegetative cells, the death rate coefficient depends significantly on environmental factors. Reichart and Mohácsi-Farkas (139) developed several kinetic models for describing the combined effect of a_w, pH, and redox potential on the thermal destruction rate of seven foodborne microorganisms (*Lactobacillus plantarum*, *Lactobacillus brevis*, *Saccharomyces cerevisiae*, *Z. bailii*, *Yarrowia lipolytica*, *Paecilomyces varioti*, and *Neosartoria fisheri*).

Recent Developments in the Thermobacteriology of Food

Previous concepts about the heat preservation of foods assumed implicitly that heat resistance values of microorganisms measured under isothermal conditions are constant and unaffected by the heating rate in the sublethal temperature range. Work during the last 2 decades has demonstrated that the rate of temperature elevation affects the subsequent isothermal death of various vegetative bacteria (95, 164). Exposure of cells to sublethal temperatures above their growth maxima (to a sublethal heat shock) induces resistance to heating at higher temperatures (96). These effects were observed in nutritionally rich media. In recent years, the induction or enhanced synthesis of a family of proteins (heat shock proteins [hsps]) in response to heat shock, and also to other environmental stresses and chemical or mechanical treatments, has been reported as a phenomenon that occurs in most, if not all, organisms (4, 10, 150). Many of the hsp inducers have in common the capacity to produce protein damage (60). It has been proposed that the signal for hsp induction is protein denaturation and an increase in the concentration of unfolded protein in the cell (123). However, the mechanism responsible for the development of thermotolerance has not been completely elucidated, and the connection between hsp induction and the acquisition of thermotolerance is unclear (4, 10). The lesions produced by various chemicals triggering thermotolerance are different from those produced by heat (10). Some investigators propose two states of tolerance to heat, one not requiring hsps and another (induced by more severe damage) requiring them (85). These observations lead to the conclusion that environmental stresses by industrial processes can induce protective response in microorganisms. Thus, the possibility that heat resistance of vegetative cells may increase during heating must be taken into account when one is establishing safe heat treatments for foods that may be heated up relatively slowly.

The traditional model of thermal inactivation kinetics, which assumes that the reduction in log numbers of survivors decreases in a linear manner with time has served the food industry and regulatory agencies well for many years. Indeed, in situations where bacterial spores are the critical targets or death is rapid, a good fit to this hypothesis is generally observed. However, as a result of the introduction of milder thermal processes that use low processing temperatures under a wide range of environmental conditions, the food industry is increasingly requiring accurate predictions for death kinetics for vegetative bacteria. Under these conditions, significant deviations from log-linear kinetics are encountered, "shoulder" and/or "tailing" of survival curves occurs frequently, and the method incorporating D and z values cannot be relied upon. The theory underlying log-linear death kinetics assumes that inactivation results from random single hits of key targets within microbial cells. Explanations offered as to why the variations in the straight-line survivor curve occur include (i) experimental errors and artifacts, such as mixed population of test organisms, flocculation and deflocculation of cells during the heating period, heat activation for spore germination, aggregation and absorption of cells to walls of the vessel, or the presence of cells in aerosol droplets above the surface of the suspension (59, 125); (ii) multiple-hit destruction mechanism (52); and (iii) variability of heat sensitivities (resistance distribution) in heated population (15, 57) and heat adaptation during treatment (57).

Recently, mathematical models were developed to describe the observed inactivation kinetics (20) which allow for variability of heat sensitivity throughout a population of cells, e.g., normal distribution of heat sensitivity. A further, improved model describes the heat sensitivity of cells as being distributed by a logistic function, which allows for accurate predictions of thermal inactivation to be made at relatively low heating rate (156) and even when parameters other than temperature, e.g., pH and NaCl concentration, are varied (20).

Microwave Heat Treatment

Electromagnetic heating of food using radio frequency (1- to 500-MHz) or microwave (500-MHz to 10-GHz) energy has the advantage of the internal heat generation. Water molecules, because of their negatively charged oxygen atoms and positively charged hydrogen atoms, are electric dipoles. When a rapidly oscillating radio frequency or microwave field is applied to a biological tissue or food rich in water, the water molecules reorient with each change in the field direction. This creates intermolecular friction, which produces heat (112).

Microwave heating modes may reduce process times and energy and water usage in some food processing areas. Microwave energy is suited for heat processing of food or pest control in the food industry (27, 144). However, the geometry of the food product, its thermal, physical, and dielectric properties, its mass, the power input, and the radiation frequency all influence the interaction of electromagnetic waves with a food (132). These factors can cause differential microwave heating of different food constituents or of different parts of a meal. This thermal nonuniformity has limited the commercial application of microwave heating for microbial inactivation, even though microwave heating has gained widespread application in home food preparation. The formation of heterogeneous temperatures within a food might create problems if it results in insufficient heating compared with conventional cooking and thereby allows survival of potentially pathogenic microorganisms (63). Therefore, metal shielding and other packaging designs which increase the heating uniformity of food are desirable (28, 61).

Regarding specific electromagnetic effects, electromagnetic fields cause ion shifts on cellular membranes, leading to changes in permeability, functional disturbances, and cell rupture (132). Theoretically, microbial cells in food may be differentially heated compared with the food matrix, which could result in specific inactivation. However, these theoretical findings need to be validated in food. So far, experimental data on the specific effect of radio frequency or microwave frequency electromagnetic energies are contradictory, but in the majority of studies, no nonthermal effects were detected (113).

A more fundamental approach in food-related research concerning microwave effects is needed, and more research has to be conducted to determine these effects and utilize advantages of various frequencies. The choice of the most widely used frequency, 2.45 GHz, was a compromise of various factors and is not necessarily the most suitable frequency for a certain application (132). Sophisticated temperature registration methods and dielectric measurements can be important tools for better understanding interactions in studies of specific effects of electromagnetic fields.

Ohmic Heating

The use of direct electrical heating, known as ohmic heating, is another development in heating technology. An electric current is passed through the material as it is pumped through a tube. The heating rate is a function of the electrical conductance of the solid and liquid phases.

If these are about the same, the two phases heat at the same rate. The design of equipment, products, and processes for ohmic heating are still in the developmental stages (148). In the limited thermobacteriological studies conducted to date, microbial death appears to be caused solely by thermal effect (121). The role of electric resistance heating in aseptic processing was discussed by Fryer (43).

ULTRASOUND

High-power ultrasound can kill microorganisms, but its use as a single variable to inactivate microbes is not practical for a number of reasons. However, sonic energy at lower power values can sensitize microbes to heat (120), and this combination with heat may be an attractive proposition for minimizing heat damage in some foods (35). Although the heat-ultrasound synergism disappears at the boiling point of the menstruum (44), this disappearance of the synergism can be prevented by overpressure, commonly no more than 1 mPa (mano-thermosonication), thereby offering the possibility of a new sterilization process for liquid foods (145).

PRESERVATION BY IRRADIATION

UV Radiation

The microbiologically most destructive wavelength range of UV radiation is between 240 and 280 nm. The great microbicidal efficiency of the so-called germicidal radiation is based on the fact that this radiation can damage nucleic acids most efficiently. Inactivation is caused by the cross-linking of thymine dimers of the DNA, thus preventing repair and reproduction. Gram-negative bacteria are most easily killed by UV, while bacterial endospores and molds are much more resistant.

Viruses are more UV resistant than bacterial cells (159). UV resistance of various microorganisms also depends on their pigment formation. Cocci that form colored colonies are less susceptible to UV than those with colorless colonies. Dark conidia of certain molds are highly UV resistant (107).

The very low penetration of UV radiation (86) and the difficulty in attaining an even exposure level over all points of the food surface are major technological problems which limit the feasibility of UV radiation in food preservation. Therefore, UV sources are used mainly for disinfection of air, e.g., in aseptic filling of liquid food, such as milk, beer, and fruit juices, in packaging sliced bread, and in the ripening rooms of cheese and dry sausages, or for disinfection of water when the use of chlorine is undesirable (16). A recent design of a treatment chamber for pasteurization of juices utilizes turbulent flow to form a "continuously renewed surface" (152). Possible uses of UV irradiation of packaging materials and preformed containers as a single treatment and in combination with hydrogen peroxide for aseptic packaging have been described by Narasimhan et al. (114) and Stannard et al. (155), respectively.

High-Intensity Pulsed Light

During the last decade, a technique has been developed into a commercially available process for decontamination of (food) surfaces with high-intensity xenon arc lamps (101). The killing effect mainly derives from the UV component of the absorbed light or heat.

Ionizing Radiation

The potential application of ionizing radiation in food products is based mainly on the fact that ionizing radiation very effectively inhibits DNA synthesis so that cell division is impaired. Given the right doses, this impairment is obtained without serious effects on the food itself. Therefore, microorganisms, insect gametes, and plant meristems can be prevented from reproducing, thus rendering the food stable (Table 28.8). There are many excellent general reviews on various aspects, prospects, and problems of food irradiation (e.g., 29, 76, 119, 165, 168–170).

In those parts of the world where the transport of food is difficult and where refrigerated storage of food is scarce and extremely expensive, the use of ionizing radiation may facilitate wider distribution of some food than would otherwise be feasible. In this way, a more

varied and possibly nutritionally superior diet may become available to the residents.

The ionizing radiation used in food irradiation is limited to high-energy electromagnetic radiation (gamma rays of ^{60}Co or X rays) with energies up to 5 MeV or electrons from electron accelerators with energies up to 10 MeV. These types of radiation are chosen because (i) they produce the desired effects with respects to the food, (ii) they do not induce radioactivity in foods or packaging materials, and (iii) they are available in quantities and at costs that allow commercial use of the process. Other kinds of ionizing radiation, in some respects, do not suit the needs of food irradiation.

The penetration of electron beams and X or gamma rays into matter occurs in different ways. The practical usable depth limit for 10 MeV of electrons in water-equivalent material is 3.9 cm, while the so-called half-thickness value for 5 MeV of X rays is 23.0 cm (165). Except for different penetration, electromagnetic radiation and electrons are equivalent in food irradiation and can be used interchangeably.

Microbiological Fundamentals

One of the main reasons for using ionizing radiation is to kill organisms that cause spoilage or are a health hazard to consumers. The biological effects of ionizing radiation on cells can be due to both direct interactions with critical cell components and indirect actions caused by radiolytic products of other molecules in the cell, particularly free radicals formed from water. The DNA in the chromosomes is the most critical target of ionizing radiation. Effects on the cytoplasmic membrane appear to play an additional important role in radiation-induced damage of cells (56). Although the changes by radiation are at the cellular level, the consequence of these changes vary with the organism. The correlation of radiation sensitivity is roughly inversely proportional to the size and complexity of an organism (Table 28.9). This correlation is related to genome size: the DNA in the nuclei of insect cells represents a target much larger than the genomes of bacteria, while the cells of mammalian organisms containing DNA

Table 28.8 Preservative effects of ionizing radiation[a]

Effect(s)	Result(s)
Inhibition of sprouting	Increased shelf life of root and bulb crops, reduction of malting losses
Decrease of after-ripening and delaying senescence of some fruits and vegetables	Increased shelf life of fruits and vegetables
Milling and sterilizing insects	Insects disinfestation of food
Prevention of growth and reproduction of parasites transmitted by food	Prevention of plastic diseases
Reduction of microbial populations	Decreased contamination of food; increased shelf life of food; prevention of food poisoning

[a] Data from reference 36.

Table 28.9 Approximate doses of radiation needed to kill various organisms[a]

Organisms	Dose (kGy)
Higher animals	0.005–0.1
Insects	0.01–1
Non-spore-forming bacteria	0.5–10
Bacterial spores	10–50
Viruses	10–200

[a] Data from reference 165.

molecules, which must provide much more genetic information than those of insects, are correspondingly larger and even more sensitive to radiation (29). Differences in radiation sensitivities within groups of similar organisms are related to differences in their chemical and physical structure and in their ability to recover from radiation injury. The amount of radiation energy required to control microorganisms in foods, therefore, varies according to the resistance of the particular species and according to the number of organisms present.

The relative radiation resistances of various foodborne microorganisms can be found in comparable forms of presentation in several reviews (e.g., 29, 39, 66, 69, 105). Summarizing data from a large number of references, Table 28.10 presents typical radiation resistances of some foodborne microorganisms in fresh and frozen foods of animal origin. In general, gram-negative bacteria, including the common spoilage organisms of many foods (e.g., pseudomonads), and particularly

enteric species including pathogens (salmonellae, shigellae, etc.), are generally more sensitive than vegetative gram-positive bacteria. The spores of the genera *Bacillus* and *Clostridium* are more resistant still. Rarely, even more highly resistant vegetable forms may be encountered, typified by *Deinococcus* (formerly *Micrococcus*) *radiodurans* (2) and the gram-negative, rod-shaped *Deinobacter* sp. (54). The radiation sensitivity of many molds is of the same order as that of vegetative bacteria. However, fungi with melanized hyphae have a radiation resistance comparable to that of bacterial spores (146). Yeasts are as resistant as the more resistant bacteria. Viruses are highly radiation resistant (165).

The death of microorganisms by ionizing radiation is the result of damage to DNA. Ionizing radiation can affect DNA directly by the ionizing ray or indirectly by the primary water radicals $\cdot$H, $\cdot$OH, and e_{aq}^-. The most important of these three radicals in DNA damage is the $\cdot$OH radical. The $\cdot$OH radicals formed in the hydration layer around the DNA molecule are responsible for 90% of DNA damage. Thus, in living cells, the indirect radiation damage is predominant. The principal lesions induced by ionizing radiation in intracellular DNA are chemical damage to the purine and pyrimidine bases, and to the deoxyribose sugar, and physicochemical damage resulting in a break in the phosphodiester backbone in one strand of the molecule (single-strand break) or in both strands at the same place (double-strand break) (108). Double-strand breaks are produced by ionizing radiation at about 5 to 10% of the rate for single-strand breaks.

Most microorganisms can repair single-strand breaks. The more sensitive organisms, such as *Escherichia coli*, cannot repair double-strand breaks, while highly resistant species (e.g., *Deinococcus* sp.) are able to repair them. The low water content of the spore protoplast is a major factor in the radiation resistance of bacterial spores. During germination, the water content of the spore protoplast increases, and therefore the radiation resistance disappears.

Radiation survival is conveniently represented with the logarithm of the number of surviving organisms plotted against radiation dose. Typical radiation survival curves are shown in Fig. 28.3. Type 1 is an exponential survival curve, quite common for the more radiation-sensitive organisms. Type 2 is characterized by an initial shoulder, indicating that equal increments of radiation are more effective at high doses than at low doses; the shoulder in the curve may be explained by a certain repair at low doses. Type 3 is characterized as concave with a resistant tail.

Similar to heat resistance, the radiation response in microbial populations can be expressed by the decimal

Table 28.10 Typical radiation resistances of some foodborne microorganisms in fresh and frozen foods of animal origin

Microorganisms	D_{10} (kGy)	
	Fresh food	Frozen food
Vibrio spp.	0.03–0.12	0.11–0.75
Yersinia enterocolitica	0.04–0.21	0.4
Pseudomonas putida	0.06–0.11	
Campylobacter jejuni	0.08–0.20	0.21–0.32
Shigella spp.		0.2–0.4
Aeromonas hydrophila	0.14–0.19	
Bacillus cereus (vegetative cells)	0.17	
Proteus vulgaris	0.2	
Escherichia coli	0.23–0.35	0.3–0.6
E. coli O157:H7	0.24–0.27	0.31–0.44
Staphylococcus aureus	0.26–0.6	0.3–0.45
Brucella abortus	0.34	
Salmonella spp.	0.3–0.8	0.4–1.3
Listeria monocytogenes	0.27–1.0	0.52–1.3
Lactobacillus spp.	0.3–0.9	
Streptococcus faecalis	0.65–1.0	
Clostridium perfringens (vegetative cells)	0.59–0.83	
Moraxella phenylpyruvica	0.63–0.88	
Clostridium botulinum type E (spores)	1.25–1.40	
B. cereus (spores)	1.6	
Clostridium sporogenes (spores)	1.5–2.2	
Clostridium botulinum types A and B (spores)	1.0–3.6	
Deinococcus radiodurans	5–3.1	
Deinobacter sp.	5.05	
Coxsackievirus		6.8–8.1

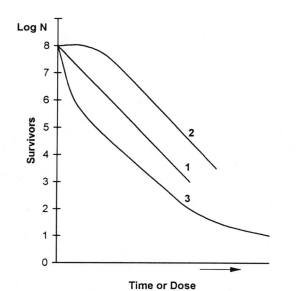

Figure 28.3 Typical radiation survival curves. N = survivors of radiation dose; N_0 = original number of viable cells. 1, exponential survival curve; 2, curve characterized by an initial shoulder; 3, concave curve with a resistant tail.

reduction dose (D_{10} value). When the dose-survival curve is a straight line, D_{10} is the reciprocal of its slope,

$$D_{10} = \frac{\text{radiation dose}}{\log N_0 - \log N}$$

where N_0 is the initial number of organisms, and N is the number of organisms surviving the radiation dose. Survival curves of types 2 and 3 are often best represented by quoting an inactivation dose, for example, the dosage to inactivate 90 or 99% of the initial viable cell count.

Several environmental factors influence radiation resistance of microorganisms (24, 51).

(i) The composition of the medium surrounding the microorganisms plays an important role in the dose requirement for a given microbiological effect. Cells irradiated in phosphate buffer are much more sensitive than those irradiated in foodstuffs. In general, the more complex the medium, the greater the competition of the medium components for the free radicals formed from water and activated molecules produced by the radiation, thus protecting the microorganisms. It is impossible to predict in which particular foods bacterial cells will be more radiation sensitive or radiation resistant.

(ii) A reduction in the moisture content (reduction of a_w) of the food protects microorganisms against the lethal effect of ionizing radiation, similar to what is known for heat treatment. In dry conditions, the yield of free radicals produced from water by radiation is lower;

thus, the indirect DNA damage that they may cause is decreased.

(iii) The temperature at which a product is treated may influence the radiation resistance of microorganisms. For vegetative cells, elevated temperature treatments, generally in the sublethal range above 45°C, synergistically enhance the lethal effect of radiation. This is thought to occur because the repair systems, which normally operate at and slightly above ambient temperatures, are damaged at higher temperatures.

(iv) Freezing causes a striking increase in radiation resistance of vegetative cells. Microbial radiation resistance in frozen foods is increased about two- or threefold higher than at ambient temperature. This increase is due to the immobilization of the free radicals and prevention of their diffusion when the medium is frozen. Thus, in the frozen state, the indirect DNA damage by ·OH radicals is nearly prevented. The change in resistance with temperature demonstrates the importance of the indirect action in high-moisture foods.

(v) The lethal effect of ionizing radiation on microbial cells increases in the presence of oxygen (160). In the total absence of oxygen, and in wet conditions, radiation resistance usually increases by a factor of 2 to 4. In dry conditions, the radiation resistance might increase by a factor of 8 to 17. Data reported for cell suspensions irradiated in sealed tubes can frequently be plotted as type 3 survival curves (Fig. 28.3). These tailing effects probably represent a shift to anaerobic conditions; that is, irradiation of aerated water will consume all the oxygen after a dose of 500 Gy.

The foregoing discussion emphasizes the necessity for specifying the environmental conditions in measuring and citing the radiation sensitivity of a microorganism.

Technological Fundamentals

Generally, there is a minimum dose requirement. Whether every mass element of a food requires irradiation will depend on the purpose of the treatment. In some cases, irradiation of the surface will suffice. In others, the entire food must receive the minimum dose. Guidelines for dose requirements are given in Table 28.11. Detailed requirements must again be considered specific for a given food.

Radurization is the term for substantial reduction in the number of spoilage microorganisms in food, thereby extending the refrigerated shelf life of a food. With relatively high doses, approaching 5 kGy, the normal spoilage organisms are often eliminated, and the more resistant and metabolically less active species, e.g., resistant *Moraxella* spp., lactic acid bacteria, and yeasts, remain. During chilled storage, this surviving biota grows less

Table 28.11 Dose requirements of various applications of food irradiation[a]

Application	Dose requirement (kGy)
Inhibition of sprouting of potatoes and onions	0.03–0.12
Insect disinfestation of seed products, flours, fresh and dried fruits, etc.	0.2–0.8
Parasite disinfestation of meat and other foods	0.1–3.0
Radurization of perishable food items (fruits, vegetables, meat, poultry, fish)	0.5–10
Radicidation of frozen meat, poultry, eggs and other foods and feeds	3.0–10
Reduction or elimination of microbial population in dry food ingredients (spices, starch, enzyme preparates, etc.)	3.0–10
Radappertization of meat, poultry, and fishery products	25–60

[a] Data from reference 36.

rapidly than the normal spoilage biota, and the storage life is correspondingly further extended to three or four times normal. As with the unirradiated foods, the nature of the spoilage biota is strongly influenced by the nature of the packaging. In anaerobic packages, lactic acid bacteria and yeasts predominate.

Radicidation is the term for reduction in number of specific viable non-spore-forming pathogenic microorganisms (other than viruses) and parasites, thereby improving the hygienic quality and reducing the risk to the public from certain pathogens.

Radappertization is the term for application of ionizing radiation to prepackaged, enzyme-inactivated foods in doses sufficient to reduce the number and/or activity of viable microorganisms to such an extent that in the absence of postprocessing contamination, no microbial spoilage or toxicity should occur, no matter how long or under what conditions the food is stored. The dose requirement for radappertization is determined by the most radiation-resistant microorganism associated with the enzyme-inactivated food. For non-acid, low-salt foods, this is the *C. botulinum* type A spore. By analogy with thermal processing, the safety of the radappertization process is based on a 12*D* reduction in the viability of botulinal spores. For low-salt, non-acid foods, the required dose is about 45 to 50 kGy (29, 170).

The usefulness of ionizing radiation in overcoming microbiological problems has been clearly demonstrated with a variety of foodstuffs and bacteria of public health significance, especially pathogenic bacteria, such as salmonellae, *L. monocytogenes,* campylobacters,

Y. enterocolitica, and *E. coli* O157:H7 (19, 39). Radiation doses of 3 to 10 kGy are sufficient to reduce viable cell counts in spices, herbs, and enzyme preparations to satisfactory levels (37).

Radiation treatment causes practically no temperature rise in the product. The largest dose likely to be employed is about 50 kGy. This amount of energy is equivalent to about 12 calories (50 J). Hence, if all the ionizing radiation degrades to heat, the temperature of a food will rise about 12°C. For this reason, irradiation has been termed a cold process.

Ionizing radiation can be applied through any type of packaging materials, including those that cannot withstand heat. Radiation can be applied after packaging, thus avoiding recontamination and reinfestation. Most standard packaging materials are satisfactory for use with irradiated foods.

One should keep in mind that the quality of the irradiated foods, as for any other preserved food, is a function of the quality of the original material and that to obtain good results, good manufacturing practice is needed. The longest shelf life may be obtained if the quality of the raw material is good and proper hygienic conditions are maintained. In certain cases, the use of more thorough inspection and sorting to remove substandard-quality raw materials before treatment (or treatments such as washing, blanching, or chemical rinse) can be used to reduce the radiation dose required to produce a high-quality product. The benefits of irradiation should never be considered a substitute for product quality or as compensation for poor handling and storage conditions.

The basic irradiation process applies a prescribed amount of ionizing radiation of either electrons or electromagnetic rays and uses certain other procedures that may be needed. Such other procedures will vary according to the food being treated and the purpose of the treatment. For radappertized foods intended to have a shelf stability without refrigeration, packaging suitable to protect the food from microbial contamination after irradiation is essential, and inactivation of the radiation-resistant food enzymes by mild heat processing may be necessary. For radurized foods, enzyme inactivation is not necessary, and protective packaging may or may not be needed. Protection to prevent reinfestation may be needed for products irradiated to control insects. Specific storage conditions could be required, depending on the nature of products treated.

Some products may require irradiation under special conditions, such as at low temperature or in an oxygen-free atmosphere, or with combination treatments such as heat and irradiation (38). Sensitization of microorganisms with the aim of lowering the dose can be obtained by physical (heat) (55) and chemical means. The microbiota surviving irradiation is more sensitive to subsequent food processing treatments than the microbiota of unirradiated products.

The actual dose employed is a balance between what is needed and what can be tolerated by the product without unwanted changes. High doses cause organoleptic changes (off flavors and/or texture changes) in high-moisture food. These undesirable effects can be limiting factors. In the case of protein foods of animal origin, the maximum dose permitted usually is set by the development of the irradiation flavor.

Radiation-induced changes in the texture of some foods, such as fresh fruits and vegetables, can be a limiting factor. Polysaccharides such as pectin can be degraded by irradiation. This is often accompanied by a release of calcium which causes softening of the product. These observations, and also the observed increase in permeability, can be explained by the fact that the natural resistance of the irradiated plant tissues is often reduced, resulting in accelerated spoilage when microorganisms are given a chance to attack the products after irradiation. However, this radiation-induced initial softening in several commodities is often compensated for by the fact that during extended postirradiation storage, this softening (after-ripening) takes place considerably more slowly than in the untreated plant tissues.

In some cases, the chemical and physical effects of irradiation can be regarded as improvements. For example, the radiation-induced softening and increase in permeability is advantageous when shortening of the cooking time (e.g., for dried vegetables) is required, and increased yield of juice can be obtained from irradiated fruits.

NEW NONTHERMAL PHYSICAL PRESERVATION TREATMENTS

During recent years, several other nonthermal physical treatments, such as high hydrostatic pressure and use of electric and magnetic fields, has attracted much interest as promising tools for food processing and preservation and for creating new types of food product (101).

HIGH HYDROSTATIC PRESSURE FOR FOOD PRESERVATION

One advantage of the application of high or ultrahigh (300 to 1,000 MPa) hydrostatic pressure is that it is transferred throughout food instantly and uniformly (Pascal principle) and thus is independent of sample size and geometry. It can be applied at ambient or moder-

ate temperatures. Because only noncovalent bonds are affected by the treatment, there is good retention of color, nutrients, and flavor (17). High hydrostatic pressure (HP)-treated fruit juices and other fruit products have been marketed in Japan since 1990, and some commercial products have been available in the United States and Europe since 1996. Several recent reviews (e.g., 7, 71, 84) summarize HP on microorganisms, foods, and food components, or developments in HP food processing technologies. The chemical principle behind the effect of ultrahigh pressure is that noncovalent bonds, such as hydrogen, ion, and hydrophobic bonds in proteins and carbohydrates, undergo changes that alter their molecular structures.

Microbial inactivation is probably due to a number of factors, including protein conformation changes and membrane perturbation, leading to cell leakage. In general, gram-positive organisms are more resistant to pressure than gram-negative organisms (93). Microbial resistances to heat and pressure are often correlated. For example, the more heat-resistant *L. monocytogenes* is also more resistant to pressure than *Salmonella* serovar Thompson, and the pressure-treated cells of these two species have marked differences in ultrastructure (97). However, the heat-resistant strain of *Salmonella* serovar Senftenberg is less resistant to pressure than other strains of *Salmonella* (102). Studies on injury and survival of *Aeromonas hydrophila* and *Y. enterocolitica* from high hydrostatic pressure suggest that post-HP growth would occur only in exaggerated temperature abuse conditions (33). HP of orange juice sensitized survivors of *E. coli* O157:H7 to the acidic conditions of orange juice, resulting in a significantly lower survival during refrigerated storage (91). Recent studies (45) showed that pressure cycling (application of consecutive, short pressure treatments interrupted by brief decompressions) strongly enhances the pressure sensitivity of *E. coli* in milk, particularly in combination with natural antimicrobial peptides and with regard to the problem of pressure-resistant mutants. Nevertheless, vegetative cells are relatively pressure sensitive, being readily killed in the range of 500 to 700 MPa, while bacterial spores are resistant to pressures far exceeding 300 mPa at ambient temperatures (50, 138). However, the combination of elevated temperatures with pressures of, e.g., 400 MPa is effective against spores (151). Spores can be inactivated by alternating pressure cycles which initiate spore germination under moderate pressure (e.g., 30 min at 80 MPa) and then killing the germinated spores by high pressure (e.g., 30 min at 350 MPa) (50). Pressure germination of spores is affected by pH and ionic environment (50). More recent data of temperature effects on pressure inactivation of bacterial spores show several log cycles of inactivation of highly heat-resistant spores of *Bacillus subtilis* and *Bacillus stearothermophilus* at 150 to 450 MPa and 70 to 95°C temperature. This suggests that combining "moderate" pressures with pasteurization temperatures can lead to "sterilization" (58, 84). Ascospores of yeasts are more pressure resistant than their vegetative cells. Nevertheless, ascospores of *S. cerevisiae* could be high-pressure-inactivated in fruit juices requiring *D* values ranging from 8 s to 10.8 min at 500 and 300 MPa, respectively (172).

High pressure on foods and food components (17) can cause protein denaturation, altered gelation properties of biopolymers, increased susceptibility to enzymatic degradation, and increased drying rates of plant tissues as a result of permeabilization of cell walls (83). Activation and inactivation of enzymes can be controlled using selected pressure and temperature conditions (82). However, various enzymes have considerable differences in barosensitivity, and further work is required to study possible regeneration of enzyme activities in foods during subsequent storage (83). The efficiency of HP, just as that of other treatments, depends on the food composition and the pH value.

Besides several batch HP systems, semicontinuous systems associated with aseptic packaging have been developed recently (152).

Electric Field Effects

High-voltage discharges (electroporation treatment) can kill microorganisms (75). An external pulsed electric field (PEF) charges cells in such a way that they behave as small dipoles and an electric potential develops over the cell membranes. When the induced electric potential exceeds some critical value, a reversible increase in membrane permeability results (8). For vegetative bacteria, this critical electric field strength is about 15 kV/cm. Much higher field strength is required for inactivation of asco- and endospores (101). When the critical electric field is greatly exceeded, the ion channels and pores of membranes are irreversibly extended to such a degree that cell contents leak out and cells die (dielectric rupture theory). The critical electric field is affected by pulse duration and resistivity of the suspension material (171). Because electric resistivity decreases as ionic strength increases, the bactericidal effect of PEF is generally inversely proportional to the ionic strength of the suspension material, whereas reduced pH increases the inactivation ratio by affecting the cytoplasm when the poration is completed (166). Microbial inactivation is proportional to the number and duration of pulses. Gram-positive bacteria are less sensitive to electric pulse treatment than gram-negative bacteria, while yeast cells, being larger than bacterial cells,

are more sensitive to PEF processing (64). PEF acts synergistically with nisin in reducing the viable count of vegetative cells of *B. cereus* (131). PEF inactivation is also a function of the treatment temperature and the microbial growth stage. Cells harvested from the logarithmic growth phase are more sensitive when compared with cells in the stationary and lag phases. Increasing the medium temperature decreases the breakdown transmembrane potential of the cell membrane in addition to causing thermal injury effects, thus resulting in a higher inactivation ratio (134).

Over the last decade, high-voltage PEF applied across electrodes between which liquid foods can be pumped has been investigated as potential nonthermal treatment for food preservation (75). Microbiological test results show that bipolar square-wave pulses are the most efficient in terms of microbial inactivation for commercial PEF pasteurization of fluid foods. Flavor and enzyme activity are not significantly diminished. The application of many rapid high-power pulses can significantly increase the temperature of the product unless this generated heat is removed.

Both electrical stimulation, as used in the meat industry, and pulsed electricity significantly reduce the viability of microorganisms on beef surfaces (8). Recent studies (79, 80) show that PEF has less pronounced effects on dry solid powders compared with those reported for liquids. Peleg (124) proposed kinetic models for PEF inactivation of microorganisms, calculating the relationship between survival rate, field strength, and the number of pulses.

A continuous PEF processing system suitable for industrial application has yet to be developed.

Magnetic Field Effects

Magnetic fields cause a change in the orientation of biomolecules and biomembranes to a direction parallel or perpendicular to the applied magnetic field and induce a change in the ionic drift across the plasma membrane (133). Thus, magnetic fields may affect growth and reproduction of microorganisms. Oscillating magnetic fields with an intensity (flux density) of 5 to 50 T with short pulses of 25 μs to 10 ms and frequencies of 5 to 500 kHz can inactivate vegetative microorganisms and pasteurize food material of high electrical resistivity, packaged in plastic packaging, without significant temperature increase (62, 133). (Microwave cooking also involves application of magnetic fields but differs in that it produces a thermal effect.) The explanation of the microbicidal effect of oscillating magnetic field pulses suggests that the oscillating magnetic field couples enough energy into the magneto-active parts (magnetic dipoles) of large biological molecules to break bonds in DNA or proteins critical for the survival and reproduction of microorganisms. As with high-voltage pulse treatment, the total exposure time is very short and produces no significant temperature increase (101). Much further research is required, however, to overcome technical difficulties, understand the mechanisms of microbial inactivation, and establish the commercial feasibility of this treatment.

Future Work Needed

Like the established physical preservation methods discussed earlier in this chapter, the nonthermal physical treatments discussed above induce sublethal injury to microbial cells surviving the treatments. The injury makes them sensitive to different physical and chemical environments to which the normal cells are resistant. Thus, both conventional and newly developed physical treatments can be used in combination with other preservative factors to increase antimicrobial efficiency and enhance safety and shelf life of foods (40, 77, 104). The various newer nonthermal physical treatments appear to be promising tools for food preservation. However, conflicting data about their effect on bacterial spores and enzymes and conflicting kinetic data need to be reconciled. The mechanistic understanding of the effects of these processes on foods, food components, and microbial cells, as well as on storage-dependent changes after nonthermal treatments, is limited. Thus, further research should concentrate on these questions.

References

1. Ababouch, L., and F. F. Busta. 1987. Effect of thermal treatments in oils on bacterial spore survival. *J. Appl. Bacteriol.* **62**:491–502.

2. Anderson, A. W., H. C. Nordan, R. F. Cain, G. Parnish, and D. Duggan. 1956. Studies on a radio-resistant micrococcus. I. The isolation, morphology, cultural characteristics and resistance to gamma radiation. *Food Technol.* **10**:575–577.

3. Anderson, J. G., and J. E. Smith. 1976. Effects of temperature on filamentous fungi, p. 191–218. *In* F. A. Skinner and W. B. Hugo (ed.), *Inhibition and Inactivation of Vegetative Microbes*. Academic Press, London, United Kingdom.

4. Auffray, Y., E. Lecesne, A. Hartky, and P. Boulibounes. 1995. Basic features of the *Streptococcus thermophilus* heat shock response. *Curr. Microbiol.* **30**:87–91.

5. Ayerst, G. 1969. The effects of moisture and temperature on growth and spore germination in some fungi. *J. Stored Prod. Res.* **5**:127–141.

6. Baird-Parker, A. C., and B. Freame. 1967. Combined effect of water activity, pH and temperature on the growth of *Clostridium botulinum* from spore and vegetative cell inocula. *J. Appl. Bacteriol.* **30**:420–429.

7. Barbosa-Canovas, G. V., V. R. Pothakamury, E. Palou, and B. Swanson. 1997. *Nonthermal Preservation of Foods*. Marcel Dekker, Inc., New York, N.Y.

8. Bawcom, D. W., L. D. Thompson, M. F. Miller, and C. B. Ramsey. 1995. Reduction of microorganisms on beef surfaces utilizing electricity. *J. Food Prot.* **58**:35–38.

9. Baxter, R. M., and N. E. Gibbons. 1962. Observations on the physiology of psychrophilism in a yeast. *Can. J. Microbiol.* **8**:511–517.

10. Boutibonnes, P., V. Bisson, B. Thammavongs, A. Hartke, J. M. Panoff, A. Benackour, and Y. Auffray. 1995. Induction of thermotolerance by chemical agents in *Lactococcus lactis* subsp. *lactis* IL 1403. *Int. J. Food Microbiol.* **25**:83–94.

11. Brown, A. D. 1974. Microbial water relations: features of the intracellular composition of sugar-tolerant yeasts. *J. Bacteriol.* **118**:769–777.

12. Brown, A. D. 1976. Microbial water stress. *Bacteriol. Rev.* **40**:803–846.

13. Brown, A. D., and J. R. Simpson. 1972. Water relations of sugar-tolerant yeasts: the role of intracellular polyols. *J. Gen. Microbiol.* **72**:589–591.

14. Brown, K. L., and C. A. Ayres. 1982. Thermobacteriology of UHT processed foods. *Dev. Food Microbiol.* **1**:119–152.

15. Cerf, O. 1977. Tailing of survival curves of bacterial spores. *J. Appl. Bacteriol.* **42**:1–19.

16. Chang, J. C. H., S. F. Ossoff, D. C. Lobe, M. H. Dorfman, C. M. Dumais, R. G. Qualls, and J. D. Johnson. 1985. UV inactivation of pathogenic and indicator microorganisms. *Appl. Environ. Microbiol.* **49**:1361–1365.

17. Cheftel, J. C. 1992. Effects of high hydrostatic pressure on food constituents: an overview, p. 195–209. *In* C. Balny, R. Hayashi, K. Heremans, and P. Masson (ed.), *High Pressure and Biotechnology*. John Libbey & Co. Ltd., London, United Kingdom.

18. Church, N. 1994. Developments in modified-atmosphere packaging and related technologies. *Trends Food Sci. Technol.* **5**:345–352.

19. Clavero, M. R. S., J. D. Monk, L. R. Beuchat, M. P. Doyle, and R. E. Brackett. 1994. Inactivation of *Escherichia coli* O157:H7, Salmonellae and *Campylobacter jejuni* in raw ground beef by gamma irradiation. *Appl. Environ. Microbiol.* **60**:2069–2075.

20. Cole, M. B., K. W. Davies, G. Munro, C. D. Holyoak, and D. C. Kilsby. 1993. A vitalistic model to describe the thermal inactivation of *Listeria monocytogenes*. *J. Ind. Microbiol.* **12**:232–239.

21. Condon, S., and F. J. Sala. 1992. Heat resistance of *Bacillus subtilis* in buffer and foods of different pH. *J. Food Prot.* **55**:605–608.

22. Corry, J. E. L. 1987. Relationship of water activity to fungal growth, p. 51–99. *In* L. R. Beuchat (ed.), *Food and Beverage Mycology*. AVI-Van Nostrand Reinhold Co., New York, N.Y.

23. Davies, A. D. 1995. Advances in modified atmosphere packaging, p. 301–320. *In* G. W. Gould (ed.), *New Methods of Food Preservation*. Blackie Academic & Professional, Glasgow, Scotland.

24. **Davies, R.** 1976. The inactivation of vegetative bacterial cells by ionizing radiation, p. 239–255. *In* F. A. Skinner and W. B. Hugo (ed.), *Inhibition and Inactivation of Vegetative Microbes.* Academic Press, London, United Kingdom.

25. **Dawes, I. W.** 1976. Inactivation of yeast, p. 279–304. *In* F. A. Skinner and W. B. Hugo (ed.), *Inhibition and Inactivation of Vegetative Microbes.* Academic Press, London, England.

26. **Deák, T.** 1991. Preservation by dehydration, p. 70–82. *In* E. Szenes and M. Oláh (ed.), *Handbook of Canning Industry.* Integra Projekt, Budapest, Hungary. (In Hungarian.)

27. **Decereau, R. V.** 1985. *Microwaves in the Food Processing Industry.* Academic Press, Inc., Orlando, Fla.

28. **Decereau, R. V.** 1992. *Microwave Foods: New Product Development.* Food & Nutrition Press, Trumbull, Conn.

29. **Diehl, J. F.** 1995. *Safety of Irradiated Foods,* 2nd ed. Marcel Dekker, Inc., New York, N.Y.

30. **Doyle, M. P., and E. H. Martihn.** 1975. Thermal inactivation of conidia from *Aspergillus flavus* and *Aspergillus parasiticus. J. Milk Food Technol.* **38:**678.

31. **Eckardt, C., and E. Ahrens.** 1977. Untersuchungen über *Byssochlamys fulva* Olliver & Smith als potentiellen Verderbniserreger in Erdbeerkonserven. II. Hitzeresistenz der Ascosporen von *Byssochlamys fulva. Chem. Mikrobiol. Technol. Lebensm.* **5:**76–80.

32. **Edgey, M., and A. D. Brown.** 1978. Response of xerotolerant and nontolerant yeasts to water stress. *J. Gen. Microbiol.* **104:**343–345.

33. **Ellenberg, L., and D. G. Hoover.** 1999. Injury and survival of *Aeromonas hydrophila* 7965 and *Yersinia enterocolitica* 9610 from high hydrostatic pressure. *J. Food Safety* **19:**263–276.

34. **Erichsen, I., and G. Molin.** 1981. Microbial flora of normal and high pH beef stored at 4°C in different gas environments. *J. Food Prot.* **44:**866.

35. **Ernshaw, R. G., J. Appleyard, and R. M. Hurst.** 1995. Understanding physical inactivation processes: combined preservation opportunities using heat, ultrasound and pressure. *Int. J. Food Microbiol.* **28:**197–219.

36. **Farkas, J.** 1980. Principles of food irradiation, p. 1–7. *Handouts of IFFIT International Training Course on Food Irradiation.* International Facility for Food Irradiation Technology, Wageningen, The Netherlands.

37. **Farkas, J.** 1988. *Irradiation of Dry Food Ingredients.* CRC Press, Inc., Boca Raton, Fla.

38. **Farkas, J.** 1990. Combination of irradiation with mild heat treatment. *Food Control* **1:**223–229.

39. **Farkas, J.** 1998. Irradiation as a method for decontaminating food. A review. *Int. J. Food Microbiol.* **44:**189–204.

40. **Fenice, M., R. Di Giambattista, J.-L. Leuba, and F. Federici.** 1999. Inactivation of *Mucor plumbeus* by combined actions of chitinase and high hydrostatic pressure. *Int. J. Food Microbiol.* **52:**109–113.

41. **Finne, G., and J. R. Matches.** 1976. Spin-labeling studies on the lipids of psychrophilic, psychrotrophic, and mesophilic clostridia. *J. Bacteriol.* **125:**211–219.

42. **Flowers, R. S., and S. E. Martin.** 1980. Ribosome assembly during recovery of heat-injured *Staphylococcus aureus. J. Bacteriol.* **141:**645–651.

43. **Fryer, P.** 1995. Electrical resistance heating of foods, p. 205–235. *In* G. W. Gould (ed.) *New Methods of Food Preservation.* Blackie Academic & Professional, Glasgow, Scotland.

44. **Garcia, M. L., J. Burgos, B. Sanz, and J. A. Ordonez.** 1989. Effect of heat and ultrasonic waves on the survival of two strains of *Bacillus subtilis. J. Appl. Bacteriol.* **67:**619–628.

45. **Garcia-Graells, C., B. Masschalek, and C. W. Michiels.** 1999. Inactivation of *Escherichia coli* in milk by high-hydrostatic-pressure treatment in combination with antimicrobial peptides. *J. Food Prot.* **62:**1248–1254.

46. **George, A. M., W. A. Cramp, and M. B. Yatwin.** 1980. The influence of membrane fluidity on the radiation induced changes in the DNA of *E. coli* K 1060. *Int. J. Radiat. Biol.* **38:**427–438.

47. **Gill, C. O., and G. Molin.** 1991. Modified atmospheres and vacuum packaging, p. 172–199. *In* N. J. Russell, and G. W. Gould (ed.), *Food Preservatives.* Blackie Academic & Professional, Glasgow, Scotland.

48. **Goldblith, S. A., L. Rey, and W. W. Rothmayr.** 1975. *Freeze Drying and Advanced Food Technology.* Academic Press, London, United Kingdom.

49. **Gorris, L. G. M.** 1994. Improvement of the safety and quality of refrigerated ready-to-eat foods using novel mild preservation techniques, p. 57–72. *In* R. P. Singh and F. A. R. Oliveira (ed.), *Minimal Processing of Foods and Process Optimization. An Interface.* CRC Press, Boca Raton, Fla.

50. **Gould, G. W.** 1973. Inactivation of spores in food by combined heat and hydrostatic pressure. *Acta Aliment.* **2:**377–383.

51. **Gould, G. W.** 1989. Drying, raised osmotic pressure and low water activity, p. 97–117. *In* G. W. Gould (ed.) *Mechanisms of Action of Food Preservation Procedures.* Elsevier Applied Science, London, United Kingdom.

52. **Gould, G. W.** 1989. Heat induced injury and inactivation, p. 11–42. *In* G. W. Gould (ed.), *Mechanisms of Action of Food Preservation Procedures.* Elsevier Applied Science, London, United Kingdom.

53. **Graham, A. F., D. R. Mason, F. J. Maxwell, and M. W. Peck.** 1997. Effect of pH and NaCl on growth from spores of non-proteolytic *Clostridium botulinum* at chill temperatures. *Lett. Appl. Microbiol.* **24:**95–100.

54. **Grant, I. R., and M. F. Patterson.** 1989. A novel radiation-resistant *Deinobacter* sp. isolated from irradiated pork. *Lett. Appl. Microbiol.* **8:**21–24.

55. **Grant, I. R., and M. F. Patterson.** 1995. Combined effect of gamma radiation and heating on the destruction of *Listeria monocytogenes* and *Salmonella typhimurium* in cook-chill roast beef and gravy. *Int. J. Food Microbiol.* **27:**117–128.

56. **Grecz, N., D. B. Rowley, and A. Matsuyama.** 1983. The action of radiation on bacteria and viruses. *In* E. S. Josephson and M. S. Peterson (ed.), *Preservation of Food*

by Ionizing Radiation, vol. 2. CRC Press, Boca Raton, Fla.

57. Han, Y. W. 1975. Death rates of bacterial spores: nonlinear survivor curves. *Can. J. Microbiol.* **21:**1464–1467.

58. Heinz, V., and O. Knorr. 1998. High pressure germination and inactivation kinetics of bacterial spores. *In* N.S. Isaacs (ed.), *High Pressure Food Science, Bioscience and Chemistry.* Royal Society of Chemistry, London, United Kingdom.

59. Hiatt, C. W. 1964. Kinetics of the inactivation of viruses. *Bacteriol. Rev.* **28:**150–163.

60. Hightower, L. E. 1991. Heat shock, stress proteins, chaperones, and proteotoxicity. *Cell* **66:**191–197.

61. Ho, Y. C., and K. L. Yam. 1993. Effect of metal shielding on microwave heating uniformity of a cylindrical food model. *J. Food Process. Preserv.* **16:**337–359.

62. Hofmann, G. A. 1985. Inactivation of microorganisms by an oscillating magnetic field. U.S. patent 4,524,079.

63. Hollywood, N. W., Y. Varabioff, and G. E. Mitchell. 1991. The effect of microwave and conventional cooking on the temperature profiles and microbial flora of minced beef. *Int. J. Food Microbiol.* **14:**67–76.

64. Hülsheger, H., J. Potel, and E. G. Neumann. 1983. Electric field effects on bacteria and yeast cells. *Rad. Environ. Biophys.* **22:**149–162.

65. Hurst, A. 1984. Reversible heat damage, p. 303–318. *In* A. Hurst and A. Nasim (ed.), *Repairable Lesions in Microorganisms.* Academic Press, London, United Kingdom.

66. Ingram, M., and J. Farkas. 1977. Microbiology of foods pasteurized by ionising radiation. *Acta Aliment.* **6:**123–185.

67. Ingram, M., and B. M. Mackey. 1976. Inactivation by cold, p. 111–151. *In* F. A. Skinner and W. B. Hugo (ed.), *Inhibition and Inactivation of Vegetative Microbes.* Society of Applied Bacteriology Symposia Series, no. 5. Academic Press, Inc., New York.

68. International Commission on Microbiological Specifications for Foods. 1980. Temperature, p. 1–37. *In* J. H. Silliker et al. (ed.), *Microbial Ecology of Foods,* vol. 1, *Factors Affecting Life and Death of Microorganisms.* Academic Press, Inc., New York, N.Y.

69. International Commission on Microbiological Specifications for Foods. 1980. Ionizing radiation, p. 46–69. *In* J. H. Silliker et al. (ed.), *Microbial Ecology of Foods,* vol. 1, *Factors Affecting Life and Death of Microorganisms.* Academic Press, Inc., New York, N.Y.

70. International Commission on Microbiological Specifications for Foods. 1980. Reduced water activity, p. 70–91. *In* J. H. Silliker et al. (ed.), *Microbial Ecology of Foods,* vol. 1, *Factors Affecting Life and Death of Microorganisms.* Academic Press, Inc., New York, N.Y.

71. Isaacs, N. S. 1998. *High Pressure Food Science, Bioscience and Chemistry.* Royal Society of Chemistry, London, United Kingdom.

72. Jakobsen, M., O. Filtenborg, and F. Bramsnaes. 1972. Germination and outgrowth of the bacterial spore in the presence of different solutes. *Lebensm. Wiss. Technol.* **5:**159–162.

73. Jay, J. M. 1992. *Modern Food Microbiology,* 4th Chapman & Hall, New York, N.Y.

74. Jermini, M. F. G., and W. Schmidt-Lorenz. 1987. Heat resistance of vegetative cells and asci of two *Zygosaccharomyces* yeasts in broth at different water activity values. *J. Food Prot.* **50:**835–841.

75. Jeyamkondan, S., D. S. Jayas, and R.A. Holley. 1999. Pulsed electric field processing of foods: a review. *J. Food Prot.* **62:**1088–1096.

76. Josephson, E. S., and M. S. Peterson (ed.). 1983. *Preservation of Foods by Ionizing Radiation.* CRC Press, Boca Raton, Fla.

77. Kalchayanand, N., T. Sikes, C. P. Dunne, and B. Ray. 1994. Hydrostatic pressure and electroporation have increased bactericidal efficiency in combination with bacteriocins. *Appl. Environ. Microbiol.* **60:**4174–4177.

78. Kang, C.K., M. Woodburn, A. Pagenkopf, and R. Cheney. 1969. Growth, sporulation and germination of *Clostridium perfringens* in media of controlled water activity. *Appl. Microbiol.* **18:**798–805.

79. Keith, W. D., L. J. Harris, and M. W. Griffiths. 1997. Reduction of bacterial levels in flour by pulsed electric fields. *J. Food Process Eng.* **21:**263–269.

80. Keith, W. D., L. J. Harris, L. Hudson, and M. W. Griffiths. 1997. Pulsed electric fields as a processing alternative for microbial reduction in spice. *Food Res. Int.* **30:**185–191.

81. Keteleer, A., and P. P. Tobback. 1994. Modified atmosphere storage of respiring product, p.59–64. *In* L. Leistner and L.G.M. Gorris (ed.), *Food Preservation by Combined Processes.* Final Report FLAIR Concerted Action No. 7, Subgroup B. EUR 15776 EN. European Commission, Luxembourg.

82. Knorr, D. 1993. Effects of high hydrostatic pressure processes on food safety and quality. *Food Technol.* **47:**156–161.

83. Knorr, D. 1994. Non-thermal processes for food preservation, p. 3–15. *In* R. P. Singh and F. A. R. Oliveira (ed.), *Minimal Processing of Foods and Process Optimization. An Interface.* CRC Press, Boca Raton, Fla.

84. Knorr, D., and V. Heinz. 1999. Recent advances in high pressure processing of foods. *New Food* **2:**15–19.

85. Laszlo, A. 1988. Evidence for two states of thermotolerance in mammalian cells. *Int. J. Hyperthermia* **4:**513–526.

86. Lee, B. H., S. Kermasha, and B. E. Baker. 1989. Thermal, ultrasonic and ultraviolet inactivation of *Salmonella* in thin films of aqueous media and chocolate. *Food Microbiol.* **6:**143–152.

87. Leistner, L. 1995. Principles and application of hurdle technology, p. 1–21. *In* G. W. Gould (ed.), *New Methods of Food Preservation.* Blackie Academic & Professional, Glasgow, Scotland.

88. Leistner, L., and N. J. Russell. 1991. Solutes and low water activity, p.111–134. *In* N. J. Russell and G. W. Gould (ed.), *Food Preservatives.* Blackie Academic & Professional, Glasgow, Scotland.

89. Le Jean, G., G. Abraham, E. Debray, Y. Candau, and G. Piar. 1994. Kinetics of thermal destruction of *Bacillus*

stearothermophilus spores using a two reaction model. *Food Microbiol.* **11**:229–241.

90. Lewis, M. 1993. UHT processing: safety and quality aspects. *Food Technol. Int. Eur.* **1993**:47–51.

91. Linton, M., J. M. J. McClements, and M. F. Patterson. 1999. Survival of *Escherichia coli* O157:H7 during storage in pressure-treated orange juice. *J. Food Prot.* **62**:1038–1040.

92. Löndahl, G., and T. E. Nilsson. 1978. Microbiological aspects of the freezing of meat and prepared foods. *Int. J. Refrig.* **1**:53–56.

93. López-Caballero, M. E., J. Carballo, and F. Jiménez-Colmenero. 1999. Microbiological changes in pressurized, prepackaged sliced cooked ham. *J. Food Prot.* **62**:1411–1415.

94. Mackey, B. M., and C. M. Derrick. 1986. Peroxide sensitivity of cold-shocked *Salmonella typhimurium* and *Escherichia coli* and its relationship to minimal medium recovery. *J. Appl. Bacteriol.* **60**:501–511.

95. Mackey, B. M., and C. M. Derrick. 1986. Elevation of heat resistance of *Salmonella typhimurium* by sublethal heat shock. *J. Appl. Bacteriol.* **61**:389–394.

96. Mackey, B. M., and C. M. Derrick. 1987. Changes in the heat resistance of *Salmonella typhimurium* during heating at rising temperatures. *Lett. Appl. Microbiol.* **4**:3–16.

97. Mackey, B. M., K. Forestiere, N. S. Isaacs, R. Stenning, and B. Brooker. 1994. The effect of high hydrostatic pressure on *Salmonella thompson* and *Listeria monocytogenes* examined by electron microscopy. *Lett. Appl. Microbiol.* **19**:429–432.

98. Macleod, R. A., and P. H. Calcott. 1976. Cold shock and freezing damage to microbes, p. 81–109. *In* T. R. G. Gray and J. R. Postgate (ed.), *The Survival of Vegetative Microbes.* Twenty-Sixth Symposium of the Society for General Microbiology. Cambridge University Press, Cambridge, United Kingdom.

99. Martens, T. (ed.). 1999. *Harmonization of Safety Criteria for Minimally Processed Foods. Rational and Harmonization Report.* FAIR Concerted Action, FAIR CT96-1020. Alma University Restaurants, Leuven, Belgium.

100. Marth, E. H. 1998. Extended shelf life refrigerated foods: microbiological quality and safety. *Food Technol.* **52**(12):57–62.

101. Mertens, B., and D. Knorr. 1992. Development of non-thermal processes for food preservation. *Food Technol.* **46**(5):124–133.

102. Metrick, C., D. Hoover, and D. Farkas. 1989. Effects of high hydrostatic pressure on heat-resistant and heat-sensitive strains of *Salmonella. J. Food Sci.* **54**:1547–1549.

103. Michels, M. J. M. 1978. Die mikrobiologische Qualität von Trockengemüse. *Lebensm. Technol. Verfahrenstech.* **29**:14–18.

104. Mohácsi-Farkas, C., G. Kiskó, J. Farkas, L. Mészáros, and T. Sáray. 1999. Assessment of antibacterial effects of essential oils by automated impedimetry and preliminary studies on their utility as biopreservatives, p. 279–281. *In* A. C. J. Tuijtelaars, R. A. Samson, F. M. Rombouts, and S. Notermans (ed.), *Food Microbiology and Food Safety into the Next Millenium.* Foundation Food Micro '99, Zeist, The Netherlands.

105. Monk, D. J., L. R. Beuchat, and M. P. Doyle. 1995. Irradiation inactivation of food-borne microorganisms. *J. Food Prot.* **58**:197–208.

106. Montville, T. J., and G. M. Sapers. 1981. Thermal resistance of spores from pH elevating strains of *Bacillus licheniformis. J. Food Sci.* **46**:1710–1712.

107. Moreno, M. A., M. del Carmen Ramos, A. Gonzalez, and G. Suarez. 1987. Effect of ultraviolet light irradiation on viability and aflatoxin production by *Aspergillus parasiticus. Can. J. Microbiol.* **33**:927–929.

108. Moseley, B. E. B. 1989. Ionizing radiation: action and repair, p. 43–70. *In* G. W. Gould (ed.), *Mechanisms of Action of Food Preservation Procedures.* Elsevier Applied Science, London, United Kingdom.

109. Mossel, D. A. A. 1975. Occurrence, prevention and monitoring of microbial quality loss of foods and dairy products. *Crit. Rev. Environ. Control* **5**:1–139.

110. Mossel, D. A. A., and M. Ingram. 1955. The physiology of the microbial spoilage of foods. *J. Appl. Bacteriol.* **18**:232–268.

111. Mossel, D. A. A., and C. B. Struijk. 1991. Public health implications of refrigerated pasteurized ("sous vide") foods. *Int. J. Food Microbiol.* **13**:187–206.

112. Mudgett, R. E. 1985. Dielectrical properties of foods, p. 15–37. *In* R. V. Decareau (ed.), *Microwaves in the Food Processing Industry.* Academic Press, Inc., Orlando, Fla.

113. Mullin, J. 1995. Microwave processing, p. 112–134. *In* G. W. Gould (ed.), *New Methods of Food Preservation.* Blackie Academic & Professional, Glasgow, Scotland.

114. Narasimhan, R., M. M. Habibullah-Khan, J. Ernest, and K. Thangavel. 1989. Effect of ultraviolet radiation on the bacterial flora of the packaging materials of milk and milk products. *Cherion.* **18**:89–92.

115. Notermans, S., J. Dufrenne, and B. M. Lund. 1990. Botulism risk of refrigerated processed foods of extended durability. *J. Food Prot.* **53**:1020–1024.

116. Nychas, G. J. E. 1994. Modified atmosphere packaging of meats, p. 417–436. *In* R. P. Singh and F. A. R. Oliveira (ed.), *Minimal Processing of Foods and Process Optimization. An Interface.* CRC Press, Boca Raton, Fla.

117. Nychas, G. J. E., and J. S. Arkoudelos. 1990. Microbiological and physico-chemical changes in minced meat under carbon dioxide, nitrogen or air at 3°C. *Int. J. Food Sci. Technol.* **25**:389.

118. Ohye, D. F., and J. H. B. Christian. 1967. Combined effects of temperature, pH and water activity on growth and toxin production by *C. botulinum* types A, B and E, p. 217–223. *In* M. Ingram and T. A. Roberts (ed.), *Botulism 1966.* Chapman & Hall, London, United Kingdom.

119. Olson, D. G. 1998. Irradiation of food. *Food Technol.* **52**(1):56–62.

120. Ordonez, J. A., B. Sanz, P. E. Hernandez, and P. Lopez. 1984. A note on the effect of combined ultrasonic and heat treatments on the survival of thermoduric streptococci. *J. Appl. Bacteriol.* **56**:175–177.

121. Palaniappan, S., S. K. Sastry, and E. R. Richter. 1992. Effects of electroconductive heat treatment and electrical

pretreatment on thermal death kinetics of selected microorganisms. *Biotechnol Bioeng.* **39**:225–232.

122. Palumbo, S. A. 1987. Is refrigeration enough to restrain foodborne pathogens? *J. Food Prot.* **49**:1003.

123. Parsell, D. A., and R. T. Sauer. 1989. Induction of heat shock-like response by unfolded protein in *Escherichia coli*: dependence on protein level not protein degradation. *Genes Dev.* **3**:1226–1232.

124. Peleg, M. 1995. A model of microbial survival after exposure to pulsed electric fields. *J. Sci. Food Agric.* **67**:93–99.

125. Perkin, A. G., F. L. Davies, P. Neaves, B. Jarvis, C. A. Ayres, K. L. Brown, W. C. Falloon, H. Dallyn, and P. G. Bean. 1980. Determination of bacterial spore inactivation at high temperatures, p. 173–188. *In* G. W. Gould and J. E. L. Corry (ed.), *Microbial Growth and Survival in Extreme Environments.* Academic Press, London, United Kingdom.

126. Pflug, I. J. 1998. *Microbiology and Engineering of Sterilization Process*, 9th ed. Environmental Sterilization Laboratory, University of Minnesota, Minneapolis.

127. Pichhardt, K. 1993. *Lebensmittelmikrobiologie*, 3rd ed. Springer-Verlag, Berlin, Germany.

128. Pitt, J. I. 1975. Xerophilic fungi and the spoilage of foods of plant origin, p. 273–307. *In* R. B. Duckworth (ed.), *Water Relations of Foods.* Academic Press, London, United Kingdom.

129. Pitt, J. I., and A. D. Hocking. 1985. *Fungi and Food Spoilage.* Academic Press, Sydney, Australia.

130. Pitt, J. I., and B. F. Miscamble. 1995. Water relations of *Aspergillus flavus* and closely related species. *J. Food Prot.* **58**:86–90.

131. Pol, I. E., H. C. Mastwijk, P. V. Bartels, and E. J. Smid. 1999. Inactivation of *Bacillus cereus* by nisin combined with pulsed electric field, p. 287–290. *In* A. C. J. Tuijtelaars, R. A. Samson, F. M. Rombouts, and S. Notermans (ed.), *Food Microbiology and Food Safety into the Next Millenium.* Foundation Food Micro '99, Zeist, The Netherlands.

132. Ponne, C. T., and P. V. Bartels. 1995. Interaction of electromagnetic energy with biological material—relation to food processing. *Radiat. Phys. Chem.* **45**:591–607.

133. Pothakamury, U. R., U. R. Monsalve-Gonzalea, G. V. Barbosa-Canovas, and B. G. Swanson. 1993. Magnetic field inactivation of microorganisms and generation of biological changes. *Food Technol.* **47**(12):85–93.

134. Pothakmury, U. R., H. Vega, Q. Zhang, G. V. Barbosa-Canovas, and B. Swanson. 1996. Effect of growth stage and processing temperature on the inactivation of *E. coli* by pulsed electric fields. *J. Food Prot.* **59**:1167–1191.

135. Potter, N. N. 1986. *Food Science*, 4th ed. AVI-Van Nostrand Reinhold, New York, N.Y.

136. Put, H. M. C., J. de Jong, F. E. M. Sand, and A. M. van Grinsven. 1976. Heat resistance studies on yeast spp. causing spoilage in soft drinks. *J. Appl. Bacteriol.* **40**:135–152.

137. Raso, J., S. Condon, and F. J. Sala Trepat. 1994. Manothermo-sonication: a new method of food preservation, p. 37–41. *In* L. Leistner and L. G. M. Gorris (ed.), *Food Preservation by Combined Process.* Final Report FLAIR

Concerted Action No. 7, Subgroup B. EUR 15776, European Commission, Luxembourg.

138. Reddy, N. R., H. M. Solomon, G. A. Fingerhut, E. J. Rhodehamel, V. M. Balasubramaniam, and P. Palaniappan. 1999. Inactivation of *Clostridium botulinum* type E spores by high pressure processing. *J. Food Safety* **19**:277–288.

139. Reichart, O., and C. Mohácsi-Farkas. 1994. Mathematical modelling of the combined effect of water activity, pH and redox potential on the heat destruction. *Int. J. Food Microbiol.* **24**:103–112.

140. Reuter, H. 1989. *Aseptic Packaging of Food.* Technomic Publishing, Lancaster, Pa.

141. Roberts, T. A., G. Hobbs, J. H. B. Christian, and N. Skovgaard. 1981. *Psychrotrophic Microorganisms in Spoilage and Pathogenicity.* Academic Press, London, United Kingdom.

142. Rockland, L. B., and L. R. Beuchat (ed.). 1987. *Water Activity: Theory and Applications to Food.* Marcel Dekker, Inc., New York, N.Y.

143. Rose, D. 1995. Advances and potential for aseptic processing, p. 283–303. *In* G. W. Gould (ed.), *New Methods of Food Preservation.* Blackie Academic & Professional, Glasgow, Scotland.

144. Rosenberg, U., and W. Bogl. 1987. Microwave pasteurization, sterilization, blanching and pest control in the food industry. *Food Technol.* **41**(6):92–99.

145. Sala, F. J., J. Burgos, S. Condon, P. Lopez, and J. Raso. 1995. Effect of heat and ultrasound on microorganisms and enzymes, p. 176–204. *In* G. W. Gould (ed.), *New Methods of Food Preservation.* Blackie Academic & Professional, Glasgow, Scotland.

146. Saleh, Y. G., M. S. Mayo, and D. G. Ahearn. 1988. Resistance of some common fungi to gamma irradiation. *Appl. Environ. Microbiol.* **54**:2134–2135.

147. Sapru, V., A. A. Teixeira, G. H. Smerage, and J. A. Lindsay. 1992. Predicting thermophilic spore population dynamics for U.H.T. sterilization processes. *J. Food Sci.* **57**:1248–1257.

148. Sastry, S. K. 1994. Ohmic heating, p. 17–33. *In* R. P. Singh and F. A. R. Oliveira (ed.), *Minimal Processing of Foods and Process Optimization. An Interface.* CRC Press, Boca Raton, Fla.

149. Sato, M., and H. Takahashi. 1968. Cold shock of bacteria. I. General features of cold shock in *Escherichia coli*. *J. Gen. Appl. Microbiol.* **14**:417–428.

150. Schlesinger, M., M. Ashburner, and A. Tissieres. 1982. *Heat Shock from Bacteria to Man.* Cold Spring Harbor Laboratory Press, Cold Spring Harbor, N.Y.

151. Seyderhelm, I., and D. Knorr. 1992. Reduction of *Bacillus stearothermophilus* spores by combined high pressure and temperature treatments. *ZFL Int. J. Food Technol. Mark. Packag. Anal.* **43**(4):17.

152. Sizer, C. E., and V. M. Balasubramaniam. 1999. New intervention processes for minimally processed juices. *Food Technol.* **53**(10):64–67.

153. Splittstoesser, D. F., S. B. Leasor, and K. M. J. Swanson. 1986. Effect of food composition on the heat resistance of yeast ascospores. *J. Food Sci.* **51**:1265–1267.

154. **Splittstoesser, D. F., and C. M. Splittstoesser.** 1977. Ascospores of *Byssochlamys fulva* compared with those of a heat resistant *Aspergillus*. *J. Food Sci.* **42:**685–688.

155. **Stannard, C. J., J. S. Abbiss, and J. M. Wood.** 1983. Combined treatment with hydrogen peroxide and ultra-violet irradiation to reduce microbial contamination levels in pre-formed packaging cartons. *J. Food Prot.* **46:**1060–1064.

156. **Stephens, P. J., M. B. Cole, and M. V. Jones.** 1994. Effect of heating rate on the thermal inactivation of *Listeria monocytogenes*. *J. Appl. Bacteriol.* **77:**702–708.

157. **Stoforos, N. C. J. Noronha, M. Hendrickx, and P. Tobback.** 1997. A critical analysis of mathematical procedures for the evaluation and design of in-container thermal processes for foods. *Crit. Rev. Food Sci. Nutr.* **37:**411–441.

158. **Stumbo, C. R., K. S. Purohit, T. V. Ramakrishnan, D. A. Evans, and F. J. Francis.** 1983. *Handbook of Lethality Guides for Low-Acid Canned Foods*, vol. 1 and 2. CRC Press, Boca Raton, Fla.

159. **Tartera, C., A. Bosch, and J. Jofre.** 1988. The inactivation of bacteriophages infecting *Bacteroides fragilis* by chlorine treatment and UV-irradiation. *FEMS Microbiol. Lett.* **56:**313–316.

160. **Thayer, D. W., and G. Boyd.** 1999. Irradiation and modified atmosphere packaging for the control of *Listeria monocytogenes* on turkey meat. *J. Food Prot.* **62:**1136–1142.

161. **Tomlins, R. I., and Z. J. Ordal.** 1976. Thermal injury and inactivation in vegetative bacteria, p. 153–190. *In* F. A. Skinner and W. B. Hugo (ed.), *Inhibition and Inactivation of Vegetative Microbes*. Society of Applied Bacteriology Symposia Series, no. 5. Academic Press, Inc., New York, N.Y.

162. **Troller, J. A.** 1983. Effect of low moisture environments on the microbial stability of foods, p. 173–198. *In* A. H. Rose (ed.), *Economic Microbiology*, vol. 8. Academic Press, London, United Kingdom.

163. **Troller, J. A., and J. V. Stinson.** 1975. Influence of water activity on growth and enterotoxin formation by *Staphylococcus aureus* in foods. *J. Food Sci.* **40:**802–804.

164. **Tsuchido, T., M. Takano, and I. Shibasaki.** 1974. Effect of temperature-elevating process on the subsequent isothermal death of Escherichia coli K-12. *J. Ferment. Technol.* **52:**788–792.

165. **Urbain, W. M.** 1986. *Food Irradiation*. Academic Press, Orlando, Fla.

166. **Vega-Mercado, H., U. P. Pothakamury, F. J. Chang, G. V. Barbosa-Canovas, and G. G. Swanson.** 1996. Inactivation of *Escherichia coli* by combining pH, ionic strength, and pulsed electric fields hurdles. *Food Res. Int.* **29:**119–199.

167. **Wilkins, P. O.** 1973. Psychrotrophic gram-positive bacteria: temperature effects on growth and solute uptake. *Can. J. Microbiol.* **19:**909–915.

168. **Wilkinson, V. M., and G. W. Gould.** 1996. *Food Irradiation. A Reference Guide*. Butterworths-Heinemann, Oxford, United Kingdom.

169. **World Health Organization.** 1994. *Safety and Nutritional Adequacy of Irradiated Food*. World Health Organization, Geneva, Switzerland.

170. **World Health Organization.** 1999. *High-Dose Irradiation: Wholesomeness of Food Irradiated with Doses Above 10 kGy*. World Health Organization, Geneva, Switzerland.

171. **Zhang, Q., B.-L. Quin, G. V. Barbosa-Cánovas, and B. G. Swanson.** 1994. Growth stage and temperature affect the inactivation of *E. coli* by pulsed electric fields, p. 104. *In Proceedings of the 37th Annual Conference of the Canadian Institute of Food Science Technology*, Vancouver, B.C., Canada.

172. **Zook, C. D., M. E. Parish, R. J. Braddock, and M. O. Balaban.** 1999. High pressure inactivation kinetics of *Saccharomyces cerevisiae* ascospores in orange and apple juices. *J. Food Sci.* **64:**533–535.

Food Microbiology: Fundamentals and Frontiers, 2nd Ed.
Edited by M. P. Doyle et al.
© 2001 ASM Press, Washington, D.C.

P. Michael Davidson

Chemical Preservatives and Natural Antimicrobial Compounds

29

The quality of a food product decreases from the time of harvest or slaughter until it is consumed. Quality loss may be due to microbiological, enzymatic, chemical, or physical changes. The consequences of quality loss caused by microorganisms include consumer hazards, due to the presence of microbial toxins or pathogenic microorganisms, and economic losses due to spoilage. Many food preservation technologies, some in use since ancient times, protect foods from the effects of microorganisms and inherent deterioration. Microorganisms may be inhibited by chilling, freezing, water activity reduction, nutrient restriction, acidification, modification of packaging atmosphere, fermentation, or nonthermal treatments (e.g., high-pressure processing) or through addition of antimicrobial compounds.

Food antimicrobial agents are chemical compounds added to or present in foods that retard microbial growth or kill microorganisms, thereby resisting deterioration in safety or quality. The major targets for antimicrobials are food poisoning microorganisms (infective agents and toxin producers) and spoilage microorganisms whose metabolic end products or enzymes cause off odors, off flavors, texture problems, discoloration, slime, and haze. Food antimicrobials are sometimes referred to as preservatives. However, food preservatives include food additives that are antimicrobial agents, antibrowning agents (e.g., citric acid), and antioxidants (e.g., butylated hydroxyanisole [BHA]). The most widely used traditional preservatives are propionates, sorbates, and benzoates.

Most food antimicrobial agents are only bacteriostatic or fungistatic and not bactericidal or fungicidal. Because food antimicrobials are generally bacteriostatic or fungistatic, they will not preserve a food indefinitely. Depending upon storage conditions, the food product eventually spoils or becomes hazardous. Food antimicrobials are often used in combination with other food preservation procedures.

Cellular targets of food antimicrobial agents include the cell wall, cell membrane, metabolic enzymes, protein synthesis, and genetic systems. The exact mechanism(s) or target(s) for food antimicrobials are often not known or well defined. There are several possible reasons for this. Researchers often focus on a single target, such as an enzyme or the cell membrane, without determining the effect on other cellular functions (78). It is difficult to pinpoint a target when many interacting reactions take place simultaneously. For example, membrane-disrupting compounds could cause leakage, interfere with active transport or metabolic enzymes, or dissipate cellular energy in the form of ATP.

P. Michael Davidson, Department of Food Science and Technology, University of Tennessee, 2509 River Rd., Knoxville, TN 37996.

Food antimicrobials likely have multiple targets, with concentration-dependent thresholds for inactivation or inhibition. A given target is important in an inhibitor's overall mechanism only when its sensitivity is within the range of the antimicrobial concentration which inhibits growth (78).

It is a difficult and somewhat arbitrary task to classify antimicrobial compounds. This chapter divides the compounds into two classes: traditional and naturally occurring. Antimicrobials are classified as traditional when they fall into one or more of the following categories: they have been used for many years, they are approved by many countries for inclusion as antimicrobials in foods (e.g., lysozyme, which is naturally occurring, is regulatory approved and is therefore traditional), or they are produced by synthetic means or are inorganic (as opposed to natural extracts or organic compounds, respectively). Ironically, synthetic traditional antimicrobials are found in nature. These include acetic acid from vinegar, benzoic acid from cranberries, and sorbic acid from mountain ash berries (rowanberries). Classification within the broad categories, especially traditional antimicrobials, is even more difficult. This chapter discusses the physical characteristics, spectra of action, applications, and mechanisms of action of antimicrobial agents. Because many compounds have similar mechanisms of action, mechanisms will be explained for the first agent and cross-referenced back for compounds with similar mechanisms.

FACTORS AFFECTING ACTIVITY

The effectiveness of food antimicrobial agents depends on many factors associated with the food product, its storage environment, its handling, and the target microorganisms themselves. Food preservation is best achieved when the antimicrobial type and concentration, storage time and temperature, food pH and buffering capacity, and the presence of other agents that may increase shelf life are known and taken into account. Gould (105) classified the factors that affect the activity of antimicrobials into microbial, intrinsic, extrinsic, and process.

Microbial factors that affect antimicrobial activity include inherent resistance (vegetative cells versus spores; strain differences), initial number and growth rate, interaction with other microorganisms (e.g., antagonism), cellular composition (Gram reaction), and cellular status (injury). Intrinsic factors affecting activity are those associated with a food product and include nutrients, pH, buffering capacity, oxidation reduction potential, and water activity. Extrinsic factors affecting antimicrobial activity include temperature of storage, atmosphere, and relative humidity. The time of storage is a most important factor. Processing factors include changes in food composition, shifts in microflora, changes in microbial numbers, and change in microstructure. Most of the factors influence microorganisms in an interactive manner.

pH is the most universally important factor influencing the effectiveness of food antimicrobial agents. Many food antimicrobials are weak acids and are most effective in their undissociated form. This is because weak acids are able to penetrate the cytoplasmic membrane of a microorganism more effectively in the protonated form. Therefore, the pK_a of these compounds is important in selecting a particular compound for an application. The lower the pH of a food product, the greater the proportion of acid in the undissociated form and the greater the antimicrobial activity. Some researchers have suggested that only the undissociated form has activity. Eklund (75), however, demonstrated that, while the undissociated form has significantly greater activity, the anion does contribute slightly to antimicrobial activity.

Another factor affecting activity is polarity. This relates both to the ionization of the molecule and to the contribution of any alkyl side groups or hydrophobic parent molecules. Antimicrobial agents must be lipophilic to attach and pass through the cell membrane but must also be soluble in the aqueous phase.

There has been much interest in the effect of environmental stress factors (e.g., heat, cold, starvation, low pH/organic acids) on developed resistance of microorganisms to subsequent stressors. This developed resistance is termed tolerance, adaptation, or habituation, depending upon how the microorganism is exposed to the stress and the physiological conditions that lead to enhanced survival (33, 92). Pathogens can develop a tolerance or adaptation to organic acids following prior exposure to low pH. For example, acid-adapted *Salmonella* strains have increased resistance to acetic, benzoic, lactic, and propionic acids and activated lactoperoxidase system; *Escherichia coli* O157:H7 has tolerances to acetic, citric, lactic, malic, and sodium lactate; and *Listeria monocytogenes* has tolerance to lactic acid (32, 33, 92, 98, 162, 163, 197). While this increased resistance could be a problem in the application of organic acids for controlling pathogens, it has not been demonstrated to be a problem in actual food processing systems.

TRADITIONAL ANTIMICROBIAL AGENTS

Organic Acids and Esters

Many organic acids are used as food additives, but not all have antimicrobial activity. The most active are acetic, lactic, propionic, sorbic, and benzoic acids.

Citric, caprylic, malic, fumaric, and other organic acids have limited activity but are used primary for flavorings. Esters of fatty acids are discussed here because they are derivatives of organic acids and have similar mechanisms.

The antimicrobial activity of organic acids is related to pH, and the undissociated form of the acid is primarily responsible for antimicrobial activity. Therefore, in selecting an organic acid for use as an antimicrobial food additive, both the product pH and the pK_a of the acid must be taken into account. The use of organic acids is generally limited to foods with pH <5.5, since most organic acids have pK_as of pH 3.0 to 5.0 (70).

The mechanisms of action of organic acids and their esters have some common elements. There is little evidence that the organic acids and related esters influence cell wall synthesis in prokaryotes or that they significantly interfere with protein synthesis or genetic mechanisms. There are some exceptions, discussed below. As stated previously, in the undissociated form, organic acids can penetrate the cell membrane lipid bilayer more easily. Once inside the cell, the acid dissociates because the cell interior has a higher pH than the exterior (122). Bacteria maintain internal pH near neutrality to prevent conformational changes to the cell structural proteins, enzymes, nucleic acids, and phospholipids. Protons generated from intracellular dissociation of the organic acid acidify the cytoplasm and must be extruded to the exterior. According to the chemiosmotic theory (191), the cytoplasmic membrane is impermeable to protons, and they must be transported to the exterior. This proton extrusion creates an electrochemical potential across the membrane called the proton motive force (PMF). The PMF is a function of the differences in potential ($\Delta\Psi$) and pH (ΔpH) and is defined as: $\Delta p = \Delta\Psi - Z\Delta$pH, where $Z = 2.3RT/F$ (R = gas constant; T = absolute temperature; F = Faraday constant). Since protons generated by the organic acid inside the cell must be extruded using energy in the form of ATP, the constant influx of these protons will eventually deplete cellular energy (Fig. 29.1). Lambert and Stratford (156) demonstrated that yeast cells pump excess protons out of the cell, utilizing energy in the form of ATP. However, the intracellular pH is eventually raised to a point that the cell may resume growth. The time it takes to accomplish this increase in intracellular pH is dependent upon the extracellular pH and inhibitor concentration and is termed "lag time."

Organic acids also interfere with membrane permeability. Sheu and Freese (257) suggested that short-chain organic acids interfere with energy metabolism by altering the structure of the cytoplasmic membrane through interaction with membrane proteins. They further hy-

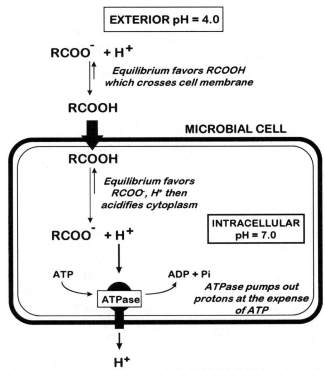

Figure 29.1 Fate of an organic acid (RCOOH) in a low-pH environment in the presence of a microbial cell.

pothesized that the interference with membrane proteins reduces ATP regeneration by uncoupling the electron transport system or by inhibiting active transport of nutrients into the cell. Sheu et al. (258) and Freese et al. (95) determined that short-chain fatty acids act as uncouplers of amino acid carrier proteins from the electron transport system. As proof, they showed that transport of L-serine, L-leucine, and malate is inhibited in membrane vesicles of *Bacillus subtilis* when exposed to acetate and other fatty acids. Later, Sheu et al. (259) found that active transport inhibition influenced energy metabolism only indirectly; cells were not necessarily ATP depleted. They suggested that inhibition of active transport was due to destruction of the PMF, which in turn caused active transport to cease. Somewhat contradictory results were obtained by Eklund (74) for both organic acids and alkyl esters of *p*-hydroxybenzoic acid (parabens). He studied the growth inhibition and uptake of amino acids and glucose for *E. coli*, *B. subtilis*, and *Pseudomonas aeruginosa* and concluded that uptake inhibition could not account for growth inhibition by propionate, benzoate, or sorbate. In a later study, Eklund (77) evaluated the effect of sorbic acid and parabens on the ΔpH and $\Delta\Psi$ components of the PMF in *E. coli* membrane vesicles. Both compounds eliminated the ΔpH but did not significantly affect the $\Delta\Psi$ component of the PMF, so active transport of amino acids continued. Eklund (77) concluded that neutralization of the PMF and subsequent transport inhibition was not the sole mechanism of action of the parabens. To summarize, the organic acids and their esters have a significant effect on bacterial cytoplasmic membranes, interfering with metabolite transport and maintenance of membrane potential. There is also considerable evidence that many organic acids and esters affect activity of microbial enzyme(s). However, because many of these studies are done with whole cells, it is not clear whether these are direct or indirect effects (77).

Acetic Acid and Acetates

Acetic acid ($pK_a = 4.75$; molecular weight [MW] 60.05; Fig. 29.2), the primary component of vinegar, and its sodium, potassium, and calcium salts, sodium and calcium diacetate, and dehydroacetic acid (methylacetopyranone) are some of the oldest food antimicrobial agents. Acetic acid is more effective against yeasts and bacteria than against molds. Only *Acetobacter* species (microorganisms involved in vinegar production), lactic acid bacteria, and butyric acid bacteria are tolerant to acetic acid (70). Bacteria inhibited by acetic acid include *Bacillus* species, *Clostridium* species, *L. monocytogenes*, *P. aeruginosa*, *Salmonella* spp., *Staphylococcus aureus*, *E. coli*, *Campylobacter jejuni*, and *Pseudomonas* spp. (70, 77, 190, 332). Molds and yeasts are more resistant to acetic acid than are bacteria. Yeasts and molds sensitive to acetic acid include *Aspergillus*, *Penicillium*, and *Rhizopus* spp. and some strains of *Saccharomyces* (70, 149).

Acetic acid and its salts have shown variable success as antimicrobial agents in food applications. Acetic acid can increase poultry shelf life when added to cut-up chicken parts in cold water at pH 2.5 (193). Addition of acetic acid at 0.1% to scald tank water used in poultry processing decreases the heat resistance of *Salmonella enterica* serovar Newport, *Salmonella enterica* serovar Typhimurium, and *C. jejuni* (201). In contrast, Lillard et al. (166) found that 0.5% acetic acid in the scald water has no significant effect on *Salmonella* spp., total aerobic bacteria, or *Enterobacteriaceae* on unpicked poultry carcasses. Acetic acid has shown variable effectiveness as an antimicrobial agent for use as a spray sanitizer on meat carcasses. Use of 2% acetic acid resulted in reductions in viable *E. coli* O157:H7 on beef after 7 days at 5°C (265). Acetic acid was the most effective antimicrobial in ground roasted beef slurries against *E. coli* O157:H7 growth in comparison with citric or lactic acid (2). Acetic acid added at 0.1% to bread dough inhibited growth of 6 log CFU of rope-forming *B. subtilis* per g in wheat bread (pH 5.14) stored at 30°C for >6 days (243). In brain heart infusion broth, 0.2% acetic acid at pH 5.1 or 0.1% acetic acid at pH 4.8 inhibited rope-forming strains of *B. subtilis* and *Bacillus licheniformis* for >6 days at 30°C (243).

Sodium acetate at 1.0% increases the shelf life of catfish fillets by 6 days when stored at 4°C compared with the control (145). Sodium acetate is also an effective inhibitor of rope-forming bacteria (*B. subtilis*) in baked goods and of the molds *Aspergillus flavus*, *Aspergillus fumigatus*, *Aspergillus niger*, *Aspergillus glaucus*, *Penicillium expansum*, and *Mucor pusillus* at pH 3.5 to 4.5 (102). It is useful in the baking industry because it has little effect on the yeast used in baking. Al-Dagal and Bazaraa (5) found that whole or peeled shrimp dipped in a 10% (wt/wt) sodium acetate solution for 2 min had extended microbiological and sensory shelf life compared with controls.

Sodium diacetate ($pK_a = 4.75$) is effective at 0.1 to 2.0% in inhibiting mold growth in cheese spread (70). At 32 mM (0.45%) in brain heart infusion broth (pH 5.4), sodium diacetate is inhibitory to *L. monocytogenes*, *E. coli*, *Pseudomonas fluorescens*, *Salmonella enterica* serovar Enteritidis, and *Shewanella putrefaciens* but not *S. aureus*, *Yersinia enterocolitica*, *Pseudomonas fragi*, *Enterococcus faecalis*, or *Lactobacillus fermentis* after 48 h at 35°C (252). In addition, 21 to 28 mM sodium diacetate suppressed growth by the natural microflora of ground beef after storage at 5°C for up to 8 days (252).

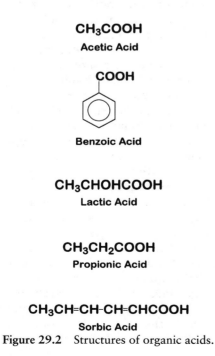

CH₃COOH

Acetic Acid

COOH

Benzoic Acid

CH₃CHOHCOOH

Lactic Acid

CH₃CH₂COOH

Propionic Acid

CH₃CH=CH-CH=CHCOOH

Sorbic Acid

Figure 29.2   Structures of organic acids.

Degnan et al. (63) showed a 2.6-log decrease in viable *L. monocytogenes* cells in blue crab meat washed with 2 M sodium diacetate after 6 days at 4°C. Dehydroacetic acid has a high pK$_a$ of 5.27 and is therefore active at higher pH values. It is inhibitory to bacteria at 0.1 to 0.4% and to fungi at 0.005 to 0.1% (70).

Acetic acid is used commercially in baked goods, cheeses, condiments and relishes, dairy product analogues, fats and oils, gravies and sauces, and meats. The sodium and calcium salts are used in breakfast cereals, cheeses, fats and oils, gelatin, hard candy, jams and jellies, meats, soft candy, snack foods, soup mixes, and sweet sauces. Sodium diacetate is used in baked goods, candy, cheese spreads, gravies, meats, sauces, and soup mixes.

The general mechanism by which acetic acid inhibits microorganisms is related to other organic acids, as discussed previously. Sheu and Freese (257) and Freese et al. (95) observed that acetic acid inhibits oxygen uptake and resultant ATP production by 76 to 77% in whole cells of *B. subtilis*. The compound does not, however, inhibit NADH oxidation by isolated membranes. Further, they found that α-glycerol phosphate- or NADH-energized uptake of serine transport in membrane vesicles of *B. subtilis* is inhibited by acetic acid. *E. coli* whole cells and vesicles give similar results. They concluded that acetate inhibits growth by uncoupling substrate transport and oxidative phosphorylation from the electron transport system. This inhibits uptake of metabolites into the cell. Later, Sheu et al. (259) determined that the short-chain fatty acids, such as acetic acid, interfere with the PMF. In addition to a demonstrated effect on the cell membrane, acetic acid may also act on cellular enzymes by reducing the intracellular pH (118).

Benzoic Acid and Benzoates

Benzoic acid (MW 122.12; Fig. 29.2) and sodium benzoate were the first antimicrobial compounds permitted in foods by the U.S. Food and Drug Administration (129). Benzoic acid occurs naturally in cranberries, plums, prunes, cinnamon, cloves, and most berries. Sodium benzoate is highly soluble in water (66.0 g/100 ml at 20°C) while benzoic acid is much less so (0.27% at 18°C). As the undissociated form of benzoic acid (pK$_a$ = 4.19) is the most effective antimicrobial agent, the most effective pH range is 2.5 to 4.5. Rahn and Conn (223) reported that the compound is 100 times as effective in acid solutions as in neutral solutions and that only the undissociated acid has antimicrobial activity. Eklund (76) demonstrated that both the dissociated and undissociated forms of benzoic acid inhibit various bacteria and a yeast but that the MIC of the undissociated acid is 15 to 290 times lower.

Benzoic acid and sodium benzoate are used primarily as antifungal agents. The inhibitory concentration of benzoic acid at pH <5.0 against most yeasts ranges from 20 to 700 μg/ml, while for molds it is 20 to 2,000 μg/ml (50). Some fungi, including *Byssochlamys nivea*, *Pichia membranaefaciens*, *Talaromyces flavus*, and *Zygosaccharomyces bailii* are resistant to benzoic acid (274). In combination with heat, 0.1% sodium benzoate reduced the time for a 3-log reduction of the ascospores of the heat-resistant mold *Neosartorya fischeri* from 205 to 210 min at 85°C to 85 to 123 min in mango and grape juice at pH 3.5 (224). *L. monocytogenes* is inhibited by 1,000 to 3,000 μg of benzoic acid per ml at pH 5.6, depending upon incubation temperature (84, 338). Benzoic acid at 0.1% is effective in reducing viable *E. coli* O157:H7 in apple cider (pH 3.6 to 4.0) by 3 to 5 logs in 7 days at 8°C (341). Sodium benzoate in combination with propionic acid delays growth of *E. coli* O157:H7 in the soft cheese queso fresco (141).

Sodium benzoate (up to 0.1%) is used as an antimicrobial agent in carbonated and still beverages, syrups, cider, margarine, olives, pickles, relishes, soy sauce, jams, jellies, preserves, pie and pastry fillings, fruit salads, and salad dressings and in the storage of vegetables (50).

There are probably multiple cellular targets for benzoic acid. Most research has focused on the cytoplasmic membrane and cellular enzymes. Freese et al. (95) hypothesized that benzoic acid uncouples both substrate transport and oxidative phosphorylation from the electron transport system. Freese (94) suggested that benzoic acid destroys the PMF by continuous transport of protons into the cell, causing disruption of the transport system.

Benzoates inhibit various microbial enzymes and enzyme systems. Acetic acid metabolism and oxidative phosphorylation (27), α-ketoglutarate and succinate dehydrogenases (27), and trimethylamine-N-oxide reductase activity of *E. coli* (154) are inhibited by benzoic acid. Aflatoxin production by *A. flavus* (51, 310, 311) and 6-phosphofructokinase activity (93) are inhibited in fungi.

Lactic Acid and Lactates

Lactic acid (pK$_a$ = 3.79; MW 90.08; Fig. 29.2) is produced naturally during fermentation of foods by lactic acid bacteria. While the acid and salts act as antimicrobials in food products, their primary uses are as pH control agents and flavorings. Lactic acid inhibits *S. aureus*, *Y. enterocolitica*, and spore-forming bacteria (29, 190, 332). Much research has been done on using lactic acid as a sanitizer on meat and poultry carcasses to reduce or eliminate pathogens. In most cases, lactic acid sprays or dips at 0.2 to 2.5% reduce contamination on beef, veal, pork, and poultry and, in some cases, improve shelf life (19, 69, 270, 271).

Sodium lactate (2.5 to 5.0%) inhibits *Clostridium botulinum*, *Clostridium sporogenes*, *L. monocytogenes*, *Salmonella* spp., *S. aureus*, *Y. enterocolitica*, and spoilage bacteria in various meat products (48, 117, 175, 188, 213, 256, 309, 323). A mixture of 2.5% sodium lactate and 0.25% sodium acetate inhibits *L. monocytogenes* growth in sliced cooked ham and a sausage product for 5 weeks at 4°C (26). Houtsma et al. (116) reported that toxin production by proteolytic *C. botulinum* is delayed at 15 and 20°C by sodium lactate concentrations of 2 and 2.5%, respectively, and that complete inhibition of toxin production at 15, 20 and 30°C occurs at 3, 4, and >4%, respectively.

Very little research has been done specifically on the mechanism of action of lactic acid against foodborne microorganisms. Presumably, it functions similarly to other organic acids and has a primary mechanism involving disruption of the cytoplasmic membrane PMF (77). Alakomi et al. (4) reported that lactic acid caused permeabilization of the outer membrane of the gram-negative bacteria *E. coli* O157:H7, *P. aeruginosa*, and *Salmonella* serovar Typhimurium. In addition, they suggested that lactic acid may act to potentiate other antimicrobial substances, such as lysozyme. There has been some speculation concerning the mechanism of lactate salts used at concentrations of 2.5% and higher. Lactate salts have minimal effects on product pH; most of the lactate remains in the less effective anionic form. There is some evidence that high concentrations of the salts reduce water activity sufficiently to inhibit microorganisms (62). However, Chen and Shelef (48) and Weaver and Shelef (323) measured the water activity of cooked meat model systems and liver sausage, respectively, containing lactate salts up to 4%, and concluded that water activity reduction is not sufficient to inhibit *L. monocytogenes*. It is most likely that at the high concentrations of lactate used, sufficient undissociated lactic acid is present, possibly in combination with a slightly reduced pH and water activity, to inhibit some microorganisms.

Propionic Acid and Propionates

Up to 1% propionic acid ($pK_a = 4.87$; Fig. 29.2) is produced naturally in Swiss cheese by *Propionibacterium freudenreichii* subsp. *shermanii*. The activity of propionates depends upon the pH of the substance to be preserved, with the undissociated acid the most active form. Eklund (76) demonstrated that undissociated propionic acid is 11 to 45 times more effective than the dissociated form.

Propionic acid and sodium, potassium, and calcium propionates are used primarily against molds; however, some yeasts and bacteria are also inhibited. The microorganism in bread dough responsible for rope formation, *B. subtilis*, is inhibited by propionic acid at pH 5.6 to 6.0 (202). Propionates (0.1 to 5.0%) retard the growth of the bacteria *E. coli*, *S. aureus*, *Sarcina lutea*, *Salmonella*, spp., *Proteus vulgaris*, *Lactobacillus plantarum*, and *L. monocytogenes*, and of the yeasts *Candida* spp. and *Saccharomyces cerevisiae* (52, 85).

Propionic acid and propionates are used as antimicrobial agents in baked goods and cheeses. Propionates may be added directly to bread dough because they have no effect on the activity of baker's yeast. Propionic acid added at 0.1% to bread dough inhibits growth of 6 log CFU of rope-forming *B. subtilis* per g in wheat bread (pH 5.30) stored at 30°C for >6 days (243). There is no limit to the concentration of propionates allowed in foods, but amounts used are generally less than 0.4%.

The primary mode of propionic acid action is probably similar to that of other organic acids, i.e., interference with cytoplasmic membrane or cellular enzymes. Propionic acid inhibits amino acid uptake and inhibits growth of various bacteria (95, 257–259). Hunter and Segel (122) also showed that amino acid transport in the mold *Penicillium chrysogenum* is inhibited by propionic acid. These researchers theorized that propionic acid neutralized the PMF of the microbial membrane by passing through the membrane as the undissociated molecule and dissociating intracellularly. Eklund (74) concluded that uptake inhibition is only partially responsible for growth inhibition, since inhibition of alanine and glucose uptake by whole cells does not correlate well with growth inhibition.

Sorbic Acid and Sorbates

Sorbic acid (Fig. 29.2) was first identified in 1859 by A. W. van Hoffman, a German chemist, from the berries of the mountain ash tree (rowanberry) (275). Sorbic acid is a *trans-trans*, unsaturated monocarboxylic fatty acid which is slightly soluble in water (0.16 g/100 ml) at 20°C. The potassium salt of sorbic acid is readily soluble in water (58.2 g/100 ml at 20°C). As with other organic acids, the antimicrobial activity of sorbic acid is greatest when the compound is in the undissociated state. With a pK_a of 4.75, activity is greatest at pH less than 6.0 to 6.5. The undissociated acid is 10 to 600 times more effective than the dissociated form (75).

Sorbates are the best characterized of all food antimicrobials as to their spectrum of action. They inhibit fungi and certain bacteria. Food-related yeasts inhibited by sorbates include species of *Brettanomyces*,

Byssochlamys, Candida, Cryptococcus, Debaryomyces, Hansenula, Pichia, Rhodotorula, Saccharomyces, Sporobolomyces, Torulaspora, and *Zygosaccharomyces* (275). Food-related mold species inhibited by sorbates belong to the genera *Alternaria, Aspergillus, Botrytis, Cephalosporium, Fusarium, Geotrichum, Helminthosporium, Mucor, Penicillium, Pullularia (Aureobasidium), Sporotrichum,* and *Trichoderma* (275). Sorbates inhibit the growth of yeasts and molds in microbiological media, cheeses, fruits, vegetables and vegetable fermentations, sauces, and meats. Sorbates inhibit growth and mycotoxin production by the mycotoxigenic molds *A. flavus, Aspergillus parasiticus, B. nivea, P. expansum,* and *P. patulum* (36, 37, 160, 240, 245). High initial mold populations can degrade sorbic acid in cheese. A number of *Penicillium, Saccharomyces,* and *Zygosaccharomyces* species can grow in the presence of and degrade potassium sorbate (24, 89, 164, 322). Sorbates may be degraded through a decarboxylation reaction resulting in the formation of 1,3-pentadiene, a compound having a kerosenelike or hydrocarbonlike odor (164).

Bacteria inhibited by sorbate include *Acinetobacter, Aeromonas, Bacillus, Campylobacter,* and *Clostridium* spp., *E. coli* O157:H7, *Lactobacillus* spp., *L. monocytogenes, Pseudomonas, Salmonella, Staphylococcus,* and *Vibrio* spp., and *Y. enterocolitica* (127, 275, 305). Sorbic acid inhibits primarily catalase-producing bacteria (336). With some exceptions (110), catalase-negative lactic acid bacteria are generally resistant to sorbates. This allows the use of sorbates in products fermented by lactic acid bacteria.

Sorbate inhibits the growth of many pathogenic bacteria in or on foods, including *Aeromonas* spp. on sundried fish with 1.5% salt, *Salmonella* spp. and *S. aureus* in sausage, *S. aureus* in bacon, *Vibrio parahaemolyticus* in seafood, *Salmonella* spp., *S. aureus,* and *E. coli* in poultry, *Y. enterocolitica* in pork, *Salmonella* serovar Typhimurium in milk and cheese, and *E. coli* O157:H7 in queso fresco cheese (108, 141, 216, 275, 300). Uljas and Ingham (307) utilized 0.1% sorbic acid in combination with freeze-thawing and storage at 25°C for 12 h or 4 h at 35°C and storage without freeze-thawing at 35°C for 6 h to achieve a 5-log reduction of *E. coli* O157:H7 in apple cider at pH 4.1. In addition, the compound inhibits growth of the spoilage bacteria *Pseudomonas putrefaciens* and *P. fluorescens,* histamine production by *Proteus morgani* and *Klebsiella pneumoniae,* and listeriolysin O production by *L. monocytogenes* (184, 275).

Sorbates are effective anticlostridial agents in cured meats and other meat and seafood products (237, 275, 276). Sorbate prevents spores of *C. botulinum* from germinating and forming toxin in beef, pork, poultry, and soy protein frankfurters and emulsions and bacon (121, 276, 277, 278). Potassium sorbate is a strong inhibitor of both *Bacillus* and *Clostridium* spore germination at pH 5.7 but much less so at pH 6.7 (269).

Lopez et al. (168) showed that 0.1% potassium sorbate inhibited unheated strains of *Bacillus stearothermophilus* spores by ca. 50% and prevented growth of spores heated at 121°C for 1 min. Against other strains, 0.1% potassium sorbate had little effect on heated spores.

Sorbate is applied to foods by direct addition, dipping, spraying, dusting, or incorporation into packaging. Baked goods can be protected from yeasts and molds through the use of 0.05 to 0.10% potassium sorbate applied either as a spray after baking or by direct addition. Food products in which sorbates are used and typical use concentrations include: beverage syrups, 0.1%; cakes and icings, 0.05–0.1%; cheese and cheese products, 0.2–0.3%; cider, 0.05–0.1%; dried fruits, 0.02–0.05%; fruit drinks, 0.025–0.075%; margarine, 0.1%; pie fillings, 0.05–0.1%; salad dressings, 0.05–0.1%; semimoist pet food, 0.1–0.3%; and wine, 0.02–0.04% (9).

One of sorbic acid's primary targets in vegetative cells appears to be the cytoplasmic membrane. Sorbic acid inhibits amino acid uptake, which in turn was theorized to be responsible for eliminating the membrane PMF through nutrient depletion (95, 257–259). Ronning and Frank (241) also showed that sorbic acid reduces the cytoplasmic membrane electrochemical gradient and consequently the PMF. They concluded that the sorbic acid-induced loss of PMF inhibits amino acid transport, which could eventually result in the inhibition of many cellular enzyme systems. In contrast, Eklund (77) showed that while low concentrations of sorbic acid reduce the ΔpH of the PMF of *E. coli* vesicles, concentrations much greater than those required for inhibition reduce, but do not eliminate, the $\Delta\Psi$ component. Since the $\Delta\Psi$ component alone could energize active uptake of amino acids, the amino acid uptake inhibition theory does not entirely explain the mechanism of inhibition by sorbic acid (77). Stratford and Anslow (284) suggest that sorbic acid does not act as a classic weak acid inhibitor, i.e., by passing through the cytoplasmic membrane as an undissociated acid and inhibiting via cytoplasmic acidification. They point out that sorbic acid is inhibitory at a relatively high pH and does not release as many protons at the MIC. They theorize that sorbic acid acts primarily through membrane disruption. In a related study, Bracey et al. (28) found little relationship between inhibition and reduction in intracellular pH of *S. cerevisiae* by sorbic acid. Since sorbic acid caused an increase in ATP consumption, which was partially attributed to increased proton pumping by membrane H^+-ATPase, they theorized that inhibition was due to induction of an energy-utilizing mechanism to compensate for a reduced intracellular pH.

The mechanism by which sorbic acid inhibits microbial growth may also be partially due to its effect on enzymes. Melnick et al. (187) theorized that sorbic acid inhibits dehydrogenases involved in fatty acid oxidation. Addition of sorbic acid results in the accumulation of β-unsaturated fatty acids that are intermediate products in the oxidation of fatty acids by fungi. This prevents the function of dehydrogenases and inhibits metabolism and growth. Sorbic acid also inhibits sulfhydryl enzymes, including fumarase, aspartase, succinic dehydrogenase, ficin, and alcohol dehydrogenase (182, 337). York and Vaughn (337) showed that sorbate reacts with the thiol group of cysteine and suggested that this is a mechanism of inactivation of sulfhydryl enzymes. Wedzicha and Brook (324) also demonstrated a 1:1 reaction between sorbate and cysteine. Martoadiprawito and Whitaker (182) proposed that sorbate inhibits the enzymes by formation of a covalent bond between the sulfhydryl or zinc hydroxide groups of the enzyme and the alpha and/or beta carbons of sorbate. Other suggested mechanisms for sorbate have involved interference with enolase, proteinase, and catalase, or the inhibition of respiration by competitive action with acetate in acetyl coenzyme A formation (10, 58, 304). Sorbate inhibits activation of the hemolytic activity of listeriolysin O of *L. monocytogenes* by reacting with cysteine (153).

The mechanism of sorbate action against bacterial spores has been studied extensively. In some of the only reports involving effects on the cell wall, Gould (104) and Seward et al. (251) demonstrated that sorbic acid inhibits cell division of germinated spores of *Bacillus* sp. and *C. botulinum* type E. Sorbic acid at pH 5.7 competitively inhibits L-alanine- and L-α-NH$_2$-*n*-butyric acid-induced germination of *Bacillus cereus* T spores and L-alanine- and L-cysteine-induced germination of *C. botulinum* 62A (269).

Miscellaneous Organic Acids

Many organic acids and their esters have been examined as potential antimicrobial agents, but most have little or no activity. They are used in foods as acidulants or flavoring agents rather than as preservatives. Fumaric acid is used to prevent the malolactic fermentation in wines (207) and as an antimicrobial agent in wines (218). Citric acid retards growth and toxin production by *A. parasiticus* and *Aspergillus versicolor* but not *P. expansum* (230). It is inhibitory to *Salmonella* sp. on poultry carcasses and in mayonnaise, to *C. botulinum* growth and toxin pro-

duction in shrimp and tomato products, and to *S. aureus* in microbiological medium (190, 221, 295, 333). Citric acid (0.156%, wt/vol) in mango juice (pH 3.5) reduced the heat resistance of *N. fischeri* spores 2.3 times at 85°C (224). Branen and Keenan (30) were the first to suggest that inhibition by citrate may be due to chelation, in studies with *Lactobacillus casei*. In contrast, Buchanan and Golden (34) found that, while undissociated citric acid is inhibitory against *L. monocytogenes*, the dissociated molecule protects the microorganism. They theorized that this protection is due to chelation by the anion.

Fatty Acid Esters

Certain fatty acid esters have antimicrobial activity in foods. One of the most effective is glyceryl monolaurate (monolaurin) (139). Monolaurin is active against gram-positive bacteria, including *Bacillus* spp., *Lactococcus* spp., *L. monocytogenes*, *Micrococcus* spp., and *S. aureus* at concentrations of ≤ 100 μg/ml but is much less effective against gram-negative bacteria (198, 199, 225, 319). Monolaurin was shown to inhibit spores and vegetative cells of *B. cereus* T, *C. botulinum* 62A, and *C. sporogenes* PA3679 at 0.075 to 0.18 mM (45). Spores were more sensitive than vegetative cells, with the exception of *C. botulinum* 62A, in which the sensitivity of each was equivalent (45). The presence of EDTA expands the activity spectrum of monolaurin to include gram-negative bacteria such as *E. coli* O157:H7 and *Pseudomonas*, *Salmonella*, and *Vibrio* spp. and decreases the MICs against gram-positive strains (139, 225). Razavi-Rohani and Griffiths (225) showed that EDTA could not be replaced by other chelators, including sodium citrate. Monolaurin is effective against *L. monocytogenes* in Camembert cheese, cottage cheese, crawfish tail homogenate, some meat products, and yogurt (199, 200, 319). Monolaurin is inhibitory to some molds and yeasts, including *Aspergillus*, *Alternaria*, *Candida*, *Cladosporium*, *Penicillium*, and *Saccharomyces* spp. (139).

Dimethyl Dicarbonate

Dimethyl dicarbonate (DMDC; Fig. 29.3) is a colorless liquid which is slightly soluble in water (3.6%). The compound is very reactive with many substances, including water, ethanol, alkyl and aromatic amines, and sulfhydryl groups (205). The primary target microorganisms for DMDC are yeasts, including *Saccharomyces*, *Zygosaccharomyces*, *Rhodotorula*, *Candida*,

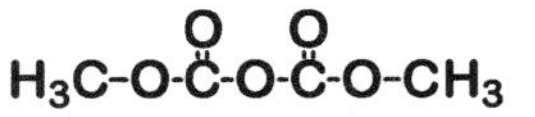

$$H_3C-O-\overset{\overset{\displaystyle O}{\|}}{C}-O-\overset{\overset{\displaystyle O}{\|}}{C}-O-CH_3$$

Figure 29.3   Structure of dimethyl dicarbonate (DMDC).

Pichia, Torulopsis, Torula, Endomyces, Kloeckera, and *Hansenula* spp. The compound is also bactericidal at 30 to 400 mg/liter to a number of species, including *Acetobacter pasteurianus, E. coli, P. aeruginosa, S. aureus,* several *Lactobacillus* species, and *Pediococcus cerevisiae* (205). The compound is bactericidal at 0.025% and more effective than either sodium bisulfite or sodium benzoate against *E. coli* O157:H7 in apple cider at 4°C (91). Molds are generally more resistant to DMDC than are yeasts or bacteria. The mechanism by which DMDC acts is most likely related to inactivation of enzymes. A related compound, diethyl dicarbonate, reacts with imidazole groups, amines, or thiols of proteins (73). In addition, diethyl dicarbonate readily reacts with histidyl groups of proteins. This can cause inactivation of the enzymes lactate dehydrogenase or alcohol dehydrogenase by reacting with the histidine in the active site (205).

Lysozyme

Lysozyme (1,4-β-N-acetylmuramidase; EC 3.2.1.17) is a 14,600-Da enzyme present in avian eggs, mammalian milk, tears and other secretions, insects, and fish. While tears contain the greatest concentration of lysozyme, dried egg white (3.5%) is the commercial source (301). Lysozyme C, the enzyme in hen's eggs, has 129 amino acids. Lysozyme is stable to heat (80°C, 2 min) at low pH but is inactivated at lower temperatures when the pH is increased. The optimum temperature for activity of the enzyme is 55 to 60°C, but it retains 50% activity at 10 to 25°C (126). The enzyme catalyzes hydrolysis of the β-1,4 glycosidic bonds between N-acetylmuramic acid and N-acetylglucosamine of the peptidoglycan of bacterial cell walls. This causes cell wall degradation and lysis in hypotonic solutions. Depending upon the enzyme source, chitin and certain esters are also susceptible to lysozyme.

Lysozyme is most active against gram-positive bacteria, probably because the peptidoglycan of the cell wall is more exposed. The enzyme inhibits foodborne bacteria, including *B. stearothermophilus, C. botulinum, Clostridium thermosaccharolyticum, Clostridium tyrobutyricum, L. monocytogenes,* and *S. aureus* (119, 120). Lysozyme is the primary antimicrobial compound in egg albumen, but its activity is enhanced by ovotransferrin, ovomucoid, and alkaline pH. Johansen et al. (133) suggested that low pH (5.5) causes increased inhibition of *L. monocytogenes* by lysozyme because the organism has a slower growth rate, which allows enzymatic hydrolysis of the cell wall to exceed the cell proliferation rate. Variation in susceptibility of gram-positive bacteria is likely due to the presence of teichoic acids and other materials that bind the enzyme and the fact that certain species have greater proportions of 1,6 or

1,3 glycosidic linkages in the peptidoglycan, which are more resistant than the 1,4 linkage (301). For example, some strains of *L. monocytogenes* are not inhibited by lysozyme alone but are inhibited when EDTA is added (119, 210). Hughey and Johnson (119) hypothesized that the peptidoglycan of the microorganism may be partially masked by other cell wall components and that EDTA enhances penetration of the lysozyme to the peptidoglycan. Lysozyme is less effective against gram-negative bacteria owing to their reduced peptidoglycan content (5 to 10%) and presence of the outer membrane of lipopolysaccharide and lipoprotein (327). Gram-negative cell susceptibility can be increased by pretreatment with chelators (e.g., EDTA) that bind Ca^{2+} or Mg^{2+}, which are essential for maintaining integrity of the lipopolysaccharide layer, or by the antibiotics polymyxin B or aminoglycosides, which disrupt the lipopolysaccharide layer (301). In addition, gram-negative cells may be sensitized to lysozyme if the cells are subjected to pH shock, heat shock, osmotic shock, drying and freeze-thaw cycling, and trisodium phosphate (39, 301, 327). Samuelson et al. (246) found that EDTA plus lysozyme inhibits *Salmonella* serovar Typhimurium on poultry. In contrast, no inhibition was demonstrated with up to 2.5 μg of EDTA per ml and 200 μg of lysozyme per ml in milk against either *Salmonella* serovar Typhimurium or *P. fluorescens* in a study by Payne et al. (210). It was theorized that there may be a significant influence of the food product on activity of lysozyme and EDTA. There is some evidence that lysozyme obtained from milk is more inhibitory to both gram-positive and gram-negative bacteria than is the enzyme from hen's egg albumen (232). Ibrahim et al. (123) found that heat denaturation of lysozyme at 80°C at pH 6.0 resulted in progressive loss of enzyme activity but increased antimicrobial activity against gram-positive and gram-negative bacteria. Their data suggested that the activity of the denatured lysozyme was due to membrane binding and subsequent perturbation.

The MIC of lysozyme against fungi, including *Candida, Sporothrix, Penicillium, Paecilomyces,* and *Aspergillus* spp., was >9,530 μg/ml in potato dextrose agar at pH 5.6 (226). Only the MICs for *Fusarium graminearum* PM162 (1,600 μg/ml) and *Aspergillus ochraceus* MM184 (3,260 μg/ml) were less than the maximum concentration evaluated. However, when combined with an equivalent concentration of EDTA, lysozyme was inhibitory to most species of fungi tested at ≤500 μg/ml.

Lysozyme is one of the few naturally occurring antimicrobial agents approved by regulatory agencies for use in foods. In Europe, lysozyme is used to prevent gas formation ("blowing") in cheeses, such as Edam and Gouda, by *C. tyrobutyricum.* Cheese manufacturers using egg white lysozyme for this purpose add a maximum of 400 mg/liter. Lysozyme is used to a great extent in Japan to preserve seafood, vegetables, pasta, and salads. The enzyme has potential for use as an antimicrobial agent with EDTA to control the growth of *L. monocytogenes* in vegetables but is less effective in refrigerated meat and soft cheese products (120). Lysozyme has been evaluated for use as a component of antimicrobial packaging (208).

Nitrites

Sodium ($NaNO_2$) and potassium (KNO_2) nitrite have a specialized use in cured meat products. Meat curing utilizes salt, sugar, spices, and ascorbate or erythorbate in addition to nitrite. Nitrite has many functions in cured meats as well as serving as an antimicrobial agent (106). As nitric oxide, it reacts with the meat pigment myoglobin to form the characteristic cured meat color, nitrosomyoglobin. It also contributes to the flavor and texture of cured meats and serves as an antioxidant. Meat curing is often combined with drying, heating, smoking, or fermentation as preservation adjuncts. At one time, sodium nitrate ($NaNO_3$) and potassium nitrate (KNO_3), also known as saltpeter, were used extensively in cured meat production. Their use was diminished when it was discovered that nitrate is converted to nitrite and that nitrite is the effective antimicrobial agent. The specific contribution of nitrite to the antimicrobial effects of curing salt was not recognized until the late 1920s, and evidence that nitrite is an effective antimicrobial agent came in the 1950s (283).

The primary antimicrobial use for sodium nitrite is to inhibit *C. botulinum* growth and toxin production in cured meats. In association with other components in the curing mix, such as salt, and reduced pH, nitrite exerts a concentration-dependent antimicrobial effect on the outgrowth of spores from *C. botulinum* and other clostridia. The use of nitrites in cured meat products to control *C. botulinum* has been studied extensively. Nitrite inhibits bacterial sporeformers by inhibiting outgrowth of the germinated spore (71). Only very high nitrite concentrations significantly inhibit spore germination.

It was first suggested in the 1920s that nitrites were more effective at a lower pH (331), and the interaction of nitrite and reduced pH against bacteria is well established. Nitrite is more inhibitory under anaerobic conditions. Ascorbate and isoascorbate enhance the antibotulinal action of nitrite, probably by acting as reducing agents (239). Storage and processing temperatures, salt concentration, and initial inoculum size also significantly influence the antimicrobial effectiveness of nitrite. Roberts and Ingram (238) and Duncan and Foster

(71) demonstrated that nitrite addition prior to heating does not increase inactivation of spores but inhibits outgrowth following heating.

Nitrite has variable effects on microorganisms other than *C. botulinum.* At 200 μg/ml and pH 5.0, nitrite completely inhibits growth of *E. coli* O157:H7 at 37°C (305). *Clostridium perfringens* growth at 20°C is inhibited by 200 μg of nitrite per ml and 3% salt or 50 μg of nitrite per ml and 4% salt at pH 6.2 in a laboratory medium (101). Nitrite is inhibitory to bacteria such as *Achromobacter, Enterobacter, Flavobacterium, Micrococcus,* and *Pseudomonas* spp. at 200 μg/g and pH 6.0 (290–292). *L. monocytogenes* growth is inhibited for 40 days at 5°C by 200 ppm of sodium nitrite with 5% NaCl in vacuum-packaged and film-wrapped smoked salmon (214). Gibson and Roberts (100, 101) found limited inhibition by nitrite (400 μg/ml) and salt (up to 6%) against fecal streptococci, *Salmonella,* and enteropathogenic *E. coli.* Certain strains of *Salmonella, Lactobacillus, C. perfringens,* and *Bacillus* are resistant to nitrite (41, 215, 234).

Meat products that may contain nitrites include bacon, bologna, corned beef, frankfurters, luncheon meats, ham, fermented sausages, shelf-stable canned cured meats, and perishable canned cured meats (e.g., ham). Nitrite is also used in a variety of fish and poultry products. The concentration used in these products is specified by governmental regulations but is generally limited to 156 ppm (mg/kg) for most products and 100 to 120 ppm (mg/kg) in bacon. Sodium erythorbate or isoascorbate is required in products containing nitrites as a cure accelerator and as an inhibitor to the formation of nitrosamines, carcinogenic compounds formed by reactions of nitrite with secondary or tertiary amines. Sodium nitrate is used in certain European cheeses to prevent spoilage by *C. tyrobutyricum* or *C. butyricum* (298).

The mechanism of nitrite inhibition of microorganisms has been studied for over 60 years. Since Ingram (124) first hypothesized that nitrite inactivated enzymes associated with respiration, nitrite has been found to inactivate a variety of enzymes or enzyme systems. However, the likely targets of clostridial inhibition by nitrite have only been elucidated in the past 20 years. Woods et al. (330) showed that nitrite caused a reduction in intracellular ATP and excretion of pyruvate in cells of *C. sporogenes.* Since these cells oxidize pyruvate to acetate to produce ATP using the phosphoroclastic system, it was theorized that this enzyme system was being inhibited by nitrite. Two enzymes in the system, pyruvate-ferredoxin oxidoreductase (PFR) and ferridoxin, were suspected to be susceptible to nitrite. Inhibition was due to reaction of nitrite in the form of nitric oxide with the nonheme iron of the proteins. PFR was most susceptible. Later, Woods and Wood (329) showed that the phosphoroclastic system of *C. botulinum* is also inhibited by nitrite. Carpenter et al. (40) confirmed that nitrite inhibits the phosphoroclastic system of *C. botulinum* and *Clostridium pasteurianum*, but that the iron-sulfur enzyme, ferredoxin, is more susceptible than PFR. McMindes and Siedler (186) reported that nitric oxide is the active antimicrobial principle of nitrite and that pyruvate decarboxylase may be an additional target for growth inhibition by nitrite. Roberts et al. (239) also confirmed inhibition of the phosphoroclastic system and found that ascorbate enhanced inhibition. In addition, they showed that other iron-containing enzymes of *C. botulinum*, including other oxidoreductases and the iron-sulfur protein hydrogenase, are inhibited. Inhibition of clostridial ferridoxin and/or pyruvate-ferredoxin oxidoreductase may be the ultimate mechanism of growth inhibition for clostridia (40, 298). These observations are supported by the fact that addition of iron to meats containing nitrite reduces the inhibitory effect of the compound (299).

The mechanism of inhibition against non-spore-forming microorganisms may be different than for spore-formers. Nitrite inhibits active transport, oxygen uptake, and oxidative phosphorylation of *P. aeruginosa* by oxidizing ferrous iron of an electron carrier, such as cytochrome oxidase, to the ferric form (244). Muhoberac and Wharton (194) and Yang (334) also found inhibition of cytochrome oxidase of *P. aeruginosa*. *E. faecalis* and *Lactococcus lactis*, which do not depend on active transport or cytochromes, are not inhibited by nitrite (244). Woods et al. (331) theorized that nitrites inhibit aerobic bacteria by binding the heme iron of cytochrome oxidase.

Parabens

Parabens (Fig. 29.4) were first reported to possess antimicrobial activity in the 1920s. Esterification of the carboxyl group of benzoic acid allows the molecule to remain undissociated up to pH 8.5, giving the parabens an effective range of pH 3.0 to 8.0 (1). In most countries, the methyl, propyl, and heptyl parabens are allowed for direct addition to foods as antimicrobial agents, while the ethyl and butyl esters are approved in some countries.

The antimicrobial activity of parabens is, in general, directly proportional to the chain length of the alkyl component. As the alkyl chain length of the parabens increases, inhibitory activity generally increases. Increasing activity with decreasing polarity is more evident against gram-positive than against gram-negative bacteria. Parabens are generally more active against molds and yeast than against bacteria (Table 29.1). Against bacteria, the parabens are more active against gram-positive genera (Table 29.2). Little research on the activity of the *n*-heptyl ester in foods seems to be available. Chan et al. (46) did show that this compound was very effective in inhibiting bacteria involved in the malolactic fermentation of wines.

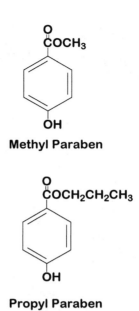

Figure 29.4 Structures of alkyl esters of *p*-hydroxybenzoic acid (parabens).

To take advantage of their respective solubility and increased activity, methyl and propyl parabens are normally used in a combination of 2–3:1 (methyl:propyl). The compounds may be incorporated into foods by dissolving in water, ethanol, propylene glycol, or the food product itself. The *n*-heptyl ester is used in fermented malt beverages (beers) and noncarbonated soft drinks and fruit-based beverages. Parabens are used in a variety of foods, including baked goods, beverages, fruit products, jams and jellies, fermented foods, syrups, salad dressings, wine, and fillings.

As parabens are phenolic derivatives, a discussion of their mechanism includes research done on phenol and related phenolic compounds, such as the phenolic antioxidants. Much of the research on mechanism of phenolic compounds has centered on their effect on cellular membranes. Judis (135) demonstrated that *E. coli* lost radioactively labeled intracellular constituents (^{14}C, ^{32}P, ^{35}S) in the presence of phenol and some of its halogenated derivatives. Judis (135) also proposed that these compounds cause physical damage to the membrane or permeability barrier. Furr and Russell (97) detected similar leakage of intracellular RNA by *Serratia marcescens* in the presence of the parabens. The amount of leakage

Table 29.1 Concentration ranges of parabens necessary for total inhibition of various fungi (pH, incubation temperature, and time vary)[a]

Fungus	Concentration (μg/ml)				
	Methyl	Ethyl	Propyl	Butyl	Heptyl
Alternaria sp.			100		50–100
Aspergillus flavus	>608	330–500	180–360	388	
Aspergillus niger	1,000	400–500	200–250	125–200	
Aspergillus parasiticus	530–>608	415–500	270–360	388	
Byssochlamys fulva			200		
Candida albicans	1,000	500–1,000	125–250	125	
Debaryomyces hansenii		400			
Fusarium graminearum	530	330	270	194	
Fusarium moniliforme	>608	330	180	290	
Fusarium oxysporum	608	330	180	290	
Penicillium citrinum	608	250	180	195	
Penicillium digitatum	500	250	63	<32	
Penicillium chrysogenum	500–>608	250–330	125–270	63–388	
Rhizopus nigricans	500	250	125	63	
Saccharomyces bayanus	930		220		
Saccharomyces cerevisiae	1,000	500	125–200	32–200	25–100
Torula utilis			200		25
Torulaspora delbrueckii		700			
Zygosaccharomyces bailii		900			
Zygosaccharomyces bisporus		400			
Zygosaccharomyces rouxii		700			

[a] Data from references 1, 132, 137, 142, 183, and 294.

is proportional to the alkyl chain length of the paraben. Davidson and Branen (60) and Degré and Sylvestre (65) found that BHA cause leakage of ^{14}C-labeled intracellular compounds from *P. fragi* and *P. fluorescens* and leakage of nucleotides from *S. aureus* Wood 46.

Other studies on the cytoplasmic membrane have evaluated nutrient uptake. Freese et al. (95) found that the parabens inhibit serine uptake as well as the oxidation of α-glycerol phosphate and NADH in membrane vesicles of *B. subtilis*. They concluded that the parabens inhibit both membrane transport and the electron transport system. Eklund (74) did a similar study using *E. coli*, *B. subtilis*, and *P. aeruginosa*. He determined the uptake of alanine by whole cells and the uptake of alanine, serine, phenylalanine, and glucose by vesicles. Parabens generally cause a decrease in amino acid uptake but not in glucose uptake. Eklund (74) postulated that, since the parabens are known to cause leakage of cellular contents, they are capable of neutralizing chemical and electrical forces which establish a normal membrane gradient. In continued work with *E. coli*, Eklund (77) found that parabens eliminated the ΔpH of the cytoplasmic membrane. In contrast, the compounds did not significantly affect the $\Delta\Psi$ component of the PMF. He concluded that neutralization of the PMF and subsequent transport inhibition could not be the only mechanism of

inhibition for the parabens. Both Eklund (74) and Freese et al. (95) stated that gram-negative bacteria are probably resistant to the parabens because of a screening effect by the cell wall lipopolysaccharide layer.

Degré et al. (64) measured the effect of BHA on the cytochrome system and its ability to oxidize various substrates by *S. aureus*. They found that BHA causes a reduction of the cytochromes and theorized that it reduces the Fe^{3+} in the cytochromes to Fe^{2+}. To measure oxidation, they assayed cell-free extracts and whole cells for oxygen uptake, with and without inhibitor. BHA inhibited respiration compared with other electron donors, such as NADH and succinate. They concluded that since BHA interferes with lipid synthesis, and lipids are involved in the respiratory process, inhibition of electron transfer reactions is due to lipid interference.

If the phenolic compounds act on the cell membrane of the microorganism, the question arises as to what portion of the membrane they attack. Kaye and Proudfoot (143) investigated the effect of phenolics on phosphatidyl ethanolamine monolayers. They found that various phenolics cause disruption of the integrity of the monolayers in direct correlation with their antimicrobial activities. The phenolic antioxidant BHT (butylated hydroxytoluene) disrupts the symmetry and increases the fluidity of lipid alkyl chains in phospholipids (81). BHA

Table 29.2 Concentration ranges of parabens necessary for total inhibition of growth of various bacteria (pH, incubation temperature and time vary)[a]

Bacterium	Concentration (μg/ml)				
	Methyl	Ethyl	Propyl	Butyl	Heptyl
Gram positive					
Bacillus cereus	1,000–2,000	830–1,000	125–400	63–400	12
Bacillus megaterium	1,000		320	100	
Bacillus subtilis	1,980–2,130	1,000–1,330	250–450	63–115	
Clostridium botulinum	1,000–1,200	800–1,000	200–400	200	
Lactococcus lactis			400		12
Listeria monocytogenes	1,430–1,600		512		
Micrococcus sp.		60–110	10–100		
Sarcina lutea	4,000	1,000	400–500	125	12
Staphylococcus aureus	1,670–4,000	1,000–2,500	350–540	120–200	12
Streptococcus faecalis		130	40		
Gram negative					
Aeromonas hydrophila	550		100		
Enterobacter aerogenes	2,000	1,000	1,000	4,000	
Escherichia coli	1,200–2,000	1,000–2,000	400–1,000	1,000	
Klebsiella pneumoniae	1,000	500	250	125	
Pseudomonas aeruginosa	4,000	4,000	8,000	8,000	
Pseudomonas fluorescens	1,310		670		
Pseudomonas fragi			4,000		
Pseudomonas putida	450				
Salmonella serovar Typhimurium			180–>300		
Vibrio parahaemolyticus			50–100		
Yersinia enterocolitica	350				

[a] Data from references 1, 16, 72, 77, 79, 136, 137, 142, 158, 161, 192, 212, 227, and 228.

causes alterations of the incorporation of [^{14}C]acetate into phospholipids, as well as changes in the ratios of tetrahymenol to polar lipids in *Tetrahymena pyriformis* (288). Al-Issa et al. (6) increased the lipid content of *L. monocytogenes* by subculturing in glycerol medium. Cells grown in this manner have increased resistance to BHA. They suggested that glycerol is involved in surface lipid synthesis, which acts as a screening mechanism against BHA. Al-Issa et al. (6) also found that strains of *L. monocytogenes* that are more resistant to BHA have lower levels of unsaturated and ante-iso fatty acids. Post and Davidson (222) also studied the effect of BHA on fatty acid composition of *B. cereus*, *C. perfringens*, *S. aureus*, *P. fluorescens*, *P. fragi*, *E. coli*, and *Salmonella* sp. They found a relationship between susceptibility to BHA and saturated/unsaturated fatty acid ratio of the gram-positive strains for the total lipid fraction. No relationships were found for the polar lipid fraction of gram-positive bacteria or either fraction of the gram-negative bacteria. Bargiota et al. (16) examined the relationship between lipid composition of *S. aureus* and resistance to parabens. Paraben-resistant strains had a greater percentage of total lipid, a higher relative percentage of phosphatidylglycerol, and decreased cyclopropane fatty

acids compared with sensitive strains. These changes could influence membrane fluidity. Juneja and Davidson (136) altered the lipid composition of *L. monocytogenes* by growth in the presence of added fatty acids. In the presence of $C_{14:0}$ or $C_{18:0}$ fatty acids, the microorganism showed increased resistance to tertiary butylhydroquinone (TBHQ) and parabens. Growth in the presence of $C_{18:1}$ increased sensitivity to the antimicrobial agents. Results indicated that for *L. monocytogenes*, a correlation existed between lipid composition of the cell membrane and susceptibility to antimicrobial compounds.

Early studies on phenol reported that high concentrations precipitated all proteins but that lower concentrations selectively inhibited essential enzymes. Chipley (49) found that 0.1 mM 2,4-dinitrophenol noncompetitively inhibits enzymes associated with the cell envelopes of both *E. coli* and *Salmonella* serovar Enteritidis. In contrast, with the same enzymes in a purified isolated state, no inhibition was demonstrated. These results suggested that a conformational change in the membrane of the vesicle causes inhibition of the enzymes and that there is no direct effect of dinitrophenol. Rico-Muñoz et al. (235) investigated the effect of BHA, BHT, TBHQ, propyl gallate, *p*-coumaric acid, ferulic acid, caffeic acid, and methyl and propyl parabens on the membrane-bound ATPase of two strains of *S. aureus*. Only BHA stimulated the activity of the enzyme. It was suggested that BHA in the membrane interferes with membrane potential, causing the ATPase to increase the hydrolysis of ATP in an attempt to regenerate membrane potential. The authors concluded that phenolic compounds probably do not have the same mechanism of action and there may be several targets which lead to inhibition of microorganisms by these compounds.

Phenolic Antioxidants

Phenolic antioxidants, including BHA, BHT, propyl gallate, and TBHQ, are used in foods primarily to delay autoxidation of unsaturated lipids. This is accomplished by interrupting the free radical chain mechanism of hydroperoxide formation during the autoxidation process. Because these compounds are phenolic in nature, it was theorized that they may have antimicrobial activity. The first study on the antibacterial effectiveness of BHA was that of Chang and Branen (47), in which *E. coli*, *Salmonella* serovar Typhimurium, and *S. aureus* were inhibited by 400, 400, and 150 μg/ml in nutrient broth, respectively. This and subsequent studies generally demonstrate that gram-positive bacteria are more susceptible to BHA than gram-negative bacteria. BHT is generally less effective than other phenolic antioxidants (59). TBHQ is an extremely effective inhibitor of gram-positive bacteria, including *S. aureus* and *L. monocytogenes*, but is less effective against gram-negative bacteria (59). Generally, BHA is a more effective antifungal agent than TBHQ, BHT, or propyl gallate. The compound inhibits *A. flavus* and *A. parasiticus*, *Byssochlamys*, *Geotrichum*, and *Penicillium* spp., and *S. cerevisiae* (59).

A number of studies have been carried out to determine the antimicrobial effectiveness of phenolic antioxidants in foods. In nearly all studies, the concentration of phenolic antioxidants required for inhibition in a food is significantly higher than that needed for in vitro inhibition, especially in meat products (59). Application studies in lower-fat and protein products or using surface applications have shown slightly more promise (3, 212).

The mechanism of inhibition by phenolic antioxidants is most likely related to other phenolics and was discussed above in the section on the parabens.

Phosphates

Some phosphate compounds, including sodium acid pyrophosphate (SAPP), tetrasodium pyrophosphate (TSPP), sodium tripolyphosphate (STPP), sodium tetrapolyphosphate, sodium hexametaphosphate (SHMP) and trisodium phosphate (TSP), have variable levels of antimicrobial activity in foods (255). Phosphates have important uses in food processing, including buffering or pH stabilization, acidification, alkalization, sequestration or precipitation of metals, formation of complexes with organic polyelectrolytes (e.g., proteins, pectin, and starch), deflocculation, dispersion, peptization, emulsification, nutrient supplementation, anticaking, antimicrobial preservation, and leavening (82).

Gram-positive bacteria are generally more susceptible to phosphates than are gram-negative bacteria. Zaika and Kim (339) found that 1% sodium polyphosphates inhibited lag and generation times of *L. monocytogenes* in brain heart infusion broth, especially in the presence of NaCl. Kelch and Bühlmann (144) tested commercial mixtures of TSPP, SAPP, STPP, and SHMP, with and without heat, against gram-positive bacteria. *S. aureus* and *E. faecalis* were completely inhibited in nutrient medium plus mild heat (50°C) while, without heat, susceptibility was variable. A mixture of TSPP, STPP, and SHMP inhibited *B. subtilis*, *C. sporogenes*, and *Clostridium bifermentans* at 0.5%. The MICs of food-grade phosphates added to early-exponential-phase cells of *S. aureus* in a synthetic medium were 0.1% for sodium ultraphosphate and sodium polyphosphate glassy and 0.5% for SAPP, STPP, and TSPP (159). Jen and Shelef (131) evaluated seven phosphate derivatives against the growth of *S. aureus* 196E. Only 0.3% SHMP (with a phosphate chain length [*n*] of 21) and 0.5% STPP or SHMP (*n* = 13 or 15) were effective growth inhibitors. Magnesium reversed the growth-inhibiting effect. Wagner and Busta (315) found that SAPP had no effect on the growth of *C. botulinum* but delayed or prevented toxicity to mice. It was theorized that this was due to binding of the toxin molecule or inactivation of the protease responsible for protoxin activation.

Phosphate derivatives also have antimicrobial activity in food products. Post et al. (219) preserved cherries against the fungal growth by *Penicillium*, *Rhizopus*, and *Botrytis* with a 10% sodium tetrapolyphosphate dip. Tanaka (289) showed that phosphate along with sodium chloride, water activity, water content, pH, and lactic acid interact to prevent the outgrowth of *C. botulinum* in pasteurized process cheese. SAPP, SHMP, or polyphosphates enhance the effect of nitrite, pH, and salt against *C. botulinum* (314). Various phosphate salts also have varying antimicrobial activity against rope-forming *Bacillus* sp. in bread and *Salmonella* sp. in pasteurized egg whites (297).

TSP is used as a sanitizer in chill water for raw poultry and other raw foods at 8 to 12%. TSP reduces viable pathogen numbers, especially *Salmonella* sp., on poultry. Lillard (165) reported a 2-log reduction of *Salmonella* serovar Typhimurium by immersing prechill chicken carcasses for 15 min in a 10% TSP solution. Inactivation was a function of the high pH (11 to 12) of the system. Kim et al. (147) found a 1.6- to 1.8-log reduction in *Salmonella* serovar Typhimurium on postchill chicken carcasses using a 10% TSP dipping treatment. Slavik et al. (267) evaluated dipping of chicken carcasses inoculated with *C. jejuni* in 10% TSP and reported a 1.5-log reduction of the pathogen. Wang et al. (320) reported that 10% TSP reduced *Salmonella* serovar Typhimurium attached to chicken skin by 1.6 to 2.3 logs. Fresh-laid eggs washed with solutions of TSP at 0.5, 1.0, or 5.0% caused changes in the shell surface, as examined using scanning electron microscopy, that might allow for increased bacterial penetration (146). Zhuang and Beuchat (342) found that dipping tomatoes for 15 s in 12% TSP (pH 11.8 to 12.6) reduced the population of *Salmonella enterica* serovar Montevideo by 4 logs on the surface and ca. 2 logs in the core. The color of the tomatoes was unaffected by TSP.

Several mechanisms have been suggested for bacterial inhibition by polyphosphates. The ability of polyphosphates to chelate metal ions appears to play an important role in their antimicrobial activity (273). Post et al. (220) found that the presence of magnesium reverses inhibition of gram-positive bacteria by polyphosphates. Jen and Shelef (131) also found that magnesium reverses inhibition of *S. aureus* by SHMP or STPP and that calcium and iron are partially effective in reversing inhibition. Maier et al. (176) demonstrated that inhibition by sodium polyphosphate glassy of *B. cereus* vegetative cells and spores was blocked or reversed by divalent cations (Mg^{2+} or Ca^{2+}). They proposed that polyphosphate inhibits cell division by blocking cell septation. They further suggested that polyphosphate inhibits the major division protein FtsZ, possibly through interaction with a required Mg^{2+}-dependent GTPase. Knabel et al. (150) stated that the chelating ability of polyphosphates is responsible for growth inhibition of *B. cereus*, *L. monocytogenes*, *S. aureus*, *Lactobacillus* sp., and *A. flavus*. Orthophosphates had no inhibitory activity against any of the microorganisms and have no chelating ability. Further, Knabel et al. (150) reported that inhibition is reduced at lower pH owing to protonation of the chelating sites on the polyphosphates. They concluded that polyphosphates inhibited gram-positive bacteria and fungi by removal of essential cations from binding sites on the cell walls of these microorganisms.

Sodium Chloride

Sodium chloride (NaCl), or common salt, is probably the oldest known food preservative. While a very few foods, such as raw meats and fish products, are preserved directly by high concentrations of sodium chloride, salt is used primarily as an adjunct to other processing methods, such as canning and pasteurization (272). In general, foodborne pathogenic bacteria are inhibited by a water activity of 0.92 or less (equivalent to an NaCl concentration of 13% [wt/vol]). The exception is *S. aureus*, which has a minimum water activity for growth of 0.83 to 0.86. Enterotoxin production by the organism is more restricted (185). Another relatively salt-tolerant foodborne pathogen is *L. monocytogenes*, which can survive in saturated salt solutions at low temperatures. Fungi are more tolerant to low water activity than are bacteria. The minimum for growth of xerotolerant fungi is 0.61 to 0.62, but most are inhibited by 0.85 or lower (57).

Sodium chloride inhibits microorganisms primarily by its plasmolytic effect. The antimicrobial activity of sodium chloride is related to its ability to reduce water activity and create unfavorable conditions for microbial growth. As the water activity of the external medium is reduced, cells are subjected to osmotic shock and rapidly lose water through plasmolysis. During plasmolysis, a cell ceases to grow and either dies or remains dormant. In order to resume growth, the cell must reduce its intracellular water activity (279). Aside from the osmotic influence on growth, other possible mechanisms include limiting oxygen solubility, alteration of pH, toxicity of sodium and chloride ions, and loss of magnesium ions (13).

Sulfites

Sulfur dioxide (SO_2) and its salts have been used as disinfectants since the time of the ancient Greeks and Romans (106, 206). The use of sulfur dioxide as a food preservative was reported in the 17th century by Evelyn (86), who suggested that casks should be filled with cider that contained sulfur dioxide (produced by burning sulfur). Salts of sulfur dioxide include potassium sulfite (K_2SO_3), sodium sulfite (Na_2SO_3), potassium bisulfite ($KHSO_3$), sodium bisulfite ($NaHSO_3$), potassium metabisulfite ($K_2S_2O_5$), and sodium metabisulfite (NaS_2O_5). While sulfites now have multiple uses as food additives, their original purpose was as an antimicrobial agent (242). As antimicrobials, sulfites are used primarily in fruit and vegetable products to control three groups of microorganisms: spoilage and fermentative yeasts and molds on fruits and fruit products (e.g., wine), acetic acid bacteria, and malolactic bacteria (206). In addition to their use as antimicrobials agents, sulfites act as antioxidants and inhibit enzymatic and nonenzymatic browning in a variety of foods. Their primary application is in fruits and vegetable products, but they are also used to a limited extent in meats.

The most important factor impacting the antimicrobial activity of sulfites is pH. Sulfur dioxide and its salts exist as a pH-dependent equilibrium mixture when dissolved in water: $SO_2 \cdot H_2O \leftrightarrow HSO_3^- + H^+ \leftrightarrow SO_3^{2-} + H^+$. Aqueous solutions of sulfur dioxide theoretically yield sulfurous acid (H_2SO_3); however, evidence indicates that the actual form is more likely $SO_2 \cdot H_2O$ (107). As the pH decreases, the proportion of $SO_2 \cdot H_2O$ increases and the bisulfite (HSO_3^-) ion concentration decreases. The pK_a values for sulfur dioxide, depending upon temperature, are 1.76 to 1.90 and 7.18 to 7.20 (107, 206, 242). The inhibitory effect of sulfites is most pronounced when the acid or $SO_2 \cdot H_2O$ is in the undissociated form. Therefore, their most effective pH range is less than 4.0. King et al. (148) demonstrated this when they found that undissociated H_2SO_3 ($SO_2 \cdot H_2O$) was the only form active against yeast and that neither HSO_3^- nor SO_3^{2-} had antimicrobial activity. Similarly, $SO_2 \cdot H_2O$ is 1,000, 500, and 100 times more active than HSO_3^- or SO_3^{2-} against *E. coli*, yeasts, and *A. niger*, respectively (229). Increased effectiveness at low pH is likely due to the ability of the unionized sulfur dioxide to pass across the cell membrane (223, 242).

Sulfites, especially as the bisulfite ion, are very reactive. These reactions not only determine the mechanism of action of the compounds, they also influence antimicrobial activity. For example, sulfites form addition compounds (α-hydroxysulfonates) with aldehydes and ketones. It is generally agreed that these bound forms have much less, or no, antimicrobial activity compared with the free forms. For example, Ough (206) reported that addition compounds with sugars completely neutralize the antimicrobial activity of sulfites against yeast. However, Stratford and Rose (286) demonstrated that pyruvate-sulfite complexes retained some activity against *S. cerevisiae*.

Sulfur dioxide is fungicidal even in low concentrations against yeasts and molds. The inhibitory concentration range of sulfur dioxide is 0.1 to 20.2 μg/ml for *Saccharomyces*, *Zygosaccharomyces*, *Pichia*, *Hansenula*, and *Candida* species (229). Roland and Beuchat (240) found that sulfur dioxide inhibits *B. nivea* growth and patulin production in apple juice.

Sulfites may be used to inhibit acetic acid-producing bacteria, lactic acid bacteria, and spoilage bacteria in meat products (106). At 100 to 200 μg/ml, sulfites inhibit *Acetobacter* sp. that cause wine spoilage (206). The concentration of sulfur dioxide required to inhibit lactic acid bacteria varies significantly, depending upon conditions, but can be 1 to 10 μg/ml in fruit products at pH 3.5 or less (326). Sulfur dioxide is more inhibitory to gram-negative rods than to gram-positive rods. Banks

and Board (12) tested several genera of *Enterobacteriaceae* isolated from sausage for their metabisulfite sensitivity. The microorganisms tested and the concentration of free sulfite necessary to inhibit their growth at pH 7.0 were: *Salmonella* sp., 15–109 μg/ml; *E. coli*, 50–195 μg/ml; *Citrobacter freundii*, 65–136 μg/ml; *Y. enterocolitica*, 67–98 μg/ml; *Enterobacter agglomerans*, 83–142 μg/ml; *S. marcescens*, 190–241 μg/ml; and *Hafnia alvei*, 200–241 μg/ml.

Sulfur dioxide is used to control the growth of undesirable microorganisms in fruits, fruit juices, wines, sausages, fresh shrimp, and acid pickles, and during extraction of starches. It is added at 50 to 100 mg/liter to expressed grape juices used for making wines to inhibit molds, bacteria, and undesirable yeasts (8). At appropriate concentrations, sulfur dioxide does not interfere with wine yeasts or with the flavor of wine. During fermentation, sulfur dioxide also serves as an antioxidant, clarifier, and dissolving agent. The optimum level of sulfur dioxide (50 to 75 mg/liter) is maintained to prevent postfermentation changes by microorganisms. In some countries, sulfites may be used to inhibit the growth of microorganisms on fresh meat and meat products. Sulfite or metabisulfite added in sausages is effective in delaying the growth of molds, yeasts, and salmonellae during storage at refrigerated or room temperature (12). Sulfur dioxide restores a bright color but may give a false impression of freshness.

Owing to their extreme reactivity, it is difficult to pinpoint the exact antimicrobial mechanism for sulfites. This reactivity is due to the ability of sulfites to act as reducing agents or take part in nucleophilic attack (107). Sulfites react with disulfide bonds of proteins and with glutathione-forming thiosulfonates: $R-S-S-R + SO_3^{2-} \rightarrow R-S-SO_3^- + RS^-$. This reaction can inactivate enzymes that have disulfide links and affect conformation of proteins. Inactivation of the cellular redox agent glutathione (γ-glutamyl-cysteinyl-glycine) can result in oxidation of protein thiol groups and oxidation of membrane lipids (107). Sulfites react with coenzymes and enzyme prosthetic groups. The coenzymes NAD and NADP are inactivated by sulfite addition. Prosthetic groups, including thiamin, folic acid, pyridoxal, flavins, and hemes, are all susceptible to sulfite inactivation. The pyrimidines cytosine and uracil, components of DNA and RNA, are susceptible to addition reactions. In vitro, bisulfite converts cytosine and 5-methylcytosine to uracil and thymine, respectively (111). In addition, the purine adenine, ATP, or NAD can form addition compounds with sulfite. These reactions may terminate DNA polymerization, transcription, and translation of RNA. Compounds containing carbonyls (aldehydes and ketones) are also subject to attack by sulfites. These compounds form α-hydroxysulfonates. Sugars that react with sulfites include arabinose, mannose, galactose, lactose, glucose, maltose, and raffinose (125). Sucrose does not react. Sulfites can also form addition products with quinones.

The most likely targets for inhibition by sulfites include disruption of the cytoplasmic membrane, inactivation of DNA replication, protein synthesis, inactivation of membrane-bound or cytoplasmic enzymes, or reaction with individual components in metabolic pathways. With *S. cerevisiae*, *Saccharomycodes ludwigii*, or *Z. bailii*, sulfites gain access to the cell by free diffusion (217, 285, 287). Other fungi have active transport systems for the compound. Sulfites dissipate the PMF and inhibit solute active transport (242). Hinze and Holzer (114) did experiments to demonstrate that the most sensitive enzyme in the Embden-Meyerhof-Parnas pathway of *S. cerevisiae* was glyceraldehyde-3-phosphate dehydrogenase. Inactivation of this enzyme led to a decrease in cellular ATP. It is still uncertain as to whether the enzyme, its cofactor NAD, or the substrate glyceraldehyde-3-phosphate is the target.

NATURALLY OCCURRING COMPOUNDS AND SYSTEMS

Many food products contain naturally occurring compounds that have antimicrobial activity. In the natural state, these compounds may play a role in extending the shelf life of a food product. In addition, many of these naturally occurring compounds have been studied for their potential as direct food antimicrobial agents. There are many problems associated with the use of natural compounds as direct food additives. An ideal naturally occurring antimicrobial would be effective enough to be added as a whole food or as an edible component (e.g., an herb or a spice). Few, if any, antimicrobial agents are present in foods at concentrations great enough to be antimicrobials without some purification. Often, even if purification of an antimicrobial is possible, adding it to another food may lead to undesirable sensory changes. The ultimate challenge is to find a naturally occurring antimicrobial agent that can be added to a "microbiologically sensitive" food product in a nonpurified form from another nonsensitive food. The nonpurified food would have to contain an antimicrobial that is completely nontoxic and is highly effective in controlling the growth of microorganisms. This may be impossible. According to Beuchat and Golden (22), the challenge is to isolate, purify, stabilize, and incorporate natural antimicrobials into foods without adversely affecting sensory, nutritional, or safety characteristics. This must be done without significant increased costs for formulation, processing, or marketing.

Animal

Lactoperoxidase System

Lactoperoxidase is an enzyme that occurs in raw milk, colostrum, saliva, and other biological secretions. The enzyme is a glycoprotein with one heme group that contains Fe^{3+}. It has 610 amino acids and a molecular mass of 78,000 (80). The enzyme is similar to other biologically significant peroxidases, such as human myeloperoxidase. Bovine milk naturally contains 10 to 60 mg of lactoperoxidase per liter (80, 327). This enzyme reacts with thiocyanate (SCN^-) in the presence of hydrogen peroxide and forms an antimicrobial compound(s); this is termed the lactoperoxidase system (LPS). Thiocyanate is found in many biological secretions, including milk. It is formed by the detoxification of thiosulfates, by metabolism of sulfur-containing amino acids, and in the diet through metabolism of various glucosides (327). Fresh milk contains 1 to 10 mg of thiocyanate per liter, which is not always sufficient to activate the LPS. Wilkins and Board (327) recommended addition of 10 to 12 mg of thiocyanate per liter. Hydrogen peroxide, the third component of the LPS, is not present in fresh milk owing to the action of natural catalase, peroxidase, or superoxide dismutase. Approximately 8 to 10 mg hydrogen peroxide per liter is required for LPS. This can be added directly, through the action of lactic acid bacteria, or through the enzymatic action of xanthine oxidase, glucose oxidase, or sulfhydryl oxidase. The amount of hydrogen peroxide required is much lower than that used in pasteurization of raw milk (ca. 300 to 800 mg/liter). In the LPS reaction, thiocyanate is oxidized to the antimicrobial hypothiocyanate ($OSCN^-$), which also exists in equilibrium with hypothiocyanous acid ($pK_a = 5.3$) (99, 233).

The LPS is more active against gram-negative bacteria, including pseudomonads, than gram-positive bacteria (25). However, it does inhibit both gram-positive and gram-negative foodborne pathogens, including salmonellae, *S. aureus*, *L. monocytogenes*, and *C. jejuni* (23, 140, 266). There is variable activity against catalase-negative microorganisms, including the lactic acid bacteria. In resistant strains, the enzyme NADH:$OSCN^-$ oxidoreductase oxidizes NADH with reduction of $OSCN^-$ (38). The LPS is theorized to be a protective mechanism against mastitis or a method for selection of beneficial microorganisms in the gut of the newborn calf. The LPS system can increase the shelf life of raw milk in countries that have a poorly developed refrigerated storage system (80). The keeping quality of milk pasteurized at 72°C for 15 s was found to be better than that of milk heated at 80°C for 15 s (18). Complete inactivation of lactoperoxidase occurred at 80°C, whereas residual lactoperoxidase activity was detected following treatment at 72°C. Since higher levels of hypothiocyanite were detected in milk treated at the lower temperature, it was theorized that lactoperoxidase may have a role in the keeping quality of pasteurized milk (18). LPS is also used as a preservation process in infant formula, ice cream, cream, cheeses, and liquid whole eggs (80).

There are several potential mechanisms for inhibition by the LPS. The thiocyanate ion can oxidize sulfhydryl groups to disulfides, sulfenylthiocyanates (-S-SCN), or sulfenic acid (-S-OH) (80). Therefore, the compound may react with cysteine side chains on proteins and inactivate enzymes with functionally important sulfhydryl groups. Enzymes that are inhibited in vitro by thiocyanate include aldolase, glyceraldehyde phosphate dehydrogenase, hexokinase, lactate dehydrogenase, and 6-phosphogluconate dehydrogenase (80). Thiocyanate may also oxidize NADH or NADPH to the corresponding NAD or NADP. Through reaction with NADH or interference with membrane proteins, thiocyanate may cause leakage of the membrane or loss of electrochemical potential. This will eventually lead to inhibition of transport of amino acids and sugars.

Lactoferrin and Other Iron-Binding Proteins

Iron-binding proteins with potential for use as food antimicrobial agents occur in milk and eggs. In milk and colostrum, the primary iron-binding protein is lactoferrin. This protein is also found in other physiological fluids and polymorphonuclear leukocytes. Milk contains low concentrations of another iron-binding protein, transferrin, that is found in larger amounts in blood. Lactoferrin is a glycoprotein with a molecular weight of around 76,500. It exists in milk primarily as a tetramer with Ca^{2+} (80). Lactoferrin has two iron-binding sites per molecule. For each Fe^{3+} bound by lactoferrin, one bicarbonate (HCO_3^-) is required. Citrate, another chelator in milk, inhibits antimicrobial activity of lactoferrin (231, 232), while the presence of bicarbonate reverses inhibition. Lactoferrin must be low in iron saturation, and bicarbonate must be present for the compound to be an effective antimicrobial. Lactoferrin is potentially active in milk owing to the low iron concentration and presence of bicarbonate (327). The exact biological role of lactoferrin is unknown; however, it may act as a barrier to infection of the nonlactating mammary gland and may help protect the gastrointestinal tract of the newborn against infection (293).

Lactoferrin is inhibitory to many microorganisms, including B. subtilis, B. stearothermophilus, L. monocytogenes, Micrococcus species, E. coli, and Klebsiella species (151, 177, 204, 211, 231). Payne et al. (211) found that bovine lactoferrin had to be reduced to 18% iron saturation using dialysis (apo-lactoferrin) to have bacteriostatic activity against four strains of L. monocytogenes and one strain of E. coli at concentrations of 15 to 30 mg/ml in ultrahigh-temperature-treated (UHT) aseptically packaged milk. At 2.5 mg/ml, the compound has no activity against Salmonella serovar Typhimurium and P. fluorescens and little activity against E. coli O157:H7 or L. monocytogenes VPHI (210). Some gram-negative bacteria may be resistant because they adapt to low-iron environments by producing siderophores, such as phenolates and hydroxamates (80). Microorganisms with a low iron requirement, such as lactic acid bacteria, would not be inhibited by lactoferrin.

Production of an iron-deficient environment by lactoferrin may be part of its mechanism of inhibition. Iron stimulates the growth of bacteria in many genera, including Clostridium, Escherichia, Listeria, Pseudomonas, Salmonella, Staphylococcus, Vibrio, and Yersinia. However, there is evidence that lactoferrin may have effects on gram-negative bacteria in addition to iron deprivation. Ellison et al. (83) studied the potential mechanisms of lactoferrin against gram-negative bacteria. They found that both lactoferrin and EDTA, another chelator, cause release of anionic lipopolysaccharides from the outer membrane of E. coli. Lactoferrin causes this loss by chelation of cations, including magnesium, calcium, and iron, that stabilize lipopolysaccharides in the membrane. This was demonstrated by a reversal of lipopolysaccharide loss by addition of iron. Since it is cationic, lactoferrin may increase the outer membrane permeability to hydrophobic compounds, including other antimicrobials. According to Naidu and Bidlack (196), lactoferrin blocks adhesion of microorganisms to mucosal surfaces, inhibits expression of fimbria and other colonizing factors of enteric pathogens, such as E. coli, and inactivates lipopolysaccharides of gram-negative bacteria.

Lactoferricin B or hydrolyzed lactoferrin (HLF) is a small peptide produced by acid-pepsin hydrolysis of bovine lactoferrin (20). It contains 25 amino acid residues and has a molecular weight of 3,000. Jones et al. (134) determined that the compound is inhibitory and cidal to Shigella and Salmonella spp., Y. enterocolitica, E. coli O157:H7, S. aureus, L. monocytogenes, and Candida sp. at concentrations ranging from 1.9 to 125 μg/ml. Proteus, Serratia, and Pseudomonas cepacia are resistant to 500 μg of lactoferricin B per ml. In contrast, while HLF was effective against L. monocytogenes, enterohemorrhagic *E. coli*, and *Salmonella* serovar Enteritidis in peptone-yeast extract-glucose broth, it was not active in a more complex medium, Trypticase soy broth (31). The addition of EDTA enhanced the activity of HLF in Trypticase soy broth, indicating that the decreased activity of HLF may have been due, in part, to excess cations in the medium. Venkitanarayanan et al. (313) found that, while 50 or 100 μg of lactoferricin B per ml reduced viable *E. coli* O157:H7 in 1% peptone, it was much less effective as an antimicrobial agent in ground beef.

Another iron-binding molecule, ovotransferrin or conalbumin, occurs in egg albumen. This 77,000- to 80,000-Da glycoprotein makes up 10 to 13% of the total egg white protein (327). Ovotransferrin has 49% sequence homology with lactoferrin (301). Each ovotransferrin molecule has two iron-binding sites and, like lactoferrin, binds an anion, such as bicarbonate or carbonate, with each ferric iron bound. The compound is 80% denatured by heating at 60°C for 5 min. The compound was first shown to be antimicrobial by Schade and Caroline (247), who demonstrated that raw egg white in nutrient broth inhibits *E. coli*, *Shigella* sp., *S. aureus*, and *S. cerevisiae*. Inhibition was reversed by iron. To be inhibitory, ovotransferrin must be in stoichiometric excess of iron, and the pH must be in the alkaline range (above 7.5) (303, 327). Ovotransferrin is inhibitory to both gram-positive and gram-negative bacteria, but the former are generally more sensitive. *Bacillus* and *Micrococcus* species are particularly sensitive (301). Some yeasts are also sensitive (312). As with lactoferrin, one of the primary mechanisms suggested for ovotransferrin is deprivation of iron from microorganisms that require the element. However, Tranter and Board (302) suggested that inhibition may be partially or completely related to the cationic nature of the protein. Certain gram-positive bacteria and yeasts are inhibited in egg albumen with or without iron (302).

Avidin

Avidin is a glycoprotein present in egg albumen. The concentration varies with the hen's age, but the mean is 0.05% of the total egg albumen protein (301). The protein has a molecular mass of 66,000 to 69,000 Da and has four identical subunits of 128 amino acids each. It is stable to heat and a wide pH range. Avidin strongly binds the cofactor biotin at a ratio of 4 molecules of biotin per molecule of avidin. Biotin is a cofactor for enzymes in the tricarboxylic acid cycle and fatty acid biosynthesis. While the biological role of the protein is unknown, it may be secreted by macrophages and may play a role in the immune system (301). Avidin inhibits growth of some

bacteria and yeasts that have a requirement for biotin, with the primary mechanism being nutrient deprivation. However, Korpela (152) investigated another potential mechanism and found that avidin can bind porin proteins in the outer membrane of *E. coli*. This finding suggested that avidin may inhibit microorganisms in vitro by interfering with transport through porins.

Plant

Spices and Their Essential Oils

Spices are roots, bark, seeds, buds, leaves, or fruit of aromatic plants that are added to foods as flavoring agents. However, it has been known since ancient times that spices and their essential oils have varying degrees of antimicrobial activity. The earliest report on the use of spices as preservatives was around 1550 B.C., when the ancient Egyptians used spices for food preservation and embalming the dead. Cloves, cinnamon, oregano, and thyme and, to a lesser extent, sage and rosemary have the strongest antimicrobial activity among spices.

The major antimicrobial components of cloves and cinnamon are eugenol [2-methoxy-4-(2-propenyl)-phenol] and cinnamic aldehyde (3-phenyl-2-propenal), respectively (Fig. 29.5). Cinnamon contains 0.5 to 10% volatile oil, of which 75% is cinnamic aldehyde and 8% is eugenol; cloves contain 14 to 21% volatile oil, 95% of which is eugenol. Fabian et al. (87) tested 10% extracts of cinnamon and clove against *B. subtilis* and *S. aureus* and found cinnamon only slightly inhibitory, while clove

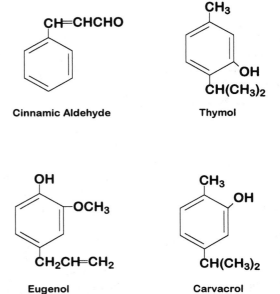

Figure 29.5 Structures of antimicrobial compounds found in spices.

extract was very inhibitory at 1:100 and 1:800, respectively. Zaika and Kissinger (340) showed that clove at 0.4% in a Lebanon bologna formulation inhibits growth and acid production by a lactic acid bacterial starter culture, while cinnamon at 0.8% inhibits growth slightly. Both spices stimulate acid production at low concentrations. Aqueous clove infusions (0.1 to 1.0% [wt/vol]) and eugenol (0.06%) inhibit outgrowth of germinated spores of *B. subtilis* in nutrient agar (7). Stecchini et al. (282) showed that essential oil of clove at 500 μg/ml inhibits *Aeromonas hydrophila* growth in microbiological media and in vacuum-packaged and air-packed cooked pork. Smith-Palmer et al. (268) determined that the 24-h MICs of cinnamon and clove essential oils against *C. jejuni, E. coli, Salmonella* serovar Enteritidis, *L. monocytogenes,* and *S. aureus* were 0.05, 0.04–0.05, 0.04–0.05, 0.03, and 0.04%, respectively, in an agar dilution assay. Azzouz and Bullerman (11) evaluated 16 ground herbs and spices at 2% (wt/vol) against nine mycotoxin-producing *Aspergillus* and *Penicillium* species. The most effective antimicrobial spice evaluated was clove, which inhibits growth initiation at 25°C by all species for over 21 days. Cinnamon is the next most effective spice, inhibiting three *Penicillium* species for over 21 days. Bullerman (35) determined that 1.0% cinnamon in raisin bread inhibits growth and aflatoxin production by *A. parasiticus.*

The antimicrobial activity of oregano and thyme has been attributed to their essential oils, which contain the terpenes carvacrol [2-methyl-5-(1-methylethyl)phenol] and thymol [5-methyl-2-(1-methylethyl)phenol], respectively (Fig. 29.5). Firouzi et al. (90) found that thyme essential oil was the most effective antimicrobial agent against *L. monocytogenes* 4b growth compared to a number of other spice and herb extracts. Similarly, thyme essential oil was the most effective antimicrobial agent among 15 spice essential oils tested by Smith-Palmer et al. (268) against *C. jejuni, E. coli, Salmonella* serovar Enteritidis, *L. monocytogenes,* and *S. aureus.* MICs of the essential oil were 0.02 to 0.05%, and minimum bactericidal concentrations were 0.03 to 0.1%. These compounds have inhibitory activity against a number of bacterial species, molds, and yeasts, including *B. subtilis, E. coli, L. plantarum, P. cerevisiae, P. aeruginosa, Proteus* species, *Salmonella* serovar Enteritidis, *S. aureus, V. parahaemolyticus,* and *A. parasiticus* (61). Ultee et al. (308) determined that carvacrol depleted intracellular ATP, reduced membrane potential, and increased permeability of the cytoplasmic membrane of *B. cereus,* leading to inhibition and eventually cell death.

Sage and rosemary also have antimicrobial activity. The active fraction of sage and rosemary may be the terpene fractions of the essential oils. Rosemary contains primarily borneol (endo-1,7,7-trimethylbicyclo[2.2.1]heptan-2-ol) along with pinene, camphene, and camphor, while sage contains thujone {4-methyl-1-(1-methylethyl)bicyclo[3.1.0]-hexan-3-one}. At 2% in growth medium, sage and rosemary are more active against gram-positive than gram-negative bacterial strains (254). The inhibitory effect of these two spices at 0.3% is bacteriostatic, while at 0.5% they are bactericidal to gram-positive strains. Of 18 spices tested against *L. monocytogenes* in culture medium, Pandit and Shelef (209) found the most effective compound to be rosemary. The most inhibitory fraction of the rosemary was α-pinene. Smith-Palmer et al. (268) demonstrated that rosemary (0.02 to 0.05%) and sage (0.02 to 0.075%) inhibited gram-positive *L. monocytogenes* and *S. aureus* but not gram-negative bacteria. Further, Pandit and Shelef (209) showed that *L. monocytogenes* Scott A growth in refrigerated fresh pork sausage was delayed by 0.5% ground rosemary or 1% rosemary essential oil. In contrast, Mangena and Muyima (179) indicated little antimicrobial activity with essential oil of rosemary against bacteria and yeasts in an agar diffusion assay. The lack of activity could be due to the assay utilized. The sensitivity of *B. cereus, S. aureus,* and *Pseudomonas* sp. to sage is greatest in microbiological medium and significantly reduced in rice and chicken and noodles (253). Loss of activity may be due to solubilization of the antimicrobial fraction in the lipid of the foods. Hefnawy et al. (112) evaluated 10 herbs and spices against two strains of *L. monocytogenes* in tryptose broth. The most effective spice was sage, which at 1% decreased viable *L. monocytogenes* by 5 to 7 logs after 1 day at 4°C. Allspice was next most effective, inactivating the microorganism in 4 days.

Sweet basil (*Ocimum basilicum* L.) essential oil has limited antimicrobial activity, with linalool and methyl chavicol the primary antimicrobial agents. Basil essential oil extract is active against certain fungi, including *Mucor* and *Penicillium* species, but has little activity against bacteria (155). A strain of *B. cereus* and *Salmonella* serovar Typhimurium were most sensitive to the basil. Wan et al. (318) screened the essential oil components of sweet basil, linalool and methyl chavicol, against 35 strains of bacteria, yeasts, and molds. Again, the compounds demonstrated limited activity against most microorganisms except *Mucor* and *Penicillium* spp. In contrast, methyl chavicol (0.1%) in filter-sterilized fresh lettuce supernatant reduced viable *A. hydrophila* by 5 logs and, as a wash for lettuce leaves, the compound was as effective as 125 μg of chlorine per ml (318). Smith-Palmer et al. (268) reported MICs for basil essential oil of 0.25, 0.25, 0.1, 0.05, and 0.1% for *C. jejuni*, *E. coli*, *Salmonella* serovar Enteritidis, *L. monocytogenes*, and *S. aureus*, respectively. Yin and Cheng (335) detected no antifungal activity against *A. flavus* or *A. niger* with water extracts of basil or ginger in a paper disk assay.

Vanillin (4-hydroxy-3-methoxybenzaldehyde) is a major constituent of vanilla beans, the fruit of an orchid (*Vanilla planifola*, *Vanilla pompona*, or *Vanilla tahitensis*). Vanillin is most active against molds and non-lactic gram-positive bacteria (130). López-Malo et al. (169) prepared fruit-based agars containing mango, papaya, pineapple, apple, and banana with up to 2,000 μg of vanillin per ml and inoculated each with *A. flavus*, *A. niger*, *A. ochraceus*, or *A. parasiticus*. Vanillin at 1,500 μg/ml significantly inhibited all strains of *Aspergillus* in all media. Cerrutti and Alzamora (43) demonstrated complete inhibition of growth for 40 days at 27°C of *Debaryomyces hansenii*, *S. cerevisiae*, *Z. bailii*, and *Zygosaccharomyces rouxii* in laboratory media and apple puree at water activity of 0.99 and 0.95 by 2,000 μg of vanillin per ml. In contrast, 2,000 μg of vanillin per ml was not effective against the yeasts in banana puree. These researchers attributed this loss of activity to binding of the phenolic vanillin by protein or lipid in banana and mango, a phenomenon demonstrated for other antimicrobial phenolic compounds (236). López-Malo et al. (170, 171) studied the effect of vanillin concentration, pH, and incubation temperature on *A. flavus*, *A. niger*, *A. ochraceus*, and *A. parasiticus* growth on potato dextrose agar. Depending upon species, vanillin in combination with reduced pH had an additive or synergistic effect on growth of the molds. The most sensitive was *A. ochraceus* (MIC = 500 μg/ml at pH 3.0, ≤25°C, or pH 4.0, ≤15°C), and the most resistant was *A. niger* (MIC = 1,000 μg/ml, pH 3.0, 15°C). Cerrutti et al. (44) utilized vanillin to produce a shelf-stable strawberry puree. A combination of 3,000 μg of vanillin per ml, 1,000 μg of calcium lactate per ml, and 500 μg of ascorbic acid per ml preserved strawberry puree acidified with citric acid to pH 3.0 and adjusted to water activity of 0.95 with sucrose for over 60 days against growth of natural microflora and inoculated yeasts.

Many other spices have been tested and shown to have limited or no activity. They include anise, bay (laurel), black pepper, cardamom, cayenne (red pepper), celery seed, chili powder, coriander, cumin, curry powder, dill, fenugreek, ginger, juniper oil, mace, marjoram, nutmeg, orris root, paprika, sesame, spearmint, tarragon, and white pepper (61, 87, 181).

Few studies have focused on the mechanism by which spices or their essential oils inhibit microorganisms. Since it has been concluded that the terpenes in essential oils

of spices are the primary antimicrobial agents, the mechanism most likely involves these compounds. Many of the most active terpenes, e.g., eugenol, thymol, and carvacrol, are phenolic in nature. Therefore, it would seem reasonable that their modes of action might be related to other phenolic compounds. As was discussed for parabens, the mode of action of phenolic compounds is generally thought to involve interference with functions of the cytoplasmic membrane, including PMF and active transport (59, 77). In addition, terpenes may have other antimicrobial mechanisms. Conner et al. (56) suggested that spice essential oils may inhibit enzyme systems in yeasts, including those involved in energy production and synthesis of structural components.

Onions and Garlic

Probably the best-characterized antimicrobial system in plants is that found in the juice and vapors of onions (*Allium cepa*) and garlic (*Allium sativum*). Walton et al. (317) and Lovell (172) reported that volatile agents from both onion and garlic inhibit *B. subtilis*, *S. marcescens*, and *Mycobacterium* sp. in microbiological media. Since those early reports, the growth and toxin production of many microorganisms have been shown to be inhibited by onion and garlic, including the bacteria *B. cereus*, *C. botulinum* type A, *E. coli*, *L. plantarum*, *Salmonella*, sp., *Shigella* sp., and *S. aureus* and the fungi *A. flavus*, *A. parasiticus*, *Candida albicans*, and species of *Cryptococcus*, *Rhodotorula*, *Saccharomyces*, *Torulopsis*, and *Trichosporon* (21, 55, 103).

Cavallito and Bailey (42) isolated the major antimicrobial compounds from garlic by using steam distillation of ethanolic extracts. They identified the antimicrobial component as allicin (diallyl thiosulfinate; thio-2-propene-1-sulfinic acid-*S*-allyl ester) (Fig. 29.6). Allicin is formed by the action of the enzyme allinase on the substrate alliin (*S*-allyl-L-cysteine sulfoxide). The reaction only occurs when cells of the garlic are disrupted, releasing the enzyme to act on the substrate. A similar reaction occurs in onion, except that the substrate is *S*-(1-propenyl)-L-cysteine sulfoxide and one of the major products is thiopropanal-*S*-oxide. The products apparently responsible for antimicrobial activity are also responsible for the flavor of onions and garlic. In addition to antimicrobial sulfur compounds, onions contain the phenolic compounds protocatechuic acid and catechol, which could contribute to their antimicrobial activity (316).

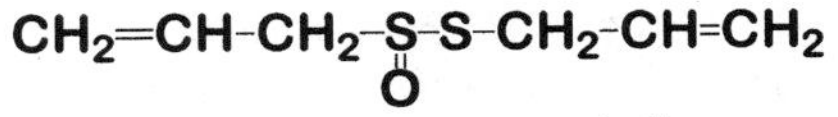

$$CH_2{=}CH{-}CH_2{-}\overset{O}{\underset{}{S}}{-}S{-}CH_2{-}CH{=}CH_2$$

Figure 29.6 Structure of allicin.

The mechanism of action of allicin is most likely inhibition of sulfhydryl-containing enzymes (21). Wills (328) found that 0.5 mM allicin inhibits many sulfhydryl enzymes, including alcohol dehydrogenase, choline esterase, choline oxidase, glyoxylase, hexokinase, papain, succinic dehydrogenase, urease, and xanthine oxidase. Carboxylase, ATPase, and β-amylase are not affected by allicin. Some nonsulfhydryl enzymes, including lactic dehydrogenase, tyrosinase, and alkaline phosphatase, are also inhibited. Barone and Tansey (17) theorized that allicin inactivates proteins by oxidizing thiols to disulfides and inhibiting the intracellular reducing activity of glutathione and cysteine. Conner et al. (56) showed that garlic oil inhibits ethanol production by *S. cerevisiae* and delays sporulation of *Hansenula anomala* and *Lodderomyces elongisporus*.

Isothiocyanates

Mustard seed has as a primary pungency component the compound allyl isothiocyanate (AIT). Isothiocyanates (R—N=C=S) are derivatives from glucosinolates in cells of plants of the *Cruciferae* or mustard family (cabbage, kohlrabi, Brussels sprouts, cauliflower, broccoli, kale, horseradish, mustard, turnips, rutabaga). These compounds are formed from the action of the enzyme myrosinase (thioglucoside glucohydrolase; EC 3.2.3.1) on the glucosinolates when the plant tissue is injured or mechanically disrupted. In addition to the allyl side group, other isothiocyanate side groups include ethyl, methyl, benzyl and phenyl. These compounds are potent antimicrobial agents (66, 67).

Isothiocyanates are inhibitory to fungi, yeasts, and bacteria in the range of 16 to 110 ng/ml in the vapor phase (128) and 10 to 600 μg/ml in liquid media (180). Inhibition against bacteria varies, but generally grampositive bacteria are less sensitive to AIT than are gramnegative bacteria. Delaquis and Mazza (66) found a 1- to 5-log decrease in viable cells of *E. coli*, *L. monocytogenes*, and *Salmonella* serovar Typhimurium in the presence of 2,000 μg of AIT per ml of air. Delaquis and Sholberg (67) examined this effect further and showed that 1,000 μg of AIT per ml of air apparently decreased viable *E. coli* O157:H7, *Salmonella* serovar Typhimurium, and *L. monocytogenes* by up to 6 logs. However, cells recovered to a large extent if they were exposed to air. *E. coli* O157:H7 was the most resistant. AIT was shown to inhibit the growth of *P. expansum*, *A. flavus*, and *Botrytis cinerea* at 100 μg of AIT per ml of air (67). Ward et al. (321) prepared horseradish essential oil distillate (~90% AIT) and applied it to the headspace of cooked roast beef or agar inoculated with *E. coli* O157:H7, *L. monocytogenes*, *Salmonella* serovar

Typhimurium, *S. aureus*, *Serratia grimeseii*, and *Lactobacillus sake*. AIT at 2 μl/liter inhibited all the microorganisms on the agar, but 20 μl/liter was required to inhibit the pathogens and spoilage microorganisms on the beef. On beef, *L. monocytogenes*, *S. grimesii*, and *L. sake* were the most resistant (321). Delaquis et al. (68) added 20 μl of horseradish essential oil per liter of air with precooked roast beef slices. The beef was stored for 28 days at 4°C, and inoculated spoilage bacteria were monitored. *Pseudomonas* sp. and members of the family *Enterobacteriaceae* were inhibited to the greatest extent, while lactic acid bacteria were more resistant. The development of off odors and off flavors was delayed, and cooked meat color was preserved in the treated roasts.

Mustard seed extract was tested in the vapor phase against *E. coli*, *Salmonella* serovar Typhimurium, *V. parahaemolyticus*, and *S. aureus* in microbiological media (250). The extract at a maximum of 180 μg/ml in the vapor phase was bacteriostatic at 30°C to *E. coli*, *Salmonella* serovar Typhimurium, and *S. aureus* at 9 h and bactericidal to *V. parahaemolyticus*. In contrast, up to 3.0% mustard seed was ineffective against *L. monocytogenes* (296).

The mechanism by which isothiocyanates inhibit cells may be due to inhibition of enzymes by direct reaction with disulfide bonds or through thiocyanate (SCN^-) anion reaction to inactivate sulfhydryl enzymes (66). Lin et al. (167) suggested that AIT inhibited *Salmonella* serovar Montevideo and *E. coli* O157:H7 through disruption of the cell membrane and leakage of intracellular material. While these compounds have very low sensory thresholds, it has been suggested that they may be useful as food antimicrobials owing to their low inhibitory concentrations.

Phenolic Compounds

A phenolic compound can be defined as any substance that possesses an aromatic ring with one or more hydroxyl groups and can include functional derivatives. Phenolic compounds occurring in foods are classified as simple phenols and phenolic acids, hydroxycinnamic acid derivatives, and the flavonoids (115). Some of the important phenolic compounds, including parabens, phenolic antioxidants (e.g., BHA and TBHQ), and certain of the terpene fraction of the essential oils of spices (e.g., thymol, carvacrol, eugenol, vanillin) have been discussed. Only the latter compounds are naturally occurring.

Much of the research on the mechanism of phenolic compounds has focused on their effect on cellular membranes. Simple phenols disrupt the cytoplasmic membrane and cause leakage of cells (59). A phenolic glycoside isolated from green olives, oleuropein, causes losses of labeled glutamate, potassium, and phosphate from *L. plantarum* cells (138). The compound also decreases cellular ATP but has no effect on the glycolysis rate. Juven et al. (138) concluded that the leakage is caused by a loss in membrane permeability. By interfering with the permeability of the cytoplasmic membrane, the phenolics inhibit cells by interfering with components of the PMF and thereby inhibiting active transport of nutrients. Phenolics may also inhibit cellular proteins directly. Rico-Muñoz et al. (235) investigated the effect of the naturally occurring propyl gallate, *p*-coumaric acid, ferulic acid, and caffeic acids on the activity of membrane-bound ATPase of *S. aureus*. All of the compounds inhibit ATPase activity. Compounds that inhibit ATPase interact directly with the enzyme, possibly through binding. The authors concluded that all phenolic compounds probably do not have the same mechanism of action and there may be several targets that lead to inhibition of microorganisms by these compounds.

Simple Phenols and Phenolic Acids

Simple phenolic compounds include monophenols (e.g., *p*-cresol), diphenols (e.g., hydroquinone), and triphenols (e.g., gallic acid). Gallic acid occurs in plants as quinic acid esters or hydrolyzable tannins (tannic acid) (115).

The only practical use of simple phenols for preservation is found in the application of wood smoke. Smoking of foods such as meats, cheeses, fish, and poultry not only imparts a desirable flavor but also has a preservative effect through drying and the deposition of chemicals. While many chemicals are deposited on smoked foods, the major contributors of flavor and antioxidant and antimicrobial effects are phenol and cresol (59, 96). Liquid smoke on the surface of cheddar cheese inhibits growth of *Aspergillus oryzae*, *Penicillium camemberti*, and *Penicillium roqueforti* (325). Of the eight major phenolic compounds in liquid smoke, isoeugenol is the most effective antifungal compound, followed by *m*-cresol and *p*-cresol.

The phenolic acids, including derivatives of *p*-hydroxybenzoic acid (protocatechuic, vanillic, gallic, syringic, ellagic) and *o*-hydroxybenzoic acid (salicylic), may be found in plants and foods. Schanderl (248) found that 100 μg of gallic, caffeic (a hydroxycinnamic acid), and quinic acids per ml halts wine yeast fermentations in the presence of 6% ethanol. Reddy et al. (228) tested gallic acid and its esters for their effectiveness against *C. botulinum* in tryptone-yeast extract-glucose medium at 37°C and found inhibition only with the esters, butyl gallate, isobutyl gallate, isoamyl gallate, *p*-octyl gallate, and *p*-dodecyl gallate. Tannic acid is inhibitory to

A. hydrophila, *E. coli*, *L. monocytogenes*, *Salmonella* serovar Enteritidis, *S. aureus*, and *S. faecalis* (53, 212). The intestinal microorganisms *Bacteroides fragilis*, *C. perfringens*, *E. coli*, and *Salmonella* serovar Typhimurium were inhibited by both tannic acid and propyl gallate, while neither *Bifidobacterium infantis* nor *Lactobacillus acidophilus* was affected (54). Chung et al. (54) theorized that tannic acid inhibited by binding iron but that propyl gallate had a different inhibitory mechanism. Stead (281) demonstrated stimulation of growth of the lactic acid spoilage bacteria *Lactobacillus collinoides* and *Lactobacillus brevis* by gallic, quinic, and chlorogenic acids. He suggested that the compounds were being metabolized by the microorganisms and concluded that the presence of these compounds in fruit-based beverages may result in antagonistic reactions with other food-grade antimicrobials agents.

Resin from the flowers of the hop vine (*Humulus lupulus* L.) is used in the brewing industry for imparting a desirable bitter flavor to beer. About 3 to 12% of the resin is composed of α-bitter acids, including humulone (humulon), cohumulone, and adhumulone, and β-bitter acids, including lupulone (lupulon), colupulone, xanthohumol, and adlupulone (203). Both types of bitter acids possess principal antimicrobial activity against gram-positive bacteria and fungi at low water activity (21). Lactic acid bacteria that spoil beer, including species of *Lactobacillus* and *Pediococcus*, were resistant to the antimicrobial effect of humulone, colupulone, and *trans*-isohumulone, while the species of the same genera that do not spoil beer were sensitive (88, 263). *E. coli* and *B. subtilis* are naturally resistant to both the iso-α-acid and β-acids, while strains of *Bacillus megaterium*, *Enterococcus salivarius*, *L. monocytogenes*, and *S. aureus* are susceptible (109, 157, 189). There is little definitive research on the antimicrobial mechanism of hop extract β-bitter acids. However, there is evidence that the mechanism may involve the cytoplasmic membrane, since the activity is selectively increased against gram-positive bacteria and fungi and is reduced by lipid (157). In addition, hop bitter acids inhibit growth of beer spoilage (gram-positive) bacteria by dissipating transmembrane pH gradient (262). However, Fernandez and Simpson (88) found that resistance to *trans*-isohumulone does not confer resistance to sorbic acid or benzoic acid, indicating a potential difference in targets for the mechanism of antimicrobial action.

Hydroxycinnamic Acids

Hydroxycinnamic acids include caffeic, *p*-coumaric, ferulic, and sinapic acids. They frequently occur as esters and less often as glucosides (115). These compounds are found in plants and foods. In tests with chlorogenic (3-caffeoyl quinic acid), isochlorogenic (dicaffeoyl quinic acid), caffeic, gallic, and ellagic acids, Sikovec (260, 261) determined that a mixture of the compounds at 800 μg/ml inhibited respiration (CO_2 production) by wine yeasts by approximately 50%. Herald and Davidson (113) demonstrated that ferulic acid at 1,000 μg/ml and p-coumaric acid at 500 and 1,000 μg/ml inhibit the growth of *B. cereus* and *S. aureus*. The compounds were much less effective against *P. fluorescens* and *E. coli*. In contrast, alkyl esters of hydroxycinnamic acids, including methyl caffeoate, ethyl caffeoate, propyl caffeoate, methyl p-coumarate, and methyl cinnamate, were effective inhibitors of the growth of *P. fluorescens* (14). Lyon and McGill (173) tested the antimicrobial activity of caffeic, cinnamic, ferulic, salicylic, sinapic, and vanillic acids on *Erwinia carotovora* and demonstrated total inhibition at pH 6.0 in nutrient broth with 0.5 μg of cinnamic, ferulic, salicylic, or vanillic acid per ml. While effective as growth inhibitors of *E. carotovora*, none of these phenolic compounds inhibit the pectolytic enzymes polygalacturonic acid lyase and polygalacturonase that are produced by the organism except caffeic acid, which only inhibits the latter (174). Stead (280) determined the effect of caffeic, coumaric, and ferulic acids against the wine spoilage lactic acid bacteria *L. collinoides* and *L. brevis*. At pH 4.8 in the presence of 5% ethanol, p-coumaric and ferulic acids were the most inhibitory compounds at 500 and 1,000 μg/ml. At 100 μg/ml, all three hydroxycinnamic acids stimulate growth of the microorganisms, suggesting that these compounds may play a role in initiating the malolactic fermentation of wines.

Many of the studies with hydroxycinnamic acids have involved their antifungal properties. Chipley and Uraih (51) showed that ferulic acid inhibited aflatoxin B_1 and G_1 production by *A. flavus* and *A. parasiticus* by up to 75%. Salicylic and *trans*-cinnamic acids totally inhibited aflatoxin production at the same level. Baranowski et al. (15) studied the effect of caffeic, chlorogenic, p-coumaric, and ferulic acids at pH 3.5 on the growth of *S. cerevisiae*. Caffeic and chlorogenic acid had little effect on the organism at 1,000 μg/ml. In the presence of p-coumaric, however, the organism was completely inhibited by 1,000 μg/ml. Ferulic acid was the most effective growth inhibitor tested. At 50 μg/ml, this compound extended the lag phase of *S. cerevisiae*, and at 250 μg/ml the growth of the organism was completely inhibited. The degree of inhibition is inversely related to the polarity of the compounds.

Furocoumarins are related to the hydroxycinnamates. These compounds, including psoralen and its derivatives, are phytoalexins in citrus fruits, parsley, carrots, celery, and parsnips. Purified psoralen and natural sources of the compound (e.g., cold-pressed lime oil, lime peel extract) have antimicrobial activity against *E. coli* O157:H7, *E. carotovora*, *L monocytogenes*, and *Micrococcus luteus* following irradiation with long-wave (365-nm) UV light (178, 306). These compounds inhibit growth by interfering with DNA replication. This occurs when the furocoumarin and DNA are exposed to UV light at 365 nm, which causes monoadducts of DNA and cross-linking of the furocoumarin and DNA (195).

Flavonoids

The flavonoids include catechins and flavons, flavonols, and their glycosides (115). Proanthocyanidins, or condensed tannins, are polymers of favan-3-ol and are found in apples, grapes, strawberries, plums, sorghum, and barley (115). Schanderl (248) reported that 2,000 μg of tannin per ml halted wine yeast fermentations earlier than a control. Singleton and Esau (264) predicted that the antibacterial effect of red and white wines would be proportional to the amount of flavonoid tannin present. Benzoic acid, proanthocyanidins, and flavonols account for 66% of cranberry microbial inhibition against *Saccharomyces bayanus*, with the latter two being the most important (183). While inhibition by benzoic acid decreases with increasing pH, there was no effect on the activity of the flavonols and increased activity of the proanthocyanidins at higher pH levels.

The flavones, flavonols, and their glycosides occur naturally in fruits and vegetables. One of the most common flavonols in the plant kingdom is quercetin. Schraufstatter (249) investigated the antimicrobial effect of chalcone, flavonone, flavone, and flavonol. All compounds except flavonol have a bacteriostatic effect on *S. aureus*. Natural sources of these compounds also have slight bacteriostatic activity against the microorganism. Payne et al. (212) found no inhibition of any of six strains of *L. monocytogenes* by quercetin at 205 μg/ml.

CONCLUSIONS

Traditional food antimicrobial agents are an important tool in preserving foods from the hazardous and detrimental effects of microorganisms. In the future, they will continue to play an important role in food preservation. However, it is doubtful that any newly discovered synthetic antimicrobial agents will be approved for use in foods by worldwide regulatory agencies. Therefore, the future of traditional antimicrobials lies in novel uses (e.g., new foods or in combination with other preservation procedures) and in combinations designed on the basis of their mechanisms so as to obtain a wide spectrum of activity against microorganisms.

Extensive research on sources and activities of natural antimicrobial agents is currently taking place. In the future, more information needs to be gathered on these natural compounds, including their activity spectra, mechanisms, targets, and interactions with environmental conditions and food components. Prior to their regulatory approval as food antimicrobials, they will have to be proven toxicologically safe. The latter point, along with cost, could be the greatest hurdle to their future use.

References

1. Aalto, T. R., M. C. Firman, and N. E. Rigler. 1953. p-Hydroxybenzoic acid esters as preservatives. I. Uses, antibacterial and antifungal studies, properties and determination. *J. Am. Pharm. Assoc.* 42:449–458.

2. Abdul-Raouf, U. M., L. R. Beuchat, and M. S. Ammar. 1993. Survival and growth of *Escherichia coli* O157:H7 in ground, roasted beef as affected by pH, acidulants, and temperature. *Appl. Environ. Microbiol.* 59:2364–2368.

3. Ahmad, S., and A. L. Branen. 1981. Inhibition of mold growth by butylated hydroxyanisole. *J. Food Sci.* 46:1059.

4. Alakomi, H.-L., E. Skyttä, M. Saarela, T. Mattila-Sandholm, K. Latva-Kala, and I. M. Helander. 1999. Lactic acid permeabilizes gram-negative bacteria by disrupting the outer membrane. *Appl. Environ. Microbiol.* 66:2001–2005.

5. Al-Dagal, M. M., and W. A. Bazaraa. 1999. Extension of shelf life of whole and peeled shrimp with organic acid salts and bifidobacteria. *J. Food Prot.* 62:51–56.

6. Al-Issa, M., D. R. Fowler, A. Seaman, and M. Woodbine. 1984. Role of lipid in butylatedhydroxyanisole resistance of *Listeria monocytogenes*. *Zentbl. Bakteriol. Hyg. A* 258:42–50.

7. Al-Khayat, M. A., and G. Blank. 1985. Phenolic spice components sporostatic to *Bacillus subtilis*. *J. Food Sci.* 50:971–974.

8. Amerine, M. A., and M. A. Joslyn. 1970. *Table Wines: the Technology of Their Production*, 2nd ed. University of California Press, Berkeley.

9. Anonymous. 1999. Sorbic acid and potassium sorbate for preserving freshness. Publ. ZS-1D. Eastman Chemical Co., Kingsport, Tenn.

10. Azukas, J. J., R. N. Costilow, and H. L. Sadoff. 1961. Inhibition of alcoholic fermentation by sorbic acid. *J. Bacteriol.* 81:189–194.

11. Azzouz, M. A., and L. B. Bullerman. 1982. Comparative antimycotic effects of selected herbs, spices, plant components and commercial fungal agents. *J. Food Sci.* 45:1298–1301.

12. Banks, J. G., and R. G. Board. 1982. Sulfite-inhibition of *Enterobacteriaceae* including *Salmonella* in British fresh sausage and in culture systems. *J. Food Prot.* 45:1292–1297.

13. Banwart, G. J. 1979. *Basic Food Microbiology*. AVI Publ., Westport, Conn.

14. Baranowski, J. D., and C. W. Nagel. 1983. Properties of alkyl hydroxycinnamates and effects on *Pseudomonas fluorescens*. *Appl. Environ. Microbiol.* 45:218–222.

15. Baranowski, J. D., P. M. Davidson, C. W. Nagel, and A. L. Branen. 1980. Inhibition of *Saccharomyces cerevisiae* by naturally occurring hydroxycinnamates. *J. Food Sci.* 45:592–594.

16. Bargiota, E. E., E. Rico-Muñoz, and P. M. Davidson. 1987. Lethal effect of methyl and propyl parabens as related to *Staphylococcus aureus* lipid composition. *Int. J. Food Microbiol.* 4:257–266.

17. Barone, F. E., and M. R. Tansey. 1977. Isolation, purification, identification, synthesis and kinetics of the activity of the anticandidal component of *Allium sativum*, and a hypothesis for its mode of action. *Mycologia* 69:793–825.

18. Barrett, N. E., A. S. Grandison, and M. J. Lewis. 1999. Contribution of the lactoperoxidase system to the keeping quality of pasteurized milk. *J. Dairy Res.* 66:73–80.

19. Bautista, D. A., N. Sylvester, S. Barbut, and M. W. Griffiths. The determination of efficacy of antimicrobial rinses on turkey carcasses using response surface designs. *Int. J. Food Microbiol.* 34:279–292.

20. Bellamy, W., M. Takase, H. Wakabayashi, K. Kawase, and M. Tomita. 1992. Antibacterial spectrum of lactoferricin B, a potent bactericidal peptide derived from the N-terminal region of bovine lactoferrin. *J. Appl. Bacteriol.* 73:472–479.

21. Beuchat, L. R. 1994. Antimicrobial properties of spices and their essential oils, p. 167–179. *In* V. M. Dillon and R. G. Board (ed.), *Natural Antimicrobial Systems and Food Preservation*. CAB Intl., Wallingford, England.

22. Beuchat, L. R., and D. A. Golden. 1989. Antimicrobials occurring naturally in foods. *Food Technol.* 43:134–142.

23. Beumer, R. R., A. Noomen, J. A. Marijs, and E. H. Kampelmacher. 1985. Antibacterial action of the lactoperoxidase system on *Campylobacter jejuni* in cow's milk. *Neth. Milk Dairy J.* 39:107–114.

24. Bills, S., L. Restaino, and L. M. Lenovich. 1982. Growth response of an osmotolerant sorbate-resistant yeast, *Saccharomyces rouxii* at different sucrose and sorbate levels. *J. Food Prot.* 45:1120–1124.

25. Björck, L. 1978. Antibacterial effect of the lactoperoxidase system on psychrotrophic bacteria in milk. *J. Dairy Res.* 45:109–118.

26. Blom, H., E. Nerbrink, R. Dainty, T. Hagtvedt, E. Borch, H. Nissen, and T. Nesbakken. 1997. Addition of 2.5% lactate and 0.25% acetate controls growth of *Listeria monocytogenes* in vacuum-packed, sensory acceptable servelat sausage and cooked ham stored at 4°C. *Int. J. Food Microbiol.* 38:71–76.

27. Bosund, L. 1962. The action of benzoic and salicylic acids on the metabolism of microorganisms. *Adv. Food Res.* 11:331–353.

28. Bracey, D., C. D. Holyoak, and P. J. Coote. 1998. Comparison of the inhibitory effect of sorbic acid and amphotericin B on *Saccharomyces cerevisiae*: is growth inhibition dependent on reduced intracellular pH? *J. Appl. Microbiol.* 85:1056–1066.

29. Brackett, R. E. 1987. Effect of various acids on growth and survival of *Yersinia enterocolitica*. *J. Food Prot.* 50:598–601.

30. Branen, A. L., and T. W. Keenan. 1970. Growth stimulation of *Lactobacillus casei* by sodium citrate. *J. Dairy Sci.* 53:593.

31. Branen, J. K., and P. M. Davidson. 2000. Activity of hydrolysed lactoferrin against foodborne pathogenic bacteria in growth media: the effect of EDTA. *Lett. Appl. Microbiol.* 30:233–237.

32. Brudzinski, L., and M. A. Harrison. 1998. Influence of incubation conditions on survival and acid tolerance response of *Escherichia coli* O157:H7 and non-O157:H7 isolates exposed to acetic acid. *J. Food Prot.* 61:542–546.

33. Buchanan, R. L., and S. G. Edelson. 1999. pH-dependent stationary-phase acid resistance response of enterohemorrhagic *Escherichia coli* in the presence of various acidulants. *J. Food Prot.* 62:211–218.

34. Buchanan, R. L., and M. H. Golden. 1994. Interaction of citric acid concentration and pH on the kinetics of *Listeria monocytogenes* inactivation. *J. Food Prot.* 57:567–570.

35. Bullerman, L. B. 1974. Inhibition of aflatoxin production by cinnamon. *J. Food Sci.* 39:1163–1165.

36. Bullerman, L. B. 1983. Effects of potassium sorbate on growth and aflatoxin production by *Aspergillus parasiticus* and *Aspergillus flavus*. *J. Food Prot.* 46:940–942.

37. Bullerman, L. B. 1984. Effects of potassium sorbate on growth and patulin production by *Penicillium patulum* and *Penicillium roqueforti*. *J. Food Prot.* 47:312–315.

38. Carlsson, J., Y. Iwami, and T. Yamada. 1983. Hydrogen peroxide excretion by oral streptococci and effect of lactoperoxidase-thiocyanate-hydrogen peroxide. *Infect. Immun.* 40:70–80.

39. Carneiro de Melo, A. M. S., C. A. Cassar, and R. J. Miles. 1998. Trisodium phosphate increases sensitivity of gram-negative bacteria to lysozyme and nisin. *J. Food Prot.* 61:839–844.

40. Carpenter, C. E., D. S. A. Reddy, and D. P. Cornforth. 1987. Inactivation of clostridial ferredoxin and pyruvate-ferredoxin oxidoreductase by sodium nitrite. *Appl. Environ. Microbiol.* 53:549–552.

41. Castellani, A. G., and C. F. Niven. 1955. Factors affecting the bacteriostatic action of sodium nitrite. *Appl. Microbiol.* 3:154–159.

42. Cavallito, C. J., and J. H. Bailey. 1944. Allicin, the antibacterial principle of *Allium sativum*. I. Isolation, physical properties and antibacterial action. *J. Am. Chem. Soc.* 16:1950–1951.

43. Cerrutti, P., and S. M. Alzamora. 1996. Inhibitory effects of vanillin on some food spoilage yeasts in laboratory media and fruit purées. *Int. J. Food Microbiol.* 29:379–386.

44. Cerrutti, P., S. M. Alzamora, and S. L. Vidales. 1997. Vanillin as an antimicrobial for producing shelf-stable strawberry puree. *J. Food Sci.* 62:608–610.

45. Chaibi, A., L. H. Ababouch, and F. F. Busta. 1996. Inhibition of bacterial spores and vegetative cells by glycerides. *J. Food Prot.* 59:716–722.

46. Chan, L., R. Weaver, and C. S. Ough. 1975. Microbial inhibition caused by p-hydroxybenzoate esters in wines. *Am. J. Enol. Vitic.* 26:201–207.

47. Chang, H. C., and A. L. Branen. 1975. Antimicrobial effects of butylated hydroxyanisole (BHA). *J. Food Sci.* 40:349–351.

48. Chen, N., and L. A. Shelef. 1992. Relationship between water activity, salts of lactic acid, and growth of *Listeria monocytogenes* in a meat model system. *J. Food Prot.* 55:574–578.

49. Chipley, J. R. 1974. Effects of 2, 4-dinitrophenol and N,N'-cyclohexylcarbodiimide on cell-envelope associated enzymes of *Escherichia coli* and *Salmonella enteritidis*. *Microbios* 10:115.

50. Chipley, J. R. 1993. Sodium benzoate and benzoic acid, p. 11–48. *In* P. M. Davidson and A. L. Branen (ed.), *Antimicrobials in Foods*, 2nd ed. Marcel Dekker, Inc., New York, N.Y.

51. Chipley, J. R., and N. Uraih. 1980. Inhibition of *Aspergillus* growth and aflatoxin release by derivatives of benzoic acid. *Appl. Environ. Microbiol.* 40:352.

52. Chung, K. C., and J. M. Goepfert. 1970. Growth of *Salmonella* at low pH. *J. Food Sci.* 35:352–357.

53. Chung, K. T., and C. A. Murdock. 1991. Natural systems for preventing contamination and growth of microorganisms in foods. *Food Microstruct.* 10:361–374.

54. Chung, K. T., Z. Lu, and M. W. Chou. 1998. Mechanism of inhibition of tannic acid and related compounds on the growth of intestinal bacteria. *Food Chem. Toxicol.* 36:1053–1060.

55. Conner, D. E., and L. R. Beuchat. 1984. Effects of essential oils from plants on growth of food spoilage yeasts. *J. Food Sci.* 49:429–434.

56. Conner, D. E., L. R. Beuchat, R. E. Worthington, and H. L. Hitchcock. 1984. Effects of essential oils and oleoresins of plants on ethanol production, respiration and sporulation of yeasts. *Int. J. Food Microbiol.* 1:63–74.

57. Corry, J. E. L. 1987. Relationships of water activity to fungal growth, p. 51–99. *In* L. R. Beuchat (ed.), *Food and Beverage Mycology*, 2nd ed. AVI/Van Nostrand Reinhold, New York, N.Y.

58. Dahl, H. K., and J. Nordal. 1972. Effect of benzoic acid and sorbic acid on the production and activities of some bacterial proteinases. *Acta Agric. Scand.* 22:29.

59. Davidson, P. M. 1993. Parabens and phenolic compounds, p. 263–306. *In* P. M. Davidson and A. L. Branen (ed.), *Antimicrobials in Foods*, 2nd ed. Marcel Dekker, Inc., New York, N.Y.

60. Davidson, P. M., and A. L. Branen. 1980. Antimicrobial mechanisms of butylated hydroxyanisole against two *Pseudomonas* species. *J. Food Sci.* 45:1607–1613.

61. Davidson, P. M., and A. S. Naidu. 2000. Phyto-phenols, p. 266–294. *In* A. S. Naidu (ed.), *Natural Food Antimicrobial Systems*. CRC Press, Inc., Boca Raton, Fla.

62. Debevere, J. M. 1989. The effect of sodium lactate on the shelflife of vacuum-packed coarse liver pate. *Fleischwirtschaft* 69:223–224.

63. Degnan, A. J., C. W. Kaspar, W. S. Otwell, M. L. Tamplin, and J. B. Luchansky. 1994. Evaluation of lactic

acid bacterium fermentation products and food-grade chemicals to control *Listeria monocytogenes* in blue crab (*Callinectes sapidus*) meat. *Appl. Environ. Microbiol.* 60:3198–3203.

64. Degré, R., M. Ishaque, and M. Sylvestre. 1983. Effect of butylated hydroxyanisole on the electron transport system of *Staphylococcus aureus* Wood 46. *Microbios* 37:7–13.

65. Degré, R., and M. Sylvestre. 1983. Effect of butylated hydroxyanisole on the cytoplasmic membrane of *Staphylococcus aureus* Wood 46. *J. Food Prot.* 46:206–209.

66. Delaquis, P. J., and G. Mazza. 1995. Antimicrobial properties of isothiocyanates in food preservation. *Food Technol.* 49:73–84.

67. Delaquis, P. J., and P. L. Sholberg. 1997. Antimicrobial activity of gaseous allyl isothiocyanate. *J. Food Prot.* 60:943–947.

68. Delaquis, P. J., S. M. Ward, R. A. Holley, M. C. Cliff, and G. Mazza. 1999. Microbiological, chemical and sensory properties of pre-cooked roast beef preserved with horseradish essential oil. *J. Food Sci.* 64:519–524.

69. Dickson, J. S., and M. E. Anderson. 1992. Microbiological decontamination of animal carcasses by washing and sanitizing systems: a review. *J. Food Prot.* 55:13–140.

70. Doores, S. 1993. Organic acids, p. 95–136. *In* P. M. Davidson and A. L. Branen (ed.), *Antimicrobials in Foods*, 2nd ed. Marcel Dekker, Inc., New York, N.Y.

71. Duncan, C. L., and E. M. Foster. 1968. Effect of sodium nitrite, sodium chloride, and sodium nitrate on germination and outgrowth of anaerobic spores. *Appl. Microbiol.* 16:406–411.

72. Dymicky, M., and C. N. Huhtanen. 1979. Inhibition of *Clostridium botulinum* by p-hydroxybenzoic acid n-alkyl esters. *Antimicrob. Agents Chemother.* 15:798–801.

73. Ehrenberg, L., I. Fedorscsak, and F. Solymosy. 1976. Diethyl pyrocarbonate in nucleic acid research. *Prog. Nucleic Acid Res. Mol. Biol.* 16:189.

74. Eklund, T. 1980. Inhibition of growth and uptake processes in bacteria by some chemical food preservatives. *J. Appl. Bacteriol.* 48:423–432.

75. Eklund, T. 1983. The antimicrobial effect of dissociated and undissociated sorbic acid at different pH levels. *J. Appl. Bacteriol.* 54:383–389.

76. Eklund, T. 1985. Inhibition of microbial growth at different pH levels by benzoic and propionic acids and esters of p-hydroxybenzoic acid. *Int. J. Food Microbiol.* 2:159–167.

77. Eklund, T. 1985. The effect of sorbic acid and esters of p-hydroxybenzoic acid on the proton motive force in *Escherichia coli* membrane vesicles. *J. Gen. Microbiol.* 131:73–76.

78. Eklund, T. 1989. Organic acids and esters, p. 161–200. *In* G. W. Gould (ed.), *Mechanisms of Action of Food Preservation Procedures*. Elsevier Applied Science, London, United Kingdom.

79. Eklund, T., I. F. Nes, and R. Skjelkvåle. 1981. Control of *Salmonella* at different temperatures by propyl paraben and butylated hydroxyanisole, p. 377. *In* T. A. Roberts, G. Hobbs, J. H. B. Christian, and N. Skovgaard (ed.), *Psy-chrotrophic Microorganisms in Spoilage and Pathogenicity*. Academic Press, London, United Kingdom.

80. Ekstrand, B. 1994. Lactoperoxidase and lactoferrin, p. 15–63. *In* V. M. Dillon and R. G. Board (ed.), *Natural Antimicrobial Systems and Food Preservation*. CAB Intl., Wallingford, United Kingdom.

81. Eletr, S., M. A. Williams, T. Watkins, and A. D. Keith. 1974. Perturbations of the dynamics of lipid alkyl chains in membrane systems: effect on the activity of membrane-bound enzymes. *Biochim. Biophys. Acta* 339:190–201.

82. Ellinger, R. H. 1972. Phosphates in food processing, p. 617–807. *In* T. E. Furia (ed.), *Handbook of Food Additives*, 2nd ed. CRC Press, Inc., Cleveland, Ohio.

83. Ellison, R. T., T. G. Giehl, and F. M. LaForce. 1988. Damage of the outer membrane of enteric gram-negative bacteria by lactoferrin and transferrin. *Infect. Immun.* 56:2774–2781.

84. El-Shenawy, M. A., and E. H. Marth. 1988. Sodium benzoate inhibits growth of or inactivates *Listeria monocytogenes*. *J. Food Prot.* 51:525–530.

85. El-Shenawy, M. A., and E. H. Marth. 1989. Behavior of *Listeria monocytogenes* in the presence of sodium propionate. *Int. J. Food Microbiol.* 8:85–94.

86. Evelyn, J. 1664. Pomona, or an appendix concerning fruit trees. *In Relation to Cider and Several Ways of Ordering It.* Supplement. *Aphorisms Concerning Cider.* P. Beale, J. Martin and J. A. Allestry, London, United Kingdom.

87. Fabian, F. W., C. F. Krehl, and N. W. Little. 1939. The role of spices in pickled food spoilage. *Food Res.* 4:269–286.

88. Fernandez, J. L., and W. J. Simpson. 1993. Aspects of the resistance of lactic acid bacteria to hop bitter acids. *J. Appl. Bacteriol.* 75:315–319.

89. Finol, M. L., E. H. Marth, and R. C. Lindsay. 1982. Depletion of sorbate from different media during growth of *Penicillium* species. *J. Food Prot.* 45:398–404.

90. Firouzi, R., M. Azadbakht, and A. Nabinedjad. 1998. Anti-listerial activity of essential oils of some plants. *J. Appl. Anim. Res.* 14:75–80.

91. Fisher, T. L., and D. A. Golden. 1998. Survival of *Escherichia coli* O157:H7 in apple cider as affected by dimethyl dicarbonate, sodium bisulfite, and sodium benzoate. *J. Food Sci.* 63:904–906.

92. Foster, J. W. 1995. Low pH adaptation and the acid tolerance response of *Salmonella typhimurium. Crit. Rev. Microbiol.* 21:215–237.

93. Francois, J., E. Vam Scjaftingen, and H. G. Hers. 1986. Effect of benzoate on the metabolism of fructose 2,6-biphosphate in yeast. *Eur. J. Biochem.* 154:141-145.

94. Freese, E. 1978. Mechanism of growth inhibition by lipophilic acids, p. 123–131. *In* J. J. Kabara (ed.), *The Pharmacological Effect of Lipids*. American Oil Chemists Society, Champaign, Il.

95. Freese, E., C. W. Sheu, and E. Galliers. 1973. Function of lipophilic acids as antimicrobial food additives. *Nature* 241:321–327.

96. Fretheim, K., P. E. Granum, and E. Vold. 1980. Influence of generation temperature on the chemical composition, antioxidative, and antimicrobial effects of wood smoke. *J. Food Sci.* 45:999–1002.

97. **Furr, J. R., and A. D. Russell.** 1972. Some factors influencing the activity of esters of p-hydroxybenzoic acid against *Serratia marcescens. Microbios* 5:189–195.

98. **Garren, D. M., M. A. Harrison, and S. M. Russell.** 1998. Acid tolerance and acid shock response of *Escherichia coli* O157:H7 and non-O157:H7 isolates provide cross protection to sodium lactate and sodium chloride. *J. Food Prot.* 61:158–161.

99. **Gaya, P., M. Medina, and M. Nunez.** 1991. Effect of the lactoperoxidase system on *Listeria monocytogenes* behavior in raw milk at refrigeration temperatures. *Appl. Environ. Microbiol.* 57:3355–3360.

100. **Gibson, A. M., and T. A. Roberts.** 1986. The effect of pH, water activity, sodium nitrite and storage temperature on the growth of enteropathogenic *Escherichia coli* and salmonellae in laboratory medium. *Int. J. Food Microbiol.* 3:183–194.

101. **Gibson, A. M., and T. A. Roberts.** 1986. The effect of pH, sodium chloride, sodium nitrite and storage temperature on the growth of *Clostridium perfringens* and faecal streptococci in laboratory medium. *Int. J. Food Microbiol.* 3:195–210.

102. **Glabe, E. F., and J. K. Maryanski.** 1981. Sodium diacetate: an effective mold inhibitor. *Cereal Foods World* 26:285–289.

103. **González-Fandos, E., M. L. García-López, M. L. Sierra, and A. Otero.** 1994. Staphylococcal growth and enterotoxins (A-D) and thermonuclease synthesis in the presence of dehydrated garlic. *J. Appl. Bacteriol.* 77:549–552.

104. **Gould, G. W.** 1964. Effect of food preservatives on the growth of bacteria from spores, p. 17. *In* N. Molin (ed.), *Microbial Inhibitors in Food.* Almqvist and Miksell, Stockholm, Sweden.

105. **Gould, G. W. (ed.).** 1989. *Mechanisms of Action of Food Preservation Procedures.* Elsevier Applied Science, London, United Kingdom.

106. **Gould, G. W.** 2000. The use of other chemical preservatives: sulfite and nitrite, p. 200–213. *In* B. M. Lund, T. C. Baird-Parker, and G. W. Gould (ed.), *The Microbiological Safety and Quality of Food.* Aspen Publ., Gaithersburg, Md.

107. **Gould, G. W., and N. J. Russell.** 1991. Sulphite, p. 72–88. *In* N. J. Russell and G. W. Gould (ed.), *Food Preservatives.* Blackie & Son Ltd., Glasgow, Scotland.

108. **Gram, L.** 1991. Inhibition of mesophilic spoilage *Aeromonas* spp. on fish by salt, potassium sorbate, liquid smoke, and chilling. *J. Food Prot.* 54:436–442.

109. **Haas, G. J., and R. Barsoumian.** 1994. Antimicrobial activity of hop resins. *J. Food Prot.* 57:59–61.

110. **Hamden, I. Y., D. D. Deane, and J. E. Kunsman.** 1971. Effect of potassium sorbate on yogurt cultures. *J. Milk Food Technol.* 34:307.

111. **Hayatsu, H., Y. Wataya, K. Kai, and S. Iida.** 1970. Reaction of sodium bisulfite with uracil, cytosine and their derivatives. *Biochemistry* 9:2858.

112. **Hefnawy, Y. A., S. I. Moustafa, and E. H. Marth.** 1993. Sensitivity of *Listeria monocytogenes* to selected spices. *J. Food Prot.* 56:876–878.

113. **Herald, P. J., and P. M. Davidson.** 1983. The antibacterial activity of selected hydroxycinnamic acids. *J. Food Sci.* 48:1378–1379.

114. **Hinze, H., and H. Holzer.** 1985. Effect of sulfite or nitrite on the ATP content and the carbohydrate metabolism in yeast. *Z. Lebensm. Unters. Forsch.* 181:87–91.

115. **Ho, C. T.** 1992. Phenolic compounds in food. An overview, p. 2–7. *In* C. T. Ho, C. Y. Lee, and M. T. Huang (ed.), *Phenolic Compounds in Food and Their Effects on Health. I. Analysis, Occurrence, and Chemistry.* American Chemical Society Symposium Series 506, American Chemical Society, Washington, D.C.

116. **Houtsma, P. C., A. Heuvelink, J. Dufrenne, and S. Notermans.** 1994. Effect of sodium lactate on toxin production, spore germination and heat resistance of proteolytic *C. botulinum* strains. *J. Food Prot.* 57:327–330.

117. **Houtsma, P. C., J. C. Wit, and F. M. Rombouts.** 1996. Minimum inhibitory concentration (MIC) of sodium lactate and sodium chloride for spoilage organisms and pathogens at different pH values and temperatures. *J. Food Prot.* 59:1300–1304.

118. **Huang, L., C. W. Forsberg, and L. N. Gibbins.** 1986. Influence of external pH and fermentation products on *Clostridium acetobutylicum* intracellular pH and cellular distribution of fermentation products. *Appl. Environ. Microbiol.* 51:1230–1234.

119. **Hughey, V. L., and E. A. Johnson.** 1987. Antimicrobial activity of lysozyme against bacteria involved in food spoilage and food-borne disease. *Appl. Environ. Microbiol.* 53:2165–2170.

120. **Hughey, V. L., R. A. Wilger, and E. A. Johnson.** 1989. Antibacterial activity of hen egg white lysozyme against *Listeria monocytogenes* Scott A in foods. *Appl. Environ. Microbiol.* 55:631–638.

121. **Huhtanen, C. M., and J. Feinberg.** 1980. Sorbic acid inhibition of *Clostridium botulinum* in nitrite-free poultry frankfurters. *J. Food Sci.* 45:453–457.

122. **Hunter, D., and I. H. Segel.** 1973. Effect of weak acids on amino acid transport by *Penicillium chrysogenum.* Evidence for a proton or charge gradient as the driving force. *J. Bacteriol.* 113:1184–1192.

123. **Ibrahim, H. R., S. Higashiguchi, M. Koketsu, L. R. Juneja, M. Kim, T. Yamamoto, Y. Sugimoto, and T. Aoki.** 1996. Partially unfolded lysozyme at neutral pH agglutinates and kills gram-negative and gram-positive bacteria through membrane damage mechanism. *J. Agric. Food Chem.* 44:3799–3806.

124. **Ingram, M.** 1939. The endogenous respiration of *Bacillus cereus.* II. The effect of salts on the rate of absorption of oxygen. *J. Bacteriol.* 24:489.

125. **Ingram, M., and K. Vas.** 1950. Combination of sulfur dioxide with concentrated orange juice. I. Equilibrium states. *J. Sci. Food Agric.* 1:21–27.

126. **Inovatech.** 2000. Inovapure product description. Canadian Inovatech, Inc., Abbotsford, B.C., Canada.

127. **International Commission on the Microbiological Specifications for Foods.** 1996. *Microorganisms in Foods 5. Microbiological Specifications of Food Pathogens.*

Blackie Academic & Professional, London, United Kingdom.

128. **Isshiki, K., K. Tokuora, R. Mori, and S. Chiba.** 1992. Preliminary examination of allyl isothiocyanate vapor for food preservation. *Biosci. Biotechnol. Biochem.* **56**:1476–1477.

129. **Jay, J. M.** 1996. Modern food microbiology, 5th ed. Chapman-Hall, New York.

130. **Jay, J. M., and G. M. Rivers.** 1984. Antimicrobial activity of some food flavoring compounds. *J. Food Safety* **6**:129–139.

131. **Jen, C. M. C., and L. A. Shelef.** 1986. Factors affecting sensitivity of *Staphylococcus aureus* 196E to polyphosphate. *Appl. Environ. Microbiol.* **52**:842–846.

132. **Jermini, M. F. G., and W. Schmidt-Lorenz.** 1987. Activity of Na-benzoate and ethyl-paraben against osmotolerant yeasts at different water activity values. *J. Food Prot.* **50**:920–927.

133. **Johansen, C., L. Gram, and A. S. Meyer.** 1994. The combined inhibitory effect of lysozyme and low pH on growth of *Listeria monocytogenes*. *J. Food Prot.* **57**:561–566.

134. **Jones, E. M., A. Smart, G. Bloomberg, L. Burgess, and M. R. Millar.** 1994. Lactoferricin, a new antimicrobial peptide. *J. Appl. Bacteriol.* **77**:208–214.

135. **Judis, J.** 1963. Studies on the mechanism of action of phenolic disinfectants. II. Patterns of release of radioactivity from *Escherichia coli* labeled by growth on various compounds. *J. Pharm. Sci.* **52**:261–264.

136. **Juneja, V. K., and P. M. Davidson.** 1992. Influence of altered fatty acid composition on resistance of *Listeria monocytogenes* to antimicrobials. *J. Food Prot.* **56**:302–305.

137. **Jurd, L., A. D. King, K. Mihara, and W. L. Stanely.** 1971. Antimicrobial properties of natural phenols and related compounds. I. Obtusastyrene. *Appl. Microbiol.* **21**:507–510.

138. **Juven, B., Y. Henis, and B. Jacoby.** 1972. Studies on the mechanism of the antimicrobial action of oleuropein. *J. Appl. Bacteriol.* **35**:559–567.

139. **Kabara, J. J.** 1993. Medium-chain fatty acids and esters, p. 307–342. *In* P. M. Davidson and A. L. Branen (ed.), *Antimicrobials in Foods*, 2nd ed. Marcel Dekker, Inc., New York, N.Y.

140. **Kamau, D. N., S. Doores, and K. M. Pruitt.** 1990. Antibacterial activity of the lactoperoxidase system against *Listeria monocytogenes* and *Staphylococcus aureus* in milk. *J. Food Prot.* **53**:1010–1014.

141. **Kasrazadeh, M., and C. Genigeorgis.** 1995. Potential growth and control of *Escherichia coli* O157:H7 in soft hispanic type cheese. *Int. J. Food Microbiol.* **25**:289–300.

142. **Kato, A., and I. Shibasaki.** 1975. Combined effect of different drugs on the antibacterial activity of fatty acids and their esters. *J. Antibacteriol. Antifung. Agents* (Japan) **8**:355–361.

143. **Kaye, R. C., and S. G. Proudfoot.** 1971. Interactions between phosphatidyl-ethanolamine monolayers and phenols in relation to antibacterial activity. *J. Pharm. Pharmacol. Suppl.* **23**:223S.

144. **Kelch, F., and X. Bühlmann.** 1958. Effect of commercial phosphates on the growth of microorganisms. *Fleischwirtschaft.* **10**:325–328.

145. **Kim, C. R., J. O. Hearnsberger, A. P. Vickery, C. H. White, and D. L. Marshall.** 1995. Extending shelf life of refrigerated catfish fillets using sodium acetate and monopotassium phosphate. *J. Food Prot.* **58**:644–647.

146. **Kim, J. W., and M. F. Slavik.** 1996. Changes in eggshell surface microstructure after washing with cetylpyridinium chloride or trisodium phosphate. *J. Food Prot.* **59**:859–863.

147. **Kim, J. W., M. F. Slavik, M. D. Pharr, D. P. Raben, C. M. Lobsinger, and S. Tsai.** 1994. Reduction of *Salmonella* on post-chill chicken carcasses by trisodium phosphate (Na$_3$PO$_4$) treatment. *J. Food Safety* **14**:9–17.

148. **King, A. D., J. D. Ponting, D. W. Sanshuck, R. Jackson, and K. Mihara.** 1981. Factors affecting death of yeast by sulfur dioxide. *J. Food Prot.* **44**:92–97.

149. **Kirby, G. W., L. Atkin, and C. N. Frey.** 1973. Further studies on the growth of bread molds as influenced by acidity. *Cereal Chem.* **14**:865.

150. **Knabel, S. J., H. W. Walker, and P. A. Hartman.** 1991. Inhibition of *Aspergillus flavus* and selected gram-positive bacteria by chelation of essential metal cations by polyphosphates. *J. Food Prot.* **54**:360–365.

151. **Korhonen, H.** 1978. Effect of lactoferrin and lysozyme in milk on the growth inhibition of *Bacillus stearothermophilus* in the Thermocult method. *Suomen Eläinlääkärilehti* **84**:255–267.

152. **Korpela, J.** 1984. Avidin, a high affinity biotin-binding protein, as a tool and subject of biological research. *Med. Biol.* **65**:5–26.

153. **Kouassi, Y., and L. A. Shelef.** 1995. Listeriolysin O secretion by *Listeria monocytogenes* in the presence of cysteine and sorbate. *Lett. Appl. Microbiol.* **20**:295–299.

154. **Kruk, M., and J. S. Lee.** 1982. Inhibition of *Escherichia coli* trimethylamine-N-oxide reductase by food preservatives. *J. Food Prot.* **45**:241–243.

155. **Lachowicz, K. J., G. P. Jones, D. R. Briggs, F. E. Bienvenu, J. Wan, A. Wilcock and M. J. Coventry.** The synergistic preservative effects of the essential oils of sweet basil (*Ocimum basilicum* L.) against acid-tolerant food microflora. *Lett. Appl. Microbiol.* **26**:209–214.

156. **Lambert, R. J., and M. Stratford.** 1999. Weak-acid preservatives: modeling microbial inhibition and response. *J. Appl. Microbiol.* **86**:157–164.

157. **Larson, A. E., R. R. Y. Yu, O. A. Lee, S. Price, G. J. Haas, and E. A. Johnson.** 1996. Antimicrobial activity of hop extracts against *Listeria monocytogenes* in media and in food. *Intl. J. Food Microbiol.* **33**:195–207.

158. **Lee, J. S.** 1973. What seafood processors should know about *Vibrio parahaemolyticus. J. Milk Food Technol.* **36**:405–408.

159. **Lee, R. M., P. A. Hartman, D. G. Olson, and F. D. Williams.** 1994. Bacteriocidal and bacteriolytic effects of selected food-grade phosphates, using *Staphylococcus aureus* as a model system. *J. Food Prot.* **57**:276–283.

160. **Lennox, J. E., and L. J. McElroy.** 1984. Inhibition of growth and patulin synthesis in *Penicillium expansum*

by potassium sorbate and sodium propionate in culture. *Appl. Environ. Microbiol.* 48:1031–1033.

161. Lewis, J. C., and L. Jurd. 1972. Sporostatic action of cinnamylphenols and related compounds on *Bacillus megaterium*, p. 384–389. *In* H. O. Halvorson, R. Hewson, and L. L. Campbell (ed.), *Spores V*. American Society for Microbiology, Washington, D.C.

162. Leyer, G. J., and E. A. Johnson. 1993. Acid adaptation induces cross-protection against environmental stress in *Salmonella typhimurium. Appl. Environ. Microbiol.* 59:1842–1847.

163. Leyer, G. J., L.-L. Wang, and E. A. Johnson. 1995. Acid adaptation of *Escherichia coli* O157:H7 increases survival in acidic foods. *Appl. Environ. Microbiol.* 61:3752–3755.

164. Liewen, M. B., and E. H. Marth. 1985. Growth of sorbate-resistant and -sensitive strains of *Penicillium roqueforti* in the presence of sorbate. *J. Food Prot.* 48:525–529.

165. Lillard, H. S. 1994. Effect of trisodium phosphate on salmonellae attached to chicken skin. *J. Food Prot.* 57:465–469.

166. Lillard, H. S., L. C. Blankenship, J. A. Dickens, S. E. Craven, and A. D. Shackelford. 1987. Effect of acetic acid on the microbiological quality of scalded picked and unpicked broiler carcasses. *J. Food Prot.* 50:112–114.

167. Lin, C.-A., J. F. Preston, and C.-I. Wei. 2000. Antibacterial mechanisms of allyl isothiocyanate. *J. Food Prot.* 63:727–734.

168. Lopez, M., S. Martinez, J. Gonzalez, R. Martin, and A. Bernardo. 1998. Sensitization of thermally injured spores of *Bacillus stearothermophilus* to sodium benzoate and potassium sorbate. *Lett. Appl. Microbiol.* 27:331–335.

169. López-Malo, A., S. M. Alzamora, and A. Argaiz. 1995. Effect of natural vanillin on germination time and radial growth of moulds in fruit-based agar systems. *Food Microbiol.* 12:213–219.

170. López-Malo, A., S. M. Alzamora, and A. Argaiz. 1997. Effect of vanillin concentration, pH and incubation temperature on *Aspergillus flavus, Aspergillus niger, Aspergillus ochraceus*, and *Aspergillus parasiticus* growth. *Food Microbiol.* 14:117–124.

171. López-Malo, A., S. M. Alzamora, and A. Argaiz. 1998. Vanillin and pH synergistic effects on mold growth. *J. Food Sci.* 63:143–146.

172. Lovell, T. H. 1937. Bactericidal effects of onion vapors. *Food Res.* 2:435.

173. Lyon, G. D., and F. M. McGill. 1988. Inhibition of growth of *Erwinia carotovora* in vitro by phenolics. *Potato Res.* 31:461–467.

174. Lyon, G. D., and F. M. McGill. 1989. Inhibition of polygalacturonase and polygalacturonic acid lyase from *Erwinia carotovora* subsp. *carotovora* by phenolics in vitro. *Potato Res.* 32:267–274.

175. Maas, M. R., K. A. Glass, and M. P. Doyle. 1989. Sodium lactate delays toxin production by *Clostridium botulinum* in cook-in-bag turkey products. *Appl. Environ. Microbiol.* 55:2226–2229.

176. Maier, S. K., S. Scherer, and M. J. Loessner. 1999. Long-chain polyphosphate causes cell lysis and inhibits *Bacillus cereus* septum formation, which is dependent on divalent cations. *Appl. Environ. Microbiol.* 65:3942–3949.

177. Mandel, I. D., and S. A. Ellison. 1985. The biological significance of the nonimmunoglobulin defense factors. *In* K. M. Pruitt and J. O. Tenovuo (ed.), *The Lactoperoxidase System: Its Chemistry and Biological Significance*. Marcel Dekker, Inc., New York, N.Y.

178. Manderfield, M. M., H. W. Schafer, P. M. Davidson, and E. A. Zottola. 1997. Isolation and identification of antimicrobial furocoumarins from parsley. *J. Food Prot.* 60:72–77.

179. Mangena, T. and N. Y. O. Muyima. 1999. Comparative evaluation of the antimicrobial activities of essential oils of *Artemisia afra, Pteronia incana* and *Rosmarinus officinalis* on selected bacteria and yeast strains. *Lett. Appl. Microbiol.* 28:291–296.

180. Mari, M., R. Iori, O. Leoni, and A. Marchi. 1993. *In vitro* activity of glucosinolate derived isothiocyanates against postharvest fruit pathogens. *Ann. Appl. Biol.* 123:155–164.

181. Marth, E. H. 1966. Antibiotics in foods—naturally occurring, developed and added. *Residue Rev.* 12:65–161.

182. Martoadiprawito, W., and J. R. Whitaker. 1963. Potassium sorbate inhibition of yeast alcohol dehydrogenase. *Biochim. Biophys. Acta* 77:536–544.

183. Marwan, A. G., and C. W. Nagel. 1986. Microbial inhibitors of cranberries. *J. Food Sci.* 51:1009–1013.

184. McKellar, R. C. 1993. Effect of preservatives and growth factors on secretion of listeriolysin O by *Listeria monocytogenes. J. Food Prot.* 56:380–384.

185. McLean, R. A., H. D. Lilly, and J. A. Alford. 1968. Effect of meat curing salts and temperature on production of staphylococcus enterotoxin B. *J. Bacteriol.* 95:1207–1211.

186. McMindes, M. K., and A. J. Siedler. 1988. Nitrite mode of action: inhibition of yeast pyruvate decarboxylase (E.C. 4.1.1.1) and clostridial pyruvate:oxidoreductase (E.C. 1.2.7.1) by nitric oxide. *J. Food Sci.* 53:917–919.

187. Melnick, D., F. H. Luckmann, and C. M. Gooding. 1954. Sorbic acid as a fungistatic agent for foods. VI. Metabolic degradation of sorbic acid in cheese by molds and the mechanism of mold inhibition. *Food Res.* 19:44–58.

188. Meng, J. H., and C. A. Genigeorgis. 1993. Modeling the lag phase of nonproteolytic *Clostridium botulinum* toxigenesis in cooked turkey and chicken breast as affected by temperature, sodium lactate, sodium chloride and spore inoculum. *Int. J. Food Microbiol.* 19:109–122.

189. Millis, R. J., and M. J. Schendel. February 1994. Inhibition of food pathogens by hop acids. U.S. patent 5,286,506.

190. Minor, T. E., and E. H. Marth. 1970. Growth of *Staphylococcus aureus* in acidified pasteurized milk. *J. Milk Food Technol.* 33:516–520.

191. Mitchell, P., and J. Moyle. 1969. Estimation of membrane potential and pH difference across the cristae membrane of rat liver mitochondria. *Eur. J. Biochem.* 7:471–484.

192. **Moir, C. J., and M. J. Eyles.** 1992. Inhibition, injury and inactivation of four psychrotrophic foodborne bacteria by the preservatives methyl p-hydroxybenzoate and potassium sorbate. *J. Food Prot.* **55:**360–366.

193. **Mountney, G. J., and J. O'Malley.** 1985. Acids as poultry meat preservatives. *Poult. Sci.* **44:**582.

194. **Muhoberac, B. B., and D. C. Wharton.** 1980. EPR study of heme-NO complexes of ascorbic acid-reduced *Pseudomonas* cytochrome oxidase and corresponding model complexes. *J. Biol. Chem.* **255:**8437–8442.

195. **Musajo, L., F. Bordin, and R. Bevilacqua.** 1967. Photoreactions at 3655A linking the 3–4 double bond of furocoumarins with pyrimidine bases. *Photochem. Photobiol.* **6:**927–931.

196. **Naidu, A. S., and W. R. Bidlack.** 1998. Milk lactoferrin—natural microbial blocking agent (MBA) for food safety. *Environ. Nutr. Interact.* **2:**35–50.

197. **O'Driscoll, B., C. G. M. Gahan, and C. Hill.** 1996. Adaptive acid tolerance response in *Listeria monocytogenes:* isolation of an acid-tolerant mutant which demonstrates increased virulence. *Appl. Environ. Microbiol.* **62:**1693–1698.

198. **Oh, D. H., and D. L. Marshall.** 1993. Influence of temperature, pH, and glycerol monolaurate on growth and survival of *Listeria monocytogenes. J. Food Prot.* **56:**744–749.

199. **Oh, D. H., and D. L. Marshall.** 1994. Enhanced inhibition of *Listeria monocytogenes* by glycerol monolaurate with organic acids. *J. Food Sci.* **59:**1258–1261.

200. **Oh, D. H., and D. L. Marshall.** 1995. Influence of packaging method, lactic acid and monolaurin on *Listeria monocytogenes* in crawfish tail meat homogenate. *Food Microbiol.* **12:**159–163.

201. **Okrend, A. J., R. W. Johnston, and A. B. Moran.** 1986. Effect of acetic acid on the death rates at 52°C of *Salmonella newport, Salmonella typhimurium,* and *Campylobacter jejuni* in poultry scald water. *J. Food Prot.* **49:**500–503.

202. **O'Leary, D. K., and R. D. Kralovec.** 1941. Development of *Bacillus mesentericus* in bread and control with calcium acid phosphate or calcium propionate. *Cereal Chem.* **18:**730–741.

203. **Omar, M. M.** 1992. Phenolic compounds in botanical extracts used in foods, flavors, cosmetics, and pharmaceuticals, p. 154–168. *In* C. T. Ho, C. Y. Lee, and M. T. Huang (ed.), *Phenolic Compounds in Food and Their Effects on Health. I. Analysis, Occurrence, and Chemistry.* American Chemical Society Symposium Series 506, American Chemical Society, Washington, D.C.

204. **Oram, J. D., and B. Reiter.** 1968. Inhibition of bacteria by lactoferrin and other iron chelating agents. *Biochim. Biophys. Acta* **170:**351–365.

205. **Ough, C. S.** 1993. Dimethyl dicarbonate and diethyl dicarbonate, p. 343–368. *In* P. M. Davidson and A. L. Branen (ed.), *Antimicrobials in Foods,* 2nd ed. Marcel Dekker, Inc., New York, N.Y.

206. **Ough, C. S.** 1993. Sulfur dioxide and sulfites, p. 137–190. *In* P. M. Davidson and A. L. Branen (ed.), *Antimicrobials in Foods,* 2nd ed. Marcel Dekker, Inc., New York, N.Y.

207. **Ough, C. S., and R. E. Kunkee.** 1974. The effect of fumaric acid on malolactic fermentation in wines from warm areas. *Am. J. Enol. Vitic.* **25:**188–190.

208. **Padgett, T., I. Y. Han, and P. L. Dawson.** 1998. Incorporation of food-grade antimicrobial compounds into biodegradable packaging films. *J. Food Prot.* **61:**1330–1335.

209. **Pandit, V. A., and L. A. Shelef.** 1994. Sensitivity of *Listeria monocytogenes* to rosemary (*Rosmarinus officianalis* L.). *Food Microbiol.* **11:**57–63.

210. **Payne, K. D., P. M. Davidson, and S. P. Oliver.** 1994. Comparison of EDTA and apo-lactoferrin with lysozyme on the growth of foodborne pathogenic and spoilage bacteria. *J. Food Prot.* **57:**62–65.

211. **Payne, K. D., P. M. Davidson, S. P. Oliver, and G. L. Christen.** 1990. Influence of bovine lactoferrin on the growth of *Listeria monocytogenes. J. Food Prot.* **53:**468–472.

212. **Payne, K. D., E. Rico-Muñoz, and P. M. Davidson.** 1989. The antimicrobial activity of phenolic compounds against *Listeria monocytogenes* and their effectiveness in a model milk system. *J. Food Prot.* **52:**151–153.

213. **Pelroy, G. A., M. E. Peterson, P. J. Holland, and M. W. Eklund.** 1994. Inhibition of *Listeria monocytogenes* in cold-process (smoked) salmon by sodium lactate. *J. Food Prot.* **57:**108–113.

214. **Pelroy, G. A., M. E. Peterson, R. Paranjpye, J. Almond, and M. W. Eklund.** 1994. Inhibition of *Listeria monocytogenes* in cold-process (smoked) salmon by sodium nitrite and packaging method. *J. Food Prot.* **57:**114–119.

215. **Perigo, J. A., and T. A. Roberts.** 1968. Inhibition of clostridia by nitrite. *J. Food Technol.* **3:**91–94.

216. **Pierson, M. D., L. A. Smoot, and N. J. Stern.** 1979. Effect of potassium sorbate on growth of *Staphylococcus aureus* in bacon. *J. Food Prot.* **42:**302–304.

217. **Pilkington, B. J., and A. H. Rose.** 1988. Reactions of *Saccharomyces cerevisiae* and *Zygosaccharomyces bailii* to sulphite. *J. Gen. Microbiol.* **134:**2823–2830.

218. **Pilone, G. J.** 1975. Control of malo-lactic fermentation in table wines by addition of fumaric acid, p. 121–138. *In* J. G. Carr, C. V. Cutting, and G. C. Whiting (ed.), *Lactic Acid Bacteria in Beverages and Foods.* Academic Press, London, United Kingdom.

219. **Post, F. J., W. S. Coblentz, T. W. Chou, and D. K. Salunhke.** 1968. Influence of phosphate compounds on certain fungi and their preservative effect on fresh cherry fruit (*Prunus cerasus* L.). *Appl. Microbiol.* **16:**138–142.

220. **Post, F. J., G. B. Krishnamurty, and M. D. Flanagan.** 1963. Influence of sodium hexametaphosphate on selected bacteria. *Appl. Microbiol.* **11:**430–435.

221. **Post, L. S., T. L. Amoroso, and M. Solberg.** 1985. Inhibition of *Clostridium botulinum* type E in model acidified food systems. *J. Food Sci.* **50:**966–968.

222. **Post, L. S., and P. M. Davidson.** 1986. Lethal effect of butylated hydroxyanisole as related to bacterial fatty acid composition. *Appl. Environ. Microbiol.* **52:**214–216.

223. **Rahn, O., and J. E. Conn.** 1944. Effect of increase in acidity on antiseptic efficiency. *Ind. Eng. Chem.* **36:**185.

224. Rajashekhara, E., E. R. Suresh, and S. Ethiraj. 1998. Thermal death rate of ascospores of *Neosartorya fischeri* ATCC 200957 in the presence of organic acids and preservatives in fruit juices. *J. Food Prot.* **61:**1358–1362.

225. Razavi-Rohani, S. M., and M. W. Griffiths. 1994. The effect of mono and polyglycerol laurate on spoilage and pathogenic bacteria associated with foods. *J. Food Safety* **14:**131–151.

226. Razavi-Rohani, S. M., and M. W. Griffiths. 1999. The antifungal activity of butylated hydroxyanisole and lysozyme. *J. Food Safety* **19:**97–108.

227. Reddy, N. R., and M. D. Pierson. 1982. Influence of pH and phosphate buffer on inhibition of *Clostridium botulinum* by antioxidants and related phenolic compounds. *J. Food Prot.* **45:**925–927.

228. Reddy, N. R., M. D. Pierson, and R. V. Lechowich. 1982. Inhibition of *Clostridium botulinum* by antioxidants, phenols and related compounds. *Appl. Environ. Microbiol.* **43:**835–839.

229. Rehm, H. J., and H. Wittman. 1962. Beitrag zur Kenntnis der antimikrobiellen Wirkung der schwefligen Saure. I. Ubersicht uber einflussnehmende Factoren auf die antimikrobielle Wirking der schwefligen Saure. *Z. Lebensm. Untersuch.-Forsch.* **118:**413–429.

230. Reiss, J. 1976. Prevention of the formation of mycotoxins in whole wheat bread by citric acid and lactic acid. *Experientia* **32:**168.

231. Reiter, B. 1978. Review of the progress of dairy science: antimicrobial systems in milk. *J. Dairy Res.* **45:**131–147.

232. Reiter, B. 1984. The biological significance and exploitation of some of the immune systems in milk: a review. *Microbiol. Aliments Nutr.* **2:**1–20.

233. Reiter, B., and B. G. Harnulv. 1984. Lactoperoxidase antibacterial system: natural occurrence, biological functions and practical applications. *J. Food Prot.* **47:**724–732.

234. Rice, K. M., and M. D. Pierson. 1982. Inhibition of *Salmonella* by sodium nitrite and potassium sorbate in frankfurters. *J. Food Sci.* **47:**1615–1617.

235. Rico-Muñoz, E., E. E. Bargiota, and P. M. Davidson. 1987. Effect of selected phenolic compounds on the membrane-bound adenosine triphosphatase of *Staphylococcus aureus*. *Food Microbiol.* **4:**239–249.

236. Rico-Muñoz, E., and P. M. Davidson. 1983. The effect of corn oil and casein on the antimicrobial activity of phenolic antioxidants. *J. Food Sci.* **48:**1284–1288.

237. Robach, M. C., and J. N. Sofos. 1982. Use of sorbate in meat products, fresh poultry and poultry products: a review. *J. Food Prot.* **45:**374–383.

238. Roberts, T. A., and M. Ingram. 1966. The effect of sodium chloride, potassium nitrate and sodium nitrite on the recovery of heated bacterial spores. *J. Food Technol.* **1:**147–163.

239. Roberts, T. A., L. F. J. Woods, M. J. Payne, and R. Cammack. 1991. Nitrite, p. 89–111. *In* N. J. Russell and G. W. Gould (ed.), *Food Preservatives*. Blackie & Son Ltd., Glasgow, Scotland.

240. Roland, J. O., and L. R. Beuchat. 1984. Biomass and patulin production by *Byssochlamys nivea* in apple juice as affected by sorbate, benzoate, SO$_2$ and temperature. *J. Food Sci.* **49:**402–406.

241. Ronning, I. E., and H. A. Frank. 1987. Growth inhibition of putrefactive anaerobe 3679 caused by stringent-type response induced by protonophoric activity of sorbic acid. *Appl. Environ. Microbiol.* **53:**1020–1027.

242. Rose, A. H., and B. J. Pilkington. 1989. Sulphite, p. 201–224. *In* G. W. Gould (ed.), *Mechanisms of Action of Food Preservation Procedures*. Elsevier Applied Science, London, United Kingdom.

243. Rosenquist, H., and Å. Hansen. 1998. The antimicrobial effect of organic acids, sour dough and nisin against *Bacillus subtilis* and *B. licheniformis* isolated from wheat bread. *J. Appl. Microbiol.* **85:**621–631.

244. Rowe, J. J., J. M. Yabrough, J. B. Rake, and R. G. Eagon. 1979. Nitrite inhibition of aerobic bacteria. *Curr. Microbiol.* **2:**51.

245. Rusul, G., and E. H. Marth. 1987. Growth and aflatoxin production by *Aspergillus parasiticus* NRRL 2999 in the presence of potassium benzoate or potassium sorbate at different initial pH values. *J. Food Prot.* **50:**820–825.

246. Samuelson, K. J., J. H. Rupnow, and G. W. Froning. 1985. The effect of lysozyme and ethylenediaminetetraacetic acid on *Salmonella* on broiler parts. *Poult. Sci.* **64:**1488–1490.

247. Schade, A. L., and L. Caroline. 1944. Raw egg white and the role of iron in growth inhibition of *Shigella dysenteriae*, *Staphylococcus aureus*, *Escherichia coli* and *Saccharomyces cerevisiae*. *Science* **100:**14–15.

248. Schanderl, H. 1962. Der Einfluss von Polyphenolen und Gerbstoffen auf die Physiologie der Weinhefe und der Wert des pH-7 Testes fur die Auswahl von Sektgrundweinen. *Mitt. Rebe Wein Ostbau Freuchtver-vert (Klosterneuburg)* **12A:**265.

249. Schraufstatter, E. 1948. Die bakteriostatische Wirking von Chalkon, Flavanon, Flavon und Flavonol. *Experientia* **4:**484–486.

250. Sekiyama, Y., Y. Mizukami, S. Dong, D. Hu, and T. Uemura. 1994. Antimicrobial activity of mustard seed extract against food poisoning bacteria. *Jpn. J. Food Microbiol.* **11:**133–136.

251. Seward, R. A., R. H. Dielbel, and R. C. Lindsay. 1982. Effects of potassium sorbate and other antibotulinal agents on germination and outgrowth of *Clostridium botulinum* Type E spores in microculture. *Appl. Environ. Microbiol.* **44:**1212–1221.

252. Shelef, L. A., and L. Addala. 1994. Inhibition of *Listeria monocytogenes* and other bacteria by sodium diacetate. *J. Food Safety* **14:**103–115.

253. Shelef, L. A., E. K. Jyothi, and M. A. Bulgarelli. 1984. Growth of enteropathogenic and spoilage bacteria in sage-containing broth and foods. *J. Food Sci.* **49:**737–740.

254. Shelef, L. A., O. A. Naglik, and D. W. Bogen. 1980. Sensitivity of some common foodborne bacteria to the spices sage, rosemary, and allspice. *J. Food Sci.* **45:**1042–1044.

255. Shelef, L. A., and J. A. Seiter. 1993. Indirect antimicrobials, p. 539–570. *In* P. M. Davidson and A. L. Branen

(ed.), *Antimicrobials in Foods*, 2nd ed. Marcel Dekker, Inc., New York, N.Y.

256. Shelef, L. A., and Q. Yang. 1991. Growth suppression of *Listeria monocytogenes* by lactates in broth, chicken, and beef. *J. Food Prot.* 54:283–287.

257. Sheu, C. W., and E. Freese. 1972. Effects of fatty acids on growth and envelope proteins of *Bacillus subtilis*. *J. Bacteriol.* 111:516–524.

258. Sheu, C. W., W. N. Konings, and E. Freese. 1972. Effects of acetate and other short-chain fatty acids on sugars and amino acid uptake of *Bacillus subtilis*. *J. Bacteriol.* 111:525–530.

259. Sheu, C. W., D. Salomon, J. L. Simmons, T. Sreevalsan, and E. Freese. 1975. Inhibitory effects of lipophilic acids and related compounds on bacteria and mammalian cells. *Antimicrob. Agents Chemother.* 7:349–363.

260. Sikovec, S. 1966. Der Einfluss einiger Polyphenole auf die Physiologie von Weinhefen. I. Der Einfluss von Polyphenolen auf den Verlauf der alkoholischen Gärung insbesondere von Emgärungen. *Mitt. (Klosterneuberg)* 16:127–138.

261. Sikovec, S. 1966. Der Einfluss von Polyphenolen auf die Physiologie von Weinhefen. II. Der Einfluss von Polyphenolen auf die Vermehrung and Atmung von Hefen. *Mitt. (Klosterneuberg)* 16:272–281.

262. Simpson, W. J. 1993. Studies on the sensitivity of lactic acid bacteria to hop bitter acids. *J. Inst. Brew.* 99:405–411.

263. Simpson, W. J., and A. R. W. Smith. 1992. Factors affecting antibacterial activity of hop compounds and their derivatives. *J. Appl. Bacteriol.* 72:327–334.

264. Singleton, V. L., and P. Esau. 1969. *Phenolic Substances in Grapes and Wine, and Their Significance*. Academic, Press, Inc., New York, N.Y.

265. Siragusa, G. R., and J. S. Dickson. 1993. Inhibition of *Listeria monocytogenes*, *Salmonella typhimurium* and *Escherichia coli* O157:H7 on beef muscle tissue by lactic or acetic acid contained in calcium alginate gels. *J. Food Safety* 13:147–158.

266. Siragusa, G. R., and M. G. Johnson. 1989. Inhibition of *Listeria monocytogenes* growth by the lactoperoxidase-thiocyanate-H$_2$O$_2$ antimicrobial system. *Appl. Environ. Microbiol.* 55:2802–2805.

267. Slavik, M. F., J. W. Kim, M. D. Pharr, D. P. Raben, S. Tsai, and C. M. Lobsinger. 1994. Effect of trisodium phosphate on *Campylobacter* attached to post-chill chicken carcasses. *J. Food Prot.* 57:324–326.

268. Smith-Palmer, A., J. Stewart, and L. Fyfe. 1998. Antimicrobial properties of plant essential oils and essences against five important food-borne pathogens. *Lett. Appl. Microbiol.* 26:118–122.

269. Smoot, L. A., and M. D. Pierson. 1981. Mechanisms of sorbate inhibition of *Bacillus cereus* T and *Clostridium botulinum* 62A spore germination. *Appl. Environ. Microbiol.* 42:477–483.

270. Smulders, F. J. M., P. Barendsen, J. G. van Logtestjin, D. A. A. Mossel, and G. M. Van der Marel. 1986. Review: lactic acid: considerations in favour of its acceptance as a meat decontaminant. *J. Food Technol.* 21:419–436.

271. Snijders, J. M. A., J. G. van Logtestjin, D. A. A. Mossel, and F. J. M. Smulders. 1985. Lactic acid as a decontaminant in slaughter and processing procedures. *Vet. Q.* 7:277–282.

272. Sofos, J. N. 1983. Antimicrobial effects of sodium and other ions in foods. *J. Food Safety* 6:45–78.

273. Sofos, J. N. 1986. Use of phosphates in low-sodium meat products. *Food Technol.* 40:52–68.

274. Sofos, J. N., L. R. Beuchat, P. M. Davidson, and E. A. Johnson. 1998. *Naturally Occurring Antimicrobials in Food*. Task Force Report No. 132, Council for Agricultural, Science, and Technology, Ames, Iowa.

275. Sofos, J. N., and F. F. Busta. 1993. Sorbic acid and sorbates, p. 49–94. *In* P. M. Davidson and A. L. Branen (ed.), *Antimicrobials in Foods*, 2nd ed. Marcel Dekker, Inc., New York, N.Y.

276. Sofos, J. N., F. F. Busta, and C. E. Allen. 1979. Botulism control by nitrite and sorbate in cured meats. A review. *J. Food Prot.* 42:739–770.

277. Sofos, J. N., F. F. Busta, and C. E. Allen. 1980. Influence of pH on *Clostridium botulinum* control by sodium nitrite and sorbic acid in chicken emulsions. *J. Food Sci.* 45:7–12.

278. Sofos, J. N., F. F. Busta, K. Bhothipaksa, C. E. Allen, M. C. Robach, and M. W. Paquette. 1980. Effects of various concentrations of sodium nitrite and potassium sorbate on *Clostridium botulinum* toxin production in commercially prepared bacon. *J. Food Sci.* 45:1285–1292.

279. Sperber, W. H. 1983. Influence of water activity on foodborne bacteria—a review. *J. Food Prot.* 46:142–150.

280. Stead, D. 1993. The effect of hydroxycinnamic acids on the growth of wine-spoilage lactic acid bacteria. *J. Appl. Bacteriol.* 75:135–141.

281. Stead, D. 1994. The effect of chlorogenic, gallic and quinic acids on the growth of spoilage strains of *Lactobacillus collinoides* and *Lactobacillus brevis*. *Lett. Appl. Microbiol.* 18:112–114.

282. Stecchini, M. L., I. Sarais, and P. Giavedoni. 1993. Effect of essential oils on *Aeromonas hydrophila* in a culture medium and in cooked pork. *J. Food Prot.* 56:406–409.

283. Steinke, P. D. W., and E. M. Foster. 1951. Botulism toxin formation in liver sausage. *Food Res.* 16:477.

284. Stratford, M., and P. A. Anslow. 1998. Evidence that sorbic acid does not inhibit yeast as a classic "weak acid preservative." *Lett. Appl. Microbiol.* 27:203–206.

285. Stratford, M., P. Morgan, and A. H. Rose. 1987. Sulphur dioxide resistance in *Saccharomyces cerevisiae* and *Saccharomycodes ludwigii*. *J. Gen. Microbiol.* 133:2173–2179.

286. Stratford, M., and A. H. Rose. 1985. Hydrogen sulphide production from sulphite by *Saccharomyces cerevisiae*. *J. Gen. Microbiol.* 131:1417–1424.

287. Stratford, M., and A. H. Rose. 1986. Transport of sulphur dioxide by *Saccharomyces cerevisiae*. *J. Gen. Microbiol.* 132:1–6.

288. Surak, J. G., and R. G. Singh. 1980. Butylated hydroxyanisole (BHA) induced changes in the synthesis of polar lipids and in the molar ratio of tetrahymenol to polar

lipids in *Tetrahymena pyriformis*. *J. Food Sci.* **45**:1251–1255.

289. Tanaka, N. 1982. Challenge of pasteurized process cheese spreads with *Clostridium botulinum* using in-process and post-process inoculation. *J. Food Prot.* **45**:1044–1050.

290. Tarr, H. L. A. 1941. Bacteriostatic action of nitrites. *Nature* (London) **147**:417–418.

291. Tarr, H. L. A. 1941. The action of nitrites on bacteria. *J. Fish Res. Board Can.* **5**:265–275.

292. Tarr, H. L. A. 1942. The action of nitrites on bacteria, further experiments. *J. Fish Res. Board Can.* **6**:74–82.

293. Teraguchi, S., K. Shin, T. Ogata, M. Kingaku, A. Kaino, H. Miyauchi, Y. Fukuwatari, and S. Shimamura. 1995. Orally administered bovine lactoferrin inhibits bacterial translocation in mice fed bovine milk. *Appl. Environ. Microbiol.* **61**:4131–4134.

294. Thompson, D. P. 1994. Minimum inhibitory concentrations of esters of p-hydroxybenzoic acid (paraben) combinations against toxigenic fungi. *J. Food Prot.* **57**:133–135.

295. Thomson, J. E., G. J. Banwart, D. H. Sanders, and A. J. Mercuri. 1967. Effect of chlorine, antibiotics, β-propiolactone, acids and washing on *Salmonella typhimurium* on eviscerated fryer chickens. *Poult. Sci.* **46**:146.

296. Ting, W. T. E., and K. E. Deibel. 1992. Sensitivity of *Listeria monocytogenes* to spices at two temperatures. *J. Food Safety* **12**:129–137.

297. Tompkin, R. B. 1983. Indirect antimicrobial effects in foods: phosphates. *J. Food Safety* **6**:13–27.

298. Tompkin, R. B. 1993. Nitrite, p. 191–262. *In* P. M. Davidson and A. L. Branen (ed.), *Antimicrobials in Foods*, 2nd ed. Marcel Dekker, Inc., New York, N.Y.

299. Tompkin, R. B., L. N. Christiansen, and A. B. Shaparis. 1978. The effect of iron on botulinal inhibition in perishable canned cured meat. *J. Food Technol.* **13**:521–527.

300. Tompkin, R. B., L. N. Christiansen, A. B. Shaparis, and H. Bolin. 1974. Effect of potassium sorbate on salmonellae, *Staphylococcus aureus*, *Clostridium perfringens*, and *Clostridium botulinum* in cooked uncured sausage. *Appl. Microbiol.* **28**:262–264.

301. Tranter, H. S. 1994. Lysozyme, ovotransferrin and avidin, p. 65–97. *In* V. M. Dillon and R. G. Board (ed.), *Natural Antimicrobial Systems and Food Preservation*. CAB Intl., Wallingford, United Kingdom.

302. Tranter, H. S., and R. G. Board. 1982. Review: the antimicrobial defense of avian eggs: biological perspective and chemical basis. *J. Appl. Biochem.* **4**:295–338.

303. Tranter, H. S., and R. G. Board. 1984. Influence of incubation temperature and pH on the antimicrobial properties of hen egg albumen. *J. Appl. Bacteriol.* **56**:53–61.

304. Troller, J. A. 1965. Catalase inhibition as a possible mechanism of the fungistatic action of sorbic acid. *Can. J. Microbiol.* **11**:611–617.

305. Tsai, S., and C. Chou. 1996. Injury, inhibition and inactivation of *Escherichia coli* O157:H7 by potassium sorbate and sodium nitrite as affected by pH and temperature. *J. Sci. Food Agric.* **71**:10–12.

306. Ulate-Rodriguez, J., H. W. Schafer, E. A. Zottola, and P. M. Davidson. 1997. Inhibition of *Listeria monocytogenes*, *Escherichia coli* O157:H7 and *Micrococcus luteus* by linear furocoumarins in a model food system. *J. Food Prot.* **60**:1050–1054.

307. Uljas, H. E., and S. C. Ingham. 1999. Combinations of intervention treatments resulting in a 5-$\log_{10}$-unit reductions in numbers of *Escherichia coli* O157:H7 and *Salmonella typhimurium* DT104 organisms in apple cider. *Appl. Environ. Microbiol.* **65**:1924–1929.

308. Ultee, A., E. P. W. Kets, and E. J. Smid. 1999. Mechanisms of action of carvacrol on the food-borne pathogen *Bacillus cereus*. *Appl. Environ. Microbiol.* **65**:4606–4610.

309. Unda, J. R., R. A. Mollins, and H. W. Walker. 1991. *Clostridium sporogenes* and *Listeria monocytogenes*: survival and inhibition in microwave-ready beef roasts containing selected antimicrobials. *J. Food Sci.* **56**:198–205.

310. Uraih, N., T. R. Cassity, and J. R. Chipley. 1977. Partial characterization of the action of benzoic acid on aflatoxin biosynthesis. *Can. J. Microbiol.* **23**:1580–1584.

311. Uraih, N., and J. R. Chipley. 1976. Effects of various acids and salts on growth and aflatoxin production by *Aspergillus flavus*. *Microbios* **17**:51–59.

312. Valenti, P., P. Visca, G. Antonini, and N. Orsi. 1985. Antifungal activity of ovotransferrin towards genus *Candida*. *Mycopathology* **89**:165–175.

313. Venkitanarayanan, K. S., T. Zhao, and M. P. Doyle. 1999. Antibacterial effect of lactoferricin B on *Escherichia coli* O157:H7 in ground beef. *J. Food Prot.* **62**:747–750.

314. Wagner, M. K. 1986. Phosphates as antibotulinal agents in cured meats. A review. *J. Food Prot.* **49**:482–487.

315. Wagner, M. K., and F. F. Busta. 1985. Inhibition of *Clostridium botulinum* 52A toxicity and protease activity by sodium acid pyrophosphate in media systems. *Appl. Environ. Microbiol.* **50**:16–20.

316. Walker, J. C., and M. A. Stahmann. 1955. Chemical nature of disease resistance in plants. *Annu. Rev. Plant Physiol.* **6**:351–366.

317. Walton, L., M. Herbold, and C. C. Lindegren. 1936. Bactericidal effects of vapors from crushed garlic. *Food Res.* **1**:163.

318. Wan, J., A. Wilcock, and M. J. Coventry. 1998. The effect of essential oils of basil on the growth of *Aeromonas hydrophila* and *Pseudomonas fluorescens*. *J. Appl. Microbiol.* **84**:152–158.

319. Wang, L.-L., and Johnson, E. A. 1997. Control of *Listeria monocytogenes* by monoglycerides in foods. *J. Food Prot.* **60**:131–138.

320. Wang, W. C., Y. Li, M. F. Slavik, and H. Xiong. 1997. Trisodium phosphate and cetylpyridinium chloride spraying on chicken skin to reduce attached Salmonella typhimurium. *J. Food Prot.* **60**:992–994.

321. Ward, S. M., P. J. Delaquis, R. A. Holley, and G. Mazza. 1998. Inhibition of spoilage and pathogenic bacteria on agar and pre-cooked roast beef by volatile horseradish distillates. *Food Res. Intl.* **31**:19–26.

322. Warth, A. D. 1977. Mechanism of resistance of *Saccharomyces bailii* to benzoic, sorbic, and other weak acids used as food preservatives. *J. Appl. Bacteriol.* **43**:215–230.

323. **Weaver, R. A., and L. A. Shelef.** 1993. Antilisterial activity of sodium, potassium or calcium lactate in pork liver sausage. *J. Food Safety* **13:**133–146.

324. **Wedzicha, B. L., and M. A. Brook.** 1989. Reaction of sorbic acid with nucleophiles: preliminary studies. *Food Chem.* **31:**29–40.

325. **Wendorff, W. L., W. E. Riha, and E. Muehlenkamp.** 1993. Growth of molds on cheese treated with heat or liquid smoke. *J. Food Prot.* **56:**963–966.

326. **Wibowo, D., R. Eschenbruch, C. R. Davis, G. H. Fleet, and T. H. Lee.** 1985. Occurrence and growth of lactic acid bacteria in wine. A review. *Am. J. Enol. Vitic.* **36:**302–313.

327. **Wilkins, K. M., and R. G. Board.** 1989. Natural antimicrobial systems, p. 285–362. *In* G. W. Gould (ed.), *Mechanisms of Action of Food Preservation Procedures.* Elsevier Applied Science, London, United Kingdom.

328. **Wills, E. D.** 1956. Enzyme inhibition by allicin, the active principle of garlic. *Biochem. J.* **63:**514–520.

329. **Woods, L. F. J., and J. M. Wood.** 1982. The effect of nitrite inhibition on the metabolism of *Clostridium botulinum.* *J. Appl. Bacteriol.* **52:**109–110.

330. **Woods, L. F. J., J. M. Wood, and P. A. Gibbs.** 1981. The involvement of nitric oxide in the inhibition of the phosphoroclastic system in *Clostridium sporogenes* by sodium nitrite. *J. Gen. Microbiol.* **125:**399–406.

331. **Woods, L. F. J., J. M. Wood, and P. A. Gibbs.** 1989. Nitrite, p. 225–246. *In* G. W. Gould (ed.), *Mechanisms of Action of Food Preservation Procedures.* Elsevier Applied Science, London, United Kingdom.

332. **Woolford, M. K.** 1975. Microbiological screening of the straight chain fatty acids (C_1-C_{12}) as potential silage additives. *J. Sci. Food Agric.* **26:**219–228.

333. **Xiong, R., G. Xie, and A. S. Edmondson.** 1999. The fate of *Salmonella enteritidis* PT4 in home-made mayonnaise prepared with citric acid. *Lett. Appl Microbiol.* **28:**36–40.

334. **Yang, T.** 1985. Mechanism of nitrite inhibition of cellular respiration in *Pseudomonas aeruginosa.* *Curr. Microbiol.* **12:**35–40.

335. **Yin, M.-C., and W.-S. Cheng.** 1998. Inhibition of *Aspergillus niger* and *Aspergillus flavus* by some herbs and spices. *J. Food Prot.* **61:**123–125.

336. **York, G. K., and R. H. Vaughn.** 1955. Resistance of *Clostridium parabotulinum* to sorbic acid. *Food Res.* **20:**60–65.

337. **York, G. K., and R. H. Vaughn.** 1964. Mechanisms in the inhibition of microorganisms by sorbic acid. *J. Bacteriol.* **88:**411–417.

338. **Yousef, A. E., M. A. El-Shenawy, and E. H. Marth.** 1989. Inactivation and injury of *Listeria monocytogenes* in a minimal medium as affected by benzoic acid and incubation temperature. *J. Food Sci.* **54:**650–652.

339. **Zaika, L. L., and A. H. Kim.** 1993. Effect of sodium polyphosphates on growth of *Listeria monocytogenes.* *J. Food Prot.* **56:**577–580.

340. **Zaika, L. L., and J. C. Kissinger.** 1981. Inhibitory and stimulatory effects of oregano on *Lactobacillus plantarum* and *Pediococcus cerevisiae.* *J. Food Sci.* **46:**1205–1210.

341. **Zhao, T., M. P. Doyle, and R. E. Besser.** 1993. Fate of enterohemorrhagic *Escherichia coli* O157:H7 in apple cider with and without preservatives. *Appl. Environ. Microbiol.* **59:**2526–2530.

342. **Zhuang, R.-Y., and L. R. Beuchat.** 1996. Effectiveness of trisodium phosphate for killing *Salmonella montevideo* on tomatoes. *Lett. Appl. Microbiol.* **22:**97–100.

Food Microbiology: Fundamentals and Frontiers, 2nd Ed.
Edited by M. P. Doyle et al.
© 2001 ASM Press, Washington, D.C.

Thomas J. Montville
Karen Winkowski
Michael L. Chikindas

30

Biologically Based Preservation Systems

The chemical and physical methods of food preservation covered in this book are well established and easily identified by consumers. Physical changes in the food or label declarations help the consumer determine how a food is processed. A little bit of knowledge can be a dangerous thing; consumers are increasingly wary of chemical preservatives and "processed" foods, even though these methods provide the unparalleled safety and diversity of our food supply. One result of these consumer trends is the increased reliance on refrigeration to ensure the safety of foods that are minimally processed and free of chemical preservatives. However, two factors make it unwise to rely too heavily on refrigeration. The first is that 20% of commercial and residential refrigerators maintain temperatures >10°C (50°F) (172). The second factor is the ability of *Listeria monocytogenes* and other pathogens to grow at or near freezing temperatures (124). Therefore, the National Food Processors Association recommends that additional barriers to microbial growth be incorporated into refrigerated foods. Such a barrier may be provided through the use of lactic acid bacteria (LAB). Their use for food preservation is accepted by consumers as "natural" and "health promoting." The fermentation of food may be the oldest form (after drying) of processing for preservation. These factors have generated tremendous interest in biologically based preservation methods.

This chapter provides an overview of the biologically based preservation technologies which can be classified as "biopreservation." Biopreservation is defined here as the use of LAB, their metabolic products, or both to improve or ensure the safety and quality of foods that are not generally considered fermented. The preservative, nutritional, and functional properties of fermented foods are covered in section VIII of this book. Acid production by LAB in temperature-abused foods ("controlled acidification") is covered in the first part of the chapter.

Some strains of LAB associated with fermented foods also produce antimicrobial proteins called bacteriocins. Bacteriocins inhibit spoilage and pathogenic bacteria without changing (e.g., through acidification or protein denaturation) the physical-chemical nature of the food being preserved. Because the use of bacteriocins is a newer and emerging area of food microbiology, the largest section of this chapter deals with this topic.

BIOPRESERVATION BY CONTROLLED ACIDIFICATION

It is widely recognized, and discussed more fully in chapter 29, that organic acids inhibit microbial growth. While this is usually done by adding organic acids to foods, LAB can produce lactic acid in situ. The controlled production of acid in situ is an important form of biopreservation.

Thomas J. Montville, Karen Winkowski, and **Michael L. Chikindas,** Department of Food Science, Food Science Bldg., 65 Dudley Rd., Rutgers–The State University of New Jersey, New Brunswick, NJ 08901-8520.

Many factors determine the effectiveness of in situ acidification, including the product's initial pH, its buffering capacity, the type and level of the challenge organism, the nature and concentration of the fermentable carbohydrate, ingredients that might influence the viability and growth rate of the LAB, and the growth rates of the LAB and target pathogen at refrigerated and abuse temperatures (74). Clearly, such applications require customization and a high level of technical support. The production of bacteriocins, diacetyl, and hydrogen peroxide may also contribute to the overall inhibition. For example, Microgard is a cultured milk product added to much of the cottage cheese in the United States as a generally regarded as safe (GRAS) food preservative. It is made by fermenting milk using *Propionibacterium shermanii* to produce acetate, propionic acid, and low-molecular-weight proteins. Although Microgard does contain a bacteriocin, the propionic acid certainly plays a major role in its activity (3, 97).

The idea of using LAB to prevent botulinal toxigenesis through in situ acid production dates back to the 1950s. This technology uses *Clostridium botulinum's* inability to grow at pH <4.8 as a defense against temperature abuse. LAB and a fermentable carbohydrate are added to the food. The LAB grow and produce acid in situ only under conditions of temperature abuse. Under proper refrigeration, the LAB cannot grow, and no acid is formed. Saleh and Ordal (141) used a "normal cheese culture," *Lactobacillus bulgaricus,* or *Lactococcus lactis* in experiments designed to prevent toxigenesis from spore inocula in chicken à la king containing a fermentable carbohydrate. When incubated at 30°C in the absence of the LAB, 16 out of 16 samples rapidly became toxic. In the presence of the "normal cheese culture," *Lb. bulgaricus,* or *Lc. lactis*, only three or fewer samples became toxic after 5 days at 30°C. In all samples to which the LAB had been added, including those positive for botulinal toxin, the pH was reduced to <4.5. The differences between the LAB cultures were not statistically significant, and the inhibition was attributed to acid production.

The discovery that carcinogenic nitrosamines are formed from nitrites used as curing agents in meats initiated a search for nitrite substitutes. Tanaka et al. (159) sought to reduce nitrite concentrations in bacon by using controlled acidification. When bacon was inoculated with 10^3 botulinal spores per gram and incubated at 28°C, toxin was produced in 58% of the conventional bacon samples prepared with 120 ppm of nitrite but no sucrose or starter culture. When similarly challenged, ≤2% bacon prepared with 80 or 40 ppm of nitrite, 0.7% sucrose, and starter cultures became toxic. The U.S. Department of Agriculture approved what is now known as the "Wisconsin process" for bacon manufacture in 1986.

BACTERIOCINS

The interest in the bacteriocins produced by LAB has grown dramatically in the last decade and is documented by many major reviews and books (e.g., 47, 65, 71, 75, 84, 116, 117, 124, 132, 153, 154, 157). One reason for this interest is that many bacteriocins inhibit food spoilage and pathogenic bacteria, such as *L. monocytogenes,* which are recalcitrant to traditional food preservation methods. In addition, bacteriocinogenic LAB are associated with, and are used as, starter cultures. The use of bacteriocins or the organisms which produce them, or both, is attractive to the food industry because the industry is facing both increasing consumer demand for natural products and increasing concern about foodborne disease. However, this interest is tempered by regulatory uncertainty and concerns that the development of bacteriocin-resistant pathogens might render the technology ineffective.

General Characteristics

The bacteriocins produced by LAB are a relatively heterogeneous group of ribosomally synthesized small proteins. They normally act against "closely related" bacteria but not against the producing organism. In some cases, the term "closely related" covers a wide range of gram-positive bacteria. Chelating agents, hydrostatic pressure, or injury can render gram-negative bacteria bacteriocin sensitive (82, 83, 151, 152). Enzymes, such as lysozyme, that act exclusively through degradative enzymatic activities are not considered bacteriocins. However, in addition to their action at the target cell's membrane, some bacteriocins, such as the colicins produced by *Escherichia coli,* inhibit protein synthesis, degrade RNA, or have other biological functions (87).

Early bacteriocin researchers often applied to bacteriocins the seven characteristics of colicins cited by Tagg's influential review (158). They would restrict the designation "bacteriocins" to those plasmid-mediated proteins produced by lethal biosynthesis which are bactericidal to a narrow range of closely related bacteria having specific binding sites for that bacteriocin. It is now clear that few LAB bacteriocins meet all of these criteria. A proposal to designate as "bacteriocins" only those proteins having all seven characteristics and to use "bacteriocinlike inhibitory substance (BLIS proteins)" to describe other antimicrobial proteins (157) has not been widely adopted. However, the BLIS designation is useful for another purpose. There are at least five separate examples in which bacteriocins have been independently isolated and named by several different investigators but were later found to be identical (based on their amino acid sequences). The resulting confusion of having many different names for the same molecule impedes research progress. In order to prevent this confusion, Jack et al. (75) proposed using "BLIS protein" followed by a strain designation as a provisional designation when proteinaceous inhibitors are first identified. A new bacteriocin name would be given only when the amino acid sequence indicates that the bacteriocin is, in fact, unique.

Bacteriocins differ in their spectra of activity, biochemical characteristics, and genetic determinants (84, 85, 117). Most bacteriocins are small (3 to 10 kDa), have a high isoelectric point, and contain both hydrophobic and hydrophilic domains. Klaenhammer (85) further classified bacteriocins into four major groups. These groupings provide a useful conceptual framework for bacteriocin researchers. This classification of bacteriocins is currently being refined to reflect recent discoveries of new molecules and their mechanism of action.

Group I bacteriocins (76) contain the unusual amino acids dehydroalanine, dehydrobutyrine, lanthionine, and β-methyllanthionine (Fig 30.1). Group I is being further subdivided into subgroups Ia and Ib on the basis of differences in secondary structure of the molecules. These unusual amino acids of group I bacteriocins are produced by posttranslational modification of serine and threonine to their dehydro forms. The dehydro amino acids react with cysteine to form thioether (single sulfhydryl)

lanthionine rings. Bacteriocins containing these lanthionine rings are commonly referred to as lantibiotics. There are many structurally similar lantibiotics. Nisin, the first and best-characterized LAB bacteriocin (73), is produced in two related forms. Nisin A contains a histidine at position 27, where nisin Z has an asparagine (115). While subtilin, produced by *Bacillus subtilis*, also contains five lanthionine rings (20) and a conformation similar to nisin, it has other amino acid substitutions, including a carboxy terminus two amino acids shorter than nisin. The 12 amino acids at the amino terminus of nisin and epidermin are similar, but epidermin lacks the central lanthionine ring common to nisin and subtilin and has a cyclized carboxy terminus. Lacticin 481 (126), lactocin S (114), and carnocin (156) are other lantibiotics produced by LAB.

Many bacteriocins belong in group II. Group II bacteriocins are a large group of small heat-stable proteins with a consensus leader sequence containing a Gly-Gly^{-1}-Xaa^{+1} cleavage site important for processing the prebacteriocin during export. Group II is subdivided into three groups. Bacteriocins active against *L. monocytogenes* and having a -Tyr-Gly-Asn-Gly-Val-Xaa-Cys amino-terminal consensus sequence are classified in the subgroup IIa. Pediocin PA-1 (whose amino acid sequence is identical to that of pediocin AcH), sakacins A and P, leucocin A, bavaricin MN, and curvacin A are members of this subgroup (47). Subgroup IIb (116) contains bacteriocins, such as lactococcin G, lactococcin M, lactacin F, and plantaricins EF and JK, which require two different peptides for activity (2, 109).

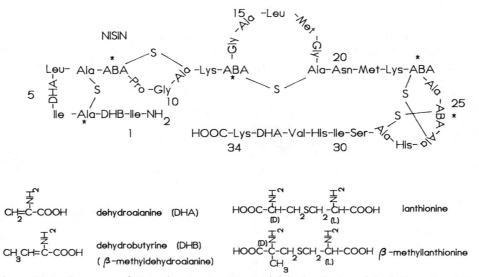

Figure 30.1 Structure of nisin showing positions of dehydroalanine, dehydrobutyrine, lanthionine, and methyl lanthionine. Reprinted from reference 94 with permission.

Bacteriocins in subgroup IIc, such as lactacin B, require reduced cysteines for activity.

Group III and IV bacteriocins differ markedly from other bacteriocins. The larger (>30 kDa) heat-labile antimicrobial proteins, such as helveticins J and V (79) and lactacins A and B (7), are classified as group III bacteriocins. Group IV bacteriocins, such as leuconocin S (154), lactocin 27 (163), and pediocin SJ-1 (142), have lipid or carbohydrate moieties. The composition and function of the nonprotein portions are largely unknown.

Classification systems and nomenclature for bacteriocins are still evolving. Klaenhammer's groupings are incorporated into a proposed numerical system analogous to the numbering system used in enzyme nomenclature. An alternate nomenclature system (75) is based on sulfhydryl chemistry. Just as bacteriocins containing lanthionine rings are called lantibiotics, those containing disulfide bonds would be called cystibiotics and those requiring reduced sulfhydryls would be called thiobiotics. Bacteriocin research is very dynamic; some time may be required before a system of definitive nomenclature is in place.

Methodological Considerations

Bacteriocin-producing bacteria are easy to isolate from foods. The methods used for their initial isolation and characterization are relatively simple (102). The most common method for demonstrating bacteriocin production (the Bac⁺ phenotype) is to overlay a colony of the putative bacteriocin producer with agar medium containing the target organism to be inhibited. An inhibition zone in the confluent growth of the target organism is presumptive evidence for bacteriocin production (Fig. 30.2). Such zones can also be produced by acid, bacteriophage, hydrogen peroxide, or other nonspecific inhibitors. Negative control experiments are required to exclude these. The positive control, confirming the inhibitor's protease sensitivity, is equally important and can often be done on the same petri dish.

Research beyond the initial isolation of the Bac⁺ bacteria and characterization of the inhibitor as a bacteriocin is more difficult. Some isolates produce large inhibition zones on agar media but have no detectable activity in broth. High (10,000 to >50,000 arbitrary units [AU]/ml) bacteriocin activities (see below for definitions) in the culture supernatants facilitate bacteriocin purification. Considerable effort to optimize the liquid medium, incubation conditions, pH, and other factors may be needed before starting purification studies. The purification usually involves salting out the protein, followed by some combination of gel filtration, ion exchange, affinity, and hydrophobic interaction

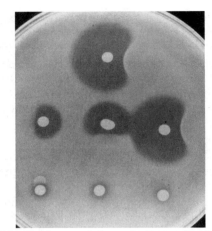

Figure 30.2 Evidence of bacteriocin production using a differed spot-on-the-lawn method. A bacteriocin-producing colony is cultured on agar and then overlaid with an agar seeded with a bacteriocin-sensitive organism. After further incubation, diffusion of bacteriocin away from the colony causes clearing zones in the lawn of sensitive bacteria. When a protease is spotted near the inhibition zone, a "dimple" of growth in the clearing zone indicates that inhibitor is a protein. The small diffuse zones around the bottom three colonies of bacteria that do not produce bacteriocins are due to acid production. Photograph courtesy of M. Stiles.

chromatography. Amino acid sequences are then determined from the electrophoretically pure protein. As more bacteriocins are being purified to the sequence level, it is becoming common to discover that independently purified bacteriocins are identical.

The lack of recognized standards for bacteriocin activity is a major impediment to progress in this field. Only for nisin has an international unit of activity been adopted. One gram of commercially available nisin preparation (Nisaplin) contains 25 mg of pure nisin and is defined as having 1 million international units (earlier known as reading units) of activity. Nisin activity is measured by the well diffusion assay of Tramer and Fowler (162) using *Micrococcus luteus* ATCC 10420 as the sensitive organism. *M. luteus* is also used to measure the activity of other peptide antimicrobial agents, such as scorpion defensins (26). Investigators frequently use other indicator strains in their assays for other bacteriocins if they are more sensitive (generate larger zone sizes) than *M. luteus*.

Bacteriocin activity can be estimated from the size of the inhibition zones produced by the diffusion of the bacteriocin in the confluent lawns of bacteriocin-sensitive bacteria. The bacteriocin can be placed in a well (such as the one made by a number 3 cork borer) in the assay medium that contains the sensitive organism. Diffusion assays can also be conducted by "spotting" a small volume (10 to 50 μl) of test material directly on a lawn of sensitive bacteria. In either case, the activity is reported as the radius or diameter of the inhibition zone, either before or after subtracting the dimension of the well or spot. The zone sizes obtained in diffusion assays are proportional to the log of the bacteriocin activity. Results are easily misinterpreted when zone sizes are reported without considering the log-linear nature of the assay.

Arbitrary units are more useful quantitative measurements of bacteriocin activity. These are usually determined by assaying twofold serial dilutions of the sample. The reciprocal of the highest dilution producing inhibition becomes the number of arbitrary units. This can be divided by the sample volume to yield arbitrary units per milliliter. The arbitrary nature of this system for measuring activity cannot be overemphasized. By virtue of the two fold dilutions, the assay is insensitive to activity differences that are less than twofold. The assumption of linearity with volume used to calculate arbitrary units per milliliter is rarely verified. The choice of indicator organism, assay technique, length of incubation time, assay medium, etc., is so idiosyncratic as to make it virtually impossible to compare the arbitrary units used by different investigators. There is no easy solution to this problem. However, by assaying a known concentration of nisin using the same experimental conditions, an approximation of the relationship between arbitrary units and equivalent nisin international units can be generated.

Some investigators use conceptually different assays to generate an arbitrary unit unrelated to those described above. For example, the ability to reduce growth rate by 50% or decrease viability by 50% can be a measure of bacteriocin activity. These assays do not require the 12- to 24-h incubation period of the diffusion assays but nonetheless are not widely used.

Recently, additional methods for bacteriocin quantification have become available. Enzyme-linked immunosorbent assays work well if antibodies to the bacteriocin of choice are available (e.g., 12, 99). A recombinant construction consisting of the *gusA* gene under the control of the nisin-regulated *nisA* promoter (e.g., 36, 125) can be used to determine nisin's concentration. Nisin induces activity of the *nisA* promoter and, subsequently, production of β-glucuronidase, which activity can be measured. Another method for nisin quantification is based on bioluminescence assay (121). This is even more sensitive than the assay based on determination of a product of the *gusA* gene expressed under the nisin-regulated promoter. This new method is based on a recombinant genetic construction, which consists of the nisin-regulated promoter and bioluminescence genes *luxAB* from *Xenorhabdus luminescens*. *Lc. lactis* MG1614 cells transformed with such a recombinant DNA can "sense" as little as 0.0125 ng of nisin per ml (176).

Bacteriocin Applications in Foods

There are several distinct applications for the use of bacteriocins in foods. Bacteriocins can be added directly to the food to inhibit spoilage or pathogenic organisms. Only nisin is commercially available for addition in pure form; however, it is mainly manufactured and used as a partially purified product of dairy or nondairy fermentation and marketed by companies under the names Nisaplin (5) and Novasin (134). The efficacy of pediocin addition has also been demonstrated. Spoilage or pathogenic organisms can also be inhibited by bacteriocinogenic cultures which are added to the food (19, 46). Bacteriocinogenic cultures can be added to nonfermented foods or used as starter cultures in fermented foods to improve safety and quality. The final application for bacteriocins is not related to spoilage or safety but rather to improving the quality of fermented foods. The use of defined starter cultures offers many benefits, such as improved quality and consistency, when compared with the use of indigenous bacteria as inocula. However, unless the indigenous microbiota can be inactivated, they usually predominate over the defined inocula. Because of this, the benefits of defined starter cultures are most pronounced in the dairy industry, where the indigenous microbiota is inactivated during pasteurization. The use of bacteriocin-producing starter cultures in fermented meats and vegetables might inactivate the indigenous microbiota and allow the use of defined starter cultures in these products.

Addition of Nisin

Nisin is added to milk, cheese, and dairy products, a variety of canned foods, mayonnaise, and baby foods throughout the world (73, 93). It is GRAS (49) as an antibotulinal agent in certain cheese spreads and is used commercially as an antimastitis teat dip (145). Nisin has potential as a treatment for ulcers, in personal hygiene, and as a general sanitizing agent. When nisin is absorbed onto surfaces, it inhibits listerial growth (30) and prevents biofilm formation (13). Nisin also sensitizes spores to heat, so that the thermal treatment can be reduced (73). This application is not approved in the United States.

Many nisin applications target *C. botulinum*. Its spores are much less nisin sensitive than are *L. monocytogenes* vegetative cells. While as little as 200 IU of nisin per ml can reduce *L. monocytogenes* viability by 6 logs, concentrations of up to 10,000 IU/ml are required to obtain similar results with botulinal spores (111).

Temperature is the major determinant of nisin's inhibitory action. Nisin is much less effective at elevated temperatures than at refrigeration temperatures against *C. botulinum* 56A spores in a model food system (137). Eventually, growth in this system occurs when nisin activity falls below some threshold level. The threshold nisin levels are lower at decreasing temperatures. For example, botulinal growth occurred when residual nisin concentrations fell below 154 IU/ml at 35°C, but not until nisin levels were <12 IU/ml at 15°C. Nisin also inhibits listeria and staphylococcal vegetative cells better at refrigerated temperatures than at elevated temperatures (24).

Many other factors influence nisin's sporostatic efficacy (144). In the studies used to support the GRAS affirmation of nisin in pasteurized process cheese, between 500 and 2,000 IU of nisin per ml inhibited botulinal spore outgrowth by 50% in broth, but levels of up to 10,000 IU/ml were ineffective in cooked meat medium (143). Nisin at 100 to 250 ppm allows some salt reduction or increased moisture levels in pasteurized process cheese spreads without elevating the risk of botulism (148). Specific phospholipids also decrease nisin's antibotulinal efficacy (137), and butterfat decreases nisin's ability to inhibit *Staphylococcus aureus* (80).

In most applications, nisin serves as one part of a multiple-barrier inhibitory system. Nisin may be a useful adjunct to modified atmosphere storage. Nisin increases the shelf life and delays toxin production by type E botulinal strains in fresh fish packaged in a carbon dioxide atmosphere. However, toxin can sometimes be detected before samples are obviously spoiled (160). The combination of nisin and modified atmosphere to prevent *L. monocytogenes* growth in pork is more effective than either used alone (48). Nisin added to liquid whole egg prior to pasteurization at 5 mg/liter extends its refrigerated shelf life from 6–11 days to 17–20 days (35).

While most of the research has been directed against botulinal spores, nisin's inhibition of spores from other bacterial species also depends on many factors. Salt antagonizes the action of nisin against *Bacillus licheniformis* spores (9). The ability of nisin to inhibit thermally stressed bacillus spores is influenced by the time-temperature combination used to affect the thermal stress, the subsequent incubation temperature, the pH, and even the type of acidulant used (123). In general, nisin is more effective at lower temperatures, against lower spore loads, and under acidic conditions.

Addition of Pediocin

While inactive against spores, pediocins inhibit *L. monocytogenes* vegetative cells. European patents cover the use of pediocin PA-1 in the form of a dried powder or culture liquid to extend the shelf life of salads and salad dressing (60) and as an antilisterial agent in foods such as cream, cottage cheese, meats, and salads (127, 169).

Pediocins are more effective than nisin in meat and are even more effective in dairy products. Dipping meat in 5,000 AU of crude pediocin PA-1 per ml decreases the viability of attached *L. monocytogenes* cells 100- to 1,000-fold. Pretreating meat with pediocin reduces subsequent *L. monocytogenes* attachment (119). Pediocin AcH at 1,350 AU/ml reduces listeria in ground beef, sausage, and other products by between 1 and 7 log cycles. Pediocin AcH is more effective at 4°C than at 25°C against *L. monocytogenes* in hot dog exudate. Emulsifiers such as Tween 80 or the entrapment of the pediocin in multilamellar vesicles increases pediocin effectiveness in fatty foods (32–34). In most cases, the bacteriocin rapidly reduces listeria viability and delays growth of the survivors.

Addition of Bacteriocin-Producing Bacteria to Nonfermented Foods

Many successful antilisterial applications of bacteriocins add the bacteriocinogenic culture rather than the pure bacteriocin. In wieners held at 4°C, *L. monocytogenes* cells grow after a 20- to 30-day lag period and increase from 10^4 to 10^6 by the end of 60 days (10). At 10^7 CFU/g, both Bac$^+$ and Bac$^-$ derivative strains of *Pediococcus acidilactici* inhibit *L. monocytogenes* growth in hot dogs for 60 days. The inhibition by the Bac$^-$ strain is not due to acidification or production of hydrogen peroxide. The control wieners have pH values similar to the wieners treated with Bac$^+$ and Bac$^-$ pediococci, and results under anaerobic conditions are similar to those under aerobic conditions. However, at low inoculum levels, the Bac$^+$ strain, but not its Bac$^-$ derivative, extends the lag period of listeria cells. The degree of inhibition increases with decreasing temperatures and, in this case, is greater under anaerobic conditions than aerobic conditions.

Lactobacillus bavaricus MN (which produces bavaricin MN) inhibits listerial growth in model gravy at 4°C, even in the absence of a fermentable carbohydrate (179). The addition of a fermentable carbohydrate, reduction of incubation temperatures, and increased lactobacillus/listeria inoculation ratios all increase the degree of inhibition. These variables also influence the success of this preservation technology in sous vide beef cubes (178). Under the most favorable conditions (4°C, high inoculation ratio, and beef packed in gravy which contains a fermentable carbohydrate), *Lb. bavaricus* causes *L. monocytogenes* viability to decline 10-fold over 6 weeks at 4°C. In the least favorable conditions (10°C, low inoculation ratio, beef cubes without gravy), listerial growth is inhibited by the bavaricin for 1 week but increases 100-fold in the beef without the *Lb. bavaricus*.

Carnobacterium piscicola LK5 is also more effective against *L. monocytogenes* at 5°C than at 19°C in ultrahigh-temperature-treated milk, dog food made from beef, pasteurized crabmeat, creamed corn, and wieners. (17). *C. piscicola* LK5 inhibits *L. monocytogenes* Scott A, even when the listeria are present at levels 100-fold higher than the *C. piscicola*.

Use of Bacteriocin-Producing Starter Cultures to Improve the Safety of Fermented Foods

If a food is going to be fermented by an LAB anyway, the use of a bacteriocin-producing starter culture can provide added value to the product. For example, the presence of a nisin producer among the strains used to make cheddar cheese provides enough nisin to increase the shelf life of pasteurized processed cheese made from it from 14 to 87 days at 22°C (135).

Bacteriocinogenic pediococci appear especially effective in fermented meats. In situ pediocin production by *P. acidilactici* PAC 1.0 during the manufacture of fermented dry sausage reduces *L. monocytogenes* viability >10-fold relative to the acid-induced decrease caused by a nonbacteriocinigenic control (53). When wild-type Bac⁺ *P. acidilactici* H is used to ferment summer sausage, 5,000 AU of pediocin per g is produced, reducing *L. monocytogenes* viability by 3.4 log units (96).

Use of Bacteriocinogenic Starter Cultures to Direct the Fermentation of Fermented Foods

The use of undefined indigenous bacteria to ferment foods compromises product quality, makes true process control difficult, and introduces an uncontrolled variable in the manufacturing of fermented foods. These problems can be overcome by the use of bacteriocinogenic starter cultures. Their ability to outgrow the indigenous microbiota (69, 129, 130, 175) and therefore to facilitate the use of defined starter cultures in unpasteurized foods is an extremely important and promising application.

There are several novel applications of nisin or nisin-producing strains in a variety of fermented foods. The use of a nisin-producing *Lc. lactis* strain paired with nisin-resistant *Leuconostoc mesenteroides* in sauerkraut fermentations retards the growth of *Lactobacillus plantarum* (69). This allows the *Ln. mesenteroides* to establish itself and results in a higher-quality product. The LAB found in wine are very sensitive to nisin and can be inhibited without inhibiting the yeast or influencing the taste (129, 130). At 100 IU/ml, nisin inhibits the bacteria that, by their malolactic fermentation, spoil wine. If the malolactic fermentation is desired, nisin-resistant *Leuconostoc oenos* can be added with nisin. This promotes the malolactic fermentation and suppresses the indigenous LAB (29).

Genetics of LAB Bacteriocins

Location of Bacteriocin Genes

The genetic information encoding bacteriocin production and immunity is located on plasmids, on the chromosome, or both. To show that bacteriocin production is plasmid mediated, researchers must provide (i) phenotypic evidence (e.g., the spontaneous loss of the Bac⁺ phenotype), (ii) physical evidence (e.g., the isolation of a plasmid from the Bac⁺ strain which is missing in Bac⁻ strains), and (iii) genetic confirmation (e.g., the introduction of the plasmid isolated from the Bac⁺ strain into a Bac⁻ strain reestablishes the Bac⁺ phenotype).

Research on lactococcins (86) provides an example of how a group of bacteriocins was determined to be plasmid mediated. Geis et al. (59) screened 280 lactococcal strains for bacteriocin production against four indicator strains. On the basis of secretion into liquid medium, sensitivity to proteolytic enzymes and precipitation by ammonium sulfate, 16 of these strains were classified as producing bacteriocinlike compounds. Some of these Bac⁺ strains easily lost their ability to produce bacteriocin and became sensitive to their own bacteriocin, thus establishing the instability of the Bac⁺ phenotype. Thirteen of these Bac⁺ strains were tested for their ability to transfer the Bac⁺ trait to plasmid-free (Bac⁻) recipient cells; four Bac⁺ transconjugants were detected (118). Plasmid analysis of these transconjugants reveal the presence of a 60-kb or 113-kb plasmid in three of the four Bac⁺ transconjugants. Additional small plasmids were observed in some isolates. Curing experiments of these transconjugants resulted in Bac⁻ variants that had lost their 60- or 113-kb plasmid. Restriction endonuclease analysis showed that two 60-kb plasmids isolated from *Lc. lactis* subsp. *cremoris* strains 9B4 and 4G6 were almost identical. Genetic analysis of the 60-kb plasmid from the 9B4 strain revealed that it carried the genes for three different bacteriocins: lactococcins A, B, and M (165, 166). In addition, pediocin A, pediocin PA-1,

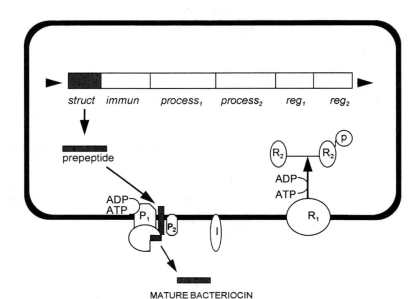

struct immun process₁ process₂ reg₁ reg₂

Figure 30.3 A generic bacteriocin operon. The structural gene (*struct*) codes for a prepropeptide which is modified and excreted by the processing gene products (P_1 and P_2) and may be regulated by a signal transduction pathway encoded by *reg₁* and *reg₂*. For additional explanations, see text.

sakacin A, lactocin, S, and the carnobacteriocins A and B are plasmid encoded.

Nisin production was initially thought to be plasmid mediated. Frequent spontaneous loss of the Nis⁺ phenotype was observed, and conjugal transfer of the Nis⁺ phenotype to Nis⁻ was reported (42, 54, 61, 149). However, no physical evidence linking the Bac⁺ phenotype to a distinctive plasmid was obtained. In addition, attempts to transform Nis⁻ strains with plasmid DNA isolated from Nis⁺ strains did not result in Nis⁺ transformants, although the plasmid pool was successfully transferred (149). By means of nucleic acid hybridization techniques, the nisin structural gene was finally located on the chromosome of *Lc. lactis* (18, 41, 131). The nisin gene resides within a 70-kb conjugative transposon and is genetically linked to the genes encoding sucrose fermentation. The integration of this transposable element into the chromosome of a Nis⁻ strain was observed after conjugation by probing the digested total DNA of transconjugants for different regions of the nisin-coding transposon (72). Other chromosomally encoded LAB bacteriocins are helveticin J (79) and lactacin B (7).

Organization of Bacteriocin Operons: a Generic Operon

The DNA sequence or amino acid composition of most bacteriocins is unknown. However, a general picture of the genetic organization of bacteriocin genes is emerging.

The structural gene encoding bacteriocin production appears to be located in an operonlike structure (85). The organization of a generic operon is shown in Fig. 30.3. The operon is organized so that it clusters the genes containing the structural information together with an array of genes involved in immunity, maturation, processing, and export of the bacteriocin molecules as well as genes encoding products involved in the regulation of bacteriocin biosynthesis. Not all operons contain all of the genes depicted, nor is the organization of the genes in each operon identical, but they share many similarities.

The *structural gene* usually encodes a prepeptide which comprises the precursor of the mature bacteriocin preceded by an N-terminal extension or "leader sequence." The secondary structure of this N-terminal extension is predicted to be an alpha helix that is cleaved during the maturation or export process. Class II bacteriocins encode a prepeptide containing a consensus sequence Gly^{-2}-Gly^{-1} at the cleavage site and present strong homology in their hydrophobicity profile. The role of the N-terminal extension is still unknown. It may be required for recognition by the maturation or export machinery. In this regard, most bacteriocin leader sequences, except for a few such as divergicin A (180), lactococcin 972 (100), propionicin T1 (51), and some enterocins (25), do not exhibit characteristics of a *sec*-dependent export proteins. Rather, they may be exported by a *sec*-independent pathway involving a transport protein encoded within the operon (see below). Alternatively, the N-terminal extension may play a role in neutralizing bacteriocin activity within the cell to protect the producer strain.

The structural genes of several LAB bacteriocins have been described and reviewed (75, 85, 140). These include the structural genes encoding nisin (nisA), pediocin PA-1 (pedA), lactococcins A and B (lcnA, lcnB), sakacin A (sakA), leucocin A (leuA), lactacin F (lafA, lafX), and lactococcin M (lcnM, lcnN).

The immunity gene confers on Bac+ cells immunity to their own bacteriocin. Immunity is specific and seems to be coordinated with bacteriocin production. Several immunity genes have been sequenced, and their ability to confer immunity to the producer strains has been proved (70, 89, 164, 170, 174). In the case of nisin, the nisI gene product is predicted to be an extracellular lipoprotein which, by anchoring to the membrane through its lipid moiety, confers immunity to the producer cells. Additional genes (nisE,-F, and -G) thought to be involved in immunity have been described in the nisin gene cluster (147).

The processing and export genes encode at least two proteins that ensure that the mature bacteriocin is formed and exported from the cytoplasm. One of these proteins is an ATP-binding cassette translocator of the hemolysin B (HlyB) subfamily. These translocators utilize ATP hydrolysis as an energy source for protein translocation and are characterized by the presence of two domains: a membrane-spanning domain and a cytoplasmic ATP-binding domain (50). Examples of genes encoding ATP-binding transporter proteins in LAB bacteriocins are nisT (nisin), lcnC (lactococcin A), and pedD (pediocin PA-1)(44, 101, 155). The second protein of the secretion apparatus is a structural homologue of the hemolysin D (HlyD), an accessory protein which may facilitate transport. The HlyD homologous protein does not seem to be present in the nisin gene cluster.

Some of the ABC transporters found in bacteria that produce class II bacteriocins have similar N-terminal amino acid sequences. This domain may contain the proteolytic cleavage site for the N-terminal extension of these bacteriocins. In contrast, the nisin gene cluster contains the nisP gene. Its gene product, a serine protease, is thought to be involved in the cleavage of nisin's N-terminal extension. In addition, the nisin operon contains the genes nisB and nisC, which encode two membrane-associated proteins involved in catalyzing the posttranslational modifications of the precursor molecules (44, 170).

The regulatory genes encode proteins homologous to proteins of the two-component regulatory systems. In this signal transduction pathway, a histidine kinase located in the membrane senses an external signal and transduces it to the cell's interior by phosphorylating a second, cytoplasmic protein referred to as a "response regulator." The phosphorylation of the response regulator activates the transcription of the bacteriocins' biosynthetic genes. Such regulatory genes occur in the nisin operon (nisKR) (45,170) and have been postulated for the sakacin A operon (sapKR)(6), the carnobacteriocins BM1 and B2 (128), and the plantaricin operon (plnBCD) (39). The bacteriocin molecule itself may be the signal molecule, playing an autoinductive role in a cell-to-cell communication pathway (37).

Genetic "Modification" of Bacteriocins

Natural variants of bacteriocins and advances in protein engineering provide insight into the role of specific amino acids in bacteriocins' biological activity. This information can be used for rational development of bacteriocins with superior activity. The two natural nisin variants, nisin A and nisin Z, differ by one amino acid (His or Asn, respectively, at position 27) but otherwise share the same structure. Nisin Z is more soluble at neutral pH values than nisin A (38). The conserved presence of modified amino residues in lantibiotics may play an important role in biological activity. However, the evidence is not conclusive: (i) while the substitution of $Dha_5 \rightarrow Dhb$ in nisin Z decreases activity 2- to 10-fold (90) and the exchange of $Dha_5 \rightarrow Ala$ in subtilin leads to the loss of inhibitory activity against outgrowth of spores, the latter mutation (both in nisin and subtilin) does not decrease activity against vegetative cells (40, 95); (ii) the exchange of $Dhb_2 \rightarrow Dha$ in nisin increases biological activity (58), but deletion of the Dha_{33} yields a molecule with similar activity to that of the wild-type nisin molecule (20). Rather, it seems that the role of dehydro residue depends on whether the action of the bacteriocin is sporicidal or bactericidal. The unusual amino acids play an important role in maintaining structural elements of the lantibiotics (140). Mutations which disrupt these structures are prone to affect the antimicrobial activity of bacteriocins. Nisin is a flexible molecule in solution except for those regions constricted by the five thioether-closed rings. The nisin molecule may be regarded as a two-fragment peptide connected by a hinge region at amino acids positions 20 to 22. The N-terminal fragment is mainly hydrophobic. The C-terminal fragment seems to adopt an alpha-helix configuration in membranelike environments (171). Mutation of $Asn_{21}/Met_{22} \rightarrow Pro$ reduces the flexibility of the molecule and decreases biological activity (56). Mutations which lead to the opening of the first (destruction of Dha_5) or third rings ($Met_{17} \rightarrow Lys$)

also decrease antimicrobial activity. On the other hand, Dha₅ in subtilin seems to be essential for sporicidal activity (95). This dehydro group may act as Michael acceptor toward sulfhydryl groups in the envelope of germinated spores.

Mechanism of Action against Vegetative Cells

LAB bacteriocins act at the cell membrane. They disrupt the integrity of the cytoplasmic membrane, thus increasing its permeability to small compounds. The addition of bacteriocins to vegetative cells results in a rapid and nonspecific efflux of preaccumulated ions, amino acids, and in some cases (nisin, las 5) but not others (pediocin PA-1), ATP molecules (21, 22, 98, 138, 167, 173, 177, 181). This increased flux of compounds across the membrane rapidly dissipates chemical and electrical gradients across the membrane. The proton motive force (PMF), an electrochemical gradient which serves as the major driving force of many vital energy-dependent processes, is dissipated within minutes of bacteriocin addition (15, 16). The chemical gradient (ΔpH) component of the PMF is dissipated faster than the electrical gradient ($\Delta\Psi$) component, probably as a result of rapid influx of protons into the cytoplasm. Bacteriocin-treated cells have decreased intracellular ATP levels. The loss of ATP can be caused by ATP efflux or ATP hydrolysis. While the efflux mechanism may be a direct result of membrane disruption, hydrolysis could result from a shift in the ATP equilibrium due to P_i efflux or as a futile attempt of the cell to regenerate the PMF (1, 177). Ultimately, these changes in permeability render the cell unable to protect its cytoplasm from the environment. This leads to cell inhibition and, possibly, death.

Given the diversity in the biochemical attributes of the LAB bacteriocin molecules and genetic determinants, it is surprising that they all act by the common mechanism of membrane permeabilization. A structural motif similar to other antimicrobial peptides occurring in nature (see below) may explain their interaction with membranes (122). In this regard, most bacteriocins with known sequences are amphiphilic cationic peptides showing alpha helix, beta sheet, or, in the case of lantibiotics, screwlike, secondary structures. Based on the amphiphilic characteristics of bacteriocins, there are at least two different mechanisms which may explain their membrane-permeabilization action (Fig. 30.4). Bacteriocins may act by a multiple-step poration complex in which bacteriocin monomers bind, insert, and oligomerize in the cytoplasmic membrane to form a pore with the hydrophilic residues facing inward and the hydrophobic ones facing outward. (The small size of bacteriocins makes it is unlikely that one molecule could form a pore.)

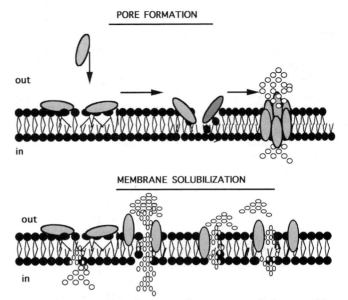

Figure 30.4 Models for pore formation and detergentlike mechanisms of bacteriocin action.

Alternatively, bacteriocins may disrupt the membrane integrity by a detergentlike membrane solubilization action. The action of nisin, pediocin PA-1, lactococcins A and B, and lactacin F in vivo is concentration and time dependent, supporting a poration complex mechanism (1, 22, 164, 173, 177). A generalized solubilization mechanism would result in an all-or-none lysis of the cells, with a sudden collapse of the bioenergetic parameters and no saturation kinetics. Further evidence supporting the mechanism of pore formation is obtained from in vitro observations. Studies in black-lipid membranes suggests that nisin, Pep 5, and other lantibiotics form transient, potential-dependent pores with a predicted diameter of 0.2 to 2 nm and lifetimes from milliseconds to seconds (139). Moreover, the addition of bacteriocins to membrane vesicles or liposomes loaded with different-size probes does not always result in leakage of the probe. Rather, leakage depends on both the size of the probe and the amount of bacteriocin added (22, 56).

Bacteriocins act similarly in that they breach the permeability barrier of the cytoplasmic membrane. However, they may require different conditions in the membrane target to establish a successful interaction. The lantibiotic bacteriocins act on energized membranes (55, 88). A recent study suggests that the N-terminal part of the nisin molecule is involved in its insertion into the lipid bilayer, while the C-terminal moiety could be involved in the initial interaction with the membrane surface (92). Prior to forming pores, nisin has to first bind to the peptidoglycan precursor lipid II, which serves as a nisin docking molecule in the membrane structure (14). Alternatively, most nonlantibiotic bacteriocins act on nonenergized membranes but seem to require a membrane receptor protein (22, 164, 173). Mutation in the *rpoN* gene, which codes for σ^{54} in *Enterococcus faecalis* JH2-2, causes resistance to group IIa bacteriocins (31). This observation allows one to speculate on a possible role for the *rpoN* gene product in the docking and/or recognition of the target cell by a IIa bacteriocin. The activity of some bacteriocins (including lactacin F, lactococcins M and G, and plantaricins EF and JK) occurs through the action of two peptides. These two-component bacteriocins are also postulated to form pores in the cytoplasmic membrane (85, 86, 109).

Mechanism of Action against Spores

Information about the mechanisms of bacteriocin action against spores deals primarily with nisin. Spore germination and outgrowth is a multistep process detailed elsewhere in this book. Nisin allows spores to germinate but inhibits outgrowth of the preemergent spore. Heat resistance and refractility under phase microscopy are lost at this point. The growth of vegetative botulinal cells is inhibited at much lower nisin concentrations than those required to inhibit outgrowth of spores. *Bacillus* species whose spore coats are opened by mechanical pressure are much more nisin sensitive than species whose spore coats are opened by lysis (93). At a molecular level, nisin modifies the sulfhydryl groups in the envelopes of germinated spores (113), presumably because the dehydro residues of nisin act as electron acceptors (63, 64).

Similarity to Other Antimicrobial Proteins

The antimicrobial proteins of LAB are not unique in structure or function. Many very different organisms produce similar "cytolytic pore-forming proteins" (122), including the colicins, cecropins, melittins, magainins, and defensins. In many cases, these have been better studied than the bacteriocins produced by LAB. Their widespread occurrence attests to the "naturalness" of bacteriocins; nature is full of similar antimicrobial proteins.

A brief review of the properties of these other antimicrobial proteins provides an important opportunity to compare and contrast them with the bacteriocins produced by LAB. Colicins made by *E. coli* have molecular masses of 35 to 70 kDa and form ion-permeable channels in the cytoplasmic membrane of sensitive bacteria (87). Cecropins are small (~25 amino acids) peptides of the moth *Hyalophora cecropia* which are active against gram-negative and gram-positive bacteria (4). Cecropins

form voltage-dependent ion channels in planar lipid membranes (23) and contain amphiphilic helices consistent with the structural requirements for membrane insertion (150). Melittin is the major toxic component of bee venom. It contains 26 amino acids, with the N-terminal region rich in hydrophobic residues and the C-terminal end predominantly hydrophilic (161). Melittin associates with lipid bilayers, causing membrane lysis, permeabilization, and inhibition of membrane-bound enzyme systems (66). Magainins are antimicrobial peptides of the frog *Xenopus laevis* and contain approximately 23 amino acids which exert their microbicidal action by permeabilizing sensitive membranes by generalized membrane destabilization, the formation of channels, or both mechanisms simultaneously (62). Defensins are small antimicrobial peptides found in cytoplasmic granules of phagocytes isolated from humans, rabbits, guinea pigs, and rats (43, 91). These small (29 to 34 amino acids) peptides have similar amino acid sequences, three-dimensional structures, and a highly conserved cysteine motif consisting of six cysteine residues forming three disulfide bonds (91). Defensins also exert their antimicrobial activity by permeabilization or disruption of cell membranes and play a major role in the oxygen-independent killing of bacteria, fungi, and certain viruses. Clearly, antimicrobial proteins are common in the biosphere and are used by many organisms to protect themselves.

Unresolved Issues

Bacteriocin Resistance

While nisin-resistant starter cultures can be used to great advantage, the appearance of nisin-resistant bacteria can undermine the use of nisin as an inhibitor of spoilage and pathogenic bacteria, Genetically stable nisin-resistant *L. monocytogenes* can be isolated at a frequency of 10^{-6} (68, 107). Nisin-resistant isolates are generated from vegetative cells of *S. aureus*, *B. licheniformis*, *B. subtilis*, *Bacillus cereus* (107), and *C. botulinum* (103) at similar frequencies. The use of multiple bacteriocins to overcome this problem has been suggested (49, 67) but will be effective only if resistance to each bacteriocin is conferred by different mechanisms. *L. monocytogenes* resistant to mesenterocin 52, curvaticin 13, or plantaricin C19 can be isolated at frequencies of 10^{-3} to 10^{-8}. Strains resistant to any of these bacteriocins are cross resistant to the other two. When all three bacteriocins are used together, resistant strains are isolated at frequencies similar to those obtained when the bacteriocins are used alone (133). Isolates resistant to all three of these bacteriocins, however, are not resistant to nisin.

Table 30.1 Parallel mechanisms of antibiotic and bacteriocin resistance[a]

Mechanism of resistance	Specificity	Example	
		Antibiotics	Bacteriocins
Destruction	Specific or general	β-Lactamases	Protease, specific "bacteriocinase," nisinase (93)
Modification	Specific	Methylation of aminoglycosides	Dehydroreductase (78)
Altered receptors	Specific	Penicillin-binding proteins	Probable, but not reported to date
Membrane composition	General	Altered membranes in resistant E. coli and bacilli	Demonstrated for nisin resistance (107)

[a] Adapted from reference 112.

Some mechanisms through which bacteria can become bacteriocin resistant parallel mechanisms of antibiotic resistance (112) (Table 30.1). This comparison is made to provide a conceptual context for bacteriocin resistance and is not meant to suggest that bacteriocin-resistant foodborne pathogens will be cross-resistant to antibiotics. Mechanisms characterized as "specific" have a biochemical specificity to a particular antimicrobial agent and cannot generate cross-resistance. General mechanisms of resistance, such as changes in membrane permeability, can be viewed as affecting the intrinsic resistance of the bacteria and might cause cross-resistance to other food preservatives. Membranes from nisin-resistant L. monocytogenes have more straight-chain fatty acids, resulting in a higher phase transition temperature (T_c) compared with the wild-type strain (107). Presumably, the lack of fluidity hinders nisin insertion into the membrane. Membrane fluidity plays an important role in listerial resistance to other antimicrobial agents (81). L. monocytogenes grown in the presence of $C_{14:0}$ or $C_{18:0}$ fatty acids have higher T_c and increased resistance to four common antimicrobial agents relative to cells grown in the presence of $C_{18:1}$, which have lower T_c and are more sensitive to other antimicrobial agents.

An important issue that must be addressed is potential cross-resistance between bacteriocins and antibiotics. Bacteriocins are different from antibiotics. They have a different mechanism of action. Therefore, cells respond to different stresses in different ways. This statement is confirmed by the sensitivity of multidrug-resistant bacteria to nisin (146). In order to survive, microorganisms can develop resistance to any antimicrobial substance used to preserve food. However, use of intelligently designed multiple-hurdle technology not only allows inhibition of bacterial growth but also makes bacteriocin-resistant cells sensitive again (108). In addition, bacteriocin-resistant bacteria may be even more sensitive to other hurdles used in food preservation (120).

The study of bacteriocin resistance is in its infancy. Under conditions closer to those of actual applications, the frequency of bacteriocin resistance may be much lower than that obtained in laboratory media optimal for growth (68, 103). Nonetheless, it is prudent to conduct additional research on bacteriocin resistance so that it might be advantageously manipulated. "Resistance management" is already a component in other antimicrobial applications, ranging from the use of antibiotics in hospitals to the use of Bacillus thuringiensis toxin as an agricultural insecticide.

Regulatory Status

The regulatory status of bacteriocins and bacteriocinogenic bacteria is unclear. LAB are generally recognized as safe (GRAS) for the production of fermented foods. GRAS status, which is conferred by the U.S. Food and Drug Administration (FDA), is especially desirable because it allows a compound to be used in a specific application without additional regulatory approval. The linkage of GRAS status to a specific application is often overlooked. Thus, the GRAS status of LAB for the production of fermented foods does not automatically make LAB, or their metabolic products, GRAS for uses such as the preservation of foods which are not fermented. Nisin is the only bacteriocin that has GRAS affirmation. The 1988 GRAS affirmation for the use of nisin in pasteurized processed cheese (52) was supported by toxicological data. This affirmation is the foundation for additional GRAS affirmations. The FDA has accepted for filing a request that nisin be affirmed as GRAS to restore the shelf life of reduced-fat pasteurized egg. A subsequent request (52a) to extend the affirmation to conventional processed eggs was also accepted for filing. Both of these requests rely on the toxicological and safety data in the 1988 affirmation and are subject to its 10,000-IU/g (250-ppm) usage limit.

Bacteriocins produced by GRAS organisms are not automatically GRAS themselves. Bacteriocins that are not GRAS are regulated as food additives (106) and require premarket approval by the FDA. A food fermented by bacteriocin-producing starters can be used as an ingredient in a second food product. Its use *as an ingredient* might coincidentally extend shelf life of the product without necessitating preservative declarations. However, if the ingredient is added *for the purpose of extending shelf life*, the FDA would probably consider it an additive and require both premarket clearance and label declaration. There is no doubt that purified bacteriocins used as preservatives require premarket approval by the FDA.

GRAS status for bacteriocins can be based on documented use prior to 1958, a consensus of scientific opinion, or a formal GRAS affirmation from the FDA. The presence of bacteriocins in foods prior to 1958 might be inferred from the ease with which bacteriocinigenic bacteria are isolated from a variety of foods. This suggests that they are long-standing members of the natural microflora of food. Furthermore, strains that produce nisin (77, 96, 136) and strains that produce pediocin PA-1 (11, 28, 57, 69) have been independently isolated from foods in different parts of the world. This demonstrates widespread occurrence of bacteriocinigenic bacteria in nature and suggests that bacteriocins have been consumed for decades. While these are reasonable arguments, the FDA might nonetheless require isolation of both the bacteriocinogenic organism and its bacteriocin from food produced prior to 1958 before accepting them as GRAS under the prior use clause.

The international regulation of bacteriocins is complex and beyond the scope of this book. Nisin is the only bacteriocin approved internationally for use in foods. The Joint Food and Agriculture/World Health Organization accepted nisin as a food additive in 1969 and set maximum intake levels as 33,000 IU per kg body weight. Based on this, many countries allow nisin in a variety of products, sometimes with no restrictions as to maximum level. In addition to milk, cheese, and dairy products, these uses include canned tomatoes, canned soups, other canned vegetables, mayonnaise, and baby food (168).

CONCLUSIONS AND OUTLOOK FOR THE FUTURE

The industrial processing and preservation of foods did not start until the late 1800s. From this time until the 1940s, the food industry was driven by physical methods for food preservation. From the 1950s until the present, chemical methods of food preservation became increasingly important, gaining a status equal to that of engineering. The current advances in nonnutritive sweeteners, fat mimetics, and functional polysaccharides are a testament to the ascendance of food chemistry. In contrast, biological methods of food preservation are still in their infancy. The first reports of genetic recombination (27), of transduction in LAB, and the importance of their plasmids occurred only in the 1970s (104, 105).

Table 30.2 Analogy betweeen the use of insecticides in production agriculture and the use of antimicrobial agents for food safety[a]

Control of insects in crops	Control of bacteria in foods
Chemical pesticides	Chemical preservatives
Integrated pest management	Microbial competition
Bioinseticides	Bacteriocins
Insecticidal plants	Antimicrobial foods

[a] Modified from reference 110.

The "shock wave" from the explosion of knowledge in molecular biology is just now hitting food microbiology. In the early 1970s, one could imagine a resurrected Pasteur entering a food microbiology laboratory and resuming his work after the briefest of orientations. This is not true of the third millenium. Our increasing reliance on the concepts and tools of molecular biology would leave Pasteur completely befuddled.

Having made its mark in analytical food microbiology, it is only a matter of time before molecular biology provides new methods for ensuring food quality and safety. The biological methods of food preservation covered in this chapter mark only the crude beginning of this new era in the life of the food industry, the era of food biology. Controlled acidification is conceptually straightforward, but its successful application depends on a variety of product-specific factors. This has limited both its commercial use and academic interest in controlled acidification. The use of antimicrobial proteins, in one form or another, is sure to increase. In production agriculture, *B. thuringiensis* insecticidal proteins have been applied to plants for the last 20 years. The genes for this protein are now being cloned into the plant itself (8). By analogy (Table 30.2), the day may come when we can genetically engineer pathogen resistance into microbially sensitive foods.

This is manuscript F10974-2-96 of the New Jersey Agricultural Experiment Station. Research in the authors' laboratory and preparation of this manuscript were supported by state appropriations, U.S. Hatch Act Funds, the U.S.-Israel Bilateral Research and Development Fund (BARD, project no. US-2113-92), and grants from the U.S. Department of Agriculture CSRS NRI Food Safety Program (nos. 91-37201-6796 and 94-37201-0994).

References

1. Abee, T., F. M. Rombouts, J. Hugenholtz, G. Guihard, and L. Letellier. 1994. Mode of action of nisin Z against *Listeria monocytogenes* Scott A grown at high and low temperatures. *Appl. Environ. Microbiol.* 60:1962–1968.

2. Allison, G., C. Fremaux, C. Ahn, and T. R. Klaenhammer. 1994. Expansion of bacteriocin activity and host range

upon complementation of two peptides encoded within the lactacin F operon. *J. Bacteriol.* **176**:2235–2241.

3. Al-Zoreky, N., J. W. Ayres, and W. Sandine. 1991. Antimicrobial activity of Microgard® against food spoilage and pathogenic microorganisms. *J. Dairy Sci.* **74**:758–763.

4. Andreu D., and R. B. Merrifield. 1985. N-terminal analogues of cecropin A: synthesis, antibacterial activity, and conformation properties. *Biochemistry* **24**:1683–1688.

5. Aplin & Barrett Ltd. http://www.aplin-barrett.co.uk/nisaplin_technical.htm.

6. Axelsson, L., and A. Holck. 1995. Molecular analysis of sakacin A gene cluster. Presented at Workshop on the Bacteriocins of Lactic Acid Bacteria—Applications and Fundamentals, Alberta, Canada, April 17–22, 1995.

7. Barefoot, S. F., and T. R. Klaenhammer. 1984. Purification and characterization of the *Lactobacillus acidophilus* bacteriocin lactacin B. *Antimicrob. Agents Chemother.* **26**:328–334.

8. Barton, B., M. Miller, M. Maffitt, J. Kofron, S. Cannon, and P. Umbeck. 1989. Development of insect resistant plants. *Dev. Ind. Microbiol.* **30**:195–202.

9. Bell, R. G., and K. M. De Lacy. 1985. The effect of nisin-sodium chloride interactions on the outgrowth of *Bacillus licheniformis* spores. *J. Appl. Bacteriol.* **59**:127–132.

10. Berry, E. D., R. W. Hutkins, and R. Mandigo. 1991. The use of bacteriocin producing *Pediococcus acidilactici* to control post processing *Listeria monocytogenes* contamination of frankfurters. *J. Food Prot.* **54**:681–686.

11. Bhunia, A. K., and M. C. Johnson. 1992. Monoclonal antibody-colony immunoblot method specific for isolation of *Pediococcus acidilactici* from foods and correlation with pediocin (bacteriocin) production. *Appl. Environ. Microbiol.* **58**:2315–2320.

12. Bouksaim, M., C. Lacroix, R. Bazin, and R. E. Simard. 1999. Production and utilization of polyclonal antibodies against nisin in an ELISA and for immunolocation of nisin in producing and sensitive bacterial strains. *J. Appl. Microbiol.* **87**:500–510.

13. Bower, C. K., J. McGuire, and M. A. Daeschel. 1995. Suppression of *Listeria monocytogenes* colonization following adsorption of nisin onto silica surfaces. *Appl. Environ. Microbiol.* **61**:992–997.

14. Breukink, E., I. Wiedemann, C. van Kraaij, O. P. Kuipers, H. Sahl, and B. de Kruijff. 1999. Use of the cell wall precursor lipid II by a pore-forming peptide antibiotic. *Science* **286**:2361–2364.

15. Bruno, M. E. C., A. Kaiser, and T. J. Montville. 1992. Depletion of proton motive force by nisin in *Listeria monocytogenes* cells. *Appl. Environ. Microbiol.* **58**:2255–2259.

16. Bruno M. E. C., and T. J. Montville. 1993. Common mechanistic action of bacteriocins from lactic acid bacteria. *Appl. Environ. Microbiol.* **59**:3003–3010.

17. Buchanan, R. L., and L. A. Klawitter. 1992. Effectiveness of *Carnobacterium piscicola* LK5 for controlling the growth of *Listeria monocytogenes* Scott A in refrigerated foods. *J. Food Safety* **12**:217–224.

18. Buchman, G. W., S. Banergee, and J. N. Hansen. 1988. Structure, expression and evolution of a gene encoding the precursor of nisin, a small protein antibiotic. *J. Biol. Chem.* **263**:16260–16266.

19. Buyong, N., J. Kok, and J. B. Luchansky. 1998. Use of a genetically enhanced, pediocin-producing starter culture, *Lactococcus lactis* subsp. *lactis* MM217, to control *Listeria monocytogenes* in cheddar cheese. *Appl. Environ. Microbiol.* **64**:4842–4845.

20. Chan, W. C., B. W. Bycroft, L. Y. Lian, and G. C. Roberts. 1989. Isolation and characterization of two degradation products derived from the peptide antibiotic nisin. *FEBS Lett.* **252**:29–36.

21. Chen, Y., and T. J. Montville. 1995. Efflux of ions and ATP depletion induced by pediocin PA-1 are concomitant with cell death in *Listeria monocytogenes* Scott A. *J. Appl. Bacteriol.* **79**:684–690.

22. Chikindas, M. L., M. J. Garcia-Garcera, A. J. M. Driessen, A. M. Ledeboer, J. Nissen-Meyer, I. F. Nes, T. Abee, W. N. Konings, and G. Venema. 1993. Pediocin PA-1, a bacteriocin from *Pediococcus acidilactici* PAC1.0, forms hydrophilic pores in the cytoplasmic membrane of target cells. *Appl. Environ. Microbiol.* **59**:3577–3584.

23. Christensen, B., J. Fink, R. B. Merrifield, and D. Mauzerall. 1988. Channel-forming properties of cecropins and related model compounds incorporated into planar lipid membranes. *Proc. Natl. Acad. Sci. USA* **85**:5072–5076.

24. Chung, K. T., J. S. Dickson, and J. D. Crouse. 1989. Effects of nisin on growth of bacteria attached to meat. *Appl. Environ. Microbiol.* **55**:1329–1333.

25. Cintas, L. M., P. Casaus, C. Herranz, L. S. Havarstein, H. Holo, P. E. Hernandez, and I. F. Nes. 2000. Biochemical and genetic evidence that *Enterococcus faecium* L50 produces enterocins L50A and L50B, the *sec*-dependent enterocin P, and a novel bacteriocin secreted without an N-terminal extension termed enterocin Q. *J. Bacteriol.* **182**:6806–6814.

26. Cociancich, S., M. Goyffon, F. Bontems, P. Bulet, F. Bouet, A. Menez, and J. Hoffman. 1993. Purification and characterization of a scorpion defensin, a 4kDa antibacterial peptide presenting structural similarity with insect defensins and scorpion toxins. *Biochem. Biophys. Res. Commun.* **194**:17–22.

27. Cohen, S. N. A. C. Chang, H. W. Boyer, and R. B. Helling. 1973. Construction of biologically functional bacterial plasmids *in vitro*. *Proc. Natl. Acad. Sci. USA* **70**:3240–3244.

28. Daba, H., C. Lacroix, J. Huang, R. E. Simard, and L. Lemieux. 1994. Simple method of purification and sequencing of a bacteriocin produced by *Pediococcus acidilactici* UL5. *J. Appl. Bacteriol.* **77**:682–698.

29. Daeschel, M. A. 1990. Controlling wine malolactic fermentation with nisin and nicin-resistant strains of *Leuconostoc oenos*. *Appl. Environ. Microbiol.* **51**:601–603.

30. Daeschel, M. A., J. McGuire, and H. Al-Makhlafi. 1992. Antimicrobial activity of nisin adsorbed to hydrophilic and hydrophobic silicon surfaces. *J. Food Prot.* **55**:731–735.

31. Dalet, K., C. Briand, Y. Cenatiempo, and Y. Hechard. 2000. The *rpoN* gene of *Enterococcus faecalis* directs sensitivity to subclass IIa bacteriocins. *Curr. Microbiol.* 41:441–443.

32. Degnan, A. J., N. Buyong, and J. B. Luchansky. 1993. Antilisterial activity of pediocin AcH in model food systems in the presence of an emulsifier or encapsulated within liposomes. *Int. J. Food Microbiol.* 18:127–138.

33. Degnan, A. J., and J. B. Luchansky. 1992. Influence of beef tallow and muscle on the antilisterial activity of pediocin AcH and liposome-encapsulated pediocin AcH. *J. Food Prot.* 55:552–554.

34. Degnan, A. J., A. E. Yousef, and J. B. Luchansky. 1992. Use of *Pediococcus acidilactici* to control *Listeria monocytogenes* in temperature-abused vacuum-packaged wieners. *J. Food Prot.* 55:98–103.

35. Delves-Broughton, J., G. C. Williams, and S. Williamson. 1992. The use of the bacteriocin, nisin, as a preservative in pasteurized white egg. *Lett. Appl. Bacteriol.* 15:133–136.

36. de Ruyter, P. G., O. P. Kuipers, and W. M. de Vos. 1996. Controlled gene expression systems for *Lactococcus lactis* with the food-grade inducer nisin. *Appl. Environ. Microbiol.* 62:3662–3667.

37. de Vos, W. M. 1995. Expression of bacteriocin genes: genetic and physiological control. Presented at Workshop on the Bacteriocins of Lactic Acid Bacteria—Applications and Fundamentals, Alberta, Canada, April 17–22, 1995.

38. de Vos, W. M., J. W. M. Mulders, R. J. Siezen, J. Hugenhoetz, and O. P. Kuipers. 1993. Properties of nisin Z and distribution of its gene, *nisZ*, in *Lactococcus lactis. Appl. Environ. Microbiol.* 59:3683–3693.

39. Diep, D. B., L. S. Havarstein, J. Nissen-Meyer, and I. F. Nes. 1994. The gene encoding plantaricin A, a bacteriocin from *Lactobacillus plantarum* C11, is located on the same transcription unit as an *agr*-like regulatory system. *Appl. Environ. Microbiol.* 60:160–166.

40. Dodd, H. M., and M. J. Gasson. 1994. p. 211–251. *In* M. J. Gasson and W. M. DeVos (ed.), *Genetics and Biotechnology of Lactic Acid Bacteria.* Chapman & Hall, Ltd., London, England.

41. Dodd, H. M., N. Horn, and M. J. Gasson. 1990. Analysis of the genetic determinant for the production of the peptide antibiotic nisin. *J. Gen. Microbiol.* 136:555–556.

42. Donkersloot, J. A., and J. Thompson. 1990. Simultaneous loss of N^5-(carboxyethyl)ornithine synthase, nisin production, and sucrose-fermenting ability by *Lactococcus lactis* K1. *J. Bacteriol.* 172:4122–4126.

43. Elsbach, P. 1990. Antibiotics from within: antibacterials from human and animal sources. *Trends Biotechnol.* 8:26–30.

44. Engelke, G., Z. Gutowski-Eckel, M. Hammelmann, and K. D. Entian. 1992. Biosynthesis of the lantibiotic nisin: genomic organization and membrane localization of the NisB protein. *Appl. Environ. Microbiol.* 58:3730–3743.

45. Engelke, G., Z. Gutowski-Eckel, P. Kiesau, K. Siegers, M. Hammelmann, and K. D. Entian. 1994. Regulation of nisin biosynthesis and immunity in *Lactococcus lactis* 6F3. *Appl. Environ. Microbiol.* 60:814–825.

46. Ennahar, S., O. Assobhel, and C. Hasselmann. 1998. Inhibition of *Listeria monocytogenes* in a smear-surface soft cheese by *Lactobacillus plantarum* WHE 92, a pediocin AcH producer. *J. Food Prot.* 61:186–191.

47. Ennahar, S., T. Sashihara, K. Sonomoto, and A. Ishizaki. 2000. Class IIa bacteriocins: biosynthesis, structure and activity. *FEMS Microbiol. Rev.* 24:85–106.

48. Fang, T. J., and L. W. Lin. 1994. Inactivation of *Listeria monocytogenes* on raw pork treated with modified atmosphere packaging and nisin. *J. Food Drug Anal.* 2:189–200.

49. Farber, J. M. 1993. Current research on *Listeria monocytogenes* in foods: an overview. *J. Food Prot.* 56:640–643.

50. Fath, M. J., and R. Kolter. 1993. ABC transporters: bacterial exporters. *Microbiol. Rev.* 57:995–1017.

51. Faye, T., T. Langsrud, I. F. Nes, and H. Holo. 2000. Biochemical and genetic characterization of propionicin T1, a new bacteriocin from *Propionibacterium thoenii. Appl. Environ. Microbiol.* 66:4230–4236.

52. Federal Register. 1988. Nisin preparation: affirmation of GRAS status as a direct human food ingredient. 21 CFR Part 184. *Fed. Regist.* 53:11247–11251.

52a. Federal Register. 1994. *Fed. Regist.* 59:42277–42278.

53. Foegeding, P. M., A. B. Thomas, D. H. Pinkerton, and T. R. Klaenhammer. 1992. Enhanced control of *Listeria monocytogenes* by in situ-produced pediocin during dry fermented sausage production. *Appl. Environ. Microbiol.* 58:884–890.

54. Fuchs, P. G., J. Zajdel, and W. T. Dobrzanski. 1975. Possible plasmid nature of the determinant for production of the antibiotic nisin in some strains of *Streptococcus lactis. J. Gen. Microbiol.* 88:189–192.

55. Gao, F. H., T. Abee, and W. N. Konings. 1991. The mechanism of action of the peptide antibiotic nisin in liposomes and cytochrome *c* oxidase proteoliposomes. *Appl. Environ. Microbiol.* 57:2164–2170.

56. Garcia-Garcera, M. J. G., G. L. Elferink, J. M. Driessen, and W. N. Konings. 1993. In vitro pore-forming activity of the lantibiotic nisin. Role of PMF force and lipid composition. *Eur. J. Biochem.* 212:417–422.

57. Garver, K. I., and P. M. Muriana. 1993. Detection, identification and characterization of bacteriocin-producing lactic acid bacteria from retail food products. *Int. J. Food Microbiol.* 19:241–258.

58. Gasson, M. J., H. M. Dodd, and N. Horn. 1995. Protein engineering of bacteriocins. Presented at Workshop on the Bacteriocins of Lactic Acid Bacteria—Applications and Fundamentals, Alberta, Canada, April 17–22, 1995.

59. Geis A., J. Singh, and M. Teuber. 1983. Potential of lactic streptococci to produce bacteriocins. *Appl. Environ. Microbiol.* 45:205–211.

60. Gonzalez, C. F. 1988. Method for inhibiting bacterial spoilage and composition for this purpose. European patent application 88101624.

61. Gonzalez, C. F., and B. S. Kunka. 1985. Transfer of sucrose-fermenting ability and nisin production phenotype among lactic streptococci. *Appl. Environ. Microbiol.* 49:627–633.

62. Grant, E., Jr., T. J. Beeler, K. M. P. Taylor, K. Gable, and M. A. Roseman. 1992. Mechanism of magainin 2a induced permeabilization of phospholipid vesicles. *Biochemistry* **31**:9912–9918.

63. Gross, E., and J. L. Morell. 1967. The presence of dehydroalanine in the antibiotic nisin and its relationship to activity. *J. Am. Chem Soc.* **89**:2791–2792.

64. Gross, E., and J. L. Morell. 1971. The structure of nisin. *J. Am. Chem. Soc.* **93**:4634–4635.

65. Guder, A., I. Wiedemann, and H. G. Sahl. 2000. Post-translationally modified bacteriocins—the lantibiotics. *Biopolymers* **55**:62–73.

66. Haberman, E. 1972. Bee and wasp venoms. *Science* **177**:314–322.

67. Hanlin, M. B., N. Kalchayan, P. Ray, and B. Ray. 1993. Bacteriocins of lactic acid bacteria in combination have greater antibacterial activity. *J. Food Prot.* **56**:252–255.

68. Harris, L. J., H. P. Fleming, and T. R. Klaenhammer. 1991. Sensitivity and resistance of *Listeria monocytogenes* ATCC 19115 Scott A and VAL 500 to nisin. *J. Food Prot.* **54**:836–840.

69. Harris, L. J., H. P. Fleming, and T. R. Klaenhammer. 1992. Novel paired starter culture system for sauerkraut, consisting of a nisin resistant *Leuconostoc mesenteroides* strain and a nisin-producing *Lactococcus lactis* strain. *Appl. Environ. Microbiol.* **58**:1484–1489.

70. Holo, H., O. Nissen, and I. F. Nes. 1991. Lactococcin A, a new bacteriocin from *Lactococcus lactis* subsp. *cremoris*: isolation and characterization of the protein and its gene. *J. Bacteriol.* **173**:3879–3887.

71. Hoover, G., and L. R. Steenson. 1993. *Bacteriocins of Lactic Acid Bacteria*. Academic Press, Inc., New York, N.Y.

72. Horn, N., S. Swindell, H. Dodd, and M. Gasson. 1991. Nisin biosyntheis genes are encoded by a novel conjugative transposon. *Mol. Gen. Genet.* **228**:129–135.

73. Hurst, A. 1981. Nisin. *Adv. Appl. Microbiol.* **27**:85–123.

74. Hutton, M. T., P. A. Chehak, and J. H. Hanlin. 1991. Inhibition of botulism toxin production by *Pediococcus acidilactici* in temperature abused refrigerated foods. *J. Food Safety* **11**:255–267.

75. Jack, R. N., J. R. Tagg, and B. Ray. 1995. Bacteriocins of gram-positive bacteria. *Microbiol. Rev.* **59**:171–200.

76. Jack, R. W., and G. Jung. 2000. Lantibiotics and microcins: polypeptides with unusual chemical diversity. *Curr. Opin. Chem. Biol.* **4**:310–317.

77. Jager, K., and S. Harlander. 1992. Characterization of a bacteriocin from *Pediococcus acidilactici* PC and comparison of bacteriocin-producing strains using molecular typing procedures. *Appl. Microbiol. Biotechnol.* **37**:631–637.

78. Jarvis, B., and J. Farr. 1971. Partial purification, specificity and mechanism of the nisin-inactivating enzyme from *Bacillus cereus*. *Biochim. Biophys. Acta* **227**:232–240.

79. Joerger M. C., and T. R. Klaenhammer. 1986. Characterization and purification of helveticin J and evidence for a chromosomally determined bacteriocin produced by *Lactobacillus helveticus* 481. *J. Bacteriol.* **167**:439–446.

80. Jones, L. W. 1974. Effect of butterfat on inhibition of *Staphylococcus aureus* by nisin. *Can. J. Microbiol.* **20**:1257–1260.

81. Juneja, V. K., and P. M. Davidson. 1993. Influence of altered fatty acid composition on resistance of *Listeria monocytogenes* to antimicrobials. *J. Food Prot.* **56**:302–305.

82. Kalchayanand, N., M. B. Hanlin, and B. Ray. 1992. Sublethal injury makes Gram-negative and resistant Gram-positive bacteria sensitive to the bacteriocins, pediocin AcH and nisin. *Lett. Appl. Microbiol.* **15**:239–243.

83. Kalchayanand, N., T. Sikes, C. P. Dunne, and B. Ray. 1994. Hydrostatic pressure and electroporation have increased bactericidal efficiency in combination with bacteriocins. *Appl. Environ. Microbiol.* **60**:4174–4177.

84. Klaenhammer, T. R. 1988. Bacteriocins of lactic acid bacteria. *Biochimie* **70**:337–349.

85. Klaenhammer, T. R. 1993. Genetics of bacteriocins produced by lactic acid bacteria. *FEMS Microbiol. Rev.* **12**:39–86.

86. Kok, J., H. Holo, M. J. van Belkum, A. J. Haandrikman, and I. F. Nes. 1993. Nonnisin bacteriocins in lactococci: biochemistry, genetics and mode of action, p. 121–150. *In* D. G. Hoover and L. R. Steenson (ed.), *Bacteriocins of Lactic Acid Bacteria*. Academic Press, Inc., New York, N.Y.

87. Konisky, J. 1982. Colicins and other bacteriocins with established modes of action. *Annu. Rev. Microbiol.* **36**:125–144.

88. Kordel, M., and H. G. Sahl. 1986. Susceptibility of bacterial, eukaryotic and artificial membranes to the disruptive action of the cationic peptide Pep 5 and nisin. *FEMS Microbiol. Lett.* **34**:139–144.

89. Kuipers, O. P., M. M. Beerthuyzen, R. J. Siezen, and W. M. de Vos. 1993. Characterization of the nisin gene cluster *nisABTCIPR* of *Lactococcus lactis*: requirement of expression of the *nisA* and *nisI* gene for producer immunity. *Eur. J. Biochem.* **216**:281–292.

90. Kuipers, O. P., H. S. Rollema, W. M. Yap, H. J. Boot, R. J. Siezen, and W. M. de Vos. 1992. Engineering dehydrated amino acid residues in the antimicrobial peptide nisin. *J. Biol. Chem.* **267**:2430–2436.

91. Lehrer, R. I., A. Barton, K. A. Daher, S. S. L. Harwig, T. Ganz, and M. E. Selsted. 1989. Interaction of human defensins with *Escherichia coli*, mechanism of bactericidal activity. *J. Clin. Invest.* **84**:553–561.

92. Lins, L., P. Ducarme, E. Breukink, and R. Brasseur. 1999. Computational study of nisin interaction with model membrane. *Biochim. Biophys. Acta* **1420**:111–120.

93. Lipinska, E. 1977. Nisin and its applications, p. 103–130. *In* M. Woodbine (ed.), *Antibiotics and Antibiosis in Agriculture*. Butterworths, London, England.

94. Liu, W., and N. Hansen. 1990. Some chemical and physical properties of nisin, a small protein antibiotic produced by *Lactococcus lactis*. *Appl. Environ. Microbiol.* **56**:2551–2558.

95. Liu, W., and N. Hansen. 1992. Enhancement of the chemical and antimicrobial properties of subtilin by site-directed mutagenesis. *J. Biol. Chem.* **267**:25078–25085.

96. **Luchansky, J. B., K. A. Glass, K. D. Harrsono, A. J. Degnan, N. G. Faith, B. Cauvin, G. Bascus-Taylor, K. Arihara, B. Bater, A. J. Maurer, and R. G. Cassers.** 1992. Genomic analysis of *Pediococcus* starter cultures used to control *Listeria monocytogenes* in turkey summer sausage. *Appl. Environ. Microbiol.* **58:**3053–3059.

97. **Lyon, W. J., J. E. Sethi, and B. A. Glatz.** 1993. Inhibition of psychrotrophic organisms by propionicin PLG-1, a bacteriocin produced by *Propionibacterium thoenii*. *J. Dairy Sci.* **76:**1506–1513.

98. **Maftah, A., D. Renault, C. Vignoles, Y. Hechard, P. Bressollier, M. H. Ratinaud, Y. Cenatiempo, and R. Julien.** 1993. Membrane permeabilization of *Listeria monocytogenes* and mitochondria by the bacteriocin mesentericin Y105. *J. Bacteriol.* **175:**3232–3235.

99. **Martinez, J. M., M. I. Martinez, C. Herranz, A. Suarez, M. F. Fernandez, L. M. Cintas, J. M. Rodriguez, and P. E. Hernandez.** 1999. Antibodies to a synthetic 1-9-N-terminal amino acid fragment of mature pediocin PA-1: sensitivity and specificity for pediocin PA-1 and cross-reactivity against Class IIa bacteriocins. *Microbiology* **145:**2777–2787.

100. **Martinez, B., M. Fernandez, J. E. Suarez, and A. Rodriguez.** 1999. Synthesis of lactococcin 972, a bacteriocin produced by *Lactococcus lactis* IPLA 972, depends on the expression of a plasmid-encoded bicistronic operon. *Microbiology* **145:**3155–3161.

101. **Marugg, J. D., C. F. Gonzalez, B. S. Kunka, A. M. Ledeboer, M. J. Pucci, M. Y. Toonen, S. A. Walker, L. C. M. Zoetmulder, and P. A. Vandenbergh.** 1992. Cloning, expression, and nucleotide sequence of genes involved in production of pediocin PA-1, a bacteriocin from *Pediococcus acidilactici* PAC 1.0. *Appl. Environ. Microbiol.* **58:**2360–2367.

102. **Mayr-Harting, A., A. J. Hedges, and R. C .W. Beerkley.** 1972. Methods for studying bacteriocins. *Methods Microbiol.* **7:**313–342.

103. **Mazzotta, A. S., A. D. Crandall, and T. J. Montville.** 1995. Resistance of *Clostridium botulinum* spores and vegetative cells to nisin, abstr. 32, p. 387, *Abstr. 95th Gen. Meet. Am. Soc. Microbiol. 1995.* American Society for Microbiology, Washington, D.C.

104. **McKay, L. L.,** 1983 Functional properties of plasmids in lactic streptococci. *Antonie Leeuwenhoek* **49:**2259–2274.

105. **McKay, L. L., B. R. Cords, and K. A. Baldwin.** 1973. Transduction of lactose metabolism in *Streptococcus lactis* C2. *J. Bacteriol.* **115:**810–815.

106. **McNamara, S.** 1986. Regulatory issues in the food biotechnology area, p. 15–28. *In* S. Harlander and T. P. Labuza (ed.), *Biotechnology in Food Processing.* Noyes Publications, Park Ridge, N.J.

107. **Ming, X., and M. A. Daeschel.** 1993. Nisin resistance of foodborne bacteria and the specific resistance responses of *Listeria monocytogenes* Scott A. *J. Food Prot.* **11:**944–948.

108. **Modi, K. D., M. L. Chikindas, and T. J. Montville.** 2000. Sensitivity of nisin-resistant *Listeria monocytogenes* to

109. **Moll, G. N., E. van den Akker, H. H. Hauge, J. Nissen-Meyer, I. F. Nes, W. N. Konings, and A. J. Driessen.** 1999. Complementary and overlapping selectivity of the two-peptide bacteriocins plantaricin EF and JK. *J. Bacteriol.* **181:**4848–4852.

110. **Montville, T. J.** 1989. The evolving impact of biotechnology on food microbiology. *J. Food Safety* **10:**87–97.

111. **Montville, T. J., A. M. Rogers, and A. Okereke.** 1992. Differential sensitivity of *Clostridium botulinum* strains to nisin. *J. Food Prot.* **56:**444–448.

112. **Montville, T. J., K. Winkowski, and R. D. Ludescher.** 1995. Models and mechanisms for bacteriocin action and application. *Int. Dairy J.* **5:**797–815.

113. **Morris, S. L., R. C. Walsh, and J. N. Hansen.** 1984. Identification and characterization of some bacterial membrane sulfhydryl groups which are targets of bacteriostatic and antibiotic action. *J. Biol. Chem.* **259:**13590–13594.

114. **Mortvedt, C. I., J. Nissen-Meyer, K. Sletten, and I. F. Nes.** 1991. Purification and amino acid sequence of lactocin S, a bacteriocin produced by *Lactobacillus sake* L45. *Appl. Environ. Microbiol.* **57:**1829–1834.

115. **Mulders, J. W., I. J. Boerrigter, H. S. Rollema, R. J. Siezen, and W. M. de Vos.** 1991. Identification and characterization of the lantibiotic nisin Z, a natural nisin variant. *Eur. J. Biochem.* **201:**581–584.

116. **Nes, I. F., and H. Holo.** 2000. Class II antimicrobial peptides from lactic acid bacteria. *Biopolymers* **55:**50–61.

117. **Nettles, C. G., and S. F. Barefoot.** 1993. Biochemical and genetic characteristics of bacteriocins of food-associated lactic acid bacteria. *J. Food Prot.* **56:**338–336.

118. **Neve, H., A. Geis, and M. Teuber.** 1984. Conjugal transfer and characterization of bacteriocin plasmids in group N (lactic acid) streptococci. *J. Bacteriol.* **157:**833–838.

119. **Nielsen, J. W., J. S. Dickson, and J. D. Crouse.** 1990. Use of a bacteriocin produced by *Pediococcus acidilactici* to inhibit *Listeria monocytogenes* associated with fresh meat. *Appl. Environ. Microbiol.* **56:**2142–2145.

120. **Nilsson, L., Y. Chen, M. L. Chikindas, H. H. Huss, L. Gram, and T. J. Montville.** 2000. Carbon dioxide and nisin act synergistically on *Listeria monocytogenes*. *Appl. Environ. Microbiol.* **66:**769–774.

121. **Nussbaum, A., and A. Cohen.** 1988. Use of a bioluminescence gene reporter for the investigation of red-dependent and gam-dependent plasmid recombination in *Escherichia coli* K12. *J. Mol. Biol.* **203:**391–402.

122. **Ojcius, D. M., and J. D. E. Young.** 1991. Cytolytic pore-forming proteins and peptides: is there a common structural motif? *Trends Biochem. Sci.* **16:**225–229.

123. **Oscroft, C. A., J. G. Banks, and S. McPhee.** 1990. Inhibition of thermally-stressed *Bacillus* spores by combinations of nisin, pH and organic acids. *Lebensm. Wiss. Technol.* **23:**538–544.

124. **Palumbo, S. A.** 1987. Can refrigeration keep our food safe? *Dairy Food Sanit.* **7:**56–60.

125. **Pavan, S., P. Hols, J. Delcour, M. C. Geoffroy, C. Grangette, M. Kleerebezem, and A. Mercenier.** 2000. Adaptation of the nisin-controlled expression system in

heat and the synergistic action of heat and nisin. *Lett. Appl. Microbiol.* **30:**249–253.

Lactobacillus plantarum: a tool to study in vivo biological effects. *Appl. Environ. Microbiol.* 66:4427–4432.

126. **Piard, J. C., O. P. Kuipers, H. S. Rollema, M. J. Desmazeaud, W. M. de Vos.** 1993. Structure, organization, and expression of the *lct* gene for lacticin 481, a novel lantibiotic produced by *Lactococcus lactis. J. Biol. Chem.* 268:16361–16368.

127. **Pucci, M. J., E. R. Vedamuthu, B. S. Kunka, and D. A. Vandenbergh.** 1988. Inhibition of *Listeria monocytogenes* by using bacteriocin PA-1 produced by *Pediococcus acidilactici* PAC 1.0. *Appl. Environ. Microbiol.* 54:2349–2353.

128. **Quadri, L. E. N., K. L. Roy, J. C. Vederos, and M. E. Stiles.** 1995. Immunity and gene organization of carnobacteriocins BM1 and B2 produced by *Carnobacterium piscicole* LV17B. Presented at Workshop on the Bacteriocins of Lactic Acid Bacteria—Applications and Fundamentals, Alberta, Canada, April 17–22, 1995.

129. **Radler, F.** 1990. Possible use of nisin in winemaking. I. Action of nisin against lactic acid bacteria and wine yeasts in solid and liquid media. *Am. J. Enol. Vitic.* 41:1–6.

130. **Radler, F.** 1990. Possible use of nisin in winemaking. II. Experiments to control lactic acid bacteria in the production of wine. *Am. J. Enol. Vitic.* 41:7–11.

131. **Rauch, P. J. G., and W. M. deVos.** 1992. Characterization of the novel nisin-sucrose conjugative transposon Tn5276 and its insertion in *Lactococcus lactis. J. Bacteriol.* 174:1280–1287.

132. **Ray, B., and M. A. Daeschel.** 1992. *Food Biopreservation of Microbial Origin.* CRC Press, Inc., Boca Raton, Fla.

133. **Rekhif, N., A. Atrih, and G. Lefebvre.** 1995. Selection and properties of spontaneous mutants of *Listeria monocytogenes* ATCC 15313 resistant to different bacteriocins produced by lactic acid bacteria strains. *Curr. Microbiol.* 230:827–853.

134. **Rhodia Food.** http://www.food.us.rhodia.com/prod_class.asp.

135. **Roberts, R. E., and E. A. Zottola.** 1993. Shelf-life of pasteurized process cheese spreads made from cheddar cheese manufactured with a nisin producing starter culture. *J. Dairy Sci.* 76:1830–1836.

136. **Rodriguez, J. M., L. M. Cintas, P. Casaus, N. Horn, H. M. Dodd, P. E. Hernandez, and M. J. Gasson.** 1995. Isolation of nisin-producing *Lactococcus lactis* strains from dry fermented sausages. *J. Appl. Bacteriol.* 78:109–115.

137. **Rogers, A. M., and T. J. Montville.** 1994. Quantification of factors influencing nisin's inhibition of *Clostridium botulinum* 56A in a model food system. *J. Food Sci.* 59:663–668, 686.

138. **Ruhr, E., and H. G. Sahl.** 1985. Mode of action of the peptide antibiotic nisin and influence of the membrane potential of whole cells and on cytoplasmic and artificial membrane vesicles. *Antimicrob. Agents Chemother.* 27:841–845.

139. **Sahl, H. G., M. Grossgarten, W. R. Widger, W. A. Cramer, and H. Brandis.** 1985. Structural similarities of the staphylococcin-like peptide Pep 5 to the antibiotic nisin. *Antimicrob. Agents Chemother.* 27:836–840.

140. **Sahl, H. G., R. W. Jack, and G. Bierbaum.** 1995. Biosynthesis and biological activities of lantibiotics with unique post-translational modifications. *Eur. J. Biochem.* 230:827–853.

141. **Saleh, M. A., and Z. J. Ordal.** 1955. Studies on growth and toxin production of *Clostridium botulinum* in precooked frozen food. II. Inhibition by lactic acid bacteria. *Food Res.* 20:340–346.

142. **Schved, F., A. Lalazar, Y. Henis, and B. J. Juven.** 1993. Purification, partial characterization and plasmid-linkage of pediocin SJ-1, a bacteriocin produced by *Pediococcus acidilactici. J. Appl. Bacteriol.* 74:67–77.

143. **Scott, V. N., and S. L. Taylor.** 1981. Effect of nisin on outgrowth of *Clostridium botulinum* spores. *J. Food Sci.* 46:117–120.

144. **Scott, V. N., and S. L. Taylor.** 1981. Temperature, pH, and spore load on the ability of nisin to prevent the outgrowth of *Clostridium botulinum* spores. *J. Food Sci.* 46:121–126.

145. **Sears, P. M., B. S. Smith, W. K. Stewart, R. Gonzalez, S. O. Rubino, S. A. Gusik, E. S. Kulisek, S. J. Projan, and P. Blackburn.** 1992. Evaluation of a nisin-based germicidal formulation on teat skin of live cows. *J. Dairy Sci.* 75:3185–3190.

146. **Severina, E., A. Severin, and A. Tomasz.** 1998. Antibacterial efficacy of nisin against multidrug-resistant Gram-positive pathogens. *J. Antimicrob. Chemother.* 41:341–347.

147. **Siegers, K., and K. D. Entian.** 1995. Genes involved in immunity to the lantibiotic nisin produced by *Lactococcus lactis* 6F3. *Appl. Environ. Microbiol.* 61:1082–1089.

148. **Somers, E. B., and S. L. Taylor.** 1987. Antibotulinal effectiveness of nisin in pasteurized process cheese spreads. *J. Food Prot.* 50:842–848.

149. **Steele, J. L., and L. L. McKay.** 1986. Partial characterization of the genetic basis for sucrose metabolism and nisin production in *Streptococcus lactis. Appl. Environ. Microbiol.* 51:57–64.

150. **Steiner, H.** 1982. Secondary structure of the cecropins: antibacterial peptides from the moth *Hyalophora cecropia. FEBS Lett.* 137:283–287.

151. **Stevens, K. A., B. W. Sheldon, N. A. Klapes, and T. R. Klaenhammer.** 1991. Nisin treatment for inactivation of *Salmonella* species and other gram-negative bacteria. *Appl. Environ. Microbiol.* 57:3613–3615.

152. **Stevens, K. A., B. W. Sheldon, N. A. Klapes, and T. R. Klaenhammer.** 1992. Effect of treatment conditions on nisin inactivation of gram-negative bacteria. *J. Food Prot.* 55:763–767.

153. **Stiles, M. E., and J. W. Hastings.** 1991. Bacteriocin production by lactic acid bacteria: potential for use in meat preservation. *Trends Food Sci. Technol.* 2:247–251.

154. **Stiles, M. E.** 1994. Bacteriocins produced by *Leuconostoc* species. *J. Dairy Sci.* 77:2718–2724.

155. **Stoddard, G. W., J. P. Petzel, M. J. van Belkum, J. Kok, and L. L. McKay.** 1992. Molecular analyses of the lactococcin A gene cluster from *Lactococcus lactis* subsp. *lactis* biovar *diacetylactis* WM4. *Appl. Environ. Microbiol.* 58:1952–1961.

156. Stoffels, G., J. Nissen-Meyer, A. Gudmundsdottir, K. Sletten, H. Holo, and I. F. Nes. 1992. Purification and characterization of a new bacteriocin isolated from a *Carnobacterium* sp. *Appl. Environ. Microbiol.* 58:1417–1422.

157. Tagg, J. R. 1991. Bacterial BLIS. *ASM News* 57:611.

158. Tagg, J. R., A. S. Dajani, and L. W. Wannamaker. 1976. Bacteriocins of gram-positive bacteria. *Bacteriol. Rev.* 40:722–756.

159. Tanaka, N. E., E. Traisman, M. H. Lee, and R. Casses. 1980. Inhibition of botulism toxin formation in bacon by acid development. *J. Food Prot.* 43:450–452.

160. Taylor, L. Y., O. O. Cann, and B. J. Welch. 1990. Antibotulinal properties of nisin in fresh fish packaged in an atmosphere of carbon dioxide. *J. Food Prot.* 53:953–957.

161. Terwilliger, T. C., and D. Eisenberg. 1982. The structure of melittin. *J. Biol. Chem.* 257:6016–6022.

162. Tramer, J., and G. G. Fowler. 1964. Estimation of nisin in foods. *J. Sci. Food Agric.* 15:522–528.

163. Upreti, G. C., and R. D. Hinsdill. 1975. Production and mode of action of lactocin 27: bacteriocin from a homofermentative *Lactobacillus*. *Antimicrob. Agents Chemother.* 7:139–145.

164. van Belkum, M. J., B. J. Hayema, R. E. Jeeninga, J. Kok, and G. Venema. 1991. Organization and nucleotide sequence of two lactococcal bacteriocin operons. Cloning of two bacteriocin genes from a lactococcal bacteriocin plasmid. *Appl. Envirion. Microbiol.* 57:492–498.

165. van Belkum, M. J., B. J. Hayema, A. Geis, J. Kok, and G. Venema. 1989. Cloning of two bacteriocin genes from lactococcal bacteriocin plasmid. *Appl. Environ. Microbiol.* 55:1187–1191.

166. van Belkum, M. J., J. Kok, and G. Venema. 1992. Cloning, sequencing, and expression in *Escherichia coli* of *lenB*, a third bacteriocin determinant from the lactococcal bacteriocin plasmid p984-6. *Appl. Environ. Microbiol.* 58:572–577.

167. van Belkum, M. J., J. Kok, G. Venema, H. Holo, I. F. Nes, W. N. Konings, and T. Abee. 1991. The bacteriocin lactococcin A specifically increases the permeability of lactococcal cytoplasmic membranes in a voltage-independent, protein-mediated manner. *J. Bacteriol.* 173:7934–7941

168. Vandenbergh, P. A. 1993. Lactic acid bacteria, their metabolic products and interference with microbial growth. *FEMS Microbiol. Rev.* 12:221–238.

169. Vandenbergh, P. A., M. J. Pucci, B. S. Kunka, and E. B. Vedamuthy. 1989. Method for inhibiting *Listeria monocytogenes* using a bacteriocin. European patent application 89101126.6.

170. van der Meer, J. R., J. Polman, M. M. Beerthuyzen, R. J. Siezen, O. P. Kuipers, and W. M. de Vos. 1993. Characterization of the *Lactococcus lactis* nisin A operon genes *nisP*, encoding a subtilisin-like serine protease involved in precursor processing, and *nisR*, encoding a regulatory protein involved in nisin biosynthesis. *J. Bacteriol.* 175:2578–2588.

171. van de Ven, F. J. M., H. W. van den Hooven, R. N. H. Konings, and C. W. Hilbers. 1991. The spatial structure of nisin in aqueous solution, p. 35–42. *In* G. Jung and H. G. Sahl (ed.), *Nisin and Novel Lantibiotics*. Escom Publishers, Leiden, The Netherlands.

172. Van Garde, S. J., and M. J. Woodburn. 1987. Food discard practices of householders. *J. Am. Diet. Assoc.* 87:322–329.

173. Venema, K., T. Abee, A. J. Haandrikman, K. J. Leenhouts, J. Kok, W. N. Konings, and G. Venema. 1993. Mode of action of lactococcin B, a thiol-activated bacteriocin from *Lactococcus lactis*. *Appl. Environ. Microbiol.* 59:1041–1048.

174. Venema, K., J. Kok, J. D. Marugg, M. Y. Toonen, A. M. Ledeboer, G. Venema, and M. L. Chikindas. 1995. Functional analysis of the pediocin operon of *Pediococcus acidilactici* PAC 1.0: PedB is the immunity protein and PedD is the precursor processing enzyme. *Mol. Microbiol.* 17:515–522.

175. Vogel, R. F., B. S. Pohle, P. S. Tichaczek, and W. Hammes. 1993. The competitive advantage of *Lactobacillus curvatus* LTH 1174 in sausage fermentations is caused by formation of curvacin. *Syst. Appl. Microbiol.* 16:457–462.

176. Wahlstrom, G., and P. E. Saris. 1999. A nisin bioassay based on bioluminescence. *Appl. Environ. Microbiol.* 65:3742–3745.

177. Winkowski, K., M. E. C. Bruno, and T. J. Montville. 1994. Correlation of bioenergetic parameters with cell death in *Listeria monocytogenes* cells exposed to nisin. *Appl. Environ. Microbiol.* 60:4186–4187.

178. Winkowski, K., A. D. Crandall, and T. J. Montville. 1993. Inhibition of *Listeria monocytogenes* by *Lactobacillus bavaricus* MN in meat systems at refrigeration temperatures. *Appl. Environ. Microbiol.* 59:2552–2557.

179. Winkowski, K., and T. J. Montville. 1992. Use of a meat isolate, *Lactobacillus bavaricus* MN, to inhibit *Listeria monocytogenes* growth in a model meat gravy system. *J. Food Safety* 13:19–31.

180. Worobo, R. W., M. J. van Belkum, H. Sailer, K. L. Roy, J. C. Vederas, and M. E. Stiles. 1995. A signal peptide secretion-dependent bacteriocin from *Carnobacterium divergens*. *J. Bacteriol.* 177:3143–3149.

181. Zajdel, J. K., P. Ceglowski, and W. T. Dobrzanski. 1985. Mechanism of action of lactostrepcin 5, a bacteriocin produced by *Streptococcus cremoris* 202. *Appl. Environ. Microbiol.* 49:969–974.

Food
Fermentations

VII

Food Microbiology: Fundamentals and Frontiers, 2nd Ed.
Edited by M. P. Doyle et al.
© 2001 ASM Press, Washington, D.C.

Mark E. Johnson
James L. Steele

Fermented Dairy Products

31

Several groups of microorganisms participate in the manufacture and ripening of fermented milk products (Table 31.1). The primary microflora used in the production of fermented milk products are the homofermentative lactic acid bacteria (LAB). Heterofermentative LAB are sometimes used but not for acid development. They are used for the production of specific flavor compounds. Raw milk naturally contains LAB, but they are mostly destroyed by pasteurization. However, this offers an opportunity to manufacturers, as they can now control the rate and extent of acid development by adding to milk specific bacteria in controlled numbers with desired metabolic activity. In turn, this allows for greater control of the manufacturing process, which ultimately leads to production of the desired end product on a more consistent basis. Traditionally, and in nonpasteurized milk, naturally occurring LAB were the acid producers, but because the numbers varied, the manufacturing processes of fermented milk products also varied. Thus, the manufacture of fermented milk products was more an art than a science.

Since the addition of the LAB starts the manufacturing process, they are commonly referred to as starter cultures. While their main function is to ferment lactose to lactic acid, it would be wrong to assume that starter bacteria are only added to produce acid. They are also involved in the development of flavor of fermented dairy products, but their importance varies with the individual product. Starter bacteria include both mesophilic (optimal growth at 25 to 30°C) and thermophilic (optimal growth at 37 to 42°C) species. Mesophilic LAB include *Lactococcus lactis* subsp. *lactis* and *Lc. lactis* subsp. *cremoris*. Thermophilic LAB include *Streptococcus thermophilus*, *Lactobacillus delbrueckii* subsp. *bulgaricus*, and *Lactobacillus helveticus*. A protocooperative relationship exists between *S. thermophilus* and starter lactobacilli, which results in an increased rate and extent of acid production. Proteolysis of casein by lactobacilli also stimulates the growth of propionibacteria (68).

In some fermented dairy products, additional bacteria, often referred to as secondary microflora (but are key to flavor development), are added to influence flavor and alter texture of the final product. Two LAB, *Leuconostoc* species and strains of *Lc. lactis* subsp. *lactis* capable of metabolizing citric acid (Cit+), are added to produce aroma compounds and carbon dioxide in cultured

Mark E. Johnson, Center for Dairy Research, Department of Food Science, University of Wisconsin-Madison, 1605 Linden Dr., Madison, Wisconsin 53706–1565.
James L. Steele, Department of Food Science, University of Wisconsin-Madison, 1605 Linden Dr., Madison, Wisconsin 53706–1565.

Table 31.1 Microorganisms involved in the manufacture of cheeses and fermented milks

Product	Principal acid producer	Intentionally introduced secondary microflora
Cheeses		
Colby, cheddar, cottage, cream	*Lactococcus lactis* subsp. *cremoris/lactis*	None
Gouda, Edam, Havarti	*Lactococcus lactis* subsp. *cremoris/lactis*	*Leuconostoc* sp., Cit⁺ *Lactococcus lactis* subsp. *lactis*
Brick, Limburger	*Lactococcus lactis* subsp. *cremoris/lactis*	*Geotrichum candidum, Brevibacterium linens, Micrococcus* sp., *Arthrobacter*
Camembert	*Lactococcus lactis* subsp. *cremoris/lactis*	*Penicillium camemberti*, sometimes *Brevibacterium linens*
Blue	*Lactococcus lactis* subsp. *cremoris/lactis*	Cit⁺ *Lactococcus lactis* subsp. *lactis, Penicillium roqueforti*
Mozzarella, provolone, Romano, Parmesan	*Streptococcus thermophilus, Lactobacillus delbrueckii* subsp. *bulgaricus, Lactobacillus helveticus*	None; animal lipases added to Romano for picante or rancid flavor
Swiss	*Streptococcus thermophilus, Lactobacillus helveticus, Lactobacillus delbrueckii* subsp. *bulgaricus*	*Propionibacterium freudenreichii* subsp. *shermanii*
Fermented milks		
Yogurt	*Streptococcus thermophilus, Lactobacillus delbrueckii* subsp. *bulgaricus*	None
Buttermilk	*Lactococcus lactis* subsp. *cremoris/lactis*	*Leuconostoc* sp., Cit⁺ *Lactococcus lactis* subsp. *lactis*
Sour cream	*Lactococcus lactis* subsp. *cremoris/lactis*	None

buttermilk and certain cheeses (Gouda, Edam, blue, and Harvarti). When used together, these bacteria make up 10 to 20% of the total starter culture, with *Leuconostoc* species being present at about three times the population of Cit⁺ *Lc. lactis* subsp. *lactis*. Heterofermentative lactobacilli (*Lactobacillus brevis, Lactobacillus fermentum*, and *Lactobacillus kefir*) are part of the varied microflora (including several yeast species) found in the more exotic cultured milks, such as kefir and koumiss, in which they produce ethanol, carbon dioxide, and lactic acid. They are not used in other fermented dairy products because of the copious quantities of carbon dioxide produced. *Propionibacterium freudenreichii* subsp. *shermanii* is added to Swiss-type cheeses, where its primary function is to metabolize L-lactic acid to propionic acid, acetic acid, and carbon dioxide. The carbon dioxide forms the "eyes" in Swiss-type cheeses. Propionibacteria also ferment citric acid to glutamic acid (17).

Other types of secondary microflora include undefined mixtures of yeasts, molds, and bacteria. These microorganisms can be added directly to the milk or are smeared, sprayed, or rubbed onto the cheese surface. This group of microorganisms has extremely varied and complex metabolic activities, their main function being to produce unique flavors. The use of these secondary microflora is usually limited to surface-ripened and mold-ripened cheeses. Yeasts (*Debaryomyces, Candida, Yersinia*, and *Geotrichum* species) and bacteria

(*Brevibacterium linens, Arthrobacter* species, and *Micrococcus* species) are employed in the aging of surface-ripened cheeses such as Limburger and Gruyère. Molds (*Penicillium camemberti* and *Penicillium roqueforti*) are used in Camembert and blue-veined cheeses, respectively.

Fermented dairy products are not commercially produced in an environment free of contaminating microorganisms. Indeed, metabolism of contaminants, e.g., lactobacilli, is essential for development of flavor in aged cheeses. Rapid acid development by starter within 4 to 8 h lowers the pH to less than 5.3 in cheese and to less than 4.6 in fermented milk products. After fermentation is complete, only acid-tolerant bacteria can grow. However, if acid development is slow or if the pH does not decrease sufficiently, contaminants which otherwise would have been inhibited may be able to grow. In some cheeses the pH can increase during ripening and permit the growth of previously inhibited bacteria. Dairy products may contain yeasts, molds, and many different genera of bacteria whose metabolic activities destroy quality. Spoilage of dairy products is described in chapter 6.

IMPORTANCE OF THE STARTER CULTURE

The key to commercial development of fermented milk products is the consistent and predictable rate of acid development by LAB. The rate and extent of pH decrease during manufacture and in the finished product are critical. pH has profound effects on moisture control during cheese manufacture, retention of coagulants, loss of minerals, hydration of proteins, and electrochemical interactions between protein molecules. These, in turn, have consequences for the development of flavor and physical properties (body and texture) of cheese and fermented milks. The reader is referred to other texts (25, 37) for discussion of these complex and interrelated phenomena.

In the past, antibiotic residues and overmature starters have been causes of inconsistent acid production. However, these problems have been overcome through monitoring of the milk supply and the use of improved starter media and preparation (76). Today, bacteriophage infection is the most common cause of inconsistent acid development, causing significant loss of revenue to the cheese industry.

In cultured milks, desired flavors are derived directly from the metabolism of starter cultures and deliberately added aroma-producing secondary microflora. Thus, the desired flavor of the product dictates the choice of microorganisms. With cultured milks and some cheeses, such as mozzarella, cream, and cottage cheese, the short time from processing to consumption (1 day to 4 weeks) and refrigerated storage precludes the development of flavor other than that produced by starter and secondary added microflora. In other cheeses, the choice of microorganism(s) depends primarily on the manufacturing protocol, such as the temperature to which the product will be subjected, the desired rate and extent of acid development, and the desired physical properties of the finished product.

There is considerable debate as to the exact contribution of the starter culture to flavor development in cheese, especially in cheeses where no secondary microflora is added. In these cheeses, nonstarter LAB, particularly lactobacilli, are the dominant adventitious microflora during ripening (59), and it is generally believed that they play the significant role in the ripening of these cheeses. The use of nonstarter LAB, especially lactobacilli, as adjunct flavor cultures is a burgeoning research area, and they are used commercially. Unfortunately, selection of appropriate cultures is a trial-and-error process since agreement on specific compounds that contribute to desired cheese flavor is often lacking. Differences in descriptions in desired flavor arise from the sheer complexity of cheese flavor as well as individual flavor perceptions and taste preferences. It should be understood that flavor development in cheese is a dynamic process and occurs in an environment that is constantly changing. It is not just that cheese develops stronger flavor with age, but that the very nature of the flavor is evolving. The development of flavor in different cheese varieties has been described (26).

LACTOSE METABOLISM

Energy transduction and fermentation pathways for carbohydrate metabolism in LAB have been described in detail (27, 60). Lactose, a disaccharide composed of glucose and galactose, is the only free-form sugar present in milk (45 to 50 g/liter). Lactococci translocate lactose into the cell by a phosphoenol-pyruvate phosphotransferase system. The lactose is phosphorylated during translocation and then cleaved by phospho-β-galactosidase into glucose and galactose 6-phosphate (Fig. 31.1). The glucose moiety enters the glycolytic pathway, and galactose 6-phosphate is converted into tagatose 6-phosphate via the tagatose pathway. Both sugars are cleaved by specific aldolases into triose phosphates, which are converted to pyruvic acid at the expense of NAD$^+$. For continued energy production, NAD$^+$ must be regenerated. This is usually accomplished by reducing pyruvic acid to lactic acid.

S. thermophilus and some thermophilic lactobacilli transport lactose via a lactose-galactose antiport system driven by an electrochemical proton gradient (60). Lactose is not phosphorylated but is cleaved by β-galactosidase to yield glucose and galactose. The glucose moiety enters the glycolytic pathway, but galactose is excreted from the cells and accumulates in milk or cheese. Thermophilic lactobacilli that do not excrete galactose and *Lb. helveticus* strains able to transport excreted galactose utilize the Leloir pathway to metabolize galactose. *Lb. delbrueckii* subsp. *bulgaricus* and most strains of *S. thermophilus* cannot metabolize galactose. This presents a problem in cheese manufacture since residual sugar can be metabolized heterofermentatively by other bacteria. Rapid production of carbon dioxide by heterofermentative bacteria causes cheese to crack and packages to swell. Residual sugar can also react with amino groups and form pink or brown pigments, i.e., Maillard-browning reaction products. Thermophiles may not metabolize the residual sugar during cold storage (4 to 7°C) of the cheese. Therefore, lactococci are sometimes included in the starter to ensure that all residual sugar is fermented. In pasta filata cheese manufacture, the curd is heated (52 to 66°C) and molded. The heat treatment may inactivate starter bacteria and prevent further sugar metabolism.

It is not known how lactose is transported in cells by *Leuconostoc* species or heterofermentative lactobacilli; however, lactose is known to be hydrolyzed by β-galactosidase (34). The galactose moiety is transformed into

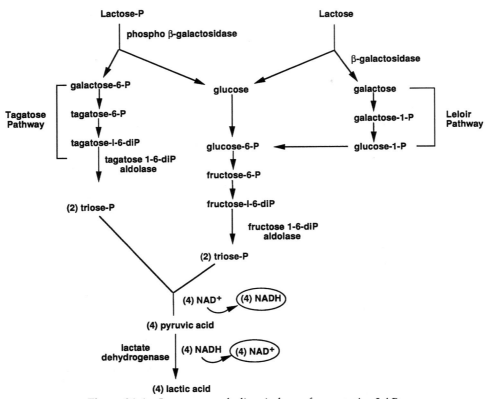

Figure 31.1   Lactose metabolism in homofermentative LAB.

glucose 6-phosphate (Leloir pathway) and, together with glucose, is metabolized through the phosphoketolase pathway (Fig. 31.2). Heterofermentative LAB lack aldolase but, through a dehydrogenation-decarboxylation

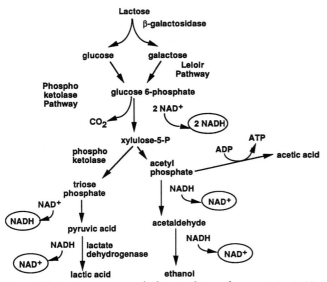

Figure 31.2   Lactose metabolism in heterofermentative LAB.

system, a pentose sugar (xylulose 5-phosphate) and carbon dioxide are formed. Xylulose 5-phosphate is then cleaved by phosphoketolase to yield glyceraldehyde and acetyl-phosphate. Lactic acid and ethanol, respectively, are formed from these intermediates, facilitating the regeneration of NAD^+. However, during cometabolism of lactose and citric acid, *Leuconostoc* species convert acetyl-phosphate into acetic acid and generate ATP.

Two enzymes, L-lactic acid dehydrogenase (L-LDH) and D-lactic acid dehydrogenase (D-LDH), are responsible for the conversion of pyruvate to L-lactic acid and D-lactic acid, respectively. Lactococci produce only L-lactic acid while *Lb. delbrueckii* subsp. *bulgaricus* and *Leuconostoc* species form only D-lactic acid. Other lactobacilli possess both enzymes and produce both D- and L-lactic acid.

The key to end-product formation from lactose metabolism is the regeneration of reducing equivalents. Lactococci have the enzymatic potential to produce compounds other than lactic acid to regenerate NAD^+, but these activities are not usually expressed under aerobic conditions (24). Oxygen in milk is used as an electron acceptor by LAB through the activity of oxidases and peroxidases (14, 66). As a consequence, hydrogen from

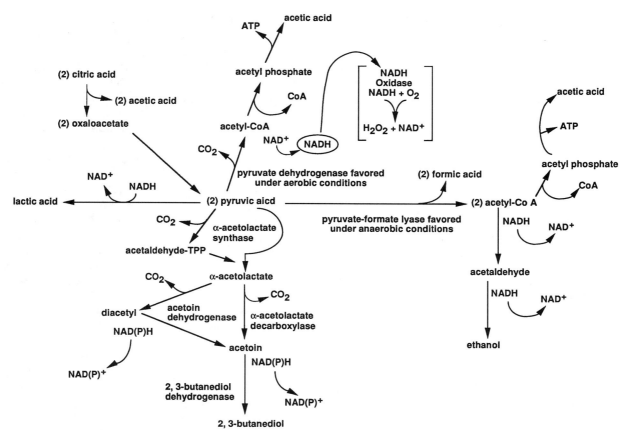

Figure 31.3 Pyruvic acid and citric acid metabolism in LAB. Abbreviations: CoA, coenzyme A; Tpp, thiamine pyrophosphate.

NADH is transferred to oxygen to produce hydrogen peroxide, and NAD$^+$ is regenerated. However, under anaerobic conditions and low sugar levels, lactococci (6, 24, 69) produce formic acid and ethanol (Fig. 31.3) to regenerate NAD$^+$. Heterofermentative LAB convert acetyl-phosphate into acetic acid rather than ethanol under aerobic conditions and regenerate NAD(P)$^+$ through NAD(P)H oxidases (51, 74).

Ethanol has a very high flavor threshold value and would not be expected to contribute directly to flavor in the amounts produced by homofermentative LAB. However, subsequent esterification of ethanol with short-chain fatty acids yields esters with very low flavor thresholds. These compounds are responsible for the fruity flavor defects of cheddar cheese (7). Short-chain fatty acids are probably generated by nonstarter LAB or exogenous sources, since starter bacteria have limited lipase activity (40).

The potential for production of diacetyl and carbon dioxide from lactose metabolism in lactococci with reduced LDH activity has been described (53). Formation

of ethanol, acetic acid, and formic acid would regenerate NAD$^+$ (Fig. 31.3).

As a result of lactose metabolism (and oxidase activity), the environment becomes anaerobic and the oxidation-reduction potential is reduced. In cheese, further metabolism by nonstarter LAB may be needed to maintain this low potential. It has been postulated that a low oxidation-reduction potential is necessary for the production and stability of reduced sulfur-containing compounds thought to be vital for the development of certain cheese flavors (32, 52). Sugar and citric acid metabolism may result in the formation of α-dicarbonyls such as glyoxal, methylglyoxal, and diacetyl. These compounds readily react with amino acids and produce a myriad of compounds that contribute to cheese flavor (33).

Although lactic acid is commonly thought of as the end of fermentation, this is not always the case. Cometabolism of citric acid and lactic acid by facultative lactobacilli will produce carbon dioxide and cause blowing of packaged cheese (28). Propionibacteria metabolize lactic acid to acetic and propionic acids. Clostridia also metabolize lactic acid to acetic acid and carbon dioxide.

In some cases, lactose is not fermented but is utilized in the production of exopolysaccharides. Under stress (low pH, low water activity), some lactococcal strains will not ferment the galactose moiety of lactose but will form exopolysaccharides containing methyl pentoses and galactose (50). *S. thermophilus* also converts lactose to an exopolysaccharide during the stationary phase of growth or at low temperatures (29).

PRODUCTION OF AROMA COMPOUNDS

Although lactic acid is the main metabolic end product of lactose metabolism in cultured dairy products and is responsible for the acid taste, it is nonvolatile and odorless and does not contribute to the aroma. The main aroma of volatile flavor components of fermented milks is produced by acetic acid, acetaldehyde, and diacetyl. In yogurt, these volatile compounds are formed by *S. thermophilus* and *Lb. delbrueckii* subsp. *bulgaricus*. *Leuconostoc* species and Cit$^+$ *Lc. lactis* subsp. *lactis* are added to produce aroma compounds in buttermilk and some cheese varieties. Literature describing the production of aroma compounds in fermented dairy products has been compiled by Imhof and Bosset (36).

Diacetyl Production

The production of diacetyl, acetic acid, and carbon dioxide from citric acid by LAB has been reviewed by Hugenholtz (35) and is shown in Fig. 31.3. The carbon dioxide produced is responsible for the holes (eyes) in Gouda and Edam cheeses and the effervescent quality of buttermilk. Milk contains 0.15 to 0.2% citric acid, but not all LAB can metabolize it. However, *Leuconostoc* species, Cit$^+$ *Lc. lactis* subsp. *lactis*, and facultative heterofermentative lactobacilli (48, 58) metabolize citric acid. *Leuconostoc* species and Cit$^+$ *Lc. lactis* subsp. *lactis* strains utilize citric acid and lactose simultaneously and under certain conditions can derive energy via metabolism of citric acid. Citric acid is transported into the cell by a citric acid permease, which is plasmid encoded in lactococci and *Leuconostoc* species (42, 71) and metabolized to pyruvic acid without generation of NADH. The result is an excess of pyruvic acid which does not have to be reduced to lactic acid to regenerate NAD$^+$; therefore, it is available for other reactions. Citrate metabolism in *Leuconostoc* sp. and *Lc. lactis* generates an electrochemical proton motive force-generating process (3). Ramos et al. (61) appear to have resolved conflicting reports on the pathway leading to the formation of diacetyl. Using ^{13}C nuclear magnetic resonance, they verified that diacetyl formation involves nonenzymatic decarboxylative

oxidation of α-acetolactate (an unstable intermediate derived from two molecules of pyruvic acid) and that the alternative suggested pathway, via a diacetyl synthase, is highly unlikely. However, Aymes et al. (2) suggested that not all the diacetyl produced by Cit$^+$ *Lc. lactis* subsp. *lactis* strains could be explained by spontaneous decarboxylation of the α-acetolactate. α-Acetolactate can only be produced when pyruvic acid accumulates within the cell. *Leuconostoc* species metabolize citric acid during growth but do not form diacetyl until the pH is below 5.4. α-Acetolactate synthase is inhibited at pH 5.4 or higher by many intermediates of lactose metabolism; however, the inhibition is relieved at lower pH values (12, 62). When diacetyl is not formed, *Leuconostoc* species form lactic acid from the pyruvic acid derived from citric acid and regenerate NAD$^+$. Since NAD$^+$ is regenerated, there is less demand to form ethanol from the acetyl-phosphate that is generated via the phosphoketolase pathway. Consequently, acetic acid is formed with the generation of ATP (Fig. 31.3), and growth is enhanced (13, 64).

Diacetyl can be reduced by 2,3-butanediol dehydrogenases (15) to acetoin and 2,3,-butanediol, both flavorless compounds. The presence of citric acid inhibits these reactions, but reduction begins when citric acid is exhausted. To ensure residual levels of citric acid in cultured milks, the ratio of starter culture to Cit$^+$ bacteria must be controlled (20). The cultured milk is stirred when the pH is reduced to 4.5 or below. This introduces oxygen and helps to increase and maintain the desired diacetyl content. The introduction of oxygen is required for nonenzymatic oxidative decarboxylation of α-acetolactate to diacetyl. In addition, LAB produce NADH oxidase, which transfers hydrogen to oxygen and regenerates NAD$^+$. The NADH oxidase activity replaces the role of the 2,3-butanediol dehydrogenases in regenerating NAD$^+$ and allows the accumulation of diacetyl, rather than its reduction to acetoin and 2,3-butanediol (4). NADH oxidase is more active at lower temperatures, while dehydrogenases are less active (5). For rapid acidification, non-citrate-metabolizing lactococci must be the dominant acid producers. If not, cometabolism of citric acid and lactose by Cit$^+$ *Lc. lactis* subsp. *lactis* would quickly consume the citric acid and result in the reduction of diacetyl to acetoin and 2,3-butanediol before pH 4.6 is reached. Hugenholtz (35) described the use of genetic engineering to construct strains of lactococci which produce elevated levels of diacetyl.

Acetaldehyde Production

There are several metabolic pathways in LAB that can lead to the formation of acetaldehyde (45, 46). This has resulted in some controversy over the primary pathway utilized by LAB. Cleavage of threonine by threonine aldolase to glycine and acetaldehyde has been suggested to be the most important mechanism for acetaldehyde production in yogurt and buttermilk (46, 78). However, using radiolabeled threonine, Wilkins et al. (77) demonstrated that only 2% of the acetaldehyde produced by mixed cultures of *Lb. delbrueckii* subsp. *bulgaricus* and *S. thermophilus* originated from threonine, even though both bacteria possess threonine aldolase. Acetaldehyde is also formed by Cit+ *Lc. lactis* subsp. *lactis* strains (41). When the ratio of diacetyl to acetaldehyde in fermented milks is lower than 3:1, a yogurt or green apple flavor defect is observed (49). The defect is due to excess metabolic activity by Cit+ *Lc. lactis* subsp. *lactis*. Excessive acetaldehyde in yogurt is the result of overripening and is always associated with high acid content. Prevention of an excessive amount of acetaldehyde may be accomplished by the use of *Leuconostoc*, which metabolizes acetaldehyde to ethanol. Obviously, to prevent overripening, the product must be cooled rapidly and stored at lower temperatures. The trend for faster acid development and larger fermentation vessels may limit the ability of the manufacturer to cool the product fast enough to prevent overripening.

PROTEOLYTIC SYSTEMS IN LAB

Proteolytic systems in LAB contribute to their ability to grow in milk and are necessary for the development of flavor in ripened cheeses. LAB are amino acid auxotrophs typically requiring several amino acids for growth. Well-characterized examples include strains of lactococci and lactobacilli which require 6 to 15 amino acids, respectively. The quantities of free amino acids present in milk are not sufficient to support the growth of these bacteria to high cell density; therefore, they require a proteolytic system capable of utilizing the peptides present in milk and hydrolyzing milk proteins (α_{S1-}, α_{S2-}, κ-, and β-caseins) to obtain essential amino acids. Peptides and amino acids formed by proteolysis may impart flavor directly or serve as flavor precursors in fermented dairy products. Additionally, the resulting flavors may have either positive or negative impacts. The production of high-quality fermented dairy products is dependent on the proteolytic systems of LAB.

Proteolytic Systems and Their Physiological Role

Since the mid-1980s, significant progress has been made in defining the biochemistry and genetics of proteolytic systems from LAB. This is especially true for the lactococcal system, in which many of the enzymes have been purified and characterized, and numerous genes encoding components of proteolytic systems have been sequenced. For extensive reviews of the proteolytic enzyme systems of LAB, see Kunji et al. (44) and Christensen et al. (11). A model of the lactococcal proteolytic enzyme system is presented in Fig. 31.4. While growing in milk, LAB obtain essential amino acids in a variety of ways. They first utilize nonprotein nitrogen sources, such as free amino acids and small peptides. Casein, which composes 80% of all proteins present in milk, becomes the primary nitrogen source after nonprotein nitrogen is depleted. Proteolytic systems in LAB can be divided into three components: enzymes outside the cytoplasmic membrane, transport systems, and intracellular enzymes.

Extensive investigations have revealed that a cell envelope-associated proteinase, designated PrtP, is the only extracellular proteolytic enzyme present in lactococci. The enzyme is a serine-protease which is expressed as a pre-pro-proteinase. A signal peptidase removes the signal peptide upon transport across the cytoplasmic membrane. Subsequently, a lipoprotein maturase (PrtM) is thought to cause a conformational change in the proproteinase, resulting in release of the pro-region, via autoproteolysis, and an active PrtP. The activated enzyme remains associated with the cell owing to the presence of a C-terminal membrane anchor sequence. Genes encoding lactococcal proteinases with different substrate specificities have been sequenced, and the amino acid residues involved in substrate binding and catalysis have been determined. A critical feature of the enzyme is its broad cleavage specificity, which results in the release of more than 100 oligopeptides from soluble β-casein, 20% of which are small enough to be transported by the oligopeptide transport system (39). Loss of PrtP, which is typically plasmid encoded in lactococcal strains, results in derivatives capable of reaching only about 10% of the final cell density of the parental strain, indicating that PrtP is essential for growth of lactococci to high cell density in milk.

Transport of nitrogenous compounds across the lactococcal cytoplasmic membrane takes place via group-specific amino acid transport systems, di/tripeptide transport systems, and an oligopeptide transport system (Opp). Of these systems, Opp is of greatest importance during growth of lactococci in milk. Growth studies with Opp derivatives have shown that Opp is essential for the uptake of PrtP-generated peptides from β-casein and oligopeptides present in the nonprotein nitrogen component of milk (39). This system, which is organized in an operon, is composed of two ATP-binding proteins, two integral membrane proteins, and a substrate-binding protein. Peptides from 4 to 18 residues can be transported with little specificity for particular side chains (18).

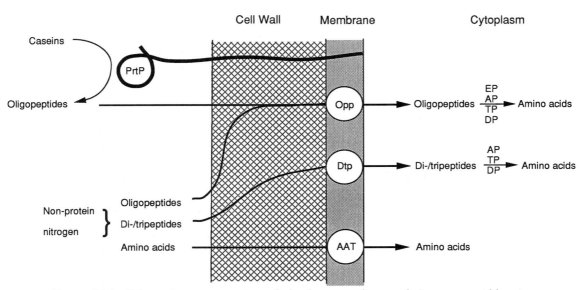

Figure 31.4 Schematic representation of the lactococcal proteolytic system. Abbreviations: PrtP, cell envelope-associated proteinase; Opp, oligopeptide transport system; Dtp, di/tripeptide transport systems; AAT, amino acid transport systems; EP, endopeptidases; AP, aminopeptidases; TP, tripeptidases; DP, dipeptidases.

Once inside the cell, peptides are hydrolyzed by peptidases. Peptidase classes that have been identified in lactococci include exopeptidases and endopeptidases. The greatest variety of enzymes are from the exopeptidase class, which includes aminopeptidases, tripeptidases, and dipeptidases. No carboxypeptidases have been detected in lactococci. Peptidases identified in lactococci and their cleavage specificities are summarized in Table 31.2. This combination of endopeptidases, aminopeptidases, tripeptidases, and dipeptidases converts the transported peptides into free amino acids, which lactococci

Table 31.2 Peptidases purified and characterized from lactococci

Peptidase	Abbreviations	Specificity[a]
X-prolyl dipeptidyl aminopeptidase	PepX	X-Pro $\updownarrow$ Y-...
Aminopeptidase N	PepN	X $\updownarrow$ Y-Z̲...
Aminopeptidase C	PepC	X$\updownarrow$Y-Z...
Aminopeptidase A	PepA	Asp(Glu)$\updownarrow$Y-Z...
Pyrrolidone carboxyl peptidase	PCP	pGlu$\updownarrow$Y-Z...
Prolyl iminopeptidase	PepI	Pro$\updownarrow$Y-Z...
Dipeptidase	PepV	X$\updownarrow$Y̲
Prolidase	PepQ	X$\updownarrow$YPro
Tripeptidase	PepT	X$\updownarrow$Y-Z
Endopeptidases	PepO, PepF	...W-X$\updownarrow$Y-Z...

[a] $\updownarrow$ indicates which peptide bond is hydrolyzed.

require for growth. Examination of the growth of strains lacking a single peptidase in milk has revealed that only the PepN single mutant grows significantly more slowly. Characterization of lactococcal derivatives lacking more than one peptidase indicated that, in general, the more peptidases that were inactivated, the slower the strain grew in milk. The only possible exception to this general rule is PepX, which has not been observed to have a significant effect on growth rate in milk (55). The most direct interpretation of these results is that the multiple peptidase mutants have a reduced ability to obtain essential amino acids from casein-derived oligopeptides, thereby limiting the availability of amino acids for new protein biosynthesis. However, it is also possible that the reduced growth rate in milk observed with multiple peptidase mutants is related to altered regulation of the proteolytic system or a reduced ability to turnover cellular proteins (11).

Proteolytic systems of other LAB have not been as extensively studied. However, numerous components of proteolytic enzyme systems of *Lb. helveticus* and *Lb. delbrueckii* have been characterized. In general, these components have homologs in the lactococcal system (11).

Proteolysis and Cheese Flavor Development

While flavor development in various types of cheese remains a poorly defined process, it is generally agreed that proteolysis is essential for flavor development in bacteria-ripened cheeses (26, 72). Proteolytic enzymes present in this group of products include chymosin, plasmin, and proteolytic enzymes from starter cultures, adjunct cultures, and nonstarter LAB. The specificities and relative activities of the proteolytic enzymes present in the cheese matrix determine which peptides and amino acids accumulate and, hence, how flavor develops. The use of isogenic strains, differing only in the specificity of activity of a single proteolytic enzyme, will have great utility in elucidating the role of individual enzymes in cheese flavor development.

Free amino acids and peptides in the cheese matrix can contribute to cheese flavor either directly or indirectly and with positive or negative effects. Cheese flavor development has been the subject of numerous comprehensive reviews (26, 67, 70). A major negative effect of proteolytic products is bitterness, which is believed to be caused by hydrophobic peptides ranging in length from 3 to 27 amino residues (47). They are believed to be generated from casein, principally by the joint action of chymosin and the LAB proteinases (9), and can be hydrolyzed to nonbitter peptides and amino acids by LAB peptidases. Therefore, the accumulation of bitter peptides is dependent on the relative rates of their formation and hydrolysis. A variety of volatile compounds can be derived from catabolism of amino acids (11, 70, 75). Numerous sulfur-containing compounds, particularly methanethiol, are thought to be important in cheese flavor. Methionine is believed to be the precursor to methanethiol; a number of enzymes and/or pathways have been identified in LAB that are capable of converting methionine to methanethiol (75). Alternatively, amino acid catabolism can give rise to compounds which have a negative impact on cheese flavor. For example, catabolism of aromatic amino acids can give rise to compounds such as indole and skatole, which contribute to "unclean" flavors in cheese (11). Overall, proteolysis is believed to be essential for development of characteristic flavor compounds in bacterial ripened cheeses; however, the mechanism(s) by which products of proteolysis give rise to beneficial flavor compounds remains unknown. Additionally, other than bitterness, the mechanisms by which proteolysis effects the development of undesirable flavor compounds are unknown.

BACTERIOPHAGES AND BACTERIOPHAGE RESISTANCE

Bacteriophage infection may lead to a decrease or complete inhibition of lactic acid production by the starter culture. This has a major impact on the manufacture of fermented dairy products, as lactic acid synthesis is required to produce these products. Additionally, slow acid production disrupts manufacturing schedules and typically results in products that are downgraded to lower economic value. The severe consequences of bacteriophage infection have led to extensive investigations into bacteriophages and the mechanisms by which LAB resist infection. Bacteriophage infection of LAB was first described in lactococcal starter cultures in the 1930s. Since then, the dairy industry has employed improved sanitation regimes, utilized sophisticated starter culture propagation vessels, developed starter culture systems to minimize the impact of phage infection, and isolated or constructed starter strains with enhanced bacteriophage resistance. However, bacteriophage infection of the starter culture has remained a significant problem in the dairy industry. Klaenhammer and Fitzgerald (43) listed four reasons why dairy product fermentations are particularly susceptible to bacteriophage infection:

1. Phage contamination can occur when fermentations are not protected from environmental contaminants and in a nonsterile fluid medium, pasteurized milk.
2. Processing efficiency is easily disrupted because batch culture fermentations occur under increasingly stringent manufacturing schedules.
3. Increasing reliance on specialized strains limits the number and diversity of available dairy starter cultures.
4. Continuous use of defined cultures provides an ever-present host for bacteriophage attack.

Historically, most bacteriophage-related problems have occurred with lactococcal starter cultures; however, problems with starter systems which employ *S. thermophilus* and *Lb. delbrueckii* subsp. *bulgaricus* have been increasing. It is principally bacteriophage infection of *S. thermophilus* strains that has been the problem (56). For more extensive reviews of bacteriophage and bacteriophage resistance in LAB, readers are referred to other reports (1, 23, 43).

Bacteriophages of LAB

Significant progress has been made on the characterization of bacteriophages from LAB. All of the bacteriophages examined contain double-stranded linear DNA genomes with either cohesive or circularly permuted, terminally redundant ends. Both lytic and temperate bacteriophages have been characterized. A major outcome of this characterization has been a clear classification, based on DNA homology and morphological studies, of lactococcal bacteriophages (38). Twelve species have been described, and type phages have been proposed for each species. Complete nucleotide sequence information is now available for seven *Lc. lactis*, three *S. thermophilus*, and two *Lactobacillus* bacteriophages (23). This information has enhanced our understanding of how bacteriophage and host interactions evolve over time and our ability to construct novel bacteriophage defense mechanisms.

Lactococcal Bacteriophage Resistance Mechanisms

Selective environmental pressure placed on lactococci by bacteriophages over thousands of years has resulted in strains which contain numerous bacteriophage defense mechanisms. The best-characterized bacteriophage-resistant strain is *Lc. lactis* ME2. This strain has been shown to contain at least five distinct phage defense loci, including one which interferes with bacteriophage adsorption, two restriction-modification systems, and two abortive infection mechanisms (43). These defense loci are encoded by plasmids capable of conjugal transfer, which suggests that genetic exchange between starter cultures has had an important role in the development of bacteriophage-resistant starter cultures. Recombinant DNA techniques have also been employed to construct lactococcal strains with enhanced bacteriophage resistance. These include the use of antisense RNA derived from conserved bacteriophage genes and the cloning of bacteriophage origins of replication on multicopy plasmids. The latter approach is thought to titrate bacteriophage replication factors and result in a bacteriophage defense mechanism similar to abortive bacteriophage infection. The remainder of this section will cover the previously mentioned three naturally occurring bacteriophage defense mechanisms.

Interference with Bacteriophage Adsorption

The adsorption of a bacteriophage to a host cell is determined by bacteriophage specificity, physicochemical properties of the cell envelope, accessibility and density of bacteriophage receptor material, and electrical potential across the cytoplasmic membrane (65). The complexity of this interaction has facilitated the isolation of bacteriophage-resistant starter cultures by exposing them to bacteriophages and isolating resistant variants. Mutants isolated in such a fashion typically have a reduced capacity to adsorb the bacteriophage(s) used in the challenge and are referred to as bacteriophage-insensitive mutants. This is a common practice for deriving starter cultures with reduced bacteriophage sensitivity.

Researchers have begun to characterize host components required for bacteriophage adsorption. It is now thought that bacteriophages initially interact reversibly with cell envelope-associated polysaccharide and then interact irreversibly with cell membrane protein(s) (57). The best-characterized example of a lactococcal mechanism which interferes with adsorption of bacteriophages has been described in *Lc. lactis* subsp. *cremoris* SK110. The reduction in the ability of bacteriophage to adsorb to cells of this strain is due to masking of the phage receptors by a galactose-containing lipoteichoic acid, rather than the absence of receptors.

Inhibition of adsorption of lactococcal bacteriophage due to the lack of a membrane-bound protein has also been reported. Characterization of phage-resistant mutants of *Lc. lactis* subsp. *lactis* C2 has revealed that these mutants bind normally with bacteriophage, but no plaques are formed. Subsequently, it was demonstrated that a 32-kDa membrane protein essential for bacteriophage infection was lacking in these mutants. Similarly, it has been suggested that a membrane protein is involved in bacteriophage adsorption to *Lc. lactis* subsp. *lactis* ML3. It remains to be determined if this protein also has a role in the injection of phage DNA into the cytoplasm (23).

Restriction and Modification Systems

Restriction/modification (R/M) systems are widely distributed in lactococci and are often plasmid encoded. In fact, it is not uncommon for strains to contain two or more R/M systems, and at least 23 plasmids that encode these systems have been identified. The two components of an R/M system are a site-specific modifying enzyme and a corresponding site-specific restriction endonuclease. This system enables the cell to differentiate bacteriophage DNA from its own DNA and inactivate foreign DNA by hydrolysis. The typical end result of bacteriophage infection of a culture containing an R/M system is a reduction in the number of progeny bacteriophage produced. The extent of reduction is dependent on both the activity of the R/M system and the number of unmodified restriction endonuclease sites on the bacteriophage genome. It is important to note that bacteriophages that escape restriction will give rise to modified progeny phage which are immune to the corresponding restriction endonuclease. Therefore, while this is an important and widely distributed mechanism of bacteriophage defense, it is also very fragile.

Two mechanisms have been identified by which bacteriophages have evolved resistance to R/M systems. Characterization of lactococcal bacteriophages has revealed that they contain far fewer restriction endonuclease sites than expected for genomes of their size, suggesting that evolutionary pressure has selected for bacteriophages with few restriction endonuclease sites. This view is supported by the observation that lactococcal bacteriophages which have recently emerged in the dairy industry are more sensitive to R/M systems. The characterization of bacteriophages which have evolved to overcome a specific R/M system has revealed that they have acquired a functional copy of the modification enzyme from that system. These examples illustrate that bacteriophage-host interactions are continually changing, with bacterial strains acquiring new defense mechanisms and bacteriophages evolving mechanisms to overcome them.

Abortive Bacteriophage Infection

Like R/M systems, abortive bacteriophage infection (Abi) systems are widely distributed in lactococci and are frequently plasmid encoded, although some are also encoded by episomes. By definition, these mechanisms inhibit bacteriophage infection following adsorption, DNA penetration, and the early stages of the bacteriophage lytic cycle. Few infections successfully release viable progeny, and those that are successful result in less progeny bacteriophage being released. The end result for the host, even those that do not release viable bacteriophages, is death.

Seventeen genes encoding Abi mechanisms have been sequenced from lactococci (23). All of the characterized Abi systems have an unusually low G+C content, suggesting that they have recently been acquired by horizontal gene transfer. Frequently, Abi genes are associated with R/M systems. These systems work well together; the R/M system reduces the number of viable bacteriophage genomes, and the Abi system reduces the number of progeny bacteriophages released from infected cells that evade the R/M system.

Some bacteriophages have the ability to overcome Abi mechanisms. These phages either have a relatively small change in their genome or have acquired a piece of chromosomal DNA from their host. Durmaz and Klaenhammer (22) proposed utilizing a culture system in which derivatives of a single strain containing different Abi and R/M systems are rotated. This strategy would reduce the selection for bacteriophages which have evolved resistance to a specific defense mechanism by rotating these mechanisms.

GENETICS OF LAB

Research on the genetics of LAB began in the early 1970s. Initially, research focused on plasmids and natural gene transfer systems in lactococci. Interest in the genetics of LAB has expanded rapidly, and there are now hundreds of researchers active in this area. This interest has resulted in the development of a relatively detailed understanding of the basic genetics of these bacteria, natural gene transfer systems, and the development of tools required for application of recombinant DNA techniques. For a more complete description of the genetics of LAB, the reader is referred to other publications (10, 19, 31, 54, 73).

General Genetics of LAB

Genetic elements of LAB which have been characterized include chromosomes, introns, transposable elements, and plasmids. Chromosomes of LAB are relatively small compared with those of other eubacteria, ranging from 1.8 to 3.4 Mbp (16). Several chromosomal genetic and physical maps have been constructed for this group of bacteria. Additionally, low-redundancy sequencing of an entire *Lc. lactis* genome was recently reported (8), and a genomic sequencing project is under way for *Lactobacillus acidophilus*. The value of genomic sequences of LAB to basic and applied research on these organisms cannot be overstated. For example, this information should allow for the development of a comprehensive view of the metabolic potential of these organisms. Group I and group II introns have been identified in LAB. These elements are ribozymes that catalyze a self-splicing reaction from mRNAs which contain the intron. To date, group I introns in LAB have only been identified in bacteriophage genomes. The only group II intron identified in LAB is associated with a plasmid-encoded gene required for conjugal transfer (21). Transposable elements, genetic elements capable of moving as discrete units from one site to another in the genome, have been described in LAB (31, 73). Insertion sequences, the simplest of transposable elements, are widely distributed in bacteria and have also been identified in numerous LAB. Their ability to mediate molecular rearrangements and affect gene regulation has had both positive and negative implications for dairy product fermentations. In one case, the incorporation of an insertion sequence into a prophage of *Lactobacillus casei* resulted in the conversion of a temperate bacteriophage into a virulent bacteriophage. Alternatively, insertion sequence-mediated cointegration has played a pivotal role in the dissemination of many beneficial characteristics via conjugation. More complex transposable elements, such as self-transmissible conjugal transposons that encode for the production of the bacteriocin nisin, the ability to metabolize sucrose, and a bacteriophage defense mechanism, have also been described in detail (31, 73). Plasmids, i.e., autonomously replicating extrachromosomal circular DNA molecules, have been identified in several LAB. These are of particular importance in lactococci, where they encode numerous characteristics essential for dairy product fermentations, including lactose metabolism, proteinase activity, oligopeptide transport, bacteriophage-resistance mechanisms, bacteriocin production and immunity, bacteriocin resistance, exopolysaccharide production, and citric acid utilization (73).

Natural Gene Transfer Mechanisms

Transduction

Transduction, the transfer of bacterial genetic material by a bacteriophage, has been demonstrated in lactococci, lactobacilli, and *S. thermophilus*. Transduction played an important role in determining that lactose metabolism and proteinase activity are plasmid encoded in lactococci. However, its use in the construction of strains to be used in industry is limited by the relatively narrow host range of transducing bacteriophages.

Conjugation

Conjugation, the transfer of genetic material from one bacterial cell to another, which requires cell-to-cell contact, has been characterized in detail in lactococci. In fact, most plasmid-encoded characteristics important in the manufacture of fermented dairy products can be transferred by conjugation. Utilizing an approach which does not require antibiotic resistance markers, conjugation has been employed to transfer plasmids that encode bacteriophage-resistance genes into commercial lactococcal strains. These strains have enhanced resistance to bacteriophage and have been used successfully in the dairy industry for several years. Conjugation is likely to be employed in the future to construct other lactococcal strains with enhanced industrial utility.

Genetic Modification of LAB Using Recombinant DNA Techniques

Experiments employing recombinant DNA techniques have led to most of the recent advances in the understanding of physiology and genetics of LAB. The power of recombinant DNA approaches is that strains can be constructed which differ in a single defined genetic alternation, e.g., inactivation of a specific gene. By comparing the wild-type culture with its isogenic derivative, the role of that gene in the phenotype being examined can be unequivocally determined. This general approach has resulted in a detailed understanding of how these bacteria utilize lactose, obtain essential amino acids, produce diacetyl, and resist bacteriophage infection. Additionally, recombinant DNA approaches have been used to construct novel bacteriophage resistance mechanisms and

overproduce enzymes of interest in dairy fermentations. In the future, it is likely that numerous commercial strains will be constructed utilizing recombinant DNA techniques. Readers interested in more comprehensive reviews on the physiology and genetics of LAB should consult other texts (30, 63).

References

1. **Allison, G. E., and T. R. Klaenhammer.** 1998. Phage defense mechanisms in lactic acid bacteria. *Int. Dairy J.* 8:207–226.

2. **Aymes, F., C. Monnet, and G. Corrieu.** 1999. Effect of alpha-acetolactate decarboxylase inactivation on alpha-acetolactate and diacetyl production by *Lactococcus lactis* subsp. *lactis* biovar *diacetylactis*. *J. Biosci. Bioeng.* 87:87–92.

3. **Bandel, M., M. E. Lhotte, C. Marty-Teysset, A. Veyrat, H. Prevost, V. Dartois, C. Divies, W. N. Konings, and J. S. Lolkema.** 1998. Mechanism of the citrate transporters in carbohydrate and citrate cometabolism in *Lactococcus* and *Leuconostoc* species. *Appl. Environ. Microbiol.* 64:1594–1600.

4. **Bassit, N., C. Y. Boquien, D. Picque, and G. Corrieu.** 1993. Effect of initial oxygen concentration on diacetyl and acetoin production by *Lactococcus lactis* subsp. *lactis* biovar *diacetylactis*. *Appl. Environ. Microbiol.* 59:1893–1897.

5. **Bassit, N., C. Y. Boquien, D. Picque, and G. Corrieu.** 1995. Effect of temperature on diacetyl and acetoin production by *Lactococcus lactis* subsp. *lactis* biovar *diacetylactis*. CNRZ 483. *J. Dairy Res.* 62:123–129.

6. **Bills, D. D., and E. A. Day.** 1966. Dehydrogenase activity of lactic streptococci. *J. Dairy Sci.* 49:1473–1477.

7. **Bills, D. D., M. E. Morgan, L. M. Libby, and E. A. Day.** 1965. Identification of compounds responsible for fruity flavor defect of experimental cheeses. *J. Dairy Sci.* 48:1168–1173.

8. **Bolotin, A., S. Mauger, K. Malarme, S. D. Erlich, and A. Sorokin.** 1999. Low-redundancy sequencing of the entire *Lactococcus lactis* IL1403 genome. *Antonie Leeuwenhoek* 76:27–76.

9. **Broadbent, J. R., M. Strickland, B. C. Weimer, M. E. Johnson, and J. L. Steele.** 1998. Small peptide accumulation and bitterness in Cheddar cheese made from single strain *Lactococcus lactis* starters with distinct proteinase specificities. *J. Dairy Sci.* 81:327–337.

10. **Chassy, B. M., and C. M. Murphy.** 1993. *Lactococcus* and *Lactobacillus*, p. 65–82. *In* A. L. Sonenshein, J. A. Hoch, and R. Losick (ed.), *Bacillus subtilis and Other Gram-Positive Bacteria.* American Society for Microbiology, Washington, D.C.

11. **Christensen, J. E., E. G. Dudley, and J. R. Pederson, and J. L. Steele.** 1999. Peptidases and amino acid catabolism in lactic acid bacteria. *Antonie Leeuwenhoek* 76:217–246.

12. **Cogan, T. M., R. J. Fitzgerald, and S. Doonan.** 1984. Acetolactate synthase of *Leuconostoc lactis* and its regulation of acetoin production. *J. Dairy Res.* 51:597–604.

13. **Cogan, T. M.** 1987. Co-metabolism of citrate and glucose by *Leuconostoc* spp.: effects on growth, substrates and products. *J. Appl. Bacteriol.* **63**:551–558.

14. **Condon, S.** 1987. Responses of lactic acid bacteria to oxygen. *FEMS Microbiol. Rev.* **46**:269–280.

15. **Crow, V. L.** 1990. Properties of 2,3-butanediol dehydrogenases from *Lactococcus lactis* subsp. *lactis* in relation to citrate fermentation. *Appl. Environ. Microbiol.* **56**:1656–1665.

16. **Davidson, B. E., N. Kordias, M. Dobos, and A. J. Hillier.** 1996. Genomic organization of lactic acid bacteria. *Antonie Leeuwenhoek* **70**:161–183.

17. **Deborde, C. D. B. Rolin, A. Bondon, J. D. D. Certaines, and P. Boyaval.** 1998. In vivo nuclear magnetic resonance study of citrate metabolism in *Propionibacterium freudenreichii* subsp. *shermanii. J. Dairy Res.* **65**:503–514.

18. **Detmers, F. J. M., E. R. S. Kunji, F. C. Lanfermeijer, B. Poolman, and W. N. Konings.** 1998. Kinetics and specificity of peptide uptake by the oligopeptide transport system of *Lactococcus lactis. Biochemistry* **37**:16671–16679.

19. **de Vos, W. M., and G. F. M. Simons.** 1994. Gene cloning and expressions system in lactococci, p. 52–105. *In* M. J. Gasson and W. M. de Vos (ed.), *Genetics and Biotechnology of Lactic Acid Bacteria.* Chapman & Hall, Ltd., London, United Kingdom.

20. **Driesson, F. M., and Z. Puhan.** 1988. Technology of mesophilic fermented milks. *Int. Dairy Fed. Bull.* **227**:75–81.

21. **Dunny, G. M., and L. L. McKay.** 1999. Group II introns and expression of conjugative transfer functions in lactic acid bacteria. *Antonie Leeuwenhoek* **76**:77–88.

22. **Durmaz, E., and T. R. Klaenhammer.** 1995. A starter culture rotation strategy incorporating paired restriction/modification and abortive infection bacteriophage defenses in a single *Lactococcus lactis* strain. *Appl. Environ. Microbiol.* **61**:1266–1273.

23. **Forde, A., and G. F. Fitzgerald.** 1999. Bacteriophage defense systems in lactic acid bacteria. *Antonie Leeuwenhoek* **76**:89–113.

24. **Fordyce, A. M., V. L. Crow, and T. D. Thomas.** 1984. Regulation of product formation during glucose or lactose limitation in nongrowing cells of *Streptococcus lactis. Appl. Environ. Microbiol.* **48**:332–337.

25. **Fox, P. F. (ed.).** 1993. *Cheese: Chemistry, Physics and Microbiology,* vol. 1 and 2. Chapman & Hall, Ltd., London, United Kingdom.

26. **Fox, P. F., J. Law, P. L. H. McSweeney, and J. Wallace.** 1993. Biochemistry of cheese ripening, p. 389–438. *In* P. F. Fox (ed.), *Cheese: Chemistry, Physics and Microbiology,* vol. 1. Chapman & Hall, Ltd., London, United Kingdom.

27. **Fox, P. F., J. A. Lucey, and T. M. Cogan.** 1990. Glycolysis and related reactions during cheese manufacture and ripening. *Food Sci. Nutr.* **29**:237–253.

28. **Fryer, T. F., M. E. Sharpe, and B. Reiter.** 1970. Utilization of milk citrate by lactic acid bacteria and "blowing" of film wrapped cheese. *J. Dairy Sci.* **37**:17–28.

29. **Gancel, F., and G. Novel.** 1994. Exopolysaccharide production by *Streptococcus salivarius* ssp. *thermophilus* cultures. 1. Conditions of production. *J. Dairy Sci.* **77**:685–688.

30. **Gasson, M. J., and W. M. de Vos (ed.).** 1994. *Genetics and Biotechnology of Lactic Acid Bacteria.* Chapman & Hall, Ltd., London, United Kingdom.

31. **Gasson, M. J., and G. F. Fitzgerald.** 1994. Gene transfer systems and transposition, p. 1–51. *In* M. J. Gasson and W. M. de Vos (ed.), *Genetics and Biotechnology of Lactic Acid Bacteria.* Chapman & Hall, Ltd., London, United Kingdom.

32. **Green, M. L., and D. J. Manning.** 1982. Development of texture and flavor in cheese and other fermented products. *J. Dairy Res.* **49**:737–748.

33. **Griffith, R., and E. G. Hammond.** 1989. Generation of Swiss cheese flavor components by the reaction of amino acids with carbonyl compounds. *J. Dairy Sci.* **72**:604–613.

34. **Huang, D. Q., H. Prevost, and C. Divies.** 1995. Principal characteristics of β-galactosidase from *Leuconostoc* spp. *Int. Dairy J.* **5**:29–43.

35. **Hugenholtz, J.** 1993. Citrate metabolism in lactic acid bacteria. *FEMS Microbiol. Rev.* **12**:165–178.

36. **Imhof, R., and J. O. Bosset.** 1994. Review: relationships between micro-organisms and formation of aroma compounds in fermented dairy products. *Z. Lebensm.-Unters.-Forsch.* **198**:267–276.

37. **International Dairy Federation.** 1988. Fermented milks: science and technology. *Int. Dairy Fed. Bull.* **227**.

38. **Jarvis, A. W., G. F. Fitzgerald, M. Mata, A. Mercenier, H. Neve, I. B. Powell, C. Ronda, M. Saxelin, and M. Teuber.** 1991. Species and type phages of lactococcal bacteriophages. *Intervirology* **32**:2–9.

39. **Juillard, V., D. Le Bars, E. R. S. Kunji, W. N. Konings, J.-C. Gripon, and J. Richard.** 1995. Oligopeptides are the main source of nitrogen for *Lactococcus lactis* during growth in milk. *Appl. Environ. Microbiol.* **61**:3024–3030.

40. **Kamaly, M. K., and E. H. Marth.** 1989. Enzyme activities of lactic streptococci and their role in maturation of cheese: a review. *J. Dairy Sci.* **72**:1945–1966.

41. **Keenan, T. W., R. C. Lindsay, M. E. Morgan, and E. A. Day.** 1966. Acetaldehyde production by single strain lactic streptococci. *J. Dairy Sci.* **49**:10–14.

42. **Kempler, G. M., and L. L. McKay.** 1981. Biochemistry and genetics of citrate utilization in *Streptococcus lactis* spp. *diacetylactis. J. Dairy Sci.* **64**:1527–1539.

43. **Klaenhammer, T. R., and G. F. Fitzgerald.** 1994. Bacteriophages and bacteriophage resistance. *In* M. J. Gasson and W. M. de Vos (ed.), *Genetics and Biotechnology of Lactic Acid Bacteria.* Chapman & Hall, Ltd., London, United Kingdom.

44. **Kunji E. R. S., I. Mierau, A. Hagting, B. Poolman, and W. N. Konings.** 1996. The proteolytic systems of lactic acid bacteria. *Antonie Leeuwenhoek* **70**:187–221.

45. **Lees, G. J., and G. R. Jago.** 1978. Role of acetaldehyde in metabolism: a review. 1. Enzymes catalyzing reactions involving acetaldehyde. *J. Dairy Sci.* **61**:1205–1215.

46. **Lees, G. J., and G. R. Jago.** 1978. Role of acetaldehyde in metabolism: a review. 2. The metabolism of acetaldehyde in cultured dairy products. *J. Dairy Sci.* **61**:1216–1224.

47. Lemieux, L., and R. E. Simard. 1992. Bitter flavour in dairy products. II. A review of bitter peptides from caseins: their formation, isolation and identification, structure masking and inhibition. *Lait* **72:**335–382.

48. Lindgren, S. E., and L. T. Axelsson. 1990. Anaerobic L-lactate degradation by *Lactobacillus plantarum*. *FEMS Microbiol. Lett.* **66:**209–214.

49. Lindsay, R. C., E. A. Day, and W. E. Sandine. 1965. Green flavor defect in lactic starter cultures. *J. Dairy Sci.* **48:**863–869.

50. Liu, S. Q., R. V. Asmundson, P. K. Gopal, R. Holland, and V. L. Crow. 1998. Influence of reduced water activity on lactose metabolism by *Lactococcus lactis* subsp. *cremoris* at different pH values. *Appl. Environ. Microbiol.* **64:**2111–2116.

51. Lucey, C. A., and S. Condon. 1986. Active role of oxygen and NADH oxidase in growth and energy metabolism of *Leuconostoc*. *J. Gen. Microbiol.* **132:**1789–1796.

52. Manning, D. J. 1979. Chemical production of essential Cheddar flavor compounds. *J. Dairy Res.* **46:**531–537.

53. McKay, L. L., and K. A. Baldwin. 1974. Altered metabolism in a *Streptococcus lactis* C2 mutant deficient in lactate dehydrogenase. *J. Dairy Sci.* **57:**181–186.

54. Mercenier, A., P. H. Pouwels, and B. M. Chassy. 1994. Genetic engineering of lactobacilli, leuconostocs and *Streptococcus thermophilous*, p. 252–294. *In* M. J. Gasson and W. M. de Vos (ed.), *Genetics and Biotechnology of Lactic Acid Bacteria*. Chapman & Hall, Ltd., London, United Kingdom.

55. Mierau I., E. R. S. Kunji, K. J. Leenhouts, M. A. Hellendoorn, A. J. Haandrikman, B. Poolman, W. N. Konings, G. Venema, and J. Kok. 1996. Multiple-peptidase mutants of *Lactococcus lactis* are severely impaired in their ability to grow in milk. *J. Bacteriol.* **178:**2794–2803.

56. Moineau, S. 1999. Applications of phage resistance in lactic acid bacteria. *Antonie Leeuwenhoek* **76:**377–382.

57. Montville, M. R., B. Ardestani, and B. L. Geller. 1994. Lactococcal bacteriophage require a host cell wall carbohydrate and a plasma membrane protein for adsorption and ejection of DNA. *Appl. Environ. Microbiol.* **60:**3204–3211.

58. Palles, T., T. Beresford, S. Condon, and T. M. Cogan. 1998. Citrate metabolism in *Lactobacillus casei* and *Lactobacillus plantarum*. *J. Appl. Microbiol.* **85:**147–154.

59. Peterson, S. D., and R. T. Marshall. 1990. Nonstarter lactobacilli in Cheddar cheese: a review. *J. Dairy Sci.* **73:**1395–1410.

60. Poolman, B. 1993. Energy transduction in lactic acid bacteria. *FEMS Microbiol. Rev.* **12:**125–148.

61. Ramos, A., K. N. Jordan, T. M. Cogan, and H. Santos. 1994. ^{13}C nuclear magnetic resonance studies of citrate and glucose cometabolism by *Lactococcus lactis*. *Appl. Environ. Microbiol.* **60:**1739–1748.

62. Ramos, A., J. S. Lolkema, W. N. Konings, and H. Santos. 1995. Enzyme basis for pH regulation of citrate and pyruvate metabolism by *Leuconostoc oenos*. *Appl. Environ. Microbiol.* **61:**1303–1310.

63. Salminen, S., and A. von Wright (ed.). 1998. *Lactic Acid Bacteria*. Marcel Dekker, Inc., New York, N.Y.

64. Schmitt, P., and C. Divies. 1991. Co-metabolism of citrate and lactose by *Leuconostoc mesenteroides* subsp. *cremoris*. *J. Ferment. Bioeng.* **71:**72–74.

65. Sijtsma, L., N. Jansen, W. C. Hazeleger, J. T. M. Wouters, and K. J. Hellingwerf. 1990. Cell surface characteristics of bacteriophage-resistant *Lactococcus lactis* subsp. *cremoris* SK110 and its bacteriophage-sensitive variant SK112. *Appl. Environ. Microbiol.* **56:**3230–3233.

66. Smart, J. B., and T. D. Thomas. 1987. Effect of oxygen of lactose metabolism in lactic streptococci. *Appl. Environ. Microbiol.* **53:**533–541.

67. Steele, J. L. 1995. Contribution of lactic acid bacteria to cheese ripening, p. 209–220. *In* E. L. Malin and M. H. Tunick (ed.), *Chemistry of Structure-Function Relationships in Cheese*. Plenum Publishing Corp., New York, N.Y.

68. Thierry, A., D. Salvat-Brunaud, and J.-L. Maubois. 1999. Influence of thermophilic lactic acid bacteria strains on propionibacteria growth and lactate consumption in an Emmental juice-like medium. *J. Dairy Res.* **66:**105–113.

69. Thomas, T. D., D. C. Ellwood, and V. M. C. Longyear. 1979. Change from homo- to heterolactic fermentation by *Streptococcus lactis* resulting from glucose limitation in anaerobic chemostat cultures. *J. Bacteriol.* **138:**109–117.

70. Urbach, G. 1995. Contribution of lactic acid bacteria to flavor compound formation in dairy products. *Int. Dairy J.* **5:**877–903.

71. Vaughan, E. E., S. David, A. Harrington, C. Daly, G. F. Fitzgerald, and W. M. De Vos. 1995. Characterization of plasmid-encoded citrate permease (*citP*) genes from *Leuconostoc* species reveals high sequence conversation with the *Lactococcus lactis citP* gene. *Appl. Environ. Microbiol.* **61:**3172–3176.

72. Visser, S. 1993. Proteolytic enzymes and their relation to cheese ripening and flavor: an overview. *J. Dairy Sci.* **76:**329–350.

73. von Wright, A., and M. Sibakov. 1998. Genetic modification of lactic acid bacteria, p. 161–210. *In* S. Salminen and A. von Wright (ed.), *Lactic Acid Bacteria*. Marcel Dekker, Inc., New York, N.Y.

74. Warriner, K. S. R., and J. G. Morris. 1995. The effects of aeration on the bioreductive abilities of some heterofermentative lactic acid bacteria. *Lett. Appl. Microbiol.* **20:**322–327.

75. Weimer, B., K. Seefeldt, and B. Dias. 1999. Sulfur metabolism in bacteria associated with cheese. *Antonie Leeuwenhoek* **76:**247–261.

76. Whitehead, W. E., J. W. Ayres, and W. E. Sandine. 1993. A review of starter media for cheese making. *J. Dairy Sci.* **76:**2344–2353.

77. Wilkins, D. W., R. H. Schmidt, R. B. Shireman, K. L. Smith, and J. J. Jezeski. 1986. Evaluating acetaldehyde synthesis from L-[^{14}C(U)] threonine by *Streptococcus thermophilus* and *Lactobacillus bulgaricus*. *J. Dairy Sci.* **69:**1219–1224.

78. Zourari, A., J. P. Accolas, and M. J. Desmazeaud. 1992. Metabolism and biochemical characteristics of yogurt bacteria. A review. *Lait* **72:**1–34.

Food Microbiology: Fundamentals and Frontiers, 2nd Ed.
Edited by M. P. Doyle et al.
© 2001 ASM Press, Washington, D.C.

Herbert J. Buckenhüskes

Fermented Vegetables

32

In vegetable processing, two preservation methods, brining (or salting) and fermentation, are closely related to each other. Depending on the amount of salt (sodium chloride) added, the raw material will be preserved owing to a decrease of water activity and high ionic activity (equilibrium salt content >10%). Otherwise, vegetables will undergo fermentation. Brining is rarely practiced today owing to problems of salt removal and decreased consumer acceptance of high-salt foods.

Fermentation is a very ancient process of preserving plant materials while retaining their nutritive value. The fermentation of vegetables may be affected by various groups of microorganisms (Table 32.1). Lactic acid bacteria and yeasts are preferentially used in the Occident (Europe and America), whereas in the Orient a great number of victuals are fermented by molds (see chapter 35). However, the most extensively used procedure for biopreservation of vegetables involves lactic acid fermentation.

Lactic acid fermentation of vegetables is believed to have first been practiced by the Chinese in prehistoric times. However, the oldest written evidence dates from the first century A.D. when Plinius described the preservation of white cabbage in earthen vessels. Today we are aware that this procedure results in lactic acid fermentation of cabbage into sauerkraut. Because of the development of efficient heat sterilizing and refrigeration systems, lactic acid fermentation has, to some extent, lost its importance as a preservation method in industrialized countries, but it is still extensively used in developing countries, where in recent years it has gained in importance. This is due to the fact that fermentations today are more than just preservation methods—they are used to

- preserve vegetables and fruits
- develop characteristic sensory properties, i.e., flavor, aroma, and texture
- destroy naturally occurring toxins and undesirable components in raw materials (Table 32.2)
- improve digestibility, especially of some legumes (Table 32.2)
- enrich products with desired microbial metabolites, e.g., L-(+)-lactic acid or amino acids (Table 32.3)
- create new products for new markets (for examples, see Table 32.3)
- enhance dietary value (26)

An up-to-date review concerning the status of vegetable fermentation in Europe was prepared by the

Herbert J. Buckenhüskes, Gewürzmüller GmbH, Klagenfurter Strasse 1-3, D-70469 Stuttgart, Germany.

Table 32.1 Types of vegetable fermentation

Type of fermentation	Product	Raw material	Major microorganisms involved
Lactic acid	Fermented vegetables	See Table 32.4	Lactic acid bacteria, e.g., *Leuconostoc mesenteroides*, *Lactobacillus plantarum*
	Fermented vegetable juices	Sauerkraut, carrots, celery, tomatoes, red beets, turnips	*Lactobacillus plantarum*, *Lactobacillus casei*, *Lactobacillus xylosus*, *Lactobacillus bavaricus*
Acetic acid	Vinegar	Grapes, potatoes, various fruits after alcoholic fermentation	*Acetobacter aceti*, *Acetobacter pasteurianus*, *Acetobacter hansenii*, *Gluconobacter oxydans*
Alcoholic	Spirits	Potatoes, horseradish (for rimanto), peas (fen-djin)	*Saccharomyces cerevisiae*, *Kluyveromyces marxianus*
	Soy sauce	Soybeans	*Aspergillus oryzae*, lactobacilli, *Pediococcus* species, yeasts

Table 32.2 Application of microorganisms in vegetable processing to achieve technological or nutritional advantages

Application	Species	Effect	References
Sauerkraut	Paired starter cultures	Aroma formation	23, 24
	Leuconostoc mesenteroides	Nisin resistant	
	Lactococcus lactis	Nisin producing	
Fermented vegetables	Various species	Increased bioavailability of iron by reducing the phytate content	2
Lupins, peas, lentils, and other legumes	Various species, e.g., *Lactobacillus acidophilus*, *Lactobacillus buchneri*, *Lactobacillus fermenti*	Improved digestibility, e.g., by reduction in flatulence-causing oligosaccharides	12, 13
Cassava	*Lactobacillus plantarum*	Reduction in linamarine	22
Delicacy salads, potato salads, ready-to-eat salads	Various	Extension of shelf life	25, 41
Salads, baby food, vegetable juices	Various	Reduction of nitrate content	44

Table 32.3 Development of new fermented vegetable products

Application	Species	Effect	Reference
Fermented fruit juices	*Lactobacillus casei*	Supplementation with L-(+)-lactic acid	43
Tempeh-like products from beans (*Vicia faba*)	*Rhizopus oligosporus*	Shelf life, aroma formation	3
Yoghurt-like product from soy proteins	Combined starter: *Streptococcus salivarius* subsp. *thermophilus* *Lactobacillus delbrueckii* subsp. *bulgaricus*	Acidification Aroma formation	4
Camembert cheese-like product from advanced soy proteins	Combined starter: *Lactococcus lactis* subsp. *cremoris*, *Lactococcus lactis* subsp. *lactis* biovar diacetylactis, *Leuconostoc mesenteroides* subsp. *cremoris*	Acidification and aroma formation	4

Table 32.4 Lactic acid-fermented vegetables and fruits produced commercially[a]

Artichoke	Melons
Capers	Olives
Carrots	Red beets
Cauliflower	Red cabbage
Celery	Silver-skinned onions (levant garlic)
Cucumbers	Swedes (*Brassica napus*)
Eggplants	Tomato-shaped paprika (green and red)
Fungi (not specified)	Turnips (*Brassica rapa*)
Green beans	Waxy paprika
Green tomatoes	White cabbage (sauerkraut, shredded and
Green paprika	whole heads)
Lupinus beans	

[a] From reference 8.

Cooperation in Science and Technology 91 Program of the European Communities (8). The proceedings list 22 vegetables that are commercially fermented (Table 32.4). In addition, an unspecified number of variably formulated vegetable blends as well as fermented vegetable juices from cabbage, carrots, celery, tomatoes, red beets, and turnips are available in the market. However, at present, only olives, cabbage for sauerkraut, and cucumbers for pickles are of major economic importance.

TECHNOLOGY OF FERMENTATION

The fermentation of vegetables represents a very complex network of independent and interactive microbiological, biochemical, enzymatic, chemical, and physical processes and reactions. Factors influencing major fermentation reactions and interactions and quality attributes affected are shown in Fig. 32.1. To complicate matters, fermentation is influenced by a multitude of exogenous factors which can be classified into four groups: technological factors, nature and amount of admixed ingredients and additives, quality of the raw material (which in turn depends on numerous agricultural factors), and the nature of microflora on raw materials. An overview of the relevant influencing factors and their important interactions is given in Fig. 32.2.

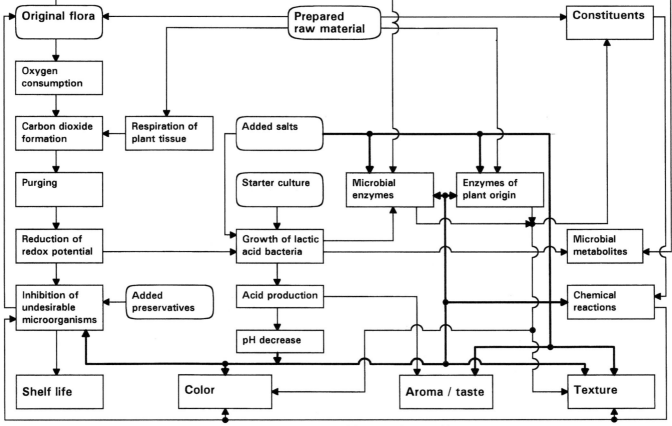

Figure 32.1 Dynamics of fermented vegetables.

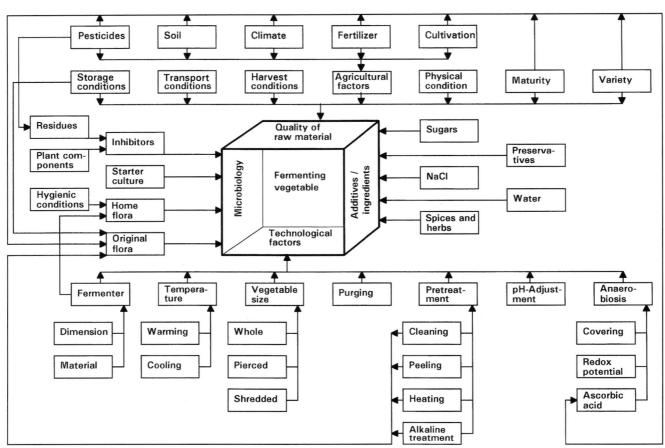

Figure 32.2   Factors influencing the quality of fermented vegetables.

Technological Factors

Botanical, physical, chemical, and textural properties of various vegetables differ widely, resulting in differences in technologies applied to fermentation. A simplified flow sheet for manufacturing lactic acid-fermented vegetables is shown in Fig. 32.3. The essential processing requirements can be described as following:

1. To produce high-quality fermented products, vegetables must be sound, undamaged, and at the proper stage of maturity. Raw materials that are to be fermented whole must be suitably size-graded to achieve a homogeneous fermentation.
2. In the first instance, mechanically damaged, bruised, or diseased fruits, dirty or withered leaves, and unripe or overripe fruit must be removed. For the production of sauerkraut, the core of the cabbage must be removed.
3. Further pretreatment depends on the particular vegetable and may include peeling (e.g., beets, carrots), blanching (green beans), cooking (beets, to

modify the texture as well as to destroy characteristic earthy flavor), or an alkaline treatment (green Spanish-style olives).

4. If the raw material is to be fermented in an altered form, it may be pierced, shredded, or sliced. Piercing of the skin or surface of plant materials facilitates access of brine to inner tissues as well as sugars and other substances from the tissues into the brine, which otherwise can only occur by relatively slow diffusion. The importance of this step in onion fermentation is illustrated in Fig. 32.4. More extensive wounding of tissue results in an accelerated decrease of pH during fermentation. Because fermentation mainly occurs in the liquid phase of the vegetable and brine system, a faster release of nutrients from pierced, shredded, or sliced onions or other vegetables results in a more rapid production of acid and, consequently, in an accelerated penetration of acid from the brine into the vegetable tissue. In cucumber fermentation, lactic acid bacteria can enter tissue from the brine

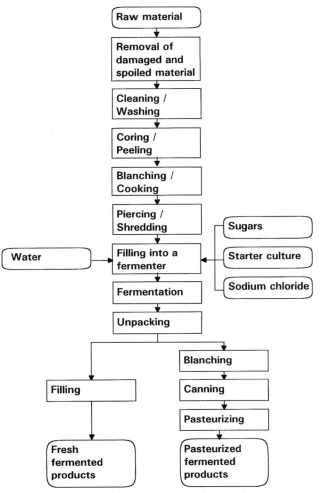

Figure 32.3 General flow sheet for the production of fermented vegetables (11).

damage. The filled vessels must be sealed in such a way that the plant material is totally covered by brine. To achieve anaerobic conditions, open tanks or silos are normally covered with wooden plates weighted with stones or by water-filled balloons. This process results in submersion of materials and, thus, exclusion of oxygen and spoilage microorganisms from the air.

6. The vast majority of vegetables are spontaneously fermented by lactic acid bacteria naturally present on their surfaces or from contact with equipment during preparation in processing facilities. Fermentative processes leading to good quality products can occur at 10°C, but the optimum temperature is between 15 and 20°C. The fermentation time depends primarily on the temperature, the type of vegetable, the degree of disintegration, and the desired kind and quality of the final product. During fermentation, tanks must be carefully monitored for acid development and microbial defects. Sometimes it is necessary to remove yeasts and molds that have grown on the surface of brines. *Debaryomyces hansenii* and *Pichia membranaefaciens* are among the most common spoilage yeasts in brine.

7. The final fermented product may either be distributed fresh, packaged or unpackaged, or pasteurized in pouches, cans, or jars. Products such as sauerkraut can be stored without any further treatment in fermentation tanks until needed for sale in the retail market. For products that easily soften, the original fermentation liquid should be replaced by a brine containing an appropriate amount of salt and lactic acid.

If fermented vegetables are to be distributed unpasteurized, it is important that all fermentable carbohydrates have been metabolized. Otherwise, secondary fermentation by yeasts may result in gaseous spoilage, brine turbidity, and potentially alcoholic fermentation. In the case of cucumbers, secondary fermentation may cause bloater formation (14). Butyric acid spoilage of fermented cucumbers can also occur as a result of secondary fermentation of lactic acid by bacteria (19).

In the case of pasteurized products, desired fermentation may be achieved as soon as the pH value is reduced to 4.1. Products are then blanched, placed in cans or jars, topped with fermentation liquid or a liquid containing salt and sometimes herbs, spices, and/or wine, and pasteurized. Until recently, pasteurization has been carried out by traditional methods of trial and error or by following the rule that "much helps much." To obtain

via stomata and grow within. Yeasts present in the brine are unable to enter cucumbers through stomata, presumably because of their larger size (15).

5. Whole, pierced, shredded, or sliced vegetables are placed in suitable fermentation vessels, which may range from 100-liter drums to tanks or silos with approximately 100-ton capacity. Filling must be done very carefully. To achieve favorable fermentation conditions, vegetables such as shredded cabbage must be salted homogeneously on a conveyor belt between shredding machines and tanks or during filling of silos. Whole fruits, such as cucumbers or tomatoes, must be placed into vessels already containing some brine to prevent mechanical

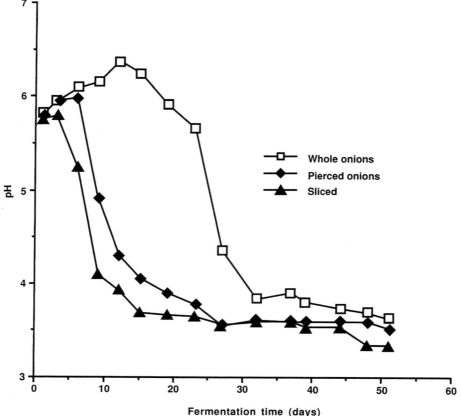

Figure 32.4   Effects of stitching and slicing on pH of fermented onions.

less heat-treated and therefore higher-quality products as well as to save energy, pasteurization processes should be evaluated and defined using *P*-value calculations. However, successful process development is difficult because of a lack of information about heat inactivation kinetics of several enzymes. Pectinolytic enzymes or lipoxygenases can cause marked changes in sensory qualities (6).

Ingredients and Additives

Sodium Chloride
In principle, vegetables can be fermented without the addition of sodium chloride (18). However, salt is often a very important ingredient for many reasons, a major one being its contribution to flavor. The amount used depends on the particular vegetable and on consumer demands. For example, the average salt content of canned sauerkraut is 11.3 g/kg in Germany and 16.7 g/kg in the United States (7).

Salt supports the development of anaerobic conditions in fermentation vessels. In sauerkraut production,

salt enhances the release of tissue fluids from shredded cabbage during filling into fermentation vessels. To increase the fermentation capacity, a part of this brine is removed while the balance takes up space between shredded cabbage pieces, thus excluding oxygen and promoting growth of lactic acid bacteria.

Salt exerts a selective effect on microorganisms naturally present on vegetables. Increasing amounts of salt inhibit the growth of undesirable bacteria and fungi and select for lactic acid bacteria. In sauerkraut fermentation, heterofermentative lactic acid bacteria are favored by low salt concentration (ca. 1%) and are greatly inhibited at 3%. Higher amounts of salt favor homofermentative species, resulting in accelerated fermentation. From a microbiological safety point of view, this may be of interest; however, an unbalanced fermentation by homofermentative lactic acid bacteria results in sauerkraut with poor flavor because of the lack of acetic acid and other desirable fermentation by-products. It is generally agreed that salt content below 0.8% often results in undesirable fermentation as well as in soft sauerkraut. The latter quality defect may be due to the fact that salt has an adverse

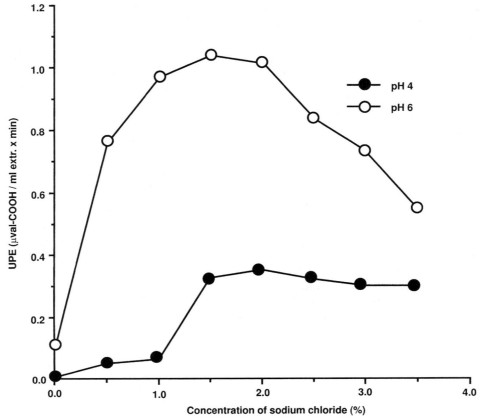

Figure 32.5 Effect of sodium chloride on the activity of cucumber (cv. Christine) pectinesterase.

effect on the texture of final products as well as on several plant and microbial enzyme systems (Fig. 32.1). For example, the effect of increasing the amount of salt at pH 4 and pH 6 on the activity of cucumber pectinesterase has been demonstrated (Fig. 32.5). Pectinesterase and other enzymes are involved in the softening of cabbage, cucumbers, olives, and other vegetables. Normally, cell wall degradation starts with enzyme-catalyzed deesterification of carboxyl groups of pectin molecules. The polysaccharide chain is then hydrolyzed by polygalacturonases which are specifically active at the deesterified positions within the chain.

While these processes may normally cause a more or less rapid disintegration of the cell wall constituents with consequent softening of the plant material, investigations have shown that deesterification may also enhance the texture of fermented vegetables. The biochemical mechanism involves linking deesterified carboxyl groups of adjoining pectin molecules with bivalent cations, e.g., calcium. In cucumber fermentation, this can be achieved by stimulation of pectinesterase by using raw materials of optimal maturity or by process control (e.g., adjust-

ment of sodium chloride and calcium chloride concentration), by inhibiting polygalacturonases through the addition of sufficient amounts of sodium chloride or by heat treatment before fermentation, and finally by the addition of calcium chloride to fermented products (34).

It is not clear whether lactic acid bacteria are extensively involved in the softening of fermented vegetables. Pectinesterase and endopolygalacturonase activity in *Lactobacillus plantarum* has been demonstrated (38, 39) but this phenomenon has not been confirmed by other researchers.

Carbohydrates

Considering the variety of fermented vegetables produced in homes and available in the marketplace, as well as the number of crops ensilaged for animal feed, it is assumed that most, if not all, plant material will undergo lactic acid fermentation if properly prepared. The major requirement is the availability of sufficient amounts of fermentable carbohydrates to be converted to lactic and acetic acid. The extent of pH decrease in fermented plant material as caused by production of acids depends on the buffering capacity of the product, which is positively correlated with protein content (5). If the buffering capacity is known, it is possible to calculate the minimum amount of fermentable carbohydrate needed to produce sufficient acid to reduce the pH to 4.1. Numerous tests with sauerkraut have shown that, in practice, approximately twice the amount of carbohydrate calculated is needed to meet the desired pH reduction. Using this as a basis for sufficient fermentation, the amount of fermentable carbohydrates such as sucrose or glucose required to achieve desired fermentation can be estimated for certain vegetables, such as cauliflower and celery.

Other Additives

The use of other food additives in fermented vegetables depends on the particular product and is restricted by law in nearly all countries. The addition of ascorbic acid to sauerkraut is commonly done to prevent gray or brown discoloration of fresh or canned products. For the same purpose, citric acid or sulfur dioxide is permitted in several countries.

To prevent microbial spoilage, and especially the growth of yeasts and molds on the surface of brines or in fermenting vegetables, sorbic acid is permitted in several countries. In some eastern European countries, the use of restricted amounts of tartaric acid, benzoic acid, or acetic acid is allowed.

Microbiology

Each particular type of vegetable provides a unique environment in terms of type, availability, and concentration of substrate, buffering capacity, competing microorganisms, and perhaps natural plant antagonists (15). As shown by characteristics compiled in Table 32.5, wounded or disintegrated plant tissues provide excellent substrates for microbial growth.

Fresh plant material harbors numerous and varied types of microorganisms. Although an extremely small population of lactic acid bacteria are present, it is assumed that plants are a natural habitat for some species. An analysis of 30 different samples of white cabbage from four growing seasons has shown that the microflora normally is dominated by aerobic bacteria (e.g., pseudomonads, enterobacteria, and coryneforms) and yeasts, while lactic acid bacteria represent 0.15 to 1.5% of the total bacterial population (40). Lactic acid bacteria mainly consist of *Leuconostoc mesenteroides* subsp. *mesenteroides*; streptococci have been isolated in only a few cases. Lactic acid bacteria reported to be associated

Table 32.5 Ecological factors affecting fermentation of vegetables

- The fermentation substrate consists of solids in a liquid environment; the brine can be circulated by a pump, distributing the microorganisms as well as the sodium chloride and the released nutrients; within the vegetable pieces, mass transfer can only occur by diffusion.
- Microorganisms originating from the raw material are distributed throughout the fermentation stock.
- *Salmonella* sp., clostridia, *Listeria* sp., and other undesirable microorganisms are present in all probability.
- Raw materials are rich in nutrients, growth factors, and minerals; however, for microbial growth, these substances must move from the plant tissue into the surrounding liquid.
- Water activity (a_w): 0.95–0.99
- pH: 5.9–6.5 for cabbage; 8.5 is maximum for olives.
- Temperature: t = 5–20°C for sauerkraut, 25–30°C for vegetable juice
- Sugar content: 25–100 g/kg of vegetable
- Buffer capacity: 0.15–0.90 g of lactic acid/100 g of vegetable
- Sodium chloride content: 0.6–2% for sauerkraut, 5–10% for brine of cucumbers

with plants are summarized in Table 32.6. Although the composition of natural microflora is affected by methods of preparing vegetables for fermentation, this does not seem to have a significant effect on the fermentation process.

The microbial population in general, as well as the population of lactic acid bacteria, undergoes considerable change during the course of vegetable fermentation. The spontaneous fermentation of cabbage, for example, has been categorized into four distinct stages:

1. Fermentation starts as soon as the cabbage is filled into vessels. When cabbage is tightly packed, the number of strictly aerobic bacteria decreases

Table 32.6 Lactic acid bacteria associated with plants[a]

Lactobacillus	*Pediococcus*
Lb. arabinosus	*P. acidilactici*
Lb. brevis	*P. pentosaceus* (formerly *P. cerevisiae*)
Lb. buchneri	*Enterococcus*
Lb. casei	*E. faecalis*
Lb. curvatus	*E. faecalis* subsp. *liquefaciens*
Lb. fermentum	*E. faecium*
Lb. plantarum	*Lactococcus*
Lb. sakei	*Lc. lactis*
Leuconostoc	
L. mesenteroides	

[a] From reference 15.

Table 32.7 Development of bacteria during sauerkraut fermentation carried out at 17°C in laminated plastic pouches[a]

Time (days)	pH	Redox potential	Aerobic bacteria			Total no.	
			Total (CFU/g)	Strictly aerobic (%)	Enterobacteria (%)	Lactic acid bacteria (CFU/g)	Yeasts (CFU/g)
0	6.48	ND[b]	1.9×10^5	95.0	1.5	6.8×10^2	3.9×10^4
1	5.72	23.0	3.0×10^5	50.0	50.0	2.8×10^4	1.1×10^4
2	5.63	22.6	1.7×10^6	7.0	93.0	6.6×10^6	1.2×10^3
3	4.38	17.5	6.9×10^6		100.0	4.4×10^8	9.5×10^2
4	4.23	15.5	1.6×10^5		100.0	9.7×10^8	1.0×10^2
5	4.04	14.1	5.0×10^3		100.0	8.3×10^8	$<10^2$
7	4.02	14.5	1.5×10^3		100.0	3.3×10^8	$<10^2$
9	4.00	15.3	$<10^2$	ND	ND	2.6×10^7	$<10^2$
11	3.93	15.4	$<10^2$	ND	ND	4.2×10^7	$<10^2$
14	3.96	15.7	$<10^2$	ND	ND	8.0×10^6	$<10^2$

[a] Data from reference 40.
[b] ND, not determined.

immediately, while facultatively anaerobic enterobacteria grow for the first 2 or 3 days (Table 32.7). During this period, oxygen dissolved in the substrate is consumed by these microorganisms and by plant respiratory processes. The change in pH is influenced by the formation of lactic, acetic, formic, and succinic acids. The production of carbon dioxide at the beginning of fermentation can result in the formation of foam.

2. Owing to a lack of oxygen, non-lactic acid bacteria are suppressed and overgrown by the facultatively anaerobic lactic acid bacteria. Lactic acid fermentation is initiated by heterofermentative lactic acid bacteria, namely, *L. mesenteroides*, followed by betabacteria, e.g., *Lactobacillus brevis*. Depending on the temperature, this succession of microorganisms is complete after 3 to 6 days, during which the concentration of lactic acid will increase to approximately 1%.

3. The third stage of fermentation is dominated by homofermentative lactic acid bacteria which are selectively favored by the complete lack of oxygen, lowered pH, and an elevated salt content. Approximately 90% of the homofermentative bacteria consist of streptobacteria, whereas streptococci and pediococci represent usually less than 10% of the total number of lactic acid bacteria. In the older literature, streptobacteria are described as a single species, namely, *Lb. plantarum* (formerly *Lactobacillus cucumeris*). Recent investigations have revealed that only 30 to 80% of the streptobacteria in sauerkraut are truly *Lb. plantarum*. During the early part of the third stage of fermentation, only 1

or 2% are true members of the species. In addition to *Lb. plantarum*, two other closely related species are of importance, namely, *Lactobacillus sakei* and *Lactobacillus curvatus*. Of special interest is the occurrence of *Lactobacillus bavaricus*, which is characterized by the exclusive formation of L-(+)-lactic acid (27). Growth of homofermentative lactic acid bacteria will increase the total acid content of the substrate to 1.5 to 2.0%. Most sauerkraut is pasteurized upon reaching pH of 3.8 to 4.1.

4. Only sauerkraut which is stored in the fermentation vessel and distributed as fresh sauerkraut undergoes the final stage of fermentation, which is dominated by *Lb. brevis* and some heterofermentative species that are able to metabolize pentoses, such as arabinose and xylose. Living plant material usually does not contain free pentoses. These sugars are liberated after harvest as a result of hydrolysis of cell wall material. The acid content may increase to 2.5%.

The initial population and growth rate of microorganisms, as well as salt and acid tolerance, are important factors that influence the sequential development of various lactic acid bacteria in most vegetable fermentations. In cucumber fermentation, excessive growth of heterofermentative *L. mesenteroides* is particularly undesirable, since the production of carbon dioxide may contribute to gaseous spoilage. The same situation occurs in the production of Spanish-style olives, which in general represent a very special type of fermentation. To destroy oleuropein, an extremely bitter-tasting glucoside in olives, freshly harvested fruits are treated with sodium hydroxide (lye) before fermentation. When the debittering process is finished, the lye must be removed by washing and neutralization, causing extensive losses in nutrients. For detailed information, see Garrido Fernández (21).

BIOGENIC AMINES

The most biogenic amines present in foods are formed by decarboxylation of corresponding amino acids through substrate-specific microbial enzymes. Biogenic amines and their precursors in fermented vegetables include ethanolamine (serine), putrescine (ornithine), cadaverine (lysine), spermidine (reaction of putrescine with a propylamine residue that is derived from methionine), phenylethylamine (phenylalanine), tyramine (tyrosine), and histamine (histidine). Concentrations of biogenic amines detected in sauerkraut are as high as 104 μg/g of histamine, 192 μg/g of tyramine, 311 μg/g of cadaverine, and 550 μg/g of putrescine (42). Vegetables are normally fermented by undefined spontaneous microflora, including decarboxylase-positive microorganisms. Genera of *Enterobacteriaceae* and *Bacillaceae* present at the beginning of fermentation, as well as species of *Lactobacillus*, *Pediococcus*, and *Streptococcus*, are known to be capable of decarboxylating one or more amino acids.

Biogenic amine formation during a spontaneous sauerkraut fermentation of 25-kg test batches of sauerkraut is shown in Fig. 32.6A. Although this investigation (29) was not attended by a detailed microbiological analysis, it is probable that the initial fermentation phase was dominated by *L. mesenteroides*. On the other hand, the formation of histamine was accompanied by a vigorous growth of *Pediococcus* species. The effect of a commercial *Lb. plantarum* starter culture (2×10^6 CFU/g) on the development of biogenic amines is illustrated in Fig. 32.6B. While the amount of histamine, cadaverine, and tyramine remained about the same as shown in Fig. 32.6A, the formation of putrescine was significantly suppressed. Despite inoculation with a starter culture, *L. mesenteroides* dominated the initial stage of fermentation; however, the growth of pediococci was not observed (29).

NITRATE REDUCTION

Depending on the plant species, the part of the plant utilized, several horticultural factors, and various processing and fermentation conditions, the nitrate content of fermented vegetables can vary considerably. Microbiological activity during the early fermentation stage converts nitrate into nitrite (1). Owing to an acidic environment, nitrite may react with secondary or tertiary amines or with amides to form undesirable nitrosamines recognized as having potent and organ-specific carcinogenicity. To improve the safety of products, attempts have been made to reduce the initial nitrate content by biotechnological means. Various microorganisms, including lactic acid bacteria, are capable of reducing nitrate. *Paracoccus denitrificans* has been reported to reduce nitrate in commercial carrot juice (28), so that nitrite accumulation was not observed.

A decrease in nitrate and nitrite content in plant products due to metabolic activity of lactic acid bacteria has been described by several researchers. Using MRS agar containing nitrate and molybdenum, it is possible to induce nitrate reductase formation by *Lactobacillus pentosus* (44). Investigations with cabbage juice have revealed that only 10% of the original nitrate is reduced and that nitrite accumulates. Several strains of *Lb. plantarum*, *Lb. pentosus*, and *L. mesenteroides* are capable of producing nitrite reductases. The final product of nitrite reduction by these bacteria is ammonia. Owing to an acidic environment, nitrate or nitrite reduction is not complete, so the process is not a major concern in the fermented vegetable industry (17).

BACTERIOPHAGES

Bacteriophages are not a problem in spontaneous fermentation of vegetables or after application of starter cultures. This is because pure-culture fermentations seldom occur. If a starter or a naturally fermented batch is infected with phages, other strains of naturally occurring lactic acid bacteria will become dominant and carry out the fermentation. Generally, the problem is much more important in natural fermentation of cucumbers and olives than in sauerkraut fermentation. Cucumbers are fermented in brine which sometimes is circulated or purged with nitrogen. In the event of an infection, both treatments cause a homogeneous dissemination of phages in the brine. On the other hand, the absence of circulating brine in sauerkraut fermentation precludes the dissemination of phages.

MICROBIAL SPOILAGE AND OTHER PROBLEMS

A summary of economically important problems which may occur during the fermentation and storage of fermented vegetables is presented in Table 32.8. Although some problems can be prevented by good manufacturing practice, their molecular or microbial background is not always clear.

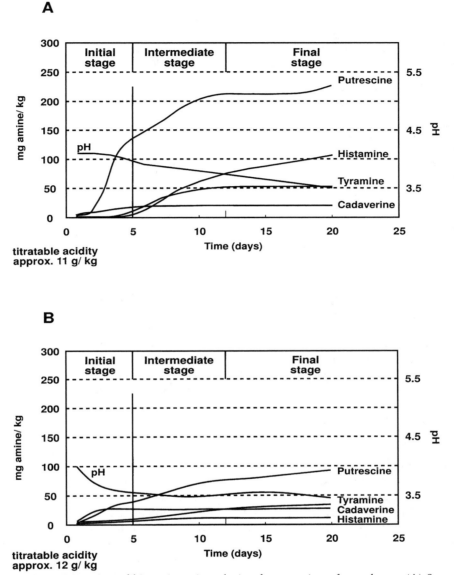

Figure 32.6 Formation of biogenic amines during fermentation of sauerkraut. (A) Spontaneous fermentation. (B) Fermentation after inoculation with *Lb. plantarum*.

Bloater formation is a serious problem in cucumber fermentation. Damage is accentuated with larger cucumbers and at higher fermentation temperatures, and the problem usually increases with an increase of dissolved carbon dioxide content in the brine. Measures should be taken to prevent or reduce the carbon dioxide content in brine and fermenting products, e.g., through the control of the initial population of bacteria and yeasts that are capable of producing carbon dioxide (30), control of excessive growth of heterofermentative lactic acid bacteria, removal of carbon dioxide by purging the brine with nitrogen gas (18), or brine circulation.

A further source of carbon dioxide sufficient to contribute to bloater damage is malolactic fermentation by certain homofermentative lactic acid bacteria, whereby malate is converted to lactate and carbon dioxide (33). Malic acid is the predominant organic acid naturally present in pickling cucumbers. The concentration ranges from 0.2 to 0.3% among various cultivars (32). The use of mutants of *Lb. plantarum*, obtained by *N*-methyl-*N*′-nitro-*N*-nitrosoguanidine mutagenesis, which lack the ability to produce carbon dioxide from malic acid has been described but not exploited commercially (16, 31).

Table 32.8 Problems which may occur during production and storage of lactic acid-fermented vegetables[a]

Vegetable	Discoloration	Softening[b]	Other problems
Cucumbers	White or gray internal spots; surface discoloration	+++	Bloater formation, off flavor, butyric acid fermentation (19)
Sauerkraut	Gray, pink, or brown	+++	Off flavor
Olives	Gray, brown, or yellow spots	+++	Off flavor, butyric acid fermentation, propionic acid fermentation
Cauliflower	Yellow or pink		
Green tomatoes	Gray	++	
Paprika	Gray	+++	Bitter taste
Carrots	Bleaching		Oxidized taste
Celery	Brown		
Green beans		++	Off flavor
Swedes	Gray or pink		

[a] Data from reference 8.
[b] ++, relevant; +++, serious problem.

STARTER CULTURES

Fermentation of the vast majority of vegetables is done by naturally occurring lactic acid microflora rather than by defined starter cultures. Economically interesting exceptions are the so-called L-(+)-sauerkraut and most fermented vegetable juices, except sauerkraut juice. L-(+)-Sauerkraut is directly fermented in small containers (300 to 400 ml) using homofermentative *Lb. bavaricus*, which mainly produces L-(+)-lactic acid, as a starter culture. The product is made in Germany and distributed directly in the fermentation container through health food stores. The *Lb. bavaricus* strain used is highly competitive so that the initial *Leuconostoc* population is suppressed and even the homofermentative lactic acid bacteria are not able to compete. In freshly made sauerkraut, the L-isomer represents more than 90% of the total lactic acid content. The suppression of the heterofermenters leads to a less developed flavor, which, however, is accepted by some consumers (27).

Most fermented vegetable juices are manufactured according to the "lactoferment-process" (10). In contrast to other vegetable products, mash or raw juice is pasteurized before fermentation so that pure culture fermentation can be achieved through the addition of a starter. Lactic acid bacteria used for this purpose include *Lactobacillus* species (*Lb. acidophilus*, *Lb. delbrueckii*, *Lb. helveticus*, *Lb. plantarum*, *Lb. salivarius*, *Lb. xylosus*, *Lb. bifidus*, *Lb. brevis*, and *Lb. casei*), *Lactococcus lactis*, and *L. mesenteroides* (11). The species most often used are *Lb. plantarum* and sometimes *Lb. casei*.

Because of requirements in some countries for certification according to ISO 9000–9004, modern consumer demands, and economic reasons, it is anticipated that the development of controlled fermentation processes will become necessary in order to consistently provide safe products with high quality. However, the fermentation of vegetables is difficult to control, primarily because of variation in shape, the large number and types of naturally occurring microorganisms, and variability in nutrient content. Since it is not possible to eliminate the natural microflora by appropriate methods, consistent products made by pure culture fermentation realistically cannot be achieved. One promising approach might be the application of defined starter cultures which are capable of growing rapidly and are highly competitive under environmental conditions used to ferment products. Depending on the particular material to be fermented as well as the desired quality of the final product, starter cultures must possess numerous characteristics. Selection criteria for starter cultures used in the fermentation of sauerkraut, cucumbers, olives, and vegetable juices are compiled in Table 32.9. However, because of modest profit margins in the vegetable fermentation industry, it is questionable whether the use of starter cultures is economically acceptable.

Extensive investigations of lactic acid bacteria used in sauerkraut fermentation have shown that strains of *Lb. plantarum* may cause a rapid decrease in pH but lead to poorly flavored products. Sauerkraut with excellent sensory characteristics has been obtained using selected strains of *L. mesenteroides* which simultaneously improve the uniformity of sauerkraut from different batches (9).

Investigations to find bacteriocin-producing strains of *L. mesenteroides* with selective advantages over

Table 32.9 Traits considered relevant in starter cultures for vegetable fermentation[a, b]

Criteria	Sauerkraut	Cucumbers	Olives	Vegetable juices[c]
Technologically relevant criteria				
Rapid and predominant growth	++	++	++	+
Homofermentative metabolism	−	++	++	++
Salt tolerance	+	++	++	0
Acid production and tolerance	++	++	++	+
Inability to metabolize organic acids	++	++	++	+
Growth at low temperatures	++	++	++	0
Few growth factors required	0	0	+	0
Tolerance of phenolic glycosides	0	0	++	0
Formation of dextrans	−	−	−	−
Pectinolytic activities	−	−	−	−
Formation of bacteriocins	+	+	+	0
Bacteriophage resistance	0	+	+	++
Sensorially relevant criteria				
Heterofermentative metabolism	++	−	−	−
Formation of flavor precursors	++	++	+	0
Nutritionally relevant criteria				
Reduction of nitrate and nitrite	+	0	0	++
Formation of L-(+)-lactate	+	+	0	++
Formation of biogenic amines	−	−	−	−

[a] Data from references 11 and 14.
[b] ++, important; +, advantageous; 0, not relevant; −, detrimental.
[c] Except sauerkraut juice.

competing but sensitive strains have not been successful. A new approach involves using a paired starter culture system consisting of a nisin-resistant *L. mesenteroides* strain and a nisin-producing *Lc. lactis* strain, both of which were isolated from commercial sauerkraut fermentations (23). Whether the presence of nisin can actually promote and extend the dominance of heterofermentative *L. mesenteroides* in a spontaneous fermentation has yet to be demonstrated (24).

PROSPECTS

Systematic scientific research on fermented vegetables dates back to the early 1900s. Several hundred research papers and reports have been published on this subject, the majority of which have dealt with microbiological, technological, and analytical aspects of sauerkraut, cucumber, and olive fermentation. Still, we are far from understanding the complete array of biochemical reactions and interactions that lead to reproducible quality of fermented vegetables. Attempts are being made to develop models for cucumber fermentation. Such models can be helpful for the prediction of microbiological safety and shelf life of pickles, detection of critical control points in production and distribution processes, optimization of production and distribution chains, and control of fermentation processes (36, 37).

A summary of the major areas of future research needs published by the Cooperation in Science and Technology program is still relevant (8):

- prevention of softening during fermentation and storage
- prevention of discoloration
- reducing the nitrate content, especially of green leafy and root vegetables
- prevention of formation of biogenic amines
- outlining controlled fermentation processes
- development and use of starter cultures
- identification and controlled formation of typical aroma components

To solve these problems, a better knowledge of ecological factors prevailing in fermenting substrates and an improved understanding of the role of various microorganisms in the development of sensory properties will be necessary. In addition, a better understanding of the genetics of lactic acid bacteria and perhaps of other microorganisms involved in the fermentation of vegetables will enhance the ability of the vegetable fermentation industry to consistently produce high-quality products.

Lactic acid fermentation is a valuable tool for the production of a wide range of vegetable products. Owing to its long tradition, lactic acid fermentation is more readily accepted by consumers than many other food preservation techniques. Given this situation and considering public opinion with regard to gene manipulation, it is questionable whether genetic engineering should be pursued to solve the problems summarized in Table 32.8. Most of the physiological properties recognized to be essential for vegetable fermentation are inherent in lactic acid bacteria, and several of the desired properties depend on plasmid-encoded traits which facilitate genetic manipulation. However, respecting consumer opinion, at present it appears expedient to take advantage of the pool of naturally available strains of lactic acid bacteria rather than to apply genetically modified microorganisms.

References

1. Andersson, R. 1984. Characteristics of the bacterial flora isolated during spontaneous lactic acid fermentation of carrots and red beets. *Lebensm. Wiss. Technol.* **17**:282–286.

2. Andersson, R., A.-S. Svanberg, and U. Svanberg. 1990. Effect of lactic acid fermentation of vegetables on the availability of iron. *FEMS Microbiol. Rev.* **9**:100.

3. Berghofer, E., and A. Werzer. 1986. Production of tempeh from domestic beans. *Chem. Mikrobiol. Technol. Lebensm.* **10**:54–62. (In German.)

4. Brückner, H., A. Salmen, A. Amar, and H. J. Buckenhüskes. 1993. Fermented dairy-like products from advanced soy proteins, p. 215–222. *In Proceedings of Euro Food Chem VII*, Valencia, Spain.

5. Buckenhüskes, H., and K. Gierschner. 1985. Determination of the buffering capacity of some vegetables. *Alimenta* **24**:83–88. (In German.)

6. Buckenhüskes, H., K. Gierschner, and W. P. Hammes. 1988. Theory and practice of pasteurization—optimization of pasteurizing of pickled vegetables. *Ind. Obst. Gemüseverwert.* **73**:315–322. (In German.)

7. Buckenhüskes, H., A. Gessler, and K. Gierschner. 1988. Analytical characterization of canned and pasteurized sauerkraut. *Ind. Obst. Gemüseverwert.* **73**:454–463. (In German.)

8. Buckenhüskes, H. J., H. Aabye Jensen, R. Andersson, A. Garrido Fernandez, and M. Rodrigo. 1990. Fermented vegetables, p. 167–187. *In P. Zeuthen, J. C. Cheftel, C. Eriksson, T. R. Gormley, P. Linko, and K. Paulus (ed.), Processing and Quality of Foods, vol. 2. Food Biotechnology: Avenues to Healthy and Nutritious Products.* Elsevier Applied Science, London, United Kingdom.

9. Buckenhüskes, H. J., and W. P. Hammes. 1990. Starter cultures in fruit and vegetable processing. *BioEngineering* **6**:34–42. (In German.)

10. Buckenhüskes, H., and K. Gierschner. 1989. Vegetable juice production—state of the art. *Flüss. Obst* **56**:751–764. (In German.)

11. Buckenhüskes, H. J. 1993. Selection criteria for lactic acid bacteria to be used as starter cultures for various food commodities. *FEMS Microbiol. Rev.* **12**:253–272.

12. Camacho, L., C. Sierra, D. Marcus, E. Guzman, M. Andrade, R. Campos, N. Diaz, M. Parraguez, and L. Trugo. 1991. Nutritional improvement of commonly consumed vegetables fermented by cultures of *Lactobacillus* spp. *Alimentos* **16**:5–11.

13. Camacho, L., C. Sierra, D. Marcus, E. Guzman, R. Campos, D. von Bäer, and L. Trugo. 1991. Nutritional quality of lupine (*Lupinus albus* cv. multolupa) as affected by lactic acid fermentation. *Int. J. Food Microbiol.* **14**:277–286.

14. Daeschel, M. A., and H. P. Fleming. 1984. Selection of lactic acid bacteria for use in vegetable fermentations. *Food Microbiol.* **1**:303–313.

15. Daeschel, M. A., R. E. Andersson, and H. P. Fleming. 1987. Microbial ecology of fermenting plant materials. *FEMS Microbiol. Rev.* **46**:357–367.

16. Daeschel, M. A., R. F. McFeeters, and H. P. Fleming. 1985. Modification of lactic acid bacteria for cucumber fermentations: elimination of carbon dioxide production from malate. *Dev. Ind. Microbiol.* **26**:339–346.

17. Emig, J. 1989. Microbial nitrate reduction and lactic acid fermentation of vegetable products. Dissertation, Hohenheim University, Stuttgart, Germany. (In German.)

18. Fleming, H. P., J. L. Etchells, R. L. Thompson, and T. A. Bell. 1975. Purging of CO_2 from cucumber brines to reduce bloater damage. *J. Food Sci.* **40**:1304–1310.

19. Fleming, H. P., M. A. Daeschel, R. F. McFeeters, and M. D. Pierson. 1989. Butyric acid spoilage of fermented cucumbers. *J. Food Sci.* **54**:636–639.

20. Fleming, H. P., L. C. McDonald, R. F. McFeeters, R. L. Thompson, and E. G. Humphries. 1995. Fermentation of cucumbers without sodium chloride. *J. Food Sci.* **60**:312–315, 319.

21. Garrido Fernández, A. 1990. Controlled fermentation of green table olives, p. 211–218. *In P. Zeuthen, J. C. Cheftel, C. Eriksson, T. R. Gormley, P. Linko, and K. Paulus (ed.), Processing and Quality of Foods, vol. 2. Food Biotechnology: Avenues to Healthy and Nutritious Products.* Elsevier Applied Science, London, United Kingdom.

22. Giraud, E., L. Gosselin, and M. Raimbault. 1992. Degradation of cassava linamarin by lactic acid bacteria. *Biotechnol. Lett.* **14**:593–598.

23. Harris, L. J., H. P. Fleming, and T. R. Klaenhammer. 1992. Characterization of two nisin-producing *Lactococcus lactis* strains isolated from a commercial sauerkraut fermentation. *Appl. Environ. Microbiol.* **58**:1477–1483.

24. Harris, L. J., H. P. Fleming, and T. R. Klaenhammer. 1992. Novel paired starter culture system for sauerkraut, consisting of a nisin-resistant *Leuconostoc mesenteroides* strain and a nisin-producing *Lactococcus lactis* strain. *Appl. Environ. Microbiol.* **58**:1484–1489.

25. Holzapfel, W. H., R. Geisen, and U. Schillinger. 1995. Protective cultures—potentiality and limitations, p. 143–161. *In H. J. Buckenhüskes (ed.), Proceedings of the 13th Filderstädter Colloquium, Fertiggerichte-Fast Food-Catering.* Gesellschaft Deutscher Lebensmitteltechnologen e.V., Bonn, Germany. (In German.)

26. Huis in't Veld, J. H. J., H. Hose, G. J. Schaafsma, H. Silla, and J. E. Smith. 1990. Health aspects of food

biotechnology, p. 273–297. *In* P. Zeuthen, J. C. Cheftel, C. Eriksson, T. R. Gormley, P. Linko, and K. Paulus (ed.), *Processing and Quality of foods*, vol. 2. *Food Biotechnology: Avenues to Healthy and Nutritious Products*. Elsevier Applied Science, London, England.

27. **Kandler, O., W. P. Hammes, M. Schneider, and K. O. Stetter.** 1987. Microbial interaction in sauerkraut fermentation, p. 302–308. *In Proceedings of the 4th International Symposium on Microbial Ecology.*

28. **Kerner, M., E. Mayer-Miebach, A. Rathjen, and H. Schubert.** 1990. Reduction of nitrate content in vegetable food using denitrifying microorganisms, p. 236–239. *In* P. Zeuthen, J. C. Cheftel, C. Eriksson, T. R. Gormley, P. Linko, and K. Paulus (ed.), *Processing and Quality of Foods*, vol. 2. *Food Biotechnology: Avenues to Healthy and Nutritious Products*. Elsevier Applied Science, London, United Kingdom.

29. **Künsch, U., H. Schärer, and A. Temperli.** 1989. Biogenic amines—a quality indicator of sauerkraut, vol. 24, p. 192–104. Tagung der Deutschen Gesellschaft für Qualitätsforschung, Kiel, Germany.

30. **McDonald, L. C., H. P. Fleming, and M. A. Daeschel.** 1991. Acidification effects on microbial populations during initiation of cucumber fermentation. *J. Food Sci.* **56:**1353–1356, 1359.

31. **McDonald, L. C., D. H. Shieh, H. P. Fleming, R. F. McFeeters, and R. L. Thompson.** 1994. Evaluation of malolactic-deficient strains of *Lactobacillus plantarum* for use in cucumber fermentation. *Food Microbiol.* **10:**489–499.

32. **McFeeters, R. F., H. P. Fleming, and R. L. Thompson.** 1982. Malic and citric acids in pickling cucumbers. *J. Food Sci.* **47:**1859–1861, 1865.

33. **McFeeters, R. F., H. P. Fleming, and R. L. Thompson.** 1982. Malic acid as a source of carbon dioxide in cucumber juice fermentation. *J. Food Sci.* **47:**1862–1865.

34. **Meurer, P.** 1991. Effect of plant origin enzymes and other factors on the texture of fermented cucumbers. Dissertation. Hohenheim University, Stuttgart, Germany. (In German.)

35. **Omran, H., H. Buckenhüskes, E. Jäckle, and K. Gierschner.** 1991. Some properties of pectinesterase, exo polygalacturonase and endo-β-1,4-gluconase of pickling cucumbers. *Dtsch. Lebensm. Rundsch.* **87:**151–156. (In German.)

36. **Passos, F. V., H. P. Fleming, D. F. Ollis, H. M. Hassan, and R. M. Felder.** 1993. Modeling the cucumber fermentation: growth of *Lactobacillus plantarum. J. Ind. Microbiol.* **12:**341–345.

37. **Passos, F. V., D. F. Ollis, H. P. Fleming, H. M. Hassan, and R. M. Felder.** 1993. Modeling the specific growth rate of *Lactobacillus plantarum* in cucumber extract. *Appl. Microbiol. Biotechnol.* **40:**143–150.

38. **Sakellaris, G., and A. E. Evangelopoulos.** 1989. Production, purification and characterization of extracellular pectinesterase from *Lactobacillus plantarum* (str. BA 11). *Biotechnol. Appl. Biochem.* **11:**503–507.

39. **Sakellaris, G., S. Nikolaropoulos, and A. E. Evangelopoulos.** 1989. Purification and characterization of an extracellular polygalacturonase from *Lactobacillus plantarum* strain BA 11. *J. Appl. Bacteriol.* **7:**77–85.

40. **Schneider, M.** 1988. Microbiology of sauerkraut fermented in small ready-to-sell containers. Dissertation, Hohenheim University, Stuttgart, Germany. (In German.)

41. **ten Brink, B.** 1993. Bacteriocins from lactic acid bacteria: natural preservatives? *Voedingsmiddelentechnologie* **26:**37–38.

42. **ten Brink, B., C. Damink, H. M. L. J. Joosten, and J. H. J. Huis in't Veld.** 1990. Occurrence and formation of biologically active amines in foods. *Int. J. Food Microbiol.* **11:**73–84.

43. **Wiesenberger, A., E. Kolb, J. A. Schildmann, and H. M. Dechent.** 1986. Lactic acid fermentation of natural substrates with low pH. *Chem. Mikrobiol. Technol. Lebensm.* **10:**32–36. (In German.)

44. **Wolf, G., and W. P. Hammes.** 1987. Lactic acid bacteria as agents for reduction of nitrate and nitrite in food. *Dechema Monogr.* **105:**271–272.

Food Microbiology: Fundamentals and Frontiers, 2nd Ed.
Edited by M. P. Doyle et al.
© 2001 ASM Press, Washington, D.C.

Steven C. Ricke
Irene Zabala Díaz
Jimmy T. Keeton

33

Fermented Meat, Poultry, and Fish Products

Fermented meat products are defined as meats that are deliberately inoculated during processing to ensure sufficient controlled microbial activity to alter the product characteristics (6). If fresh meat is not preserved or cured in some manner, it spoils rapidly owing to the growth of indigenous gram-negative bacteria and subsequent putrefaction resulting from their metabolic activities (29). Although some manufacturers still depend upon naturally occurring microflora to ferment meat, most use starter cultures consisting of a single species or multiple species combinations of lactic acid bacteria and/or micrococci that have been selected for metabolic activities especially suited for fermentation in meat ecosystems. Understanding the technological, microbiological, and biochemical processes that occur during meat, poultry, and fish fermentation is essential to ensure safe, palatable products.

MANUFACTURE OF FERMENTED MEAT AND POULTRY PRODUCTS

Sausage Categories and Meat Fermentation
Dry and semidry sausages represent the largest category of fermented meat products, with many of the present-day processing practices having their origin in the Mediterranean region. Traditionally, dry sausages acquired their particular sensory characteristics from exposure to salt and the rapid drying conditions existing in the warm, dry Mediterranean climate. These products were heavily seasoned and typically not smoked, and they derived their name, sausage, from the Latin term *salsus*, meaning salted. Sausage processing practices later spread to northern Europe and, by the Middle Ages, hundreds of varieties of dry and semidry sausages were manufactured across the continent. In contrast to the Mediterranean variety, northern European sausages were prepared during the cold winter months and stored until summer, and thus were called summer sausages. These sausages contained more water than their Mediterranean counterparts and were lightly spiced, heavily smoked at cool temperatures, and less susceptible to spoilage owing to colder ambient temperatures. Summer sausages are similar to present-day semidry sausages. Examples of the most common dry and semidry sausage varieties are listed in Table 33.1 and are categorized as salamis or cervelats.

Product Categories and Compositional Endpoint Characteristics
Compositional characteristics of dry and semidry sausage categories produced in the United States, and some processing criteria, have been defined by the U.S.

Steven C. Ricke, Irene Zabala Díaz, and Jimmy T. Keeton, Department of Poultry Science, Kleberg Animal and Food Science Center, Texas A&M University, College Station, TX 77843-2472.

Table 33.1 Categories and origins of selected dry and semidry sausages[a]

Category	Origin	Description and unique characteristics
Dry sausages		
Salamis (Genoa, Milano, Siciliano)	Italy	Lean pork (coarse), some with wine or cured beef (fine), garlic, large-diameter casing (beef or hog bungs) with flax twine, some with coating of white mold
Lombardia salami	Italy	Coarse cut, high fat content, brandy added, twine wrap over casing
Cappicola	Italy	Boneless pork shoulder butt combined with red hot or sweet peppers, mildly cured
Mortadella	Italy	Finely chopped, cured beef and pork with added cubes of backfat, mildly spiced, smoked, encased in beef bladders
Pepperoni	Italy	Cured pork, some beef, cubed fat, red peppers, small-diameter casing
Chorizos	Spain or Portugal	Pork (coarse), highly spiced, hot, small-diameter casing
D'Arles	France	Similar to Italian salamis, coarse, large-diameter casings (hog bungs)
Lyons	France	All-pork (fine), diced fat (fine), spices and garlic, cured, large-diameter casing
Alesandri and Alpino	United States	Similar to Italian salami
Katenrauchwurst, Dauerwurst, Plockwurst, Zerevelat	Germany	Dry sausages, smoked, air dried
Semidry sausages		
Summer sausages (cervelat, farmer cervelat)	Generic	Mildly seasoned soft cervelat, beef and pork (coarse), no garlic, cured, small-diameter casing
Holsteiner cervelat	Germany	Similar to farmer cervelat, packed in ring-shaped casing
Thuringer cervelat	Germany	Medium dry to soft, tangy flavor, mildly spiced, smoked, some beef and veal added
Gothaer cervelat	Germany	Very lean pork, fine chopped, cured, soft texture, mild seasoning
Goteborg cervelat, medwurst	Sweden	Coarse, salty, soaked in brine before heavy smoke
Landjaeger cervelat	Switzerland	Small diameter (frankfurter size), pressed flat, smoked, flavored with garlic and caraway seeds
Teewurst, frische Mettwurst	Germany	Undried, spreadable, smoked
Lebanon bologna	United States	Smoked, coarse, large diameter, very acid

[a] From references 6, 74, 82, and 84.

Department of Agriculture-Food Safety Inspection Service (USDA-FSIS) under 9 CFR part 319, subpart I, "Semi-Dry Fermented Sausage," and subpart J, "Dry Fermented Sausage." FSIS policy memo 056 (dated January 12, 1983) requires that these "shelf-stable" products be nitrited, cured, fermented, reach a final pH of 5.0 or less, and have an M/P ratio (moisture/protein) of 3.1:1.0 or less (75). Examples of the compositional characteristics of dry and semidry fermented meat products are given in Table 33.2.

Manufacturing Procedures and Processing Conditions

Dry and semidry sausage manufacture using starter cultures involves the following basic steps: (i) reducing the particle size of high-quality raw meat trimmings; (ii) incorporation of salt, nitrate (Europe, mostly) or nitrite (United States and Europe), glucose, spices, seasonings, and a specific inoculum selected on the basis of incubation temperature optimum and level of lactic acid desired; (iii) uniformly blending all ingredients and further

reducing of particle size; (iv) vacuum stuffing into a semipermeable casing to minimize the presence of oxygen; (v) incubation (ripening) at or near the temperature optimum of the starter culture until a specific pH endpoint is achieved or until carbohydrate utilization is complete; (vi) heating (usually, but product dependent) of the product to inactivate the inoculum and ensure pathogen destruction; and (vii) drying (aging) the product to

Table 33.2 Composition of two types of fermented sausages[a]

Parameter	Summer sausage/ Thuringer cervelat	Pepperoni/ hard salami
Moisture	50	30
Fat	24	39
Protein	21	21
Salt	3.4	4.2
pH	4.9	4.7
Total acidity	1.0	1.3
Yield	90	64

[a] Values (except pH) are expressed as percentage (wt/wt). Adapted from reference 99.

Table 33.3 Generic manufacturing scheme for dry and semidry fermented sausages with starter cultures[a]

Processing sequence	Semidry sausage	Dry sausage (North America)	Dry sausage (Europe)
Meat tissue selection	Fresh/frozen meats with low microbial populations, no discoloration, no off odors, limited age, no dark, firm and dry (DFD) tissue, trimmed free of blood clots, glands, sinews, gristle, bruises, refrigerated to <4.5°C (<40°F)	Same as semidry	Same as semidry
Comminution, grinding, blending ingredients; inoculum level at ca. 10[7] CFU/g	Fresh/tempered −3°C (26°F) meats, coarse grind/chop/mince lean (1/4–1/2 in., 6.35–12.7 mm) and fat (1/2–1 in., 12.7–25.4 mm) meats separately, combine each to a specified fat endpoint; blend with seasonings and cure ingredients to uniformly distribute ingredients; avoid overmixing and excessive protein extraction or fat smearing	Same as semidry	Same as semidry
Ingredients used Salt (2.5–3%) Glucose (0.4–0.8%) Nitrite (<150 mg/kg) Sodium erythorbate (550 mg/kg) Antioxidants (natural or synthetic) Spices (sterilized)	Rehydrate frozen or lyophilized culture with nonchlorinated, distilled water at ambient temperature <1 h before use. Tap water is usually acceptable. If antimicrobial compounds are present (i.e., chlorine) and are in excess, distilled water should be used. Add inoculum (high-temperature optimum for smokehouse incubation) to meat batch; blend, but avoid excessive mixing (causes fat smearing and coating of lean particles); fine grind/chop/mince (1/8–3/16 in., 3.2–4.8 mm) to specified particle size	Same as semidry. Exceptions: Use high-temperature inoculum (>32.5°C, 90°F) for smokehouse incubation, low-temperature inoculum (21.3°C, 70°F) for "green" or "ripening" room incubation	Same as semidry Exceptions: Use nitrate (200–600 mg/kg) alone or in combination with nitrite (nitrate used mostly in Europe; only for Lebanon bologna and country-style hams in U.S.); low-temperature fermentation (21.3°C, 70°F) used most often
Vacuum encasing (stuffing)	Keep at 2°C (ca. 34°F); vacuumize to remove oxygen and encase in fibrous or natural casing. Oxygen exclusion accelerates anaerobic fermentation, favors lactic acid bacteria growth, color, and flavor development	Same as semidry	Same as semidry
Incubation and fermentation (ripening)	High-temperature smokehouse incubation at 32.5–38.1°C (90–100°F) and 90% relative humidity for ≥18 h (dependent upon sausage diameter) to an endpoint pH of <4.7. Air movement >1 m/s; smoke at the end of fermentation	Low-temperature smokehouse incubation or "ripening" room at 15–26°C (60–78°F) and 90% relative humidity for ca. 72 h to an endpoint pH of <4.7. Air movement >1 m/s; smoke application at the end of fermentation if desired	Low-temperature "ripening" at 26°C (78°F) and 88% relative humidity for 3 days to an endpoint pH of 4.7–4.8. Chamber relative humidity held at 5–10% lower than relative humidity within the sausage or use the following schedule: a. 22.2–23.9°C (72–75°F), 94–95% relative humidity for 24 h b. 20.0–22.2°C (68–72°F), 90–92% relative humidity for 24 h c. 18.3–20.0°C (65–68°F), 85–88% relative humidity for 24 h

(Continued)

Table 33.3 Generic manufacturing scheme for dry and semidry fermented sausages with starter cultures[a] *(Continued)*

Processing sequence	Semidry sausage	Dry sausage (North America)	Dry sausage (Europe)
Drying (aging)	Drying chamber at 12.9–15.7°C (55–60°F) and 65–70% relative humidity for ≥12 days (dependent upon sausage diameter) to specified moisture/protein ratio	Drying chamber at 10–11.2°C (50–52°F) and 68–72% relative humidity for ≥21 days (dependent upon sausage diameter) to specified moisture/protein ratio	Remains in the ripening room or moved to drying chamber at 20°C (68°F) and 88% relative humidity for 10 days, then 15.7°C (60°F) and 82% relative humidity for 14 days to a specified endpoint or hold at 11.7–15.0°C) (53–59°F) at 75–80% relative humidity to specified endpoint

[a] Adapted from reference 6 and 59, and based on comments from reference 75.

the required moisture or moisture/protein endpoint. A generic scheme for manufacturing dry and semidry sausages using starter cultures is given in Table 33.3.

Factors Affecting Color, Texture, Flavor, and Appearance of Fermented Meats

Raw Meat Tissues

Fresh or frozen raw meats to be used for fermented sausages should be chilled to <4.5°C, have low microbial populations, be free of physical and chemical defects, and meet the compositional specifications for the product being manufactured. The predominant bacteria that develop in meats not held under vacuum are typically gram-negative, oxidase-positive, aerophilic rods composed of psychrotrophic pseudomonads along with psychrotrophic *Enterobacteriaceae* (59). Small numbers of lactic acid bacteria and other gram-positive microorganisms are initially present in meat and become the dominant microflora if oxygen is excluded, as in the case of vacuum-packaged or vacuum-encased meat products.

Postrigor pH and residual glycogen concentration of muscle tissues influence the quality of fermented sausages. ATP levels in muscle tissues average 1 μm/g 24 h after slaughter, while pH values range from 5.5 to 5.7 for beef, 5.7 to 5.9 for pork, and 5.8 to 6.0 for poultry. Pork meat sometimes exhibits pale, soft, and exudative (PSE) characteristics and has tissue pH values of 5.3 to 5.5, a wet or watery meat surface, and a pale pink color. However, these tissues may be incorporated into dry sausages at levels up to 50% of the meat block without impairing sensory qualities (100). Use of 100% PSE meats in sausages, however, will likely result in products with a pale, yellowish, cured color, poor water-holding capacity, lower water activity (a_w), increased susceptibility to oxidative rancidity, and poor (soft, grainy, noncohesive) textural characteristics. Meats exhibiting a

dark, firm, and dry (DFD) condition are characterized by a dry surface, dark red color, and high pH (>6.0 to 6.2). This condition occurs more frequently in beef than in pork. DFD trimmings are not suitable for dry sausage because of their excessive water-binding capacity and potential for accelerated microbial spoilage. Dark red meat from more mature animals, however, may be desirable because of its contribution to product appearance.

Use of lamb and mutton meats in fermented sausages is limited (2), but studies indicate that acceptable sausage products can be produced when combined with appropriate seasonings and limited amounts of mutton fat (11, 109). Fat tissues from beef, lamb, or pork have a high proportion of saturated fatty acids and yield products that are firmer and have a more desirable texture than products containing poultry and turkey fats, which have a predominance of polyunsaturated fatty acids. Polyunsaturated fatty acids are more susceptible to autoxidation and rancidity, which can lead to the development of off flavors. Thus, poultry meats may be a less desirable source of fat for fermented sausage formulations, because they contain higher levels of polyunsaturated fatty acids.

Lean poultry meat, if used, is often supplemented with pork or beef meat to avoid sausages that appear too light in color and to ensure appropriate textural attributes. Poultry sausages are usually lower in fat (15%) than red meat sausages (23 to 45%), initially have a higher pH, and contain more moisture that can affect product uniformity (diameter) during drying. Incorporation of turkey thigh meat into sausages can result in darker red products owing to higher myoglobin content in the lean tissue, but turkey, as with poultry, has a higher proportion of polyunsaturated fatty acids and moisture compared with red meats. Slightly higher levels of fermentable carbohydrate should be used in formulas containing higher-pH meats, such as poultry. Heating >68.9°C (155°F), applying heating schedules as outlined by the USDA (101), or heating for the same time and temperature combinations as specified for roast beef (102) may be required to ensure destruction of *Salmonella* sp. if present on poultry tissues. Mechanically deboned poultry meat is an acceptable meat source in fermented dry sausages when limited to 10% of the meat block (38), as sausages tend to become soft when higher amounts are used.

Ingredients

Incorporation of sodium chloride, sodium or potassium nitrite and/or nitrate, glucose, and homofermentative lactic acid starter cultures (Table 33.3) in sausage formulas dramatically alters the ecology of the culture environment and chemical characteristics of finished products (Table 33.4). Incubation of sausages at the optimum inoculum growth temperature under a reduced-oxygen environment causes a reduction in pH by the rapid conversion of glucose to lactic acid as a consequence of the exponential growth of lactic acid bacteria and sub-

sequent suppression of indigenous microflora, such as psychrotrophic pseudomonads, *Enterobacteriaceae*, and most pathogens. *Pseudomonas* species are sensitive to salt, nitrite, elevated incubation temperatures (>37°C), and reduced oxygen tension, while competitiveness of the *Enterobacteriaceae* is restricted at reduced oxygen levels, low pH, and the presence of salt.

The glucose content of postrigor meats (4.5 to 7 μmol/g) is not sufficient to significantly reduce pH; therefore, 0.4 to 0.8% fermentable carbohydrate in the form of glucose, sucrose, or maltose is added to sausage formulas to enable reduction of the pH to 4.6 to 5.0. About 1 oz of glucose per 100 lb of meat (0.62 g of glucose/kg of meat) is required to reduce the pH by 0.1 unit. Carbohydrates such as lactose, raffinose, trehalose, dextrins, and maltodextrins, when used in place of glucose, yield less lactic acid owing to incomplete utilization (50). In the United States, fermented sausages having final pH values of 4.8 to 5.0 contain approximately 25 g of lactic acid per kg (dry weight), while some Italian and Hungarian salamis that are fermented with limited amounts of carbohydrate may show a decrease in pH of only 0.5 unit.

Table 33.4 Chemical characteristics of selected fermented sausage products

Category	Final pH	Lactic acid (%)	Moisture/ protein ratio	Moisture loss (%)	Moisture[a] (%)	Comments
Dry sausages[b]	5.0–5.3 (<5.3)	0.5–1.0	<2.3:1	25–50	<35	Heat processed (optional)[c]; dried or aged after fermentation for moisture loss; may be smoked
Cervelat			1.9:1		32–38	Shelf stable
Cappicola			1.3:1		23–29	
German "Dauerwurst"	4.7–4.8		1.1:1		25–27	
German salami	4.7–4.8		1.6:1		34–35	
Pepperoni	4.5–4.8	0.8–1.2	1.6:1	35	25–32	
Italian salami, hard or dry			1.9:1	30	32–38	
Genoa salami	4.9	0.79	2.3:1	28	33–39	
Thuringer, dry	4.9	1.0	2.3:1	28	46–50	
Semidry sausages[b]	4.7–5.1 (<5.3)	0.5–1.3	2.3–3.7:1	8–15	45–50	Heat processed[c]; typically smoked; packaged after processing and chilling
Lebanon bologna	4.7	1.0–1.3	2.5:1	10–15	56–62	Keep refrigerated
Cervelat, soft			2.6:1	10–15		
Salami, soft			2.3–3.7:1	10–15	41–51	
Summer sausage	<5.0	1.0	3.1:1	10–15	41–52	
Thuringer, soft			3.7:1		46–50	
Other (for comparison)						
Dried beef			2.04:1	29		
Beef jerky			0.75:1	>50	28–30	
Air-dried sausage			2.1:1			

[a] Water activity ranges for dry and semidry sausages are <0.85–0.91 and 0.90–0.94, respectively. The European Economic Directive 77/99 requests a_w of <0.91 or pH<4.5 for dry sausages to be shelf stable or a combined a_w and pH of <0.95 and <5.2, respectively.
[b] Data from references 1, 6, 52, 72, 81, 84, 97, and 103.
[c] USDA–FSIS Title 9 CFR may be amended to require specified time and temperature heating combinations after fermentation or verification that processing conditions destroy all pathogenic microorganisms.

Nitrates and nitrites (40 to 50 mg/kg, minimum) in fermented sausages react with the heme moiety of myoglobin to facilitate the development of cured color, retard lipid oxidation, inhibit the growth of *Clostridium botulinum* (through a synergistic relationship with salt), and enhance cured flavor. In the United States, sodium chloride (2.5 to 3.0%) or a combination of sodium and potassium chloride is used with sodium or potassium nitrite (156 mg/kg maximum ingoing into the new product, not to exceed 200 mg/kg in the final product). Nitrite serves as the primary curing agent in fermented sausages, but sodium or potassium nitrate may be legally added to dry sausages at a maximum level of 1,718 mg/kg, calculated as sodium nitrate. In Europe, sodium or potassium nitrate (200 to 600 mg/kg), in combination with nitrate-reducing microorganisms, such as *Micrococcus varians* and *Staphylococcus carnosus*, are utilized during low-temperature fermentation (ripening) to enable metabolism by these acid-sensitive, nitrate-reducing bacteria. Sodium ascorbate is often used in combination with nitrates and nitrites in dry sausages at concentrations up to 550 and 600 mg/kg in the United States and Europe, respectively. Ascorbates, isoascorbates, or erythorbates accelerate the reduction of nitrous acid to nitric oxide, thus enhancing color development, reducing residual nitrite, and retarding the formation of *N*-nitrosamines, a class of potent carcinogens.

Ground pepper, paprika, garlic, mace, pimento, cardamon, red pepper, and mustard are commonly used in fermented sausages, but as with all spices, they should be sterilized to avoid wild fermentations. Red pepper and mustard are known to stimulate lactic acid formation, possibly because of the available manganese in these spices, which enhances the glycolytic enzyme fructose-1,6-diphosphate aldolase (59). Garlic, rosemary, and sage contain antioxidant and antimicrobial compounds and may assist in preserving flavor, color, and microbial shelf life of fermented sausages.

Glucono-delta-lactone (GDL), a chemical acidulant, is hydrolyzed to gluconic acid, which is then converted to lactic acid and acetic acid by indigenous lactobacilli. The acidulant is sometimes used at concentrations ranging from 0.25 to 0.5%, although up to 1% is allowed in the United States. At levels of 0.25% GDL (38), dry sausage color and consistency are important in conjunction wih a rapid decline in the pH caused by natural starter culture fermentation. At higher concentrations, GDL can inhibit growth of lactobacilli, produce undesirable aromas, and impart a sweet flavor from the unfermented sugar.

Smoke contains phenols, carbonyls, and organic acids, which may act to preserve sausage products. Phenols are effective antioxidants and microbial inhibitors, while carbonyls contribute a desirable amber color by combining with free amino groups to form brown furfural compounds. Organic acids in smoke, such as formic, acetic, propionic, butyric, and isobutyric acids, inhibit the growth of microorganisms on the surface of sausages and promote coagulation of surface proteins.

MANUFACTURE OF FERMENTED FISH PRODUCTS

Fermented fish products include a variety of fish sauces, fish pastes, and fish/vegetable blends that have been salted, packed whole in layers, or ground into small particles and then fermented in their own "pickle." These products are eaten as a proteinaceous staple or condiment in Southeast Asia but are consumed as a condiment in northern Europe (12). Fish fermentation involves minimal bacterial conversion of carbohydrates to lactic acid but entails extensive tissue degradation by proteolytic and lipolytic enzymes derived from viscera and muscle tissues. Low-molecular-weight compounds from fish tissue degradation are the primary contributors to aroma and flavor, characteristics of sauces. Indigenous microorganisms, however, do contribute to aroma and flavor but are limited to species tolerant of high salt concentrations (10 to 20%) in the curing brine. Partial tissue hydrolysis is responsible for the unique textural attributes of pastes and fish/vegetable blends.

Fish Sauces

Fish sauces, such as nuoc-mam (Vietnam), patis (Philippines), nam-pla (Thailand), budu (Malaysia), nuoc-mam-nuoc (Thailand), and shottsuru (Japan), are liquids consumed as a condiment with rice and vary in color from clear brown to yellow-brown. Sauces have a predominantly salty taste and are derived from decanting or pressing fermented fish or shrimp after a 9-month to 1-year fermentation (67). Products fermented over a 1- to 2-year period have a distinctive sharp, meaty aroma and may range in protein content from 9.6 to 15.2%. Commercial fish sauce production begins with layering seine-netted fish, shrimp, or shellfish with salt in concrete vats in an approximate ratio of 3:1 (fish/salt), sealing the vat, allowing supernatant liquor to develop, and carefully decanting this liquid. Enzymatic digestion or fermentation may range from 6 months for small fish to 18 months for larger species. The first liquid removed from the fermenting fish contains an abundance of peptides, amino acids, ammonia, and volatile fatty acids and is considered the highest quality. Extracts may be supplemented with caramel, caramelized sugar, molasses, roasted corn, or roasted barley to enhance the color and keeping qualities. For some products, the sauce is ripened in the sun for 1 to 3 months and blended with bacterial by-products from the manufacture of monosodium glutamate.

Nitrogen content of the supernatant liquor increases as a result of proteolysis during fermentation. Initially, salt penetrates the tissues by osmosis (0 to 25 days), a protein-rich liquid develops through autolysis (80 to 120 days), and, ultimately, the fish tissue is transformed into a nitrogen-containing liquid (140 to 200 days). Proteases such as cathepsin B and trypsinlike enzymes have been shown to increase soluble protein content of the liquor during the first 2 months, but their activity gradually decreases through an inhibition feedback mechanism with the buildup of amino acids and polypeptides. New polypeptide formation occurs during the last fermentation stage as the level of free amino acids increases. Aseptically produced fish sauces do not have a typical aroma (13), suggesting that some microbial involvement is required for aroma development. Bacterial populations in raw fish, predominantly facultative anaerobes, have been shown to be high (2.7×10^4 CFU/g) initially but decline (2×10^3 CFU/g) over a 6- to 9-month period. *Bacillus*, *Lactococcus*, *Micrococcus*, and *Staphylococcus* sp. have been confirmed as indigenous microflora, but as fermentation progresses, populations and number of species decline with the changing brine environment. Crisan and Sands (21) identified *Bacillus cereus* and *Bacillus licheniformis* as the dominant bacteria in nam-pla, but after 7 months' fermentation, another strain of *B. licheniformis*, *Bacillus megaterium*, and *Bacillus subtilis* were the dominant species. *Micrococcus copoyenes*, *M. varians*, *Bacillus pumilus*, and *Candida clausenii* are other halotolerant microorganisms, i.e., those capable of growth in 10% brine but not 20%, that have been isolated from sauces.

The characteristic aroma and flavor of fish sauce is complex and cannot be attributed to specific volatile fatty acids, peptides, or amino acids derived from bacterial fermentation alone. One theory suggests a complex interaction of enzymatic activity and oxidation during the fermentation, however, some evidence exists for bacterial production of volatile fatty acids in fresh fish before salting and in the brief period following salting (14). Dougan and Howard (28) characterized nam-pla aroma as being ammoniacal-trimethylamine, cheesy-ethanoic, and *n*-butanoic, and having other aromas attributable to low-molecular-weight volatile fatty acids, meaty ketones, keto acids, and amino acids, especially glutamic acid. The flavor of fish sauce has a strong salt component with contributions from a combination of mono-amino acids and possibly aspartic and glutamic acids. Other

factors, such as pH of the brine, fermentation temperature, and salt concentration, also affect flavor.

Fish Paste

Fish pastes, more widely produced than sauces, are consumed raw or cooked as a condiment with rice and vegetables. In some countries, fish pastes may be the primary source of dietary protein for low-income families. A wide variety of fermented fish pastes are produced and require shorter processing periods than sauces. These include bagoong, tinabal, and balbakwa (Philippines), pra-hoc (Cambodia), padec (Laos), blachan or bleachon, trassi-udang (Malaysia), sidal (Pakistan), and shiokara (Japan) (12). Paste production consists of mixing cleaned, eviscerated, whole or ground fish, shrimp, plankton, or squid with salt in a ratio of 3:1 (fish/salt) and then placing them in vats to ferment. Proteolytic enzymes from viscera tissue, and to some extent bacteria, break down the tissue until it attains a pasty consistency. Aging in hermetically sealed containers may follow, or the paste may be hand kneaded and then aged. Pickle (liquid exudate), which forms as a result of the osmotic differential of the brine solution and the fish tissue, is decanted, aged, and consumed as fish sauce. When pickle no longer forms, the fish paste is ready for use or aged further. Typical paste has a salty, cheeselike aroma, but other characteristics vary depending upon the method and region of production.

Bacteria do not appear to play a major role in the proteolysis of fish paste but may contribute to aroma and flavor or to spoilage. Populations of 6.5×10^3 CFU/g, representing 40 bacterial species, have been reported to occur in bagoong (20% salt), of which *Bacillus*, *Micrococcus*, and *Moraxella* species were dominant (32). Eighteen strains of bacteria, some of which were halophilic, were found in shiokara (13, 110), and *Micrococcus* species appeared to be responsible for ripening. Halophilic *Vibrio* and *Achromobacter* species were also isolated and were likely responsible for spoilage.

Proteolytic enzymes from fish viscera, stomach, pancreas, and intestine are more active than muscle enzymes and are most often responsible for release of free amino acids and polypeptides, which are believed to undergo further microbial synthesis to yield specific flavor compounds. Some of the compounds derived from proteolytic degradation include volatile fatty acids (formic, ethanoic, propanoic, isobutanoic, and *n*-pentanoic acids), ammonia, trimethylamine, and mono- and dimethylamines. Aminobutane and 2-methylpropylamine are thought to result from microbial action. If carbohydrates are added to the raw materials and fermented to alcohol, this may suppress proteolytic and microbial activity. Bogoong is the residue of partially hydrolyzed fish or shrimp (67) having a pH of 5.2 and containing 65 to 68% moisture, 13 to 15% protein, 2% fat, 20 to 30% salt, 1.45% nitrogen, 0.18% volatile nitrogen, 0.015% trimethylamine, 0.011% hydrogen sulfide, and a 35% solids base. Pra-hoc, used in dishes such as soups, may contain as much as 24% protein and 17% salt. Shiokara, a Japanese fish or squid product with or without malted rice, is texturally between a sauce and paste and is characterized as a dark brown liquid containing lumps of solid tissue. In its final form, the squid product contains 74.2% water, 7.8% salt, 11.6% protein, and 8.7% ash.

Fermented Rice and Shrimp or Rice and Fish

Balao balao, a traditional food of the Philippines, is a cooked rice and shrimp product which is fermented at room temperature for 7 to 10 days in a 20% salt brine (67). A similar Filipino fermented product in which fish is substituted for shrimp is known as burong isda, while the Japanese version is called naresushi or funasushi. Bacterial isolates involved in the fermentation of burong have been demonstrated to be capable of starch hydrolysis and are identified (67) as having characteristics similar to *Lactobacillus plantarum* and *Lactobacillus coryneformis* subsp. *coryneformis* but with the capacity to convert oligosaccharides and reducing sugars to lactic acid. An enzyme with a pH optimum of 4.0 has been isolated from these bacteria and found to hydrolyze amylose to oligosaccharides. However, during the initial stages of fermentation, other bacteria may hydrolyze starch for use by lactic acid bacteria.

STARTER CULTURES IN MEATS

The use of starter cultures in fermented meat products is a relatively recent practice compared with their use in fermented dairy foods and alcoholic beverages (55). The rationale for the use of a starter culture in meat fermentation is similar in concept to the use of starter cultures in dairy products. In the United States, lactobacilli or pediococci are the predominant culture microorganisms, while in Europe, these genera are most often used in combination with micrococci and staphylococci (8, 55). Proper inoculation and incubation procedures are among the most critical steps for the production of safe, flavorful, uniform, and wholesome fermented sausages. Inoculation is accomplished by one of three methods: natural fermentation, which relies on indigenous microflora in the meat to serve as the inoculum; back inoculation, which involves transfer of a portion of raw meat from a previous batch of sausage to the present batch, i.e., transfer

from a mother batch of raw sausage; or the use of starter cultures, i.e., inoculation of unfermented meat with a pure strain or strains of lactic acid bacteria. Use of commercial starter cultures is the predominant method of inoculation in the United States.

Development of Commercial Starter Cultures

In fresh meat, lactic acid bacteria are a minor component of the microflora, but when meat is packaged and stored under vacuum, the resulting microenvironment facilitates the growth of lactic acid bacteria (49). Commercial application of starter cultures in the cheese industry during the 1930s led investigators to identify potential starter cultures for fermented meats in the 1940s. Use of pure cultures in sausages began after several studies in the 1950s demonstrated that lactic acid bacteria are responsible for lactic acid production (23, 69–71). An essential requirement for starter cultures is that they can be produced and preserved in a viable and metabolically active form suitable for commercial distribution. The additional requirements for selection of starter cultures include producing adequate quantities of lactic acid, being able to grow at a salt concentration of at least 6%, being homofermentative, catalase positive, and able to enhance flavor in the finished sausage, and not producing biogenic amines and slimes (75). In the United States, *Lactobacillus* species were first used as starter cultures for meat fermentations in the temperature range of 20 to 25°C. Original starter cultures were logically derived from the predominant microflora of fermented meat products, but when attempts were made to preserve these cultures via lyophilization, as was routinely done with dairy starter cultures, these strains invariably died. Deibel et al. (24) were able to surmount this problem by using *Pediococcus cerevisiae* (now *Pediococcus acidilactici*), the first commercially available meat starter culture. Although *P. cerevisiae* was not a predominant bacterium in naturally fermented meats, it did, in addition to surviving lyophilization, possess characteristic lactic acid fermentation ability, have a higher optimal growth temperature, and tolerate salt concentrations up to at least 6.5% (6, 8, 24). When Deibel et al. (24) developed a lyophilization procedure for maintaining starter cultures consisting of pediococci, it was considered the best method for distributing a viable culture in a reliable and economical manner (10). Eventually, lyophilization proved commercially problematic because rehydration procedures were time-consuming and introduced unacceptable variation (6). Also, lyophilized cultures exhibited inordinately long lag phases, thus lengthening fermentation times. Implementation of frozen culture technology in the late 1960s, along with improvements in conditions for storing, handling, and shipping, led to commercial acceptance of frozen culture concentrates (10, 30). This approach eliminated the need for a rehydration step and provided higher numbers of viable cells than did lyophilization (79). Widespread use of pure starter culture strains in the United States occurred during the late 1970s and was a consequence of the development of efficient, high-volume dry sausage production technologies that required short ripening times and produced consistently uniform products with low defect levels.

Foodborne illness outbreaks associated with coagulase-positive staphylococci also focused attention on the need to control the fermentation process and ensure the production of safe products. In addition to solving these problems, the introduction of frozen culture concentrates also renewed interest in lactobacilli as starter cultures (6, 9). In 1974, *Lb. plantarum* was patented as a starter culture to be used alone or in combination with *P. acidilactici* for dry and semidry sausages (6, 31).

Currently, the predominant genera of bacteria used either singly or as mixed starter cultures in the United States and Europe are *Pediococcus*, *Lactobacillus*, *Micrococcus*, and *Staphylococcus* (6, 39). These cultures are available either fresh, frozen, freeze-dried, or in a low-temperature stabilized liquid form, with the frozen culture being used most often (7). *Lb. plantarum*, *P. acidilactici*, and *Pediococcus pentosaceus* (7, 22) are favored in the United States for their rapid and nearly complete (>90%) conversion of glucose to lactic acid (pH 4.6 to 5.1) at high temperatures (32°C). Culture type selection in the United States is based on fermentation temperature and final pH of the product (75). Final desired pH of fermented meat products depends on the "activity" of meat culture, which generally refers to the ability of a culture to reduce the pH of the meat sample under a set of defined conditions, especially the desired temperature of the fermentation (75). The rate of pH decline increases with increasing culture "activity"(75). Highly active meat starter cultures are able to decrease pH in a designed meat system from pH of 5.6 to 4.8 within 8 h of fermentation. *P. acidilactici* is used in high-temperature fermentation (35 to 46.1°C) and rapid pH decline (75). For low-temperature fermentation (21.1 to 35°C) and rapid pH decline, *P. pentosaceus* is preferred, while for low temperatures (21.1 to 35°C), and slow pH decline (slow fermentation), *Lb. plantarum* strains are preferred (75). In addition, selected *Micrococcus* spp. are added for color and flavor development. To improve textural properties of fermented sausages, control pathogens, and expand the temperature range for

fermentation, mixtures of some meat starter cultures can be used (75).

In Europe, less fermentative (pH 5.2 to 5.6) organisms—*Staphylococcus xylosus*, *S. carnosus*, *Staphylococcus simulans*, *Staphylococcus saprophyticus*, and, to a lesser extent, *Micrococcus* sp. (62), which has a lower temperature optimum (<24°C)—are preferred for flavor development and red color enhancement. Micrococci, which are commonly employed as starter cultures in the manufacture of European dry sausage, reduce nitrate and nitrite via reductase enzymes. Chemical reduction of nitrate/nitrite to nitric oxide via nitrate and nitrite reductase, followed by reaction with the singular heme moiety of myoglobin, forms dinitrosylhemochrome, which develops into a characteristic pink cure color when heated (73). Some lactobacilli (*Lactobacillus delbrueckii* subsp. *lactis*, *Lb. sake*, *Lb. farciminis*, *Lb. brevis*, *Lb. buchneri*, and *Lb. suebicus*) reduce nitrite to nitric oxide in vitro (75, 108). Red color development in meat can also be attributed to bacterial consumption of oxygen. Increased oxygen consumption by rapidly growing facultative anaerobes such as lactobacilli could reduce oxygen tension on meat surfaces (75). *Lactobacillus fermentum* forms a physical barrier which would limit access of oxygen to the meat surface underneath. In theory, bacteria initially consume oxygen and reduce the oxygen pressure to levels that allow for metmyoglobin formation (3, 75). However, further consumption of oxygen establishes a low oxygen environment and allows for metmyoglobin reduction to bright red myoglobin derivatives. Bacterial metabolites or intracellular components from bacterial cells are presumed responsible for metmyoglobin conversion, but the exact mechanism of the oxymyoglobin oxidation is not fully understood (75). *P. pentosaceus* and *P. acidilactici* have temperature optima of 32 and 35°C, respectively, while lactobacilli have lower optima, in the range of 21 to 24°C. Rapidly fermented products, such as beef sticks and pepperoni, utilize *P. acidilactici* and *P. pentosaceus* alone or in combination with *Lactobacillus* species or *Micrococcus* spp., which permits the use of elevated fermentation temperatures ranging from 32 to 46°C. *Lactobacillus curvatus*, *Lb. sake*, *Lb. plantarum*, *P. pentosaceus*, and *P acidilactici* produce bacteriocins which may find broader application in starter cultures for meat fermentations in the future.

Characteristics of Commercial Meat Starter Cultures

Bacterial starter cultures for meat and poultry products in the United States have been selected on the basis of being homofermentative, capable of rapidly converting glucose or sucrose to DL-lactic acid anaerobically with sustained growth to pH 4.5, tolerant of salt brines (NaCl, KCl) up to 6%, and capable of growth in the presence of sodium or potassium nitrate or nitrite in the range of 600 mg/kg (nitrate) and 150 mg/kg (nitrite). These cultures are aciduric, capable of growth in the range of 21 to 43°C, nonproteolytic, nonlipolytic, inactivated at 60.5 to 63.2°C, resistant to phage infection and mutation, and able to outgrow and/or suppress pathogens. Specific advantages favoring the use of commercial starter cultures over natural fermentations, i.e., back inoculation, are consistency of the inoculum, reduced risk of bacterial cross-contamination, uniformity of lactic acid development, production of desirable flavor components, predictability of pH endpoint (as regulated by carbohydrate level and incubation temperature), and reduced risk of proteolytic and pathogenic bacterial outgrowth. Other benefits include the acceleration of fermentation time to increase commercial plant throughput and the reduction of product defects, such as off flavors, lack of tangy flavor, excessive softness, crumbly texture, gas pockets, and pinholes, all of which can be attributed to heterofermentative bacteria.

An inoculum population of 10^7 CFU/g of raw product is sufficient for rapid lactic acid development within 6 to 18 h under controlled temperature, air flow, and relative humidity conditions (7), but incubation time is also dependent upon carbohydrate type and concentration, spice composition, exclusion of oxygen, and product diameter or thickness. In the United States, commercial processors inoculate meats with homofermentative, gram-positive bacteria that have been isolated from and adapted to specific meat products. These inocula (starter cultures) may consist of a single species of Lb. plantarum, Lactobacillus pentosus, Lb. sake, Lb. curvatus, M. varians, P. acidilactici, or P. pentosaceus, or, more commonly, a combination of these bacteria. Use of starter cultures ensures dominance of desirable microorganisms, production of acids consisting of >90% lactic acid, and modulation of fermentation based on combinations of inoculum level, glucose concentration, and incubation time/temperature intervals (6, 7). Thus, the combined effects of low pH, increased acidity, concomitant loss of moisture during drying, reduction of a_w, concentration of curing salts, such as sodium chloride and sodium nitrite, bacterial inhibition of spoilage or pathogenic microorganisms, and heat processing (if applied) preserve fermented meat and poultry products against spoilage by inactivating indigenous tissue and bacterial enzymes.

Mold and yeast starter cultures are not widely used in the United States, with the exception of dry sausages produced in the San Francisco area. White or gray molds are typical on casing surfaces of sausages produced in Hungary, Italy, Spain, Greece, Yugoslavia, Romania, Slovakia, and the Czech Republic. Their effect is primarily cosmetic, but it has been reported that the mycelial coat can reduce moisture loss and facilitate uniform drying (6). Catalase produced by molds may serve as an antioxidant by reacting with surface oxygen to prevent it from entering the product, while nitrate reductase promotes the development of red surface color (62). Green mold on the surface of sausage is typically the result of excessive humidity and sporulation, but Penicillium chrysogenum on French sausage is noted for its ability to produce a preferred ebony color. Molds are capable of decomposing lactic acid which increases pH and results in a milder flavor (36). Penicillium nalgiovense is most commonly used for this purpose, but application of P. chrysogenum and Penicillium camemberti may also be used. In Germany, only Penicillium candidum, P. nalgiovense, and Penicillium roqueforti are approved for application to sausages. The use of nontoxigenic molds may reduce the risk of mycotoxin production by other molds (36, 54).

Yeasts such as Candida famata and Debaryomyces hansenii are used alone or in combination with bacterial cultures to produce a powdery surface or a fruity/alcoholic aroma which may be construed as spoiled if uncontrolled (6). These cultures are added at a population of 10^6 CFU/g and are characterized by a high salt tolerance and growth at low a_w, e.g., as low as 0.87 (40). Streptomyces griseus subsp. hutter produces a cellar-ripened sausage aroma and enhances color because of its nitrate reductase and catalase activities (62).

Classification of Bacterial Starter Cultures for Meat

The genera most commonly used by meat starter cultures are Lactobacillus, Pediococcus, Micrococcus, and Staphylococcus (6, 39, 97); Pediococcus, Lactobacillus, and other lactic acid bacteria are preferred when acid production is of primary importance. Specific strains of Micrococcus and coagulase-negative staphylococci are used in meat curing when lower incubation temperatures and less acid production are required for flavor development, as found in many European sausages (6, 97). Eight genera (Lactobacillus, Leuconostoc, Pediococcus, Streptococcus, Carnobacterium [formerly Lactobacillus], Enterococcus, Lactococcus, and Vagococcus [the latter three genera were formerly Streptococcus]) are most commonly used as starter cultures (44). Lactobacillus hordniae and Lactobacillus xylosus are now in the genus Lactobacillus, while

Streptococcus diacetilactis has been classified as a citrate-utilizing strain of *Lactococcus lactis* subsp. *lactis*.

Lactobacilli and pediococci are gram-positive, nonsporing rods or cocci which produce lactic acid as a major end product during the fermentation of carbohydrates (5, 97). Originally classified on the basis of morphology, glucose fermentation pathway, optimal growth temperature, and stereoisomer of lactic acid produced, this phenotypic grouping has remained largely intact but may change considerably as molecular characterization is completed (5, 97). Ongoing research on oligonucleotide cataloging and 16S and 23S rRNA sequencing indicates that all gram-positive bacteria cluster in one of the 11 major eubacterial phyla and can be further divided into two main groups or clusters (5, 97, 107). One cluster, designated the *Actinomycetes* subdivision, is composed of bacteria possessing a mol% G+C of the DNA above 55% and would include meat fermentation genera such as *Micrococcus*, while the low-mol% G+C cluster, designated the *Clostridium* subdivision, would include meat fermentation genera such as the staphylococci (5, 96, 97, 107). Lactic acid bacteria are thought to form a large related cluster which phylogenetically lies between strictly anaerobic species such as the clostridia and facultative or strictly aerobic staphylococci and bacilli (5, 46, 47, 97).

At the species level, lactobacilli appear to fall into three clusters which do not appear to correlate with current classification schemes (5, 97). Most of the species involved or associated with meat fermentations fall within the *Lactobacillus casei*/*Pediococcus* subgroup, composed of obligately homofermentative, some heterofermentative, and all of the facultative heterofermentative lactobacilli (5). The pediococci are gram-positive cocci and are the only lactic acid bacteria capable of dividing in two planes. Consequently, they can appear as pairs, tetrads, or other formations (33, 79, 97). The latest taxonomic classification places pediococci as members of the gram-positive facultative anaerobic phylogenetic cluster that contains 15 genera, including the staphylococci, streptococci, micrococci, and leuconostocs (88). This designation will probably change, because comparison studies using 16S rRNA and nucleotide sequencing have indicated that pediococci are more closely aligned with the phylogenetic cluster that includes the lactobacilli and leuconostocs (20, 95, 105). Most of the strains designated *P. cerevisiae* that were used as meat starters have been reclassified as *P. acidilactici*, and now *P. acidilactici*, *Pediococcus damnosus*, *Pediococcus parvulus*, and *P. pentosaceus* form an evolutionary group with the *Lb. pentosus*, *Lb. brevis*, and *Lb. buchneri* complex (97).

The lactobacilli associated with meat fermentations are members of the genus *Lactobacillus*. They are characterized as gram-positive, non-spore-forming rods which are catalase negative on media not containing blood, are usually nonmotile, usually reduce nitrate, and ferment glucose (41, 48). The genus *Lactobacillus* currently consists of more than 50 species (97). DNA hybridization and sequencing comparisons indicate that despite the numerous species, as a whole, they compose a well-defined group of bacteria (41). Part of the reason for the large number of species is that the genus *Lactobacillus* has been exhaustively studied, not only from the perspective of taxonomy and identification, but also from the standpoint of nutritional requirements for application in studies on biochemistry and metabolism (89). Most research on meat fermentations has emphasized *Lb. plantarum* as the predominant species (6, 97). However, attempts to identify lactobacilli isolated from meat are usually less than successful, because most of the documented descriptions and schemes of identification are based on isolates from other food sources (49, 81, 86, 90). A series of investigations on atypical lactic acid streptococci isolated from fermented meats resulted in their identification as *Lb. sake* and *Lb. curvatus* (reference 43 as cited in reference 39). These species outnumbered the typical lactobacilli identified as *Lb. plantarum*, *Lb. brevis*, *Lb. alimentarius*, *Lb. casei*, *Lb. farciminis*, *Weissella viridescens*, unspecified leuconostocs, and pediococci by 1,000-fold and were typically psychrotrophic and less acid tolerant (39, 80, 97). Efforts to accurately classify and identify these strains are becoming more important as various isolates of lactic acid bacteria become more commonly used as starter cultures.

Starter Culture Metabolism

The term "fermentation" was first defined by Pasteur as life in environments devoid of oxygen. However, meat fermentations are more accurately defined as bacterial or microbial metabolic processes in which carbohydrates and related compounds are oxidized, with the release of energy in the absence of oxygen as a final electron acceptor. Thus, by definition, fermentative microorganisms cannot use oxygen as a terminal electron acceptor to generate ATP and must either conduct anaerobic respiration using more electronegative electron acceptors (carbon dioxide, sulfate, nitrate, or fumarate) or ferment intermediate metabolites formed from the substrate as an electron acceptor (35). The bacteria responsible for fermentation are either facultative or obligate anaerobes.

As a collective group, the gram-positive acidogenic (predominantly lactic acid) bacteria include the genera *Lactobacillus*, *Streptococcus*, *Pediococcus*,

Leuconostoc, *Lactococcus*, and *Enterococcus*, which can metabolize a large number of mono- and oligosaccharides, polyalcohols, aliphatic compounds, mono-, di-, and tricarboxylic acids, and some amino acids, although individual species characteristically have a limited range of carbon and energy sources (57). During fermentation of sausages, members of this group of bacteria are responsible for two basic microbiological processes that occur simultaneously and are interdependent, viz., the production of nitric oxide by nitrate- and nitrite-reducing bacteria and decrease in pH as a result of anaerobic glycolysis. These two activities are synergistic owing to the pH dependency of nitrite and nitrate reduction. The following discussion details reactions that are important in meat, poultry, and fish fermentations and is based on several excellent reviews (5, 46, 55).

Carbohydrate Fermentation Pathways

Acidogenic bacteria ferment indigenous and added carbohydrates primarily to form lactic acid. The formation of lactic acid, which is an anaerobic process, helps to create a reduced pH environment in the meat, and this in turn contributes to the development of cured meat color. The formation of lactic acid also causes coagulation of meat proteins and, in combination with drying, gives sausages their characteristic firm texture. There are three potential pathways that the lactic acid bacteria may use to form lactic acid from carbohydrates. The homofermentative pathway yields two moles of lactic acid per mole of glucose; the heterofermentative pathway yields one mole of lactic acid, one mole of ethanol, one mole of acetic acid, and one mole of carbon dioxide per mole of glucose; and the bifidum pathway (which will not be discussed at length here, since it is generally not found in the bacteria important in meat starter cultures) yields three moles of acetic acid and two moles of lactate per two moles of glucose (35). The homofermentative pathway or glycolysis is the primary means of generating lactic acid in meat fermentations. Both *P. acidilactici* and *P. pentosaceus* are microaerophilic and, under anaerobic growth conditions, homofermentative, yielding DL-lactic acid (33, 79). Glucose and most other monosaccharides are fermented, and, unlike other pediococci, both species can ferment pentoses (33). It is probable that *P. acidilactici* and *P. pentosaceus* transport glucose via the phosphoenolpyruvate phosphotransferase system and derive ATP using the Embden-Meyerhof pathway (glycolysis). The Embden-Meyerhof pathway features the formation of fructose 1,6-diphosphate (FDP), which is cleaved by FDP aldolase to form dihydroxyacetonephosphate and glyceraldehyde 3-phosphate. These three-carbon intermediates are converted to pyruvate, which also energetically favors ATP formation via substrate-level phosphorylation at two separate metabolic transformation steps (two ATPs per glucose). Because the resulting NADH formed during glycolysis requires oxidizing to regenerate NAD^+ and maintain oxidation-reduction balance, pyruvate is reduced to lactate by a NAD^+-dependent lactate dehydrogenase. Since lactic acid is virtually the only end product, the fermentation is referred to as a homolactic fermentation (5).

Heterofermentation is also characterized by dehydrogenation steps yielding 6-phosphogluconate, followed by decarboxylation and formation of a pentose 5-phosphate. The pentose is cleaved by phosphoketolase to form glyceraldehyde 3-phosphate, which eventually becomes lactic acid through glycolysis, and acetyl phosphate (which is required to maintain electron balance) is reduced to ethanol (5, 35). In theory, homofermentative and heterofermentative bacteria can be distinguished by the pattern of fermentation end products and the presence or absence of FDP aldolase and phosphoketolase (46). Thus, obligately homofermentative species possess a constitutive FDP aldolase and lack phosphoketolase, while the reverse is true of obligately heterofermentative species (5, 46, 48). Species of lactic acid bacteria used to ferment meat can be regarded essentially as facultative heterofermenters because they not only possess a constitutive FDP aldolase, and consequently use glycolysis for hexose fermentation, but they also possess a pentose-inducible phosphoketolase. The type of fermentation is dependent on growth conditions. In an ecosystem such as meats, with a wide variety of complex substrates serving as sources of pentoses, organic acids, and other fermentable compounds in addition to hexoses it is possible for lactic acid bacteria to be homofermentative when using hexose but heterofermentative, when using pentoses and other substrates. Lactic acid is the primary metabolite derived from both homo- and heterofermentation, but during heterofermentation production of mixtures of lactic acid, acetic acid, ethanol, and carbon dioxide also occurs.

ATP Generation, Pyruvate Metabolism, and Energy Recycling

The most important energy requirements of the bacterial cell are macromolecular synthesis and the transport of essential solutes against a concentration gradient (5). Fermentation that is characteristic of lactic acid bacteria is essentially the oxidation of a substrate to generate energy-rich intermediates that in turn are used for ATP production by substrate-level phosphorylation. Although lactic acid bacteria can tolerate a more minimal and broad range of internal pHs, there still must be a substantial amount of ATP generated to keep the cytoplasmic pH above a threshold level to offset the massive external acid production (5).

Bacteria have considerable flexibility in the pathways they can use to generate ATP and regulate its distribution. In bacteria which rely on an electron transport chain, large quantities of ATP can be generated via a proton motive force across the electron transport chain coupled to a H^+-translocating ATP synthase (5). Even though lactic acid bacteria do not possess ATP-generating capabilities via electron transport, lactobacilli do have some metabolic flexibility by altering pyruvate pathways and conserving energy by end-product efflux. Some lactic acid bacteria have alternative pathways of pyruvate metabolism other than direct reduction to lactic acid. Oxidation involves the formation of NADH from NAD^+, which requires regeneration back to NADH, and pyruvate serves as the key intermediate for this by acting as the electron acceptor (5). Pyruvate-formate lyase generates formate and acetyl coenzyme A (acetyl-CoA) from pyruvate and CoA, and the acetyl-CoA can be used as either a precursor for substrate-level phosphorylation or direct reduction to ethanol or both. This alteration in pyruvate metabolism has been demonstrated in studies in which lactobacilli have been shown to shift fermentation patterns and concomitantly the amount of potential ATP formed as the growth rate changes (27). In these studies, mixed acid-forming strains of lactobacilli produced less lactic acid and more acetic acid with decreasing growth rate, and for every mole of acetic acid, an extra mole of ATP (two versus one mole ATP produced per mole of lactic acid) was potentially formed (27). During fermentation, lactic acid bacteria form sufficient end products in the cytoplasm to result in a very high internal to external cell gradient (76). It has been suggested that when a fermentation product such as lactic acid exceeds the electrochemical proton gradient, additional metabolic energy is gained by product efflux through carrier-mediated transport in symport with protons (61, 76, 98). Metabolic energy is conserved, because a product gradient is essentially a proton gradient which can be used to directly produce ATP by proton-driven ATPase, and it helps to maintain a proton motive force without consumption of ATP (5, 76).

Nitrate/Nitrite Reduction

Cured meat derives its characteristic pink color from the reaction of myoglobin (Fe^{2+}) or metmyoglobin (Fe^{3+}) with nitric oxide to ultimately form nitric oxide myoglobin, an unstable, bright pink compound (73). Nitric oxide myoglobin is formed directly from the reaction of purple-red myoglobin with nitric oxide. Another proposed route of formation may be through the oxygenation of myoglobin to bright red oxymyoglobin (Fe^{2+}) and subsequent oxidation to brown metmyoglobin (Fe^{3+}). Nitric oxide reduces metmyoglobin to gray-brown nitric oxide metmyoglobin (Fe^{2+}), which transmutates to nitric oxide myoglobin with the loss of oxygen. Nitric oxide myoglobin, when heated, forms nitrosohemochrome (Fe^{2+}), the stable pink pigment present in commercially processed cured meats. Nitrate/nitrite-reducing bacteria transform added nitrates into nitrites and eventually into more reduced forms for reaction with the dominant myoglobin or metmyoglobin pigment forms. *Micrococcaceae* possess nitrate reductase to convert nitrate to nitrite, which in turn is converted to nitric oxide via nitrous acid under acidic conditions. Nitric oxide is a strong electron donor and rapidly reacts with the heme moiety of myoglobin to form the characteristic pink meat pigment, dinitrosylhemochrome. Some strains of lactic acid bacteria may produce metabolites, such as carbon dioxide, acetate, formate, succinate, and acetoin, during fermentation that cause off flavors; however, nitrate and nitrite prevent the formation of certain off flavors from compounds such as formate by their inhibition of pyruvate/formate lyase activity in lactobacilli (40). The formation of nitric oxide from the reduction of nitrates and nitrites is crucial, because it reduces nitrosometmyoglobin to nitrosomyoglobin and induces red color formation (55). The advantage of microbial activity is the reduction of nitrates, thus removing excess nitrate/nitrite from the meat.

Metabolic Activities Important in Commercial Starter Cultures

Strains of *P. acidilactici* were initially selected for use as starter cultures because they ferment carbohydrates rapidly to lactic acid at higher temperatures and effectively lower the pH of the meat from a range of 5.6 to 6.2 down to 4.7 to 5.2 (10). Traditionally, semidry sausages (summer sausage, thuringer, beef sticks) and some types of pepperoni are allowed to undergo greening, i.e., fermentation at 16 to 38°C, which favors the development of indigenous microflora and extends fermentation times (10). Although indigenous microflora have the advantage of imparting unique flavor(s) and other desirable sensory properties to sausages, they also have sufficient variation in lactic acid production to be at a considerable disadvantage for large-scale production of sausage. The use of starter cultures consisting of pediococci allows for the inoculation of large numbers of one microorganism into the raw meat followed by a uniform fermentation.

More important, meat inoculated with *P. acidilactici* can be incubated at higher temperatures (43 to 50°C), which precludes the growth of most indigenous microorganisms so that flavor and pH characteristics are predictable (10). Consequently, commercial production is enhanced, since flavor development can be controlled and the fermentation process can be hastened.

Strains of *P. acidilactici* that have been selected and developed for use as commercial starter cultures are well suited for the production of semidry sausages, which are fermented and/or smoked at higher temperatures (26 to 50°C). However, when used for production of dry sausages, which require lower fermentation temperatures (15 to 27°C), these strains generally produce lactic acid at a much slower rate (10). Since *P. pentosaceus* has a lower optimum temperature for growth than *P. acidilactici* (28 to 32°C versus 40°C) (33), it has been promoted as an effective meat starter culture. *P. pentosaceus* is also more attractive because it has a 25% lower Arrhenius energy of activation for fermentation than *P. acidilactici* (78). Thus, *P. pentosaceus* has a lower optimum growth temperature and a higher capacity for rapid production of lactic acid than *P. acidilactici*.

Metabolic Contributions of Starter Cultures to Sausage Sensory Qualities

Fermentation of sausages most often involves inoculation of ground meats with a starter culture, followed by an incubation period to allow enzymatic conversion of available carbohydrate to approximately equimolar concentrations of DL-lactic acid. Fermentable carbohydrates in the form of residual muscle glucose (0.1 mg/g) and added glucose or sucrose are the primary sources of energy available for metabolism. The decline in pH to 4.5 to 4.7 with the buildup of lactic acid results in a dramatic loss of water-holding capacity as the pH nears the isoelectric point (pI ~ 5.1) of myofibrillar proteins. Product texture becomes firmer, density increases with dehydration during aging, and partial denaturation of the myofibrillar proteins occurs owing to the presence of lactic acid. Thus, the tangy flavor and chewy texture of fermented sausages are consequences of the dominant fermentation metabolite, lactic acid, and dehydration.

Lactic acid is the dominant flavor component in fermented sausage, but spices, salt, sugar, sodium nitrite reaction products, smoke, and meat components such as fatty acids, amino acids, and peptides also contribute to the flavor profile (74). Secondary flavor contributions from metabolites of lactobacilli and pediococci are minor in fermented sausages produced in the United States; however, natural heterofermentative lactobacilli (*Lb. brevis*, *Lb. buchneri*) from back inocula-

tions, when combined with staphylococci and micrococci, give many European sausages their characteristic flavor as a result of volatile acids, alcohols, and carbon dioxide. *Micrococcaceae*, for example, possess lipolytic and proteolytic enzymes, which generate aldehydes, ketones, and short-chain fatty acids that give characteristic aromas and flavors to sausages (59). They also produce catalase, which decomposes hydrogen peroxide. *Leuconostoc mesenteroides* and *Lb. brevis* produce ethanol, acetic acid, lactic acid, pyruvic acid, acetoin (which imparts a nutty flavor and aroma), and carbon dioxide to give an effervescent sensation, but fermentation conditions must be controlled to avoid excessive pinholes, gas pockets, and off flavors.

In natural fermentations as well as in sausages inoculated with lactobacilli, *Lb. sake* and *Lb. curvatus* have been observed to be the dominant bacteria at temperatures of 25°C and below, while *Lb. plantarum* tends to dominate at higher fermentation temperatures (45). *Lb. sake*, *Lb. curvatus*, and *Lb. plantarum*, to a lesser extent, form DL-lactic acid and have flavin-dependent oxidases capable of forming hydrogen peroxide. Tolerance to peroxide may be one of the factors contributing to their dominance during fermentation. Hydrogen peroxide, if not decomposed, can induce oxidation of unsaturated fatty acids, leading to rancid flavor development, and/or oxidize the heme component of myoglobin, leading to fading or the formation of green, yellow, gray, or other off-color pigments (62). *Lb. curvatus* does not possess pseudocatalase or true catalase and may permit the accumulation of hydrogen peroxide. *Lb. sake*, *Lb. plantarum*, and pediococci, on the other hand, exhibit catalase activity and can therefore decompose hydrogen peroxide formed during fermentation. Sensory analysis of unspiced fermented dry sausages produced using various starter culture combinations (15) has revealed that sausage flavor is influenced by culture composition. Berdague et al. (15) reported that the butter odor of dry sausages was largely dependent upon degradation of sugars by way of pyruvic acid and that curing and rancid odors were correlated with compounds resulting from lipid oxidation. Their work indicated that *Staphylococcus saprophyticus* and *Staphylococcus warneri* were most associated with butter odor, which in turn was correlated to the presence of acetoin, diacetyl, 1,3-butanediol, and 2,3-butanediol. Distinctive curing odors were associated with combinations of *S. carnosus* and *P. acidilactici*, *S. carnosus* and *Lb. sake*, and *S. carnosus* and *P. pentosaceus* and were correlated with 2-pentanone, 2-hexanone, 2-heptanone, and an unknown compound. A combination of *S. saprophyticus* and *Lb. sake*, which produced the most acetic acid, was associated with fruity odors derived from esters and a less intense rancid odor. Meat proteins are hydrolyzed by endogenous and microbial proteases to yield peptides and amino acids, which are in turn degraded to ammonia and amines, causing a slight rise in the pH. Demeyer and Samejima (25), however, suggested that the major protease activity in fermented meat is derived from the meat enzymes. Amino acid degradation products and nucleotide inosine monophosphate intensify meat flavors and contribute to the overall sausage flavor. Thus, it appears that compounds produced via endogenous proteolysis in combination with starter culture fermentation play a significant role in the nonacidic flavor and aroma of fermented sausages.

Genetics and Biotechnology of Meat Starter Cultures

To utilize biotechnological approaches with meat starter cultures, genetic transfer systems and mobile genetic elements for carrying a potentially important gene(s) must be identified. The standard approach to gene cloning in lactic acid bacteria is to develop plasmid vectors based either on indigenous cryptic plasmids or on heterologous plasmids resistant to a broad range of antibiotics (34). Since Chassy et al. (19) first demonstrated the presence of plasmids in lactobacilli, significant progress has been made, particularly in elucidating genetic systems in dairy lactobacilli and *Lc. lactis* (34, 41).

Plasmids have been detected in many strains of *P. pentosaceus*, *P. cerevisiae*, and *P. acidilactici*, and some may be transferable between strains. Some properties that are characteristic of a particular species of *Pediococcus* have been shown to be plasmid linked. The production of bacteriocin and bacteriocinlike substances and the ability to ferment some sugars have been shown to be linked to or encoded by plasmids in strains of *P. cerevisiae*, *P. pentosaceus*, and *P. acidilactici*. Numerous plasmids have been isolated from meat lactobacilli and some have been functionally identified. Nes (68) observed that 8 of the 10 strains of *Lb. plantarum* examined for the presence of extrachromosomal DNA contained from one to six plasmids. Six strains containing plasmids were commonly used as starter cultures in dry sausage. Metabolic functions in *Lb. plantarum* and four atypical *Lactobacillus* species isolated from fresh meat (56) that have been shown to be plasmid linked include maltose utilization and cysteine metabolism (91). When Schillinger and Lücke (87) screened 221 strains of lactobacilli from meat and meat products for antibacterial compounds, they linked immunity and production of a bacteriocin in a strain of *Lb. sake* with an 18-MDa plasmid by showing that cured variants no longer possessed either trait. *Lb. sake* also produces a bacteriocin, lactocin S, which inhibits species of *Lactobacillus*, *Pediococcus*, and *Leuconostoc*. Both bacteriocin production by *Lb. sake* and its immunity factor have been shown to be associated with an unstable 50-kb plasmid (64–66).

Genetic transfer has been demonstrated in several species of lactobacilli and lactococci, but limited studies have been performed with species associated directly with meat fermentations. In vivo gene transfer systems have involved either conjugation, which requires cell-to-cell contact (bacterial mating) for transfer of genetic material, or transduction (transfection or phage-mediated genetic exchange), while in vitro physiological transformation requires uptake of naked DNA by the recipient cell (42, 104). Intergenic conjugation of an antibiotic-resistant streptococcal plasmid was demonstrated in *Lb. plantarum* (106), and successful inter- and intragenic conjugation of this same plasmid has also been reported for *Lb. plantarum* (92) as well as a strain of *Lactobacillus* isolated from fermented sausage (83). Conjugative transfer of an *Escherichia-Streptococcus* nonconjugative shuttle plasmid has also been achieved in *Lb. plantarum* by cointegration formation (93). Further genetic analysis and manipulation of meat lactobacilli have been handicapped by a lack of natural competence and transformation procedures (41, 63). This will change with the emergence of electroporation methods that have been successively used to transform plasmids in *Lb. plantarum* (4, 17, 58, 77).

Although shuttle vectors have been used to construct and introduce heterologous genes into lactobacilli, they generally contain antibiotic resistance markers which greatly facilitate the process but may not receive regulatory approval for use in foods (41). Furthermore, considerable instability and loss of function occur during replication, because these plasmids for the most part belong to a group which replicates via single-stranded DNA intermediates or rolling-circle replication (34, 37). This form of replication is a potential source of plasmid segregational instability. With meat lactobacilli, vectors containing replicons from *Lactobacillus*, *Staphylococcus*, or *Escherichia coli* sequences have all been shown to contribute to this instability (53, 77). Therefore, plasmids containing a *Lactobacillus* replicon can be stabilized under selective conditions and used to express genes from multiple copies of the plasmid but under nonselective conditions are frequently segregationally unstable and lost (53, 77). Consequently, efforts have focused on gene cloning strategies that enable heterologous genes to integrate in a stable fashion into the bacterial chromosome (34, 43). Recombination in chromosomes can occur as general recombination when chromosomal DNA is transferred from one bacterium to the next via conjugation or transduction, which necessitates extensive sequence homology between externally introduced DNA and chromosomal DNA as well as specific bacterial recombination factors (104). Recombination can also be mediated by transposable genetic elements which involve short target sequences of 10 nucleotide base pairs or fewer and occur independent of bacterial host function (104). Transposons have been introduced into *Lb. curvatus* (51) and *Lb. plantarum* (4), and integration of plasmids by single crossover events within regions of homology can be accomplished by constructing plasmid suicide vectors that lack the ability to replicate in lactic acid bacteria (34). Scheirlinck et al. (85) used a suicide plasmid to integrate *Bacillus stearothermophilus* α-amylase genes and *Clostridium thermocellum* endoglucanase genes into *Lb. plantarum*. This homologous insertion into the chromosome was made possible by generating plasmid constructs containing an unknown portion of a *Lactobacillus* chromosome as part of the suicide vector to facilitate site-specific integration. Complete replacement of a gene(s) can be achieved with homologous double-crossover recombination events. An alternative to using homologous sequences is to locate insertion sequences already present on the chromosome that can

transpose chromosomal DNA to plasmids and use these in the construction of integrative vectors (42). Only a few such elements have been identified in meat lactobacilli, and only one, IS*1163* from *Lb. sake*, has been isolated, sequenced, and described in any detail (94).

Development of genetically engineered strains of lactobacilli that have large-scale use in fermented meats or other foods has remained elusive. Part of the problem is that transfer of structural genes to a new host does not guarantee that the genes will be expressed (42). Optimal expression of cloned genes requires not just cloning of the structural gene but also efficient promoters, ribosome-binding sites, and termination sites, all of which must be identified, isolated, cloned, and sequenced (42). Integration systems that have been used do not allow high enough levels of heterologous gene expression for practical application to fermented meats and give an unsatisfactory stabilization of the foreign gene (43). One approach to solving this problem essentially involves using the structural gene (*amyL*) that encodes production and secretion of α-amylase from *B. licheniformis* as a reporter gene for the cloning of expression-secretion sequences from *Lb. plantarum* (43). The key requirement for this approach was that the *amyL* gene did not possess a specific and/or regulated promoter that was operational in any gram-positive bacterium outside its genus

Table 33.5 Research needs for meat starter cultures[a]

Improving current starter culture technology
 Substrate specificity of starter culture microorganisms
 Immobilized cells for meat fermentation
 Increased rate of lactic acid production (reduction of lag phase)
 Control of pathogens without heating
 Use of starter cultures in combination with acidulants, e.g., GDL
 Identification and testing of nitrite substitutes (development of cured color with *C. botulinum* outgrowth protection)
 New cryoprotectant (antifreeze) solutions for frozen starter cultures
 Development of new pediocins that are more soluble, are not bound by fat, and are subject to proteolysis or inactivated by other meat components
 Production of biogenic amine-free (negative) meat starter culture
 Development of meat cultures to be grown at higher salt concentration
 Production of starter cultures for pathogen and yeast control
 Rapid detection and identification using molecular detection techniques
 Improvement of starter cultures for higher level of nitrite reduction and better red color development
 Probiotic lactic acid bacteria (*Lactobacillus acidophilus*) applied to meat fermentation
Use of cultures to enhance meat product nutrition and quality
 Enhancement of nutritional quality by in situ production of critical dietary nutrients for selected populations
 Greater utilization of nontraditional meat sources via fermentation
 Accelerated curing of traditional meat products (nonnitrite hams, bacon, comminuted meats, meat snacks, natural acidification against *C. botulinum*)
 Generation of natural antioxidants in situ (α-tocopherol production by starter cultures)
 Production of antimycotic agents for mold inhibition
 Development of starter cultures unique to fish pastes and sauces

[a] Based on comments in references 22 and 75.

of origin, *Bacillus*. Thus, the high expression and secretion of over 90% of the *Bacillus* extracellular amylase in *Lb. plantarum* serves as an expression reporter to indicate that replacement of the *Bacillus* promoter by an *Lb. plantarum* promoter had taken place. The plasmid containing the silent *amyL* coding frame could subsequently be used as a probe to locate expression (transcription and translation) or expression-secretion regions on the chromosome of *Lb. plantarum* strains expressing and secreting α-amylase.

Newly developed techniques for genetic manipulations in lactic acid bacteria have been introduced. Inducible systems for gene expression that utilize inducible promoters linked to promoterless reporter genes have been described in *Lactococcus* sp. Recent applications have shown the applicability of such expression systems in other lactic acid bacteria (26). The recent publication of the entire genome sequence of *Lc. lactis* will provide novel genetic tools for the genetic manipulation of lactic acid bacteria, including new insertion elements, prophages, and competence genes that have been found after analysis of the entire genome sequence (16). These findings should lead to the development of procedures that will enhance the understanding of gene regulation and genetic engineering of lactic acid bacteria. Numerous opportunities are available for advances in meat starter culture research. Potential research areas are summarized in Table 33.5. The number of groups conducting research on genetic systems, particularly in lactobacilli, has multiplied severalfold in recent years. Earlier predictions (18, 60) for biotechnological approaches to optimize fermentations are nearing reality.

References

1. **Acton, J. C.** 1977. The chemistry of dry sausages. *Proc. Recip. Meat Conf.* **30:**49–62.
2. **Al-Sheddy, I. A., D. Y. C. Fung, and C. L. Kastner.** 1995. Microbiology of fresh and restructured lamb meat: a review. *Crit. Rev. Microbiol.* **21:**31–52.
3. **Arihara, K., H. Kushida, Y. Kondo, M. Itoh, J. B. Luchansky, and R. G. Cassens.** 1993. Conversion of metmyoglobin to bright red myoglobin derivatives by *Chromobacterium violaceum*, *Kurthia* sp., and *Lactobacillus fermentum* JCM1173. *J. Food Sci.* **58:**38–42.
4. **Aukrist, T., and I. F. Nes.** 1988. Transformation of *Lactobacillus plantarum* with the plasmid pTV1 by electroporation. *FEMS Microbiol. Lett.* **52:**127–132.
5. **Axelsson, L. T.** 1993. Lactic acid bacteria: classification and physiology, p. 127–159. *In* S. Salminen and A. von Wright (ed.), *Lactic Acid Bacteria*. Marcel Dekker, Inc., New York, N.Y.
6. **Bacus, J. N.** 1986. Fermented meat and poultry products. *Adv. Meat Res.* **2:**123–164.
7. **Bacus, J. N.** 1995. Personal communication.
8. **Bacus, J. N., and W. L. Brown.** 1981. Use of microbial cultures: meat products. *Food Technol.* **35:**74–78, 83.
9. **Bacus, J. N., and W. L. Brown.** 1985. The lactobacilli: meat products, p. 58–71. *In* S. E. Gilliland (ed.), *Bacterial Starter Cultures for Foods*. CRC Press, Inc., Boca Raton, Fla.
10. **Bacus, J. N., and W. L. Brown.** 1985. The pediococci: meat products, p. 86–95. *In* S. E. Gilliland (ed.), *Bacterial Starter Cultures for Foods*. CRC Press, Inc., Boca Raton, Fla.
11. **Bartholomew, D. R., and C. I. Osuala.** 1986. Acceptability of flavor, texture, and appearance of mutton processed meat products made by smoking, curing, spicing, adding starter cultures and modifying fat source. *J. Food Sci.* **51:**1560–1562.
12. **Beddows, C. G.** 1985. Fermented fish and fish products, p. 1–39. *In* B. J. B. Wood (ed.), *Microbiology of Fermented Foods*, vol. II. Elsevier Applied Science Publishers, London, United Kingdom.
13. **Beddows, C. G., A. G. Ardeshir, and W. Johari bin Daud.** 1979. Biochemical changes occurring during the manufacture of Budu. *J. Sci. Food Agric.* **30:** 1097–1103.
14. **Beddows, C. G., A. G. Ardeshir, and W. Johari bin Daud.** 1980. Development and origin of the volatile fatty acids in Budu. *J. Sci. Food Agric.* **31:**86–92.
15. **Berdague, J. L., P. Monteil, M. C. Montel, and R. Talon.** 1993. Effects of starter cultures on the formation of flavor compounds in dry sausage. *Meat Sci.* **35:**275–287.
16. **Bolotin, A., S. Mauger, K. Malarme, S. D. Ehrlich, and A. Sorokin.** 1999. Low-redundancy sequencing of the entire *Lactococcus lactis* IL 1403 genome. *Antonie Leeuwenhoek* **76:**27–76.
17. **Bringle, F., and J.-C. Hubert.** 1990. Optimized transformation by electroporation of *Lactobacillus plantarum* strains by plasmid vectors. *Appl. Microbiol. Biotechnol.* **33:**664–670.
18. **Chassy, B. M.** 1987. Prospect for the genetic manipulation of lactobacilli. *FEMS Microbiol. Lett.* **46:**297–312.
19. **Chassy, B. M., E. Gibson, and A. Giuffrida.** 1976. Evidence for extrachromosomal elements in *Lactobacillus*. *J. Bacteriol.* **127:**1576–1578.
20. **Collins, M. D., A. M. Williams, and S. Wallbanks.** 1990. The phylogeny of *Aerococcus* and *Pediococcus* as determined by 16 rRNA sequence analysis: description of *Tetragenococcus* gen. nov. *FEMS Microbiol. Lett.* **70:**255–262.
21. **Crisan, E. V., and A. Sands.** 1975. The microbiology of four fermented fish sauces. *Appl. Microbiol.* **29:**106.
22. **Curtis, S. I.** 1995. Personal communication.
23. **Deibel, R. H., and C. F. Niven, Jr.** 1957. *Pediococcus cerevisiae*: a starter culture for summer sausage. *Bacteriol. Proc.*, p. 14–15.
24. **Deibel, R. H., G. D. Wilson, and C. F. Niven, Jr.** 1961. Microbiology of meat curing. IV. A lyophilized *Pediococcus cerevisiae* starter culture for fermented sausages. *Appl. Microbiol.* **9:**239–243.
25. **Demeyer, D., and K. Samejima.** 1991. Animal biotechnology and meat processing, p. 127–143. *In* L. O. Fiems,

B. G.Cottyn, and D. I. Demeyer (ed.), *Animal Biotechnology and the Quality of Meat Production.* Elsevier, Amsterdam, The Netherlands.

26. de Vos, W. M. 1999. Gene expression systems for lactic acid bacteria. *Curr. Opin. Microbiol.* **2**:289–295.

27. de Vries, W., W. M. C. Kapteijn, E. G. van der Beek, and A. H. Stouthamer. 1970. Molar growth yields and fermentation balance of *Lactobacillus casei* L3 in batch cultures and continuous cultures. *J. Gen. Microbiol.* **63**:333–345.

28. Dougan, J., and G. E. Howard. 1975. Some flavouring constituents of fermented fish sauces. *J. Sci. Food Agric.* **26**:887–894.

29. Egan, A. F. 1983. Lactic acid bacteria of meat and meat products. *Antonie Leeuwenhoek* **49**:327–336.

30. Everson, C. W., W. E. Danner, and P. A. Hammes. 1970. Bacterial starter cultures in sausage products. *J. Agric. Food Chem.* **18**:570–571.

31. Everson, C. W., W. E. Danner, and P. A. Hammes. June 1974. Process for curing dry and semidry sausages. U.S. patent 3,814,817.

32. Fujii, T., S. D. Basuki, and H. Tozawa. 1980. Microbiological studies on the ageing of fish sauce; chemical composition and microflora of fish sauce produced in the Philippines. *Nippon Suissan Gakkaishi* **46**:1235–1240.

33. Garvie, E. I. 1986. Genus *Pediococcus*, p. 1075–1079. *In* P. H. A. Sneath, N. S. Mair, M. E. Sharpe, and J. G. Holt (ed.), *Bergey's Manual of Systematic Bacteriology*, vol. 2. The Williams & Wilkins Co., Baltimore, Md.

34. Gasson, M. J. 1993. Progress and potential in the biotechnology of the lactic acid bacteria. *FEMS Microbiol. Rev.* **12**:3–20.

35. Gottschalk, G. 1986. *Bacterial Metabolism*, 2nd ed. Springer-Verlag, New York, N.Y.

36. Grazia, L., P. Romano, A. Bagni, D. Roggiani, and G. Guglielmi. 1986. The role of moulds in the ripening process of salami. *Food Microbiol.* **3**:19–25.

37. Gruss, A., and S. D. Ehrlich. 1989. The family of highly interrelated single-stranded deoxyribonucleic acid plasmids. *Microbiol. Rev.* **53**:231–241.

38. Hammer, G. F. 1987. Meat processing: ripened products. *Fleischwirtschaft* **67**:71–74.

39. Hammes, W. P., A. Bantleon, and S. Min. 1990. Lactic acid bacteria in meat fermentation. *FEMS Microbiol. Lett.* **87**:165–174.

40. Hammes, W. P., and H. J. Knauf. 1994. Starters in the processing of meat products. *Meat Sci.* **36**:155–168.

41. Hammes, W. P., N. Weiss, and W. Holzapfel. 1992. The genera *Lactobacillus* and *Carnobacterium*, p. 1535–1594. *In* A. Balows, H. G. Trüper, M. Dworkin, W. Harder, and K.-H. Schleifer (ed.), *The Prokaryotes: A Handbook on the Biology of Bacteria: Ecophysiology, Isolation, Identification, Applications*, 2nd ed. Springer-Verlag, New York, N.Y.

42. Harlander, S. K. 1992. Genetic improvement of microbial starter cultures, p. 20–26. *In Applications of Biotechnology to Traditional Fermented Foods.* National Academy Press, Washington, D.C.

43. Hols, P., T. Ferain, D. Garmyn, N. Bernard, and J. Delcour. 1994. Use of homologous expression-secretion signals and vector-free stable chromosomal integration in engineering of *Lactobacillus plantarum* for α-amylase and levanase expression. *Appl. Environ. Microbiol.* **60**:1401–1413.

44. Jay, J. M. 1992. Fermented foods and related products, p. 371–409. *In Modern Food Microbiology*, 4th ed. Chapman & Hall, New York, N.Y.

45. Kagermeier, A. 1981. Taxonomie und Vorkommen von Milchsaurebakterien in Fleischprodukten. Dissertation. Fakultat fur Biologie, Ludwig-Maximilian-Universität München, Munich, Germany.

46. Kandler, O. 1983. Carbohydrate metabolism in lactic acid bacteria. *Antonie Leeuwenhoek* **49**:209–224.

47. Kandler, O. 1984. Current taxonomy of lactobacilli. *Dev. Ind. Microbiol.* **25**:109–123.

48. Kandler, O., and N. Weiss. 1986. Regular, non-sporing Gram-positive rods, p. 1208–1234. *In* P. H. A. Sneath, N. S. Mair, M. E. Sharpe, and J. G. Holt (ed.), *Bergey's Manual of Systematic Bacteriology*, vol. 2. The Williams & Wilkins Co., Baltimore, Md.

49. Kitchell, A. G., and B. G. Shaw. 1975. Lactic acid bacteria in fresh and cured meat, p. 209–220. *In* J. G. Carr, C. V. Cutting, and G. C. Whiting (ed.), *Lactic Acid Bacteria in Beverages and Food.* Academic Press, Inc., New York, N.Y.

50. Klettner, P.-G., and D. List. 1980. Beitrag zum Einfluss der Kohlenhydratart auf den Verlauf der Rohwurstreifung. *Fleischwirtschaft* **60**:1589–1593.

51. Knauf, H. J., R. F. Vogel, and W. P. Hammes. 1989. Introduction of the transposon Tn919 into *Lactobacillus curvatus*. *FEMS Microbiol. Lett.* **65**:101–104.

52. Languer, H. J. 1972. Aromastoffe in der Rohwurst. *Fleischwirtschaft* **52**:1299–1306.

53. Leer, R. J., N. van Luijk, M. Posno, and P. H. Pouwels. 1992. Structural and functional analysis of two cryptic plasmids from *Lactobacillus pentosus* MD353 and *Lactobacillus plantarum* ATCC 8014. *Mol. Gen. Genet.* **234**:265–274.

54. Leistner, L. 1986. Mould-ripened foods. *Fleischwirtschaft* **66**:1385–1388.

55. Liepe, H. U. 1983. Starter cultures in meat production, p. 400–424. *In* H.-J. Rehm and G. Reed (ed.), *Biotechnology, Food and Feed Production with Microorganisms*, vol. 5. Verlag Chemie, Weinheim, Germany.

56. Liu, M.-L., J. K. Kondo, M. B. Barnes, and D. T. Bartholomew. 1988. Plasmid-linked maltose utilization in *Lactobacillus* spp. *Biochimie* **70**:351–355.

57. London, J. 1990. Uncommon pathways of metabolism among lactic acid bacteria. *FEMS Microbiol. Lett.* **87**:103–112.

58. Luchansky, J. B., P. M. Muriana, and T. R. Klaenhammer. 1988. Application of electroporation for transfer of plasmid DNA to *Lactobacillus, Lactococcus, Leuconostoc, Listeria, Pediococcus, Bacillus, Staphylococcus, Enterococcus* and *Propionibacterium. Mol. Microbiol.* **2**:637–646.

59. Lücke, F.-K. 1985. Fermented sausages, p. 41–83. *In*

B. J. B. Wood (ed.), *Microbiology of Fermented Foods*, vol. 2. Elsevier Applied Science Publishing Co., Inc., London, United Kingdom.

60. **McKay, L. L., and K. A. Baldwin.** 1990. Applications for biotechnology: present and future improvements in lactic acid bacteria. *FEMS Microbiol. Lett.* **87:**3–14.

61. **Michels, P. A. M., J. P. J. Michels, J. Boonstra, and W. N. Konings.** 1979. Generation of electrochemical proton gradient in bacteria by the extrusion of metabolic end products. *FEMS Microbiol. Lett.* **5:**357–364.

62. **Mogensen, G.** 1993. Starter cultures, p. 1–22. *In* J. Smith (ed.), *Technology of Reduced-Additive Foods.* Blackie Academic and Professional, New York, N.Y.

63. **Morelli, L., P. S. Cocconcelli, V. Bottazzi, G. Damiani, L. Ferretti, and V. Sgaramella.** 1987. *Lactobacillus* protoplast transformation. *Plasmid* **17:**73–75.

64. **Mortvedt, C. L., and I. F. Nes.** 1989. Bacteriocin production by a *Lactobacillus* strain isolated from fermented meat. *Eur. Food Chem. Proc.* **1:**336–341.

65. **Mortvedt, C. I., and I. F. Nes.** 1990. Plasmid-associated bacteriocin production by a *Lactobacillus sake* strain. *J. Gen. Microbiol.* **136:**1601–1607.

66. **Mortvedt, C. I., J. Nissen-Meyer, K. Sletten, and I. F. Nes.** 1991. Purification and amino acid sequence of lactocin S, a bacteriocin produced by *Lactobacillus sake* L45. *Appl. Environ. Microbiol.* **57:**1829–1834.

67. **National Research Council.** 1992. *Applications of Biotechnology to Traditional Fermented Foods*, p. 121–149. National Academy Press, Washington, D.C.

68. **Nes, I. F.** 1984. Plasmid profiles of ten strains of *Lactobacillus plantarum*. *FEMS Microbiol. Lett.* **21:**359–361.

69. **Niinivaara, F. P.** 1955. The influence of pure cultures of bacteria on the maturing and reddening of raw sausage. *Acta Agric. Fenn.* **85:**95–101.

70. **Niven, C. F., Jr.** 1951. Sausage discolorations of bacterial origin. *Am. Meat Inst. Found. Bull.*, no. 13.

71. **Niven, C. F., Jr., R. H. Deibel, and G. D. Wilson.** 1958. The AMIF Sausage Starter Culture. Circular no. 41. American Meat Institute Foundation, Chicago, Ill.

72. **Palumbo, S. A., and J. L. Smith.** 1977. Lebanon bologna processing. *Proc. Recip. Meat Conf.* **30:**63–68.

73. **Pearson, A. M., and W. F. Tauber.** 1984. *Processed Meats*, 2nd ed. AVI Publishing Co., Inc. Westport, Conn.

74. **Pederson, C. S.** 1979. Fermented sausage, p. 210–234. *In Microbiology of Food Fermentations*. AVI Publishing Co., Inc. Westport, Conn.

75. **Podolak, R.** Personal communication.

76. **Poolman, B.** 1993. Energy transduction in lactic acid bacteria. *FEMS Microbiol. Rev.* **12:**125–148.

77. **Posno, M., R. J. Leer, N. Van Luijk, M. J. F. van Giezen, P. T. H. M. Heuvelmans, B. C. Lokman, and P. H. Pouwels.** 1991. Incompatibility of *Lactobacillus* vectors with replicons derived from small cryptic *Lactobacillus* plasmids and segregational instability of the introduced vectors. *Appl. Environ. Microbiol.* **57:**1822–1828.

78. **Raccach, M.** 1984. Method for selection of lactic acid bacteria and determination of minimum temperature for meat fermentations. *J. Food Prot.* **47:**670–671.

79. **Raccach, M.** 1987. Pediococci and biotechnology. *Crit. Rev. Microbiol.* **14:**291–309.

80. **Reuter, G.** 1975. Classification problems, ecology and some biochemical activities of lactobacilli in meat products, p. 221–229. *In* J. G. Carr, C. V. Cutting, and G. C. Whiting (ed.), *Lactic Acid Bacteria in Beverages and Food*. Academic Press, Inc., New York, N.Y.

81. **Rogosa, M., and M. E. Sharpe.** 1959. An approach to the classification of the lactobacilli. *J. Appl. Bacteriol.* **22:**329–340.

82. **Romans, J. R., W. J. Costello, C. W. Carlson, M. L. Greaser, and K. W. Jones.** 1994. Sausages, p. 773–886. *In The Meat We Eat*. Interstate Publishers, Inc., Danville, Ill.

83. **Romero, D. A., and L. L. McKay.** 1986. Isolation and plasmid characterization of a *Lactobacillus* species involved in the manufacture of fermented sausage. *J. Food Prot.* **48:**1028–1035.

84. **Rust, R. E.** 1976. *Sausage and Processed Meats Manufacturing*. American Meat Institute, Washington, D.C.

85. **Scheirlinck, T., J. Mahillon, H. Joos, P. Dhaese, and F. Michiels.** 1989. Integration and expression of α-amylase and endoglucanase genes in the *Lactobacillus plantarum* chromosome. *Appl. Environ. Microbiol.* **55:**2130–2137.

86. **Schillinger, U., and F.-K. Lücke.** 1987. Identification of lactobacilli from meat and meat products. *Food Microbiol.* **4:**199–208.

87. **Schillinger, U., and F.-K. Lücke.** 1989. Antibacterial activity of *Lactobacillus sake* isolated from meat. *Appl. Environ. Microbiol.* **55:**1901–1906.

88. **Schleifer, K. H.** 1986. Gram-positive cocci, p. 999–1002. *In* P. H. A. Sneath, N. S. Mair, M. E. Sharpe, and J. G. Holt (ed.), *Bergey's Manual of Systematic Bacteriology*, vol. 2. The Williams & Wilkins Co., Baltimore, Md.

89. **Sharpe, M. E.** 1981. The genus *Lactobacillus*, p. 1653–1679. *In* M. P. Starr, H. G. Trüper, A. Balows, and H. G. Schlegel (ed.), *The Prokaryotes—a Handbook on Habitats, Isolation, and Identification of Bacteria*, vol. II. Springer-Verlag, New York, N.Y.

90. **Sharpe, M. E. T. F. Fryer, and D. G. Smith.** 1966. Identification of the lactic acid bacteria. *In* B. M. Gibbs and F. A. Skinner (ed.), *Identification Methods for Microbiologists*, part A. Academic Press Ltd., London, England.

91. **Shay, B. J., A. F. Egan, M. Wright, and P. J. Rogers.** 1988. Cysteine metabolism in an isolate of *Lactobacillus sake*: plasmid composition and cysteine transport. *FEMS Microbiol. Lett.* **56:**183–188.

92. **Shrago, A. W., B. M. Chassy, and W. J. Dobrogosz.** 1986. Conjugal plasmid transfer (pAMb1) in *Lactobacillus plantarum*. *Appl. Environ. Microbiol.* **52:**574–576.

93. **Shrago, A. W., and W. J. Dobrogosz.** 1988. Conjugal transfer of group B streptococcal plasmids and comobilization of *Escherichia coli-Streptococcus* shuttle plasmids to *Lactobacillus plantarum*. *Appl. Environ. Microbiol.* **54:**824–826.

94. **Skaugen, M., and I. F. Nes.** 1994. Transposition in

Lactobacilli sake and its abolition of lactocin S production by insertion of IS*1163*, a new member of the IS*3* family. *Appl. Environ. Microbiol.* **60**:2818–2825.

95. **Stackebrandt, E., V. J. Fowler, and C. R. Woese.** 1983. A phylogenetic analysis of lactobacilli, *Pediococcus pentosaceus* and *Leuconostoc mesenteroides. Syst. Appl. Microbiol.* **4**:326–337.

96. **Stackebrandt, E., and M. Teuber.** 1988. Molecular taxonomy and phylogenetic position of lactic acid bacteria. *Biochimie* **70**:317–324.

97. **Stiles, M. E., and W. H. Holzapfel.** 1997. Lactic acid bacteria of foods and their current taxonomy. *Int. J. Food Microbiol.* **36**:1–29.

98. **ten Brink, R. Otto, U. P. Hansen, and W. N. Konings.** 1985. Energy recycling by lactate efflux in growing and nongrowing cells of *Streptococcus cremoris. J. Bacteriol.* **162**:383–390.

99. **Terrell, R. N., G. C. Smith, and Z. L. Carpenter.** 1977. Practical manufacturing technology for dry and semi-dry sausage. *Proc. Recip. Meat Conf.* **30**:39–44.

100. **Townsend, W. E., C. E. Davis, and C. E. Lyon.** 1978. Some properties of fermented dry sausage prepared from PSE and normal pork. *In Kongressdokumentation, 24th Europäischer Fleischforscher-Kongress,* Kulmbach, Germany.

101. **U.S. Department of Agriculture, Food Safety Inspection Service.** 1995. Prescribed treatment for pork and products containing pork to destroy trichinae, Part 318.10. *Code of Federal Regulations, Title 9.* Office of the Federal Register, Washington, D.C.

102. **U.S. Department of Agriculture, Food Safety Inspection Service.** 1995. Requirements for the production of cooked beef, roast beef, and cooked corn beef, Part 318.17. *Code of Federal Regulations, Title 9.* Office of the Federal Register, Washington, D.C.

103. **Vandekerckhove, P., and D. Demeyer.** 1975. Die Zusammernstzung belgischer Rohwurst (Salami). *Fleischwirtschaft* **55**:680–682.

104. **von Wright, A., and M. Sibakov.** 1993. Genetic modification of lactic acid bacteria, p. 161–198. *In* S. Salminen and A. von Wright (ed.), *Lactic Acid Bacteria.* Marcel Dekker, Inc., New York, N.Y.

105. **Weiss, N.** 1992. The genera *Pediococcus* and *Aerococcus,* p. 1502–1507. *In* A. Balows, H. G. Trüper, M. Dworkin, W. Harder, and K.-H. Schleifer (ed.), *The Prokaryotes— a Handbook on the Biology of Bacteria: Ecophysiogy, Isolation, Identification, Applications,* 2nd ed. Springer-Verlag, New York, N.Y.

106. **West, C. A., and P. J. Warner.** 1985. Plasmid profiles and transfer of plasmid-encoded antibiotic resistance in *Lactobacillus plantarum. Appl. Environ. Microbiol.* **50**:1319–1321.

107. **Woese, C. R.** 1987. Bacterial evolution. *Microbiol. Rev.* **51**:221–271.

108. **Wolf, G., E. K. Aarendt, U. Pfahler, and W. P. Hammes.** 1990. Heme-dependent and heme-independent nitrite reduction by lactic acid bacteria results in different N-containing products. *Int. J. Food Microbiol.* **10**:323–330.

109. **Wu, W. J., D. C. Rule, J. R. Busboom, R. A. Field, and B. Ray.** 1991. Starter culture and time/temperature of storage influences on quality of fermented mutton sausage. *J. Food Sci.* **56**:919–925.

110. **Zenitani, B.** 1955. Studies on fermented fish products. I. On the aerobic bacteria in "Shiokara." *Bull. Jpn. Soc. Sci. Fish.* **21**:280–283.

Food Microbiology: Fundamentals and Frontiers, 2nd Ed.
Edited by M. P. Doyle et al.
© 2001 ASM Press, Washington, D.C.

Larry R. Beuchat

Traditional Fermented Foods

34

Fermented foods of plant and animal origin are an integral part of the diet of people in many countries. While the preparation of many traditional or "indigenous" fermented foods and beverages is a household art, the preparation of others, e.g., soy sauce, has evolved to a biotechnological process on a commercial scale. Several books (24, 75, 85, 104) and reviews (9, 11, 12, 18, 19, 32, 34, 36, 54, 55, 62, 63, 66, 68, 78, 82–84, 90, 105, 108, 111) have been published on the subject of traditional fermented foods. A book describing potential applications of biotechnology to traditional fermented foods was published by the U.S. National Research Council (76). A dictionary and guide to fermented foods of the world (17) and a glossary of indigenous fermented foods (101) provide excellent descriptions of known biochemical and microbiological processes associated with fermented foods.

Table 34.1 lists some of the more common traditional fermented foods prepared and consumed in various countries. The information summarized in this table was compiled from several reviews of the subject but is not necessarily complete or even accurate in all aspects. However, the diversity of types of fermented foods consumed by various cultures is evident. Only a few

traditional fermented foods will be covered here, mainly to illustrate the complexity of biochemical, sensory, and nutritional changes that can result from more-or-less controlled microbial activity in a range of raw materials.

BAKERY PRODUCTS

One of the oldest traditional fermented foods is bread, which, in various forms, has been a staple in the diets of many population groups for several centuries. The history of bread can be traced back about 6 millennia. The production of loaf bread supposes the development of the oven and the discovery of dough fermentation. The Egyptians developed a baking oven approaching the design of a modern oven about 2700 B.C. (82). Since 1750 B.C. there were professional bakers in Egypt. The development of cereal foods has proceeded through several stages, from roasted grain to gruels to flat breads, and finally to leavened bread loaves. All stages are practiced today. About 60% of the world population, mainly in Central America, parts of South America, Africa, the Middle East, and the Far East, eat gruels made from grains and flat breads. The production of bakery products consists of five major steps, i.e., the

Larry R. Beuchat, Center for Food Safety and Quality Enhancement, Department of Food Science and Technology, University of Georgia, 1109 Experiment St., Griffin, GA 30223-1797.

Table 34.1 Traditional fermented foods[a]

Product	Geographic region	Substrate	Microorganism(s)	Nature of product	Product use
Ang-kak (anka, red rice)	China, Southeast Asia	Rice	*Monascus purpureus*	Dry red powder	Colorant
Bagni	Caucasus	Millet	Unknown	Liquid	Drink
Bagoong	Philippines	Fish	Unknown	Paste	Seasoning agent
Banku	Ghana	Maize, cassava	Lactic acid bacteria, yeasts	Dough	Staple
Bonkrek	Central Java (Indonesia)	Coconut press cake	*Rhizopus oligosporus*	Solid	Roasted or fried in oil, used as a meat substitute
Bouza	Egypt	Wheat	Yeasts	Liquid	Thick acidic beverage
Braga	Romania	Millet	Unknown	Liquid	Drink
Bread	International	Wheat, rye, other grains	*Saccharomyces cerevisiae*, other yeasts, lactic acid bacteria	Solid	Staple
Burukutu	Savannah regions of Nigeria	Sorghum and cassava	Lactic acid bacteria, *Candida* spp., *Saccharomyces cerevisiae*	Liquid	Creamy drink with suspended solids
Busa	Tartars of Krim, Turkestan, Egypt	Rice or millet, sugar	*Lactobacillus* and *Saccharomyces* spp.	Liquid	Drink
Chee-fan	China	Soybean wheat curd	*Mucor* sp., *Aspergillus glaucus*	Solid	Eaten fresh, cheeselike
Chicha	Peru	Maize	*Aspergillus*, *Penicillium* spp., yeasts, bacteria	Liquid	Alcoholic beverage
Chickwangue	Congo	Cassava roots	Bacteria	Paste	Staple
Chinese yeast	China	Soybeans	Mucoraceous molds and yeasts	Solid	Eaten fresh or canned, used as a side dish with rice
Darassum	Mongolia	Millet	Unknown	Liquid	Drink
Dawadawa (daddowa, iru, kpalugu, kinda)	West Africa, Nigeria	African locust bean	*Bacillus subtilis*, *Bacillus licheniformis*	Solid, sun-dried	Eaten fresh, supplement to soups, stews
Dhokla	India	Bengal gram and wheat	Lactic acid bacteria	Spongy	Staple
Dosai (doza)	India	Black gram and rice	*Leuconostoc mesenteroides*, yeasts	Spongy, pancakelike	Breakfast food
Fish sauce (nuoc-mam, patis, mampla, ngam-pya-ye)	Southeast Asia	Fish	Bacteria	Liquid	Seasoning agent
Gari	West Africa	Cassava root	*Corynebacterium manihot*, *Geotrichum candidum*	Flour	Eaten boiled as staple with stews, vegetables
Hamanatto	Japan	Whole soybeans, wheat flour	*Aspergillus oryzae*, *Streptococcus*, *Pediococcus* spp.	Beans retain individual form, raisinlike, soft	Flavoring agent for meat and fish, eaten as snack
Idli	Southern India	Rice and black gram	Lactic bacteria (*Leuconostoc mesenteroides*), *Torulopsis*, *Candida* spp., *Trichosporon pullulans*	Spongy, moist, steamed bread	Bread substitute

(Continued)

Table 34.1 *(Continued)*

Product	Geographic region	Substrate	Microorganism(s)	Nature of product	Product use
Injera	Ethiopia	Teff, or maize wheat, barley, sorghum	*Candida guilliermondii*	Moist, breadlike pancake	Bread substitute
Inyu	Taiwan, China, Hong Kong	Black soybeans	*Aspergillus oryzae*	Liquid	Flavor enhancer
Jamin-bang	Brazil	Maize	Yeasts and bacteria	Bread- or cakelike	Bread substitute
Kaanga-kopuwai	New Zealand	Maize	Bacteria and yeasts	Soft, slimy	Eaten as vegetable
Kanji	India	Rice and carrots	*Hansenula anomala*	Liquid	Sour, added to vegetables
Katsuobushi	Japan	Whole fish	*Aspergillus glaucus*	Solid, dry	Seasoning agent
Kecap	Indonesia and vicinity	Soybeans, wheat	*Aspergillus oryzae, Lactobacillus, Hansenula, Saccharomyces* spp.	Liquid	Condiment, seasoning agent
Kenkey	Ghana	Maize	Unknown	Mush	Steamed, eaten with vegetables
Ketjap	Indonesia	Black soybeans	*Aspergillus oryzae*	Syrup	Seasoning agent
Khaman	India	Bengal gram	Lactic acid bacteria	Solid, cakelike	Breakfast food
Kimchi (kim-chee)	Korea	Cabbage, vegetables, sometimes seafood, nuts	Lactic acid bacteria	Solid and liquid	Condiment
Kinema	Nepal, Sikkim, Darjeeling district of India	Soybeans	*Bacillus subtilis, Enterococcus faecium,* yeasts	Solid	Snack
Kishk (kushuk, kushik)	Egypt, Syria, Arab world	Wheat, milk	Lactic acid bacteria, *Bacillus* spp.	Solid	Dried balls dispersed rapidly in soups
Lafun	West Africa, Nigeria	Cassava root	Lactic acid bacteria	Paste	Staple food
Lao-chao	China, Indonesia	Rice	*Rhizopus oryzae, Rhizopus chinensis, Chlamydomucor oryzae, Saccharomycopsis* sp.	Soft, juicy, glutinous	Eaten as such as dessert or combined with eggs, seafood
Mahewu (magou)	South Africa	Maize	Lactic acid bacteria (*Lactobacillus delbrueckii*)	Liquid	Drink, sour and nonalcoholic
Meitauza	China, Taiwan	Soybean cake	*Actinomucor elegans, Mucor meitauza*	Solid	Fried in oil or cooked with vegetables
Meju	Korea	Black soybeans	*Aspergillus oryzae, Rhizopus* spp.	Paste	Seasoning agent
Merissa	Sudan	Sorghum	*Saccharomyces* sp.	Liquid	Drink
Minchin	China	Wheat gluten	*Paecilomyces, Aspergillus, Cladosporium, Fusarium, Syncephalastum, Penicilium, Tricothecium* spp.	Solid	Condiment
Miso (chiang, jang, doenjang, tauco, tao chieo)	Japan, China	Rice and soybeans or rice and other cereals, such as barley	*Aspergillus oryzae, Torulopsis etchellsii, Lactobacillus* sp.	Paste	Soup base, seasoning
Munkoyo	Africa	Millet, maize, or kaffir corn plus roots of munkoyo	Unknown	Liquid	Drink
Natto	Northern Japan	Soybeans	*Bacillus natto*	Moist, mucilagenous	Meat substitute

(Continued)

Table 34.1 Traditional fermented foods[a] *(Continued)*

Product	Geographic region	Substrate	Microorganism(s)	Nature of product	Product use
Ogi	Nigeria, West Africa	Maize	Lactic bacteria (*Cephalosporium, Fusarium, Aspergillus, Penicillium* spp., *Saccharomyces cerevisiae, Candida mycoderma* (*C. valida* or *C. vini*)	Paste, porridge	Staple, eaten for breakfast, weaning babies
Oncom (ontjom, lontjom)	Indonesia	Peanut press cake	*Neurospora intermedia,* less often *Rhizopus oligosporus*	Solid	Roasted or fried in oil, used as meat substitute
Papadam	India	Black gram	*Saccharomyces* spp.	Solid, crisp	Condiment
Peujeum	Java	Banana, plantain	Unknown	Solid	Eaten fresh or fried
Pito	Nigeria	Guinea corn or maize or both	Yeasts, lactic acid bacteria	Liquid	Drink
Poi	Hawaii	Taro corms	*Lactobacillus* bacteria, *Candida vini* (*Mycoderma vini*), *Geotrichum candidum*	Semisolid	Side dish with fish, meat
Pozol	Southeastern Mexico	Maize	Molds, yeasts, bacteria	Dough, spongy	Diluted with water, drunk as basic food
Prahoc	Cambodia	Fish	Unknown	Paste	Seasoning agent
Puto	Philippines	Rice	Lactic acid bacteria, *Saccharomyces cerevisiae*	Solid	Snack
Rabadi	India	Maize and buttermilk	Lactic acid bacteria	Semisolid	Mush, eaten with vegetables
Sierra rice	Ecuador	Unhusked rice	*Aspergillus flavus, Aspergillus candidus, Bacillus subtilis*	Solid	Brownish-yellow, seasoning
Sorghum beer (Ibantu beer, kaffir beer, leting, joala, utshivala, mqom-boti, igwelel)	South Africa	Sorghum, maize	Lactic acid bacteria, yeasts	Liquid	Drink, acidic and weakly alcoholic
Soybean milk yogurt	China, Japan	Soybeans	Lactic acid bacteria	Liquid	Drink
Soy sauce (chaing-yu, shoyu, toyo, kanjang, kecap, seeieu)	Japan, China, Philippines, other parts of Asia	Soybeans and wheat	*Aspergillus oryzae* or *Aspergillus soyae,* *Lactobacillus* bacteria, *Zygosaccharomyces rouxii*	Liquid	Seasoning for meat, fish, cereals, vegetables
Sufu (tahur, taokaoan, tao-hu-yi)	China, Taiwan	Soybean whey curd	*Actinomucor elegans, Mucor hiemalis, M. silvaticus, M. subtilissimus*	Paste	Soybean cheese, condiment
Tao-si	Philippines	Soybeans plus wheat flour	*Aspergillus oryzae*	Semisolid	Seasoning agent, side dish
Taotjo	East Indies	Soybeans plus roasted wheat meal or glutinous rice	*Aspergillus oryzae*	Semisolid	Condiment

(Continued)

Table 34.1 (Continued)

Product	Geographic region	Substrate	Microorganism(s)	Nature of product	Product use
Tapé	Indonesia and vicinity	Cassava or rice	*Saccharomyces cerevisiae, Hansenula anomala, Rhizopus oryzae, Chlamydomucor oryzae, Mucor* sp., *Endomycopsis fibuliger (Saccharomycopsis* sp.)	Soft solid	Eaten fresh as staple
Tarhana	Turkey	Parboiled wheat meal and yogurt (2:1)	Lactic acid bacteria	Biscuits or powder	Dried seasoning for soups
Tauco	West Java (Indonesia)	Soybeans, cereals	*Rhizopus oligosporus, Aspergillus oryzae*	Paste	Condiment
Tempeh (tempe kedeke)	Indonesia and vicinity, Surinam	Soybeans	*Rhizopus* spp., principally *R. oligosporus*	Solid	Fried in oil, roasted, or used as meat substitute in soup
Thumba (bojah)	West Bengal	Millet	*Endomycopsis fibuliger*	Liquid	Drink, mildly alcoholic
Torani	India	Rice	*Hansenula anomala, Candida guilliermondii, C. tropicalis, Geotrichum candidum*	Liquid	Seasoning for vegetables
Ugba	Nigeria, West Central Africa	African oil bean	*Bacillus subtilis, Micrococcus luteus, Micrococcus roseus, Bacillus* sp.	Solid	Condiment, flavoring agent
Waries	India, Pakistan	Black gram flour, Bengal gram flour	*Candida* spp., *Saccharomyces* spp.	Spongy	Spicy condiment eaten with vegetables, legumes, rice

[a] Compiled from references 12, 17, 30, 31, 35, 36, 68, 75, 80, 82, 83, 85, and 91.

preparation of raw materials, dough formation, dough processing and fermentation, baking, and preservation. Attention will be given here to breads owing their sensorial properties, at least in part, to fermentative activities of microorganisms.

Preparation of Raw Materials

Wheat or rye flours are the major ingredients in basic, leavened bread recipes. Wheat breads are leavened with yeasts, while rye breads require, in addition to yeasts, acidification either by the use of sourdough starter or by the addition of acid. The function of flours is to absorb water and to form a cohesive, viscoelastic mass, i.e., dough. In wheat breads, crackers, and other bakery products consisting largely of wheat flour, functionality is complemented by starch. However, the most important functional component is gluten, upon which the texture of bakery products is most dependent. Other ingredients include water, fat, and salt, which greatly influence the texture of bakery products; sugar to promote fermentation, color development, and flavor; and optional ingredients such as eggs, milk, skim milk powder, spices, cocoa, fruits, oil seeds, and dough conditioners (68, 82). The latter ingredients impart desired flavor, aroma, color, and texture characteristics to specific types of products.

Top fermenting strains of *Saccharomyces cerevisiae*, known as baker's yeast, are widely used for commercial production of baked goods. Leavening of doughs requires the addition of 1 to 6% yeast, based on the weight of the flour; the optimum temperatures for growth and fermentation are 28 and 32°C, respectively (82). The optimum pH is between 4 and 5. Baker's yeast is available in the form of yeast cakes, bulk or crumbled yeast, yeast cream, active dry baker's yeast, and instant active dry baker's yeast. Yeast cakes are the most popular form and are similar to bulk or crumbled yeast (30 to 32% yeast solids), the main difference being that bulk yeast is not extruded to form uniform blocks but rather broken into irregular pieces after removal from a filter. Yeast cream or liquid yeast is essentially a water suspension of baker's yeast with about 18% yeast solids. Active dry baker's yeast (92 to 96% yeast solids) is available in granular or powder form in hermetically sealed containers under vacuum or an inert gas atmosphere and can be stored without substantial loss of activity for up to 1 year. Its use is mainly in bakeries in tropical and subtropical countries and for home use. Rehydration at 35 to 42°C is required before use in bakery products. Instant active dry yeast (95% yeast solids) may be added directly to flour or dough during mixing.

Sourdough starters contain 10^7 to 10^9 lactic acid bacteria per gram and 10^5 to 10^7 yeasts per gram. Both homofermentative and heterofermentative *Lactobacillus* species, in various combinations, as well as *S. cerevisiae*, *Pichia saitoi*, *Issatchchenka orientalis*, and *Torulopsis holmii* may be present. Homofermentative lactobacilli are less likely to produce the desired sensory qualities of sourdough breads than are heterofermentative species, such as *Lactobacillus brevis*, *Lactobacillus fermentum*, and *Lactobacillus sanfranciscensis*. The predominant lactic acid bacterium in San Francisco sourdough French bread is *Lb. sanfranciscensis*, which ferments maltose but not sucrose, glucose, xylose, arabinose, galactose, rhamninose, or raffinose (40). Yeasts include *T. holmii*, which ferments glucose, sucrose, galactose, and raffinose, but not maltose; *Saccharomyces inusitus*, which ferments maltose, glucose, sucrose, and galactose; and *Saccharomyces exiguus*, which grows in the presence of lactic acid bacteria (88).

Dough Formation and Processing

Raw materials are mixed to form the base for development of the dough to eventually produce bakery products. The baking quality of the flour depends on the variety of grain and milling processes (82). Conditions under which flour is stored and sieved before using in doughs can also affect the quality of bakery products.

Doughs containing largely flour and water are prepared in mixers. Incorporation of air in small bubbles is essential for the leavening of dough and the grain of the bread. The number of bubbles does not increase during fermentation, but the size does. The choice of mixer and the conditions during mixing largely determine the final volume of bakery products and the structure of the crumb (82). Conditions such as pressure, speed and intensity of mixing, temperature, and time of mixing all affect the dough characteristics and subsequent baked product quality (68).

Principal methods of processing and fermenting doughs to produce bread include those described as sponge dough, straight dough, liquid ferment, continuous mix, and short-time dough. Various methods to produce sourdoughs also exist. The sponge dough process involves mixing part of the flour, water, and yeast to produce a sponge or predough. After the sponge has fully fermented at 25°C, it is mixed with the remaining flour, water, and other ingredients to form the final dough formulation and is held at 26 to 28°C. The time required for preparation of the sponge and final dough vary depending on the formulation and desired sensorial qualities of the final baked product. However, sponge fermentation time is generally in the range of 3.5 to 5 h.

In the straight dough process, flour, water, yeast, salt, and all other ingredients are combined to form a dough. After mixing, the dough is fermented for 2 to 4 h at 26 to 32°C. If an overnight fermentation (8 to 12 h) is desired, the amount of yeast added to the dough is reduced, and the dough is held at 18 to 20°C. Straight doughs require less labor, time, and equipment to prepare than sponge doughs. However, breads made from straight doughs have a characteristically blander flavor than breads made from sponge doughs, which may be less acceptable to some consumers.

A liquid ferment or brew using yeast, sugar, salt, and up to 70% of the flour of the ingredient formula is prepared using the liquid ferment process. The temperature of fermentation is 23 to 31°C, with the temperature rising 2 to 5°C during a 2- to 3-h fermentation period. The pH of the fermented liquid is 4.6 to 5.2, depending upon the type of dough and bread desired. Lactic acid bacteria are responsible for the reduction in pH. Liquid ferments have the same function as sponge doughs. They are mixed with the remaining formulation ingredients to produce the final dough.

Continuous mix processes for preparing dough use a liquid ferment, with flour (Am Flow process) or without flour (Do-Maker process), a premixer in which all of the ingredients of the formula are combined, and a developer in which the dough is formed under pressure and with intensive mixing. The dough is softer than that produced by the sponge dough process and contains too much water to be worked by hand. Continuous mix processes are timesaving compared with conventional processes.

Short-time or no-time doughs are obtained by mechanical or chemical means rather than by fermentation. In the Chorleywood bread process, dough is made by combining all ingredients and mixing intensively in a batch, high-speed mixer (68, 82). Mechanical dough development is enhanced by the addition of oxidants, such as potassium iodate, potassium bromate, or ascorbic acid. The process may be carried out with continuous extrusion of dough directly into baking pans, producing bread with very fine grain. In the Brimec process, dough mixing is carried out with high- or low-speed mixers. The addition of 1% shortening and high levels of oxidants is required to produce good bread.

Bread doughs containing more than 20% rye flour require acidification. Otherwise, the development of desirable flavor, aroma, and texture cannot be achieved. Commercial sourdough cultures, spontaneous (natural) souring, or a portion of sourdough used in a preceding batch can be used to achieve acidification. Modifications or processes to make sourdoughs are numerous (82), but biological modifications are brought about largely by the growth of lactic acid bacteria. In addition to acid production, protease and peptidase activity result in the production of amino acids and peptides, which are used by yeasts during sourdough fermentation.

Regardless of the method used to process doughs, a fermentation period is essential. Enzymatic activity of *S. cerevisiae* increases, and the production of carbon dioxide causes the dough to rise. Fermentation is complete when the formation of a colloidal dough structure and the intensity of yeast activity have reached their optima. If fermentation is extended, the gluten absorbs too much water and the dough becomes too bulky. Fully fermented dough is divided into pieces, rounded, and permitted to rest for 5 to 30 min, the intermediate proof. Doughs containing rye flour may require only a 1- to 7-min intermediate proof. The dough is then placed into a molder for final proofing, which is done in a proof box or cabinet for periods ranging from 30 to 60 min, depending on the size of the dough piece and the activity of the yeast. The temperature of the proof box is maintained at 30 to 40°C. A relative humidity of 65 to 85% is required to promote the development of desired surface appearance of baked products.

Baking

Baking stabilizes dough structure and results in the formation of the characteristic aroma, flavor, and color of baked products. The production of gas causes a 40% increase in volume of dough, although this can be controlled within limits by the temperature of baking, type of baking oven, relative humidity, and time of baking (82).

Preservation

Aside from preserving sensorial qualities of baked products, steps must be taken to prevent microbial spoilage. A spoilage defect called rope results from the growth of *Bacillus* species, principally in wheat breads that have not been acidified or in breads with high concentrations of sugar, fat, or fruits. A slightly bitter taste and fruity aroma are followed by yellow, brown, and finally red discoloration and a repulsive aroma. Molds associated with spoilage include species of *Aspergillus*, *Penicillium*, *Mucor*, and *Rhizopus*. A defect called red bread results from the growth of *Neurospora intermedia*. Yeasts responsible for white, chalky spots include *Pichi burtonii* and *Geotrichum candidum*.

Prevention or retardation of microbial spoilage can be achieved by the use of appropriate packaging, application of chemical preservatives, or freezing (68). Propionic acid and its calcium and sodium salts are effective against molds as well as *Bacillus* species that cause rope. The development of rope can be prevented by addition of acetic acid or calcium acetate to dough. Sorbic acid and its potassium or sodium salts are highly effective in controlling the growth of yeasts and molds.

SOY SAUCE

The principles of soy sauce fermentation have been reviewed (32, 107, 111). There are five main types of soy sauce recognized in Japan (28). Koikuchi soy sauce is an all-purpose seasoning characterized by a strong aroma and dark reddish brown color. Usukuchi soy sauce is lighter in color, milder in flavor, and used mainly for cooking when preservation of the original color and flavor of foods is desired. Tamari soy sauce has a strong flavor and is dark brown in color. Saishikomi soy sauce contains a trace of alcohol, and Shiro soy sauce is characterized by a high level of reducing sugars and a yellowish tan color. All of these types of soy sauce contain relatively high levels of salt, in the range of 17 to 19%, and all are used as seasoning agents to enhance the flavor of meats, seafood, and vegetables. Typical ranges in other characteristics are: pH 4.6 to 4.8, 2.2 to 30 Bé, 0.5 to 2.5 g of total nitrogen per ml, 0.2 to 1.1 g of formal nitrogen per 100 ml, 3.8 to 2.0 g of reducing sugar per 100 ml, and from a trace to 2.2 ml of ethanol per 100 ml.

Preparation of Soybeans and Wheat

Soaking and cooking of soybeans and roasting and crushing (cracking) of wheat are separate processes in the early stages of soy sauce production. Whole soybeans or defatted soybean meal or flakes can be used. If whole beans are used, oil must eventually be removed from the fermented mash; otherwise, an inferior product will result. Utilization of nitrogen is higher and fermentation time is shorter with defatted beans than with whole soybeans. This may be due to a lower surface/volume ratio in whole beans versus meal, hence a more pronounced physical restraint in whole beans with regard to access of microbial enzymes to soybean components during fermentation. Whole beans or meal are soaked for 12 to 15 h at ambient temperature or, preferably, at about 30°C until the weight is doubled. Soaking is done either by changing still water every 2 or 3 h or by constantly running water over the beans. If water is not changed, spore-forming *Bacillus* species may proliferate and eventually be deleterious to end product quality. Depending upon the depth of soybeans in the water, the temperature of those in the bottom layer of tanks may increase if water is not changed or circulated during soaking. Hydrated soybeans or meal are then drained, covered with water again, and steamed to achieve further softening and pasteurization. If pressure is used during steaming, the soybeans can be sterilized. The conditions for cooking soybeans influence enzyme activity during subsequent fermentation and may affect the turbidity of the final product (64). Turbidity is caused by insoluble undenatured protein which is dispersed in concentrated salt solution. With increased moisture or pressure during steaming, the soy protein tends to denature more readily; excessively denatured soy protein, on the other hand, is less accessible for enzyme reaction. Consequently, the yield of soluble nitrogen and other soluble compounds will be reduced. Rapid cooling on an industrial scale is done by spreading the soybeans in about a 30-cm layer on perforated traylike platforms and forcing air through them (111). It is important to reduce the temperature to less than 40°C within a few hours. Otherwise, proliferation of naturally occurring microorganisms may spoil the soybeans before controlled fermentation can be initiated.

Concurrent with the preparation of soybeans is the roasting and crushing (cracking) of wheat. The roasting of wheat contributes to the aroma and flavor of soy sauce. Wheat flour or wheat bran may be used in place of whole wheat kernels. Breakdown and conversion products resulting from the roasting process include guaiacyl series compounds, such as vanillin, vanillic acid, ferulic acid, and 4-ethylguaiacol (5). The free phenolic compound content increases with heating owing to degradation of lignin and glycosides. Roasting causes the formation of several brown reaction products that contribute to the desired sensory properties of soy sauce.

Koji Process

Koji is an enzyme preparation produced from cereals or sometimes pulses that is used as a starter for larger batches of traditional fermented foods. In the case of soy sauce, seed (tane) koji is produced by culturing single or mixed strains of *Aspergillus oryzae* or *Aspergillus sojae* on either steamed, polished rice or a mixture of wheat bran and soybean flour (111). Seed koji is added to a soybean/wheat mixture at 0.1 to 0.2% to produce what is then simply called koji.

Strains of *A. oryzae* or *A. sojae* used to prepare koji must have high proteolytic and amylolytic activities and should contribute to the characteristic aroma and flavor of soy sauce. Lipase (113), cellulase (29), and several acid, neutral, and alkaline proteases as well as peptidases (50–52) may also be produced by *A. oryzae* and *A. sojae*. On a commercial scale, a mixture of equal weights of cooked soybeans and roasted wheat are spread in 5-cm layers in trays made from bamboo strips or in stainless steel trays, inoculated with the seed koji, and stacked in such a way as to allow good air circulation. Temperature (25 to 35°C) and moisture (27 to 37%) control are important to the development of good koji; a temperature of 30°C for 2 to 3 days is desirable. A high moisture content is required at the beginning of the incubation period when mycelial growth occurs, followed by a lower moisture content in later stages when spores are being formed (111). The incubation period should be long enough to achieve adequate production and accumulation of enzymes but not so long as to encourage excess sporulation, which may impart undesirable flavors to the finished product. Good-quality koji is clear yellow to yellowish green.

Enzymes produced by *A. oryzae* that contribute to desirable soybean fermentation have been described (29, 50–52, 112). The effects of diisopropyl-phosphorofluoridate, SH reagents such as *p*-chloromercuribenzoate and monoiodoacetate, and metal-chelating agents such as EDTA, α,α'-dipyridyl, and *p*-phenanthroline as well as carboxypeptidases, alkaline and neutral proteinases, leucine aminopeptidases, and lipases on soy sauce fermentation have been studied. Koji infected with molds such as *Rhizopus* or *Mucor* species should be discarded because of the undesirable flavor and aroma it will impart to soy sauce.

Mash (Moromi) Stage

Mature koji is mixed with an equal amount or more (up to 120% by volume) of brine to form the mash (moromi). The sodium chloride content of the mash should be 17 to 19%. Concentrations less than 16% salt may enable growth of undesirable putrefactive bacteria during subsequent fermentation and aging. Higher salt concentrations retard the growth of desirable osmophilic yeasts, namely, *Zygosaccharomyces rouxii*, and halophilic bacteria. The mycelium of the koji mold is not tolerant of the high concentration of salt and, consequently, dies during the very early stage of mash preparation.

If mash fermentation is allowed to proceed naturally without controlling temperature, as would be the situation in the family home, 12 to 14 months is required for the fermentation and aging process. If the mash is kept in large wooden, concrete, or metal containers such as those used by commercial manufacturers, the temperature is usually maintained at 35 to 40°C, thus reducing the fermentation and aging period to 2 to 4 months. Regardless of temperature controls, it is important to stir the mash intermittently with a wooden stick on a small scale or with metal paddles or compressed air in modern commercial facilities. Stirring must be correlated to a certain extent with the rate of carbon dioxide production. Elevated levels of carbon dioxide will enhance the growth of certain anaerobic microorganisms which may impart undesirable flavor and aroma to the finished product. Excessive aeration, on the other hand, will hinder proper fermentation. Koji enzymes hydrolyze proteins to yield peptides and free amino acids during the early stages of fermentation. Starch is converted to simple sugars, which in turn are fermented by microorganisms to yield lactic, glutamic, and other acids as well as alcohols and carbon dioxide. As a consequence, the pH of the mash drops from near neutrality to 4.5 to 4.8.

Various groups of bacteria and yeasts predominate in sequence during mash fermentation and aging (112). *Pediococcus halophilus*, a salt-tolerant bacterium, grows readily in the first stage of fermentation, converting simple sugars to lactic acid and causing a decrease in pH. Later, *Z. rouxii*, *Torulopsis* species, and other yeasts dominate. Molds that may grow on the surface of the mash are believed to have no relation to proper fermentation or aging (108). Soy sauce owes its pleasant aroma and flavor largely to the enzymatic activities of microorganisms. A partial list of flavor components identified in soy sauce is shown in Table 34.2. *P. halophilus* and, perhaps, *Lactobacillus* species produce lactic and other organic acids, which contribute to aroma and flavor. However, yeasts probably make the greatest contribution to the characteristic sensory qualities of soy sauce. By-products of fermentation, such as 4-ethylguaiacol, 4-ethylphenol and 2-phenylethanol (109), furfuryl alcohol (45), pyrazines, furanones (58–61), and ethyl acetate (110), are among the main flavor-contributing compounds. The amounts and types of by-products may differ, depending on the conditions used to process soybeans before fermentation (20).

The extent of aging can be determined by measuring the glutamic acid content. Liquid (sauce) is removed from the mash with a press or by siphoning off the top of the mash. Fresh brine is sometimes added to the residue, and a second, lower-quality fermentation is allowed to proceed for 1 or 2 months before a second drawing is made. Oil is removed from the filtrate by decantation.

Table 34.2 Some flavor components in soy sauce[a]

Acetaldehyde
Acetic acid
Acetone
2-Acetyl furan
2-Acetylpyrrole
Benzaldehyde
Benzoic acid
Benzyl alcohol
Borneol
Bornyl acetate
Butanoic acid
1-Butanol
2,6-Dimethoxyphenol
2,3-Dimethylpyrazine
2,6-Dimethylpyrazine
Diethyl succinate
Ethanol
Ethyl acetate
Ethyl benzoate
3-Ethyl-2,5-dimethylpyrazine
Ethyl-2-hydroxypropanoate (ethyl lactate)
2-Ethyl-6-menthylpyrazine
Ethyl myristate
4-Ethylphenol
Ethyl phenylacetate
Furfural
Furfuryl acetate
Furfuryl alcohol
2,3-Hexanedione
2-Hexanone
3-Hydroxyl-2-butanone (acetoin)
4-Hydroxy-2-ethyl-5-methyl-3(2H)-furanone
4-Hydroxy-5-ethyl-2-methyl-3(2H)-furanone
4-Hydroxy-5-methyl-3(2H)-furanone
3-Hydroxyl-2 methyl-4-pyrone (maltol)
2-Methoxy-4-ethylphenol (4-ethylguaracol)
2-Methoxyphenol (guaiacol)
3-Methylbutanal
3-Methylbutanoic acid
3-Methyl-1-butanol
3-Methylbutylacetate
2-Methylpropanal
2-Methyl propanoic acid
2-Methyl-1-propanol
2-Methylpyrazine
3-Methyl-3-tetrahydrofuranone
3-Methylthio-1-propanol (methional)
4-Pentanolide
Phenyl acetaldehyde
2-Phenylethanol
2-Phenylethyl acetate
Propanal
2-Propanol

[a] Adapted from Nunomura et al. (58–61).

Pasteurization

Raw soy sauce is pasteurized at 70 to 80°C, thus killing the vegetative cells of microorganisms and denaturing most enzymes and other proteins. Alum or kaolin may be added to enhance clarification, after which the sauce is filtered and bottled. Preservatives may be added to prevent growth of yeasts during storage (111). Butyl-*p*-hydroxybenzoate and sodium benzoate are most widely used.

MISO

Fermented soybean pastes are known as miso in Japan, chiang in China, jang or doenjang in Korea, tauco in Indonesia, and tao chieo in Thailand. In addition to soybeans and salt, most of these products also contain cereals such as rice or barley. In Japan, miso is mainly used as a base for soups.

Methods for preparing fermented soybean pastes differ somewhat, but the basic process is the same (12). Rice miso is made from rice, soybeans, and salt; barley miso is made from barley, soybeans, and salt; and soybean miso is made from soybeans and salt. These major types of miso are further classified on the basis of degree of sweetness and saltiness. The procedure for making miso consists of first preparing the koji and the soybeans (simultaneous processes), then brining or fermentation, and finally aging. The preparation of rice miso is considered here to illustrate the general procedure for making miso.

Koji Preparation

Polished rice is washed and soaked in water overnight at about 15°C to bring the moisture content to about 35%. Excess water is removed, and the rice is steamed at atmospheric pressure for 40 min to 1 h. The cooked rice is then deposited in trays or on platforms and cooled to about 35°C before seed koji, prepared as described for use in soy sauce manufacturing, is added at a ratio of 1 g/kg of rice based on a viable spore count of 10^9/g of seed koji (25). The mixture is placed in a rotating fermentor drum in which the temperature, air circulation, and atmospheric relative humidity are controlled. The temperature is maintained at 30 to 35°C to promote growth of *A. oryzae* and maximum production of proteases and carbohydrases. Overheating is usually caused by rapid growth of undesirable bacteria which may be present on the uncooked rice and survive the steaming process. Ventilation should be adequate to supply sufficient oxygen and to remove carbon dioxide, and humidity should be such that neither drying nor sticking of the rice occurs. Fermentation is complete after 40 to 50 h or when koji is characterized by a sweet aroma and flavor; musty odors indicate an inferior-quality koji. The addition of salt to koji as it is removed from fermentors or trays retards further growth of *A. oryzae*.

Preparation of Soybeans

Whole soybeans are prepared for fermentation concurrently with the preparation of koji. The soybeans should also be large and uniform in size and have an ability to absorb water and cook very rapidly. After extraneous materials are removed by mechanical equipment or by hand, soybeans are washed and soaked in water for 18 to 22 h. Water should be changed during the soaking period, especially in the summer months when temperatures are elevated, to control the proliferation of bacteria. At the end of the soaking period, soybeans have increased in volume by about 240% and in weight by 220 to 260%. Drained soybeans are cooked in water or steamed at 115°C for about 20 min or until they are sufficiently soft to be easily pressed flat between the thumb and finger. Flavor and color development can be achieved by varying the heating temperature and time.

Fermentation and Aging

Cooked, cooled soybeans are then mixed with salted koji and an inoculum consisting of a portion of miso from a previous batch or pure cultures of osmophilic yeasts and bacteria. Strains of *Z. rouxii*, *Torulopsis* sp., and *P. halophilus* are the most important microorganisms in miso fermentation (100). The mixture, known as green miso, is packed into vats or tanks to undergo anaerobic fermentation and aging at 25 to 30°C. White miso takes about 1 week, salty miso 1 to 3 months, and soybean miso over 1 year. White miso contains 4 to 8% salt, which permits rapid fermentation, and yellow or brown misos contain 11 to 13% salt. Moisture content ranges from 44 to 52%, protein from 8 to 19%, carbohydrate from 6 to 30%, and fat from 2 to 10%, depending on the ratio of soybeans, rice, and barley used as ingredients.

During fermentation and aging, soybean protein is digested by proteases produced by *A. oryzae* in the koji. Amino acids and their salts, particularly sodium glutamate, contribute to flavor. The addition of commercial enzyme preparations to enhance fermentation has met with some success. The relative amount of carbohydrates in miso is a reflection of the amount of rice in the product. Starch is extensively saccharified by koji amylases to yield glucose and maltose, some of which is utilized as a source of energy by the microorganisms responsible for fermentation. Miso contains 0.6 to 1.5% acids, mainly lactic, succinic, and acetic. Esters that are formed with ethyl and higher alcohols, together with fatty acid esters derived from fatty acids of soybean lipid, are important in giving miso its characteristic aroma (81). Total tocopherol content is decreased in the cooking process but is unaffected during aging (114). Substantial hydrolysis of triglycerides may occur during the early stages of fermentation. Antioxidative activity in miso is attributed in part to the existence of isoflavones, tocopherols, lecithin, compounds of amino-carbonyl reactions, and living microbial cells which tend to have a reducing activity (25). Cooking and steaming reduce thiamin and riboflavin content of soybeans.

NATTO

Natto is a Japanese name given to fermented whole soybeans, but related products are known as tou-shih by the Chinese, tao-tjo by the East Indians, and tao-si by the Filipinos (75, 89). Color, aroma, and flavor of these products vary, depending upon the microorganisms used to ferment the soybeans; however, products are generally dark in color and have a pungent but pleasant aroma and often a harsh flavor due to their relatively high free fatty acid and low-molecular-weight protein content. Fermented whole soybeans are eaten with boiled rice or as a seasoning agent with cooked meats, seafood, and vegetables.

Three major types of natto are prepared in Japan (39, 75). Itohiki-natto, produced in large quantities in eastern Japan, is referred to simply as natto. Washed soybeans are soaked overnight or until they are approximately doubled in weight, then steamed for about 15 min and inoculated with *Bacillus natto*, a variant strain of *Bacillus subtilis*. The beans are packaged in approximately 150-g quantities and allowed to ferment at 40 to 45°C for 18 to 20 h (30). Production of polymers of glutamic acid by *B. natto* causes the surface of the final product to have a viscous appearance and texture. A second type of fermented whole soybean is known as yuki-wari-natto. This product is made by mixing itohiki-natto with salt and rice koji and then aging at 25 to 30°C for about 2 weeks. Hama-natto is a third major type of whole fermented soybean. Soybeans are soaked in water for about 4 h, steamed without pressure for 1 h, cooled, inoculated with a koji prepared from roasted wheat and barley, and fermented for about 20 h or until covered with the green mycelium of *A. oryzae* (75). After drying to a moisture content of about 12%, soybeans are submerged in a salt brine along with strips of ginger and allowed to age under pressure for 6 to 12 months. Breakdown of proteins, carbohydrates, and lipids during aging contributes to desirable sensory qualities. In addition to hydrolytic enzymes originating from *A. oryzae* in the koji, enzymes produced by bacteria such as *Micrococcus*, *Lactococcus*, and *Pediococcus* spp. may also contribute to hydrolysis of soybean components.

The proximate composition of natto varies greatly, but the ranges (as percentage of wet weight) (75) are: water, 55.0 to 60.8; protein, 16.7 to 22.7; fat 0.7 to 8.5; carbohydrate, 5.4 to 6.6; and ash, 2.1 to 3.0.

SUFU

Sufu is a mold-fermented soybean curd produced and consumed largely in the Orient. Preparation consists of three major steps, namely, making a soybean milk curd, fermenting the curd with an appropriate mold(s), and finally brining the fermented curd. Soybeans are washed, soaked overnight in water, and ground in a fashion similar to that followed to make tofu. After boiling or steaming for 20 to 30 min, the ground mass is strained through a fine sieve to separate the soybean milk from the insoluble residue. Alternatively, soybeans may be soaked, ground, and strained without heating. The milk is then heated to boiling to inactivate trypsin inhibitors and to reduce some of the undesirable beany flavor. Coagulation of the milk is achieved by adding calcium sulfate or magnesium sulfate and, occasionally, acid. The curd is then transferred to a cloth-lined wooden box and pressed to remove the whey. The resulting curd (tofu) contains 80 to 85% water, 10% protein, and 4% lipid. Tofu is prepared for fermentation by cutting into 3-cm cubes and soaking for 1 h in a brine containing about 6% sodium chloride and 2.5% citric acid. This treatment retards or prevents the growth of bacterial contaminants but has little effect on growth of desired molds during subsequent stages of sufu preparation. The cubes are boiled in brine for about 15 min, cooled, placed in perforated trays, surface-inoculated with a selected mold, and incubated at 12 to 25°C for 2 to 7 days, depending upon the type and rate of growth of the mold and the desired flavor characteristics of the final product. White to yellowish white mycelium covers each cube, known as pehtze, at the end of the incubation period. Pehtze contain about 74% water, 12% protein, and 4.3% lipid (97).

Molds isolated from sufu these include *Mucor corticolus*, *Mucor hiemalis* (*dispersus*), *Mucor praini*, *Mucor racemosus*, *Mucor silvaticus*, *Mucor subtilissimus*, *Actinomucor elegans*, and *Rhizopus chinensis* (28, 30, 32), all of which secrete proteases to hydrolyze soybean protein and yield peptides and amino acids, thus contributing to flavor development.

The last step of making sufu involves brining and aging. Depending upon the desired flavor and color, pehtzes may be submerged in salted, fermented rice or soybean mash, fermented soybean paste, or a solution containing 5 to 12% sodium chloride, red rice, and 10% ethanol. Red rice and soybean mash impart a red color to sufu. Use of brine containing high levels of ethanol results in sufu with a marked alcoholic bouquet. In addition to imparting taste, sodium chloride also enhances the release of mycelial enzymes which penetrate the molded cubes and hydrolyze soybean components. The aging period ranges from 1 to 12 months, at which time the sufu is consumed as a condiment or used to season vegetables or meat.

MEITAUZA

Meitauza is a fermented product made from the waste from ground, steeped, strained soybeans resulting from preparation of tofu and sufu. Soybean cakes approximately 10 to 14 cm in diameter and 2 to 3 cm thick are fermented for 10 to 15 days with moderate aeration (30). During fermentation, cakes become covered with white mycelium of *Mucor meitauza* (*A. elegans*), a principal mold involved in sufu fermentation. At the end of fermentation, cakes are partially sundried. Meitauza is cooked in vegetable oil or with vegetables as a flavoring agent.

LAO-CHAO

Lao-chao, also known as chiu-niang or tien-chiu-niang by the Chinese, is a fermented rice product. Glutinous rice is first steamed and cooked, then mixed with a small amount of commercial starter known as chiu-yueh or peh-yueh (99). The mass is incubated at ambient temperature for 2 to 3 days during which yeasts hydrolyze the starch, rendering the product soft, juicy, sweet, fruity, and slightly alcoholic (1 to 2%). Mucoraceous molds, including *Rhizopus oryzae*, *R. chinensis*, and *Chlamydomucor oryzae*, can be consistently isolated from lao-chao. *Endomycopsis* sp., one of the few yeasts capable of producing amylases and utilizing starch, is also an integral part of the necessary microflora. Lao-chao is consumed as such, or it may be cooked with eggs and served as a dessert.

ANG-KAK

Red rice (ang-kak, ankak, anka, ang-quac, beni-koji, aga-koji) has been used in the fermentation industry for preparing red rice wine and foods such as sufu, fish sauce, fish paste, and red soybean curd. Pigments produced by *Monascus purpureus* and *Monascus anka* on a rice substrate are used as household and industrial food colorants in many Oriental countries. The major pigments produced by *Monascus* species are monascorubrin, rubropunctatin, and monascorubramine (103). Monascorubrin (red) and monascin (yellow) pigments produced by *M. purpureus*

have been studied most extensively. The optimum cultural conditions for the production of pigments by a *Monascus* species isolated from the solid koji of Kaoliang liquor are reported to be pH 6.0 for a 3-day incubation at 32°C (43). Among the carbon sources tested, starch, maltose, and galactose were suitable for pigment production; a starch content of 3.5% (5% rice powder) and sodium nitrate or potassium nitrate content of 0.5% gave maximum yield of pigment in laboratory media. Zinc may act as a growth inhibitor of *M. purpureus* and concomitantly as a stimulant for glucose uptake and the synthesis of secondary metabolites such as pigments (7).

PUTO

Puto is a fermented rice cake prepared in the Philippines from 1-year-old rice that is ground with sufficient water to allow fermentation before steaming (77). The product is similar to idli prepared from rice and black gram mungo (*Phaseolus mungo*) in India. The quality of puto depends on the microflora present in the milled rice as well as the variety of rice used as a starting material. There is a high degree of correlation between amylose content (within limits of 20 to 27%) and general acceptability.

RAGI

The Indonesian word "ragi" connotes the starter or inoculum used to initiate various kinds of fermentations (24). Thus, for example, Indonesian bread (roti) is made with a baker's yeast preparation (ragiroti), and fermented glutinous rice or cassava products called tapé ketan and tapé ketella, respectively, are prepared using ragi-tapé. Each type of ragi has a distinct mycological profile and is a source of enzymes necessary for the breakdown of carbohydrates and proteins in grains, legumes, and roots used as main fermentation substrates. Of particular importance is amylase produced by *Endomycopsis fibuliger* in ragi-tapé. Yeasts isolated from ragi-tapé are known to produce α-D-1,4-gluconoglucohydrolase, which releases the β form of glucose by hydrolysis and has high specific activities toward maltodextrins with four degrees of polymerization, amylose, amylopectin, and glycogen but little or no activity toward α-methyl- or p-nitrophenyl-α-glucoside (37).

TAPÉ

Indonesian tapé ketan is a fermented, partially liquefied, sweet-sour, mildly alcoholic rice paste (4, 22). In the traditional process, fermentation is initiated by the addition of powdered ragi made from rice flour containing the desired fungi. Tapé ketan (tepej) is prepared by fermenting glutinous rice, whereas tapé ketella (Indonesian), tapé télo (Javanese), and peujeum (Sudanese) are prepared by fermenting cassava roots. Rice or peeled, chopped cassava is steamed or boiled until soft, spread in thin layers in bamboo trays, inoculated with powdered ragi, covered with a banana or other suitable leaf, and allowed to ferment for 1 to 2 days. The product will take on a white appearance, soft texture, and pleasant, sweet, alcoholic aroma and flavor. Amylolytic molds such as *Amylomyces rouxii* and alcohol-producing yeasts, particularly *Endomycopsis burtonii*, appears to be necessary for preparing good tapé.

TEMPEH

Tempeh is a fermented soybean product consumed largely in Indonesia, New Guinea, and Surinam. Kedelee or kedele, meaning soybean, is used to differentiate tempeh made using soybeans from tempeh bongkrek, a product prepared from coconut press cake (copra). Other beans, peas, and cereals are also used to make tempeh (57). Tempeh, as described in the following text, is synonymous with tempeh kedelee.

Preparation of Soybeans and Fermentation

Soybeans are soaked in water at ambient temperature overnight or until hulls (testae) can be easily removed by hand. Lactic acid bacteria and yeasts are predominant in water in which soybeans have been soaked, suggesting their involvement in the fermentation (56). A comprehensive, quantitative study of the ecology of soybean soaking for tempeh fermentation has been made (47). *Lactobacillus casei* and *Lactococcus* species dominate the fermentation, but significant contributions are made by other bacteria and yeasts, e.g., *Pichia burtonii*, *Candida diddensiae*, and *Rhodotorula mucilaginosa*.

After removing the hulls from the soaked soybeans, cotyledons may be pressed slightly to remove more water and then mixed with a small amount of tempeh from a previous batch or a commercial starter. The inoculated beans are then spread onto bamboo frames, wrapped in banana leaves, and allowed to ferment at ambient temperature for 1 to 2 days. At this point, the soybeans are covered with white *Rhizopus oligosporus* mycelium and bound together as a cake (33).

The temperature and moisture content of the fermenting substrate are critical if a good-quality tempeh is to be obtained. The most desirable temperature is between 30 and 38°C. The fermenting beans should be kept covered during fermentation to retard the rate of loss of moisture.

However, slow diffusion of air to and release of gas from the product are essential to promote proper growth and metabolic activities of the *R. oligosporus*. Production of black sporangia and spores is undesirable and usually indicates inadequate environmental conditions during fermentation or a product which has been kept beyond its normal shelf life of about 1 day. Fermented, yeasty off odors are often produced during storage (102). The production of ammonia as a result of enzymatic breakdown of mycelia and soybean protein causes the tempeh to be inedible within a very short period of time. Tempeh is either sliced, dipped in salt solution and deep-fried in coconut oil, or cut into pieces and used in soups.

Biochemical Changes

While other genera of molds are occasionally found in tempeh, only *Rhizopus* can produce acceptable tempeh (30). *Rhizopus* species are known to produce carbohydrases, lipases, proteases, phytases, and other enzymes (58); *R. oligosporus*, *R. stolonifer*, *R. arrhizus*, *R. oryzae*, *R. formosaensis*, and *R. achlamydosporus* are among the species. The principal species used in Indonesia is *R. microsporus*. Among the carbohydrases produced by *R. oligosporus* are endocellulase, xylanase, and arabanase.

The hemicellulose content (as glucose) is reduced from 2.8% in raw soybeans, to 2.0% as the beans are cleaned and cooked, and to 1.1% after fermentation (94). The fiber content, however, may increase upon fermentation owing to the production of mold mycelium (70). Protease production by *R. oligosporus* is substantial and may play an important role in developing good-quality tempeh. Two proteolytic enzyme systems, one with an optimum pH at 3.0 and the other at 5.5, have been described (98). The pH 5.5 system dominates in tempeh fermentation. Amino acid profiles are not changed significantly upon fermentation; however, free amino acids may increase as much as 85-fold (49). Prolonged fermentation may result in losses of lysine (102).

R. oligosporus possesses strong lipase activity (14) and has been shown to hydrolyze over one-third of the neutral fat in soybeans during a 3-day fermentation period (96). Except for the depletion of as much as 40% of the linolenic acid in the late stages of fermentation, there apparently is no preferential utilization of any fatty acid.

ONCOM

Oncom is a fermented peanut press cake product prepared and consumed largely in Indonesia (8, 10, 12). According to Hesseltine (30), the flavor of fermented peanut press cake is fruitlike and somewhat alcoholic but takes on a mincemeat or almond character if the product is deep-fried.

Preparation of Peanuts and Fermentation

After extraction of oil from peanuts, the press cake is broken up and soaked in water for about 24 h (8, 30, 32). Oil that rises to the surface of the water during the soaking period is removed, and the press cake is then steamed for 1 to 2 h and pressed into a layer about 3 cm deep in a bamboo frame. The mass is inoculated with a portion of a previous batch of oncom containing either *N. intermedia* (*N. sitophila*) or, less often, *R. oligosporus*, covered with banana leaves, and allowed to ferment for 1 to 2 days, at which time the internal portion of the mass has been invaded by mycelia. Aeration is important in the production of oncom, as are temperature, moisture content, and degree of press cake granulation. A temperature range of 25 to 30°C is suitable for producing the best oncom. Sporulation may occur, resulting in an orange to apricot-colored product if *N. intermedia* is used or a gray to black product if *R. oligosporus* is used. Only the surface of the fermented press cake is covered with colored conidia or spores.

Cassava (tapioca), potato peels, or other high-carbohydrate materials may be added to the press cake before inoculation to enhance fermentation. The addition of cassava to peanuts has been shown to promote the growth of *Rhizopus* sp. (30), and the addition of citric acid (1.25% by weight), cassava (1%), and sodium chloride (0.63%) to defatted peanut flour has been demonstrated to enhance the growth of *N. intermedia*, *R. oligosporus*, and *Rhizopus delemar* (14).

Biochemical Changes

N. intermedia and *R. oligosporus* produce hydrolytic enzymes that act upon peanut constituents during fermentation. Crude protein is elevated slightly while true protein is decreased from 94% of the crude protein in peanut press cake to 74% of the crude protein in oncom (93). High-molecular-weight globulins are hydrolyzed to smaller components, and the percentages of specific amino acids and proportions of specific amino acids within the free amino acid fraction change during fermentation (15). The nitrogen solubility of the peanut substrate is increased from 5% to 24% and 19% when fermented with *N. intermedia* and *R. oligosporus*, respectively (69). Maximal protease activity of *N. intermedia* is at pH 6.5 (13). The protein efficiency ratio of fermented peanuts apparently is not increased over that of heated unfermented peanuts (70, 93, 95).

N. intermedia has strong α-galactosidase activity (106). The increased digestibility of oncom compared with unfermented peanuts may be attributable in part to the decrease in levels of these sugars. Lipid is hydrolyzed as a consequence of fermentation (14). The free fatty acid fraction of fermented peanuts contains a significantly higher level of saturated fatty acids, particularly palmitic and stearic acids, and lower amounts of linoleic acid than does the total lipid of oncom. The phytic acid content of peanut press cake is reduced by fermentation (27).

IDLI

Idli is a steamed fermented dough made in India from various proportions (1:4 to 4:1) of rice and black gram flours (75). Other ingredients, such as cashew nuts, ghee, chili peppers, ginger, fried cumin seeds, or curry leaves, may be added to the dough in small quantities to impart additional flavor. To prepare idli, dehulled black gram and rice are washed and soaked in water separately for 5 to 10 h at ambient temperature (31). After soaking, the black gram is ground with water to give a coarse paste, whereas the rice is ground to give a smooth gelatinous paste. Salt (about 0.8%) is added to a mixture of pastes, and fermentation is allowed to proceed for 15 to 24 h. The fermented mass is then steamed to yield a soft, spongy product. The open texture is attributed to the protein (globulin) and polysaccharide (arabino-galactan) in black gram (89).

Idli batter volume increases 1.6- to 3.1-fold, and the pH decreases from an initial 6.0 to 4.3 during fermentation (86). Bacteria identified as part of the microflora responsible for production of good idli include *Leuconostoc mesenteroides*, *Lactobacillus delbrueckii*, *Lb. fermentum*, *Lactobacillus lactis*, *Streptococcus faecalis*, and *Pediococcus cerevisiae* (6, 42, 46, 71, 72). Yeasts that may be involved in idli fermentation include *Oidium lactis* (*Geotrichum candidum*), *T. holmii*, *Torulopsis candida*, and *Trichosporon pullulans* (6, 23, 42). Lactic acid bacteria are responsible for pH reduction and may also contribute to improvement of the nutritional value of unfermented black gram and rice. An increase in thiamin and riboflavin contents occurs as a result of fermentation (41). Bacteria may also play a role in the breakdown of phytate present in black gram. *L. mesenteroides* isolated from soybean idli secretes β-N-acetylglucosaminidase and α-D-mannosidase, which are involved in the hydrolysis of hemagglutinin (73).

WARIES AND PAPADAM

Waries is a spicy condiment shaped in the form of a ball 3 to 8 cm in diameter which is used in cooking with vegetables, beans, or rice in India (6). Dehulled beans are soaked in water, ground into a coarse paste, and mixed with spices such as asafoetida, caraway, cardamom, cloves, fenugreek, ginger, red pepper, and salt, and a small amount of paste from a previously fermented batch. The paste is fermented for 4 to 10 days at ambient temperature, then formed into balls and air dried in the sun. The surface of the balls becomes sealed with a mucilaginous coating during the drying process, thus entrapping gases produced by *Candida* species and *S. cerevisiae* (75).

Papadam is similar to waries but does not contain fenugreek or ginger. Circular wafers are prepared from a mixture of black gram paste and spices which have been fermented for 4 to 6 h (6). Yeasts responsible for fermentation are the same as those found in waries. Papadams are served roasted or deep-fried and consumed as a condiment.

DAWADAWA

Locust beans (*Parkia filicoidea*) are fermented to produce a seasoning agent in West Africa. The product is known as daddawa in Nigeria, kpalugu in northern Ghana, kinda in Sierra Leone, and neteton in Gambia (63, 75). To prepare dawadawa, the yellow powdery pulp is removed from the dark brown to black locust bean seeds, and the seeds are boiled in water with the possible addition of potash until slightly soft. The beans are then stored overnight in earthenware, metallic pots, or baskets to further soften the seed coat. The black seed coats are then removed, and the swollen cotyledons are washed in water, boiled for about 30 min, and deposited in a tray, pot, or basket. The preparation is covered with leaves or sheets of polyethylene and left to ferment for 2 to 3 days at ambient temperature. Microorganisms responsible for fermentation undoubtedly include spore-forming bacilli, lactic acid bacteria, and yeasts (62). Metabolic activities of the microflora result in a mucilaginous, strongly proteolytic, ammoniacal smelling substance which covers and binds the individual beans (62). Moisture is partially removed from the fermented, dark brown mass by sun drying before it is pounded into flattened cakes and further dried to prolong shelf life. Darkening during sun drying is due to polyphenol oxidation (16).

GARI

Fermented root of the cassava plant (also known as manioc, mandioca, apiun, yuca, cassada, or tapioca) is known as gari in the rain forest belt of West Africa. Preparation of gari consists of the following stages (60). First, the corky outer peel and the thick cortex are removed, and the inner portion of the root is grated. The pulp is then packed into jute bags, and weights are applied to express some of the juice. After 3 to 4 days of fermentation, cassava is sieved and heated while constantly being turned over a hot steel pan or in an oven. This process has been termed "garifying" (65). The final product contains 10 to 15% moisture, 80 to 85% starch, 0.1% fat, 1 to 1.5% crude protein, and 1.5 to 2.5% crude fiber. Palm oil may be added as a colorant just before or after drying.

Fresh cassava roots contain cyanogenic glucosides, viz., linamarin and lotaustralin, that decompose during the fermentation of gari with the liberation of gaseous hydrocyanic acid (21). *Lactobacillus plantarum* and other lactic acid bacteria also contribute significantly to decreasing the pH (67), which causes hydrolysis of cyanogenic glucosides. The acid condition favors the growth of *G. candidum*, which produces aldehydes and esters that give gari its characteristic aroma and flavor. Other yeasts and molds (26, 79) and lactic acid bacteria (53) undoubtedly contribute to flavor development. The fermentation of gari is reported to be self-sterilizing, exothermic, and anaerobic and to proceed in two stages at an optimum temperature of about 35°C (1). Lactic and formic acids are produced with a trace of gallic acid.

OGI

Corn is eaten in West Africa principally as a porridge known as ogi (Nigeria) or kenkey (Ghana). The Bantu equivalent to ogi is called mahewu in southern Africa. To prepare ogi, kernels of corn are soaked in warm water for 1 to 3 days, after which they are wet-milled and sieved with water through a screen to remove fiber, hulls, and much of the germ (2). The filtrate is fermented to yield a sour, white, starchy sediment. Ogi may be diluted in water to 8 to 10% solids and boiled into a pap or cooked and turned into a stiff gel (eke) before eating. Ogi is a major breakfast cereal for adults and a traditional food for weaning babies. Lactic acid bacteria are largely responsible for fermentation of ogi, although other bacteria, yeasts, and molds may be involved.

INJERA

Injera is an Ethiopian bread made from teff, sorghum, wheat, barley, corn, or a mixture of these grains (87). Teff flour and water are combined with irsho, a fermented yellow fluid saved from a previous batch. The resultant thin, watery paste is generally incubated for 1 to 3 days. A portion of the fermented paste is then mixed with three parts water and boiled to give a product called absit, which is in turn mixed with a portion of the original fermented flour to yield a thin injera. Thick injera (aflegna) is teff paste which has undergone only minimal fermentation (12 to 24 h) and is characterized by a sweet flavor and a reddish color. A third type of injera (komtata) is made from overfermented paste and, consequently, has a sour taste, probably owing to extensive growth of lactic acid bacteria. While the microflora responsible for fermentation of the sweeter types of injera have not been fully determined, *Candida guilliermondii* apparently is a primary yeast in this process. Regardless of the method used to prepare injera, the paste is baked or grilled to result in a breadlike product similar in appearance to pancakes.

POI

Corms of the taro plant are the principal material used to prepare poi in Hawaii and islands in the South Pacific (3). Cooked corms are peeled and ground or pounded to a fine consistency. The addition of water at this point results in fresh poi. The second phase of poi preparation involves fermentation at ambient temperature for 1 to 3 days or longer. As fermentation progresses, texture changes from a sticky mass to one having a more watery and fluffy consistency. Lactic acid bacteria are the predominant microflora during early stages of fermentation. *Lb. delbrueckii*, *Lactobacillus pastorianus*, *Lactobacillus pentacetius*, *Lactococcus lactis*, and *Saccharomyces kefir* produce large amounts of lactic acid and moderate amounts of acetic, propionic, succinic, and formic acids. *Candida vini* and *G. candidum* are prevalent in later stages of fermentation and are thought to be responsible for imparting the pleasant fruity aroma and flavor to older poi.

FERMENTED FISH PRODUCTS

Fermented fish sauce and paste are popular condiments in the Orient. Whole small fish, with or without entrails, or shrimp are heavily salted (up to 30% sodium chloride), packed into containers, and fermented from periods ranging from a few days to over a year (100). Roasted cereals, glutinous or red rice flour, or bran may be added in varying amounts to prepare fish pastes. The fermentation is essentially anaerobic, involving bacterial and autolytic breakdown of proteins and lipids in fish tissue to result in highly flavored products. Because of their high salt content, consumption of large quantities of fish sauce and paste is limited. However, these products represent an important dietary source of calcium.

Fish sauce and paste are known as nuoc-mam and mams, respectively, in Cambodia and Vietnam, ngampya-ye and ngapi in Myanmar, patis and bagoong in

the Philippines, and mampla and kapi in Thailand (92, 100). Traditional methods for preparing fish sauce or paste are similar, but variations result in differences in appearance, aroma, texture, and flavor characteristics in the final products. Preparation of various fermented fish sauces and pastes is described in chapter 33.

SAFETY AND NUTRITIONAL ASPECTS

Of interest to food microbiologists and sanitarians is the possibility of microorganisms producing toxic substances during fermentation or storage of indigenous fermented foods. An investigation of the aflatoxin-forming ability of 238 strains of *Aspergillus* used in the Japanese food industry has revealed that 52 strains produced fluorescent compounds, but none produced aflatoxin (44). None of the 46 domestic rice, 11 imported rice, 108 miso, and 28 rice-koji samples examined contained aflatoxin. Over 200 strains of industrial koji molds have been reported to be nonaflatoxigenic (48). Molds isolated from 24 samples of miso, katuobushi, and tane koji are capable of producing koji acid and β-nitropropionic acid, but none produced aflatoxin (38). While opportunistic mycotoxigenic molds and pathogenic bacteria may occasionally grow in improperly fermented or stored products, traditional fermented foods must be considered microbiologically safe. Otherwise, diseases and illnesses implicating the presence of these microorganisms in traditional fermented foods would be more evident.

Many traditional fermented foods are staples in the diets of vast populations of people who would otherwise have less than minimum intakes of protein and/or calories. While the quality or quantity of proteins in vegetable-based fermented foods generally is not dramatically increased over that of raw substrates, the digestibility may be improved. Antinutritional and toxic components in plant materials may actually be reduced by fermentation (74). The positive contribution of traditional fermented foods to the nutritional well-being of those who consume them on a regular basis is recognized.

References

1. **Akinrele, I. A.** 1964. Fermentation of cassava. *J. Sci. Food Agric.* 15:589–594.
2. **Akinrele, I. A.** 1970. Fermentation studies on maize during the preparation of a traditional African starch-cake food. *J. Sci. Food Agric.* 21:619–625.
3. **Allen, O. N., and E. K. Allen.** 1933. The manufacture of poi from taro in Hawaii: with special emphasis upon its fermentation. *Hawaii Agric. Exp. Stn. Bull.* 70.
4. **Ardhana, M. M., and G. H. Fleet.** 1989. The microbial ecology of tape ketan fermentation. *Int. J. Food Microbiol.* 9:157–165.

5. **Asao, Y., and T. Yokotsuka.** 1958. Studies on flavorous substances in soy sauce. Part XVI. Flavorous substances in raw soy sauce. *J. Agric. Chem. Soc. Jpn.* **32:**617–623.

6. **Batra, L. R., and P. D. Millner.** 1974. Some Asian fermented foods and beverages and associated fungi. *Mycologia* **66:**942–950.

7. **Bau, Y.-S., and H.-C. Wong.** 1979. Zinc effects on growth, pigmentation and antibacterial activity of *Monascus purpureus* fungi. *Physiol. Plant* **46:**63–67.

8. **Beuchat, L. R.** 1976. Fungal fermentation of peanut presscake. *Econ. Bot.* **30:**227–234.

9. **Beuchat, L. R.** 1978. Microbial alterations of grains, legumes and oilseeds. *Food Technol.* **32**(5):193–198.

10. **Beuchat, L. R.** 1982. Flavor chemistry of fermented peanuts. *Ind. Eng. Chem. Res. Dev.* **21:**533–536.

11. **Beuchat, L. R.** 1987. Traditional fermented food products, p. 209–306. *In* L. R. Beuchat (ed.), *Food and Beverage Mycology*, 2nd ed. Van Nostrand Reinhold, New York, N.Y.

12. **Beuchat, L. R.** 1995. Indigenous fermented foods, p. 505–559. *In* H.-J. Rehm, G. Reed, A. Puhler, and P. Stadler (ed.), *Biotechnology, a Multi-Volume Comprehensive Treatise*, 2nd ed., vol. 9. VCH, Weinheim, Germany.

13. **Beuchat, L. R., and S. M. M. Basha.** 1976. Protease production by the ontjom fungus, *Neurospora sitophila*. *Eur. J. Appl. Microbiol.* **2:**195–203.

14. **Beuchat, L. R., and R. E. Worthington.** 1974. Changes in lipid content of fermented peanuts. *J. Agric. Food Chem.* **22:**509–512.

15. **Beuchat, L. R., C. T. Young, and J. P. Cherry.** 1975. Electrophoretic patterns and free amino acid composition of peanut meal fermented with fungi. *Can. Inst. Food Sci. Technol. J.* **8:**40–45.

16. **Campbell-Platt, G.** 1980. African locus bean (*Parkia* species) and its West African fermented food product, dawadawa. *Ecol. Food Nutr.* **9:**123–132.

17. **Campbell-Platt, G.** 1987. *Fermented Foods of the World: a Dictionary and Guide.* Butterworths, London, United Kingdom.

18. **Campbell-Platt, G., and P. E. Cook.** 1989. Fungi in the production of foods and ingredients. *J. Bacteriol. Symp. Suppl. Ser.* **18 67:**117S–131S.

19. **Chavan, J. K., and S. S. Kadam.** 1989. Nutritional improvement of cereals by fermentation. *Crit. Rev. Food Sci. Nutr.* **28:**349–400.

20. **Chou, C.-C., and M.-Y. Ling.** 1988. Biochemical changes in soy sauce prepared with extruded and traditional raw materials. *Food Res. Int.* **31:**487–492.

21. **Collard, P., and S. Levi.** 1959. A two-stage fermentation of cassava. *Nature* (London) **183:**620–621.

22. **Cronk, T. C., K. H. Steinkraus, L. R. Hackler, and L. R. Mattick.** 1977. Indonesian tapé ketan fermentation. *Appl. Environ. Microbiol.* **33:**1067–1073.

23. **Desikachar, H. S. R., R. Radhakrishnamurthy, G. Ramarao, S. B. Kadkol, M. Srinivasan, and V. Subrahmanyan.** 1960. Studies on idli fermentation. I. Some accompanying changes in the batter. *J. Sci. Ind. Res.* **19:**168–172.

24. **Dwidjoseputro, D., and F. T. Wolf.** 1970. Microbiological studies of Indonesian fermented foodstuffs. *Mycopath. Mycol. Appl.* **41:**211–222.

25. **Ebine, H.** 1971. Miso, p. 127. *In* T. Kawabata, M. Fujimaki, and H. Mitsuda (ed.), *Conversion and Manufacture of Foodstuffs by Microorganisms.* Saikon Publishing Co., Tokyo, Japan.

26. **Ekunsanmi, T. J., and S. A Odunfa.** 1990. Ethanol tolerance, sugar tolerance and invertase activities of some yeast strains isolated from steep water of fermenting cassava tubers. *J. Appl. Bacteriol.* **69:**672–675.

27. **Fardiaz, D., and P. Markakis.** 1981. Degradation of phytic acid in oncom (fermented peanut press cake). *J. Food Sci.* **46:**523–525.

28. **Fukushima, D.** 1979. Fermented vegetable (soybean) protein and related foods of Japan and China. *J. Am. Oil Chem. Soc.* **56:**357–362.

29. **Goel, S. K., and B. J. B. Wood.** 1978. Cellulose and exoamylase in experimental soy sauce fermentations. *J. Food Technol.* **13:**243–248.

30. **Hesseltine, C. W.** 1965. A millennium of fungi, food, and fermentation. *Mycologia* **57:**149–197.

31. **Hesseltine, C. W.** 1979. Some important fermented foods of Mid–Asia, the Middle East, and Africa. *J. Am. Oil Chem. Soc.* **56:**367–374.

32. **Hesseltine, C. W.** 1983. Microbiology of oriental fermented foods. *Annu. Rev. Microbiol.* **37:**575–601.

33. **Hesseltine, C. W., M. Smith, B. Bradle, and K. S. Djien.** 1963. Investigation of tempeh, an Indonesian food. *Dev. Ind. Microbiol.* **4:**275–287.

34. **Hesseltine, C. W., M. Smith, and H. L. Wang.** 1967. New fermented cereal products. *Dev. Ind. Microbiol.* **8:**179–186.

35. **Hesseltine, C. W., and H. L. Wang.** 1980. The importance of traditional fermented foods. *BioScience* **30:**402–404.

36. **Hesseltine, C. W., and H. L. Wang.** 1986. *Indigenous Fermented Food of Non-Western Origin.* Mycologia Memoir no. 11. Cramer, Berlin, Germany.

37. **Kato, K., K. Kuswanto, L. Banno, and T. Harada.** 1976. Identification of *Endomycopsis fibuligera* isolated from ragi in Indonesia and properties of its crystalline glucoamylase. *J. Ferment. Technol.* **54:**831–837.

38. **Kinosita, R., T. Ishiko, S. Sugiyama, T. Seto, S. Igarasi, and I. E. Goetz.** 1968. Mycotoxins in fermented food. *Cancer Res.* **28:**2296–2311.

39. **Kiuchi, K., T. Ohta, H. Itoh, T. Takabayashi, and H. Ebine.** 1976. Studies on lipids of natto. *J. Agric. Food Chem.* **24:**404–407.

40. **Kline, L., and T. F. Sugahara.** 1971. Microorganisms of the San Francisco sour dough bread process. II. Isolation and characterization of undescribed bacterial species responsible for the souring activity. *Appl. Microbiol.* **21:**459–465.

41. **Lakshmi, I.** 1978. Studies on fermented foods, M.S. thesis. University of Baroda, Baroda, India.

42. **Lewis, Y. S., and D. S. Johar.** 1953. Microorganisms in fermenting grain mashes used for food preparations. *Cent. Food Technol. Res. Inst.* **2:**228.

43. **Lin, C.-F.** 1973. Isolation and cultural conditions of *Monascus* sp. for the production of pigment in a submerged culture. *J. Ferment. Technol.* **51:**407–414.

44. Matsuura, S., M. Manabe, and T. Sato. 1970. Surveillance for aflatoxins of rice and fermented-rice products in Japan, p. 48–55. *In* M. Herzberg (ed.), *Proceedings of the First U.S.-Japan Conference on Toxic Microorganisms, Mycotoxins, Botulism.* U.S. Government Printing Office, Washington, D.C.

45. Morimoto, S., and N. Matsutani. 1969. Studies on the flavor components of soy sauce; isolation of furfuryl alcohol and the formation of furfuryl alcohol by yeasts and molds. *J. Ferment. Technol.* 47:518–525.

46. Mukherjee, S. K., M. N. Albury, C. S. Pederson, A. G. Van Veen, and K. H. Steinkraus. 1965. Role of *Leuconostoc mesenteroides* in leavening the batter of idli, a fermented food of India. *Appl. Microbiol.* 13:227–231.

47. Mulyowidarso, R. K., G. H. Fleet, and K. A. Buckle. 1989. The microbial ecolog of soybean soaking for tempe production. *Int. J. Food Microbiol.* 8:35–46.

48. Murakami, H. S., S. Takase, and T. Ishii. 1967. Production of fluorescent substances in rice koji, and their identification by absorption spectrum, *J. Gen. Appl. Microbiol.* 14:97–110.

49. Murata, K., H. Ikehata, and T. Miyamoto. 1967. Studies on the nutritional value of tempeh. *J. Food Sci.* 32:580–584.

50. Nakadai, T., S. Nasuno, and N. Iguchi. 1972. Purification and properties of acid carboxypeptidase I from *Aspergillus oryzae. Agric. Biol. Chem.* 36:1343–1352.

51. Nakadai, T., S. Nasuno, and Iguchi, N. 1973. Purification and properties of alkaline proteinase from *Aspergillus oryzae. Agric. Biol. Chem.* 37:2685–2694.

52. Nakadai, T., S. Nasuno, and N. Iguchi. 1973. Purification and properties of neutral proteinase from *Aspergillus oryzae. Agric. Biol. Chem.* 37:2695–2701.

53. Ngaba, P. R., and J. S. Lee. 1979. Fermentation of cassava (*Manihot escuelenta* Crantz). *J. Food Sci.* 44:1570–1571.

54. Nout, M. J. R. 1994. Fermented foods and food safety. *Food Res. Int.* 27:291–298.

55. Nout, M. J. R. 1995. Useful role of fungi in food processing, p. 295–303. *In* R. S. Samson, E. S. Hoekstra, J. C. Frisvad and O. Filtenberg (ed.), *Introduction to Food-Borne Fungi*, 4th ed. Centraalbureau voor Schimmelculture Baarn, The Netherlands.

56. Nout, M. J. R., M. A. De Dreu, A. M. Zuurbier, and T. M. G. Bonants-van Laarhoven. 1987. Ecology of controlled soyabean acidification for tempeh manufacture. *Food Microbiol.* 4:165–172.

57. Nout, M. J. R., and F. M. Rombouts. 1990. Recent developments in tempe research. *J. Appl. Bacteriol.* 69:609–633.

58. Nunomura, N., M., Sasaki, Y. Asao, and T. Yokotsuka. 1976. Identification of volatile components in shoyu (soy sauce) by gas chromatography. *Agric. Biol. Chem.* 40:485–491.

59. Nunomura, N., M. Sasaki, Y. Asao, and T. Yokotsuka. 1976. Isolation and identification of 4-hydroxy-2(or 5)ethyl-5(or 2)methyl-3(2H)furanone as a flavor component in shoyu (soy sauce). *Agric. Biol. Chem.* 40:491–496.

60. Nunomura, N., M. Sasaki, and T. Yokotsuka. 1979. Isolation of 4-hydroxy-5-methyl-3(2H)furanone, a flavor component in shoyu (soy sauce). *Agric. Biol. Chem.* 43:1361–1367.

61. Nunomura, N., M. Sasaki, and T. Yokotsuka. 1980. Shoyu (soy sauce) flavor components: acidic fractions and the characteristic flavor component. *Agric. Biol. Chem.* 44:339–345.

62. Odunfa, S. A. 1985. African fermented foods, p. 155–191. *In* B. J. B. Wood (ed.), *Microbiology of Fermented Foods.* Elsevier, London, United Kingdom.

63. Odunfa, S. A. 1988. Review: African fermented foods. From art to science. *Mircen J.* 4:259–273.

64. Ogawa, G., and A. Fujita. 1980. Recent progress in soy sauce production in Japan, p. 381. *In* G. E. Inglett and G. E. Munck (ed.), *Recent Progress in Cereal Chemistry.* Academic Press, Inc., New York, N.Y.

65. Ogunsua, A. O. 1980. Changes in some chemical constituents during the fermentation of cassava tubers (*Manihot esculenta*, Crantz). *Food Chem.* 5:249–255.

66. Onishi, H. 1990. Yeasts in fermented foods, p. 167–198. *In* J. F. T. Spencer and D. M. Spencer (ed.), *Yeast Technology.* Springer-Verlag, Berlin, Germany.

67. Oyewole, O. B., and S. A. Odunfa. 1990. Characterization and distribution of lactic acid bacteria in cassava fermentation during fufu production. *J. Appl. Bacteriol.* 68:145–152.

68. Ponte, J. G., and C. C. Tsen. 1987. Bakery products, p. 233–267. *In* L. R. Beuchat (ed.), *Food and Beverage Mycology*, 2nd ed. Van Nostrand Reinhold, New York, N.Y.

69. Quinn, M. R., and L. R. Beuchat. 1975. Functional property changes resulting from fungal fermentation of peanut flour. *J. Food Sci.* 40:475–478.

70. Quinn, M. R., L. R. Beuchat, J. Miller, C.T. Young, and R.E. Worthington. 1975. Fungal fermentation of peanut flour: effects on chemical composition and nutritive value. *J. Food Sci.* 40:470–474.

71. Rajalakshmi, R., and K. Vanaja. 1967. Chemical and biological evaluation of the effects of fermentation on the nutritive value of foods prepared from rice and grams. *Br. J. Nutr.* 21:467–473.

72. Ramakrishnan, C. V. 1976. Preschool child malnutrition. Pattern, prevalence and prevention. *Baroda J. Nutr.* 3:1–39.

73. Rao, G. S. 1978. Studies on fermented foods with reference to hemagglutin in hydrolyzing bacteria isolated from rice–soy idli batter. Ph.D. thesis. University of Baroda, Baroda, India.

74. Reddy, N. R., and M. D. Pierson. 1994. Reduction in antinutritional and toxic components in plant foods by fermentation. *Food Res. Int.* 27:281–290.

75. Reddy, N. R., M. D. Pierson, and D. K. Salunkhe. 1986. *Legume-Based Fermented Foods.* CRC Press, Inc., Boca Raton, Fla.

76. Ruskin, F. R. 1992. *Applications of Biotechnology to Traditional Fermented Foods.* National Academy Press, Washington, D.C.

77. Sanchez, P. C. 1975. Varietal influence on the quality of Philippine rice cake (puto). *Philippine Agric. J.* 58:376–382.

78. **Sanni, A. I.** 1993. The need for process optimization of African fermented foods and beverages. *Int. J. Food Microbiol.* **18:**85–95.

79. **Sanni, M. O.** 1989. The mycoflora of gari. *J. Appl. Bacteriol.* **67:**239–242.

80. **Sarkar, P. K., J. P. Tamang, P. E. Cook, and J. D. Owens.** 1994. Kinema—a traditional soybean fermented food: proximate composition and microflora. *Food Microbiol.* **11:**47–55.

81. **Shibasaki, K., and C. W. Hesseltine.** 1962. Miso fermentation. *Econ. Bot.* **16:**180–195.

82. **Spicher, G., and J.-M. Brummer.** 1995. Baked goods, p. 241–319. *In* H.-J. Rehm, G. Reed, A. Puhler, and P. Stadler (ed.), *Biochemistry, a Multi-Volume Comprehensive Treatise*, 2nd ed., vol. 9. VCH, Weinheim, Germany.

83. **Stanton, W. R., and A. Wallbridge.** 1969. Fermented food processes. *Process Biochem.* **4**(4):45–51.

84. **Steinkraus, K. H.** 1994. Nutritional significance of fermented foods. *Food Res. Int.* **27:**259–267.

85. **Steinkraus, K. H.** 1983. Handbook of indigenous fermented foods. Marcel Dekker, New York, N.Y.

86. **Steinkraus, K. H., A. G. Van Veen, and D. B. Thiebeau.** 1967. Studies on idli—an Indian fermented black gram-rice food. *Food Technol.* **21:**110–113.

87. **Stewart, R. B., and A. Getachew.** 1962. Investigations of the nature of injera. *Econ. Bot.* **16:**127–130.

88. **Sugihara, T. F., L. Kline, and M. W. Miller.** 1971. Microorganisms of the San Francisco sourdough bread process. I. Yeasts responsible for the leavening action. *Appl. Microbiol.* **21:**456–458.

89. **Susheelamma, N. S., and M. V. L. Rao.** 1979. Functional role of the arabinogalactan of black gram (*Phaseolus mungo*) in the texture of leavened foods (steamed puddings). *J. Food Sci.* **44:**1309–1312, 1316.

90. **Tamang, J. P., P. K. Sarkar, and C. W. Hesseltine.** 1988. Traditional fermented foods and beverages of Darjeeling and Sikkim—a review. *J. Sci. Food Agric.* **44:**375–385.

91. **Ulloa-Sosa, M.** 1974. Mycofloral succession in pozol from Tabasco, Mexico. *Biol. Soc. Mex. Microbiol.* **8:**17–48.

92. **van Veen, A. G.** 1953. Fish preservation in Southeast Asia. *Adv. Food Res.* **4:**209–231.

93. **van Veen, A. G., D. C. W. Graham, and K. H. Steinkraus.** 1968. Fermented peanut press cake. *Cereal Sci. Today* **13:**96–98.

94. **van Veen, A. G., and G. Schaefer.** 1950. The influence of the tempeh fungus on the soya bean. *Doc. Neerl. Indones. Morbis Trop.* **2:**270–281.

95. **van Veen, A. G., and K. H. Steinkraus.** 1970. Nutritive value and wholesomeness of fermented foods. *J. Agric. Food Chem.* **18:**576–578.

96. **Wagenknecht, A. C., L. R. Mattick, L. M. Lewin, D. B. Hand, and K. H. Steinkraus.** 1961. Changes in soybean lipids during tempeh fermentation. *J. Food Res.* **26:**373–376.

97. **Wai, N. S.** 1968. Investigation of the various processes used in preparing Chinese cheese by the fermentation of soybean curd with *Mucor* and other fungi, p. 89. Tech. Rept., USDA, Public Law 480 Proj. UR-A6-(40)-1.

98. **Wang, H. L., and C. W. Hesseltine.** 1965. Studies on the extracellular proteolytic enzymes of *Rhizopus oligosporus. Can. J. Microbiol.* **11:**727–732.

99. **Wang, H. L., and C. W. Hesseltine.** 1970. Sufu and laochao. *J. Agric. Food Chem.* **18:**572–575.

100. **Wang, H. L., and C. W. Hesseltine.** 1982. Oriental fermented foods, p. 492. *In* G. Reed (ed.), *Prescott and Dunn's Industrial Microbiology*, 4th ed. AVI Publishing Co., Westport, Conn.

101. **Wang, H. L., and C. W. Hesseltine.** 1986. Glossary of indigenous fermented foods, p. 317–344. *In* C. W. Hesseltine and H. L. Wang (ed.), *Indigenous Fermented Food of Non-Western Origin.* Cramer, Berlin, Germany.

102. **Winarno, F. G., and N. R. Reddy.** 1986. Tempe, p. 95–117. *In* N. R. Reddy, M. D. Pierson, and D. K. Salunke (ed.), *Legume-Based Fermented Foods.* CRC Press, Inc., Boca Raton, Fla.

103. **Wong, H.-C.** 1982. Antibiotic and pigment production by *Monascus purpureus*. Ph.D. dissertation, University of Georgia, Athens.

104. **Wood, B. J. B.** 1965. *Microbiology of Fermented Foods*, vol. 2. Elsevier, London, United Kingdom.

105. **Wood, B. J. B.** 1994. Technology transfer and indigenous fermented foods. *Food Res. Int.* **27:**269–280.

106. **Worthington, R. E., and L. R. Beuchat.** 1974. α-Galactosidase activity of fungi on intestinal gas-forming peanut oligosaccharides. *J. Agric. Food Chem.* **22:**1063–1066.

107. **Yokotsuka, T.** 1960. Aroma and flavor of Japanese soy sauce. *Adv. Food Res.* **10:**75.

108. **Yokotsuka, T.** 1971. Shoyu, p. 117–125. *In* M. Fujimaki and H. Mitsuda (ed.), *Conversion and Manufacture of Foodstuffs by Microorganisms.* Saikon Publishing Co., Tokyo, Japan.

109. **Yokotsuka, T., T. Sakasai, and Y. Asao.** 1967. Studies on flavorous substances in shoyu. Part 35. Flavorous compounds produced by yeast fermentation. *J. Agric. Chem. Soc. Jpn.* **41:**428–433.

110. **Yong, F. M., K. H. Lee, and H. A. Wong.** 1981. The production of ethyl acetate by soy yeast (*Saccharomyces rouxii* Y-1096). *J. Food Technol.* **16:**177–185.

111. **Yong, F. M., and B. J. B. Wood.** 1974. Microbiology and biochemistry of the soy sauce fermentation. *Adv. Appl. Microbiol.* **17:**157–194.

112. **Yong, F. M., and B. J. B. Wood.** 1976. Microbial succession in experimental soy sauce fermentation. *J. Food Technol.* **11:**525–536.

113. **Yong, F. M., and B. J. B. Wood.** 1977. Biochemical changes in experimental soy sauce koji. *J. Food Technol.* **12:**163–175.

114. **Yoshida, H., and G. Kajimoto.** 1972. Changes in lipid components during miso making. Studies on the lipids of fermented foodstuffs (Part I). *J. Jpn. Soc. Food Nutr.* **25:**415–421.

Food Microbiology: Fundamentals and Frontiers, 2nd Ed.
Edited by M. P. Doyle et al.
© 2001 ASM Press, Washington, D.C.

Sterling S. Thompson
Kenneth B. Miller
Alex S. Lopez

35

Cocoa and Coffee

Cocoa and coffee are two of the many foods that rely on a microbial curing process or fermentation for flavor development. The popularity and worldwide appeal of these products are due primarily to their unique flavor and aroma. Although a primary curing process is conducted in the preparation of each product before marketing, fermentation of cocoa is absolutely essential for flavor development, whereas with coffee, the curing process is less crucial to flavor and more important for the removal of pulp. Consequently, this chapter will focus mainly on the more comprehensive role of fermentation in cocoa curing and to a lesser extent in the production of coffee.

COCOA PROCESSING

Commercial cocoa is derived from the seeds (beans) of the ripe fruit (pods) of the plant *Theobroma cacao*, which is native to the Amazon region of South America. It has been used by the Amerindians to produce a beverage since time immemorial and was introduced to Europe in the 15th century by Cortés during the period of discovery and colonization of the Americas. Its popularity and

demand led to the establishment and spread of rootstock to virtually all of the European colonies located between 15 degrees north and south of the equator with climates that could support cocoa production.

Of the *Theobroma* species, only *T. cacao* produces beans suitable for chocolate manufacturing. Immediately following the harvest of ripe fruit, or following a brief storage period, the seeds are removed and subjected sequentially to a fermentation and drying process often referred to as "curing," which is carried out on farms, estates, or cooperatives in the producing countries. The origin of this process has been lost in antiquity, but it was believed at one time that fermentation was conducted simply to aid in removing the mucilaginous pulp surrounding the seed so as to facilitate drying and storage, as in the case of coffee. This in fact is true, but the main reason for fermentation of cocoa is to induce biochemical transformations within the beans that lead to formation of the color, aroma, and flavor precursors of chocolate. Without this treatment, cocoa beans are excessively bitter and astringent and, when processed, do not develop the flavor that is characteristic of chocolate. The character and strength of chocolate flavor are

Sterling S. Thompson and Kenneth B. Miller, Microbiology Research, Technical Center, Hershey Foods Corporation, 1025 Reese Ave., Hershey, PA 17033-0805.
Alex S. Lopez, c/o Malaysian Cocoa Board, Jen. Tunka Abd. Rahman, Beg Berkunci 211, Kota Kimbalu 88999, Sabah, Malaysia.

governed primarily by the genetic constitution of the co-coa variety, while the fermentation process releases and develops this flavor potential (30). The inherited char-acteristics of the bean therefore set a limit to what can be achieved by fermentation. It is impossible to improve genetically inferior material by superior processing tech-niques, yet, on the other hand, it is quite easy to ruin good-quality cocoa by careless or inadequate curing.

The cocoa fruit varies among varieties in size, shape, external color, and appearance. These characters have of-ten been used in classifying cocoa, but as far as the flavor quality is concerned, the only really important morpho-logical differences are those that distinguish between the white-seeded Criollo variety of South and Central Amer-ica and the purple-seeded Forastero variety of the Ama-zon. The former type is the source of the original "fine" cocoa which has almost disappeared from the market because of its susceptibility to disease, its lower produc-tivity, and its replacement by the hardier, more prolific Forastero varieties and their varietal crosses, which now account for over 95% of the world production. Hence, the following discussion refers primarily to the process-ing of Forastero cocoa.

Flowers are produced seasonally from cushions that emerge on the bark of the trunk and stems. Fertilized flowers bear fruit 170 days from pollination, a period during which the fruit grows to maturity and changes color from green or dark red-purple to yellow, orange, or red, depending on the variety. The mature fruits are thick walled and contain 30 to 40 beans, each enveloped in a sweet, white, mucilaginous pulp and loosely attached to an axial placenta. Only the beans are used in chocolate manufacturing. For the purpose of describing the curing process, the bean may be envisaged as comprising two main parts, namely, the testa (seed coat) together with the attached sugary, mucilaginous pulp that surrounds it, and the embryo or the cotyledons contained within. The mucilage, containing sugars and citric acid (58), serves as a substrate for microorganisms that are involved in the natural fermentation process; the cotyledons, referred to as the "nib" in the cured bean, are used in chocolate manufacturing.

Processing begins with the harvesting of healthy ripe fruits, an operation carried out over a period of 3 to 4 days at a frequency which varies according to the size of the farm and yield. Fermentation is a batch-type process, and harvesting is conducted to allow for the accumula-tion of sufficient material for each batch while taking pre-cautions that, in the process, pods do not overripen and the seeds within do not germinate. The pods are usually collected in piles in the field and broken open on site or at the processing plant (fermentary) at the end of the har-vesting operation. The beans are removed manually or mechanically on some large estates in West African coun-tries, Mexico, and Brazil, where pod breaking and bean extraction are mechanized (35). Once removed from the pods, the seeds are aggregated in heaps or in receptacles of various types and left to ferment for a period of 2 to 8 days. During this interval, microorganisms which are transferred to the seeds from laborers' hands, fruit sur-faces, and containers used in transport and fermentation degrade the cells of the mucilage that surround the bean (39). The collective microbial activity resulting from the accidental inoculation by a multitude of microorganisms is referred to as "fermentation" or "sweating." This pro-cess results in the liberation of pulp juices from which al-cohols and acids are produced with the evolution of heat. During fermentation, concentrations of ethanol and lac-tic and acetic acids sequentially increase and decrease. This can result in an excess of acetic acid that remains at the end of fermentation (59). Together, these factors bring about changes and affect the curing of the bean. The two principal objectives of fermentation are to re-move mucilage, thus allowing aeration during fermenta-tion of the beans and facilitating drying later on, and to provide the heat and acetic acid necessary to inhibit ger-mination, which assures proper curing of the beans (31).

Methods of Fermentation

The manner of fermenting cocoa varies considerably from country to country, and in many instances, even ad-jacent farms may adopt different curing methods. Much effort has been made to standardize fermentation prac-tices. Today, many of the primitive methods such as fer-mentation in banana leaf-lined holes in the ground, in derelict canoes, and in makeshift banana and bamboo frames are the exception rather than the rule. In general, large farms with adequate cocoa fruit production will have permanent facilities specifically constructed for this purpose. In such instances, fermentation is carried out in batteries of wooden or fiberglass boxes. However, most of the world's cocoa is produced on small holdings un-der very rural conditions; here, relatively small volumes are produced and do not always merit permanent pro-cessing facilities. In this case, cocoa is fermented in any convenient receptacle such as fruit boxes, baskets, plastic buckets, or fertilizer bags, or when these are not readily available, the beans are simply piled on a sheet and cov-ered with any handy material. On the whole, however, the majority of the world's cocoa is fermented on drying platforms, in heaps covered with banana leaves, in bas-kets, or in an assortment of wooden boxes (13, 26, 50).

Approximately one-half of the world's crop is fermented in some type of box, and the remaining half is fermented in heaps or by using other primitive methods.

Fermentation on Drying Platforms

Fermentation on drying platforms is practiced in parts of Central America where Criollo cocoa was once grown. Wet cocoa beans are spread directly on drying platforms where they ferment and dry during the day and are heaped into piles each night to conserve heat and retard the growth of surface molds. Criollo cocoa requires only a short period of fermentation (about 2 to 3 days) for flavor development. Forastero varieties require a fermentation time of 5 to 8 days for the development of flavor (14). Although Criollo cocoa has been largely replaced by Forastero hybrids in Central American countries, in many instances, the old method of fermentation still persists (46). This practice preserves the fine flavor characteristics of Criollo beans; however, it is inappropriate for Forastero varieties, which require longer fermentation times for optimal flavor development. Fermenting cocoa on the drying floor is convenient, but unless properly managed, the process tends to produce underfermented cocoa with the added danger of undesirable mold growth and its consequences of off flavor development.

Fermentation in Heaps

Fermentation in heaps is a popular method among smallholding farmers in Ghana and many other African cocoa-producing countries. It also has been observed sporadically in the Amazon region of Brazil. This method does not require a permanent structure and is well suited to family holdings with a small production. Judging from Ghanaian cocoa, fermentation in heaps can produce good-quality products. Varying quantities of cocoa beans from 25 to 1,000 kg are heaped in the field on plantain leaves and covered with the same material. The beans are mixed (turned) periodically to ensure even fermentation and to decrease the potential for mold growth. This is often done daily or every other day by forming another heap. Mixing is laborious and small heaps may not be turned at all. The duration of fermentation is from 4 to 7 days.

Fermentation in Baskets

Fermentation in baskets is practiced principally by small-scale producers in Nigeria, the Amazon region, the Philippines, and some parts of Ghana. Small lots of cocoa are placed in woven baskets lined with plantain leaves. The surface is covered with plantain leaves and weighted down. The turning procedure and fermentation process are similar to that used for small heaps. Basket fermentation is often used when cocoa fermented in heaps is vulnerable to predial larceny.

Fermentation in Boxes

Fermentation in boxes is considered to be an improvement over other methods. This batch process requires a fixed volume of cocoa and is the method of choice on large estates. The size of the containers varies from region to region, but the design and function are standard. The container or "sweat box" may be a single unit or one of a number of compartments within a large box created by subdividing the space into units measuring approximately 1 by 1 by 1 m, with either fixed or movable internal partitions. These boxes hold between 600 and 700 kg of freshly harvested (wet) cocoa beans. The box is always raised above ground level, over a drain which carries away the pulp juices (sweatings) liberated by the degradation of the mucilage during fermentation. The wooden floor of the box generally has holes or spaces between the boards or slats to facilitate drainage and aeration. Sweat boxes vary considerably in size from that of a small fruit box (0.4 by 0.4 by 0.5 m) to some measuring 7 by 5 by 1 m, used on some Malaysian estates (24). Large estates and cooperatives often have batteries of 20 to 30 sweat boxes arranged in tiers in three to seven rows, one below the other to facilitate mixing or turning. Mixing is achieved by simply removing a dividing wall and shoveling the beans into the next box or, in the case of the tier design, into the box below. On some Malaysian estates, boxes are built on pallets, and a forklift is used to transfer the contents into an empty box. Variations occur not only in the size of sweat boxes but also in the type of wood used in their construction and in drainage and aeration methods and the duration of fermentation. The recommendation is to ferment a 1-m^3 volume for 6 to 7 days with two to three mixings during this period. In the majority of cases, boxes are filled to within 10 cm of the top, and the surface is covered with a padding of banana leaves or jute sacking to help retain the heat and prevent the surface beans from drying.

In some countries, fermentation norms have been modified in an attempt to overcome problems such as acidity by varying the prefermentation treatment and the depth of beans in the sweat boxes (30). In Malaysia, for instance, harvested cocoa pods are stored up to 15 days before breaking to remove the beans, or the beans may be pressed or predried to reduce the pulp volume before fermentation (4, 10).

Fermentation progess is assessed by the odor and the external and internal color changes in the beans. When the process is judged complete, the beans are dried in the sun or in mechanical dryers.

Microbiology of Cocoa Fermentation

Fermentation begins immediately after the beans are removed from the pods, as they become inoculated with a variety of microorganisms from the pod surface, knives, laborers' hands, containers used to transport the beans to the fermentary, dried mucilage on surfaces of the fermentation box (tray, platform, or basket) from the previous fermentation, insects, and banana or plantain leaves (18, 19, 26, 36, 39, 48, 55). It is the pulp surrounding the beans, not the cocoa bean, that undergoes microbial fermentation. Chemical changes take place within the bean as a result of the fermentation of the pulp. The testa of the bean acts as a natural barrier between microbial fermentation activities outside the bean and chemical reactions within the bean. However, there is a migration of ethanol, acetic acid, and water of microbial origin from the outside to the inside of the bean. After the bean dies, soluble bean components are leached through the skin and lost in the drainings. The pulp consists of about 85% water, 2.7% pentosans, 0.7% sucrose, 10% glucose and fructose, 0.6% protein, 0.7% acids, and 0.8% inorganic salts (21), making it a rich substrate for microbial growth. The concentration of sucrose, glucose, and fructose is influenced by the age of the pod (56).

The initial microbial population is variable in number and type; however, the key groups active during fermentation of beans are yeasts, lactic acid bacteria, and acetic acid bacteria (28, 39, 48, 61). Climatic conditions may influence the sequence of microorganisms involved in the fermentation (61). It is theorized that *Bacillus* species play an important role during the latter stages of the fermentation and become the dominant group during drying (61). More research is needed to confirm this theory. Over 100 aerobic spore-forming bacteria were isolated from cocoa bean fermentations in Bahia. Bacteria were identified as *Bacillus subtilis, B. licheniformis, B. firmus, B. coagulans, B. pumilus, B. macerans, B. polymyxa, B. laterosporus, B. stearothermophilus, B. circulans, B. pasteurii, B. megaterium, B. brevis,* and *B. cereus. B. subtilis, B. circulans,* and *B. licheniformis* were encountered more frequently than the other *Bacillus* species during fermentation (61).

During fermentation, yeasts, lactic acid bacteria, and acetic acid bacteria develop in succession. Species of microorganisms that have been detected in cocoa during fermentation in Ghana, Malaysia, and Belize are listed in Table 35.1. At the onset of fermentation, a pH of 3.4 to 4.0, a sugar content of 10 to 12%, and a low oxygen tension favor the growth of yeasts (56, 60). Yeasts utilize the carbohydrates in the pulp under aerobic and anaerobic conditions and may form 40 to 65% of the microflora when the fermentation begins (9, 40). The yeast phase lasts 24 to 48 h, during which populations may increase to 90% of the total microflora. Yeast populations have been determined in several investigations (28).

Some yeasts produce various pectinolytic enzymes that degrade the cocoa pulp, thereby aiding in the drainage of juices (16, 48). In addition to metabolizing sugar to produce ethanol, yeasts utilize citric acid, causing the pH to increase (48). All yeast species that contribute to fermentation are not present simultaneously, but follow a succession which is influenced by the turning step (aeration) and the fact that fermenting bean masses are not homogeneous (28). Several genera of yeasts are involved in fermentation (Table 35.1). In one study (9), of the 142 yeast genera detected in fermenting bean masses, 105 were asporogenous and 37 were ascosporogenous. Between 48 and 72 h of fermentation, the yeast population begins to decrease so that by day 3, it is reduced to 10% of the total microbial population (4). Three factors are responsible for the rapid decline in the dominance of yeasts. First, yeasts rapidly metabolize sucrose, glucose, and fructose in the pulp to form carbon dioxide and ethanol, causing a reduction in energy source. Second, the production of ethanol produces a toxic environment that suppresses yeast growth. For example, Schwan et al. (61) reported that a decline in the opulation of *Kloeckera apiculata* was associated with an ethanol concentration greater than 4%. A small amount of heat is developed simultaneously with ethanol production (2). Third, acetic acid, which is produced from ethanol by the acetic acid bacteria, is also toxic to yeasts. The acetic acid concentration may reach 1 to 2% (6).

The anaerobic conditions created by yeasts make the environment suitable for lactic acid bacteria (40). Lactic acid bacteria prefer a low oxygen concentration or, if oxygen is present, a high concentration of carbon dioxide (28). Such an environment develops as the pulp collapses and the yeast population decreases. The population of lactic acid bacteria increases rapidly, but large numbers may be present for only a brief period (30, 48, 50). The lactic acid bacteria population has been observed to reach 10^6 to 10^7 CFU/g in a typical fermentation in Belize and a high of 20% of the total microflora after 1.5 days in fermentations in Trinidad (55). In Brazil, the lactic acid bacteria population is about 65% of the total microflora after 14 h of fermentation. The lactic acid bacteria population remained high up to 3 days, at which time it decreased to less than 10% of the total microflora (40).

Table 35.1 Microorganisms isolated from fermenting cocoa beans

Microorganism	Country		
	Ghana[a]	Malaysia[a]	Belize[b]
Lactic acid bacteria	Lactobacillus plantarum	Lactobacillus plantarum	Lactobacillus plantarum
	Lactobacillus mali	Lactobacillus collinoides	Lactobacillus fermentum
	Lactobacillus collinoides		Lactobacillus brevis
	Lactobacillus fermentum		Lactobacillus buchneri
			Lactobacillus cellobiosus
			Lactobacillus casei subsp. pseudoplantarum
			Lactobacillus delbrueckii
			Lactobacillus fructivorans
			Lactobacillus kandleri
			Lactobacillus gasseri
			Leuconostoc mesenteroides
			Leuconostoc paramesenteroides
			Leuconostoc oenos
Acetic acid bacteria	Acetobacter rancens	Acetobacter rancens	Acetobacter spp.
	Acetobacter ascendens	Acetobacter lovaniensis	Gluconobacter oxydans
	Acetobacter xylinum	Acetobacter xylinum	
	Gluconobacter oxydans	Gluconobacter oxydans	
Yeasts	Candida spp.	Candida spp.	Brettanomyces claussenii
	Hansenula spp.	Debaryomyces spp.	Candida spp.
	Kloeckera spp.	Hanseniaspora spp.	Candida boidinii
	Pichia spp.	Hansenula spp.	Candida cocoai
	Saccharomyces spp.	Kloeckera spp.	Candida intermedia
	Saccharomycopsis spp.	Rhodotorula spp.	Candida guilliermondii
	Schizosaccharomyces spp.	Saccharomyces spp.	Candida krusei
	Torulopsis spp.	Torulopsis spp.	Candida reukaufii
			Kloeckera apis
			Kloeckera javanica
			Pichia membranaefaciens
			Saccharomyces cerevisiae
			Saccharomyces chevalieri
			Schizosaccharomyces spp.
			Schizosaccharomyces malidevorans

[a] From Carr et al. (7).
[b] S. S. Thompson and J. Pfeifer, unpublished data.

Both homofermentative and heterofermentative lactic acid bacteria occur in cocoa fermentations; however, the majority are homofermentative (61). Lactic acid bacteria detected in traditional box fermentation of cocoa beans in Brazil were isolated and characterized as being homofermentative and heterofermentative. The homofermentative species included *Lactobacillus plantarum*, *Lactobacillus casei*, *Lactobacillus delbrueckii*, *Lactobacillus acidophilus*, *Pediococcus cerevisiae*, *Pediococcus acidilactici*, and *Streptococcus* (*Lactococcus*) *lactis*. The heterofermentative species included *Leuconostoc mesenteroides* and *Lactobacillus brevis* (41). Citric acid is metabolized either to acetic acid, carbon dioxide, and lactic acid by heterofermentative species or to acetylmethylcarbinol and carbon dioxide by homofermentative species. Lactic acid bacteria may be more important in cocoa fermentation in Brazil where, during the first 48 h of fermentation, their population is consistently larger than the yeast population (40). This differs from most other fermentations where yeasts are the dominant microorganism during the first 48 h. If lactic acid bacteria remain as a high percentage of the total microbial population during the fermentation, high concentrations of lactic acid will be produced. Since lactic acid is not volatile, it will remain in the chocolate after manufacturing, producing an undesirable chocolate (52, 64, 72).

As the beans are turned to aerate the mass, more of the pulp is metabolized, and conditions become aerobic. The population of lactic acid bacteria decreases and the population of acetic acid bacteria increases with increased aeration. The population of acetic acid bacteria generally reached 10^5 to 10^6 CFU/g in a typical fermentation in Belize. Two genera of acetic acid bacteria, *Acetobacter* and *Gluconobacter*, have been isolated from fermenting cocoa beans. *Acetobacter* species occur more frequently than *Gluconobacter* species (6, 7, 39). The acetic acid bacteria population has been observed to make up 80 to 90% of the total microflora after 2 days in fermentations in Trinidad (55). Acetic acid bacteria oxidize ethanol to acetic acid exothermally (2), causing the temperature of the bean mass to rise to 45 to 50°C. Turning the beans periodically facilitates oxidation of the ethanol to acetic acid and conserves the high temperature of the bean mass. When all of the ethanol is oxidized to acetic acid and then carbon dioxide and water, fermentation subsides and the temperature of the bean mass decreases quickly.

During the later stages of fermentation and while drying, aerobic spore-forming *Bacillus* species develop and may become dominant (6, 39). *Bacillus* species are present during the first 72 h of the fermentation, but during this early stage, their population remains constant. They become dominant later in the fermentation, making up over 80% of the microbial population (60, 62). Development of *Bacillus* species in the bean mass is favored by increased aeration, an increased pH (3.5 to 5.0) of the pulp, and an increase in temperature to 45 to 50°C (6, 39, 62).

Bacillus species can produce several compounds that may contribute to the acidity and off flavors of fermented cocoa. The C_3, C_4, and C_5 free fatty acids that are present in the bean mass during the aerobic phase of fermentation may contribute to the development of some of the off flavors of chocolate (33, 44). The importance of *Bacillus* species in cocoa bean fermentation is not well established, but they are reputed to produce acetic and lactic acids, 2,3-butanediol, and tetramethylpyrazine, which can affect the flavor of chocolate (33, 62, 73).

A key factor that must be considered when deciding on a fermentation scheme is when to remove the beans from their fermentation environment and begin drying. Extending the fermentation can result in undesirable microbial activity, leading to putrefaction and the production of compounds such as butyric and valeric acids that contribute to off flavors (34). Forsyth and Quesnel (13) suggested that the following factors may collectively indicate when fermentation is optimum: (i) external color of the beans; (ii) time schedule; (iii) decrease in temperature; (iv) bean cut test and the internal color used as a criterion; (v) aroma of the fermenting mass; (vi) plumping or swelling of the beans.

A more desirable measurement of optimum fermentation would be a chemical method that is relatively rapid, inexpensive, and easy to perform and interpret. We have observed that the end point of fermentation can be determined by the pH of the beans, provided a normal temperature curve is established. The minimum pH that gives acceptable cocoa liquor is 5.2; however, the actual fermentation pH may be slightly lower. Other methods to measure the qualitative and quantitative changes that occur during cocoa fermentation have been reviewed by Shamsuddin and Dimick (63).

Fermentation Using Pure Culture Seeding

Fermentation of cocoa beans continues to be a "natural process"; however, studies have been conducted to determine the potential application of a defined microbial inoculum to improve the natural fermentation technique (11, 57, 59). A seeding fermentation study was conducted using *Saccharomyces chevalieri*, *Candida zeylanoides*, and *Kluyveromyces fragilis*, isolates obtained from previous fermentations. Of the three yeasts used, the fermentation using *S. chevalieri* produced the better quality chocolate (57). *Torulopsis candida* (ATCC 20031), *Candida norvegensis* (ATCC 22971), *K. fragilis* (ATCC 8601), and *S. chevalieri* were evaluated for their ability to increase the yield of the cocoa sweatings for use in the development of products such as soft drinks, jams, and marmalades. These controlled fermentations did not have a negative effect on the physicochemical properties of the cocoa sweatings or the quality of the chocolate (11). An inoculum cocktail consisting of *Saccharomyces cerevisiae* var. *chevalieri*, *Lactobacillus lactis*, *Lactobacillus plantarum*, *Acetobacter aceti*, and *Gluconobacter oxydans* subsp. *suboxydans* was successfully used to produce an acceptable quality chocolate (59). We conducted bench-top seeding studies using an inoculum mixture isolated from natural fermentation studies in Belize. The mixture contained 11 microorganisms consisting of yeasts and lactic and acetic acid bacteria. The final chocolate was judged to be acceptable by a trained taste panel (S. S Johnson and J. Pfeifer, unpublished data). Several issues and questions must be resolved to determine whether a starter culture approach can supplement or replace the traditional fermentation.

Biochemistry of Cocoa Fermentation

The actual production of chocolate flavor precursors occurs within the cocoa bean and is primarily the result of biochemical changes that take place during fermentation and drying. The mode of fermentation and the microbial environment during these stages of cocoa production provide the necessary conditions for complex biochemical reactions to occur. Although flavor compounds such as lactic and acetic acids are produced external to the bean by microbial activity, chocolate flavor development is largely dependent on the enzymatic formation of flavor precursors within the cotyledon that are unique to cocoa. Such classes of compounds include free amino acids, peptides, reducing sugars, and polyphenols. When fermented dried cocoa beans containing these flavor precursors are subjected to roasting during chocolate manufacture, a necessary step in flavor development, a series of complex nonenzymatic browning reactions occurs to produce flavor and colored compounds characteristic of chocolate (23, 29, 51, 53, 54). However, if unfermented cocoa beans lacking these precursor compounds are roasted, very little chocolate flavor is produced. It is, therefore, important that these flavor precursor compounds be formed inside the cocoa bean during fermentation.

The initiation of fermentation also corresponds to an incipient germination phase which is necessary for mobilization of the enzymes and hydration of bean components in preparation for growth. However, the germination phase is undesirable in cocoa beans used to make chocolate.

Bean Death

Bean death is a critical event during cocoa fermentation which allows the biochemical reactions responsible for flavor development to occur within the cocoa bean. Although rising temperatures and increasing acetic acid concentrations during fermentation have been implicated in causing seed death (42), more recent data (27) indicate that the production of ethanol during the anaerobic yeast growth phase correlates very closely with death of the seed. Total inability of the seed to germinate occurs about 24 h after maximum concentrations of ethanol are attained within the cotyledon (Fig. 35.1). As a result, events associated with germination and certain quality defects, e.g., the utilization of valuable seed components such as cocoa butter and the opening of the testa by hypocotyl extension, will not occur. This produces a more stable, desirable end product.

From a flavor perspective, events associated with the death of the seed also cause cellular membranes to leak and permit enzymes and substrates to react to form flavor precursor compounds important to chocolate flavor development. Activity of these enzymes in the cotyledon results in significant increases in free amino acid and reducing sugar contents (glucose and fructose) (Fig. 35.1). While the temperature of the beans increases, the concentration of organic acids increases, causing a decrease in pH. All of these factors influence the biochemistry within the bean and have an impact on cocoa flavor and quality.

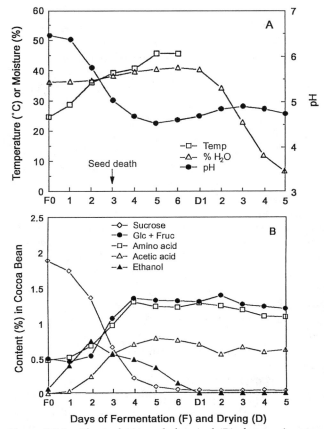

Figure 35.1 Physical (A) and chemical (B) changes in cocoa beans during fermentation and drying in Belize. Fermentation was conducted with 2,000 lb of wet cocoa beans from ripe pods in wooden boxes that were turned daily. Drying was conducted in flat-bed dryers indirectly heated with hot air. Data represent an average of 11 fermentation trials using composite samples collected daily. (A) Temperature was measured in the whole bean mass. Moisture (%) and pH analyses are based on shell-free cotyledons. (B) Sucrose, glucose (Glc), fructose (Fruc), total amino acid, acetic acid, and ethanol contents (%) were determined by analysis of water extracts of shell-free cotyledon samples. Data from Lehrian (27).

Environmental Factors

There are several environmental factors, viz., pH, temperature, and moisture, in the fermenting mass that influence cocoa bean enzyme reactions. Each enzyme has an optimum pH at which it is most active, and within a defined range, an enzyme reaction accelerates as temperature increases. In addition, a certain amount of moisture is necessary to allow enzymes and their substrates to react to form products. Significant changes in pH, temperature, and moisture occur during cocoa fermentation and drying processes (Fig. 35.1) that influence the type and quantity of flavor precursor compounds produced by enzymatic action.

Moisture content within the cotyledon during fermentation is usually more than 35% and will permit adequate migration of enzymes and substrates for enzymatic activity. However, once the drying process begins, moisture content gradually decreases, making it increasingly difficult for enzymes and substrates to react. When a moisture content of 6 to 8% is achieved, virtually all enzyme activity ceases.

The pH of the unfermented cotyledon is about 6.5 and may decrease to as low as 4.5 by the end of the fermentation. This lowering of pH occurs after seed death and is primarily due to the diffusion into the bean of organic acids produced by lactic and acetic acid bacteria. It is the growth of these and other microorganisms that also contributes to the increasing temperature of the mass of fermenting beans. Typically, bean mass temperature will rise from 25°C to about 50°C, followed by a slight decrease as bacterial growth subsides. An increase in temperature of more than 20°C during fermentation can have a profound impact on enzyme activity. If very little change in temperature occurs, enzyme activity is reduced, resulting in fewer flavor precursors and poor chocolate flavor. Likewise, if appropriate amounts of organic acids are not produced during fermentation, the pH of the cotyledon will not be suitable for optimal enzyme activity, and the flavor profile of the resulting cocoa will be affected. However, too much acid will produce excessive sourness that can mask the chocolate flavor.

Consequently, there is a delicate balance among the length of fermentation, environmental factors, and microbial activity that influences enzyme activity within the cotyledon. Hydrolytic and oxidative enzymes play a major role in reactions that produce flavor precursors. A summary of cocoa bean enzymes, their substrates, and pH optima is given in Table 35.2 (30). More recently, Hansen et al. have studied cocoa bean enzymes and their activities during the fermentation and drying process in an effort to understand their impact on cocoa flavor and quality (20).

Hydrolytic Enzyme Reactions

Hydrolytic enzymes such as invertase, glycosidases, and proteases have highest activity during the anaerobic phase of cocoa fermentation. The products of these enzyme activities during cocoa fermentation fall into three basic categories: sugars, amino acids/peptides, and cyanidins. Sugars and amino acids/peptides participate in nonenzymatic browning reactions during roasting to form important chocolate flavor precursors, whereas the cyanidins have more of an impact on color development and some minor flavor components.

Sucrose is the major sugar in unfermented cocoa beans. It is not a reducing sugar and therefore does not participate in nonenzymatic browning reactions that occur during roasting to contribute to chocolate flavor. However, sucrose is converted to glucose and fructose by invertase during the fermentation process. These reducing sugars represent more than 95% of the total reducing monosaccharides in cocoa beans, and their concentrations increase almost three fold during fermentation, while sucrose is depleted (Fig. 35.1).

Another class of hydrolytic enzymes within the cocoa bean that contributes to both flavor and color during fermentation is the glycosidases. The substrates for these enzymes are the purple-colored anthocyanins located in specialized vacuoles within the cotyledon and are responsible for the characteristic deep purple color of the unfermented bean. The actions of specific glycosidase enzymes begin at seed death and are responsible for cleaving the sugar moieties, galactose, and arabinose attached to the anthocyanins. This results in a bleaching of the purple color of the beans as well as the release of reducing sugars that can participate in flavor precursor reactions during roasting (12). Pigments themselves do not carry any flavor potential (12, 33, 49). Cocoa beans that still contain significant purple color are considered to have been poorly fermented and are less desirable.

Although there are small amounts of free amino acids present in unfermented cocoa beans, the total free amino

Table 35.2 Characteristics of the principal enzymes active during the curing of the cocoa bean[a]

Enzyme	Location	Substrate	Product	pH	Temp (°C)	Reference(s)
Invertase	Testa	Sucrose	Glucose and fructose	4.0	52	32
				5.25	37	
Glycosidases (β-galactosidase)	Bean	Glycosides (3-β-D-galactosidyl cyanidin and 3-α-L-arabinosidyl cyanidin)	Cyanidin and sugars	3.8–4.5	45	12
Proteases	Bean	Proteins	Peptides and amino acids	4.7	55	5, 43
Polyphenol oxidases	Bean	Polyphenols (epicatechin)	σ-Quinones and σ-diquinones	6.0	31.5, 34.5	45

[a] Information taken from Lopez (30).

acid pool increases significantly during fermentation due to the action of both endo- and exoproteases on cocoa bean proteins. After seed death occurs, these proteolytic enzymes are free to act on protein substrates within the bean, and their activity becomes dependeant on pH and temperature. A vicilin-like globular storage protein within the cotyledon is the primary target of these proteolytic enzymes, and ratios of free amino acids and peptides that are unique to cocoa are produced (70). These flavor precursor compounds contribute to the development of cocoa flavor when roasted in the presence of reducing sugars.

Oxidative Enzyme Reactions

Significant oxidative enzyme activity also occurs, being most prevalent late in the aerobic phase of fermentation but continuing well into the drying of cocoa. Polyphenol oxidase is the major oxidase in cocoa and is responsible for much of the brown color that occurs during fermentation as well as some flavor modifications. This enzyme becomes active during the aerobic phase of the fermentation as a result of oxygen permeating the cotyledon. Events that contribute to activity include seed death, subsequent breakdown of cellular membranes, reduction in the amount of seed pulp, and aeration of the bean mass by agitation. Oxygen continues to penetrate the beans during the drying process, enabling polyphenol oxidase activity to continue until rising temperatures and insufficient moisture become inhibiting factors.

Catechins and leucocyanidins are the major classes of polyphenols that are subject to oxidation in cocoa beans. Epicatechin makes up more than 90% of the total catechin fraction and is the major substrate of polyphenol oxidase (17). Oxidation of epicatechin during the aerobic phase of fermentation and drying is largely responsible for the characteristic brown color of fermented cocoa beans. Polyphenols in the dihydroxy configuration are oxidized to form quinones which in turn can polymerize with other polyphenols or complex with amino acids and proteins to yield characteristic colored compounds and high-molecular-weight insoluble material. This complexation also has an impact on flavor. The formation of these less soluble polyphenolic complexes reduces astringency and bitterness associated with native polyphenols present in unfermented cocoa (14, 47). In addition, the ability of polyphenols to complex with proteins results in the reduction of off flavors associated with the roasting of peptide and protein material (22, 71).

Flavor and Quality Implications

The ultimate goal of biochemical changes during fermentation is to produce cocoa beans with desirable flavor and color characteristics. Good chocolate flavor potential is achieved by the production of specific amino acids, peptides, and sugars through the action of proteases and invertases on cocoa bean substrates. Proper control of fermentation conditions and microflora assures that concentrations of organic acids are maintained at reasonable levels to minimize sour and putrid off flavors while still developing the pH and temperature environment for enzyme-substrate reactions that produce chocolate flavor precursors. Enzyme-mediated conversion of polyphenol materials during fermentation and drying processes reduces astringency and bitterness and produces the desirable brown color typical of properly fermented cocoa beans. The drying process will then preserve the flavor and color characteristics of the beans until they are made into chocolate.

Although cocoa has been successfully produced for centuries, flavor characteristics of specific varieties of cocoa are becoming diluted due to the prevalence of genetic hybrids. While these hybrids are being selected for high crop yield and disease resistance, certain flavor attributes are being lost. The actual identity of the flavor precursors responsible for chocolate flavor has not yet been confirmed. However, recent work has focused on characterizing cocoa bean enzymes and proteins that yield breakdown products unique to cocoa during the fermentation process (69). This work needs to continue in order to understand flavor development and maintain the high quality of chocolate flavor the consumer expects.

Drying

After the beans are fermented, they have a moisture content of about 40 to 50%, which must be reduced to 6 to 8% for safe storage. A higher final moisture content will result in mold growth during storage. The drying process relies on air movement to remove water. This environment favors aerobic microorganisms which proliferate at rates that decrease with moisture loss. Sun drying is the preferred method, but in regions where harvesting coincides with frequent rainfall, some form of artificial drying is necessary and desirable. In general, sun drying is employed on small farms, whereas large estates may resort to both natural and artificial drying. Sun drying allows a slow migration of moisture throughout the bean, which transports flavor precursors that had been formed during fermentation.

During sun drying, beans are placed on wooden platforms, mats, polypropylene sheets, or concrete floors in layers ranging from 5 to 7 cm thick. The beans are constantly mixed to promote uniform drying, to break agglomerates that may form, and to discourage mold growth. Under sunny conditions, the beans dry in about a week, but under cloudy or rainy conditions, drying times may be prolonged to 3 or 4 weeks, increasing the risk of mold development and spoilage.

Various types of artificial dryers employed to overcome the dependence on weather conditions have been described by McDonald et al. (37). Hot air dryers of one form or another, fueled by wood or oil as a source of cheap, readily available energy, are generally employed. The beans may be heated by direct contact with the flue gases; however, the preferred method relies on indirect heating via heat exchangers. Improperly used or poorly maintained heating systems present the danger of contamination with smoke which results in smoky or hammy off-flavors characteristic of beans from some countries. Platforms, trays, and rotary dryers of various designs, coupled to furnaces, are used, but in every instance, the initial drying must be slow and with frequent mixing to obtain uniform removal of water. This results in volatilization of acids and sufficient time for oxidative, biochemical reactions to occur. For this reason, temperatures should not exceed 60°C and drying times should take at least 48 h. Elevated temperatures also tend to produce cocoa with brittle shells and cotyledons which crumble during handling. In short, the drying rate should be controlled so as to remove moisture at a rate that will avoid case hardening (rapid drying on the bean surface with moisture retention inside the bean) or excessive mold growth, while still allowing sufficient time for biochemical oxidative reactions and loss of acid to occur.

Storage

Due to marketing practices and manufacturing procedures, fermented, cured, dried beans are stored for periods of 3 to 12 months in warehouses on farms, at wharfs in exporting and receiving countries, and at factories before being processed into chocolate. The efficiency of the drying process will determine the shelf life of the product. Uniformly dried beans with a moisture content of 7 to 8%, when stored at a relative humidity of 65 to 70%, will generally maintain that moisture, resist mold growth and insect infestation, and not require repeated fumigation. The cocoa quality can change during storage depending on temperature, relative humidity, and ventilation conditions. Slow oxidation and acid loss continue to enhance product quality somewhat, but prolonged storage results in a noticeable staling (31).

COFFEE PROCESSING

Coffee beans may also undergo a fermentation step to prepare the fruit for commercial use (3, 8, 66). Like cocoa, the fermentation of coffee has the important goal of breaking down the pulp layer surrounding the beans in order to aid in the processing of the fruit into a desirable finished product. Unlike cocoa, however, the role of fermentation in coffee is less critical in the development of coffee flavor, although improperly fermented fruit can result in undesirable off flavors. Coffee beans are produced by the genus *Coffea*. Over 40 species are known, but only a few are used to produce coffee. *Coffea arabica*, *C. canephora*, *C. robusta*, *C. liberica*, and *C. excelsa* are the most important species. The coffee fruit is a fleshy berry approximately the size of a small cherry (8).

It takes approximately 1 year from the flowering stage for a fruit, or coffee cherry as it is commonly called, to reach maturity. As the fruit ripens, it changes color from green to cherry red. The ripe, red coffee cherry consists of two green beans surrounded by a pulp layer and enclosed in an outer skin. It is necessary to first remove the outer skin and pulp layer to obtain the green coffee beans, which are then dried, roasted, milled, and used for making coffee beverages. The removal of the pulp layer can be accomplished by either a natural drying step yielding "natural coffee" or a wet fermentation step yielding "washed coffee." The percent yield of dried coffee from ripe cherries varies among species: Arabica produces 12 to 18%, Robusta produces 17 to 22%, and Liberica produces about 10%.

Natural Coffee

The "natural" drying process relies on partial dehydration of the pulp layer while the coffee fruit is ripening on the tree. The ripe fruit is then harvested and subjected to additional drying, and the dried pulp layer and outer skin are removed by hulling. Some mucilage will penetrate the coffee bean during drying, and for this reason, the natural process produces a light-brown-colored bean instead of the blue-green color of wet-processed coffee. Drying must be carefully controlled to avoid excessive fermentation during the initial stages which can result in reduced product quality. Molds that may develop on slowly drying coffee can produce off flavors typical of a butyric fermentation. Coffee is dried to a moisture content below 13%.

Washed Coffee

The wet process is accomplished by harvesting ripe coffee fruit and immediately removing the majority of pulp by mechanical squeezing or depulping. This is followed by a fermentation step which is used to convert the remaining mucilage to water-soluble products that can be removed by washing. In general, wooden or concrete bins are used for this fermentation step. The white, sticky, partially depulped beans are held under water for 12 to

60 h, depending on environmental conditions, ripeness of the cherries, and the variety being processed. Microbial by-products are periodically washed away during the fermentation step to avoid the development of excessive off flavors. Once the sticky pulp layer is converted to water-soluble products, the beans are washed with water and dried to a moisture content of about 12%.

Microbiology and Biochemistry of Coffee Fermentation

As stated above, the purpose of fermentation is removal of the mucilage from around the seed. A natural fermentation, which involves several different microorganisms, is the primary method used to remove the mucilage. Three nonmicrobial methods, namely, the addition of enzymes, chemical treatment (sodium hydroxide), and mechanical force, may also be used to remove mucilage (25). However, only the natural fermentation used in the wet process to produce washed coffee will be discussed.

Coffee is generally fermented in wooden or concrete tanks. Mucilage is easily metabolized by most of the microorganisms that have been identified in coffee fermentation studies. Mucilage is composed of pectin, pectic acids, reducing sugars, sucrose, caffeine, chlorogenic acid, amino acids, and hydrolytic and oxidative enzymes (25).

Many coffee fermentation studies have been conducted over the years, but researchers do not agree on which microorganism(s) is responsible for digestion of the mucilage. However, there is agreement that several microorganisms are present during the fermentation. Molds, yeasts, several species of lactic acid bacteria, coliforms, and other gram-negative bacteria have been isolated. These microorganisms originate from the surface of the fruit and the soil (1, 15, 67).

Since the mucilage composition consists largely of pectic substances (25), the microorganisms responsible for colonization and utilization of this material must be capable of producing pectinases. Coffee fermentation studies have demonstrated that the highest microbial activity occurs during the first 12 to 24 h after the beans are harvested (68). Most of the microorganisms detected during this early stage belong to the genera *Aerobacter* (*Enterobacter*) and *Escherichia*. Populations of these bacteria increase from 10^2 to 10^9 CFU/g during that first 24-h period. Pectinolytic species of *Bacillus*, *Fusarium*, *Penicillium*, and *Aspergillus* have also been detected. Only a few yeast species have been identified. Fermentation studies conducted on Kona coffee beans demonstrated that *Erwinia dissolvens* was the main cause of mucilage decomposition (15).

The pH of unfermented coffee beans can range from 5.4 to 6.4. During fermentation, the pH may decrease to 3.7 (1, 38). Low levels of ethanol have been detected under both aerobic and anaerobic conditions (38). However, ethanol may reach slightly higher levels under anaerobic fermentation conditions (25).

Duration of a natural fermentation will depend on climate conditions, regional factors, coffee variety, degree of anaerobiosis of the ferment, and the microorganisms present (25). An optimal fermentation time is considered to be 16 to 24 h (66). Extending the fermentation will result in development of flavor defects and bean discoloration (38, 46). Once the fermentation is complete, any remaining mucilage is removed by washing. Washing the beans is required to avoid excessive fermentation and enhance the drying process.

CONCLUSION

Although serving somewhat different purposes, microbial fermentation plays a critical role in the production of both cocoa and coffee. A common goal of fermentation of both of these products is the breakdown and removal of fruit pulp. This process aids in the proper drying of coffee and, in the case of cocoa, provides a suitable environment for flavor and color development. As in most natural curing processes, there is a delicate balance between environmental factors and conditions enabling microbial activities, all of which influence the biochemical changes that take place in processed foods. Coffee and cocoa are no exceptions, and it is the proper control of the fermentation process that largely determines the color and flavor quality of final products. Consequently, understanding the microbiology and biochemistry of cocoa and coffee production, as well as the factors that influence them, is critical to quality control.

References

1. Agate, A. D., and J. V. Bhat. 1966. Role of pectinolytic yeasts in the degradation of the mucilage layer of *Coffee* and *Robusta* cherries. *Appl. Microbiol.* 14:256–260.
2. Anonymous. 1979. *The Fermentation of Cocoa.* National Union of Cocoa Producers, Villahermosa, Mexico.
3. Arunga, R. 1982. Coffee, p. 259–274. *In* A. H. Rose (ed.), *Economic Microbiology*, vol. 7, *Fermented Foods*. Academic Press, Inc., New York, N.Y.
4. Biehl, B., B. Meyer, G. Crone, L. Pollman, and M. B. Said. 1984. Chemical and physical changes in the pulp during ripening and post-harvest storage of cocoa pods. *J. Sci. Food Agric.* 48:189–208.
5. Biehl, B., U. Passern, and D. Passern. 1977. Subcellular structures in fermenting cocoa beans—effect of aeration and temperature during seed and fragment incubation. *J. Sci. Food Agric.* 28:41–52.

6. **Carr, J. G., P. A. Davies, and J. Dougan.** 1979. *Cocoa Fermentation in Ghana and Malaysia*. University of Bristol Research Station, Long Ashton, Bristol, and Tropical Products Institute, London, United Kingdom.

7. **Carr, J. G., P. A. Davies, and J. Dougan.** 1980. *Cocoa Fermentation in Ghana and Malaysia. II*. University of Bristol Research Station, Long Ashton, Bristol, and Tropical Products Institute, London, United Kingdom.

8. **Castelein, J., and H. Verachtert.** 1983. Coffee fermentation, p. 587–615. *In* H. J. Rehm and G. Reed (ed.), *Biotechnology*, vol. 5. Chemie-Verlag, Weinheim, Germany.

9. **de Camargo, R. J., J. Leme, and A. M. Filho.** 1963. General observations on the microflora of fermenting cocoa beans (*Theobroma cacao*) in Bahia (Brazil). *Food Technol.* **17:**116–118.

10. **Duncan, R. J. E., G. Godfrey, T. N. Yap, G. L. Pettipher, and T. Tharumarajah.** 1989. Improvement of Malaysian cocoa bean flavor by modification of harvesting, fermentation and drying methods—the Sime-Cadbury process. *Planter* **65:**157–173.

11. **Dzogbefia, V. P., R. Buamah, and J.H. Oldham.** 1999. The controlled fermentation of cocoa (*Theobroma cacao* L.) using yeasts: enzymatic process and associated physico-chemical changes in cocoa sweatings. *Food Biotechnol.* **13:**1–12.

12. **Forsyth, W. G. C., and V. C. Quesnel.** 1957. Cocoa glycosidase and colour changes during fermentation. *J. Sci. Food Agric.* **8:**505–509.

13. **Forsyth, W. G. C., and V. C. Quesnel.** 1957. Variations in cocoa preparation, p. 157–168. *In Sixth Meeting of the Interamerican Cocoa Committee*, Bahia, Brazil.

14. **Forsyth, W. G. C., and V. C. Quesnel.** 1963. Mechanisms of cocoa curing. *Adv. Enzymol.* **25:**457–492.

15. **Frank, H. A., N. A. Jum, and A. S. Dela Cruz.** 1965. Bacteria responsible for mucilage-layer deposition in Kona coffee cherries. *Appl. Microbiol.* **13:**201–207.

16. **Gauthier, B., J. Guiraud, J. C. Vincent, J. P. Porvais, and P. Galzy.** 1977. Components on yeast flora from the traditional fermentation of cocoa in the Ivory Coast. *Rev. Ferment. Ind. Aliment.* **32:**160–163.

17. **Griffiths, L. A.** 1957. Detection of the substrate of enzymatic browning in cocoa by a post-chromatographic enzymatic technique. *Nature* **180:**1373–1374.

18. **Grimaldi, J.** 1954. *The Cocoa Fermentation Process in the Cameroon*. V. Reunion del Comite Técnico Interamericano del Cocoa, 5RTIC/DOC 24. Bioq.12.

19. **Grimaldi, J.** 1978. The possibilities of improving techniques of pod breaking and fermentation in the traditional process of the preparation of cocoa. *Cafe Cacao The* **22:**303–316.

20. **Hansen, C. E., M. del Olmo, and C. Burri.** 1998. Enzyme activities in cocoa beans during fermentation. *J. Sci. Food Agric.* **77:**273–281.

21. **Hardy, F.,** 1960. *Cocoa Manual*, p. 350. Inter American Institute of Agricultural Science, Turrialba, Costa Rica.

22. **Hardy, F., and C. Rodrigues.** 1953. Quantitative variations in nitrogenous components of the cocoa bean: effects of genetic type, and soil type, p. 89–91. *In A Report on Cocoa Research 1945–1951*. The Imperial College of Tropical Agriculture, St. Augustine, Trinidad, British West Indies.

23. **Hodge, J.** 1953. Chemistry of browning reactions in model systems. *J. Agric. Food Chem.* **1:**928–943.

24. **Hoi, O. K.** 1977. Cocoa bean processing—a review. *Planter* **53:**507–530.

25. **Jones, K. L., and S. E. Jones.** 1984. Fermentations involved in the production of cocoa, coffee and tea, p. 433–446. *In* M. E. Bushell (ed.), *Modern Applications of Traditional Biotechnologies*. Elsevier, New York, N.Y.

26. **Knapp, A. W.** 1937. *Cocoa Fermentation. A Critical Survey of Its Scientific Aspects*. John Bale Sons and Curnow, London, United Kingdom.

27. **Lehrian, D. W.** 1989. Recent developments in the chemistry and technology of cocoa processing, p. 22–33. *In* Y. B. Che Man, M. N. B. Abdul Karim, and B. A. Amabi (ed.), *Food Processing Issues and Prospects*. Faculty of Food Science and Biotechnology, Universiti Pertanian Malaysia.

28. **Lehrian, D. W., and G. R. Patterson.** 1983. Cocoa fermentation, p. 529–575. *In* H. J. Rehm and G. Reed (ed.), *Biotechnology*, vol. 5. Chemie–Verlag, Weinheim, Germany.

29. **Lopez, A. S.** 1972. The development of chocolate aroma from non-volatile precursors, p. 640–646. *In Fourth International Conference on Cocoa Research (Trinidad and Tobago)*.

30. **Lopez, A. S.** 1986. Chemical changes occurring during the processing of cocoa, p. 19–53. *In* P. S. Dimick (ed.), *Proceedings of the Cocoa Biotechnology Symposium*. Department of Food Science. The Pennsylvania State University, University Park, Pa.

31. **Lopez, A. S., and P. S. Dimick.** 1995. Cocoa fermentation, p. 562–577. *In* H. J. Rehm and G. Reed (ed.), *Biotechnology*, vol. 9. Chemie-Verlag, Weinheim, Germany.

32. **Lopez, A. S., D. W. Lehrian, and L. V. Lehrian.** 1978. Optimum temperature and pH of invertase of the seeds of *Theobroma cacao* L. *Rev. Theobromo* (Brazil) **8:**105–112.

33. **Lopez, A. S., and V. C. Quesnel.** 1971. An assessment of some claims relating to the production and composition of chocolate aroma. *Int. Choc. Rev.* **26:**19–24.

34. **Lopez, A. S., and V. C. Quesnel.** 1973. Volatile fatty acid production in cocoa fermentation and the effect on chocolate flavor. *J. Sci. Food Agric.* **24:**319–326.

35. **Lozano, A. R.** 1958. Mechanical pod breakers. *Cocoa* (Turrialba) **3(14):**35.

36. **Martelli, H. L., and H. F. K. Dittmar.** 1961. Cocoa fermentation. V. Yeasts isolated from cocoa beans during the curing process. *Appl. Microbiol.* **9:**370–371.

37. **McDonald, C. R., R. A. Lass, and A. S. Lopez.** 1981. Cocoa drying—a review. *Cocoa Growers' Bull.* **31:**5–39.

38. **Menchu, J. F., and C. Rolz.** 1973. Coffee fermentation technology. *Cafe Cacao The* **17:**53–61.

39. **Ostovar, K., and P. G. Keeney.** 1973. Isolation and characterization of microorganisms involved in the fermentation of Trinidad's cocoa beans. *J. Food Sci.* **38:**611–617.

40. **Passos, M. F. L., A. S. Lopez, and D. O. Silva.** 1984. Aeration and its influence on the microbial sequence in cocoa fermentations in Bahia with emphasis on lactic acid bacteria. *J. Food Sci.* **49:**1470–1474.

41. **Passos, M. F. L., D. O. Silva, A. S. Lopez, C. L. L. F. Ferreira, and W. V. Guimaraes.** 1984. Characterization and distribution of lactic acid bacteria from traditional cocoa bean fermentations in Bahia. *J. Food Sci.* **49:**205–208.

42. **Quesnel, V. C.** 1965. Agents inducing the death of the cocoa seeds during fermentation. *J. Sci. Food Agric.* **16:**441–447.

43. **Quesnel, V. C.** 1972. Biochemistry and processing, p. 48. *In Annual Report of Cocoa Research.* University of the West Indies, St. Augustine, Trinidad British West Indies.

44. **Quesnel, V. C.** 1972. Cacao curing in retrospect and prospect, p. 602–606. *In Fourth International Conference on Cocoa Research (Trinidad and Tobago).*

45. **Quesnel, V. C., and K. Jugmohunsingh.** 1970. Browning reaction in drying cocoa. *J. Sci. Food Agric.* **21:**537.

46. **Rodriquez, D. B., and H. A. Frank.** 1969. Acetaldehyde as a possible indicator of spoilage in coffee Kona (Hawaiian coffee) *J. Sci. Food Agric.* **20:**15–19.

47. **Roelofsen, P. A.** 1953. Polygalacturonase activity of yeast, *Neurospora* and tomato extract. *Biochim. Biophys. Acta* **10:**410–413.

48. **Roelofsen, P. A.** 1958. Fermentation, drying, and storage of cocoa beans. *Adv. Food Res.* **8:**225–296.

49. **Rohan, T. A.** 1957. Cocoa preparation and quality. *West Afr. Cocoa Res. Inst.* **1956–57:**76.

50. **Rohan, T. A.** 1963. *Processing of Raw Cocoa from the Market.* FAO Agriculture studies no. 60. Food and Agriculture Organization, Rome, Italy.

51. **Rohan, T. A.** 1969. The flavor of chocolate; its precursors and a study of their reactions. *Gordian* **69:**443–447, 500–501, 542–544, 587–589.

52. **Rohan, T. A., and T. Stewart.** 1965. The precursors of chocolate aroma: the distribution of free amino acids in different commercial varieties of cocoa beans. *J. Food Sci.* **30:**416–419.

53. **Rohan, T. A., and T. S. Stewart.** 1967. The precursors of chocolate aroma: production of free amino acids during fermentation of cocoa beans. *J. Food Sci.* **32:**396–398.

54. **Rohan, T. A., and T. S. Stewart.** 1967. The precursors of chocolate aroma: production of reducing sugars during fermentation of cocoa beans. *J. Food Sci.* **32:**399–402.

55. **Rombouts, J. E.** 1952. Observations on the microflora of fermenting cocoa beans in Trinidad. *Trinidad Proc. Soc. Appl. Bacteriol.* **15:**103–111.

56. **Rombouts, J. E.** 1953. Critical review of the yeast species previously described from cocoa. *Trop. Agric.* (Trinidad) **30:**34–41.

57. **Sanchez, J., G. Daguenet, J. P. Guiraud, J. C. Vincent, and P. Galzy.** 1985. A study of the yeast flora and the effect of pure culture seeding during the fermentation process of cocoa beans. *Lebensm.-Wiss. Technol.* **18:**69–76.

58. **Saposhnikova, K.** 1952. Changes in the acidity and carbohydrates during ripening of the cocoa fruit: variations of acidity and weight of seeds during fermentation of cocoa in Venezuela. *Agron. Trop.* (Maracay) **2:**185–195.

59. **Schwan, R. F.** 1998. Cocoa fermentations conducted with a defined microbial cocktail inoculum. *Appl. Environ. Microbiol.* **64:**1477–1483.

60. **Schwan, R. F., A. S. Lopez, D. O. Silva, and M. C. D. Vanetti.** 1990. Influência da frequência e intervalos de revolvimentos sobre a fermentacao de cocoa e qualidade do chocolate. *Rev. Agratrop.* **2:**22–31.

61. **Schwan, R. F., A. H. Rose, and R. G. Board.** 1995. Microbial fermentation of cocoa beans, with emphasis on enzymatic degradation of the pulp. *J. Appl. Bacteriol.* **79:**96–107.

62. **Schwan, R. F., M. C. D. Vanetti, D. O. Silva, A. S. Lopez, and C. A. deMoraes.** 1986. Characterization and distribution of aerobic spore-forming bacteria from cocoa fermentations in Bahia. *J. Food Sci.* **51:**1583–1584.

63. **Shamsuddin, S. B., and P. S. Dimick.** 1986. Qualitative and quantitative measurements of cocoa bean fermentation, p. 55–78. *In* P. S. Dimick (ed.), *Proceedings of the Cacao Biotechnology Symposium.* Department of Food Science, The Pennsylvania State University, University Park, Pa.

64. **Sieki, K.** 1973. Chemical changes during cocoa fermentation using the tray method in Nigeria. *Rev. Int. Choc.* **28:**38–42.

65. **Sivetz, M., and N. Desrosier.** 1979. Harvesting and handling green coffee beans, p. 74–116. *In Coffee Technology.* AVI Publishing Company Inc., Westport, Conn.

66. **Suryakantha Raju, K., S. Vishveshwara, and C. S. Srinivasan.** 1978. Association of some characters with cup quality in *Coffea canephora* × *Coffea arabica* hybrids. *Indian Coffee* 195–197.

67. **Van Pee, W., and J. M. Castelein.** 1972. Study of the pectinolytic microflora, particularly the Enterobacteriaceae, from fermenting coffee in the Congo. *J. Food Sci.* **37:**171–174.

68. **Vaughn, R. H., R. DeCamargo, H. Falanghe, G. Mello-Ayres, and A. Serzedello.** 1958. Observations on the microbiology of the coffee fermentation in Brazil. *Food Technol.* **12:**57.

69. **Voigt, J., and B. Beihl.** 1995. Precursors of the cocoa specific aroma components are derived from the vicilin class (7S) globulin of the cocoa seeds by proteolytic processing. *Bot. Acta* **108:**283–289.

70. **Voigt, J., D. Wrann, H. Heinrichs, and B. Biehl.** 1994. The proteolytic formation of essential cocoa-specific aroma precursors depends on particular chemical structures of the vicilin-class globulin of the cocoa seeds lacking in the globular storage proteins of coconuts, hazelnuts and sunflower seeds. *Food Chem.* **51:**197–205.

71. **Wadsworth, R. V.** 1955. *The Preparation of Cocoa,* p. 131–142. *In 1955 Cocoa Conference.* London, United Kingdom.

72. **Weissberger, W., T. E., Kavanagh, and P. G. Keeney.** 1971. Identification and quantification of several non-volatile organic acids of cocoa beans. *J. Food Sci.* **36:**877–879.

73. **Zak, D. K., K. Ostovar, and P. G. Keeney.** 1972. Implication of *Bacillus subtilis* in the synthesis of tetramethylpyrazine during fermentation of cocoa beans. *J. Food Sci.* **37:**967–968.

Food Microbiology: Fundamentals and Frontiers, 2nd Ed.
Edited by M. P. Doyle et al.
© 2001 ASM Press, Washington, D.C.

Iain Campbell

Beer

36

This chapter is a general overview of the scientific principles of the brewing industry. More detailed information on the science and technology of beer production is available in textbooks by Lewis and Young (9), Hough et al. (2, 7), and Moll (11).

The legal definition of beer varies among countries. The strictest definition, as in Germany, limits the ingredients to hops, yeast, water, and malt, not necessarily of barley, although barley is understood if no other cereal is specified. In many other countries, sugars or unmalted cereals are permitted as adjuncts, providing up to 30% of the fermentable sugar. Shown in Fig. 36.1 is a general outline of the brewing process.

Archeological evidence has shown that beer was produced as early as 3000 B.C., perhaps as a fortuitous discovery from the baking of bread. Yeast is unable to ferment the starch of cereal grains, so a preliminary germination is required, in which the starch and protein are hydrolyzed enzymically to simple sugars and amino acids which provide the main nutrients for *Saccharomyces cerevisiae* to ferment. This could have happened accidentally with moist grain, and naturally occurring fermentative yeasts would have produced a primitive beer. In subsequent developments over the millennia, barley became the principal cereal for beer production because

of its husk, which provided an excellent natural filter for clarification of the extracted wort.

MALTING

The first stage of the brewing process is the production of malt, now almost always carried out by specialist maltsters. It is a common mistake to think that during malting the grains convert starch to simple sugars which the yeast can metabolize. In fact, that happens during mashing, and the changes during malting are limited to a modification of the starch to facilitate subsequent hydrolysis.

Not all barley is suitable for malting. One important property is the nitrogen content. Too low a level of nitrogen will restrict yeast growth in the subsequent fermentation, but normally the problem is too high a level of nitrogen in the barley, more than is necessary for yeast growth, with the resulting excess nitrogen encouraging microbial, particularly bacterial, spoilage of the final beer. Also, grain intended for malting must be carefully dried and stored to avoid risk of either dormancy (delayed germination) or death of the barley (5, 12).

The malting process occurs in three stages: steeping, germination, and kilning. Steeping in water over 24 to

Iain Campbell, Department of Biological Sciences, Heriot-Watt University, Riccarton, Edinburgh EH14 4AS, Scotland.

Malting

↓

Conversion of barley starch to fermentable sugars

(glucose, maltose, maltotriose) and protein to free amino nitrogen

Milling, Mashing

↓

Extraction of sugars, amino acids, and other yeast nutrients and

enzymes with hot water to yield sweet wort

Wort Boiling

↓

Boiling with hops to extract aroma and bittering compounds,

then sterilization to yield hopped wort

Fermentation

↓

Conversion by *Saccharomyces cerevisiae* of fermentable sugars

to ethanol and carbon dioxide

Post-Fermentation Treatments

↓

Maturation (improvement of flavor), clarification, packaging,

pasteurization

Figure 36.1 The brewing process.

48 h, usually in two stages with an "air rest" between, stimulates the growth of the embryo plant, as would occur if it had been planted in soil. Germination is controlled by temperature, aeration, and humidity to the point where the stem is just about to emerge from the grain, by which time rootlets have already formed. By that stage, the zone of modification of the starch has extended almost to the opposite end of the grain from the embryo. In the traditional malting process, grain transferred from the steeping vessel was spread on the floor as a layer about 0.8 m deep and manually turned daily

to prevent the developing rootlets binding together over the 2 weeks of germination. Since late in the 19th century, numerous types of mechanical malting systems have been developed to speed up the process and reduce the labor requirements. Figure 36.2 shows one of the first, invented by Saladin, which is still operated using essentially the same design. The rotating helical turners carried on a gantry run along the full length of the germination vessel to turn the germinating grain, which is aerated by moist air passing through the slotted floor.

At the end of germination the grain has a moisture content of about 50%, so drying to <6% moisture is required for storage. However, kilning is more complex than simply drying the grain. Malt is not only a source of fermentable sugars and other nutrients for yeasts, but also a source of amylolytic and proteolytic enzymes for hydrolysis of any additional unmalted cereal in the recipe. Although these enzymes are moderately heat resistant, they are increasingly inactivated by drying temperatures above 70°C. Also, higher temperatures result in darker-colored malt by enhancing browning reactions between the sugars and nitrogen components of the grain. For some beers, darker malts are desirable, but with the penalty of reduced enzymatic activity for hydrolysis of cereal adjuncts.

The production of malt is more the concern of the botanist and biochemist, but there are important microbiological aspects. First, malt or malt extract is a food product, and high standards of hygiene are required. Although some bacterial contamination is inevitable, potentially the most troublesome contaminants are molds. Certain molds have specific spoilage effects or produce mycotoxins (see chapters 21 through 23 of this volume). In the case of some species of *Fusarium*, a polypeptide "gushing factor" creates nuclei for development of CO_2 bubbles, which can result in violent frothing of the beer when the bottle or can is opened. However, since growth

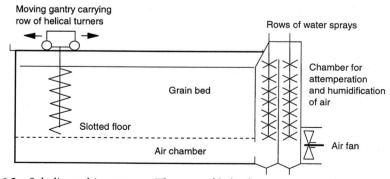

Figure 36.2 Saladin malting system. The row of helical turners moves from one end of the grain bed to the other at intervals during germination to turn the malt and prevent matting of the rootlets.

of any mold on barley gives an obvious weathered appearance and a moldy aroma which persists to the final beer, simple sensory assessment is sufficient for the maltster to decide on acceptance or rejection of a barley sample. In fact, the maltster normally has no alternative to such immediate assessment when deciding to accept a load of grain from a waiting truck, since conventional microbiological analysis would require at least 1 week.

MASHING AND WORT PRODUCTION

Wort is the sugary solution prepared from malt, either alone or with sugar or unmalted cereal adjunct if appropriate, when the resulting grist is extracted with warm water. Details of the process vary between breweries, and it is impractical to deal with all of the possibilities here.

In traditional milling, although the contents of the grains must be ground finely to maximize the yield of fermentable extract, it is important that the outermost layer, the husk, is only cracked rather than ground to powder. In a later stage of the process the husk functions as a natural filter. Barley has become the dominant cereal of the brewing industry largely because its husk structure is ideal for this purpose.

The traditional decoction mashing process for Bavarian and Czech beers has origins predating the invention of the thermometer. Consistent brewing requires reproducible conditions, and consistent temperatures are achieved by mixing the grist with a measured volume of well water, heating a measured proportion to boiling, and returning the boiled slurry to the mash tun. Well water is of constant temperature throughout the year, and water always boils at the same temperature at a given location, so with the same volumes each time, mashing temperatures would always be the same. Experience has shown that several steps of increasing temperature are required; we now know the optima of α- and β-amylases and protease of the malt. This was achieved by a succession of accurately measured decoctions. While double- or triple-decoction mashing undoubtedly produces a good-quality wort, even from poorly modified malts, the process is expensive and now is used only for prestigious products, for which the traditional process has sales appeal. Infusion mashing is also a traditional process, associated particularly with British ales and similar beers of Belgium and northern Germany. At its simplest, the mash tun contains a floating bed of grist, suspended on entrapped air, with wort drawn off at the base and continuously replaced by a top spray of hot sparge water until further extraction of sugar is impractical. Originally,

infusion mashing operated at a constant temperature of about 70°C (traditionally the temperature at which the brewer's reflection was just obscured by the steam beginning to rise from the brewing liquor), and well-modified malt was essential (2).

The majority of breweries use a mashing vessel in which the grist is mixed with warm water, heated in steps through the optimum temperatures of the various mashing enzymes, and finally filtered in the lauter tun (German: lauter = clear), a tank with a horizontal slotted metal base on which the bed of husk material functions as a filter, since all cereal solids must be removed before the next stage of the process. When unmalted cereal forms part of the recipe, the ground cereal is heated in a cereal cooker, at least to the gelatinization temperature of the starch of the cereal (70 to 75°C in the case of corn, a common ingredient), but often to boiling, and then transferred to the mashing vessel to allow hydrolysis of its starch and protein by the malt enzymes.

In recent years, the most modern breweries have replaced the lauter tun with a mash filter, in reality a type of filter press (11). As in the lauter tun, the cereal solids constitute the filter bed, held within a row of vertical polypropylene sheets with numerous fine perforations, much smaller than the slots in the bottom plates of the lauter tun. The malt can therefore be more finely ground than for a lauter tun, giving better extraction of nutrients, but the main advantage of the mash filter is that in each of the filter chambers the mash can be squeezed to expel the wort. The process is faster than filtration in a lauter tun and a much smaller amount of sparge liquor is required, so that stronger worts can be produced, and more quickly.

HOPS AND HOP BOILING

Over brewing history, many different herbs have been used as flavoring for beer, but for the past five centuries hops have been preferred, almost certainly for microbiological reasons. It is recognized that hops have an important antimicrobial, particularly antibacterial, effect and it is presumed that the medieval brewers realized that hopped beers maintained their quality for longer than beers with other flavorings.

Hops require a temperate climate, but in practice their cultivation is restricted largely to certain areas, e.g., the county of Kent in Britain and Washington State in the United States. The different hop varieties in common use vary in their content of the bitter acids, resins, and oils which contribute to beer flavor and aroma, and whether they contain seeds. The selection of a particular variety, or blend of varieties, is an important part of a beer recipe. Flavor is extracted by boiling, with the incidental advantages of sterilizing and concentrating the wort, isomerizing hop α- and β-acids, and purging the wort of harsh grainy flavors. Also, if glucose, sucrose, or maltose crystals or syrup are used, this is the appropriate stage for addition (7).

In appearance, hops resemble small pine cones but have a softer texture. Cone hops are still used by some traditional breweries, but the majority of breweries now use processed hops, either pellets prepared from ground cone material, or hop extract. Modern hop extraction technology uses liquid CO_2 to avoid potential problems of residual organic solvent. Also, various preisomerized hop products are available, which can be added in the last few minutes of boiling to reduce losses. So it is interesting that the boiling stage, originally intended for extraction of flavors, is now in some breweries simply a process of biochemical and microbiological stabilization of the wort.

To the brewer, the most important hop components are the resins, tannins, and essential oils. Typically, these constitute approximately 14, 4, and 0.5%, respectively, of the weight of the mature hop cones. Bitterness of beer is due to the α-acid (humulone) and β-acid (lupulone) components of the resins, but not directly. These complex acids are isomerized during hop boiling to the bitter iso-α-acids and iso-β-acids, isohumulone and hulupone, respectively (7, 9, 20). Within these groups of isoacids are numerous analogs according to the acyl side chains on the resin molecule. Essential oils contribute aroma to the beer, but most of the oil is lost during boiling. Therefore, aroma hops, as distinct from high-α-acid bitterness hops, are added late in the boil to ensure maximum aroma effect. Hop tannins have little direct influence on flavor, but react with malt protein to form hot break, a precipitate of protein/tannin complexes and insoluble calcium salts and phosphates. This must be removed before fermentation since the particulate material would adversely affect fermentation and flavor development.

After traditional hop boiling, i.e., with cone hops, the wort is clarified by filtration through the settled bed of spent hops. Pellets or extract do not provide a suitable filter medium, so after their introduction, the wort has to be clarified by centrifugation. A better system is used in modern breweries, in which the hopped wort is run tangentially into a circular tank, the whirlpool hop separator. Hop debris and hot break collect at the center of the resulting vortex, while clear wort is drawn from the side of the vessel.

The wort, still at approximately 100°C, has to be cooled to 20°C or less before "pitching" with yeast inoculum. Less obvious, but nevertheless essential, is the requirement, explained below, to aerate the wort to 6 to 8 ppm of dissolved O_2 (DO). After cooling and aeration, the wort is ready for the fermentation stage of the process.

FERMENTATION

Brewing Yeast (*S. cerevisiae*)

Formerly, actively fermenting yeasts of the fermentation industries, both culture yeasts and common contaminant "wild yeasts," were classified as different species of *Saccharomyces*. *Saccharomyces cerevisiae*, used for traditional northern European ales and similar beers, was collected (skimmed) as "yeast head" from the surface of an active fermentation, and *Saccharomyces carlsbergensis*, "bottom yeast" which did not form such a head, had to be collected from the base of the vessel at the end of fermentation (8, 21). Originally, bottom yeast was used only for Czech, Bavarian, and Danish lager beers, but now non-head-forming yeasts are more widely used to maximize the useful capacity of enclosed fermentation vessels. Although most important as wine yeasts, *Saccharomyces bayanus*, *Saccharomyces ellipsoideus*, and *Saccharomyces uvarum* are also possible brewery contaminants, as is *Saccharomyces diastaticus*, an amylolytic yeast. All are now classified officially by yeast taxonomists as a single species, *S. cerevisiae*, but still it is convenient to distinguish the different types by their former specific names.

S. cerevisiae is not a facultative anaerobe like *Escherichia coli* and several other bacteria (16, 21). Although brewing yeast changes between oxidative and fermentative metabolism according to aerobic or anaerobic conditions, it cannot grow anaerobically indefinitely. As with all eukaryotic cells, yeast cell membranes contain unsaturated fatty acids (UFA) and sterols, which can be synthesized only under aerobic conditions. The amounts of UFA and sterols from malt that are naturally present in wort are too low to support yeast growth, hence the requirement for initial aeration of the wort to allow the yeast to synthesize these compounds.

"Pitching yeast" in satisfactory condition must be added in the correct amount to the fermentation vessel; usually 1×10^7 to 2×10^7 cells/ml is added, although measurement of yeast cake or slurry by weight is more convenient on a production scale. In the course of a typical fermentation, the yeast population increases by a factor of about 8, i.e., only three successive cell divisions. Subsequent multiplication is inhibited by the lack of O_2, UFA, and sterols, and since aeration late in the fermentation would cause flavor problems, no further cell growth occurs.

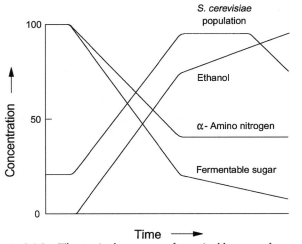

Figure 36.3 Theoretical progress of a typical brewery fermentation, showing changes in population of *S. cerevisiae* and concentration of fermentable sugar, amino nitrogen, and ethanol. The graph shows yeasts in suspension and the start of settling of cells from the beer at the end of fermentation. The time axis is not calibrated, since fermentation rates differ widely between breweries. Other variables are shown as percentages of the initial or final value, expressed as 100%.

A brewery fermentation is essentially the lag, log, and early stationary phases of the microbial growth curve (Fig. 36.3) (7, 9). Brewing is unusual in modern biotechnology in reusing the culture from the previous fermentation, although a new culture is prepared at regular intervals, e.g., after 15 successive fermentations. In the lag phase of 6 to 12 h, *S. cerevisiae* utilizes the DO in the wort to restore its UFA/sterol supply and adjusts from the anaerobic, acid, alcoholic conditions at the end of the previous fermentation to the very different environment of fresh wort.

Nutrients, including sugars, are taken up from the wort to generate new yeast cells; fermentable sugars also provide the energy for this process. A wort of 10° Plato, i.e., 10% solute, has a specific gravity of 1.040, which is largely due to the sugars in solution. Amino acids, vitamins, and inorganic salts account for less than 1% of the dissolved solids. Typically, the approximate sugar composition of such a wort would be 4% maltose, 2% glucose, 2% maltotriose, and 2% higher dextrins, which are not utilized by brewing yeast. However, the various fermentable sugars are not utilized simultaneously. This is related to the transport of nutrients into yeast cells and varies with different strains. Normally, brewing yeasts transport and metabolize the simplest sugar, i.e., glucose, first, and maltose transport begins only after much of the glucose has been metabolized. For a similar reason, utilization of maltotriose begins late in the fermentation, when most of the maltose has been consumed. Note also in Fig. 36.3 that fermentation of sugars continues after yeast growth ceases, due to the continuing action of the relevant enzymes, but more slowly than before. However, without yeast growth and protein synthesis, there is no further requirement for amino nitrogen.

Ethanol and CO_2 are the main products of yeast metabolism, but small amounts of other compounds make an important contribution to the flavor of the beer (16, 22). By-products of the Embden-Meyerhof pathway, the major metabolic route in anaerobic fermentation, contribute to flavor, but a major effect is the nitrogen metabolism of yeast. *S. cerevisiae* has a limited range of amino acid permeases, and most amino acids used for protein synthesis have to be synthesized by the yeast cell. Keto acids, which are intermediates in the biosynthetic pathway, may be converted to higher alcohols (Fig. 36.4) which make a significant contribution to the flavor of

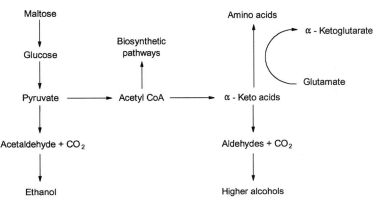

Figure 36.4 Formation of higher alcohols (fusel alcohols) as by-products of amino acid biosynthesis. Note the similarity of the reactions pyruvate → acetaldehyde → ethanol and α-keto acid → aldehyde → higher alcohols; both are important for redox balance under anaerobic conditions.

beer, and indeed most fermented beverages. Also related to transport is the decrease in pH during the fermentation by excretion of H^+ during uptake of nutrients. The pH of wort is normally 5.1 to 5.2 and decreases during fermentation to 3.8 to 4.0, mainly during the period of yeast growth.

Progress of Fermentation

Traditional fermentation vessels are open rectangular tanks of 2 to 3 m in depth (7). Such vessels are particularly useful for top yeasts, those brewing strains which rise to the surface of the fermenting beer, from where they are skimmed off and used as an inoculum for the next fermentation. Since about 1970, the cylindroconical fermentation vessel (Fig. 36.5) has become the preferred type (10, 11, 22). The cylindroconical fermentation vessel is enclosed to reduce the risk of contamination and facilitate recovery of CO_2 and has a conical lower section to facilitate recovery of bottom yeast, now almost universally used, which settles out late in the fermentation. The shape of the vessel also encourages a vigorous mixing of yeast cells in the fermenting wort, giving a fast fermentation. Central upward movement and peripheral downward movement of bubbles of CO_2 are encouraged by cooling panels on the walls.

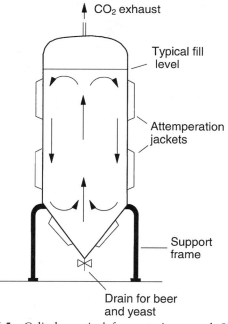

Figure 36.5 Cylindroconical fermentation vessel. Vigorous circulation of fermenting wort is created by central upward flow of CO_2 bubbles and downward flow in contact with the cooling jackets. Cooling of the cone section is optional, depending on whether the brewery stores settled yeast before the next fermentation.

In recent years, the larger brewery companies have installed very large cylindroconical fermentation vessels, of capacity up to 6,000 hl, for large-scale production of their major brands. Although all cylindroconical vessels operate in a similar way, an important difference is that the very large vessels could be four to five times the size of the brewhouse and must be filled in instalments. The first batch of wort must be inoculated immediately, as it is dangerously poor microbiological practice to hold uninoculated wort. The wort is aerated, the yeast begins to grow, and the second, third, fourth, and possibly fifth batches of wort are added as available. Unfortunately, there is a biochemical problem. The wort must be aerated since yeast needs oxygen to grow efficiently. An intermediate state of oxygen deficiency causes excessive production of esters, especially ethyl acetate. However, oxygen added late in the fermentation encourages the production of unwanted diacetyl. Therefore the solution is to add with the first batch of wort most of the yeast and DO required, perhaps three times as much as that volume of wort would normally require; the rest of the pitching yeast and DO are added with the second batch of wort, and the later wort additions are fermented by the yeast already present.

An important property of brewing yeasts is the ability to flocculate, i.e., spontaneously aggregate into clumps late in the fermentation. Too early flocculation stops the fermentation as the yeasts cells settle out of partly fermented beer. Nonflocculent yeasts are also troublesome; they have to be removed by filtration or centrifugation. Currently the preferred explanation of flocculation is a lectin-like activity of the cell wall which develops during fermentation, but other factors are also involved, e.g., divalent ions, particularly Ca^{2+}, decreasing amounts of flocculation-inhibiting fermentable sugar, and increasing amounts of stimulatory ethanol (16, 17).

POSTFERMENTATION TREATMENTS

Conditioning

Green beer from the fermentation vessel contains acetaldehyde, diacetyl, and other unwanted by-products of yeast metabolism, which must be removed. Although present in only small amounts, these compounds have a major effect on the flavor, or more correctly, the aroma of the beer. In traditional cask conditioning, the beer, still with about 1% fermentable sugar, undergoes a secondary fermentation in casks, absorbing these undesirable flavor compounds and generating sufficient CO_2 for dispense. Similar changes are associated with traditional low-temperature (0 to 2°C) secondary fermentation of pilsner and other lager beers (German, lager = store) over several months of storage. It is now known that such storage at low temperature is not essential; beneficial flavor changes can be achieved more rapidly by a few days' storage of the beer containing 10^5 to 10^6 yeast cells/ml at 15°C, followed by cooling to 0°C only long enough to precipitate chill-haze material and yeast sediment. It is important to avoid accidental access of air at this stage, since diacetyl is formed from acetolactate by spontaneous chemical reactions in the presence of O_2, and yeast is no longer present to remove that diacetyl.

Filtration

With the exception of cask-conditioned beers, which are clarified by addition of fining agents, it is normal practice to filter beer to achieve clarity. Various designs of filters are in common use, most using either cellulose fibers or diatomite or pumice powder as filter media (11). Such materials, which adsorb microbial and inert haze-forming material within the filter, do not sterilize the beer, since sufficiently fine filters would cause unacceptably slow flow rates. Typically the yeast count is reduced to 10^2 cells/liter. At least 10^5 yeast cells/ml are required to form a visible haze.

Membrane filtration, which does sterilize beer, is becoming increasingly popular. Pasteurization is unnecessary for sterile-filtered beer, so membrane filtration avoids substantial energy cost and potential flavor defects that can result from heating the beer. With two filters used in series, one a rough prefilter, to remove as much particulate material as possible, and the second a membrane filter to sterilize, an acceptable flow rate and long filter life can be achieved.

Pasteurization and Packaging

Draft beer is pasteurized by passage through a heat exchanger at 70°C before filling into already cleaned and sterilized kegs. Beer for sale in bottles or cans is pasteurized after filling. In both systems the heat treatment is equal in terms of pasteurization units (1 PU = 60°C for 1 min), but tunnel pasteurizers for bottles and cans use a lower temperature (typically 62°C) for a longer time. Individual breweries have their own preferred pasteurization treatment. In theory, 5 PU is sufficient to kill the small numbers of *S. cerevisiae* cells likely to pass through a cellulose filter, but to eliminate the slightly more heat-resistant bacterial or wild yeast contaminants which could occasionally be present, up to 30 PU may be applied. Since increasing heat treatment could adversely affect flavor, choice of PU value is a compromise between potential risks of oxidized flavors and microbial spoilage (7, 9).

HIGH-GRAVITY BREWING

The basic principle of high-gravity (HG) brewing is that it is theoretically possible to double the production of a brewery, without the expense of additional brewhouse or fermentation capacity, by fermenting double-strength wort. The resulting double-strength beer would be diluted to normal strength at the final stage of the process. The economic advantages are obvious. At first sight the addition of water to beer might seem rather questionable, but that water would have been used anyway, in the first stage of the standard process. Unfortunately, the idea is very difficult to put into practice. In order to match the flavor of HG beer with the normal product, the concentration of malt, hops, yeast, and DO in the wort must be doubled (16).

Dealing with the nonmicrobiological aspects first, with standard mash and lauter tuns, doubling the weight of malt does not yield double the extract. Extraction becomes less efficient at higher concentrations, and efficient sparging of the grain bed in the lauter tun dilutes the wort excessively. Therefore, until the development of the mash filter, which can produce HG wort from an all-malt grist, it was necessary to use sugar syrup adjuncts to increase the gravity, which is illegal in some countries. There is a similar problem with hops, since extraction becomes progressively less efficient from larger quantities of hops or hop pellets. That particular problem was solved by the use of hop extracts, especially preisomerized extracts which may be added either late in hop boiling or even at the stage of dilution of HG beer to normal strength.

Although the brewer can easily add twice the number of yeast cells to the double-strength wort, it is difficult to achieve twice the DO concentration. A maximum of 6 to 8 mg of DO per liter is routinely achieved by injecting air into wort between the cooler (after hop boiling) and the fermentation vessel. In the HG process, twice the number of yeast cells would require 12 to 16 mg of DO per liter, achieved only by injecting pure oxygen rather than air. Reduced oxygen levels cause excess ester production.

A standard wort contains 10% sugar. Double-strength wort containing 20% sugar is likely to retard yeast growth by osmotic stress. A more serious situation occurs at the end of fermentation, when the yeast is recovered from beer containing 8% instead of 4% ethanol by volume, and its viability may be too low for successful repitching in the next fermentation. A brewery that produces its own culture yeast to replace the yeast culture occasionally could easily prepare a new yeast culture for each fermentation. Since the pure culture plant is already available, the additional expense is small in comparison with the savings from HG operation. Finally, the dilution water must be deoxygenated to <0.1 mg of DO per liter to prevent any flavor problems from staling reactions or development of diacetyl. A deaeration system is the only equipment which is required specifically for HG operation.

The example of brewing to double strength described above has only recently been successfully accomplished in practice. Previously, 1.5 times normal strength was the practical limit, because of flavor problems. Without sufficiently efficient oxygenation of the wort in the first stages of the fermentation, *S. cerevisiae* is unable to grow normally and produces excessive amounts of esters, as indicated in the section above on very large cylindroconical vessels. With recent developments in oxygenation, however, two times normal strength is now practicable.

MICROBIAL CONTAMINANTS OF THE BREWING PROCESS

Beer production carries a risk of microbial contamination at all stages of the process. In barley during malting, the most important contaminants are molds. Field fungi which develop on barley during growth are seldom a problem. *Fusarium* spp. may develop sufficiently to cause "gushing" only after a particularly wet growing season (4). Storage fungi, however, principally *Aspergillus* and *Penicillium* spp., can have serious effects, mentioned earlier, in forming mycotoxins.

In the brewery itself, yeasts and bacteria are potentially the troublesome contaminants. In beer, its acidic (usually about pH 3.9), alcoholic, anaerobic characteristics, with the additional antibacterial effects of CO_2 and hop acids and oils, restrict the range of potential spoilage organisms. Yeasts on grain are mainly aerobes that are unable to grow in fermenting wort or beer, but some species of the genera *Brettanomyces*, *Candida*, *Debaryomyces*, *Pichia*, *Saccharomyces*, *Torulaspora*, and *Zygosaccharomyces* are potential spoilage yeasts (3). Presumably their natural habitat and original source are plants, but once established in a brewery they are difficult to eradicate and are often transferred from one fermentation to the next in the pitching yeast, in increasing numbers each time. The fermentative genera *Saccharomyces*, *Torulaspora*, and *Zygosaccharomyces* and equivalent nonsporing *Candida* spp. may contaminate the fermentation or persist through filtration and pasteurization to produce turbidity and off flavors in the beer. Most spoilage organisms are smaller than the 5- to 7-μm diameter of culture yeasts, so filters designed to retain brewing yeasts are less efficient in removing contaminants. *Dekkera* and its nonsporing anamorph *Brettanomyces* cause acetic acid spoilage and turbidity. *Debaryomyces*, *Pichia*, and equivalent aerobic *Candida* spp. are oxidative yeasts and so are limited to the early

stages of fermentation, or to beer to which air has gained access postfermentation. These yeasts cause turbidity and yeasty or estery off flavor, and in bottled or canned beer often form a surface film that fragments to flaky particles or sediment.

Only a few species of the lactic acid bacteria, i.e., *Lactobacillus* and *Pediococcus*, are sufficiently resistant to the antibacterial properties of beer to grow at any stage of the process, causing turbidity and an off flavor, often attributable to diacetyl (13). Diacetyl is a strongly flavored minor by-product of their metabolism which is highly valued in the dairy industry as the flavor of butter but considered by many people to be a spoilage fault of beer. Superattenuation by starch-fermenting lactobacilli and slime formation by pediococci are other possible faults. *Zymomonas* can also grow at all stages, but its most usual effect is in finished beer, causing further fermentation with turbidity and off flavors, including that reminiscent of rotten apples, or dimethyl sulfide aroma.

Other possible spoilage bacteria are limited by their properties to specific stages of the process (19). *Acetobacter* and *Gluconobacter* spp. are strict aerobes and grow, with characteristic acetic acid production, only on beer accidentally exposed to air. Enterobacteria (including *Obesumbacterium*, the most important of that group in the brewery environment) cause turbidity and off flavor, often indole, phenols, diacetyl, H_2S, and dimethyl sulfide. They grow well in the early stages of fermentation until inhibited by the decreasing pH and increasing ethanol content. Even so, they survive to be transferred with the yeast recovered for the next fermentation. Finally, *Megasphera* (cocci) and *Pectinatus* (rods) are recently discovered strictly anaerobic gram-negative bacteria that form acetic, butyric, and propionic acids, H_2S, dimethyl sulfide, and turbidity; these have become troublesome only because of modern advances in maintaining very low DO levels in beer.

Other bacteria are occasionally isolated, e.g., *Bacillus* and *Micrococcus* spp., but are unable to grow in wort or beer. Their main nuisance value is the necessary laboratory testing to confirm that they are not the more troublesome *Lactobacillus* or *Pediococcus*, which are of similar appearance (13). A simple distinguishing test is that the lactic acid bacteria do not form catalase (Fig. 36.6).

Isolation of Microbial Contaminants

Culture media for microbiological quality control can be either nonselective or selective. All brewing yeasts and bacteria should grow on nonselective media, usually either malt extract agar or Wallerstein Laboratories nutrient (WLN) agar, a semisynthetic medium of more consistent composition. These media are useful for examination, by plating or membrane filtration, of samples

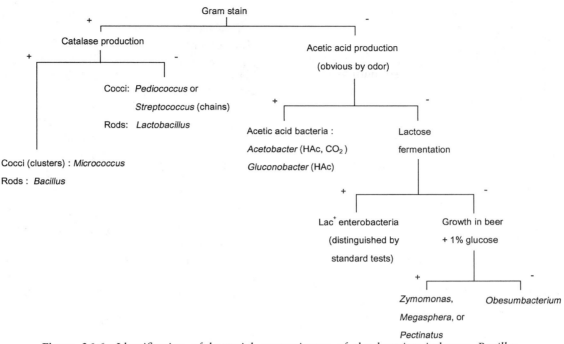

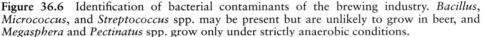

Figure 36.6 Identification of bacterial contaminants of the brewing industry. *Bacillus*, *Micrococcus*, and *Streptococcus* spp. may be present but are unlikely to grow in beer, and *Megasphera* and *Pectinatus* spp. grow only under strictly anaerobic conditions.

where no microorganisms should be found or should be present only in low numbers, e.g., pasteurized beer or swabs or rinsings of recently cleaned and sterilized equipment. Any microorganism, whether bacteria, culture yeast, or wild yeast, can be regarded as a contaminant in these situations.

Selective media are intended to suppress the growth of culture yeast but allow the growth of any microorganisms that may be present. Such media should normally be used to examine samples of pitching yeast, or samples taken during or immediately after fermentation, where culture yeast is present. A commonly used selective medium is lysine agar, a synthetic medium containing glucose, inorganic salts, vitamins, and L-lysine as sole nitrogen source. *Saccharomyces* spp. are unable to grow, while other yeast genera do grow, using the $-NH_2$ groups of lysine for all nitrogen requirements. Unfortunately, wild strains of *S. cerevisiae* and other *Saccharomyces* spp. are particularly likely contaminants, but cannot be detected on lysine agar. However, diastatic yeasts, including the former *S. diastaticus*, can be isolated on a medium consisting of starch, salts, vitamins, and ammonium sulfate, which all yeasts can utilize as a nitrogen source.

The most useful selective agent for detecting or enumerating of bacterial contaminants is the antifungal antibiotic cycloheximide (Acti-Dione), which can be added to any suitable culture medium, but usually to WLN as actidione agar. Some wild yeasts are sufficiently resistant to grow but, in general, a more effective way to recover wild *Saccharomyces* contaminants relies on their sporulation. Sporulation, which is stimulated by starvation conditions, is presumably advantageous in nature, but most culture yeasts, after innumerable generations in rich medium, have lost the ability to form viable spores. Yeast spores, unlike bacterial spores, are only marginally more heat resistant than the vegetative cells, but the difference is sufficient to recover contaminants on nonselective media after heat treatment. Practical details of microbiological analyses are fully explained in analytical manuals (1, 18).

In general, full identification of recovered contaminants is unnecessary. Bacteria and non-*Saccharomyces* wild yeasts need be identified to genus only, as indicated in Fig. 36.6 and 36.7. Only fermentative yeasts need further identification to determine whether or not they are *Saccharomyces* and, if so, *S. cerevisiae*. Even identification to species in this case is insufficient. It is important to determine whether it is the culture yeast itself, possibly persisting on an improperly cleaned fermentor or other equipment surfaces or surviving through filtration and pasteurization. Remedial action for that situation would obviously be different from the discovery of a wild *S. cerevisiae* strain as a contaminant. A quick and simple

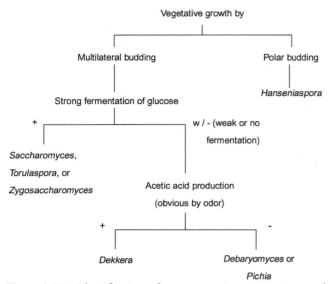

Figure 36.7 Identification of common yeast contaminants of the brewing industry. All genera listed are teleomorphic, i.e., form spores. Anamorphic (nonsporing) forms of *Dekkera* and *Hanseniaspora* are *Brettanomyces* and *Kloeckera*, respectively. The anamorph of all other genera listed is *Candida*. *S. cerevisiae* ferments glucose, sucrose, maltose, and raffinose but not lactose.

test to distinguish industrial yeast strains is to use the range of antifungal compounds on a multidisk, as used in medical mycology. In theory, a small-scale fermentation can distinguish the culture yeast from others, but the delay in obtaining useful results is unacceptably long. With recent advances in genetic methods, DNA fingerprinting is a potentially useful rapid method to distinguish different strains, but at present is not sufficiently developed for routine use (6, 14).

Cleaning and Sanitizing

In most industrial fermentation processes, steam is the usual method of sanitizing. In the brewery, however, chemical sterilants are preferred, mainly because they can be used at ambient temperature and avoid attemperation problems with adjacent working fermentors (15). Modern breweries use automatic in-place cleaning and sanitizing equipment to remove deposits of yeast and organic soil derived from beer foam by a powerful jet of water and detergent, followed by a spray of chemical sanitizers, followed by a spray of sterile rinse water. Until recently, caustic sanitizers containing 2% NaOH and additives were widely used, but acid sanitizers based on phosphoric or peracetic ($CH_3 \cdot COOOH$) acids have become more popular, not least because they are unaffected by CO_2 and can be used in closed vessels without the long delay and expense of draining off valuable gas.

Chlorine-based sanitizers are also used, but not widely because of the possibility of flavor problems. Residual chlorine can react with unknown compounds in beer to produce a strong medicinal flavor of organohalogen compounds. Quaternary ammonium compounds and biguanides are also regarded with suspicion. These compounds are widely used in other food industries because of their protective persistence on treated surfaces, even after rinsing, but this is unacceptable in the brewing industry because of the possible adverse effects of residual sanitizer on beer foam.

MODERN DEVELOPMENTS

For most of its history, brewing was either a domestic enterprise or associated with monasteries. Developments over the past 300 years have coincided with increasingly large-scale commercial production and with general improvements in scientific knowledge and technology. Research in the brewing industry has improved our understanding of the processes involved, and better quality malts and beers can be produced more rapidly and efficiently. For example, in 1900 it took at least 14 days to produce a batch of malt, mashing required up to 8 h, ale and lager fermentations could last 7 and 14 days, respectively, and subsequent lager maturation took many weeks longer. By 1980 these times had been halved, with improvement in consistency and quality, although some traditionalist consumers might disagree.

New products offer possibilities for increasing market share. Recent examples include diet beer or light beer, which contain lower carbohydrate content and, by implication, lower calorie content. Up to 20% of the carbohydrate in wort can be dextrins, which *S. cerevisiae* does not utilize, and so remain in the beer. Even in countries where addition of hydrolytic enzymes is not permitted, dextrin-fermenting yeasts can legally be used to ferment dextrin. One possible source of such strains is hybridization between *S. diastaticus* and brewing yeasts, and successful results have indeed been achieved in that way. Genetic modification has also produced potentially useful diastatic yeasts, but at present commercial brewers, in Europe at any rate, are reluctant to risk the public disapproval of their use. Another obvious use of diastatic yeasts is for beers of higher alcohol content, by fermentation of the dextrins (6, 16). In the currently popular ice beer, the alcohol content is increased by selectively freezing out part of the water content of the beer, with the incidental benefit of improved flavor and stability due to the chilling process.

Continuous fermentation is not actually a modern development, but recently there has been renewed interest in continuous fermentation of beer. Between about 1955 and 1970 there was considerable research on continuous operation of industrial microbiological processes in general. At that time many breweries built continuous fermentation systems based on a succession of two or three stirred vessels in which the homogeneous contents replicated the successive stages of batch fermentation: aerobic growth of the yeast, anaerobic fermentation, and, in some breweries, a third anaerobic vessel for flavor development and yeast separation. Other breweries preferred a single-vessel system, an unstirred tower in which different levels represented different stages of batch fermentation, and fermented beer overflowed from a yeast separator at the top. Although undoubtedly faster and more economical than previous batch operation, both types of continuous systems had the serious disadvantage of high cost of construction and operation in comparison with the cylindroconical vessels which were introduced in their modern form during the 1960s. Another problem was the difficulty of replicating in a different type of fermentation the flavors of traditional batch beers.

Recently, there has been renewed interest in continuous fermentation, now using immobilized-cell technology. Although perfectly successful on a laboratory scale, full-size fermentors suffer from damage to the immobilized-cell support by the expansion of bubbles of CO_2 as they rise up the necessarily tall column. It is possible that new types of support currently under development may avoid this difficulty, but meanwhile immobilized-cell systems are restricted to processes with less vigorous evolution of gas, postfermentation conditioning, or the production of low-alcohol beer. Moll (11) has provided a list of attempts over the years to operate continuous fermentation processes.

There is no doubt that yet more novel fermentation methods and products will be developed in the brewing industry in the future. We can also be sure that there will be a continuing demand for traditional beers, and the small breweries specializing in such products are certainly enjoying much success at the present time.

References

1. **Baker, C. D. (ed.).** 1991. *Recommended Methods of Analysis.* Institute of Brewing, London, United Kingdom.
2. **Briggs, D., J. S. Hough, R. Stevens, and T. W. Young.** 1981. *Malting and Brewing Science,* 2nd ed., vol. I, *Malt and Sweet Wort.* Chapman and Hall, London, United Kingdom.
3. **Campbell, I.** 1996. Wild yeasts in brewing and distilling, p. 193–208. *In* F. G. Priest and I. Campbell (ed.), *Brewing Microbiology,* 2nd ed. Chapman and Hall, London, United Kingdom.
4. **Flannigan, B.** 1996. The microflora of barley and malt, p. 83–125. *In* F. G. Priest and I. Campbell (ed.), *Brewing Microbiology,* 2nd ed. Chapman and Hall, London, United Kingdom.
5. **Gibson, G.** 1989. Malting plant technology, p. 279–325. *In* G. H. Palmer (ed.), *Cereal Science and Technology.* Aberdeen University Press, Aberdeen, United Kingdom.
6. **Hammond, J. R. M.** 1996. Yeast genetics, p. 43–82. *In* F. G. Priest and I. Campbell (ed.), *Brewing Microbiology,* 2nd ed. Chapman and Hall, London, United Kingdom.
7. **Hough, J. S., D. Briggs, R. Stevens, and T. W. Young.** 1982. *Malting and Brewing Science,* vol. II, *Hopped Wort and Beer.* Chapman and Hall, London, United Kingdom.
8. **Kreger-van Rij, N. J. W.** 1987. Classification of yeasts, p. 5–61. *In* A. H. Rose and J. S. Harrison (ed.), *The Yeasts,* 2nd ed., vol. 1. Academic Press, London, United Kingdom.
9. **Lewis, M. J., and T. W. Young.** 1995. *Brewing.* Chapman and Hall, London, United Kingdom.
10. **Maule, D. R.** 1986. A century of fermenter design. *J. Inst. Brewing* **92:**137–145.
11. **Moll, M.** 1991. *Beers & Coolers.* Lavoisier/Springer, Syracuse, NJ.
12. **Palmer, G. H.** 1989. Cereals in malting and brewing, p. 61–242. *In* G. H. Palmer (ed.), *Cereal Science and Technology.* Aberdeen University Press, Aberdeen, United Kingdom.
13. **Priest, F. G.** 1996. Gram-positive brewery bacteria, p. 127–161. *In* F. G. Priest and I. Campbell (ed.), *Brewing Microbiology,* 2nd ed. Chapman and Hall, London, United Kingdom.
14. **Schofield, M. A., S. M. Rowe, J. R. M. Hammond, S. W. Molzahn, and D. R. Quain.** 1995. Differentiation of yeast strains by DNA fingerprinting. *J. Inst. Brewing* **101:**75–78.
15. **Singh, M., and J. Fisher.** 1996. Cleaning and disinfection in the brewing industry, p. 271–300. *In* F. G. Priest and I. Campbell (ed.), *Brewing Microbiology,* 2nd ed. Chapman and Hall, London, United Kingdom.
16. **Stewart, G. G., and I. Russell.** 1986. One hundred years of yeast research and development in the brewing industry. *J. Inst. Brewing* **92:**537–558.
17. **Stratford, M.** 1992. Yeast flocculation: a new perspective. *Adv. Microb. Physiol.* **33:**1–71.
18. **Thorn, J. A.** 1992. *ASBC Methods and Analysis,* 8th ed. American Society for Brewing Chemists, St. Paul, Minn.
19. **Van Vuuren, H. J. J.** 1996. Gram-negative spoilage bacteria, p. 163–191. *In* F. G. Priest and I. Campbell (ed.), *Brewing Microbiology,* 2nd ed. Chapman and Hall, London, United Kingdom.
20. **Verzele, M.** 1986. 100 years of hop chemistry and its relevance to brewing. *J. Inst. Brewing* **92:**32–48.
21. **Yarrow, D.** 1984. Genus 22. *Saccharomyces* Meyen ex Reess, p. 379–395. *In* N. J. W. Kreger-van Rij (ed.), *The Yeasts, a Taxonomic Study,* 3rd ed. Elsevier, Amsterdam, The Netherlands.
22. **Young, T. W.** 1996. The biochemistry and physiology of yeast growth, p. 13–42. *In* F. G. Priest and I. Campbell (ed.), *Brewing Microbiology.* Chapman and Hall, London, United Kingdom.

Food Microbiology: Fundamentals and Frontiers, 2nd Ed.
Edited by M. P. Doyle et al.
© 2001 ASM Press, Washington, D.C.

Graham H. Fleet

Wine

37

Winemaking is a bioprocess that has its origins in antiquity. Scientific understanding of the process commenced with the studies of Louis Pasteur, who demonstrated that wines were the product of an alcoholic fermentation of grape juice by yeasts (1). Since the time of Pasteur, winemaking has developed into a modern, multinational industry with a strong research and development base in the disciplines of viticulture and enology. Viticulture concerns the study of grapes and grape cultivation, while enology covers postharvest processing of the grapes, from crushing through fermentation to packaging and retailing of the wine.

Microorganisms are fundamental to the winemaking process. To understand their contribution, it is necessary to know (i) the taxonomic identity of each species associated with the process, (ii) the kinetics of growth and survival of each species throughout the process, (iii) the biochemical activities of these species and how such activities determine the physicochemical properties of the wine, (iv) the influence of winemaking practices upon microbial growth and activity, and (v) the ultimate impact of microbial action upon sensory quality and consumer acceptability of the wine. This chapter will focus on the occurrence, growth, and significance of microorganisms in winemaking. It covers wines produced only

from grapes and includes table wines, sparkling wines, and fortified wines.

THE PROCESS OF WINEMAKING

Details of the process of winemaking are beyond the scope of this chapter and are described elsewhere (13, 132, 139). Figure 37.1 outlines the main steps in the production of white and red table wines.

Grapes

Numerous varieties of the grape, *Vitis vinifera*, are used in winemaking, with the particular variety determining the fruity or floral characteristics of the final product. Some main varieties used in white winemaking are riesling, traminer, Muller-Thurgau, chardonnay, Semillon, and sauvignon blanc, while those used in red winemaking include cabernet sauvignon, merlot, cabernet franc, pinot noir, shiraz, gamay, grenache, and barbera. The grapes are harvested at an appropriate stage of maturity, which determines the chemical composition of the juice extracted from them. Particularly important are the concentrations of sugars and acids which are the major constituents of the juice (Table 37.1) and have an important impact on its fermentation properties. Other preharvest

Graham H. Fleet, Department of Food Science and Technology, The University of New South Wales, Sydney, New South Wales 2052, Australia.

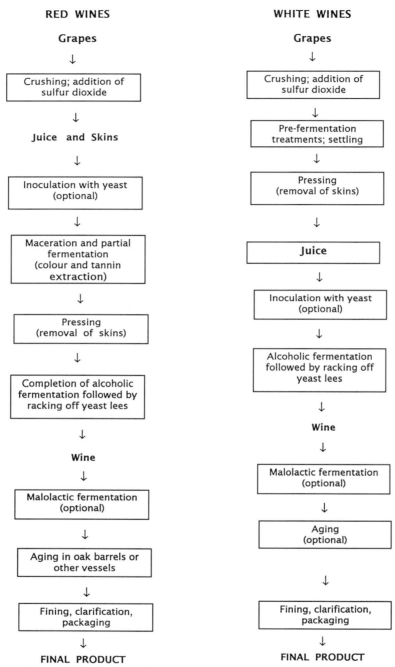

Figure 37.1 Outline of processes for making red and white wines.

conditions that affect the chemical composition of the grape and its juice include climate, sunlight exposure, soil, use of fertilizers, availability of water, vine age, and use of fungicides and insecticides. Traditionally, grapes were harvested manually, but now there is increasing use of mechanical harvesters, which are often operated at night to minimize the temperature of the berries at the time of crushing.

Crushing and Prefermentation Treatments

For white wines, the grapes are mechanically destemmed and crushed, and the juice is drained away from the skins. If required, clarification of the juice is done by cold settling, filtration, centrifugation, or combinations of these methods. Cold settling is generally done at 5 to 10°C for 24 to 48 h with the addition of pectolytic enzymes to help break down grape material. The juice is then transferred

Table 37.1 Main components of grape juice[a]

Substance	Concn
Glucose	75–150 mg/ml
Fructose	75–150 mg/ml
Pentose sugars	0.8–2 mg/ml
Pectin	0.1–1 mg/ml
Tartaric acid	2–10 mg/ml
Malic acid	1–8 mg/ml
Citric acid	0.1–0.5 mg/ml
Ammonia	5–150 μg/ml
Amino acids (total)	50–2500 μg/ml
Protein	10–100 μg/ml
Vitamins (varies with vitamin)	μg-mg/ml
Anthocyanins	0.5 mg/ml
Flavonoids, nonflavonoids	0.1–1.0 mg/ml

[a] Data from references 5, 13, 86, 87, and 132.

to the fermentation tank, where the fermentation commences naturally or it may be initiated by inoculation with selected yeasts (50).

Red grapes are mechanically destemmed and crushed, and the juice plus skins (must) are directly transferred to the fermentation tank. Fermentation begins either naturally or after inoculation, and during the first few days, the skins rise to the top of the juice to form a cap. Throughout this early stage, often described as maceration, juice is regularly pumped over the cap. The purpose of this step is to extract purple and red anthocyanin pigments, as well as other phenolic substances, from the grape skins to give color, tannic, and astringent character to the wine. The extraction process is assisted by the production of ethanol during this preliminary fermentation. When sufficient extraction has been achieved, the partially fermented wine is drained and pressed from the skins to another tank for completion of the fermentation. Some variations to this process include thermovinification and carbonic maceration. In thermovinification, the juice plus skins are heated to 45 to 55°C with pumping over to accelerate color and tannin extraction, after which the juice is separated from the skins and transferred to the fermentation tank. In carbonic maceration, uncrushed grapes are placed in a tank, which is gassed with carbon dioxide to remove oxygen. The temperature is maintained at 25 to 35°C for several days during which the grapes undergo endogenous metabolism that extracts color and phenolics from the skin. After about 8 to 10 days, the grapes are pressed to yield a partially fermented juice (1 to 1.5% ethanol) that is transferred to a tank for subsequent fermentation (12, 13).

Other pretreatments to the juice or must include the adjustment of pH and sugar concentration (where permitted), addition of diammonium phosphate or other nutrients to assist yeast growth, addition of sulfur dioxide (50 to 75 μg/ml) as an antioxidant, antimicrobial agent, and inhibitor of phenol oxidase activity, and addition of ascorbic acid or erythorbic acid as an antioxidant.

Alcoholic Fermentation

Traditionally, fermentation of the juice was conducted in large wooden barrels or concrete tanks, but most modern wineries now use relatively sophisticated stainless steel tanks with facilities for temperature control (principally cooling), cleaning in place, and other features for process management (13, 41). However, some premium-quality white wines (e.g., chardonnay) may be fermented in wooden (oak) barrels. White wines are generally fermented at 10 to 18°C for 7 to 14 days or more, where the lower temperature and slower fermentation rate favor the retention of desirable volatile flavor compounds. Red wines are fermented for about 7 days at 20 to 30°C, where the higher temperature is necessary to extract color from the grape skins.

The alcoholic fermentation can be conducted either as an indigenous or wild fermentation or as an induced or seeded fermentation. With indigenous fermentation, yeasts resident in the grape juice initiate and complete the fermentation. With seeded fermentation, selected stains of yeasts, generally species of *Saccharomyces cerevisiae*, are inoculated into the juice at initial populations of 10^6 to 10^7 cells/ml. Such yeasts have been commercially available as active dry preparations for the last 30 to 40 years and are now used extensively throughout the world, especially in the newer wine-producing countries, the United States, Australia, and South Africa (35). The advantages and disadvantages of indigenous and seeded fermentations have been well discussed (1, 5, 40, 68, 83, 135, 136, 150). Essentially, seeded fermentations are more rapid and predictable, while indigenous fermentations have a more varied outcome, with the potential of failures but with the prospect of wines with more interesting character due to contributions from a greater range of yeast species.

Alcoholic fermentation is considered complete when the fermentable sugars, glucose and fructose, of the juice are completely utilized. The wine is then drained or pumped (racked) from the sediment of yeast and grape material (lees) and transferred to stainless steel tanks or wooden barrels for malolactic fermentation, if desired, and aging. Clarification by filtration or centrifugation may be done at this stage. Leaving the wine in contact with the lees for long periods is not encouraged because the yeast cells autolyze, with the potential of adversely affecting wine flavor and providing nutrients for the subsequent growth of spoilage bacteria.

Malolactic Fermentation

It has been known since the early 1900s that, after alcoholic fermentation, wines frequently undergo another fermentation, which has been termed the malolactic fermentation (13, 29, 71, 97, 179). The malolactic fermentation commences naturally, about 2 to 3 weeks after completion of the alcoholic fermentation, and lasts about 2 to 4 weeks. Lactic acid bacteria resident in the wine are responsible for the malolactic fermentation, but many winemakers now choose to encourage this reaction by inoculation of commercial cultures of *Oenococcus oeni*, formerly known as *Leuconostoc oenos* (39). The main reaction is decarboxylation of L-malic acid to L-lactic acid, giving a decrease in acidity of the wine and an increase in its pH by about 0.3 to 0.5 unit. Wines produced from grapes cultivated in cool climates tend to have higher concentrations of malic acid (pH 3.0 to 3.5), which can mask their varietal character. A decrease in acidity by malolactic fermentation gives a wine with a softer, mellower taste. Also, growth of malolactic bacteria in wine contributes additional metabolites that may confer complex and interesting flavor characteristics. Apart from flavor considerations, there are practical reasons for having wines complete malolactic fermentation. Wines that have not undergone malolactic fermentation before bottling risk this reaction occurring at some later stage in the bottle. If this happens, the wine becomes gassy and cloudy and is considered to be spoiled. There is also a view that wines with completed malolactic fermentation have greater microbiological stability and are less prone to spoilage by other species of lactic acid bacteria. After malolactic fermentation by O. oeni, it is believed that fewer nutrients are available for microbial growth and that bacteriocin production may be a further inhibitory factor.

Malolactic fermentation is not necessarily beneficial to all wines. Wines produced from grapes grown in warmer climates tend to be less acid (pH >3.5). Further reduction in acidity by malolactic fermentation is deleterious to overall sensory balance, and also it increases this pH to values at which spoilage bacteria are more likely to grow. However, preventing the natural occurrence of malolactic fermentation in these wines (as might occur after bottling) is an extra technical burden. Consequently, many winemakers prefer to encourage the malolactic fermentation and later adjust wine acidity, if necessary. Nevertheless, there are winemakers who prefer not to have the malolactic fermentation occur in their wines.

Postfermentation Processes

Most white wines are not stored for lengthy periods after completion of the alcoholic fermentation or malolactic fermentation. If storage is necessary, it is generally done in stainless steel tanks. Some white wines (e.g., chardonnay) may be aged in wooden barrels. Most red wines are aged for periods of 1 to 2 years by storage in wooden (generally oak) barrels. During this time, chemical reactions that contribute to flavor development occur between wine constituents and components extracted from the wood of the barrels (13, 132). Critical points for control during storage and aging are exclusion of oxygen and addition of sulfur dioxide to free levels of 20 to 25 μg/ml. These controls are necessary to prevent the growth of spoilage bacteria and yeasts and to prevent unwanted oxidation reactions.

Just before bottling, the wines may be cold stored at 5 to 10°C to precipitate excess tartarate and then clarified by application of one or more processes, which include addition of fining agents (bentonite, albumen, isinglass, gelatin), centrifugation, pad filtration, and membrane filtration. For some white wines with residual sugar, potassium sorbate up to 100 to 200 μg/ml may be added to control yeast growth (50, 132).

Wine Flavor

The distinctive flavors of wine originate from the grape and the processing operations, which include alcoholic fermentation, malolactic fermentation, and aging (24, 116, 139, 153). The grapes contribute trace amounts of many volatile components (e.g., terpenes) that give wines their distinctive varietal, fruity character. In addition, they contribute nonvolatile acids (tartaric and malic) that impact on flavor, and tannins (flavanoid phenols) that give bitterness and astringency. The fermentation steps, especially alcoholic fermentation, increase the chemical and flavor complexity by assisting extraction of compounds from the grapes, modifying some grape-derived substances and producing a vast array of volatile and nonvolatile metabolic end products. Further chemical alterations occur during aging; also, enzymes derived from the grapes and excreted by yeasts and malolactic bacteria, as well as those added at prefermentation, might be expected to participate in chemical-flavor transformations. Thus, the final flavor represents contributions from many compounds and cannot be attributed to any one "impact substance."

YEASTS

Yeasts are significant in winemaking because they carry out the alcoholic fermentation, they can cause spoilage of the wine, and their autolytic products may affect sensory quality and influence the growth of malolactic and spoilage bacteria. Procedures for their isolation, enumeration, and identification have been reviewed (53, 57). Molecular methods for their identification have been developed (33, 34, 49).

Origin

Wine yeasts originate from any of three sources, namely, the surfaces of grapes, the surfaces of winery equipment (crushers, presses, fermenters, tanks, pipes, pumps, barrels, filtration units), and inoculum cultures (100, 101, 176).

The main yeasts found on mature, sound grapes are species of *Kloeckera* and *Hanseniaspora*, with lesser representations from species of *Candida*, *Metschnikowia*, *Cryptococcus*, *Pichia*, *Kluyveromyces*, and *Hansenula*. *S. cerevisiae*, the principal wine yeast, occurs at very low populations (<50 CFU/ml) on sound grapes and is rarely isolated from this source by direct plating methods. Freshly crushed grapes generally yield a must or juice with a yeast population of 10^3 to 10^5 CFU/ml, of which *Kloeckera* and *Hanseniaspora* species compose 50 to 70%. However, numerous factors affect the total yeast population and relative proportions of individual species on grapes. These include temperature, rainfall, and other climatic conditions, degree of maturity at harvest, physical damage due to mold or insect or bird attack, grape variety, and application of fungicides (10, 53, 86, 101). The chemical composition and physical properties of the waxy, outer surface of the skin (67) are likely to have an important influence on the yeast ecology of the grape, but this relationship remains unstudied.

The surfaces of winery equipment that come into contact with the grape juice and wine are locations for development of the so-called residential or winery yeast flora. The extent of this development depends upon the nature of the surface and the effectiveness of cleaning and sanitizing operations. *S. cerevisiae* is prevalent on these surfaces, which are considered to be the main source of this species in wine fermentations. Such sources are likely to harbor multiple strains of *S. cerevisiae* that have accumulated from the grapes and starter cultures used in previous vintages (25, 53, 66, 100).

Starter cultures, if used to inoculate the juice, will be a principal source of yeasts. Presently, various strains of *S. cerevisiae* are used, but the future could see the development and use of other species (68, 160). Commercial yeast preparations used for inoculation are not necessarily pure and could contain a proportion of species other than *S. cerevisiae* (35, 131, 136).

Growth during Fermentation

Over the last 100 years, many studies have described the yeast populations that grow during alcoholic fermenta-

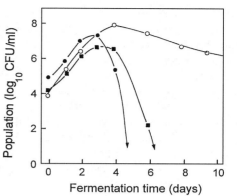

Figure 37.2 Generalized growth of yeast species during alcoholic fermentation of wine. ○, *S. cerevisiae*; ●, *Kloeckera* and *Hanseniaspora* species; ■, *Candida* species. Variations will occur in the initial and maximum populations for each species; for fermentations inoculated with *S. cerevisiae*, the initial population is approximately 10^6 CFU/ml (56).

tion. Early studies were essentially qualitative descriptions of the main species isolated from different stages (early, mid, final) of fermentation. Subsequent studies followed the growth of individual species throughout the entire course of fermentation. Recently, molecular techniques have made it possible to follow the development of particular strains throughout fermentation. Virtually all ecological studies to date have shown that *S. cerevisiae* is the principal wine yeast. Invariably, this species predominates during the middle to final stages of fermentation, but there are important contributions of other yeast species that must be considered (5, 10, 40, 53, 85, 87).

Figure 37.2 gives a general representation of the growth of yeasts during the fermentation of grape juice, whether it be conducted by indigenous or inoculated processes. As mentioned, freshly extracted grape juice harbors a yeast population of 10^3 to 10^5 CFU/ml, composed mostly of *Kloeckera* and *Hanseniaspora* species, but species of *Candida*, *Metschnikowia*, *Pichia*, *Hansenula*, *Kluyveromyces*, and *Rhodotorula* also occur. These species are often referred to as the non-*Saccharomyces* yeasts or wild flora. The juice will also contain low populations of *S. cerevisiae*, depending on the extent of its occurrence on grapes and contamination from equipment used to process the juice. Fermentation is initiated by the growth of various species of *Kloeckera*, *Hanseniaspora*, *Candida*, and *Metschnikowia* (e.g., *Kloeckera apiculata*, *Hanseniaspora valbyensis*, *Candida stellata*, *Candida colliculosa*, and *Metschnikowia pulcherrima*) as well as *S. cerevisiae*. The growth of the non-*Saccharomyces* species is generally limited to the first 2 to 4 days of fermentation, after which they die off. Nevertheless, they achieve maximum populations of 10^6 to 10^7 CFU/ml before death, and such growth is metabolically significant in terms of substrates utilized (hence not available to *S. cerevisiae*) and end products released into the wine. Also, the dead cells become part of the yeast pool for subsequent autolysis. Their death is attributed to an inability to tolerate the increasing concentrations of ethanol, which is largely produced by *S. cerevisiae*. After 4 days or so, the fermentation is continued and completed by *S. cerevisiae*, which reaches final populations of about 10^8 CFU/ml. Wine strains of *S. cerevisiae* are particularly noted for their strong production and tolerance of ethanol (up to 15% [vol/vol] or more). By comparison, strains of *K. apiculata* or *C. stellata* rarely tolerate ethanol concentrations greater than 5 or 8%, respectively, at 20 to 25°C (58).

Application of molecular techniques to the study of wine fermentations has revealed a greater ecological complexity than that shown in Figure 37.1. These techniques include karyotyping by pulsed-field gel electrophoresis, restriction fragment length polymorphism analysis of mitochondrial DNA and ribosomal DNA subunits, amplified fragment length polymorphism analysis of restriction nuclease digests of total DNA, and other PCR methods (33, 34, 66, 129, 154). Essentially, there can be an evolution of different strains within any one species throughout the course of the fermentation. Different phases (early, mid, final) of the alcoholic fermentation may be dominated by different strains of *S. cerevisiae*; also, strain variation can occur within the non-*Saccharomyces* yeasts (25, 66, 109, 127–129, 154, 155). Changes in colony morphology (56), secondary end-product formation (123, 144), and killer phenotype (111) of isolates are other indicators of strain variation throughout fermentation. It is thought that this variation is due to the natural diversity that occurs in winery ecosystems, the selective pressures of the changing environment (e.g., increasing ethanol concentration and other end products), or rearrangement of chromosomal DNA within the one strain as fermentation progresses (95, 102, 109, 112). Molecular techniques have been very useful also in demonstrating the carryover of particular strains of *S. cerevisiae* from one vintage to the next, the geographical specificity of particular strains, and the

dominance or nondominance of inoculated strains (25, 66, 127, 177).

Despite significant research, understanding of the yeast ecology of wine fermentations is probably incomplete. To date, all studies on this subject have used conventional cultural methods for isolation of the yeasts. Experience in other fields of microbial ecology suggests that this basic cultural approach can grossly underestimate the species diversity in ecosystems. In other words, all of the species associated with wine fermentations may not yet be known. Molecular studies based on extraction and analyses of the total microbial DNA from ecosystems are needed to more reliably characterize the associated microflora (54). Moreover, viable but nonculturable species may occur in wine ecosystems and escape detection (104). Another ecological issue is the potential formation of biofilms during wine fermentation. In red wine fermentations, for example, both juice and grape skins constitute the ecosystem. It is possible that a biofilm of yeasts will associate with the skins and present a population and species and strain profile which may be different from that in the juice (54).

Factors Affecting Yeast Growth during Fermentation

The process of winemaking involves many intrinsic and extrinsic variables that determine the duration and completeness of fermentation (Table 37.2) (10, 15, 40, 56, 85, 137). These factors influence the rate and extent of growth of individual yeast species and strains, but detailed information on these effects remains limited. Yeast growth is best measured by plate counts (52, 53, 57), but carbon dioxide production, as measured by loss of culture weight, and utilization of juice sugars are also used to monitor growth and fermentation kinetics (45).

Juice Composition

In most circumstances, grape juices provide all the nutrients and conditions necessary for a vigorous and complete fermentation. However, chemical and physical properties of the juice vary according to grape variety, climatic influences, viticulture practices, and maturity

Table 37.2 Factors affecting the growth of yeasts during alcoholic fermentation

Composition of grape juice	Processing factors	Ecological factors
Sugar concentration	Addition of sulfur dioxide	Influences of grape fungi and bacteria
Assimilable nitrogen	Extent of settling and clarification of juice	Population and composition of indigenous yeasts
pH	Addition of yeast nutrients	Killer yeasts
Fungicide and pesticide residues	Inoculation with selected yeasts	
Content of dissolved oxygen	Temperature control	
Accumulation of toxic metabolites	"Pumping over"	

at harvest. Relevant properties include sugar concentration, amount of nitrogenous substances, concentrations of vitamins, dissolved oxygen content, amount of soluble solids, fungicide and pesticide residues, pH, and presence of any yeast-inhibitory or -stimulatory substances produced by growth of fungi and bacteria on the grapes.

The concentration of fermentable hexoses in grape juice varies between 150 and 300 mg/ml (Table 37.1) and may be as high as 400 mg/ml in juice prepared from grapes infected with *Botrytis cinerea* (*pourriture noble*, noble rot) (42, 87). Initial sugar concentration might be expected to affect growth rates of the different species and strains of wine yeasts and, consequently, the extent to which they contribute to the overall fermentation. The growth of *C. stellata* and *Torulaspora delbrueckii* may be favored in juices with higher sugar concentrations (5, 18, 87).

Free amino acids and ammonium ions (Table 37.1) are the principal nitrogen sources used by yeasts during fermentation. For a long time, it was thought that most juices contained sufficient nitrogen substrates (>150 mg/liter of assimilable nitrogen) to allow rapid and complete fermentation. However, it is now known that some juices, especially those that have been heavily processed or clarified, may not have sufficient nitrogen nutrients to allow maximum yeast growth and complete fermentation (75). Moreover, nitrogen availability in the vineyard and use of nitrogen fertilizers can significantly affect the concentration of assimilable nitrogen in the juice and subsequent yeast growth (164). The nitrogen demand by yeasts increases significantly with increasing sugar concentration in the juice and also varies with the strain of *S. cerevisiae*. Consequently, supplementation of juices with various yeast foods or diammonium phosphate is now common practice to ensure that nitrogen availability is not a factor that limits yeast growth. Most studies on the nitrogen requirements of wine yeasts have been conducted with *S. cerevisiae*, and little or nothing is known about the nitrogen demands of non-*Saccharomyces* species or their ability to remove specific nitrogen substrates from the juice before the growth of *S. cerevisiae*. The ability of wine yeasts to utilize grape juice proteins as a source of nitrogen requires further consideration. Strains of *S. cerevisiae* generally do not produce extracellular proteolytic enzymes (52). However, some strains of the non-*Saccharomyces* wine yeasts, e.g., *K. apiculata* and *M. pulcherrima*, are proteolytic (19, 52).

Grape juices generally contain sufficient concentrations of vitamins (inositol, thiamine, biotin, pantothenic acid, and nicotinic acid) to permit maximum growth of *S. cerevisiae* (118). Vitamin losses may occur in heavily processed juices, in which case supplementation can improve yeast growth. Non-*Saccharomyces* species are more demanding of vitamins than *S. cerevisiae*, and vitamin availability could be a factor which limits their contribution to fermentation. The pH of grape juice varies between 2.8 and 4.0, depending on the concentration of tartaric and malic acids. Although growth and fermentation rates by *S. cerevisiae* are decreased as the pH is decreased from 3.5 to 3.0 (18, 70), it is not fully known how juice pH affects the relative growth rates of the non-*Saccharomyces* yeasts and their potential to influence alcoholic fermentation.

Treatment of grapes with fungicides and pesticides before harvest can give juices that contain residues of these substances. Depending on their concentration and chemical nature, these residues may decrease yeast growth and even change the ecology, thereby leading to slow or incomplete fermentations (14, 56, 106).

Conditions which stimulate yeast growth and fermentation include aeration of the juice before or during the early stages of fermentation and the presence of grape solids and particulate materials (9, 36, 119). Different yeast species may be selectively adsorbed to such particles to form a biofilm of immobilized biomass, as noted already.

Clarification of Grape Juice

The procedures used to clarify juices, especially for white wine fermentations, will influence the populations of indigenous yeasts in the juice and their potential contribution to the fermentation. Centrifugation and filtration remove yeast cells, thereby decreasing or eliminating the contribution of indigenous species to the fermentation. In contrast, clarification by cold-settling presents opportunities for the growth of indigenous yeasts, especially those species or strains that grow well at low temperatures (e.g., *K. apiculata*) (9, 18, 107).

Sulfur Dioxide

The addition of sulfur dioxide to grapes or juice for controlling oxidation reactions and restricting the growth of indigenous microflora is a well-established practice (117, 145). However, a general view that addition of sulfur dioxide (50 to 100 μg/ml, total) to juice will suppress the growth of non-*Saccharomyces* yeasts relative to *S. cerevisiae* is not supported by recent studies. Strong growth of *K. apiculata* and various *Candida* species occurs during the early stages of most commercial fermentations to which the usual amounts of sulfur dioxide have been added. In one study, growth of *K. apiculata* was not inhibited by total sulfur dioxide concentrations of 100 to 250 μg/ml (56). These findings question the efficacy of sulfur dioxide in controlling indigenous yeasts and challenge one of the basic reasons for using sulfur dioxide in winemaking. Despite the importance of sulfur dioxide in winemaking, good comparative data on the responses of wine yeasts to this agent are lacking.

Temperature of Fermentation

Temperature control has become an important practice in modern winemaking. Temperature affects the rate of growth and metabolic activities of yeasts (18). It also affects the ability of individual yeast species to tolerate ethanol (58), thereby determining their survival and contribution to the fermentation. Fastest yeast growth and fermentation occur at 25 to 30°C, and the ecology of the fermentation follows the pattern outlined in Fig. 37.2. However, when the temperature is decreased below 20°C, there is an increased contribution of the non-Saccharomyces species to the fermentation. Species such as *K. apiculata* and *C. stellata* exhibit increased tolerance to ethanol and do not die off, as shown in Fig. 37.2. They can produce maximum populations of 10^7 to 10^8 CFU/ml which remain viable until the end of fermentation (70). Moreover, they may have faster growth rates than *S. cerevisiae* at low temperatures (18). The impact of such ecological shifts on the chemical composition and sensory quality of wines has yet to be determined.

Inoculation of the Juice

Perhaps the most significant technological innovation in winemaking during the last 20 to 30 years has been the seeding (inoculation) of the juice with selected strains of *S. cerevisiae*. These strains have been selected according to criteria that facilitate the process of winemaking and will yield a product of desired quality (Table 37.3) (23, 35, 68, 74). Seeding of the fermentation is undertaken with the assumption and expectation that the inoculated strain will outcompete and dominate over indigenous strains of *S. cerevisiae* and the non-*Saccharomyces* yeasts. While much evidence shows that inoculated strains dominate at the end of fermentation, the view that early growth of the indigenous species is suppressed or insignificant cannot be supported. Growth of the indigenous non-*Saccharomyces* species, according to Fig. 37.2, still occurs (69); moreover, indigenous strains of *S. cerevisiae* may grow despite massive competition from the seeded strain. Indeed, if conditions in the juice do not favor growth of the seeded strain, indigenous *S. cerevisiae* may dominate the fermentation; this can be verified by molecular techniques that allow differentiation of yeast strains. Although there is a high probability that inoculated *S. cerevisiae* will dominate the fermentation, seeding will not necessarily guarantee the dominance of any particular strain or its exclusive contribution to the fermentation (25, 66, 108, 128, 154, 155). Significant factors that affect this outcome will be the population of indigenous yeasts already in the juice and the extent to which they have adapted to grow in that juice (150).

Interactions with Other Microorganisms

Various species of molds, acetic acid bacteria, and lactic acid bacteria naturally occur on grapes and on winery equipment. Conditions which allow their proliferation on the grape or in the juice before yeast growth have the potential to affect the ecology and success of the alcoholic fermentation and are discussed in later sections.

Killer yeasts are certain strains that produce extracellular proteins or glycoproteins, termed killer toxins, that can destroy other yeasts (158, 173). Usually, strains of one species only kill strains within that species, but, as more yeasts are examined, it is becoming evident that

Table 37.3 Some desirable and undesirable characteristics of wine yeasts[a]

Desirable	Undesirable
Complete and rapid fermentation of sugars	Production of sulfur dioxide
High tolerance of alcohol	Production of hydrogen sulfide
Resistance to sulfur dioxide	Production of volatile acidity
Fermentation at low temperatures	High formation of acetaldehyde, pyruvate, and esters
Production of good flavor and aroma profiles	
Production of glycerol	Foaming properties
Production of β-glycosidases	Formation of ethyl carbamate precursors
Malic acid degradation	Production of polyphenol oxidase; affects wine color
Killer phenomenon	
Good sedimentation properties	Inhibition of malolactic fermentation
Tolerance of pesticides and fungicides	Production of mousy and other taints
Ferment under pressure	
Suitability for mass culture, freeze-drying distribution, and rehydration	

[a] From references 23, 35, and 74.

killer interactions between different species may occur. Killer toxin-producing strains of *S. cerevisiae* and killer-sensitive strains of *S. cerevisiae* occur as part of the natural flora of wine fermentations. In some wineries, killer strains may predominate at the end of fermentation, suggesting that they have asserted their killer property and taken over the fermentation. Killer strains have also been found within wine species of *Candida*, *Pichia*, *Hanseniaspora*, and *Hansenula*, and, indeed, some of these strains can assert their killer action against wine strains of *S. cerevisiae*. Expression of the killer phenomenon during wine fermentations is affected by many factors, including ethanol concentration, pH, temperature, assimilable nitrogen, presence of fining agents, and the relative populations of killer and killer-sensitive strains (56, 111). There are several implications of killer yeasts in winemaking. First, inoculated strains of *S. cerevisiae* could be destroyed by indigenous killer strains of *S. cerevisiae* or non-*Saccharomyces* species, leading to premature cessation of the fermentation, slower fermentation, or completion of the fermentation by a less desirable species. Second, there may be an advantage in conducting the fermentation with selected or genetically engineered killer strains of *S. cerevisiae* for the purposes of controlling the growth of less desired indigenous species. Moreover, strains could be selected or constructed to produce stable, broad-spectrum killer toxins that would protect the wine from infection by spoilage yeasts. Finally, strains might be selected to have immunity against the killing action of indigenous yeasts, thereby giving them a greater chance of dominating the fermentation (158, 173).

Stuck, Sluggish Fermentations

A sporadic but serious problem is the premature cessation of yeast growth and alcoholic fermentation, giving wine with residual, unfermented sugar (>2 to 4 g/liter) and a lower than expected concentration of ethanol. Such fermentations are commonly referred to as being stuck or sluggish if they take longer than normal to give low residual sugar (9, 75, 168). Factors considered to cause this problem include excessive clarification and processing of the juice, fermentation temperature too high, juice deficiency in nutrients or growth factors, presence of fungicide residues, influences from other microorganisms such as molds, acetic acid bacteria, and killer yeasts, ethanol toxicity, and accumulation of medium-chain-length fatty acids such as octanoic and decanoic acids that can become toxic to yeast growth. An overriding consideration is failure in the transport of grape juice sugars into the yeast cell and the multitude of factors that affect expression of the various genes responsible for this transporter activity (8, 9). Initiatives to overcome stuck fermentations include the addition of nitrogen-containing yeast foods, controlled aeration of the juice or wine, and the addition of yeast cell wall hulls or other bioadsorbents to remove toxic substances (9, 110).

Biochemistry

Yeasts utilize grape juice constituents as substrates for their growth, thereby generating metabolic end products that are excreted into the wine (13, 40, 74). The main products are carbon dioxide and ethanol and, to a lesser extent, glycerol and succinic acid. In addition, many hundreds of volatile and nonvolatile secondary metabolites are produced in small amounts that, collectively, contribute to the sensory quality of the wine. These substances include a vast range of organic acids, higher alcohols, esters, aldehydes, ketones, sulfur compounds, and amines. The chemical identities of individual substances, their flavor or aroma sensation, their flavor threshold, and their concentrations in wines are well documented (24, 116, 153), and the biochemistry of their formation in *S. cerevisiae*, at least, is reasonably well known (6, 40, 87). However, further studies are needed to determine the metabolic characteristics of the non-*Saccharomyces* yeasts. The production of these metabolite varies considerably depending on the yeast strain, yeast species, and conditions of fermentation (144). Tables 37.4 and 37.5

Table 37.4 Concentrations of volatile compounds (μg/ml) produced by different species of wine yeasts[a]

Yeast species	Propanol	Isobutanol	Isoamyl alcohol	2-Phenyl ethanol	Ethyl acetate	Isoamyl acetate	Acetoin
Saccharomyces cerevisiae	0.4–170	5–666	17–769	5–83	10–205	0.1–16	0–29
Kloeckera apiculata	4–25	3–60	10–117	10–35	40–870	0.04–1.1	56–187
Candida stellata	4–8	13–21		6–11	7–25	0.1–0.4	
Pichia anomala	3–15	18–29	11–25	27	137–2,150	1–11	
Metschnikowia pulcherrima	1–43	37–123	21–243	22	150–382	0.1–0.8	
Zygosaccharomyces bailii	18–25	20–30	48–85	13–22	23–53	0.1–0.5	17–24
Pichia membranifaciens	<1	1–9	0.5–9.5		16–21	1–6	
Brettanomyces bruxellensis	2–3	19	46	26	36–860	<1	

[a] Data from references 5, 24, 27, 44, 52, 68, 78, 87, 105, 116, 144–149, 153–155, 159, 163, 175, and 186.

Table 37.5 Concentrations of ethanol, glycerol, acetaldehyde, acetic acid, and succinic acid produced by different species of wine yeasts[a]

Yeast	Ethanol (%)	Glycerol (mg/ml)	Acetaldehyde (μg/ml)	Acetic acid (mg/ml)	Succinic acid (mg/ml)
Saccharomyces cerevisiae	6–23	3.7–6.8	15–30	0.1–2.0	0.6–1.7
Kloeckera apiculata	2–7	5.5–8.2	8–54	0.1–1.2	0.3
Candida stellata	6–7			1.08–1.3	
Pichia anomala	0.5–5	0.2–2.2	3.2–8.1	1.6	0.2
Metschnikowia pulcherrima	2–4	2.7–4.2	23–40	0.1–0.2	
Zygosaccharomyces bailii	5–13			0.1–0.3	1.6
Pichia membranifaciens	0.1–0.5	4.1–5.4	2.9	0.3	
Brettanomyces bruxellensis	9–12			1–7	

[a] Data from references 5, 24, 27, 44, 52, 68, 78, 87, 105, 116, 144–149, 153–155, 157, 159, 163, 175, and 186.

show some of the main metabolites produced by yeasts associated with wine fermentations.

Research in this field is now moving to understand the expression of metabolic genes under winemaking conditions (140). The first stage of the alcoholic fermentation involves gene expression under conditions of exponential growth in an environment of low pH and relatively high sugar concentration. The second stage involves significant metabolic activity of stationary-phase yeast cells under the added stress of ethanol presence. Many of the secondary metabolites that impact on wine flavor are produced during this second stage (74).

Carbohydrates

Glucose and fructose in juice are metabolized by the glycolytic pathway to pyruvate, which is decarboxylated by pyruvate decarboxylase to acetaldehyde. Acetaldehyde is reduced to ethanol by alcohol dehydrogenase (8). Although most of the pyruvate is converted to ethanol and carbon dioxide, small proportions are converted to secondary metabolites (Fig. 37.3). Glycerol, which imparts desirable smoothness and viscosity to the wine, is derived

from the glycolytic intermediate dihydroxy acetone. Its production is increased by the presence of sulfur dioxide, higher incubation temperature, and increased sugar concentration (62), but there are significant strain and species influences (138). Transport of sugars into the cell and factors which affect this activity are important rate-limiting steps in the fermentation process (8, 9).

The potential for wine yeasts to degrade grape pectins and enhance juice extraction needs more consideration, given that some strains of *S. cerevisiae* may produce pectolytic enzymes (11, 19, 52). Yeast glycosidase activities are attracting increasing interest since these enzymes have the potential to enhance wine flavor through the hydrolysis of glycosylated terpenoids of the grape. Species of *Candida*, *Kloeckera* and *Pichia* found in grape juice may be stronger producers of these enzymes than *S. cerevisiae* (19, 103).

Nitrogen Compounds

Wine yeasts take up and metabolize ammonium ions and amino acids in the juice as sources of nitrogen (75). However, genes for the transport and assimilation of amino

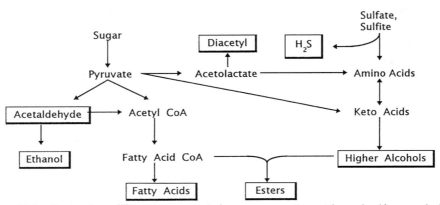

Figure 37.3 Derivation of flavor compounds from sugar, amino acids, and sulfur metabolism by yeasts (6, 75).

acids are not expressed until after the ammonium ions are utilized (151). The mechanisms of transport of these compounds into the cells and biochemical pathways for their metabolism, involving decarboxylation, transamination, reduction, and deamination reactions, have been reviewed (89). There is little evidence that *S. cerevisiae* produces extracellular proteolytic enzymes to degrade grape juice proteins. However, some non-*Saccharomyces* species such as *K. apiculata* and *M. pulcherrima* produce extracellular proteases that could be involved in the breakdown of juice proteins (19, 52).

The availability of nitrogen in the juice and its metabolism by yeasts have important implications in winemaking (9, 75). Juices that are limiting in nitrogen content can give stuck or sluggish fermentations and wines with unacceptably high contents of hydrogen sulfide. Metabolism of arginine (the most predominant amino acid in grape juices) by *S. cerevisiae* can lead to the production of urea, which is able to react with ethanol to form ethyl carbamate, a suspected carcinogen. The amount of urea produced depends on many factors, including the concentration of arginine relative to other nitrogen substances in the juice and the strain of *S. cerevisiae* (76). Juice nitrogen can have a significant impact on the amounts of higher alcohols, fatty acids, esters, and carbonyl compounds, such as aldehydes and diacetyl, produced by yeasts (75).

In addition to utilization, yeasts can release amino acids into the wine. This occurs during the final stages of alcoholic fermentation, by mechanisms not fully understood, and later when the cells have died and there is autolytic degradation of yeast proteins (20, 77). These amino acids can serve as nutrients for the growth of malolactic bacteria or spoilage bacteria.

Sulfur Compounds

Yeasts produce a range of volatile sulfur compounds which, above certain threshold concentrations, have a negative impact on wine flavor (120, 133). The most predominant compounds are sulfur dioxide (sulfite), hydrogen sulfide, and dimethyl sulfide, with lesser amounts of other organic sulfites, mercaptans, and thioesters. The production of sulfite and hydrogen sulfide in *S. cerevisiae* is linked to the biosynthesis of cysteine and methionine by the sulfate reduction pathway, (64, 80, 165). These sulfur-containing amino acids exert feedback control over this pathway and the production of sulfite and hydrogen sulfide. The formation of sulfite by *S. cerevisiae* depends on the strain. Most strains produce less than 10 μg/ml, but some give levels up to 100 μg/ml (145). High sulfite-producing strains are avoided in winemaking because of the negative effect of this metabolite on

wine quality, its potential to cause allergic reactions in some consumers, and its potential to inhibit malolactic bacteria.

At concentrations exceeding 50 μg/liter, hydrogen sulfide gives an unpleasant "rotten egg" aroma to wine. Many chemical and biological factors affect the production of hydrogen sulfide during wine fermentation, the most significant of which is the strain of *S. cerevisiae*. Some strains produce hydrogen sulfide at concentrations exceeding 1 μg/ml. Hydrogen sulfide production is genetically based, but it is also influenced by factors such as the composition of the grape juice and fermentation conditions (165). Elemental sulfur used as a fungicide on grapes prior to harvest, metabisulfite added to the grapes and juice at crushing, and sulfate that occurs naturally in the juice are all significant precursors of hydrogen sulfide in the wine. Deficiencies in assimilable nitrogen in the juice can cause hydrogen sulfide production by *S. cerevisiae* (75). Under these conditions, the intracellular pool of cysteine and methionine is low, allowing the sulfate reduction pathway to operate with consequent production of hydrogen sulfide. If the juice contains an adequate supply of assimilable nitrogen (e.g., ammonium ions and amino acids), cysteine and methionine are produced at concentrations which, through feedback inhibition, decrease the activity of the sulfate reduction pathway and production of hydrogen sulfide. There is also a view that, under nitrogen limiting conditions, *S. cerevisiae* degrades intracellular proteins to provide essential amino acids, including cysteine and methionine from which hydrogen sulfide is formed (75, 80). The potential for the non-*Saccharomyces* yeasts to produce sulfur compounds during wine fermentations requires study.

Organic Acids

Of the numerous organic acids produced in wine by yeasts, succinic and acetic acids are the most significant (130). Succinic acid has a bitter, salty taste and is produced by *S. cerevisiae* at concentrations up to 2.0 mg/ml, depending on strain. Lower concentrations are produced by non-*Saccharomyces* species (157). The production of succinic acid is not associated with any major defects in wine quality. In contrast, acetic acid becomes detrimental to wine flavor at concentrations exceeding 1.5 mg/ml and may lead to stuck fermentations (48). Most strains of *S. cerevisiae* produce only small amounts (<0.75 mg/ml) of this acid, but some can produce greater than 1.0 mg/ml and are unsuitable for winemaking (74). Factors that limit yeast growth, such as low temperature, high sugar concentrations, low pH, deficiency in available nitrogen, and excessive clarification, cause increased acetic acid production by *S. cerevisiae* (36, 157).

Candida, Kloeckera, and *Hanseniaspora* species may produce higher amounts of acetic acid than *S. cerevisiae* (Table 37.5), but there is substantial strain variation in this property (147).

The production of lactic acid by wine yeasts is considered insignificant (<0.1 mg/ml), but there are species of *Saccharomyces* (130) and some strains of *Kluyveromyces thermotolerans* (108) and *C. colliculosa* (17) that can produce this acid in amounts of 5 to 10 mg/ml. Such strains could be used to increase the acidity of some wines. Although tartaric acid is prevalent in grape juice and wine, it is not metabolized by wine yeasts (130). However, malic acid is partially (5 to 50%) metabolized by *S. cerevisiae* and other wine yeasts. It is completely degraded by some species of *Schizosaccharomyces* and some strains of *Zygosaccharomyces bailii* (61, 130) and is oxidatively decarboxylated to pyruvate, which is then converted to ethanol. The possibility of using species of *Schizosaccharomyces* to deacidify wines in place of the malolactic fermentation has attracted considerable interest, but such use must be carefully controlled, as these yeasts can produce off flavors (55, 61, 182).

Yeasts produce small amounts (1 to 15 μg/ml) of free fatty acids in wines (87, 130). Of special note are hexanoic, octanoic, and decanoic acids which, on accumulation, may become toxic to *S. cerevisiae* and contribute to stuck fermentations (9).

Autolysis

The autolytic degradation of yeast cells at the end of alcoholic fermentation and during cellar storage of the wines is often underestimated as a significant biochemical event. During autolysis, yeast proteins, nucleic acids, and lipids are extensively degraded, releasing peptides, amino acids, nucleotides, bases, and free fatty acids, into the wine. These products affect wine flavor and serve as nutrients for the growth of bacteria. In addition to *S. cerevisiae,* the non-*Saccharomyces* species are involved in autolytic reactions (20, 77, 126).

Flavor Compounds

Yeasts produce many organic acids, higher alcohols, esters, aldehydes, ketones, sulfur compounds, and amines that, in conjunction with grape constituents, determine wine flavor (24, 74, 116, 153). Wines produced by indigenous fermentations are often perceived as having more complex and interesting flavors than those produced by inoculation with selected strains of *S. cerevisiae* (5, 68, 86, 135, 136). Such views imply distinctive contributions by indigenous, non-*Saccharomyces* yeasts to the flavor profile, but definitive studies that connect their growth to a chemical profile and a particular sensory out-

come remain wanting. Nevertheless, there are numerous studies which show that non-*Saccharomyces* species produce some compounds in greater or lesser amounts than *S. cerevisiae* (68, 144) (Tables 37.4 and 37.5) and that mixed culture fermentations involving these species as well as *S. cerevisiae* have different chemical profiles than fermentation conducted solely by *S. cerevisiae* (78, 102, 154, 160, 186). However, these conclusions are complicated by the fact that metabolite production by particular yeasts varies considerably, depending on the strain and conditions of fermentation (74, 144, 147–149).

Spoilage Yeasts

Growth of inappropriate species or strains of yeasts during alcoholic fermentation can give an inferior wine (e.g., one high in content of esters, acetic acid, hydrogen sulfide), and the product is considered spoiled. Spoilage can also occur during storage of wine in the cellar and after bottling (13, 40, 57, 166, 170). Wine that is exposed to air (e.g., as in incompletely filled barrels or tanks) quickly develops a film or surface flora of weakly fermentative or oxidative yeasts in the genera *Candida, Pichia,* and *Hansenula.* Particularly significant is *Pichia membranifaciens.* These species oxidize ethanol, glycerol, and acids, giving wines with unacceptably high levels of aetaldehyde, esters, and acetic acid. Fermentative species that grow in bottled wines and in wines during cellar storage include *Z. bailii, Brettanomyces* and *Dekkera* species (*Brettanomyces intermedia*), and *Saccharomycodes ludwigii.* In addition to causing excessive carbonation, sediments, and haze, these species produce acid and estery off flavors. The growth of *Brettanomyces* species is also associated with the production of unpleasant mousy or medicinal taints because of the formation of tetrahydropyridines and volatile phenolic substances, such as 4-ethylguaiacol and 4-ethylphenol (166, 170). Wines that contain residual sugars are prone to refermentation unless properly managed. In such cases, the growth of *S. cerevisiae* constitutes a form of spoilage.

Genetic Improvement of Wine Yeasts

A vast pool of wine yeasts naturally occurs in vineyards and wineries, and selection from these reservoirs has generally met the needs of winemaking. Nevertheless, the process of strain selection and development can be accelerated and more specifically targeted through the use of classical and modern genetic improvement technologies (2, 74, 125). Classical mutagenesis has yielded strains of *S. cerevisiae* that give decreased levels of higher alcohols and better fermentation performance at low temperatures. Traditional mating and hybridization methods have given strains with decreased production of hydrogen sulfide and improved flocculation characteristics. Protoplast fusion has been used to introduce the killer property into desirable strains of *S. cerevisiae* and to introduce the malic acid-degrading properties of *Schizosaccharomyces* species into *S. cerevisiae*. Recombinant DNA technologies are being used to produce strains of *S. cerevisiae* with a range of desirable properties, such as increased malic acid degradation, broad-spectrum killer activity, improved flocculation and foaming characteristics, decreased production of unwanted metabolites such as H_2S and acetic acid, secretion of enologically advantageous enzymes such as glycosidases, pectinases and proteases, improved utilization of amino acids, increased lactic acid production for low-acid wines, increased levels of desirable flavor components (e.g., glycerol), and production of bacteriocins active against spoilage bacteria (37, 74, 151, 152, 178). In addition to giving a pleasing product and an efficient fermentation, any new strain must be acceptable to legislative authorities and consumers on safety and ethical criteria.

LACTIC ACID BACTERIA

Lactic acid bacteria are significant in winemaking because they cause spoilage and are responsible for malolactic fermentation. In addition, they will release autolytic products into the wine. Wine lactic acid bacteria have the unique ability to tolerate the stresses of the wine environment, namely, low pH, presence of ethanol and sulfur dioxide, low temperature, and dilute concentrations of nutrients. Their occurrence, growth, and significance in wines have been reviewed (29, 71, 72, 84, 96, 97, 179). The main species of concern occur in the genera *Oenococcus, Leuconostoc, Pediococcus,* and *Lactobacillus* and include *O. oeni, Leuconostoc mesenteroides, Pediococcus parvulus, Pediococcus pentosaceus, Pediococcus damnosus* (formerly *Pediococcus cerevisiae*), and various species of *Lactobacillus* (e.g., *L. brevis, L. plantarum, L. fermentum, L. buchneri, L. hilgardii,* and *L. trichodes*). *O. oeni* (formerly known as *L. oenos* [39]) is uniquely found in wine ecosystems and is intimately associated with the occurrence of malolactic fermentation in wines produced in many countries. Consequently, the taxonomy, biochemistry, and physiology of this species have been specifically studied (97, 174). Molecular methods for the species and strain differentiation of wine lactic acid bacteria have been developed (3, 161, 184).

Ecology

The lactic acid bacteria of wines originate from the grapes and winery equipment, but inoculation of selected

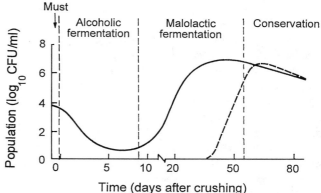

Figure 37.4 Growth of lactic acid bacteria during vinification of red wines, pH 3.0 to 3.5. The solid line shows the growth of *O. oeni*, often the only species present. Occasionally, species of *Lactobacillus* and *Pediococcus* develop toward the end of malolactic fermentation or at later stages during conservation (broken line). For wines of pH 3.5 to 4.0, a similar growth curve is obtained, but there may be slight growth and death of lactic acid bacteria during the early stages of alcoholic fermentation. Also, there is a greater chance that species of *Lactobacillus* and *Pediococcus* will grow and conduct malolactic fermentation.

species to conduct the malolactic fermentation is widely practiced. Freshly extracted grape juice contains lactic acid bacteria at populations of 10^3 to 10^4 CFU/ml, but they undergo little or no growth during the alcoholic fermentation and tend to die off because of competition from yeasts (Fig. 37.4). Nevertheless, these bacteria are capable of abundant growth in the juice and, if yeast growth is delayed, they could grow and spoil the juice or cause stuck alcoholic fermentation (9, 46, 97). About 1 to 3 weeks after completion of the alcoholic fermentation, the surviving lactic acid bacteria commence vigorous growth to conduct the malolactic fermentation. Final populations of 10^6 to 10^8 CFU/ml are produced. The onset, duration, and ecology of this growth are determined by many factors, including the properties of the wine, vinification variables, and influences of other microorganisms. Consequently, the natural occurrence of malolactic fermentation and its completion by the preferred species, *O. oeni*, can be unpredictable.

Of the wine properties, pH, concentration of ethanol, and concentration of sulfur dioxide have strong influences on the growth of lactic acid bacteria. Different species, and even strains within species, show different responses to these properties (30, 57, 179). Wines of low pH (e.g., pH 3.0), high ethanol content (>12% [vol/vol]), and high total sulfur dioxide (>50 μg/ml) are less likely to support the growth of lactic acid bacteria and may not undergo successful malolactic fermentation. Strains of *O. oeni* are more tolerant of low pH than those of *Leuconostoc*, *Pediococcus*, and *Lactobacillus* species and generally predominate in wines of pH 3.0 to 3.5. Wines with pH values exceeding 3.5 tend to have a mixed microflora, consisting of *O. oeni* and various species of *Pediococcus* and *Lactobacillus*. Species of *Pediococcus* and *Lactobacillus* are more tolerant of higher concentrations of sulfur dioxide than *O. oeni* and are more likely to occur in wines with higher amounts of this substance (30, 31). Thus, winemaker management of pH and sulfur dioxide content is important if it is desired to have malolactic fermentation conducted solely by *O. oeni*.

Microbiological factors that affect the growth of *O. oeni* in wine and successful completion of malolactic fermentation include excessive growth of molds and acetic acid bacteria on grapes, yeast species, and strains responsible for the alcoholic fermentation, and bacteriophages. Substances produced by the growth of fungi or acetic acid bacteria on damaged grapes could either stimulate or inhibit malolactic fermentation (82, 97, 179). During alcoholic fermentation and subsequent autolysis, yeasts release nutrients that are believed to encourage the growth of lactic acid bacteria. However, the growth of some strains of *S. cerevisiae* during alcoholic fermentation can be inhibitory to the subsequent growth of *O. oeni* (72, 180). Production of high concentrations of sulfur dioxide, inhibitory proteins, or toxic fatty acids (hexanoic, octanoic, decanoic, and dodecanoic) by these strains can be inhibitory to the bacteria (16, 73, 97). It is not known how the growth of non-*Saccharomyces* species impacts on development of the malolactic fermentation. Bacteriophages active against *O. oeni* have been isolated from wines and can interrupt and delay the malolactic fermentation (32, 97). Lysogeny of *O. oeni* is common, and bacteriophage-resistant strains have been described (30, 122).

The fate of lactic acid bacteria after malolactic fermentation depends on the wine and winemaking practices, but they may survive in wine for long periods. Because wine pH is increased by malolactic fermentation, it becomes a more favorable environment for bacterial growth, and spoilage microorganisms such as *Pediococcus* and *Lactobacillus* species could develop, especially in wines of pH 3.5 to 4.0. As noted already, wines after malolactic fermentation may have better microbiological stability. Nevertheless, there are reports that *O. oeni* and various species of *Lactobacillus* and *Pediococcus* can reestablish growth in such wines (47, 180). The significance of bacteriocin production by the growth of lactic acid bacteria in wines requires more study. The death of *O. oeni* in wines subsequent to the growth of *Pediococcus* and *Lactobacillus* species can occur (31, 141). Wine

isolates of *P. pentosaceus* and *L. plantarum* that have bacteriocin activity against wine strains of *Oenococcus*, *Lactobacillus*, and *Pediococcus* have been reported (114, 167). Edwards et al. (47) noted the production of substances by *O. oeni* that are inhibitory to the growth of *P. parvulus*. Strains of *O. oeni* that produce bacteriocins with broad-spectrum action against pediococci and lactobacilli would have obvious value in controlling spoilage by these bacteria.

Biochemistry

Lactic acid bacteria change the chemical composition of wine by utilizing some constituents for growth and generating metabolic end products. Depending upon the species and strains that develop, such changes will have a positive or negative impact on wine quality. Detailed studies that correlate the growth of lactic acid bacteria with chemical changes in the composition of wine are few (31). Consequently, the biochemical mechanisms by which lactic acid bacteria grow in wines and impact on flavor are still poorly understood (71, 97).

Carbohydrates

Wines contain residual amounts of glucose and fructose (0.5 to 1 mg/ml) and smaller amounts of other hexose and pentose sugars (<0.5 mg/ml). Concentrations of many of these sugars decrease with the growth of lactic acid bacteria, but consistent trends have not emerged (31). This finding agrees with the ability of these bacteria to ferment a wide range of hexose and pentose sugars and the substantial variation between species and strains in conducting these reactions (30, 93, 121, 142, 174). The pathways utilized by lactic acid bacteria for sugar metabolism are well known (13, 22). *Pediococcus* species as well as some species of *Lactobacillus* ferment hexoses by the Embden-Meyerhof-Parnas pathway (homofermenters), while *O. oeni* and some species of *Lactobacillus* use the hexose monophosphate or phosphoketolase pathway (heterofermenters). This latter pathway is used by all species to metabolize pentose sugars. It is not known how the wine conditions of low pH and high ethanol concentration affect the kinetics of transport of different hexose and pentose sugars into the cells and their subsequent metabolism. We have observed that the sugar fermentation profiles of *O. oeni*, *P. parvulus*, *L. plantarum*, and *L. brevis* are substantially different at pH 6.0 and 4.0 (30).

Some strains of *O. oeni* as well as other lactic acid bacteria can produce extracellular viscous polysaccharide materials that have the potential to affect the filterability of wines (97, 174, 179). It is not known if species of wine lactic acid bacteria produce enzymes that break down grape juice pectins. The occurrence of gly-cosidase activity in these bacteria requires further study since these enzymes have the potential to release aroma and flavor volatiles from glycosylated terpenes and phe-nols of grapes (103).

Nitrogen Compounds

The concentrations of some wine amino acids (e.g., argi-nine and histidine) decrease with the growth of lactic acid bacteria and are consistent with their utilization as a nitrogen source (31, 71). Extracellular protease pro-duction by O. oeni has been reported; this could be of nutritional advantage to the organism as well as min-imizing the risk of protein haze and sediments in wine (143). Concentrations of some amino acids are increased after malolactic fermentation, but it is not clear if this is due to enzymatic hydrolysis of wine proteins and pep-tides or to bacterial autolysis (174). Decarboxylation of amino acids to produce amines of potential public health significance has been reported for wine lactic acid bac-teria, especially for species of Pediococcus, but it is now recognized that many strains of O. oeni also have this property (84, 87, 97, 179). Some wine lactic acid bac-teria, including strains of O. oeni, metabolize arginine to citrulline, which is a precursor of ethyl carbamate, a potential carcinogen (94).

Organic Acids

The decarboxylation of L-malic acid to L-lactic acid is one of the most significant metabolic reactions conducted by lactic acid bacteria in wines. These bacteria possess mechanisms for the transport of malic acid into the cell, efflux of lactic acid, and the malolactic enzyme for decar-boxylation. Purification, properties, and gene sequence of the malolactic enzyme, as well as the bioenergetics of the reaction, have been reported for several species (71, 84, 96, 97). However, factors which regulate the expression of malolactic activity are not completely un-derstood. Glucose-induced inhibition has been reported in some strains of O. oeni (105). The metabolism of tartaric acid is not usually encountered during or after malolactic fermentation, but when it occurs, the wine is spoiled (166). Citric acid can be completely or par-tially metabolized during malolactic fermentation, de-pending on the wine pH and species of lactic acid bac-teria. Its degradation is frequently correlated with small increases in the concentration of acetic acid and diacetyl; this observation is consistent with the action of citrate lyase, which is produced by some but not all species of wine lactic acid bacteria (31, 97, 98). The concentra-tions of fumaric, gluconic, and pyruvic acids can decrease during malolactic fermentation. Gluconic acid, which is

significantly increased in wines made from grapes in-fected with B. cinerea, is metabolized by most lactic acid bacteria except the pediococci (179). Sorbic acid, which may be added to wines in some countries to control yeast growth, can be metabolized by O. oeni to form 3,5-hexadien-2-ol and 2-ethoxyhexa-3-diene, which cause geraniumlike off flavors (57, 179).

Flavor Compounds

Much has been written about the impact of malolactic fermentation on wine flavor (29, 71, 72, 97). Malolactic fermentation not only affects the taste of wine through deacidification, but it also contributes other flavor char-acteristics (often described as buttery, nutty, fruity, or vegetative) that may enhance or detract from overall ac-ceptability (38). Such changes are related to the wine con-stituents metabolized by the malolactic bacteria and the nature, concentration, and flavor threshold of products generated. Autolysis of malolactic bacteria could also af-fect flavor. Attempts to connect sensory impression to fla-vor substances produced by particular species or strains of malolactic bacteria have proved elusive because of confounding influences of grape variety, yeasts, and vini-fication variables. However, diacetyl production during malolactic fermentation has been linked to the evolution of buttery aromas that could be considered desirable or undesirable depending on their concentration and the wine style (90, 97, 98). Metabolism of citrate gives rise to diacetyl, but it is also a by-product of sugar metabolism. Other flavor constituents that increase in concentration during malolactic fermentation include acetic acid, di-ethyl succinate, and ethyl lactate, as well as other volatile acids, esters, and higher alcohols (29).

Spoilage Reactions

Uncontrolled growth of lactic acid bacteria during or after malolactic fermentation results in wine spoilage, the type of spoilage depending upon the particular wine and species of lactic acid bacteria which grow (13, 40, 57, 166, 179). Wines containing high concentrations of residual glucose and fructose are more likely to support bacterial growth, with the production of unacceptable amounts of acetic acid and D-lactic acid. Mannitol taint is caused by some strains of heterofermentative lacto-bacilli (e.g., L. brevis) and is due to enzymatic reduction of fructose to mannitol. Excessive production of diacetyl (>1 to 5 μg/ml) can give overpowering buttery flavors. L. hilgardii, L. brevis, and Lactobacillus cellobiosus have been implicated in the formation of mousy taints owing to their production of acetyl-tetrahydropyridines (26). Degradation of glycerol, especially by species of Pedio-coccus and Lactobacillus, can lead to the production

Table 37.6 Desirable properties of malolactic bacteria

Strong malolactic activity under wine conditions
Strong ability to grow in wines, including those of low pH (3.0 to 3.2) and high ethanol (14%) and those containing sulfur dioxide (50 μg/ml total), and at low temperatures (15 to 20°C)
Resistance to destruction by bacteriophages; nonlysogenic
Resistance to fungicide and pesticide residues
Production of desirable malolactic flavors and no off flavors
Production of bacteriocins effective against spoilage bacteria
Nonproduction of biogenic amines and precursors of ethyl carbamate
Nonproduction of yeast inhibitory factors if used before alcoholic fermentation
Suitability for mass culture, freeze-drying, distribution, and rehydration

of acrolein and associated development of bitterness. Spoilage arising from the degradation of tartaric and sorbic acids has been mentioned already. However, the ability to metabolize glycerol and tartaric acid is not widespread among the lactic acid bacteria. The production of extracellular polysaccharides by some species (e.g., *P. damnosus*) leads to ropiness (97).

Control of Malolactic Fermentation

Control of malolactic fermentation has emerged as one of the challenges of modern winemaking (29, 72). Because the natural occurrence of malolactic fermentation can be unpredictable, commercial cultures of malolactic bacteria have become available for inoculation into the wine to induce this reaction. Generally, these are strains of *O. oeni* that have been selected for a range of desirable criteria (Table 37.6) (35, 71, 83). Many factors affect the successful induction of malolactic fermentation by inocula, the most important of which are selection of the appropriate strain, proper reactivation and preculture of the freeze-dried concentrate, level of inoculum, and timing of inoculation (73, 84, 115). Generally, wines are inoculated to give 10^6 to 10^7 CFU/ml of bacteria just after alcoholic fermentation is completed. Arguments have been advanced for inoculating malolactic bacteria into the juice either before or simultaneously with the yeast culture. Under these conditions, bacterial cells are not exposed to the stresses of high ethanol concentration, giving a higher probability of successful growth and malic acid degradation. Risks associated with early inoculation of malolactic bacteria include their metabolism of grape juice sugars to yield unacceptable concentrations of acetic acid and interference with yeast growth and alcoholic fermentation.

Even under "optimum" conditions, inoculation with malolactic bacteria does not ensure successful comple-

tion of malolactic fermentation. In some instances, the particular properties of the wine may not be suitable for growth of the bacterial strain or mixture of strains inoculated. Several biotechnological innovations have been considered to overcome this problem. Bioreactor systems consisting of high concentrations of malolactic bacteria immobilized in beads of alginate or carrageenan have been developed (23, 29, 41, 55). An alternative system uses high densities (10^9 to 10^{10} CFU/ml) of *O. oeni* retained within a cell recycle membrane bioreactor (59, 60). Under these conditions, cells of malolactic bacteria act as biocatalysts and, in the absence of growth, rapidly convert malic acid to lactic acid in wine that is passed through the reactor on a continuous basis. Such technologies give rapid, continuous deacidification of wines but have not proved commercially successful, mainly because of instability of the malolactic activity of cells in the reactor.

The yeast *Schizosaccharomyces pombe* rapidly metabolizes malic acid under anaerobic conditions. Malic acid is decarboxylated to pyruvate by the malic enzyme, and then pyruvate is decarboxylated to acetaldehyde, which is reduced to ethanol (130). Many attempts have been made to exploit this property to deacidify wines in place of the bacterial malolactic fermentation. In one approach, these yeasts are cultured in grape juice either with or before inoculation with *S. cerevisiae*. In another strategy, juices or wines are passed through reactors containing immobilized cells of *S. pombe*. While successful malic acid degradation has been achieved with this yeast, it may produce undesirable off flavors (41, 55, 61, 182). *Z. bailii* is another yeast that gives good degradation of malic acid and may have potential for wine deacidification (130, 146).

The availability of strains of *S. cerevisiae* that could carry out malolactic fermentation simultaneously with alcoholic fermentation would be most attractive to winemakers. Using recombinant DNA technology, the malolactic gene from lactic acid bacteria has been incorporated into wine strains of *S. cerevisiae*. However, the constructs showed poor uptake of malic acid and instability of malolactic expression, revealing the need for increased understanding of the molecular biology of the malolactic reaction and its regulation, including transport of malic acid into the cell (2, 84, 96). A construct of *S. cerevisiae* expressing the malolactic gene of *Lactococcus lactis* and the malate transport gene of *S. pombe* was subsequently developed and reported to give good malolactic activity in wine, but is not yet commercialized (178).

Complete prevention of malolactic fermentation is an option preferred by some winemakers. Wines of low pH (<3.2), high ethanol (>14%), and high sulfur dioxide (>50 μg/ml) are less prone to malolactic fermentation. The commercially available bacteriocin nisin, at 100 units/ml, inhibits the growth of malolactic and spoilage lactic acid bacteria in wines (28, 51). The addition of lysozyme to wine is another possibility for controlling the growth of these bacteria (63).

ACETIC ACID BACTERIA

Acetic acid bacteria cause the vinegary spoilage of wines through the oxidation of ethanol to acetaldehyde and acetic acid. In addition, their growth on grapes before fermentation can produce substances that not only affect wine flavor but can interfere with the growth of yeasts, leading to stuck fermentations (43, 57, 88, 166). The main species of concern are *Gluconobacter oxydans*, *Acetobacter pasteurianus*, and *Acetobacter aceti* in which strains have evolved that can tolerate the wine environment. However, the taxonomy of these bacteria continues to evolve (162), and additional species are likely to be found in wine ecosystems.

Ecology

Sound, unspoiled grapes harbor low populations of acetic acid bacteria, generally less than 10^2 CFU/g, with *G. oxydans* being the predominant species. Damaged, spoiled grapes and those infected with the mold *B. cinerea* have higher populations (>10^6 CFU/g) that are characterized by a dominance of *A. pasteurianus* and *A. aceti*. The populations of acetic acid bacteria in freshly extracted grape juice reflect those present on the grapes.

In the absence of yeasts, acetic acid bacteria quickly grow in grape juice, reaching populations of 10^6 to 10^8 CFU/ml. The extent of their growth during alcoholic fermentation depends on their initial populations, relative to yeasts, in the juice. In juice prepared from sound grapes on which the yeast population is 10 to 100 times greater than that of acetic acid bacteria, there appears to be little growth or influence of these bacteria on the alcoholic fermentation (44, 82). However, when initial populations of these bacteria exceed about 10^4 CFU/ml, they may grow in conjunction with yeasts during the early stages of fermentation. Populations as high as 10^8 CFU/ml can develop but die off owing to the combined influences of ethanol and anaerobosis caused by yeast growth. Nevertheless, acetic acid and other substances produced by acetic acid bacteria become inhibitory to the yeast, causing premature cessation of the alcoholic fermentation (43, 44).

At the end of alcoholic fermentation, the population of acetic acid bacteria is generally less than 10^2 CFU/ml. *G. oxydans* is rarely isolated from wines at this stage,

and *A. pasteurianus* and *A. aceti* are the main species found. Subsequent transfer of the wine from fermentation tanks to other storage vessels may produce sufficient agitation and aeration to encourage the growth of surviving acetic acid bacteria to populations of 10^5 CFU/ml or higher. It is not uncommon to isolate *A. pasteurianus* or *A. aceti* from wines during bulk storage in barrels or tanks kept under anaerobic conditions (43, 53, 88, 181). The mechanisms by which these aerobic bacteria survive for long periods under apparently anaerobic or semi-anaerobic conditions require explanation. Their growth is activated by exposure of the wine to air, and the wine is quickly spoiled.

Biochemistry

Oxidation of ethanol to acetic acid is a key reaction of acetic acid bacteria. Ethanol is first oxidized by ethanol dehydrogenase to acetaldehyde, which is then oxidized by acetaldehyde dehydrogenase to acetic acid. There is substantial variation between strains of *G. oxydans*, *A. aceti*, and *A. pasteurianus* in the strengths of these reactions. Species of *Acetobacter* further oxidize acetic acid to carbon dioxide and water by the tricarboxylic acid (TCA) cycle, but this reaction is inhibited in the presence of ethanol. Species of *Gluconobacter* do not have a fully functional TCA cycle and are unable to completely oxidize acetic acid. Sulfur dioxide, which is generally present in wine, can chemically trap acetaldehyde, causing an accumulation of this intermediate at the expense of its further oxidation to acetic acid. Ethanol concentrations above 10% become increasingly inhibitory to the growth of acetic acid bacteria and their ability to oxidize ethanol. Aldehyde dehydrogenase is less stable than ethanol dehydrogenase at high concentrations of ethanol, and such conditions give an increased accumulation of acetaldehyde. Lower oxygen concentrations also favor the accumulation of acetaldehyde. In addition to acetic acid and acetaldehyde, ethyl acetate is another end product that is significant in the vinegary spoilage of wines (43, 44, 57, 88).

Acetobacter and *Gluconobacter* lack a functional Embden-Meyerhof-Parnas pathway and metabolize hexose and pentose sugars by the hexose monophosphate pathway to acetic and lactic acids. However, *Acetobacter* species give only weak metabolism of sugars, which appears to be associated with their decreased ability to phosphorylate these substrates. At pH values below 3.5 or glucose concentrations above 5 to 15 mM, as occur in grapes or grape juice, metabolism of glucose by the hexose monophosphate pathway is inhibited, and glucose is directly oxidized to gluconic and ketogluconic acids. Consequently, grapes heavily infected with *G. oxydans* give juices with high concentrations (50 to 70 mg/ml) of gluconic acid. Strains of *A. aceti* and *A. pasteurianus* also produce gluconic acid in grape juice but to a lesser extent than *G. oxydans* (44).

Strains of *G. oxydans* and *A. aceti* oxidize glycerol to dihydroxyacetone. Glycerol is not normally present in grape juice, but its concentration may be significant (up to 20 mg/ml) in juices prepared from grapes infected with *B. cinerea* or other molds. Metabolism of glycerol by acetic acid bacteria, either on grapes, during the early stages of alcoholic fermentation, or in wine, leads to significant production of dihydroxyacetone that could affect wine quality (43, 166).

Species of *Acetobacter* use the TCA cycle for metabolism of organic acids, causing decreases in concentrations of citric, succinic malic, tartaric, and lactic acids in wines. Moreover, lactate may be oxidized to acetoin. The metabolism of amino acids and protein by wine acetic acid bacteria has not been examined. Another property relevant in wine production is their ability to produce extracellular fibrils of cellulose and other polysaccharides (43).

OTHER BACTERIA

There are scattered reports that *Bacillus* and other bacterial species can survive and grow in wines and contribute to spoilage (13, 57). Species of *Bacillus* implicated include *B. coagulans*, *B. circulans*, and *B. subtilis* isolated from spoiled sweet wines; *B. megaterium* isolated from spoiled brandy; and *B. coagulans* and *B. badius*, which were isolated from wine corks and had limited ability to grow in wines (53, 91, 166). Juices and wines of high pH (e.g., 4.0) have, on rare occasions, been spoiled by the growth of *Clostridium butyricum* and have elevated concentrations of butyric, isobutyric, propionic, and acetic acids (166). Species of *Actinomyces* and *Streptomyces* have been isolated from wines, corks, and wooden barrels and could contribute to the musty, earthy, or corky taints sometimes found in wines (53, 91).

MOLDS

Molds (filamentous fungi) significantly impact on wine production and quality by affecting grapes in the vineyard and by causing off flavors and taints in wines by contamination of corks and wooden barrels (42, 53). Moldy grapes may contain mycotoxins that could carry over into the wine. Low levels of ochratoxin A have been reported in some wines (185). Although molds are usual contaminants of grape juice, the conditions of anaerobiosis, increasing ethanol concentration, and presence of sulfur dioxide are strong deterrents to their growth during wine fermentation and conservation.

Grapes and vines harbor a range of mold species, the population and diversity of which depend on grape variety, degree of berry maturity and physical damage, climatic conditions, and viticultural practices. Infections of the vine by species such as *Plasmopara viticola* (downy mildew), *Uncinula necator* (powdery mildew), *Phomopsis viticola* (cane and leaf spot), and *Eutypa lata* (dieback) have major effects on the quality and yield of grapes and, eventually, can kill the vine (156). Rotting and spoilage of grape berries before harvest can be caused by a variety of fungal species, the principal one of which is *B. cinerea*, which causes bunch rot (113). Other species include those of *Penicillium, Aspergillus, Rhizopus, Mucor, Alternaria*, and *Cladosporum*. Grapes that are heavily infected with molds have altered chemical composition and mold enzymes that adversely affect wine flavor and color and the growth of yeasts during alcoholic fermentation (4, 42). Fungal contamination of vines and grapes is controlled by the application of fungicides (113). However, their use must be carefully managed to minimize residues in the juice and their potential inhibitory effects on the alcoholic and malolactic fermentations (14, 106). The biocontrol of grape fungi with appropriate yeasts could be a future direction for minimizing or eliminating the use of chemical fungicides (169).

Although *B. cinerea* is well known for causing bunch rot, its controlled development on grapes is the basis for producing the distinctive and highly prized, botrytized sweet wines—sauternes, Trockenbeerenauslese, and Tokay (42, 139). Under certain climatic and viticultural conditions, this species can parasitize healthy grape berries without significantly disrupting the general integrity of their skin, causing the so-called *pourriture noble*. Mold growth on and in the berry leads to its dehydration, with consequent concentration of its chemical constituents. This concentration effect, along with fungal metabolism of grape sugars and acids (especially tartaric) gives a juice that, typically, has increased sugar concentration (300 to 400 mg/ml), high concentration of glycerol (3 mg/ml) and other polyols, high concentration of gluconic acid, and less tartaric acid. The fermentation of such juices requires particular attention as they are prone to become stuck, possibly owing to nitrogen deficiency combined with the higher sugar content; also, there is evidence that the mold secretes antiyeast substances. *B. cinerea* also produces various phenolic oxidases and glycosidases that can affect wine color and flavor, and it produces extracellular soluble glucans that block membranes during filtration processes (4).

Mold contamination of wine corks and wooden barrels can cause earthy, moldy, or corky taints in wines and rejection of the product (21, 91). Molds use the cork or wood as growth substrates, generating potent aroma compounds, such as trichloroanisoles, 1-octen-3-one, geosmin, and guaiacol, that are subsequently leached into the wine to cause the taint. Molds isolated from these sources include species of *Penicillium, Aspergillus, Trichoderma, Cladosporium, Paecilomyces*, and *Monilia*.

SPARKLING WINES OR CHAMPAGNE

Sparkling wines contain dissolved carbon dioxide, which is produced by a secondary fermentation of a base wine in a closed system, such as a bottle or a tank (13, 81, 132). The microbial ecology and biochemistry of base wine production is essentially the same as for table wines. The carbon dioxide emerges as bubbles (sparkling) when the pressure is released at the time of consumption. Three types of secondary fermentation processes are used. The traditional *méthode champenoise* requires fermentation in the bottle in which the wine is sold. A base wine (usually white wine) is selected to which sugar or sugar syrup is added to give a final concentration of about 2.4 mg/ml. Yeast nutrients (diammonium phosphate and vitamins) may also be added. A suitable strain of *S. cerevisiae* is inoculated into the wine, which is thoroughly mixed and transferred to the bottles (*levurage* and *tirage*). The bottles are sealed with crown caps or corks and then stored at 12 to 15°C for fermentation. The bottles are specially made to withstand the high pressure (about 600 kPa) of carbon dioxide produced during fermentation. Fermentation is completed in 3 to 6 months, after which the wine is aged in the bottles in contact with the yeast lees for at least 6 to 12 months, or longer for premium-quality wines. After storage, the bottles are restacked with their neck downward and periodically twisted or shaken to facilitate settling of the yeast sediment into the neck. This process is called riddling, or *remuage*. Finally, the sedimented yeast plug is removed from the bottle neck by a process termed *dégorgement*. The yeast plug is frozen by dipping the neck of the bottle into calcium chloride solution at −24°C, and then the cork or cap is removed, allowing the internal pressure to force out the yeast plug with little loss of wine. The lost wine is replaced by the addition of base wine (*dosage*), which may contain a small amount of sugar to give a desired amount of sweetness, and sulfur dioxide. *Remuage, dégorgement*, and *dosage* are automated processes in the modern winery. Following these operations, the bottles are corked, and an *agraffe* (wire clamp) is applied.

Variations in the *méthode champenoise* include the transfer method and bulk fermentation or Charmat process. In the transfer process, fermentation is conducted in the bottle as described, followed by a short maturation on lees. The bottle contents are chilled and emptied into a pressurized tank. *Dosage* is added to the wine, which is filtered to remove the yeasts, and bottled. In the Charmat process, the base wine, with added sugar and yeast, is fermented in a pressurized tank. After fermentation, the wine is chilled and clarified by centrifugation before transfer to a second tank (pressurized) containing the *dosage*. The wine is finally filtered and bottled.

Special strains of *S. cerevisiae* are required for the secondary fermentation. Criteria for these strains include giving complete sugar fermentation under conditions of low temperature (10 to 15°C), relatively high ethanol concentration (8 to 12%), low pH (as low as 3.0), presence of up to 20 μg of free sulfur dioxide per ml, low nutrient availability, and increasing pressure of carbon dioxide (up to 600 kPa); flocculate and sediment to facilitate the riddling process; giving good flavor; and undergoing autolysis during aging on the lees. Usually, the yeast is inoculated into the wine base at a population of 1×10^6 to 5×10^6 CFU/ml. The secondary fermentation is completed over the next 30 to 40 days, during which time the yeast grows to a maximum population of about 10^7 CFU/ml. Subsequently, yeast cells slowly die and, generally, viable cells cannot be detected after 100 days. Autolysis of the yeast then commences (7, 20, 171).

Chemical changes occur in two phases, namely, during the secondary fermentation and during the period of aging and yeast autolysis. During secondary fermentation, the added sugar is utilized with the production of carbon dioxide and about 1% ethanol. In the early stages, amino nitrogen is consumed, but after exhaustion of sugar, yeast cells give a small efflux of amino nitrogen into the wine. Minor changes in the concentrations of glycerol, some organic acids, and some esters have been noted, but the data are not consistent (7). The yeast cells are operating under extreme environmental conditions, especially increasing concentration of carbon dioxide and physical pressure, and these circumstances are likely to affect their metabolic behavior.

Yeast autolysis can have a distinctive (mostly positive) impact upon sparkling wine quality in terms of aroma, flavor, bubble size, and bubble persistence. As mentioned earlier, the autolysis of yeast cells is characterized by the degradation of cellular macromolecules and release of their degradation products, as well as other cell constituents, into the external medium (20, 77, 126). Consistent with these reactions, the aging process has been correlated with gradual increases in concentrations of proteins, amino acids, nucleic acids, lipids, free fatty acids, and mannoproteins (originating from the yeast cell wall) in the wine. In addition, changes occur in the concentrations of various esters, higher alcohols, carbonyl compounds, terpenes, and lactones, which may or may not be related to yeast autolysis. By mechanisms not fully understood, these changes influence sparkling wine aroma, flavor, and bubble properties (20, 92, 124, 172).

An innovation in sparkling wine production is the use of yeasts immobilized in beads of alginate, principally for the purposes of accelerating removal of yeasts by the riddling process (23, 183). The entrapped yeast cells perform satisfactorily during the secondary fermentation and subsequent aging, giving products with sensory quality comparable to that produced by the traditional process. Strategies to accelerate yeast autolysis and to decrease the time and cost of aging would be worthy of development and could exploit the use of killer yeasts (171).

FORTIFIED WINES

Fortified wines, such as sherry, port, and Madeira, have ethanol concentrations of 15 to 22%. This higher ethanol concentration is achieved by the addition of ethanol (usually derived from the distillation of wine products) at certain stages during the process. Details of fortified wine production are given elsewhere (13, 65, 134).

With sherry-style wines, an essentially dry white wine base is produced from particular grape varieties. The yeast ecology and biochemistry of the fermentation are similar to those described already. At the completion of alcoholic fermentation, the wine is fortified and transferred to oak casks for aging and maturation. In the case of oloroso sherries, the wine is fortified to 18 to 19% ethanol; this stops any further microbiological processes. Consequently, subsequent maturation is a chemical process. With the fino sherries, the wine is fortified to 15 to 16% ethanol and aged in a series of oak casks by the solera system, whereby portions of the wine in each cask are systematically removed and replaced with an amount of younger wine of the same style so that the wine is continuously blended and emerges with a consistent character. The frequency of transfers and amount of wine removed from each cask vary according to the producer. This process encourages the natural formation of a film or velum of yeast growth (flor) at the air-wine interface. Essentially, the velum is a wrinkled layer of yeast biomass about 5 to 10 mm thick. It represents a unique ecological niche where yeasts (and probably bacteria as well) have adapted to survive and metabolize in the presence of high concentrations of ethanol. Based on the analysis of mitochondrial DNA and chromosomal DNA profiles, the velum mainly consists of four hydrophobic strains of *S. cerevisiae* (79, 99), but strains of *T. delbrueckii* (*C. colliculosa*) and *Z. rouxii* may also be present. There appears to be a definitive evolution of species and strains during the process of velum development, and the ecology may be different at the top (aerobic) and bottom (anaerobic) locations. The oxidative metabolism of the velum decreases the concentrations of wine acids, glycerol, and alcohol and substantially increases the concentration of acetaldehyde (99). Amino acids (including proline) are assimilated with the consequent production of higher alcohols. These changes, as well as many other less quantitative reactions conducted by the flor yeasts contribute to the unique flavor of fino sherries.

Port-style wines are prepared from a red wine base, while the Madeira-style wines are produced from a blend of red and white wines. In these processes, fortification with ethanol is done at an appropriate stage during the alcoholic fermentation so that the fermentation is arrested to give a wine with residual unfermented sugar and a desired amount of sweetness. These wines are then subjected to particular processes for aging and maturation that do not involve any secondary fermentations because of the high concentration of ethanol (134).

Despite their high ethanol concentrations, fortified wines may undergo spoilage from ethanol-tolerant strains of yeasts (*Zygosaccharomyces bisporus*, *S. cerevisiae*, and *Rhodotorula*, *Candida*, and *Brettanomyces* species) and lactic acid bacteria (*L. hilgardii* and other species) (166, 170).

References

1. **Amerine, M. A.** 1985. Winemaking, p. 67–81. *In* H. Koprowski and S. A. Plotin (ed.), *World's Debt to Pasteur*. Alan R. Liss, Inc., New York, N.Y.

2. **Barre, P., F. Vezinhet, S. Dequin, and B. Blondin.** 1993. Genetic improvement of wine yeasts, p. 265–287. *In* G. H. Fleet (ed.), *Wine Microbiology and Biotechnology*. Harwood Academic Publishers, Chur, Switzerland.

3. **Bartowsky, E. J., and P. A. Henschke.** 1999. Use of polymerase chain reaction for specific detection of the malolactic fermentation bacterium *Oenococcus oeni* (formerly *Leuconostoc oenos*) in grape juice and wine samples. *Aust. J. Wine Res.* **5:**39–44.

4. **Bayonove, C.** 1989. Incidences des attaques parasitaires fongiques sur la composition qualitative du raisin et des vins. *Rev. Franc. Oenol.* **116:**29–39.

5. **Benda, I.** 1982. Wine and brandy, p. 293–402. *In* G. Reed (ed.), *Prescott and Dunn's Industrial Microbiology*, 4th ed. AVI Publishing Co., Westport, Conn.

6. **Berry, D. R., and D. C. Watson.** 1987. Production of organoleptic compounds, p. 345–366. *In* D. R. Berry, I. Russell, and G. Stewart (ed.), *Yeast Biotechnology*. Allen and Unwin, London, United Kingdom.

7. **Bidan, P., M. Feuillat, and J. Moulin.** 1986. Rapport de la France. Les vins Mousseux. *Bull. Off. Int. Vin* **59:**563–626.

8. **Bisson, L. F.** 1993. Yeasts—metabolism of sugars, p. 55–75. *In* G. H. Fleet (ed.), *Wine Microbiology and Biotechnology*. Harwood Academic Publishers, Chur, Switzerland.

9. **Bisson, L. F.** 1999. Stuck and sluggish fermentations. *Am. J. Enol. Viticult.* **50:**107–119.

10. **Bisson, L. F., and R. E. Kunkee.** 1991. Microbial interactions during wine production, p. 37–68. *In* J. G. Zeikus and E. A. Johnson (ed.), *Mixed Cultures in Biotechnology*. McGraw-Hill, Inc., New York, N.Y.

11. **Blanco, P., C. Sierro, A. Diaz, N. M. Reboredo, and T. G. Villa.** 1997. Grape juice biodegradation by polygalacturonases from *Saccharomyces cerevisiae*. *Int. Biodeterior. Biodegrad.* **40:**115–118.

12. **Boulton, R.** 1995. Red wines, p. 121–158. *In* A. G. H. Lea and J. R. Piggott (ed.), *Fermented Beverage Production*. Blackie Academic & Professional, Glasgow, Scotland.

13. **Boulton, R. B., V. L. Singleton, L. F. Bisson, and R. E. Kunkee.** 1995. *Principles and Practices of Winemaking*. Chapman & Hall, New York, N.Y.

14. **Cabras, P., A. Angrioni, V. L. Garau, F. M. Pirisi, G. A. Farris, G., Madau, and G. Emonti.** 1999. Pesticides in fermentative processes of wine. *J. Agric. Food Chem.* **47:**3854–3857.

15. **Cantarelli, C.** 1989. Factors affecting the behaviour of yeast in wine fermentation, p. 127–151. *In* C. Cantarelli and G. Lanzarini (ed.), *Biotechnology Applications in Beverage Production*. Elsevier Applied Science, London, United Kingdom.

16. **Capucho, I., and M. V. San Ramão.** 1994. Effect of ethanol and fatty acids on malolactic activity of *Leuconostoc oenos*. *Appl. Microbiol. Biotechnol.* **42:**391–395.

17. **Charoenchai, C., and G. H. Fleet.** Unpublished data.

18. **Charoenchai, C., G. H. Fleet, and P. Henschke.** 1998. Effects of temperature, pH and sugar concentration on the growth rates and cell biomass of wine yeast. *Am. J. Enol. Viticult.* **49:**283–288.

19. **Charoenchai, C., G. H. Fleet, P. Henschke, and B. E. N. Todd.** 1997. Screening of non-*Saccharomyces* wine yeasts for the presence of extracellular hydrolytic enzymes. *Aust. J. Grape Wine Res.* **3:**2–8.

20. **Charpentier, C., and M. Feuillat.** 1993. Yeast autolysis, p. 225–242. *In* G. H. Fleet (ed.), *Wine Microbiology and Biotechnology*. Harwood Publishers, Chur, Switzerland.

21. **Chatonnet, P., N. J. Boidron, and D. Dubourdieu.** 1994. The microflora of cork wood and their development during the seasoning and ageing in the open air. *J. Int. Sci. Vigne Vin* **28:**185–201.

22. **Cogan, T. M.** 1995. Flavor production by dairy starter cultures. *J. Appl. Bacteriol. Symp. Suppl.* **79:**49S–64S.

23. **Colagrande, O., A. Silva, and M. D. Fumi.** 1994. Recent applications of biotechnology in wine production. *Biotechnol. Prog.* **10:**2–18.

24. **Cole, V. C., and A. C. Noble.** 1995. Flavor chemistry and assessment, p. 361–385. *In* A. G. H. Lea and J. R. Piggott (ed.), *Fermented Beverage Production*. Blackie Academic & Professional, Glasgow, Scotland.

25. **Constanti, M., M. Poblet, L. Arola, A. Mas, and J. M. Guillamón.** 1997. Analysis of yeast populations during alcoholic fermentation in a newly established winery. *Am. J. Enol. Viticult.* **48:**339–344.

26. **Costello, P., T. H. Lee, and P. A. Henschke.** 2000. Mousy off-flavour spoilage of wine by lactic acid bacteria, p. 226–230. *In* A. Lonvaud-Funel (ed.), *Oenologie 99*. Sixth Symposium International D'Oenologie. Technique & Documentation, Paris, France.

27. **Cottrell, T. H. E., and M. R. McLellan.** 1986. The effect of fermentation temperature on chemical and sensory characteristics of wine from seven white grape cultivars grown in New York State. *Am. J. Enol. Viticult.* **37:**190–194.

28. **Daeschel, M. A., D. S. Jung, and B. T. Watson.** 1991. Controlling malolactic fermentation with nisin and nisin-resistant strains of *Leuconostoc oenos*. *Appl. Environ. Microbiol.* **57:**601–603.

29. **Davis, C., D. Wibowo, R. Eschenbruch, T. H. Lee, and G. H. Fleet.** 1985. Practical implications of malolactic fermentation—a review. *Am. J. Enol. Viticult.* **36:**209–301.

30. **Davis, C. R., D. Wibowo, G. H. Fleet, and T. H. Lee.** 1988. Properties of wine lactic acid bacteria: their potential enological significance. *Am. J. Enol. Viticult.* **39:**137–142.

31. **Davis, C. R., D. Wibowo, T. H. Lee, and G. H. Fleet.** 1986. Growth and metabolism of lactic acid bacteria during and after malolactic fermentation of wines at different pH. *Appl. Environ. Microbiol.* **51:**539–545.

32. **Davis, C. R., N. F. A. Silveira, and G. H. Fleet.** 1985. Occurrence and properties of bacteriophages of *Leuconostoc oenos* in Australian wines. *Appl. Environ. Microbiol.* **50:**872–876.

33. **de Barros Lopes, M., A. Soden, P. A. Henschke, and P. Langridge.** 1996. PCR identification of commercial yeast strains using intron splice site primers. *Appl. Environ. Microbiol.* **62:**4514–4520.

34. **de Barros Lopes, M., S. Rainieri, P. A. Henschke, and P. Langridge.** 1999. AFLP fingerprinting for analysis of yeast genetic variation. *Int. J. Syst. Bacteriol.* **49:**915–924.

35. **Degré, R.** 1993. Selection and cultivation of wine yeast and bacteria, p. 421–447. *In* G. H. Fleet (ed.), *Wine Microbiology and Biotechnology*. Harwood Academic Publishers, Chur, Switzerland.

36. **Delfini, G., and A. Costa.** 1993. Effects of grape must lees and insoluble materials on the alcoholic fermentation rate and the production of acetic acid, pyruvic acid and acetaldehyde. *Am. J. Enol. Viticult.* **44:**86–92.

37. **Dequin, S., E. Baptista, and P. Barre.** 1999. Acidification of grape musts by *Saccharomyces cerevisiae* wine yeast strains genetically engineered to produce lactic acid. *Am. J. Enol. Viticult.* **50:**45–50.

38. **de Revel, G., N. Martin, L. Pripis-Nicolau, A. Lonvaud-Funel, and A. Bertrand.** 1999. Contribution to the knowledge of malolactic fermentation influence on wine aroma. *J. Agric. Food Chem.* **47:**4003–4008.

39. **Dicks, L. M. T., F. Dellaglio, and M. D. Collins.** 1995. Proposal to reclassify *Leuconostoc oenos* as *Oenococcus*

oeni (corrig) gen. nov. *Int. J. Syst. Bacteriol.* **45**:395–397.

40. **Dittrich, H. H.** 1995. Wine and brandy, p. 464–503. *In* G. Reed and T. W. Nagodawithana (ed.), *Biotechnology*, 2nd ed., vol. 9. *Enzymes, Biomass, Food and Feed*. VCH, Weinheim, Germany.

41. **Divies, C.** 1993. Bioreactor technology and wine fermentation, p. 449–475. *In* G. H. Fleet (ed.), *Wine Microbiology and Biotechnology*. Harwood Academic Publishers, Chur, Switzerland.

42. **Doneche, B.** 1993. Botrytized wines, p. 327–351. *In* G. H. Fleet (ed.), *Wine Microbiology and Biotechnology*. Harwood Academic Publishers, Chur, Switzerland.

43. **Drysdale, G. S., and G. H. Fleet.** 1988. Acetic acid bacteria in winemaking—a review. *Am. J. Enol. Viticult.* **39**:143–154.

44. **Drysdale, G. S., and G. H. Fleet.** 1989. The effect of acetic acid bacteria upon the growth and metabolism of yeasts during the fermentation of grape juice. *J. Appl. Bacteriol.* **67**:471–481.

45. **Dubois, C., C. Manginot, J.-L. Roustan, J.-M. Sablayrolles, and P. Barre.** 1996. Effect of variety, year and grape maturity on the kinetics of alcoholic fermentation. *Am. J. Enol. Viticult.* **47**:363–368.

46. **Edwards, C. G., K. M. Haag, M. D. Collins, R. A. Huston, and Y. C. Huang.** 1998. *Lactobacillus kunkeei* sp. nov: a spoilage organism associated with grape juice fermentations. *J. Appl. Microbiol.* **84**:698–702.

47. **Edwards, C. G., J. C. Peterson, T. D. Boylston, and J. D. Vasile.** 1994. Interactions between *Leuconostoc oenos* and *Pediococcus* spp. during vinification of red wines. *Am. J. Enol. Viticult.* **45**:49–55.

48. **Eglinton, J. M., and P. A. Henschke.** 1999. Restarting incomplete fermentations: the effect of high concentrations of acetic acid. *Aust. J. Grape Wine Res.* **5**:71–78.

49. **Esteve-Zarzoso, B., C. Belloch, F. Uruburu, and A. Querol.** 1999. Identification of yeasts by RFLP analysis of the 5.8S rRNA gene and the two ribosomal internal transcribed spacers. *Int. J. Syst. Bacteriol.* **49**:329–337.

50. **Ewart, A.** 1995. White wines, p. 95–120. *In* A. G. H. Lea and J. R. Piggott (ed.), *Fermented Beverage Production*. Blackie Academic & Professional, Glasgow, Scotland.

51. **Faia, A. M., and F. Radler.** 1990. Investigation of the bactericidal effect of nisin on lactic acid bacteria of wine. *Vitis* **29**:233–238.

52. **Fleet, G. H.** 1992. Spoilage yeasts. *Crit. Rev. Biotechnol.* **12**:1–44.

53. **Fleet, G. H.** 1993. The microorganisms of winemaking—isolation, enumeration and identification, p. 1–26. *In* G. H. Fleet (ed.), *Wine Microbiology and Biotechnology*. Harwood Academic Publishers, Chur, Switzerland.

54. **Fleet, G. H.** 1999. Microorganisms in food ecosystems. *Int. J. Food Microbiol.* **50**:101–117.

55. **Fleet, G. H.** 1999. Alternative fermentation technology, p. 172–177. *In* R. J. Blair, A. N. Sas, P. F. Hayes, and P. J. Hoj (ed.), *Proceedings of the Tenth Australian Wine Industry Technical Conference*, Winetitles, Adelaide, Australia.

56. **Fleet, G. H., and G. M. Heard.** 1993. Yeasts—growth during fermentation, p. 27–54. *In* G. H. Fleet (ed.), *Wine Microbiology and Biotechnology*. Harwood Academic Publishers, Chur, Switzerland.

57. **Fugelsang, K. C.** 1997. *Wine Microbiology*. Chapman & Hall, New York, N.Y.

58. **Gao, C., and G. H. Fleet.** 1988. The effects of temperature and pH on the ethanol tolerance of the wine yeasts, *Saccharomyces cerevisiae, Candida stellata* and *Kloeckera apiculata*. *J. Appl. Bacteriol.* **65**:405–410.

59. **Gao, C., and G. H. Fleet.** 1994. Degradation of malic acid by high density cell suspensions of *Leuconostoc oenos*. *J. Appl. Bacteriol.* **76**:632–637.

60. **Gao, C., and G. H. Fleet.** 1995. Cell-recycle membrane bioreactor for conducting continuous malolactic fermentation. *Aust. J. Grape Wine Res.* **1**:32–38.

61. **Gao, C., and G. H. Fleet.** 1995. Degradation of malic and tartaric acids by high density cell suspensions of wine yeast. *Food Microbiol.* **12**:65–71.

62. **Gardner, N., N. Rodrigue, and C. P. Champagne.** 1993. Combined effects of sulfites, temperature and agitation time on production of glycerol in grape juice by *Saccharomyces cerevisiae*. *Appl. Environ. Microbiol.* **59**:2022–2028.

63. **Gerbaux, V., A. Villa, C. Monamy, and A. Bertrand.** 1997. Use of lysozyme to inhibit malolactic fermentation and to stabilise wine after malolactic fermentation. *Am. J. Enol. Viticult.* **48**:49–53.

64. **Giudici, P., and R. E. Kunkee.** 1994. The effect of nitrogen deficiency and sulfur-containing amino acids on the reduction of sulfate to hydrogen sulfide by wine yeasts. *Am. J. Enol. Viticult.* **45**:107–112.

65. **Goswell, R. W., and R. E. Kunkee.** 1977. Fortified wines, p. 478–533. *In* A. H. Rose (ed.), *Economic Microbiology*, vol. 1. Academic Press Ltd., London, United Kingdom.

66. **Gutiérrez, A. R., P. Santamaria, S. Epifania, P. Garijo, and R. Lopez.** 1999. Ecology of spontaneous fermentation in one winery during 5 consecutive years. *Lett. Appl. Microbiol.* **29**:411–415.

67. **Hardie, W. J., T. P. O. O'Brien, and V. G. Jaudzems.** 1996. Morphology, anatomy and development of the pericarp after anthesis in grape, *Vitis vinifera*. *Aust. J. Grape Wine Res.* **2**:97–142.

68. **Heard, G. M.** 1999. Novel yeasts in winemaking—looking to the future. *Food Aust.* **51**:347–352.

69. **Heard, G. M., and G. H. Fleet.** 1985. Growth of natural yeast flora during the fermentation of inoculated wines. *Appl. Environ. Microbiol.* **50**:727–728.

70. **Heard, G. M., and G. H. Fleet.** 1988. The effects of temperature and pH on the growth of yeast species during the fermentation of grape juice. *J. Appl. Bacteriol.* **65**:23–28.

71. **Henick-Kling, T.** 1993. Malolactic fermentation, p. 289–326. *In* G. H. Fleet (ed.), *Wine Microbiology and Biotechnology*. Harwood Academic Publishers, Chur, Switzerland.

72. **Henick-Kling, T.** 1995. Control of malolactic fermentation in wine: energetics, flavor modification and methods of starter culture preparation. *J. Appl. Bacteriol. Symp. Suppl.* **79**:29S–37S.

73. **Henick-Kling, T., and Y. H. Park.** 1994. Considerations for the use of yeast and bacterial starter cultures: sulfur dioxide and timing of inoculation. *Am. J. Enol. Viticult.* **45**:464–469.

74. **Henschke, P. A.** 1997. Wine yeast, p. 527–560. *In* F. K. Zimmermann and K. D. Entian (ed.), *Yeast Sugar Metabolism.* Technomic Publishing Co., Lancaster, Pa.

75. **Henschke, P., and V. Jiranek.** 1993. Yeasts—metabolism of nitrogen compounds, p. 77–164. *In* G. H. Fleet (ed.), *Wine Microbiology and Biotechnology.* Harwood Academic Publishers, Chur, Switzerland.

76. **Henschke, P. A., and C. S. Ough.** 1991. Urea accumulation in fermenting grape juice. *Am. J. Enol. Viticult.* **42**:317–321.

77. **Hernawan, T., and G. H. Fleet.** 1995. Chemical and cytological changes during the autolysis of yeasts. *J. Ind. Microbiol.* **14**:440–450.

78. **Herraiz, T., G. Reglero, M. Herraiz, P. J. Martin-Alvarez, and M. D. Cabezudo.** 1990. The influence of the yeast and type of culture on the volatile composition of wines fermented without sulfur dioxide. *Am. J. Enol. Viticult.* **41**:313–318.

79. **Ibeas, J. I., I. Lozano, F. Perdiganes, and J. Jimenez.** 1997. Dynamics of flor yeast populations during the biological ageing of sherry wines. *Am. J. Enol. Viticult.* **48**:75–79.

80. **Jiranek, V., P. Langridge, and P. A. Henschke.** 1995. Regulation of hydrogen sulfide liberation in wine-producing *Saccharomyces cerevisiae* strains by assimilable acid nitrogen. *Appl. Environ. Microbiol.* **61**:461–467.

81. **Jordan, A. D.** 1994. Style and quality of Australian sparkling wines over the past 20 years. *Food Aust.* **46**:184–188.

82. **Joyeux, A., S. Lafon-Lafourcade, and P. Ribéreau-Gayon.** 1984. Metabolism of acetic acid bacteria in grape must. Consequences on alcoholic and malolactic fermentation. *Sci. Aliments.* **4**:247–255.

83. **Kunkee, R. E.** 1984. Selection and modification of yeasts and lactic acid bacteria for wine fermentation. *Food Microbiol.* **1**:315–332.

84. **Kunkee, R. E.** 1991. Some roles of malic acid in the malolactic fermentation in winemaking. *FEMS Microbiol. Rev.* **88**:55–72.

85. **Kunkee, R. E., and L. Bisson.** 1993. Wine-making yeasts, p. 69–128. *In* A. H. Rose and J. S. Harrison (ed.), *The Yeasts,* 2nd ed., vol. 5, *Yeast Technology.* Academic Press Ltd., London, United Kingdom.

86. **Kunkee, R. E., and M. A. Amerine.** 1970. Yeasts in winemaking, p. 50–71. *In* A. H. Rose and J. S. Harrison (ed.), *The Yeasts,* 1st ed., vol. 3, *Yeast Technology.* Academic Press, Ltd., London, United Kingdom.

87. **Lafon-Lafourcade, S.** 1983. Wine and brandy, p. 81–63. *In* H. J. Rehm and G. Reed (ed.), *Biotechnology,* vol. 5, *Food and Feed Production with Microorganisms.* Verlag Chemie, Weinheim, Germany.

88. **Lafon-Lafourcade, S., and A. Joyeux.** 1981. Les bacteries acetique du vin. *Bull. Off. Int. Vin* **54**:803–829.

89. **Large, P. J.** 1986. Degradation of organic nitrogen compounds by yeasts. *Yeast* **2**:1–34.

90. **Laurent, M. H., T. Henick-Kling, and T. Acree.** 1994. Changes in the aroma and odour of Chardonnay wine due to malolactic fermentation. *Wein Wiss.* **49**:3–10.

91. **Lee, T. H., and R. F. Simpson.** 1993. Microbiology and chemistry of cork taints in wine, p. 353–372. *In* G. H. Fleet (ed.), *Wine Microbiology and Biotechnology.* Harwood Academic Publishers, Chur, Switzerland.

92. **Liger-Belair, G., R. Marchal, B. Robillard, M. Vignes-Adler, A. Maujean, and P. Jeandet.** 1999. Study of effervescence in a glass of champagne. Frequencies of bubble formation, growth rates, and velocities of rising bubbles. *Am. J. Enol. Viticult.* **50**:317–323.

93. **Liu, S. Q., C. R. Davis, and J. D. Brooks.** 1995. Growth and metabolism of selected lactic acid bacteria in synthetic wines. *Am. J. Enol. Viticult.* **46**:166–174.

94. **Liu, S. Q., and G. J. Pilone.** 1998. A review: arginine metabolism in wine lactic acid bacteria and its practical significance. *J. Appl. Microbiol.* **84**:315–327.

95. **Longo, E., and F. Vezinhet.** 1993. Chromosomal rearrangements during vegetative growth of a wild strain of *Saccharomyces cerevisiae. Appl. Environ. Microbiol.* **59**:322–326.

96. **Lonvaud-Funel, A.** 1995. Microbiology of the malolactic fermentation: molecular aspects. *FEMS Microbiol. Lett.* **126**:209–214.

97. **Lonvaud-Funel, A.** 1999. Lactic acid bacteria in the quality improvement and depreciation of wine. *Antonie Leeuwenhoek* **76**:317–331.

98. **Martineau, B., and T. Henick-Kling.** 1995. Performance and diacetyl production of commercial strains of malolactic bacteria in wine. *J. Appl. Bacteriol.* **78**:526–536.

99. **Martinez, P., M. J. Valcarcel, L. Perez, and T. Benitez.** 1995. Metabolism of *Saccharomyces cerevisiae* flor yeasts during fermentation and biological ageing of fino sherry: by-products and aroma compounds. *Am. J. Enol. Viticult.* **49**:240–250.

100. **Martini, A.** 1993. Origin and domestication of the wine yeast *Saccharomyces cerevisiae. J. Wine Res.* **4**:165–176.

101. **Martini, A., M. Ciani, and G. Scorzetti.** 1996. Direct enumeration and isolation of wine yeasts from grape surfaces. *Am. J. Enol. Viticult.* **47**:435–440.

102. **Mateo, J. J., M. Jimenez, T. Huerta, and A. Pastor.** 1991. Contribution of different yeasts isolated from musts of monastrell grapes to the aroma of wine. *Int. J. Food Microbiol.* **14**:153–160.

103. **McMahon, H., B. W. Zoecklein, K. Fugelsang, and Y. Jasinski.** 1999. Quantification of glycosidase activities in selected yeasts and lactic acid bacteria. *J. Ind. Microbiol. Biotechnol.* **23**:198–203.

104. **Millet, V., and A. Lonvaud-Funel.** 2000. The viable but non-culturable state of wine microorganisms during storage. *Lett. Appl. Microbiol.* **30**:136–141.

105. **Miranda, M., A. Ramos, M. Veiga-da-Cunha, C. Loureiro-Dias, and H. Santos.** 1997. Biochemical basis for glucose-induced inhibition of malolactic fermentation in *Leuconostoc oenos. J. Bacteriol.* **179**:5345–5354.

106. **Mlikota, F., P. Males, and B. Cvjetkovic.** 1996. Effectiveness of five fungicides on grape vine grey mould and their effects on must fermentation *J. Wine Res.* **7**:103–110.

107. **Mora, J., and A. Mulet.** 1991. Effects of some treatments of grape juice on the population and growth of yeast species during fermentation. *Am. J. Enol. Viticult.* 42:133–136.

108. **Mora, J., J. I. Barbas, and A. Mulet.** 1990. Growth of yeast species during the fermentation of musts inoculated with *Kluyveromyces thermotolerans* and *Saccharomyces cerevisiae. Am. J. Enol. Viticult.* 41:156–159.

109. **Mortimer, R. K., P. Romano, G. Suzzi, and M. Polsnelli.** 1994. Genome renewal: a phenomenon revealed from genetic study of 43 strains of *Saccharomyces cerevisiae* derived from natural fermentation of grape musts. *Yeast* 10:1543–1552.

110. **Munoz, E., and M. Ingeldew.** 1990. Yeast hulls in wine fermentations—a review. *J. Wine Res.* 1:197–210.

111. **Musmanno, R. A., T. di Maggio, and G. Coratza.** 1999. Studies on strong and weak killer phenotypes of wine yeasts: production, activity of toxin in must, and its effect in mixed culture fermentation. *J. Appl. Microbiol.* 87:932–938.

112. **Nadal, D., D. Carro, J. Fernandez-Larrea, and B. Pina.** 1999. Analysis and dynamics of the chromosomal complements of wild sparkling-wine yeast strains. *Appl. Environ. Microbiol.* 65:1688–1695.

113. **Nair, N. G.** 1990. Strategies for fungicidal control of bunch rot of grapes caused by *Botrytis cinerea* in the Hunter Valley. *Aust. N. Z. Wine Ind. J.* May:218–220.

114. **Navarro, L., M. Zarazaga, J. Saenz, F. Ruiz-Larrea, and C. Torres.** 2000. Bacteriocin production by lactic acid bacteria isolated from Rioja wines. *J. Appl. Microbiol.* 88:44–51.

115. **Nielsen, J. C., C. Prahl, and A. Lonvaud-Funel.** 1996. Malolactic fermentation in wine by direct inoculation with freeze-dried *Leuconostoc oenos* cultures. *Am. J. Enol. Viticult.* 47:42–48.

116. **Nykanen, L.** 1986. Formation and occurrence of flavor compounds in wine and distilled alcoholic beverages. *Am. J. Enol. Viticult.* 37:84–96.

117. **Ough, C. S., and E. A. Crowell.** 1987. Use of sulfur dioxide in winemaking. *J. Food Sci.* 52:386–388, 393.

118. **Ough, C. S., M. Davenport, and K. Joseph.** 1989. Effects of certain vitamins on growth and fermentation rate of several commercial active dry wine yeasts. *Am. J. Enol. Viticult.* 40:208–213.

119. **Ough, C. S., and M. L. Groat.** 1978. Particle nature, yeast strain and temperature interactions on the fermentation rates of grape juice. *Appl. Environ. Microbiol.* 35:881–885.

120. **Park, S. K., R. B. Boulton, E. Bartra, and A. C. Noble.** 1994. Incidence of volatile sulfur compounds in California wines. A preliminary survey. *Am. J. Enol. Viticult.* 45:341–344.

121. **Pimentel, M. S., M. H. Silva, I. Cortes, and A. M. Faia.** 1994. Growth and metabolism of sugar and acids of *Leuconostoc oenos* under different conditions of temperature and pH. *J. Appl. Bacteriol.* 76:42–48.

122. **Poblet-Icart, M., A. Bordons, and A. Lonvaud-Funel.** 1998. Lysogeny of *Oenococcus oeni* (syn. *Leuconostoc oenos*) and study of their induced bacteriophages. *Curr. Microbiol.* 36:365–369.

123. **Polsinelli, M., P. Romano, G. Suzzi, and R. Mortimer.** 1996. Multiple strains of *Saccharomyces* on a single grape vine. *Lett. Appl. Microbiol.* 23:110–114.

124. **Presa-Owens, C. D. L., P. Schlich, H. D. Davies, and A. C. Noble.** 1998. Effect of *Methode Champenoise* process on aroma of four *V. vinifera* varieties. *Am. J. Enol. Viticult.* 49:289–294.

125. **Pretorius, I.** 2000. Tailoring wine yeast for the new millennium: novel approaches to the ancient art of winemaking. *Yeast* 16:675–729.

126. **Pueyo, E., A. Martinez-Rodriguez, M. Polo, G. Santa-Maria, and B. Bartolome.** 2000. Release of lipids during yeast autolysis in a model wine system. *J. Agric. Food Chem.* 48:116–122.

127. **Querol, A., E. Barrio, T. Huerta, and D. Ramon.** 1992. Molecular monitoring of wine fermentations conducted by active dry yeast strains. *Appl. Environ. Microbiol.* 58:2948–2953.

128. **Querol, A., E. Barrio, and D. Ramon.** 1994. Population dynamics of natural *Saccharomyces* strains during wine fermentation. *Int. J. Food Microbiol.* 21:315–323.

129. **Querol, A., and D. Ramón.** 1996. The application of molecular techniques in wine microbiology. *Trends Food Sci. Technol.* 7:73–77.

130. **Radler, F.** 1993. Yeasts—metabolism of organic acids, p. 165–182. *In* G. H. Fleet (ed.), *Wine Microbiology and Biotechnology.* Harwood Academic Publishers, Chur, Switzerland.

131. **Radler, F., and B. Lotz.** 1990. The microflora of active dry yeast and the quantitative changes during fermentation. *Wien Wiss.* 45:114–122.

132. **Rankine, B. L.** 1989. *Making Good Wine. A Manual of Winemaking Practices for Australia and New Zealand.* Sun Books, Melbourne., Australia.

133. **Rauhut, D.** 1993. Yeasts—production of sulfur compounds, p. 183–223. *In* G. H. Fleet (ed.), *Wine Microbiology and Biotechnology.* Harwood Academic Publishers, Chur, Switzerland.

134. **Reader, H. P., and M. Dominguez.** 1995. Fortified wines: sherry, port and madeira, p. 159–207. *In* A. G. H. Lea and J. R. Piggott (ed.), *Fermented Beverage Production.* Blackie Academic & Professional, Glasgow, Scotland.

135. **Reed, G., and T. W. Nagodawithana (ed.).** 1992. *Yeast Technology,* 2nd ed. AVI Publishing Co./Van Nostrand Reinhold, New York, N.Y.

136. **Reed, G., and T. W. Nagodawithana.** 1988. Technology of yeast usage in wine making. *Am. J. Enol. Viticult.* 39:83–90.

137. **Regueiro, L. A., C. L. Costas, and J. E. L. Rubio.** 1993. Influence of viticultural and enological practices on the development of yeast populations during winemaking. *Am. J. Enol. Viticult.* 44:405–408.

138. **Remize, F., J. M. Sablayrolles, and S. Dequin.** 2000. Reassessment of the influence of yeast strain and environmental factors on glycerol production in wine. *J. Appl. Microbiol.* 88:371–378.

139. **Ribéreau-Gayon, P., D. Dubourdieu, B. Donéche, and A. Lonvaud.** 2000. *Handbook of Enology*, vol. 1. *The Microbiology of Wine and Vinifications*. John Wiley & Sons, Chichester, United Kingdom.

140. **Riou, C., J. M. Nicaud, P. Barre, and C. Gaillardin.** 1997. Stationary-phase gene expression in *Saccharomyces cerevisiae* during wine fermentation. *Yeast* **13**:903–915.

141. **Rodriguez, A. V., and M. C. Manca de Nadra.** 1995. Mixed culture of *Lactobacillus hilgardii* and *Leuconostoc oenos* isolated from Argentine wine. *J. Appl. Bacteriol.* **78**:521–525.

142. **Rodriguez, A. V., and M. C. Marca de Nadra.** 1994. Sugar and organic acid metabolism in mixed cultures of *Pediococcus pentosaceus* and *Leuconostoc oenos* isolated from wine. *J. Appl. Bacteriol.* **77**:61–66.

143. **Rollán, G. C., M. E. Farias, A. M. Strasser de Saad, and M. C. Manca de Nadra.** 1998. Exoprotease activity of *Leuconostoc oenos* in stress conditions. *J. Appl. Microbiol.* **85**:219–223.

144. **Romano, P.** 1997. Metabolic characteristics of wine strains during spontaneous and inoculated fermentation. *Food Technol. Biotechnol.* **35**:255–260.

145. **Romano, P., and G. Suzzi.** 1993. Sulfur dioxide and wine microorganisms, p. 373–394. *In* G. H. Fleet (ed.), *Wine Microbiology and Biotechnology*. Harwood Academic Publishers, Chur, Switzerland.

146. **Romano, P., and G. Suzzi.** 1993. Potential use for *Zygosaccharomyces* species in winemaking. *J. Wine Res.* **4**:87–94.

147. **Romano, P., G. Suzzi, G. Comi, and R. Zironi.** 1992. Higher alcohol and acetic acid production by apiculate wine yeasts. *J. Appl. Bacteriol.* **73**:126–130.

148. **Romano, P., G Suzzi, L. Turbanti, and M. Polsinelli.** 1994. Acetaldehyde production in *Saccharomyces cerevisiae* wine yeasts. *FEMS Microbiol. Lett.* **118**:213–218.

149. **Romano, P., G. Suzzi, R. Zironi, and G. Comi.** 1993. Biometric study of acetoin production in *Hanseniaspora guilliermondii* and *Kloeckera apiculata*. *Appl. Environ. Microbiol.* **59**:1838–1841.

150. **Ross, J. P.** 1997. Going wild: wild yeast in winemaking. *Wines Vines* Sept.:16–21.

151. **Salmon, J. M., and P. Barre.** 1998. Improvement of nitrogen assimilation and fermentation kinetics under enological conditions by derepression of alternative nitrogen-assimilatory pathways in an industrial *Saccharomyces cerevisiae* strain. *Appl. Environ. Microbiol.* **64**:3831–3837.

152. **Schoeman, H., M. A. Vivier, M. du Toit, L. M. T. Dicks, and I. Pretorius.** 1999. The development of bactericidal yeast strains by expressing the *Pediococcus acidilactici* pediocin gene (pedA) in *Saccharomyces cerevisiae*. *Yeast* **15**:647–656.

153. **Schreier, P.** 1979. Flavor composition of wines: a review. *Crit. Rev. Food Sci. Nutr.* **12**:59–111.

154. **Schutz, M., and J. Gafner.** 1993. Analysis of yeast diversity during spontaneous and induced alcoholic fermentations. *J. Appl. Bacteriol.* **75**:551–558.

155. **Schutz, M., and J. Gafner.** 1994. Dynamics of the yeast strain population during spontaneous alcoholic

156. **Scott, E., B. E. Stummer, D. L. Whisson, A. M. Corrie, and R. van Heeswijck.** 1999. Molecular approaches in the study of grapevine pathogens and pests, p. 146–149. *In* R. J. Blair, A. N. Sas, P. F. Hayes, and P. B. Hoj (ed.), *Proceedings of the Tenth Australian Wine Industry Conference, Winetitles*, Adelaide, Australia.

157. **Shimazu, Y., and M. Watanabe.** 1981. Effects of yeast strains and environmental conditions on formation of organic acids in must during fermentation. *J. Ferment. Technol.* **59**:27–32.

158. **Shimizu, K.** 1993. Killer yeasts, p. 243–264. *In* G. H. Fleet (ed.), *Wine Microbiology and Biotechnology*. Harwood Academic Publishers, Chur, Switzerland.

159. **Shinohara, T.** 1984. L'importance des substances volatiles dur vin. Formation et effets sur la qualité. *Bull. Off. Int. Vin* **57**:606–618.

160. **Soden, A., I. L. Francis, H. Oakey, and P. A. Henschke.** 2000. Effects of co-fermentation with *Candida stellata* and *Saccharomyces cerevisiae* on the aroma and composition of Chardonnay wine. *Aust. J. Grape Wine Res.* **6**:21–30.

161. **Sohier, D., and A. Lonvaud-Funel.** 1998. Rapid and sensitive *in situ* hybridisation method for detecting and identifying lactic acid bacteria in wine. *Food Microbiol.* **15**:391–397.

162. **Sokollek, S. J., C. Hertel, and W. P. Hammes.** 1998. Cultivation and preservation of vinegar bacteria. *J. Biotechnol.* **60**:195–206.

163. **Soufleros, E., and A. Bertrand.** 1979. Rôle de la souche de levure dans la production des substances volatiles au cours de la fermentation du jus de raisin. *Conn. Vigne Vin* **13**:181–198.

164. **Spayd, S. E., C. W. Nagel, and C. G. Edwards.** 1995. Yeast growth in riesling juice as affected by vineyard nitrogen fertilisation. *Am. J. Enol. Viticult.* **46**:49–55.

165. **Spiropoulos, A., J. Tanaka, I. Flerianos, and L. F. Bisson.** 2000. Characterisation of hydrogen sulfide fermentation in commercial and natural wine isolates of *Saccharomyces*. *Am. J. Enol. Viticult.* **51**:233–248.

166. **Sponholz, W. R.** 1993. Wine spoilage by microorganisms, p. 395–420. *In* G. H. Fleet (ed.), *Wine Microbiology and Biotechnology*. Harwood Academic Publishers, Chur, Switzerland.

167. **Strasser de Saad, A. M., S. E. Pasteris, and M. C. Manca de Nadra.** 1995. Production and stability of pediocin N5p in grape juice medium. *J. Appl. Bacteriol.* **78**:473–476.

168. **Strehaiano, P.** 1993. Stuck fermentations: causes and physiological factors. *Rev. Fr. Oenol.* **140**:41–48.

169. **Suzzi, G., P. Romano, I. Ponti, and C. Montuschi.** 1995. Natural wine yeasts as biocontrol agents. *J. Appl. Bacteriol.* **78**:304–309.

170. **Thomas, S. D.** 1993. Yeasts as spoilage organisms in beverages, p. 517–562. *In* A. H. Rose and J. S. Harrison (ed.), *The Yeasts*, 2nd ed. vol. 5, *Yeast Technology*. Academic Press Ltd., London, United Kingdom.

171. **Todd, B. E. N., G. H. Fleet, and P. A. Henschke.** 2000. Promotion of autolysis through the interaction of killer

and sensitive yeasts: potential application in sparkling wine production. *Am. J. Enol. Viticult.* **51**:65–72.

172. Tvetanov, O. S., and G. K. Bambalov. 1994. Effect of yeast strain on the content of nitrogenous compounds and the quality of red sparkling wines stored with yeasts. *J. Wine Res.* **5**:41–52.

173. van Vuuren, H. J., and C. J. Jacobs. 1992. Killer yeasts in the wine industry. A review. *Am. J. Enol. Viticult.* **43**:119–128.

174. van Vuuren, H. J. J., and L. M. T. Dicks. 1993. *Leuconostoc oenos*—a review. *Am. J. Enol. Viticult.* **44**:99–112.

175. van Zyl, J. A., M. J. De Vries, and A. S. Zeeman. 1963. The microbiology of South African winemaking. III. The effect of different yeasts on the composition of fermented musts. *S. Afr. J. Agric. Sci.* **6**:165–180.

176. Vaughn-Martini, A., and A. Martini. 1995. Facts, myths and legends on the prime industrial microorganism. *J. Ind. Microbiol.* **14**:514–522.

177. Versavaud, A., P. Courcoux, C. Roulland, L. Dulau, and J. N. Hallet. 1995. Genetic diversity and geographical distribution of wild *Saccharomyces cerevisiae* strains from the wine-producing area of Charentes, France. *Appl. Environ. Microbiol.* **61**:3521–3529.

178. Volschenk, H., M. Viljoen, J. Grobler, F. Bauer, A. Lonvaud-Funel, M. Denayrolles, R. E. Subden, and H. J. J. van Vuuren. 1997. Malolactic fermentation in grape must by a genetically engineered strain of *Saccharomyces cerevisiae*. *Am. J. Enol. Viticult.* **48**:193–197.

179. Wibowo, D., R. Eschenbruch, C. Davis, G. H. Fleet, and T. H. Lee. 1985. Occurrence and growth of lactic acid

bacteria in wine—a review. *Am. J. Enol. Viticult.* **36**:302–313.

180. Wibowo, D., G. H. Fleet, T. H. Lee, and R. E. Eschenbruch. 1988. Factors affecting the induction of malolactic fermentation in red wines with *Leuconostoc oenos*. *J. Appl. Bacteriol.* **64**:421–428.

181. Wilkes, K. L., and M. R. Dharmadhikari. 1997. Treatment of barrel wood infected with acetic acid bacteria. *Am. J. Enol. Viticult.* **48**:516–520.

182. Yokotsuka, K., A. Otaki, A. Naitoh, and H. Tanaka. 1993. Controlled deacidification and alcohol fermentation of a high-acid grape must using two immobilized yeasts, *Schizosaccharomyces pombe* and *Saccharomyces cerevisiae*. *Am. J. Enol. Viticult.* **44**:371–377.

183. Yokotsuka, K., M. Yajima, and T. Matsudo. 1997. Production of bottle-fermented sparkling wine using yeast immobilised in double-layer gel beads or strands. *Am. J. Enol. Viticult.* **48**:471–481.

184. Zavaleta, A. I., A. J. Martinez-Murcia, and F. Rodriguez-Valera. 1997. Intraspecific genetic diversity of *Oenococcus oeni* as derived from DNA fingerprinting and sequence analysis. *Appl. Environ. Microbiol.* **63**:1261–1267.

185. Zimmerli, B., and R. Dick. 1996. Ochratoxin A in table wine and grape juice—occurrence and risk assessment. *Food Addit. Contam.* **13**:655–668.

186. Zironi, R., P. Romano, G. Suzzi, F. Battistutta, and G. Comi. 1993. Volatile metabolites produced in wine by mixed and sequential cultures of *Haneniaspora guiliermondii* or *Kloeckera apiculata* and *Saccharomyces cerevisiae*. *Biotechnol. Lett.* **15**:235–238.

Advanced Techniques in Food Microbiology

IX

Food Microbiology: Fundamentals and Frontiers, 2nd Ed.
Edited by M. P. Doyle et al.
© 2001 ASM Press, Washington, D.C.

Peter Feng

Development and Impact of Rapid Methods for Detection of Foodborne Pathogens

38

INTRODUCTION

Microbiological Testing of Foods

Analysis of foods for the presence of bacteria is a common practice today for ensuring food quality and safety. Prior to the 1900s, however, there were no national regulations for food safety in the United States (67). As a result, foods were seldom tested for microbial contamination, and consumers relied solely on food manufacturers to ensure that foods were unadulterated and, therefore, safe for consumption. But as society became industrialized, the rapid growth in populations in the cities increased the demand for food production, which also increased consumer concerns for the safety of foods. For instance, growth in the dairy industry in response to demand exceeded the handling and transportation capacity of many farms, and milk became a dangerous vehicle for disseminating infections such as diphtheria, typhoid, and scarlet fever (52). Unscrupulous producers, seeking to meet demand but also to increase profits by reducing production costs, began to adulterate foods. Dilution of milk with water or addition of charcoal dust or sawdust to ground coffee and cocoa became common practice. Producers sometimes added borax and formaldehyde to butter and milk to prolong the shelf life of these and other perishable foods.

In response to the increased public concerns for the safety of foods, the U.S. Congress in 1906 passed the Federal Food and Drug Act to prohibit interstate commerce of foods known to be adulterated or misbranded or to contain substances injurious to health (52, 67). However, the burden of proof of food adulteration or contamination rested with the federal government and not with the food industry; hence, only the federal regulatory agencies were actively developing methods and testing foods for contamination. That situation changed in 1938 with the passage of the Food, Drug, and Cosmetic Act, which revised the authority of the federal government and redefined the requirements for food safety. The 1938 act established that it was the manufacturer's responsibility not only to know the law but also to comply with its requirements. Hence, producers now had to test their products to ensure that the ingredients and the foods produced were safe for consumption (67). This stimulated research and development of better analytical methods to test for the presence of bacteria, toxins, or other contaminants in foods.

Today, microbial analyses of food samples are routinely performed. However, the introduction and gradual implementation of the Hazard Analysis and Critical Control Point (HACCP) concept to safeguard foods "from farm to fork" will probably reduce our

Peter Feng, Division of Microbiological Studies, U.S. Food and Drug Administration, Washington, DC 20204.

dependence on end product analysis. End product testing is generally regarded as ineffective in ensuring the safety of foods. But microbiological testing and the demand for better testing methods will continue, as these assays still play critical roles in food production, in the establishment and validation of HACCP procedures, and in the monitoring of food quality to ensure the distribution of safe foods to the consumer.

Currently, there are many microbiological methods to test for most of the pathogens commonly found in foods. For the most part, these are conventional methods that rely extensively on culture and agar media to grow and enumerate viable bacterial cells. However, the advent of biotechnology has greatly altered food testing methods, and there are numerous companies that are actively developing assays that are specific, faster, and often more sensitive than conventional methods in testing for microbial contaminants in foods. These assays, often referred to as "rapid methods," use technologies such as special fluorogenic or chromogenic enzyme substrates, monoclonal or antibody cocktails, nucleic acids, and PCR to develop more efficient diagnostic assays. Some of these are automated. This chapter examines the impact of biotechnology on the emergence of rapid methods; the continued problems of testing for pathogens in foods; the major types of rapid assay formats; the impact and limitations of rapid methods; and the importance of validating these methods before application. Since it is impossible to discuss all the technologies and formats that are still under development, this chapter will focus only on formats used in commercially available tests. In the last section of the chapter, however, a few novel technologies are presented which may provide extremely rapid and sensitive assays for testing for pathogens in foods.

Problems in Food Analysis

Microbiological analysis of foods, especially for particular pathogen species, remains a challenging task for virtually all assays and technologies. The problem may be due in part to the fact that bacteria are not uniformly distributed in foods, but analysis is also complicated by the heterogeneity of food matrices. Foods are composed of an infinite array of ingredients that include proteins, carbohydrates, fats, oils, chemicals, preservatives, and many other compounds. The physical form of foods is also highly variable: foods come in powder, liquid, gel, solid, semisolid, or other forms. These differences in viscosity, compounded by the presence of ingredients such as fats and oils can interfere with proper mixing. This results in nonuniform homogenates which do not allow reproducible analysis or consistent isolation of specific pathogens.

In addition to the problems of variable matrices, testing of foods is further complicated by the presence of indigenous microbiota. In some raw foods, such as sprouts, populations are as high as 10^8 cells/g. Normal microbiota generally pose no significant health risks, but their presence often interferes with the selective isolation and identification of specific pathogens, which are usually found in much lower numbers (12). This interference becomes especially critical when foods are being analyzed for pathogens such as *Shigella* sp. and enterohemorrhagic *Escherichia coli* (EHEC) of serotype O157:H7. These pathogens have low infectious dosages, making the consumption of low numbers of cells dangerous (44).

The problems caused by normal microbiota and matrix interference are further exacerbated if the specific pathogen of interest is injured. Many processes or treatments used in food manufacturing, such as heat, cold, drying, freezing, osmotic activity, chemical additives, preservatives, and other factors, can sublethally injure bacterial cells (53). These injured or stressed cells are sensitive to many of the compounds or antibiotics used in selective enrichment media; hence, unless injured cells are allowed to resuscitate, they may easily be outgrown by other bacteria present in the sample (53). Injury is explained in greater detail in chapter 2.

To overcome these difficulties, conventional microbiological procedures had to be constantly modified or adapted to test for specific bacteria or food types. One of the more effective means implemented over the years is the stepwise cultural enrichment of food samples to enhance the detection of specific pathogens (76). The process usually begins with a "preenrichment" step, whereby foods are incubated in nonselective media to allow repair of injured or stressed cells. Aliquots are then transferred into "selective enrichment," whereby the pathogen of interest is allowed to grow while the growth of normal biota is suppressed. Occasionally, a "postenrichment" step is also included to further increase the number of pathogens to detectable levels. Culture-enriched samples are then plated onto selective and/or differential media, in which colonies are presumptively identified by distinct phenotypes. Definitive identification requires a battery of biochemical tests and, sometimes, serological typing as well (85).

Although the procedure for conventional methods described above is effective and has been used routinely for isolating and identifying bacterial pathogens in foods, the method is labor-intensive and time-consuming. As a result, conventional methods usually take several days to perform and are inadequate for making quick assessments of the microbiological safety of foods.

Development and Origin of Rapid Methods

Scientific advances have introduced a number of technologies that are being used to detect foodborne pathogens and their toxins (29). These assays are collectively referred to as rapid methods, a term that has steadily gained in popularity and interest worldwide. Many reviews have been published on the use of rapid methods to test for pathogens in foods (7, 14, 21, 29, 30, 37, 41, 55, 85), seafood (57), and the environment (74) and for foodborne yeasts (20).

Definitions of Rapid Detection and Identification

To define exactly what constitutes a rapid method is not simple, as "rapid" is subjective to interpretation. As a result, a rapid method can be an assay that gives instant or real-time results, but such ideal rapid methods do not yet exist. On the other hand, a rapid method can also be a simple modification of a procedure that shortens the assay time. Since conventional methods, generally regarded as "the gold standard," often take days if not a week to complete, any modification that reduces this analysis time can technically be called a rapid method. Between these two extremes of definitions, therefore, virtually all new methods that have been developed or are being developed can be classified as rapid method. So the term "rapid methods" encompasses a large group of diverse technologies and assay formats.

"Identification" and "detection" are two terms often used in conjunction with rapid methods that also warrant clarification. Although these terms have been used synonymously, identification usually refers to determining the identity of a pure culture isolate, whereas detection refers to detecting the presence of an organism directly in foods, without the need to isolate the organism. Although other investigators may differ in their interpretation of these terms, in this chapter the above distinction will be used so that rapid detection and rapid identification refer to different applications, with the main difference being that identification requires a pure culture isolate of the organism.

The term rapid method was virtually nonexistent in the literature 20 years ago; hence, it gives the impression of being a newly formed concept. However, within the broad definition of rapid method, developments in this area have been ongoing for years as scientists have found easier and simpler means to do microbial analysis. One notable example is the work of Weaver and colleagues in the late 1940s (14, 37); they revolutionized the concept of biochemical identification by using concentrated inoculum of cells on smaller volumes of reagents to speed up reaction time. This gave rise to the miniaturized biochemical identification kits that became popular in clinical testing in the mid-1970s and later in food applications in the 1980s. But aside from this and a few other innovations, development of rapid methods did not accelerate until major advances were made in biotechnology.

Impact of Biotechnology—Delphi Forecast

Basic scientific research gave rise to biotechnology (59), which in turn stimulated developments in many areas, including rapid diagnostic methods. This evolution in technology is neatly illustrated in the pyramid of biotechnology shown in Fig. 38.1 (48). The beginning or apex of the pyramid can probably be attributed to the research or discovery of the DNA in the 1950s, which stimulated a flurry of research in molecular biology. After decades of intensive efforts in this golden era of basic research, a wealth of knowledge was generated, including the development of innovative reagents and laboratory techniques in the 1970s: the discovery of restriction enzymes that enabled manipulation and cloning of specific DNA; highly specific monoclonal antibodies that recognized a single antigenic epitope; better tissue culture techniques that allowed cultivation of fastidious organisms and viruses; and advancements in fermentation technology, which made specialty enzymes and reagents readily available for laboratory research. This era of research and technology development, shown in the top three tiers of the pyramid (Fig. 38.1), produced numerous Nobel laureates. Initially, these new technologies were restricted to laboratory research. But in the 1980s, many concepts began to be adapted for practical application, including the detection of bacterial pathogens.

The swiftness with which these technologies impacted on the field of food diagnostics is astounding. For instance, an information survey known as the Delphi Forecast was conducted in 1981, when 29 renowned microbiologists around the world were invited to predict diagnostic technologies that would be used in food microbiology in the future (46). In the area of cell enumeration, they forecasted the use of automated colony counting and instruments that determined cell density by measuring conductance or specific bacterial metabolites. In bacterial identification, they predicted the use of miniaturized biochemical kits and systems that identified bacteria based on the utilization of special substrates. Most of these predictions turned out to be accurate, including the projected year of implementation. Many of those systems and instruments that were predicted exist today. Obviously missing from the predictions, however, were assays based on nucleic acids or

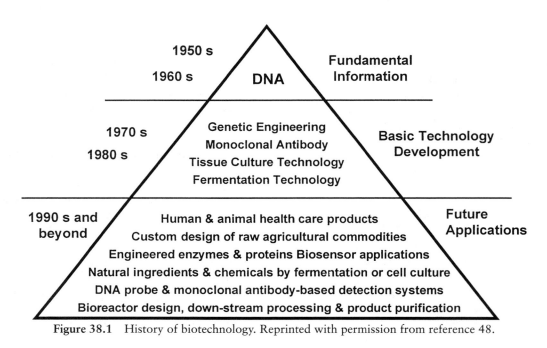

Figure 38.1 History of biotechnology. Reprinted with permission from reference 48.

antibodies. At the time the Delphi Forecast was conducted, both of these molecular technologies were basic laboratory research tools. Few foresaw that they would dominate pathogen testing within a decade after the forecast.

RAPID METHOD FORMATS

Rapid methods are usually more sensitive and specific than microbiological methods for identification or detection of pathogens in foods. The common formats of commercial rapid methods are (i) miniaturized biochemical identification kits, (ii) modification of conventional microbiological methods, and (iii) antibody-based and (iv) nucleic acid-based detection assays. Within each group, there are also some automated identification and detection assays. Under the broader definition of "rapid," all of these are rapid methods, since they do expedite the identification or detection process. But if the more stringent definition is applied, the use of the term rapid methods to describe these assays may be misleading. For instance, in identification tests, the actual assay may be rapid, but since a pure culture is required, it still takes time to grow and plate that isolate. Likewise, the complexity of food matrices and the other analysis problems discussed above have precluded the use of most detection methods directly on food samples. As a result, almost all rapid methods designed to detect a specific pathogen still require enrichment of the food samples in media prior to testing. In other words, the performance of the assays themselves may be very rapid, ranging from a few minutes to a few hours, but the total time for analysis of a food sample requires a few days or longer because of cultural enrichment. Despite the need for enrichment, however, many rapid methods use abbreviated enrichment steps and eliminate the need for differential plating for colony isolation. So there are still significant reductions in overall assay time as compared with conventional methods. The technologies and the many unique designs used in rapid methods are discussed in the following sections.

Miniaturized Biochemical and Other Identification Assays

Bacteria isolated from foods are identified by their biochemical characteristics. To generate a biochemical profile is a labor-intensive, media-consuming process, in which pure cultures of bacteria are subjected to a battery of tests to determine various enzymatic activities and the utilization of assorted carbohydrates. The initial concept to miniaturize biochemical testing began in the 1940s, when the use of small vials greatly economized media usage, and the use of concentrated inocula considerably reduced incubation time (14, 49). Since inception, miniaturized biochemical tests steadily gained in popularity in clinical microbiology, and later, the benefits of using these kits in food analysis became apparent.

Miniaturized biochemical test kits are used for the identification of pure cultures of bacteria isolated from food (49). Most kits consist of a disposable, multichamber device, containing 15 to 30 media or substrates, designed to identify a specific group or species. With the exception of a few systems, in which results can be read in 4 h, most kits require 18 to 24 h of incubation (14). In general, miniaturized biochemical tests show 90 to 99% accuracy and are comparable to conventional identification methods (14, 37, 49). Although most kits are designed for enteric bacteria, biochemical kits for the identification of non-*Enterobacteriaceae* are also available, including kits for *Campylobacter, Listeria*, anaerobes, nonfermenting gram-negative organisms and gram-positive bacteria.

Miniaturized biochemical identification kits are all very similar in procedure, cost, and performance and are certainly much simpler and more economical than conventional methods (15). However, there may be some differences in the size of the database: the older kits have been used longer and may have a more extensive identification database. The concept of the minisystems has basically remained the same, but the newer assays are designed to be more user friendly, with less hands-on manipulation or need for extraneous reagents, compared with the earlier kits. Some miniaturized kits are listed in Table 38.1.

Consistent with this shift toward simplicity, advances in instrumentation have enabled automation of the miniaturized biochemical identification tests. In clinical

Table 38.1 Partial list of miniaturized biochemical kits and automated systems for identifying foodborne bacteria[a]

System	Format	Manufacturer	Organisms
API[b]	Biochemical	bioMérieux	*Enterobacteriaceae, Listeria, Staphylococcus, Campylobacter,* nonfermenters, anaerobes
Cobas IDA	Biochemical	Hoffmann-La Roche	*Enterobacteriaceae*
Micro-ID[b]	Biochemical	REMEL	*Enterobacteriaceae, Listeria*
EnterotubeII	Biochemical	Roche	*Enterobacteriaceae*
Spectrum 10	Biochemical	Austin Biological	*Enterobacteriaceae*
RapID	Biochemical	Innovative Diag.	*Enterobacteriaceae*
BBL Crystal	Biochemical	Becton Dickinson	*Enterobacteriaceae, Vibrionaceae,* nonfermenters, anaerobes
Minitek	Biochemical	Becton Dickinson	*Enterobacteriaceae*
Microbact	Biochemical	Microgen	*Enterobacteriaceae,* gram-negative organisms, nonfermenters, *Listeria*
Vitek[b]	Biochemical[c]	bioMérieux	*Enterobacteriaceae,* gram-negative and gram-positive organisms
Microlog	C oxidation[c]	Biolog	*Enterobacteriaceae,* gram-negative and gram-positive organisms
MIS[b]	Fatty acid[c]	Microbial-ID	*Enterobacteriaceae, Listeria, Bacillus, Staphylococcus, Campylobacter*
Walk/Away	Biochemical[c]	MicroScan	*Enterobacteriaceae, Listeria, Bacillus, Staphylococcus, Campylobacter*
Replianalyzer	Biochemical[c]	Oxoid	*Enterobacteriaceae, Listeria, Bacillus Staphylococcus, Campylobacter*
Riboprinter	Nucleic acid[c]	Qualicon	*Salmonella, Staphylococcus, Listeria, Escherichia coli*
Cobas Micro-ID	Biochemical[c]	Becton Dickinson	*Enterobacteriaceae,* gram-negative organisms, nonfermenters
Malthus[b]	Conductance[c]	Malthus	*Salmonella, Listeria, Campylobacter, E. coli, Pseudomonas,* coliforms
Bactometer	Impedance	bioMérieux	*Salmonella*

[a] Modified from reference 31.
[b] Selected systems adopted by the Association of Official Analytical Chemists International (AOAC) as an official first or final action.
[c] Automated systems.

analysis, there are already many automated instruments that can identify and determine the antibiotic susceptibility of the organisms (83). Most automated systems make identifications based on biochemical profiles, and some, such as the Vitek instrument, have been approved by the Association of Official Analytical Chemists International (AOAC) (see later section) for the identification of various foodborne bacteria. In this test, a disposable, credit card-sized plastic device containing up to 30 biochemical tests is inoculated with a pure culture suspension of cells. The instrument incubates and monitors changes in biochemical reactions to generate a phenotypic profile, which is then compared with the computer database to provide identification.

A few other automated systems identify bacteria based on compositional or metabolic properties. The Microbial Identification System uses gas chromatography to analyze cell extracts of pure bacterial cultures to generate a fatty acid profile (21). The computer then analyzes and calculates the peak data on the chromatograms and compares them with the database for best-match identification. The Biolog Identification System generates identification profiles based on the ability of the bacteria to oxidize a panel of 95 different carbon sources (60). The respiration of viable bacterial cells on these carbon sources causes changes in the redox potential, which are detected colorimetrically using a tetrazolium-based compound. The color changes are measured by a spectrophotometer, and the resulting profile is compared with the reference database in the computer. Some of these automated identification systems are listed on Table 38.1.

Modifications and Specialized Media

Another category of rapid methods includes a large variety of assays, ranging from specialized media to simple modifications of conventional assays, that result in savings in labor, time, and materials. The group includes tests like Petrifilm, which uses disposable cardboard containing dehydrated media, designed for enumerating total bacteria, specific bacterial species, or mold and yeasts (18). This test eliminates the need for preparing media and agar plates, economizes in storage and incubation space, and also simplifies disposal of materials after analysis. The Isogrid tests uses specially designed hydrophobic grid membrane filters (HGMF) that can handle larger cell densities. This reduces the number of dilutions needed prior to filtration. HGMF increases counting capacity from 80 CFU/ml, allowed on normal membranes, to up to 1,600 CFU/ml (24). There are two other developments that have had a big impact in this area. One was the introduction of specialized substrates that are used in media or in substrate-impregnated disks

to enable rapid detection of trait enzymatic activities (41, 49); the other is an assay that enumerates total bacterial counts based on bacterial ATP.

Color pH indicators have been used in media for years to measure changes in acidity due to bacterial metabolic activities. However, the introduction of chromogenic and fluorogenic substrates greatly stimulated the development of special microbiological media. These substrates provide a quick measure of specific enzyme activities, some of which are trait characteristics of certain bacteria or bacterial groups, thus providing a quick presumptive identification of the organism. Some chromogenic substrates, like o-nitrophenyl-β-D-galactoside (ONPG) and 5-bromo-4-chloro-3-indolyl-β-D-galactoside (X-Gal), which was developed for use in molecular biology to detect promoter activities of DNA fragments cloned into expression vectors, have been incorporated into culture media to measure β-galactosidase (GAL) activity in enteric bacteria. Similarly, the fluorogenic substrate 4-methylumbelliferyl-β-D-glucuronide (MUG) has been added to various media to specifically detect E. coli β-glucuronidase (GUD) activity (32). Today, many specialized media, including substrate-saturated paper discs, use MUG or other chromogenic GUD substrates for detection of E. coli (27, 38). One particular area that has seen a great increase in specialized media is in the presumptive isolation of E. coli O157:H7 or other EHEC. These media often use combinations of chromogenic substrates to measure GAL, GUD, or sorbitol fermentation, so that bacterial species exhibiting these phenotypes or combinations of them will produce different-colored colonies. Several of these assays and specialized media are listed in Table 38.2.

Bioluminescence assays that detect bacterial ATP have also become common in food analysis. The key components of these assays are the enzyme luciferase, the organic compound luciferin, and magnesium ions. In the presence of ATP, luciferin is oxidized to oxyluciferin and, in the process, photons are released. By using ATP as the limiting reagent, the number of photons emitted, which can be quantified with a luminometer, becomes directly proportional to the estimated number of bacterial cells initially present in the sample (21, 49). Luciferase reactions are rapid and, unlike other methods, which require culture enrichment, ATP assays can detect 10^4 CFU/ml in 5 to 10 min without the need for enrichment. These assays are useful for monitoring the sanitary quality of foods or food processing surfaces; however, since all living organisms produce ATP, they can only be used to estimate the number of the total bacterial population, not the presence of specific bacteria or pathogens.

Table 38.2 Partial list of other commercially available rapid methods and specialty substrate media for detection of foodborne bacteria[a]

Organism	Trade name	Assay format[b]	Manufacturer
Bacteria	Redigel[c]	Media	RCR Scientific
	Isogrid[c]	HGMF	QA Labs
	Enliten	ATP	Promega
	Profile-1	ATP	New Horizon
	Biotrace	ATP	Biotrace
	Lightning	ATP	Idexx
	Petrifilm[c]	Media-film	3M
	SimPlate	Media	Idexx
Coliform/*E. coli*	Isogrid[c]	HGMF/MUG	QA Labs
	Petrifilm[c]	Media-film	3M
	SimPlate	Media	Idexx
	Redigel	Media	RCR Scientific
	ColiQuik[d]	MUG/ONPG	Hach
	ColiBlue[d]	Media	Hach
	Colilert[c,d]	MUG/ONPG	Idexx
	LST-MUG[c]	MPN media	Difco, GIBCO
	ColiComplete[c]	MUG-X-Gal	BioControl
	Colitrak	MPN-MUG	BioControl
	ColiGel, E*Colite[d]	MUG-X-Gal	Charm Sciences
	CHROMagar	Medium	CHROMagar
E. coli	MUG disc	MUG	REMEL
	CHROMagar	Medium	CHROMagar
EHEC[e]	Rainbow Agar	Medium	Biolog
	BCMO157:H7	Medium	Biosynth
	Fluorocult O157:H7	Medium	Merck
Listeria monocytogenes	BCM	Medium	Biosynth
Salmonella sp.	Isogrid[c]	HGMF	QA Labs
	OSRT	Medium/motility	Unipath (Oxoid)
	Rambach	Medium	CHROMagar
	MUCAP	C_8 esterase	Biolife
	XLT-4	Medium	Difco
	MSRV[c]	Medium	
Yersinia sp.	Crystal violet	Dye binding	Polysciences

[a] Modified from reference 31.
[b] Abbreviations: HGMF, hydrophobic grid membrane filter; MPN, most probable number; MUG, 4-methyl-umbelliferyl-β-D-glucuronide; ONPG, o-nitrophenyl-β-D-galactoside.
[c] Adopted by the AOAC as an official first or final action.
[d] Application for water analysis.
[e] EHEC, enterohemorrhagic *E. coli*.

Nucleic Acid-Based Assays

There are many nucleic acid-based assays, but only DNA probe, PCR, and bacteriophage have been developed commercially for detecting foodborne pathogens. Recently, a number of DNA-based molecular typing methods, including pulsed field gel electrophoresis, restriction fragment length polymorphism, and ribotyping, have also been developed. Typing methods are extremely useful for epidemiological investigations and for characterizing and fingerprinting foodborne pathogens (22, 43); however, molecular typing methods are not within the scope of this chapter and therefore will not be discussed.

DNA Probes

DNA probes for detecting pathogens in foods were introduced in the early 1980s (36). At a 1987 meeting on the future market opportunities in diagnostics, many experts had the optimistic outlook that probes would become a dominant diagnostic tool in the food industries worldwide. However, the actual commercial growth of probe assays has been unexpectedly slow, perhaps because of

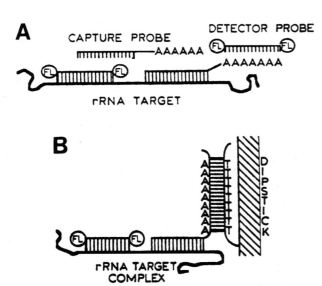

Figure 38.2 Diagrams illustrating the DNA probe format of the GENE-TRAK assay. (A) Hybridization of capture and detector probes to the rRNA target. (B) Dipstick capture of the hybridized complex. Reprinted with permission from reference 28.

Table 38.3 Partial list of commercially available nucleic acid-based assays used in the detection of foodborne bacterial pathogens[a]

Organism	Trade name	Format	Manufacturer
Clostridium botulinum	Probelia	PCR	BioControl
Campylobacter sp.	AccuProbe	Probe	GEN-PROBE
	GENE-TRAK	Probe	GENE-TRAK
Escherichia coli	GENE-TRAK	Probe	GENE-TRAK
E. coli O157:H7	BAX	PCR	Qualicon
	Probelia	PCR	BioControl
Listeria sp.	GENE-TRAK[b]	Probe	GENE-TRAK
	AccuProbe	Probe	GEN-PROBE
	BAX	PCR	Qualicon
	Probelia	PCR	BioControl
Salmonella sp.	GENE-TRAK[b]	Probe	GENE-TRAK
	BAX	PCR	Qualicon
	BIND[c]	Phage	BioControl
	Probelia	PCR	BioControl
Staphylococcus aureus	AccuProbe	Probe	GEN-PROBE
	GENE-TRAK	Probe	GENE-TRAK
Yersinia enterocolitica	GENE-TRAK	Probe	GENE-TRAK

[a] Modified from reference 31.
[b] Adopted by the AOAC as an official first or final action.
[c] Bacterial Ice Nucleation Diagnostics.

the initial use of radioactive markers in these assays. So, despite the fact that DNA probes have been developed for most of the major foodborne pathogens (34, 62), only a few companies are actively pursuing this format for food testing. Probe assays generally target rRNA, taking advantage of the fact that the higher copy number of bacterial rRNA provides a naturally amplified target and affords greater sensitivity.

The GENE-TRAK assay uses a combination of DNA hybridization and enzyme immunoassay for the detection of foodborne pathogens. The key components are a polydeoxyadenosine-tailed capture probe and a fluorescein-labeled detector probe, both of which are specific for adjacent regions of the target rRNA (Fig. 38.2A). After cultural enrichment of the food sample and chemical cell lysis, the rRNA target is hybridized simultaneously with both probes, and the hybrid is subsequently immobilized with a polydeoxythymidine-coated dipstick (Fig. 38.2B). Unbound or nonspecific binding is removed by multiple washing steps, and the bound complex is detected by enzyme immunoassay with peroxidase-conjugated anti-fluorescein antibody and a colorimetric peroxidase substrate (17).

The AccuProbe assay by GEN-PROBE is a hybridization protection assay in which a single-stranded probe, tagged with a chemiluminescent acridium ester label, is hybridized to specific sequences in the 16S rRNA. Hybridization protects the probe and label, as unbound

probes are inactivated by differential hydrolysis. Addition of alkaline H_2O_2 activates the label to emit light at 430 nm, which can be quantified with a luminometer. The AccuProbe assay is used mostly for identification of pure culture isolates of bacteria; however, it has also been explored as a test for pathogens in food enrichment samples (69, 73). Selected AccuProbe and GENE-TRAK assays are listed in Table 38.3.

PCR

Although the use of DNA probes for food analysis has not gained the projected commercial dominance, the basic principle of hybridization used in probe assays is being utilized by other technologies. In the PCR assay, short fragments of DNA (probes) or primers are hybridized to a specific target sequence or template, which is then enzymatically amplified by *Taq* polymerase using a thermocycler. PCR is an extremely powerful tool that enables exponential amplification of a specific sequence in a short time. PCR assays can be designed in a multiplex format to simultaneously detect several targets. There are many reviews of the extensive efforts to develop PCR assays for detecting foodborne pathogens (5, 50, 51, 56). Theoretically, PCR can amplify a single copy of DNA a million-fold in less than 2 h; hence, it has the potential to eliminate or greatly reduce the dependence on cultural enrichment. But despite the vast numbers of publications on the use of PCR in tests for foodborne pathogens, there are only a few commercial assay kits (Table 38.3). This may in part be attributed to the fact that, like other assays, PCR assays are affected by the complex composition of foods. Because the extreme sensitivity achievable by PCR using pure cultures is reduced when testing foods, some cultural enrichment is still required prior to PCR analysis.

PCR interference lies mostly with the presence of inhibitors in foods, which can affect primer binding or amplification efficiency or cause the occurrence of false-negative results (78, 91). In clinical testing, the presence of PCR inhibitors in fecal (96), serum (63), and blood (1, 33) samples is well documented. Also, humic substances in the soil and the presence of compounds such as iron affect primer binding or inhibit PCR amplification when environmental samples are tested (87). In the application of PCR for testing foodborne pathogens, foods with high protein and fat content, such as chicken, meat, soft cheeses, and oyster tissue, tend to interfere most with PCR amplification. This may be due to the presence of proteinases in these samples, which can degrade the enzyme (63, 78). There are exceptions, however; samples of bean sprout tissue, which are low in fat, are inhibitory to PCR (2). In addition to food components, excess levels of some ingredients used in enrichment or media and compounds used in the extraction of DNA from food samples can also interfere with PCR (2, 40, 78). As a result, food samples or culture-enriched samples often need to be treated or extracted to produce suitable template DNA for PCR analysis. Furthermore, because of the complexity of foods, a preparation method developed for a particular food type may not be usable for other foods. For example, extraction using alkaline conditions and detergent can reduce PCR inhibitory effects from high fat- and protein-containing foods (78). But other foods may require dilutions, washings, filtrations, or centrifugation steps to remove inhibitory effects. Some pretreatment steps effective in preparing selected food, soil, and clinical samples for PCR analysis are summarized by Lantz et al. (63). Also, immunomagnetic separation (see later section), in which antibodies are used to capture and concentrate specific bacteria, is effective in reducing the inhibitory effects of foods and has facilitated the use of PCR in testing foods and environmental and clinical samples (72, 79).

Aside from PCR, there are other methods for DNA amplification (99). Some of these, such as ligase chain reaction and self-sustained sequence replication (3SR), are being explored for testing for foodborne pathogens (90, 95, 99).

Bacteriophage

The highly specific interaction of phage with its bacterial host can be used to detect specific pathogens in foods. Two examples are bioluminescence and ice nucleation. As discussed above, assays that measure ATP cannot differentiate bacterial species, so they are used mainly to quantify total bacterial load in foods or to monitor sanitation. But bioluminescence assays can be used to identify bacteria if specific bacteriophages are engineered to carry detectable markers. Several species in the genera *Alteromonas*, *Photobacterium*, and *Vibrio* carry the *lux* operon that encodes bacterial luciferase and therefore are naturally bioluminescent (4, 84). The *lux* genes from *Vibrio fischeri* were cloned into Charon 30 phage to develop an assay that is specific for *E. coli* (89). Since phages do not have mechanisms for gene expression, the *lux* genes are silent. But once the recombinant phage infects its host, the inserted *lux* genes are expressed within a short time, resulting in a bioluminescent host. The amount of luminescence emitted can be quantified with a luminometer and calibrated to reflect the original bacterial load present in the sample. The recombinant Charon phage detected as few as 10 *E. coli* cells in 100 min after infection and detected 10^2 to 10^3 cells in milk or urine in 1 h, without cultural enrichment (61, 89). Similar sensitivities were reported for a *lux* recombinant phage using pure cultures of *Salmonella* sp. (88). However, since the required detection limit for *Salmonella* is 1 cell in 25 g of food, a short cultural enrichment is still required in bioluminescence assays.

The Bacterial Ice Nucleation Diagnostic (BIND) assay for *Salmonella* sp. is another bacteriophage assay. It uses the *ina* gene for ice nucleation and is based on the physical properties of freezing. Small, homogeneous volumes of water can be supercooled to $-40°C$ and remain liquid; but if nucleating agents are present, ice crystals will form at $-2°C$ (94, 97). Most ice nuclei are inorganic molecules, but some phytopathogenic bacteria, such as *Erwinia*, *Pseudomonas*, and *Xanthomonas* spp., produce proteins that also nucleate ice (94). Ice nucleation proteins, encoded by the *ina* gene, seem to bind water molecules to simulate configuration of ice crystals, thereby promoting freezing (97). By cloning the *ina* gene into a *Salmonella*-specific phage, the presence of this pathogen in foods can be detected by using a freeze-indicator dye (97, 98). The detection limit of the BIND assay is reported to be as low as 20 CFU/ml but ranges between 10^2 and 10^4 CFU/ml, depending on the *Salmonella* serovar being detected; hence, enrichment of foods is still needed for optimal detection. Aside from detection assays, ice nucleation technology is also being explored by industry to raise freezing temperatures for foods in order to reduce freezing times and to improve the quality, flavor, and texture of frozen foods (64).

Antibody-Based Assays

Antibody has been used for many years to serotype bacteria. One of the earliest typing schemes was devised in the 1920s and 1930s to identify *Salmonella* serotypes (25). Development of the radioimmunoassay expanded the use of antibodies to include quantitative determination of antigens, but it was not until the introduction of highly sensitive tests, such as the enzyme-linked immunosorbent assay (ELISA) (23), that the potential of antibodies in detection assays was realized. The highly specific binding of antibody, especially monoclonal antibody, and the simplicity and versatility of antigen-antibody reactions have facilitated the design of a variety of assays and formats (39). In general, antibody assays are classed into five formats (28), and they compose the largest group of rapid methods used in food testing.

Latex Agglutination and Reverse Passive Latex Agglutination

The simplest antibody test is latex agglutination (LA), whereby antibody-coated colored latex beads or colloidal gold particles are used to test bacterial cell suspensions. If the specific antigen is present, visible clumping or a precipitate is formed (Fig. 38.3). Although agglutination reactions are extremely simple and occur almost instantly, they are not very sensitive and require about 10^7 CFU for a reaction. This, however, still represents a 10- to 100-fold increase in sensitivity over the traditional heme or cell agglutination assays (10). For analysis of

bacterial toxins, the assay is known as reverse passive latex agglutination (RPLA), and it is usually done in a microplate (58). The main difference is that the antigens (bacterial cells) in LA are insoluble, whereas in RPLA, the antigens (toxins) are soluble so that, instead of clumping, the formation of a diffuse lattice work is considered positive. In food analysis, LA assays have been evaluated for detecting *Salmonella* sp.; however, multistep enrichment is needed prior to testing, and in some cases, suspended food particles affect the interpretation of agglutination reactions (19). LA is most useful for quick serological identification or confirmation of isolates from foods (65), and RPLA is used mostly for analysis of toxins in food extracts or for toxin production by pure culture isolates.

Immunodiffusion

Immunodiffusion is another simple antibody assay format used in testing for pathogens in foods, but the *Salmonella* 1-2 Test is the only commercial assay that uses this design (93). It is a modification of the Ouchterlony test, but instead of a gel matrix, it uses a two-chamber, L-shaped device (Fig. 38.4). The vertical or motility chamber contains semisolid agar, and the top of this chamber is inoculated with an antibody specific for

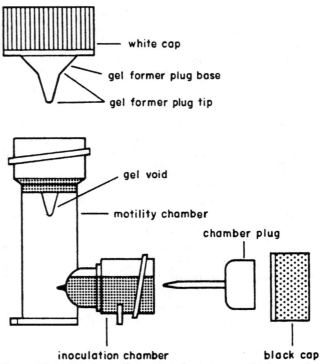

Figure 38.4 Diagram illustrating the 1-2 test, which uses the immunodiffusion assay format for detecting antigen-antibody interactions. Reprinted with permission from reference 28.

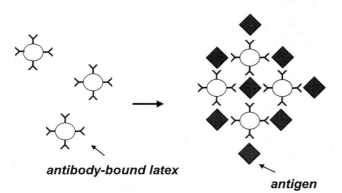

Figure 38.3 Diagram illustrating the latex agglutination assay format, in which agglutination or clumping results from specific antigen-antibody-latex interaction.

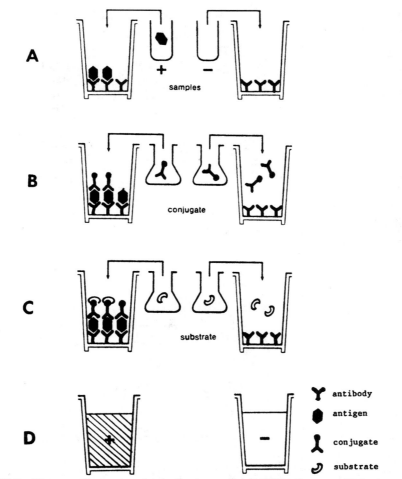

Figure 38.5 Diagram illustrating the antibody sandwich ELISA format. (A) Antigen capture by antibody-coated support. (B) Antigen binding by a second antibody conjugated with an enzyme. (C) Enzymatic activity on colorimetric substrates. (D) Positive and negative reactions as determined by color development. Reprinted with permission from reference 28.

Salmonella flagellar antigen. The horizontal or inoculation chamber contains selective enrichment and is inoculated with an aliquot of the preenrichment broth. Motile salmonellae, if present, are growth-amplified in the selective enrichment chamber and migrate into the motility chamber. When the diffusing antibody contacts the flagellar antigen, a visible line of precipitation is formed. Nonmotile salmonellae are not detected in this assay.

Enzyme-Linked Immunosorbent Assay (ELISA)

The ELISA is the earliest and probably the most prevalent format of antibody assay used for pathogen detection in foods. Commercially available ELISAs are usually designed as a "sandwich" assay, whereby antibody-coated solid matrices are used to capture the antigen (bacteria or toxin) from enrichment cultures, and a second antibody conjugated with an enzyme is added to form an antibody-antigen-conjugate "sandwich." Enzymes commonly used are alkaline phosphatase or horseradish peroxidase. A colorimetric substrate is then added, which is cleaved by the enzyme to produce a colored product that can be recorded visually or with a spectrophotometer (Fig. 38.5). Although the microtiter plate is the most commonly used solid matrix, variations of ELISA have been designed using dipsticks, paddles, membranes (9), pipette tips (35), or other supports. The detection sensitivity of ELISA is usually about 10^4 to 10^5 CFU/ml for whole bacterial cells and a few nanograms per milliliter or less for toxins or proteins (10); hence, cultural enrichment or extraction is

Table 38.4 Partial list of commercially available, antibody-based assays for the detection of foodborne pathogens and toxins[a]

Organism/toxin	Trade name	Assay format[b]	Manufacturer
Bacillus cereus diarrheal toxin	TECRA	ELISA	TECRA
	BCET	RPLA	Unipath
Campylobacter sp.	Campyslide	LA	Becton Dickinson
	Meritec-campy	LA	Meridian
	MicroScreen	LA	Mercia
	VIDAS	ELFA[c]	bioMérieux
	EiaFOSS	ELISA[c]	Foss
	TECRA	ELISA	TECRA
Clostridium botulinum toxin	ELCA	ELISA	Elcatech
Clostridium perfringens enterotoxin	PET	RPLA	Unipath
EHEC[d] O157	RIM O157&H7	LA	REMEL
	E. coli O157	LA	Unipath
	Prolex	LA	PRO-LAB
	EcolexO157	LA	Orion Diagnostica
	Wellcolex O157	LA	Murex
	E. coli O157	LA	TechLab
	O157&H7	Sera	Difco
	PetrifilmHEC	Ab-blot	3M
	EZ COLI	Tube-EIA	Difco
	Dynabeads	Ab-beads	Dynal
	EHEC-TEK	ELISA	Organon-Teknika
	Assurance[e]	ELISA	BioControl
	HECO157	ELISA	3M Canada
	TECRA	ELISA	TECRA
	E. coli O157	ELISA	LMD Lab
	Premier O157	ELISA	Meridian
	E. coli O157:H7	ELISA	Binax
	E. coli Rapitest	ELISA	Microgen
	Transia card	ELISA	Transia
	E. coli O157	EIA/capture	TECRA
	VIP[e]	Ab-ppt	BioControl
	Reveal	Ab-ppt	Neogen
	NOW	Ab-ppt	Binax
	Quix Rapid O157	Ab-ppt	Universal HealthWatch
	Immuno*Card*STAT	Ab-ppt	Meridian
	VIDAS	ELFA[c]	bioMérieux
	EiaFOSS	ELISA[c]	Foss
Shiga toxin (Stx)	VEROTEST	ELISA	MicroCarb
	Premier EHEC	ELISA	Meridian
	Verotox-F	RPLA	Denka Seiken
ETEC[d]			
Labile toxin (LT)	VET-RPLA	RPLA	Oxoid
Stabile toxin (ST)	*E. coli* ST	ELISA	Oxoid
Listeria sp.	Microscreen	LA	Microgen
	Listeria Latex	LA	Microgen
	Listeria-TEK[e]	ELISA	Organon Teknika
	TECRA[e]	ELISA	TECRA
	Assurance[e]	ELISA	BioControl
	Transia Listeria	ELISA	Transia
	Pathalert	ELISA	Merck
	Listertest	Ab-beads	VICAM
	Dynabeads	Ab-beads	Dynal
	VIP[e]	Ab-ppt	BioControl

(Continued)

Table 38.4 *(Continued)*

Organism/toxin	Trade name	Assay format[b]	Manufacturer
	Clearview	Ab-ppt	Unipath
	RAPIDTEST	Ab-ppt	Unipath
	VIDAS[e]	ELFA[c]	bioMérieux
	EiaFOSS	ELISA[c]	Foss
	UNIQUE	Capture-EIA	TECRA
Salmonella sp.	Bactigen	LA	Wampole Labs
	Spectate	LA	Rhone-Poulenc
	Microscreen	LA	Mercia
	Wellcolex	LA	Laboratoire Wellcome
	Serobact	LA	REMEL
	RAPIDTEST	LA	Unipath
	Dynabeads	Ab-beads	Dynal
	Screen	Ab-beads	VICAM
	CHECKPOINT	Ab-blot	KPL
	1-2 Test[e]	Diffusion	BioControl
	SalmonellaTEK[e]	ELISA	Organon Teknika
	TECRA[e]	ELISA	TECRA
	EQUATE	ELISA	Binax
	BacTrace	ELISA	KPL
	LOCATE	ELISA	Rhone-Poulenc
	Assurance[e]	ELISA	BioControl
	Salmonella	ELISA	GEM Biomedical
	Transia	ELISA	Transia
	Bioline	ELISA	Bioline
	VIDAS[e]	ELFA[c]	bioMérieux
	OPUS	ELISA[c]	TECRA
	PATH-STIK	Ab-ppt	LUMAC
	Reveal	Ab-ppt	Neogen
	Clearview	Ab-ppt	Unipath
	UNIQUE[e]	Capture-EIA	TECRA
Salmonella enterica serovar enteritidis	1-2 Test	Diffusion	BioControl
Shigella sp.	Bactigen	LA	Wampole Labs
	Wellcolex	LA	Laboratoire Wellcome
Staphylococcus aureus	Staphyloslide	LA	Becton Dickinson
	AureusTest[e]	LA	Trisum
	Staph Latex	LA	Difco
	S. aureus VIA	ELISA	TECRA
Enterotoxin	SET-EIA	ELISA	Toxin Technology
	SET-RPLA	RPLA	Unipath
	TECRA[e]	ELISA	TECRA
	Transia SE	ELISA	Transia
	RIDASCREEN	ELISA	R-Biopharm
	VIDAS	ELFA[c]	bioMérieux
	OPUS	ELISA[c]	TECRA
Vibrio cholerae	CholeraSMART	Ab-ppt	New Horizon
	BengalSMART	Ab-ppt	New Horizon
	CholeraScreen	Agglutination	New Horizon
	BengalScreen	Agglutination	New Horizon
Enterotoxin	VET-RPLA[f]	RPLA	Unipath

[a] Modified from reference 31.
[b] Abbreviations: Ab-ppt, immunoprecipitation; EIA, enzyme immunoassay; ELFA, enzyme-linked fluorescent assay; ELISA, enzyme-linked immunosorbent assay; LA, latex agglutination; RPLA, reverse passive latex agglutination.
[c] Automated ELISA.
[d] EHEC, enterohemorrhagic *E. coli*; ETEC, enterotoxigenic *E. coli*.
[e] Adopted by the AOAC as an official first or final action.
[f] Also detects *E. coli* LT enterotoxin.

needed before detection. Some ELISAs have been modified to include signal amplification to improve sensitivity. One of these, the ELISA ELCA (enzyme-linked coagulation assay) for *Clostridium botulinum* toxins, uses a coagulation-based amplification for detecting bound toxin-antibody complexes (77). Manipulations are involved in ELISAs to pipette samples and reagents, and the solid matrices also need to be washed thoroughly to avoid false reactions. For some assays, there are programmed autoanalyzers that perform ELISA procedures and reduce hands-on time. These automated ELISA instruments are commercially available, and most still use the sandwich antibody format. Some, such as the enzyme-linked fluorescence assay (ELFA), have been modified to use a fluorogenic substrate to enhance sensitivity (13, 16). After enrichment, automated systems can perform an ELISA in 45 min, as compared with a manual ELISA, which requires a few hours. Some of these ELISAs and automated systems are listed in Table 38.4.

Immunomagnetic Separation

In addition to their application in detection assays, antibodies can also be used in food analysis to selectively capture bacteria and shorten cultural enrichment time. In the immunomagnetic separation (IMS) technology, specific antibodies coupled to magnetic particles or beads are used to capture pathogens of interest from preenrichment media. IMS is analogous to selective cultural enrichment, whereby the growth of other bacteria is suppressed while the pathogen of interest is allowed to grow. The difference between the two processes, however, is that in selective enrichment, chemical reagents or antibiotics are used to select for the pathogens, while in IMS, pathogens are selectively captured by antibodies. Since reagents can be harsh and may cause cell stress or injury, IMS is a milder alternative for target enrichment; also, the elimination of selective enrichment steps shortens sample analysis time. Bacteria purified by IMS can be further tested by plating onto selective media or can be identified by serological, genetic, or other methods. IMS has been used to select for *Listeria* sp. (68), EHEC of the O157:H7 serotype (6, 86), *Salmonella* sp. (92), and other pathogens from foods, as well as from environmental and clinical samples (72, 79).

Immunoprecipitation

Recently, assays using immunoprecipitation or immunochromatography have also been introduced for detecting pathogens in foods. These assays are based on the technology developed for clinical diagnostics and are commonly used in home pregnancy tests. The assay is basically a sandwich antibody test, but instead of

enzyme conjugates, the detection antibodies are labeled colored latex beads or colloidal gold. Most of these assays consist of small, disposable plastic devices or dipsticks, with a sample inoculation chamber, a test chamber, and a control chamber where the results are read. After enrichment, a 0.1-ml sample is added to the inoculation chamber, where it is wicked across to the test and control chambers in a few minutes and the results are generated. Conceptually, the device consists of a series of absorbent pads that are saturated, in sequence, with latex-bound detection antibody (inoculation chamber) and capture antibody (test chamber), both of which are specific for the target antigen. In the control chamber there is another capture antibody, but it is specific not to the antigen but rather to the detection antibody. Hence, as the sample wicks across the pads, the antigen, if present, reacts first with the labeled detection antibody; this complex is then bound by the captured antibody (test chamber) to form a visible band of precipitation (8, 26). As the excess complex diffuses across to the control chamber, it is retained by the second capture antibody to form another line of precipitation (Fig. 38.6). The control chamber ensures that the sample has flowed across the entire device, so that the appearance of precipitation bands in both the test and control chambers indicates a valid positive test. Immunoprecipitation assays are extremely simple, requiring no washing or manipulations, and can be completed within 10 min after cultural enrichment. Many antibody-based assays for detecting foodborne pathogens now use this format (Table 38.4).

LIMITATIONS AND IMPACT

The number of rapid methods available for detecting bacterial pathogens and toxins in foods is impressive. However, as discussed earlier, food analysis remains challenging because food matrices interfere with microbiological methods, including rapid methods. As a result, despite the great impact rapid methods have had on food testing, there are also limitations.

LIMITATIONS

Enrichment

By themselves, most rapid methods can be done in a few minutes to a few hours; hence, they are more rapid than traditional microbiological procedures. But, in the analysis of foods, the problems inherent in food testing do not allow the direct application of rapid methods to testing for pathogens in foods. As a result, samples still need to be enriched in growth media before analysis. This continued reliance on culturing is the time-limiting step and

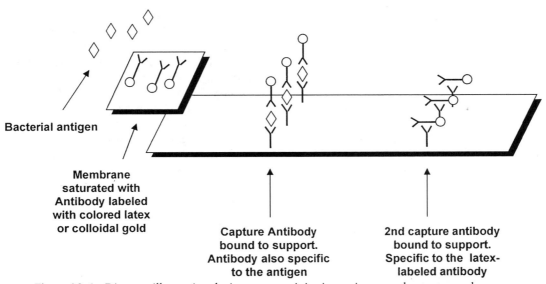

Bacterial antigen

Membrane saturated with Antibody labeled with colored latex or colloidal gold

Capture Antibody bound to support. Antibody also specific to the antigen

2nd capture antibody bound to support. Specific to the latex-labeled antibody

Figure 38.6 Diagram illustrating the immnuoprecipitation or immunochromatography assay format, in which samples wick across a series of membranes containing specific reagents.

somewhat diminishes the impact of rapid methods. Still, enrichment is a limitation only in terms of assay speed. Enrichment also provides essential benefits, such as dilution of the interfering effects of inhibitors, differentiation of viable from nonviable cells, and repair of stressed or injured cells.

Confirmation

An ideal application for rapid methods is for quick screening of foods for the presence of a particular pathogen or toxin. AOAC International-approved rapid methods are mostly designated for preliminary screening—negative results are regarded as definitive, but positive results are considered presumptive and must be confirmed by standard microbiological methods. Unlike culture tests, which almost always yield a viable isolate, many rapid methods require cell lysis or heat denaturation to expose antigens or targets; hence, the viable cells needed for confirmation can only be obtained by repeat analysis of the original enrichment medium using culture techniques. Although confirmation extends analysis time by several days, it may not be an imposing limitation, since most results encountered in the analysis of food samples are negative; there are few positive results to be confirmed.

Interference

Evaluation of rapid methods shows that some methods may perform better in some foods than others (47, 70). This may be attributed mostly to interference by normal biota and by food components, some of which can be especially troublesome for the technologies used in these assays. For example, ingredients in infant formula and intrinsic peroxidase in some vegetables cause false reactions in ELISAs that use peroxidase conjugates. Simple procedural variations, such as insufficient washing of the microtiter plate wells, can produce variable results in ELISAs, especially if the food contains highly "sticky" ingredients. Similarly, an ingredient in foods that inhibits an antibody assay by interfering with antigen-antibody interactions may not affect a DNA-based test. Other components can inhibit DNA hybridization or *Taq* polymerase but have no effect on antibodies. Since method efficiencies may be food dependent, it is advisable to perform some comparative studies to ensure that a particular assay will be effective in the analysis of that food type.

Gene Expression

DNA-based assays are highly specific, but the specificity is often dictated by short oligonucleotides. Hence, a positive result, for instance with a probe or PCR primers specific for a toxin gene, only indicates that bacteria with those gene sequences are present and that the bacteria have the potential to be toxigenic. It does not indicate that the cell is necessarily viable in the case of PCR, or that the gene is actually expressed and that the toxin is made. Bacteria carry genetic sequences that are not expressed owing to genetic mutations or physiological factors. Similarly, in cases of clostridial and staphylococcal intoxication, in which illnesses are caused by the ingestion of preformed toxins, DNA probes and PCR assays can be used to detect the presence of cells, but they are of limited use in detecting the presence of preformed toxins in foods.

Single-Target Design

Almost all rapid methods are designed to detect a single target, which makes them well suited for use in quality control programs to screen foods for the presence of specific targets. But in surveillance programs, a food may be tested simultaneously for the presence of several pathogens. Likewise, in illness investigations, the food implicated may be suspected to contain a particular pathogen based on clinical symptoms, but the actual pathogen is unknown. In these situations, when multiple pathogen analyses may be needed, the requirement for several single-target tests to analyze one food sample may make the procedure complex and costly.

Overabundance of Assays

Currently, there are at least 30 different commercial assays for testing for *Salmonella* sp. or EHEC O157:H7 (31). Such a large number of options can be confusing and overwhelming to the user; more important, this has limited the ability of the scientific community to effectively evaluate and validate these assays (see later section).

Impact on Regulations

The microbiological safety of foods is of primary concern to public health. The food industry and regulatory agencies have implemented programs to monitor foods for the presence of pathogens and microbial toxins. Many of the methods used in these compliance, surveillance, and quality control programs are time-consuming microbiological methods. Moreover, most of the food samples analyzed are negative for the presence of pathogens. This is an ideal situation in which rapid methods can be used to quickly screen large numbers of samples to eliminate those that are negative, thereby facilitating the performance of these programs. Before implementation, however, rapid methods should be carefully evaluated or, preferably, validated to ensure that they can effectively detect the pathogen of interest in that food type. Another issue of concern is that, currently, a positive result by a rapid method is regarded as presumptive and needs to be confirmed, but negative results are accepted without confirmation. This creates situations in which false-positive results are caught during confirmation but false-negative results are not detected unless other assays are used simultaneously during analysis. False-negative reactions are of great concern in terms of food safety and have serious implications, as they may result in foodborne infections.

As detection methods improve, levels of sensitivity also increase. While this may be good news for the consumer, the greater detection sensitivity can also create interesting challenges for the food industry and regulatory agencies. For example, current specifications for *Salmonella* sp. in ready-to-eat foods are "zero tolerance" or "absence." Likewise, microbial toxins are not permitted in foods, as the presence of any amount of these toxins is regarded as adulterating the food. But these "zero tolerance" limits are flexible, because the sensitivity of the methods varies, and "absence" may really mean "not detected." The problem with increased detection sensitivity is that it may give rise to situations in which foods previously analyzed by conventional methods and found to conform with the requirement of "absence" may no longer meet the same specifications if a more sensitive method is used. Such cases have already occurred; e.g., canned foods contaminated with low levels of bacterial toxins were not detected by traditional methods but were detected by rapid methods. In these situations, will food manufacturers have to modify their processing facilities or upgrade their quality control programs to comply with the greater test sensitivity? What is the health risk of the lower levels of pathogen and toxin detected? Will there be a need to implement changes every time a more sensitive method is introduced? Since guidelines are method dependent, will sterile foods be the ultimate limit as methods get more and more sensitive?

VALIDATION OF RAPID METHODS

All new methods developed for detecting foodborne pathogens or toxins should be validated by comparative testing with standard microbiological methods before their routine use in food analysis. In the United States, assays used in testing clinical specimens are reviewed and approved by the Center for Devices and Radiological Health of the Food and Drug Administration (FDA). However, in the analysis of foods, there are no such requirements for government approval. Although the federal agencies that regulate food safety—the Center for Food Safety and Applied Nutrition of the FDA and the Food Safety Inspection Service (FSIS) of the U.S. Department of Agriculture (USDA)—do have internal processes for evaluating methods used in food testing, most of the validations are done by outside, independent organizations. There are many national and international bodies that validate the effectiveness of testing methods for foods. In the United States, the principal organization is the AOAC, and methods approved by the AOAC are generally regarded as "official" or "standard" methods.

The International Standards Organization and the International Dairy Federation are prominent organizations dealing with validation, and the Codex Alimentarius of the United Nations accepts validation by all of the above bodies.

FDA's BAM and USDA's MLG

The FDA's *Bacteriological Analytical Manual* (*BAM*), published by AOAC, began in 1965 as a loose-leaf document intended to standardize methods internally within FDA testing labs. However, the reputation and usefulness of *BAM* in food analysis quickly spread, and it became generally available. *BAM* has since undergone eight revisions. The current 8A edition, 1998, is over 500 pages and is a compilation of media and methods preferred by the FDA for testing for bacterial pathogens, toxins, and other contaminants in foods and cosmetics. Many of the methods in *BAM* are AOAC-approved methods. But *BAM* also contains methods developed by the FDA and others that have been peer reviewed and evaluated by FDA scientists but have not been fully validated by collaborative studies (54). *BAM* is frequently cited and often requested by the food industry, nationally and internationally, and has been translated into several foreign languages, including Chinese and Spanish.

Similarly, the *Microbiology Laboratory Guidebook* (*MLG*), published by the FSIS of the USDA, is a manual of methods preferred by the FSIS for the microbiological analysis of meat, poultry, and egg products that are regulated by the USDA. Currently in its third edition, 1998, *MLG* contains many methods that have been internally evaluated for testing these products for microbial pathogens and other contaminants.

Both *BAM* and *MLG* also include some commercially available methods in their procedures or for specific applications; however, both manuals have clearly stated positions disclaiming the endorsement or official approval of these tests.

AOAC International Validation Programs

The most widely accepted method validation programs in the United States are those of AOAC International. There are currently three programs: the collaborative study and peer-verified programs are administered by AOAC International. The test kit Performance Tested Program is administered by AOAC Research Institute, a subsidiary of AOAC (3).

For many years, the AOAC used only the collaborative study program to validate microbiological and chemical assays used for testing foods. This is a rigorous, multistep process beginning with (i) ruggedness testing, to determine the effects of minor changes in parameters on assay performance; (ii) a precollaborative study, to establish the applicability of the method to multiple foods or to selected foods; and (iii) a collaborative study, which is a comparative testing of the method with standard methods by at least 15 collaborating labs, using multiple food samples that have been seeded at three levels (unseeded, low, and high) of the specific pathogen in question. Each sample is tested in five replicates, so that over 1,000 data points are generated for each method. This process is managed by an associated referee and closely monitored by a general referee, a methods committee, and statistical and safety advisors, who critique and approve experimental designs and protocols as well as review the final results. Methods that pass scrutiny are presented before the AOAC Official Methods Board for approval as a first action method. These methods become eligible for voting for final action status, if after 3 years of use by the scientific community, no major flaws were uncovered that could not be satisfactorily addressed or resolved before the general referee and the methods committee. Only methods that have been approved through the collaborative study process are regarded as standard methods and become listed in the *Official Methods of Analysis of the AOAC International* (3).

The peer-verified program of AOAC was established in 1993 and is a much less stringent validation program than the collaborative study program. However, it is designed to meet the requirements of some analysts or assay developers who wish to evaluate a new method quickly or to revise a validated method to extend its applicability or improve its performance. In this program, the analyst generates the performance data, writes an experimental protocol for approval by AOAC, and then selects an independent laboratory to repeat the approved protocol. Both sets of performance data are then reviewed by two technical reviewers, selected by the AOAC technical referee, who also grants acceptance or rejection of the method. Although it is a much simpler program, peer-verified methods are not regarded as official or standard methods (3).

The test kit performance testing program is administered by the AOAC Research Institute (AOAC RI), which was established as a subsidiary of the AOAC in 1992 to provide a third-party review of performance claims of manufactured test kits. Its sole purpose, therefore, is to validate commercially available rapid method kits. In this process, the manufacturer submits kit performance claims and data in support of those claims to AOAC RI. Two experts retained by AOAC RI review the material to see if the claims are substantiated by the data. If they are, the reviewers develop a protocol and use an independent laboratory for additional performance testing. If the newly generated data are in concurrence with the data submitted by the manufacturer, the kit is granted performance-tested status. Like the peer-verified methods, performance-tested methods are not regarded as official or standard methods.

Since implementation, the Performance Tested Program of the AOAC RI has often been confused with the collaborative study validation of the AOAC; both programs are also mutually competitive. As a result, the AOAC has recently modified the process by incorporating performance-tested validation into the collaborative study program, so that all kits with proprietary technologies must first be submitted for validation to the AOAC RI. Once it has attained performance-tested status, the kit may remain as such; but if the manufacturer opts to seek official status, the performance-tested status is accepted as equivalent to a precollaborative study so that it can proceed directly to the collaborative study phase of the official validation program.

In Tables 38.1 through 38.4, the methods that have attained official AOAC first or final action status are indicated. However, since these methods continue to be revised, the information shown may not be the most current.

FUTURE DEVELOPMENTS

The technologies predicted for food diagnostics in the Delphi Forecast of 1981 (46) have all become realities and the next generation of technologies is already being explored for use in food analysis. Some of these, like biosensors and DNA microarray (chips), have great potential in providing versatile assays that can detect analytes with extreme sensitivity and in near real time.

Immunosensors or Biosensors

First introduced in the 1970s, a biosensor consists of a biological or biologically derived sensing component, such as an antibody, enzyme, or receptor ligand, that is coupled to a physicochemical transducer, which can convert the biological signals, whether optical (light), electrochemical, thermometric (heat), or peizoelectric (mass), to generate measurable, digital electronic readings (66, 82). In clinical applications, compact, pocket-sized biosensors for testing for glucose levels in diabetics have been available since the 1980s (82). The use of biosensors in food analysis has progressed from the experimental to the developmental stage. An immunosensor, using antibody-coupled piezoelectric crystals, can rapidly detect 10^6 E. coli cells/ml in buffer (71). Other sensors have reported similar sensitivities, and some have even detected 10^4 Salmonella cells/ml (66). The sensitivity, however, decreases greatly when testing foods, owing to interference by food components. Detection sensitivity can be improved by coupling biosensor detection with IMS capture. One IMS-electrochemiluminescence biosensor was able to detect, within 1 h, 10^2 to 10^3 CFU/ml of E. coli O157 or Salmonella enterica serovar Typhimurium cells in buffer and 2×10^2 CFU/ml of these pathogens seeded in milk, juices, surface waters, and various suspensions of ground beef, ground chicken, and fish (100). A recent review by Goldschmidt (42) examines the scope of sensor application to microbiological analysis.

Other types of biosensors include the electronic nose that can detect and differentiate products or contaminants based on the emissions of volatile organic compounds, and artificial membranes that change color with the binding of the specific analytes. One sensor membrane is made of a polymer of diacetylenic acid to which has been incorporated ganglioside receptors of various bacterial toxins. The membrane has a characteristic blue color, which turns red when toxin is bound. The reaction is rapid and specific, and the color change can be quantified with a spectrophotometer (11).

The potential application of biosensor technology to food testing offers several attractive features. Many of these systems are portable, so can be used for field testing or on-the-spot analysis. Most are very rapid tests that can be done within an hour or less and are capable of testing multiple analytes simultaneously. However, the lack of adequate sampling and sample preparation techniques and interference by food components have hampered the application of these tests to food analysis. If these problems can be resolved, biosensors may be able to provide near real-time analysis for pathogens in foods, with potential for online monitoring and control of food processing (80).

DNA Microarray (Chips)

The basic concept of complementary hybridization used in DNA probe assays is undergoing a resurgence with the development of the DNA microarray technology. Also known as gene chips, this technology used photolithography, which was developed by computer chip makers to embed microcircuitry onto silicon wafers and adapted to embedding DNA oligonucleotides (75, 81). A gene chip can be made of glass or nylon membranes, and there are two basic variants. In one format, target DNA is amplified by PCR and spotted onto a membrane, which is then probed either singly or simultaneously by hundreds of labeled probes to determine specific hybridization. In the other format, an array of oligonucleotides are either synthesized directly on a glass chip or embedded after synthesis, and then exposed to labeled target

DNA (75). Fluorescent labels are commonly used in gene chip assays, and these can be labeled chemically or during PCR amplification; however, some systems have had problems distinguishing weak fluorescence from background signals. As a result, electrical signals instead of fluorescence patterns are also being explored for detecting specific hybridization (81). Some gene chips can hold up to 400,000 or more distinct testing spots. These can work as a miniaturized molecular biology laboratory to allow simultaneous testing of thousands of targets or analytes. Microarray assays are used to study gene expression, detection of gene mutations or polymorphism, and gene sequencing and mapping. The advantage to their potential application to diagnostics is that they not only have the capacity for multiple pathogen testing but can also be used to detect the presence of specific virulence genes for each pathogen detected in order to provide virulence profiles for the organisms. A gene chip developed by immobilizing rRNA-specific oligonucleotides on a matrix of polyacrylamide gel pads on a glass slide was used to enumerate nitrifying bacteria in soil. By labeling the target DNA or RNA with fluorescein isothiocyanate, specific hybridization reactions were detected by fluorescence (45). The assay can be modified to quantify individual populations by using a multicolor detection system. Despite the tremendous potential of microarray technology, only a few products have been developed—so far, none of them for foodborne pathogens. Aside from the difficulties inherent in food testing, another factor that may be hindering its general application may be disputes over patented genes, proprietary technologies, or licensing agreements that are involved in designing DNA microarray assays (81).

CONCLUSIONS

Traditionally, methods to detect foodborne pathogens have relied primarily on culture media to select and propagate viable cells in foods. Although effective, these assays are labor-intensive and time-consuming, often requiring several days to complete. Advances in biotechnology have introduced new techniques to food testing that are faster and more sensitive than traditional methods. These assays, collectively known as rapid methods, comprise a large, diverse group of tests, including miniaturized biochemical tests, specialized media and substrates, and DNA- and antibody-based detection assays. Most of these tests are single-target assays that are designed to expedite analysis; hence, they are well suited for use in preliminary screening of large numbers of food samples for the presence of a specific pathogen. However, since these assays remain susceptible to interference by normal bacterial biota and by the complexity of food matrices, it is essential that they be thoroughly evaluated, preferably by comparative or official collaborative studies, before use in routine analysis. Rapid methods are often more sensitive than conventional microbiological tests, but as assays become more and more sensitive, they may present interesting challenges to the food industry and regulatory agencies. Last, technology continues to advance at a great pace. The next generation of assays currently being developed potentially has the capability for near real-time and online monitoring of multiple pathogens.

I thank G. J. Jackson for helpful comments and critical reading of this manuscript.

References

1. **Akane, A., K. Matsubara, H. Nakamura, S. Takahashi, and K. Kimura.** 1992. Identification of the heme compound copurified with deoxyribonucleic acid (DNA) from bloodstains, a major inhibitor of polymerase chain reaction (PCR) amplification. *J. Forensic Sci.* 39:362–372.

2. **Andersen, M. R., and C. J. Omiecinski.** 1992. Direct extraction of bacterial plasmids from food for polymerase chain reaction. *Appl. Environ. Microbiol.* 58:4080–4082.

3. **Andrews, W. H.** 1996. AOAC International's three validation programs for methods used in the microbiological analysis of foods. *Trends Food Sci. Technol.* 7:147–151.

4. **Baker, J. M, M. W. Griffiths, and D. L. Collins-Thompson.** 1992. Bacterial bioluminescence: application in food microbiology. *J. Food Prot.* 55:62–70.

5. **Barrett, T., P. Feng, and B. Swaminathan.** 1997. Amplification methods for detection of foodborne pathogens, p. 171–181. *In* H. H. Lee, S. A. Morse, and O. Olsvik (ed.), *Nucleic Acid Amplification Techniques: Application to Disease Diagnosis.* Eaton Publishing, Boston, Mass.

6. **Bennett, A. R., S. MacPhee, and R. P. Bett.** 1996. The isolation and detection of *Escherichia coli* O157 by use of immunomagnetic separation and immunoassay procedures. *Lett. Appl. Microbiol.* 22:237–243.

7. **Blackburn, C. W.** 1993. Rapid and alternative methods for the detection of salmonellas in foods. *J. Appl. Bacteriol.* 75:199–214.

8. **Brinkman, E., R. Van Beurden, R. Mackintosh, and R. Beumer.** 1995. Evaluation of a new dip-stick test for the rapid detection of Salmonella in foods. *J. Food Prot.* 58:1023–1027.

9. **Calicchia, M. L., J. D. Reger, C. I. N. Wang, and D. W. Osato.** 1994. Direct enumeration of *Escherichia coli* O157:H7 from Petrifilm EC count plates using the Petrifilm Test kit—HEC without sample pre-enrichment. *J. Food Prot.* 57:859–864.

10. **Candish, A. A. G.** 1991. Immunological methods in food microbiology. *Food Microbiol.* 8:1–14.

11. Charych, D., Q. Cheng, A. Reichert, G. Kuziemko, M. Stroh, J. O. Nagy, W. Spevak, and R. C. Stevens. 1996. A "litmus test" for molecular recognition using artificial membranes. *Chem. Biol.* 3:113–120.

12. Clark, J. A. 1980. The influence of increasing numbers of non-indicator organisms upon the detection of indicator organisms by the membrane filter and presence-absence tests. *Can. J. Microbiol.* 26:827–832.

13. Cohen, A. E., and K. F. Kerdahi. 1996. Evaluation of a rapid and automated enzyme-linked fluorescent immunoassay for detecting *Escherichia coli* serogroup O157 in cheese. *J. AOAC Int.* 79:858–860.

14. Cox, N. A., D. Y. C. Fung, J. S. Bailey, P. A. Hartman, and P. C. Vasavada. 1987. Miniaturized kits, immunoassays and DNA hybridization for recognition and identification of foodborne bacteria. *Dairy Food Sanit.* 7:628–631.

15. Cox, N. A., D. Y. C. Fung, M. C. Goldschmidt, J. S. Bailey, and J. E. Thomson. 1984. Selecting a miniaturized system for identification of Enterobacteriaceae. *J. Food Prot.* 47:74–77.

16. Curiale, M. S., V. Gangar, and C. Gravens. 1997. VIDAS enzyme-linked fluorescent immunoassay for detection of *Salmonella* in foods: collaborative study. *J. AOAC Int.* 80:491–504.

17. Curiale, M. S., M. J. Klatt, and M. A. Mozola. 1990. Colorimetric deoxyribonucleic acid hybridization assay for rapid screening of *Salmonella* in foods: collaborative study. *J. Assoc. Off. Anal. Chem.* 73:248–256.

18. Curiale, M. S., T. Sons, D. McIver, J. S. McAllister, B. Halsey, D. Robles, and T. L. Fox. 1991. Dry rehydratable film for enumeration of total coliforms and *Escherichia coli* in foods: collaborative study. *J. Assoc. Off. Anal. Chem.* 74:635–648.

19. D'Aoust, J.-Y., A. M. Sewell, and P. Greco. 1991. Commercial latex agglutination kits for the detection of foodborne *Salmonella.* *J. Food Prot.* 54:725–730.

20. Deak, T. 1995. Methods for the rapid detection and identification of yeasts in foods. *Trends Food Sci. Technol.* 6:287–292.

21. Dziezak, J. D. 1987. Rapid methods for microbiological analysis of food. *Food Technol.* 41:56–73.

22. Earnshaw, R., and J. Gidley. 1992. Molecular methods for typing food pathogens. *Trends Food Sci. Technol.* 3:39–43.

23. Engvall, E., and P. Perlmann. 1971. Enzyme-linked immunosorbent assay (ELISA) quantitative assay of immunoglobulin G. *Immunochemistry* 8:871–874.

24. Entis, P. 1989. Hydrophobic grid membrane filter/MUG method for total coliform and *Escherichia coli* enumeration in foods: collaborative study. *J. Assoc. Off. Anal. Chem.* 72:936–950.

25. Ewing, W. H. 1986. *Edward's and Ewing's Identification of Enterobacteriaceae*, 4th ed. Elsevier, New York, N.Y.

26. Feldsine, P. T., R. L. Forgey, M. T. Falbo-Nelson, and S. L. Brunelle. 1997. *Escherichia coli* O157:H7 visual immunoprecipitate assay: a comparative validation study. *J. AOAC. Int.* 80:43–48.

27. Feldsine, P. T., M. T. Falbo-Nelson, and D. L. Hustead. 1993. Substrate supporting disc method for confirmed detection of total coliforms and *E. coli* in all foods: comparative study. *J. AOAC Int.* 76:988–1005.

28. Feng, P. 1992 Commercial assay systems for detecting foodborne *Salmonella*: a review. *J. Food Prot.* 55:927–934.

29. Feng, P. 1996. Emergence of rapid methods for identifying microbial pathogens in foods. *J. AOAC Int.* 79:809–812.

30. Feng, P. 1997. Impact of molecular biology on the detection of foodborne pathogens. *Mol. Biotechnol.* 7:267–278.

31. Feng, P. 1998. Rapid methods for detecting foodborne pathogens, p. App.1.01–App.1.16. *In FDA Bacteriological Analytical Manual*, 8A ed. AOAC International, Gaithersburg, Md.

32. Feng, P., and P. A. Hartman. 1982. Fluorogenic assays for immediate confirmation of *Escherichia coli*. *Appl. Environ. Microbiol.* 43:1320–1329.

33. Feng, P., S. P. Keasler, and W. E. Hill. 1992. Direct identification of *Yersinia enterocolitica* in blood by polymerase chain reaction. *Transfusion* 32:850–854.

34. Feng, P., K. A. Lampel, and W. E. Hill. 1996. Developments in food technology: applications and economic and regulatory considerations, p. 203–229. *In* C. A. Dangler and B. Osburn (ed.), *Nucleic Acid Analysis: Principles and Bioapplications*. Wiley & Sons, New York, N.Y.

35. Firstenberg-Eden, R., and N. M. Sullivan. 1997. EZ Coli rapid detection system: a rapid method for the detection of *Escherichia coli* O157 in meat and other foods. *J. Food Prot.* 60:219–225.

36. Fitts, R., M. Diamond, C. Hamilton, and M. Neri. 1983. DNA-DNA hybridization assay for detection of *Salmonella* spp. in foods. *Appl. Environ. Microbiol.* 46:1146–1151.

37. Fung, D. Y. C., N. A. Cox, and J. S. Bailey. 1988. Rapid methods and automation in the microbiological examination of foods. *Dairy Food Sanit.* 8:292–296.

38. Gaillot, O., P. D. Camillo, P. Berche, R. Courcol, and C. Savage. 1999. Comparison of CHROMagar salmonella medium and Hektoen Enteric agar for isolation of salmonellae from stool samples. *J. Clin. Microbiol.* 37:762–765.

39. Gazzaz, S. S., B. A. Rasco, and F. M. Dong. 1992. Application of immunochemical assays to food analysis. *Crit. Rev. Food Sci. Nutr.* 32:197–229.

40. Gibb, A. P., and S. Wong. 1998. Inhibition of PCR by agar from bacteriological transport media. *J. Clin. Microbiol.* 36:275–276.

41. Giese, J. 1995. Rapid microbiological testing kits and instruments. *Food Technol.* 49:64–71.

42. Goldschmidt, M. 1999. Biosensors—scope in microbiological analysis, p. 268–278. *In* R. K. Robinson, C. A. Batt, and P. Patel (ed.), *Encyclopedia of Food Microbiology*. Academic Press Ltd., London, United Kingdom.

43. Grif, K., H. Karch, C. Schneider, F. D. Daschner, L. Beutin, T. Cheasty, H. Smith, B. Rowe, M. P. Dierich, and F. Allerberger. 1998. Comparative study of five different techniques for epidemiological typing of *Escherichia coli* O157. *Diagn. Microbiol. Infect. Dis.* 32:165–176.

44. **Griffin, P. M., B. P. Bell, P. R. Cieslak, J. Tuttle, T. J. Barrett, M. P. Doyle, A. M. McNamara, A. M. Shefer, and J. G. Wells.** 1994. Large outbreak of *Escherichia coli* O157:H7 infections in the Western United States: the big picture, p. 7–12. *In* M. A. Karmali and A. G. Goglio (ed.), *Recent Advances in Verocytotoxin-Producing Escherichia coli Infections.* Elsevier Science B.V., Amsterdam, The Netherlands.

45. **Guschin, D. Y., B. K. Mobarry, D. Proudnikov, D. A. Stahl, B. E. Rittmann, and A. D. Mirzabekov.** 1997. Oligonucleotide microchip as genosensors for determinative and environmental studies in microbiology. *Appl. Environ. Microbiol.* **63:**2397–2402.

46. **Gutteridge, C. S., and M. L. Arnott.** 1989. Rapid methods: an over the horizon view, p. 297–319. *In* M.R. Adams and C.F.A. Hope (ed.), *Rapid Methods in Food Microbiology: Progress in Industrial Microbiology*, Elsevier, New York, N.Y.

47. **Hanai, K., M. Satake, H. Nakanishi, and K. Venkateswaran.** 1997. Comparison of commercially available kits with standard methods for detection of *Salmonella* strains in foods. *Appl. Environ. Microbiol.* **63:**775–778.

48. **Harlander, S.** 1989. Food biotechnology: yesterday, today and tomorrow. *Food Technol.* **43:**196–206.

49. **Hartman, P. A., B. Swaminathan, M. S. Curiale, R. Firstenberg-Eden, A. N. Sharpe, N. A. Cox, D. Y. C. Fung, and M. C. Goldschmidt.** 1992. Rapid methods and automation, p. 665–746. *In* C. Vanderzant and D. F. Splittstoesser (ed.), *Compendium of Methods for the Microbiological Examination of Foods.* American Public Health Association, Washington, D.C.

50. **Hill, W. E.** 1996. The polymerase chain reaction: application for the detection of foodborne pathogens. *Crit. Rev. Food Sci. Nutr.* **36:**123–173.

51. **Hill, W. E., and O. Olsvik.** 1994. Detection and identification of foodborne pathogens by the polymerase chain reaction: food safety applications, p. 268–289. *In* P. D. Patel (ed.), *Rapid Analysis Techniques in Food Microbiology.* Chapman & Hall, Ltd., London, United Kingdom.

52. **Hunt, P. R.** 1989. Development and growth of the Food and Drug Administration. *Food Technol.* **43:**280–286.

53. **Hurst, A.** 1977. Bacterial injury: a review. *Can. J. Microbiol.* **23:**935–944.

54. **Jackson, G. J., and I. K. Wachsmuth.** 1998. The US Food and Drug Administration's selection and validation of tests for foodborne microbes and microbial toxins. *Food Control* **7:**37–39.

55. **Jay, J. M.** 1985. Analysis of food products for microorganisms or their products by nonculture methods, p. 87–126. *In* D. W. Gruenwedel and J. R. Whitaker (ed.), *Food Analysis Principles and Techniques.* Marcel Dekker, Inc., New York, N.Y.

56. **Jones, D. D., and A. K. Bej.** 1994. Detection of foodborne microbial pathogens using polymerase chain reaction methods, p. 341–365. *In* H. G. Griffin and A. M. Griffin (ed.), *PCR Technology: Current Innovations.* CRC Press, Boca Raton, Fla.

57. **Kalamaki, M., R., J. Prioce, and D. Y. C. Fung.** 1997. Rapid methods for identifying seafood microbial pathogens and toxins. *J. Rapid Methods Automation Microbiol.* **5:**87–137.

58. **Karmali, M. A., M. Petric, and M. Bielaszewska.** 1999. Evaluation of a microplate latex agglutination method (Verotox–F) for detecting and characterizing verotoxin (Shiga toxin) in *Escherichia coli. J. Clin. Microbiol.* **37:**396–399.

59. **Kennedy, M. J.** 1992. The evolution of the word "biotechnology." *Trends Food Sci. Technol.* **3:**154–156.

60. **Klinger, J. M., R. P. Stowe, D. C. Obenhuber, T. O. Groves, S. K. Mishra, and D. L. Pierson.** 1992. Evaluation of the Biolog automated microbial identification system. *Appl. Environ. Microbiol.* **58:**2089–2092.

61. **Kodikara, C. P., H. H. Crew, and G. S. A. B. Stewart.** 1991. Near on-line detection of enteric bacteria using *lux* recombinant bacteriophage. *FEMS Microbiol. Lett.* **83:**261–266.

62. **Lampel, K. A., P. Feng, and W. E. Hill.** 1992. Gene probes used in food microbiology, p. 151–188. *In* D. Bhatnagar and T. E. Cleveland (ed.), *Molecular Approaches to Improving Food Safety.* Van Nostrand Reinhold, New York, N.Y.

63. **Lantz, P.-G., B. Hahn-Hagerdal, and P. Radstrom.** 1994. Sample preparation methods in PCR-based detection of food pathogens. *Trends Food Sci. Technol.* **5:**384–389.

64. **Li, J., and T.-C. Lee.** 1995. Bacterial ice nucleation and its potential application in the food industry. *Trends Food Sci. Technol.* **6:**259–265.

65. **March S. B., and S. Ratnam.** 1989. Latex agglutination test for detection of *Escherichia coli* O157. *J. Clin. Microbiol.* **27:**1675–1677.

66. **Medina, M. B.** 1997. SPR Biosensor: food science applications. *Food Testing Analysis* **3:**14–16.

67. **Middlekauff, R. D.** 1989. Regulating the safety of foods. *Food Technol.* **43:**296–306.

68. **Mitchell, B. A., J. A. Milbury, A. M. Brookins, and B. J. Jackson.** 1994. Use of immunomagnetic capture on beads to recover *Listeria* from environmental samples. *J. Food Prot.* **57:**743–745.

69. **Ninet, B., E. Bannerman, and J. Bille.** 1992. Assessment of the Accuprobe *Listeria monocytogenes* culture identification reagent kit for rapid colony confirmation and its application in various enrichment broths. *Appl. Environ. Microbiol.* **58:**4055–4059.

70. **Niroomand, F., and C. Ford.** 1994. Comparison of rapid techniques for detection of *Escherichia coli* O157:H7. *J. Rapid Methods Automation Microbiol.* **3:**85–96.

71. **Oh, S.** 1993. Immunosensors for food safety. *Trends Food Sci. Technol.* **4:**98–103.

72. **Olsvik, O., T. Popovic, E. Skjerve, K. S. Cudjoe, E. Hornes, J. Ugelstad, and M. Uhlen.** 1994. Magnetic separation techniques in diagnostic microbiology. *Clin. Microbiol. Rev.* **7:**43–54.

73. **Partis, L., K. Newton, J. Murby, and R. J. Wells.** 1994. Inhibitory effects of enrichment media on the Accuprobe test for *Listeria monocytogenes. Appl. Environ. Microbiol.* **60:**1693–1694.

74. Pickup, R. W. 1991. Development of molecular methods for the detection of specific bacteria in the environment. *J. Gen. Microbiol.* **137**:1009–1019.

75. Ramsay, G. 1998. DNA chips: state-of-the art. *Nat. Biotechnol.* **16**:40–44.

76. Ray, B. 1979. Methods to detect stressed microorganisms. *J. Food Prot.* **42**:346–355.

77. Roman, M. G., J. Y. Humber, P. A. Hall, N. R. Reddy, H. A. Solomon, M. X. Triscott, G. A. Beard, J. D. Bottoms, T. Cheng, and G. J. Doellgast. 1994. Amplified immunoassay ELISA-ELCA for measuring *Clostridium botulinum* type E neurotoxin in fish fillets. *J. Food Prot.* **57**:985–990.

78. Rossen, L., P. Norskov, K. Holmstrom, and O. F. Rasmussen. 1992. Inhibition of PCR by components of food samples, and DNA-extraction solutions. *Int. J. Food Microbiol.* **17**:37–45.

79. Safarik, I., M. Safarikova, and S. J. Forsythe. 1995. The application of magnetic separation in applied microbiology. *J. Appl. Bacteriol.* **78**:575–585.

80. Schugerl, K., B. Hitzmann, H. Jurgens, T. Kullick, R. Ulber, and B. Weigal. 1994. Challenges in integrating biosensors and FIA for on-line monitoring and control. *Trends Biotechnol.* **14**:21–31.

81. Service, R. S. 1998. Microchip arrays put DNA on the spot. *Science* **282**:396–399.

82. Severs, A. H. 1994. Biosensors for food analysis. *Trends Food Sci. Technol.* **5**:230–232.

83. Stager, C. E., and J. R. Davis. 1992. Automated systems for identification of microorganisms. *Clin. Microbiol. Rev.* **5**:302–327.

84. Stewart, G. S. A. B., and P. Williams. 1992. *lux* genes and the application of bacterial bioluminescence. *J. Gen. Microbiol.* **138**:1289–1300.

85. Swaminathan, B., and P. Feng. 1994. Rapid detection of food-borne pathogenic bacteria. *Annu. Rev. Microbiol.* **48**:401–426.

86. Tomoyasu, T. 1998. Improvement of the immunomagnetic separation method selective for *Escherichia coli* O157 strains. *Appl. Environ. Microbiol.* **64**:376–382.

87. Tsai, Y.-L., and B. H. Olson. 1992. Detection of low numbers of bacterial cells in soil and sediments by polymerase chain reaction. *Appl. Environ. Microbiol.* **58**:754–757.

88. Turpin, P. E., K. A. Maycroft, J., Bedford, C. L. Rowlands, and E. M. H. Wellington. 1993. A rapid luminescent-phage based MPN method for the enumeration of *Salmonella typhimurium* in environmental samples. *Lett. Appl. Microbiol.* **16**:24–27.

89. Ulizer, S., and J. Kuhn. 1987. Introduction of *lux* genes into bacteria, a new approach for specific determination of bacteria and their antibiotic susceptibility, p. 463–472. *In* J. Scholmerich, R. Andreesen, A. Kapp, M. Ernst, and W. G. Wood (ed.), *Bioluminescence and Chemiluminescence New Perspectives.* John Wiley & Sons, Toronto, Ontario, Canada.

90. Uyttendaele, M., R. Schukkink, B. van Gemen, and J. Debevere. 1995. Development of NASBA, a nucleic acid amplification system, for identification of *Listeria monocytogenes* and comparison to ELISA and a modified FDA method. *Int. J. Food Microbiol.* **27**:77–89.

91. Vaneechoute, M., and J. Van Eldere. 1997. The possibilities and limitations of nucleic acid amplification technology in diagnostic microbiology. *J. Med. Microbiol.* **46**:188–194.

92. Varmut, A. E. M., A. A. J. M. Franken, and R. P. Beumer. 1992. Isolation of salmonellas by immunomagnetic separation. *J. Appl. Bacteriol.* **72**:112–118.

93. Warburton, D. W., P. T. Feldsine, and M. T. Falbo-Nelson. 1995. Modified immunodiffusion method for detection of *Salmonella* in raw flesh and highly contaminated foods: collaborative study. *J. AOAC Int.* **78**:59–68.

94. Warren G., and P. Wolber. 1991. Molecular aspects of microbial ice nucleation. *Mol. Microbiol.* **5**:239–243.

95. Wiedmann, M., F. Barany, and C. A. Batt. 1993. Detection of *Listeria monocytogenes* with a nonisotopic polymerase chain reaction-coupled ligase chain reaction assay. *Appl. Environ. Microbiol.* **59**:2743–2745.

96. Wilde, J., J. Eiden, and R. Yolken. 1990. Removal of inhibitory substances from human fecal specimens for detection of group A rotaviruses by reverse transcriptase and polymerase chain reactions. *J. Clin. Microbiol.* **28**:1300–1307.

97. Wolber, P. K., and R. L. Green. 1990. New rapid method for the detection of *Salmonella* in foods. *Trends Food Sci. Technol.* **1**:80–82.

98. Wolber, P. K., P. L. Green, W. T. Tucker, N. M. Watanabe, C. A. Vance, R. A. Fallon, C. Lindhardt, and A. J. Smith. 1995. Transduction of *ina* genes for bacterial identification, p. 283–298. *In* R. E. Lee, G. J. Warren, and L. V. Gusta (ed.), *Biological Ice Nucleation and Its Applications.* American Phytopathological Society, St. Paul, Minn.

99. Wolcott, M. J. 1992. Advances in nucleic acid-based detection methods. *Clin. Microbiol. Rev.* **5**:370–386.

100. Yu, H., and J. G. Bruno. 1996. Immunomagnetic-electrochemiluminescent detection of *Escherichia coli* O157:H7 and *Salmonella typhimurium* in foods and environmental water samples. *Appl. Environ. Microbiol.* **62**:587–592.

Food Microbiology: *Fundamentals and Frontiers*, 2nd Ed.
Edited by M. P. Doyle et al.
© 2001 ASM Press, Washington, D.C.

Todd R. Klaenhammer

Probiotics and Prebiotics

39

THE PROBIOTIC CONCEPT

Many beneficial bacteria within our food supply inhabit the gastrointestinal tract (GIT) of humans, animals, and birds. Probiotic (which means "for life") bacteria have long been considered to influence general health and well-being via their commensal association with the GIT and its normal microflora. The concept of probiotics was first popularized at the turn of the 20th century by the Russian Nobel laureate Elie Metchnikoff (Fig. 39.1). Metchnikoff proposed that a normal, healthy gastrointestinal microflora in humans and animals provided resistance against "putrefactive" intestinal pathogens (3). He theorized that the intestinal flora influences the incidence and severity of enteric infections and either enhances or slows atrophy and aging processes. During that period, it was reported that infants subsisting on breast milk develop a characteristic intestinal flora of rod- and bifid-shaped bacteria, then called *Bacillus bifidus* (now known to represent *Lactobacillus* and *Bifidobacterium* species). The fermentative bacteria constituting this flora produced lactic acid, were associated with fermented milk products, and did not produce putrefactive compounds or toxins. Metchnikoff proceeded to isolate a *Lactobacillus* culture from fermented milk consumed by Bulgarian peasants, who were renowned

for living long and healthy lives. The culture produced large amounts of lactic acid and survived during intestinal implantation studies. From these and many other observations made during his work on the intestinal microflora, Metchnikoff suggested that by transforming the "wild population of the intestine into a cultured population...the pathological symptoms may be removed from old age, and...in all probability, the duration of life of man may be considerably increased" (as related by Bibel in reference 3). The value of consuming fermented dairy products (yogurt, kefir, sour cream) became popular in Europe, with the Pasteur Institute supporting the manufacture of a product called Le Ferment, based on the Bulgarian *Lactobacillus* sp. Later, Minoru Shirota (ca. 1930; related in reference 45) selected from human feces a *Lactobacillus* culture, *Lactobacillus casei* strain Shirota, that survived passage through the GIT. The fermented milk Yakult is manufactured using a pure culture of strain Shirota. The product remains a dietary staple in Japan and Korea and is now available in Europe.

THE NORMAL MICROFLORA

Hundreds of microbial species live as commensals, largely on mucosal tissues of the nose, mouth, intestinal

Todd R. Klaenhammer, Department of Food Science, Schaub Hall, Box 7624, North Carolina State University, Raleigh, NC 27695-7624.

Figure 39.1 Probiotic pioneers: Eli Mechnikoff (1845–1916) (from reference 45 with permission); (top) *Lactobacillus casei* Shirota (from reference 45 with permission); (bottom) *Bifidobacterium* sp. (from reference 7 [Danone/INRA Versailles], with permission).

tract and vagina (Fig. 39.2). Bacterial populations found throughout the human body are estimated at 10^{14} cells, 10-fold more cells than the 10^{13} composing the human body itself. It is well established that the composition of the microflora is complex, dynamic, and specific to each host and can change markedly with diet, age, and lifestyle (Fig. 39.3) (32, 51). Comparative studies with colonized and germfree gnotobiotic animals have established that the normal gastrointestinal microflora is responsible for many important properties that affect nutrition (digestion, feed conversion), pathogen sensitivity, enterohepatic circulation (bile salt hydrolases and glucuronidases), and immunomodulation via the muscosal immunity system (51). Microbial interference and colonization resistance are properties of the normal microflora that reflect the ability of the existing flora to competitively occupy available sites and establish an immunity profile through which host and nonhost microorganisms are recognized. The collective microfloras of humans, animals, and fowl vary considerably with the architecture of their GIT (51). Groups of microorganisms are located at various

locations throughout the GIT and can reflect those that are either harmful or beneficial or those that may act positively or negatively, depending on the circumstances and specific strains involved (Fig. 39.4) (12). For example, there is a clear distinction between *Escherichia coli* and the fecal coliforms composing the normal flora and those varieties that are pathogenic and/or enterotoxigenic, such as *E. coli* O157:H7. General impacts on the host can depend on the many factors that direct the composition of the normal microflora, including diet, age, exposure to various microorganisms, and the genetic or physiological conditions of the host tissues themselves. The digestive tract is composed four of major microbial population categories (8, 52, 54), defined as:

- Autochthonous microflora—populations of microbes that are present at high levels and permanently colonize the host
- Normal microflora—microorganisms that are frequently present, but can vary in number and be sporadically absent

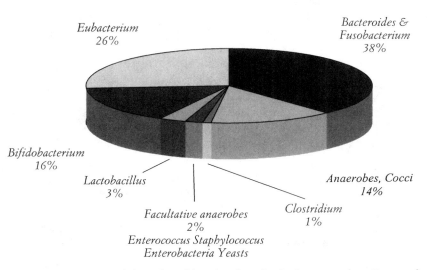

Figure 39.2 Composition of the culturable microflora in the human colon. From reference 36a with permission.

- True pathogens—microorganisms that are periodically acquired, but can persist, causing infection and disease
- Allochthonous microflora—microbes of another origin and present temporarily (most probiotic cultures are allochthonous [54])

Molecular technologies (55) have revealed that individual humans carry their own unique collection of autochthonous strains. While the exact microorganisms vary considerably between individuals, it is now clear that each person exhibits a persistent flora that can be recovered repeatedly over extended periods of time. Severe disturbances to the GIT, such as antibiotic therapy, enteric infections, or dietary stress, can temporarily disrupt the autochthonous and normal microflora. However, these populations appear to recover quickly as the host returns to normalcy. Of the predominant colonic microorganisms, both *Lactobacillus* and *Bifidobacterium* species are considered exclusively beneficial and elicit no harmful or pathogenic responses within their normal ecological niches (Fig. 39.4). As a result, they are the two major species upon which the probiotic and prebiotic concepts have evolved. From this platform, the question of probiotics and their ability to positively or negatively impact the autochthonous, normal, or pathogenic microflora of the GIT can be discussed.

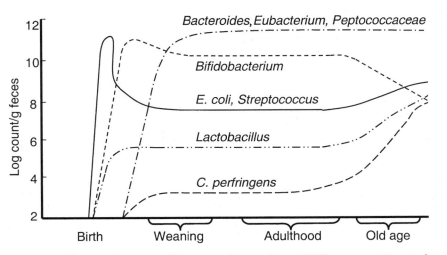

Figure 39.3 Differences in species of bacteria in human feces of different ages. From reference 32 with permission.

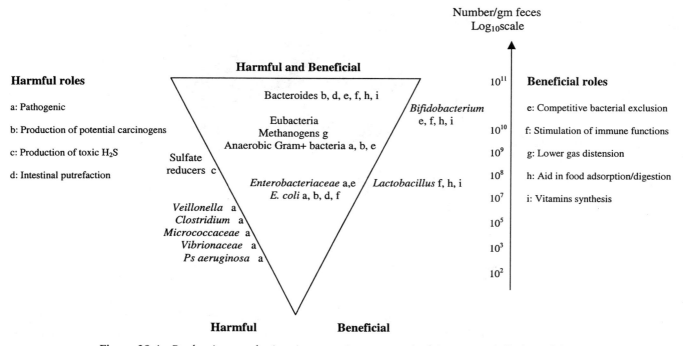

Figure 39.4 Predominant colonic microorganisms categorized into potentially harmful or beneficial groups. Adapted from reference 12.

PROBIOTICS, ABIOTICS, AND PREBIOTICS

In contrast to "antibiotic," the term "probiotic" was first coined to describe a substance produced by one microorganism that stimulates the growth of another microorganism (27). Fuller (10) defined probiotics as "a live microbial feed supplement which beneficially affects the host animal by improving its intestinal balance." The definition has evolved in recent years to include the many types of probiotic cultures (mono- and mixed-strain cultures, multiple probiotic species), applications (gastrointestinal versus topical), and mechanisms of probiotic activity (live cells, dead cells, and cellular components). Therefore, the following definitions will be applied in this text:

Probiotics: Live microbial cell preparations that when applied or ingested in certain numbers exert a beneficial effect on health and well-being (adapted from definitions in references 9, 14, and 38)

Abiotics: Nonviable probiotic organisms or cellular components thereof that exert beneficial effects on health or well-being (adapted from reference 46)

In addition to the multiple impacts of probiotics and abiotics, benefits to intestinal health can be realized by feeding ingredients that selectively promote the growth of the existing beneficial microflora. Growth factors in human breast milk, which stimulate the growth of *Bifidobacterium* sp., have long been recognized to change the composition of the intestinal microflora. Bifidogenic factors are typically complex carbohydrates (e.g., galactosyllactose in breast milk) which are not metabolized by the host or microflora residing in the upper GIT. As a consequence, these factors reach the colon, where they are preferentially metabolized by the residing bifidobacteria. Stimulation of the natural population levels of beneficial bacteria in this manner decreases colonic pH, stimulates mucosal immunity, and retards enteric infections (6). Expansion of this core concept toward the development of food ingredients, metabolized selectively by one or more groups of beneficial intestinal bacteria, has been championed by Gibson and Roberfroid (12), leading to the following definitions:

Prebiotic: A nondigestible food ingredient that beneficially affects the host by selectively stimulating the growth and/or activity of one or a limited number of bacteria in the colon

Synbiotics: The combination of a prebiotic ingredient with a probiotic culture

Table 39.1 Prebiotic compounds influencing members of the intestinal microflora[a]

Prebiotic factor	Origin	Microbes stimulated	Effects
Oligosaccharides	Onion, garlic, chicory root, burdock, asparagus, Jerusalem artichoke, soybean, wheat bran	*Bifidobacterium* species	Increase in bifidobacteria, suppression of putrefactive bacteria, prevention of constipation and diarrhea
Fructooligosaccharides (inulin, oligofructose)	Same as for oligosaccharides	*Bifidobacterium* species, *Lactobacillus acidophilus*, *Lactobacillus casei*, *Lactobacillus plantarum*	Growth of bifidobacteria and acid promotion
Fructans	Ash-free white powder from tubers of Jerusalem artichoke	*Bifidobacterium* species	Growth of bifidobacteria
Human kappa casein and derived glycomacropeptide	Human milk: chymotrypsin and pepsin hydrolysate	*Bifidobacterium bifidum*	Growth promotion
Stachyose and raffinose	Soybean extract	*Bifidobacterium* species	Growth factor
Casein macropeptide	Bovine milk	*Bifidobacterium* species	Growth promotion
Lactitol (4-*O*-β-D-galactopyranosyl) D-glucitol	Synthetic sugar alcohol of lactose	*Bifidobacterium* species	Growth promotion
Lactulose (4-*O*-β-D-galactopyranosyl) D-fructose	Synthetic derivative of lactose	*Bifidobacterium* species	Growth promotion

[a] Compiled from references 13, 39, and 50.

PREBIOTICS

Considering the specificity and uniqueness of the autochthonous and normal microflora, the use of prebiotics to stimulate the growth and activity of beneficial bacteria in an individual's intestinal microflora is a logical and effective approach to extend probiotic benefits. Food ingredients classified as prebiotics generally exhibit the following characteristics:

- limited hydrolysis and absorption in the upper GIT
- selective growth stimulation of beneficial bacteria in the colon
- potential to repress pathogens and limit virulence via a number of processes, including attenuation of virulence, immunostimulation, and stimulation of a beneficial flora that promotes colonization resistance

The best-known prebiotics are fructooligosaccharides derived from food sources. The largest natural source is inulin, recovered by water extraction from the chicory root. Inulin is composed of a glucose-fructose(n) polymer, with a degree of polymerization (DP; the number of fructose residues) that varies from 2 to 60; the average DP is 10. The bonds are beta-1-2-osidic linkages. Inulin can also be found in edible plants, such as onions, asparagus, bananas, wheat, and Jerusalem artichokes. Oligofructoses that contain glucose-fructose(n) moieties, plus polymeric fructose chains with DP values from 2 to 10, can be synthesized or generated as a hydrolysis product of inulin. Most prebiotic compounds are bifidogenic

in nature (Table 39.1). However, there are some exceptions. Notable among these is the finding by Kaplan and Hutkins (19) that some lactobacilli residing in the small intestine also use fructo-oligosaccharides. Studies with a variety of bifidogenic factors and prebiotics are providing evidence of health-related effects occurring via the prebiotic and its stimulated microflora in the areas of colonization resistance, reduction of colon cancer markers in animals (enzymes and aberrant crypt foci), reduction of serum triglycerides, and enhanced mineral adsorption (calcium, magnesium, iron, and zinc) (reviewed in reference 6).

The current potential for prebiotics rests largely with the investigation of those compounds that can be extracted from foods and their impact on the residing intestinal microflora and health-based markers. One area of exciting research is the generation of designer prebiotics that may offer multiple activities in retarding undesirable microorganisms, better promoting the native desirable microflora, or stimulating the growth or activity of synbiotic cultures (11). The following are good targets for the development of prebiotics:

- expand avenues for incorporation into appropriate food vehicles
- improved stimulation of beneficial floras
- exhibit antipathogenic properties; antiadhesive and attenuation
- identify low-dosage forms

- derived from dietary polysaccharides
- noncariogenic
- good preservative and drying characteristics
- low caloric value
- controllable viscosity

PROBIOTIC CULTURES

Probiotic microorganisms designed for delivery in food or dairy products, via supplementation or fermentation, are usually members of the genus *Lactobacillus* or *Bifidobacterium* (Table 39.2). Nonpathogenic microorganisms that occupy important niches in the host gut or tissues, such as yeasts, enterococci, and *Enterobacteriaceae*, are also used as human and animal probiotics.

Considerable effort has been directed to the development of probiotic cultures for animals. Antibiotics have been used extensively in animal feeds since the 1950s to improve growth and feed conversion. Concern over the development, transmission, and spread of antibiotic resistance determinants through the common practice of feeding animals antibiotics has revived interest in developing probiotic cultures for animals and poultry (10). Nearly 30 years ago, Nurmi and Rantala (34) discovered that newly hatched chicks could be protected from *Salmonella* infection by exposure to a suspension of gut contents from the adult chicken. This competitive exclusion concept has been revisited in a probiotic mixture called Preempt, which is composed of 29 species of nonpathogenic bacteria isolated from the chicken gut. The probiotic mixture is sprayed over the chicks so that, as they preen, their intestinal tract is seeded with microbes which occupy this ecological niche to provide

Table 39.2 Examples of human probiotic species and strains with research documentation[a]

Bifidobacterium breve Yakult
Bifidobacterium lactis (BB12)
Bifidobacterium longum (SBT-2928, BB536)
Lactobacillus acidophilus (NCFM, SBT-2062, DDS1)
Lactobacillus casei (Shirota, CRL431, DN014 001, immunitas)
Lactobacillus delbrueckii subsp. *bulgaricus* (2038)
Lactobacillus fermentum
Lactobacillus johnsonii (La1, Lj1)
Lactobacillus paracasei (CLR431, F19)
Lactobacillus plantarum (299V)
Lactobacillus reuteri (SD2112)
Lactobacillus rhamnosus (GG, 271, GR1)
Lactobacillus salivarius (UCC118)
Saccharomyces boulardii
Streptococcus thermophilus (1131)

[a] Compiled from reference 42.

Table 39.3 Probiotic cultures used in farm animals[a]

Genus	Probiotic species or strains
Bacillus species....	*B. cereus, B. licheniformis, B. mesentericus, B. natto, B. toyoi*
Bifidobacterium species.........	*B. animalis, B. bifidum, B. breve, B. pseudolongum, B. subtile, B. thermophilium*
Lactobacillus species.........	*L. acidophilus, L. brevis, L. casei, L. delbrueckii* subsp. *bulgaricus, L. fermentum, L. helveticus, L. plantarum, L. reuteri, L. rhamnosus*
Other............	*Aspergillus, oryzae, Candida pintolopesii, Clostridium botulinum, Lactococcus lactis, Pediococcus pentosaceus, Saccharomyces cerevisiae*; Preempt, 29 species; cocktail of *E. coli* (17 strains) and *Proteus mirabilis* (1 strain)

[a] Adapted from reference 10.

colonization resistance against *Salmonella, Campylobacter,* and *Listeria* spp. (16, 33, 37). Ironically, this "seeding" occurs naturally in small flocks where mothers are in close contact with hatchlings. Similarly, Zhao et al. (60) specifically targeted colonization of *E. coli* O157:H7 in cattle by creating a probiotic cocktail of 17 isolates of *E. coli* and one *Proteus mirabilis* strain. These were isolated from cattle on the basis of their ability to inhibit *E. coli* O157:H7 in vitro. In challenge studies with *E. coli* O157:H7, the "probiotic mixture" reduced the level of pathogen carriage in most of the probiotic-treated animals. The primary health targets for animal probiotics are enhancing animal growth, weight gain, and reduction of enteric pathogens. There is significant interest in food microbiology, because control of enteric pathogens in foods at the farm level can greatly alter the risk of foodborne illness. Microorganisms used as probiotics for farm animals (Table 39.3) are pure cultures, such as *Lactobacillus reuteri* (4), or multiple-strain cultures that act more broadly in multiple hosts under varied conditions (10).

Bacterial probiotics are generally effective in chickens, pigs, and preruminant calves, whereas fungal probiotics have shown better results in adult ruminants. In the animal probiotic field, host specificity, age of the animal, and targeted benefit are critically important in selecting cultures for specific probiotic applications.

HEALTH BENEFITS

The important role of the intestinal microflora in health and resistance to disease is well recognized. Clinical investigators with well-characterized cultures are slowly accumulating evidence to support the primary claim that

Table 39.4 Proposed health benefits and mechanisms of probiotics[a]

Proposed health outcome	Suspected mechanisms
Promote lactose digestion in lactose-intolerant individuals	Lactase activity from probiotic cultures
	Lactase released from transient bile-sensitive lactic acid bacteria that lyse in the small intestine
Resistance to enteric pathogens	Colonization resistance
	Systemic immunity
	Shortened duration of and enhanced recovery from diarrhea
	Adjuvant increasing secretory antibody production
	Alteration to unfavorable or antagonistic conditions for pathogens (lowering pH, bacteriocins, and production of short-chain fatty acids)
Anticarcinogenic	Antimutagenic activity—binding of mutagens
	Lowering the procarcinogenic enzyme activities (nitroreductase, β-glucuronidase, azoreductase) of colonic bacteria in humans and animals
	Reduction in aberrant crypt foci in colon of animals
	Stimulation of immune function
	Influence on secondary bile salt concentrations
Reduce toxic impact of small-bowel bacterial overgrowth	Lower production of toxic metabolites, e.g., dimethyamine, by colonic microflora
Immune system modulation	Strengthen nonspecific and antigen-specific defenses against infections and tumors
	Adjuvant effect on antigen-specific immune responses
	Lowering of inflammatory responses
	Amelioration of atopic eczema in infants
	Influencing Th1/Th2 cells and cytokine production
Blood lipids, heart disease	Alteration of bile salt hydrolase activity
	Reduction of cholesterol—mechanism?
Antihypertensive effects	Bacterial peptidase action on milk proteins produces a tripeptide that acts as an angiotensin-1-converting enzyme inhibitor to lower blood pressure in hypertensive animals
Reduced activity of ulcerative *Helicobacter pylori*	Production of inhibitors by lactic acid bacteria, in fermented milks, against *H. pylori*
Hepatic encephalopathy	Competitive exclusion or inhibition of urease-producing gut flora
Limit urogenital infections	Adhesion to urinary and vaginal cells
	Competitive exclusion
	Production of inhibitors (biosurfactants, hydrogen peroxide)
Alteration of gut motility	Unknown

[a] Compiled from references 41–43 and 52.

probiotics exert a beneficial influence on the intestinal ecosystem and may also offer some protection against gastrointestinal infections and inflammatory bowel disease. Proposed health outcomes and their suspected mechanisms are compiled in Table 39.4. For details of the research supporting each these above clinical areas, the reader is referred to the Institute of Food Technologists' Scientific Status Summary (41) and Tannock's book on this topic (52).

Some of the health claims for probiotics are controversial. However, well-designed studies using carefully selected and defined probiotic cultures are increasingly supporting these health outcomes (52). General examination of the clinical studies suggests three major avenues through which probiotic cultures appear to carry out beneficial activities in the GIT via probiotic and/or abiotic mechanisms (Fig. 39.5).

1. Probiotic cultures appear to affect the microbial composition and the associated metabolic and enzymatic activities of resident harmful or developing pathogenic bacteria. Alterations of the microflora are likely prerequisites to lowering procarcinogenic enzymes, limiting production of carcinogens, and

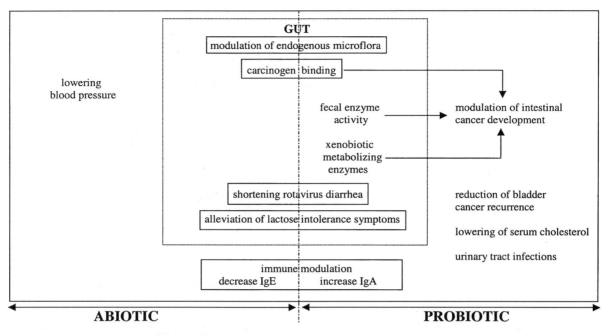

Figure 39.5   Health benefits and suspected mechanisms of probiotics versus abiotics.

altering in the concentrations of secondary bile acids.

2. Probiotic cultures, by their intimate association with the intestinal mucosa, their cellular components, and their effects on the associated microflora, appear to improve immunological function in the GIT and the integrity of the mucosal barrier. Modulation and enhancement of the immune system appear to be focal mechanisms that impact many of the effects shown in Table 39.2, such as inflammatory responses, anticarcinogenic activity, and colonization resistance (17, 31).

3. Abiotic components, which are dead cells, and their cellular constituents (enzymes, cell wall components) or nonviable growth byproducts (bioactive peptides) play important roles in eliciting probiotic-type benefits. Clear examples are delivery of lactase by dead cells (47, 56), immune system modulation by cellular components (15, 29, 36), and antihypertensive effects by angiotensin-converting enzyme tripeptides (28).

Overarching conclusions on the benefits of probiotics are difficult to draw because the study results can vary with the strains used, the population level of probiotic cells delivered, the health marker targeted, and the number, age, and condition of the subjects evaluated. There is also little known about the minimal dose of probiotic

and the frequency of consumption required to elicit a physiological effect. Levels of cells delivered can range from 10^6 to 10^9 CFU per ml or per g depending on the food carrier and the method of elevating the probiotic bacterial concentration. For example, in the manufacture of Sweet Acidophilus milk, *Lactobacillus acidophilus* NCFM cells are concentrated to 10^{10} CFU/ml in frozen culture concentrates and then added directly to 2% fluid milk to yield a target concentration approaching 10^7 CFU/ml. Alternatively, as is the case in some yogurt fermentations, probiotic cultures are added as inoculants and allowed to grow during the fermentation phase, reaching levels of 10^8 to 10^9 CFU/ml. The proposed target level for daily consumption of probiotic cultures is 10^8 to 10^9 CFU per day, or up to 10^9 to 10^{10} CFU/day if losses are expected during stomach transit (42). Achieving these levels of viable probiotic cultures in foods can be a challenging technological task. Substantial effort is required, on a strain-to-strain basis, to design the fermentation, concentration, and storage (dried or frozen) conditions that will enhance the viability and functional activity of probiotic cultures. Stress preconditioning is one way to enhance the storage stability of probiotic cultures in harsh environments (21).

MOLECULAR TAXONOMY

There are a multitude of potential bacterial strains that can be used as probiotics, and not all strains elicit the

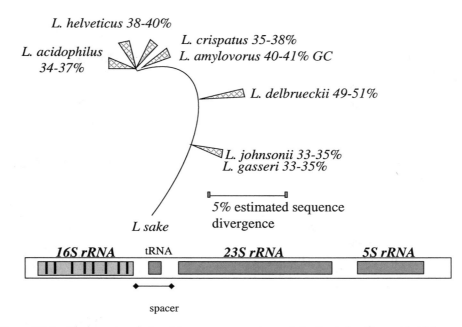

Figure 39.6 Phylogenic relationships among members of the *Lactobacillus acidophilus* complex. Vertical bars represent nine variable regions in the 16S rRNA gene used for phylogenetic identification and analysis. Compiled from references 21, 44, and 55.

same technological properties or probiotic characteristics. Proper identification of probiotic strains and cocktails of multiple cultures is the first critical step in establishing the cause-and-effect relationships responsible for the health benefits under investigation. This has been a historical problem for the probiotic field. Strains under investigation were often misidentified and incorrectly named, and health claims were inappropriately inferred on the basis of name recognition. Underlying these problems were two basic issues: (i) reliance on phenotypic criteria for taxonomic identification and (ii) the existence of multiple closely related species that could not be distinguished by phenotypic criteria or gross genetic analysis (e.g., percent GC content). Therefore, the genus and species names used in the probiotic literature prior to 1995 should be viewed carefully.

Phylogenetic analysis is now widely recognized as the most powerful tool for taxonomic classification of bacterial cultures (59). The availability of molecular tools to properly identify probiotic species and individual strains over the last decade has helped dispel confusion over strain identity and ancestry. The accumulating sequence information on rRNAs provides a growing resource for comparative identification of probiotic cultures, both established candidates and potentially new candidates. At this juncture, phylogenetic analysis has recognized 54 species of lactobacilli, 18 of which are considered to be of some interest in probiotics, and 31 species of *Bifidobacterium,* 11 of which have been

detected in human feces (53). Phylogenetic analysis can be conducted in varying degrees and combined with other characteristics (phenotypes) as needed to make definitive taxonomic classifications. A number of DNA sequence-based typing systems have been used to analyze conserved regions of the rRNA operon or other conserved genes in probiotic cultures (35):

1. PCR amplification and sequencing of ~1,500 bp of the 16S rRNA gene
2. PCR amplification and sequencing of ~450 bp of the internal transcribed spacer region between the 16S and 23S rRNA genes (Fig. 39.6)
3. PCR amplification and sequencing of ~50 bp of the variable region of 16S rRNA to identify members of the *L. acidophilus* complex (25)
4. PCR amplification and sequencing of alternative genes that are universally present and highly conserved, e.g., the *recA* gene of bifidobacteria (23)

The availability of efficient and cost-effective commercial sequencing facilities has made bacterial species identification by phylogenetic analysis commonplace. The ability to properly identify probiotic species provides the first giant step toward eliminating any confusion over strain identity and ancestry. The practice is also expected to uncover new probiotic species that have been hidden below the surface of traditional taxonomic descriptions. DNA sequence analysis of the rRNA operons of strains loosely classified as *L. acidophilus* reveal six closely related species that compose the "acidophilus" complex: *L. acidophilus, L. crispatus, L. amylovorus, L. gallinarum, L. gasseri,* and *L. johnsonii* (Fig. 39.6) (18, 22, 26). All of these species are considered to have probiotic potential, but it remains to be determined what specific roles and benefits are exerted by each species exclusively, or collectively by the group.

MOLECULAR APPROACHES TO INVESTIGATE PROBIOTIC BACTERIA

The primary scientific barrier to the acceptance of probiotics is the inability to explain how a probiotic influences an individual's complex GIT microflora and then to determine the impact of those specific changes on factors that could impact general health and well-being. This has led to scientific criticism of the probiotic field and a charge to demonstrate benefits with the rigor of pharmaceutical studies, applying new molecular technologies to investigate direct cause-and-effect relationships (52). The recent introduction of molecular technologies to microbial ecology provides a set of powerful tools for the analysis of the complex, variable, and dynamic communities of the GIT and the impact of probiotic cultures on those communities. These technologies do not rely on cultural techniques and therefore provide a more accurate picture of the microbial composition, diversity, and fluctuation within a community (55). The power of this technology was illustrated by Suau et al. (49), who sequenced 284 PCR-generated 16S rRNA amplicons cloned from one human fecal sample and reported that only 24% of the sequences correlated with known (culturable) organisms. This means that 76% of the sequences represent previously unidentified microorganisms. Understanding the impact of probiotics on the intestinal microflora will be enhanced considerably by the use of molecular methods that can more accurately reflect those changes occurring within the entire gastrointestinal microbiota.

Among these technologies are powerful genetics-based molecular techniques used to fingerprint specific DNA patterns that are characteristic for a single strain. Methods applied to probiotic cultures over the past 5 years have been reviewed by O'Sullivan (35):

- ribotyping
- analysis of restriction fragment length polymorphisms of genomic DNA using pulsed-field gel electrophoresis (Fig. 39.7)
- randomly amplified polymorphic DNA
- 16S rDNA sequencing
- in situ PCR and fluorescent in situ hybridization allowing in situ visualization of a specific probiotic

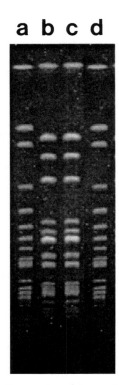

Figure 39.7 DNA fingerprint of the predominant *Lactobacillus* culture isolated from human feces (a) before feeding a probiotic, (b and c) after feeding *Lactobacillus acidophilus,* and (d) 2 weeks after feeding was halted. *Sma*I-digested DNA fragments prepared from individual *Lactobacillus* colonies were separated by pulsed-field electrophoresis.

culture, or indicator flora, within mixed microbial populations

The molecular tools available for genomic fingerprinting can unequivocally link a fed probiotic culture with the strain recovered in the GIT or feces (2, 54, 57). These technologies have demonstrated that when feeding is stopped, the probiotic strain is no longer recovered after about 14 to 25 days. It is thought that probiotics (allochthonous flora) are not likely to permanently colonize the GIT. They would require continuous delivery to maintain their presence (52).

Analysis of complex microbial communities has been revolutionized by the analysis of rRNA genes and the use of in situ probes or denaturing gradient gel electrophoresis (DGGE) to distinguish between different species that may be present. A sample from the GIT is used in a PCR with universal primers designed to amplify a variable region in the 16S rRNA that is flanked by sequences conserved among all bacteria (Fig. 39.6). The amplicons are identical in length but vary in their melting properties because of variations in the internal 16S rRNA sequence and percent GC content of the organism's DNA. Analysis under denaturing conditions using temperature or chemicals yields a specific banding profile, with each band representing a 16S rRNA amplicon from a different bacterium that was originally present in the GIT sample. These DNA amplicon bands can be extracted from the gel and sequenced to reveal the identity of the organisms present. The presence and intensity of the bands may also reflect the relative population levels of various species in the sample. Profiles on human fecal samples show some bands common among different individuals, reflecting universal and dominant species. There are also unique bands characteristic of an individual's microflora at that moment in time (61). In a recent study, DGGE analysis of human fecal samples after feeding *Lactobacillus salivarius* revealed a markedly stable flora that was not modified by probiotic treatment (55). Additional work of this type is needed and may require targeting of specific microbial populations in the GIT that are expected to be impacted by the introduction of probiotic cultures. The documented ability of foods and food-bearing probiotics to modify the intestinal flora will be instrumental in proving or refuting probiotic claims on a mechanistic basis.

The developments in molecular techniques over the past decade have removed many of the key issues that had previously hindered scientific progress in probiotics. Exacting methods for the identification, tracking, and analysis of probiotic cultures within complex microbial ecosystems now promise to revolutionize our understanding of the functional roles and in vivo effects associated with probiotic bacteria.

CRITERIA FOR SELECTION OF PROBIOTIC CULTURES

Different probiotic species and even different strains within a species exhibit distinctive properties that can markedly affect their survival in foods, fermentation characteristics, and other probiotic properties. Strain selection becomes, therefore, a critical parameter to ensure the culture's fermentation or probiotic performance. Desirable traits for selection of functional probiotics are summarized in Table 39.5 under four major categories (21).

These criteria are based on past experience with microbial selection in vitro, propagation (viability, technological suitability), and safe use of lactic acid bacteria in foods (nonpathogenic, nontoxic, genetically stable, normal inhabitant of target species, viability). Some criteria (i.e., survival during transit, retention in microflora, anti-inflammatory, immunostimulation) require in vivo analysis in human or animal studies. These studies can be

Table 39.5 Desirable selection criteria for probiotic strains[a]

Appropriateness
- taxonomic identification known by phylogenetic analysis and rRNA sequencing
- origin—normal inhabitant of the species targeted and isolated from a healthy individual
- safety—nontoxic, nonpathogenic, "generally recognized as safe" status

Technological suitability
- amenable to mass production and storage: adequate growth, recovery, concentration, freezing, dehydration, storage, and distribution
- viability at high populations (preferred at 10^7 to 10^9 CFU/g)
- stability of desired characteristics during culture preparation, storage, and delivery
- provides desirable organoleptic qualities (or no undesirable qualities) when included in foods or fermentation processes
- genetically stable to maintain phenotypic properties
- genetically accessible for potential modification

Competitiveness
- capable of survival, proliferation, and metabolic activity at the target site in vivo
- resistant to bile
- resistant to acid
- able to compete with the normal microflora, including the same or closely related species; potentially resistant to bacteriocins, acid, and other antimicrobial agents produced by residing microflora
- adherence, colonization, and retention evaluated

Performance and functionality
- able to exert one or more clinically documented health benefits
- antagonistic toward pathogenic/cariogenic bacteria
- production of antimicrobial substances (bacteriocins, hydrogen peroxide, organic acids, or other inhibitory compounds)
- immunostimulatory
- anti-inflammatory
- antimutagenic
- anticarcinogenic
- production of bioactive compounds (enzymes, vaccines, peptides)

[a] Adapted from reference 21.

lengthy and expensive and can become more complicated as multiple probiotic strains, singly or in cocktails, are evaluated for functional properties. As a result, additional efforts are needed to develop model systems in animals, or improve in vitro systems, that better predict probiotic performance in vivo. Some recent examples include the development of a transit tolerance test (5) that evaluates probiotic survival in simulated gastric juice (pH 2) containing pepsin and sodium chloride. The majority (14 of 15 strains) of probiotic cultures lost 90% viability during the gastric portion of the test but survived well under conditions mimicking those of the small intestine. Gastric transit can be detrimental to delivery of viable probiotic cells to the small intestine (42). Efforts to protect cultures with dairy foods or components that act as buffering agents can improve survival (5). A more dynamic in vitro system for screening probiotic cultures has been developed by Marteau et al. (30) with a simulated GIT model composed of major components (stomach, small intestine, colon), chemical constituents (pH, bile, pepsin, nutrients, water), and peristaltic movement. In this model in vitro system, the survival of lactobacilli and bifidobacteria strains was similar to the that seen in human in vivo survival studies.

Selection criteria that address competitiveness (e.g., adherence, antimicrobial agents, bacteriocins) and performance (e.g., immunostimulation, anticarcinogenic) remain more complicated and controversial, because the underlying mechanisms by which probiotics exert functional roles in vivo remain to be elucidated. Therefore, it is impossible to precisely define the microbial features, or collection of characteristics, that dictate some of the above selection criteria. Establishing the mechanisms that link in vitro phenotypic criteria to in vivo functionality will present one of the major scientific challenges for probiotics in the coming decade. In this regard, this field is poised perfectly to exploit the recent progress in sequencing capacity and functional genomics toward the investigation of these bacteria and their probiotic capabilities (20). There are a myriad of possible probiotic strains, representing a diverse set of phenotypes, which are being linked increasingly to a variety of benefits. Defining and screening genetic traits important for functional probiotic activities promises to identify superior strains and allow the construction of strain combinations that elicit unique or multiple effects.

DELIVERY VEHICLES FOR BIOACTIVE COMPOUNDS

Lactic acid bacteria, such as lactobacilli and lactococci, have been consumed at concentrations of 10^8 to 10^9 CFU/g for centuries in fermented foods, notably in dairy products. Their generally recognized as safe status, acid tolerance, and capacity for safe consumption at elevated levels provides a unique opportunity to exploit these beneficial bacteria as live vehicles for delivery of biological molecules to the small intestine. Oral ingestion of proteins (vaccines or enzymes) results in denaturation, degradation, and loss of biological activity. In contrast, bioactive molecules can be protected by viable or nonviable cells during passage through the stomach and then released into the GIT. Bioactive molecules targeted for delivery by probiotic lactic acid bacteria include vaccine antigens against viral and bacterial pathogens, digestive

enzymes for humans, and growth-promoting enzymes in animals. Wells et al. (58) have pioneered the development and use of lactic acid bacteria as vaccine delivery vehicles. Recently, it was demonstrated that *Lactococcus lactis* secreting the cytokine interleukin-10 could decrease the GIT inflammatory responses of mice with colitis (48). There are many exciting opportunities to employ probiotic bacteria as vehicles to deliver bioactive molecules to targeted locations in the GIT. This is likely to be one of the most important areas for practical application of probiotic cultures that are derived through recombinant DNA technology (24).

SAFETY

Probiotic lactic acid bacteria have been consumed in various forms, at high concentrations, for centuries. As a consequence, they enjoy a generally recognized as safe status in U.S. regulatory statutes. They are not considered pathogenic in any capacity, and recent reviews conclude that the infection potential of lactobacilli and bifidobacteria is low (1, 40). This conclusion is based on the wide presence of these probiotic bacteria in foods, their commensal relationship with the host, and an extremely low correlation with infections (42). However, rare incidents of lactic acid bacteria associated with infected tissues are documented, but the bacteria occur only as opportunists (often in endocarditis) in immunocompromised and elderly individuals. Some intestinal isolates, primarily *Lactobacillus rhamnosus*, have warranted increased surveillance because of a higher correlation rate with secondary infections. When considered in context, these incidents are extremely rare and speak more to the susceptibility of the individual than to the infective potential of the probiotic. However, it remains an important and critical question whether or not *any* lactic acid bacterium isolated from the GIT or feces can be considered inherently safe. The history of safe use suggests that the answer is yes, but it is widely accepted that new isolates should be evaluated for pathogenicity and toxicity in animal studies before inclusion in any commercial probiotic products for humans. Any pathogenic potential of probiotic cultures will await the results of ongoing genome sequencing projects that are expected to uncover the genetic content of many key probiotic species and reveal the presence or, more likely, the absence of infective or pathogenic determinants.

CONCLUSIONS

The probiotic field has offered considerable promise since the early observations by Metchnikoff about the importance of the GIT microflora in limiting the development of putrefactive organisms and the potential role of lactic acid bacteria in establishing a healthy microflora. Today, microbiological and molecular methods are providing critical new insights into probiotic bacteria, the normal microflora, and the interactions between the two that may be responsible for eliciting beneficial outcomes realized in health and well-being. Proving or refuting the probiotic concept will be a worthy challenge to the most talented scientists because of the complexity of the host, dynamic interactions within microbial ecosystems, and the multifaceted impacts of our associated microorganisms on health and well-being. Some of the key issues and important challenges for food microbiologists working on probiotics in the years ahead will be the following.

- taxonomic identification of all probiotic cultures used in research, clinical trials, and commercial products
- definition of the active principles responsible for probiotic and abiotic activities
- correlation of genus, species, strain, phenotype, and genotype to specific probiotic functions
- determination of the impact of probiotics on the normal microflora and associated host tissues
- stabilization of cultures and components for delivery in probiotic applications
- exploiting of probiotic lactic acid bacteria for targeted delivery of novel bioactive compounds
- science-driven implementation of findings in commercial development
- ensured safety of probiotic cultures, components, and food carriers
- identification of physiologically relevant biomarkers that can assess parameters of probiotic effectiveness (strain, dose, growth, colonization potential)
- epidemiological and long-term studies in humans and animals to investigate the impact of probiotic cultures on diseases of longevity, e.g., colon cancer, heart disease, and inflammatory bowel diseases

Paper no. FSR 01–10 of the Journal Series of the Dept. of Food Science, NCSU, Raleigh, N.C. I thank Rodolphe Barrangou-Poueys for his creative input into Figures 39.4 and 39.5 and W. Michael Russell and Michael Callanan for their critical review of the text.

References

1. Adams, M. R., and P. Marteau. 1995. On the safety of lactic acid bacteria from food. *Int. J. Food Microbiol.* 27:263–264.

2. Alander, M., R. Satokari, R. Korpela, M. Saxelin, T. Vilpponen-Salmela, T. Mattila-Sandholm, and A. von Wright. 1999. Persistence of colonization of human colonic mucosa by a probiotic strain, *Lactobacillus rhamnosus* GG, after oral consumption. *Appl. Environ. Microbiol.* 65:351–354.

3. Bibel, D. J. 1988. Elie Metchnikoff's bacillus of long life. *ASM News* 54:661–665.

4. Cacas, I. A. 1998. *Lactobacillus reuteri*: an effective probiotic for poultry and other animals, p. 475–516. *In* S. Salminen and A. von Wright (ed.), *Lactic Acid Bacteria: Microbiology and Functional Aspects*. Marcel Dekker, Inc., New York, N.Y.

5. Charteris, W. P., P. M. Kelly, L. Morelli, and J. K. Collins. 1998. Development and application of an in vitro methodology to determine the transit tolerance of potentially probiotic *Lactobacillus* and *Bifidobacterium* species in the upper human gastrointestinal tract. *J. Appl. Bacteriol.* 84:759–768.

6. Crittenden, R. G. 1999. Prebiotics, p. 141–156. *In* G. W. Tannock (ed.), *Probiotics: a Critical Review*. Horizon Scientific Press, Norfolk, United Kingdom.

7. Danone World Newsletter. 1997. Bifidobacteria. http://www.danonenewsletter.fr. *Danone World Newsletter*, no. 16.

8. Dubos, R., R. W. Schaedler, R. Costello, and P. Hoet. 1965. Indigenous, normal and autochthonous flora of the gastrointestinal tract. *J. Exp. Med.* 122:67–76.

9. Fuller, R. 1989. Probiotics in man and animals. *J. Appl. Bacteriol.* 66:365–378.

10. Fuller, R. 1999. Probiotics for farm animals, p. 15–22. *In* G. W. Tannock (ed.), *Probiotics: a Critical Review*. Horizon Scientific Press, Norfolk, United Kingdom.

11. Gibson, G. R. 1998. Dietary modulation of the human gut microflora using prebiotics. *Br. J. Nutr.* 80(Suppl. 2):S209–S212.

12. Gibson, G. R., and M. B. Roberfroid. 1995. Dietary modulation of the human colonic microbiota: introducing the concept of prebiotics. *J. Nutr.* 125:1401–1412.

13. Gomes, A. M. P., and F. X. Malcata. 1999. *Bifidobacterium* spp. and *Lactobacillus acidophilus*: biological, biochemical, technological and therapeutical properties relevant for use as probiotics. *Trends Food Sci. Technol.* 10:139–157.

14. Guarner, F., and G. J. Schaafsma. 1998. Probiotics. *Int. J. Food Microbiol.* 39:237–238.

15. Hosono, A., J. Lee, A. Ametani, M. Natsume, M. Hirayama, T. Adachi, and S. Kaminogawa. 1997. Characterization of a water-soluble polysaccharide fraction with immunopotentiating activity from *Bifidobacterium adolescentis* M101-4. *Biosci. Biotechnol. Biochem.* 61:312–316.

16. Hume, M. E., D. E. Corrier, D. J. Nisbet, and J. R. DeLoach. 1998. Early *Salmonella* challenge time and reduction in chick cecal colonization following treatment with a characterized competitive exclusion culture. *J. Food Prot.* 61:673–676.

17. Isolauri, E., E. Salminen, and S. Salminen. 1998. Lactic acid bacteria and immune modulation. *In* S. Salminen and A. von Wright (ed.), *Lactic Acid Bacteria: Microbiology and Functional Aspects*. Marcel Dekker, Inc., New York, N.Y.

18. Johnson, J. L., C. F. Phelps, C. S. Cummins, J. London, and F. Gasser. 1980. Taxonomy of the *Lactobacillus acidophilus* group. *Int. J. Syst. Bacteriol.* 30:53–68.

19. **Kaplan, H., and R. W. Hutkins.** 2000. Fermentation of fructooligosaccharides by lactic acid bacteria and bifidobacteria. *Appl. Environ. Microbiol.* **66:**2682–2684.

20. **Klaenhammer, T. R.** 1998. Functional activities of *Lactobacillus* probiotics: genetic mandate. *Int. Dairy J.* **8:**497–506.

21. **Klaenhammer, T. R., and M. J. Kullen.** 1999. Selection and design of probiotics. *Int. J. Food Microbiol.* **50:**45–58.

22. **Klaenhammer, T. R., and W. M. Russell.** 2000. Species of the *Lactobacillus acidophilus* complex, p. 1151–1157. *In* R. K. Robinson, C. Batt, and P. D. Patel (ed.), *Encyclopedia of Food Microbiology*, vol. 2. Academic Press, Inc., San Diego, Calif.

23. **Kullen, M. J., L. J. Brady, and D. J. O'Sullivan.** 1997. Evaluation of using a short region of the *recA* gene for rapid and sensitive speciation of dominant bifidobacteria in the human large intestine. *FEMS Microbiol. Lett.* **154:**377–383.

24. **Kullen, M. J., and T. R. Klaenhammer.** 1999. Genetic modification of intestinal lactobacilli and bifidobacteria, p. 65–84. *In* G. W. Tannock (ed.), *Probiotics: a Critical Review.* Horizon Scientific Press, Norfolk, United Kingdom.

25. **Kullen, M. J., R. B. Sanozky-Dawes, D. C. Crowell, and T. R. Klaenhammer.** 2000. Use of DNA sequence of variable regions of the 16S rRNA gene for rapid and accurate identification of bacteria in the *Lactobacillus acidophilus* complex. *J. Appl. Microbiol.* **89:**511–516.

26. **Lauer, E., C. Helming, and O. Kandler.** 1980. Heterogeneity of the species *Lactobacillus acidophilus* (Moro) Hansen and Moquot as revealed by biochemical characteristics and DNA-DNA hybridization. *Zentbl. Bakteriol. Mikrobiol. Hyg. 1 Abt. Orig. C* **1:**150–168.

27. **Lilly, D. M., and R. H. Stillwell.** 1965. Probiotics: growth promoting factors produced by microorganisms. *Science* **147:**747–748.

28. **Maeno, M., N. Tamamoto, and T. Takano.** 1996. Identification of antihypertensive peptides from casein hydrolysate produced by a proteinase from *Lactobacillus helveticus* CP790. *J. Dairy Sci.* **73:**1316–1321.

29. **Marin, M. L., J. H. Lee, J. Murtha, Z. Ustunol, and J. J. Pestka.** 1997. Differential cytokine production in clonal macrophage and T-cell lines cultured with bifidobacteria. *J. Dairy Sci.* **80:**2713–2720.

30. **Marteau, P., M. Minekus, R. Havenaar, and J. H. H. Huis in't Veld.** 1997. Survival of lactic acid bacteria in a dynamic model of the stomach and small intestine: validation and the effect of bile. *J. Dairy Sci.* **80:**1031–1037.

31. **McCracken, V. J., and H. R. Gaskins.** 1999. Probiotics and the immune system, p. 85–111. *In* G. W. Tannock (ed.), *Probiotics: a Critical Review.* Horizon Scientific Press, Norfolk, United Kingdom.

32. **Mitsuoka, T.** 1992. The human gastrointestinal tract, p. 69–114. *In* B. J. B. Wood (ed.), *The Lactic Acid Bacteria*, vol. 1. *The Lactic Acid Bacteria in Health and Disease.* Elsevier Applied Science, London, United Kingdom.

33. **Nisbet, D. J., G. I. Tellez, V. K. Lowry, R. C. Anderson, G. Garcia, G. Nava, M. H. Kogut, D. E. Corrier, and L. H. Stanker.** 1998. Effect of a commercial competitive exclusion culture (Preempt) on mortality and horizontal transmission of *Salmonella gallinarum* in broiler chickens. *Avian Dis.* **42:**651–656.

34. **Nurmi, E., and M. Rantala.** 1973. New aspects of *Salmonella* infection in broiler production. *Nature* (London) **241:**210–211.

35. **O'Sullivan, D. J.** 1999. Methods for the analysis of the intestinal microflora, p. 23–44. *In* G. W. Tannock (ed.), *Probiotics: a Critical Review.* Horizon Scientific Press, Norfolk, United Kingdom.

36. **Perdigon, G., M. E. Nader de Macias, S. Alvarez, G. Oliver, and A. A. Pesce de Ruiz Holgado.** 1986. Effect of perorally administered lactobacilli on macrophage activation in mice. *Infect. Immun.* **53:**404–410.

36a. **Puhan, Z.** 1999. Effect of probiotic fermented dairy products in human nutrition. *Ind. Latte* **35**(3–4):3–11.

37. **Radloff, J.** 1998. Spray guards chicks from infections. *Sci. News* **153:**196.

38. **Salminen, S., A. Ouwehand, Y. Benno, and Y. K. Lee.** 1999. Probiotics: how should they be defined? *Trends Food Sci. Technol.* **10:**107–110.

39. **Salminen, S., M. Roberfroid, P. Ramos, and R. Fonden.** 1998. Prebiotic substrates and lactic acid bacteria, p. 343–358. *In* S. Salminen and A. von Wright (ed.), *Lactic Acid Bacteria: Microbiology and Functional Aspects.* Marcel Dekker, Inc., New York, N.Y.

40. **Salminen, S., A. Von Wright, L. Morelli, P. Marteau, D. Brassart, W. M. de Vos, R. Fonden, M. Saxelin, K. Collins, G. Mogensen, S.-E. Birkeland, and T. Mattila-Sandholm.** 1998. Demonstration of safety of probiotics—a review. *Int. J. Food Microbiol.* **44:**93–106.

41. **Sanders, M. E.** 1999. Probiotics—Scientific Status Summary. *Food Technol.* **53:**67–77.

42. **Sanders, M. E., and J. Huis in't Veld.** 1999. Bringing a probiotic-containing functional food to market: microbiological, product, regulatory, and labeling issues. *Antonie Leeuwenhoek* **76:**293–315.

43. **Sandholm, T. M., S. Blum, J. K. Collins, R. Crittenden, W. de Vos, C. Dunne, R. Fonden, G. Grenov, E. Isolauri, B. Kiely, P. Marteau, L. Morelli, A. Ouwehand, R. Reniero, M. Saarela, S. Salminen, M. Saxelin, E. Schiffrin, F. Shanahan, E. Vaughan, and A. von Wright.** Probiotics: towards demonstrating efficacy. *Trends Food Sci. Technol.* **10:**393–399.

44. **Schleifer, K. H., and W. Ludwig.** 1995. Phylogenetics for the genus *Lactobacillus* and related genera. *Syst. Appl. Microbiol.* **18:**461–467.

45. **Shortt, C.** 1998. Living it up for dinner. *Chem. Ind.* **20:**300–303.

46. **Shortt, C.** 1999. The probiotic century: historical and current perspectives. *Trends Food Sci. Technol.* **10:**411–417.

47. **Somkuti, G. A., M. E. Dominiecki, and D. H. Steinberg.** 1998. Permeabilization of *Streptococcus thermophilus* and *Lactobacillus delbrueckii* subsp. *bulgaricus* with ethanol. *Curr. Microbiol.* **36:**202–206.

48. **Steidler, L., W. Hans, L. Schotte, S. Neirynck, F. Obermeier, W. Falk, W. Fiers, and E. Remaut.** 2000. Treatment of murine colitis by *Lactococcus lactis* secreting IL-10. *Science* **289:**1352–1355.

49. Suau, A., R. Bonnet, M. Sutren, J.-J. Godon, G. R. Gibson, M. D. Collins, and J. Dore. 1999. Direct analysis of genes encoding 16S rRNA from complex communities reveals many novel molecular species. *Appl. Environ. Microbiol.* **65**:4799–4807.

50. Tamime, A. Y., V. M. E. Marshall, and R. K. Robinson. 1995. Microbiological and technical aspects of milks fermented by bifidobacteria. *J. Dairy Res.* **62**:151–187.

51. Tannock, G. W. 1995. *The Normal Microflora.* Chapman & Hall, Ltd., London, United Kingdom.

52. Tannock, G. W. 1999. *Probiotics: a Critical Review.* Horizon Scientific Press, Norfolk, United Kingdom.

53. Tannock, G. W. 1999. A fresh look at the intestinal microflora, p. 5–14. *In* G. W. Tannock (ed.), *Probiotics: a Critical Review.* Horizon Scientific Press, Norfolk, United Kingdom.

54. Tannock, G. W. 1999. Analysis of the intestinal microflora: a renaissance. *Antonie Leeuwenhoek* **76**:265–278.

55. Vaughan, E. E., H. G. H. J. Heilig, E. G. Zoetendal, R. Satokari, K. Collins, A. D. L. Ackermans, and W. M. de Vos. 1999. Molecular approaches to study probiotic bacteria. *Trends Food Sci. Technol.* **10**:400–404.

56. Vesa, T. H., P. Parteau, S. Zidi, F. Briet, P. Pochart, and J. C. Rambaud. 1996. Digestion and tolerance of lactose from yoghurt and different semi-solid fermented dairy products containing *Lactobacillus acidophilus* and bifidobacteria in lactose maldigesters—is bacterial lactase important? *Eur. J. Clin. Nutr.* **50**:730–733.

57. Walter, J., G. W. Tannock, A. Tilsala-Timisjarvi, S. Rodtong, D. M. Loach, K. Munro, and T. Alatossava. 2000. Detection and identification of gastrointestinal *Lactobacillus* species by using denaturing gradient gel electrophoresis and species-specific PCR primers. *Appl. Environ. Microbiol.* **66**:297–303.

58. Wells, J. M., K. Robinson, L. M. Chamberlain, K. M. Schofield, and R. W. LePage. 1996. Lactic acid bacteria as vaccine delivery vehicles. *Antonie Leeuwenhoek* **70**:317–330.

59. Woese, C. R. 1987. Bacterial evolution. *Microbiol. Rev.* **51**:221–271.

60. Zhao, T., M. P. Doyle, B. G. Harmon, C. A. Brown, P. O. Mueller, and A. H. Parks. 1998. Reduction of carriage of enterohemorrhagic *Escherichia coli* O157:H7 in cattle by inoculation with probiotic bacteria. *J. Clin. Microbiol.* **36**:641–647.

61. Zoetendal, E. G., A. D. L. Akkermans, and W. M. de Vos. 1998. Temperature gradient gel electrophoresis analysis of 16S rRNA from human fecal samples reveals stable and host-specific communities of active bacteria. *Appl. Environ. Microbiol.* **64**:3854–3859.

Food Microbiology: Fundamentals and Frontiers, 2nd Ed.
Edited by M. P. Doyle et al.
© 2001 ASM Press, Washington, D.C.

Richard C. Whiting
Robert L. Buchanan

Predictive Modeling and Risk Assessment

40

Modeling the behavior of food microorganisms began about 1920 with the development of methods for calculating thermal death times. This revolutionized the canning industry (43). A resurgence in predictive modeling began in the 1980s that was driven by a proliferation of refrigerated and limited-shelf-life foods, the development of multiple-hurdle preservation systems, and the advent of personal computers. The microbiological, mathematical, and statistical tools existed prior to this expansion in modeling efforts. However, without the ability to disseminate cumbersome equations and to solve them rapidly and repeatedly at one's desk, the extensive effort necessary to create and use microbial models would not be worthwhile (13). Since the first edition of this book was written (97), modeling techniques have become standard tools for designing experimentation and describing the results. Because microbial models now appear regularly in the food literature, this chapter will summarize the concepts and most frequently used modeling methods, focusing on literature published since the first edition. The literature on modeling specific microorganisms or foods is now too large to review in this chapter. It is hoped that this chapter and its citations will be a good starting point for developers and users of microbial models. Other reviews of microbial modeling can be found in references 21, 48, 60, 79, 93, and 95.

Another advance since the first edition of this book is the linking of a series of food processing and storage steps into a continuous process model, potentially extending from farm to fork. This permits comparisons between the efficacy of individual control steps to limit the numbers of pathogens consumed. For example, the overall impact of steps to limit raw product contamination at the farm or processing establishment could be contrasted to the ability to prevent growth during storage. Awareness has also evolved that the number of pathogens consumed per se is not of primary concern. Instead, reducing the likelihood of illness is recognized as the objective in controlling pathogens. This joining of the process calculations to the likelihood of illness brings predictive modeling into the arena of risk assessment and Hazard Analysis and Critical Control Point system (HACCP), the food industry's system of risk management. Subjective evaluations of a food product's risk to consumers have a long history in the food industry and regulatory agencies. Modeling techniques provide the tools to make these evaluations quantitative. Risk assessment also introduced the concept of using the entire distribution for a parameter value in the analysis, not just the mean or "worst-case" value.

Richard C. Whiting and Robert L. Buchanan, FDA, Center for Food Safety and Applied Nutrition, 200 C St. S.W., Washington, DC 20204.

Evaluating the impact of the distributions from variations and uncertainties in data provides a much more complete understanding of the process being analyzed than single point estimations.

MODELS

In classic microbial research, data handling merely presented graphs that illustrated microbial growth or inactivation for several levels of one or perhaps two factors. The readers had to subjectively interpolate these results to their situation. This became increasingly difficult when three or more changing factors were involved. Standard deviations or other estimates of error that inform the user about the precision of the estimates were frequently not provided.

Modeling in food microbiology assumes that the growth or inactivation of microorganisms in a food or model system is predictable within the limits of normal biological variability and can be described by mathematical equations. It commonly assumes that the measured behavior in broth cultures or other model systems predicts the microorganism's behavior in foods with corresponding levels of environmental factors (pH, temperature) (59, 79). Exceptions do occur, however, if a food contains another significant factor not present in the model. This possibility requires that a user validate a specific model for individual foods with a few selected tests. When a food and pathogen are frequently associated together, specific models are developed for them. Usually temperature is the most important environmental variable. Product-specific models will be more accurate than models based on broth or other systems.

Because of the complexity of microbial biochemistry, particularly in relation to non-steady-state kinetics, and a lack of knowledge of the limiting factors in foods, modeling in food microbiology has taken a more empirical approach than the modeling techniques used for fermentation and other biochemical processes. Empirical though these models may be, they are based on established linear and nonlinear regression techniques. Growth data or model parameters are fitted to equations by using interactive least-squares computer algorithms. Assumptions about randomness, normal distributions, interpolations with tested ranges rather than extrapolations outside the ranges, parsimony, and stochastic specifications must be made in microbial modeling as they are for any statistical application of regression.

All models, both for individual steps and multipart processes, are simplifications that represent the complex biochemical processes controlling microbial growth or inactivation. They must be simplified to a reasonable number of input variables and steps. These variables need to be easily measured (e.g., temperature and pH) or known for a food (e.g., added salt level). An input variable of glucose or peptide concentration would necessitate an analysis of the food before the model could be used. Therefore, a model for an individual step or a complete food process must simultaneously be sufficiently detailed to provide a useful prediction, but simple enough that data are available. This balance between simplicity and complexity means that no model will be "best" for all situations. When process models are created, the opportunities for additional detail are numerous; the modeler must judge when sufficient detail has been included in the model to provide an appropriate answer.

Levels of Models

Models can be conceived as having three levels (97). The first or primary level is a mathematical expression that describes the changes in microbial numbers with time. An example is a growth model that estimates the change in log CFU/ml with time. Another primary-level model is one that describes the decreasing counts with time during thermal processing, such as the widely used decimal reduction time, or D value. Alternatively, microbial numbers may be estimated by turbidity, conductance, or biochemical assay such as PCR (or enzyme-linked immunosorbent assay). Describing the formation of a microbial toxin or other metabolic product with time constitutes another type of primary-level model.

The secondary level of modeling is an equation describing how the parameters of the primary model change with changes in environmental or other factors. These equations might be based on Arrhenius or square root relationships (62), particularly if temperature is the primary factor of concern, as is often the case if a specific group of foods is being modeled. When other factors such as organic acid or nitrite concentrations are included in the model, polynomial regression equations are used. They are very flexible, with squared, cubic, and cross-product terms, but have less mechanistic interpretation than other secondary models. The z value in thermal process calculations is a familiar secondary-level model. It describes the change in the D value with changing temperature.

In the tertiary-level models, this process is reversed to obtain a prediction or estimation. Environmental values of interest are entered into the secondary-level models to obtain specific parameter values for the primary model. The primary model is then solved for increasing periods of time to obtain the growth or inactivation curve expected from that combination of environmental values. Because of the unwieldiness of these equations, both the secondary- and primary-level models are used in conjunction with spreadsheet or other software programs. This avoids the reentering of equations, takes advantage of graphics capabilities of the software, and allows performance of other calculations. Such applications software is designated as the tertiary level. Tertiary systems vary in complexity from an equation on a spreadsheet to expert systems or risk assessment simulations. Two widely available tertiary systems that contain a variety of growth and other models are the Food MicroModel (Leatherhead Food Research Association, Surrey, United Kingdom; tel: +44 [0] 372 376761) (59) and the Pathogen Modeling Program (U.S. Department of Agriculture, Wyndmoor, Pa. tel: 215-836-3794) (13).

Microbial models are also classified as primary- or secondary-level growth models or inactivation/survival models. Within growth models there are kinetic models to estimate lag times and exponential growth rates. The growth/no growth boundary and the probability of growth are also modeled. Microbial inactivation and survival are modeled with the same techniques. The classic thermal death model assumes the log number of surviving microorganisms declines linearly with time of treatment. Nonlinear population declines are observed, and several approaches to modeling these situations are proposed. For these situations, a variety of mathematical approaches are used for the primary- and secondary-level models. The next section describes the most frequently used and recently proposed models.

Growth Models

The primary level of growth modeling consists of equations that estimate increasing numbers of a microorganism with time (i.e., a growth curve). Simple estimates of the exponential growth rate and the Gompertz and Baranyi equations are the most frequently used models. However, users of these models note the increasing width of the confidence intervals surrounding the models' predictions as the conditions become increasingly unfavorable for growth. These increased confidence intervals arise from the increased difficulty in obtaining precise data and the inherent increase in microbial variability at these conditions. Often kinetic data for the primary-level parameters cannot be effectively collected near these conditions of greatest interest. Alternative forms of modeling, such as the growth/no growth boundary and the probability of growth models, can be applied at these conditions to help clarify whether growth may be expected.

Growth/No Growth Boundary

This form of modeling describes the relatively narrow range where multiple environmental parameters begin to interactively prevent growth. This description provides important information for formulating products with minimal risk to consumers and food processors (63). The model's capability to designate a combination of environmental factors that does not allow any growth is particularly important when developing means for controlling pathogens when even small amounts of growth are unacceptable. A highly infectious microorganism such as *Escherichia coli* O157:H7 or *Shigella flexneri* or a toxin producer such as *Clostridium botulinum* are examples of such a pathogen.

This type of model is developed by inoculating, storing, and observing growth in substrates such as microbiological broth media with specified environmental and compositional characteristics (e.g., pH, salt concentration, temperature) over an extended period of time. A secondary-level, logit (p) model with parameters for the environmental and compositional conditions is used to describe the boundary (73). The logit (p) is defined by

$$\text{logit}(p) = \ln[p/(1 - p)] \qquad (1)$$

where p is the probability that growth occurs. Each medium having no growth is given the probability value, logit (p), of 0, and those with growth the value of 1. If replicates exist for a condition, p can range from 0 to 1. This model may be fitted by standard logistic regression procedures such as PROC LOGISTIC (SAS Inst., Cary, N.C.). An example of a growth/no growth model based on the square root growth model (see below) for *S. flexneri* had the form (73):

$$\begin{aligned}
\text{logit}(p) = {} & b_0 + b_1 \ln(T - T_{\min}) + b_2 \ln(\text{pH} - \text{pH}_{\min}) \\
& + b_3 \ln(a_w - a_{w\min}) + b_4 \ln(\text{NO}_{2\max} - \text{NO}_2)
\end{aligned}$$

$$(2)$$

where the coefficients b_x are parameters for the model estimated by the regression analyses. The other parameters, $T_{\min}$, $\text{pH}_{\min}$, $a_{w\min}$ and $\text{NO}_{\max}$, can be estimated independently and are constant for a microorganism. There are models for Shiga toxin-producing *E. coli* that extend to temperatures above the optimum (80) and *Listeria monocytogenes* with factors of temperature, pH, NaCl, and lactic acid (87). After the model is fitted to the data to determine the parameter values, the position of the growth/no growth boundary can be estimated by letting $p = 0.5$, a 50% chance of growth (logit [0.5] = 0). Other probability levels could be chosen, e.g., $p = 0.1$, which would provide a more conservative estimate. Once the probability level and the parameter values for the other factors are chosen, the boundary value for the remaining factor can be calculated.

A time-to-growth (TTG) model was created for the yeast *Zygosaccharomyces bailii* at various pH, NaCl, fructose, and acetic acid levels using the LIFEREG (SAS Inst.) procedure to accommodate the time-censored data. (Data are considered "time censored" if some samples do not have growth at the end of the experiment, in this case, 29 days, but might have if the sampling period had been longer [54].) The natural log of TTG was modeled by a second-order, two-factor interaction regression model.

Probability of Growth

In addition to growth/no growth, the probability of growth within 29 days was estimated for the *Z. bailii* data (54). The probability of growth is modeled by

$$P_{(29 \text{ days})} = e^w/(1 + e^w) \qquad (3)$$

where $w = -3.46 - 8.99$ (fructose) -3.25 (NaCl) $+ 3.77$ pH $- 6.25$ (acetic acid). Fructose and NaCl are in molar concentrations, and acetic acid is the percentage of undissociated acid.

Models 2 and 3 do not use time as a factor because the model is for a specified storage time. Other situations may show the probability of growth changing more gradually over time or never achieving complete growth; an example of this model was developed for the germination and growth of *C. botulinum* spores (100). When a set of tubes containing spores was stored, a period without growth was followed by a period of increasing numbers of turbid samples and then no additional turbid samples for the remainder of the storage period. Graphs of the fraction of tubes showing growth with time had sigmoid shapes with varying placements, slopes, and heights. This increasing probability of growth with time was modeled with a logistic function

$$P = P_{\max}/[1 + e^{k(\tau - t)}] \qquad (4)$$

where P is the fraction of tubes with growth (0 to 1) at time t, $P_{\max}$ is the maximum fraction of tubes with growth at the end of the modeling time, k is the slope term for the rate of increasing tubes with growth, and τ is the time of the inflection point where P equals 0.5 $P_{\max}$. Polynomial regression equations were then calculated for each of the three parameters ($P_{\max}$, k, and τ) to describe their changes with temperature, pH, NaCl concentration, and spore numbers. Under favorable environmental conditions, the value for τ is small and the values for the k and $P_{\max}$ parameters are high. As conditions become less favorable, τ increases and the k and $P_{\max}$ parameters decrease. The model estimates that the mean τ time increases as the spore numbers approach small values, but the confidence interval about τ increases. Confidence limits for τ also increase as the value increases; i.e., the variation about the mean time for growth increases as that value increases.

Lag-Phase Duration

Many growth models incorporate lag phases into their overall structure. A correlation between the generation time and the lag-phase duration is often presumed (33). This is apparent when the inoculum cells are in the same physiological state and have the same cultural history (77).

The lag-phase duration can be modeled independently of generation time (growth rate) with a modified Arrhenius relationship. The Arrhenius relationship assumes the log of the rate (1/lag-phase duration) is inversely proportional to the reciprocal of the absolute temperature. This relationship is based on the control of the rate by a single rate-limiting enzymatic reaction. Typical Arrhenius plots have been observed at suboptimal temperatures for some bacteria; unfortunately, plots with two distinct slopes were more frequently found (76). An expanded version of the Arrhenius equation that includes curvilinearity was fitted to a variety of food microorganisms (32). The lag time was related to water activity and the temperature by the relationship

$$\ln(1/\text{lag time}) = C_0 + C_1/T + C_2/T^2 + C_3\, a_w \qquad (5)$$

where C_x are specific parameter values for a microorganism.

Several models are proposed for the lag times of *Listeria* species (36). The square root model is adapted by inverting and taking the natural logarithm; one form of this model is

$$\ln \lambda = \ln\{1/[b(T - T_{\min})]^2\} \qquad (6)$$

where λ is the lag time and $T_{\min}$ is the notational temperature where growth ceases. Other variations of the model change the exponent

$$\ln \lambda = p/(T - T_{\min}) \qquad (7)$$

and

$$\ln \lambda = \exp[a + (b/K^o)] \qquad (8)$$

Another approach to modeling of the lag-phase duration is based on the concept that the transition from nongrowth to growth depends on two factors, the magnitude of the biochemical changes that the cell must undergo and the rate that a cell can adapt. A two-dimensional matrix of lag times of *Aeromonas hydrophila* cells transferred from a prior incubation temperature to a new incubation temperature showed that both temperatures have a major effect on the lag-phase duration (50). Temperature transitions are modeled for cells in different physiological states (94). For the same transition, cells in the exponential growth phase have shorter lag times than cells in the stationary phase, stationary-phase cells starved in dilute medium, or frozen exponential-phase cells. Desiccated exponential cells have the longest lag times. For exponential growth, stationary, and desiccated cells, the longest lag times are for cells grown at higher temperatures (28 to 37°C) and transferred to lower temperatures (4 to 8°C). Cells transferred to the higher temperatures have relatively short lag times, regardless of physiological state or prior temperature. For each physiological state, quadratic regression equations for the logarithm of the lag time against the two incubation temperatures are calculated. For cells in the starved or frozen states, the prior temperature has less influence on the lag time; the amount of needed biological change is more related to recovering from the stress than the prior temperature. The rate the cells adjusted from these two states, determined by the growth temperature, is more significant than the prior temperature.

The commonly observed lag phase and transition to exponential growth phase are a phenomenon of a culture (i.e., population), not an individual cell (22). Each individual cell makes the metabolic changes and begins to replicate. The curvature in the plotted growth is the result of the distribution of individual lag times and the summation of the growth of the initial fraction of cells that makes the transition. Cultures with large numbers of cells have a high probability that cells with the shortest lag phase will be present to initiate growth. If the numbers of cells are small, the observed lag phases will reflect the distribution in lag-phase duration.

Injury

Sublethal heating or other stressful treatments can injure bacterial cells. To be detected and enumerated, these cells must have a nutritionally and environmentally favorable medium, usually without selective agents. These cells need time in this medium to make biochemical repairs before they can resume growth and regain normal characteristics. This repair or recovery period can be modeled as a variation of the lag phase. A multicompartment kinetic model was presented that describes sublethal injury, resuscitation, induced lag, and exponential growth (47). Nonlinear models for injury and resuscitation are proposed. This approach to modeling repair is applied to *L. monocytogenes* (61), and a table of fitted parameters after various heating temperatures and times

is presented. The model is

$$\log_{10}(\% \text{ noninjured cells}) = k - (R_r/a_r)(\exp[-a_r t] - 1) \tag{9}$$

where k is the initial level of $\log_{10}(\%$ noninjured cells), R_r is the maximum rate of recovery, and a_r describes the decreasing rate of recovery as the $\log_{10}(\%$ noninjured cells) approaches complete recovery, i.e., $\log_{10}(\%$ noninjured cells) approaches 2.0.

The repair of heat-injured *L. monocytogenes* is followed by enumeration on selective and nonselective media (26). The increasing number of recovered cells follows a first-order relationship for all of the temperature-pH-NaCl conditions tested.

Primary-Level Growth Models

Microbial growth is an exponential process, and the specific growth rate, $\mu(t)$, is the slope of the natural logarithm of cell concentration with time (6)

$$\mu(t) = d[\ln x(t)]/dt \tag{10}$$

where $x(t)$ is the concentration of cells at a given time, t. Microbial data are typically calculated and plotted as the base-10 logarithm, and the observed exponential growth rate is the specific growth rate divided by 2.3. The doubling time, Td, for a culture is

$$Td = \ln 2/\mu(t) \approx 0.69/\mu(t) \tag{11}$$

This is different from the average generation time, GT,

$$GT = 1/\mu(t) \tag{12}$$

Before the development of formal models, an estimation of the exponential growth rate was made from the linear portion of the plotted growth curve. This process essentially can be replicated by computer curve-fitting software. Buchanan et al. (22) proposed this approach with a discontinuous three-phase linear model:

Lag phase $\qquad N_t = N_0 \qquad t \leq t_{\text{lag}} \tag{13}$

Exponential-growth

phase $\qquad N_t = N_0 + \mu(t - t_{\text{lag}}) t_{\text{lag}} < t < t_{\text{max}} \tag{14}$

Stationary phase $\qquad N_t = N_{\text{max}} \qquad t \geq t_{\text{max}} \tag{15}$

where N_t is the log number of cells/ml at time t, N_0 is the log initial number of cells, N_{max} is the log stationary-phase number, μ is the growth rate, t_{lag} is the lag-phase duration, and t_{max} is the time N_{max} is reached. The stationary phase of the model may be omitted if desired. The curvature of a plotted growth curve between the lag and exponential-growth phases is described as a characteristic of a culture as the first cells end their lag phase and begin doubling. By assuming individual cells in a culture have variances for t_{lag} and μ, the curvilinear shape of a traditional growth curve can be reproduced. When single values for these parameters are assumed, the model has an angle between lag and growth. The model does not assume that cells have a submaximal growth rate during the transition from lag to exponential-growth phases.

Roberts and colleagues (42) first used the Gompertz equation for food microbiology. A series of sigmoid-shaped functions were evaluated, and the Gompertz equation with its sharper curvature between the lag and exponential-growth phases than for the transition to stationary phase fit microbial data better than other sigmoid functions (105). The Gompertz equation is widely used to model microbial growth. Its basic form is

$$Y_t = A + C \exp\{-\exp[-B(t - M)]\} \quad (16)$$

where Y_t is the $\log_{10}$ cell concentration at time t, A is inoculum level, C is the growth from inoculum to stationary phase, B is the relative growth rate, and M is the time when the maximum growth rate is achieved. The lag time and exponential-growth rate are estimated by the following equations:

$$\text{lag time} \qquad = M - 1/B \quad (17)$$
$$\text{exponential growth rate} = BC/e \quad (18)$$

Alternatively, the exponential-growth rate and lag time can be calculated by taking the first and second derivatives of the Gompertz equation, respectively (16).

Although frequently used, this model has several undesirable characteristics. The fit of a data set often does not have A equal the inoculum, and there is no linear exponential-growth phase. Consequently, the exponential growth rate at time M is usually greater than that estimated by models having linear exponential-growth phases.

These limitations can be overcome by providing a more mechanistic or biological basis and having a differentiable model. Baranyi and colleagues (5, 7) proposed a model that includes a linear exponential-growth phase, $\mu(x)$, and a lag phase that is determined by an adjustment function. The adjustment function, $\alpha(t)$, is multiplied by the maximum specific growth rate to give the growth rate at a specified time,

$$dx/dt = \alpha(t)\mu(x)x \quad (19)$$

The cell populations, x, are in arithmetic values, not their logarithm.

Initially, the adjustment function has a value of zero because the cells are not growing. With time, the cells adjust to their new environment, the adjustment factor increases to 1, and the cells are at the maximum growth rate for that combination of environmental conditions.

The decreasing growth rate as the cell population approaches the stationary phase is modeled by

$$\mu(x) = \mu_{\max}[1 - (x/x_{\max})] \quad (20)$$

Combining these equations, rearranging and converting to the $\log_{10}$ CFU/ml yields

$$Y_t = Y_0 + \mu_{\max} A(t)$$
$$- \ln\{1 + [\exp(\mu_{\max}\alpha(t) - 1) \exp(Y_{\max} - Y_0)]\} \quad (21)$$

where $\mu_{\max}$ is the maximum exponential growth rate, $A(t)$ is the integral of the adjustment function $\alpha(t)$, and Y_t, Y_0, and $Y_{\max}$ are the numbers at time t, at inoculation, and at maximum cell density (stationary phase), respectively.

An important concept presented by this model is that the observed lag phase is a combination of the physiological state of the cell (q_0) and the adjustment to the new environment (ν). One form of the adjustment function is

$$\alpha(t) = q_0/[q_0 + e^{-\nu t}] \quad (22)$$

The lag-phase duration (λ) is calculated by

$$\lambda = \ln(1 + 1/q_0)/\nu \quad (23)$$

If the cells are not ready to grow (small q_0) or the rate of adjustment is slow (small ν), the lag phase will be lengthy.

Secondary-Level Growth Models

The parameters of the primary-level models presented above determine the way microbial populations change with time in a specific and constant environment. The secondary-level models calculate how the primary-level parameters change with changes in the environment. The Arrhenius model presented for modeling the lag phase is a secondary level model describing the change in the lag phase with changes in temperature.

The Arrhenius relationship has not received much use for growth modeling because the reciprocal plots of absolute temperature versus log growth rate are frequently nonlinear. However, the effect of temperature and CO_2 levels on the growth of spoilage microorganisms in fish with a modified Arrhenius was successfully modeled (55). The model is

$$\ln(\mu_{\max}) = \ln(\mu_{\text{ref}} - d_{CO_2} \cdot [CO_2])$$
$$+ (E_A/R)[(1/T_{\text{ref}}) - (1/T)] \quad (24)$$

where μ_{ref} is the $\mu_{\max}$ under reference storage conditions, T is in absolute temperature (K), T_{ref} is the temperature under the reference storage conditions, d_{CO_2} is a constant expressing the effect of CO_2 on the $\mu_{\max}$, $[CO_2]$ is the concentration of CO_2 in the package, E_A is the activation energy, and R the universal gas constant.

Polynomial regression equations are frequently used to model the effect of several factors on the lag phases and exponential-growth rates or generation times (21). The approach makes no assumptions about the relationship between independent and modeled variables. The equation is the best fit to the particular data set. The more complex the equation, with interactions and quadratic or cubic terms, the more "flex" in the multidimensional surface and the closer to the data it would be. Whether this complexity is warranted or whether the model is overparameterized can be determined by testing the fit of the equation with a new data set. These equations can be simplified by removing terms that are not statistically significant. Because the variation usually increases with increasing growth rate or lag time, the logarithm or other transformation of the values is frequently used to normalize the variances.

The square root or Bèlerádek model is based on the linear relationship between the square root of the exponential-growth rate below the optimum growth rate and the temperature (62, 64). The basic form of the model is

$$\mu^{0.5} = b(T - T_0) \tag{25}$$

where μ is the growth rate, b is the parameter value for a specific condition, and T_0 is the temperature where the extrapolated growth rate would be zero. The T_0 is characteristic of a microorganism, and a group of lines of growth rates in different media will extrapolate to the same T_0. This allows expansion of the model to

$$\mu^{0.5} = b(T - T_0)(pH - pH_0)^{0.5}(a_w - a_{w0})^{0.5} \tag{26}$$

The pH_0 and a_{w0} are the extrapolated values when the growth rate is notational zero. Temperatures above the optimum can be modeled by

$$\mu^{0.5} = b(T - T_{min})\{1 - \exp[b_2(T - T_{max})]\} \tag{27}$$

where T_{min} is the extrapolated zero temperature for the low-temperature range and T_{max} is the extrapolated zero temperature for the high-temperature range. Carbon dioxide concentrations inside modified atmosphere packages were added to this model (34, 55).

The gamma model is based on the concept that the observed growth rate is the product of all of the environmental factors affecting growth, a mathematical expression of the hurdle concept (103). The observed growth rate (γ is a fraction of the optimal growth rate μ_{opt}.

$$\gamma = \mu/\mu_{opt} \tag{28}$$

A dimensionless gamma term that ranges from 0 to 1 can

be added for each factor that affects growth, for example,

$$\gamma(T) = [(T - T_{min})/(T_{opt} - T_{min})]^2 \tag{29}$$

$$\gamma(pH) = [(pH - pH_{min})(pH_{max} - pH)]/ \\ [(pH_{opt} - pH_{min})(pH_{max} - pH_{opt})] \tag{30}$$

$$\gamma(aw) = (a_w - a_{w min})/(1 - a_{w min}) \tag{31}$$

The gamma values for the specific environmental condition are determined with the respective minimum, optimum, and maximum parameter values, which remain constant for a microorganism, and the parameter values of interest. The observed growth rate for a specific set of conditions would be the product of these gamma terms:

$$\mu = \mu_{opt} \, \gamma(T)\gamma(pH)\gamma(a_w) \tag{32}$$

An approach to modeling salt concentrations with the gamma model is based upon the Monod equation (49):

$$\mu = \mu_{opt}[S/(K_s + S)] = \mu_{opt'}[(S_{MIC} - S)/(K_s + (S_{MIC} - S))] \tag{33}$$

where S is the NaCl concentration, S_{MIC} is the minimum inhibitory concentration, and K_s is the salt concentration that results in half-maximum growth rate.

The gamma model has the advantage of being very adaptable, additional factors can easily be added or deleted and the T_{min}-type parameter values remain constant despite adding factors or changes within a gamma factor. If a new strain of a pathogen emerges with different characteristics, the T_{min}-type parameter can readily be determined with minimal experimental data.

Artificial Neural Networks

A new technique for modeling the response of a variable to a set of input parameters is the artificial neural network (ANN). The theory and application of ANN were described for predictive microbial modeling (45, 66). An ANN was developed as the secondary-level model to estimate the Gompertz parameters for a data set of *S. flexneri* growth with environmental factors of temperature, pH, NaCl, and $NaNO_2$ concentrations. The feasibility of ANN to model the parameters for microbial growth was further demonstrated (27, 41).

Modeling Dynamic Growth Conditions

The models discussed to this point are for constant, single-combination environments. Many foods have changing environmental conditions, frequently changing temperatures. One modeling approach is to divide the growth period into appropriately small intervals and calculate the fraction of the lag phase that is achieved in each interval based on the average environmental condition during that interval. When the fractions sum to

1.0, the lag phase is finished. Exponential growth can be similarly treated; intervals are designated which have sufficiently uniform conditions; growth is calculated for each interval and summed.

This approach has obvious limitations, particularly when changes are very rapid. Van Impe et al. (88, 89) used a time-implicit model with a continuously differential equation based upon the Gompertz equation. The growth model developed by Baranyi and Roberts (5) is also suitable for modeling fluctuating conditions (3, 11). A finite element analysis to model the temperature transition gradients in packaged fluid milk was coupled to a dynamic growth model for *L. monocytogenes* (2). The finite element analysis calculates the temperature distributions within the food using the food's thermodynamic characteristics and the external temperatures. Microbial growth is modeled for each small portion of the food according to its specific temperature profile. The overall growth is then determined by summing growth for the entire food volume. This avoids basing all of the growth on the center or the location of lowest temperature in the food, a worst-case approach.

Modeling Competition

The effect of competition between two microorganisms has been noted but has received little modeling. It is a difficult area to find the principal factors and parameters to model. One microorganism may affect another by lowering the pH (fermented foods) or producing a bacteriocin. These actions may be modeled as the dynamic production of a pH change or a metabolite by the first microorganism and inhibition on the second. The inhibition by lactic acid-producing *Lactococcus lactis* on the growth of *L. monocytogenes* was modeled with this approach (12). Other mechanisms of microbial competition are less understood. The Gompertz equation was fitted to growth curves for *E. coli* O157:H7 grown in broth in the presence of competing microbiota (35). The effects on lag phase, growth rates, and maximum population varied depending upon the competitor. The Baranyi model (equations 19–23) was used to quantify differences in lag time and growth rate between cultures of single species and multiple species of spoilage microorganisms from refrigerated meat (71).

Inactivation/Survival Models

The inactivation and survival of microbes in processes leading to a decline in their numbers can be modeled. Inactivation refers to an active process of killing microorganisms, frequently using heat. Survival refers to microorganisms in environmental conditions that do not permit growth wherein the microorganisms decrease in number with time. Survival is a passive and slow process. Conditions with high acidity or an antimicrobial could be considered as either inactivation or survival. From a modeling perspective, however, there is little difference between the two processes and they will be discussed together.

The classic log-linear thermal death time model assumes first-order kinetics: a constant proportion of organisms is inactivated in each successive time period. This model assumes that all cells have the same heat resistance and the death of an individual cell results from a random inactivation of a critical molecule. The D value is the time for a 1-log decrease in viability at a defined temperature in a specific matrix. The z value is the change in the log of the D value with changing temperature, a secondary-level model. This model has a long history of successfully modeling the heat inactivation in retort and other thermal processes.

First-order primary-level model
$$N = N_0 e^{-kt} \tag{34}$$

where N is the concentration of surviving cells, N_0 is the initial concentration, k is the inactivation rate, and t is the time.

D-value primary-level model
$$N = N_0 e^{-t/D} \tag{35}$$

where D is the time for 1 log decrease.
The secondary-level model parameter for the D value is the z value.

$$\log (D_1/D_2) = (T_2 - T_1)/z \tag{36}$$

Arrhenius and Eyring relationships are less empirical, having been based on chemical and thermodynamic theory. For this secondary-level model, the log of the inactivation rate is inversely related to the reciprocal of the absolute temperature (75).

$$\ln k = -E_A/RT \tag{37}$$

where E_A is the inactivation energy, R is the universal gas constant, and T is the temperature in Kelvin. Both inactivation rate (k) and D-value approaches can satisfactorily model inactivation of microorganisms under constant and changing temperature (40).

Increasingly, researchers are reporting data with nonlinear inactivation kinetics. The nonlinearity can appear as a lag or shoulder period before any death begins or a tailing of an apparently more resistant portion of the population. Such derivations from linear kinetics are particularly evident when inactivation conditions are relatively mild, e.g., heating of vegetative bacterial cells in the range from 50 to 55°C. The simplest approach when a shoulder appears is to use a discontinuous model with two linear sections (17):

$$N = N_0 \qquad \text{when } t < t_{\text{lag}} \qquad (38)$$

$$N = -b(t - t_{\text{lag}}) \qquad \text{when } t > t_{\text{lag}} \qquad (39)$$

Biphasic thermal inactivation of *Salmonella enterica* serovar Enteritidis was modeled by having two populations, each with their respective *D* values (51). First-order kinetics models that fit data with shoulders and tails were developed (102). Thermal inactivation of spores was modeled with a population dynamics theory that links first-order activation and inactivation processes (84). The shoulder was ascribed to the need to activate spores before they become susceptible to thermal destruction.

Curved inactivation data were fitted by an exponentially damped model (31):

$$\log_{10}(N/N_0) = -kt \exp(-\sigma t) \qquad (40)$$

where σ represents the damping coefficient. The greater the value of σ the more curvature in the inactivation plot. An undesirable characteristic of this model is that after the population declines, the curvature eventually has the population increasing.

Semilogarithmic thermal inactivation curves of *C. botulinum* spores were described (70):

$$\log_{10}(N/N_0) = -b(T)t^{n(T)} \qquad (41)$$

where $b(T)$ and $n(T)$ are temperature- and medium-dependent coefficients. These two terms are modeled with empirical exponential equations.

Whiting (92) used a logistic model for lethality curves exhibiting a shoulder and tails for *Salmonella* sp. and *Staphylococcus aureus*:

$$N = N_0 + \log[F_1([1 + \exp(-b_1 t_{\text{lag}})]/\{1 + \exp[b_1(t - t_{\text{lag}})]\})$$
$$+ (1 - F_1)([1 + \exp(-b_2 t_{\text{lag}})]/\{1 + \exp[b_2(t - t_{\text{lag}})]\})] \qquad (42)$$

where F_1 is the fraction of the cells in the major population and $(1 - F_1)$ is the fraction in the subpopulation, b_1 and b_2 are the rate of declines for the major and subpopulation, respectively, and t_{lag} is the shoulder or lag period. The *D* values are approximated by $D = 2.3/b$. Logistic algorithms to model the effect of heating rate on thermal inactivation were presented (85).

The Baranyi model (equations 19 to 23) was used to fit thermal inactivation survival curves for *L. monocytogenes* that has linear curves, lag phases, tailing, and sigmoid inactivation (101). The Baranyi model was also used for survival of *E. coli* O157:H7 in eggplant salad and oregano essential oil (83). The Gompertz model was used to model the inactivation of *L. monocytogenes* in sausage products (10).

Modeling Dynamic Inactivation

An extensive analysis of the structure of most of the inactivation/survival models was presented by Geeraerd et al. (41) using data from five different pathogens. A new four-parameter model with shoulder, tail, and log-linear inactivation was developed:

$$dN/dt = -kN$$
$$dC_c/dt = -k_{\max} C_c \qquad (43)$$
$$k = k_{\max}\{[1/(1 + C_c)][1 - (N_{\text{res}}/N_0)]\}$$

where C_c is a measure of the physiological state of the population, $k_{\max}$ is the first-order inactivation rate, and N_{res} is the residual population responsible for tailing. Differentiating the static model with respect to time derives a dynamic version of this model. It can account for the changes in temperature with time and integrate the total inactivation. This model is related to both the Baranyi (equations 19 to 23) and Whiting (equation 42) models. The Baranyi model is also differentiable and suitable for modeling dynamic conditions.

Statistical Considerations for Modeling

It is widely recognized by users of simple linear regression that extrapolation beyond the data range is inappropriate and yields potentially erroneous estimations. This is equally applicable to polynomial regression used in microbial modeling where the multidimensional space that defines the interpolation space or the polyhedron that encloses all of the tested combinations can be very complex (8). The unreliability of the estimation becomes greater closer to the boundary of the interpolation region. Overparameterization (the inclusion of too many parameters) of a descriptive model can result in a model that closely fits the particular data set used to develop the model but is illusory in terms of its real predictive ability. In other words, the additional parameters are modeling the errors within the data set, not the overall relationship between the factors being modeled.

Various indices have been proposed to indicate the quality or "goodness of fit" of a model. The error mean square, *F*, and R^2 values from regression analyses are frequently reported that use the absolute difference between the predicted and observed values. Simultaneously plotting data and estimates from the model is necessary to observe whether there are any specific areas within the interpolation space where the model is poorly fitted. Two complementary indices of the quality of predictive models are the bias and accuracy factors (78). The bias factor is the average of the $\log_{10}$ ratio of predicted/observed values, thereby reducing the influence of large values. The calculation for generation time (GT) is

$$\text{Bias factor} = 10^{(\Sigma \log(\text{GT predicted}/\text{GT observed})/n)} \quad (44)$$

Perfect agreement between predicted and observed values will have a value of 1.0. Values greater than 1.0 indicate that longer GT were predicted than observed. Because over- and underestimates tend to average out in the bias factor, an estimate of the average accuracy of the estimates can be calculated with the accuracy factor using the absolute value of the log ratios.

$$\text{Accuracy factor} = 10^{(\Sigma |\log(\text{GT predicted}/\text{GT observed})|/n)}$$
$$(45)$$

The larger the accuracy factor value, the less accurate the average estimate. These factors were further developed to compare different models (4). Defining X = (difference between predicted and observed values), the accuracy factor is

$$A_f = \exp\left[\sqrt{\left(\sum (X^2)\right)}\right] \quad (46)$$

and the bias factor is

$$B_f = \exp\left[\sum (X)\right] \quad (47)$$

Regression analysis assumes the data are normally distributed. If they are not, the data should be transformed before modeling. This requirement also applies to nonlinear curve-fitting procedures. Determining whether a data set is normally distributed, either before or after a transformation, is not always easy, particularly for typical data sets for microbiology with few replicates. Because variation clearly increases with increasing value for microbiological parameters such as generation time or lag-phase duration, the logarithms of these values are typically taken. The consequences of different transformations, such as logarithm, square root, and inverse, on the variances for the distributions for modeling growth rate, lag phase, and other microbial parameters are described (1, 18, 67, 74).

Validation

It is important to validate a model for the specific food to ensure that the model is sufficiently accurate for its intended use (86). Most modelers internally validate their models by comparing the data to the corresponding predicted values. External validation uses new data from additional replicates to assess the quality of the predictions. The adequacy of the model can be assessed by the statistical measures described above and by graphing. However, individual foods may have additional factors that affect microbial growth that are not included in the model. Therefore, users of a model should conduct trials

with a few selected inoculated food samples to confirm that the predictions by the model are accurate for their specific circumstances.

Strain Variation

Different strains of a foodborne microbial pathogen do not have identical characteristics. However, microbiological studies have rarely measured this natural diversity, and it has not been incorporated into most models. This diversity can be large; representative data for thermal inactivation D values for strains of *L. monocytogenes*, *Salmonella* serovar Enteritidis, and *E. coli* O157:H7 range from 6.5 to 26 min, 11.8 to 31.3 s, and 17.4 to 35.4 min, respectively (28, 58, 82). The ratio of standard deviation to the mean between strains of *E. coli* O157:H7 for the exponential-growth rate, time for 4 logs of decline, and thermal D values at 55 and 60°C were 0.16, 0.4, 0.4, and 0.3, respectively (98). These variations between strains are probably greater than the uncertainties from the experimental procedures and statistical modeling.

HACCP and Modeling

The modeling previously discussed focuses on describing a single processing step with constant environmental conditions although dynamic models, such as the Baranyi growth model, are suitable for changing conditions. These growth, survival, and inactivation models can be linked to model an entire process from the input of raw ingredients to the final product. An example of a model for a frozen meat patty process by Zwietering and Hasting (104) estimated the microbial counts at each step. Changes in initial contamination or temperature can be made and the expected consequences for the final product calculated. These models can, therefore, be useful in estimating growth or survival of a pathogen, determining the shelf life, assisting product development, optimizing processes, predicting the effects of changing a processing step or whole process, and educating production personnel and others on the consequences of process control/out-of-control (64). Predictive models can help set the parameters for critical control points (37).

RISK ANALYSES

Thus far in this chapter our discussion on modeling has focused on the estimated or predicted mean value. The mean value for growth rate is multiplied by the mean storage time to predict the amount of growth. That there are distributions in these mean values is widely recognized, often standard deviations or confidence ranges are provided, but the algebraic calculations only consider the means for both input parameters and the calculated values. This approach is a "deterministic" or "point-estimate" analysis. Making a "worst-case" estimate is another example of a deterministic analysis.

Quantitative risk analysis incorporates the distributions of the model parameters into the calculations. The sources of the distributions about the mean value are from variability and uncertainty (91). Variability refers to the differences that actually exist in the population, different packages of a food are on the retail shelf for different periods of time before purchase or the temperature cycles within an oven. To change variability, a change in that step or processing parameter must be made. Uncertainty refers to a lack of knowledge about the true value for a parameter. This includes experimental errors and statistical confidence intervals. It also includes the lack of data. Additional research and better methodology can reduce uncertainty. An estimate of the amount of *Cryptosporidium* sp. consumed in the United States, for example, has a high level of uncertainty because the methodology is imprecise and relatively few foods have been analyzed for this pathogen.

These distributions can be described in a variety of ways, such as normal, lognormal, beta, binomial, pert, triangle, or histogram distributions (91). If the input parameters of a model or a food process are distributions, the calculated output will be a distribution (25). To make these calculations, simulation or Monte Carlo modeling is used. In this method, a value is randomly selected from each distribution according to the characteristics of that distribution and the model is calculated. A new set of values from each distribution is selected and the process is recalculated (iterated). This continues until a distribution of calculated values is created. Its mean or median can describe this distribution, but the entire shape and spread of the distribution are important in the interpretation of the process. For a thermal inactivation calculation, for example, the inactivation received by the lowest 5% or 1% of the iterations of the model may be more relevant to setting critical control point parameters than the average amount of inactivation. This type of analysis is termed "probabilistic" or "stochastic" and provides a much more complete description of the process being assessed than the deterministic analysis.

Risk analysis is the overall process of defining a problem, collecting information, calculating results, risks, and uncertainties, evaluating potential mitigations, and deciding on a course of action. Risk analysis has three components: risk management, risk communication, and risk assessment (38). Members of the risk management group control the risk analysis: the problem is identified, the scope is defined, the risk assessment is commissioned,

management options and mitigations are identified, and the preferred management option is selected. Risk management considers societal values and preferences as well as scientific information, technical feasibility, and costs. Risk communication involves an interactive exchange of information and opinions. This includes communication with the public and stakeholders as well as communication between individuals involved with the risk analysis process. Risk assessment is the scientific analysis of the problem. It collects and organizes the available scientific information, develops any models and calculates outputs, and presents results with attendant uncertainties to the risk managers. Risk assessors evaluate the effect of any mitigations that the risk managers may request, including establishing the boundaries for options in relation to what actions are supported by the science available. A risk analysis is an interactive process among these three components. In particular, the risk managers and risk assessors need to be in regular contact as decisions are made during the development of the structure of the risk assessment to ensure the risk assessment provides the most appropriate information pertaining to the problems facing the risk managers. Nevertheless, maintaining the scientific integrity of the risk assessment is of utmost importance.

The risk assessment process is a tool for situations where available data are less than ideal. It determines the state of knowledge, including estimates of the impact of the quality of the data. The purpose of a risk assessment may be to identify critical data deficiencies, to target specific future research, to provide insight into the general or relative relationships between components of the process being analyzed, or to estimate quantitative values for risks or other outputs. The quality of the available data must become progressively better for the latter purposes. Data quality refers to the completeness, relevancy, and validity of data, design and purpose of the study, year data were collected, season, geographic area, laboratory versus food plant, in vitro versus inoculated pack, same microbial species or genus, agreement with related studies, peer review, and expert opinion versus systematic collection.

Quantitative Risk Assessment

The risk assessment is the measurement of the risk and an identification of factors that influence it. The basic principles of microbial risk assessment have been published by U.S. and international bodies and will be summarized in this chapter (14, 15, 29, 52, 53, 72). Several of the key definitions are: hazard, the agent that causes harm; risk, the likelihood of adverse health effect from a hazard; and transparency, having all data, assumptions, logic, value judgments, limitations, and uncertainties in the risk assessment fully stated, documented, and accessible for review.

The first step in the risk assessment is to identify the purpose and scope of the risk assessment and define the goals and expected outputs. This designates the parameters to be incorporated and establishes the latitudes of the assumptions that may be necessary. An important characteristic of the risk assessment is its potential to evaluate the entire farm-to-fork-to-illness continuum and allow comparisons of factors or mitigations on the farm to those in processing or food preparation. However, the scope of an individual risk assessment does not need to include the entire continuum, only what is necessary to provide information for the problems facing the risk managers.

The risk assessment itself has four components: hazard identification, exposure assessment, hazard characterization, and risk characterization.

Hazard identification is the collection of epidemiological, biological, and other pertinent information and expert knowledge on microorganisms and their sources in foods. It includes information on the occurrence and severity of illness in consumers. The amounts, frequencies, and sources of microorganisms should be estimated. Epidemiological, microbiological, and food consumption data are utilized. This step is a qualitative evaluation of the information (or lack of information) and is presented to the risk managers to refine the structure and goals of the risk assessment.

Exposure assessment determines the number of microorganisms or amount of microbial toxin consumed. An exposure assessment uses information on contamination of raw ingredients, growth/inactivation during processing, storage and preparation, and consumption patterns (frequency, serving size). Both the frequency of occurrence and levels of contaminating microorganisms are needed. The predictive microbial models are an essential part of the exposure assessment. The outputs from this component are typically the number of pathogens consumed per serving or the frequency of consuming different numbers of a pathogen.

Hazard characterization, often termed the dose-response relationship, is the evaluation of the effect on human health from consuming the microbial pathogen. The nature, severity, and duration

of the adverse effects and the interaction of the host (individual) are considered. The immunological status of the person is an important factor in the response to many foodborne pathogens. The food matrix in which the pathogen is found may also affect the response. Microbial factors that affect virulence are also components of the hazard characterization. The typical output of this component is a relationship between number of pathogens consumed and the probability of infection, illness, or death. Often a separate dose-response relationship is determined for different susceptible populations (children, elderly, immunocompromised, pregnant women).

Risk characterization is the integration of the exposure assessment and the hazard characterization. The overall probability of occurrence and severity of illness in specific populations from consuming the foods is calculated. The risk characterization includes a description of statistical and biological uncertainties that accompany any estimates that are made.

The final step is to produce a full and systematic report of the risk assessment. As necessary, this includes an interpretation of the findings so that it is useful to the risk managers, the intended recipients of the report. Transparency to both technical and nontechnical audiences is important. The risk analysis process is also iterative, the risk managers may specify additional questions to be answered or mitigations to be tested on the basis of the initial risk assessment.

Microbial exposure assessments are a critical component of the risk analysis and, compared to chemical risk assessments, a more complex process. The number of microorganisms consumed can change by orders of magnitude depending on the characteristics and parameters of the food process. At the present time, microbial risk assessments are concerned with acute illness following a single exposure to the pathogen with short onset times in contrast to many environmental chemicals. Chronic exposure to low levels of a microbial pathogen can stimulate an immune response. But this is not understood sufficiently to assess. Similarly, chronic effects or sequelae are also not usually considered. The dose-response relationships are not well characterized at the present time. However, human feeding studies and epidemiological data do exist that provide more precise data to estimate the response of humans to microbial pathogens than is frequently available for chemical hazards.

Exposure Assessment

The microbial models discussed previously are necessary components of the exposure assessment (56). However, microbial pathogens are often present in low numbers or at low frequencies in foods. This requires the exposure assessment to evaluate both concentration and prevalence. For the latter, Poisson and binomial distributions are used to determine the probabilities of contamination or number of packages with low levels of contamination. Low-level contamination in a holding tank, for example, may result in most packages that are subsequently filled from that tank being uncontaminated. A certain frequency of packages will have one pathogen, a lower frequency of packages will have two pathogens, and so forth.

The process of assembling the data for a risk assessment typically reveals a lack of scientific data. Survey data on the prevalence of contamination in our food supply are often incomplete. Little information is collected on many process variables such as time in transit or at retail, and there is a severe lack of knowledge on food preparation practices in both food service and homes. Expert opinion is often the only source of data for important parameters in the process modeling.

The structure of the risk assessment in itself can be a major component in the uncertainty. The level of complexity and detail necessary for the risk assessment to provide adequate answers for the risk managers is as much an art as a science. The number of factors in a food process or the number of routes that a food may take after it leaves the processor can be numerous. Adding additional steps, e.g., including the transit from grocery store to home refrigerator, may improve the quality of the estimate of the number of pathogens at consumption. However, every added parameter has its uncertainty; this may overwhelm the gain from the added step. Parsimony in individual growth and inactivation models is a virtue; simplicity in modeling a food process is also a virtue. A model should be only as encompassing and detailed as required. In practice, it is often necessary to assemble and calculate a risk assessment and analyze the output to determine whether sufficient detail (or too much) has been incorporated. Sensitivity analysis, which shows the impact of the uncertainty or variation about a parameter on the output probability distribution, provides important clues on the impact of that parameter in the risk assessment. Scenario trials, where a series of values for a parameter are tested, are also valuable in determining the appropriateness of the risk assessment's structure.

Hazard Characterization

The goal of the dose-response relationship is to provide an estimate of the probability of illness after consumption of a specified number of pathogens (19, 30). This relationship depends on the characteristics of the pathogen, the susceptibility of the host, and the food matrix. The human health endpoint must be specified. Infection is generally used to refer to a colonization of the intestinal tract, and morbidity to refer to signs of illness in an individual. In some individuals, infection and morbidity lead to mortality, or death.

Microbial pathogens may be infectious, toxico-infectious, or toxin producers. An infectious microorganism (*L. monocytogenes*) must survive the digestive process, attach in the intestinal tract, and invade the intestinal epithelium (gastroenteritis) or the body (sepsis). A toxico-infectious pathogen has a similar process, except that the toxin is produced (*E. coli* O157:H7) or released (*Clostridium perfringens*) in the intestinal tract, which then affects the intestinal epithelium or the body. A toxigenic pathogen, such as *S. aureus* and *C. botulinum*, produces the toxin in the food before consumption. Different strains of a pathogen may have widely different abilities to cause illness, depending on the genes for virulence factors the strain possesses and how they are expressed.

Individuals vary in their susceptibilities to microbial pathogens, depending on their age, general health and nutrition, immune status, and other factors. Segments of the population with increased risk have depressed immune capability, particularly for the infectious and toxico-infectious pathogens. These individuals are children, the elderly, and those with immune system diseases or taking immune system-depressing drugs for other medical conditions. Individuals with achlorhydria or who use antacids, which reduce stomach acidity, may have increased risk.

In recent years there is increasing realization that the food matrix affects the probability of illness. Microorganisms may adapt to acidic environments and thereby more effectively survive passage through the stomach. The food environment may affect expression of various virulence factors before the pathogen is consumed. Foods with high buffering capacity may reduce stomach acidity, and fats may protect pathogens from stomach acids. The transit time through the stomach depends on whether the food was eaten with a meal or not.

Data for developing dose-response relationships come from human feeding trials, animal feeding studies, in vitro biochemical experiments, and epidemiological investigations. Human feeding trials can only be done with pathogens unlikely to cause serious illness or death and must expose healthy adults, not the immunocompromised populations of greatest interest. The number of individuals in a trial is very limited, making the detection of illness at a low dose and frequency very unlikely. Animal trials can use larger numbers of animals at a dose and allow studies with different host susceptibilities. Studies of virulence factors are also possible. However, the relationship between the susceptibility of animals and humans to the pathogen is often poorly understood. Laboratory mice are several orders of magnitude more susceptible to infection by *L. monocytogenes* than are humans, for example. Epidemiological investigations have as their primary objective the identification and prevention of further consumption of the contaminated food. Usually by the time the food is identified, there is none remaining for microbial testing or it has little relevance to its microbial condition when consumed. Frequently in outbreaks the number of individuals who became ill is known, but the number of people who consumed the food and did not become ill is unknown. Follow-up to determine the amount of food consumed and the characteristics of individuals who did and did not become ill is only occasionally done. A few outbreaks have been investigated thoroughly and provided valuable information regarding susceptible populations and provided data to relate human responses to animal feeding studies.

Dose-Response Modeling

When the logarithm of ingested pathogens and the probability of illness are plotted, a sigmoid relationship is typically observed. The precise shape at the lower dose range is of great concern and very difficult to collect data for. For infectious microorganisms, a single cell is considered to have the potential for causing illness under appropriate circumstances. Several models have been proposed for modeling the dose-response relationship; the three most frequently used models are the exponential, beta-Poisson, and Weibull-gamma (19, 30, 44, 48).

The exponential model is

$$P = 1 - \exp(-rd) \tag{48}$$

where P is the probability of infection, d is the dose, and r is the parameter specific for each pathogen and human population. It assumes that the host susceptibility and pathogen virulence are constant for a population.

The beta-Poisson model is

$$P = 1 - (1 + d/\beta)^{-\alpha} \tag{49}$$

where α is the infectivity parameter and β is the curve shape parameter. The beta-Poisson model is derived from the exponential model but assumes a beta distribution for the host-pathogen interaction.

The Weibull-gamma model is

$$P = 1 - [1 + (d^b)/\beta]^{-\alpha} \tag{50}$$

where b is the curve shape parameter and α and β are infectivity parameters. Derived from the Weibull model, this model assumes the host-pathogen interaction is described by a gamma distribution. With appropriate parameter values, the Weibull-gamma model simplifies to either the beta-Poisson ($b = 1$) or exponential model ($\alpha = 1$).

In general, increasing the number of parameters in the model (i.e., exponential = 1, beta-Poisson = 2, Weibull-gamma = 3) increases the likelihood that the model will adequately fit a data set, but this can be offset by increasing uncertainty from overparameterization.

Risk Characterization

This step is the integration of the exposure assessment and the hazard characterization to estimate the likelihood of illness from consuming the food. Depending on the needs of the risk managers, the estimated parameter may be a risk per serving or total number of cases. It may be for the entire population or a specific susceptible group. Other outputs may be requested, such as a ranking or comparison of different scenarios. Because simulation modeling was used to generate distributions, another purpose of the risk characterization is to communicate the uncertainties and to provide an estimate of the confidence in any value calculated by the analysis. The risk characterization may describe the impact of individual steps in the model and evaluate different mitigations. Because simulation modeling is a new technique, effectively explaining, interpreting, and communicating the results to risk managers and stakeholders are very important. Because risk assessors frequently make choices in their use of data, the risk characterization must be transparent for its conclusions to be accepted.

Explaining where variation or uncertainty in the data affects the resulting outcome is also important. Unacceptably high risks from variation will need to be corrected by better process design and control, whereas control of apparently high risks from uncertainty should be addressed by additional data collection and research. In the latter situation, other controls may be put in place until the additional data are obtained. In many cases, the separation of variation and uncertainty is difficult, and the data will contain both sources of variation. The risk assessment can provide a qualitative evaluation of which source predominates that is sufficient for the risk managers to base their actions on.

The hazard characterization may quantify the frequency of occurrence of different degrees of severity of the hazard. However, the risk characterization must still provide guidance to the risk managers of the impact on public health of the different outcomes. Whether cost-benefit analyses and other measures of the quality of life should be part of the risk characterization or should follow completion of the risk assessment is currently debated by risk assessors. However, epidemiological records can estimate the proportion of infections that result in illness and lead to treatment by physicians, hospitalizations, or death (23). Costs of treating a patient, lost salary, and other economic consequences can be estimated. Economic and risk analyses may affect choice of mitigations by determining the most effective and efficient allocation of resources (65). A plant with a stable workforce, for example, may realize greater benefits from worker training because it has a lower likelihood of losing training value to worker attrition than a plant with high worker turnover. Larger plants are more likely to profit from purchasing new expensive technologies that have significant economies of scale.

A recent example of microbial risk assessment comparing relative risks from different foods is by U.S. Food and Drug administration (FDA) and the U.S. Department of Agriculture Food Safety and Inspection Service (USDA FSIS) for *L. monocytogenes* (www.foodsafety.gov/~dms/lmrisk.html). Examples of farm-to-fork process microbial risk assessments are those by the USDA FSIS on *E. coli* O157:H7 in ground beef (www.fsis.usda.gov/ophs/ecolrisk/pubmeet/sld001.htm), by the FDA on *Vibrio parahaemolyticus* in oysters (www.foodsafety.gov/~dms/vprisk.html), and by the USDA/FSIS on *Salmonella* serovar Enteritidis in shell eggs (www.fsis.usda.gov/ophs/risk/contents.htm). Process risk assessments published in the scientific literature include assessments of *Salmonella* serovar Enteritidis in eggs (96, 99), *L. monocytogenes* in soft cheese from raw milk (9), *E. coli* O157:H7 in ground beef (24), and *L. monocytogenes* in smoked fish (57). The World Health Organization and the Food and Agriculture Organization Codex Alimentarius Commission Committee on Food Hygiene commissioned risk assessments for *L.monocytogenes*, *Salmonella* serovar Enteritidis in eggs, and *Salmonella* spp. in broilers to help reach international consensus on management of these pathogens (www.who.int/fsf/mbriskassess).

Risk Assessment in HACCP

HACCP is a risk management system and is accommodated within the risk analysis paradigm (39). The risk assessment helps analyze the entire food process, identify the critical points, and set critical limits. It also evaluates alternative processes and out-of-process events, such as a breakdown in a refrigeration system. The risk assessment provides the underlying support to the HACCP plan and is the means to quantitatively link the food process to public health (20, 68, 69, 81). Determining equivalent levels of safety for foods in international trade is another developing role for risk assessments (46).

Several terms have been proposed by the International Commission on Microbiological Specifications for Foods (90) for controlling the risks from foodborne illnesses through risk assessments and HACCP systems. First, there must be a target level of safety termed the tolerable level of risk, the probability of illness in a specified individual from consuming a serving of the food. This level of risk is a societal value judgment, not just a scientific question. It involves the risk individuals will tolerate in foods they enjoy. It also involves technical feasibility, costs, and sensory quality. Once an acceptable level of risk is determined, the dose-response relationship estimates the frequency and extent of exposure to a pathogen that constitutes that risk. This exposure is the food safety objective and becomes the microbiological threshold that the food processor controls his process to achieve. The food technologist, food engineer, and microbiologist collaborate to design the food process by evaluating combinations of process steps and within-step parameters and selecting the combination that most efficiently produces the best product that meets the food safety objective. For each critical step in the process, the objective of that step is specified, e.g., a 5 $\log_{10}$ inactivation. This specification is the performance criteria, and there may be several technologies and combinations of operating parameters that can achieve it. Microbial inactivation, for example, can be attained by thermal pasteurization, pulsed electrical fields, or ultrahigh pressure. For thermal pasteurization there are many time-temperature combinations that result in a specified inactivation. The specific process and parameters chosen are the process criteria and become an integral part of the HACCP plan.

CONCLUSION

The techniques of predictive modeling and risk assessment can link all of the processing steps to estimate an incidence of illness from consuming that food. Evaluation of the safety of a food is now a quantitative, instead of qualitative, procedure. Safety is no longer a poorly defined, apparent absence of illness or a below-detectable level of contamination. Instead, the probability or rate of occurrence of a specific illness in an individual must be considered, and there is rarely a complete absence of risk. These rapidly advancing tools will require more quantitative microbial data collection and effective communication to make this new paradigm effective and accepted by consumers and stakeholders.

References

1. **Alber, S. A., and D. W. Schaffner.** 1992. Evaluation of data transformations used with the square root and Schoolfield models for predicting bacterial growth rate. *Appl. Environ. Microbiol.* **58:**3337–3342.

2. **Alavi, S. H.** 1997. Development of an integrated dynamic growth-finite element model for predicting the growth of *Listeria monocytogenes* in packaged fluid whole milk. M.Sc. thesis. Department of Agricultural Biological Engineering, Pennsylvania State University, University Park, Pa.

3. **Alavi, S. H., V. M., Puri, S. J. Knabel, R. H. Mohtar, and R. C. Whiting.** 1999. Development and validation of a dynamic growth model for *Listeria monocytogenes* in fluid whole milk. *J. Food Prot.* **62:**170–176.

4. **Baranyi, J., Pin, C., and T. Ross.** 1999. Validating and comparing predictive models. *Int. J. Food Microbiol.* **48:**159–166.

5. **Baranyi, J., and T. A. Roberts.** 1994. A dynamic approach to predicting bacterial growth in food. *Int. J. Food Microbiol.* **23:**277–294.

6. **Baranyi, J., and T. A. Roberts.** 1995. Mathematics of predictive food microbiology. *Int. J. Food Microbiol.* **26:**199–218.

7. **Baranyi, J., T. A. Roberts, and P. McClure.** 1993. A non-autonomous differential equation to model bacterial growth. *Food Microbiol.* **10:**43–59.

8. **Baranyi, J., T. Ross, T. A. McMeekin, and T. A. Roberts.** 1996. Effects of parameterization on the performance of empirical models used in 'predictive microbiology.' *Food Microbiol.* **13:**83–91.

9. **Bemrah, N., M. Sanaa, M. H. Cassin, M. W. Griffiths, and O. Cerf.** 1998. Quantitative risk assessment of human listeriosis from consumption of soft cheese made from raw milk. *Prev. Vet. Med.* **37:**129–145.

10. **Bhaduri, S., P. W. Smith, S. A. Palumbo, C. O. Turner-Jones, J. L. Smith, B. S. Marmer, R. L. Buchanan, L. L. Zaika, and A. C. Williams.** 1991. Thermal destruction of *Listeria monocytogenes* in liver sausage slurry. *Food Microbiol.* **8:**75–78.

11. **Bovill, R., J. Bew, N. Cook, M. D'Agostino, N. Wilkinson, and J. Baranyi.** 2000. Predictions of growth for *Listeria monocytogenes* and *Salmonella* during fluctuating temperature. *Int. J. Food Microbiol.* **59:**157–165.

12. **Breidt, F., and H. P. Flemming.** 1998. Modeling of the competitive growth of *Listeria monocytogenes* and *Lactococcus lactis* in vegetable broth. *Appl. Environ. Microbiol.* **64:**3159–3165.

13. **Buchanan, R. L.** 1993. Developing and distributing user-friendly application software. *J. Ind. Microbiol.* **12:**251–255.

14. **Buchanan, R. L.** 1997. National Advisory Committee on Microbiological Criteria for Foods "Principles of risk assessment for illnesses caused by foodborne biological agents." *J. Food Prot.* **60:**1417–1419.

15. **Buchanan, R. L.** 1998. Principles of risk assessment for illness caused by foodborne biological agents. *J. Food Prot.* **61:**1071–1074.

16. **Buchanan, R. L., and M. L. Cygnarowicz.** 1990. A mathematical approach toward defining and calculating the duration of the lag phase. *Food Microbiol.* **7:**237–240.

17. **Buchanan, R. L., M. H. Golden, and R. C. Whiting.** 1993. Differentiation of the effects of pH and lactic or acetic acid concentration on the kinetics of *Listeria monocytogenes* inactivation. *J. Food Prot.* **56:**474–478, 484.

18. **Buchanan, R. L., and J. G. Phillips.** Updated models for the effects of temperature, initial pH, NaCl, and NaNO$_2$ on the aerobic and anaerobic growth of *Listeria monocytogenes.* *J. Quant. Microbiol.,* in press.

19. **Buchanan, R. L., J. L. Smith, and W. Long.** 2000. Microbial risk assessment: dose-response relations and risk characterization. *Int. J. Food Microbiol.* **58:**159–172.

20. **Buchanan, R. L., and R. C. Whiting.** 1998. Risk assessment: a means for linking HACCP plans and public health. *J. Food Prot.* **61:**1531–1534.

21. **Buchanan, R. L., R. C. Whiting, and S. A. Palumbo.** 1993. Proceedings of workshop on the application of predictive microbiology and computer modeling techniques to the food industry. *J. Ind. Microbiol.* **12:**137–359.

22. **Buchanan, R. L., R. C. Whiting, and W. C. Damert.** 1997. When is simple good enough: a comparison of the Gompertz, Baranyi, and three-phase linear models for fitting bacterial growth curves. *Food Microbiol.* **14:**313–326.

23. **Buzby, J. C., T. Roberts, and B. M. Allos.** 1997. Estimating annual costs of *Campylobacter*-associated Guillain-Barré syndrome. Agricultural Economic Report Number 756, Economic Research Service, U.S. Department of Agriculture, Washington, D. C.

24. **Cassin, M. H., A. M. Lammerding, E. C. D. Todd, W. Ross, and R. S. McColl.** 1998. Quantitative risk assessment for *Escherichia coli* O157:H7 in ground beef hamburgers. *Int. J. Food Microbiol.* **41:**21–44.

25. **Cassin, M. H., G. M. Paoli, and A. M. Lammerding.** 1998. Simulation modeling for microbial risk assessment. *J. Food Prot.* **61:**1560–1566.

26. **Chawla, C. S., H. Chen, and C. W. Donnelly.** 1996. Mathematically modeling the repair of heat-injured *Listeria monocytogenes* as affected by temperature, pH, and salt concentration. *Int. J. Food Microbiol.* **30:**231–242.

27. **Cheroutre-Vialette, M., and A. Lebert.** 2000. Modelling the growth of *Listeria monocytogenes* in dynamic conditions. *Int. J. Food Microbiol.* **55:**201–207.

28. **Clavero, M. R. S., L. R. Beuchat, and M. P. Doyle.** 1998. Thermal inactivation of *Escherichia coli* O157:H7 isolated from ground beef and bovine feces, and suitability of media for enumeration. *J. Food Prot.* **61:**285–289.

29. **Codex Alimentarius Commission.** 1996. Principles and guidelines for the application of microbiological risk assessment. CX/FH 96/10, August 1996.

30. **Coleman, M., and H. Marks.** 1998. Topics in dose-response modeling. *J. Food Prot.* **61:**1550–1559.

31. **Daughtry, B. J., K. R. Davey, C. J. Thomas, and A. P. Verbyla.** 1997. Food processing—a new model for the thermal destruction of contaminating bacteria, p. A113–A116. Seventh International Congress on Engineering and Food—ICEF7. April 14–17, 1997, Brighton, United Kingdom.

32. **Davey, K. R.** 1991. Applicability of the Davey (linear Arrhenius) predictive model to the lag phase of microbial growth. *J. Appl. Bacteriol.* **70:**253–257.

33. **Delignette-Muller, M. L.** 1998. Relation between the generation time and the lag time of bacterial growth kinetics. *Int. J. Food Microbiol.* **43:**97–104.

34. **Devlieghere, F., B. van Belle, and J. Debevere.** 1999. Shelf life of modified atmosphere packed cooked meat products: a predictive model. *Int. J. Food Microbiol.* **46:**57–70.

35. **Duffy, G., R. C. Whiting, and J. J. Sheridan.** 1999. The effect of pH, temperature and competitive microflora on the growth kinetics of *Escherichia coli* O157:H7. *Food Microbiol.* **16:**299–307.

36. **Duh, Y.-H., and D. W. Schaffner.** 1993. Modeling the effect of temperature on the growth rate and lag time of *Listeria innocua* and *Listeria monocytogenes. J. Food Prot.* **56:**205–210.

37. **Elliott, P. H.** 1996. Predictive microbiology and HACCP. *J. Food Prot.* 1996(Suppl.):48–53.

38. **FAO/WHO.** 1997. Risk management and food safety. FAO Food and Nutrition Paper number 65. Food and Agriculture Organization, Rome, Italy.

39. **Foegeding, P. M.** 1997. Driving predictive modelling on a risk assessment path for enhanced food safety. *Int. J. Food Microbiol.* **36:**87–95.

40. **Fujikawa, H., and T. Itoh.** 1998. Thermal inactivation analysis of mesophiles using the Arrhenius and z-value models. *J. Food Prot.* **61:**910–912.

41. **Geeraerd, A. H., C. H. Herremans, C. Cenens, and J. F. van Impe.** 1998. Application of artificial neural networks as a non-linear modular modeling technique to describe bacterial growth in chilled food products. *Int. J. Food Microbiol.* **44:**49–68.

42. **Gibson, A. M., N. Bratchell, and T. A. Roberts.** 1988. Predicting microbial growth: growth responses of salmonellae in a laboratory medium as affected by pH, sodium chloride and storage temperature. *Int. J. Food Microbiol.* **6:**155–178.

43. **Goldblith, S. A., M. A. Joslyn, and J. T. R. Nickerson.** 1961. *An Introduction to the Thermal Processing of Foods,* vol. 1, p. 1128. AVI Pub. Co., Westport, Conn.

44. **Haas, C. N., A. Thayyar-Madabusi, J. B. Rose, and C. P. Gerba.** 1999. Development and validation of dose-response relationship for *Listeria monocytogenes. Quant. Microbiol.* **1:**89–102.

45. **Hajmeer, M. N., l. A Basheer, and Y. M. Najjar.** 1997. Computational neural networks for predictive microbiology. II. Application to microbial growth. *Int. J. Food Microbiol.* **34:**51–66.

46. **Hathaway, S. C.** 1997. Development of food safety risk assessment guidelines for foods of animal origin in international trade. *J. Food Prot.* **60:**1432–1438.

47. **Hills, B. P., and B. M. Mackey.** 1995. Multi-compartment kinetic models for injury, resuscitation, induced lag and growth in bacterial cell populations. *Food Microbiol.* **12:**333–346.

48. **Holcomb, D. L., M. A. Smith, G. O. Ware, Y.-C. Hung, R. E. Brackett, and M. P. Doyle.** 1999. Comparison of six dose-response models for use with food-borne pathogens. *Risk Anal.* **19:**1091–1100.

49. **Houtsma, P. C., B. J. M. Kusters, J. C. de Wit, F. M. Rombouts, and M. H. Zwietering.** 1994. Modelling growth rates of *Listeria monocytogenes* as a function of lactate concentration. *Int. J. Food Microbiol.* **24:**113–123.

50. **Hudson, J. A.** 1993. Effect of pre-incubation temperature on the lag time of *Aeromonas hydrophila. Lett. Appl. Microbiol.* **16:**274–276.

51. **Humpheson, L., M. R. Adams, W. A. Anderson, and M. B. Cole.** 1998. Biphasic thermal inactivation kinetics in *Salmonella enteritidis* PT4. *Appl. Environ. Microbiol.* **64:**459–464.

52. **ICMSF.** 1998. Potential application of risk assessment techniques to microbiological issues related to international trade in food and food products. *J. Food Prot.* **61:**1075–1086.

53. **ILSI.** 1996. A conceptual framework to assess the risks of human disease following exposure to pathogens. *Risk Anal.* **16:**841–847.

54. **Jenkins, P., P. G. Poulos, M. B. Cole, M. H. Vandeven, and J. D. Legan.** 2000. The boundary for growth of *Zygosaccharomyces bailii* in acidified products described by models for time to growth and probability of growth. *J. Food Prot.* **63:**222–230.

55. **Koutsoumanis, P. K, P. S. Taoukis, E. H. Drosinos, and G. E. Nychas.** 2000. Applicability of an Arrhenius model for the combined effect of temperature and CO_2 packaging on the spoilage microflora of fish. *Appl. Environ. Microbiol.* **66:**3528–3534.

56. **Lammerding, A. M., and A. Fazil.** 2000. Hazard identification and exposure assessment for microbial food safety risk assessments. *Int. J. Food Microbiol.* **58:**147–157.

57. **Lindqvist, R., and A. Westöö.** 2000. Quantitative risk assessment for *Listeria monocytogenes* in smoked or gravid salmon and rainbow trout in Sweden. *Int. J. Food Microbiol.* **58:**181–196.

58. **Mackey, B. M., C. Pritchet, A. Norris, and G. C. Mead.** 1990. Heat resistance of *Listeria:* strain differences and effects of meat type and curing salts. *Lett. Appl. Microbiol.* **10:**251–255.

59. **McClure, P. J., C. De W. Blackburn, M. B. Cole, P. S. Curtis, J. E. Jones, J. D. Legan, I. D. Ogden, M. W. Peck, T. A. Roberts, J. P. Sutherland, and S. J. Walker.** 1994. Modelling the growth, survival and death of microorganisms in foods: the UK Food Micromodel approach. *Int. J. Food Microbiol.* **23:**265–276.

60. **McDonald, K., and D.-W. Sun.** 1999. Predictive food microbiology for the meat industry: a review. *Int. J. Food Microbiol.* **52:**1–27.

61. **McKellar, R. C., G. Butler, and K. Stanich.** 1997. Modelling the influence of temperature on the recovery of

Listeria monocytogenes from heat injury. *Food Microbiol.* **14**:617–625.

62. **McMeekin, T. A., J. Olley, T. Ross, and D. A. Ratkowsky.** 1993. *Predictive Microbiology: Theory and Application.* John Wiley & Sons, New York, N.Y.

63. **McMeekin, T. A., K. Presser, D. Ratkowsky, T. Ross, M. Salter, and S. Tienungoon.** 2000. Quantifying the hurdle concept by modelling the bacterial growth/no growth interface. *Int. J. Food Microbiol.* **55**:93–98.

64. **McMeekin, T. A., and T. Ross.** 1996. Modeling applications. *J. Food Prot.* 1996(Suppl.):37–42.

65. **Morales, R. A., and R. M. McDowell.** 1998. Risk assessment and economic analysis for managing risks to human health from pathogenic microorganisms in the food supply. *J. Food Prot.* **61**:1567–1570.

66. **Najjar, Y. M., I. A. Basheer, and M. N. Hajmeer.** 1997. Computational neural networks for predictive microbiology. I. Methodology. *Int. J. Food Microbiol.* **34**:27–49.

67. **Ng, T. M., and D. W. Schaffner.** 1997. Mathematical models for the effects of pH, temperature, and sodium chloride on the growth of *Bacillus stearothermophilus* in salty carrots. *Appl. Environ. Microbiol.* **63**:1237–1243.

68. **Notermans, S., G. Gallhoff, M. H. Zwietering, and G. C. Mead.** 1995. The HACCP concept: specification of criteria using quantitative risk assessment. *Food Microbiol.* **12**:81–90.

69. **Notermans, S., and G. C. Mead.** 1996. Incorporation of elements of quantitative risk analysis in the HACCP system. *Int. J. Food Microbiol.* **30**:157–173.

70. **Peleg, M., and M. B. Cole.** 2000. Estimating the survival of *Clostridium botulinum* spores during heat treatments. *J. Food Prot.* **63**:190–195.

71. **Pin, C., and J. Baranyi.** 1998. Predictive models as means to quantify the interactions of spoilage organisms. *Int. J. Food Microbiol.* **41**:59–72.

72. **Potter, M. E.** 1996. Risk assessment terms and definitions. *J. Food Prot.* 1996(Suppl.):6–9.

73. **Ratkowsky, D. A., and T. Ross.** 1995. Modelling the bacterial growth/no growth interface. *Lett. Appl. Microbiol.* **20**:29–33.

74. **Ratkowsky, D. A., T. Ross, M. Macario, T. W. Dommett, and L. Kamperman.** 1996. Choosing probability distributions for modelling generation time variability. *J. Appl. Bacteriol.* **80**:131–137.

75. **Reichardt, W.** 1994. Modelling the destruction of *Escherichia coli* on the basis of reaction kinetics. *Int. J. Food Microbiol.* **23**:449–465.

76. **Reichardt, W., and R. Y. Morita.** 1982. Temperature characteristics of psychrotrophic and psychrophilic bacteria. *J. Gen. Microbiol.* **128**:565–568.

77. **Robinson, T. P., M. J. Ocio, A. Kaloti, and B. M. Mackey.** 1998. The effect of the growth environment on the lag phase of *Listeria monocytogenes. Int. J. Food Microbiol.* **44**:83–92.

78. **Ross, T.** 1996. Indices for performance evaluation of predictive models in food microbiology. *J. Appl. Bacteriol.* **81**:501–508.

79. **Ross, T., and T. A. McMeekin.** 1994. Predictive microbiology. *Int. J. Food Microbiol.* **23**:241–264.

80. **Salter, M. A., D. A. Ratkowsky, T. Ross, and T. A. McMeekin.** 2000. Modelling the combined temperature and salt (NaCl) limits for growth of a pathogenic *Escherichia coli* strain using nonlinear logistic regression. *Int. J. Food Microbiol.* **61**:159–167.

81. **Serra, J. A., E. Domenech, I. Escriche, and S. Martorell.** 1999. Risk assessment and critical control points from the production perspective. *Int. J. Food Microbiol.* **46**:9–26.

82. **Shah, D. B., J. G. Bradshaw, and J. T. Peeler.** 1991. Thermal resistance of egg-associated epidemic strains of *Salmonella enteritidis. J. Food Sci.* **56**:391–393.

83. **Skandamis, P. N., and G. E. Nychas.** 2000. Development and evaluation of a model predicting the survival of *Escherichia coli* O157:H7 NCTC 12900 in homemade eggplant salad at various temperatures, pHs, and oregano essential oil concentrations. *Appl. Environ. Microbiol.* **66**:1646–1653.

84. **Smerage, G. H., and A. A. Teixeira.** 1993. A new view on the dynamics of heat destruction of microbial spores. *J. Ind. Microbiol.* **12**:211–220.

85. **Stephens, P. J., M. B. Cole, and M. V. Jones.** 1994. Effect of heating rate on the thermal inactivation of *Listeria monocytogenes. J. Appl. Bacteriol.* **77**:702–708.

86. **te Giffel, M. C., and M. H. Zwietering.** 1999. Validation of predictive models describing the growth of *Listeria monocytogenes. Int. J. Food Microbiol.* **46**:135–149.

87. **Tienungoon, S., D. A. Ratkowsky, T. A. McMeekin, and T. Ross.** 2000. Growth limits of *Listeria monocytogenes* as a function of temperature, pH, NaCl, and lactic acid. *Appl. Environ. Microbiol.* **66**:4979–4987.

88. **Van Impe, J. F., B. M. Nicolaï, T. Martens, J. De Baerdemaeker, and J. Vandewalle.** 1992. Dynamic mathematical model to predict microbial growth and inactivation during food processing. *Appl. Environ. Microbiol.* **58**:2901–2909.

89. **Van Impe, J. F., B. M. Nicolaï, M. Schellekens, T. Martens, and J. De Baerdemaeker.** 1995. Predictive microbiology in a dynamic environment: a system theory approach. *Int. J. Food Microbiol.* **25**:227–249.

90. **Van Schothorst, M.** 1998. Principles for the establishment of microbiological food safety objectives and related control measures. *Food Control* **9**:379–384.

91. **Vose, D. J.** 1998. The application of quantitative risk assessment to microbial food safety. *J. Food Prot.* **61**:640–648.

92. **Whiting, R. C.** 1993. Modeling bacterial survival in unfavorable environments. *J. Ind. Microbiol.* **12**:240–246.

93. **Whiting, R. C.** 1995. Microbial modeling in foods. *Crit. Rev. Food Sci. Technol.* **35**:467–494.

94. **Whiting, R. C., and L. K. Bagi.** 2000. Modelling the lag phase of *Listeria monocytogenes,* p. 148–156. *In* J. F. M. van Impe and K. Bernaerts (ed.) *Proceeding 3rd International Conference on Predictive Modelling in Foods.* Katholieke Universiteit Leuven, Leuven, Belgium.

95. **Whiting, R. C., and R. L. Buchanan.** 1994. Microbial modeling—IFT scientific status summary. *Food Technol.* **48**:113–120.

96. **Whiting, R. C., and R. L. Buchanan.** 1997. Development of a quantitative risk assessment model for *Salmonella*

enteritidis in pasteurized liquid eggs. *Int. J. Food Microbiol.* **36:**111–125.

97. **Whiting, R. C., and R. L. Buchanan.** 1997. Predictive modeling, p. 728–739. *In* M. P. Doyle, L. R. Beuchat, and T. J. Montville (ed.), *Food Microbiology, Fundamentals and Frontiers.* ASM Press, Washington, D. C.

98. **Whiting, R. C., and M. H. Golden.** Variation among *E. coli* O157:H7 strains relative to their growth, survival, thermal inactivation and toxin production in broth. *Int. J. Food Microbiol.*

99. **Whiting, R. C., A. Hogue, W. D. Schlosser, E. D. Ebel, R. Morales, A. Baker, and R. McDowell.** 2000. A risk assessment for *Salmonella enteritidis* in shell eggs. *J. Food Sci.* **65:**864–869.

100. **Whiting, R. C., and T. P. Strobaugh.** 1998. Expansion of the time-to-turbidity model for proteolytic *Clostridium botulinum* to include spore numbers. *Food Microbiol.* **15:**449–453.

101. **Xiong, R., G. Xie, A. S. Edmondson, R. H. Linton, and M. A. Sheard.** 1999. Comparison of the Baranyi model with the modified Gompertz equation for modelling thermal inactivation of *Listeria monocytogenes* Scott A. *Food Microbiol.* **16:**269–279.

102. **Xiong, R., G. Xie, A. S. Edmondson, and M. A. Sheard.** 1999. A mathematical model for bacterial inactivation. *Int. J. Food Microbiol.* **46:**45–55.

103. **Zwietering, M. H., J. C. de Wit, and S. Notermans.** 1996. Application of predictive microbiology to estimate the number of *Bacillus cereus* in pasteurised milk at the point of consumption. *Int. J. Food Microbiol.* **30:**55–70.

104. **Zwietering, M. H., and A. P. M. Hasting.** 1997. Modelling the hygienic processing of foods—a global process overview. *Trans IChemE.* **75**(Part C):159–167.

105. **Zwietering, M. H., I. Jongenburger, F. M. Rombouts, and K. van't Riet.** 1990. Modeling of the bacterial growth curve. *Appl. Environ. Microbiol.* **56:**1875–1881.

Food Microbiology: Fundamentals and Frontiers, 2nd Ed.
Edited by M. P. Doyle et al.
© 2001 ASM Press, Washington, D.C.

Dane Bernard

41

Hazard Analysis and Critical Control Point System:
Use in Controlling Microbiological Hazards

Many of the preceding chapters of this book present current information regarding consumer health problems associated with food or water. These problems can often be successfully addressed by applying a management system, the Hazard Analysis and Critical Control Point (HACCP) system. It is specifically designed to address food safety risks.

It is the responsibility of the food industry to raise, transport, process, and prepare foods that present a minimum level of risk from foodborne hazards. To accomplish this, industry must identify and control the potential risks posed by pathogenic microorganisms, toxic chemicals, and physical agents. The controls applied must be both practical and achievable within the limits of the available technology. To assist in identifying hazards and managing risks, the HACCP system has been developed.

HACCP is currently regarded as the best system available for designing programs to assist food firms in producing foods that are safe to consume. Recognizing this, the U.S. Food and Drug Administration (FDA) and the U.S. Department of Agriculture's Food Safety and Inspection Service (USDA FSIS) have published final rules (7–9) mandating the development and implementation of HACCP plans to cover a variety of domestic and imported seafoods, juices, and meat and poultry products.

Paralleling these U.S. government initiatives is an international movement toward a food safety assurance and regulatory scheme based on the principles of HACCP. For example, the European Union (E.U.) has adopted several product-specific ("vertical") directives that will require specific foods introduced into commerce within the E.U. (including imports) to be produced in accordance with HACCP principles. Examples include the directive on fishery products (91/493/EEC), the directive on milk, heat-treated milk, and milk-based products (92/46/EEC), and the directive on meat products (92/5/EEC). In addition to these, the E.U. has adopted a "horizontal" directive (93/43/EEC) that requires consistency with HACCP principles for a wide range of food items. The E.U. is reportedly considering a new horizontal directive that will mandate HACCP for a wide range of foods made or imported into the E.U. Countries outside the E.U. such as Australia, New Zealand, Canada, Japan, Egypt, South Africa, and many others have also adopted or are considering HACCP-based food safety control systems.

Dane Bernard, National Food Processors Association, 1350 I St., NW, Suite 300, Washington, DC 20005-3305.

The agreements and treaties that support the World Trade Organization (WTO) illustrate the growing importance of HACCP as a tool to judge the acceptability of foods traded internationally. These agreements give guidance to WTO member nations about how to establish sanitary and phytosanitary requirements without creating technical trade barriers (an illegal practice that may result in legal sanctions against a country that is member of the WTO). For the establishment of sanitary measures addressing food safety, these agreements reference the standards, guidelines, and recommendations developed and adopted by the Codex Alimentarius Commission as an acceptable scientific baseline for consumer protection (4).

The 20th Session of the Codex Alimentarius Commission (July 1993) adopted the document "Guidelines for the Application of the Hazard Analysis Critical Control Point (HACCP) System." The Commission noted that the text was "urgently needed" in order to incorporate it into the "Draft Code of Practice—General Principles of Food Hygiene." To illustrate the rapid development of our understanding of HACCP and its worldwide application, the Codex Committee on Food Hygiene undertook to revise the guideline almost immediately. The Codex Alimentarius Commission in June 1997 adopted the revised document "Hazard Analysis and Critical Control Point (HACCP) System and Guidelines for Its Application." The HACCP guidelines are appended to the Code of Hygienic Practice (5), which serves to emphasize that, for all its merits, HACCP does not stand alone but depends on a foundation of general good hygienic practices, a point that will be discussed later.

ORIGIN OF HACCP

The acronym HACCP was first coined during collaborative work between the Pillsbury Company and the National Aeronautics and Space Administration to describe the systematic approach to food safety developed to assure the safety of foods being produced for manned space flights. To approach the goal of 100% assurance that these foods would be safe to consume, a system had to be developed that went well beyond the limited effectiveness of sampling and analyzing finished goods for the presence of hazardous microorganisms, potentially injurious foreign materials, or toxic chemicals. Methods used by others, including "Haz-Ops" studies and "Modes of Failure Analysis," were considered in the search for new ways to assure that foods were safe. Although the HACCP concept reflects much of this earlier thinking, these programs did not take into account the special needs of the food industry. From these ap-

proaches, the HACCP concept was distilled. The essential concept is that if knowledge can be ascertained about how a product (in this case a food) may become unsafe if consumed, then control measures can be developed to prevent and/or detect such failures and keep food that presents an unacceptable risk from reaching consumers. This concept formed the foundation for development of the HACCP system of food safety control.

Dr. Howard Bauman presented the HACCP concept, along with the original three principles, to the public in 1971 at the first National Conference on Food Protection (17). Although the concept created some interest in the early 1970s, initial attempts at incorporating HACCP into food operations were not very successful. It is likely that a major contributor to these early unsuccessful attempts was a lack of description and understanding of HACCP concepts. Early HACCP plans typically had far more critical control points than were needed to assure production of a safe food, were not "plant friendly," and were too cumbersome to be sustained. However, in 1985, a subcommittee of the Food Protection Committee of the National Academy of Sciences issued a report on microbiological criteria that included a strong endorsement of HACCP (14). On the basis of this recommendation, the National Advisory Committee on Microbiological Criteria for Foods (NACMCF) was formed in 1988 as an advisory group to the Secretaries of Agriculture, Commerce, Defense, and Health and Human Services. Part of the mission of the NACMCF was to develop material to promote an understanding of HACCP and to encourage adoption of the HACCP approach to food safety. The NACMCF has been a leader in elaborating HACCP principles and in providing guidance for their application to food processing operations and to the sponsoring agencies. In recognition that the ideas underlying the successful application of HACCP continue to evolve, the NACMCF has updated its 1989 and 1992 guidance documents. The committee's latest guidance document on HACCP, completed in 1997, was published in 1998 (16).

HACCP OVERVIEW

The approach taken in this chapter is that HACCP plans should only address significant food safety hazards. It is these safety hazards that warrant the extensive monitoring, record keeping, and verification activities required by the HACCP system of food safety management. FDA, in its final rule on seafood safety, and USDA's FSIS, in its regulation on HACCP and pathogen reduction, referenced above, also limit HACCP to issues of food *safety*. There is also growing international acceptance of the "safety-only" philosophy of HACCP. This may be viewed as an unnecessarily narrow interpretation by some, as the concepts embodied in the seven principles for developing a HACCP plan can be applied to the control of quality factors and economic adulteration as well. However, the safety-only approach is increasingly viewed as a functional necessity in order for HACCP to be globally successful. HACCP is likely to be used to judge the acceptability of products in domestic and international trade. While quality is a negotiable item (especially when viewed on a global scale), safety should be nonnegotiable. That said, recent experience shows that even HACCP "experts" still have difficulty coming to agreement on the safety implications of certain hazards. This results in a situation in which it is sometimes difficult to obtain agreement on what is a good HACCP plan. This situation becomes even worse when quality factors are included in the mix. Thus, if the seven principles are applied to develop a control strategy for *quality* factors, this should not be labeled as a HACCP plan.

The HACCP principles describe a format for identifying and assuring control of those factors (hazards) that are *reasonably likely* to cause a food to be unsafe for consumption. HACCP is based on a commonsense application of technical and scientific principles to the food production process from field to table. The process of developing a HACCP plan is analogous to conducting a qualitative assessment of risk posed by a particular hazard, coupled with selection and implementation of appropriate control measures.

The concept of a qualitative risk assessment was provided a measure of international recognition through its elaboration during an expert consultation sponsored by the Food and Agriculture Organization (FAO) of the United Nations and the World Health Organization (21). Once the hazards that present significant consumer risk are determined, a management plan is developed to control those hazards. The results of the risk assessment are used to determine which risks are significant, e.g., reasonably likely to present consumer safety problems if not properly controlled.

In the hazard analysis step (principle 1), hazards of potential significance are identified and evaluated. Then decisions are made as to which hazards are significant. The remaining six HACCP principles describe the steps necessary to develop a management strategy to control the significant hazards.

Most contemporary references on HACCP point to its applicability to all segments of the food industry, from basic agriculture through food preparation and handling, food processing, food packaging, food service, distribution systems, and consumer handling and use. In a broad sense this may be true, but the way in which the HACCP principles are interpreted and applied will differ, depending on which segment in the food chain is being addressed. This subject is discussed further in a paper published by FAO (10). The major focus of this chapter is on the application of HACCP to the food manufacturing segment, with other segments noted where appropriate.

The basic premise underlying HACCP is to prevent hazards during production rather than to rely on inspection of finished goods in hopes of detecting and correcting problems. Although testing of finished products usually plays a role in a HACCP system (as a verification tool), the main focus is on prevention through process control. As most food microbiologists will attest, results derived from sampling and testing only support a particular claim about the units tested. When interpreting test results, the limits of the testing methodology and the sampling plan must be considered, because statements about the nature of units not tested can only be made by inference within the context of these limitations. By contrast, if the hazards of concern are properly identified and effective control measures are available and followed, HACCP provides greater confidence in the safety of each unit produced than can be obtained through any practical sampling and testing program. Thus, firms that produce foods in accord with a properly designed HACCP plan will typically have a reduced need for finished product testing and analysis.

A CAUTIONARY NOTE

In the opinion of many HACCP authorities, the most important (and least understood) activity in the development of a HACCP plan is the hazard analysis step. Although this process will be discussed in more depth below, it must be emphasized that a HACCP plan will not be effective in minimizing risks associated with a food unless the hazard analysis is done correctly and effective control measures are adopted. A note of caution, however: HACCP is not a "zero" defects program, even though this should always be the ultimate goal in food safety. When applied to processes with a well-defined and controlled microbial kill step, such as canning, we closely approach the zero risk threshold. However, application of HACCP to raw or raw transformed products for which control steps are less effective and less well defined is a risk-reduction exercise that cannot offer the same level of protections as in other processing areas. Unless and until more effective control measures are available and adopted in these parts of the industry, gains available through adoption of HACCP alone will be marginal.

In addition, any hazard analysis will be limited by the information available to the experts conducting the analysis. Some risk factors, such as a pathogen that adapts to a new environment, cannot be anticipated and thus cannot be addressed in a HACCP plan. HACCP is not a magic bullet that will cure all food safety problems. However, adoption of HACCP will put a system in place that has the potential to be more responsive to newly identified problems. Because the natural world is always presenting us with new hazards, it is vital to the success of a HACCP system that plans are reviewed in light of new scientific information. To be effective over time, the hazard analysis and resulting HACCP plans should be reviewed frequently.

USING HACCP TO IDENTIFY RISKS AND CONTROL HAZARDS

The systematic approach to food safety embodied by HACCP is based on seven principles. The principles adopted by the NACMCF are used in this chapter. Although there are variations of the principles adopted and published by other groups and official bodies, there is surprising agreement on the basic concepts. However, in reviewing most of the publications describing HACCP, e.g., those by the International Life Sciences Institute (12), the Food Processors Institute (20), Campden and Chorleywood Food Research Association (13), etc., some variation is noted in the interpretation of the principles and in the resulting guidance about how they are to be applied during development and implementation of HACCP plans. Nevertheless, there is surprising agreement in the overarching message that users of HACCP should not focus so tightly on the "doctrine" that the intended commonsense approach to food safety management is lost. In other words, one should use common sense in applying the seven principles.

As noted earlier, HACCP continues to evolve as a concept. As additional real-world experience is gained with the system, more techniques will be developed that will make it easier to understand and apply. The guidance documents about HACCP will also continue to be modified. Thus, each firm, besides following the legal requirements that pertain to its operation, should do what it believed is appropriate to develop a workable, plant-friendly, and effective HACCP plan. Remember that many of the substantial benefits of HACCP accrue because plans will be operation and product (or product category) specific, therefore fitting the unique needs of each food facility. Each operation therefore should view the HACCP guidance available today as just that, "guidance." Having taken this relatively traditional stand re-

garding the individuality of HACCP plans, the imposition of HACCP as a regulatory program brings with it unique challenges to this notion. In the opinion of the author, the challenge is whether individually developed (and often highly variable) HACCP plans can ultimately be supported in a government-regulated system.

PRELIMINARY STEPS

Certain preliminary steps must be undertaken before beginning the hazard analysis. The recommended steps that precede the hazard analysis include: (i) gain management support, (ii) assemble the HACCP team, (iii) describe the food and the method of its distribution, (iv) identify the intended use and consumers of the food, (v) develop a flow diagram, and (vi) verify the flow diagram.

Regarding point one from the above list, most who have attempted to develop HACCP programs agree that it is very difficult to put HACCP systems in place without firm support by company management. HACCP often requires basic changes in management styles and requires the commitment of resources to the process. Assembling a HACCP team (point ii) can be especially challenging for small firms that do not have a large technical staff. Even though it may be a challenge for some firms, most now agree that individuals with the necessary expertise must be involved in the process. However, they need not necessarily be employed at the firm. All firms developing HACCP plans must find a way to have the requisite expertise reflected in their plans. The team should be multidisciplinary and include individuals from areas such as engineering, production, sanitation, quality assurance, and food microbiology. The team should also include local personnel, who are involved in the operation, as they are more familiar with the variability and limitations of the operation. Points iii through ix from the list are necessary to allow the development of HACCP plans that are tailored to specific products and food operations. To address these preliminary steps, the food firm must have sufficient information about the ingredients, processing methods, the distribution system, and the expected consumers so it can determine where and how significant food safety hazards may be introduced or aggravated.

ESTABLISHING PLAN-SPECIFIC OBJECTIVES AND PERFORMANCE CRITERIA

The topics of food safety objectives and performance criteria have recently been introduced into discussions of the application of risk analysis in food safety systems (11). The ultimate role of these concepts in development and implementation of HACCP plans will be an item for discussion for some time in venues such as Codex. Without a doubt, these terms will become more fully defined and accepted at some point. But until the exact definitions of food safety objectives are developed and agreed to, it would be premature to recommend how these terms should be applied to the development of HACCP plans. However, at present it has proven useful for the HACCP team to identify plan-specific public health-related objectives as the plan is being developed. These may be in the form of a statement of the food safety-related results expected from implementation of the HACCP plan. These objectives will typically become clear as the team conducts its hazard analysis and determines the significant hazards that will be addressed in the plan. Where possible, performance criteria should be identified that lead to meeting the objective. An example of such an objective is exemplified by the qualitative statement that canned foods should pose only a negligible risk from *Clostridium botulinum*. This can be accomplished by achieving or exceeding the performance criterion of a 12-log reduction (or equivalent) of the target organism. A similar qualitative objective would be to produce a pasteurized product that poses a negligible risk from vegetative pathogens such as *Salmonella* sp. In the production of pasteurized egg products, a 9-log reduction of *Salmonella* sp. is thought to adequately protect consumers from this hazard (1). Process steps to inactivate pathogenic organisms should be designed to meet criteria that have been developed to be protective of public health.

For raw products, defined and agreed-to objectives are more difficult to determine, and associated performance criteria are even more challenging. In this case, the objective may be to produce a product that presents a risk of illness from vegetative pathogens that is as low as reasonably achievable. This objective is usually accomplished by application of treatments designed to reduce numbers of microorganisms present or reduce the potential for cross contamination and by control of conditions (e.g., preservatives) specifically designed to reduce or prevent the proliferation of pathogens.

Similar public health-related objectives and criteria should be identified for chemical and physical hazards as well.

PRINCIPLE 1: CONDUCT A HAZARD ANALYSIS

After addressing the preliminary points discussed above, the HACCP team conducts a hazard analysis or ensures that an analysis is conducted. The purpose of the hazard analysis is to develop a list of hazards that are of such significance that they are reasonably likely to cause injury or illness if not effectively controlled. Hazards that are not reasonably likely to occur would not require further consideration within a HACCP plan. The NACMCF (16) defines a hazard (for HACCP) as "a biological, chemical, or physical agent that is reasonably likely to cause illness or injury in the absence of its control." The NACMCF (16) also provides a definition for a hazard analysis as "the process of collecting and evaluating information on hazards associated with the food under consideration to decide which are significant and must be addressed in the HACCP plan." The results of the hazard analysis identify the hazards that should be addressed in the HACCP plan along with control measures for each hazard. For a hazard to be listed among the significant hazards addressed in the HACCP plan, the hazard should be of such a nature that its prevention, elimination, or reduction to acceptable levels is essential to the production of a safe food (15).

On this last point, note that HACCP does not provide an excuse for food operators to ignore low-risk hazards or to ignore existing legal requirements, including those that define adulteration. Operations must be in compliance with all legal requirements. Low-risk hazards should not be dismissed as insignificant. It is expected that these hazards will usually be addressed within other management systems, such as Good Manufacturing Practice (GMP) programs, which are prerequisites for the HACCP system. Most authorities agree that if a food operation does not have sound prerequisite programs that comply with GMPs and other regulatory requirements, HACCP will have little chance for success. It is clear, however, that the NACMCF recognized that all biological, chemical, or physical properties that *may* cause a food to be unsafe for consumption are not *significant* enough to warrant being addressed within a HACCP plan. Making the decision as to whether a hazard is reasonably likely to present a significant consumer safety problem is at the heart of the hazard assessment.

Conducting a Hazard Analysis

The NACMCF (16) advises that the process of conducting a hazard analysis involves two stages: hazard identification and hazard evaluation.

Hazard Identification

Hazard identification is sometimes described as a brainstorming session. During this stage, the HACCP team reviews the ingredients used in the product, the activities conducted at each step in the process, and the equipment used to make the product. The team also considers the method of storage, distribution, the intended use, and consumers of the product.

On the basis of its review of the information just described, the team develops a list of potential biological, chemical, and physical hazards that may be introduced, increased (e.g., pathogen growth), or controlled at each step described on the flow diagram. During hazard identification, the HACCP team should focus on developing a list of potential hazards associated with each ingredient, each process step under direct control of the food operation, and the final product. Knowledge of any adverse health-related events historically associated with the product is important in this exercise.

Potential Biological Hazards

A biological hazard is one that, if not properly controlled, is reasonably likely to result in foodborne illness. The primary organisms of concern are pathogenic bacteria, such as *C. botulinum*, *Listeria monocytogenes*, *Salmonella* species, and *Staphylococcus aureus*. It is also important to identify potential hazards from other biological agents, such as pathogenic viruses and parasites. The potential biological hazards associated with a product and process should be listed in preparation for evaluation.

Potential Chemical Hazards

Potential chemical hazards include toxic substances and any other compounds that may render a food unsafe for consumption not only by the general public, but also by the small percentage of the population that may be particularly sensitive to a specific chemical. For example, sulfiting agents used to preserve fresh leafy vegetables, dried fruits, and wines have caused allergic-type reactions in sensitive individuals. Other chemical hazards to be considered include aflatoxin and other mycotoxins, fish and shellfish toxins, scombrotoxin (histamine) from the decomposition of certain types of fish, and allergenic ingredients such as tree nuts or shellfish.

As in the case of biological hazards, the HACCP team must identify all potential chemical hazards associated with the production of the food commodity before evaluating the significance of each.

Potential Physical Hazards

The HACCP team must identify any potential physical hazards associated with the finished product. Foreign objects that are capable of injuring the consumer represent potential physical hazards. Examples of such potential hazards include glass fragments, pieces of wood or metal, plastic, etc. During the hazard evaluation stage, differentiation may be made between foreign objects that are aesthetically unpleasant and those that are capable of causing injury. The HACCP team should define the per-

formance criteria (size, shape, etc.) of physical hazards capable of causing injury. For example, a Public Health Hazard Analysis Board at FSIS concluded that bone particles less than 10 mm in size do not pose a significant hazard to consumers (3). In addition, a thorough review of data by FDA indicated that objects less than 7 mm in size pose little risk unless a food is intended for a special risk group such as infants or the elderly (18). FDA also notes in its current Compliance Policy Guide that products containing a hard or sharp foreign object that measures 7 mm to 25 mm in length should be considered adulterated (2).

Hazard Evaluation

The second stage of the hazard analysis, the hazard evaluation, is conducted after the list of potential biological, chemical, or physical hazards is assembled. During the hazard evaluation, the HACCP team decides which of the potential hazards present a significant risk to consumers. Based on this evaluation, a determination is made as to which hazards must be addressed in the HACCP plan. During the evaluation of each potential hazard, the food, its method of preparation, transportation, storage, and the likely consumers of the product should be considered. The team must consider the influence of the likely food preparation practices and methods of storage and whether the intended consumers are particularly susceptible to a potential hazard.

According to the NACMCF (16), each potential hazard should be evaluated on the basis of the severity of the potential hazard and its likely occurrence. Severity is the seriousness of the consequences (potential illness or injury) resulting from exposure to the hazard. Considerations of severity (e.g., the magnitude and duration of illness or injury and the potential for sequelae) can be helpful in understanding the public health impact of the hazard. Consideration of the likely occurrence of the hazard in the food as eaten is usually based on a combination of experience, epidemiological data, and information in the technical literature. When conducting the hazard evaluation, it is helpful to consider the likelihood of occurrence and the severity of the potential consequences if the hazard is not properly controlled. Because likelihood of occurrence and severity of consequences of a hazard are the two main factors to be considered in a risk assessment, many have compared a hazard analysis to a qualitative assessment of risk to consumers as a result of a hazard in food. Although the process and output of a risk assessment are different from those of a hazard analysis, the identification of potential hazards and the hazard evaluation may be facilitated by using the same systematic approach.

One tool developed to help visualize the concept of evaluating hazards according to a qualitative estimate of risk is contained in Table 41.1. This table is an adaptation of a table developed by the NACMCF in 1997 (16; Appendix D) to help explain the two stages of conducting a hazard analysis. This table gives three examples using a logic sequence that parallels a qualitative risk assessment scheme for conducting a hazard analysis. Although the examples in this table relate to biological hazards, chemical and physical hazards are equally important to consider.

Hazards identified as significant in one operation or facility may not be significant in another operation producing the same or a similar product. This may happen because of differences in sources of supply, product formulation, differences in production methods, etc. For example, because of differences in equipment and/or an effective maintenance program, the probability of metal contamination may be high in one facility but remote in another. The key to this determination, however, is the proper conduct and documentation of a hazard analysis. Vast differences in HACCP plans should not be observed between two facilities that manufacture the same items in a very similar manner.

When a team has determined that a potential hazard is not reasonably likely to occur, this decision should be based on all available evidence from literature and, when available, from the plant's own historical records. A summary of the HACCP team's deliberations and the rationale developed during the hazard analysis must be kept for future reference. This summary should include the conclusions developed during the hazard evaluation regarding the likelihood of occurrence and the severity of hazards. This information will be useful during validation, future reviews, and updates of the hazard analysis and the HACCP plan.

IDENTIFICATION OF CONTROL MEASURES

Upon completion of the hazard analysis, the hazards associated with each step in the production of the food should be listed along with any measure(s) that are used to control the hazard(s). The term control measure was substituted for preventive measure in the 1997 NACMCF HACCP document (16) because not all hazards can be prevented, but control is usually possible. The NACMCF (16) defines control measures as "any action or activity that can be used to prevent, eliminate or reduce a significant hazard." More than one control measure may be required to control a specific hazard. More than one hazard may be addressed by a specified control measure.

As an example, consider the identification of an appropriate control strategy for *Salmonella* sp. in the production of a product formulated with egg, such as that described in Table 41.1. When selecting a control measure, the HACCP team considers the intrinsic properties of the food and its method of preparation compared with those physical parameters that permit growth and survival of *Salmonella* sp. and with those conditions needed to inactivate it. On the basis of the HACCP team's review of control options for *Salmonella* sp., the team may determine that sufficient heat must be applied to kill the organisms and that, to serve this purpose, only pasteurized liquid eggs will be used in formulating this product. If done correctly, this step will reduce the likelihood of *Salmonella* sp. occurrence in finished goods to an acceptable level. Thus, the control measure selected is use of pasteurized eggs.

INFLUENCE OF PREREQUISITE PROGRAMS ON THE HAZARD ANALYSIS

As noted earlier in this chapter, each operation must have a firm foundation in prerequisite programs before considering the implementation of a HACCP plan. One of the most difficult decisions that a HACCP team will face is determining whether a potential hazard can be managed under existing programs, such as GMP compliance, or whether the hazard should be managed in the HACCP plan. Prerequisite programs typically include objectives other than food safety, and it may not be easy to associate performance of a prerequisite program element, e.g., pest control or chemical storage programs, with specific production lots or batches (19). Consequently, it is usually more effective to manage non-food safety issues and low-risk hazards within a quality system rather than including their performance and control as part of the HACCP plan. This is appropriate provided that uninterrupted adherence to the prerequisite program is not essential for food safety.

Nevertheless, prerequisite programs play an important role in controlling potential health hazards. Occasional deviation from a prerequisite program requirement would alone not be expected to create a food safety hazard or concern. For example, supplier control programs and chemical control programs can be used to minimize potential chronic health hazards such as those from mycotoxins or pesticides. Similarly, foreign material contamination in many food processes can be minimized by preventive maintenance programs and by upstream control devices such as sifters or magnets. Deviations from compliance in a prerequisite program usually do not result in action against the product. In

Table 41.1 Examples of how the stages of hazard analysis are used to identify and evaluate hazards

Hazard analysis stage		Product		
		Frozen cooked beef patties produced in a manufacturing plant	Product containing eggs prepared for food service	Commercial frozen precooked, boned chicken for further processing
Stage 1 hazard identification	Determine potential hazards associated with product.	Enteric pathogens (i.e., E. coli O157:H7 and Salmonella sp.)	Salmonella sp. in finished product	Staphylococcus aureus in finished product
Stage 2 hazard evaluation	Assess severity of health consequences if potential hazard is not properly controlled.	Epidemiological evidence indicates that these pathogens cause severe health effects, including death, among children and elderly. Undercooked beef patties have been linked to disease from these pathogens.	Salmonella sp. causes a foodborne infection that is a moderate to severe illness; salmonellosis can be caused by ingestion of only a few cells of Salmonella sp.	Certain strains of S. aureus produce an enterotoxin, which can cause a moderate foodborne illness.
	Determine likelihood of occurrence of potential hazard if not properly controlled.	Likely occurrence of E. coli O157:H7 is low to remote while the likelihood of salmonellae is moderate in raw beef trimmings.	Product is made with liquid eggs, which have been associated with past outbreaks of salmonellosis. Recent problems with Salmonella enterica serotype Enteritidis in eggs cause increased concern. Probability of Salmonella sp. in raw eggs cannot be ruled out. If not effectively controlled, some consumers are likely to be exposed to Salmonella sp. from this food.	Product may be contaminated with S. aureus due to human handling during boning of cooked chicken. Enterotoxin capable of causing illness will only occur as S. aureus multiplies to about 10^6/g. Operating procedures during boning and subsequent freezing prevent growth of S. aureus; thus, the potential for enterotoxin formation is very low.
	Using information above, determine if this potential hazard is to be addressed in the HACCP plan.	The HACCP team decides that enteric pathogens are hazards for this product.	HACCP team determines that if the potential hazard is not properly controlled, consumption of product is likely to result in an unacceptable health risk.	The HACCP team determines that the potential for enterotoxin formation is very low. However, it is still desirable to keep the initial number of S. aureus organisms low. Employee practices that minimize contamination, rapid carbon dioxide freezing, and handling instructions have been adequate to control this potential hazard.
		Hazards must be addressed in the plan.	Hazard must be addressed in the plan.	Potential hazard does not need to be addressed in HACCP plan.

contrast, potential acute health hazards such as the presence of *Salmonella* sp. are usually controlled in a HACCP system in which definitive critical control points (CCPs) are available to eliminate the hazard. Deviations from compliance in a HACCP system normally result in action against the product, such as evaluation of product to determine appropriate action (rework, destroy, etc.). This is a key consideration that can aid in distinguishing between control points within prerequisite programs and CCPs that should be included in a HACCP plan.

The decision as to whether a hazard warrants control within a HACCP program will depend for the most part on the HACCP team's evaluation of the risk to consumers that might result from a failure to control the hazard. Use of the logical evaluation of potential hazards, as presented in Table 41.1 following the same systematic approach as a qualitative risk assessment, should help in this task.

Other considerations when deciding on GMPs versus HACCP involve those hazards that are controlled, in part, by systemwide adherence to GMPs or sanitation standard operating procedures. Recontamination of cooked products by *L. monocytogenes* is an example of this type of situation. Control measures to prevent recontamination of cooked product by *L. monocytogenes* may be so diverse that management within a HACCP program will be extremely difficult. In addition, if the plant chooses to address the possibility of postprocessing contamination by *L. monocytogenes* within a HACCP plan, the plant may find it difficult to identify discrete CCPs and critical limits that are meaningful. More importantly, failure to conduct a single environmental control activity at the prescribed time is not likely to result in contamination by *L. monocytogenes*. This would call into question whether this is really a CCP. Additionally, many in the scientific community remain skeptical about the potential for small numbers of *L. monocytogenes* organisms to present a health hazard even to immunocompromised individuals. In this example, as with all hazard analysis exercises, it is important that the HACCP team develop a scientific rationale for its decisions.

A summary of the HACCP team deliberations and the rationale developed during identification of the control measures should be kept for future reference.

PRINCIPLE 2: DETERMINE THE CCPs

Once the significant hazards and control measures are identified, CCPs within the process scheme are identified. A control point is any point, step, or procedure at which biological, physical, or chemical factors can be controlled. A CCP is defined by the NACMCF (16) as "a step at which control can be applied and [where control] is essential to prevent or eliminate a food safety hazard or reduce it to an acceptable level." Each significant hazard identified during the hazard analysis should be addressed by at least one CCP. Examples of CCPs include cooking, chilling, product formulation control, application of a bactericidal rinse, a decontamination step, etc.

Whereas measures needed for adequate control of a hazard should be reflected in the HACCP plan, to keep HACCP programs plant-friendly and sustainable, CCPs should not be redundant. Redundant CCPs typically add little to the margin of safety but will add to the record keeping and administrative burden of a firm's management structure. They will also add significant cost without concomitant benefit. Experience has shown that HACCP plans that are unnecessarily cumbersome will not be supported over extended periods.

As an example of this concept, a firm may have multiple magnets and metal detectors in a line to protect production equipment from damage and consumers from harm. If the product is passed through a metal detection/reject device after it is in its final package, the last detector will typically be regarded as the CCP for this hazard, while up-line metal detectors or magnets will be considered control points.

For the product referred to above that was formulated with pasteurized liquid eggs, the pasteurization step occurs at a supplier's facility. In this case, the firm purchasing the eggs should confirm that the product received has been pasteurized. Alternately, if the firm was producing an egg-containing item that is cooked in its facility, the CCP for *Salmonella* sp. control could be located at the cooking step.

CONSIDERATIONS FOR RAW PRODUCTS

The objective of a HACCP plan will be different for raw products than for products that are processed by cooking or applying some other bactericidal treatment. For products that receive a defined bactericidal treatment, the hazard from microbial pathogens can be reduced, often to negligible levels, through proper design and application of the treatment. For raw products, however, the risk associated with the product can, at best, be only minimized.

The degree to which risk can be minimized for raw products will often be limited by the technology available. The HACCP team should consider this and determine whether the available interventions are adequate or whether some redesign of the line or process might be in order to reduce risk appropriately. This type of evaluation is typically one of the hidden values of conducting the hazard analysis. Many firms will find that simple adjustments in a line configuration or conduct of a particular operation may pay large dividends in terms of product safety.

Processors of raw products should establish CCPs that are within their control. For foods intended to be cooked after leaving the manufacturer's control, the firm should evaluate the potential for inadequate cooking, either intentionally or unintentionally. If a significant possibility exists that the product may be inadequately cooked, producers should minimize their reliance on the consumer's cooking step as a CCP.

DECISION TREES AND THEIR USE

CCP decision trees have been developed to assist firms in identifying CCPs in the process; both the NACMCF and Codex documents include various versions of this tool. The HACCP team may use a decision tree to evaluate each of the points where food safety hazards can be prevented, eliminated, or reduced to acceptable levels. Each point should then be categorized as either a control point or a CCP. The results of this evaluation should be summarized and added to the material developed in principle 1.

The most common problem with using a decision tree is trying to apply the questions in a tree prior to the completion of the hazard analysis. By applying the questions to potential hazards that are not reasonably likely to cause illness or injury, the HACCP team will unintentionally identify CCPs that are not related to controlling product safety. Another common problem is that no decision tree appears to give correct answers all the time. According to the NACMCF (16), "Although application of the CCP decision tree can be useful in determining if a particular step is a CCP for a previously identified hazard, it is merely a tool and not a mandatory element of HACCP. A CCP decision tree is not a substitute for expert knowledge."

PRINCIPLE 3: ESTABLISH CRITICAL LIMITS

A critical limit is defined as "a maximum and/or minimum value to which a biological, chemical or physical parameter must be controlled at a CCP to prevent, eliminate or reduce to an acceptable level the occurrence of a significant food safety hazard" (16). Each CCP will have one or more control measures that must be properly applied to assure prevention, elimination, or reduction of significant hazards to acceptable levels. Each control measure will have an associated critical limit(s) that serves as a safety boundary for the control measure(s) applied at each CCP and should be viewed as the dividing line between product that is unacceptable from

that which is acceptable. Critical limits may be set for such factors as temperature, time, physical dimension, humidity, moisture level, water activity, pH, titratable acidity, salt concentration, available chlorine, viscosity, and presence or concentration of preservatives. Sensory information such as texture, aroma, and visual appearance is rarely a critical limit. Although product condition is often important, it is typically related more to product quality than to its potential for producing illness or injury. Remember a point made earlier in this chapter that HACCP does not change existing laws regarding adulteration. It is illegal to process a food that "...consists in whole or in part of any filthy, putrid or decomposed substance..." (6). Processors are bound by existing statute to meet this mandate. However, food quality should not be addressed in a HACCP plan unless it is directly linked to the safety of a product.

Setting Critical Limits

Each firm must assure that the critical limits identified in its HACCP plan are adequate for the intended purpose. Thus the firm or its process authority should base critical limits on a benchmark or performance standard which the treatment should achieve. There are many possible sources for these benchmarks. They may be derived from sources such as regulatory standards and guidelines (when they relate directly to a health hazard), literature surveys, experimental studies, and experts such as process authorities. Referring again to our earlier example of pasteurized liquid eggs, U.S. regulatory authorities require a treatment targeted to achieve a 9-log cycle reduction of *Salmonella* enterica serotype Typhimurium. Thus, a cooking process could be judged adequate if it achieves this performance standard. The firm should have adequate scientific studies to validate the critical limits it has selected and the proper delivery of the control measure to each unit of product. The normal variation in process delivery should also be considered when establishing critical limits, and a firm should consider setting target or operational values such that critical limits are not routinely violated. Validation, which includes assurance that the critical limits are well founded as well as effectively and consistently delivered, will be discussed in more detail in the verification section of this chapter.

Microbiological Criteria as Critical Limits

Just as microbiological testing is not a good tool for monitoring, microbiological criteria are typically not useful as critical limits in a HACCP program. Because HACCP targets process control, factors that lend themselves to "real-time" monitoring and quick feedback should be identified as control measures. Thus, the critical limits used in a HACCP program do not typically relate to a criterion that is directly associated with microbiological testing of products or ingredients. However, there will be instances where the only option open to a processor is to hold lots of an ingredient and perform microbiological testing before release. This may arise when a particular ingredient can only be obtained from sources where little control is exercised over factors that affect the contaminants or pathogens associated with the ingredient. In this case, a CCP may be located at receiving, where the incoming ingredient would be sampled for analysis and placed in controlled storage until the results of the analysis are available. In this instance, microbiological testing is the monitoring activity and a microbiological criterion would be the critical limit. When using this approach, no product should be released until the results of all testing are finalized.

PRINCIPLE 4: ESTABLISH MONITORING PROCEDURES

Monitoring is a planned sequence of observations or measurements to assess whether a CCP is under control and to produce an accurate record for future use in verification. What is monitored will be a measurement of the physical factor identified as the control (preventive) measure for a significant hazard. Examples of monitoring activities include:

- measuring temperature
- tracking elapsed time, e.g., time at a specific temperature
- sampling product and determining pH
- determining moisture level or water activity

Monitoring serves three main purposes. First, monitoring tracks the system's operation in a manner essential to food safety management. If monitoring detects a trend toward loss of control, that is, approaching a target level, then action can be taken to bring the process back into control before a deviation occurs. Second, monitoring is used to determine when a deviation occurs at a CCP, i.e., violation of the critical limit, which will initiate an appropriate corrective action. Third, it provides the basis for written documentation of process control for verifying that the HACCP plan has been followed.

When establishing monitoring activities, the firm should specify:

- what control measure(s) is being monitored
- how often the monitoring needs to be conducted
- what procedures will be followed to collect data and what methods and equipment will be used
- who will be responsible for performing each monitoring activity

The frequency at which activities are monitored must be consistent with the needs of the operation in relation to the variation inherent in the control step; it must also be adequate to assure proper control. Frequency of monitoring should be integrated with a product coding system that is designed to prevent excessive amounts of product from being involved in a corrective action if problems arise and a critical limit is violated. For example, if a product must reach a certain temperature during a continuous cooking step and that temperature is only checked once per hour, then if the temperature drops below the critical limit, all product produced since the last check must be reworked or destroyed. Less product would be involved if checks were more frequent or if a method of continuous monitoring with automatic line shutdown or diversion were used.

In most instances, control measures that can be monitored on a continuous basis are preferred over those that do not provide this level of security. Those control measures that can be monitored continuously are typically amenable to continual recording and electronic monitoring with automated trend analysis. Such monitoring can provide advanced warning to allow correction of a problem before violation of a critical limit occurs. In the example above, constant monitoring of temperature with an alarm to warn of a trend toward violation of a critical limit may have avoided a deviation or at least would minimize the amount of product subject to corrective action.

Certain characteristics, however, cannot be measured on a continuous basis, or they may not need continuous monitoring. Products that are mixed in a batch, such as some acidified products, do not need continuous monitoring of pH. A control measure like water activity in a food cannot easily be monitored continually. Such items will be monitored through a planned sequence of sampling and testing of appropriate frequency to document control.

One of the underlying concepts of HACCP is to promote awareness of food safety and individual responsibility to the line level in an operation. One way to promote this is to involve line operators in monitoring activities. In many operations, monitoring activities are assigned to quality control personnel when they could be accomplished just as effectively by line operators, with the extra benefit of line worker involvement. The role of quality control personnel may be more appropriate in verification, or "checking the checker."

Those responsible for monitoring a control measure at a CCP must:

- be trained in the appropriate monitoring techniques
- understand the importance of monitoring

- accurately report/record results of the monitoring activity
- immediately report deviations from critical limits so that corrective actions can be taken

Returning again to the example of a baked item formulated with liquid eggs, it was determined that the control measure for *Salmonella* sp. would be heat applied during cooking and that the CCP would be a cooking step. The parameters being monitored by the firm are temperature of the oven and the duration of the cooking cycle. Critical limits for these physical factors must be set such that the performance criterion identified earlier (minimum 9-log cycle inactivation of *Salmonella* sp.) is met. If this firm chooses to use equipment that providess continuous monitoring (recording) of time and temperature, then the operator's duty is to assure that the monitoring equipment is working and to make a manual reading periodically as a verification activity. If there is a deviation from either the time or temperature component of the cooking step, the operator should initiate the proper corrective action.

PRINCIPLE 5: ESTABLISH CORRECTIVE ACTIONS

The HACCP system is designed to identify situations where health hazards are reasonably likely to occur and to establish strategies to prevent or minimize their occurrence. However, ideal circumstances do not always prevail. When there is a deviation from established critical limits, corrective actions must be taken. Corrective actions, in the HACCP context, are procedures to be followed when a deviation or failure to meet a critical limit occurs. Corrective actions should be developed and adopted that address how the firm will:

- fix or correct the cause of noncompliance to assure that the process is brought under control and the critical limits are being met
- determine the disposition of noncompliant product
- determine whether adjustments in the HACCP plan are needed

Because of the variations in CCPs for different foods and the diversity of possible deviations, corrective action plans must be developed that are specific to the particular operation. A corrective action plan must be developed for deviations that may occur at each CCP. Individuals who have a thorough understanding of the process, product, and HACCP plan are to be assigned responsibility for assuring that the appropriate corrective actions have been implemented. Records of the corrective actions that have been taken must be maintained. Long-term solutions should be sought when a particular critical limit appears to be violated routinely. Such repeated events should trigger a review of the plan, the CCP, the identified critical limit, and the method of process control to determine what improvements are needed to reduce the deviations.

Corrective actions for inadequately cooking the egg-containing product considered earlier would include correcting the equipment problem that caused the temperature or time error. In addition, any product involved in the deviation must not be released until it is evaluated to determine whether it is safe for consumption. In this case, microbiological testing is seldom appropriate; the number of samples that would need analysis to detect a low-level contaminant may be excessive. Instead, the preferred method would be to obtain an expert evaluation of the lethality actually delivered to the product during the deviation period. If this is determined to be adequate, the product may be safely consumed. Alternately, the product may be reprocessed, diverted for another use in which it will be rendered safe, or destroyed.

PRINCIPLE 6: ESTABLISH VERIFICATION PROCEDURES

There are several HACCP-related activities that will be conducted under the label of verification. Verification is the use of methods, procedures, or tests, in addition to those used in monitoring, to determine whether the HACCP plan is being followed and whether the records of monitoring activities are accurate. Verification is a second level of review, beyond the primary review of line personnel who are actually conducting the monitoring. This secondary level of review is typically the responsibility of quality control personnel and will be aimed at verifying that records are being kept accurately and that verification sampling is conducted when appropriate. Verification often involves a third level of review in which outside audit teams review all aspects of the HACCP plan, monitoring procedures, record keeping practices, etc.

In addition, the concept of validation has evolved as a separate and distinct activity to be conducted as a subset of verification. In essence, validation is described as that element of verification focused on collecting and evaluating scientific and technical information to determine whether the HACCP plan, when properly implemented, will effectively control the significant hazards. Validation usually focuses on the process of assuring that the plan reflects identification of all significant hazards, that the control measures identified are appropriate, that the designated critical limits are adequate, and that the rest of the features of the plan are satisfactory. Most now believe that validation should be viewed as a distinct function but one that is carried out in parallel with other activities. Plans may need to be revalidated whenever significant changes occur in product formulations or equipment.

Validation of the control procedures for *Salmonella* sp. identified in the earlier example for a cooked product would include a determination of whether the time and temperature of the cook process were adequate to inactivate an appropriate number of *Salmonella* cells. In addition, validation information should include the rationale for the microbiological criterion that the cook process is supposed to achieve. In this case, reference to the USDA *Egg Pasteurization Manual* (1) could be used. The determination of the appropriate time and temperature to achieve the appropriate level of kill would involve reviewing appropriate literature or consulting with experts in the field to determine approximate thermal inactivation kinetics of the organism in the product. It will also be necessary to determine the heating rate of the product in the oven to be used for cooking so that the cooking time and temperature needed to achieve an adequate kill can be established. Uniform delivery of the cook step in the oven must also be validated so that each unit of product passing through the oven is assured of receiving at least the minimum treatment. Alternatively, the cook step can be validated microbiologically by using "inoculated packs"—product inoculated with a nonpathogenic simulator organism—if these are statistically designed to assure that "worst-case" conditions are encountered and enough sample units are run.

Verification activities for the cook step would include calibrations necessary to assure that the time and temperature measurements and other data produced during monitoring are accurate. The firm should also review all HACCP records, especially the records for any lethal step, such as a heat treatment, before final release of any lot of product. Some may decide to include occasional microbiological testing of finished product as an additional verification step. Verification activities should take place on a regular and predetermined schedule. Some will be conducted daily, others will be weekly, monthly, or even less frequently. The need for accuracy and precision, as well as equipment variation, should be considered when establishing verification schedules.

PRINCIPLE 7: ESTABLISH RECORD-KEEPING AND DOCUMENTATION PROCEDURES

Record keeping is integral to maintaining a HACCP system. Without effective record keeping and review, the HACCP system will not be maintained as an ongoing practice. Records provide the basis for management assessment of the safety of products produced and for documenting the safety of products to customers. Records provide a means to trace the production history of foods produced, to document that critical limits were met, and to prove that any needed corrective actions were carried out.

The HACCP plan and associated support documents should be on file at the food establishment. Generally, the records used in the HACCP system will include the following:

1. The HACCP plan

   - listing of the HACCP team and assigned responsibilities
   - description of the product and its intended use
   - block flow diagram for the entire manufacturing process indicating CCPs
   - hazards associated with each CCP and preventive measures
   - critical limits for each CCP
   - monitoring procedures (who, what, when, where, how)
   - corrective action plans for deviations from critical limits
   - record keeping procedures
   - procedures for verification of the HACCP system
   - records of validation studies

2. Records obtained during the operation of the plan

   - records of data collected by monitoring of control measures at CCPs
   - corrective action records
   - records of certain verification activities

Examples of operational records for a cooked product might include a computerized log of the operational details of the oven used for cooking. Such records would document the time the unit was turned on, when it reached temperature, when product began to move through, the record of temperature and time during the production run, etc. A record of operator checks of a separate temperature-monitoring device, as well as confirmation of throughput or residence time in the oven, would also be expected. If any deviations occurred, a record of the time and extent of the deviation should be available, along with a description of actions needed to bring the cooker back into control and the disposition of any product produced while the deviation occurred. In this case, the processor may choose to stop the line automatically if a time or temperature parameter is violated. This eliminates the need to determine product disposition. Other operational records would include the calibration records for the temperature-monitoring equipment, time measurements, and any other practices needed to assure the accuracy of the primary monitoring activities.

In addition to the above, there should be a HACCP master file that will contain a record of the deliberations of the HACCP team and documentation for various aspects of the plan. This should include justification for the critical limits identified, details on sampling plans and methods of analysis, standard operating procedures for monitoring activities and corrective actions, and other details that may need further review and modification to support the plan. The forms used, format, practices to be followed, and review procedures should be explained in the HACCP master file.

CONCLUSIONS

The HACCP concept provides a systematic, structured approach to assuring the safety of food products. However, there is no blueprint or universal formula for putting together the specific details of a HACCP plan. The plan must be dynamic. It must allow for modifications to production, substitution of new materials and ingredients, and development of new products. For HACCP to work well, it should be a participatory endeavor at all levels of an organization, both in formulating and managing the plan.

References

1. **Anonymous.** 1968. *Egg Pasteurization Manual*. U.S. Department of Agriculture, Agricultural Research Service, Washington, D.C.

2. **Anonymous.** 1999. Foods—adulteration involving hard or sharp foreign objects. *Food and Drug Administration Compliance Policy Guides Manual—Update No. 12.* Compliance Policy Guide, Chapter 5, Section 555.425. http://www.fda.gov/ora/compliance_ref/cpg/cpgfod/cpg555-425.htm

3. **Anonymous.** 1995. Bone particles and foreign material in meat and poultry products (a report to the Food Safety and Inspection Service). Public Health Hazard Analysis Board, USDA, FSIS.

4. **Anonymous.** World Trade Organization agreement on the application of sanitary and phytosanitary measures. http://www.wto.org/english/docs_e/legal_e/ursum_e.htm # bAgreement

5. **Codex Alimentarius Commission.** 1997. Recommended International Code of Practice General Principles of Food Hygiene. CAC/RCP 1–1969, Revision 3. http://www.fao.org/waicent/faoinfo/ECONOMIC/esn/codex/STANDARD/standard.htm

6. **Federal Food, Drug, and Cosmetic Act.** Title 21, U.S.Code, Section 342 (a) (3).

7. **Federal Register.** 1995. Procedures for the safe and sanitary processing and importing of fish and fishery prod-

ucts; final rule. *Fed. Regist.* 60:65096–65202 (December 18).

8. **Federal Register.** 1996. Pathogen reduction; hazard analysis and critical control point (HACCP) systems; final rule. *Fed. Regist.* 61:38806–38989 (July 25).

9. **Federal Register.** 2001. Hazard analysis and critical control point (HACCP); procedures for the safe and sanitary processing and importing of juice. *Fed. Regist.* 66:6137–6202 (January 19).

10. **Food and Agriculture Organization of the United Nations.** 1995. The use of hazard analysis and critical control point (HACCP) principles in food control. Report of an FAO Expert Technical Meeting, Vancouver, Canada. Food and Nutrition Paper 58. Food and Agriculture Organization of the United Nations, Rome, Italy.

11. **International Commission on Microbiological Specifications for Foods, M. Van Schothorst, Secretary.** 1998. Principles for the establishment of microbiological food safety objectives and related control measures. *Food Control* 9:379–384.

12. **International Life Sciences Institute.** 1997. *A Simple Guide to Understanding and Applying the Hazard Analysis Critical Control Point Concept*, 2nd ed. International Life Sciences Institute, Washington, D.C.

13. **Leaper, S. (ed.).** 1997. *HACCP: A Practical Guide.* Technical Manual No. 38. Campden and Chorleywood Food Research Association, Chipping Camden, Gloucestershire, United Kingdom.

14. **National Academy of Sciences/National Research Council.** 1985. *An Evaluation of the Role of Microbiological Criteria for Foods and Food Ingredients.* National Academy Press, Washington, D.C.

15. **National Advisory Committee on Microbiological Criteria for Foods.** 1992. Hazard analysis and critical control point system. *Int. J. Food Microbiol.* 16:1–23.

16. **National Advisory Committee on Microbiological Criteria for Foods.** 1998. Hazard analysis and critical control point principles and applications guidelines. *J. Food Protect.* 61:762–775.

17. **National Conference on Food Protection.** 1971. *Proceedings of the National Conference on Food Protection.* Department of Health, Education, and Welfare, Public Health Service, Washington, D.C.

18. **Olsen, A. R.** 1998. Regulatory action criteria for filth and other extraneous materials. I. Review of hard or sharp foreign objects as physical hazards in food. *Reg. Toxicol. Pharmacol.* 28:181–189.

19. **Sperber, W. H., K. E. Stevenson, D. T. Bernard, K. E. Deibel, L. J. Moberg, L. R. Hontz, and V. N. Scott.** 1998. The role of prerequisite programs in managing a HACCP system. *Dairy Food Environment. Sanit.* 18:418–423.

20. **Stevenson, K. E., and D. T. Bernard (ed.).** 1999. *HACCP: A Systematic Approach to Food Safety*, 3rd ed. The Food Processors Institute, Washington, D.C.

21. **World Health Organization.** 1995. *Application of Risk Analysis to Food Issues.* Report of the Joint FAO/WHO Expert Consultation, 13–17 March 1995. World Health Organization, Geneva, Switzerland.

Index